CIVIL ENGINEERING
(OBJECTIVE TYPE)

LIST OF CONTRIBUTORS

Authors   Chapters

★ **Dr. P. Jaya Rami Reddy**   4, 5, 8, 11, 12, 17, 18
Prof. & Head, Deptt. of Civil Engg.
G. Pulla Reddy Engg. College, Kurnool.

★ **Sri A. Kameswara Rao**   6, 7, 9, 10, 16, 17
Asst. Professor in Civil Engineering
NBKR Institute of Science & Technology, Vidyanagar.

★ **Sri S. Sriramam**   14, 15
Professor and Head, Deptt. of Civil Engineering
NBKR Institute of Science & Technology, Vidyanagar.

★ **Sri K.S.V. Radhakrishna**   3, 13
Prof. of Civil Engineering
NBKR Institute of Science & Technology, Vidyanagar.

★ **Dr. R.K. Bansal**   1, 2
Deptt. of Mechanical Engineering
Delhi College of Engineering, Delhi.

CIVIL ENGINEERING
(OBJECTIVE TYPE)

Edited by

Dr. P. Jaya Rami Reddy

*Prof. & Head, Deptt. of Civil Engg.
G. Pulla Reddy Engg. College, Kurnool
Andhra Pradesh*

LAXMI PUBLICATIONS (P) LTD
(An ISO 9001:2015 Company)

BENGALURU • CHENNAI • GUWAHATI • HYDERABAD • JALANDHAR
KOCHI • KOLKATA • LUCKNOW • MUMBAI • RANCHI
NEW DELHI

CIVIL ENGINEERING (OBJECTIVE TYPE)

© by Laxmi Publications (P) Ltd.

All rights reserved including those of translation into other languages. In accordance with the Copyright (Amendment) Act, 2012, no part of this publication may be reproduced, stored in a retrieval system, translated into any other language or transmitted in any form or by any means, electronic, mechanical, photocopying, recording or otherwise. Any such act or scanning, uploading, and or electronic sharing of any part of this book without the permission of the publisher constitutes unlawful piracy and theft of the copyright holder's intellectual property. If you would like to use material from the book (other than for review purposes), prior written permission must be obtained from the publishers.

Printed and bound in India
Typeset at : Goswami Associates, Delhi
First Edition : 2007; Reprint : 2008, 2009, 2010, 2011; Fourth Edition : 2015; Reprint : 2016, 2017, 2018, Edition : 2019
ISBN 978-81-318-0878-8

Limits of Liability/Disclaimer of Warranty: The publisher and the author make no representation or warranties with respect to the accuracy or completeness of the contents of this work and specifically disclaim all warranties. The advice, strategies, and activities contained herein may not be suitable for every situation. In performing activities adult supervision must be sought. Likewise, common sense and care are essential to the conduct of any and all activities, whether described in this book or otherwise. Neither the publisher nor the author shall be liable or assumes any responsibility for any injuries or damages arising herefrom. The fact that an organization or Website if referred to in this work as a citation and/or a potential source of further information does not mean that the author or the publisher endorses the information the organization or Website may provide or recommendations it may make. Further, readers must be aware that the Internet Websites listed in this work may have changed or disappeared between when this work was written and when it is read.

All trademarks, logos or any other mark such as Vibgyor, USP, Amanda, Golden Bells, Firewall Media, Mercury, Trinity, Laxmi appearing in this work are trademarks and intellectual property owned by or licensed to Laxmi Publications, its subsidiaries or affiliates. Notwithstanding this disclaimer, all other names and marks mentioned in this work are the trade names, trademarks or service marks of their respective owners.

Branches		
✆ Bengaluru	080-26 75 69 30	
✆ Chennai	044-24 34 47 26,	24 35 95 07
✆ Guwahati	0361-254 36 69,	251 38 81
✆ Hyderabad	040-27 55 53 83,	27 55 53 93
✆ Jalandhar	0181-222 12 72	
✆ Kochi	0484-237 70 04,	405 13 03
✆ Kolkata	033-22 27 43 84	
✆ Lucknow	0522-220 99 16	
✆ Mumbai	022-24 93 12 61	
✆ Ranchi	0651-220 44 64	

Published in India by

Laxmi Publications (P) Ltd.
(An ISO 9001:2015 Company)
113, GOLDEN HOUSE, DARYAGANJ,
NEW DELHI - 110002, INDIA
Telephone : 91-11-4353 2500, 4353 2501
Fax : 91-11-2325 2572, 4353 2528
www.laxmipublications.com   info@laxmipublications.com

C—00344/019/06
Printed at: Ajit Printing Press, Delhi

CONTENTS

Chapters	Pages
1. Fluid Mechanics and Hydraulic Machines	1–80
2. Engineering Mechanics	81–140
3. Strength of Materials	141–190
4. Hydrology	191–223
5. Water Resources Engineering	224–280
(Including Irrigation, Open Channel Flow and Water Power)	
6. Water Supply Engineering	281–332
7. Sanitary Engineering	333–373
8. Soil Mechanics	374–513
9. Highway Engineering	514–559
10. Railway Engineering	560–599
11. Building Materials	600–634
12. Building Construction	635–677
13. Theory of Structures	678–714
14. Reinforced Comment Concrete Structures	715–798
15. Steel Structures	799–854
16. Surveying	855–896
17. Construction Planning, PERT, CPM	897–911
18. Estimating and Quantity Surveying	912–925
19. Air Pollution	926–969
20. Watershed Management	970–975

PREFACE TO THE FOURTH EDITION

Another revision only shows the popularity the book enjoys. We are happy to bring out the fourth edition. The opportunity has been utilized to add a new chapter on Watershed Management, to incorporate the suggestions received from various readers, to revise the jist of the subject matter presented at the beginning of every chapter and to increase the number of objective type questions in each chapter. It is hoped that these revisions and additions will enhance its usefulness further.

–Authors

PREFACE TO THE FIRST EDITION

Objective type testing has become necessary to screen out and rank large number of candidates competing for employment or post-graduate admission. Therefore, there is an increasing need for this type of books.

This book has been prepared keeping in view the syllabii prescribed for various competitive examinations like GATE, UPSC (I.E.S.), I.A.S. etc. All the subjects of Civil Engineering have been covered. To make the reader familiar with the topic quickly, a gist is given at the beginning of each chapter. It is hoped that the book will be found useful by all those who are preparing for the competitive examinations.

I am grateful to all my colleagues who have readily agreed and contributed to this venture. Special thanks are due to Dr. R.K. Bansal, Delhi College of Engineering, Delhi, for his contribution to the first and second chapters.

Suggestions for improving the usefulness of the book will be greatly appreciated and duly incorporated in the next edition.

–Authors

Chapter 1: Fluid Mechanics and Hydraulic Machines

1. INTRODUCTION

DEFINITIONS AND FLUID PROPERTIES

Fluid mechanics is that branch of science which deals with the behaviour of the fluid (*i.e.* liquids or gases) when they are at rest or in motion. When the fluids are at rest, there will be no relative motion between adjacent fluid layers and hence velocity gradient $\left(\dfrac{du}{dy}\right)$, which is defined as the change of velocity between two adjacent fluid layers divided by the distance between the layers, will be zero. Also the shear stress $\tau = \mu \dfrac{du}{dy}$ will be zero in which $\dfrac{du}{dy}$ is the velocity gradient or **rate of shear strain**.

The law, which states that the shear stress (τ) is directly proportional to the rate of shear strain $\left(\dfrac{du}{dy}\right)$, is called **Newton's Law of viscosity**. Fluids which obey Netwon's law of viscosity are known as **Newtonian fluids** and the fluids which do not obey this law are called **Non-Newtonian fluids**.

(*i*) **Density** or **mass density**. It is defined as the mass per unit volume of a fluid and is denoted by the symbol ρ (rho).

(*ii*) **Weight density** or **specific weight**. It is defined as the weight per unit volume of a fluid and is denoted by the symbol w.

Mathematically, $$\rho = \frac{\text{Mass of fluid}}{\text{Volume of fluid}}$$

and $$w = \frac{\text{Weight of fluid}}{\text{Volume of fluid}} = \frac{\text{Mass of fluid} \times g}{\text{Volume}} = \rho \times g.$$

The value of density (ρ) for water is 1000 kg/m^3 and of specific weight or weight density (w) is 1000×9.81 N/m^3 or 9810 N/m^3 in S.I. units.

(*iii*) **Specific volume.** It is defined as volume per unit mass and hence it is the reciprocal of mass density. **Specific gravity** is the ratio of weight density or mass density of the fluid to the weight density or mass density of a standard fluid at a standard temperature. For liquids, water is taken as a standard fluid at 4°C and for gases, air is taken as standard fluid.

(*iv*) **Viscosity.** It is defined as the property of a fluid which offers resistance to the movement of one layer of fluid over another adjacent layer of the fluid. Unit of viscosity in MKS is expressed as $\frac{\text{kgf-sec}}{\text{m}^2}$, in SI system as $\frac{\text{Ns}}{\text{m}^2}$ and in CGS as $\frac{\text{dyne-sec}}{\text{m}^2}$. The unit of viscosity in CGS is also called Poise.

The equivalent numerical value of one poise in MKS units is obtained by dividing 98.1 and in SI units is obtained by dividing 10.

Kinematic viscosity is defined as the ratio of dynamic viscosity to density of fluid. It is denoted by the Greek symbol (ν) called 'nu'. Unit of kinematic viscosity in MKS is m^2/sec and in CGS is cm^2/sec which is also called stoke. The viscosity of a liquid decreases with the increase of temperature while the viscosity of the gas increases.

(*v*) **Compressibility.** It is the reciprocal of the bulk modulus of elasticity, which is defined as the ratio of compressive stress to volumetric strain. Mathematically,

$$\text{Bulk Modulus} = \frac{\text{Increase of pressure}}{\text{Volumetric strain}} = \frac{dp}{-\left(\frac{dV}{V}\right)}$$

$$\therefore \quad \text{Compressibility} = \frac{1}{\text{Bulk Modulus}} = \frac{-\left(\frac{dV}{V}\right)}{dp}$$

(*vi*) **Surface tension.** It is defined as the tensile force acting on the surface of a liquid in contact with a gas such that the contact surface behaves like a membrane under tension. It is expressed as force per unit length and is denoted by σ (called sigma). Hence unit of surface tension in MKS is kgf/m while in SI it is N/m.

The relation between surface tension (σ) and difference of pressure (p) between inside and outside of a liquid drop is given by $p = \frac{4\sigma}{d}$

For a soap bubble, $$p = \frac{8\sigma}{d}$$

For a liquid jet, $$p = \frac{2\sigma}{d}.$$

(*vii*) **Capillarity.** It is defined as a phenomenon of rise or fall of a liquid surface in a small vertical tube held in a liquid relative to general level of the liquid. The rise or fall of liquid is given by

$$h = \frac{4\sigma \cos\theta}{wd}$$

where d = Dia. of tube

θ = Angle of contact between liquid and glass tube.

(*viii*) **Ideal fluid** is a fluid which offers no resistance to flow and is incompressible. Hence for ideal fluid viscosity (μ) is zero and density (ρ) is constant.

(*ix*) **Real fluid** is a fluid which offers resistance to flow. Hence viscosity for real fluid is not zero.

PRESSURE AND ITS MEASUREMENT

Pressure at a point is defined as the force per unit area. The Pascal's law states that intensity of pressure for a fluid at rest is equal in all directions. The pressure at any point in a incompressible fluid (*i.e.* liquid) at rest is equal to the product of weight density of fluid and vertical height from free surface of the liquid.

Mathematically, $p = wz = \rho g z$.

(*i*) **Hydrostatic law** states that the rate of increase of pressure in the vertically downward direction is equal to the specific weight of the fluid *i.e.* $\frac{dp}{dz} = w = \rho g$.

(*ii*) **Absolute pressure** is the pressure measured with reference to absolute zero pressure while gauge pressure is the pressure measured with reference to atmospheric pressure. Thus the pressure above the atmospheric pressure is called gauge pressure. Vacuum pressure is the pressure below the atmospheric pressure. Mathematically,

Gauge pressure = Absolute pressure – Atmospheric pressure

Vacuum pressure = Atmospheric pressure – Absolute pressure.

(*iii*) **Manometers** are defined as the devices used for measuring the pressure at a point in a fluid. They are classified as:

1. Simple Manometers, and
2. Differential Manometers.

Simple manometers are used for measuring pressure at a point while differential manometers are used for measuring the difference of pressures between the two points in a pipe or two different pipes.

(iv) The pressure at a point in a static compressible fluid is obtained by combining two equations i.e., equation of state for a gas $\left(\dfrac{p}{\rho} = RT\right)$ and the equation given by hydrostatic law $\left(\dfrac{dp}{dz} = -\rho g\right)$. For isothermal process, the pressure at a height Z in a static compressible fluid is given as $p = p_0 e^{-gZ/RT}$

(v) For adiabatic process the pressure and temperature at a height Z are

$$p = p_0\left[1 - \dfrac{\gamma-1}{\gamma}\dfrac{gZ}{RT_0}\right]^{\dfrac{\gamma-1}{\gamma}} \text{ and } T = T_0\left[1 - \dfrac{\gamma-1}{\gamma}\dfrac{gZ}{RT_0}\right]$$

where p_0 = Absolute pressure at ground or sea-level

R = Gas constant, γ = Ratio of specific heats

T_0 = Temperature at ground or sea-level.

HYDROSTATIC FORCES ON PLANE SURFACES

The **force** exerted by a static liquid on a vertical, horizontal and inclined surface immersed in the liquid is given by

$$F = \rho g A \bar{h}$$

where ρ = Density of the liquid

A = Area of the immersed surface

$\bar{h}$ = Depth of the centre of gravity of the immersed surface from free surface of the liquid.

(i) **Centre of pressure** is defined as the point of application of the resultant pressure on the surface. The depth of centre of pressure (h^*) from free surface of the liquid is given by

$$h^* = \dfrac{I_G \sin^2\theta}{A\bar{h}} + \bar{h} \quad \text{for inclined surface}$$

$$= \dfrac{I_G}{A\bar{h}} + \bar{h} \quad \text{for vertical surface}$$

The centre of pressure for a plane vertical surface lies at a depth of two-third the total height of the immersed surface from free surface.

(ii) **The total force on a curved surface** is given by $F = \sqrt{F_x^2 + F_y^2}$

where F_x = Horizontal force on a curved surface and is equal to total pressure force on the projected area of the curved surface on the vertical plane

and F_y = Vertical force on the curved surface and is equal to the weight of the liquid actually or virtually supported by the curved surface.

The inclination of the resultant force on curved surface with horizontal is given by

$$\tan \theta = \frac{F_y}{F_x}$$

(iii) **The resultant force on a sluice gate** is given by

$$F = F_1 - F_2$$

where F_1 = Pressure force on the upstream side of the sluice gate
F_2 = Pressure force on the downstream side of the sluice gate.

(iv) **Lock-gates.** For a lock-gate, the reaction between the two gates (P) is equal to the reaction at the hinge (R), i.e., $R = P$ and the reaction between the two gates (P) is given by $P = \dfrac{F}{2 \sin \theta}$

where F = Resultant water pressure on the lock-gate = $F_1 - F_2$
and θ = Inclination of the gate with the normal to the side of the lock.

BUOYANCY AND FLOTATION

Buoyant force is the upward force or thrust exerted by a liquid on body when the body is immersed in the liquid. The point through which the buoyant force is supposed to act is called **centre of buoyancy**. It is denoted by B. The point, about which a floating body starts oscillating when the body is given a small angular displacement, is known as **Metacentre**. It is denoted by M. The distance between the meta-centre (M) and centre of gravity (G) of a floating body is known as **meta-centric height**. This is denoted by GM and mathematically it is given as

$$GM = \frac{I}{V} - BG$$

where I = Moment of Inertia of the plan of the floating body at the water surface
V = Volume of the body submerged in water
BG = Distance between the centre of gravity (G) and centre of buoyancy (B).

(i) **Conditions of equilibrium** of a floating and submerged body are:

Equilibrium	Floating body	Submerged body
(i) Stable	M should be above G	B should be above G
(ii) Unstable	M should be below G	B should be below G
(iii) Neutral	M and G coincide	B and G coincide

(ii) **The meta-centric height (GM)** experimentally is given by

$$GM = \frac{wx}{W \tan \theta}$$

where w = Movable weight
x = Distance through which w is moved
W = Weight of floating body including w
θ = Angle through which floating body is tilted

(iii) **The time period of oscillation** of a floating body is given by $T = 2\pi \sqrt{\dfrac{k^2}{GM \times g}}$

where k = Radius of gyration, GM = Meta-centric height.

KINEMATICS OF FLUID

Kinematics is defined as that branch of science which deals with the study of fluid in motion without considering the forces causing the motion. The fluids flow may be compressible or incompressible; steady or unsteady; uniform or non-uniform; laminar or turbulent; rotational or irrotational; one, two or three dimensional.

(i) If the density (ρ) changes from point to point during fluid flow, it is known **compressible flow**. But if density (ρ) is constant during fluid flow, it is called **incompressible flow**. Mathematically,

$\rho \neq$ Constant for compressible flow

$\rho =$ Constant for incompressible flow.

(ii) If the fluid characteristic like velocity, pressure, density etc. do not change at a point with respect to time, the fluid flow is known as **steady flow**. If these fluid characteristic change with respect to time, the fluid flow is known as unsteady flow. Mathematically,

$\left(\dfrac{\partial v}{\partial t}\right) = 0, \left(\dfrac{\partial p}{\partial t}\right) = 0$ or $\left(\dfrac{\partial \rho}{\partial t}\right) = 0$ for steady flow, and

$\left(\dfrac{\partial v}{\partial t}\right) \neq 0, \left(\dfrac{\partial p}{\partial t}\right) \neq 0$ or $\left(\dfrac{\partial \rho}{\partial t}\right) \neq 0$ for unsteady flow.

(iii) If the velocity in a fluid flow does not change with respect to the length of direction of flow, the flow is said **uniform** and if the velocity change it is known **non-uniform** flow. Mathematically,

$\left(\dfrac{\partial v}{\partial s}\right) = 0$ for uniform, and $\left(\dfrac{\partial v}{\partial s}\right) \neq 0$ for non-uniform flow.

(iv) If the Reynold number (R_e) in a pipe is less than 2000, the flow is said to be **laminar** and if the Reynold number is more than 4000, the flow is said to be **turbulent**.

Reynolds number (R_e) is given by $R_e = \dfrac{\rho V D}{\mu}$ or $\dfrac{VD}{\nu}$

where V = Velocity of fluid, D = Dia. of pipe

μ = Viscosity of fluid, ν = Kinematic viscosity of fluid.

(v) If the fluid particles while flowing along stream lines also rotate about their own axis, that flow is known as **rotational flow** and if the fluid particles, while flowing along stream lines, do not rotate about their own axis, that type of flow is called **irrotational flow**.

(vi) The rate of discharge for incompressible fluid is given by
$$Q = A \times V$$

(vii) **Continuity equation** is written is general form as
$$\rho A V = \text{constant}$$

and in differential form as $\dfrac{\partial u}{\partial x} + \dfrac{\partial v}{\partial y} + \dfrac{\partial w}{\partial z} = 0\quad$ for three-dimensional flow

and $\qquad\dfrac{\partial u}{\partial x} + \dfrac{\partial v}{\partial y} = 0\quad$ for two-dimensional flow

(viii) The components of acceleration in x, y and z direction are

$$a_x = u\frac{\partial u}{\partial x} + v\frac{\partial u}{\partial y} + w\frac{\partial u}{\partial z} + \frac{\partial u}{\partial t}$$

$$a_y = u\frac{\partial v}{\partial x} + v\frac{\partial v}{\partial y} + w\frac{\partial v}{\partial z} + \frac{\partial v}{\partial t}$$

$$a_z = u\frac{\partial w}{\partial x} + v\frac{\partial w}{\partial y} + w\frac{\partial w}{\partial z} + \frac{\partial w}{\partial t}$$

(ix) **Local acceleration** is defined as the rate of change of velocity at a given point. In the above components of acceleration the expressions $\dfrac{\partial u}{\partial t}, \dfrac{\partial v}{\partial x}$ and $\dfrac{\partial w}{\partial t}$ are called local acceleration.

(x) **Convective acceleration** is defined as the rate of change of velocity due to change of position of fluid particles in a fluid flow.

(xi) **Velocity potential function** (ϕ) is defined as the scalar function of space and time such that its negative derivative with respect to any direction gives the fluid velocity in that direction. Hence the components of velocity in x, y and z direction in terms of velocity potential are

$$u = -\frac{\partial \phi}{\partial x},\ v = -\frac{\partial \phi}{\partial y}\ \text{and}\ w = -\frac{\partial \phi}{\partial z}.$$

(xii) **Stream function** (ψ) is defined as the scalar function of space and time, such that its partial derivative with respect to any direction gives the velocity component at right angles to that direction. It is defined only for two-dimensional flow. The velocity components in x and y directions in terms of stream function are

$$u = -\frac{\partial \psi}{\partial y}\ \text{and}\ v = \frac{\partial \psi}{\partial x}.$$

(xiii) **Equipotential line** is a line along which the velocity potential (ϕ) is constant. A grid obtained by drawing a series of equipotential lines and stream lines is called a flow net.

(*xiv*) **Angular deformation** or shear deformation is defined as the average change in the angle contained by two adjacent sides. It is also called shear strain rate and is given by

$$\text{Shear strain rate} = \frac{1}{2}\left[\frac{\partial v}{\partial x} + \frac{\partial u}{\partial y}\right]$$

Rotational components of a fluid particle are given as

$$\omega_z = \frac{1}{2}\left[\frac{\partial v}{\partial x} - \frac{\partial u}{\partial y}\right], \quad \omega_x = \frac{1}{2}\left[\frac{\partial w}{\partial y} - \frac{\partial v}{\partial z}\right], \quad \omega_y = \frac{1}{2}\left[\frac{\partial u}{\partial z} - \frac{\partial w}{\partial x}\right]$$

Vorticity is equal to two times the value of rotation.

(*xv*) **Vortex flow** is defined as the flow of a fluid along a curved path. It is of two types namely (*i*) Forced vortex flow and (*ii*) Free vortex flow. If the fluid particles are moving round a curved path with the help of some external torque the flow is called **forced vortex flow**. And if no external torque is acquired to rotate the fluid particles, the flow is called **free-vortex flow**. The relation between tangential velocity and radius for vortex flow is given by

$$r = \omega \times r \qquad \text{for forced vortex}$$
$$v \times r = \text{constant} \qquad \text{for free vortex.}$$

The pressure variation along the radial direction for vortex flow along a horizontal plane,

$$\frac{\partial p}{\partial r} = \frac{\rho v^2}{r}$$

For forced vortex flow, $\quad z = \dfrac{v^2}{2g}$

For free vortex flow the equation is $\dfrac{p_1}{\rho g} + \dfrac{v_1^2}{2g} + z_1 = \dfrac{p_2}{\rho g} + \dfrac{v_2^2}{2g} + z_2$.

DYNAMICS OF FLUID

Dynamics of fluid flow is defined that branch of science which deals with the study of fluids in motion considering the forces which cause the flow.

(*i*) **Euler's equation** of motion is obtained by considering forces due to pressure and gravity. **Navier-Strokes equations** are obtained by considering pressure force, gravity force and viscous force. **Reynold's equation of motion** are obtained by considering pressure force, gravity force, viscous force and force due to turbulence.

(*ii*) **Bernoulli's equation** is obtained by integrating the Euler's equation of motion. It states that, "For a steady, ideal flow of an incompressible fluid, the total energy which consists of pressure energy $\left(\dfrac{p}{\rho g}\right)$, kinetic energy $\left(\dfrac{v^2}{2g}\right)$ and datum energy (z) at any point of the fluid is constant."

Mathematically, it is written as

$$\frac{p}{\rho g} + \frac{v^2}{2g} + z = \text{constant}$$

or
$$\frac{p_1}{\rho g} + \frac{v_1^2}{2g} + z_1 = \frac{p_2}{\rho g} + \frac{v_2^2}{2g} + z_2$$

Bernoulli's equation for **real fluids** is written as

$$\frac{p_1}{\rho g} + \frac{v_1^2}{2g} + z_1 = \frac{p_2}{\rho g} + \frac{v_2^2}{2g} + z_2 + h_L$$

where h_L = Loss of energy between section 1 and 2.

(iii) **Applications of Bernoulli's equation** are:

(i) Venturimeter, (ii) Orificemeter, and (iii) Pitot-tube.

The discharge through a venturimeter is given by

$$Q = C_d \frac{A_1 A_2}{\sqrt{A_1^2 - A_2^2}} \times \sqrt{2gh}$$

where h = Difference of pressure head in terms of fluid head flowing through venturimeter for a horizontal venturimeter

C_d = Co-efficient of venturimeter

A_1 = Area at the inlet of venturimeter

A_2 = Area at the throat of the venturimeter.

The value of 'h' is given by the differential U-tube manometer. For a horizontal venturimeter or inclined venturimeter

$$h = x \left[\frac{S_h}{S_f} - 1\right] \text{ for manometer with heavier liquid}$$

$$= x \left[1 - \frac{S_l}{S_f}\right] \text{ for manometer with lighter liquid}$$

where x = Difference in the readings of the differential manometer

S_h = Specific gravity of heavier liquid in manometer

S_f = Specific gravity of liquid flowing through venturimeter

S_l = Specific gravity of lighter liquid in manometer.

(iv) **Pitot tube** is used to find the velocity of a flowing fluid at any point in a pipe or a channel. The velocity is given by the relation,

$$V = C_v \sqrt{2gh}$$

where C_v = Co-efficient of pitot-tube,

and h = Difference of the pressure head.

(v) **Momentum equation** states that the net force acting on a fluid mass is equal to the change in momentum per second (or rate of change of momentum) in that direction. Mathematically, it is written as

$$F = \frac{d}{dt}(mv)$$

where mv = Momentum.

(vi) **The impulse-momentum equation** is given by

$$F \times dt = d(mv)$$

and it states that the impulse of a force (F) acting on a fluid mass (m) in a short interval of time (dt) is equal to the change of momentum $d(mv)$ in the direction of force.

(vii) **Force on a bend,** exerted by a flowing fluid in the direction of x and y, are given by

$$F_x = \rho Q[v_{1x} - v_{2x}] + (p_1 A_1)_x + (p_2 A_2)_x$$
$$F_y = \rho Q[v_{1y} - v_{2y}] + (p_1 A_1)_y + (p_2 A_2)_y$$

where
v_{1x} = Initial velocity in the x-direction
v_{2x} = Final velocity in the x-direction
$(p_1 A_1)_x$ = Initial pressure force in x-direction
$(p_2 A_2)_x$ = Final pressure force in x-direction and so on

Resultant force on the bend is $F_R = \sqrt{F_x^2 + F_y^2}$.

ORIFICE AND MOUTHPIECE

Orifice is a small opening of any cross-section on the side or at the bottom of a tank, through which a fluid is flowing. **A mouthpiece** is a short length of a pipe which is two or three times its diameter in length, fitted in a tank or vessel containing the fluid.

Hydraulic Co-efficients

(i) There are three *hydraulic co-efficients* namely,

(a) Co-efficient of velocity, C_v (b) Co-efficient of contraction, C_c

(c) Co-efficient of discharge, C_d.

(ii) The expression for co-efficient of velocity in terms of x, y co-ordinates from vena-contracta is

$$C_v = \frac{x}{\sqrt{4yH}}$$

where H = Height of water from the centre of orifice.

(iii) Co-efficient of discharge for different types of mouthpieces are

(a) $C_d = 0.855$ ...For external mouthpiece
(b) $C_d = 0.707$ ...For internal mouthpiece running full
(c) $C_d = 0.50$ ...For internal mouthpiece running free
(d) $C_d = 1.0$ ...For convergent or convergent divergent.

NOTCH AND WEIR

Notch is a device used for measuring the rate of flow of a liquid through a small channel. A **weir** is a concrete or masonry structure placed in the open channel over which the flow occurs.

(*i*) The **discharge** through the following notches or weirs is given by

$$Q = \tfrac{2}{3} C_d \times L \times H^{3/2} \qquad \text{...For rectangular notch or weir}$$

$$= \tfrac{8}{15} C_d \times \tan\tfrac{\theta}{2} \times \sqrt{2g} \times H^{5/2} \qquad \text{...For a triangle notch or weir.}$$

(*ii*) **The discharge** through a trapezoidal notch or weir is equal to the sum of discharge through a rectangular notch and the discharge through a triangular notch.

(*iii*) **The error** in discharge due to error in measurement of head over a notch is given by

$$\frac{dQ}{Q} = \frac{3}{2}\frac{dH}{H} \qquad \text{...For a rectangular notch}$$

$$= \frac{5}{2}\frac{dH}{H} \qquad \text{...For a triangular notch}$$

where Q = Discharge through notch and H = head over the notch.

An error of 1% in measuring H will produce 1.5% error in discharge over a rectangular notch and 2.5% error in discharge over a triangular notch.

(*iv*) **Velocity of approach** (V_a) is defined as the velocity with which the water approaches the notch or weir. This is given by

$$V_a = \frac{\text{Discharge over the notch}}{\text{Cross-section area of channel}}$$

The head due to velocity of approach is given by $h_a = \dfrac{V_a^2}{2g}$

Discharge over a rectangular weir, considering velocity of approach is given by

$$Q = \tfrac{2}{3} C_d L \sqrt{2g}\,[(H_1 + h_a)^{3/2} - h_a^{3/2}]$$

VISCOUS FLOW

Viscous flow is the flow for which Reynold number is less than 2000 or the fluid flows in layers.

(*i*) For the viscous flow through **circular pipes,** the shear stress distribution, velocity distribution, ratio of maximum velocity to average velocity and difference of pressure head are given by

(*a*) $\tau = -\dfrac{\partial p}{\partial x}\dfrac{r}{2}$ ...(Linear)

(*b*) $u = -\dfrac{1}{4\mu}\dfrac{\partial p}{\partial x}[R^2 - r^2] = U_{max}\left[1 - \left(\dfrac{r}{R}\right)^2\right]$ ...(Parabolic)

(c) $\dfrac{U_{max}}{\bar{u}} = 2.0$

(d) $h_f = \dfrac{32\mu \bar{u} L}{\rho g D^2}$

where $\dfrac{\partial p}{\partial x}$ = Pressure gradient, τ = Shear stress

r = Radius at any point, R = Radius of the pipe

U_{max} = Maximum velocity, h_f = Loss of pressure head

$\bar{u}$ = Average velocity, D = Diameter of pipe.

(ii) For viscous flow between **two parallel plates,** the shear stress distribution, velocity distribution, ratio of maximum velocity to average velocity and difference of pressure head are given by

(a) $\tau = -\dfrac{1}{2}\dfrac{\partial p}{\partial x}[t - 2y]$ ...(Linear)

(b) $u = -\dfrac{1}{2\mu}\dfrac{\partial p}{\partial x}[ty - y^2]$ ...(Parabolic)

(c) $\dfrac{U_{max}}{\bar{u}} = 1.5$

(d) $h_f = \dfrac{12\mu \bar{u} L}{\rho g t^2}$.

where t = Distance between two plates,

y = Distance from the plates.

(iii) **Kinetic energy correction factor** (α) is defined as the ratio of kinetic energy per second based on actual velocity to kinetic energy per second based on average velocity. For a circular pipe through which viscous flow is taking place, $\alpha = 2.0$.

(iv) **Momentum correction factor** (β) is defined as the ratio of momentum of a fluid based on actual velocity to the momentum of the fluid based on average velocity. For a circular pipe, having viscous flow, $\beta = \dfrac{4}{3}$.

(v) The **co-efficient of friction** (f) which is a function of Reynold number is given by

$f = \dfrac{16}{R_e}$ ...for viscous flow or for $R_e < 2000$

$= \dfrac{0.079}{R_e^{1/4}}$ ...for R_e varying from 4000 to 10^5

$= 0.0008 + \dfrac{0.05525}{R_e^{0.237}}$ ...for $R_e \geq 10^5$ but $\leq 4 \times 10^7$.

TURBULENT FLOW

(i) **Smooth and rough boundaries.** If the average height (k) of the irregularities projecting from the surface of the boundary is small compared with the thickness of the laminar sub-layer (δ'), the boundary is known as **smooth**. But if k is large in comparison to δ', the boundary is known as **rough**. Mathematically,

$$\frac{k}{\delta'} < 0.25 \qquad \text{...for smooth boundary}$$
$$> 6.0 \qquad \text{...for rough boundary.}$$

And if $\frac{k}{\delta'}$ lies between 0.25 to 6.0, the boundary is in transition.

Darcy formula is given by $h_f = \dfrac{4fLV^2}{d \times 2g}$

where h_f = Head loss due to friction and is known as **Major,** Head loss.

Chezy's formula is given by $\quad V = C\sqrt{m \times i}$

where C = Chezy's constant,

m = Hydraulic mean depth = $\dfrac{d}{4}$ for circular pipe (running full),

i = Loss of head per unit length = $\dfrac{h_f}{L}$.

FLOW THROUGH PIPES

(*i*) **Minor losses**

(*a*) Loss of head due to *sudden expansion* (h_e) is given by $h_e = \dfrac{(V_1 - V_2)^2}{2g}$.

(*b*) Loss of head due to *sudden contraction* (h_c) is given by

$$h_c = \left(\frac{1}{C_c} - 1\right)^2 \frac{V_2^2}{2g}, \qquad \text{where } C_c = \text{Co-efficient of contraction}$$

(*c*) Loss of head at the inlet of a pipe, $h_i = 0.5 \dfrac{V^2}{2g}$.

(*d*) Loss of head at the outlet of a pipe, $h_o = \dfrac{V^2}{2g}$.

(*ii*) **Hydraulic gradient and total energy lines.** The line representing the sum of pressure head and datum head with respect to some reference line is called hydraulic gradient line (H.G.L.) while the line representing the sum of pressure head, datum head and velocity head with respect to some reference line is known as total energy line (T.E.L.).

(*iii*) The **equivalent size** of the pipes connected in series is given by

$$\frac{L}{d^5} = \frac{L_1}{d_1^5} + \frac{L_2}{d_2^5} + \frac{L_3}{d_3^5} + \ldots\ldots$$

where $\quad L$ = Equivalent length of pipe = $L_1 + L_2 + L_3$

d = Equivalent size of the pipe

d_1, d_2, d_3 = Diameters of pipes connected in series.

(iv) For **parallel pipes,** the loss of head in each pipe is same and rate of flow in main pipe is equal to the sum of the rate of flow in each pipe, connected in parallel i.e.,

(a) $Q = Q_1 + Q_2 + \ldots$ (b) $h_{f_1} = h_{f_2} = \ldots$

(v) **Power transmitted** through a pipe is given by,

$$\text{H.P.} = \frac{w \times Q \times [H - h_f]}{75}$$

where H = Total head at the inlet of pipe

h_f = Head lost due to friction

Efficiency of power transmission through pipes, $\eta = \dfrac{[H - h_f]}{H}$

This efficiency will be maximum when $h_f = \dfrac{H}{3}$.

Diameter of **nozzle** for maximum power transmission through nozzle is

$$d = \left(\frac{D^5}{8fL}\right)^{1/4}$$

where d = Diameter of nozzle at outlet, D = Diameter of pipe

L = Length of pipe, f = Co-efficient of friction.

(vi) **Water hammer.** When a liquid is flowing through a long pipe fitted with a valve at the end of the pipe and the valve is closed suddenly, a pressure wave of high intensity is produced behind the valve. This pressure wave of high intensity is having the effect of hammering action of the walls of the pipe. This phenomenon is known as water hammer. The intensity of pressure rise (p_i) due to water hammer is given by

$$p_i = \frac{\rho L V}{t} \quad \text{...If valve is closed gradually}$$

$$= V\sqrt{K\rho} \quad \text{...If valve is closed suddenly}$$

$$= V\sqrt{\frac{\rho}{\frac{1}{K} + \frac{D}{Et}}} \quad \text{...If valve is closed suddenly and pipe is elastic.}$$

where L = Length of pipe, V = Velocity of flow

K = Bulk modulus of fluid, D = Diameter of pipe

t = Time for closing valve

E = Modulus of elasticity for pipe material

The valve of closure is said to be gradual if $t > \dfrac{2L}{C}$

The valve closure is said to be sudden if $t < \dfrac{2L}{C}$

where C = Velocity of pressure wave produced due to water hammer = $\sqrt{\dfrac{K}{\rho}}$.

DIMENSIONAL AND MODEL ANALYSIS

(a) **Hydraulic Similarities.** There are three types of similarities that must exist between the model and prototype. They are: (i) Geometric similarity, (ii) Kinematic similarity, and (iii) Dynamic similarity.

Geometric similarity means the similarity of all linear dimensions of model and prototype. **Kinematic** similarity means the similarity of motion between model and prototype. **Dynamic similarity** means the similarity of forces between the model and prototype.

(b) **Dimensionless Parameters.** They are five dimensionless parameters, namely: (i) Reynold's number, (ii) Froude number, (iii) Euler number, (iv) Weber number and (v) Mach number.

Reynold's number is the ratio of inertia force to viscous force and is given by

$$R_e = \frac{\rho V D}{\mu} \text{ or } \frac{V \times D}{\nu} \text{ for pipe flow.}$$

Froude number is the ratio of square root of the inertia force to gravity force and is given by

$$F_e = \sqrt{\frac{F_i}{F_g}} = \frac{V}{\sqrt{Lg}}.$$

Euler's number is the ratio of square root of inertia force to pressure force and is given by

$$E = \sqrt{\frac{F_e}{F_p}} = \frac{V}{\sqrt{p/\rho}}$$

Weber number is the ratio of square root of inertia force to surface tension force and is given by

$$W = \sqrt{\frac{F_i}{F_s}} = \frac{W}{\sqrt{\sigma/L\rho}}$$

Mach number (M) is the ratio of square root of inertia force to elastic force and is given by

$$M = \sqrt{\frac{F_i}{F_e}} = \frac{V}{\sqrt{K/\rho}} = \frac{V}{C}.$$

where C = Velocity of sound wave in air.

(c) **Models** are of two types namely (i) Undistorted and (ii) Distorted Model. If the models are geometrically similar to its proto-type the models are known as **undistorted model.** And if the models are having different scale ratios for horizontal and vertical dimensions, the models are known as **distorted model.**

BOUNDARY LAYERS

(i) **Boundary layer.** When a solid body is immersed in a flowing fluid, there is a narrow region to the fluid in the neighbourhood of the solid body, where the velocity of the fluid varies from zero to free-stream velocity. This narrow region of fluid is called boundary layer.

(ii) Boundary layer may be laminar or turbulent. If the Reynold number of the flow defined as

$$R_e = \frac{U \times x}{\nu}$$

is less than 5×10^5, the boundary layer is called laminary boundary layer. And if R_e is more than 5×10^5, the boundary layer is called turbulent boundary layer.

In the above expression, U = Free stream velocity,
x = Distance from leading edge,
ν = Kinematic viscosity.

(iii) **Displacement thickness (δ^*)** is given by $\delta^* = \int_0^\delta \left(1 - \frac{u}{U}\right) dy$

(iv) **Momentum thickness (θ)** is given by, $\theta = \int_0^\delta \frac{u}{U}\left(1 - \frac{u}{U}\right) dy$

(v) **Energy thickness** $= \int_0^\delta \frac{u}{U}\left[1 - \frac{u^2}{U^2}\right] dy$

(vi) **Von Karman momentum** integral equation is given

$$\frac{\tau_0}{\rho U^2} = \frac{\partial \theta}{\partial x}, \text{ where } \theta = \int_0^\delta \frac{u}{U}\left(1 - \frac{u}{U}\right) dy$$

(vii) Velocity profile for turbulent boundary is given by $\frac{u}{U} = \left(\frac{y}{\delta}\right)^{1/7}$

where u = Velocity within boundary layer
U = Free stream velocity
δ = Boundary layer thickness.

DRAG AND LIFT FORCES

(i) **Drag and Lift Forces.** The force, exerted by a flowing fluid on a solid body in the direction of motion, is called drag force while the force perpendicular to the direction of motion is called lift force. Mathematically, they are given as

$$F_D = C_D A \frac{\rho U^2}{2}, \quad F_L = C_L A \frac{\rho U^2}{2}$$

where F_D = Drag force, $\quad F_L$ = Lift force
A = Projected area of body, $\quad U$ = Free-stream velocity
C_D = Co-efficient of drag, $\quad C_L$ = Co-efficient of lift.

(ii) The drag force on a sphere for $R_e \angle 0.2$ is given by

$$F_D = 3\pi\mu DU.$$

COMPRESSIBLE FLOW

(i) Bernoulli's equation for **compressible fluid** is given by

$$\frac{p}{\rho g}\log_e p + \frac{V^2}{2g} + Z = \text{constant for iso-thermal process}$$

$$\frac{\gamma}{\gamma-1}\frac{p}{\rho} + \frac{V^2}{2g} + Z = \text{constant for adiabatic process.}$$

(ii) **Velocity of sound wave** is given by

$$C = \sqrt{\frac{dp}{d\rho}} = \sqrt{\frac{k}{\rho}} \quad \text{...In terms of Bulk Modulus } k$$

$$= \sqrt{\frac{p}{\rho}} = \sqrt{RT} \quad \text{...For isothermal process}$$

$$= \sqrt{\frac{\gamma p}{\rho}} = \sqrt{\gamma RT} \quad \text{...For adiabatic process}$$

(iii) If $M < 1$ flow is called **sub-sonic flow**

$M > 1$ flow is called **super-sonic flow**

$M = 1$ flow is called **sonic flow.**

In sub-sonic flow, the disturbance always moves ahead of the projectile. In sonic flow, the disturbance moves along the projectile while in super-sonic flow that disturbance lags behind the projectile.

(iv) **The Mach angle** (α) is given by $\sin\alpha = \frac{C}{V} = \frac{1}{M}$

(v) **Area velocity relationship** for compressible fluid is given as $\frac{dA}{A} = \frac{dV}{V}[M^2 - 1]$

For $M < 1$, the velocity decreases with increase of area.

For $M > 1$, the velocity decreases with the decrease of area.

(vi) For maximum flow through an orifice of nozzle fitted to the tank the pressure ratio $\frac{p_2}{p_1}$ $= n = 0.528$.

(vii) The **compressibility correction factor** is given by

$$\text{C.C.F.} = \left[1 + \frac{M_1^2}{4} + \frac{2-\gamma}{4}M_1^4 + \ldots\ldots\right].$$

CHANNEL FLOW

(*i*) **Channel flow.** Channel flow means the flow of a liquid through a passage at atmospheric pressure. The flow of liquid in a channel takes place under the force of gravity which means the flow takes place due to the slope of the bed of the channel. The flow in a channel is classified as:

1. Steady and unsteady flow
2. Uniform and non-uniform flow
3. Laminar and turbulent flow
4. Sub-critical, critical and super-critical flow.

For the *steady flow*, the velocity at a point in a channel with respect to time should be constant. And for *unsteady flow*, the velocity at a point with respect to time should be variable.

The flow in a channel will be *uniform* if the velocity with respect to direction of flow is constant. The velocity will be constant if area of flow is constant. Area of flow will be constant if depth of flow is constant. Hence the flow in a channel will be uniform if the depth of flow is constant. But if the depth of flow is variable then the flow in a channel is known as *non-uniform*.

The non-uniform flow is divided into gradually varied flow (G.V.F.) and rapidly varied flow (R.V.F.). If the depth of flow varies gradually, then the non-uniform flow is known as gradually varied flow. And if the depth of flow varies rapidly, then non-uniform flow is called rapidly varied flow.

The flow in a channel is said to be *laminar* if the Reynolds number (R_e) is less than 500 or 600. Reynolds number is case of open channel flow is given by

$$R_e = \frac{\rho V R}{\mu}$$

where ρ = Density of liquid, V = Velocity of liquid, μ = Viscosity

R = Hydraulic radius or hydraulic mean depth

$$= \frac{A}{P} = \frac{\text{Area of flow}}{\text{Wetted Perimeter}}.$$

If the Reynolds number is more than 2000, then the flow is open channel is known as *turbulent*.

(*ii*) The flow in open channel is said to be **sub-critical** if the Froude Number (F_e) is less than 1.0.

The Froude number is given by, $F_e = \dfrac{V}{\sqrt{gD}}$

where V = Velocity of flow

D = Hydraulic depth of channel

$$= \frac{A}{T}$$

where A = Wetted area, T = Top width of the channel.

(*iii*) The flow in a channel is known as **critical flow** if the Froude number is equal to 1.0. But if the Froude number is more than 1.0, then flow in the channel is known as **super-critical flow**.

(*iv*) **The velocity through a channel** is given by

(*a*) Chezy's Formula

(*b*) Manning's Formula.

Chezy's Formula

The velocity by Chezy's formula is given by, $V = C\sqrt{mi}$

where C = Chezy's constant and depends upon the surface of the channel

m = Hydraulic mean depth

$$= \frac{A}{P} = \frac{\text{Area of Flow}}{\text{Wetted Perimeter}}$$

i = Slope of the bed of the channel.

Manning's Formula

The velocity through a channel according to Manning's Formula is given by, $V = \dfrac{1}{N} m^{2/3} i^{1/2}$

where N = Manning's constant and depends upon the surface of the channel

m = Hydraulic mean depth

i = Slope of the bed of the channel.

(*v*) **The relation between Chezy's constant** (C) and Manning's constant (N) is given by

$$C = \frac{1}{N} m^{1/6}.$$

The Chezy's constant is not a dimensionless co-efficients. The dimension of C is given by $L^{1/2} T^{-1}$.

(*vi*) **Different Types of Channels**

The different types of channels are

1. Rectangular channel
2. Trapezoidal channel
3. Circular channel
4. Triangular channel.

(*vii*) **Values of A (area of flow) and P (Wetted perimeter) for different channels.**

1. *Rectangular channel* [See Fig. 1.1 (*a*)]

Let b = Width of channel, d = Depth of flow

then $A = b \times d$

$P = b + 2d$

2. *Trapezoidal channel* [See Fig. 1.1 (b)]

Let b = Width of flow, d = Depth of flow

n = Side slope which is expressed as 1 vertical to n horizontal

then $A = (b + nd) \times d$

$P = b + 2d\sqrt{n^2 + 1}$

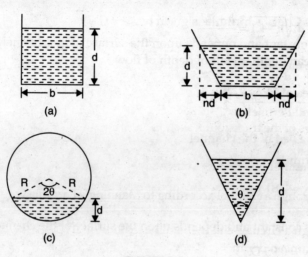

FIGURE 1.1

3. *Circular channel* [Refer Fig. 1.1 (c)]

Let R = Radius of the channel, d = Depth of water in circular channel

2θ = Angle subtended at the centre of the channel by the free surface of water

then $A = R^2 \left(\theta - \dfrac{\sin 2\theta}{2} \right)$

and $P = 2R\theta$

4. *Triangular channel* [Refer Fig. 1.1 (d)]

Let d = depth of flow, θ = Angle of triangular channel

then $A = d^2 \tan\left(\dfrac{\theta}{2}\right)$

and $P = 2d \sec\left(\dfrac{\theta}{2}\right)$

(*viii*) **Most Efficient Section of a Channel.** The section of a channel is said to be most efficient if the discharge through the channel is maximum for a given area of flow, given surface resistance and given slope of the bed of the channel. The most efficient section is also known as most economical section of the channel.

(a) **A rectangular channel** will be most efficient, if

1. Depth of flow = Half of width of channel i.e., $d = \frac{1}{2} \times b$

2. Hydraulic mean depth = Half of depth of flow i.e., $m = \frac{1}{2} \times d$.

(b) **A trapezoidal channel** will be most efficient if

1. Length of sloping side = Half of top width

i.e., $$d\sqrt{n^2 + 1} = \frac{b + 2d}{2}.$$

2. Hydraulic mean depth = Half of depth of flow

i.e., $$m = \frac{d}{2}$$

3. The three sides of the trapezoidal section are the tangential to the semi-circle described on the water-line.

(c) **A circular channel** will be most efficient for maximum velocity, if

1. Depth of flow = 0.81 times the dia. of the circular channel.

2. Hydraulic mean depth = 0.3 times the dia. of the channel.

The flow through a circular channel will be maximum if

1. Depth of flow = 0.95 D.

(ix) **Specific Energy.** The total energy of a flowing liquid in a channel with respect to the bed of channel is known as specific energy. It is given by

$$E = y + \frac{V^2}{2g}$$

where y = Depth of flow, V = Velocity of flow.

The curve which represents the variation of specific energy with the depth of flow is known as specific energy curve.

(x) **Critical Depth.** The depth of flow of water at which the specific energy is minimum, is known as critical depth. For a rectangular channel, the critical depth (y_c) is given by

$$y_c = \left(\frac{q^2}{g}\right)^{1/3}$$

where q = Rate of flow per unit width of channel.

The velocity of flow, corresponding to critical depth is known as critical velocity.

(xi) **Hydraulic Jump.** The sudden rise of water level which takes place due to the transformation of super-critical flow to the subcritical flow, is known as hydraulic jump. And when hydraulic jump takes place, a loss of energy due to eddy formation and turbulence occurs. The depth of water after hydraulic jump is given by

$$y_2 = -\frac{y_1}{2} + \sqrt{\left(\frac{y_1}{2}\right)^2 + \frac{2q^2}{gy_1}}$$

where y_2 = Depth of water after hydraulic jump
y_1 = Depth of water before hydraulic jump
q = Discharge per unit width.

The loss of energy (h_L) during hydraulic jump is given by, $h_L = \dfrac{(y_2 - y_1)^3}{4y_1 y_2}$.

HYDRAULIC TURBINES

(i) **Hydraulic Turbines.** The hydraulic machines which convert the hydraulic energy into the mechanical energy, are called turbines.

(ii) **The force exerted by a jet of water** on a stationary plate in the direction of jet is given by

$F_x = \rho A V^2$ ...For a vertical plate
$\quad = \rho A V^2 \sin^2 \theta$ ...For an inclined plate
$\quad = \rho A V^2 [1 + \cos \theta]$ ...For a curved plate

(iii) For force exerted by the jet of water having velocity V on a plate moving with a velocity u is given by

$F_x = \rho A (V - u)^2$ ...For a vertical plate
$\quad = \rho A (V - u)^2 \sin^2 \theta$ ...For an inclined plate
$\quad = \rho A (V - u)^2 [1 + \cos \theta]$ ...For a curved plate

(iv) **Efficiency of a series** of vanes is given as $\eta = \dfrac{2u(V - u)}{V^2}$

Efficiency will be maximum, when $u = \dfrac{V}{2}$. Maximum efficiency = 50%.

(v) **For a series of curved** radial vanes, the work done per unit weight per second

$$= \frac{1}{g}[V_{w_1} u_1 \pm V_{w_2} u_2]$$

where V_{w_1}, V_{w_2} = Velocity of whirl at inlet and outlet.

(vi) **The net head** on the turbine is given by $H = H_g - h_f$
where H = Net head, H_g = Gross head, h_f = Head loss due to friction.

(vii) **The efficiencies** of a turbine are: (a) Hydraulic efficiency (η_h), (b) Mechanical efficiency (η_m) and (c) Overall efficiency (η_0).

(a) Hydraulic efficiency (η_h) is given by

$$\eta_h = \frac{\text{Power given by water of the number}}{\text{Power supplied at inlet}} = \frac{\text{R.P.}}{\text{W.P.}}$$

$$= \frac{W}{g} \frac{[V_{w_1}u_1 \pm V_{w_2}u_2]}{75} \bigg/ \left(\frac{W \times H}{75}\right) = \frac{(V_{w_1}u_1 \pm V_{w_2}u_2)}{gH}$$

(b) Mechanical efficiency (η_m) is given by $\eta_n = \dfrac{\text{S.P.}}{\text{R.P.}}$

(c) Overall efficiency is given by $\eta_0 = \dfrac{\text{S.P.}}{\text{W.P.}} = \eta_n \times \eta_h$.

(viii) **Impulse Turbine.** If at the inlet of a turbine, total energy is only kinetic energy, the turbine is called impulse turbine. Pelton wheel is an impulse turbine.

(ix) **Reaction Turbine.** If at the inlet of a turbine, the total energy is kinetic energy as well as pressure energy, the turbine is called reaction turbine. Francis and Kaplan turbines are reaction turbines.

(x) **Jet Ratio (m)** is defined as the ratio of diameter (D) of Pelton wheel to the diameter (d) of the jet or

$$\text{Jet ratio, } m = \frac{D}{d}$$

(xi) Pelton wheel is a tangential flow impulse turbine, Francis is an inward flow reaction turbine and Kaplan is an axial flow reaction turbine. The rate of flow (Q) through the turbine is given by

$$Q = \frac{\pi}{4} d^2 \times \sqrt{2gH} \qquad \text{...For Pelton Turbine}$$

$$= \pi D_1 B_1 V_{f1} \qquad \text{...For Francis Turbine}$$

$$= \frac{\pi}{4} [D_1^2 - D_0^2] \times V_{f1} \qquad \text{...For Kaplan Turbine}$$

where H = Net head, $\quad V_{f1}$ = Velocity of flow at inlet

D_0 = Dia. of Kaplan Turbine, D_b = Hub diameter.

(xii) **Draft-tube.** It is a pipe of gradually increasing area used for discharging water from the exit of a reaction turbine.

(xiii) **Specific Speed** of a turbine is defined as the speed at which a turbine runs when it is working under a unit head and develops unit (i.e. 1 kW power). It is given by $N_s = \dfrac{N\sqrt{P}}{H^{5/4}}$, where P = Shaft power, H = Net head on turbine.

(a) **Unit speed (N_u)** is the speed of a turbine, when the head on the turbine is one metre. It is given by $N_u = \dfrac{N}{\sqrt{H}}$.

(b) **Unit discharge (Q_u)** is the discharge through a turbine when the head (H) on the turbine is unity. It is given by $Q_u = \dfrac{Q}{\sqrt{H}}$.

(c) **Unit power (P_u)** is the power developed by a turbine when the head on the turbine is unity. It is given by $P_u = \dfrac{P}{H^{3/2}}$.

(xiv) **Characteristic Curves.** The following three are the important characteristic curves of a turbine:

1. Main characteristic curve, 2. Operating characteristic curves, and 3. Constant efficiency curves.

(xv) **Governing of a Turbine** is defined as the operation by which the speed of the turbine is kept constant under all conditions of working.

CENTRIFUGAL PUMPS

(i) The hydraulic machine which converts the mechanical energy into pressure energy by means of centrifugal force is called **centrifugal pump**. The work done by impeller on water per second per unit weight of water

$$= \frac{1}{g} V_{w_2} \times u_2$$

where V_{w_2} = Velocity of whirl at outlet, u_2 = Tangential velocity of wheel at outlet.

(ii) **The manometric head (H_m)** is the head against which a centrifugal pump has to work. It is given as

$$H_m = \frac{V_{w_2} \times u_2}{g} - \text{Loss of head in impeller and casing}$$

$$= \frac{V_{w_2} \times u_2}{g} \quad \text{If losses in pump are zero}$$

$$= \text{Total head at outlet} - \text{Total head at inlet}$$

$$= \left(\frac{p_0}{w} + \frac{V_0^2}{2g} + Z_0\right) - \left(\frac{p_i}{w} + \frac{V_i^2}{w} + Z_i\right)$$

$$= h_s + h_d + h_{fd} + \frac{V_d^2}{2g}$$

(iii) **The efficiencies of a pump** are: (a) Manometric efficiency (η_{man}), (b) Mechanical (η_m) and (c) overall efficiency (η_0). They are expressed as

$$\eta_{man} = \frac{gH_m}{V_{m_2} \times u_2}$$

$$\eta_m = \frac{P}{S.P.} = \left(\frac{W}{g} \times \frac{V_{w_2} \times u_2}{75}\right) \Big/ S.P.$$

$$\eta_0 = \eta_{man} \times \eta_m$$

where P = Power on impeller, S.P. = Shaft power.

(iv) **Multistage centrifugal** pumps are used to produce a high head or to discharge a large quantity of water. The produce a high head, the impellers are connected in series while to discharge a large quantity of liquid, the impellers are connected in parallel.

(v) **Specific speed of a pump** is defined as the speed at which a pump runs when the head developed is one metre and discharge is one cubic metre. It is given as $N_s = \dfrac{N\sqrt{Q}}{H_m^{3/4}}$, where H_m = Manometric head.

(vi) **Cavitation** is defined as the phenomenon of formation of vapour bubbles and sudden collapsing of the vapour bubbles.

RECIPROCATING PUMP

(i) The discharge through a reciprocating pump per second is given by

$$Q = \frac{ALN}{60} \qquad \text{...for a single acting}$$

$$= \frac{2ALN}{60} \qquad \text{...for a double acting}$$

where A = Area of position, L = Length of stroke.

(ii) **Work done by reciprocating** pump per second

$$= \frac{wALN}{60}(h_s + h_d) \qquad \text{...for a single acting}$$

$$= \frac{2wALN}{60}(h_s + h_d) \qquad \text{...for a double acting.}$$

(iii) **The pressure head due** to acceleration is given by

$$h_a = \frac{l}{g} \times \frac{A}{a} \times \omega^2 r \cos\theta$$

where l = Length of suction or delivery pipe
       a = Area of suction or delivery pipe
       r = Radius of crank = L/2
       ω = Angular speed = $\dfrac{2\pi N}{60}$.

(*iv*) **Indicator diagram** is a graph between the pressure head in the cylinder and the distance travelled by the piston from inner dead centre for one complete revolution of the crank. Work done by the pump is proportional to the area of indicator diagram. Area of ideal indicator diagram is the same as the area of indicator diagram due to acceleration in suction and delivery pipes.

(*v*) **Air vessel** is a device used: (*a*) to obtain a continuous supply of water at uniform rate, (*b*) to save a considerable amount of work and (*c*) to run the reciprocating pump at a high speed without separation.

The mean velocity ($\overline{V}$) for a single acting pump is $\overline{V} = \dfrac{A}{a} \dfrac{\omega r}{\pi}$.

The work saved by fitting air vessels in a single acting reciprocating pump is 84.8% while in a double acting, the work saved is 39.2%.

MISCELLANEOUS HYDRAULIC DEVICES

The miscellaneous hydraulic devices are hydraulic press, hydraulic accumulator, hydraulic intensifier, hydraulic ram, hydraulic lift, hydraulic cranes, hydraulic coupling and hydraulic torque converter.

(*i*) **Hydraulic press** is a device used for lifting heavy weights by the application of a much smaller force. **Hydraulic accumulator** is a device used for storing the energy of a fluid in the form of pressure energy. **Hydraulic intensifier** is a device used for increasing the pressure intensity of a liquid. **Hydraulic ram** is a pump which raises water without any external power (such as electricity) for its operation. **Hydraulic lift** is a device used for carrying persons or goods from one floor to another floor in a multi-storeyed building. **Hydraulic crane** is a device used for raising or transferring heavy weights. **Hydraulic coupling** is a device, in which power is transmitted from driving shaft to driven shaft without any change of torque while **torque convertor** is a device in which arrangement is provided for getting increased or decreased torque at the driven shaft.

(*ii*) **Capacity of a hydraulic accumulator** = $p \times A \times L$

where p = Liquid pressure supplied by pump

A = Area of sliding ram, and L = Stroke of the ram.

(*iii*) **Hydraulic ram** has two efficiencies namely D' Aubuisson's efficiency and Rankine efficiency. They are given by

$$\text{D' Aubuisson's, } \eta = \frac{wH}{Wh} \text{ and Rankine's, } \eta = \frac{w(H-h)}{(W-w) \times h}$$

where w = Weight of water raised/sec, W = Weight of water supplied raised/sec,

h = Height of water in supply tank, and H = Height of water raised.

II. OBJECTIVE TYPE QUESTIONS

Fluid Properties

1. An ideal fluid is defined as the fluid which
   (a) is compressible
   (b) is incompressible
   (c) is incompressible and non-viscous (inviscid)
   (d) has negligible surface tension.

2. Newton's law of viscosity states that
   (a) shear stress is directly proportional to the velocity
   (b) shear stress is directly proportional to velocity gradient
   (c) shear stress is directly proportional to shear strain
   (d) shear stress is directly proportional to the viscosity.

3. A Newtonian fluid is defined as the fluid which
   (a) is incompressible and non-viscous
   (b) obeys Newton's law of viscosity
   (c) is highly viscous
   (d) is compressible and non-viscous.

4. Kinematic viscosity is defined as equal to
   (a) dynamic viscosity × density
   (b) dynamic viscosity / density
   (c) dynamic viscosity × pressure
   (d) pressure × density.

5. Dynamic viscosity (μ) has the dimensions as
   (a) MLT^{-2}
   (b) $ML^{-1}T^{-1}$
   (c) $ML^{-1}T^{-2}$
   (d) $M^{-1}L^{-1}T^{-1}$.

6. Poise is the unit of
   (a) mass density
   (b) kinematic viscosity
   (c) viscosity
   (d) velocity gradient.

7. The increase of temperature
   (a) increases the viscosity of a liquid
   (b) decreases the viscosity of a liquid
   (c) decreases the viscosity of a gas
   (d) increases the viscosity of a gas.

8. Stoke is the unit of
   (a) surface tension
   (b) viscosity
   (c) kinematic viscosity
   (d) none of the above.

9. The multiplying factor for converting one poise into MKS unit of dynamic viscosity is
   (a) 9.81
   (b) 98.1
   (c) 981
   (d) 0.981.

10. Surface tension has the units of
    (a) force per unit area
    (b) force per unit length
    (c) force per unit volume
    (d) none of the above.
11. The gases are considered incompressible when mach number
    (a) is equal to 1.0
    (b) is equal to 0.50
    (c) is more than 0.3
    (d) is less than 0.2.
12. Kinematic viscosity (ν) is equal to
    (a) $\mu \times \rho$
    (b) $\dfrac{\mu}{\rho}$
    (c) $\dfrac{\rho}{\mu}$
    (d) none.
13. Compressibility is equal to
    (a) $\dfrac{-\left(\dfrac{dV}{V}\right)}{dp}$
    (b) $\dfrac{dp}{-\left(\dfrac{dV}{V}\right)}$
    (c) $\dfrac{dp}{d\rho}$
    (d) $\sqrt{\dfrac{dp}{d\rho}}$.
14. Hydrostatic law of pressure is given as
    (a) $\dfrac{\partial p}{\partial z} = \rho g$
    (b) $\dfrac{\partial p}{\partial z} = 0$
    (c) $\dfrac{\partial p}{\partial z} = z$
    (d) $\dfrac{\partial p}{\partial z} = $ constant.
15. Four curves are shown in Fig. 1.2 with velocity gradient $\left(\dfrac{\partial u}{\partial y}\right)$ along x-axis and viscous shear stress (τ) along y-axis. Curve A corresponds to
    (a) ideal fluid
    (b) newtonian fluid
    (c) non-newtonian fluid
    (d) ideal solid.
16. Curve B in Fig. 1.2 corresponds to
    (a) ideal fluid
    (b) newtonian fluid
    (c) non-newtonian fluid
    (d) ideal-solid.

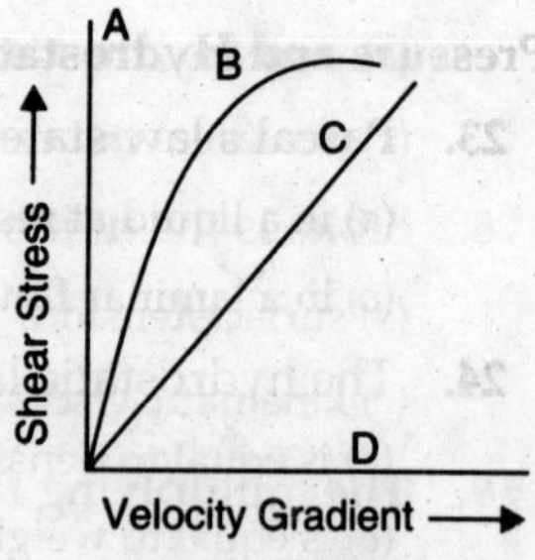

FIGURE 1.2

17. Curve C in Fig. 1.2 corresponds to
    (a) ideal fluid
    (b) newtonian fluid
    (c) non-newtonian fluid
    (d) ideal solid.
18. Curve D in Fig. 1.2 corresponds to
    (a) ideal fluid
    (b) newtonian fluid
    (c) non-newtonian fluid
    (d) ideal solid.
19. The relation between surface tension (σ) and difference of pressure (Δp) between the inside and outside of a liquid droplet is given as
    (a) $\Delta p = \dfrac{\sigma}{4d}$
    (b) $\Delta p = \dfrac{\sigma}{2d}$
    (c) $\Delta p = \dfrac{4\sigma}{d}$
    (d) $\Delta p = \dfrac{\sigma}{d}$.
20. For a soap bubble, the surface tension (σ) and difference of pressure (Δp) are related as
    (a) $\Delta p = \dfrac{\sigma}{4d}$
    (b) $\Delta p = \dfrac{\sigma}{2d}$
    (c) $\Delta p = \dfrac{4\sigma}{d}$
    (d) $\Delta p = \dfrac{8\sigma}{d}$.
21. For a liquid jet, the surface tension (σ) and difference of pressure (Δp) are related as
    (a) $\Delta p = \dfrac{\sigma}{4d}$
    (b) $\Delta p = \dfrac{\sigma}{2d}$
    (c) $\Delta p = \dfrac{4\sigma}{d}$
    (d) $\Delta p = \dfrac{2\sigma}{d}$.
22. The capillary rise or fall of a liquid is given by
    (a) $h = \dfrac{\sigma \cos\theta}{4\rho g d}$
    (b) $h = \dfrac{4\sigma \cos\theta}{\rho g d}$
    (c) $h = \dfrac{8\sigma \cos\theta}{\rho g d}$
    (d) none of the above.

Pressure and Hydrostatic Forces on Surfaces

23. Pascal's law states that pressure at a point is equal in all direction
    (a) in a liquid at rest
    (b) in a fluid at rest
    (c) in a laminar flow
    (d) in a turbulent flow.
24. The hydrostatic law states that rate of increase of pressure in a vertical direction
    (a) is equal to density of the fluid
    (b) is equal to specific weight of the fluid
    (c) is equal to weight of the fluid
    (d) none of the above.
25. Fluid statics deals with the following forces
    (a) viscous and gravity forces
    (b) viscous and gravity forces
    (c) gravity and pressure forces
    (d) surface tension and gravity forces.

26. Gauge pressure at a point is equal to
    (a) absolute pressure plus atmospheric pressure
    (b) absolute pressure minus atmospheric pressure
    (c) vacuum pressure plus absolute pressure
    (d) none of the above.

27. Atmospheric pressure head in terms of water column is
    (a) 7.5 m
    (b) 8.5 m
    (c) 9.81 m
    (d) 10.30 m.

28. The hydrostatic pressure on a plane surface is equal to
    (a) $\rho g A \bar{h}$
    (b) $\rho g A \bar{h} \sin^2 \theta$
    (c) $\frac{1}{4} \rho g A \bar{h}$
    (d) $\rho g A \bar{h} \sin \theta$

    where A = Area of plane surface and $\bar{h}$ = Depth of centroid of the plane area below the liquid free surface.

29. Centre of pressure of a plane surface immersed in a liquid is
    (a) above the centre of gravity of the plane surface
    (b) at the centre of gravity of the plane surface
    (c) below the centre of gravity of the plane surface
    (d) none of the above.

30. The resultant hydrostatic force acts through a point known as
    (a) centre of gravity
    (b) centre of buoyancy
    (c) centre of pressure
    (d) none of the above.

31. For submerged curved surface, the vertical component of the hydrostatic force is
    (a) mass of the liquid supported by the curved surface
    (b) weight of the liquid supported by the curved surface
    (c) the force of the projected area of the curved surface on vertical plane
    (d) none of the above.

32. Manometer is a device used for measuring
    (a) velocity at a point in a fluid
    (b) pressure at a point in a fluid
    (c) discharge of a fluid
    (d) none of the above.

33. Differential manometers are used for measuring
    (a) velocity at a point in a fluid
    (b) pressure at a point in a fluid
    (c) difference of pressure between two points
    (d) none of the above.

34. The pressure at a height Z in a static compressible fluid undergoing isothermal compression is given as

(a) $p = p_0 e^{-\frac{gR}{ZT}}$ □ (b) $p = p_0 e^{-\frac{gT}{RZ}}$ □

(c) $p = p_0 e^{-\frac{RT}{gZ}}$ □ (d) $p = p_0 e^{-\frac{gZ}{RT}}$ □

where p_0 = Pressure at ground level, R = Gas constant, T = Absolute temperature.

35. The pressure at a height Z in a static compressible fluid undergoing adiabatic compression is given by

(a) $p = p_0 \left[1 - \frac{\gamma-1}{\gamma} \frac{RT_0}{gZ}\right]^{\frac{\gamma}{\gamma-1}}$ □ (b) $p = p_0 \left[1 - \frac{\gamma}{\gamma-1} \frac{RT_0}{gZ}\right]^{\frac{\gamma}{\gamma-1}}$ □

(c) $p = p_0 \left[1 - \frac{\gamma-1}{\gamma} \frac{gZ}{RT_0}\right]^{\frac{\gamma}{\gamma-1}}$ □ (d) none of the above. □

36. The temperature at a height Z in a static compressible fluid undergoing adiabatic compression is given as

(a) $T = T_0 \left[1 - \frac{\gamma-1}{\gamma} \frac{RT_0}{gZ}\right]$ □ (b) $T = T_0 \left[1 - \frac{\gamma-1}{\gamma} \frac{gZ}{RT_0}\right]$ □

(c) $T = T_0 \left[1 - \frac{\gamma}{\gamma-1} \frac{RT_0}{gZ}\right]$ □ (d) none of the above. □

37. Temperature lapse-rate is given by

(a) $L = -\frac{R}{g}\left[\frac{\gamma-1}{\gamma}\right]$ □ (b) $L = -\frac{R}{g}\left[\frac{\gamma}{\gamma-1}\right]$ □

(c) $L = -\frac{g}{R}\left[\frac{\gamma-1}{\gamma}\right]$ □ (d) none of the above. □

38. When the fluid is at rest, the shear stress is
   (a) maximum □   (b) zero □
   (c) unpredictable □   (d) none of the above. □

39. The depth of centre of pressure of an inclined immersed surface from free surface of liquid is equal to

(a) $\frac{I_G}{A\bar{h}} + \bar{h}$ □ (b) $\frac{I_G A \sin^2\theta}{\bar{h}} + \bar{h}$ □

(c) $\frac{I_G \sin^2\theta}{A\bar{h}} + \bar{h}$ □ (d) $\frac{I_G \bar{h}}{A \sin^2\theta} + \bar{h}$. □

40. The depth of centre of pressure of a vertical immersed surface from free surface of liquid is equal to

(a) $\frac{I_G}{A\bar{h}} + \bar{h}$ □ (b) $\frac{I_G A}{\bar{h}} + \bar{h}$ □

(c) $\frac{I_G \bar{h}}{A} + \bar{h}$ □ (d) $\frac{A\bar{h}}{I_G} + \bar{h}$. □

41. The centre of pressure for a plane vertical surface lies at a depth of
    (a) half the height of the immersed surface
    (b) one-third the height of the immersed surface
    (c) two-third the height of the immersed surface
    (d) none of the above.

42. The inlet length of a venturimeter
    (a) is equal to the outlet length
    (b) is more than the outlet length
    (c) is less than the outlet length
    (d) none of the above.

43. Flow of a fluid in a pipe takes place from
    (a) higher level to lower level
    (b) higher pressure to lower pressure
    (c) higher energy to lower energy
    (d) none of the above.

Buoyancy and Flotation

44. For a floating body, the buoyant force passes through the
    (a) centre of gravity of the body
    (b) centre of gravity of the submerged part of the body
    (c) metacentre of the body
    (d) centroid of the liquid displaced by the body.

45. The condition of stable equilibrium for a floating body is
    (a) the metacentre M coincides with the centre of gravity G
    (b) the metacentre M is below centre of gravity G
    (c) the metacentre M is above centre of gravity G
    (d) the centre of buoyancy B is above centre of gravity G.

46. A submerged body will be in stable equilibrium if
    (a) the centre of buoyancy B is below the centre of gravity G
    (b) the centre of buoyancy B coincides with G
    (c) the centre of buoyancy B is above the metacentre M
    (d) the centre of buoyancy B is above G.

47. The metacentric height of a floating body is
    (a) the distance between metacentre and centre of buoyancy
    (b) the distance between the centre of buoyancy and centre of gravity
    (c) the distance between metacentre and centre of gravity
    (d) none of the above.

48. The point, through which the buoyant force is acting, is called
    (a) centre of pressure
    (b) centre of gravity
    (c) centre of buoyancy
    (d) none of the above.

49. The point, through which the weight is acting, is called
    (a) centre of pressure
    (b) centre of gravity
    (c) centre of buoyancy
    (d) none of the above.

50. The point, about which a floating body, starts oscillating when the body is tilted is called
    (a) centre of pressure   (b) centre of buoyancy
    (c) centre of gravity   (d) meta-centre.

51. The meta-centric height (GM) is given by
    (a) $GM = BG - \dfrac{I}{V}$ (b) $GM = \dfrac{V}{I} - BG$
    (c) $GM = \dfrac{I}{V} - BG$ (d) none of the above.

52. For floating body, if the meta-centre is above the centre of gravity, the equilibrium is called
    (a) stable   (b) unstable
    (c) neutral   (d) none of the above.

53. For a floating body, if the meta-centre is below the centre of gravity, the equilibrium is called
    (a) stable   (b) unstable
    (c) neutral   (d) none of the above.

54. For a floating body, if the meta-centre coincides with the centre of gravity, the equilibrium is called
    (a) stable   (b) unstable
    (c) neutral   (d) none of the above.

55. For a floating body, if centre of buoyancy is above the centre of gravity, the equilibrium is called
    (a) stable   (b) unstable
    (c) neutral   (d) none of the above.

56. For a submerged body, if the centre of buoyancy is above the centre of gravity, the equilibrium is called
    (a) stable   (b) unstable
    (c) neutral   (d) none of the above.

57. For a submerged body, if the centre of buoyancy is below the centre of gravity, the equilibrium is called
    (a) stable   (b) unstable
    (c) neutral   (d) none of the above.

58. For a submerged body, if the centre of buoyancy coincides with the centre of gravity, the equilibrium is called
    (a) stable   (b) unstable
    (c) neutral   (d) none of the above.

59. For a submerged body, if the meta-centre is below the centre of gravity, the equilibrium is called
    (a) stable   (b) unstable
    (c) neutral   (d) none of the above.

60. The meta-centric height (GM) experimentally is given as

   (a) $GM = \dfrac{W \tan \theta}{wx}$
   (b) $GM = \dfrac{w \tan \theta}{W \times x}$
   (c) $GM = \dfrac{wx}{W \tan \theta}$
   (d) $GM = \dfrac{Wx}{w \tan \theta}$

   where w = Movable weight, W = Weight of floating body including w, θ = Angle of tilt.

61. The time period of oscillation of a floating body is given by

   (a) $T = 2\pi \sqrt{\dfrac{GM \times g}{k^2}}$
   (b) $T = 2\pi \sqrt{\dfrac{k^2}{GM \times g}}$
   (c) $T = 2\pi \sqrt{\dfrac{GM}{gk^2}}$
   (d) $T = 2\pi \sqrt{\dfrac{gk^2}{GM}}$

   where k = Radius of gyration, GM = Meta-centric height and T = Time period.

Kinematics and Dynamics of Flow

62. The necessary condition for the flow to be steady is that
   (a) the velocity does not change from place to place
   (b) the velocity is constant at a point with respect to time
   (c) the velocity changes at a point with respect to time
   (d) none of the above.

63. The necessary condition for the flow to be uniform is that
   (a) the velocity is constant at a point with respect to time
   (b) the velocity is constant in the flow field with respect to space
   (c) the velocity changes at a point with respect to time
   (d) none of the above.

64. The flow in the pipe is laminar if
   (a) Reynold number is equal to 2500
   (b) Reynold number is equal to 4000
   (c) Reynold number is more than 2500
   (d) None of the above.

65. A stream line is a line
   (a) which is along the path of a particle
   (b) which is always parallel to the main direction of flow
   (c) across which there is no flow
   (d) on which tangent drawn at any point gives the direction of velocity.

66. Continuity equation can take the form
   (a) $A_1 V_1 = A_2 V_2$
   (b) $\rho_1 A_1 = \rho_2 A_2$
   (c) $\rho_1 A_1 V_1 = \rho_2 A_2 V_2$
   (d) $p_1 A_1 V_1 = p_2 A_2 V_2$.

67. Pitot-tube is used for measurement of
   (a) pressure
   (b) flow
   (c) velocity at a point
   (d) discharge.

68. Bernoulli's theorem deals with the law of conservation of
    (a) mass
    (b) momentum
    (c) energy
    (d) none of the above.
69. Continuity equation deals with the law of conservation of
    (a) mass
    (b) momentum
    (c) energy
    (d) none of the above.
70. Irrotational flow means
    (a) the fluid does not rotate while moving
    (b) the fluid moves in straight lines
    (c) the net rotation of fluid-particles about their mass centres is zero
    (d) none of the above.
71. The velocity components in x and y-directions in terms of velocity potential (ϕ) are
    (a) $u = -\dfrac{\partial \phi}{\partial x}, v = \dfrac{\partial \phi}{\partial y}$
    (b) $u = \dfrac{\partial \phi}{\partial y}, v = \dfrac{\partial \phi}{\partial x}$
    (c) $u = -\dfrac{\partial \phi}{\partial y}, v = -\dfrac{\partial \phi}{\partial x}$
    (d) $u = -\dfrac{\partial \phi}{\partial x}, v = -\dfrac{\partial \phi}{\partial y}$.
72. The velocity components in x and y-directions in terms of stream function (ψ) are
    (a) $u = \dfrac{\partial \psi}{\partial x}, v = \dfrac{\partial \psi}{\partial y}$
    (b) $u = -\dfrac{\partial \psi}{\partial x}, v = \dfrac{\partial \psi}{\partial y}$
    (c) $u = \dfrac{\partial \psi}{\partial y}, v = \dfrac{\partial \psi}{\partial x}$
    (d) $u = -\dfrac{\partial \psi}{\partial y}, v = \dfrac{\partial \psi}{\partial x}$.
73. The relation between tangential velocity (v) and radius (r) is given by
    (a) $V \times r$ = constant for forced vortex
    (b) V/r = constant for forced vortex
    (c) $V \times r$ = constant for free vortex
    (d) V/r = constant for free vortex.
74. The pressure variation along the radial direction for vortex flow along a horizontal plane is given as
    (a) $\dfrac{\partial p}{\partial r} = -\rho \dfrac{V^2}{r}$
    (b) $\dfrac{\partial p}{\partial r} = \rho \dfrac{V}{r^2}$
    (c) $\dfrac{\partial p}{\partial r} = \rho \dfrac{V^2}{r}$
    (d) none of the above.
75. For a forced vortex flow the height of paraboloid formed is equal to
    (a) $\dfrac{p}{w} + \dfrac{V^2}{2g}$
    (b) $\dfrac{V^2}{2g}$
    (c) $\dfrac{V^2}{r^2 \times 2g}$
    (d) $\dfrac{\omega r^2}{2g}$.

76. Bernoulli's equation is derived making assumptions that
    (a) the flow is uniform, steady and incompressible
    (b) the flow is non-viscous, uniform and steady
    (c) the flow is steady, non-viscous, incompressible and irrotational
    (d) none of the above.

77. The Bernoulli's equation can take the form
    (a) $\dfrac{p_1}{\rho_1} + \dfrac{V_1^2}{2g} + Z_1 = \dfrac{p_2}{\rho_2} + \dfrac{V_2^2}{2g} + Z_2$
    (b) $\dfrac{p_1}{\rho_2 g} + \dfrac{V_1^2}{2} + Z_1 = \dfrac{p_2}{\rho_2 g} + \dfrac{V_2^2}{2} + Z_2$
    (c) $\dfrac{p_1}{\rho_1 g} + \dfrac{V_1^2}{2g} + Z_1 = \dfrac{p_2}{\rho_2 g} + \dfrac{V_2^2}{2g} + gZ_2$
    (d) $\dfrac{p_1}{\rho_1 g} + \dfrac{V_1^2}{2g} + Z_1 = \dfrac{p_2}{\rho_2 g} + \dfrac{V_2^2}{2g} + Z_2$.

78. The flow rate through a circular pipe is measured by
    (a) Pitot-tube
    (b) Venturi-meter
    (c) Orifice-meter
    (d) None of the above.

79. If the velocity, pressure, density etc., do not change at a point with respect to time, the flow is called
    (a) uniform
    (b) incompressible
    (c) non-uniform
    (d) steady.

80. If the velocity, pressure, density etc., change at a point with respect to time, the flow is called
    (a) uniform
    (b) compressible
    (c) unsteady
    (d) incompressible.

81. If the velocity in a fluid flow does not change with respect to length of direction of flow, it is called
    (a) steady flow
    (b) uniform flow
    (c) incompressible flow
    (d) rotational flow.

82. If the velocity in a fluid flow changes with respect to length of direction of flow, it is called
    (a) unsteady flow
    (b) compressible flow
    (c) irrotational flow
    (d) none of the above.

83. If the density of a fluid is constant from point to point in a flow region, it is called
    (a) steady flow
    (b) incompressible flow
    (c) uniform flow
    (d) rotational flow.

84. If the density of a fluid changes from point to point in a flow region, it is called
    (a) steady flow
    (b) unsteady flow
    (c) non-uniform flow
    (d) compressible flow.

85. If the fluid particles move in straight lines and all the lines are parallel to the surface, the flow is called
    (a) steady
    (b) uniform
    (c) compressible
    (d) laminar.

86. If the fluid particles move in a zig-zag way, the flow is called
    (a) unsteady (b) non-uniform
    (c) turbulent (d) incompressible.

87. The acceleration of a fluid particle in the direction of x is given by
    (a) $A_x = u \dfrac{\partial}{\partial x} + v \dfrac{\partial v}{\partial y} + w \dfrac{\partial w}{\partial z} + \dfrac{\partial u}{\partial t}$
    (b) $A_x = u \dfrac{\partial u}{\partial x} + u \dfrac{\partial v}{\partial y} + v \dfrac{\partial w}{\partial z} + \dfrac{\partial u}{\partial t}$
    (c) $A_x = u \dfrac{\partial u}{\partial x} + v \dfrac{\partial u}{\partial y} + w \dfrac{\partial u}{\partial z} + \dfrac{\partial u}{\partial t}$
    (d) none of the above.

88. The local acceleration in the direction of x is given by
    (a) $u \dfrac{\partial u}{\partial x} + \dfrac{\partial u}{\partial t}$
    (b) $\dfrac{\partial u}{\partial t}$
    (c) $u \dfrac{\partial u}{\partial x}$
    (d) none of the above.

89. The convective acceleration in the direction of x is given by
    (a) $u \dfrac{\partial u}{\partial x} + v \dfrac{\partial v}{\partial y} + w \dfrac{\partial w}{\partial z}$
    (b) $u \dfrac{\partial u}{\partial x} + u \dfrac{\partial u}{\partial y} + u \dfrac{\partial u}{\partial z}$
    (c) $u \dfrac{\partial u}{\partial x} + u \dfrac{\partial v}{\partial y} + u \dfrac{\partial w}{\partial z}$
    (d) $u \dfrac{\partial u}{\partial x} + v \dfrac{\partial u}{\partial y} + w \dfrac{\partial u}{\partial z}$.

90. Shear strain rate is given by
    (a) $\dfrac{1}{2}\left(\dfrac{\partial u}{\partial x} + \dfrac{\partial v}{\partial y}\right)$
    (b) $\dfrac{1}{2}\dfrac{\partial u}{\partial x} + \dfrac{\partial v}{\partial y}$
    (c) $\dfrac{1}{2}\left(\dfrac{\partial v}{\partial x} + \dfrac{\partial u}{\partial y}\right)$
    (d) $\dfrac{1}{2}\dfrac{\partial v}{\partial x} + \dfrac{\partial u}{\partial y}$.

91. For a two-dimensional fluid element in x-y plane, the rotational component is given as
    (a) $\omega_z = \dfrac{1}{2}\left(\dfrac{\partial v}{\partial x} + \dfrac{\partial u}{\partial y}\right)$
    (b) $\omega_z = \dfrac{1}{2}\left(\dfrac{\partial u}{\partial x} - \dfrac{\partial v}{\partial y}\right)$
    (c) $\omega_z = \dfrac{1}{2}\left(\dfrac{\partial u}{\partial x} + \dfrac{\partial v}{\partial y}\right)$
    (d) $\omega_z = \dfrac{1}{2}\left(\dfrac{\partial v}{\partial x} - \dfrac{\partial u}{\partial y}\right)$.

92. Vorticity is given by
    (a) two times the rotation (b) 1.5 times the rotation
    (c) three times the rotation (d) equal to the rotation.

93. Study of fluid motion with the forces causing the flow is known as
    (a) kinematics of fluid flow (b) dynamics of fluid flow
    (c) statics of fluid flow (d) none of the above.

94. Study of fluid motion without considering the forces causing the flow is known as
    (a) kinematics of fluid flow   □   (b) dynamics of fluid flow   □
    (c) statics of fluid flow   □   (d) none of the above.   □

95. Study of fluid at rest, is known as
    (a) kinematics   □   (b) dynamics   □
    (c) statics   □   (d) none of the above.   □

96. The term $V^2/2g$ is known as
    (a) kinetic energy   □   (b) pressure energy   □
    (c) kinetic energy per unit weight   □   (d) none of the above.   □

97. The term $p/\rho g$ is known as
    (a) kinetic energy per unit weight   □   (b) pressure energy   □
    (c) pressure energy per unit weight   □   (d) none of the above.   □

98. The term Z is known as
    (a) potential energy   □   (b) pressure energy   □
    (c) potential energy per unit weight   □   (d) none of the above.   □

99. The discharge through a venturimeter is given as

    (a) $Q = \dfrac{A_1^2 A_2^2}{\sqrt{A_1^2 - A_2^2}} \times \sqrt{2gh}$ □   (b) $Q = \dfrac{A_1 A_2}{\sqrt{2A_1^2 - A_2^2}} \times \sqrt{2gh}$ □

    (c) $Q = \dfrac{A_1 A_2}{\sqrt{A_1^2 - A_2^2}} \times \sqrt{2gh}$ □   (d) none of the above.   □

100. The difference of pressure head (h) measured by a mercury-oil differential manometer is given as

    (a) $h = x \left[1 - \dfrac{S_g}{S_0} \right]$ □   (b) $h = x [S_g - S_0]$ □

    (c) $h = x [S_0 - S_g]$ □   (d) $h = x \left[\dfrac{S_g}{S_0} - 1 \right]$ □

    where x = Difference of mercury level, S_g = Specific gravity of mercury, and S_0 = Specific gravity of oil.

101. The difference of pressure head (h) measured by a differential manometer containing lighter liquid is

    (a) $h = x \left[1 - \dfrac{S_l}{S_0} \right]$ □   (b) $h = x \left[\dfrac{S_l}{S_0} - 1 \right]$ □

    (c) $h = x [S_0 - S_l]$ □   (d) none of the above   □

    where S_l = Specific gravity of lighter liquid in manometer
    S_0 = Specific gravity of fluid flowing
    x = Difference of lighter liquid levels in differential manometer.

102. Pitot-tube is used to measure
    (a) discharge
    (b) average velocity
    (c) velocity at a point
    (d) pressure at a point.

103. Venturimeter is used to measure
    (a) discharge
    (b) average velocity
    (c) velocity at a point
    (d) pressure at a point.

104. Orifice-meter is used to measure
    (a) discharge
    (b) average velocity
    (c) velocity at a point
    (d) pressure at a point.

105. For a sub-merged curved surface, the horizontal component of force due to static liquid is equal to
    (a) weight of liquid supported by the curved surface
    (b) force on a projection of the curved surface on a vertical plane
    (c) area of curved surface × pressure at the centroid of the submerged area
    (d) none of the above.

106. For a sub-merged curved surface, the component of force due to static liquid is equal to
    (a) weight of the liquid supported by curved surface
    (b) force on a projection of the curved surface on a vertical plane
    (c) area of curved surface × pressure at the centroid of the sub-merged area
    (d) none of the above.

107. An oil of specific gravity 0.7 and pressure 0.14 kgf/cm² will have the height of oil as
    (a) 70 cm of oil
    (b) 2 m of oil
    (c) 20 cm of oil
    (d) 10 cm of oil.

108. The difference in pressure head, measured by a mercury water differential manometer for a 20 m difference of mercury level will be
    (a) 2.72 m
    (b) 2.52 m
    (c) 2.0 m
    (d) 0.2 m.

109. The difference in pressure head, measured by a mercury-oil differential manometer for a 20 cm difference of mercury level will be (sp. gr. of oil = 0.8)
    (a) 2.72 m of oil
    (b) 2.52 m of oil
    (c) 3.20 m of oil
    (d) 2.0 m of oil.

110. The rate of flow through a venturimeter varies as
    (a) H
    (b) $\sqrt{H}$
    (c) $H^{3/2}$
    (d) $H^{5/2}$.

111. The rate of flow through a V-notch varies as
    (a) H
    (b) $\sqrt{H}$
    (c) $H^{3/2}$
    (d) $H^{5/2}$.

Orifices and Mouthpieces

112. The range for co-efficient of discharge (C_d) for a venturimeter is
 (a) 0.6 to 0.7
 (b) 0.7 to 0.8
 (c) 0.8 to 0.9
 (d) 0.95 to 0.99.

113. The co-efficient of velocity (C_v) for an orifice is
 (a) $C_v = \sqrt{\dfrac{4x^2}{yH}}$
 (b) $C_v = \sqrt{\dfrac{2x}{4yH}}$
 (c) $C_v = \sqrt{\dfrac{x^2}{4yH}}$
 (d) none of the above.

114. The co-efficient of discharge (C_d) in terms of C_v and C_c is
 (a) $C_d = \dfrac{C_v}{C_c}$
 (b) $C_d = C_v \times C_c$
 (c) $C_d = \dfrac{C_c}{C_v}$
 (d) none of the above.

115. An orifice is known as large orifice when the head of liquid from the centre of orifice is
 (a) more than 10 times the depth of orifice
 (b) less than 10 times the depth of orifice
 (c) less than 5 times the depth of orifice
 (d) none of the above.

116. Which mouthpiece is having maximum co-efficient of discharge?
 (a) external mouthpiece
 (b) convergent divergent mouthpiece
 (c) internal mouthpiece
 (d) none of the above.

117. The co-efficient of discharge (C_d)
 (a) for an orifice is more than that for a mouthpiece
 (b) for internal mouthpiece is more than that external mouthpiece
 (c) for a mouthpiece is more than that for an orifice
 (d) none of the above.

118. Orifices are used to measure
 (a) velocity
 (b) pressure
 (c) rate of flow
 (d) none of the above.

119. Mouthpieces are used to measure
 (a) velocity
 (b) pressure
 (c) viscosity
 (d) rate of flow.

120. The ratio of actual velocity of a jet of water at veena-contracta to the theoretical velocity is known as
 (a) co-efficient of discharge
 (b) co-efficient of velocity
 (c) co-efficient of contraction
 (d) co-efficient of viscosity.

121. The ratio of actual discharge of a jet of water to its theoretical discharge is known as
    (a) co-efficient of discharge □   (b) co-efficient of velocity □
    (c) co-efficient of contraction □   (d) co-efficient of viscosity. □

122. The ratio of the area of the jet of water at veena-contracta to the area of orifice is known as
    (a) co-efficient of discharge □   (b) co-efficient of velocity □
    (c) co-efficient of contraction □   (d) co-efficient of viscosity. □

123. The discharge through a large rectangular orifice is
    (a) $\frac{2}{3} C_d \times b \times \frac{2}{3}\sqrt{2g}(\sqrt{H_2} - \sqrt{H_1})$ □   (b) $\frac{8}{15} C_d \times b \times \sqrt{2g}(H_2^{3/2} - H_1^{3/2})$ □
    (c) $\frac{2}{3} C_d \times b \times \sqrt{2g}(H_2^{3/2} - H_1^{3/2})$ □   (d) none of the above □

    where b = Width of orifice, H_1 = Height of liquid above top edge of the orifice, H_2 = Height of liquid above bottom edge of orifice.

124. The discharge through fully submerged orifice is
    (a) $C_d \times b \times (H_2 - H_1) \times \sqrt{2g} \times H^{3/2}$ □   (b) $C_d \times b \times (H_2 - H_1) \times \sqrt{2g}$ □
    (c) $C_d \times b \times (H_2^{3/2} - H_1^{3/2}) \times \sqrt{2gH}$ □   (d) none of the above □

    where H = Difference of liquid levels on both sides of the orifice
    H_1 = Height of liquid above top edge orifice of upstream side
    H_2 = Height of liquid above bottom edge of orifice on upstream side.

Notches and Weirs

125. Notch is a device used for measuring
    (a) rate of flow through pipes □   (b) rate of flow through a small channel □
    (c) velocity through a pipe □   (d) velocity through a small channel. □

126. The discharge through a rectangular notch is given by
    (a) $Q = \frac{2}{3} C_d \times L \times H^{5/2}$ □   (b) $Q = 2/3\, C_d \times L \times H^{3/2}$ □
    (c) $Q = \frac{2}{3} C_d \times L \times H^{5/2}$ □   (d) $Q = 8/15\, C_d \times L \times H^{3/2}$. □

127. The discharge through a triangular notch is given by
    (a) $Q = 2/3\, C_d \times \tan\frac{\theta}{2} \times \sqrt{2gH}$ □   (b) $Q = 2/3\, C_d \times \tan\frac{\theta}{2} \times \sqrt{2g} \times H^{3/2}$ □
    (c) $Q = 8/15\, C_d \times \tan\frac{\theta}{2} \times \sqrt{2g}\, H^{5/2}$ □   (d) none of the above □

    where θ = Total angle of triangular notch, H = Head over notch.

128. The discharge through a trapezoidal notch is given as
 (a) $Q = 2/3 \, C_{d1} \times L \times H^{3/2} + 8/15 \, C_{d2} \times \tan \theta/2 \times \sqrt{2g} \times H^{3/2}$
 (b) $Q = 2/3 \, C_{d1} \times L \times H^{5/2} + 8/15 \, C_{d2} \times \tan \theta/2 \times \sqrt{2g} \, H^{3/2}$
 (c) $Q = 2/3 \, C_{d1} \times L \times H^{3/2} + 8/15 \, C_{d2} \times \tan \theta/2 \times \sqrt{2g} \, H^{5/2}$
 (d) none of the above
 where $\theta/2$ = Slope of the side of the trapezoidal notch.

129. The error in discharge due to the error in the measurement of head over a rectangular notch is given by
 (a) $\dfrac{dQ}{Q} = \dfrac{5}{2} \dfrac{dH}{H}$
 (b) $\dfrac{dQ}{Q} = \dfrac{3}{2} \dfrac{dH}{H}$
 (c) $\dfrac{dQ}{Q} = \dfrac{7}{2} \dfrac{dH}{H}$
 (d) $\dfrac{dQ}{Q} = \dfrac{1}{2} \dfrac{dH}{H}$.

130. The error in discharge due to the error in the measurement of head over a triangular notch is given by
 (a) $\dfrac{dQ}{Q} = \dfrac{5}{2} \dfrac{dH}{H}$
 (b) $\dfrac{dQ}{Q} = \dfrac{3}{2} \dfrac{dH}{H}$
 (c) $\dfrac{dQ}{Q} = \dfrac{7}{2} \dfrac{dH}{H}$
 (d) $\dfrac{dQ}{Q} = \dfrac{1}{2} \dfrac{dH}{H}$.

131. The velocity with which the water approaches a notch is called
 (a) velocity of flow
 (b) velocity of approach
 (c) velocity of whirl
 (d) none of the above.

132. The discharge over a rectangular notch considering velocity of approach is given as
 (a) $Q = \dfrac{3}{2} C_d L \sqrt{2g} \, (H^{3/2} - h_a^{3/2})$
 (b) $Q = \dfrac{2}{3} C_d L \sqrt{2g} \, (H - h_a)^{3/2}$
 (c) $Q = \dfrac{2}{3} C_d L \sqrt{2g} \, [(H + h_a)^{3/2} - h_a^{3/2}]$
 (d) none of the above
 where H = Head over notch, and h_a = Head due to velocity of approach.

133. The velocity of approach (V_a) is given by
 (a) $V_a = \dfrac{\text{Discharged over notch}}{\text{Area of notch}}$
 (b) $V_a = \dfrac{\text{Discharged over notch}}{\text{Area of channel}}$
 (c) $V_a = \dfrac{\text{Discharged over notch}}{\text{Heat over notch} \times \text{Width of channel}}$
 (d) none of the above.

134. Francis's formula for a rectangular weir with end contraction suppressed is given as
 (a) $Q = 1.84 \, LH^{5/2}$
 (b) $Q = 2/3 \, L \times H^{3/2}$
 (c) $Q = 1.84 \, LH^{3/2}$
 (d) $Q = 2/3 \, L \times H^{5/2}$.

135. Francis's formula for a rectangular weir for two end contractions is given by
    (a) $Q = 1.84[L - 0.2 \times 2H]H^{5/2}$
    (b) $Q = 1.84[L - 0.2H]H^{3/2}$
    (c) $Q = 1.84[L - 0.2H]H^{5/2}$
    (d) none of the above.

136. Bazin's formula for discharge over a rectangular weir without velocity of approach is given by
    (a) $Q = mL \times \sqrt{2gH^{5/2}}$
    (b) $Q = mL \times \sqrt{2g} \times H^{3/2}$
    (c) $Q = m \times L \times \sqrt{2gH}$
    (d) none of the above

    where $m = 0.405 + \dfrac{0.003}{H}$ and H = Head over weir.

137. Cipolletti weir is a trapezoidal weir having side slope of
    (a) 1 horizontal to 2 vertical
    (b) 4 horizontal to 1 vertical
    (c) 1 horizontal to 4 vertical
    (d) 1 horizontal to 3 vertical.

Laminar and Turbulent Flow Through Pipes

138. A flow is said to be laminar when
    (a) the fluid particles move in a zig-zag way
    (b) the Reynold number is high
    (c) the fluid particles move in layers parallel to the boundary
    (d) none of the above.

139. For the laminar flow through a circular pipe
    (a) the maximum velocity = 1.5 times the average velocity
    (b) the maximum velocity = 2.0 times the average velocity
    (c) the maximum velocity = 2.5 times the average velocity
    (d) none of the above.

140. The loss of pressure head for the laminar flow through pipes varies
    (a) as the square of velocity
    (b) directly as the velocity
    (c) as the inverse of the velocity
    (d) none of the above.

141. For the laminar flow through a pipe, the shear stress over the cross-section
    (a) varies inversely as the distance from the centre of the pipe
    (b) varies directly as the distance from the surface of the pipe
    (c) varies directly as the distance from the centre of the pipe
    (d) remains constant over the cross-section.

142. For the laminar flow between two parallel plates
    (a) the maximum velocity = 2.0 times the average velocity
    (b) the maximum velocity = 2.5 times of the average velocity
    (c) the maximum velocity = 1.33 times the average velocity
    (d) none of the above.

143. The valve of the kinetic energy correction factor (α) for the viscous flow through a circular pipe is
    (a) 1.33
    (b) 1.50
    (c) 2.0
    (d) 1.25.

144. The valve of the momentum correction factor (β) for the viscous flow through a circular pipe is
    (a) 1.33
    (b) 1.50
    (c) 2.0
    (d) 1.25.

145. The pressure drop per unit length of a pipe for laminar flow is
    (a) equal to $\dfrac{12\mu \overline{U} L}{\rho g D^2}$
    (b) equal to $\dfrac{12\mu \overline{U}}{\rho g D^2}$
    (c) equal to $\dfrac{32\mu \overline{U} L}{\rho g D^2}$
    (d) none of the above.

146. For viscous flow between two parallel plates, the pressure drop per unit length is equal to
    (a) $\dfrac{12\mu \overline{U} L}{\rho g D^2}$
    (b) $\dfrac{12\mu \overline{U} L}{D^2}$
    (c) $\dfrac{32\mu \overline{U} L}{D^2}$
    (d) $\dfrac{12\mu \overline{U}}{D^2}$.

147. The velocity distribution in laminar flow through a circular pipe follows the
    (a) parabolic law
    (b) linear law
    (c) logarithmic law
    (d) none of the above.

148. A boundary is known as hydrodynamically smooth, if
    (a) $\dfrac{k}{\delta'} = 0.3$
    (b) $\dfrac{k}{\delta'} > 0.3$
    (c) $\dfrac{k}{\delta'} < 0.25$
    (d) $\dfrac{k}{\delta'} = 6.0$

    where k = Average height of the irregularities from the boundary and δ' = Thickness of laminar sub-layer.

149. The co-efficient of friction for laminar flow through a circular pipe is given by
    (a) $f = \dfrac{0.0791}{(R_e)^{1/4}}$
    (b) $f = \dfrac{16}{R_e}$
    (c) $f = \dfrac{64}{R_e}$
    (d) none of the above.

150. The loss of head due to sudden expansion of a pipe is given by
    (a) $h_L = \dfrac{V_1^2 - V_2^2}{2g}$
    (b) $h_L = \dfrac{0.5 V_1^2}{2g}$
    (c) $h_L = \dfrac{(V_1 - V_2)^2}{2g}$
    (d) none of the above.

151. The loss of head due to sudden contraction of a pipe is equal to
    (a) $\left(\dfrac{1}{C_c}-1\right)^2 \dfrac{V_2}{2g}$
    (b) $\left(1-\dfrac{1}{C_c}\right)^2 \dfrac{V_2}{2g}$
    (c) $\dfrac{1}{C_c}\left(1-\dfrac{V_2^2}{2g}\right)$
    (d) none of the above.

152. Hydraulic gradient line (H.G.L.) represents the sum of
    (a) pressure head and kinetic head
    (b) kinetic head and datum head
    (c) pressure head, kinetic head and datum head
    (d) pressure head and datum head.

153. Total Energy Line (T.E.L.) represents the sum of
    (a) pressure head and kinetic head
    (b) kinetic head and datum head
    (c) pressure head and datum head
    (d) pressure head, kinetic head and datum head.

154. When the pipes are connected in series, the total rate of flow
    (a) is equal to the sum of the rate of flow in each pipe
    (b) is equal to the reciprocal of the sum of rate of flow in each pipe
    (c) is the same as flowing through each pipe
    (d) none of the above.

155. Power transmitted through pipes will be maximum when
    (a) Head lost due to friction = $\frac{1}{2}$ total head at inlet of the pipe
    (b) Head lost due to friction = $\frac{1}{4}$ total head at inlet of the pipe
    (c) Head lost due to friction = total head at the inlet of the pipe
    (d) Head lost due to friction = $\frac{1}{3}$ total head at the inlet of the pipe.

156. The valve closure is said to be gradual if the time required to close the valve
    (a) $t = \dfrac{2L}{C}$
    (b) $t \leq \dfrac{2L}{C}$
    (c) $t < \dfrac{4L}{C}$
    (d) $t > \dfrac{2L}{C}$.
    where L = Length of pipe, C = Velocity of pressure wave.

157. The velocity of pressure wave in terms of bulk modulus (K) and density (ρ) is given by
    (a) $C = \sqrt{\dfrac{\rho}{K}}$
    (b) $C = \sqrt{K\rho}$
    (c) $C = \sqrt{\dfrac{K}{\rho}}$
    (d) none of the above.

158. The co-efficient of friction in terms of shear stress is given by

(a) $f = \dfrac{2\tau V^2}{\tau_0}$ (b) $f = \dfrac{2\tau_0}{\rho V^2}$

(c) $f = \dfrac{\tau_0}{2\rho V^2}$ (d) $f = \dfrac{\rho V^2}{2\tau_0}$.

159. Reynold shear stress for turbulent flow is given by

(a) $\tau = \overline{\rho u v'}$ (b) $\overline{\tau} = \mu \dfrac{\partial u}{\partial y}$

(c) $\overline{\tau} = \eta \dfrac{du}{dy}$ (d) none of the above

where u', v' = Fluctuating component of velocity in the direction x and y and η = Eddy viscosity.

160. The shear stress in turbulent flow due to Prandtl is given by

(a) $\overline{\tau} = \rho l^2 \left(\dfrac{du}{dy}\right)^2$ (b) $\overline{\tau} = \rho^2 l \left(\dfrac{du}{dy}\right)^2$

(c) $\overline{\tau} = \rho^2 l^2 \left(\dfrac{du}{dy}\right)$ (d) none of the above

where l = Mixing length.

161. Shear velocity (u_*) is equal to

(a) $\sqrt{\rho \tau_0}$ (b) $\sqrt{\dfrac{\tau_0}{\rho}}$

(c) $\sqrt{\dfrac{\rho}{\tau_0}}$ (d) $\dfrac{1}{\sqrt{\rho \tau_0}}$

where τ_0 = Shear stress at the surface.

162. The velocity distribution in turbulent flow for pipes is given by

(a) $u = U_{max} + 5.5 u_* \log_e (y/R)$ (b) $u = 2.5 u_* \log_e (y/R)$

(c) $u = U_{max} + 2.5 u_* \log_e (y/R)$ (d) none of the above.

where u_* = Shear velocity, R = Radius of pipe, y = Distance from pipe wall, U_{max} = Centre-line velocity.

163. When the pipes are connected in parallel, the total loss of head

(a) is equal to the sum of the loss of head in each pipe

(b) is same as in each pipe

(c) is equal to the reciprocal of the sum of loss of head in each pipe

(d) none of the above.

164. L_1, L_2, L_3 are the length of three pipes, connected in series. If d_1, d_2 and d_3 are their diameters, then the equivalent size of the pipe is given by

(a) $\dfrac{L}{d^5} = \dfrac{L_1}{d_1^5} + \dfrac{L_2}{d_2^5} + \dfrac{L_3}{d_3^5}$ □   (b) $\dfrac{d^5}{L} = \dfrac{d_1^5}{L_1} + \dfrac{d_2^5}{L_2} + \dfrac{d_3^5}{L_3}$ □

(c) $Ld^5 = L_1 d_1^5 + L_2 d_2^5 + L_3 d_3^5$ □   (d) none of the above   □

where $L = L_1 + L_2 + L_3$.

165. The power transmitted in kW through pipe is given by

(a) $\dfrac{\rho \times g \times Q \times H}{75}$ □   (b) $\dfrac{\rho \times g \times Q \times h_f}{1000}$ □

(c) $\dfrac{\rho \times g \times Q \times (H - h_f)}{4500}$ □   (d) $\dfrac{\rho \times g \times Q \times (H - h_f)}{1000}$ □

where H = Total head at the inlet of pipe, h_f = Head lost due to friction in pipe and Q = Discharge per second.

166. Efficiency of power transmission through pipe is given by

(a) $\dfrac{H - h_f}{H}$ □   (b) $\dfrac{H}{H + h_f}$ □

(c) $\dfrac{H - h_f}{H + h_f}$ □   (d) none of the above   □

where H = Total head at inlet, h_f = Head lost due to friction.

167. Maximum efficiency of power transmission through pipe is

(a) 50%   □   (b) 66.67%   □
(c) 75%   □   (d) 100%.   □

168. Diameter of nozzle (d) for maximum power transmission is given by

(a) $d = \left(\dfrac{D^4}{8fL}\right)^{1/5}$ □   (b) $d = \left(\dfrac{D^5}{8fL}\right)^{1/5}$ □

(c) $d = \left(\dfrac{D^5}{8fL}\right)^{1/4}$ □   (d) none of the above   □

where D = Dia. of pipe, L = Length of pipe.

169. Water-hammer in pipes takes place when
    (a) fluid is flowing with high velocity   □
    (b) fluid is flowing with high pressure   □
    (c) flowing fluid is suddenly brought to rest by closing the valve   □
    (d) flowing fluid is gradually brought to rest.   □

170. The pressure rise (p_i) due to water hammer, when the valve is closed suddenly and pipe is assumed rigid, is equal to

(a) $V\sqrt{\dfrac{k}{\rho}}$ ☐ (b) $V\sqrt{k\rho}$ ☐

(c) $V\sqrt{\dfrac{\rho}{k}}$ ☐ (d) $Vk\rho$ ☐

where V = Velocity of flow, k = Bulk modulus of water and ρ = Density of fluid.

171. The pressure rise (p_i) due to water hammer, when valve is closed gradually is equal to

(a) ρLV ☐ (b) $\dfrac{\rho LV}{t}$ ☐

(c) $\dfrac{\rho t}{VL}$ ☐ (d) $\dfrac{\rho}{LVt}$ ☐

where t = Time required to close the valve.

172. The pressure rise (p_i) due to water hammer, when valve is closed suddenly and pipe is elastic, is equal to

(a) $V \times \sqrt{\dfrac{kEt}{\rho D}}$ ☐ (b) $V \times \sqrt{\dfrac{\frac{1}{k}+\frac{D}{Et}}{\rho}}$ ☐

(c) $V \times \sqrt{\dfrac{\rho}{\frac{1}{k}+\frac{D}{Et}}}$ ☐ (d) none of the above ☐

where E = Modulus of elasticity for pipe material, D = Diameter of pipe, t = Time required to close valve and k = Bulk modulus of water.

173. The pressure rise (p_i) due to water hammer depends on
   (a) the diameter of pipe only ☐
   (b) the length of pipe only ☐
   (c) the required to close the valve only ☐
   (d) elastic properties of the pipe material only ☐
   (e) elastic properties of liquid flowing through pipe only ☐
   (f) all of the above. ☐

174. The valve closure is said to be sudden if the time required to close the valve

(a) $t = \dfrac{2L}{C}$ ☐ (b) $t < \dfrac{2L}{C}$ ☐

(c) $t > \dfrac{2L}{C}$ ☐ (d) none of the above ☐

where C = Velocity of pressure wave produced and L = Length of pipe.

175. For a viscous flow through circular pipes, certain curves are shown in Fig. 1.3. Curve A is for
   (a) shear stress distribution ☐
   (b) velocity distribution ☐
   (c) pressure distribution ☐
   (d) none of the above. ☐

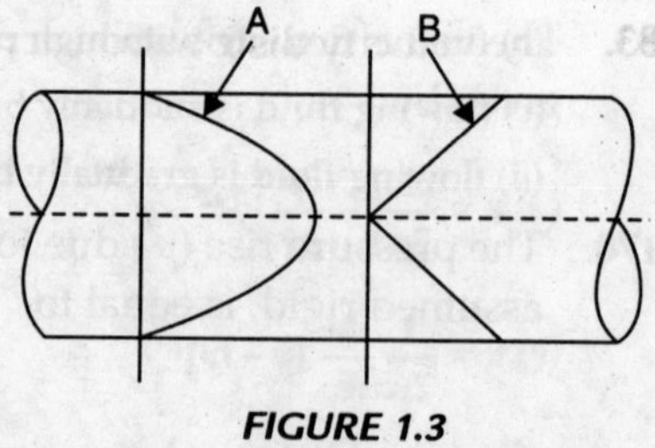

FIGURE 1.3

176. Curve B in Fig. 1.3 is for
   (a) shear stress distribution ☐
   (b) velocity distribution ☐
   (c) pressure distribution ☐
   (d) none of the above. ☐

177. Figure 1.4 shows four curves for velocity distribution across a section for Reynolds number equal to 1000, 4000, 6000 and 10000. Curve A corresponds to Reynold number equal to
   (a) 1000 ☐ 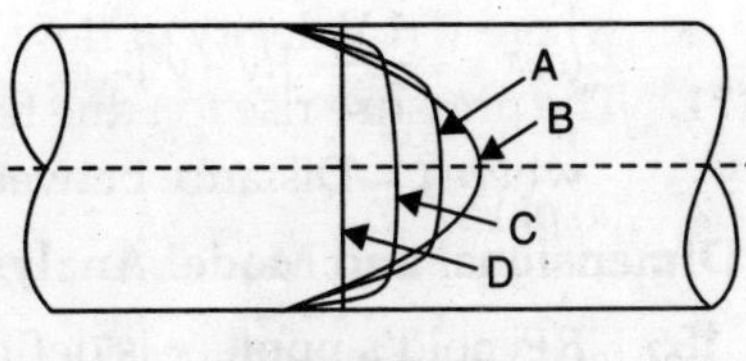
   (b) 4000 ☐
   (c) 6000 ☐
   (d) 10000. ☐

FIGURE 1.4

178. Curve B in Fig. 1.4 corresponds to Reynolds number
   (a) 1000 ☐
   (b) 4000 ☐
   (c) 6000 ☐
   (d) 10000. ☐

179. Curve C in Fig. 1.4 corresponds to the Reynold number
   (a) 1000 ☐
   (b) 4000 ☐
   (c) 6000 ☐
   (d) 10000. ☐

180. Curve D in Fig. 1.4 corresponds to the Reynold number
   (a) 1000 ☐
   (b) 4000 ☐
   (c) 6000 ☐
   (d) 10000. ☐

181. The shear stress distribution across a section of a circular pipe having viscous flow is given by
   (a) $\tau = \frac{\partial p}{\partial x} r^2$ ☐
   (b) $\tau = \frac{\partial p}{\partial x} \frac{r}{2}$ ☐
   (c) $\tau = -\frac{\partial p}{\partial x} \frac{r}{2}$ ☐
   (d) $\tau = -\frac{\partial p}{\partial x} \times 2r$. ☐

182. The velocity distribution across a section of a circular pipe having viscous flow is given by
   (a) $u = U_{max} \left[1 - \left(\frac{r}{R}\right)^2\right]$ ☐
   (b) $u = U_{max} [R^2 - r^2]$ ☐
   (c) $u = U_{max} \left[1 - \frac{r}{R}\right]^2$ ☐
   (d) none of the above. ☐

183. The velocity distribution across a section of two fixed parallel plates having viscous flow is given by
   (a) $u = \frac{1}{2\mu}\left(-\frac{\partial p}{\partial x}\right)(t^2 - y^2)$ ☐
   (b) $u = -\frac{1}{2\mu}\frac{\partial p}{\partial x}[ty - y^2]$ ☐
   (c) $u = \frac{1}{2\mu}\frac{\partial p}{\partial x}[y - ty]$ ☐
   (d) $u = -\frac{1}{2\mu}\frac{\partial p}{\partial x}[t - y^2]$ ☐

where t = Distance between two plates and y is measured from the lower plate.

184. The shear stress distribution across a section of two fixed parallel plates having viscous flow is given by

(a) $\tau = -\dfrac{1}{2}\dfrac{\partial p}{\partial x}[t^2 - y^2]$ ☐
(b) $\tau = -\dfrac{1}{2}\dfrac{\partial p}{\partial x}[t - 2y]$ ☐
(c) $\tau = -\dfrac{1}{2}\dfrac{\partial p}{\partial x}[ty - y^2]$ ☐
(d) $\tau = \dfrac{1}{2}\dfrac{\partial p}{\partial x}[y - ty]$ ☐

where t = Distance between two parallel plates and y is measured from the plate.

Dimensional and Model Analysis

185. Reynold's number is defined as the
    (a) ratio of inertia force to gravity force ☐
    (b) ratio of viscous force to gravity force ☐
    (c) ratio of viscous force to elastic force ☐
    (d) ratio of inertia force to viscous force. ☐

186. Froude's number is defined as the ratio of
    (a) inertia force to viscous force ☐
    (b) inertia force to gravity force ☐
    (c) inertia force to elastic force ☐
    (d) inertia force to pressure force. ☐

187. Mach number is defined as the ratio of
    (a) inertia force to viscous force ☐
    (b) viscous force to surface tension force ☐
    (c) viscous force to elastic force ☐
    (d) inertia force to elastic force. ☐

188. Euler's number is the ratio of
    (a) inertia force to pressure force ☐
    (b) inertia force to elastic force ☐
    (c) inertia force to gravity force ☐
    (d) none of the above. ☐

189. Models are known undistorted model, if
    (a) the prototype and model are having different scale ratios ☐
    (b) the prototype and model are having same scale ratio ☐
    (c) model and prototype are kinematically similar ☐
    (d) none of the above. ☐

190. Geometric similarity between model and prototype means
    (a) the similarity of discharge ☐
    (b) the similarity of linear dimensions ☐
    (c) the similarity of motion ☐
    (d) the similarity of forces. ☐

191. Kinematic similarity between model and prototype means
    (a) the similarity of forces ☐
    (b) the similarity of shape ☐
    (c) the similarity of motion ☐
    (d) the similarity of discharge. ☐

192. Dynamic similarity between model and prototype means
    (a) the similarity of forces ☐
    (b) the similarity of motion ☐
    (c) the similarity of shape ☐
    (d) none of the above. ☐

193. Reynold number is expressed as
    (a) $R_e = \dfrac{\rho \mu L}{V}$ ☐
    (b) $R_e = \dfrac{V \mu L}{\rho}$ ☐
    (c) $R_e = \dfrac{\rho V L}{\mu}$ ☐
    (d) $R_e = \dfrac{V \times L}{\nu}$ ☐

194. Froude's number (F_e) is given by

   (a) $F_e = V\sqrt{\dfrac{L}{g}}$
   (b) $F_e = V\sqrt{\dfrac{g}{L}}$
   (c) $F_e = \dfrac{V}{\sqrt{L \cdot g}}$
   (d) none of the above.

195. Mach number (M) is given by

   (a) $M = \dfrac{C}{V}$
   (b) $M = V \times C$
   (c) $M = \dfrac{V}{C}$
   (d) none of the above.

196. The ratio of inertia force to viscous force is known as
   (a) Reynold number
   (b) Froude number
   (c) Mach number
   (d) Euler number.

197. The square root of the ratio of inertia force to gravity force is called
   (a) Reynold number
   (b) Froude number
   (c) Mach number
   (d) Euler number.

198. The square root of the ratio of inertia force to force due to compressibility is known as
   (a) Reynold number
   (b) Froude number
   (c) Mach number
   (d) Euler number.

199. The square root of the ratio of inertia force to pressure force is known as
   (a) Reynold number
   (b) Froude number
   (c) Mach number
   (d) Euler number.

200. Model analysis of pipes flow are based on
   (a) Reynold number
   (b) Froude number
   (c) Mach number
   (d) Euler number.

201. Model analysis of free surface flows are based on
   (a) Reynolds number
   (b) Froude number
   (c) Mach number
   (d) Euler number.

202. Model analysis of aeroplanes and projectile moving at super-sonic speed are based on
   (a) Reynold number
   (b) Froude number
   (c) Mach number
   (d) Euler number.

Boundary Layer Flow

203. Boundary layer on a flat plate is called laminar boundary layer if
   (a) Reynold number is less than 2000
   (b) Reynold number is less than 4000
   (c) Reynold number is less than 5×10^5
   (d) None of the above.

204. Boundary layer thickness (δ) is the distance from the surface of the solid body in the direction perpendicular to flow, where the velocity of fluid is equal to
    (a) Free stream velocity
    (b) 0.9 times the free-stream velocity
    (c) 0.99 times the free stream velocity
    (d) None of the above.

205. Displacement thickness (δ^*) is given by
    (a) $\delta^* = \int_0^\delta \left(1 - \dfrac{U}{u}\right) dy$
    (b) $\delta^* = \int_0^\delta \dfrac{U}{u}\left(1 - \dfrac{U}{u}\right) dy$
    (c) $\delta^* = \int_0^\delta \dfrac{u}{U}\left(1 - \dfrac{u^2}{U^2}\right) dy$
    (d) none of the above.

206. Momentum thickness (θ) is given by
    (a) $\theta = \int_0^\delta \dfrac{u}{U}\left(1 - \dfrac{u}{U}\right) dy$
    (b) $\theta = \int_0^\delta \left(1 - \dfrac{u}{U}\right) dy$
    (c) $\theta = \int_0^\delta \dfrac{u}{U}\left(1 - \dfrac{u^2}{U^2}\right) dy$
    (d) none of the above.

207. Energy thickness (δ^{**}) is equal to
    (a) $\int_0^\delta \dfrac{u}{U}\left[1 - \dfrac{u}{U}\right] dy$
    (b) $\int_0^\delta \dfrac{u}{U}\left(1 - \dfrac{u^2}{U^2}\right) dy$
    (c) $\int_0^\delta \dfrac{u}{U}\left(1 - \dfrac{u}{U}\right)^2 dy$
    (d) none of the above.

208. Von-Karman momentum integral equation is given as
    (a) $\dfrac{\tau_0}{\frac{1}{2}\rho U^2} = \dfrac{\partial \theta}{\partial x}$
    (b) $\dfrac{\tau_0}{\rho U^2} = \dfrac{\partial \theta}{\partial x}$
    (c) $\dfrac{\tau_0}{2\rho U^2} = \dfrac{\partial \theta}{\partial x}$
    (d) none of the above.

209. The boundary layer separation takes place if
    (a) pressure gradient is zero
    (b) pressure gradient is + ve
    (c) pressure gradient is negative
    (d) none of the above.

210. The condition for boundary layer separation is
    (a) $\left(\dfrac{\partial u}{\partial y}\right)_{y=0} = +\text{ve}$
    (b) $\left(\dfrac{\partial u}{\partial y}\right)_{y=0} = -\text{ve}$
    (c) $\left(\dfrac{\partial u}{\partial y}\right)_{y=0} = 0$
    (d) none of the above.

211. The boundary layer flow will be attached to the surface if

    (a) $\left(\dfrac{\partial u}{\partial y}\right)_{y=0} = 0$ ☐ (b) $\left(\dfrac{\delta u}{\delta y}\right)_{y=0} = +\text{ve}$ ☐

    (c) $\left(\dfrac{\partial u}{\partial y}\right)_{y=0} = -\text{ve}$ ☐ (d) none of the above. ☐

212. The condition for detached flow is

    (a) $\left(\dfrac{\partial u}{\partial y}\right)_{y=0} = 0$ ☐ (b) $\left(\dfrac{\partial u}{\partial y}\right)_{y=0} = +\text{ve}$ ☐

    (c) $\left(\dfrac{\partial u}{\partial y}\right)_{y=0} = -\text{ve}$ ☐ (d) none of the above. ☐

213. Drag is defined as the force exerted by a flowing fluid on a solid body
    (a) in the direction of flow ☐
    (b) perpendicular to the direction of flow ☐
    (c) in the direction which is at an angle of 45° to the direction of flow ☐
    (d) none of the above. ☐

214. Lift force is defined as the force exerted by a flowing fluid on a solid body
    (a) in the direction of flow ☐ (b) perpendicular to the direction of flow ☐
    (c) at an angle of 45° to the direction of flow ☐ (d) none of the above. ☐

215. Drag force is expressed mathematically as
    (a) $F_D = \tfrac{1}{2}\rho U^2 \times C_D \times A$ ☐ (b) $F_D = \tfrac{1}{2}\rho U^2 \times C_D \times A$ ☐
    (c) $F_D = 2\rho U^2 \times C_D \times A$ ☐ (d) none of the above. ☐

216. Lift force (F_L) is expressed mathematically as
    (a) $F_L = \tfrac{1}{2}\rho U^2 \times C_L$ ☐ (b) $F_L = \tfrac{1}{2}\rho U^2 \times C_L \times A$ ☐
    (c) $F_L = 2\rho U^2 \times C_L \times A$ ☐ (d) $F_L = \rho U^2 \times C_L \times A$. ☐

217. Total drag on a body is the sum of
    (a) pressure drag and velocity drag ☐ (b) pressure drag and friction drag ☐
    (c) friction drag and velocity drag ☐ (d) none of the above. ☐

218. A body is called stream-lined body when it is placed in a flow and the surface of the body
    (a) coincides with the streamlines ☐ (b) does not coincide with the streamlines ☐
    (c) is perpendicular to the streamlines ☐ (d) none of the above. ☐

219. A body is called bluff body if the surface of the body
    (a) coincides with the streamlines ☐ (b) does not coincide with the streamlines ☐
    (c) is very smooth ☐ (d) none of the above. ☐

220. The drag on a sphere (F_D) for Reynolds number less than 0.2 is given by
 (a) $F_D = 5\pi\mu DU$
 (b) $F_D = 3\pi\mu DU$
 (c) $F_D = 2\pi\mu DU$
 (d) $F_D = \pi\mu DU$.

221. The skin friction drag on a sphere (for Reynolds number less than 0.2) is equal to
 (a) one-third of the total drag
 (b) half of the total drag
 (c) two-third of the total drag
 (d) none of the above.

222. The pressure drag on a sphere (for Reynolds number less than 0.2) is equal to
 (a) one-third of the total drag
 (b) half of the total drag
 (c) two-third of the total drag
 (d) none of the above.

223. Terminal velocity of a falling body is equal to
 (a) the maximum velocity with which body will fall
 (b) the maximum constant velocity with which body will fall
 (c) half of the maximum velocity
 (d) none of the above.

224. When a falling body has attained terminal velocity, the weight of the body is equal to
 (a) drag force minus buoyant force
 (b) buoyant force minus drag force
 (c) drag force plus the buoyant force
 (d) none of the above.

225. The tangential velocity of ideal fluid at any point on the surface of the cylinder is given by
 (a) $u_\theta = \frac{1}{2} U \sin \theta$
 (b) $u_\theta = U \sin \theta$
 (c) $u_\theta = \frac{1}{2} U \sin \theta$
 (d) none of the above.

226. The lift force (F_L) produced on a rotating circular cylinder in a uniform flow is given by
 (a) $F_L = \dfrac{LUT}{\rho}$
 (b) $F_L = \rho LUT$
 (c) $F_L = \dfrac{\rho UT}{\rho}$
 (d) $F_L = \dfrac{\rho LU}{\Gamma}$

 where L = Length of the cylinder, U = Free stream velocity, Γ = Circulation.

227. The lift co-efficient (C_L) for a rotating cylinder in a uniform flow is given by
 (a) $C_L = \dfrac{\Gamma U}{R}$
 (b) $C_L = \dfrac{\Gamma R}{U}$
 (c) $C_L = \dfrac{\Gamma}{RU}$
 (d) $C_L = \dfrac{RU}{\Gamma}$.

228. The circulation developed on an airfoil is given by
 (a) $\Gamma = \dfrac{CU \sin \alpha}{\pi}$
 (b) $\Gamma = \pi CU \sin \alpha$
 (c) $\Gamma = \dfrac{\pi CU}{\sin \alpha}$
 (d) $\Gamma = \dfrac{\pi \sin \alpha}{CU}$.

 where C = Chord length, U = Velocity of airfoil, α = Angle of attack.

Boundary Layer Flow

229. The boundary-layer takes place
    - (a) for ideal fluids
    - (b) for pipe-flow only
    - (c) for real fluids
    - (d) for flow over flat plate only.

230. The boundary layer is called turbulent boundary layer if
    - (a) Reynold number is more than 2000
    - (b) Reynold number is more than 4000
    - (c) Reynold number is more than 5×10^5
    - (d) None of the above.

231. Laminar sub-layer exists in
    - (a) laminar boundary layer region
    - (b) turbulent boundary layer region
    - (c) transition zone
    - (d) none of the above.

232. The thickness of laminar boundary layer at a distance x from the leading edge over a flat plate varies as
    - (a) $x^{4/5}$
    - (b) $x^{1/2}$
    - (c) $x^{1/5}$
    - (d) $x^{3/5}$.

233. The thickness of turbulent boundary layer at a distance x from the leading edge over a flat plate varies as
    - (a) $x^{4/5}$
    - (b) $x^{1/2}$
    - (c) $x^{1/5}$
    - (d) $x^{3/5}$.

234. The separation of boundary layer takes place in case of
    - (a) negative pressure gradient
    - (b) positive pressure gradient
    - (c) zero pressure gradient
    - (d) none of the above.

235. The velocity profile for turbulent boundary layer is
    - (a) $\dfrac{u}{U} = \sin\left(\dfrac{\pi}{2}\dfrac{y}{\delta}\right)$
    - (b) $\dfrac{u}{U} = \left(\dfrac{y}{\delta}\right)^{1/7}$
    - (c) $\dfrac{u}{U} = 2\left(\dfrac{y}{\delta}\right) - \left(\dfrac{y}{\delta}\right)^2$
    - (d) $\dfrac{u}{U} = \dfrac{3}{2}\left(\dfrac{y}{\delta}\right) - \dfrac{1}{2}\left(\dfrac{y}{\delta}\right)^3$.

236. The drag force exerted by a fluid on a body immersed in the fluid is due to
    - (a) pressure and viscous force
    - (b) pressure and gravity forces
    - (c) pressure and turbulence forces
    - (d) none of the above.

Compressible Flow

237. Equation of state is expressed as
    - (a) $p\rho = RT$
    - (b) $\dfrac{p}{\rho} = RT$
    - (c) $\dfrac{\rho}{p} = RT$
    - (d) $\dfrac{p}{\rho} = \dfrac{R}{T}$

    where p = Absolute pressure, T = Absolute temperature, R = Gas constant, ρ = Density of gas.

238. The continuity equation in differential form is

(a) $\dfrac{dA}{A} + \dfrac{dV}{V} + \dfrac{d\rho}{\rho} = 0$ ☐

(b) $AdA + VdV + \rho d\rho = 0$ ☐

(c) $\dfrac{A}{dA} + \dfrac{V}{dV} + \dfrac{\rho}{d\rho} = $ Constant ☐

(d) $\dfrac{dA}{A} + \dfrac{dV}{V} + \dfrac{d\rho}{\rho} = $ Constant. ☐

239. Velocity of sound wave (C) for isothermal process is given by

(a) $C = \sqrt{\dfrac{p}{\rho}}$ ☐

(b) $C = \sqrt{\dfrac{k p}{\rho}}$ ☐

(c) $C = \sqrt{p\rho}$ ☐

(d) $C = \sqrt{\dfrac{p}{k\rho}}$ ☐

where k = Ratio of specific heats.

240. Super-sonic flow means

(a) Mach number = 1.0 ☐
(b) Mach number = 1.0 ☐
(c) Mach number > 1.0 ☐
(d) None of the above. ☐

241. Sonic-flow means

(a) Mach number < 1.0 ☐
(b) Mach number = 1.0 ☐
(c) Mach number > 1.0 ☐
(d) None of the above. ☐

242. In sonic-flow, the disturbances, created by a projectile, moves

(a) along the projectile ☐
(b) ahead of the projectile ☐
(c) behind the projectile ☐
(d) none of the above. ☐

243. In super-sonic flow, the projectile (which creates disturbances) moves

(a) ahead of the disturbances ☐
(b) along the disturbances ☐
(c) behind the disturbances ☐
(d) none of the above. ☐

244. Mach angle (α) is given by

(a) $\sin \alpha = \dfrac{V}{C}$ ☐

(b) $\sin \alpha = VC$ ☐

(c) $\sin \alpha = \dfrac{C}{V}$ ☐

(d) $\sin \alpha = \dfrac{1}{VC}$. ☐

245. For sub-sonic flow, if the area of flow increases

(a) velocity is constant ☐
(b) velocity increases ☐
(c) velocity decreases ☐
(d) none of the above. ☐

246. For super-sonic flow, if the area of flow increases, then

(a) velocity decreases ☐
(b) velocity increases ☐
(c) velocity is constant ☐
(d) none of the above. ☐

247. The area velocity relationship for compressible fluids is

(a) $\dfrac{dA}{A} = \dfrac{dV}{V} [1 - M^2]$
(b) $\dfrac{dA}{A} = \dfrac{dV}{V} [M^2 - 1]$
(c) $\dfrac{dA}{A} = \dfrac{dV}{V} [1 - V^2]$
(d) $\dfrac{dA}{A} = \dfrac{dV}{V} [C^2 - 1]$.

Channel Flow I

248. The flow in open channel is laminar if the Reynold number is
(a) 2000
(b) less than 2000
(c) less than 500
(d) none of the above.

249. The flow in open channel is turbulent if the Reynold number is
(a) 2000
(b) more than 2000
(c) more than 4000
(d) 4000.

250. If the Froude number in open channel flow is less than 1.0, the flow is called
(a) critical flow
(b) super-critical
(c) sub-critical
(d) none of the above.

251. If the Froude number in open channel flow is equal to 1.0, the flow is called
(a) critical flow
(b) streaming flow
(c) shooting flow
(d) none of the above.

252. If the Froude number in open channel flow is more than 1.0, the flow is called
(a) critical flow
(b) streaming flow
(c) shooting flow
(d) none of the above.

253. Chezy's formula is given as
(a) $V = i\sqrt{mC}$
(b) $V = C\sqrt{mi}$
(c) $V = m\sqrt{Ci}$
(d) none of the above.

254. The discharge through a rectangular channel is maximum when
(a) $m = \dfrac{d}{3}$
(b) $m = \dfrac{d}{2}$
(c) $m = 2d$
(d) $m = \dfrac{3d}{2}$

where m = Hydraulic mean depth, d = Depth of flow.

255. The discharge through a trapezoidal channel is maximum when
(a) half of top width = sloping side
(b) top width = half of sloping side
(c) top width = 1.5 × sloping side
(d) none of the above.

256. The maximum velocity through a circular channel takes place when depth of flow is equal to
(a) 0.95 times the diameter
(b) 0.5 times the diameter
(c) 0.81 times the diameter
(d) 0.3 times the diameter.

257. The maximum discharge through a circular channel takes place when depth of flow is equal to
    (a) 0.95 times the diameter
    (b) 0.3 times the diameter
    (c) 0.81 times the diameter
    (d) 0.5 times the diameter.

258. Specific energy of a flowing fluid per unit weight is equal to
    (a) $\dfrac{p}{\rho \times g} + \dfrac{V^2}{2g}$
    (b) $\dfrac{p}{\rho \times g} + h$
    (c) $\dfrac{V^2}{2g} + h$
    (d) $\dfrac{p}{\rho \times g} + \dfrac{V^2}{2g} + h$.

259. The depth of flow after hydraulic jump is
    (a) $d_2 = \dfrac{d_1}{2}[\sqrt{1+8(F_e)_1^2} - 1]$
    (b) $d_2 = \dfrac{d_1}{2}[1 + \sqrt{1+8(F_e)_1^2 - 1}]$
    (c) $d_2 = \dfrac{d_1}{2} + \sqrt{\dfrac{d_1^2}{4} + 8(F_e)_1}$
    (d) none of the above.

260. The depth of flow at which specific energy is minimum is called
    (a) normal depth
    (b) critical depth
    (c) alternate depth
    (d) none of the above.

261. The critical depth (h_c) is given by
    (a) $\left(\dfrac{q^2}{g}\right)^{1/2}$
    (b) $\left(\dfrac{q}{g}\right)^{1/3}$
    (c) $\left(\dfrac{q^2}{g}\right)^{1/3}$
    (d) $\left(\dfrac{q^2}{g}\right)^{2/3}$

    where q = Rate of flow per unit width of channel.

262. For a circular channel, the wetted perimeter is given by
    (a) $\dfrac{R\theta}{2}$
    (b) $3R\theta$
    (c) $2R\theta$
    (d) $R\theta$

    where R = Radius of circular channel and θ = Half the angle subtended by the water surface at the centre.

263. For a circular channel, the area of flow is given by
    (a) $R^2\left(2\theta - \dfrac{\sin 2\theta}{2}\right)$
    (b) $R^2\left(\theta - \dfrac{\sin 2\theta}{2}\right)$
    (c) $R^2(\theta - \sin 2\theta)$
    (d) none of the above

    where θ = Half the angle subtended by water surface at the centre and R = Radius of circular channel.

264. The hydraulic mean depth is given by

   (a) $\dfrac{P}{A}$ □
   (b) $\dfrac{P^2}{A}$ □
   (c) $\dfrac{A}{P}$ □
   (d) $\sqrt{\dfrac{A}{P}}$ □

   where A = Area and P = Wetted perimeter.

265. A most economical section is one which for a given cross-sectional area, slope of bed (i) and co-efficient of resistance has

   (a) maximum wetted perimeter □
   (b) maximum discharge □
   (c) maximum depth of flow □
   (d) none of the above. □

Hydraulic Turbines

266. Specific speed of a turbine is defined as the speed of the turbine which

   (a) produces unit power at unit head □
   (b) produces unit power at unit discharge □
   (c) delivers unit discharge at unit head □
   (d) delivers unit discharge at unit power. □

267. A pump is defined as a device which converts

   (a) hydraulic energy into mechanical energy □
   (b) mechanical energy into hydraulic energy □
   (c) kinetic energy into mechanical energy □
   (d) none of the above. □

268. A turbine is a device which converts

   (a) hydraulic energy into mechanical energy □
   (b) mechanical energy into hydraulic energy □
   (c) kinetic energy into mechanical energy □
   (d) electrical energy into mechanical energy. □

269. The force exerted by a jet of water on a stationary vertical plate in the direction of jet is given by

   (a) $F_x = \rho A V^2 \sin^2 \theta$ □
   (b) $F_x = \rho A V^2 [1 + \cos \theta]$ □
   (c) $F_x = \rho A V^2$ □
   (d) none of the above. □

270. The force exerted by a jet of water on a stationary inclined plate in the direction of jet is given by

   (a) $F_x = \rho A V^2$ □
   (b) $F_x = \rho A V^2 \sin^2 \theta$ □
   (c) $F_x = \rho A V^2 [1 + \cos \theta]$ □
   (d) $F_x = \rho A V^2 [1 + \sin \theta]$. □

271. The force exerted by a jet of water on a stationary curved plate in the direction of jet is equal to

   (a) $\rho A V^2$ □
   (b) $\rho A V^2 \sin^2 \theta$ □
   (c) $\rho A V^2 (1 + \cos \theta)$ □
   (d) $\rho A V^2 [1 + \sin \theta]$. □

272. The force exerted by a jet of water having velocity V on a vertical plate, moving with a velocity u is given by

   (a) $F_x = \rho A (V - u)^2 \sin^2 \theta$ □
   (b) $F_x = \rho A (V - u)^2$ □
   (c) $F_x = \rho A (V - u)^2 [1 + \cos \theta]$ □
   (d) none of the above. □

273. The force exerted by a jet of water having velocity V on a series of vertical plates moving with velocity u is given by
    (a) $P_x = \rho A V^2$
    (b) $F_x = \rho A (V-u)^2$
    (c) $F_x = \rho A V u$
    (d) none of the above.

274. Efficiency of the jet of water having velocity V striking a series of vertical plates moving with a velocity u is given by
    (a) $\eta = \dfrac{2V(V-u)}{u^2}$
    (b) $\eta = \dfrac{2u(V-u)}{V^2}$
    (c) $\eta = \dfrac{u^2}{V^2(V-u)}$
    (d) none of the above.

275. Efficiency, of the jet of water having velocity V and striking a series of vertical plates moving with a velocity u, is maximum when
    (a) $u = 2V$
    (b) $u = \dfrac{V}{2}$
    (c) $u = \dfrac{3V}{2}$
    (b) $u = \dfrac{4V}{3}$.

276. Maximum efficiency of a series of vertical plates is
    (a) 66.67%
    (b) 33.33%
    (c) 50%
    (d) 80%.

277. For a series of curved radial vanes, the work done per second per unit weight is equal to
    (a) $\dfrac{1}{g} V w_1 u_1 + V w_2 u_2$
    (b) $\dfrac{1}{g} [V_1 u_1 + V_2 u_2]$
    (c) $\dfrac{1}{g} [V w_1 u_2 \pm V w_2 u_2]$
    (d) none of the above.

278. The net head (H) on the turbine is given by
    (a) H = Gross Head + Head lost due to friction
    (b) H = Gross Head − Head lost due to friction
    (c) H = Gross Head + $\dfrac{V^2}{2g}$ − Head lost due to friction.

279. Hydraulic efficiency of a turbine is defined as the ratio of
    (a) power available at the inlet of turbine to power given by water to the runner
    (b) power at the shaft of the turbine to power given by water to the runner
    (c) power at the shaft of the turbine to the power at the inlet of turbine
    (d) none of the above.

280. Mechanical efficiency of a turbine is the ratio of
    (a) power at the inlet to the power at the shaft of turbine
    (b) power at the shaft to the power given to the runner

(c) power at the shaft to power at the inlet of turbine
(d) none of the above.

281. The overall efficiency of a turbine is the ratio of
    (a) power at the inlet of turbine to the power at the shaft
    (b) power at the shaft to the power given to the runner
    (c) power at the shaft to the power at the inlet of turbine
    (d) none of the above.

282. The relation between hydraulic efficiency (η_h), mechanical efficiency (η_m) and overall efficiency (η_0) is
    (a) $\eta_h = \eta_0 \times \eta_m$
    (b) $\eta_0 = \eta_h \times \eta_m$
    (c) $\eta_0 = \dfrac{\eta_m}{\eta_h}$
    (d) none of the above.

283. A turbine is called impulse if at the inlet of the turbine
    (a) total energy is only kinetic energy
    (b) total energy is only pressure energy
    (c) total energy is the sum of kinetic energy and pressure energy
    (d) none of the above.

284. A turbine is called reaction turbine if at the inlet of the turbine the total energy is
    (a) kinematic energy only
    (b) kinetic energy and pressure energy
    (c) pressure energy only
    (d) none of the above.

285. Tick mark the correct statement
    (a) Pelton wheel is a reaction turbine
    (b) Pelton wheel is a radial flow turbine
    (c) Pelton wheel is an impulse turbine
    (d) None of the above.

286. Francis turbine is
    (a) an impulse turbine
    (b) a radial flow impulse turbine
    (c) an axial flow turbine
    (d) a reaction radial flow turbine.

287. Kaplan turbine is
    (a) an impulse turbine
    (b) a radial flow impulse turbine
    (c) an axial flow reaction turbine
    (d) a radial flow reaction turbine.

288. Jet ratio (m) is defined as the ratio of
    (a) diameter of jet of water to diameter of Pelton wheel
    (b) velocity of vane to the velocity of jet of water
    (c) velocity of flow of the jet of water
    (d) diameter of Pelton wheel to diameter of the jet of water.

289. Flow ratio is defined as the ratio of
    (a) velocity of flow at inlet to the velocity given by $\sqrt{2gH}$
    (b) velocity of runner at inlet to the velocity of flow at inlet

(c) velocity of runner to the velocity given by $\sqrt{2gH}$

(d) none of the above.

290. Speed ratio is given by

(a) $\dfrac{u}{\sqrt{2gh}}$

(b) $\dfrac{V_f}{\sqrt{2gh}}$

(c) $\dfrac{\sqrt{2gH}}{V_f}$

(d) $\dfrac{V_w}{\sqrt{2gH}}$.

291. The speed ratio for Pelton wheel varies from

(a) 0.45 to 0.50

(b) 0.6 to 0.7

(c) 0.3 to 0.4

(d) 0.8 to 0.9.

292. The discharge through Pelton Turbine is given by

(a) $Q = \pi DBV_f$

(b) $Q = \dfrac{\pi}{4} d^2 \times \sqrt{2gH}$

(c) $Q = \dfrac{\pi}{4} [D_0^2 - D_0^2] \times V_f$

(d) none of the above.

293. The discharge through Francis Turbine is given by

(a) $Q = \pi DBV_f$

(b) $Q = \dfrac{\pi}{4} d^2 \times \sqrt{2gh}$

(c) $Q = \dfrac{\pi}{4} [D_0^2 - D_0^2]$

(d) none of the above.

294. The discharge through Kaplan turbine is given by

(a) $Q = \pi DBV_f$

(b) $Q = \dfrac{\pi}{4} d^2 \times \sqrt{2gH}$

(c) $Q = \dfrac{\pi}{4} [D_0^2 - D_0^2]$

(d) $Q = 0.9 \, \pi DBV_f$

295. Draft tube is used for discharging water from the exit of

(a) an impulse turbine

(b) a Francis turbine

(c) a Kaplan turbine

(d) a Pelton wheel.

296. Specific speed of a turbine is defined as the speed at which the turbine runs when

(a) working under unit head and discharging one litre per second

(b) working under unit head and develops unit horse power

(c) develops unit horse power and discharges on litre per second

(d) none of the above.

297. The specific speed (N_s) of a turbine is given by

(a) $N_s = \dfrac{N\sqrt{P}}{H^{3/4}}$

(b) $N_s = \dfrac{N\sqrt{Q}}{H^{3/4}}$

(c) $N_s = \dfrac{N\sqrt{P}}{H^{5/4}}$

(d) $N_s = \dfrac{NP^{5/4}}{\sqrt{H}}$.

298. Unit speed is the speed of a turbine when it is working
(a) under unit head and develops unit power
(b) under unit head and discharge one m³/sec
(c) under unit head
(d) none of the above.

299. Unit discharge is the discharge of a turbine when
(a) the head on turbine is unity and it develops unit power
(b) the head on turbine is unity and it moves at unit speed
(c) the head on the turbine is unity
(d) none of the above.

300. Unit power is the power developed by a turbine when
(a) head on turbine is unity and discharge is also unity
(b) head = one metre and speed is unity
(c) head on turbine is unity
(d) none of the above.

301. The unit speed (N_u) is given by the expression

(a) $N_u = \dfrac{N}{H^{3/2}}$

(b) $N_u = \dfrac{N}{H^{3/4}}$

(c) $N_u = \dfrac{N}{\sqrt{H}}$

(d) $N_u = \dfrac{N}{H^{5/4}}$.

302. The unit discharge (Q_u) is given by the expression

(a) $Q_u = \dfrac{Q}{\sqrt{H}}$

(b) $Q_u = \dfrac{Q}{H^{3/2}}$

(c) $Q_u = \dfrac{Q}{H^{3/4}}$

(d) $Q_u = \dfrac{Q}{H^{5/4}}$.

303. Unit power (P_u) is given by the expression

(a) $P_u = \dfrac{P}{\sqrt{H}}$

(b) $P_u = \dfrac{Q}{H^{3/2}}$

(c) $P_u = \dfrac{Q}{H^{3/4}}$

(d) $P_u = \dfrac{Q}{H^{5/4}}$.

304. The unit discharge (Q_u) and unit speed (N_u) curves for different turbines are shown in Fig. 1.5. Curve A is for
    (a) Francis Turbine ☐
    (b) Kaplan Turbine ☐
    (c) Pelton Turbine ☐
    (d) Propeller Turbine. ☐

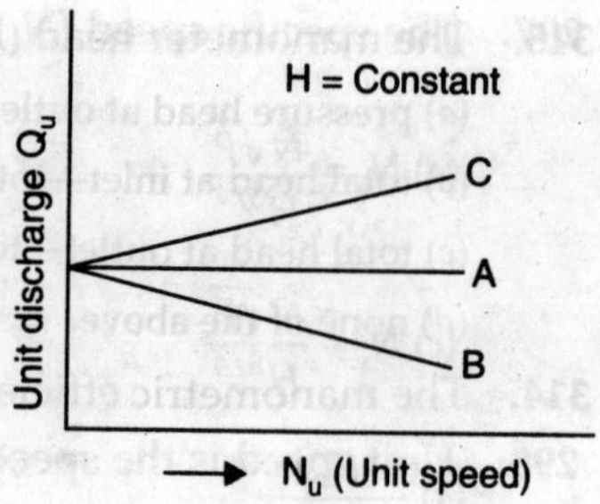

305. Curve B in Fig. 1.5 is for
    (a) Francis Turbine ☐
    (b) Kaplan Turbine ☐
    (c) Pelton Turbine ☐
    (d) Propeller Turbine. ☐

FIGURE 1.5

306. Curve C in Fig. 1.5 is for
    (a) Francis Turbine ☐   (b) Kaplan Turbine ☐
    (c) Pelton Turbine ☐   (d) Propeller Turbine. ☐

307. Tick mark the correct statement
    (a) Curves at constant speed are called main characteristic curves. ☐
    (b) Curves at constant head are called main characteristic curves. ☐
    (c) Curves at constant efficiency are called operating characteristic curves. ☐
    (d) Curves at constant efficiency are called main characteristic curves. ☐

308. Main characteristic curve of a turbine means
    (a) curves at constant speed ☐   (b) curves at constant efficiency ☐
    (c) curves at constant head ☐   (d) none of the above. ☐

309. Operating characteristic curves of a turbine means
    (a) curves drawn at constant speed ☐   (b) curves drawn at constant efficiency ☐
    (c) curves drawn at constant head ☐   (d) none of the above. ☐

310. Muschel curves means
    (a) curves at constant head ☐   (b) curves at constant speed ☐
    (c) curves at constant efficiency ☐   (d) none of the above. ☐

311. Governing of a turbine means
    (a) the head is kept constant under all condition of working ☐
    (b) the speed is kept constant under all conditions ☐
    (c) the discharge is kept constant under all conditions ☐
    (d) none of the above. ☐

Centrifugal and Reciprocating Pumps

312. The work done by impeller of a centrifugal pump on water per second per unit weight of water is given by
    (a) $\frac{1}{g} Vw_1 u_1$ ☐   (b) $\frac{1}{g} Vw_2 u_2$ ☐
    (c) $\frac{1}{g} (Vw_1 u_2 - Vw_1 u_1)$ ☐   (d) none of the above. ☐

313. The manometer head (H_m) of a centrifugal pump is given by
    (a) pressure head at outlet of pump—pressure head at inlet
    (b) total head at inlet—total head at outlet
    (c) total head at outlet—total head at inlet
    (d) none of the above.

314. The manometric efficiency (η_{man}) of a centrifugal pump is given by
    (a) $\dfrac{H_m}{gVw_2u_2}$
    (b) $\dfrac{gH_m}{Vw_2u_2}$
    (c) $\dfrac{Vw_2u_2}{gH_m}$
    (d) $\dfrac{g \times Vw_2u_2}{H_m}$.

315. Mechanical efficiency ($\eta_{mech.}$) of a centrifugal pump is given by
    (a) (power at the impeller)/shaft power
    (b) shaft power/power at the impeller
    (c) power possessed by water/power at the impeller
    (d) power possessed by water/shaft power.

316. To produce a high head by multi-stage centrifugal pumps, the impellers are connected
    (a) in parallel
    (b) in series
    (c) in parallel and in series both
    (d) none of the above.

317. To discharge a large quantity of liquid by multi-stage centrifugal pump, the impellers are connected
    (a) in parallel
    (b) in series
    (c) in parallel and in series
    (d) none of the above.

318. Specific speed of a pump is the speed at which a pump runs when
    (a) head developed is unity and discharge is one cubic metre
    (b) head developed is unity and shaft horse power is also unity
    (c) discharge is one cubic metre and shaft horse power is unity
    (d) none of the above.

319. The specific speed (N_s) of a pump is given by the expression
    (a) $N_s = \dfrac{N\sqrt{Q}}{H_m^{5/4}}$
    (b) $N_s = \dfrac{N\sqrt{P}}{H_m^{3/4}}$
    (c) $N_s = \dfrac{N\sqrt{Q}}{H_m^{3/4}}$
    (d) $N_s = \dfrac{N\sqrt{P}}{H_m^{5/4}}$.

320. The operating characteristic curves of a centrifugal pump are shown in Fig. 1.6.
    Curve A is for
    (a) head
    (b) efficiency
    (c) power
    (d) none of the above.

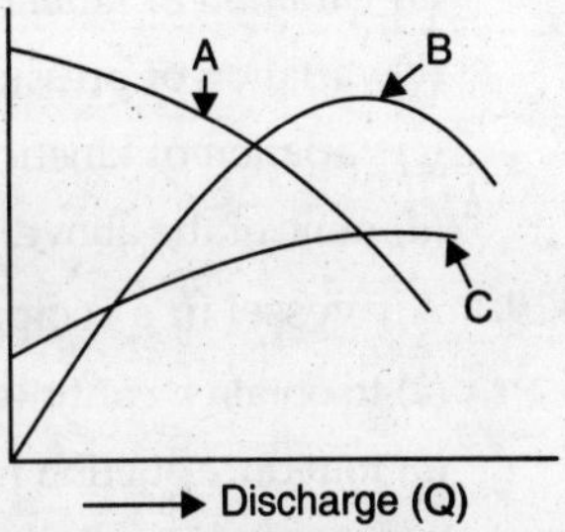

FIGURE 1.6

321. Curve B in Fig. 1.6 is for
    (a) head □
    (b) efficiency □
    (c) power □
    (d) none of the above. □

322. Curve C in Fig. 1.6 is for
    (a) head □
    (b) efficiency □
    (c) power □
    (d) none of the above. □

323. Cavitation will take place if the pressure of the flowing fluid at any point is
    (a) more than vapour pressure of the fluid □
    (b) equal to vapour pressure of the fluid □
    (c) is less than vapour pressure of the fluid □
    (d) none of the above. □

324. Cavitation can take place in case of
    (a) Pelton Wheel □
    (b) Francis Turbine □
    (c) Reciprocating pump □
    (d) Centrifugal pump. □

325. Tick mark the correct statement
    (a) Centrifugal pump convert mechanical energy into hydraulic energy by sucking liquid into chamber. □
    (b) Reciprocating pumps convert mechanical energy into hydraulic energy by means of centrifugal force. □
    (c) Centrifugal pumps convert mechanical energy into hydraulic energy by means of centrifugal force. □
    (d) Reciprocating pumps convert hydraulic energy into mechanical energy. □

326. The discharge through a single acting reciprocating pump is
    (a) $Q = \dfrac{ALN}{60}$ □
    (b) $Q = \dfrac{2ALN}{60}$ □
    (c) $Q = ALN$ □
    (d) $Q = 2ALN$. □

327. The pressure head due to acceleration (h_a) is reciprocating pump is given by
    (a) $h_a = \dfrac{l}{g} \times \dfrac{a}{A} \times \omega^2 r \sin\theta$ □
    (b) $h_a = \dfrac{l}{g} \times \dfrac{A}{a} \times \omega^2 r \sin\theta$ □
    (c) $h_a = \dfrac{l}{g} \times \dfrac{A}{a} \times \omega^2 r \cos\theta$ □
    (d) $h_a = \dfrac{A}{a} \omega^2 r \sin\theta$ □

    where A = Area of cylinder, a = Area of pipe and r = Radius of crank.

328. Indicator diagram shows for one complete revolution of crank the
    (a) variation of kinetic head in the cylinder □
    (b) variation of pressure head in the cylinder □
    (c) variation of kinetic and pressure head in the cylinder □
    (d) none of the above. □

329. Air vessel in a reciprocating pump is used
    (a) to obtain a continuous supply of water at uniform rate □
    (b) to reduce suction head □

(c) to increase the delivery head
(d) none of the above.

330. The work saved by fitting as air vessel to a single acting reciprocating pump is
    (a) 39.2%
    (b) 84.8%
    (c) 48.8%
    (d) 92.3%.

331. The work saved by fitting an air vessel to a double acting reciprocating pump is
    (a) 39.2%
    (b) 84.8%
    (c) 48.8%
    (d) 92.3%.

332. The pressure, at which separation takes place, is known separation pressure or separation pressure head. For water, the limiting value of separation pressure head is
    (a) 2.5 m (abs.)
    (b) 7.5 m (abs.)
    (c) 10.3 m (abs.)
    (d) 5 m (abs.).

333. During suction stroke of a reciprocating pump, the separation may take place
    (a) at the end of suction stroke
    (b) in the middle of suction stroke
    (c) in the beginning of suction stroke
    (d) none of the above.

334. During delivery stroke of a reciprocating pump, the separation may take place
    (a) at the end of delivery stroke
    (b) in the middle of delivery stroke
    (c) in the beginning of the delivery stroke
    (d) none of the above.

Miscellaneous Hydraulic Devices

335. Hydraulic accumulator is a device used for
    (a) lifting heavy weights
    (b) storing the energy of a fluid in the form of pressure energy
    (c) increasing the pressure intensity of a fluid
    (d) none of the above.

336. Hydraulic intensifier is a device used for
    (a) storing energy of a fluid in the form of pressure energy
    (b) increasing pressure intensity of a fluid
    (c) transmitting power from one shaft to another
    (d) none of the above.

337. Hydraulic ram is a pump which works
    (a) on the principle of water-hammer
    (b) on the principle of centrifugal action
    (c) on the principle of reciprocating action
    (d) none of the above.

338. Hydraulic coupling is a device used for
    (a) transmitting same torque to the driven shaft
    (b) transmitting increased torque to the driven shaft
    (c) transmitting decreased torque to the driven shaft
    (d) none of the above.

339. Torque converter is a device used for
 (a) transmitting same torque to the driven shaft
 (b) transmitting increased torque to the driven shaft
 (c) transmitting decreased torque to the driven shaft
 (d) transmitting increased or decreased torque to the driven shaft.

340. Capacity of a hydraulic accumulator is given as equal to
 (a) pressure of water supplied by pump × volume of accumulator
 (b) pressure of water × area of accumulator
 (c) pressure of water × stroke of the ram of accumulator
 (d) none of the above.

341. Kaplan turbine is a propeller turbine in which the vanes fixed on the bub are
 (a) non-adjustable
 (b) adjustable
 (c) fixed
 (d) none of the above.

342. If the head on the turbine is more than 300 m, the type of turbine used should be
 (a) Kaplan
 (b) Francis
 (c) Pelton
 (d) Propeller.

343. If the specific speed of a turbine is more than 300, the type of turbine is
 (a) Pelton
 (b) Kaplan
 (c) Francis
 (d) Pelton with more jets.

344. Run-away speed of a Pelton wheel means
 (a) full load speed
 (b) no load speed
 (c) no load speed with no governor mechanism
 (d) none of the above.

345. Spouting velocity means
 (a) actual velocity of jet
 (b) ideal velocity of jet
 (c) half of ideal velocity of jet
 (d) none of the above.

346. Surge tank in a pipe line is used to
 (a) reduce the loss of head due to friction in pipe
 (b) make the flow uniform in pipe
 (c) relieve the pressure due to water hammer
 (d) none of the above.

347. Hydraulic ram is a device used for
 (a) storing energy of a water in the form of pressure energy
 (b) increasing pressure intensity of water
 (c) lifting small quantity of water to a greater height by means of large quantity of water falling through small height
 (d) none of the above.

348. For low head and high discharge, the suitable turbine is
    (a) Pelton
    (b) Francis
    (c) Kaplan
    (d) None of the above.

349. For high head and low discharge, the suitable turbine is
    (a) Pelton
    (b) Francis
    (c) Kaplan
    (d) None of the above.

350. The flow of water, leaving the impeller, in a centrifugal pump casing is
    (a) forced vortex flow
    (b) free vortex flow
    (c) centrifugal flow
    (d) none of the above.

351. Rotameter is used for measuring
    (a) density of fluids
    (b) velocity of fluids in pipes
    (c) discharge of fluids
    (d) viscosity of fluids.

352. A current meter is a device used for measuring
    (a) velocity
    (b) viscosity
    (c) current
    (d) pressure.

353. A hot wire anemometer is a device used for measuring
    (a) viscosity
    (b) velocity of gases
    (c) pressure of gases
    (d) none of the above.

354. D' Aubuissons efficiency of a Hydraulic Ram as compared to Rankine's efficiency is
    (a) less
    (b) more
    (c) equal
    (d) none of the above.

355. The value of specific weight for water in S.I. units is equal to
    (a) 981 N/m³
    (b) 98.1 N/m³
    (c) 9810 N/m³
    (d) 1000 N/m³.

356. The angle of contact (θ) between water and glass tube in case of capillary rise is equal to
    (a) 0°
    (b) 90°
    (c) 128°
    (d) 150°.

357. The angle of contact (θ) between mercury and glass tube in case of capillary depression is
    (a) 0°
    (b) 90°
    (c) 128°
    (d) 150°.

358. Numerical value of gauge pressure is
    (a) more than absolute pressure
    (b) less than absolute pressure
    (c) equal to the absolute pressure
    (d) none of the above.

359. Hydraulic mean depth is given by
    (a) $\dfrac{P}{A}$
    (b) $\dfrac{A}{P}$
    (c) $A \times P$
    (d) $\dfrac{1}{AP}$

    where A = Area and P = Wetted perimeter.

360. For a superonic flow, velocity increases
    (a) with the decrease of area of flow
    (b) with the increase of area of flow
    (c) when area of flow is constant
    (d) none of the above.

Channel Flow II

361. A flow in a channel will be laminar, if Reynolds number of the flow
    (a) is less than 2000
    (b) is less than 1500
    (c) is less than 1000
    (d) is less than 500.

362. The flow in the channel will be turbulent if Reynolds number is more than
    (a) 500
    (b) 1000
    (c) 1500
    (d) 2000.

363. If the depth of flow along the length of the channel is constant, then flow is known as
    (a) laminar
    (b) uniform
    (c) non-uniform
    (d) steady.

364. If the depth of flow along the length of the channel is variable, then flow is known as
    (a) laminar
    (b) uniform
    (c) non-uniform
    (d) steady.

365. If the velocity at a section of a channel is constant with respect to time, then the flow in the channel is known as
    (a) uniform
    (b) laminar
    (c) steady
    (d) unsteady.

366. If the velocity at a section of a channel is variable with respect to time, then the flow in the channel is known as
    (a) uniform
    (b) laminar
    (c) steady
    (d) unsteady.

367. If the depth of flow in a channel is varying slowly, then the flow is known as
    (a) G.V.F.
    (b) R.V.F.
    (c) P.V.F.
    (d) M.V.F.

368. If the depth of flow in a channel is varying rapidly, then the flow is known as
    (a) G.V.F.
    (b) R.V.F.
    (c) P.V.F.
    (d) M.V.F.

369. For sub-critical flow in a channel
    (a) $F_e = 1$
    (b) $F_e > 1$
    (c) $F_e < 1$
    (d) none of the above

    where F_e = Froude number.

370. Hydraulic mean depth is given by
    (a) $m = \dfrac{P}{A}$
    (b) $\dfrac{A}{P}$
    (c) $m = A \times P$
    (d) $\dfrac{1}{AP}$

    where A = Area of flow, P = Wetted perimeter, and m = Hydraulic mean depth.

371. The Froude number is given by

(a) $F_e = \sqrt{\dfrac{V}{gD}}$ ☐ (b) $F_e = \dfrac{V}{\sqrt{gD}}$ ☐

(c) $F_e = \dfrac{V \times g}{\sqrt{D}}$ ☐ (d) $F_e = \dfrac{g}{\sqrt{VD}}$ ☐

where V = Velocity of flow, D = Hydraulic depth of channel, and
g = Acceleration due to gravity.

372. The hydraulic depth of a channel is equal to

(a) $D = A \times T$ ☐ (b) $D = \dfrac{A}{T}$ ☐

(c) $D = \dfrac{1}{AT}$ ☐ (d) $D = \dfrac{T}{A}$ ☐

where A = Area of flow and T = Top width.

373. Chezy's formula is used for finding ...... in a channel.

(a) velocity ☐ (b) discharge ☐
(c) slope ☐ (d) hydraulic mean depth. ☐

374. Chezy's formula is given by,

(a) $V = \sqrt{C\,mi}$ ☐ (b) $Q = AC\sqrt{mi}$ ☐

(c) $V = C\sqrt{mi}$ ☐ (d) $m = \dfrac{A}{P}$ ☐

where V = Velocity of flow, C = Chezy's constant,
m = Hydraulic mean depth, and i = Slope of the bed of channel.

375. Manning's formula is used for finding ...... in a channel.

(a) rate of flow ☐ (b) velocity ☐
(c) hydraulic mean depth ☐ (d) slope. ☐

376. Manning's formula is given by

(a) $\dfrac{1}{N}\sqrt{mi}$ ☐ (b) $\dfrac{1}{N}\,m^{2/3}\,i^{1/2}$ ☐

(c) $\dfrac{1}{N}\,m^{1/6}\,i^{1/2}$ ☐ (d) $\dfrac{1}{N}\,m^{2/5}\,i^{3/4}$ ☐

where N = Manning's constant, m = Hydraulic mean depth, and
i = Slope of the bed of channel.

377. The relation between Chezy's constant and Manning's constant is given by

(a) $C = \dfrac{1}{N}\,m^{1/2}$ ☐ (b) $C = \dfrac{1}{N}\,m^{1/3}$ ☐

(c) $C = \dfrac{1}{N}\,m^{1/6}$ ☐ (d) $C = \dfrac{1}{N}\,m^{1/9}$. ☐

378. Which one of the following is a dimensionless constant?
    (a) Chezy's constant
    (b) Manning's constant
    (c) Reynolds number
    (d) none of the above.

379. The dimensions of the Chezy's constant is given by
    (a) $L^{1/2} T^{1/2}$
    (b) $L^{1/2} T^{3/4}$
    (c) $L^{1/2} T^{-1}$
    (d) $LT^{1/2}$.

380. The expression $\left(\dfrac{b \times d}{b + 2d}\right)$ represents the hydraulic mean depth for
    (a) triangular channel
    (b) rectangular channel
    (c) trapezoidal channel
    (d) circular channel
    where b = Width of channel and d = Depth of flow.

381. The area of flow of a trapezoidal channel is given by
    (a) $A = (b + 2nd) \times d$
    (b) $A = (b + nd) \times d$
    (c) $A = (2b + nd) \times d$
    (d) $A = (b + nd) \times 2d$
    where b = Width, d = Depth, and n = Side slope of the channel.

382. The wetted perimeter in a trapezoidal is given by
    (a) $P = b + d\sqrt{n^2 + 1}$
    (b) $P = 2b + d\sqrt{n^2 + 1}$
    (c) $P = b + 2d\sqrt{n^2 + 1}$
    (d) $P = 2b + d\sqrt{n^2 + 1}$.

383. If angle subtended at the centre of the circular channel by free surface of water is '2θ', then wetted perimeter is given by,
    (a) $P = R \times \theta$
    (b) $P = 2R\theta$
    (c) $P = 4R\theta$
    (d) $P = 3R\theta$.

384. The area of flow for a circular channel is given by
    (a) $A = R^2 (\theta - 2 \sin 2\theta)$
    (b) $A = 2R^2 (2\theta - \sin 2\theta)$
    (c) $A = R^2 \left(\theta - \dfrac{\sin 2\theta}{2}\right)$
    (d) $A = R^2 \left(\dfrac{\theta}{2} - \sin 2\theta\right)$
    where R = Radius and 2θ = Angle at the centre.

385. If the angle of the triangular channel is θ, then area of flow is given by
    (a) $A = 2d^2 \tan 2\theta$
    (b) $A = d^2 \tan \theta$
    (c) $A = 3d^2 \tan \dfrac{\theta}{2}$
    (d) $A = d^2 \tan \left(\dfrac{\theta}{2}\right)$
    where d = Depth of flow.

386. The wetted perimeter P in the above question is given by
    (a) $P = 2d \sec \left(\dfrac{\theta}{2}\right)$
    (b) $P = \dfrac{d}{2} \sec \theta$
    (c) $P = d \sec 2\theta$
    (d) $P = 2d \sec \theta$.

387. A section of a channel is most efficient if the discharge through the channel is maximum for a given
  (a) area of flow
  (b) surface resistance
  (c) slope of the bed of channel
  (d) all of the above.

388. A rectangular channel will be most efficient if depth of flow is equal to
  (a) width of the channel
  (b) twice the width of the channel
  (c) half the width of channel
  (d) one-quarter width of channel.

389. For the most efficient rectangular channel, the hydraulic mean depth should be equal to
  (a) depth of flow
  (b) half of depth of flow
  (c) twice the depth of flow
  (d) one-quarter depth of flow.

390. A trapezoidal channel will be most efficient if length of sloping side is equal to
  (a) twice the top width
  (b) top width
  (c) half of the top width
  (d) one-quarter of top width.

391. If d = depth of flow, b = width of trapezoidal channel and n = slope of the sides of the channel, then for most efficient trapezoidal channel, the condition is
  (a) $d\sqrt{n^2+1} = b + 2nd$
  (b) $2d\sqrt{n^2+1} = b + nd$
  (c) $d\sqrt{n^2+1} = b + 2nd$
  (d) $2d\sqrt{n^2+1} = b + 2nd$.

392. For the most efficient trapezoidal channel, the hydraulic mean depth should be
  (a) twice the depth of flow
  (b) 1.5 times the depth of flow
  (c) equal to the depth of flow
  (d) half the depth of flow.

393. A trapezoidal channel will be most efficient if
  (a) three sides of trapezoidal are equal
  (b) three sides of trapezoidal are unequal
  (c) three sides of trapezoidal section are tangential to the semi-circle described on the water line
  (d) none of the above.

394. A circular channel will have maximum velocity if depth of flow is equal to
  (a) diameter of circular channel
  (b) 0.81 times the diameter of circular channel
  (c) 0.75 times the diameter of circular channel
  (d) 0.625 times the diameter of circular channel.

395. The rate of flow through a circular channel will be maximum if depth of flow is equal to
  (a) diameter of circular channel
  (b) 0.95 times the dia. of circular channel
  (c) 0.81 times the dia. of circular channel
  (d) 0.75 times the dia. of circular channel.

396. The total energy of flowing water in a channel with respect to bed of channel is known as
  (a) total energy
  (b) specific energy
  (c) hydraulic gradient
  (d) mechanical energy.

397. The specific energy is given by
    (a) $y + \dfrac{V^2}{2g}$
    (b) $2y + \dfrac{V^2}{2g}$
    (c) $y + \dfrac{V^2}{8}$
    (d) $\dfrac{P}{w} + z + \dfrac{V^2}{2g}$
    where y = Depth of flow and V = Velocity of flow.

398. Critical depth is the depth of water at which
    (a) specific energy is maximum
    (b) specific energy is minimum
    (c) specific energy is zero
    (d) specific energy is one plus minimum specific energy.

399. For a rectangular channel, the critical depth (y_c) is given by
    (a) $y_c = \dfrac{q^2}{g}$
    (b) $y_c = \left(\dfrac{3q^2}{g}\right)^{1/3}$
    (c) $y_c = \left(\dfrac{q^2}{g}\right)^{1/3}$
    (d) $y_c = \left(\dfrac{q^2}{g}\right)^{1/4}$
    where q = Rate of flow per unit width.

400. The velocity of flow, corresponding to critical depth is known as
    (a) maximum velocity
    (b) minimum velocity
    (c) critical velocity
    (d) average velocity.

401. Hydraulic jump will take place if depth of flow is
    (a) more than critical depth
    (b) equal to critical depth
    (c) less than critical depth
    (d) two times the critical depth.

402. The loss of energy during hydraulic jump is given by
    (a) $h_L = \dfrac{(y_1 + y_2)}{4y_1 y_2}$
    (b) $h_L = \dfrac{(y_1 - y_2)^2}{2y_1 y_2}$
    (c) $h_L = \dfrac{(y_2 - y_1)^2}{2y_1 y_2}$
    (d) $h_L = \dfrac{(y_2 - y_1)^3}{4y_1 y_2}$
    where y_1 = Depth of water before hydraulic jump
    y_2 = Depth of water after hydraulic jump.

III. TRUE/FALSE

1. Newton's law of viscosity states that the shear stress is directly proportional to rate of shear strain.
    (a) True
    (b) False.

2. Kinematic viscosity is equal to viscosity multiplied by density.
   (a) True   □   (b) False.   □
3. Surface tension has the unit of force per unit length.
   (a) True   □   (b) False.   □
4. Gauge pressure at a point is equal to absolute pressure minus atmospheric pressure.
   (a) True   □   (b) False.   □
5. The increase of temperature decreases the viscosity of a gas.
   (a) True   □   (b) False.   □
6. The gases are considered incompressible when Mach number is less than 0.2.
   (a) True   □   (b) False.   □
7. Atmospheric pressure head in terms of meter of water is 7.5 m.
   (a) True   □   (b) False.   □
8. The centre of pressure of a plane horizontal surface immersed in a liquid coincides with the centre of gravity.
   (a) True   □   (b) False.   □
9. For a floating body, the buoyant force passes through the centre of gravity of the body.
   (a) True   □   (b) False.   □
10. A submerged body will be in stable equilibrium if the centre of buoyancy is above the centre of gravity.
    (a) True   □   (b) False.   □
11. The necessary condition for the flow to be steady is that the velocity changes at a point with respect to time.
    (a) True   □   (b) False.   □
12. The flow in the pipe is laminar if the Reynolds number is equal or less than 2000.
    (a) True   □   (b) False.   □
13. Continuity equation is based on law of conservation of energy.
    (a) True   □   (b) False.   □
14. The pressure variation along the radial direction for vertex flow in a horizontal plane is expressed as $\dfrac{dp}{dr} = \rho \dfrac{V^2}{r}$.
    (a) True   □   (b) False.   □
15. The co-efficient of discharge for an orifice is more than that of for a mouthpiece.
    (a) True   □   (b) False.   □
16. For the laminar flow through a circular pipe, the variation of shear stress across the cross-section of the pipe is parabolic.
    (a) True   □   (b) False.   □

17. The value of momentum correction factor for the viscous flow through a circular pipe is equal to two.
    (a) True   (b) False.

18. For a hydrodynamically smooth boundary, the term $\frac{k}{\delta'}$ should be less than 0.25 (where k = average height of the irregularities from the boundary and δ' is the thickness of laminar sub-layer).
    (a) True   (b) False.

19. The relation between co-efficient of friction (f) and Reynolds number (R_e) for laminar flow through a pipe is given by $f = \frac{16}{R_e}$.
    (a) True   (b) False.

20. Power transmitted through a pipe will be maximum when the head lost due to friction in the pipe is one-third of the total at the inlet of the pipe.
    (a) True   (b) False.

21. The ratio of inertia force to viscous force is known as Reynolds number.
    (a) True   (b) False.

22. Boundary layer on a flat plate is known as laminar boundary layer if Reynolds number is less than 2000.
    (a) True   (b) False.

23. The boundary layer separation takes place if pressure gradient is positive.
    (a) True   (b) False.

24. The force exerted by a flowing fluid on a solid body in the direction of flow is known as drag force.
    (a) True   (b) False.

25. A body is called stream-lined body when it is placed in a flow and the surface of the body coincides with the stream-lines.
    (a) True   (b) False.

26. The terminal velocity of a falling body is equal to the maximum constant velocity with which body will fall.
    (a) True   (b) False.

27. For isothermal process, the velocity of sound wave is equal to $\sqrt{\frac{p}{\rho}}$, where p = pressure and ρ = density.
    (a) True   (b) False.

28. Flow of a fluid in a pipe takes place from higher pressure to lower pressure and not from higher energy to lower energy.
    (a) True   (b) False.

29. The shear velocity (u_*) is equal to $\sqrt{\dfrac{\tau_0}{\rho}}$, where τ_0 = shear stress at the surface and ρ = density of fluid.
    (a) True ☐   (b) False. ☐

30. When pipes are connected in parallel, the total loss of head is equal to the sum of the loss of head in each pipe.
    (a) True ☐   (b) False. ☐

31. The maximum efficiency of power transmission through pipe is 50%.
    (a) True ☐   (b) False. ☐

32. Water-hammer in pipes takes place when the flowing fluid is suddenly brought to rest by closing the valve.
    (a) True ☐   (b) False. ☐

33. For super-sonic flow, if the area of flow increases, then the velocity also increases.
    (a) True ☐   (b) False. ☐

34. The velocity through a channel for uniform flow by Chezy's formula is given by, $V = C\sqrt{m \times i}$, where C = Chezy's constant, m = hydraulic mean depth and i = slope of the bed of channel.
    (a) True ☐   (b) False. ☐

35. Specific speed of a turbine is defined as the speed of the turbine which produces unit power at unit discharge.
    (a) True ☐   (b) False. ☐

36. A turbine converts hydraulic energy into mechanical energy whereas a pump converts mechanical energy into hydraulic energy.
    (a) True ☐   (b) False. ☐

37. A jet of water is having a velocity V and strikes a series of vertical plates which are moving with velocity u. The efficiency of the system will be maximum if $u = \dfrac{V}{2}$.
    (a) True ☐   (b) False. ☐

38. Jet ratio is defined as the ratio of the diameter of the jet to the diameter of Pelton turbine.
    (a) True ☐   (b) False. ☐

IV. FILL IN THE BLANKS

1. Kaplan tubine is …… turbine.
   (a) a reaction ☐   (b) an impulse ☐

2. The term $\left(\dfrac{N}{\sqrt{H}}\right)$ represents …… of a turbine.
   (a) specific speed ☐   (b) unit speed. ☐
   where N = Speed of the turbine and H = Net head on the turbine.

78 CIVIL ENGINEERING (OBJECTIVE TYPE)

3. Stoke is the unit of ....
   (a) viscosity ☐   (b) kinematic viscosity. ☐
4. Absolute pressure ...... atmospheric pressure is equal to gauge pressure.
   (a) plus ☐   (b) minus. ☐
5. If the Reynolds number is equal to or less than ......, the flow in pipe is known as laminar.
   (a) 4000 ☐   (b) 2000. ☐
6. The pitot-tube is used to measure ...... at a point in a flowing liquid.
   (a) discharge ☐   (b) velocity. ☐
7. The velocity profile, across the section of a pipe for laminar flow, follows ...... law.
   (a) linear ☐   (b) parabolic. ☐
8. The ratio of inertia force to ...... force is known as Reynolds number.
   (a) gravity ☐   (b) viscous. ☐
9. Kinematic similarity means similarity of ...... between model and prototype.
   (a) motion ☐   (b) forces. ☐
10. The force exerted by a flowing fluid on a solid body ...... is known as drag force.
    (a) in the direction of flow ☐   (b) perpendicular to the direction of flow. ☐

ANSWERS
Objective Type Questions

1. (c)	2. (b)	3. (b)	4. (b)	5. (b)	6. (c)
7. (b), (d)	8. (c)	9. (b)	10. (b)	11. (d)	12. (b)
13. (a)	14. (a)	15. (d)	16. (c)	17. (b)	18. (a)
19. (c)	20. (d)	21. (d)	22. (b)	23. (b)	24. (b)
25. (c)	26. (b)	27. (d)	28. (a)	29. (c)	30. (c)
31. (b)	32. (b)	33. (c)	34. (d)	35. (c)	36. (b)
37. (c)	38. (b)	39. (c)	40. (a)	41. (c)	42. (c)
43. (c)	44. (c)	45. (b)	46. (d)	47. (c)	48. (c)
49. (b)	50. (d)	51. (c)	52. (a)	53. (b)	54. (c)
55. (d)	56. (a)	57. (b)	58. (c)	59. (d)	60. (c)
61. (b)	62. (b)	63. (b)	64. (d)	65. (c)	66. (c)
67. (c)	68. (c)	69. (a)	70. (c)	71. (d)	72. (d)
73. (b), (c)	74. (c)	75. (b)	76. (c)	77. (d)	78. (c), (b)
79. (d)	80. (c)	81. (c)	82. (d)	83. (b)	84. (d)
85. (d)	86. (c)	87. (c)	88. (b)	89. (d)	90. (c)
91. (d)	92. (a)	93. (a)	94. (a)	95. (c)	96. (c)
97. (c)	98. (c)	99. (c)	100. (d)	101. (a)	102. (c)
103. (a)	104. (a)	105. (b)	106. (a)	107. (b)	108. (c)
109. (c)	110. (b)	111. (d)	112. (d)	113. (c)	114. (b)
115. (b)	116. (b)	117. (c)	118. (c)	119. (d)	120. (b)

121. (a)	122. (c)	123. (c)	124. (b)	125. (b)	126. (b)
127. (c)	128. (c)	129. (b)	130. (a)	131. (b)	132. (c)
133. (b)	134. (c)	135. (b)	136. (b)	137. (c)	138. (c)
139. (b)	140. (b)	141. (c)	142. (d)	143. (c)	144. (a)
145. (d)	146. (d)	147. (a)	148. (c)	149. (b)	150. (c)
151. (d)	152. (d)	153. (d)	154. (c)	155. (d)	156. (d)
157. (c)	158. (b)	159. (a)	160. (a)	161. (b)	162. (c)
163. (b)	164. (a)	165. (d)	166. (a)	167. (b)	168. (c)
169. (c)	170. (b)	171. (b)	172. (c)	173. (f)	174. (b)
175. (b)	176. (a)	177. (a)	178. (b)	179. (c)	180. (d)
181. (c)	182. (a)	183. (b)	184. (b)	185. (d)	186. (b)
187. (d)	188. (a)	189. (b)	190. (b)	191. (c)	192. (c)
193. (c)	194. (c)	195. (c)	196. (a)	197. (b)	198. (c)
199. (d)	200. (a)	201. (b)	202. (c)	203. (c)	204. (c)
205. (d)	206. (a)	207. (b)	208. (b)	209. (b)	210. (c)
211. (b)	212. (c)	213. (a)	214. (b)	215. (a)	216. (b)
217. (b)	218. (a)	219. (b)	220. (b)	221. (c)	222. (a)
223. (b)	224. (c)	225. (c)	226. (b)	227. (c)	228. (b)
229. (c)	230. (c)	231. (b)	232. (b)	233. (a)	234. (b)
235. (b)	236. (a)	237. (b)	238. (a)	239. (a)	240. (c)
241. (b)	242. (a)	243. (a)	244. (c)	245. (c)	246. (b)
247. (b)	248. (c)	249. (b)	250. (c)	251. (a)	252. (c)
253. (b)	254. (b)	255. (a)	256. (c)	257. (a)	258. (c)
259. (a)	260. (c)	261. (c)	262. (c)	263. (b)	264. (c)
265. (b)	266. (a)	267. (b)	268. (a)	269. (c)	270. (d)
271. (c)	272. (b)	273. (a)	274. (b)	275. (b)	276. (c)
277. (c)	278. (b)	279. (d)	280. (b)	281. (c)	282. (b)
283. (a)	284. (b)	285. (c)	286. (d)	287. (c)	288. (d)
289. (a)	290. (a)	291. (a)	292. (b)	293. (a)	294. (c)
295. (b), (c)	296. (b)	297. (c)	298. (c)	299. (c)	300. (c)
301. (c)	302. (a)	303. (b)	304. (c)	305. (a)	306. (b)
307. (b)	308. (c)	309. (a)	310. (c)	311. (b)	312. (b)
313. (c)	314. (b)	315. (a)	316. (b)	317. (a)	318. (a)
319. (c)	320. (a)	321. (b)	322. (c)	323. (c)	324. (b), (d)
325. (c)	326. (a)	327. (c)	328. (b)	329. (a)	330. (b)
331. (a)	332. (a)	333. (c)	334. (a)	335. (b)	336. (b)
337. (a)	338. (a)	339. (d)	340. (a)	341. (b)	342. (c)
343. (b)	344. (c)	345. (b)	346. (c)	347. (c)	348. (c)
349. (a)	350. (b)	351. (c)	352. (a)	353. (b)	354. (b)

355. (c)	356. (a)	357. (c)	358. (b)	359. (b)	360. (b)
361. (d)	362. (d)	363. (b)	364. (c)	365. (c)	366. (d)
367. (a)	368. (b)	369. (c)	370. (b)	371. (b)	372. (b)
373. (a)	374. (c)	375. (b)	376. (b)	377. (c)	378. (c)
379. (c)	380. (b)	381. (b)	382. (c)	383. (b)	384. (c)
385. (d)	386. (a)	387. (d)	388. (c)	389. (b)	390. (c)
391. (d)	392. (d)	393. (c)	394. (b)	395. (b)	396. (b)
397. (a)	398. (b)	399. (c)	400. (c)	401. (c)	402. (d)

True/False

1. (a)	2. (b)	3. (a)	4. (a)	5. (b)	6. (a)
7. (b)	8. (a)	9. (b)	10. (a)	11. (b)	12. (a)
13. (b)	14. (a)	15. (b)	16. (b)	17. (b)	18. (a)
19. (a)	20. (a)	21. (a)	22. (a)	23. (a)	24. (a)
25. (a)	26. (a)	27. (a)	28. (b)	29. (a)	30. (b)
31. (b)	32. (a)	33. (a)	34. (a)	35. (b)	36. (a)
37. (a)	38. (b)				

Fill in the Blanks

1. (a)	2. (b)	3. (b)	4. (b)	5. (b)	6. (b)
7. (b)	8. (b)	9. (a)	10. (a)		

Chapter 2 ENGINEERING MECHANICS

I. INTRODUCTION

INTRODUCTION AND DEFINITION

Mechanics is that branch of science which deals with the bodies when they are at rest or in motion. When the bodies are at rest, the branch of mechanics is known as *Statics* and if the bodies are in motion, the branch of mechanics is known as '*Dynamics*. Dynamics is further divided into two parts namely (*i*) Kinematics and (*ii*) Kinetics. *Kinematics* is the branch of mechanics which deals with the study of rigid bodies in motion without considering the forces, which cause motion. *Kinetics* is the branch of mechanics which deals with the study of rigid bodies in motion, taking into consideration the forces.

1. Force is that action which moves or tends to move a body. The units of force are (*i*) Newton (N) in S.I. units, (*ii*) Kilogram force (kgf) in M.K.S. units, and (*iii*) Dyne in C.G.S. units.

2. Newton is a force which acts on a mass of one kilogram and produces an acceleration of one metre per second square. **Dyne** is force which acts on a mass of one gram and produces an acceleration of one centimetre per second square. The relation between newton (N) and dyne is given by

$$1 \text{ N} = 10^5 \text{ dyne}$$

Force is a vector quantity which means it is having magnitude and direction. A single force which produces the same effect as a number of forces acting together is called the *resultant* of these forces. If the forces are acting in a straight line, their resultant is equal to the algebraical sum of the forces. If the forces are acting in different directions, their resultant is obtained by: (*a*) Law of triangle of forces, (*b*) Law of parallelogram of forces, and (*c*) Law of polygon of forces.

(*a*) **Law of triangle of forces** states that if two forces acting on a body are represented in magnitude and direction by the two sides of a triangle taken in order, then their resultant is given by the third side of the triangle taken in the opposite order.

(*b*) **Law of parallelogram of forces** states that if two forces, acting at a point of a body, be represented in magnitude and direction by the two adjacent sides of a parallelogram, their resultant may be represented in magnitude and direction by the diagonal of the parallelogram, which passes through their point of intersection.

(*c*) **Law of polygon of forces** states that if a number of forces acting on a point of a body are represented in magnitude and direction by the sides of a polygon, taken in order, then their resultant is represented in magnitude and direction by the closing side of the polygon taken in the opposite direction. Conversely, if any number of forces acting at a point can be represented in magnitude and direction by the sides of a polygon taken in order, the forces are in equilibrium.

3. The forces acting on a body may be: (*a*) Coplanar, (*b*) Non-coplanar, (*c*) Concurrent, (*d*) Non-concurrent, (*e*) Coplanar concurrent and (*f*) Collinear etc.

4. **Coplanar forces** are those forces, whose lines of action lie on the same plane. *Non-coplanar* forces are those forces whose lines of action do not lie in the same plane. *Concurrent forces* are those forces, which meet at a point and if the forces do not meet at a point, the forces are called *non-concurrent*. If the lines of action of the forces lie in the same plane and they meet at a point, those forces are called *coplanar concurrent forces*. *Collinear forces* are those forces, whose lines of action lie on the same line.

5. **Lami's theorem** states that if three coplanar forces acting at a point be in equilibrium, then each force is proportional to the sine of the angle between the other two.

6. **Moment of a force** about a point is the product of the magnitude of the force and perpendicular distance of its line of action from the point.

7. When a number of forces acting on a rigid body are in equilibrium, then the sum of moments of the forces which tend to turn the body in one direction about any given axis is equal to the sum of the moments of the forces which tend to turn the body in the opposite direction about the same axis. This is known as *Principle of Moments*.

8. When a number of coplanar forces are acting on a particle, the algebraic sum of the moments of all the forces about any point is equal to the moment of their resultant force about the same point. This is known as **Varignon's theorem of moments.**

9. A system of coplanar forces will be in equilibrium if the sum of the resolved components of the forces of the system in any two perpendicular directions is zero separately and the sum of the moments of the forces about a point in their plane is zero. Conversely, if a system of coplanar forces is in equilibrium, the sum of the resolved components of the forces of the system in any two perpendicular directions must be separately zero and also the algebraic sum of their moments about any point in their plane must be zero.

10. **A couple** consists of two equal, opposite and parallel forces acting on a body. The perpendicular distance between the two parallel forces is called the *arm* of the couple. The *moment of a couple* is equal to the product of the magnitude of one of the forces and the sum of the couple. The couple tends to rotate a body. If two couples are acting on a body, the body will be in equilibrium

ENGINEERING MECHANICS

if both the couples have equal moments, are acting in the same plane and their directions of rotation are opposite.

EQUATIONS OF MOTIONS

The rate of change of displacement of a body is called velocity or linear velocity (V) while the rate of change of angular displacement is called angular velocity (ω). The relation between linear velocity and angular velocity is given by

$$V = \omega r.$$

The rate of change of velocity is called acceleration (a) while the rate of change of angular velocity is called angular acceleration (x). The relation between acceleration (a) and angular acceleration is given by

$$a = \alpha r.$$

The acceleration (a) in differential form is represented by $a = \dfrac{dv}{dt}$ in terms of velocity

$$= \dfrac{d^2 s}{dt^2} \text{ in terms of displacement}$$

$$= \dfrac{V dv}{ds} \text{ in terms of velocity and displacement.}$$

(a) **Equations for Linear Motions.** For a body moving with a uniform acceleration (a) the equations are

(i) $V = u + at$ 	(ii) $S = ut + \frac{1}{2} at^2$

(iii) $V^2 = u^2 + 2aS$

where u = Initial velocity, V = Final velocity

t = Time taken in sec, S = Distance travelled.

The distance travelled by a body in nth second is given by

$$S_n = u + \dfrac{a}{2}(2n - 1).$$

(b) **Equations for Angular Motions.** For a body rotating with a uniform angular acceleration (α), the equations are

(i) $\omega = \omega_0 + \alpha t$ 	(ii) $\theta = \omega_0 t + \frac{1}{2} \alpha t^2$

(iii) $\omega^2 = \omega_0^2 + 2\alpha\theta$

where ω_0 = Initial angular velocity in rad/sec

ω = Final velocity

t = Time in sec

θ = Angle traversed in radians.

The angle traversed by a body in nth second, $\theta_n = \omega_0 + \dfrac{\alpha}{2}(2n - 1).$

KINETICS

Kinetics deals with the study of rigid bodies in motion, where the forces which couse motion are considered.

When two bodies are connected by a light inextensible string and pass over a smooth pulley, both the bodies will have equal acceleration (*a*) and tension (*T*) in both sides of the string will be equal.

(*i*) Let W_1 and W_2 are the weights attached at the two ends of the string as shown in Fig. 2.1. If $W_1 > W_2$ or W_1 is moving downwards, then

Acceleration, $\quad a = \left(\dfrac{W_1 - W_2}{W_1 + W_2}\right) g \text{ m/s}^2,$

Tension, $\quad T = \dfrac{2W_1 W_2}{W_2 + W_2}$

and Pressure on the pulley, $2T = \dfrac{2W_1 W_2}{W_2 + W_2}$.

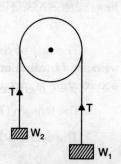

FIGURE 2.1

(*ii*) If the weight W_2 is placed on a smooth horizontal plane and weight W_1 is hanging free, as shown in Fig. 2.2 ($W_1 > W_2$), then

Acceleration, $\quad a = \dfrac{W_1 \times g}{W_1 + W_2}$

and Tension, $\quad T = \dfrac{W_1 W_2}{W_1 + W_2}$.

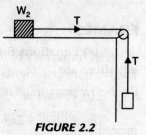

FIGURE 2.2

(*iii*) If the weight W_2 is placed on a rough horizontal plane (having co-efficient of friction, μ) and weight W_1 hangs freely. When W_1 moves downward.

Acceleration, $\quad a = \dfrac{(W_1 - \mu W_2)g}{W_1 + W_2}$

and Tension, $\quad T = \dfrac{(1+\mu)W_1 W_2}{(W_1 + W_2)}$.

(*iv*) If the weight W_2 is lying on a smooth inclined plane (having inclination α) and W_1 hangs freely as shown in Fig. 2.3. When W_1 moves downward, then

Acceleration, $\quad a = \dfrac{(W_1 - W_2 \sin \alpha)g}{(W_1 + W_2)}$

and Tension, $\quad T = \dfrac{W_1 W_2 (1 + \sin \alpha)}{(W_1 + W_2)}$.

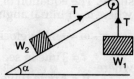

FIGURE 2.3

(v) Weight W_2 is lying on a rough inclined plane (having inclination α and co-efficient of friction μ) and W_1 hangs freely. When W_1 moves downwards, then

Acceleration, $$a = \frac{(W_1 - W_2 \sin \alpha - \mu W_2 \cos \alpha)}{(W_1 + W_2)} \times g$$

and Tension, $$T = \frac{W_1 W_2 (1 + \sin \alpha + \mu \cos \alpha)}{(W_1 + W_2)}$$

The acceleration of a body moving down an inclined plane is given by

$a = g \sin \theta$ ..... when inclined plane is smooth

$= g[\sin \theta - \mu \cos \theta]$ ..... when inclined plane is rough

where θ = Inclination of plane with horizontal and

μ = Co-efficient of friction between body and inclined plane.

The tension (T) in the cables supporting a lift is given by

$$T = W\left(1 + \frac{a}{g}\right) \text{ when lift is moving up}$$

$$= W\left(1 + \frac{a}{g}\right) \text{ when lift is moving down.}$$

FRAMES

A frame may be defined as a structure, made up of several bars, rivetted or welded together. The bars are also known as the members of the frame. There are three types of frames namely (i) Perfect frames, (ii) Deficient frame and (iii) Redundant frames.

A frame is called a *perfect frame*, which has got the number of members as given by the formula

$$n = (2j - 3)$$

where n = Number of members and j = Number of joints.

A frame is called *deficient frame* if it has got less number of members than given by $n = 2j - 3$. A frame is called *redundant frame* if it has got more number of members than given by $n = 2j - 3$.

PROJECTILES

Projectile is defined as an object which is projected in air with certain initial velocity at a certain angle. The object traces some path in air and falls on the ground at a point, other than the point of projection. The path traced by the projectile is known as *trajectory*, which is parabolic in shape. The equation of trajectory is

$$y = x \tan \alpha - \frac{gx^2}{2u^2 \cos^2 x}$$

where α = Angle of projection

u = Initial velocity of projectile

x and y = Horizontal and vertical distances covered by the projectile during the time 't'.

The velocity with which a projectile is thrown in air is known as *velocity of projection*. And the angle to the horizontal at which a projectile is thrown in air is called *angle of projection*.

The horizontal range (R), which is the distance between the point of projection and the point where the projectile strikes the ground, is given as

$$R = \frac{u^2 \sin 2\alpha}{g}$$

R will be maximum for a given value of u, when $\sin 2\alpha = 1 = \sin 90°$ or $\alpha = 45°$. Then maximum value of R will be as

$$R_{max} = \frac{u^2}{g}$$

Time of flight (t), which is the time taken by the projectile to reach maximum height and to return back to the ground is given as,

$$t = \frac{2u \sin \alpha}{g}$$

Maximum height (H), reached by projectile is given as

$$H = \frac{u^2 \sin^2 \alpha}{2g}$$

where u = Velocity of projection, and α = Angle of projection.

FRICTION

The property of bodies, by virtue of which a force is exerted by a stationary body on the other moving body to resist its motion, is called friction and the force exerted is called force of friction.

(*i*) **Static friction.** Static friction is the force called into play, when one body rests upon another and a force is applied to make one slide over the other but the body is not sliding. If the body is moving (or sliding) then the force of friction that exists between the moving body and stationary body is called *dynamic friction*. The force of friction, which exists when the body is just on the point of moving (or sliding), is called *limiting friction*. Limiting friction is the maximum amount of friction that comes into play. Dynamic friction is always less than the limiting friction.

(*ii*) **The coefficient of friction (μ).** It is the ratio of force of limiting friction to normal reaction between two bodies. And angle of friction (φ) is the angle between the normal reaction and resultant of limiting friction and normal reaction. The relation between coefficient of friction (μ) and angle of friction (φ) is given by

$$\mu = \tan \phi.$$

(*iii*) **Angle of repose.** It is the inclination of a plane with the horizontal at which the body placed on the inclined plane is just on the point of moving down.

Or

The angle of repose is defined as the maximum inclination of a plane at which a body remains in equilibrium over the inclined plane by the assistance of friction only.

(iv) **Friction on a horizontal plane.** The force P required to move the body of weight W on a rough horizontal plane as shown in Fig. 2.4 is given by

$$P = W \tan \phi$$

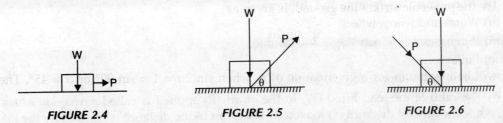

FIGURE 2.4 **FIGURE 2.5** **FIGURE 2.6**

If P is inclined at an angle θ with the horizontal plane as shown in Fig. 2.5, then the value of P necessary to move the body is given by

$$P = \frac{W \sin \phi}{\cos (\theta - \phi)}$$

where ϕ = Angle of friction and P will be minimum when $\theta = \phi$

$P_{min} = W \sin \phi$.

For Fig. 2.6, The necessary value of P to move that body is given by

$$P = \frac{W \sin \phi}{\cos (\theta + \phi)}.$$

(v) **Friction on an inclined plane.** Figure 2.7 shows a body of weight W, placed on an inclined plane of inclination α and force P is applied at an angle θ with the plane. The necessary value of P to move the body up is given by

$$P = W \frac{\sin (\alpha - \phi)}{\cos (\theta - \phi)}$$

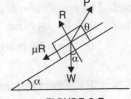

FIGURE 2.7

where ϕ = Angle of friction.

If the body is on the point of moving down, then the value of P is given by

$$P = \frac{W(\alpha - \phi)}{\cos (\theta + \phi)}.$$

In Fig. 2.8, force P is applied horizontally and the body is on the point of moving up the plane. The necessary value of P is given by

$$P = W \tan (\alpha + \phi).$$

If the body (Fig. 2.8) is on the point of moving down the plane, then the necessary horizontal force P is given by

$$P = W \tan (\alpha - \phi).$$

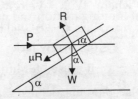

FIGURE 2.8

LIFTING MACHINES

Lifting machines are the machines used for lifting heavy loads or weights by applying comparatively small force. The important lifting machines are:

(i) Wheel and axle—simple and differential.
(ii) Worm and worm-wheel.
(iii) Purchase crab-winch-single and double.
(iv) Pulleys.
(v) Screw jack—simple and differential.

(i) The ratio of the load lifted (W) to the effort (P) applied is called *mechanical advantage* (M.A.), while the ratio of distance (x) moved by the effort to the distance (y) moved by the load is known as *velocity ratio* (V.R.). Mathematically

$$\text{M.A.} = \frac{W}{P} \text{ and V.R.} = \frac{x}{y}.$$

(ii) **Efficiency of a machine** is given by,

$$\eta = \frac{\text{Output}}{\text{Input}} = \frac{W \times \text{Distance moved by } W}{P \times \text{Distance moved by } P} = \frac{W \times y}{P \times x}$$

$$= \frac{\frac{W}{P}}{\frac{x}{y}} = \frac{\text{M.A.}}{\text{V.R.}}$$

If the efficiency of a machine is 100%, the machine is known as **ideal machine.** If the efficiency of a machine is less than 50%, it is known as **self-locking machine**. Self-locking machines are also called irreversible machine. If the efficiency of a machine is more than 50%, the machine is known as reversible machine.

(iii) **Law of machine** is expressed as

$$P = mW + c$$

where P = Effort applied, W = Load lifted and m and c = Constants.

The amount of friction, present in a machine, may be expressed in terms of effort or load. Additional effort is required to overcome this friction. Additional load is to be lifted if friction is not there.

Additional effort $= P - \dfrac{W}{\text{V.R.}}$

and additional load $= P \times \text{V.R.} - W$

where V.R. = Velocity ratio, P = Actual effort applied
  W = Actual load lifted.

(iv) **Screw jack**

It is used for lifting heavy weights. The angle of screw (α) in terms of pitch (p) is given by

$$\tan \alpha = \frac{p}{\pi d}$$

where d = Mean diameter of screw.

(a) The effort applied horizontally at the mean radius of the screw jack to lift a load of W is given by

$$P = W \tan(\alpha + \phi)$$

and to lower the load W, the effort is given by $P = W \tan(\alpha - \phi)$
where α = Angle of screw, and ϕ = Angle of friction.

(b) The effort applied at the end of the handle of a screw jack is given by

$$P = \frac{Wd}{2L} \tan(\alpha + \phi) \text{ to lift a load } W$$

$$= \frac{Wd}{2L} \tan(\alpha - \phi) \text{ to lower a load } W$$

where d = Mean dia. of the screw, L = Length of the handle
α = Angle of screw, ϕ = Angle of friction.

(c) **Efficiency of a screw-jack** is independent of the weight lifted or effort applied. Efficiency of a screw-jack for raising a load W is given by

$$\eta = \frac{\tan \alpha}{\tan(\alpha + \phi)}$$

The efficiency of a screw-jack will be maximum if

$$\alpha = 45° - \frac{\phi}{2}.$$

The maximum efficiency of a screw-jack is given by

$$\eta_{max} = \frac{1 - \sin \phi}{1 + \sin \phi}$$

where ϕ = Angle of friction.

(v) **Simple wheel and axle**

Let W = Load lifted.
   P = Effort applied,
   D = Diameter of wheel, and
   d = Diameter of axle.

Then velocity ratio (V.R.) of this system is given by,

$$\text{V.R.} = \frac{D}{d} \text{ and } \eta = \frac{M.A.}{V.R.}.$$

(vi) **Differential wheel and axle**

There are two axles and one wheel in this system.

Let D = Diameter of wheel,
   d_1 = Diameter of axle B, and
   d_2 = Diameter of axle C.

The diameter of axle B is more than the diameter of axle C.

Then, $\quad$ V.R. $= \dfrac{2D}{(d_1 - d_2)}$ and $\eta = \dfrac{\text{M.A.}}{\text{V.R.}}$.

(vii) Worm and worm wheel

Let $\quad L$ = Radius of the wheel (or length of handle)

$\quad r$ = Radius of load drum

$\quad T$ = Number of teeth on the worm wheel.

(a) For a single-threaded worm

If the effort wheel completes one revolution, then the worm will also complete one revolution. But the worm wheel will move through one teeth. Hence for one revolution of the effort wheel:

Distance moved by effort $= 2\pi L$

Distance moved by load $=$ Distance of one teeth

$= \dfrac{\text{Distance of one revolution of load drum}}{\text{Number of teeth on worn wheel}}$

$= \dfrac{2\pi r}{T}$

$\therefore \quad$ V.R. $= \dfrac{\text{Distance moved by effort}}{\text{Distance moved by load}}$

$= \dfrac{2\pi L}{\left(\dfrac{2\pi r}{T}\right)} = \dfrac{L \times T}{r}$.

(b) For a double-threaded worm

Here the worm-wheel will move by two teeths and load will move by $\dfrac{2\pi r \times 2}{T}$

$\therefore \quad$ V.R. $= \dfrac{2\pi L}{\left(\dfrac{2\pi r \times 2}{T}\right)} = \dfrac{L \times T}{2r}$

If worm is n-threaded, then V.R. $= \dfrac{L \times T}{n \times r}$.

(viii) Single Purchase crab Winch

It consists of an effort axle and a load axle. On effort axle, a small toothed wheel known as pinion is mounted whereas on load axle, a large toothed wheel known as spur wheel is mounted. The pinion meshes with spur wheel.

Let $\quad T_1$ = Number of teeth on the pinion,

$\quad T_2$ = Number of teeth on the spur wheel,

$\quad L$ = Length of lever arm, and

$\quad D$ = Diameter of the load axle.

When lever arm makes one revolution, the pinion also makes one revolution. The spur wheel makes $\left(\dfrac{T_1}{T_2}\right)$ revolution. Also the load axle makes $\left(\dfrac{T_1}{T_2}\right)$ revolution.

Distance moved by effort $= 2\pi L$

Distance moved by load $= (\pi D) \times \dfrac{T_1}{T_2}$

$\therefore\quad$ V.R. $= \dfrac{2\pi L}{\pi D \times \dfrac{T_1}{T_2}} = 2 \times \dfrac{L}{D} \times \dfrac{T_2}{T_1}$.

(ix) Double Purchase crab Winch

It consists of effort axle, load axle and intermediate axle. On the intermediate axle a pinion and a spur wheel is mounted. The pinion of intermediate axle gears with the spur wheel of the load axle. And the spur wheel of the intermediate axle gears with the pinion of the effort axle.

Let $\;T_1$ = No. of teeth on pinion of effort axle

T_2 = No. of teeth on the spur wheel of intermediate axle

T_3 = No. of teeth on the pinion of intermediate axle

T_4 = No. of teeth on the spur wheel of load axle

L = Length of lever, and

D = Diameter of load axle.

Distance moved by load $= \pi D \times \dfrac{T_1}{T_2} \times \dfrac{T_3}{T_4}$.

Distance moved by effort $= 2\pi L$

$\therefore\quad$ V.R. $= \dfrac{2\pi L}{\pi D \times \dfrac{T_1}{T_2} \times \dfrac{T_3}{T_4}} = \dfrac{2L}{D} \times \dfrac{T_2}{T_1} \times \dfrac{T_4}{T_3}$.

(x) Pulleys

A pulley may be a fixed pulley or a movable pulley. For a single fixed pulley, mechanical advantage is equal to one, whereas in case of a single movable pulley, the mechanical advantage is more than one.

The pulleys are generally used in certain combinations to obtain a higher mechanical advantage and efficiency. The important system of pulleys are:

(a) First system of pulley,

(b) Second system of pulley, and

(c) Third system of pulley.

For these system of pulleys, the velocity ratio is given by:

(a) *For first system*
$$V.R. = 2^n$$
where n = No. of movable pulleys.

(b) *For second system*
$$V.R. = n$$
where n = No. of segments supporting the load
= Total number of pulleys in two blocks.

If the weight of lower block is considered, than also
$$V.R. = n.$$

(c) *For third system*
$$V.R. = 2^n - 1$$
where n = No. of pulleys.

For Weston's Differential Pulley Block
$$V.R. = \frac{2D}{(D-d)}$$
where D = Dia. of large pulley of upper block
d = Dia. of smaller pulley of upper block.

Sometimes, this velocity ratio is also expressed as
$$V.R. = \frac{2T_1}{T_1 - T_2}$$
where T_1 = No. of teeth of larger pulley and
T_2 = No. of teeth of smaller pulley.

VIRTUAL WORK

Virtual work is the work done by a force on a body due to small virtual (*i.e.* imaginary) displacement. Principle of Virtual work states, "If a system of forces acting on a body or system of bodies be in equilibrium and if the system is imagined to undergo a small displacement consistent with the geometrical conditions, then the algebraic sum of the virtual work done by the forces of the system is zero".

For most of the problems solved by virtual work method, it is assumed that there is no friction in hinges, bearing or along sliding surfaces. And if all the forces act in the same direction, the system is given virtual displacement in the direction of forces.

CENTRE OF GRAVITY (C.G.)

The point, through which the whole weight of a body is supposed to act, is called the centre of gravity of the body.

The location of the C.G. of the following cases are:

(i) The C.G. of a rectangle or parallelogram is at the point of intersection of its diagonals.

(ii) The C.G. of a semi-circular arc [Fig. 2.9 (a)] is at $2r/\pi$ above base AB.

(iii) The C.G. of a semi-circle [Fig. 2.9 (b)] is at $4r/3\pi$ above base AB.

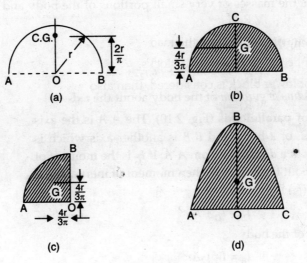

FIGURE 2.9

(iv) The C.G. of a quadrant of a circle [Fig. 2.9 (c)] is at $4r/3\pi$ from AO and OB.

(v) The C.G. of a triangle lies at the point of concurrence of the medians of the triangle.

(vi) C.G. of a solid right circular cone is at $\frac{1}{4}$ of the total height above the base on axis.

(vii) C.G. of a thin hollow right circular cone is at $\frac{1}{3}$ of the total height above the base on axis.

(viii) C.G. of a solid hemisphere is at $3r/8$ from base.

(ix) C.G. of a thin hollow hemisphere is at $r/2$ from base.

(x) C.G. of solid pyramid or cone is at $\frac{1}{4}$ th of the total height above base.

(xi) C.G. of hollow pyramid or cone is at $\frac{1}{3}$ of the total height above base.

(xii) C.G. of prism or cylinder hollow or solid is at $\frac{1}{2}$ of the total height.

(xiii) C.G. of a parabola [Fig. 2.9 (d)] is at $\frac{2}{5}$ of the axis OB above AC.

The C.G. of a body consisting of different areas is given by

$$\bar{x} = \frac{a_1 x_1 + a_2 x_2 + a_3 x_3 + \ldots\ldots}{a_1 + a_2 + a_3 + \ldots\ldots}$$

and

$$\bar{y} = \frac{a_1 y_1 + a_2 y_2 + a_3 y_3 + \ldots\ldots}{a_1 + a_2 + a_3 + \ldots\ldots}$$

If a given section is symmetrical about X-X-axis or Y-Y-axis, the C.G. of the section will lie on the axis of symmetry.

MOMENT OF INERTIA (M.O.I.)

The moment of inertia of a body of mass M about an axis is given by

$$m_1 r_1^2 + m_2 r_2^2 + \ldots = \Sigma m_1 r_1^2 = I$$

where $m_1, m_2, \ldots$ are the masses of very small portions of the body and $r_1, r_2, \ldots$ heir distances from the axis.

If k is such a length from the axis that

$$Mk^2 = I = m_1 r_1^2 + m_2 r_2^2 + \ldots$$

then k is called the *radius of gyration* of the body about the axis.

(a) **Theorem of parallel axis** (Fig. 2.10). The A-A is the axis passing through C.G. of a body and B-B is another axis which is parallel to A-A and is at a distance r from A-A. If I_G is the moment of inertia of the body about the axis A-A, then moment of inertia of the body about axis B-B (I_B) is given by

$$I_B = I_G + Mr^2$$

where M is the mass of the body

$$I_B = I_G + Ar^2$$

where A = Area of the body.

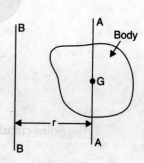

FIGURE 2.10

(b) **Theorem of perpendicular axis** (Fig. 2.11). Let A is the lamina, lying in the plane x-y, where OX and OY are two axes at right angles to one another. If OZ is the axis at right angles to the plane x-y, then moment of inertia of the lamina about OZ is equal to the sum of moments of inertia about axes OX and OY. Mathematically,

$$I_z = I_x + I_y$$

where I_z = M.O.I. about axis OZ
I_x = M.O.I. about axis OX
I_y = M.O.I. about axis OY.

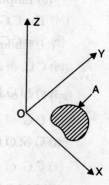

FIGURE 2.11

(c) **M.O.I. of Various Sections**

(i) Rectangular section $I = \dfrac{1}{12} BD^3$.

(ii) Hollow rectangular section $I = \dfrac{BD^3 - bd^3}{12}$.

(iii) $I = \dfrac{B}{12}(D^3 - d^3)$.

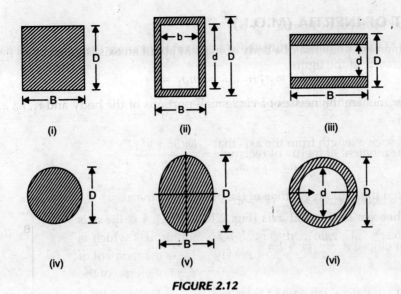

FIGURE 2.12

(*iv*) Solid circular section $I = \dfrac{\pi}{64} D^4$.

(*v*) Elliptical section $I = \dfrac{\pi}{64} BD^3$.

(*vi*) Hollow circular section $I = \dfrac{\pi}{64} (D^4 - d^4)$.

(*vii*) M.O.I. of a triangle about base = $\dfrac{bh^3}{12}$, where b = Base and h = Height.

(*viii*) M.O.I. of a triangle about an axis passing through its C.G. = $\dfrac{bh^3}{36}$.

(*ix*) M.O.I. of a solid cone about $Y\text{-}Y = \dfrac{3}{10} Mr^2$.

M.O.I. of a solid cone about axis $X\text{-}X$ passing through C.G. of the cone

$$= \dfrac{3}{20} M \left(r^2 + \dfrac{h^2}{4} \right).$$

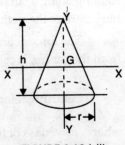

FIGURE 2.12 (vii)

(*x*) M.O.I. of solid sphere = $\dfrac{2}{5} Mr^2$.

(*xi*) M.O.I. of solid circular cylinder about vertical axis = $\dfrac{Mr^2}{2}$.

(d) **Polar Moment of Inertia, (I_0)**

It is the M.O.I. of a plane figure about an axis at right angle to its plane and passing through the centre of gravity of the figure.

(i) Polar moment of inertia of a circle = $\dfrac{\pi d^4}{32}$.

(ii) Polar moment of inertia of rectangle = $\dfrac{bh(b^2 + h^2)}{12}$.

(iii) I_0 of a hollow circle = $\dfrac{\pi}{32}(D^4 - d^4)$.

(iv) I_0 for ellipse = $\dfrac{\pi ab}{64}(a^2 + b^2)$.

(v) I_0 for equilateral triangle of sides $x = \dfrac{x^4}{16\sqrt{3}}$.

(e) **Momentum.** The product of mass of a body and its velocity is known as momentum. *Impulse* is the product of force and time, when time interval is very small and force is large one. The large force acting for a very small interval of time is called the *impulsive* force. Mathematically,

Impulse = $F \times t$, where F is large and t is very small.

Force F is known as impulsive force.

The angular momentum of a rotating body, having moment of inertia I and rotating with angular velocity ω, is equal to $I \times \omega$. If a torque T is acting on the body, then

$$T = I\alpha, \text{ where } \alpha = \text{Angular acceleration.}$$

The unit of torque will be in Newton-metre as I is in kg m². If torque is to be obtained in kgf-m, then it should be divided by 'g' (i.e., 9.81).

(f) **Centripetal and Centrifugal Forces.** Centripetal force is the force which acts on a body, rotating along a circular path, along the radius of the circular path and is always directed towards the centre of the path.

The force acting on a body, rotating in a circular or curved path along the radius and away from the centre, is called *centrifugal force*. This is given by

$$\text{Centrifugal force} = m\dfrac{V^2}{r} \text{ or } m\omega^2 r$$

where m = Mass of the body,

r = Radius of the circular path,

ω = Angular velocity of the body, and

V = Linear velocity of body.

(g) **Banking of Roads.** Banking or super-elevation of roads is the process of raising the outer edge of the roads above the inner edge. The angle of super-elevation or banking (θ) is given by

$$\theta = \tan^{-1}\left(\frac{V^2}{gr}\right)$$

where V = Velocity of vehicle and r = Radius of circular path.

The maximum velocity (V_{max}) of a vehicle on a level circular path to avoid

(a) Skidding is $\quad V_{max} = \sqrt{\mu g r}$

and (b) Over turning is $\quad V_{max} = \sqrt{\dfrac{d\,gr}{2h}}$

where h = Height of C.G. of vehicle from ground level,

d = Distance between the centre lines of the wheels, and

r = Radius of circular path.

SIMPLE HARMONIC MOTION (S.H.M.)

A body is said to describe simple harmonic motion (S.H.M.) if it moves in a straight line such that its acceleration is proportional to its distance from a fixed point and is always directed towards the fixed point.

(i) For a simple harmonic motion, *periodic time (T)* is given by

$$T = \frac{2\pi}{\omega}$$, where ω = Angular velocity of the particle.

(ii) Number of cycles per second is called *frequency* and it is equal to $\dfrac{1}{T}$, where T is the periodic time.

∴ $\quad$ Frequency $= \dfrac{1}{T} = \dfrac{\omega}{2\pi}$.

(iii) *The velocity of* the particle moving with S.H.M. at a distance y from the mean position is

$$V = \omega\sqrt{r^2 - y^2}$$, where r = Amplitude of the motion

and acceleration is given by $a = -\omega^2 y$.

(iv) The velocity of the particle moving with S.H.M. is maximum, when it is passing through mean position. Hence when $y = 0$, velocity is maximum and is given by, $V_{max} = \omega r$.

(v) The acceleration of the particle moving with S.H.M. is maximum, when it is at its extreme position, *i.e.*, when $y = r$ = amplitude.

∴ Maximum acceleration $a_{max} = -\omega^2 r$.

When particle is at the mean position, acceleration is zero and when particle is at its extreme position, velocity is zero.

(vi) Time period of a **simple pendulum** is given by

$$T = 2\pi \sqrt{\frac{L}{g}}, \text{ where } L = \text{Length of pendulum.}$$

When the bob of the pendulum moves from one extremity to the other, half an oscillation is said to be completed. This constitutes **one Beat**. A pendulum, which executes one beat per second, is called **seconds pendulum**. Hence for a seconds pendulum, the time period is equal to two seconds. Length of a second pendulum is obtained from

$$T = 2\pi \sqrt{\frac{L}{g}} = 2 \quad \therefore \quad L = \left(\frac{2}{2\pi} \times \sqrt{g}\right)^2 = \left(\frac{1}{\pi}\sqrt{9.81}\right)^2$$

$$= 0.994 \text{ m or } 99.40 \text{ cm}.$$

(vii) The time period of an **elastic string or spring** is given by

$$T = 2\pi \sqrt{\frac{\delta}{g}}, \qquad \text{where } \delta = \text{Static extension.}$$

(viii) **Gain or loss of oscillations due to** change in length or g of a simple pendulum is given by

$$\frac{dn}{n} = \frac{dg}{2g} - \frac{dl}{2l}$$

where
$\begin{cases} \dfrac{dn}{n} = \text{Change in number of oscillation} \\ \dfrac{dg}{g} = \text{Change of gravity} \\ \dfrac{dl}{l} = \text{Change of length of pendulum.} \end{cases}$

(a) If L is constant, then $\dfrac{dn}{n} = \dfrac{dg}{2g}$

(b) If g is constant, then $\dfrac{dn}{n} = -\dfrac{dl}{2l}$

$-$ve sign means that $\dfrac{dl}{l}$ decreases, as $\dfrac{dn}{n}$ increases. Or with the increase of length, the clock goes slow.

(ix) **Compound Pendulum.** A body of any shape or size which is made to oscillate about a fixed axis is called a compound pendulum. The body shown in Fig. 2.13 oscillates about an axis Y-Y or through point of suspension O. G is the centre of gravity of the body and $OG = h$. $OP = L$, the length of a simple pendulum which has the same time period as compound pendulum. The time period of compound pendulum is

$$T = 2\pi \sqrt{\frac{k^2 + h^2}{gh}}$$

where k = Radius of gyration and $h = OG$ or length between the point of suspension (O) and centre of gravity (G).

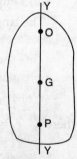

FIGURE 2.13

The point P is called *centre of oscillation*. If P be made centre of suspension, O will become centre of oscillation. Hence centre of oscillation ans suspension are interchangable.

The point P is also called *centre of percussion* which is the point on the body at which if a blow is given, no reaction will be felt at the point of suspension.

BELTS AND ROPES DRIVES

The belts and ropes are used for transmitting power from one shaft to another shaft. The belts are of the following important types:

(*i*) Open belt drive,

(*ii*) Cross belt drive, and

(*iii*) Compound belt drive.

(*i*) **The velocity ratio** of a belt is given by

$$\frac{N_2}{N_1} = \frac{d_1}{d_2} \qquad \text{...... if thickness of belt is neglected}$$

$$= \frac{d_1 + t}{d_2 + t} \qquad \text{...... if thickness of belt is considered}$$

$$= \frac{d_1}{d_2} \times \left[\frac{E \times \sqrt{f_2}}{E + \sqrt{f_1}}\right] \qquad \text{...... if creep of belt is considered}$$

where N_1 = Speed of driver pulley,

N_2 = Speed of driven pulley,

d_1 = Dia. of driver pulley,

d_2 = Dia. of driven pulley,

t = Thickness of belt,

E = Young's modulus,

f_1 = Stress in the belt on tight side, and

f_2 = Stress in the belt on slack side.

(*ii*) **The velocity ratio of a compound belt drive is given by**

$$\frac{\text{Speed of last follower}}{\text{Speed of first driver}} = \frac{\text{Product of dia. of drivers}}{\text{Product of dia. of follower}}.$$

(*iii*) **Length of a Belt**

The length of a belt is given by,

$$L = \pi (r_1 + r_2) + \frac{(r_1 - r_2)^2}{x} + 2x \qquad \text{...... For an open belt}$$

$$= \pi (r_1 + r_2) + \frac{(r_1 + r_2)^2}{x} + 2x \qquad \text{...... For a crossed belt}$$

where r_1 = Radius of larger pulley,

r_2 = Radius of smaller pulley,

x = Distance between the centres of two pulleys, and

L = Length of the belt.

(iv) Ratio of Tensions

The ratio of the tensions on the two sides of a belt is given by,

$$\frac{T_1}{T_2} = e^{\mu \times \theta}$$

where T_1 = Tension on the tight side,
 T_2 = Tension on the slack side,
 μ = Co-efficient of friction between belt and pulley,
 θ = Angle of contact in radians
  = $(180 - 2\alpha)$ ...... For an open belt
  = $(180 + 2\alpha)$ ...... For a crossed belt

where $\alpha = \sin^{-1} \dfrac{(r_1 - r_2)}{x}$ ...... For an open belt

  $= \sin^{-1} \dfrac{(r_1 + r_2)}{x}$ ...... For a crossed belt.

(v) Power Transmitted by a belt

Power transmitted is given by

$$P = \frac{(T_1 - T_2) \times v}{75},$$ where T_1 and T_2 are in kgf

$$= (T_1 - T_2) \times v \text{ watt},$$ where T_1 and T_2 are in Newton.

(A) The centrifugal tension (T_c) is given by

$$T_c = \frac{w}{g} \times v^2 \text{ kgf} \quad \ldots \text{M.K.S. units}$$

$$= m \times v^2 \text{ Newton} \quad \ldots \text{S.I. units}$$

where w = Weight of belt per metre length in kgf,
 m = Mass of belt per metre length in kg.

(B) The maximum tension (T_{max}) is given by

$$T_{max} = f \times \text{Area of belt}$$
$$= f \times (b \times t)$$
$$= T_1 + T_c \quad \ldots \text{if centrifugal tension is considered}$$
$$= T_1 \quad \ldots \text{if centrifugal tension is neglected.}$$

(C) *For maximum power transmission:*

(a) Velocity of belt (v) is given by

$$v = \sqrt{\frac{g \times T_{max}}{3w}} \quad \ldots \text{in M.K.S. Units}$$

$$= \sqrt{\frac{T_{max}}{3 \times m}} \quad \ldots \text{in S.I. Units}$$

(b) $T_{max} = 3T_c$ or $T_c = \frac{1}{3} T_{max}$

(c) $T_1 = \frac{2}{3} T_{max}$.

(D) *Initial tension* (T_0) in a belt is given by

$$T_0 = \frac{T_1 + T_2}{2} \quad \text{... if } T_c \text{ is neglected}$$

$$= \frac{T_1 + T_2 + 2T_e}{2} \quad \text{... if } T_c \text{ is considered.}$$

WORK, POWER AND ENERGY

Work. Work is the product of force and distance. Mathematically the work done is given by

Work done = $F \times S$ ...When force and distance are in the same direction

= $F \cos \theta \times S$ ...When force acts at an angle θ with the direction of displacement.

(*i*) **The unit of work** is N m (or Joule). Hence one **Joule** is the work done by a force of 1 N when the displacement is 1 m.

(*ii*) The work done is also represented by the **area of force-displacement curve**.

(*iii*) The work done on a **rotating body** by a torque (T) is given by

Work done = $T \times \theta$

where θ = Angular displacement is radians.

Power. The rate of doing work is known as power. Hence the power is the work done per second. The unit of power is N m/s or Watt (W). Hence

Power = Work done per second

$$= \frac{\text{Work done}}{\text{Time}} = \frac{\text{Force} \times \text{Distance}}{\text{Time}}$$

$$= \text{Force} \times \frac{\text{Distance}}{\text{Time}} = \text{Force} \times \text{Velocity}$$

The force and velocity should be in the same direction.

The power required to rotate a body is given by,

Power = Torque × Angular velocity

$= T \times \omega$

$$= T \times \frac{2\pi N}{60} = \frac{2\pi NT}{60} \text{ watt.}$$

II. OBJECTIVE TYPE QUESTIONS

Basic Definitions

1. Triangle law of forces states that if two forces acting at a point are represented in magnitude and direction by the two sides of the triangle taken in order, then their resultant is given by the
    (*a*) third side of the triangle taken in the same order ☐
    (*b*) third side of the triangle taken in the opposite order ☐
    (*c*) sum of the two forces acting ☐
    (*d*) none of the above. ☐

2. Law of polygon of forces states that
   (a) if a number of forces acting at a point are represented by the sides of a polygon taken in order, then their resultant is represented in magnitude and direction by the closing side of the polygon, taken in the same order. ☐
   (b) if a number of forces acting at a point are represented in magnitude and direction by the sides of a polygon taken in order, then their resultant is represented in magnitude and direction by the closing side of the polygon, taken in the opposite order. ☐
   (c) the resultant of a number of forces acting on a point is the sum of all forces. ☐
   (d) none of the above. ☐

3. Two forces P and Q are acting at an angle θ, their resultant (R) is given by
   (a) $R = \sqrt{P^2 + Q^2 + 2AB \sin 2\theta}$ ☐
   (b) $R = \sqrt{P^2 + Q^2 + 2AB \cos \theta}$ ☐
   (c) $R = \sqrt{P^2 + Q^2 - 2AB \cos \theta}$ ☐
   (d) $R = \sqrt{P^2 + Q^2 + 2AB \cos 2\theta}$. ☐

4. Two forces A and B are acting at an angle θ and their resultant B makes an angle α with the force A, then
   (a) $\tan \alpha = \dfrac{B \sin \theta}{B + A \cos \theta}$ ☐
   (b) $\tan \alpha = \dfrac{A \sin \theta}{A + B \cos \theta}$ ☐
   (c) $\tan \alpha = \dfrac{B \sin \theta}{A + B \cos \theta}$ ☐
   (d) $\tan \alpha = \dfrac{A \cos \theta}{B + A \sin \theta}$. ☐

5. Two forces A and B are acting at an angle θ and their resultant R makes an angle α with the force A, then
   (a) $\cos \alpha = \dfrac{A + B \sin \theta}{\sqrt{A^2 + B^2 - 2AB \cos \theta}}$ ☐
   (b) $\cos \alpha = \dfrac{A + B \sin \theta}{\sqrt{A^2 + B^2 + 2AB \cos \theta}}$ ☐
   (c) $\cos \alpha = \dfrac{B \sin \theta}{\sqrt{A^2 + B^2 + 2AB \cos \theta}}$ ☐
   (d) none of the above. ☐

6. Lami's theorem states that if
   (a) three forces acting at a point are in equilibrium, they can be represented by the three sides of a triangle. ☐
   (b) the three forces acting at a point can be represented in magnitude and direction by the sides of a triangle, the forces are in equilibrium. ☐
   (c) three forces acting at a point are in equilibrium, each force is proportional to the sine of the angle between the other two. ☐
   (d) none of the above. ☐

7. The forces which do not meet at a point are called
   (a) non-coplanar forces ☐
   (b) coplanar forces ☐
   (c) non-concurrent forces ☐
   (d) concurrent forces. ☐

8. The forces whose lines of action do not lie in the same plane, are called
   (a) non-coplanar forces ☐
   (b) coplanar forces ☐
   (c) non-concurrent forces ☐
   (d) none of the above. ☐

9. The forces, whose line of action lie on the same line, are known as
   (a) coplanar forces ☐
   (b) concurrent forces ☐
   (c) collinear forces ☐
   (d) none of the above. ☐

10. The forces, whose lines of action does not lie in the same plane but are meeting at one point, are known as
    (a) coplanar concurrent forces
    (b) non-coplanar concurrent forces
    (c) non-coplanar non-concurrent forces
    (d) none of the above.

11. The forces, whose lines of action lie in the same plane and are meeting at one point, are known as
    (a) coplanar concurrent forces
    (b) coplanar non-concurrent forces
    (c) non-coplanar concurrent forces
    (d) none of the above.

12. Coplanar concurrent forces means the lines of action of forces
    (a) lie in the same plane
    (b) lie in the same plane but the forces are not meeting at one point
    (c) lie in the same plane and forces are meeting at one point
    (d) none of the above.

13. The forces, which meet at a point, are known as
    (a) collinear forces
    (b) coplanar forces
    (c) concurrent forces
    (d) none of the above.

14. The forces, which lie in the same plane, are called
    (a) collinear forces
    (b) coplanar forces
    (c) concurrent forces
    (d) none of the above.

15. Tick mark the correct statement
    (a) The algebraic sum of the resolved parts of a number of force in a given direction is equal to their resultant.
    (b) The algebraic sum of the resolved parts of a number of force in a given direction is equal to two times their resultant.
    (c) The algebraic sum of the resolved parts of a number of forces in a given direction is equal to the resolved component of the resultant in that direction.
    (d) None of the above.

16. The statement—if three forces acting at a point can be represented in magnitude and direction by the sides of a triangle taken in order, the forces are in equilibrium—is known as
    (a) Lami's theorem
    (b) Law of polygon of forces
    (c) Law of triangle of forces
    (d) Newton's law of forces.

17. The statement—the algebraic sum of the moments taken about any point in the plane of forces is zero—is known as
    (a) Law of polygon of forces
    (b) Lami's theorem
    (c) Newton's law of forces
    (d) Law of moments.

18. Two couples will balance one another when they are in the same plane and
    (a) have equal moments and their direction of rotation is same
    (b) have unequal moments and their direction of rotation is opposite
    (c) have equal moments and their direction of rotation is opposite
    (d) none of the above.

19. The number of members (n) and number of joints (j) in a perfect frame is given by
    (a) $n = (3j - 2)$
    (b) $n = (2j - 3)$
    (c) $j = (2n - 3)$
    (d) $j = (3n - 2)$.

20. A frame, which has got less number of members than given by the formula $n = (2j - 3)$, is called a
    (a) perfect frame
    (b) deficient frame
    (c) redundant frame
    (d) none of the above.

21. A frame, which has got more number of members than given by the formula $n = (2j - 3)$, is called a
    (a) perfect frame
    (b) deficient frame
    (c) redundant frame
    (d) none of the above.

22. A frame, which has got the number of members equal to the number of members given by $n = (2j - 3)$, is called a
    (a) perfect frame
    (b) deficient frame
    (c) redundant frame
    (d) none of the above.

23. The resultant of two forces each equal to $\dfrac{P}{4}$ and acting at right angles is
    (a) $\dfrac{P}{2}$
    (b) $\dfrac{P}{2\sqrt{2}}$
    (c) $\sqrt{2}P$
    (d) $\dfrac{P}{\sqrt{2}}$.

24. Two forces of magnitudes 4 and 5 N act at an angle of 60°, the resultant force is equal to
    (a) 6 N
    (b) $\sqrt{61}$
    (c) 7 N
    (d) 9 N.

25. A body will be in equilibrium when
    (a) the algebraic sum of vertical components of all forces is zero
    (b) the algebraic sum of horizontal components of all forces is zero
    (c) the algebraic sum of moments of all forces about a point is zero
    (d) all the above.

Equations of Motion

26. Rate of change of displacement of a body is called
    (a) acceleration
    (b) velocity
    (c) momentum
    (d) none of the above.

27. Rate of change of velocity of body is called
    (a) acceleration
    (b) velocity
    (c) momentum
    (d) none of the above.

28. The product of mass and velocity of a body is called
    (a) acceleration
    (b) velocity
    (c) momentum
    (d) none of the above.

29. If a body is moving with a uniform acceleration (a), then final velocity (V) of the body after time 't' is equal to

   (a) $ut + \frac{1}{2}at^2$ ☐   (b) $u + at$ ☐
   (c) $u^2 + 2aS$ ☐   (d) none of the above.   ☐

   where u = Initial velocity, S = Distance travelled in t seconds.

30. If a body is moving with a uniform acceleration (a), then the distance travelled by a body after time 't' is equal to

   (a) $ut + \frac{1}{2}at^2$ ☐   (b) $u + at$ ☐
   (c) $u^2 + 2aV$ ☐   (d) none of the above.   ☐

31. If a body is moving with a uniform acceleration (a), then the distance travelled by the body in nth second is given by

   (a) $\frac{u+a}{2}(1-2n)$ ☐   (b) $\frac{u+a}{2}(n-2)$ ☐
   (c) $u + \frac{a}{2}(2n-1)$ ☐   (d) none of the above.   ☐

32. If a body is moving in a curved path, the motion of the body is called

   (a) rectilinear   ☐   (b) rotational   ☐
   (c) curvilinear   ☐   (d) none of the above.   ☐

33. If a body is moving in a straight line, the motion of the body is called

   (a) rectilinear   ☐   (b) rotational   ☐
   (c) curvilinear   ☐   (d) none of the above.   ☐

34. If a body is moving in a circular path, the motion of the body is called

   (a) rectilinear   ☐   (b) rotational   ☐
   (c) curvilinear   ☐   (d) none of the above.   ☐

35. Rate of change of angular velocity is called

   (a) acceleration   ☐   (b) angular acceleration   ☐
   (c) kinetic energy   ☐   (d) none of the above.   ☐

36. The relation between linear acceleration (a) and angular acceleration (α) is given by

   (a) $\alpha = a \times r$ ☐   (b) $\alpha = \frac{a}{r}$ ☐
   (c) $\alpha = \frac{1}{a \times r}$ ☐   (d) $\alpha = \frac{r}{a}$.   ☐

37. The angular displacement by a rotating body in the nth second is equal to

   (a) $\omega_0 + \left(\frac{n-2}{2}\right)\alpha$ ☐   (b) $\left(\frac{\omega_0 \times n}{2}\right) \times \alpha$ ☐
   (c) $\omega_0 + \left(\frac{2n-1}{2}\right)\alpha$ ☐   (d) $\omega_0 + \left(\frac{1-2n}{2}\right)\alpha$.   ☐

   where ω_0 = Initial angular velocity of the body moving in a circle
   α = Uniform angular acceleration.

38. The linear velocity (V) of a rotating body is given by

    (a) $V = \dfrac{\omega}{r}$ ☐
    (b) $V = \omega \times r$ ☐
    (c) $V = \dfrac{1}{\omega r}$ ☐
    (d) none of the above. ☐

39. The expression ($\tfrac{1}{2} mV^2$) denotes

    (a) centrifugal force ☐
    (b) kinetic energy ☐
    (c) potential energy ☐
    (d) none of the above. ☐

40. The expression $\left(\dfrac{mV^2}{r}\right)$ represents

    (a) centrifugal force ☐
    (b) kinetic energy ☐
    (c) potential energy ☐
    (d) none of the above. ☐

41. The expression ($\tfrac{1}{2} I\omega^2$) represents

    (a) centrifugal force ☐
    (b) kinetic energy ☐
    (c) kinetic energy of rotation ☐
    (d) potential energy. ☐

42. A force P of high magnitude acts on a body for a small interval of time (Δt). The product of P and Δt is called

    (a) impulsive force ☐
    (b) kinetic energy of the body ☐
    (c) impulse ☐
    (d) none of the above. ☐

43. The force P in question 42 is called

    (a) impluse ☐
    (b) impulsive force ☐
    (c) propelling force ☐
    (d) none of the above. ☐

44. Energy lost by a body (of mass m and moving with a velocity V) when it strikes another body (of mass M at rest) due to impact is equal to

    (a) $\dfrac{mV^2}{2g}\left(1 + \dfrac{m}{m+M}\right)$ ☐
    (b) $\dfrac{mV^2}{2g}(m + M - 1)$ ☐
    (c) $\dfrac{mV^2}{2g}\left(1 - \dfrac{m}{m+M}\right)$ ☐
    (d) none of the above. ☐

45. Tension in a cable supporting a lift, when left is going up is equal to

    (a) $W\left(1 - \dfrac{a}{g}\right)$ ☐
    (b) $W\left(1 + \dfrac{a}{g}\right)$ ☐
    (c) $W\left(W - \dfrac{W}{g}\right)$ ☐
    (d) $W\left(g + \dfrac{a}{g}\right)$ ☐

    where a = Uniform acceleration of lift; and W = Weight carried by lift.

46. Tension a cable supporting a lift, when lift is going down is equal to

(a) $W\left(1 - \dfrac{a}{g}\right)$ □ (b) $W\left(1 + \dfrac{a}{g}\right)$ □

(c) $a\left(W - \dfrac{W}{g}\right)$ □ (d) $W\left(g + \dfrac{a}{g}\right)$ □

where a = Uniform acceleration of lift; and W = Weight carried by lift.

47. When two bodies of mass (m and $2m$) are connected by a light inextensible string and pass over a smooth pulley, then acceleration of one body is

(a) equal to the acceleration of the other body □
(b) two time the acceleration of the other body □
(c) half the acceleration of the other body □
(d) none of the above. □

48. When two bodies of mass (M and $2M$) are connected by a light inextensible string and pass over a smooth pulley, then

(a) tension in both sides of the string will be equal □
(b) tension in one side of the string is two times the tension in the other side of the string □
(c) tension in one side of the string is half the tension in the other side of the string □
(d) none of the above. □

49. Two bodies of masses m_1 and m_2 are connected by a light in extensible string and pass over a smooth pulley. If the mass m_1 is coming down, then the acceleration of both the bodies is equal to

(a) $\dfrac{g(m_1 + m_2)}{(m_1 - m_2)}$ □ (b) $\dfrac{g(m_1 - m_2)}{(m_1 + m_2)}$ □

(c) $\dfrac{g(m_1 m_2)}{m_1 - m_2}$ □ (d) $\dfrac{g(m_1 m_2)}{m_1 + m_2}$ □

50. For question 49, the tension in the string will be equal to

(a) $\dfrac{2(m_1 + m_2)}{m_1 - m_2}$ □ (b) $\dfrac{2 m_1 m_2}{m_1 - m_2}$ □

(c) $\dfrac{2 m_1 m_2}{m_1 + m_2}$ □ (d) $\dfrac{2(m_1 + m_2)}{m_1 - m_2}$ □

51. Figure 2.14 shows the two bodies of masses m_1 and m_2 connected by a light inextensible string and a passing over a smooth pulley. Mass m_2 lies on a smooth horizontal plane. When mass m_1 is moving downward the acceleration of the two bodies is equal to

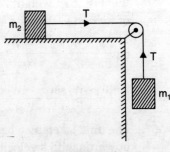

FIGURE 2.14

(a) $\dfrac{m_1 g}{m_1 - m_2}$ m/s² □

(b) $\dfrac{m_1 g}{m_1 + m_2}$ m/s² □

(c) $\dfrac{m_2 g}{m_1 + m_2}$ □

(d) $\dfrac{m_2 g}{m_1 - m_2}$ □

108 Civil Engineering (Objective Type)

52. Refer to Fig. 2.14, the tension (T) will be equal to

   (a) $\dfrac{m_1 m_2}{m_1 - m_2}$ N ☐   (b) $\dfrac{m_1 m_2}{m_1 + m_2}$ ☐

   (c) $\dfrac{m_1 - m_2}{m_1 m_2}$ ☐   (d) $\dfrac{m_1 + m_2}{m_1 m_2}$. ☐

53. If the weight W_2 in Fig. 2.14 is resting on a rough horizontal plane (having co-efficient of friction as μ), then the acceleration is equal to

   (a) $\dfrac{(W_1 + \mu W_2)g}{(W_1 + W_2)}$ ☐   (b) $\dfrac{(\mu W_1 + W_2)g}{(W_1 + W_2)}$ ☐

   (c) $\dfrac{(W_1 - \mu W_2)g}{(W_1 + W_2)}$ ☐   (d) $\dfrac{(\mu W_1 + W_2)g}{W_1 - W_2}$. ☐

54. If the weight W_2 in Fig. 2.14 is resting on a rough horizontal plane (having co-efficient of friction as μ), then tension in the string is equal to

   (a) $\dfrac{(1+\mu) W_1 W_2}{W_1 + W_2}$ ☐   (b) $\dfrac{(1+\mu) W_1 W_2}{W_1 - W_2}$ ☐

   (c) $\dfrac{\mu W_1 W_2}{W_1 + W_2}$ ☐   (d) $\dfrac{\mu W_1 W_2}{W_1 - W_2}$. ☐

55. Two weights W_1 and W_2 are connected by a light inextensible string. Weight W_2 is placed on a smooth inclined plane of inclination α and W_1 hangs freely as shown in Fig. 2.15. If W_1 moves downwards then acceleration is equal to

   (a) $\dfrac{W_1 - W_2 \sin \alpha}{(W_1 + W_2)}$ ☐

   (b) $\dfrac{(W_1 - W_2 \sin \alpha)g}{W_1 + W_2}$ ☐

   (c) $\dfrac{W_1 + W_2}{W_1 - W_2 \sin \alpha}$ ☐

   (d) $\dfrac{(W_1 + W_2)g}{W_1 - W_2 \sin \alpha}$. ☐

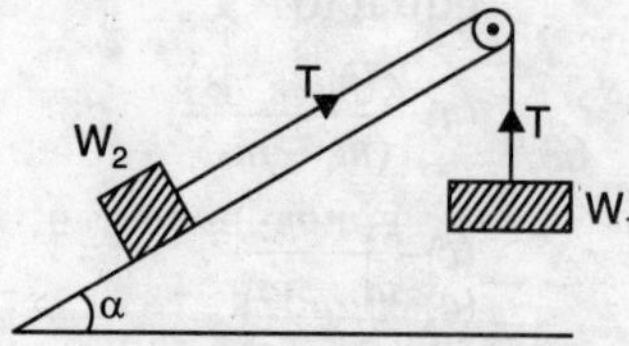

   FIGURE 2.15

56. If in Fig. 2.15, weight W_2 is placed on a rough inclined plane of inclination α and co-efficient of friction μ, then the acceleration is equal to

   (a) $\dfrac{(W_1 - W_2 \sin \alpha - \mu W_2 \cos \alpha)}{(W_1 + W_2)}$ ☐   (b) $\dfrac{(W_1 + W_2 \sin \alpha - \mu W_2 \cos \alpha)}{(W_1 + W_2)}$ ☐

   (c) $\dfrac{(W_1 - W_2 \sin \alpha - \mu W_2 \cos \alpha)g}{(W_1 + W_2)}$ ☐   (d) $\dfrac{(W_1 - W_2 \sin \alpha - \mu)g}{(W_1 + W_2)}$. ☐

57. The time taken by a ball (of weight 500 N) to return back to earth, if it is thrown vertically upwards with a velocity 4.9 m/s is equal to

   (a) $\tfrac{1}{2}$ s ☐   (b) 1 s ☐

   (c) 2 s ☐   (d) 3 s. ☐

58. The maximum height attained by a ball (of weight 500 N) which is thrown vertically upwards with a velocity 4.9 m/s is equal to
    (a) 100 cm
    (b) 245 cm
    (c) 122.5 cm
    (d) 980 cm.

59. A tower is of height 100 m. A stone is thrown up from the foot of water with a velocity of 20 m/s and at the same time another stone is dropped from the top of the tower. The two stones will meet after
    (a) 10 s
    (b) 5 s
    (c) 2 s
    (d) 7.5 s.

60. The maximum height reached by a stone (of weight 50 N) which is thrown vertically upward with an initial velocity 19.6 m/s would be
    (a) 20 m
    (b) 19.6 m
    (c) 30 m
    (d) 25 m.

61. A body is moving with a velocity of 2 m/s. If the velocity of body becomes 5 m/s after 4 seconds, the acceleration of the body would be
    (a) 1 m/s^2
    (b) 0.75 m/s^2
    (c) 1.5 m/s^2
    (d) 0.375 m/s^2.

62. A body is rotating with an angular velocity of 5 radians/s. After 4 seconds, the angular velocity of the body becomes 1.3 radians per sec. The angular acceleration of the body would be
    (a) 3 rad/s^2
    (b) 2 rad/s^2
    (c) 1 rad/s^2
    (d) 1.5 rad/s^2.

63. A flywheel starting from rest and accelerating uniformly performs 20 revolution in 4 seconds. The angular velocity of flywheel after 8 seconds would be
    (a) 30 rad/s
    (b) 35 rad/s
    (c) 40 rad/s
    (d) 55 rad/s.

64. A body is moving in a straight line with an initial velocity of 4 m/s. After 5 seconds the velocity of the body becomes 9 m/s. The distance travelled by the body in third second would be
    (a) 6 m
    (b) 5.5 m
    (c) 6.5 m
    (d) 4 m.

65. A body is rotating with an angular velocity of 5 radians/s. After 4 seconds, the angular acceleration of the body becomes 13 radians/s. If the body is rotating with uniform acceleration, the angle covered by the body in the third second would be
    (a) 20 radians
    (b) 25 radians
    (c) 15 radians
    (d) 10 radians.

66. A body is moving with a velocity of 10 m/s. The time required, to stop the body within a distance of 5 m, is equal to
    (a) 3 second
    (b) 5 second
    (c) 1 second
    (d) 0.5 second.

67. A stone dropped into a well is heard to strike the water after 4 seconds. If the velocity of sound is 350 m/s, the depth of well would be
    (a) 150 m
    (b) 70.75 m
    (c) 100 m
    (d) 35.375 m.
68. A light string passes over a smooth, weightless pulley and has weights 40 N and 60 N attached to its end as shown in Fig. 2.16. The tension in string will be
    (a) 60 N
    (b) 50 N
    (c) 48 N
    (d) 20 N.
69. Refer to Fig. 2.16, the acceleration, with which the weight 60 N descends, is
    (a) $\frac{g}{5}$
    (b) $\frac{g}{4}$
    (c) $2g$
    (d) $5g$.

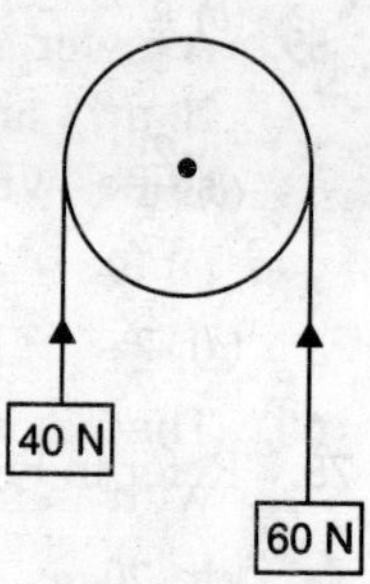

FIGURE 2.16

70. Refer to Fig. 2.16, the pressure on pulley would be
    (a) 100 N
    (b) 96 N
    (c) 20 N
    (d) 50 N.
71. Figure 2.17 shows two weights 40 N and 60 N connected by a light inextensible string and passes over a smooth weightless pulley. The weight 40 N is resting on a rough horizontal plane with μ = 0.3 and weight 60 N hangs freely and is moving downward. The tension in the string would be
    (a) 20 N
    (b) 100 N
    (c) 31.2 N
    (d) 50 N.
72. Refer to Fig. 2.17, the acceleration with which weight 60 N descends would be
    (a) 0.08 g
    (b) 20 g
    (c) 100 g
    (d) 50 g.

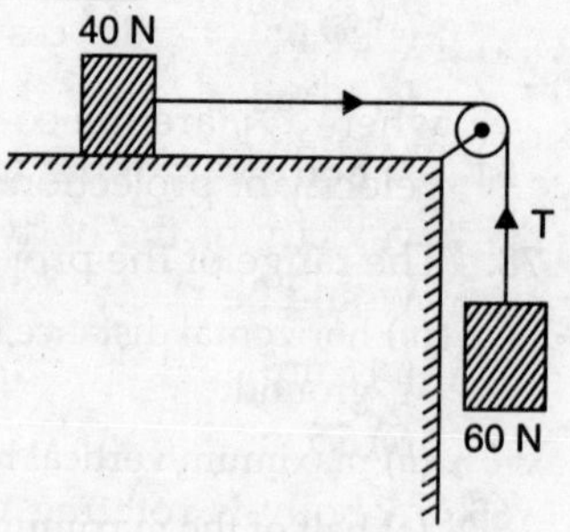

FIGURE 2.17

73. Two weights of 50 N and 150 N (of two blocks A and B respectively) are connected by a string and frictionless and weightless pulleys as shown in Fig. 2.18. The tension in the string would be
    (a) 100 N
    (b) 200 N
    (c) 64.3 N
    (d) 50 N.

74. Refer to Fig. 2.18, the acceleration to block A would be

   (a) $\dfrac{g}{7}$ □

   (b) $\dfrac{g}{5}$ □

   (c) $\dfrac{2g}{5}$ □

   (d) $\dfrac{2g}{7}$.   □

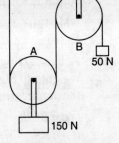

FIGURE 2.18

75. Refer to Fig. 2.18, the acceleration of block B would be

   (a) $\dfrac{g}{7}$ □   (b) $\dfrac{g}{5}$ □

   (c) $\dfrac{2g}{5}$ □   (d) $\dfrac{2g}{7}$.   □

Projectiles

76. The path traced by a projectile in the space is

   (a) hyperbolic   □   (b) parabolic   □

   (c) linear   □   (d) none of the above.   □

77. The equation of the path travelled by a projectile in space is given as

   (a) $y = x \tan \alpha - \dfrac{2gx^2}{u^2 \cos^2 \alpha}$ □   (b) $y = x \cos \alpha - \dfrac{2gx^2}{u^2 \tan^2 \alpha}$ □

   (c) $y = x \tan \alpha - \dfrac{gx^2}{2u^2 \cos^2 \alpha}$ □   (d) $y = x \sin \alpha - \dfrac{gx^2}{2u^2 \cos^2 \alpha}$.   □

   where x, y are the co-ordinates of any point on the path, from point of projection and u = velocity of projection.

78. The range of the projectile is

   (a) horizontal distance between the point of projection and the point where the projectile strikes the ground   □

   (b) maximum vertical height attained by the projectile   □

   (c) half of the maximum vertical height attained by the projectile   □

   (d) none of the above.   □

79. The range (R) is given by

   (a) $\dfrac{u^2 \sin^2 2\alpha}{2g}$ □   (b) $\dfrac{u^2 \sin^2 2x}{g}$ □

   (c) $\dfrac{u^2 \sin^2 2\alpha}{g}$ □   (d) none of the above.   □

   where α = Angle of projection and u = Velocity of projection.

80. Time of flight of projectile is defined as the time
    (a) taken by the projectile to reach maximum height
    (b) taken by the projectile to reach maximum height and to return back to the ground
    (c) taken by the projectile to return from maximum height to the ground
    (d) none of the above.

81. The time of flight (t) is equal to
    (a) $\dfrac{u \sin \alpha}{2g}$
    (b) $\dfrac{2u \sin \alpha}{g}$
    (c) $\dfrac{2g}{u \sin \alpha}$
    (d) $\dfrac{g}{2u \sin \alpha}$.

82. Maximum height attained by a projectile is equal to
    (a) $\dfrac{u^2 \sin \alpha}{2g}$
    (b) $\dfrac{u \sin^2 \alpha}{2g}$
    (c) $\dfrac{u^2 \sin^2 \alpha}{2g}$
    (d) $\dfrac{2u^2 \sin^2 \alpha}{g}$.

83. The velocity of projectile at a height k is equal to
    (a) $\sqrt{u^2 + 2gh}$
    (b) $\sqrt{u - 2gh}$
    (c) $\sqrt{2u^2 - gh}$
    (d) $\sqrt{gh - 2u^2}$.

84. Velocity of projectile after an interval 't' is given by
    (a) $\sqrt{u^2 + g^2t^2 - (2 \sin \alpha) \times gt}$
    (b) $\sqrt{u^2 - (2 \sin \alpha) \times gt}$
    (c) $\sqrt{u^2 + g^2t^2 - (2u \sin \alpha) \times gt}$
    (d) none of the above.

85. Maximum horizontal range of a projectile, having velocity of projection u and angle of projection as α, is equal to
    (a) $\dfrac{u^2 \sin 2\alpha}{2g}$
    (b) $\dfrac{u^2}{2g}$
    (c) $\dfrac{2g}{u^2}$
    (d) $\dfrac{g}{2u^2}$.

86. If R_{max} is the maximum range of a projectile, then the range (R) of the projectile when fixed at an angle of 30°, with the same initial velocity would be
    (a) $\sqrt{2} \, R_{max}$
    (b) $\dfrac{\sqrt{3}}{2} R_{max}$
    (c) $\dfrac{1}{8} R_{max}$
    (d) $\dfrac{1}{4} R_{max}$.

87. If R_{max} is the maximum range of a projectile, then the maximum height of the projectile when fired at angle 30°, with the same velocity would be
    (a) $\frac{1}{4} R_{max}$ (b) $\frac{1}{2} R_{max}$
    (c) $\frac{1}{8} R_{max}$ (d) none of the above.

88. The angle of projection for maximum range of a projectile is
    (a) 90°   (b) 30°
    (c) 60°   (d) 45°.

89. One newton is equal to
    (a) 10^3 dyne   (b) 10^2 dyne
    (c) 10^5 dyne   (d) 10^4 dyne.

90. Dyne is the force acting on a mass of
    (a) one kilogram to produce an acceleration of one metre per second square
    (b) one gm to produce an acceleration of one m/sec^2
    (c) one kg to produce an acceleration of one cm/sec^2
    (d) one gm to produce an acceleration of one cm/sec^2.

91. One newton is a force acting on a mass of
    (a) one gm to produce an acceleration of one m/sec^2
    (b) one kg to produce an acceleration of one m/sec^2
    (c) one kg to produce an acceleration of one cm/sec^2
    (d) none of the above.

92. Joule is the unit of
    (a) velocity   (b) force
    (c) work   (d) acceleration.

93. Joule is expressed in S.I. units as
    (a) N m/s   (b) N m^2
    (c) N m   (d) m N.

94. Watt is the unit of
    (a) force   (b) velocity
    (c) work   (d) power.

95. Watt is expressed in S.I. units as
    (a) N m/s   (b) N m^2/s
    (c) J/s   (d) N m.

96. Pressure in S.I. units is expressed as
    (a) N/m   (b) N/m^2
    (c) N m^2 (d) N m.

97. One metric horse power is equal to
    (a) 746 watts   (b) 736 watts
    (c) 550 watts   (d) 75 watts.

98. Momentum of a body is given by
    (a) mass × velocity
    (b) mass × change of velocity
    (c) moment × distance
    (d) velocity × acceleration.

99. The expression $I \times \omega$ (where I = moment of inertia and ω = angular velocity) represents
    (a) power
    (b) angular momentum
    (c) moment (angular)
    (d) none of the above.

100. The torque (T) acting on a rotating body is given as $T = I\alpha$, where I = M.O.I. in kg-m² and α = angular acceleration in rad/sec². The units of T will be as
    (a) kgf-m
    (b) N m
    (c) dyne metre
    (d) none of the above.

Friction

101. Limiting force of friction is defined as the frictional force which exists when a body
    (a) is moving with maximum velocity
    (b) is stationary
    (c) just begins to slide over the surface
    (d) none of the above.

102. Co-efficient of friction is the ratio of
    (a) force of friction to reaction between two bodies
    (b) force of friction to normal reaction between two bodies
    (c) force of limiting friction to reaction between two bodies
    (d) force of limiting friction to normal reaction between two bodies.

103. Angle of friction (ϕ) is the angle between the
    (a) limiting friction and normal reaction
    (b) limiting friction and the resultant of limiting friction and normal reaction
    (c) normal reaction and the resultant of limiting friction and normal reaction
    (d) none of the above.

104. The co-efficient of friction (μ) in terms of angle of friction (ϕ) is given by
    (a) $\phi = \tan \mu$
    (b) $\mu = \sin \phi$
    (c) $\mu = \tan \phi$
    (d) $\mu = \dfrac{1}{\tan \phi}$.

105. The force of friction which exists when the body is in motion is called
    (a) static friction
    (b) limiting friction
    (c) dynamic friction
    (d) none of the above.

106. Dynamic friction is always
    (a) more than static friction
    (b) more than limiting friction
    (c) less than limiting friction
    (d) none of the above.

107. A body of weight W is resting on a horizontal plane. A force P is applied parallel to the plane to move the body. The value of P, necessary to move the body against the resistance of friction is
    (a) $W/\tan \phi$
    (b) $W \sin \phi$
    (c) $W \tan \phi$
    (d) $W \cos \phi$
    where ϕ = Angle of friction.

108. The force P is applied at an angle θ with the horizontal plane on which a body of weight W is placed as shown in Fig. 2.19. The value of P, necessary to move the body is equal to

   (a) $\dfrac{W \cos \phi}{\cos(\theta - \phi)}$

   (b) $W \dfrac{\sin \phi}{\cos(\theta - \phi)}$

   (c) $\dfrac{W \tan \phi}{\sin(\theta - \phi)}$

   (d) none of the above

   where φ = Angle of friction.

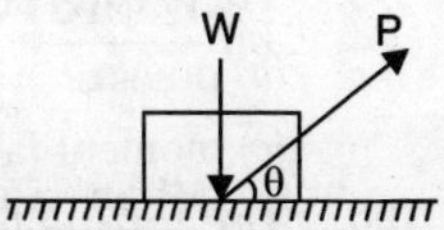

   FIGURE 2.19

109. Refer to Fig. 2.19, the necessary force P to move the body will be minimum, when

   (a) θ = 2φ   (b) θ = φ
   (c) θ = φ/2   (d) none of the above

   where φ = Angle of friction.

110. Refer to Fig. 2.19, the minimum force P to move the body is

   (a) W tan φ   (b) W cos φ
   (c) W sin φ   (d) W/sin φ

   where φ = Angle of friction.

111. The necessary force P, applied at an angle θ with the horizontal plane, on which a body of weight W is placed, to move the body is (see Fig. 2.20)

   (a) $\dfrac{W \cos \phi}{\cos(\theta - \phi)}$

   (b) $\dfrac{W \sin \phi}{\cos(\theta - \phi)}$

   (c) $\dfrac{W \tan \phi}{\cos(\theta + \phi)}$

   (d) $\dfrac{W \tan \phi}{\cos(\theta - \phi)}$

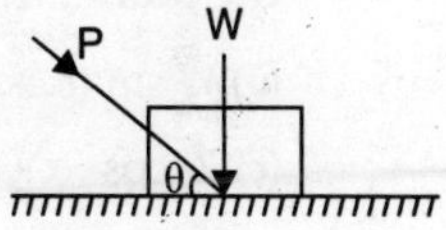

   FIGURE 2.20

   where φ = Angle of friction.

112. Refer to Fig. 2.20, the minimum force P to move the body is equal to

   (a) W tan φ   (b) W sin φ
   (c) W cos φ   (d) W/tan φ

   where φ = Angle of friction.

113. Refer to Fig. 2.20, the force P will be minimum when

   (a) θ = 2φ   (b) θ = φ
   (c) θ = (90 − φ)   (d) θ = (90 − 2φ).

114. The necessary force P, applied parallel to an inclined plane having inclination α with horizontal to move a body of weight W, up the plane is equal to (Refer to Fig. 2.21).

   (a) $\dfrac{W \sin \alpha}{\cos \phi}$ ☐

   (b) $\dfrac{W \sin (\alpha + \phi)}{\cos \phi}$ ☐

   (c) $\dfrac{W \cos (\alpha + \phi)}{\cos \phi}$ ☐

   (d) $\dfrac{W \tan (\alpha + \phi)}{\cos \phi}$ ☐

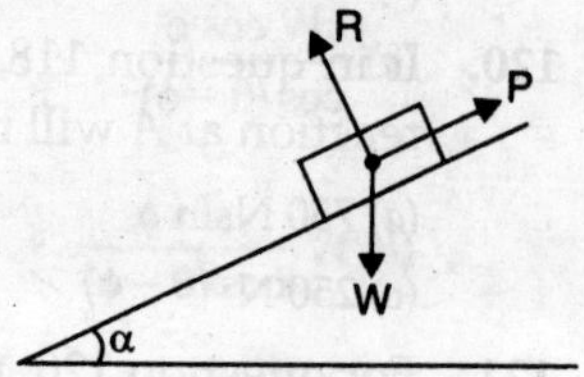

   FIGURE 2.21

   where φ = Angle of friction.

115. Refer to Fig. 2.21, if the body is on the point of moving down the plane, then necessary force P will be equal to

   (a) $\dfrac{W \sin (\alpha + \phi)}{\cos \phi}$ ☐ (b) $\dfrac{W \sin (\alpha - \phi)}{\tan \phi}$ ☐

   (c) $\dfrac{W \sin (\alpha - \phi)}{\cos \phi}$ ☐ (d) $\dfrac{W \sin (\alpha - \phi)}{\tan \phi}$ ☐

116. Figure 2.22 shows a body of weight W placed on an inclined plane having inclination α with the horizontal. Force P is applied horizontally. When the body is on the point of moving up the plane, the necessary value of P is

   (a) W sin (α + φ) ☐ (b) W tan (α + φ) ☐
   (c) W cos (α + φ) ☐ (d) W sin (α − φ). ☐

117. Refer to Fig. 2.22, if the body is on the point of moving down the plane, the necessary force P would be

   (a) W sin (α − φ) ☐
   (b) W tan (α + φ) ☐
   (c) W tan (α − φ) ☐
   (d) W sin (α + φ). ☐

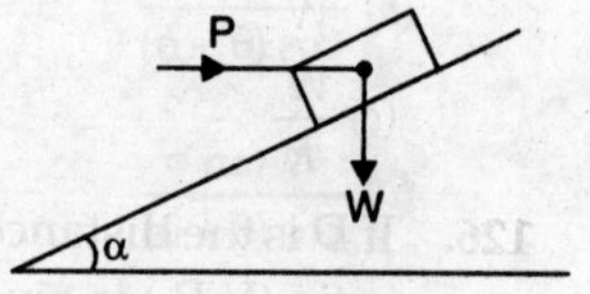

   FIGURE 2.22

118. A ladder of weight 250 N is placed against a smooth vertical wall and a rough horizontal floor (μ = 0.3) as shown in Fig. 2.23. If the ladder is on the point of sliding, the reaction at A will be

   (a) 250 N ☐
   (b) 261 N ☐
   (c) 125 N ☐
   (d) 500 N. ☐

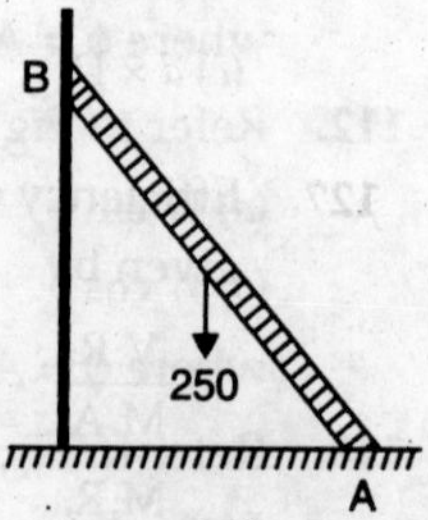

   FIGURE 2.23

119. Refer to Fig. 2.23, the reaction at B will be
   (a) 250 N
   (b) 125 N
   (c) 75 N
   (d) 500 N.

120. If in question 118, a man of weight 500 N stands on the ladder at its middle point. The reaction at A will be (when ladder is on the point of sliding)
   (a) 750 N
   (b) 783 N
   (c) 250 N
   (d) 1000 N.

121. For question 120, the reaction at B will be
   (a) 750 N
   (b) 783 N
   (c) 225 N
   (d) 1000 N.

122. A horizontal force of 400 N is applied on a body of weight 1200 N, placed on a horizontal plane. If the body is just on the point of motion, the angle of friction would be
   (a) 20°
   (b) 18° 26′
   (c) 10°
   (d) 25°.

123. A body of weight W is placed on an inclined plane. The angle made by the inclined plane with horizontal, when the body is on the point of moving down is called
   (a) angle of inclination
   (b) angle of repose
   (c) angle of friction
   (d) angle of limiting friction.

124. If angle of friction is zero
   (a) force of friction will act normal to the plane
   (b) force of friction will act opposite to the direction of motion
   (c) force of friction will be zero
   (d) force of friction will be infinite.

125. If P is the effort required to lift a load (W), then mechanical advantage (M.A.) is given by
   (a) $\dfrac{P}{W}$
   (b) $P \times W$
   (c) $\dfrac{W}{P}$
   (d) $\dfrac{1}{P \times W}$.

126. If D is the distance moved by the effort and d is the distance moved by the load, then velocity ratio (V.R.) is given by
   (a) $\dfrac{d}{D}$
   (b) $\dfrac{D}{d}$
   (c) $d \times D$
   (d) $\dfrac{1}{D \times d}$.

127. Efficiency of a machine in terms of mechanical advantage (M.A.) and velocity ratio (V.R.) is given by
   (a) $\dfrac{V.R.}{M.A.}$
   (b) $\dfrac{1}{(V.R.) \times (M.A.)}$
   (c) $\dfrac{M.R.}{V.R.}$
   (d) $(V.R.) \times (M.A.)$.

128. Self-locking machine is one which has efficiency
    (a) 100%
    (b) less than 50%
    (c) more than 50%
    (d) none of the above.

129. A reversible machine is one which have efficiency
    (a) 100%
    (b) less than 50%
    (c) more than 50%
    (d) none of the above.

130. Non-reversible machine is also called
    (a) ideal machine
    (b) self-locking machine
    (c) actual machine
    (d) none of the above.

131. The law of the machine is given by
    (a) $W = mP + C$
    (b) $P = mW + C$
    (c) $C = W + mP$
    (d) $C = mW + P$

    where P = Effort applied, W = Load lifted,
    m = Constant and equal to the slope of the line,
    C = Another constant.

132. For an ideal machine
    (a) M.A. is less than V.R.
    (b) M.A. is greater than V.R.
    (c) M.A. is equal to V.R.
    (d) None of the above.

133. In an actual machine, the amount of friction present may be expressed in terms of effort. Additional effort is required to overcome this friction. The value of additional effort required is equal to
    (a) $\dfrac{W}{V.R.} P$
    (b) $P - \dfrac{W}{V.R.}$
    (c) $P + \dfrac{W}{V.R.}$
    (d) $P + \dfrac{V.R.}{W}$

    where V.R. = Velocity ratio and W = Load applied.

134. The amount of friction, present in a machine, may be expressed in terms of load. Additional load is to be lifted if the friction is not present. The value of this additional load is equal to
    (a) $W \times P \times V.R.$
    (b) $\dfrac{P}{V.R.} + W$
    (c) $P \times V.R. - W$
    (d) $\dfrac{V.R.}{P} + W.$

135. The efficiency of a screw-jack for raising a load W is equal to
    (a) $\dfrac{\cos \alpha}{\tan (\alpha + \phi)}$
    (b) $\dfrac{\sin \alpha}{\tan (\alpha + \phi)}$
    (c) $\dfrac{\tan (\alpha + \phi)}{\tan \alpha}$
    (d) $\dfrac{\tan \alpha}{\tan (\alpha + \phi)}$

    where α = Angle of screw (Helix angle); and ϕ = Angle of friction.

136. The efficiency of a screw-jack will be maximum, if angle of screw or helix angle (α) is equal to

   (a) $\dfrac{\pi}{2} - \phi$ ☐ (b) $\dfrac{\pi}{4} - \phi$ ☐

   (c) $\dfrac{\pi}{4} + \phi$ ☐ (d) $\dfrac{\pi}{4} - \dfrac{\phi}{2}$ ☐

   where φ = Angle of friction.

137. The maximum value of efficiency of a screw-jack for raising a load W is equal to

   (a) $\dfrac{1 - \tan\phi}{1 + \sin\phi}$ ☐ (b) $\dfrac{\sin\phi}{1 + \cos\phi}$ ☐

   (c) $\dfrac{1 - \sin\phi}{1 + \sin\phi}$ ☐ (d) $\dfrac{\cos\phi}{1 + \sin\phi}$ ☐

138. The efficiency of a screw-jack for a given value of angle of friction.

   (a) depends upon the weight lifted only ☐
   (b) depends upon the effort applied only ☐
   (c) depends upon weight and effort only ☐
   (d) independent of weight lifted or effort applied. ☐

139. For a differential wheel and axle having diameter of effort wheel as D and diameters of two axles as d_1 and d_2 ($d_1 > d_2$), the velocity ratio (V.R.) is equal to

   (a) $\dfrac{2D}{d_1 + d_2}$ ☐ (b) $\dfrac{2D}{2d_1 - d_2}$ ☐

   (c) $\dfrac{2D}{d_1 - d_2}$ ☐ (d) $\dfrac{D}{d_1 - d_2}$ ☐

Centre of Gravity

140. The C.G. of a triangle lies at the point of concurrence of

   (a) the right bisectors of the angle of the triangle ☐
   (b) the medians of the triangle ☐
   (c) the altitudes from the vertices on the opposite side ☐
   (d) none of the above. ☐

141. The C.G. of solid hemisphere lies on the central radius at a distance

   (a) $\dfrac{3r}{4}$ from the plane base ☐ (b) $\dfrac{3r}{8}$ from the plane base ☐

   (c) $\dfrac{8r}{3}$ from the plane base ☐ (d) none of the above. ☐

142. The C.G. of a semi-circular lamina lies on the central radius at a distance of

   (a) $\dfrac{4r}{3\pi}$ from base diameter ☐ (b) $\dfrac{3r}{8}$ from base diameter ☐

   (c) $\dfrac{8r}{3}$ from base diameter ☐ (d) none of the above. ☐

143. The C.G. of a solid right circular cone lies on the axis at a height
    (a) half of the total height above the base
    (b) one-third of the total height above the base
    (c) one-fourth of the total height above the base
    (d) none of the above.

144. The C.G. of a thin hollow right circular cone lies on the axis at a height
    (a) half of the total height
    (b) one-third of the total height
    (c) one-fourth of the total height
    (d) none of the above.

145. The C.G. of a semi-circular arc is at the central radius at a distance of
    (a) $\dfrac{3r}{4}$ from base diameter
    (b) $\dfrac{3r}{8}$ from base diameter
    (c) $\dfrac{2r}{\pi}$ above base diameter
    (d) $\dfrac{r}{2\pi}$ above base diameter.

146. The C.G. of a quadrant of a circle is at a distance of
    (a) $\dfrac{3r}{4\pi}$ from the axis
    (b) $\dfrac{4r}{3\pi}$ from the axis
    (c) $\dfrac{3r}{8}$ from the axis
    (d) $\dfrac{8r}{3}$ from the axis.

147. The C.G. of a thin hollow hemisphere is at a distance of
    (a) $\dfrac{r}{3}$ from base
    (b) $\dfrac{r}{2}$ from base
    (c) $\dfrac{r}{4}$ from base
    (d) none of the above.

148. The C.G. of a solid cone lies on the central axis at a distance of
    (a) $\frac{1}{3}$ of the total height above base
    (b) half the total height above base
    (c) $\frac{1}{4}$ of the total height above base
    (d) 2/5 of total height above base.

149. The C.G. of a hollow cone lies on the central axis above base at a distance of
    (a) $\frac{1}{3}$ of the total height
    (b) half of the total height
    (c) $\frac{1}{4}$ of the total height
    (d) 2/5 of the total height.

150. The C.G. of a hollow pyramid lies on the central axis above base at a distance of
    (a) $\frac{1}{4}$ of the total height
    (b) $\frac{1}{3}$ of the total height
    (c) $\frac{1}{2}$ of the total height
    (d) 2/5 of the total height.

151. The C.G. of a hollow cylinder lies on the vertical axis above base at a distance of
    (a) $\frac{1}{4}$ th of the total height
    (b) $\frac{1}{3}$ rd of the total height
    (c) half of the total height
    (d) 2/5th of the total height.

152. The C.G. of a parabola as shown in Fig. 2.24 lies on the axis OB at a distance of
   (a) $\frac{1}{5}$ th of OB
   (b) $\frac{1}{4}$ th of OB
   (c) 2/5th of OB
   (d) $\frac{1}{3}$ rd of OB.

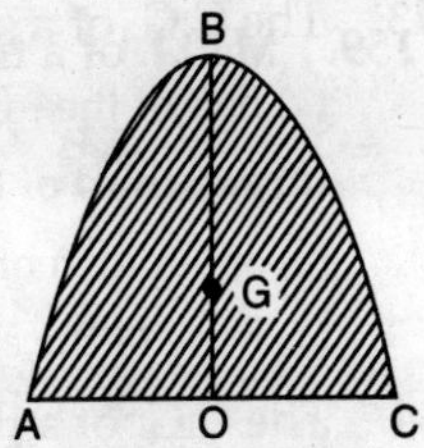

FIGURE 2.24

Moment of Inertia

153. If A = area of the body, I_G = moment of inertia of body about an axis passing through its C.G. and I_0 = moment of inertia of the body about an axis parallel to the axis passing through C.G. and at a distance x, then according to the theorem of parallel axis
   (a) $I_G = I_0 + Ax^2$
   (b) $I_0 = I_G + Ax^2$
   (c) $I_0 = I_G - Ax^2$
   (d) none of the above.

154. If I_x = M.O.I. about x-axis and I_y = M.O.I. about y-axis, then moment of inertia about z-axis is given by
   (a) $I_z = I_x - I_y$
   (b) $I_z = I_y - I_x$
   (c) $I_z = I_x \times I_y$
   (d) none of the above.

155. A thin rod of length L and mass M will have moment of inertia about an axis passing through one of its edge and perpendicular to the rod,
   (a) $\dfrac{Ml^2}{12}$
   (b) $\dfrac{ML^2}{6}$
   (c) $\dfrac{ML^3}{3}$
   (d) $\dfrac{ML^2}{3}$.

156. Moment of inertia of a rectangular section having b = Width and d = Depth about x-axis is given by
   (a) $I_x = \dfrac{bd^3}{12}$
   (b) $I_x = \dfrac{b^3 d}{12}$
   (c) $I_x = \dfrac{b^2 d^2}{6}$
   (d) none of the above.

157. M.O.I. of a circular section of diameter d about an axis passing through its C.G. lying in the plane of the section is given by
   (a) $I_x = \dfrac{\pi d^4}{32}$
   (b) $I_x = \dfrac{\pi d^4}{64}$
   (c) $I_x = \dfrac{\pi d^4}{16}$
   (d) none of the above.

158. M.O.I. of a triangular section about an axis passing through its base is given by
   (a) $I = \dfrac{bh^3}{12}$
   (b) $I = \dfrac{bh^3}{32}$
   (c) $I = \dfrac{bh^2}{36}$
   (d) none of the above

where b = Width at a base and h = Height of triangle.

159. M.O.I. of a triangular section, about an axis passing through its C.G. is

   (a) $I = \dfrac{bh^3}{12}$
   (b) $I = \dfrac{bh^3}{32}$
   (c) $I = \dfrac{bh^2}{36}$
   (d) none of the above

160. M.O.I. of a solid sphere of mass M and radius R is given by

   (a) $I = \dfrac{MR^2}{12}$
   (b) $I = \dfrac{2}{5} MR^2$
   (c) $I = \dfrac{MR^2}{36}$
   (d) $I = \dfrac{3}{5} MR^2$

161. M.O.I. of a thin spherical shell of mass M and radius r is given as

   (a) $\dfrac{2}{5} Mr^2$
   (b) $\dfrac{2}{3} Mr^2$
   (c) $\dfrac{3}{2} Mr^2$
   (d) $\dfrac{4}{3} Mr^2$.

162. M.O.I. of a solid cone about its vertical axis is

   (a) $\dfrac{10}{3} Mr^2$
   (b) $\dfrac{5}{3} Mr^2$
   (c) $\dfrac{3}{10} Mr^2$
   (d) $\dfrac{3}{5} Mr^2$.

163. The units of moment of inertia of mass are

   (a) kg-m³
   (b) kg-m²
   (c) kg-m
   (d) kg-m⁴.

164. The units of moment of inertia of area are

   (a) kg-m³
   (b) kg-m²
   (c) m⁴
   (d) none of the above.

165. M.O.I. of a solid cone about an axis passing through its C.G. and parallel to base is

   (a) $\dfrac{3}{10} Mr^2$
   (b) $\dfrac{3}{5} Mr^2$
   (c) $\dfrac{3}{20} M\left(r^2 + \dfrac{h^2}{4}\right)$
   (d) $\dfrac{10}{3} Mr^2$.

166. M.O.I. of a solid circular cylinder about vertical axis is

   (a) $\dfrac{3}{10} Mr^2$
   (b) $\dfrac{3}{5} Mr^2$
   (c) $\dfrac{3}{2} Mr^2$
   (d) $\dfrac{Mr^2}{2}$

167. M.O.I. of elliptical section, of major axis = D and minor axis = B, is

   (a) $\dfrac{\pi}{64} D^4$
   (b) $\dfrac{\pi}{64} d^4$
   (c) $\dfrac{\pi}{64} BD^3$
   (d) $\dfrac{\pi}{64} B^2 D^2$.

168. Polar moment of inertia of a circle (I_0) is given by

   (a) $\dfrac{\pi}{64} d^4$
   (b) $\dfrac{\pi}{32} d^4$
   (c) $\dfrac{\pi}{16} d^4$
   (d) $\dfrac{\pi}{32} d^3$.

169. Polar moment of inertia of a rectangle ($b \times h$) is given by

   (a) $\dfrac{bh^3}{12}$
   (b) $\dfrac{hb^3}{12}$
   (c) $\dfrac{bh(b^2 + h^2)}{12}$
   (d) $\dfrac{b^2 h^2 (b + h)}{12}$.

170. Polar moment of inertia of an equilateral triangle of sides x is given by

   (a) $\dfrac{x^4}{16}$
   (b) $\dfrac{x^4}{16\sqrt{3}}$
   (c) $\dfrac{x^4}{32}$
   (d) $\dfrac{x^4}{64}$.

171. Polar moment of inertia is

   (a) the moment of inertia of an area about an axis parallel to centroidal axis
   (b) equal to moment of inertia
   (c) the moment of an area about an axis which is not lying in the plane of the area
   (d) the moment of inertia of an area about a line or axis perpendicular to the plane of the area.

172. The moment of inertia of a triangle (having base = b and height = h) with respect to an axis through the apex and parallel to the base is

   (a) $\dfrac{bh^3}{12}$
   (b) $\dfrac{bh^3}{36}$
   (c) $\dfrac{bh^3}{4}$
   (d) $\dfrac{bh^3}{10}$.

173. Newton's second law for rotary motion states that

   (a) rate of change of rotation of a body about a fixed axis is directly proportional to the impressed external force and takes place in the direction of force.
   (b) rate of change of momentum is directly proportional to the impressed force and takes plane in the direction of force.
   (c) rate of change of rotation (angular momentum) is directly proportional to the impressed external torque and takes place in the direction of force.
   (d) none of the above.

174. The relation between external torque (T) acting on a body and the angular acceleration (α) is given by

(a) $T = \dfrac{I}{\alpha}$ □ (b) $T = \dfrac{\alpha}{I}$ □

(c) $T = I\alpha$ □ (d) none of the above □

where T is in newton-metres.

Circular Motion

175. If a body is moving in a circular path, a force comes into play which acts along the radius of circular path and is directed towards the centre of the path. This force is called

(a) centrifugal force □ (b) centripetal force □

(c) shear force □ (d) none of the above. □

176. The force, acting on a body moving in a circular path along the radius away from the centre of the path, is called

(a) centrifugal force □ (b) centripetal force □

(c) shear force □ (d) none of the above. □

177. Centrifugal force is given by

(a) $\dfrac{W}{g} \dfrac{\omega^2}{r}$ □ (b) $\dfrac{W}{g} \omega^2 r^2$ □

(c) $\dfrac{W}{g} V^2 r$ □ (d) $\dfrac{W}{g} \dfrac{V^2}{r}$ □

where W = Weight of body in kgf, V = Linear velocity,
ω = Angular velocity, and r = Radius of circular path.

178. The expression $T \times \omega$ (where T = torque in N m and ω angular velocity) gives

(a) work done due to rotation □ (b) horse power □

(c) power in watts □ (d) force. □

179. Kinetic energy due to rotation is expressed as

(a) K.E. = $\dfrac{I\omega}{2g}$ □ (b) K.E. = $\dfrac{I^2 \times \omega}{2g}$ □

(c) K.E. = $\dfrac{I\omega^2}{2g}$ □ (d) none of the above. □

180. Radius of gyration is expressed as

(a) $k = \sqrt{\dfrac{m}{I}}$ □ (b) $\sqrt{\dfrac{I}{m}}$ □

(c) $k = \sqrt{mI}$ □ (d) $\dfrac{I}{\sqrt{mI}}$. □

181. Super-elevation (or banking) of roads is the process of
    (a) raising the inner edge of the road above the outer edge
    (b) raising the outer edge of the roads above the inner edge
    (c) keeping both the edges at the same level
    (d) none of the above.

182. The angle of super-elevation (or banking) is given by
    (a) $\theta = \tan^{-1} \dfrac{V^2}{g}$
    (b) $\theta = \tan^{-1} \dfrac{V^2}{gr}$
    (c) $\theta = \tan^{-1} \dfrac{gr}{V^2}$
    (d) $\theta = \tan^{-1} \dfrac{r}{gV^2}$

    where V = Velocity of the vehicle, r = Radius of circular path.

183. The maximum velocity of a vehicle on a level circular path to avoid skidding is given by
    (a) $\sqrt{\dfrac{\mu}{gr}}$
    (b) $\sqrt{\mu gr}$
    (c) $\sqrt{\dfrac{\mu r}{g}}$
    (d) $\dfrac{1}{\sqrt{\mu gr}}$

    where μ = Co-efficient of friction between the wheels of the vehicle and the ground
    r = Radius of circular path.

184. The maximum velocity (V_{max}) of a vehicle on a level circular path to avoid outer-turning is given by
    (a) $\sqrt{\dfrac{dgr}{2h}}$
    (b) $\sqrt{\dfrac{gr}{2hr}}$
    (c) $\sqrt{\dfrac{2h}{dgr}}$
    (d) $\sqrt{\dfrac{dg}{2hr}}$

    where h = Height of C.G. of the vehicle from the ground level
    d = Distance between the centre lines of the wheels
    r = Radius of circular path.

185. The gravitational acceleration at a place is 6 times the value of gravitational acceleration at earth, the weight of the body at that place will be
    (a) one-sixth of the weight at earth
    (b) same as at earth
    (c) 6 times the weight at earth
    (d) none of the above.

186. A flywheel, mounted on a shaft, weighs 500 kg and has a radius of gyration 50 cm about its axis of rotation. If the flywheel starts from rest and shaft is subjected to a moment of 625 Nm, the speed of shaft after 4 seconds is equal to
    (a) 30 rad per sec
    (b) 20 rad/sec
    (c) 10 rad per sec
    (d) 40 rad/sec.

187. A curved road is generally provided a slope, which is known as
    (a) angle of banking
    (b) angle of repose
    (c) angle of friction
    (d) angle of reaction.

188. The angle of banking provided on the curved roads depends upon
    (a) the velocity of vehicle only
    (b) the square of velocity of vehicle only
    (c) the square or velocity of vehicle and radius of circular path
    (d) co-efficient of friction between the vehicle and road contact point.

189. To prevent side thrust on the wheel flanges of a train, moving round a curve
    (a) outside rails are raised
    (b) inner rails are raised
    (c) rails are kept at the same level
    (d) thrust eliminators are provided.

Simple Harmonic Motion

190. A body is said to describe simple harmonic motion if it moves in a straight line such that
    (a) its acceleration is proportional to its distance from a fixed point on the straight line and is directed away from the fixed point
    (b) its velocity is proportional to its distance from a fixed point and is directed towards the fixed point
    (c) its velocity is proportional to its distance from a fixed point and is directed away from the fixed point
    (d) none of the above.

191. The time period of one oscillation of a simple pendulum is given by
    (a) $\pi \sqrt{\dfrac{L}{g}}$
    (b) $2\pi \sqrt{\dfrac{L}{g}}$
    (c) $2\pi \sqrt{\dfrac{L}{2g}}$
    (d) $\pi \sqrt{\dfrac{2L}{g}}$.

192. Beat is equal to
    (a) one oscillation
    (b) twice the oscillation
    (c) half the oscillation
    (d) none of the above.

193. If a pendulum executes one beat per second, it is called
    (a) simple pendulum
    (b) second's pendulum
    (c) compound pendulum
    (d) none of the above.

194. Length of a second's pendulum is equal to
    (a) 50 cm
    (b) 99.4 cm
    (c) 150 cm
    (d) 80 cm.

195. The periodic time (T) of a S.H.M. is given by
    (a) $T = \dfrac{\omega}{2\pi}$
    (b) $T = \dfrac{\pi}{2\omega}$
    (c) $T = \dfrac{2\pi}{\omega}$
    (d) $T = \dfrac{2\omega}{\pi}$.

196. The velocity of a particle moving with S.H.M. is maximum, when it is at
   (a) extreme position
   (b) its mean position
   (c) a point between its mean and extreme position
   (d) none of the above.

197. The acceleration of a particle moving with S.H.M. is maximum, when it is at
   (a) its extreme position
   (b) its mean position
   (c) a point between its mean and extreme positions
   (d) none of the above.

198. The velocity of a particle moving with S.H.M. is zero, when it is at
   (a) its extreme position
   (b) its mean position
   (c) a point between its mean and extreme position
   (d) none of the above.

199. The acceleration of a particle moving with S.H.M. is zero, when it is at
   (a) its extreme position
   (b) its mean position
   (c) a point between its mean and extreme position
   (d) none of the above.

200. The period of oscillation of a helical spring is given by
   (a) $T = 2\pi \sqrt{\dfrac{g}{\delta}}$
   (b) $T = 2\pi \sqrt{\dfrac{\delta}{g}}$
   (c) $T = \sqrt{\dfrac{g}{2\delta}}$
   (d) $T = \pi \sqrt{\dfrac{2\delta}{g}}$

   where δ = Static extension.

201. For a compound pendulum, the time period is given as
   (a) $T = 2\pi \sqrt{\dfrac{K_G^2}{gh}}$
   (b) $T = 2\pi \sqrt{\dfrac{K_G^2 + h^2}{gh}}$
   (c) $T = 2\pi \sqrt{\dfrac{gh}{K_G^2}}$
   (d) $T = 2\pi \sqrt{\dfrac{gh}{K_G^2 + h^2}}$

   where K_G = Radius of gyration
   h = Distance from the point of suspension and centre of the gravity of the body.

202. Centre of percussion is a point at which
   (a) resultant pressure force is supposed to act
   (b) weight of the body is acting
   (c) if a blow is given, no reaction will be felt at the point of suspension of a body
   (d) resultant force of buoyancy is supposed to act.

203. Centre of oscillation and centre of suspension for a compound pendulum are
 (a) acting at the same point
 (b) interchangable
 (c) not-interchangable
 (d) none of the above.

204. The velocity of a particle moving with S.H.M. at a distance from mean position is given by
 (a) $\sqrt{\omega(r^2 - y^2)}$
 (b) $\omega \sqrt{r^2 - y^2}$
 (c) $\omega \sqrt{1 - \dfrac{y^2}{r^2}}$
 (d) $\sqrt{\omega \left(1 - \dfrac{y^2}{r^2}\right)}$

where r = Amplitude of the motion, and ω = Angular velocity.

205. The acceleration of a particle moving with S.H.M. at a distance y from mean position is given by
 (a) $-\omega y^2$
 (b) ωy^2
 (c) $-\omega^2 y$
 (d) $-\omega^2/y$.

206. The gain or loss of number of oscillations are given by
 (a) $\dfrac{dn}{n} = \dfrac{dg}{g} - \dfrac{dl}{2l}$
 (b) $\dfrac{dn}{n} = \dfrac{dl}{l} - \dfrac{dg}{2g}$
 (c) $\dfrac{dn}{n} = \dfrac{dg}{g} - \dfrac{dl}{l}$
 (d) $\dfrac{dn}{n} = \dfrac{dg}{2g} - \dfrac{dl}{2l}$.

207. At a place, the clock will go slow if
 (a) length of pendulum is increased
 (b) length of pendulum is decreased
 (c) mass of its bob is increased
 (d) none of the above.

208. The length of a simple pendulum which has the same time period as compound pendulum is given by
 (a) $h + \dfrac{k^2}{h^2}$
 (b) $h^2 + \dfrac{k^2}{h}$
 (c) $h + \dfrac{k^2}{h}$
 (d) $h + \dfrac{k}{h^2}$

where h = Length between the point of suspension and centre of gravity, and
 k = Radius of gyration.

209. The time period of a simple pendulum will be doubled if
 (a) its length is doubled
 (b) its length is halved
 (c) its length is increased four times
 (d) its length is increased eight times.

210. At a place, a clock will go fast if
 (a) its length is increased
 (b) its length is decreased
 (c) mass of its bob is increased
 (d) mass of its bob is decreased.

211. The maximum velocity of a particle moving with S.H.M. is 2 m/sec and maximum acceleration is 20 m/sec². The frequency of the motion is equal to

   (a) $\dfrac{20}{\pi}$ (b) $\dfrac{10}{\pi}$

   (c) $\dfrac{5}{\pi}$ (d) $\dfrac{1}{\pi}$.

212. A body moving with S.H.M. is having frequency as three vibrations per second and amplitude as 20 cm. The maximum velocity of the body is equal to

   (a) 0.377 m   (b) 3.77 m

   (c) 37.7 m   (d) 0.00377 m.

213. The periodic time (T) of a particle with S.H.M. is

   (a) directly proportional to the mass of the particle
   (b) directly proportional to the angular velocity
   (c) directly proportional to the square of angular velocity
   (d) inversely proportional to the angular velocity.

Basic Definitions

214. A force is acting on a mass of one kilogram and produces an acceleration of one metre per second square. Then the force is known as

   (a) dyne   (b) newton
   (c) kg-weight   (d) kgf.

215. A force is acting on a mass of one gram and produces an acceleration of one cm per second square. Then the force is known as

   (a) dyne   (b) newton
   (c) gm-weight   (d) gmf.

216. One newton is equal to

   (a) 10^3 dyne   (b) 10^4 dyne
   (c) 10^5 dyne   (d) 10^6 dyne.

217. The linear acceleration a is equal to

   (a) $\dfrac{dv}{dt}$ (b) $v\dfrac{dv}{ds}$

   (c) $\dfrac{d^2s}{dt^2}$ (d) any one of the above

   (e) none of the above.

218. The acceleration of a body moving down an inclined smooth plane is equal to

   (a) $g \cos \theta$ (b) $-g \sin \theta$
   (c) $g \sin \theta$ (d) $g \tan \theta$

   where θ = Inclination of plane with horizontal.

219. The acceleration of a body moving up an inclined smooth plane is equal to
    (a) $g \cos \theta$
    (b) $- g \sin \theta$
    (c) $g \sin \theta$
    (d) $g \tan \theta$
    where θ = Inclination of plane with horizontal.

220. The acceleration of a body moving down a rough inclined plane is equal to
    (a) $g \sin \theta$
    (b) $g [\sin \theta - \mu \cos \theta]$
    (c) $g \tan \theta$
    (d) $g [\sin \theta - \mu \sin \theta]$
    where θ = Inclination of plane with the horizontal.

221. The radius of gyration (k) for a circular lamina is equal to
    (a) $\sqrt{2} R$
    (b) $\dfrac{R}{\sqrt{2}}$
    (c) $0.6324 R$
    (d) $0.5 R$
    where R = Radius.

222. The radius of gyration (k) for a solid cylinder is equal to
    (a) $\sqrt{2} R$
    (b) $\dfrac{R}{\sqrt{2}}$
    (c) $0.6324 R$
    (d) $0.5 R$.

223. The radius of gyration (k) for a solid sphere is equal to
    (a) $\sqrt{2} R$
    (b) $\dfrac{R}{\sqrt{2}}$
    (c) $0.6324 R$
    (d) $0.5 R$

224. The kinetic energy due to rotation of a body is equal to
    (a) $\frac{1}{2} I \omega^2$
    (b) $\frac{1}{2} m V^2$
    (c) $2 I \omega^3$
    (d) $\frac{1}{2} I^2 \omega$.

225. If a body is having motion of translation as well as motion of rotation, then total kinetic energy is equal to
    (a) $\frac{1}{2} m V^2$
    (b) $\frac{1}{2} m V^2 + \frac{1}{2} I \omega^2$
    (c) $\frac{1}{2} m V^2 + \frac{1}{2} \omega I^2$
    (d) none of the above.

226. The angular acceleration (α) of a rotating body is equal to
    (a) $\dfrac{d\omega}{dt}$
    (b) $\dfrac{d^2 \theta}{dt^2}$
    (c) $\omega \dfrac{d\omega}{d\theta}$
    (d) any one of the above.

227. The radius of gyration (k) is equal to
 (a) $\sqrt{\dfrac{A}{I}}$ (b) $\sqrt{\dfrac{I}{A}}$
 (c) $\sqrt{AI}$ (d) $\sqrt{\dfrac{I}{AI}}$

 where I = Moment of inertia, and A = Area.

228. When a projectile is projected on an inclined plane, the range on the inclined plane will be maximum if
 (a) $2\alpha - \beta = 90°$ (b) $2\alpha + \beta = 90°$
 (c) $\alpha + 2\beta = 90°$ (d) $\alpha - 2\beta = 90°$

 where α = Angle of projection with the horizontal, and
   β = Inclination of the inclined plane with the horizontal.

229. If the projectile is projected down an inclined plane, the range on the inclined plane will be maximum if
 (a) $2\alpha - \beta = 90°$ (b) $2\alpha + \beta = 90°$
 (c) $\alpha + 2\beta = 90°$ (d) $\alpha - 2\beta = 90°$

 where α = Angle of projection with the horizontal, and
   β = Inclination of the inclined plane with the horizontal.

230. Angle of repose is
 (a) less than angle of friction   (b) more than angle of friction
 (c) equal to angle of friction   (d) none of the above.

Lifting Machines

231. For a simple wheel and axle, the velocity ratio is given by
 (a) $D/(d+1)$ (b) d/D
 (c) D/d (d) $(D+1)/d$

 where D = Diameter of wheel, and d = Diameter of axle.

232. A weight of 48 N is to be raised by a wheel and axle. The diameters of wheel and axle are 400 mm and 100 mm respectively. The force required at the wheel is 16 N. Then the velocity ratio of this machine will be
 (a) $\dfrac{1}{4}$ (b) $\dfrac{2}{4}$
 (c) $\dfrac{3}{4}$ (d) 4.

233. In the above question, the mechanical advantage will be
 (a) $\dfrac{1}{3}$ (b) 3
 (c) 4   (d) 5.

234. In question 232, the efficiency of the machine will be
 (a) 90%   (b) 80%
 (c) 75%   (d) 50%.

235. For a differential wheel and axle, the velocity ratio is given by
    (a) $D/(d_1 - d_2)$
    (b) $2D/(d_1 - d_2)$
    (c) $D/2(d_1 - d_2)$
    (d) $2D/(2d_1 - d_2)$
    where D = Dia. of wheel, d_1 = Dia. of bigger axle, d_2 = Dia. of smaller axle.

236. For a differential wheel and axle, the diameter of wheel is 25 cm. The diameters of differential axles are 10 cm and 9 cm. An effort of 30 N is applied to lift a load of 900 N. Then the velocity ratio of the machine will be
    (a) 10
    (b) 25
    (c) 50
    (d) 60.

237. In question 236, the mechanical advantage will be
    (a) 10
    (b) 20
    (c) 30
    (d) 40.

238. In question 236, the efficiency of the machine will be
    (a) 40%
    (b) 60%
    (c) 75%
    (d) 80%.

239. The differential wheel and axle is having large value of .... as compared to simple wheel and axle.
    (a) mechanical advantage
    (b) velocity ratio
    (c) efficiency
    (d) all of the above.

240. The velocity ratio of worm and worm-wheel, when the worm is single threaded, is given by
    (a) $\dfrac{2LT}{r}$
    (b) $\dfrac{LT}{r}$
    (c) $\dfrac{L \times T}{2r}$
    (d) $\dfrac{L \times T}{3r}$
    where L = Length of the handle, T = Number of teeth on the worm wheel, and
    r = Radius of load drum.

241. When the worm is double-threaded, then the velocity ratio of worm and worm wheel is
    (a) $\dfrac{2LT}{r}$
    (b) $\dfrac{LT}{r}$
    (c) $\dfrac{L \times T}{2r}$
    (d) $\dfrac{L \times T}{3r}$

242. The number of teeth on the worm-wheel of a single-threaded worm and worm wheel is 60. Calculate the velocity ratio if the diameter of the effort wheel is 25 cm and that of load drum is 12.5 cm. The load lifted is 600 N when a force of 20 N is applied to this machine
    (a) 60
    (b) 120
    (c) 180
    (d) 240.

243. In question 242, the mechanical advantage would be
    (a) 30
    (b) 45
    (c) 60
    (d) 75.

244. In question 242, the efficiency of the machine would be
    (a) 75% (b) 60%
    (c) 50% (d) 25%.

245. The velocity ratio of single purchase crab winch is given by
    (a) $\dfrac{L}{D} \times \dfrac{T_2}{T_1}$ (b) $\dfrac{2L}{D} \times \dfrac{T_2}{T_1}$
    (c) $\dfrac{L}{2D} \times \dfrac{T_2}{T_1}$ (d) $\dfrac{L}{3D} \times \dfrac{T_2}{T_1}$

    where L = Length of lever arm, D = Diameter of the load axle,
    T_1 = Number of teeth on the pinion, and T_2 = Number of teeth on the spur wheel.

246. The number of teeth on pinion and spur wheel of a single purchase crab winch are 10 and 100 respectively. The diameter of load axle is 30 cm and length of lever arm is also 30 cm. The velocity ratio of this machine will be
    (a) 10 (b) 20
    (c) 30 (d) 40.

247. For a single fixed pulley, the mechanical advantage is
    (a) less than one (b) more than one
    (c) equal to one (d) none of the above.

248. For a single movable pulley, the mechanical advantage is
    (a) less than one (b) more than one
    (c) equal to one (d) none of the above.

249. The velocity ratio for the first system of pulleys is equal to
    (a) $2 \times n$ (b) 2^n
    (c) $2/n$ (d) $\dfrac{1}{2^n}$

    where n = No. of movable pulleys.

250. For the first system of pulleys, there are four movable pulleys. A load of 1440 N is lifted by an effort of 100 N. Then the velocity ratio of this pulley will be
    (a) 4 (b) 8
    (c) 16 (d) 32.

251. In question 250, the efficiency of the pulley will be
    (a) 90% (b) 80%
    (c) 60% (d) 50%.

252. In case of second system of pulley, the velocity ratio is equal to
    (a) $n + 1$ (b) $n - 1$
    (c) n (d) $n + 2$

    where n = No. of segments supporting the movable block or load.

253. A weight of 200 N is to be lifted by an effort of 60 N, by second system of pulleys having three pulleys in the upper block and two pulleys in the lower block. The velocity ratio of the system will be
(a) 2   (b) 3
(c) 1   (d) 5.

254. In question 253, the efficiency of the system will be
(a) 80%   (b) 66.67%
(c) 50%   (d) 33.33%.

255. In third system of pulley, the velocity ratio is equal to
(a) 2^n (b) $2^n - 1$
(c) $2^n + 1$ (d) $2^n + 2$

where n = No. of pulleys in the system.

256. There are four pulleys in a third system of pulley. The load lifted is 1800 N by an effort of 160 N. Then the velocity ratio of the pulley will be
(a) 16   (b) 15
(c) 17   (d) 18.

257. In question 256, the efficiency of the machine will be
(a) 50%   (b) 60%
(c) 70%   (d) 75%.

258. In a Weston differential pulley block, the upper block has two pulleys of diameters 25 cm and 20 cm. A load of 100 N is lifted by an effort of 20 N, by this machine. The velocity ratio of this machine is
(a) 40   (b) 30
(c) 20   (d) 10.

259. The efficiency of the machine given in question 258, is
(a) 50%   (b) 60%
(c) 70%   (d) 75%.

Belt Drive

260. The velocity ratio of a belt is given by (if thickness of belt is neglected)
(a) $\frac{N_2}{N_1} = \frac{d_2}{d_1}$ (b) $\frac{N_2}{N_1} = \frac{d_1}{d_2}$
(c) $\frac{N_2}{N_1} = \frac{d_2 + 1}{d_1 + 1}$ (d) $\frac{N_2}{N_1} = \frac{d_1 + 1}{d_2 + 1}$

where N_1 = speed of driver pulley, N_2 = speed of driven pulley,

d_1 = Dia. of driver pulley, and d_2 = Dia. of driven pulley.

261. The velocity ratio of a belt (if thickness of belt is considered) is given by
(a) $\frac{N_2}{N_1} = \frac{d_2 + t}{d_1 + t}$ (b) $\frac{N_2}{N_1} = \frac{d_1 + 2}{d_2 + 2}$
(c) $\frac{N_2}{N_1} = \frac{d_1 + t}{d_2 + t}$ (d) $\frac{N_2}{N_1} = \frac{d_2 + 1}{d_1 + 1}$.

262. The velocity ratio of a compound belt drive is given by

(a) $\dfrac{\text{Speed of last follower}}{\text{Speed of first driver}} = \dfrac{\text{Product dia. of follower}}{\text{Product of dia. driver}}$

(b) $\dfrac{\text{Speed of last follower}}{\text{Speed of first driver}} = \dfrac{\text{Dia. of last follower}}{\text{Dia. of first driver}}$

(c) $\dfrac{\text{Speed of last follower}}{\text{Speed of first driver}} = \dfrac{\text{Product of dia. of drivers}}{\text{Product of dia. of follower}}$

(d) None of the above.

263. A shaft is driven with the help of a belt which is passing over the engine and shaft. The engine is running at 200 r.p.m. The diameters of engine pulley is 51 cm and that of shaft is 30 cm. The speed of the shaft will be

(a) 200 r.p.m. (b) 300 r.p.m.
(c) 340 r.p.m. (d) 400 r.p.m.

264. In the question 263, the speed ratio would be

(a) 2 (b) 1.7
(c) 1.5 (d) 1.3.

265. In question 263, what should be dia. of the pulley of engine, so that the speed of shaft is 400 r.p.m.?

(a) 100 cm (b) 75 cm
(c) 60 cm (d) 40 cm.

266. A shaft is driven by a belt of thickness 1 cm. The belt is passing over the engine and the shaft. The engine is running at 310 r.p.m. The diameter of engine pulley is 50 cm and that of shaft is 30 cm. The speed of the shaft will be

(a) 400 r.p.m. (b) 500 r.p.m.
(c) 510 r.p.m. (d) 520 r.p.m.

267. The speed ratio in question 266, would be

(a) 2 (b) $\dfrac{50}{30}$

(c) $\dfrac{51}{31}$ (d) $\dfrac{52}{32}$.

268. An engine is running at 200 r.p.m. With the help of belt, the engine drives a line shaft. The diameters of pulleys on engine and on line shafts are 80 cm and 40 cm respectively. A 100 cm dia. pulley one line shaft drives a 20 cm dia. pulley keyed to a dynamo shaft. The speed of dynamo shaft would be

(a) 1000 r.p.m. (b) 2000 r.p.m.
(c) 1500 r.p.m. (d) 4000 r.p.m.

269. In question 268, the speed ratio of the compound belt drive would be

(a) 10 (b) 15
(c) 20 (d) 25.

270. For an open belt drive, the length (L) of the belt is equal to

(a) $\pi(r_1 + r_2) + \dfrac{(r_1 + r_2)^2}{2x} + x$

(b) $\pi(r_1 - r_2) + \dfrac{(r_1 - r_2)^2}{2x} + 2x$

(c) $\pi(r_1 + r_2)^2 + \dfrac{(r_1 - r_2)^2}{2x} + 2x$

(d) $\pi(r_1 + r_2) + \dfrac{(r_1 + r_2)^2}{2x} + 2x$

where r_1 = Radius of larger pulley, r_2 = Radius of smaller pulley, and
x = Distance between the centres of two pulleys.

271. For a crossed belt-drive, the length (L) of the belt is equal to

(a) $\pi(r_1 + r_2) + \dfrac{(r_1 + r_2)^2}{2x} + x$

(b) $\pi(r_1 - r_2) + \dfrac{(r_1 - r_2)^2}{2x} + 2x$

(c) $\pi(r_1 + r_2)^2 + \dfrac{(r_1 - r_2)^2}{2x} + 2x$

(d) $\pi(r_1 + r_2) + \dfrac{(r_1 + r_2)^2}{2x} + 2x$

272. The length of the open belt drive depends upon

(a) sum of the radii
(b) difference of the radii
(c) both the sum and difference of the radii
(d) none of the above.

273. The length of the crossed belt drive depends upon

(a) sum of radii
(b) difference of radii
(c) both sum and difference of the radii
(d) none of the above.

274. The ratio of tensions on the two sides of a belt is given by

(a) $\dfrac{T_1}{T_2} = e \times \mu \times \theta$

(b) $\dfrac{T_1}{T_2} = e^{\mu \times \theta}$

(c) $\dfrac{T_1}{T_2} = \mu^{e \times \theta}$

(d) $\dfrac{T_1}{T_2} = e^{(\mu/\theta)}$

where T_1 = Tension on the tight side, T_2 = Tension on the slack side
μ = Co-efficient of friction between belt and pulley
θ = Angle of contact in radians.

275. The angle of contact for an open belt-drive is given by

(a) $\theta = (180 + 2\alpha)$
(b) $\theta = (180 - 2\alpha)$
(c) $\theta = \dfrac{180 + \alpha}{2}$
(d) $\theta = \dfrac{180 - \alpha}{2}$.

276. The angle of contact for a crossed belt is given by

(a) $\theta = (180 + 2\alpha)$
(b) $\theta = (180 - 2\alpha)$
(c) $\theta = \dfrac{180 + \alpha}{2}$
(d) $\theta = \dfrac{180 - \alpha}{2}$.

277. The value of α for an open-belt drive is given by

   (a) $\alpha = \sin^{-1}\left(\dfrac{r_1 - r_2}{2x}\right)$
   (b) $\alpha = \sin^{-1}\left(\dfrac{r_1 + r_2}{2x}\right)$
   (c) $\alpha = \sin^{-1}\left(\dfrac{r_1 - r_2}{x}\right)$
   (d) $\alpha = \sin^{-1}\left(\dfrac{r_1 + r_2}{x}\right)$.

278. The value of α for a crossed-belt drive is given by

   (a) $\alpha = \sin^{-1}\left(\dfrac{r_1 - r_2}{2x}\right)$
   (b) $\alpha = \sin^{-1}\left(\dfrac{r_1 + r_2}{2x}\right)$
   (c) $\alpha = \sin^{-1}\left(\dfrac{r_1 - r_2}{x}\right)$
   (d) $\alpha = \sin^{-1}\left(\dfrac{r_1 + r_2}{x}\right)$.

279. The power transmitted by a belt is given by

   (a) $P = (T_1 - T_2) \times v$
   (b) $P = \dfrac{(T_1 - T_2) \times v}{4500}$
   (c) $P = \dfrac{(T_1 - T_2) \times v}{75}$
   (d) $P = \dfrac{(T_1 + T_2) \times v}{75}$

   where T_1 and T_2 are tension in N.

280. The power in kilo-watt transmitted by a belt is given by

   (a) $P = (T_1 - T_2) \times v$
   (b) $P = \dfrac{(T_1 - T_2) \times v}{1000}$
   (c) $P = \dfrac{(T_1 - T_2) \times v}{75}$
   (d) $P = \dfrac{(T_1 + t_2) \times v}{75}$

   where T_1 and T_2 are in Newton.

281. The centrifugal tension (T_c) in belt is given by

   (a) $T_c = \frac{1}{2} mv^2$ Newton
   (b) $T_c = m \times v^2$ Newton
   (c) $T_c = 2 \times m \times v^2$ Newton
   (d) $T_c = 3 \times m \times v^2$ Newton.

282. For maximum power transmission, the velocity of belt is given by

   (a) $v = \sqrt{\dfrac{T_{max}}{m}}$
   (b) $v = \sqrt{\dfrac{3T_{max}}{m}}$
   (c) $v = \sqrt{\dfrac{T_{max}}{3m}}$
   (d) $v = \sqrt{\dfrac{2T_{max}}{m}}$

   where T_{max} = Maximum tension in the belt, m = Mass of belt per metre length in kg.

283. For maximum power transmission, maximum tension in the belt is given by

   (a) $T_{max} = 2T_c$
   (b) $T_{max} = 3T_c$
   (c) $T_{max} = 4T_c$
   (d) $T_{max} = \frac{1}{3} T_c$

   where T_c = Centrifugal tension in the belt.

284. For maximum power transmission, the tension on the tight side should be
    (a) one-third of maximum tension
    (b) two-third of maximum tension
    (c) one-quarter of maximum tension
    (d) three fourth of maximum tension.

285. The initial tension in the belt is given by (if centrifugal tension is neglected)
    (a) $T_0 = \dfrac{T_1 + T_2}{3}$
    (b) $T_0 = \dfrac{T_1 + T_2}{2}$
    (c) $T_0 = \dfrac{T_1 + T_2}{4}$
    (d) $T_0 = \dfrac{2T_1 + T_2}{3}$.

286. If centrifugal tension is considered, then initial tension in the belt is given by
    (a) $T_0 = \dfrac{T_1 + T_2 + T_c}{3}$
    (b) $T_0 = \dfrac{T_1 + T_2 + 2T_c}{3}$
    (c) $T_0 = \dfrac{T_1 + T_2 + 2T_c}{3}$
    (d) $T_0 = \dfrac{T_1 + T_2 + T_c}{2}$.

287. The tensions on the tight and slack sides of a belt are 200 N and 100 N. If centrifugal tension is neglected, then initial tension in the belt will be
    (a) 25 N
    (b) 300 N
    (c) 150 N
    (d) 175 N.

288. If in question 287, the belt is running at 4 m/sec, then the power transmitted by the belt will be
    (a) 1200 W
    (b) 600 W
    (c) 400 W
    (d) 200 W.

289. If in question 287, the centrifugal tension is considered and mass of belt is 0.5 kg per metre length when belt is running at 4 m/sec then centrifugal tension will be equal to
    (a) 16 N
    (b) 8 N
    (c) 4 N
    (d) 2 N.

290. For the above question, the initial tension will be equal to
    (a) 150 N
    (b) 158 N
    (c) 164 N
    (d) 172 N.

III. TRUE/FALSE

1. The resultant of two forces A and B (which are coplaner and are acting at an angle θ) is given by $R = \sqrt{P^2 + Q^2 + 2AB \sin \theta}$
    (a) True
    (b) False.

2. For a perfect frame, the number of members (n) and number of joints (j) is given by $n = 2j - 3$
    (a) True
    (b) False.

ANSWERS
Objective Type Questions

1. (b)	2. (b)	3. (b)	4. (c)	5. (b)	6. (c)
7. (c)	8. (a)	9. (c)	10. (b)	11. (a)	12. (c)
13. (c)	14. (b)	15. (c)	16. (c)	17. (d)	18. (c)
19. (b)	20. (b)	21. (c)	22. (a)	23. (b)	24. (b)
25. (d)	26. (b)	27. (a)	28. (c)	29. (b)	30. (a)
31. (c)	32. (c)	33. (a)	34. (b)	35. (b)	36. (b)
37. (c)	38. (b)	39. (b)	40. (a)	41. (c)	42. (c)
43. (b)	44. (c)	45. (b)	46. (a)	47. (a)	48. (a)
49. (b)	50. (c)	51. (b)	52. (b)	53. (c)	54. (a)
55. (b)	56. (c)	57. (b)	58. (c)	59. (b)	60. (b)
61. (b)	62. (b)	63. (c)	64. (c)	65. (d)	66. (c)
67. (b)	68. (c)	69. (a)	70. (b)	71. (c)	72. (a)
73. (c)	74. (a)	75. (d)	76. (b)	77. (c)	78. (a)
79. (c)	80. (b)	81. (b)	82. (c)	83. (b)	84. (c)
85. (b)	86. (b)	87. (c)	88. (d)	89. (c)	90. (d)
91. (b)	92. (c)	93. (c)	94. (d)	95. (a, c)	96. (b)
97. (b)	98. (a)	99. (b)	100. (b)	101. (c)	102. (d)
103. (c)	104. (c)	105. (c)	106. (c)	107. (c)	108. (b)
109. (b)	110. (c)	111. (c)	112. (b)	113. (c)	114. (b)
115. (c)	116. (b)	117. (c)	118. (b)	119. (c)	120. (b)
121. (c)	122. (b)	123. (b)	124. (c)	125. (c)	126. (b)
127. (c)	128. (b)	129. (c)	130. (b)	131. (b)	132. (c)
133. (b)	134. (c)	135. (d)	136. (d)	137. (c)	138. (d)
139. (c)	140. (b)	141. (b)	142. (a)	143. (c)	144. (b)
145. (c)	146. (b)	147. (b)	148. (c)	149. (a)	150. (b)
151. (c)	152. (c)	153. (b)	154. (c)	155. (d)	156. (a)
157. (b)	158. (c)	159. (a)	160. (b)	161. (b)	162. (c)
163. (b)	164. (c)	165. (c)	166. (d)	167. (c)	168. (b)
169. (c)	170. (b)	171. (d)	172. (c)	173. (c)	174. (c)
175. (b)	176. (a)	177. (d)	178. (c)	179. (c)	180. (b)
181. (b)	182. (b)	183. (b)	184. (a)	185. (c)	186. (c)
187. (a)	188. (c)	189. (a)	190. (d)	191. (b)	192. (c)
193. (b)	194. (b)	195. (c)	196. (b)	197. (a)	198. (a)
199. (b)	200. (b)	201. (b)	202. (c)	203. (b)	204. (b)
205. (c)	206. (d)	207. (a)	208. (c)	209. (c)	210. (b)
211. (c)	212. (b)	213. (d)	214. (b)	215. (a)	216. (c)
217. (d)	218. (c)	219. (b)	220. (b)	221. (b)	222. (b)

223. (c)	224. (a)	225. (b)	226. (d)	227. (b)	228. (a)
229. (b)	230. (c)	231. (c)	232. (d)	233. (b)	234. (c)
235. (b)	236. (c)	237. (c)	238. (b)	239. (d)	240. (b)
241. (c)	242. (b)	243. (a)	244. (d)	245. (b)	246. (b)
247. (c)	248. (b)	249. (b)	250. (c)	251. (a)	252. (c)
253. (d)	254. (b)	255. (b)	256. (b)	257. (d)	258. (d)
259. (a)	260. (b)	261. (c)	262. (c)	263. (c)	264. (b)
265. (c)	266. (c)	267. (c)	268. (b)	269. (a)	270. (c)
271. (d)	272. (c)	273. (a)	274. (b)	275. (b)	276. (a)
277. (c)	278. (d)	279. (a)	280. (b)	281. (b)	282. (c)
283. (b)	284. (b)	285. (b)	286. (c)	287. (c)	288. (c)
289. (b)	290. (b)				

True/False

1. (b) 2. (a).

Chapter 3 STRENGTH OF MATERIALS

I. INTRODUCTION

Strength of Materials extends the study of forces that was begun in Engineering Mechanics. The field of Mechanics covers the relations between forces acting on rigid bodies. Strength of Materials deals with the relations between externally applied loads and their internal effects on bodies. The deformations in the bodies, however small they may be, are of major interest.

Every body offers resistance against any disturbance to its natural state of formation. This resistance, against deformation, per unit area is known as the stress. It is measured in terms of force exerted per unit area.

The stresses acting on any surface can be resolved into two components one normal to the surface and the other tangential to the surface. These are the two basic stresses, namely **Normal stresses** and **Shearing stresses.** All other stresses are either similar to these or a combination of these.

The axial stresses resulting form axial forces which tend to elongate a member are called **Tensile stresses.** Members which are subjected to predominantly axial tensile stresses are known as **ties.** Such axial stresses which tend to shorten or compress a member are called **Compressive stresses.** Members which are subjected to predominantly axial compressive stresses are called **Struts.** As per Saint Venant's principle, these axial stresses will be uniformly distributed over the entire area of cross-section. If a force is parallel to the surface on which it acts, it is called a **shear force.** The direction of shear stress on any plane will be parallel to the plane of shear.

Ultimate strength is the greatest stress the material can withstand without rupture. For obvious reasons of safety, the full strength is never utilised. Allowable stress is that part of the ultimate strength which will be used when maximum permissible force is applied. In the straight line or Elastic theory, the Factor of Safety is, the stress at which the material yields (called as yield stress) divided by allowable stress. The selection of factor of safety will be based on 'degree of safety' required, economy, reliability of the material, required life of the structure, loading conditions, method of assessment of qualities of the material, method of assessment of loads, access into the structure for maintenance and environmental conditions.

Strain is the ratio of deformation δl, to the length l over which it occurred. Thus, $\varepsilon = \delta l/l$. Strain is a non-dimensional quantity whereas stress is expressed in N/mm² etc.

Generally almost all the structural materials will be capable of recovering original shape and size upon the removal of the deforming forces, provided they are not excessive. This property is called the elasticity. The limit of stress upto which a material can exhibit Elasticity is called the Elastic limit. If the material is stressed beyond this limit full recovery is not possible. Some permanent deformations will be left unrecovered. That part, which is irrecoverable is called **Permanent Set.** Robert Hooke formulated that "Stress is proportional to strain within certain limits". The limit upto which Hooke's law holds good is called Proportional limit or limit of Proportionality. Within the Elastic Zone, stress is proportional to strain upto proportional limit and beyond this it is not linearly related. Thus the zone upto Elastic limit (Elastic zone) can be divided as Linearly Elastic zone and Non-linearly Elastic zone. The constant of proportionality between stress and strain, which gives a measure of stiffness or elasticity of the material is called 'Modulus of Elasticity' or Young's Modulus (after Dr. Thomas Young) and is denoted by E.

Plasticity of a material is the property of the material by virtue of which it continues to deform without any considerable increase in stress.

The ratio of shear stress to shear strain within the shearing proportional limit is known as Shear Modulus, or Modulus of Rigidity, denoted by N, C or G.

Any normal stress is accompanied by a strain in its own direction and an opposite kind of strain in every direction at right angles to it. The ratio of unrestrained lateral linear strain to longitudinal linear strain is called **Poisson's Ratio**, μ, $1/m$, or ν. Usually, this varies from about 0.15 to 0.4.

The change in volume per unit volume is called the volumetric strain. If the strains in the three principal directions are ε_x, ε_y and ε_z, the volumetric strain will be the sum of these strain, i.e. $\dfrac{\delta V}{V} = \varepsilon_x + \varepsilon_y + \varepsilon_z$. If a body is subjected to three stresses σ_x, σ_y and σ_z all tensile in the three principal directions x, y and z, respectively, strains in the three directions will be given by

$$\varepsilon_x = \frac{\sigma_x}{E} - \frac{1}{mE}(\sigma_y + \sigma_z)$$

$$\varepsilon_y = \frac{\sigma_y}{E} - \frac{1}{mE}(\sigma_z + \sigma_x)$$

$$\varepsilon_z = \frac{\sigma_z}{E} - \frac{1}{mE}(\sigma_x + \sigma_y)$$

When a body is subjected to three mutually perpendicular like direct stresses of same intensity, the ratio of this stress to the corresponding volumetric strain in the body is known as Bulk Modulus, K.

The three elastic constants E, N and K together with Poisson's Ratio are connected by the following equations.

$$E = 2N\left(1 + \frac{1}{m}\right), \quad E = 3K\left(1 - \frac{2}{m}\right), \quad E = \frac{9NK}{N + 3K}.$$

It can be inferred that the Modulus of Rigidity for a material will be less than Modulus of Elasticity Or, generally the materials will be weak in shear in comparison to normal stress *e.g.*, for mild steel, $N \cong 0.4\ E$.

Strength is the ability of a material to resist forces by developing the stress based on the nature of the force without failure.

Stiffness is that property of the material due to which a material can resist deformation.

Tenacity is the ultimate tensile strength.

Ductility is the property of the material that enables it to be drawn permanently through great changes for shape without rupture. This is indicated by percentage elongation and percentage reduction in area from a tension test.

Brittleness is the opposite of ductility. Brittle materials fail suddenly without warning when stressed beyond their strength. They can not accommodate much change in shape without rupture.

Hardness is the ability of a material to resist indentation or abrasion or cutting, or scratching etc. It is given as force divided by the surface area (of contact) of the spherical indentation (*e.g.*, BHN).

Toughness is the property of a material which enables it to absorb energy at high stress without fracture. The measure of toughness is the amount of energy that a unit volume of the material has absorbed after being stressed upto the point of fracture.

Creep is the property by which a material continues to deform with time under sustained loading.

Resilience is the amount of energy stored in a strained body, also known as **Strain Energy**. The maximum amount of strain energy which can be stored in a unit volume without permanent set is called **Proof Resilience** or **Modulus of Resilience**.

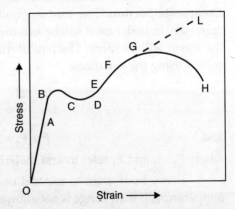

If a ductile material like mild steel is stressed upto fracture, the stress strain curve will be obtained as shown in figure aside in which O is starting point, *i.e.*, zero stress and zero strain. A is proportional limit, B is Elastic limit. C is upper yield point. D is lower yield point. Point G is the ultimate strength. H is nominal breaking stress and L is actual breaking stress. Solid curve is based on original area of cross-section. Dotted curve is based on actual area of cross-section (possibly) from time to time. OA is linearly elastic zone and AB is non-linearly elastic zone. DF is plastic zone. FG is strain hardening zone.

Members with varying cross-sections are called Non-prismatic members. The total elongation or shortening of the member will be the algebraic sum of the changes in lengths of individual portions. If the cross-sections suddenly vary, $\delta l = \Sigma \delta l_i$

where
$$\delta l_i = \frac{P_i l_i}{A_i E_i}$$

If the cross-section is uniformly varying, the change in an element is to be integrated. For a circular rod tapering uniformly from a diameter d_1 to d_2 over a length of l and subjected to a load P, the overall change in length, δl will be given by $\delta l = \dfrac{4Pl}{E\pi d_1 d_2}$

The strain energy stored in a body due to axial stress is given by

$$U = \frac{1}{2} \times \text{Stress} \times \text{Strain} \times \text{Volume} \quad \text{or} \quad U = \frac{\sigma^2}{2E} \times \text{Volume}.$$

If an axial load P is applied gradually, corresponding stress will be, P/A. If the axial load P is suddenly applied, max. instantaneous stress will be $2P/A$, which will gradually adjust to P/A. If the axial load P is dropped through a height 'h' or applied with equivalent impact, max. instantaneous stress is given by

$$\sigma = \frac{P}{A}\left\{1 + \sqrt{1 + \frac{2EAh}{Pl}}\right\}$$

If the loading is through a shock, and if the shock transmits U units of energy,

$$\sigma = \sqrt{\frac{2EU}{Al}}$$

Whenever a member is made up of more than one material, the section is known as **Composite Section**. The load applied on the member will be distributed among the various materials. In such case it will be assumed that there will not be relative slip among themselves, *i.e.*, the strain will be same. The equilibrium condition gives one equation and the problem can be solved using the equations

$$E = \frac{\delta l}{l} = \frac{P_1}{A_1 E_1} = \frac{P_2}{A_2 E_2} = \frac{P_3}{A_3 E_3} \ldots\ldots \text{etc.,}$$

and
$$P = P_1 + P_2 + P_3 + \ldots\ldots \text{etc.}$$

where P_1, A_1 and E_1 refer to first material and so on.

Every body tries to expand or contract depending on the changes in the surrounding temperature. If this change is not allowed the stresses will be developed in the body corresponding to such prevented changes.

If prevented change $= l\alpha t$, strain $= \alpha t$

Corresponding stress $= E\alpha t$

Force developed $= AE\alpha t$

where A is area of cross-section, E is Young's modulus, α is coefficient of linear expansion and t is the change in temperature.

If a cylindrical bar of length l is subjected to its own weight, ρ being the unit weight of material, (By integrating extension in an element)

$$\delta l = \rho l^2 / 2E$$

If a conical bar is subjected to its own weight,

similarly, $$\delta l = \frac{\rho l^2}{6E}.$$

Any structure whether a bridge or building etc. will be subjected to various types of loads. Dead load includes all loads which permanently act on the structure including the self weight of the member under consideration. During the course of design of the structure only the weight of that element under consideration is to be assumed, all the other elements supported by it must have been designed already and the dead load from them known.

Live load is that part which comes on to the structure based on the purpose when the structure is put to use.

The loads may be concentrated at some points (known as concentrated loads) or may be distributed (known as distributed loads).

The internal forces induced in various members of a structure will depend on how the structure is supported.

If the ends of the structure are simply placed over supports, due to bending, the ends of the member get lifted up, thereby reducing the contact area which results in high bearing stresses. There is no connection between member and support. This will not be stable for high horizontal components.

The end of the member may be supported over a hinge on rollers through a pin. This type of support cannot offer reaction in the plane of rollers. This will permit the end of the member to rotate about the centre of pin and hence offers only one reaction component normal to the plane of the rollers.

A hinged support will offer two reaction components since it prevents vertical and horizontal movements of the ends but permits rotation.

A built in support or fixed support, when the end of the member is either built into the support or rigidly connected, will offer three reaction components, as it will not permit the end of the member to move or to rotate.

A member is said to be overhanging when the member does not stop over a support but continues beyond. Sometimes the end of the member may be left unsupported.

By the application of equations of statics, namely $\Sigma H = 0$, $\Sigma V = 0$ and ΣM about any point lying in the plane of the structure is equal to zero, only three unknown external reaction components can be determined. If the total number of reaction components, based on the nature of supporting exceeds the number of equations which statics can offer, the member is said to be statically indeterminate. Degree of indeterminacy is total number of reaction components minus three.

A force is completely known when all the four elements, magnitude, direction, sense and point of application, are known. Space diagram is the diagram which indicates the structure along with system of forces that act on it to some scale clearly indicating the direction, sense and points of application of various forces. The Force Polygon is the Vector diagram showing the forces in magnitude, direction and sense. The closing line of the force polygon gives the equilibrant of the force system. If the polygon of forces closes by itself, including the reactions, the body will be in equilibrium. Sometimes the force polygon fails to give the idea about the point of application of resultant of a force system, when it is not a concurrent one. In such cases a Link Diagram or also known as Funicular Polygon is drawn linking the Space Diagram and Force Polygon.

The Shear Force at any section is equal to the net transverse force (*i.e.*, the algebraic sum of the transverse components of force) either to the right or to the left of the section including reactions. Obviously, shear force to the left or right of the section will be same.

The Bending moment at any section is equal to the algebraic sum of the moments about this section of various forces either to the right or to the left of the section including reactions.

If y is the deflection of the beam at a section x distant from origin,

$$EI \frac{d^4y}{dx^4} \propto W_x, \text{ the load intensity at } x.$$

$$EI \frac{d^3y}{dx^3} \propto F_x, \text{ the shear force at } x$$

$$EI \frac{d^2y}{dx^2} \propto M_x, \text{ the bending moment at } x$$

$$EI \frac{dy}{dx} \propto \theta_x, \text{ the slope of the elastic curve at } x$$

It can be observed from above that by the integration of function of load intensity the shear force will be obtained and etc. Thus the following Laws of Beam Diagrams can be established.

First Law of Beam Diagrams

Slope of curve at any point in any diagram

= Length of ordinate at the corresponding point in the next higher diagram

The order of diagrams is (1) Load intensity diagram (2) Shear force diagram (3) Bending moment diagram. Higher diagram means the just previous diagram.

Second Law of Beam Diagrams

Difference in lengths of any two ordinates in any diagram

= Area between the corresponding ordinates of the next higher diagram.

Simple bending means flexure by pure couples applied to a beam without shearing force. The assumption that Plane transverse sections remain plane and normal to the axis of the beam even after bending means the strains in the various fibres will be proportional to their distances from neutral axis. Neutral axis is the line of intersection of Neutral plane and transverse section of the beam. This will pass through the centre of gravity of cross-section. The assumptions that material obeys Hooke's law, limit of proportionality not exceeded and Young modulus is same in compression and tension mean that the stresses will be proportional to the strains in all the fibres.

The flexure formula is $\dfrac{M}{I} = \dfrac{\sigma}{y} = \dfrac{E}{R}$, where M is the bending moment acting at the section, I is the second moment of the area about an axis passing through centroid, σ is the stress in a fibre which is at a distance of y from neutral axis. E is Young's modulus and R is the radius of curvature (change).

This flexure formula is applicable with following limitations :

(a) Loads must be static loads.

(b) The beam must be free from initial or residual stresses.

(c) The beam must fail by bending only and not by twisting, or buckling etc.

(d) There must be a plane of symmetry for the cross-section and all loads must be applied in this plane.

Modulus of section with reference to any axis is the moment of inertia divided by the distance of extreme fibre from neutral axis.

It can be observed from flexure formula that $\dfrac{1}{R} = \dfrac{M}{EI}$, or the bent shape for a given moment will depend on the value of EI. E is a material property and I is a cross-sectional property. Hence the value EI is combined property of the beam section giving its stiffness. It is called Flexural rigidity.

The variation in bending moment along the axis causes differential horizontal stresses to develop. These stresses will be proportional to rate of change of bending moment or shear force at the section.

The shear stress on a fibre at any level is given by $\tau = \dfrac{FA\bar{y}}{Ib}$,

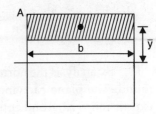

where F is the shear force at the section, $A\bar{y}$ is the moment of the area between the fibre under consideration and corresponding extreme fibre taken about the neutral axis I is the moment of inertia about the centroidal axis, b is the width of the section at the level of the fibre under consideration

Generally, unless the width at neutral axis is large, maximum shear stress occurs near the neutral axis. The distribution of shear stress is of parabolic variation. The ratio of maximum shear stress to average shear stress will depend thus on the shape of the cross-section. When a body is subjected to bending moment M, the strain energy stored will be $\displaystyle\int_0^l \dfrac{M^2 dx}{2EI}$. From the flexure formula, $\dfrac{1}{R} = \dfrac{M}{EI}$, or when M is constant (*i.e.*, pure bending moment) and EI is constant (*i.e.*, prismatic member), the bent shape (or elastic curve) will be of constant curvature. So a prismatic beam subjected to pure bending moment will bend into circular arc.

In a simply supported beam subjected to pure bending moment, the maximum deflection will occur at centre and of magnitude $\dfrac{Ml^2}{8EI}$.

As can be seen from the previous equations, the deflections can be calculated by integrating the bending moment expression twice in succession. The constants of integration can be determined using the boundary conditions. If slope and deflection are calculated at a section, the slope and deflection at other section can be calculated by drawing the qualitative picture of elastic curve. Especially when there is no load in between the two sections.

$$\theta_C = \theta_B, \quad \Delta_C = \Delta_B + \theta_B \cdot \dfrac{l}{2}$$

It can be observed that even through the stresses act on a plane, the cracks will form on a different plane altogether. So there may be planes over which the stresses may be of critical nature and magnitude. On any plane the stresses may be resolved into two components normal and tangential. Such planes on which no tangential stresses act will be known as Principal planes. The only normal stresses acting on these principal planes are called Principal Stresses. The larger one is called Major Principal Stress and the smaller one is called minor principal stress. In any stressed member the maximum and minimum normal stresses will be Principal stresses. The two principal planes will be at right angles to each other. If the principal stresses are σ_1 and σ_2 both of same nature maximum shear stress will be given by $\dfrac{\sigma_1 - \sigma_2}{2}$.

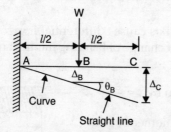
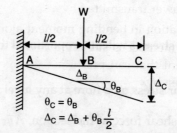

By analysis the normal and tangential stresses on any plane inclined at a given angle can be found. The plane carrying maximum shear can be determined by differentiating shear stress expression w.r.t. θ and equating to zero.

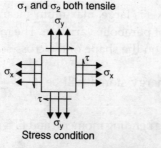
Stress condition

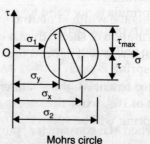
Mohrs circle

Ellipse of stress and Mohr's circle of stress can be used. If no shear stress acts on these two planes and σ_x and σ_y are equal and of same nature, then Mohr's circle will be a point. The radius of any Mohr's circle will give the magnitude of the corresponding maximum shear stress.

A moment acting on a member about the longitudinal axis trying to twist one end of the member with respect to the other is called twisting moment or Torque. A bending moment bends the member whereas a twisting moment twists the member. A beam supporting a balcony or portico, itself being supported rigidly at the two ends or one end by columns is an example for a member under combined bending and torsion. Shafts in automobiles, beams in grid works will be subjected to torque.

The assumption that a plane cross-section perpendicular to the longitudinal axis remains same even after the application of torque is valid only for circular sections. The torsional formula is $\dfrac{T}{J} = \dfrac{\tau}{r} = \dfrac{N\theta}{l}$, where T is the twisting moment or torque in Nm. J is polar moment of inertia of the cross-section in m^4. τ is the shear stress in the material at a radial distance of r, in N/m^2, r is the radial distance of fibre in m. N is the modulus of rigidity in N/m^2. θ is the relative twist between two sections, l apart, in radians. l is the distance between the two section in m.

The stresses generated in the material will be shear stresses since the applied twisting moment will force each cross-section to slip over the other. The material on the periphery will be subjected to maximum shear stress. The ductile materials will allow cross-sections to slip easily and they may fail by shear slip between cross-sections. Brittle materials will fail by diagonal tension the crack occurring at 45° to the longitudinal axis. The maximum twist will be small when compared to that in a ductile material.

Horse power transmitted by a shaft $= \dfrac{2\pi n T}{0.45 \times 10^6}$, where n is r.p.m., T is torque in kg cm.

When a member is subjected to a bending moment M and twisting moment T, a single bending moment that can produce the same major principal stress called, equivalent Bending moment, is given by,

$$M_{eq} = \frac{1}{2}\{M + \sqrt{M^2 + T^2}\}$$

Similarly, equivalent twisting moment producing same max. shear stress is

$$T_{eq} = \sqrt{M^2 + T^2}.$$

The torsional formula is not valid for impact loads. This is valid only for circular sections. If a crack develops it starts on the periphery and develops towards the centre, if it is a pure torsion.

A helical spring will be close coiled or open coiled based on the angle of helix. If the angle is more and hence the spring is open coiled, it will develop bending stresses in addition to shear stresses when subjected to axial loads. If the mean coil radius is R, every cross-section is subjected to a twisting moment of WR, when the spring is subjected to axial force W. Thus the material of a close coiled helical spring will be subjected to shear stress when the spring is subjected to axial load.

If the spring is subjected to axial twist, the number of coils in the spring will increase or decrease depending on the nature of the twisting moment. Because the overall length of the spring is same, when number of coils changes, the radius of curvature of the spring coils changes. When the radius of curvature changes for a member, the straining action will be bending moment. Therefore when a close coiled helical spring is subjected to axial twist, bending stresses will develop in the material.

When an axial load of W acts on a close coiled helical spring, the spring undergoes a compression (or deflection) which is given by $\delta = \dfrac{64WR^3n}{Cd^4}$, where W is the load, R is the mean radius of the coil of the spring, n is the number of coils in the spring, C is the modulus of rigidity and d is the diameter of the spring wire.

The load required to cause unit deflection is called **Stiffness** $\left(K = \dfrac{W}{\delta}\right)$. If two springs are in series, and their Stiffness are K_1 and K_2 the stiffness of the combined spring is $K = \dfrac{W}{\delta_1 + \delta_2}$, where $\delta_1 = \dfrac{W}{K_1}$ and $\delta_2 = \dfrac{W}{K_2}$ or $K = \dfrac{K_1 K_2}{K_1 + K_2}$.

If two springs with stiffnesses K_1 and K_2 are in parallel, the stiffness of the combined spring is $K = \dfrac{W}{\delta}$, where $W = W_1 + W_2$.

$$W_1 = K_1\delta, \; W_2 = K_2\delta, \text{ and } W = \delta(K_1 + K_2)$$
$$\therefore \quad K = K_1 + K_2$$

If the helical springs crack, the crack will be initiated on the **inner side of curvature**.

The stiffness of spring is $\dfrac{Cd^4}{64R^3n}$

The energy stored in a compressing spring is

$$U = \dfrac{1}{2}W\delta = \dfrac{W}{2} \times \dfrac{64WR^3n}{Cd^4} = \dfrac{32W^2R^3n}{Cd^4}$$

If the diameter to wall thickness ratio in a shell is less than 20 $\left(i.e., \dfrac{d}{t} < 20\right)$, the cylinder can be considered as thick cylinder. If $\dfrac{d}{t} > 20$ the cylinder can be considered as a thin cylinder. The hoop stress developed to withstand the internal fluid pressure in a thin cylinder will not vary along the thickness of the wall to a considerable extent and hence variation can be ignored. In a thick cylinder the hoop stress varies considerably with maximum at inner radius and minimum at outer radius for an internal fluid pressure.

To improve the capacity of thin cylinders, often they will be wound with wire under tension, which will develop hoop compression in the cylinder and thus the latter can withstand higher pressure.

To make thick cylinders more efficient and to have most favourable distribution of hoop stress, another thick cylinder will be jacketed on to this. Instead of having a heavy thick cylinder, number of (laminated) thin cylinders forming a compound cylinder fitted by shrinking on will serve more efficiently. In a cylindrical shell subjected to internal pressure only, the element on the surface will be subjected to two principal stresses. A longitudinal crack will develop if it fails due to σ_h. An element of an elemental ring of a thick cylinder will be subjected to three principal stresses, σ_h, σ_l and σ_r.

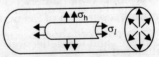

The contents of Lame's equations can be calculated for a given problem based on the conditions of pressure.

Some structures like masonry dams, chimneys etc. will be subjected to combined bending and axial stresses. The masonry structures will be weak in tension. Therefore to avoid tension, the resultant of forces at any level should pass through, middle third of the base at that level.

That portion of the area of cross-section of a member within which if a compressive force is applied no tension develops anywhere in the cross-section is called *Core*. In other words, to avoid tension at any fibre in the cross-section, the compressive force should be applied within the core. The eccentricity of a compressive load is thus limited.

A column is a vertical member subjected to compressive force predominantly.

The short columns will fail by crushing. The long columns will fail due to instability, which will be usually earlier if the column is long. The load which can cause buckling in columns will depend on E, I and L and end conditions. Euler analysed columns with different end conditions.

For both ends hinged $\qquad P_{cr} = \dfrac{\pi^2 EI}{L^2}$

For both ends fixed $\qquad P_{cr} = \dfrac{4\pi^2 EI}{L^2}$

For one end fixed and the other free $\qquad P_{cr} = \dfrac{\pi^2 EI}{4L^2}$

For one end fixed and the other hinged $\qquad P_{cr} = \dfrac{2\pi^2 EI}{L^2}$

Effective length of a column is that length of the same column but with such a modified length with which the column will have same strength as that of a hinged ended column. Thus the effective length will be

for a column with both ends hinged $L_e = L$

for a column with both ends fixed $L_e = L/2$

for a column with one end fixed and the other free $L_e = 2L$

for a column with one end fixed and the other hinged $L_e = L/\sqrt{2}$.

A column will have maximum strength when its both ends are fixed.

Radius of gyration of a section is $r = \sqrt{I/A}$

The radius of gyration will depend on the axis about which bending takes place. The ratio, L_e/r is known as Slenderness ratio. The Eulers formulae are valid only when $L_e/r > 120$.

The failure of load resisting members, under static loading usually consists of one of two types of action, namely (1) inelastic deformation (yielding) or (2) brittle fracture. Brittle fracture means separation of the material without accompanying measurable yielding. Which one of these two modes of failure occurs depends on the inherent internal characteristics and structure of the material and also on the external conditions such as temperature, state of stress, type of loading, rate of loading etc.

A shell is a structure which will have a curved surface. The curvature of the surface will be so designed to avoid development of bending moment and shear force. Because the material is subjected only to direct stresses, the material in a shell structure will be put to more efficient use.

II. OBJECTIVE TYPE QUESTIONS

1. Elasticity of a body is
    (a) the property by which a body returns to its original shape after removal of the load ☐
    (b) the ratio of stress to strain ☐
    (c) the resistance to the force acting ☐
    (d) large deformability as in case of rubber. ☐

2. The stress in a member subjected to a force is
    (a) continued deformation under sustained loading ☐
    (b) load per unit area ☐
    (c) the resistance offered by the material per unit area to a force ☐
    (d) the strain per unit length. ☐

3. A perfectly elastic body is
    (a) that body which recovers its original shape completely after removal of force ☐
    (b) a body of such a material with a lot of extensibility ☐
    (c) a body made of rubber only ☐
    (d) a body whose cross-sectional dimensions are very small. ☐

4. Permanent set is
    (a) the force which acts permanently on the body ☐
    (b) irrecoverable deformation in the body ☐
    (c) the shape of the member just after completion of construction ☐
    (d) ratio of Poisson's Ratio to Young's Modulus. ☐

5. In the case of a partially elastic body that part of the work done by the external forces during deformation is dissipated in the form of heat, which is developed in the body during
    (a) non-elastic deformation ☐ (b) elastic deformation ☐
    (c) 50% of total deformation ☐ (d) last 25% of deformation. ☐

STRENGTH OF MATERIALS   153

6. The law "Stress is proportional to strain within certain limits" is formulated by
   (a) Thomas Young   (b) Poisson
   (c) Mohr   (d) Robert Hook.

7. A member with a cross-section of A mm² is subjected to a force of PN. It is L mm long and of Young's Modulus E N/mm². The linear strain will be
   (a) $\frac{PL}{AE}$ N/mm   (b) $\frac{PA}{LE}$ N/mm²
   (c) $\frac{P}{AE}$ mm/mm   (d) $\frac{AP}{LE}$ mm/mm.

8. Young's Modulus is the ratio of the normal stress to the
   (a) normal strain within elastic limit
   (b) reciprocal of normal strain within elastic limit
   (c) normal strain within proportional limit
   (d) normal strain at yield point.

9. As per elastic theory of design the factor of safety is the ratio of
   (a) working stress to yield stress   (b) yield stress to working stress
   (c) ultimate strength to yield stress   (d) ultimate load to load at yield.

10. In a composite bar the load distribution among different materials of which it is made is based on the assumption that all the materials will have
    (a) equal areas   (b) same Young's Modulus
    (c) same strain   (d) same stress.

11. The stress due to temperature change in a member depends on
    (a) length of the member   (b) area of cross-section
    (c) supporting conditions at the two ends   (d) none of the above.

12. In a Uniaxial tension test on a mild steel bar, the Lueders' lines will be
    (a) inclined at 45° to the direction of tensile stress applied
    (b) perpendicular to the direction of tensile stress applied
    (c) along the direction of tensile stress
    (d) none of the above.

13. The percentage elongation of a material from a direct tensile test indicates
    (a) ductility   (b) strength
    (c) yield stress   (d) ultimate strength.

14. The percentage reduction in area of a member from a direct tension test indicates
    (a) ductility   (b) elasticity
    (c) malleability   (d) brittleness.

15. The standard gauge length over which the extension is to be measured to determine percentage elongation of a specimen with initial cross-section of a_0 is
    (a) $6.56 \sqrt{a_0}$ (b) $5.65 \sqrt{a_0}$
    (c) $6.65 \sqrt{a_0}$ (d) $a_0 \sqrt{6.56}$.

16. The Poisson's Ratio is the ratio of
    (a) lateral elongation to linear elongation
    (b) lateral stress to linear stress
    (c) lateral strain to longitudinal strain
    (d) Young's Modulus of elasticity to Modulus of Rigidity.
17. The equation connecting Young's Modulus (E) Poisson's ratio (1/m) and Modulus of Rigidity (C), is
    (a) $C = 2E (1 + 1/m)$
    (b) $E = 2C(1 - 2/m)$
    (c) $E = 3C(1 + 1/m)$
    (d) $E = 2C(1 + 1/m)$.
18. Modulus of Rigidity is the ratio of
    (a) shear stress to shear strain
    (b) normal stress to normal strain
    (c) Poisson's ratio to ultimate strength in compression
    (d) lateral stress to lateral strain.
19. A tie is a member which
    (a) connects two joints
    (b) is subjected to axial tension primarily
    (c) does not suffer any stress irrespective of loading conditions
    (d) suffers two equal and opposite forces at the two ends.
20. A strut is a member which
    (a) connects two joints
    (b) is subjected to shear force predominantly
    (c) is subjected to axial compressive force predominantly
    (d) is subjected to bending moment and shear force along with any axial force.
21. A principal plane is a plane which carries
    (a) maximum shear stress
    (b) the given stresses of higher magnitude acting
    (c) no shear stress
    (d) plane inclined at 45° to x-axis.
22. The angle between the two principal planes is
    (a) 45°
    (b) 90°
    (c) 30°
    (d) 60°.
23. If an element of a specimen of brittle material is subjected to shear stress, the crack propagation if occurs, will be inclined to the sides of the element at
    (a) 45°
    (b) 90°
    (c) 30°
    (d) 60°.
24. Bulk Modulus, Young's Modulus and Poisson's Ratio are connected by the relation
    (a) $K = E/3 (1 - 2/m)$
    (b) $E = K/3 (1 + 2/m)$
    (c) $E = 2K (1 + 1/m)$
    (d) $E = 3K (1 - 2/m)$.

25. A prismatic bar is supported at top and is subjected to its own weight. The area of cross-section is A, length is L, density of the materials is ρ and Young's Modulus is E. The total elongation of the bar is given by
    (a) $\rho l^2/E$
    (b) $\rho l^2/2E$
    (c) $\rho AL/2E$
    (d) $\rho l^2/2EA$.

26. The elongation of a conical bar, supported at top under the action of its own weight if the length of the bar is l, the diameter of the base is d and the density of materials is ρ, is given by
    (a) $\rho l^2/6E$
    (b) $\pi \rho d^2 l/4E$
    (c) $4\rho l/\pi d^2 E$
    (d) $6\rho l^2/d^2 E$.

27. In the case of uniaxial tension the angle between the plane carrying maximum shear stress and direction of tensile force will be
    (a) 90°
    (b) 0°
    (c) 45°
    (d) 60°.

28. An element is subjected to two normal stresses σ_1 and σ_2, both tensile on two mutually perpendicular planes. There are no shear stresses on these two planes. The max. shear stress is given by
    (a) $(\sigma_1 + \sigma_2)/2$
    (b) $(\sigma_1 - \sigma_2)/2$
    (c) $\sigma_1 + \sigma_2$
    (d) $\sigma_1 - \sigma_2$.

29. The maximum shear stress from a Mohr's circle is given by
    (a) the diameter of the circle
    (b) the distance of centre from the origin
    (c) the distance of farthest point on the Mohr's circle from origin
    (d) the radius of the circle.

30. An element is subjected to two equal and like stresses σ, on two mutually perpendicular planes. The shape of the Mohr's circle will be
    (a) a circle of radius 2σ
    (b) a circle of radius σ
    (c) a circle of radius $\sigma/2$
    (d) a point.

31. If the strains in the three principal directions are ε_x, ε_y and ε_z, the volumetric strain will be
    (a) $1/3\,(\varepsilon_x + \varepsilon_y + \varepsilon_z)$
    (b) $(\varepsilon_x + \varepsilon_y + \varepsilon_z)$
    (c) $3(\varepsilon_x + \varepsilon_y + \varepsilon_z)$
    (d) $(\varepsilon_x + \varepsilon_y + \varepsilon_z)^3$.

32. The residual stress in a member is
    (a) the stress due to the loading on the member
    (b) the average of initial and final stresses
    (c) deformation stress
    (d) instantaneous stress due to sudden loading.

33. The stress along the contact surface of a rivet and the member is
    (a) bearing stress
    (b) compressive stress
    (c) shearing stress
    (d) axial tensile stress.

34. A plate 100 mm wide, 10 mm thick is having a hole of diameter 10 mm, symmetrical about the axis of the plate. The plate is subjected to a force of 9 kN. The maximum stress on a section passing through centre of the hole will be
    (a) 10 N/mm²
    (b) > 10 N/mm²
    (c) < 10 N/mm²
    (d) 9 N/mm².

35. In a stressed body the maximum normal stress at any point is always
    (a) a principal stress
    (b) average of maximum and minimum shear stresses
    (c) sum of the two normal stresses acting in two principal directions
    (d) none of these.

36. In a stressed body the minimum normal stress at any point is always
    (a) average of two normal stresses acting in the two principal directions
    (b) a principal stress
    (c) difference of two normal stresses in the two principal directions plus shear stresses on those planes
    (d) none of these.

37. The maximum shear stress will be equal to
    (a) one half of the algebraic difference of the maximum and minimum principal stresses at the point
    (b) the intensity of shear stress acting on the major principal plane
    (c) the shear stress acting on the planes in the direction of two principal directions
    (d) none of these.

38. The ratio of strengths of a rivet in double shear to that in a single shear will be
    (a) 1
    (b) $\frac{1}{2}$
    (c) 2
    (d) none of these.

39. When a column is resting on a base plate, the stresses along the surfaces of contact are
    (a) compressive stresses
    (b) shear stresses
    (c) tensile stresses
    (d) bearing stresses.

40. An isotropic material is the one which
    (a) has same structure at all the points
    (b) has Youngs Modulus equal to Modulus of Rigidity
    (c) has the elastic constants, identical in all the directions
    (d) obeys Hooke's law upto failure.

41. If the Poisson's ratio of a material is 0.25, the ratio of Modulus of Rigidity to the Young's Modulus is
    (a) 2
    (b) 0.4
    (c) 2.5
    (d) 4.

42. The actual breaking stress of a ductile material from a tension test will be
    (a) greater than ultimate strength
    (b) equal to ultimate strength
    (c) equal to nominal breaking stress
    (d) less than the ultimate strength but greater than nominal breaking stress.

43. The difference in placing the end of a beam simply over a support and the supporting end through a hinge on rollers is that the roller support
    (a) can offer reaction in the plane of rollers
    (b) can offer moment reaction
    (c) will not allow the end to lift up due to deflection
    (d) will not offer reaction normal to the plane of rollers.

44. The number of reaction components possible at a hinge on rollers support is
    (a) 1   (b) 2
    (c) 0   (d) 3.

45. The number of reaction components possible at a hinged end for a general loading is
    (a) 1   (b) 0
    (c) 2   (d) 3.

46. The number of reaction components possible at a fixed or built in end for a general loading is
    (a) 1   (b) 3
    (c) 2   (d) 0.

47. A simply supported beam is subjected to a pure moment. This will be resisted through
    (a) a moment reaction at hinged end
    (b) a moment reaction at hinge on rollers end
    (c) a couple formed by the reactions from the two supports
    (d) external support capable of resisting moment which is necessarily to be provided.

48. A cantilever beam is the one which is supported with
    (a) one end hinge and other on rollers   (b) one end fixed and the other on rollers
    (c) both ends on rollers   (d) one end fixed and the other free.

49. A horizontal beam with both the ends hinged will be statically determinate for this type of loading
    (a) purely vertical loading   (b) purely inclined loads
    (c) any general loading   (d) inclined loads with moments.

50. A beam is said to be, in general, stable and statically determinate for general loading when the number of reaction components is
    (a) greater than 3   (b) 0
    (c) less than 3   (d) 3.

51. A beam is supported over three rollers lying in the same plane. The beam is stable for
    (a) any general loading
    (b) loading with no component in the direction of the beam
    (c) loading with no component perpendicular to the direction of beam
    (d) only when no load except self weight acts.

52. A simply supported beam is subjected to a udl of intensity w/m throughout the length of the span. The B.M. diagram will be a
    (a) triangle with $wl^2/8$ max. ordinate
    (b) rectangle with uniform ordinate $wl^2/8$
    (c) parabola with $wl^2/8$ max. ordinate
    (d) parabola with $wl/4$ max. ordinate.

53. The bending moment in a beam will be maximum where
    (a) the S.F. is uniform
    (b) the S.F. is maximum
    (c) the S.F. is zero
    (d) none of these.

54. The slope of curve of S.F. diagram at any section will be equal to
    (a) the slope of loading at that section
    (b) the ordinate of loading diagram at that section
    (c) the area of loading diagram from end to that section
    (d) the bending moment at that section.

55. The slope of curve of B.M. diagram at any section will be equal to
    (a) the slope of loading diagram at that section
    (b) the slope of S.F. diagram at that section
    (c) the ordinate of S.F. diagram at that section
    (d) the area of S.F. diagram starting from any one end.

56. To difference between B.M. values at any two sections will be equal to
    (a) the area of S.F. diagram between those two sections
    (b) the difference in slopes of S.F. diagram at the same sections
    (c) the area of loading diagram between the two sections
    (d) the moment of area of loading diagram between the two sections taken about mid-point between the two sections.

57. The difference between S.F. values at any two sections will be equal to
    (a) the area of B.M. diagram between the two sections
    (b) the difference between the slopes of the curve of loading diagram at the two sections
    (c) the ordinate of S.F. diagram at one section plus the slope of the loading diagram multiplied by the distance between the two sections
    (d) the area of the loading diagram between those two sections.

58. A cantilever beam is subjected to a uniformly varying load with zero intensity at the free end and with the maximum intensity of w/m at the fixed end. The maximum bending moment will be
    (a) $wl^2/6$
    (b) $wl^3/3$
    (c) $wl^2/1.5$
    (d) $2wl^2/3$.

59. A simply supported beam is subjected to a uniformly varying load with zero intensity at the two ends increasing to w/m at the centre. The maximum S.F. is
    (a) wl/2
    (b) wl/4
    (c) wl²/2
    (d) wl²/4.

60. A simply supported beam is subjected to a uniformly varying load with zero intensity at the two ends increasing to w/m at the centre. The max. B.M. will occur at
    (a) l/3 from centre
    (b) l/2 from ends
    (c) at the two ends
    (d) none of these.

61. A simply supported beam is subjected to a uniformly varying load with zero intensity at the two ends increasing to w/m at the centre. The maximum B.M. will be equal to
    (a) wl²/12
    (b) wl²/24
    (c) wl²/6
    (d) w²l/8.

62. Which of the following end conditions permits the displacement in any direction and also rotation ?
    (a) fixed end
    (b) hinged end
    (c) free end
    (d) roller end.

63. Points of contraflexure are the points where
    (a) the S.F. is zero
    (b) the B.M. is zero
    (c) the beam is supported
    (d) the B.M. changes its sign.

64. Which of the following beams will have points of contraflexure ? (self weight is negligible)
    (a) a simply supported beam with udl over entire span
    (b) a over hanging beam with loading only over supported span and not on overhangs
    (c) fixed beam subjected to concentrated load
    (d) cantilever subjected to uniformly varying loads.

65. A cantilever of span l is subjected to bending moment M at the free end. The S.F. diagram will be
    (a) a triangle with maximum ordinate at the fixed end
    (b) a rectangle with ordinate M
    (c) a parabola with maximum ordinate of $Ml^2/2$ at the fixed end
    (d) no shear force at all.

66. A beam is said to have been subjected to a pure bending moment, when
    (a) S.F. is maximum
    (b) the load is applied as a udl throughout the span
    (c) S.F. in a length is zero
    (d) the load is applied at the mid-span section only.

67. A mild steel flat of cross-section 10 mm × 120 mm is bent into a circular arc of radius 10 m. Assuming Young's Modulus as 2×10^5 N/mm², the required B.M. is
    (a) 2 kNm
    (b) 20 kNm
    (c) 200 Nmm
    (d) 200 Nm.

68. A rectangular section 100 × 200 mm is subjected to a moment of 20 kNm. The maximum bending stress is
    (a) 30 N/mm² □   (b) 5/6 N/mm² □
    (c) 10000 N/mm² □   (d) 300 N/mm² □

69. The slope of the B.M. diagram changes its sign when
    (a) B.M. changes its sign □   (b) S.F. is maximum □
    (c) S.F. changes its sign □   (d) rate of loading changes suddenly. □

70. The minimum moment of resistance of a beam with material of allowable stress σ, is given by
    (a) $\dfrac{\sigma \times I}{\text{Distance of farthest fibre from neutral axis}}$
    (b) $\sigma \times I$
    (c) $\dfrac{\sigma \times I}{\text{distance of nearer extreme fibre}}$
    (d) σ × I × ratio of distances of the two extreme fibres.

71. The assumption that the cross-sections plane before bending remain plane even after bending means
    (a) the strains in the fibres are proportional to their distances from neutral axis □
    (b) the bending moment will be resisted by the central core of the section □
    (c) the stresses in the fibres are proportional to their distances from neutral axis □
    (d) the neutral axis lies at mid height. □

72. The assumption in the theory of bending that the material is homogeneous and isotropic and has the same value of Young's Modulus in tension and compression implies that
    (a) the stresses are proportional to strains at all fibres □
    (b) the strains are proportional to distances of the fibres from neutral axis □
    (c) the beam bends into a circular arc □
    (d) the neutral axis lies at the centre. □

73. The moment of inertia of a rectangular section $b \times d$ about the base is
    (a) $1/12\ bd^3$ □   (b) $1/3\ bd^3$ □
    (c) $1/36\ bd^3$ □   (d) $bd^2/6$. □

74. For a circular section, $I_{zz} = I_{xx} + I_{yy}$. Then zz-axis
    (a) coincides with xx-axis □
    (b) coincides with yy-axis □
    (c) will be tangential to the circle □
    (d) will be passing through point of intersection of xx and yy-axes and perpendicular to the plane of cross-section. □

75. The moment of inertia of triangular section $b \times h$, about the base is
    (a) $bh^3/12$ □   (b) $b^2h^2/2$ □
    (c) $bh^3/36$ □   (d) $b^2h^2/36$. □

76. In the case of T section, the maximum bending stress will occur at
    (a) neutral axis
    (b) extreme fibre in the flange
    (c) extreme fibre in the web
    (d) junction of web and flange.
77. The neutral axis of any section is
    (a) the axis passing through middle point of the height
    (b) the line of intersection of neutral plane with cross-section
    (c) the axis about which the moment of inertia is minimum
    (d) longitudinal axis of the member.
78. A rectangular section with b/d ratio of 0.5 and a circular section have same area of cross-section 10,000 mm². The ratio of moment of resistance of rectangle to that of circle is
    (a) 1.67
    (b) 0.6
    (c) 1
    (d) 0.5.
79. Which of the following sections is the most efficient in carrying bending moments ?
    (a) rectangular section
    (b) circular sections
    (c) I-section
    (d) T-section.
80. The maximum shear stress will always occur at
    (a) neutral axis
    (b) the top extreme fibre
    (c) the bottom extreme fibre
    (d) a fibre in the cross-section depending on the configuration.
81. The ratio of maximum shear stress to average shear stress in the case of a rectangular section is
    (a) 4/3
    (b) 3/2
    (c) 1
    (d) 2.
82. In an I section almost all the shear force is taken by
    (a) web
    (b) top flange
    (c) bottom flange
    (d) half the depth of each flange.
83. In the case of H section the maximum shear stress will occur at
    (a) top fibres
    (b) bottom fibres
    (c) neutral axis
    (d) at the junction of the web and flanges.
84. A simply supported beam is subjected to a udl of w/m over the entire span of 2 m. The cross-section is 24 × 50 cm. The maximum permissible shear stress is 10 kg/cm². What is the maximum intensity of load that is permissible on the beam ?
    (a) 8000 kg
    (b) 4000 kg
    (c) 16000 kg
    (d) 12000 kg.
85. A beam of length $(l + 2a)$ is supported on supports l apart with equal overhangs on each side. Two concentrated loads W each are applied one at each end. The shear stress at the neutral axis at a section in between the supports is
    (a) $WA\overline{Y}/Ib$
    (b) $2WA\overline{Y}/Ib$
    (c) zero
    (d) $WA\overline{Y}/2Ib(L + 2W)$.

86. A straight beam subjected to pure bending will be bending in the shape of
    (a) circular arc
    (b) parabolic arc
    (c) straight line
    (d) triangle.

87. Longitudinal cracks observed in timber beams are due to
    (a) high bending stresses
    (b) application of concentrated loads over the beams
    (c) shear failure between layers
    (d) timber not being strong in compression.

88. In the case of beams of brittle materials, and with square cross-section and if they are supported with one diagonal as neutral axis, by removing small triangular portions from top and bottom corners, the section modulus will
    (a) not change
    (b) decrease
    (c) increase
    (d) depend on bending moment.

89. The flexure formula is valid only for
    (a) static loads
    (b) dynamic loads
    (c) static loads with no residual stresses
    (d) dynamic loads with no residual stresses.

90. For the materials which do not have a pronounced yield point from a tension test the yield point is taken to be
    (a) stress corresponding to 0.2% strain from the stress strain curve
    (b) the stress at which the permanent set reaches the value 0.2%
    (c) ultimate strength
    (d) average of ultimate strength and actual breaking stress.

91. The product EI is called
    (a) flexural rigidity
    (b) torsional rigidity
    (c) second moment of area
    (d) none of these.

92. A cantilever is subjected to a concentrated load W at the mid-point of the span. The slope at the free end will be
    (a) $WL^2/6EI$
    (b) $WL^2/2EI$
    (c) $WL^2/3EI$
    (d) $WL^2/8EI$.

93. When a cantilever AB is subjected to a concentrated load at the free end, the slope and deflection at the free end are $WL^2/2EI$ and $WL^3/3EI$. If the same load is applied at mid span point, the deflection at the free end will be
    (a) $\dfrac{5}{384}\dfrac{WL^3}{EI}$
    (b) $\dfrac{5}{48}\dfrac{WL^3}{EI}$
    (c) $\dfrac{WL^3}{6EI}$
    (d) $\dfrac{WL^3}{16EI}$.

94. A cantilever is subjected to moment M at the free end. The deflection at free end is
    (a) $\dfrac{ML^2}{2EI}$
    (b) $\dfrac{ML^2}{3EI}$
    (c) $\dfrac{ML^3}{2EI}$
    (d) $\dfrac{ML^3}{3EI}$.

95. A simply supported beam is subjected to a central concentrated load. The slope at the two ends is given by

   (a) $\dfrac{WL^2}{6EI}$ (b) $\dfrac{WL^3}{48EI}$

   (c) $\dfrac{WL^2}{48EI}$ (d) $\dfrac{WL^2}{16EI}$.

96. A beam is supported with equal overhangs over the two supports. Each overhang is a and supported span is L. The deflection of the mid point of the beam due to two concentrated loads W each acting on the two ends is

   (a) $\dfrac{WaL^2}{6EI}$ (b) $\dfrac{WaL^2}{8EI}$

   (c) $\dfrac{WaL^2}{8EI}$ (d) $\dfrac{Wa^2L^2}{8EI}$.

97. The expression $EI d^4y/dx^4$ at any section for a beam is equal to
   (a) load intensity at the section
   (b) S.F. at the section
   (c) B.M. at the section
   (d) slope at the section.

98. What is the maximum number of unknown reaction components that can be determined using only statics
   (a) 0 (b) 1
   (c) 2 (d) 3.

99. Statically indeterminate structures are opted inspite of being indeterminate because
   (a) they can be made of good material
   (b) they can be designed to have larger dimensions
   (c) higher loads can be supported with less consumption of material
   (d) they can be easily analysed.

100. In carriage springs in addition to changing the number of plates, to have uniform strength
   (a) the thickness of each plate will be changing
   (b) the width of the plate will be gradually decreased
   (c) additional rivets are provided
   (d) welding is done.

101. Twisting moment is a moment applied in the plane of cross-section acting about
   (a) longitudinal axis (b) neutral axis
   (c) yy-axis (d) xx-axis.

102. When a member is subjected to a twisting moment, the material will be subjected to
   (a) axial tension (b) shear stresses
   (c) bending stresses (d) axial compressive stress.

103. If a shaft is subjected to pure twisting moment, an element on the surface is subjected to
   (a) normal tensile stress (b) normal compressive stress
   (c) pure shear stress (d) bending stress.

104. When a member is subjected to twisting moment the stress in any elementral ring will be proportional to
   (a) its radial distance from centre ☐   (b) area of cross-section ☐
   (c) Young's Modulus of material ☐   (d) Modulus of rigidity of the material. ☐

105. When a shaft made up of a brittle material is subjected to a twisting moment as shown, the possible crack propagation will be
   (a) 45° clockwise w.r.t. axis ☐
   (b) perpendicular to axis ☐
   (c) 45° anti-clockwise w.r.t. axis ☐
   (d) parallel to axis. ☐

106. Torsional rigidity is
   (a) EI ☐   (b) Z ☐
   (c) NJ ☐   (d) $\dfrac{N\theta}{l}$. ☐

107. Polar moment of inertia of a circular section with radius R is
   (a) $\pi R^4/32$ ☐   (b) $\pi R^3/32$ ☐
   (c) $\pi R^4/16$ ☐   (d) $\pi R^4/2$. ☐

108. In a shaft subjected to pure torsion, maximum shear stress will occur at
   (a) centre of shaft ☐   (b) periphery ☐
   (c) a distance of semi-radius from centre ☐   (d) none of these. ☐

109. What are the units of torque ?
   (a) Nm ☐   (b) N ☐
   (c) N/m ☐   (d) N/m². ☐

110. The maximum shear stress produced in a shaft is 5 N/mm². The shaft is of 40 mm diameter. The value of twisting moment is
   (a) 628 Nm ☐   (b) 62.8 Nm ☐
   (c) 125.6 Nm ☐   (d) 1256 Nm. ☐

111. Polar moment of inertia of a hollow circular section with external diameter D and internal diameter d is
   (a) $\dfrac{\pi}{64}(D^4 - d^4)$ ☐   (b) $\dfrac{\pi D}{32}(D^3 - d^3)$ ☐
   (c) $\dfrac{\pi D}{32}(D^4 - d^4)$ ☐   (d) $\dfrac{\pi D^2}{32}(D^2 - d^2)$. ☐

112. A brittle material will
   (a) fail after giving ample warning ☐   (b) fail suddenly ☐
   (c) never fail ☐   (d) never be used for structural purposes. ☐

113. Which of the following is a relatively ductile material ?
   (a) mild steel ☐   (b) cast iron ☐
   (c) high carbon steel ☐   (d) bronze. ☐

114. By increasing the carbon content in steel, the ultimate tensile strength will
    (a) decrease
    (b) not be affected
    (c) increase
    (d) become zero.

115. With the increase in the carbon content, the ductility of the steel will
    (a) increase
    (b) decrease
    (c) not be affected
    (d) difficult to tell.

116. Which of the following materials will have the highest Young's Modulus ?
    (a) brass
    (b) copper
    (c) mild steel
    (d) timber.

117. Creep of a material is
    (a) continued deformation with time under sustained loading
    (b) disappearance of deformation on removal of load
    (c) not being ductile
    (d) to become brittle.

118. What is tenacity ?
    (a) ultimate strength in tension
    (b) ultimate strength in compression
    (c) ultimate shear strength
    (d) ultimate impact strength.

119. At ordinary temperatures, how is the yield point affected with rate of loading ?
    (a) not greatly influenced
    (b) very greatly influenced
    (c) not influenced at all
    (d) difficult to tell.

120. How does the elastic limit vary with increase in temperature ?
    (a) will increase
    (b) will not change
    (c) will decrease
    (d) none of the above.

121. How does the Young's Modulus vary with increase in temperature ?
    (a) will not be affected
    (b) will decrease
    (c) will increase
    (d) difficult to tell.

122. The moment of resistance developed in a beam subjected to bending moment will, at any instant, be
    (a) equal to bending moment applied at that instant
    (b) greater than bending moment applied
    (c) less than bending moment applied
    (d) always the full strength of the beam.

123. Which of the following materials is suitable for transverse test for modulus of rupture ?
    (a) mild steel
    (b) timber
    (c) aluminium
    (d) rubber.

124. What is the number of basic elements of a force ?
    (a) 3
    (b) 1
    (c) 2
    (d) 4.

125. What are the units of twist in torsional formula, $\frac{T}{J} = \frac{\tau}{r} = \frac{N\theta}{l}$?
    (a) degrees
    (b) N/mm²
    (c) radians
    (d) Nm.

126. A cylindrical member made of ductile metal subjected to a static torsional moment will generally fail
    (a) by yielding due to circumferential shearing stress
    (b) by sudden fracture
    (c) by bending
    (d) due to diagonal tensile stress.

127. A plane section before twisting will remain plane after twisting if the cross-section is
    (a) rectangular
    (b) circular
    (c) square
    (d) triangular.

128. The torsional formula is valid if the shearing stresses
    (a) are uniform throughout the cross-section
    (b) are within the shearing proportional limit
    (c) are zero throughout the cross-section
    (d) are caused due to twisting moment applied only at the free end.

129. The power transmitted in kW by a shaft rotating at a speed of n rpm transmitting a mean torque of T (in kg m) is given by
    (a) $\pi nT/3060$
    (b) $2\pi nT/4500$
    (c) πnT
    (d) $3060\, \pi nT$.

130. When a shaft of diameter d is subjected to a bending moment M and torque T, the equivalent B.M. is given by
    (a) $\dfrac{M + \sqrt{M^2 + T^2}}{2}$
    (b) $\dfrac{M - \sqrt{M^2 + T^2}}{2}$
    (c) $\dfrac{16}{\pi d^3}(M + \sqrt{M^2 + T^2})$
    (d) $\dfrac{32}{\pi d^2}(M + \sqrt{M^2 + T^2})$.

131. When a shaft of diameter d is subjected to a bending moment M and a torque T, the equivalent torque is given by
    (a) $\dfrac{M + \sqrt{M^2 + T^2}}{2}$
    (b) $\sqrt{M^2 + T^2}$
    (c) $\dfrac{16}{\pi d^3}(M^2 + T^2)^{1/2}$
    (d) $\dfrac{M - \sqrt{M^2 + T^2}}{2}$.

132. As per maximum principal stress theory, when a shaft is subjected to a bending moment M and torque T, and if p is allowable stress in axial tension, the diameter d of shaft is given by
    (a) $d^4 = \dfrac{16}{\pi p}(M + \sqrt{M^2 + T^2})$
    (b) $d^3 = \dfrac{16}{\pi p}(M + \sqrt{M^2 + T^2})$
    (c) $d^2 = \dfrac{4}{\pi p}(M + \sqrt{M^2 + T^2})$
    (d) $d^3 = \dfrac{32}{\pi p}(M + \sqrt{M^2 + T^2})$.

133. When a close coiled helical spring is subjected to an axial compressive load the material will be subjected to
   (a) axial compressive stress
   (b) axial tensile stress
   (c) shear stress
   (d) bending stress.

134. When a close coiled helical spring with n coils, with mean radius R and wire diameter d is subjected to axial load, the compression in the spring is given by
   (a) $\dfrac{64\,WR^3 n}{Cd^3}$
   (b) $\dfrac{64\,WR^3 n}{Cd^4}$
   (c) $\dfrac{CWR^3 n}{32 d^4}$
   (d) $\dfrac{32\,WR^3 n}{Cd^4}$.

135. Two close coiled helical springs of stiffnesses K_1 and K_2 are placed in series. The stiffness of the combined spring is
   (a) $K_1 + K_2$
   (b) $\dfrac{K_1 K_2}{K_1 + K_2}$
   (c) $\dfrac{K_1 + K_2}{K_1 K_2}$
   (d) greater of K_1 and K_2.

136. Two close coiled helical springs of stiffnesses K_1 and K_2 are placed in parallel. The set is placed in between two rigid plates and subjected to axial load. The stiffness of the combination will be
   (a) $K_1 + K_2$
   (b) $\dfrac{K_1 + K_2}{K_1 K_2}$
   (c) greater of K_1 and K_2
   (d) $\dfrac{K_1 K_2}{K_1 + K_2}$.

137. The stiffness of a spring is
   (a) the load required for breaking the spring
   (b) load required to compress the spring upto shearing proportional limit
   (c) load required to produce unit deflection
   (d) load per coil of the spring.

138. Two close coiled helical springs are arranged in parallel. A load W is applied on to them through a set of rigid plates. The ratio of loads shared by the first and the second springs is
   (a) $K_1 : K_2$
   (b) $K_2 : K_1$
   (c) $\dfrac{K_1}{K_2} : \dfrac{K_2}{K_1}$
   (d) $\dfrac{K_1 K_2}{K_1 + K_2} : \dfrac{K_2}{K_1 + K_2}$.

139. The diamond cone indentor is used in
   (a) Rockwell hardness test
   (b) Brinell hardness test
   (c) Vicker's hardness test
   (d) Direct shear test.

140. The diamond quadrilateral pyramid indentor is used in
    (a) Rockwell hardness test
    (b) Brinell hardness test
    (c) Vicker's hardness test
    (d) Direct shear test.

141. In Brinell Hardness test, the type of indentor used is
    (a) hard steel cone
    (b) hard steel ball
    (c) mild steel ball
    (d) diamond cone.

142. Glass will obey Hooke's law upto
    (a) yield point
    (b) proof stress
    (c) fracture
    (d) 50% of stress at fracture.

143. The effect of the size of the specimen on ultimate strength will be more serious for
    (a) ductile materials
    (b) brittle materials
    (c) hard materials
    (d) none of the above.

144. The proportional limit of mild steel specimen is taken as the stress corresponding to a permanent set of
    (a) 0.2%
    (b) 0.1%
    (c) 0.25%
    (d) 0.01%.

145. For structural carbon steel the ratio of yield stress to ultimate strength will be about
    (a) 0.55 to 0.6
    (b) 0.2 to 0.3
    (c) 1.0
    (d) 1.2 to 1.5.

146. Within the elastic range of a tensile test the deviation from Hooke's law and some after effects may be noticed. This is due to
    (a) thermo elastic effect
    (b) material does not at all obey Hooke's law
    (c) yield stress is crossed
    (d) elastic limit is greater than ultimate strength.

147. Who introduced the terms upper and lower yield points?
    (a) Robert Hooke
    (b) Thomas Young
    (c) Mohr
    (d) Bach.

148. The lower yield point is more significant than upper yield point because
    (a) it is less than upper yield point
    (b) it is less influenced by shape of the specimen
    (c) it occurs before upper yield point
    (d) strain is more at this point.

149. The compression test is commonly used for testing
    (a) ductile materials
    (b) rubber
    (c) brittle materials
    (d) none of these.

150. When a cylindrical specimen of a ductile material is subjected to compression
    (a) it ultimately assumes the shape of a flat disc
    (b) it fails by fracture like a brittle material
    (c) it fails immediately after starting of loading
    (d) diameter of the specimen decreases.

151. The impact tests are used to determine
    (a) ultimate crushing strength
    (b) toughness
    (c) ductility
    (d) tenacity.

152. Toughness is
    (a) ability to absorb energy during plastic deformation
    (b) higher ultimate strength
    (c) stress at yield
    (d) strain energy at yield.

153. The brittle materials have low toughness because
    (a) they have large plastic deformation before fracture
    (b) they have only small plastic deformation before fracture
    (c) they can absorb high impacts
    (d) they fail very progressively.

154. How is Izod test specimen supported ?
    (a) vertical cantilever
    (b) horizontal cantilever
    (c) simply supported
    (d) none of these.

155. The notch provided on the specimen for impact test should be placed on
    (a) compression side
    (b) tension side
    (c) all along the perimeter of the specimen
    (d) notch should face the hammer.

156. An open coiled helical spring when subjected to an axial load, the material is subjected to
    (a) bending moment
    (b) twisting moment
    (c) bending moment and twisting moment
    (d) direct normal stress.

157. A thin cylinder is the one in which
    (a) the variation is hoop stress along the thickness can be neglected
    (b) the hoop stress can be neglected
    (c) the thickness is less than 20 mm
    (d) only internal fluid pressure acts.

158. A cylinder can be assumed as a thin cylinder when the diameter to thickness ratio is
    (a) < 20
    (b) > 20
    (c) 10
    (d) negligible.

159. A thin cylinder of diameter 100 mm and thickness 5 mm is subjected to a internal fluid pressure of 10 N/mm². The hoop stress is
    (a) 100 N/mm²
    (b) 10 N/mm²
    (c) 50 N/mm²
    (d) 5000 N/mm².

160. A thin cylinder is subjected to an external fluid pressure. The hoop stress will be
    (a) compressive
    (b) tensile
    (c) bending stress
    (d) zero.

161. A cylinder is to be designed as a thick cylinder when d/t ratio is
    (a) < 20
    (b) > 20
    (c) 10
    (d) negligible.

162. In a thick cylinder the hoop stress, along the thickness
    (a) varies uniformly
    (b) is constant
    (c) zero
    (d) is of parabolic variation.

163. In a cylinder wound with wire under tension, the cylinder will be before the introduction of fluid, under
    (a) tension
    (b) bending
    (c) hoop compression
    (d) will not be stressed at all.

164. The longitudinal stress in a thin cylinder subjected to internal fluid pressure of σ will be
    (a) $\dfrac{\sigma d}{4t}$
    (b) $\dfrac{\sigma d}{2t}$
    (c) $\dfrac{\sigma}{t}$
    (d) $\dfrac{\sigma \times d \times t}{4}$.

165. The volumetric strain in the case of a thin cylinder subjected to internal pressure σ, is
    (a) $\dfrac{\sigma d}{4tE}\left(\dfrac{5}{2} - \dfrac{2}{E}\right)$
    (b) $\dfrac{\sigma d}{2tE}\left(\dfrac{5}{2} - \dfrac{1}{m}\right)$
    (c) $\dfrac{\sigma d}{4tE}\left(3 - \dfrac{2}{m}\right)$
    (d) $\dfrac{\sigma d}{2tE}\left(\dfrac{5}{2} - \dfrac{2}{m}\right)$.

166. The hoop stress in the case of a thin spherical shell will be
    (a) $\dfrac{\sigma d}{4t}$
    (b) $\dfrac{\sigma d}{2t}$
    (c) σtd
    (d) $\dfrac{4\sigma}{td}$.

167. The volumetric strain in a thin spherical shell will be
    (a) $\dfrac{3\sigma d}{4tE}\left(1 - \dfrac{1}{m}\right)$
    (b) $\dfrac{4\sigma d}{3tE}\left(1 - \dfrac{1}{m}\right)$
    (c) $\dfrac{3\sigma d}{4tE}\left(1 + \dfrac{1}{m}\right)$
    (d) $\dfrac{3\sigma d}{4tE}\left(1 - \dfrac{2}{m}\right)$.

168. The initial tension in the wire wound on a thin cylinder, upon introduction of fluid under pressure, will
    (a) decrease
    (b) become zero
    (c) increase
    (d) not change.

169. In a thick cylinder the variation in hoop stress along the thickness is given by
    (a) $\dfrac{b}{x^2} + a$
    (b) $\dfrac{b}{x^2} - a$
    (c) $\dfrac{b}{a} + x^2$
    (d) $\dfrac{b}{a} - x^2$.

170. In a thick cylinder for internal fluid pressure, the hoop stress will be maximum at
    (a) outer surface
    (b) inner surface
    (c) centre of the thickness
    (d) at the centre of hollow portion.

171. A close coiled helical spring has mean coil radius R and diameter of wire d. It has n coils. The effect of direct shearing stress due to axial load, besides developing the twisting moment, will depend on
    (a) n
    (b) $\dfrac{d}{R}$ ratio
    (c) material of spring
    (d) none of these.

172. For heavy springs the cracks usually start on
    (a) the inner side of the coil
    (b) outer side of the coil
    (c) centre of the coil wire
    (d) none of these.

173. When a thick cylinder is subjected to external fluid pressure only, hoop stress will be maximum at
    (a) outer surface
    (b) inner surface
    (c) centre of thickness
    (d) centre of the shell.

174. When a thick cylinder is subjected to internal pressure only, maximum radial pressure will occur at
    (a) inner surface
    (b) outer surface
    (c) centre of thickness
    (d) none of these.

175. When a thick cylinder is subjected to external pressure only, maximum radial pressure will occur at
    (a) inner surface
    (b) outer surface
    (c) centre of thickness
    (d) none of these.

176. A thin cylindrical pipe is tested for internal fluid pressure. If it has failed due to hoop tension mainly, the crack will be
    (a) longitudinal
    (b) circumferential
    (c) at 45° to longitudinal axis
    (d) none of these.

177. When a hollow thick cylinder (jacket) is shrunk on to another thick cylinder, the inner cylinder will be subjected to
    (a) hoop tension with maximum at outer surface
    (b) hoop tension with maximum at inner surface
    (c) hoop compression with maximum at outer surface
    (d) hoop compression with maximum at inner surface.

178. A jacket is shrunk on to another cylinder. After the introduction of the fluid, the jacket will be subjected to
    (a) hoop compression with maximum at inner fibre
    (b) hoop tension with maximum at inner fibre
    (c) hoop compression with maximum at outer fibre
    (d) hoop tension with maximum at outer fibre.

179. If a cylinder is made up of thin-walled cylindrical sheets or laminations, the hoop stress will be of
    (a) the most unfavourable distribution with very high stress at outermost fibre ☐
    (b) the most favourable distribution with more or less uniform distribution over the entire thickness of compound shell ☐
    (c) same distribution as that of ordinary thick cylinder of total thickness ☐
    (d) similar distribution as that of ordinary thick cylinder of total thickness but with hoop compression. ☐

180. What is strain energy ?
    (a) the energy stored in a body because of being strained ☐
    (b) the energy that is spent for straining a body ☐
    (c) stress × strain ☐
    (d) volume × stress × strain. ☐

181. The ratio of maximum instantaneous stress due to sudden loading to the maximum stress due to gradual loading will be
    (a) 1 ☐ (b) 3 ☐
    (c) 2 ☐ (d) $\frac{1}{2}$. ☐

182. The strain energy stored in a body due to an axial stress is
    (a) $\frac{stress^2 \times volume}{2E}$ ☐ (b) stress × strain × area ☐
    (c) $\frac{stress^2}{2E} \times area$ ☐ (d) stress × strain × volume. ☐

183. Strain energy due to bending stresses is
    (a) $\int \frac{M^2 dx}{2EI}$ ☐ (b) $\int \frac{M dx}{2EI}$ ☐
    (c) $\int \frac{M^2 dx}{4EI}$ ☐ (d) $\int \frac{M dx}{4E^2 I^2}$. ☐

184. The core of a circular section of radius R is
    (a) a square of diagonal $2R$ ☐ (b) a circle of radius $R/2$ ☐
    (c) a circle of radius $R/4$ ☐ (d) a circle of radius R. ☐

185. The core of a rectangular section $b \times d$ is
    (a) a rhombus with diagonals $b/3$ and $d/3$ ☐ (b) a rectangle with sides $b/3$ and $d/3$ ☐
    (c) a rectangle with sides $b/6$ and $d/6$ ☐ (d) a rhombus with diagonals $b/6$ and $d/6$. ☐

186. What is the core of a section ?
    (a) that portion of the cross-section within which if a compressive force is applied, tension is not produced any where in the section ☐
    (b) the central portion of the cross-section of similar shape but with half the area ☐
    (c) the area within which a compressive force if applied will not cause any compressive stress ☐
    (d) none of these. ☐

187. To avoid any possibility of tension occurring in masonry structures, the resultant of various forces at any level must pass through
    (a) the section
    (b) the centre of the section
    (c) middle third of the width or depth of the section
    (d) a corner of the section.

188. The Euler crippling load for a column with both ends hinged is
    (a) $\pi^2 EI/l^2$
    (b) $4\pi^2 EI/l^2$
    (c) $\pi^2 EI/4l^2$
    (d) $2\pi^2 EI/l^2$.

189. The Euler crippling load for a column with both ends fixed is
    (a) $\pi^2 EI/l^2$
    (b) $4\pi^2 EI/l^2$
    (c) $\pi^2 EI/4l^2$
    (d) $2\pi^2 EI/l^2$.

190. The Euler crippling load for a column with one end fixed and the other free is
    (a) $\dfrac{\pi^2 EI}{L^2}$
    (b) $\dfrac{4\pi^2 EI}{L^2}$
    (c) $\dfrac{\pi^2 EI}{4L^2}$
    (d) $\dfrac{2\pi^2 EI}{L^2}$.

191. The Euler crippling load for a column with one end fixed and the other hinged is
    (a) $\dfrac{\pi^2 EI}{L^2}$
    (b) $\dfrac{4\pi^2 EI}{L^2}$
    (c) $\dfrac{\pi^2 EI}{4L^2}$
    (d) $\dfrac{2\pi^2 EI}{L^2}$.

192. The effective length of a column with both ends fixed is
    (a) $l/2$
    (b) $2l$
    (c) $l/\sqrt{2}$
    (d) l.

193. The effective length of a column with both ends hinged is
    (a) $l/2$
    (b) $2l$
    (c) $l/\sqrt{2}$
    (d) l.

194. The effective length of a column with one end fixed and the other end free is
    (a) $l/2$
    (b) $2l$
    (c) $l/\sqrt{2}$
    (d) l.

195. The effective length of column with one end fixed and the other end hinged is
    (a) $l/2$
    (b) $2l$
    (c) $l/\sqrt{2}$
    (d) l.

196. In the case of long columns the maximum permissible stress depends on
    (a) the ultimate crushing strength of the material
    (b) the maximum slenderness ratio
    (c) radius of gyration only
    (d) effective length only.

197. When a closed coiled helical spring is subjected to an axial tensile load the material of the spring will be subjected to
    (a) axial compressive stress
    (b) bending stress
    (c) axial tensile stress
    (d) shear stress.

198. When a close coiled helical spring is subjected to an axial twist, the material of the spring will be subjected to
    (a) bending stresses
    (b) axial compressive stress
    (c) axial tensile stress
    (d) shear stress.

199. In the Euler's theory for long columns it is assumed that
    (a) the failure occurs by crushing of the material
    (b) the failure occurs by buckling only
    (c) the column shortens so much that it will become a short column
    (d) the column is having initial curvature and the load is eccentrically applied.

200. Euler's formulae for columns can be used if the slenderness ratio is
    (a) < 100
    (b) 0
    (c) > 120
    (d) > 120 but less than 180.

201. The strain energy stored in a member due to shear stress τ is,
    (a) $\dfrac{\tau^2}{2N} \times \text{volume}$
    (b) $\dfrac{\tau^2}{2E} \times \text{volume}$
    (c) $\dfrac{\tau^2 \times \text{area}}{2E}$
    (d) $\dfrac{\tau^2}{2N}$.

202. A thin cylinder designed as per the theory of thick cylinders, will be
    (a) very unsafe
    (b) just unsafe
    (c) on safer side
    (d) very much safe.

203. A thick cylinder designed as per the theory of thin cylinders, will be
    (a) on unsafe side
    (b) just safe
    (c) very much on safer side
    (d) difficult to tell.

204. If a shaft subjected to pure torsion cracks, the crack will start
    (a) at centre of cross section
    (b) on the periphery
    (c) throughout the cross-section simultaneously
    (d) none of the above.

205. Which of the following gives Modulus of Elasticity ?
    (a) ratio of linear stress to linear strain
    (b) ratio of shear stress to shear strain
    (c) ratio of lateral strain to longitudinal strain
    (d) ratio of the normal stress (of equal magnitude on all six faces) on a solid cube to the volumetric strain.

206. In problem 205, which would give the modulus of rigidity ?
207. In problem 205, which would give the bulk modulus ?
208. In problem 205, which would give the Poisson's ratio ?
209. The property that is responsible for the body to return to its original shape after the removal of the external load is known as
    (a) resilience ☐ (b) plasticity ☐
    (c) ductility ☐ (d) elasticity. ☐
210. The stress-strain curve obtained by gradually increasing axially applied load on a ductile material to the point of failure is given below.
    Identify the elastic limit
    (a) p ☐
    (b) e ☐
    (c) y ☐
    (d) y'. ☐
211. In the figure of problem 210, identify the limit of proportionality.
212. In the figure of problem 210, identify the upper yield point.
213. In the figure of problem 210, identify the lower yield point.
214. In the figure of problem 210, the stress corresponding to point u, is called
    (a) breaking stress ☐ (b) yield stress ☐
    (c) nominal stress ☐ (d) ultimate stress. ☐
215. In the figure of problem 210, the stress corresponding to point f, is known as
    (a) breaking stress ☐ (b) yield stress ☐
    (c) nominal stress ☐ (d) ultimate stress. ☐
216. In the figure of problem 210, which is the breaking point
    (a) e ☐ (b) y' ☐
    (c) u ☐ (d) f. ☐
217. The S.I. unit of modulus of elasticity is
    (a) N/cm^2 ☐ (b) N/m^2 ☐
    (c) dyne/cm^2 ☐ (d) no units. ☐
218. A rod of length L and uniform cross-sectional area A is rigidly fixed at its top and is hanging. At any section which is at a distance x from the lower end, the stress due to its own self-weight is proportional to
    (a) x ☐ (b) $\dfrac{1}{x}$ ☐
    (c) x^2 ☐ (d) $\dfrac{1}{x^2}$. ☐

219. Identify the correct relationship that exists between the modulus of elasticity E, modulus of rigidity C, and bulk modulus K.

    (a) $E = \dfrac{3K+C}{9KC}$
    (b) $E = \dfrac{3KC}{3K+C}$
    (c) $E = \dfrac{9KC}{3K+C}$
    (d) $E = \dfrac{K}{C}$.

220. For a certain material Poisson's ratio is 0.25. Then the ratio of modulus of elasticity to the modulus of rigidity for the material is

    (a) 0.4
    (b) 2.5
    (c) 4
    (d) 0.5.

221. For a certain material Poisson's ratio is 0.25. Then the ratio of modulus of elasticity to the bulk modulus for the material is

    (a) 1.5
    (b) 1.0
    (c) 0.5
    (d) 0.

222. The work done to strain a material within elastic limits is known as

    (a) resistance
    (b) virtual work
    (c) resilience
    (d) work modulus.

223. A rod uniformly reduces its diameter from D to d over a length L. If an axial load P is applied, the corresponding elongation produced is given by

    (a) $\dfrac{\pi E D d}{4PL}$
    (b) $\dfrac{8PL}{\pi E D d}$
    (c) $\dfrac{\pi E D d}{8PL}$
    (d) $\dfrac{4PL}{\pi E D d}$.

224. In the problem 223, if the rod has uniform diameter D, the elongation is given by

    (a) $\dfrac{\pi E D^2}{4PL}$
    (b) $\dfrac{8PL}{\pi E D^2}$
    (c) $\dfrac{\pi E D^2}{8PL}$
    (d) $\dfrac{4PL}{\pi E D^2}$.

225. A rod of square section of side D at one end tapers to a square section of side d at the other end. If its length is L, the elongation produced by an axial load P is given by

    (a) $\dfrac{EdD}{4PL}$
    (b) $\dfrac{4PL}{EDd}$
    (c) $\dfrac{EL}{EDd}$
    (d) $\dfrac{EDd}{PL}$.

226. A bar is made of different materials to have a composite section and carries an external load. Then

    (a) strain in all the material is same
    (b) the sum of the individual loads carried by different materials is equal to the external load
    (c) both (a) and (b)
    (d) none of the above.

227. The modular ratio of two materials is defined as

(a) $\dfrac{K_1}{K_2}$ ☐   (b) $\dfrac{E_1}{E_2}$ ☐

(c) $\dfrac{C_1}{C_2}$ ☐   (d) $\left(\dfrac{E_1}{E_2}\right)^2$. ☐

228. If α denotes the co-efficient of linear expansion, T the rise in temperature the thermal stress is given by

(a) $ET\alpha$ ☐   (b) $\dfrac{E\alpha}{T}$ ☐

(c) $\dfrac{ET}{\alpha}$ ☐   (d) $\dfrac{\alpha}{ET}$. ☐

229. The circumferential tensile stress induced when one cylinder is shrunk over the other is known as

(a) longitudinal stress ☐   (b) ultimate stress ☐
(c) shear stress ☐   (d) hoop stress. ☐

230. If the stress at elastic limit is p_e, the modulus of resilience is given by

(a) $\dfrac{2E}{p_e^2}$ ☐   (b) $\dfrac{p_e^2}{2E}$ ☐

(c) $\dfrac{4E}{p_e^2}$ ☐   (d) $\dfrac{p_e^2}{4E}$. ☐

231. The volumetric strain produced in a sphere is ...... times the strain in its diameter

(a) two ☐   (b) three ☐
(c) four ☐   (d) one and a half. ☐

232. A pull of 20 t is suddenly applied to a rod of cross-sectional area 40 cm². The stress produced in the rod is equal to

(a) 0.5 t/cm² ☐   (b) 1.0 t/cm² ☐
(c) 2.0 t/cm² ☐   (d) 4 t/cm². ☐

233. Let the strains produced in length and diameter of a cylindrical rod be α and β. Then the volumetric strain is given by

(a) α + 2β ☐   (b) α + β ☐
(c) α − β ☐   (d) α − 2β. ☐

234. A rod of length L is hanging vertically and carries a load P at the bottom. Let the weight per unit length of the rod be w. Then the tensile force in the rod at a distance y from the support is given by

(a) P ☐   (b) P − wy ☐
(c) P + wy ☐   (d) P + w(L − y). ☐

235. A body is subjected to an axial tensile stress p. Then the normal stress on any oblique plane inclined at an angle θ to the cross-section of the body is given by
    (a) $p_n = p \sin \theta$
    (b) $p_n = p \cos \theta$
    (c) $p_n = p \sin^2 \theta$
    (d) $p_n = p \cos^2 \theta$.

236. In the problem 235, what would be the tangential shear stress on the oblique plane?
    (a) $p_t = \dfrac{p}{2} \sin 2\theta$
    (b) $p_t = \dfrac{p}{2} \sin \theta$
    (c) $p_t = \dfrac{p}{2} \cos 2\theta$
    (d) $p_t = 0$.

237. For what angle θ, will the normal stress be maximum?
    (a) 90°
    (b) 45°
    (c) 30°
    (d) 0°.

238. For what angle θ, will the tangential shear stress be maximum?
    (a) 90°
    (b) 45°
    (c) 30°
    (d) 0°.

239. When a member is subjected to an axial tensile load, the plane normal to the axis experiences
    (a) maximum shear stress
    (b) maximum normal stress
    (c) minimum normal stress
    (d) none of the above.

240. In the problem of 239 a plane inclined at 45° to the axis carries
    (a) maximum shear stress
    (b) minimum shear stress
    (c) maximum normal stress
    (d) minimum normal stress.

241. A member is subjected to an axial tensile stress of p. Then the maximum shear stress induced in the member is equal to
    (a) $\dfrac{p}{2}$
    (b) p
    (c) $2p$
    (d) p^2.

242. A body is subjected to normal stresses p_1 and p_2 in two mutually perpendicular directions alongwith simple shear q. Then the maximum principal stress is given by
    (a) $\dfrac{p_1 + p_2}{2} + \dfrac{\sqrt{(p_1 - p_2)^2 + q^2}}{2}$
    (b) $\dfrac{p_1 + p_2}{2} - \dfrac{\sqrt{(p_1 - p_2)^2 + q^2}}{2}$
    (c) $\dfrac{\sqrt{(p_1 - p_2)^2 + 4q^2}}{2}$
    (d) $\dfrac{\sqrt{(p_1 - p_2)^2 - 4q^2}}{2}$.

243. In problem 242, minimum principal stress is given by.

244. In problem 242, minimum shear stress is given by.

245. The angle between the plane of principal stress and the plane of maximum shear will be
    (a) 30°
    (b) 60°
    (c) 45°
    (d) 90°.

246. A circle is marked on a mild steel plate and then it is subjected to two normal stresses in mutually perpendicular directions alongwith simple shear. After the loading the circle assumes the shape of
   (a) ellipse
   (b) cycloid
   (c) remains as circle
   (d) square.

247. A conical bar with base diameter D and length L has a unit weight of γ. The elongation due to its weight is given by
   (a) $\dfrac{6E}{\gamma L^2}$
   (b) $\dfrac{\gamma L^2}{6E}$
   (c) $\dfrac{3E}{\gamma L^2}$
   (d) $\dfrac{\gamma L^2}{3E}$.

248. A simply supported beam AB carries a u.d.l. of ω throughout the span. What concentrated load should be applied at the centre to cause the same bending moment as u.d.l.?
   (a) $\dfrac{\omega l}{4}$
   (b) ωl
   (c) $\dfrac{\omega l}{2}$
   (d) $\dfrac{\omega l}{8}$.

249. The moment of inertia of a rectangular section about the base is
   (a) twice the moment of inertia about the centroidal axis
   (b) three times the moment of inertia about the centroidal axis
   (c) four times the moment of inertia about the centroidal axis
   (d) none of the above.

250. If M, I and R denote the bending moment, the moment of inertia and the radius of curvature of the beam, which of the following equations is correct
   (a) $\dfrac{1}{R} = \dfrac{d^2y}{dx^2} = \dfrac{EI}{M}$
   (b) $\dfrac{1}{R} = \dfrac{d^2y}{dx^2} = \dfrac{M}{EI}$
   (c) $R = \dfrac{d^2y}{dx^2} = \dfrac{M}{EI}$
   (d) $R = \dfrac{d^2y}{dx^2} = \dfrac{EI}{M}$.

251. Which of the following represents the shear force at a section of the beam ?
   (a) $EI \dfrac{d^4y}{dx^4}$
   (b) $EI \dfrac{d^3y}{dx^3}$
   (c) $EI \dfrac{d^2y}{dx^2}$
   (d) $EI \dfrac{dy}{dx}$.

252. In problem 251, which will represent the rate of loading ?

253. In problem 251, which will represent the bending moment ?

254. A rectangular beam carries a maximum bending moment of M. If its depth is doubled, its moment carrying capacity will be
   (a) M
   (b) $2M$
   (c) $3M$
   (d) $4M$.

255. A simply supported beam of span L carrying a u.d.l. registers a deflection of y cm at the centre. If the span of the beam is doubled, the deflection at the centre for the same u.d.l. would be
   (a) $2y$
   (b) $4y$
   (c) $8y$
   (d) $16y$.

256. In problem 255, if the u.d.l. is double instead of the span, the corresponding deflection would be
   (a) $2y$
   (b) $4y$
   (c) $8y$
   (d) $16y$.

257. In a cantilever beam made of reinforced cement concrete, the reinforcement steel should be at the
   (a) top of the section
   (b) centre of the section
   (c) bottom of the section
   (d) any where.

258. The load on a circular short column of diameter D, in order to keep the stress wholly compressive, may be applied anywhere within a concentric circle of diameter
   (a) $\dfrac{D}{2}$
   (b) $\dfrac{D}{3}$
   (c) $\dfrac{D}{4}$
   (d) $\dfrac{D}{8}$.

259. The maximum eccentricity of the vertical load on a short column of external diameter D and internal diameter d, for no tension to occur, is
   (a) $\dfrac{8D}{D^2+d^2}$
   (b) $\dfrac{D^2+d^2}{8D}$
   (c) $\dfrac{D^2+d^2}{D}$
   (d) $\dfrac{D^2-d^2}{8D}$.

260. If the bending moment changes from positive to negative, the section where such change occurs is known as
   (a) point of inflation
   (b) crest point
   (c) point of contraflexure
   (d) neutral point.

261. In which of the following cases will the shear force as well as bending moment at the free end of a cantilever beam be equal to zero ?
   (a) when it carries a concentrated load at mid span
   (b) when it carries a concentrated load at the end
   (c) when it carries a u.d.l. over the entire span
   (d) none of the above.

262. A solid circular shaft of diameter D is subjected to an axial load to produce a stress of p N/cm². If the same axial load is applied to a hollow circular shaft of external diameter D and internal diameter $D/2$, the corresponding stress produced would be

(a) $\dfrac{p}{2}$ N/cm²   □   (b) $\dfrac{p}{4}$ N/cm²   □

(c) $\dfrac{3p}{2}$ N/cm²   □   (d) $\dfrac{4p}{3}$ N/cm².   □

263. In problem 262, if the internal diameter is $D/3$ what is the stress in the second shaft?

(a) $\dfrac{p}{2}$ N/cm²   □   (b) $\dfrac{p}{4}$ N/cm²   □

(c) $\dfrac{8}{9}$ N/cm²   □   (d) $\dfrac{9p}{8}$ N/cm².   □

264. The statement that "the deflection at any point in a beam subjected to any load system is given by the partial derivative of the total strain energy stored w.r.t. the load acting at that point in the direction in which the deflection is desired" is known as

(a) the first theorem of Castigliano   □   (b) Maxwell's theorem   □

(c) Clapeyron's theorem   □   (d) Pappu's theorem.   □

265. In a fixed beam, the slopes at the ends are

(a) minimum   □   (b) maximum   □

(c) same as at the centre   □   (d) zero.   □

266. Which of the following relations is correct?

(a) Factor of safety = crippling load – safe load   □

(b) Factor of safety = crippling load/safe load   □

(c) Factor of safety = crippling load × safe load   □

(d) Factor of safety = safe load/crippling load.   □

267. As per the Rankine's formula, the crippling load is given by

(a) $\dfrac{f_c A}{1 + a\left(\dfrac{l}{K}\right)^2}$ □   (b) $\dfrac{f_c A}{1 + a\left(\dfrac{l_e}{K}\right)^2}$ □

(c) $\dfrac{f_c A}{1 - a\left(\dfrac{l}{K}\right)^2}$ □   (d) $\dfrac{f_c A}{1 - a\left(\dfrac{l_e}{K}\right)^2}$.   □

268. In the Rankine's formula the constant a is equal to

(a) $\dfrac{\pi^2 E}{f_c}$ □   (b) $\dfrac{\pi^2}{E f_c}$ □

(c) $\dfrac{E f_c}{\pi^2}$ □   (d) $\dfrac{f_c}{\pi^2 E}$.   □

269. For a column of given material, the Rankine's constant depends on

(a) length of the column   □   (b) diameter of the column   □

(c) moment of inertia of the column   □   (d) none of the above.   □

270. Which of the following is Gordon's formula for columns ?

(a) $\dfrac{f_c A}{1 + b\left(\dfrac{l_e}{d}\right)^2}$ ☐

(b) $\dfrac{f_c A}{1 + a\left(\dfrac{l_e}{K}\right)^2}$ ☐

(c) $f_c A - BA\left(\dfrac{l_e}{K}\right)^2$ ☐

(d) $\dfrac{f_c A}{1 - b\left(\dfrac{l_e}{d}\right)^2}$ ☐

where K = Least radius of gyration ; d = Least lateral dimension ;
l_e = Effective length ; a, b and B = Constant.

271. In problem 270, which is the Johnson's parabolic formula ?
272. In problem 270, which is the Rankine's formula ?
273. Margin in the case of riveted joints is defined as the distance between the
    (a) centre of rivets in adjacent rows ☐
    (b) centre of rivet hole to the nearest edge of the plate ☐
    (c) centres of two consecutive rivets in a row ☐
    (d) any of the above. ☐
274. Tearing off the plate between the rivet hole and the edge of the plate will not happen if the margin distance is at least
    (a) 0.5d ☐
    (b) 0.75d ☐
    (c) d ☐
    (d) 1.5d. ☐
275. In a riveted joint, the quantity ($p \times t \times f_t$) gives, (p being the pitch)
    (a) strength of solid plate ☐
    (b) bearing strength ☐
    (c) crushing strength ☐
    (d) tearing strength. ☐
276. In a riveted joint, let P_t, P_s and P_c denote the maximum load per pitch length from the view point of tearing, shear and crushing respectively, and let F be the strength of the solid plate. Then the efficiency of the joint is given by
    (a) $\dfrac{P_t}{F}$ ☐
    (b) $\dfrac{P_s}{F}$ ☐
    (c) $\dfrac{P_c}{F}$ ☐
    (d) $\dfrac{\text{Least of } P_t, P_s \text{ and } P_c}{F}$. ☐
277. The tensile, shearing and crushing stresses in a riveted joint are computed based on
    (a) diameter of the rivet ☐
    (b) diameter of drilled hole ☐
    (c) average of (a) and (b) ☐
    (d) none of the above. ☐
278. A welded joint, compared to a riveted joint, has
    (a) more strength ☐
    (b) less strength ☐
    (c) same strength ☐
    (d) difficult to predict. ☐
279. The Unwin's formula for the diameter of rivet in terms of the thickness of the plate is given by
    (a) $d = 16\sqrt{t}$ ☐
    (b) $d = 9\sqrt{t}$ ☐
    (c) $d = 6\sqrt{t}$ ☐
    (d) $d = 3\sqrt{t}$. ☐

280. If d, n and f_s denote the diameter of the rivet, number of rivets per pitch length and permissible shear stress respectively, then the shear strength pitch length of a lap joint is given by

(a) $4n \times \dfrac{\pi}{4} d^2 \times f_s$ 　　　(b) $2n \times \dfrac{\pi}{4} d^2 \times f_s$

(c) $3n \times \dfrac{\pi}{4} d^2 \times f_s$ 　　　(d) $n \times \dfrac{\pi}{4} d^2 \times f_s$.

281. In problem 280, which expression gives the shear strength of a butt joint?

282. If f_c denotes the safe crushing stress, the crushing strength of a riveted joint per pitch length is equal to

(a) $\dfrac{\pi}{4} d^2 \times f_c \times n$ 　　　(b) $\pi d \times t \times f_c \times n$

(c) $d \times t \times f_c \times n$ 　　　(d) $p \times t \times f_c \times n$.

283. A spring has a stiffness of k. If it is cut into two equal portions, the stiffness of the cut portion of the spring is

(a) k 　　　(b) $\dfrac{k}{2}$

(c) $2k$ 　　　(d) $\dfrac{k}{4}$.

284. In the case of thick cylinders, the longitudinal stress
    (a) varies with maximum at outer surface to minimum at inner surface
    (b) varies with maximum at inner surface to maximum at outer surface
    (c) is uniform throughout the thickness
    (d) is zero everywhere.

285. The radial, circumferential and longitudinal stresses at any point in a thick cylinder are all
    (a) tensile 　　　(b) compressive
    (c) shear 　　　(d) none of the above.

286. A thick cylinder of internal radius R_i and the external radius R_o is subjected to internal fluid pressure p_i. Then the minimum radial stress is given by

(a) $\dfrac{2 p_i R_i^2}{R_o^2 - R_i^2}$ 　　　(b) p_i

(c) 0 　　　(d) $\left(\dfrac{R_o^2 + R_i^2}{R_o^2 - R_i^2} \right) \times p_i$.

287. In problem 286, which is the maximum radial stress?
288. In problem 286, which is the maximum circumferential stress?
289. In problem 286, which is the minimum circumferential stress?

290. A thick spherical shell is subjected to an internal fluid pressure p. Then the radial stress at any radius x is given by

(a) $\dfrac{2b}{x^3} - a$ ☐   (b) $\dfrac{b}{x^3} + a$ ☐

(c) $\dfrac{x^3}{2b} - a$ ☐   (d) $\dfrac{x^3}{b} + a$ ☐

where a and b are constants.

291. In problem 290, what is the circumferential stress at any radius x?

292. What is the maximum shear stress induced in a thin cylindrical shell subjected to an internal fluid pressure p?

(a) $\dfrac{pD}{2t}$ ☐   (b) $\dfrac{pD}{4t}$ ☐

(c) $\dfrac{pD}{8t}$ ☐   (d) $\dfrac{pD}{t}$ ☐

293. What is the maximum shear stress induced in a thin spherical shell subjected to an internal fluid pressure p?

(a) $\dfrac{pD}{2t}$ ☐   (b) zero ☐

(c) $\dfrac{pD}{4t}$ ☐   (d) $\dfrac{pD}{8t}$ ☐

294. A helical spring is subjected to an axial load W. If d, n, R and C denote the diameter of the spring, number of coils, mean radius of the coil and modulus of rigidity respectively, the deflection of the spring is given by

(a) $\dfrac{64 W^2 R^2 n}{Cd^3}$ ☐   (b) $\dfrac{64 W^2 R^3 n}{Cd^2}$ ☐

(c) $\dfrac{64 WR^3 n}{Cd^4}$ ☐   (d) $\dfrac{32 W^2 R^3 n}{Cd^4}$ ☐

295. The stiffness of the close-coiled helical spring is given by

(a) $\dfrac{Cd^3}{64 W^2 R^2 n}$ ☐   (b) $\dfrac{Cd^3}{64 W^2 R^3 n}$ ☐

(c) $\dfrac{Cd^4}{32 W^2 R^2 n}$ ☐   (d) $\dfrac{Cd^4}{64 R^3 n}$ ☐

296. The maximum shear stress in a close-coiled helical spring is given by

(a) $\dfrac{16WR}{\pi d^3}$ ☐   (b) $\dfrac{WR}{16\pi d^3}$ ☐

(c) $\dfrac{16WR}{\pi d^2}$ ☐   (d) $\dfrac{64WR}{\pi d^3}$ ☐

297. Three beams made of same material (but of different sections) of same span are subjected to same maximum bending moment. Then which section will have the maximum weight per unit length ?

   (a) circular
   (b) square
   (c) rectangular
   (d) none of the above.

298. A simply supported beam of span l carrier a u.d.l. of w kg/m. What is the magnitude of concentrated load to be applied at the centre of this beam which would produce the same deflection as the u.d.l. ?

   (a) $\dfrac{3}{8} wl$
   (b) $\dfrac{wl}{2}$
   (c) $\dfrac{5}{8} wl$
   (d) $\dfrac{7}{8} wl$.

299. A solid circular cylinder has a diameter D. A hollow circular cylinder has an internal diameter D and has the same cross-sectional area as the solid cylinder. Then the ratio of the moment of inertia of the hollow cylinder to that of solid cylinder is equal to

   (a) 2
   (b) 3
   (c) 4
   (d) 8.

300. A rectangular section has dimensions of 10 cm × 20 cm. The ratio of the moment of inertia about x-axis passing through its centroid to the moment of inertia about y-axis passing through its centriod is equal to

   (a) 8
   (b) 4
   (c) 6
   (d) 2.

301. The CGS unit of stress is

   (a) erg
   (b) dyne/cm²
   (c) N/cm²
   (d) kg/cm².

302. The SI unit of stress is

   (a) N/m²
   (b) N/cm²
   (c) Pascal
   (d) Both (a) and (c).

303. When a tensile load of P newtons is applied on a circular rod of diameter D and length L, it produces an elongation of x units. What is the elongation produced by the same tenside load on a hollow circular rod with extermed diameter D and internal diameter $0.5D$ and made up of the same material and of same length?

   (a) $\dfrac{4}{3} x$
   (b) $\dfrac{3}{4} x$
   (c) $\dfrac{1}{2} x$
   (d) $2x$.

304. A compound bar consists of steel and bronze bars with areas of 10 cm² and 20 cm² and Youngs modulus of Elasticity 2×10^5 N/mm² and 1×10^5 N/mm² respectively. If the loads shared by them are P_s and P_b, then $P_s : P_b$ will be

   (a) 1 : 1
   (b) 1 : 2
   (c) 1 : 3
   (d) 1 : 4.

305. A solid circular bar changes its diameter from $(D - a)$ to $(D + a)$ over a length L. What is the percentage error in estimating the Youngs modulus E if the mean diameter is used instead of varying diameter in the calculations

   (a) $\dfrac{D^2}{D^2 + a^2} \times 100$ ☐
   (b) $\left(\dfrac{D - a}{D}\right)^2 \times 100$ ☐
   (c) $\left(\dfrac{10a}{D}\right)^2$ ☐
   (d) $\left(\dfrac{D + a}{D}\right)^2 \times 100$. ☐

306. Arrange the values of Youngs, Bulk and Rigidity modulus of mild steel in the ascending order of magnitude

   (a) $E - K - C$ ☐
   (b) $K - C - E$ ☐
   (c) $C - E - K$ ☐
   (d) $C - K - E$. ☐

307. The relationship among the three elastic constants is

   (a) $E = \dfrac{3K + C}{9KC}$ ☐
   (b) $\dfrac{3K + C}{KC}$ ☐
   (c) $E = \dfrac{9KC}{3K + C}$ ☐
   (d) $\dfrac{KC}{3K + C}$. ☐

308. The volumetric strain in a sphere is equal to ...... times the linear strain in the diameter

   (a) 1 ☐
   (b) 2 ☐
   (c) 3 ☐
   (d) 4. ☐

309. If e_v is the volumetric strain in a circular rod, e_d is the strain in diameter and e_l is the strain in length, then

   (a) $e_v = e_d + e_l$ ☐
   (b) $e_v = 2e_d + e_l$ ☐
   (c) $e_v = e_d + 2e_l$ ☐
   (d) $e_v = 2e_d + 2e_l$. ☐

310. When a cube is subjected to a pressure p on all faces, the strain energy stored is given by

   (a) $\dfrac{p^2}{2K} \times$ Volume of cube ☐
   (b) $\dfrac{p^2}{4K} \times$ Volume of cube ☐
   (c) $\dfrac{p^2}{6K} \times$ Volume of cube ☐
   (d) $\dfrac{p^2}{8K} \times$ Volume of cube. ☐

311. When a rectangular block is subjected to a shear stress q, the strain energy stored is given by

   (a) $\dfrac{q^2}{8C} \times$ Volume of the block ☐
   (b) $\dfrac{q^2}{6C} \times$ Volume of the block ☐
   (c) $\dfrac{q^2}{4C} \times$ Volume of the block ☐
   (d) $\dfrac{q^2}{2C} \times$ Volume of the block. ☐

312. What is the radius of gyration of rectangular section of base b and depth d about one of its vertical edges?

   (a) $d/\sqrt{12}$ ☐
   (b) $b/\sqrt{12}$ ☐
   (c) $b/\sqrt{3}$ ☐
   (d) $d/\sqrt{3}$. ☐

313. In the case of rectangular section of base b and depth d, what is the ratio of radius of gyration about vertical edge to the radius of gyration about the base

    (a) $\dfrac{d}{b}$
    (b) $\dfrac{b}{d}$
    (c) $\left(\dfrac{d}{b}\right)^2$
    (d) $\left(\dfrac{b}{d}\right)^2$.

314. The cross-section of a structural member has an area of 256 cm² with $I_{xx} = 48 \times 10^4$ cm⁴ and $I_{yy} = 16 \times 10^4$ cm⁴. What is its radius of gyration about z-z axis

    (a) 50 cm
    (b) 60 cm
    (c) 70 cm
    (d) 80 cm.

315. From a circle of diameter D, a smaller circle of diameter d is cut out concentrically. What is the ratio of I_{xx} of the original circle to the I_{xx} of the circle after taking out the inner circle if d/D is 1/2 ?

    (a) $\dfrac{16}{15}$
    (b) $\dfrac{15}{16}$
    (c) $\dfrac{8}{7}$
    (d) $\dfrac{7}{8}$.

316. A beam RACBS with its mid-point at C is simply supported at A and B while the overhangs RA and BS are equal in length. Two concentrated loads equal in magnitude act at R and S respectively. Which portion of the beam is subjected to pure bending ?

    (a) RA
    (b) AB
    (c) BS
    (d) AC.

317. Which portion of the beam in question 316 experiences zero shear ?

    (a) BS
    (b) AR
    (c) AB
    (d) CB.

318. Which portion of the beam in question 316 experiences a constant bending moment ?

    (a) RS
    (b) AS
    (c) AB
    (d) BS.

319. Which point of the beam in question 316 experiences maximum sagging moment ?

    (a) C
    (b) R
    (c) S
    (d) no sagging moment in the entire beam.

320. In the beam in question 316 if the portion AB is subjected to u.d.l. and the two concentrated loads are removed, which portion of the beam experiences zero shear

    (a) AC and CB
    (b) RA and BS
    (c) RA and AC
    (d) CB and BS.

321. If the downward deflection is considered positive, then in the beam of question 320 over which portions of the beam the deflections will be negative ?

    (a) RA and BS
    (b) AB
    (c) RA and AC
    (d) CB and BS.

322. In the beam in question 320, at what points the slope is zero ?
   (a) A
   (b) C
   (c) B
   (d) R and S.

323. In the beam of question 320 at what points the deflection is zero?
   (a) R and S
   (b) C
   (c) R, C and S
   (d) A and B.

324. A beam RACBS is simply supports at A and B. The spans are AB = l, RA = BS = a, and C is mid-point of AB. The beam is subjectes to u.d.l. over its entire length that is from R to S. The beam will experience the hogging moment over its entire length when
   (a) $a \geq l/2$
   (b) $a = l/2$
   (c) $a < l/2$
   (d) difficult to guess.

325. In the beam of question 324, the sagging moment becomes zero wen
   (a) $a < l/2$
   (b) $a = l/2$
   (c) $a > l/2$
   (d) there is no relation between a and l.

326. In the beam in question 324 for what value of $\frac{a}{l}$, will the sagging and hogging moments be equal ?
   (a) $\frac{1}{\sqrt{2}}$
   (b) $\frac{1}{\sqrt{4}}$
   (c) $\frac{1}{\sqrt{8}}$
   (d) $\frac{1}{\sqrt{16}}$.

327. A cantilever beam AB of length l carries u.d.l. of w_1 kg/m over its entire length. Another identical cantilever beam carries u.d.l. of w_2 kg/m over half the span nearer to the fixed end A. What shall be $\frac{w_2}{w_1}$ for the B.M. at the fixed end to be same in both the beams
   (a) 2
   (b) 4
   (c) 8
   (d) 16.

328. In question 327 what shall be $\frac{w_2}{w_1}$, if the S.F. at the fixed end has to be same in both the beams
   (a) 2
   (b) 4
   (c) 8
   (d) 16.

329. A cantilever beam of length l is subjected to a concentrated load of P newtons at its free end. The reaction at the fixed end will be
   (a) Anti-clockwise moment Pl and an upward vertical force P
   (b) Anti-clockwise moment Pl and an downward vertical force P
   (c) clockwise moment Pl and an upward vertical force P
   (d) clockwise moment Pl and an downward vertical force P.

330. In the case of rectangular beams the ratio of max shear stress to the average shear stress is equal to
  (a) 0.50
  (b) 1.0
  (c) 1.25
  (d) 1.5.

ANSWERS
Objective Type Questions

1. (a)	2. (c)	3. (a)	4. (b)	5. (a)	6. (d)
7. (c)	8. (c)	9. (b)	10. (c)	11. (c)	12. (a)
13. (a)	14. (a)	15. (b)	16. (c)	17. (d)	18. (a)
19. (b)	20. (c)	21. (c)	22. (b)	23. (a)	24. (d)
25. (b)	26. (a)	27. (c)	28. (b)	29. (d)	30. (d)
31. (b)	32. (c)	33. (a)	34. (b)	35. (a)	36. (b)
37. (a)	38. (c)	39. (d)	40. (c)	41. (b)	42. (a)
43. (c)	44. (a)	45. (c)	46. (b)	47. (c)	48. (d)
49. (a)	50. (d)	51. (b)	52. (c)	53. (c)	54. (b)
55. (c)	56. (a)	57. (d)	58. (a)	59. (b)	60. (b)
61. (a)	62. (c)	63. (d)	64. (c)	65. (d)	66. (c)
67. (d)	68. (a)	69. (c)	70. (a)	71. (a)	72. (a)
73. (b)	74. (d)	75. (a)	76. (c)	77. (b)	78. (a)
79. (c)	80. (d)	81. (b)	82. (a)	83. (d)	84. (a)
85. (c)	86. (a)	87. (c)	88. (c)	89. (c)	90. (b)
91. (a)	92. (d)	93. (b)	94. (a)	95. (d)	96. (c)
97. (a)	98. (d)	99. (c)	100. (b)	101. (a)	102. (b)
103. (c)	104. (a)	105. (a)	106. (c)	107. (d)	108. (b)
109. (a)	110. (b)	111. (c)	112. (b)	113. (a)	114. (c)
115. (b)	116. (c)	117. (a)	118. (a)	119. (a)	120. (c)
121. (b)	122. (a)	123. (b)	124. (d)	125. (c)	126. (a)
127. (b)	128. (b)	129. (a)	130. (a)	131. (b)	132. (b)
133. (c)	134. (b)	135. (b)	136. (a)	137. (c)	138. (a)
139. (a)	140. (c)	141. (b)	142. (c)	143. (b)	144. (d)
145. (a)	146. (a)	147. (d)	148. (b)	149. (c)	150. (a)
151. (b)	152. (a)	153. (b)	154. (a)	155. (b)	156. (c)
157. (a)	158. (b)	159. (a)	160. (a)	161. (a)	162. (d)
163. (c)	164. (a)	165. (d)	166. (a)	167. (a)	168. (c)
169. (a)	170. (b)	171. (b)	172. (a)	173. (b)	174. (a)
175. (b)	176. (a)	177. (d)	178. (b)	179. (b)	180. (a)
181. (c)	182. (a)	183. (a)	184. (c)	185. (b)	186. (a)
187. (c)	188. (a)	189. (b)	190. (c)	191. (d)	192. (a)

193. (d)	194. (b)	195. (c)	196. (b)	197. (d)	198. (a)
199. (b)	200. (c)	201. (a)	202. (c)	203. (a)	204. (b)
205. (a)	206. (b)	207. (d)	208. (c)	209. (d)	210. (b)
211. (a)	212. (c)	213. (d)	214. (d)	215. (a)	216. (d)
217. (b)	218. (a)	219. (c)	220. (b)	221. (a)	222. (c)
223. (d)	224. (d)	225. (c)	226. (c)	227. (b)	228. (a)
229. (d)	230. (b)	231. (b)	232. (b)	233. (a)	234. (d)
235. (d)	236. (a)	237. (d)	238. (d)	239. (b)	240. (a)
241. (a)	242. (a)	243. (b)	244. (c)	245. (d)	246. (a)
247. (b)	248. (c)	249. (c)	250. (b)	251. (b)	252. (a)
253. (c)	254. (d)	255. (d)	256. (a)	257. (a)	258. (c)
259. (b)	260. (c)	261. (c)	262. (d)	263. (d)	264. (a)
265. (d)	266. (b)	267. (b)	268. (d)	269. (d)	270. (a)
271. (c)	272. (b)	273. (b)	274. (d)	275. (a)	276. (d)
277. (b)	278. (a)	279. (c)	280. (d)	281. (b)	282. (c)
283. (c)	284. (c)	285. (d)	286. (c)	287. (b)	288. (d)
289. (a)	290. (a)	291. (b)	292. (c)	293. (b)	294. (c)
295. (d)	296. (a)	297. (a)	298. (c)	299. (b)	300. (b)
301. (b)	302. (d)	303. (a)	304. (a)	305. (c)	306. (d)
307. (c)	308. (c)	309. (b)	310. (a)	311. (b)	312. (c)
313. (b)	314. (a)	315. (a)	316. (b)	317. (c)	318. (c)
319. (d)	320. (b)	321. (a)	322. (b)	323. (d)	324. (a)
325. (b)	326. (c)	327. (b)	328. (a)	329. (a)	330. (d)

Chapter 4 HYDROLOGY

I. INTRODUCTION

Hydrology is the science that deals with the occurrence, movement and circulation of water. It is a highly interdisciplinary science which draws, for its investigation, many principles from other sciences like Fluid mechanics, geology, geohydrology, hydrogeology, hydrometeorology, limnology (science of lakes), cryology (science of snow and ice), ecology etc.

Water exists in three forms, namely, gaseous, liquid and solid forms. It is kept in circulation by the energy provided by the sun. The group of various arcs such as precipitation, runoff, interception, evaporation, transpiration, infiltration etc., which represent different paths through which the water in nature circulates is called the hydrologic cycle.

The gaseous envelope around the globe earth is divided into number of layers such as troposphere, stratosphere, chemosphere, ionosphere etc. In the troposphere, the temperature decreases with altitude while in the stratosphere it more or less remains constant with elevation. All the atmospheric activities are confined to troposphere.

The main constituents of the atmosphere are nitrogen, oxygen and the inert gases. The variable constituents are carbon dioxide, ozone and water vapour.

The atmosphere always contains water vapour. The partial pressure exerted by water vapour is called the vapour pressure denoted by e. The maximum amount of vapour that can be accommodated by the atmosphere depends on the temperature. When atmosphere possesses the maximum vapour that it can hold, then it is said to be saturated. The partial pressure exerted by water vapour at saturated condition is called the saturation vapour pressure denoted by e_s. The ratio of $\dfrac{e}{e_s}$ is called the relative humidity. The relative humidity can be measured by psychrometer and hygrograph.

The solar energy is measured in calories per square cm one cal/cm² is called one langley. It can be measured by Pyrheliograph which is also known as Pyranometer.

The temperature are measured by thermometers. The continuous recording of temperature is done by thermograph.

A vast and deep body of air which acquired homogeneous temperature and humidity characteristics at any elevation is called air mass. The region between two distinct air masses is called air front. A disturbance created at the air front develops into cyclone leading to cyclonic precipitation. In cyclones of northern hemisphere the pressure is low at the centre and wind blows spirally inward in counter clockwise direction and in clockwise direction in Southern hemisphere.

The high pressure areas covering large areas where wind blows spirally outward are called anti-cyclones.

For the formation of precipitation, availability of sufficient moisture, its cooling to saturation condition, condensation and the growth of droplets are the basic conditions required.

If the cooling occurs as the air mass is lifted up a slope, the resulting precipitation is called orographic. As the air mass yields rainfall on the windward side, the rainfall on the leeward side is negligible or scanty. Such a region is called rain shadow region.

If the cooling occurs due to lifting of air by heating at the earth surface and results in precipitation then it is called convective precipitation.

The growth of droplets is hastened by the presence of condensation nuclei. Dry ice and silver iodide are sprayed into clouds to act as condensation nuclei and hasten the formation of precipitation. Such a process is called cloud seeding or artificial rain making.

The various forms of precipitation are drizzle, glaze, rain, sleet, snow, hail, dew etc.

The precipitation is measured by rain gauges. They are of two types: ordinary rain gauges and recording rain gauges.

Symons rain gauge with a receiving area of 127 mm diameter is the most commonly used ordinary rain gauge in India.

Tipping bucket type, weighing type, float type (or syphon type) are the recording gauges. They usually have the receiving area of 203 mm. They give the record of cumulative depth of rainfall against time which is nothing but a rainfall mass curve.

The rate at which the depth of rainfall is accumulating is called the rainfall intensity expressed in mm/h. It can be computed only from the record obtained from recording gauges. A graph showing variation of rainfall intensity against time is called rainfall hyetograph.

Rainfall can also be measured by radar. The rainfall intensity depends on the intensity of the echoes.

Rain gauges should be located on plain ground well-protected from the gusty winds. The nearest objective should be at least at a distance of twice its height. As per ISI recommendations there should be one gauge for 520 km² in plain areas and one gauge per 130 km² in hilly areas.

Intensity duration analysis of storms indicate that as the duration of storm increases the intensity decreases.

Isohyets are the lines drawn on a map which connect all the points receiving equal depth of rainfall.

The average depth of rainfall is determined by any of the three methods. Arithmetic mean, Thiessen polygon and Isohyetal. The isohyetal method gives the most accurate value.

In the arithmetic mean method the average rainfall P is given by

$$P = \frac{P_1 + P_2 + \ldots + P_n}{n}$$

where $P_1, P_2, \ldots, P_n$ are the rainfall recorded at the n gauges respectively which lie within the area.

In Thiessen polygon method, the average rainfall depth is given by $P = \dfrac{A_1 P_1 + A_2 P_2 + \ldots + A_n P_n}{A}$, where $A_1, A_2, \ldots$ are the areas of the polygons around rain gauge stations and $A = (A_1 + A_2 + \ldots + A_n)$ is the total area. $\dfrac{A_1}{A}, \dfrac{A_2}{A}, \ldots, \dfrac{A_n}{A}$ are called the Thiessen weights of the corresponding gauges.

The depth-area-duration analysis show that the average depth of rainfall decreases as the area increases.

The double mass curve is a graph drawn between the cumulative annual rainfall at any station against the cumulative average annual rainfall of certain number of nearby rain gauge stations. This curve is used to check the consistency (and to correct if found necessary) of the record at any suspected gauge.

The height of water surface in the river above an arbitrary datum is called the stage. The variation of stage with time is called the stage hydrograph. The graph showing the variation of discharge with time is called the discharge hydrograph or simply the hydrograph. The discharge hydrograph is obtained from stage hydrograph through the stage-discharge relationship, which is also known as the rating curve.

The discharge in the stream is measured by first finding the average velocity and then multiplying it with the area of flow. The velocity itself is measured by the current meter. The velocity of flow is directly proportional to the speed N of the current meter in revolutions per second. That is, $V = a + bN$, where a and b are the constants of the meter. The velocity across any vertical approximately varies as parabola, with the maximum velocity occurring at or near the free surface.

In the single point method the velocity measured at 0.6 depth from the free surface is taken as the average velocity. Whereas in two point method, the arithmetic mean of the velocities measured at 0.2 depth and 0.8 depth from the free surface is taken as the average velocity.

The discharge is expressed in m^3/s. The larger units of volume are hectare-metre, million m^3, comec-day, cumec-month etc. The relevant conversions are: one hectare-metre = 10000 m^3; one million m^3 = 100 ha-m, one cumec-day = 86400 m^3 = 8.64 ha-m and one cumec month = 2592000 m^3 = 259.2 ha-m etc.

The graph showing the cumulative volume of water against time is called a flow mass curve or simply a mass curve. Mass curve can be very conveniently used to find the storage capacity of a reservoir to meet a given uniform demand.

The graph showing the discharge in the stream against the percentage time such discharge is equalled or exceed is called a flow duration curve. It is mainly used in hydropower studies to ascertain the firm power and secondary power available at a given site.

Infiltration is the process by which water enters the surface strata of the earth. The maximum rate at which a given soil absorbs water is called its infiltration capacity and is denoted by f. The infiltration capacity depends on many factors such as depth of surface detention, initial soil moisture, compaction, surface vegetation, temperature etc.

Infiltration capacity decreases continuously within the storm and it is well-described by the Horton's equation

$$f = f_c + (f_o - f_c)e^{-kt}$$

where f_o and f_c are the initial and minimum infiltration rates and k is a constant.

Infiltration is generally measured by double ring infiltrometers. But the actual infiltration during a storm may or may not be equal to the infiltration capacity depending on whether the rainfall intensity is more than the infiltration capacity or not.

However, for many design purposes the average infiltration indices such as φ-index and ω-index are sufficient.

φ-index is defined as the average rainfall intensity above which the rainfall volume equals the observed run-off volume.

ω-index is a refined version of φ-index. The depression storage is excluded while computing it. Therefore, it will be always less than φ-index.

Evaporation in the process by which water from liquid or solid state passes into the vapour state and is diffused into atmosphere. Dalton has first shown that evaporation is proportional to the vapour pressure deficit. This is known as the Dalton's law and is written as $E = c(e_s - e)$.

The value of the proportionality constant c depends on the other factors affecting the evaporation such as radiation, temperature, wind, atmospheric pressure, quality of water, size of water body, nature of evaporating surface etc.

It is observed that the salinity in water decreases the evaporation rate.

Evaporation is estimated using empirical equations and it is measured using the evaporation pans (also called evaporimeters).

The three types of evaporation pans are land pan (or surface pan), floating pan and sunken pan.

Because of exposure conditions, the evaporation from a pan is always more than that from an adjacent large water body like a lake or reservoir. The ratio of lake evaporation to the pan evaporation is called the pan co-efficient. Obviously, pan co-efficient is always less than unity. The sunken pan has the largest pan co-efficient.

Evaporation from water bodies can be reduced by applying certain chemical compounds which form a layer of one molecule thickness at the surface. The commonly used chemical is cetyl alcohol.

Transpiration is the process by which water vapour escapes from the living plant, principally through the leaves and enters the atmosphere.

Evapotranspiration is the sum of water used by plants in a given area in transpiration and the water evaporated from the adjacent soil in the area. It is approximately taken as the consumptive use.

Lysimeter is the instrument used to measure the evapotranspiration. There are some empirical equations to estimate the evapotranspiration from meteorological data.

The water available below the surface of earth is called the ground water. It has been an important water resource through the ages.

Geological formations which contain ground water and at the same time which are sufficiently permeable to transmit and yield water in usable quantities are called aquifers.

An aquifer is called unconfined if it contains the water table in it. An aquifer which is sandwiched between two relatively impervious layers is called a confined aquifer.

A formation which may contain water, but cannot transmit water because of its poor permeability is called an aquiclude.

If the confining layer is sufficiently permeable to transmit water vertically but not permeable enough to transport water laterally, it is called an aquitard.

If one of the confining layers is an aquitard then the aquifer is known as a leak aquifer (semi-confined aquifer).

A geological formation which does not contain ground water is called an aquifuge.

The surface obtained by connecting the equilibrium water levels in tubes or piezometers is called the piezometric surface. The piezometric surface of a confined aquifer is above the upper confining layer.

If a well penetrates into the confined aquifer, water in the well rises to the piezometric level. If the piezometric surface at the place of the well is above the ground level, it will yield a flowing well.

A well penetrating into an unconfined aquifer is called the water table well.

When the water table intersects the stream, the aquifer contributes run-off to the stream and such a stream is called an effluent stream.

When the water table is below the bed of the stream, water flows from the stream to the aquifer. Such a stream is called an "influent stream".

Darcy postulated that the velocity through the aquifers being very small, is directly proportion to the hydraulic gradient. Darcy's law is written as $V = K \dfrac{dh}{dx}$, where the constant of proportionality K is called the co-efficient of permeability or the hydraulic conductivity. The product of K and the thickness of the aquifer B is called the transmissibility (or transmissivity) and is denoted by T. That is, $T = KB$.

Specific yield of an aquifer is the ratio of the water which will drain freely from a saturated formation to the total volume of the formation.

Storage co-efficient is defined as the volume of water that an aquifer releases from or takes into storage per unit surface area of aquifer per unit drop of water table or piezometric surface.

For unconfined aquifers the storage co-efficient and the specific yield are same.

The steady state yield from a tube-well in an unconfined aquifer is given as

$$Q = \dfrac{\pi K(h_1^2 - h_2^2)}{\ln\left(\dfrac{r_1}{r_2}\right)}$$

where h_1 and h_2 are the water levels in the two observation wells located at the radial distances r_1 and r_2 from the main well.

The yield can also be written as $Q = \dfrac{\pi K(H^2 - h_w^2)}{\ln(R/r_w)}$

where h_w and r_w are the water level in the well and radius of the well, H is the thickness of the aquifer and R is the radius of influence. The radius of influence is the distance from the main well where the cone of depression just commences or where the drawdown is zero.

The yield from the confined aquifer is given as

$$Q = \frac{2\pi KB(h_1 - h_2)}{\ln(r_1/r_2)} \text{ or } Q = \frac{2\pi T(H - h_w)}{\ln(R/r_w)}.$$

The steady state yield equation was first developed by Dupuit and modified later by Thiem.

In open wells the diameter is very large.

The specific yield or the specific capacity of an open well is the ratio of K/A, where A is the area of the well.

From the recuperation test we can obtain the specific capacity

$$\frac{K}{A} = \frac{2.303}{t} \log_{10} \frac{h_1}{h_2}$$

where t is the time required for the water level to recupe from h_1 to h_2.

The precipitation falling on the earth's surface becomes surface run-off after meeting the infiltration. The infiltrated water may become ground water after deep percolation or it may percolate laterally which is known as subsurface run-off. The subsurface run-off which reaches the streams rather quickly is called prompt subsurface run-off.

The surface run off plus the prompt subsurface run-off is called the direct run-off.

When water table rises, it may contribute run-off to the stream which is known as ground water run-off.

The sum of ground water run-off and the delayed subsurface run-off is called base flow.

The sum of direct run-off and base flow constitute the total run-off.

However, some investigators treat the total run-off as consisting of surface run-off and ground water run-off only.

Out of total precipitation, that portion which produces the direct run-off is called the rainfall excess or precipitation excess.

Unit hydrograph, which was conceived first by Sherman, relates the rainfall excess and the direct run-off (including its time distribution) produced by it. Therefore, the first step in the derivation of unit hydrograph is to separate the base flow from total run-off.

Unit hydrograph is defined as the direct run-off hydrograph produced by 1 cm of rainfall excess uniformly distributed over the entire basin area and also uniformly distributed over a specified period of time known as the unit duration.

Unit hydrograph is based on the principle of linearity and principle of linearity and principle superposition.

If a D-hr unit hydrograph is available, unit hydrographs with other durations such as $2D$, $3D$ etc. can be obtained easily from the principle of superposition. As the duration of unit hydrograph increases, the base period also increases and consequently the peak ordinate decreases.

From a given unit hydrograph, if other unit hydrographs whose durations are not integral multiples of the original duration are required, they can be obtained using the S-curve hydrograph.

A S-curve hydrograph is a hydrograph of direct run-off resulting from a rainfall excess of constant intensity occurring for a very long period. The name is derived from the deformed S-shape. It attains an equilibrium ordinate at the end of the base period of the unit hydrograph from which it is derived. It is nothing but the summation of the given unity hydrograph.

Unit hydrograph may be used to predict the direct run-off of any storm with any intensity provided the duration of the storm is same as the duration of unit hydrograph.

The word "Unit" in the unit hydrograph is meant to denote the unit duration, though it is sometimes wrongly interpreted to denote unit depth of run-off.

When the duration of unit hydrograph approaches zero, it becomes instantaneous unit hydrograph.

There may be basins which do not have rainfall and run-off records to derive unit hydrograph and yet unit hydrograph may be required for such basins for flood estimation etc. In such cases the synthetic unit hydrographs are developed.

Synthetic unit hydrograph is not a unit hydrograph derived from the rainfall run-off records of the basin but it is derived from the unit hydrographs available for the nearby basins. It is derived from the relationships established between the unit hydrograph parameters and the basin parameters as proposed by Snyder.

$$t_p = C_t(LL_c)^{0.3} \text{ and } D = \frac{t_p}{5.5}.$$

$$Q_p = \frac{C_p A}{t_p}$$

$$T = 3 + 3\frac{t_p}{24}$$

In the above equations t_p is the basin lag which is the time difference between the peak of the hydrograph and the centroid of the rainfall. L is the length of main stream, L_c is the distance between basin outlet and a point on the stream which is nearest to the centroid of the basin, D is the standard duration, Q_p is the peak ordinate, A is the basin area, T is the base period, and C_p and C_t are the empirical constants.

The factors affecting run-off may be grouped under two heads. The climatic factors which include: type of precipitation, intensity of rainfall, duration of rainfall, areal distribution of rainfall, direction of storm movement, antecedent precipitation etc.

The physiographic factors are: Land use, type of soil, area of the basin, shape of the basin, elevation, slope, orientation, type of drainage network, indirect drainage, artificial drainage etc.

According to principle of time invariance, the base periods of all the direct run-off hydrographs produced by storm of same duration will also be same.

According to principle of linearity, the ordinates of the direct run-off hydrographs of common base period will be proportional to the volumes of run off represented by them.

Flood routing may be defined as the procedure whereby the shape of a flood hydrograph at a particular location on the stream is determined from the known or assumed flood hydrograph at some other location upstream.

Flood routing is based on the continuity equation and another equation which relates storage to outflow/inflow.

In the case of channels Muskingum method is used. According to this method, the outflow is given as

$$Q_2 = C_0 I_2 + C_1 I_1 + C_2 Q_1$$

where $\quad C_0 = \dfrac{0.5\Delta t - Kx}{K(1-x) + 0.5\Delta t}$, $C_1 = \dfrac{0.5\Delta t + Kx}{K(1-x) + 0.5\Delta t}$, $C_2 = \dfrac{K(1-x) + 0.5\Delta t}{K(1-x) + 0.5\Delta t}$

K and x are the storage constants of the reach, and Δt is the routing interval. The constants C_0, C_1 and C_2 are called the routing constants. It can be easily verified that they add up to unity.

In the case of reservoir routing I.G. Pul's method is commonly used.

Since the processes like rainfall and run-off are essentially random in nature, their future occurrence can be best predicted by analysing the data collected on them in the past.

Let X be the random variable denoting the event such as the annual peak discharge. Then its value observed for n years in the past $x_1, x_2, ..., x_n$ is known as its sample. The statistics of this sample like mean, and standard deviation are computed from the following equations:

$$\text{Mean, } \bar{x} = \dfrac{x_1 + x_2 + ... x_n}{n} = \dfrac{1}{n}\sum x_i$$

$$\text{Standards deviation, } S_x = \sqrt{\dfrac{1}{n-1}\sum (x_i - \bar{x})^2}$$

The co-efficient of variation is defined as the ratio of standard deviation to the mean.

The observed values of annual floods are arranged in the descending order of magnitude. Then ranks are assigned starting with 1 for the highest flood, 2 for the next highest and so on, and n for the least. If we consider a flood with a rank m, it has been equalled or exceeded m years out of n years. So on the average such flood can be expected to occur once in $\dfrac{n}{m}$ years. This is what is known as its return period or the recurrence interval.

$$T = \dfrac{n}{m} \qquad \text{California formula}$$

$$T = \dfrac{n+1}{m} \qquad \text{Weibul's formula}$$

$$T = \frac{2n}{2m-1}$$ Hazen's formula

A graph plotted between the flood magnitude against its return period is called the probability plotting. This many be used to obtain the flood magnitude for any given return period or the return period of any given flood peak by interpolation and extrapolation. Results from such interpolation or extrapolation would be in error since the analysis does not consider the theoretical probability distributions of the variable.

The theoretical distribution widely used to fit the flood data is the Gumbel's extreme value distribution. Log-Pearson Type III distribution is popular in United States.

According to Gumbel's distribution, the cumulative probability of occurrence of any flood with magnitude x is given by $P = 1 - \frac{1}{T} = e^{-e^{-y}}$

where y is called the reduced variate $y = a(x - x_f)$. The parameters a and x_f can be obtained from the sample data.

From the interpretation of probability, a T year flood may occur in any year with a probability of $\frac{1}{T}$. The probability that the structure does not fail in any year is, therefore, $\left(1 - \frac{1}{T}\right)$. Assuming that annual flood events are independent, the probability that the structure will not fail in the next N years is $\left(1 - \frac{1}{T}\right)^N$.

Hence, the probability that the structure may fail in any one of the next N years, which is nothing but the risk in the design, is equal to $\left[1 - \left(1 - \frac{1}{T}\right)^N\right]$.

When soil is moved from its present location erosion is said to have occurred. The soil particles which are detached and moved are called sediments.

Normal erosion is mainly due to natural processes like rainfall, wind etc. and is very slow. Accelerated erosion is human induced (deforestation, forest fires, etc.) and is very fast.

The factors affecting erosion are rainfall regime, vegetal cover, soil type, land slope and land use.

Erosion is more in intense storms since the size of the rain drop and velocity of the rain drop are more.

Vegetal cover reduces erosion.

Cohesive soil resists splash erosion more than the cohesionless soil.

Erosion is more on steep soils.

Proper soil conservation methods such as contour bunding will reduce erosion. Rill is a very small stream. Rill erosion is the removal of soil by small concentration of flowing water.

Gully erosion is the removal of soil from rivulets which are formed due to sufficient accumulation of overland flow on slope grounds.

Sheet erosion is the wearing away of a thin layer of soil on the land surface, especially between rills, mainly by overland flow

Channel erosion refers to the erosion occurring in the stream channels in the form of stream bank erosion or streambed degradation. Gross erosion is the summation of erosion from all sources within the catchment area. Sediment transport refers to the mechanism by which sediment is moved downstream by flowing water.

Sediment carried by water in suspension is known as suspended load.

Sediment carried by water by rolling or sliding along the river bottom is called bed load and by bouncing along the bed is known as saltation.

Minute particles of colloidal sizes which always remain in suspension are called wash load.

The bottom shear stress τ_0 also known as tractive stress developed in the flow is responsible for bed load. For wide channels it is given by

$$\tau_0 = \gamma d S_0$$

where γ = Specific weight of water, d = Depth of flow and S_0 = Slope of the channel.

When τ_0 exceeds the threshold or critical tractive stress τ_c the particles on the bed begin to move.

The bed load is given by

$$q_s = C_s \frac{\tau_0}{\gamma} (\tau_0 - \tau_c)$$

in which q_s is in kg/s/metre width of channel and C_s is an empirical constant.

The amount of sediment eroded and removed from the sources is known as gross sediment production. This is usually expressed in tons /km²/year.

Sediment yield refers to the actual delivery of eroded soil particles to a given downstream point. It is usually expressed as tons/years.

Sediment delivery ratio is defined as the ratio of sediment yield to the gross sediment production. It varies from 0 to 1.

Since the eroded particles may be deposited before they reach the downstream point sediment yield ≤ gross sediment production.

When the sediment laden water enters the reservoir sediment deposition takes place.

If the inflowing sediment laden water is denser than the surface water in the reservoir, it moves slowly below the surface in the form of a density current (or turbidity current).

The five sediments which are deposited near the dam and the density currents carrying sediments may be flushed out by opening the scour sluices of the dam.

The trap efficiency of the reservoir is defined as the ratio of the sediment trapped to the incoming sediment and is expressed in per cent. It is a function of the ratio of the reservoir capacity to mean annual run off volume. Trap efficiency will be less for reservoirs with small capacity-inflow ratio.

II. OBJECTIVE TYPE QUESTIONS

1. The science with deals with occurrence, movement and circulation of water is called
   (a) hydrogeology
   (b) geohydrology
   (c) hydrology
   (d) hydrography.
2. Limnology is the science which deals with
   (a) surface streams
   (b) lakes
   (c) glaciers
   (d) snow and ice.
3. Hydrometeorology is the science which deals with
   (a) water in the atmosphere
   (b) water below of surface of the earth
   (c) water in the surface streams
   (d) water in oceans.
4. Relative humidity of the atmosphere is defined as the ratio of
   (a) actual vapour pressure to the vapour pressure at 0°C
   (b) actual vapour pressure to the atmospheric pressure
   (c) weight of water to the weight of air
   (d) actual vapour pressure to the saturation vapour pressure.
5. A pressure of one millibar is equal to
   (a) 100 N/m^2
   (b) 1000 N/m^2
   (c) 10000 N/m^2
   (d) 100000 N/m^2.
6. A pressure of 1 N/m^2 is equal to
   (a) 0.1 millibar
   (b) 0.01 millibar
   (c) 0.001 millibar
   (d) 1 millibar.
7. The ratio of the radiation reflected back to the radiation received by the surface is called its
   (a) radiation coefficient
   (b) Plank's constant
   (c) albedo
   (d) Bowen's ratio.
8. In which of the following does the temperature remain constant with elevation
   (a) troposphere
   (b) mesosphere
   (c) ionosphere
   (d) stratosphere.
9. Langley is the unit which measures
   (a) infiltration
   (b) permissibility
   (c) radiation
   (d) albedo.
10. Isobar is a line which joins points of equal
    (a) rainfall depth
    (b) temperature
    (c) humidity
    (d) atmospheric pressure.
11. In cyclones of Northern hemisphere wind blows
    (a) clockwise inward
    (b) anti-clockwise inward
    (c) clockwise outward
    (d) anti-clockwise outward.

12. In anti-cyclones of Northern hemisphere wind blows
    (a) clockwise inward
    (b) anti-clockwise inward
    (c) clockwise outward
    (d) anti-clockwise outward.
13. The instrument used to measure the wind velocity in the atmosphere is
    (a) current meter
    (b) atmometer
    (c) pyranometer
    (d) anemometer.
14. In the following, identify the one which is different from the rest
    (a) rain
    (b) drizzle
    (c) hail
    (d) fog.
15. Rain shadow region is formed on the
    (a) windward side of mountain when rain yielding mass passes over it
    (b) Leeward side of mountain when rain yielding air mass passes over it
    (c) Plains when rain yielding air mass passes over it
    (d) none of the above.
16. The albedo of solid surface is in the range
    (a) 0.95 to 1
    (b) 0.5 to 0.75
    (c) 0.1 to 0.3
    (d) 0.001 to 0.01.
17. The albedo of the water surface is nearer to
    (a) 0.5
    (b) 0.05
    (c) 0.25
    (d) 0.75.
18. Which is the odd one in the following?
    (a) snow
    (b) sleet
    (c) rain
    (d) hail.
19. The convective precipitation is caused when
    (a) vertical instability of moist air is produced by surface heating
    (b) the disturbance on the air front develops into cyclone
    (c) the colder air rises into warm air
    (d) all of the above.
20. Rain shadow region in India is found
    (a) to the west of western ghats
    (b) to the west of eastern ghats
    (c) to the south of Himalayas
    (d) to the east of western ghats.
21. The cyclonic precipitation is caused due to
    (a) disturbance caused on the frontal surface between cold and warm air masses
    (b) the thermal convective currents
    (c) the orographic cooling when air mass is lifted up a slope
    (d) none of the above.
22. The instrument used to measure the humidity of the atmosphere continuously with time is called
    (a) Barograph
    (b) Thermograph
    (c) Hygrograph
    (d) Thermo-hygrograph.

23. The instrument which records the variation of temperature with time is called
    (a) Barograph
    (b) Thermograph
    (c) Hygrograph
    (d) Thermo-hygrograph.

24. The instrument which records the variation of both temperature and humidity with time is called
    (a) Barograph
    (b) Thermograph
    (c) Hygrograph
    (d) Thermo-hygrograph.

25. Pyranometer is the instrument which measures
    (a) the duration of sunshine
    (b) radiation
    (c) evaporation
    (d) none of the above.

26. Rainfall hyetograph shows the variation of
    (a) cumulative rainfall with time
    (b) rainfall intensity with time
    (c) rainfall depth over an area
    (d) rainfall intensity with the cumulative rainfall.

27. Fainfall mass curve shows the variation of
    (a) rainfall intensity with time
    (b) rainfall intensity with cumulative rainfall
    (c) rainfall excess with time
    (d) cumulative rainfall with time.

28. The diameter of the receiving area of Syphon type recording raingauge is equal to
    (a) 127 mm
    (b) 300 mm
    (c) 203 mm
    (d) 500 mm.

29. The diameter of the receiving area of Symons ordinary rain gauge is equal to
    (a) 127 mm
    (b) 300 mm
    (c) 203 mm
    (d) 500 mm.

30. In selecting a site for a rain gauge the nearest object should be at a minimum distance of
    (a) twice its height
    (b) three times its height
    (c) equal to its height
    (d) anywhere.

31. Double mass curve technique is used
    (a) to prepare rainfall hyetograph from rainfall mass curve
    (b) to check the consistency of record at a suspected rain gauge station
    (c) to derive the hydrograph
    (d) to derive the S-curve hydrograph.

32. The maximum rainfall intensity at a given location
    (a) increases with increase in duration of rainfall
    (b) decreases with increase in duration of rainfall
    (c) is independent of the duration of rainfall
    (d) difficult to predict.

33. The most accurate method of finding the average depth of rainfall over an area is
    (a) Thiessen polygon method
    (b) Isohyetal method
    (c) Arithmetic mean method
    (d) Any of the above.

34. The Thiessen polygonal areas of the four rain gauge stations A, B, C and D in a catchment are 75, 125, 150 and 150 km² respectively. If the average depth of rainfall for the catchment is given as 5 cm and the rainfall recorded at B, C and D are 5 cm, 4 cm and 5 cm. What is the rainfall at A?
    (a) 8 cm
    (b) 7 cm
    (c) 6 cm
    (d) 6 cm.

35. A major river basin is divided into four sub-basin with areas of 900, 700, 1000 and 1400 km² respectively. If the average annual rainfalls on the sub-basins are 100, 80, 100 and 110 cm respectively what is the rainfall for the basin as a whole?
    (a) 85 cm
    (b) 95 cm
    (c) 100 cm
    (d) 105 cm.

36. Four rain-gauge stations A, B, C and D in a catchment area have recorded 20, 25, 22 and 15 cm respectively. If their Thiessen weights are 0.3, 0.4, 0.1 and 0.2. What is the average depth of rainfall on the catchment?
    (a) 21.2 cm
    (b) 22.2 cm
    (c) 20.2 cm
    (d) 19.2 cm.

37. For a given storm the average depth of rainfall over an area
    (a) increases with increase in area
    (b) decreases with increase in area
    (c) has no relation with area
    (d) none of the above.

38. An isohyet is a line joining points of
    (a) equal rainfall intensity
    (b) equal rainfall depth
    (c) equal evaporation
    (d) equal humidity.

39. The chart removed from a recording type raingauge gives
    (a) the rainfall mass curve
    (b) to rainfall hyetrograph
    (c) the isohyetal map
    (d) the double mass curve.

40. As per Indian standards how many raingauges should be installed in a catchment with an area of 1000 km² lying in plains?
    (a) 6
    (b) 4
    (c) 2
    (d) 8.

41. Intensity of rainfall means
    (a) total rainfall during a storm
    (b) rainfall per unit area
    (c) the rate at which the rainfall depth is accumulating
    (d) volume of rain water per unit area.

42. The typical characteristic of convective showers is that they are of
    (a) high intensity and long duration
    (b) high intensity and short duration
    (c) low intensity and long duration
    (d) low intensity and short duration.

43. The double mass curve is prepared by plotting
    (a) cumulative annual rainfall at a station against cumulative annual rainfall at a neighbouring station
    (b) cumulative annual rainfall at a station against time in years
    (c) cumulative annual rainfall at a station against the cumulative average annual rainfall of a certain number of neighbouring stations
    (d) the annual rainfall of this year against the annual rainfall of previous year.

44. In which of the following the water equivalent would be minimum
    (a) fresh powder snow
    (b) virgin snow
    (c) coarse snow
    (d) granular ice.

45. In which of the following the water equivalent would be maximum.
    (a) fresh powder snow
    (b) virgin snow
    (c) coarse snow
    (d) packed snow.

46. The snow fall is generally measured in terms of
    (a) weight of snow per unit area
    (b) equivalent depth of water
    (c) depth of snow fallen
    (d) any of the above.

47. In radar measurement of rainfall, the energy of echo waves depends upon
    (a) the solar radiation
    (b) wind velocity
    (c) the size of the drop
    (d) the inclination of rainfall.

48. Thiessen polygon method is used
    (a) to determine the parameters of the aquifer
    (b) to locate the depth of water table
    (c) to compute the average depth of rainfall
    (d) to drive the ordinates of unit hydrograph.

49. The precipitation formation is hastened by sending
    (a) dry ice into clouds
    (b) silver iodide into clouds
    (c) both (a) and (b)
    (d) none of the above.

50. In the two point method of finding the average velocity using the current water across a vertical in a open channel, the velocities are measured below the free surface at
    (a) 0.25 and 0.75 depths
    (b) 0.20 and 0.80 depths
    (c) 0.40 and 0.60 depths
    (d) 0.15 and 0.85 depths.

51. In the single point method of finding the average velocity using the current meter across a vertical in a open channel, the velocity is measured below the free surface at
    (a) 0.8 depth
    (b) 0.7 depth
    (c) 0.6 depth
    (d) 0.5 depth.

52. In N is the speed of the current meter in revolutions per second, the velocity measured by it is proportional to
    (a) $N^{1/2}$
    (b) $N^{3/2}$
    (c) N^2
    (d) N.

53. The rating curve of a stream gauging station given the variation of
    (a) discharge in the stream with the variation of
    (b) discharge in the stream with the stage
    (c) discharge in the stream with water surface slope
    (d) discharge in the stream with the velocity of flow.

54. The crest gauge is used to record
    (a) the lowest stage in the river
    (b) the average stage in the river
    (c) the peak stage in the river
    (d) none of the above.

55. The stage in the river is defined as
    (a) the elevation of water surface with reference to an arbitrary datum
    (b) the average depth of flow in the stream
    (c) the radius of a semi-circle whose area is equal to the area of flow
    (d) none of the above.

56. A hydrograph is the graph drawn between
    (a) discharge in the river and the stage in the river
    (b) discharge and time
    (c) stage and time
    (d) none of the above.

57. One cumec-day is equal to
    (a) 8.64 hectare metres
    (b) 86.4 hectare metres
    (c) 864 hectare metres
    (d) 0.864 hectare metres.

58. One hectare-metre is equal to
    (a) 10000 m^3
    (b) 1000 m^2
    (c) 1000000 m^3
    (d) 100000 m^3.

59. Flow mass curve is the graph drawn between
    (a) the rate of flow and time
    (b) cumulative volume of flow and time
    (c) the cumulative discharge and time
    (d) cumulative volume of flow and the discharge.

60. Flow mass curve is used
    (a) to determine the storage capacity of the reservoir to meet a given uniform demand
    (b) to check the consistency of the flow record at a given site
    (c) to derive the unit hydrograph
    (d) to develop synthetic unit hydrograph.

61. Flow duration curve is the graph drawn between
    (a) the discharge and time
    (b) the discharge and the percentage of time such discharge is exceeded
    (c) cumulative rate of flow and time
    (d) cumulative volume of flow and time.

62. The concept of unit hydrograph was first introduced by
    (a) Dalton
    (b) Sherman
    (c) Darcy
    (d) Gumbel.
63. The word 'unit' in the unit hydrograph denotes
    (a) the unit depth of runoff
    (b) unit duration of the storm
    (c) unit base period of the hydrograph
    (d) arbitrary.
64. The unit hydrograph is the graphical relation between
    (a) total rainfall and the total runoff
    (b) total rainfall and the direct runoff
    (c) effective rainfall and the total runoff
    (d) effective rainfall and the direct runoff.
65. The peak ordinate of a 4 h unit hydrograph of a basin is 270 m^3/s. Then, the peak ordinate of a 8 h unit hydrograph of same will be basin
    (a) equal to 270 m^3/s
    (b) less than 270 m^3/s
    (c) more than 270 m^3/s
    (d) difficult to tell.
66. The base period of a 6 h unit hydrograph of a basin is 84 h. Then, the base period of a 12 h unit hydrograph of the same basin will be
    (a) 90 h
    (b) 84 h
    (c) 72 h
    (d) 168 h.
67. A storm with a uniform intensity occurring over a basin for a period of 6 h produced an effective rainfall of 15 cm and the peak discharge in the corresponding direct runoff hydrograph is 930 m^3/s. Over the same basin, if another storm of same duration but with an effective rainfall of 7.5 cm occurs what is the peak ordinate of the direct rainfall hydrograph produced by it
    (a) 930 m^3/s
    (b) 1860 m^3/s
    (c) 2790 m^3/s
    (d) 465 m^3/s.
68. The concentration time of the basin is
    (a) the time between the centre of the rainfall and the peak discharge
    (b) the base period of the hydrograph
    (c) the time taken by the water particle at the remotest point of the basin to reach the basin outlet
    (d) the duration of rainfall.
69. Direct runoff is the sum of
    (a) the surface runoff and the base flow
    (b) the base flow and the ground water runoff
    (c) the delayed subsurface runoff and deep percolation
    (d) the surfaces runoff and the prompt sub-surface runoff.

70. The base flow is
   (a) the difference between total runoff and direct runoff
   (b) the sum of surface runoff and delayed sub-surface runoff
   (c) the difference between prompt sub-surface runoff and delayed sub-surface runoff
   (d) none of the above.

71. The S-curve hydrograph is
   (a) the summation of the unity hydrograph
   (b) the summation of the total runoff hydrograph
   (c) the summation of the rainfall hyetograph
   (d) none of the above.

72. The S-curve hydrograph is used
   (a) to estimate the peak flood from a basin due to a given storm
   (b) to convert the unit hydrograph of given duration into a unit hydrograph of any other duration
   (c) to develop synthetic unit hydrograph
   (d) to estimate the infiltration losses.

73. The lag time of the basin is
   (a) the time between the centroid of rainfall diagram and the peak ordinate of the hydrograph
   (b) the time between the beginning and ending of direct runoff
   (c) the time between the beginning and ending of effective rainfall
   (d) the time taken for the remotest particle to reach the basin outlet.

74. Synthetic unit hydrograph of a basin is the unit hydrograph derived from
   (a) the available rainfall and runoff records of the basin
   (b) the rainfall and runoff records of the nearby basins
   (c) the arbitrary fixation of its shape
   (d) none of the above.

75. In the synthetic unit hydrograph proposed by Snyder the peak ordinate is given in terms of basin area A and basin lag t_p as
   (a) $Q_p = \dfrac{2C_p A}{t_p}$
   (b) $Q_p = \dfrac{C_p A}{t_p}$
   (c) $Q_p = \dfrac{C_p A^2}{t_p}$
   (d) $Q_p = \dfrac{C_p t_p}{A}$.

76. The basin lag of Snyders synthetic unit hydrograph is given by
   (a) $t_p = C_t (LL_c)^{0.3}$
   (b) $t_p = C_t \left(\dfrac{LL_c}{\sqrt{5}}\right)^{0.3}$
   (c) $t_p = C_t (LL_c)^{0.6}$
   (d) $t_p = C_t \left(\dfrac{L}{L_c}\right)^{0.3}$.

77. The base period of the Snyders synthetic unit hydrograph is given by

   (a) $T = 1 + 3t_p$
   (b) $T = 1 + \dfrac{t_p}{8}$
   (c) $T = 3 + 3 \cdot \dfrac{t_p}{24}$
   (d) $T = 1 + \dfrac{t_p}{24}$.

78. The 4 h unit hydrograph of a basin can be approximated as a triangle with base period of 48 h and a peak ordinate of 200 m^3/s. The area of the basin would be
   (a) 1728 km^2
   (b) 3456 km^2
   (c) 864 km^2
   (d) 5184 km^2.

79. If the duration of a unit hydrograph approaches zero, the resulting unit hydrograph is known as
   (a) S-curve hydrograph
   (b) synthetic unit hydrograph
   (c) constant unit hydrograph
   (d) instantaneous unit hydrograph.

80. According to the principle of linearity of unity hydrograph theory
   (a) the base periods of direct runoff hydrographs produced by storm of same duration will also be same
   (b) the ordinates of the direct runoff hydrographs of a common base period are directly proportional to the volume of run-off represented by each hydrograph
   (c) the base periods of direct runoff hydrograph is proportional to the depth of direct runoff
   (d) the base period of direct runoff hydrograph is proportional to the duration of effective rainfall.

81. Unit hydrograph method is generally used to estimate the peak flood when the catchment area does not exceed
   (a) 1000 km^2
   (b) 1500 km^2
   (c) 5000 km^2
   (d) 10000 km^2.

82. Infiltration capacity of the soil is defined as
   (a) the depth of water absorbed by the soil during the storm
   (b) the maximum rate at which the soil absorbs water
   (c) the average rainfall intensity during the storm
   (d) the rate at which the runoff is generated.

83. The infiltration capacity of the given soil
   (a) increases with increase in the initial soil moisture
   (b) decreases with increase in the initial soil moisture
   (c) is independent of the initial soil moisture
   (d) difficult to tell.

84. All other factors remaining same, the infiltration capacity in winter
   (a) is less than that in summer
   (b) is more than that in summer
   (c) is same as that in summer
   (d) is difficult to tell.

85. In the standard notation, the Horton's infiltration equation is given by
    (a) $f = f_c + (f_o - f_c) e^{-kt}$
    (b) $f = f_o + (f_c - f_o) e^{-kt}$
    (c) $f = f_c + (f_o + f_c) e^{-kt}$
    (d) $f = f_c + (f_o - f_c) e^{kt}$.

86. The observed runoff during a 6 h storm with a uniform intensity of 15 mm/h over a basin of area 300 km² is 21.6 million m³. The average infiltration rate during the storm is
    (a) 3 mm/h
    (b) 6 mm/h
    (c) 12 mm/h
    (d) 18 mm/h.

87. A 6 h storm with hourly intensities of 7, 18, 25, 12, 10 and 3 mm/h produced a runoff of 33 mm. Then the φ-index is
    (a) 7 mm/h
    (b) 3 mm/h
    (c) 10 mm/h
    (d) 8 mm/h.

88. φ-index is defined as
    (a) the difference between maximum and minimum infiltration capacities
    (b) difference between the total rainfall and total runoff divided by the duration of the storm
    (c) the rainfall intensity above which the rainfall volume equals the observed runoff volume
    (d) minimum infiltration rate during the storm.

89. W-index will be always
    (a) equal to φ-index
    (b) more than φ-index
    (c) less than φ-index
    (d) difficult to tell.

90. According to Dalton's law evaporation is directly proportional to
    (a) the vapour pressure gradient
    (b) the difference between saturation vapour pressure at 100°C and the actual vapour pressure
    (c) the difference between the actual vapour pressure and the saturation vapour pressure at 0°C
    (d) the difference between the saturation vapour pressure at given temperature and the saturation vapour pressure at 0°C.

91. The saturation deficit of the atmosphere is the difference between
    (a) e_s at given temperature and e
    (b) e_s at 100° and e_s at 0°
    (c) e_s at given temperature and e_s at 0°C
    (d) e and e_s at 0°C.
    where e_s is the saturation vapour pressure and e is the actual vapour pressure.

92. The pan co-efficient is defined as the
    (a) ratio of lake evaporation to pan evaporation
    (b) ratio of pan evaporation to the lake evaporation
    (c) product of pan evaporation and lake evaporation
    (d) difference between pan and lake evaporation.

93. The pan evaporation is
    (a) always less than lake evaporation
    (b) always more than lake evaporation
    (c) always equal to the lake evaporation
    (d) sometimes less and sometimes more than lake evaporation.

94. Which of the following has the largest pan co-efficient?
    (a) land pan
    (b) suken pan
    (c) floating pan
    (d) surface pan.

95. The chemical compound which is generally used to reduce the evaporation from water bodies is
    (a) D.D.T
    (b) alum
    (c) cetyl alcohol
    (d) potassium dichromate.

96. The salinity in water
    (a) reduces the evaporation
    (b) does not affect evaporation
    (c) increases the evaporation
    (d) difficult to say.

97. Lysimeter is the instrument used to measure
    (a) evaporation
    (b) transpiration
    (c) infiltration
    (d) evapotranspiration.

98. The abscissa in a psychrometric chart is
    (a) relative humidity
    (b) dry bulb temperature
    (c) wet bulb temperature
    (d) wet bulb depression.

99. The evaporation through plants and from the surrounding soil together is called
    (a) hydration
    (b) vapourisation
    (c) transpiration
    (d) evapotranspiration.

100. An aquifer is a geological formation which
    (a) does not contain water
    (b) contains water but does not transmit
    (c) contains water and also transmits water
    (d) is an out crop ozzing out water.

101. Which one of the following formations does not contain ground water?
    (a) aquifer
    (b) aquifuge
    (c) aquitard
    (d) aquiclude.

102. In the case of a flowing well, the piezometric surface is
    (a) below the ground level
    (b) above the ground level
    (c) between ground level and the water surface in the well
    (d) below the water surface in the well.

103. In the case of a water table well, the piezometric surface
    (a) is above the ground level
    (b) is below the water level in the well
    (c) coincides with water level in the well
    (d) is between the water level in the well and ground level.

104. Darcy's law for ground water motion states that the velocity
    (a) is proportional to hydraulic gradient
    (b) is proportional to the square of hydraulic gradient
    (c) is inversely proportional to hydraulic gradient
    (d) is proportional to the logarithm of hydraulic gradient.

105. An influent stream is one which
    (a) contributes runoff to the ground water
    (b) derives runoff from ground water
    (c) neither contributes nor derives runoff from ground water
    (d) flows only below the ground.

106. In the above question which is an effluent stream.

107. Specific yield of an aquifer is defined as the ratio of the
    (a) volume of pore space to the total volume of soil
    (b) volume of water freely drained from a saturated soil to the volume of soil
    (c) volume of water retained when a saturated soil is freely drained to the volume of soil
    (d) volume of prove space to volume of soil grains.

108. The steady discharge from a well in an unconfined aquifer is given by the equation

    (a) $Q = \dfrac{\pi K (h_1^2 - h_2^2)}{\ln\left(\dfrac{r_1}{r_2}\right)}$
    (b) $Q = \dfrac{\pi KB (h_1 - h_2)}{\ln\left(\dfrac{r_1}{r_2}\right)}$
    (c) $Q = \dfrac{\ln (r_1/r_2)}{\pi K(h_1^2 - h_2^2)}$
    (d) $Q = \dfrac{\ln (r_1/r_2)}{\pi B(h_1^2 - h_2^2)}$.

    where h_1 and h_2 are the water levels in the two observation wells at radial distances of r_1 and r_2, K is the permeability of the aquifer and B is the thickness of the aquifer.

109. Which of the equations in the above problem gives the discharge from a confined aquifer under steady state conditions?

110. The equation for steady state yield from a tube-well first developed by
    (a) Darcy
    (b) Dupuit
    (c) Jacob
    (d) Chow.

111. Another name for an unconfined aquifer is
    (a) leak aquifer
    (b) perched aquifer
    (c) artesian aquifer
    (d) water table aquifer.

112. Radius of influence is the horizontal distance between
    (a) the centre of the well and a point on the cone of depression of maximum drawdown
    (b) the centre of the well and a point on the cone of depression of zero drawdown
    (c) the centre of the well and the outer edge of the well
    (d) the centre of the well and the first observation well.

113. The transmissivity of the aquifer is defined as the product of
    (a) radius of well and radius of influence
    (b) thickness of aquifer and radius of influence
    (c) radius of the well and the permeability of the aquifer
    (d) permeability of the aquifer and the thickness of the aquifer.

114. Specific capacity of an open well is the ratio of
    (a) area of the well to the permeability of the aquifer
    (b) permeability of the aquifer to the area of the well
    (c) area of the well to steady state drawdown
    (d) permeability of the aquifer to steady state drawdown.

115. In the recuperation test of an open well if T is the time for recuperation of water level from h_1 to h_2, then the specific capacity of the open well is given by
    (a) $\dfrac{2.303}{T} \log_{10}\left(\dfrac{h_1}{h_2}\right)$
    (b) $\dfrac{2.303}{T} \log_{10}\left(\dfrac{h_2}{h_1}\right)$
    (c) $\dfrac{T}{2.303} \log_{10}\left(\dfrac{h_2}{h_1}\right)$
    (d) $\dfrac{\log_{10}(h_2/h_1)}{2.303\, T}$.

116. The difference between the maximum and minimum values of a sample is called its
    (a) standard deviation
    (b) median
    (c) range
    (d) mode.

117. The unbiased standard deviation of a sample data is given by
    (a) $s = \sqrt{\dfrac{\Sigma(x_i - \bar{x})^2}{n}}$
    (b) $s = \sqrt{\dfrac{\Sigma(x_i - \bar{x})^2}{(n-1)}}$
    (c) $s = \sqrt{\dfrac{\Sigma(x_i - \bar{x})}{n}}$
    (d) $s = \sqrt{\dfrac{\Sigma(x_i - \bar{x})^2}{n(n-1)}}$.

118. The co-efficient of variation is defined as the ratio of
    (a) standard deviation to skewness co-efficient
    (b) mean to standard deviation
    (c) standard deviation to mean
    (d) square of standard deviation to mean.

119. If m is the rank of a flood in n years record the Weibul's formula for computing the return period is
    (a) $T = \dfrac{n}{m}$
    (b) $T = \dfrac{n}{m+1}$
    (c) $T = \dfrac{n}{m-1}$
    (d) $T = \dfrac{n+1}{m}$.

120. The Hazen's formula for computing the return period is
    (a) $T = \dfrac{n}{m}$
    (b) $T = \dfrac{2n}{2m-1}$
    (c) $T = \dfrac{2n+1}{2m}$
    (d) $T = \dfrac{2n}{2m+1}$.

121. The California formula for return period is
    (a) $T = \dfrac{n}{m}$
    (b) $T = \dfrac{2n}{2m-1}$
    (c) $T = \dfrac{n+1}{m}$
    (d) $T = \dfrac{n}{m-1}$.

122. The most commonly used formula for computing return period is
    (a) California
    (b) Hazen's
    (c) Werbul's
    (d) Beard's.

123. The most commonly used probability distribution to fit the flood data is
    (a) Normal distribution
    (b) Gumbel's distribution
    (c) Log-normal distribution
    (d) Log-Pearson distribution.

124. A flood with a return period of 100 years is the flood which occurs
    (a) every 100th year
    (b) the maximum observed flood in the past 100 years
    (c) once in every 100 years on the average
    (d) only after 100 years in the immediate future.

125. A spillway is designed for T year flood and has an estimated useful life period of N years. Then the probability that it will not fail in the next N years is given by
    (a) $\left(1 - \dfrac{1}{T}\right)^N$
    (b) $1 - \left(1 - \dfrac{1}{T}\right)^N$
    (c) $1 - e^{-N/T}$
    (d) e^{-NT}.

126. The Ryve's formula for maximum flood from a catchment of area A is given by
    (a) $Q = CA^{2/3}$
    (b) $Q = CA^{3/2}$
    (c) $Q = CA^{1/3}$
    (d) $Q = CA^{3/5}$.

127. Which is the Dicken's formula for flood peak in the above problem.

128. In problem 125, what is the probability that the T year flood may occur in any one of the next N years.

129. The probability that a T year flood occurs in any year is
    (a) $\dfrac{1}{T}$
    (b) $\left(\dfrac{1}{T}\right)^2$
    (c) $\log\left(\dfrac{1}{T}\right)$
    (d) e^{-T}.

130. In the channel routing by the Muskingum method, the values of the routing constants c_0 and c_1 are –0.2 and 0.5 respectively. The value of the third routing constant c_2 is
    (a) 0.5
    (b) –0.2
    (c) –0.5
    (d) 0.7.

131. Which of the following is not one of the physiographic factors affecting the runoff from a basin?
    (a) size of the basin
    (b) slope of the main stream
    (c) antecedent precipitation
    (d) drainage net work.

132. The relationship between the reduced variate y and the return period T is given by
    (a) $1 - \dfrac{1}{T} = e^{-e^{-y}}$
    (b) $\dfrac{1}{T} = e^{-e^{-y}}$
    (c) $1 + \dfrac{1}{T} = e^{-e^{-y}}$
    (d) $\dfrac{1}{T} = 1 + e^{-e^{-y}}$.

133. The most commonly used method for reservoir routing is
    (a) Muskingum method
    (b) Snyders method
    (c) Chow's method
    (d) Pul's method.

134. The place which records highest annual rainfall in India
    (a) Trivandrum
    (b) Bombay
    (c) Chirrapunji
    (d) Goa.

135. Enveloping curve is a method
    (a) to determine the peak flood
    (b) to determine the infiltration capacity
    (c) to estimate the interception
    (d) to route the flood through a reservoir.

136. The average atmospheric pressure expressed in millibars is
    (a) 10.13
    (b) 101.3
    (c) 1013
    (d) 10130.

137. The average atmospheric pressure expressed in dynes/cm^2 is
    (a) 100
    (b) 1000
    (c) 10^5
    (d) 10^6.

138. Relative humidity may be defined as the ratio of
    (a) the actual vapour pressure to the saturation vapour pressure
    (b) the saturation vapour pressure to the actual vapour pressure
    (c) the vapour pressure deficit to the saturation vapour pressure
    (d) the saturation vapour pressure to the vapour pressure deficit.

139. Isohyets are
    (a) lines joining points of equal rainfall intensity
    (b) lines joining points of equal storm duration
    (c) lines joining points of equal relative humidity
    (d) lines joining points of equal rainfall depth.

140. The moisture useful for plant growth is
    (a) capillary water
    (b) gravity water
    (c) hygroscopic water
    (d) all the above.

141. Darcy's law gives the velocity of flow in
    (a) openchannels
    (b) pipes
    (c) porous medium
    (d) pumps.

142. Which of the following cannot be used as a humidity measuring device?
    (a) Hydrometer
    (b) Dry and wet bulb thermometers
    (c) Phychrometer
    (d) Hygrometer.

143. The number of rain gauges required in a given area to estimate the average depth of rainfall with a given accuracy
    (a) is more if the rainfall is non-uniformly distributed
    (b) is less if the rainfall is non-uniformly distributed
    (c) is more if the rainfall is uniformly distributed
    (d) is independent of the variability of rainfall.

144. The site selected for measurement of snowfall should be
    (a) horizontal
    (b) open to snowfall and in isolation
    (c) sheltered against winds and drifting snow
    (d) all the above.

145. The ordinate of the instantaneous unit hydrograph is proportional to
    (a) the ordinate of S-curve hydrograph
    (b) the slope of S-curve hydrograph
    (c) inverse of the ordinate of S-curve hydrograph
    (d) inverse of the slope of S-curve hydrograph.

146. S-curve hydrograph derived its name because
    (a) it has the deformed S-shape
    (b) it is proposed by Snyder
    (c) it gives the storage of runoff
    (d) none of the above.

147. S-curve hydrograph can be used only
    (a) to obtain unit hydrograph of longer duration from unit hydrograph of shorter duration
    (b) to obtain unit hydrograph of shorter duration from unit hydrograph of longer duration
    (c) to obtain unit hydrograph of any duration from unit hydrograph of any given duration
    (d) to obtain a synthetic unit hydrograph.

148. A S-curve hydrograph is derived from a D-hour unit hydrograph of a basin with a drainage area of A sq km. The equilibrium discharge ordinate in the S-curve is given by
    (a) $\dfrac{27.8A}{D}$ m³/s
    (b) $\dfrac{2.78A}{D}$ m³/s
    (c) $\dfrac{278A}{D}$ m³/s
    (d) $\dfrac{0.278A}{D}$ m³/s.

149. The ordinate of a direct runoff hydrograph are measured at Δt-hour intervals and summed up as ΣQ. The area of the catchment is A sq km. Then the depth of runoff d in cm is given by

   (a) $\dfrac{0.36 \, \Delta t \, \Sigma Q}{A}$ ☐
   (b) $\dfrac{0.036 \, \Delta t \, \Sigma Q}{A}$ ☐
   (c) $\dfrac{0.36 \, A \, \Sigma Q}{\Delta t}$ ☐
   (d) $\dfrac{0.036 \, A \, \Sigma Q}{\Delta t}$. ☐

150. The trap efficiency of a reservoir is a function of
   (a) age of the reservoir ☐
   (b) the reservoir capacity ☐
   (c) total inflow ☐
   (d) (reservoir capacity/total inflow). ☐

151. The standard time at which the daily rainfall is recorded in India is
   (a) 7.30 A.M. ☐
   (b) 8.30 A.M. ☐
   (c) 9.30 A.M. ☐
   (d) 5.30 P.M. ☐

152. A spillway is designed for a 5 year flood. What is the probability that the design flood occurs at least once in the next 5 years?
   (a) 0.2 ☐
   (b) 0.8 ☐
   (c) 0.672 ☐
   (d) 0.5. ☐

153. The two parameters sufficient to describe the symmetrical normal distribution are
   (a) mean and standard deviation ☐
   (b) mean and range ☐
   (c) range and variance ☐
   (d) mean and kurtosis co-efficient. ☐

154. If the permeability of a porous medium is not same in all the directions then it is known as
   (a) thixotropic ☐
   (b) isoentropic ☐
   (c) isotropic ☐
   (d) anisotropic. ☐

155. Which of the following is preferred for measuring the velocity of flow in a natural stream?
   (a) Pitot tube ☐
   (b) Hot-wire anemometer ☐
   (c) Current meter ☐
   (d) Rod float. ☐

156. The theory of infiltration was enunciated by
   (a) Sherman ☐
   (b) Dalton ☐
   (c) Darcy ☐
   (d) Horton. ☐

157. Rainfall simulators are used for the determination of
   (a) rainfall ☐
   (b) interception ☐
   (c) evaporation ☐
   (d) infiltration. ☐

158. To which category does the Symon's rain gauge belong?
   (a) tipping-bucket ☐
   (b) ordinary rain gauge ☐
   (c) syphon ☐
   (d) weighing. ☐

159. The type of recording rain gauge used in India is
   (a) weighing type ☐
   (b) float type ☐
   (c) tipping-bucket type ☐
   (d) none of the above. ☐

160. The Thiessen weights of the four influencing rain gauge stations of a catchment area are 0.1, 0.2, 0.3 and 0.4. If the rainfalls recorded at these gauges are 40, 30, 20 and 10 mm respectively what is the average depth of rainfall over the basin?
    (a) 20 mm
    (b) 30 mm
    (c) 40 mm
    (d) 10 mm.

161. The isohyets drawn for a storm yielded the following data:

Isohyet in mm	45–55	55–65	65–75	75–85
Area between isohyets in km^2	100	200	300	400

The average depth of rainfall is equal to
    (a) 50 mm
    (b) 60 mm
    (c) 70 mm
    (d) 80 mm.

162. It is predicted that a storm yielding a maximum daily rainfall of 20 cm has a return period of 5 years. Then the place will receive a maximum daily rainfall of 20 cm
    (a) one in every five years
    (b) on the average once in 5 years
    (c) four times in every five years
    (d) none of the above.

163. The depth of flow in a stream at a vertical is 10 m. The velocities measured at 2 m and 8 m depth are 0.6 m/s and 0.4 m/s respectively. What is the average velocity for the vertical?
    (a) 1 m/s
    (b) 0.25 m/s
    (c) 0.5 m/s
    (d) 0.75 m/s.

164. The water balance equation for a catchment area in terms of rainfall (P), runoff (R), evaporation (E) and storage (S) is written as
    (a) $R = P - E \pm \Delta S$
    (b) $R = P + E \pm \Delta S$
    (c) $R = E - P \pm \Delta S$
    (d) $P = E - R \pm \Delta S$.

165. In the Muskingum channel routing equation, the sum of the three routing constants C_0, C_1 and C_2 should be equal to
    (a) 1/3
    (b) 2/3
    (c) 1
    (d) 3.

166. The units of the constant K in the Muskingum storage equation are
    (a) m/s
    (b) m^3/s
    (c) hr
    (d) m^3.

167. The units of the constant x in the Muskingum storage equation are
    (a) m
    (b) m/s
    (c) m^3/s
    (d) no units.

168. For channel routing, the Muskingum storage equation is given by
    (a) $K[xI + (1 - x)Q]$
    (b) $K[xQ + (1 - x)I]$
    (c) $K[xQ + (1 - x)Q]$
    (d) $K[xI + (1 + x)Q]$.

169. The ratio of the total number of streams draining the basin to the basin area is known as
    (a) drainage density
    (b) stream density
    (c) drainage efficiency
    (d) all the above.

170. The ratio of the total lengths of the streams draining the basin to the basin area is known as
    (a) drainage density
    (b) stream density
    (c) drainage efficiency
    (d) stream efficiency.

171. If A and L denote the area of the basin and the length of the main stream, the form factor of the basin is given by
    (a) $\dfrac{A}{L}$
    (b) $\dfrac{A}{L^2}$
    (c) $\dfrac{L}{A}$
    (d) $\dfrac{L}{A^2}$.

172. A storm produced an intensity of 30 mm/hr for a period of 6 hours. If the area of the basin is 800 km² and average infiltration rate is 5 mm/hour, what is the volume of runoff produced by the storm in hectare-metres?
    (a) 12000
    (b) 120
    (c) 12
    (d) 1.2.

173. I.M.D. stand for
    (a) Indian Mining Department
    (b) Indian Mineral Deposits
    (c) India Meteorological Department
    (d) International Monetary Debt.

174. The strange's table gives the relationship between
    (a) temperature and evaporation
    (b) rainfall and infiltration
    (c) rainfall and runoff
    (d) runoff and area of the basin.

175. The responsibility of gauging the major rivers in the country lies with
    (a) central ground water board
    (b) central water commission
    (c) central board of irrigation and power
    (d) central water agency.

176. Removal of soil particles from the present location is known as
    (a) sedimentation
    (b) saltation
    (c) erosion
    (d) excavation.

177. If d and v are the diameter and velocity of rain drop, then the erosive power of the rain drop is proportional to
    (a) d^2v^3
    (b) d^3v^2
    (c) d^2v^2
    (d) d^3v^3.

178. Removal of soil by small concentration of flowing water is known as
    (a) rill erosion
    (b) gully erosion
    (c) sheet erosion
    (d) channel erosion.

179. Removal of soil from rivulets which are formed due to sufficient accumulation of overland flow on sloped grounds is known as
    (a) rill erosion   (b) gully erosion
    (c) sheet erosion   (d) channel erosion.

180. Wearing away of a thin layer of soil on the land surface, especially between rills mainly by overland flow is known as
    (a) rill erosion   (b) gully erosion
    (c) sheet erosion   (d) channel erosion.

181. Erosion occurring in the stream channels in the form of stream bank erosion and stream bed degradation is known as
    (a) rill erosion   (b) gully erosion
    (c) sheet erosion   (d) channel erosion.

182. Soil particles detached from the present location are called
    (a) gravel   (b) sediments
    (c) sand   (d) none of the above.

183. Sediment carried by flowing water in suspension is known as
    (a) suspended load   (b) bed load
    (c) saltation   (d) wash load.

184. Sediment carried by water by rolling and sliding along the stream bottom is known as
    (a) suspended load   (b) bed load
    (c) saltation   (d) wash load.

185. Sediment carried by water by bouncing along the stream bed is known as
    (a) suspended load   (b) bed load
    (c) saltation   (d) wash load.

186. Minute particles of colloidal sizes which always remain in suspension in water and carried by water are called
    (a) suspended load   (b) bed load
    (c) saltation   (d) wash load.

187. If γ is the specific weight of water, d is the depth of flow and S_0 is the bed slope of the channel, then the tractive stress τ_0 developed in water at the bottom of the channel is given by
    (a) $\tau_0 = \gamma d \sqrt{S_0}$ (b) $\tau_0 = \gamma \sqrt{dS_0}$
    (c) $\tau_0 = \gamma dS_0$ (d) $\tau_0 = \gamma/dS_0$.

188. If τ_c is the critical tractive stress, the bed load q_s per metre width of the stream is given by (C_s is an empirical constant)
    (a) $q_s = C_s \dfrac{\tau_0}{\gamma} \sqrt{(\tau_0 - \tau_c)}$ (b) $q_s = C_s \dfrac{\tau_0}{\gamma}(\tau_0 - \tau_c)$
    (c) $q_s = C_s \dfrac{\tau_0}{\gamma}(\tau_0 - \tau_c)^{1.5}$ (d) $q_s = C_s \dfrac{\tau_0}{\gamma}(\tau_0 - \tau_c)^2$.

189. The ratio of the sediment deposited in the reservoir to the incoming sediment is called
    (a) Sediment efficiency of the reservoir   (b) Trap efficiency of the reservoir
    (c) Erosion efficiency of the catchment   (d) None of the above.

190. The sediment laden water with higher density than the surface water in the reservoir flowing underneath the surface water is called
    (a) Eddy current   (b) Sediment current
    (c) Density current   (d) Flush current.

191. The trap efficiency of a reservoir is a function of
    (a) ratio of reservoir storage capacity to mean annual runoff volume
    (b) ratio of reservoir storage capacity to the square root of the mean annual runoff volume
    (c) ratio of reservoir storage capacity to the square of the mean annual runoff volume
    (d) ratio of reservoir storage capacity to the logarithm of the mean annual runoff volume.

192. The trap efficiency of a reservoir, after commissioning will
    (a) increase with time
    (b) decrease with time
    (c) increase initially for some time and decrease later
    (d) decrease initially for some time and increase later.

193. If the initial specific weight of sediments depositing in a reservoir is γ_1, the specific weight after t years is γ_t, and K is the consolidation coefficient, then
    (a) $\gamma_t = \gamma_1 + K \log t$ (b) $\gamma_t = \gamma_1 + K \cdot t$
    (c) $\gamma_t = \gamma_1 + K/t$ (d) $\gamma_t = \gamma_1 + K \sqrt{t}$.

194. The annual flood peak of a stream is estimated to have 50 years and 200 years flood of 2400 m³/s and 3060 m³/s. The 100 years flood of this stream will be close to
    (a) 2730 m³/s   (b) 2000 m³/s
    (c) 3500 m³/s   (d) 4000 m³/s.

195. Evaporation may also be viewed as
    (a) convection   (b) indirect radiation
    (c) indirect cooling   (d) indirect sublimation.

196. The idea of the synthetic unit hydrograph was first introduced by
    (a) Sherman   (b) Darcy
    (c) Dalton   (d) Snyder.

197. Distribution graph shows the variation of
    (a) discharge against time
    (b) the percentage of runoff volume over a time interval against time
    (c) cumulative infiltration depth with time
    (d) cumulative rainfall depth with time.

198. When the temperature is equal to dew point temperature, the existing vapour pressure of the air e and the saturation vapour pressure e_s are related as

(a) $e < e_s$ ☐   (b) $e = e_s$ ☐
(c) $e > e_s$ ☐   (d) $e = 0.5\, e_s$. ☐

199. In reservoir routing discharge is expressed in m³/s and the time is measured in hours. Then the storage in the reservoir, for the purpose of constructing storage-discharge curve shall be expressed in
   (a) m³ ☐   (b) cumec-days ☐
   (c) cumec-hours ☐   (d) million m³. ☐

200. If the annual withdrawl of water from a groundwater basin exceeds the average annual recharge, then
   (a) overdraft is said to occur ☐   (b) inferior recharge is said to occur ☐
   (c) basin is said to give higher yield ☐   (d) none of the above. ☐

ANSWERS
Objective Type Questions

1. (c)	2. (b)	3. (a)	4. (d)	5. (a)	6. (b)
7. (c)	8. (d)	9. (c)	10. (d)	11. (a)	12. (d)
13. (d)	14. (d)	15. (a)	16. (c)	17. (b)	18. (c)
19. (a)	20. (d)	21. (a)	22. (c)	23. (b)	24. (d)
25. (b)	26. (b)	27. (d)	28. (c)	29. (a)	30. (a)
31. (b)	32. (b)	33. (b)	34. (b)	35. (c)	36. (a)
37. (b)	38. (b)	39. (a)	40. (d)	41. (c)	42. (b)
43. (c)	44. (a)	45. (d)	46. (b)	47. (c)	48. (c)
49. (c)	50. (b)	51. (c)	52. (d)	53. (b)	54. (c)
55. (a)	56. (b)	57. (a)	58. (a)	59. (b)	60. (a)
61. (b)	62. (b)	63. (b)	64. (d)	65. (b)	66. (a)
67. (d)	68. (c)	69. (d)	70. (a)	71. (a)	72. (b)
73. (a)	74. (b)	75. (b)	76. (a)	77. (c)	78. (a)
79. (d)	80. (b)	81. (c)	82. (b)	83. (b)	84. (a)
85. (a)	86. (a)	87. (d)	88. (c)	89. (c)	90. (a)
91. (a)	92. (a)	93. (b)	94. (b)	95. (c)	96. (a)
97. (d)	98. (b)	99. (d)	100. (c)	101. (b)	102. (b)
103. (c)	104. (a)	105. (a)	106. (b)	107. (b)	108. (a)
109. (b)	110. (b)	111. (d)	112. (b)	113. (d)	114. (b)
115. (a)	116. (c)	117. (b)	118. (c)	119. (d)	120. (b)
121. (a)	122. (c)	123. (b)	124. (c)	125. (a)	126. (a)
127. (b)	128. (b)	129. (a)	130. (d)	131. (c)	132. (a)
133. (d)	134. (c)	135. (a)	136. (c)	137. (d)	138. (a)
139. (d)	140. (a)	141. (c)	142. (a)	143. (a)	144. (d)
145. (b)	146. (a)	147. (c)	148. (b)	149. (a)	150. (d)
151. (b)	152. (c)	153. (a)	154. (d)	155. (c)	156. (d)

157. (d)	158. (b)	159. (b)	160. (a)	161. (c)	162. (b)
163. (c)	164. (a)	165. (c)	166. (c)	167. (d)	168. (a)
169. (b)	170. (a)	171. (b)	172. (a)	173. (c)	174. (c)
175. (b)	176. (c)	177. (b)	178. (a)	179. (b)	180. (c)
181. (d)	182. (b)	183. (a)	184. (b)	185. (c)	186. (d)
187. (c)	188. (b)	189. (b)	190. (a)	191. (a)	192. (b)
193. (a)	194. (a)	195. (c)	196. (d)	197. (b)	198. (b)
199. (c)	200. (a)				

Chapter 5

WATER RESOURCES ENGINEERING

(Including Irrigation, Open Channel Flow and Water Power)

I. INTRODUCTION

The process of artificially supplying water to soil for raising the crops is called Irrigation. As the rainfall is highly non-uniformly distributed in space and is confined to about 4 months in monsoon at many places and as it is basically an agricultural country, Irrigation is essential in India.

There are two types of irrigation: Flow irrigation where the water is supplied to the fields by gravity and lift irrigation, where water is lifted up to ground level and then it is made to flow by gravity.

The methods of applying water to the crops are basically of three types. They are surface irrigation, sub-surface irrigation and sprinkler irrigation.

In the surface irrigation water is applied on the ground surface which infiltrates and then it is absorbed by plants.

In the sub-surface irrigation water is supplied directly to the root zone of the crops.

In the sprinkler irrigation water is applied in the form of spray (resembling rain).

In the surface irrigation methods we have flooding, furrows and contour farming. In the controlled flooding the various methods are free flooding, border strips, basin flooding, zig-zag method etc.

Irrigation water should have acceptable quality. Presence of excessive salts make water saline and unsuitable for irrigation.

If total salt content in water exceeds 2000 ppm, it is not suitable for irrigation.

Boron is the most toxic element. If its concentration exceeds 2 ppm it is harmful to the crops.

The ratio of weight of water present in the soil to the weight of dry soil is called the moisture content.

When all the pores in the soil are occupied by water, the soil is said to be saturated. The moisture content at saturation condition is called the saturation capacity.

When a saturated sample is drained some water will move out under gravity and the remaining water is held in the soil against gravity by the capillary forces.

The moisture content of the soil after it is drained under gravity is called the field capacity. The difference between saturation capacity and field capacity indicates the superfluous water or the gravity water. The gravity water is not available for plant growth.

The plant roots extract moisture from soil and the moisture in the soil will be continuously depleted. The moisture content at which plants can no longer extract sufficient water from the soil for their growth is called permanent wilting point or wilting co-efficient. The difference between the field capacity and the permanent wilting point is the available water. For healthy growth of the crop the moisture in the soil should be maintained at or slightly below the field capacity.

When a soil sample is ovendried its moisture content will be zero. If it is now kept in atmosphere, it will absorb some moisture from the atmosphere which is held in soil due to chemical forces. This water is known as hygroscopic water and it can not be removed from the soil unless by heating.

The consumptive use or the evapotranspiration is the amount of water required (usually expressed as the depth over a given area) to meet the transpiration needs for the healthy growth of plants and also the evaporation from the surrounding soil.

Not all the rain which falls on the ground can be utilised by the plants because most of it runs off. The portion of rainfall which is effectively used by the plant to meet its consumptive need is called the effective rainfall.

Net irrigation requirement = Consumptive use – effective rainfall

$$NIR = CU - ER$$

Field irrigation requirement = Net irrigation requirement + Field application losses such as runoff percolation and evaporation

$$FIR = NIR + \text{field application losses}$$

Gross irrigation requirement = Field irrigation requirement + conveyance losses

$$GIR = FIR + \text{Conveyance losses}$$

$$\text{Water application efficiency} = \frac{NIR}{FIR} \times 100$$

$$\text{Water conveyance efficiency} = \frac{FIR}{GIR} \times 100.$$

It is obvious that the sprinkler irrigation has the maximum water application efficiency and also maximum water conveyance efficiency.

Sprinkler irrigation, though involves large initial expenditure, is suitable when the land is highly undulating. In such cases, contour farming is also adopted when the irrigation is done by flooding.

Duty may be defined as the number of hectares of a particular crop which can be irrigated by continuous supply of 1 cumec of water throughout the base period of the crop. Duty is denoted by D while the base period in days is denoted by B.

If the volume of water supplied to the crop throughout its base period is spread over the area of the crop the depth it would occupy is called delta denoted by Δ, expressed in Ha-m/Ha or simply m.

The relationship between duty and delta is given by $\Delta = \dfrac{8.64\,B}{D}$.

The value of duty depends on the place where it is measured. The duty measured on the field will be high, while the duty measured at the head of the canal is less owing to conveyance losses.

The total area which can be irrigated by a canal system is called the gross commanded area (G.C.A.). This G.C.A. less the unfertile barren land and the other areas of habitation represents the culturable commanded area (C.C.A.). The area on which the crop is actually grown is called culturable cultivated area.

C.C.A. = G.C.A. − the unfertile and inhabited area

$$\text{Intensity of irrigation} = \dfrac{\text{Culturable cultivated area}}{\text{G.C.A.}}$$

Crops which are sown at the beginning of south-west monsoon are called Khariff crops. Crops which are sown in autumn are called Rabi crops.

Crop ratio is defined as the ratio of the area irrigated in Rabi season to the area irrigated in Khariff season.

The crops which increase the nitrogen content of the soils and hence the fertility are called leguminious crops.

Rice, maize, jowar, pulses and groundnut come under Khariff crops.

Gram, wheat, tobacco and potato come under Rabi crops. Cotton is a long duration crop with a base period of about 8 months.

Sugar cane is a perennial crop whose base period is spread over almost the entire year.

Reservoirs are formed by damming the rivers to store water for using the same in dry periods. Certain storage in the reservoir is earmarked to accommodate the silt that is likely to be trapped after the reservoir is formed. Such storage is called the dead storage. All the sluices which draw water from the reservoir for various uses should have their sill levels above the dead storage level.

Suitable sites for locating reservoir are selected after thorough topographic, geological and Hydrologic investigations.

The storage available at a potential site between any two successive contours with areas A_1 and A_2 and with an elevation difference h can be found out from one of the following equations:

$$V = \dfrac{h}{2}(A_1 + A_2) \qquad \text{(Trapezoidal formula)}$$

$$V = \dfrac{h}{3}(A_1 + A_2 + \sqrt{A_1 A_2}) \qquad \text{(Cone formula)}$$

The trap efficiency of the reservoir is a function of the ratio of the capacity of the reservoir to the inflow rate. As this ratio increases the trap efficiency also increases.

The structure built across a stream to form a reservoir is called a dam.

A dam is said to be overflow dam when water is allowed to flow over it otherwise it is called non-overflow dam.

Depending on the material used, the dams are known as either rigid dams or non-rigid dams.

Gravity dams, arch dams, buttress dams, steel dams and timber dam come under the category of rigid dams.

Earth dams and rockfill dams come under the category of non-rigid dams.

A gravity dam resists all the external forces by its own self weight. It is more permanent than others and requires least maintenance but requires good foundations.

An archdam resists the water thrust by arch action and transmits the reaction force to the two abuttments. It is preferred in narrow gorges where the hillocks of the valley are very strong. Idukki dam in Kerala is an example of the Arch dam in India.

Earth dams are cheap to construct and require no skilled labour. They generally utilise the material available at the site. They can be built on any type of foundation with proper design.

The water pressure on a gravity dam is given by

$$P = \frac{WH^2}{2},$$ where H is the height of water stored and it acts horizontally at a height of $\frac{H}{3}$ above the base.

The uplift pressure on the base will have an intensity of WH at u/s face and WH' at the d/s face, where H' is the tail water. If a drainage gallery is provided the intensity at the line of the gallery will be taken as

$$w[H' + \tfrac{1}{3}(H - H')].$$

The height of the wave h_w is computed using Molitor's formula. Then the wave pressure is given by

$$P_w = 2w\, h_w$$ and acts at a height of $\tfrac{3}{8} h_w$ above the still water surface.

The other forces include self-weight ice pressure, silt pressure, wind pressure and earthquake pressure.

The factor of safety against overturning is defined as the ratio of the sum of all the restoring moments to, the sum of all the overturning moments, both taken about the toe. It should always be more than 1.5.

The shear friction factor ensures stability against sliding. It is given by

$$\text{S.F.F.} = \frac{\mu \Sigma (V - U) + bq}{\Sigma H}$$

where q is the shear strength of the joint, V is the total vertical downward forces, U is the total up lift force, H is the total horizontal force and μ is the co-efficient of friction. Its value should be between 1.5 and 4.0 depending on the loading condition.

To prevent cracking, the design should ensure that no tension develops. This is guaranteed if the resultant falls within the middle third of the base or eccentricity is less than one-sixth of the base. That is, $e \leq \dfrac{b}{6}$.

The maximum normal stress on the base is given by

$$p_n = \dfrac{V}{b}\left(1 + \dfrac{6e}{b}\right).$$ This occurs at the toe.

The maximum principal stress is given as

$$\sigma = p_n \sec^2 \phi,$$ where ϕ is the angle made by the d/s face with the vertical.

The elementary profile of gravity dam is a right-angled triangle. It is modified to suit to the particular conditions.

If the height of the dam is large, the stresses developed on the base will also increase and therefore the base width may be increased (by making the u/s face inclined instead of vertical) to keep the stress within permissible limits.

A low gravity dam is the dam in which the maximum compressive stress is less than the allowable stress when the u/s face is vertical. The limiting height of low gravity dam is given by

$$H = \dfrac{f}{w(S+1)}$$

where f is the allowable stress, w is the unit weight of water and S is the specific gravity of the dam material.

Drainage galleries in gravity dams are provided to intercept the seepage through the body of the dam and drain off, to facilitate the drilling and grouting of foundations, to give access to the instrumentation installed and to relieve the uplift pressures.

Construction joints in gravity dams are provided to facilitate the construction work to be carried out in stages.

Construction joints are provided to prevent cracks due to shrinkage produced by temperature changes. The construction joints are properly keyed to permit the transfer of shearing stresses and are properly sealed by water stops to prevent leakages.

Arch dams are two types: constant radius arch dam in which the radius of the arch is constant and constant angle arch dams (a special variety of variable arch dam) in which the central angle of the arch is constant.

For constant angle arch dam the best central angle is 133° 34'. At this angle the volume of concrete is minimum.

The dams which combine the features of arch dams and buttress dams are called multiple arch buttress dams. The Meer Alam dam near Hyderabad, A.P. is of this type.

Based on method of construction, the earth dams are known as rolled fill dams or hydraulic fill dams.

Based on materials used, the rolled fill dams are further divided into homogeneous embankment type, zoned embankment type and Diaphragm type.

In homogeneous type, the top flow line intersects the d/s face which is not at all desirable. To overcome this difficulty a horizontal filter is provided on the d/s side.

In the zoned embankment type the central portion is made of highly impervious material like clay which is called the core and to give stability to the core previous material is dumped on either side which are called shells.

In the diaphragm type dam the central impervious portion is a very thin wall.

The split-up of earth dam failures is as follows:

Hydraulic failures = 40%
Seepage failures = 30%
Structural failures = 30%

Toe drain, horizontal drain and chimney drain are the devices which are used to relieve the pore pressure in the body of the dam and increase the stability.

Impervious cutoff, cutoff trench, upstream impervious blanket and relief wall are the devices which are used to control the seepage through the foundation of the earth dams.

The top flow line, also known as the phreatic line in the earth dam section has the shape of a parabola. If its focal distance is s, the discharge per unit length of the dam is then given as $q = Ks$, where K is the permeability of the dam material. Alternatively if a flow net is drawn.

$$q = KH \frac{N_f}{N_d}$$

where N_f is the number of flow channels, N_d is the number of equipotential drops and H is the height of water stored. If the dam material is anisotropic with K_x and K_y as the permeabilities in the x and y directions. The equivalent permeability for the transformed isotropic section is given as

$$K' = \sqrt{K_x K_y}.$$

Flownets may also be constructed using the electrical analogy method. The upstream face is an equipotential line.

The failure of the slopes in an earth dam occurs along a circle known as the slip circle. The stability analysis of slopes is done by Sweedish slip circle method or by Bishop's method, the former being the most generally accepted.

Since the base of the earth dam is large because of flat side slopes normally the stability analysis of foundation against shear is not required. But when the foundation consists of fine, loose cohesionless material or unconsolidated clays and silts it may be necessary to carry out the stability analysis of the foundations.

Any storage reservoir should be provided with arrangements to discharge the surplus water to the d/s side safely. The component in the storage headworks which serves this purpose is called the spillway.

Based on the utility the spillway may be called a main spillway or an emergency spillway.

Emergency spillway is an additional safety measure which comes into operation only when unprecedented floods arrive. Its sill is normally kept at or above the Full Reservoir level.

The various types of spillways are ogee spillway, chute spillway, side channel spillway, tunnel or conduit spillway, morning glory spillway, or shaft spillway, and siphon spillway.

Ogee spillway is most commonly used. Its shape conforms to the nappe over a rectangular weir. Its discharge equation is also same as that of rectangular weir. The pressures on the ogee spillway are atmospheric only when the head of the flow is equal to the design head. If the head is more than design head negative pressures will be developed on the ogee spillway and the discharge increases. In the ogee spillway the discharge is proportional to $H^{3/2}$.

In the siphon spillway the discharge takes place under the siphonic action. The discharge is proportional to $H^{1/2}$. But the head in this case is the difference in head race and tail race water levels.

The water discharged over the spillway will have a lot of kinetic energy. Unless this energy is dissipated and its velocity reduced it may erode away the river bed at the foot of spillway and cause danger to the structure.

The commonly used energy dissipators are the hydraulic jump type, the rollers bucket, and the ski-jump bucket.

For a given discharge the normal depth of flow in the river d/s of the spillway may not be same as the depth of flow required for the formations of the jump. If it is less than required the jump will be pushed d/s and if it is more, the jump will be pushed u/s.

The stilling basin where the hydraulic jump occurs is therefore, designed to suit to the tail water conditions at the site.

For example, the floor may be depressed or the sloping glacis may be provided etc.

In order to contain the jump within the stilling basin and to reduce the length of the jump some appurtenance like end sill, chute block and baffle piers may be used.

When the river bed consists of hard rock and if the tail water depth is less than the depth required for the formation of the jump, the ideal choice would be the ski-jump energy dissipator.

Gates are provided over the spillways to increase the useful storage without much of additional cost. Judicious operation of gates can increase the usefulness of the reservoir manifold.

Various types of gates are Radial gates, needle, flash boards, stop logs, vertical lift gates, bear trap gates, rolling gates and drum gates.

Radial gate is also known as a tainter gate.

Stoney gate is a type of vertical lift gate.

Weir is an obstruction of small height built across a river to raise water level and divert water into the canal.

Barrage also has the same purpose but its crest will be almost at the river bed level and the raising of water level is done by the gates.

The other components of diversion headworks are divide wall, scouring sluices fish ladder, head regulator, and silt excluder.

The divide wall prevents the cross currents and the flow parallel to the weir and thereby eliminates the formation of vortices and deep scour. It also provides a still packet of water in front of the head regulator.

The purpose of the fish ladder is to allow the migration of fish from u/s to d/s and *vice versa*.

The scour sluices (also known as under sluices) maintain deep channel in front of the head regulator and dispose of heavy silt from time to time. They also carry a part of the flood.

Head regulator allows water into canal under controlled and regulated conditions.

Silt excluder is meant to prevent the entry of silt into canal.

The weirs are to be generally founded on permeable river beds. Therefore the foundation floor thickness of weirs should be sufficient to resist the uplift pressures.

If H is the percolation head and L is the length of seepage path (also known as the percolation length) the hydraulic gradient is H/L.

According to Bligh's theory the safe creep length is given by $L = CH$, where C is called the co-efficient of creep.

In the Bligh's creep theory the vertical and horizontal creeps are given the same weightage and the hydraulic gradient is uniform everywhere throughout the creep length. Therefore in this theory the d/s cutoff has no special significance except to increase the creep length.

In the Lane's weighted creep theory the horizontal creep is given a weightage of 1 while the vertical creep is given a weightage of 1.5.

Both Bligh's theory and Lane's weighted creep theory fail to recognize the importance of exit gradient.

According to Khosla's potential flow theory the hydraulic gradient of the seepage flow is not same throughout. In the case of a simple floor with negligible thickness the seepage pressure head varies as a sine curve and the hydraulic gradient at the exit is infinity. Also the outer faces of end piles are more effective than the inner faces in dissipating the uplift pressure.

The d/s sheet pile is essential to contain the exit gradient and prevent undermining.

There will be mutual interference of the piles on the uplift pressures. The effect of d/s pile is to increase the uplift pressure at the u/s pile.

At the end of solid apron on the d/s of the foundation an inverted filter and a launching apron (also known as talus, which is made of rough stones in two or three layers) are provided.

The inverted filter relieves the uplift pressure while the launching apron protects the d/s pile from scour holes. Talus may also be provided at the u/s end of the solid apron.

Whatever, care we may take, some silt may still find its way into the canal at head regulator.

Silt extractor or silt ejector is the device which is constructed at distance away from the head regulator to remove the silt from the canal.

Based on the alignment the canals are classified as contour canal, ridge (or watershed) canal and side slope canal.

As the name suggests the ridge canal runs along the ridge line of the watershed and it has command area on both sides. It is not having any cross drainage works.

Contour canal runs along a contour and it has commad area only on one side.

Side slope canal is normal to the contours and it has steep bed slope.

A canal which is designed to carry water round the year is called a perennial canal.

A canal which feeds two or more canals is called a feeder canal.

The order of the network of an irrigation canal system is Main canal—Branch canal—Major distributory—Minor distributory—Water course.

The discharge in the minor distributory is less than $\frac{1}{4}$ m³/s.

Canals draw a fair share of silt from the river. When these canals are not lined and when they run in alluvium soils, they must be so designed that they do not either silt or scour.

Kennedy and Lacey have proposed silt theories to design the canals for non-silting and non-scouring conditions.

According to Kennedy's theory the silt is kept in suspension only due to the eddies generated from the bed. The critical velocity is given by $V_0 = 0.55\, D^{0.64}$, where D is the depth of flow. He made use of the Kutter's equation for finding the mean velocity. Design becomes unique only when B/D ratio is known. Otherwise different designs exist for different bed slopes. The silt supporting capacity is proportional to $V_0^{0.25}$.

According to Lacey's theory the silt is kept in suspension due to the eddies generated from both bed and also sides. The relevant equations are

$$V^2 = \frac{2}{5} f R$$

$$V = 10.8\, R^{2/3}\, S^{1/3}$$

$$P = 4.75\, \sqrt{Q}$$

$$R = 0.47 \left(\frac{Q}{f}\right)^{1/3}$$

$$f = 1.76\, \sqrt{m_r}$$

where V is the velocity, R is the scour depth, f is the silt factor, P is the perimeter, Q is the discharge, S is the slope, all under regime condition, and m_r is the mean diameter of the silt particles in mm.

Generally the canal section will be such that a part of it will be in excavation and the remaining in embankment. If the volume of excavation is just equal to the volume of soil required to form the embankment on either side, it will be most economical. Accordingly, the depth which satisfies the above condition is called the balanced depth of cutting.

Berm is a narrow strip of land at the ground level between the inner toe of the bank and top edge of the cutting.

Free board is the level difference between top of the bank and full supply level (F.S.L.).

Borrow pits are required when the volume of filling exceeds the volume of cutting.

When the volume of excavation is in excess of filling, spoil banks are formed.

Counter berm (or back berm) is provided on the outside of the canal bank to contain the hydraulic gradient line within the bank.

When the water table rises very near to the root zone of the crop affecting the fertility of the soil and the yield of the crop, the soil is said to be water logged. The land will be water logged when the water table is within 1.5 m to 2.1 m below ground level.

The reasons for water logging are excessive irrigation, inadequate surface drainage, seepage from canal system, obstruction to drainage etc.

It can be remedied by providing efficient surface drainage and subsurface drainage, restricting the irrigation, reducing the seepage from canals etc.

Seepage loss is the major loss in the irrigation canals. This loss can be minimised by lining the canals. It also acts as anti-water logging measure besides reducing the evaporation losses and eliminating the weed growth. The various types of lining are: concrete lining, Brick lining, soil cement lining, shotcrete lining, Precast concrete lining, cement motar lining asphaltic lining etc.

An outlet is a small structure which allows water from the distributing channel to a water course or field channel.

If the discharge in the outlet depends on the difference in water levels in the distributing channel and field channel, it is called a non-modular outlet.

If the discharge in the outlet depends only on the fluctuations of the water level in the distributing channel and is independent of the fluctuations in water levels of the field channel, it is called a semi-modular or flexible outlet.

If the discharge in the outlet remains constant irrespective of the fluctuations in the water level in distributing channel and field channel, it is called rigid module or flexible outlet.

$$F = \text{Flexibility} = \frac{dq/q}{dQ/Q}$$

q = The discharge through the outlet
Q = The discharge in the distributing channel

when $F = 1$, it is called the proportional outlet
$F > 1$, it is called the hyper proportional outlet
$F < 1$, it is called the sub-proportional outlet

$$S = \text{Sensitivity} = \frac{\left(\dfrac{dq}{q}\right)}{\left(\dfrac{dG}{D}\right)}$$

where G is the gauge reading such that $G = 0$, when $q = 0$
D is the depth of water in the distribution channel.

For rigid module sensitivity is zero.

Submerged pipe outlet is an example of non-modular outlet.

Pipe outlet discharging freely into atmosphere, Kennedy's gauge outlet, open flume outlet come under the category of semi-modules.

Gibb's module is an example of rigid module.

When the natural ground slope is steeper than the design slope canal falls are provided. Canal fall would lower the water level and dissipate the energy associated with the drop.

Types of falls: ogee fall, rapid fall, stepped fall, Notch fall Vertical drop fall, Glacis type of fall, cylinder fall (or syphon well drop).

If the discharge can be measured at the fall it is called a meter-fall otherwise non-meter fall.

Sarda fall is a vertical drop fall. It uses rectangular crest when $Q < 14 \text{ m}^3/\text{s}$ and trapezoidal crest when $Q > 14 \text{ m}^3/\text{s}$.

Montague fall is a glacis type fall.

Canal regulators are required to direct water from main canal to branch canal or branch canal to major distributory etc. and also to maintain proper levels in the canals to achieve this objective.

The regulator at the head of the off-taking canal is called the head regulator. The regulator on the parent channel just below the off-take point is called cross regulator.

Sometimes a bridge can be combined with regulator.

A canal escape is a structure constructed on an irrigation canal for the purpose of wasting some of its water.

Cross drainage works are required whenever a canal crosses a natural drain during its course.

When the canal runs above the drain it is called an aqueduct. When there is a sufficient head room between bottom of the canal trough and the high flood level (H.F.L.) in the drain it is called a simple aqueduct. When the bottom of canal trough is below the H.F.L. it is called a syphon aqueduct.

When the drain runs above the canal it is called a superpassage. If the F.S.L. is above the river bed level but below H.F.L. it is called a canal syphon.

When canal and drain cross each other at same level it is called a level crossing.

When the section of the canal including the earthen banks is not altered while it passes over the drain, it is called type I aqueduct.

When the outer slopes of the banks are supported by walls and the flow section is not altered it is called type II aqueduct. The type II syphon aqueduct is also called an under tunnel.

When the canal section is flumed it is called the type III aqueduct.

River training works are required to direct and guide the river flow, to make the river course stable and reduced bank erosion and to pass the flood discharges safely.

The important river training works are guide banks, Groynes or spurs, Levees or embankments and pitched islands etc.

Guide banks are provided to guide the stream near a structure so as to confine it in a reasonable width of the rivers. Guide bank is also known as Bell's bund.

Groynes, which are also known as spurs, are the structures constructed in a transverse direction to the river flow and extend into the river. They are provided to protect the river bank and train the flow along a certain course.

In a repelling groyne the axis of the groyne makes an obtuse angle with the direction of flow.

The axis of an attracting groyne makes an acute angle with the river flow direction.

The axis of a deflecting groyne is normal to the river flow.

In the case of Denehy's groyne, the head has the shape of T, while the shape of Hockey spur resembles the hockey stick.

Levee is an earthen dike constructed parallel to the river. It is also known as marginal bund.

If the depth of flow in a open channel at a section does not change with time the flow is said to be steady. Otherwise, it is called unsteady flow.

The steady flow in open channels is of two types: Uniform flow and Varied (or non-uniform) flow.

The varied flow can be divided into three categories: gradually varied flow, rapidly varied flow, spatially varied flow.

In uniform flow the depth of flow is same along the channel.

In a gradually varied flow the depth of flow either increases or decreases along the channel very gradually.

In a rapidly varied flow the depth of flow changes very rapidly along the channel.

In a spatially varied flow, the discharge either increases of decreases along the channel.

The hydraulic jump is an example of rapidly varied flow, while the flow in a side channel spillway is an example of spatially varied flow.

The velocity in uniform flow in open channels is given by

$$V = C\sqrt{RS} \quad \text{Chezy's}$$

$$V = \frac{1}{n} R^{2/3} S^{1/2} \quad \text{Manning's}$$

where C is the Chezy's co-efficient, n is the Manning's rugosity co-efficient

S is the bed slope

$$R = \frac{A}{P} = \text{the hydraulic radius}$$

A = Area of flow

P = Wetted perimeter

The hydraulic depth D is given as the ratio $\frac{A}{T}$, where T is the top width.

The specific energy is the energy measured w.r.t. channel bottom as the datum

$$E = y + \frac{V^2}{2g}.$$

For a given discharge, the specific energy will be minimum when the flow is critical. The corresponding depth of flow is called the critical depth, denoted by y_c.

When the depth of flow y is less than y_c the flow is super-critical and when $y > y_c$ it is subcritical.

The specific energy increases as the depth of flow increases in sub-critical flow whereas in super-critical flow it decreases.

The Froude number of the flow, F is given by $F = \dfrac{V}{\sqrt{gD}}$

where V is the velocity and D is the hydraulic depth. For critical flow; $F = 1$, when $F < 1$ the flow is sub-critical and if $F > 1$, the flow is supercritical.

For a given specific energy there are two possible depth of flow one in super-critical regime and the other in sub-critical regime. These two depths are called the alternate depths.

For a given specific energy, the discharge will be maximum when the flow is critical.

In a sub-critical flow, if the width of the channel is reduced, the depth of flow decreases and the reverse happens in super-critical flow.

In a sub-critical flow, if a hump is placed the depth of flow over the hump is less than the u/s depth of flow and the reverse is true in super-critical flow.

The specific force at a section is given by

$$\text{Specific force} = \frac{Q^2}{gA} + A\bar{z}$$

where $\bar{z}$ is the depth to the centroid of the section from the free surface.

The two depths of flow one in super-critical regime and the other in sub-critical regime for which the specific force is same are called the conjugate depths.

For a given discharge, the specific force will be minimum when the flow is critical.

For a given specific force, the discharge is maximum when the flow is critical.

When the flow is critical, $\frac{V^2}{2g} = \frac{D}{2}$. For rectangular channels since $D = y$, the velocity head equals half the depth of critical flow. In triangular channels at critical flow the velocity head is one-fourth the depth of flow.

The section of the open channel is the most efficient when the wetted perimeter is minimum for the given area.

In the best rectangular section $B = 2y$.

In the best triangular section the bottom angle is 90°.

In other words, the best rectangular and triangular sections are half of a square.

In the best trapezoidal section the top width is twice the length of the inclined side, the side slope is 60°, and it is half of a regular hexagon.

If y_1 and y_2 are the depth before and after the jump, and F_1 is the Froude number of the flow before the jump,

$$\frac{y_2}{y_1} = \frac{1}{2}[-1 + \sqrt{1 + 8F_1^2}].$$

Loss in jump, $\quad \Delta E = \dfrac{(y_2 - y_1)^3}{4y_1y_2} = \dfrac{(V_1 - V_2)^3}{2g(V_1 + V_2)}$

$$y_1 y_2 (y_1 + y_2) = 2y_c^3$$

Height of jump, $h_j = y_2 - y_1$
Length of jump $L_j = (5 \text{ to } 7) h_j$
$F_1 = 1$ to 1.7 undular jump
$F_1 = 1.7$ to 2.5 weak jump
$F_1 = 2.5$ to 4.5 oscillating jump
$F_1 = 4.5$ to 9.0 steady jump
$F_1 > 9.0$ strong jump.

Efficiency of the jump $= \dfrac{E_2}{E_1} = \dfrac{E_1 - \Delta E}{E_1}$.

The depth of flow to carry a given discharge under uniform flow conditions over a given slope is called the normal depth y_n.

If $y_n < y_c$ it is steep slope, $y_n = y_c$ it is critical slope, $y_n > y_c$ it is mild slope.

The gradually varied flow in open channels is created by placing obstructions in the uniform flow, by terminating the channel abruptly, by changing the bed slope suddenly etc.

The dynamic equation of gradually varied flow is given by

$$\dfrac{dy}{dx} = \dfrac{S_o - S_f}{1 - \dfrac{Q^2 T}{g A^3}}, \text{ where } S_f \text{ is the energy slope.}$$

when $\dfrac{dy}{dx}$ is positive the profile is called back water curve and

when $\dfrac{dy}{dx}$ is negative the profile is called drawdown curve.

Three types of profiles are possible on any slope. For example, M_1, M_2, M_3 and S_1, S_2, S_3 on mild and steep slopes respectively.

Drawdown curves can occur only in second zone.

Backwater curves can occur only in first and third zones.

A rectangular channel is said to be wide if $B > 10y$. For wide rectangular channels, the hydraulic radius is approximately equal to the depth of flow. That is, $R \approx y$. With this condition, if Manning's equation is used for velocity.

$$\dfrac{dy}{dx} = S_o \dfrac{1 - \left(\dfrac{y_n}{y}\right)^{10/3}}{1 - \left(\dfrac{y_c}{y}\right)^3}$$

and if Chezy's equation is used

$$\dfrac{dy}{dx} = S_o \dfrac{1 - \left(\dfrac{y_n}{y_c}\right)^3}{1 - \left(\dfrac{y_c}{y}\right)^3}.$$

The section factor for uniform flow is $AR^{2/3}$, or $A\sqrt{R}$ depending on whether Manning's or Chezy's equation is used.

The section factor for critical flow is $A\sqrt{D}$.

In circular channels, maximum velocity occurs if $y = 0.81\,d$, where d is the diameter of the channel.

Maximum discharge according to Chezy's equation occurs when $y = 0.95\,d$ and according to Manning's equation when $y = 0.938\,d$.

The hydraulic jump is also known as a standing wave. The depth of flow after the jump is called the sequent depth. For a given initial depth y_1, the sequent depth is always less than the alternate depth y_2. This is because of losses in the hydraulic jump.

The flow over a chute spillway is generally supercritical.

The potential energy possessed by water when it is stored in reservoirs can be utilized to run the turbines which in turn activate generators and produce electricity. Such an arrangement is called a hydroelectric plant. The power produced in such plants is known as water power or hydro power.

The load (that is the demand for power) on a hydroplant is not uniform. It is a variable with the peak load occurring somewhere in the evening hours.

The graph showing the variation of load with time is called a load curve.

The ratio of the average load to peak load is called load factor.

The installed capacity of the plant will be in excess of the peak load.

The difference between installed capacity and peak load is called the reserve capacity.

The ratio of the average load to the installed capacity is called the capacity factor.

The capacity factor is always less than load factor.

Higher load factor and higher capacity factor indicate the better utilization of the plant capacity.

A graph plotted between the load and the percentage time such load is equalled or exceeded is called a load duration curve.

The storage provided to take care of the hourly fluctuations is called pondage. The enlarged water body above the intake which serves this purpose is called a forebay.

Booms are provided to deflect and divert the ice and debris from intake to spillway.

Trash racks are provided to prevent the entry of trash into intakes.

Water hammer is produced in penstock pipes due to sudden changes in the discharge.

Surge tanks are provided to reduce the water hammer pressures in penstock pipes. They should be located as nearer to the turbine as possible.

II. OBJECTIVE TYPE QUESTIONS

1. Contour bunding is practiced in
   (a) plain areas ☐
   (b) hilly areas ☐
   (c) water logged areas ☐
   (d) dried-up tanks. ☐

2. Which of the following has the maximum water application efficiency?
   (a) Surface irrigation ☐   (b) Lift irrigation ☐
   (c) Sprinkler irrigation ☐   (d) Furrow irrigation. ☐

3. The soil moisture useful for plant growth is
   (a) gravity water ☐   (b) hygroscopic water ☐
   (c) capillary water ☐   (d) all the above. ☐

4. The field capacity is the moisture content present in the soil
   (a) when it is completely saturated ☐
   (b) when all the gravity water is removed from it after saturation ☐
   (c) when the oven dry sample absorbs moisture from atmosphere ☐
   (d) none of the above. ☐

5. The field capacity of an irrigation soil depends on
   (a) both porosity and pore size ☐   (b) only on porosity ☐
   (c) only on pore size ☐   (d) porosity and depth of root zone. ☐

6. Available soil moisture is the difference between
   (a) saturation capacity and field capacity ☐
   (b) saturation capacity and permanent wilting point ☐
   (c) field capacity and permanent wilting point ☐
   (d) saturation capacity and temporary wilting point. ☐

7. The moisture content of the soil below which plants cannot extract sufficient water for their requirements is called
   (a) field capacity ☐   (b) saturation capacity ☐
   (c) temporary wilting point ☐   (d) permanent wilting point. ☐

8. Soil moisture deficiency is the difference between
   (a) saturation capacity and the existing soil moisture content ☐
   (b) field capacity and the existing soil moisture content ☐
   (c) permanent wilting point and the existing moisture content ☐
   (d) temporary wilting point and the existing moisture content. ☐

9. Basin irrigation is a method of irrigation in which
   (a) water is applied to straight ditches parallel to a row of plants ☐
   (b) sewage effluent is used instead of fresh water ☐
   (c) water lifted by pumps is stored in large basins and then applied to fields ☐
   (d) a basin is formed around each tree or a group of trees and water is applied to these basins. ☐

10. The field capacity and dry unit weight of an irrigation soil are 25% and 1.5 g/cc respectively. If the root zone depth is 0.8 m. What is the depth of water required to bring the existing soil moisture of 15% to the field capacity?
    (a) 6 cm ☐   (b) 12 cm ☐
    (c) 18 cm ☐   (d) 24 cm. ☐

11. For an irrigation field lying in a sandy undulating terrain, the most desirable method of applying water is
    (a) basin flooding
    (b) furrow irrigation
    (c) free flooding
    (d) sprinkler irrigation.

12. When the water table is within the root zone depth and is detrimental to the plant life, the land is said to be
    (a) super saturated
    (b) water logged
    (c) over nourished
    (d) none of the above.

13. Which of the following may lead to water logging of the fields?
    (a) poor drainage
    (b) excessive seepage from nearby reservoirs and canals
    (c) over irrigation
    (d) all the above.

14. Water present in the soil which cannot be removed except by heating is called
    (a) gravity water
    (b) capillary water
    (c) hygroscopic water
    (d) free water.

15. Effective rainfall for a crop may be defined as
    (a) the portion of the rainfall which is utilized by crops
    (b) the total rainfall
    (c) the total rainfall minus the total run off
    (d) none of the above.

16. Consumptive use of a crop is defined as the
    (a) total amount of water applied to the crop during its life period
    (b) total amount of water applied minus the total rainfall during its life period
    (c) total amount of water utilised by the crop for its evapo-transpiration requirements
    (d) total amount of water used in the plant metabolism.

17. A climatic region lacking enough water for agriculture without artificial irrigation is called
    (a) arid zone
    (b) dry zone
    (c) desert zone
    (d) none of the above.

18. If B is the base period in days, D is the duty in hectares/cumec and Δ is the delta of the crop in m, the relation between them is given by
    (a) $D = 8.64 B\Delta$
    (b) $\Delta = 8.65 BD$
    (c) $\Delta = \dfrac{0.864 B}{D}$
    (d) $\Delta = \dfrac{8.64 B}{D}$.

19. The duty at the field of a crop is 100 hectares/cumec. If the canal losses are 25%, what is the duty at the head of the canal?
    (a) 750
    (b) 1250
    (c) 250
    (d) 800.

20. The time factor of a canal is defined as the ratio of
    (a) the number of days of irrigation period to the number of days the canal has run
    (b) the number of days the canal has run to the number of days of irrigation period
    (c) the duty at the head of canal to the duty at the field
    (d) the number of days the canal has run at its capacity.

21. The capacity factor of a canal is defined as the ratio of
    (a) the mean discharge in the canal to the peak discharge
    (b) peak discharge to the average discharge
    (c) the peak discharge to the ayacut irrigated by the canal
    (d) the ayacut irrigated to the peak discharge.

22. The duty of water at the outlet is also known as
    (a) time factor
    (b) capacity factor
    (c) full supply co-efficient
    (d) outlet factor.

23. The canal has to irrigate 12000 hectares of rice with a duty of 1000 hectares/cumec. For what discharge should the canal be designed if the capacity factor is 0.8 and the time factor is 0.75
    (a) 9.6 m³/s
    (b) 9 m³/s
    (c) 20 m³/s
    (d) 12.8 m³/s.

24. Crop ratio is defined as the ratio of area irrigated
    (a) in Rabi season to Kharif season
    (b) in Kharif season to Rabi season
    (c) under perennial crop to non-perennial crops
    (d) under perennial crop to total area.

25. The Kharif crop is sown
    (a) at the end of south-west monsoon
    (b) at the end of north-east monsoon
    (c) the beginning of south-west monsoon
    (d) in mid summer.

26. Which of the following is not a Rabi crop
    (a) sugar cane
    (b) groundnut
    (c) wheat
    (d) potato.

27. Nitrogen content in the soil can be increased by raising one of the following crops in crop rotation
    (a) garden crop
    (b) aquatic crop
    (c) leguminous-crop
    (d) perennial crop.

28. The average Δ of rice crop is nearer to
    (a) 40 cm
    (b) 80 cm
    (c) 120 cm
    (d) 160 cm.

29. Net irrigation requirement of a crop is given as
    (a) consumptive use + field losses
    (b) consumptive use + conveyance losses
    (c) consumptive use + field losses + conveyance losses
    (d) consumptive use – effective rainfall.

30. The field irrigation requirement is computed as
    (a) consumptive use + field application losses
    (b) net irrigation requirement + field application losses
    (c) net irrigation requirement + conveyance losses
    (d) consumptive use + conveyance losses.

31. The gross irrigation requirement is given by
    (a) consumptive use + conveyance losses
    (b) field irrigation requirement + conveyance losses
    (c) net irrigation requirement + conveyance losses
    (d) consumptive use + field application losses.

32. The depth of root zone for rice is generally about
    (a) 8 cm   (b) 12 cm
    (c) 16 cm   (d) 10 cm.

33. The most commonly adopted method of irrigation for cereal crops is
    (a) furrow   (b) basin flooding
    (c) check flooding   (d) sub-surface irrigation.

34. The intensity of irrigation is defined as the ratio of
    (a) culturable commanded area to gross commanded area
    (b) gross commanded area to culturable commanded area
    (c) culturable cultivated area to culturable commanded area
    (d) culturable cultivated area to gross commanded area.

35. The heavy crop among the following is
    (a) hemp   (b) sugar cane
    (c) tobacco   (d) cotton.

36. Identify the correct pair of crop and harvesting time from the following:
    (a) Tobacco—December   (b) Rice—July
    (c) Potato—February   (d) Gram—April.

37. Which of the following is leguminous crop?
    (a) rice   (b) sugar cane
    (c) groundnut   (d) hemp.

38. What type of crop is the sugar cane
    (a) Kharif   (b) Rabi
    (c) Hot weather   (d) Perennial.

39. The most desirable alignment of an irrigation canal is along
    (a) the ridge line   (b) a contour line
    (c) the valley line   (d) a straight line.

40. The type of canal meant for diversion of flood waters of river is
    (a) ridge canal   (b) inundation canal
    (c) perennial canal   (d) permanent canal.

41. An irrigation canal which is designed to irrigate all round the year may be called
    - (a) permanent canal
    - (b) continuous canal
    - (c) perennial canal
    - (d) all weather canal.

42. Canals which are excavated directly from the rivers with or without head regulator are called
    - (a) natural canals
    - (b) ditch canals
    - (c) seasonal canals
    - (d) innudation canals.

43. The canals meant for the purpose of draining off water from water logged areas are called
    - (a) seepage canals
    - (b) percolation canals
    - (c) drains
    - (d) ditch canals.

44. The cone formula to comput the storage volume V between two contours with an elevation difference of h m is given by
    - (a) $V = \dfrac{h}{2}(A_1 + A_2)$
    - (b) $V = \dfrac{h}{3}(A_1 + A_2 + \sqrt{A_1 A_2})$
    - (c) $\dfrac{h}{3}(A_1 + 2A_2)$
    - (d) $V = \dfrac{h}{3}(2A_1 + A_2)$.

45. Dead storage in a reservoir is provided
    - (a) to meet the emergency needs
    - (b) to mitigate the floods
    - (c) to accommodated the silt trapped in the reservoir
    - (d) to increase the useful life period.

46. Which of the following is a non-rigid dam?
    - (a) gravity dam
    - (b) earth dam
    - (c) arch dam
    - (d) buttress dam.

47. Which of the following is a rigid dam
    - (a) gravity dam
    - (b) earth dam
    - (c) rockfill dam
    - (d) coffer dam.

48. The external forces acting on a gravity dam are resisted by
    - (a) transferring the thrust to the abuttments
    - (b) the cantilever beam bending action
    - (c) the self weight of the dam
    - (d) none of the above.

49. Arch dams are generally preferred in
    - (a) wide rivers with good foundation at shallow depth
    - (b) wide rivers with weak foundation
    - (c) narrow gorges with strong abuttments
    - (d) reservoirs where provision is made for future increase in capacity.

50. The famous arch dam in India is at
    - (a) Bhakra
    - (b) Khadakvasla
    - (c) Nagarjuna Sagar
    - (d) Idikki.

51. Beaver dam is a type of
    (a) earth dam
    (b) steel dam
    (c) timber dam
    (d) buttress dam.

52. The type of dam which requires least maintenance is
    (a) steel dam
    (b) gravity dam
    (c) timber dam
    (d) rockfill dam.

53. If h_w is the height of the wave, the maximum intensity of wave pressure occurs at a height of (measured above still water surface)
    (a) $\dfrac{h_w}{2}$
    (b) $\dfrac{h_w}{4}$
    (c) $\dfrac{h_w}{8}$
    (d) $\dfrac{h_w}{16}$.

54. If w is the specific weight of water and h_w is the height of the wave, the total wave pressure is equal to
    (a) $2w\, h_w^2$
    (b) $4w\, h_w^2$
    (c) $8w\, h_w^2$
    (d) $16w\, h_w^2$.

55. The head water and tail water depths in a gravity dam are H and H'. The intensity of up lift pressure at the line of drainage gallery is then given by
    (a) $w[H' + \frac{1}{3}(H - H')]$
    (b) $w/3\,(H + H')$
    (c) $w[H - \frac{1}{3}(H - H')]$
    (d) $w[H + \frac{2}{3}(H - H')]$.

56. The lower limit of factor of safety against overturning in a gravity dam is
    (a) 1.25
    (b) 1.75
    (c) 2.0
    (d) 1.5.

57. If p_n is the normal stress at the toe, and ϕ is the angle made by the downstream face with the vertical the maximum principal stress at the toe of the gravity dam for no tail water condition is given by
    (a) $p_n \tan^2 \phi$
    (b) $p_n \sec^2 \phi$
    (c) $p_n \cot^2 \phi$
    (d) $p_n \csc^2 \phi$.

58. A low gravity dam is one in which
    (a) the height of water stored is less than 30 m
    (b) the resultant just passes through the down stream middle third point
    (c) the maximum principal stress is less than the allowable crushing strength and the upstream face is entirely vertical
    (d) the height of the dam is less than 5 times the top width.

59. If the allowable crushing strength is f, w is the unit weight of water and S is the specific gravity of the dam material, the limiting height of the low gravity dam is given by

   (a) $H = \dfrac{f}{w(S+1)}$
   (b) $H = \dfrac{f}{w(S-1)}$
   (c) $H = \dfrac{f}{(w+S+1)}$
   (d) $H = \dfrac{f}{(w+S-1)}$.

60. The type of dam which can be raised easily, if required, is
   (a) gravity dam
   (b) earth dam
   (c) arch dam
   (d) none of the above.

61. For no tension to develop in the gravity dam the resultant of all the external forces should always lie
   (a) at the centre of the base
   (b) within the middle third portion of the base
   (c) within the d/s third portion
   (d) with the u/s third portion.

62. For no tension to develop in the gravity dam, the eccentricity of the resultant force should be
   (a) $< \dfrac{b}{3}$
   (b) $< \dfrac{b}{4}$
   (c) $< \dfrac{b}{6}$
   (d) $< \dfrac{b}{12}$.

63. If H is the height of water to be stored, S is the specific gravity of the dam material and μ is the co-efficient of friction the base width of the elementary profile satisfying the condition of no tension is given by
   (a) $\dfrac{H}{\mu.S}$
   (b) $\dfrac{H}{\mu(S-1)}$
   (d) $\dfrac{H}{\sqrt{S}}$
   (d) $\dfrac{H}{\sqrt{S-1}}$.

64. If the resultant falls outside the middle third for reservoir full condition the gravity dam may fail due to
   (a) sliding
   (b) crushing
   (c) tension
   (d) over turning.

65. The contraction joints in a gravity dam are provided
   (a) to ensure proper transfer of stresses
   (b) to eliminate stress concentrations
   (c) to prevent cracks in the dam that may develop due to temperature changes
   (d) to facilitate the construction of dam in stages.

66. The leakage through the joints in a gravity dam is prevented by providing
   (a) keys
   (b) drainage galleries
   (c) water stops
   (d) all of the above.

67. Which of the following is not a purpose of the drainage gallery
    (a) intercepts the seepage through the dam and drains it off
    (b) facilitates the drilling and grouting operations
    (c) reduces the cost of structure
    (d) provides access for inspection.

68. The most economical central angle of an arch dam is
    (a) 33°
    (b) 93°
    (c) 183°
    (d) 133°.

69. The example of multiple arch type Butress dam in India is
    (a) Mir-Alam dam
    (b) Koyna dam
    (c) Idikki dam
    (d) Tunga Bhadra dam.

70. In a homogeneous embankment type earth dam, the phreatic line is kept well within the body of the dam by
    (a) providing proper u/s slope protection
    (b) providing proper d/s slope protection
    (c) suitably increasing the top width
    (d) providing horizontal drainage filter at the d/s face.

71. Hydraulic failures of earth dams account for
    (a) 40% of the total failures
    (b) 30% of the total failures
    (c) 50% of the total failures
    (d) 60% of the total failures.

72. The purpose of the outer shells in a zoned embankment type earth dam is
    (a) to reduce the seepage through the dam
    (b) to provide stability to the central core
    (c) to permit steep slopes for the u/s and d/s faces
    (d) to avoid cut off trench.

73. The cut off trench in earth dams is provided
    (a) to reduce the seepage through the dam
    (b) to reduce the seepage through the foundation
    (c) to increase the stability of the slopes
    (d) to reduce the dam section and economise.

74. For reservoir full condition the upstream face of an earth dam is
    (a) a stream line
    (b) an equipotential line
    (c) neither a stream line nor an equipotential line
    (d) an interface.

75. If s is the focal length of the basic parabola describing the phreatic line and k is the permeability of the dam material, the seepage discharge per unit length of the earth dam is given by
    (a) $q = ks^2$
    (b) $q = k/s$
    (c) $q = s/k$
    (d) $q = ks$.

76. If the number of flow channels is N_f and the number of equipotential drops is N_d, in a flow net constructed for an earth dam storing water to a height of H metres, the seepage discharge per unit length of the dam is then given by

   (a) $q = KH \dfrac{N_f}{N_d}$ (b) $q = KH \dfrac{N_d}{N_f}$

   (c) $q = \dfrac{K \cdot N_f}{H \cdot N_d}$ (d) $q = \dfrac{H \cdot N_f}{K \cdot N_d}$.

77. A flow net constructed for an earth dam storing water to a height of 20 m, the number of flow channels and the number of potential drops are found to be 4 and 10 respectively. If the permeability of the dam material is 3 m/day, the seepage per metre length of the dam is equal to

   (a) 48 m³/day   (b) 96 m³/day
   (c) 24 m³/day   (d) 12 m³/day.

78. The permeabilities in x and y directions of an anisotropic soil are K_x and K_y. The equivalent permeability of the transformed isotropic soil is given by

   (a) $\sqrt{\dfrac{K_x}{K_y}}$ (b) $\sqrt{K_x K_y}$

   (c) $\dfrac{K_x}{\sqrt{K_y}}$ (d) $\dfrac{K_y}{\sqrt{K_x}}$.

79. Which of the following is different from the rest?
   (a) horizontal drainage filter   (b) chimney drain
   (c) rock toe   (d) cut off trench.

80. Which of the following is different from the rest?
   (a) rock toe   (b) u/s impervious blanket
   (c) relief well   (d) impervious cut off.

81. Identify the suitable section to be adopted when the available material is fine gravel and coarse sand and the foundation is largely impervious.

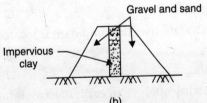

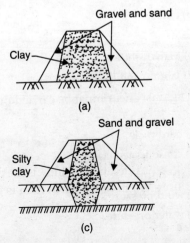

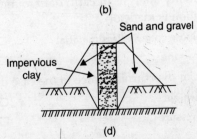

82. The most widely used method of slope stability analysis of earth dam is
    (a) British slip circle method
    (b) Bishop's slip circle method
    (c) Sweedish slip circle method
    (d) Nordic slip circle method.

83. The stability of the up-stream slope of an earth dam becomes critical for the condition of
    (a) steady state seepage when the reservoir is full
    (b) reservoir empty
    (c) sudden draw down
    (d) none of the above.

84. The maximum failures of earthen dams have occurred due to
    (a) the erosion caused by burrowing animals
    (b) the piping under excessive hydraulic gradient
    (c) overtopping caused by insufficient spillway capacity
    (d) sloughing of d/s sloping.

85. In which state the famous Idikki Arch dam is located?
    (a) Uttar Pradesh
    (b) Punjab
    (c) West Bengal
    (d) Kerala.

86. In which state the Koyna dam is located?
    (a) Gujarat
    (b) Karnataka
    (c) Maharashtra
    (d) Orissa.

87. In which state the largest gravity dam (Nagarjuna Sagar) is located?
    (a) Andhra Pradesh
    (b) Madhya Pradesh
    (c) Uttar Pradesh
    (d) Arunachal Pradesh.

88. In which state the highest gravity dam is located?
    (a) West Bengal
    (b) Punjab
    (c) Rajasthan
    (d) Tamilnadu.

89. In an earth dam the top flow line has the shape of
    (a) straight line
    (b) ellipse
    (c) parabola
    (d) circle.

90. In which state the Hirakund dam is located?
    (a) Orissa
    (b) Maharashtra
    (c) Kerala
    (d) Jammu and Kashmir.

91. The stability analysis of earth dam foundation against shear must be carried out when the foundation is made of
    (a) fine loose cohesionless material
    (b) hard rock
    (c) coarse sand
    (d) consolidated clay.

92. The major overturning force in the case of gravity dam is
    (a) self-weight
    (b) wind pressure
    (c) silt pressure
    (d) water pressure.

93. The most common type of spillway used in gravity dams is
    (a) ogee spillway
    (b) siphon spillway
    (c) side channel spillway
    (d) chute spillway.

94. The suction pressure on an ogee spillway is caused when the head on the spillway is
    (a) equal to the design head
    (b) < the design head
    (c) > the design head
    (d) ≥ the design head.

95. In ogee spillway the discharge is proportional to
    (a) $H^{1/2}$
    (b) $H^{5/2}$
    (c) $H^{3/2}$
    (d) $H^{5/4}$.

96. The crest level of an emergency spillway is kept at
    (a) the same level as the main spillway
    (b) slightly below the crest of the main spillway
    (c) at the full reservoir level
    (d) slightly above the main spillway.

97. Another name for shaft spillway is
    (a) conduit spillway
    (b) morning glory spillway
    (c) tunnel spillway
    (d) trough spillway.

98. In the siphon spillway the discharge is proportional to
    (a) $H^{1/2}$
    (b) $H^{5/2}$
    (c) $H^{3/2}$
    (d) $H^{5/4}$.

99. Under siphonic action the head acting on the siphon spillway is equal to
    (a) head water level minus the crest level
    (b) crest level minus the tail water level
    (c) head water level minus the tail water level
    (d) none of the above.

100. In a chute spillway the flow is generally
    (a) sub-critical flow
    (b) laminar flow
    (c) critical flow
    (d) super critical flow.

101. The flow over a side channel spillway is the best example of
    (a) rapidly varying flow
    (b) critical flow
    (c) gradually varied flow
    (d) spatially varied flow.

102. The energy dissipated in a hydraulic jump with initial and sequent depth of y_1 and y_2 is given as
    (a) $\dfrac{(y_2 - y_1)^3}{y_1 y_2}$
    (b) $\dfrac{(y_2 - y_1)^3}{2 y_1 y_2}$
    (c) $\dfrac{(y_2 - y_1)^3}{3 y_1 y_2}$
    (d) $\dfrac{(y_2 - y_1)^3}{4 y_1 y_2}$.

103. The floor of the stilling basin of a hydraulic jump type energy dissipator is depressed below the bed of the d/s channel when tail water rating curve
    (a) lies below the jump rating curve □ (b) lies above the jump rating curve □
    (c) coincides with the jump rating curve □ (d) none of the above. □

104. The sheet of water flowing over a ogee spillway is called
    (a) jet □ (b) jet stream □
    (c) nappe □ (d) wake. □

105. The purpose of the end sill in the stilling basin of a hydraulic jump type energy dissipator is
    (a) to increase the tail water depth □
    (b) to reduced the length of jump and control scour □
    (c) to counteract the uplift on the floor □
    (d) to dissipate the energy by impact action. □

106. The purpose of the baffle pier in the stilling basin of a hydraulic jump type energy dissipator is
    (a) to dissipate the energy by friction □ (b) to counteract the uplift on the floor □
    (c) to reduce the height of the jump □ (d) to increase the length of the jump. □

107. The length of hydraulic jump will generally be
    (a) 2 to 3 times the height of the jump □ (b) 3 to 5 times the height of the jump □
    (c) 5 to 7 times the height of the jump □ (d) 7 to 9 times the height of the jump. □

108. When the sound rock is available in the river bed and tail water rating curve lies below the jump rating curve, the energy dissipator that is recommended is
    (a) Roller bucket □ (b) Depressed floor □
    (c) sloping glacis below the river bed □ (d) ski-jump bucket. □

109. Radial gate is also known as
    (a) tainter gate □ (b) stoney gate □
    (c) drum gate □ (d) rolling gate. □

110. Stoney gate is a type of
    (a) radial gate □ (b) drum gate □
    (c) vertical lift gate □ (d) rolling gate. □

111. The entry of debris into intakes is prevented by providing
    (a) a gate □ (b) trash racks □
    (c) diversion tunnel □ (d) bell-mouthed shape for the entrance. □

112. The crest level of a barrage is generally kept
    (a) almost at the river bed level □ (b) above the river bed level □
    (c) below the river bed level □ (d) at the maximum flood level. □

113. The divide wall in a diversion headwork is provided
    (a) to increase the head of flow through the head regulator □
    (b) to prevent the formation of vortices in front of head regulator □
    (c) to control the silt entry into channel □
    (d) to reduce the uplift pressure on the apron. □

114. Which of the following is not a component of the diversion headwork?
    (a) fish ladder (b) divide wall
    (c) head regulator (d) spillway.

115. The foundation of a weir consists of a horizontal floor of length 30 m an u/s pile of depth 8 m and a d/s pile of depth 12 m. The creep length according to Bligh's theory is
    (a) 50 m (b) 70 m
    (c) 90 m (d) 110 m.

116. The creep length in the above problem according to Lane's weighted creep theory is
    (a) 50 m (b) 70 m
    (c) 90 m (d) 110 m.

117. According to Khosla's theory the exit gradient in the case of a floor with negligible floor thickness and without d/s pile is
    (a) 0 (b) ∞
    (c) 1 (d) H/L.

118. The effect of d/s pile on the u/s pile, due to mutual interference, is
    (a) to increase the uplift pressure (b) to decrease the uplift pressure
    (c) to eliminate the uplift pressure (d) none of the above.

119. In the case of permeable foundations the d/s pile is provided
    (a) to reduce the uplift pressure on floor
    (b) to prevent undermining
    (c) to modify the uplift pressure on the floor
    (d) to increase the vertical creep.

120. The launching apron on the d/s side of permeable foundation is provided
    (a) to increase the effective creep length (b) to hasten the energy dissipation
    (c) to protect the d/s pile from scour holes (d) to reduce the depth of d/s pile.

121. The inverted filter immediately after the impervious floor on the d/s side of a permeable foundation is provided
    (a) to relieve the uplift pressure (b) to increase the effective creep length
    (c) to hasten the energy dissipation (d) to protect the d/s pile from scour holes.

122. If the percolation head is 5 m and the safe hydraulic gradient is 1 in 10, the length of the impervious floor from Bligh's theory is
    (a) 25 m (b) 100 m
    (c) 75 m (d) 50 m.

123. In the above problem, if an upstream pile of depth 7.5 m is provided, then what is the length of impervious floor?
    (a) 42.5 (b) 50 m
    (c) 35 m (d) 57.5 m.

124. The entry of silt into the canal can be controlled by
    (a) silt extractor (b) silt excluder
    (c) silt ejector (d) the head regulator.

125. For effective control of silt entry into the canal the sill of the head regulator should be
  (a) below the sill of the under sluices
  (b) above the sill of the under sluices
  (c) at the same level as the sill of under sluices
  (d) at the maximum flood level.
126. Cross drainage works are not required when the canal is completely
  (a) a ridge canal   (b) a contour canal
  (c) side slope canal   (d) carrier canal.
127. Which of the following canal has the command area only on one side?
  (a) ridge canal   (b) side slope canal
  (c) contour canal   (d) feeder canal.
128. In the above problem, in which of the canals may one bank be avoided?
129. The discharge in the minor distributories will be generally less than
  (a) $\frac{1}{8}$ m³/s   (b) $\frac{1}{4}$ m³/s
  (c) $\frac{3}{8}$ m³/s   (d) $\frac{1}{2}$ m³/s.
130. According to Kennedy's theory the critical velocity is proportional to (if D is the depth of flow)
  (a) $D^{0.6}$ (b) $D^{0.64}$
  (c) $D^{0.68}$ (d) $D^{1.64}$.
131. If V_0 is the critical velocity, according to Kennedy's theory the silt supporting velocity of a canal is proportional to
  (a) $V_0^{0.64}$ (b) $V_0^{1.5}$
  (c) $V_0^{2.5}$ (d) V_0.
132. If Q is the discharge, the wetted perimeter of a regime channel according to Lacey's theory is given by
  (a) $P = 47.5\sqrt{Q}$ (b) $P = 475\sqrt{Q}$
  (c) $P = 0.475\sqrt{Q}$ (d) $P = 4.75\sqrt{Q}$.
133. If Q is the discharge and f is the silt factor, the regime scour depth according to Lacey's theory is given by
  (a) $R = 4.75\sqrt{Q}$ (b) $R = 135\dfrac{Q^2}{f}$
  (c) $R = 0.47\left(\dfrac{Q}{f}\right)^{1/3}$ (d) $R = 1.35\left(\dfrac{Q}{f}\right)^{2/3}$.

134. If d is the mean diameter of the silt particles the Lacey's silt factor f is proportional to
    (a) $d^{1/2}$
    (b) $d^{-1/2}$
    (c) d
    (d) d^2.

135. According to Kennedy's theory the silt is kept in suspension due to the eddies generated from
    (a) the entire perimeter
    (b) from the sides only
    (c) from the bed only
    (d) none of the above.

136. In the above problem, according to Lacey's theory the silt is kept in suspension due to the eddies generated from

137. If w is the unit weight of water, S is the bed slope and R is the hydraulic radius, the average tractive force on the canal bed is given by
    (a) $\tau_0 = wR\sqrt{S}$
    (b) $\tau_0 = wRS^{1.5}$
    (c) $\tau_0 = wRS^{2.5}$
    (d) $\tau_0 = wRS$.

138. The balanced depth of cutting of a canal is one in which
    (a) the volume of cutting is equal to the volume of embankment
    (b) the volume of cutting is less than the volume of embankment
    (c) there is no cutting but only embankment
    (d) there is only cutting and no embankment.

139. The counter berm for a canal is provided
    (a) to increase the stability of the bank
    (b) to increase the cross-sectional area of the canal
    (c) to contain the saturation gradient line within the bank
    (d) to be used as an inspection track.

140. A spoil bank is formed when
    (a) the canal section is too large
    (b) the canal has steep bed slope
    (c) the canal alignment is meandrous
    (d) the volume of excavation is in excess of embankment filling

141. Water logging is the state of the soil where
    (a) the water table is brought very near to the ground surface
    (b) the water table is at deep depth
    (c) the moisture in the soil is beyond the reach of plant roots
    (d) none of the above.

142. The water is not suitable for irrigation if the total salt concentration exceeds
    (a) 1000 ppm
    (b) 2000 ppm
    (c) 3000 ppm
    (d) 4000 ppm.

143. The water is unsuitable for irrigation when the boron content exceeds the limit of
    (a) 0.5 ppm
    (b) 1 ppm
    (c) 1.5 ppm
    (d) 2 ppm.

144. Leaching is the process where by
    (a) nutrients are added to the soil in the form of solution
    (b) excess salts are dissolved in water and removed through infiltration
    (c) weeds are removed through excessive tillage
    (d) none of the above.

145. The acidity in the soils is indicated by a pH value of
    (a) more than 7
    (b) 7
    (c) less than 7
    (d) 0.

146. Which one of the following is different from the rest
    (a) Asphaltic lining
    (b) shotcrete lining
    (c) cement concrete lining
    (d) precast concrete lining.

147. In non-modular outlet the discharge is affected by
    (a) the fluctuations both in the distributory and field channels
    (b) the fluctuations only in the distribution channel
    (c) the fluctuations only in the field channel
    (d) is independent of fluctuations in the distribution and field channels.

148. In the above problem, the discharge in the flexible outlet is affected by

149. In problem 147, the discharge in the rigid module is affected by

150. Gibb's module is a type of
    (a) non-modular outlet
    (b) semi-module outlet
    (c) rigid module outlet
    (d) open flume outlet.

151. Kennedy's gauge outlet comes under the category of
    (a) non-modular outlet
    (b) semi-module outlet
    (c) rigid module outlet
    (d) open flume outlet.

152. The purpose of the cross-regulator is
    (a) to control the discharge into the off-take canal
    (b) to maintain proper levels in the main canal
    (c) to control the silt entry into the off-take canal
    (d) none of the above.

153. Falls in the canals are provided when
    (a) natural ground slope is same as the design slope
    (b) natural ground slope is more than the design slope
    (c) natural ground slope is less than the design slope
    (d) the canal runs in deep cutting.

154. Syphon well drop (or cylindrical fall) in the canals is provided when the discharge in the canal
    (a) is large and the drop is small
    (b) is large and the drop is also large
    (c) is small and the drop is also small
    (d) is small but the drop is large.

155. In the case of a notch type of canal fall, the surplus energy is dissipated
    (a) over the glacis
    (b) in the cistern
    (c) in the hydraulic jump
    (d) none of the above.

156. A canal escape is a structure constructed for the purpose of
    (a) dissipating excess energy
    (b) acting as a forebay
    (c) discharging waste water from the canal
    (d) none of the above.

157. When canal runs above the drain, the cross drainage work provided is
    (a) aqueduct
    (b) super passage
    (c) canal syphon
    (d) level crossing.

158. When the canal runs below the drain, the cross drainage work provided is called
    (a) aqueduct
    (b) super passage
    (c) level crossing
    (d) syphon aqueduct.

159. In the case of an aqueduct if the canal section (including banks) is not changed, then it is called
    (a) type I aqueduct
    (b) type II aqueduct
    (c) type III aqueduct
    (d) none of the above.

160. In the case of an aqueduct, if the canal section is flumed then it is called
    (a) type I aqueduct
    (b) type II aqueduct
    (c) type III aqueduct
    (d) none of the above.

161. A type II syphon aqueduct is also called
    (a) canal syphon
    (b) simple aqueduct
    (c) under tunnel
    (d) under canal.

162. Syphon aqueduct is selected as the cross drainage work when the canal bed level
    (a) is below the maximum flood level in the drain
    (b) is above the maximum flood level in the drain
    (c) is below the bed level of the drain
    (d) none of the above.

163. Canal syphon is selected as the suitable cross drainage work when the full supply level in the canal is
    (a) below the bed level of the drain
    (b) above the bed level of the drain but below the high flood level
    (c) above the high flood level
    (d) none of the above.

164. The length of water way required for the drain at a cross drainage work is obtained as
    (a) $L = 475\sqrt{Q}$
    (b) $L = 47.5\sqrt{Q}$
    (c) $L = 4.75\sqrt{Q}$
    (d) $L = 0.475\sqrt{Q}$.

165. In syphon aqueduct the depth of d/s cut-off below high flood level is taken as
    (a) the regime scour depth
    (b) twice the regime scour depth
    (c) 3 time the regime scour depth
    (d) 1.5 times the scour depth.

166. In the type III aqueduct, the splay of the upstream transition canal is generally taken as
    (a) 4 : 1   (b) 3 : 1
    (c) 2 : 1   (d) 1 : 1.

167. In the type III aqueduct, the splay of the d/s transition canal is
    (a) 4 : 1   (b) 3 : 1
    (c) 2 : 1   (d) 1 : 1.

168. Which of the following is the aggrading river?
    (a) loosing its bed
    (b) building up its bed
    (c) neither loosing nor building its bed
    (d) developing a high degree of sinuosity.

169. In the above problem, which is the degrading river

170. Guide banks are provided
    (a) to train the flow of a river along a specified course
    (b) to confine the width of the river
    (c) to reduce the flood peak
    (d) none of the above.

171. Spurs are provided
    (a) to train the flow of a river along a specified course
    (b) to confine the width of the river
    (c) to reduce the flood peak
    (d) none of the above.

172. Guide bank is also known as
    (a) groyne   (b) spur
    (c) marginal bund   (d) Bell's bund.

173. Identify the attracting groyne from the following:

174. In problem 173, identify the repelling groyne.
175. In problem 173, identify the deflecting groyne.
176. In problem 173, identify the Denehy's groyne.
177. In a repelling groyne, the axis of the groyne (w.r.t. the river flow direction) makes
    (a) an acute angle   (b) an obtuse angle
    (c) a right angle   (d) an angle of 180°.

178. In an attracting groyne the axis of the groyne (w.r.t. the river flow direction) makes
    (a) an acute angle
    (b) an obtuse angle
    (c) a right angle
    (d) an angle of 180°.
179. In a deflecting groyne, the axis of the groyne (w.r.t. the river flow direction) makes
    (a) an acute angle
    (b) an obtuse angle
    (c) a right angle
    (d) an angle of 180°.
180. The trap efficiency of the reservoir is a function of
    (a) the inflow into reservoir
    (b) ratio of inflow to reservoir capacity
    (c) ratio of reservoir capacity to inflow
    (d) reservoir capacity.
181. With age, the trap efficiency of a reservoir is likely to
    (a) increase
    (b) decrease
    (c) remain constant
    (d) difficult to tell.
182. For an economical design of a gravity dam, the shear friction factor should lie between (depending on loading condition)
    (a) 1 and 2
    (b) 1.5 and 4
    (c) 5 and 7
    (d) 0 and 1.
183. Dry stone pitching on a sloping face of an earth dam is known as
    (a) reinforcement
    (b) lining
    (c) revetment
    (d) cushion.
184. The type of dam recommended for a site with narrow gorge and steep and strong side slopes is
    (a) gravity dam
    (b) earth dam
    (c) steel dam
    (d) arch dam.
185. The water held in pores of soil by capillary forces against gravity pull is called
    (a) Gravity water
    (b) Hygroscopic water
    (c) Chemical water
    (d) Capillary water.
186. The laboratory estimate of field capacity is known as
    (a) saturation capacity
    (b) moisture equivalent
    (c) wilting range
    (d) percolation co-efficient.
187. The process of washing out of the salt from the upper zone of the soil by flooding is called
    (a) desaltation
    (b) separation
    (c) leaching
    (d) wild flooding.
188. A canal meant to convey water from one source to the other is known as
    (a) feeder canal
    (b) perennial canal
    (c) commuter canal
    (d) link canal.
189. The weight of silt carried by the river per unit volume of water is termed as
    (a) silt grade
    (b) silt factor
    (c) silt ratio
    (d) silt charge.

190. For a given depth of flow in a canal the velocity which keeps the canal free from silting and scouring is called
    (a) optimum velocity
    (b) economic velocity
    (c) ideal velocity
    (d) critical velocity.

191. The ratio of total quantity of water supplied to a crop during its base period to the area is called
    (a) duty
    (b) delta
    (c) consumptive use
    (d) base depth.

192. The area irrigated by one m³/s of water flowing throughout the base period of the crop is called
    (a) duty
    (b) delta
    (c) crop factor
    (d) crop ratio.

193. According to Lacey's theory the regime velocity is proportional to
    (a) $R^{2/3}S^{1/2}$
    (b) $R^{1/3}S^{1/2}$
    (c) $R^{2/3}S^{1/3}$
    (d) $R^{1/3}S^{1/3}$.

194. The weed growth in a canal leads to
    (a) reduction in silting
    (b) reduction in discharge
    (c) increase in velocity of flow
    (d) reduction in depths of flow.

195. The shape of a lined canal recommended by ISI is
    (a) triangular
    (b) parabolic
    (c) trapezoidal
    (d) semi-circular.

196. At what fraction of depth below the free surface does the average velocity occur in open channel flow
    (a) 0.2
    (b) 0.4
    (c) 0.6
    (d) 0.8.

197. The difference in elevations of top of bank and full supply level of a canal is called
    (a) berm
    (b) free board
    (c) critical depth
    (d) none of the above.

198. The component of a diversion head work which facilitates the migration of fish from u/s or d/s to the other side is known as
    (a) fish net
    (b) fish channel
    (c) fish ladder
    (d) fish pond.

199. An impervious wall inside an earthen dam to reduce the seepage is called
    (a) core wall
    (b) diaphragm wall
    (c) pug wall
    (d) all of the above.

200. The retaining wall in continuation of abutment both upstream and downstream is called
    (a) a flared wall
    (b) flank wall
    (c) wing wall
    (d) both (b) and (c).

201. An obstruction of small height built across a stream to divert water into an off take channel is known as
    (a) storage head work
    (b) diversion head work
    (c) bifurcation channel
    (d) forebay.

202. An earthen embankment built on each side of a river for some distance as a flood control measure is called
    (a) groyne
    (b) spur
    (c) dyke
    (d) retaining wall.

203. The ratio of rate of change of discharge in the outlet to the rate of change of the water level in the distributory w.r.t. the normal depth of flow is called
    (a) setting
    (b) sensitivity
    (c) efficiency
    (d) drowning ratio.

204. The top of the weir or spillway is called
    (a) ridge
    (b) head
    (c) crest
    (d) peak.

205. If H is the percolation head, L is the length of impervious apron, d_1 and d_2 are the depths of u/s and d/s piles then according to Khosla's theory the exist gradient depends on
    (a) H, L, d_1, d_2
    (b) H, L, d_1
    (c) H, L, d_2
    (d) L, d_1, d_2.

206. The ratio of rate of change of discharge in the outlet to the rate of change of discharge in the distributory is known as
    (a) flexibility
    (b) rigidity
    (c) proportionality
    (d) efficiency.

207. In montague type canal fall the glacis is
    (a) straight
    (b) parabolic
    (c) cycloidal
    (d) none of the above.

208. The net irrigation requirement of a crop is 64 cms. If the field application losses and the conveyance losses are each 20% what is the depth of water to be applied?
    (a) 80 cm
    (b) 100 mm
    (c) 128 cm
    (d) 76.8 mm.

209. The unique design of a channel by Kennedy's silt theory requires that
    (a) the bed width of canal must be given
    (b) the depth of flow must be given
    (c) the ratio of breadth to depth must be given
    (d) the slope of the channel must be given.

210. According to Lacey's regime theory the relation between V, R and f is
    (a) $f = \dfrac{2}{5}\dfrac{V^2}{R}$
    (b) $f = \dfrac{5}{2}\dfrac{V^2}{R}$
    (c) $f = \dfrac{2}{\sqrt{5}}\dfrac{V^2}{R}$
    (d) $f = \dfrac{5}{\sqrt{2}}\dfrac{V^2}{R}$.

211. Garrett's diagrams are used for the graphical solution of design of canal by
    (a) Lacey's theory
    (b) Kennedy's theory
    (c) Lane's theory
    (d) Lindlay's theory.

212. On the d/s side of the weir if h is the height of the hydraulic gradient line above the top of the floor and S is the specific gravity of the floor material then the required thickness of the floor is given by
    (a) $t = \dfrac{h+t}{S+1}$
    (b) $t = \dfrac{h}{S+1}$
    (c) $t = \dfrac{h}{S-1}$
    (d) $t = \dfrac{h-t}{S-1}$.

213. The velocity permitted in the barrels of syphon aqueduct is of the order
    (a) 4 to 5 m/s
    (b) 3 to 4 m/s
    (c) 2 to 3 m/s
    (d) 1 to 2 m/s.

214. The Unwin's formula for the head loss through the syphon barrel is given by $h = \left(1 + f_1 + f_2 \dfrac{1}{R}\right)\dfrac{V^2}{2g}$. The factor f_1 in the above formula is to account for
    (a) the head loss at entry
    (b) head loss at exit
    (c) the head loss due to friction
    (d) head loss in the transition.

215. Talus is a different name for
    (a) groyne
    (b) spur
    (c) launching apron
    (d) none of the above.

216. The permissible velocity in concrete lined canals is
    (a) 0.5 m/s
    (b) 1 m/s
    (c) 2 m/s
    (d) 4 m/s.

217. The pH value of soil which makes the soil unsuitable for irrigation
    (a) 6.5
    (b) 7
    (c) 8
    (d) 11.

218. The dimensions of Manning's rougosity co-efficient n
    (a) $L^{2/3}T^{-1}$
    (b) $LT^{-1/3}$
    (c) $L^{1/3}T$
    (d) LT^{-1}.

219. The dimensions of Chezy's C
    (a) $L^{1/2}T^{-1}$
    (b) LT^{-1}
    (c) L^2T^{-1}
    (d) LT^{-2}.

220. The relation between Chezy's C and Manning's n
    (a) $C = \dfrac{R^{1/6}}{n}$
    (b) $C = nR^{1/6}$
    (c) $C = \dfrac{n}{R^{1/6}}$
    (d) $C = \dfrac{R^{1/6}}{n^2}$.

221. Hydraulic jump may also be called
   (a) positive surge
   (b) negative surge
   (c) standing wave
   (d) none of the above.

222. Hydraulic jump is an example of
   (a) gradually varied flow
   (b) spatially varied flow
   (c) rapidly varied flow
   (d) unsteady flow.

223. The flow in a open channel is said to be super critical when
   (a) $F < 1$
   (b) $F > 1$
   (c) $F = 1$
   (d) $F > 10$.

224. If A is the area of flow, P is the wetted perimeter and T is the top width of flow, the hydraulic radius is defined as
   (a) $\dfrac{A}{P}$
   (b) $\dfrac{A}{T}$
   (c) $\dfrac{P}{A}$
   (d) $\dfrac{T}{A}$.

225. The hydraulic depth is defined as
   (a) $\dfrac{A}{P}$
   (b) $\dfrac{A}{T}$
   (c) $\dfrac{P}{A}$
   (d) $\dfrac{T}{A}$.

226. In the most efficient rectangular section the width is equal to
   (a) depth
   (b) $\dfrac{\text{depth}}{2}$
   (c) 2 depth
   (d) 1.5 depth.

227. In the most efficient triangular section, the bottom angle is
   (a) 45°
   (b) 60°
   (c) 90°
   (d) 120°.

228. In the most efficient trapezoidal section which of the following is true
   (a) the top width is twice the length of sloping side
   (b) the hydraulic radius is half the depth of flow
   (c) the shape is half of regular hexagon
   (d) all the above.

229. If D is the hydraulic depth, in a critical flow
   (a) $\dfrac{V^2}{2g} = \dfrac{D}{4}$
   (b) $\dfrac{V^2}{2g} = \dfrac{D}{2}$
   (c) $\dfrac{V^2}{2g} = \dfrac{3}{4}D$
   (d) $\dfrac{V^2}{2g} = D$.

230. In a rectangular channel of depth of flow d when the flow is critical the specific energy is equal to
    (a) d
    (b) 1.5 d
    (c) 2d
    (d) 2.5 d.

231. In a triangular channel of depth of flow d, when the flow is critical, the specific energy is equal to
    (a) d
    (b) 2.25 d
    (c) 1.5 d
    (d) 1.75 d.

232. In a super critical flow, as the depth of flow increases, the specific energy
    (a) also increases
    (b) decreases
    (c) does not change
    (d) none of the above.

233. In a subcritical flow, as the depth of flow increases, the specific energy
    (a) also increases
    (b) decreases
    (c) does not change
    (d) none of the above.

234. The supercritical flow is also known as
    (a) tranquil flow
    (b) shooting flow
    (c) rapid flow
    (d) both (b) and (c).

235. The subcritical flow is also known as
    (a) tranquil flow
    (b) shooting flow
    (c) rapid flow
    (d) both (a) and (b).

236. If q is the discharge per unit width in a rectangular channel, the critical depth is given by
    (a) $y_c = \left(\dfrac{q^2}{g}\right)$
    (b) $y_c = \left(\dfrac{q}{g}\right)^{1/3}$
    (c) $y_c = \left(\dfrac{q^2}{g}\right)^{1/3}$
    (d) $y_c = \left(\dfrac{q}{g^2}\right)^{1/3}$.

237. If Q is the discharge in a triangular channel of given side slope, the critical depth is proportional to
    (a) $Q^{1/3}$
    (b) $(Q^2)^{1/3}$
    (c) $(Q^2)^{1/5}$
    (d) $Q^{1/5}$.

238. Which of the following is a correct statement
    (a) For a given discharge, the specific energy is minimum when the flow is critical
    (b) For a given specific energy, the discharge will be maximum when the flow is critical
    (c) In a critical flow, the velocity of flow and wave celerity are equal
    (d) all the above.

239. If B is the width and d is the depth of flow, a rectangular channel is said to be wide if
    (a) $B > d$
    (b) $B > 2d$
    (c) $B > 5d$
    (d) $B > 10d$.

240. In a wide rectangular channel the section factor for uniform flow is proportional to
   (a) $d^{3/2}$
   (b) $d^{2/3}$
   (c) $d^{5/3}$
   (d) $d^{10/3}$.

241. In a rectangular channel carrying subcritical, flow, if the bottom width is reduced, the depth of flow
   (a) increases
   (b) decreases
   (c) remains constant
   (d) difficult to tell.

242. The depth of flow over hump in a rectangular channel carrying subcritical flow
   (a) is less than depth of flow u/s
   (b) is more than depth of flow u/s
   (c) is equal to the depth of flow u/s minus the height of hump
   (d) none of the above.

243. The depth of flow over a hump in a rectangular channel carrying super critical flow
   (a) is less than depth of flow u/s
   (b) is more than depth of flow u/s
   (c) is equal to the depth of flow u/s minus the height of hump
   (d) none of the above.

244. The maximum velocity in a circular channel of diameter D occurs when the depth of flow is
   (a) 0.81 D
   (b) 0.5 D
   (c) 0.75 D
   (d) 0.98 D.

245. According to Chezy's equation, the maximum discharge in a circular channel of diameter D occurs when the depth of flow is
   (a) 0.95 D
   (b) 0.5 D
   (c) 0.75 D
   (d) 0.81 D.

246. According to Manning's equation, the maximum discharge in a circular channel of diameter D occurs when the depth of flow is
   (a) 0.469 D
   (b) 0.7035 D
   (c) 0.95 D
   (d) 0.938 D.

247. The two depths of flow for which the specific energy is same are called
   (a) alternate depths
   (b) sequent depths
   (c) conjugate depths
   (d) none of the above.

248. The depths of flow for which the specific force is same are called
   (a) alternate depths
   (b) conjugate depths
   (c) initial depths
   (d) none of the above.

249. If y_n and y_c are normal and critical depths of flow, the channel is said to be mild sloped channel, if
   (a) $y_n < y_c$
   (b) $y_n = y_c$
   (c) $y_n > y_c$
   (d) $y_n = 2y_c$.

250. In the above problem which is a steep slope channel.
251. In gradually varied flows all back water curves can occur
    (a) only in zone 1
    (b) only in zone 2
    (c) only in zone 3
    (d) either in zone 1 or in zone 3.
252. In gradually varied flows all drawdown curves can occur
    (a) only in zone 1
    (b) only in zone 2
    (c) only in zone 3
    (d) either in zone 1 or in zone 3.
253. The profile found u/s of an obstruction placed in a channel with $y_n > y_c$ is
    (a) M_1
    (b) S_1
    (c) M_2
    (d) S_2.
254. The profile found u/s of an abrupt end in a channel with $y_n > y_c$ is
    (a) M_1
    (b) S_1
    (c) M_2
    (d) S_2.
255. The alternate depths in a channel are 1 m and 2 m. Then the critical depth is
    (a) 0.534 m
    (b) 0.756 m
    (c) 1.387 m
    (d) 2.252 m.
256. In a hydraulic jump on a horizontal floor of a rectangular channel, the relationship between y_1, y_2 and y_c is given by
    (a) $y_1 y_2 (y_1 + y_c) = 2 y_1^2$
    (b) $y_1 y_2 (y_c + y_2) = 2 y_2^3$
    (c) $y_1 y_2 (y_1 + y_2) = 2 y_c^3$
    (d) $y_1 y_2 (y_1 + y_c) = \dfrac{y_c^3}{2}$.
257. E_1 is the energy before the jump and E_2 is the energy after the jump. The efficiency of the jump is then defined as
    (a) $\dfrac{E_1}{E_2}$
    (b) $\dfrac{(E_1 - E_2)}{E_1}$
    (c) $\dfrac{E_2}{E_1}$
    (d) $\dfrac{(E_1 - E_2)}{E_2}$.
258. $F_1 > 9$, the hydraulic jump is known as
    (a) steady jump
    (b) strong jump
    (c) efficient jump
    (d) weak jump.
259. If V_1 and V_2 are the velocities before and after the jump, the energy loss in the jump is given by
    (a) $\Delta E = \dfrac{(V_1 - V_2)}{4 V_1 V_2}$
    (b) $\Delta E = \dfrac{(V_1 - V_2)^3}{4(V_1 + V_2)}$
    (c) $\Delta E = \dfrac{(V_1 - V_2)^3}{2g}$
    (d) $\Delta E = \dfrac{(V_1 - V_2)^3}{2g(V_1 + V_2)}$.

260. In a wide rectangular channel, using Manning's equation, the dynamic equation for gradually varied flow may be written as

(a) $\dfrac{dy}{dx} = S_0 \dfrac{1-\left(\dfrac{y_n}{y}\right)^{10/3}}{1-\left(\dfrac{y_c}{y}\right)^{10/3}}$ □   (b) $\dfrac{dy}{dx} = S_0 \dfrac{1-\left(\dfrac{y_n}{y}\right)^{10/3}}{1-\left(\dfrac{y_c}{y}\right)^{3}}$ □

(c) $\dfrac{dy}{dx} = S_0 \dfrac{1-\left(\dfrac{y_n}{y}\right)^{3}}{1-\left(\dfrac{y_c}{y}\right)^{3}}$ □   (d) $\dfrac{dy}{dx} = S_0 \dfrac{1-\left(\dfrac{y_n}{y}\right)^{3}}{1-\left(\dfrac{y_c}{y}\right)^{10/3}}$ □

261. In the above problem, identify the dynamic equation for gradually varied flow in wide rectangular channels when Chezy's equation is used.

262. What is an adverse slope
    (a) very steep slope where uniform flow is not possible □
    (b) slope which maintains critical flow for all discharges □
    (c) slope which increases the bed elevation in the direction of flow □
    (d) slope which decreases the bed elevation in the direction of flow. □

263. The section factor for uniform flow is
    (a) $AR^{2/3}$ □   (b) $AD^{2/3}$ □
    (c) $\dfrac{A}{\sqrt{D}}$ □   (d) $A\sqrt{D}$. □

264. In the above problem, identify the section factor for critical flow.

265. Discharge in laboratory channels can be measured by
    (a) venturiflume □   (b) standing wave flume □
    (c) parshall flume □   (d) all the above. □

266. A channel of bed slope 0.0001 carries a discharge of 10 m³/s when the depth of flow is 1.2 m. What is the discharge carried by this channel at the same depth of flow if the slope is increased to 0.0009?
    (a) 90 m³/s □   (b) 30 m³/s □
    (c) 60 m³/s □   (d) 15 m³/s. □

267. At the foot of the spillway if the available tail water depth is less than the depth required for the formation of the jump, the location of the jump is
    (a) shifted up stream □
    (b) shifted down stream □
    (c) shifted both ways depending on the discharge □
    (d) none of the above. □

268. Which of the following hydro electric plants is provided with an underground power station
    (a) Hirakud □   (b) Bhakra □
    (c) Sharavati □   (d) Koyna. □

269. One horse power is equal to
 (a) 1.36 kW
 (b) 0.736 kW
 (c) 0.75 kW
 (d) 1.736 kW.

270. Load factor is defined as the ratio of
 (a) average load to the installed capacity
 (b) average load to the reserve capacity
 (c) reserve capacity to the installed capacity
 (d) average load to peak load.

271. In the above problem which is the capacity factor?

272. Reserve capacity of a hydroelectric plant is the difference between
 (a) the peak load and the average load
 (b) the installed capacity and the average load
 (c) the installed capacity and the peak load
 (d) the peak load and the minimum load.

273. If N, Q, P and H denote the rotational speed, discharge, power developed and head on the turbine respectively, the specific speed is given by
 (a) $N_s = \dfrac{N\sqrt{P}}{H^{3/4}}$
 (b) $N_s = \dfrac{N\sqrt{P}}{H^{5/4}}$
 (c) $N_s = \dfrac{N\sqrt{Q}}{H^{3/4}}$
 (d) $N_s = \dfrac{N\sqrt{Q}}{H^{5/4}}$.

274. If p is the number of pairs of poles in a generator coupled to a turbine, and f is the frequency of the current produced by it, the synchronous speed of the turbine is equal to
 (a) $\dfrac{120f}{p}$
 (b) $\dfrac{12f}{p}$
 (c) $\dfrac{6f}{p}$
 (d) $\dfrac{60f}{p}$.

275. Load duration curve represents
 (a) load against time
 (b) cumulative energy against time
 (c) load against percentage of time such load is exceeded
 (d) none of the above.

276. Pondage is the storage provided
 (a) to take care of the seasonal fluctuations in the stream flow
 (b) to take care of the annual fluctuations in the stream flow
 (c) to take care of hourly fluctuations in load demand for a small period
 (d) any of the above.

277. Run of the river plants are designed so that they can use the water
    (a) from a large storage
    (b) as it comes in the river with some pondage
    (c) which is pumped back from tail race to head race
    (d) none of the above.
278. Which of the following is generally used as a base load plant?
    (a) hydroelectric plant
    (b) pumped storage plant
    (c) thermal plant
    (d) any of the above.
279. The enlarged body of water above the intake of a hydroelectric plant is called
    (a) Reservoir
    (b) Pond
    (c) Boom
    (d) Forebay.
280. Boom in the intake of a hydroelectric plant serves the purpose of
    (a) controlling the velocity of a flow in the penstock
    (b) preventing the debris or trash from entering the penstock
    (c) closing the gate of penstock in emergency
    (d) deflecting or diverting the ice or trash from intake to spillway.
281. Surge tanks in hydroelectric plants are provided
    (a) to protect the penstock pipe from water hammer pressure
    (b) to maintain the uniform discharge in penstock pipes
    (c) to reduce the frictional loss in penstock pipes
    (d) none of the above.
282. Water hammer in penstock pipes is caused by
    (a) sudden changes in water level of the reservoir
    (b) sudden changes in discharge
    (c) gradual changes in discharge
    (d) sudden changes in temperature.
283. A surge tank in a hydroelectric plant should be located
    (a) as nearer to the reservoir as possible
    (b) as nearer to the turbine as possible
    (c) exactly midway between reservoir and turbine
    (d) anywhere.
284. Which of the following is an impulse turbine?
    (a) Kaplan
    (b) Francis
    (c) Turgo wheel
    (d) Deriaz.
285. Which of the following is a reaction turbine?
    (a) Pelton wheel
    (b) Turgo wheel
    (c) Banki
    (d) Deriaz.

286. At a hydroelectric plant 1600 h.p. of power can be generated under a head of 81 m. If the synchronous speed of the turbine is 2430 rpm, what is the suitable type of turbine
    (a) Pelton wheel ☐ (b) Kaplan ☐
    (c) Francis ☐ (d) Propeller. ☐

287. Which of the following turbines is different from the rest?
    (a) Fournyron ☐ (b) Kaplan ☐
    (c) Pelton wheel ☐ (d) Francis. ☐

288. The bulb turbine comes under the category of
    (a) high head turbine ☐ (b) medium head turbines ☐
    (c) low head turbines ☐ (d) high head and high speed turbine. ☐

289. Which dimensionless number is important in the study of cavitation in turbines?
    (a) Webers number ☐ (b) Mach number ☐
    (c) Thomas number ☐ (d) Eulers number. ☐

290. A pelton wheel is
    (a) a tangential flow impulse turbine ☐ (b) an inward flow impulse turbine ☐
    (c) an inward flow reaction turbine ☐ (d) an outward flow reaction turbine. ☐

291. The maximum continuous power that can be generated at a hydroelectric plant throughout the year is called
    (a) firm power ☐ (b) installed capacity ☐
    (c) base power ☐ (d) assured power. ☐

292. The runaway speed of a turbine is
    (a) the actual running speed at design load ☐
    (b) the synchronous speed of the generator ☐
    (c) the speed attained by the turbine under no load condition ☐
    (d) the speed of the wheel when Governor fails. ☐

293. When the load on the turbine is uniform, the type of turbine generally selected is
    (a) Pelton wheel ☐ (b) Francis ☐
    (c) Kaplan ☐ (d) Propeller. ☐

294. When the load on the turbine is varying, the type of turbine selected is
    (a) Pelton wheel ☐ (b) Francis ☐
    (c) Kaplan ☐ (d) Propeller. ☐

295. In which of the following turbine, the draft tube is not required
    (a) Pelton ☐ (b) Francis ☐
    (c) Kaplan ☐ (d) Propeller. ☐

296. A draft tube in a reaction turbine is provided
    (a) to prevent cavitation ☐
    (b) to increase the effective head on the turbine by an amount equal to height of setting ☐
    (c) to convert part of kinetic energy at the exit of turbine into working head ☐
    (d) both (b) and (c). ☐

297. Cavitation is possible only in
    (a) impulse turbines
    (b) reaction turbines and pumps
    (c) reaction turbines and impulse turbines
    (d) impulse turbines and pumps.

298. In which of the following situations a surge tank is not required
    (a) power house located immediately at the toe of the dam
    (b) power house located far away from the dam
    (c) power house located far away from the dam with pressure relief value
    (d) none of the above.

299. In which of the following the running cost will be least
    (a) thermal plant
    (b) nuclear plant
    (c) hydroplant
    (d) both in (a) and (b).

300. The pipe which carries water under pressure from reservoir to turbine is called
    (a) scroll case pipe
    (b) draft tube
    (c) pen stock
    (d) none of the above.

301. Which of the following statements is true?
    (a) capillary water, held by surface tension forces against gravity is useful for the plant growth
    (b) gravity water is unavailable for plants
    (c) sandy soil contain more capillary water than the clays
    (d) none of the above.

302. The consumptive use of a crop is the depth of water equal to
    (a) that transpired by the crop
    (b) that evaporated by the crop
    (c) that evaporated from adjacent soil, and that used by the crop and transpired
    (d) none of the above.

303. The irrigation canals are generally aligned along
    (a) ridge line
    (b) straight line
    (c) contour line
    (d) valley line.

304. A canal meant to drain off water from the waterlogged areas is known as
    (a) valley canal
    (b) drain
    (c) auxiliary canal
    (d) surplus canal.

305. Canals which take off from ice-fed perennial rivers are known as
    (a) permanent canals
    (b) ice canals
    (c) fixed canals
    (d) perennial canals.

306. If the water table is nearer to the ground surface, an unlined irrigation canal may become useless, because
    (a) seepage loss will be considerable
    (b) ayacut area becomes waterlogged soon
    (c) spread of malaria
    (d) all the above.

307. The optimum *k* or depth for rice is
    (a) 19 cm
    (b) 13 cm
    (c) 9 cm
    (d) 30 cm.

308. Lacey's regime condition will be realised, if
    (a) silt grade in the channel is variable
    (b) channel flows in unlimited, incoherent alluvium of the same character as that of transported material
    (c) discharge in the channel is variable
    (d) silt charge in the channel is variable.

309. The optimum *k* or depth of wheat is
    (a) 13 cm
    (b) 9 cm
    (c) 30 cm
    (d) 19 cm.

310. The main reason for the silting up of a canal is
    (a) inadequate slope
    (b) non-regime section
    (c) defective head regulator
    (d) all the above.

311. The full supply level of a canal at its head regulator should be kept
    (a) at the same F.S.L. of parent canal
    (b) 15 cm below the F.S.L. of parent canal
    (c) 15 cm above the F.S.L. of parent canal
    (d) none of the above.

312. The full supply level (F.S.L.) of a gravity canal is
    (a) slightly above the ground level
    (b) exactly at the ground level
    (c) always below the ground level
    (d) about 4 to 5 m above the ground level.

313. The berms formed finally in the canals will serve the purpose of
    (a) strengthening of banks
    (b) controlling seepage loss
    (c) protecting the banks from erosion
    (d) all the above.

314. The difference between the level of the top of the bank and the F.S.L. is known as
    (a) safe margin depths
    (b) berm
    (c) free board
    (d) none of the above.

315. The minimum free board to be provided should be 90 cm if the discharge in the canal is between
    (a) 100 and 150 m^3/s
    (b) 60 and 100 m^3/s
    (c) 30 and 60 m^3/s
    (d) 30 and 33 m^3/s.

316. Borrow pits should preferably be taken from the
    (a) field on the left side of the canal
    (b) field on the right side of the canal
    (c) field on the left as well as right of the canal
    (d) central half width of the section of the canal.

317. Disposal of excess excavated earth is used to make spoil bank on
    (a) left side
    (b) right side
    (c) both the sides
    (d) all the above.

318. A land is said to be waterlogged if the pores in the soil are saturated within
    (a) a depth of 40 cm
    (b) a depth of 50 cm
    (c) root zone depth
    (d) all the above.

319. The top of the capillary zone
    (a) lies above the piezometric surface at all points
    (b) lies below the piezometric surface at all points
    (c) coincides with the piezometric surface
    (d) none of the above.

320. Waterlogging can be prevented by
    (a) reducing the percolation from canals and water courses
    (b) increasing the ground water withdrawls
    (c) providing efficient subsurface drainage system
    (d) all the above.

321. A canal has a length of L km and the perimeter of its cross-section is P m. The amount required to line the canal, if the cost of lining per sq. m is ₹ C, will be
    (a) 10000 ALC
    (b) $\dfrac{ALC}{10000}$
    (c) 1000 PLC
    (d) $\dfrac{ALC}{1000}$.

322. A triangular channel section has side slopes of 1 vertical: z horizontal. If the depth of flow is y, what is the hydraulic depth?
    (a) y
    (b) $\dfrac{y}{3}$
    (c) $\dfrac{y}{4}$
    (d) $\dfrac{y}{2}$.

323. The most desirable section for a lined canal is
    (a) trapezoidal with rounded corners for large sections
    (b) triangular with round bottom for small sections
    (c) both (a) and (b)
    (d) none of the above.

324. The vertical component of the hydrostatic force on an inclined surface is obtained as
    (a) half of the horizontal component
    (b) zero
    (c) equal to horizontal component
    (d) as the weight of water resting on the inclined surface.

325. The design of a major hydraulic structure founded on a permeable foundation is generally based on
    (a) Khosla's theory of independent variables
    (b) Bligh's creep theory
    (c) Lane's weighted creep theory
    (d) none of the above.

326. The momentum equation as applied to a hydraulic jump in a rectangular channel of negligible bed slope may be written as

   (a) $y_1^2 - y_2^2 = \dfrac{2q}{g}(v_1 - v_2)$ ☐
   (b) $y_2^2 - y_1^2 = \dfrac{2q}{g}(v_1 - v_2)$ ☐
   (c) $y_1^2 - y_2^2 = \dfrac{2q}{g}(v_2 - v_1)$ ☐
   (d) $y_1^2 + y_2^2 = \dfrac{2q}{g}(v_1 + v_2)$ ☐

327. For a given discharge in a channel, Blench curves give the relationship between the loss of head and
   (a) specific energy u/s ☐
   (b) specific energy d/s ☐
   (c) critical depth of water d/s ☐
   (d) depth of water d/s. ☐

328. The purpose of a cross regulator is
   (a) to increase the depth of flow u/s when the main canal is running with low flow ☐
   (b) to regulate the flow in the distributories ☐
   (c) to pass excess flood water ☐
   (d) none of the above. ☐

329. Pick up the correct statement from the following:
   (a) The escape must lead the surplus water to natural drains. ☐
   (b) The capacity of escape should not be less than the capacity of canal at its location. ☐
   (c) Escapes are essential safety valves in canal system. ☐
   (d) All the above. ☐

330. Bed bars are provided in a canal
   (a) to control the silt entry ☐
   (b) to gauge the discharge ☐
   (c) to rise the supply level ☐
   (d) to watch the general behaviour of the canal. ☐

331. The flow in a canal syphon is
   (a) pressure flow ☐
   (b) under atmospheric ☐
   (c) critical flow ☐
   (d) none of the above. ☐

332. When a canal and natural drain approach each other at a same elevation, the cross drainage work provided is
   (a) an aqueduct ☐
   (b) a syphon ☐
   (c) a level crossing ☐
   (d) any of the above. ☐

333. The uplift pressure on the roof of an inverted syphon will be maximum when
   (a) drain is running with H.F.L. ☐
   (b) drain is running dry ☐
   (c) canal is running with F.S.L. ☐
   (d) canal is running dry. ☐

334. The meandering of a river is most likely in
   (a) boulder stage of the river ☐
   (b) delta stage of the river ☐
   (c) rocky stage of the river ☐
   (d) trough stage of the river. ☐

335. The crest level of a barrage is kept
   (a) almost at the river bed level with large gates ☐
   (b) high with no gates ☐

(c) high with large gates
(d) low with no gates.

336. Meandering of the river generally takes place, in
   (a) rocky stage
   (b) delta stage
   (c) boulder stage
   (d) trough stage.

337. The width of a meander belt is the transverse distance between
   (a) apex point of one curve and the apex point of the reverse curve
   (b) apex point and the crossing
   (c) two banks of meandering river
   (d) none of the above.

338. The sinuosity of a meander may be defined as the ratio of
   (a) meander length to width of meander
   (b) meander length to half the width of the meander
   (c) curved length to the straight distance
   (d) none of the above.

339. A deficit in the sediment load of flowing water may cause a river to become
   (a) meandering type
   (b) aggrading type
   (c) degrading type
   (d) none of the above.

340. In general, the length of meander, the width of meander and the width of the river vary as
   (a) Q
   (b) $Q^{1/2}$
   (c) Q^2
   (d) Q^3.

341. A Francis turbine is
   (a) an inward flow reaction turbine
   (b) outward flow reaction turbine
   (c) axial flow reaction turbine
   (d) none of the above.

342. A Kaplan turbine is
   (a) low head axial flow turbine
   (b) high head reaction turbine
   (c) medium head reaction turbine
   (d) high head impulse turbine.

343. In two similar turbines of different size, which of the following quantities should be equal
   (a) $\left(\dfrac{H}{N^2D^2}\right)$
   (b) $\left(\dfrac{Q}{ND^3}\right)$
   (c) $\dfrac{N\sqrt{P}}{H^{5/4}}$
   (d) all the above.

344. A turbine runs at 500 r.p.m. when working under a head of 16 m. If the head falls to 4 m what is the new speed in r.p.m.?
   (a) 125
   (b) 200
   (c) 250
   (d) 400.

345. Runaway speed of a turbine is
   (a) the normal running speed
   (b) speed at no load when governor fails
   (c) synchronous speed
   (d) none of the above.

346. A pumped storage plant is generally used as
    (a) a base load plant
    (b) a peak load plant
    (c) an emergency plant
    (d) none of the above.

347. The inclination of the deck with the horizontal in the case of a buttress dam is usually kept between
    (a) 10° and 15°
    (b) 15° and 20°
    (c) 20° and 25°
    (d) 35° and 45°.

348. In the case of a buttress dam too steep a slope may lead to
    (a) sliding
    (b) increased inclined stress
    (c) over turning
    (d) none of the above.

349. In the case of a buttress dam too flat a slope may lead to
    (a) excessive tensile stress on u/s face
    (b) high inclined stress at the heel
    (c) over turning
    (d) over stressing of buttress.

350. The slenderness ratio of a buttress is given by
    (a) $\dfrac{\text{Height of butters}}{\text{Thickness of buttress}}$
    (b) $\dfrac{\text{Height of butters}}{\text{Width of buttress}}$
    (c) $\dfrac{\text{Height of butters}}{\text{Spacing of buttress}}$
    (d) any of the above.

351. The advantage of burried penstock
    (a) less accessible to inspection
    (b) tendency to slide along steep slopes
    (c) no expansion joints needed
    (d) location difficult.

352. The function of an anchor block is
    (a) it prevents the pipe from sliding down the hill
    (b) it controls the movement of pipeline due to vibration of water hammer
    (c) it resists the unbalanced hydrodynamic force at a change in direction
    (d) all the above.

353. Which one of the following is an integral part of an electric generator?
    (a) transformer
    (b) exciter
    (c) circuit breaker
    (d) bus bar.

354. The Froude number may also be defined as the ratio of the velocity of flow in the channel to
    (a) the velocity of sound in water
    (b) the velocity of sound in air
    (c) the velocity of wave celerity in shallow depths
    (d) any of the above.

355. The maximum power which can be produced throughout the year at a hydroelectric plant is known as
    (a) Firm power
    (b) Primary power
    (c) Base power
    (d) any of the above.

356. The power which is available on and above the firm power is known as
    (a) surplus power
    (b) secondary power
    (c) grace power
    (d) none of the above.

357. The movement of water under or around a structure built on a permeable foundation is called
    (a) infiltration
    (b) deep percolation
    (c) seepage
    (d) creep.

358. The irrigation with sewage, (instead of natural water) where the disposal of sewage is the primary object is called
    (a) sewage irrigation
    (b) wealth from waste irrigation
    (c) broad irrigation
    (d) none of the above.

359. The proportion of silt in water is called
    (a) silt density
    (b) silt factor
    (c) silt ratio
    (d) silt charge.

360. If W and D denote the bed-width and depth of flow in a channel the combined seepage and evaporation losses per km length are proportional to
    (a) $(W + D)^{1/6}$
    (b) $(W + D)^{1/3}$
    (c) $(W + D)^{1/2}$
    (d) $(W + D)^{2/3}$.

361. Stone pitching, or any other material laid on a sloping face of an earthen bank to maintain its slope or to protect it from erosion is called
    (a) filter
    (b) revetment
    (c) shrouding
    (d) none of the above.

362. An excavation filled with impervious material to reduce seepage is called
    (a) cut of trench
    (b) key trench
    (c) both (a) and (b)
    (d) none of the above.

363. A weir in which the tail water level is more than the crest level is called
    (a) stalled weir
    (b) submerged weir
    (c) flooded weir
    (d) all the above.

364. A weir in which the tail water level is less than the crest level is called
    (a) free weir
    (b) normal weir
    (c) aerated weir
    (d) natural weir.

365. A fall in which the crest is kept at or near the canal bed without any glacis, is
    (a) natural fall
    (b) notch fall
    (c) gravity fall
    (d) none of the above.

366. The difference in levels between the top of capillary zone and the water table is
    (a) zero
    (b) negative
    (c) positive
    (d) difficult to tell.

367. In Montague type fall
    (a) a straight glacis is used
    (b) a parabolic glacis is used
    (c) a cycloidal glacis is used
    (d) no glacis is used.

368. The ratio of inertial forces to gravity forces is called
   (a) Gravity number
   (b) Reynolds number
   (c) Weber number
   (d) Froude number.

369. The expression for Froude number is
   (a) $\dfrac{V}{\sqrt{gd}}$
   (b) $\dfrac{V^2}{\sqrt{gd}}$
   (c) $\sqrt{\dfrac{V}{gd}}$
   (d) $\dfrac{\sqrt{V}}{gd}$.

370. In the equation $F = \dfrac{V}{\sqrt{gd}}$, d denotes
   (a) the depth of flow
   (b) hydraulic depth
   (c) hydraulic radius
   (d) any of the above.

371. In a M_2 profile at a section the depth of flow is 1.5 m. Another section with a depth of 1.8 m will be
   (a) u/s of the first section
   (b) d/s of the first section
   (c) sometimes u/s and sometimes d/s
   (d) none of the above.

372. The syphon well drop is ideal for
   (a) low discharges and low drops
   (b) low discharges and large drops
   (c) high discharges and low drops
   (d) high discharges and high drops.

373. In a homogeneous earth dam without d/s horizontal filter, portion of the d/s face through which seepage flow comes out is
   (a) stream line
   (b) equipotential line
   (c) neither a stream line nor an equipotential line
   (d) both (a) and (b).

374. A canal carries a discharge of 5 m³/s with a slope of $\dfrac{1}{100}$. For the same depth of flow its discharge carrying capacity will be doubled by increasing its slope to
   (a) $\dfrac{1}{50}$
   (b) $\dfrac{1}{25}$
   (c) $\dfrac{1}{10}$
   (d) $\dfrac{1}{75}$.

375. The profile formed between the sluice gate and the hydraulic jump is
   (a) M_3
   (b) C_3
   (c) S_3
   (d) A_3.

376. Crop rotation means
   (a) not growing any crop during alternate year
   (b) growing simultaneously two different crops in alternate rows

(c) growing different crops in successive seasons
(d) none of the above

377. Natural fertility is generally found in
(a) black soils
(b) alluvium soils
(c) red soils
(d) sandy soils.

378. The duty of well water is more than the duty of canal water because
(a) well water is less turbid than canal water
(b) well water is used more economically
(c) well water irrigation incurs less conveyance losses
(d) both (b) and (c).

379. Earthquake forces acting on a gravity dam
(a) reduce the self weight of the dam
(b) increase the uplift forces on the base of the dam
(c) increase the horizontal water pressure acting on the dam
(d) none of the above.

380. The source from which the soil is obtained to make up for the difference between the soil required for embankment and the soil available from the canal excavation is known as
(a) Borrow pit
(b) spoil bank
(c) dowel
(d) soil extractor.

381. The canal breach may occur when
(a) piping takes place through the canal bank
(b) canal banks are over-topped by excess flows
(c) cultivator cuts the embankment to derive additional supply
(d) all the above.

382. Fluming of a canal in the cross-drainage work will
(a) increase the duty
(b) avoid uplift pressure
(c) reduce the cost of cross drainage work
(d) enable the canal to draw water from drain.

383. When the length of the body wall of the fall is less than the normal width of the canal, the fall is then called
(a) Notch fall
(b) flumed fall
(c) Sarda fall
(d) ogee fall.

384. Which of the following soils has the least permeability
(a) clay
(b) silt
(c) gravel
(d) sand.

385. If the discharge over a weir is expressed as $Q = KH^n$ the value of n in the case of proportional weir is
(a) 2.5
(b) 2.0
(c) 1.5
(d) 1.0.

386. In the above problem, what is the value of n in the case of cippoletti weir
    (a) 1.0
    (b) 1.5
    (c) 2.0
    (d) 2.5.
387. The proportional weir is also called as
    (a) Sutro weir
    (b) cippoletti weir
    (c) contractionless weir
    (d) suppressed weir.
388. If m is the average particle size of silt in mm, then the silt factor is proportional to
    (a) m
    (b) $\sqrt{m}$
    (c) $m^{2/3}$
    (d) $m^{1/5}$.
389. In the case of flow over a rectangular suppressed weir, the pressure underneath the nappe will be
    (a) very high
    (b) moderately high
    (c) negative
    (d) atmospheric.
390. The slope of the sides of a cippoletti weir is
    (a) $1H : 6V$
    (b) $1H : 4V$
    (c) $1H : 2V$
    (d) $1H : 1V$.
391. Which of the following is also known as superfluous water?
    (a) gravitational water
    (b) capillary water
    (c) hygroscopic water
    (d) water in overland flow.
392. According to Khosla's potential flow theory, the undermining of the floor commences at
    (a) intermediate point
    (b) starting point
    (c) tail end
    (d) none of the above.
393. The degree of sinuosity is the ratio between the
    (a) meander length and width of meander
    (b) meander length and width of river
    (c) curved length and straight air distance
    (d) none of the above.
394. Tortuosity of meandering river is the ratio of
    (a) meandering length to width of meander
    (b) curved length along the river to the direct axial length
    (c) inverse of (b)
    (d) meander length to width of river.
395. The bed of the canal is lowered in the case of
    (a) canal syphon
    (b) level crossing
    (c) syphon aqueduct
    (d) all the above.
396. Flow mass curve is used to determine
    (a) the storage capacity of reservoir for a given demand
    (b) the average demand that can be met by the given storage capacity of the reservoir
    (c) the evaporation losses from the reservoir
    (d) both (a) and (b).

397. Flow duration curve is a convenient tool to assess the ...... available at the site.
    (a) firm power
    (b) secondary power
    (c) tertiary power
    (d) average power.

398. In Lanes weighted creep theory the weights proposed for horizontal and vertical creeps are in the ratio of
    (a) $1H:3V$
    (b) $3H:1V$
    (c) $2H:1V$
    (d) $1H:2V$.

399. The critical hydraulic gradient, according to Khoslas theory, in alluvial soils is approximately
    (a) 2.5
    (b) 2.0
    (c) 1.5
    (d) 1.0.

400. Which of the following is a rigid dam?
    (a) Earthfill dam
    (b) Rockfill dam
    (c) Buttress dam
    (d) none of the above.

ANSWERS
Objective Type Questions

1. (b)	2. (c)	3. (c)	4. (b)	5. (a)	6. (c)
7. (d)	8. (b)	9. (d)	10. (b)	11. (d)	12. (b)
13. (d)	14. (c)	15. (a)	16. (c)	17. (a)	18. (d)
19. (d)	20. (b)	21. (a)	22. (d)	23. (c)	24. (a)
25. (c)	26. (b)	27. (c)	28. (c)	29. (d)	30. (b)
31. (b)	32. (b)	33. (c)	34. (c)	35. (b)	36. (c)
37. (d)	38. (d)	39. (a)	40. (b)	41. (c)	42. (d)
43. (c)	44. (b)	45. (c)	46. (b)	47. (a)	48. (c)
49. (c)	50. (d)	51. (c)	52. (b)	53. (c)	54. (a)
55. (a)	56. (d)	57. (b)	58. (c)	59. (a)	60. (b)
61. (b)	62. (c)	63. (d)	64. (c)	65. (c)	66. (c)
67. (c)	68. (d)	69. (a)	70. (d)	71. (a)	72. (b)
73. (b)	74. (b)	75. (d)	76. (a)	77. (c)	78. (b)
79. (d)	80. (a)	81. (b)	82. (c)	83. (c)	84. (c)
85. (d)	86. (c)	87. (a)	88. (b)	89. (c)	90. (a)
91. (a)	92. (d)	93. (a)	94. (c)	95. (c)	96. (c)
97. (b)	98. (a)	99. (c)	100. (d)	101. (d)	102. (d)
103. (a)	104. (c)	105. (b)	106. (a)	107. (c)	108. (d)
109. (a)	110. (c)	111. (b)	112. (a)	113. (b)	114. (d)
115. (b)	116. (c)	117. (b)	118. (a)	119. (b)	120. (c)
121. (a)	122. (d)	123. (c)	124. (b)	125. (b)	126. (a)
127. (c)	128. (a)	129. (b)	130. (b)	131. (c)	132. (d)
133. (c)	134. (a)	135. (c)	136. (a)	137. (d)	138. (a)
139. (c)	140. (d)	141. (a)	142. (b)	143. (d)	144. (b)
145. (c)	146. (a)	147. (a)	148. (b)	149. (d)	150. (c)

151. (b)	152. (b)	153. (b)	154. (d)	155. (b)	156. (c)
157. (a)	158. (b)	159. (a)	160. (c)	161. (c)	162. (a)
163. (b)	164. (c)	165. (d)	166. (c)	167. (b)	168. (b)
169. (a)	170. (b)	171. (a)	172. (d)	173. (c)	174. (a)
175. (b)	176. (d)	177. (b)	178. (a)	179. (c)	180. (c)
181. (b)	182. (b)	183. (c)	184. (d)	185. (d)	186. (b)
187. (c)	188. (a)	189. (d)	190. (d)	191. (b)	192. (a)
193. (c)	194. (b)	195. (c)	196. (c)	197. (b)	198. (c)
199. (d)	200. (d)	201. (b)	202. (c)	203. (b)	204. (c)
205. (c)	206. (a)	207. (b)	208. (b)	209. (c)	210. (b)
211. (b)	212. (c)	213. (c)	214. (a)	215. (c)	216. (c)
217. (d)	218. (c)	219. (a)	220. (a)	221. (c)	222. (c)
223. (b)	224. (a)	225. (b)	226. (c)	227. (c)	228. (d)
229. (b)	230. (b)	231. (b)	232. (b)	233. (a)	234. (d)
235. (a)	236. (c)	237. (c)	238. (d)	239. (d)	240. (c)
241. (b)	242. (a)	243. (b)	244. (a)	245. (a)	246. (d)
247. (a)	248. (b)	249. (c)	250. (a)	251. (d)	252. (b)
253. (a)	254. (c)	255. (c)	256. (c)	257. (c)	258. (b)
259. (d)	260. (b)	261. (c)	262. (c)	263. (a)	264. (d)
265. (d)	266. (b)	267. (b)	268. (d)	269. (b)	270. (d)
271. (a)	272. (c)	273. (b)	274. (d)	275. (c)	276. (c)
277. (b)	278. (c)	279. (d)	280. (d)	281. (a)	282. (b)
283. (b)	284. (c)	285. (d)	286. (c)	287. (c)	288. (c)
289. (c)	290. (a)	291. (a)	292. (c)	293. (b)	294. (c)
295. (a)	296. (d)	297. (b)	298. (a)	299. (c)	300. (c)
301. (a)	302. (c)	303. (a)	304. (b)	305. (d)	306. (d)
307. (a)	308. (b)	309. (a)	310. (d)	311. (b)	312. (a)
313. (d)	314. (b)	315. (b)	316. (d)	317. (d)	318. (c)
319. (a)	320. (d)	321. (c)	322. (d)	323. (c)	324. (d)
325. (a)	326. (b)	327. (b)	328. (a)	329. (d)	330. (d)
331. (a)	332. (c)	333. (a)	334. (b)	335. (a)	336. (d)
337. (a)	338. (c)	339. (c)	340. (b)	341. (a)	342. (a)
343. (d)	344. (c)	345. (b)	346. (b)	347. (d)	348. (a)
349. (b)	350. (a)	351. (c)	352. (d)	353. (b)	354. (c)
355. (a)	356. (b)	357. (d)	358. (c)	359. (d)	360. (d)
361. (b)	362. (c)	363. (b)	364. (a)	365. (b)	366. (c)
367. (b)	368. (d)	369. (a)	370. (b)	371. (a)	372. (a)
373. (c)	374. (b)	375. (a)	376. (c)	377. (b)	378. (d)
379. (c)	380. (a)	381. (d)	382. (c)	383. (c)	384. (a)
385. (d)	386. (b)	387. (a)	388. (b)	389. (c)	390. (b)
391. (a)	392. (c)	393. (c)	394. (b)	395. (a)	396. (d)
397. (a)	398. (a)	399. (d)	400. (c)		

Chapter 6 WATER SUPPLY ENGINEERING

I. INTRODUCTION

WATER BORNE DISEASES

Polluted water spreads certain communicable diseases.

1. **Bacterial origin:** Cholera, Typhoid, Paratyphoid, Bacillary dysentery
2. **Protozoal origin:** Amoebic dysentery
3. **Viral origin:** Poliomyelities Infectious Hepatitis (Yellow Jaundice)
4. **Helminthal origin:** Hookworm, Roundworm and Tapeworm diseases
5. **Chemical origin:**

    (*a*) Methemoglobinemia—(Blue Baby Syndrome)—Occurs only in infants less than 6 months old due to NITRITES in drinking water

    (*b*) Fluorosis: Dental fluorosis: Enamel of teeth is destroyed

    Skeletal fluorosis: Bones loose elasticity and become stiff. Blood coagulation and tissue respiration are also affected

    (*c*) (Endemic) Goitre: Thyroid gland affected due to deficiency of Iodine in drinking water

    (*d*) Lead: Cumulative poison.

Water Related Diseases

They spread because of insect vectors breeding on water: Malaria, Filaria, Yellow fever, Dengue, Chicken guinea

Protected Water: Water is said to be protected when it is
1. Aesthetically attractive
2. Palatable (tasty)
3. Hygienically safe and free from diseases causing germs.
4. Possessing concentration of minerals just at optimum level.
5. Free from odours and colours.

DEMAND OF WATER

1. Domestic demand—for drinking, bathing, cooking, washing, flushing, cleaning and ablution. It depends on (*i*) customs and habits of people, (*ii*) climate, (*iii*) system and supply, (*iv*) quality of water, (*v*) method of sewage disposal, (*vi*) status of consumer, (*vii*) distribution pressure, (*viii*) method of charging, and (*ix*) alternate sources.

   As per Indian Standards, it is 135 litres/capita/day.
2. Industrial and commercial demand.
3. Public demand
4. Fire demand

(*a*) **Based on Population**

   (*i*) National Board of Fire Underwriters $Q = 3860 \sqrt{P}[1 - 0.01\sqrt{P}]$.

   (*ii*) Kuichling $\quad Q = 3180\sqrt{P}$

   (*iii*) Freeman $\quad Q = 1135\left(\dfrac{P}{5} + 10\right)$

   (*iv*) Burton $\quad Q = 5660\sqrt{P}$

   Q = Quantity of water in litres/minute

   P = Population in thousands

(*b*) **Based on the "Nature" of Materials**

Insurance Services Office Formula:

$$F = 3.7 \times C \times \sqrt{A}$$

F = Fire flow in litres/second

C = Co-efficient depending on the inflammability of the type of construction

   1.5 for inflammable materials as wood

   1.0 for ordinary construction

   0.6 for fire resistant construction

A = Total floor area in all the storeys of a building in sq. m.

A minimum of 4 streams to supply water jets incessantly for a minimum of 4 hours is required.

5. Leakages, losses and thefts.
   Leakages are due to bad quality pipes, appurtenances and bad workmanship.
   Losses are due to bad management.
   Thefts are illegal tappings.

METHOD OF FORECASTING FUTURE POPULATION

1. **Arithmetical increase method,** $P_n = P + n \times a$

   P = Present population

   P_n = Population after 'n' decades

   a = Average increase per decade.

   It gives too low results for young cities.

   Best suited for old towns with sluggish growth.

2. **Incremental increase method,** $P_n = P + n \cdot a + \dfrac{n(n+1)}{2} \cdot b$

   where b = Incremental increase per decade.

3. **Geometrical increase method,** $P_n = P\left[1 + \dfrac{r}{100}\right]^n$

   where r = Per cent growth per decade.

   It gives very high rates.

   Best suited for cities in their initial stages of growth.

4. **Decreasing rate of increase method:**

   (Declining growth method)

   $$P_n = P\left[1 + \dfrac{r-s}{100}\right]\left[1 + \dfrac{r-2s}{100}\right]\left[1 + \dfrac{r-3s}{100}\right]\ldots\left[1 + \dfrac{r-ns}{100}\right]$$

   where s = decrease in % growth.

5. **Graphical method.** A neat curve of population versus Decades is drawn with the existing data and it is skillfully extended for the future decades.

6. **Comparison method.** The city under question is compared with a well developed city having similar features and future population computed.

DESIGN PERIOD

It is the period of posterity for which treatment units as pumps and pipe lines are to be designed.

10 to 60 years are adopted in India

If
  (*i*) Inflation rate is high
  (*ii*) Rate of interest on the loan for the projected is low
  (*iii*) City has slow growth rate
  (*iv*) Superior quality pipes and fittings are available
  (*v*) Difficult to relay the pipeline then longer design periods are adopted.

QUALITY OF WATER

Impurity	Causes
Suspended Impurities	
(a) Silt, Clay	Turbidity
(b) Bacteria	
Pathogenic	Typhoid, Paratyphoid, Cholera, Bascillary, Dysentery.
Non-Pathogenic	1. Slime forming bacteria
	2. Iron bacteria—Cause pitting, turbidity.
	3. Sulphur bacteria—Acidity, corrosion.
(c) Virus	1. Infectious hepatities (Jaundice)
	2. Poliomyelities.
(d) Algae and Protozoa	—Colour, Turbidity, Odour, Taste, Acidity and Amoebic dysentery.
(e) Helminths	—Tape worm, Round worm
Dissolved Impurities	
(a) $Ca(HCO_3)_2$, $Mg(HCO_3)_2$,	Alkalinity, Hardness
$CaSO_4$, $MgSO_4$,	Hardness, Laxative effect
$CaCl_2$, $MgCl_2$	Corrosion
(b) Na_2SO_4	Foaming in boilers
(c) NaF	Fluorosis
(d) Iron and Manganese	Red, Brown/Black Colour, Taste, Corrosion.
(e) Lead	Cumulative Poison, Plumbo solvency
(f) Nitrates	Methamoglobinemia (Blue babies)
(g) CO_2	Acidity, Corrosion
(h) H_2S	Acidity, Corrosion and Foul-smell.

TESTS ON WATER

ALKALINITY

It is the ability to neutralize H^+ ions.

Sources. Salts of weak acids and strong bases, removal of CO_2 by algae during photosynthesis.

    Types. Bicarbonate alkalinity,

              Carbonate alkalinity,

              Hydroxide alkalinity,

Phenolpthalein end point (at a pH of 8.3) indicates complete neutralization of OH^- alkalinity + half of CO_3^{--} alkalinity.

Methyl orange end point (at a pH of 4.5) indicates the total alkalinity due to OH^-, CO_3^{--} and HCO_3^-.

Bicarbonate alkalinity and hydroxide alkalinity do not coexist.

Hydroxide alkalinity is also called caustic alkalinity

While Bicarbonate Alkalinity is almost harmless, Hydroxide Alkalinity is highly undesirable.

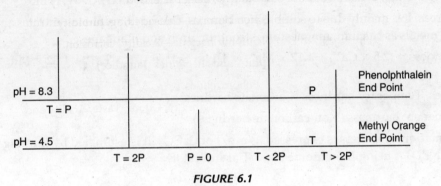

FIGURE 6.1

When $T = P$ only Hydroxide alkalinity exists

When $T = 2P$ only Carbonate alkalinity exists

When $P = 0$ only Bicarbonate alkalinity exists

When $T < 2P$, Hydroxide alkalinity = $2P - T$, Carbonate Alkalinity = $2(T - P)$

When $T > 2P$, Carbonate Alkalinity = $2P$, Bicarbonate Alkalinity = $T - 2P$

Estimation. Alkalinity is expressed as mg/l of $CaCO_3$.

$$= \frac{\text{Vol. of acid} \times \text{Normality of acid } (0.02 \text{ N}) \times 1000 \times \text{Eq. wt. of } CaCO_3 \,(= 50)}{\text{Vol. of sample}}$$

Effects. Natural waters are slightly alkaline. Alkalinity imparts bitter taste.

Excess alkalinity is harmful for irrigation.

Excess bicarbonate alkalinity induces hardness and scale formation.

ACIDITY

It is the capacity to donate H^+ ions.

Sources. Hydrolysis of salts of strong acids and weak bases (as $AlSO_4$, $FeSO_4$), Carbon dioxide, Industrial wastes as those from Rayon mills, Metal finishing operations.

Types. Mineral acidity (pH < 4.5) is given by methyl orange end point.

Phenolphthalien end point (pH > 8.3) gives total acidity *i.e.*, both Mineral acidity and Carbon dioxide acidity.

Estimation. Acidity is expressed as mg/l of $CaCO_3$.

Acidity can be obtained by titrating the water sample with 0.02 N Sodium Hydroxide using methyl orange and phenolphthalein as indicators.

Effects. Acidity interferes with water treatment operations as softening.

It corrodes steel and zinc coating of pipe material.

It affects aquatic life.

It neutralises the binding power of cement when used in RCC.

HARDNESS

The water readily gives lather with soap is called SOFT WATER.

The water that consumes more soap to give lather is called HARD WATER

Sources. It is mainly due to soluble bicarbonates, chlorides, sulphates, nitrates and silicates of divalent metals as calcium, magnesium, strontium, iron and manganese.

Hardness = $(2.5 \times Ca^{++} + 4.12 \times Mg^{++} + 1.14 \times Sr^{++} + 1.79 \times Fe^{++} + 1.82 \times Mn^{++})$ mg/l as $CaCO_3$

Types. Temporary hardness (Carbonate hardness)

Permanent hardness (Non-carbonate hardness).

Estimation. Hardness is expressed as mg/l of $CaCO_3$. It is estimated by titrating the water sample with EDTA using Eriochrome black **T** as indicator.

Or

By titrating the water sample with standard soap solution.

When alkalinity is less than total hardness,

Carbonate hardness = Alkalinity

Non-carbonate hardness = Total hardness–Carbonate hardness

When alkalinity is greater than total hardness.

Carbonate hardness = Total hardness

Non-carbonate hardness = 0.

Classification:

0 to 75 mg/l	—Soft
75 to 150 mg/l	—Moderately hard
150 to 300 mg/l	—Hard
more than 300 mg/l	—Very hard

Effects. Scales formed in boilers because of hard water are insulators of heat. Hence more fuel is consumed. More soap consumed in laundry. Vegetables are toughened.

Advantages of hard waters:

Taste is induced to water because of dissolved salts.

Moderately hard water is preferred to absolutely soft water for irrigation.

Absolutely soft waters are "aggressive" *i.e.*, dissolve metals in them.

Cardiovascular diseases are reported with soft water as the calcium of hard waters replaces Cadmium that causes heart troubles. Calcium also strengthens bones.

CHLORIDES

Sources. Waste waters, ground sources, industrial wastes as tanneries.

    Estimation. Chlorides are estimated by titrating the water sample with silver nitrate using potassium chromate as indicator.

    i.e., all the chlorides are converted to silver chloride which is insoluble in water and forms white precipitate. After all the chloride is precipitated, Silver of silver Nitrate reacts with chromate of Potassium chromate forming silver chromate, brick red in colour.

    Effects. Excess chlorides are injurious to people suffering from heart and kidney troubles.

    Sodium chloride alone gives the saline taste. Chloride concentration < 250 mg/l is desirable for domestic consumption.

    Chloride anion corrodes the pipe material and increases conductivity.

RESIDUAL CHLORINE

It is the amount of chlorine remaining in water 15 to 30 minutes after the application of chlorine.

    Residual chlorine can be estimated by adding "Orthotolidine" to the sample and comparing the intensity of yellow colour developed which directly gives the amount of residual chlorine.

    Generally an amount of 0.05 to 0.2 mg/l of residual chlorine is required for municipal water supply.

DISSOLVED OXYGEN

It is the amount of oxygen soluble in a water sample at a particular temperature and pressure.

    Its solubility increases with pressure but reduces with increase in temperature. As impurities gather its concentration gets reduced.

    Dissolved oxygen for pure water at 0°C and at atmospheric pressure is 14.6 mg/l

    Dissolved oxygen for pure water at 20°C and at atmospheric pressure is 9 mg/l

    Dissolved oxygen for pure water at 35°C and at atmospheric pressure is 7 mg/l

    Sources. D.O. may shoot up to 30 mg/l momentarily because of photosynthesis of algae.

    Estimation. Winkler's (Iodimetric) Test:

    Oxygen present in water sample oxidizes Mn^{++} to Mn^{++++} when NaOH and KI are added. On acidification Mn^{++++} reverts to Mn^{++} liberating free iodine from KI. The liberated iodine is titrated against 0.025 N sodium thiosulphate using starch as indicator.

    Effects. A minimum D.O. of 2 to 4 mg/l is required for the survival of aquatic life.

    Higher D.O. causes corrosion.

SUSPENDED SOLIDS

They are insoluble particles remaining suspended in water because of velocity currents.

    Size–between 0.1 mm to 0.001 mm.

    Sources. Oils, greases, clay, silt, plant fibres, algae, bacteria.

    Estimation. One litre of sample filtered.

    Non-filterable solids represent suspended solids.

Effects. They give unpleasant appearance.

Because of them chemicals are adsorbed, pathogens are shielded.

DISSOLVED SOLIDS

Size 10^{-6} to 10^{-8} mm.

Sources. Inorganic and organic solids, decay products of vegetation.

Estimation. Filtered water when heated to 104°C the residue left over givens out. Total Dissolved Solids (TDS). When the residue is further ignited at 600°C for 1 hour—the content still remaining is called 'Fixed Solids" and it represents—"Inorganic fraction" of TDS. The content got volatilised is called 'volatile matter' and represent organic fraction of TDS.

Effects. Dissolved salts cause taste, odour and colour to water. They may alter the water quality and even render the water toxic. However 'taste' is attributed to water because of TDS only.

For domestic consumption concentration of TDS ≯ 1000 mg/l and preferably less than 500 mg/l.

CONDUCTIVITY

It is the capacity of a water sample to conduct electricity through it.

Sources. Pure water is a poor conductor of electricity.

More is TDS in a water sample more is its specific conductance.

Estimation. Specific conductance is the conductivity in a cubic centimetre field as 25°C.

It is measured as millisiemens per metre.

Ionised salts readily register more specific conductance whereas salts that do not ionise readily will not register any.

TURBIDITY

It is the interference due to suspended and colloidal matter for the passage of light.

Amount of turbidity depends on: number, size, shape and refractive index of suspended particles.

Sources. Particles of clay, silt, finely divided organic matter, micro-organisms, soaps, detergents.

Estimation. 1 mg of silica dissolved in 1 litre of distilled water is one unit of turbidity. Instead of silica when 1 mg of Formazin is dissolved in 1 litre of distilled water it is called Formazin Turbidity Unit (FTU).

Effects. Turbidity presents unpleasant appearance. It adsorbs chemicals and causes odours and tastes, interferes with light penetration and hence photosynthesis. Sediments deposited may harm aquatic life.

It shields bacteria and hence more amount of the disinfectant is required to kill them. It also increases load on filters.

SCHMUTZE DECKE (Zoogleal layer) of slow sand filter gets coated with silt when turbidity is very high. The slow sand fitter may go out of operation.

COLOUR

Colour is a visible pollutant. Apparent colour is because of suspended solids.

True colour is due to dissolved solids.

Sources. Iron, manganese, industrial wastes as mining, refineries, pulp and paper, chemicals, textiles and slaughter houses, decaying organic matter.

Estimation. 1 mg of platinum (in the form of chloroplatinate ions) + $\frac{1}{2}$ mg of metallic cobalt dissolved of 1 litre of distilled water gives 500 hazen units of colour.

Pure water is supposed to be COLOURLESS. Presence of colour degrades the quality of water.

Effects. Colour is aesthetically objectionable. It prevents photosynthesis of aquatic plants. Stains clothes. plumbing fixtures.

Not suited to laundering, dyeing, paper, dairy and beverage manufacture.

Water for domestic consumption should not have colour greater than 5 units.

TEMPERATURE

It is a catalyst, a depressant, an activator, a restrictor, a stimulator, a controller, a killer–is the most influential parameter.

Temperature affects self purification of stream, rate of chemical reactions, biodegradation of organic matter and solubility of gases.

Rise in temperature enhances toxicity of poisons and intensity of odours.

Low temperature affects coagulation, filtration and efficiency of chlorination.

Taste gets affected with rise in temperature.

Sudden increases in temperature indicates thermal pollution.

TASTES AND ODOURS

Water of agreeable taste is desired for domestic consumption.

Taste of a water sample can be sweet, sour, bitter or salty.

While odourless water is required for domestic consumption, offensive odours reduce appetite for food, impaired respiration, nausea, vomiting, mental perturbation and hence lowered water consumption.

Flavour Profile may be analysed as Aromatic, Bitter, Cucumber, Earthy, Fishy, Geranium, Goaty, Grassy, Mouldy, Putrefactive and Vile.

Sources 1. Microorganisms as 'Actinomycetes' (bacteria) and blue green algae

2. Decomposing organic matter

3. Dissolved gases

4. Chlorine (due to the formation of Chlorophenols)

5. Industrial wastes

6. Minerals.

Estimation. Threshold Odour Number (TON)

$$= \frac{x \text{ ml of odourous water sample} + (200 - x) \text{ of distilled water}}{x}$$

It is the 'DILUTION FACTOR' required to produce odour just perceptible.

TON greater than 3 is not recommended from aesthetic sense.

Effect. Tastes and odours may be objectionable from aesthetic point of view. Drinking water should have pleasant taste but without any odour. Salt waters cannot quench the thirst and on the other hand aggravate the thirst.

pH (potential Hydrogen)

It is the Hydrogen ion concentration or Hydrogen ion activity.

pH is the negative logorthm of the Hydrogen ion concentration.

A sample is said to be acidic when its pH is greater than '0' but less than '8.3'.

A sample is said to be alkaline when its pH is greater than '4.5' but less than '14'.

Thus in a pH range of 4.5 to 8.3, the sample can be acidic due to Carbon dioxide acidity (H_2CO_3) or alkaline due to Bicarbonate (HCO_3^-) Alkalinity.

Thus both Acidity and Alkalinity may coexist in a sample.

IRON AND MANGANESE

Ferrous Iron—(Fe^{++}) and Manganous salts—(Mn^{++}) are highly soluble in water and cause—colour (Iron—red colour, Mn—black colour), turbidity and metallic taste.

Consumption of Iron rich water may not pose problems, but over a prolonged period may cause HEMOSIDEROSIS—leading to liver disorder.

Eggs easily get spoiled when washed in waters rich in iron concentration.

Clothes are stained when washed in Iron and Manganese rich water.

Cooking utensils and plumbing fixtures (as wash basins and commodes) are also stained.

In the presence of oxygen, gets converted to Fe^{+++} and this ferric iron is precipitated and deposited as rusty, gelatinous, slimy lumps of TUBERCLES of Ferric Hydroxides on the walls of the carrying pipe.

Tubercles roughen the surface of flow and reduce the carrying capacity of the pipe line.

While Ferrous and Manganous salts are soluble in water, Ferric (Fe^{+++}) and Manganic (Mn^{++++}) salts are insoluble and get separated.

INDICATOR ORGANISM

ESCHERCHIA COLI (E. Coli) are Aerobic, Non spore forming, Gram stain, Negative rods those ferment lactose with gas production within 48 h at 37°C (Human body temperature).

E. Coli are bacteria those stay in the intestine of human beings and come out along with excreta. Hence its presence in water indicates of contamination. It is called as indicator organism as its presence indicates bacterial contamination.

Standards for drinking water

Turbidity	≯	5 Units (NTU)
Colour	≯	5 Units (Platinum Cobalt scale)
Taste		None objectionable
Odour	≯	3 (Threshold Odour Number)
Total solids	≯	500 mg/l
pH	≮	7 ≯ 8.5
Total Hardness (as $CaCO_3$)	≯	200 mg/l
Chlorides	≯	200 mg/l
Sulphates	≯	200 mg/l
Fluorides	≯	1 mg/l
Nitrates	≯	45 mg/l
Iron	≯	0.1 mg/l
Manganese	≯	0.05 mg/l
Zinc	≯	5 mg/l
Lead	≯	0.1 mg/l
Cadmium	≯	0.01 mg/l
Coliform bacteria (E. Coli or B. Coli)		Nil/100 ml

SOURCES OF WATER

1. Surface Sources
2. Ground Sources

Surface Sources

Lakes. An elevated lake (formed on mountains slopes) gives the purest water as chances for its pollution are less. A lake is said to be **Oligotrophic,** when its waters are clear, transparent and nutrient poor.

Slight increase in the nutrient content makes the lake **Mesotrophic.** Waters of mesotrophic lakes are green coloured and rich in aquatic plants.

When a lake is turbid, nutrient (Nitrates and Phosphate) rich, rich in blue green algae and coarse fish, it is called **Eutrophic** lake.

Ponds and tanks are smaller lakes. A reservoir is artificially created lake by constructing a dam across the flow of a river or a stream.

Lake, pond, tank or reservoir water need not be subjected to sedimentation as they had little or no velocity of flow and are the stilling basins themselves.

Rivers. Rivers are large bodies of flowing water. Stream is a small river. Many stream merge to form a river.

Because of velocity of flow they gather suspended and colloidal solids. Rivers are the mostly polluted bodies because both domestic and industrial waste waters are let off into them and as they flow over a vast catchment area.

GROUND SOURCES

Wells. (*a*) Shallow well, (*b*) Deep well.

Shallow well is the well-formed by tapping the aquifer nearest to ground level. It is known for delivering impure water. It may get dried up during summer. A deep well is formed by tapping an aquifer below an impervious layer. Hence its quality may not reflect the conditions nearer to ground level.

It is less prone for pollution but may contain more dissolved solids.

Infiltration well is one provided nearer to river banks with an idea of getting pure water. Pure water requiring only disinfection shall get collected.

Infiltration Gallery is a perforated open jointed tunnel like pipe laid along or across a stream course at shallow depth to tap relatively pure water.

An Artesian well is a deep well with its aquifer confined between two impervious layers— and is under pressure.

Intake is a device to draw clean, safe and palatable water of sufficient quantities in all seasons from surface sources.

Treatment of Water

Impurity	Treatment
1. Floating solids	Screens
2. Suspended solids	Sedimentation
3. Colloidal solids	Sedimentation aided by coagulation
4. Micro organisms	Filtration, Disinfection
5. Dissolved gases	Aeration
6. Hardness	Chemical Treatment
7. Colour	Coagulation Filtration through activated carbon bed, chlorination
8. Turbidity	Coagulation, Filtration
9. Tastes and odours	Adsorption, Coagulation and Filtration, Chlorination, Aeration

Flow diagram

Raw water ⟶ Screens ⟶ Presedimentation ⟶ Aeration ⟶ Coagulation ⟶ Sedimentation ⟶ Filtration ⟶ Disinfection ⟶ pH Correction ⟶ Distribution

Screens

They are mechanical devices to exclude (floating) debris, (leaves, twigs etc.), fish and eels. Mesh screens with opening size of 1 mm to 25 mm are used.

AERATION

Aeration is the mechanical dispersion of water in air or injection of air into water to

1. expel colour and taste producing substances as H_2S and volatile organic compounds.
2. oxidise colour causing Fe^{++} and Mn^{++} compounds to insoluble Fe^{+++} and Mn^{++++} compounds.
3. expel acidity causing CO_2.
4. inject oxygen into water which improves taste, reduces BOD or nuisance potential, oxidises impurities and also disinfects water.

Cascades: It is the water in thin sheets falling through a head of 1 to 3 m at the rate of surface loading of 100 to 200 $m^2/m^3/s$ at a velocity of 0.3 m/s over sloping floor or that with a flight of steps.

Perforated or Packing tower is a cylindrical container of less than 3 m diameter through which water falls or trickles over a number of perforated plates while air current is sent up the column in the reverse direction of flow. Some times instead of perforated plates packed bed of broken stone is provided above air supply and below water jets.

It is designed at the rate of 2000 $m^3/m^2/d$

Nozzles: They are 20 to 40 mm in diameter horizontally spaced at intervals of 0.5 to 3.5 m, Spraying 5 to 10 l/s of water as fine jets operating at a pressure of 70 kPa.

Stand Pipe: It is a cylindrical pipe operating at a head of 1 to 9 m through which water flows up and slides along its perimeter during which time it is getting aerated.

It is designed at 100 to 300 $m^3/m^2/s$

Spray Tower: A round and cylindrical pipe of a number of circular trays are arranged at a vertical spacing of 250 to 750 mm.

The lower most tray is of bigger diameter and the diameters of trays decrease with increase in height. Some times the trays are filled with coke coated with a strong oxidising agent as potassium permanganate or dichromate the hasten oxidation.

The rate of loading is 1 to 1.5 m^3/m^2 per minute.

The trays are designed at 50 to 150 $m^3/m^2/s$.

Diffused Aerators: A rectangular tank is provided with perforated floor and sides is designed to have a width of 3 to 9 m, water depth of 2.5 to 5 m and the volume of the tank should not be greater than 150 m^3.

Through the diffused holes compressed air is injected into the water at the rate of 0.1 to 1 m^3 of air per m^3 of water.

SEDIMENTATION

It is the separation of suspended solids from water by stilling the water.

Sedimentation is the settling of solid particles from a liquid medium due to

(*i*) difference in the specific gravities of the solid particles and the medium

(*ii*) size and shape of the particles and mass action

(*iii*) viscosity of the medium.

Types of Sedimentation Tanks

(*i*) Conventional type

(*ii*) Shallow depth type or tube settlers

(a) *Fill and draw type.* Tanks are filled and kept stand still for the period of detention and then emptied. They are rarely used except for industrial use.

(b) *Continuous flow type.* Water shall be flowing at a velocity of 2.5 mm to 5 mm/s (15 cm to 30 cm per minute).

Based on shape they are mainly classified as

(i) Rectangular tanks (ii) Circular tanks

(iii) Hopper bottomed tanks

Over flow rate $\left(i.e., \dfrac{\text{Quantity treated per day}}{\text{Surface area}}\right)$ governs the efficiency rather than the detention period.

Rectangular Tanks

(a) Plain type (b) Baffled type

The purpose of the baffles is to increase the length of flow.

Design factors:

$$\dfrac{l}{b} \not> 4$$

$$b \not> 12 \text{ m}$$

Liquid depth	= 3 to 4 m
Free board	= 0.5 m
Surface loading	= 20 to 33 m³/m²/d
Weir loading	$\not>$ 250 m³/m/d
Detention period	= $\dfrac{24V}{Q}$ hours = 1 to 8 hours usually
where	V = Volume of tank (m³)
	Q = Discharge (m³/day)

Horizontal velocity of flow $\not>$ 2.5 mm/s.

Circular Tanks

(a) Radial flow type

(b) Peripheral flow type

Circular tanks have a diameter upto 60 m but 30 m is most common due to operational problems of sludge scrapper.

Hopper bottomed Tanks

They are circular or square in plan with steep sloping (about 60° with horizontal bottoms).

They are ideally suited for small quantities of highly turbid waters.

Tube Settlers

They are shallow depth sedimentation tanks. They consist of steeply inclined tubes 50 mm diameter/side circular, square or hexagonal in cross-section and 600 mm long. Water flows upwards—more is the contact area and sludge settles at bottom.

COAGULATION

Colloids are very fine particles of 1 to 500 nanometres (or millimicrometres) in size. They possess like charge and hence repel each other and move helter skelter called Brownian Movement. They are either hydrophobic (hate water) or hydrophilic (love water).

Coagulation is the process of destabilizing (by reducing repulsive force between them) these colloids by the addition of a coagulant. The inorganic and organic colloidal particles causing turbidity are hydrophobic and can be easily separated from water by the addition of Aluminium salts (alum) and iron salts (Copperas, Ferric Chloride etc). Coagulant is an electrolyte which neutralises the electric charge of colloids and hence they get agglomerated (come close to each other) as a FLOC.

Flocculation is the process of grouping of the decharged colloids by gentle mixing or stirring so that they become bulky and settle.

A coagulant aid is a substance that stimulates floc formation, reduces the amount of coagulant and widens the pH range of the coagulant. Activated carbon, activated silica and polyelectrolites are some coagulant aids.

Solutions of starch, soap and synthetic detergents contain hydrophilic colloids which are more stable.

Flash mixing (rapid mixing of coagulant for 1 minute), Flocculation (promoting the growth of floc by gentle stirring for about 30 minutes) and then sedimentation for 2 to 4 hours is the ideal treatment before filtration when the turbidity exceeds 25 NTU.

Flocculator—clarifier or clariflocculator is a single unit where "mixing of coagulant", "flocculation" and "sedimentation" are performed in a single compartment. Its overflow rate may be anywhere in between 20 and 100 $m^3/m^2/day$, depending on the quality of water.

FILTRATION

Filtration is the process of straining the water through a grannular bed.

Colour, turbidity, colloidal solids, taste and odour causing substances, iron, manganese and bacteria are removed by filtration.

Theory of Filtration

1. **Mechanical Straining:** Particles of smaller size pass through their voids and bigger particles are retained on sand.
2. **Sedimentation and Adsorption:** Voids between sand particles may act as miniature sedimentation tanks where very fine solids may settle down.
   Impurities may get adsorbed on to the sand particles.
3. **Electrolytic Action:** Media of filter have an electrolytic charge and impurities of opposite polarity are readily attracted.
4. **Biological Action:** Biological impurities as Bacteria, Protozoa, Algae and higher organisms get collected over the filter bed as a zoogleal layer, called SCHMUTZE DECKE. This zoogleal film consumes the impurties of water.

Based on the rate of filtration sand filters are classified as slow sand filters and rapid sand filters.

Slow Sand Filter

It is a water tight rectangular bed of 1000 to 5000 m² in area.

0.25 to 0.35 mm sand of 900 to 1500 mm depth is laid over 30 to 280 mm sized gravel of 300 to 500 mm depth.

A zoogleal mat **Schmutze** (dirty) **Decke** (skin) comprising of bacteria, algae and protozoa–formed on the top of the filter removes most of the impurities of the raw water.

Though the rate of filtration is very low, it removes 99.9% of bacteria. It does not require any pretreatment other than pre-sedimentation.

But when the turbidity of the raw water is more than 40 units the bed easily gets clogged.

Rapid Sand Filter

It is 8 to 200 m² square or rectangular tank with an enclosure tank.

0.4 mm to 1 mm sand of 600 to 750 mm depth is laid over 450 mm thick gravel layer.

Perforated pipes or perforated plates provided below the gravel bed collect filtered water and lead to the main.

When head loss increases and rate of filtration considerably gets reduced, wash water flowing upwards in the reverse direction of filtration cleans the media. This process is called Back washing.

Item	S.S.F.	R.S.F.
1. Rate of filtration	1 to 10 m³/m²/d (0.01 to 0.1 lit/m²/s)	100 to 200 m³/m²/d (1 to 2.3 lit/m² s)
2. Size	2000 m²	40 m² to 400 m²
3. Bed composition	300 mm deep gravel layer below a sand layer of 1 m reduced to 600 mm due to scrappings	450 mm deep gravel below a sand layer of 750 mm.
4. Size of sand	0.25 mm to 0.35 mm	0.45 mm and higher
Uniformity co-eff.	2 to 3	≤ 1.5.
5. Arrangement of sand grains	Unstratified	Finest at top. Size increases with depth.
6. Under drainage	Half round tile laterals draining into main	Perforated lateral pipes draining into main pipe
7. Loss of head	60 mm initially 1200 mm finally	300 mm initially 2.7 m finally.
8. Penetration of suspended solids	Superficial	Very deep.
9. Interval of cleaning	20 to 60 days	$\frac{1}{2}$ day to 2 days.
10. Method of cleaning	(i) Scrapping of top 12 mm layer	(i) Agitating the media by compressed air.
	(ii) Cleaning the exposed bed by jets of water	(ii) Lifting the impurities by the flow in the reverse direction of filtration.

11. Quantity of wash water	0.2% to 0.6% of quantity filtered	1% to 6% of quantity filtered.
12. Pre-treatment	Aeration, Sedimentation with or without coagulation (optional)	Aeration, Sedimentation aided by coagulation.
13. Post-treatment	Chlorination (optional)	Chlorination (compulsory)
Major process of treatment	Entrapment of impurities in the zoogleal film (SCHMUTZE DECKE)	Interception and interparticle settling of sand media.
14. Cost of construction	Higher	Lower
15. Cost of operation	Lower	Higher
16. Depreciation of plant	Lower	Higher
17. Efficiency	98 to 99% of bacteria removed	90% bacteria removed
	Colour fully removed if flocculated	Colour removed if flocculated
	Turbidity removed without coagulation	Turbidity removed by coagulation.
	Bed gets clogged if turbidity exceeds 40 units	Can take almost any amount of turbidity of natural waters.
18. Skill in operation	Highly skilled operation not required.	Efficiency depends on skilled supervision.

Pressure Filter. It is an enclosed cylindrical filter 1.5 to 3 m in diameter and 3 to 8 m high. Its composition of media is same as R.S. Filters. But a pressure of 3×10^5 to 7×10^5 Pa is applied for rapid filtration. Its rate of filtration is about 120 m^3/m^2/d (1.4 lit/m^2/s) to 250 m^3/m^2/d (2.8 litre/m^2/s).

However their use is confined to swimming pools and industries. Also back washing it not so effective in pressure filters.

DISINFECTION

Disinfection is the opposite of "infection" *i.e.*, it is the removal of causative organisms responsible for infection *i.e.*, pathogen removal.

Sterilization is the total destruction of all life *i.e.*, flora and fauna both pathogenic and non-pathogenic.

Factors influencing disinfection
1. Turbidity—shields the organism from the influence of disinfectant.
2. Colloids and particulates—adsorb the disinfectant.
3. Viruses, cysts and ova—are more resistant to disinfection. They need higher doses of disinfectant for their elimination.
4. Higher temperatures and longer time of contact increase the efficiency of disinfection.

A disinfectant should—
1. be easy of prepare, handle and store,
2. not alter the quality of water,
3. not corrode the pipes nor containers.

Chlorine: Chlorine is a yellowish green coloured gas with pungent odour. It is 2.5 times heavier than air and is moderately soluble in water. It is highly toxic and the toxicity increases with temperature.

Chlorine is the mostly used disinfectant. Besides killing pathogenic and nuisance micro-organisms, it precipitates iron and manganese in water, destroys odour and taste causing elements.

$$Cl_2 + H_2O \xrightarrow{pH > 4} HOCl + HCl$$

$$HOCl \xrightleftharpoons{pH\ (6\ to\ 10)} H^+ + OCl^-$$

Chlorine existing in water as "Hypochlorous acid" (HOCl) and "Hypochlorite ion" (OCl$^-$) is defined as "Free available chlorine".

'HOCl' is more effective than OCl$^-$ and hence disinfecting power of chlorine decreases with rise in pH.

At 20°C–Percentages of free available chlorine.

pH	% of HOCl	% of OCl$^-$
6	96.8	3.2
7	75.2	24.8
7.5	49.1	50.9
8	23.2	76.8
9	2.9	97.1

When ammonia present in water chlorine reacts with it to form Monochloramine (NH$_2$Cl) and Dichloramine (NHCl$_2$) called Combined Available Chlorine'. The disinfecting ability of Combined Available Chlorine is inferior to that of 'Free Available Chlorine' and takes more time.

Break Point Chlorination

When chlorine is a continuously applied for disinfection and 'Residual Chlorine' is gauged the following curve OACBD is obtained:

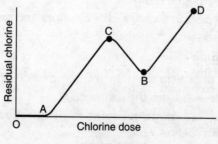

FIGURE 6.2

O to A—Free chlorine oxidizes reducing agents as nitrites, hydrogen sulphide, ferrous and manganous ions etc., and gets reduced to chloride which is not a disinfectant.

A to C—Organic matter reacts with chlorine to form chloro-organic compounds. Ammonia present in water reacts with chlorine to form monochloramine and dichloramine ($NH_2Cl + NHCl_2$) called "Combined Residual Chlorine". When its concentration exceeds 2 to 3 mg/l, it destroys pathogens after a contact period of 30 minutes.

C to B—Further addition of chlorine produces trichloramine (NCl_3), N, N_2O—none of which are disinfectants. Hence further addition of chlorine reduces available chlorine.

B—Break Point.

B to D—All reactions are completed and ammonia is completely oxidized. Any subsequent addition of chlorine shall remain as free available chlorine (HOCl).

A minimum of 0.2 mg/l of residual chlorine is desirable to guard against post-contamination.

To improve efficiency of coagulation, reduce tastes and odours and to reduce the load on filters—chlorine may be applied before filtration. It is called **Prechlorination**.

During epidemics a chlorine residual as high as 1 or 2 mg/l is left in drinking waters to prevent the spread of waterborne diseases. It is **Super Chlorination**.

Advantages of Chlorine: Chlorine is a very cheap disinfectant. It is highly bactericidal at low concentrations and leaves considerable residual. It is non toxic to higher forms of life. It controls Algae, Iron fixing and slime forming bacteria in pipe lines, Filter flies on Trickling filters, controls H_2S, anaerobic conditions besides reducing BOD.

Higher doses of chlorine induce odour, taste, cough and throat infection. Chlorinated hydrocarbons as

Trihalomethanes: Chlorinated hydrocarbons as Trihalomethanes (THMs) are suspected to be carcinogenous. Super chlorinated waters are also corrosive.

Chlorine is toxic to fish and causes eye irritation to swimmers. Hence **Dechlorination** is done by reducing agents such as sodium sulphite, sodium bisulphate, SO_2 or by adsorption on Activated carbon or by Aeration or boiling.

Residual chlorine in a water sample can be measured by adding "orthotolidine" to the water. Chlorine in water reacts with orthotolidine to form "Holoquinone", a yellow coloured compound. The intensity of the yellow colour is proportional to the amount of residual chlorine present.

Bromine is used to disinfect swimming pool water. It causes less eye irritation compared to chlorine.

Iodine is used to disinfect small quantities of water.

Ozonation

Ozone is an unstable isotope of oxygen of pungent smell.

It requires less contact time (30 s), less sensitive to pH, kills viruses and spores more effectively, oxidises other impurities causing tastes and odours, oxidises Iron and Manganese as well, the end products are non toxic and on the other hand enrich the water quality and is not carcinogeneous.

But as it is highly unstable, at least 3 times as costly as chlorine, leaves no residual—not universally adopted.

Ultra Violet Rays

Special mercury lamps enclosed in quartz globes emit ultraviolet rays. Clear waters (free from turbidity) in layers not deeper than 120 mm are exposed to the rays. In a contact time of less than 15 s, bacteria, viruses and even spores are killed.

Excess Lime Treatment

Too high or too low a pH causes pathogen kill. Excess lime raises pH to greater than 11, kills 99.9% of coliforms.

Ions of Heavy Metals

Silver, Copper and Mercury exert bactericidal properties. About 0.05 mg/l of Silver is highly effective. It is more efficient at higher temperatures.

Ultra Sonic Waves at frequencies of 20 to 400 kHz kill bacteria over contact period of 2 s to 60 min.

SOFTENING

Hard waters do not readily give lather with soap. They form scales in boilers.

However, extremely soft waters are tasteless, and cause pipe corrosion.

Hardness is mainly because of

1. $Ca(HCO_3)_2$
2. $Mg(HCO_3)_2$

Temporary hardness or Carbonate hardness

3. $CaCl_2$
4. $MgCl_2$
5. $CaSO_4$
6. $MgSO_4$

Permanent hardness or Non-carbonate hardness

Lime–Soda Process. Calcium carbonate and magnesium hydroxide are almost insoluble in water. Hence salts of calcium are precipitated out by the addition of soda (Na_2CO_3) as $CaCO_3$ and those of magnesium by the addition of lime [$Ca(OH)_2$] as $Mg(OH)_2$. $CaCO_3$ is precipitated at a pH of 9 to 9.5 while $Mg(OH)_2$ is precipitated at a pH of 11. Lime soda process is reasonably effective in reducing Turbidity, Colour, Bacteria, Viruses and Fluorides.

Zeolite or ion exchange process. Hard water is passed through sodium zeolite which absorbs Ca and Mg getting converted to Ca and Mg zeolites, releasing sodium salts. The rate of (hydraulic) loading is 230 to 460 $m^3/m^2/d$.

After some days of operation the entire sodium zeolite bed may get converted to Ca and Mg Zeolite. When the effluent is harder than the influent, it is called **Breakthrough**. Regeneration with 10% NaCl brine solution rejuvenates the bed.

Limitations of Zeolite Process

1. Raw water rich in turbidity, oil and H_2S coat the zeolite bed and make it ineffective.

2. Concentrations of Fe^{++} and Mn^{++} greater than 2 mg/l oxidize and precipitate on grains of zeolite.

3. Hardness greater than 800 mg/l require frequent regeneration.

4. Effluent is rich in Na salts–continuous consumption of which may affect people suffering from heart, kidney or circulatory ailments.

5. Waters are not to be chlorinated before the zeolite process treatment.

DEFLUORIDATION

Fluorine is the most active element known. It is not found free in nature.

It is present in more than 100 minerals as Fluorite, Cryolite, Apatite and Topaz, and in igneous rocks as Granites, Pegmatites and in volcanic flows.

More is the non-carbonate hardness (of Ca and Mg in particular) less is the fluoride concentration.

Effects of Fluorides

During the state of formation of permanent teeth in young children, fluorides combine chemically with tooth enamel forming harder and stronger teeth more resistant to 'Cavitation in teeth' called **Dental Caries.**

Fluorides in water reduce the incidence of 'Osteoporosis' (bones easily getting weakened) and hardening of arteries in old people.

Excess concentration of fluorides causes 'Hypoplosia of teeth' (mottled enamel or Dental fluorosis).

Skeletal fluorosis is the other disease where bones loose their strength, joints become stiff, chest loses its mobility and breathing becomes abdominal. Pain develops over the entire body with tingling sensation because of the acute poisoning of the central nervous system. 'Jeenovolgam' is skeletal fluorosis with bent leg bones.

Animals also are affected. Molars show abnormal wear, their teeth do not oppose each other in upper and lower jaws and mastic action is rendered difficult. Hence they refuse to eat. Because of stiff joints they develop dragging gait.

Crops grown on fluoride rich belts contribute fluorides.

Threshold Limit

Higher is the temperature, more is the water consumption and less is the tolerable limit.

$$\text{ISI—0.6 to 1.2 mg/l}$$
$$\text{ICMR—1 to 2 mg/l}$$
$$\text{WHO—1 to 1.5 mg/l.}$$

Removal Methodologies

1. **Bone charcoal.** Degreased bones are heated to 400°C to 600°C for 10 minutes and powdered. Bones contain Tricalcium Phosphate for which fluorides have good affinity.

2. Alum along with Activated silica clay and lime can remove fluorides.

3. Dolomite Lime (lime rich in Magnesium) can remove fluorides.

4. Activated Carbons as burnt paddy husk, Bentonite clay, Fuller's earth, Silica gel, Bauxite and Sodium Silicate remove fluorides by 'adsorption'.

REMOVAL OF IRON AND MANGANESE

Fe^{++} and Mn^{++} are soluble in water.

Effects:

1. They cause turbid, yellow, red, brown or black waters causing stains in laundering and plumbing fixtures.

2. They support the growth of micro organisms in distribution system (Iron bacteria as **Leptothrix, Crenothrix & Gallionella**) clogging pipes.

3. Metallic taste is attributed to water.

4. A disease called 'Hemosiderosis' is caused because of consumption of excess Iron.

Removal:

Fe^{+++} and Mn^{++} are removed along with hardness.

Fe^{++} and Mn^{++++} are insoluble in water.

Fe^{++} alone can be removed by Aeration, Sedimentation and Filtration. But when both Fe^{++} and Mn^{++} along with organic matter are present, strong oxidising agents are required.

1. Raw water $\longrightarrow$ Lime $\longrightarrow$ Aeration $\longrightarrow$ Chlorination $\longrightarrow$ Clarification $\longrightarrow$ Filtration $\longrightarrow$ Chlorination $\longrightarrow$ Distribution.

2. Raw water $\longrightarrow$ Potassium Permanganate (Oxidation) $\longrightarrow$ Clarification $\longrightarrow$ Filtration $\longrightarrow$ Chlorination $\longrightarrow$ Distribution

3. Raw water $\longrightarrow$ MgO $\longrightarrow$ Diatomateous earth (Calcined $MgCO_3$) $\longrightarrow$ Rapid mix $\longrightarrow$ Diatomateous (earth filter) $\longrightarrow$ Chlorination $\longrightarrow$ Distribution (5' to 10')

4. $Fe^{++}/Mn^{++} + Na_2Z$ (Zeolite) $\longrightarrow FeZ/MnZ + 2Na^+$

5. Adsorption on a bed of Pyrolusite (MnO_2).

DESALINATION

It is reduction of the mineral content of water.

Fresh water shall be having a total dissolved solids of less than 1000 mg/l whereas sea water around 35000 mg/l.

The salt content can be minimised by the following methods:

1. **Distillation.** When water boils steam of water vapour with volatile impurities emerges out leaving dissolved salts behind. Condensation of the water vapour gives out salt free water.

In multi-staged distillation 10 to 50 compartments of boilers are kept in series—each one operating at a pressure less than the preceding one—so that water boils and steam generates at low temperature. The condensing vapour of one heats the contents of the next unit.

Distillation is quite costly but is indepedent of total dissolved solids concentration.

The treated water has a sodium content of concentration less than 100 mg/l and chloride concentration less than 200 mg/l.

2. Freezing. When a refrigerant is added to salt water and made to freeze—pure water forms ice crystals whereas salty matter remains in solution. The ice can be melted to yield pure water.

3. Reverse osmosis. When a thin layer (0.1 to 0.15 mm thick semi permeable membrane) separates two salt solutions—water from dilute sample flows into that of higher concentration and is called osmosis. But when a pressure greater than osmotic pressure is applied water flows from a denser liquid to the dilute liquid.

It is relatively a cheaper process.

DISTRIBUTION OF WATER

Methods of Distribution 1. Gravitational System. When the source of water is at a higher level (as a reservoir formed on mountainous slopes) than the area of distribution, it is adopted.

2. Direct Pumping System. It is pumping of the water from the source to the consumer.

3. Combined Pumping and Gravity System. It is pumping of the treated water to an overhead tank from where water reaches the consumer by gravity.

Systems of Supply 1. Continuous System. Water is supplied to the consumer round the clock. It is the best suited system provided the city had less undulations.

2. Intermittent System. It is supply of the water during a part of the day.

When a city is full of undulations, water may not rise to the summits unless other distribution is cut-off. Also repairs can be easily done during non-supply hours.

But in case of any leakages exfiltration takes place during supply hours and infiltration takes place during non-supply hours. Further if a fire accident occurs during non-supply hours no water will be available to fight the fire.

Layout of Distribution Net Works

1. Dead end system. If had only one main from which sub-mains and laterals branch off but nowhere any two pipes merge again and hence full of "dead ends". Stagnant water columns are formed at dead ends. "Blow off valves" are to be provided to throw off the stagnant water.

Computations of discharges can be done easily and this system is the only system suited to irregularly grown cities.

2. Ring system. It is an improved dead-end system where the dead ends are interconnected.

3. Grid iron system. It is a closely knit network of mains, submains and laterals mutually at right angles or parallel to each other. It is free from dead ends and water is under perfect circulation. A planned city with rectangular layout of roads either perfectly levelled or on a gentle slope alone is suited to it.

4. Radial system. An elevated reservoir is created for every zone provided with radial network of roads.

Storage Capacity of a Reservoir

∴ Min. storage capacity of the reservoir = max. surplus + max. deficit = $a + b$

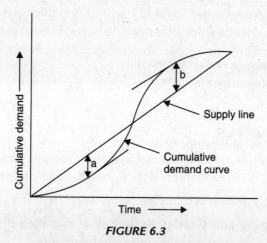

FIGURE 6.3

Types of Pipes

Cast iron pipes. C.I. Pipes are resistant to corrosion and have a long life (upto 100 years).

Steel pipes. They are affected by "pitting" of exterior and "turberculation" of interior surfaces because of highly acidic waters.

They are coated with bitumen on external face and cement mortar on the interior surfaces for longevity.

They are quite common for rising mains.

R.C.C. pipes. They are durable (life span ≈ 75 years) water tight and have low maintenance cost. They are heavy and hence can be easily laid under water (*i.e.*, the pipe will not set afloat even when it is empty).

But they are brittle and heavy.

Asbestos cement pipes. They are light, easy to handle and easily jointed. They offer smooth surfaces and resist corrosion better.

They are brittle and easily punctured by tools.

Galvanised iron pipes. They are the mostly used pipes for plumbing. Because of the zinc coating they last longer (≈ 40 years).

Acidic waters may corrode the zinc coating.

Plastic pipes. P.V.C. pipes are commonly used where the pipe need be bent as in case of a goose neck etc. They are smooth and corrosion resistant but do not have enough rigidity.

II. OBJECTIVE TYPE QUESTIONS

1. Of the total content of water on globe the available quantity for use is less than
   - (*a*) 20%
   - (*b*) 2%
   - (*c*) 0.1%
   - (*d*) 0.03%.

2. Water for domestic consumption should be
   - (a) colourless, odourless and tasteless
   - (b) free from dissolved salts
   - (c) hygienically safe
   - (d) attractive for looks.
3. A pollutant is
   - (a) any foreign element in excess concentration
   - (b) an element causing change in quality
   - (c) a harmful element
   - (d) an adultrant.
4. For good taste a desirable temperature of water is
   - (a) less than 11°C
   - (b) less than 20°C
   - (c) 30°C
   - (d) room temperature.
5. A hot water or waste water sample may have
   - (a) more bacteria
   - (b) greater biological activity
   - (c) no bacteria
   - (d) less odours and tastes.
6. True colour of water is due to
   - (a) suspended solids
   - (b) colloidal solids
   - (c) volatile solids
   - (d) acids in solution.
7. A source of colour in water is due to
   - (a) silt
   - (b) clay
   - (c) organic debris
   - (d) inorganic inert matter.
8. Purest water may have
   - (a) no colour
   - (b) faint bluse green colour
   - (c) dark blue colour
   - (d) brownish yellow colour.
9. Highly coloured waters are
   - (a) unaesthetic
   - (b) highly polluted
   - (c) require elaborate treatment
   - (d) rich in iron and manganese.
10. An industry that insists on colour free water is
    - (a) fertilisers
    - (b) dairy
    - (c) tannery
    - (d) steel.
11. Coloured waters may affect the following water treatment unit
    - (a) plain sedimenation
    - (b) sedimenation aided by coagulation
    - (c) filtration
    - (d) chlorination.
12. Permissible colour for drinking water is not more than
    - (a) 5 units
    - (b) 10 units
    - (c) 25 units
    - (d) 50 units.
13. If treated water when reaches the consumer is coloured, it is due to
    - (a) colouring pigments in water
    - (b) iron and manganese in water
    - (c) bacteria and algae in water
    - (d) corrosion of pipe line.

14. An equivalent of mg/l is
    - (a) g/m³
    - (b) kg/m³
    - (c) T/MI
    - (d) ppm.
15. Turbidity for domestic water is undesirable because it
    - (a) is unaesthetic
    - (b) causes change of taste
    - (c) give apparent colour
    - (d) prevents light penetration and hence photosynthesis.
16. For a river in flash flood, turbidity is mainly due to
    - (a) inorganic soil
    - (b) silica
    - (c) iron, manganese and zinc
    - (d) organic wastes of plants and animals.
17. Turbidity depends on the presence of the following in water
    - (a) suspended solids
    - (b) suspended solids and colloidal solids
    - (c) any material capable of preventing light
    - (d) intensity of colouration.
18. Turbidity is mainly due to
    - (a) floating solids
    - (b) suspended solids
    - (c) colloidal solids
    - (d) dissolved solids.
19. Turbidity is the ability of water to
    - (a) scatter light
    - (b) retain suspended solids
    - (c) retain colloidal solids in suspension
    - (d) detain dissolved solids.
20. Turbidity is measured by the
    - (a) concentration of suspended solids collected after centrifuging
    - (b) concentration of colloidal and suspended solids after sedimentation
    - (c) intensity of light scattered when light is passed through
    - (d) difference in suspended solid concentration before and after filtration.
21. Turbidity of a water sample depends on
    - (a) concentration and density of suspended and colloidal solids
    - (b) concentration of suspended, colloidal and dissolved solids
    - (c) concentration of organic and inorganic matter
    - (d) concentration, size, shape and refractive index of suspended solids.
22. The standard unit of turbidity is produced when
    - (a) 1 mg of silicon dioxide is dissolved in 1 litre of distilled water
    - (b) 1 mg of platinum is dissolved in 1 litre of distilled water
    - (c) 1 mg of sodium chloride dissolved in 1litre of distilled water
    - (d) 1 mg of cobalt being dissolved in 1 litre of distilled water.
23. Units of turbidity are
    - (a) mg/l
    - (b) NTU
    - (c) potassium chloroplatinate units
    - (d) no units.

24. Turbidity is undesirable because it
    (a) renders waters unpalatable
    (b) affects photosynthesis of aquatic life
    (c) causes milky white colouration to water
    (d) causes red and brown stains.
25. Turbidity can be removed by
    (a) coagulation and filtration
    (b) aeration and sedimentation
    (c) activated carbon adsorption
    (d) disinfection.
26. Due to highly turbid waters efficiency of the following operation gets affected
    (a) sedimentation
    (b) sedimentation aided by coagulation
    (c) filtration
    (d) disinfection.
27. Tastes along with odours are produced in natural waters due to
    (a) organic matter
    (b) inorganic matter
    (c) total dissolved solids
    (d) alkaline solids.
28. If 20 ml of an odourous water sample is needed 180 ml of odour free distilled water to produce 200 ml of odour free mixture, then the threshold odour number (TON) is
    (a) 10
    (b) 9
    (c) 1
    (d) 0.9.
29. Water for domestic use should have
    (a) sweet smell
    (b) faint smell
    (c) inoffensive smell
    (d) no smell.
30. Conductivity represents the concentration of the following in the water sample
    (a) total solids
    (b) dissolved solids
    (c) volatile solids
    (d) fixed solids.
31. Conductivity directly depends on
    (a) total solids
    (b) total dissolved solids
    (c) ionised dissolved solids
    (d) volatile organic solids.
32. Conductivity reflects
    (a) organic solid concentration
    (b) inorganic solid concentration
    (c) total dissolved solids
    (d) compounds which readily dissociate in solution.
33. Conductivity is standardised at
    (a) 20°C
    (b) 37.5°C
    (c) 15°C
    (d) 25°C.
34. Suspended solids are less in ground water because
    (a) they are absorbed by the soil
    (b) they are filtered through the soil layers
    (c) they are set floating on water
    (d) they readily settle to the bottom.

35. The usual size of suspended solids is
    (a) > 100 μm
    (b) < 1 mm and > 1 μm
    (c) < 1 μm and > 10^{-3} μm
    (d) > 1 μm and < 100 μm.

36. Suspended solids are undersirable in domestic waters because they
    (a) induce taste
    (b) cause disease
    (c) produce obnoxious odour
    (d) are unaesthetic.

37. Non-filterable residue of a water sample represent
    (a) floating solids
    (b) suspended solids
    (c) colloidal solids
    (d) dissolved solids.

38. Volatile solids represent
    (a) total dissolved solids (TDS)
    (b) suspended and colloidal solids
    (c) organic dissolved solids
    (d) inorganic dissolved solids.

39. A water sample is termed turbid when it
    (a) fails to transmit light through it
    (b) is rich in suspended solids
    (c) is rich in both suspended and colloidal solids
    (d) is rich in total solids.

40. The content of total solids in drinking water shall not be greater than
    (a) 50 mg/l
    (b) 100 mg/l
    (c) 500 mg/l
    (d) 2000 mg/l.

41. If a sample is heated to 600°C, the fraction getting evaporated represents
    (a) fixed solids
    (b) volatile solids
    (c) total dissolved solids
    (d) settleable solids.

42. pH of water sample containing 0.1008 g of H ion per litre is
    (a) 1
    (b) 10
    (c) 4
    (d) 13.

43. pH of a water sample is 7. pH of another water sample of same quantity is 8, then pH of the mixture is
    (a) 7.50
    (b) 7.74
    (c) 7.26
    (d) 7.86.

44. Higher pH for water is undesirable because
    (a) it corrodes zinc, copper and lead pipes
    (b) it induces sour taste
    (c) it renders chlorination less effective
    (d) it promotes growth of iron and sulphur bacteria.

45. A desirable pH value for domestic water is
    (a) 7
    (b) 6 to 8
    (c) 5 to 9
    (d) 7 to 8.5.

46. Alkalinity in natural waters is due to
    (a) salts of weak bases strong acids
    (b) drainages from abandoned mines
    (c) industrial wastes from rayon mills and steel mills
    (d) photosynthesis of algae in water.

47. The two alkalinities which do not co-exist are
    (a) carbonate and bicarbonate
    (b) bicarbonate and hydroxide
    (c) hydroxide and carbonate
    (d) any two.

48. Caustic alkalinity is because of
    (a) Caustic soda
    (b) Hydroxides
    (c) Carbonates
    (d) Bicarbonates.

49. If 20 ml of 0.02 H_2SO_4 is needed to produce methyl orange end point with 200 ml of water sample, then total alkalinity of the water sample is
    (a) 0.1 mg/l
    (b) 10 mg/l
    (c) 100 mg/l
    (d) 400 mg/l.

50. If 20 ml of 0.02 H_2SO_4 could produce phenolpthalein end point and 30 ml the methly orange end point with a 100 ml water sample then its hydroxide alkalinity is
    (a) 0
    (b) 100 mg/l
    (c) 200 mg/l
    (d) 300 mg/l.

51. Phenolpthalein end point represents
    (a) bicarbonate alkalinity
    (b) full carbonate + half bicarbonate alkalinity
    (c) half bicarbonate + full hydroxide alkalinity
    (d) full hydroxide + half carbonate alkalinity.

52. If 50 ml of a water sample titrated against 0.02 H_2SO_4 could register a phenolphthalein end point of 10 ml and methyl orange end point of 30 ml then the alkalinities are
    (a) 10 mg/l of hydroxide and 20 mg/l of carbonate
    (b) 10 mg/l of carbonate and 20 mg/l of bicarbonate
    (c) 10 mg/l of bicarbonate and 20 mg/l of carbonate
    (d) 10 mg/l of bicarbonate and 20 mg/l of hydroxide.

53. A minimum amount of .... mg/l of alkalinity is desirable in any water.
    (a) 10
    (b) 20
    (c) 50
    (d) 100.

54. Excess alkalinity is undesirable in swimming pools because
    (a) Nervous system is affected
    (b) Skin infections caused
    (c) Lacrimal fluid around the eye is altered causing eye irritation
    (d) Swimmer's itch is caused.

55. Excess alkalinity causes CHLOROSIS in plants which is the loss of green pigment 'chlorophyl' because of
    (a) precipitation of Iron as Iron hydroxide ☐
    (b) leaf not getting nutrients ☐
    (c) photosynthesis is prevented ☐
    (d) chlorine bleaches all the colours. ☐

56. Because of excess alkalinity
    (a) corrosion is caused ☐
    (b) pitting and tuberculation is induced ☐
    (c) water is coloured ☐
    (d) incrustations develop. ☐

57. Due to the photosynthetic activity of algae pH of pools of water
    (a) will not get changed ☐
    (b) increases during the day time ☐
    (c) increases during the night time ☐
    (d) progressively increases with time. ☐

58. pH range of mineral acidity is
    (a) < 4.5 ☐
    (b) > 4.5 but < 8.3 ☐
    (c) < 7 ☐
    (d) > 8.3. ☐

59. pH range of CO_2 acidity is
    (a) < 4.5 ☐
    (b) > 4.5 but < 8.3 ☐
    (c) < 7 ☐
    (d) > 8.3. ☐

60. Excess acidity causes
    (a) eye irritation of swimmers ☐
    (b) incrustations in pipes ☐
    (c) chlorosis in plants ☐
    (d) corrosion. ☐

61. Chloride that is insoluble in water is
    (a) Sodium chloride ☐
    (b) Magnesium chloride ☐
    (c) Calcium chloride ☐
    (d) Silver chloride. ☐

62. Salty taste is mainly due to
    (a) NaCl ☐
    (b) $CaCl_2$ ☐
    (c) $MgCl_2$ ☐
    (d) AgCl. ☐

63. Laxative effect is because of
    (a) $CaSO_4$ ☐
    (b) $CaCl_2$ ☐
    (c) $MgCl_2$ ☐
    (d) $MgSO_4$. ☐

64. A buffer is that
    (a) shock absorber of pollutants ☐
    (b) maintains same temperature ☐
    (c) maintains same pH ☐
    (d) maintains same dissolved oxygen concentration. ☐

65. A minimum amount of D.O. desirable in any water body is
    (a) 1 m/l ☐
    (b) 2 mg/l ☐
    (c) 3 mg/l ☐
    (d) 5 mg/l. ☐

66. In a D.O. test oxygen is replaced by equivalent in
    (a) iodine
    (b) sodium thiosulphate
    (c) manganous sulphate
    (d) sulphuric acid.
67. Sodium Azide is added to the D.O. test to neutralise the effect of
    (a) iodine
    (b) sulphuric acid
    (c) manganous sulphate
    (d) nitrites.
68. Greater D.O. is undesirable in
    (a) boilers
    (b) aerobic treatment
    (c) natural water bodies
    (d) anaerobic treatement.
69. Supersaturation of any water with D.O. causes
    (a) Eutrophication
    (b) Gas bubble disease of fish
    (c) Methamoglobinemia
    (d) Endemic Goitre.
70. BOD represents
    (a) pollutional strength of a waste
    (b) pollutional strength of an organic fraction of wastes
    (c) pollutional strength of inorganic fraction of wastes
    (d) pollutional strength of bio-degradable organic fraction of wastes.
71. BOD equation is
    (a) $y = L_i (1 - 10^{-kt})$
    (b) $x = L_T(1 - 10^{-kt})$
    (c) $K_T = K_{20} (1.047)^{T-20}$
    (d) $x = K(1 - 10^{-kt})$.
72. In the BOD equation K is called
    (a) oxygenation constant
    (b) deoxygenation constant
    (c) reaeration constant
    (d) combined deoxygenation and reaeration constant.
73. In the figure shown for BOD, AB represents
    (a) carbonaceous stage
    (b) Nitrogenous stage
    (c) proteinous stage
    (d) first phase.
74. For ordinary domestic sewage BOD reaction is expected to get completed in about (at 20°C)
    (a) 5 days
    (b) 10 days
    (c) 20 days
    (d) 30 days.

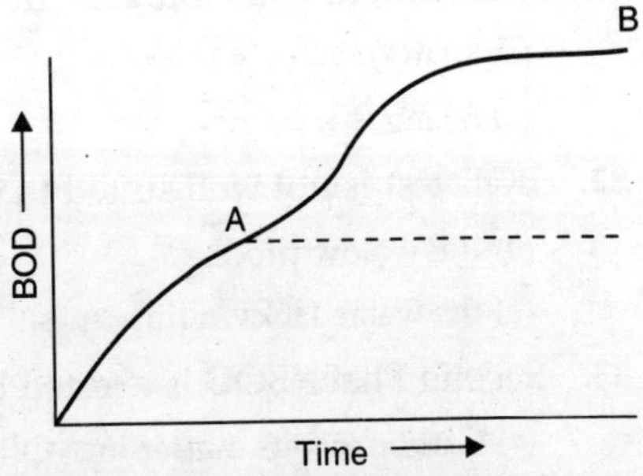

FIGURE 6.4

75. Usual population equivalent of BOD is
    (a) 10 g/capita/day
    (b) 20 g/capita/day
    (c) 50 g/capita/day
    (d) 80 g/capita/day.
76. BOD test is standardised at
    (a) 10°C and 10 day
    (b) 20°C and 5 day
    (c) 37°C and 3 day
    (d) 50°C and 2 day.
77. Light is excluded in BOD incubator to
    (a) encourage anaerobic conditions
    (b) prevent growth of algae
    (c) control aerobes
    (d) encourage growth of fungi.
78. A waste water sample of 2 ml is made upto 300 ml in a BOD bottle with distilled water. Initial D.O. of the sample is 8.0 and after 5 days it is 2.0. Its BOD is
    (a) 6 mg/l
    (b) 894 mg/l
    (c) 900 mg/l
    (d) 2400 mg/l.
79. A waste water sample of 5 ml is made upto 300 ml with distilled water. The sample had an initial D.O. of 8.0 mg/l and after 5 days the D.O. is O. BOD of the sample
    (a) 8 mg/l
    (b) 472 mg/l
    (c) 480 mg/l
    (d) test is invalid.
80. In a BOD test, three samples gave the following readings.

	Initial D.O.	D.O. after 5 days of incubation
P	7.8	6.6
Q	7.8	4.0
R	7.8	0.5

    If the dilution ratio is 50, then the exact BOD of the sample is
    (a) 50 mg/l
    (b) 60 mg/l
    (c) 190 mg/l
    (d) 365 mg/l.
81. The effluent of a biological treatment unit should have a BOD less than
    (a) 10 mg/l
    (b) 30 mg/l
    (c) 50 mg/l
    (d) 0 mg/l.
82. BOD test is not well suited to industrial wastes because
    (a) it is a slow process
    (b) toxic chemicals produce wrong results
    (c) the waste lacks in nutrients
    (d) oxidation is incomplete.
83. Second Phase BOD is exerted because of the oxidation of
    (a) Carbonaceous matter into CO_2 and H_2O
    (b) Lipids into CO_2 and H_2O
    (c) Ammonia to nitrates
    (d) Both carbohydrates and proteins.

84. For any waste
    (a) COD may be less than BOD
    (b) BOD and COD are equal
    (c) COD is always greater than BOD
    (d) depending upon the percentage of biodegradable matter, it may be either way.
85. Result of COD is generally higher than that of BOD because
    (a) the strong oxidising conditions in the test
    (b) complete oxidation of organic and inorganic compounds
    (c) oxidation of gases like ammonia
    (d) the stablest compounds are being oxidized.
86. In a COD test the most common interference is caused by the ions of
    (a) chlorides   (b) ferrous iron
    (c) sulphates   (d) nitrites.
87. Time taken for COD test is about
    (a) 3 hrs.   (b) 24 hr.
    (c) 3 days   (d) 30 days.
88. An advantage of COD test over BOD test is
    (a) it oxidized all matter including toxins
    (b) it cannot oxidize benzene, pyridine etc.
    (c) it is relatively a quick process
    (d) biodegradable matter cannot be distinguished from the rest.
89. Presence of the following indicates recent pollution
    (a) ammonical nitrogen   (b) albuminoidal nitrogen
    (c) nitrites   (d) nitrates.
90. Presence of nitrogen in a waste water sample is due to the decomposition of
    (a) carbohydrates   (b) proteins
    (c) fats   (d) vitamins.
91. Eutrophication of a lake is
    (a) organic pollution
    (b) inorganic pollution
    (c) enrichment of water quality with nutrients
    (d) deterioration of water bodies with biological growths.
92. One of the important elements responsible for eutrophication is
    (a) nitrates   (b) phosphates
    (c) sulphates   (d) chlorides.
93. Combined available chlorine means
    (a) HCl + HOCl   (b) HOCl + OCl
    (c) $NH_2Cl + NHCl_2$ (d) $NH_2Cl + NHCl_2 + NCl_3$.

94. Methemoglobinemia (Blue baby syndrome) is due to
    (a) Oxidation of haemoglobin
    (b) Nitrates replacing oxygen in haemoglobin
    (c) Nitrites replacing oxygen in haemoglobin
    (d) Carbon monoxide replacing oxygen in haemoglobin.

95. Gastro intestinal irritation is produced because of excess of the following in water
    (a) chlorides
    (b) sulphates
    (c) nitrates
    (d) phosphates.

96. "Dental Caries" will be absent where the following is present
    (a) mottled enamel of teeth
    (b) itai-itai
    (c) methamoglobinemia
    (d) cancer.

97. Concentration of fluorides desirable in water is
    (a) ≯1 mg/l
    (b) 1 to 2 mg/l
    (c) 10 to 20 mg/l
    (d) not more than 250 mg/l.

98. More than 1 mg/l of fluorides
    (a) cause dental caries
    (b) cause mottled enamel of teeth
    (c) cause skeletal fluorosis
    (d) prevent dental cavities and tooth decay in children.

99. Fluoridation is done to
    (a) protect children's teeth
    (b) protect from dental fluorosis
    (c) protect from both dental and skeletal fluorosis
    (d) add taste to water.

100. Sudden increase in chloride concentration in water indicates
    (a) sewage contamination
    (b) ground water mixed with surface water
    (c) fresh water being mixed with sea water
    (d) increase in hardness of water.

101. Kidney damage may be caused due to the excess concentration of
    (a) fluorides
    (b) nitrates
    (c) chlorides
    (d) nitrites.

102. Absolutely soft waters are required for
    (a) drinking
    (b) boilers
    (c) prevention of corrosion in pipes
    (d) washing with synthetic detergent soap.

103. Besides Ca and Mg, the other metallic ions responsible for hardness are
    (a) Na and K
    (b) Fe and Mn
    (c) Cl and SO_4
    (d) Si and C.

104. Hardness is desirable for
    (a) laundering
    (b) boiler feed water
    (c) wash water of sinks and porcelain tubs
    (d) cardiovascular diseases.

105. A desirable feature of hardness is
    (a) more soap consumed
    (b) skin roughened
    (c) vegetables toughened
    (d) scales formed inside pipes protect them from corrosion.

106. If the total alkalinity of a water sample is 100 mg/l and its total hardness is 300 mg/l, then its carbonate and non-carbonate hardness are respectively
    (a) 100 mg/l and 200 mg/l
    (b) 300 mg/l and zero
    (c) 200 mg/l and 100 mg/l
    (d) 100 mg/l and zero.

107. If the total alkalinity of a water sample is 300 mg/l and total hardness is 100 mg/l, then the carbonate and non-carbonate hardnesses are respectively
    (a) 100 mg/l and 200 mg/l
    (b) 300 mg/l and 0
    (c) 200 mg/l and 100 mg/l
    (d) 100 mg/l and 0.

108. Excess concentration of sodium salts in water causes
    (a) hardness
    (b) bitter taste
    (c) alkalinity
    (d) laxative effect.

109. Excess concentration of magnesium salts in water causes
    (a) hardness
    (b) bitter taste
    (c) alkalinity
    (d) laxative effect.

110. Iron in water causes
    (a) stains on clothes
    (b) eutrophication
    (c) rust
    (d) corrosion.

111. Lead in water causes
    (a) plumbo solvency
    (b) stains on clothes and paper
    (c) damages nervous system
    (d) permanent bluish skin.

112. A communicable disease is
    (a) cancer
    (b) fluorosis
    (c) tuberculosis
    (d) goitre.

113. A water borne disease is
    (a) malaria
    (b) cancer
    (c) dysentery
    (d) encephalitis.

114. Due to improper storing of water the following disease spreads
    (a) tuberculosis
    (b) filaria
    (c) methamoglobinemia
    (d) fluorosis.

115. For health, presence of traces of the following elements is a must in drinking water
    (a) fluorides and iodine
    (b) chlorides and nitrates
    (c) phosphates and carbonates
    (d) chlorides and sulphates
116. A water borne virus infection is
    (a) cholera
    (b) swimmer's itch
    (c) jaundice
    (d) cancer.
117. An indicator organism is
    (a) pathogenic bacteria
    (b) non-pathogenic bacteria
    (c) facultative bacteria
    (d) non-pathogenic bacteria of the same family of pathogens.
118. An indicator organism
    (a) should be a pathogen
    (b) is of human origin but not found in water
    (c) harmless bacterium found in some types of waters
    (d) should produce more gas when incubated.
119. Bacteria utilise
    (a) liquid food
    (b) soluble food
    (c) solid food
    (d) inorganic solid food.
120. Autotrophic bacteria derive energy from
    (a) inorganic compounds
    (b) organic ompounds
    (c) ultraviolet rays of the sun
    (d) both organic and inorganic compounds.
121. Oily odour and taste producing substances are produced by
    (a) bacteria
    (b) protozoa
    (c) algae
    (d) rotifers.
122. An indicator organism for bacteriological quality of water is
    (a) *salmonella typhi*
    (b) *escherchia coil*
    (c) *mycobacterium tuberculosis*
    (d) *shigella dysenteria*.
123. In drinking water number of coliform bacteria should not exceed
    (a) 0/100 ml
    (b) 1/100 ml
    (c) 10/100 ml
    (d) 100/100 ml.
124. The average domestic demand for an Indian city is
    (a) 135 lpcd
    (b) 270 lpcd
    (c) 500 lpcd
    (d) 750 lpcd.
125. Kuichling's formula for fire demand is $Q =$

    (a) $3182 \sqrt{P}$
    (b) $1136 \left[\dfrac{P}{10} + 10\right]$
    (c) $4640 \sqrt{P} (1 - 0.01 \sqrt{P})$
    (d) $5660 \sqrt{P}$.

126. A factor affecting domestic demand is
   (a) dead end system or recticulation system of network of pipes
   (b) timing of water supply
   (c) water supplied is soft or hard
   (d) climatic conditions.
127. For an old city with constraints for growth the best method of forecasting future population is
   (a) arithmetical increase method
   (b) geometrical increase method
   (c) graphical method
   (d) incremental increase method.
128. If the present population of a city is 'a', rate of increase per decade is 'b', incremental increase per decade is 'c', percentage growth per decade is 'r', then population after 'n' decades by incremental increase method is
   (a) $a + nb$
   (b) $a + n(b + c)$
   (c) $a + nb + \dfrac{n(n+1)}{2} c$
   (d) $a \left[1 + \dfrac{r}{100}\right]^n$.
129. Design period mainly depends on
   (a) percentage interest at which the loan is taken
   (b) capacity of the municipality to repay
   (c) quality of fittings used
   (d) rate of growth of population.
130. Ground water is
   (a) free from suspended solids but contain dissolved solids
   (b) free from harmful bacteria but may contain harmless bacteria
   (c) free from floating impurities but may have suspended solids
   (d) free from dissolved solids but may have colloidal solids.
131. A shallow well is one tapping an aquifer
   (a) nearest to ground level
   (b) below an impervious layer
   (c) sandwiched in between two impervious layers
   (d) runs through a number of strata.
132. A deep well
   (a) easily gets dried up during summer
   (b) may yield constant discharge
   (c) is not deeper than a shallow well
   (d) is formed by just tapping the nearest aquifer to the ground.
133. A river intake may be situated on
   (a) a straight reach
   (b) the convex side of a curved course
   (c) downstream side of outfall sewer
   (d) proximity to industries.

134. River water may have
   (a) more suspended solids and less dissolved solids
   (b) more dissolved solids and less suspended solids
   (c) more dissolved solids and less floating solids
   (d) more colloidal and dissolved solids.

135. A good source of water requiring practically the least treatment is
   (a) a perennial river
   (b) an impounded reservoir
   (c) a deep well
   (d) an elevated lake.

136. Sedimentation may not be required for water from a
   (a) shallow well
   (b) deep well
   (c) river
   (d) canal.

137. Waters from the following source is likely to be hard
   (a) river
   (b) lake
   (c) deep well
   (d) shallow well.

138. Permissible velocity through coarse screens is not greater than
   (a) 1 m/s
   (b) 1 m/min
   (c) 0.3 m/s
   (d) 0.3 m/min.

139. A discrete particle is one
   (a) whose settling is unaffected by the neighbouring particles
   (b) whose settling influences neighbouring particles, i.e., due to mass action
   (c) for which hindered settling takes place
   (d) which settles as compressed settling.

140. The main factor responsible for sedimentation of a particle is
   (a) specific gravity of the particle
   (b) specific gravity of the medium
   (c) difference of sp. gr. of particle and medium
   (d) sum of sp. gr. of particle and medium.

141. If Q = Quantity of water to be treated per day, V = Volume of sedimentation basin then detention period in hours is
   (a) $\dfrac{24V}{Q}$
   (b) $\dfrac{24Q}{V}$
   (c) $\dfrac{Q}{24V}$
   (d) $\dfrac{V}{24Q}$.

142. Generally "flow through period" of a sedimentation tank is
   (a) less than detention period
   (b) more than detention period
   (c) exactly equal to the detention period
   (d) some time more and some time less.

143. Over flow rate of a sedimentation tank is
    (a) $\dfrac{Q}{\text{Plan area}}$
    (b) $\dfrac{Q}{\text{Area of longitudinal section}}$
    (c) $\dfrac{Q}{\text{Cross-sectional area}}$
    (d) $\dfrac{Q}{\text{Plan area} \times \text{Liquid depth}}$.

144. Coagulant should be used for sedimentation when turbidity of raw water exceeds
    (a) 5 units
    (b) 10 units
    (c) 50 units
    (d) 100 units.

145. Coagulation is affected because of higher concentration of
    (a) phosphates
    (b) nitrates
    (c) bicarbonates
    (d) sulphates.

146. Coagulating power of a chemical increases with its
    (a) atomic weight
    (b) molecular weight
    (c) density
    (d) valency.

147. Colloidal particles are associated with
    (a) tyndal effect
    (b) discrete particle
    (c) Schmutzedecke
    (d) Brownian movement.

148. Lime is added to water or waste samples to
    (a) destabilise colloidal oil and organic emulsion suspensions
    (b) to assist precipitation of hydroxides of metals by lowering their pH
    (c) remove calcium hardness
    (d) convert calcium carbonate to calcium bicarbonate.

149. Optimum dose of coagulant means the least dose that produces
    (a) maximum amount of floc
    (b) floc whose density is very high
    (c) maximum amount of floc within the least time
    (d) readily settleable floc.

150. The jar test is to be carried out immediately after the collection of the water sample so that
    (a) concentration of dissolved gases as oxygen, nitrogen, carbon dioxide shall not change
    (b) sedimentation of Colloids do not take place before the test
    (c) quality of water may not get deteriorated
    (d) least amount of coagulant can be used.

151. Methamoglobinemia is due to
    (a) Fluorides
    (b) Chlorides
    (c) Nitrites
    (d) Nitrates.

152. When Alumina is used as a coagulant for neutral water the end products obtained are
    (a) $Al(OH)_3$
    (b) $Al(OH)_3 + H_2SO_4$
    (c) $Al(OH)_3 + CaSO_4$
    (d) $Al(OH)_3 + CaSO_4 + CO_2$.

153. A coagulant aid is one which
  (a) promotes coagulation
  (b) stimulates floc formation
  (c) increases dosage of coagulant
  (d) narrows down the pH range.

154. Copperas is
  (a) ferric chloride
  (b) ferric sulphate
  (c) ferrous chloride
  (d) ferrous sulphate.

155. Chlorinated copperas is
  (a) chlorine and ashes of copper
  (b) chlorine and ferrous sulphate
  (c) ferrous sulphate and ferrous chloride
  (d) ferric sulphate and ferric chloride.

156. Detention time in a "flocculator" is
  (a) 10 s to 20 s
  (b) 20 min to 30 min
  (c) 1 hour to 8 hours
  (d) 2 hours to 3 hours.

157. Detention time in a Flash Mixer is
  (a) 30 s to 60 s
  (b) 20 min to 30 min.
  (c) 1 h to 8 h
  (d) 2 h to 3 h.

158. *Schmutzedecke* will be formed in
  (a) slow sand filters
  (b) rapids sand filters
  (c) pressure filters
  (d) both slow sand filters and pressure filters.

159. Back washing is highly effective in case of
  (a) slow sand filter
  (b) rapid sand filter
  (c) pressure filter
  (d) rapid sand filters and pressure filters.

160. Fine sand is used as media in case of
  (a) slow sand filter
  (b) rapids and filter
  (c) pressure filter
  (d) both rapid and slow sand filters.

161. Effluent from this unit requires no other treatment other than disinfection
  (a) slow sand filter
  (b) rapid sand filter
  (c) pressure filter
  (d) both rapid and slow sand filters.

162. Coagulant may not be added for treatment before
  (a) slow sand filter
  (b) rapids and filter
  (c) pressure filter
  (d) slow and rapid sand filters.

163. Iso-electric point is the pH value at which
  (a) the electrical charge of a colloid is zero
  (b) the colloidal charge of a particle is zero
  (c) coagulation takes place quite effectively
  (d) ideal concentration of coagulant is present for rapid coagulation.

164. Coagulant is a must before letting water into
  (a) slow sand filter
  (b) rapid sand filter
  (c) pressure filter
  (d) both rapid and sand pressure filters.

165. A 10 cm deep filters removes 90% suspended solids present in water. If 99% removal is needed, the depth of filter and filter performance coefficient are
    (a) 20 cm and 0.909/cm
    (b) 11 cm and 9/cm
    (c) 11 cm and 0.909/cm
    (d) 20 cm and 9/cm.

166. A highly flexible filter is
    (a) slow sand filter
    (b) rapid sand filter
    (c) pressure filter
    (d) both rapid and slow sand filters.

167. This is used to treat swimming pool water
    (a) slow sand filter
    (b) rapid sand filter
    (c) pressure filter
    (d) both rapid and slow sand filters.

168. Disinfection is the process of
    (a) killing all the bacteria
    (b) killing only pathogenic bacteria
    (c) removal of causative organism for disease
    (d) complete destruction of life.

169. One of the physical agents responsible for disinfection is
    (a) ultra violet rays
    (b) ozone
    (c) heat
    (d) lime.

170. A desirable property of a disinfectant is that it should
    (a) act instantaneously
    (b) kill all the bacteria
    (c) not alter the quality of water
    (d) be very cheap.

171. Efficiency of disinfection is
    (a) $\dfrac{\text{No. of pathogens killed}}{\text{No. of micro-organisms present initially}} \times 100$

    (b) $\dfrac{\text{No. of pathogens killed}}{\text{No. of pathogens present initially}} \times 100$

    (c) $\dfrac{\text{No. of micro-organisms killed}}{\text{No. of pathogens present initially}} \times 100$

    (d) $\dfrac{\text{No. of micro-organisms killed}}{\text{No. of micro-organisms present initially}} \times 100$.

172. Mud balls are formed in a filter because of
    (a) raw water rolling over the filter surface
    (b) quick filtration of turbid water
    (c) frequent back washing
    (d) bond developed between impurities and sand grains due to inefficient back washing.

173. Sterilization is
    (a) boiling
    (b) disinfection
    (c) eliminating all life
    (d) killing disease causing organisms.
174. A heavy metal capable of disinfecting is
    (a) lead
    (b) iron
    (c) zinc
    (d) silver.
175. To disinfect swimming pool water, we use
    (a) chlorine
    (b) bromine
    (c) iodine
    (d) fluorine.
176. Iodine is used to disinfect
    (a) swimming pool
    (b) small pools of drinking water
    (c) municipal water supply
    (d) raw water wihtout any other treatment.
177. The only disinfectant available in the three states (solid, liquid and gas) is
    (a) chlorine
    (b) bromine
    (c) iodine
    (d) fluorine.
178. Pick up a disinfectant from the following
    (a) $Ca(OH)_2$
    (b) $KMnO_4$
    (c) HCl
    (d) DDT.
179. Residual Chlorine is given out when
    (a) chlorine required for disinfection and oxidation of organic matter is less than applied chlorine
    (b) chlorine required is more than that supplied
    (c) chlorine applied is just sufficient to that required
    (d) none of the above.
180. In chlorination the most effective kill is due to
    (a) HCl
    (b) HOCl
    (c) OCl
    (d) Cl.
181. Disinfection ability of a chemical depends on
    (a) concentration of disinfectant
    (b) time of concentration
    (c) concentration of disinfectant × time of concentration
    (d) $\dfrac{\text{concentration of disinfectant}}{\text{time of concentration}}$.
182. In the curve shown "Break Point" is
    (a) A
    (b) B
    (c) C
    (d) D.

183. The usual amount of residual chlorine required after 10 minutes of contact is
    (a) 0.1 mg/l
    (b) 0.2 mg/l
    (c) 1 mg/l
    (d) more than 3 mg/l.
184. "Free chlorine Residual" means
    (a) HOCl + HCl
    (b) HCl + OCl
    (c) HOCl + OCl
    (d) chloramines.

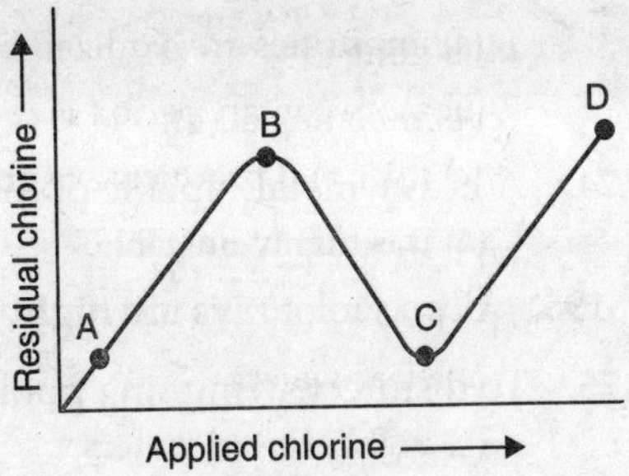

FIGURE 6.5

185. "Prechlorination" is done to
    (a) kill bacteria
    (b) inactivate viruses
    (c) reduce tastes and odours
    (d) reduce quantity of chlorine used.
186. Plain chlorination is done after
    (a) aeration
    (b) sedimentation
    (c) filtration
    (d) softening.
187. Super chlorination is done
    (a) in day to day practice
    (b) during summer
    (c) during an epidemic
    (d) when no other treatment is resorted to.
188. Dechlorination is done with
    (a) chloramines
    (b) sulphur dioxide
    (c) heating
    (d) potassium dichromate.
189. Before commissioning a new water supply line it is filled with a concentrated solution of chlorine and kept for 24 hours. The initial and later concentrations of chlorine are
    (a) 3 and 0.2 mg/l
    (b) 10 and 5 mg/l
    (c) 20 and 10 mg/l
    (d) 50 and 25 mg/l.
190. A main disadvantage of chlorine as disinfectant is
    (a) it is costly
    (b) it corrodes pipe lines
    (c) it is difficult to handle
    (d) it may cause cancer.
191. Presence of (residual) chlorine is readily detected by
    (a) methyl orange
    (b) chrome black T
    (c) orthotolidine
    (d) holoquinone.
192. Besides disinfection chlorination is effective in the removal of
    (a) turbidity
    (b) hardness
    (c) colour
    (d) carbon dioxide.
193. An advantage of ozone over chlorine for disinfection is it (ozone)
    (a) is easy to apply
    (b) leaves sufficient chemical to guard against post contamination
    (c) is cheaper than chlorine
    (d) gives good taste to water.

194. A disadvantage of ozone is
    (a) impurities are oxidized before disinfection commences
    (b) more contact period is required
    (c) toxic end products are given out
    (d) it is highly unstable.

195. Ultra violet rays are highly effective for disinfection for
    (a) clear waters
    (b) hard waters
    (c) highly turbid waters
    (d) waters rich in suspended solids.

196. An advantage of ultraviolet radiation is
    (a) it is cheap
    (b) it can be used for any type of water
    (c) post contamination is impossible
    (d) no chemical is added to water.

197. An undesirable gas in domestic water is
    (a) CH_4
    (b) NH_3
    (c) O_2
    (d) H_2S.

198. Fluorides are removed from water by
    (a) activated carbon
    (b) activated silica
    (c) activated alumina
    (d) caustic soda.

199. Iron and manganese can be removed from water by
    (a) sedimentaion
    (b) sedimentation aided by coagulation
    (c) filtration
    (d) oxidation.

200. Absolutely soft waters
    (a) cause no damage to pipe line
    (b) give good taste
    (c) give lather with synthetic detergent soap
    (d) form no scales in boilers.

201. Besides calcium and magnesium, hardness is also caused because of salts of
    (a) sodium
    (b) potassium
    (c) strontium
    (d) aluminium.

202. Soft waters have a hardness of
    (a) zero
    (b) less than 75 mg/l
    (c) less than 150 mg/l
    (d) less than 300 mg/l.

203. Hardness is removed from water based on the relative insolubility of the following
    (a) $CaCO_3$ and $MgCO_3$
    (b) $Ca(OH)_2$ and $MgCO_3$
    (c) $Ca(OH)_2$ and $Mg(OH)_2$
    (d) $CaCO_3$ and $Mg(OH)_2$.

204. The solubility of $CaCO_3$ in pure water at 20°C is
    (a) 0
    (b) 9 mg/l
    (c) 17 mg/l
    (d) 50 mg/l.

205. The solubility of $Mg(OH)_2$ in pure water at 20°C is
 (a) 0   (b) 9 mg/l
 (c) 17 mg/l   (d) 50 mg/l.
206. $CaCO_3$ gets precipitated at a pH of
 (a) 6.5   (b) 7
 (c) 9 to 9.5   (d) 11.
207. $Mg(OH)_2$ gets precipitated at a pH of
 (a) 6.5   (b) 7
 (c) 9 to 9.5   (d) 11.
208. Carbonate hardness is due to
 (a) $Ca(HCO_3)_2$, $Mg(HCO_3)_2$ and $MgCO_3$
 (b) $Ca(HCO_3)_2$, $Mg(HCO_3)_2$, $CaCO_3$ and $MgCO_3$
 (c) $CaCO_3$
 (d) $CaCO_3$ and $MgCO_3$.
209. The effluent from the lime soda process contains a hardness of
 (a) 0   (b) 10 mg/l
 (c) 50 mg/l   (d) 100 mg/l.
210. Zeolite process can lower the hardness upto
 (a) 0   (b) 10 mg/l
 (c) 50 mg/l   (d) 100 mg/l.
211. To activate the zeolite bed the following is done
 (a) recycling   (b) back washing
 (c) base exchange   (d) regeneration.
212. When the source of water is a reservoir on a mountain slope and the city is in plains then the system of distribution is
 (a) gravity system
 (b) pumping system
 (c) combined pumping and gravity system
 (d) gravity system for a part and pumping for the remaining.
213. Pumping system is best suited when
 (a) fire accidents occur frequently
 (b) density of population is high and space available is less
 (c) source of water is at low level
 (d) power failures are more common.
214. Combined pumping and gravity flow system is best suited, where
 (a) the city is in plains and source is fairly elevated
 (b) the city had a gentle slope and source is elevated
 (c) the city is on steep slopes and source is below
 (d) any type of topography.
215. A disadvantage of combined pumping and gravity flow is that
 (a) it needs pumping round the clock
 (b) elevated storage reservoirs required
 (c) adequate pressure may not exist in pipes
 (d) we may not get enough water to fight fire.

216. An advantage of intermittent system of supply of water is
    (a) it is economical
    (b) supply is assured during a fire accident
    (c) pumping is for limited hours
    (d) repairs can be carried out during non-supply hours.

217. A major disadvantage of intermittent system of supply of water is
    (a) bigger pipes and pumps required
    (b) more number of valves required
    (c) consumers should store water
    (d) infiltration of impurities may occur through leaky joints.

218. Intermittent system is more popular in India because
    (a) supply hours can be staggard for different zones of different elevations
    (b) less quantity of water shall be sufficient
    (c) wastage is quite less
    (d) it is highly economical in the long run.

219. The purpose of stand pipe is that it
    (a) increases storage capacity of water
    (b) helps fire fighting
    (c) is of great help in intermittent supply system
    (d) boosts pressures in pipes.

220. Losses in distribution system are due to
    (a) poor quality of pipes and fittings
    (b) unauthorised connections
    (c) dead end system of net work of pipes
    (d) low morale of people.

221. Storage capacity of a distribution reservoir should be a minimum of
    (a) a or b whichever is greater
    (b) a + b
    (c) y_1
    (d) $y_1 + y_2$.

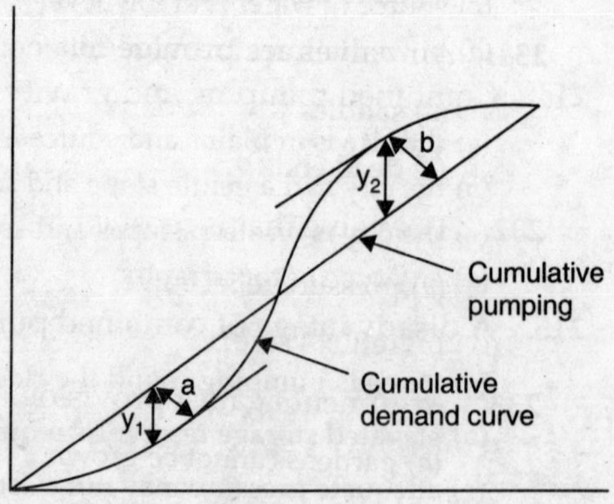

222. The best material for pipes to be laid under water is
    (a) cast iron
    (b) steel
    (c) cement concrete
    (d) asbestos cement

FIGURE 6.6

223. Mostly used pipe for plumbing is
    (a) steel
    (b) cast iron
    (c) cement concrete
    (d) galvanised iron.

224. The best pipe of water mains known for long life is
    (a) cast iron
    (b) steel
    (c) cement concrete
    (d) asbestos cement.

225. The pipe that is light, having low coefficient of expansion and resistant to corrosion is
    (a) cement concrete
    (b) hume pipe
    (c) asbestos cement
    (d) cast iron.

226. The only advantage of dead end system is
    (a) it had more number of dead ends
    (b) blow off valves empty stagnant pockets of water
    (c) less pipe length
    (d) suited for any town.

227. Grid iron system is best suited to
    (a) an irregularly grown old town
    (b) undulating topography
    (c) radial roads
    (d) planned city on a gentle slope.

228. A disadvantage of interlaced system is
    (a) not suited for any town
    (b) elevated reservoirs required
    (c) more length of pipes required
    (d) repairs cannot be attended easily.

229. Computations of discharges is difficult for
    (a) dead end system
    (b) reticulation system
    (c) radial system
    (d) ring system.

230. Blow off valves are a must when
    (a) undulations exist
    (b) saddles exist
    (c) dead ends exist
    (d) fire hydrants exist.

231. Air valves are provided at
    (a) saddles
    (b) summits
    (c) dead ends
    (d) regularly at 1 km intervals.

232. The valve that is most commonly provided on rising mains is
    (a) pressure relief valve
    (b) blow off valves
    (c) reflux valve
    (d) sluice valve.

233. Argument against provision of water meters is
    (a) gardens cannot be grown
    (b) swimming pools have to pay more
    (c) cleaning of roads is costly
    (d) sanitation of poor may be neglected.

234. A goose neck is
    (a) a bent flexible pipe provided between ferrule and stop cock ☐
    (b) a T-shaped brass length between water meter and ferrule ☐
    (c) a straight G.I. pipe between service pipe and stop cock ☐
    (d) a bent rigid pipe between service pipe and water meter. ☐

235. Colouration is caused by iron and manganese when their concentrations are respectively greater than
    (a) 0.3 mg/l and 0.1 mg/l   ☐ (b) 0.5 mg/l and 0.3 mg/l ☐
    (c) 0.8 mg/l and 0.5 mg/l   ☐ (d) 0.1 mg/l and 0.05 mg/l. ☐

236. Floatation is a process where
    (a) lighter particles settle ☐
    (b) lighter particles are separated ☐
    (c) lighter particles are set to float unaided ☐
    (d) lighter particles float due to the attachment of gas bubbles. ☐

III. TRUE/FALSE

1. Resistivity = Resistance × Distance between electrodes.
2. TOC is generally less than BOD because it measures carbon but not the oxygen consumption.
3. Turbidity is because of suspended solids but the measured turbidity cannot be taken as proportionate to the suspended solid concentration.
4. The concentration of settleable solids in an Imhoff cone is read as mg/l.
5. Compounds which dissociate in solution to give separate ions will have high conductivity and which do not dissociate have low conductivity.
6. Grannular filters remove large particles by mechanical straining and sedimentation and very small particles by diffusion but most of intermediate sized particles may escape out.
7. Except in case of highly turbid waters the efficiency of the removal of impurities is independent of applied concentration in case of grannular filters.
8. A coagulant should produce water soluble hydroxides.
9. Chlorine is capable of reducing odours due to biological origin but produces pungent odours with phenols.
10. Chlorine produces trihalomethane with organic matter which is carcinogenous.
11. To counteract post contamination with ozone is not possible, chlorine dioxide is added after disinfection.
12. A concentration greater than 0.4 mg/l of ozone is required for effective kill.
13. The percentage kill is always directly proportional to the concentration of ozone.
14. Bromine is used as a disinfectant in swimming pools as it causes less eye irritation.
15. Higher alkalinity imparts bitter taste to water.

16. While calcium hardness poses no problem to public health, magnesium hardness may induce laxative effect if people are not accustomed to.
17. Fluoride concentrations less than 5 mg/l may not cause bone fluorosis.
18. Silver is a toxic metal.
19. When the demand of oxygen is more than the supply, aerobic conditions are created.
20. No organic material is resistant to biological degradiations.
21. Pseudohardness is caused because of high concentration of sodium salts.
22. D.O. of a lake may be higher during the night time.
23. The specific gravity of a discrete particle changes with time.
24. Terminal settling velocity for discrete particles of laminar flow is given by, $v_t = \dfrac{g(\rho_p - \rho_m)d^2}{18\mu}$.
25. As the flocculating particles retain the same size, and shape throughout, Stoke's law is applicable to them.
26. Stoke's law gives the horizontal component of velocity of flow which in turn decides the length of the sedimentation basin.
27. Sedimentation tanks designed for flocculating particles are generally deeper than those designed for discrete particles.
28. Horizontal velocity of flow is not the consideration in the design of circular sedimentation tanks whereas it is for rectangular sedimentation tanks.
29. Peripheral flow type of circular sedimentation tank has greater detention period for the same capacity of radial flow type tanks.
30. Most of the colloids in surface waters are negatively charged.
31. Over dosing of the coagulant results in non-removal of colloidal particles.
32. Less turbidity requires more coagulant than high turbidity.
33. Bentonite clay (a coagulant aid) addition to water reduces its turbidity.
34. Flocculation is the process where the colloids agglomerate by the turbulence created.
35. Temporary hardness is caused because of Alkaline Anions and permanent hardness because of Acidic Anions.
36. The least water soluble salts are calcium carbonate and magnesium hydroxide.
37. Calcium carbonate causes hardness.
38. Lime-soda process of removal of hardness is independent of pH.
39. Ideal pH for the removal of calcium hardness is 9.5 and magnesium hardness is 10.5.
40. Complete removal of hardness is impossible by lime-soda process and about 10 mg/l of magnesium hardness and 40 mg/l of calcium hardness are retained by the treated water.
41. Recarbonation is done to convert insoluble calcium carbonate into soluble calcium bicarbonate.

42. Before ion exchange process for the removal of hardness, sedimentation, iron and manganese removal is to be done and chlorination should not be done.
43. Fluorides may not be present in hard waters.
44. Slow sand filters can accept any amount of turbidity, while rapid sand filters cannot operate when turbidity is greater than 40 units.
45. While impurities get penetrated to the full depth in case of rapid sand filters, they are confined to the top 30 to 60 mm in slow sand filters.
46. Effective size of filter sand is the size of the sieve opening which would permit the passage of 10% of the total weight of sand.
47. Uniformity co-efficient of sand is the ratio between size of sieve opening permitting 60% passage to that permitting 10% by weight.
48. Slow sand filter has a high colour removal efficiency compared to rapid sand filter.
49. While the sand for rapid sand filter is stratified, the sand in slow sand filter is unstratified.
50. Fluoridation helps only the strengthening of children's teeth but not of adults.
51. More is the non-carbonate hardness, less is the fluoride concentration.
52. Turbidity has no effect on the amount of disinfectant.
53. Particulate matter may shield pathogens and may adsorb disinfectants.
54. Ferric and manganic salts are water soluble whereas ferrous and manganous salts are water insoluble.
55. Odours of ground waters may be easily removed by aeration whereas it may not be quite effective for surface waters.
56. Aeration expels hydrogen sulphide, carbon dioxide and other volatile matter but increases dissolved oxygen concentration.
57. Disinfection is not usually done for waters of pH less than 6 and more than 8.
58. Hydrogen peroxide is not only a strong oxidizing agent but also a good disinfectant too.
59. Copper ions are not only algicidal but strongly bactericidal also.
60. Among the igneous rocks Basalt is a good aquifer.
61. While Quartzite may harbour some water, sand stones are impervious.
62. None of the metamorphic rocks are good aquifers.
63. While sands and gravels retain water, clays and silts do not retain water.
64. Shale is a good example of aquiclude.
65. Floatation is the removal of lighter suspended impurities from water and is the converse of sedimentation.

ANSWERS

Objective Type Questions

1. (d)	2. (c)	3. (b)	4. (a)	5. (b)	6. (b)
7. (c)	8. (a)	9. (a)	10. (b)	11. (d)	12. (a)

13. (b)	14. (a)	15. (a)	16. (a)	17. (b)	18. (c)
19. (a)	20. (c)	21. (a)	22. (d)	23. (b)	24. (b)
25. (a)	26. (c)	27. (a)	28. (a)	29. (d)	30. (b)
31. (c)	32. (d)	33. (d)	34. (b)	35. (d)	36. (d)
37. (d)	38. (c)	39. (a)	40. (c)	41. (b)	42. (a)
43. (c)	44. (c)	45. (d)	46. (d)	47. (b)	48. (b)
49. (c)	50. (b)	51. (d)	52. (c)	53. (b)	54. (c)
55. (a)	56. (c)	57. (b)	58. (a)	59. (b)	60. (a)
61. (d)	62. (a)	63. (d)	64. (c)	65. (c)	66. (a)
67. (d)	68. (a)	69. (b)	70. (d)	71. (a)	72. (b)
73. (b)	74. (d)	75. (d)	76. (b)	77. (d)	78. (c)
79. (d)	80. (c)	81. (b)	82. (b)	83. (c)	84. (c)
85. (a)	86. (a)	87. (a)	88. (c)	89. (a)	90. (b)
91. (c)	92. (b)	93. (c)	94. (c)	95. (b)	96. (a)
97. (a)	98. (b)	99. (a)	100. (a)	101. (c)	102. (b)
103. (b)	104. (d)	105. (a)	106. (a)	107. (d)	108. (b)
109. (d)	110. (a)	111. (a)	112. (c)	113. (c)	114. (b)
115. (a)	116. (c)	117. (d)	118. (b)	119. (b)	120. (a)
121. (c)	122. (b)	123. (a)	124. (a)	125. (a)	126. (d)
127. (a)	128. (c)	129. (d)	130. (a)	131. (a)	132. (b)
133. (a)	134. (a)	135. (d)	136. (b)	137. (c)	138. (a)
139. (a)	140. (c)	141. (a)	142. (a)	143. (a)	144. (c)
145. (a)	146. (d)	147. (d)	148. (a)	149. (d)	150. (a)
151. (c)	152. (b)	153. (b)	154. (d)	155. (d)	156. (b)
157. (a)	158. (a)	159. (b)	160. (a)	161. (a)	162. (a)
163. (b)	164. (c)	165. (b)	166. (c)	167. (c)	168. (c)
169. (c)	170. (c)	171. (d)	172. (d)	173. (c)	174. (d)
175. (b)	176. (b)	177. (a)	178. (b)	179. (a)	180. (b)
181. (c)	182. (c)	183. (b)	184. (c)	185. (c)	186. (c)
187. (c)	188. (b)	189. (d)	190. (d)	191. (c)	192. (c)
193. (d)	194. (d)	195. (a)	196. (d)	197. (d)	198. (c)
199. (d)	200. (d)	201. (c)	202. (b)	203. (d)	204. (c)
205. (b)	206. (c)	207. (d)	208. (a)	209. (c)	210. (a)
211. (d)	212. (a)	213. (b)	214. (d)	215. (b)	216. (d)
217. (d)	218. (a)	219. (d)	220. (a)	221. (d)	222. (c)
223. (d)	224. (a)	225. (c)	226. (d)	227. (d)	228. (c)
229. (b)	230. (c)	231. (b)	232. (c)	233. (d)	234. (a)
235. (a)	236. (d)				

True/False

1. T	2. T	3. T	4. F	5. T	6. T
7. T	8. F	9. T	10. T	11. T	12. T
13. F	14. T	15. T	16. T	17. T	18. T
19. F	20. F	21. T	22. F	23. F	24. T
25. F	26. F	27. T	28. T	29. T	30. T
31. T	32. T	33. F	34. T	35. T	36. T
37. F	38. F	39. T	40. T	41. T	42. T
43. T	44. F	45. T	46. T	47. T	48. F
49. T	50. T	51. T	52. F	53. T	54. F
55. T	56. T	57. T	58. F	59. F	60. T
61. F	62. T	63. T	64. T	65. T	

Chapter 7 SANITARY ENGINEERING

1. INTRODUCTION

The sanitary engineering aims at creating an attractive, comfortable and clean environment by providing (*i*) protected water supply; (*ii*) quick and hygienic disposal of solid, liquid and gaseous wastes; (*iii*) optimum light and sound by eliminating smoke, odour, noise, and glare and agents of disease transmission (such as flies).

Conservancy system (Dry system). In this system, different wastes (including human excreta) are collected, conveyed and disposed off by human beings.

It is cheap where man power available is plenty.

Water carriage system. Human body wastes including urine, and night soil are collected in impermeable receptacles carried by force of water to a far off place through underground pipes to the place of disposal.

Sewage. It is a liquid waste (of foul nature) of domestic or industrial origin. Waste water from toilets is called sanitary sewage.

Sullage. It is waste water less foul in nature such as the one from baths.

Storm water. It is the run off from roads, buildings and other catchment areas. Unless it is quickly disposed off:

(1) it causes hinderance to traffic.

(2) it becomes breeding centre for mosquitoes spreading malaria and filaria.

(3) it creates unsightly conditions.

(4) it spreads disease causing elements and filth over vast areas.

(5) it soaks the roads weakening the bond between aggregates.

D.W.F. It is dry weather flow—*i.e.*, flow available in any season *i.e.*, due to "Sanitary Sewage".

Sewer. It is the pipe carrying sewage.

Sewerage. It is the "Collection + Conveyance" of sewage.

Systems of Sewerage

Separate system. Sanitary sewage and storm water are collected separately, conveyed separately and disposed off separately.

It is best suited to (1) areas of uneven rainfall *i.e.*, rainfall is not uniformly distributed throughout the year, (2) in hilly areas with steep slopes, (3) where deep excavations cannot be easily done.

Combined system. Both sanitary sewage and storm water are collected into the same pipe.

It is best suited to areas, where

(1) rainfall is evenly distributed throughout the year.

(2) storm water flow $\ngtr 10 \times$ D.W.F.

(3) area is in plains.

(4) excavation is easy and less costly.

(5) area is congested to harbour more number of pipes.

(6) both storm water and sewage need be pumped.

Partly Separate System. When rain falls, the few initials showers contribute to run off full of impurities. The subsequent run off is more in volume and less foul. The initial discharges are collected into the sanitary sewer and the subsequent discharge is diverted to storm water drain by providing a "leaping weir" etc.

Hydraulics of Sewers

Max. flow thro' a sewer $= 3 \times$ Av. flow $\quad$ [Bigger the sewer, less is the

Min. flow thor' a sewer $= \frac{1}{2} \times$ Av. flow $\quad$ multiplying factor]

(assumed)

Sanitary sewers are to run $\frac{2}{3}$ full to allow for

(*i*) fluctuations of flow.

(*ii*) gases of decomposition to get accumulated at top.

and (*iii*) keeping the pressure atmospheric and for gravity flow.

Velocity of flow should be more than 'Self cleansing velocity' (0.6 m/s) and at the same time less than 'Scouring Velocity' (3 m/s).

$$v = \frac{1}{n} R^{2/3} \cdot S^{1/2}$$

where n = Co-eff. of roughness, R = Hyd. mean depth, S = Slope.

Quantity of storm water $Q = \dfrac{1}{360} AIR$

Q = Discharge (Cumecs)
A = Area of catchment (hectares)
I = Intensity of rainfall (mm/hr)
R = Co-eff. of run off (depends on "imperviousness" and may be anywhere between 0.01 and 0.95).

Design Period. Design period for a city depends on

1. Rate of growth — fast / slow
2. Economic status and availability of funds — afford to pay more / cannot pay more
3. Topography of the land.
4. Expected life of sewers and appurtenances.
5. Ease of relaying the sewer line.

Longer design periods are to be adopted when slow rate of growth, higher rate of inflation, better quality pipes exist. Also when lower rate of interest is charged and difficult to extend the existing line.

A value of 5 to 50 years is commonly adopted.

Sections of Sewers. Circular sections are common upto 1.5 m dia.

But velocity drops considerably when they run less than 1/4 full.

Hence circular sections are mainly used in separate system.

Egg-shaped or Ovoid sewers offer greater velocities of flow even when the discharge is less than 1/4 the full. Hence they are ideally suited for combined system.

But their construction is difficult.

A box type sewer is either a square or rectangular sewer.

It is ideally suited for an outfall sewer draining into sea as the friction at the corners help against tides.

Lateral. The first sewer to receive sewage from houses.

Branch Sewer. It receives discharges from laterals.

Main Sewer. Branch sewers drain into a main sewer.

A sewerage district may have only one main sewer.

Trunk Sewer. Two or more main sewers drain into a trunk sewer.

Outfall Sewer. It is that part of the trunk sewer leading to disposal units.

SEWER APPURTENANCES

Manhole. It is an inspection chamber to inspect, clean and repair sewers.

Working chamber of the manhole should have a min. dimensions of 1.2 m (in the direction of flow) × 0.9 m (at right angle to flow).

Manholes are to be provided at
1. every change in direction
2. every change in grade
3. every change in elevation
   (when the drop > 90 cm it is called a drop manhole)
4. when more than two pipes meet
   (then their 'tops' or '0.8 d' line should be at the same level)
5. when the dia of a sewer changes
6. at 50 to 200 m intervals on straight lengths ('Rodding distance for cleaning is the governing factor).

Lamphole. It is a 20 to 30 cm dia. circular shaft over a sewer whose main function is to permit a lamp to be lowered—thus illuminating the sewer line. It may be also used for flushing of sewer with water.

However a lamphole can never be a substitute for a manhole.

Leaping Weir. When rainfall occurs the initial run off is quite foul in nature and is collected into the sanitary sewer because of its low velocity of flow. The subsequent run off is less foul and more in volume jumps over the weir and reaches a storm water drain.

Syphon spillway also serves the same purpose *i.e.*, diverts excess spill over into an open channel.

Inverted Syphon. It is a depressed sewer taken below a road, canal or any other obstruction to a sewer line. It runs full and its pressure is more than atmospheric pressure. A velocity > 0.9 m/s is desirable in it to prevent any deposition.

CHARACTERISTICS OF SEWAGE

Colour. Fresh sewage is either grey or light yellow in colour.

Stale sewage is dark brown in colour.

Odour. Fresh sewage has inoffensive, mild earthy odour or even odourless.

Septic sewage produces odours because of ammonia, hydrogen sulphide, methane, carbon disulphide, indol and skatol.

Solids. The weight of solids left after evaporating to 103°–105°C is called "Dissolved Solids".

When it is further ignited to 600°C (in a muffle furnace) the weight lost is due to 'Volatile Solids" which are nothing but organic solids. The remaining dissolved solids are called "fixed solids".

Total solids = Suspended solids + Volatile solids + Dissolved solids.

Nitrogen. Presence of Ammonical introgen in sewage indicates very recent pollution.

Nitrite is a highly unstable form of nitrogen. It is either oxidized to Nitrates or reduced to Ammonia. Its presence indicates of pollution sometime back.

Nitrate is the stablest form of Nitrogen.

Too much of Nitrates promote 'Eutrophication' of lakes.

EUTROPHICATION

Eutrophication is the enrichment of lakes with nutrients (phosphates and nitrates) which in turn promote algal growths. Small concentrations of nutrients promote growth of "Green Algae" which is the food for zooplankton and fish in turn. With excessive nutrients 'Blue green' algae flourish which is not utilised by zooplankton and fish. Blue green algae blooms spreading as a mat at top prevent light penetration. Decaying algae settle to the bottom reducing dissolved oxygen. Malodours are produced. Shores are infested with aquatic weed growths that add to obnoxious conditions existing. Waters become turbid.

Eutrophication can be reduced by

(*i*) minimising nutrients (in particular phosphates),

(*ii*) replacing or diluting the waters with nutrient free water,

(*iii*) injecting cold waters at the bottom,

(*iv*) cutting weeds and removing algae, and

(*v*) application of algicides as copper sulphate.

Chlorides. Every human contributes 8 to 15 g of NaCl per day.

Hence sudden increase in chloride concentration of any water supply indicates sewage contamination.

Sulphides. Sulphates are the end products of aerobic decomposition whereas hydrogen sulphide is the end product of anaerobic decomposition.

pH. Fresh sewage is slightly alkaline. pH decreases with time *i.e.*, as the sewage becomes stale.

However, a pH range between 5 and 10 is the best suited for most of biochemical reactions.

Dissolved oxygen. Pure water with no impurities at the room temperature may have a D.O. of 7 to 8 mg/l. Addition of sewage lowers it.

D.O. in water depends on:

1. Temperature,

2. Atmospheric pressure,

3. Percentage oxygen present in atmosphere,

4. Area of surface exposed,

5. Purity of water,

6. Oxygen deficiency of water,

7. Turbulence at surface, and

8. Presence of algae.

A min. D.O. of 3 mg/l is to be maintained in any waterbody to prevent nuisance conditions.

Bacteria

Anaerobic Bacteria. They fluorish in the absence of (dissolved) oxygen. These bacteria derive oxygen in the combined state for their metabolism *i.e.*, they decompose radicals as nitrates and sulphates and consume the (combined) oxygen in them. They prefer darkness and stagnant media.

Their end products are odourous and are unstable. Also the reactions take place quite slowly.

Aerobic bacteria. They fluorish in the presence of free (Dissolved) oxygen for their metabolism. They prefer sunlight and moving medium.

They deliver inoffensive and stable end products. The reactions take place quite fast.

Aerobic metabolism yields 20 to 30 times the energy of that of anaerobic metabolism.

Facultative aerobic bacteria. They fluorish both under aerobic and anaerobic conditions, though they prefer Aerobic conditions to anaerobic conditions.

Similarly Facultative Anaerobic bacteria prefer Anaerobic conditions to Aerobic conditions.

Psychrophilic or *cryptophilic* bacteria fluorish at less than 20°C.

Mesophilic bacteria fluorish in a temperature range of 20°C – 45°C.

Thermophilic bacteria fluorish in a temperature range of 45°C – 60°C.

A pH range of 5 to 9 is the best suited for bacterial growth though extreme cases of pH < 1 can be tolerated by Sulphur bacteria.

Escherischia Coli—a non-pathogenic intestinal bacterium is used as "indicator organism" of sewage contamination because an average adult excretes 2000 million *E. Coli* per day.

Algae. They are chlorophyl bearing microscopic, multi-cellular, photo-synthetic plants without roots, stems or leaves.

They may range in size from microscopic unicellular forms smaller than bacteria to sea weeds a few metres long !

Green algae because of their photosynthesis may supersaturate any water sample with D.O.

Protozoa. Unicellular asexual animals which consume bacteria.

Viruses. They are the tiniest parasitic pathogens of highly specific reactions. They pass through the pores of filters which retain bacteria. They require a living cell for reproduction. They can infect humans, animals, plants, fish, insects and micro-organisms as bacteria.

Control of viruses through chemicals is not possible. It is achieved by producing antibodies in the host cell.

Smallpox, Chickenpox, Rabies, Yellow fever, Mumps, Poliomyelitis, Dengue, Influenza, Cancer and common cold are some of the virus infections.

Bio-Chemical Oxygen Demand. B.O.D. at a time and temperature is the amount of oxygen required by the micro-organisms for the oxidation of organic matter to yield stable end products. It is the amount of oxygen required by micro-organisms for the oxidation of waste to yield stable end products as CO_2 and H_2O at a particular temperature and time.

$$y = L_i[1 - 10^{-kt}]$$

where L_i = Ultimate BOD, or initial Oxygen equivalent
k = Deoxygenation constant/day
t = No. of days
y = BOD exerted or oxygen absorbed (demand satisfied) in t days.
or BOD at any time t = ultimate BOD $(1 - 10^{-kt})$

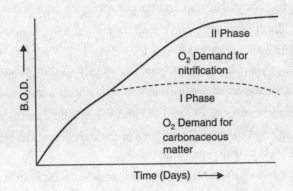

FIGURE 7.1

Rate constant 'K' varies with temperature $K_T = K_{20}[1.047^{T-20}]$

Similarly, ultimate BOD L_i also varies with temperature.

$$[L_i]_T = [L_i]_{20} (0.02T + 0.6).$$

BOD test is carried out by collecting the sample and diluting it with sufficient quantity of pure D.O. rich water. The sample being filled in BOD bottles (of capacity 300 ml) is incubated as 20°C and tested after 5 days for their D.O. concentration.

$$BOD = \begin{bmatrix} \text{Initial D.O.} \\ \text{of the sample} \\ \text{(after diluting)} \end{bmatrix} - \begin{bmatrix} \text{Final} \\ \text{D.O.} \end{bmatrix} \times \text{Dilution factor.}$$

$$\text{Diluting factor} = \frac{\text{Capacity of the BOD bottle}}{\text{Volume of the sample}} = \frac{1}{\text{Dilution ratio}}$$

The final D.O. should be a min. of 1 mg/l and if it falls below the test is invalid.

Theoretically, infinite time is required for the complete exertion of BOD, but in practice it is considered to be complete after 20 days. (for domestic sewage)

It was found that 65 to 70% of the total first stage BOD (Carbonaceous oxygen demand) is exerted in the first 5 days. Hence the test is carried out for 5 days only and unless otherwise mentioned BOD of any waste is 5 days BOD (the sample being incubated at 20°C).

In general Nitrification (oxygen demand by the **Nitrogen** bacteria) commence after 5th, **6th** or 7th day for domestic sewage (II stage BOD) **and is not taken** into account.

Limitations of BOD Test

1. Pollutional strength of wastes toxic to bacteria cannot be assessed.
2. Only the pollutional strength of biodegradable organics is given out.
3. High concentration of active, acclimated bacterial seed is required.
4. The test is specified as 5 d and 20°C. It does not include Nitrogenous or ultimate carbonaceous stage, nor the ambient temperature.

Population equivalent. A value of 80 g/day of [5 days, 20°C] BOD is taken as population equivalent (*i.e.*, BOD contribution per head/day).

Chemical Oxygen Demand (COD). COD is the amount of oxygen consumed by the oxidizable matter from a strong oxidizing agent.

It oxidizes all organic matter to $CO_2 + H_2O$ and takes only 3 hours for completion.

BOD test may be affected due to the presence of toxic chemicals as copper, phenols, chlorine, cyanides etc., while COD test had no such drawbacks.

But it is not possible to differentiate between relatively inert and slowly bio-degradable matter as fats and lignins from the rest.

For any waste COD > BOD.

$\dfrac{BOD}{COD}$ gives an idea of readily biodegradable matter. The higher the ratio, more is the percentage of bio-degradable matter in the sample.

Natural Methods of Disposal of Sewage

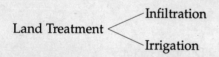

Infiltration. It is the application of raw sewage or sewage given primary treatment on to a prepared bed of soil.

Loams, sandy soils and other porous beds are the best suited for infiltration.

Marshy soils, closely compacted clays, hard rocks and loose and hungry soils are not suited.

The area is levelled and given a gentle slope. Application of sewage is intermittently done.

Sewage Irrigation. Sewage is a perennial source of water.

It is rich in nutrients as nitrates, phosphates, potash, calcium, magnesium, sodium and sulphur. It is also rich in trace elements as zinc, molybdenum, copper, boron and vitamins and hormones needed for plant growth.

Organic fraction of sewage is the best soil conditioner.

Problem	Solution
1. Sanitary sewage will be having less phosphates.	Supplement phosphates.
2. Sodium salt accumulation.	Flood the sewage farm with fresh water of low salt content or apply gypsum.
3. Toxic substances as synthetic detergent (ABS) and Boron.	Dilution.
4. Health hazards.	Sewage farm workers provided with gumboots and gloves. Consumers should not eat the raw products.
5. Sewage sickness It is the property of the land getting clogged due to the continual application of sewage and its inability to take any more load though the loading rate is quite reduced.	1. Deep ploughing every year to increase permeability. 2. Tilling for effective soil aeration. 3. Flooding with pure water to leech out salts. 4. Resting 2 to 3 (rainy) months of year and every alternate summer. 5. Lowering GWT below 2 m from G.L.

Dilution. Sewage given preliminary treatment may be let off into flowing bodies of water. Before being discharged into, the sewage

1. should be free from unsightly floating matter, oils and greases.
2. should not have a suspended solid concentration greater than 150 mg/l.
3. should not have a BOD greater than 150 mg/l.

The receiving stream should have

1. a min. flow of 100 l/s/1000 people.
2. enough initial D.O. and even after the addition of sewage the D.O. should not fall below 3 mg/l.

Condition	Dilution ratio
1. Domestic sewage with no treatment	> 500
2. Preliminary treatment given. Suspended solids $\not>$ 150 mg/l	300 – 500
3. Preliminary treatment + chemical precipitation to remove suspended solids. ($\not>$ 60 mg/l)	150 – 300
4. Suspended solids $\not>$ 30 mg/l BOD $\not>$ 20 mg/l i.e., completely treated	$\not<$ 150

A stream with turbulent flow, water falls and rapids have more capacity to purify themselves than a slow and sluggish stream.

SELF PURIFICATION OF STREAMS

Self purification of a stream is its ability to regain its normal level of purity when polluted.

It depends on dilution, sedimentation, oxidation, sunlight, temperature, initial D.O., and currents. Higher they are less is the time taken for self-purification and the stream can take more pollutional loads.

Zones of Pollution

1. **Zone of degradation.** Indicates recent pollution, D.O. falls to 40%, water is turbid, algae die.

2. **Zone of active decomposition.** D.O. goes down from 40% to zero and again comes back to 40%. Water is dark coloured. Fish die. Foul smells emanate. Fungi appear.

3. **Zone of recovery.** D.O. rises to 40%. Water becomes clear. Algae reappear.

4. **Zone of clear waters.** D.O. gets saturated. Normal stream conditions restored.

SEWAGE TREATMENT

Preliminary treatment. Removal of large floating solids, by screens Grit by Grit chamber and grease by skimming tanks and grease traps to mainly improve flow characteristics.

Primary treatment. Plain sedimentation, Chemical coagulation and Fine screening to marginally improve the quality of sewage.

Secondary treatment. For domestic sewage it is a biological treatment removing most of the pollutional load as BOD and suspended solids. It is almost the final treatment of sewage.

Aerobic Units	Anaerobic Units
Trickling filter	Septic Tank
R.B.C.	Imhoff Tank
Activated Sludge Process	Anaerobic Lagoon
Oxidation Pond	Upward flow Anaerobic filter.
Aerated lagoon	
Oxidation ditch	

Tertiary treatment. Necessary only in specific cases as industrial wastes, Removal of Nitrogen, Phosphorus, Dissolved solids, etc.

Physical operations as Screening, Sedimentation are called **Unit Operations.**

Biological operations as Trickling Filtration, Activated sludge processing are called **Unit Processes.**

Screens. Fixed screens

    (a) Fine screens (< 25 mm holes)—Rarely used.

    (b) Medium screens (25 to 75 mm) not commonly used.

    (c) Coarse screens (> 75 mm clearances) used before pumps.

SANITARY ENGINEERING

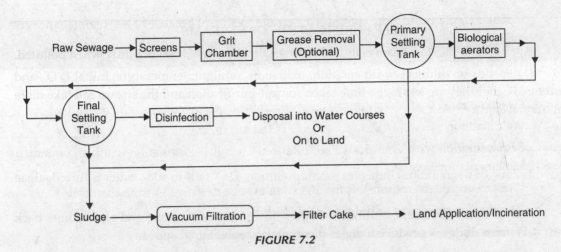

FIGURE 7.2

Rotating screens—Cut the floating solids to a size less than 6 mm.

Velocity of flow $\ngtr$ 1 m/s and preferably 300 mm/s.

Grit Chamber. Grit is inert matter as sand, gravel, egg shells, bone chips, seeds, cinders, and large organic particles of size greater than 0.2 mm and specific gravity greater than 2.65.

Grit chambers are provided to remove them so that

1. mechanical equipment can be protected from heavy wear.
2. to prevent clogging of sewers and to protect pumps.
3. frequent digester cleaning is avoided.

Settling velocity = 20 mm/s.

It controls the detention period.

Grit chamber is a rectangular tank designed for a detention period of 45 seconds to 90 seconds and a horizontal velocity of flow of 250 to 450 mm/s.

Sedimentation Tanks. Sedimentation is the process of separation of suspended solids from water or waste water by the force of gravity. Based on their location they are classified as

1. Primary settling tanks: before biological treatment units
2. Intermediate settling tanks: in between two high rate biological units
3. Secondary settling tanks: after the biological treatment unit.

Primary clarifiers. It is rectangular or circular in plan.

Rectangular tanks have a width less than 6 m.

Length	= 3 to 5 times width
Liquid depth	= 2 to 2.5 m

Circular tank have a dia. of 10 to 40 m.

Over flow rate	= 20 to 30 m³/m²/day
BOD removal	= 30 to 40%
Weir loading for circular tanks	= 180 m³/m/day
Detention period	= 1 to $1\frac{1}{2}$ hrs.

(Greater detention periods promote septic conditions and hence are avoided).

Circular clarifiers are more common in sewage treatment because of less maintenance cost.

Diameter	= 10 to 40 m.
Liquid depth	= 2 to 3 m
Bottom slope	= 8%
Overflow rate	= 20 to 30 m^3/m^2/d
Weir loading	= 180 m^3/m/d.

Sedimentation with Chemicals Precipitation. Chemical coagulation is not so common for sewage sedimentation because

1. micro-organisms responsible for digestion may be destroyed by the chemical.
2. settling in secondary settling tanks is affected.
3. more sludge is produced whose disposal is a problem.

Copperas, chlorinated copperas, ferric sulphate, alum, lime, ferrous chloride and ferric chloride are some coagulants used.

Sewage coagulation is quite useful for

1. industrial wastes rich in chemicals.
2. phosphate removal (reducing eutrophication).

Biological Treatment. It is mostly favoured when the waste water is predominantly biodegradable.

When $\frac{BOD}{COD} > 0.6$, biological treatment is best suited.

When $\frac{BOD}{COD} > 0.3$, but < 0.6 biological treatment is recommended after seeding.

When $\frac{BOD}{COD} < 0.3$, biological treatment is not possible.

Micro-organisms under favourable conditions remove dissolved organic solids and colloidal solids and get themselves removed. Large number of organisms in a small reactor decompose the organic matter in a smaller interval of time under ideal conditions with high efficiency.

Aerobic Biological reactors are classified as:

1. Attached film growth reactors and 2. Suspended film growth reactors.

Attached film growth reactors are those where bacterial growth occurs over porus medium. Oxygen is supplied by AIR DIFFUSION through pores.

Attached film growth reactors are

1. Trickling filter
2. Rotating Biological Contactor (RBC).

Trickling Filter. It is circular, rectangular or of any other shape but circular tanks 30 to 60 m in dia. are quite common.

Media is 25 mm to 100 mm in size.

Media may be crushed rock, gravel, anthracite coal, coke, cinders, blast furnace slag, broken bricks, clinkers, wooden lath, ceramics and plastics or any other of porus clean, durable, insoluble, weather resistant and strong enough to support its own weight, suitable for the growth of active microbial film.

Synthetic media *i.e.*, plastics have high porosity (95% whereas other media cannot have a porosity more than 50%), low weight, greater specific surface area (200 m²/m³ against 50 to 65 m²/m³ for conventional media).

The trickling filter with synthetic media and depth more than 12 m is called **Bio Tower**.

Trickling filter works on the principle of "Sorption + Biological oxidation".

A zoogleal film of bacteria, protozoa, and fungi forms over the media. They support rotifers, sludge worms, insect larvae and snails.

Standard Rate Trickling Filter is 3 to 6 m deep with no recirculation.

High Rate Trickling Filters have a depth of 1 to 1.5 m and the effluent is recycled.

Recirculation factor
$$F = \frac{1+r}{(1+0.1r)^2}$$

N.R.C. formula,
$$\frac{C_i - C_e}{C_i} = \frac{1}{1 + 0.014\sqrt{\frac{Q \cdot C_i}{VF}}}$$

where C_i = Influent BOD (mg/l), $\quad C_e$ = Effluent BOD (mg/l)

Q = Load (m³/d), $\quad V$ = Volume of filter bed (m³)

$$\frac{C_e}{C_i} = e^{-\frac{KD}{Q^{0.67}}}$$

where K = Rate constant (1.89/d), D = Depth of filter (m)

Q = Hyd. loading rate (m³/m²/d)

Velz formula:
$$C_e = \left(\frac{C_i + rC_e}{1+r}\right) e^{-KD}$$

K = Experimental constant (0.57 for low rate, 0.49 for high rate)

r = Recirculation ratio

N.R.C. formula:

$$\text{Efficiency} = \frac{100}{1 + 0.0044\sqrt{10U}}, \quad \text{where } U = \text{Organic loading (g/m³/d)}$$

Item	Standard rate trickling filter	High rate trickling filter
Hydraulic loading	2 to 5 m^3/m^2/day	10 to 30 m^3/m^2/day
Organic (BOD) loading	100 to 400 g/m^3/day	500 to 1500 g/m^3/day
Dosing	Intermittent (5 to 15 minutes)	Continuous (15 seconds)
Sloughing	Occasionally	Regularly
Recirculation ratio	Nil	0.5 to 3
Depth	2 to 3 m (most common)	1 to 1.5 m
Effluent	Highly nitrified	Not fully nitrified
BOD of effluent	Less than 20 mg/l	More than 30 mg/l
Secondary sludge	Black, highly oxidized, fine particles	Brown, not fully oxidized particles tending to septacity
Odours	More	Less
Psychoda populations	More	Less

ROTATING BIOLOGICAL CONTACTOR (R.B.C.)

(Attached Film Growth Reactor)

An RBC works on the same principle of a Trickling Filter. While waste water moves relative to media in T.F., media moves relative to waste water in RBC. In both the cases media is loaded intermittently, gets aerated and again loaded.

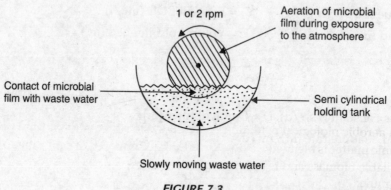

FIGURE 7.3

RBC consists of a series of closely spaced circular discs (3 to 4 m in dia. and 10 mm thick placed 30 to 40 mm centre to centre) of PVC. A module 3.7 m in dia and 7.6 m long offers around 10,000 sq m of surface area for biological slime layer growth.

The discs rotate at 1 or 2 rpm with 35% to 40% (of cross-sectional area) submergence.

Biological growth gets attached to the surface of the discs to a thickness of 1 to 3 mm. The rotation alternately brings the biomass in contact with the organic matter of the waste water and free atmosphere for the adsorption of oxygen.

RBC has low power demand, greater process stability, relatively higher organic loading rates. It gives less amount of sludge compared to Activated Sludge Process.

Hydraulic loading rate = 0.04 to 0.06 m³/m²/d

Organic loading rate = 50 to 60 g/m²/d

Sewage is retained in the semicircular holding tank for 45 to 90 minutes.

$$Q(C_i - C_e) = \text{P.A.} \frac{C_e}{K_s + C_e}$$

where P = Org. loading rate (g/m²/d), K_s = Experimental constant

A = Total area of the discs (m²)

Advantages and Disadvantages:

No foaming, bulking or insect nuisance.

But sensitive to temperature and shock loading.

Suspended Film Growth Reactor: It is a reactor where the zoogleal film of bacteria, protozoa, rotifers and algae remain in suspension along with waste water during the period of aeration.

ACTIVATED SLUDGE PROCESS

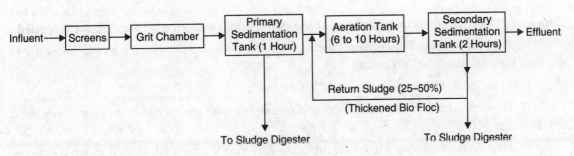

FIGURE 7.4

ASP is an aerobic biological oxidation process. Waste water effluent from primary setting tank rich in organic matter is blended with 'return sludge' composed of active micro-organisms to increase the available biomass and to speed up the reaction.

The biomass involves bacteria, protozoa, rotifers and algae.

$$\text{Food} + \text{Microbes} + \text{Oxygen} \xrightarrow{\text{Nutrients}}$$

(Organic waste) (Sludge) (Air)
[BOD] [MLSS]

$$\text{New cells} + \text{Energy} + CO_2 + H_2O + NH_3$$

(Surplus sludge) (End products)

F/M – 0.7 to 0.2 in conventional ASP. M.L.S.S. of 2000 to 2500 mg/l is required. About 60 to 95 m³ of air/kg of BOD is required.

About 8 m³ of air is required/m³ of domestic sewage.

Organic (BOD) loading = 500 to 650 g/m³/day

BOD removal desired = 20 (T + 1), where T = Aeration period in hours.

Efficiency = 90 to 95%.

$$\text{Volume of Aeration Tank} = \frac{QL_i}{F \cdot X_v}$$

Q = Sewage flow (m³/d), L_i = BOD conc. (mg/l)

F = F/M ratio (usually taken as 0.5), X_v = MLVSS concentration (mg/l)

Aeration tank:

Volume	≯ 150 m³
Liquid depth	= 3 to 4 m
Breadth/Depth	= 1 to 2.5

Different Types of ASP

Depending upon the mode of application of the mixed liquor and aeration the ASP is classified as:

1. Plug Flow Type. The conventional type. Aeration is done uniformly along the length of the tank between the inlet and outlet.

It is highly sensitive and cannot take shock loads.

2. Tapered Aeration Type. Spacing of diffusers is closer near the inlet and increases along the direction of flow.

3. Step Aeration Type. Mixed liquor is applied into the aeration tank through 3 or more inlets located along the length of the tank.

4. Continuously Mixed Activated Sludge Process. It has a number of inlets and the sewage is continuously mixed so that the tank contents are homogeneous. Return sludge is 35 to 100%.

While F/M ratio varies from inlet to outlet in other methods, it remains constant here. It can also take shock loads.

5. Extended Aeration System. It is with more aeration period (18 h to 36 h as against 6 to $7\frac{1}{2}$ h for other processes), less F/M ratio (0.2 to 0.05 against 0.7 to 0.3 for other methods) and more amount of return sludge (50 to 200%).

Activated sludge process	Trickling filter
1. Suspend culture	Attached culture media
2. Return sludge supplies the required number of microbes to treat sewage	Microbes resting on media treat sewage
3. Less area required	More area required
4. Less loss of head	More loss of head
5. Conventional ASP cannot accept shock loading	Can take shock loads
6. Volume of sludge is more	Volume of sludge is relatively less
7. Uniform quality of effluent given	Sloughing takes place occasionally
8. Skilled operation essential System is fully aerobic	Skilled operation is not required Thin layer attached to media is anaerobic and then a thick aerobic layer over it
9. Efficiency is 90 to 95%	Efficiency is 60 to 98%

SLUDGE TREATMENT

Sludge is the sediment collected in various treatment units.

It is rich in organic matter, moisture and bacteria.

Depending on the clarifier sludge is classified as

1. Primary sludge and 2. Secondary sludge.

1. Primary sludge: Primary sludge is from primary sedimentation tanks—which is mainly composed of inorganic solids and coarser organic solids—which are grannular and denser (sp. gr. = 2 to 2.5).

2. Secondary sludge: Secondary sludge is from secondary settling tanks—which is finely divided and predominantly organic in nature and is relatively lighter (Sp. gr. = 1.2 to 1.3).

Sludge contains 70 to 75% of "free water" that can be separated by sedimentation and 20 to 25% "floc water" trapped in the interstices of the floc particles which can only be separated by mechanical dewatering.

Flow dia. for sludge treatment:

Sludge $\longrightarrow$ Thickening $\longrightarrow$ Digestions $\longrightarrow$ Conditioning $\longrightarrow$ Dewatering $\longrightarrow$ Disposal

Sludge Thickening

It is the process of concentrating the solids by separating the water.

The sludge thickener is a circular sedimentation tank with a rotating deep truss provided with vertical pickets which cut the sludge blanket and release the entrapped water. The sludge thickener has a detention time of 24 h, Hydraulic loading rate of 10 to 30 $m^3/m^2/d$ and a (side water) depth of 3 to 4 m.

Dissolved air floatation is a sludge thickening process where chemicals as Alum and copperas are added to the sludge and fine air bubbles under a pressure of many atmospheres are introduced to cause floatation and subsequent thickening of the sludge.

Dissolved air floatation rapidly separates the solids which settle very slowly by gravity thickening.

Centrifugation of the sludge also causes sludge thickening but operation and maintenance are costly.

Sludge Digestion

Sludge digestion is the process of
1. stabilising organic matter and lowering BOD.
2. elimination of pathogens.
3. removal of the food for rodents and flies
4. recovery of Methane, a useful gas
5. reduction in volume of sludge and ready dewatering of it.

That is during digestion

1. Liquifaction. Complex organic matter is broken down by Acid forming saprophytic bacteria into short chain volatile fatty acids as Acetic acid, Butyric acid and Propionic acid.

2. Gasification. The acids so formed are converted to gases as Methane and Carbon dioxide

$$Acids \longrightarrow Alcohols \longrightarrow CH_4 + CO_2$$

Factors Controlling Digestion

1. Temperature. Bacteria which prefer 10°–15°C are called **Cryptophilic** or **Psychrophilic** bacteria, these which prefer 33°–37°C are **Mesophilic** and 45°–60°C are **Thermophilic** bacteria.

Bacterial activity gets doubled with 10°C rise in temperature. *i.e.*, when a waste gets digested in 10 days at 20°C, the same may get digested in just 2 days at 35°C.

Most digesters operate in Mesophilic range while high rate digesters operate in Thermophilic range.

2. pH. 6.5 to 8 is ideal for most of the bacteria.

3. Nutrients. COHNPS—while domestic sewage is rich in carbohydrates (COH), it is poor in N P and S and they are to be supplemented.

4. Intimate contact. Being microscopic bacteria easily get segregated because of their low density.

Intimate mixing well distributes bacteria and nutrients and efficiency improves.

5. Inoculation. Seeding of digester with microbes from digested sludge improves efficiency.

6. Type of operation. (*a*) **Batch operation**—leaving the contents of the digester to undergo stratification *i.e.*, Digested sludge, Active layer of digested solids, Tank liquor, Scum and Gas formed one over the other. It is less efficient.

(*b*) **Continuous Mixing operation.** Where the contents of the tank are intimately mixed. Bacteria and nutrients are intimately mixed and there are no dead pockets or stratification.

Item	Standard rate digester	High rate digester
1. Stratification	Stratification occurs as gas at top. Scum Supernatant tank liquor. Active layer of digesting solids Digested sludge below.	Contents are intimately mixed.
2. Effects of stratification	1. Micro-organisms responsible for digestion may be segregated due to density currents.	Because of intimate mixing inoculation of sludge is being done. Hence capacity is more. Efficiency is also more.
	2. Toxins produced remain in position and inhibit further reaction.	Toxins get diluted.
	3. 50% of volume of digester remains idle.	Full volume is utilised.
3. Detention period	30 to 90 days.	10 to 20 days.
4. Solids loading	0.5 to 1.6 kg/m³/day	1.6 to 6.4 kg/m³/day
5. Temperature	Mesophilic	Thermophilic.

Digester volume $V = \left[V_f - \dfrac{2}{3}(V_f - V_d) \right] t$

where V_f = Vol. of fresh sludge added (m³/d).

V_d = Vol. of digested sludge with drawn (m³/day).

t = Period of digestion (days).

Digestion tanks are 6 m to 35 m in dia.

Sludge depths 7.5 to 14 m.

Digested sludge is dried on porus beds in 20 to 30 cm thick layers.

Dried sludge is used as a manure. It is a good soil builder.

Septic Tank. It is an underground water tight tank where settling of solids, floatation of grease, anaerobic decomposition of accumulated organic matter and of sludge take place.

Carbon dioxide, methane and hydrogen sulphide are given off.

Generally it is rectangular with a min. breadth of 1 m and length = 2 to 3 times breadth.

Liquid depth is 1.2 to 2.5 m.

A detention period of 24 hrs is more common for Indian conditions.

Design Particulars:

1. Minimum breadth = 0.75 m

2. Minimum length = 1.5 m

3. $\dfrac{\text{Length}}{\text{Breadth}}$ = 2, 3 or 4

4. Liquid depth = 1.5 ± 0.2 m
5. Minimum volume = 3 m^3
6. Detention period = 24 h
7. Sludge contribution = 75 l/capita/year
8. Free board = 0.5 m
9. Bottom slope = 5 to 10%
10. Ventilation stack = 50 mm in dia.

Septic tank effluent may have a BOD of even 200 mg/l and hence requires further treatment. The effluent of the septic tank is to be further disposed of into:

1. Tile drains or Dispersion trenches

or 2. Soak pit

The ideal soil for dispersion trenches should have a rate of percolation not greater than 180 s/mm (should not take more than 180 seconds for a fall in level of 1 mm.)

It is ideal if 50 s/mm.

(When it is less than 2 s/mm it is "loose and hungry soil"–undesirable)

Ground Water Table must be at a minimum depth of 1.8 m.

Sludge is cleared once in every 3 to 5 years.

Waste Stabilisation Pond or Oxidation Pond

It is a large basin (> 1.2 hectares in area) with earthen embankments. Its depth ≯ 0.6 m to prevent the growth of rooted aquatic weeds—which in turn become breeding centres for mosquitoes. If the liquid depth exceeds 1.5 m anaerobiosis may be developed in bottom layers.

It works on the principle of Symbiosis of Algae and Aerobic Bacteria.

Aerobic bacteria decompose organic matter of sewage giving out CO_2, Ammonical Nitrogen and Phosphates.

Algae utilise these nutrients during their photosynthesis and inturn release oxygen—raising the D.O. level of sewage.

Depending on solar insolation, a detention period of 10 to 30 days is adopted.

Oxidation ponds do not require electric power. Where land is inexpensive and more number of solar days do occur in a year (*i.e.*, less number of cloudy days) to promote photosynthesis—it is best suited. No skill is required to operate it.

Its only disadvantages are the effluent had more suspended solids and anaerobic conditions may be developed in temperate zones on cloudy days occurring on consecutive days.

Aerated Lagoons

They are also large basins having a depth of 2.4 m to 4 m.

Floating aerators create turbulence and mix the contents of the lagoon and supply oxygen for bio-oxidation of waste.

Detention period is relatively less (3 to 5 days) as the lagoons function round the clock.

Aerated lagoon is regarded as Activated sludge unit without recycling of return sludge as their microbial flora are similar to the ASP.

Aerated lagoons have the advantage of accepting shock loads and treating biodegradable toxins as phenols.

Oxidation Ditch

It works on the principle of extended aeration.

It is an elongated oval shaped continuous flow channel of 1 to 1.5 m liquid depth and 0.5 m free board.

It is provided with one or two cage rotors which

1. supply necessary propulsion for complete mixing and
2. supply oxygen required for biological activity.

The sewage free from floating matter and grit is mixed with 50 to 100% return sludge to maintain a MLSS concentration of 3000 to 5000 mg/l.

Rotors 700 to 1000 mm in dia. rotate at 70 rpm and have a usual depth of immersion of 150 mm. A min. vel. of 0.3 m/s is maintained.

A usual detention time of 10 to 16 h is adopted.

Oxidation ditch can take shock loads.

SOLID WASTE DISPOSAL

Refuse comprises of garbage, rubbish, ashes, dead animals, street sweepings, and industrial wastes.

Garbage. It is the putrescible waste resulting from growing, handling, preparation, cooking and consumption of food.

Rubbish. It is the non-putrescible waste other than ashes. *e.g.*, papers, leaves, cans, glass, crockery, cardboard, wood, scrap metals.

Ashes. Product of incineration.

Smaller dead animals. Rats, cats, dogs.

Street Sweepings. Mostly inert.

Refuse either mixed or separate as garbage, rubbish etc., is collected into fire-proof, rodent proof metallic containers with tight lids. It is to be emptied twice in a week.

Refuse is disposed off by

1. Sanitary Land Filling. (Controlled tipping).

Natural depressions of ground are made use off.

Refuse is dumped, compacted, a thin layer of soil is laid over it, and this bed is ready to receive another layer of refuse.

2. Salvaging is extraction of hog feed etc., from garbage, scrap metals etc., from rubbish, coarse aggregate from street sweepings etc.

3. Incineration. Incineration of any type of refuse is associated with fuel or auxiliary fuel and hence is costly. It is quite commonly adopted by individual residences.

HOUSE FITTINGS

Water Closet

1. Squatting type, or Indian Type. It is nothing but a saddle like porcelain basin of varying depth the maximum depth being about 500 mm when fitted with trap. Length of the commode lies between 45 cm to 65 cm and it is dependent on the length of the urinating jet. It is so fixed in position such that its flushing rim is at the floor level.

2. Pedestal type or European type. It is relatively a higher one compared to the squatting type. It rests over the floor. It is designed to incorporate a minimum of 50 mm of water seal in it. It is ideally suited to people accustomed for sitting on chairs.

The pipe which drains off the contents from a water closet is called "Soil Pipe".

3. Lavatory Basin or Wash Basin. It is a receptacle about 50 cm long placed at a height of about 0.75 m above floor level. It is to receive the waste water during face or hand washings.

4. Urinals. They are designed to receive urine.

(i) **Bowl type.** This type are best suited for the people of the same height as the students of a specific class.

(ii) **Stall type.** Stall type urinals are best suited for people of different heights as in a public place.

P, Q or S traps. A trap is a bent pipe to contain water seal. The water seal prevents the backward travel of the foul smelling gases from the soil pipe into the commode.

Q or P traps are quite common after a squatting type of water closet whereas S trap is usually provided after a wash basin.

Floor traps are provided for kitchens and baths.

An intercepting trap is the last trap from a house drain. It contains a minimum water seal of 75 mm. Its main object is to prevent the travel of foul smelling gases from the branch sewer into the house drain (lateral).

Soil pipe collects the discharges from toilets and urinal (sewage).

Waste pipe collects discharges from baths and wash basins (sullage).

Ventilation pipe is a pipe provided to dispel foul smelling gases into the atmosphere. It is provided with a conical crown with slits for the easy escape of gases into the atmosphere and to provide an unstable base so that birds cannot build their nests. It is called a "Cowl".

Antisyphonage pipe is provided in multi-storeyed buildings. Its purpose is the prevention of sucking of the water seal in the lower floors when sewage runs down.

ENVIRONMENTAL POLLUTION

Air, water, food, heat and light are the basic elements to sustain life.

Pollutant is any foreign element that is finding its way into air, water and food and endangering its quality.

Also excessive or irregularly distributed heat, light and sound causes discomfort.

Air Pollutants. Suspended Particulate Matter (SPM).

(a) Dusts (b) Fumes
(c) Smokes (d) Bioparticles.

Mists and Gases

(a) Carbon monoxide (b) Oxides of Sulphur
(c) Oxides of Nitrogen (d) Hydrogen Sulphide.

Water Pollutants. Water is a universal solvent. Almost everything is soluble in to a certain extent.

Domestic waste water and industrial waste water are polluting natural bodies of water.

Thermal Pollution. Hot waters from thermal power stations, steel mills, petroleum refineries and paper mills raise the temperature of natural bodies of water. This rise decreases D.O., accelerates biological activity and growth of aquatic plants which contribute odours and undesirable tastes to water.

Noise. Noise is unwanted sound. The sound that is either too loud or that is of varying intensity causes annoyance.

Similarly too much concentration of light causes glare, the most unpleasant one for anybody.

Ecology. Nature consists of "Living (Biotic) organisms" and "Non-living (abiotic) environment".

Any living thing can flourish in its selected environment *i.e.*, the environment controls the species of life that is called its 'habitat'.

Again, the environmental factors are being controlled by the life (trees and animals) existing there.

In other words, in a closely knit ecosystem (Biotic + Abiotic factors) every thing influences all others, and all are interdependent.

i.e., just as the land and climate determine the types of trees present so also plants growing on sand dunes build up a soil radically different from one the original substratum.

Hence if an existing species is wiped out or a strange species was introduced into an ecosystem its impact shall be felt everywhere.

A food chain is the transfer of energy from one source to another.

Plant ⟶ Herbivore ⟶ Carnivore ⟶ Still bigger carnivore

A nectar sipping butterfly ⟶ Dragon fly ⟶ Hawk ⟶ Snake ⟶ Frog.

Types of food chains

1. **Predator food chain.** Starts from a plant base and goes from smaller to larger animals.

2. **Parasite chain.** Goes from larger to smaller organisms.

3. **Saprophytic chain.** Dead matter to micro-organisms.

Food chains are interconnected with each other. With operations as "Commensalism" (at the same dining table), "Mutualism" and "Parasitism".

Biological Magnification. While energy decreases as the trophic level increases, it becomes more and more concentrated with each link in a process and this is called "Biological Magnification".

e.g., 1 part/g wt of DDT in water

800	"	"	"	Plankton
12000	"	"	"	Minnow (fish)
35000	"	"	"	Predatory fish
92000	"	"	"	Fish eating birds.

Man is the fundamental culprit in upsetting the ecological balance. His most common activities as growing crops, killing animals and reptiles, cutting down of forests, using pesticides, and starting of industries culminated in complete disruption and destruction of nature's cycle.

II. OBJECTIVE TYPE QUESTIONS

1. Conservation system is best suited
   - (a) in slums ☐
   - (b) in densely populated area ☐
   - (c) for people of low morale ☐
   - (d) when man power is cheap. ☐

2. A deplorable aspect of conservancy system is
   - (a) recurring cost is high ☐
   - (b) vehicles are required to carry night soil ☐
   - (c) vast areas for disposal are necessary ☐
   - (d) human elements is involved in collection and transportation of human waste. ☐

3. Water carriage system needs more
   - (a) men ☐
   - (b) area ☐
   - (c) water ☐
   - (d) money. ☐

4. Combined system may be favoured where
   - (a) rain fall is concentrated in a season of the year and DWF is often fluctuating from day to day ☐
   - (b) rain fall is scattered throughout the year and DWF is too small compared to storm water ☐
   - (c) rain fall is distributed throughout the year such that it is ≤ 10 × DWF ☐
   - (d) city is on steep rocky slopes. ☐

5. Separate system is adopted in
   - (a) plains with a gentle slope ☐
   - (b) thickly populous districts ☐
   - (c) areas where rain fall is distributed throughout the year ☐
   - (d) steap rocky slopes. ☐

6. Partly separate system is
   - (a) collecting DWF and storm water in the same sewer ☐
   - (b) collecting DWF and drainage of initial showers into the same sewer ☐

(c) collection of sewage and sullage into the same sewer
(d) collection of storm water and a part of DWF into the same sewer.

7. Leaping weir is provided only in
   (a) conservancy system
   (b) separate system
   (c) partly separate system
   (d) combined system.

8. Sewage is
   (a) waste water from bathrooms
   (b) drainage from roads
   (c) waste water from kitchen
   (d) any waste water of domestic or industrial origin.

9. Sullage is
   (a) waste water from baths
   (b) drainage from roads
   (c) industrial liquid waste
   (d) waste water from toilets.

10. Sewerage is
    (a) collection of sewage
    (b) collection + conveyance of sewage
    (c) collection + conveyance + treatment of sewage
    (d) collection + conveyance + treatment + ultimate disposal of sewage.

11. A sanitary sewer is expected to run
    (a) full
    (b) half full
    (c) $\frac{2}{3}$ full
    (d) 90% full.

12. Top space is left free in a sanitary sewer for
    (a) gases to accumulate
    (b) fluctuations to be met
    (c) allowing shock loading
    (d) maintaining uniform pressure.

13. Max. flow in sewers may be taken as
    (a) $1\frac{1}{2}$ × av. flow
    (b) 2 × av. flow
    (c) 3 × av. flow
    (d) 5 × av. flow.

14. In general "Imperviousness" of any catchment shall be between
    (a) 1% to 95%
    (b) 0.1% to 10%
    (c) 0.01% to 95%
    (d) 0.0001% to 100%.

15. Time of concentration depends on
    (a) intensity of rain fall
    (b) duration of rainfall
    (c) slope of catchment
    (d) evaporation and percolation.

16. Velocity of flow in a sewer should be between
    (a) 0.6 m/s and 3 m/s
    (b) 2 m/s and 10 m/s
    (c) 30 cm/s and 90 cm/s
    (d) 30 m/s and 90 m/s.

17. For a partly running circular sewer as the depth of flow increases
    (a) wetted perimeter increases
    (b) velocity of flow increases
    (c) velocity of flow decreases
    (d) discharge increases.
18. Ovoid sewers are best suited for
    (a) separate system
    (b) combined system with less fluctuations
    (c) combined system with very wide fluctuations
    (d) outfall sewer.
19. A sewer that collects sewage from toilets is called
    (a) lateral
    (b) branch sewer
    (c) main sewer
    (d) outfall sewer.
20. That length of the main sewer leading to treatment or disposal units is called
    (a) lateral
    (b) branch sewer
    (c) trunk sewer
    (d) outfall sewer.
21. A manhole is provided
    (a) at every 500 m intervals
    (b) at every corner
    (c) when flow gets divided
    (d) when direction or grade changes.
22. When fall in elevation is greater than 90 cm the manhole provided is called
    (a) lamp hole
    (b) flight manhole
    (c) drop manhole
    (d) fall manhole.
23. A flight sewer is provided when the gradient is
    (a) zero
    (b) low
    (c) steep
    (d) maximum (*i.e.*, vertical).
24. When more than 2 sewers join in a manhole
    (a) their tops should be at the same level
    (b) their centres should be at the same level
    (c) their bottoms should be at the same level
    (d) they can be either way.
25. The working chamber of a manhole, along the direction of flow should have a minimum length of
    (a) 0.6 m
    (b) 0.9 m
    (c) 1.2 m
    (d) 2.4 m.
26. "Never discharge the contents of a bigger sewer into a smaller though the capacity of the later may be more due to steep slope" because
    (a) depth of flow may increase
    (b) gases generated may explode
    (c) back flow may occur
    (d) sewer may get clogged.
27. Before entering a manhole a candle is lowered into the manhole
    (a) to illuminate it
    (b) to detect toxic gases
    (c) to give a signal to the adjacent manhole
    (d) to find out presence of oxygen.

28. A lamphole is helpful in
    (a) illuminating sewer line
    (b) cleaning sewer line
    (c) repairing
    (d) testing sewers.
29. A syphon spillway is an arrangement to
    (a) divert DWF into open drain
    (b) divert excessive DWF into open drain
    (c) divert excessive storm water into open drain
    (d) divert excessive combined flow into open drain.
30. Colour of fresh sewage is
    (a) green
    (b) brown
    (c) pink
    (d) grey.
31. The best pH range suited for most of bacteria is
    (a) 1 to 2
    (b) 5 to 9
    (c) 7
    (d) 7 to 10.
32. The bacteria that flourish at high temperature (greater than 45°C) are called
    (a) cryptophilic
    (b) psychrophilic
    (c) mesophilic
    (d) thermophilic.
33. Aerobic bacteria
    (a) consume oxygen in combined state
    (b) prefer darkness to light
    (c) prefer movement to stagnation
    (d) produce less energy and more end products.
34. Anaerobes give out
    (a) more energy
    (b) less unmetabolised organics
    (c) less odourous products
    (d) unstable end products.
35. The common end products for both aerobiosis and anaerobiosis is
    (a) H_2S
    (b) CH_4
    (c) NO_3
    (d) CO_2.
36. A unit working purely on anaerobiosis is
    (a) septic tank
    (b) activated sludge process
    (c) trickling filter
    (d) contact bed.
37. Anaerobic treatment is best suited for
    (a) high efficiency
    (b) toxic wastes
    (c) dilute inorganic wastes
    (d) strong organic wastes.
38. A natural method of disposal of sewage is
    (a) sewage irrigation
    (b) septic tank treatment
    (c) composting
    (d) aerated lagooning.

39. Sewage farming is best suited for India because
    (a) it is abundantly available
    (b) it is a cheap method of sewage disposal
    (c) it is a hygienic method
    (d) nutrients and trace elements required by plants are available in sewage.
40. Sewage sickness is a sickness developed by
    (a) humans due to consumption of crops grown on sewage
    (b) plants because of harmful elements of sewage irrigation
    (c) land unable to accept any more loading of sewage
    (d) sewage itself due to pathogens within.
41. An ideal crop of sewage irrigation is
    (a) potato
    (b) beans
    (c) sugarcane
    (d) banana.
42. When raw sewage is to be disposed off by dilution sewage should
    (a) be free from suspended solids
    (b) be free from floating impurities
    (c) be having a BOD less than 150 mg/l
    (d) not be very hot.
43. Self-purification of a stream is quite rapid when
    (a) stream had sluggish flow
    (b) straight course with uniform cross-section
    (c) course is full of rapids and water falls
    (d) more bacterial pollution exists.
44. In the figure zone II represents
    (a) zone of degradation
    (b) zone of active decomposition
    (c) zone of recovery
    (d) zone of clear waters.

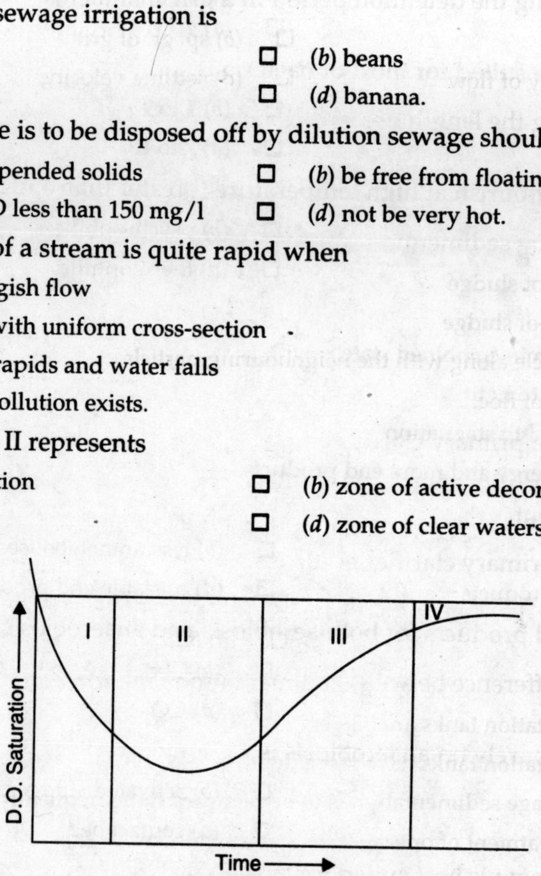

FIGURE 7.5

45. Coarse screens are iron bars with a clearance of
    (a) less than 25 mm
    (b) 25 to 75 mm
    (c) 75 mm
    (d) 100 mm.
46. Grit is
    (a) inert matter of sp. gr. > 2.65
    (b) organic matter of sp. gr. nearer to 1
    (c) organic and inert matter combined
    (d) colloidal matter of heavy sp. gr.

47. Grit chamber had a detention period of
    (a) 45 s to 90 s
    (b) 2 min to 5 min
    (c) 20 min to 30 min
    (d) 2 hours to 4 hours.

48. Velocity of flow through a grit chamber is
    (a) 30 to 60 cm/s
    (b) 30 to 60 cm/min
    (c) 15 to 45 cm/s
    (d) 15 to 45 cm/min.

49. Grit obtained in a grit chamber is used
    (a) as a manure
    (b) as a soil conditioner
    (c) as a coarse aggregate
    (d) as a substitute for sand.

50. The factor governing the detention period in a grit chamber is
    (a) size of grit
    (b) sp. gr. of grit
    (c) horizontal velocity of flow
    (d) settling velocity.

51. The factor deciding the length of a grit chamber is
    (a) detention period
    (b) horizontal velocity of flow
    (c) settling velocity
    (d) surface loading.

52. 'Mass action' during sedimention is the property of
    (a) increase in mass of sludge
    (b) decrease in mass of sludge
    (c) settling of a particle along with the neighbouring particle
    (d) quick spreading of floc.

53. Overflow rate for a primary clarifier is
    (a) 20–30 m^3/m^2/day
    (b) 5–10 m^3/m^2/day
    (c) 30–100 m^3/m^2/day
    (d) 200–300 m^3/m^2/day.

54. BOD removal of primary clarifier is
    (a) 50%–60%
    (b) 30%–40%
    (c) 10%–20%
    (d) Nil.

55. A fundamental difference between sedimentation tank for water and sewage is
    (a) sewage sedimentation tanks are bigger
    (b) sewage sedimentation tanks have more depth
    (c) sludge from sewage sedimentation is to be removed more frequently
    (d) it can be final treatment of operation in water treatment.

56. Sedimentation aided by coagulation is rarely done for sanitary sewage because
    (a) chemicals cost more
    (b) time of treatment is further prolonged
    (c) sludge cannot be disposed off easily
    (d) chemicals may be harmful to living organisms of sewage.

57. Sedimentation aided by coagulation is adopted
    (a) almost everywhere
    (b) where biological treatment is to follow next

(c) when industrial wastes are blended with sanitary sewage
(d) when phosphates are less in sewage.

58. A septic tank is a water tight tank where the following operation(s) take place
    (a) sedimentation
    (b) sedimentation + digestion (anaerobic)
    (c) digestion (aerobic)
    (d) decomposition of organic and inorganic matter by bacteria.

59. Mostly adopted detention period for a septic tank under Indian conditions is
    (a) 12 hours
    (b) 24 hours
    (c) 48 hours
    (d) 72 hours.

60. The gases given out of a septic tank are
    (a) $CO_2 + SO_2 + N$
    (b) $CO_2 + PH_3 + NH_3$
    (c) $CO_2 + CH_4 + H_2S$
    (d) $CH_4 + O_2 + H_2$.

61. Septic tanks are best suited for
    (a) municipalities
    (b) industries
    (c) scattered residences
    (d) congested areas.

62. Septic tank effluent is
    (a) quite clear and sparkling
    (b) always over septicized
    (c) having fine organic solids with high BOD
    (d) black tarry liquor with earthen odour.

63. Media used in Trickling Filter is
    (a) organic matter
    (b) inorganic matter
    (c) broken lime stone
    (d) any inert, insoluble and tough material.

64. In a Trickling filter the thickness of the aerobic biofilm is
    (a) 0.1 to 0.2 mm
    (b) > 10 mm
    (c) varies from time to time being very less initially and increasing uptill sloughing
    (d) depends on the rate of hydraulic loading.

65. Usual size of media in Standard Rate Trickling Filter is
    (a) 6 mm to 30 mm
    (b) 10 mm to 20 mm
    (c) 25 mm to 100 mm
    (d) 100 mm to 250 mm.

66. Depth of Standard Rate Trickling Filter is
    (a) 0.5 to 1 m
    (b) 1 to 2 m
    (c) 1.5 to 4.8 m
    (d) 1 to 8 m.

67. Biological layer stickling to the trickling filter media consists of
    (a) bacteria
    (b) protozoa
    (c) bacteria + algae
    (d) bacteria + protozoa + fungii.

68. The usual rate of organic loading on a Standard Rate Trickling Filter is
    (a) 10 to 100 g/m^3/day
    (b) 100 to 400 g/m^3/day
    (c) 400 to 1000 g/m^3/day
    (d) 1000 to 3000 g/m^3/day.

69. An extension of Trickling Filter is
    (a) rotating biological contactor
    (b) intermittent sand filter
    (c) bio tower
    (d) contact bed.

70. The primary operations in a Trickling Filter are
    (a) sorption + biological oxidation
    (b) filtration + aeration
    (c) oxidation + nitrification
    (d) aerobic + anaerobic oxidation.

71. A very important character is media of a trickling Filter is
    (a) greater specific surface area
    (b) less specific gravity
    (c) capability of supporting heavy loads
    (d) bigger size with angular edges.

72. The usual rate of hydraulic loading on a High Rate Trickling Filter is
    (a) 1 to 2 m^3/m^2/day
    (b) 2 to 5 m^3/m^2/day
    (c) 5 to 10 m^3/m^2/day
    (d) 10 to 30 m^3/m^2/day.

73. Recirculation ratio for a Standard Rate Trickling Filter is
    (a) 0
    (b) 0.5 to 3
    (c) 3 to 6
    (d) 20.

74. Effluent BOD of a Standard Rate Trickling Filter is less than
    (a) 10 mg/l
    (b) 20 m/l
    (c) 30 mg/l
    (d) 50 mg/l.

75. A High Rate Biological treatment unit has a $\frac{F}{M}$ ratio of
    (a) less than 0.1
    (b) 0.2 to 0.5
    (c) 0.1 to 1
    (d) greater than 1.

76. Effluent from a High Rate Trickling Filter is
    (a) black and highly oxidized
    (b) light grey and oxidized
    (c) brown and not fully oxidized
    (d) well nitrified and free from odours.

77. *Psychoda alternata* is found in
    (a) septic tanks
    (b) standard rate trickling filters
    (c) high rate trickling filters
    (d) activated sludge process.

78. Filter unloading is
    (a) downward trickling of sewage
    (b) removal of sludge
    (c) emptying of trickling filter media
    (d) detachment of zoogleal slime.

79. To improve efficiency recirculation of the following is done in HR trickling filters
    (a) sludge
    (b) zoogleal slime
    (c) effluent
    (d) raw sewage.

80. The immediate layer sticking to the media of a trickling filter is of
    (a) aerobic bacteria
    (b) anaerobic bacteria
    (c) facultative aerobic bacteria
    (d) bacteria + algae.

81. An equivalent term of "Mean Cell Residence Time" is
    (a) solids retention time
    (b) solids age
    (c) sludge detention time
    (d) average cell time.

82. Approximate percentage of the RBC disc submerged under sewage is
    (a) 20%
    (b) 40%
    (c) 50%
    (d) 75%.

83. In operation a RBC is nearer to
    (a) aerated lagoon
    (b) oxidation ditch
    (c) activated sludge process
    (d) trickling filter.

84. Organic loading of RBC is expressed as
    (a) $g/m^3/d$
    (b) $m^3/m^2/d$
    (c) $g/m^2/d$
    (d) $m^3/m/d$.

85. Activated sludge process is a biological process involving
    (a) aerobic + anaerobic bacteria
    (b) aerobic bacteria + protozoa + algae
    (c) anaerobic bacteria + fungii
    (d) facultative bacteria + algae.

86. Media provided in ASP is
    (a) broken stone
    (b) anthracite coal
    (c) plastics
    (d) none.

87. Aeration period in the conventional aeration tank of ASP is
    (a) 1 to 2 hours
    (b) 2 to 4 hours
    (c) 6 to 10 hours
    (d) 10 to 20 hours.

88. Amount of return sludge in conventional ASP is
    (a) 10 to 20%
    (b) 25 to 50%
    (c) 50 to 100%
    (d) 100 to 300%.

89. Food/Microbes [F/M] ratio in ASP is
    (a) 1 initially and 2 finally
    (b) 2 initially and 1 finally
    (c) 0.7 initially and 0.3 finally
    (d) 0.5 initially and 0 finally.

90. F/M is accurately given by
    (a) $\dfrac{BOD}{Total\ Solids}$
    (b) $\dfrac{COD}{MLSS}$
    (c) $\dfrac{BOD}{MLVESS}$
    (d) $\dfrac{BOD}{MLSS}$.

91. Organic loading of ASP is
    (a) 100 to 200 $g/m^3/day$
    (b) 200 to 400 $g/m^3/day$
    (c) 500 to 650 $g/m^3/day$
    (d) 650 to 800 $g/m^3/day$.

92. Efficiency of activated sludge process is
    (a) 30 to 40%
    (b) 50 to 60%
    (c) 60 to 98%
    (d) 90 to 95%.

93. Conventional ASP is not suited to
    (a) strong organic wastes
    (b) shock loading and varying quality of influent
    (c) very dilute wastes
    (d) high recirculation.

94. Sludge from a Biological treatment unit is
    (a) harmless inorganic sediment
    (b) rich in organic matter and bacteria
    (c) rich in inorganic matter and micro organisms
    (d) rich in bigger inert solids.

95. A treatment unit developed on algal-bacteria mutual symbiosis is
    (a) activated sludge process
    (b) rotating biological contactor
    (c) waste stabilization pond
    (d) aerated lagoon.

96. Sludge digestion is
    (a) disposal of sludge
    (b) dilution of sludge
    (c) stabilization of sludge
    (d) removal of waste products from sludge.

97. Anaerobic sludge digestion mainly yields
    (a) methane
    (b) hydrogen sulphide
    (c) ammonia
    (d) carbon dioxide.

98. The detention period for a standard rate digester is
    (a) 10 to 20 days
    (b) 20 to 30 days
    (c) 30 to 90 days
    (d) 90 to 180 days.

99. In a standard rate digester the main bacteria are
    (a) cryptophilic
    (b) psychrophilic
    (c) mesophilic
    (d) thermophilic.

100. Quantity of methane produced in a sludge digester is
    (a) 1 to 5 m^3/1000 people
    (b) 10 to 20 m^3/100 people
    (c) 15 to 30 m^3/1000 people
    (d) 25 to 50 m^3/1000 people.

101. Usually the digesters have a dia of
    (a) 2 to 6 m
    (b) 6 to 35 m
    (c) 10 to 60 m
    (d) 50 to 100 m.

102. Supernatant from a digester will be having a BOD of
    (a) less than 20 mg/l
    (b) greater than 200 mg/l
    (c) greater than 500 mg/l
    (d) greater than 5000 mg/l.

103. Mixing of contents of a digester
    (a) reduces its capacity
    (b) produces more foam and scum
    (c) toxic products produced are diluted
    (d) stratification takes place.

104. Dried sludge is a good
    - (a) fuel
    - (b) soil builder
    - (c) manure
    - (d) filler of a low lying area.

105. The length of an Indian type squating pan is governed by
    - (a) depth of water seal
    - (b) number of users
    - (c) length of urinating jet
    - (d) system of supply of water.

106. Pedestal type commode is
    - (a) completely above floor level
    - (b) partly above and partly below floor level
    - (c) completely below floor level
    - (d) completely above ground level.

107. Any trap is intended to trap
    - (a) water
    - (b) sewage of sullage
    - (c) any liquid waste
    - (d) fowl gases.

108. Any trap should necessarily have
    - (a) water seal
    - (b) sewage or sullage
    - (c) a bent shape
    - (d) grating.

109. The last trap provided in a house drainage system is
    - (a) Q-trap
    - (b) floor trap
    - (c) Nahani trap
    - (d) intercepting trap.

110. Stall type urinals are best suited to
    - (a) high schools
    - (b) residental houses
    - (c) people of same height
    - (d) public places.

111. Bowl type of urinals are ideal for
    - (a) people of the same height
    - (b) commercial areas
    - (c) public places
    - (d) people of different heights.

112. Soil stack collects waste waters from
    - (a) baths
    - (b) sinks and wash basins
    - (c) urinals
    - (d) toilets.

113. Waste pipe collects liquid waste from
    - (a) toilets and urinals
    - (b) wash basins and baths
    - (c) urinals and kitchens
    - (d) from any room.

114. The pipe having min. dia. in plumbing system is
    - (a) waste pipe
    - (b) soil pipe
    - (c) ventilation pipe
    - (d) anti-syphonage pipe.

115. Anti-syphonage pipe need not be provided in
    - (a) single-storeyed buildings
    - (b) multi-storeyed buildings
    - (c) residential flats
    - (d) industrial buildings.

116. The object of cowl is to
    - (a) give ornamental look
    - (b) prevent entry of rainfall into the waste pipe

(c) quickly dispel fowl gases
(d) prevent nest building.

117. The solid waste resulting from growing, handling, preparation, cooking and consumption of putrescible organic matter is called
    (a) refuse
    (b) rubbish
    (c) garbage
    (d) cinders.

118. Pieces of papers, crockery, scrap metals etc., are classified as
    (a) refuse
    (b) rubbish
    (c) garbage
    (d) ashes and cinders.

119. Salvaging is
    (a) dumping of solid waste
    (b) sanitary land filling of solid waste
    (c) composting and soil conditioning
    (d) extraction of essence from waste.

120. Pick out the odd statement
    (a) excessive Chlorides in water, increase conductance
    (b) iron in water causes spots and stains on textile and paper
    (c) scales formed in boilers because of hard water, are good conductors of heat
    (d) fishermen nets are deteriorated because of acidic and alkaline waters.

121. Pick up the odd statement
    (a) hot water kills bacteria
    (b) hot water induces stratification and fish retreat to bottom layers
    (c) suspended solids settle and cover spawning grounds of fish
    (d) floating solids and liquids interfere with natural reaeration.

122. Agent responsible for ecological imbalance
    (a) carnivores
    (b) herbivores
    (c) man
    (d) omnivores.

123. The activity that may not affect ecological balance is
    (a) growing crops
    (b) developing gardens
    (c) industries
    (d) use of pesticides

124. Eco system is
    (a) community and environment
    (b) species and habitat
    (c) communities and habitat
    (d) biosphere and habitat.

125. The relationship between "Biotic" and "Abiotic" substances of an ecosystem is
    (a) reciprocity
    (b) cannibalism
    (c) commensalism
    (d) parasitism.

126. "Ecological Niche" is
    (a) habitat of an organism
    (b) community and environment of an organism
    (c) organisms profession (position and status)
    (d) confinement of a species to a specific region.

127. A food chain
   (a) is transfer of energy from one organism to another ☐
   (b) is relationship between producers, herbivores and carnivores ☐
   (c) is comprising of any three or more organisms ☐
   (d) is combination of organisms at different trophic levels. ☐

128. "A nectar sipping butterfly → Dragon fly → Frog → Snake → Hawk" is
   (a) Predator food chain ☐   (b) Parasitic food chain ☐
   (c) Saprophytic food chain ☐   (d) Food web. ☐

129. "Biological Magnification" is
   (a) decrease in concentration with trophic level ☐
   (b) increase in concentration with trophic level ☐
   (c) equilibrium in concentration ☐
   (d) magnification of poisons with time. ☐

130. Pick up the odd statement. Eutrophication is
   (a) due to flow of nitrates and phosphates into stagnant bodies of water ☐
   (b) growth of blue-green algae ☐
   (c) supersaturation of water with dissolved oxygen due to green algae mats ☐
   (d) decomposition of organic bottom muds. ☐

131. An effective method of controlling eutrophication of lake water is
   (a) weed cutting ☐   (b) thorough mixing ☐
   (c) copper sulphate application ☐   (d) diluting with nutrient free water very often. ☐

III. TRUE/FALSE

1. Sewerage is the collection, conveyance and treatment of sewage.
2. When the rainfall is evenly distributed throughout the year combined system is favoured.
3. When the drainage area is steep separate system is preferred.
4. Dry weather flow represents the sanitary sewage discharge.
5. The resting floor should be depressed by about 50 cm to accommodate for trap attached to squatting type water basin where as the pedestal type can directly rest over the floor.
6. The pipe carrying sullage is called waste pipe.
7. Soil pipe carries the discharges from lavatory basin.
8. Intercepting traps maintain greater water seal than any other trap.
9. When the rate of inflation is very high smaller design periods are adopted.
10. Relaying of the pipe lines should be done well before the expected life span.
11. Time of concentration is directly proportional to imperviousness of the catchment area.
12. Self cleansing velocity for most of the sewers is not less than 60 cm/s.
13. Generally scouring velocity for design of earthenware sewers is not greater than 3 m/s.

14. In partly flowing circular sewer hydraulic mean depth increases with depth of flow.
15. Circular sewers are the best suited for separate system where the fluctuations between max. and min. flow are less.
16. An ovoid section almost gives the same hydraulic mean depth irrespective of the depth of flow.
17. Box sectioned sewers are mostly used as outfall sewers draining into sea.
18. A manhole is a must when two or more sewers meet.
19. When more than three sewers of different diameters meet in a manhole, their bottom are kept at the same level.
20. A drop manhole is provided when the drop in elevation is greater than 0.8 m.
21. A lamphole serves the same purpose as a manhole.
22. Catch basins are usually provided in separate system.
23. Leaping weir is provided in partly separate system of sewerage.
24. An inverted siphon is a depressed sewer below a river, underbridge or sub-way.
25. Before a person enters the manhole a bare flame is lowered into the manhole and when it is withdrawn after 3 minutes the flame should not be put off.
26. By cutting off oxygen supply for considerable time, fungus growths can be controlled.
27. The length of the Indian type or squatting pan varies from 45 cm to 63 cm. Higher values are adopted in public places because the factor governing the length is the length of urinating jet.
28. Because of lightning nitrates are formed in nature.
29. Higher temperatures of waste waters cause the growth of undesirable water plants and waste water fungus.
30. Anaerobic reactions prefer darkness and stagnant media.
31. Methane is also produced during aerobic reactions.
32. There are no common end products for both aerobic and anaerobic reactions.
33. Aerobic metabolism yields 20 to 30 times as much energy as that anaerobic metabolism.
34. Algae are microscopic multicellular photosynthetic plants without roots, stems or leaves.
35. Clear, cold, mountain lakes support only a few algae whereas warm polluted lakes enriched with nitrates and phosphates harbour dense algal blooms.
36. Eutrophication is the enrichment of water with nutrients.
37. Chemical oxygen demand cannot distinguish between readily bio-degradable matter and slow bio-degradable matter.
38. Anaerobic bacteria do not require any type of oxygen and they fluorish in the absence of oxygen.
39. Aerobic processes synthesize 0.4 to 0.8 kg biomass per kg BOD whereas anaerobic processes synthesize 0.08 to 0.2 kg biomass per kg BOD.
40. Anaerobic treatment is relatively cheap and slow and ideally suited for strong wastes of high BOD, low sulphur content and less settleable solids.

41. During the endogenous phase of bacterial growth, the external food supply is cut off and hence bacteria have to use the stored energy within them.
42. Permissible velocity of flow through the screen chamber is 30 cm/min.
43. Grit is inert organic and inorganic matter of specific gravity > 2.65.
44. The general settling velocity of sand particles is 21 mm/s.
45. Preaeration (*i.e.*, before sedimentation) of sewage helps in flocculation.
46. Sludge scraper in a primary clarifier shall be rotating at 0.02 rpm.
47. Efficiency of a sedimentation tank has nothing to do with the location of inlet and outlets.
48. A hopper-bottomed sedimentation tank is pyramidal or conical in elevation its bottom slope being 45° to 60° to the horizontal.
49. Any septic tank should have a minimum volume of 3 m^3.
50. Septic tanks are not economical for a greater than 120 users.
51. Septic tank effluent is highly stable has a BOD **less** than 30 mg/l and require no other treatment.
52. Imhoff tank is a two storeyed septic **tank with** separate zones for sedimentation and sludge digestion.
53. An imhoff tank can be used for 1000 users.
54. An intermittent sand filter was almost the same as a rapid sand filter.
55. Intermittent sand filter when improved for higher rate of filtration gave birth to Trickling Filter.
56. In a trickling filter the thickness of the aerobic layer at all times remains more or less uniform *i.e.*, being 0.1 to 0.2 mm thick.
57. In a trickling filter the biological oxidation remains uniform throughout its depth.
58. In a high rate trickling filter the load is applied at a time for an interval not exceeding 15 seconds.
59. Recirculation ratio should be taken into account while computing organic loading on a high rate trickling filter.
60. The effluent from a high rate trickling filter consists of brown colloids that undergo putrefaction.
61. Recirculation do not exist in standard rate trickling filter.
62. The biofilm of a trickling filter consists of anaerobic layer in addition to the aerobic film.
63. Sloughing takes place continuously in a standard rate trickling filter.
64. A bio-tower is a tall trickling filter but provided with synthetic media of high porosity, low weight and more specific surface area.
65. A trickling filter is a purely aerobic process with suspended culture media.
66. In a trickling filter sloughing takes place because of the thickening of the biofilm and the inner anaerobic bacteria getting no food and hence losing the grip over media.
67. High rate trickling filters are free from psychoda as their larvae are continuously washed away.

68. A rotating biological contractor is a fixed film system.
69. In a rotating biological contactor the media is relatively stationery to the load.
70. In a T.F. media is stationery while the sewage moves down, whereas in a R.B.C. the media moves as the sewage remains relatively stationery.
71. In a conventional Activated Sludge Process F/M ratio is greater than 1.
72. Any high rate unit has a F/M less than 0.2.
73. Extended Aeration systems have a F/M ratio of about 0.2 to 0.1.
74. About 95 m^3 of air is required for every kg of BOD in conventional activated sludge system.
75. Mixed liquor is influent + 25% to 50% of effluent.
76. A waste stabilization pond is a waste water pond best suited for tropics with considerable wind currents.
77. In a shallow ponds of depth less than 60 cm rooted aquatic weeds may grow.
78. Aerated lagoon can be regarded as Activated Sludge Process but without return sludge.
79. Usual detention periods in aerated lagoons are 4 to 5 day.
80. Aerated lagoon mainly functions on photosynthesis.
81. Aerated lagoon can treat bio-degradable toxins as phenols whereas conventional activated sludge cannot.
82. An oxidation ditch is a high rate unit.
83. The liquid depth in a oxidation ditch is 1 to 1.5 m.
84. Sludge digestion is the process of decomposition of organic matter into relatively stable end products.
85. Digested sludge attracts flies and is a good food for rodents.
86. Most sludge digesters prefer thermophilic zone.
87. High rate sludge digesters operate at detention periods of 30–90 days.
88. Due to stratification about 50% of the volume of the digester remains idle.
89. During aerobic sludge digestion auto-oxidation of bacteria takes place with no further cell synthesis.
90. Effluent from aerobic sludge digesters have a BOD greater than 5000 mg/l.
91. Elutriation is the washing of sludge with water to get rid of very fine suspended matter and ammonia compounds.
92. Garbage is an inert solid waste.
93. Rubbish is a putrescible solid waste.
94. All solid wastes both bio-degradable as well as inert are termed as Refuse.
95. Ashes can also be called as rubbish.
96. In India the rate of generation of municipal solid waste is about 400 g/capita/day.
97. Ignitable waste is a hazardous solid waste.
98. 70% of municipal solid wastes are of organic nature.

99. Garbage is to be disposed off once in three or four days to prevent breeding of flies.
100. Rubbish is to be daily collected.
101. Sanitary land fill contaminates nearby sources of water.
102. Predominant gases generated from land fill are hydrogen, oxygen and nitrogen.
103. Marshy lands are best suited for sanitary land fill.
104. While carbon dioxide escapes into the atmosphere, methane is retained in the lower layers of sanitary land fill.
105. Incineration method of disposal of solid wastes is a cheap process.
106. Sifting is fine solid material as ash obtained after incinerating the solid waste.
107. The end products remaining after dissimilatory and assimilatory bacterial activity of a municipal solid waste is called compost.
108. Compost is a good soil conditioner.
109. Less the amount of moisture quicker is the composting operation.
110. When the temperature of the compost is kept between 60°C and 70°C for 24 hours, pathogens, seeds and weeds get destroyed.
111. Bacteria activity ceases at temperatures greater than 66°C.
112. Too low a C/N ratio in compost gives off ammonia and too high value of it robs nitrogen from the soil.
113. Turning is regularly done in aerobic composting to prevent drying and caking.
114. pH less than 8.5 is to be maintained in a compost so as to minimise the loss of ammonia gas.
115. Pyrolysis is the same as combustion.

ANSWERS
Objective Type Questions

1. (d)	2. (d)	3. (c)	4. (c)	5. (d)	6. (b)
7. (c)	8. (d)	9. (a)	10. (b)	11. (c)	12. (a)
13. (c)	14. (c)	15. (c)	16. (a)	17. (d)	18. (c)
19. (a)	20. (d)	21. (d)	22. (c)	23. (c)	24. (a)
25. (c)	26. (a)	27. (d)	28. (a)	29. (c)	30. (d)
31. (b)	32. (d)	33. (c)	34. (d)	35. (d)	36. (a)
37. (d)	38. (a)	39. (d)	40. (c)	41. (d)	42. (c)
43. (c)	44. (b)	45. (c)	46. (a)	47. (a)	48. (c)
49. (c)	50. (d)	51. (b)	52. (c)	53. (a)	54. (b)
55. (c)	56. (d)	57. (c)	58. (b)	59. (b)	60. (c)
61. (c)	62. (c)	63. (d)	64. (a)	65. (c)	66. (c)
67. (d)	68. (b)	69. (c)	70. (a)	71. (a)	72. (d)
73. (a)	74. (c)	75. (d)	76. (c)	77. (b)	78. (d)
79. (c)	80. (b)	81. (a)	82. (b)	83. (d)	84. (c)

85. (b)	86. (d)	87. (c)	88. (b)	89. (c)	90. (c)
91. (c)	92. (d)	93. (b)	94. (b)	95. (c)	96. (c)
97. (a)	98. (c)	99. (c)	100. (c)	101. (b)	102. (d)
103. (c)	104. (b)	105. (c)	106. (a)	107. (d)	108. (a)
109. (d)	110. (d)	111. (a)	112. (d)	113. (b)	114. (d)
115. (a)	116. (d)	117. (c)	118. (b)	119. (d)	120. (c)
121. (a)	122. (c)	123. (b)	124. (b)	125. (a)	126. (a)
127. (a)	128. (a)	129. (b)	130. (b)	131. (d)	

True/False

1. F	2. F	3. T	4. T	5. T	6. T
7. F	8. T	9. F	10. F	11. F	12. T
13. T	14. F	15. T	16. T	17. T	18. T
19. F	20. F	21. F	22. F	23. T	24. T
25. T	26. T	27. T	28. T	29. T	30. T
31. F	32. F	33. T	34. T	35. T	36. T
37. T	38. F	39. T	40. T	41. T	42. F
43. T	44. T	45. T	46. T	47. F	48. T
49. T	50. T	51. F	52. T	53. T	54. T
55. T	56. T	57. F	58. T	59. F	60. T
61. T	62. T	63. F	64. T	65. F	66. T
67. T	68. T	69. F	70. T	71. F	72. F
73. T	74. T	75. F	76. T	77. T	78. T
79. T	80. F	81. T	82. F	83. T	84. T
85. F	86. T	87. T	88. T	89. T	90. F
91. T	92. F	93. F	94. T	95. F	96. T
97. T	98. T	99. T	100. F	101. T	102. T
103. F	104. T	105. T	106. T	107. T	108. T
109. F	110. T	111. T	112. T	113. T	114. F
115. F					

Chapter 8 SOIL MECHANICS

I. INTRODUCTION

ORIGIN OF SOILS

For engineering purposes soil may be defined as an assemblage of discrete solid particles of organic or inorganic composition. Soils are formed by the process of physical and/or chemical weathering.

Soils resulting from disintegration of rock may stay at the place of their formation, known as residual soils. If the soils are carried away by forces of gravity, water, wind and ice and deposited at another location, they are known as transported soils.

Transported soils may be further subdivided based on the transporting agency and the place of deposition as listed below:

1. *Glacial soils (or Till)*. Soils that are transported by glaciers.
2. *Aeoline deposits (or Loess)*. Soil deposits formed by wind.
3. *Alluvial deposits*. Soil deposits formed by rivers and streams.
4. *Marine deposits*. Soil deposits formed by sea water.
5. *Lacustrine deposits*. Soil deposited in lake beds.
6. *Talus*. Soils that are transported by gravity.

STRUCTURE OF SOILS

The structure of a soil may be defined as the manner of arrangement of soil particles and the electrical forces acting between adjacent particles.

The following types of structures are commonly studied:

(*i*) *Single grained structure.* It is characteristic of coarse grained soils.

(*ii*) *Honey-comb structure.* It is characteristic of fine grained soils, especially silts.

(*iii*) *Flocculent structure.* This structure is characteristic of fine grained soils such as clays.

(*iv*) *Dispersed structure.* Develops in clayey soils which are remoulded.

THE THREE PHASE SYSTEM

Bulk soil as it exists in nature is more or less a random assembly of soil particles with **air** and/or water occupying the void space amongst the particles. In most situations the liquid in the soil is water and the gas in the soil is air. Therefore, soil will be considered as an aggregation of solid particles, water and air. A schematic diagram of the three phase system is shown below.

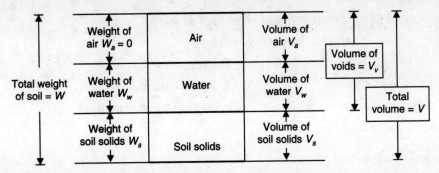

In a dry soil or in a saturated soil the three phase system reduces to two phases, *viz.*, **soil solids and air, and soil solids and water** respectively. Thus soil exists in either two phase or **three phase composition**.

BASIC RELATIONS

Volume relationships

$$\text{Void ratio } e = \frac{\text{Volume of voids } (V_v)}{\text{Volume of soil solids } (V_s)}$$

Typical values of void ratio in soils may range from 0.50 to 1.50.

$$\text{Porosity } n = \frac{\text{Volume of voids}}{\text{Total volume}} \times 100$$

Porosity is expressed as a percentage. Range of porosity is $0 \leq n \leq 100$

$$\text{Degree of saturation } S = \frac{\text{Volume of water } (V_w)}{\text{Volume of voids } (V_v)} \times 100$$

It is expressed as a percentage.

Thus for dry soil, $\qquad S = 0\%$

For saturated soil, $\qquad S = 100\%$

For partially saturated soil $0 \leq S \leq 100\%$

The parameter porosity and void ratio are related to each other as follows:

$$e = \frac{n}{1-n}$$

$$n = \frac{e}{1+e}$$

Weight relationship

$$\text{Water content } (w) = \frac{\text{Weight of water } (W_w)}{\text{Weight of soil solids } (W_s)} \times 100$$

This is expressed as a percentage.

For dry soil $w = 0\%$

The value of the water content can be as high as 400%.

Water content of a soil can be determined by

1. Oven drying method
2. Pycnometer method
3. Sand bath method
4. Alcohol method.
5. Calcium carbide method
6. Torsion balance method
7. Radiation method.

Unit weight of water $\quad\quad\quad\quad\quad\quad\quad\quad \gamma_w = \dfrac{W_w}{V_w}$

Unit weight of soil solids $\quad\quad\quad\quad\quad\quad \gamma_s = \dfrac{W_s}{V_s}$

Total or Bulk or Moist unit weight of soil $\quad \gamma = \dfrac{W}{V}$

Dry unit weight of soil $\quad\quad\quad\quad\quad\quad\quad \gamma_d = \dfrac{W_s}{V}$

Saturated unit weight of soil $\quad\quad\quad\quad\quad \gamma_{sat} = \dfrac{W_{sat}}{V}$

Buoyant (submerged) unit weight of soil $\quad \gamma = \gamma_{sat} - \gamma_w$

Given G, γ_w, and e then dry density $\quad\quad (\gamma_d) = \dfrac{G\gamma_w}{1+e}$

Saturated density $\quad\quad\quad\quad\quad\quad\quad\quad\quad \gamma_{sat} = \dfrac{(G+e)}{1+e}\gamma_w$

Submerged density $\quad\quad\quad\quad\quad\quad\quad\quad \gamma_{sub} = \dfrac{(G-1)}{1+e}\gamma_w$

Bulk density $\quad\quad\quad\quad\quad\quad\quad\quad\quad\quad \gamma = \dfrac{(G+e \cdot S)}{1+e}\gamma_w \text{ or } \gamma = \gamma_d(1+w)$.

Specific gravity. The specific gravity of solid particles is defined as the ratio of the mass of a given volume of solids to the mass of an equal volume of water at 4°C.

Specific gravity of soil solids, $G = \dfrac{\text{Density of soil solids } (\rho_s)}{\text{Density of water } (\rho_w)}$

Generally G value varies from 2.0 to 3.0. The smaller values are for coarse grained soils.

Mass specific gravity G_m. It is defined as the ratio of the mass density of the soil to the mass density of water

$$G_m = \dfrac{\rho}{\rho_w}$$

Obviously $\quad G_m < G$

This is also called as apparent specific gravity or bulk specific gravity.

Absolute specific gravity (G_a). If both the permeable and impermeable voids are excluded from volume of solids, the remaining volume is the true or absolute volume of the solids.

$$G_a \dfrac{(\rho_s)_a}{\rho_w} = \dfrac{\text{Mass density of absolute solids}}{\text{Density of water}}$$

This is also called as "Grain specific gravity" or "specific gravity of solids".

Specific gravity of soil solids is determined by

1. Pycnometer method
2. Density bottle method
3. Measuring flask method
4. Gas Jar method
5. Shrinkage limit method.

Density index. It is the ratio of the difference between the void ratio of a cohesionless soil in the loosest state and any given void ratio to the difference between its voids ratio in the loosest and the densest states.

$$I_D = \dfrac{e_{max} - e}{e_{max} - e_{min}} \times 100$$

where $e_{max} = e$ of the soil in loosest state

$e_{min} = e$ of the soil in densest state

$e = e$ at the in-situ state.

Density index is a measure of degree of compaction and the stability of a stratum is indirectly reflected by the compactness.

Designation	I_D (%)
Very loose	0 to 15
Loose	15 to 35
Medium	35 to 65
Dense	65 to 85
Very dense	85 to 100

Relationship between w, e and G and S

$$e = \frac{wG}{S}$$

Percentage of air voids $\quad n_a = \dfrac{\text{Volume of air voids }(V_a)}{\text{Total volume of soil }(V)} \times 100$

Air content $(a_c) = \dfrac{\text{Volume of air voids }(V_a)}{\text{Volume of voids }(V_v)}$

$a_c = 1 - S$, where S is the degree of saturation

Field density of a soil is determined by
1. Sand replacement method
2. Water displacement method
3. Core cutter method
4. Water balloon method
5. Submerged weight method
6. Radiation method.

CLASSIFICATION OF SOILS

Need for classification

For understanding the nature, the items are grouped together, otherwise very little progress would be made in understanding this environment.

A chemist will group together elements which have something in common.

A biologist groups together animals that are in some way alike.

So the attempt of classification is to choose those criteria which are relevant to the purpose in view. Soils that are grouped in order of performance for one type of physical conditions will not necessarily have the same order of performance under some other physical conditions.

For a proper evaluation of the suitability of soil, for use as foundation or construction material, information about its properties, in addition to classification, is frequently necessary. Those properties which help to assess the engineering behaviour of a soil and which assist in determining its classification accurately are termed "Index properties". The tests required to determine the index properties are in fact "classification tests".

Therefore soil classification tests have proved to be a very useful tool to the soil engineer. It gives general guide lines about some of the engineering properties of soils like permeability, compressibility and strength.

CLASSIFICATION ON THE BASIS OF GRAIN SIZE

The basis of using size of particles as a criterion for classifying soil is that the engineering behaviour of a soil which has predominently small particles would differ from the behaviour of a soil which has predominently larger particles.

Natural soils occur in many varieties and types. For engineering purposes the soils may be broadly grouped into two categories, namely

(*i*) Coarse grained soils

(*ii*) Fine grained soils.

Obviously, the basis of such grouping is the size of the individual soil grains.

Gravels and sands are examples of coarse grained soils, and silts and clays are examples of fine grained soils.

Clay size	Silt size	Sand size			Gravel size		Cobble size	Boulders
		Fine	Medium	Coarse	Fine	Coarse		
0.002	0.075	0.425	2.0	4.75	20	80	300	

Diameter in mm

This type of classification test determines the range of sizes of particles in the soil and the percentage of particles in each of these size ranges. This is also called "grain size distribution" or "Mechanical analysis" meaning the separation of a soil into its different size fractions.

The particle size distribution is found in two stages:

(*i*) Sieve analysis

(*ii*) Sedimentation analysis.

Sieve analysis is useful for coarse grained soils. Silt and clay particle sizes cannot be separated by sieving. For these soil types, sedimentation tests are used. Hydrometer test and pipette tests are the common sedimentation tests.

The results of these grain size analysis are presented in a semi-logarithmic plot taking particle size on *x*-axis (log scale) and percentage finer on *y*-axis (ordinary scale). The primary reason for representing grain size to log scale in the grain size distribution is that a very wide range of grain sizes can be represented in one plot.

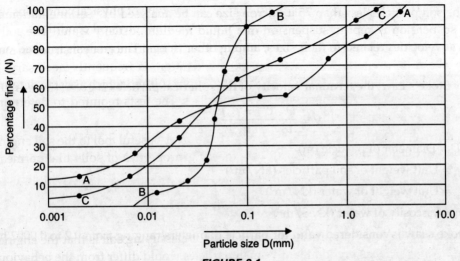

FIGURE 8.1

For convenience, the size of soil particle is expressed in equivalent diameter which corresponds to the size of the sieve opening or the size of a spherical particle falling through a soil suspension in sedimentation test.

In the above figure, it can be observed that curve C—C spreads over a large range of particle sizes. Such soils are called well graded soils. Curve B—B is confined to a narrow range of particles and hence called uniformly graded or poorly graded soils. In the soils corresponding to curve A—A, some of the particles are missing. There is a gap in gradation, hence it is called gap graded.

The sizes corresponding to 10% finer, 30% finer and 60% finer have been designated as D_{10}, D_{30} and D_{60} respectively. Size D_{10} is known as "effective diameter".

With these sizes the following quantities are defined:

Uniformity coefficient $$C_u = \frac{D_{60}}{D_{10}}$$

Coefficient of curvature
or
Coefficient of gradation
$$C_c = \frac{D_{30}^2}{D_{60} \cdot D_{10}}$$

The range of values of C_u and C_c for well and poorly graded soils is given below:

Soil type	Gradation	C_u	C_c
Gravel	Well graded	> 4	$1 < C_c < 3$
	Poorly graded	Not possessing the above values	
Sand	Well graded	> 6	$1 < C_c < 3$
	Poorly graded	Not possessing the above values	

SEDIMENTATION ANALYSIS

The soil particles less than 75 μ micron size can be analysed by "sedimentation analysis". The soil suspension is kept in suspension in a liquid medium, usually water. The soil particles descend at velocities related to their sizes, among other things. The analysis is based on Stokes's Law.

By Stoke's Law, the terminal velocity of the spherical particles is given (in m/s) by:

$$V = \frac{D^2}{18} \frac{(\gamma_s - \gamma_w)}{\eta}$$

where D = Diameter of particle (m)
γ_s = Unit weight of soil particles (kN/m³)
γ_w = Unit weight of water (kN/m³)
η = Viscosity of water (kN.S/m²)

Stoke's law is considered valid for particle diameters ranging from 0.2 to 0.0002 mm.

CONSISTENCY OF SOILS

Consistency is that property of a soil which is manifested by its resistance to flow. It can also be looked upon as the degree of firmness of a soil and is often directly related to strength. The consistency of a fine grained soil is largely influenced by the water content of the soil. Consistency is generally described as soft, medium stiff, stiff or hard.

A gradual decrease in water content of a fine grained soil causes the soil to pass from a liquid state to a plastic state, from plastic state to a semi solid state and finally to a solid state. The water contents at these changes of state are different for different soils.

The water contents that correspond to these changes of state are called Atterberg Limits. These four consistency states are shown in Fig. 8.2.

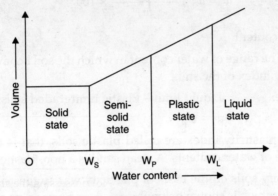

FIGURE 8.2

The water contents corresponding to transition from one state to the next are known as the Liquid limit (W_L), the Plastic limit (W_P) and the Shrinkage limit (W_S).

Liquid Limit. It is defined as the arbitrary limit of water content that represents the boundary between the liquid and plastic states of the soil. At liquid limit, the soil possesses a small value of shear strength. The liquid limit is the minimum water content at which the soil tends to flow as a liquid.

The liquid limit is determined in the laboratory as the water content at which a part of soil, cut by a groove of standard dimensions, will flow together for a distance of 12 mm under the impact of 25 blows. The drop of the cup used for this test is 10 mm.

The compressibility of a soil generally increases with an increase in liquid limit.

Plastic Limit. It is the arbitrary limit of water content that represents the boundary between the plastic and semi solid states of soil. The plastic limit is the minimum water content at which the change in shape of the soil is accompanied by visible cracks.

In the laboratory, the plastic limit is determined as the water content at which a soil will just begin to crumble when rolled into a thread of approximately 3 mm in diameter.

Shrinkage Limit. It is the arbitrary limit of water content that represents the boundary between the semi-solid and solid states of soil. The shrinkage limit may also be defined as the lowest water content at which the soil is fully saturated. Reduction in water content beyond shrinkage limit does not cause any reduction in total volume of soil mass.

In the laboratory, the shrinkage limit is determined by completely drying out a lump of soil and measuring its final volume and mass.

$$\text{Shrinkage limit, } w_s = \frac{(w_i - w_d) - (v_i - v_d)\gamma_w}{w_d} \times 100$$

where w_i = Initial wet weight of the soil
w_d = Final dry weight of the soil
v_i = Initial volume of the soil pat
v_d = Dry volume of the soil pat

or
$$w_s = w_1 - \frac{(v_i - v_d)\gamma_w}{w_d} \times 100$$

where w_1 = Initial water content.

Plasticity Index. The range of water content in which the soil behaves like a plastic material is known as the plasticity index of the soil.

∴ Plasticity index   = Liquid limit – Plastic limit

$$I_P = w_L - w_P$$

Soils with a high plasticity index are called plastic soils, that is they behave as a plastic material for a large range of water contents. A clean sand is a non-plastic material.

A method to classify soils on the basis of plasticity was suggested by Cassagrande in the form of a plasticity chart which is shown below:

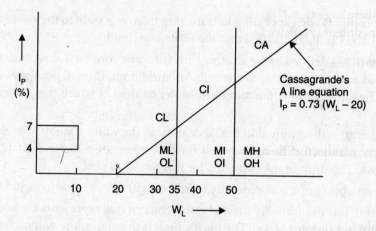

FIGURE 8.3

C = Clay M = Silt O = Organic L = Low Compressibility
I = Medium Compressibility H = High Compressibility

P.I. of a soil is a measure of the amount of clay in the soil, in other words, the fineness of the particles. Soils with high organic content have low plasticity Index.

Shrinkage Index. The range of water content within which a soil is in a semi-solid state of consistency.

∴ Shrinkage Index = Plastic limit – Shrinkage limit
$$= w_p - w_s$$

S.I. can be used as an indicator for the amount of clay.

Shrinkage ratio. It is defined as the ratio of given volume change, expressed as percentage of dry volume to the corresponding change in water content above shrinkage limit.

$$S.R. = \frac{\frac{V_1 - V_2}{V_d}}{(w_1 - w_2)} \times 100$$

V_1 = Volume of soil mass at water content w_1
V_2 = Volume of soil mass at water content w_2
V_d = Volume of dry soil mass.

Volumetric shrinkage (V.S.). It is defined as the decrease in volume expressed as a percentage of the dry volume of the soil mass, when the moisture content is decreased from given percentage to shrinkage limit.

$$V.S. = \frac{V_1 - V_d}{V_d} \times 100$$

∴ $V.S. = S.R. (w_1 - w_2)$.

Linear Shrinkage (L.S.). It is defined as the change in length divided by the initial length, when the water content is reduced to the shrinkage limit.

Liquidity Index $$I_L = \frac{\text{Water content} - \text{Plastic limit}}{\text{Plasticity index}}$$

$I_L > 1$—Soil is in liquid state.

$I_L < 0$—Soil is in semi-solid state.

Consistency Index (or) Relative consistency $$I_C = \frac{\text{Liquid limit} - \text{Water content}}{\text{Plasticity index}}$$

$I_C > 1$ Soil is in semi-solid state.

$I_C < 0$ Soil behaves like a liquid.

Obviously $I_L + I_C = 1$.

ACTIVITY OF CLAYS

The presence of even small amounts of clay particles can be found with the help of Activity ratio.

Activity ratio $$(A_k) = \frac{\text{Plasticity Index}}{\text{Clay fraction (or) percentage of clay sizes}}$$

Activity	Classification
< 0.75	Inactive
0.75 to 1.25	Normal
> 1.25	Active

Flow index (I_f) is the slope of the flow curve obtained from a liquid limit test.

$$I_f = \frac{W_1 - W_2}{\log(N_2/N_1)}$$

Toughness Index = I_t = $\dfrac{\text{Plasticity Index}}{\text{Flow Index}}$.

SENSITIVITY OF CLAYS

Sensitivity (s) of a clay is defined as the ratio of its unconfined compression strength (q_u) in the natural or undisturbed state to that in the remoulded state, without any change in water content.

$$s = \frac{q_u(\text{undisturbed})}{q_u(\text{remoulded})}.$$

Thixotropy of Clays. Clayey soils may loose some strength as a result of remoulding. With the passage of time, however, the strength increases, though not back to the original value. This phenomenon of strength gain, with no change in volume or water content is called **Thixotropy**.

PERMEABILITY

The ability of any porous medium such as soil to transmit fluids through the inter-connected voids is known as the Permeability. The Permeability is a very important engineering property of the soils, because it has a dominating influence on the total engineering behaviour of soil.

Concept of Heads. To define the water head at a point, the following three heads must be considered.

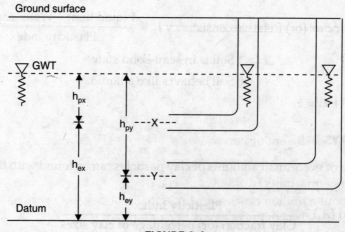

FIGURE 8.4

1. *Pressure head* (h_p). It is the pressure at any point by the unit weight of water. In other words, it is the height to which water raises in the stand pipe from the point of insertion. In the Fig. 8.4, h_{px} is pressure head at X and h_{py} is the pressure head at point Y.

2. *Elevation head* (h_e). It is the height of the point from the selected datum level. In Fig. 8.4, h_{ex} is the elevation head at X and h_{ey} is elevation head at point Y.

3. *Total head* (h). It is the sum of pressure head (h_p) and elevation head (h_e).

Therefore, Total Head at (h_x) = $h_{px} + h_{ex}$ and at Y (h_y) = $h_{py} + h_{ey}$

Darcy's law. Knowing the value of permeability of a soil enables the engineer to determine the quantity of water which can flow through the soil. Based on his experiments he found that the flow velocity is proportional to hydraulic gradient ($V \propto i$). It is also called as discharge velocity; Darcian velocity (or) superficial velocity.

According to Darcy's law, the rate of flow through a soil can be calculated using the equation:

$$q = K.i.A$$

where K is coefficient of permeability (mm/sec or cm/sec) or m/sec

i is hydraulic gradient = h/L (a dimensionless parameter)

A is the cross-sectional area.

The flow velocity can be obtained using the equation:

$$V = Ki$$

The flow velocity is different from the velocity inside the soil pores which is known as the seepage velocity. Because the flow is continuous, discharge must be the same throughout the system. Thus

$$q = A \times V = A_v \times V_s$$

where A_v = Cross-sectional area of the voids.

$$\therefore \quad V_s = \left(\frac{A}{A_v}\right) \times V = \frac{1}{n} \times V, \quad \text{where } n = \text{Porosity}$$

or

$$V = n . V_s = \left(\frac{e}{1+e}\right) V_s$$

From the above relation, it is clear that the seepage velocity is greater than the flow velocity.

Seepage velocity = $V_s = \dfrac{V}{n} = \dfrac{K.i}{n} = \dfrac{K}{n} i = K_p . i$

where K_p is the constant of proportionality, and is called as "the coefficient of percolation".

$$K_p = K/n$$

Factors influencing Coefficient of Permeability

1. *The size and shape of the soil particles i.e., Soil Type*

Difference in permeability of different soils can be enormous. For example, the permeability of a sandy soil is about a million times as much as that of a clay. Following table presents typical values of coefficient of permeability (K) for different types of soils.

Soil	Coefficient of permeability cm/s
Gravel	10^0
Coarse sand	10^0 to 10^{-1}
Medium sand	10^{-1} to 10^{-2}
Fine sand	10^{-2} to 10^{-3}
Very fine sands and silts	10^{-3} to 10^{-7}
Clays	10^{-7} to 10^{-9}

Hazen gives the following equation for clean sands in loose state $K(\text{cm/s}) = C \cdot D_{10}^2$ where C is empirical coefficient which varies from 90 to 120; often assumed as 100.

2. *The function of void ratio (or) Density*

The variation of K with void ratio can be mathematically expressed as

$$K \propto \frac{e^3}{1+e} \text{ (or) } K \propto e^2 \text{ (or) } \log K \propto e$$

The relation $e \propto \log K$ is close to a straight line for nearly all types of soils, as shown in Fig. 8.5.

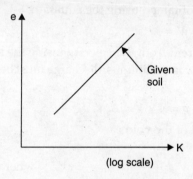

FIGURE 8.5

3. *The detailed arrangement of the individual soil grains i.e.,* the structure (or) soil fabric.

If all particles were aligned in one direction, one can readily visualize water being able to move with ease along the resulting longitudinal voids. On the other hand, a random arrangement of soil particles would make the water movement relatively difficult.

K for soil placed in a relatively dry state is greater than K for soil placed in a relatively moist state.

4. *Function of the Permeant.* The ease with which a liquid can flow through a soil also depends on the liquid itself.

For example, a thick viscous oil will move at a slower rate than water. Therefore, greater the viscosity the lower is the permeability.

It is customary to standardise the value of K at 20°C

$$K_{20} = \frac{\mu_T}{\mu_{20}} \times \frac{\gamma_{20}}{\gamma_T} K_T$$

where K_{20} = Value of coefficient of Permeability at 20°C.

K_T = Value of coefficient of Permeability at T°C.

μ_{20} = Viscosity of Permeant at 20°C.

γ_T = Viscosity of Permeant at T°C.

γ_{20} = Unit weight of Permeant at 20°C.

γ_T = Unit weight of Permeant at T°C.

5. *Presence of Discontinuities*. Presence of air in soil makes it unsaturated and

$$K_{unsat.\ soil} < K_{sat.\ soil}.$$

Generally $\left(\dfrac{K_{unsat.}}{K_{sat.}}\right)_{\text{at same void ratio}} = \left(\dfrac{S}{100}\right)^{3.5}$

where S is degree of saturation (varies from 0 to 100%).

Permeability of Stratified Soils

Flow along the direction of stratification

$$K_H = \dfrac{\sum_{i=1}^{n} K_i H_i}{\sum_{i=1}^{n} H_i} \text{ or } \dfrac{K_1 H_1 + K_2 H_2 + \ldots + K_n H_n}{H_1 + H_2 + \ldots + H_n}$$

where K_H = Equivalent coefficient of permeability along direction of stratification for an equivalent layer of thickness equal to sum of thickness of all layers

K_n = K of nth layer

H_n = Thickness of the nth layer

n = Total number of soil layers.

Flow perpendicular to the direction of stratification

$$K_V = \dfrac{\sum_{i=1}^{n} H_i}{\sum_{i=1}^{n} \dfrac{H_i}{K_i}} = \dfrac{H_1 + H_2 + \ldots + H_n}{\dfrac{H_1}{K_1} + \dfrac{H_2}{K_2} + \ldots + \dfrac{H_n}{K_n}}$$

where K_V = Equivalent coefficient of permeability perpendicular to the direction of stratification for an equivalent layer of thickness equal to the sum of thickness of all the layers.

The above equations are valid only when one-dimensional flow takes place in horizontal or vertical direction. The approximate relationships indicate that K_H is governed by the most pervious layer and K_V is governed by the most impervious layer.

In general $K_H > K_V$.

Determination of coefficient of permeability. The coefficient of permeability can be determined by both laboratory tests and field tests.

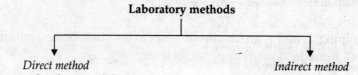

Direct method
1. Constant head permeability test.
2. Falling head permeability test.

Indirect method
1. Consolidation test.
2. Computation from the grain size.
3. Horizontal capillary test.

Constant heat permeability test. This test is more suitable for coarse grained soils where reasonable quantity of water can be collected in a given period. The given soil sample can be remoulded by compacting the given soil at the given density in the permeability mould. The length (L) and cross-section of sample (A) are measured. From an overhead tank water is allowed to pass through the soil sample, the head (h) causing the flow is maintained constantly throughout the test period. When a steady flow is established through the soil sample, a reasonable quality of water (Q) is collected in a measuring jar in a given period of time (t). Then from Darcy's law

$$q = \frac{Q}{t} = K.i.A$$

or

$$\frac{Q}{t} = K \cdot \frac{h}{L} A$$

or

$$K = \frac{Q \cdot L}{t.h.A}$$

Falling head permeability test. This test is used for fine grained soils which are relatively less permeable. In this method the amount of water entering into the soil sample is measured instead of the amount of water which is flowing out of sample. The soil sample is prepared in the same way as that for constant head test. The water is allowed to enter into the sample and the amount of water entered into the sample in a certain period of time (t) is measured using a burette of certain cross-sectional area (a).

Let h_1 and h_2 are the heads at times t_1 and t_2 respectively. Then the coefficient of permeability can be calculated from the equation

$$K = \frac{aL}{A(t_2 - t_1)} \log_e (h_1/h_2)$$

Field methods for determination of coefficient of permeability
- Pumping out test
  - Pumping out test in a confined aquifer
  - Pumping out test in an unconfined aquifer
- Pumping in test
  - Open end test
  - Packer test

Pump-out test

(i) *Unconfined Aquifer*: $K = \dfrac{q}{1.36(z_2^2 - z_1^2)} \log\left(\dfrac{r_2}{r_1}\right)$

where q is rate of pumping; z_1 and z_2 are depths of water in the observation wells distant, r_1 and r_2 from the pumped well from the bottom impervious layer.

(ii) *Confined Aquifer*: $K = \dfrac{q}{2.72\, H(z_2 - z_1)} \log\left(\dfrac{r_2}{r_1}\right)$

where q is rate of pumping; H is thickness of the aquifer; z_1 and z_2 are depths of water in the observation wells.

SEEPAGE

Ground water is frequently encountered at some of the construction sites. The movement of water through the soil under a hydraulic gradient is called as "Seepage". A hydraulic gradient is supposed to exist between two points if there exist a difference in the hydraulic head at the two points. The seepage of water will pose some problems like:

(i) loss of stored water through an earth dam or foundation.

(ii) settlement of structures that are constructed on compressible soils due to consolidation effect.

(iii) instability for soil slopes.

(iv) forces on hydraulic structures due to the pressure exerted by the percolating water.

For studying the seepage forces, Laplace equation is best suited. The solution of two Laplace equations for the potential functions and flow functions takes the form of two sets of orthogonal curves.

Flow Lines. These set of curves represent the trajectories of seepage. The space between the two adjacent flow lines is assumed to be a flow channel with an impervious boundary at both ends such that water does not cross the flow lines.

Equipotential Lines. These set of curves represent lines of equal head. The head loss caused by water crossing two adjacent equipotential lines is termed the potential drop.

Flow Net. The entire pattern of flow lines and equipotential lines is referred to as a flow net. Thus, a flow net is a graphical representation of head and direction of seepage at every point.

The properties of flow net are as follows:

(i) Equipotential and flow lines intersect or meet orthogonally.

(ii) The potential drop between any two adjacent equipotential lines is the same.

(iii) The quantity of water flowing through two adjacent flow channels is the same.

(iv) Velocity of flow is more in small size square figures, so as to keep the discharge same.

Applications of flow nets

1. Quantity of seepage water

The total discharge through the complete flow net per unit length can be calculated using the equation

$$q = KH \frac{N_f}{N_d}$$

where N_f is the number of flow channels.

N_d is the number of potential drops.

H is the total head loss.

The ratio of N_f/N_d is called as the shape factor of the flow net.

2. Amount of seepage pressure

Seepage pressure = $(H - N_d \Delta h) \gamma_w$

where N_d is the number of potential drops (by a value of Δh). This seepage pressure acts in the direction of flow.

3. Amount of uplift pressure

Uplift pressure = $h_w \gamma_w$

where h_w = Piezometric head.

4. Exit gradient

The maximum hydraulic gradient at downstream end of flow lines is termed the exit gradient.

In other words, it is the hydraulic gradient where the percolating water leaves the soil mass and emerges into free water.

This is given by $\quad i_e = \dfrac{\Delta h}{L}$

Δh is head loss over the last field.

L is length of smallest square in the last field.

Piping. Because of the instability caused by a high value of hydraulic gradient at the end face of percolating soil mass, the soil grains are dislodged and are eroded. Such erosions at the exit cause gradually a pipe-shaped discharge channel. The width of this pipe-shaped channel and the hydraulic gradient will increase with time and will lead to a failure of the structure which is constructed on the soil. Such a type of failure is called piping failure.

Filters or drain materials are used for preventing piping failures. Such materials should satisfy the following two requirements:

(i) The gradation of the material should be such that it should allow a rapid drainage without developing the seepage forces.

(ii) The gradation of filter material should be capable of forming small size pores such that the migration of adjacent particles through the pores is prevented.

(iii) It should add some self weight to the structure.

The criterion for the design of a graded filter as per Terzaghi is given as

$$\frac{(D_{15}) \text{ filter material}}{(D_{85}) \text{ protected soil material}} < 5 < \frac{(D_{15}) \text{ filter material}}{(D_{15}) \text{ protected soil material}}.$$

QUICK CONDITION

When water flows upwards in soil, at a certain hydraulic gradient, the upward forces neutralise the downward gravitational forces. This particular gradient of flow is called critical hydraulic gradient (i_{cr}).

The critical hydraulic gradient is given by the equation:

$$i_{cr} = \frac{G-1}{1+e}$$

We know

$$\gamma_{sub} = \left(\frac{G-1}{1+e}\right) \gamma_w$$

$$\therefore \quad \frac{\gamma_{sub}}{\gamma_w} = \frac{G-1}{1+e}$$

$$\therefore \quad i_{cr} = \frac{G-1}{1+e} = \frac{\gamma_{sub}}{\gamma_w} \simeq 1.$$

The value of critical hydraulic gradient is approximately equal to unity. This means that an upward hydraulic gradient of magnitude i_{cr} will be just sufficient to start the phenomenon of "boiling" in sand. A saturated sand becomes "Quick" or "Alive" at this critical hydraulic gradient. Contrary to the popular belief, "quick sand", is neither a material nor does it have any capacity to suck anything into itself. Quick sand is a type of hydraulic condition and not a type of sand, and many a soil can become "Quick".

Thus, at this "Quick condition", flow will be occurring in the upward direction under critical hydraulic gradient and soil loses all its effective stress. Shear strength of all soils is a function of effective stress. Some soils lose their shear strength when effective stress reduces to zero. In such soils the "Quick" condition is dangerous because that soil can no longer support anything.

CONSOLIDATION

When a structure is placed on a foundation which consists of soil stratum, the loads from the structure will be transferred to soil, in turn the soil will be stressed. The foundation that is not designed properly may fail by either shear failure of the soil or by excessive settlement of the soil.

The settlement of the foundation is defined as the vertical deformation from the original level. The settlement of soil will occur because of the re-arrangement of soil particles themselves in a closer state to enable the soil to sustain the applied stress. In saturated soils, closer state of particles is possible if the water is pushed out of the soil.

Therefore this process by which a decrease in water content of saturated soil takes place because of the expulsion of water from the voids of the soil is known as consolidation.

Compressibility behaviour of cohesionless and cohesive soils are not one and the same. A static load on a cohesionless soil may compress the soil immediately but the same load on a cohesive soil may compress the soil very slowly.

The compression consists of (*i*) Elastic or immediate compression, (*ii*) Primary compression, and (*iii*) Secondary compression.

Elastic compression is immediate and it is due to elastic bending and re-orientation of the soil particles.

Primary compression is due to volume change of soil mass. The application of load on soil mass creates hydrostatic pressures in a saturated soil. This excess pore water pressure is dissipated by the gradual expulsion of water through the voids of the soil, which results in a volume change and it depends on the time. This volume change results in the compression of soil mass. This compression of the soil under static load is called "**Consolidation**".

The decrease in volume per unit increase of pressure is defined as the COMPRESSIBILITY of the soil.

SPRING ANALOGY

The compressibility of soils can be well understood by means of spring analogy which is a mechanical model. The model is as shown in Fig. 8.6.

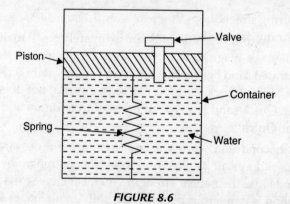

FIGURE 8.6

The model consists of a piston sliding in a cylindrical container. This piston is connected to the bottom of container through a spring as shown in above figure. The piston is provided with a valve which acts as a controlled outlet.

The response of soils to consolidation is similar to Spring analogy. The similarity is as shown below.

Spring analogy	Soil
(*i*) Spring	(*i*) The skeleton of soil solids
(*ii*) Piston	(*ii*) Soil surface
(*iii*) Valve (outlet)	(*iii*) Voids in the soils
(*iv*) Water	(*iv*) Pore water in the soils
(*v*) Discharge capacity of outlet.	(*v*) Coefficient of permeability.

Hydrodynamic time lag. The time required for the total expulsion of pore water and it depends on permeability of soils.

Plastic time lag. The time required for the transmission of the applied stress to the grains and the effective stress to reach a constant value.

The e-log p curve plotted for an experiment with loading and unloading conditions will be as shown in Fig. 8.7.

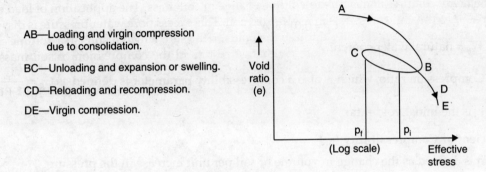

AB—Loading and virgin compression due to consolidation.
BC—Unloading and expansion or swelling.
CD—Reloading and recompression.
DE—Virgin compression.

FIGURE 8.7

The behaviour during unloading of the soil sample represented by the portion BC of the curve can be used for defining the swelling characteristics of the soil. The slope of this line indicates the swelling index.

Swelling index $C_s = \dfrac{e_i - e_f}{\log p_i - \log p_f}$.

The time taken for the volume change to occur is a function of

(i) Amount of water that has to be expelled out from the soil which again depends on

    (a) loading conditions (*i.e.*, how much stress is applied),

    (b) boundary conditions (*i.e.*, how much soil is effected) and

    (c) an engineering property of soil (*i.e.*, compressibility of soil),

(ii) Permeability of soil.

(iii) Location and the number of draining surfaces.

Compression Index (C_c)

From Fig. 8.8, $\quad C_c = \dfrac{e_0 - e_1}{\log_{10}\left(\dfrac{p_1}{p_0}\right)}$

where e_0 = Void ratio corresponding to eff. stress p_0

e_1 = Void ratio corresponding to eff. stress p_1.

In other words C_c is defined as the slope of the linear portion of the curve drawn between effective stress (on log scale) and the void ratio (on ordinary scale) and it is a dimensionless parameter.

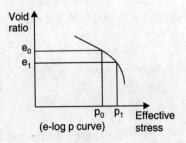

(e-log p curve)

FIGURE 8.8

C_c can be determined from the following equation:

$$C_c = 0.09 (W_L - 10) \text{ for undisturbed sample.}$$
$$= 0.007 (W_L - 10) \text{ for remoulded sample.}$$

where W_L is liquid limit of soil in per cent.

(C_c) undisturbed or field sample $= 1.3 (C_c)$ remoulded sample.

The expression for organic soils and peats in terms of insitu (natural) water content is as given below:

$$C_c = 0.0115 W_n$$

where W_n = natural water content.

Compression ratio, which is also a compressibility parameter is defined as $C_R = \dfrac{C_c}{1+e_0}$

where e_0 is the initial void ratio.

Coefficient of Compressibility (a_v)

It is defined as the change in volume of soil per unit increase in the pressure.

$a_v = -\dfrac{\Delta_v}{\Delta_p}$. The negative sign indicates that with increase in pressure, there is decrease in volume.

a_v is also defined as the secant slope of the effective stress and void ratio curve.

$$a_v = -\dfrac{\Delta_e}{\Delta_p}$$

Coefficient of Volume Change (m_v)

It is defined as the change in volume of soil per unit of initial volume due to unit increase in pressure. It is also called coefficient of volume compressibility or Modulus of volume change.

$$m_v = -\dfrac{\Delta v / v_0}{\Delta_p} = -\dfrac{\Delta e}{1+e_0} \cdot \dfrac{1}{\Delta_p}$$

In the above equation Δ_e term represents the change in void ratio and represents the change in volume of the saturated soil occurring through expulsion of pore water, and $(1 + e_0)$ represents initial volume, both for unit volume of soil solids.

We know
$$a_v = -\dfrac{\Delta_e}{\Delta_p}$$

$\therefore$
$$m_v = \dfrac{a_v}{1+e_0}$$

It can also be defined as the compression of a soil layer per unit of original thickness due to a given unit increase in pressure.

$$\therefore \quad m_v = -\frac{\Delta_H}{H} \cdot \frac{1}{\Delta_p}$$

or
$$\Delta_H = m_v H . \Delta_p$$

From the above equations we can say

$$\frac{\Delta_H}{H} = \frac{\Delta e}{1+e_0} \Rightarrow \Delta_H = \frac{\Delta e}{1+e_0} \cdot H$$

$\therefore$ Consolidation settlement $S_c = \Delta_H = \dfrac{\Delta e}{1+e_0} \cdot H$

We know
$$\Delta_e = C_c = \log\left(\frac{p_i}{p_0}\right)$$

$$\therefore S_c = \Delta_H = H \cdot \frac{C_c}{1+e_0} \cdot \log\left(\frac{p_i}{p_0}\right) = H \cdot \frac{C_c}{1+e_0} \log\left(\frac{p_0 + \Delta_p}{p_0}\right)$$

The coefficient of consolidation, $C_v = \dfrac{K}{m_v \gamma_w} = \dfrac{K(1+e)}{a_v \gamma_w}$

Degree of Consolidation (U_z). It is defined as the ratio, expressed as a percentage of the amount of consolidation at a given time, within a soil mass to the total amount of consolidation obtainable under a given stress condition.

$$U_z = \frac{e_0 - e_t}{e_0 - e_f} = \frac{p - p_0}{p_f - p_0} = 1 - \frac{U}{U_i}$$

where e_f = Void ratio at the end of consolidation.

e_t = Void ratio during consolidation at time, t.

e_0 = Void ratio at the start of consolidation.

p = Effective stress at time 't'.

p_0 = Pressure at starting of consolidation.

p_f = Pressure at end of consolidation.

U = Pore pressure excess at time 't'.

U_i = Pore pressure excess at the start of consolidation.

Assuming that the compression takes place in the vertical direction only, then the degree of consolidation also represents the percentage settlement.

$$\frac{S}{S_f} = 1 - \frac{U}{U_i}$$

where S = Settlement completed at any time 't'

S_f = Final settlement (at 100% consolidation).

$$\therefore \quad S = S_f\left(1 - \frac{U}{U_i}\right)$$

Time factor (T_v). It is a dimensionless constant and is defined by the equation

$$T_v = \frac{C_v t}{d^2}$$

where C_v = Coefficient of consolidation
t = Time (variable)
d = Drainage path (For double drainage, the drainage path is equal to half the thickness).

The time factor T_v for different percentage of consolidation can be determined using the following equation:

$$T_v = \frac{\pi}{4}\left(\frac{U}{100}\right)^2 \qquad \text{when } U \le 60\%$$

$$= -0.933 \log_{10}\left(1 - \frac{U}{100}\right) - 0.0851 \qquad \text{when } U > 60\%$$

Determination of coefficient of consolidation. The results of the consolidation test on the undisturbed soil samples are utilised to find the coefficient of consolidation (C_v). The characteristics of the theoretical relationship between the time factor and degree of consolidation were utilised to fit the laboratory curve to determine well defined point on the time compression curve corresponding to given degree of consolidation.

The commonly used methods for the curve fitting are:
(i) Square root time fitting method.
(ii) Logarithm of time fitting method.

Normally consolidated soils. If a soil has never been subjected to an effective pressure greater than the existing over burden pressure and which is also completely consolidated by the existing over burden is said to be normally consolidated.

Over consolidation soils. If a soil layer has been subjected in the past to a pressure more than the present over burden pressure, it is said to be over consolidated. Over consolidation might be due to the weight of an ice sheet or glacier which has melted away or by other geologic over burden and structural loads which no longer exist now.

Over consolidation ratio (OCR) is defined as:

$$\text{OCR} = \frac{\text{Pre-consolidation pressure}}{\text{Present effective overburden stress}} = \frac{p_c}{p_0}$$

Under consolidated soils. A soil which is not fully consolidated under the existing over burden pressure is called an under consolidated soil.

COMPACTION

Introduction

Most of the earthen constructions like highways, airways, embankments and dams require soil fill which is laid in layers and then compacted. Based on cost considerations generally it will be economical to use the local soil for fills and for foundations rather than to import large quantities of suitable materials. The properties of these locally available soils must be improved by some means. Compacting of soil is the most obvious and simple way of increasing the supporting capacity and of course the stability of the soil.

SOIL MECHANICS

Compaction of soil is defined as the process by which the soil particles are artificially rearranged and packed together by some mechanical means in order to increase the unit weight of soil. This is achieved by forcing the soil solids to move closer due to the expulsion of air from the voids and this is usually accomplished by rolling or tamping.

COMPACTION vs. CONSOLIDATION

Similarities. At the end of both processes:

(1) A closer packing of soil grains results.

(2) Shear strength increases

(3) Compressibility and permeability decreases.

Differences

Compaction	Consolidation
1. It is almost instantaneous	1. It is time dependent.
2. It occurs due to reduction of air voids.	2. It occurs due to expulsion of pore water from voids.
3. Soil is unsaturated.	3. Soil is saturated.
4. For a specified compaction energy, the compaction of a solid takes place only upto a certain limiting water content.	4. No limiting value of moisture content for the consolidation.

Proctor developed the relationship between soil density water content and compactive effort by conducting tests on soil specimens. He varied the water content by keeping the compaction energy the same.

The relationship between dry density and water content shows a peak at some water content. The addition of water to soil facilitates compaction by reducing the surface tension.

The relationship between dry density and water content is as shown in Fig. 8.9 and is called as compaction curve.

The maximum density obtained is called maximum-dry density (MDD). The water content corresponding to this MDD is called the optimum moisture content (OMC)

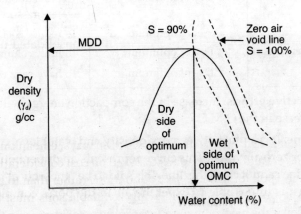

FIGURE 8.9

From the figure, the following points can be noted:

1. For a given compactive effort the dry density of soil first increases with increase in water content. Beyond a certain value of water content the trend is reversed.

2. The degree of saturation is always less than 100% even at high value of water content.

The reason for the increase and then decreases in dry density can be explained as follows:

Addition of water to begin with facilitates easier movement of particles and their closer packing, therefore an increase in the soil density. But, beyond a certain limit the water level becomes excessive and tends to occupy space which otherwise would have been occupied by solid particles resulting in a decrease in the dry density.

The following are the generally found important effects of compaction:

(*i*) Compaction increases the dry density of the soil, in turn increasing its shear strength and bearing capacity.

(*ii*) Compaction reduces the permeability of the soil.

(*iii*) Compaction decreases the settlement of the soil.

The characteristics of a soil which is compacted depends on whether it is dry side of optimum or wet side of optimum as given below:

Characteristics of soil	Dry side of optimum	Wet side of optimum
Structure	flocculated	dispersed
Shear strength	high	low
Compressibility	more	less
Permeability	more	less

On account of the differences in soil characteristics the soil may be compacted on dry side of optimum or on wet side of optimum depending on the performance required from the soil. The following table gives the information under various circumstances.

To be used in	Compacted soil at	Reasons
Homogeneous earth dams.	Dry of optimum	To prevent building up of high pore water pressure.
Core of earth dams	Wet of optimum	To reduce coefficient of permeability.
As subgrade for pavements.	Dry of optimum	To limit volume change in sub-grade.
Fills	Dry of optimum	To have easy working conditions.

Effect of compactive efforts. Increase in the compaction energy will result in an increase in the M.D.D. and a decrease in the O.M.C.

Zero air void line. It is a plot between dry density and water level corresponding to degree of saturation of 100% or zero air voids. This curve represents an upper bound for dry unit weight. The moisture density line cannot cross this line. It is said to be "theoretical" because it can never be reached in practice as it is impossible to expel the air in the pores completely by the process of compaction.

The zero air void density for any moisture content can be calculated from the expression

$$\gamma_d = \frac{G\gamma_w}{1+e}$$

where $e = \dfrac{wG}{S}$, G = Specific gravity,

γ_w = Density of water,

e = Void ratio, w = Water content of compacted soil,

S = Degree of saturation = 100% in this case.

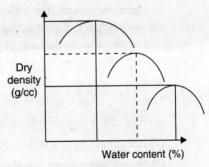

FIGURE 8.10

TYPES OF LABORATORY COMPACTION TESTS

The compaction characteristics of the soil are determined in the laboratory by any one of the following methods:

1. Standard proctor's test (light compaction).
2. Modified proctor's test (Heavy compaction).
3. Harvard miniature compaction test.
4. Abbot's compaction test.
5. Jodhpur mini compactor test.
6. Indian standard compaction tests.

Specification for compaction tests (Standard and Modified proctor's test).

Test	Mould			Hammer		Layers	Blows per layer	Energy kg. mm/cm³
	Dia (mm)	Height (mm)	Volume (mm³)	Mass (kg)	Height of fall (cm)			
Standard	100	127.3	1000	2.60	31	3	25	60.45
Proctor's test	150	127.3	2250	2.60	31	3	56	60.45
Modified	100	127.3	1000	4.89	45	5	25	272.60
Proctor's test	150	127.3	2250	4.89	45	5	56	272.60

Harvard Miniature Compaction Test

The compaction in this test is achieved by "kneading action" of a cylindrical tamper 12.7 mm. in dia. The mould has a volume of 62.4 cm³. The tamper operates through a compression spring so that the tamping force is controlled not to exceed a certain predetermined value. For different soils and different compactive efforts desired, the no. of layers, no. of tampings and the tamping force may be varied.

In-Situ or Field compaction

For any construction job which requires soil to be used as a foundation material, compaction in-situ or in the field is necessary.

Soil compaction can be achieved by different means such as tamping action, kneading action, vibration and impact. Compaction equipment operating on the tamping, kneading or impact principle are effective in the case of cohesive soils, while those operating on the kneading, tamping or vibratory principle are effective in the case of cohesionless soils.

The primary types of compaction equipment are (*i*) Rollers (*ii*) Rammers and (*iii*) Vibrators. Out of these, the most common are Rollers. Rollers are further classified as follows:

(*a*) Smooth-wheeled rollers (compaction by static compression to the soil)

(*b*) Pneumatic tyred rollers (compaction by kneading action)

(*c*) Sheep foot rollers (compaction is achieved by a combination of tamping and kneading).

Relative compaction

The results of laboratory compaction are not directly applicable to field compaction because the equipment used in field is different as well as no lateral confinement is possible in field compaction. The relative compaction is defined as:

$$\text{Relative compaction} = \frac{\text{Field dry density } (\gamma_d)}{\text{Maximum dry density in the lab test}} \times 100$$

Usually 90 to 97% relative compaction can be obtained.

Compaction of sands

The compaction characteristics of pure sandy soils are different from those of cohesive soils.

In case of purely sandy soils, the dry density-W/C relation-ship will be as shown in the Fig. 8.11.

Initially there is some decrease in dry density due to bulking of sand. The surface tension forces due to capillarity water resist closer arrangements of sand particles. The trend reverses when sufficient quantity of water is available.

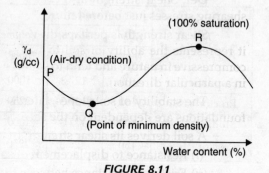

FIGURE 8.11

The following observation can be obtained from the graph:

M.D.D. results when the soil is either dry or saturated. So, to achieve maximum compaction in sands, they must be compacted either in dry or saturated state by flooding with water.

SHEAR STRENGTH

STRESSES

Stress in a Soil Mass

Stress is generally defined as the force per unit area. Stress will develop within the soil mass due to external load on the soil or due to its own weight.

If any load is applied on a soil mass which may be dry, then the stresses will be transmitted through the points of contact between the soil grains. It is not possible to determine the stresses at the points of contact of soil grains. So, generally the stress is computed as the applied load divided by contact area.

Types of Stresses

The stresses in a soil mass can be divided into three categories as follows:

1. *Total stress* (σ). It is defined as the total load per unit area of cross section. The total stress can be either due to external applied load or due to self-weight of the soil or due to both.

2. *Neutral stress* (u). It is defined as the stress carried by the pore water due to the applied load. This is also called as "Neutral pressure" or "Pore water pressure". The amount of neutral stress at a depth 'Z' below the water surface $u = \gamma_w \cdot Z$ where γ_w is unit weight of water.

3. *Effective stress* (σ'). It is defined as the difference between the total stress (σ) at that point and the pore water pressure (u) at that location.

$$\sigma' = \sigma - u$$

It is also called Inter grannular pressure and is also defined as the load transmitted through the soil grains divided by the area of cross section including both solid particles and voids.

Unlike total stress and pore water pressure, effective stress is not a physical parameter. It cannot be measured in the field by any instrument. It can be found by an arithematical expression, *i.e.*, by substracting one physical parameter (u) from another physical parameter (σ). Thus, the effective stress is only a mathematical concept which is found to be very useful for studying the engineering behaviour of soil.

SHEAR STRENGTH

Def. Shear strength also called as shearing strength may be defined as the resistance to shearing stresses just before failure.

Shear strength is perhaps the most important engineering property of soil. In other words, it represents the ability of soil to withstand shear stresses. Unlike normal stresses, which are compressive in nature and tend to squeeze soil, the shear stresses tend to displace the soil particles in a particular direction.

The stability of soil slopes, lateral earth pressures of soils and also the bearing capacity of foundations are dependent on the shear strength of the soil.

A soil derives its shear strength from the following:

(*i*) Resistance to displacement of soil because of interlocking of soil particles.

(*ii*) Frictional resistance between the individual soil particles, which may be sliding friction, rolling friction, or both.

(*iii*) Adhesion between the soil particles or cohesion.

A material in general has three different types of strength:

(1) Compressive strength   (2) Tensile strength   (3) Shear strength.

Soil has zero tensile strength. Under compressive forces soil fails in shear. Hence the shear strength is critical in case of soils.

Friction. It is the primary source of shear strength in most of the soils.

Angle of internal friction (ϕ)

In case of soil mass there is tendency of the soil particles to slide on each other under load. Generally, the planes through a point are acted by normal stresses and shear stresses.

The angle of friction between soil particles which try to slide on each other is called "angle of internal friction" or "angle of shearing resistance".

The surface or plane along which there is possibility of sliding of soil mass is called "Potential sliding surface".

If a dry sand is allowed to slump on a horizontal ground, it will make a slope with horizontal (ϕ), which is nothing but the angle of internal friction. This angle of slope is also called as "Angle of Repose".

Cohesion. It is the mutual attraction that exists between fine particles and tends to hold them together in a solid mass.

Mohr-Coulomb Failure Theory

Among the several theories available to define shear strength behaviour, Mohr-Coulomb's strength theory has been found to be the most practical method. According to this theory, the failure of soil mass will not occur on a plane where shear stress is maximum but on a plane where there is combination of normal and shear stresses.

At failure conditions of the soil element there will be one particular plane called "failure plane" along which the shear stress reaches its failure value. This failure of shear stress (τ_f) along the failure plane is known as shear strength.

The shear strength is a function of the normal stress on the failure plane and is given by the relationship.

$$S = \tau_f = C + \sigma \tan \phi$$

where $S = \tau_f$: Shear strength (Shear stress at failure)

C = Cohesion

ϕ = Angle of shearing resistance

σ = Normal stress on the failure plane.

From the point of view of shear strength three types of soil can exist.

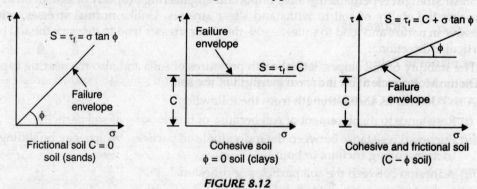

FIGURE 8.12

But the shear strength of soil is always controlled by the effective stress and the basic equation can be modified as

$$S = \tau_f = C' + \sigma' \tan \phi' = C' + (\sigma - u) \tan \phi'$$

where C' = Effective cohesion.

ϕ = Effective angle of internal friction.

σ = Total normal stress.

u = Pore water pressure.

Soil Mechanics

C' and ϕ' are called effective stress parameters, whereas C and ϕ are known as total stress parameters.

Principal Planes

In general, for any system of forces at a point there will be three mutually perpendicular planes on which the stress is wholly normal and shear stress is zero. The three planes are called principal planes and are mutually perpendicular to each other. The normal stress acting on these planes is called principal stress. The plane on which stress is maximum is called as major principal plane, and the plane on which stress is minimum is called as minor principal plane. The third plane on which the stress is neither maximum nor minimum is called intermediate principal plane.

Mohr's Circle

When a body is acted upon by an externally applied stress system, the state of stress on any plane through the body can be determined using the concept of the Mohr circle.

Let OP and OQ be the directions of major and Minor principal planes at any point 'O' in a medium. Let σ_1 and σ_3 be the intensities of the major and minor principal stresses respectively.

Then according to Mohr, the stress condition on any plane at 'O' can be determined by drawing a stress circle with $(\sigma_1 - \sigma_3)$ as diameter.

The stress circle is drawn on normal stress and shear stress axes. Now we can determine the stress ϕ on any plane which is passing through point 'O'.

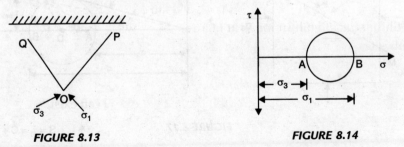

FIGURE 8.13   **FIGURE 8.14**

Origin of planes or Pole

On the Mohr circle a line is drawn parallel to the plane on which the known stresses act. The line intersects the Mohr's circle at a point 'P'. This point is known as Pole or origin of planes.

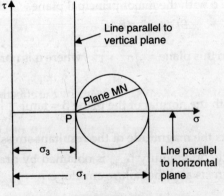

FIGURE 8.15

Limit conditions

We know that for cohesionless soils, the maximum shear strength is developed when the angle of obliquity equals its limitting value φ. For this condition the failure envelope becomes a tangent to the stress circle inclined at an angle φ to the σ axis.

From figure it can be noticed that the failure plane is not the plane subjected to the maximum value of shear stress. The criterion for failure is maximum obliquity but not the maximum shear stress.

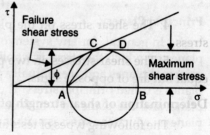

FIGURE 8.16

Hence in Fig. 8.16, although the plane AD is subjected to a greater shear stress than the plane AC, plane AD is not the plane of failure but the plane AC is plane of failure.

Important characteristics of Mohr's circle

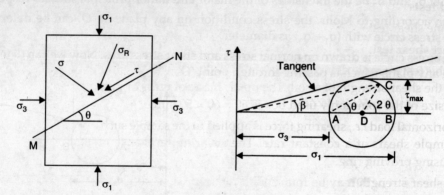

FIGURE 8.17

1. The maximum shear stress $\tau_{max} = \dfrac{\sigma_1 - \sigma_3}{2}$ i.e., the radius of Mohr's circle.

τ_{max} occurs on a plane inclined at 45° to the principal plane.

2. The point C on the Mohr's Circle indicates the stresses (normal and shear) on a plane (MN) which makes an angle φ with the major principal plane.

$$\angle BDC = 2 \angle BAC$$

The resultant stress on this plane = $\sqrt{\sigma^2 + \tau^2}$, where σ is normal stress and τ is shear stress.

Angle of obliquity with the normal of the plane $\beta = \tan\left(\dfrac{\tau}{\sigma}\right)$.

The line OC represents the magnitude of the resultant stress on this plane MN.

3. The maximum angle of obliquity β_{max} is obtained by drawing a tangent to the Mohr's circle from the origin. This line is also the failure envelope

$$\tan\beta_{max} = \frac{(\sigma_1-\sigma_3)/2}{(\sigma_1+\sigma_3)/2} \implies \beta_{max} = \tan^{-1}\left(\frac{\sigma_1-\sigma_3}{\sigma_1+\sigma_3}\right)$$

4. The shear stress on the plane of maximum obliquity is less than the maximum shear stress.

5. The shear stresses on two planes which are at right angles to each other are numerically equal but are of opposite signs.

Determination of shear strength of soil

The following types of tests are the most commonly used to determine the shear strength of soils.

Laboratory tests
1. Direct shear test
2. Triaxial compression test
3. Unconfined compression test
4. Laboratory vane shear test.

Field tests
1. Vane shear test
2. Penetration test.

Direct shear test

For this test a simple device called "shear box apparatus" is used for finding the shear strength of a soil. The box is made of brass or gun metal. The size of the commonly used box is 60 × 60 × 50 mm.

A horizontal load *i.e.*, shearing force is applied to the sample such that the sample shears at a constant rate. The shearing resistance is measured using proving ring.

The shear strength may be found with the sample subjected to varying compressive loads and a graph of shear stress against compressive stress is plotted.

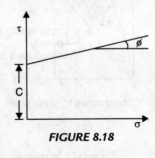

FIGURE 8.18

From the graph, the shear strength parameters C and ϕ can be determined.

This test is normally used for cohesionless soils since this gives good results for grannular soils.

Size of soil specimen that is tested: 60 × 60 × 25 mm.

Advantages of Direct Shear Test
1. The sample preparation is easy.
2. The test is simple.
3. It is ideally suited for drained tests on cohesionless soils.
4. As the thickness of sample is small, the drainage is quick and the pore pressure dissipates very rapidly. Therefore consolidated drained and undrained tests take relatively small time.

Disadvantages of Direct Shear Test
1. The distribution of normal and shear stress along the predetermined failure plane is very complex.

2. Drainage condition during this test cannot be controlled as the water content of a saturated soil changes rapidly with stress.
3. The area of the failure plane is not constant during the direct shear test. This area will be less than the original area of the soil specimen.
4. The soil is sheared on a predetermined horizontal plane. This forced plane is not necessarily the weakest plane. This is the most important limitation of the Direct shear test.
5. The effect of lateral restraint by the side walls of the shear box is likely to affect the results.
6. The measurement of pore water pressure is not possible.

Stress-strain curves in case of Direct Shear Test

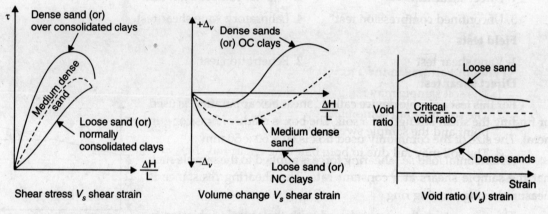

FIGURE 8.19

Triaxial Compression Test

This is the most popular and extensively used test. As the name suggests, the soil specimen is subjected to three compressive stresses in mutually perpendicular directions, one of the three stresses being increased until the specimen fails in shear.

Advantages of Triaxial Compression Test

1. The stress distribution is uniform on the shear plane during the progress of testing.
2. The soil specimen fail in shear on the weakest shear plane.
3. Complete control of the drainage is possible.
4. The pore water pressure can be measured accurately during the test.
5. The possibility to vary the cell pressure is possible in this test to simulate the field conditions for the sample.

Disadvantages of Triaxial Compression Test

1. The apparatus is elaborate, costly and bulky.
2. The drained tests takes more time.
3. It is not possible to determine the cross-sectional area of the specimen at large strains, as the assumption that the specimen remains cylindrical does not hold good.

4. The consolidation of the specimen is isotropic which is contrary to field where it is anisotropic.

Principal stresses in Triaxial test

The triaxial test can be considered as happening in two stages, the first bring the application of the cell water pressure (σ_3). While the second is the application of a deviator stress σ_d i.e., $\sigma_1 - \sigma_3$.

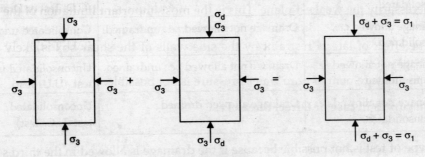

FIGURE 8.20

Types of Failures in the Triaxial Test

1. The soil sample may fail in pure shear.
2. The soil may fail completely by barelling and in this failure there is no definite failure point and the sample swells.
3. The soil may fail due to both barelling effect and shear.

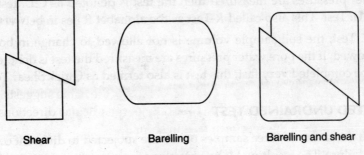

Shear   Barelling   Barelling and shear

FIGURE 8.21

Types of Triaxial Compression Tests

In laboratory, out of the four possible types, three types of triaxial compression tests can be conducted. Each type represents a particular combination of drainage conditions of the soil sample in the two stages of the test.

Generally there are three different and successive stages in a triaxial shear test. These three stages are:

Stage 1. Preparation of soil sample.

Stage 2. Application of cell pressure.

Stage 3. Application of axial stress *i.e.*, deviator stress at constant cell pressure.

Sl. No.	Drainage conditions during the two stages of triaxial testing		Type of test
	Second Stage (application of cell pressure only)	Third Stage (application of additional axial stress at constant cell pressure)	
1.	Drainage allowed i.e., consolidated.	Drainage allowed i.e., drained.	Consolidated drained test (C.D. Test or Drained test).
2.	Drainage allowed i.e., consolidated.	Drainage not allowed i.e., undrained.	Consolidated undrained test (C.U. Test).
3.	Drainage not allowed i.e., unconsolidated	Drainage not allowed i.e., undrained.	Unconsolidated undrained test. (U.U. Test).
4.	Drainage not allowed i.e., unconsolidated.	Drainage allowed drained.	Unconsolidated test (U.C. Test).

This type of test is not possible because if the drainage is allowed in the third stage, it will also drain off water on account of pore water pressure developed in the second stage and the conditions will become those of the C.D. Test.

In the C.D. Test, the soil sample volume changes in both the stages of the triaxial test and no pore water is allowed to remain in the sample. As the complete drainage takes very long time, the test is termed as Slow Shear Test or *S*-Test.

In the C.U. test, the soil sample volume changes in the first stage but not in the second stage; so, in the second stage pore water pressure is induced in the sample and it can be measured.

If pore water pressures are measured then the test is denoted as C.U. test, and if it is not measured then U.U. Test. This also called *R*-Test as the alphabet R lies in between *Q* and *S*.

In the U.U. Test, the soil sample volume is not allowed to change in both stages as the drainage is not allowed. If the pore water pressures are measured the test is denoted as UU test. As the U.U. test can be completed very fast, this test is also termed as Quick Shear Test or *Q*-Test.

UNCONSOLIDATED UNDRAINED TEST

In this test a minimum of three samples of soil are subjected to different cell pressures and then are loaded to failure. The resulting Mohr's envelope in terms of total stresses will be as shown in Fig. 8.22.

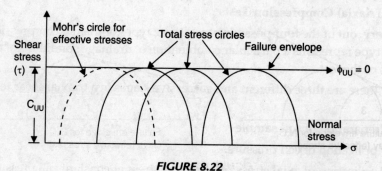

FIGURE 8.22

As seen from the Fig. 8.22 the failure envelope is horizontal and the shear strength is given by

$$\tau_f = C_{uu}$$

where C_{uu} = Cohesion intercept in U.U. test. In this test angle of shearing resistance ϕ_{uu} is zero.

For this test, the failure envelope cannot be drawn in terms of effective stresses, because for all the tests the effective stress remains the same. This is due to the fact that an increase in the confining pressure results in an equal increase in pore water pressure for a saturated sample under undrained conditions.

The Deviator stress for all the three specimens is same.

CONSOLIDATED UNDRAINED TEST

The failure envelope in terms of total stresses can be drawn from the results of this test.

In the case of normally consolidated clays, shear strength in total stress terms is $\tau_f = \sigma \tan \phi$

For the normally consolidated clays, the shear strength in terms of effective stress terms is

$$\tau_f = \sigma' \tan \phi'$$

The failure envelope has an intercept of cohesion (C') for over consolidated clays. So, the shear strength in case of over consolidated clays is

$$\tau_f = C' + \sigma' \tan \phi'.$$

CONSOLIDATED DRAINED TEST OR DRAINED TEST

Because the drainage is allowed in both stages, the pore water pressure at all stages of the test is zero, and always the effective stress prevails Mohr's Circles in terms of effective stresses can be drawn to get effective stresses parameters.

The shear strength equation

$$\tau_f = C_{CD'} + \sigma' \tan \phi_{CD'}$$

For N.C. Soils $C_{CD'} = 0$

∴ Theoretically, for N.C. soils $\phi_{CU'} = C_{CD'}$

Similarly $\phi_{CU'} = \phi_{CD'}$

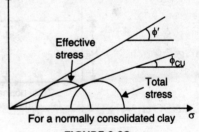

For a normally consolidated clay

FIGURE 8.23

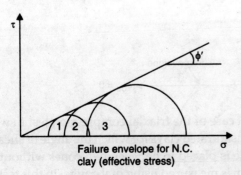

Failure envelope for N.C. clay (effective stress)

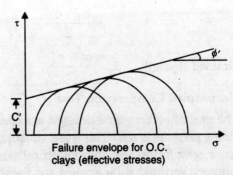

Failure envelope for O.C. clays (effective stresses)

FIGURE 8.24

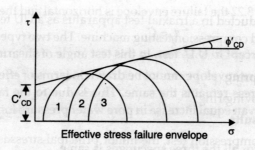

FIGURE 8.25

Hence, for N.C. soils the effective stress parameters C' and ϕ' obtained from consolidated drained test and that from consolidated undrained tests are approximately equal and do not depend upon the manner of testing.

Whereas total stress parameters C and ϕ depend upon the manner of testing.

UNDRAINED AND DRAINED SHEAR STRENGTH

Consider two identical soil samples. The two samples will have same effective stress parameters C' and ϕ'. Let both soil samples be consolidated under the same confining pressure (σ'_c).

After this let one of the sample be tested under U.U. condition and the other be tested under drained condition i.e., under C.D. condition.

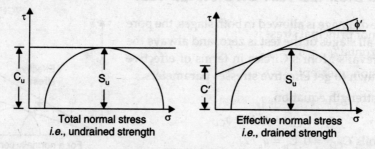

FIGURE 8.26

Undrained strength, S_u = Undrained cohesion C_u

$$= \frac{\sigma_1 - \sigma_3}{2}$$

Drained strength, $S_d = \dfrac{\sigma'_1 - \sigma'_C}{2}$.

Unconfined Compression Test

The unconfined compression test is a special case of the triaxial compression test in which the confining pressure is zero. This test is suitable for saturated clays for which the angle of shearing resistance is zero. In this test, the cylindrical sample is placed in between two cones without any lateral support. Vertical load is applied and the strain is measured using dial gauge. In this test, the sample is sheared rapidly and no drainage takes place.

This test can be conducted in a triaxial test apparatus as a UU test, it is more convenient to perform it in an unconfined compression testing machine. The two types of machines that are used for this tests.

1. Machine with a spring.
2. Machine with proving ring.

Presentation of Results

In an unconfined compression test, the minor principal stress is zero. The major principal stress is equal to the deviator stress and is calculated as

$$\sigma_1 = Q/A$$

where Q = Axial load, A = Area of cross-section.

The axial stress at which the specimen fails is known as the unconfined compressive strength (q_u).

While calculating the axial stress, the area of cross-section of the specimen at the axial strain should be used. The corrected area can be obtained from

$$A = \frac{A_0}{1-\varepsilon}$$

where A_0 = Original area of cross-section, ε = Axial strain.

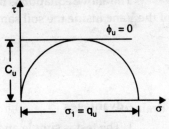

FIGURE 8.27

Mohr's Circle

As the minor principal stress is zero, the Mohr's circle passes through the origin. The failure envelope is horizontal ($\phi_u = 0$)

The cohesion intercept = Radius of Mohr's circle

$$= \frac{q_u}{2}$$

$$\therefore \quad C_u = q_u/2.$$

Advantages of the test

1. The test is convenient, simple and quick.
2. It is ideally suited for measuring the unconsolidated undrained shear strength of saturated clays.

Disadvantages

1. The test cannot be conducted on fissured clays.
2. The test may be a misleading one for soils for which the angle of shearing resistance is not zero.

VANE SHEAR TEST

The Vane shear test is used to determine the in-situ shear strength of clayey soil. The undrained shear strength of soft clays can be determined in a laboratory by this test.

This apparatus consists of four blades at the end of the rod i.e., at the bottom end. IS code recommends that the height of the vane should be equal to twice the overall diameter.

The vane is pushed slowly into the soil and rotated at uniform speed. The torque required to rotate the vane is measured and that the shear strength of soil is evaluated by relating the maximum torque to the shear strength provided by the soil around the perimeter and at the top and bottom of the cylinder.

The expression for shear strength from the results of Vane shear is given by

$$\tau_f = \frac{T}{\pi D^2 \left(\frac{H}{2} + \frac{D}{6}\right)}$$

where τ_f = Shear strength, T = Maximum torque
       H = Height of vane, D = Diameter of vane.

The above equation is modified if the top of the Vane is above the soil surface and the depth of the Vane inside the soil sample is H_1. In such a case

$$\tau_f = \frac{T}{\pi D^2 \left(\frac{H_1}{2} + \frac{D}{12}\right)}$$

Advantages
1. The test is simple and quick.
2. It is ideally suited for the determination of the in-situ undrained shear strength of non-fissured, fully saturated clay.

Disadvantages
1. The test will not give good results when the failure envelope is not horizontal.
2. The test cannot be conducted on the clayey soil which contains sand or silt.

Sensitivity of Clays

Changes in the stress and environment with time may make the soil to have a higher strength in the undisturbed state than in the remoulded state. The term sensitivity (S_t) is generally used to describe this difference in strength.

Sensitivity is defined as the ratio of undrained strength of the in-situ soil (undisturbed soil) to the undrained strength of that soil determined after remoulding it at the in-situ water content i.e.,

$$S_t = \frac{(\tau_f) \text{ undisturbed}}{(\tau_f) \text{ remoulded}}.$$

The value of sensitivity may vary from 1 to 100, and thus accordingly are classified as insensitive, medium sensitive, sensitive, extra sensitive and quick.

Description	Sensitivity	
Insensitive	1	
Low Sensitivity	1–2	
Medium Sensitivity	2–4	Highly over consolidated clays are insensitive.
Sensitive	4–8	
Very Sensitive	8–16	
Quick	> 16	

LIQUEFACTION

Generally the dry sand has zero cohesion but moist sand may have apparent cohesion due to surface tension.

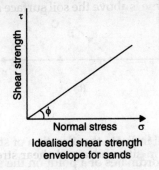 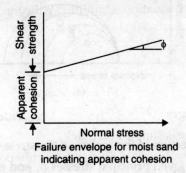

FIGURE 8.28

If a sample of dense sand is subjected to shear, it tends to expand. This is due to the soil particles trying to ride on each other during sliding. The expansion under shear is termed as "Dilatancy".

In case of saturated dense sands if there is sudden application of stress in the field, the soil does not get time to expand and negative pore water pressures develop. This results in increased effective stresses and greater shear strength. In case of loose sands, this is reverse. The loose sands try to compress by bringing their grains closer together. This results in reduced effective stress and reduction in corresponding shear strength.

If fine sands in the saturated condition are subjected to sudden loading, the tendency towards volume reduction causes pore water pressure to increase instantaneously. This causes a large decrease in the shear resistance, as the shear strength is given by

$$\tau_f = \sigma' \tan \phi' = (\sigma - u) \tan \phi$$

where $\sigma = u \rightarrow$ the soil looses all its shear strength and thus behaves like liquid. This phenomenon is known as Liquefaction.

Stress-Path

Stress-path is defined as a line or curve which indicates the progressive changes in the stresses as the load acting on the soil changes. The commonly used stress-path is that given by Lambe.

Lambe's stress-path is a line drawn through the points representing the maximum shear stresses acting on the specimen as the load is changed.

Fig. 8.29 (a) represents successive states as σ_1 is increased with a constant value of σ_3. A diagram with a number of Mohr's circles will look odd. It will be convenient if we plot only the points of maximum shear stresses. Thus, the locus of the points on the Mohr's diagram whose coordinates represents the maximum value of shear stresses is called as **Stress-path**.

It will be more convenient to draw the stress-path on p–q plot as shown in Fig. 8.29 (b). As seen from the figure there is no need to draw complete circles, but we can plot only the points corresponding to maximum shear stress.

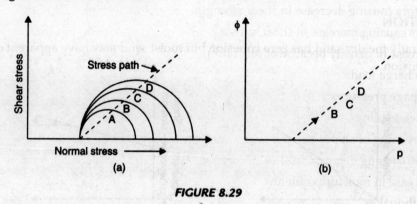

FIGURE 8.29

The direction of the arrow on the stress-path will indicate the direction of stress changes. For the given principal stresses σ_1 and σ_3, the coordinates of a point on the stress path are

$$p = \frac{\sigma_1 + \sigma_3}{2} \quad \text{and} \quad q = \frac{\sigma_1 - \sigma_3}{2}$$

Type of Stress-Paths

1. **Effective stress-path.** It is generally denoted as E.S.P. and is plotted between effective stresses.

$$\left(\frac{\sigma_1' + \sigma_3'}{2}\right) \text{ and } \left(\frac{\sigma_1' - \sigma_3'}{2}\right)$$

2. **Total stress-path.** It is generally denoted as T.S.P. and is plotted between total stresses.

$$\left(\frac{\sigma_1 + \sigma_3}{2}\right) \text{ and } \left(\frac{\sigma_1 - \sigma_3}{2}\right).$$

STABILITY OF SLOPES

The slope is an inclined boundary surface between air and the body of an earthwork. Earth slopes may be natural or man-made. These are invariably required in the construction of highways, railways, earth dams etc.

When the ground surface is sloping, forces exist which tend to cause the soil to move from higher elevations to lower elevations.

The important forces which are responsible for the instability of earthen slopes are:

1. Force of gravity (because of self weight of soil).
2. Seepage force (because of seeping water).

These forces induce the shear stresses in the soil. If the shear strength along any plane is not sufficient to take the shear stresses, failure occurs in the form of sliding of soil masses. So, the stability of these earth slopes is important for an engineer.

The factors leading to the failure of slopes may be classified as:

1. Factors causing increase in shear stresses.
2. Factors causing decrease in shear strength.

Factors causing increase in shear stress:

1. Increase of density because of saturation.
2. Surcharge loads.
3. Seepage pressures.
4. Shock loadings.
5. Steepening of slopes either by excavation or by natural erosion.

Factors causing decrease in shear strength:

1. Increase in moisture content.
2. Earthquakes.
3. Increase in pore water pressure.
4. Loss of cementing medium.

Slopes may be classified as:

1. Infinite slopes.
2. Finite slopes.

1. Infinite slope. It represents the boundary surface of a semi-infinite soil mass inclined to the horizontal. This type of slope is just hypothetical in nature.

2. Finite slope. This type of slope will be having a base and top surface with a finite value of height.

Types of Slope Failures

1. Rotational failure. Failure by rotation along a slip surface.

(a) *Toe failures*: Failure along the surface which passes through the toe.

(b) *Slope failure*: Failure occurs along a surface which intersects the slope above the toe of slope.

(c) *Base failure*: Failure occurs along a surface which passes below the toe.

2. Transitional failure. These types of failures occur in infinite slopes along a failure surface which is parallel to the slope.

3. Compound failure. It is a combination of the rotational and transitional failures.

Stability Analysis of Infinite Slopes

(a) **Cohesionless soils.** The factor of safety of an infinite slope is defined as the ratio of shear strength of soil to the required shear stress along the surface of failure.

Factor of safety against sliding $F = \dfrac{\tau_s}{\tau_f} = \dfrac{\tan\phi}{\tan i}$

τ_s = Shear strength along the failure surface.
τ_f = Shear stress along the failure surface.
ϕ = Angle of shearing resistance.
i = Angle of slope of earth with horizontal.

For the limiting equilibrium condition F.S. = 1

$\therefore \qquad \tan\phi = \tan i$

Thus, the maximum inclination of an infinite slope in a cohesionless soil by considering stability considerations is equal to angle of shearing resistance of soil.

(b) **Cohesive soils**

$$\text{F.S.} = \dfrac{c' + (\gamma H \cos^2 i)\tan\phi'}{\gamma H \cos i \sin i}$$

So, for a C-ϕ soil, the slope is stable as long as the slope angle $i \leq \phi$. If $i > \phi$, the slope will be stable only upto a limited height known as critical height which is given by the expression:

$$H_c = \dfrac{C'}{\gamma(\tan i - \tan\phi')\cos^2 i}$$

where C' = Cohesion; γ = Unit weight of soil.

A dimensionless parameter is used for the analysis of stability of C-ϕ soil and is called stability number (S_n).

Stability number is defined as the ratio of mobilised cohesion and γH

$$S_n = \dfrac{C_m}{\gamma H}$$

If the factor of safety with respect to cohesion is F_c, then $F_c = C/C_m$

Mobilised cohesion at a depth H, $C_m = \dfrac{C}{F_c}$

$\therefore \qquad S_n = \dfrac{C}{F_c \gamma H} = \cos^2 i (\tan i - \tan\phi)$

Stability factor: Reciprocal of stability number is called as stability factor.

Stability Analysis of Finite Slopes

For finite slopes, generally failure occurs because of rotational failure.

The methods that are commonly used for the stability analysis are:

(*i*) Swedish method or Slip circle method.

(*ii*) Friction circle method.

Swedish method or Method or slices

This method assumes the surface of sliding to be an arc of a circle. The analysis is done by dividing the area within the slip circle into a number of vertical slices.

(*i*) **Purely Cohesive Soils ($\phi = 0$ soils)**

θ = Angle made by ship circle at the point of motion

R = Radius of slip circle

e = Eccentricity

W = Weight of soil mass ABD

Factor of safety against sliding $= \dfrac{\text{Resisting moment}}{\text{Disturbing moment}}$

$= \dfrac{CR^2\theta}{We}$

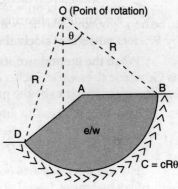

FIGURE 8.30

C is cohesion along slip surface.

A series of slip circles are selected and the circle along which the factor of safety is minimum is the likely failure plane.

Cohesive-frictional Soil

Factor of safety against sliding $= \dfrac{C.L + \Sigma N . \tan \phi}{\Sigma T}$

L = Total arc length

ΣN = Sum of normal components of the weight W.

ΣT = Sum of tangential components of the weight W

(*ii*) **Frictional Circle Method**

— Generally used for a $C\text{-}\phi$ soil. r = Radius of failure circle

— This method also assumes the failure surface as an arc of a circle.

In this method of analysis, the resultant force R will be tangential to the friction circle of radius $Kr \sin \phi$. The friction circle is also called as ϕ circle, where K is a factor whose value is greater than unity. Force triangle is drawn and values required are evaluated.

A number of slip circles are analysed and the surface which gives a low factor of safety is the likely failure plane.

Stress Distribution in Soils

Stresses are induced in the soil mass due to (1) weight of the overburden or (2) by structural loads or applied loads.

These stresses are required for the (1) determination of earth pressures (2) stability analysis of a soil mass and (3) settlement analysis of foundations.

The stresses due to overburden increases with depth but the stresses due to the applied loads decreases with depth.

The knowledge of stress distribution is essential for the analysis of deformation behaviour of foundation of any structure. The stress strain behaviour of soils is highly complex and it depends upon physical properties of soil like water content, void ratio etc., rigidity of the foundation, the type of loading on foundation and the type of foundation.

Generally, all theories for the determination of stress-distribution, the soil mass is assumed to be homogeneous, elastic, isotropic and semi-infinite. The theories developed by Boussinesq and by Westergaard's are widely used in practice.

When the ground surface is horizontal, the stresses due to self weight of the soil are known as Geostatic stresses. In such a case, the stresses are normal to the horizontal and vertical planes, and these are found to be the principal planes.

The vertical stress on any horizontal plane at a depth Z

$$\sigma_z = \frac{\text{Weight of overburden soil}}{\text{Area}} = \frac{\gamma(Z \cdot A)}{A} = \gamma Z$$

Horizontal stress acts on a vertical plane. The horizontal stress can be calculated with the help of coefficient of lateral stress which is defined as the ratio of horizontal stress (σ_h) to the vertical stress (σ_v)

$$K = \sigma_h/\sigma_v \implies \sigma_h = K\sigma_v$$

Boussinesq Theory

(*a*) Vertical stress due to concentrated load:

Vertical stress at a depth of 'z' $\sigma_z = K_B \cdot \dfrac{Q}{z^2}$, where K_B is Boussinesq's coefficient.

$$K_B = \frac{3}{2\pi[1 + (r/z)^2]^{5/2}}$$

where Q is the magnitude of the concentrated load, z is the depth; r is the radial distance.

(*b*) Horizontal stress due to a concentrated load:

$$\sigma_h = \frac{3Q}{2\pi} \cdot \frac{rz^2}{R^5}, \text{ where } R = \sqrt{x^2 + y^2 + z^2}$$

where x, y and z are the coordinates of the point at which the horizontal stress is determined.

Vertical stress distribution on a horizontal plane

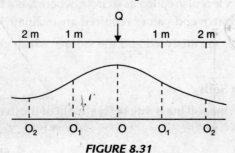

FIGURE 8.31

Vertical stress distribution on a vertical plane

The stress distribution is as shown in Fig. 8.32. The maximum vertical stress occurs at $r/Z = 0.817$. This corresponds to the point on the vertical plane with the line drawn at 39° 15′ to the vertical axis of the load.

Vertical stress due to a line load:

$$\sigma_z = \frac{2q'}{\pi z}\left[\frac{1}{1+(x/z)^2}\right]^2$$

where q' = Line load, x = Horizontal distance,
z = Depth

Vertical stress under a strip load:

$$f = \frac{q}{\pi}(\theta + \sin\theta)$$

where q = Uniformly distributed load intensity per unit area.
B = Width of the strip.

$$\sigma_z = \frac{q}{\pi}(2\theta + \sin 2\theta \cos 2\phi)$$

where $2\theta = \alpha_2 - \alpha_1$
$2\phi = \alpha_2 + \alpha_1$

Vertical stress under a uniformly circular loaded area:

$$\sigma_z = q\left[\left\{1 - \frac{1}{1+(R/z)^2}\right\}^{3/2}\right]$$

where q = Intensity of load, R = Radius of foundation, z = Depth.

$$\sigma_z = q\left[1 - \left\{\frac{1}{1+\tan^2\theta/2}\right\}^{3/2}\right] = 1 - \cos^3\theta$$

Vertical stress under corner of a rectangular area:

$$(\sigma_z) \text{ at } p = \frac{q}{4\pi}\cdot\frac{2mn(m^2+n^2+1)^{1/2}}{m^2+n^2+m^2n^2+1}\cdot\frac{m^2+n^2+2}{m^2+n^2+1} + \tan^{-1}\left[\frac{2mn\sqrt{m^2+n^2+1}}{m^2+n^2-m^2n^2+1}\right]$$

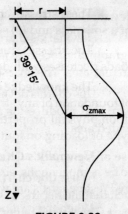

FIGURE 8.32

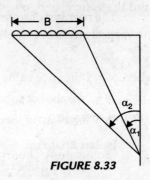

FIGURE 8.33

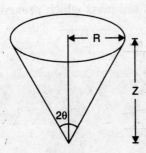

FIGURE 8.34

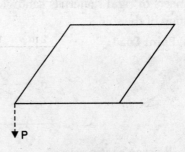

FIGURE 8.35

Newmark's Influence Chart

Newmark's chart is a graphical representation of Boussinesq's theory. Newmark developed a graphical procedure for determining the vertical stress below the centre of the uniformly loaded circular area. The chart developed by Newmark's is also called as "Influence Chart".

This chart is developed in such a way that each area on the chart when loaded with a uniform load, will produce the same increment of vertical stress at a depth z beneath the centre of the loaded area.

Use of Newmark's Chart

To use this chart, a plan of the uniformly loaded area is drawn on tracing sheet to a scale such that the depth z at which the stress is to be found equals to the length which is marked on the chart. The tracing paper on which the loaded area is drawn is then placed on the Newmark's chart with the point where the stress is to be found is located over the centre of the influence chart.

The number of squares that are covered by the plan diagram are counted. The intensity of the vertical stress at a depth z will be given by the equation

$$\sigma_z = I.N.q$$

where I = Influence value of the Newmark's chart.

N = Number of squares covered by the plan of loaded area.

q = Intensity of loading.

Isobar Diagram

An Isobar is a line which connects all the points below the ground surface where the vertical pressure is same. For any load, many isobars corresponding to various vertical pressure intensities can be drawn. It is generally found that an isobar resembles the shape of a bulb or that of an onion because the vertical pressure at all the points in a horizontal plane at equal radial distances from the centre of the load will be same. Thus, the stress isobar is called the "bulb of pressure" or Simply "Pressure bulb". According to one concept, the zone within which the stresses have a significant effect on the settlement of structures is known as the "pressure bulb" and it is generally treated that an isobar of 0.1 Q forms a pressure bulb. The isobars of higher intensities 0.2 Q, 0.3 Q etc., will lie within the isobar of 0.1 Q.

Westergaard's Analysis

Westergaard developed a solution to determine distribution of stress due to a point load in soils which are generally anisotropic in nature. He made the assumption that the soil is composed of thin sheets of rigid materials sandwiched in a homogeneous soil mass which prevent lateral deformation of the soil.

(a) **Point Load:**

$$\sigma_z = \frac{Q}{2\pi z^2} \frac{[(1-2\mu)/(2-2\mu)]}{\left[\frac{(1-2\mu)}{(2-2\mu)} + \left(\frac{r}{z}\right)^2\right]^{3/2}}$$

where μ = Poisson's ratio

When $\mu = 0$,
$$\sigma_z = \frac{Q}{\pi z^2} \frac{1}{[1 + 2(r/z)^2]^{3/2}} = \frac{Q}{z^2} \cdot K_w$$

where K_w = Westergaard's influence coefficient.

(b) **Uniformly loaded Circular area:**

$$\sigma_z = q \left[1 - \left\{ \frac{1}{1 + \left(\frac{R}{C^n Z}\right)^2} \right\}^{1/2} \right]$$

q = Intensity of load; R = Radius of circular footing;

z = Depth below the G.S.; C^n = Constant = $\sqrt{\dfrac{1 - 2\mu}{2 - 2\mu}}$

where μ = Poisson's ratio.

CONTACT PRESSURE

Contact pressure is the actual pressure that will be transmitted from the foundation to the soil. From the observations it is found that the pressure distribution beneath the footing, symmetrically loaded is not uniform. The pressure intensity depends upon type of soil and the rigidity of footing. Contact pressure distribution diagrams that are observed in flexible and rigid footings are as follows:

Flexible footings

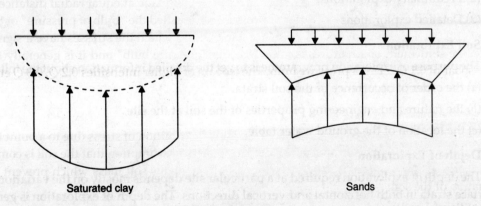

FIGURE 8.36

For flexible footings the contact pressure distribution is uniform. The settlement will be varying from center to the end.

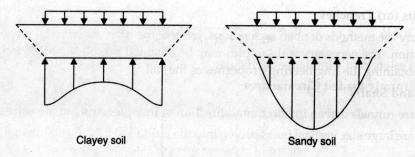

FIGURE 8.37

For rigid footings, the settlement is uniform. The pressure intensity varies from centre to the end.

SOIL EXPLORATION

For judging the suitability of the site for any Engineering work, site investigations are necessary. Soil exploration programme is necessary for preparing the designs and for analysing the safety considerations of the concerned project. The investigations that are conducted either in the laboratory or in the field for obtaining the necessary data are called as soil exploration programme. The purpose of the site investigation is to get the information about the various conditions at the site.

Stages in Sub-surface Exploration

Generally the sub-surface exploration is carried out in three stages as mentioned:

(1) Reconnaissance

(2) Preliminary explorations.

(3) Detailed explorations.

Soil Exploration

- The purpose of exploration programme is to get the detailed information about the following:

(*a*) the order of occurrence of the soil strata.

(*b*) the nature and engineering properties of the soil at the site.

(*c*) the location of the ground water table.

Depth of Exploration

The depth of exploration required at a particular site depends mostly on the variation of the sub-surface strata in both horizontal and vertical directions. The depth of exploration is generally governed by the depth of the influence zone, which will be generally given by an isobar diagram.

Method of Soil Exploration

(1) *Direct Method*: Pits (or) trenches; Drifts and Shafts.

(2) *Semi-direct methods*: Borings.

(3) *Indirect methods*: Penetration tests; Geophysical methods.

Test Pits (or) Trenches

These are the methods of open excavations. In these methods, soils can be inspected in their natural condition. The necessary soil samples may be obtained by soil sampling techniques and are used for obtaining the Engineering properties of the soil.

Drifts and Shafts

Drifts are tunnels driven in horizontal direction to find the nature of the soil strata.

Shafts are large size vertical holes driven into the ground to find the nature of the soil strata.

Borings

These are vertical bore holes drilled into the ground to get the information about the soil strata.

The methods that are used generally for drilling the bore holes:

(1) Auger boring (2) Auger and Shell borings (3) Wash boring (4) Percussion drilling (5) Rotary drilling.

Auger boring. An auger is a device that is used for advancing a bore hole into the ground. The two types of augers that are used are (1) hand augers (2) mechanical augers.

Auger and Shell boring. The sides of the bore hole cannot remain unsupported, the soil is prevented from falling with the help of casing (or) shell. Casings may be used for sandy soils (or) stiff clays.

Wash borings. It is a simple method used for making holes in all types of soil except those which are mixed with gravel and boulders. This method used for exploration below the ground water table.

Percussion drilling. This method is used for making holes in rocks, boulders and other hard strata. For this method, a heavy drill bit is driven into the ground by repeated blows.

Rotary drilling. In this method, the bore hole is advanced by a rotating drill rod. It is driven by a rotary drive mechanism by the application of downward pressure. This method is generally used to obtain the rock core samples.

Types of soil samples. The soil samples are generally are of two types:

(1) Disturbed samples   (2) Undisturbed samples.

(1) **Disturbed samples.** This is a sample in which the natural structure of the soil gets disturbed either fully or partly. However, these samples will represent the composition of the soil. These can be used to determine the index properties of the soil.

(2) **Undisturbed samples.** This is a sample in which the natural structure as well as the water content of the soil are retained. These samples can be used for determining both the index properties and engineering properties of the soil.

Sample Disturbance. The disturbance to the soil specimen depends upon the type of samples and the method of sampling. The factors that will be governing the disturbance of a sample are:

(1) Cutting edge

(2) Inside wall friction and

(3) Non-return valve.

The following are defined with respect to the diameters as follows:

(a) Area ratio

$$= \frac{\text{Max. cross-sectional area of the cutting edge}}{\text{Area of the soil sample}}$$

$$= \frac{D_2^2 - D_1^2}{D_1^2} \times 100$$

where D_1 = Inner diameter of the cutting edge.

D_2 = Outer diameter of the cutting edge.

(b) Inside clearance $= \dfrac{D_3 - D_1}{D_1} \times 100$

where D_3 = Inner diameter of the sampling tube.

(c) Outside clearance $= \dfrac{D_2 - D_4}{D_4} \times 100$

where D_4 = Outside diameter of the sampling tube.

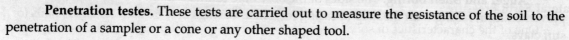

FIGURE 8.38

Penetration testes. These tests are carried out to measure the resistance of the soil to the penetration of a sampler or a cone or any other shaped tool.

There are two important types of tests that are ganerally used: (1) Standard penetration test (2) Cone penetration test.

Standard penetration test. For this test a split-spoon sampler with a 50.8 mm outer diameter and 35 mm inner diameter is used. This sampler is generally driven with the help of a 63.5 kg hammer with 75 cm free fall. The number of blows that are required for a penetration of 300 mm is designated as the standard penetration value or number denoted as 'N'.

Cone penetration test. This is also called as Dutch Cone test. There are two types of Cone penetration tests depending upon the type of loading:

(a) Static cone penetration (b) Dynamic cone penetration test.

These tests are carried out to get a continuous record of the resistance of soil by penetrating a dutch cone with a base of 10 cm² and an apex angle of 60°. For obtaining the resistance, the cone is pushed downwards at a steady rate to a depth of 35 mm each time. The approxi. relationship between cone resistance (q_c) and SPT NO. (N) are as follows:

(a) Gravels $q_c = 8N$ to $10 N$
(b) Sands $q_c = 5N$ to $6 N$
(c) Silty sands $q_c = 3N$ to $4 N$
(d) Silts and clays $q_c = 2 N$

where q_c is in kg/cm²

Geophysical Methods

These methods involve the technique of finding the sub-surface details by means of measuring some physical properties of the sub-surface and correlating them. These methods are of course approximate ones. These methods determine the conditions over large distances. These are suitable for investigating large areas in a short time period.

The two geophysical methods which are mostly used are:

(1) Electrical resistivity method and   (2) Seismic refraction method.

1. Electrical resistivity method. In this method the sub-surface details are obtained by measuring the electrical resistivity of the sub-surface strata. The types of electrical resistivity methods that are generally used are: (a) Profiling method (b) Sounding method.

In the profiling method, the distance between the electrodes is kept constant, so the depth of investigation is fixed. This method is preferred when investigation has to be done over large areas at a particular depth.

In the sounding method, the spacing between the electrodes is not same. The spacing between the electrodes is gradually increased till the exploration depth is reached. This method is useful in studying the changes in the strata with the increase in depth.

2. Seismic refraction method. These methods are on the principle that the shock waves have different velocities in different soils. This method is very fast in finding the profiles of different strata provided the deeper layers have increasingly greater density and higher velocities.

Lateral Earth Pressure

One of the characteristics of soil is its tendency to exert a lateral pressure as that of liquids on the structures which are in contact with soils.

Lateral earth pressure is the force exerted by a soil mass on the side retained by a structure, for example, a retaining wall.

A retaining structure is a temporary or permanent structure which is used to provide lateral support to soil or other material.

Soil mass will be stable if it is laid at a slope flatter than the safe slope. But at some locations where the space is limited, the soil has to be retained at a slope steeper than the safe slope. In such circumstances, a retaining structure is necessary to provide a lateral support to the soil mass.

Retaining wall helps in maintaining the ground surface at different elevations on either side of it.

Water Pressure at Rest

The pressure at any point in water is same in all directions as shown in the Fig. 8.39. Both the lateral pressure and the vertical pressure at any depth 'H' will be given by $\gamma_w \cdot H$.

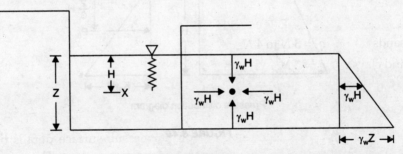

FIGURE 8.39

∴ Total force per unit length of wall

$$P = \text{Area of pressure distribution diagram} = (1/2)\,\gamma_w z \cdot z = \frac{\gamma_w z^2}{2}$$

This force will act at a distance of (2/3) z from top or (1/3) z from bottom *i.e.*, it acts at the centroid of the pressure distribution diagram.

Earth Pressures

In case of soil, the lateral pressure at any point will not be equal to the vertical pressure at that point, because soil possess shear strength but water has no shear strength.

Based on the deformation of the retaining wall, the lateral pressure can be grouped into three types:

1. At-rest earth pressure
2. Active earth pressure
3. Passive earth pressure.

1. At-rest earth pressure. The earth pressure at rest is the lateral pressure that exists in the soil deposits which have not been subjected to lateral yielding *i.e.*, when the soil is at rest. This case occurs when the retaining wall is firmly fixed at its top and is not allowed to move in lateral direction.

The at rest condition is also known as "elastic equilibrium condition", as no part of the soil mass has failed.

The lateral earth pressure intensity (p_0) at any depth in a soil is given by $p_0 = K_0 \cdot \gamma z$
where K_0 is coefficient of earth pressure at rest.

γ is unit weight of soil.

z is the depth at which the pressure is required. The expression for K_0 is $K_0 = \dfrac{\mu}{1-\mu}$

where μ is Poisson's ratio of the soil.

K_0 value can also be calculated from the expression $K_0 = (1 - \sin\theta)$, where θ angle of shearing resistance of soil.

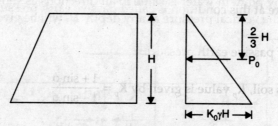

Pressure distribution diagram

FIGURE 8.40

Total earth pressure at rest $P_0 = (1/2) K_0 \gamma H.H = \dfrac{K_0 \gamma H^2}{2}$

and it acts at a distance of $(2/3) H$ from top as shown in Fig. 8.41.

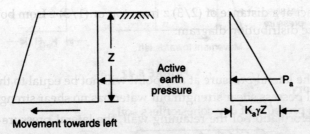

FIGURE 8.41

Active and passive earth pressures are the limiting conditions and represents the states of plastic equilibrium. A state of plastic equilibrium exists in a soil mass when every part of it is just on the point of failing in shear *i.e.*, they are just on the verge of failure.

Active earth pressure. This state occurs when the soil mass yields in such a way that it tends to stretch the soil mass horizontally. When a retaining wall moves away from the backfill, there will be stretching of the soil mass and the active state of earth pressure exists and it is a state of plastic equilibrium.

The lateral earth pressure at this condition is called active earth pressure and is given by

$$p_a = K_a \gamma Z$$

where K_a is coefficient of active earth pressure.

For a cohesionless soil, K_a value is given by $K_a = \dfrac{1 - \sin \phi}{1 + \sin \phi}$

Total active thrust $P_a = (1/2) K_a \gamma H^2$ and acts at a distance of $(2/3) H$ from top as shown in Fig. 8.41.

Passive earth pressure. This state occurs when the wall tends to compress the soil horizontally. When retaining wall moves towards the backfill, there will be a compression of the soil mass and a state of passive earth pressure exists and it is a state of plastic equilibrium.

The lateral pressure at this condition is called passive earth pressure and is given by

$$p_p = K_p \gamma H$$

where K_p is coefficient of passive earth pressure.

For a cohesionless soil, K_p value is given by $K_p = \dfrac{1 + \sin \phi}{1 - \sin \phi}$

Total passive pressure $P_p = \dfrac{1}{2} K_p \gamma H^2$ and acts at a distance of $\dfrac{2}{3} H$ from top as shown in Fig. 8.42.

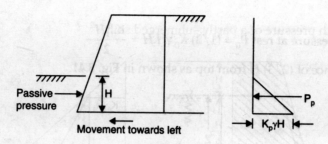

FIGURE 8.42

Rankine's Theory

(a) Active earth pressure of dry cohesionless soil.

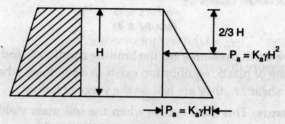

FIGURE 8.43

K_a = Coefficient of active earth pressure.
γ = Unit weight of the soil.
H = Height of soil mass.

(b) Active earth pressure of cohesionless soil carrying uniform surcharge:

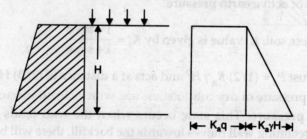

FIGURE 8.44

q = Surcharge intensity.

(c) Active earth pressure of a submerged soil:

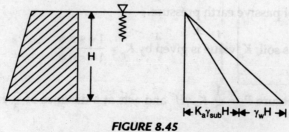

FIGURE 8.45

γ_{sub} = Submerged unit weight of soil.

(d) Active earth pressure of a partly submerged soil:

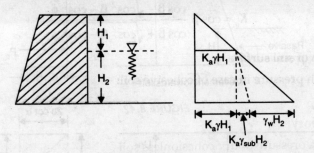

FIGURE 8.46

γ_w = Unit weight of water.

(e) Active earth pressure in case of a stratified soil:

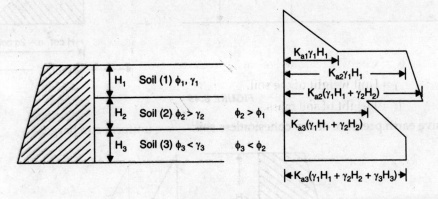

FIGURE 8.47

(f) Active earth pressure of dry cohesionless soil with inclined surface:

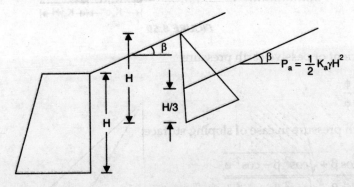

FIGURE 8.48

Pressure distribution diagram:

$$K_a = \cos\beta \, \frac{\cos\beta - \sqrt{\cos^2\beta - \cos^2\phi}}{\cos\beta + \sqrt{\cos^2\beta - \cos^2\phi}}$$

where β = Inclination of soil surface.

(g) Active earth pressure in case of cohesive soil:

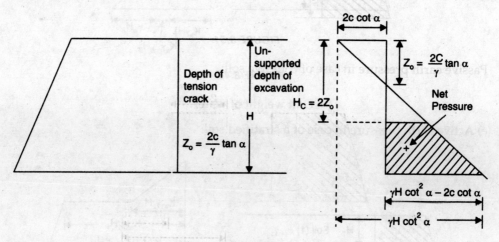

FIGURE 8.49

Passive earth pressure of dry cohesionless soil:

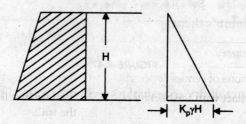

FIGURE 8.50

where K_p = Coefficient of passive earth pressure

$$= \frac{1 + \sin\phi}{1 - \sin\phi}$$

Passive earth pressure in case of sloping surface:

where $K_p = \cos\beta \, \dfrac{\cos\beta + \sqrt{\cos^2\beta - \cos^2\phi}}{\cos\beta - \sqrt{\cos^2\beta - \cos^2\phi}}$

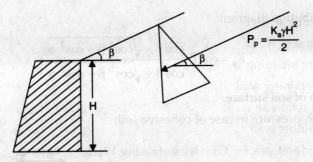

FIGURE 8.51

Passive earth pressure in case of cohesive soil:

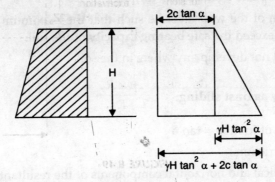

FIGURE 8.52

Rankine's theory Vs. Coulomb's theory:

Rankine theory	Coulomb theory
1. It is based on ideal conditions of semi-infinite soil.	1. Need not be such an ideal condition.
2. The face of the wall in contact with the backfill is smooth.	2. Wall friction is allowed on the face of the wall.
3. Yielding about a base point is necessary for both active and passive states to develop.	3. Yielding should ensure only a critical condition for a soil wedge to slide (or) torn off from the rest of the soil mass.
4. Potential sliding surfaces are planes making angles of $(45 - \phi/2)$ with the direction of is major principal stress.	4. Potential sliding surfaces are planes which make some angle with horizontal and dependent on angle of shearing resistance and angle of wall friction.

Retaining walls

A retaining wall is one of the important variety of retaining structures. These are relatively rigid walls which are used for supporting the soil mass laterally so that the soil can be retained at different elevations on both sides of the wall.

Types of retaining walls

(a) Gravity retaining walls.

(b) Semi-Gravity retaining walls.

(c) Cantilever retaining wall.

(d) Counterfort retaining wall.

(e) Buttress retaining wall.

Stability Considerations for Gravity Retaining Walls

The basic criteria for a satisfactory design of a retaining wall can be summarised as follows:

(a) The wall must be safe against sliding.

(b) The wall must be safe against over-turning.

(c) The base width of the wall must be such that the maximum pressure exerted on the foundation soil does not exceed the safe bearing capacity of the soil.

(d) Tension should not develop any where in the soil.

(a) **Factor of safety against sliding:** $\dfrac{\mu \times R_V}{R_H}$

where μ is the coefficient of friction = $\tan \delta$

δ = Angle of wall friction.

R_V and R_H are vertical and horizontal components of the resultant force.

A minimum factor of safety of 1.5 against sliding is generally preferred for the designs.

(b) **Factor of Safety against overturning**

$$= \dfrac{\text{Restoring moments about toe}}{\text{Overturning moments about toe}}$$

A minimum value of 1.5 for cohesionless soils and a value of 2.0 for cohesive soils is generally recommended for retaining walls.

(c) **Safety against bearing capacity failure**

The vertical component R_V of the resultant force must not exceed the bearing capacity of the soil.

The maximum pressure, $p_{max} = \dfrac{R_V}{b}\left(1 + \dfrac{6e}{b}\right)$

and the minimum pressure, $p_{min} = \dfrac{R_V}{b}\left(1 - \dfrac{6e}{b}\right)$

where b is the width of base and e is eccentricity.

Factor of safety against bearing capacity failure

$$= \dfrac{\text{Allowable bearing capacity}}{\text{Maximum pressure}}$$

A factor of safety of 3 is generally recommended for retaining walls.

(d) **No tension should develop**

When eccentricity is greater than $b/6$, then tension develops at the heel. Tensional forces are not desirable for soils which are of particulate medium, because soils have very less value of tensile strength.

BEARING CAPACITY

Foundation. It is a part of a structure which is in contact with the ground and it transmits the loads from the superstructure to the soil beneath it.

Footing. This is a portion of the foundation, which transmits loads directly to the foundation soil.

Foundations may be classified into two types:

(a) Shallow foundations   (b) Deep foundations.

A foundation is termed as a shallow foundation, if it is laid at a depth equal to or less than its width and it is termed as a deep foundation, if it is laid at a depth greater than its width. Any foundation must satisfy the following criteria:

(a) The foundation must be stable against shear failure.

(b) The settlement of the foundation must be within the safe limits.

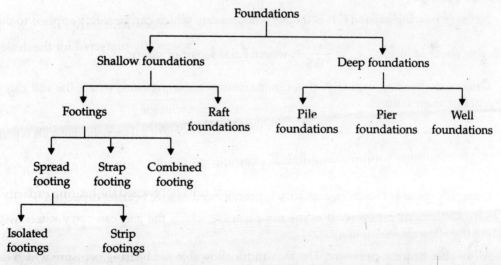

Isolated footings. These are a type of spread footings, where in they will be used to "Spread out" the load from footings.

Strip footings. These are the types of foundations which are used to "Spread out" the load from the walls.

Strap footings. It consists of two or more isolated footings connected by a beam called "Strap". This strap acts as a connecting beam and makes the two footings to behave as a single unit but does not take any soil reaction. This type of footings are required when the footing of an exterior column cannot extend into the adjoining property.

Combined footings. It supports two or more columns when the areas required for individual footings would overlap. It is also opted when the property line is so close to one column that a spread footing would be eccentrically loaded when kept entirely within the property line. By joining this column with another interior column, the load will be evenly distributed. The shape of this footing may be rectangular or trapezoidal in plan.

Raft foundations. These are also called as Mat foundations. It is a large footing (or) slab supporting several columns as well as walls under the structure. This type of foundations preferred when the allowable pressure of the soil is low (or) when the columns and walls of the structure are such that their individual footings would overlap.

Bearing capacity. The load carrying capacity of the foundation soil, which enables it to bear the loads of the structure.

Ultimate bearing capacity (q_{ult}). It is the maximum gross pressure at the base of the foundation at which the soil fails in shear.

Net bearing capacity ($q_{net\,ult}$). It is the maximum net pressure at which the soil fails in shear.

$$q_{net\,ult} = q_{ult} - \gamma D$$

where γ = Unit weight of the soil
 D = Depth of the foundation.

Net ultimate bearing capacity is the difference of ultimate bearing capacity and the overburden pressure.

Net safe bearing capacity. It is the net soil pressure which can be safely applied to the soil considering shear failure. $q_{net\,safe} = \dfrac{q_{net\,ult}}{F.S}$, where F.S is factor of safety.

Gross safe bearing capacity. It is the maximum gross pressure which the soil can carry safely without shear failure.

$$q_{safe} = q_{net\,safe} + \text{overburden pressure.}$$

$$= \dfrac{q_{net\,ult}}{F.S} + \gamma D$$

Generally gross safe bearing capacity is pronounced as safe bearing capacity.

Safe settlement pressure. It is the net pressure which the soil can carry safely without exceeding the allowable settlement.

Allowable bearing pressure. The maximum allowable net bearing pressure which can be applied on soil at which the soil neither fails in shear nor exceeds the allowable settlement. It is also called as allowable bearing capacity (or) allowable soil pressure.

Factors affecting bearing capacity. The ultimate bearing capacity of a foundation depends upon:

(1) Physical and engineering properties of the soil.
(2) Size, shape, depth and roughness of the footing.
(3) Location of the ground water table.
(4) Initial stresses, if any.

Determination of the ultimate bearing capacity

Ranakine's theory. According to Rankine, the ultimate bearing capacity is given by the equation:

$$q_{ult} = \gamma D_f \left(\frac{1 + \sin \phi}{1 - \sin \phi} \right)^2.$$

This equation is generally not preferred. This is rewritten to get the value of the minimum depth of the foundation.

$$(D_f)_{min} = \frac{q_{ult}}{\gamma} = \left(\frac{1 - \sin \phi}{1 + \sin \phi} \right)^2.$$

Terzaghi's bearing capacity theory

Terzaghi's bearing capacity equation for a strip footing

$$ult\ B.C. = q_{ult} = CN_c + \gamma D N_q + 0.5\ \gamma\ BN_\gamma$$

where C is the cohesion; γ is the unit weight of soil

D is the depth of the footing; B is the width

N_c, N_q and N_γ are the Terzaghi's bearing capacity factors.

The net ultimate bearing capacity will be given by

$$q_{net\ ult} = q_{ult} - \gamma D$$
$$= CN_c + (N_q - 1)\ \gamma D + 0.5\ \gamma\ BN_\gamma$$

Types of Shear failures

The three modes of shear failures under a footing are: (1) General shear failure, (2) local shear failure, and (3) Punching shear failure.

In the Terzaghi's B.C. equation N_c, N_q and N_γ refers to the condition of General shear failure.

For local shear failure condition, Terzaghi has proposed the following parameters for the soil.

Mobilised cohesion $C_m = \frac{2}{3} C$

Mobilised angle of shearing resistance: $\phi_m = \tan^{-1}\left(\frac{2}{3} \tan \phi \right)$

Bearing capacity equation for local shear failure in the case of a strip footing becomes:

$$q_{ult} = \frac{2}{3} C N_c' + \gamma D_f N_q' + 0.5\ \gamma\ BN_\gamma'$$

where N_c', N_q' and $N_{\gamma'}$ are the bearing capacity factors corresponding to local shear failure and these will depend upon mobilised angle of shearing resistance (ϕ_m).

Effect of water table on ultimate bearing capacity

The bearing capacity equation by Terzaghi is derived by the assumption that the **water table is located at a great depth**. If the water table is located near to the base of the foundation, the bearing capacity equation needs a small modification as shown below:

Ult. B.C. for a... $q_{ult} = CN_c + \gamma D_f N_q R_{W_1} + 0.5 \gamma BN_\gamma R_{W_2}$ strip footing

where R_{W_1} and R_{W_2} are the water table correction factors

which are given by $R_{W_1} = 0.5\left(1 + \dfrac{Z_{W_1}}{D_f}\right)$, $R_{W_2} = 0.5\left(1 + \dfrac{Z_{W_2}}{B}\right)$

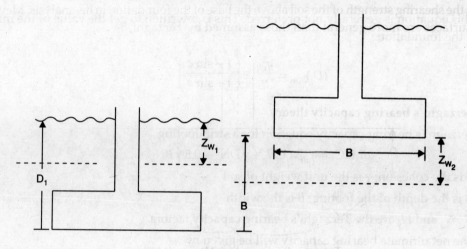

FIGURE 8.53

When the water table is at Ground surface

$Z_{W_1} = 0 \Rightarrow R_{W_1} = 0.5$

$Z_{W_2} = 0 \Rightarrow R_{W_2} = 0.5$

When the water tables is at the base of the foundation:

$Z_{W_1} = D_f \Rightarrow R_{W_1} = 1.0$

$Z_{W_2} = 0 \Rightarrow R_{W_2} = 0.5$

The maximum value of f_{W_1} is equal to depth of footing.

When the water table is below the base of the footing:

$Z_{W_1} = D_f = R_{W_1} = 1.0$

Z_{W_2} = Depth of GWT below base of footing.

R_{W_2} = Corresponding value from equation.

The maximum value of Z_{W_2} is the base of the foundation.

Bearing Capacity of Circular and Square Footing

Based on the experimental results, Terzaghi has proposed the following equation for determining the ultimate bearing capacity of circular and square footings:

q_{ult} for circular footing $= 1.3\, CN_c + \gamma D_f N_q + 0.3\, \gamma BN_\gamma$, where B refers to the diameter of the footing.

q_{ult} for square footing $= 1.3\, CN_c + \gamma D_f N_q + 0.4\, \gamma BN_\gamma$, where B refers to the side of the footing.

Meyerhof's Bearing Capacity Theory

Meyerhof considered the failure mechanism similar to that assumed by Terzaghi, but considers the shearing strength of the soil above the base of the foundation in his analysis. Meyerhof's rupture surfaces are more general than those assumed by Terzaghi.

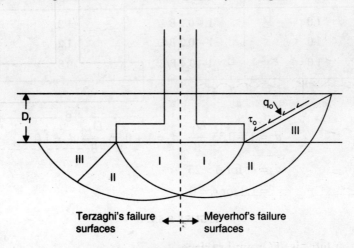

FIGURE 8.54

	Terzaghi's theory	Meyerhof's theory
Zone I	Elastic zone	Elastic zone
Zone II	Radial shear zone	Radial shear zone
Zone III	Rankine's passive zone	Mixed shear zone

According to Meyerhof, the equation for the bearing capacity will be

$$q_{ult} = CN_c + q_o N_q + 0.5\, \gamma BN_\gamma$$

where N_c, N_q and N_γ are Meyerhof's bearing capacity factors.

q_o is the normal stress on the equivalent free surface.

The value of N_c according to Meyerhof for a strip footing

$$N_c = 5.5 \cdot (1 + 0.25\, D_f/B)$$

For square or circular footing $N_c = 6.2(1 + 0.32\, D_f/B)$, where b is the side of a square or diameter of circular footing.

Brinch Hansen's Method

Brinch Hansen has proposed the following equation for the determination of ultimate bearing capacity.

$$q_{ult} = CN_c S_c d_c i_c + q\, N_q S_q d_q i_q + 0.5\, \gamma B\, N_\gamma S_\gamma d_\gamma i_\gamma$$

where S refers to the shape factor
d refers to the depth factor
i refers to the inclination factor
N_c, N_q and N_γ are the bearing capacity factors.

Shape factors

	Continuous footing strip footing (width = b)	Rectangular footing (size B × L)	Square footing (Size = B)	Circular footing (Diameter B)
S_c	1.0	1 + 0.2 B/L	1.3	1.3
S_q	1.0	1 + 0.2 B/L	1.2	1.2
S_γ	1.0	1 − 0.4 B/L	0.8	0.6

Depth factors

$$d_c = 1 + 0.35 \frac{D_f}{B} \quad d_q = 1 + 0.35 \frac{D_f}{B} \quad d_\gamma = 1.0$$

$d_q = d_c$ for $\phi > 25°$
    $= 1.0$ for $\phi = 0$.

Inclination factors

$$i_c = 1 - \frac{H}{2C_a BL} \quad i_q = 1 - 1.5\, H/V \quad i_\gamma = (i_q)^2$$

Limitation $\quad H \leq V \tan \delta + C_a BL$

where H and V are Horizontal and vertical components of the total load.

δ is the angle of friction between soil and foundation.

C_a is adhesion between soil and footing.

B is the size of the footing.

L is length of the footing parallel to H.

Eccentrically loaded foundations

In practice, foundations are some times subjected to moments in addition to the vertical loads, which in turn leads to eccentricity to the foundation. The maximum and minimum pressures in this case are

$$q_{max} = \frac{Q}{BL}\left(1 + \frac{6e}{B}\right) \text{ and } q_{min} = \frac{Q}{BL}\left(1 - \frac{6e}{B}\right)$$

where e refers to eccentricity $= \frac{M}{Q}$.

When $= B/6$, then $q_{min} = 0$; when $e > B/6$ – tension develops and that is not preferred for soil mass.

The maximum pressure q_{max} should be less than the safe bearing capacity.

Some times the vertical load will not be acting at the center of gravity of foundation. This in turn develops a moment. When the eccentricity is in both directions, the concept of useful width introduced by Meyerhof is used. According to him, the effective footing dimensions are

Effective length $\qquad L' = L - 2e_x$
Effective width $\qquad B' = B - 2e_y$
Effective area $\qquad A' = B' \times L'$

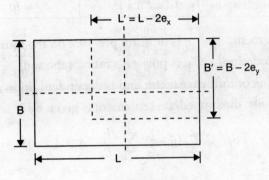

FIGURE 8.55

By Meyerhof's concept, the area of the footing which is symmetrical about the load is considered as useful.

Settlement Analysis

The settlement of footing refers to the sinking of a foundation due to vertical deformation of the underlying soil.

Settlements may be classified as uniform or total settlement, and non-uniform or differential settlement. If the different elements of a foundation structure undergo a constant settlement, then the foundation is said to have undergone a uniform settlement. If the different elements of a foundation undergo varied settlements then the foundation is said to be under non-uniform (or) differential settlement.

The factors which may result in a uniform settlement are (*i*) Elastic compression of soil (*ii*) Primary and secondary consolidation of soil (*iii*) Lowering of ground water table (*iv*) Underground erosion of soil.

The factors which will lead to a differential settlement are:

(*i*) Unequal loading, in other words, non-uniform pressure distribution on soil.

(*ii*) Overloading of adjacent land by heavy loads.

(*iii*) Non-homogeneous sub-soil conditions beneath the structure.

(*iv*) Variation in the amount of load transfer from columns to foundations.

Settlements of foundations

Analytical methods are available for computing the settlements of shallow foundations under a symmetrical vertical load.

Foundation settlement under a load can be classified into (a) Immediate settlement (b) Consolidation settlement (c) Secondary consolidation settlement.

Immediate settlement (S_i). Also called as elastic settlement and it occurs immediately (or) within a short period after application of the load.

In the case of cohesionless soils, both the elastic as well as primary compression settlements will occur almost together because of high permeabilities.

The immediate settlement of a flexible foundation in case of a cohesive soil is given by

$$S_i = q \cdot B \cdot \left(\frac{1-\mu^2}{E_s}\right) I_f$$

where S_i is immediate settlement, q is uniform pressure on the foundation,
B is width of the foundation, μ is poisson's ratio of the soil,
I_f is the influence value or influence factor and is dependent upon ratio of L/B.

In the cohesionless soils, the immediate settlement is given by

$$S_i = C_1 C_2 (\bar{q} - q) \sum_{Z>0}^{2B} \frac{I_z}{E_s} \cdot \Delta_z$$

C_1 = Correction factor for the depth of the foundation embodiement = $1 - 0.5 \left(\frac{q}{(\bar{q}-q)}\right)$

C_2 = Correction factor for creep in soils = $1 + 0.2 \log_{10}$ (time in years/0.1)

$\bar{q}$ = Pressure at the level of the foundation,

q = Surcharge, E_s = Modulus of elasticity,

I_f = Influence factor, Δ_z = Incremental thickness of soil.

The immediate settlement for cohesionless soils can be calculated with the help of either standard penetration test or dutch cone penetration test.

$$S_i = \frac{H}{C_s} \log_e \left(\frac{p + \Delta_p}{p}\right)$$

where H = Thickness of the layer

C_s = Compressibility constant = $1.5 \dfrac{C_r}{p}$

C_r = Static cone resistance (kN/m²)
   = 430 N to 1930 N (kN/m²)

where N is SPT number.

p = Effective overburden pressure at the centre of the layer before any excavation or application of load.

Δ_p = Increase in the pressure at the centre of the layer due to foundation load.

Consolidation settlement (S_c). This is the settlement due to the primary consolidation of the soil and it occurs due to the dissipation of excess pore water pressure in the soil. This settlement is generally determined by the expression given by Terzaghi as follows:

$$S_c = C_c \frac{H}{1+e_0} \log_{10}\left(\frac{p + \Delta_p}{p}\right)$$

where C_c = Compression index, p = Initial effective overburden pressure,

e_0 = Initial void ratio, H = Thickness of compressible layer,

Δ_p = Increase in the pressure due to the load.

Secondary consolidation settlement (S_s). This is the settlement due to secondary consolidation of the soil, which is believed to occur after the completion of primary consolidation.

Secondary consolidation settlement, $S_s = C_\alpha H_p \log\left(\dfrac{t_s}{t_p}\right)$

where C_α is coefficient of Secondary consolidation.

H_p = Thickness of soil mass after completion of primary consolidation.

t_s = Any time after completion of primary consolidation at which the intension of secondary consolidation is resorted.

t_p = Time to complete primary consolidation.

Total settlement of a footing = $S = S_i + S_c + S_s$.

PILE FOUNDATIONS

When the soil immediately beneath the foundation is too weak to support the structure, the depth of the foundation is increased till more suitable soil is met with, and these types of foundations are called as deep foundations. Among deep foundations, pile foundations, Pier foundations and Well foundations are common.

A Pile is a relatively small diameter shaft, which is driven or installed in the ground by suitable means.

Pier foundations and well foundations are relatively larger in cross-section than a Pile.

Pile foundations are commonly used in the following circumstances:

(a) When the structure is expected to carry large uplift loads.

(b) When the soil near the ground surface is weak to support the load from super-structure.

(c) When the foundation is expected to take lateral loads.

(d) To reduce differential settlements.

(e) When the foundation is expected to take eccentric loads and inclined loads.

(f) When the scouring of soil which is there immediately below a shallow foundation is expected to occur.

(g) When the foundation has to transmit the loads below the active zone in case of expansive soils.

Classification of Pile foundations

(a) **Based on the material used**

(i) Timber Piles   (ii) Concrete Piles

(iii) Steel Piles   (iv) Composite Piles.

(b) **Based on the method of load transfer**
  (i) End bearing Piles
  (ii) Friction Piles
  (iii) Combined end bearing and friction piles.

(c) **Based on the method of forming**
  (i) Cast-in-situ Piles
  (ii) Pre-cast Piles
  (iii) Prestressed Piles.

(d) **Based on the method of installation**
  (i) Driven Piles
  (ii) Bored Piles
  (iii) Jacked Piles
  (iv) Jetted Piles
  (v) Vibrated Piles.

(e) **Based on the Use**
  (i) Load bearing Piles
  (ii) Compaction Piles
  (iii) Tension Piles
  (iv) Load Test Piles
  (v) Sheet Piles
  (vi) Anchor Piles
  (vii) Laterally Loaded Piles.

(f) **Based on cross-section of Pile**
  (i) Circular Piles
  (ii) Square Piles
  (iii) Rectangular Piles
  (iv) Hexagonal Piles
  (v) Octagonal Piles.

(g) **Based on displacement of the soil**
  (i) Displacement Piles
  (ii) Non-displacement Piles.

Pile Driving. The hammers that are commonly used are:
  (i) Drop Hammer
  (ii) Single-acting hammer
  (iii) Double acting hammer
  (iv) Diesel hammer
  (v) Vibratory hammer.

Load Carrying Capacity of Piles

The ultimate bearing capacity of a pile is the maximum load which it can carry without failure of the foundation. The allowable load of a pile is the load which can be safely applied on the pile at which the soil neither fails in shear nor undergoes excessive settlement.

The following are the methods for determining the Pile Capacity:
  (i) Static methods
  (ii) Dynamic methods
  (iii) Penetration tests
  (iv) Load tests on Piles.

The first 2 are the theoretical approaches and the last 2 are practical or field approaches.

The ultimate load capacity of piles can be estimated by calculating the resistance derived from the end bearing and friction component of the total pile capacity.

Static Methods

$$\begin{Bmatrix} \text{Ultimate capacity} \\ \text{of a pile} \end{Bmatrix} = \begin{Bmatrix} \text{End-bearing} \\ \text{resistance} \end{Bmatrix} + \begin{Bmatrix} \text{Skin-frictional} \\ \text{resistance} \end{Bmatrix}$$

$$Q_{ult} = Q_p + Q_s$$
$$= q_p \cdot A_p + f_s \cdot A_s$$

q_p = Bearing capacity in point bearing for the pile.
A_p = Bearing area of the base of the pile.
f_s = Unit skin friction for the pile-soil system.
A_s = Surface area of the pile in contact with the soil.

The value of q_p can be calculated using the expression

$$q_p = CN_c + \gamma DN_q + 0.5 \gamma BN_r$$

which is the same form as the bearing capacity of shallow foundation.

Unit skin friction. The general form for the unit skin friction resistance is given by $f_s = C_a + \sigma_h \tan \delta$, where C_a = adhesion; σ_h = average lateral pressure of the soil against pile surface; δ = angle of wall friction.

For sandy soils .... $C_a = 0$

∴ Unit skin friction:

$$f_s = \sigma_h \tan \delta$$

where σ_h = Soil pressure normal to the pile surface.
$\tan \delta$ = Coefficient of friction between soil and pile material.

∴ $f_s = K \sigma_v \tan \delta$; since $\sigma_h = Kv$

Values of tan δ

Material	tan δ	(δ) value
Wood	0.4	0.67φ
Concrete	0.45	0.75φ
Steel	0.2 to 0.4	20°

∴ Ultimate load capacity for Piles in sands:

$$Q_{ult} = (\gamma DN_q + 0.5 \gamma BN_r) A_p + (K \sigma_v \tan \delta) A_s$$

Since the value of $0.5 \gamma BN_r$ will be very small value when compared to the second term γDN_q, the equation can be written as $\gamma DN_q \cdot A_p + K \sigma_v \tan \delta A_s$.

For cohesive soils:

$$f_s = C_a \quad \text{Since } \tan \delta = 0$$

The adhesion value is given by $C_a = \alpha \cdot C$
where α is the adhesion factor
C is cohesion of the soil.

∴ $$Q_{ult} = (CN_c + \gamma D) A_p + C_a A_s$$

(For $\phi = 0$, $N_q = 1$ and $N_r = 0$)

Dynamic Methods

The load carrying capacity of a pile is estimated from the resistance of the pile to the penetration during driving. These formulae are derived based on the principle that the energy input from the pile hammer is equal to the energy used to drive the pile plus the energy losses. The most widely used dynamic formulae are:

(1) **Engineering News Record formula (ENR formula).**

Ultimate load
$$Q_{ult} = \frac{W.H.\eta_h}{S+C}$$

where W is the weight of hammer, h is height of fall of hammer

η_h is efficiency of hammer.

For drop hammers:
$$\eta_h = 0.7 \text{ to } 0.9$$

For steam hammers:
$$\eta_h = 0.75 \text{ to } 0.85$$

S is penetration of Pile per hammer blow

C is constant (for drop hammer $C = 2.54$ cm and for steam hammers $C = 0.254$ cm)

A factor of safety of 6.00 is generally used to calculate the allowable load capacity.

$$Q_{all} = \frac{Q_{ult}}{FS} = \frac{W_h \, nh}{F.S.(S+C)}$$

The product $W.h$ can be replaced by the rated energy of the hammer (E)

$$Q_{all} = \frac{E.\eta_h}{FS(S+C)}$$

Hiley's formula

Ultimate load
$$Q_{ult} = \frac{W.h.\eta_h.\eta_b}{S+C/2}$$

η_b = Efficiency of hammer blow

S = Penetration per hammer blow (cm)

C = Constant = $C_1 + C_2 + C_3$

C_1 = Temporary compression of dolly and packing

= $1.77 \, Q/A$, where the driving is without dolly or helmet

= $9.05 \, Q/A$, where the driving is with a short dolly and helmet

C_2 = Temporary compression of pile = $0.0657 \, QL/A$

C_3 = Temporary compression of soil = $3.55 \, Q/A$

where Q is ultimate driving resistance

L is length of Pile in metres

A is cross-sectional area of pile in cm^2.

The efficiency of hammer blow (η_b) depends upon the weight of hammer (W); weight of pile and helmet (p) and coefficient of resitution (e).

When $W > ep$ $\quad \eta_b = \dfrac{W + e^2 p}{W + p}$

When $W < ep$ $\quad \eta_b = \dfrac{W + e^2 p}{W + p} - \left(\dfrac{W - ep}{W + P}\right)^2$

Danish Formula

$$\text{Ultimate load} = Q_{ult} = \dfrac{W_h\, \eta_h}{S + C/2}$$

where C is constant and is elastic compression of pile.

$$C = \sqrt{\dfrac{2\,\eta_h (WhL)}{AE_p}}$$

where L is length of Pile.

W and h are weight and height of fall of hammer.

A is cross-sectional area of pile.

E_p is modulus of elasticity of pile material.

The allowable load is found using a factor of safety of 3 to 4.

Disadvantages of dynamic formulae are:

(A) The static resistance is not necessarily equal to the dynamic resistance.

(B) These formulae does not give any estimation about the settlements of soil.

(C) The assumption of free impact between the two bodies is not justified, since the pile is not a free body.

In-situ Penetration Tests

Meyerhof has proposed the following equation for calculating the carrying capacity of a pile in granular soils using SPT and cone resistance values.

Point bearing resistance $\quad q_p = q_c$ (static cone test)

$\hspace{8em} = 4\,N$ (SPT value test)

$\therefore \hspace{4em} q_p = q_c = 4\,N$ (q_p and q_c will be in kg/cm^2)

q_c = Static cone resistance value at bearing stratum.

N = Standard penetration test value at the bearing stratum.

Skin friction resistance, $f_s = q\,\dfrac{q_{c\,\text{ave.}}}{200}$ (static cone test)

$\hspace{6em} = \dfrac{N_{\text{ave.}}}{50}$ (SPT value test)

$$\therefore \quad f_s = \frac{q_c}{200} = \frac{N_{ave.}}{50} \quad (q_c \text{ and } f_s \text{ will be in kg/cm}^2)$$

$q_{c\ ave.}$ = Average cone resistance in kg/cm^2

$N_{ave.}$ = Average value of N along the length of Pile.

Negative Skin friction

When the soil layer surrounding a portion of the pile shaft settles more than the pile, a downward drag occurs on the pile. This downward drag is called as negative skin friction.

Negative skin friction develops when a soft or loose soil surrounding the pile settles after the pile has been installed as shown in Fig. 8.56.

Piles installed in freshly placed fills of soft compressible deposits as shown in Fig. 8.56 are subjected to a downward drag which is called as negative skin friction.

The magnitude of negative skin friction for a single pile in a filledup soil deposit can be calculated from the expression.

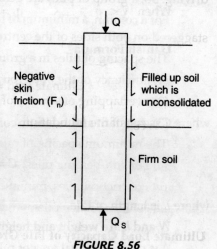

FIGURE 8.56

Negative skin friction $F_n = C_a \times L_c \times P$

where C_a is unit adhesion = αC_u

C_u is undrained cohesion for the compressible layer.

L_c is length of pile in the compressible layer

P is perimeter of pile.

In cohesionless soils
$$F_n = 1/2\ k \cdot \gamma \cdot L_c^2 \tan \delta\ P$$

K is coefficient of earth pressure.

f is angle of wall friction.

When a pile group passes through a soft, unconsolidated stratum, the higher of the following value is used in the design.

$$F_{ng} = F_n \times n$$
$$= C_u L_c P_g + \gamma L_c A_g$$

where F_n = Negative skin friction value of a single pile

P_g = Perimeter of the group

γ = Unit weight of the compressible soil.

A_g = Area of the pile group.

The effect of negative skin friction on the factor of safety against ultimate pile load capacity of a pile can be found by equation.

$$\text{Factor of Safety} = \frac{\text{Ultimate pile load capacity}}{\text{Working load} + \text{Negative skin friction}}$$

GROUP ACTION OF PILES

Single piles are rarely used as foundation members. Generally a group of piles, connected together at the top either by a beam or a slab called a Pile Cap, is used.

Generally the driven piles are not used as single piles, because of the chance for the occurrence of the eccentricity of piles due to the tendency of pile to wander in the lateral direction during driving. So, a group of piles are generally preferred.

For a column, a minimum of three number of piles are used and for walls, piles are generally staggered on both sides of the centre line of the wall in a uniform way.

The spacing of piles in a group depends generally on

(1) efficiency of the pile group
(2) overlapping of stresses of piles in the group
(3) cost of the foundation.

The minimum spacing of piles as recommended by IS Code is

(a) For end-bearing piles; Minimum spacing = $2.5\,d$

(b) For friction piles, minimum spacing = $3.0\,d$

(c) For compaction piles which are driven in loose sands or fillings, minimum spacing = $2.0\,d$.

Ultimate Load Capacity of Pile Groups

In the case of pile groups in compression, the ultimate load capacity of a pile group need not necessarily equal to the sum of the individual load capacities of the individual piles in the group.

If piles are spaced sufficient distance apart, then the capacity of the pile group is the sum of the individual capacities of the pile.

The ratio of the ultimate load capacity of the pile group $Q_{ult}(g)$ to the sum of the individual load capacities of the piles in the group, is called the group efficiency (η_g)

$$\eta_g = \frac{Q_{ult}(g)}{N\, Q_{ult}(i)}$$

where N is number of piles in the group

$Q_{ult}(i)$ is the ultimate load capacity of a single pile.

The Converse-Labarre formula that is used for the determination of group efficiency of piles is

$$\eta_g = 1 - \frac{\theta}{90}\left(\frac{m(n-1)+n(m-1)}{mn}\right)$$

where m = Number of rows of piles

n = Number of piles in a row

$\theta = \tan^{-1}(d/s)$ in degrees

where d is diameter of pile, s is spacing of piles.

Settlement of a Pile group

The settlement of a pile group is more than the settlement of a single pile even when the loads on the single pile and on each individual pile of the pile group are the same. It is due to overlapping of stresses due to the piles in the group. The zone of influence for a pile group is much larger and deeper than for a single pile.

Pile groups in Sand

The settlement of a pile group in sand can be determined by using skempton's equation which relates the settlement of pile group with that of an individual pile as follows:

$$\frac{S_g}{S_i} = \left(\frac{4B + 2.7}{B + 3.6}\right)^2$$

where S_g = Settlement of the pile group at an average load of Q per pile
S_i = Settlement of a single pile at a load Q
B = Width of pile group in metres.

Meyerhof expressed the ratio of settlement of a pile group to that of a single pile for square pile groups driven in sand as

$$\frac{S_g}{S_i} = \frac{S(5 - S/3)}{(1 + 1/\gamma)^2}$$

where S = Ratio of pile spacing to pile diameter.
r = Number of rows for square pile group

The settlement of a single pile can be obtained from Pile load test results.

Settlement of pile groups in clay

The common approach to calculate the settlement of pile groups in clays in the "Equivalent footing approach". In this approach, a fictitious footing is assumed at some arbitrary depth in the soil and the settlement of this footing is calculated. The settlement of this equivalent footing is assumed to be equal to the settlement of the pile group.

$$\text{Consolidation settlement} = C_c \cdot \frac{H}{1 + C_0} \log_{10}\left(\frac{p + \Delta p}{p}\right)$$

where H is the thickness of the compressible stratum. Initial overburden pressure (p) and the stress increase (Δp) are calculated at the middle of the compressible soil.

Pile Groups in Clay

Experimental results have indicated that a group of piles may fail in one of following two ways:

(a) By individual pile failure.
(b) By block failure.

The ultimate load capacity of the pile group $[Q_{ult}(g)]$ based on individual pile failure is given by

$$Q_{ult}(g) = N * Q_{ult}(1)$$

where $Q_{ult}(1)$ is ultimate load capacity of a single pile.

In block failure, the soil bound by the perimeter of the pile group and the embedded length of the pile acts as one unit or a block.

$$Q_{ult}(g) = C N_c A_p(g) + \alpha C A_s(g)$$

$A_p(g)$ = Cross-sectional area of block

$A_s(g)$ = Surface area of block which is in contact with soil.

Pile Groups in Sand

The group efficiency of a friction pile group in sand is generally believed to be more than one at closer spacings, due to compaction of sand by pile driving operation. The efficiency value approaches unity when the pile spacing is increased to 5 to 6 times the diameter. This is true only in loose sands. In dense sands the group efficiency is less than one. An efficiency value of unity is generally taken for design purposes, in other words, the load capacity of the pile group is equal to the sum of individual pile load capacities.

II. OBJECTIVE TYPE QUESTIONS

1. Who coined the term Soil Mechanics?
   - (a) Terzaghi
   - (b) Cassagrande
   - (c) Newmark
   - (d) Rankine.

2. To what category do the gravel and sand belong?
   - (a) cohesionless soils
   - (b) cohesive soils
   - (c) marine soils
   - (d) expansive soils.

3. Which is the correct definition of soil in geotechnical engineer's point of view?
   - (a) Soil is the top surface of the earth where plants can grow.
   - (b) Soil is the unconsolidated material consisting particles produced by disintegration of rock which may or may not contain organic matter.
   - (c) Soil is the relatively thin surface zone which can contain moisture.
   - (d) None of the above.

4. Pick up the correct sequence of geological cycle of formation of soil
   - (a) transportation—upheaval—deposition—weathering
   - (b) transportation—deposition—weathering—upheaval
   - (c) weathering—upheaval—deposition—transportation
   - (d) weathering—transportation—deposition—upheaval.

5. Chemical weathering occurs because of
   - (a) oxidation
   - (b) carbonation
   - (c) hydration
   - (d) all the above.

6. Physical weathering occurs due to
   (a) temperature changes
   (b) wedging action of ice
   (c) spreading of roots of plants
   (d) all the above.
7. If the soil stays at a place above the parent rock where it is produced, then it is called
   (a) stationary soil
   (b) static soil
   (c) residual soil
   (d) immobile soil.
8. The soil which contains 'finest particles' is
   (a) Silt
   (b) Gravel
   (c) Clay
   (d) Sand.
9. Generally soil refers to
   (a) solid medium
   (b) liquid medium
   (c) particulate medium
   (d) gaseous medium.
10. An agent responsible for the transportation of soil is
    (a) wind
    (b) water
    (c) gravity
    (d) all the above.
11. Cohesionless soils are formed due to
    (a) oxidation of rocks
    (b) leaching action of water on rocks
    (c) blowing of hot and cold wind
    (d) physical disintegration of rocks.
12. Loess is
    (a) over consolidated clay
    (b) fine sand
    (c) wind borne soil
    (d) marine soil.
13. Peat is composed of
    (a) clay and sand
    (b) decayed vegetable matter
    (c) inorganic silt and silty clay
    (d) synthetic chemicals.
14. Kaolinite, Illite and Montmorillonite represent
    (a) residual soils
    (b) cohesionless soils
    (c) structural arrangement of soils
    (d) clay minerals.
15. China clay is also called
    (a) Kaolin
    (b) Illite
    (c) Montmorillonite
    (d) none of the above.
16. Bentonite contains predominantly
    (a) Kaolinite
    (b) Illite
    (c) Montmorillonite
    (d) none of the above.
17. Loam means
    (a) silt with little sand
    (b) sand with little clay
    (c) clay with little silt
    (d) mixture of sand, silt and clay sized particles in equal porportions.

18. Honey combed structure is found in
    (a) gravels
    (b) coarse sands
    (c) fine silts and clays
    (d) highly plastic clays.
19. Soil transported by water and deposited at the bottom of the lake is known as
    (a) alluvial soil
    (b) lacustrine soil
    (c) loess
    (d) dune sand.
20. Agent of transportation for alluvial soils is
    (a) water
    (b) gravity
    (c) wind
    (d) ice.
21. Soil is considered as
    (a) single phase system
    (b) two phase system
    (c) three phase system
    (d) none of the above.
22. Soil can exist in
    (a) three phase system
    (b) two phase system
    (c) single phase system
    (d) all the above.
23. In the unit phase diagram for a soil mass
    (a) total volume is taken as unity
    (b) volume of water is taken as unity
    (c) volume of soil solids is taken as unity
    (d) none of the above.
24. Which of the following is a glacial soil?
    (a) Transported by running water
    (b) Deposited at the bottom of the lakes
    (c) Deposited in sea water
    (d) None of the above.
25. Talus is the soil transported by
    (a) gravitational force
    (b) water
    (c) glacier
    (d) wind.
26. If a soil suffers a change in volume by the application of external loads but recovers its volume immediately after the load gets removed, such a soil is termed as
    (a) compressible soil
    (b) plastic soil
    (c) elastic soil
    (d) perfect soil.
27. Shape of clay particles is
    (a) rounded
    (b) rectangular
    (c) needle like
    (d) none.
28. For a cohesive soil, the modulus of elasticity
    (a) varies linearly with depth
    (b) is constant with depth
    (c) can't say
    (d) none of the above.
29. For a cohesionless soil, the modulus of elasticity
    (a) varies linearly with depth
    (b) is constant with depth
    (c) can't say
    (d) none of the above.

30. Which of the following is a cohesionless soil?
    (a) sand
    (b) silt
    (c) clay
    (d) silt-clay.

31. Finest grain particles are contained in
    (a) fine sand
    (b) silt
    (c) clay
    (d) none of the above.

32. Which of the following statements is true?
    (a) In a dry soil all the voids are filled with air.
    (b) In a saturated soil all the voids are filled with water.
    (c) In a partly saturated soil the voids are occupied by both air and water.
    (d) All the above.

33. The ratio of volume of voids to the total volume of a given soil is
    (a) void ratio
    (b) porosity
    (c) air content
    (d) air ratio.

34. The ratio of total weight of saturated soil to its total volume is called
    (a) saturated unit weight
    (b) submerged unit weight
    (c) buoyant unit weight
    (d) wet unit weight.

35. Submerged unit weight of soil is equal to the saturated density
    (a) plus unit weight of water
    (b) minus unit weight of water
    (c) multiplied by unit weight of water
    (d) divided by unit weight of water.

36. The specific gravity of a soil is the ratio of the unit weight of the soil solids to that of water at a temperature of
    (a) 27°C
    (b) 36°C
    (c) 4°C
    (d) 17°C.

37. The principle involved in the relation $\gamma_{sub} = \gamma_{sat} - \gamma_w$ is
    (a) Stoke's law
    (b) Archimedes principle
    (c) Darcy's law
    (d) All the above.

38. The most commonly used specific gravity of a soil in soil mechanics is
    (a) mass sp. gr.
    (b) apparent sp. gr.
    (c) grain sp. gr.
    (d) bulk sp. gr.

39. If kerosene is used instead of water in the Pycnometer method of determination of grain sp.gr., the expression for the grain sp. gr.
    (a) should be added to the sp. gr. of kerosene
    (b) should be multiplied by the sp. gr. of kerosene
    (c) should be divided by the sp. gr. of kerosene
    (d) should be subtracted by the sp. gr. of kerosene.

40. Specific gravity of soil is
    (a) same for clays and sands
    (b) determined by hydrometer
    (c) less than 2.0 for most soils
    (d) more than 2.5 for most soils.

41. The ratio of weight of water to the total weight of the soil is known as
    (a) water content
    (b) degree of saturation
    (c) saturated density
    (d) none of the above.

42. Any change in moisture content of a soil changes
    (a) value of angle of shearing resistance
    (b) strength of soil
    (c) amount of compaction required
    (d) all the above.

43. The trace moisture absorbed by a dry **soil from** the atmosphere is called
    (a) moisture content
    (b) water content
    (c) hygroscopic moisture
    (d) none of the above.

44. The maximum temperature recommended in the oven dry method for water content determination is
    (a) 80°C
    (b) 100°C
    (c) 110°C
    (d) 130°C

45. Water content of a soil can
    (a) take a value only from 0% to 100%
    (b) less than 0%
    (c) be greater than 100%
    (d) never be greater than 100%.

46. The ratio of weight of water to the weight of solids is called
    (a) degree of saturation
    (b) water content
    (c) void ratio
    (d) porosity.

47. Valid range for degree of saturations of a soil in percentage
    (a) $0 < S < 100$
    (b) $0 \leq S \leq 100$
    (c) $0 > S > 100$
    (d) $S \leq 0$.

48. The ratio of volume of water present in a given soil mass to the volume of voids in it is known as
    (a) percentage voids
    (b) porosity
    (c) degree of saturation
    (d) water content.

49. Density index for a natural soil is used to express
    (a) percentage voids
    (b) relative compactness
    (c) shear strength of clays
    (d) specific gravity.

50. Core-cutter method is used for
    (a) determining density of soil
    (b) obtaining samples for direct shear test
    (c) determining bearing capacity of soil
    (d) compacting soil.

51. Theoretically, the void ratio in a soil can have a value of
    (a) more than one only
    (b) less than one only
    (c) equal to one
    (d) can be less (or) more than one.

52. Relationship between void ratio e and porosity n is given by
    (a) $e = n(1 + n)$
    (b) $e = n(1 + e)$
    (c) $e = n(1 - e)$
    (d) $e = n(1 + n)$.

53. In hydrometer analysis the principle used is
    (a) Newton's law
    (b) Darcy's law
    (c) Stoke's law
    (d) Reynold's law.
54. Stoke's law gives the following velocity of a spherical particle falling freely in an infinite liquid medium
    (a) falling velocity
    (b) actual velocity
    (c) terminal velocity
    (d) all the above.
55. The soil sample for the hydrometer analysis is pretreated with the following agent
    (a) Oxidising agent
    (b) Reduction agent
    (c) Sulphuric agent
    (d) All the above.
56. In a hydrometer analysis the reason for addition of hydrogen peroxide and heating it is to remove
    (a) water present
    (b) air present
    (c) organic matter
    (d) calcium compounds.
57. In a hydrometer analysis for removing calcium compounds the following acid is used
    (a) 0.1 N HCl
    (b) 0.2 N HCl
    (c) 0.1 N H_2SO_4
    (d) 0.2 N H_2SO_4.
58. The deflocculating agent which is used in the sedimentation analysis is
    (a) Hydrochloric acid
    (b) Hydrogen peroxide
    (c) Sodium hexa meta phosphate
    (d) Sodium chlorite.
59. Hydrometer is a device which is used to measure
    (a) temp. of liquids
    (b) density of liquids
    (c) sp. gr. of liquids
    (d) all the above.
60. The range of values printed on a hydrometer are
    (a) 1.0 to 1.04
    (b) 0.995 to 1.04
    (c) 0.9 to 1.04
    (d) none.
61. In the hydrometer analysis, the effective depth will be
    (a) constant
    (b) goes on increasing
    (c) goes on decreasing
    (d) none.
62. In the pipettle analysis, the sampling depth will be
    (a) constant
    (b) goes on increasing
    (c) goes on decreasing
    (d) none.
63. The correction to be applied to hydrometer reading is
    (a) Meniscus correction
    (b) Temperature correction
    (c) Deflocculating agent correction
    (d) all the above.
64. The composite correction to be applied for in a hydrometer reading is
    (a) $C = C_t - C_m \pm C_d$
    (b) $C = C_t \pm C_m - C_d$
    (c) $C = C_m - C_d \pm C_t$
    (d) none
    where C_t = Temp. correction, C_m = Miniscus correction
           C_d = Deflocculating correction.

65. Hydrometers are usually calibrated at a temperature of
    (a) 4°C
    (b) 20°C
    (c) 24°C
    (d) 27°C.

66. If the temperature at the time of test is more than that of calibration of the hydrometer, the hydrometer reading will be
    (a) less
    (b) more
    (c) same
    (d) none.

67. With the increase of deflocculating agent, the density of soil suspension will
    (a) decrease
    (b) increase
    (c) be constant
    (d) none.

68. Stoke's law is applicable when the effective diameter of the particles is less than
    (a) 0.0002 mm
    (b) 0.002 mm
    (c) 0.02 mm
    (d) 0.2 mm.

69. Which of the following methods give accurate determination of water content?
    (a) sand bath method
    (b) alcohol method
    (c) calcium carbide method
    (d) oven drying method.

70. Equation for bulk unit weight γ in terms of specific gravity G, specific weight of water γ_w, voids ratio e, and degree of saturation S is written as
    (a) $\gamma = \dfrac{\gamma_w(G+eS)}{(1+e)}$
    (b) $\gamma = \dfrac{\gamma_w(G-eS)}{(1+e)}$
    (c) $\gamma = \dfrac{\gamma_w(G+eS)}{(1-e)}$
    (d) $\gamma = \dfrac{\gamma_w(G-eS)}{(1-e)}$.

71. The relation between void ratio e, specific gravity G, water content w, and degree of saturation S is given by
    (a) $e = \dfrac{wG}{S}$
    (b) $e = \dfrac{SG}{w}$
    (c) $e = \dfrac{wS}{g}$
    (d) $e = wGS$.

72. The water content of a saturated soil is 50%. If the specific gravity of the solids is 2.4, what is its void ratio?
    (a) 0.6
    (b) 1.2
    (c) 1.8
    (d) 2.4.

73. A moist sample of soil weighs 24 g in a tin lid. The tin lid above weighs 14 g. Oven dry weight of tin and sample is 22 g. What is the water content of the soil?
    (a) 12.5%
    (b) 37.5%
    (c) 25%
    (d) 50%.

74. Using standard notation, the degree of saturation is given by
    (a) $S = \dfrac{V_w}{V_v}$
    (b) $\dfrac{V_v}{V_w}$
    (c) $\dfrac{V_w}{V}$
    (d) $\dfrac{V}{V_w}$.

75. If W and W_s denote the total weight and weight of water in the sample, the water content is given by
   (a) $1 - \dfrac{W}{W_s}$
   (b) $1 + \dfrac{W}{W_s}$
   (c) $\dfrac{W}{W_s} - 1$
   (d) $\dfrac{W}{W_s} + 1$.

76. The ratio of volume of air V_a to the volume of voids V_v is called
   (a) degree of saturation
   (b) degree of aeration
   (c) air content
   (d) none of the above.

77. Air content and degree of saturation are related by
   (a) $a_c + S = 0$
   (b) $a_c = 1 + S$
   (c) $a_c + S = 1$
   (d) $a_c = S$.

78. Percentage air voids and porosity are related by
   (a) $n = n_a \times a_c$
   (b) $n_a = n \times a_c$
   (c) $a_c = n + n_a$
   (d) $a_c = n - n_a$.

79. A soil has a bulk density of 2.4 g/cm³ and water content of 20%. What is its dry density?
   (a) 2 g/cm³
   (b) 1.2 g/cm³
   (c) 1.4 g/cm³
   (d) 1.6 g/cm³.

80. The basic relationship between water content w, bulk density γ, and dry density γ_d is given by
   (a) $\gamma_d = \dfrac{\gamma}{1-w}$
   (b) $\gamma_d = \dfrac{1+w}{\gamma}$
   (c) $\gamma_d = \dfrac{\gamma}{1+w}$
   (d) $\gamma_d = \dfrac{1-w}{\gamma}$.

81. To obtain grain size distribution, hydrometer analysis is appropriate for
   (a) peats
   (b) sands and gravels
   (c) silts and clays
   (d) all soils.

82. The minimum size of silt particles is
   (a) 0.002 mm
   (b) 0.04 mm
   (c) 0.06 mm
   (d) 0.03 mm.

83. The assumption in sedimentation analysis is as follows
   (a) soil particles are spherical
   (b) particles settle independent of other particles
   (c) walls of the jar where in settlement analysis is done do not affect the settlement
   (d) all the above.

84. The maximum size of clay particles is
   (a) 0.002 mm
   (b) 0.04 mm
   (c) 0.06 mm
   (d) 0.08 mm.

SOIL MECHANICS

85. The relative density of the soil is given by
    (a) $\dfrac{e_{max} - e}{e_{max} + e_{min}}$
    (b) $\dfrac{e_{max} - e}{e_{max} - e_{min}}$
    (c) $\dfrac{e_{max} + e}{e_{max} + e_{min}}$
    (d) $\dfrac{e_{max}}{e_{min}}$.

86. Textural classification is merely based on
    (a) grain size
    (b) plasticity index
    (c) shape of particles
    (d) consistency limits.

87. The effective size of the soil is
    (a) D_{15}
    (b) D_{85}
    (c) D_{10}
    (d) D_{90}.

88. Uniformity co-efficient is the ratio of
    (a) D_{10} to D_{60}
    (b) D_{60} to D_{10}
    (c) D_{30} to D_{60}
    (d) D_{60} to D_{30}.

89. $(D_{30})^2 / (D_{60} \times D_{10})$ is known as
    (a) uniformity co-efficient
    (b) co-efficient of curvature
    (c) gradation co-efficient
    (d) none of the above.

90. The symbol SP indicates
    (a) sands
    (b) poorly graded sand
    (c) well graded sands
    (d) none.

91. The symbol GM indicates
    (a) gravels
    (b) gravelly silts
    (c) silty gravels
    (d) silts.

92. The symbol SC indicates
    (a) sands
    (b) sandy clay
    (c) clay sands
    (d) clays.

93. The symbol CL indicates
    (a) silts
    (b) clays
    (c) inorganic clays
    (d) organic clays.

94. The symbol OH indicates
    (a) other soils
    (b) clay of high plasticity
    (c) silt of high plasticity
    (d) organic silts and organic silts of high plasticity.

95. Peat is nothing but
    (a) organic silts
    (b) organic clay
    (c) inorganic silt
    (d) inorganic clay.

96. Generally, when compacted the GW type of soils will be
    (a) pervious
    (b) very pervious
    (c) non-pervious
    (d) cannot say.

97. The workability of the following soil will be excellent as a construction material
    (a) GW
    (b) GP
    (c) GM
    (d) GC.
98. The compressibility of the following soil will be excellent when compacted and saturated
    (a) GW
    (b) GP
    (c) GM
    (d) GC.
99. The shearing strength of the following soil will be excellent when compacted and saturated
    (a) SW
    (b) SP
    (c) SM
    (d) SC.
100. The permeability of the following soil will be almost zero
    (a) GW
    (b) SP
    (c) SW
    (d) GC.
101. The shearing strength of the following soil will be poor
    (a) CL
    (b) OL
    (c) SC
    (d) GC.
102. The compressibility of the following soil is some what less
    (a) OL
    (b) MH
    (c) CH
    (d) OH.
103. The workability of the following soil is better
    (a) OL
    (b) MH
    (c) CH
    (d) OH.
104. Soil in which some of the intermediate size particles are missing is known as
    (a) poorly graded soil
    (b) non-uniform soil
    (c) ill proportioned soil
    (d) skip graded soil.
105. Soil which contains the particles of different sizes in good proportion is called
    (a) uniform soil
    (b) well graded soil
    (c) consistent soil
    (d) none of the above.
106. For a well graded soil the co-efficient of curvature will be between
    (a) 1 and 10
    (b) 2 and 8
    (c) 3 and 7
    (d) 1 and 3.
107. D_{10} of the soil is the diameter in mm such that
    (a) 10% of the soil is coarser than this value
    (b) 10% of the soil is finer than this value
    (c) this value has no bearing on particle size distribution
    (d) none of the above.
108. The groove which is cut for the determination of liquid limit has to flow a distance of
    (a) 10 mm
    (b) 11 mm
    (c) 12 mm
    (d) 13 mm.

109. The drop of the cup for the liquid limit determination is
    (a) 10 mm
    (b) 11 mm
    (c) 12 mm
    (d) 13 mm.

110. At liquid limit a soil has
    (a) high shear strength
    (b) no shear strength
    (c) negligible (or) very small shear strength
    (d) nothing to do with shear strength.

111. When the plastic limit cannot be determined, the soil is reported to be
    (a) unplastic
    (b) plastic
    (c) semi-plastic
    (d) non-plastic.

112. A soil consisting of approximately equal percentage of sand, silt and clay is referred to as
    (a) well soil
    (b) perfect soil
    (c) poor soil
    (d) loam.

113. In the field, coarse grained soils and fine grained soils are distinguished based on
    (a) the water content
    (b) the sp. gr.
    (c) the colour
    (d) appearance of individual grains to the naked eye.

114. In the field, sand can be distinguished from silt by
    (a) Shaking test
    (b) Dilatancy test
    (c) Dispersion test
    (d) Rolling test.

115. Shaking test is also called as
    (a) Dilatancy test
    (b) Dispersion test
    (c) Rolling test
    (d) all.

116. In the field, dilatancy test is used to differenciate clay from
    (a) Gravels
    (b) Sands
    (c) Silts
    (d) none.

117. Toughness test is also called as
    (a) Shaking test
    (b) Rolling test
    (c) Strength test
    (d) Dispersion test.

118. In the field, Rolling test is used to differentiate clay from
    (a) Gravels
    (b) Sands
    (c) Silts
    (d) None.

119. For classification purposes, coarse grained soils shall be divided into following number of sub-divisions
    (a) One
    (b) Two
    (c) Three
    (d) Four.

120. For classification purposes, fine grained soils shall be divided into following number of sub-divisions based on liquid limit
    (a) One
    (b) Two
    (c) Three
    (d) Four.

121. The coarse grained soils will lie on boundary classification when they have this percentage of fines
    (a) 1 to 10%
    (b) 10 to 20%
    (c) 5 to 12%
    (d) 12 to 20%.

122. The basis for the classification of the grained soils is
    (a) liquid limit
    (b) grain size
    (c) water content
    (d) plasticity chart.

123. A reduction in liquid limit after oven drying to a value less than three-fourth of the liquid limit before oven-drying is an indication of
    (a) coarse grained soils
    (b) fine grained soils
    (c) organic soils
    (d) none.

124. Soils which plot above the A-line, and which have a plasticity Index between 4 to 7 are classified as
    (a) CL
    (b) ML
    (c) CL-ML
    (d) all.

125. According to IS classification, the value of liquid limit for fine grained soils of high compressibility is
    (a) 15%
    (b) 35%
    (c) 50%
    (d) 75%.

126. Specific surface of the following soil is more
    (a) gravels
    (b) sands
    (c) silts
    (d) clays.

127. More hygroscopicity will be observed in
    (a) gravels
    (b) sands
    (c) silts
    (d) clays.

128. Hydroscopic water can be removed in the lab. by oven-drying at a temp. of
    (a) 100°C
    (b) less than 100°C
    (c) 105 to 110°C
    (d) Greater than 110°C.

129. The minimum water content at which the soil just begins to crumble when rolled into threads 3 mm in diameter is known as
    (a) shrinkage limit
    (b) plastic limit
    (c) liquid limit
    (d) consistency limit.

130. When the plastic limit cannot be determined, the soil is reported to be
    (a) unplastic
    (b) plastic
    (c) semi-plastic
    (d) non-plastic.

131. The maximum water content at which a reduction in water content does not cause a decrease in volume of soil mass is known as
    (a) liquid limit
    (b) plastic limit
    (c) shrinkage limit
    (d) ductile limit.

132. In the laboratory determination of SL the volume of the dry soil pat is determined by
    (a) oven dry method
    (b) sand bath method
    (c) radiation method
    (d) mercury replacement method.

133. Shrinkage ratio will be equal to
    (a) mass sp. gr.
    (b) absolute sp. gr.
    (c) grain sp. gr.
    (c) sp. gr. of water.

134. Plasticity index is obtained as the difference between
    (a) liquid limit and shrinkage limit
    (b) shrinkage limit and plastic limit
    (c) liquid limit and plastic limit
    (d) none of the above.

135. The ratio of plasticity index to clay fraction is known as
    (a) activity ratio
    (b) flow index
    (c) liquidity index
    (d) toughness index.

136. Two soils A and B are tested in the lab. for the consistency limits. The results are

	A	B
Liquid limit	40%	60%
Plastic limit	20%	25%

    Which soil is more plastic?
    (a) A
    (b) B
    (c) both
    (d) none.

137. The slope of the flow curve obtained in liquid limit test is called
    (a) liquidity index
    (b) plasticity index
    (c) toughness index
    (d) flow index.

138. Two soils A and B are having a flow index of 9 and 6 respectively. Then which soil has a better strength as a function of WC
    (a) A
    (b) B
    (c) both
    (d) none.

139. The ratio of plasticity index to flow index is called
    (a) activity ratio
    (b) liquidity index
    (c) toughness index
    (d) none of the above.

140. Two soils P and Q are having a toughness index of 4 and 10 respectively. Which soil has a better strength at plastic limit
    (a) P
    (b) Q
    (c) both
    (d) none.

141. Rock flour is the name of
    (a) clays
    (b) organic clays
    (c) organic silts
    (d) inorganic silts.

142. With reference to shrinkage and swelling, the most active clay mineral is
    (a) Kaolinite
    (b) Montmorillonite
    (c) Illite
    (d) None.

143. In a soil if the water content is equal to the liquid limit, its relative consistency is
    (a) 1
    (b) 0
    (c) 10
    (d) 5.
144. The difference between plastic limit and shrinkage limit is called
    (a) fluidity index
    (b) relative consistency
    (c) plasticity index
    (d) shrinkage index.
145. The water content at which the soil changes from liquid state to plastic state is known as
    (a) plastic limit
    (b) liquid limit
    (c) shrinkage limit
    (d) none of the above.
146. The property of the soil which permits it to be deformed rapidly without volume change, rupture and elastic rebound is termed as
    (a) ductility
    (b) malleability
    (c) elasticity
    (d) plasticity.
147. The liquidity index is given as
    (a) $\dfrac{w - \text{P.L.}}{\text{P.I.}}$
    (b) $\dfrac{w + \text{P.L.}}{\text{P.I.}}$
    (c) $\dfrac{\text{P.I.}}{w - \text{P.L.}}$
    (d) $\dfrac{\text{P.I.}}{w + \text{P.L.}}$
    where w is the water content, P.L. is the plastic limit and P.I. is the plasticity index.
148. Which of the following statements is true?
    (a) The soil is acidic when its pH < 7
    (b) The soil is acidic when its pH > 7
    (c) The soil is alkaline when its pH = 7
    (d) The organic matter in the soil is responsible for hygroscopic water.
149. When other factors remaining constant, the velocity with which a soil grain settle out of suspension depends upon
    (a) shape of grain
    (b) size of grain
    (c) weight of grain
    (d) all the above.
150. Degree of saturation of a soil is generally
    (a) above 100%
    (b) below zero
    (c) between 0% and 100%
    (d) none of the above.
151. Thixotropy of soils refers to
    (a) gain of strength of soil with passage of time after it has been remoulded
    (b) loss of strength of soil with passage of time after it had been remoulded
    (c) thickening of soil particles with water
    (d) none of the above.
152. The ratio of the compressive strength of unconfined clay in an undisturbed state to that in a remoulded state is
    (a) more than one
    (b) less than one
    (c) equal to one
    (d) equal to zero.

153. The basis for all soil classification system is
    (a) permeability characteristics
    (b) specific gravity of solids
    (c) grain size and plasticity characteristics
    (d) none of the above.

154. The property of soil which allows water to flow through the soil is known as
    (a) capillarity
    (b) permeability
    (c) fluidity
    (d) viscosity.

155. The law which states that the velocity of flow through porous medium is directly proportional to the hydraulic gradient is
    (a) Newton's law
    (b) Stoke's law
    (c) Darcy's law
    (d) Reynold's law.

156. Darcy's law is applicable if a soil is
    (a) homogeneous
    (b) incompressible
    (c) isotropic
    (d) all the above.

157. According to Darcy's law, the flow velocity can be obtained by
    (a) Ki/A
    (b) $Ki\,A$
    (c) Ki
    (d) K/iA.

158. Coefficient of percolation and coefficient of permeability are related by
    (a) $K = Kp/n$
    (b) $K = Kpn$
    (c) $Kp = K/n$
    (d) None.

159. Coefficient of percolation has a value
    (a) less than the cofficient of permeability
    (b) equal to the coefficient of permeability
    (c) greater than coefficient of permeability
    (d) none of the above.

160. The constant of proportionality between seepage velocity and hydraulic gradient is called
    (a) coefficient of permeability
    (b) coefficient of percolation
    (c) coefficient of transmissibility
    (d) seepage coefficient.

161. Capillary force is dependent on
    (a) pore pressure
    (b) water content
    (c) depth to water table
    (d) surface tension of water.

162. The zone of soil mass above the water table which is saturated by capillary moisture is called
    (a) capillary fringe
    (b) capillary soil
    (c) submerged soil
    (d) none.

163. The number of heads that are used for defining the water head at a point in a soil mass are
    (a) one
    (b) two
    (c) three
    (d) four.

164. Velocity head in soils will be
    (a) very high
    (b) high
    (c) negligible
    (d) can't say.

165. Find out the correct statement from the following:
    (a) coefficient of percolation will be lesser than coefficient of permeability
    (b) gravels will be having very small coefficient of permeability
    (c) the equation given by Hazen is $K(cm/sec) = C.D_{10}$
    (d) negative pore pressure can exist.

166. The pore water pressure in the capillary zone is
    (a) positive
    (b) negative
    (c) zero
    (d) can't say.

167. Water which the soil particles freely adsorb from atmosphere by the physical forces of attraction, and is held by the force of adhesion is known as
    (a) Contact moisture
    (b) Hygroscopic water
    (c) Adsorbed water
    (d) All the above.

168. Capillary water is held in the interstices of soil due to capillarity forces. The capillary forces depends upon
    (a) surface tension of water
    (b) pressure in water
    (c) size of soil pores
    (d) all the above.

169. Out of the following, which will have the maximum value of surface tension force
    (a) mercury
    (b) petrol
    (c) water
    (d) acetone.

170. A soil which does not permit the passage or seepage of any permeant through its voids, is known as
    (a) Solid soil
    (b) Hard soil
    (c) Impermeable soil
    (d) Honey comb soil.

171. In most of the practical flow problems in soil mechanics, the flow is
    (a) Laminar
    (b) Turbulent
    (c) Supersonic
    (d) Subsonic.

172. The value of Reynold's number for laminar flow through soil is
    (a) less than 20,000
    (b) less than 2,000
    (c) less than 200
    (d) less than 20.

173. The velocity of percolation is defined as
    (a) the discharge per unit of gross c/s area
    (b) the discharge per unit of net c/s area
    (c) the discharge per unit of total c/s area
    (d) none of the above.

174. Magnitude of capillary rise is more in
    (a) silts
    (b) sands
    (c) clays
    (d) gravels.

175. The height to which water can be lifted by capillarity is
    (a) independent of the atmospheric pressure value
    (b) depends upon the atmospheric pressure value
    (c) increases with decreasing diameter of the passage and is independent of the atmospheric pressure
    (d) none.

176. Piping in soils is due to
    (a) low exit gradient
    (b) erosion of subsoil by high velocity of seepage flow
    (c) leakage of water through pipes laid in dams
    (d) passage of water through well-connected pores in soil.
177. Permeability of soil varies
    (a) inversely as square of grain size
    (b) as square of grain size
    (c) as grain size
    (d) inversely as grain size.
178. Physical properties of a permeant which influences the permeability are
    (a) viscosity only
    (b) unit weight only
    (c) both viscosity and unit weight
    (d) none.
179. Select the correct statement. The permeability is
    (a) Directly proportional to unit weight, inversely proportional to viscosity
    (b) Directly proportional to viscosity, inversely proportional to unit weight
    (c) Directly proportional to both unit weight and viscosity
    (d) Inversely proportional to both unit weight and viscosity.
180. Due to rise in temperature, the unit weight and viscosity of the permeant fluid are reduced to 90% and 60% respectively. If other things remain constant, the coefficient of permeability
    (a) increases by 33.3%
    (b) decreases by 33.3%
    (c) increases by 50%
    (d) increases by 25%.
181. Units of co-efficient of permeability
    (a) cm/s
    (b) s/cm
    (c) cm/s^2
    (d) s^2/cm.
182. Coefficient of permeability of a soil
    (a) increases with the decrease in temperature
    (b) increases with the increase in temperature
    (c) does not depem upon temperature
    (d) none of the above.
183. Void ratio of soil 'A' = 0.80
    Void ratio of soil 'B' = 1.20
    Which soil is more pervious?
    (a) soil A
    (b) soil B
    (c) both A and B
    (d) none.
184. A sand has e = 0.8 and another clay has e = 1.2; which is more pervious?
    (a) sand
    (b) clay
    (c) both
    (d) none.
185. The coefficient of permeability is proportional to void ratio as
    (a) e
    (b) $1/e^3$
    (c) $e^3/1+e$
    (d) $e/1+e^2$.

186. Which of the following soils has largest permeability?
    (a) Sand
    (b) Gravel
    (c) Silt
    (d) Clay.

187. The coefficient grain size of soil (D_{10}) is 0.3 mm, the value of coefficient of permeability as per Hazen is (coefficient of Hazen $C = 10$)
    (a) 3 mm/sec
    (b) 9 mm/sec
    (c) 0.09 mm/sec
    (d) 0.9 mm/sec.

188. Falling head permeameter is preferable when soil sample is
    (a) clayey
    (b) silty sand
    (c) sandy
    (d) sandy gravel.

189. The coefficient of permeability of fine grained soils can be determined indirectly from the data obtained from a consolidation test on a sample. It's value is given by
    (a) $K = C_v m_v r_w$
    (b) $K = \dfrac{C_v}{m_v r_w}$
    (c) $K = \dfrac{1}{C_v m_v r_w}$
    (d) $K = \dfrac{m_v r_w}{C_v}$.

190. In the case of stratified soils, the permeabilities K_x and K_z along and across stratification are related as
    (a) $K_x < K_z$
    (b) $K_x = K_z$
    (c) $K_x > K_z$
    (d) no relation between K_x and K_z.

191. The range of coefficient of permeability of sands is about
    (a) $< 10^{-6}$
    (b) 10^{-4} to 10^{-6}
    (c) 10^{-2} to 10^{-4}
    (d) 10^{-2} to 1.

192. The value of coefficient of permeability of clays is
    (a) $< 10^{-6}$
    (b) 10^{-4} to 10^{-6}
    (c) 10^{-2} to 10^{-4}
    (d) 10^{-2} to 1.

193. Soils with a value for coefficient of permeability ranging between 10^{-4} to 10^{-6} are classified as
    (a) pervious soils
    (b) semi-pervious soils
    (c) impervious soils
    (d) all the above.

194. The expression for critical gradient is
    (a) $i_c = \dfrac{G-1}{1+e}$
    (b) $i_c = \dfrac{G+1}{1+e}$
    (c) $i_c = \dfrac{1-e}{G-1}$
    (d) $i_c = \dfrac{1+e}{G+1}$.

195. The presence of entrapped air in a soil will
    (a) increase the permeability
    (b) decrease the permeability
    (c) no effect on permeability
    (d) can't say.

196. The presence of organic matter in a soil will
   (a) increase the permeability
   (b) decrease the permeability
   (c) no effect on permeability
   (d) difficult to guess.
197. The hydraulic head that would produce a quick condition in a sand stratum of thickness 1.5 m and of sp. gr. 2.67 and void ratio of 0.67 is equal to
   (a) 1.0 m
   (b) 2.0 m
   (c) 1.5 m
   (d) 3.0 m.
198. A critical hydraulic gradient occurs when
   (a) flow is in upward direction
   (b) effective stress is zero
   (c) seepage pressure is in upward direction
   (d) all the above.
199. The phenomenon in which a cohesionless soil loses all its shear strength and the soil particles have a tendency to move up in the direction of flow, is known as
   (a) boiling condition
   (b) quick sand
   (c) quick condition
   (d) all the above.
200. Quick sand is
   (a) pure silica sand
   (b) a condition in which cohesion is decreased quickly
   (c) a sand which can act as a quick filter
   (d) a condition in which cohesionless soil looses its shear strength due to the upward flow of water.
201. The quantity of seepage in a soil is
   (a) directly proportional to the coeff. of permeability
   (b) inversely proportional to the coeff. of permeability
   (c) directly proportional to the uniformity coeff. of the soil
   (d) none of the above.
202. Flow net consists of a number of stream lines and equipotential lines which are
   (a) parallel to each other
   (b) perpendicular to each other
   (c) orthogonal to each other
   (d) none of the above.
203. The seepage force in a soil is
   (a) parallel to the equipotential lines
   (b) perpendicular to the equipotential lines
   (c) perpendicular to the flow lines
   (d) none.
204. The seepage force in soils at exit is proportional to
   (a) the head of water at d/s
   (b) the head of water at u/s
   (c) the head causing flow
   (d) the exit gradient.
205. Flow net is used for the determination of
   (a) exit gradient
   (b) seepage
   (c) hydrostatic pressure
   (d) all the above.
206. Electrical analogy method is used to
   (a) find electrical conductivity of soil
   (b) draw flownet
   (c) find the depth to water table
   (d) find the water content in a soil.

207. The flow net can be obtained by
    (a) electrical analogy method
    (b) graphical method
    (c) solution of laplace equations
    (d) all the above.
208. The shape factor of a flownet is defined as
    (a) N_d/N_f
    (b) N_f/N_d
    (c) both
    (d) none.
209. Find out the correct statement about shape factor
    (a) dependent on permeability
    (b) independent of permeability
    (c) not related to permeability
    (d) none.
210. The u/s face of an earth dam is
    (a) an equipotential line
    (b) phreatic line
    (c) flow line
    (d) none of the above.
211. The top flow line in the case of seepage flow through an earth dam has the shape of
    (a) parabola
    (b) ellipse
    (c) hyperbola
    (d) none of the above.
212. Flow through an earth dam is a case of
    (a) unconfined flow
    (b) confined flow
    (c) both
    (d) none.
213. The phreatic line is
    (a) the u/s face of the earth dam
    (b) d/s face of the earth dam
    (c) the top flow line
    (d) none of the above.
214. The hydrostatic pressure on the phreatic line with in a dam section is
    (a) greater than atmospheric pressure
    (b) less than atmospheric pressure
    (c) equal to the atmospheric pressure
    (d) none.
215. If s is the focal length of the Kozeny's parabola and k is the coefficient of permeability, the seepage flow rate per unit length of the dam is given as
    (a) $q = k/s$
    (b) $q = s/k$
    (c) $q = k\sqrt{s}$
    (d) $q = k.s$.
216. If h is the head of water stored, N_f is the number of flow lines and N_d is the number of equipotential drops in a flownet, then the seepage flow rate is given by
    (a) $q = k.h \dfrac{N_f}{N_d}$
    (b) $q = k.h. \dfrac{N_f}{N_d}$
    (c) $q = k.h \left(\dfrac{N_f}{N_d}\right)^2$
    (d) none of the above.
217. The discharge 'q' through an anisotropic soil mass can be obtained from the expression
    (a) $q = k.h. \dfrac{N_f}{N_d}$
    (b) $q = k'.h. N_f/N_d$
    (c) $q = k'h\, N_f N_d$
    (d) $q = k.h.N_f N_d$
    where k' is modified coeff. of permeability
    k is coeff. of permeability.

218. The modified co-efficient of permeability k' in the case of an anisotropic soil is given by
    (a) $k' = k_x - k_z$ ☐   (b) $k' = \sqrt{k_x \cdot k_z}$ ☐
    (c) $k' = (k_x \cdot k_z)^{1.5}$ ☐   (d) $k' = (k_x \cdot k_z)^2$. ☐

219. In a flownet, which of the following statement is true?
    (a) Flow lines and equipotential lines intersect at oblique angles ☐
    (b) Flow lines and equipotential lines intersect orthogonally ☐
    (c) Flow rate through each flow channel is different ☐
    (d) The head along each flow line is constant. ☐

220. The permeability of a soil deposit-in-situ can be best obtained by
    (a) falling head permeameter ☐   (b) constant head permeameter ☐
    (c) pumping test ☐   (d) none of the above. ☐

221. The minimum number of observation wells required to determine the permeability of a stratum in the field by pumping test is
    (a) one ☐   (b) two ☐
    (c) three ☐   (d) none of the above. ☐

222. The total discharge from two wells situated near to each other is
    (a) sum of the discharges from individual wells ☐
    (b) greater than the sum of the discharges from the wells ☐
    (c) less than the sum of the discharges from the wells ☐
    (d) none. ☐

223. The Laplace equation governing the two dimensional seepage flow is given by
    (a) $\dfrac{\partial h}{\partial x} + \dfrac{\partial h}{\partial z} = 0$ ☐   (b) $\dfrac{\partial h}{\partial x} = \dfrac{\partial h}{\partial z}$ ☐
    (c) $\dfrac{\partial^2 h}{\partial x^2} = \dfrac{\partial^2 h}{\partial z^2}$ ☐   (d) $\dfrac{\partial^2 h}{\partial x^2} + \dfrac{\partial^2 h}{\partial z^2} = 0$. ☐

224. When the steady seepage occurs in an isotropic soil, the head causing the flow satisfies the following equation
    (a) Darcy's equation ☐   (b) Hazen's equation ☐
    (c) Lapace's equation ☐   (d) none. ☐

225. Identify the incorrect statement
    (a) The only water which can be obtained from the aquifer is that which will flow by gravity. ☐
    (b) A high porosity indicates that an aquifer will yield large volumes of water to well. ☐
    (c) Aquifuge is an impermeable formation which neither contains water nor transmits any water. ☐
    (d) All the above. ☐

226. The rate of flow of water through a vertical strip of aquifer of unit width and extending the full saturation height under unit hydraulic gradient is known as
    (a) Coefficient of permeability ☐   (b) Storage coefficient ☐
    (c) Specific retension ☐   (d) Coefficient of transmissibility. ☐

227. In order to increase the FS against piping
   (a) a blanket of impermeable material is laid for same length along the downstream ground surface ☐
   (b) pump out the water periodically, downstream ☐
   (c) both ☐
   (d) none. ☐

228. In a stratified soil deposit with number of layer having different permeabilities, the average permeability along the bedding planes is given by
   (a) $k_x = \dfrac{z_1 + z_2 + \ldots + z_n}{k_1 + k_2 + k_3 + \ldots + k_n}$ ☐
   (b) $k_x = \dfrac{k_1 z_1 + k_2 z_2 + \ldots + k_n z_n}{z_1 + z_2 + \ldots + z_n}$ ☐
   (c) $k_x = \dfrac{k_1 + k_2 + \ldots + k_n}{z_1 + z_2 + \ldots + z_n}$ ☐
   (d) $k_x = \dfrac{z_1 + z_2 + \ldots + z_n}{k_1 z_1 + k_2 z_2 + \ldots + k_n z_n}$. ☐

229. In a stratified soil deposit the average permeability normal to the bedding planes is given by
   (a) $k_z = \dfrac{z_1 + z_2 + \ldots + z_n}{k_1 + k_2 + \ldots + k_n}$ ☐
   (b) $k_z = \dfrac{k_1 + k_2 + \ldots + k_n}{z_1 + z_2 + \ldots + z_n}$ ☐
   (c) $k_z = \dfrac{z_1 + z_2 + \ldots + z_n}{\dfrac{z_1}{k_1} + \dfrac{z_2}{k_2} + \ldots + \dfrac{z_n}{k_n}}$ ☐
   (d) $k_z = \dfrac{z_1 + z_2 + \ldots + z_n}{k_1 z_1 + k_2 z_2 + \ldots + k_n z_n}$. ☐

230. Which of the following is correct?
   (a) $k_x = k_z$ ☐
   (b) $k_x < k_z$ ☐
   (c) $k_x > k_z$ ☐
   (d) none ☐
   where k_x = Permeability along the bedding planes
   k_z = Permeability normal to the bedding planes.

231. If k_x and k_z are the permeability values in x and z-directions respectively in a two dimensional flow situation, the equivalent coefficient of permeability k_e is given by
   (a) $k_e = k_x + k_z$ ☐
   (b) $k_e = k_x - k_z$ ☐
   (c) $k_e = k_x/k_z$ ☐
   (d) $k_e = \sqrt{k_x \cdot k_z}$. ☐

232. A coarse grained soil has a void ratio of 0.7 and specific gravity of 2.7. The critical gradient at which the quick sand condition occurs is
   (a) 1.00 ☐
   (b) 0.75 ☐
   (c) 0.5 ☐
   (d) 0.25. ☐

233. The seepage flow through a porous medium is generally
   (a) Turbulent ☐
   (b) Super critical ☐
   (c) Transitional ☐
   (d) Laminar. ☐

234. Criterion for the design of an inverted filter is
   (a) $\dfrac{D_{15} \text{ of filter}}{D_{85} \text{ of base}} < 5 < \dfrac{D_{15} \text{ of filter}}{D_{15} \text{ of base}}$ ☐
   (b) $\dfrac{D_{85} \text{ of filter}}{D_{15} \text{ of base}} < 5 < \dfrac{D_{15} \text{ of filter}}{D_{15} \text{ of base}}$ ☐
   (c) $\dfrac{D_{15} \text{ of filter}}{D_{15} \text{ of base}} < 5 < \dfrac{D_{85} \text{ of filter}}{D_{15} \text{ of base}}$ ☐
   (d) $\dfrac{D_{15} \text{ of filter}}{D_{15} \text{ of base}} < 5 < \dfrac{D_{85} \text{ of filter}}{D_{85} \text{ of base}}$. ☐

235. Identify the incorrect statement. A practical filter
    (a) prevents the movement of soil particles due to the flowing water
    (b) is designed in such a way so as to provide quick drainage
    (c) has subsequent layers increasingly coarser than the previous one
    (d) none of the above.

236. The coefficient of electro-osmotic permeability
    (a) is the same as the coefficient of hydraulic permeability
    (b) varies in the value over a wide range depending on the size of voids in the soil
    (c) varies in value over a wide range depending on the grain size of the sample
    (d) is almost independent of the grain size of the soil sample.

237. Bleeder wells are required to
    (a) increase discharge from main well
    (b) store drainage water
    (c) collect seepage water in dams
    (d) relieve pressure in impervious layers.

238. Neutral stress refers to
    (a) major principal stress
    (b) minor principal stress
    (c) pore water pressure
    (d) total stress.

239. The difference between total pressure and pore water pressure is known as
    (a) major principal stress
    (b) minor principal stress
    (c) effective stress
    (d) none of the above.

240. A flownet may be used for the determination of
    (a) exit gradient
    (b) seepage flow rate
    (c) seepage pressure
    (d) all the above.

241. Which of the following is responsible for shear resistance in soils?
    (a) intergranular friction
    (b) cohesion and adhesion between the soil particles
    (c) both (a) and (b)
    (d) none of the above.

242. In an undrained plastic clay the shear strength is due to
    (a) internal friction
    (b) cohesion
    (c) inter-granular friction
    (d) none of the above.

243. Through a point in a loaded soil, the normal stress is maximum on
    (a) major principal plane
    (b) minor principal plane
    (c) plane making an angle of 45 with principal planes
    (d) none of the above.

244. The maximum shear stress occurs on the filament which makes an angle with the horizontal plane equal to
    (a) $22\frac{1}{2}°$
    (b) $45°$
    (c) $60°$
    (d) $67\frac{1}{2}°$.

245. The seepage velocity v_s and the Darcy's velocity v are related as
   (a) $v = \dfrac{v_s}{n}$
   (b) $v_s = \dfrac{v}{n}$
   (c) $v_s = v \cdot n$
   (d) $v_s = \dfrac{n}{v}$.

246. When analogy is drawn between Darcy's law for flow of water and Ohm's law for flow of current, what is analogous to the quantity of seepage?
   (a) Resistance
   (b) Voltage
   (c) Current
   (d) Conductivity.

247. The relationship between the hydraulic potential h, the position head above the datum z and the piezometric head h_w is given by
   (a) $h_w = h - z$
   (b) $z = h + h_w$
   (c) $h = z - h_w$
   (d) $z = h_w - h$.

248. An isobar is
   (a) a curve connecting points of equal vertical pressure
   (b) a point at which the net pressure is zero
   (c) a curve connecting points of equal total pressure
   (d) a curve connecting points of equal pore water pressure.

249. Consolidation theory was enunciated by
   (a) Rankine
   (b) Westergaard
   (c) Skempton
   (d) Terzaghi.

250. Consolidation is a process involving
   (a) sudden compression of soil
   (b) tilting and failure of structure
   (c) abnormal sinking of foundation
   (d) gradual expulsion of pore water.

251. Consolidation is generally considered to be a function of
   (a) total stress
   (b) neutral stress
   (c) effective stress
   (d) none.

252. In the Terzaghi's theory of one dimensional consolidation it is assumed that
   (a) soil particles are compressible
   (b) soil is non-homogeneous
   (c) coefficient of volume change is not constant
   (d) compression is one-dimensional.

253. The unit of the coefficient of consolidation is
   (a) cm^2/cm
   (b) cm/sec
   (c) cm^2/sec
   (d) gm-cm^2/sec.

254. When a static load is applied, the consolidation is fast in the case of
   (a) clays
   (b) silty clays
   (c) sandy silts
   (d) sands.

255. The aim of doubling the pressure each time in the consolidation test is to see that the soil is always
   (a) over consolidated condition ☐ (b) normally consolidated condition ☐
   (c) under consolidation condition ☐ (d) none. ☐

256. The maximum over consolidation ratio of normally consolidated soil is
   (a) one ☐ (b) two ☐
   (c) three ☐ (d) four. ☐

257. In Terzaghi's theory of one-dimensional consolidation
   (a) plastic lag alone is considered and hydrodynamic lag is ignored ☐
   (b) hydrodynamic lag alone is considered and plastic lag is ignored ☐
   (c) both hydrodynamic and plastic lag are ignored ☐
   (d) both hydrodynamic and plastic lag are considered. ☐

258. Direct measurement of permeability of the soil specimen can be made at any stage of loading in
   (a) fixed ring type consolidometer ☐ (b) floating ring type consolidometer ☐
   (c) both ☐ (d) none. ☐

259. Square root of time fitting method is used for calculating
   (a) compression index ☐ (b) co-efficient of consolidation ☐
   (c) co-efficient of compressibility ☐ (d) co-efficient of volume compressibility. ☐

260. Empirical relationship between the compression index C_c and liquid limit L.L. of normally loaded clays of low to medium sensitivity is
   (a) $C_c = 0.009$ (L.L. − 10%) ☐ (b) $C_c = 0.009$ (10% − L.L.) ☐
   (c) $C_c = 0.09 \left(\dfrac{1}{L.L.} - 10\%\right)$ ☐ (d) $C_c = 0.1$ (L.L. − 25%). ☐

261. The expression for organic soils and peats is
   (a) $C_c = 0.009$ (L.L. − 10%) ☐ (b) $C_c = 0.007$ (L.L. − 10%) ☐
   (c) $0.0115\, w_n$ ☐ (d) none. ☐

262. If the consolidation pressure at any point is denoted by Δ_p, $\overline{\Delta}_p$ represents that portion of consolidation stress which, at a given time, is transmitted from grain to grain and u is the corresponding excess hydrostatic pressure
   (a) $\Delta_p = \overline{\Delta}_p + u$ ☐ (b) $\overline{\Delta}_p = \Delta_p + u$ ☐
   (c) $\overline{\Delta}_p = \Delta_p - u$ ☐ (d) $\Delta_p = \overline{\Delta}_p - u$. ☐

263. The value of compression index for a remoulded sample whose liquid limit is 30% is
   (a) 0.021 ☐ (b) 0.027 ☐
   (c) 0.21 ☐ (d) 0.27. ☐

264. The coefficient of consolidation of a soil is affected by
   (a) compressibility only ☐ (b) permeability only ☐
   (c) both compressibility and permeability ☐ (d) none. ☐

265. Compressibility of a clayey soils will be
    (a) equal to that of sandy soils
    (b) greater than sandy soils
    (c) less than clayee soils
    (d) can't say.
266. The co-efficient of compressibility of an over consolidated clay will be
    (a) equal to that of a normally consolidated clay
    (b) greater than that of a normally consolidated clay
    (c) less than that of a normally consolidated clay
    (d) can't say.
267. The time required for full dissipation of pore water pressure will depend on
    (a) thickness of soil sample only
    (b) co-efficient of permeability of soil only
    (c) both thickness and co-efficient of permeability only
    (d) none of the above.
268. When compared to deformation of thin soil sample because of consolidation, the deformation of thick soil sample will be
    (a) less
    (b) more
    (c) equal
    (d) can't say.
269. The time required for the completion of consolidation for a thin soil sample when compared with that of a thick soil sample will be
    (a) less
    (b) more
    (c) equal
    (d) can't say.
270. As the value of drainage path increases the time required for consolidation will
    (a) decrease
    (b) increase
    (c) constant
    (d) can't say.
271. The change in the thickness of a soil sample may be due to
    (a) expulsion of pore water or pore air from the voids
    (b) compression of the soil solid particles
    (c) compression of water (or) air in the voids of soil
    (d) all the above.
272. Primary consolidation of soil is due to the expulsion of
    (a) absorbed water only
    (b) adsorbed water only
    (c) both (a) and (b)
    (d) none.
273. Secondary consolidation of soil is due to the expulsion of
    (a) absorbed water only
    (b) adsorbed water only
    (c) both (a) and (b)
    (d) none.
274. The relationship between the time factor T_v, coefficient of consolidation C_v, the length of drainage path d, and time t is given by
    (a) $T_v = \dfrac{C_v \cdot d^2}{t}$
    (b) $T_v = \dfrac{C_v \cdot t^2}{d}$
    (c) $T_v = \dfrac{C_v \cdot t}{d^2}$
    (d) $T_v = \dfrac{C_v \cdot t^2}{d^2}$.

SOIL MECHANICS 475

275. The time factor for a particular degree of consolidation
   (a) depends upon the coefficient of consolidation
   (b) depends upon the drainage path
   (c) depends upon the distribution of initial excess hydrostatic pressure
   (d) none of the above.

276. The compression resulting from long term static load and resulting expulsion of water is known as
   (a) compaction
   (b) inverse swelling
   (c) consolidation
   (d) none of the above.

277. Compression of soils occurs rapidly if voids are occupied by
   (a) air
   (b) water
   (c) partly air and partly water
   (d) none of the above.

278. The compressibility of a field deposit is
   (a) the same as that shown by a laboratory sample
   (b) greater than that shown by a laboratory sample
   (c) smaller than that shown by a laboratory sample
   (d) none.

279. The void ratio reduced from e_0 to e when the effective pressure is increased to $(p_0 + \Delta_p)$ from p_0. The co-efficient of compressibility is given by
   (a) $\dfrac{\Delta_p}{e_0 - e}$
   (b) $\dfrac{e_0 - e}{\Delta_p}$
   (c) $\dfrac{\Delta_p}{e}$
   (d) $\dfrac{e}{p_0}$.

280. When the effective pressure is increased by p, the soil is reducing its volume by ΔV from its original volume V_0. The co-efficient of volume change is given by
   (a) $m_v = \dfrac{-(\Delta_v/V_0)}{1+\Delta_p}$
   (b) $m_v = \dfrac{1+\Delta_p}{-(\Delta_v/V_0)}$
   (c) $m_v = \dfrac{1+\Delta_p}{-(\Delta_v/V_0)}$
   (d) $m_v = \dfrac{-(\Delta_v/V_0)}{\Delta_p}$.

281. The co-efficient of consolidation C is given by
   (a) $C = \dfrac{K}{\gamma_w m_v}$
   (b) $C = \dfrac{\gamma_w m_v}{K}$
   (c) $C = \dfrac{K}{m_v}$
   (d) $C = \dfrac{m_v}{K}$.

282. The value of percentage consolidation U_z if the initial excess pore pressure is U_i; and the pore pressure at that particular time is U, is
   (a) $U_z = \left(1 - \dfrac{U}{U_i}\right)$
   (b) $U_z = \left(1 - \dfrac{U}{U_i}\right) \times 100$
   (c) $U_z = \dfrac{U}{U_i} \times 100$
   (d) $U_z = \left(1 + \dfrac{U}{U_i}\right)$.

283. Mathematically speaking, the time taken for 100% consolidation is
    (a) 5 years
    (b) 10 years
    (c) zero
    (d) infinite.

284. The degree of consolidation $U\%$ for a given time t is obtained from the relation
    (a) $U = f(C_v)$
    (b) $U = f(m_v)$
    (c) $U = f(T_v)$
    (d) $U = C_v \dfrac{d^2U}{df^2}$.

285. If consolidation experiments are conducted on two soil samples of the same soil but having different thickness (thick and thin) then the deformations will be higher in
    (a) thick
    (b) thin
    (c) constant
    (d) none.

286. As the length of drainage path increases the time required for consolidation
    (a) decreases
    (b) increases
    (c) remains constant
    (d) can't say.

287. The ratio of settlement at any time t, to the final settlement is called as
    (a) percentage settlement
    (b) partial settlement ratio
    (c) degree of consolidation
    (d) residual consolidation.

288. A clay deposit subjected to pressure in the past which is more than the present over burden pressure is known as
    (a) normally consolidated soil
    (b) over-consolidated soil
    (c) under-consolidated soil
    (d) none of the above.

289. A clay deposit which is not fully consolidated under the existing over burden pressure is known as
    (a) normally consolidated soil
    (b) over-consolidated soil
    (c) under-consolidated soil
    (d) none of the above.

290. A soil which is fully consolidated under the existing over burden pressure is known as
    (a) normally consolidated soil
    (b) over consolidated soil
    (c) under-consolidated soil
    (d) none of the above.

291. Over-consolidation may be due to
    (a) the weight of ice sheet or glacier which is melted away
    (b) weight of the land slide which is removed
    (c) both (a) and (b)
    (d) none of the above.

292. Even after the complete dissipation of excess pore pressure, a little more consolidation is possible. This is called as
    (a) Primary consolidation
    (b) Secondary consolidation
    (c) Compressibility
    (d) Compaction.

293. Secondary consolidation is
    (a) caused by hydrodynamic lag
    (b) caused by hydrostatic pressure
    (c) caused by Creep
    (d) none.

294. Which of the following clays behave like a dense sand
    (a) normally consolidated clay
    (b) under consolidated clay
    (c) over consolidated clay with a low OCR
    (d) over consolidated clay with a high OCR.
295. For a soil layer with double drainage and thickness H, the drainage path is equal to
    (a) $2H$
    (b) $H/2$
    (c) H^2
    (d) none of the above.
296. The final settlement of a soil of thickness H_0, initial void ratio e_0, initial effective pressure σ^1 subject to an increase in effective pressure of $\Delta\sigma^1$ is given by
    (a) $\dfrac{C_c}{1+e_0} H_0 \cdot \log\left(\dfrac{\sigma^1 + \Delta\sigma^1}{\sigma^1}\right)$
    (b) $\dfrac{C_c \cdot H_0}{1+e_0} \cdot \log\left(\dfrac{\sigma^1}{\sigma^1 + \Delta\sigma^1}\right)$
    (c) $\dfrac{1+e_0}{C_c \cdot H_0} \cdot \log\left(\dfrac{\sigma^1 + \Delta\sigma^1}{\sigma^1}\right)$
    (d) $\dfrac{1+e_0}{C_c \cdot H_0} \cdot \log\left(\dfrac{\sigma^1}{\sigma^1 + \Delta\sigma^1}\right)$.
297. A clay deposit suffers a total settlement of 10 cm with one way drainage, then with two way drainage, it suffers a total settlement of
    (a) 20 cm
    (b) 5 cm
    (c) 10 cm
    (d) Nil.
298. The consolidation settlement of a soil is
    (a) directly proportional to compression index
    (b) inversely proportional to compression index
    (c) directly proportional to the voids ratio
    (d) none of the above.
299. The differential equation governing one dimensional consolidation is written as
    (a) $\dfrac{\partial u}{\partial t} = C_v \dfrac{\partial z}{\partial u}$
    (b) $\dfrac{\partial u}{\partial t} = C_v \dfrac{\partial u}{\partial z}$
    (c) $\dfrac{\partial u}{\partial t} = C_v \dfrac{\partial^2 z}{\partial u^2}$
    (d) $\dfrac{\partial u}{\partial t} = C_v \dfrac{\partial^2 u}{\partial z^2}$.
300. If the time required for 60% consolidation of a remoulded sample of clay with single drainage is p, then the time required to consolidate the same soil with same degree of consolidation but with double drainage is
    (a) p
    (b) $2p$
    (c) $p/4$
    (d) $4p$.
301. The pressure-void ratio curve will be
    (a) linear when plotted on ordinary paper
    (b) linear when plotted on semi-log paper
    (c) linear when plotted on log-log paper
    (d) none of the above.
302. Standard proctor test is used for determining
    (a) optimum moisture content (OMC)
    (b) void ratio
    (c) co-efficient of consolidation
    (d) pavement thickness.
303. Compaction of soil is aimed at
    (a) decreasing dry density
    (b) increasing porosity
    (c) decreasing void ratio
    (d) decreasing shear strength.

304. The process of compaction of a soil involves
    (a) expulsion of pore water
    (b) expulsion of air voids
    (c) expulsion of both pore water and air voids
    (d) none.
305. Compaction of a soil
    (a) increases dry density
    (b) decreases porosity
    (c) both (a) and (b)
    (d) none.
306. For the same soil, increase in compaction effort
    (a) does not affect OMC
    (b) increases OMC
    (c) decreases OMC
    (d) decreases dry density.
307. Select the correct statement.
    (a) Increase in the compactive effort will decrease the dry density of the soil
    (b) Irrespective of the compactive effort, a soil cannot reach the zero-air voids condition
    (c) both (a) and (b)
    (d) none.
308. The most effective method for compacting sand is by using
    (a) pneumatic rollers
    (b) sheep foot rollers
    (c) steel tyred rollers
    (d) vibration.
309. The ratio of dry density obtained in the field to the proctor's maximum dry density is called
    (a) Compaction energy
    (b) Compactive effort
    (c) Relative compaction
    (d) None.
310. The soil involved in the compaction is
    (a) perfectly saturated soil
    (b) partly saturated soil
    (c) submerged soil
    (d) none.
311. The maximum dry density upto which any soil can be compacted depends upon
    (a) water content only
    (b) amount of compactive effort only
    (c) both water content and compactive effort
    (d) none.
312. For better strength and stability, the coarse grained and fine grained soils are respectively compacted as
    (a) dry of OMC and wet of OMC
    (b) wet of OMC and wet of OMC
    (c) wet of OMC and dry of OMC
    (d) dry of OMC and dry of OMC.
313. Compaction process may be accompanied by
    (a) Rolling
    (b) Tamping
    (c) Vibration
    (d) Any of the above.
314. The maximum compaction for any given water content corresponds to a per cent air-voids equal to
    (a) 0
    (b) 1.0
    (c) 0.1
    (d) can't say.

315. The zero air void density (r_d) for any moisture content can be calculated by the equation

    (a) $r_d = \dfrac{Gr_w}{1 + \dfrac{wG}{S}}$

    (b) $r_d = \dfrac{Gr_w}{1 - \dfrac{wG}{S}}$

    (c) $r_d = \dfrac{Gr_w}{1 + wG}$

    (d) none.

316. The coefficient of sub-grade reaction of a soil is
    (a) load per unit deformation
    (b) pressure per unit deformation
    (c) density per unit deformation
    (d) none of the above.

317. Plasticity needle is used to determine
    (a) plastic limit of the soil used in compaction
    (b) penetration resistance to control field compaction
    (c) penetration value of bitumen used in road construction
    (d) swelling index of black cotton soil.

318. In the modified proctor test the drop height of the rammer is
    (a) 30 cm
    (b) 45 cm
    (c) 60 cm
    (d) 75 cm.

319. The admixture used in soil stabilisation is
    (a) cement
    (b) lime
    (c) bitumen
    (d) any of the above.

320. Clays containing organic matter are made suitable for stabilisation, by first adding a small percentage of
    (a) hydrated lime
    (b) cement
    (c) calcium chloride
    (d) resins.

321. Cut back bitumen refers to bitumen
    (a) cut to the required size for filling gaps
    (b) mixed with a solvent
    (c) heated to decrease viscosity
    (d) mixed with water before adding the aggregate.

322. For calculating the stress distribution in soils, Boussinesq assumed the point load to exist
    (a) below the ground level
    (b) below water table
    (c) at the ground level
    (d) at water table.

323. The stress developed at a point in a soil mass due to a concentrated load which is applied at the ground surface is inversely proportional to
    (a) depth of the concerned point from ground surface
    (b) square of the depth of the point from ground surface
    (c) cube of the depth of the point from ground surface
    (d) square root of the depth of the point from ground surface.

324. Total number of stress components at a particular point within a soil mass because of the load is
    (a) 3
    (b) 6
    (c) 9
    (d) 12.

325. The value of Poisson's ratio for an elastic material generally varies from
    (a) 0 to 1.0
    (b) 0.0 to 0.5
    (c) 1 to 1.5
    (d) 1.5 to 2.0.

326. Geostatic stresses are due to
    (a) static loads
    (b) dynamic loads
    (c) self weight of soil
    (d) none of above.

327. The intensity of vertical pressure at a point which at a depth of 10 m exactly below a concentrated load of 10 t is
    (a) $\dfrac{0.5}{\pi}$ kN/m²
    (b) $\dfrac{0.15}{\pi}$ kN/m²
    (c) $\dfrac{0.5}{\pi}$ t/m²
    (d) $\dfrac{0.15}{\pi}$ t/m².

328. Select the incorrect statement.
    (a) The Boussinesq theory assumes that the soil has a constant value of modulus of elasticity
    (b) Westergaard's theory takes into account the anisotropy of the soils.
    (c) The vertical stress at a point by Boussinesq's theory depends upon modulus of elasticity and Poisson's ratio
    (d) None of the above.

329. The intensity of vertical stress just below the load point is given by
    (a) $\dfrac{3\phi}{\pi z^2}$
    (b) $\dfrac{3\phi}{2z^2}$
    (c) $\dfrac{3\phi}{2\pi z^2}$
    (d) none.

330. The value of vertical stress
    (a) decreases with an increase in (r/f) ratio
    (b) decreases with a decrease in r/f ratio
    (c) increase with an increase in (r/f) ratio
    (d) difficult to guess.

331. Select the incorrect statement:
    An isobar is a curve
    (a) points of equal stress intensity
    (b) which approaches leminslate curve
    (c) which will be in the shape of an Onion (or) a bulb
    (d) none of the above.

332. Select the incorrect statement.
    (a) Pressure bulb is nothing but an isobar
    (b) Isobar is a contour of equal stress
    (c) Isobars of higher intensities will lie out side
    (d) None of the above.

333. Vertical stress on a vertical plane which is at a particular radial distance from the axis of a vertical concentrated load
    (a) is same at all depth
    (b) decreases with depth constantly
    (c) increases first, attains a maximum value and then decreases
    (d) decreases first, attains a maximum value and then increases.

334. The maximum value of stress on a vertical plane which is at a particular radial distance from the axis of load corresponds to the point of intersection of the vertical plane with the line which is drawn at an angle of
    (a) 39° 15'
    (b) 45°
    (c) 90°
    (d) none.

335. The expression for the vertical stress at a depth z due to a line load of intensity q' is given by
    (a) $\sigma_z = \dfrac{3q'}{2\pi z^2} \left[\dfrac{1}{1+(r/z)^2}\right]^{5/2}$
    (b) $\sigma_z = \dfrac{2q'}{2z} \left[\dfrac{1}{1+(r/z)^2}\right]^{2}$
    (c) $\sigma_z = \dfrac{q'}{\pi} [2\theta + \sin 2\theta]$
    (d) none of the above

    where r is radius, x is coordinate of the concerned point in x-direction, θ is angle.

336. The expression for the vertical stress at a depth z under the center of circular area of diameter 'R' carrying a uniformly distributed load of intensity q is given by
    (a) $\sigma_z = 1 - \left\{\dfrac{1}{1+(R/z)^2}\right\}^{3/2}$
    (b) $\sigma_z = 1 - \left\{\dfrac{1}{1+(z/R)^2}\right\}^{3/2}$
    (c) $\sigma_z = q \left\{1 - \dfrac{1}{1+\left(\dfrac{2R}{z}\right)^2}\right\}^{3/2}$
    (d) $\sigma_z = q \left\{1 - \dfrac{1}{1+\left(\dfrac{R}{z}\right)^2}\right\}^{3/2}$

337. If the angle subtended by the circular area at the point where the vertical stress is required is 2θ, then the influence factor can be obtained from the equation
    (a) $I = 1 - \sin^2 \theta$
    (b) $I = 1 - \cos^2 \theta$
    (c) $I = 1 - \sin^3 \theta$
    (d) $I = 1 - \cos^3 \theta$.

338. The expression for vertical stress at a point below the corner of a rectangular loaded area was derived by
    (a) Boussinesq
    (b) Westergaard
    (c) Newmark
    (d) Fenske.

339. Select the incorrect statement
    (a) The stresses increases with depth because of over burden pressure
    (b) The stresses decreases with depth because of the applied load
    (c) The stresses decreases with depth in case of both overburden and applied loads
    (d) None of the above.

340. The speciality of the Newmark's diagram is that it can be used for finding the stress below
    (a) rectangular loaded areas, at any point
    (b) circular loaded areas, below the centre
    (c) rectangular loaded area, below the corner
    (d) any shape of loaded area at any point.

341. Newmark's chart is based on the concept of vertical stress under
    (a) a concentrated load
    (b) a line load
    (c) a strip load
    (d) a uniformly loaded circular area

342. The units for the influence factor of a Newmark's chart is
    (a) kN/m²
    (b) kN/m
    (c) m/sec
    (d) none.

343. What is the ratio of R_2/z for a Newmark's chart, when the circles are divided into 20 sectors. Take the value of influence factor as 0.005
    R_2 = radius of second circle, z = depth of the point at which stress is required
    (a) 0.270
    (b) 0.37
    (c) 0.40
    (d) 0.47.

344. The radius of tenth circle for a Newmark's chart, when it is divided into 20 sectors and if the influence factor is 0.005 is
    (a) 2.54
    (b) 3.54
    (c) unity
    (d) infinite.

345. The expression for the vertical stress at a point p using Newmark's chart is
    (a) $\sigma_z = I \times n \times q$
    (b) $\sigma_z = \dfrac{I \times n}{q}$
    (c) $\sigma_z = \dfrac{I \times q}{n}$
    (d) none.

346. Westergaard's theory is more appropriate for
    (a) layered soils
    (b) homogeneous deposits
    (c) anisotropic soils
    (d) normally consolidated homogeneous soils.

347. When compared with Boussinesq coefficient values, the values of Westergaard's coefficient will be
    (a) higher
    (b) smaller
    (c) same
    (d) can't say.

348. Fanske's charts are based on
    (a) Boussinesq's theory
    (b) Newmark's theory
    (c) Westergaard's theory
    (d) none.

349. The vertical stress at any point can be calculated approximated by
    (a) Equivalent point load method
    (b) 2 : 1 distribution method
    (c) Sixty degrees distribution
    (d) All the above.

350. The upward pressure due to soil on the underside of a footing is generally called as
    (a) vertical stress
    (b) tangential stress
    (c) contact pressure
    (d) none of the above.

351. Studies on the contact pressure distribution are done by
    (a) Boussinesq
    (b) Westergaard
    (c) Newmark
    (d) Borowicka.

352. Select the correct statement
    (a) Contact pressure distribution will be different for sands and clays
    (b) When the footing is flexible, contact pressure distribution is uniform
    (c) When the footing is rigid, the settlement of the footing is uniform
    (d) All the above.

353. The maximum contact pressure for a rigid footing on a cohesionless soil will be at
    (a) edges
    (b) center
    (c) between centre and edge
    (d) none.

354. Generally hand augers are used when the depth is about
    (a) 6 m
    (b) 12 m
    (c) 25 m
    (d) 37 m.

355. Select the correct statement.
    The depth of exploration
    (a) should be atleast equal to the significant depth
    (b) depends upon the degree of variation of sub-surface details
    (c) is governed by the influence zone
    (d) all the above.

356. The depth of exploration for a square footing should be at least
    (a) width of footing
    (b) 1.5 times width of footing
    (c) twice the width of footing
    (d) 3 times width of footing.

357. The depth of exploration of a strip footing should be at least
    (a) width of footing
    (b) 1.5 times width of footing
    (c) twice the width of footing
    (d) 3 times width of footing.

358. The depth upto which the stress increment due to the applied loads can produce significant settlement is called as
    (a) depth of settlement
    (b) depth of foundation
    (c) significant depth
    (d) none.

359. The minimum depth of exploration in case of gravity dam is
    (a) base of dam
    (b) twice the base of dam
    (c) height of the dam
    (d) twice the height of dam.

360. For an area of about 0.4 hectares, the minimum number of bore holes is
    (a) 1
    (b) 3
    (c) 5
    (d) 7.

361. The minimum clear working space for a pit as recommended by IS code is
    (a) 1 m × 1 m
    (b) 1.2 m × 1.2 m
    (c) 1.4 m × 1.4 m
    (d) 2 m × 2 m.

362. Bore holes are found to be economical than the test pits when the depth of exploration is around
    (a) 2 m
    (b) 4 m
    (c) 6 m
    (d) 10 m.

363. Adits are also known as
    (a) Test pits
    (b) Trenches
    (c) Shafts
    (d) Drifts.

364. Drifts are naturally made in
    (a) horizontal direction
    (b) vertical direction
    (c) inclined direction
    (d) none.

365. Shafts are naturally made in
    (a) horizontal direction
    (b) vertical direction
    (c) inclined direction
    (d) none.

366. The mechanical augers will become inconvenient when the depth is greater than
    (a) 6 m
    (b) 12 m
    (c) 25 m
    (d) 37 m.

367. For a sandy soil below the water table, the presence of a casing is
    (a) required
    (b) not required
    (c) not at all necessary
    (d) none.

368. Disadvantage with the auger boring is
    (a) difficult to locate the changes in soil strata
    (b) not useful for higher depths
    (c) soil samples are disturbed
    (d) all the above.

369. An undisturbed sample is that in which
    (a) neither the natural structure nor the properties are preserved
    (b) both the natural structure and the properties are preserved
    (c) natural structure is disturbed but the water content is retained
    (d) none.

370. Select the incorrect statement:
    In case of a wash boring
    (a) the equipment is relatively light and inexpensive
    (b) water content of a soil sample is retained
    (c) undisturbed samples are obtained
    (d) process will be slow in hard soils.

371. For making holes in hard strata the preferred one is
    (a) Hand auger
    (b) Mechanical auger
    (c) Wash boring
    (d) Percussion drilling.

372. Samples from auger boring are
    (a) undisturbed samples
    (b) representative samples
    (c) non-representative samples
    (d) none.

373. For undisturbed samples, the area ratio should be less than
    (a) 1%
    (b) 5%
    (c) 7%
    (d) 10%.

374. For undisturbed samples, the inside clearance should be between
    (a) 1 to 3%
    (b) 3 to 5%
    (c) 5 to 7%
    (d) 7 to 10%.

375. For undisturbed samples, the outside clearance should be between
    (a) 0 to 2%
    (b) 2 to 4%
    (c) 4 to 6%
    (d) 6 to 8%.

376. The most critical factor which affects the sample disturbance
    (a) area ratio
    (b) inside clearance
    (c) outside clearance
    (d) none of the above.

377. The samples are called thick wall samplers when the area ratio is about
    (a) 10 to 25%
    (b) 25 to 30%
    (c) 30 to 40%
    (d) 40 to 50%.

378. The internal and external diameters of a split spoon sampler is
    (a) 25 mm and 50 mm
    (b) 30 mm to 50.8 mm
    (c) 35 mm and 50.8 mm
    (d) 40 mm and 50.8 mm.

379. Chunk samples are
    (a) Splits spoon samples
    (b) Scraper bucket samples
    (c) Hand carved samples
    (d) None.

380. Standard penetration test is best suited for
    (a) cohesionless soils
    (b) clayey soils
    (c) silty soils
    (d) none.

381. In case of standard penetration test, if the number of blows exceeds this value for 150 mm penetration, it is generally refused
    (a) 20
    (b) 35
    (c) 50
    (d) 70.

382. Generally the SPT number is corrected for
    (a) dilatancy only
    (b) overburden pressure only
    (c) both for dilatancy and overburden pressure
    (d) none.

383. The apex angle of a Dutch cone is
    (a) 30°
    (b) 45°
    (c) 60°
    (d) 90°.

384. For very soft and sensitive clayee soils, the best exploration method is
    (a) Hand auger
    (b) Mechanical auger
    (c) Wash boring
    (d) Vane shear test.

385. In standard penetration test, when the SPT number (N) value is greater than 15, the equivalent corrected value (N_c) according to Terzaghi and Peck is
    (a) $N_c = 15 - \frac{1}{2}(N - 15)$
    (b) $N_c = 15 + \frac{1}{2}(N - 15)$
    (c) $N_c = 15 - \frac{1}{2}(N + 15)$
    (d) $N_c = 15 + \frac{1}{2}(N + 15)$.

386. The standard penetration test is performed in a hole of diameter about
    (a) 25 mm
    (b) 25 mm to 55 mm
    (c) 55 to 150 mm
    (d) None.

387. The area ratio of the split spoon sampler is
    (a) 100%
    (b) 112%
    (c) 125%
    (d) 150%.

388. Seismic methods are based on the principle that
    (a) soils will have different resistivity values with a variation in their structure
    (b) soils will have different conductivity values of current with a variation in the their structure
    (c) soils will have different velocities of shock waves with a variation in their structure
    (d) none of the above.

389. The Geophones will
    (a) measure the electrical resistivity of a soil
    (b) convert the ground vibration into electrical impulses
    (c) be useful for drawing a bore log
    (d) none of the above.

390. The electrical resistivity (ρ) of any conductor is given by the equation
    (a) $\rho = RA/L$
    (b) $\rho = LA/R$
    (c) $\rho = RAL$
    (d) none.
    where R = Resistance in ohms, A = Cross-sectional area in cm^2, L = Length of sample in cm.

391. The conductance of a material will be
    (a) equal to electrical resistivity
    (b) twice the value of electrical resistivity
    (c) reciprocal of electrical resistivity
    (d) square root of electrical resistivity.

392. Seismic methods are developed based on the assumption that
    (a) the subsoil layers will have equal thicknesses
    (b) the velocity of the shock waves decreases with depth
    (c) the velocity of the shock waves increases with depth
    (d) none.

393. Select the correct statement
    (a) Electrical sounding method studies the changes in the strata with depth at a particular point
    (b) Electrical profiling method studies the boundaries of different strata

(c) Resistivity mapping method is nothing but the electrical profiling method
(d) All the above.

394. In electrical resistivity method of exploration, the depth of exploration is roughly proportional to
 (a) spacing of electrodes
 (b) size of electrodes
 (c) number of electrodes
 (d) none of the above.

395. The mean resistivity value (p) in electrical resistivity method is calculated by
 (a) $p = \dfrac{2\pi DI}{V}$
 (b) $p = \dfrac{2\pi DV}{I}$
 (c) $p = 2\rho\, DVI$
 (d) none

 where D = Distance between electrodes, I = Current supplies, V = Voltage drop.

396. Vane shear test is used for
 (a) measuring shear strength of cohesive soil
 (b) measuring void ratio of sandy soils
 (c) measuring bearing capacity of soils
 (d) all the above.

397. Shearing strength of cohesionless soil depends upon
 (a) dry density
 (b) void ratio
 (c) loading rate
 (d) normal stress.

398. The strength theory applicable to a C-ϕ soil is
 (a) Darcy's law
 (b) Terzaghi's theory
 (c) Mohr-Coulomb theory
 (d) all.

399. The effective angle of internal friction will be
 (a) zero
 (b) 90
 (c) limited to a maximum of 45
 (d) all the above.

400. Effective stress is
 (a) the stress at the particles contact
 (b) governing factor for the engineering properties of the soil
 (c) a physical parameter
 (d) all the above.

401. Rise in the water table position upto the ground surface causes
 (a) increase in effective stress
 (b) decrease in effective stress
 (c) increase in total stress
 (d) decrease in total stress.

402. In Mohr's stress circle, the co-ordinates of the centre of the circle are
 (a) (0, 0)
 (b) (σ_1, σ_3)
 (c) $\left(\dfrac{\sigma_1 + \sigma_3}{2}, 0\right)$
 (d) $\left(\dfrac{\sigma_1 + \sigma_3}{2}, \dfrac{\sigma_1 - \sigma_3}{2}\right)$.

403. The angle made by the failure envelope with horizontal in the Mohr's circle gives
 (a) cohesion
 (b) angle of internal friction
 (c) angle of wall friction
 (d) surcharge angle.

404. Unconfined compression test is generally done on saturated clays for which the apparent angle of shearing resistance is
 (a) 0
 (b) 15°
 (c) $22\frac{1}{2}°$
 (d) 30°.

405. Unconfined compression strength is obtained from
 (a) undrained test
 (b) drained test
 (c) slow test
 (d) consolidated drained test.

406. In the unconfined compressive strength, the corrected area of cross-section (A_c) at any strain can be calculated by
 (a) $A_c = A_0$
 (b) $A_c = \dfrac{A_0}{1+\varepsilon}$
 (c) $A_c = A_0(1-\varepsilon)$
 (d) $A = \dfrac{A_0}{1-\varepsilon}$

 where A_0 = Original area of cross-section, ε = Strain.

407. In the unconfined compression test on a sample of initial volume V and length L, the area of cross-section at failure is taken as
 (a) $\dfrac{V}{L+\Delta L}$
 (b) $\dfrac{V}{L-\Delta L}$
 (c) $\dfrac{V+\Delta V}{L+\Delta L}$
 (d) $\dfrac{V-\Delta V}{L-\Delta L}$

 where ΔV = Change in volume and ΔL = Change in length.

408. If a clayey soil specimen is subjected to a vertical compressive load, the angle by the tracks with the horizontal is
 (a) zero
 (b) 45°
 (c) 90°
 (d) 180.

409. The unconfined compression test can be conducted on
 (a) sandy soils
 (b) clayey soils only
 (c) both sandy and clayey soils
 (d) none.

410. The angle between the two planes on which the shearing stress is zero is
 (a) zero
 (b) 30°
 (c) 45°
 (d) 90°.

411. In Mohr's stress circle the radius of circle when σ_1 and σ_3 are major principal stress and τ is the shear stress on these planes is
 (a) $\sqrt{\dfrac{\sigma_1-\sigma_3}{2}+\tau^2}$
 (b) $\dfrac{\sigma_1-\sigma_3}{2}+\tau$
 (c) $\sqrt{\dfrac{\sigma_1-\sigma_3}{2}+\tau}$
 (d) $\sqrt{\left(\dfrac{\sigma_1-\sigma_3}{2}\right)^2+\tau^2}$.

412. In the Mohr's circle for the above condition the centre of circle has coordinates as
    (a) $\left(\dfrac{\sigma_1 - \sigma_3}{2}, 0\right)$ (b) $\left(\dfrac{\sigma_1 + \sigma_3}{2}, 0\right)$
    (c) $\left(\dfrac{\sigma_1 + \sigma_3}{2}, \dfrac{\sigma_1 - \sigma_3}{2}\right)$ (d) $\left(0, \dfrac{\sigma_1 + \sigma_3}{2}\right)$.

413. In the triaxial test, when the soil sample is subjected to only the cell pressure, then the Mohr's circle will be

    (a) 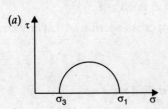   (b)

    (c) 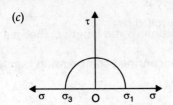   (d)

414. In Mohr's circle the angle made by the plane which consists of maximum shear stress with horizontal at origin of planes is
    (a) zero   (b) 45°
    (c) 90°   (d) 180°.

415. The maximum shear stress in case of Mohr's circle will be numerically equal to
    (a) 1   (b) 3
    (c) $\dfrac{\sigma_1 + \sigma_3}{2}$ (d) $\dfrac{\sigma_1 - \sigma_3}{2}$.

416. The shear stresses on two planes which are at right angles to each other are
    (a) numerically equal and of the same sign
    (b) numerically equal and are of the opposite sign
    (c) not equal
    (d) can't say.

417. The shape of plot between shear and normal stresses according to Mohr's theory is
    (a) straight line   (b) curve
    (c) elliptical   (d) all.

418. Expansion of soils under shear is known as
    (a) liquefaction   (b) volumetric deformation
    (c) critical expansion   (d) dilatancy.

419. In a strain-controlled shear test
    (a) the shear force is increased at a constant rate
    (b) the shearing strain increases at a given rate
    (c) both (a) and (b)
    (d) none.

420. In a stress-controlled shear test
    (a) the shear force is increased at a constant rate
    (b) the shearing strain increases at a given rate
    (c) both (a) and (b)
    (d) none.

421. The stress-strain characteristics will be better observed in
    (a) stress-controlled test
    (b) strain-controlled test
    (c) both (a) and (b)
    (d) none.

422. The test which represents the field conditions more closely is
    (a) stress-controlled test
    (b) strain-controlled test
    (c) both (a) and (b)
    (d) none.

423. Undrained shear strength S_u of saturated clay tested in unconfined compression is given in terms of unconfined compressive strength q_u as
    (a) $S_u = 1/2\, q_u$
    (b) $S_u = q_u$
    (c) $S_u = 2\, q_u$
    (d) $S_u = 2/3\, q_u$.

424. The value $N\phi = \tan^2(45 + \phi/2)$ is called
    (a) bearing capacity factor
    (b) flow value
    (c) friction factor
    (d) stability number.

425. Shear strength of soil is determined by the equation
    (a) $C = s + \sigma \tan \phi$
    (b) $\sigma = C + s \tan \phi$
    (c) $s = C + \sigma \tan \phi$
    (d) $s = \sigma + C \tan \phi$.

426. The stress that is responsible for the mobilisation of shearing strength of a soil
    (a) total stress
    (b) effective stress
    (c) neutral stress
    (d) none.

427. Rise of water table above the ground surface causes
    (a) equal increase in pore water pressure and total stress
    (b) equal increase in pore water pressure and effective stress
    (c) equal increase in total stress and effective stress
    (d) none of the above.

428. Rise in water table upto the ground surface results in
    (a) reduction of pore water pressure but no change in total stress
    (b) increase of pore water pressure and increase in total stress
    (c) increase of pore water pressure and increase in total stress
    (d) increase of pore water pressure and decrease in total stress.

429. Drainage conditions during test can be controlled best in
   (a) direct shear test
   (b) vane shear test
   (c) unconfined compression test
   (d) triaxial shear test.

430. The type of test in which no significant volume changes are expected is
   (a) consolidated drainage test
   (b) consolidated undrained test
   (c) unconsolidated undrained test
   (d) all the above.

431. If the stress-strain curve of a clayey soil showed a peak, it can be
   (a) normally consolidated clay
   (b) under consolidated clay
   (c) over consolidated clay
   (d) none.

432. The type of shearing test in which there is a pre-determined failure plane
   (a) Direct shear test
   (b) Trianial test
   (c) Vane shear test
   (d) Unconfined compression test.

433. The shear failure exhibited by loose sands is known as
   (a) Elastic failure
   (b) Plastic failure
   (c) Brittle failure
   (d) None of the above.

434. The shear failure exhibited by dense sands is known as
   (a) Elastic failure
   (b) Plastic failure
   (c) Brittle failure
   (d) None of the above.

435. Select the correct statement
   (a) A well graded sandy soil will exhibit more strength than a uniform sandy soil
   (b) The sand may appear to have cohesion due to capillary moisture
   (c) The angle of shearing resistance of sand is independent of rate of loading
   (d) All the above.

436. Clays generally exhibit plasticity property when they are mixed with
   (a) Kerosene
   (b) Oil
   (c) HCl
   (d) Water.

437. Select the correct statement.
   (a) The shear strength in cohesionless soils is mainly due to function between particles.
   (b) In a drained test, the loose sand shows a decrease in volume.
   (c) In dense sands, interlocking between particles also contribute to the strength
   (d) All the above.

438. Select the incorrect statement.
   (a) The value of angle of shearing resistance is related to the density index
   (b) The shear strength decreases with increase in the confining pressure
   (c) The interlocking effect increases with an increase in the density
   (d) none of the above.

439. The shear strength of sands with the following shape will be more
   (a) sharp edged particles
   (b) rounded edged particles
   (c) flat particles
   (d) none.

440. The ultimate strength of a loose and dense sand samples which are consolidated to the same value of effective stress will be
    (a) same □   (b) more for dense sands □
    (c) more for loose sends □   (d) none. □

441. In a consolidated drained test on a normally consolidated clayey soils, the volume of the soil during the shearing will
    (a) remain same □   (b) increase □
    (c) decrease □   (d) can't say. □

442. The stress-strain curve of an over consolidated clay is similar to that of
    (a) normal consolidated clay □   (b) loose sands □
    (c) silts □   (d) dense sands. □

443. The stress-strain curve of a normal consolidated clay is similar to that of
    (a) gravels □   (b) loose sands □
    (c) silts □   (d) dense sands. □

444. In the triaxial test, the intermediate principal stress will be equal to
    (a) Major principal stress □
    (b) Deviator stress □
    (c) Minor principal stress □
    (d) Average of major and minor principal stresses. □

445. Select the correct statement.
    (a) When compared to triaxial test, direct shear test is a slow test □
    (b) In a Direct shear test, the plane of shear failure is random □
    (c) In a triaxial test, there will be better control on the drainage of the soil when compared to direct shear test □
    (d) None of the above. □

446. Vane shear test is a
    (a) field test □   (b) laboratory test □
    (c) can be done both in field and laboratory □   (d) none. □

447. Select the incorrect statement.
    (a) The failure envelope of normally consolidated clays passes through the origin. □
    (b) The failure envelope of over consolidated clays passes through the origin. □
    (c) The failure envelope of a moist sands shows a cohesion intercept on shear stress axis. □
    (d) None of the above. □

448. The failure envelope for a clayey soil will be horizontal in the case of
    (a) unconsolidated undrained test □   (b) consolidated undrained test □
    (c) consolidated drained test □   (d) none. □

449. Stress distribution on the failure plane in the case of a triaxial test is
    (a) zig-zag □   (b) non-uniform □
    (c) uniform □   (d) can't say. □

450. Select the incorrect statement.
    (a) The shear strength of clayey soil will depend on the drainage condition. ☐
    (b) The angle of shearing resistance decreases with an increase in plasticity index of clay. ☐
    (c) As the clay content in a cohesive soil increases, the angle of shearing resistance value decreases. ☐
    (d) The shear strength of a disturbed sample is more than that of the undisturbed sample. ☐
451. In unconfined compression test, alround stress is
    (a) equal to major principal stress ☐   (b) half the major principal stress ☐
    (c) equal to zero ☐   (d) equal to intermediate principal stress. ☐
452. The value of cohesion of a saturated clay will be ...... times the value of unconfined compression strength.
    (a) 2 ☐   (b) 1.0 ☐
    (c) 0.5 ☐   (d) zero. ☐
453. For saturated clays under undrained conditions
    (a) $S = q_u$ ☐   (b) $S = q_u/2$ ☐
    (c) $S = (-u) \tan \phi$ ☐   (d) $S = c + \tan \phi$. ☐
454. The neutral stress on the soil is due to
    (a) weight of water present in soil pores ☐   (b) deformation of soil grains ☐
    (c) external load acting on the soil ☐   (d) weight of the soil grains. ☐
455. When a shear stress is applied on a dense sand, then the volume of sand will
    (a) remain constant ☐   (b) increase ☐
    (c) decrease ☐   (d) can't say. ☐
456. The application of deviator stress on a specimen in case of a triaxial test will produce shear stress on
    (a) vertical plane only ☐   (b) horizontal plane only ☐
    (c) both (a) and (b) ☐   (d) none. ☐
457. Sensitivity of a claycy soil is defined as
    (a) ratio of strength of an undisturbed soil sample to that of a remoulded sample ☐
    (b) ratio of strength of a remoulded soil sample to that of an undisturbed soil sample ☐
    (c) ratio of deviator load to cross sectional area ☐
    (d) none. ☐
458. The test that will be preferred for testing of shear strength of a saturated clay is
    (a) Direct shear test ☐   (b) Vane shear test ☐
    (c) Triaxial test ☐   (d) Unconfined compression test. ☐
459. The line joining the points of maximum shear stresses as the load is changing is called
    (a) Isobar diagram ☐   (b) Stress circle ☐
    (c) Mohr's circle ☐   (d) Stress-path. ☐
460. The coordinates of a point on the stress path are
    (a) $\dfrac{\sigma_1+\sigma_3}{2}, \dfrac{\sigma_1-\sigma_3}{2}$ ☐   (b) $\dfrac{\sigma_1-\sigma_3}{2}, \dfrac{\sigma_1+\sigma_3}{2}$ ☐
    (c) $\dfrac{\sigma_1-\sigma_3}{2}, \dfrac{\sigma_1-\sigma_3}{2}$ ☐   (d) none. ☐

461. Select the incorrect statement.
    (a) The stress-strain curve for dense sands exhibits relatively high initial tangent modulus ☐
    (b) The stress-strain curve for loose sands exhibits low initial tangent modulus ☐
    (c) In a drained test, the dense sand shows a volume decrease, but as the strain increases, the volume starts increasing ☐
    (d) None of the above. ☐
462. Most of the shear tests are done in equipment which are
    (a) stress controlled ☐   (b) strain controlled ☐
    (c) drainage controlled ☐   (d) volume controlled. ☐
463. For a saturated soil, Skempton's constant 'B' value is
    (a) zero ☐   (b) 1.0 ☐
    (c) 10.0 ☐   (d) Any value. ☐
464. Pile foundation are usually provided for
    (a) bridges ☐   (b) high rise multistoreyed buildings ☐
    (c) runways ☐   (d) residential buildings. ☐
465. Number of piles required to support a column
    (a) 1 ☐   (b) 2 ☐
    (c) 3 ☐   (d) 4. ☐
466. Underreamed piles are generally
    (a) driven piles ☐   (b) bored piles ☐
    (c) precast piles ☐   (d) all the above. ☐
467. Bitumen, when added to soil, mainly
    (a) increases cohesion ☐   (b) accelerates compaction ☐
    (c) acts as water proofing agent ☐   (d) fills voids in soils. ☐
468. For a foundation of depth D in a cohesionless soil, the ultimate bearing capacity q_u is given by
    (a) $\gamma.D \left(\dfrac{1-\sin\phi}{1+\sin\phi} \right)$ ☐
    (b) $\gamma.D \left(\dfrac{1+\sin\phi}{1-\sin\phi} \right)$ ☐
    (c) $\gamma.D \left(\dfrac{1-\sin\phi}{1+\sin\phi} \right)^2$ ☐
    (d) $\gamma.D \left(\dfrac{1+\sin\phi}{1-\sin\phi} \right)^2$. ☐
469. Terzaghi's equation for ultimate bearing capacity of a circular footing of diameter B is
    (a) $q_u = 1.3\, cN_c + \gamma DN_v + 0.3\, \gamma BN_r$ ☐
    (b) $q_u = 1.3\, cN_c + \gamma DN_q + 0.5\, \gamma BN_r$ ☐
    (c) $q_u = 1.3\, cN_c + \gamma DN_q + 0.4\, \gamma BN_r$ ☐
    (d) $q_u = cN_c + \gamma DN_q + 0.5\, \gamma BN_r$. ☐
470. In question 469, which is the equation for ultimate bearing capacity of a square footing of size B?
471. In question 469, which equation gives the ultimate bearing capacity of a strip footing of width B and infinite length?

472. According to Terzaghi in case of presence of water table the ultimate bearing capacity equation has to be modified as
    (a) $q_{ult} = CN_c + \gamma DN_q + 0.5\gamma BN_r$
    (b) $q_{ult} = CN_c - \gamma DN_q - 0.5\gamma BN_r$
    (c) $q_{ult} = CN_c - \gamma DN_q R_{w_1} - 0.5\gamma BN_r R_{w_2}$
    (d) $q_{ult} = CN_c + \gamma DN_q R_{w_1} + 0.5\gamma BN_r R_{w_2}$
    where R_{w_1} and R_{w_2} are correction factors.

473. Any foundation must satisfy that
    (a) the soil below does not fail in shear
    (b) the settlement of footing is within safe limits
    (c) the pressure enerted by footing should not exceed the allowable bearing pressure
    (d) all the above.

474. The safe bearing capacity of a soil can be defined as
    (a) maximum pressure which the soil can carry safely without shear failure
    (b) ultimate load on the bearing area
    (c) pressure at which the settlement will not exceed the allowable settlement
    (d) none.

475. The expression for the gross safe bearing capacity is
    (a) $\dfrac{q_{ult}}{F.S.}$
    (b) $\dfrac{q_{net\ ult}}{F.S.}$
    (c) $\dfrac{q_{net\ ult}}{F.S. + \gamma.D}$
    (d) none.

476. The bearing capacity of a soil does not depend on
    (a) applied load
    (b) size of the footing
    (c) shear parameters of soil
    (d) none.

477. The net bearing pressure which can be used for the design of foundations
    (a) gross safe bearing capacity
    (b) ultimate bearing capacity
    (c) net safe settlement pressure
    (d) net allowable bearing pressure.

478. The value of net allowable bearing pressure is
    (a) net safe bearing capacity
    (b) net safe settlement pressure
    (c) minimum of net safe bearing capacity and net safe settlement pressure
    (d) none.

479. The analysis made from the theory of penetration of punches into metals is
    (a) Terzaghi's analysis
    (b) Prandtt's analysis
    (c) Meyerhof's analysis
    (d) Rankine's analysis.

480. Select the incorrect statement:
    The assumptions made by Terzaghi are
    (a) the base of the footing is smooth
    (b) shear strength of the soil is governed by Mohr-Coulomb equation
    (c) the footing is a shallow foundation
    (d) the shear strength of the soil above the base of the footing is neglected.

481. The resultant passive pressure on the surfaces of the elastic wedge in Terzaghi's theory consists of
    (a) Passive pressure because of cohesion
    (b) Passive pressure because of surcharge
    (c) Passive pressure because of unit weight
    (d) All the above.

482. The Terzaghi's bearing capacity factors are functions of
    (a) internal friction angle
    (b) cohesion
    (c) friction angle between footing and soil
    (d) all the above.

483. The type of shear failure that is expected for a dense sand (or) a stiff clay is
    (a) general shear failure
    (b) local shear failure
    (c) punching shear failure
    (d) all the above.

484. The type of shear failure that can be expected for a loose sand (or) a soft clay is
    (a) general shear failure
    (b) local shear failure
    (c) punching shear failure
    (d) all the above.

485. The shear failure for which a peak can be observed for the load-settlement curve is
    (a) general shear failure
    (b) local shear failure
    (c) punching shear failure
    (d) all the above.

486. The shear failure for which no heave will be observed is
    (a) general shear failure
    (b) local shear failure
    (c) punching shear failure
    (d) all the above.

487. Terzaghi has suggested mobilised cohesion and mobilised angle of internal friction for
    (a) general shear failure
    (b) local shear failure
    (c) punching shear failure
    (d) all the above.

488. The value of mobilised cohesion will be
    (a) equal to C
    (b) 1/3 C
    (c) 2/3 C
    (d) 2 C
    where C is cohesion of the soil.

489. The type of shear failure that can be expected for a cohesionless soil whose angle of internal friction is less than 29° is
    (a) general shear failure
    (b) local shear failure
    (c) punching shear failure
    (d) all the above.

490. The type of shear failure that can be expected for a soil whose relative density is less than 35% is
    (a) general shear failure
    (b) local shear failure
    (c) punching shear failure
    (d) all the above.

491. The type of shear failure that can be expected for a soil whose void ratio is less than 0.55 is
    (a) general shear failure
    (b) local shear failure
    (c) punching shear failure
    (d) all the above.

492. The value of reduction factors R_{w_1} and R_{w_2} when the water table is at the ground surface are (where R_{w_1}: correction for surcharge term and R_{w_2}: correction for weight term)
    (a) 1.0 and 1.0
    (b) 0.5 and 1.0
    (c) 1.0 and 0.05
    (d) 0.5 and 0.5.

493. The values of reduction factors R_{w_1} and R_{w_2} when the water table is at a depth equal to or greater than the width of the footing are
    (a) 1.0 and 1.0
    (b) 0.5 and 1.0
    (c) 1.0 and 0.5
    (d) 0.5 and 0.5.

494. The value of reduction factors R_{w_1} and R_{w_2} when the water table is at the base of the footing are
    (a) 1.0 and 1.0
    (b) 0.5 and 1.0
    (c) 1.0 and 0.5
    (d) 0.5 and 0.5.

495. The deformations of soils under a strip footing are
    (a) one dimensional
    (b) two dimensional
    (c) three dimensional
    (d) none.

496. The deformations of soils under a circular or square footing are
    (a) one dimensional
    (b) two dimensional
    (c) three dimensional
    (d) none.

497. Main difference between Terzaghi and Meyerhof's theories is based on
    (a) roughness of the footing
    (b) type of loading
    (c) shear strength of soil above the base of the footing
    (d) all the above.

498. According to Meyerhof, the foundation depth parameter is
    (a) normal stress on the equivalent free surface
    (b) shear stress on the equivalent free surface
    (c) angle made by the equivalent free surface with horizontal
    (d) all the above.

499. According to Meyerhof, the value of the angle made by the equivalent free surface with horizontal will be
    (a) zero
    (b) 30°
    (c) 45°
    (d) 90°.

500. The maximum value of N_c for a strip footing according to Meyerhof when $D_{f/b}$ ratio is greater than 2.5 is
    (a) 5.14
    (b) 5.7
    (c) 8.25
    (d) 9.0.

501. Skempton found that for a cohesive soil the bearing capacity factor N_c in Terzaghi's equation
    (a) decreases with depth
    (b) increases with depth
    (c) is constant with depth
    (d) none.

502. The minimum value of N_c suggested by Skempton for a strip footing is
    (a) 5.14
    (b) 5.7
    (c) 7.50
    (d) 6.20.

503. The minimum value of N_c suggested by Skempton for a square or circular footing is
    (a) 5.14
    (b) 5.7
    (c) 7.50
    (d) 6.2.

504. According to Skempton, N_c reaches the maximum value when the ratio of $D_{f/b}$ is
    (a) less than 4.5
    (b) greater than 4.5
    (c) greater than 2.0
    (d) none.

505. A raft foundation is preferred for
    (a) columns of industrial buildings
    (b) load bearing walls of multistoreyed buildings
    (c) columns of a building which are closely spaced
    (d) none.

506. The permissible value of tension for soil when the footings are subjected to both compression and bending is
    (a) zero
    (b) 1 N/cm^2
    (c) 1 t/m^2
    (d) depends on soil.

507. According to Terzaghi's theory, the value of the ultimate bearing capacity for a strip footing which is at the ground surface on a purely cohesive soil is
    (a) 5.7 C
    (b) 5.14 C
    (c) 7.5 C
    (d) 6.2 C.

508. Using a plate load test, we can determine
    (a) the allowable bearing pressure
    (b) the ultimate bearing capacity
    (c) the settlements for a given intensity of loading
    (d) all the above.

509. The size of the plate for a plate load test will be
    (a) less than 0.2 m
    (b) between 0.3 m and 0.75 m
    (c) greater than 1.0 m
    (d) any size.

510. In a cohesionless soil, the settlement of a 30 cm plate in a plate load test is 2 cm, then the settlement of square footing of 90 cm side under the same load intensity will be
    (a) 2 cm
    (b) 4 cm
    (c) 6 cm
    (d) 4.5 cm.

511. In a cohesive soil, the settlement of a 30 cm plate in a plate load test is 2 cm, then the settlement of a square footing of 90 cm side under the same load intensity will be
    (a) 2 cm
    (b) 4 cm
    (c) 6 cm
    (d) 4.5 cm.

512. The seating load that will be generally applied in a plate load test is
    (a) 2 kN/m²
    (b) 5 kN/m²
    (c) 7 kN/m²
    (d) 10 kN/m².

513. In the plate load test, the load will be applied in the increments of
    (a) 20% of estimated safe load (or) 10% of estimated ultimate load
    (b) 10% of estimated safe load (or) 20% of estimated ultimate load
    (c) 30% of estimated safe load (or) 40% of estimated ultimate load
    (d) 40% of estimated safe load (or) 30% of estimated ultimate load.

514. The minimum settlement that is to be observed for ending the plate load test
    (a) 10 mm
    (b) 15 mm
    (c) 20 mm
    (d) 25 mm.

515. The zero correction for the plate load test results is proposed by
    (a) Terzaghi
    (b) Meyerhof
    (c) Abbet
    (d) Rankine.

516. According to Housel, for the design of a shallow foundation for a given safe settlement, the minimum number of plate load tests to be conducted is
    (a) 1
    (b) 2
    (c) 3
    (d) 4.

517. The generally observed limitations for the plate load test are
    (a) size
    (b) time
    (c) scale (or) size
    (d) all the above.

518. Rise of water table upto the ground surface in case of a cohesionless soil will reduce the ultimate bearing by (approximately)
    (a) 10%
    (b) 25%
    (c) 50%
    (d) 75%.

519. Two footings, one with circular and the other of square are founded in purely saturated clay soil. The ratio of their ultimate bearing capacities according to Terzaghi is (assume diameter of circular footing = width of square footing)
    (a) 1.0
    (b) 1.5
    (c) 1.3
    (d) 0.75.

520. Two footings, one with circular and the other of square are founded on the surface of cohesionless soil. The ratio of their ultimate bearing capacities according to Terzaghi is (assume diameter of circular footing is equal to the width of square footing)
    (a) 1.0
    (b) 1.5
    (c) 1.3
    (d) 0.75.

521. Two or more footings connected by a beam is called as
    (a) Strap footings
    (b) Cantilever footing
    (c) Pump-handle foundation
    (d) All the above.

522. When the area of all the footings covers more than 50% of the total area of the structure, then the foundation that is preferrable is
   (a) isolated foundation
   (b) combined footings
   (c) raft foundation
   (d) none of the above.

523. Mat foundations are generally preferred when
   (a) the allowable soil pressure is low
   (b) the expected differential settlement for individual footings is high
   (c) Area covered by individual footing is more than 50% of total area of structure
   (d) none of the above.

524. California bearing ratio test is used
   (a) to find the bearing capacity of the soil
   (b) to find the thickness of a flexible pavement
   (c) to find the ratio of ultimate bearing capacity to net bearing capacity
   (d) none of the above.

525. The process of compaction may involve
   (a) rolling
   (b) tamping
   (c) vibration
   (d) all of the above.

526. If ϕ is the angle of shearing resistance, the angle between the direction of failure and the direction of major principal plane is given by
   (a) $45 + \phi$
   (b) $45 - \phi$
   (c) $(45 + \phi)/2$
   (d) $45 + \phi/2$.

527. The bearing capacity of a weak soil may be improved by
   (a) increasing the depth of foundation
   (b) compacting the soil by ramming
   (c) removing the poor soil and filling the gap with sand, rubble etc.
   (d) any of the above.

528. If the co-efficient of active earth pressure is 1/2, what is the value of co-efficient of passive earth pressure?
   (a) 1/4
   (b) 1/2
   (c) 1
   (d) 2.

529. A soil mass is said to be in inelastic equilibrium if it is in
   (a) Active state
   (b) Passive state
   (c) At-rest state
   (d) All the above.

530. A soil mass is said to be in plastic equilibrium if it is
   (a) in the plastic stage
   (b) stressed to maximum value
   (c) on the verge of failure
   (d) all the above.

531. The base width of retaining wall is B. What is the maximum permissible eccentricity if the wall should not fail in tension?
   (a) $B/3$
   (b) $2B/3$
   (c) $B/6$
   (d) $B/2$.

532. The material that is retained by a retaining structure is generally called as
    (a) surcharge
    (b) backfill
    (c) soil slope
    (d) all the above.

533. The portion of the back fill material lying above a horizontal plane at the elevation of the top of the retaining wall is called as
    (a) surcharge
    (b) backfill
    (c) soil slope
    (d) all the above.

534. Retaining walls are predominantly subjected to
    (a) vertical loads
    (b) lateral loads
    (c) upward loads
    (d) none of the above.

535. The coefficient of earth pressure is calculated using the
    (a) theory of plasticity
    (b) theory of elasticity
    (c) theory of shear strength
    (d) none of the above.

536. For calculating the value of the coefficient of earth pressure at rest the soil is assumed to be
    (a) elastic
    (b) isotropic and homogeneous
    (c) semi-infinite
    (d) all the above.

537. The earth pressure at rest corresponds to a condition of lateral strain equal to
    (a) infinite
    (b) unity
    (c) zero
    (d) can't say.

538. The value of coefficient of earth pressure for soils will lie within the range of
    (a) 0 to 0.3
    (b) 0.4 to 0.8
    (c) 0.8 to 1.0
    (d) 1 to 1.4.

539. In active state of plastic equilibrium in a cohesionless soil with a horizontal ground surface
    (a) major principal stress is vertical
    (b) major principal stress is horizontal
    (c) minor principal stress is vertical
    (d) none of the above.

540. In the passive state of plastic equilibrium in a cohesionless soil with a horizontal ground surface
    (a) major principal stress is vertical
    (b) major principal stress is horizontal
    (c) minor principal stress is vertical
    (d) none of the above.

541. For the passive state of earth pressure to develop, the retaining wall will
    (a) move away from the backfill
    (b) move towards the back fill
    (c) remains at-rest
    (d) none of the above.

542. In the active state of earth pressure
    (a) vertical stress is constant and horizontal stress decreases
    (b) vertical stress is constant and horizontal stress increases
    (c) vertical stress and horizontal stress decreases
    (d) vertical stress and horizontal stress increases.

543. In the passive state of earth pressure
   (a) vertical stress is constant and horizontal stress decreases
   (b) vertical stress is constant and horizontal stress increases
   (c) vertical stress and horizontal stress decrease
   (d) vertical stress and horizontal stress increases.

544. The state of shear failure corresponding to the minimum earth pressure condition is called
   (a) Active state
   (b) Passive state
   (c) At-rest state
   (d) None of the above.

545. The value of the earth pressure will be maximum in
   (a) Active state
   (b) Passive state
   (c) At-rest state
   (d) None of the above.

546. For the full active pressure to reach in case of dense sands, the horizontal strain that is required is
   (a) greater than 2%
   (b) greater than 10%
   (c) less than 0.5%
   (d) less than − 0.5%.

547. For full passive resistance to develop in case of dense sands, the horizontal strain that is required is
   (a) about 2%
   (b) about 10%
   (c) about 0.5%
   (d) none of the above.

548. If the backfill is having a uniform surcharge of intensity q per unit area, the lateral pressure everywhere will increase by a quantity of
   (a) intensity q
   (b) $\gamma . q$
   (c) $K_a q$
   (d) none of above
   where γ is unit weight, K_a is coefficient of active earth pressure.

549. The equivalent height of the fill in case of retaining wall with a surcharge is
   (a) γ/q
   (b) $\gamma . q$
   (c) q/γ
   (d) none of above
   where γ is unit weight, q is surcharge intensity.

550. The equation for the coefficient of passive earth pressure as given by Rankine for an inclined fill is
   (a) $K_p = \cos \beta \dfrac{\cos \beta - \sqrt{\cos^2 \beta - \cos^2 \phi}}{\cos \beta + \sqrt{\cos^2 \beta - \cos^2 \phi}}$
   (b) $K_p = \cos \beta \dfrac{\cos \beta - \sqrt{\cos^2 \beta - \cos^2 \phi}}{\cos \beta - \sqrt{\cos^2 \beta + \cos^2 \phi}}$
   (c) $K_p = \cos \beta \dfrac{\cos \beta + \sqrt{\cos^2 \beta - \cos^2 \phi}}{\cos \beta - \sqrt{\cos^2 \beta - \cos^2 \phi}}$
   (d) none of the above
   where β is angle of surcharge of the fill, ϕ is angle of internal friction of the soil.

551. The expression for the total active earth pressure in case of a horizontal sandy soil is
   (a) $K_a \gamma H$
   (b) $K_a q$
   (c) $K_a q H$
   (d) $1/2 \, K_a \gamma H^2$.

552. The expression for coefficient of earth pressure at rest is proportional to
    (a) $\dfrac{\mu}{1-\mu}$
    (b) $1 + \sin \phi$
    (c) $1 - \cos \phi$
    (d) $1 - \sin \phi$.

553. In the case of a backfill with a sloping surface, the total active pressure on the wall of height H acts at
    (a) $H/3$ above the base parallel to the base
    (b) $H/3$ above the base parallel to the sloping surface
    (c) $H/2$ above the base parallel to the base
    (d) $H/2$ above the base parallel to the sloping surface.

554. In Rankine's earth pressure theory, the following assumption is made
    (a) wall face is rough and vertical
    (b) back fill is cohesive soil
    (c) wall face is smooth and vertical
    (d) elastic equilibrium is satisfied.

555. Distribution of earth pressure with depth is
    (a) parabolic
    (b) hydrostatic
    (c) non-linear but increasing
    (d) decreasing and increasing.

556. The active earth pressure is proportional to
    (a) $\tan^2 (45 + \phi/2)$
    (b) $\tan (45 - \phi/2)$
    (c) $\tan^2 (45 - \phi/2)$
    (d) none of the above.

557. In the active state the angle made by the failure plane with major principal plane which is horizontal is
    (a) $45 - \phi/2$
    (b) $45 + \phi/2$
    (c) 45
    (d) zero.

558. In the passive state the angle made by the failure plane with the major principal plane which is vertical is
    (a) $45 - \phi/2$
    (b) $45 + \phi/2$
    (c) 45
    (d) zero.

559. When the failure wedge of a soil moves upwards and inwards, it indicates
    (a) active state
    (b) passive state
    (c) at rest condition
    (d) none of the above.

560. The condition in which the failure envelope will not touch the Mohr's circle is
    (a) active state condition
    (b) passive state condition
    (c) at rest condition
    (d) none of the above.

561. Select the correct statement:
    In Coulomb's theory
    (a) equilibrium of the sliding wedge is taken into consideration
    (b) the slip surface is a plane surface which passes through the heel of the wall
    (c) for the active case, the trial surface which gives the largest force to maintain the equilibrium of wedge is taken into consideration
    (d) all the above.

562. Select the incorrect statement:
 In Coulomb's theory
 (a) the surface of the wall is assumed to be rough
 (b) the sliding wedge itself acts as a rigid body
 (c) the resultant earth pressure on the wall is inclined at an angle to the normal of the wall, which is the angle of wall friction
 (d) none of the above.

563. The value of angle of wall friction for concrete walls is generally taken as
 (a) $\frac{1}{3}\phi$
 (b) $\frac{1}{2}\phi$
 (c) $\frac{2}{3}\phi$
 (d) ϕ.

564. The expression for the active earth pressure coefficient as given by Coulomb is

 (a) $K_a = \dfrac{\sin^2(\beta+\phi)}{\sin^2\beta \sin(\beta+\delta)\left[1 + \sqrt{\dfrac{\sin(\phi+\delta)\sin(\phi-i)}{\sin(\beta-\delta)\sin(\beta+i)}}\right]^2}$

 (b) $K_a = \dfrac{\sin^2(\beta+\phi)}{\sin^2\beta \sin(\beta-\delta)\left[1 - \sqrt{\dfrac{\sin(\phi+\delta)\sin(\phi-i)}{\sin(\beta-\delta)\sin(\beta+i)}}\right]^2}$

 (c) $K_a = \dfrac{\sin^2(\beta+\phi)}{\sin^2\beta \sin(\beta-\delta)\left[1 + \sqrt{\dfrac{\sin(\phi+\delta)\sin(\phi-i)}{\sin(\beta-\delta)\sin(\beta+i)}}\right]^2}$

 (d) none of the above

 where ϕ is angle of internal friction, δ is angle of wall friction, i is angle of surcharge, β is angle made by the back of the wall with the horizontal.

565. The resultant active earth pressure in the case of Coulomb's theory will act at a height of
 (a) $H/3$ from the top of the wall
 (b) $H/3$ above the base of the wall
 (c) $H/2$ above the base of the wall
 (d) none of the above.

566. Coefficient of earth pressure at rest is
 (a) greater than the active earth pressure coefficient but less than the passive earth pressure coefficient
 (b) less than the active earth pressure coefficient but greater than the passive earth pressure coefficient
 (c) less than both the active and passive earth pressure coefficients
 (d) greater than both the active and passive earth pressure coefficients.

567. Wedge shape failure is assumed by
 (a) Coulomb
 (b) Rankine
 (c) Poncelet
 (d) Terzaghi.

568. Compared to dry back fill submerged back fill will exert
    (a) same earth pressure
    (b) less earth pressure
    (c) more earth pressure
    (d) difficult to tell.

569. A plane inclined at an angle ϕ to the horizontal at which the soil is expected to stay without sliding
    (a) ϕ-line
    (b) repose line
    (c) natural slope line
    (d) all the above.

570. If the resultant force at the bottom of retaining wall lies outside the middle third, the failure will be due to
    (a) crushing
    (b) sliding
    (c) upthrust
    (d) overturning.

571. In the case of retaining walls, surcharge is
    (a) excess moisture in wall
    (b) extra load on the wall
    (c) additional load carrying capacity of wall
    (d) extra load on the horizontal backfill.

572. Earth pressure at rest is given by the equation
    (a) $K_0 = \mu/(1-\mu)$
    (b) $K_0 = (1-\mu)/\mu$
    (c) $K_0 = \mu/(1+\mu)$
    (d) $K_0 = (1+\mu)/\mu$.

573. In cohesive soils, depth of vertical cut upto which no lateral support is required is given by
    (a) $2c/\gamma$
    (b) $4c/\gamma$
    (c) $2\gamma/c$
    (d) $4\gamma/c$.

574. The critical vertical depth (H_c) of free standing soil will be
    (a) $2C/\gamma \tan(45 + \phi/2)$
    (b) $4C/\gamma \tan^2(45 + \phi/2)$
    (c) $4C/\gamma \tan(45 + \phi/2)$
    (d) $4C/\gamma \tan^3(45 + \phi/2)$
    where ϕ = angle of internal friction, C = cohesion; γ = unit weight, δ = angle of wall friction.

575. Cohesion in a soil
    (a) increases active pressure and decreases passive pressure
    (b) decreases active pressure and increases passive pressure
    (c) decreases both active and passive pressures
    (d) increases both active and passive pressures.

576. Rebhann's graphical method is based upon
    (a) Rankine's theory
    (b) Westergaard's theory
    (c) Culmann's theory
    (d) Coulomb's theory.

577. Cohesive soils are
    (a) poor for back fill because of large lateral pressure
    (b) good for back fill because of large lateral pressure
    (c) good for back fill because of small lateral pressure
    (d) none of the above.

578. Factor of safety against sliding shall be
    (a) at least 1.5
    (b) more than 3.0
    (c) between 2.5 and 3.0
    (d) equal to 1.0.

579. Active earth pressure is
    (a) always less than passive earth pressure
    (b) always greater than passive earth pressure
    (c) sometimes greater than passive earth pressure
    (d) equal to passive earth pressure.

580. One of the graphical methods for earth pressure determination is
    (a) Newmark's influence chart method
    (b) Mohr diagram method
    (c) Culmann's method
    (d) Taylor's method.

581. Sheet pile walls are used as
    (a) retaining walls for water front construction
    (b) load bearing pile foundations
    (c) seepage preventing devices
    (d) uplift preventing devices.

582. As sheet pile walls are embedded in soil, for the design
    (a) active pressure only is considered
    (b) passive pressure only is considered
    (c) active and passive pressures are considered
    (d) at rest pressures are considered.

583. Sheet piles are held in position by
    (a) self weight of sheet pile.
    (b) tie rods which are anchored
    (c) embedding bottom of sheet pile
    (d) adjusting water level on one side.

584. Caisons are structures used for
    (a) underpinning
    (b) lifting and transporting foundation parts
    (c) dewatering while laying foundations.
    (d) constructing foundations in proper position.

585. The structure which derives its stability due to self weight is
    (a) sheet pile wall
    (b) bulk head
    (c) cantilever retaining wall
    (d) masonry retaining wall.

586. For normally consolidated clay deposits stability analysis of slopes by the following method is appropriate
    (a) Friction circle method
    (b) Swedish circular arc method
    (c) Slices method
    (d) None of the above.

587. If a slope represents the boundary surface of a semi-infinite soil mass inclined to the horizontal and the soil properties for all identical depths below the surface are constant, it is called as
    (a) Infinite slope
    (b) Finite slope
    (c) Critical slope
    (d) None of the above.

588. For slopes of limited extent, the surface of slippage, is usually along
    (a) parabolic arc
    (b) an elliptical arc
    (c) circular arc
    (d) a straight line.

589. Base failure refers to failure surface which
    (a) is above the toe of the slope
    (b) includes toe of the slope
    (c) is below the toe of the slope
    (d) none of the above.

590. Toe failure is most likely in the case of
    (a) steep slopes
    (b) gentle slopes
    (c) all inclinations of slopes
    (d) very steep slopes.

591. When a weak plane exists above the toe, then the probable type of failure that can be expected for the stability of the slope is
    (a) slope failure
    (b) base failure
    (c) toe failure
    (d) transitional failure.

592. When a weak plane exists below the toe, then the probable type of failure that can be expected for the stability of the slope is
    (a) slope failure
    (b) base failure
    (c) toe of failure
    (d) transitional failure.

593. The type of slope failure that can be expected for an infinite slope is
    (a) slope failure
    (b) base failure
    (c) toe failure
    (d) transitional failure.

594. Stability number is given by
    (a) $\gamma H/C$
    (b) $\gamma C/H$
    (c) $H/\gamma C$
    (d) $C/\gamma H$.

595. For a submerged slope, the stability number is given by
    (a) $C/\gamma H$
    (b) $\gamma H/C$
    (c) $C/\gamma_{sub} H$
    (d) $C/\gamma_{sat} H$.

596. For the sudden draw down case, the stability number is given by
    (a) $C/\gamma H$
    (b) $\gamma H/C$
    (c) $C/\gamma_{sub} H$
    (d) $C/\gamma_{sat} H$.

597. The stability number will be zero for
    (a) cohesive soils
    (b) C-ϕ soils
    (c) frictional soils
    (d) none.

598. For a clay slope of height 20 m, the stability number is 0.05, $\gamma = 25$ kN/m^3, $C = 30$ kN/m^2, the critical height of slope is
    (a) 30 m
    (b) 20 m
    (c) 24 m
    (d) 26 m.

599. Critical failure surface is the surface along which factor of safety is
    (a) maximum
    (b) constant
    (c) minimum
    (d) variable.

600. The factor of safety of the slope against sliding due to shear is given by
    (a) $FS = \tau/S$
    (b) $FS = S/\tau$
    (c) $FS = S \times \tau$
    (d) none.

601. The slope of a soil is greater than the slope of Mohr-Coulomb strength envelope, then the slope will be
    (a) stable slope
    (b) perfectly stable
    (c) unstable
    (d) can't say.

602. For a base failure of a slope, depth factor (D_f) is
    (a) equal to 1
    (b) zero
    (c) < 1
    (d) > 1.

603. The method of slices is applicable to
    (a) stratified soils
    (b) uniform soil slopes
    (c) non-uniform slopes
    (d) homogeneous soils.

604. Rapid drawdown is a case when stability analysis is made considering
    (a) downstream slope with seepage forces
    (b) upstream slope with lowered reservoir level
    (c) both slopes with seepage forces
    (d) both slopes with pore pressures.

605. Stability of slopes can be increased by
    (a) adopting gentle slopes
    (b) by stabilising the soil
    (c) consolidating and densifying the soil
    (d) all the above.

606. Berms are used to
    (a) increase the weight of dam
    (b) reduce seepage losses
    (c) increase shear strength
    (d) increase factor of safety.

607. Select the incorrect statement
    (a) sheet piles and retaining walls can be installed for increasing the stability of soil slopes
    (b) transitional failures may occur along slopes of layered materials
    (c) for increasing the stability of a slope, densification can be done in case of cohesionless soils
    (d) proper drainage along a slope reduces the stability of a soil slope.

608. Stability analysis shall be made considering
    (a) total stresses
    (b) normal stresses
    (c) effective stresses
    (d) shear stresses.

609. The factors which are responsible for land slides are
    (a) topography and climate
    (b) geological and hydrological conditions
    (c) weathering
    (d) all the above.

610. Based on pore pressure the critical stage at which the stability of the embankment should be checked is
    (a) rapid draw down situation
    (b) at the end of construction
    (c) steady-state seepage situation
    (d) all the above.

611. Factor of safety of slopes is defined as
    (a) F.S. = $\dfrac{\text{total stress}}{\text{effective stress}}$
    (b) F.S. = $\dfrac{\text{resisting moment}}{\text{overturning moment}}$
    (c) F.S. = $\dfrac{\text{overturning moment}}{\text{resisting moment}}$
    (d) F.S. = $\dfrac{\text{shear stress}}{\text{normal stress}}$.

612. Factor of safety of embankments shall be
   (a) at least 1.0
   (b) at least 1.5
   (c) at least 2.0
   (d) at least 2.0.

613. Shallow footing is one whose depth is
   (a) always equal to width
   (b) less than the width
   (c) more than the width
   (d) none of the above.

614. For foundation at shallow depth, Terzaghi assumed that at failure
   (a) failure surface extends upto ground level
   (b) failure surface terminates at base level of foundation
   (c) bottom of footing is smooth
   (d) elastic conditions exist.

615. Which of the following requires greatest deformation?
   (a) local shear failure
   (b) general shear failure
   (c) composite shear failure
   (d) rigid failure.

616. For $\phi = 0$ case, N_c value according to Terzaghi is
   (a) 9.5
   (b) 5.7
   (c) 5.14
   (d) 5.52.

617. Value of factor of safety adopted in foundation design is
   (a) 1.0 to 1.5
   (b) 1.5 to 2.0
   (c) 2.0 to 2.5.
   (d) 2.5 to 3.00.

618. Correlations and design charts are available for sands, relating allowable soil pressure and
   (a) cone resistance
   (b) maximum dry density
   (c) unconfined compressive strength
   (d) standard penetration test value.

619. Raising of water table in shallow foundations, near foundation level
   (a) reduces bearing capacity
   (b) increases bearing capacity
   (c) does not affect bearing capacity
   (d) increases and then decreases bearing capacity.

620. Total settlement that can be observed for a footing will consist of
   (a) immediate (or elastic compression)
   (b) primary consolidation settlement
   (c) secondary consolidation settlement
   (d) all the above.

621. The compression of the soil because of expulsion of gases and due to re-arrangement of particles is called as
   (a) Immediate settlement
   (b) Distortion settlement
   (c) Contact settlement
   (d) All the above.

622. The magnitude of immediate settlement will be generally more in case of
   (a) highly permeable soils
   (b) soils of low permeability
   (c) soils with no permeability
   (d) none of the above.

623. The immediate settlement in a cohesionless soils can be calculated with the help of
    (a) Standard penetration test
    (b) Dutch cone penetration test
    (c) Charts
    (d) All the above.

624. The primary consolidation settlement (S_c) can be found from
    (a) $S_c = C_c \cdot \dfrac{H}{1+e_0} \log_e \left(\dfrac{p+\Delta_p}{p}\right)$
    (b) $S_c = C_c \cdot \dfrac{1+e_0}{H} \log_e \left(\dfrac{p+\Delta_p}{p}\right)$
    (c) $S_c = C_c \cdot \dfrac{H}{1+e_0} \log_{10} \left(\dfrac{p+\Delta_p}{p}\right)$
    (d) $S_c = C_c \cdot \dfrac{1+e_0}{H} \log_{10} \left(\dfrac{p+\Delta_p}{p}\right)$

    where C_c = Compression index, p = Initial overburden pressure, Δ_p = Increment in the pressure.

625. It is considered that secondary consolidation in a clayey soils will occur because of dissipation of
    (a) absorbed water
    (b) adsorbed water
    (c) capillary water
    (d) none.

626. According to IS code, the permissible values of total settlement in case of isolated footings on clayeys and sands are
    (a) 65 mm and 40 mm
    (b) 40 mm and 65 mm
    (c) 65 mm and 100 mm
    (d) 100 mm and 65 mm.

627. According to IS code, the permissible values of total settlements of rafts on clays and sands are
    (a) 65 mm and 40 mm
    (b) 40 mm and 65 mm
    (c) 65 mm and 100 mm
    (d) 100 mm and 65 mm.

628. The type of settlement which causes more trouble for the stability of structure is
    (a) consolidation settlement
    (b) uniform settlement
    (c) differential settlement
    (d) none of the above.

629. The damage will be severe if the angular distortion is about
    (a) 1/50
    (b) 1/90
    (c) 1/120
    (d) 1/180.

630. According to IS code, the permissible values of different settlements in case of clays and sands are
    (a) 25 mm and 40 mm
    (b) 40 mm and 25 mm
    (c) 25 mm and 100 mm
    (d) none.

631. There will be no damage for the structure if the angular distortion is about
    (a) 1/50
    (b) 1/90
    (c) 1/120
    (d) 1/180.

632. The concept of useful width for the determination of bearing capacity is generally used for
    (a) Isolated footings
    (b) Strap footings
    (c) Axially loaded footings
    (d) Eccentrically loaded footings.

633. Which of the following soils will have least value of safe load?
   (a) sand stone
   (b) lime stone
   (c) moorum
   (d) soft chalk.

634. Black cotton soil is not suitable for foundations because of its
   (a) black colour
   (b) low bearing capacity
   (c) cohesive particles
   (d) swelling and shrinkage nature.

ANSWERS
Objective Type Questions

1. (a)	2. (a)	3. (b)	4. (d)	5. (d)	6. (d)
7. (c)	8. (b)	9. (c)	10. (d)	11. (d)	12. (c)
13. (b)	14. (d)	15. (a)	16. (c)	17. (d)	18. (c)
19. (b)	20. (a)	21. (c)	22. (d)	23. (b)	24. (d)
25. (a)	26. (c)	27. (c)	28. (b)	29. (a)	30. (a)
31. (c)	32. (d)	33. (b)	34. (a)	35. (b)	36. (c)
37. (b)	38. (c)	39. (d)	40. (d)	41. (d)	42. (d)
43. (c)	44. (c)	45. (c)	46. (b)	47. (b)	48. (c)
49. (b)	50. (a)	51. (d)	52. (b)	53. (c)	54. (c)
55. (a)	56. (c)	57. (b)	58. (c)	59. (c)	60. (b)
61. (b)	62. (a)	63. (d)	64. (c)	65. (d)	66. (a)
67. (b)	68. (a)	69. (d)	70. (a)	71. (a)	72. (b)
73. (c)	74. (c)	75. (c)	76. (c)	77. (b)	78. (b)
79. (a)	80. (c)	81. (c)	82. (a)	83. (d)	84. (a)
85. (b)	86. (a)	87. (c)	88. (b)	89. (b)	90. (b)
91. (c)	92. (c)	93. (b)	94. (d)	95. (b)	96. (a)
97. (a)	98. (d)	99. (a)	100. (c)	101. (b)	102. (a)
103. (a)	104. (d)	105. (b)	106. (d)	107. (b)	108. (c)
109. (a)	110. (c)	111. (d)	112. (d)	113. (d)	114. (c)
115. (a)	116. (c)	117. (b)	118. (c)	119. (b)	120. (c)
121. (c)	122. (d)	123. (d)	124. (c)	125. (c)	126. (d)
127. (d)	128. (c)	129. (b)	130. (d)	131. (c)	132. (d)
133. (a)	134. (c)	135. (a)	136. (b)	137. (d)	138. (b)
139. (c)	140. (b)	141. (d)	142. (b)	143. (b)	144. (d)
145. (b)	146. (d)	147. (a)	148. (a)	149. (d)	150. (c)
151. (a)	152. (a)	153. (c)	154. (b)	155. (c)	156. (d)
157. (b)	158. (c)	159. (c)	160. (b)	161. (d)	162. (a)
163. (c)	164. (c)	165. (d)	166. (b)	167. (d)	168. (d)
169. (a)	170. (c)	171. (a)	172. (b)	173. (b)	174. (c)
175. (c)	176. (b)	177. (b)	178. (c)	179. (a)	180. (c)
181. (a)	182. (b)	183. (b)	184. (a)	185. (c)	186. (b)
187. (d)	188. (a)	189. (a)	190. (c)	191. (d)	192. (a)
193. (b)	194. (a)	195. (b)	196. (b)	197. (c)	198. (d)

199. (d)	200. (d)	201. (a)	202. (c)	203. (b)	204. (d)
205. (d)	206. (b)	207. (d)	208. (b)	209. (b)	210. (a)
211. (a)	212. (a)	213. (c)	214. (c)	215. (d)	216. (a)
217. (b)	218. (b)	219. (b)	220. (c)	221. (b)	222. (c)
223. (d)	224. (c)	225. (b)	226. (d)	227. (a)	228. (b)
229. (c)	230. (c)	231. (d)	232. (a)	233. (d)	234. (a)
235. (d)	236. (b)	237. (d)	238. (c)	239. (c)	240. (d)
241. (c)	242. (b)	243. (a)	244. (b)	245. (b)	246. (c)
247. (a)	248. (a)	249. (d)	250. (d)	251. (c)	252. (d)
253. (c)	254. (d)	255. (b)	256. (a)	257. (b)	258. (a)
259. (b)	260. (a)	261. (c)	262. (d)	263. (c)	264. (c)
265. (b)	266. (c)	267. (c)	268. (b)	269. (a)	270. (b)
271. (d)	272. (a)	273. (b)	274. (a)	275. (c)	276. (c)
277. (a)	278. (b)	279. (b)	280. (d)	281. (a)	282. (b)
283. (d)	284. (c)	285. (a)	286. (b)	287. (c)	288. (b)
289. (c)	290. (a)	291. (c)	292. (d)	293. (c)	294. (d)
295. (b)	296. (a)	297. (c)	298. (a)	299. (b)	300. (c)
301. (b)	302. (a)	303. (c)	304. (b)	305. (c)	306. (c)
307. (b)	308. (d)	309. (c)	310. (b)	311. (c)	312. (a)
313. (d)	314. (a)	315. (c)	316. (b)	317. (b)	318. (b)
319. (d)	320. (d)	321. (b)	322. (c)	323. (b)	324. (c)
325. (b)	326. (c)	327. (d)	328. (c)	329. (c)	330. (a)
331. (b)	332. (c)	333. (c)	334. (a)	335. (b)	336. (d)
337. (d)	338. (c)	339. (c)	340. (d)	341. (d)	342. (d)
343. (c)	344. (d)	345. (a)	346. (a)	347. (b)	348. (c)
349. (d)	350. (c)	351. (d)	352. (d)	353. (b)	354. (a)
355. (d)	356. (b)	357. (d)	358. (c)	359. (c)	360. (c)
361. (b)	362. (c)	363. (d)	364. (a)	365. (b)	366. (b)
367. (a)	368. (d)	369. (b)	370. (c)	371. (d)	372. (c)
373. (d)	374. (a)	375. (a)	376. (a)	377. (a)	378. (c)
379. (c)	380. (a)	381. (c)	382. (c)	383. (c)	384. (d)
385. (b)	386. (c)	387. (b)	388. (c)	389. (b)	390. (a)
391. (c)	392. (c)	393. (d)	394. (a)	395. (b)	396. (a)
397. (d)	398. (c)	399. (c)	400. (d)	401. (a)	402. (c)
403. (b)	404. (a)	405. (a)	406. (d)	407. (b)	408. (b)
409. (b)	410. (d)	411. (d)	412. (b)	413. (d)	414. (b)
415. (d)	416. (b)	417. (b)	418. (d)	419. (b)	420. (a)
421. (b)	422. (a)	423. (a)	424. (b)	425. (c)	426. (b)
427. (a)	428. (c)	429. (d)	430. (c)	431. (c)	432. (a)
433. (b)	434. (c)	435. (d)	436. (d)	437. (d)	438. (b)
439. (a)	440. (a)	441. (c)	442. (d)	443. (b)	444. (c)
445. (c)	446. (c)	447. (b)	448. (a)	449. (c)	450. (d)

Soil Mechanics

451. (c)	452. (c)	453. (b)	454. (a)	455. (b)	456. (d)
457. (a)	458. (d)	459. (d)	460. (a)	461. (d)	462. (b)
463. (b)	464. (b)	465. (c)	466. (b)	467. (c)	468. (d)
469. (a)	470. (c)	471. (d)	472. (d)	473. (d)	474. (a)
475. (c)	476. (a)	477. (d)	478. (c)	479. (b)	480. (a)
481. (d)	482. (a)	483. (a)	484. (c)	485. (a)	486. (c)
487. (b)	488. (c)	489. (b)	490. (b)	491. (a)	492. (d)
493. (a)	494. (c)	495. (b)	496. (c)	497. (c)	498. (d)
499. (d)	500. (c)	501. (b)	502. (a)	503. (d)	504. (b)
505. (c)	506. (a)	507. (b)	508. (d)	509. (b)	510. (d)
511. (c)	512. (c)	513. (a)	514. (d)	515. (c)	516. (b)
517. (d)	518. (c)	519. (a)	520. (d)	521. (d)	522. (c)
523. (d)	524. (b)	525. (d)	526. (d)	527. (d)	528. (d)
529. (c)	530. (c)	531. (c)	532. (b)	533. (a)	534. (b)
535. (b)	536. (d)	537. (c)	538. (b)	539. (a)	540. (b)
541. (b)	542. (a)	543. (b)	544. (a)	545. (b)	546. (d)
547. (a)	548. (c)	549. (c)	550. (c)	551. (d)	552. (d)
553. (b)	554. (c)	555. (b)	556. (c)	557. (b)	558. (b)
559. (b)	560. (c)	561. (d)	562. (d)	563. (c)	564. (c)
565. (b)	566. (a)	567. (a)	568. (c)	569. (d)	570. (d)
571. (d)	572. (a)	573. (b)	574. (c)	575. (b)	576. (d)
577. (a)	578. (a)	579. (a)	580. (c)	581. (a)	582. (c)
583. (b)	584. (d)	585. (d)	586. (b)	587. (a)	588. (c)
589. (c)	590. (a)	591. (a)	592. (b)	593. (d)	594. (d)
595. (c)	596. (d)	597. (c)	598. (c)	599. (c)	600. (b)
601. (c)	602. (d)	603. (d)	604. (b)	605. (d)	606. (d)
607. (d)	608. (c)	609. (d)	610. (d)	611. (b)	612. (b)
613. (b)	614. (b)	615. (a)	616. (b)	617. (d)	618. (a)
619. (d)	620. (d)	621. (d)	622. (a)	623. (d)	624. (a)
625. (b)	626. (a)	627. (d)	628. (c)	629. (a)	630. (b)
631. (d)	632. (d)	633. (d)	634. (d)		

Chapter 9 HIGHWAY ENGINEERING

1. INTRODUCTION

'The Community pays for good roads whether it has them or not—it pays more if it has not got them'.

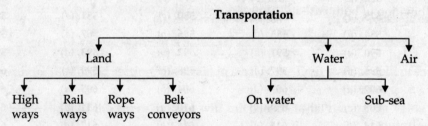

Highway	Railway
1. From starting point to destination.	From nearest Rly. stn. to the nearest Rly. stn. of destination.
2. Practically no need of packing of goods.	More care in packing required.
3. Takes less time.	Takes more time.
4. Steep gradients no problem.	Suited to flat gradients.
5. Suited to light traffic.	Suited to heavy traffic.
6. Suited to small distances.	Suited to long distances.

Highway Engineering

Necessity of Transport

1. For movement of people and goods as from

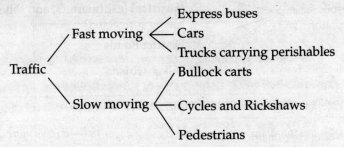

To keep price line stable.

2. For Cultural and Social advancement and National integration and good links with neighbouring countries.

3. For Security, Law and order and defence.

4. For promotion of tourism.

Classification of Roads

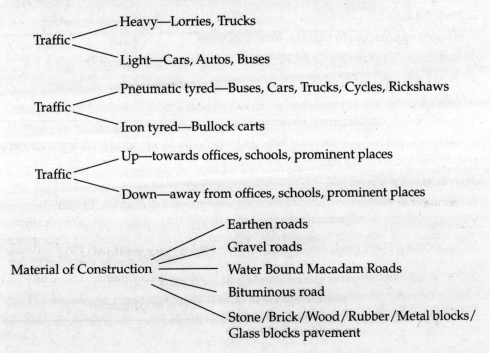

Combination of both Fast and Slow traffic is called "Mixed traffic".

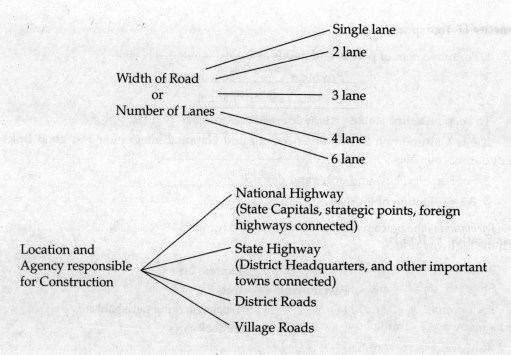

Planning

Factors Governing Alignment

1. Straight alignment—shortest—best.

2. Easy gradients.

3. Straight approaches to bridges, level crossings.

To be avoided

1. Deep Cuttings.

2. Valleys.

3. Sharp curves, blind corners, steep grades.

4. Sensitive places as places of worship, security spots as defence production units, neighbouring foreign territory etc.

Surveys to be Conducted

1. Economic survey.

2. Traffic survey.

3. Engineering survey : (*a*) Reconnaissance (*b*) Preliminary survey (*c*) Final location survey.

Geometric Design

```
_____
            Wearing Course
_____
             Base Coat
_____
       Base Course/Soling/Foundation
_____
              Sub-Base
_____
              Sub-Grade
_____
  ///₹        ///₹         ///₹
        Embankment/G.L./Cutting
```

Formation is the bottom most prepared ground (ground level, embankment or cutting) over which the highway rests.

Subgrade is the first artificial course provided.

Strength, stability and bearing power of the road depend on it.

Its top must be 600 to 1000 mm above HFL (high flood level) and that is why the road is called a highway.

Subbase: It is a grannular layer (of gravel, sand, coal or ashes) provided to improve drainage and hence strength of *base course* above it.

Base course, foundation or soling is the important course which protects the layers below it from rain and snow (pumping and frosting).

A pavement is classified as flexible or rigid based on it.

Base coat is the layer above base course but below the wearing course.

It is an optional layer whose function is to add to the properties of the wearing course.

Wearing course is the topmost layer of a highway which directly comes in contact with traffic.

It should be strong, stable, impervious, abrassion and shock resisting.

It should offer reasonable longitudinal and lateral friction, should be even and provide good visibility.

Skid is the advancement of the wheel without any rotation.

Longitudinal skid is because of smooth road surface, steep gradients, and wornout tyres.

Lateral skid is because of steep camber, less super elevation for fast moving vehicles and more super elevation for slow moving traffic and wornout tyres.

Slip is the rotation of wheel but without any advancement.

It is because of smooth and soft pavements.

Unevenness, roughness or cumulative vertical deformation is the ups and downs of the top of the wearing course causing uneasiness for the driver and passengers, more wear and tear of tyre and other parts of automobile, more fuel consumption, more time of travel and more number of accidents.

It is because of
1. Poor or uneven bearing capacity of formation,
2. Inadequate compaction of subgrade soil,
3. Less thickness of base course,
4. Inadequate surface and subsurface drainage,
5. Improper construction and compaction of wearing course.

Unevenness Index	Comfortable Speed
Less than 150 cm/km	No limit
Between 150 and 250 cm/km	Less than 100 kmph
More than 350 cm/km	Less than 50 kmph

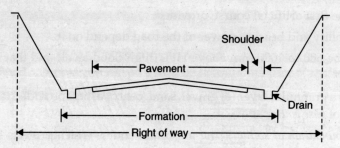

Camber: Camber is crossfall. It is the transverse slope of a highway. It is provided for
1. Improved appearance.
2. Easy drainage.
3. Easy separation of up and down traffic.

$$\text{Camber} = \frac{2a}{b}$$

Amount of Camber to be provided depends on :
1. Rainfall.
2. Permeability of road surface.
3. Smoothness of Wearing Course.

Road	Camber recommended
Earth	1 in 20
Gravel	1 in 24
WBM	1 in 36 to 1 in 48
Bituminous	1 in 48 to 1 in 60
Concrete	1 in 70 to 1 in 80

Gradient

Gradient is the longitudinal slope of a highway. It is to be provided for
1. easy drainage,
2. minimising the cost of forming high embankments and deep cuttings

Ruling gradient is the normal gradient provided for the safe and comfortable journey.
Limiting gradient is slightly steeper provided to reduce horizontal length of travel.

Exceptional gradient is the steepest gradient provided in exceptional cases for a very short length. Minimum gradient is the least gradient to be adopted to fecilitate easy drainage.

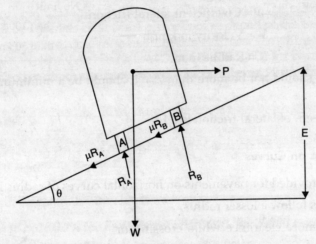

R_A and R_B are to be perpendicular to the inclined surface.

Ruling gradient	1 in 30 in plains
Limiting gradient	1 in 20 in plains
Exceptional gradient	1 in 15 in plains
Minimum gradient	1 in 500 for cement concrete roads,
	1 in 200 for Bituminous roads and
	1 in 100 for Gravel roads.

Curves

A horizontal curve is provided whenever there is change in direction.
A vertical curve is provided whenever there is change in gradient.
A curve is said to be sharp when its radius is less than 50 m.
Gradient may not be adopted on a horizontal curve *i.e.*, gradient can be zero on a horizontal curve.

But when the gradient is to be provided on a horizontal curve, it is less compared to straights on either side, and this reduction is called 'Grade Compensation'.

Grade Compensation on Curves $\ngtr \dfrac{76}{R}\%$

Super Elevation

It is the raising of the outer edge of a road on a horizontal curve over its inner edge to compensate the centrifugal force $P\left(=\dfrac{WV^2}{gR}\right)$

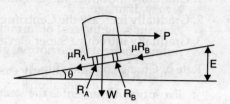

$$\tan\theta + \mu = \dfrac{V^2}{127 R},$$

$\tan\theta$ = Super elevation.
μ = Co-efficient lateral friction
V = Velocity in kmph
R = Radius in m.

Super elevation should not be more than 7%. It should be a minimum of 1.4% (1 in 70 to fecititate drainage)

Safe limit for co-eff. of lateral friction is 0.15.

Widening of Pavement on Curves

Extra width is provided for pavements on horizontal curves of radius less than 300 m

1. As rear wheels follow a lesser radius,
2. To maintain more clearance while crossing or overtaking,
3. For easy manoeuvring of overhanging length,
4. To get greater visibility.

Mechanical Widening Width $\left.\right\}$ $W_m \simeq \dfrac{l^2}{2R}$

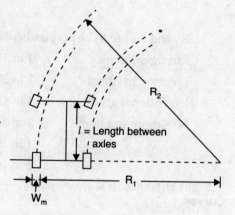

l = Length between axles

Psychological Widening Width $\left.\right\}$ $W_{ps} = \dfrac{V}{9.5\sqrt{R}}$

$\therefore$ Extra width $\quad W_e = \dfrac{n.l^2}{2R} + \dfrac{V}{9.5\sqrt{R}}$

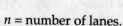

n = number of lanes.

1. Provide 'W_e' on the inner side for sharp curves.
2. Provide $\dfrac{W_e}{2}$ on either side of broad curves.

Transition Curves. Transition curve is a curve of continuously changing radius. They are introduced in between the straight portion and Circular Curve to

1. Gradually introduce the Curvature.
2. Gradually increase the Centrifugal force.
3. Gradually introduce Super elevation and extra widening.

$$\text{Rate of Change of Centrifugal acceleration} \quad C = \frac{v^3}{LR} = \frac{0.021V^3}{LR}$$

v = Velocity in m/s, $\quad L$ = Length of transition curve in m

R = Radius in m, $\quad C = \dfrac{80}{75 + V}, \quad v$ = Velocity in kmph

$$0.5 \leq C \leq 0.8$$

MINIMUM LENGTH OF A TRANSITION CURVE

Minimum length of a transition curve is the maximum value of:

1. $\quad L = \dfrac{0.021 V^3}{CR}$

2. (a) $\quad L = nE$

where $\dfrac{1}{n}$ = Rate of change of super elevation,

E = Max. super elevation provided

When the pavement is rotated about the inner edge and

(b) $\quad L = \dfrac{nE}{2}$ when the pavement is rotated about the crown.

3. $\quad L = \dfrac{2.7 V^2}{R}$ in plains.

Sight Distance. It is clear visible distance ahead of the driver.
It depends on

1. Time of perception ⎫ Alertness of driver ⎫
                      ⎬ Care and skill   ⎬ 2 to 2.5 s
2. Time of reaction   ⎭ Vision and weather ⎭

3. Time of brake application

(a) Speed   (b) Brakes condition   (c) Type of pavement

(d) Level, down slope or up slope   (e) Climate

∴ Total distance = $vt + \dfrac{v^2}{2g(\mu \pm i)\eta}$

t = Time of perception and reaction, μ = Co-eff. of longitudinal friction
vt_2 = Lag distance
i = Slope + for ↑ and – ve for ↓, η = Efficiency of brakes.

$\dfrac{v^2}{2g(\mu \pm i)\eta}$ = Braking distance

Overtaking sight distance

$$d = 0.7 \times v_s + 6 \text{ (6 m being the length of vehicle)}$$

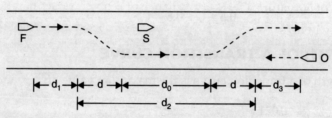

Length of overtaking = $v_s \times t_p + \{2d + v_s \times t_0\} + v_f \times t_0$

v_s = Velocity of slow moving vehicle, $\quad v_f$ = Velocity of fast moving vehicle,
t_p = Time of perception, $\quad\quad\quad\quad\quad t_0$ = Time of overtaking

Vertical Curves

Summit Curves: A summit curve occurs when a rising steep slope joins rising flatter slope, up slope is followed by down slope or falling flatter slope is followed by falling steep slope such that the net change is –ve and deflection angle is always downwards.

An advantage of a summit curve is that stress on the tyre is the least as the centrifugal force acts upwards.

Length of summit curves

$$L = \dfrac{(i_1 - i_2)D_s^2}{[\sqrt{2H} + \sqrt{2h}]^2} \quad\quad \text{where } L > D_s$$

i_1 = Initial slope, i_2 = Later slope, H = Height of driver's eye above pavement (= 1 : 2 m),
h = Height of obstruction above pavement (0.15 m), D_s = Stopping sight distance.

∴ $\quad\quad L = \dfrac{(i_1 - i_2)D_s^2}{4.4}$

when $\quad L < D_s$

$$L = 2D_s - \dfrac{4.4}{(i_1 - i_2)}$$

Minimum radius of a vertical curve $\bigg\}$ $R = \dfrac{L}{i_1 - i_2}$

Valley Curve

A valley curve occurs when a falling gradient is followed by a rising gradient, when a flatter rising gradient leads to a steep rising gradient and when a steeper falling gradient is followed by flatter falling gradient. Net change in slope is + ve and the deflection angle is always upwards.

In a valley curve the centrifugal force vertically acts downwards increasing the pressure on tyre. To minimise the time for which the peak down thrust exerted on tyres the valley curve is designed as a composition of two transition curves with no circular curve in between.

Length of Valley Curves

$$L = \frac{2v^3}{CR} = 2 \times \frac{v^3(i_1 - i_2)}{C}$$

where R = Radius of curvature, C = Rate of change of centrifugal acceleration.

v = Velocity in m/s.

Minimum length of a valley curve $L_{min} = 0.5 \times V$ metres.

Design of Flexible and Rigid Pavements

Flexible	Rigid
Low flexural strength.	High flexural strength.
Wearing course deflects if sub-grade deflects.	Top slab bridges over small irregularities of sub-grade.
Design—empirical rules followed.	Design is quite precise.
Less impervious.	More impervious.

Design

1. Max. wheel load

Max. axle load (IRC) = 8160 kg

∴ Max. equivalent single wheel load ($MESWL$) = 4080 kg.

2. Contact Pressure

Influence of tyre pressure is very high in the upper layers—It diminishes considerably as depth increases

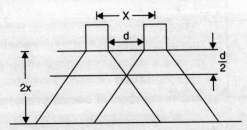

Rigidity factor = $\dfrac{\text{Contact pressure}}{\text{Tyre pressure}}$

It is 'l' for a contact of pressure of 7 kg/cm^2.

EWL can be determined from the graph drawn for a deflection $\frac{'d'}{2}$ under a load 'P', and deflection '$2x$' under a load '$2P$'.

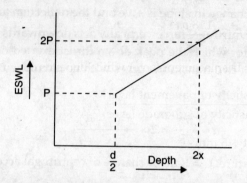

Repetition of loads

Due to repeated applications of loads elastic and plastic deformation occur.

McLeod assumed 100% thickness of a slab to withstand 10^6 repetitions of the load, while 25% of its thickness withstands application of load only ONCE.

If a 25 cm thick pavement fails under a load P applied for 'x' number of repetitions and under a load Q for 'y' number of repetitions, then

$$\left.\begin{array}{r}\text{Equivalent wheel load}\\ \text{factor for the load } Q\end{array}\right\} = \frac{x}{y}$$

Structural damage caused by an axle varies as the Fourth Power of its ratio to the standard axle load. (80 kN or 8160 kg)

$$\therefore \quad \text{Equivalence factor } F = \left(\frac{L}{L_s}\right)^4 = \left(\frac{L}{80}\right)^4.$$

Elastic Modulus

Vertical deformation $\quad \Delta = \dfrac{1.5\, pa}{E_s} \quad$ $\begin{bmatrix}\text{for flexible plate}\\ \text{as a rubber tyre}\end{bmatrix}$

p = Surface pressure, a = Radius of loaded area

E_s = Modulus of elasticity of soil, $\Delta = \dfrac{1.18\, pa}{E_s} \quad$ $\begin{bmatrix}\text{for rigid plate}\\ \text{as mild steel plate}\end{bmatrix}$

$$\left.\begin{array}{r}\text{Vertical stress}\\ \text{at a depth } z\end{array}\right\} = \sigma_s = p\left[1 - \frac{z^3}{(a^2 + z^2)^{3/2}}\right]$$

(Boussinesq's Theory)

Burmister's Theory

$$\Delta = F_w \times \frac{1.5\, pa}{E_2} \quad \text{(flexible pavements)}$$

$$\Delta = F_w \times \frac{1.18\, pa}{E_2} \quad \text{(rigid pavements)}$$

F_w = Displacement factor which depends on E_1/E_2 and pavement thickness.

where $\dfrac{E_1 (= \text{Modulus of elasticity of pavement layer})}{E_2 (= \text{Modulus of elasticity of subgrade layer})}$

Allowable settlement ≯ 5 mm.

Rigid pavements

$$\left. \begin{array}{l} \text{Radius of relative stiffness} \\ \text{(Westergaard)} \end{array} \right\} l = \left[\frac{Eh^3}{12(1-\mu^2)K} \right]^{1/4}$$

l = Radius of relative stiffness (cm)
E = Modulus of elasticity of cement concrete (kg/cm^2)
h = Slab thickness (cm)
μ = Poisson's ratio for cement concrete
K = Modulus of subgrade reaction (kg/cm^3).

Temperature stress

$$f = \frac{E \propto (\delta t)\, C}{2}$$

f = Temperature stress in the edge (kg/cm^2)
E = Modulus of elasticity (kg/cm^2)
$\propto$ = Co-eff. of thermal expn. of cement concrete
δt = Change in temperature
C = Bradbury's co-eff.

Earth Roads. Mostly used roads in India.

Camber	1 in 20 to 1 in 12
Min. gradient	1 in 120 desirable
Min. width	8 m (2 lanes).

Gravel Road

Camber	1 in 30 to 1 in 25.

WBM Road. Broken aggregates is bound together by wet stone dust produced during wear and tear.

1. Subgrade cleared, cleaned rolled to a flatter camber.

2. 120 to 180 mm stone or broken stone are packed so that their pointed ends are upwards. Hollow spaces at top are filled with broken stone or gravel and rolled with a 10 ton roller.

3. Metal spread over the prepared base, rolled with 10 ton roller.

4. Screenings are spread to fill voids—dry rolled—wet rolled.

5. After 24 hours, 6 mm thick sandy—clay is spread and heavy dose of water applied—Rolled.

6. After 24 hours, the surface is rolled and fine sand is sprinkled.

7. Curing—for 10 days.

8. Road opened for traffic.

WBM road can bear 450 ton of pneumatic traffic and 450 ton of Iron wheeled traffic/day.

Bituminous Pavements

1. Provide a good top layer for low cost roads
2. Smooth, impervious riding surface
3. Skid resistance
4. Resistance against wear and deformation.

Surface Painting

WBM road—subjected to traffic atleast for 6 months—irregularities corrected—cleaned to 5 to 10 mm into joints—Bitumen applied @ 2 kg/sq m as first coat—1 kg/sq m for 2nd coat—covered with clean, dry chips—rolled with 8 ton roller.

Full grouted Macadam Road

A min. thickness of 200 mm WBM road—Base

1. Dry aggregates spread to 50 to 100 mm thickness Dry rolled—8 to 10 ton roller.

2. Hot/Cold Bitumen applied @ 8–14 kg/sq m—to go to full depth.

3. 10 mm stone chips applied @ $1\frac{1}{2}$ to 2 m³/100 m².

4. Rolled—10 to 12 ton roller. Opened for traffic.

5. After 7 days—road cleaned—seal coat applied @ $\frac{1}{2}$ to $1\frac{1}{2}$ kg/m².

6. Chips (6 mm) applied @ 1 m³/100 sq m rolled with 10–12 ton roller. Road is opened for traffic after 24 hrs.

Semi-grouted Macadam Pavement

1. WBM surface—cleaned.

2. Hoggin material (stone powder/moorum/coarse sand) laid to 10 to 20 mm—rolled—8 to 10 ton roller.

3. Water sprinkled.

4. Bitumen applied @ 4 to 7 kg/m².

5. 10 mm stone chips applied @ $1\frac{1}{2}$ to 2 m³/100 m².

6. Rolled with 10–12 ton roller.

The remaining is same as for full grouted Macadam Road.

Bitumen Bound Macadam Pavement

1. Base prepared.
2. Prime coat @ $\frac{1}{2}$ to 1 kg/sq m—applied.
3. A homogeneous premix at 150 to 180° C—placed to the desired grade and camber.
4. Rolled—8 to 10 ton.

Finished thickness—50 mm/75 mm, where GWT is too nearer to G.L. and Freezing and thawing so common—this road is best suited.

Premixed Bitumen Carpet

1. Base prepared.
2. Priming coat applied.
3. Aggregate and bitumen—separately heated and mixed to form homogeneous mix—spread uniformly.
4. Rolled—6 to 9 ton roller—kept damp.
5. Sealing coat $\left(\text{Bitumen} + \dfrac{6 \text{ mm stone chips}}{\text{Coarse sand}}\right)$ applied.
6. Rolling—6 to 9 ton roller.

Opened for traffic after 24 hrs.

Bituminous Concrete Pavement

1. Base prepared.
2. Bituminous levelling coarse laid.
3. Metal + Sand + Mineral filler—heated separately mixed with hot bitumen and laid to a uniform thickness.
4. Rolling 8 to 10 ton roller and then with 15 to 30 ton roller.
5. Sealing coat provided.
6. Opened for traffic after 24 hours.

Cement Concrete Roads

1. Can take Heavy Conc. loads as iron tyred traffic.
2. Can bear high temperatures.
3. Can span over defective subgrade and bridges over minor defects.
4. Resists oil spillages.
5. Requires very little maintenance.

Construction

1. Base prepared—given camber.
2. Steel or wooden form—provided.
3. Subgrade kept moist at least for 12 hours before concreting.
4. Coarse aggregate + Sand + Cement + Water mixed, transported, and placed in position :

> **3.5 m–4.5 m alternate bays**
> **or still wider and long strips—Concreted**

5. Surface compacted.

6. Canvas belt 15 to 30 cm wide and 60 cm longer than the width of pavement provided with wooden handles is applied longitudinally.

7. Grade, camber and alignment checked.

8. Cured—28 days.

9. Expansion joints, contraction joints warping joints, and construction joints and longitudinal joints are filled with an impervious elastic membrane.

TRAFFIC ENGINEERING

It is efficient planning, designing of pavement and operating smooth, quick, convenient, comfortable and safe streams of traffic without any conflicts.

1. **Physical features of road user.** Vision, Hearing. Physical strength to steer and apply brakes.

2. **Mental characteristics.** Ready wit, ability to understand traffic signals, traffic behaviour, and psychology of road user. Patience and perseverance are the desirable qualities of a driver.

3. **Environmental characteristics:** Snow, Rain, fog, mist, lighting, glare

Traffic Study. Traffic volume is expressed as

Average Annual Daily Traffic (AADT)

Enoscope is used to study "Spot Speed".

Traffic density = No. of vehicles/unit length (km).

Vehicles are converted to equivalent PCUs (Passenger car units) as follows :

Vehicle	Running	Waiting for signal	Parked
Car	1	1	1
Bus	2.2	2.8	3.4
Scooter	0.4	0.3	0.2
Pedal cycle	0.7	0.4	0.1
Hand Cart	4.6	3.2	0.3

$$AADT = \frac{\text{Total traffic per year as PCUs}}{365 \text{ (366 in case of leap year)} \times 24}$$

30th Highest Hourly Volume :

It is the traffic volume/hour exceeding AADT only 29 times in an year.

Origin and Destination (O and D) studies. They are conducted to get information of

1. Existing traffic flow along different routes

2. Lines (directions) along which the traffic prefers to travel

3. Changes desired as

(*a*) One way

(*b*) By passes (by skipping some intermediate stations)

(c) Inclusion of more intermediate stations
(d) Grade separated intersections
(e) Parking areas to be provided/eliminated

Traffic signs

Regulatory : (orders) : One way traffic, Dead slow, speed limit etc.

CIRCULAR DISCS

Warning : (Conditions ahead) : Right turn, left turn, 'S' curves etc.

RECTANGULAR BLOCK BELOW A RED TRIANGLE

Informatory : They provide information about bus stop, parking places, roads ahead etc.–

RECTANGULAR BOARDS
Traffic signals

Traffic signals are time based devices at a grade intersection to segregate traffic along different channels with no scope for any conflict.

They are provided when

1. Number of vehicles exceed 250/lane/h
2. A busy main road (> 1000 veh/h) rarely permits, crossing of another stream (> 100 veh/h) to cross it.
3. Number pedestrian crossing the road exceeds 150/h
4. In an year more than 5 accidents take place involving death, injury or loss of property worth ₹ 2000 or more.

Road Markings

They segregate and streamline and regulate traffic.

Traffic Islands

1. Divisional islands—separate up and down traffic.
2. Channelizing islands—guide traffic into proper channels. The most common one is the one provided for guiding left turning traffic.
3. Pedestrian loading island—provided between cycle lane and Bus lane for the sake of bus passengers.
4. Rotary—a big island around which traffic moves in clockwise direction, without coming to stand still on any of its legs. The shape of the central island is a square, diamond or circular when traffic on all the legs are of equal importance.

It is elliptical when traffic along its major diameter is prominent compared to that in other directions.

Traffic is supposed to maintain a speed of 30 to 40 kmph on the rotary.

Super elevation is not provided for the curved road but instead the road is made rough to present a coefficient of friction of 0.47 at 30 kmph and 0.43 at 40 kmph.

An entry radius of 15 to 25 m at 30 kmph and 20 to 35 m at 40 kmph are adopted.

A weaving angle of 15° is desirable as a lesser angle increases the diameter of the central island.

A weaving length of 30 to 60 m for 30 kmph rotary and 45 to 90 m for 40 kmph rotary are adopted.

Width of the carriageway is a minimum of 9.5 m.

Advantages:

It is an intersection at grade where traffic from 4 to 7 directions merge, move clockwise at 30 or 40 kmph and diverge without any conflicts. It is suited for a traffic density of 500 to 3000 vehicles per hour.

Intersections

I. Intersection at Grade

1. Crossing angle preferably 90°, but not less than 60°.
2. Enough clear line of sight *i.e.*, no blind corners.

Any vehicle should safely travel for 8 seconds after sighting the other vehicle.

3. Area of conflict at the intersection should be as small as possible.
4. Relative speed and angle of approach should be small.
5. Area of entry is converged (funnelling) to compel speed reduction and area of exit is widened to encourage quick exit.

(Radius of exit ≃ 2 × radius of entry).

II. Grade Separated Intersection

To avoid conflicts during crossing over 2 level or 3 level flyovers are provided so that straight, left turning and right turning traffic move along different channels assigned for them. Diamond and clover leaf intersections are examples.

II. OBJECTIVE TYPE QUESTIONS

1. In a country road length in km is taken to a population of
    (a) a hundred
    (b) a thousand
    (c) a lakh
    (d) a million.

2. Economic survey of a proposed road project includes a detailed survey of
    (a) agricultural and industrial products available in the area
    (b) resources of income to local bodies as toll tax
    (c) origin and destination of traffic
    (d) soil characteristics at various places.

3. An undesirable element to be avoided while fixing the alignment of the highway is
    (a) straight and short route
    (b) right angled crossing for bridges, culverts and level crossing
    (c) rising ground and high embankments
    (d) proximity to a place of worship.

4. A precise survey is
    (a) Reconnaissance
    (b) Preliminary survey
    (c) Final location survey
    (d) Economic survey.

5. Mixed traffic means
   (a) both up and down traffic
   (b) light traffic as of cycles to that of heavy traffic as of trucks
   (c) pedestrians + animal drawn coaches + lorries
   (d) slow moving and fast moving traffic.

6. The road connecting district head quarters of a state is
   (a) National Highway
   (b) State Highway
   (c) District road (Major)
   (d) Minor district road.

7. The road connecting a district headquarters of one state to the district headquarters of another bordering state is called
   (a) National Highway
   (b) State Highway
   (c) Major District Road
   (d) Expressway.

8. 2000 M 60 means
   (a) Number of vehicles per day is 2000, traffic is mixed and design speed is 60 kmph
   (b) Designed for 60 vehicles/hour (minimum) and 2000 vehicles/hour (maximum)
   (c) Length of the highway is 2000 km and design speed is 60 kmph
   (d) After travelling 2000 metres a 60 m long rough road is going to be met.

9. Total kilometrage for NH, SH and major district roads is given by
   (a) $\dfrac{A}{8} + \dfrac{B}{32} + 1.6N + 8T + D - R$
   (b) $0.32N + 0.8Q + 1.6R + 3.2S + D$
   (c) $\dfrac{N}{2} + \dfrac{S}{5} + \dfrac{D}{6} + \dfrac{OD}{8} + \dfrac{V}{10} + D'$
   (d) $\dfrac{P_0}{10000} + \dfrac{P_r}{1000} - E.$

10. An old worn out tyre may offer more friction on a dry surface than a new tyre with treads because
    (a) old surface got more accustomed for the surface of road
    (b) new tyre had a smooth surface
    (c) old surface had more contact area
    (d) hollows of treads of tyre had more friction.

11. The min. (desirable) value of co-efficient of friction along longitudinal direction is
    (a) 1
    (b) 0.5
    (c) 0.4
    (d) 0.15.

12. The minimum co-efficient of lateral friction for a highway is
    (a) 1
    (b) 0.5
    (c) 0.4
    (d) 0.15.

13. Unevenness of a pavement should be preferably less than
    (a) 100 cm/km
    (b) 150 cm/km
    (c) 250 cm/km
    (d) 329 cm/km.

14. If you have to choose between an alignment of highway through cutting, embankment, pavement at ground level itself and a tunnel, the best choice is
    (a) road nearer to ground level ☐   (b) embankment ☐
    (c) cutting ☐   (d) tunnel. ☐

15. Right of way in the figure below is
    (a) a ☐   (b) b ☐
    (c) c ☐   (d) d. ☐

16. In the figure above width of formation is
    (a) a ☐   (b) b ☐
    (c) c ☐   (d) d. ☐

17. Width of a traffic lane is
    (a) 3.75 m ☐   (b) 5.50 m ☐
    (c) 7.00 m ☐   (d) 7.50 m. ☐

18. Camber is
    (a) $\dfrac{b}{a}$ ☐
    (b) $\dfrac{2b}{b}$ ☐
    (c) $\dfrac{b}{2a}$ ☐
    (d) $\dfrac{a}{b}$. ☐

19. One of the natural factors influencing Camber is
    (a) type of material used for wearing coarse ☐   (b) topography of the area ☐
    (c) nature of subsoil met with ☐   (d) amount of rainfall. ☐

20. The primary object of providing Camber is
    (a) easy drainage ☐   (b) improved appearance ☐
    (c) easy separation of up and down traffic ☐   (d) easy overtaking facility. ☐

21. Camber depends on
    (a) smoothness of base coarse ☐   (b) permeability of subgrade ☐
    (c) amount of rainfall ☐   (d) grade of wearing coarse. ☐

22. For earthen roads the most Common Camber is
    (a) 1 in 20
    (b) 1 in 24
    (c) 1 in 36
    (d) 1 in 48.
23. Ruling gradient in plains is
    (a) 1 in 20
    (b) 1 in 30
    (c) 1 in 40
    (d) 1 in 50.
24. Limiting gradient in plains is
    (a) 1 in 15
    (b) 1 in 20
    (c) 1 in 30
    (d) 1 in 40.
25. Exceptional gradient in plains is
    (a) 1 in 15
    (b) 1 in 20
    (c) 1 in 30
    (d) 1 in 40.
26. Minimum gradient to be adopted for a black top road is
    (a) 0
    (b) 1 in 200
    (c) 1 in 20
    (d) 1 in 50.
27. Grade compensation on curves is a maximum of
    (a) $\dfrac{56}{R}$
    (b) $\dfrac{76}{R}$
    (c) $\dfrac{100}{R}$
    (d) $\dfrac{156}{R}$.
28. Super elevation + lateral friction should not be greater than
    (a) $\dfrac{V^2}{127R}$
    (b) $\dfrac{V^2}{225R}$
    (c) $\dfrac{V^2}{7.4R}$
    (d) $\dfrac{V^2}{14.28R}$.
29. Minimum super elevation provided is
    (a) 7%
    (b) 10%
    (c) not less than the grade of the road
    (d) not less than camber at the section.
30. Max. amount of super elevation should not be greater than
    (a) 2%
    (b) 3%
    (c) 5%
    (d) 7%.
31. Minimum super elevation on a horizontal curve is
    (a) 0
    (b) 7%
    (c) 1.4%
    (d) gradient on either side.
32. Higher value of super elevation is highly undesirable for
    (a) fast moving vehicles
    (b) slow moving vehicles
    (c) mixed traffic
    (d) non-snow falling areas.

33. Higher values of super elevation is dangerous in case of
    (a) a fast moving motor cycle
    (b) heavily loaded (but with a light material as cotton) bullock cart
    (c) a road that is always dry
    (d) a long truck with a trailer.
34. While negotiating a curve
    (a) wheels of both axles tread the same path
    (b) front wheels follow a less radius than rear wheels
    (c) rear wheels follow a less radius than the front wheels
    (d) depending on right hand or left hand curve it varies.
35. When the distance between the axles is 'l' and radius of the curve is R then mechanical widening width is
    (a) $\dfrac{l^2}{R}$
    (b) $\dfrac{l^2}{2R}$
    (c) $\dfrac{2l^2}{R}$
    (d) $\dfrac{l}{2R}$.
36. Psychological widening width is
    (a) $\dfrac{V^2}{9.5R}$
    (b) $\dfrac{V}{9.5\sqrt{R}}$
    (c) $\dfrac{V}{9.8R}$
    (d) $\dfrac{V^2}{9.8\sqrt{R}}$.
37. On sharp curves widening of the carriage way is done by
    (a) providing more width on the inner curve
    (b) providing more width on the outer curve
    (c) distributing half on inner and half on the outer
    (d) distributing $\frac{3}{4}$ on the outer and $\frac{1}{4}$ on the inner.
38. In plains the minimum length of transition curve is
    (a) $\dfrac{V^2}{R}$
    (b) $\dfrac{V^2}{1.5R}$
    (c) $\dfrac{2.7V^2}{R}$
    (d) $\dfrac{V^2}{24R}$.
39. For sight distance calculation 'Time of perception' and reaction depends on
    (a) speed of the vehicle
    (b) gradient of road
    (c) alertness of driver
    (d) nature of pavement.
40. Generally the time of perception in simple cases is taken as
    (a) 2 seconds
    (b) 30 seconds
    (c) 60 seconds
    (d) 120 seconds.

41. Stopping distance is given by
    (a) $vt + \dfrac{v^2}{2g(\mu \pm i)}$
    (b) $vt \pm \dfrac{v^2}{2gi}$
    (c) $vt\mu + \dfrac{v^2}{2g}$
    (d) $\left(vt + \dfrac{v^2}{2g}\right)\mu$.

42. The normal elevation of driver's eye is
    (a) 100 cm above road level
    (b) 120 cm above road level
    (c) 150 cm above road level
    (d) 200 cm above road level.

43. For non-passing sight distance a stationary object of this height is considered
    (a) 10 cm
    (b) 15 cm
    (c) 50 cm
    (d) 100 cm.

44. In the figure below the over taking sight distance is
    (a) BC
    (b) BD
    (c) AC
    (d) AD.

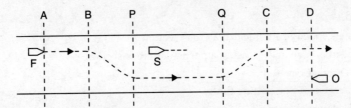

45. In the above figure 'distance of over taking' is
    (a) PQ
    (b) BQ
    (c) PC
    (d) BC.

46. Length of summit curve L in the adjacent figure is
    (a) $\dfrac{(i_1 + i_2) D_s^2}{\sqrt{2H} + \sqrt{2h}}$
    (b) $\dfrac{(i_1 - i_2) D_s^2}{[\sqrt{2H} + \sqrt{2h}]^2}$
    (c) $\dfrac{(i_1 \pm i_2) D_s^2}{\sqrt{2H} + \sqrt{2h}}$
    (d) $\left[\dfrac{i_1 + i_2}{\sqrt{2H} + \sqrt{2h}}\right]^2 D_s.$

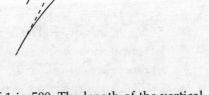

47. A rising gradient of 1 in 50 meets a falling gradient of 1 in 500. The length of the vertical curve if the rate of change of gradient is 1% per 100 m
    (a) 45.45 m
    (b) 180 m
    (c) 200 m
    (d) 220 m.

48. A valley curve is composed of just two transition curves without any circular curve in between to
    (a) minimise the length of valley curve
    (b) increase sight distance
    (c) minimise the time of acting of down thrust on tyres
    (d) nullify the centrifugal force.
49. Length of a vehicle controls the design of
    (a) vertical profile of the road
    (b) overtaking distance
    (c) geometrics and cross-sectional characteristics
    (d) axle and wheel load.
50. Weight of a vehicle affects the design of
    (a) camber and gradient of a road
    (b) pavement thickness and gradient of a road
    (c) cross drainage works and tunnels
    (d) permissible speed of vehicle.
51. Travel speed is
    (a) instantaneous speed of a vehicle at a cross-section
    (b) average speed of a vehicle crossing a particular cross-section
    (c) $\dfrac{\text{Distance covered}}{\text{Time of travel excluding halting time}}$
    (d) $\dfrac{\text{Distance covered}}{\text{Time of travel including halting time}}$.
52. The layer that is directly coming in contact with the traffic is
    (a) Wearing course   (b) Base course
    (c) Sub base   (d) Sub grade.
53. A pavement is classified as flexible pavement or rigid pavement based on its
    (a) Wearing course   (b) Base course
    (c) Sub base   (d) Sub grade.
54. The first artificial course provided on a highway
    (a) Sub grade   (b) Sub base
    (c) Base course   (d) Base coat.
55. Strength, stability and bearing power of a highway depend on
    (a) Formation   (b) Sub grade
    (c) Base course   (d) Wearing course
56. Skidding occurs on _____ pavement
    (a) dry   (b) wet
    (c) smooth   (d) soft.
57. The layer of the road that may not contribute to unevenness is
    (a) Base coat   (b) Formation
    (c) Base course   (d) Sub base.

58. Greater unevenness causes
    (a) slipping
    (b) skidding
    (c) glare
    (d) accidents.
59. When unevenness index is more than 350 cm/km length, then the comfortable speed is less than
    (a) 25 kmph
    (b) 50 kmph
    (c) 100 kmph
    (d) 150 kmph.
60. An instrument used to measure Roughness index is
    (a) Enoscope
    (b) Deflectometer
    (c) Seismograph
    (d) Bump integrator.
61. A pavement that offers poor visibility is
    (a) Gravel road
    (b) WBM road
    (c) Bituminous road
    (d) Cement concrete road.
62. A drawback of a rigid pavement is that
    (a) no crack occurs even if local settlement takes place
    (b) it acts as a bridge to cover minor depression like irregularity
    (c) any small rift further widens
    (d) its ability to withstand iron wheeled traffic is less.
63. An example of a rigid pavement is
    (a) earthen road
    (b) water bound Macadam road
    (c) bitumen road
    (d) concrete road.
64. For an earthen road the min. desirable gradient is
    (a) 1 in 12
    (b) 1 in 20
    (c) 1 in 120
    (d) 1 in 200.
65. Mostly used road in India is
    (a) earthen road
    (b) water Bound Macadam road
    (c) bitumen road
    (d) cement concrete road.
66. Base course of a WBM road consists of
    (a) 120 to 180 mm stone placed on their ends upwards
    (b) 120 to 180 mm stone placed on their ends upwards + 50 mm stone Macadam + Water
    (c) 120 to 180 mm stone placed on their ends upwards + 50 mm stone Macadam + Water + 2 layers of metal each 150 thick
    (d) 120 to 180 mm stone placed on their ends upwards + 50 mm stone Macadam + Water + 2 layers of metal each 150 mm thick + Rolled + Screenings applied.
67. A well prepared WBM road can accept
    (a) 450t of pneumatic load/day
    (b) 450t of iron wheeled traffic/day
    (c) 450t of pneumatic traffic + 450t of iron wheeled traffic/day
    (d) 900t of mixed traffic/day.

68. Soil stabilizer is used for
    (a) improving properties of low cost roads
    (b) improving skid resistance of a surface
    (c) growing vegetation on the side slopes of embankment
    (d) flattening of slope providing drainage and prevention of rock slides.

69. Cement stabilizations is best suited for
    (a) sandy soils
    (b) clayey soils
    (c) clays rich in sulphates
    (d) soils rich in lime.

70. When lime is added to a soil, it
    (a) increases grain size due to electrolytic and chemical action
    (b) lowers plastic limit and increases liquid limit
    (c) renders soil brittle
    (d) reduces binding action.

71. Bitumen stabilization acts as a
    (a) hydrophilic medium
    (b) adhesive for coarse grains and water proofing agent for fine grains
    (c) destroyer of organic matter and hence more strength
    (d) thin cover that gets oxidized in a few days.

72. A hygroscopic salt
    (a) Calcium hydroxide
    (b) Calcium silicate
    (c) Calcium chloride
    (d) Calcium carbonate.

73. The best stabilizer for black cotton soil is
    (a) Lime
    (b) Bitumen
    (c) Cement
    (d) Calcium chloride.

74. Surface painting is
    (a) marking of white and yellow lines on pavement
    (b) providing a thin Bituminous layer over WBM roads
    (c) application of Bitumen concrete layer over WBM road
    (d) application of a hot mix of bitumen and then stone chips over it.

75. For surface painting the approximate quantity of Bitumen required is
    (a) 1 kg/m² for first coat and 2 kg/m² for second coat
    (b) 2 kg/m² for first coat and 1 kg/m² for second coat
    (c) 5 kg/m² for first coat and 2 kg/m² for second coat
    (d) 8 kg/m² for first coat and 14 kg/m² for second coat.

76. Rolling should be
    (a) from centre (middle of pavement) to edge
    (b) from edge to centre
    (c) from edge to centre with an overlap of a min. of 30 cm
    (d) from one edge to another edge with a min. overlap of one half the width or roller.

77. If full grouted method of construction of Bituminous roads, Bitumen is applied at the rate of
    (a) 1 to 2 kg/m²
    (b) 4 to 8 kg/m²
    (c) 8 to 14 kg/m²
    (d) 10 to 20 kg/m².
78. In full grouted method of construction of Bituminous roads, the second coat of Bitumen is applied at the rate of
    (a) $\frac{1}{2}$ to $1\frac{1}{2}$ kg/m²
    (b) 1 to 2 kg/m²
    (c) 4 to 8 kg/m²
    (d) 8 to 14 kg/m².
79. In full grouted method of construction of Bituminous roads, rolling is done with
    (a) initially 8t roller and 10t roller later
    (b) initially 10t roller and 20t roller later
    (c) initially 1t roller and 2t roller later
    (d) initially 15t roller and 30t roller later.
80. In semigrouted Macadam Pavement the hoggin material is
    (a) Bitumen
    (b) Water
    (c) Stone powder
    (d) Fine sand.
81. In Bitumen Bound Macadam roads a Bituminous premix at the following temperature is applied
    (a) 0°C
    (b) 100°C
    (c) 150°C
    (d) 200°C.
82. A Bitumen Bound Macadam Pavement is best suited where
    (a) rainfall is heavy
    (b) ground water table is at shallow depths
    (c) the climate is hot and humid
    (d) rainfall is scant.
83. In the premix method of Bitumen road construction aggregate is also heated
    (a) for easy workability
    (b) for easy spreading
    (c) to get a homogeneous mix
    (d) to economise the quantity of Bitumen.
84. Seal coat is a layer of
    (a) cement concrete
    (b) coarse sand + Bitumen
    (c) water repellant agent
    (d) adhesive to improve bond between aggregates.
85. Purpose of the seal coat is to provide
    (a) an even surface
    (b) required grade
    (c) camber
    (d) an impervious layer.
86. The best road suited to pneumatic and iron wheeled traffic is
    (a) Farthen road
    (b) Water Bound Macadam road
    (c) Bitumen Bound Macadam road
    (d) Cement Concrete road.
87. Cement concrete road can be laid over
    (a) any surface of ground
    (b) an earthen road subjected to traffic for 1 year
    (c) a WBM road of thickness greater than 150 mm and subjected to traffic for 1 month
    (d) any black top road in use at least for 2 weeks.

88. Subgrade preparation for a cement concrete road includes
    (a) levelling and compacting
    (b) levelling + compacting + camber is given
    (c) levelling + compacting to the camber + kept moist for 12 hours
    (d) levelling + compacting to the camber + kept moist for 12 hours + cement concrete layer laid.
89. Permissible tolerance in grade, camber and alignment of a cement concrete road is
    (a) 0.01%   (b) 0.1%
    (c) 1%   (d) 2%.
90. In a cement concrete road expansion joints are provided at intervals of
    (a) 4 m   (b) 10 m
    (c) 20 m   (d) 50 m.
91. Longitudinal ruts are formed because of
    (a) iron wheeled traffic
    (b) combined iron wheeled and pneumatic traffic
    (c) heavy rainfall
    (d) heavy axle loads of vehicles.
92. A deep and big depression on a road is called
    (a) longitudinal rut   (b) cross rut
    (c) pot hole   (d) crack.
93. For filling the pot hole of an earthen road, the hardness of the filling material
    (a) should have the same hardness of neighbouring material
    (b) should be harder than the neighbouring material
    (c) should be softer than the neighbouring material
    (d) may be harder or softer.
94. To fill pot holes of a Bituminous surface
    (a) heated stone chips or coarse sand and Bitumen are applied
    (b) primer is applied and filled with gravel and rammed
    (c) a thin layer of Bitumen is spread and then premixed material is placed
    (d) 60 mm thick layers of premixed Bituminous concrete is placed in position and rammed.
95. Bleeding can be controlled by the application of
    (a) dust   (b) heated stone chips
    (c) metal   (d) hard rolling.
96. Wave and corrugations may be formed because of
    (a) lack of bond for aggregate   (b) wearing coat being more elastic
    (c) defective rolling   (d) excessive tangential compression.
97. Maximum equivalent single wheel load as per IRC is
    (a) 8160 kg   (b) 4080 kg
    (c) 2040 kg   (d) 1020 kg.

98. As per Boussinesq's theory vertical stress at a depth z is
    (a) $p\left[1-\dfrac{z^3}{(a^2+z^2)}\right]$
    (b) $p\left[1-\dfrac{z^2}{(a^3+z^3)}\right]$
    (c) $p\left[1-\dfrac{z^3}{(a^2+z^2)^{3/2}}\right]$
    (d) $p\left[1-\dfrac{z^2}{(a^2+z^2)^{3/2}}\right]$.

99. Rigidity factor is '1' when contact pressure is
    (a) 5 kg/cm^2
    (b) 7 kg/cm^2
    (c) 9 kg/cm^2
    (d) 11 kg/cm^2.

100. If the thickness of a flexible pavement is 't' for a load 'p' then according to McLeod, pavement thickness for 10^6 repetitions of the same load is
    (a) t
    (b) $2t$
    (c) $4t$
    (d) $10t$.

101. Equivalence factor =
    (a) $\dfrac{\text{axle load}}{\text{standard axle load}}$
    (b) $\left[\dfrac{\text{axle load}}{\text{standard axle load}}\right]^2$
    (c) $\left[\dfrac{\text{axle load}}{\text{standard axle load}}\right]^4$
    (d) $\left[\dfrac{\text{axle load}}{\text{standard axle load}}\right]^{1/2}$.

102. Vertical deflection for a flexible plate is
    (a) $\dfrac{1.5pa}{E_s}$
    (b) $\dfrac{1.18pa}{E_s}$
    (c) $\dfrac{pa}{E_s}$
    (d) $\dfrac{pa}{1.5E_s}$.

103. In Burmister's theory displacement factor depends on
    (a) ratio of modulii of elasticity of pavement and subgrade layers
    (b) ratio of vertical deflection of pavement and subgrade layers
    (c) Poisson's ratio of the pavement material
    (d) thickness of wearing course.

104. As per IRC the max. width of a vehicle is
    (a) 1.75 m
    (b) 2.20 m
    (c) 2.44 m
    (d) 3.12 m.

105. PUC equivalent for a bus is
    (a) 1.00
    (b) 1.75
    (c) 2.25
    (d) 6.00.

106. PUC equivalent for a cycle is
    (a) 0.2
    (b) 1.00
    (c) 2.25
    (d) 6.00.

107. The safe speed on a highway is
    (a) 50th percentile speed
    (b) 75th percentile speed
    (c) 85th percentile speed
    (d) 98th percentile speed.

108. 'Weaving' is
    (a) merging
    (b) diverging
    (c) crossing
    (d) merging, travelling and diverging.

109. AADT is
    (a) Total traffic in a day
    (b) Total traffic of a month/28, 29, 30 or 31
    (c) Total traffic in an year/365 or 366
    (d) Total traffic in an year/365 (or 366) × 24

110. 30th highest hourly volume means
    (a) AADT exceeded it 29 times in an hour
    (b) AADT exceeded it 29 times in a day
    (c) AADT exceeded it 29 times in a month
    (d) AADT exceeded it 29 times in an year.

111. Traffic density is
    (a) no. of vehicles moving in a specific direction per lane per day
    (b) no. of vehicles moving in a specific direction per hour
    (c) no. of vehicles per unit length
    (d) max. no. of vehicles passing a given point in one hour.

112. Accident may occur because of
    (a) alert driver
    (b) disciplined travellers
    (c) sun and wind
    (d) stray cattle.

113. 66.67% of the accidents take place because of
    (a) traffic moving in opposite directions
    (b) traffic changing lanes
    (c) right turning traffic
    (d) left turning traffic.

114. Minimum lateral clearance desirable from the pavement edge is
    (a) 0.50 m
    (b) 0.75 m
    (c) 1.55 m
    (d) 1.85 m.

115. Capacity of a 3.75 m traffic lane is
    (a) 1000 PCU
    (b) 2500 PCU
    (c) 5000 PCU
    (d) 10000 PCU.

116. An advantage of manual counting of traffic is
    (a) permits traffic classification by the type of vehicle
    (b) suited in any climate
    (c) highly accurate
    (d) it can be carried out for any length of time.

117. A disadvantage of mechanical counting of traffic is
    (a) it is quite expensive
    (b) it is unpopular in inclement weather
    (c) it cannot make classified counts
    (d) it may not accurate.

118. An effective way of conducting 'Origin and destination studies' to extract more information is
    (a) road side interview
    (b) licence plate method
    (c) return post card method
    (d) tag on car method.

119. An instrument used to study 'Spot Speeds' in traffic engineering is
    (a) speedometer
    (b) enoscope
    (c) speed recorder
    (d) enometer.
120. 'Fixed delay' in a highway is due to
    (a) pedestrians crossing the rod
    (b) parked vehicles
    (c) traffic signals
    (d) road repairs.
121. A lamp post at the edge of the pavement reduces the capacity of the lane to
    (a) 92%
    (b) 83%
    (c) 72%
    (d) 61%.
122. Pick up the odd statement
    (a) shopkeepers desire to have more parking space nearer to their shops
    (b) through traffic prefer a wide road with no parkings
    (c) if no official parking space is provided commercial vehicle driver will be having more number of haltings
    (d) the car driver ALWAYS wants free parking everywhere.
123. The area of most acute vision of a driver is a cone of
    (a) 3°
    (b) 10°
    (c) 15°
    (d) 20°.
124. Traffic signs and devices should be placed within a cone of
    (a) 3°
    (b) 10°
    (c) 15°
    (d) 20°.
125. An advantage of 'one way traffic' system is
    (a) reduced number of points of conflicts
    (b) saving of fuel
    (c) short joining distances
    (d) quality of improvement in environment.
126. In the figure below if the road is open to 'one way traffic only' then the number of points of conflicts is
    (a) 1
    (b) 3
    (c) 5
    (d) 10.

127. A basic requirement of 'Intersection at grade' is
    (a) the area of conflict should be small
    (b) the relative speed should be high
    (c) the relative angle of approach of the vehicles should be high
    (d) it should clearly indicate what the driver should do.

128. The minimum radius for intersection curve when the speed is 35 kmph is
   (a) 15 m
   (b) 25 m
   (c) 35 m
   (d) 50 m.

129. A separate provision for a right turn lane is to be provided when right turning traffic is more than
   (a) 25% of total traffic
   (b) $33\frac{1}{3}$% of total traffic
   (c) $37\frac{1}{2}$% of total traffic
   (d) 50% of total traffic.

130. The purpose of a 'divisional island' is to eliminate
   (a) nose to tail collision
   (b) head on collision
   (c) side swipe
   (d) tail to tail collision.

131. A channelization island provides
   (a) equal entry and exit widths
   (b) funnel shaped entry and wider exit
   (c) wider entrance and funnel shaped exit
   (d) high relative speed at entry and low speed at exit.

132. A channelization island should have
   (a) small entry radius and large exit radius
   (b) large entry radius and small exit radius
   (c) equal radii for entry and exit
   (d) large entry and exit radii.

133. The traffic island that segregates the left turning from the rest is
   (a) channelization island
   (b) divisional island
   (c) pedestrian loading island
   (d) rotary island.

134. Pedestrian loading island is located in between
   (a) up and down lanes
   (b) left turning and other streams
   (c) foot path and cycle tracks
   (d) cycle lane and motor vehicle vane.

135. The advantage of a rotary is
   (a) traffic is in continuous motion
   (b) no waiting by traffic
   (c) vehicles move in the same direction
   (d) left turn is relatively easy.

136. The crossing angle should be a min. of
   (a) 20°
   (b) 30°
   (c) 60°
   (d) 80°.

137. Rotary is ideally suited
   (a) when traffic is more than 4 streams join at the junction
   (b) when traffic is very heavy
   (c) when the pedestrian traffic is heavy
   (d) for congested areas.

138. An elliptical rotary island is provided when
   (a) most of the traffic is along its major axis
   (b) most of the traffic is along its minor axis
   (c) right turning traffic is more than 25%
   (d) traffic density is more than 3000 vehicles/h.

139. The number of points of conflicts in the figure below when 'Right turn' is prohibited in
    (a) 1 (b) 4
    (c) 8 (d) 10.

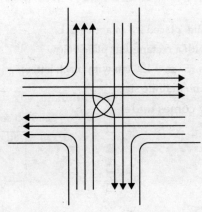

140. Grade separation had the advantage of
    (a) easy right turn (b) no speed restriction
    (c) number of points of conflicts is nil (d) occupying the least space.

141. A disadvantage of 'Diamond Junction' is
    (a) no provision for easy right turn (b) two over bridges are required
    (c) two parallel roads are required (d) crossing angle is 90°.

142. A three level round about is preferred to a clover leaf junction when
    (a) less area is available
    (b) left turn traffic has a direct path
    (c) heavy traffic is to be handled
    (d) construction of an over bridge is very difficult.

143. For the most effective traffic control, adopt
    (a) more number of traffic signs (b) limited number of traffic signs
    (c) more varieties of traffic signs (d) more 'warning' and less 'informatory' signs.

144. The traffic sign in the figure given below is a
    (a) warning sign (b) informatory sign
    (c) regulatory sign (d) route marking sign.

Q. 144   Q. 145

145. The road sign given above is a
    (a) warning sign (b) informatory sign
    (c) regulatory sign (d) route marker sign.

146. Regulatory signs are
    (a) red circular discs 60 cm dia. placed 2.8 m above G.L.
    (b) red triangle 45 cm side with a rectangular plate below
    (c) rectangular yellow plates 75 cm × 120 cm with black letters
    (d) rectangular in shape but no definite size.

147. Unguarded level crossing comes under
    (a) warning sign
    (b) informatory sign
    (c) regulator sign
    (d) route marker sign.

148. The above figure indicates
    (a) warning sign (b) informatory sign
    (c) regulatory sign (d) route marker sign.

149. An advantage of traffic signal is
    (a) no rear end collision (b) easy segregation of traffic
    (c) quick movement of vehicles (d) orderly movement of vehicles.

150. A disadvantage of Traffic Signals is
    (a) traffic along one or two directions is permitted while the remaining has to wait
    (b) waiting vehicles let off exhaust gases into the environment
    (c) head on collision may occur
    (d) pedestrians may face problems.

151. Yellow colour of a 'Coloured light traffic signal' indicates
    (a) go (b) stop
    (c) be ready to go (d) clearance time.

152. Centre line for an urban road of more than 4 lanes is
    (a) broken line
    (b) 3 m long broken line with a gap of 4.5 m in between lines
    (c) continuous thick line
    (d) two thick parallel lines with a gap of 75 mm in between.

153. Thick white and black lines of 2 m to 4 m long provided along the width of a highway indicates
    (a) lane line (b) centre line
    (c) cycle track (d) pedestrian crossing.

III. TRUE/FALSE

1. Haphazard growth of suburban dwelling centres not bound by any city regulations is called *ribbon growth*.
2. Revenue obtained by the road traffic is too meagre compared to the amount spent for its development.
3. Central road fund was formed by levying a rupee per eight gallons of diesel oil.
4. Indian Road Congress was formed in the year 1943.
5. The road connecting foreign highways with the state counterparts is called a state highway.
6. Straight alignment of a highway is the shortest but may not be the cheapest.
7. Absolutely straight alignment may not be desirable at all the times because it makes the driver to relax.
8. While on the highway the driver of a vehicle is free to change lanes and even the direction of the vehicle. The driver of a railway coach cannot take independent decisions but this freedom of the former is a factor responsible for more number of accidents on a highway.
9. Alignment of a highway is so chosen to have as many cross drainage works as possible so that quick drainage takes place.
10. The rise or fall in excess of the floating gradient is called ineffective rise or ineffective fall.
11. A reconnaissance survey may be less accurate but furnishes all the details related to the alignment and along the various alternate routes proposed.
12. Alignment of a highway should be finalised before conducting the preliminary survey.
13. When number of gate Closures × Intensity of traffic in tonnes per day is greater than 50,000 an over bridge is preferred to a level crossing.
14. When more than 50% of the traffic is terminating at the station, a bypass road is desirable.
15. Design of cross-section, gradient, sight distance, radius of horizontal curves, amount of super elevation to be provided and characteristics of wearing course and layers below it depend on "design speed".
16. In India the maximum permissible design speed on National Highways in plains is 100 kmph.
17. On busy roads a minimum limit for design speed is also required as low speed may block normal movements of traffic.
18. The maximum forward skid co-efficient is greater than maximum lateral skid co-efficient for almost all the vehicles during their braking test in a plain.
19. When the longitudinal displacement is less than the circumferential movement of the wheels it is called "slipping" whereas when the longitudinal displacement is more than the circumferential movement it is called "skidding".
20. Skidding is because of less friction between the tyre of the vehicle and surface of the pavement.
21. When the braking force exceeds frictional force between the tyre and pavement skidding takes place.

22. "Slip" results in because of loose and slippery surface of pavement.
23. Slipping may take place due to sudden increase in acceleration of a vehicle.
24. Co-efficient of friction between tyre and pavement increases with load and temperature.
25. Co-efficient of friction increases with the speed of the vehicle.
26. Worn out and smoothened tyres with no identity of treads offer higher co-efficient of friction during dry weather whereas new tyres with treads in perfect condition offer higher co-efficient of friction during wet weather.
27. 0.15 and 0.4 are the values recommended by IRC for lateral and longitudinal co-efficients of friction against lateral skidding and longitudinal skidding.
28. "Unevenness" is the cumulative (either upwards or downwards measured in cm) vertical displacement from the proposed even longitudinal profile per kilometre horizontal length of the road.
29. The highly desirable unevenness index in less than 150 cm/km for any speed of vehicle for the comfortable ride of passengers.
30. Unevenness index not exceeding 320 cm/km gives comfortable ride at 100 kmph.
31. Retardation is limited to 3.92 m/sec^2 for the comfort of passengers.
32. While black top roads give poor visibility, light coloured pavements give more glare.
33. A dry pavement gives more glare than a wet pavement.
34. The maximum width of a vehicle using the road should not be greater than 2.44 m.
35. The maximum height of a single decked vehicle should not exceed 3.81 m.
36. The maximum height of a double decker vehicle should not exceed 4.72 m.
37. The maximum length of a single unit with two axles is 10.67 m.
38. The maximum length of any vehicle along with a trailer combination should not exceed 18.29 m.
39. No combination of vehicles should consist of more than two units and their total length is restricted to 18.99 m.
40. As per IRC axle load is not to exceed 8200 kg.
41. A terrain is said to be "rolling" when it is not very plain, full of undulations but at the same time is not mountainous.
42. A minimum clearance width of 1.2 m is required between the vehicles during overtaking.
43. Flatter cambers are desirable for WBM roads.
44. Steeper cambers are the required for slow moving traffic as a bullock cart.
45. Parabolic type of camber is ideal for fast vehicles which use the relatively flatter central width of the road while overtaking.
46. Camber should never be less than 50% of the gradient.
47. Width of a single lane carriage way is 3.75 m. While it is 7 m for a two lane road.
48. Median is a longitudinal strip provided in between two traffic lanes to separate up and down traffic and to prevent head on collision between the vehicles.

49. Shoulders are the extension of pavement width beyond pavement edges to accommodate stationary vehicles but not to be used as a regular traffic lane by the moving vehicles.
50. Guard rails are needed on embankments when the height of filling is more than 3 m.
51. Roadway includes width of carriage way + width of shoulders on either side.
52. Width of roadway for National Highway is 15 m.
53. Recommended "Land width" for a National Highway is 45 m.
54. For stopping sight distance, the height of the eye level of driver is 1.2 m and the height of the stationary object is 15 cm above the road surface.
55. Intellection time is the time required by the driver to take a decision.
56. PIEV time varies from 10 seconds to 30 seconds.
57. Braking distance = $\dfrac{v^2}{2g\left(f \pm \dfrac{n}{100}\right)}$.
58. Safe passing sight distance is the minimum distance ahead of the driver for overtaking a slow moving vehicle and at the same time causing the least inconvenience to the traffic in the opposite direction.
59. Overtaking sight distance is depended on the relative speed of the overtaking vehicle and independent of the speed of the vehicle coming in the opposite direction.
60. Overtaking sight distance along an upward gradient or downward gradient is more than the corresponding on a level road.
61. Overtaking sight distance down the slope is more compared to up the grade.
62. The minimum length of an overtaking zone is a minimum of 3 × overtaking sight distance.
63. At an uncontrolled intersection the time interval between sighting the vehicle on a leg approximately at right angles and approaching the junction at the design speed should be more than 2 seconds.
64. "Centrifugal Ratio" = $\dfrac{\text{Centrifugal force}}{\text{Weight of the vehicle}}$.
65. 'Overturning' of a vehicle takes place while negotiating a horizontal curve when the centrifugal ratio = b/h

    where b = Distance between the two wheels of the axle

    and h = Height above the pavement where the centrifugal force acts.
66. Centrifugal force is always greater than the frictional force.
67. Centrifugal force = $\dfrac{Wv^2}{9R}$ always acts parallel to the surface of the pavement.
68. Frictional force F always acts parallel to the surface of contact and always in the opposite direction of motion.
69. Equilibrium super elevation = $\dfrac{V^2}{127R}$.

70. Passengers do not feel jolting while negotiating curves when the super elevation is greater than $\dfrac{V^2}{127R}$.
71. Super elevation is limited to 7% in plains.
72. Negative super elevation results in case of the outer ring of horizontal curves of bigger radius where normal camber is retained.
73. When super elevation is to be provided it is not possible to provide camber.
74. The super elevation is zero at the beginning of the transition curve and it gradually reaches its full value at the middle point of the curve.
75. Extra width provided on curves is generally kept on the outer side of the horizontal curve.
76. In case of plain circular curves without any transition curves extra width is provided on the inner side of the curve.
77. Ruling gradient is steeper than limiting gradient.
78. Exceptional gradients are provided near hair pin bends and approaches to cause ways.
79. Exceptional gradient should not be provided for more than 100 m at a stretch.
80. Exceptional gradient should not be provided for lengths greater than 60 m/km of road length.
81. The gradient at any section should not be flatter than the ruling gradient.
82. The minimum gradient can be zero when the road is in a perfect plane country.
83. Percent grade compensation on horizontal curves is given out by the equation

    Per cent grade compensation = $\dfrac{30 + 5}{R} \times 100$.
84. Grade compensation is not required for gradients flatter than 1 in 25.
85. A valley curve is formed when a flatter downward gradient is followed by a steeper downgrade.
86. A summit curve is a parabola while a valley curve is two transition curves with no circular curve in between.
87. When a flat downward slope follows a level road a summit curve is obtained.
88. The question of sight distance does not arise in case of valley curves.
89. In valley curves the centrifugal force adds to the pressure on springs while in summit curves the force pressure on springs relieves.
90. PCU of a bullock cart is greater than that of a truck with a trailer.
91. In the cumulative frequency curve 85th percentile is the governing factor for overtaking distances and imposition of speed limits.
92. For the driver the field of very clear vision is a cone whose apex angle is 10°.
93. The maximum permissible load on an axle is 8165 kg.
94. Traffic volume is the maximum number of vehicles crossing a specific cross-section of a road at the peak time.

95. 30th highest hourly volume means the peak volume of traffic that will be exceeded only 29 times in a year.
96. Spot speed is measured by Enoscope.
97. "Head way" decreases with increase in speed to the vehicle.
98. Highest traffic density occurs when the traffic volume is very small.
99. The capacity of a 3.75 m wide traffic lane with earthen shoulders is 1000 PCU/day (in both the directions) whereas it is 2500 PCU/day if hard 1 m wide shoulders are provided on either side.
100. Besides the motorist, the vehicle, road conditions, and weather, a non-motorist road user without road sense and stray cattle are also responsible for accidents.
101. Of all the manoeuvres left turn involves the maximum number of conflicts while a right turn had the minimum number of conflicts in India.
102. At an intersection a straight going vehicle has the minimum number of conflicts.
103. At an intersection provided with traffic signals the number of rear end collisions may increase considerably.
104. Amber coloured light provided in between red and green lights for a traffic signal to provide "Clearance time" of the vehicles.
105. Longitudinal solid yellow line should never be crossed by a driver.
106. As relative speed increases, always the severity of the accidents increase.
107. Channelisation should be always followed by traffic signals.
108. At any rotary intersection in India the traffic moves only in clockwise direction.
109. Under normal conditions no traffic signals are required at a rotary intersection.
110. An urban rotary intersection is generally designed for a speed of 30 kmph.
111. An elliptical rotary takes more traffic along its minor axis and least traffic along the major axis.
112. Weaving angle should be smaller for smooth flow of traffic but an angle less than 15° increases the length of axis of a rotary.
113. Weaving length should be a minimum of 4 × width of weaving section.
114. For accident free manoeuvres provide as large a weaving length as possible.
115. Vehicles entering a rotary should accelerate, move along the weaving length with uniform speed and decelerate at the exit.
116. The radius at the exit of a rotary is generally $1\frac{1}{2}$ to 2 times at the entry.
117. Channelizing islands must be provided for every leg leading to a rotary.
118. A rotary is to be provided on a level ground or on a slope not steeper than 1 in 50.
119. A rotary may operate well for a combined traffic volume of 500 to 5000 PCU per hour.
120. A rotary is the best suited when the right turning traffic is less than 30% of the total traffic.
121. Fuel consumption for travelling 270 m on a level ground is less than that consumed for a stop at the traffic signal.

122. Rotaries can be had when the number of legs at an intersection is between 4 and 7.
123. Rotary is the best suited when the pedestrian traffic is high on all the legs.
124. Rotaries become ineffective to handle slow moving traffic.
125. A rotary should never be provided at a summit.
126. A grade separated intersection is mainly designed to offer the maximum facility for left turning traffic (in India).
127. Grade separation is ideally suited to a flat terrain.
128. An underpass poses the problem of drainage.
129. A diamond crossing is advantageous over the clover leaf junction when the straight, left and right turning traffic is almost equal at the intersection.
130. Parking at 45° or 60° to the centre line of the road is easy but taking out a parked vehicle is difficult.
131. Right angled parking is used for steep grades.
132. Regulatory road signs are provided with 60 cm dia. red discs, warning signs with an equilateral triangle of 45 cm side whereas informatory signs are just rectangular boards.
133. Two different signs should not be placed on the same post and the minimum distance between two neighbouring traffic signs is 30 m.
134. "Loading islands" are provided to receive bus/tram passengers.
135. Traffic streams should merge at an angle greater than 15°.
136. "Bending" of the traffic stream controls the speed of vehicles.
137. Acceleration and deceleration lanes need not have the full width of a traffic lane.
138. Coarse sand on pavement increases abrasing action.
139. Metal with angular edges is required for WBM and Bituminous roads whereas rounded aggregate is preferred for concrete pavements.
140. The maximum value of Los Angels Abrasion test permitted for good aggregate used for concrete construction is 30%.
141. The cylinders of Deval's Attrition Machine rotate at 33 r.p.m. to make 10,000 revolutions during the test.
142. The aggregate impact value should not exceed 45% when it is used as wearing course.
143. Flakiness index = $\left[\dfrac{\text{Weight of aggregate whose least dimension} \leq \frac{3}{5} \text{ their mean dimension}}{\text{Total weight of aggregate}}\right]$.
144. The desirable angularity number for aggregate used in construction is less than 11.
145. Rock should have a water absorption ratio of less than 0.6%.
146. The penetration value should be 80/100 for road tar.
147. Ductility value of bitumen should not exceed 50.
148. Ductility test is conducted at 27°C and the rate of pull is 50 mm/min.

149. Viscosity test determines the time in seconds taken for 50 ml of the sample to flow through a standard orifice at a given testing temperature.
150. Bitumen is having more specific gravity than tar.
151. The lowest temperature at which a Bituminous material momentarily catches fire is called its fire point.
152. In the flash point test the test flame is applied from atleast 17°C below the expected flash point and then at intervals of 1° to 3°C rise in temperature.
153. When the Bituminous vapour gets ignited and burns for 5 seconds it is called as flash point.
154. The bitumen used for pavement construction should have a flash point greater than 175°C.
155. For good bitumen the solubility should be 90%.
156. Percentage loss on heating of bitumen at 163°C for 5 hours should not exceed 1 or 2%.
157. The maximum content of water in bitumen should not exceed 0.2% by weight.
158. In cut backs RC—5 would contain the highest proportion of solvents whereas RC—0 has the lowest content.
159. RC—3, MC—3 and SC—3 have the same initial viscosity.
160. An emulsion consists of two immiscible liquids the globules of one homogeneously dispersed in the other.
161. Bituminous emulsions are used for patch repair works and in soil stabilisation.
162. Bitumen coats the aggregates more easily than tar.
163. Tar has more free carbon compared to bitumen.
164. Bigger the size of the aggregate, more is the stability.
165. As the height of the road layer increases size of the aggregate also increases.
166. About 5 cm aggregate is used in the base course while it is 1.25 cm in the wearing course.
167. A pavement is said to be a rigid pavement when the load is transferred from particle to particle and from layer to layer and any void created is readily filled by the material at top or sides.
168. When lean cement concrete is used in the base course the type of pavement is called "semi-rigid" pavement.
169. The lowest layer of a pavement is the base course.
170. Sub-grade is the bottom most layer of a highway over which the embankment is built in.
171. Compaction is always advantageous for clayey soils in cuttings.
172. Sub-grade should be atleast 1.2 m above GWT in any season.
173. The purpose of sub-base is to form a layer of separation between the base and the underlying soil which has undesirable characteristics.
174. The depth of sub-base may not be uniform but varies considerably and may be even zero at some sections.
175. Base course is a very important layer which resists vertical settlement or lateral (horizontal) shear of wearing course and acts as a protective layer for sub-base and sub-grade.

176. The minimum thickness of base course is 75 mm.
177. "Pumping" occurs not only in rigid pavements, but in flexible pavements as well.
178. "Base coat" is an optional layer provided above base course but below the wearing course.
179. The weight of the vehicle is more important in the pavement design rather than the axle or wheel load.
180. Total depth of pavement is always influenced by the tyre pressure.
181. Pneumatic tyre demands strong and hard aggregate as wearing course whereas steel tyre demands strong and hard material for base course.
182. Rigidity Factor = $\dfrac{\text{Contact Pressure}}{\text{Tyre Pressure}}$.
183. Duel wheels are provided for the rear assembly when the load on the axle is greater than 8170 kg.
184. Stresses start overlapping from a depth $\dfrac{d}{2}$ to $2s$ in case of duel wheel where

    d = Clear spacing between wheels ; and s = Centre to centre spacing of wheels.
185. McLeod assumes the design load to get repeated 10^6 times before the pavement fails.
186. If a pavement structure fails for 10^6 repetitions of 2200 kg and 5×10^5 repetitions of 2700 kg, then the equivalent wheel load factor is 0.5.
187. Frost action creates voids in the sub-grade.
188. Frost action can be temporarily rectified by the addition of Calcium and Sodium Chloride to the sub-grade which lower freezing temperature.
189. As per Steel's hypothesis as the value of Group Index increases its strength also increases.
190. The pressure (applied at a uniform rate) needed to force out water from a compacted sub-grade soil sample is called "exudation pressure".
191. Stiffness factor (in Triaxial test) = $\dfrac{\text{Modulus of elasticity of sub-grade}}{\text{Modulus of elasticity of pavement}}$.
192. Burrmister's method assumes the elastic modulus of each layer of the pavement increases with depth.
193. Stress due to corner load = $\dfrac{3 \times \text{Corner load}}{(\text{Thickness of slab})^2}$.
194. Friction stress induced is more in bottom fibres of a long concrete slab.
195. The expansion gaps in cement concrete roads should not be greater than 25 mm under any circumstances.
196. The maximum spacing of contraction joints in case of non-reinforced cement concrete slabs is at 13 m.
197. Dowel bars are not provided when the thickness of the slab is less than 15 cm.
198. Organic soil as peat is not undesirable for embankment construction of a highway.
199. Pneumatic tyred rollers are ideally suited to compact fine sandy soils.

200. Sheep foot rollers are more suited to compact clayey soils.
201. For an earthen road a camber of 1 in 20 is provided.
202. Gravel or earthen roads may last for considerable length of time if exclusively subjected to pneumatic tyred traffic or iron tyred traffic but not to mixed traffic.
203. Rolling is to be done by starting from one edge and proceeding towards the other and by providing enough overlap.
204. Big boulders or full bricks are to be used for the lower layers of a WBM road to get an even surface.
205. Screenings completely fill the voids left by coarse aggregate during the construction of WBM road.
206. While excess binder in case of cement concrete roads adds to the strength of the pavement, bituminous binder in excess of the optimum is not conducive for the satisfactory performance of the pavement.
207. Bituminous roads need more time for construction than cement concrete roads.
208. Surface dressing is the application of a less viscous bitumen as binder on to a low cost road as a WBM road.
209. A tack coat is the same as a prime coat but applied over a relatively impervious pavement as a black top or cement concrete road.
210. In the Penetration Macadam type of construction, the bitumen is applied as a thin sheet and then the aggregates are spread over it and compacted by rolling.
211. Full Grout Macadam type of construction is employed where the temperature is very high.
212. A thin section of Bitumen Bound Macadam is as goods as a thick layer of untreated WBM road.
213. For Bitumen Bound Macadam Construction the aggregate need not be very tough even to take heavy loads.
214. A bituminous premix consisting of 12 mm chips, coarse sand in addition to the bitumen with a overall thickness of 25 mm is called a bituminous carpet.
215. In addition to the coarse aggregate, fine aggregate and hot bitumen, bituminous concrete consists of an inert mineral filler.
216. A bituminous concrete road is stronger than a cement concrete road.
217. Sheet asphalt road is the same as a bituminous concrete road but contains no coarse aggregate.
218. Mastic asphalt layer requires no rolling *i.e.*, compaction.
219. Surface dressing can be carried out in any weather *i.e.*, even when it is slightly raining or in an extremely cold weather.
220. In a concrete slab tensile stresses are developed at the bottom layer during mid-night.
221. Expansion joints are provided in cement concrete roads at 20 ± 2 m intervals.
222. An expansion joint in a cement concrete road may serve as a construction joint as well but the converse is not true.
223. Lime is the mostly used modifier in case of highly plastic soils.

224. Non-cohesive soils need treatment with water-proofing or water repellant agents as bitumen and resins to impart cohesion between the particles.
225. Use of excess fines than the optimum quantity to fill the voids left by the coarse aggregate renders the mix more strong and less prone for frost susceptibility.
226. According to Fuller at the maximum density

    Percentage finer material of dia. less than $d = 100 \sqrt{\dfrac{d}{D}}$

    where D = Size of the largest particle.
227. Sulphates in the soil and $MgSO_4$ in particular are detrimental to soil stabilizing admixtures as cement.
228. When plastic soils are treated with lime the soil becomes finely divided with less affinity for water and hence swelling is reduced.
229. Lime stabilized soil is the best suited at sub-zero temperature climates but not that efficient in hot climates.
230. Hydrated lime is mostly used for stabilization than quick lime because the latter causes skin burns.
231. The most commonly used bituminous stabilizers are emulsions and cut backs.
232. The best type of stabilization for black cotton soils is the addition of portland cement to the wearing course.
233. Desert sand can be stabilized by adding Bitumen and kankar dust as filler.
234. When a part of deformation of a pavement surface due to heavy loading is left over even after the removal of the load, the deformation is called "consolidation deformation".
235. Loss of base course material results in mainly because of iron tyred traffic.
236. Alligator cracking occurs in bituminous pavement because of fatigue.
237. A localised heaving up of the pavement in a temperate country in cold weather is called "Frost Heaving".
238. "Reflection Cracks" occur in bituminous layers provided over cement concrete pavements.
239. Excessive use of vibrator, deficiency of cement in the cement concrete or presence of chemicals in the cement causes in the pavement rough and ugly appearance known as "scaling".
240. "Cross ruts" are formed along an earthen road because of heavy rainfall or due to steep camber while "Longitudinal ruts" are formed due to intensive traffic.
241. Pot holes formed are enclosed by a rectangular boundary the material inside which is to be taken out and the pit formed is carefully filled with aggregate and binder such that it projects 10 mm above the existing level.
242. For filling pot holes and other maintenance work hot tar coal is usually applied as binder but not cold mix as emulsion or cut back.
243. Excess bitumen when applied may 'bleed' which increases the life of seal coat of wearing course.
244. Corrugations when get developed due to defective rolling will spread indefinitely.

245. Corrugations are easily developed in marshy soils, clays and when bigger stones are used for sub-grade.
246. Wetness, plastic clay, grease and coarse sand particles of round shape induce skidding in the pavement.
247. Bituminous roads have more skid resistance than any other pavement.
248. Sideway skidding is due to insufficient super elevation on curve and lack of grip.
249. Impending skidding is mainly due to too smooth a pavement or worn out tyre lacking in adequate grip.
250. Straight skidding may take place because of abrupt destruction of momentum of a vehicle particularly on steep downward slopes.
251. Cement concrete joints are to be renewed soon after summer and before the onset of monsoon. Also they are to be attended in winter.
252. Highest ground water table level in any season should be 1.2 m below the subgrade level.
253. A grannular or impermeable bituminous layer just below the wearing course shall be effective in cutting off capillary rise of water.

ANSWERS
Objective Type Questions

1. (c)	2. (a)	3. (d)	4. (c)	5. (b)	6. (b)
7. (b)	8. (a)	9. (a)	10. (c)	11. (c)	12. (d)
13. (b)	14. (b)	15. (d)	16. (c)	17. (a)	18. (b)
19. (d)	20. (a)	21. (c)	22. (a)	23. (b)	24. (b)
25. (a)	26. (b)	27. (b)	28. (a)	29. (d)	30. (d)
31. (c)	32. (c)	33. (b)	34. (c)	35. (b)	36. (b)
37. (a)	38. (c)	39. (c)	40. (a)	41. (a)	42. (b)
43. (b)	44. (d)	45. (d)	46. (b)	47. (b)	48. (c)
49. (b)	50. (b)	51. (d)	52. (a)	53. (b)	54. (a)
55. (b)	56. (c)	57. (a)	58. (d)	59. (b)	60. (d)
61. (c)	62. (c)	63. (d)	64. (c)	65. (a)	66. (b)
67. (c)	68. (a)	69. (a)	70. (a)	71. (b)	72. (c)
73. (a)	74. (b)	75. (d)	76. (c)	77. (c)	78. (a)
79. (a)	80. (c)	81. (c)	82. (b)	83. (c)	84. (b)
85. (d)	86. (d)	87. (c)	88. (c)	89. (b)	90. (c)
91. (b)	92. (c)	93. (a)	94. (d)	95. (b)	96. (c)
97. (b)	98. (c)	99. (b)	100. (c)	101. (c)	102. (a)
103. (a)	104. (c)	105. (c)	106. (a)	107. (b)	108. (d)
109. (d)	110. (d)	111. (c)	112. (d)	113. (c)	114. (d)
115. (a)	116. (a)	117. (c)	118. (c)	119. (b)	120. (c)
121. (c)	122. (d)	123. (a)	124. (b)	125. (a)	126. (a)

127. (a) 128. (c) 129. (b) 130. (b) 131. (b) 132. (a)
133. (a) 134. (d) 135. (b) 136. (c) 137. (a) 138. (a)
139. (b) 140. (a) 141. (a) 142. (a) 143. (b) 144. (a)
145. (c) 146. (a) 147. (a) 148. (b) 149. (d) 150. (b)
151. (d) 152. (d) 153. (d).

True/False

1. T   2. F   3. F   4. F   5. F   6. T
7. T   8. T   9. F   10. T   11. T   12. F
13. T   14. F   15. T   16. T   17. T   18. T
19. T   20. T   21. T   22. T   23. T   24. F
25. F   26. T   27. T   28. T   29. T   30. F
31. T   32. T   33. F   34. T   35. T   36. T
37. T   38. T   39. T   40. T   41. T   42. T
43. F   44. F   45. T   46. T   47. T   48. T
49. T   50. T   51. T   52. F   53. T   54. T
55. F   56. F   57. T   58. T   59. F   60. T
61. F   62. T   63. T   64. T   65. F   66. F
67. F   68. T   69. T   70. T   71. T   72. T
73. T   74. F   75. F   76. T   77. F   78. T
79. T   80. T   81. F   82. F   83. T   84. T
85. F   86. T   87. T   88. F   89. T   90. T
91. T   92. F   93. T   94. F   95. T   96. T
97. F   98. T   99. T   100. T 101. F 102. F
103. T 104. T 105. T 106. F 107. T 108. T
109. T 110. T 111. F 112. T 113. T 114. F
115. F 116. T 117. T 118. T 119. T 120. F
121. T 122. T 123. F 124. T 125. F 126. T
127. F 128. T 129. F 130. T 131. T 132. T
133. T 134. T 135. F 136. T 137. T 138. T
139. T 140. F 141. T 142. F 143. T 144. T
145. T 146. F 147. T 148. T 149. T 150. F
151. F 152. T 153. F 154. T 155. F 156. T
157. T 158. F 159. T 160. T 161. T 162. T
163. T 164. T 165. F 166. T 167. F 168. T
169. F 170. F 171. F 172. T 173. T 174. T
175. T 176. T 177. F 178. T 179. F 180. F
181. F 182. T 183. T 184. T 185. T 186. T
187. T 188. T 189. F 190. T 191. F 192. T
193. T 194. T 195. T 196. F 197. T 198. F

199. F	200. T	201. T	202. T	203. F	204. F
205. F	206. T	207. F	208. T	209. T	210. F
211. F	212. T	213. T	214. T	215. T	216. F
217. T	218. T	219. F	220. F	221. T	222. T
223. T	224. F	225. F	226. T	227. T	228. T
229. F	230. T	231. T	232. F	233. T	234. T
235. F	236. T	237. T	238. T	239. T	240. T
241. T	242. F	243. F	244. T	245. T	246. T
247. F	248. T	249. T	250. T	251. T	252. T
253. F					

Chapter 10 RAILWAY ENGINEERING

1. INTRODUCTION

Railway is a smooth but rigid base for a smooth wheel to roll on.

The first train ran between Stockton and Darlington in the year 1825.

In India the first train rolled on the track between Boribunder and Thana in the year 1853.

Gauge. It is the clear inner distance between the two running faces of the rails.

Economic, political and strategic consideration decide the "gauge".

Wider gauges can handle more traffic, with greater speeds and with greater safety. But they need flatter gradients, and flatter curves. Cost of construction increases as $(Gauge)^2$.

Today the broadest gauge in the world is Broad Gauge (1676 mm) while the narrowest gauge is Light Gauge (610 mm). 62% of the total rail tack is laid to the Standard Gauge (1435 mm).

Break of Gauge. When more than one gauge is adopted in a country it is called "break of gauge".

5 different gauges are adopted in India:

Broad gauge	1676 mm (5' 6")	
Metre gauge	1000 mm	
Narrow gauge	762 mm (2' 6")	
Light gauge	610 mm (2' 0")	
Standard gauge	1435 mm ($4'\ 8\frac{1}{2}"$)	
Adopted for Kolkata Metro		

In order to connect remote areas where the earnings by the Railways are less, they adopted less gauge widths as it had less initial cost of construction.

But in the long run 'Break of Gauge' had no advantages. Further it is causing:

1. Inconvenience to passengers.

2. Duplications of facilities as signals, platforms, waiting halls etc.

3. Inefficient operation of rolling stock *i.e.*, shortage for one gauge when those belonging to another gauge are lying idle.

4. Delay, loss, theft and breakages of goods during transhipment.

Formation. 1. It is the prepared bed of soil receiving ballast over it. It may be an embankment, cutting or a level ground. It provides stability to the track by supporting ballast over it and preventing ballast penetration through it.

2. It helps in quick drainage (the formation is given a camber or cross fall of 1 in 40).

In case of weak soils as Black Cotton Soils—formation is covered with a thin blanket of coarse sand or moorum to improve drainage and prevent softening. Otherwise the levels fluctuate forming 'Heaved track'—in monsoon.

Formation need be strengthened in case of weak soils.

Ballast. It is a granular bed resting on formation and supporting sleepers over it.

1. It holds the sleepers in position by resisting longitudinal and lateral displacement and preventing vertical settlement.

2. It forms resilient and elastic track by providing sufficient cushioning action. (Depth of cushion should be a minimum of 200 mm for BG).

3. It causes quick drainage thus protecting formation.

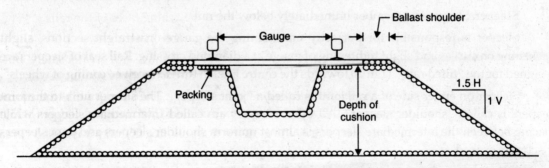

Broken stone (Quartzite, trap, Basalt, Granite, Sand Stone and lime stone) of size 25 mm (to be used under points and crossings) to 50 mm (under concrete and wooden sleepers) is the best type of ballast.

Angular stones are desirable because of their interlocking properly but they damage wooden sleepers while packing. Packing is the compact ballast mass under the rail seat to maintain levels, alignment and elasticity of the track.

Gravel due to its smooth and regular shape had little interlocking property and hence more depth of cushion is required with it. Because of its easy spreading nature it forms 'Centre Bound sleepers' with wooden and steel sleepers. It maintains good track with C.I. sleepers.

Diamond shaped (over burnt) brick ballast is used when stone ballast is not readily available. But it easily get powdered under heavy loads forming a dusty track. Also metal sleepers are corroded because of it while Rails develop Corrugations.

Ashes and Cinders (from steam locomotives) drain well and form a non-slippery track during wet weather. But they easily get powdered and form a dusty track in dry weather. Steel gets corroded because of them.

Coarse sand free from clay forms a silent track.

But sand is easily blown off by wind and hence need be renewed periodically. It enters moving parts of rolling stock increasing their wear.

It is covered with a layer of gravel to prevent blow off. Sand can be used for light weight and slow moving trains of MG, NG and LG.

The Ballast section is trapezoidal in cross section with the top surface projecting beyond the ends of sleepers called "ballast shoulders".

Side slopes of 1.5 H to 1 V are adopted.

Depth of cushion is the depth of ballast at the middle length of Gauge, below the sleeper but above formation. A minimum of 150 mm depth is followed for NG (including LG) and 250 mm for BG on trunk routes.

On curves the super elevation is given by adding more ballast under the outer rail. Also length of shoulders is increased to improve lateral stability of the track.

Packing is well compacted (rammed) ballast under the rail seat (below sleeper) to withstand heavy loading and dispersal of load over a wider area.

Sleeper. Sleeper is a member immediately below the rail.

Sleeper is responsible for perfectly maintaining the gauge on straight sections, slight slackening on curves and slight tightening of gauge at a diamond crossing. Rail seat of sleepers are designed for the "tilted rails" (1 in 20 towards the centre line) in turn to receive "coning of wheels".

Sleeper on either side of a rail joint is called a "joint sleeper". The sleeper next to the joint sleepers is called "shoulder sleeper". All other sleepers are called Intermediate sleepers while spacing between the intermediate sleepers is almost uniform shoulder sleepers and joint sleepers are closely spaced.

The number of sleepers per rail length is called "Sleeper density". If the length of rail is taken as 13 m, a sleeper density of M + 7 means, 13 + 7 = 20 sleepers being provided in that rail length.

1. Wooden sleepers
2. Steel sleepers
3. Cast Iron sleepers
4. Concrete sleepers

are the four varieties of sleepers that are common on Indian Railways.

While wooden sleepers are costly and Cast Iron sleepers (which can be recast even if broken) were more common at one time, concrete is fast replacing them.

Item	Wooden Sleeper	Steel Sleeper	C.I. Sleeper	Concrete Sleeper
Life	About 15 year	About 35 years	About 50 years	About 60 years
Cost	₹ 800 or more	₹ 400	₹ 300	₹ 450
Weight (B G)	60 kg (depends on the tree origin)	80 kg	87 kg	270 kg
Gauge	May slacken	Perfectly maintained	May slacken or may be tightened if the tie bar is bent	Perfectly maintained
Causes of damage	Easily affected by white ants	Easily gets corroded	Gets corroded	Not affected by white ants or corrosion
Choice of Ballast	Can bear with any ballast	Gravel cause centre bound track, Ashes, cinders, brick ballast cause corrosion	Ashes, cinders, brick ballast cause corrosion	Can bear any type of ballast
High speeds	Suitable	Suitable	Not suitable	Suitable
Creep	The highest	Low	Low	A bare minimum
Scrap value	Low	Moderate	Highest	Nil
Rigidity of track	Least	Better longitudinally lesser laterally	Better	Best
Bearing area	Maximum	Less	Least	Moderate

Rail. Rail is a horizontal continuous beam to support rolling wheels over it. It offers a hard and smooth surface laid to a gentle slope and it is tilted inwards.

A double headed rail which is symmetric about the horizontal axis was used first with the idea of using both the flanges for gauge faces. But the bottom surface got dented and became rough and hence could not be used as a gauge face.

Bull headed rail with a smaller foot flange and bigger head flange evolved as a result. But the base is quite unstable and needs chairs to hold it. Flat footed rail came into existence as a result.

Standard length of a rail for BG is taken as 13 m, and for MG and NG (including LG) it is 12 m.

But when the track is welded the length of the rail can be 200 m to 4000 m.

The rail section is defined by its weight per metre length. 52 kg/m is the section generally adopted for BG. 60 kg/m is recommended for trunk routes.

The axle load a rail can bear $\simeq 510 \times$ weight of rail per metre length.

Points and Crossings. Points and Crossing is an arrangement for a vehicle to transfer itself from one track to another.

A pair of split switches (consisting of tapering rail head) kept at a constant distance and always one keeping contact with the stock rail (toe of switch) while the other away from the stock rail and creating enough flangeway (throw of switch) are responsible for the cross over.

When one rail meets another at an angle it is called crossing. The angle between the two rails is called "crossing angle".

$$\text{Crossing Number} = \frac{\text{Spread at the leg of crossing}}{\text{Length of point rail from TNC}}$$

Crossing No.	Speed on BG
1 in $8\frac{1}{2}$	25 kmph
1 in 12	40 kmph
1 in 16	50 kmph
1 in 20	60 kmph
1 in 24	100 kmph

Maintenance of Track. Due to continuous running of trains 'Gauge, Alignment, levels and grades, Points and Crossings and Elasticity of the track' are affected.

About 70% of the damage to the track is caused by goods traffic. (Heavy axle load, close wheel spacing, unequal distribution of load on the two wheels of the axle and absence of good springs).

Good maintenance. Enhances life span of track and rolling stock, offers better comfort to passengers, averts accidents, and reduces consumption of fuel.

Daily maintenance. Comprises of a team of PWI, APWI, Mate, Keyman and Gang. Track length of about 4 km is attached to 1 mate, 1 Keyman and 8 gangmen.

Each Gang attends "GANG BEAT" comprising of duties as

        Packing,

        Tightening of joints

        Driving in spikes and keys

        Keeping good drainage

        Observing alignment

        Slewing

        Recording of creep

        Proper maintenance of points and crossings.

Annual maintenance. It is taken up immediately after the rainy reason.

        Removal of growth of vegetation on the track

        Opening out of ballast, lifting and packing,

        Replacement of damaged sleepers, Removal, grease application for fishplates, application and refixing of Fishplates

Checking of Alignment and gauge, checking of super elevation on curves. Longitudinal levelling of track and Boxing of ballast.

High joint or a riding joint is formed due to overpacking of joint and shoulder sleepers or sinkage and settlement of intermediate sleepers. It produces galloping effect.

Blowing joint is formed because of thin layer of dust coated ballast and loose packing. Battered rail ends, wide expansion gaps and weak sleepers aggravate the situation. Dust is blown out as the train passes over. During rainy season jets of water may spurt out.

Hogged Rails. They are rails bent down at ends because of:

1. battering action of wheels
2. loose packing under joints
3. loose fish plates

The track can be remedied by:

1. cutting and removing the hogged ends
2. replacing the hogged rail by a new one
3. dehogging the bend down rail
4. welding the track.

Battered Rail End. It is permanent deformation (depression) occurring at the beginning of a rail length due to impact loads of the rolling wheel jumping from the previous rail length. It occurs when the (expansion) joint is a bit wider. Battering is measured as difference in elevations at the end (*i.e.,* beginning) and at 300 mm away from it on the same rail.

While a battering of 2 mm or less can be ignored, the rail must be replaced when it is 3 mm or more.

Buckling of Rails. It is the widening of the gauge width because of ( ) shape developed for the rails which were straight and parallel earlier.

Buckling of rails is because of:

1. inadequate expansion gap
2. inadequate ballast cushion
3. less numbers of sleepers provided
4. no control over creep
5. longer lengths of welded track
6. improper slewing* of track on curves.

It is the most serious defect as it results in derailment.

Corrugated Rails. When depressions of 0.1 mm or more are formed at regular intervals of 20 to 3000 mm (wavelength), corrugations are said to be developed. They cause discomfort as the running of wheels will be rough and produce enormous roar (sound intensity shall be more when wavelength is less).

They occur

1. when brick is used as ballast
2. when track runs through long tunnels
3. on electrified sections of track
4. when steel of rails contain excess phosphorus

---

*Slewing is bringing the track to the original alignment on horizontal curves which moved away due to 1. Excessive centrifugal forces, 2. Less super elevation, and 3. Poor anchorage between sleeper and ballast.

5. when cushiong effect is drastically changing over a smaller length because of varying bearing capacity and varying depth of ballast and

6. because of bad joints.

Centre Bound Sleepers. Due to continuous application of compressive stresses and when the ballast lacks in interlocking action, ballast under the rail seats spread either way and the track becomes centre bound, *i.e.*, being supported at the centre only. This is quite common when gravel was used as ballast under steel sleepers (which are of inverted channel section depressed at the middle of the track) push away the ballast under the rail seat which are channellised to the middle of the track because of compressive forces.

Creep. It is the 'permanent stretching' of the rails along the direction of predominant motion due to rolling loads, braking force and impact loads.

Creep is more on curves and less on straight lengths.

Creep is more on steep slopes and down grades in particular.

Creep is more at starting and stopping points, at accelerating and decelerating spots.

Creep is more on unidirectional tracks than on two way tracks.

Creep is more for weak joints, loose packing, wider expansion gaps, lighter rails, light sleepers, less sleeper density, light and rounded ballast and heavy loads.

Creep may not be uniform being very less for one rail and quite high for the other rail of the same track.

Effects of Creep:

1. Sleepers get out of square (at right angles to rails) and out of position.
2. Gauge is affected.
3. Joints move out of support
4. Shearing and breaking of fishplates and bolts takes place
5. Points and crossings get distorted
6. Buckling of rails may result in because of closing of expansion gaps.

It is measured as "mm/rail length/month".

It is either prevented or checked by pulling the rail backward, particularly when it exceeds 150 mm per rail length.

Gradient. Gradient is the longitudinal slope of the rail track.

"Ruling gradient" is the steepest permissible slope.

It is 1 in 150 in plains and 1 in 100 in hilly regions.

In a valley curve if a train acquires enough momentum during it downward travel to easily ascend the following rising gradient without additional energy, then this rising gradient (succeeding the falling gradient) is called 'Momentum Gradient'.

'Pusher gradient' is steeper than the Ruling gradient (laid only for limited length) where another engine from the rear pushes the train. It may be even 1 in 37 for BG and is still steeper for smaller gauges.

Though "level sections" for limited lengths are not uncommon in railways a 'Minimum Gradient' of 1 in 1000 is adopted to fecilitate drainage.

Stations and yards are preferably provided on a level surface and in case a gradient is unavoidable it should not be steeper than 1 in 400 though 1 in 1000 is preferred.

Otherwise stationary vehicles may roll over causing accidents.

On curves ruling gradient is reduced and this reduction in gradient is called 'Grade Compensation'.

It is   0.04% per degree of curve for BG
          0.03% per degree of curve for MG
and   0.02% per degree of curve for NG.

Curves. Degree of a curve is the angle subtended by it at the centre by a '30 m' arc.

$$\frac{D}{30} = \frac{360°}{2\pi R}$$

$$\therefore D = \frac{1720}{R}$$

Max. degree of curve for BG = 10°

Max. degree of curve for MG = 16°

Max. degree of curve for NG = 40°.

Super elevation. It is the raising of the outer rail over the inner rail to over come the centrifugal force on curves.

$$P = \frac{WV^2}{gR}$$

To counteract this excessive horizontal thrust acting horizontally and away from the centre of the curve the outer rail is raised over the inner rail by an amount 'E'

$$E = \frac{GV^2}{127\,R}$$

All the trains running on the track may not run at this maximum speed and hence super elevation is limited to 10% of gauge.

Max. super elevation $\approx \frac{1}{10} \times$ Gauge.

A max. value of 165 mm for BG is adopted as limiting super elevation or limiting Cant.

If a train travels at a velocity and the super elevation provided is just sufficient to it, it is called "equilibrium super elevation".

If the super elevation provided is not adequate to the speeds permitted—it is called "Cant deficiency".

$\therefore$ Cant deficiency = Cant required – Cant provided

Cant deficiency $\not> $ 75 mm for BG

Speed on Curves:

Old values. Safe speed when transition curves are provided on either side of circular curves:

For BG $V = 4.35\sqrt{R - 67}$ $\Big\}$ for $V \not> 50$ kmph
For NG $V = 3.65\sqrt{R - 6}$

When transition curves are not provided 80% of the values are to be adopted.
For high speed trains

$$V = 4.58\sqrt{R}$$

New, Values:

For BG $\quad V = 0.27 (C_a + C_d) R$

where C_a = Actual cant provided in mm

C_d = Cant deficiency permitted in mm

For MG $\quad V = 0.347 (C_a + C_d) R$.

Transition Curves: Transition curves are curves of varying radius generally introduced between a straight line (of infinite radius) and a circular curve (of radius 'R') to:

1. gradually reduce the radius (from ∞ on straight line to R of circular curve)

2. gradually introduce superelevation ('O' on straight line to 'E' of circular curve—it remains as a constant of 'E' along the circular curve)

3. gradully widen the Gauge (Being 'G' on straight line to '$G + d$' on circular curve)

4. gradually reduce the tilting of rails (from '$\frac{1}{20}$' on straight line to 'O' on circular curve).

Length of the Transition curve:

Maximum of

1. $L = 7.20 \, E$ (E = Superelevation in cm)
2. $L = 0.073 \, DV_{max}$ (D = Cant deficiency in cm)
3. $L = 0.073 \, EV_{max}$ (V_{max} = Max. velocity)
4. $L = 4.4\sqrt{R}$ (R = Radius of circular curve)
5. $L = 0.07 \dfrac{V_{max}^3}{R}$

Widening of Track on Curves:

The Gauge which fairly remains uniform on straight sections is widened on horizontal curves. It is because the smaller diameter of coned wheel coming in contact with inner rail travels less distance compared to the outer rail for the same number of revolutions of the axle and wheel assembly.

$$d = \frac{13(L + l_f)^2}{R} \text{ cm}$$

where L = Rigid wheel base

= 6 m for BG and 4.88 m for MG

l_f = Lap of flange in m.

$$l_f = 0.02\sqrt{h^2 + D \cdot h}$$

where h = Depth of wheel flange below rail top level (cm)

D = Dia. of wheel (cm)

Tolerance:

	Variation
Gauge	≯ 2 mm from sleeper to sleeper
Depression of joints	≯ 5 mm
Cross levels	≯ 4 mm

Stations and Yards. Station is a definite place meant for passengers to board the train or alight from it.

Yard is an annexure of a railway station meant for loading and unloading of goods.

1. Terminal station: Terminal station is either the first or the last station *i.e.*, having flow of traffic on one side of it only. In general these are very big stations provided with all the passenger and goods facilities.

2. Junction: Junction is a place where traffic flow takes place in more than two directions. Two or more main lines or branch lines may join at this place. Hence all the facilities for passengers, goods, as well as for rolling stock are to be provided here.

3. Way side station: Way side station is any other station any where in between two junctions or two terminal stations.

Way side stations are further classified as "Block Stations"— where movements of trains are controlled by telegraphic messages and various fixed signals.

'Flag station' is a station where no fixed signals nor block instruments exist and movements of the trains are controlled by red and green signal only.

Halt is either a temporary or permanent stop for some passenger trains and is an inferior flag station.

	Item	Particulars
1.	Length of platform	Length of the longest train with a regular halt [≮ 180 m]
2.	Width	≯ 3.67 m
3.	Height	0.75 to 0.85 m for BG ⎫ 0.30 to 0.40 m for MG ⎬ High level 0.25 to 0.40 m for NG ⎭ 0.45 m for BG—low level At rail level for flag stations 1.05 m for BG ⎫ 0.70 m for MG ⎬ for goods platform 0.60 m for NG ⎭

Passenger platforms is one where passengers wait. Coaching yard is the place where maintenance and repairs to the coaches are carried out. Goods platform is meant for loading and unloading of goods.

Marshalling yard. It is an integral part of a goods yard provided at junctions to receive goods trains, to sort the wagons by shunting operations destinationwise and to reassemble into new trains.

These may be: 1. *Flat yards*. Where the engine continuously in contact with bogies shall be doing sorting.

2. *Gravitational yards*. Where natural topography permits tracks to be laid to a downward slope so that bogies can roll on their own.

3. *Hump yard*. Where an artificial hump (2.5 to 3 m high) is created and the wagons are pushed on to this hump from where they roll down on to the sorting lines due to the acquired momentum.

Locoyard is a place where examination, over hauling and repairs are carried out to the locomotives.

Signalling. Signal is a permanent or temporary arrangement to control the movement of trains safety, quickly and efficiently.

Outer signal. Outer signal is the first signal sighted by an approaching train. It conveys three aspects

i.e.,
	On	—Stop dead
	Off Caution	—Proceed with caution and be ready to stop
	Off Clear	—Proceed with confidence at the speed of driver's choice.

Home signal. Home signal is provided at the boundary of the station limit. Incidentally it acts as a routing signal also. An elevated home signal indicates 'main line' and has the three aspects while all the others meant for loop lines have only two aspects (on and off caution).

Starter signal is the first departure signal.

Advance starter signal is provided beyond all the points and crossings. Its sole purpose is to protect the shunting operations within the station yard.

A block is the section of track in between the Advance Starter signal and the immediate outer signal (of the neighbouring station) met with.

Two aspect lower quadrant signals convey cnly two aspects *i.e.* 'on' an 'off-Caution' and hence a 'warner' with a fish tail is introduced to convey one more aspect that "off—clear". This warner signal cannot be set to "off" unless:

1. The points are set for the main track and

2. Starter (of main line), Home (of main line), Advance starter and outer stop signal are set "off".

Three aspect upper quadrant signals convey three aspects. Also "drooping" is prevented in this case.

Interlocking. It is a safety arrangement to:

1. Set the signal off only when the points are set for it and impossible to disturb the point unless the signal is set to "on".

2. Prevent conflicting signals being given simultaneously.

Level Crossing. When a highway and Railway cross each other at the same level it is called a level crossing.

1. Special type   : When traffic is exceptionally heavy
2. A Class   : For National Highways
3. B Class   : For Metalled roads
4. C Class   : For Unmetalled roads
5. D Class   : For cattle or pedestrians only.

Crossing Angle between Highway and Rail road be preferably 90° and under any circumstance ≮ 45°. The Road should be straight at least for a length of 30 m on either side of the Rail Crossing.

At least a length of 7.5 m on either side should be levelled for the highway vehicles to wait.

II. OBJECTIVE TYPE QUESTIONS

1. The first railway passenger train ran between
    (a) Stockton and Darlington   (b) Mohawk and Hudson
    (c) Nuremberg and Furth   (d) Boribunder and Thana.

2. The first railway train of the world rolled on the rails in the year
    (a) 1825   (b) 1833
    (c) 1835   (d) 1853.

3. The first railway train in India had its first run between Boribunder and Thana in the year
    (a) 1825   (b) 1833
    (c) 1835   (d) 1853.

4. The first railway train in India ran between
    (a) Howrah and Raniganj   (b) Mumbai and Thana
    (c) Chennai and Bangalore   (d) Kolkata and Delhi.

5. Gauge is the
    (a) clear inner distance between the two rails of track
    (b) clear outer distance between the two rails of track
    (c) centre to centre distance between rails of track
    (d) inner distance between the two rails—2 × flange thickness of the running wheel.

6. Perfect gauge is maintained because of
    (a) ballast   (b) formation
    (c) rails   (d) sleepers.

7. "Gauge" on Indian Railways is the
    (a) minimum distance between running faces of the two inner rails
    (b) distance between the running faces measured 14 mm below the rail table

(c) distance between the running faces measured 15.88 mm below the rail table ☐
(d) distance between the running faces measure 16 mm below the rail table. ☐

8. More 'Gauge Width' is recommended
   (a) in hilly areas ☐
   (b) for cheap plate laying ☐
   (c) to attain greater speeds ☐
   (d) when the traffic flow is highly fluctuating. ☐

9. Standard Gauge is
   (a) 1676 mm ☐
   (b) 1524 mm ☐
   (c) 1435 mm ☐
   (d) 1000 m. ☐

10. 62% of the total rail track of the world is laid to
    (a) broad gauge ☐
    (b) standard gauge ☐
    (c) metre gauge ☐
    (d) narrow gauge. ☐

11. World's widest Gauge is
    (a) 1524 mm ☐
    (b) 1676 mm ☐
    (c) 1829 mm ☐
    (d) 2286 mm. ☐

12. World's narrowest Gauge is
    (a) 1000 mm ☐
    (b) 914 mm ☐
    (c) 762 mm ☐
    (d) 610 mm. ☐

13. Light Gauge (610 mm) is adopted
    (a) in plains ☐
    (b) where the revenue earnings are low ☐
    (c) where there is no other type of transport ☐
    (d) where acquisition of load is a problem. ☐

14. For a wider Gauge
    (a) curves can be sharp ☐
    (b) gradients can be steep ☐
    (c) speed can be greater ☐
    (d) more can be the weight of the train. ☐

15. Cost of construction of a railway track is proportional to
    (a) $\sqrt{Gauge}$ ☐
    (b) Gauge ☐
    (c) $Gauge^2$ ☐
    (d) $Gauge^3$. ☐

16. A good reason in favour of 'Break of Gauge' is
    (a) different gauges suit different regions ☐
    (b) alternate routes are available between two important stations ☐
    (c) less important routes are connected by MG, NG and LG ☐
    (d) separate platforms are provided to segregate passengers. ☐

17. Gauge is slightly widened on
    (a) points ☐
    (b) diamond crossing ☐
    (c) curves ☐
    (d) tracks for fast trains. ☐

18. Gauge is slightly tightened
    (a) on curves ☐
    (b) at diamond crossing ☐
    (c) when creep is more ☐
    (d) at level crossings. ☐

19. Cost of construction of a new railway line per kilometre length is about
    (a) 1 lakh
    (b) 10 lakhs
    (c) 40 lakhs
    (d) 1 crore.

20. Tractive resistance of a railway line is only $\frac{1}{5}$ that of the pneumatic tyre on metalled road is because of
    (a) flat slopes on railway track
    (b) steel wheel to steel rail has the least friction
    (c) vacuum braking system
    (d) cushioning developed in the track.

21. Units of track modulus are
    (a) kg cm
    (b) kg cm/cm
    (c) kg/cm
    (d) kg/cm/cm.

22. The prepared bed of ground to receive ballast is called
    (a) packing
    (b) boxing
    (c) embankment or cutting
    (d) formation.

23. A good formation
    (a) bears heavier loads
    (b) offers comfortable riding
    (c) gives easy maintenance
    (d) supports ballast.

24. The top of formation is usually given a camber of
    (a) 1 in 10
    (b) 1 in 20
    (c) 1 in 40
    (d) 1 in 100.

25. Heaved track is formed because of
    (a) formation of poor drainage properties
    (b) ballast with no interlocking action
    (c) sleeper that cannot keep the gauge
    (d) rails that developed corrugations.

26. The side slope that is quite common in cutting is
    (a) $2H : 1V$
    (b) $1\frac{1}{2}H : 1V$
    (c) $1 : 1$
    (d) $1H : 2V$.

27. A track is elastic mainly because of
    (a) rails
    (b) sleepers
    (c) ballast
    (d) formation.

28. The primary function of a ballast is to
    (a) maintain the gauge
    (b) provide elasticity
    (c) conceal and make up irregularities of sleepers
    (d) prevent growth of vegetation on formation.

29. Strength of the ballast should be enough to resist
    (a) abrasion and crushing
    (b) impact loads and beater blows
    (c) weathering and discharges from toilets
    (d) load from wheels.
30. Durability of ballast is its ability to resist
    (a) crushing loads
    (b) abrasion and discharges from toilets
    (c) beaters blows
    (d) compressive and lateral thrusts.
31. Ballast should not be
    (a) hard
    (b) non-porous
    (c) smooth
    (d) non-flaky.
32. The best type of ballast is
    (a) granite
    (b) sand stone
    (c) lime stone
    (d) quartzite.
33. A draw back of gravel as ballast is
    (a) packing is very difficult
    (b) it is full of impurities
    (c) it is costly
    (d) it forms Centre Bound Track.
34. An advantage of sand as ballast is
    (a) sand is easily blown out
    (b) it offers smooth track
    (c) it is cheap
    (d) vegetation may grow.
35. Ashes and cinders are not to be used as ballast along with steel with sleepers because
    (a) they are porous and easily get powdered under heavy loads
    (b) they corrode steel
    (c) they do not provide good grip for steel sleeper
    (d) their dust rises high under traffic and spreads affecting sanitation of track.
36. Brick bats when used as ballast
    (a) forms Heaved track
    (b) forms centre bound track
    (c) corrode rails
    (d) create corrugation on rails.
37. Dusty track is created because of the following ballast
    (a) lime stone
    (b) brick bats
    (c) sand
    (d) blast furnace slag.
38. The ballast that is best suited to a steel sleeper is
    (a) gravel
    (b) brick bats
    (c) ashes and cinders
    (d) quartzite.
39. The minimum depth of ballast cushion recommended for trunk routes of Broad Gauge is
    (a) 150 mm
    (b) 200 mm
    (c) 250 mm
    (d) 300 mm.
40. The ballast in between two adjacent sleepers is called
    (a) ballast cushion
    (b) shoulder ballast
    (c) packing
    (d) crib ballast.

41. Primary function of a sleeper is
    - (a) to maintain gauge
    - (b) to take loads from rails
    - (c) to give cushioning action
    - (d) to give stability to the track.

42. A sleeper should
    - (a) be easy to carry
    - (b) be elastic
    - (c) be less affected by derailment
    - (d) permit easy lifting and packing.

43. The usual depth of a wooden sleeper to be provided for broad gauge is
    - (a) 10 cm
    - (b) 13 cm
    - (c) 16 cm
    - (d) 20.32 cm.

44. Treating of wooden sleepers with 60% ASCU to prevent infection before seasoning is called
    - (a) Creosoting treatment
    - (b) Impregnation treatment
    - (c) Prophylactic treatment
    - (d) Boutton treatment.

45. The ideal preservative for wooden sleepers is
    - (a) ASCU
    - (b) chromated zinc chloride
    - (c) pentachlorophenol
    - (d) coal tar creosote.

46. If S = strength of timber and H = hardness of timber at 12% moisture content for a timber sleeper, then its "composite sleeper index" is
    - (a) $\dfrac{S+H}{2}$
    - (b) $\dfrac{S+9H}{10}$
    - (c) $\dfrac{S+10H}{11}$
    - (d) $\dfrac{S+10H}{20}$.

47. An ideal sleeper in many ways is
    - (a) wooden sleeper
    - (b) steel sleeper
    - (c) cast iron sleeper
    - (d) concrete sleeper.

48. A main disadvantage of timber sleeper is
    - (a) light weight
    - (b) attacked by white ants
    - (c) pilferage is maximum
    - (d) does not maintain gauge.

49. Check sleeper is a sleeper
    - (a) provided at an angle other than right angles to the rail
    - (b) of separate blocks of timber connected by a tie bar
    - (c) provided vertically at buffer stops
    - (d) laid longitudinally below rails.

50. The only type of sleeper to be provided over bridges is
    - (a) wooden sleeper
    - (b) steel sleeper
    - (c) cast iron sleeper
    - (d) concrete sleeper.

51. Adzing of sleepers is
    - (a) providing tapering depth of sleeper to suit super elevation on curves
    - (b) separate wooden blocks under rails being connected by a tie bar

(c) cutting of sleeper to receive the foot flange to suit tilting of rails
(d) impregnation of the wooden sleeper with creosote.

52. Steel sleeper had the advantage of
    (a) keeping perfect gauge
    (b) being ideally suited to receive points and crossings
    (c) being used for any type of rail
    (d) being fairly withstanding corrosion.

53. Life of a steel sleeper is about
    (a) 15 years
    (b) 50 years
    (c) 70 years
    (d) depends on design.

54. The cross-section of a steel sleeper is
    (a) angle
    (b) rectangle
    (c) channel
    (d) I.

55. Ends of the steel sleeper are
    (a) bulb shaped
    (b) convex shaped
    (c) sinuous
    (d) perfectly flat and rectangular.

56. The usual failure of steel sleeper is along
    (a) edges
    (b) centre
    (c) rail seat
    (d) longitudinally.

57. Bearing plates when provided at the rail seat of a steel sleeper
    (a) prolong life
    (b) accept more loads
    (c) increase resilience of the track
    (d) prevent the track from being centre bound.

58. C.I. sleeper had the advantage of
    (a) being in three parts and hence easy to repair
    (b) having less lateral stability
    (c) being easy to maintain gauge
    (d) being less corroded.

59. A drawback of C.I. sleeper is
    (a) it is in 3 parts
    (b) its scrap value is less
    (c) the tie when bent affects gauge
    (d) its grip is less and track easily becomes corrugated.

60. An advantage of concrete sleeper is
    (a) it is heavy with a flat bottom
    (b) packing is done very easily
    (c) they are the cheapest sleepers
    (d) it is neither corroded nor eaten by white ants.

61. The cheapest of all the sleepers is
    (a) wooden sleeper
    (b) steel sleeper
    (c) cast iron sleeper
    (d) concrete sleeper.

62. The lightest sleeper is
    (a) wooden sleeper
    (b) steel sleeper
    (c) cast iron sleeper
    (d) concrete sleeper.

63. The sleeper that requires the least maintenance is
    (a) wooden sleeper
    (b) steel sleeper
    (c) cast iron sleeper
    (d) concrete sleeper.

64. The best sleeper suited for track circuiting is
    (a) wooden sleeper
    (b) steel sleeper
    (c) cast iron sleeper
    (d) concrete sleeper.

65. The sleeper not suited to high speeds
    (a) wooden sleeper
    (b) steel sleeper
    (c) cast iron sleeper
    (d) concrete sleeper.

66. The sleeper of highest scrap value is
    (a) wooden sleeper
    (b) steel sleeper
    (c) cast iron sleeper
    (d) concrete sleeper.

67. The sleeper that had the least scrap value is
    (a) wooden sleeper
    (b) steel sleeper
    (c) cast iron sleeper
    (d) concrete sleeper.

68. The sleeper that can withstand appreciable amount of bending at its centre is
    (a) wooden sleeper
    (b) cast iron sleeper
    (c) steel sleeper
    (d) concrete sleeper.

69. The sleeper to be used when the formation is new or it is treacherous or it is black cotton soil whose bearing power is very less is
    (a) wooden sleeper
    (b) cast iron sleeper
    (c) steel sleeper
    (d) concrete sleeper.

70. The sleeper best suited to be provided at a level crossing is
    (a) wooden sleeper
    (b) cast iron sleeper
    (c) steel sleeper
    (d) concrete sleeper.

71. Lateral resistance is very high in case of
    (a) wooden sleeper
    (b) cast iron sleeper
    (c) steel sleeper
    (d) concrete sleeper.

72. The most suitable type of sleeper for a long welded rail is
    (a) wooden sleeper
    (b) cast iron sleeper
    (c) steel sleeper
    (d) concrete sleeper.

73. The heaviest of all the sleepers is
    (a) wooden sleeper
    (b) cast iron sleeper
    (c) steel sleeper
    (d) concrete sleeper.

74. The sleeper that is to be only fixed by elastic fastenings is
    (a) wooden sleeper
    (b) cast iron sleeper
    (c) steel sleeper
    (d) concrete sleeper.

75. The sleepers worst hit because of corrugations developed on running table are
    (a) wooden
    (b) steel
    (c) C 1
    (d) concrete.

76. Sleeper density is
    (a) number of sleeper per km length
    (b) number of sleeper per rail length
    (c) density of the material of sleeper
    (d) minimum distance between any two neighbouring sleepers.

77. Joint sleeper is
    (a) a common sleeper for two tracks
    (b) a sleeper of more than one material
    (c) a sleeper on either side of a rail joint
    (d) two separate blocks being connected by a tie bar.

78. Shoulder sleeper is a sleeper
    (a) between any two intermediate sleepers
    (b) between a joint sleeper and an intermediate sleeper
    (c) on either side of a rail joint
    (d) nearer to the abuttment of a bridge.

79. Sleeper density $M + 7$ means
    (a) no. of sleepers required per kilometre length is 1007.
    (b) no. of sleeper between any two stations = distance in km + 7
    (c) no. of sleepers per rail length = length of rail in metres + 7
    (d) spacing of sleepers = 107 cm c/c.

80. A rail should possess
    (a) vertical stiffness
    (b) lateral stability on curves
    (c) horizontal moment of resistance = 3 × vertical moment of resistance
    (d) centre of gravity in the middle $\frac{1}{3}$ height.

81. For long welded rails sleeper density should be a minimum of
    (a) $M + 3$
    (b) $M + 4$
    (c) $M + 7$
    (d) $M + 9$.

82. For short welded rails the minimum sleeper density is
    (a) M + 3
    (b) M + 4
    (c) M + 7
    (d) M + 9.

83. Rails of track are similar to
    (a) girders of a bridge
    (b) lining of a tunnel
    (c) base course of a highway
    (d) levelling course of a building.

84. For a good rail section $\dfrac{\text{Horizontal moment of resistance}}{\text{Vertical moment of resistance}}$ in any type of tractions is
    (a) 1 in 2
    (b) 1 in 5
    (c) 1 in 10
    (d) 1 in 20.

85. The object of double headed rail is to
    (a) provided symmetrical section about both vertical and horizontal axes
    (b) use both the flanges for riding
    (c) employ chairs to hold the rail
    (d) gain more vertical stiffness.

86. The standard length of rail for BG is
    (a) 10 m
    (b) 13 m
    (c) 15 m
    (d) 20 m.

87. The maximum length of a long welded rail adopted by the Indian Railways is
    (a) 13 m
    (b) 39 m
    (c) 1000 m
    (d) from one station to the neighbouring station.

88. Continuous welded rails have a length of
    (a) 13 m
    (b) 39 m
    (c) 1000 m
    (d) from one station to the neighbouring station.

89. Excessive thermal stresses were reported in short welded rails when the rail lengths are more than
    (a) 3
    (b) 5
    (c) 7
    (d) 9.

90. The standard length of a rail is decided depending on
    (a) its weight per metre length
    (b) sleeper density
    (c) longest bogie
    (d) permissible high speed.

91. Canting rails adopted by Indian Railways is
    (a) 1 in 10
    (b) 1 in 20
    (c) 1 in 40
    (d) 1 in 80.

92. Coning of wheels is provided
    (a) to give dynamic stability of the rolling stock
    (b) to prevent lateral slip
    (c) to save material of wheels
    (d) to suit super elevation on curves.

93. The outer and inner wheels though cast monolithic cover different distances on a curve for the same number of revolutions because of
    (a) adzing of sleepers
    (b) tilting of rails
    (c) coning of wheels
    (d) widening of gauge.

94. The gauge is widened on curves so that
    (a) tilting of outer rail is averted
    (b) centrifugal force is reduced
    (c) different diameters are in contact with inner and outer rails
    (d) super elevation of the outer rail is fully implemented.

95. The weight of rail section to be adopted depends on
    (a) heaviest axle load
    (b) max. length of bogie
    (c) nature of haul, whether steam, diesel or electric
    (d) gauge.

96. The usual standard rail section weighs
    (a) 60 kg/m
    (b) 52 kg/m
    (c) 44 kg/m
    (d) 37 kg/m.

97. The rails used today on trunk routes of Broad Gauge are
    (a) 44.41 kg
    (b) 52 kg
    (c) 60 kg
    (d) 75 kg.

98. The maximum axle load borne by a 60 kg rail of Broad Gauge is
    (a) 260 × 60 kg
    (b) 320 × 60 kg
    (c) 440 × 60 kg
    (d) 560 × 60 kg.

99. Excess content of the following in steel for rail manufacture may be responsible for the development of corrugations of rails
    (a) Nickel
    (b) Phosphorus
    (c) Carbon
    (d) Chromium.

100. Tongue rail is made up of
    (a) low manganese steel
    (b) medium manganese steel
    (c) high manganese steel
    (d) medium carbon steel.

101. The thin edge of the tongue rail is called
    (a) toe
    (b) heel
    (c) switch
    (d) throw.

102. Switch is
    (a) a tongue rail
    (b) a stock rail
    (c) combination of both tongue and stock rails
    (d) tongue and stock rail combination but separated by flange way.

103. Number of crossing is

    (a) $\dfrac{\text{spread at the leg of crossing}}{\text{length of point rail from TNC}}$ (b) $\dfrac{\text{spread at the leg of crossing}}{\text{length of point rail from ANC}}$

    (c) $\dfrac{\text{spread at the leg of crossing}}{\text{length of splice rail}}$ (d) $\dfrac{\text{spread at the leg of crossing}}{\text{length of lead rail from throat to tongue rail}}$.

104. Square crossing is to be avoided because

    (a) it requires more space
    (b) only one track can be used at a time
    (c) both wheels of the axle have to simultaneously jump over
    (d) wear is more.

105. The crossing number adopted by Indian Railways for high speeds is

    (a) 1 in $8\dfrac{1}{2}$ (b) 1 in 12
    (c) 1 in 16   (d) 1 in 20.

106. Gap is the

    (a) distance between throat and TNC
    (b) distance between throat and ANC
    (c) distance between toe of crossing to heel of crossing
    (d) flange way between point rail and wing rail.

107. Obtuse angle crossing had

    (a) a pair of splice rail and point rail + running rail + wing rail on either side of throat
    (b) two pairs of splice rails and point rails + running rail + wing rail on either side of throat
    (c) a pair of splice and point rails + running rail
    (d) a pair of point and splice rails + a wing rail.

108. The minimum thickness of ANC is

    (a) zero   (b) 2 mm
    (c) 3 mm   (d) 6 mm.

109. The purpose of a check rail is to

    (a) provide flange way as in case of a level crossing
    (b) guide the wheels before jumping over the gap
    (c) hold and guide one wheel of the axle when the other is jumping over the gap
    (d) lead the wheels on to a turnout

110. Heel divergence is

    (a) distance between running faces of stock rail and tongue rail at the heel of switch
    (b) distance between the inner faces of stock rail and tongue rail at the heel of switch
    (c) flange way + 3 mm to 6 mm for wear
    (d) length of the stock rail per unit spread.

111. Switch angle =
    (a) $\dfrac{\text{heel divergence}}{\text{the vertical length of tongue rail}}$ ☐
    (b) $\dfrac{\text{heel divergence}}{\text{length of stock rail}}$ ☐
    (c) $\dfrac{\text{throw of switch}}{\text{actual length of tongue rail}}$ ☐
    (d) same as the crossing angle. ☐

112. Flange way is a minimum of
    (a) 60 mm + 3 mm to 6 mm for wear ☐
    (b) 50 mm + 3 mm to 6 mm for wear ☐
    (c) 40 mm + 3 mm to 6 mm for wear ☐
    (d) 30 mm + 3 mm to 6 mm for wear. ☐

113. Throw of switch for BG is
    (a) 89 mm ☐
    (b) 91 mm ☐
    (c) 93 mm ☐
    (d) 95 mm. ☐

114. Length of the tongue rails depends on
    (a) crossing number ☐
    (b) switch angle ☐
    (c) heel divergence ☐
    (d) length of stock rail. ☐

115. Flattest diamond crossing on BG is
    (a) 1 in $8\dfrac{1}{2}$ ☐
    (b) 1 in 10 ☐
    (c) 1 in 12 ☐
    (d) 1 in 16. ☐

116. When one line is straight and the other is to its left then it is a
    (a) right hand turnout ☐
    (b) left hand turnout ☐
    (c) similar flexure ☐
    (d) contrary flexure. ☐

117. A scissor cross over is provided where
    (a) space is restricted but shunting operations are brisk ☐
    (b) traffic is less ☐
    (c) land is cheap ☐
    (d) passenger traffic is heavy. ☐

118. A temporary arrangement of a double line track where a small length of track is common for up and down tracks is called
    (a) ladder track ☐
    (b) gauntlet track ☐
    (c) herring bone grid ☐
    (d) triangle. ☐

119. When two parallel tracks one of Broad Gauge and other of Metre Gauge are provided with a common bridge across a wide river, the arrangement of the track is called
    (a) pinion track ☐
    (b) gauntlet track ☐
    (c) ladder line ☐
    (d) herring bond grid. ☐

120. Max. damage is done to the track because of
    (a) fast moving but light weight vehicles ☐
    (b) slow moving but heavy trains ☐
    (c) heavy axle load that too unequally distributed ☐
    (d) frequency acceleration and deceleration. ☐

121. A basic necessity for regular track maintenance is
    (a) it gives max. comfort to passengers
    (b) it prolongs life of rolling stock
    (c) it saves a lot of fuel
    (d) it averts accidents.
122. The tolerance limit permissible for gauge is
    (a) + 3 mm, – 3 mm
    (b) + 6 mm, – 3 mm
    (c) + 13 mm, – 3 mm
    (d) + 19 mm.
123. The sleeper that is worst hit due to derailment is
    (a) wooden sleeper
    (b) steel sleeper
    (c) cast iron sleeper
    (d) concrete sleeper.
124. The sleeper that completely loses shape because of derailment is
    (a) wooden sleeper
    (b) steel sleeper
    (c) cast iron sleeper
    (d) concrete sleeper.
125. The sleepers that can accept severe shock loading are
    (a) wooden and steel sleepers
    (b) steel and cast iron sleepers
    (c) cast iron and concrete sleepers
    (d) concrete and wooden sleepers.
126. If the ballast is dirty
    (a) it may not permit drainage
    (b) it may cause rough riding
    (c) cushioning action may get lost
    (d) ballast may get consolidated leading to high formation pressures.
127. Daily maintenance includes
    (a) tightening of joints
    (b) removal of vegetation from ballast
    (c) removal of damaged sleepers
    (d) removal of fishplates, grease application and refixing.
128. Daily maintenance organisation consists of
    (a) SM + ASM + Cabin men
    (b) Controller + Mechanical Chief Engineer + Civil Chief Engineer + Bridges Chief Engineer
    (c) General Manager + Chief Operating Superintendent + I.O.W. + A.I.O.W.
    (d) PWI + A.P.W.I. + Mate + Keyman.
129. The length of track attached to each gang is
    (a) 1 km
    (b) 2 km
    (c) 3 km
    (d) 4 km.
130. Slewing operation is
    (a) squaring of sleepers
    (b) checking of gauge
    (c) boxing and dressing of ballast
    (d) correcting alignment.
131. Cant and Level board is provided to check
    (a) tilting of rails
    (b) gradients of track
    (c) super elevation on curves
    (d) widening of gauge.

132. Packing is
    (a) loose ballast responsible for cushion
    (b) closely compacted ballast under rail seat
    (c) giving a trapezoidal cross-section to ballast
    (d) extra length of ballast provided beyond the end of sleeper.
133. Packing is done by
    (a) two people standing face to face on either side of rail
    (b) four people, two on either rail facing the same direction
    (c) four people, two working diagonally back to back on each rail
    (d) eight, four attacking the four corners of each rail.
134. High joint is
    (a) a very wide joint formed due to extreme contraction during winter
    (b) a raised joint due to over packing
    (c) an unsupported joint due to the elevation of intermediate sleepers
    (d) a joint provided on hilly tracks to prevent downward skidding.
135. A high joint is formed because of
    (a) less packing under rail seat
    (b) support sleeper
    (c) sinking of joint sleeper
    (d) sinking of intermediate sleepers.
136. Rail ends may become hogged because of
    (a) loose packing
    (b) no sleeper under the joint
    (c) long welded lengths
    (d) excess tightening of fish botts.
137. The maximum value of kink or battering permissible per rail length is
    (a) 1 mm
    (b) 2 mm
    (c) 3 mm
    (d) 4 mm.
138. Buckling of rails is
    (a) vising of the rail convexly upwards
    (b) sinking of the rail at its mid length
    (c) widening of gauge
    (d) tightening of gauge.
139. Buckling is more common when
    (a) track is welded over long lengths
    (d) greater amount of expansion gap is provided at rail joints
    (c) the alignment has more summit and valley curves
    (d) slewing is frequently done.
140. The most dangerous defect of a rail track that requires urgent rectification is
    (a) buckling of rails
    (b) corrugations of rails
    (c) crippling of rails
    (d) hogging of rails.
141. A ballastless, sleeperless track can be prepared when rails rest over
    (a) bituminous pavement
    (b) compacted and consolidated natural ground
    (c) hard rocky soil
    (d) rubben pads on cement concrete pavement.
142. Blowing joint is formed because of
    (a) loose packing and deficiency of ballast
    (b) over packing and more cushion
    (c) concrete sleepers and high carbon rails
    (d) expansion joint being closed.

143. A remedy for Heaved track is to cover the formation with
    (a) a thin blanket of ashes or moorum
    (b) a thin layer of clay
    (c) a layer of fine sand
    (d) a thick layer of broken stone.

144. Centre Bound Sleeper means
    (a) sleeper is supported only at its centre
    (b) sleeper with firm grip at the edges
    (c) sleeper provided with a longitudinal support at their middle length
    (d) longitudinal sleeper.

145. Centre bound track gets formed when steel sleepers are used because of
    (a) flow of ballast takes place along the (inverted) channel
    (b) the depth of ballast at the centre of gauge is the minimum
    (c) packing easily gets disturbed pushing the ballast
    (d) the sleeper being bent down at its centre.

146. Centre Bound track is often formed when the ballast is
    (a) gravel
    (b) granite
    (c) brick bats
    (d) blast furnace slag.

147. "Centre bound sleepers" are mostly formed with
    (a) C.I. pot sleeper
    (b) C.I. plate sleeper
    (c) steel trough sleepers
    (d) concrete sleepers.

148. Longitudinal sag in track takes place
    (a) on bridges
    (b) at level crossings
    (c) at approaches to rigid structures like bridges
    (d) when the formation is strong but ballast is insufficient.

149. Tolerance limit of Gauge on Indian Railways is
    (a) 3 mm tight to 6 mm slack
    (b) 8 mm tight to 8 mm slack
    (c) not more than 10 mm from sleeper to sleeper
    (d) not more than 10 mm for alternate sleepers.

150. Due to slipping of the wheels the rail forms
    (a) crushed head
    (b) battered ends
    (c) split head
    (d) horizontal fissure.

151. Corrugations were found mostly on
    (a) soft rails than hard rails
    (b) rails receiving heavy and slow moving rains
    (c) rails on concrete sleepers resting on sand ballast
    (d) rails on steel sleepers and broken brick ballast.

152. Corrugations are relatively less on
    (a) electric traction
    (b) less spacing between axles
    (c) wheel slipping
    (d) welded track.

153. Corrugations are formed on rails due to the excess content of
    (a) carbon
    (b) magnanese
    (c) silicon
    (d) phosphorus.

154. Creep is
    (a) temporary elongation of rail in the direction of motion
    (b) permanent elongation in the longitudinal direction
    (c) buckling of rails across the gauge
    (d) skidding of the wheels.

155. Creep will be less
    (a) in hot weather than in cold weather
    (b) on double track than on single track
    (c) in cuttings than on embankments
    (d) in swampy soils than on dry soils.

156. One of the most undesirable effects of creep is
    (a) the rails get elongated
    (b) points and crossing are disturbed
    (c) expansion joints get closed
    (d) fish plates are pulled.

157. Creep is principally due to
    (a) wave motion of rails due to moving trains
    (b) rigid holding of track
    (c) motions in either direction as on a single track
    (d) longer lengths of rails.

158. Adjustment of rails is needed when the creep exceeds
    (a) 2 mm
    (b) 10 mm
    (c) 15 mm
    (d) 150 mm.

159. At a stretch the maximum number of expansion joints which can be jammed due to creep before notification is to be done is
    (a) 3
    (b) 4
    (c) 6
    (d) 10.

160. Creep anchors should not be provided at
    (a) level crossings
    (b) points and crossings
    (c) approaches to stations and yards
    (d) bridges.

161. One method of creep prevention is
    (a) use wooden sleepers
    (b) provide creep anchors near joints
    (c) pull back when the creep exceeds 75 mm
    (d) provide concrete sleepers.

162. To prevent buckling
    (a) increase shoulder ballast
    (b) do not provide creep anchors
    (c) increase the spacing between sleepers
    (d) increase the gap of expansion joint.

163. The defect caused because of the slipping of driving wheel of the engine on the rail surface is called
    (a) hogging
    (b) scabing
    (c) wheel burns
    (d) shelling.
164. If the angle of wear of a rail head exceeds the following value derailment may take place
    (a) 10°
    (b) 25°
    (c) 40°
    (d) 60°.
165. The object of providing super elevation on curves is
    (a) to limit the lateral thrust on the outer rail
    (b) to create comfort for passengers
    (c) to minimise wear of sleepers
    (d) to increase speeds of rolling stock.
166. On Indian Railways Super elevation is provided by
    (a) keeping inner rail normal and raising the outer rail
    (b) keeping the outer rail normal and depressing the inner rail
    (c) elevating outer rail half way and depressing the inner for the remaining half
    (d) adopting Special Rail section of varying web depth.
167. Super elevation =
    (a) $\dfrac{GV^2}{1.25 R}$
    (b) $\dfrac{GV^2}{1750 R}$
    (c) $\dfrac{GV^2}{127 R}$
    (d) $\dfrac{GV^2}{1719 R}$.
168. The max. value of super elevation provided on Indian Railways is
    (a) 165 mm
    (b) 140 mm
    (c) 90 mm
    (d) 65 mm.
169. Cant excess causes
    (a) discomfort to riders
    (b) more wear for rolling stock
    (c) greater lateral stress
    (d) more pressure on the inner rail.
170. Max. cant deficiency permitted is
    (a) 40 mm
    (b) 50 mm
    (c) 60 mm
    (d) 75 mm.
171. Negative super elevation arises in case of
    (a) main and loop lines
    (b) transition curves
    (c) similar flexure
    (d) contrary flexure.
172. Length of a vertical curve between two gradients of 1 in 100 (rising) and 1 in 50 (falling) meeting in a summit is
    (a) 300 m
    (b) 600 m
    (c) 900 m
    (d) 1800 m.

173. The length of a vertical summit curve provided between two slopes is 500 m. If the slopes are the other way the length of the valley curve to be provided between them is
    (a) 250 m
    (b) 500 m
    (c) 750 m
    (d) 1000 m.

174. Max. degree of curve on BG is
    (a) 10°
    (b) 16°
    (c) 20°
    (d) 40°.

175. Max. ruling gradient permitted in Indian Railways in plains is
    (a) 1 in 50
    (b) 1 in 100
    (c) 1 in 150
    (d) 1 in 500.

176. Mountainous railways are those which have a grade steeper than
    (a) 1%
    (b) 3%
    (c) 6%
    (d) 8%.

177. A central rack is provided for mountainous railway tracks when the gradient is steeper than
    (a) 1%
    (b) 3%
    (c) 6%
    (d) 8%.

178. If the ruling gradient on BG is 1 in 200, the grade compensation on a 3° curve is
    (a) 0.06%
    (b) 0.09%
    (c) 0.12%
    (d) 0.38%.

179. Safe speed on BG curve is V =
    (a) $4.4\sqrt{R-70}$ kmph
    (b) $3.6\sqrt{R-6}$ kmph
    (c) $4.4\sqrt{R-60}$ kmph
    (d) $3.6\sqrt{R-20}$ kmph.

180. The desirable radial acceleration over curves for a comfortable travelling of passengers is
    (a) 100 mm/sec^3
    (b) 305 mm/sec^3
    (c) 500 mm/sec^3
    (d) 1000 mm/sec^3.

181. The resistance offered by the rolling stock is called
    (a) Track resistance
    (b) Frictional resistance
    (c) Journal resistance
    (d) Tractive resistance.

182. The curve resistance for a 100 t load on BG for a 2° curve is
    (a) 15 kg
    (b) 30 kg
    (c) 45 kg
    (d) 60 kg.

183. Co-efficient of friction is very low in case of
    (a) dry surface
    (b) greasy surface
    (c) wet surface
    (d) frosty weather inside a tunnel.

184. Maximum permissible gradient for a station yard is
    (a) 1 in 1000
    (b) 1 in 500
    (c) 1 in 400
    (d) 1 in 300.

185. The desirable gradient for a station yard is not to exceed
    (a) 1 in 1000
    (b) 1 in 500
    (c) 1 in 400
    (d) 1 in 300.
186. Fast moving trains should be received on
    (a) main line
    (b) loop line
    (c) trap siding
    (d) sick siding.
187. The maximum permissible speed on a branch line is
    (a) 12 kmph
    (b) 24 kmph
    (c) 36 kmph
    (d) 48 kmph.
188. A flag station is one where
    (a) no train is scheduled to stop but provided only for operational convenience
    (b) only passenger trains halt and express and goods trains may not halt
    (c) movement of trains is controlled by flags only
    (d) everything is temporary including the station.
189. A siding is a line connected to the main line by a cross over
    (a) at one end and the other end is a dead end
    (b) at either end and there are no dead ends
    (c) and detached from the main line so that shunting operations are independently done
    (d) takes the trains away from main line.
190. A sick siding is the line meant for
    (a) passenger trains to come near the platform
    (b) damaged siding that needs repairs
    (c) siding meant for sick people and wounded soldiers
    (d) bogies declared unfit for travel to wait.
191. A refuse line is a
    (a) line to receive goods bogies in marshalling yard
    (b) idle line rarely used that too at junctions and terminal stations
    (c) busy line to receive incoming trains at terminals
    (d) line to harbour a slow moving trains for crossing or over taking by a fast moving train at way side stations.
192. A double line had the advantage of
    (a) reducing creep
    (b) reducing maintenance cost
    (c) permitting easy over taking and crossing
    (d) avoiding loss of time due to crossing.
193. The line provided for a passenger train to come nearer to the platform is
    (a) refuse line
    (b) trap siding
    (c) loop line
    (d) branch line.
194. The ends of passenger platform are provided with a slope of
    (a) 1 in 2
    (b) 1 in 4
    (c) 1 in 6
    (d) 1 in 8.

195. The siding provided on steep slopes so that a wagon at rest will not enter the main line is called
    (a) trap siding
    (b) catch siding
    (c) sick siding
    (d) refuge siding.

196. The purpose of a marshalling yard is to
    (a) receive goods trains and to reorient their destinations
    (b) receive and split the trains bound towards different destinations
    (c) receive, break rearrange and despatch the goods trains
    (d) receive unload and load the goods wagons bound to different destinations.

197. Flat type of marshalling yard is preferred to the others when
    (a) space available is small and restricted
    (b) less number of engines are available
    (c) goods to be transported include perishables
    (d) pneumatic retarders are provided to bring the bogies to rest quickly.

198. In the hump yard the steepest gradient is provided at
    (a) pushing end
    (b) hump itself
    (c) first falling gradient
    (d) final falling gradient.

199. Marshalling yards are a must at every
    (a) important way side station
    (b) important junction
    (c) important flag station
    (d) terminal.

200. An arrangement provided to transfer a bogie from one track to another is called a
    (a) turn table
    (b) traverser
    (c) transfer table
    (d) cross over.

201. The function of a derailing switch is to
    (a) derail wagons at the time of necessity
    (b) prevent derailment of a train by accident
    (c) connect shunting lines with running lines
    (d) isolate shunting lines from running lines.

202. A fouling mark is provided
    (a) on the main line to mark the entry into station yard
    (b) on the loop line indicating the beginning of the track leading nearer to platform
    (c) on converging lines indicating where a bogie should rest
    (d) between parallel lines indicating the minimum distance to be maintained between them.

203. A dead end is provided with
    (a) level track embedded in sand
    (b) falling gradient with a split switch
    (c) rising gradient with a vertical cross sleeper
    (d) rising gradient and vertical post with a buffing beam.

204. The very first signal met by a train before leaving a station is
    (a) outer signal ☐ (b) home signal ☐
    (c) starter signal ☐ (d) advance starter signal. ☐

205. Green over red signal spotted before approaching a station indicates
    (a) on ☐ (b) off caution ☐
    (c) off clear ☐ (d) no meaning. ☐

206. When a train is approaching a station if red over Green signal is spotted in the two aspect upper quadrant system it indicates
    (a) on ☐ (b) off caution ☐
    (c) off clear ☐ (d) no meaning. ☐

207. The 'warner' signal can be set to 'off' position only when
    (a) the block ahead is clear ☐
    (b) home and starter signals are set off ☐
    (c) home, starter and advance starter are set off ☐
    (d) starter and home of main line and advance starter and stop signal of outer are also set off. ☐

208. Off-caution' means
    (a) proceed very slowly ☐
    (b) proceed slowly and watch for the next signal ☐
    (c) proceed at your desired speed ☐
    (d) do not enter the station yard. ☐

209. The signal that also acts as a routing signal
    (a) outer signal ☐ (b) home signal ☐
    (c) starter signal ☐ (d) advance starter signal. ☐

210. A stop signal with a black circular ring around the white hand is a
    (a) warner signal ☐ (b) goods signal ☐
    (c) loco (shed) signal ☐ (d) calling on signal. ☐

211. The outer signal is provided at a minimum distance of
    (a) 1 km from home signal ☐ (b) 580 metres from home signal ☐
    (c) 860 metres from home signal ☐ (d) 180 metres from home signal. ☐

212. If the outer signal is 'on' then the train should stop dead
    (a) just at the signal post *i.e.*, should not cross it ☐
    (b) 10 m infront of signal post ☐
    (c) 90 m before signal post ☐
    (d) 100 m before signal post. ☐

213. The signal located exactly at the boundary of the station limit is
    (a) outer signal ☐ (b) home signal ☐
    (c) starter signal ☐ (d) advance starter signal. ☐

214. Detonator is an audible signal provided in a foggy weather to instruct the driver to
    (a) stop dead
    (b) proceed with caution
    (c) proceed with confidence
    (d) look for the signal ahead.
215. The minimum distance between the home signal and the points and crossings of the approaching station is
    (a) 90 m
    (b) 100 m
    (c) 180 m
    (d) 540 m.
216. The purpose of the advance starter signal is to
    (a) just repeat the message of starter
    (b) permit the train into the next block
    (c) act as a check even if the starter is set off
    (d) protect the station yard from shunting operations.
217. A 'block' is the zone between
    (a) outer signal to outer signal between two neighbouring block stations
    (b) advance starter to advance starter of two neighbouring block stations
    (c) outer signal to advance starter signal
    (d) advance starter and outer signal.
218. A temporary dwarf signal provided below a stop signal, the dwarf signal being 'on' while the top stop signals off is
    (a) calling on signal
    (b) repeater signal
    (c) co-acting signal
    (d) shunt signal.
219. A dwarf slave signal provided to the same post of a stop signal and just repeats the on and off positions of the matter stop signal is
    (a) calling on signal
    (b) repeater signal
    (c) co-acting signal
    (d) routing signal.
220. The purpose of a repeater signal is to
    (a) repeat the functions of any of starter, advance starter, home or outer signal
    (b) give advance warning to the driver
    (c) caution the passengers on the platform
    (d) reflect the conditions of the next block.
221. The primary purpose of a signal is to
    (a) avert accidents
    (b) regulate speeds
    (c) convey messages
    (d) create understanding between passengers and train operators.
222. The signal that regulates the speeds is
    (a) starter signal
    (b) advance starter signal
    (c) outer signal
    (d) home signal.

223. Three aspects are primarily required for
   (a) starter signal
   (b) advance starter signal
   (c) home signal
   (d) outer signal.

224. The signal given in side indicates
   (a) stop dead
   (b) control the speed
   (c) proceed at 10 kmph
   (d) end of speed restriction.

225. This signal indicates
   (a) stop dead
   (b) control the speed
   (c) proceed at 10 kmph
   (d) end of speed restriction.

226. The signal given in side conveys
   (a) stop dead
   (b) control the speed
   (c) speed limited is 10 kmph
   (d) end of speed limit.

227. The enclosed signal is
   (a) sighting board for passenger trains
   (b) sighting board for goods trains
   (c) track under repair, slow down
   (d) track under repair, stop dead.

228. An arrangement provided to prevent disturbing the points set is
   (a) lock bar
   (b) detector
   (c) compensator
   (d) warner.

229. Disc signals are provided for controlling
   (a) passenger trains
   (b) goods trains
   (c) shunting operations
   (d) marshalling operations.

230. The pit where wheels of locomotive can be removed is called
   (a) ash pit
   (b) drop pit
   (c) examination pit
   (d) locoshed.

231. Interlocking is mostly a
   (a) Mechanical device
   (b) Electrical device
   (c) Safety device
   (d) Operational device.

232. The signals provided at the lowest level are
   (a) Stop signals
   (b) Coacting signals
   (c) Caution indicators
   (d) Point indicators.

233. The system of train operation followed in emergencies is
   (a) Absolute block system
   (b) One engine system
   (c) Time interval system
   (d) Pilot guard system.

234. One important principle of interlocking is
   (a) points should be capable of being disturbed only when the signal is set 'on' ☐
   (b) signal should not be set 'off' unless all the points are set for it ☐
   (c) two different signals when set 'off' should not cause confusion ☐
   (d) up signals should interlock down signals and *vice versa.* ☐
235. Central control for the operation of signals of a station shall be with
   (a) station master ☐ (b) cabin men ☐
   (c) controller ☐ (d) divisional mechanical engineer. ☐
236. The best system of a railway high way crossing is
   (a) level crossing ☐ (b) road over rail track ☐
   (c) road under rail track ☐ (d) both road over and road under bridges. ☐
237. The angle of crossing between the Railway and highway should be
   (a) 90° ☐ (b) 60° to 90° ☐
   (c) 45° to 90° ☐ (d) 0° to 90°. ☐
238. Highway meeting the Rail track should be straight for atleast a length of
   (a) 10 metres from the centre of a rail track ☐ (b) 20 metres from the centre of a rail track ☐
   (c) 30 metres from the centre of a rail track ☐ (d) 40 metres from the centre of a rail track. ☐
239. The desirable maximum gradients on approaches on either side of a level crossing is
   (a) 1 in 10 ☐ (b) 1 in 30 ☐
   (c) 1 in 50 ☐ (d) 1 in 70. ☐
240. The minimum length of a level ground provided on either side of a level crossing for waiting of (highway) vehicles is
   (a) 5 m ☐ (b) 7.5 m ☐
   (c) 10 m ☐ (d) 12.5 m. ☐
241. The additional rail or similar straight piece provided parallel to the rail at a level crossing to maintain flange way is called
   (a) check rail ☐ (b) wing rail ☐
   (c) guard rail ☐ (d) split rail. ☐

III. TRUE/FALSE

1. Indian Railways have the world's broadest gauge (1676 mm) and narrowest gauge (610 mm) which respectively constitute 6% and 6% of the world's total length of rail track. But it has no standard gauge (1435 mm) which is 62% of world's total rail track.
2. Though higher speeds are possible with a wider gauge, strength of track, power of engine, weight of train, nature of formation, sharpness of curves and steepness of gradients govern the maximum speed attainable.
3. The cost of rolling stock is based on traffic volume but not on gauge width.
4. The lateral force exerted by steam locomotives on rails is less than that exerted by diesel and electric engines.
5. Track modulus is the resistance offered against deformation.
6. Improper drainage of ballast causes the decay of wooden sleepers, corrosion of steel sleepers and leakage of current between adjoining rails in track circuiting.
7. Shoulder is the extrawidth of ballast beyond the ends of sleeper.

8. Puncturing of formation and penetration of ballast into it causes mixing of ballast with earth and thereby considerably increases the bearing capacity of the formation.
9. Ash is a conductor and hence not suited as ballast for track circuiting.
10. Higher stresses result in on the formation if the extent of compaction of the ballast is very less.
11. A highly elastic sleeper may carry 44% of the axle load whereas it is 100% when the track is rigid.
12. When the sleepers are of different materials and ages uneven distribution of load occurs even when the other factors are the same.
13. As in America in India also the spacing between sleepers is always kept uniform.
14. Freshly creosoted sleeper catches fire easily than an untreated sleeper whereas with time the ignition temperature for a creosoted sleeper rises fairly above untreated sleeper.
15. Impregnation of wooden sleepers with creosote not only enhances its life but also improves it strength.
16. Hard wood sleepers are laid with sap wood on top and heart wood below whereas for soft wood sleepers heart wood is on top and sap wood is below.
17. Maintenance cost of a wooden sleeper increases with its age.
18. Any spike on a wooden sleeper offers maximum resistance to lateral deflection and little to the longitudinal creep.
19. When water enters the spike hole it causes corrosion of spike and hence the grip increases.
20. A "rail screw" offers less resistance to lateral deflection and more resistance to longitudinal movement than a "dog spike".
21. Where anti-creep bearing plates are fixed "round spikes" and "screw spikes" are used instead of "dog spikes".
22. For a track sleeper the minimum composite sleeper index is 780.
23. As the altitude increases rain fall increases, but pollutants decrease and this increases the life of a steel sleeper.
24. In general sleepers of the same material last for the same period whereas steel sleepers have varying life from one to another.
25. A steel sleeper is condemned when it loses 25% of it original weight.
26. The cost of reconditioning of a steel sleeper is less than that of a wooden sleeper.
27. When CSTI sleepers are used, wooden sleepers are provided under joints to prevent hogging of rail ends.
28. CSTI sleeper with fastenings give a resilient track capable of absorbing vibrations and thus suited for heavy traffic at high speeds.
29. Concrete sleepers should not be used on curves of radius less than 500 m.
30. Concrete sleepers are the best suited for long welded rails but not under the joints.
31. For the same weight a Flat Footed Rail is stronger than a Bull Headed Rail.
32. Bull Headed Rails offer smoother riding compared to Flat Footed Rails.

33. Flat Footed Rails keep better alignment of track than Bull Headed Rails.
34. The fastenings of Bull Headed Rails have a greater tendency than Flat Footed Rails to get loose.
35. Heavier is the rail less is the maintenance cost.
36. Any rail is designated by its weight per one length of rail.
37. Canting of 1 in 20 is advantageous over that of 1 in 40.
38. On any section of rail track—be it be straight, curved, over switches or crossings—rails are always tilted at 1 in 20.
39. If the tilting is flatter than 1 in 20 wheels move laterally and swaying of bogies results in.
40. Manufacture of longer lengths of the rail is uneconomical.
41. When the running wheel moves over, a heavier rail depresses more than a lighter rail and hence more is the wastage of power.
42. Corrosion of rails is more inside tunnels particularly when hauled by steam locomotives.
43. For a track well-packed and provided with adequate ballast and with a regular and periodical maintenance a square joint is advantageous over a staggered joint.
44. Increase in percentage of carbon and phosphorus in rail steel increases resistance to corrosion, whereas increase in the percentage of sulphur reduces it.
45. Long welded rails develop corrugations easily than short rail lengths jointed by bolts and fish plates.
46. More the number of joints, more is the wear and tear, more is the noise produced and rough is the running.
47. Welded track presents no noise.
48. Thermal expansion for a single rail length and that of 5 welded rail lengths is almost the same.
49. In a long welded rail the central portion is not effected by thermal changes but only the end 100 m on either side called "breathing length" alone undergo expansion and contraction.
50. Long welded rails are not to be provided when either the rail, sleeper, ballast or formation need frequent repairs or renewals.
51. When long welded rails are provided, level crossings should not fall within the breathing lengths.
52. "Switch expansion joints" are provided at the ends of long welded rails.
53. In a "short welded rail" its entire length undergoes expansion or contraction unlike long welded rail.
54. Tightening of the gauge by about 4 mm reduces corrugations.
55. Swaying of the rolling stock is controlled because of the coning of wheels.
56. "End batter of a rail" is measured as the difference between the rail at the end and at a point 1 matre away from the end.
57. "Rail end batter" is said to be severe when it exceeds 3 mm.
58. Larger amount of depression and less wavelength of corrugations cause greater 'roar'.

59. Riding is rough on corrugated rails.
60. Corrugations have no impact on the life of sleepers.
61. Excess slipping and skidding of wheels over rails may induce corrugations.
62. Sleepers are always laid at right angles to the direction of rails.
63. Tongue rails are manfactured out of medium manganese steel.
64. The purpose of stretcher bar is to maintain the fixed distance between the tongue rail and stock rail.
65. Crossing number is the cotangent of the angle made by the turnout with the straight track.
66. Crossing number is the length to be traversed along the point rail from TNC to get unit spread from the splice rail measured at right angles to it.
67. The speed, permitted on 1 in $8\frac{1}{2}$ turnout is more than that of 1 is 12 turnout.
68. As the crossing angle decreases the 'gap' at the crossing increases.
69. When a wheel of the axle is jumping over the gap the other wheel's flange (of the same axle) shall be confined in the 41 mm clearance between the stock rail and check rail.
70. A gauntlet track has no switches.
71. Creep is more on loose grounds such as new embankments and swampy areas.
72. Creep develops rapidly in winter.
73. Due to creep suspended joints may become supported joints causing rough track.
74. Track subjected to heavy creep should not be opened in the mid-day of summer as the rails may buckle due to loosened grip of ballast.
75. Creep is more along ascending gradients and less along descending gradients.
76. When buckling of rails occurs speed of rolling stock should be controlled not to exceed 15 kmph.
77. Slipping of wheels burns the top of rail whereas skidding produces flat spots on the wheel tread.
78. Slightly tight gauge gives better running than a slightly slack gauge as the former arrests the lateral play of wheels.
79. Approaches of girder bridges may sink during rainy season in black cotton soil.
80. Measured shovel packing is not suited for soft formations.
81. Measured shovel packing can be adopted for any type of sleepers and for any type of ballast.
82. Curves of less than 3° are said to be flat curves.
83. Gauge is not widened when the degree of curve is less than 4° 30'.
84. Super elevation at the beginning of the transition curves is zero, it progressively increases to a maximum at the end of the transition curves and it remains uniform throughout the following circular curve.
85. "Cant excess" occurs when a train on a curve has higher speed than the one for which the super elevation is designed.

86. What a train moves at a speed more than the equilibrium speed on a curve the wear on the outer rail is more while if it has a speed less than the equilibrium speed the wear on the inner rail is more.
87. For curves sharper than 8° on broad gauge a check rail is provided along the inner curve with a clearance of about 43 mm so that the wear on the side of the outer rail is controlled.
88. An additional rail provided to maintain flange way at level crossings and on bridges is called 'guard rail' and the same when provided at a crossing is called 'check rail'.
89. 'Pusher grade' is a steep gradient requiring an additional engine for hauling.
90. While the highway has a convex shaped cross section to provide camber, railway sleepers are depressed at their middle to form concave shape to hold rails to the tilted position.

ANSWERS
Objective Type Questions

1. (a)	2. (a)	3. (d)	4. (b)	5. (a)	6. (d)
7. (a)	8. (c)	9. (c)	10. (b)	11. (b)	12. (d)
13. (b)	14. (c)	15. (c)	16. (c)	17. (c)	18. (b)
19. (c)	20. (b)	21. (d)	22. (d)	23. (c)	24. (c)
25. (a)	26. (c)	27. (c)	28. (b)	29. (b)	30. (b)
31. (c)	32. (d)	33. (d)	34. (b)	35. (b)	36. (d)
37. (b)	38. (d)	39. (d)	40. (d)	41. (a)	42. (d)
43. (b)	44. (c)	45. (d)	46. (d)	47. (a)	48. (d)
49. (b)	50. (a)	51. (c)	52. (a)	53. (b)	54. (c)
55. (a)	56. (c)	57. (d)	58. (a)	59. (c)	60. (d)
61. (c)	62. (b)	63. (d)	64. (a)	65. (c)	66. (c)
67. (d)	68. (a)	69. (a)	70. (a)	71. (c)	72. (d)
73. (d)	74. (d)	75. (d)	76. (b)	77. (c)	78. (b)
79. (c)	80. (a)	81. (c)	82. (b)	83. (a)	84. (b)
85. (b)	86. (b)	87. (c)	88. (d)	89. (a)	90. (c)
91. (b)	92. (b)	93. (c)	94. (c)	95. (a)	96. (b)
97. (c)	98. (d)	99. (b)	100. (c)	101. (a)	102. (c)
103. (a)	104. (c)	105. (d)	106. (b)	107. (b)	108. (d)
109. (c)	110. (a)	111. (a)	112. (a)	113. (d)	114. (a)
115. (b)	116. (b)	117. (a)	118. (b)	119. (b)	120. (c)
121. (d)	122. (b)	123. (d)	124. (c)	125. (a)	126. (d)
127. (a)	128. (d)	129. (d)	130. (d)	131. (c)	132. (b)
133. (c)	134. (b)	135. (d)	136. (a)	137. (b)	138. (c)
139. (a)	140. (a)	141. (d)	142. (a)	143. (a)	144. (a)
145. (a)	146. (a)	147. (c)	148. (c)	149. (a)	150. (a)
151. (d)	152. (d)	153. (d)	154. (b)	155. (c)	156. (b)
157. (a)	158. (d)	159. (c)	160. (d)	161. (d)	162. (a)

163. (c)	164. (b)	165. (a)	166. (a)	167. (c)	168. (a)
169. (d)	170. (d)	171. (d)	172. (c)	173. (d)	174. (a)
175. (c)	176. (b)	177. (c)	178. (c)	179. (a)	180. (b)
181. (c)	182. (d)	183. (b)	184. (c)	185. (a)	186. (a)
187. (b)	188. (c)	189. (a)	190. (d)	191. (d)	192. (d)
193. (c)	194. (c)	195. (b)	196. (c)	197. (a)	198. (c)
199. (b)	200. (b)	201. (d)	202. (c)	203. (d)	204. (c)
205. (b)	206. (d)	207. (d)	208. (b)	209. (b)	210. (b)
211. (b)	212. (a)	213. (b)	214. (d)	215. (c)	216. (d)
217. (d)	218. (a)	219. (a)	220. (a)	221. (d)	222. (c)
223. (d)	224. (b)	225. (d)	226. (a)	227. (b)	228. (a)
229. (c)	230. (b)	231. (c)	232. (d)	233. (d)	234. (a)
235. (a)	236. (b)	237. (c)	238. (c)	239. (b)	240. (b)
241. (c)					

True/False

1. F	2. T	3. T	4. F	5. T	6. T
7. T	8. F	9. T	10. T	11. T	12. T
13. F	14. T	15. F	16. T	17. T	18. T
19. F	20. T	21. T	22. T	23. T	24. F
25. T	26. F	27. T	28. T	29. T	30. T
31. T	32. T	33. F	34. T	35. T	36. F
37. F	38. F	39. T	40. T	41. F	42. T
43. F	44. T	45. F	46. T	47. T	48. T
49. T	50. T	51. T	52. T	53. T	54. T
55. T	56. F	57. T	58. T	59. T	60. F
61. T	62. F	63. F	64. F	65. T	66. T
67. F	68. T	69. T	70. T	71. T	72. F
73. T	74. T	75. F	76. F	77. T	78. T
79. F	80. T	81. F	82. T	83. T	84. T
85. F	86. T	87. F	88. T	89. T	90. T

Chapter 11 BUILDING MATERIALS

I. INTRODUCTION

Stones are used in the construction of buildings since prehistoric times. They are obtained from rocks. They are loosing their universal use for the following reasons: (*i*) Steel R.C.C. are taking the place of stones since they are less bulky, stronger and more durable, (*ii*) Strength of stones cannot be rationally assessed, (*iii*) Dressing of stones is time consuming, (*iv*) They are not conveniently and cheaply available in plains, (*v*) Good quality stones are not available at all places.

Geologically rocks are classified as igneous, sedimentary and metamorphic rocks. Physically they are classified as stratified, and unstratified: chemically they are classified as Agrillaceous, silicious and calcareous.

When molten magma of silicates is forced up, it solidifies and the igneous rocks are formed. Solidification at the surface of earth gives Basalts and traps. Solidification below the earth gives granite.

When the layers of deposited debris, sand and silt are subjected to enormous over burden pressures for millions of years, the sedimentary rocks are formed, examples, sand stones and lime stones.

Due to structural changes, the igneous and sedimentary rocks find their way deep into earth where they are subjected to high temperatures and heavy pressures and this causes change in texture and mineral composition resulting in the formation of metamorphic rocks. Examples: (*i*) granite changes to gneiss, (*ii*) Sand stone changes to quartzite, (*iii*) Lime stone changes to marble stone, and (*iv*) Shale changes to slate.

If the rocks can be split easily along distinct layer they are called stratified (ex. slate, marble, lime stone and all sedimentary rocks).

If rocks show no stratification and cannot be split into thin layers, they are called unstratified (ex. granite, basalt and trap).

If the principal constituent is clay, the rocks are called agrillaceous, if it is sand they are called silicious and if it is lime they are called calcareous. Slate and laterite—agrillaceous rocks, Quartzite and granite—silicious rocks. Lime stone and marble stone—calcareous.

The art of taking out stones of various sizes from natural rocks is called the quarrying. Quarrying may be done by different methods such as excavating, wedging, heating and blasting.

Explosives used in blasting are gun powder, dynamite, gun cotton and cordite. Dynamite and gun cotton are exploded by detonation whereas gun powder and cordite can be ignited by means of a fuse.

The process of giving a definite and regular shape with smooth faces to the stones is called dressing. Cast stone or artificial stone is made with cement and natural aggregate to the required shape and size.

In masonry, stones should always be kept in position such that the pressure acting on them is at right angles to their natural bedding plane, if they have.

The common building stones of India are granite, gneiss, basalt and trap, lime stone, marble, sand stone, laterite, chalk etc.

Brick is one of the oldest and most extensively used material of construction. Bricks are made by moulding the tampered clay to suitable size and shape, and drying and burning them later. If burning is not done they are called sun-dried bricks which are very weak and should not be used where they are exposed to rain.

Iron oxide present in the raw materials gives the colour to bricks and lime plays the role of binding the particles and reduces shrinkage. Excessive lime melts the bricks and they may loose the shape.

Iron pyrites, pebbles of stone and gravel, excessive alkalies are harmful ingredients.

Digging—weathering—blending—tempering moulding—drying—and burning is the sequence of operations in brick manufacturing.

Pug mill is sometimes used for tempering the earth. Moulding may be either by hand or machine. In India, as the climate is favourable the bricks are sun-dried whereas in the western countries artificial methods are employed for drying the bricks. Bricks are burnt in clamps or kilns.

Bricks burnt in clamps are of poor quality compared to kiln burnt bricks. Bull's trench kiln and Hoffmans kiln are semi-continuous and continuous kilns respectively.

Good quality bricks should be well burnt with uniform red colour, free from cracks, give clear ringing noise when struck against each other. They should not absorb more than 15% of water by weight when kept in water for 24 h. They should have a crushing strength not less than 55 kg/cm^2.

The bricks specially manufactured to withstand high temperatures (used in chimneys and furnaces etc.,) are called refractory bricks. The alumina content is increased in these bricks.

The standard size of the brick as per Indian standards is 19 cm × 9 cm × 9 cm so that with 1 cm thick mortar joint the size becomes 20 cm × 10 cm × 10 cm. But the size adopted by public works departments of different states are different.

To meet the special requirements, perforated bricks, hollow bricks, coping bricks, King closer bricks, Queen closer bricks and cornice bricks are manufactured.

Hollow bricks are light and provide good insulation against heat and sound.

Queen closer is half of the regular brick cut longitudinally.

Tiles are thin slabs which are made by burning the brick-earth in kilns or by using concrete.

The brick earth used for manufacture of tiles is purer than that of the bricks.

Tiles are generally used for covering roofs, for floorings or for making drains etc.

Terra cotta is a type of earthenware which finds its use as a substitute for stones on the ornamental parts of the building.

Lime has been extensively used for all types of construction purposes. Cement is a substitute for lime. But since lime is cheap and locally available it is still used in constructions of buildings.

Lime is classified into Fat lime, poor or lean lime and Hydraulic lime.

Fat lime contains high calcium oxide (about 93%) and impurities less than 5%. When the purest calcium carbonate is calcined, carbon dioxide is driven off leaving quick lime. When water is added to quick lime, the lumps are broken, heat is generated and it swells in its volume giving fat lime. It is used for plastering and white-washing and not for mortar because it has poor strength and is slow hardening.

The clayey impurities in poor lime exceed 5%. It takes longer time to slake and hardens slowly. It is used for both plastering and mortar.

Hydraulic lime sets even under water (hence its name) unlike fat and poor limes which can set only by absorbing carbon dioxide from atmosphere. Lime stone which contains silica and alumina can give hydraulic lime.

When silica and alumina content in lime is less than 15% it is called feeble hydraulic lime, between 15 and 25% it is called moderately hydraulic lime, and between 25 and 30% it is called eminently hydraulic lime. Complex silicates and aluminates present in hydraulic lime start chemical reaction in the presence of water and this results in setting of lime. Hydraulic lime gives very good strength to mortar and therefore it is used in the construction of thick walls and also structures below ground level.

A dry powder obtained after adding just sufficient water to quick lime is called hydrated lime.

Quick lime as it comes from the kilns is called lump lime.

A thin pourable suspension of slaked lime in water is called milk of lime.

Lime is obtained by burning lime stone either in clamps or kilns. Kiln may be of intermittent or continuous type.

The process of adding water to quick lime to form calcium hydroxide is called slaking. When the slaked lime is required in the form of paste or putty, tank slaking is adopted, otherwise platform slaking is done.

Cement is obtained by mixing lime stone and clay and burning them and then grinding to fine powder. Approximate composition of raw materials used in the manufacturing of ordinary portland cement is

Calcium oxide (CaO)—60 to 65%, Silica (SiO_2)—20 to 25%,

Aluminium oxide (Al_2O_3)—4 to 8%, Ferrous oxide (Fe_2O_3)—2 to 4% and Magnesium oxide (MgO)—1 to 3%.

The above compounds undergo some chemical combinations during the manufacturing process. The constituents of the end product are mainly Tri-calcium silicate, Di-calcium silicate and Tri-calcium Aluminate.

The constituent responsible for the cementing property in cement is the Tri-calcium silicate and therefore the more it is present in cement the better the cement is.

The first to set and harden is the aluminate, next is tri-silicate and the slowest is di-silicate. Therefore initial strength of cement is due to aluminate. All the three give out heat during reaction with water and it is aluminate which gives maximum heat and is responsible for undesirable properties in concrete.

A cement with less aluminate will have low initial strength but high ultimate strength.

Mixing of raw materials may be done in ball mill and the burning is done in rotary kilns.

When one gram of cement is heated for one hour at 900 to 1000°C, the loss on ignition should not exceed 4% for good cement and insoluble residue should not exceed 1.5%.

The residue retained on No. 9 IS Sieve should not exceed 10% by weight.

The physical tests on cement are conducted in the temperature range of 25°–29°C.

Consistency test, to find the proper amount of water to be added to the cement, is done in Vicat apparatus. The penetration between 33 and 35 mm in the vicat test indicate the normal consistency.

Le-chatelier's apparatus is used to conduct the soundness test. If the concrete undergoes too much change in volume after setting, it results in distortion and cracks and this is known as unsoundness.

In the process of hardening, the time at which the cracks that appear in the concrete do not reunite is called initial setting time.

The time at which the concrete attains sufficient strength and hardness is called final setting time.

Both initial and final setting times of cement can be found out using the Vicat apparatus.

Compressive strength of cement is judged by determining the compressive strength of cubes made of cement-sand mortar.

Tensile strength of cement is judged by testing the tensile strength of briquettes made of cement-sand mortar.

For good cement the compressive strength after 7 days should not fall below 175 kg/cm^2 and tensile strength should not be less than 25 kg/cm^2.

The initial setting time of good ordinary port land cement should not be less than 30 minutes and final setting time should not be more than 10 h.

Pozzolona portland cement contain 20 to 30% of pozzolona materials. The pozzolona materials are not cementitious by themselves, but they react with calcium hydroxide in presence of water at ordinary temperatures and form compounds possessing cementitious properties.

Pozzolona cement is produced either by grinding together the portland cement clinker and pozzolona, or by intimately or uniformly blending portland cement and fine pozzolona.

Compared to ordinary portland cement, the pozzolona portland cement gains strength rather slowly and needs the formwork to be kept for longer periods and has higher ultimate strength. It has same initial and final setting times.

For constructions under water, the quick setting cement is used. It has initial setting time of 5 minutes and final setting time of 30 minutes.

The inert materials like gravel, sand, brick bats etc., which are used along with cement, lime or mud in the preparation of mortar and concretes are called the aggregates.

The aggregate which passes through 4.75 mm sieve and which is completely retained on 0.15 mm sieve is called fine aggregate.

The aggregates which do not pass through 4.75 mm sieve are called coarse aggregates. If the size of the aggregate is between 75 mm and 150 mm it is called cyclopean aggregate.

Sand, crushed stone, ash or cinder are examples of fine aggregates.

Stone ballast, broken bricks and gravel are examples of coarse aggregates.

When water is added to sand, a thin film of water is formed around sand particles due to surface tension which keeps them apart and causes increase in volume of sand. This increase in volume is known as bulking of sand.

The bulking of sand is maximum when moisture content is around 4%.

A paste formed by mixing fine aggregate such as sand and binding material like cement or lime with water in specified proportions is called mortar.

Lime mortars are used in masonry joints to bind stones or bricks, for plastering of surfaces, for pointing and in lime concrete. Cement mortar is also used for all the above purposes and it is much stronger.

Mortars and concretes may be prepared by hand mixing or machine mixing. The quality of them will be better and more uniform if it is done in mixer machines.

The proportions of ingredients in the mortar depend on the purpose for which it is meant. The ratio of cement to sand in commonly used mortars are: For masonry work 1 : 6; for internal plastering 1 : 5 and for external plastering and plastering of R.C.C. work 1 : 3.

The cement mortar having fluid consistency is called grout. The process of applying mortar on concrete under pressure through the nozzle of a cement gun is called guniting.

The joints of masonry work are raked to a depth of about 1.25 cm and then they are filled with mortar. This is known as pointing. Pointing is done to close any crevices left in the mortar joints, (so that the entry of moisture into the wall is prevented) to improve the appearance of the pointed surface, and to give protection to the walls constructed with mud mortar.

A composite material made of coarse aggregate, fine aggregate and binding material in specified proportions mixed with water may be called the concrete. Freshly prepared concrete is called wet concrete and when it is set and hardened it is called set concrete or simply concrete.

Lime concrete is generally used as a levelling course for foundations as base concrete under floors and for roof finish.

Cement concrete is much stronger and more versatile. It is used in all reinforced works. Concrete without reinforcement is called plain concrete.

The proportion of ingredients in a concrete mix depend on the purpose for which the concrete is used. For general R.C.C. work the proportion of cement, fine aggregate and coarse aggregate in cement concrete are 1 : 2 : 4. When high strength is required the mix may be of 1 : 1 : 2 or 1 : 1.5 : 3. For mass concrete works the lean mix such as 1 : 3 : 6 or 1 : 4 : 8 may by used.

The ingredients in concrete are mixed in specified proportion by weight. But in practice it is usually done by volume.

A 50 kg cement bag has a volume of 0.0345 m^3. But if it is taken out of bag it becomes loose and increases its volume. It will be very convenient if fine and coarse aggregates are measured by a container whose volume is 34.5 litres.

To obtain uniform and thorough mixing it is preferable that the mixing be done in mixer machines called concrete mixers.

For a given proportion of ingredients it is water-cement ratio which governs the strength of the concrete. While higher water-cement ratio increases the workability it reduces the strength of the concrete.

Slump test is carried out for controlling the water content in concrete. The apparatus of slump test is very simple and consists of a cone with both ends open, the diameter at the base 20 cms, at the top 10 cms and height 30 cms. After filling the cone with wet concrete, it is lifted. The extent by which the concrete drops is called the slump. It is measured from the top of the cone.

Workability of concrete is better determined by compaction factor test. A compaction factor of 0.95 indicates good workability.

The mean size of the aggregate is provided by fineness modulus. It is found by taking the cumulative percentage of aggregates retained on a set of ten IS Sieves and dividing the sum by 100.

If fineness modulus of fine aggregate is F_1, and coarse aggregate F_2, then the ratio of the fine aggregate to coarse aggregate in the mix to obtain a fineness modulus of F for the combined aggregate is given by

$$X = \frac{F_2 - F}{F - F_1}.$$

All works executed using cement or lime are to be cured. Curing is the process wherein water is made available for concrete to attain its full strength. Curing may be done by spraying water, covering the surface with wet cloth or wet sand.

Wood obtained from trees and used for engineering purposes is called timber. Trees which grow inwards in a longitudinal fibrous mass are called endogenous trees. *Examples*—Coconut, bamboo, palm, cane etc. Trees which grow outward by adding one concentric ring every year are called exogenous trees. *Examples*—Teak, sal, deodar, shisham etc.

The outermost protective covering layer is called bark. The first formed portion of the stem is called pith. The rings arranged in concentric circles around the pith are known as annual rings. These rings help in estimating the age of the tree. Innermost rings surrounding the pith is the heart wood. Thin horizontal veins radiating outward from the pith are called medullary rays.

The attack of fungus on wood reducing it to a dry powder is called dry rot. The decay of timber due to alternate wetting and drying is called wet rot.

The wooden logs are cut into pieces by different methods of sawing which include: flat sawing, quarter sawing, radial sawing and tangential sawing. The most economical of them is the flat sawing.

Freshly felled trees contain lot of moisture. If this is not removed the wood is likely to warp, crack and shrink. The removal of moisture under controlled conditions at a uniform rate is called seasoning of timber.

Seasoning makes timber resistant to decay, lighter, stronger and stable. It is easier to paint and polish seasoned wood. Seasoned timber has got better electrical resistance.

Air seasoning (or natural seasoning) of wood is simple and economical but it is a slow process. It requires more stacking space and gives relatively stronger wood.

Kiln seasoning (or artificial seasoning) of wood is quite technical and expensive but it is a quick process. It requires less stacking space and it gives relatively weak timber.

Timber can be made fire resistant (not completely) by soaking it in ammonium sulphate, ammonium chloride and ammonium phosphate or by spraying sodium silicate, potassium silicate and ammonia phosphate.

Applying creosote oil under pressure makes the timber resistant to rot and attack from white ant.

Tarring and painting of wood preserves the wood.

Thin sheets of wood which are peeled of or sliced or sawn from a log of wood are called veneers. Rose and teak wood are commonly used for making veneers. Veneers used for making plywoods are known as plies. Odd number of plies, arranged such that the grain of one layer is at right angles to the other, are glued together to form plywood. The plywood is identified by its thickness in mm.

Paints and varnishes are used with the purpose of providing protection to metal, timber and plastered surfaces from the corrosive effects of weather, heat moisture or gases etc. and also improving their appearance.

Paints are classified as oil paints, cement paints, water paints and bituminous paints. The special paints include fireproof paints, luminous paints etc.

An oily liquid (usually linseed oil) in which the base and the pigments are dissolved to obtain paint is called the vehicle. The material (usually white lead, zinc oxide, or metallic powders of aluminium copper etc.,) which provides body to the paint is called base. The ingredient added to get the desired colour is called the pigment.

Depending on the base, the paints may be called lead paints, zinc paints, aluminium paints etc.

A thinner (usually turpentine oil) is the liquid added to the paint to obtain desired consistency.

The material added to the paint to hasten the drying of vehicle is called a drier.

The material used in place of base to reduce the cost of paint (such as silicon, charcoal etc.,) is called inert filler.

Improper seasoning of wood, excessive use of drier or too many coats of paint result in cracks extending throughout the thickness of the paint. This results in scaling.

If hair cracks developed on a painted surface enclose small areas, it is called crazing. If they enclose large areas it is called crocodiling.

Blisters are formed when the surface to be painted is oily or greasy or when moisture is still present in the pores of the wood.

Cellulose paints (with a trade name Duco) are used for painting automobiles, aircrafts and other costly things.

Distemper is a water paint. It is used for painting masonary walls.

Varnish is a solution of resin in turpentine or alcohol. It is applied to the painted surface to increase the brilliance or to the unpainted wood surface to brighten the ornamental appearance of the grains of wood. Spirit varnish is also called French polish.

Common ferrous metals used in construction of buildings are: cast iron, wrought iron and different forms of steel.

Common non-ferrous metals are: aluminium, copper, zinc, lead and tin.

The variation in carbon content gives different form of iron. Pig iron is the crudest form. Wrought iron is the purest form. Cast iron is in between. The carbon content in cast iron varies between 2 and 4.5%. Whereas in wrought iron it is less than 0.25%.

In mild steel the carbon content is between 0.15% and 0.3%. High carbon steel has a carbon content of 0.55 to 1.5%. In high tension steel it is kept below 0.15%. Steel goes on becoming harder and tougher with the increase in its carbon content.

Chromium is added to make the steel corrosion resistent. When the chromium content is more than 16% it is called stainless steel.

Other varieties of steel are: Nickel steel, vanadium steel, tungsten steel and manganese steel.

Steel or iron is galvanised to protect it from the rusting and corrosion. Depositing a fine film of zinc on steel or iron by dipping them in molten zinc is called galvanising.

In electroplating a thin film of nickel, chromium, cadmium copper or zinc is deposited on the surface to be protected by the process of electrolysis.

Metals can be joined by welding, soldering or brazing. Solder is an alloy of lead and tin in the proportion of 1 : 2. Brazing solder is a mixture of tin, zinc and copper in the proportion of 1 : 3 : 4.

Brass is an alloy of copper (60 to 70%) and zinc. It is used for making pumps certain machine parts and household utensils.

Bronze is an alloy of 90% copper and 10% tin.

The base metal of the alloys duralumin and Y-alloy is aluminium. Duralumin is used for making cables, aeroplane parts etc.

Plastics are synthetic materials that are increasingly used in the constructions. Many electrical and sanitary fittings, insulators, floats etc., are made of plastics. PVC is a kind of thermoplastic. It stands for Polyvinyl chloride.

Glass is obtained by fusion of silica, and potash or soda at a temperature of more than 1000°C. It has been extensively used in building construction for glazing doors and windows.

Glasses are usually classified into 3 categories. Soda lime glass, lead glass and boro-silicate glass.

Soda lime glass is used for glazing purpose and to make ordinary glassware.

Lead glass is used for electrical bulbs and optical glasses. It is also called flint glass.

Boro-silicate glass withstands high temperatures and is therefore used for making laboratory equipment and cooking utensils.

When one face of the glass is made rough by grinding it is called ground glass. It is used for windows of bath rooms and toilets.

When glass is reinforced with wire mesh it is called wired glass. It is used for glazing north light trusses.

Tar, bitumen and asphalt are used in damp proofing the buildings, water proofing roofs and constructing metalled roads.

Tar is obtained as a by product in the distillation of coal, resinous wood, or bituminous shales.

Bitumen is obtained as a by product in the refining of petroleum.

Asphalt is a natural or artificial mixture of some inert matter and bitumen.

Asbestos and portland cement are used in making a variety of asbestos cement (A.C.) products like corrugated sheets, pipes, tiles etc. These products are very popular in building construction. Asbestos is available in nature in the form of a fibrous material.

II. OBJECTIVE TYPE QUESTIONS

1. Which of the following is the reason for the decrease in the use of stones as building material
   (a) steel and R.C.C. are less bulky and more durable
   (b) strength of stones cannot be rationally analysed
   (c) stones are not conveniently available in plains
   (d) all the above.

2. The solidification of molten magma when it reaches the surface of earth results in the formation of
   (a) sedimentary rocks
   (b) metamorphic rocks
   (c) basalts and traps
   (d) granite.

3. The solidification of molten magma within the earth's crust results in the formation of
   (a) sedimentary rocks
   (b) metamorphic rocks
   (c) basalts and traps
   (d) granite.

4. Identify the process responsible for the formation of sedimentary rocks
   (a) solidification of molten mass of silicates below or at the surface of earth
   (b) changes in texture or mineral composition or both of igneous and sedimentary rocks due to high temperature and heavy pressures

(c) deposited layers of sand and silt subjected to enormous overburden pressures over geological times ☐
   (d) none of the above. ☐
5. In problem 4, identify the process responsible for the formation of igneous rocks.
6. In problem 4, identify the process responsible for the formation of metamorphic rocks.
7. Under metamorphism, which of the following changes is correct
   (a) granite changes into gneiss ☐
   (b) sand stone changes into quartzite ☐
   (c) lime stone changes into marble ☐
   (d) all the above. ☐
8. The agrillaceous rocks have their principal constituents as
   (a) lime ☐
   (b) clay ☐
   (c) sand ☐
   (d) none of the above. ☐
9. In problem 8, what is the principal constituent of calcareous rocks.
10. In problem 8, what is the principal constituent of silicious rocks.
11. The process of taking out stones of various sizes from natural rocks is known as
    (a) dressing ☐
    (b) seasoning ☐
    (c) quarrying ☐
    (d) none of the above. ☐
12. The process of giving definite and regular shape to stones with smooth faces is known as
    (a) pitching ☐
    (b) dressing ☐
    (c) seasoning ☐
    (d) none of the above. ☐
13. In stone masonry, the stones are placed in position such that the natural bedding plane is
    (a) normal to the direction of pressure they carry ☐
    (b) parallel to the direction of pressure they carry ☐
    (c) at 45° to the direction of pressure they carry ☐
    (d) at 60° to the direction of pressure they carry. ☐
14. If the molten magma forces itself into an already existing rock in the earth's crust and solidifies there, such a rock is known as
    (a) metamorphic rock ☐
    (b) extrusive rock ☐
    (c) intrusive rock ☐
    (d) igneous rock. ☐
15. Pegmatite is an example of
    (a) sedimentary rock ☐
    (b) extrusive igneous rock ☐
    (c) intrusive igneous rock ☐
    (d) metamorphic rock. ☐
16. Lime stone comes under the category of
    (a) aqueous rock ☐
    (b) stratified rock ☐
    (c) sedimentary rock ☐
    (d) all the above. ☐
17. Laterite is an example of
    (a) agrillaceous rock ☐
    (b) volcanic rock ☐
    (c) organic rock ☐
    (d) silicious rock. ☐

18. Slate and marble stone belong to
    (a) igneous rocks
    (b) metamorphic rocks
    (c) sedimentary rocks
    (d) foliated rocks.
19. In problem 18, to which category does the sand stone belong.
20. The physical classification divides the rocks into
    (a) calcareous, agrillaceous and silicious
    (b) organic, semi-organic and inorganic
    (c) igneous, sedimentary and metamorphic
    (d) stratified, unstratified and foliated.
21. In problem 20, identify the chemical classification of rocks.
22. In problem 20, identify the geological classification of rocks.
23. Mica, mainly, is composed of
    (a) calcium carbonate
    (b) magnesium and calcium silicate
    (c) silica with oxygen
    (d) potassium and aluminium silicate.
24. The moisture absorption of a good stone should be less than
    (a) 1%
    (b) 5%
    (c) 10%
    (d) 15%.
25. Most of the stones possess the specific gravity in the range of
    (a) 1 to 1.5
    (b) 1.5 to 2.0
    (c) 2.4 to 2.8
    (d) 3 to 4.
26. Which of the following has highest crushing strength
    (a) lime stone
    (b) granite
    (c) gneiss
    (d) laterite.
27. Which of the following has the lowest crushing strength
    (a) basalt
    (b) granite
    (c) diorite
    (d) laterite.
28. Find the one which is not used in quarrying
    (a) gun powder
    (b) gun cotton
    (c) marble powder
    (d) dynamite.
29. Quartzite is a
    (a) sandy rock
    (b) silicious rock
    (c) organic rock
    (d) calcareous rock.
30. Basalt can be classified as
    (a) sedimentary rock
    (b) metamorphic rock
    (c) intrusive igneous
    (d) extrusive igneous.
31. The reason for the popularity of bricks as construction material is that
    (a) they are cheap and available locally at all places
    (b) they are durable and possess fairly good strength and lighter than stones
    (c) they have very good insulating property against heat and sound
    (d) all the above.

32. In the composition of good bricks, the total content of silt and clay, by weight, should not be less than
    (a) 20%
    (b) 30%
    (c) 50%
    (d) 75%.

33. Pug mill is the device used for
    (a) excavating the soil for preparation of bricks
    (b) tempering the earth needed for manufacture of bricks
    (c) burning the bricks
    (d) none of the above.

34. The indentation provided in a face of the brick is called
    (a) frog
    (b) pallet
    (c) strike
    (d) none of the above.

35. The most widely used kiln in India is
    (a) Hoffman's kiln
    (b) Bull's trench kiln
    (c) clamp kiln
    (d) none of the above.

36. The standard size of brick as per Indian standards is
    (a) 20 × 10 × 10 cm
    (b) 23 × 12 × 8 cm.
    (c) 19 × 9 × 9 cm
    (d) 18 × 9 × 9 cm.

37. Refractory bricks are specially manufactured
    (a) to withstand high temperature
    (b) to withstand high crushing pressure
    (c) to have high insulation against sound
    (d) none of the above.

38. When a brick is cut into two halves longitudinally, one part is called
    (a) king closer
    (b) cornice brick
    (c) queen closer
    (d) voussoir.

39. A brick which is given a wedge like shape to be used in the construction of arches is called
    (a) king closer
    (b) cornice brick
    (c) queen closer
    (d) voussoir.

40. The red colour obtained by the bricks is due to the presence of
    (a) lime
    (b) silica
    (c) manganese
    (d) iron oxide.

41. The soil used for the manufacture of bricks should preferably not contain the following material
    (a) alkalies
    (b) pebbles
    (c) kankar
    (d) all the above.

42. The number of standard bricks required for one cubic metre of brick masonry is
    (a) 400
    (b) 500
    (c) 750
    (d) 250.

43. Formation of whitish deposit on the bricks due to the presence of excess salts is called
    (a) efflorescence
    (b) disintegration
    (c) warping
    (d) floating.

44. The water absorption of a good brick after 25 h immersion should be less than
    (a) 25%
    (b) 20%
    (c) 15%
    (d) 10%.

45. Hollow bricks are generally used with the purpose of
    (a) reducing the cost of construction
    (b) providing insulation against heat
    (c) increasing the bearing area
    (d) ornamental look.

46. Tiles are used for
    (a) covering the roofs and floorings
    (b) making drains
    (c) both (a) and (b)
    (d) none of the above.

47. Terra cotta, in buildings, is used for
    (a) insulation
    (b) ornamental work
    (c) sewage lines
    (d) sanitary services.

48. The lime which has high calcium oxide content and which sets only in the presence of carbon dioxide is called
    (a) fat lime
    (b) hydraulic lime
    (c) magnesium lime
    (d) lean lime.

49. The lime which has the property of setting in water is known as
    (a) fat lime
    (b) hydraulic lime
    (c) hydrated lime
    (d) quick lime.

50. The quick lime as it comes from kilns is called
    (a) milk lime
    (b) hydraulic lime
    (c) lump lime
    (d) hydrated lime.

51. A dry powder obtained on treating quick lime with just enough water to satisfy its chemical affinity for water under the condition of its hydration is called
    (a) hydraulic lime
    (b) hydrated lime
    (c) milk of lime
    (d) none of the above.

52. The process of adding water to lime to convert it into a hydrated lime is termed as
    (a) watering
    (b) baking
    (c) hydration
    (d) slaking.

53. The constituent responsible for setting of hydraulic lime under water is
    (a) silica
    (b) clay
    (c) calcium oxide
    (d) carbon dioxide.

54. A thin pourable suspension of slaked lime in water is known as
    (a) lime water
    (b) lime paint
    (c) milk of lime
    (d) lime lotion.

55. Lime suitable for making mortar of good strength
    (a) hydraulic lime
    (b) fat lime
    (c) lean lime
    (d) none of the above.

56. The calcination of pure lime result in
    (a) quick lime
    (b) hydraulic lime
    (c) hydrated lime
    (d) fat lime.

57. Plaster of Paris can be obtained from the calcination of
    (a) lime stone
    (b) gypsum
    (c) dolomite
    (d) bauxite.

58. The silicious and aluminous minerals, which do not have cementitious qualities by themselves but which react with lime in the presence of water at normal temperature to form cementitious compounds, are known as
    (a) glazed materials
    (b) procelains
    (c) pozzolonic materials
    (d) carbonacious materials.

59. The process of heating the lime stone to redness in contact with air is termed
    (a) carbonation
    (b) oxidation
    (c) hydration
    (d) calcination.

60. The raw materials having more than 10% proportion in the manufacture of cement are
    (a) calcium oxide and silica
    (b) calcium oxide and magnesium oxide
    (c) magnesium oxide and ferrous oxide
    (d) silica and ferrous oxide.

61. After the addition of water to it, the cement sets and hardens due to
    (a) the heat produced by the chemical action
    (b) the hydration and hydrolysis of the constituent compounds of cement
    (c) binding action of water
    (d) none of the above.

62. Which constituent of the cement, upon addition of water, sets and hardness first
    (a) tri-calcium silicate
    (b) tri-calcium aluminate
    (c) di-calcium silicate
    (d) free lime.

63. In problem 62, which constituent is the best cementing material ?

64. The quality of cement is good if it has more of
    (a) di-calcium silicate
    (b) tri-calcium aluminate
    (c) tri-calcium silicate
    (d) free lime.

65. The ingredient added in the manufacturing process to control the setting time of cement is
    (a) magnesium sulphate
    (b) free lime
    (c) gypsum
    (d) calcium sulphate.

66. In the sieve analysis of fineness test, the residue on No. 9 sieve after 15 minutes of sieving should not be more than
    (a) 5%
    (b) 7.5%
    (c) 10%
    (d) 15%.

67. The temperature range at which the consistency test is conducted
    (a) 25°C to 29°C
    (b) 20°C to 25°C
    (c) 15°C to 20°C
    (d) 30°C to 35°C.

68. The consistency test is performed to find
    (a) the correct water-cement ratio
    (b) the fineness of the cement
    (c) the compressive strength
    (d) tensile strength.

69. For normal consistency, the penetration in Vicat apparatus should be between
    (a) 20 to 30 mm
    (b) 33 to 35 mm
    (c) 35 to 38 mm
    (d) > 40 mm.

70. Le-Chatallier's apparatus is used to carry out
    (a) consistency test
    (b) soundness test
    (c) compressive strength test
    (d) tensile strength.

71. In briquette test, the seven day tensile strength of good portland cement should not be less than
    (a) 20 kg/cm^2
    (b) 30 kg/cm^2
    (c) 25 kg/cm^2
    (d) 35 kg/cm^2.

72. The seven day compressive strength of a good portland cement, as obtained from the compressive test on cement-sand mortar cubes, should not be less than
    (a) 125 kg/cm^2
    (b) 150 kg/cm^2
    (c) 175 kg/cm^2
    (d) 200 kg/cm^2.

73. The initial setting time of ordinary portland cement should not be less than
    (a) 15 minutes
    (b) 30 minutes
    (c) 45 minutes
    (d) one hour.

74. The final setting time of ordinary portland cement should not be more than
    (a) 5 h
    (b) 7.5 h
    (c) 10 h
    (d) 12.5 h.

75. Quick setting cement is used
    (a) for the construction of structures under water
    (b) to obtain very high strength
    (c) where resistance to acidic water is required
    (d) none of the above.

76. The initial and final setting times of quick setting cement are
    (a) 15 minutes and 45 minutes
    (b) 20 minutes and 1 hour
    (c) 10 minutes and 30 minutes
    (d) 5 minutes and 30 minutes.

77. The pozzolona protland cement gains the strength
    (a) in the same time as ordinary portland cement
    (b) in less time than ordinary portland cement
    (c) in more time than ordinary portland cement
    (d) difficult to tell.

78. Compared to the ordinary portland cement, the ultimate strength of the pozzolona portland cement is
    (a) same
    (b) more
    (c) less
    (d) very much less.

79. In low heat cement, the proportion of the following compound is kept at low value
    (a) tricalcium aluminate
    (b) tricalcium silicate
    (c) both (a) and (b)
    (d) none of the above.

80. In the chemical composition test for loss on ignition, the cement is heated to a temperature of
    (a) 100°C
    (b) 200°C
    (c) 500°C
    (d) 1000°C.

81. Loss on ignition in cement should not exceed
    (a) 1%
    (b) 4%
    (c) 8%
    (d) 12%.

82. The insoluble residues in good cement should be
    (a) between 4 and 8%
    (b) less than 8%
    (c) between 8 and 10%
    (d) less than 1.5%.

83. Excessive free lime and excessive magnesia present in the cement make the cement
    (a) unsound
    (b) to have very low initial setting time
    (c) to have very low compressive strength
    (d) gain strength faster.

84. The dry process of manufacturing cement has become obsolete, because, in comparison to wet process
    (a) it is slow and costly
    (b) the quality of cement produced by it is inferior
    (c) it is difficult to maintain the correct proportions of constituents
    (d) all the above.

85. Snowcem is
    (a) mixture of lime and pigment
    (b) chalk powder
    (c) coloured cement
    (d) none of the above.

86. The aggregate is called fine aggregate if it is completely retained on
    (a) 0.15 mm sieve
    (b) 0.30 mm sieve
    (c) 4.75 mm sieve
    (d) none of the above.

87. The aggregate is called coarse aggregate if it is completely retained on
    (a) 10 mm sieve
    (b) 15 mm sieve
    (c) 29 mm sieve
    (d) 4.75 mm sieve.

88. The bulking of sand is due to
    (a) the increase in space between the particles caused by the surface tension effect of moisture
    (b) the swelling of air in voids
    (c) the viscous effect of moisture
    (d) none of the above.

89. The increase in volume of dry sand when water is added is called
    (a) honey combing
    (b) bulking
    (c) segregation
    (d) bleeding.

90. The sand in mortar
    (a) increases the volume of mortar
    (b) reduces the shrinkage and cracking
    (c) helps the pure lime to set by allowing penetration of air which provides the needed carbon dioxide
    (d) all the above.

91. The aggregate is called the cyclopean aggregate if its size is
    (a) between 0.15 mm and 4.5 mm
    (b) 4.75 mm to 40 mm
    (c) 75 mm to 150 mm
    (d) more than 150.

92. The cement mortar mix generally used for masonry work is
    (a) 1 : 3
    (b) 1 : 5
    (c) 1 : 6
    (d) 1 : 10.

93. The cement mortar mix generally used for internal plastering is
    (a) 1 : 3
    (b) 1 : 5
    (c) 1 : 6
    (d) 1 : 10.

94. The cement mortar mix commonly used for external plastering and plastering of reinforced cement concrete works is
    (a) 1 : 3
    (b) 1 : 5
    (c) 1 : 6
    (d) 1 : 10.

95. Pointing is the process whereby
    (a) the masonry joints are filled up with mortar after raking out for small depth
    (b) the grooves arc cut on a plastered surface to give a look of masonry
    (c) small circles looking like points are cut out at random on a plastered surface
    (d) none of the above.

96. The purpose of pointing is
    (a) to seal off any crevices left in the mortar joint and there by prevent the entry of moisture into walls
    (b) to improve the appearance of a wall when it is not plastered
    (c) to protect the masonry joints laid in mud mortar
    (d) all the above.

97. The process of applying cement mortar under pressure through a nozzle is called
    (a) pressurising
    (b) prestressing
    (c) guniting
    (d) none of the above.

98. The volume of one bag of cement weighing 50 kg is
    (a) 0.05 m^3
    (b) 0.0345 m^3
    (c) 0.025 m^3
    (d) 0.04 m^3.

99. The strength of cement concrete for a given mix depends on
    (a) water-cement ratio
    (b) final setting time
    (c) initial setting time
    (d) none of the above.
100. The minimum water-cement ratio to obtain workable concrete is
    (a) 0.6
    (b) 0.55
    (c) 0.5
    (d) 0.4.
101. Compaction factor for good workability of concrete is
    (a) 0.7
    (b) 0.80
    (c) 4.85
    (d) 0.95.
102. Slump test facilitates
    (a) controlling of water-cement ratio of concrete during construction
    (b) the determination of initial and final setting times of cement
    (c) the determination of workability of concrete
    (d) none of the above.
103. The concrete mix used for general R.C.C. work is
    (a) 1 : 2 : 4
    (b) 1 : 1 : 2
    (c) 1 : 4 : 8
    (d) 1 : 5 : 10.
104. Fineness modulus is
    (a) the diameter of the sieve on which 50% of coarse aggregate is retained
    (b) the diameter of the sieve on which 50% of fine aggregate is retained
    (c) an index which gives the mean size of the aggregates used in the mix
    (d) none of the above.
105. The fineness modules of coarse and fine aggregates are F_2 and F_1. What is the ratio X of fine aggregate to coarse aggregate in a mix whose desired fineness modulus is F
    (a) $X = \dfrac{F_2 - F_1}{F - F_2}$
    (b) $X = \dfrac{F_2 - F}{F - F_1}$
    (c) $X = \dfrac{F_1 - F_2}{F_1 - F}$
    (d) $X = \dfrac{F - F_2}{F - F_1}$.
106. The process of keeping concrete wet to enable it to attain full strength is known as
    (a) curing
    (b) wetting
    (c) drenching
    (d) quenching.
107. Curing of concrete can be done by
    (a) spraying
    (b) ponding
    (c) covering with moist cloth
    (d) any of the above.
108. Which one of the following does not belong to endogenous trees
    (a) palm
    (b) bamboo
    (c) teak
    (d) cane.
109. Which one of the following does not belong to exogenous trees
    (a) coconut
    (b) teak
    (c) shisham
    (d) sal.

110. The solution of salts from the soil absorbed by the trees which becomes a viscous solution due to loss of moisture and action of carbon dioxide is known as
    (a) pith
    (b) cambium
    (c) bark
    (d) sap.

111. The age of the tree can be judged from
    (a) the height
    (b) the diameter
    (c) annual rings
    (d) piths.

112. The layer between the bark of the tree and the sap wood which is not yet converted into wood is called
    (a) heart wood
    (b) cambium layer
    (c) soft wood layer
    (d) pith.

113. Which of the following is not a hard wood?
    (a) deodar
    (b) sal
    (c) teak
    (d) oak.

114. Which of the following is not a soft wood
    (a) deodar
    (b) walnut
    (c) shisham
    (d) kail.

115. When the timber is attacked by fungus and reduced to powder, it is called
    (a) wet rot
    (b) dry rot
    (c) druxiness
    (d) doatiness.

116. Which of the following wood has the maximum resistance to white ants
    (a) deodar
    (b) sal
    (c) walnut
    (d) teak.

117. Creosote oil is used to preserve the wood from
    (a) rot and white ant
    (b) fire hazards
    (c) cracking
    (d) none of the above.

118. Timber can be made reasonably fire resistant
    (a) by soaking it in ammonium sulphate
    (b) by applying tar paint
    (c) by pumping creosote oil into timber under high pressure
    (d) none of the above.

119. The main purpose of seasoning is
    (a) to make the timber fire resistant
    (b) to remove the moisture from the timber at uniform rate
    (c) to make the timber water proof
    (d) none of the above.

120. The seasoning of timber
    (a) makes the timber light, strong and stable
    (b) prevents warping, cracking and shrinkage in timber

(c) makes timber resistant to decay by fungi, termites etc., and also resistant to electricity
(d) all the above.

121. Which of the following statements is not correct
    (a) kiln seasoned timber is stronger than natural seasoned timber
    (b) moisture content in the timber can be reduced to any desired level in kiln seasoning
    (c) kiln seasoning is quicker than natural seasoning
    (d) kiln seasoning requires less stacking space but more expensive than the natural seasoning.

122. In a well-seasoned timber, the moisture content will be in the range of
    (a) 20–25% (b) 15–20%
    (c) 10–12% (d) 5–7%.

123. Most economical of the methods of sawing wood is
    (a) radial sawing (b) tangential sawing
    (c) quarter sawing (d) flat sawing.

124. A thin sheet of wood sliced from a log of wood is called
    (a) plywood (b) lamin board
    (c) veneer (d) none of the above.

125. In the manufacture of plywoods the veneers are placed such that the grains of one layer are
    (a) at 45° with the grains of the other
    (b) at right angles with the grains of the other
    (c) at 60° with grains of the other
    (d) parallel to the grains of the other.

126. Plywood is identified by
    (a) volume (b) weight
    (c) thickness (d) area.

127. The ingredient which gives the desired colour to a paint is called
    (a) base (b) pigment
    (c) vehicle (d) solvent.

128. The most commonly used substance as a vehicle in the oil paints is
    (a) zinc oxide (b) turpentine oil
    (c) white lead (d) linseed oil.

129. The oil/liquid in which base and pigment are dissolved to form a paint is called
    (a) thinner (b) filler
    (c) vehicle (d) none of the above.

130. Distemper is a type of
    (a) oil paint (b) enamel paint
    (c) water paint (d) varnish.

131. The base in a paint has the following function
    (a) it forms the body of the paint
    (b) it reduces the shrinkage cracks

(c) it reinforces the films of the paint after it has dried and prevents the penetration of paint to lower surfaces

(d) all the above.

132. Turpentine oil is used in paints as
    (a) thinner
    (b) base
    (c) carries
    (d) drier.

133. The paint used for automobiles is
    (a) distemper
    (b) emulsion paint
    (c) oil paint
    (d) any of the above.

134. Duco is the trade name for
    (a) bituminous paint
    (b) oil paint
    (c) cellulose paint
    (d) water paint.

135. The paint which gives illumination during nights is called
    (a) fluorescent paint
    (b) cellulose paint
    (c) enamel paint
    (d) none of the above.

136. The function of a paint is
    (a) to give a clean, colourful and pleasing surface
    (b) to increase the life of the painted surface
    (c) to protect the surface from corrosion and other weather effects
    (d) all the above.

137. The painting work is generally specified by
    (a) weight of the paint applied
    (b) labour used in the painting
    (c) area of the painted surface
    (d) any of the above.

138. The cracks in the painted surface extending throughout the thickness of the paint are caused due to
    (a) improper seasoning of the painted wood
    (b) excessive use of drier
    (c) too many coats of paint resulting in excessive thickness
    (d) all the above.

139. The small areas on painted surface enclosed by hair line cracks are called
    (a) crazing
    (b) crocodiling
    (c) chalking
    (d) blistering.

140. The paints used in aircrafts are
    (a) dry paints
    (b) cellulose paints
    (c) water paints
    (d) emulsion paints.

141. The defect in painting caused due to sliding of one layer of paint over another, is known as
    (a) wrinkling
    (b) peeling
    (c) alligatoring
    (d) none of the above.

142. The varnish is essentially made of
    (a) resin
    (b) solvent
    (c) driver
    (d) both (a) and (b).

143. French polish is made by dissolving the resin
    (a) in oil
    (b) in water
    (c) in spirit
    (d) in turpentine.

144. Snowcem is
    (a) a cement paint
    (b) an oil paint
    (c) an enamel paint
    (d) a cellulose paint.

145. The property of material enabling it to be drawn into thin wires is called
    (a) toughness
    (b) hardness
    (c) ductility
    (d) malleability.

146. The property of metals because of which they can be transformed into different shapes by heating is called
    (a) malleability
    (b) ductility
    (c) flowability
    (d) resilience.

147. The property due to which steel can withstand hammer blows is called
    (a) toughness
    (b) hardness
    (c) resilience
    (d) plasticity.

148. The carbon content in cast iron is in the range of
    (a) 0.1 to 0.2%
    (b) 0.15 to 0.30%
    (c) 1 to 1.5%
    (d) 2 to 4.5%.

149. The carbon content in mild steel is in the range of
    (a) 0.1 to 0.2%
    (b) 0.15 to 0.30%
    (c) 1 to 1.5%
    (d) 2 to 4.5%.

150. The carbon content in high tension steel is
    (a) < 0.15%
    (b) < 0.30%
    (c) < 0.45%
    (d) < 2%.

151. The chromium content in stainless steel is more than
    (a) 6%
    (b) 11%
    (c) 16%
    (d) 20%.

152. Invar is a nickel steel with nickel content between
    (a) 30–40%
    (b) 20–30%
    (c) 10–20%
    (d) 40–50%.

153. The steel used for making rail tracks is
    (a) tungsten steel
    (b) manganese steel
    (c) chromium steel
    (d) none of the above.

154. Galvanising means
    (a) depositing a fine film of zinc on iron or steel by dipping it in molten zinc
    (b) depositing a thin layer of tin by dipping steel in molten tin
    (c) depositing a fine film of nickel or zinc on iron by the process of electrolysis
    (d) none of the above.

155. Which of the following metals is poisonous ?
    (a) aluminium
    (b) copper
    (c) lead
    (d) tin.

156. The descending order of specific gravities of Zinc (Z), Tin (T), Lead (L), and Aluminium (A) is
    (a) L—T—Z—A
    (b) T—Z—L—A
    (c) L—A—T—Z
    (d) L—Z—A—T.

157. The constituents metals of brass are
    (a) copper and tin
    (b) copper and zinc
    (c) copper and aluminium
    (d) copper and manganese.

158. The constituent metals of Bronze alloy are
    (a) copper and tin
    (b) copper and zinc
    (c) copper and aluminium
    (d) copper and manganese.

159. The composition of copper and tin in the alloy bronze is
    (a) 90% and 10%
    (b) 80% and 20%
    (c) 70% and 30%
    (d) 60% and 40%.

160. The constituent metals of the solder are
    (a) lead and tin
    (b) lead and zinc
    (c) lead and copper
    (d) lead and iron.

161. The constituent metals of brazing solders are
    (a) tin-lead-copper
    (b) tin-lead-zinc
    (c) tin-zinc-copper
    (d) tin-zinc-iron.

162. In the heat treatment of steel, the process of heating steel to red hot and dipping it in a bath of cold water or oil is called
    (a) annealing
    (b) normalising
    (c) hardening
    (d) none of the above.

163. PVC stands for
    (a) polythene vanadium carbide
    (b) polyvinyl chloride
    (c) polyvinyl carbide
    (d) polythene vinyl chloride.

164. Monel metal is an alloy of
    (a) nickel and lead
    (b) copper and zinc
    (c) zinc and tin
    (d) nickel and copper.

165. The proportions of copper and zinc in Muntz metal are
    (a) 60%–40%
    (b) 70%–30%
    (c) 80%–20%
    (d) 90%–20%.

166. The constituent metals of German silver are
    (a) zinc-copper-nickel
    (b) zinc-copper-chromium
    (c) zinc-copper-tin
    (d) zinc-copper-lead.
167. The main ingredient in the manufacture of glass is
    (a) silica
    (b) potash
    (c) soda
    (d) lime.
168. The type of glass used for the window of bath room and toilets is
    (a) coloured glass
    (b) ground glass
    (c) block glass
    (d) flint glass.
169. The type of glass used in glazing the north light trusses is
    (a) ground glass
    (b) wired glass
    (c) flint glass
    (d) block glass.
170. Asphalt is a mixture of
    (a) bitumen and inert mineral matter
    (b) bitumen and asbestos
    (c) bitumen and cement
    (d) tar and asbestos.
171. The asbestos content in asbestos cement sheets is roughly about
    (a) 5%
    (b) 10%
    (c) 15%
    (d) 30%.
172. The type of glass used to make laboratory equipment and cooking utensils is
    (a) soda lime glass
    (b) lead glass
    (c) boro-silicate glass
    (d) wired glass.
173. Water glass is
    (a) water formed by condensation on glass
    (b) water free from impurities
    (c) glass made from pure silica
    (d) solution of sodium silicate in water.
174. The height of the cone in slump test is
    (a) 20 cm
    (b) 30 cm
    (c) 45 m
    (d) 50 cm.
175. The melting point of glass is in the range of
    (a) 1200 to 1400°C
    (b) 1400 to 2000°C
    (c) 800 to 950°C
    (d) 400 to 500°C.
176. The base metal in Y-alloy is
    (a) aluminium
    (b) copper
    (c) iron
    (d) zinc.
177. Shingle is
    (a) water bound pebbles
    (b) disintegrated laterite
    (c) crushed granite
    (d) none of the above.
178. Sewer pipes are made of
    (a) earthenware
    (b) stoneware
    (c) terracotta
    (d) all the above.

179. Glazing of clay products is done
    (a) to improve their appearance
    (b) to protect them from atmospheric effect
    (c) to protect them from corrosive action
    (d) all the above.

180. Rapid hardening cement attains early strength due to
    (a) larger proportion of lime grounded finer than in ordinary cement
    (b) small proportion of lime grounded finer than in ordinary cement
    (c) presence of excess percentage of gypsum
    (d) none of the above.

181. Portland cement manufactured from pure white chalk and clay but free from iron oxide, is known as
    (a) low heat portland cement
    (b) quick setting cement
    (c) white cement
    (d) none of the above.

182. Strength of cement, with storage
    (a) increases
    (b) decreases
    (c) same
    (d) difficult to tell.

183. Good quality sand is never obtained from the following source
    (a) riverbed
    (b) nala
    (c) sea
    (d) gravel powder.

184. According to Indian standards, at what percentage of moisture content will the weight of timber be specified
    (a) 12%
    (b) 10%
    (c) 8%
    (d) 6%.

185. Plastic asphalt is
    (a) a mixture of cement and asphalt
    (b) a natural asphalt
    (c) a refinery product
    (d) none of the above.

186. Chemical formula for quick lime is
    (a) $CaCO_3$
    (b) $Ca(OH)_2$
    (c) CO_3CO_2
    (d) none of the above.

187. In paints, pigment is responsible for
    (a) glossy face
    (b) durability
    (c) smoothness
    (d) colour.

188. Fibre glass has the characteristic that
    (a) it retains heat longer
    (b) it is shockproof and fire retardant
    (c) it has a higher strength to weight ratio
    (d) all the above.

189. The pigment used in paints for corrosive resistance is
    (a) red lead
    (b) white lead
    (c) gypsum
    (d) ferrous oxide.

190. The pigment commonly used in the manufacture of paints is
    (a) ambers
    (b) iron oxide
    (c) lamp black
    (d) all the above.
191. Lacquer paints are
    (a) generally applied on structural steel
    (b) are more durable compared to enamel paints
    (c) consisting of resin and nitro-cellucose
    (d) all the above.
192. Paints most resistant to fire are
    (a) enamel paints
    (b) aluminium paints
    (c) asbestos paints
    (d) cement paints.
193. Lacquer is
    (a) distemper
    (b) oil paint
    (c) spirit varnish
    (d) none of these.
194. Bullet proof glass is made of thick glass sheet and a sandwiched layer of
    (a) steel
    (b) stainless steel
    (c) high strength plastic
    (d) chromium plate.
195. Dextrine is
    (a) rubber based adhesive
    (b) animal glue
    (c) starch glue
    (d) none of the above.
196. Pig iron made from haemitite ore, free from sulphur, phosphorus and copper, is known as
    (a) Bessemer pig
    (b) grey or foundry pig
    (c) white or forge pig
    (d) mottled pig.
197. In question 196, which one represents the pig iron obtained from the furnace that is properly provided with fuel at very high temperature ?
198. In question 196, which one represents the pig iron obtained from the furnace with insufficient fuel at low temperature ?
199. The type of pig iron used for the manufacture of steel by Bessemer process is
    (a) Bessemer pig
    (b) grey pig
    (c) white pig
    (d) Mottled pig.
200. In question 199, which type of pig iron is used for the manufacture of wrought iron ?
201. In question 199, which is the most unsuitable pig iron for manufacture of light and ornamental castings ?
202. For melting one tonne of cast iron, the requirement of materials is
    (a) 700 m^3 of air
    (b) 20 kg of lime stone
    (c) 100 kg of coke
    (d) all the above.
203. The process involved in the manufacture of wrought iron from pig iron is
    (a) refining
    (b) pudding
    (c) rolling
    (d) all the above.

204. Vanadium steel is generally used for
    (a) railway switches and crossing
    (b) bearing balls
    (c) magnets
    (d) axles and springs.

205. The percentage of chromium and nickel in stainless steel, respectively, are
    (a) 18% and 8%
    (b) 8% and 18%
    (c) 12% and 36%
    (d) 36% and 12%.

206. The percentage of cobalt in high carbon steel to make permanent magnets is
    (a) 15%
    (b) 20%
    (c) 25%
    (d) 35%.

207. The ingredient which makes the stainless steel corrosion resistant is
    (a) carbon
    (b) sulphur
    (c) chromium
    (d) manganese.

208. Rock formed by the process of gradual deposition are known as
    (a) igneous rocks
    (b) metamorphic rocks
    (c) sedimentary rocks
    (d) volcanic rocks.

209. The sub-classification of sedimentary rocks
    (a) volcanic and plutonic
    (b) mechanical, chemical, organic
    (c) intrusive, extrusive
    (d) stratified and unstratified.

210. Sand stone comes under the category of
    (a) sedimentary
    (b) metamorphic
    (c) igneous
    (d) volcanic.

211. Marble stone comes under the category of
    (a) igneous
    (b) metamorphic
    (c) stratified
    (d) foliated.

212. If the rocks are formed due to alteration of original structure under heat and excessive pressure, then they are known as
    (a) igneous
    (b) sedimentary
    (c) volcanic
    (d) metamorphic.

213. Rocks in which the main constituent is silica are known as
    (a) calcareous
    (b) argillaceous
    (c) silicious
    (d) sandy.

214. The classification of Kaolin is
    (a) calcareous
    (b) argillaceous
    (c) silicious
    (d) organic.

215. The tendency of the minerals to split along a certain plane is called
    (a) lusture
    (b) softness
    (c) fractive
    (d) cleavage.

216. Which of the following does not show good cleavage ?
    (a) dolomite (b) calcite
    (c) silica (d) mica.
217. The fracture in asbestos is
    (a) fibrous (b) regular
    (c) irregular (d) none of the above.
218. The common type of stone used for railway ballast is
    (a) basalt or trap granite (b) marble
    (c) slate (d) sand stone.
219. Pick up the odd one from the rest
    (a) gun powder (b) marble powder
    (c) dynamite (d) gun-cotton.
220. Which property of stones is determined using Smith's test?
    (a) hardness (b) specific gravity
    (c) soluble and clayey matter (d) durability.
221. Which of the following impurities is undesirable in the soil used for brick making ?
    (a) kankar (b) alkali
    (c) iron oxide (d) both (a) and (b).
222. To improve the quality of bricks sometimes sand is added to the powdered soil. This process is known as
    (a) mixing (b) blending
    (c) treating (d) none of the above.
223. The shape of the brick gets deformed due to rain water falling on hot brick. This defect is known as
    (a) chuffs (b) bloating
    (c) nodules (d) lamination.
224. Bricks are likely to get discoloured by the formation of white deposit when they contain large proportion of soluble salts. This phenomenon is called
    (a) lamination (b) efflorescence
    (c) bloating (d) underburning.
225. Bloating of bricks is due to
    (a) presence of excess carbonaceous matter in the clay
    (b) presence of gas forming material in the clay
    (c) bad or rapid burning
    (d) all the above.
226. Swelling of bricks is known as
    (a) bulking (b) bladdening
    (c) bloating (d) none of the above.

227. To which category do the chromite bricks belong ?
   (a) ordinary fire bricks
   (b) acid refractory bricks
   (c) basic refractory bricks
   (d) neutral refractory bricks.

228. In question 227, to which category do the bauxite bricks belong ?

229. In question 227, to which category do the dolomite bricks belong ?

230. For what purpose are fire bricks used
   (a) to increase the heat flow
   (b) to decrease the heat flow
   (c) to reflect heat
   (d) to protect the building against lightning.

231. The advantage of adding pozzolona to lime is
   (a) to reduce the shrinkage
   (b) to increase resistance to cracking
   (c) to impart greater strength
   (d) all the above.

232. The type of lime used for the constructions under water is
   (a) hydraulic lime
   (b) fat lime
   (c) quick lime
   (d) pure lime.

233. The ingredient which accounts for the most in cement is
   (a) silica
   (b) iron oxide
   (c) lime
   (d) aluminium.

234. The ingredient which accounts for the least in cement is
   (a) silica
   (b) iron oxide
   (c) lime
   (d) aluminium.

235. The separation of water on the fresh concrete is known as
   (a) segregation
   (b) hydration
   (c) bleeding
   (d) none of the above.

236. The fractured surface of a stone generally indicates its
   (a) grain size
   (b) texture
   (c) toughness
   (d) strength.

237. The structure of the stones obtained from the rocks of igneous origin is generally
   (a) foliated
   (b) stratified
   (c) regular
   (d) unstratified.

238. The fracture of a stone may be
   (a) even
   (b) uneven
   (c) concoidal
   (d) any of the above.

239. The arrangement, size and shape of the constituent mineral grains of a stone gives the information about its
   (a) texture
   (b) toughness
   (c) lusture
   (d) none of the above.

240. If a blasting explosive like gun powder is to be destroyed, it must be
   (a) packed in a container and buried
   (b) burnt in open fire
   (c) thrown into water
   (d) any of the above.

241. On Mohr's scale, a scratch with the aid of a finger nail indicates a hardness of
    (a) 1 (b) 2
    (c) 3 (d) 4.
242. Hard silicious rocks which cannot be scratched by a knife represent a hardness of
    (a) 7 (b) 6
    (c) 5 (d) 4.
243. Which of the following can be used as a preservative for building stones ?
    (a) raw and boiled linseed oil (b) coal tar
    (c) alum soap solution (d) any of the above.
244. A mortar prepared by mixing wood powder or saw dust with the cement or lime mortar is known as
    (a) cinder mortar (b) wood mortar
    (c) light weight mortar (d) gauged mortar.
245. A badly mixed cement concrete may lead to
    (a) cracks (b) bleeding
    (c) segregation (d) honey-combing.
246. The explosive not used for blasting the rocks under water is
    (a) gun powder (b) gun cotton
    (c) dynamite (d) cordite.
247. To which grade do the bricks possessing a compressive strength of not less than 140 kg/cm^2 belong ?
    (a) A (b) AA
    (c) B (d) C.
248. When a bitumen is graded 75/15, the figure 75 represents
    (a) viscosity in centipoise (b) softening point in °C
    (c) fire point in °C (d) flash point in °C.
249. The operation of removal of impurities of clay adhering to iron ore is known as
    (a) calcination (b) purification
    (c) dressing (d) refining.
250. Bitumen paint renders
    (a) protective surface (b) shining surface
    (c) smooth surface (d) hard surface.
251. The purpose of the soundness test is
    (a) to determine the presence of free lime
    (b) to determine the setting time
    (c) to determine the sound proof quality of cement
    (d) to determine the fineness.

252. The time required in hours for the bricks burnt in kilns to cool down, so that they can be unloaded, is
   (a) 4
   (b) 10
   (c) 8
   (d) 12.

253. The specific surface expressed in sq cm/gm of a good portland cement should not be less than
   (a) 1750
   (b) 2000
   (c) 2250
   (d) 2500.

254. The coefficient of linear expansion of concrete is
   (a) less than that of steel
   (b) more than that of steel
   (c) almost same as that of steel
   (d) many times more than that of steel.

255. The defect caused due to over-maturity and unventilated storage during transit is called
   (a) heart shake
   (b) cup shake
   (c) foxiness
   (d) rind gall.

256. The solvent used in cement paints is
   (a) turpentine
   (b) water
   (c) spirit
   (d) kerosene.

257. Distemper is used on
   (a) plastered surface not exposed to weather
   (b) plastered surface exposed to weather
   (c) roof tops
   (d) unplastered brick wall.

258. The best primer used for structural steel work is
   (a) zinc oxide
   (b) red lead
   (c) white lead
   (d) iron oxide.

259. The commonly used cement in making cement paints is
   (a) portland cement
   (b) alumina cement
   (c) white cement
   (d) any of the above.

260. The crushing strength of a stone depends upon its
   (a) texture
   (b) specific gravity
   (c) lusture
   (d) both (a) and (b).

261. Which of the following is a volcanic type of rock ?
   (a) Granite
   (b) Shale
   (c) Slate
   (d) Basalt.

262. Which of the following is the parent rock of marble ?
   (a) Sandstone
   (b) Limestone
   (c) Shale
   (d) Basalt.

263. Which of the following rocks accepts good polishing ?
   (a) Shale
   (b) Laterite
   (c) Marble
   (d) Breccia.

264. Into how many units are all the rocks geologically classified ?
    (a) 3 units
    (b) 4 units
    (c) 5 units
    (d) 6 units.
265. Which of the following rocks is used in cement manufacturing ?
    (a) Granite
    (b) Limestone
    (c) Sandstone
    (d) Quartzite.
266. The rock with poorest strength is
    (a) Shale
    (b) Dolerite
    (c) Pegmatite
    (d) Granite.
267. When the molten magma from the earth's crust comes onto the surface of the earth, it is called
    (a) Larva
    (b) Zava
    (c) Lava
    (d) Pumice.
268. The FROG is provided in the bricks for
    (a) advertisement
    (b) making a good key
    (c) reducing the weight
    (d) improving the crushing strength.
269. Bricks produced by washing the inside surface of the brick mould with water are known as
    (a) sand moulded bricks
    (b) hollow bricks
    (c) reinforced bricks
    (d) slop moulded bricks.
270. Bricks manufactured by clearing the inside surface of the brick mould with sand are known as
    (a) sand moulded bricks
    (b) hollow bricks
    (c) reinforced bricks
    (d) slop moulded bricks.
271. The size of the brick adopted by the Public Work Department (PWD) of Punjab Government is
    (a) 20 cm × 10 cm × 10 cm
    (b) 23 cm × 11 cm × 7 cm
    (c) 23 cm × 10.8 cm × 7 cm
    (d) 19 cm × 9 cm × 9 cm.
272. Machine moulding of bricks can be done by
    (a) Plastic clay method
    (b) Dry pressed clay method
    (c) Both (a) and (b)
    (d) None of these.
273. How many bricks can be moulded per day by a moulder and a helper?
    (a) 1000
    (b) 200
    (c) 2000
    (d) 500.
274. Identify the odd man out
    (a) Bulls Trench kiln
    (b) Hoffman kiln
    (c) Tunnel kiln
    (d) Allahabad kiln.
275. In which part of our country Black Cotton Soil Bricks are entensively used?
    (a) Central Western
    (b) North Eastern
    (c) Southern
    (d) Northern.

276. In lime manufacturing, if the fuel is not allowed to have direct contact with lime, then the kiln is called
    (a) Flare kiln
    (b) Flame kiln
    (c) Hybrid kiln
    (d) Clamp kiln.

277. In lime manufacturing, if the fuel is allowed to have direct contact with lime, then the kiln is called
    (a) Flare kiln
    (b) Flame kiln
    (c) Hybrid kiln
    (d) Clamp kiln.

278. In continuous flame type kiln, the maximum diameter of the kiln is at
    (a) Bottom
    (b) Mid-height
    (c) Top
    (d) Quarter height from top.

279. Lime is used for
    (a) Cementation
    (b) Fertiliser
    (c) Grouting
    (d) Pointing.

280. Oilwell cement is generally used for
    (a) road works
    (b) plastening
    (c) cementing oil wells
    (d) none of the above.

281. The fluxing agent in the manufacturing of white cement is
    (a) Bauxite
    (b) Gypsum
    (c) Borax
    (d) Gryolite.

282. Find the odd man out
    (a) C_3S
    (b) C_2S
    (c) C_3A
    (d) CO_2.

283. The mass of cement present in a standard cement bag is
    (a) 25 kg
    (b) 50 kg
    (c) 75 kg
    (d) 100 kg.

284. In the manufacturing of cement the nodules from the rotary kiln are called
    (a) pallets
    (b) fines
    (c) balls
    (d) clinkers.

285. The glass used for security purposes is
    (a) wired glass
    (b) ground glass
    (c) bullet proof glass
    (d) coloured glass.

286. If a timber is having three annual rings, it's age is about
    (a) 9 years
    (b) 6 years
    (c) 3 years
    (d) 1½ years.

287. As the number of modullary rays increases, generally the strength of the timber
    (a) decreases
    (b) increases
    (c) may increase or may decrease
    (d) difficult to guess.

288. Good timber must emit
    (a) sonorous sound   (b) dull sound
    (c) no sound   (d) noisy sound.
289. Find the odd man out
    (a) Bark   (b) Pith
    (c) Sap   (d) Draxiness.
290. Pick up the bogue compound from the following
    (a) C_3S (b) C_2S
    (c) C_3A (d) all the above.

ANSWERS
Objective Type Questions

1. (d)	2. (c)	3. (d)	4. (c)	5. (a)	6. (b)
7. (d)	8. (b)	9. (a)	10. (c)	11. (c)	12. (b)
13. (a)	14. (c)	15. (c)	16. (d)	17. (a)	18. (b)
19. (c)	20. (d)	21. (a)	22. (c)	23. (d)	24. (b)
25. (c)	26. (c)	27. (d)	28. (c)	29. (b)	30. (d)
31. (d)	32. (c)	33. (b)	34. (a)	35. (b)	36. (c)
37. (a)	38. (c)	39. (d)	40. (d)	41. (d)	42. (b)
43. (a)	44. (b)	45. (b)	46. (c)	47. (b)	48. (a)
49. (b)	50. (c)	51. (b)	52. (d)	53. (c)	54. (c)
55. (a)	56. (a)	57. (b)	58. (c)	59. (d)	60. (a)
61. (b)	62. (b)	63. (a)	64. (b)	65. (c)	66. (c)
67. (a)	68. (a)	69. (b)	70. (b)	71. (c)	72. (c)
73. (b)	74. (c)	75. (a)	76. (d)	77. (c)	78. (b)
79. (c)	80. (d)	81. (b)	82. (d)	83. (a)	84. (d)
85. (c)	86. (a)	87. (d)	88. (a)	89. (b)	90. (d)
91. (c)	92. (c)	93. (b)	94. (a)	95. (a)	96. (d)
97. (c)	98. (b)	99. (a)	100. (d)	101. (d)	102. (a)
103. (a)	104. (c)	105. (b)	106. (a)	107. (d)	108. (c)
109. (a)	110. (d)	111. (c)	112. (b)	113. (a)	114. (c)
115. (b)	116. (d)	117. (a)	118. (a)	119. (b)	120. (d)
121. (a)	122. (c)	123. (d)	124. (c)	125. (b)	126. (c)
127. (b)	128. (d)	129. (c)	130. (c)	131. (d)	132. (a)
133. (c)	134. (c)	135. (a)	136. (d)	137. (c)	138. (d)
139. (a)	140. (b)	141. (c)	142. (d)	143. (c)	144. (a)
145. (c)	146. (a)	147. (a)	148. (d)	149. (b)	150. (a)
151. (c)	152. (a)	153. (b)	154. (a)	155. (c)	156. (a)
157. (b)	158. (a)	159. (a)	160. (a)	161. (c)	162. (c)

163. (b)	164. (d)	165. (a)	166. (a)	167. (a)	168. (b)
169. (b)	170. (a)	171. (c)	172. (c)	173. (d)	174. (b)
175. (c)	176. (a)	177. (b)	178. (b)	179. (d)	180. (a)
181. (c)	182. (b)	183. (c)	184. (a)	185. (a)	186. (a)
187. (d)	188. (d)	189. (a)	190. (d)	191. (d)	192. (c)
193. (c)	194. (c)	195. (a)	196. (a)	197. (b)	198. (c)
199. (a)	200. (c)	201. (d)	202. (d)	203. (b)	204. (d)
205. (a)	206. (d)	207. (c)	208. (c)	209. (b)	210. (a)
211. (b)	212. (d)	213. (c)	214. (b)	215. (d)	216. (c)
217. (a)	218. (a)	219. (b)	220. (c)	221. (d)	222. (b)
223. (a)	224. (b)	225. (d)	226. (c)	227. (d)	228. (c)
229. (c)	230. (b)	231. (d)	232. (a)	233. (c)	234. (b)
235. (c)	236. (b)	237. (d)	238. (d)	239. (a)	240. (c)
241. (b)	242. (a)	243. (d)	244. (c)	245. (d)	246. (a)
247. (b)	248. (b)	249. (c)	250. (a)	251. (a)	252. (d)
253. (c)	254. (c)	255. (c)	256. (b)	257. (a)	258. (b)
259. (c)	260. (d)	261. (d)	262. (b)	263. (c)	264. (a)
265. (b)	266. (a)	267. (c)	268. (b)	269. (d)	270. (a)
271. (c)	272. (c)	273. (a)	274. (d)	275. (a)	276. (a)
277. (b)	278. (b)	279. (a)	280. (c)	281. (d)	282. (d)
283. (b)	284. (d)	285. (c)	286. (c)	287. (b)	288. (a)
289. (d)	290. (d).				

Chapter 12 BUILDING CONSTRUCTION

I. INTRODUCTION

A structure constructed to house any activity of man may be called a building. Basically any building has two parts : sub-structure or foundations and super-structure.

Foundation, which is in direct contact with the ground and located below the ground level, transmits all the loads from super-structure to the supporting soil. Hence, foundation is the most important part of the building. It evenly distributes the loads over large area thereby reducing the load intensity, provides even surface for the structure to rest, gives lateral stability to the structure and ensures safety against undermining and protection against soil movements. Thorough soil investigations are to be done before deciding on the type of foundation and its design.

The investigation methods include : (*i*) Open pit excavations, (*ii*) Borings, (*iii*) Sub-surface soundings and (*iv*) Geophysical methods such as seismic refraction and electrical resistivity.

Open pit excavation is suitable for shallow depths. The other methods are used when the soil is to be explored for deeper depths.

Static penetration and standard penetration methods in which the resistance to penetration is correlated with the soil properties, belong to the category of sub-surface soundings. The record of samples obtained from boring and arranged according to their depth of occurrence is called a bore-log.

When the angle of internal fraction ϕ of cohesionless soil is known the Rankines formula to obtain minimum depth of foundation may be used.

$$D = \frac{q}{\gamma}\left[\frac{1-\sin\phi}{1+\sin\phi}\right]^2$$, where q is the intensity of loading which is less than or equal to the safe bearing capacity of the soil and γ is the unit weight of soil.

If the depth of foundation is less than its width it is termed a shallow foundation (also known as open foundation) otherwise it is known as deep foundation. The structural unit constructed in masonry or concrete under the base of a wall or a column which distributes the load over a large area is called the footing. The term footing is generally associated with shallow foundations.

A footing which provides a continuous longitudinal bearing (as in the case of walls) is known as strip footing.

A footing which supports two columns is called a combined footing and if it supports more than two columns it is called a continuous footing. When two footings are joined by a beam it is called a strap footing.

Grillage and Raft foundations are shallow foundations. A grillage foundation is provided for heavily loaded steel stanchions in soils of poor bearing capacity. The depth of foundation is generally limited to about 1.5 m. The load from stanchion is spread over large area by means of layers (tiers) of steel or timber joists placed at right angles to the next layer. The steel joists are embedded in concrete so that they are kept in position and are protected from corrosion. The distance between the flanges is kept at $1\frac{1}{2}$ times the width of flange or 30 cm whichever is less. However, the minimum spacing is 8 cm so that concrete can be easily poured. Timber grillages are useful in water logged areas where bearing capacity of soil is low and where steel grillages are likely to get corroded.

Raft foundation is shallow foundation which covers the entire area underneath the structure and supports the walls and columns. When the allowable bearing capacity is low, the individual foundations are so large that the gap left between them is small that it will be economical to use raft foundation. Raft foundations are also useful where the settlements are difficult to control.

Machine foundations are to be treated carefully as they are subjected to dynamic loads due to vibrations. The permissible bearing pressure is generally taken as $\frac{1}{2}$ to $\frac{1}{4}$ of that under static loads.

Pile foundations, pier foundations and well foundations (or caissons) are various types of deep foundations. Out of these pile foundation is more widely used in building construction.

A pile which transmits the load coming on to it to a suitable bearing stratum are called end-bearing piles or simply bearing piles. That means they are taken to such depth till they meet bearing stratum.

A pile which transmits the load coming on to it through the friction developed between its surface area and the surrounding soil is called a friction pile.

Compaction piles are those which do not take any load by themselves, but they improve the bearing capacity of the surrounding soil through compaction.

Depending on the material used, the piles may be classified as concrete piles (both precast and *cast-in-situ*), timber piles, steel piles and composite piles (made of concrete and timber or concrete and steel).

Steel piles may be *H*-piles, pipe piles or sheet piles.

Sheet piles are used to reduce the seepage and uplift under hydraulic structure.

The batter piles are used to resist large horizontal or inclined forces.

A simplex pile has a metal shoe. Franki pile has enlarged base of mushroom shape, which gives the effect of spread footing. Under reamed piles will have one or more bulbs on the pile stem and are generally used in expansive soils.

Cantilever sheet piles are very common in the construction of coffer dams.

From scour depth consideration or due to low bearing pressure, if the foundation of bridge has to be taken to more than 5 to 7 m depth, open excavation becomes costly and uneconomical as heavy timbering (method of preventing the caving in of foundations in loose soils by wooden planks and struts) is required. Also the excavated earth is loose when refilled and is susceptible for easy scouring. In all these cases well foundations are the only solution.

When the well foundations is completed, its bottom is plugged with concrete, the well is filled with sand and the top also is plugged and the pier is then constructed over it. In circular well foundations the steining of the well (*i.e.*, the thickness of the well) is about $\frac{D}{4}$, where D is the diameter of the well. In sinking the well, care must be taken to see that it is truly vertical. The tilt should not exceed 1 in 20.

Superstructure of a building consists of masonry walls which suport the roof and they contain doors and window meant for passage and ventilation and aeration. A building unit bonded with mortar may be called masonry. Masonry may be called as brick masonry, stone masonry and composite masonry depending on whether they are made of bricks or stones or a mixture of bricks stones and concrete blocks respectively.

A horizontal layer of masonry unit is called a course. A brick or stone placed in position such that its length is parallel to the face of the wall is called stretcher and when it is perpendicular to the face of the wall it is called header. In the case of stone masonry stretcher is also known as through stone. A course showing only stretchers on the exposed face is called stretcher course and similarly a course showing only headers is called header course. The lower face of brick or stone is called bed. In the case of stones, they must be placed in position such that their natural bed is always perpendicular to the direction of pressure. The overlapping of stones or bricks in masonry such that no continuous vertical joints are formed and they are tied together is known as bond. The exterior corner or angle of a wall is known as quoin. The units forming the corner are called quoins, *i.e.*, quoin stone or quoin brick. If the length of quoin is parallel to the wall it is called a quoin stretcher and when it is perpendicular it is called quoin header. The surface of the wall exposed to weather is called face and which is not exposed is called back. The portion of the wall between facing and backing is called hearting.

If a brick is cut into two halves longitudinally two queen closers are obtained. The portion of the brick in which the width at one end is full and the width at the other end is half of that in full brick is called king closer. Bevelled closer is a brick in which the width tapers from full width at one end to half width at the other end. In a mitred closer one long face is full and the other long face is small in length. The portion of the brick cut across the width is called a bat. Depending on its size it may be called half-bat or three-fourth bat.

The bottom surface of door or a window opening is a sill. The horizontal member spanning the window or door opening in a wall and supporting the masonry above the opening is called a lintel. The horizontal projection at the base of the wall is called plinth. Jambs are the vertical sides of a finished opening for the door or window, and the exposed vertical surfaces left out after the door and window has been fixed in position are called reveals.

A projected stone meant to support joist or truss is called corbel. A projection of ornamental course near the top of the wall or near the junction of wall and roof is called cornice. The covering over an exposed top of a wall is known as coping. The sloping face provided on the surface of sills, cornices and copings is called weathering. The groove made on the underside a projecting element such as sill, cornice etc., so that rain water can be drained clear of the wall is known as throating.

The bricks left projecting in alternate courses to provided bed for the future extension of the wall is called toothing. The termination of a wall in stepped fashion is known as racking back.

The stone masonry in which the stones used are either not dressed at all or very roughly dressed is called a random rubble masonry. The random rubble masonry may be coursed or uncoursed. The stone masonry in which the stones are having straight bed and sides is called square rubble masonry. In ashlar masonry accurately dressed stones in courses and in cement mortar are used.

First class bricks should not absorb water more than one-sixth of their weight when immersed for one hour whereas the second class bricks should not absorb more than one-fourth of their weight.

An imaginary vertical line which includes the vertical joints in brick masonry is called perpend. If the brick is rounded at one or two edges it is called bull nose brick.

A bond in which all the bricks are laid as stretcher is known as a stretcher bond. This is possible only in half-brick walls. A bond in which all the bricks are laid as headers is called a header bond. This is very weak bond which can be adopted only for one brick walls. A bond which consists of alternate courses of stretchers and headers is called the English bond. English bond is the strongest bond. If each course consists of alternate headers and stretchers such a bond is called Flemish bond. Though Flemish bond is weaker than English bond, it gives more pleasing appearance. An English bond in which queen closers are placed next to quoin headers and a header is introduced next to quoin stretcher in every alternate stretcher course is called English cross bond. This has an improved appearance compared to ordinary English bond. An English bond in which every stretcher course starts with a three-fourth bat at the quoin, and in every alternate stretcher course a header is placed next to the three-fourth bat provided at the quoin is called a dutch bond. In this bond the corners of the wall are very strong. A bond in which stretcher bricks are laid on edges and headers are placed on bed in alternate layers forming a continuous cavity is called soldier's (or Silverlock's) bond. A bond in which the bonding bricks are kept at inclination is called a raking bond. A bond in which a header course is provided only after a few stretcher courses is called a garden wall bond.

A junction at which two walls meet at 90° is called a Tee-junction, and at an angle other than 90° it is called squint junction. When two internal walls intersect a cross junction is formed.

The brick masonry providing one or more steps outside a door in the external wall is called thresholds. Brickwork built within wooden frames is called brick nogging.

In cavity walls, two walls separated by a small gap of 4 to 10 cms are connected together by metal pins or bonding bricks at suitable intervals. The inner and outer skin walls in a cavity wall should have a minimun thickness of 10 cm each. Cavity walls prevent external moisture into the building, provide better insulation against heat and sound and they are cheap. Also their self-weight is less compared to solid walls. They are also called hollow walls. A wall of considerable less thickness which divides the space into rooms may be called a partition wall. A load bearing partition wall is called an inner wall. A brick wall built up within the frame work of wooden members is called the brick nogging. The vertical members of the frame are called studs, while the horizontals are called nogging pieces. There are also clay block partitions, glass partitions, timber partitions asbestos sheet partitions etc.

Different types of flooring are (*i*) flag stone flooring, (*ii*) concrete flooring, (*iii*) terrazo flooring, (*iv*) mosaic flooring, (*v*) tile flooring, (*vi*) cork flooring, (*vii*) glass flooring, (*viii*) granolithic flooring, (*ix*) wooden or timber flooring, (*x*) brick flooring, (*xi*) linoleum flooring. The choice of flooring depends on economy appearance of flooring required, special purpose of the flooring, if any, like sound and heat considerations and whether it is ground floor or upper floor etc.

In flag stone flooring laminated stones of rectangular or square shape are laid over a concrete base. It is not very elegant but durable if properly done. In brick flooring a layer of well burnt uniform shaped bricks are laid over concrete base. It is cheap and can be done fast but it suffers from the water absorbing quality of bricks. Concrete flooring is very common in residential and other types of buildings. In granolithic flooring granite chips of size of 13 mm or less are used as coarse aggregate in concrete. In terrazzo flooring top surface is made of concrete consisting of marble chips and cement in rich proportion of 1 : 1.25 to 1 : 2. The marble chips may have size of 3 mm to 6 mm. When the concrete is set the chips are exposed by grinding to give special appearance. In mosaic flooring, the mosaic pieces (mainly of broken glazed tiles or marble stone) are inserted into the top layer of flooring while it is still wet. In bathrooms and kitchens of residential buildings in hospitals and sanitoriums and in temples where cleanliness is utmost important marble flooring is preferred. Timber flooring is used for dancing halls, sanitoriums etc. Rubber flooring and cork flooring make the room noiseless and hence they are used in office or public buildings, theatres, galleries, broadcasting stations etc. Linoleum is available in rolls and flooring is done by spreading these rolls. The rolls may be available in 2 to 4 m widths and 2 to 6 mm thickness. Glass flooring is a special purpose flooring which is used when lighting is to be transmitted from upper floor to lower floor.

For upper floors, the common types of flooring are R.C.C. flooring, jack arch flooring, timber flooring, steel joist and stone or precast concrete flooring etc. In R.C.C. flooring sometimes flat slabs are used without any beams to gain more head room.

The horizontal member spanning an opening such as door or window in a masonry wall and which support the load above the opening is called a lintel. In some cases, an arch may be used for the same purpose. Based on materials used lintels are classified into timber, stone, brick, steel and R.C.C. lintels.

When the wall length on either side of the opening is more than the effective span L, and the height of the wall above the opening is more than $L \sin 60°$ because of the arch action developed over the opening the load coming on the lintel is equal to weight of an equilateral triangle of length L and not the weight of the rectangular portion above the opening. In the other cases where the length on either side is less than $\dfrac{L}{2}$ and the height of the wall above the opening is also less than $L \sin 60°$, the load will be more than triangle.

In the case of an arch, the inner curve is called intrados and the outer curve is called extrados. The inner surface may be called soffit or intrados. The wedge-shaped bricks or stones forming the arch are called voussoirs. The highest point of arch is called crown and the voussoir at the crown is known as key. The curved triangular portion between the extrados and the horizontal line through the arch is called spandril. The inclined surface on the abutment which receives the arch and from which the arch springs is called skewback. The points at which the curve of the arch springs are called the springings and the horizontal line joining the springings is known as the springing line. The first voussoir at springing level immediately adjacent to skewback is called springer. The clear vertical distance between springing line and the highest point on the intrados is called the rise. Horizontal distance between supports is the span. The perpendicular distance between the intrados and extrados is the depth of the arch.

In a flat arch there is no rise and skew backs make an angle of 60° with the horizontal. A circular arch in which the centre lies below the springing line is a segmental arch, and if the centre lies on the springing line it is semi-circular, and if the centre is above it is a horse-shoe arch. The arch made of two arcs of a circle meeting at apex point is called a pointed arch or gothic arch. In this arch if the triangle formed by the springing points and the apex is an isosceles triangle, it is called a lancet arch. The arch in which the depth is more at crown than at springings is called venetian arch. The venetian arch with intrados as a circle is called the florentine arch. The arch constructed either on a flat arch or on a wooden lintel is known as the relieving arch. A semi-circular arch with two vertical portions at the springings is called the stilted arch.

The lower half of the arch between the crown and the skew-back is known as haunch. A continuous row of arches with intermediate supports as piers is called an arcade.

An arrangement of steps provided for vertical transportation between the floors of a building may be called a stair. The alternative to stair is the lift which is usually thought of when the number of floors in the building exceeds 3. Lifts may be operated either manually (which is very rare now-a-days) or electrically. Lifts are very fast and comfortable and the persons using lift need not exert themselves. Even in buildings provided with lift, stair is a must as an emergency measure. For small height ramps may be used in place of stairs. Ladder is a very crude substitute for stairs. Escalators are used in special circumstances (airports, shopping malls etc.).

In a stair, the upper horizontal portion of a step upon which the foot is placed while ascending and descending is called a tread. The vertical portion of the step between two successive treads is called the riser. Normally all the steps in a stair (or at least in a flight) will have same tread and same rise. A continuous series of steps between landings is called the flight. The level platform at the bottom or end of a flight is known as a landing, which facilitates change in the direction of stair, breaks the monotony in straight stair and provides respite for the users.

The vertical distance between two successive tread faces is known as the rise, while the horizontal distance between two successive riser faces is called the going. For comfortable ascent and descent, the sum of going and rise expressed in cms should be between 40 to 45 while their product should be between 400 to 450. The rounded off projected part of the tread beyond the face of the riser is called nosing. Scotia is the moulding provided under the nosing to improve the elevation of the step and to provide strength to nosing. The sloping member which supports the stair is called string or stringer. The imaginary line tangential to nosings and parallel to the stringers is called line of nosing. The angle made by the line of nosing with the horizontal is the pitch or slope of the stair. The pitch should be limited to 30° to 45°. Newel post is a vertical member at ends of flights connecting the string and hand rail. A vertical support to the hand rail is known as Baluster. A row of balusters is called balustrade. The total horizontal length of stair in plan including landings is called the Run.

In domestic buildings, the width of stair is about 90 cm. In public buildings it may be 150 to 180 cms. From view point of comfort, the number of steps in a flight should not exceed 12 and should not be less than 3.

An ordinary step rectangular in plan is called a flier. A group of tapering steps radiating from same point are called winders. Tapering steps which do not radiate from same point are called dancing steps are balancing steps.

Stairs in which there is only one turn either to the left or right are called quarter turn stairs, which change the direction by 180° are called half turn, and which change direction three times are called three-quarter turn stairs. Dog legged and open well stairs come under half-turn stairs. In straight stairs there is no change in direction and they run straight between two floors. Bifurcated stair has a wider flight at the bottom which bifurcates into two narrower flights one turning to the left and the other turning to the right. They are generally used in public buildings at their entrance halls. Helical, circular and spiral stairs come under the category of continuous stairs. Spiral stair will have all winders.

Roof is the uppermost portion of the building. It gives protection to the building from sun, rain and wind. The roof may be flat or inclined. The roofs with inclined surfaces are also known as pitched roofs. To obtain architectual effects some times the roof surfaces are given special shapes such as cylindrical and parabolic shells or hyperbolic paraboloids and hyperboloids of revolution. In such cases they are called curved roofs.

Pitched roofs are further classified into lean-to, Hip, Gable, Deck, Mansard and Gambrel roofs. In lean-to roof there is only one sloping side. It is used for small spans such as in verandahs. A roof which slopes in two directions is called gable roof. At the end face of such roofs a vertical triangle is formed and the wall covering this triangle is called gable wall. A roof sloping in two directions with a break in each slope is called gambrel roof. A roof formed by four sloping surfaces is called hip roof. At the end faces of hip roof sloping triangles are formed. A roof which slopes in four directions and with a break in each slope is called Mansard roof. A hip roof with flat top is known as deck roof.

The pitched roofs are generally supported by either wooden or steel trusses. The top line of the sloping roof is called the ridge. Hip is the intersection of two sloping surfaces where the exterior angle is more than 180° and if it is less than 180°, it is called valley. Ridge piece is the horizontal wooden member along the ridge which supports the common rafters. The rafters along the hip and valley are respectively called hip rafter and valley rafter. Common rafters, principal rafters, purlins and boardings form the required structural frame to support the roof material.

Roofs with only common rafters are single roofs and those with rafters and purlins are called double roof. Lean to roof, couple roof, couple close roofs and collar tie roof come under the single roofs. These roofs are used when the span is limited to 5 m. Lean-to roof for spans of less than 2.5 m, couple roof for less than 3.6 m, and collar tie roof up to 5 m. When the span exceeds 5 m, trusses are used. King post truss has only one vertical members called king post and is used for spans between 5 to 8 m. The spacing of trusses along the wall is generally limited to 3 m. In queen post truss, there will be two vertical posts called queen posts and is employed for spans of 8 to 12 m. When the span exceeds 12 m a combination of king and queen post trusses is used.

When span is more than about 10 m, the timber trusses prove to be heavy and uneconomical. In such cases, steel trusses made of rolled steel structural sections such as angles, channels, Tees and plates are used. As the timber has become prohibitively costly, in India steel trusses are being used for all spans. Fink truss, Howe truss, North-light truss etc., are some common steel trusses used for industrial buildings.

Roofing material may be of thatched covering, wood shingles, tiles, galvanised corrugated sheets, asbestos cement sheets etc. Again due to economy, versatility and other properties like light weight, water proof, fire resistant, and availability in fairly big sizes asbestos cement sheets are very popular compared to other materials.

Roofs which are perfectly horizontal or with a slope less than 10° are called flat roofs. They need terracing so that they are protected from the adverse effects of heat, rain snow in cold climates, etc. The flat roofs are normally given about 7.5 to 10 cm thick brick bat lime concrete over which two or three layers of flat tiles are laid. Just below the brickbat concrete the damp proof course may be inserted.

The joints in carpentry works are classified into lengthening, widening, bearing, framing, angle or corner and oblique shouldered joints.

Lapped joint, fished joint, scarfed or spliced joint and tabled joint are the lengthening joints and last one is used when the member is subjected to both compression and tension.

Dove tailed joint may be used for widening and bearing joints. Rebate joint and tongued and grooved joints are very common widening joints.

In mortoise and tenon joint a projected piece of one member (known as tongue or tenon) fits into a hole (known as mortoise) of the other member. Bridle joint is generally used in wooden trusses at juctions of struts and ties. In oblique tenon joint the tenon of the inclined member is oblique.

Nails, screws, bolts and nuts, dogs etc., are the common fastenings used by a carpenters. A screw with square or hexagonal head is called coach screw. A U-shaped wrought irons fastening with pointed ends which is used to connect the members is called dog.

Bead plane, jack plane and rebate plane are the planing tools which are used to plane the surfaces. Gimlet, auger are used to make holes. Saw and chisel are used for cutting wood and shaping joints. Hammer, mallet, screw driver etc., are used for driving nails and screws.

Doors and windows are essential parts of building and indispensable. They consist of a frame and shutters which can be swung around a hinge. According to I.S. notation, the doors are designated as 10 DS 20, 12 DT 21 etc. For example, 10 DS 20 means a single shuttered door of width 100 cm and height 200 cm. Similarly 12 DT 21 means a double shuttered door of width 120 cm and height 210 cm. The size of the door is measured outer to outer of the frame. Doors may be made of steel or timber.

In a door or window frame, the top horizontal member is called the head and the bottom one is called the sill. The projections of head and sill beyond two outer vertical edge of the frame are called horns. The frames are provided with two or three hold fasts on the vertical members so that the frame is held in proper position without any warping or bends.

In a shutter, the vertical outer member is called the style and the horizontal members are called rails (top, bottom and lock rails). The shutter may be panelled or glazed.

The horizontal member which sub-divides the frame horizontally is called the transome while the vertical member which sub-divides the frame vertically is called mullion. A recess made inside the frame to receive the shutter is known as rebate.

The various types of doors are battened-ledged-braced-framed doors, panelled doors, glazed or sash doors, flush doors, louvered doors, wire-gauged doors, revolving doors, sliding doors, swing doors, collapsible doors, rolling shutter doors, etc.

The various types of windows are : fixed, pivoted, double hung, sliding, casement, glazed, louvered, bay, corner, dormer, gable, lantern, sky light, etc. A small window fixed at a greater height than window is called ventilator. Sometimes a window and ventilator may be combined in the same frame with necessary transome and mullion.

A thin coat of mortar applied to wall surfaces may be called plastering. Plastering gives protection to the wall against rain and other atmospheric agencies, gives decorative effect to the walls, conceals the inferior material and workmanship, gives protection against vermin, and eliminates the gathering of dirt and dust by the walls as it provides smooth surface. It also provides a plane surface for painting, colour washing or white-washing. Plastering may be done either with cement mortar or lime mortar or lime-cement mortar.

The surface to which the first coat of plastering is to be applied is called the background. If the coating is done in three coats, the first coat is known as rendering coat, the second as the floating coat and the third is called the finishing or setting coat.

Development of hair cracks in an irregular pattern of a plastered surface is known as crazing. Development of local swellings is termed the blisterings. Patches of plaster not adhering to the previous coat is called the flaking. Dislodgement of plaster from the background surface is known as peeling.

Raking the joints in masonry to a depth of about 2 cms and filling them with rich mortar is known as pointing. The various types of pointing are flush, struck, recessed, V pointing etc.

The dampness in buildings may be caused due to movement of moisture from ground into the walls due to capillary action, rain from uncovered tops of walls, rain beating on external walls, poor drainage, defective construction, imperfect roof slope, etc.

Dampness in buildings has many bad effects and hence buildings are made damp proof D.P.C. stands for damp proof course. There are many ways of damp proofing. Hot bitumen, mastic asphalt, metal sheets and plastic sheets are some of the materials used for this purpose.

Cement concrete is the most widely used non-homogeneous material in building construction. It is made of cement, fine aggregate and coarse aggregate mixed in a predetermined proportion. To take care of its weakness in resisting tension, it is reinforced with steel in tension zones. Then, it is called reinforced cement concrete abbreviated as R.C.C. The strength of concrete depends on the proportions of constituent materials. For example, M 15 grade concrete (with 1 : 2 : 4 mix) has a 28 days characteristic strength of 15 N/mm^2.

The removal of impure air from a building by replacing it with fresh air from outside may be called ventilation. Natural ventilation of buildings is provided by keeping sufficient number of windows, doors and ventilators. Mechanical or artificial ventilation is done by exhaust fans, fans, air coolers and air-conditioners.

Air-conditioning is the process of treating air so as to control simultaneously its temperature, humidity, purity and distribution to meet the requirements of the conditioned space. The temperature range within which the comfort is obtained for the majority of the people is called comfortable zone. The comfort also depends on humidity and air velocity. Effective temperature comfortable zone varies from 20 to 23°C in summer and 18 to 22°C in winter. For Indian conditions the comfortable zone varies from 25°C with 60% relative humidity to 30°C with 45% relative humidity.

Acoustics is the science which deals with various aspects of sound. The sound which produces displeasing effect to the ear is called noise. The intensity of sound is measured in decibels (db). The persistence of sound even after the source of sound has ceased is called reverberation. An echo is produced when the reflected sound wave reaches the ear just when the original sound has already been heard. The points where the reflected sound waves are concentrated are called sound foci. Dead spots are those where the sound intensity is so low that it is insufficient for hearing. Using good sound absorbing materials may eliminate echoing and proper geometrical design is necessary to avoid sound foci and dead spots.

II. OBJECTIVE TYPE QUESTIONS

1. Reconnaissance survey means
   (a) conducting the surveys to get the plan of the site ☐
   (b) conducting the surveys to get the contours of site ☐
   (c) visual inspection of the site and study of the topographical features ☐
   (d) none of the above. ☐

2. Soil investigation at a site are carried out to obtain the information regarding
   (a) the engineering and physical properties of the soils ☐
   (b) the details of the soil profile to a considerable depth ☐
   (c) the depth to ground water-table and its seasonal fluctuations ☐
   (d) all the above. ☐

3. The investigation method suitable for shallow depths is
   (a) borings ☐   (b) open excavation ☐
   (c) geophysical ☐   (d) none of the above. ☐

4. The soil investigation method using the penetrometers belongs to the category of
   (a) open excavation ☐   (b) borings ☐
   (c) sub-surface soundings ☐   (d) geophysical. ☐

5. The number of blows recorded in a standard penetration test is for a penetration of
   (a) 10 cm ☐   (b) 20 cm ☐
   (c) 30 cm ☐   (d) 40 cm. ☐

6. In a static penetrometer test the angle of the penetrating cone is
   (a) 30°
   (b) 45°
   (c) 60°
   (d) 75°.
7. The weight and fall used in standard penetrometer test to drive the cone are
   (a) 50 kg and 50 cm
   (b) 60 kg and 60 cm
   (c) 70 kg and 70 cm
   (d) 65 kg and 75 cm.
8. In the case of soil investigation by boring, the record of soil samples arranged according to the depth where they are found is known as
   (a) bore-log
   (b) bore-sounding
   (c) bore-map
   (d) none of the above.
9. The plate loading test gives
   (a) the ultimate bearing capacity of the soil
   (b) safe bearing capacity of the soil
   (c) the depth of underlying rock
   (d) none of the above.
10. The relationship between the ultimate bearing capacity q_f, the net ultimate bearing capacity q_{nf} and the depth of footing D is (γ being the unit weight of the soil)
    (a) $q_{nf} = q_f + D\gamma$
    (b) $q_f = q_{nf} + \gamma D$
    (c) $q_{nf} = q_f - D\gamma$
    (d) $q_f = q_{nf} - \gamma D$.
11. According to Rankine's theory, the minimum depth of foundation is given by
    (a) $D = \dfrac{q_s}{\gamma}\left(\dfrac{1-\sin\phi}{1+\sin\phi}\right)^2$
    (b) $D = \dfrac{q_s}{\gamma}\left(\dfrac{1+\sin\phi}{1-\sin\phi}\right)^2$
    (c) $D = \dfrac{\gamma}{q_s}\left(\dfrac{1-\sin\phi}{1+\sin\phi}\right)^2$
    (d) $D = \dfrac{\gamma}{q_s}\left(\dfrac{1+\sin\phi}{1-\sin\phi}\right)^2$.
12. The safe bearing capacity of the soil can be improved by
    (a) increasing the depth of foundation
    (b) grouting the soil
    (c) draining the soil if water table is very near the base of the footing
    (d) all the above.
13. The minimum depth of foundation below the ground level in any soil is
    (a) 120 cm
    (b) 100 cm
    (c) 80 cm
    (d) 60 cm.
14. The method of protecting the foundations in loose soils from caving in, is called
    (a) timbering
    (b) shoring
    (c) both (a) and (b)
    (d) none of the above.
15. Osmosis is
    (a) an electrical process where fine grained cohesive soils are drained using electricity
    (b) a method of chemical stabilisation
    (c) a method to determine the depth to water table
    (d) a method of shoring.

16. The approximate bearing capacity of a sound rock without any defects may be taken as
    (a) 5 to 10 t/m²   (b) 10 to 15 t/m²
    (c) 100 to 200 t/m²   (d) 300 to 350 t/m².
17. Which of the following has the lowest bearing capacity
    (a) fine sand loose and dry   (b) moist clay
    (c) coarse sand loose and dry   (d) coarse gravel.
18. In the problem 17 which has the highest bearing capacity.
19. The type of foundation recommended for heavily loaded steel stanchions in soils with poor bearing capacity
    (a) grillage foundation   (b) raft foundation
    (c) pile foundation   (d) well foundation.
20. In problem 19, which is the type of foundation recommended for buildings with heavy loads on soils with poor bearing capacity and in which it is difficult to control the settlement.
21. In problem 19, which foundation is recommended for bridges when the depth of foundation is large.
22. Well foundation is also known as
    (a) caisson   (b) swage pile
    (c) under ream   (d) none of the above.
23. The most commonly used deep foundation in buildings
    (a) well foundation   (b) pile foundation
    (c) raft foundation   (d) grillage foundation.
24. If the load coming on to a pile is transferred to a suitable bearing stratum at its bottom end, then it is called
    (a) sheet pile   (b) friction pile
    (c) bearing pile   (d) compaction pile.
25. If the load coming on to a pile is resisted by the friction developed between the surface area of the pile and the surrounding material, then it is called
    (a) sheet pile   (b) friction pile
    (c) bearing pile   (d) compaction pile.
26. A pile, which by itself does not carry any load but improves the bearing capacity of the soil is called
    (a) sheet pile   (b) friction pile
    (c) bearing pile   (d) compaction pile.
27. A pile commonly used to reduce seepage and control the uplift pressures under hydraulic structures is known as
    (a) sheet pile   (b) friction pile
    (c) bearing pile   (d) compaction pile.
28. A pile which has an enlarged shape at the base is called
    (a) vibro pile   (b) franki pile
    (c) under reamed pile (d) none of the above.

29. A *cast-in-situ* concrete pile with one or more enlarged bulbs on its stem is called
    (a) vibro pile
    (b) franki pile
    (c) under reamed pile
    (d) simplex pile.

30. The distance between the flanges of the beams in any tier of a grillage foundation should be equal to
    (a) half the flange width
    (b) flange width
    (c) one and a half times the flange width or 30 cm whichever is less
    (d) minimum of 60 cm.

31. The maximum permissible differential settlement in the case of foundations on sandy soils is generally about
    (a) 5 mm
    (b) 10 mm
    (c) 20 mm
    (d) 25 mm.

32. The maximum permissible differential settlement in the case of foundations on clayee soils is generally about
    (a) 10 mm
    (b) 20 mm
    (c) 40 mm
    (d) 100 mm.

33. The maximum permissible total settlement in the case of foundations on clayee soils is generally about
    (a) 50 mm
    (b) 100 mm
    (c) 150 mm
    (d) 200 mm.

34. The maximum permissible total settlement in the case of raft foundations on sandy soils is generally about
    (a) 25 mm
    (b) 45 mm
    (c) 65 mm
    (d) 85 mm.

35. Uniform distribution of loads from walls on to piles is achieved by connecting the group of piles supporting the structure
    (a) by tie rods at the bottom of the piles
    (b) by tie rods at the top of the piles
    (c) by a concrete beam
    (d) isolating the piles without any connection.

36. The offsets in the stepped strip footing in brick masonry is
    (a) 5 cm
    (b) 10 cm
    (c) 15 cm
    (d) 20 cm.

37. The maximum permissible eccentricity of load on a rectangular foundation with width B is equal to
    (a) $\dfrac{B}{3}$
    (b) $\dfrac{B}{6}$
    (c) $\dfrac{B}{2}$
    (d) $\dfrac{B}{4}$.

38. The safety of the structure can be disturbed by
    (a) lowering or rising of water table
    (b) excavations in the immediate vicinity of the structure
    (c) mining or tunnelling operations in the neighbourhood
    (d) any of the above.
39. A foundation could fail due to
    (a) unequal settlement of foundations
    (b) unequal distribution of weight of the structure
    (c) horizontal movement of earth adjoining the structure
    (d) any of the above.
40. The advantage of timber piles is that they are
    (a) flexible and light
    (b) stronger and durable
    (c) free from weathering effects
    (d) any of the above.
41. The safe bearing capacity under dynamic loads such as in machine foundations is taken to be
    (a) more than that for static loads
    (b) same as that for static loads
    (c) less than that for static loads
    (d) depends on the weight of machine.
42. To improve the bearing capacity of the foundation, the cement grouting may be used if the foundation consists of
    (a) clay
    (b) compacted sand
    (c) gravel
    (d) rock.
43. If the soil is water logged, its bearing capacity may be improved by
    (a) compaction
    (b) grouting
    (c) drainage
    (d) chemical treatment.
44. When two or more individual column footings are joined by a beam, it is called
    (a) strip footing
    (b) step footing
    (c) combined footing
    (d) strap footing.
45. A footing which supports two columns is known as
    (a) continuous footing
    (b) combined footing
    (c) strip footing
    (d) step footing.
46. If footing supports more than two columns, it is called
    (a) continuous footing
    (b) combined footing
    (c) strip footing
    (d) strap footing.
47. The safe bearing capacity of medium clay which can be readily indented with a thumb nail is taken as
    (a) 10 t/m^2
    (b) 15 t/m^2
    (c) 25 t/m^2
    (d) 45 t/m^2.

48. The thickness of steining in well foundations is generally about (where D is the diameter of the well)

    (a) $\dfrac{D}{2}$ (b) $\dfrac{D}{3}$

    (c) $\dfrac{D}{4}$ (d) $\dfrac{3}{4}D$.

49. The tilt in the well foundations should not exceed

    (a) $\dfrac{1}{150}$ (b) $\dfrac{1}{200}$

    (c) $\dfrac{1}{100}$ (d) $\dfrac{1}{50}$.

50. The piles used in the construction of coffer dams

    (a) bearing piles   (b) sheet piles

    (c) friction piles   (d) compaction piles.

51. The pile used to resist large horizontal or inclined force is

    (a) batten pile   (b) bearing pile

    (c) anchor pile   (d) tension pile.

52. The exterior corner of a wall is known as

    (a) perpend   (b) jamb

    (c) quoin   (d) frog.

53. The surface of the wall exposed to weather is called

    (a) the face   (b) the back

    (c) the side   (d) perpend.

54. In the above problem which is the surface of the wall not exposed to weather ?

55. The portion of the wall between the facing and backing of the wall is called

    (a) jamb   (b) frog

    (c) throating   (d) hearting.

56. The portion of the brick obtained by cutting into two portions longitudinally is called

    (a) bat   (b) king closer

    (c) queen closer   (d) mitred closer.

57. The portion of the brick obtained by cutting it in transverse direction is called

    (a) bat   (b) king closer

    (c) queen closer   (d) bevelled closer.

58. The portion of a brick which is cut at one corner such that at one end its width is half of that of a full brick is called

    (a) king closer   (b) queen closer

    (c) bevelled closer   (d) mitred closer.

59. The brick which is cut such that at one end half the width is maintained and at the other end full width is maintained is called
    (a) king closer
    (b) queen closer
    (c) mitred closer
    (d) bevelled closer.

60. The brick which is cut such that one longer face is full and the other longer face is small in length is called
    (a) king closer
    (b) queen closer
    (c) mitred closer
    (d) bevelled closer.

61. The overlapping of bricks or stones in masonry is called
    (a) frog
    (b) joint
    (c) jamb
    (d) bond.

62. A groove provided on the underside of projecting elements such as cornices, copings such that rain water can be discharged clear of the wall surface is called
    (a) corbel
    (b) throating
    (c) hearting
    (d) weathering.

63. A brick or stone put in position in the masonry such that its length is perpendicular to the face of the wall is known as
    (a) header
    (b) stretcher
    (c) closer
    (d) bull nose.

64. A brick or stone put in position in the masonry such that its length is parallel to the face of the wall is known as
    (a) header
    (b) stretcher
    (c) closer
    (d) bull nose.

65. In random rubble masonry
    (a) the stones are neatly dressed and laid in courses
    (b) the stones are neatly dressed but not built to courses
    (c) the stones are either not dressed or roughly dressed and may be coursed or uncoursed
    (d) the stones are not dressed and are laid without mortar.

66. In the above problem which is ashlar masonry.

67. The safe compressive strength of ashlar masonry of granite stone in 1 : 6 cement mortar may be taken as
    (a) 100 t/m^2
    (b) 130 t/m^2
    (c) 160 t/m^2
    (d) 190 t/m^2.

68. The tool used by the masons to check the verticality of walls is
    (a) square
    (b) spirit level
    (c) nicker
    (d) plumb bob.

69. The vertical sides of openings for doors and window are called
    (a) jambs
    (b) reveals
    (c) buttresses
    (d) pilasters.

70. The horizontal member spanning the window and door openings in the masonry walls is called
    (a) sill
    (b) bed
    (c) lintel
    (d) arch.

71. The bricks arranged projecting in alternate courses for the purpose of bonding future masonry work is called
    (a) stoolings
    (b) lacing
    (c) throating
    (d) toothing.

72. A projected stone in masonry wall provided to serve as a support to joist or truss is known as
    (a) cornice
    (b) corbel
    (c) coping
    (d) cramp.

73. A projected ornamental course near the top of the wall or near the junction of the wall and ceiling is known as
    (a) cornice
    (b) corbel
    (c) coping
    (d) cramp.

74. The covering on the exposed top of a wall is called
    (a) cornice
    (b) corbel
    (c) coping
    (d) cramp.

75. A low height wall constructed along the edge of the roof is called
    (a) gable
    (b) plinths
    (c) string course
    (d) parapet.

76. A brick which has one or two rounded edges and used to obtain rounded corners in the wall is called
    (a) curved brick
    (b) bull nose brick
    (c) rounded brick
    (d) circular brick.

77. Stretcher bond is possible only in
    (a) half brick wall
    (b) one brick wall
    (c) one and half brick wall
    (d) wall of any thickness.

78. The bond which consists of alternate courses of headers and stretchers is known as
    (a) English bond
    (b) Flemish bond
    (c) Raking bond
    (d) Dutch bond.

79. The bond in which each course consists of alternate stretchers and headers is known as
    (a) English bond
    (b) Flemish bond
    (c) Raking bond
    (d) Dutch bond.

80. The bond in which the bonding bricks are at an inclination to the direction of the wall is known as
    (a) English bond
    (b) Flemish bond
    (c) Raking bond
    (d) Dutch bond.

81. The English bond in which the corners of the wall are strengthened by starting every stretcher course with a three-fourth bat at the quoin and a header is placed next to the three-fourth bat at the quoin in alternate stretcher course is called
    (a) English bond   (b) Flemish bond
    (c) Raking bond   (d) Dutch bond.

82. The English bond in which queen closers are placed next to quoin headers and a header is introduced next to the quoin stretcher in every alternate stretcher course
    (a) Raking bond   (b) Dutch bond
    (c) English cross bond   (d) Facing bond.

83. A bond in which a header course is provided only after few stretcher courses is known as
    (a) Dutch bond   (b) English cross bond
    (c) Facing bond   (d) Garden wall bond.

84. A junction at which two walls meet at an angle other than 90° is called
    (a) oblique junction   (b) squint junction
    (c) irregular junction   (d) obtuse junction.

85. A tool used by masons both for lifting and spreading mortar and also to cut the bricks is
    (a) trowel   (b) scutch
    (c) bolster   (d) brick hammer.

86. The steps provided in the form of brick masonry outside the door in external walls is called
    (a) capping   (b) thresholds
    (c) pedestals   (d) stairs.

87. A brick masonry could fail due to
    (a) rupture along a vertical joint in poorly bonded walls
    (b) shearing along a horizontal plane
    (c) crushing due to overloading
    (d) any of the above.

88. Good bonding in the brick masonry is ensured when
    (a) all bricks used are of uniform size
    (b) the vertical joints in alternate courses are in plumb from top to base
    (c) bats are used as sparingly as possible
    (d) all the above.

89. A bonding in which all courses are laid as stretchers is known as
    (a) Flemish bond   (b) Header bond
    (c) Stretcher bond   (d) English bond.

90. A bond in which stretcher bricks are laid on edge in alternate courses forming continuous cavity is known as
    (a) Dutch bond   (b) Soldier bond
    (c) Herringbone bond   (d) Cavity bond.

91. A temporary structure erected with a purpose of providing a safe working platform is known as
    (a) centering
    (b) shore
    (c) rake
    (d) scaffolding.

92. An arrangement for supporting an unsafe structure temporarily, till it is rendered safe or dismantled, is known as
    (a) scaffolding
    (b) hauling
    (c) shoring
    (d) jacking.

93. An arrangement supporting an existing structure by providing supports underneath is known as
    (a) underpinning
    (b) scaffolding
    (c) lifting
    (d) jacking.

94. The indentation marks made on the top of the brick to provide key for holding the mortar is known as
    (a) key
    (b) depression
    (c) frog
    (d) valley.

95. The compressive strength of first class and second class bricks should not be less than
    (a) 40 kg/cm^2
    (b) 50 kg/cm^2
    (c) 80 kg/cm^2
    (d) 100 kg/cm^2.

96. The pressure due to wind of velocity V is proportional to
    (a) V
    (b) V^2
    (c) $\frac{1}{V}$
    (d) $\frac{1}{V^2}$.

97. When two walls separated by a small gap, are connected together by metal pins, it is called
    (a) partition wall
    (b) double wall
    (c) cavity wall
    (d) normal wall.

98. Cavity walls
    (a) prevent dampness from entering the building
    (b) have lesser dead load for given wall thickness
    (c) are better insulated against heat and sound
    (d) all the above.

99. A wall which is constructed to divide the space within the building into rooms is called
    (a) partition wall
    (b) cavity wall
    (c) normal wall
    (d) plain wall.

100. A load bearing partition wall is called
    (a) external wall
    (b) internal wall
    (c) main wall
    (d) none of the above.

101. Brick masonry constructed within a framework of wooden member to act as a partition wall is called
    (a) brick nogging
    (b) block partition
    (c) trussed partition
    (d) lath partition.

102. When first class bricks are immersed in water for one hour, the weight of water absorbed by them should not exceed
    (a) $\frac{1}{5}$ of weight of brick
    (b) $\frac{1}{4}$ of weight of brick
    (c) $\frac{1}{6}$ of weight of brick
    (d) $\frac{1}{2}$ of weight of brick.

103. A flooring which is made of thin laminated stone slabs of regular geometric shape laid on concrete bedding is called
    (a) flag stone flooring
    (b) tiled flooring
    (c) granolothic flooring
    (d) stone block flooring.

104. Cork flooring is essentially used in
    (a) all residential buildings
    (b) industrial buildings
    (c) bath rooms and kitchens
    (d) buildings such as libraries, theatres and broadcasting stations which are to be noiseless.

105. Timber flooring is preferred in
    (a) all residential buildings
    (b) only bath rooms
    (c) auditoriums and dancing halls
    (d) public buildings.

106. The most commoly used flooring in residential and public buildings is
    (a) mosaic flooring
    (b) concrete flooring
    (c) marble flooring
    (d) granolithic flooring.

107. In flat slab floor, the flared portion of the columns just below the slab is called
    (a) capital
    (b) pedestal
    (c) pilaster
    (d) cleat.

108. The biggest disadvantage of brick flooring is that
    (a) it is very expensive
    (b) it requires skilled workmen
    (c) it is water absorbent
    (d) it is not fire resistant.

109. Glass flooring is generally preferred when
    (a) extra cleanliness is required
    (b) light is to be transmitted from upper to lower floor
    (c) aesthetic appearance is required
    (d) none of the above.

110. Flooring used in bath rooms and kitchens of posh residential buildings is
    (a) mosaic flooring
    (b) glass flooring
    (c) granolithic flooring
    (d) marble flooring.

111. The disadvantage of jack arch floor is
    (a) the ceiling of the floor is not plain from below
    (b) extra tie rods are required in the end spans to withstand lateral thrust due to arch action

(c) the joints are susceptible for corrosion
(d) all the above.

112. Terrazzo flooring is obtained
    (a) by using marble chips as aggregate in concrete
    (b) by spreading the marble chips over ordinary wet concrete
    (c) by mixing marble powder in ordinary concrete
    (d) none of the above.

113. The oxalic acid is spread and rubbed over the floor which has been ground with machine
    (a) to make the floor surface look smooth and uniform
    (b) to fill up any dents that are left on the floor surface
    (c) to make the surface appear glossy
    (d) to make the surface durable.

114. The joints provided in wooden floors are of
    (a) dove tail type
    (b) tongue and groove type
    (c) butt type
    (d) none of the above.

115. The bearing of the lintel should be
    (a) 10 cm
    (b) height of lintel
    (c) $\frac{1}{10}$ to $\frac{1}{12}$ of span
    (d) minimum of above values.

116. Arches over openings are preferred to lintel beams when
    (a) special architectural appearance is required
    (b) loads are heavy and span is large
    (c) strong abutments are available
    (d) all the above.

117. Brick lintels are used when the span is
    (a) less than 2 m
    (b) less than 1 m
    (c) less than 3 m
    (d) anything since they are cheap.

118. The advantage of R.C.C. lintel is that
    (a) the sunshade can be cast monolithic along with lintel
    (b) there is no limitation on span
    (c) they are strong and fire resistant
    (d) all the above.

119. If b is the width of the wall, L is the effective span of the lintel, γ is the unit weight of masonry, H is the height of wall above the lintel and if the length of wall on one side of the lintel is less than half the effective span the load on the lintel W is given by
    (a) $W = \gamma bL^2 \frac{\sqrt{3}}{4}$
    (b) $W = \gamma bLH$
    (c) $W = \gamma bL^2$
    (d) none of the above.

120. In the above problem, if the length of the wall on either side of the lintel is less than half the effective span, the load on the lintel is given by

121. In problem 119, if $H > L \sin 60°$ and length of the wall on either side of the lintel is more than half the effective span, the load on the lintel is given by.
122. In the problem 119 if $H < L \sin 60°$, the load on the lintel is given by
123. The inner curve of an arch is known as
    - (a) spandril
    - (b) extrados
    - (c) intrados or soffit
    - (d) arcade.
124. In the above problem, which is known as the outer curve of the arch.
125. The wedge shaped units of masonry forming an arch are called
    - (a) closers
    - (b) bevelled units
    - (c) keys
    - (d) voussoirs.
126. Spandril is
    - (a) the curved space formed between the extrados and the horizontal line through the crown
    - (b) the space in the form of a sector between the intrados and the centre of the arch
    - (c) the curved space between the intrados and springing line
    - (d) none of the above.
127. The highest point on the extrados of the arch is known as
    - (a) summit
    - (b) ridge
    - (c) crown
    - (d) peak.
128. The voussoir fixed at the crown of the arch is called
    - (a) closer
    - (b) springer
    - (c) haunch
    - (d) key.
129. A row of arches in continuation is called
    - (a) multiple arch
    - (b) arch line
    - (c) arcade
    - (d) conclave.
130. The points from which the curve of arch springs at the supports are called
    - (a) starting points
    - (b) base points
    - (c) span points
    - (d) springing points.
131. The inclined surface on the abutment which acts as a seat for the arch is known as
    - (a) skew back
    - (b) springer
    - (c) impost
    - (d) haunch.
132. The rise of the arch is defined as the clear vertical distance between
    - (a) the springing line and the highest point on extrados
    - (b) the springing line and the highest point on intrados
    - (c) the centre of the arch and the crown of the arch
    - (d) none of the above.
133. An arch which is formed by two arcs of circles meeting at the apex point is known as
    - (a) pointed arch
    - (b) stilted arch
    - (c) segmental arch
    - (d) sector arch.

134. In a pointed arch if the apex point and the springing points become corners of an isosceles triangle, then it is known as
    (a) florentine arch □   (b) lancet arch □
    (c) horse-shoe arch □   (d) Bull's eye arch. □
135. The pointed arch in which the depth is more at the crown than at the springings is known as
    (a) stilted arch □   (b) variable depth arch □
    (c) relieving arch □   (d) venetian arch. □
136. If the centre of the arch lies on the springing line, it is
    (a) segmental arch □   (b) semi-circular arch □
    (c) Bull's eye arch □   (d) horse-shoe arch. □
137. If the centre of the arch lies above the springing line, it is
    (a) segmental arch □   (b) semi-circular arch □
    (c) Bull's eye arch □   (d) horse-shoe arch. □
138. If the centre of arch lies below the springing line, it is
    (a) segmental arch □   (b) semi-circular arch □
    (c) Bull's eye arch □   (d) horse shoe arch. □
139. An arch in the form of a perfect circle used for circular windows is called
    (a) segmental arch □   (b) semi-circular arch □
    (c) Bull's eye arch □   (d) horse-shoe arch. □
140. An arch which is constructed either on a flat arch or on a wooden lintel to provide greater strength is called
    (a) stilted arch □   (b) florentine arch □
    (c) venetian arch □   (d) relieving arch. □
141. The angle of the skew-back with the horizontal in the case of flat arches is usually
    (a) 75° □   (b) 60° □
    (c) 45° □   (d) 30°. □
142. The portion of the wall which supports an arch is termed as
    (a) pier □   (b) column □
    (c) abutment □   (d) bearer. □
143. The lower half of the arch between the crown and the skew-back is known as
    (a) haunch □   (b) spandril □
    (c) rise □   (d) soffit. □
144. The upper horizontal portion of the step where the foot is placed while ascending or descending is known as
    (a) base □   (b) tread □
    (c) flight □   (d) soffit. □
145. The vertical distance between the tops of two successive treads is called
    (a) flight □   (b) soffit □
    (c) run □   (d) rise. □

146. The projected portion of the tread (which is usually rounded) beyond the face of the riser is
    (a) nosing
    (b) scotia
    (c) soffit
    (d) newel.

147. A vertical member supporting the hand rail is known as
    (a) newel
    (b) strut
    (c) stud
    (d) baluster.

148. A continuous series of steps between landings is called
    (a) going
    (b) stretch
    (c) flight
    (d) run.

149. A level platform at the top or bottom of a flight between the floors is known as
    (a) resting
    (b) landing
    (c) breaking
    (d) passing.

150. A vertical member placed at the ends of flights connecting the hand rail is termed as
    (a) balustrade
    (b) starting
    (c) newel post
    (d) winder.

151. A group of tapering steps radiating from same point is known as
    (a) dancing steps
    (b) splayed steps
    (c) winders
    (d) fliers.

152. In domestic buildings, the sufficient width of the stair is about
    (a) 0.45 m
    (b) 0.9 m
    (c) 1.2 m
    (d) 1.5 m.

153. In public buildings the width of the steps may be kept between
    (a) 0.45 to 0.9 m
    (b) 1.0 to 1.5 m
    (c) 2.0 to 2.5 m
    (d) 1.5 to 1.8 m.

154. Normally the rise of the steps would be between
    (a) 10 cm to 15 cm
    (b) 20 cm to 25 cm
    (c) 5 cm to 10 cm
    (d) 25 cm to 30 cm.

155. The normal tread length varies between
    (a) 30 cm to 50 cm
    (b) 15 cm to 20 cm
    (c) 25 cm to 30 cm
    (d) 20 cm to 25 cm.

156. Spiral stair is used when
    (a) there is space restriction
    (b) the cost of the stair is to be minimised
    (c) architectural appearance is to be improved
    (d) none of the above.

157. A stair in which all the steps are winders is-
    (a) circular stair
    (b) helical stair
    (c) bifurcated stair
    (d) spiral stair.

158. A stair which changes its direction by 180° is
    (a) three-quarter turn stair   (b) half-turn stair
    (c) quarter-turn stair   (d) straight stair.
159. Bifurcated stair is commonly used
    (a) at the entrance of public buildings
    (b) as an emergency stair at the back of the building
    (c) in small residential buildings
    (d) none of the above.
160. When stone slabs are used for risers and treads they are connected by
    (a) scotia   (b) dowel
    (c) baluster   (d) stringer.
161. Dog-legged stair is a
    (a) half-turn stair   (b) quarter-turn stair
    (c) three-quarter turn stair   (d) continuous stair.
162. In a stair, when rise and going are expressed in cms, their sum should be between
    (a) 40 to 45 cm   (b) 30 to 45 cm
    (c) 20 to 45 cm   (d) 40 to 55 cm.
163. When expressed in cms, the product of rise and going in a stair should be between
    (a) 100 to 150   (b) 200 to 250
    (c) 300 to 350   (d) 400 to 450.
164. Lift becomes essential in a building when the number of floors exceeds
    (a) 2   (b) 3
    (c) 4   (d) 6.
165. The pitch or slope of the stair, for comfortable ascent and descent, should be between
    (a) 45 to 60°   (b) 30 to 45°
    (c) 20 to 30°   (d) 25 to 30°.
166. Approximately triangular-shaped stones used as steps are called
    (a) spandrils   (b) stone steps
    (c) triangular steps   (d) monoblock steps.
167. How many treads would be there in a straight stair connecting two floors with height difference of 3.6 m ? The rise is 15 cm
    (a) 23   (b) 24
    (c) 22   (d) 26.
168. How many treads would be there in a dog-legged stair connecting two floor with a height difference of 3.6 m ? The rise of the steps in 15 cm
    (a) 26   (b) 24
    (c) 23   (d) 22.
169. A roof which slopes in all the four directions is
    (a) gable roof   (b) hip roof
    (c) gambrel roof   (d) mansard roof.

170. A roof which slopes in two directions is known as
    (a) gable roof
    (b) hip roof
    (c) gambrel roof
    (d) mansard roof.

171. A roof which slopes in two directions and with a break in each slope is known as
    (a) gumbel roof
    (b) gerard roof
    (c) gambrel roof
    (d) gilbert roof.

172. A roof which slopes in all four directions and having flat top is known as
    (a) deck roof
    (b) duck roof
    (c) dual roof
    (d) durable roof.

173. A roof which slopes in all the four directions and with a break in each slope is known as
    (a) mat roof
    (b) mansard roof
    (c) miller roof
    (d) mixed roof.

174. The horizontal wooden member at the apex of the roof supporting the common rafters is known as
    (a) ridge piece
    (b) purlin
    (c) batten
    (d) cleat.

175. The horizontal wooden or steel members which support the common rafter in a truss are called
    (a) purlins
    (b) battens
    (c) cleats
    (d) posts.

176. The limiting span of a couple roof is about
    (a) 2.5 m
    (b) 3.5 m
    (c) 4.5 m
    (d) 5.5 m.

177. In king post truss there will be
    (a) only one vertical post
    (b) only two vertical posts
    (c) more than two vertical posts
    (d) any number of vertical posts.

178. The number of vertical posts in a queen post truss is
    (a) 1
    (b) 2
    (c) more than 2
    (d) any number.

179. Lean to roof is used generally
    (a) for small sheds
    (b) in verandahs
    (c) outhouses
    (d) any of the above.

180. Queen-post trusses are used when the span is
    (a) 8 to 12 m
    (b) 5 to 8 m
    (c) < 5 m
    (d) > 15 m.

181. Combination of king-post truss and queen post truss is used when the span is
    (a) < 5 m
    (b) 5 to 8 m
    (c) 8 to 12 m
    (d) > 12 m.

182. A truss is called a composite truss when it
    (a) has two slopes on each side
    (b) is a combination of king and queen post-trusses
    (c) is made of two different materials
    (d) has both valleys and hips.

183. Steel strusses have almost replaced the timber trusses because
    (a) they can be used for large spans
    (b) they are more economical than timber and easy to fabricate
    (c) they are fire-proof, more strong and permanent
    (d) all the above.

184. When more lighting is required inside the buildings such as workshops and factories with fairly large spans
    (a) north-light trusses are preferred
    (b) fink trusses are preferred
    (c) howe trusses are preferred
    (d) cambered French trusses are preferred.

185. The material commonly used to cover the pitched roofs with steel trusses is
    (a) thatched covering of reed or straw
    (b) tiles
    (c) wood shringles
    (d) asbestos cement sheets.

186. The desired slope in Madras-terrace roofing is obtained by
    (a) using varying thickness of brick bat concreting
    (b) using varying mortar thickness while laying flat tiles
    (c) adjusting the height of the furring piece at the centre
    (d) providing camber to the joists.

187. Joining two wooden members at an angle is known as
    (a) mitring
    (b) housing
    (c) grooving
    (d) studding.

188. A semi-circular moulding provided on the edges of wood is known as
    (a) veneering
    (b) rebating
    (c) studding
    (d) beading.

189. When a member is subjected to both compression and tension, the type of lengthening joint to be used is
    (a) lapped joint
    (b) fished joint
    (c) tabled joint
    (d) spliced joint.

190. The joint commonly used in wooden trusses where struts and ties meet
    (a) dove-tailed joint
    (b) cogged joint
    (c) notched joint
    (d) bridle joint.

191. The tool normally used by carpenters to plane the surface
    (a) jack plane
    (b) bead plane
    (c) rebate plane
    (d) gimlet.

192. The tool used for cutting small moulding along the edges
    (a) jack plane
    (b) bead plane
    (c) rebate plane
    (d) gimlet.

193. In doors and windows, style is
    (a) the vertical member of the frame
    (b) the vertical member of the shutter
    (c) the member which divides the frame into two vertical portions
    (d) none of the above.

194. The horizontal members of the shutter of a door or window are called
    (a) sills
    (b) horns
    (c) rails
    (d) panels.

195. A horizontal member of the frame which divides the window horizontally is
    (a) rail
    (b) transit
    (c) transome
    (d) divider.

196. A vertical member of the frame which divides the window vertically is
    (a) mullion
    (b) pillion
    (c) cross style
    (d) rail.

197. The dimensions of a door or window frame are measured
    (a) outer to outer on both sides
    (b) outer to outer on vertical side and inner to inner on the other side
    (c) the reverse of (b)
    (d) inner to inner on both sides.

198. The normal height of the door in residential buildings
    (a) 1.5 m
    (b) 1.75 m
    (c) 2.0 m
    (d) 2.5 m.

199. As per Indian Standards, the door designated as 10 DT 20 means
    (a) a single shuttered door of 1 m × 2 m
    (b) a double shuttered door of 1 m × 2 m
    (c) a single shuttered door of 2 m × 1 m
    (d) a double shuttered door of 2 m × 1 m.

200. The thickness of shutter of the door generally does not exceed
    (a) 10 mm
    (b) 20 mm
    (c) 30 mm
    (d) 50 mm.

201. The depression on the recess made inside the door frame to receive the shutter is called
    (a) reveal
    (b) rebate
    (c) jamb
    (d) run.

202. The sill of window in residential buildings is generally kept above the floor level by
    (a) 40 cm
    (b) 80 cm
    (c) 120 cm
    (d) 160 cm.

203. A door in which the shutter is formed by joining the vertical wooden boards side by side with tongue and grooved joints and which are fixed together by horizonal supports is called

   (a) framed and panelled door □   (b) flush door □
   (c) sash door □   (d) battened and ledged door. □

204. In the battened and ledged door, the vertical wooden boards are called

   (a) battens □   (b) ledges □
   (c) rails □   (d) braces. □

205. In the battened and ledged door, the horizontal wooden boards holding the battens together are called

   (a) rails □   (b) braces □
   (c) ledges □   (d) panels. □

206. The inclined wooden boards which provide additional strength to battens in the battened and ledged door are called

   (a) ledges □   (b) braces □
   (c) rails □   (d) panels. □

207. The type of door used when good lighting is to be permitted into the room

   (a) panelled door □   (b) louvered door □
   (c) wire gauged door □   (d) glazed door. □

208. The type of door used when good ventilation is required in the room and at the same time privacy is maintained

   (a) panelled door □   (b) louvered door □
   (c) wire gauged door □   (d) glazed door. □

209. The type of door commonly used for garrages, shops, and godowns

   (a) swing door □   (b) revolving door □
   (c) rolling shutter door □   (d) flush door. □

210. The window in which the shutters can rotate either horizontally or vertically about pivots is called

   (a) revolving window □   (b) rotating window □
   (c) flexible window □   (d) pivoted window. □

211. A small window fixed at a greater height than the regular window (generally 30 to 50 cm below the roof level) is called

   (a) anciliary window □   (b) ventilator □
   (c) roof window □   (d) casement window. □

212. A vertical window provided on the sloping roof is known as

   (a) dormer window □   (b) gable window □
   (c) lantern window □   (d) bay window. □

213. The fixture fixed on the external doors where pad locks are to be used is known as

   (a) barrel bolt □   (b) latch □
   (c) aldrop □   (d) hook. □

214. Joining two wooden boards at an oblique angle is termed
    (a) beading
    (b) rebating
    (c) grooving
    (d) mitring.

215. The usual way of specifying the screws used in wood work is by
    (a) diameter
    (b) length
    (c) weight
    (d) any of the above.

216. U-shaped pieces made of wrought iron with pointed ends which are used to connect the members are known as
    (a) dogs
    (b) clamps
    (c) connectors
    (d) warps.

217. A carpentry tool used for making holes is
    (a) drilling machine
    (b) boring machine
    (c) reamber
    (d) gimlet.

218. A window provided on a pitched roof parallel and slightly above the sloping roof surface is known as
    (a) top ventilator
    (b) sky light
    (c) dormer window
    (d) gable window.

219. A window with a fixed glass provided on the flat roof is known as
    (a) sky light
    (b) lantern window
    (c) dormer window
    (d) top ventilator.

220. A small wooden block hinged on the post outside the door is known as
    (a) eye
    (b) horn
    (c) cleat
    (d) stop.

221. The window projecting beyond the main external wall is
    (a) anciliary window
    (b) auxiliary window
    (c) bay window
    (d) overhanging window.

222. The maximum allowable deflection of a timber beam supporting a roof of span L is
    (a) $\dfrac{L}{160}$
    (b) $\dfrac{L}{200}$
    (c) $\dfrac{L}{360}$
    (d) $\dfrac{L}{400}$.

223. The timber sections of width more than 5 cm and thickness less than 5 cm are known as
    (a) logs
    (b) planks
    (c) scantlings
    (d) sheets.

224. The projections of the heads or sills of door or window are called
    (a) horns
    (b) keys
    (c) stops
    (d) bells.

225. A wall constructed to withstand the pressure of an earthfilling is known as
    (a) parapet wall
    (b) retaining wall
    (c) butting wall
    (d) cavity wall.

226. If single coat plastering is adopted what is its minimum thickness
    (a) 18 mm
    (b) 15 mm
    (c) 12 mm
    (d) 6 mm.
227. In single coat plastering the thickness of plastering should not exceed
    (a) 20 mm
    (b) 15 mm
    (c) 12 mm
    (d) 10 mm.
228. A vertical strip of mortar formed on the surface to be plastered before first coat, is known as
    (a) dot
    (b) screed
    (c) dado
    (d) flaking.
229. In a three coat plastering the first coat is known as
    (a) rendering coat
    (b) floating coat
    (c) setting coat
    (d) base coat.
230. In a three coat plastering the second coat is known as
    (a) floating coat
    (b) intermediate coat
    (c) setting coat
    (d) bonding coat.
231. In a three coat plastering the third coat is known as
    (a) floating coat
    (b) setting coat
    (c) bonding coat
    (d) peripheral coat.
232. In which of the following pointings do the dust and water not accumulate
    (a) recessed pointing
    (b) V-pointing
    (c) struck pointing
    (d) flush pointing.
233. As per latest I.S. code the characteristic strength of M 15 grade concrete at 28 days is
    (a) 150 N/cm^2
    (b) 15 N/cm^2
    (c) 15 N/mm^2
    (d) 150 N/m^2.
234. The characteristic strength of a material is defined as the strength below which
    (a) not more than 5% of the test results fall
    (b) not less than 5% of the test results fall
    (c) not more than 50% of the test results fall
    (d) not more than 25% of the test results fall.
235. The props of from work of the beams with span exceeding 6 m can be removed after
    (a) 7 days
    (b) 10 days
    (c) 14 days
    (d) 21 days.
236. D.P.C. stands for
    (a) double plastered column
    (b) double pinned cleat
    (c) damp proof course
    (d) damp protection cover.
237. The dampness in the buildings may be caused due to
    (a) moisture rise in the walls through capillary action
    (b) rain beating against external walls and rain travel from wall tops
    (c) defective construction such as poor joints in masonry, defective throating, imperfect roof slope etc.
    (d) all of the above.

238. A semi-rigid material ideally suited to make damp proof course is
    (a) bitumen
    (b) metal sheet
    (c) mastic asphalt
    (d) none of the above.

239. To facilitate quick flow of rain water on R.C.C. flat roof towards spouts, it is usually given a slope of
    (a) not exceeding 2 to 3°
    (b) about 15°
    (c) 20°
    (d) $22\frac{1}{2}°$.

240. The surface to which the first coat of plaster is to be applied is called
    (a) base surface
    (b) reference surface
    (c) back ground surface
    (d) first surface.

241. A series of hair cracks developed on a finished plastered surface is known as
    (a) zig-zagging
    (b) random cracking
    (c) crocodiling
    (d) crazing.

242. Few swellings developed locally here and there on a finished plastered surface are called
    (a) bubbling
    (b) boiling
    (c) blistering
    (d) honey combing.

243. The roughening of the back ground surface (when they are smooth as in the case of R.C.C. surface obtained over smooth centering) to provide necessary key for plastering is called
    (a) hacking
    (b) pinning
    (c) dotting
    (d) hindering.

244. The formation of loose patches of plaster due to poor bond between successive coats is called
    (a) scaling
    (b) peeling
    (c) flaking
    (d) popping.

245. The complete dislocation of some portion of plastered surface due to imperfect bond resulting in patches is known as
    (a) scaling
    (b) peeling
    (c) flaking
    (d) popping.

246. The intensity of sound is measured in
    (a) pitches
    (b) tones
    (c) decibels
    (d) audels.

247. The points of higher sound intensity where the reflected sound waves meet is called
    (a) sound foci
    (b) dead spots
    (c) echoes
    (d) none of the above.

248. The points of low sound intensity causing unsatisfactory hearing are called
    (a) sound foci
    (b) dead spots
    (c) echoes
    (d) weak spots.

249. The suitable remedy for the acoustic defect of sound foci in the new design
   (a) add sound absorbers
   (b) introduce suitable diffusers
   (c) avoid curvilinear interiors
   (d) none of the above.

250. In cinema theatres, to avoid reverberation, the longitudinal walls should be
   (a) perfectly parallel
   (b) converging towards screen
   (c) converging towards rear
   (d) should be curvilinear.

251. The volume of space per person to be provided in cinema theatres for good acoustics is
   (a) 2 to 3 m^3
   (b) 3.7 to 4.2 m^3
   (c) 5 to 6 m^3
   (d) > 7 m^3.

252. The noise level in rustle of leaves is
   (a) 100 db
   (b) 50 db
   (c) 10 db
   (d) 1 db.

253. A screw with a square head is known as
   (a) coach screw
   (b) cushion screw
   (c) cord screw
   (d) cone screw.

254. Ventilation means
   (a) cooling the air
   (b) heating the air
   (c) allowing bright light
   (d) replacing the impure air in the building by fresh air from outside.

255. Air conditioning means
   (a) cooling the air to 15°C
   (b) cooling, purifying and controlling humidity of air
   (c) adding moisture to air
   (d) none of the above.

256. The comfortable temperature range in summer may be generally taken as
   (a) 10 to 15°C
   (b) 15 to 20°C
   (c) 20 to 23°C
   (d) 30 to 35°C.

257. According to IS, a window designated as 10 WS 15 means
   (a) a fixed window of size 1 m × 1.5 m
   (b) a fixed window of size 0.5 m × 0.75 m
   (c) a single shuttered window of size 1 m × 1.5 m
   (d) a double shuttered window of size 1 m × 1.5 m

258. Reverberation means
   (a) sound produced uninterruptedly by a source
   (b) sound produced by a source intermittently
   (c) persistence of sound even after the source of sound has ceased
   (d) acoustic illusion.

259. The sound will be painful to the ear when its intensity exceeds
    (a) 50 db
    (b) 100 db
    (c) 130 db
    (d) 200 db.
260. Which of the following has higher fire resistance
    (a) timber
    (b) concrete
    (c) glass
    (d) brick.
261. The maximum area served by one test pit, in the case of open test pit method of investigation, is about
    (a) 15 m × 15 m
    (b) 30 m × 30 m
    (c) 50 m × 50 m
    (d) 100 m × 100 m.
262. Which of the following materials is not generally used in making piles ?
    (a) timber
    (b) steel
    (c) stainless steel
    (d) R.C.C.
263. Quick sand
    (a) is a sand which readily absorbs moisture
    (b) is a sand which allows flow of water through it at rapid rate
    (c) is pure silica used in glass industry
    (d) none of the above.
264. Cement grouting is considered as a useful measure for improving foundations on
    (a) rock
    (b) clay
    (c) compact sand
    (d) moist sand.
265. Vacuum system is used for dewatering soils when the permeability of soil is less than
    (a) 0.0001 cm/s
    (b) 0.00001 cm/s
    (c) 0.01 cm/s
    (d) 0.1 cm/s.
266. Minute cracks in stone containing calcite which form hard veins, are known as
    (a) vents
    (b) shakes
    (c) sand holes
    (d) mottle.
267. Stones having spotted appearance due to the presence of chalky substances are known as
    (a) shakes
    (b) chalk holes
    (c) mottle
    (d) vents.
268. Intensities of acceleration due to earthquakes are represented on
    (a) Mohr's scale
    (b) Dalton scale
    (c) Reid scale
    (d) Kossi-Forrel scale.
269. When a wall is not laterally supported, its effective height is taken as
    (a) 1.5 times its actual height
    (b) 1.25 times its actual height
    (c) 0.5 times its actual height
    (d) 0.75 times its actual height.
270. A member resembling a column which is an integral part of a load bearing wall is known as
    (a) Lateral support
    (b) Pier
    (c) Buttress
    (d) abutment.

271. The ventilating top of a sewage pipe, or a hood shaped top of a chimney is called
    (a) cap
    (b) hat
    (c) corbel
    (d) cowl.
272. A recess made inside of a wall to accommodate pipes or electric wiring etc. is called
    (a) recess
    (b) rebate
    (c) drip
    (d) chases.
273. Which of the following has the highest calorific value?
    (a) marshy ground
    (b) rocky area
    (c) clay
    (d) sand.
274. Which of the following has highest calorific value ?
    (a) diesel oil
    (b) bitumen
    (c) timber
    (d) paper.
275. Which of the following has lowest unit weight ?
    (a) cork
    (b) glass
    (c) saw dust
    (d) bamboo.
276. The limiting value of the ratio of plinth area to plot area in the case of industrial buildings
    (a) 1.0
    (b) 0.8
    (c) 0.6
    (d) 0.4.
277. In the case of residential plots of area 200 sq m, the ratio of plinth area to plot area is restricted to
    (a) 0.9
    (b) 0.66
    (c) 0.5
    (d) 0.33.
278. The permissible area in the case of residential plots of size 1000 sq. m or more is generally
    (a) 666 m^2
    (b) 555 m^2
    (c) 333 m^2
    (d) 222 m^2.
279. The height of rooms meant for human habitation must not be less than
    (a) 5 m
    (b) 2.75 m
    (c) 5.5 m
    (d) 1.5 m.
280. The height of bathroom in residential building must not be less than
    (a) 1.5 m
    (b) 2.8 m
    (c) 2.4 m
    (d) 3.5 m.
281. The area required for parking of a car
    (a) 2.8 m^2
    (b) 28 m^2
    (c) 14 m^2
    (d) 140 m^2.
282. The area required for parking of scooters and motorcycles
    (a) 2.8 m^2
    (b) 1.4 m^2
    (c) 8.2 m^2
    (d) 4.1 m^2.
283. The parking area of a bicycle
    (a) 1.4 m^2
    (b) 2.8 m^2
    (c) 4.2 m^2
    (d) 0.7 m^2.

284. In the case of overhead electric line of 220 V, the minimum distance of verandah or balcony from the live conductor must be more than
    (a) 1.0 m
    (b) 1.4 m
    (c) 2.4 m
    (d) 3.0 m.

285. In case of 33 kV lines the minimum distance of verandah etc., from the live conductor must be more than
    (a) 1.2 m
    (b) 2.4 m
    (c) 3.6 m
    (d) 4.8 m.

286. Stone chips or broken bricks are also known as
    (a) scrap
    (b) waste
    (c) dust
    (d) spall.

287. Mastic is
    (a) a rich lime mortar
    (b) a rich cement mortar
    (c) a putty used in paint work
    (d) bitumen based material used for water proofing damp proofing etc.

288. A facing and shelf, usually ornamental above the fire place is called
    (a) chimney
    (b) arris
    (c) mantel
    (d) cross-cavity.

289. Which of the following is not the standard wattage of electrical lamps ?
    (a) 40
    (b) 50
    (c) 60
    (d) 100.

290. Which of the following is not the usual size of the ceiling fans ?
    (a) 1400 mm
    (b) 1200 mm
    (c) 1000 mm
    (d) 900 mm.

291. Which of the following has the highest bearing capacity ?
    (a) black cotton soil
    (b) soft clay
    (c) fine sandy soil
    (d) hard rock.

292. Which of the following has the lowest bearing capacity ?
    (a) black cotton soil
    (b) coarse sandy soil
    (c) laminated rock
    (d) hard rock.

293. Depth of lean concrete bed at the bottom of the footing is
    (a) 10 cm
    (b) 15 cm
    (c) equal to its projection beyond wall base
    (d) equal to half the width of wall in the super structure.

294. Black cotton soil is often termed as treacherous soil. The reason being
    (a) it swells excessively when wet
    (b) it shrinks excessively when dry
    (c) undergoes large volumetric changes
    (d) all the above.

295. The black cotton soil is unsuitable for foundations because
  (a) it is difficult to excavate it ☐   (b) permeability is uncertain ☐
  (c) bearing capacity is low ☐   (d) none of the above. ☐

296. Raft foundations are preferred when the area required for all the individual footings is more than
  (a) 20% of the total area ☐   (b) 30% of the total area ☐
  (c) 50% of the total area ☐   (d) 10% of the total area. ☐

297. The type of steel pile which is generally sunk in soft clay and loose sand of low bearing capacity is
  (a) screw pile ☐   (b) disc pile ☐
  (c) H-pile ☐   (d) none of the above. ☐

298. The instrument generally used to measure the depth of foundation is
  (a) steel tape ☐   (b) boning rod ☐
  (c) levelling staff ☐   (d) ranging rod. ☐

299. The piles driven to increase the capacity of supporting loads on vertical piles are known as
  (a) sinking piles ☐   (b) raking piles ☐
  (c) eccentric piles ☐   (d) none of the above. ☐

300. Which of the following is true ?
  (a) D.P.C. should be continuous. ☐   (b) D.P.C. should be impervious. ☐
  (c) D.P.C. may be horizontal or vertical. ☐   (d) all the above. ☐

301. In ordinary residential buildings the D.P.C. may be provided at
  (a) between ground level and water table level ☐   (b) at ground level ☐
  (c) at plinth level ☐   (d) at water table level. ☐

302. For the construction of 10 cm thick brick wall, the bond preferred is
  (a) english bond ☐   (b) flemish bond ☐
  (c) header bond ☐   (d) stretcher bond. ☐

303. If the mortar contains both cement and lime as the binding materials, then it is called
  (a) mixed mortar ☐   (b) heterogeneous mortar ☐
  (c) gauged mortar ☐   (d) ungauged mortar. ☐

304. The advantage of air entrained concrete in lining walls and roofs
  (a) it is heat insulating ☐   (b) it is sound insulating ☐
  (c) both (a) and (b) ☐   (d) none of the above. ☐

305. A cut made in the frame of a door to receive the shutter is known as
  (a) rebate ☐   (b) groove ☐
  (c) louver ☐   (d) stop. ☐

306. A flooring made of 4 to 6 mm marble chips is known as
  (a) marble flooring ☐   (b) terrazo flooring ☐
  (c) mossaic flooring ☐   (d) all the above. ☐

307. The dadoing in a bath room should be upto
 (a) ceiling
 (b) 15 cm from floor revel
 (c) 200 cm from floor level
 (d) level of tap.
308. The ceiling height of a building is the height
 (a) of the ceiling above ground level
 (b) between flooring and ceiling
 (c) of the roof above ground level
 (d) none of the above.
309. Expansion joints in masonry walls are provided if length exceeds
 (a) 10 m
 (b) 25 m
 (c) 40 m
 (d) 50 m.
310. In the case of R.C.C. roof slabs covered by insulating layers, the maximum spacing of expansion joints is
 (a) 20 to 30 m
 (b) 30 to 40 m
 (c) 10 to 15 m
 (d) more than 60 m.
311. The lower part of a structure which transmits the load to the soil is known as
 (a) Plinth
 (b) Foundation
 (c) Basement
 (d) Super-structure.
312. The foundation in a building is provided to
 (a) distribute the load over a large area
 (b) increase the overall stability of the structure
 (c) transmit the load to the bearing surface at uniform rate
 (d) perform all the above functions.
313. The minimum load which causes the failure of a foundation is called .... of the soil.
 (a) threshold strength
 (b) nominal strength
 (c) ultimate bearing strength
 (d) ultimate compressive strength.
314. The safe bearing capacity of a soil is given by
 (a) nominal strength/factor of safety
 (b) ultimate bearing strength/factor of safety
 (c) nominal strength × factor of safety
 (d) ultimate bearing strength × factor of safety.
315. For buildings on clayee soils, the minimum depth of foundation is
 (a) 0.2 to 0.4 m
 (b) 0.9 to 1.6 m
 (c) 0.6 to 0.9 m
 (d) 0.4 to 0.6 m.
316. Which of the following statements is false
 (a) The foundation in a building will increase the bearing capacity of the soil
 (b) The foundation in a building will distribute the load uniformly over the bearing surface
 (c) The foundation in a building will distribute the load over a large area
 (d) The foundation in a building will increase the overall stability of the structure.
317. The total horizontal force P per metre length of a retaining wall of height H holding a soil of density γ and angle of internal friction ϕ is given by
 (a) $P = \gamma H \left(\dfrac{1 - \sin \phi}{1 + \sin \phi} \right)$
 (b) $P = \gamma H \left(\dfrac{1 + \sin \phi}{1 - \sin \phi} \right)$
 (c) $P = \dfrac{\gamma H^2}{2} \left(\dfrac{1 - \sin \phi}{1 + \sin \phi} \right)$
 (d) $P = \dfrac{\gamma H^2}{2} \left(\dfrac{1 + \sin \phi}{1 - \sin \phi} \right)$.

318. The most extensively used type of pointing in brick and stone masonaries is
    (a) tuck pointing
    (b) struck pointing
    (c) V-groove pointing
    (d) flush pointing.
319. The type of pointing in which the mortar is first pressed into the raked joint and then finished off flush with the edges of the brick or stone is called
    (a) tuck pointing
    (b) struck pointing
    (c) V-groove pointing
    (d) flush pointing.
320. The type of pointing in which the mortar is first pressed into the raked joint and the face of the pointing is kept vertical and inside the plane of the wall by means of a suitable tool is called
    (a) Recessed pointing
    (b) Rubbed pointing
    (c) Beaded pointing
    (d) Tuck pointing.
321. The process of filling up all nail holes, cracks etc., with putty is known as
    (a) knotting
    (b) priming
    (c) finishing
    (d) stopping.
322. The width of jamb is
    (a) 57 mm to 76 mm
    (b) 76 mm to 114 mm
    (c) 114 mm to 138 mm
    (d) 138 mm to 152 mm.
323. The depth of jamb is
    (a) 52 mm to 76 mm
    (b) 76 mm to 114 mm
    (c) 114 mm to 138 mm
    (d) 138 mm to 152 mm.
324. In the road works the slump of concrete commonly adopted is
    (a) 12 to 20
    (b) 20 to 28
    (c) 35 to 50
    (d) 50 to 100.
325. In wooden stairs the thickness of the member used as tread is generally
    (a) 25 mm
    (b) 38 mm
    (c) 48 mm
    (d) 55 mm.
326. The bricks used for face work shall be
    (a) second class
    (b) brick tiles
    (c) first class
    (d) refractory bricks.
327. Which of the following criteria shall be satisfied by a partition wall?
    (a) It should be provide adequate privacy in rooms in respect of light and sound
    (b) It should be rigid enough to withstand the vibrations
    (c) It should be constructed from light and durable material
    (d) All the above.
328. The weight of the furniture stored in a building is termed as
    (a) Dead load
    (b) Live load
    (c) both (a) and (b)
    (d) none of the above.

329. While calculating the loads, the specific weight of brick masonry is generally taken as
    (a) 1200 kg/m³   (b) 1600 kg/m³
    (c) 1920 kg/m³   (d) 2300 kg/m³.
330. The size of the plate used in plate loading test in clayee soils is
    (a) 30 cm × 30 cm   (b) 45 cm × 45 cm
    (c) 60 cm × 60 cm   (d) 75 cm × 75 cm.
331. The size of the plate used in plate loading test in gravelly or sandy soils is
    (a) 30 cm × 30 cm   (b) 45 cm × 45 cm
    (c) 60 cm × 60 cm   (d) 75 cm × 75 cm.
332. Columns in a workshop supporting Gantry Cranes will have
    (a) Pile foundations   (b) Grillage foundations
    (c) Raft foundations   (d) Well foundation.
333. The least thickness of a stone masonry wall is
    (a) 100 mm   (b) 230 mm
    (c) 350 mm   (d) 500 mm.
334. The angle of repose for dry sand may be taken as
    (a) 15°   (b) 30°
    (c) 45°   (d) 10°.
335. The wall constructed across the slope to protect the slope from the adverse effects is called
    (a) Buttresses   (b) Breast wall
    (c) Retaining wall   (d) Counterfort retaining wall.
336. While the depth of the rebate equals the thickness of the door shutter, the width of the rebate is usually
    (a) 38 mm   (b) 25 mm
    (c) 12 mm   (d) 8 mm.
337. In a panelled door, the rails fixed in between the lock rail and top rail are known as
    (a) Frieze rail   (b) mullion
    (c) lock rail   (d) top rail.
338. The glass one side of which is patterned while rolling and which obscures the direct vision but does not obstruct light is known as
    (a) obscured glass   (b) figured glass
    (c) roughcast glass   (d) all the above.
339. The most suitable type of door for air-conditioned rooms
    (a) swinging door   (b) sliding door
    (c) revolving door   (d) rolling shutter door.
340. The number of steps in a flight should not be less than
    (a) 12   (b) 3
    (c) 6   (d) 9.

341. The number of steps in an ordinary flight shall not be more than
 (a) 12
 (b) 3
 (c) 6
 (d) 9.

342. The extreme vertical support to the hand railing provided at the top and bottom of a flight is known as
 (a) baluster
 (b) newel post
 (c) balustrade
 (d) barrister.

343. The under surface of a stair is known as
 (a) soffit
 (b) ceiling
 (c) waist
 (d) all the above.

344. The additional support given to the nosing in a wooden stair is known as
 (a) soffit
 (b) cornice
 (c) barrister
 (d) scotia.

345. In a stair, the ratio of the vertical distance covered to the horizontal distance occupied by the flight is called
 (a) pitch
 (b) gradient
 (c) slope
 (d) camber.

346. The vertical intermediate support to the hand railing is known as
 (a) balustrade
 (b) barrister
 (c) baluster
 (d) newel post.

347. In a stair the most suitable position to provide winders is
 (a) at the top
 (b) at the bottom
 (c) in the middle
 (d) near the landing.

348. The height of hand railing above the tread should be generally between
 (a) 30 and 50 cm
 (b) 75 and 80 cm
 (c) 80 and 120 cm
 (d) 50 and 75 cm.

349. The narrow strips of wood or bands of plaster laid on the floor at intervals to act as guides for bringing the whole of the work to a true and even surface are termed as
 (a) battens
 (b) screeds
 (c) markers
 (d) helpers.

350. The cheapest floor which remains warm in winter and cool in summer is
 (a) Mud floor
 (b) Rubber floor
 (c) Timber floor
 (d) Asphalt floor.

ANSWERS
Objective Type Questions

1. (c)	2. (d)	3. (b)	4. (c)	5. (c)	6. (c)
7. (d)	8. (a)	9. (a)	10. (b)	11. (a)	12. (d)
13. (c)	14. (c)	15. (a)	16. (d)	17. (b)	18. (d)
19. (a)	20. (b)	21. (d)	22. (a)	23. (b)	24. (c)

25. (b)	26. (d)	27. (a)	28. (b)	29. (c)	30. (c)
31. (d)	32. (c)	33. (b)	34. (c)	35. (c)	36. (a)
37. (b)	38. (d)	39. (d)	40. (a)	41. (c)	42. (d)
43. (c)	44. (d)	45. (b)	46. (a)	47. (c)	48. (c)
49. (b)	50. (b)	51. (a)	52. (c)	53. (a)	54. (b)
55. (d)	56. (c)	57. (a)	58. (a)	59. (d)	60. (c)
61. (d)	62. (b)	63. (a)	64. (b)	65. (c)	66. (a)
67. (c)	68. (d)	69. (a)	70. (c)	71. (d)	72. (b)
73. (a)	74. (c)	75. (d)	76. (b)	77. (a)	78. (a)
79. (b)	80. (c)	81. (d)	82. (c)	83. (d)	84. (b)
85. (a)	86. (b)	87. (d)	88. (d)	89. (c)	90. (b)
91. (d)	92. (c)	93. (a)	94. (c)	95. (c)	96. (b)
97. (c)	98. (d)	99. (a)	100. (b)	101. (a)	102. (c)
103. (a)	104. (d)	105. (c)	106. (b)	107. (a)	108. (c)
109. (b)	110. (d)	111. (d)	112. (a)	113. (c)	114. (b)
115. (d)	116. (d)	117. (b)	118. (d)	119. (c)	120. (b)
121. (a)	122. (b)	123. (c)	124. (b)	125. (d)	126. (a)
127. (c)	128. (d)	129. (c)	130. (d)	131. (a)	132. (b)
133. (a)	134. (b)	135. (d)	136. (b)	137. (d)	138. (a)
139. (c)	140. (d)	141. (b)	142. (c)	143. (a)	144. (b)
145. (d)	146. (a)	147. (d)	148. (c)	149. (b)	150. (c)
151. (c)	152. (b)	153. (d)	154. (a)	155. (c)	156. (a)
157. (d)	158. (b)	159. (a)	160. (b)	161. (a)	162. (a)
163. (d)	164. (c)	165. (b)	166. (a)	167. (a)	168. (d)
169. (b)	170. (a)	171. (c)	172. (a)	173. (b)	174. (a)
175. (a)	176. (b)	177. (a)	178. (b)	179. (d)	180. (a)
181. (d)	182. (c)	183. (d)	184. (a)	185. (d)	186. (c)
187. (a)	188. (d)	189. (c)	190. (d)	191. (a)	192. (c)
193. (b)	194. (c)	195. (c)	196. (a)	197. (a)	198. (c)
199. (b)	200. (c)	201. (b)	202. (b)	203. (d)	204. (a)
205. (c)	206. (b)	207. (d)	208. (b)	209. (c)	210. (d)
211. (b)	212. (a)	213. (c)	214. (d)	215. (b)	216. (a)
217. (d)	218. (b)	219. (b)	220. (d)	221. (c)	222. (c)
223. (c)	224. (a)	225. (b)	226. (d)	227. (a)	228. (b)
229. (a)	230. (a)	231. (b)	232. (d)	233. (c)	234. (a)
235. (d)	236. (c)	237. (d)	238. (c)	239. (a)	240. (c)
241. (d)	242. (c)	243. (a)	244. (c)	245. (b)	246. (c)
247. (a)	248. (b)	249. (c)	250. (b)	251. (b)	252. (c)
253. (a)	254. (d)	255. (b)	256. (c)	257. (c)	258. (c)

259. (c)	260. (b)	261. (a)	262. (c)	263. (d)	264. (a)
265. (b)	266. (b)	267. (c)	268. (d)	269. (d)	270. (b)
271. (d)	272. (d)	273. (b)	274. (a)	275. (a)	276. (c)
277. (b)	278. (c)	279. (b)	280. (c)	281. (b)	282. (a)
283. (a)	284. (c)	285. (c)	286. (d)	287. (d)	288. (c)
289. (b)	290. (c)	291. (d)	292. (a)	293. (c)	294. (d)
295. (b)	296. (c)	297. (a)	298. (b)	299. (b)	300. (d)
301. (c)	302. (d)	303. (c)	304. (c)	305. (a)	306. (c)
307. (c)	308. (b)	309. (c)	310. (a)	311. (b)	312. (d)
313. (c)	314. (b)	315. (b)	316. (a)	317. (c)	318. (d)
319. (d)	320. (a)	321. (d)	322. (b)	323. (a)	324. (b)
325. (b)	326. (c)	327. (d)	328. (b)	329. (c)	330. (c)
331. (a)	332. (b)	333. (c)	334. (b)	335. (b)	336. (c)
337. (a)	338. (d)	339. (c)	340. (b)	341. (a)	342. (b)
343. (a)	344. (d)	345. (a)	346. (c)	347. (b)	348. (b)
349. (b)	350. (a)				

Chapter 13 THEORY OF STRUCTURES

I. INTRODUCTION

1. Propped Cantilevers and Fixed Beams

A cantilever is a beam which derives the three reaction components necessary for stability from the only support *i.e.*, the fixed end. As a result, the maximum bending moment and maximum shear force will be higher than in the case of any other beam. In planning, very often, it would be possible to provide a simple support near the free end of a cantilever or, one of the two simple supports at the two ends could be converted as a built in end. In such cases the resulting structure will be a cantilever supported at the free end. This will be equivalent to a cantilever with a condition that the free end is not permitted to have any linear displacement in a direction perpendicular to the plane of the additional simple support at the free end. In other words, the reaction at that support will be just sufficient to eliminate the deflection at that point. This cantilever, called Propped Cantilever, will be obviously Statically Indeterminate. The condition of compatibility of deflection near that end will provide additional equation or equations for solving the redundant reaction components. Thus either the maximum shear force or the maximum bending moment or both will be less than the respective values in the case of a simple cantilever.

The consistent deformation method adopts the equation of compatibility of deformation consistent with the deformation of the final structure. This is also known as Maxwell's method. The General method or the Superposition Equation method. If the excess reaction components or the excess supports are removed, a basic structure which is statically determinate is obtained. The excess supports removed can be replaced by unknown reaction elements. The deformations of this basic determinate structure, under the action of given loading and the unknown reaction elements in lieu of the excess supports, should be consistent with those of the original indeterminate structure. The unknown reaction components can be determined with the help of such equations of deformation compatibility.

If a roller support has been removed at a certain point, the deformation compatibility is that the deflection perpendicular to the plane of rollers must be zero. If a hinged support is removed, the two requirements are that the horizontal and vertical deflections at that point are zero. Similarly if a fixed support is removed, the three requirements are that the rotation, vertical and horizontal deflections must be zero. The simultaneous equations, equal in number to the unknown reaction components, can be solved to determine the reaction components and with that the structure can be completely analyzed. It is obvious that one point of contraflexure occurs in the case of a propped cantilever and two in the case of a fixed beam for any loading. Fig. 13.1 shows S.F. and B.M. diagrams for a propped cantilever subjected to u.d.l. throughout and Fig. 13.2 shows the same for a central concentrated load.

The point of contraflexure will be towards the fixed support.

Similarly Fig. 13.3 and Fig. 13.4 shows S.F. and B.M. diagrams due to u.d.l. over the entire span and concentrated load at the centre of the span respectively over a fixed beam.

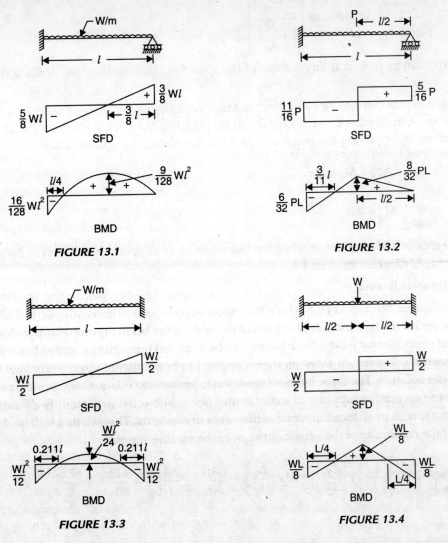

FIGURE 13.1

FIGURE 13.2

FIGURE 13.3

FIGURE 13.4

The propped cantilever and fixed beams can be solved by Successive Integration of load intensity function or B.M. equation and imposing boundary conditions accordingly to solve for integration constants. Similarly a simple supported beam propped at any point can be solved using appropriate deformation compatibility.

The prop is assumed to be rigid and at the same level as that of the other support. Sometimes the prop may be above or below the level of the other supports. Then the deflection due to unknown reaction element will be equated to otherwise free deflection plus or minus the level difference of prop accordingly.

Sometimes the prop can be an elastic prop. The reaction in the prop will be proportional to the deflection that occurs in the prop e.g., a spring used as a prop. If the stiffness of a prop is K, in the case of a cantilever propped at the free end and subjected to u.d.l. throughout the span the reaction from the prop is given by

$$P = wl \left[\frac{\frac{3}{8}}{1 + \frac{3EI}{Kl^3}} \right]$$

The prop can be placed at any section of the span thus making the final deflection at that point zero.

The fixed beams can also be solved by Moment Area theorems. The end moments M_A and M_B of any beam with fixed ends at A and B can be obtained by solving the two simultaneous equations derived by appliying the two Moment Area theorems. The equations are

$$M_A + M_B = \mu \frac{2}{l} \qquad \qquad ...(1)$$

$$M_A + 2M_B = \frac{6\mu \cdot \bar{x}}{l^2} \qquad \qquad ...(2)$$

where μ is the area of B.M. diagram treating the beam as simply supported and $\bar{x}$ is the distance to the c.g. of the M/EI diagram from end A.

2. Continuous Beams

The continuous beam is a beam which is continuous over one or more supports in addition to being supported at the two ends. Any or both of the ends of the beam may be unsupported if the total number of supports is at least 3. Such beams can be analysed through Consistent Deformation method. Clapeyron's theorem of three moments applied to two adjacent spans successively will provide an easier solution. The three moment equation expresses the relation between the bending moments at three successive supports of a continuous beam, subjected to any applied loading on the various spans, with or without unequal settlements of supports. This relation will be derived on the basis of the continuity of the elastic curve over the middle support.

$$M_A \left(\frac{L_1}{I_1} \right) + 2M_B \left[\left(\frac{L_1}{I_1} \right) + \left(\frac{L_2}{I_2} \right) \right] + M_C \left(\frac{L_2}{I_2} \right) = \frac{6A_1 a_1}{I_1 L_1} - \frac{6A_2 a_2}{I_2 L_2} + \frac{6Eh_A}{L_1} + \frac{6Eh_C}{L_2}$$

where the two spans are AB and BC and M_A, M_B and M_C are the moments at the respective supports. L_1 and L_2 are the two spans AB and BC. I_1 and I_2 are the respective moments of inertia. $A_1 a_1$ is the moment of the bending moment diagram of span AB about the end A treating the span AB as simply supported. $A_2 a_2$ is for span BC similarly. h_A and h_C are the amounts by which the supports A and C are above the intermediate support B. If A and C, or any of them, lie (or lies) below the level of B, h_A and h_C are given –ve sign.

After determining the moments at the sections over the supports, the support reactions can be calculated thus completing the analysis of the beam.

3. Pin Jointed Frames

A pin jointed frame is a structure composed of a number of straight bars, of same or different materials, fastened together at their ends by pins or hinges which are frictionless. If loaded only at the joints, the displacements will be so adjusted that all the members will be subjected to only direct stresses and no shear force and no bending moment.

When the axes of all the members of the frame, the loads acting and the reaction components from supports lie in the same plane, the frame is called a plane frame. To avoid bending moment and resulting tension on the masonry supports, there must be a tie connecting the two joints supported on masonry. Thus to support one load, there must be three members with three joints. Every additional joint (under every load) will be formed by adding two more members. So the number of members, 'n' required to have 'j' number of joints will be $(2j - 3)$. Any frame which satisfies the equation $n = (2j - 3)$ will be a perfect frame. If $n \neq (2j - 3)$, it will be known as an imperfect frame. If $n < (2j - 3)$, it will be deficient and will be unstable for a general loading. If $n > (2j - 3)$ in any frame, it will be an internally redundant frame. It may not be possible to analyze this frame by simple methods. If the total number of reaction components from the supports is more than 3, the frame will be externally redundant. The method of joints for analysis provides only two equations of static equilibrium namely $\Sigma V = 0$ and $\Sigma H = 0$ at every joint. The axes of all members at a joint meet at a point, or in other words the forces in various members at a joint will be concurrent. Hence the third condition of equilibrium, $\Sigma M = 0$ will not yield any additional equation. Therefore, for analysis, the joints must be so selected in succession that at the joint under consideration only two forces be unknown and they can be determined. In the method of sections the truss is divided into two portions, the section cutting only three members with unknown forces, at a time. The free body diagram of any portion is considered and the three equations of statics be applied to solve the forces.

To solve an internally redundant truss, a basic truss is derived by removing the redundant members. They will be replaced by pairs of forces and then the conditions that "the change in distance between the displaced end joints of each redundant member must be equal to the change in the length of the said redundant" are applied. Change in the length of the redundant member is FL/AE.

4. Energy Methods

Castigliano, in 1879, published the results of an elaborate research on statically indeterminate structures in which he used two theorems. The two theorems are

"In any structure the material of which is elastic and follows Hooke's law and in which the temperature is constant and the supports unyielding, the first partial derivative of the strain energy with respect to any particular force is equal to the displacement of the point of application of that force in the direction of its line of action."

"In any structure, the material of which is linearly elastic or non-linearly elastic and in which the temperature is constant and the supports are unyielding, the first partial derivative of the strain energy with respect to any particular deflection component is equal to the force applied at the point and in the direction corresponding to that deflection component."

These two theorems can be used to determine the displacement (both linear and rotational) at a section of a beam or frame and the unknown force (redundant force) acting on a structure at a section respectively.

Betti's law states "In any structure the material of which is elastic and follows Hooke's law and in which the supports are unyielding and the temperature is constant, the external virtual work done by a system of forces P during the deformation caused by a system of forces Q is equal to the external virtual work done by the Q system during the deformation caused by the P system".

Maxwell's law of Reciprocal Deflections states "In any structure the material of which is elastic and follows Hooke's law and in which the supports are unyielding and the temperature constant the deflection of point 1 in a direction AB due to a load P at point 2 acting in a direction CD is numerically equal to the deflection of point 2 in the direction CD due to a load P at point 1 acting in the direction AB".

In case of pin jointed frames, the members carry only axial forces. The total strain energy of the system will be the sum of the strain energy in all the members. The deflection of any joint in any direction will be given by partial derivatives of total strain energy with respect to the force acting at that joint in the direction in which the deflection is desired. If required, an imaginary load is to be assumed and to be made equal to zero after partial differentiation. When the method of virtual work is used to compute truss deflections, only one component of the deflection of one joint can be calculated at a time. To obtain the magnitude and direction of the true absolute movement of one joint, two separate applications of the method are usually required. One graphical solution using Williot diagram with Mohr's correction (known as Williot-Mohr diagram) will determine the resultant deflection of all joints of truss simultaneously.

5. Influence Lines and Moving Loads

Influence line for any parameter at a section (such as reaction, shear force, bending moment or force in a member of a truss) is a diagram showing the variation in the parameter at that section as a unit load crosses the span, usually from left to right. The ordinate at any point will give the value of the parameter at the desired section, when the unit load occupies position of ordinate.

Fig. 13.5 shows influence line for reaction at A for a cantilever.

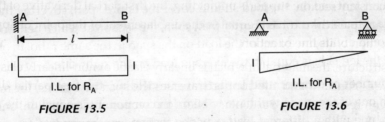

I.L. for R_A

FIGURE 13.5

I.L. for R_A

FIGURE 13.6

Fig. 13.6 shows influence line (I.L.) for reaction at A for a simply supported beam.

Fig. 13.7 shows I.L. for shear force at P for a cantilever and Fig. 13.8 for a simply supported beam respectively.

I.L. for S.F. at P

FIGURE 13.7

I.L. for S.F. at P

FIGURE 13.8

Fig. 13.9 and Fig. 13.10 show the influence lines for B.M. at P for a cantilever and a simply supported beam respectively.

I.L. for B.M. at P

FIGURE 13.9

I.L. for B.M. at P

FIGURE 13.10

When a concentrated load acts on a beam at any section, the value of a certain parameter at a given section will be equal to the value of the load multiplied by the ordinate of the respective influence line diagram, under the load. If a number of concentrated loads act, the total value of the parameter will be $\Sigma W_i Y_i$, where W_i is ith concentrated load any Y_i is the ith ordinate of influence line diagram under the ith load.

When a uniform distributed load occupies a certain position on the beam, the total value of any parameter will be equal to the product of load intensity and the area of respective influence line diagram covered under the load.

If a u.d.l. longer than span traverses a simply supported beam, for max +ve or –ve shear force to occur at a given section, the load should completely cover only that portion, +ve or –ve accordingly, of the respective influence line diagram. If a u.d.l. shorter than span traverses a simply supported beam, for max +ve or –ve shear force to occur at a given section, the load should cover only that portion, +ve or –ve, of the respective influence line diagram and extend from the section towards the end leaving the end portion uncovered, if it comes to.

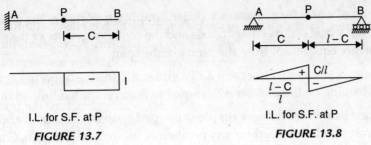

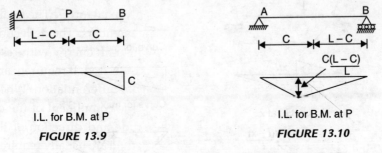

If a u.d.l. longer than span traverses a simply supported beam, for maximum bending moment to occur at any section, the load should cover the entire span, covering the entire area of influence line diagram. If the u.d.l. is shorter than span, the section at which bending moment is to be maximum should divide the extent of the load on the span in the same ratio. In other words, the average load on the left side is equal to the average load on the right side.

When a number of concentrated loads, traverses the span in a series, for maximum shear force at a section one of the loads will have to be on the section. By inspection, one or more load positions, every time with a different load over the section, can be studied for maximum shear force. The condition for maximum B.M. is that movement of one load from one side to another side of the section changes the sign of the difference between the average loadings on two sides. Maximum B.M. at that section occurs when that load is on the section.

When a series of wheel loads crosses a girder, simply supported at the ends the maximum B.M. under any given wheel occurs when its axis and the centre of gravity of the series of the load system on the span are equidistant from the centre of the span.

The maximum B.M. at a given section X of a girder AB, freely supported at the ends, occurs when the average loading on the portion AX is equal to the average loading on the portion XB.

When a single load W traverses a simple supported span maximum +ve S.F. or –ve S.F. or B.M. will occur always under the load for any position of the load on the span. Consider a section at a distance x from left support, as shown in Fig. 13.11.

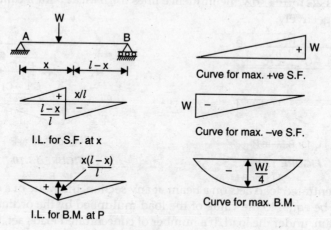

FIGURE 13.11

Maximum +ve S.F. at $\quad x = \dfrac{Wx}{l} \quad$ Straight line variation

Maximum –ve S.F. at $\quad x = \dfrac{W(l-x)}{l} \quad$ Straight line variation

Maximum +ve B.M. at $\quad x = \dfrac{Wx(l-x)}{l} \quad$ Parabolic variation.

When a u.d.l. longer than span traverses a simply supported span, considering a section at a distance x from left support, maximum values will occur for the following positions of load. Refer to Fig. 13.12.

Maximum +ve S.F. ... u.d.l. covering entire left side portion of span from the section only.

Maximum –ve S.F. ... u.d.l. covering entire right side portion of span from the section only.

Maximum B.M. ... u.d.l. covering entire span both on left and right side portions of the section.

Therefore, the equations for the parameters will be

Maximum +ve S.F. ... $\dfrac{wx^2}{2l}$ ...Parabolic

Maximum –ve S.F. .. $\dfrac{w(l-x)^2}{2l}$ ...Parabolic

Maximum +ve B.M. ... $\dfrac{wx(l-x)}{2}$ ...Parabolic

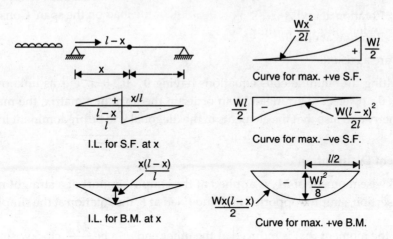

FIGURE 13.12

When a u.d.l. shorter than span transverses a simply supported beam of span l, the parameters will be maximum as detailed below :

Maximum +ve S.F. ... u.d.l. with nose on the section and extending towards left from the section.

Maximum –ve S.F. ... u.d.l. with tail on the section and extending towards right from the section.

Maximum +ve B.M. ... u.d.l. located near the section such that the section divides the span as well as length of u.d.l. in the same ratio.

6. Slope Deflection Method

In this method, all joints are considered rigid, *i.e.*, the angles between members at the joints are considered not to change in value as loads are applied.

In the slope deflection method, the rotations of the joints are treated as unknowns.

For any one member, the end moments can be expressed in terms of the end rotations. But, to satisfy the conditions of equilibrium, the sum of the end moments which any joint exerts on the ends of members meeting there must be zero. This equation of equilibrium furnishes the necessary condition to cope with the unknown rotation of the joint and when these unknown joint rotations are found, the end moments can be computed from the slope deflection equations.

The slope deflection equations are

$$M_{AB} = M_{FAB} + \frac{2EI}{l}(-2\theta_A - \theta_B)$$

$$M_{BA} = M_{FBA} + \frac{2EI}{l}(-\theta_A - 2\theta_B),$$

where M_{FAB} = Fixed end moment at end A
M_{FBA} = Fixed end moment at end B
θ_A = Rotation at end A
θ_B = Rotation at end B
EI and l are as usual.

After getting the simultaneous equations having $\theta_A, \theta_B, \theta_C,...$ etc. as unknowns, if the co-efficients of $\theta_A, \theta_B, \theta_C$ etc. are arranged in an order in the form of a matrix, the matrix will have diagonal symmetry and also that the elements in the diagonal will be predominant in the particular equation.

7. Moment Distribution

If a clockwise moment of M_A is applied at the simple support of a straight member AB, of constant cross-section, simply supported at A and fixed at B, the rotation at the simple support will be $\frac{M_A \cdot L}{4EI}$ and the moment that is induced at the other end will be $\frac{M_A}{2}$ clockwise.

In other words, for a span AB which is simply supported at A and fixed at B, a clockwise rotation of θ_A can be effected by applying a clockwise moment of $M_A = \left(\frac{4EI}{L}\right)\theta_A$ at A which in turn induces a clockwise moment of $M_B = \frac{M_A}{2}$ on the member at B. The expression $\frac{4EI}{l}$ is usually called the stiffness factor.

Stiffness factor is defined as the moment required to be applied at A to cause a rotation of 1 radian at A of a span AB simply supported at A and fixed at B.

The induced moment at B is known as Carry Over Moment. The ratio of $\dfrac{M_B}{M_A}$ (i.e., $\dfrac{1}{2}$) is known as Carry Over Factor. Thus the Carry Over Factor is the ratio of the moment induced at the end B to the moment applied at A.

If a joint consist in several members meeting, any applied moment will be distributed among various members in proportion to their stiffnesses. Therefore, a Distribution Factor can be defined as the factor that decides the share of moment applied at the joint to be borne by a particular member.

A joint will be assumed to be a rigid one. Therefore, the rotation of various members meeting at the joint will be same. Each member requires a particular moment to have that rotation. In other words, for a given rotation of the joint, each of the members offers a moment resistance in proportion to its stiffness. The sum of such resistances equals the applied or induced moments.

Distribution Factor, D.F. = $\dfrac{K_i}{\Sigma K_i}$, where K_i is the stiffness of ith member.

The stiffness of a member when the far end is hinged is $\dfrac{3EI}{L}$. On comparison, it can be observed that the stiffness of a member when far end is hinged is $\dfrac{3}{4}$ of the stiffness when the far end is fixed.

8. Rotation Contribution Method (KANI's Method)

The general Slope Deflection equations are

$$M_{AB} = M_{FAB} + \dfrac{6EI\delta}{l^2} + \dfrac{2EI}{l}(2\theta_A + \theta_B)$$

and

$$M_{BA} = M_{FBA} + \dfrac{6EI\delta}{l^2} + \dfrac{2EI}{l}(2\theta_B + \theta_A)$$

where δ is the lateral displacement of one end with respect to the other. The slope deflection equations relate the values of the deformations with those of the end moments. Thus, for every member there will be two slope deflection equations corresponding to it. The slope deflection method, therefore, involves the solution of numerous simultaneous equations. Depending upon the number of members, a computer may be needed for the solution of the formulated simultaneous equations. The Rotation Contribution method presented by Gas par Kani offers a well-organised iterative procedure.

In Kani's method, certain trial values are assumed for the rotation contributions and Displacement Contributions. These are progressively corrected by repeated application of certain simple equations. Even if an arithmetic error creeps in, it gets automatically corrected in the subsequent cycles. The results converge very rapidly. Moreover, sway calculations and non-sway claculations are carried out in single operation. The procedure can be modified and adopted fast if the dimensions of the sections are changed.

The terms $\dfrac{2EI\theta_A}{l}$ and $\dfrac{2EI\theta_B}{l}$ denoted by M'_{AB} and M'_{BA} will be referred to as Rotation Contributions, at A and B respectively.

$$\left.\begin{array}{l} M_{AB} = M_{FAB} + 2M'_{AB} + M'_{BA} \\ M_{BA} = M_{FBA} + 2M'_{BA} + M'_{AB} \end{array}\right\} \qquad \ldots(A)$$

The equations are for the moments at A and B of AB. For joint A, the end A of the member is referred to as near end and the end B as far end (and *vice-versa* at joint B).

Thus the moment at the near end of member is given by the algebraic sum of the following values :

(*i*) the fixed end moment at the near end due to external loading

(*ii*) twice the rotation contribution at the near end

(*iii*) the rotation contribution at the far end.

Denoting the $\dfrac{I}{l}$ value by K, $\qquad \dfrac{M'_{AB}}{\Sigma M'_{AB}} = \dfrac{K}{\Sigma K}$

$$M'_{AB} = \dfrac{K}{\Sigma K}\left[-\tfrac{1}{2}(\Sigma M_{FAB} + \Sigma M'_{BA})\right]$$

The factor $\dfrac{-K}{2\Sigma K}$ will be called Rotation Factor. The rotation factors for members meeting at a joint are obtained by distributing the value $-\dfrac{1}{2}$ among members in proportion to their K values.

$$M'_{AB} = \text{R.F.}\,(\Sigma M_{FAB} + \Sigma M'_{BA}) \qquad \ldots(B)$$

The fixed end moments for the members can be calculated directly from the loading. If trial values are assumed for the rotation contribution at far ends, the individual rotation contributions at the near end can be calculated from equation (B).

Equation (B) is repeatedly applied to the various joints one after the other in an order. The rotation contributions calculated at a joint (when it is considered as near joint) are taken as the trial values when the joint becomes a far one to some other near joint, to which the equation (B) will be applied.

For the assumption of trial values for rotation contributions, it is customary to take zero as the trial value in the first cycle of operation.

After completing sufficient number of cycles of application of equation (B) the results will be substituted in equations (A) to get final moments.

9. Three Hinged Arches

Arches are constructed to smooth geometrical shapes. The centre line of the arch is called arch axis. The highest point on the arch axis is called Crown of the Arch. The lowest end points are called springing points or springings of the arch. The vertical distance of the crown above the springings called the rise of the arch. An arch can be treated as a curved beam whose ends are restrained against horizontal movement.

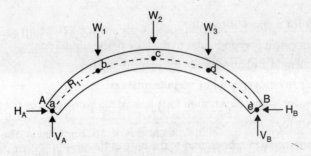

FIGURE 13.13

Fig. 13.13 shows an arch with some concentrated loads acting.

If H_A and V_A are known, the resultant of H_A and V_A, R_1, will act along ab. The resultant R_2 of W_1 and R_1 will act along bc and so on. Therefore, the dotted line $abcde$ will represent the line of action of the internal forces created in the arch. This line $abcde$ will correspond to the link polygon or theoretical arch, and keeps, W_1, W_2,......etc. in equilibrium. Thus, an imaginary link work of thrust is created inside the arch. This is called the line of thrust or the Linear Arch for a given set of loads.

It is the funicular polygon and will consists of straight lines (giving directions of resultants) intersecting on the lines of action of the given loads.

If the loading is continuous instead of point loads, the line of thrust becomes a continuous curve.

Eddy's theorem states that "The bending moment at any point of the arch axis is proportional to the vertical intercept between the arch axis and the line of thrust".

Therefore, if the arch axis coincides with the line of thrust the moment in the arch will be zero. The ideal arch axis is one which coincides with the line of thrust or funicular polygon for a given set of loads, which are same as the B.M. diagram for a given system of loads. When the load is uniformly distributed, the B.M. diagram will be parabolic. Hence arches should have parabolic axes if they are used to support uniformly distributed loads.

Eddy's theorem is valid only for vertical loads.

The three hinged arches will have hinged supports at the two ends and also have a third hinge anywhere between the two ends, usually at the crown. The line of thrust in such arches must pass through these hinges.

II. OBJECTIVE TYPE QUESTIONS

1. A statically indeterminate structure is the one which
   (a) cannot be analysed using the equations of statics alone ☐
   (b) cannot be analysed at all ☐
   (c) is not stable for general loading ☐
   (d) can be analysed with the equations of statics alone. ☐

2. A roller support for a space structure
   (a) permits the movement perpendicular to the base of the support only ☐
   (b) permits movement in any direction in the plane of the base ☐
   (c) does not permit any movement but permits rotation ☐
   (d) permits movement in any direction in the plane of the base but does not permit any rotation. ☐

3. A hinged support for a space structure
   (a) permits the movement perpendicular to the base of the support only ☐
   (b) permits movement in any direction in the plane of the base ☐
   (c) does not permit any movement at all but permits rotation about all three mutually perpendicular axes through the joint ☐
   (d) does not permit movement as well as rotation in any direction. ☐

4. A fixed support for a space structure
   (a) permits the translation only in all directions ☐
   (b) permits rotation only about all three mutually perpendicular axes through the joint ☐
   (c) permits rotation about all the three mutually perpendicular axes and movement in any direction in the plane of the base ☐
   (d) permits neither rotation about nor translation along, any of the three principal axes. ☐

5. Free body diagram is
   (a) the diagram of the body or a part of the body in isolated equilibrium ☐
   (b) the diagram of a body freed from all the forces that have been acting ☐
   (c) the diagram of a body with no supports at all ☐
   (d) none of the above. ☐

6. The free body diagram of a portion of a body will be in equilibrium under the action of
   (a) only external loading acting on that part without consideration of support reactions ☐
   (b) all super-imposed loads, self-weight, support reactions, if any, and the internal reactions exposed at the cuts ☐
   (c) only support reactions and the internal reactions exposed at the cuts ☐
   (d) external loading acting on that part and internal reactions exposed at the cuts. ☐

7. The analyis of a structure is
   (a) deciding the material of the member ☐
   (b) deciding the dimensions of the member ☐
   (c) calculating the magnitude and nature of various straining actions at salient points of the structure ☐
   (d) planning of the structure. ☐

8. The design of a structure is
   (a) the planning of the structure ☐
   (b) the calculation of straining actions at salient points ☐
   (c) deciding the material and proportions of the various members of the structure ☐
   (d) none of the above. ☐

9. The statically indeterminate structures can be solved by
   (a) using equations of statics alone
   (b) equations of compatibility alone
   (c) ignoring all deformations and assuming the structure to be rigid
   (d) using the equations of statics and the necessary number of equations of compatibility.

10. The equations of compatibility are written based on
    (a) the geometry of the deformed structure under the action of several forces acting
    (b) the nature of external forces
    (c) the duration of external forces
    (d) none of the above.

11. A plane structure is a structure
    (a) the various members of which lie in a plane
    (b) the thickness of various members of which will be very small
    (c) in which there will not be any bending moment
    (d) none of the above.

12. A one dimensional structure is one in which
    (a) one dimension is very small and other two are large
    (b) one dimension is much larger than the other two dimensions
    (c) all the three dimensions are equal
    (d) none of the above.

13. A two dimensional structure is one in which
    (a) two dimensions are very much larger than the third
    (b) all the dimensions are equal
    (c) two dimensions are very much smaller than the third
    (d) none of the above.

14. A dam is a
    (a) one dimensional structure
    (b) two dimensional structure
    (c) three dimensional structure
    (d) none of the above.

15. A beam is a
    (a) one dimensional structure
    (b) two dimensional structure
    (c) three dimensional structure
    (d) none of the above.

16. A slab is a
    (a) one dimensional structure
    (b) two dimensional structure
    (c) three dimensional structure
    (d) none of the above.

17. A plane structure when subjected to a force lying outside the plane will be
    (a) stable
    (b) statically determinate
    (c) unstable
    (d) statically indeterminate.

18. An externally indeterminate structure is the one in which
    (a) the total number of external reaction components is more than the equations of statics applicable to the structure as a whole ☐
    (b) the total number of external reaction components is more than the degree of internal indeterminacy ☐
    (c) the sum of the external reaction components and the degree of internal redundancy is greater than three ☐
    (d) none of the above. ☐

19. An internally indeterminate structure
    (a) must be externally indeterminate ☐
    (b) must be externally determinate ☐
    (c) may be an unstable structure based on supports ☐
    (d) none of the above. ☐

20. A beam is a structural member predominantly subjected to
    (a) transverse loads ☐   (b) axial forces ☐
    (c) twisting moment ☐   (d) none of the above. ☐

21. A beam is completely analysed, when
    (a) support reactions are determined ☐
    (b) shear and moment diagrams are found ☐
    (c) the moment of inertia is uniform throughout the length ☐
    (d) none of the above. ☐

22. A rigid frame is a structure composed of members which are connected by
    (a) rigid joints ☐   (b) simple bearing ☐
    (c) a single rivet ☐   (d) none of the above. ☐

23. A frame is completely analysed when
    (a) the variation in shear is found throughout the frame ☐
    (b) the variation in direct stress, shear and moment is found throughout the frame ☐
    (c) the shear and moment reaction from member to member are known at every joint ☐
    (d) none of the above. ☐

24. A truss is completely analysed, when
    (a) the direct stresses in all the members are found ☐
    (b) all the external reactions components are determined ☐
    (c) the equilibrium is satisfied ☐
    (d) none of the above. ☐

25. In a co-planar parallel force system, the number of unknown forces that can be found by the principles of statics is
    (a) 3 ☐   (b) 2 ☐
    (c) 1 ☐   (d) 0. ☐

26. In a general co-planar force system, the number of unknown forces that can be found by the principle of statics is
    (a) 1 ☐ (b) 2 ☐
    (c) 3 ☐ (d) 0. ☐

27. A prop is
    (a) an additional support provided to a stable structure to avoid any displacement at the desired point in a direction perpendicular to the plane of the prop ☐
    (b) the support which gives the maximum number of reactions for the stability of the structure ☐
    (c) a support which will not affect the geometry of the deformed structure ☐
    (d) a dummy support provided for architectural purposes. ☐

28. A rigid prop is one which
    (a) permits 50% of free deflection, that would have occurred if the prop were not there ☐
    (b) does not permit any displacement perpendicular to the plane of prop ☐
    (c) does not offer any reaction ☐
    (d) supports the entire load and relieves all other supports completely. ☐

29. An elastic prop is one which
    (a) does not offer any reaction ☐
    (b) supports the entire load and relieves all other supports completely ☐
    (c) develops reaction propotional to the compression in itself ☐
    (d) none of the above. ☐

30. A sinking prop is one which
    (a) permits any amount of deflection ☐
    (b) does not permit any deflection at all ☐
    (c) is provided below the level of regular supports and becomes effective after the respective deflection occurs ☐
    (d) none of the above. ☐

31. A cantilever subjected to a uniformly distributed load of intensity $\dfrac{w}{m}$ is propped by a rigid prop to the same level of fixed support. The reaction in the prop is
    (a) $\dfrac{3}{8}wl$ ☐ (b) $\dfrac{wl}{2}$ ☐
    (c) wl ☐ (d) $\dfrac{5}{8}wl$. ☐

32. A cantilever is subjected to a concentration load, W at the free end and is propped at the free end to the same level as that of the fixed support. The reaction in the prop (rigid) will be
    (a) $\dfrac{W}{2}$ ☐ (b) W ☐
    (c) 2W ☐ (d) $\dfrac{3}{8}W$. ☐

33. A cantilever is subjected to a uniformly distributed load, $\dfrac{w}{m}$. It is propped by a rigid prop to a level δ higher than the fixed end. The reaction in prop will be

   (a) $\dfrac{3}{8}wl$

   (b) $\dfrac{wl}{2}$

   (c) $3\left(\dfrac{wl}{8} + \dfrac{EI\delta}{l^3}\right)$

   (d) $3\left(\dfrac{wl}{8} - \dfrac{EI\delta}{l^3}\right)$.

34. A cantilever is subjected to a uniformly distributed load, $\dfrac{w}{m}$. It is propped by a spring of stiffness K to the same level, as that of the fixed end, before loading. The reaction in prop will be

   (a) $\dfrac{3}{8}wl + K$

   (b) $\dfrac{3}{8}wl - K$

   (c) $wl\left[\dfrac{\dfrac{3}{8}}{1 - \dfrac{3EI}{Kl^3}}\right]$

   (d) $wl\left[\dfrac{\dfrac{3}{8}}{1 + \dfrac{3EI}{Kl^3}}\right]$.

35. For the application of moment area method, for finding deflection at a section in a beam
   (a) the position of at least one tangent to the elastic curve, at any section, should be known
   (b) the $\dfrac{M}{EI}$ diagram must be a triangle
   (c) the beam must be of uniform moment of inertia
   (d) the B.M. diagram if known is sufficient.

36. If the span of a real beam is l, the span of the corresponding conjugate beam is
   (a) $\dfrac{l}{2}$
   (b) l
   (c) $2l$
   (d) l × number of supports.

37. The loading on the conjugate beam will be
   (a) loading on the real beam divided by EI
   (b) B.M. diagram multiplied by EI
   (c) B.M. diagram divided by S.F. diagram
   (d) B.M. diagram divided by EI.

38. The shear force at a section in the conjugate beam corresponds to
   (a) shear force multiplied by EI at that section in real beam
   (b) deflection at that section multiplied by EI in real beam
   (c) EI times slope at that section in real beam
   (d) slope at that section in real beam.

39. The deflection at a section in the real beam is equal to
   (a) the bending moment at that section in the conjugate beam
   (b) EI times the bending moment at that section in the conjugate beam
   (c) the shear force at that section in the conjugate beam
   (d) the moment of the bending moment diagram of conjugate beam about that section.

40. For a conjugate beam, the fixed end of a real beam corresponds to
    (a) fixed end
    (b) free end
    (c) hinged end
    (d) hinged end on rollers.

41. The free end of a cantilever corresponds to
    (a) free end of the corresponding conjugate beam
    (b) hinged end of the correspondig conjugate beam
    (c) fixed end of the corresponding conjugate beam
    (d) none of the above.

42. A support over which the real beam is continuous will correspond to
    (a) an internal hinge in the conjugate beam
    (b) a hinged support in the conjugate beam
    (c) a fixed support in the conjugate beam
    (d) a discontinuity in the conjugate beam.

43. The displacements of joints of a truss can be obtained directly from
    (a) space diagram
    (b) force diagram
    (c) Williot Mohr diagram
    (d) funicular polygon.

44. If a basic structure is obtained by removing the roller support of an indeterminate structure the requirement the basic structure has to satisfy is that
    (a) the deflection in the direction perpendicular to the supporting surface must be zero
    (b) the displacement in the direction of supporting surface must be zero
    (c) the displacement in any direction at that point must be zero
    (d) none of the above.

45. If a hinged support of an indeterminate structure is removed to obtain a basic structure the resulting basic structure shall satisfy the requirement at that point
    (a) the vertical displacement must be zero
    (b) the horizontal displacement must be zero
    (c) the vertical and horizontal displacements must be zero
    (d) none of the above.

46. If a fixed support of an indeterminate structure is removed to obtain a basic structure the resulting basic structure shall satisfy the requirement at that point
    (a) the rotation must be zero
    (b) the vertical displacement must be zero
    (c) the horizontal displacement to zero
    (d) the rotation, the horizontal and vertical displacements must be zero.

47. The basic form of a pin jointed frame is a
    (a) triangle
    (b) rectangle
    (c) trapezium
    (d) parallelogram.

48. Two inclined struts connected by a hinge at the ends and supporting a load at the joint cannot be supported over two masonry walls because
    (a) the walls cannot bear the compressive stress
    (b) the walls cannot take up horizontal component since tension develops due to bending
    (c) the walls gets crushed due to lateral thrust
    (d) the struts will get crushed.

49. In a pin jointed plane frame all the loads are assumed to act
    (a) in the plane of the frame
    (b) perpendicular to the plane of the frame
    (c) in a plane inclined at 45° to the plane of the frame
    (d) none of the above.

50. In a pin jointed plane frame all the loads will be assumed to act at
    (a) the centre of the members only
    (b) the joints of members only
    (c) only the top chord joints
    (d) none of the above.

51. By assuming that all the forces acting on a pin connected truss are co-planar and act at joints only it can be expected that all the member will be subjected to
    (a) direct compressive stress only
    (b) direct tensile stress only
    (c) direct stresses only
    (d) direct stresses and shear stress only.

52. A pin jointed frame with number of joints j, and number of members n will be a perfect frame, if
    (a) $n = (2j + 3)$
    (b) $n > (2j - 3)$
    (c) $n < (2j - 3)$
    (d) $n = (2j - 3)$.

53. A pin jointed plane frame with j number of joints and n number of members will be internally redundant, if
    (a) $n > (2j - 3)$
    (b) $n < (2j - 3)$
    (c) $n = (2j - 3)$
    (d) $n > (2j + 3)$.

54. In a pin jointed frame it is sufficient if the forces in all the members meeting at a joint are
    (a) co-planar
    (b) co-planar and concurrent
    (c) equal in magnitude
    (d) none of the above.

55. In a pin jointed frame the members meeting at a joint must be so arranged that
    (a) the axes of all the members are concurrent and coplanar
    (b) not more than two axes meet at a point
    (c) the axes must be parallel to each other
    (d) at least three axes meet at a common point.

56. For analyzing pin jointed frames by the method of joints, the joints must be selected in succession such that
    (a) only two members meet at that point
    (b) only three members meet at that joint
    (c) only two unknown forces exist at that joint
    (d) there in no external force acting in the joint.

57. For analyzing pin jointed frames by the method of joints, the number of equations of static equilibrium available is
    (a) 3
    (b) 1
    (c) 0
    (d) 2.

58. For analyzing pin jointed frames by the method of sections, the section should be so chosen that it cuts
    (a) only three members at a time
    (b) any number of members but only two members with unknown forces
    (c) any number of members but only three members with unknown forces
    (d) not more than one member.

59. The ratio of strength of a fixed beam to that of a simply supported beam of same span under u.d.l. throughout with regards to shear is
    (a) 3
    (b) 2
    (c) 0.5
    (d) 1.0.

60. The ratio of load carrying capacity of a fixed beam to that of a simply supported beam having same maximum bending moment under u.d.l. throughout the span is
    (a) 1.5
    (b) 1.0
    (c) 0.6667
    (d) 3.0.

61. The ratio of maximum – ve bending moment of a cantilever to that of a cantilever propped at the free end to the same level for same span and same u.d.l. throughout the span is
    (a) 2
    (b) 3
    (c) 4
    (d) 5.

62. The ratio of load carrying capacity of a fixed beam to that of a cantilever of same span, having same maximum bending moment under u.d.l. throughout the span is
    (a) 6
    (b) 3
    (c) 4
    (d) 2.

63. A fixed beam of span l is subjected to a u.d.l. w/m. The ratio of sum of maximum – ve B.M. and + ve B.M. to maximum + ve B.M. in the case of a simply supported beam of same span under same loading
    (a) 3
    (b) 2
    (c) 1
    (d) 0.5.

64. A simply supported beam is propped at the centre by a rigid prop. It is subjected to a central concentrated load W. The decrease in each end support reaction because of provision of prop is
    (a) 100%
    (b) 50%
    (c) 75%
    (d) 25%.

65. For finding forces in the members of a truss by the graphical method, the force diagram is suggested by
    (a) Mohr
    (b) Timoshenko
    (c) Maxwell
    (d) Betti.

66. The theorem of three moments for the solution of continuous beams is formulated by
    (a) Young
    (b) Maxwell
    (c) Williot-Mohr
    (d) Clapeyron.

67. If one of the supports of a continuous beam sinks by δ, the induced moments will be

    (a) $\dfrac{6EI\delta}{l^2}$ (b) $\dfrac{EI\delta}{6l^2}$

    (c) $\dfrac{6EIl}{\delta}$ (d) $\dfrac{6E\delta}{Il^2}$.

68. If a continuous beam has a fixed end, with unknown moment reaction there, for applying theorem of three moments
    (a) the fixed support is replaced by a unit moment
    (b) an imaginary span of any lengh but with infinite moment of inertia is added beyond the support
    (c) it is not possible to analyze
    (d) none of the above.

69. In a propped cantilever subjected to u.d.l. throughout the span, the point of contraflexure will occur at
    (a) $l/2$ (b) $l/4$ from propped end
    (c) $l/4$ from fixed end (d) propped end.

70. The ratio of section moduli required for a fixed beam and a propped cantilever of same span with same material if both are subjected to u.d.l. throughout the span is
    (a) 0.667 (b) 1.5
    (c) 1.0 (d) 2.5.

71. In a fixed beam is subjected to u.d.l. throughout the span, the point of contraflexure will occur at
    (a) $l/2$ (b) at the two fixed supports
    (c) $0.21\,l$ from each of the supports (d) $0.667\,l$ from each of the supports.

72. The term centre of gravity was introduced by
    (a) Archimedes (b) Otto Mohr
    (c) Saint Venant (d) Coulomb.

73. The influence line diagram for S.F. or B.M. at a section is
    (a) the value of S.F. or B.M. at that section when the unit load is placed over that section only
    (b) the value of S.F. or B.M. at that section when the unit load is at the centre of the span
    (c) the variation in the value of S.F. or B.M. at that section as the unit load transverses the span from left to right
    (d) the S.F. or B.M. diagram.

74. The influence line diagram for reaction at a support of a simply supported beam is a
    (a) triangle with ordinate 1 at that support
    (b) triangle with ordinate 1 at the other support
    (c) rectangle with ordinate of 1
    (d) rectangle with ordinate of $\dfrac{1}{2}$.

75. The influence line diagram for reaction at the support of a cantilever will be a
    (a) triangle with zero ordinate at fixed end and unit ordinate at free end
    (b) triangle with unit ordinate at fixed end and zero ordinate at free end
    (c) rectangle with ordinate $\frac{1}{2}$
    (d) rectangle with unit ordinate.

76. The influence line diagram for shear force at a section on a cantilever will be
    (a) a rectangle of length equal to full length of span with unit ordinate
    (b) a rectangle extending between free end and the section with unit ordinate
    (c) a triangle extending for whole length of span with unit ordinate at fixed support
    (d) a triangle extending between the section and the fixed end with unit ordinate at fixed support.

77. In the influence line diagram for S.F. at a section in a simply supported beam, the sum of maximum –ve ordinate and maximum +ve ordinate will be
    (a) 0   (b) 2
    (c) 1   (d) none of the above.

78. The influence line diagram for B.M. at a section in a cantilever will be a triangle extending between the section and the
    (a) fixed end with maximum ordinate under the section
    (b) fixed end with maximum ordinate under the fixed end
    (c) unsupported end with maximum ordinate at the section
    (d) unsupported end with maximum ordinate at the unsupported end.

79. The influence line diagram for B.M. at a section of a simply supported beam will be a
    (a) triangle with maximum ordinate at the section and zero at the two supports
    (b) triangle with maximum ordinate at the support nearer to the section and zero at the other support
    (c) rectangle with unit ordinate
    (d) triangle with maximum ordinate of unity at the section and zero at both the supports.

80. If a section divides the span of a simply supported beam as c and $(l - c)$, the ordinate of influence line diagram, for B.M. at the section, will be
    (a) maximum at the section and equal to $c(l - c)$
    (b) maximum at the section and equal to $\frac{c(l-c)}{l}$
    (c) maximum at the supports and equal to unity
    (d) none of the above.

81. If a u.d.l. longer than span l, traverses the span, the +ve shear force at a section will be equal to
    (a) net area of influence line diagram multiplied by intensity of u.d.l.
    (b) +ve area of influence line diagram multiplied by intensity of u.d.l.
    (c) maximum +ve ordinate of influence line diagram multiplied by intensity of u.d.l. multiplied by span
    (d) none of the above.

82. A simply supported beam is traversed by a u.d.l. shorter than span. For maximum shear force to occur at a given section, the u.d.l. should cover the span such that
    (a) the u.d.l. occupies equal lengths on each side of the section ☐
    (b) the u.d.l. extends only from the section towards nearer support ☐
    (c) the u.d.l. extends from the section towards either of the sides based on nature of shear force covering only +ve or –ve area ☐
    (d) the section divides the span and u.d.l. in the same ratio. ☐

83. The shear force due to dead load developed at a section will be given by
    (a) dead load intensity multiplied by net area of inuflence line diagram for shear force at the section ☐
    (b) dead load intensity multiplied by +ve or –ve area whichever is greater, of influence line diagram ☐
    (c) product of influence line diagram ordinate, +ve or –ve whichever is greater, at that section and dead load intensity ☐
    (d) none of the above. ☐

84. A simply supported beam is traversed by a u.d.l. longer than span, the maximum B.M. at a section will be given by the product of
    (a) the area of influence line diagram for B.M. at that section and average load on the span when the load covers the longer portion either to left or to right of section ☐
    (b) the area of influence line diagram and intensity of u.d.l. ☐
    (c) the maximum ordinate and total load on the span ☐
    (d) none of the above. ☐

85. A beam is subjected at a number of concentrated loads. The B.M. at a section of the beam is given by
    (a) the sum of the ordinates under all loads multiplied by the sum of all the loads ☐
    (b) the average of all ordinates multiplied by the average of all loads ☐
    (c) the sum of the products of each concentrated load and the ordinate under it ☐
    (d) none of the above. ☐

86. A simply supported beam is traversed by a u.d.l. shorter than span. For maximum B.M. to occur at a given section
    (a) the section should divide the span and the u.d.l. in the same ratio ☐
    (b) the u.d.l. should extend from the section onto longer portion of the span ☐
    (c) the u.d.l. should be symmetrically placed about the section ☐
    (d) none of the above. ☐

87. A series of concentrated loads traverses a simply supported beam. The maximum B.M. under any given load occurs when
    (a) the centre of gravity of the load system and that load are equidistant from the centre of the span ☐
    (b) that load is placed over the centre of the span and the rest of the load system acts accordingly ☐
    (c) average load on the span should be higher with the particular load anywhere on the span ☐
    (d) none of the above. ☐

88. A series of concentrated loads traverses a simply supported beam. For maximum B.M. at a given section to occur, the load system must be placed such that
   (a) the heaviest load is placed over the section   ☐
   (b) the average loading on the two sides of the section is same   ☐
   (c) maximum number of loads will cover the span   ☐
   (d) none of the above.   ☐

89. A simply supported beam is traversed by a series of concentrated loads. The maximum B.M. at a section will occur when that load acts on this section, which
   (a) when moved from one side of the section to another, the difference between the average loadings on the two sides changes its sign   ☐
   (b) is the heaviest of all loads   ☐
   (c) is the first load   ☐
   (d) none of the above.   ☐

90. When a single load W moves over a simply supported beam, the maximum B.M. at a section will occur when the load is placed
   (a) over the section   ☐   (b) at centre of span   ☐
   (c) over the nearer support   ☐   (d) none of the above.   ☐

91. When a single load W moves over a simply supported beam, the maximum S.F. at a section will occur when the load is placed
   (a) over the nearer support   ☐   (b) over the farther support   ☐
   (c) over the section   ☐   (d) at centre of span.   ☐

92. When a single load W moves over a simply supported beam, the curve for maximum shear force, +ve or –ve will be a
   (a) parabola with maximum ordinate W at centre of span and zero at supports   ☐
   (b) parabola with maximum ordinate W at a support   ☐
   (c) triangle with maximum ordinate W at a support   ☐
   (d) a triangle with maximum ordinate W at centre and zero at the two supports.   ☐

93. When a single concentrated load W traverses a simply supported beam, the curve for maximum +ve B.M. will be a
   (a) parabola with maximum ordinate at centre   ☐
   (b) parabola with maximum ordinate at one of the supports   ☐
   (c) triangle with maximum ordinate at centre   ☐
   (d) triangle with maximum ordinate at one of the supports.   ☐

94. A simply supported beam is traversed by a single concentrated load W. The absolute maximum B.M. will be occurring
   (a) at centre with a value of $\dfrac{Wl}{4}$ ☐
   (b) at $\dfrac{l}{4}$ from each of the supports with a maximum of $\dfrac{Wl}{8}$ ☐
   (c) at the two supports with zero at centre   ☐
   (d) none of the above.   ☐

95. A u.d.l. of intensity $\frac{w}{m}$ longer than span traverses a simply supported beam. The curve for maximum shear force, +ve or –ve, will be

   (a) a triangle with maximum ordinate $\frac{wl}{2}$ at centre and zero at the two supports ☐

   (b) a triangle with maximum ordinate wl at one of supports ☐

   (c) a parabola with maximum ordinate $\frac{wl}{2}$ at both of the supports simultaneously ☐

   (d) a parabola with maximum ordinate $\frac{wl}{2}$ at one of the supports with zero at the other support. ☐

96. A u.d.l. of intensity $\frac{w}{m}$ longer than span traverses a simply supported beam. The curve for maximum bending moment will be

   (a) a parabola with maximum ordinate of $\frac{wl^2}{8}$ at centre and zero at ends ☐

   (b) a parabola with maximum ordinate of $\frac{wl^2}{8}$ at ends with zero at centre ☐

   (c) a triangle with $\frac{wl}{4}$ ordinate at centre and zero at ends ☐

   (d) a triangle with $\frac{wl^2}{8}$ ordinate at centre and zero at ends. ☐

97. A continuous beam of two equal spans l, each, is simply supported at the two ends. It is subjected to u.d.l., $\frac{w}{m}$, on both the spans. The reaction at the central support will be

   (a) wl ☐ (b) $\frac{3}{2} wl$ ☐

   (c) $\frac{5}{4} wl$ ☐ (d) $2wl$. ☐

98. A continuous beam has two equal spans, l each, and is simply supported at the two ends. If it is subjected to u.d.l. $\frac{w}{m}$, on both the spans. The bending moment at the central support will be

   (a) $\frac{wl^2}{4}$ ☐ (b) $\frac{wl^2}{2}$ ☐

   (c) $\frac{wl^2}{12}$ ☐ (d) $\frac{wl^2}{8}$. ☐

99. A beam carrying u.d.l. of w/m throughout is simply supported at the two ends and continuous over a hinge on roller support. The moment reaction, the central support has to provide to the beam for equilibrium, is

   (a) 0
   (b) $\dfrac{wl^2}{2}$
   (c) $\dfrac{wl^2}{8}$
   (d) $\dfrac{wl^2}{4}$.

100. A multi-span continuous beam is simply supported at the two ends. One of the intermediate supports sinks with respect to other supports. Due to this settlement, the bending moment will get modified at
   (a) only the two supports adjacent to the one that sinks
   (b) only the support that sinks
   (c) all the supports
   (d) all the supports except the two end supports.

101. In a rigid jointed frame, the joints are considered
   (a) to rotate only as a whole
   (b) not to rotate at all
   (c) that 50% of members rotate in clockwise direction and 50% in anti-clockwise direction
   (d) none of the above.

102. In slope deflection method, the end moments for any member are expressed
   (a) as zero
   (b) in terms of the unknown end rotations
   (c) as equal
   (d) none of the above.

103. In a rigid jointed frame, the rotating of various members meeting at the joint will be
   (a) equal
   (b) proportional to the length of the member
   (c) proportional to the stiffness
   (d) proportional to the respective moment of inertia.

104. In slope deflection method, the unknown rotations at various joints are determined by considering
   (a) the equilibrium of the joint
   (b) the rigidity of the joint
   (c) the equilibrium of the structure
   (d) none of the above.

105. In the simultaneous equations derived from slope deflection method the matrix formed by the co-efficients of various rotations will have
   (a) all elements equal
   (b) all elements zero
   (c) all elements along the diagonal zero
   (d) diagonal symmetry with diagonal element predominant in that respective equation.

106. The moment distribution method is formulated by
   (a) Hardy Cross
   (b) Thomas Young
   (c) Mohr
   (d) Kani.

107. The simultaneous equations of slope deflection method can be solved by iteration in
 (a) moment distribution method
 (b) consistent deformation method
 (c) conjugate beam method
 (d) Williot Mohr method.

108. A beam is simply supported at end A and fixed at B. If a moment M is applied at the simple end, the moment developed at the fixed end will be
 (a) $-M$
 (b) $+M$
 (c) $\dfrac{+M}{2}$
 (d) $\dfrac{-M}{2}$.

109. The stiffness factor of a member is the moment required to be applied at the simply supported end to produce
 (a) a unit rotation of one radian at fixed end
 (b) a unit rotation of one radian at simply supported end
 (c) a unit deflection at the simply supported end
 (d) a unit rotation of one radian at both the ends.

110. The stiffness factor at the near end of a member with far end fixed is
 (a) $\dfrac{4EI}{l}$
 (b) $\dfrac{3EI}{l}$
 (c) $\dfrac{EI}{l}$
 (d) EI.

111. The stiffness factor at the near end of a member with far end hinged is
 (a) $\dfrac{4EI}{l}$
 (b) $\dfrac{3EI}{l}$
 (c) $\dfrac{EI}{l}$
 (d) EI.

112. The ratio of stiffness of a member when far end is hinged to that of the member when far end is fixed is
 (a) 1
 (b) 2
 (c) $\dfrac{3}{4}$
 (d) $\dfrac{4}{3}$.

113. At a joint where several members meet, any applied moment will be distributed among the various members
 (a) equally
 (b) in proportion to their stiffnesses
 (c) equally among such members whose far ends are fixed
 (d) equally among such members whose far ends are hinged.

114. The distribution factor of a member at a joint is
 (a) the ratio of the moment borne by the member to the total moment applied at the joint
 (b) the ratio of the area of the member to the sum of the areas of several members
 (c) the ratio of the moment induced at the far end to the moment applied at the near end
 (d) none of the above.

115. If K_i the stiffness of ith member at a joint, the distribution factor for the member is

   (a) $\dfrac{K_i}{\Sigma K_i}$ □   (b) ΣK_i □

   (c) K_i □   (d) $(\Sigma K_i - K_i)$. □

116. The sway correction for a frame may be required because we make an assumption that
   (a) the joints do not rotate at all □
   (b) the applied moment is distributed as per relative stiffnesses □
   (c) the joints do not move □
   (d) none of the above. □

117. The sway correction for frame will be necessary
   (a) only when the frame is unsymmetrical □
   (b) only when the loading is unsymmetrical □
   (c) only when both loading and frame are unsymmetrical □
   (d) when either the loading or the frame is unsymmetrical. □

118. Kani's 'Rotation Contribution' method is advantageous over Moment Distribution method since
   (a) Kani's method is iterative □
   (b) any arithmetic error that creeps in will automatically get corrected □
   (c) it involves actual solution of simultaneous equations □
   (d) none of the above. □

119. Sway calculations and non-sway calculations are carried out in a single operation in
   (a) Kani's method □   (b) moment distribution method □
   (c) unit load method □   (d) none of the above. □

120. If the preliminary dimensions of the sections are changed relatively, the analysis can be modified fast in
   (a) moment distribution method □   (b) Kani's method □
   (c) double integration method □   (d) consistent deformation method. □

121. When an end of continuous beam is fixed, in Kani's method, the rotation contribution will be

   (a) zero □   (b) $\dfrac{EI}{l}$ □

   (c) $\dfrac{2EI}{l}$ □   (d) EI. □

122. In Kani's method an overhang can be conveniently dealt with by regarding it as a member with
   (a) infinite length □   (b) zero length □
   (c) unit length □   (d) none of the above. □

123. In Kani's method, the displacement contribution of a member with a sway of δ, is

   (a) $EI\delta$ □   (b) $\dfrac{6EI\delta}{l^2}$ □

   (c) $\dfrac{4EI\delta}{l}$ □   (d) $\dfrac{3EI}{l}$. □

124. The vertical distance of the crown of the arch above the springings is called
    (a) span of the arch
    (b) rise of the arch
    (c) thickness of the arch
    (d) none of the above.
125. An arch can be treated as a curved beam
    (a) whose ends are restrained against horizontal movement
    (b) whose ends do not provide any reaction
    (c) whose ends are unsupported
    (d) none of the above.
126. The line of thrust of the linear arch is
    (a) the axis of the arch
    (b) the imaginary link work of thrust for a given set of loads
    (c) the springing line
    (d) none of the above.
127. The moments in the arch will be zero, if
    (a) ends are hinged
    (b) ends are fixed
    (c) the arch axis coincides with the line of thrust
    (d) the arch axis is parallel to the line of thrust.
128. The arches meant for supporting uniformly distributed loads, to avoid any bending moment, must be
    (a) circular
    (b) elliptical
    (c) parabolic
    (d) none of the above.
129. The Eddy's theorem is valid for
    (a) vertical loads only
    (b) horizontal loads only
    (c) dynamic loads only
    (d) all loads.
130. A three hinged arch will have three hinges
    (a) two at the two ends and one any where in between the two ends
    (b) two at the two ends and the third at the crown only
    (c) one hinge at the crown essentially and the other two any where
    (d) none of the above.
131. In a three hinged arch, the line of thrust
    (a) must pass through all the three hinges
    (b) must pass through two end hinges and may or may not pass through the third hinge
    (c) must not pass through any of the hinges
    (d) must pass through two end hinges only.
132. A suspension cable, supporting loads, will be under
    (a) tension
    (b) bending
    (c) compression
    (d) compression and bending.

133. An arch and a simply supported beam are of same span and are subjected to same loading. Denoting the maximum bending moment in the case of arch as M_a and the maximum bending moment in the case of beam as M_b which of the following is true ?
    (a) $M_a > M_b$
    (b) $M_b > M_a$
    (c) $M_a = M_b$
    (d) none of the above.

134. If L is the span of a 3 hinged arch, h is the rise and w the u.d.l. per unit length over the entire span, the horizontal reaction at each support is given by
    (a) $\dfrac{wL^2}{4h}$
    (b) $\dfrac{wL^2}{6h}$
    (c) $\dfrac{wL^2}{8h}$
    (d) $\dfrac{wL^2}{10h}$.

135. The intercept between a given arch and a linear arch at a section is proportional to
    (a) the horizontal thrust
    (b) the shear force at the section
    (c) the vertical reaction at the support
    (d) the bending moment at the section.

136. A semicircular arch of radius R is hinged at the two supports which are at the same level and also at the crown. It is subjected to a u.d.l. of w per metre length over the entire span. The distance of the section with maximum bending moment measured from the centre is equal to
    (a) $\dfrac{R}{2}$
    (b) 0
    (c) $\dfrac{R\sqrt{3}}{2}$
    (d) $\dfrac{3R}{2}$.

137. A three hinged parabolic arch has its abutments at depth of h_1 and h_2 below the crown. It is subjected to a concentrated load of W at the crown. If the span of the arch is L, the horizontal thrust at each support is given by
    (a) $\dfrac{WL}{\sqrt{h_1^2 + h_2^2}}$
    (b) $\dfrac{WL}{(\sqrt{h_1} + \sqrt{h_2})^2}$
    (c) $\dfrac{WL}{(h_1 + h_2)}$
    (d) $\dfrac{WL}{(h_1 + h_2)^2}$.

138. What is the horizontal thrust at each support of a two hinged parabolic arch subjected to u.d.l. over the entire span ?
    (a) $\dfrac{wL^2}{8h}$
    (b) $\dfrac{wL^2}{6h}$
    (c) $\dfrac{wL^2}{4h}$
    (d) $\dfrac{wL^2}{2h}$.

139. The reaction locus for a two hinged parabolic arch is
    (a) a horizontal line parallel to the span
    (b) a parabolic curve
    (c) a cycloid
    (d) none of the above.

140. The reaction locus for a two hinged parabolic arch is
   (a) a horizontal line parallel to the span   □   (b) a parabolic curve   □
   (c) a cycloid   □   (d) none of the above.   □

141. A three hinged arch ACB consists of two quandrantal parts AC and CB of radii R_1 and R_2 respectively (with $R_2 > R_1$) hinged at A, C and B. It is subjected to a vertical concentrated load W at the crown C. If the vertical reactions at the two abutments A and B are denoted by V_A and V_B, then
   (a) $V_A > V_B$ □   (b) $V_A < V_B$ □
   (c) $V_A = V_B$ □   (d) none of the above.   □

142. The maximum tension in a cable occurs
   (a) at the highest point in the cable   □   (b) at the lowest point in the cable   □
   (c) at the centre of the cable   □   (d) at all points in the cable.   □

143. A parabolic cable is subjected to a rise of $t°$ in temperature. Then the increase in the dip of the cable is proportional to
   (a) span length   □   (b) square of span length   □
   (c) cube of span length   □   (d) reciprocal of span length.   □

144. A three-hinged girder is used to stiffen a cable of a suspension bridge. Then the bending moment in the girder due to dead load will be
   (a) maximum at mid span   □   (b) maximum at quarter span   □
   (c) maximum at one-eighth span   □   (d) zero over the entire span.   □

145. A cable subjected to u.d.l. over its entire span assumes a shape of
   (a) semi-circle   □   (b) an isosceles triangle   □
   (c) parabola   □   (d) none of the above.   □

146. The span and dip of a parabolic cable are L and h respectively. Then the length of the cable is approximately equal to
   (a) $L + \dfrac{3}{8} h$ □   (b) $L + \dfrac{8}{3} h$ □
   (c) $L + \dfrac{3}{8} \dfrac{h^2}{L}$ □   (d) $L + \dfrac{8}{3} \dfrac{h^2}{L}$.   □

147. In a cable subjected to a u.d.l. over the entire span, the minimum tension is given by
   (a) $\dfrac{wL^3}{16h}$ □   (b) $\dfrac{wL^2}{8h}$ □
   (c) $\dfrac{wL^2}{4h}$ □   (d) zero.   □

148. The maximum tension in a cable subjected to a u.d.l. over the entire span is given by
   (a) $\dfrac{wL}{2}\sqrt{\dfrac{L^2}{16h^2}+1}$ □   (b) $\dfrac{wL}{2}\sqrt{\dfrac{16h^2}{L^2}+1}$ □
   (c) $\dfrac{wL^2}{16h}$ □   (d) $\dfrac{16h}{wL^2}$.   □

149. The bending moment in a cable carrying a system of loads will be
    (a) maximum at the centre
    (b) minimum at the centre
    (c) zero at all points
    (d) none of the above.

150. The influence line diagram for the horizontal thrust at the support of a three hinged arch when a concentrated load passes over the span is
    (a) a triangle with maximum ordinate at the centre
    (b) a triangle with maximum ordinate at one of the supports
    (c) a parabola with maximum ordinate at the centre
    (d) none of the above.

151. A fixed beam is subjected to a couple of moment M_0 at the centre. The corresponding fixed moment at each end is equal to
    (a) M_0
    (b) $\dfrac{M_0}{2}$
    (c) $\dfrac{M^2}{4}$
    (d) $\dfrac{M_0}{8}$.

152. A simply supported beam and a fixed beam are of same span and same uniform flexural rigidity. If they are subjected to same u.d.l. over the entire span, the deflection at the centre in the case of fixed beam will be
    (a) half of the deflection at the centre of the simply supported beam
    (b) one-fourth of the deflection at centre of the simply supported beam
    (c) three-fourths of the deflection at the centre of the simply supported beam
    (d) one-fifth of the deflection at the centre of the simply supported beam.

153. What is the ratio of the bending moment at the centre of a simply supported beam to the bending moment at the centre of a fixed beam, when both are of same span and both are subjected to same u.d.l.
    (a) 1.5
    (b) 3
    (c) 6
    (d) 9.

154. A fixed beam is subjected to a u.d.l. over its entire span. The points of contraflexure will occur on either side of the centre at a distance of.......from the centre.
    (a) $\dfrac{L}{\sqrt{3}}$
    (b) $\dfrac{L}{3}$
    (c) $\dfrac{L}{2\sqrt{3}}$
    (d) $\dfrac{L}{4\sqrt{3}}$.

155. "The first partial derivative of the strain energy in a structure with respect to any particular deflection component is equal to the force applied at the point and in the direction corresponding to that deflection component" is the statement of
    (a) the Castigliano's theorem
    (b) Maxwell's theorem
    (c) Betti's theorem
    (d) none of the above.

156. A simply supported beam of span L carries a u.d.l. of w per unit length over the entire span. The strain energy stored by the beam is given by

(a) $\dfrac{w^2 L^5}{84 EI}$ ☐ (b) $\dfrac{w^2 L^5}{18 EI}$ ☐

(c) $\dfrac{w^2 L^5}{240 EI}$ ☐ (d) $\dfrac{w^2 L^5}{284 EI}$. ☐

157. A simply supported beam of span L carries a concentrated load W at the centre. The strain energy stored in the beam is

(a) $\dfrac{W^2 L^3}{48 EI}$ ☐ (b) $\dfrac{W^2 L^3}{96 EI}$ ☐

(c) $\dfrac{W^2 L^3}{24 EI}$ ☐ (d) $\dfrac{W^2 L^3}{192 EI}$. ☐

158. Two wires AO and BO support a vertical load W at O as shown in the adjoing figure. The wires are of equal length and equal cross-sectional area. The tension in each wire is equal to

(a) $\dfrac{W}{2}$ ☐

(b) W ☐

(c) $W \sqrt{2}$ ☐

(d) $\dfrac{W}{\sqrt{2}}$. ☐

159. In the question 158, what is the vertical deflection of O

(a) $\dfrac{WL}{AE}$ ☐ (b) $\dfrac{WL}{2AE}$ ☐

(c) $\dfrac{WL}{4AE}$ ☐ (d) $\dfrac{WL}{8AE}$. ☐

160. The maximum strain energy stored in a material at elastic limit per unit volume is known as
(a) resilience ☐ (b) proof resilience ☐
(c) modulus of resilience ☐ (d) modulus of rigidity. ☐

161. Beam $ABCDE$ is fixed at A, roller supported at C and hinged at E. A concentrated inclined load W_1 acts at B and another concentrated inclined load W_2 acts at D. The external statical indeterminacy for this continuous beam is
(a) 1 ☐ (b) 2 ☐
(c) 3 ☐ (d) 4. ☐

162. For the continuous beam in question (161) if the concentrated loads W_1 and W_2 act vertically instead of in the inclined direction, what is the external statical indeterminancy?
(a) 1 ☐ (b) 2 ☐
(c) 3 ☐ (d) 4. ☐

163. Beam ABCD is simply supported at A, roller supported at B and D and hinged at C. The spans being AB = 10 m, BC = CD = 5 m. It is subjected to a uniformly distributed load of 1 kN/m. What is the reaction at B for this compound beam?

   (a) 5 kN
   (b) 10 kN
   (c) 12.5 kN
   (d) 15 kN.

164. For the compound beam in question 163, what is the bending moment at C?

   (a) maximum
   (b) cannot be determined
   (c) zero
   (d) none of the above.

165. What is the ratio of the section modulus of a circular section of diameter D to the section modulus of a square section of side D?

   (a) $\dfrac{3\pi}{4}$
   (b) $\dfrac{3\pi}{8}$
   (c) $\dfrac{3\pi}{16}$
   (d) $\dfrac{3\pi}{32}$.

166. A beam of length L is subjects to bending moment M and the moment of inertia of its section is I. Both M and I are varying over its length. The total strain energy of the beam is given by

   (a) $\int_0^L \dfrac{M^2}{EI} dx$
   (b) $\int_0^L \dfrac{2M^2}{EI} dx$
   (c) $\int_0^L \dfrac{4M^2}{EI} dx$
   (d) $\int_0^L \dfrac{M^2}{2EI} dx$.

167. For the truss shown in the figure aside what is the force in the member RS

   (a) 50 kN compression
   (b) 50 kN tension
   (c) 25 kN compression
   (d) zero.

168. A square column with its side of b metres is subjected to a vertical load P acting at the centroid of one of the quarters of the square. What is the maximum bending stress?

   (a) $\dfrac{3P}{b^2}$
   (b) $\dfrac{2P}{b^2}$
   (c) $\dfrac{P}{b^2}$
   (d) $\dfrac{4P}{b^2}$.

169. Which of the following has no influence on the shear stress developed in a shaft while transmitting the power?

   (a) radius of shaft
   (b) length of shaft
   (c) angle of twist
   (d) modulus of rigidity.

170. A three hinged arch with a span 2L and rise H is hinged at the crown. An isolated concentrated load of P is acting at a distance of a from the left support. What is the horizontal reaction at the left support ?

 (a) $\dfrac{Pa}{H}$ 　　　 (b) $\dfrac{Pa}{2H}$

 (c) $\dfrac{2P}{aH}$ 　　　 (d) $\dfrac{2H}{Pa}$.

171. In question 170, what is vertical reaction ?

 (a) $\dfrac{Pa}{2L}$ 　　　 (b) $\dfrac{P.L}{a}$

 (c) $\dfrac{Pa}{L}$ 　　　 (d) $\dfrac{Pa^2}{2L}$.

172. The external indeterminancy in a two hinged arch is

 (a) 4 　　　 (b) 3
 (c) 2 　　　 (d) 1.

173. The shape of a freely suspended cable is

 (a) elliptical 　　　 (b) parabolic
 (c) hyperbolic 　　　 (d) none of the above.

174. Slenderness ratio is defined as the ratio of the effective length of the column to the

 (a) least radius of gyration 　　　 (b) least lateral dimension
 (c) radius of gyration about polar axis 　　　 (d) none of the above.

175. In a three hinged arch, two hinges are provided at supports. The third hinge is generally provided at

 (a) one quarter span 　　　 (b) at the crown
 (c) any where 　　　 (d) one-eighth span.

176. A function whose value at a point represents some structural property such as shear force, bending moment etc., as a unit load is placed at that point is called

 (a) structural function 　　　 (b) analysis function
 (c) influence line 　　　 (d) none of the above.

177. In a simply supported beam of span L, a unit load traversers from left to right. The influence line for bending moment at a section 'a' metre from left support is a triangle with a maximum ordinate at 'a' as

 (a) $a\left(1-\dfrac{a}{L}\right)$ 　　　 (b) $a\left(1-\dfrac{L}{a}\right)$

 (c) $a\left(\dfrac{a}{L}-1\right)$ 　　　 (d) $a\left(\dfrac{L}{a}-1\right)$.

178. The ratio of the stiffness factor of a member when the far end is hinged to the stiffness factor of the same member when the far end is fixed is

(a) $\dfrac{1}{4}$ ☐   (b) $\dfrac{1}{3}$ ☐

(c) $\dfrac{1}{2}$ ☐   (d) $\dfrac{3}{4}$ ☐

179. In moment distribution method, the stiffness factor of a member whose far end is fixed is

(a) $\dfrac{EI}{L}$ ☐   (b) $\dfrac{2EI}{L}$ ☐

(c) $\dfrac{3EI}{L}$ ☐   (d) $\dfrac{4EI}{L}$. ☐

180. In question 179, what is the stiffness factor of the member if its far end is hinged?

(a) $\dfrac{3EI}{L}$ ☐   (b) $\dfrac{4EI}{L}$ ☐

(c) $\dfrac{EI}{L}$ ☐   (d) $\dfrac{2EI}{L}$. ☐

ANSWERS
Objective Type Questions

1. (a)	2. (b)	3. (c)	4. (d)	5. (a)	6. (b)
7. (c)	8. (c)	9. (d)	10. (a)	11. (a)	12. (b)
13. (a)	14. (c)	15. (a)	16. (b)	17. (c)	18. (a)
19. (c)	20. (a)	21. (b)	22. (a)	23. (b)	24. (a)
25. (b)	26. (c)	27. (a)	28. (b)	29. (c)	30. (c)
31. (a)	32. (b)	33. (c)	34. (d)	35. (a)	36. (b)
37. (d)	38. (d)	39. (a)	40. (b)	41. (c)	42. (a)
43. (c)	44. (a)	45. (c)	46. (d)	47. (a)	48. (b)
49. (a)	50. (b)	51. (c)	52. (d)	53. (a)	54. (b)
55. (a)	56. (c)	57. (d)	58. (c)	59. (d)	60. (a)
61. (c)	62. (a)	63. (c)	64. (a)	65. (c)	66. (d)
67. (a)	68. (b)	69. (c)	70. (a)	71. (c)	72. (a)
73. (c)	74. (a)	75. (d)	76. (b)	77. (c)	78. (d)
79. (a)	80. (b)	81. (b)	82. (c)	83. (a)	84. (b)
85. (c)	86. (a)	87. (a)	88. (b)	89. (a)	90. (a)
91. (c)	92. (c)	93. (a)	94. (a)	95. (d)	96. (a)
97. (c)	98. (d)	99. (a)	100. (d)	101. (a)	102. (b)
103. (a)	104. (a)	105. (d)	106. (a)	107. (a)	108. (c)
109. (b)	110. (a)	111. (b)	112. (c)	113. (b)	114. (a)
115. (a)	116. (c)	117. (d)	118. (b)	119. (a)	120. (b)

121. (a)	122. (a)	123. (b)	124. (b)	125. (a)	126. (b)
127. (c)	128. (c)	129. (a)	130. (a)	131. (a)	132. (a)
133. (b)	134. (c)	135. (d)	136. (c)	137. (b)	138. (a)
139. (b)	140. (a)	141. (c)	142. (a)	143. (b)	144. (d)
145. (c)	146. (d)	147. (b)	148. (a)	149. (c)	150. (a)
151. (c)	152. (d)	153. (b)	154. (c)	155. (a)	156. (c)
157. (b)	158. (d)	159. (a)	160. (c)	161. (c)	162. (b)
163. (d)	164. (c)	165. (c)	166. (b)	167. (d)	168. (a)
169. (b)	170. (b)	171. (a)	172. (d)	173. (b)	174. (a)
175. (b)	176. (c)	177. (a)	178. (d)	179. (d)	180. (a).

Chapter 14: Reinforced Cement Concrete Structures

I. INTRODUCTION

GENERAL

Cement concrete is a composite material obtained by mixing its three ingredients, namely, coarse aggregate, fine aggregate and cement in predetermined proportions (which depend on the strength to be possessed by the concrete) with specified amount of water. The coarse aggregate may sometimes be referred as gravel or metal and fine aggregate is sand. Initially the mixture will be a plastic mass which can be poured in suitable moulds, called *forms* and becomes hard progressively. This process is known as *setting*. The setting time can be divided into 3 distinct phases. The *initial set* requires 30 to 60 minutes and during this phase the concrete decreases its plasticity and develops resistance to flow. The second phase, known as *final set*, may vary between 5 to 6 hours after mixing. During the third phase, known as *progressive hardening* the concrete increases its strength.

The advantage of concrete is that it can be cast to any shape and size with an appropriate *form work*. With proper curing it attains most of its strength by the end of one month after mixing. The strength and hardness of concrete depend on the quality and proportions of the ingredients used and the properties of concrete vary almost as widely as different kinds of stones.

The concrete described above is known as *plain concrete*. Concrete is fairly strong in compression and weak in tension and it can be used where the tensile stresses are absent or negligibly small. However, the concrete used in beams, slabs etc, reinforcement bars (usually mild steel bars) have to be embedded in concrete at the tensile zones. The concrete is then called the *reinforced cement concrete* abbreviated as R.C.C.

Merits of R.C.C.:

(a) The *coefficient of linear expansion* of concrete is almost equal to that of steel.

(b) Concrete can be moulded into any shape and size and its ingredient materials are easily available.

(c) Concrete constructions are economical and their maintenance cost is almost nil

(d) Concrete is durable and is not easily affected by the atmospheric agencies

(e) Concrete is fire resistant and its construction is superior to steel and timber construction

(f) Monolithic construction is possible with concrete and this provides a greater flexibility in planning and design.

Types of Cement. There are many types of cement such as Ordinary Portland Cement (OPC), Low Heat Portland Cement, Portland – *pozzolona* Cement (PPC), High alumina cement, natural cements and special cements like masonry cement, expansive cement etc. The type of cement must be chosen such that it is the most appropriate to the work. The specifications to be satisfied by the cement can be found in the relevant IS-code.

Aggregates. Aggregates used for concrete must comply with the norms laid down in IS: 383–1970.

Measurement of Materials. All the three ingredient materials have to be measured by weight. In our country cement is supplied in bags weighing 50 kg each. The volume of cement in one bag may be taken as 34.5 litres that is 0.0345 m^3.

Water Cement Ratio. Water cement ratio is defined as the ratio of the volume of water used in making concrete to the volume of cement used. The workability and strength of concrete depend on water cement ratio. For a given proportion of materials there is one optimum value of water cement. If the actual water cement is less than this optimum it will not only reduce the strength but may be also insufficient to ensure complete setting of cement. Likewise, if the actual water cement ratio is more than optimum, it will increase the workability but decrease the strength. Some practical values of water cement for R.C.C. are about: 0.45 for 1 : 1 : 2 concrete, 0.50 for 1 : 1.5 : 3 and 0.60 for 1 : 2 : 4.

Durability of Concrete. The property of concrete by virtue of which it resists the disintegration and decay is called the durability. The disintegration and decay in concrete may be due to:

(a) Use of unsound cement which produces changes in hardened concrete due to delayed chemical reactions.

(b) Use of less durable aggregate which is acted upon by cement and atmospheric gases.

(c) Excessive pores formed while making concrete, which permits harmful gases causing disintegration.

(d) Freezing and thawing of water sucked through the cracks causing disintegration.

(e) Expansion and contraction occurring due to temperature changes, or alternate wetting and drying.

One of the main characteristic of the concrete influencing the durability is its *permeability*. Higher permeability permits rather the free flow of potentially deleterious substances like water, oxygen, carbon dioxide, chloride, sulphate and others. To ensure durability care should be taken to see that proper ingredient materials are used, mixing and compaction is properly done and sufficient cover is provided for the embedded reinforcement bars.

Workability of Concrete. Workability may be defined as the ease with which the concrete may be mixed, handled, transported, placed in position and compacted. The major factor influencing the workability is the amount of water present in the mix. The concrete mix proportions chosen should be such that the concrete is of adequate workability for the placing conditions of concrete and can be properly compacted.

Grades of Concrete. IS 456 : 2000 specifies 15 grades of concrete which are designated as $M10, M15, M20, M25, M30, M35, M40, M45, M50, M55, M60, M65, M70, M75$ and $M80$ in which the letter M refers to the mix and the number that follows the letter M refers to the *characteristic strength* in N/mm² (f_{ck}) of 150 mm concrete cube at 28 days. Thus for example $M20$ grade refers to a concrete mix whose characteristic strength is 20 N/mm².

The characteristic strength is defined as the strength of the material below which not more than 5% of the test results are expected to fall. Suppose n (say 80) cubes of $M20$ grade concrete are tested for their compressive strength. Then at the maximum 0.05 n (that is 4) cubes can have strength of less than 20 N/mm² and the remaining 0.95 n (that is 76) cubes shall have strength more than or equal to 20 N/mm².

From the definition of characteristic strength it is implied that there is only 5% chance or a probability of 0.05 for the actual strength to be less than the characteristic strength. To put it in other words the characteristic strength has 95% reliability.

The characteristic cube compressive strength of concrete is denoted by f_{ck}.

An estimate of tensile strength of concrete in flexure denoted by f_{cr} may be obtained from the following equation:

$$f_{cr} = \sqrt{f_{ck}} \qquad \qquad ...(14.1)$$

The modulus of elasticity of concrete denoted by E_c can be assumed as follows:

$$E_c = 5000 \sqrt{f_{ck}} \qquad \qquad ...(14.2)$$

IS 456:2000 recommends that minimum grade of concrete used in reinforced cement concrete works shall not be less than $M20$.

Steel Reinforcement. The steel reinforcement used in concrete are generally of the following types:

(a) Mild steel bars conforming to IS 432 (part-I): 1966 and Hot rolled mild steel deformed bars conforming to IS 1139 : 1966.

These bars have yield strength of 250 N/mm². Hence they are referred to as Fe 250 steel.

(b) Hot rolled high yield strength deformed bars conforming to IS 1139 : 1966 and Cold worked steel high strength deformed bars conforming to IS 1786 : 1979 (Grade Fe 415 and Fe 500) having 0.2% proof stress as 415 N/mm² and 500 N/mm² respectively. These are also known as CTD (cold twisted deformed) bars.

The idealized stress-strain curve for mild steel bars, and the stress-strain curves for CTD bars are shown in Fig. 14.1.

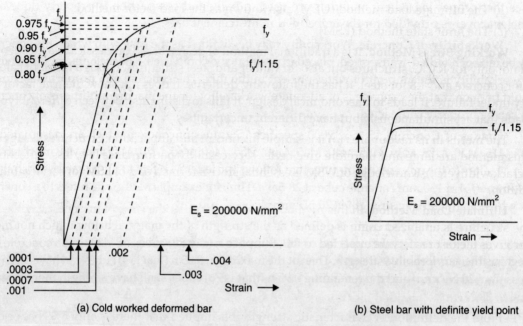

FIGURE 14.1 Stress-strain curves for mild steel and CTD bar.

The characteristic strength of steel is denoted by f_y and is defined as the strength below which not more than 5% of the test results are expected to fall. Following the similar interpretation as that of concrete grades the Fe 250, Fe415 and Fe500 grade steels will have characteristic strengths of 250 N/mm², 415 N/mm² and 500 N/mm² respectively. Fe415 is also referred to as Tor 40 and similarly Fe500 is also known as Tor 50.

When strain (x-axis) *versus* stress (y-axis) curve is prepared, we notice that the strain increases at a yield stress and the curve exhibits horizontal portion which is called *plateau*.

The steel bar can be strained beyond yield plateau by twisting or stretching and then unloading. The process is known as cold working and the tor-steel bars available in India are of this type. A twisted bar has considerable increased yield stress and it has yield stress 50 to 100% more than ordinary steel bars. Thus designs using twisted bars lead to saving in steel.

One can expect improved *bond* between concrete and steel when deformed bars are used. A deformed bar is obtained by providing lugs or ribs deformation on the surface of the bar.

High yield strength deformed bars (HYSD) are in common use. The use of HYSD bars also requires the simultaneous use of grade of concrete of $M20$ or higher.

The modulus of elasticity of steel denoted by E_s is generally taken as 2×10^5 N/mm². The stress-strain relation for steel is assumed to be same both in tension and compression.

DESIGN CONCEPTS

The reinforced cement concrete structures may be designed using any one of the following three methods:

(*a*) The working stress method (WSM) also called the Modular Ratio Method.

(b) The ultimate load method (ULM) also known as the load factor method.

(c) The limit state method (LSM).

Working Stress Method. It is a traditional method of design based on classical elastic theory used not only for R.C.C. structures but also for timber and steel structures. It uses a factor of safety of 3 for concrete and 1.8 for steel. It has the following demerits. It does not show the true factor of safety under failure. It leads to uneconomical design. It fails to distinguish between different types of loads that act simultaneously but have different uncertainties.

The merits in its favour are : (a) it is simple in concept and application. Structures designed by this method are large and therefore give better serviceability performance (i.e., less deflection, less crack width etc.,). Knowledge of WSM is essential since it forms a part of LSM for serviceability condition.

Ultimate Load Method. In this method stress condition at the state of impending collapse of the structure is analyzed using the non-linear stress-strain curves of concrete and steel. Load factor gives factor of safety. In brief, the ultimate load method ensures safety at ultimate loads but disregards the serviceability at service loads.

Limit State Method. A limit state is a state of impending failure beyond which the structure ceases to perform its intended function satisfactorily, in terms of safety or serviceability. Two types of limit states are considered in design. They are limit state of collapse and limit state of serviceability. The limit state of collapse include the limit state of collapse in flexure, in compression, in compression and uniaxial bending, in compression and biaxial bending, in shear, in bond, in torsion and in tension. The limit state of serviceability include the limit state of deflection, the limit state of cracking and the other limit states such vibration, fire resistance, durability etc.

In general, the structure shall be designed based on the most critical limit state and it shall be checked for other limit states.

Characteristic Strength and Partial Safety Factors. Suppose the tests conducted on n samples have given strengths $f_1, f_2, ..., f_n$. Then the mean strength f_m is given by

$$f_m = (f_1 + f_2 + ... + f_n)/n = \frac{1}{n} \sum_{i=1}^{n} f_i \qquad ...(14.3)$$

Similarly the standard deviation of the strength f_s is given by

$$f_s = \sqrt{\frac{\sum_{i=1}^{n} (f_i - f_m)^2}{n-1}} \qquad ...(14.4)$$

Now, if we assume that the strength follows normal probability distribution, the characteristic strengths f_k can be obtained from the relation

$$f_k = f_m - 1.645 f_s \qquad ...(14.5)$$

because the area of the normal curve between $-\infty$ and -1.645 is 0.05. The design strength of the material f_d is given by

$$f_d = f_k / \gamma_m \qquad ...(14.6)$$

where γ_m is the partial safety factor appropriate to the material and the limit state being considered. For concrete $\gamma_m = 1.5$ and for steel $\gamma_m = 1.15$

Thus for steel the design stress is given as $f_{ds} = f_y/1.15 = 0.87 f_y$. The IS 456 : 2000 code suggests that for design purpose the compressive strength of concrete in the structure shall be assumed to be 0.67 times the characteristic strength of concrete in cube and in addition the partial safety factor of 1.15 shall be applied. Thus the design stress in concrete f_{dc} is given by $f_{dc} = (0.67 f_{ck})/1.5 = 0.45 f_{ck}$. This is reflected in stress-strain curves

Characteristic Loads. A characteristic load is defined as that value of load which has a 95 per cent probability of not being exceeded during the life of structure. If F_m is the mean load, F_s is the standard deviation of the load, then the characteristic load F_k is given by

$$F_k = F_m + 1.645 F_s \qquad ...(14.7)$$

Equation (14.7) is again based on the assumption that the load F is a random variable following normal probability distribution and the normal curve has 95% of its area between $-\infty$ and $+ 1.645$. However, since adequate data is not available, IS 456 : 2000 suggests that

      the dead loads (DL) given in IS 875 (part-I)

      the imposed loads (IL) or the live loads (LL) given in IS 875 (part-II)

      the wind loads (WL) given in IS 875 (part-III)

      the snow loads (SL) given in IS 875 (part-IV)

      the seismic or earthquake loads (EL) given in IS 1893

shall be assumed as the characteristic loads

The design load is given by

$$F_d = \gamma_f F_k \qquad ...(14.8)$$

where γ_f is the partial safety factor appropriate to the nature of loading and the limit state being considered. The design load F_d given by eqn. (14.8) is also known as *factored load*.

Partial safety factors for loads may be taken from Table 18 of the code IS 456 : 2000.

Stress-Strain curves for concrete and steel. The most important characteristic of the materials in R.C.C. (that is concrete and steel) required in the design is the stress-strain curve.

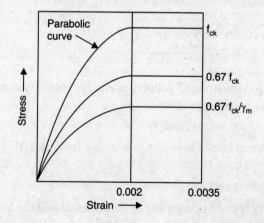

FIGURE 14.2 **Stress-strain curve for concrete.**

A typical stress strain curve for concrete is shown in Fig. 14.2. It may be noted that each curve is parabolic in the initial portion up to a strain of 0.002. At a strain of 0.002 (at 0.2% strain) the stress remains constant with increasing load until a strain of 0.35% is reached, when the concrete is said to have failed.

The stress-strain curve for steel is shown in Fig. 14.1. It may be observed from this figure that for mild steel the stress is proportional to strain up to yield point and thereafter the strain increases at constant stress.

For the stress-strain curves of Fe 415 and Fe 500, there is no definite yield point. Hence yield stress in taken as 0.2 per cent proof stress. The stress-strain curves for these two types of steel are linear up to a stress of $0.8 f_y$ and strains are elastic. Thereafter, they are not linear. The salient points on the stress-strain curves of Fe 415 and Fe 500 in the non linear portion are given in Table 14.1

Table 14.1. Salient Points on the Design Stress-Strain Curve for Cold Worked Bars

Stress level	Fe 415 : f_y = 415 N/mm²		Fe500 : f_y = 500 N/mm²	
	Strain	Stress N/mm²	Strain	Stress N/mm²
$0.8 f_{yd}$	0.00144	288.7	0.00174	347.8
$0.85 f_{yd}$	0.00163	306.7	0.00195	369.6
$0.90 f_{yd}$	0.00192	324.8	0.00226	391.3
$0.95 f_{yd}$	0.00241	342.8	0.00277	413.0
$0.975 f_{yd}$	0.00276	351.8	0.00312	423.9
$1.0 f_{yd}$	0.00380	360.9	0.00417	434.8

LIMIT STATE OF COLLAPSE IN FLEXURE (SINGLE REINFORCED RECTANGULAR SECTIONS)

Design for the limit state of collapse in flexure is based on the following assumptions:

(*i*) Plane sections normal to the axis remain plane after bending.

(*ii*) The maximum strain in concrete at the outermost compression fiber is taken as 00.0035.

(*iii*) The relationship between the compressive stress distribution in concrete and the strain in concrete may be assumed to be rectangular, trapezoidal, parabolic or any other shape which results in prediction of strength in substantial agreement with the results of the test. An acceptable stress-strain curve is given in Fig. 14.3. For design purpose the compressive strength of concrete in the structure shall be assumed to be 0.67 times the characteristic strength. The partial safety factor of $\gamma_m = 1.5$ shall be applied in addition

(*iv*) The tensile strength of concrete is ignored

(*v*) The stresses in the steel reinforcement are derived from the representative stress-strain curve for the type of steel used. For design purpose, the partial safety factor for steel $\gamma_m = 1.15$ shall be applied.

(vi) The maximum strain in the tension reinforcement in the section at failure shall not be less than $[(f_y/115 E_s) + 0.002]$.

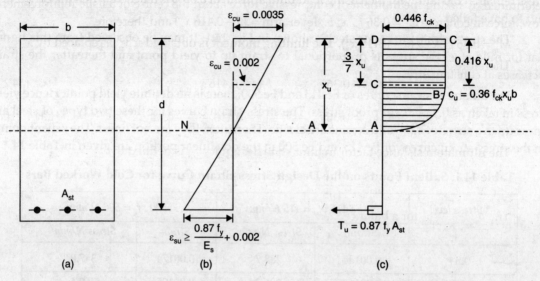

FIGURE 14.3 Stress-strain distribution across the depth of rectangular section.

The stress and strain distribution across the depth of a rectangular section is shown in Fig. 14.3. The notation is self explanatory. If x_u is the depth to Neutral Axis (NA), then from the similarity of triangles in the strain diagram, we have

$$\frac{0.0035}{x_u} = \frac{0.0035 + (0.002 + 0.87 f_y/E_s)}{d}$$

where d is the effective depth. The effective depth d of the section is the vertical distance between the tension reinforcement and the maximum compression fiber excluding the thickness of the finishing material not placed monolithically with the member. Thus we have $d = (D - \varphi/2 - clear\ cover)$ where D is the gross or overall depth and φ is the diameter of the reinforcement bars, the reinforcement being provided in only one layer. The simplification of the above expression replacing E_s by 2×10 N/mm^2 gives

$$\frac{x_u}{d} = \frac{700}{1100 + 0.87 f_y} \qquad \ldots(14.9)$$

The value of x_u given by eqn. (14.9) is the max because both concrete and steel are taken to fail simultaneously. Therefore we can write

$$\frac{x_{u,max}}{d} = \frac{700}{1100 + 0.87 f_y} \qquad \ldots(14.10)$$

It may be noted here that $x_{u,max}$ is dependent on the grade of steel only. Substituting appropriate value of f_y, we obtain that, for Fe 250 $x_{u,max}/d = 0.53$, for Fe450 it is 0.48 and for Fe 500 it is 0.46.

In a balanced section the steel reinforcement reaches its yield stress at the same instant when the ultimate strain is reached in concrete. The limiting moment of resistance $M_{u,\text{lim}}$ can be determined by taking the moment of compressive force about the centre of tensile reinforcement. Now, force in concrete $C_u = 0.36 f_{ck} x_u b$; lever arm $= (d - 0.416 x_u)$ and therefore

$M_u = 0.36 f_{ck} x_u b (d - 0.416 x_u)$. The limiting moment is obtained if x_u is replaced by $x_{u,\text{max}}$. Therefore

$$M_{u,\text{lim}} = 0.36 f_{ck} x_{u,\text{max}} b (d - 0.416 x_{u,\text{max}}). \qquad \ldots(14.11)$$

$$M_{u,\text{lim}} = 0.36 f_{ck} \frac{x_{u,\text{max}}}{d}\left(1 - 0.416 \frac{x_{u,\text{max}}}{d}\right) bd^2 \qquad \ldots[14.11(a)]$$

The ultimate resistance factor in limit state design $R_{u,\text{lim}}$ is defined by

$$R_{u,\text{lim}} = \frac{M_{u,\text{lim}}}{bd^2} \qquad \ldots(14.12)$$

$R_{u,\text{lim}}$ for a balanced section is given by

$$R_{u,\text{lim}} = 0.36 f_{ck} \frac{x_{u,\text{max}}}{d}\left(1 - 0.416 \frac{x_{u,\text{max}}}{d}\right) \qquad \ldots(14.13)$$

Taking the values of $\left(\dfrac{x_{u,\text{max}}}{d}\right)$ from eqn. (14.10) and substituting them in eqn. (14.13), we get

for Fe 250, $M_{u,\text{lim}} = 0.1489 f_{ck} bd^2$ and $R_{u,\text{lim}} = 0.1489 f_{ck}$
for Fe 415, $M_{u,\text{lim}} = 0.3810 f_{ck} bd^2$ and $R_{u,\text{lim}} = 0.3810 f_{ck}$
for Fe 500, $M_{u,\text{lim}} = 0.1330 f_{ck} bd^2$ and $R_{u,\text{lim}} = 0.1330 f_{ck}$

If A_{st} is the area of reinforcement steel provided, then the reinforcement ratio p_t is given by

$$p_t = \frac{A_{st}}{bd} \qquad \ldots(14.14)$$

The reinforcement required for a balanced section can be obtained by equating the forces of compression and tension, that is forces in steel and concrete

$$A_{st}(0.87 f_y) = 0.36 f_{ck} b x_{u,\text{max}}$$

$$p_{t,\text{lim}} = \frac{A_{st,\text{max}}}{bd} = 0.414 \frac{f_{ck}}{f_y} \frac{x_{u,\text{max}}}{d} \qquad \ldots(14.15)$$

The actual design may not always adopt a balanced section. When the strain in extreme compression member reaches 0.0035, the actual strain, ε_s, in steel can have the following values:

(i) ε_s = failure strain, corresponding to balanced section
(ii) ε_s > failure strain, corresponding to under-reinforced section
(iii) ε_s < failure strain, corresponding to over reinforced section

We know that the failure strain in steel $\varepsilon_s \geq \left[\dfrac{f_y}{1.15 E_s} + 0.002\right]$.

Over-reinforced section. In an over reinforced section $p_t > p_{t,\lim}$ and $x_u > x_{u,\max}$. Since concrete fails first, the failure will be sudden and therefore is not desirable. The code recommends that if $\left(\dfrac{x_u}{d}\right)$, is found to be greater than $\left(\dfrac{x_{u,\max}}{d}\right)$, the section should be redesigned.

Under-reinforced section. In an under reinforced section the steel provided $p_t < p_{t,\lim}$ and hence depth to NA will be less than $x_{u,\max}$. That is $x_u < x_{u,\lim}$. The variation of strain across the depth is shown in Fig. 14.4 both for under reinforced and over reinforced sections.

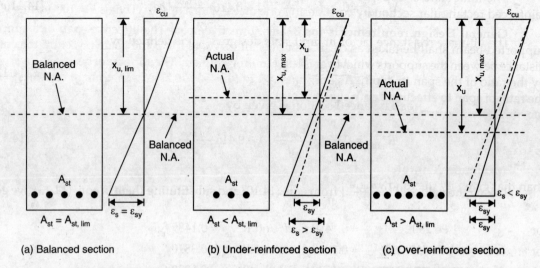

(a) Balanced section   (b) Under-reinforced section   (c) Over-reinforced section

FIGURE 14.4 Stress variation in balanced, under and over reinforced sections.

In the under reinforced section, the strain in steel at the limit state of collapse will be more than $\left(0.002 + \dfrac{0.87 f_x}{E_s}\right)$ and the stress in steel will fully reach to its maximum value $0.87 f_y$. The actual value of $\left(\dfrac{x_u}{d}\right)$ can be determined by equating the compressive and tensile forces, that is $C_u = T_u$, which gives

$$\dfrac{x_u}{d} = 2.417\, p_t\, \dfrac{f_y}{f_{ck}} \qquad \ldots(14.16)$$

The moment of resistance of an under reinforced section can be obtained by multiplying the tensile force by the lever arm

$$M_u = 0.87 f_y A_{st} (d - 0.416\, x_u) = 0.87 f_y A_{st} d \left(1 - 0.416\, \dfrac{x_u}{d}\right)$$

Using eqn. (14.16) for $\left(\dfrac{x_u}{d}\right)$ and then simplifying the above expression, we get

$$M_u = 0.87 f_y A_{st} d \left(1 - \frac{f_y}{f_{ck}} \frac{A_{st}}{bd}\right) \quad \ldots(14.17)$$

or

$$\frac{M_u}{bd^2} = 0.87 f_y p_t \left(1 - \frac{f_y}{f_{ck}} p_t\right) \quad \ldots(14.18)$$

The balanced design gives smallest concrete area and maximum reinforcement area. Since the cost of steel is quite high compared to concrete the balanced section need not necessarily be economical. Under reinforced design is always desirable.

The limit state of collapse in flexure in the case of slabs may be treated similar to singly reinforced rectangular section by considering one metre width of slab. That is $b = 1000$ mm.

General Design requirements for Beams. Effective span: The effective span of a simply supported beam shall be taken as clear span plus effective depth of the beam or centre to centre distance between the supports whichever is less Limiting stiffness: The stiffness of beams is governed by the ratio of the span to depth. As per clause 23.2 of IS 456 : 2000 for spans not exceeding 10 m, the ratio of span to effective depth shall not exceed 7 for cantilever, 20 for simply supported and 26 for continuous beams.

However, for spans exceeding 10 m, the above limits may be multiplied by $(10/L)$, where L is span in m.

Minimum Reinforcement. The minimum area of tension reinforcement should not be less than that given by the following expression.

$$A_{st} \geq \frac{0.85bd}{f_y} \quad \ldots(14.19)$$

This work out to only 0.2% for Fe 415 steel and 0.34% for Fe250 steel.

Maximum Reinforcement. The maximum area of tension reinforcement should not exceed 4% of the gross cross-sectional area. That is, $A_{st} < 0.04\ bD$, where D is the gross depth of beam.

Spacing of Bars: The horizontal distance between two parallel main reinforcing bars shall usually be not less than the greatest of the following.

(a) Diameter of the bar if the diameters are equal

(b) diameter of the largest bar if the bars are unequal

(c) 5 mm more than the nominal maximum size of the aggregate

When there are two or more rows of bars, the bars shall be vertically in line and the minimum vertical distance between the bars shall be 15 mm or two thirds of nominal maximum size of aggregate, or the maximum size of the bar whichever is greater.

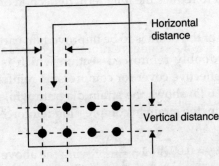

FIGURE 14.5 Reinforcement particulars for a 2 layer case.

Cover to Reinforcement. The reinforcement in concrete shall have cover of thickness as given below.

(a) At each end of reinforcement bar not less than 25 mm nor less than twice the diameter of the bar

(b) For longitudinal reinforcing bar in beam not less than 25 mm nor less than the diameter of such bar.

Side Face Reinforcement. Where the depth of the beam exceeds 750 mm, side face reinforcement shall be provided equally along the two faces with the area of such reinforcement being not less than 0.1% of the beam area at a spacing not exceeding 300 mm or width of the beam whichever is less.

The ISI special publication SP 16 titled "Design aids for Reinforced concrete to IS 456" contains a number of charts and tables which will be highly useful in determining the area of steel required.

Tables 1 to 4 give the percentage steel required for various values of $\left(\dfrac{M_u}{bd^2}\right)$ and f_y and for concrete grades of f_{ck} = 15, 20, 25 and 30 N/mm². Charts 1 to 18 give the moment of resistance per metre width for different depths (ranging from 5 to 80 cm) and varying percentage of steel and for various values of f_{ck} = 15 and 20 using steel of grades f_y = 250, 415 and 500.

DOUBLY REINFORCED RECTANGULAR SECTIONS

The case of $M_u > M_{u,\lim}$ can be addressed in two ways: Either by increasing the depth of the section till $M_u = M_{u,\lim}$ or by providing compression reinforcement. It may not be possible to increase the depth due to architectural and other considerations. Then the only alternative available would be to provide the reinforcement in the compression zone, that is making the section doubly reinforced.

The doubly reinforced sections are adopted under the following situations:

(i) When the depth is restricted due to architectural and other considerations

(ii) At supports of continuous beams where the bending moment changes its sign

(iii) In precast members which may be subjected to reversal of bending moment during handling.

(iv) In bracing members of a frame due to changes in the direction of wind loads.

(v) When it is required to reduce the long term deflection or to increase the stiffness of the beam.

(vi) When the ductility of the beam is to be improved. In earthquake zones, structures.

Figure 14.6 shows a doubly reinforced section which is provided with compression reinforcement at a depth d' (effective cover for compression reinforcement) below the outermost compression fiber. Figure 14.6 (b) shows the strain diagram while Fig. 14.6 (c) shows the stress block. From the strain diagram, the strain in compression reinforcement ε_{sc} is given by

$$\varepsilon_{sc} = 0.0035 \left(1 - \dfrac{d'}{x_a}\right) \qquad \ldots(14.20)$$

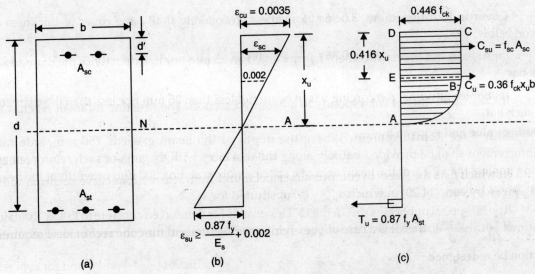

FIGURE 14.6 Variation of stress and strain in a doubly reinforced sections.

If f_{sc} is the stress developed in compression steel which depends on ε_s and which can be read off from the stress-strain curve of the steel used in the structure or from Table 14.1, the compressive force in compression steel C_s can be written as

$$C_s = f_{sc} A_{sc} \qquad \ldots(14.21)$$

where A_{sc} is the area of compression steel. Similarly the compression force in concrete C_c is written as

$$C_c = 0.36 f_{ck} b x_u \qquad \ldots(14.22)$$

Therefore the total compressive force C_u is given by

$$C_u = C_s + C_c = f_{sc} A_{sc} + 0.36 f_{ck} b x_u \qquad \ldots(14.23)$$

The total tensile force in the tension reinforcement T_u is obtained from

$$T_u = 0.87 f_y b A_{st} \qquad \ldots(14.24)$$

where A_{st} is the total tension reinforcement. The location of NA can be computed by equating C_u and T_u. That is

$$0.36 f_{ck} b x_u + f_{sc} A_{sc} = 0.87 f_y b A_{st} \qquad \ldots(14.25)$$

Solving for x_u from eqn. (14.25) required an interactive procedure since f_{sc} depends on ε_{sc} which in turn depends on x_u. If the loss of compression area occupied by compression steel is taken into account eqn. (14.25) is modified as

$$0.36 f_{ck} x_u b + (f_{sc} - 0.446 f_{ck}) A_{sc} = 0.87 f_y b A_{st} \qquad \ldots(14.26)$$

However, the term $0.446 f_{ck} A_{sc}$ will be very small and hence can be neglected. Generally, eqn. (14.25) only is used to find x_u.

The limiting value of $\dfrac{x_{u,\lim}}{d}$ can be obtained using the expression

$$\dfrac{x_{u,\lim}}{d} = \dfrac{700}{1100 + 0.87 f_y}. \text{ If } \dfrac{x_u}{d} < \dfrac{x_{u,\lim}}{d}, \text{ then the moment of resistance } M_u \text{ is given by}$$

$$M_u = 0.36 f_{ck} x_u b (d - 0.416 x_u) + A_{sc} f_{sc} (d - d')$$

$$M_u = 0.36 f_{ck} \frac{x_u}{d}\left(1 - 0.416 \frac{x_u}{d}\right) b d^2 + A_{sc} f_{sc} (d - d') \qquad ...(14.27)$$

If $\dfrac{x_u}{d} = \dfrac{x_{u,\text{lim}}}{d}$ then it corresponds to a balanced section. The resisting moment reaches its limiting value and is given by

$$M_{u,\text{lim}} = 0.36 f_{ck} x_{u,\text{lim}} b(d - 0.416 x_{u,\text{lim}}) + A_{sc} f_{sc} (d - d') \qquad ...(14.28)$$

In which f_{sc} is the stress in compression steel found from stress strain curve corresponding to ε_{sc} given by eqn. (14.20) in which $x_{u,\max}$ is substituted for x_u

If $\dfrac{x_u}{d} > \dfrac{x_{u,\max}}{d}$, then it is a case of over-reinforced section and the code recommends that the section be redesigned.

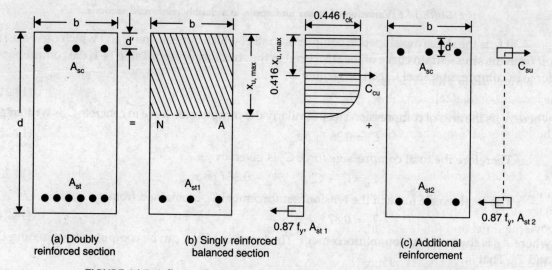

FIGURE 14.7 Split up of moments of resistance in doubly reinforced beam.

Design of doubly reinforced section to resist the applied moment $M_{u,a}$: From Fig. 14.7, the moment taken care of by the balanced single reinforced section is $M_{u,\text{lim}}$ and $(M_{u,a} - M_{u,\text{lim}})$ denoted by $M_{u,2}$ has to be taken care of by the additional tension reinforcement A_{st2} and the compression reinforcement A_{sc}. Therefore, a doubly reinforced section can be taken as equivalent to a singly reinforced balanced section and a section with additional tension and compression reinforcement. Thus we have A_{st} the total tension reinforcement to consist of A_{st1} the tension reinforcement needed for the balanced section and A_{st2} the additional tension reinforcement. That is $A_{st} = A_{st1} + A_{st2}$

$A_{st1} = p_{t,\text{lim}} bd$, where $p_{t,\text{lim}}$ is given by eqn. (14.15)

$$A_{st2} = \frac{M_u - M_{u,\text{lim}}}{0.87 f_y (d - d')} = \frac{A_{sc} f_{sc}}{0.87 f_y} \qquad ...(14.29)$$

The compression reinforcement A_{sc} can be obtained from

$$A_{sc} = \frac{M_u - M_{u,\lim}}{f_{sc}(d-d')} = \frac{M_2}{f_{sc}(d-d')} \qquad \ldots(14.30)$$

A_{sc} can also be obtained from

$$A_{sc} = \frac{0.87 f_y A_{st2}}{(f_{sc} - 0.45 f_{ck})} \qquad \ldots(14.31)$$

Also, $\quad p_c = \dfrac{A_{sc}}{bd} \qquad \ldots(14.32)$

where p_c is the per cent compression reinforcement

Tables 45 to 56 in SP 16 give the percentage tension and compression reinforcements, that is, p_t and p_c for different combinations of $\dfrac{d'}{d}$ (in the range 0.05 to 0.20), grades of concrete (f_{ck} = 15, 20, 25, and 30 N/mm^2), grades of steel (f_y = 250, 415, and 500 N/mm^2) and the moment of resistance factor $\dfrac{M_u}{bd^2}$ (in the range 2.24 to 8.3) which provide quick solution to the design problems.

Steel Beam Theory. If the amount of compression reinforcement required equals or exceeds the amount of total tension reinforcement in the above analysis the beam section may be designed using the steel beam theory. In this theory the compressive resistance provided by the concrete is wholly ignored. Equal areas of compression and tension reinforcement are provided. That is

$$A_{sc} = A_{st} = M_u / [0.87 f_y (d-d')]$$

Lintels. Lintel is an ordinary beam provided over an opening in the wall such as door or window. Load coming on to the lintel depends on (*i*) height of the masonry above the lintel (*ii*) length of the supporting walls at the ends of the lintel (*iii*) positions of the openings in the wall above the lintel and (*iv*) load transmitted by the roof slab on to the wall. Once the moment to be carried by the lintel is arrived at the lintel can be designed as a singly reinforced section. The width of the lintel beam is taken equal to the width of the walls.

T-BEAMS AND L-BEAMS

When slabs are running over more than one adjacent spans, subjecting the beams on which they are resting to uniformly distributed load, since beam and slab are cast monolithically, the slab acts as a flange to the beam and participates in taking compressive forces. In such situations the beam is designed as either T-beam (when the slab is on both sides of the beam) or L-beam (when the slab is only on one side of the beam). Figure 14.8 shows the location of T and L beams, where JK and MN are beams and PJKQ, JMNK and MRSN are the slabs.

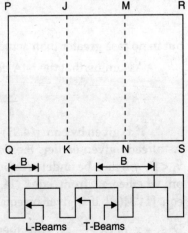

FIGURE 14.8 T-Beams and L-Beams.

In Fig. 14.9 which shows the cross-section of a typical T-beam, b_w = breadth of the rib, b_f = breadth of the flange, d_w = depth of the rib, d = effective depth and D_f = thickness of the flange

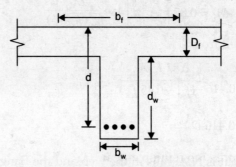

FIGURE 14.9 Cross-section of a T-Beam.

The effective width b_f of the flange is taken as the least of the following:

(a) $b_f = \dfrac{l_0}{6} + 6D_f + b_w$, where l_0 is the distance between points of zero moment in the beam

and (b) $b_t = b_w + \frac{1}{2}$ (sum of the clear distances to the adjacent beams on either side)

However, for isolated beams, the effective flange width shall be obtained as given below but in no case greater than the actual width.

$$b_l = \dfrac{l_0}{\dfrac{l_0}{b} + 4} + b_w \qquad ...(14.33)$$

For continuous beams and frames l_0 may be taken as 0.7 times the effective span. For L-beams

$$b_f = \dfrac{l_0}{12} + b_w + 3D_f \qquad ...(14.34)$$

For an isolated L-beam

$$b_l = \dfrac{l_0}{\dfrac{l_0}{b} + 4} + b_w \qquad ...[14.34\,(a)]$$

but in no case greater than actual width

Assuming that the N.A. falls within the flange we can show that

$$\dfrac{x_u}{d} = 2.417\, p_t\, \dfrac{f_y}{f_{ck}} \qquad ...(14.35)$$

If x_u given by eqn. (14.35) is less than D_f, it will be treated as rectangular section of width b_f. As already given earlier, the depth of critical NA is given by eqn. [14.34(a)]. The T beam with $x_u < D_f$ will then be under reinforced or balanced or over reinforced rectangular section depending on whether x_u from eqn. (14.35) is less than or equal to or more $x_{u,\max}$ calculated from eqn. [14.34(a)]. If x_u given by eqn. (14.35) is greater than D_f, again there will be two cases : (i) whether $D_f \leq \dfrac{3}{7} x_u$ or $D_f > \dfrac{3}{7} x_u$. Thus there are total 3 cases in all.

Case (i): $D_f > x_u$ that is N.A. is entirely within the flange, the moment of resistance is given by

$$M_u = 0.87 f_y A_{st} d \left(1 - \frac{f_y}{f_{ck}} \frac{A_{st}}{b_f d}\right) \qquad \text{...(14.36)}$$

when $D_f = x_u$, then

$$M_u = 0.87 f_y A_{st} d \left(1 - 0.416 \frac{D_f}{D}\right) \qquad \text{...(14.37)}$$

or $\quad M_u = 0.36 f_{ck} b_f D_f (d - 0.416 D_f) \qquad \text{...[14.37(a)]}$

Case (ii): $D_f \leq \frac{3}{7} x_u$ that is N.A. falls in the web and the flange lies completely within the rectangular portion of the stress block. Then M_u is given by

$$M_u = 0.36 \frac{x_u}{d}\left(1 - 0.416 \frac{x_u}{d}\right) f_{ck} b_w d^2 + 0.446 f_{ck} (b_f - b_w) D_f (d - 0.5 D_f) \qquad \text{...(14.38)}$$

when $D_f = \frac{3}{7} x_u$, then M_u is given by

$$M_u = 0.36 f_{ck} b_w \left(\frac{7}{3} D_f\right)\left(d - 0.416 \frac{7}{3} D_f\right) + 0.446 f_{ck} (b_f - b_w) D_f (d - 0.5 D_f) \qquad \text{...(14.39)}$$

Case (iii): $D_f > \frac{3}{7} x_u$. In this case also NA falls in the web, but the distribution of compression stress in the flange will not be totally rectangular. Then M_u for this case is given by

$$M_u = 0.36 \frac{x_u}{d}\left(1 - 0.416 \frac{x_u}{d}\right) f_{ck} b_w d^2 + 0.446 f_{ck} (b_f - b_w) y_f (d - 0.5 y_f) \qquad \text{...(14.40)}$$

where y_f is the depth of equivalent rectangular block given by

$$y_f = 0.15 x_u + 0.65 D_f \text{ (but not greater than } D_f\text{)} \qquad \text{...(14.41)}$$

In most of the cases where T-beams are designed for practical situations, the neutral axis falls within the flange and the area of steel to be provided can be obtained using Table 1 to 4 in SP:16 as in the design of rectangular beam

Design Tables 57 to 59 are useful to compute the limiting moment of resistance factor $\frac{M_{u,\lim}}{b_w d^2 f_{ck}}$ for singly reinforced T-sections. These table cover different grades of steel (250, 415 and 500) $\frac{D_f}{d}$ ratio varying from 0.06 to 0.45 and $\frac{b_f}{b_w}$ ratio varying from 1 to 10.

When $M_{u,\lim} < M_u$, then T-beam has to be designed as doubly reinforced section providing required compression reinforcement. IS: 456–2000 specifies that the area of the tension reinforcement shall not be less than that given by

$$\frac{A_{st}}{b_w d} = \frac{0.85}{f_y} \quad \text{or} \quad A_{st} \geq \frac{0.85\, b_w d}{f_y} \qquad \ldots(14.42)$$

Similarly the maximum reinforcement shall not exceed 4% of the gross sectional area (based on web width). That is

$$A_{st} \leq 0.04\, b_w D \qquad \ldots(14.43)$$

LIMIT STATE DESIGN FOR SHEAR

The total external shear force V_u is jointly resisted by the concrete as well as shear reinforcement. That is $V_u = V_{uc} + V_{us}$. Where V_{uc} is the shear strength of concrete and is given by $V_{uc} = \tau_c\, bd$, where τ_c is the design shear strength of concrete which can be read from table 19 in IS 456 : 2000 or can be computed using the following expression

$$\tau_c = \frac{0.85\sqrt{0.8 f_{ck}}\,(\sqrt{1+5\beta}-1)}{6\beta} \qquad \ldots(14.44)$$

where $\beta = \dfrac{0.8 f_{ck}}{6.89\, p_t}$ (but not less than 1) $\qquad \ldots[14.44\,(a)]$

The nominal shear stress in beam of uniform depth is given by

$$\tau_v = \frac{V_u}{bd} \qquad \ldots(14.45)$$

For solid slabs, the design shear strength for concrete shall be taken as $\tau_c k$, where the value of k is tabulated on page 72 of IS : 456 : 2000.

For members subject to axial compression P_{uc}, the design shear strength of concrete obtained from eqn. (14.44) shall be multiplied by the following factor

$$\delta = 1 + \frac{3\, P_{uc}}{A_g f_{ck}} \quad \text{but not more than 1.5} \qquad \ldots(14.46)$$

where A_g = Gross area of concrete in mm^2. The case of member subjected to axial tension is not covered by IS 456 : 2000

The distance between the support and the load is known as *shear span* and is denoted by a_v. The shear failure in beams without shear reinforcement is influenced by the ratio a_v/d. The three types of failure that may be experienced by a beam are : (*i*) Splitting and compression failure in deep beam when $\dfrac{a_v}{d} < 1$, (*ii*) Shear compression or shear tension failure when $1 < \dfrac{a_v}{d} < 2.8$, and (*iii*) Diagonal tension failure when $2.8 < \dfrac{a_v}{d} < 6$.

Grade of concrete, p_t and grade of steel, ratio of shear span to effective depth $\frac{a_v}{d}$, compressive force, compression reinforcement, axial tensile force and shear reinforcement provided are the factors which affect the shear resistance of a R.C. member.

Shear reinforcement may be provided in the form of vertical stirrups, bent up bars along with stirrups, and inclined stirrups. The shear force to be resisted by the shear reinforcement $V_{us} = V_u - V_{cu}$.

When vertical stirrups are used with total cross-section area A_{sv} of each set of bar or link, the spacing of stirrups s_v is given by

$$s_v = \frac{0.87 f_y A_{sv} d}{V_{us}} = \frac{0.87 f_y A_{sv}}{(\tau_v - \tau_c) b} \qquad ...(14.47)$$

When bars inclined at 45° are provided, the spacing of inclined bars is given by

$$s_v = \frac{0.87 f_y A_{sv} d\sqrt{2}}{V_{us}} \qquad ...(14.48)$$

As per IS : 456 – 2000, minimum reinforcement in the form of stirrups shall be provided such that

$$\frac{A_{sv}}{b s_v} \geq \frac{0.4}{0.87 f_g} \quad \text{or} \quad \frac{A_{sv}}{b s_v} \geq \frac{0.46}{f_y} \qquad ...(14.49)$$

This corresponds to spacing of stirrups as under.

$$s_v \leq \frac{0.87 f_y A_{sv}}{0.4 b} \quad \text{or} \quad s_v \leq \frac{2.175 f_y A_{sv}}{b} \qquad ...(14.50)$$

Under no circumstances, even with shear reinforcement, shall the nominal shear stress τ_v in beams exceed $\tau_{c,max}$ given by eqn. (14.44).

However, where the calculated maximum shear stress is less than $0.5\ \tau_c$ and also in the members of minor structural importance such as lintel beams this provision need not be complied with.

The shear reinforcement provided in R.C.C. members has the following functions.

(a) It prevents sudden shear failure and imparts ductility to the member thereby giving sufficient warning impending failure. It prevents the brittle shear failure.

(b) It guards against concrete cover bursts and loss of bond to tension steel.

(c) It holds the main reinforcement in place and helps in maintaining the requirements of cover for concrete and spacing of reinforcement.

(d) It acts as necessary tie for compression reinforcement.

(e) It confines concrete and thereby increasing its strength.

(f) It prevents failure that can be caused due to shrinkage and thermal stresses.

LIMIT STATE DESIGN FOR BOND

The force which acts parallel to the axis of the bar on the interface between the reinforcement bar and the surrounding concrete and which prevents the slippage between the concrete and the steel bar is known as bond. The bond force developed is due to the combined effect of adhesion between concrete and steel, friction due to shrinkage of concrete, and interlocking of ribs of the steel bar with concrete.

Table 14.2. Design Bond stress for Plain Bars in Tension (IS 456 : 2000)

Grade of concrete	M20	M25	M30	M35	M40 and above
Design bond stress τ_{bd} in N/mm²	1.2	1.4	1.5	1.7	1.9

The design bond stress τ_{bd} in the limit state method for plain bars is tension is as given in Table 14.2. For deformed bars these values are increased by 60% and for bars in compression, the value of design bond stress in tension shall be increased by 25%.

Thus the bond stress in compression for plain bars in obtained by multiplying the value from Table 14.2 by 1.25 whereas the bond stress in compression for the deformed bars is obtained by multiplying the value from Table 14.2 by (1.6 × 1.25).

The development length which ensures enough bond force to prevent the slippage is given by

$$L_d = \frac{0.87 f_y \phi}{4 \tau_{bd}} \qquad ...(14.51)$$

where ϕ is the nominal diameter of the bar

The anchorage value of a standard U type hook (180° bend) shall be taken equal to 16 ϕ.

The anchorage value of a standard bend shall be taken as 4 times ϕ for each 45° in the bend subjected to a maximum of 16 times ϕ.

The development length L_d at simple supports shall be limited to

$$L_d \leq \frac{M_1}{V} + L_0 \qquad ...(14.52)$$

and at a simple support where the compressive reaction confines the ends of reinforcing it shall be confined to

$$L_d \leq 1.3 \frac{M_1}{V} + L_0 \qquad ...[14.52\ (a)]$$

where

M_1 = Moment of resistance of the section

V = Shear force at the section due to design loads

L_0 = Extended length of the bar beyond point of zero bending which is limited to the effective depth or 12 ϕ whichever is greater

In eqn. (14.52), M_1 is computed from

$$M_1 = 0.87 f_y A_{st1} (d - 0.416 x_u) \qquad \ldots(14.53)$$

$$x_u = \frac{0.87 f_y A_{st1}}{(0.36 f_{ck} b)} \qquad \ldots[14.53(a)]$$

In the case of splicing of tension reinforcement, splices should not be used for $\phi \geq 36$ mm instead bars may be welded. In bars of $\phi < 36$ mm the lap length in flexural tension shall be L_d or 30ϕ whichever is larger and in direct tension $2L_d$ or 30ϕ whichever is larger. The lap length in compression shall be L_d or 24ϕ whichever is larger.

LIMIT STATE DESIGN FOR TORSION

Equivalent shear V_e is computed as

$$V_e = V_u + 1.6 \frac{T_u}{b} \qquad \ldots(14.54)$$

The nominal shear stress τ_{ve} is given by

$$\tau_{ve} = \frac{V_e}{bd} \qquad \ldots(14.55)$$

which shall not exceed $\tau_{c,\max}$ given in Table 14.2.

The equivalent bending moment M_{e1} is given by

$$M_{e1} = M_u + M_t \qquad \ldots(14.56)$$

where M_u is the bending moment at the section, and

$$M_t = T_u \left(\frac{(1 + D/b)}{1.7} \right) \qquad \ldots(14.57)$$

where
T_u = The torsional moment
d = Effective depth
b = Breadth of the beam
D = Overall depth of the beam

The longitudinal reinforcement must be designed for M_{e1}. If $M_t > M_u$ then the longitudinal reinforcement shall be provided on the flexural compression face such that the beam can also withstand an equivalent moment M_{e2} given by

$$M_{e2} = M_t - M_u \qquad \ldots(14.58)$$

where M_{e2} is taken to act in direction opposite to M_u.

When the depth of beam exceeds 450 mm, the side face reinforcement shall be provided along the two faces. The total area of such reinforcement shall not be less than 0.1 percent of the web area and shall be distributed equally on two faces at a spacing not exceeding 300 mm or web thickness whichever is less.

LIMIT STATE OF SERVICEABILITY IN DEFLECTION AND CRACKING

Various considerations that come under the purview of serviceability are : (*i*) deflection (*ii*) cracking (*iii*) vibrations (*iv*) slenderness (*v*) impermeability (*vi*) acoustic and (*vii*) thermal insulation. The most important are deflection and cracking. It should be noted that the safety requirements against flexure, shear, bond etc. need not automatically satisfy the serviceability requirements. The two most important serviceability conditions are : (1) the member should not undergo excessive deformation and (2) the crack width at the surface of the R.C.C. member should not be more than that specified by the code. As per IS 456 : 2000, the combinations of loads for serviceability conditions should be the largest of the following:

(*i*) 1.0 DL + 1.0 LL (*ii*) 1.0 DL + 1.0 WL (*iii*) 1.0 DL + 0.80 LL + 0.8 WL (or EL)

where DL = Dead load, LL = Live load, WL = Wind load and EL = Earthquake load.

SLABS

Slabs are simple plane structural members whose thickness is smaller compared to its length and breadth. They are most commonly used as roof coverings and floors. They can be in various shapes like square, rectangle, circle, triangle etc. Slabs support transverse loads and transfer them to the supports by bending action in one or more directions. They are generally supported by beams or walls. When supported on all four edges, the slabs are classified into *one way slab* or *two way slab* depending on the ratio of longer span l_y to shorter span l_x. When $\left(\dfrac{l_y}{l_x}\right)$ is greater than 2 they are called *one way slabs*. Bending in one way slab occurs only in one direction that is along shorter span. Minimum reinforcement is to be provided along the longer span to distribute the load uniformly and also to resist the temperature and shrinkage stresses. In slabs supported on all the four edges if $\dfrac{l_y}{l_x} < 2$ the bending takes place both along l_x and l_y and hence the reinforcement has to be provided in both the directions. Obviously, when the slab is supported only on two opposite edges it behaves like one way slab.

A slab supported directly on the columns without any intermediate beams is called a *flat slab*.

Circular slabs are used to provide (*i*) roof to a room circular in plan (*ii*) floor for a circular tank (*iii*) roof of a pump house constructed over circular well and (*iv*) roof of a traffic control post etc. When a circular slab simply supported at the edge is loaded with u.d.l., it bends in the form of a *saucer* due to which the stresses develop in both the radial and circumferential directions. Reinforcement needs to be provided in these two directions, which is not practical as this leads to congestion near the centre of the slab. Alternative method is to provide reinforcement in the form of mesh of bars.

Basic Rules for Design of slabs. In the case of simply supported slabs the effective span is taken as the clear distance between the supports plus the effective depth of the slab or the distance between centre to centre of supports whichever is less.

The value of ratio of span to effective depth for spans upto 10 m are as follows: cantilever – 7, simply supported – 20, continuous – 26.

The minimum reinforcement, in the case of mild steel, shall not be less than 0.15% of the total cross-sectional area. In the case of high strength deformed bars or welded wire fabric, however, this limit can be reduced to 0.12%

The maximum diameter of the reinforcing bars shall not exceed 1/8 of the total thickness of the slab.

The horizontal distance between parallel main reinforcement bars shall not be more than 3 times the effective depth of a solid slab or 300 mm whichever is less.

The horizontal distance between parallel reinforcement bar provided against shrinkage and temperature shall be more than 5 times the effective depth of the slab or 450 mm whichever is less.

The horizontal distance between two parallel main reinforcing bars shall not be less than the largest of the following: (*i*) the diameter of the bar if the diameters are equal (*ii*) the dia of the largest bar if the diameters are unequal, and (*iii*) 5 mm more than the nominal size of the coarse aggregate.

At each end of reinforcement bar the cover shall not be less than 25 mm or twice the diameter of such bar whichever is more.

The bottom cover for reinforcement shall not be less than 20 mm or the diameter of the bar whichever is more.

One way Slab. The analysis and design of one way slab is same as that of a beam of width one metre.

Suitable depth for the slab is assumed from stiffness consideration taking an appropriate value for the ratio (span/effective depth). Loads acting on the slab are computed per one metre width of slab. The factored moment and shear force are then computed. In the case of a simply supported slab $M_u = \dfrac{w_u l^2}{8}$ and $V_u = \dfrac{w_u l}{2}$, where l is the shorter span. The minimum depth required is computed using the expression

$$M_u = M_{u,\text{lim}} = k f_{ck} b d^2 \qquad \ldots(14.59)$$

where $b = 1000$ mm, $k = 0.138$ for Fe 415. The assumed depth shall be more than this. Otherwise, increase the assumed depth. Compute the area of steel per metre width of the slab from the expression.

$$M_u = 0.87 f_y A_{st} \left(1 - \dfrac{f_y A_{st}}{f_{ck} b d}\right) \qquad \ldots(14.60)$$

Spacing is then given by $S = \dfrac{a_{st}}{A_{st}} \times 1000$, where a_{st} is the area of bar used, and A_{st} total area of steel required. Spacing should not be more than $3d$ or 300 mm whichever is less. Distribution steel is provided at 0.15% of gross area if it is for HYSD bars. Spacing of distributing steel should not be more than $5d$ or 450 mm whichever is less. The nominal shear stress τ_v computed for the critical section shall be less than $k\tau_c$, where τ_c is obtained from eqn. (14.49) or Table 19 on page 73 of IS: 456–2000 and k is taken from Table 18 on page 72 of IS: 456–2000. Slabs are normally found to be safe in shear and do not require any special shear reinforcement.

The development length provided must be more than that given by eqn. (14.52). Normally the slabs satisfy the development length also.

Two way slabs. When slabs are supported on all four edges and the ratio of longer span to shorter span $\left(\dfrac{l_y}{l_x}\right)$ is less than or equal to 2, the slabs are likely to bend along the two spans and such slabs are called two way slabs. The load is transferred in both the directions to the four supporting edges and hence the main reinforcement is to be provided in both the directions to resist the two way bending. In the case of two way slabs the bending moment and deflections are less compared to one way slabs for similar loading and hence the thickness required will be less.

Two way slabs may have their corners held down or free to lift. When the slabs simply supported on all the four sides and are subjected to transverse loads, the bending of the slab in both the directions causes the corners to lift. Therefore torsion reinforcement is required in order that the corners are held down.

Depending on support conditions the two way labs are divided into the following three categories: (*i*) slabs simply supported on all the four edges and the corners are free to lift (*ii*) restrained slabs; that is slabs with fixed or continuous edges and (*iii*) the slabs simply supported on all the four edges with corners held down.

Simply supported two way slabs with corners not held down. The maximum bending moments per unit width along the two directions are calculated as

$$M_x = \alpha_x \, w l_x^2 \qquad \ldots(14.61)$$
$$M_y = \alpha_y \, w \, l_y^2 \qquad \ldots(14.62)$$

where w is the uniformly distributed load on the slab, α_x and α_y are the moment coefficients given in Table 27 of IS : 456–2000. At least 50% of the tension reinforcement provided at the mid span should be extended up to the supports. The remaining 50% should extend to within $0.1 \, l_x$ or $0.1 \, l_y$ of the support as appropriate.

Restrained two way slabs. Slabs with their corners prevented from lifting are termed as the restrained slabs. At the discontinuous edges torsion reinforcement needs to be provided at the corners if the corners are to be held down. Moments developed in restrained two way slabs (either continuous or discontinuous) depend on $\dfrac{l_y}{l_x}$ ratio and the edge conditions. For uniformly distributed loads maximum positive moment will be developed at mid span and maximum negative moment will be developed at the supports. The design moments for restrained slabs are calculated using the eqn. (14.61) and eqn. (14.62) again but with the moment coefficients taken from Table 26 of IS: 456–2000 for the appropriate edge conditions.

The depth of the slab is determined from stiffness consideration. For two way slabs with $l_x < 3.5$ m and $LL \leq 3$ kN/m² the allowable $\dfrac{l_x}{d}$ ratio is as given below:

	Fe 250	Fe 415
Simply supported slabs	35	28
Fixed continuous slabs	40	32

If $l_x > 3.5$ m and $LL > 3$ kN/m² the $\dfrac{l_x}{d}$ may be taken to be same as in the case of one way slabs.

For the design moments the area of steel at the mid span in both the directions is obtained from eqn. (14.59). As the short span bars are provided in the bottom layers and the long span bars are provided above them, effective depth d for short span $d = D - \dfrac{\phi}{2}$ whereas for the longer span the effective depth $d_1 = D - \phi$.

DESIGN OF STAIR CASES

Stair cases are essential and important component of buildings which enable the vertical movement between different floors. They may be broadly classified as given below.

1. Straight stair
2. Quarter turn stair
3. Half turn stair
4. Dog legged stair
5. Open newel stair with quarter space landing
6. Geometrical stairs such as circular stair, spiral stair etc.,

The width of stair is determined by the purpose of the stair and may generally vary between 1 m for residential buildings to 2 m in public buildings.

The flight of the stair is the length of the stair between two landings and the number of steps in a flight may vary between 3 and 12.

The rise and tread should be so proportional to provide comfortable access. Generally the sum of the tread and twice the rise is about 500 mm and the product of rise the thread is in the range 40000 to 42000.

In residential buildings rise may be 150 – 180 mm and the thread may be 200 – 210 mm. While in public buildings the rise is 120 to 150 mm and the tread is between 200 to 300 mm.

When overcrowding is not expected, the live load may be taken as 3 kN/m². If overcrowding is likely then it is taken as 5 kN/m².

When the stair slab (*waist lab*) is supported by side walls or a stringer beam on both the sides the stair is said to be spanning horizontally. Each step would then be structurally equivalent to an individual beam.

When the inclined stair slab together with the landing is supported at the top and bottom of the flight and without support on the sides, then the stair is said to span longitudinally. Dog legged, open well and quarter turn stairs belong to this category.

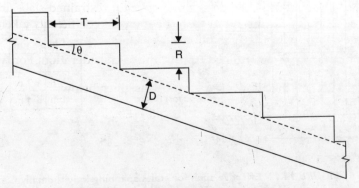

FIGURE 14.10 Tread and rise in a stair.

Design of Stairs spanning longitudinally. The depth of the section shall be taken as the minimum thickness perpendicular to the soffit of the stair as shown in Fig. 14.10.

Effective span L is taken, for different cases as shown in Fig. 14.11.

Dead load DL is taken as the sum of w_1, w_2 and (0.5 to 1 kN/m²) towards finishing) where the weight of waist slab per unit horizontal area w_1 is given by

$$w_1 = \frac{D\sqrt{R^2 + T^2}}{T} \times 25 = 25 D \sqrt{1 + \left(\frac{R}{T}\right)^2} \qquad \ldots(14.63)$$

and weight of steps per unit horizontal area w_2 is given by

$$w_2 = \frac{25}{2} R \qquad \ldots(14.64)$$

R being the rise in metres. The design is similar to w rectangular beam.

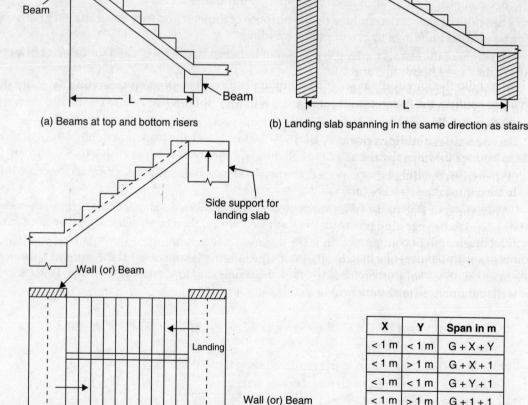

FIGURE 14.11 Effective span for stairs spanning longitudinally.

COLUMNS

A vertical member whose effective length is greater than 3 times its least lateral dimension carrying compressive loads is called a column.

A vertical compression member whose effective length is less than 3 times its least lateral dimension is called a *pedestal*

An inclined member carrying compressive load is called a *strut*.

Columns are generally square, rectangular or circular in shape. For architectural purpose they may be sometimes hexagonal or octagonal in shape.

Though a concrete column is supposed to carry axial compression load and though concrete is strong in compression, both longitudinal and transverse reinforcements are provided in a column to serve the following functions.

(*i*) to increase the load carrying capacities of columns and to reduce the size of the column

(*ii*) to resist the tensile stress produced in the column due to eccentric loads, due to moment coming on to the column and the transverse loads

(*iii*) to prevent sudden and brittle failures

(*iv*) to provide ductility to the column

(*v*) to reduce the effects of creep and shrinkage due to sustained loading

(*vi*) to prevent longitudinal buckling of longitudinal reinforcement

(*vii*) to hold the longitudinal reinforcement in position at the time of concreting

(*viii*) to confirm the concrete and prevent longitudinal splitting

(*ix*) to resist diagonal tension caused due to transverse shear

Based on type of reinforcement columns are classified as tied, spiral and composite columns. Based on loading they are termed as axially loaded and eccentrically loaded (uniaxial or biaxial). Based on slenderness ratio they are known as short or long columns.

If the columns are to take lateral loads they are called unbraced columns, when the lateral supports provided at the ends of the columns to take lateral loads then they are called braced columns.

Effective length of column, which is the distance between points of zero bending moments (or points of contraflexure) of a buckled column, depends on the unsupported length and the end condition can be obtained from Table 28 or IS: 456–2000.

If the column is to fail by compression and not by buckling, then it must satisfy the following limits.

(*a*) The unsupported length shall not exceed 60 times the least lateral dimension

That is $L < 60\,b$

(*b*) If one end of the column is unrestrained then $L < 100\,b^2/D$

where b = width and D = depth of cross-section

No column can have perfectly axial load. Eccentricity can be induced due to improper construction or actual conditions of loading. Hence all columns must be designed for a minimum eccentricity given by

$$e_{min} = \frac{l}{500} + \frac{\text{Lateral dimension}}{30}$$

subject to a minimum of 20 mm.

When the eccentricity does not exceed 0.05 times the lateral dimensions, the ultimate load on the column is given by

$$P_u = 0.4 f_{ck} A_c + 0.67 f_y A_{sc} \qquad \ldots(14.65)$$

where P_u = The factored axial load
A_{sc} = Area of longitudinal reinforcement
A_g = Gross-section area
A_c = Area of concrete = $A_g - A_{sc}$
f_{ck} = Characteristic compression strength of concrete
f_y = Characteristic compression strength of steal

$A_g = b\,D$ for rectangular columns and $A_g = \frac{\pi}{4} D^2$ for circular columns. Also the percentage steel p is given by $(100\, A_{sc}/A_g)$. Eqn. (14.65) can be rearranged in terms of p to assume the form

$$\frac{P_u}{A_g} = 0.4 f_{ck} + (0.67 f_y - 0.4 f_{ck}) \frac{p}{100} \qquad \ldots(14.66)$$

Eqn. (14.66) facilitates construction of design charts with $\dfrac{P_u}{A_g}$, f_{ck} and p as variables for different types of steels (that is for different f_y). Such charts are given in the design aids SP : 16 (chart nos. 24, 25 and 26).

The load carrying capacities of a short column can be increased by 5 percent if it is having transverse reinforcement which is helical instead of ties provided the ratio of the volume of helical reinforcement v_{hr} to the core volume V_k is not less than $0.36 \left(\dfrac{A_g}{A_k} - 1 \right) \dfrac{f_{ck}}{f_{sh}}$

That is $\qquad (P_u)_{\text{helical}} = 1.05\,(P_u)_{\text{ties}}$

While calculating the diameter of the core it is taken as $(D - 2 \times \text{clear cover})$

Design requirements for columns (clause 26.5.3 of IS 456 : 2000)

Longitudinal Reinforcement:

(*i*) The cross-sectional area of longitudinal reinforcement shall not be less than 0.8% and not more than 6% of gross cross-sectional area of the column.

(*ii*) If any column has larger cross-sectional than actually required (for other reasons), the minimum reinforcement shall be 0.8% of the required area and not the area provided.

(*iii*) The minimum no. of longitudinal rods shall be 4 in the case of rectangular column and 6 in the case of circular column.

(*iv*) Minimum diameter of longitudinal bars is 12 mm.

(*v*) Spacing of longitudinal bars measured along the periphery shall not exceed 300 mm.

(*vi*) In the case of pedestals nominal longitudinal reinforcement shall not be less than 0.15% of the cross-sectional area.

Transverse Reinforcement:

(*i*) The diameter of lateral ties shall not be less than 1/4th of diameter of the largest longitudinal bar and in no case less than 6 mm.

(*ii*) The pitch of the ties shall not be more than the least of lateral dimension of the column, 16 times the diameter of the smallest longitudinal bar and 300 mm.

(*iii*) The pitch of the helical reinforcement shall not be more than 75 mm, shall not be more than 1/6th of the core diameter, shall not be less than 25 mm and shall not be less than 3 times the diameter of the helical bar.

Columns subjected to axial load and biaxial bending

Some eccentricity is inherent in the concrete columns due to non-homogenity of the material, inaccuracies in loading, inaccuracies in construction etc. Therefore the code recommends that all columns shall be designed for minimum eccentricity equal to the unsupported length of column/500 plus lateral dimension/30 subject to a minimum of 20 mm. Also there are many occasions a column is subjected to uniaxial or biaxial moment in addition to xial load. If this e_{min} is less than $0.05\ D$, the column may be designed as axially loaded column. On the other hand if $e_{min} > 0.05\ D$ the column has to be designed for axial load P_u and moment $M_u = P_u\ e_{min}$.

The design of column subject to axial load and uniaxial bending will involve lengthy calculations by trial and error. The design charts given in SP: 16 may be used conveniently. According to the method suggested by IS: 456–2000 for the design of columns subjected to combined axial load and biaxial bending, the column is safe if

$$\left[\frac{M_{ux}}{M_{ux1}}\right]^{\alpha_n} + \left[\frac{M_{uy}}{M_{uy1}}\right]^{\alpha_n} \leq 1.0 \qquad \text{...(14.67)}$$

where M_{ux} and M_{uy} are the moments about x and y-axes due to design load

M_{ux1} is the maximum uni-axial moment capacity for bending about x-axis with axial load P_u when $M_{uy} = 0$

M_{uy1} is the maximum uni-axial moment capacity for bending about y-axis with axial load P_u when $M_{ux} = 0$

P_u is factored axial load

P_{ur} is axial load capacity of the section under pure axial load and is given by
$P_{uz} = 0.45 f_{ck} A_c + 0.75 f_2 A_s$

α_n is a factor related to the ratio of P_u/P_{uz}. When the ratio is ≤ 0.2, α_n is 1.0, when the ratio is ≥ 0.8, $\alpha_n = 2.0$ and in between it varies linearly.

A column is considered long (or *slender*) when the slenderness ratios $\dfrac{l_{ex}}{D}$ and $\dfrac{l_{ey}}{b}$ are more than 12. The maximum possible stress in long R.C.C. column is obtained by applying a reduction coefficient to the permissible stresses in concrete and steel. The reduction coefficient C_r is given by

$$C_r = 1.25 - \frac{l_e}{48b} \qquad \text{...(14.68)}$$

or
$$C_r = 1.25 - \frac{l_e}{160 \, r_{min}} \qquad \text{...[14.68(a)]}$$

where b = Least lateral dimension of the column, and

r_{min} = Least radius of gyration of the column.

FOOTINGS

The substructure (the portion of the structure below the ground level) which transmits the load of structure to the supporting soil is commonly known as foundation. Footing is that portion of the foundation which ultimately delivers the load to the soil. Foundations are provided to ensure that (i) the pressure intensity on the underlying soil does not exceed the bearing capacity of the soil and (ii) the settlement of the structure is uniform and is within the permissible limits.

The bearing capacity of the soil can be obtained from the laboratory tests on samples collected from the site. The minimum depth of foundation, according to Rankines theory, is given by

$$D = \frac{P}{\gamma} \left(\frac{1 - \sin \phi}{1 + \sin \phi} \right)^2 \qquad \text{...(14.69)}$$

where P is the beaming capacity, γ is the unit weight and ϕ is the angle of internal friction.

In footings made of plain and R.C.C. and resting on soils the thickness at the edge shall not be less than 150 mm and in the case of pile cap slabs it shall not be less than 300 mm.

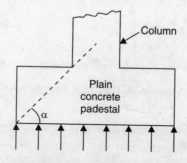

FIGURE 14.12 Pedestal footing.

In the case of plain concrete pedestals, the angle α between the plane passing through the bottom edge of the pedestal and the corresponding junction edge of the column with pedestal and the horizontal base (Fig. 14.12) shall satisfy the condition that

$$\tan \alpha \not< 0.9 \sqrt{\frac{100 \, q_0}{f_{ck}} + 1} \qquad \text{...(14.70)}$$

where q_0 = calculated maximum bearing pressure at the base of the pedestal in N/mm².

The bending moment at any section shall be determined by passing a vertical plane through the section which extends completely across the footing and computing the moment of force acting on the centre area of the footing on one side of the said plane. For the footings supporting columns the vertical section is at the face of the column. For example the max moment in the case of an isolated square footing is given by

$$M_u = \frac{pl(l-a)^2}{8} \qquad ...(14.71)$$

where l is the size of the footing, a is the size of the column and p is the soil pressure.

The critical section for checking shear, in the case of a footing acting essentially as a wide beam with a potential diagonal crack extending in a plane across the entire width shall be assumed as a vertical section located at a distance equal to the effective depth of footing from the face of the column in the case of footing on soils and at a distance equal to half the effective depth in the case of footing on the piles.

In case of two way action of the footing with potential diagonal cracking along the surface of a truncated pyramid or cone, the critical section shall be taken at a distance equal to the half of the effective depth from the periphery of the column.

The critical section for checking the development length shall be same as that for bending moment, and also at all other vertical planes where abrupt change of section occur. If the reinforcement is curtailed the anchorage requirement shall be checked.

The bearing pressure σ_{cbr} on the loaded area is given by

$$\sigma_{cbr} = \frac{W}{a \times b} \qquad ...(14.72)$$

where W is the load on the column and a and b are the sides of the column.

If A_1 is supporting area of the footing and A_2 is the loaded area at the column base then,

$$\sigma_{cbr} \leq 0.45 f_{ck} \sqrt{\frac{A_1}{A_2}} \qquad ...(14.73)$$

For footings minimum normal cover shall be 50 mm.

The nominal reinforcement for concrete sections of thickness greater than 1 m shall be 360 mm² per metre length in each direction on each face. This provision does not supercede the requirement of minimum tensile reinforcement based on the depth of the section.

RETAINING WALLS

A structure used for maintaining the ground surfaces at different elevations on either side of it is called a retaining wall. The material retained or supported by the retaining wall is called *backfill*. The backfill may have its top surface either horizontal or inclined. That portion of the backfill lying above the horizontal plane at the elevation of top of the wall is called *surcharge* and the inclination of the top surface of the surcharge with the horizontal is called the *surcharge angle* usually denoted by β.

Retaining walls are commonly classified into following types

(*i*) Gravity retaining walls

(*ii*) Cantilever retaining walls: ⊥ (inverted *T*) shaped or *L* shaped

(*iii*) Counterfort retaining walls

(*iv*) Buttressed retaining walls

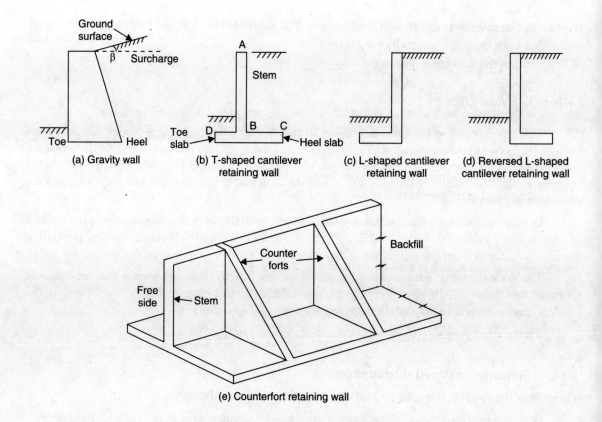

FIGURE 14.13 Types of retaining walls.

In the gravity wall [Fig. 14.13(a)] the earth pressure exerted by the backfill is resisted by the self weight of the wall. The wall can be of masonary or concrete and will be designed such that no negative (tensile) stresses are developed anywhere in the wall.

The cantilever retaining wall shown in Fig. [14.13 (b), (c) and (d)], usually of R.C.C., resists the earth pressure and the other vertical pressures by way of bending.

In a counterfort retaining wall as shown in Fig. [14.13 (e)] the vertical stem and the heel slab are strengthened by counterforts at suitable distances. This makes the stem as well as heel slab to act as continuous slabs. However the toe slab acts as cantilever bending upwards.

In a buttress type retaining wall the counterforts which are also called buttress are provided on the other side of the back full. They are not commonly used.

The active earth pressure acting at the base of a retaining wall of height H is given by

$$p_a = K_a \gamma H \qquad \text{...(14.74)}$$

where γ is unit weight of backfill soil, K_a coefficient of active earth pressure given by

$$K_a = \frac{1 - \sin \phi}{1 + \sin \phi} \qquad \text{...(14.75)}$$

ϕ is the angle of internal friction.

When there is uniform surcharge of ω per unit area on the top of backfill
$$p_a = K_a \gamma H + K_a \omega \qquad \text{...(14.76)}$$
When the backfill is sloping at an angle β, then
$$p_a = K_a \gamma H$$
where K_a is now given by
$$K_a = \cos \beta \, \frac{\cos \beta - \sqrt{\cos^2 \beta - \cos^2 \phi}}{\cos \beta + \sqrt{\cos^2 \beta - \cos^2 \phi}} \qquad \text{...(14.77)}$$
When $\beta = 0°$, eqn. (14.77) reduces to eqn. (14.75)
The total earth pressure P is calculated from
$$P = \frac{1}{2} K_a \gamma H^2 \qquad \text{...(14.78)}$$
with the resultant acting at a distance of $\frac{H}{3}$ from the base, its direction being horizontal in the case of uniform surcharge and parallel to the surface of the back full in the case of sloping surcharge.

For the case of passive earth pressure, the same equations are used with K_p replacing K_a, where
$$K_p = \frac{1 + \sin \phi}{1 - \sin \phi} \qquad \text{...(14.79)}$$

Similarly for sloped surcharge case
$$K_p = \cos \beta \, \frac{\cos \beta + \sqrt{\cos^2 \beta - \cos^2 \phi}}{\cos \beta - \sqrt{\cos^2 \beta - \cos^2 \phi}} \qquad \text{...(14.80)}$$

Factor of safety against sliding given by $F_2 = \frac{\mu \Sigma W}{P_a}$ shall be more than 1.5 where μ is coefficient of friction between concrete and soil, ΣW is the sum of all net downward forces and P_a is horizontal earth pressure.

Factor of safety against overturning given by $F_1 = \frac{M_R}{M_O}$ shall be more than 2.0, where M_R is total resisting moment and M_O is total overturing moment both taken about the toe.

The overturning moment M_O is calculated using
$$M_O = \frac{K_a \gamma H^3}{6}$$

The distance of point of application of the resultant force is given by $\bar{x} = \frac{\Sigma M}{\Sigma W}$

where ΣM is the net moment at the toe. Then the eccentricity is given by $e = \frac{b}{2} - \bar{x}$

The design must ensure that resultant falls always in the middle third of the base.

That means
$$e \not> \frac{b}{6}$$

The top width of stem shall be more than 200 mm. The width of base slab $b = 0.5\,H$ to $0.6\,H$ for walls without surcharge and $b = 0.6\,H$ to $0.7\,H$ for walls with surcharge. Toe projection may be $\dfrac{b}{3}$ to $\dfrac{b}{4}$. Thickness of the base slab shall be same as the thickness of stem.

Heel slab is designed for maximum B.M. due to earth pressure from bottom, weight of the earth from top and also the weight heel slab.

Toe slab is designed for maximum B.M. due to earth pressure from the bottom and the weight of toe slab from top.

When the height of the retaining wall is more then 5.5 m, the counterforts are provided.

Economical spacing of counterforts is $\dfrac{H}{3}$ to $\dfrac{H}{2}$, where H is the height of retaining wall.

Generally it is kept between 3 to 3.5 m.

The thickness of the base slab in centimetres is given as $2\,L\,H$, where L is the span of conterfort in m and H is height of retaining wall in m.

Base width $= 0.6\,H$ to $0.7\,H$ and the toe projection is one forth of base width.

WATER TANKS

There are 3 types namely: (i) Underground (ii) Resting on ground and (iii) Elevated water tanks. Watertightness is an important criterion. Usually more than M20 are used. The permissible tensile stresses in concrete and the permissible stresses in steel are reduced to control cracking, and are governed by IS : 370 (Part-II).

Circular Tanks

(a) **Spherical Domes.** Large circular tanks are covered with spherical domes which are supported by a ring beam. The thickness of R.C.C. spherical dome is generally not less than $\dfrac{1}{500}$ of the diameter with values of 50 to 100 mm for domes in the range of 25 to 50 m respectively. The reinforcement in the dome is made up of wire mesh. Domes are designed for Meridional thrust and hoop stress. If the concentrated load on the dome is W and the distributed load on the dome is w per unit area.

Meridional thrust
$$T = \frac{wR}{1+\sin\theta} + \frac{W}{2\pi R \sin^2\theta} \qquad \ldots(14.81)$$

and hoop stress
$$f = \frac{wR}{t}\left[\cos\theta - \frac{1}{1+\cos\theta}\right] - \frac{W}{2\pi R\,t\,\sin^2\theta} \qquad \ldots(14.82)$$

where R = Radius of the dome of sphere

θ = Angle between radius vector and vertical

t = Thickness of the dome

(b) **Circular Tanks with Flexible or Free Base.** The intensity of water pressure at any depth h is equal to wh. The corresponding hoop tension per metre height at this level is $T = \dfrac{whD}{2}$, where D = the internal diameter of the tank. Area of steel per metre height $A_{st} = \dfrac{wHD}{2 \times 1000}$ in cm², where w is unit weight if water in kg/m³.

Thickness of tank wall is calculated from cracking consideration using the formula,

$$\sigma_{ct} = \dfrac{\dfrac{wHD}{2}}{100t + (m-1)A_{st}} \qquad \ldots(14.83)$$

where σ_{ct} = Allowable tensile stress in concrete in kg/cm²

t = Thickness of tank wall in cm.

The thickness of the wall should not be less than 15 cm.

(c) **Circular Tank with the Wall Restrained at the Base.** For hinged and fixed bases the co-effecients for moments and ring tension compiled in Tables of IS : 3370 (Part IV) 1967 are to be used. These co-efficients are expressed as a function of a the non-dimensional parameter $\dfrac{H^2}{Dt}$.

Hoop tension per metre height = co-efficient × $wh\dfrac{D}{2}$

B.M. per metre run = co-efficient × wH^3

Shear force at the base of the wall = co-efficient × wH^2

Rectangular Water Tanks

Moments and shear co-efficients are given in IS : 3370 (part IV) 1967 for different edge conditions. Alternatively the approximate design method given below can be adopted:

(a) **For Tanks with Ratio of $\dfrac{L}{B} < 2$.** The bending in the walls is predominant in the vertical direction in the bottom of $\dfrac{H}{4}$ or 1 m (whichever is greater, and is designed as a cantilever.

The corners are designed for the maximum moment obtained after moment distribution with the intensity of pressure $p = w(H - h)$.

In the absence of moment distribution the bending moments may be computed by the following approximate expressions:

B.M. at centre of span $= \dfrac{pB^2}{12}$ (producing tension on outer face)

B.M. at ends of span $= \dfrac{pB^2}{12}$ (producing tension on water face)

In addition to the B.M., the walls are subjected to direct tension given by

Direct tension on long walls $= T_l = w(H-h)\dfrac{B}{2}$

Direct tension on short walls $= T_b = w(H-h)\dfrac{L}{2}$

Design moment $= M - Tx$

where x is the distance of c.g. of reinforcement from c.g. of the section.

Area of steel $A_{st} = \dfrac{M - Tx}{\sigma_{st}\, jd} + \dfrac{T}{\sigma_{st}}$.

(b) **For Tanks with Ratio of** $\dfrac{L}{B} > 2$. In this case long walls are assumed to bend vertically and hence are designed as cantilevers. Short walls are assumed to bend horizontally supported on long walls above $\dfrac{H}{4}$ or 1 m from bottom (whichever is greater).

B.M. for long walls $= \dfrac{wH^2}{6}$

B.M. for short walls above $\dfrac{H}{4}$ or 1 m $= \dfrac{w(H-h)B^2}{16}$

Maximum cantilever moment for short wall $= \dfrac{wH^2}{6}$.

In addition direct pulls are considered for long and short walls.

INTZE Type Elevated Tank

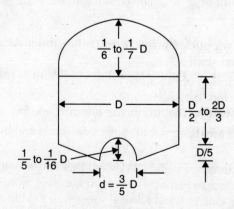

FIGURE 14.14 INTZE tank.

(a) **Dimensions.** Rise of spherical dome $= \dfrac{D}{6}$ to $\dfrac{D}{7}$

Height of cylindrical walls $= \dfrac{D}{6}$ to $\dfrac{2}{3}D$

Height of conical dome $= \dfrac{D}{5}$

Inclination of conical dome $= 45°$

Rise of domed floor $= \dfrac{3}{25}D$ to $\dfrac{D}{10}$

D = Diameter of Cylindrical wall portion.

(b) **Spherical Dome.** It is designed to take

(i) Self-weight of dome, (ii) weight of latern, (iii) weight of water proof covering, (iv) live load.

The design is based on the formulae given under domes. Usually 75 mm to 100 mm thick slab with minimum reinforcement of 0.2% is provided.

(c) **Top Ring Beam.** It is provided to take the hoop tension produced by the horizontal component of the meridional thrust induced in the dome at the junction with cylindrical wall.

$$\text{Hoop tension in ring beam} = T \text{ and } \sigma_{ct} = \dfrac{T}{A_c + (m-1)A_{st}} \qquad ...(14.84)$$

(d) **Cylindrical Wall.** It is designed for hoop tension due to water pressure. If a centre shaft is provided, the shaft is designed for hoop compression. Hoop tension in cylindrical wall $= \dfrac{whD}{2}$, where h is depth of water.

(e) **Conical or Inclined Slab.** The slab is designed to span between the two ring beams and also for hoop tension due to water pressure as a conical dome.

Horizontal thrust at any height $= H_x$

Hoop tension $= H_x \dfrac{D}{2}$

(f) **Domed Floor.** It is supported on the ring girder. This floor slab is designed to carry the water load above it and central shaft above, if provided.

(g) **Ring Girder.** The design of this girder depends on the number if supports on which it rests. The ring girder is subjected to bending and torsional moments. The moments are computed using co-efficients which depend on the number of supports.

(h) **Columns.** Columns are designed for both axial loads and bending moment due to wind loads on the tank.

(i) **Braces.** Braces are designed using steel beam theory and equal reinforcement is provided in tension and compression zones to resist positive and negative moments produced by the column moments above and below it, due to wind loads. Moment in the brace is twice the moment in the columns if braces are placed at regular intervals.

(j) **Foundation.** The foundation normally consists of a circular ring on a circular column.

ULTIMATE LOAD THEORY

Assumptions in Ultimate Load Design

(a) The distribution strain across the depth of a member in bending remains linear right up to failure.

(b) The beam will fail when the maximum compressive strain in the concrete reaches a particular value depending on the quality of concrete.

(c) The shape of the stress diagram for the concrete at failure can be fully defined by means of co-efficients.

(d) The concrete does not resist tension.

(e) The strain in the steel is a certain proportion of the strain in the concrete at the same level.

Design of Beams

(a) **For Primary Tension Failure**

$$M_u = A_{st} \cdot \sigma_{sy} d \left[1 - \frac{K_2}{K_1 K_3} \cdot \frac{A_3 \sigma_{sy}}{bd f_c}\right] \qquad \ldots(14.85)$$

where A_{st} = Area of steel provided, σ_{sy} = Yield in steel, b = Breadth of the beam

d = Effective depth, f_c = Cylinder compressive strength K_1, K_2, K_3 = Stress block parameters.

As per IS : 456 – 1964

$$M_u = A_{st} \cdot \sigma_{sy} d \left[1 - 0.91 \frac{A_{3t} \sigma_{sy}}{bd \sigma_{cu}}\right].$$

For primary compression failure

$$M_u = b K_u d^2 K_1 K_3 f_c (1 - K_2 K_u)$$

where $K_u = \sqrt{\{pm_p + (0.5 pm_p)^2\}} - 0.5 pm_p$

$p = \dfrac{A_u}{bd}$; $m_p = \dfrac{E_s E_u}{K_1 K_3 f_c}$ = plastic modular ratio.

(b) **For Balanced Failure**

$$M_u = A_{st} \cdot \sigma_{sy} d \left[1 - 0.59 \frac{A_{3t} \sigma_{sy}}{bd f_c}\right]$$

As per Whitney's theory $M_u = \dfrac{1}{3} b d^2 f_c$ and $A_u = 0.425 \dfrac{bd f_c}{\sigma_{sy}}$

As per IS : 456 – 1964, $M_u = 0.185 b d^2 \sigma_{cu}$

where σ_{cu} = Ultimate cube compressive strength

$$A_{st} = 0.236 \, bd \, \frac{\sigma_{cu}}{\sigma_{sy}}$$

(c) For Primary Compression Failure:

Is : 456 does not permit over reinforced beams.

As per IS : 456 – 1964, Ultimate Factored Load = 1.5 D.L + 2.2 L.L.

Eccentrically Loaded Columns and Struts

For balanced failure

$$P_{ub} = \frac{2}{3} \sigma_{cu} bxd + A_{sc} \left(\sigma_{scu} - \frac{2}{3} \sigma_{cu} \right) - A_{st} \sigma_{sy} \qquad ...(14.86)$$

where $\quad x = \dfrac{5950}{7000 + \sigma_{sy}}$

σ_{scu} = Stress in the steel in the compression face.

$$P_{ub} e_{ub} + P_{ub} \frac{1}{2} (d - d_c) = \frac{2}{3} \sigma_{cu} bxd^2 (1 - 0.5x) + A_{sc} \left(\sigma_{scu} - \frac{2}{3} \sigma_{cv} \right)(d - d_c)$$

where $\quad d_c$ = Depth to steel in face adjacent to direct load

e_{ub} = Eccentricity of load corresponding to a failure load of P_{ub}.

Ultimate Shear Strength

As per ACI, the permissible nominal ultimate shear stress

$$q_c = 0.45 \sqrt{0.8 \, \sigma_{cu}}$$

PRE STRESSED CONCRETE

General

Prestressed concrete is basically concrete in which internal stresses of a suitable magnitude and distribution are introduced so that the stress resulting from external loads are counteracted to a desired degree.

Advantages of Prestressed Concrete

    (a) Prestressed members remain free from cracks.

    (b) Efficient utilization of high compressive strength of concrete and high tensile strength of steel.

    (c) Section of the member is reduced resulting in the economy of construction.

    (d) Diagonal tension in concrete is reduced.

    (e) Effect of bending moment due to dead load on members is eliminated to a certain extent.

Disadvantages

    (a) Special equipment is required and prestressing cannot be carried out under all circumstances.

    (b) It requires expert supervision and good technical knowledge.

Materials Required

Minimum compressive strength of concrete for pretensioned members = 42 N/mm².
Minimum compressive strength of concrete for post-tensioned members = 35 N/mm².
Minimum cement content in concrete for pretensioned members = 350 kg/m³.
Minimum cement content in concrete for post tensioned members = 400 kg/m³.
Ultimate tensile strength of high tensile wires = 1600 N/mm².
Ultimate tensile strength of high tensile bars = 1000 N/mm².

Losses of Prestress

(a) **Loss Due to Elastic Deformation of Concrete:**

$$f_c = \frac{E_s}{E_c} \quad \ldots(14.87)$$

where f_c = Prestress in concrete at the level of steel.

In case of post tensioning if all the tendons are tensioned at a time, the loss of prestress due to elastic deformation.

(b) **Loss Due to Shrinkage of Concrete:**

(i) For pretensioning = $\dfrac{300 \times 10^{-6}}{\log(t+2)} E_s$

(ii) For post tensioning in humid conditions = $\dfrac{200 \times 10^{-6}}{\log(t+2)} E_s$

(iii) For post tensioning in dry conditions = $\dfrac{300 \times 10^{-6}}{\log(t+2)} E_s$

where t = Age of the concrete in days at the time of stress transfer.

(c) **Lose Due to Creep of Concrete:**

(i) *Ultimate Creep Strain Method:*

Loss of stress in steel due to creep of concrete = $\in_{cc} f_c E_s$

where $\in_{cc}$ = Ultimate creep strain for a sustained unit stress

$$= 17.5 \frac{(1.25 - h_a)}{0.35} \sqrt{\frac{40}{f_{cu}}} \cdot \frac{f_{cu}}{f_{ci}} \times 10^{-6} \text{ mm/mm per N/mm}^2$$

h_a = Ambient humidity of the environment
f_{cu} = 28 days cube strength of concrete
f_{ci} = Strength of concrete at transfer
f_{ci} = Compressive stress in concrete at the level of steel.

(ii) *Creep Co-efficient method:*

Loss of stress in steel = $\phi f_c \dfrac{E_s}{E_c}$

where ϕ = Creep co-efficient = $\dfrac{\text{Creep strain}}{\text{Elastic Strain}}$

= 1.5 for watery situations to 4.0 for dry condition.

(d) **Loss Due to Relaxation of Stress in Steel.** It is expressed as a percentage of the initial stress in steel. IS : 1343 recommends a value varying from 0 to 90 N/mm² for stress in wires varying from $0.5 f_p$ to $0.8 f_p$, where f_p is the characteristic strength of prestressing steel.

(e) **Loss of Stress Due to Friction:**

$$P_x = P_0 e^{-(\mu\alpha + kx)}$$

where P_x = Prestressing force in the cable at a distance x from the tensioning end
P_0 = Prestressing force at the jacking end
μ = Co-efficient of friction between cable and duct = 0.25 to 0.55
α = The cumulative angle in radians through the tangent to the cable profile has turned between any two points under consideration
k = Friction co-efficient for wave effect = 15×10^{-4} to 50×10^{-4} per metre.

(f) **Losses due to Anchorage Slip.** Loss in prestress for post tensioned members = $\dfrac{E_s \Delta}{L}$

where E_s = Modulus of elasticity of steek
Δ = Slip of anchorage
L = Length of cable.

(g) **Total Losses Allowed in design:**
For pre-tensioned members = 18% approximately
For post-tensioned members = 15% approximately.

Shear

The design of prestressed concrete beams for shear is similar to that of R.C.C. beams.

Deflection

Deflections in prestressed concrete beams are calculated by the usual formulae of elementary strength of materials. The forces and moments contributing to the deflection of prestressed concrete beam are

(a) Dead load w_d and live load w_l acting downwards.
(b) Upward load w_p caused by curvature of tensioned cables.
(c) Moment $M_p = \pm P_c$ due to anchorage of cables at the end sections. This moment will cause upward deflection when 'e' is below the centroidal axis.

For equal eccentricity at the two ends

$$\delta_{\text{down}} = \dfrac{5}{384}[w_d + w_l - w_p]\dfrac{L^4}{EI} \pm \dfrac{M_p L^2}{8EI} \qquad ...(14.88)$$

The upward deflection under load condition is similarly calculated.

II. OBJECTIVE TYPE QUESTIONS

1. M 10 grade of concrete approximates
   (a) 1 : 3 : 6 mix
   (b) 1 : 1 : 2 mix
   (c) 1 : 2 : 4 mix
   (d) 1 : 1½ : 3 mix.

2. M 25 grade of concrete approximates
   (a) 1 : 3 : 6 mix
   (b) 1 : 1 : 2 mix
   (c) 1 : 2 : 4 mix
   (d) 1 : 1½ : 3 mix.

3. The normal grades of concrete used in reinforced concrete buildings are
   (a) M15, M20, M25
   (b) M7.5, M10, M15
   (c) M5, M7.5, M10
   (d) M20, M25, M30.

4. The normal grades of concrete used in reinforced concrete water tanks are
   (a) M15, M20, M25
   (b) M7.5, M10, M15
   (c) M5, M7.5, M10
   (d) M20, M25, M30.

5. The approximate allowable stress in bending compression in reinforced concrete is
   (a) $0.256 f_{ck}$
   (b) $0.446 f_{ck}$
   (c) $0.336 f_{ck}$
   (d) $0.306 f_{ck}$.

6. The approximate allowable stress in axial compression in reinforced concrete is
   (a) $0.25 f_{ck}$
   (b) $0.44 f_{ck}$
   (c) $0.33 f_{ck}$
   (d) $0.30 f_{ck}$

   where f_{ck} is the characteristic strength of concrete.

7. The tensile strength of concrete to be used in the design of reinforced concrete members is
   (a) $0.2 f_{ck}$
   (b) $0.1 f_{ck}$
   (c) $0.7 \sqrt{f_{ck}}$
   (d) zero.

8. The permissible bearing stress on full area of concrete shall be taken as
   (a) $0.20 f_{ck}$
   (b) $0.30 f_{ck}$
   (c) $0.25 f_{ck}$
   (d) $0.45 f_{ck}$.

9. The minimum quantity of cement content needed in one m³ of a reinforced concrete which is not directly exposed to weather is about (in kg)
   (a) 200
   (b) 250
   (c) 300
   (d) 350.

10. The minimum quantity of cement content needed in one m³ of a reinforced concrete which is exposed to sea weather conditions (in kg)
    (a) 200
    (b) 250
    (c) 300
    (d) 350.

11. The design shear stress in reinforced cement concrete depends on
    (a) characteristic strength of concrete
    (b) percentage of longitudinal tensile reinforcement
    (c) characteristic strength of steel
    (d) both (a) and (b).
12. The maximum shear stress permitted in reinforced concrete members depends upon
    (a) grade of concrete
    (b) percentage of longitudinal tensile reinforcement
    (c) shear reinforcement provided
    (d) all the above.
13. The ratio of the allowable bond stress in deformed bars to that of plain bars is about
    (a) 1.2
    (b) 1.3
    (c) 1.4
    (d) 1.6.
14. The ultimate tensile stress in mild steel bars of grade I should be greater than (in N/mm^2)
    (a) 415
    (b) 260
    (c) 140
    (d) 100.
15. The ultimate tensile stress in mild steel bars of grade II should be greater than (in N/mm^2)
    (a) 430
    (b) 375
    (c) 260
    (d) 140.
16. The ultimate tensile stress of medium tensile deformed bars should be (in N/mm^2)
    (a) less than 420
    (b) greater than 420
    (c) greater than 450
    (d) greater than 500.
17. The yield stress of 40 mm diameter mild steel bars of grade I should be greater than (in N/mm^2)
    (a) 260
    (b) 240
    (c) 200
    (d) 130.
18. The proof stress of mild steel deformed bars of 20 mm diameter should be greater than (in N/mm^2)
    (a) 425
    (b) 360
    (c) 250
    (d) 200.
19. The proof stress of cold-twisted deformed bars of 20 mm diameter is about (in N/mm^2)
    (a) 500
    (b) 415
    (c) 360
    (d) 250.
20. When mild steel conforming to Grade II is to be used with the design details having been worked out on the basis of Grade I Steel, the area of reinforcement should be
    (a) decreased by 10%
    (b) increased by 10%
    (c) increased by 20%
    (d) remained as it is.
21. The allowable tensile stress in mild steel stirrups in reinforced cement concrete is (in N mm^2)
    (a) 140
    (b) 190
    (c) 230
    (d) 260.

22. The allowable tensile stress in high yield strength deformed steel stirrups used in reinforced cement concrete is (in N/mm²)
    (a) 140
    (b) 190
    (c) 230
    (d) 260.

23. The allowable compressive stress in mild steel bars in doubly reinforced cement concrete beams when compressive resistance of the concrete is taken into account is (in N/mm²)
    (a) 130
    (b) 140
    (c) $m\sigma_{cb}$
    (d) $1.5\, m\sigma_{cb}$.

24. The allowable compressive stress in mild steel reinforcement bars in reinforced concrete columns is (in N/mm²)
    (a) 130
    (b) 190
    (c) $m\sigma_{cb}$
    (d) $1.5\, m\sigma_{cb}$.

25. The main reinforcement in a simply supported R.C.C. member is placed at
    (a) top fibre
    (b) side fibres
    (c) bottom fibre
    (d) top and bottom fibres.

26. The main reinforcement in R.C.C. cantilever members is placed at
    (a) top fibre
    (b) side fibres
    (c) bottom fibre
    (d) top and bottom fibres.

27. The main reinforcement in R.C.C. continuous members is placed at
    (a) top fibre
    (b) side fibres
    (c) bottom fibre
    (d) top and bottom fibres.

28. An additional longitudinal reinforcement in webs of beams should be placed if the depth of the web is more than 750 mm at
    (a) top fibre
    (b) side fibres
    (c) bottom fibres
    (d) top and bottom fibres.

29. The maximum strain in the tension reinforcement at failure shall not be less than
    (a) $\dfrac{f_y}{1.15\, E_s} - 0.002$
    (b) $\dfrac{f_y}{1.15\, E_s} + 0.002$
    (c) $\dfrac{1.15\, E_s}{f_y} - 0.002$
    (d) $\dfrac{1.15\, E_s}{f_y} + 0.002$.

30. The depths to N.A. in a singly reinforced section is given by
    (a) $\dfrac{x_u}{d} = \dfrac{700}{1100 + 0.87 f_y}$
    (b) $\dfrac{x_u}{d} = \dfrac{1100 + 0.87 f_y}{700}$
    (c) $\dfrac{x_4}{d} = \dfrac{700}{1100 - 0.87 f_y}$
    (d) $\dfrac{x_4}{d} = \dfrac{1100 - 0.87 f_y}{700}$.

31. The ratio of the compressive strength of concrete is structure to the characteristic strength of concrete in cube is equal to
    (a) 1.67
    (b) 1/0.67
    (c) 0.67
    (d) 1/1.67.

32. The shape of the stress-strain curve for compression in concrete is generally taken to be
    (a) hyperbolic
    (b) straight line
    (c) rectangular
    (d) parabolic.
33. By over-reinforcing the section in tension, the moment of resistance of beam can be increased by not more than
    (a) 10%
    (b) 15%
    (c) 18%
    (d) 25%.
34. The steel beam theory of doubly reinforced beams assumes that
    (a) tension is resisted by tension steel only
    (b) compression is resisted by compression steel only
    (c) stress in tension steel equals the stress in compression steel
    (d) all the above.
35. The reinforcement ratio p_t in a balanced section of a R.C.C. beams is given by
    (a) $0.414 \dfrac{f_{ck}}{f_y} \dfrac{x_{u,max}}{d}$
    (b) $\dfrac{1}{0.414} \dfrac{f_{ck}}{f_y} \dfrac{x_{u,min}}{d}$
    (c) 1.419
    (d) $\dfrac{1}{1.419}$.
36. The partial factor of safety for steel in limit state design is
    (a) 1.65
    (b) 1.75
    (c) 1.50
    (d) 1.15.
37. The partial factor of safety for concrete in limit state design is
    (a) 1.65
    (b) 1.75
    (c) 1.50
    (d) 1.15.
38. The maximum allowable percentage of tension reinforcement in R.C.C. beams is
    (a) 14%
    (b) 0.4%
    (c) 4%
    (d) 0.09%.
39. The minimum percentage of high yield strength deformed reinforcement bars (HYSD) provided in R.C.C. slabs is
    (a) 0.4
    (b) 0.15
    (c) 0.12
    (d) 0.2.
40. The maximum diameter of the reinforcement bars in R.C.C. slabs is
    (a) 20 mm
    (b) 16 mm
    (c) span/100
    (d) thickness of slab/8.
41. The maximum diameter of the reinforcement bars in R.C.C. beams is limited to
    (a) 28 mm
    (b) 40 mm
    (c) one-eighth of the least dimension of the beams
    (d) one-tenth of the depth of beams.

42. The maximum spacing of the torsional reinforcement in R.C.C. beams must not exceed
    (a) 300 mm
    (b) b_1
    (c) $\dfrac{b_1 + d_1}{4}$
    (d) all the above.

43. The minimum number of torsional longitudinal reinforcement bars to be provided in R.C.C. beams is
    (a) 2
    (b) 4
    (c) 8
    (d) 6.

44. The effective width of flange in R.C.C. T-beams should be restricted to
    (a) $\dfrac{l_0}{6} + b_w + 6D_f$
    (b) $\dfrac{l_0}{12} + b_w + 3D_f$
    (c) $\dfrac{l_0}{3}$
    (d) centre to centre distance of the beams.

45. The probability that the actual strengths of material is more than or equal to its characteristic strength is
    (a) 0.75
    (b) 0.85
    (c) 0.95
    (d) 0.65.

46. A flat slab is supported on
    (a) beams
    (b) columns
    (c) beams and columns
    (d) columns monolithically built with slab.

47. Enlarged head of a supporting column of a flat slab, is known as
    (a) column head
    (b) drop panel
    (c) capital
    (d) column strip.

48. Thickened part of the flat slab over its supporting column is known as
    (a) column head
    (b) drop panel
    (c) capital
    (d) column strip.

49. The diameter of the column head supporting a flat slab is about
    (a) 0.25 times the span
    (b) 2.5 times the diameter of the column
    (c) 4 times the diameter of the column
    (d) 0.5 times the span.

50. The effective width of a column strip of a flat slab is taken as
    (a) one-fourth the width of the panel
    (b) half the width of the panel
    (c) half the diameter of the column
    (d) the diameter of the column.

51. If W is the load per unit area on a circular slab of radius R, then the maximum radial moment at the centre of a simply supported slab is equal to
    (a) $\dfrac{WR^2}{16}$
    (b) $\dfrac{2WR^2}{16}$
    (c) $\dfrac{3WR^2}{16}$
    (d) $\dfrac{5WR^2}{16}$.

52. If W is the load per unit area on a circular slab of radius R, then the maximum circumferential moment at the centre of the slab is

   (a) $\dfrac{WR^2}{16}$ ☐
   (b) $\dfrac{2WR^2}{16}$ ☐
   (c) $\dfrac{3WR^2}{16}$ ☐
   (d) $\dfrac{5WR^2}{16}$. ☐

53. In a circular slab with fixed ends subjected to a uniformly distributed load, the ratio of the maximum negative and maximum positive radial moments is
   (a) 1 ☐
   (b) 2 ☐
   (c) 3 ☐
   (d) 4. ☐

54. The minimum distance over which a bar should be extended (curtailment) beyond which it is no longer needed for bending in R.C.C. beams is
   (a) Effective depth of beam or 12 times the diameter of the bar whichever is greater ☐
   (b) 48 times the diameter of the bar ☐
   (c) 1.5 times the depth or 24 times the diameter of the bar whichever is greater ☐
   (d) $\dfrac{\phi \sigma_s}{4\tau_{bd}}$. ☐

55. The tension reinforcement in a R.C.C. beam can be cut off at a point when it is no longer needed if
   (a) enough bond length is available ☐
   (b) the shear at the cut-off point does not exceed two-thirds ☐
   (c) bending moment is zero ☐
   (d) all the above. ☐

56. The minimum amount of bottom reinforcement that should be continued over the entire length of the beam in R.C.C. continuous beam is
   (a) 50% ☐
   (b) 40% ☐
   (c) $33\frac{1}{3}$% ☐
   (d) 25%. ☐

57. The minimum amount of negative moment reinforcement in R.C.C. beams that should extend beyond the point of inflection is
   (a) one-third of the total ☐
   (b) one-fourth of the total ☐
   (c) zero ☐
   (d) half of the total. ☐

58. The minimum length beyond which at least one-third of the negative moment reinforcement should extend beyond the inflection point is
   (a) 12 times the diameter of the bar ☐
   (b) one-sixteenth of the clear span ☐
   (c) effective depth of the member ☐
   (d) greater of the above. ☐

59. The splicing of the reinforcement bars in R.C.C. beams can be done at a section where
   (a) shear force is zero ☐
   (b) bending moment is zero ☐
   (c) bending moment is less than half the maximum B.M. on the beam ☐
   (d) bending moment is more than half the maximum B.M. on the beam. ☐

60. Lap splices in tension reinforcement are permitted for bars
    (a) larger than 36 mm
    (b) less than 36 mm
    (c) less than 25 mm
    (d) all diameter bars.

61. The lap length of a direct, tension reinforcement bar in a R.C.C. beams should be more than
    (a) 30 times the diameter of the bar
    (b) twice the development length or 30 times the diameter of the bar
    (c) 48 times the diameter of the bar
    (d) 24 times the diameter of the bar or development length.

62. The minimum lap length at the splice of compression reinforcement in a R.C.C. member should be more than
    (a) 30 times the diameter of the bar
    (b) twice the development length or 30 times the diameter of the bar
    (c) 48 times the diameter of the bar
    (d) 24 times the diameter of the bar or development length.

63. When bars of two different diameters are to be spliced the lap length is calculated on the basis of
    (a) diameter of the larger bar
    (b) diameter of the smaller bar
    (c) mean diameter of the bars
    (d) three-fourths diameter of the larger bar.

64. The minimum horizontal distance between two main reinforcement bars should be
    (a) diameter of the larger bar or 5 mm more than the nominal maximum size of coarse aggregate
    (b) 5 mm more than the nominal size of the aggregate
    (c) 5 mm more than the diameter of the bar
    (d) size of the aggregate or 5 mm more than the diameter of the bars.

65. The minimum vertical space of the main reinforcement in R.C.C. beams is
    (a) diameter of the larger bar or 5 mm more than the nominal maximum size of coarse aggregate
    (b) diameter of the larger bar or two-third the nominal maximum size of coarse aggregate
    (c) 5 mm more than the diameter of the bar
    (d) 15 mm.

66. The maximum horizontal distance between parallel main deformed reinforcement in beams is
    (a) 215 mm to 300 mm
    (b) 125 mm to 235 mm
    (c) 105 mm to 195 mm
    (d) 50 mm to 165 mm.

67. The maximum horizontal distance between parallel main reinforcement bars in solid slabs is
    (a) 5 times the effective depth or 450 mm whichever is smaller
    (b) 3 times the effective depth or 450 mm whichever is less
    (c) 200 mm or 3 times the effective depth whichever is more
    (d) 200 mm or 5 times the effective depth whichever is more.

68. The minimum cover to the main bars in R.C.C. beam must be
    (a) 15 mm or diameter of the bar
    (b) 25 mm or diameter of the bar
    (c) 40 mm or diameter of the bar
    (d) 25 mm or size of the coarse aggregate.

69. The minimum cover to the main reinforcement in R.C.C. solid slabs is
    (a) 15 mm or diameter of the bar
    (b) 25 mm or diameter of the bar
    (c) 40 mm or diameter of the bar
    (d) 25 mm or size of the coarse aggregate.

70. The final deflection due to all loads including the effects of temperature, creep and shrinkage should not normally exceed
    (a) $\frac{\text{span}}{250}$ or 20 mm whichever is less
    (b) $\frac{\text{span}}{250}$
    (c) $\frac{\text{span}}{350}$ or 20 mm whichever is less
    (d) $\frac{\text{span}}{350}$.

71. The deflection including the effects of temperature, creep and shrinkage occurring after erection of partitions and the application of finishes should not normally exceed
    (a) $\frac{\text{span}}{250}$ or 20 mm whichever is less
    (b) $\frac{\text{span}}{250}$
    (c) $\frac{\text{span}}{350}$ or 20 mm whichever is less
    (d) $\frac{\text{span}}{350}$.

72. For cantilever beams and slabs, the basic value of span to effective depth ratio is
    (a) 7
    (b) 10
    (c) 20
    (d) 26.

73. For the simply supported beams and slabs the basic value of span to effective depth ratio is
    (a) 7
    (b) 10
    (c) 20
    (d) 26.

74. For simply supported two way slabs of small spans with HYSD bars of grade Fe 415 the span to over all depth ratio is limited to
    (a) 20
    (b) 26
    (c) 28
    (d) 35.

75. A compression member is termed as column or strut if the ratio of its effective length to the least lateral dimension is more than
    (a) 1
    (b) 2
    (c) 3
    (d) no limit.

76. A compression member is considered as short when both the slenderness ratio $\frac{l_{ex}}{D}$ and $\frac{l_{ey}}{b}$ are less than
    (a) 12
    (b) 10
    (c) 8
    (d) 16.

77. The effective length of column with one and effectively held in position and other restrained against rotation at both directions is
    (a) 2 l
    (b) 1.2 l
    (c) l
    (d) 0.65 l.

78. The effective length of a column with both ends effectively held in position, but not restrained against rotation is
    (a) 2 l
    (b) 1.2 l
    (c) l
    (d) 0.65 l.

79. A column is regarded as long column if the ratio of its effective length to least lateral radius of gyration is more than
    (a) 150
    (b) 100
    (c) 60
    (d) 40.

80. The minimum eccentricity that should be used in the design of any R.C.C. column is given by
    (a) 50 mm
    (b) $\dfrac{\text{unsupported length}}{500}$
    (c) $\dfrac{\text{lateral dimension}}{30}$
    (d) $\dfrac{\text{unsupported length}}{500} + \dfrac{\text{lateral dimension}}{30}$.

81. The minimum percentage of longitudinal reinforcement in R.C.C. columns is
    (a) 0.6
    (b) 0.8
    (c) 1.2
    (d) 4.0.

82. The maximum percentage of longitudinal reinforcement in R.C.C. columns is
    (a) 0.8
    (b) 2.0
    (c) 4.0
    (d) 6.0.

83. The minimum diameter of longitudinal reinforcement in R.C.C. column should not be less than
    (a) 6 mm
    (b) 10 mm
    (c) 8 mm
    (d) 12 mm.

84. The allowable axial load on a short R.C.C. column is given by
    (a) $0.4 f_{ck} A_g + (0.67 f_y - 0.4 f_{ck}) A_s$
    (b) $0.446 f_{ck} A_g + (0.75 f_y - 0.446 f_{ck}) A_s$
    (c) 1.05 times that given by (b)
    (d) 1.05 times that given by (a)

85. The allowable axial load on a short R.C.C column with helical reinforcement is given by
    (a) $0.4 f_{ck} A_g + (0.67 f_y - 0.4 f_{ck}) A_s$
    (b) $0.446 f_{ck} A_g + (0.75 f_y - 0.446 f_{ck}) A_s$
    (c) 1.05 times that given by (b)
    (d) 1.05 times that given by (a)

86. The spacing of longitudinal reinforcement in columns measured along the perimeter of the section shall not be more than
    (a) 200 mm
    (b) 250 mm
    (c) 275 mm
    (d) 300 m.

87. An inclined structural member carrying axial compression is called
    (a) pedestal
    (b) strut
    (c) stay
    (d) cord.

88. The design of uncracked R.C.C. columns subjected to axial load and bending is governed by
    (a) $\left(\dfrac{M_{ux}}{M_{ux1}}\right)^{\alpha_n} + \left(\dfrac{M_{uy}}{M_{uy1}}\right)^{\alpha_n} = 1$
    (b) $\left(\dfrac{M_{ux}}{M_{ux1}}\right)^{\alpha_n} + \left(\dfrac{M_{uy}}{M_{uy1}}\right)^{\alpha_n} > 1$
    (c) $\left(\dfrac{M_{ux}}{M_{ux1}}\right)^{\alpha_n} + \left(\dfrac{M_{uy}}{M_{uy1}}\right)^{\alpha_n} < 1$
    (d) none of the above.

89. The strength of short column with helical reinforcement can be taken as 1.05 times the strength of similar column with ties, it the ratio of volume of helical reinforcement to volume of the core is not less than
    (a) $0.36\left(\dfrac{A_k}{A_g} - 1\right)\dfrac{f_{ck}}{f_{yh}}$
    (b) $0.36\left(\dfrac{A_g}{A_k} - 1\right)\dfrac{f_{yh}}{f_{ck}}$
    (c) $0.36\left(\dfrac{A_g}{A_k} - 1\right)\dfrac{f_{ck}}{f_{yh}}$
    (d) $0.36\left(\dfrac{A_k}{A_g} - 1\right)\dfrac{f_{yh}}{f_{ck}}$.

90. The minimum number of longitudinal steel bars in R.C.C. rectangular columns must be
    (a) 2
    (b) 4
    (c) 6
    (d) 8.

91. The minimum number of longitudinal steel bars in R.C.C. circular columns must be
    (a) 2
    (b) 4
    (c) 6
    (d) 8.

92. The minimum number of longitudinal steel bars in helically reinforced R.C.C. columns must be
    (a) 2
    (b) 4
    (c) 6
    (d) 8.

93. Spacing of longitudinal bars measured along the priphery of a circular R.C.C. column shall not exceed
    (a) 150 mm
    (b) 75 mm
    (c) 250 mm
    (d) 300 mm.

94. The maximum spacing of longitudinal bars with a tie bent around alternate bars in R.C.C. columns is
    (a) 50 mm
    (b) 75 mm
    (c) 100 mm
    (d) 48 ϕ_{tr}.

95. The spacing of the longitudinal bars effectively tied should not exceed the following if additional longitudinal bars provided in between are tied in one direction by open ties
    (a) 50 mm
    (b) 7.5 mm
    (c) 180 mm
    (d) 48 ϕ_{tr}.

96. The diameter of transverse reinforcement of columns should be equal to one fourth of the diameter of the main rods but not less than
    (a) 5 mm
    (b) 6 mm
    (c) 8 mm
    (d) 10 mm.

97. The spacing of transverse reinforcement of a column shall not be more than
    (a) the least lateral dimension of a column
    (b) sixteen times the smallest diameter of the longitudinal reinforcement bar to be tied
    (c) forty-eight times the diameter of the transverse reinforcement
    (d) all of the above.

98. If the size of a column is reduced above the floor, the main bars of the column are
    (a) continued up
    (b) bent inward at the floor level
    (c) stopped just below the floor level and separate lap bars are provided
    (d) both (b) and (c).

99. The allowable compressive stress in a reinforcement concrete wall compared to that of R.C.C. columns
    (a) is equal
    (b) is less
    (c) is more
    (d) bears no relation.

100. With the increase in the ratio of the height to length of the wall, after certain value, the allowable compressive stress in R.C.C. walls
    (a) decreases
    (b) increases
    (c) does not change
    (d) oscillates.

101. The minimum percentage of main reinforcement in R.C.C. walls when it is not carrying any load is
    (a) 0.4
    (b) 0.15
    (c) 0.1
    (d) no such limit.

102. The minimum percentage of the total mild steel reinforcement in R.C.C. walls is about
    (a) 0.4
    (b) 0.2
    (c) 0.32
    (d) no such limit.

103. The design criterion in R.C.C. walls is very similar to
    (a) slabs
    (b) retaining walls
    (c) plates
    (d) columns.

104. The percentage increase in the allowable compressive stress in R.C.C. walls whose height is more than twice the length of the wall is
    (a) 0%
    (b) 10%
    (c) 20%
    (d) 30%.

105. A R.C.C. wall may be considered long for design purposes to give reduction in the load capacity if
    (a) (height/thickness) is more than 12
    (b) (length/thickness) is more than 16
    (c) (height/thickness) is more than 20
    (d) no fixed relation.

106. The minimum percentage of vertical reinforcement in R.C.C. walls with HYSD bars is
    (a) 0.12
    (b) 0.15
    (c) 0.20
    (d) 0.25.

107. The minimum percentage of horizontal reinforcement in R.C.C. walls with mild steel bars is
    (a) 0.12
    (b) 0.15
    (c) 0.20
    (d) 0.25.

108. The maximum spacing of the vertical reinforcement in R.C.C. walls is
    (a) 450 mm
    (b) 600 mm
    (c) five times the thickness of the wall
    (d) three times thickness of the wall.

109. The maximum spacing of the horizontal reinforcement in R.C.C. walls is
    (a) 450 mm
    (b) 600 mm
    (c) five times thickness of the wall
    (d) three times the thickness of the wall.

110. The maximum diameter of the nominal reinforcement in R.C.C. walls up to 150 mm thick should be
    (a) 8 mm
    (b) 12 mm
    (c) 16 mm
    (d) 20 mm.

111. In a cantilever retaining wall of height "H", the horizontal pressure of earth will act at a distance of
    (a) $\dfrac{H}{3}$ from the top
    (b) $\dfrac{H}{3}$ from the base
    (c) $\dfrac{H}{2}$ from the top
    (d) $\dfrac{H}{4}$ from the base.

112. If 'w' is the weight per unit volume, "p" the safe bearing capacity, and 'φ' is the angle of repose of the soil retained by the retaining wall, the minimum depth of foundation to be provided for the retaining wall is
    (a) $\dfrac{p}{w}\left(\dfrac{1-\sin\phi}{1+\sin\phi}\right)^2$
    (b) $\dfrac{p}{w}\left(\dfrac{1+\sin\phi}{1-\sin\phi}\right)^2$
    (c) $\dfrac{p}{w}\left(\dfrac{1-\sin\phi}{1+\sin\phi}\right)$
    (d) $\dfrac{p}{w}\left(\dfrac{1+\sin\phi}{1-\sin\phi}\right)$.

113. In the case mentioned in problem 112, if the soil is surcharged at an angle of "α", then the earth pressure exerted by the soil will be
    (a) $\dfrac{wh^2}{2}\sin\phi$
    (b) $\dfrac{wh^2}{2}\left(\dfrac{1-\sin\phi}{1+\sin\phi}\right)$
    (c) $\dfrac{wh^2}{2}\cos\alpha\left(\dfrac{1-\sin\phi}{1+\sin\phi}\right)$
    (d) $\dfrac{wh^2}{2}\cos\alpha\left(\dfrac{\cos\alpha-\sqrt{\cos^2\alpha-\cos^2\phi}}{\cos\alpha+\sqrt{\cos^2\alpha+\cos^2\phi}}\right)$.

114. If the retained earth of a retaining wall is surcharged uniformly by "w_1/m^2", then the height of retaining wall assumed as imaginarily is

   (a) decreased by $\dfrac{w_1}{w}$
   (b) increased by $\dfrac{w_1}{w}$
   (c) decreased by $\dfrac{w}{w_1}$
   (d) increased by $\dfrac{w}{w_1}$.

115. In order to make a retaining wall safe against sliding, its horizontal thrust should be less than
   (a) total vertical load of the wall
   (b) co-efficient of friction between soil and the base slab
   (c) $\dfrac{\text{total vertical load of the wall}}{\text{co-efficient of friction between soil and base slab}}$
   (d) total vertical load of the wall × co-efficient of friction between soil and the base slab.

116. The minimum thickness of stem at the top in a cantilever retaining wall should be
   (a) 100 mm
   (b) 150 mm
   (c) 200 mm
   (d) 300 mm.

117. The minimum percentage of mild steel reinforcement to be provided in any direction in a retaining wall as
   (a) 0.12
   (b) 0.15
   (c) 0.20
   (d) 0.22.

118. The maximum spacing of the reinforcement being provided in a retaining wall is
   (a) 100 mm
   (b) 250 mm
   (c) 450 mm
   (d) 600 mm.

119. Normally counterforts in a retaining wall are spaced at an interval of
   (a) 1.5 m to 2.5 m
   (b) 2.5 m to 3.5 m
   (c) 3.5 m to 4.5 m
   (d) 4.5 m to 5.5 m.

120. Bending moment in vertical stem and heel slab near counterforts is taken as
   (a) $\dfrac{wl^2}{10}$
   (b) $\dfrac{wl^2}{12}$
   (c) $\dfrac{wl^2}{12}$
   (d) $\dfrac{wl^2}{8}$

121. Bending moment in vertical stem and heel slab at centre (in between counterforts) is taken as
   (a) $\dfrac{wl^2}{8}$
   (b) $\dfrac{wl^2}{10}$
   (c) $\dfrac{wl^2}{12}$
   (d) $\dfrac{wl^2}{16}$.

122. In a cantilever retaining wall with the horizontal backfill the earth pressure exerted by the soil can be calculated by

   (a) $\dfrac{wh^2}{2}\sin\phi$ ☐
   (b) $\dfrac{wh^2}{2}\left(\dfrac{1-\sin\phi}{1+\sin\phi}\right)$ ☐

   (c) $\dfrac{wh^2}{2}\left(\dfrac{1+\sin\phi}{1-\sin\phi}\right)$ ☐
   (d) $\dfrac{wh^2}{2}\left(\dfrac{1-\sin\phi}{1+\sin\phi}\right)^2$. ☐

123. The minimum factor of safety against sliding for a retaining wall is
   (a) 1.0 ☐
   (b) 1.75 ☐
   (c) 2.0 ☐
   (d) 3.0. ☐

124. The minimum factor of safety against overturning for a retaining wall is
   (a) 3.0 ☐
   (b) 2.0 ☐
   (c) 1.5 ☐
   (d) 1.0. ☐

125. A foundation is classified as shallow foundation when depth of the foundation is
   (a) more than its width ☐
   (b) equal or less than its width ☐
   (c) less than its length ☐
   (d) more than its length. ☐

126. A foundation is classified as deep foundation when depth of the foundation is
   (a) more than its width ☐
   (b) equal or less than its width ☐
   (c) less than its length ☐
   (d) more than its length. ☐

127. When a number of columns are supported by beams and slab, then the foundation provided is known as
   (a) spread foundation ☐
   (b) strap and strip foundation ☐
   (c) mat or raft foundation ☐
   (d) pile foundation. ☐

128. The safe bearing capacity of an ordinary soil generally varies in between
   (a) 50 kN/m² to 100 kN/m² ☐
   (b) 100 kN/m² to 200 kN/m² ☐
   (c) 200 kN/m² to 300 kN/m² ☐
   (d) 300 kN/m² to 400 kN/m². ☐

129. According to IS : 456, the thickness of R.C.C. footing on soils at its edges is kept not less than
   (a) 100 mm ☐
   (b) 150 mm ☐
   (c) 200 mm ☐
   (d) 300 mm. ☐

130. According to IS : 456, the thickness at the edge on top of piles for a footing on the piles should not be less than
   (a) 100 mm ☐
   (b) 150 mm ☐
   (c) 200 mm ☐
   (d) 300 mm ☐

131. If $'p'$ is the net upward pressure on a square footing of side "b" for a square column of side "a", the maximum bending moment is given by

   (a) $\dfrac{pb^2}{8}$ ☐
   (b) $\dfrac{pb(b-a)^2}{8}$ ☐

   (c) $\dfrac{p.b.a^2}{8}$ ☐
   (d) $\dfrac{p(b-a)^2}{8}$. ☐

132. For calculating the depth of an isolated sloped footing of width "b" for a column of size "a", the width of the footing is taken as

   (a) $b + \frac{(b-a)}{8}$ ☐
   (b) $(a + 150)$ mm ☐
   (c) "a" mm ☐
   (d) $\frac{(b-a)}{8}$. ☐

133. If the width of the foundation for two columns is restricted, the shape of the footing generally adopted is
   (a) square ☐
   (b) rectangle ☐
   (c) trapezoidal ☐
   (d) circular. ☐

134. In a combined footing for two columns carrying unequal loads, the maximum hogging bending moment will occur
   (a) under the lighter column ☐
   (b) under the heavier column ☐
   (c) in between the two columns ☐
   (d) at the edges. ☐

135. One way shear strength in a R.C.C. footing is checked at a distance equal to.......of the footing from the face of the column
   (a) one-fourth of the effective depth ☐
   (b) one-half of the effective depth ☐
   (c) three-fourth of the effective depth ☐
   (d) the effective depth. ☐

136. Two way shear in a R.C.C. footing is checked at a distance equal to.....of the footing from the face of the column along the surface of truncated cone.
   (a) one-fourth of the effective depth ☐
   (b) one-half of the effective depth ☐
   (c) three-fourth of the effective depth ☐
   (d) the effective depth. ☐

137. The critical section for checking the development length in a footing shall be at the same planes as for
   (a) bending moment ☐
   (b) one way shear ☐
   (c) two way shear ☐
   (d) one way and two way shear. ☐

138. The zone in which transverse bending is likely to occur may be obtained by drawing a line from the faces to the column making an angle 'θ' with the horizontal where "θ" may be
   (a) 30° ☐
   (b) 45° ☐
   (c) 60° ☐
   (d) 90°. ☐

139. A raft foundation is provided if the areas of the foundation exceeds the plan area of the building by
   (a) 50% ☐
   (b) 40% ☐
   (c) 30% ☐
   (d) 20%. ☐

140. In a combined footing, the bottom bars under a column should be extended into the interior of the slab. This distance should be greater of the distance up to the point of contraflexure and
   (a) 30 times diameter of the main reinforcement from the outer face of the column ☐
   (b) 42 times diameter of the main reinforcement from the outer face of the column ☐
   (c) 30 times diameter of the main reinforcement from the inner face of the column ☐
   (d) 42 times diameter of the main reinforcement from the inner face of the column. ☐

141. Pile foundations are normally used
    - (a) in soft clayey soils
    - (b) in heavy load situations
    - (c) when the bearing area required is not available
    - (d) in loose sandy soils.

142. Pile foundations are invariably used in
    - (a) bridge foundations
    - (b) machine foundations
    - (c) tall buildings
    - (d) towers.

143. Generally for piles, the concrete used is of grade
    - (a) M 10
    - (b) M 15
    - (c) M 20
    - (d) M 30.

144. Untreated timber piles......for foundations.
    - (a) can be used in the ground below the water table
    - (b) cannot be used in the ground below the water table
    - (c) can be used above the water level
    - (d) should not be used

145. One of the preservatives used in timber piles for foundations is
    - (a) bitumen
    - (b) creosote oil
    - (c) olive oil
    - (d) petroleum resin.

146. The diameter of the timber piles can be about
    - (a) 120 to 350 mm
    - (b) 100 to 250 mm
    - (c) 50 to 150 mm
    - (d) 200 to 500 mm.

147. Timber piles of long lengths can be
    - (a) spliced
    - (b) not spliced
    - (c) of only one length
    - (d) none of the above.

148. The advantage of a concrete pile over a timber pile is
    - (a) no decay due to termites
    - (b) no restriction on the length
    - (c) not necessary to cut below the water mark
    - (d) all of the above.

149. The amount of reinforcement provided in precast concrete piles is usually governed by
    - (a) hammer force
    - (b) direct load on the pile
    - (c) frictional resistance
    - (d) handling forces.

150. The minimum cover to the reinforcement of precast concrete piles is
    - (a) 15 mm
    - (b) 25 mm
    - (c) 40 mm
    - (d) 60 mm.

151. Reinforcement is normally used in *cast-in-situ* concrete to
    - (a) resist impact stresses
    - (b) resist temperature, shrinkage forces etc.
    - (c) primarily resist compression
    - (d) resist tension during uplift.

152. The minimum number of piles needed in a group of piles to support a column is
    (a) 1
    (b) 2
    (c) 3
    (d) 4.

153. When the length of a pile is up to 34 times their least width, then the area of the steel should not be less than
    (a) 1.25% of gross cross-sectional area of the pile
    (b) 1.50% of gross cross-sectional area of the pile
    (c) 1.75% of gross cross-sectional area of the pile
    (d) 2.0% of gross cross-sectional area of the pile.

154. The spacing of the lateral reinforcement in a pile shall not be
    (a) less than half the least width of the pile
    (b) greater than half the least width of the pile
    (c) less than three-fourth the least width of the pile
    (d) less than three-fourth the maximum width of the pile.

155. The clear cover to all the reinforcements including binding wires in a pile should not be less than
    (a) 12 mm
    (b) 25 mm
    (c) 40 mm
    (d) 50 mm.

156. To ensure that the hogging bending moment at two points of suspension of a pile of length L, equals the sagging moment at its centre, the distances of points of suspension from either end should be equal to
    (a) $0.107 L$
    (b) $0.207 L$
    (c) $0.307 L$
    (d) $0.407 L$.

157. A pile of length l, carrying a u.d.l. "w" per metre length is suspended at two points, the maximum bending moment at the centre of the pile or at the points of suspension is
    (a) $\dfrac{wl^2}{16}$
    (b) $\dfrac{wl^2}{24}$
    (c) $\dfrac{wl^2}{36}$
    (d) $\dfrac{wl^2}{47}$.

158. A pile of length l, carrying a u.d.l. "w" per metre length is suspended at the centre and from two other points $0.15\ l$ from either end, the maximum hogging moment will be
    (a) $\dfrac{wl^2}{30}$
    (b) $\dfrac{wl^2}{60}$
    (c) $\dfrac{wl^2}{90}$
    (d) $\dfrac{wl^2}{120}$.

159. To reduce the possibilities of cracking, generally the minimum grade of concrete used in R.C.C. water tanks is
    (a) M 15
    (b) M 20
    (c) M 25
    (d) M 10.

160. The allowable direct tensile stress for M 20 concrete in R.C.C. water tanks is (in N/mm^2)
    (a) 1.2
    (b) 1.5
    (c) 1.7
    (d) 2.0.

161. The allowable bending tensile stress for M 30 concrete in R.C.C. water tanks is (in N/mm^2)
    (a) 1.2
    (b) 1.5
    (c) 1.7
    (d) 2.0.

162. In water tanks, slabs should not be concreted in lengths more than
    (a) 5 m
    (b) 7.5 m
    (c) 10 m
    (d) 20 m.

163. In water tanks, to accommodate any additional thermal displacement, expansion joints should be provided at an interval of
    (a) 5 m
    (b) 7.5 m
    (c) 10 m
    (d) 30 m.

164. For sections upto a thickness of 100 mm the minimum area of mild steel reinforcement to be provided in walls, floors, and roofs of water tanks in two perpendicular directions should be
    (a) 0.2% of the concrete section
    (b) 0.3% of the concrete section
    (c) 0.5% of the concrete section
    (d) 0.8% of the concrete section.

165. The minimum percentage of HYSD bars to be provided in R.C.C. Walls, floors and roofs of 450 mm thick in water tanks is
    (a) 0.16
    (b) 0.24
    (c) 0.4
    (d) 0.3.

166. The hoop tension in R.C.C. water tanks must be resisted by
    (a) steel only
    (b) partly by concrete and partly by steel
    (c) steel alone or along with the concrete
    (d) concrete alone.

167. The allowable direct tensile stress in mild steel bars placed near the liquid face up to 225 mm thick walls is (in MPa)
    (a) 100
    (b) 110
    (c) 125
    (d) 140.

168. The allowable direct tensile stress in mild steel bars placed 225 mm beyond the water face in R.C.C. walls is (in MPa)
    (a) 100
    (b) 110
    (c) 125
    (d) 140.

169. The allowable direct tensile stress in HYSD bars placed near the liquid face upto 225 mm thick walls is (in MPa)
    (a) 150
    (b) 190
    (c) 230
    (d) 140.

170. As per IS : 3370, when shrinkage stresses are allowed, the permissible tensile stress in concrete can be
    (a) decreased by 33.33%
    (b) increased by 33.33%
    (c) decreased by 25%
    (d) increased by 25%.

171. While designing the wall of a circular water tank restrained at the base, the moments and shear are calculated based on $K = \dfrac{48H^4}{D^2T^2}$ (Dr. Reissner's method) and as per ISI, the value of K depends upon the ratio

   (a) $\dfrac{12H^4}{D^2T^2}$ ☐ (b) $\dfrac{H^2}{DT}$ ☐

   (c) $\dfrac{16H^4}{D^2T^2}$ ☐ (d) $\dfrac{H^2}{\sqrt{DT}}$. ☐

172. R.C.C. structural members in pure bending submerged in water are designed by the following criterion

   (a) $z = \dfrac{M}{\sigma_{bt}}$ ☐ (b) $M = Qbd^2$ ☐

   (c) $A = \dfrac{P}{\sigma_{at}}$ ☐ (d) $\dfrac{M}{Z\sigma_{bt}} + \dfrac{P}{A\sigma_{at}} \leq 1$. ☐

173. R.C.C. members in water tanks subjected to hoop tension are designed based on the following criterion

   (a) $Z = \dfrac{M}{\sigma_{bt}}$ ☐ (b) $M = Qbd^2$ ☐

   (c) $A = \dfrac{P}{\sigma_{at}}$ ☐ (d) $\dfrac{M}{Z\sigma_{bt}} + \dfrac{P}{A\sigma_{at}} \leq 1$. ☐

174. R.C.C. walls of water tanks subjected to bending moment and hoop tension are designed by the following criterion

   (a) $Z = \dfrac{M}{\sigma_{bt}}$ ☐ (b) $M = Qbd^2$ ☐

   (c) $A = \dfrac{P}{\sigma_{at}}$ ☐ (d) $\dfrac{M}{Z\sigma_{bt}} + \dfrac{P}{A\sigma_{at}} \leq 1$. ☐

   where Z = Modulus of the transformed section,

   A = Transformed cross-sectional area,

   b and d = Width and effective depth of the section respectively,

   σ_{bt} and σ_{at} = Allowable bending and direct tensile stresses in concrete respectively,

   P and M = Direct force and bending moment on the section,

   Q = Constant associated with moment of resistance of a section.

175. The rise of the dome provided in Intz tank is

   (a) $\dfrac{1}{4}$ of the span ☐ (b) $\dfrac{1}{6}$ of the span ☐

   (c) 300 mm ☐ (d) $\dfrac{1}{8}$ of the span. ☐

176. The top ring beam of a Intz tank is designed for
    (a) hoop tension
    (b) hoop compression
    (c) diagonal tension
    (d) bending moment.

177. The vertical wall of the Intz tank is designed for
    (a) hoop tension
    (b) hoop compression
    (c) diagonal tension
    (d) bending moment.

178. The ratio of the diameter of the bottom dome to the diameter of the Intz should be
    (a) 0.4 to 0.5
    (b) 0.6 to 0.7
    (c) 0.7 to 0.8
    (d) 0.9 to 1.0.

179. From economy point of view, the inclination of conical dome of an Intz tank with horizontal should be in between
    (a) 30° to 40°
    (b) 40° to 45°
    (c) 50° to 55°
    (d) 60° to 75°.

180. The stress-strain curve for concrete up-to elastic yield point will be in the shape of
    (a) a zig-zag line
    (b) parabolic line
    (c) nearly a straight line
    (d) semi-circular line.

181. The structural analysis used in the ultimate strength design of R.C.C. structures is
    (a) elastic analysis
    (b) plastic analysis
    (c) ultimate analysis
    (d) elasto-plastic analysis.

182. The ultimate load used in the ultimate strength design of R.C.C. structures refers to
    (a) collapse load
    (b) working loads multiplied by load factors
    (c) total or partial collapse load
    (d) plastic hinge load.

183. The ultimate strength of a balanced reinforced concrete beam is proportional to
    (a) crushing strength of the concrete
    (b) tensile strength of the concrete
    (c) tensile strength of the steel
    (d) crushing strength of the concrete and/or tensile strength of the steel.

184. The ultimate strength of an under reinforced R.C.C. beam is proportional to
    (a) crushing strength of the concrete
    (b) crushing strength of the concrete and/or tensile strength of the steel
    (c) tensile strength of the concrete
    (d) tensile strength of the steel.

185. The ultimate strength of an over reinforced R.C.C. beam is proportional to
    (a) crushing strength of the concrete
    (b) tensile strength of the concrete
    (c) tensile strength of the steel
    (d) crushing strength of the concrete and/or tensile strength of the steel.

186. The ultimate bending compressive strain allowed in R.C.C. beam is
    (a) 0.0035
    (b) 0.002
    (c) 0.003
    (d) 0.0033.

187. The primary compression failure in R.C.C. beam is caused in
    (a) under-reinforced beams
    (b) balanced beams
    (c) over-reinforced beams
    (d) all types of beams.

188. The secondary compression failure in R.C.C. beam is caused in
    (a) under-reinforced beams
    (b) balanced beams
    (c) over-reinforced beams
    (d) all types of beams.

189. The ratio of the allowable ultimate crushing stress of R.C.C. beams in bending to that of 150 mm cube strength is
    (a) 0.40
    (b) 0.67
    (c) 0.85
    (d) 1.0.

190. The ratio of the allowable ultimate crushing stress of R.C.C. columns in bending to that of 150 mm cube strength is
    (a) 0.40
    (b) 0.67
    (c) 0.85
    (d) 1.0.

191. For a balanced design, according to Whitney, the maximum depth of concrete stress block in R.C.C. beams should be
    (a) 0.43 d
    (b) 0.30 d
    (c) 0.59 d
    (d) 0.537 d.

192. For a balanced design, according to IS : 456, the maximum depth of concrete stress block in R.C.C. beams should be
    (a) 0.43 d
    (b) 0.30 d
    (c) 0.59 d
    (d) 0.537 d.

193. As per IS : 456–1964 the ultimate bending moment capacity of an under reinforced R.C.C. beam section is
    (a) $0.185\, bd^2 f_{ck}$
    (b) $0.87 \left(1 - \dfrac{n}{2}\right) d A_{st} f_y$
    (c) $0.25\, bd^2 f_{ck}$
    (d) $0.4\, bd^2 f_{ck} + 0.67\, d\, A_{st} f_y.$

194. As per IS : 456–1964, the ultimate bending moment capacity of a balanced R.C.C. beam section is
    (a) $0.185\, bd^2 f_{ck}$
    (b) $0.87 \left(1 - \dfrac{n}{2}\right) d A_{st} f_y$
    (c) $0.25\, bd^2 f_{ck}$
    (d) $0.4\, bd^2 f_{ck} + 0.67\, d A_{st} f_y.$

195. As per IS : 456–1964, the ultimate bending moment capacity of an over reinforced R.C.C. beam section is
    (a) $0.185\, bd^2 f_{ck}$
    (b) $0.87 \left(1 - \dfrac{n}{2}\right) d A_{st} f_y$
    (c) $0.25\, bd^2 f_{ck}$
    (d) $0.4\, bd^2 f_{ck} + 0.67\, d A_{st} f_y.$

196. As per Whitney's theory, the ultimate bending moment capacity of an under-reinforced R.C.C. beam section is
    (a) $A_{st} f_{ck} d (1 - 0.59q)$ □
    (b) $A_{st} f_y d (1 - 0.59q)$ □
    (c) $A_{st} f_y d \times 0.59q$ □
    (d) $A_{st} f_y d$. □

197. As per Whitney's theory, the ultimate bending moment capacity of a balanced R.C.C. beam section is
    (a) $0.185 \, bd^2 f_{ck}$ □
    (b) $0.87 \left(1 - \dfrac{n}{2}\right) dA_{st} f_y$ □
    (c) $0.25 f_{ck}$ □
    (d) $0.4 \, bd^2 f_{ck} + 0.67 \, dA_{st} f_y$. □

198. As per Whitney's theory, the ultimate moment copacity of an over-reinforced R.C.C. section is
    (a) same as that of balanced section □
    (b) more than that of balanced section □
    (c) less than that of balanced section □
    (d) none of the above. □

199. As per IS : 456–1964, the ultimate load in structures in which effect of wind and earthquake load is neglected, should be
    (a) 1.2 D.L + 2.5 L.L □
    (b) 1.5 D.L + 1.8 L.L □
    (c) 1.5 D.L + 2.2 L.L □
    (d) 1.5 D.L + 0.5 L.L. □

200. As per IS : 456–1964, the ultimate load in structures in which effect of wind or earthquake load is considered, should be
    (a) 1.2 D.L + 2.5 L.L + 0.5 W.L □
    (b) 1.5 D.L + 1.8 L.L + 2.2 W.L □
    (c) 1.5 D.L + 2.2 L.L + 0.5 W.L □
    (d) 1.5 D.L + 2.1 L.L + 2.2 W.L. □

201. The combined load factor applied in the ultimate strength design of R.C.C. members is
    (a) 1.5 □
    (b) 1.2 □
    (c) 1.8 □
    (d) 2.2. □

202. As per IS : 456–1964, the area of steel reinforcement to be provided for balanced R.C.C. beams using ultimate load theory should be
    (a) $0.185 \, bd \, \dfrac{f_{ck}}{f_y}$ □
    (b) $0.185 \, bd^2 \, \dfrac{f_{ck}}{f_y}$ □
    (c) $0.236 \, bd \, \dfrac{f_{ck}}{f_y}$ □
    (d) $0.236 \, bd^2 \, \dfrac{f_{ck}}{f_y}$. □

203. The maximum compressive strain permitted in R.C.C. columns at failure load is
    (a) 0.002 □
    (b) 0.003 □
    (c) 0.0035 □
    (d) 0.004. □

204. The maximum compressive strain permitted in uncracked R.C.C. columns under bending and compression is
    (a) 0.002 □
    (b) 0.003 □
    (c) 0.0035 □
    (d) 0.004. □

205. The ultimate load carrying capacity of R.C.C. columns under pure axial load is

    (a) $0.185\, bd\, f_{ck}$
    (b) $0.87\left(1-\dfrac{n}{2}\right)A_s f_y$
    (c) $0.25\, bd\, f_{ck}$
    (d) $0.4\, bd\, f_{ck} + 0.67\, A_s f_y$.

206. The ratio of the ultimate load capacity of helically-reinforced concrete slender columns to that of tied columns is

    (a) 0.85
    (b) 1.00
    (c) 1.05
    (d) 1.15.

207. The maximum unsupported length of a R.C.C. column unrestrained in a plane should be

    (a) 15 m
    (b) 60 D
    (c) $\dfrac{100 b^2}{D}$
    (d) $\dfrac{100 D^2}{b}$.

208. The reduction co-efficient applied to long columns designed by ultimate strength design as compared to that designed by working stress design is

    (a) less
    (b) same
    (c) more
    (d) no relation at all.

209. The reduction factor to be applied to long R.C.C. columns designed by ultimate strength design is

    (a) $1 - \dfrac{L}{160r}$
    (b) $1.5 - \dfrac{L}{160r}$
    (c) $1.25 - \dfrac{L}{160r}$
    (d) $1.15 - \dfrac{L}{160r}$.

    where L = Effective length, r = Minimum radius of gyration.

210. In prestressed concrete structures the prestressing of the concrete is done to compensate the stresses caused by

    (a) dead load
    (b) working loads
    (c) live loads
    (d) dynamic loads.

211. The net effect due to prestressing in prestressed concrete beams is usually

    (a) tension
    (b) compression
    (c) bending and tension
    (d) bending and compression.

212. Prestressing in concrete beams is usually obtained by

    (a) external prestressing
    (b) chemical reaction
    (c) internal prestressing
    (d) stream curing.

213. A tendon with two straight line segments and an intermediate kink introduces in a prestressed concrete beam

    (a) compression
    (b) bending and compression
    (c) compression, bending and shear
    (d) tension and shear.

214. An eccentric tendon anchored perpendicular to the plane of concrete at the end section in prestressed beams introduces
    (a) compression
    (b) bending and compression
    (c) compression, bending and shear
    (d) tension and shear.
215. A parabolic tendon in prestressed beam causes an equivalent balancing
    (a) concentrated shear force
    (b) constant moment
    (c) distributed force
    (d) uniformly distributed force.
216. The prestress at the time of transfers of the tendon force to the concrete is called
    (a) anchor prestress
    (b) initial prestress
    (c) final prestress
    (d) partial prestress.
217. The prestress transferred at the time of working load is called
    (a) anchor prestress
    (b) initial prestress
    (c) final prestress
    (d) partial prestress.
218. The decrease in stress caused in a prestressed beam at constant strain is called
    (a) creep loss
    (b) relaxation
    (c) shrinkage
    (d) transfer stress.
219. An increase in strain in a concrete member at constant stress is called
    (a) creep loss
    (b) relaxation
    (c) shrinkage
    (d) transfer stress.
220. Strain in concrete at zero stress is called
    (a) creep loss
    (b) relaxation
    (c) shrinkage
    (d) transfer stress.
221. Fully prestressed concrete beam means
    (a) non-tension is permitted in the beams
    (b) no cracking is permitted in the beams
    (c) working loads are completely resisted by the prestressing force
    (d) full prestressing is applied to start with.
222. Prestressing can be efficiently used for the following members
    (a) columns and struts
    (b) ties
    (c) beams and pipes
    (d) wall panels.
223. The quality of concrete used in pre-tensioned prestressed concrete is about
    (a) M 30 to M 45
    (b) M 35 to M 50
    (c) M 40 to M 60
    (d) M 45 to M 65.
224. The quality of concrete used in post-tensioned prestressed concrete is about
    (a) M 30 to M 50
    (b) M 40 to M 60
    (c) M 25 to M 45
    (d) M 45 to M 65.
225. If the minimum age of member when full design stress expected is 6 months, the age factor to be used is
    (a) 1.0
    (b) 1.10
    (c) 1.15
    (d) 1.20.

226. The approximate value of shrinkage strain for pre-tensioned concrete is
    (a) 0.0003
    (b) 0.0002
    (c) 0.0004
    (d) $\dfrac{0.0002}{\log_{10}(t+2)}$.

    where t is the age of concrete at transfer in days.

227. In the age at loading for concrete is 28 days, the ultimate creep strain is estimated using the value of creep co-efficient as
    (a) 2.2
    (b) 1.8
    (c) 1.6
    (d) 1.1.

228. The Youngs modulus of strands conforming to IS : 6006–1970 is (kN/mm^2)
    (a) 210
    (b) 205
    (c) 200
    (d) 195.

229. The permissible relaxation of stress for prestressing steel after 1000 h is (in MPa)
    (a) 10 to 15
    (b) 35 to 90
    (c) 100 to 150
    (d) 150 to 200.

230. The approximate ratio of the permissible stress at transfer in concrete to the 150 mm cube crushing stress in prestressed concrete is about
    (a) 0.3 to 0.4
    (b) 0.35 to 0.45
    (c) 0.4 to 0.5
    (d) 0.45 to 0.55.

231. The approximate ratio of permissible stress at working load to the ultimate stress in high tensile steel in prestressed concrete is about
    (a) 0.45 to 0.55
    (b) 0.55 to 0.65
    (c) 0.65 to 0.75
    (d) 0.75 to 0.85.

232. The approximate ratio of the permissible stress at the time of jacking to the ultimate stress of high tensile steel in prestressed concrete is about
    (a) 0.4 to 0.5
    (b) 0.5 to 0.6
    (c) 0.6 to 0.7
    (d) 0.7 to 0.8.

233. The approximate ratio of the permissible bearing stress at anchor to the 150 mm cube crushing stress is about
    (a) 0.4 to 0.5
    (b) 0.5 to 0.6
    (c) 0.6 to 0.7
    (d) 0.7 to 0.8.

234. The maximum permissible tensile stress in pretensioned prestressed concrete in buildings is about (in MPa)
    (a) 0
    (b) 0 to 1
    (c) 1 to 2
    (d) 2.0 to 2.5.

235. The maximum tensile stress allowed in prestressed concrete in water tanks is about (in MPa)
    (a) 0
    (b) 0 to 1
    (c) 1 to 2
    (d) 2.0 to 2.5.

236. The approximate ultimate tensile stress of high tensile steel bars in prestressed concrete is (in MPa)
   (a) 900 to 1200 ☐ (b) 1200 to 1500 ☐
   (c) 1500 to 2000 ☐ (d) 2000 to 2500. ☐

237. The approximate ultimate tensile stress of high tensile steel wires used in prestressed concrete is (in MPa)
   (a) 900 to 1200 ☐ (b) 1200 to 1500 ☐
   (c) 1500 to 2000 ☐ (d) 2000 to 2500. ☐

238. The approximate ultimate tensile stress of high tensile cables used in prestressed concrete is (in MPa)
   (a) 900 to 1200 ☐ (b) 1200 to 1500 ☐
   (c) 1500 to 2000 ☐ (d) 2000 to 2500. ☐

239. Working stress design assumes
   (a) elastic stress-strain relation ☐ (b) linear stress-strain relation ☐
   (c) elastic and linear stress-strain relation ☐ (d) isotropic material. ☐

240. Working stress design assumes
   (a) partial recovery of deflections caused during some load conditions ☐
   (b) full recovery of deflections caused during some load conditions ☐
   (c) partial recovery of deflections caused in all types of load conditions ☐
   (d) full recovery of deflections caused in all types of load conditions. ☐

241. The stresses caused by different loads can be superimposed in
   (a) working stress design ☐ (b) ultimate strength design ☐
   (c) limit state design ☐ (d) plastic design. ☐

242. The critical load conditions for which a design is to be made in prestressed concrete structures are
   (a) dead and live loads ☐ (b) transfer and live loads ☐
   (c) different working loads ☐ (d) transfer and working loads. ☐

243. Working loads may be defined as
   (a) maximum loads consisting of dead and live loads ☐
   (b) a set of ciritical loads in service conditions ☐
   (c) all loads that occur during service conditions ☐
   (d) a set of maximum loads for which no failure will occur. ☐

244. The minimum cover in any pretensioned prestressed concrete member protected from direct weather is
   (a) 15 mm ☐ (b) 20 mm ☐
   (c) 25 mm ☐ (d) 30 mm. ☐

245. The minimum cover in any post-tensioned prestressed concrete member protected from direct weather is the greater of the cable or
   (a) 15 mm ☐ (b) 20 mm ☐
   (c) 25 mm ☐ (d) 30 mm. ☐

246. The minimum cover in any pretensioned prestressed concrete member located in aggressive environment is
    (a) 20 mm
    (b) 25 mm
    (c) 30 mm
    (d) 40 mm.

247. The minimum clear spacing for single wires used in pre-tension system is
    (a) 2 times the diameter of the wire
    (b) $1\frac{1}{2}$ times the maximum size of aggregate
    (c) 3 times the diameter of the wire and that given in (b) whichever is greater
    (d) combination of 'a' and 'b' whichever is greater.

248. The minimum clear spacing of non-grouped cables or large bars shall be
    (a) 40 mm
    (b) maximum size of the cable or bar
    (c) 5 mm plus maximum size of aggregate
    (d) greater of a, b and c.

249. The minimum clear horizontal spacing between groups of cables or ducts of grouped cables shall be greater of
    (a) 30 mm or 5 mm more than maximum size of aggregate
    (b) 40 mm or 5 mm more than maximum size of aggregate
    (c) 50 mm or maximum size of aggregate
    (d) 40 mm or maximum size of aggregate.

250. The minimum clear vertical spacing between groups of cables or ducts of grouped cables shall be
    (a) 50 mm
    (b) 40 mm
    (c) 60 mm
    (d) maximum size of aggregate.

251. The mix proportions of cement mortar to be placed in the joints in precast prestressed member is
    (a) 1 : 1
    (b) 1 : 2.5
    (c) 1 : 2
    (d) 1 : 1.5.

252. The thickness of cement mortar to be used at the butted joints of precast prestressed concrete segments should be about
    (a) 30 mm to 38 mm
    (b) 25 mm to 32 mm
    (c) 12 mm to 18 mm
    (d) 18 mm to 25 mm.

253. The thickness of cement grout to be used at the butt joints of precast prestressed concrete members should be
    (a) 10 mm
    (b) 12 mn
    (c) 18 mm
    (d) 25 mm.

254. The grouting of ducts in post-tensioned prestressed concrete causes
    (a) reduction in deflections
    (b) reduction in crack-width
    (c) reduction in corrosion
    (d) all the above.

255. The mix used in the grouting of ducts in post-tensioned prestressed concrete is
    (a) cement grout only
    (b) cement-sand mortar grout
    (c) cement with resin
    (d) cement or cement-sand mortar grout.

256. The grouting of ducts in post-tensioned prestressed concrete works should be done under a pressure of about (in MPa)
    (a) 0.5
    (b) 0.7
    (c) 1.2
    (d) 1.5.

257. Long-line prestressing in prestressed concrete is used in
    (a) *cast-in situ* pre-tensioned construction
    (b) *cast-in-situ* post-tensioned construction
    (c) precast post-tensioned construction
    (d) precast pretensioned construction.

258. The unit mould method of prestressing is commonly used in
    (a) precast construction
    (b) post-tensioned construction
    (c) *cast-in-situ* construction
    (d) cantilever construction.

259. The diameter of the high tensile steel tendons used is
    (a) 3 mm to 5 mm
    (b) 5 mm to 8 mm
    (c) 6 mm to 10 mm
    (d) 8 mm to 12 mm.

260. In the Magnel Blaton system of prestressing, wires in the same layer and the wires in adjacement layers are separated with a clearance of
    (a) 2 mm
    (b) 4 mm
    (c) 8 mm
    (d) 10 mm.

261. In the Magnel Blaton system of prestressing, the wires are anchored by wedging two at a time into sandwich plates. These plates are provided with two wedge shaped grooves on its two faces and the thickness of plate is generally taken as
    (a) 10 mm
    (b) 16 mm
    (c) 20 mm
    (d) 25 mm.

262. In the Magnel Blaton system each sandwich plate can generally anchor
    (a) 8 wires
    (b) 12 wires
    (c) 16 wires
    (d) 20 wires.

263. In the Freyssinet system, the group of wires arranged to form cable with spiral ring inside, consists of
    (a) 4 wires
    (b) 8 wires
    (c) 12 wires
    (d) 16 wires.

264. In the Gifford Udall system, in prestressed concrete a tendon of 28 mm diameter is used. This tendon is stronger than
    (a) 12 × 6 mm wire
    (b) 16 × 6 mm wires
    (c) 16 × 8 mm wires
    (d) 12 × 10 mm wires.

265. In the Lee-McCall system, the diameter of the high tensile alloy steel wires (silico-manganese steel) used is
    (a) 5 mm to 10 mm
    (b) 10 mm to 16 mm
    (c) 16 mm to 22 mm
    (d) 22 mm to 30 mm.

266. The amount of slip from friction wedges holding the wires may be taken on an average as
    (a) 1 mm
    (b) 2.5 mm
    (c) 5 mm
    (d) 10 mm.

267. The approximate loss of prestress due to creep in steel in prestressed concrete is about
    (a) 1%
    (b) 3%
    (c) 6%
    (d) 10%.
268. The approximate loss of prestress in prestressed concrete due to creep in concrete is about
    (a) 1%
    (b) 3%
    (c) 6%
    (d) 10%.
269. The approximate loss of prestress due to shrinkage of concrete in prestressed concrete is about
    (a) 1%
    (b) 3%
    (c) 6%
    (d) 10%.
270. The approximate loss of prestress due to the slippage of anchorage in long span prestressed concrete beams is about
    (a) 0 to 5%
    (b) 5 to 8%
    (c) 8 to 12%
    (d) 10 to 15%.
271. The approximate loss of prestress caused by friction in shallow curved tendons is about
    (a) 0 to 5%
    (b) 5 to 8%
    (c) 8 to 12%
    (d) 10 to 15%.
272. The approximate total percentage loss of prestress in pretensioned concrete beams is about
    (a) 0 to 10%
    (b) 10 to 15%
    (c) 15 to 20%
    (d) 20 to 25%.
273. The approximate total percentage loss of prestress in post-tensioned concrete beams is about
    (a) 0 to 10%
    (b) 10 to 15%
    (c) 15 to 20%
    (d) 20 to 25%.
274. The net shear force on concrete in prestressed concrete members is less when compared with that in R.C.C. because of
    (a) pre-compression
    (b) transverse component of the cable force
    (c) anchorage force
    (d) the statement is wrong.
275. The shear force on prestressed concrete element depends on
    (a) distributed loads only
    (b) concentrated loads only
    (c) variation in net bending moment
    (d) torsion also.
276. The shear force on a prestressed concrete element is proportional to
    (a) rate of change of moment
    (b) uniformly distributed load
    (c) concentrated loads
    (d) distributed and concentrated loads.
277. The shear stress at any level caused by transverse shear force in prestressed concrete beams is given by
    (a) $\dfrac{VQ}{Ib}$
    (b) $\dfrac{V}{bjd}$
    (c) $\dfrac{V_e}{bjd}$
    (d) $\dfrac{V_e Q}{Ib}$.
    where V = Shear force, V_e = Effective shear force, b = Width, I = Moment of inertia,
            Q = First moment of the area of the section above that level taken about centroidal axis
            jd = Lever arm.

278. Anchorage plates in post-tensioned prestressed concrete structures are placed
    (a) perpendicular to the plane of the concrete ☐
    (b) perpendicular to the axis of the tendon ☐
    (c) inclined to the plane of the concrete ☐
    (d) no special relation. ☐
279. The size of the anchor plate in post-tensioned prestressed concrete structures depends upon
    (a) bearing capacity of the concrete ☐
    (b) prestressing force ☐
    (c) stress in the cable and bearing stress of the concrete ☐
    (d) prestressing force bearing capacity of the concrete. ☐
280. The tendons in post-tensioned prestressed concrete structures
    (a) must be anchored on steel and anchor plates ☐
    (b) can be anchored through cones bearing directly on concrete ☐
    (c) can be anchored through special buttoning without steel plates ☐
    (d) can be anchored at the ends only. ☐
281. The bursting stress in concrete at the anchorage zone is generated on planes approximately
    (a) parallel to the axis of the cable ☐
    (b) perpendicular to the axis of the cable ☐
    (c) perpendicular to the end planes of the beam ☐
    (d) parallel to the end planes of the beam. ☐
282. The bursting stresses in prestressed concrete members are developed at
    (a) maximum bending moment zone ☐ (b) maximum shear zone ☐
    (c) anchorage zone ☐ (d) bond zone. ☐
283. Spalling stresses are produced in post-tensioned prestressed concrete members at
    (a) maximum bending moment zone ☐ (b) anchorage zone ☐
    (c) maximum shear zone ☐ (d) bond zone. ☐
284. Spalling stresses are produced because of
    (a) bursting force ☐ (b) inadequate anchor block ☐
    (c) insufficient bond length ☐ (d) high concentrated tendon force. ☐
285. The bonding of wires in post-tensioned construction is done to
    (a) bond between cables and concrete ☐ (b) minimize anchorage ☐
    (c) transfer the tendon force to the concrete ☐ (d) prevent corrosion. ☐
286. The anchor cones in post-tensioned prestressed concrete beams are designed
    (a) primarily for hoop tension ☐ (b) primarily for hoop compression ☐
    (c) for torsion ☐ (d) for bearing compression. ☐
287. Anchor cones are used in pretensioned prestressed concrete beams to
    (a) transfer prestress ☐ (b) protect the wires from slipping ☐
    (c) bond the wires ☐ (d) there are no such cones. ☐

288. The bond stress along the length of the wire in pretensioned prestressed concrete construction
    (a) is almost constant ☐
    (b) decreases from maximum to zero in a short length ☐
    (c) varies proportional to the shear force ☐
    (d) varies proportional to the bending moment. ☐

289. A load bearing wall is designed to carry
    (a) vertical loads ☐
    (b) live load plus self-weight ☐
    (c) imposed loads ☐
    (d) wind load plus self-weight. ☐

290. A partition wall is designed to carry
    (a) live loads ☐
    (b) imposed loads ☐
    (c) vertical loads ☐
    (d) no external loads. ☐

291. A shear wall is designed to carry
    (a) axial shear and bending forces ☐
    (b) shear force ☐
    (c) shear and bending ☐
    (d) axial and bending. ☐

292. A cavity wall is designed to carry
    (a) axial, shear and bending forces ☐
    (b) shear force ☐
    (c) shear and bending ☐
    (d) axial and bending. ☐

293. A curtain wall is designed to carry
    (a) live, dead and wind loads ☐
    (b) wind load plus self-weight ☐
    (c) shear force ☐
    (d) imposed loads. ☐

294. The slenderness ratio of masonry walls should be limited to
    (a) 40 ☐
    (b) 30 ☐
    (c) 20 ☐
    (d) 10. ☐

295. A masonry wall is said to be short wall if the slenderness ratio is less than
    (a) 20 ☐
    (b) 15 ☐
    (c) 12 ☐
    (d) 8. ☐

296. The allowable compressive strength of brick masonry is about......of the compressive strength of the brick.
    (a) one-twelfth or less ☐
    (b) one-eighth or less ☐
    (c) one-sixth or less ☐
    (d) one-fourth or less. ☐

297. The allowable compressive strength of brick masonry is about.......of the compressive strength of the mortar.
    (a) 30% or less ☐
    (b) 40% or less ☐
    (c) 50% or less ☐
    (d) 60% or less. ☐

298. The ratio of the allowable stress in bending to that in direct compression in masonry is about
    (a) 1.5 ☐
    (b) 1.33 ☐
    (c) 1.25 ☐
    (d) 0.8. ☐

299. The basic allowable compressive stress in masonry is......with an increase in the area of the masonry.
   (a) unchanged
   (b) increased
   (c) decreased
   (d) no relation at all.

300. The allowable compressive stress under concentrated load in brick masonry is.....that under uniform load conditions.
   (a) same as
   (b) less than
   (c) 25% more than
   (d) 50% more than.

301. In the limit state method, the maximum strain in concrete in bending compression is taken to be
   (a) 0.35
   (b) 0.035
   (c) 0.0035
   (d) 0.00035.

302. The design strength of the materials in the limit state method is given by
   (a) $f_d = \dfrac{f}{\gamma_m}$
   (b) $f_d = \dfrac{1.5f}{\gamma_m}$
   (c) $f_d = \dfrac{\gamma_m}{f}$
   (d) none of the above

   where f is the characteristic strength of the material and γ_m is the partial safety factor.

303. In the limit state method, the partial safety factor for concrete is taken as
   (a) 3
   (b) 2
   (c) 1
   (d) 1.5.

304. In the limit state method, the partial safety factor for steel is taken as
   (a) 1.05
   (b) 1.15
   (c) 1.25
   (d) 1.35.

305. The maximum strain in the tension reinforcement in the section (designed as per limit state method) at failure shall not be less than
   (a) $\dfrac{1.15 E_s}{f_y} + 0.002$
   (b) $\dfrac{f_y}{1.15 E_s} + 0.002$
   (c) $\dfrac{1.15 E_s}{f_y}$
   (d) none of the above

   where f_y is the characteristic strength and E_s is the modulus of elasticity of steel.

306. The limiting moment of resistance of a balanced rectangular section of size b and d, when steel of Fe 415 grade is used, is given by
   (a) $0.0133 f_{ck} bd^2$
   (b) $0.033 f_{ck} bd^2$
   (c) $0.133 f_{ck} bd^2$
   (d) $1.33 f_{ck} bd^2$.

307. The limiting reinforcement percentage p_t, for a singly reinforced balanced rectangular section is given by (when Fe 415 grade steel is used)
   (a) $\dfrac{p_t \lim \times f_y}{f_{ck}} = 19.87$
   (b) $\dfrac{p_t \lim \times f_y}{f_{ck}} = 21.88$
   (c) $\dfrac{p_t \lim \times f_y}{f_{ck}} = 18.95$
   (d) none of the above.

308. In the limit state method, the over-reinforced sections
    (a) are not permitted
    (b) are permitted
    (c) are permitted only in extreme cases
    (d) none of the above.

309. If the factored moment to be carried by a rectangular section is more than the limiting moment of the section
    (a) the section can be designed as a over-reinforced section
    (b) the section has to be redesigned increasing the dimensions or design the section as doubly reinforced
    (c) either (a) and (b)
    (d) none of the above.

310. The distance to the neutral axis from the top compression fibre of a singly reinforced rectangular section is given by
    (a) $x_u = \dfrac{0.36 f_{ck} b}{0.87 f_y A_{st}}$
    (b) $x_u = \dfrac{0.36 f_y A_{st}}{0.87 f_{ck} b}$
    (c) $x_u = \dfrac{0.87 f_y A_{st}}{0.36 f_{ck} b}$
    (d) $x_u = \dfrac{0.87 f_{ck} b}{0.36 f_y A_{st}}$.

    where A_{st} is the area of the steel.

311. If A_{sv} and s_v denote the area and spacing of vertical stirrups, and d the effective depth, the strength of the vertical stirrups is given by
    (a) $\dfrac{0.87 f_y A_{sv}}{s_v \cdot d}$
    (b) $\dfrac{0.87 A_{sv}}{f_y \cdot s_v \cdot d}$
    (c) $\dfrac{0.87 f_y}{A_{sv} \cdot s_v \cdot d}$
    (d) $\dfrac{0.87 f_y \cdot A_{sv} \cdot d}{s_v}$.

312. The allowable bond stress in tension for plain bars in the limit state method for M 15 grade concrete is
    (a) 0.6 N/mm²
    (b) 1.2 N/mm²
    (c) 1.0 N/mm²
    (d) 15 N/mm².

313. For deformed bars, the allowable bond stress in tension is increased by......per cent over the corresponding value of plain bars.
    (a) 20
    (b) 40
    (c) 60
    (d) 80.

314. For bars in compression, the values of bond stress for bars in compression shall be increased by......per cent.
    (a) 25
    (b) 50
    (c) 75
    (d) 100.

315. When the factored shear force and factored torsional moment acting on a beam are denoted by V_u and T_u and the breadth of the beam by b the equivalent shear V_e is given by
    (a) $T_u + 1.6 \dfrac{V_u}{b}$
    (b) $V_u + \dfrac{1.6 T_u}{b}$
    (c) $T_u + \dfrac{b}{1.6 V_u}$
    (d) $V_u + \dfrac{b}{1.6 T_u}$.

316. In the limit state method, the maximum compressive strain in axial compression in concrete is taken as
    (a) 0.0035
    (b) 0.035
    (c) 0.002
    (d) 0.02.

317. When the minimum eccentricity does not exceed 5 per cent of the lateral dimension, the load carrying capacity of a reinforced concrete column as per the limit state method is given by
    (a) $P_u = 0.4 f_{ck} \cdot A_c + 0.75 f_y A_{sc}$
    (b) $P_u = 0.4 f_{ck} \cdot A_c + 0.6 f_y A_{sc}$
    (c) $P_u = 0.4 f_{ck} \cdot A_c + f_y A_{sc}$
    (d) $P_u = f_{ck} \cdot A_c + f_y A_{sc}$.

318. All columns shall be designed for an eccentricity not less than
    (a) 10 mm
    (b) 20 mm
    (c) 30 mm
    (d) 40 mm.

319. In the limit state method, the compressive force in concrete is taken to act at a distance of.....times the neutral axis measured from the extreme compression fibre.
    (a) 0.22
    (b) 0.32
    (c) 0.42
    (d) 0.52.

320. The surface width of cracks shall not generally exceed
    (a) 0.1 mm
    (b) 0.2 mm
    (c) 0.3 mm
    (d) 0.4 mm.

321. The nominal shear stress in beams of M 15 grade concrete (even with shear reinforcement) under no circumstances shall exceed
    (a) 2.5 N/mm²
    (b) 1.5 N/mm²
    (c) 0.5 N/mm²
    (d) no limit.

322. The increase in strength of concrete after six months is generally
    (a) 5%
    (b) 10%
    (c) 15%
    (d) 20%.

323. Modular ratio in reinforced concrete is defined as
    (a) $\dfrac{\text{Modulus of elasticity of steel}}{\text{Modulus of elasticity of concrete}}$
    (b) $\dfrac{\text{Modulus of elasticity of concrete}}{\text{Modulus of elasticity of steel}}$
    (c) $\dfrac{\text{Coefficient of linear expansion of steel}}{\text{Coefficient of linear expansion of concrete}}$
    (d) none of the above.

324. A slight upward curvature provided at the time of construction so that it will straighten out and attain correct shape on loading is called
    (a) camber
    (b) starter
    (c) preset
    (d) super elevation.

325. The entrained air in concrete
    (a) increases the strength
    (b) decreases the workability
    (c) increases the workability
    (d) none of the above.
326. The drying of concrete after casting results in
    (a) expansion
    (b) shrinkage
    (c) neither shrinkage nor expansion
    (d) difficult to predict.
327. During transportation of concrete the separation of coarse aggregate from mortar is called
    (a) separation
    (b) creeping
    (c) bleeding
    (d) segregation.
328. The separation of water from a freshly mixed concrete is known as
    (a) separation
    (b) creeping
    (c) bleeding
    (d) segregation.
329. The type of cement preferred in the construction of massive concrete dam is
    (a) ordinary portland cement
    (b) rapid harding cement
    (c) low heat cement
    (d) white cement.
330. Which of the constituent materials of concrete is inert ?
    (a) water
    (b) cement
    (c) aggregates
    (d) all the above.
331. For an aggregate to be called as a cyclopean aggregate its size must be larger than
    (a) 35 mm
    (b) 55 mm
    (c) 75 mm
    (d) 95 mm.
332. A corarse aggregate passes through 7.5 mm sieve and retained on 60 mm sieve. It will be termed flaky if its minimum dimension is less than
    (a) 40.5 mm
    (b) 67.5 mm
    (c) 50.5 mm
    (d) 20.5 mm.
333. In problem 332, the aggregate will be known as elongated if its length is more than
    (a) 111.5 mm
    (b) 121.5 mm
    (c) 101.5 mm
    (d) 91.5 mm.
334. An aggregate is said to be flaky if its least dimension is less than.......of the mean dimension.
    (a) three-fifths
    (b) two-thirds
    (c) half
    (d) one-fourth.
335. Los Angels machine is used to test the aggregate for
    (a) impact value
    (b) abrasion resistance
    (c) water absorption
    (d) crushing strength.
336. If the fineness modulus of sand is 2.5 it is graded as
    (a) coarse sand
    (b) medium sand
    (c) fine sand
    (d) very fine sand.
337. Water cement ratio may be defined as
    (a) $\dfrac{\text{Volume of water}}{\text{Volume of cement}}$
    (b) $\dfrac{\text{Weight of water}}{\text{Weight of cement}}$
    (c) $\dfrac{\text{Weight of water}}{\text{Weight of concrete}}$
    (d) any of the above.

338. As per I.S. specifications full strength of concrete is attained......days after casting.
    (a) 7
    (b) 14
    (c) 21
    (d) 28.

339. High increase in temperature......the strength of concrete.
    (a) increases
    (b) has no effect on
    (c) decreases
    (d) none of the above.

340. The concrete mix which is not amicable for smooth finish is said to possess
    (a) segregation
    (b) internal friction
    (c) bleeding
    (d) hardness.

341. Workability of concrete mix having very low water-cement ratio has to be ascertained by
    (a) slump test
    (b) bending test
    (c) compaction factor test
    (d) compression test.

342. The grade of concrete not recommended by IS : 456 is
    (a) M 100
    (b) M 400
    (c) M 500
    (d) none.

343. A moist sand occupies a depth of 15 cm and when fully saturated it occupies 12 cm. Then the bulking of the moist sand is
    (a) 10%
    (b) 15%
    (c) 20%
    (d) 25%.

344. The process of proper and accurate measurement of concrete ingedients for uniformity of proportion is known as
    (a) batching
    (b) grading
    (c) mixing
    (d) none of the above.

345. Expansion joints are provided if the lengths of the concrete structure exceeds
    (a) 35 m
    (b) 45 m
    (c) 15 m
    (d) 25 m.

346. At a given water content, workability of concrete is more if the aggregates used are
    (a) rounded aggregates
    (b) flaky aggregates
    (c) angular aggregates
    (d) none of the above.

347. Which of the following tests gives the workability of concrete ?
    (a) compression test
    (b) bending test
    (c) slump test
    (d) none of the above.

348. Which of the following statements is correct ? As the area of steel increases
    (a) depth of neutral axis decreases
    (b) depth of neutral axis increases
    (c) lever arm increases
    (d) none of the above.

349. In a singly reinforced beam, if the concrete is stressed to its allowable limit earlier than steel, the section is said to be
    (a) over reinfored section
    (b) under reinforced section
    (c) balanced section
    (d) economical section.

350. A circular slab, when subjected to external loading, deforms to assume as shape of
   (a) hemi-sphere
   (b) ellipsoid
   (c) semi-hemisphere
   (d) paraboloid.

351. A square column of side a is founded on a square footing of side b. The maximum bending moment on the footing corresponding to a net upward pressure p is given by
   (a) $\dfrac{pb(b-a)^2}{8}$
   (b) $\dfrac{pb(b+a)^2}{8}$
   (c) $\dfrac{pb(b-a)}{4}$
   (d) $\dfrac{pb(b+a)}{4}$.

352. If the allowable punching shear stress is q, the depth of footing from punching shear criterion to carry a load W is given by
   (a) $D = \dfrac{W(a-b)}{4a^2bq}$
   (b) $D = \dfrac{W(a^2-b^2)}{8abq}$
   (c) $D = \dfrac{W(a^2+b^2)}{8abq}$
   (d) $D = \dfrac{(a^2-b^2)}{8a^2bq}$.

353. While driving the piles, the drop must be at least
   (a) 140 cm
   (b) 120 cm
   (c) 100 cm
   (d) 80 cm.

354. A retaining wall of height h retains earth up to its top without any surcharge. The resultant pressure acts parallel to the top surface and from the base at a distance of
   (a) $\dfrac{h}{2}$
   (b) $\dfrac{h}{4}$
   (c) $\dfrac{h}{3}$
   (d) $\dfrac{2h}{2}$.

355. The live load adopted in the design of an inaccessible roof slab is
   (a) 0 kg/m^2
   (b) 150 kg/m^2
   (c) 100 kg/m^2
   (d) 75 kg/m^2.

356. The maintenance cost of concrete structures is
   (a) practically nil
   (b) exhorbitant
   (c) moderate
   (d) as much as for timber structures.

357. The reason for choosing mild steel bars as reinforcement in tensile zones of concrete is that
   (a) The modulus of elasticity of mild steel is about the same as that of concrete
   (b) The poissons ratio of mild steel is almost the same as that of concrete
   (c) The coefficient of linear expansion is about the same as that of concrete
   (d) The stress-strain plots of both mild steel and concrete are identical in shape.

358. The advantage of concrete is that
   (a) it can be moulded into any shape and size
   (b) it is durable and maintenance cost is practically nil.
   (c) it is fire resistant
   (d) all the above.

359. For concrete, the initial set generally requires
    (a) 30 to 60 days
    (b) 30 to 60 hours
    (c) 30 to 60 minutes
    (d) 30 to 60 seconds.

360. For concrete, the final set generally requires
    (a) 5 to 6 minute
    (b) 5 to 6 hours
    (c) 5 to 6 days
    (d) 5 to 60 seconds.

361. The optimum water-cement ratio for 1 : 2 : 4 concrete is around
    (a) 0.75
    (b) 0.60
    (c) 0.45
    (d) 0.30.

362. The weight of cement contained in one standard bag is
    (a) 30 kg
    (b) 40 kg
    (c) 50 kg
    (d) 60 kg.

363. The probability that the actual strength of the material is more than its characteristic strength is
    (a) 0.05
    (b) 0.10
    (c) 0.90
    (d) 0.95.

364. In the limit state design, the strength of concrete as well as the strength of steel is taken to be as random variable that follows
    (a) normal probability distribution
    (b) log-normal probability distribution
    (c) Gamma probability distribution
    (d) exponential probability distribution.

365. The chances for the actual strength of concrete to be less than its characteristic strength are
    (a) 95%
    (b) 90%
    (c) 10%
    (d) 5%.

366. Compared to plain bars, the deformed bars will have
    (a) better bond with concrete
    (b) improved yield stress
    (c) economical design
    (d) all the above.

367. The modulus of elasticity of steel is generally taken as
    (a) 2×10^3 N/mm^2
    (b) 2×10^4 N/mm^2
    (c) 2×10^5 N/mm^2
    (d) 2×10^{11} N/mm^2.

368. If f_m is the mean compressive strength and f_s is the standard deviation of the compressive strength of concrete obtained after testing a large number of samples, then the characteristic strength of concrete is given by
    (a) $f_k = f_m + 1.645 f_s$
    (b) $f_k = f_m - 1.645 f_s$
    (c) $f_k = f_m + 0.645 f_s$
    (d) $f_k = f_m - 0.645 f_s$.

369. The design load obtained after multiplying the characteristic load by the partial safety factor is also known as
    (a) factored load
    (b) probable load
    (c) possible load
    (d) plausible load.

370. In terms of the mean load F_m and the standard deviation of the load F_s the characteristic load F_k is given by
   (a) $F_k = F_m + 1.645 F_s$
   (b) $F_k = F_m + 0.5 F_s$
   (c) $F_k = F_m - 1.645 F_s$
   (d) $F_k = F_m - 0.5 F_s$.

371. Concrete fails at a strain of
   (a) 0.2
   (b) 0.002
   (c) 0.0035
   (d) 0.35.

372. The shape of the stress-strain curve of concrete up to a strain of 0.002 is
   (a) parabolic
   (b) linear
   (c) hyperbolic
   (d) none of the above.

373. The shape stress-strain curve of mild steel upto yield stress is
   (a) parabolic
   (b) linear
   (c) hyperbolic
   (d) none of the above.

374. In the case of Fe 415 and Fe 500 steels, the yield stress is taken as ...... per cent proof stress.
   (a) 0.02
   (b) 0.035
   (c) 0.2
   (d) 0.35.

375. In the limit state designs, the strain in the outermost compression fibre at the time of failure is taken to be
   (a) 0.0025
   (b) 0.0030
   (c) 0.0035
   (d) 0.0040.

376. The moment of resistance factor R_u is given as
   (a) $R_u = \dfrac{M_u}{b}$
   (b) $R_u = \dfrac{M_u}{bd}$
   (c) $R_u = \dfrac{M_u}{b^2 d}$
   (d) $R_u = \dfrac{M_u}{bd^2}$.

377. Over reinforced section is not desirable because
   (a) heavy reinforcement makes it uneconomical
   (b) failure is sudden as concrete fails first
   (c) its design is highly complicated
   (d) none of the above.

378. The balanced design gives
   (a) smallest concrete and maximum A_{st}
   (b) smallest concrete and minimum A_{st}
   (c) largest concrete and maximum A_{st}
   (d) largest concrete and minimum A_{st}.

379. In a simply supported beam carrying uniformly distributed load, the spacing of vertical stumps near supports would be
   (a) uniform
   (b) closer
   (c) wider
   (d) no specific pattern.

380. For a fixed beam of span l carrying uniformly distributed load the bending moment at the critical section is given by
   (a) $\dfrac{wl^2}{8}$
   (b) $\dfrac{wl^2}{4}$
   (c) $\dfrac{wl^2}{12}$
   (d) $\dfrac{wl^2}{24}$.

381. In the limit state design of a rectangular section the ratio of the distance to neutral axis x_u to the effective depth $d \left(\dfrac{x_u}{d} \right)$ depends only on

   (a) f_y
   (b) f_{ck}
   (c) both f_y and f_{ck}
   (d) none of the above.

382. Moment of resistance in an under reinforced section is

   (a) $0.87 f_y A_{st} d \left(1 - \dfrac{A_{st}}{bd} \cdot \dfrac{f_y}{f_{ck}}\right)$
   (b) $0.97 f_y A_{st} d \left(1 - \dfrac{A_{st}}{bd} \cdot \dfrac{f_y}{f_{ck}}\right)$
   (c) $0.87 f_y A_{st} d \left(1 + \dfrac{A_{st}}{bd} \cdot \dfrac{f_y}{f_{ck}}\right)$
   (d) $0.97 f_y A_{st} d \left(1 + \dfrac{A_{st}}{bd} \cdot \dfrac{f_y}{f_{ck}}\right)$.

383. The combination of loads for serviceability conditions should be the ...... of (i) DL + LL (ii) DL + WL and (iii) DL + 0.8LL + 0.8 WL (or EL).

   (a) the largest
   (b) the average
   (c) the smallest
   (d) 1.5 times the average.

384. In general the formulae for short terms deflection at the mid span can be expressed as

   (a) $K \dfrac{Wl}{EI}$
   (b) $K \dfrac{Wl^2}{EI}$
   (c) $K \dfrac{Wl^3}{EI}$
   (d) $K \dfrac{Wl^4}{EI}$.

385. As per IS 456 : 2000, the vertical deflection limits in a cantilever beam may generally be assumed to be satisfied provided that (span/depth) ratio does not exceed

   (a) 26
   (b) 20
   (c) 7
   (d) 3.

386. The moment of resistance factor $R_{u,\lim}$, when Fe 415 steel is used, is given by

   (a) $0.1489 f_{ck}$
   (b) $0.1330 f_{ck}$
   (c) $0.1399 f_{ck}$
   (d) $0.1381 f_{ck}$.

387. Which of the following is not a measure adopted for reducing deflection?

   (a) Decrease steel reinforcement
   (b) Increase the depth
   (c) Increase the camber
   (d) use rich concrete.

388. Cracks will be within the allowable limits if code specification for

   (a) maximum and minimum spacing of bars are followed
   (b) maximum and minimum areas of steel are followed
   (c) curtailment and anchorage of bars are followed
   (d) all the above are followed.

389. The assessed surface width of crack should not generally exceed

   (a) 0.2 mm
   (b) 0.1 mm
   (c) 0.02 mm
   (d) 0.01 mm.

390. Which of the following is not a method for analysing two way slabs with fixed edges?
    (a) Hardy-cross method
    (b) Pigeauds method
    (c) Marcus's method
    (d) IS code method.

391. The drop of a flat slab may have thickness ...... more than the rest of the slab
    (a) 5 to 10%
    (b) 10 to 15%
    (c) 20 to 30%
    (d) 25 to 50%.

392. According to IS : 456–2000, in the case of slender columns, the permissible stresses in steel and concrete are reduced by multiplying with a coefficient C_r given by
    (a) $1.25 - \dfrac{160 r_{min}}{l_e}$
    (b) $1.25 - \dfrac{l_e}{160 r_{min}}$
    (c) $2.25 - \dfrac{160 r_{min}}{l_e}$
    (d) $2.25 - \dfrac{l_e}{160 r_{min}}$.

393. A combined footing is needed when
    (a) the columns are too near and their footings overlap
    (b) the bearing capacity of the soil is very less
    (c) the end column is near the property line
    (d) all the above.

394. When all the columns and walls of a structure are supported by a single footing then it is called
    (a) an individual footing
    (b) combined footing
    (c) raft footing
    (d) none of the above.

395. When the pile is suspended from two points, the minimum B.M. will occur if the overhang from each support is
    (a) 0.106 L
    (b) 0.206 L
    (c) 0.306 L
    (d) 0.406 L.

396. The minimum moment in the case of pile suspended from two points is given by
    (a) $\dfrac{wL^2}{37}$
    (b) $\dfrac{wL^2}{27}$
    (c) $\dfrac{wL^2}{47}$
    (d) $\dfrac{wL^2}{17}$.

397. The minimum moment in the case of a pile suspended from three points is given by
    (a) $\dfrac{wL^2}{80}$
    (b) $\dfrac{wL^2}{90}$
    (c) $\dfrac{wL^2}{70}$
    (d) $\dfrac{wL^2}{60}$.

398. A thumb rule for spacing of counterports is given by
    (a) $l = 1.5 \left(\dfrac{H}{\gamma}\right)^{1/2}$
    (b) $l = 2.5 \left(\dfrac{H}{\gamma}\right)^{1/3}$
    (c) $l = 3.5 \left(\dfrac{H}{\gamma}\right)^{1/4}$
    (d) $l = 4.5 \left(\dfrac{H}{\gamma}\right)^{1/5}$.

399. In steel beam theory the compression reinforcement A_{sc} and the tension reinforcement A_{st} are related as

(a) $A_{sc} = A_{st}$ ☐   (b) $A_{st} > A_{sc}$ ☐
(c) $A_{st} < A_{sc}$ ☐   (d) no relation. ☐

400. The compression force C_u resisted by the concrete above neutral axis will be acting at a distance of ...... from the extreme compression fibre.

(a) $\dfrac{1}{7} x_u$ ☐   (b) $\dfrac{2}{7} x_u$ ☐
(c) $\dfrac{3}{7} x_u$ ☐   (d) $\dfrac{4}{7} x_u$. ☐

ANSWERS
Objective Type Questions

1. (a)	2. (b)	3. (a)	4. (d)	5. (c)	6. (a)
7. (c)	8. (c)	9. (b)	10. (d)	11. (d)	12. (a)
13. (c)	14. (a)	15. (b)	16. (c)	17. (b)	18. (c)
19. (b)	20. (b)	21. (a)	22. (c)	23. (d)	24. (a)
25. (c)	26. (a)	27. (d)	28. (b)	29. (b)	30. (a)
31. (c)	32. (d)	33. (d)	34. (d)	35. (a)	36. (d)
37. (c)	38. (c)	39. (c)	40. (d)	41. (c)	42. (d)
43. (b)	44. (a)	45. (c)	46. (d)	47. (c)	48. (b)
49. (a)	50. (b)	51. (c)	52. (c)	53. (b)	54. (a)
55. (b)	56. (d)	57. (a)	58. (d)	59. (c)	60. (b)
61. (b)	62. (b)	63. (a)	64. (a)	65. (b)	66. (b)
67. (b)	68. (b)	69. (a)	70. (b)	71. (c)	72. (a)
73. (c)	74. (d)	75. (c)	76. (a)	77. (d)	78. (c)
79. (c)	80. (d)	81. (b)	82. (d)	83. (d)	84. (a)
85. (c)	86. (d)	87. (b)	88. (a)	89. (c)	90. (b)
91. (c)	92. (c)	93. (d)	94. (b)	95. (d)	96. (a)
97. (d)	98. (d)	99. (c)	100. (a)	101. (b)	102. (c)
103. (d)	104. (c)	105. (a)	106. (a)	107. (d)	108. (a)
109. (a)	110. (c)	111. (b)	112. (a)	113. (d)	114. (b)
115. (d)	116. (c)	117. (b)	118. (c)	119. (b)	120. (b)
121. (d)	122. (b)	123. (b)	124. (c)	125. (b)	126. (a)
127. (c)	128. (b)	129. (b)	130. (d)	131. (b)	132. (a)
133. (b)	134. (c)	135. (d)	136. (b)	137. (a)	138. (b)
139. (a)	140. (b)	141. (c)	142. (c)	143. (c)	144. (a)
145. (b)	146. (a)	147. (a)	148. (d)	149. (d)	150. (c)
151. (b)	152. (c)	153. (a)	154. (b)	155. (c)	156. (b)
157. (d)	158. (c)	159. (b)	160. (a)	161. (d)	162. (b)

163. (d)	164. (b)	165. (a)	166. (a)	167. (b)	168. (c)
169. (a)	170. (b)	171. (b)	172. (a)	173. (c)	174. (d)
175. (b)	176. (b)	177. (a)	178. (b)	179. (c)	180. (c)
181. (a)	182. (b)	183. (d)	184. (d)	185. (a)	186. (a)
187. (c)	188. (a)	189. (b)	190. (a)	191. (d)	192. (a)
193. (b)	194. (a)	195. (a)	196. (b)	197. (c)	198. (a)
199. (c)	200. (c)	201. (c)	202. (c)	203. (a)	204. (c)
205. (d)	206. (c)	207. (c)	208. (b)	209. (c)	210. (b)
211. (d)	212. (c)	213. (c)	214. (b)	215. (d)	216. (b)
217. (c)	218. (b)	219. (a)	220. (c)	221. (c)	222. (c)
223. (c)	224. (a)	225. (c)	226. (a)	227. (c)	228. (d)
229. (b)	230. (c)	231. (b)	232. (d)	233. (d)	334. (c)
235. (a)	236. (a)	237. (c)	238. (c)	239. (c)	240. (d)
241. (a)	242. (d)	243. (b)	244. (b)	245. (d)	246. (c)
247. (c)	248. (d)	249. (b)	250. (a)	251. (d)	252. (d)
253. (b)	254. (d)	255. (a)	256. (b)	257. (d)	258. (a)
259. (b)	260. (b)	261. (d)	262. (a)	263. (c)	264. (b)
265. (d)	266. (b)	267. (b)	268. (c)	269. (b)	270. (b)
271. (b)	272. (d)	273. (c)	274. (b)	275. (c)	276. (a)
277. (d)	278. (d)	279. (c)	280. (b)	281. (a)	282. (c)
283. (b)	284. (d)	285. (d)	286. (a)	287. (d)	288. (b)
289. (c)	290. (d)	291. (a)	292. (d)	293. (b)	294. (c)
295. (d)	296. (c)	297. (b)	298. (c)	299. (a)	300. (d)
301. (c)	302. (a)	303. (d)	304. (b)	305. (b)	306. (c)
307. (a)	308. (a)	309. (b)	310. (c)	311. (d)	312. (c)
313. (c)	314. (a)	315. (b)	316. (c)	317. (b)	318. (b)
319. (c)	320. (c)	321. (a)	322. (b)	323. (a)	324. (a)
325. (c)	326. (b)	327. (d)	328. (c)	329. (c)	330. (c)
331. (c)	332. (a)	333. (b)	334. (a)	335. (b)	336. (c)
337. (b)	338. (d)	339. (c)	340. (d)	341. (c)	342. (c)
343. (d)	344. (a)	345. (b)	346. (a)	347. (c)	348. (b)
349. (a)	350. (d)	351. (a)	352. (d)	353. (b)	354. (c)
355. (d)	356. (a)	357. (c)	358. (d)	359. (c)	360. (b)
361. (b)	362. (c)	363. (d)	364. (a)	365. (d)	366. (d)
367. (c)	368. (b)	369. (a)	370. (a)	371. (c)	372. (a)
373. (b)	374. (c)	375. (c)	376. (d)	377. (b)	378. (a)
379. (b)	380. (d)	381. (a)	382. (a)	383. (a)	384. (c)
385 (c)	386. (d)	387. (a)	388. (d)	389. (b)	390. (a)
391. (d)	392. (b)	393. (d)	394. (c)	395. (b)	396. (c)
397. (b)	398. (c)	399. (a)	400. (c)		

Chapter 15 STEEL STRUCTURES

I. INTRODUCTION

BOLTED CONNECTIONS

Uses. Turned and fitted bolts and high strength friction slip bolts are used for avoiding slip in the connections. Bolts are used in place of rivets in such connections where rivets cannot be driven and also where the rivets are subjected to tensile stresses. Ordinary bolts are used for connections in temporary structures.

Stresses. Bolted connection is subjected to bending, bearing, shearing and tension.

Bending stress are neglected in the case of short bolts and may exist in long bolts when the grip exceeds 5 times the diameter and should be estimated as usual.

Bearing stresses are of no significance if the loading is not excessive. The shear is more or less uniform and is distributed over its gross area. When tensile stresses are to be estimated in the bolt, the area of cross-section at the root of the thread shall be taken.

Allowable Stresses in Bolts for Mild Steel

In Tension:

Over 38 mm diameter	= 1260 kg/cm^2
More than 20 mm and less than 38 mm	= 945 kg/cm^2
Less than 20 mm	= 785 kg/cm^2.

In Shear:

Turned and fitted	= 1025 kg/cm^2
Black bolts	= 865 kg/cm^2.

In Bearing :

Turned and fitted   = 2360 kg/cm²

Black bolts   = 2045 kg/cm².

When the effect of wind or seismic load is taken into account the above stressess may be increased by 25%.

The diameter of the bolt hole shall be taken as the nominal diameter of bolts plus 1.6 mm.

RIVETED CONNECTIONS

Uses. Rivets are widely used in bridges and buildings. In normal steel work, riveted connections are called upon to resist shear, bearing and sometimes tension.

The Allowable Stresses. The allowable stresses in rivets depend upon the type of steel and how they are driven *viz.* (*a*) Power driven shop rivets; (*b*) Power driven field rivets; and (*c*) Hand driven rivets.

Pitch and Edge Distances. The minimum pitch is 2.5 times the diameter of the hole.

The maximum pitch varies for tension and compression members.

The edge distance increases with the diameter of the hole and depends upon the type of edge.

Rivet Size. It is common practice to provide only one size of a rivet in any particular connection. The diameter of the rivet is given by

$$d_1 = 6\sqrt{t}$$

where d_1 = Nominal diameter of the rivet, t = Thickness of the plate in mm.

Strength of a Riveted Joint:

(*a*) Strength in shear

Strength of joint in single shear $= n \cdot \dfrac{\pi}{4} d^2 f_s$

Strength of joint in double shear $= n \cdot 2 \cdot \dfrac{\pi}{4} d^2 f_s$.

(*b*) Strength in bearing

Strength of joint in bearing $= n f_b \cdot td$.

(*c*) Strength in tearing

Strength of joint in tearing $= n(p - d) t f_t$

where n = Number of rivets per pitch length

         f_s = Allowable shear stress in rivet

         f_b = Allowable bearing stress in rivet

         f_t = Allowable tensile stress in plate.

Eccentrically Loaded Riveted Joint. A riveted joint may be eccentrically loaded in two ways, namely, the line of action of the applied load lies in the plane of rivets or the line of action of load may be in a plane perpendicular to the plane in which the rivets lie.

(a) **Load line being in the plane of rivets.**

The rivets are subjected to shear only and the resultant shear force in the most stressed rivet is given by

$$F = \sqrt{F_a^2 + F_m^2 + 2F_a \cdot F_m \cos \theta}$$

where F_a = Axial force = $\dfrac{P}{n} = \dfrac{\text{Total load}}{\text{Number of rivets}}$

F_m = Force due to moment = $\dfrac{M \cdot r}{\Sigma r^2}$, n = Number of rivets

r = Distance of the rivet from C.G. of rivet system

P = Eccentric load, θ = Angle between the forces F_a and F_m.

(b) **Axis of load not lying in the plane of rivets.**

Rivets will be subjected to tension and shear. The rivets will be so proportioned that

$$\left(\dfrac{f_s}{P_s}\right)^2 + \left(\dfrac{f_t}{P_t}\right)^2 \leq 1$$

where f_s = Actual shear stress in rivet

f_t = Actual tensile stress in rivet

P_s = Permissible shear stress in rivet

P_t = Permissible tensile stress in rivet.

As per IS : 800–1984 the above equation is modified as

$$\dfrac{\tau_{vf,cal}}{\tau_{vf}} + \dfrac{\sigma_{tf,cal}}{\sigma_{tf}} \leq 1.4$$

where $\tau_{vf\,cal}$ = Calculated shear stress in the rivet

$\sigma_{tf\,cal}$ = Calculated tensile stress in the rivet

τ_{vf} = Maximum permissible shear stress in rivet

σ_{tf} = Maximum permissible tensile stress.

(i) **Rivets in Initial Tension**

$$f_t = \dfrac{T_{max}}{A_r}, \quad f_s = \dfrac{V}{NA_r} \quad \text{and} \quad n = \sqrt{\dfrac{6M}{PmR}}$$

where $T_{max} = \dfrac{M \cdot Y_{max}}{\Sigma y^2}$

Y_{max} = Distance of farthest rivet from centre of gravity of the rivets

A_r = Area of the rivet hole

N = Number of rivets

P = Pitch of rivets

n = Number of rivets per line

m = Number of rivet lines

R = Rivet value in tension.

(ii) **Rivets with no Initial Tension**

$$n = 0.8 \sqrt{\frac{6M}{pmR}}.$$

In this case the neutral axis is assumed to lie about $\frac{1}{6}$ to $\frac{1}{7}$ of the depth of the connection from the lower end of the connecting angles.

WELDED CONNECTIONS

Advantages of Welded Joint:

(a) Economical
(b) Connections are rigid
(c) 100 per cent efficient joint
(d) Clean and neat joint.

Disadvantages of Welded Joint:

(a) Over rigidity
(b) Joints can develop brittle fracture
(c) Welded metal has less fatigue strength
(d) Warping of parts may take place
(e) Highly skilled labour is required.

Types of Welds:

(a) Butt welds
(b) Fillet welds } Single and double V, U, J

(c) Slot welds
(d) Plug welds.

Size. The size of a fillet weld is the minimum leg length in the cross-section.

Effective Throat Thickness. In case of butt weld, the effective throat thickness shall be taken as the thickness of the thinner part joined.

In case of fillet weld, the effective throat thickness 't' is given by $t = ks$

where k = A constant which depends upon the angle between the meeting surface (varies between 0.7 to 0.5)

s = Size of the weld.

Minimum Size. Depends on the thickness of the plate to be welded

Up to and including 9.5 mm thick	3 mm size
Over 9.5 mm and including 19.0 mm thick	5 mm size
Over 19.0 mm and including 32 mm thick	6 mm size
Over 32 mm and including 50 mm thick	8 mm size
Over 50 mm	Special precautions are necessary.

Permissible Stresses in Butt Welds

In direct tension and direct compression	1420 kg/cm^2
In bending tension and bending compression	1575 kg/cm^2
In shear	1025 kg/cm^2

Permissible Stress in Filled Weld

In shear   1025 kg/cm².

Site Welds. The permissible stresses are 80% of the values given in 3.7 and 3.8, above.

TENSION MEMBERS

Tension members whether single section or built-up ones, are reduced in their sectional areas because of rivets used in fabrication.

Net Sectional Area. The net sectional areas to be used in the design are:

(a) For plates connected by chain riveting $A_{net} = (b - nd)\,t$

where b = Width of the plate
   n = Number of holes at each section
   d = Diameter of hole
   t = Thickness of the plate.

(b) For plates connected by *zig-zag* or staggered riveting

$$A_{net} = \left(b - nd + n_1 \frac{p^2}{4g}\right) t$$

where n_1 = Number of gauge spaces
   n = The number of rivet holes
   p = Staggered pitch *i.e.*, pitch parallel to the applied load
   g = Gauge, *i.e.*, distance between two consecutive rivets in the same chain measured at right angle to the direction of stress in the member.

(c) Single angle whose both legs are connected.

Formulae given in (a) and (b) above depending on the condition.

(d) Single angle section connected by one leg only

$$A_{net} = a + bk$$

where a = Net sectional area of the connected leg
   b = Area of the outstanding leg
   $k = \dfrac{1}{1 + 0.35 \dfrac{b}{a}}$
   $k = \dfrac{3a}{3a + b}$ as per IS : 800–1984.

(e) Pair of angles placed back to back (or a single tee) conneced by only one leg of each angle (or by the flange of tee) to the same side of a gusset plate

$$A_{net} = a + bk$$

where a = Net sectional area of the connected legs (or flange of the tee)
   b = Area of the outstanding legs (or web of the tee)
   $k = \dfrac{1}{1 + 0.2 \dfrac{b}{a}}$ and $k = \dfrac{5a}{5a + b}$ as per IS : 800–1984.

(f) Double angles or tees placed back to back and connected to each side of gusset or to each side of a part of rolled section.

$$A_{net} = \text{Gross area} - \text{Deduction of area for holes.}$$

Design. Load carried by axially loaded member

$$P = A_{net} \times \text{Allowable stress in axial tension } (P_t).$$

Members subjected to axial and bending tension must be so proportioned that the quantity

$$\frac{f_t}{P_t} + \frac{f_{bt}}{P_{bt}} \le 1.$$

where f_t = Calculated axial tensile stress = $\dfrac{\text{Axial force}}{\text{Net area}}$

f_{bt} = Calculated bending tensile stress in the extreme fibre = $\dfrac{\text{Bending moment}}{\text{Section modulus}}$

P_t = Permissible axial tensile stress

P_{bt} = Permissible bending tensile stress.

COMPRESSION MEMBERS

Columns. A column is defined by its slenderness ratio. Broadly columns can be classified as:

(a) Short columns $\dfrac{l}{r} < 60$

(b) Medium columns $60 < \dfrac{l}{r} < 100$

(c) Long columns $\dfrac{l}{r} > 100$.

Effective Length. Depending on the end conditions, the effective length '*l*' of a compression member in terms of its length '*L*' can be obtained from Table 15 of IS : 800–1962 or Table 5.2 of IS : 800–1984.

Allowable Stress. Allowable stress for axial compression depends on $\dfrac{l}{r}$ ratio and can be obtained from Table 2 of IS : 800–1962. These values are based on the secant formula assuming a minimum eccentricity. Depending upon the type of loads carried by a compression member, slenderness ratio must not exceed the values given below:

(a) Member carrying loads resulting from dead loads and superimposed loads: 180.

(b) Member carrying loads resulting from wind or seismic forces: 250.

(c) Member normally acting as a tie in a roof truss or a bracing system but subject to possible reversal of stress resulting from the action of wind or seismic forces: 350.

As per IS : 800–1984.

Permissible stress in axial compression = $\dfrac{0.6\, f_{cc} f_y}{[(f_{cc})^n + (f_y)^n]^{1/n}}$

where $f_{cc} = \dfrac{\pi^2 E r^2}{l^2}$, f_y = Yield stress, $n = 1.4$.

The values based on above formula are given in Table 5.1 for yield stresses varying between 220 to 540 MPa.

Built-up-Members. Built-up-sections such as double channel sections, double I sections etc., are to be connected by single lacing, double lacing or batten system so that the components act together.

(a) **Specifications for Lacing System**

1. Lacing system is designed for a shear force (S) equal to 2.5 per cent of the axial load.

Force in member in single lacing system = $\dfrac{S}{n} \operatorname{cosec} \theta$

Force in member in double lacing system = $\dfrac{S}{2n} \operatorname{cosec} \theta$.

2. The angle between the lacing bars and the axis of the member is kept between 40° and 70°.

3. The maximum spacing of lacing bars, *i.e.*, centre to centre of their end fastenings, should be such that the minimum slenderness ratio of the main components of the compression member between the consecutive connections should not be greater than 0.7 times the most unfavourable slenderness ratio of the member as a whole or 50 whichever is less.

4. The slenderness ratio of lacing bar should not exceed 145.

5. The minimum width of the lacing bars depends upon the size of rivets and is approximately equal to 3 times the nominal rivet diameter.

(b) **Specifications for Batten System**

1. Batten system is designed for a shear force (S) equal to 2.5 per cent of the axial load.

2. They are subjected to

$$\text{Moment} = \dfrac{Sd}{2n} \text{ and shear force} = \dfrac{Sd}{na}$$

where d = Distance between centre to centre of battens longitudinally

n = Number of parallel planes of battens

a = Minimum distance across, *i.e.*, between the centroids of the rivet groups or welding.

3. (a) If $\dfrac{l}{r_{yy}} \leq 0.8 \dfrac{l}{r_{xx}}$ of the compression member then $\left(\dfrac{d}{r}\right)_{max} \leq 50$ or $0.7 \dfrac{l}{r_{xx}}$ of the member as a whole.

(b) If $\dfrac{l}{r_{yy}} > 0.8 \dfrac{l}{r_{xx}}$, then $\left(\dfrac{d}{r}\right)_{max} \leq 40$ or $0.6 \dfrac{l}{r_{xx}}$ of the member as a whole

where r_{yy} = Radius of gyration about an axis perpendicular to the plane of battens

r_{xx} = Radius of gyration about an axis parallel to the plane of battens.

4. (a) Depth of intermediate battens $\geq \frac{3}{4}$ th distance between the end connections joining the batten with the main member.

(b) Depth of end battens ≥ distance between the end connections joining the batten with the main member.

5. Thickness of batten plate $\geq \frac{1}{50}$ th of the distance between the inner most connecting lines.

Column Splice. A column should preferably be spliced 300 mm or 150 mm above the floor line. If the ends are not milled, the splice plates and their connections to the column are designed to transmit all the forces. If the ends are milled, the splice plates and their connections are designed to carry 50% of the direct load, in addition to the moments carried by the members.

Column Bases. Three types of column bases are used for the axially loaded columns. (a) Slab base (b) Gussetted base (c) Grillage foundation.

(a) **Slab base.** Thickness of a rectangular slab base for an I-section.

$$t = \sqrt{\dfrac{3w}{P_{bct}}\left(A^2 - \dfrac{B^2}{4}\right)}$$

where w = The pressure or loading on the underside of the base

A = The greater projection of the plate beyond the column

B = The lesser projection of the plate beyond the column

P_{bct} = The permissible bending stress in slab bases.

Thickness of a square slab base for a solid round column

$$t = \sqrt{\dfrac{9W}{16P_{bct}} \cdot \dfrac{D}{(D-d)}}$$

where W = The total axial load, D = Length of side of base

d = The diameter of the reduced end, if any, of the column.

(b) **Gusseted base.** When the end of the column is machined for complete bearing on the base plate, 50% of the axial column load is assumed to be transferred to the base plate by direct bearing and the remaining 50% of axial load will be transferred through the fastenings including gusset plates, cleat angles, stiffeners etc.

Where the ends of column shaft and the gusset plates are not faced for complete bearing, the fastenings connecting them to the base plate will be designed to transmit all forces to which the base is subjected.

(c) **Grillage foundation.** The grillage foundation consists of two or more tiers of steel beams laid at right angles to distribute load over a larger area. The grillage is generally encased in concrete.

The permissible stresses in grillage beams may exceed allowable stresses by $33\frac{1}{3}\%$, in case the following conditions are fulfilled:

1. The beam is unpainted and solidly encased in ordinary dense concrete with 10 mm aggregate and of a works cube strength not less than 160 kg/cm² at 28 days.

2. Pipe separators or equivalent are used to keep the beams properly spaced so that the distance between the edge of adjacent flanges is not less than 75 mm.

3. The thickness of the concrete cover on the top of the upper flanges, at the ends, and at the outer edges of the sides of the outermost beams is not less than 100 mm.

The area of the grillage will be determined by the bearing capacity of concrete or soil on which it is supported.

For top tier beams:

$$\text{Maximum bending moment} = \frac{W}{8}(L-l)$$

$$\text{Maximum shear force} = \frac{W}{2L}(L-l)$$

where l is length of the base plate parallel to the beams

L is length of top tier beams.

For bottom tier beams:

$$\text{Maximum bending moment} = \frac{W}{8}(B-b)$$

$$\text{Maximum shear force} = \frac{W}{2B}(B-b)$$

where b is width of the base plate parallel to the beams; B is length of bottom tier beams.

BEAMS

Definition. A beam is a structural element or member subjected to transverse loads and reactions.

A large beam supporting a smaller beam is a girder.

Allowable Bending Stresses. For rolled I-beams with or without cover plates and double channel beams with cover plates when compression flange is laterally supported, the allowable stresses in tension and compression is same and are given below:

Rolled I-beams and channels = 1650 kg/cm²
Other sections, thickness ≤ 20 mm = 1650 kg/cm²
Other sections, thickness > 20 mm = 1575 kg/cm²
Compound beam I or channels with cover plates, thickness ≤ 20 mm = 1650 kg/cm²
Compound beam I or channels with cover plates, thickness > 20 mm = 1575 kg/cm².

Allowable Shear Stresses. For parts other than rolled steel sections maximum shear stress = 1100 kg/cm² ≈ $0.45 f_y$.

For unstiffened webs with $\dfrac{d}{t_w}$ not greater than 85,

Average shear stress = 945 kg/cm² ≈ $0.4 f_y$.

Deflection. The maximum deflection in beams should not exceed $\dfrac{1}{325}$ of span for simply supported beams and $\dfrac{2}{325}$ of span for cantilever beams.

Generally the deflection criterion is satisfied if the detpth of I-beams is more than $\dfrac{\text{Span}}{19}$ for simply supported beams, $\dfrac{\text{Span}}{12}$ for cantilevers and $\dfrac{\text{Span}}{50}$ for fixed beams.

Design Considerations

(a) **Modulus of sections**

Modulus of section $Z = \dfrac{\text{Maximum bending moment}}{\text{Allowable bending stress}}$.

For the required section modulus, the deepest section which is also lightest in weight should be chosen.

(b) **Shear stress.** Shear stress at any section is given by $\dfrac{FA\bar{y}}{Ib}$.

Secondary Design Considerations

(a) **Local buckling.** To prevent local buckling of flange plates the thickness of outstanding compression flanges or cover plates beyond the web or the first line of rivets should be at least $\dfrac{1}{16}$ of the width of projection. Also the thickness of the compression plate between two parallel lines of rivets or webs should be at least $\dfrac{1}{50}$ of the distance between parallel lines.

(b) **Web crippling.** At points of concentrated loads and reactions the web buckles in the vicinity of the fillet toe of the web due to transmission of the compressive load from the wider flange to the narrower web. The stresses produced are localized bearing stresses and should not exceed the value P_b equal to 1890 kg/cm².

The actual bearing stress $P_b = \dfrac{P}{(b + 2h_2\sqrt{3})t_w} < P_b$ at intermediate load point and

$P_b = \dfrac{P}{(b + h_2\sqrt{3})t_w} < P_b$ at the reaction point.

where b = Stiff portion of bearing
h_2 = Distance from the outer face of the flange to the web toe of the fillet
t_w = Thickness of web.

(c) **Vertical buckling or web crippling.** Vertical buckling is a tendency for the web to buckle laterally due to heavy compressive stress, where concentrated loads are distributed by bearing plates. Bearing stiffeners should be provided at points of concentrated load and points of support, where concentrated load or reaction exceeds the value of $f_c.t_w.B$.

$$B = b + \frac{D}{2} \text{ at support}$$

$$B = b + D \text{ under concentrated load}$$

b = Length of stiff portion of the bearing
D = Depth of the girder
f_c = Axial stress for column for a slenderness ratio of

(i) $\dfrac{h_1\sqrt{3}}{t_w}$ when both the flanges are held against lateral deflection and rotation,

(ii) $\dfrac{h_1\sqrt{6}}{t_w}$ when bottom flange is assumed restrained against movement and rotation, while the top flange is held in position only, $h_1 = D - 2h_2$.

(d) **Diagonal buckling.** The web of a beam buckles in the diagonal direction due to shear in the web. The allowable compressive stress corresponding to the slenderness ratio $\dfrac{d}{t_w}\sqrt{6}$ should not be less than the shear stress in web, where d is the depth of web.

PLATE GIRDERS

Plate Girders. Plate girder is a built-up beam using plates and angles and connecting them by rivets or welding. These are more economical for short spans and heavy loads.

Effective Span. The effective span for a plate girder is usually the centre to centre of end bearings.

Depth of Girder
For lighter loads   Span/12
For medium loads   Span/10
For heavy loads   Span/8.

The economical depth of the girder can be obtained from the expressions

$$d = k\sqrt[3]{\frac{M}{f_b}} \quad \text{or} \quad d = 1.1\sqrt{\frac{M}{f_b t_w}}$$

where M = Maximum bending moment
f_b = Maximum permissible bending stress
k = Parameter equal to 5 for welded girder and 5.5 for riveted girder
t_w = Thickness of the web.

Width of Flange. Width of flange should be within the range of $\frac{L}{40}$ to $\frac{L}{50}$.

Self Weight.

For riveted plate girder = $\frac{W}{300}$ per metre run.

For welded plate girder = $\frac{W}{400}$ per metre run.

W = Total superimposed load on the girder.

Allowable Stresses.

Up to 20 mm web thickness

Bending stress	= 1575 kg/cm²
Average shear stress	= 945 kg/cm²

Over 20 mm web thickness

Bending stress	= 1500 kg/cm²
Average shear stress	= 865 kg/cm².

Minimum Thickness of Plates

For girders exposed to weather but accessible for painting	= 6 mm
For girders exposed to weather but not accessible for painting	= 8 mm
For heavy bridges	= 8 mm
For railway bridges	= 10 mm.

Moment of Resistance. Moment of resistance of a plate girder = $m = f_b \left(A_n + \frac{A_w}{8} \right) D$

where f_b = Maximum allowable bending stress
A_n = Net area of the flange
A_w = Area of web
D = Effective girder depth or web depth.

Specifications

(a) **Web plate**

Minimum thickness of web plate for unstiffened webs = $\frac{d}{85}$

For vertically stiffened web = $\frac{1}{180}$ of the smallest clear panel dimension or $\frac{d}{180}$, whichever is greater.

For webs with longitudinal stiffeners at $\frac{2}{5} d$ from compression flange $= \frac{d}{250}$.

For webs with additional horizontal stiffener at the neutral axis $= \frac{d}{400}$.

The greater unsupported clear dimension of web panel should not exceed $270 \, t_w$.

(b) **Flange plates**

Maximum outstand for compression flange $= 16t$

Maximum outstand for tension flange $= 20t$

Area of flange angles in riveted girder shall be at least $\frac{1}{3}$ of the gross flange area.

Flange Plate Curtailment

(a) **Riveted girders.** Distance at which the flange plate can be curtailed from centre

$$x = \frac{L}{2}\sqrt{\frac{A - A_x}{A}}$$

where L = Effective span, A = Flange area required at the centre,

A_x = Flange area required at a distance x from centre.

The first cover plate should not be curtailed.

Each flange plate should be extended beyond its theoretical cut-off point and the extension shall contain sufficient rivets to develop in the plate, the load calculated for the bending moment in the girder at the theoretical cut-off point.

(b) **Welded girders.** The change in flange plate is accomplished by using various length plates of different thickness. These plates are butt welded to form a continuous flange.

If the difference in thickness of the two plates exceeds 25% of the thickness of the thinner plates or 3.8 mm, whichever is greater, the dimensions of the thicker part should be reduced at the butt joint to those of the smaller part, the slope being not steeper than 1 in 5.

Connections

(a) **Rivets connecting flange plate to flange angles.**

$$\text{Pitch of rivets} = \frac{nRd\left(A_f + \dfrac{A_w}{6}\right)}{VA_p} \text{ for compression flange}$$

$$= \frac{nRd\left(A_f' + \dfrac{A_w}{8}\right)}{VA_p'} \text{ for tension flange}$$

where n = Number of rivets in one pitch length, R = Rivet value

d = Depth of web, A_f = Area of each flange

A_w = Area of web, A_p = Area of cover plates

A_f' = Net area of each flange, A_p' = Net area of cover plates

V = Vertical shear force at the section considered.

(b) **Rivets connecting flange angles to web**

$$\text{Pitch of rivets} = \frac{R}{\left[\left\{\frac{VA_f}{d\left(A_f + \frac{A_w}{6}\right)}\right\} + w^2\right]^{1/2}} \text{ for compression flange}$$

$$= \frac{R \cdot d\left(A_f' + \frac{A_w}{8}\right)}{VA_f} \text{ for tension flange}$$

w = Uniformly distributed load on the girder per unit length.

(c) **Welded girder-connection flange plate and web**

(i) Size of continuous fillet weld = $\dfrac{VA\bar{y}}{1450\, I}$ cm.

(ii) If intermittent fillet weld is used, clear distance between adjacent intermittent fillet welds

$$= d = l_i \left[\frac{2S_w I}{VA\bar{y}} - 1\right]$$

where l_i = Effective length of intermittent fillet welds
S_w = Strength of fillet weld per cm length.

l_i should not be less than four times the weld size with a minimum of 38 mm. The value of 'd' should not exceed 12 times the thickness of thinner plate for compression flange or 200 mm and 16 times the thickness of thinner plate for tension flange or 200 mm whichever is less.

Maximum size of fillet weld = $\frac{2}{3}$ the web thickness approximately.

Stiffeners

(a) **Intermediate Stiffeners**

These are provided when $\dfrac{d}{t_w} > 85$

Spacing of stiffeners $\not< 0.33d$
$\not> 1.5d.$

Moment of inertia of stiffeners $\not< \dfrac{1.5 d^3 t^3}{C^2}.$

For flats outstand should not be greater than $12t$.

For other sections outstand should not be greater than $16t$.

(b) **Bearing stiffeners.** The function of bearing stiffener is to transmit concentration of load so as to avoid local bending failure of the flange and local crippling or buckling of the web.

Area of the stiffeners is found on the basis of local permissible contact bearing pressure of 1890 kg/cm^2.

The bearing stiffeners together with the web plate shall be designed as a column with an equivalent reduced slenderness ratio.

Effective length = 0.7 actual length.

The area of the section which resists compression is the area of pair of stiffeners together with a length of web on each side of the centre line of the stiffeners equal to 20 times of the web thickness where available.

In case of riveted bearing stiffeners filler plates shall be used.

(c) **Longitudinal stiffeners.** In addition to the vertical intermediate stiffeners, longitudinal stiffeners shall be provided if $t_w < \dfrac{d}{200}$. One horizontal stiffener shall be placed on the web at a distance from the compression flange = $\tfrac{2}{5}$ of the distance from the compression flange to the neutral axis. The moment of inertia of these stiffeners shall not be less than $4 C_1 t_w^3$, where C_1 is the actual distance between stiffeners.

If the thickness of web is less than $\dfrac{d}{250}$, a second horizontal stiffener shall be placed at the neutral axis of the girder for which $I \not< dt_w^3$.

(d) **Stiffener connections.** The stiffener connection to web plate shall be designed to develop a shear force in tonnes per cm run of not less than $\dfrac{t_w^2}{2h}$, where 'h' is the outstand leg width of stiffener.

If intermittent welds are used, the distance between the effective lengths of any two weld, even if staggered on opposite sides of the stiffeners should not exceed 16 times the thickness of the stiffener nor 300 mm.

If the welding is done on one side of stiffener or on two sides but staggered, the effective length of each weld should not be less than four times the thickness of the stiffener where rivets are used

Minimum pitch $\not<$ 2.5 times the diameter of hole, and

Maximum pitch $\not>$ 32 times the thickness of the thinner outside plate or 300 mm which ever is less.

Web Splice. Web splices are designed to resist the shears and moments at the section. Three types of web splices commonly used are

(a) **Rational splice.** In this splice, two cover plates extend from edge of top flange angles to the edge of bottom flange angles. The combined thickness of cover plates is made equal to the thickness of the web.

$$\text{Pitch of rivets} = p = \dfrac{nR}{\sqrt{\left(f_b t_w\right)^2 + \left(\dfrac{V}{d_e}\right)^2}}$$

where R = Rivet value

f_b = Bending stress at the level of rivets connecting flange angles

V = Shear force at the section

d_e = Effective depth of the plate girder

n = Number of rows of rivets.

Four cover plates on the vertical legs of the flange angles are provided. Number of rivets required towards decreasing moment side from splice

$$m = \frac{P_2}{R - P_1}$$

where $P_2 = \frac{My}{I} \times$ Area of portion of web beneath flange angles

$$P_1 = \frac{V}{d_e} \cdot \frac{A_f}{A_f + \frac{A_w}{6}} \cdot p$$

p = Pitch of these rivets.

(b) **Moment splice.** This splice is used for deep girders. In this splice the moment is assumed to be taken by four moment plates, two at the top and two at the bottom, and shear is assumed to be taken by shear plates two in number placed on either side of the web.

For Moment Plates

$$A_s = \frac{MI_w d_w}{If d_1^2}$$

$$t_s = \frac{A_s}{2(d_s - n_1 d)}$$

$$n = \frac{A_s f d_1}{R d_w}$$

where A_s = Net area of two moment plates

t_s = Thickness of the moment plates

n = Number of rivets required on each side of web splice

M = Bending moment at the section

I_w = Gross moment of inertia of the web

I = Gross moment of inertia of the girder

d_w = Depth of web plate

f = Bending stress at the extreme fibre of web plate

d_1 = Distance between c/c of moment plates

d_s = Depth of moment plate

d = Diameter of rivets

n_1 = Number of rivets in one vertical row

R = Rivet value.

For Shear Plates

The combined thickness of shear plates resist shear at the section. The width of the plates is kept sufficient to accommodate rivets given by $\dfrac{V}{R}$.

(c) **Shear Splice.** Splice plates are provided between flange angles on either side only. They are designed to take the full shear and moment originally taken by web

Moment carried by web $\qquad M_w = \dfrac{MI_w}{I}$

Thickness of the splice plates $\qquad t = \dfrac{3M_w}{h_s^2 P_{bt}}$

Number of horizontal rows of rivets $\quad n = \sqrt{\dfrac{6M_w}{lpR}}$

Number of vertical rows of rivets $\qquad l = \dfrac{6M_w P}{h_s^2 R}$

where h_s = Depth of splice plates, P_{bt} = Permissible bending stress,

p = Pitch of rivets, R = Rivet value, $n = \dfrac{h_s}{p}$.

GANTRY GIRDER

End Carriage Wheel Base. The wheel base may be taken as given below:

Span of crane in m	Up to 18 m	18 m to 22 m	above
Minimum wheel base	$\dfrac{\text{Span}}{5}$	3.6 m	$\dfrac{\text{Span}}{6}$

Vertical Loads. Allow an impact allowance of 25% for E.O.T. cranes and 10% for hand-operated cranes.

Lateral Thrust. Total lateral thrust = 10% of combined weight of crab and lift load for E.O.T. cranes and 5% of combined weight of crab and lift load for hand-operated cranes.

Longitudinal Thrust. 5% of the actual wheel loads for E.O.T. as well as hand operated cranes.

Weight of Crab. $\frac{1}{5}$ th lift load + 500 kg.

Approximate Dead Load of the Girder and Rail. 800 kg to 1000 kg for a span of 6 m and 10t crane capacity.

Depth of Girder.

$$\frac{\text{Span}}{10} \text{ to } \frac{\text{Span}}{12}.$$

Allowable Stresses. The allowable tensile and shear stresses are same as that for ordinary beams.

The allowable compressive stress depends on the critical stress C_s when the girder is laterally unsupported. The sum of the compressive stresses due to vertical loads and lateral loads should not exceed the value of P_{bc} corresponding to the calculated value of the critical stress C_s.

If the moments of inertia of compression and tension flanges about y-y axis is different,

$$C_s = 10.1 \times 10^6 \frac{I_y' h}{Z_x l^2} \left[\sqrt{1 + \frac{0.162 K l^2}{I_y' h^2}}\right] + K_2 \cdot 10.1 \times 10^6 \frac{I_y' h}{Z_x l^2} \text{ kg/cm}^2$$

where $I_y' = K_1 \times$ Moment of inertia of the section about the y-y axis at the point of maximum bending moment

K_1 = Co-efficient depends on N, the ratio on the total area of both flanges at the point of minimum bending moment to the corresponding area at the point of maximum bending moment (for flanges of constant area, $K_1 = 1$)

h = Distance between the centre of gravity of the compression flange and the centre of gravity of the tension flange (approximately depth of the section minus thickness of the flange)

K = Torsional constant = $\sum \dfrac{b' t'^3}{3}$, b' and t' are the width and average thickness of the rectangular components of which the section is made of

l = Effective length of compression flange

Z_x = Section modulus about x-x axis with reference to the compression flange

K_2 = Co-efficient depends on M, the ratio of moment of intertia of compression flange alone about its axis parallel to the y-y axis of the girder as a whole to that of the whole section about the y-y axis of the girder, both at the point of maximum bending moment (for flanges of equal moment of inertia, $M = 0.5$ and $K_2 = 0$).

Deflection. The maximum vertical deflection under dead and imposed loads shall not exceed the following values:

(a) Hand operated cranes $\dfrac{L}{500}$

(b) E.O.T. cranes of capacity up to 50t $\dfrac{L}{750}$

(c) E.O.T. cranes of capacity over 50t $\dfrac{L}{1000}$

where L = Span of crane runway girder.

ROOF TRUSSES

Pitch. The pitch of the roof truss is the ratio of the height of the truss to the span. The slope of the truss depends on the type of sheeting.

Corrugated iron $26\frac{1}{2}°$

Asbestos sheets $20°$

Slates $35°$

Tiles $40°$.

Spacing of Truss. Common spacing of trusses ranges from 3 m to 5 m. Economical spacing varies from $\frac{1}{3}$ to $\frac{1}{5}$ of span.

Spacing of Purlins. Purlins are spaced such that they exit at each node of the truss to avoid bending in the main rafter. The spacing of the purlins also depends on the safe span of the sheeting material.

Loads

1. Dead Loads

(a) *Sheeting*

G.I. sheeting   15 kg/m²

A.C. sheeting   18 kg/m²

Glazing 6 mm thick   25 to 30 kg/m².

(b) *Purlins*. Self-weight of purlins with corrugated sheets varies from 6 to 9 kg/m² area covered by purlin.

(c) *Trusses*

$$\text{Weight of truss} = W = \frac{L}{3} + 5 \text{ kg}$$

where L = Span of truss, W = Weight of truss per m² in kg for a spacing of 4 m.

For other spacing proportional values may be taken.

Weight of wind bracing = 1.5 kg/m² of plan area.

2. Live Loads. For sloping roofs of greater than 10° slope the live load is 75 kg/m² less 1 kg/m² for every degree increase in the slope upto 20° and less 2 kg/m² for every degree increase in slope over 20° with a minimum of 40 kg/m².

3. Snow Loads. No snow load is considered if slope is greater than 50°. For other slopes 2.5 kg/m² per cm depth of snow may be taken.

4. Wind Loads. The external wind pressure on roofs will depend on the slope of the roof and internal pressure on the degree of permeability of building.

For small permeability, internal pressure   = 0

For normal permeability, internal pressure = $\pm\, 0.2\, p$

For large openings, internal pressure   = $\pm\, 0.5\, p$

where 'p' is the pressure intensity which depends on the zone in which the structure is located.

Design of Purlins

Depth of purlin $\not< \dfrac{L}{45}$

Width of purlin $\not< \dfrac{L}{60}$.

Maximum bending moment in purlin = $\dfrac{WL}{10}$

where L = Spacing of trusses or span of purlins

W = Total distributed load on the purlin due to all loads.

Design of Tension Members

(a) Double angle sections shall be used for main tie and single angle sections for other members.

(b) Minimum size of angle is ISA 50 × 50 × 6 mm for main tie and for others ISA 50 × 50 × 5 mm.

(c) Minimum thickness of gusset plate is 6 mm

(d) Minimum diameter of rivet is 16 mm.

(e) The design of members shall be based on net area.

(f) Maximum permissible slenderness ratio of tension members when subjected to reversal of stress is 350.

Design of Compression Members

(a) Double angle sections are used for main rafter and for other compression members single angle sections are used.

(b) Minimum size of angle is ISA 50 × 50 × 6 mm for main rafter and for other ISA 50 × 50 × 5 mm.

(c) Maximum slenderness ratio of member is 180.

(d) Effective length of a member depends on continuity of the member and the number of rivets used to connect it. It varies between 0.7 to 1 of the actual length.

(e) When single rivet or bolt is used to connect the member the allowable compressive stress is taken as 0.8 times the normal allowable stress depending on the slenderness ratio. When member is connected by two or more rivets, it is designed for normal allowable stress.

II. OBJECTIVE TYPE QUESTIONS

1. The diameter of rivet hole with respect to the nominal diameter of the rivet should be
    (a) 1.5 to 2 mm more ☐   (b) 2.5 to 3 mm more ☐
    (c) 0.5 to 1 mm more ☐   (d) 3.0 to 3.5 mm more. ☐

2. The minimum cross pitch should be more than.....times the diameter of the hole.
    (a) 5 ☐   (b) 3 ☐
    (c) 2.5 ☐   (d) 4.0. ☐

3. The ratio of the maximum cross pitch to the thickness of the plates connected should be
   (a) 12
   (b) 16
   (c) 4
   (d) 32.

4. The maximum cross pitch allowed in riveted connection is about
   (a) 100 mm
   (b) 200 mm
   (c) 300 mm
   (d) 400 mm.

5. The maximum longitudinal pitch allowed in a tension riveted joint is
   (a) 16t or 200 mm whichever is less
   (b) 12t or 200 mm whichever is less
   (c) 4.5t or 200 mm whichever is less
   (d) 4t + 100 mm.

6. The maximum longitudinal pitch allowed in a compression riveted joint is
   (a) 16t or 200 mm whichever is less
   (b) 12t or 200 mm whichever is less
   (c) 4.5t or 200 mm whichever is less
   (d) 4t + 100 mm.

7. The maximum longitudinal pitch allowed in a butt riveted joint is
   (a) 12 D
   (b) 8 D
   (c) 4.5 D
   (d) 3.5 D.

8. The maximum longitudinal pitch allowed in staggered riveted tension member is
   (a) 24t or 300 mm whichever is less
   (b) 18t or 300 mm whichever less
   (c) 6 D or 300 mm whichever is more
   (d) 6t + 200 mm.

9. The maximum longitudinal pitch allowed in staggered riveted compression member is
   (a) 24t or 300 mm whichever is less
   (b) 18t or 300 mm whichever is less
   (c) 6 D or 300 mm whichever is more
   (d) 6t + 200 mm.

10. The approximate minimum edge distance in sheared plates from the centre of hole in a riveted connection is
    (a) 4t + 100 mm
    (b) 2.5 D
    (c) 4t + 37 mm
    (d) 2t + 37 mm.

11. The maximum edge distance in sheared plates from the centre of hole in a riveted connection is
    (a) 4t + 100 mm
    (b) 4t + 37 mm
    (c) 4t
    (d) 100 mm.

12. The maximum pitch of tacking rivets in the case of tension members separated by solid spacers should not be more than
    (a) 600 mm
    (b) 1000 mm
    (c) 1500 mm
    (d) 1800 mm.

13. The diameter of a bolt hole with respect to the nominal diameter of the bolt should be
    (a) 1.4 mm more
    (b) 1.0 mm more
    (c) 1.2 mm more
    (d) 1.6 mm more.

14. The requirements of pitch, edge distances etc. for bolts shall be......than in the case of rivets.
    (a) more
    (b) less
    (c) same
    (d) no bearing with.

15. The permissible axial tensile stress in mild steel power driven shop rivet is (in MPa)
    (a) 126 □   (b) 78.5 □
    (c) 63.0 □   (d) 94.5. □

16. The ratio of the permissible tensile stress in power driven shop rivets to the yield stress of mild steel is
    (a) 0.53 □   (b) 0.4 □
    (c) 0.33 □   (d) 0.27. □

17. The ratio of the permissible tensile stress in bolts with diameter less than 20 mm to the yield stress of mild steel is about
    (a) 0.53 □   (b) 0.4 □
    (c) 0.33 □   (d) 0.27. □

18. The ratio of the permissible tensile stress in tension rods to the yield stress of mild steel is
    (a) 0.53 □   (b) 0.4 □
    (c) 0.33 □   (d) 0.27. □

19. The ratio of the permissible shear stress of power driven field rivets to the yield stress of mild steel is
    (a) 0.43 □   (b) 0.40 □
    (c) 0.33 □   (d) 0.37. □

20. The ratio of the permissible shear stress of power driven shop rivets to the yield stress of mild steel is
    (a) 0.43 □   (b) 0.40 □
    (c) 0.33 □   (d) 0.37. □

21. The ratio of the permissible shear stress of Hand driven rivets to the yield stress of mild steel is
    (a) 0.43 □   (b) 0.4 □
    (c) 0.33 □   (d) 0.37. □

22. The ratio of the permissible bearing stress of power driven shop rivets to the yield stress of mild steel is
    (a) 1.0 □   (b) 0.9 □
    (c) 0.67 □   (d) 0.87. □

23. The ratio of the permissible bearing stress of power driven field rivets to the yield stress of mild steel is
    (a) 1.0 □   (b) 0.9 □
    (c) 0.67 □   (d) 0.87. □

24. The ratio of the permissible bearing stress of hand driven rivets to the yield stress of mild steel is
    (a) 1.0 □   (b) 0.9 □
    (c) 0.67 □   (d) 0.87. □

25. The ratio of the permissible bearing stress of turned and fitted bolts to the yield stress of mild steel is
    (a) 1.0
    (b) 0.9
    (c) 0.67
    (d) 0.87.

26. The ratio of the permissible bearing stress of black bolts to the yield stress of high tensile steel is
    (a) 1.00
    (b) 0.9
    (c) 0.87
    (d) 0.67.

27. The permissible stresses in rivets or bolts under wind or seismic load can be increased by.......times the normal permissible stresses.
    (a) 1.1
    (b) 1.25
    (c) 1.33
    (d) 1.5.

28. When two plates are placed end to end and are joined by cover plates, the joint is known as
    (a) lap joint
    (b) butt joint
    (c) chain riveted lap joint
    (d) double cover butt joint.

29. The strength of a riveted joint is equal to
    (a) shearing strength of the rivets
    (b) bearing strength of the rivets
    (c) tearing strength of the plates
    (d) least of (a), (b) and (c).

30. Efficiency of a riveted joint is defined as the ratio of
    (a) least strength of a riveted joint to the strength of the solid plate
    (b) greatest strength of a riveted joint to the strength of the solid plate
    (c) least strength of a riveted plate to the greatest strength of the joint
    (d) all the above.

31. Number of rivets required in a joint is equal to
    (a) $\dfrac{\text{load}}{\text{shear strength of a rivet}}$
    (b) $\dfrac{\text{load}}{\text{rivet value}}$
    (c) $\dfrac{\text{load}}{\text{bearing strength of a rivet}}$
    (d) $\dfrac{\text{load}}{\text{tearing strength of plate}}$.

32. If P is the load applied to a bracket with an eccentricity e, the resisting moment M offered by a rivet at a distance r from the centre of gravity is
    (a) $\dfrac{Pe^2 r}{\Sigma r^2}$
    (b) $\dfrac{P.e.r}{\Sigma r^2}$
    (c) $\dfrac{\Sigma r^2}{Per}$
    (d) $\dfrac{\Sigma r^3}{Per}$.

33. Rivets and bolts subjected to shear and tension should be so proportioned that
    (a) $\left(\dfrac{f_s}{p_s}\right)^2 + \left(\dfrac{f_t}{p_t}\right)^2 > 1$
    (b) $\left(\dfrac{f_s}{p_s}\right)^2 + \left(\dfrac{f_t}{p_t}\right)^2 = 1$
    (c) $\left(\dfrac{f_s}{p_t}\right)^2 + \left(\dfrac{f_t}{p_t}\right)^2 < 1$
    (d) $\left(\dfrac{f_s}{p_s}\right)^2 + \left(\dfrac{f_t}{p_t}\right)^2 \leq 1$.

34. Tacking rivets in compression plates exposed to weather have a pitch not exceeding 200 mm or
    (a) 32 times the thickness of outside plate ☐
    (b) 24 times the thickness of outside plate ☐
    (c) 16 times the thickness of outside plate ☐
    (d) 8 times the thickness of outside plate. ☐

35. If p and d are the pitch and the gross diameter of rivets, the efficiency (η) of a riveted joint used in the design is
    (a) $\dfrac{p-d}{p}$ ☐
    (b) $\dfrac{p+d}{p}$ ☐
    (c) $\dfrac{p}{p-d}$ ☐
    (d) $\dfrac{p}{p+d}$. ☐

36. Generally the size of a butt weld is indicated by the effective throat thickness, but in the case of incomplete penetration, it is taken as
    (a) half the thickness of the thicker part joined ☐
    (b) three-fourths thickness of the thicker part joined ☐
    (c) three-fourths thickness of the thinner part joined ☐
    (d) seven-eights thickness of the thinner part joined. ☐

37. The strength of a butt weld is
    (a) about 70 to 90 per cent of the main member ☐
    (b) equal to that of the main member ☐
    (c) equal to or more than that of the main member ☐
    (d) more than that of the main member. ☐

38. The strength of a fillet weld is
    (a) about 80 to 95 per cent of the main member ☐
    (b) equal to that of the main member ☐
    (c) more than that of the main member ☐
    (d) equal to or more than that of the main member. ☐

39. Fillet welding is associated with cover plates
    (a) not at all true ☐
    (b) sometimes ☐
    (c) all the times ☐
    (d) none of the above. ☐

40. The cross-section of a standard fillet weld is a triangle with base angles of
    (a) 45° and 45° ☐
    (b) 30° and 60° ☐
    (c) 40° and 50° ☐
    (d) 35° and 55°. ☐

41. The size of the fillet weld is given by
    (a) smaller side of the triangle ☐
    (b) throat of the fillet ☐
    (c) smaller size of the plate welded ☐
    (d) hypotenuse of the triangle. ☐

42. A fillet weld whose axis is parallel to the direction of the applied load is known as
    (a) end fillet weld ☐
    (b) diagonal fillet weld ☐
    (c) side fillet weld ☐
    (d) flat fillet weld. ☐

43. The throat in a fillet weld is
    (a) smaller side of the triangle of the fillet
    (b) larger side of the triangle of the fillet
    (c) hypotenuse of the triangle of the fillet
    (d) perpendicular distance from the root to the hypotenuse.
44. The maximum size of the fillet weld that can be made in a single pass is
    (a) 6 mm   (b) 8 mm
    (c) 10 mm   (d) 12 mm.
45. The maximum size of the fillet weld that can be applied to the edge of the plate should be
    (a) 1.5 mm less than the thickness of the plate
    (b) equal to the thickness of the plate
    (c) 1.5 mm more than the thickness of the plate
    (d) 0.75 times the thickness of the plate.
46. The maximum size of the fillet weld that can be applied to the toe of an angle or the round edge should be
    (a) 1.5 mm less than the thickness of the plate
    (b) equal to the thickness of the plate
    (c) 1.5 mm more than the thickness of the plate
    (d) 0.75 times the thickness of the plate.
47. The minimum size of the fillet weld that can be used is
    (a) 1 mm   (b) 2 mm
    (c) 3 mm   (d) 5 mm.
48. The length of side/longitudinal fillet welds should not be less than
    (a) 16 times the thickness of the thinner part connected
    (b) the perpendicular distance between the welds
    (c) twice the perpendicular distance between the welds
    (d) 4 times the size of weld or 40 mm whichever is greater.
49. The effective length of an intermittent fillet weld should be
    (a) 16 times the thickness of the thinner part connected
    (b) the perpendicular distance between the welds
    (c) twice the perpendicular distance between the welds
    (d) 4 times the size of weld or 40 mm whichever is greater.
50. Transverse spacing of the side fillet welds should not be less than
    (a) 16 times the thickness of the thinner part connected
    (b) 12 times the thickness of the thinner part connected
    (c) 20 times the thickness of the thinner part connected
    (d) 4 times the size of the weld or 40 mm whichever is greater.

51. The effective length of a fillet weld is
    (a) $l - 2s$
    (b) $l - 4s$
    (c) $0.8 l$
    (d) $0.9 l$
    where l = Total length of the weld and s = Size of the weld.

52. The effective throat size of a fillet weld is
    (a) 0.707 times the size of the weld
    (b) equal to the size of the weld
    (c) function of the angle between the fusion sides
    (d) hypotenuse of the triangle.

53. The ratio of the strength of the weld material to that of the parent body is
    (a) more than one
    (b) equal to one
    (c) less than one
    (d) less or more than one.

54. The weakest plane in a fillet weld is
    (a) smaller of the sides
    (b) side parallel to the force
    (c) side normal to the force
    (d) throat.

55. Continuous weld of constant thickness is called
    (a) fillet weld
    (b) seam weld
    (c) stitch weld
    (d) butt weld.

56. A series of continuous spot welds is called
    (a) fillet weld
    (b) seam weld
    (c) stitch weld
    (d) butt weld.

57. The actual thickness of a butt weld when compared with the thickness of the plate is
    (a) more
    (b) less
    (c) equal
    (d) more or less.

58. The permissible shear stress in mild steel weld material is about (in MPa)
    (a) 100
    (b) 110
    (c) 125
    (d) 94.5.

59. The percentage reduction in permissible stresses in field welds when compared with those of shop welds is
    (a) zero
    (b) 10
    (c) 20
    (d) 33.

60. The approximate percentage increase in permissible stresses under wind and seismic conditions when compared with normal live load conditions in welded joints is about
    (a) 50
    (b) 33
    (c) 25
    (d) 10.

61. The minimum thickness of steel members exposed to weather and accessible for painting is
    (a) 10 mm
    (b) 8 mm
    (c) 6 mm
    (d) 4 mm.

62. The minimum thickness of steel members exposed to weather and not accessible for painting is
    (a) 10 mm ☐   (b) 8 mm ☐
    (c) 6 mm ☐   (d) 4 mm. ☐

63. The allowable axial tensile stress in rolled mild steel I-beams and channels is (in kg/cm^2)
    (a) 1420 ☐   (b) 1500 ☐
    (c) 1810 ☐   (d) 2125. ☐

64. The allowable axial tensile stress in mild steel plates of over 20 mm thick is (in kg/cm^2)
    (a) 1420 ☐   (b) 1500 ☐
    (c) 1810 ☐   (d) 2125. ☐

65. The allowable axial tensile stress in high tension steel plates upto and including 45 mm thick is (in kg/cm^2)
    (a) 1420 ☐   (b) 1500 ☐
    (c) 1810 ☐   (d) 2125. ☐

66. The allowable tensile bending stress in mild steel plates up to and including 20 mm thick is (in kg/cm^2)
    (a) 1575 ☐   (b) 1970 ☐
    (c) 1500 ☐   (d) 1650. ☐

67. When the effect of wind or seismic load is taken into account, the permissible tensile stresses specified under live load conditions may be exceeded by
    (a) 10% ☐   (b) 25% ☐
    (c) $33\frac{1}{3}$% ☐   (d) 40%. ☐

68. The maximum permissible slenderness ratio of steel ties is
    (a) 180 ☐   (b) 250 ☐
    (c) 350 ☐   (d) no limit. ☐

69. The maximum permissible slenderness ratio of steel ties likely to be subjected to possible reversal of stress due to wind or seismic forces is
    (a) 180 ☐   (b) 250 ☐
    (c) 350 ☐   (d) no limit. ☐

70. The net area of round bars to resist tension is the area of cross-section
    (a) at the mid section ☐
    (b) at the root of the thread ☐
    (c) is based on the average diameter of the rod ☐
    (d) is the average of (a) and (b). ☐

71. Net sectional area of a tension member is equal to its gross sectional area
    (a) plus the area of the rivet holes ☐   (b) divided by the area of the rivet holes ☐
    (c) minus twice the area of the rivet holes ☐   (d) minus the area of the rivet holes. ☐

72. In a tension member with *zig-zag* riveting, the net width of the plate is given by

    (a) $b - nd + n_1 \dfrac{p^2}{4g}$ □   (b) $b + nd - n_1 \dfrac{p^2}{4g}$ □

    (c) $b - nd - n_1 \dfrac{p^2}{4g}$ □   (d) $b + nd + n_1 \dfrac{p^2}{4g}$ □

    where b = Width of the plate, n = Number of rivet holes
    n_1 = Number of gauge spaces, d = Diameter of rivet holes
    g = The gauge, p = The staggered pitch.

73. If a single angle in tension is connected by one leg only, the net effective sectional area of the angles shall be taken as

    (a) $a - \dfrac{b}{1 + 0.35 \dfrac{b}{a}}$ □   (b) $a + \dfrac{b}{1 + 0.35 \dfrac{b}{a}}$ □

    (c) $a - \dfrac{b}{1 + 0.2 \dfrac{b}{a}}$ □   (d) $a + \dfrac{b}{1 + 0.2 \dfrac{b}{a}}$. □

    where a = Net sectional area of the connected leg
    b = Area of the outstanding leg.

74. If a pair of angles back-to-back in tension are connected by only one leg of each angle to the same side of a gusset, the net effective area shall be taken as

    (a) $a - \dfrac{b}{1 + 0.35 \dfrac{b}{a}}$ □   (b) $a + \dfrac{b}{1 + 0.35 \dfrac{b}{a}}$ □

    (c) $a - \dfrac{b}{1 + 0.2 \dfrac{b}{a}}$ □   (d) $a + \dfrac{b}{1 + 0.2 \dfrac{b}{a}}$. □

75. For double angles carrying direct tension placed back-to-back and connected to each side of a gusset, the effective sectional area is equal to gross sectional area of

    (a) the section □
    (b) the section plus the area of the rivet holes □
    (c) the section minus the area of the rivet holes □
    (d) the section minus one-half of the area of the rivet holes. □

76. Members subject to both bending and axial tension shall be so proportioned that

    (a) $\dfrac{f_t}{P_t} + \dfrac{f_{bt}}{P_{bt}} > 1$ □   (b) $\dfrac{f_t}{P_t} + \dfrac{f_{bt}}{P_{bt}} \leq 1$ □

    (c) $\dfrac{f_t}{P_t} + \dfrac{f_{bt}}{P_{bt}} \geq 1$ □   (d) $\dfrac{f_t}{P_t} + \dfrac{f_{bt}}{P_{bt}} < 1$ □

    where f_t = The calculated axial tensile stress
    f_{bt} = The calculated bending tensile stress in the extreme fibre
    P_t = The permissible axial tensile stress
    P_{bt} = The permissible bending tensile stress in the extreme fibre.

77. In a tension splice the number of rivets carrying calculated shear stress through a packing greater than 6 mm thick is to be increased by 2.5% for each.....thickness of packing.
    (a) 1 mm
    (b) 1.5 mm
    (c) 2.0 mm
    (d) 2.5 mm.

78. The maximum deflection allowed in steel ties is
    (a) $\dfrac{L}{450}$
    (b) $\dfrac{L}{400}$
    (c) $\dfrac{L}{300}$
    (d) $\dfrac{L}{350}$.

79. The allowable direct tensile stress in structural steel is
    (a) $0.6 f_y$
    (b) $0.66 f_y$
    (c) $0.7 f_y$
    (d) $0.75 f_y$.

80. The minimum thickness of steel in main members not directly exposed to weather shall be
    (a) 10 mm
    (b) 8 mm
    (c) 6 mm
    (d) 4.5 mm.

81. The minimum thickness of steel in secondary members not directly exposed to weather shall be
    (a) 8 mm
    (b) 6 mm
    (c) 5 mm
    (d) 4.5 mm.

82. Effective length of a compression member effectively held in position and restrained in direction at both ends is
    (a) 0.67 L
    (b) 0.85 L
    (c) L
    (d) 1.5 L

    where L is the actual length of the member measured between the centres of effective lateral supports.

83. Effective length of a compression member effectively held in position at both ends and restrained in direction at one end is
    (a) 0.67 L
    (b) 0.85 L
    (c) L
    (d) 1.5 L.

84. Effective length of a compression member effectively held in position at both ends but not restrained in direction is
    (a) 0.67 L
    (b) 0.85 L
    (c) L
    (d) 1.5 L.

85. Effective length of a compression member effectively held in position and restrained in direction at one end and at the other end partially restrained in direction but not held in position is
    (a) 0.67 L
    (b) 0.85 L
    (c) L
    (d) 1.5 L.

86. The maximum slenderness ratio of compression members carrying loads resulting from dead and superimposed loads should be
    (a) 180 (b) 250
    (c) 350 (d) 400.

87. The maximum slenderness ratio of compression members carrying loads resulting from wind or seismic loads should be
    (a) 180 (b) 250
    (c) 350 (d) 400.

88. The maximum slenderness ratio of steel members acting as wind bracings should be
    (a) 180 (b) 250
    (c) 350 (d) 400.

89. The maximum deflection allowed in steel columns should be (where L is the actual length of the column)
    (a) $\dfrac{L}{250}$ (b) $\dfrac{L}{300}$
    (c) $\dfrac{L}{400}$ (d) $\dfrac{L}{350}$.

90. The minimum thickness of unstiffened outstanding legs of mild-steel compression member should be
    (a) 10 mm (b) 16 mm
    (c) $\dfrac{B}{14}$ (d) $\dfrac{B}{16}$.

    where B = Width of outstand.

91. Outstanding length of a compression member consisting of an angle or channel is measured as
    (a) nominal width of the section
    (b) one-half of the nominal width of the section
    (c) distance from the free edge to the first row of rivets
    (d) slenderness ratio of the section.

92. Outstanding length of a compression member consisting of flanges of beams and tee sections is measured as
    (a) nominal width of the section
    (b) one-half of the nominal width of the section
    (c) distance from the free edge to the first row of rivets
    (d) slenderness ratio of the section.

93. The unsupported width of unstiffened mild steel plate subjected to compressive load measured between adjacent lines of rivets should be less than
    (a) 50t (b) 80t
    (c) 90t (d) 45t

    where t is thickness of the plate.

94. For design purposes the unsupported width of unstiffened mild-steel plate subjected to compression is taken as
    (a) 50t
    (b) 80t
    (c) 90t
    (d) 45t.

95. For determination of allowable stress in axial compression, IS : 800—1962 has adopted
    (a) Euler's formula
    (b) Rankines formula
    (c) Perry Robertson formula
    (d) Secant formula.

96. The allowable compressive stress in steel columns is
    (a) directly proportional to the slenderness ratio
    (b) inversely proportional to the slenderness ratio
    (c) proportional to the square of the slenderness ratio
    (d) non-linearly related to the slenderness ratio.

97. The most economical section for a column is
    (a) I-section
    (b) Tubular section
    (c) Solid round section
    (d) Rectangular section.

98. When large value of radius of gyration is not required
    (a) two channels are placed back to back
    (b) flanges are kept outward
    (c) flanges are kept inward
    (d) two I-sections are used.

99. The effective length of a double angle strut with angles placed back-to-back and connected to both the sides of a gusset plate, by not less than two rivets is
    (a) 0.67 L
    (b) 0.85 L
    (c) L
    (d) 1.5 L.

100. Allowable permissible stress corresponding to the slenderness ratio of double angles placed back-to-back, connected to one side of a gusset plate is reduced by
    (a) 20%
    (b) 25%
    (c) 30%
    (d) 40%.

101. The slenderness ratio of single angle discontinuous struts connected by a single rivet or bolt should not exceed
    (a) 120
    (b) 180
    (c) 350
    (d) 400.

102. The effective length of a single angle discontinuous strut connected by two or more rivets at each end is taken as
    (a) L
    (b) 0.8 L
    (c) 0.85 L
    (d) 0.7 L

    where L is the length centre-to-centre of inter-section at each end.

103. Lacing or battening of compound steel columns
    (a) increases the capacity of the column
    (b) decreases the buckling of the columns
    (c) decreases local buckling of member
    (d) is a must.

104. Battens in compound steel columns are provided mainly to
    (a) ensure unified behaviour
    (b) increase the column capacity
    (c) decrease the buckling in members
    (d) prevent buckling.

105. The most commonly used sections for lateral system to carry shear force in built up columns are
    (a) rolled steel plates ☐
    (b) rolled steel angles ☐
    (c) rolled steel channels ☐
    (d) all the above. ☐

106. Lacing bars and battens in compound steel columns are designed to resist a transverse shear force of
    (a) 1.5% of the axial load ☐
    (b) 2.0% of the axial load ☐
    (c) 2.5% of the axial load ☐
    (d) 3.0% of the axial load. ☐

107. The slenderness ratio of lacing bars is limited to
    (a) 200 ☐
    (b) 145 ☐
    (c) 350 ☐
    (d) 400. ☐

108. The effective length of a bar in double lacing system is taken as......times the distance between the inner end rivets
    (a) 0.7 ☐
    (b) 0.8 ☐
    (c) 0.85 ☐
    (d) 1.0. ☐

109. The minimum width of a lacing bar depends on
    (a) nominal diameter of rivet ☐
    (b) thickness of lacing bar ☐
    (c) length of the member ☐
    (d) no relation at all. ☐

110. In a single lacing system, the thickness of the lacing bar should be
    (a) $\frac{1}{30}$ th length between inner end rivets ☐
    (b) $\frac{1}{40}$ th length between inner end rivets ☐
    (c) $\frac{1}{5}$ th length between inner end rivets ☐
    (d) $\frac{1}{60}$ th length between inner end rivets. ☐

111. Compression members composed of two channels placed back-to-back and separated by a small distance are connected together by riveting so that the minimum slenderness ratio of each member between the connections does not exceed
    (a) 70 ☐
    (b) 60 ☐
    (c) 50 ☐
    (d) 40. ☐

112. Members subjected to both bending and axial compression shall be so proportioned that
    (a) $\frac{f_c}{P_c} + \frac{f_{bc}}{P_{bc}} > 1$ ☐
    (b) $\frac{f_c}{P_c} + \frac{f_{bc}}{P_{bc}} \geq 1$ ☐
    (c) $\frac{f_c}{P_c} + \frac{f_{bc}}{P_{bc}} \leq 1$ ☐
    (d) $\frac{f_c}{P_c} + \frac{f_{bc}}{P_{bc}} < 1$ ☐

    where f_c = Calculated average axial compressive stress
    f_{bc} = Calculated bending compressive stress in extreme fibre
    P_c = Permissible compressive stress on the member subjected to axial compressive load only
    P_{bc} = Permissible bending compressive stress on the extreme fibre.

113. The column splice is used to increase
    (a) strength of the column   (b) cross-sectional area of the column
    (c) length of the column   (d) all the above.

114. If the depths of a column splices are equal, then the column splice is provided
    (a) with filler plates   (b) with bearing plates
    (c) with filler and bearing   (d) none of the above.

115. If the depth of the section of an upper column is much smaller than the lower column
    (a) filler plates are provided with column splice
    (b) bearing plates are provided with column splice
    (c) neither filler nor bearing plates are provided with column splice
    (d) filler and bearing plates are provided with column splice.

116. Column bases are mainly subjected to and designed for
    (a) bending and compression   (b) bearing and tension
    (c) compression and tension   (d) bearing and compression.

117. The minimum thickness of a rectangular slab base subjected to uniformly distributed load is given by

    (a) $\sqrt{\dfrac{3w}{P_{bct}}\left(A^2 + \dfrac{B^2}{4}\right)}$ (b) $\sqrt{\dfrac{3w}{P_{bct}}\left(A^2 - \dfrac{B^2}{4}\right)}$

    (c) $\sqrt{\dfrac{3w}{P_{bct}}\left(B^2 + \dfrac{A^2}{4}\right)}$ (d) $\sqrt{\dfrac{3w}{P_{bct}}\left(B^2 - \dfrac{A^2}{4}\right)}$.

    where A = The greater projection of the plate beyond the column
            B = The lesser projection of the plate beyond the column
            w = The pressure or loading on the underside of the base
            P_{bct} = The permissible bending stress in slab base.

118. The minimum thickness of a square base D provided for a circular column of diameter d and carrying a total axial load W is given by

    (a) $\sqrt{\dfrac{9W}{16 P_{bct}}\left(\dfrac{D}{D+d}\right)}$ (b) $\sqrt{\dfrac{16W}{9 P_{bct}}\left(\dfrac{D}{D+d}\right)}$

    (c) $\sqrt{\dfrac{9W}{16 P_{bct}}\left(\dfrac{D}{D-d}\right)}$ (d) $\sqrt{\dfrac{16W}{1 P_{bct}}\left(\dfrac{D-d}{D}\right)}$.

    where P_{bct} = the permissible bending stress in the steel.

119. A column footing is provided
    (a) to spread the column load over a large area
    (b) to ensure that intensity of bearing pressure between the column footing and soil does not exceed permissible bearing capacity of the soil
    (c) to distribute the column load over the soil through the column footing
    (d) all the above.

120. In a grillage foundation maximum bending moment occurring at the centre of grillage beams is given by

(a) $\dfrac{P}{4}(L+a)$ ☐ (b) $\dfrac{P}{4}(L-a)$ ☐

(c) $\dfrac{P}{8}(L+a)$ ☐ (d) $\dfrac{P}{2}(L+a)$ ☐

where P, L and a are axial load, length of the beam and length of the column base respectively.

121. In a grillage foundation the maximum shear force occurs
   (a) at the edge of grillage beam ☐ (b) at the centre of grillage beam ☐
   (c) at the edge of base plate ☐ (d) at the centre of base plate. ☐

122. For an economical design of a combined footing to support two equal column loads, the projections of beams in lower tier are kept such that bending moments under columns are equal to
   (a) bending moment at the centre of the beam ☐
   (b) half the bending moment at the centre of the beam ☐
   (c) twice the bending moment at the centre of the beam ☐
   (d) bending moment at the edge of the beam. ☐

123. The moment carrying capacity of steel structural sections is governed by the following stresses
   (a) tensile stress ☐ (b) bending tensile stress ☐
   (c) bending compressive stress ☐ (d) bending compressive stress or tensile stress. ☐

124. The rolled steel *I*-sections are most commonly used as the beams because they provide
   (a) large moment of interia with less cross-sectional area ☐
   (b) greater moment of resistance as compared to other sections ☐
   (c) greater lateral stability ☐
   (d) all the above. ☐

125. In rolled steel beams, shear force is resisted by
   (a) flanges only ☐ (b) web only ☐
   (c) web and flanges together ☐ (d) top flange only. ☐

126. The effective length (*l*) of a simply supported beam with ends restrained against torsion, and the ends of compression flange partially restrained against lateral bending is given by
   (a) l = span ☐ (b) $l = 1.20 \times$ span ☐
   (c) $l = 0.85 \times$ span ☐ (d) $l = 0.7 \times$ span. ☐

127. The effective length (*l*) of a simply supported beam with ends restrained against torsion, and the ends of compression flange fully restrained against lateral bending is given by
   (a) l = span ☐ (b) $l = 1.20 \times$ span ☐
   (c) $l = 0.85 \times$ span ☐ (d) $l = 0.7 \times$ span. ☐

128. The effective length (*l*) of a simply supported beam with ends unrestrained against torsion and the ends of compression flange unrestrained against lateral bending is given by
   (a) l = span ☐ (b) $l = 1.20 \times$ span ☐
   (c) $l = 0.85 \times$ span ☐ (d) $l = 0.7 \times$ span. ☐

129. For cantilever beams built in at the support and restrained against torsion at the free end, the effective projecting length (l) is given by
    (a) $l = 0.75 L$ ☐   (b) $l = L$ ☐
    (c) $l = 2L$ ☐   (d) $l = 0.85 L$ ☐
    where L = Projecting length of the cantilever.

130. For cantilever beams continuous and partially restrained against torsion at the support and free at the end the effective length (l) is given by
    (a) $l = 0.75 L$ ☐   (b) $l = L$ ☐
    (c) $l = 2L$ ☐   (d) $l = 0.85 L$. ☐

131. For a simply supported beam, the maximum deflection permitted is
    (a) $\dfrac{1}{300}$ of the span ☐   (b) $\dfrac{1}{325}$ of the span ☐
    (c) $\dfrac{1}{350}$ of the span ☐   (d) $\dfrac{1}{400}$ of the span. ☐

132. A simply supported beam carrying uniformly distributed load will be safe in deflection if the ratio of its span and depth is
    (a) < 19 ☐   (b) > 19 ☐
    (c) < 24 ☐   (d) > 24. ☐

133. The thickness of an outstanding flange in steel beams is subjected to a minimum values because of
    (a) durability ☐   (b) local bending ☐
    (c) local buckling ☐
    (d) compressive stress and lateral buckling of the beam. ☐

134. The minimum thickness of unstiffened compression outstands of mild steel plates in beams is limited to
    (a) 8 mm ☐   (b) $\dfrac{b_f}{25}$ ☐
    (c) $\dfrac{D}{40}$ ☐   (d) $\dfrac{b_s}{16}$. ☐
    where b_f = Flange width, D = Depth of the beam, b_s = Flange outstand.

135. The lateral buckling of I-beams in steel is governed mostly by
    (a) boundary conditions of the flanges ☐   (b) maximum moment of inertia ☐
    (c) torsional constant ☐   (d) width of the flanges. ☐

136. The width of unstiffened mild steel plates in tension flanges of a steel beam should be limited to
    (a) $16 t$ ☐   (b) $20 t$ ☐
    (c) $\dfrac{D}{2}$ ☐   (d) $\dfrac{b_f}{2}$. ☐
    where D = Depth of the beam, b_f = Width of the flange, and
    t = Thickness of the outstanding leg.

137. The minimum thickness of unstiffened web plates in mild steel beams should be

   (a) 6 mm
   (b) $\dfrac{d}{85}$
   (c) $\dfrac{D}{100}$
   (d) $\dfrac{L}{325}$.

   where d = Clear distance between the flange angles, D = Depth of the beam and L = Span.

138. The minimum thickness of vertically stiffened web plates in mild steel beams should be

   (a) 6 mm
   (b) $\dfrac{D}{200}$
   (c) $\dfrac{D}{325}$
   (d) $\dfrac{d}{200}$.

139. Web crippling in beams generally occurs at the points where

   (a) bending moment is maximum
   (b) shear force is maximum
   (c) concentrated loads act
   (d) deflection is maximum.

140. The bearing stress produced under a concentrated load in a beam is given by

   (a) $\dfrac{W}{(b+h\sqrt{3})t_w}$
   (b) $\dfrac{W}{(b+2h\sqrt{3})t_w}$
   (c) $\dfrac{W}{(b+2h\sqrt{2})t_w}$
   (d) $\dfrac{W}{(b+h\sqrt{2})t_w}$.

   where b = Length of the bearing plate, h = Depth of the root of the fillet, t_w = Thickness of the web plate.

141. To prevent web buckling, the allowable compressive stress is found corresponding to a slenderness ratio of

   (a) $\dfrac{d_w\sqrt{2}}{t_w}$
   (b) $\dfrac{d_w\sqrt{2}}{t_w}$
   (c) $\dfrac{d_w\sqrt{6}}{t_w}$
   (d) $\dfrac{d_w\sqrt{3}}{t_w}$.

   where d_w = Clear depth of web between root fillets, t_w = Thickness of web.

142. Web buckling at a continuous support can be prevented if

   (a) $\dfrac{f_c}{f_{cr}} + \dfrac{f_s}{f_{scr}} \leq 1$
   (b) $\dfrac{d}{t_w} < 180$
   (c) $\left(\dfrac{f_c}{f_{cr}}\right)^2 + \left(\dfrac{f_s}{f_{scn}}\right)^2 \leq 1$
   (d) $\dfrac{d}{t_w} > 180$

   where f_c and f_s are the actual longitudinal compression and shear stresses and f_{cr} and f_{scr} are the critical longitudinal compression and shear stresses which will buckle the web plate individually.

143. The allowable shear stress in stiffened webs of mild steel beams decreases with
    (a) decrease in the spacing of the stiffeners ☐ (b) increase in the spacing of the stiffeners ☐
    (c) decrease in effective depth ☐ (d) all the above. ☐

144. The approximate allowable bending compressive stress in mild steel beams in which the flange is continuously laterally supported is (in kg/cm^2)
    (a) 1400 ☐ (b) 1500 ☐
    (c) 1650 ☐ (d) 2285. ☐

145. The approximate allowable bending compressive stress in mild steel plate girders having continuous lateral support to the flanges is (in kg/cm^2)
    (a) 1400 ☐ (b) 1500 ☐
    (c) 1650 ☐ (d) 1575. ☐

146. The permissible bending stresses in slab bases is (in kg/cm^2)
    (a) 1890 ☐ (b) 1650 ☐
    (c) 1575 ☐ (d) 1500. ☐

147. The allowable maximum shear stress in rolled mild steel sections is (in kg/cm^2)
    (a) 945 ☐ (b) 1100 ☐
    (c) 1340 ☐ (d) 1575. ☐

148. The allowable average shear stress in unstiffened webs of rolled mild steel sections is (in kg/cm^2)
    (a) 945 ☐ (b) 1100 ☐
    (c) 1340 ☐ (d) 1575. ☐

149. The allowable average shear stress is kg/cm^2 in stiffened mild steel plate girders of $\dfrac{d}{t_w} = 200$ is
    (a) 945 ☐ (b) 1100 ☐
    (c) 745 ☐ (d) 700. ☐

150. In the case of steel sections subjected to combined bending and compression, the following will govern the working stress design
    (a) principal stress criterion ☐ (b) failure criterion ☐
    (c) maximum strain criterion ☐ (d) none of the above. ☐

151. The minimum spacing of vertical stiffeners in a plate girder is given by
    (a) 0.33 d ☐ (b) 0.4 d ☐
    (c) 0.7 d ☐ (d) d ☐
    where d is the distance between the flange angles.

152. The maximum spacing of vertical stiffeners in a plate girder is limited to
    (a) 0.7 d ☐ (b) d ☐
    (c) 1.5 d ☐ (d) 2d. ☐

153. Intermediate vertical stiffeners are provided if the thickness of the mild steel web is less than
    (a) $\dfrac{d}{85}$
    (b) $\dfrac{d}{100}$
    (c) $\dfrac{d}{180}$
    (d) $\dfrac{d}{200}$.

154. Intermediate vertical stiffeners are provided in plate girders to
    (a) transfer concentrated loads
    (b) prevent excessive deflection
    (c) eliminate web buckling
    (d) eliminate local buckling.

155. The length of an outstanding leg of a vertical stiffener for mild steel sections in terms of its thickness "t" is
    (a) $12t$
    (b) $14t$
    (c) $16t$
    (d) $20t$.

156. If d is the clear depth of a plate girder, t is the minimum required thickness of the web, c is the maximum clear distance between vertical stiffeners, the moment of inertia of a stiffener about the face of the web should not be less than
    (a) $\dfrac{1.5 d^4 t}{c}$
    (b) $\dfrac{1.5 d^3 t^3}{c^2}$
    (c) $\dfrac{1.5 d^2 t^3}{c}$
    (d) $\dfrac{1.5 d^4 t^3}{c^2}$.

157. Bearing stiffeners are provided in plate girders to
    (a) eliminate web buckling
    (b) transfer concentrated loads
    (c) prevent excessive buckling
    (d) eliminate local buckling.

158. Bearing stiffeners in plate girders are provided at
    (a) mid span
    (b) quarter points
    (c) supports
    (d) equal intervals.

159. Bearing stiffeners are designed as
    (a) columns
    (b) ties
    (c) beams
    (d) beam-ties.

160. The ratio of the effective length to its actual length of bearing stiffeners is
    (a) 1
    (b) 1.5
    (c) 2
    (d) 0.7.

161. When bearing stiffeners are required to provide restraint against torsion the moment of inertia of the stiffener about the centre line of the web plate should be not less than
    (a) $\dfrac{D^2 T^2}{250} \dfrac{R}{W}$
    (b) $\dfrac{D^3 T}{250} \dfrac{R}{W}$
    (c) $\dfrac{D T^2}{250} \dfrac{R}{W}$
    (d) $\dfrac{D^3 T}{300} \dfrac{R}{W}$.

    where D = Overall depth of the girder, T = Maximum thickness of compression flange, R = Reaction on the bearing, and W = Total load on the girder.

162. Horizontal stiffeners are provided in plate girders if the thickness of the web is
    (a) 8 mm
    (b) less than $\dfrac{d}{200}$
    (c) less than $\dfrac{L}{200}$
    (d) equal to that of the flange

    where d = The clear distance between flange angles, L = Span of the girder.

163. The smaller permissible clear dimension of the web in the panel of a plate girder is
    (a) $85t$
    (b) $180t$
    (c) $200t$
    (d) $270t$.

164. The larger permissible clear dimension of the web in the panel of a plate girder is
    (a) $85t$
    (b) $180t$
    (c) $200t$
    (d) $270t$.

165. The flange-web joint in plate girder is designed to resist
    (a) axial stress
    (b) normal stress
    (c) shear stress
    (d) bending stress.

166. The joint connecting the cover plate and flange angle in plate girder is designed to resist
    (a) axial stress
    (b) normal stress
    (c) shear stress
    (d) bending stress.

167. The web splice in plate girders is primarily designed to withstand
    (a) bending moment
    (b) axial force
    (c) shear force
    (d) all the above.

168. The web splice in plate girder is subjected to
    (a) axial force only
    (b) shear and axial force
    (c) bending movement and axial force
    (d) shear force and bending moment.

169. The flange splice in plate girder is subjected to
    (a) axial force only
    (b) shear and axial force
    (c) bending moment and axial force
    (d) shear force and bending moment.

170. In a plate girder the web splicing is done at
    (a) minimum shear location
    (b) minimum moment location
    (c) maximum shear location
    (d) maximum moment location.

171. In a plate girder the flange splicing is done at
    (a) minimum shear location
    (b) minimum moment location
    (c) maximum shear location
    (d) maximum moment location.

172. The load on a lintel is assumed as uniformly distributed if the masonry above it, is upto a height of
    (a) the effective span
    (b) 1.25 times the effective span
    (c) 1.5 times the effective span
    (d) 2.0 times the effective span.

173. The horizontal thrust on the tie rods provided at the end beam of a Jack Arch is given by
    (a) $\dfrac{WL}{4R}$
    (b) $\dfrac{WL}{8L}$
    (c) $\dfrac{WL}{8R}$
    (d) $\dfrac{WL}{2R}$.

    where W, L and R are total load on the arch, its span and its rise respectively.

174. The minimum diameter of tie rods to be used in jack arch is
    (a) 10 mm
    (b) 12 mm
    (c) 16 mm
    (d) 8 mm.

175. The maximum deflection permitted in the beam of jack arch is
    (a) $\dfrac{\text{Span}}{300}$
    (b) $\dfrac{\text{Span}}{350}$
    (c) $\dfrac{\text{Span}}{400}$
    (d) $\dfrac{\text{Span}}{480}$.

176. The rise of a Jack arch is kept about
    (a) $\dfrac{1}{3}$ to $\dfrac{1}{4}$ of the span
    (b) $\dfrac{1}{4}$ to $\dfrac{1}{6}$ of the span
    (c) $\dfrac{1}{6}$ to $\dfrac{1}{8}$ of the span
    (d) $\dfrac{1}{8}$ to $\dfrac{1}{12}$ of the span.

177. Span of continuous fillers are considered approximately equal if the longest span does not exceed the shortest span by more than
    (a) 5%
    (b) 10%
    (c) 15%
    (d) 20%.

178. The bending moment for filler joists at the interior support is
    (a) $\dfrac{WL^2}{10}$
    (b) $-\dfrac{WL^2}{10}$
    (c) $-\dfrac{WL^2}{12}$
    (d) $\dfrac{WL^2}{12}$.

    where W = The dead plus live load per unit length of span, L = The span.

179. The ratio of the span of filler joists (centre to centre of supports) to the depth from the underside of the joist to the top of the structural concrete should not exceed
    (a) 35
    (b) 60
    (c) 12
    (d) 45.

180. In case of cantilever fillers, the ratio of span to the depth should not exceed
    (a) 35
    (b) 20
    (c) 12
    (d) 8.

181. The minimum width of a solid casing for a cased beam is equal to
    (a) $b + 125$ mm
    (b) $b + 100$ mm
    (c) $b + 75$ mm
    (d) $b + 50$ mm.

182. Modified moment of inertia of sections with a single web is equal to moment of inertia of the section about the y-y axis at the point of maximum bending moment multiplied by the ratio of
    (a) area of compression flange at the minimum bending moment to the corresponding area at point of maximum bending moment
    (b) area of tension flange at the minimum bending moment to the corresponding area at point of maximum bending moment

(c) total area of flanges at the minimum bending moment to the corresponding area at the point of maximum bending moment ☐

(d) area of compression flange at the maximum bending moment to the area of tension flange at the maximum bending moment. ☐

183. Gantry girders have to be designed to resist
   (a) transverse loads ☐
   (b) transverse and lateral loads ☐
   (c) lateral and axial loads ☐
   (d) transverse lateral and axial loads. ☐

184. The drag force caused on the gantry girder in hand-operated cranes is....than that in the electrically operated ones.
   (a) more ☐
   (b) less ☐
   (c) equal ☐
   (d) no relation. ☐

185. The ratio of the impact factor used in hand-operated cranes to that of the electrically-operated one in gantry beams is
   (a) more than 1 ☐
   (b) less than 1 ☐
   (c) equal to 1 ☐
   (d) no relation. ☐

186. The percentage of the impact load with respect to the wheel load of an electrically-operated gantry girder is
   (a) 5% ☐
   (b) 10% ☐
   (c) 20% ☐
   (d) 25%. ☐

187. The lateral force on the rail head from a hand-operated crane girder when compared with the crab and lifting weight is
   (a) 20% ☐
   (b) 15% ☐
   (c) 10% ☐
   (d) 2.5%. ☐

188. The maximum deflection permitted in gantry girders is
   (a) $\dfrac{L}{250}$ ☐
   (b) $\dfrac{L}{500}$ ☐
   (c) $\dfrac{L}{1000}$ ☐
   (d) $\dfrac{L}{750}$, where L is the span. ☐

189. The member of a roof truss which supports the purlins is called as
   (a) principal rafter ☐
   (b) principal tie ☐
   (c) sag tie ☐
   (d) main strut. ☐

190. The member of a roof truss which is parallel to the span of the truss and primarily under tension is called as
   (a) principal rafter ☐
   (b) principal tie ☐
   (c) sag tie ☐
   (d) main strut. ☐

191. The sag tie in a truss is mainly used to reduce
   (a) tension ☐
   (b) compression ☐
   (c) moment and deflection ☐
   (d) weight of the truss. ☐

192. The nature of the main force in a minor sling of a roof truss is
   (a) bending ☐
   (b) compression ☐
   (c) compression and bending ☐
   (d) tension. ☐

193. The plate used as a connecting piece at the intersection of two or more structural members in a roof truss is called as
    (a) template
    (b) gusset plate
    (c) base plate
    (d) shoe plate.

194. The plate attached to the top of a column or base of a roof truss to provide a connecting link between the column and truss is called as
    (a) template
    (b) gusset plate
    (c) base plate
    (d) shoe plate.

195. The plate used at the bottom of a steel column to connect it with the foundation is called as
    (a) template
    (b) gusset plate
    (c) base plate
    (d) shoe plate.

196. The pitch of a truss is defined as the ratio of
    (a) height of the truss to the span
    (b) height of the truss to one-half of the span
    (c) height of the truss to length of principal rafter
    (d) none of the above.

197. The economical pitch of a roof truss subjected to snow loads and heavy winds is
    (a) $\frac{1}{6}$
    (b) $\frac{1}{5}$
    (c) $\frac{1}{4}$
    (d) $\frac{1}{3}$.

198. The economic range of spacing of roof trusses is
    (a) $\frac{1}{2}$ to $\frac{1}{3}$ of span
    (b) $\frac{1}{2}$ to $\frac{1}{4}$ of span
    (c) $\frac{1}{4}$ to $\frac{1}{6}$ of span
    (d) $\frac{1}{3}$ to $\frac{1}{5}$ of span.

199. Sag rods in a roof truss are used for connecting
    (a) principal rafters
    (b) purlins
    (c) main ties
    (d) all the above.

200. By cambering the truss the forces in the members are
    (a) increased
    (b) decreased
    (c) not altered
    (d) none of the above..

201. The shape factor used for circular sections is
    (a) 0.7
    (b) 0.8
    (c) 0.9
    (d) 1.0.

202. The minimum thickness of the plates used in pressed steel tanks is
    (a) 3 mm
    (b) 4 mm
    (c) 5 mm
    (d) 6 mm.

203. The segmental bottom of a circular tank is subjected to
    (a) hoop tension
    (b) hoop compression
    (c) bending moment
    (d) hoop tension and hoop compression.

204. The minimum thickness of the plates used in water tanks other than pressed tanks is
    (a) 3 mm
    (b) 4 mm
    (c) 5 mm
    (d) 6 mm.

205. A structural element used to support a vertical cladding is called
    (a) cleat
    (b) runner
    (c) ferrule
    (d) diaphragm.

206. A rigid structural element provided in beams to prevent the distortion of shape is called
    (a) cleat
    (b) runner
    (c) ferrule
    (d) diaphragm.

207. A structural member provided at the level of the top of the columns around an industrial building to resist the bending force on the total building is called
    (a) eave girder
    (b) baluster
    (c) eave board
    (d) tie girder.

208. A structural member connecting columns at or below the basement level to transmit horizontal forces is called
    (a) eave girder
    (b) baluster
    (c) eave board
    (d) tie girder.

209. The load factor is defind as
    (a) $\dfrac{\text{ultimate load}}{\text{working load}}$
    (b) $\dfrac{\text{plastic load}}{\text{working load}}$
    (c) $\dfrac{\text{limit load}}{\text{working load}}$
    (d) $\dfrac{\text{yield load}}{\text{working load}}$.

210. The load factor applied to dead loads in the design of steel structures under live load conditions is
    (a) 2.0
    (b) 1.7
    (c) 1.5
    (d) 1.3.

211. The load factor applied to wind and seismic loads in design of steel structures is
    (a) 2.2
    (b) 1.8
    (c) 1.5
    (d) 1.3.

212. The load factor applied to live loads in the design of steel structures is
    (a) 2.0
    (b) 1.7
    (c) 1.5
    (d) 1.3.

213. The plastic section modulus is defined as
    (a) the ratio of plastic moment capacity of a section to the yield moment
    (b) the ratio of moment of inertia to radius of gyration
    (c) radius of gyration multiplied by the area
    (d) the ratio of plastic moment capacity to the yield stress.

214. The design criterion for the plastic design of steel beams is
    (a) $M_p \geq M_u$ ☐ (b) $M_p \geq$ load factor $\times M_u$ ☐
    (c) $Zf_y \geq M_u$ ☐ (d) $Zf_y \leq M_u$. ☐
    where M_u = The ultimate moment, M_p = Plastic moments capacity
    Z = Section modulus, f_y = Yield stress.

215. The design criterion for the design of steel columns is
    (a) $P_y \geq 1.7 Af_y$ ☐ (b) $P_{yp} \geq P_u$ ☐
    (c) $P_{yp} \geq P_{cr}$ ☐ (d) $P_u \geq P_{cr}$ ☐
    where P_u = Ultimate axial load, P_{yp} = Yield capacity of the cross-section
    P_{cr} = Critical load, A = Area of cross-section, f_y = Yield stress.

216. The design criterion for the plastic design of steel ties is
    (a) $P_u \geq 1.7 Af_y$ ☐ (b) $P_u < 1.7 A_{fy}$ ☐
    (c) $1.7 A_{fy} > P_{cr}$ ☐ (d) $1.7 Af_y \geq P_u$. ☐

217. The plastic hinge in a section is caused
    (a) when the material at a section reach plastic state ☐
    (b) when the extreme fibres at a section reach the yield state ☐
    (c) when all the fibres at a section reach the yield state ☐
    (d) none of the above. ☐

218. The moment-curvature relation at a plastic hinge is
    (a) linear ☐ (b) parabolic ☐
    (c) constant curvature for all moments ☐ (d) constant moment for different curvatures. ☐

219. The ratio of the plastic moment capacity to yield moment capacity of a section is
    (a) more than one ☐ (b) less than one ☐
    (c) equal to one ☐ (d) none of the above. ☐

220. The ratio of the plastic moment capacity to the yield moment capacity of a rolled steel beam section is about
    (a) 1.7 ☐ (b) 1.5 ☐
    (c) 1.15 ☐ (d) 0.85. ☐

221. The ratio of the plastic moment capacity to the yield moment capacity of a rectangular section is
    (a) 1.7 ☐ (b) 1.5 ☐
    (c) 1.15 ☐ (d) 0.85. ☐

222. The ratio of the plastic moment capacity to the yield moment capacity of a circular section is
    (a) 1.7 ☐ (b) 1.5 ☐
    (c) 1.15 ☐ (d) 0.85. ☐

223. The ratio of the ultimate moment capacity to the plastic moment capacity is
    (a) equal to one ☐ (b) more than one ☐
    (c) less than one ☐ (d) more than three. ☐

224. The ultimate strength design of steel structures makes use of
   (a) plastic analysis of structures   □   (b) elastic structural analysis   □
   (c) ultimate analysis   □   (d) elastic and plastic analysis.   □

225. The limit design of steel structures makes use of
   (a) plastic analysis of structures   □   (b) elastic structural analysis   □
   (c) ultimate analysis   □   (d) elastic and plastic analysis.   □

226. The plastic design of a structure which is based on the mechanical method of plastic analysis
   (a) is safe   □   (b) is unsafe   □
   (c) gives no indication of safety   □   (d) none of the above.   □

227. The plastic design of a steel structure which is based on the kinematic method of plastic analysis
   (a) is safe   □   (b) is unsafe   □
   (c) gives no indication of safety   □   (d) none of the above.   □

228. The plastic design of steel structures based on the statical method of plastic analysis
   (a) is safe   □   (b) is unsafe   □
   (c) gives no indication of safety   □   (d) none of the above.   □

229. The plastic design of steel structures which is based on the lower bound theorem of plastic analysis
   (a) is safe   □   (b) is unsafe   □
   (c) gives no indication of safety   □   (d) none of the above.   □

230. The plastic design of steel structures which is based on the upper bound theorem of plastic analysis
   (a) is safe   □   (b) is unsafe   □
   (c) gives no indication of safety   □   (d) none of the above.   □

231. Rolled steel beams are classified into how many series as per Indian standards?
   (a) 2   □   (b) 3   □
   (c) 5   □   (d) 4.   □

232. Rolled steel beams are used to resist
   (a) bending stress   □
   (b) tensile stress in independent sections   □
   (c) compressive stress in independent sections   □
   (d) all the above.   □

233. The twisting of channel sections may be due to the assymetry with respect to the axis
   (a) perpendicular to the web   □   (b) parellel to the web   □
   (c) parallel to flanges   □   (d) perpendicular to flanges.   □

234. Increase in percentage carbon content in steel causes decrease in
   (a) strength   □   (b) hardness   □
   (c) ductility   □   (d) brittleness.   □

235. If V is the velocity of wind and K is a constant, the wind pressure p is given by
    (a) $p = KV^2$
    (b) $p = \dfrac{K}{V^2}$
    (c) $p = KV$
    (d) $p = \dfrac{K}{V}$.

236. Permissible stress may also be known as
    (a) working stress
    (b) yield stress
    (c) ultimate stress
    (d) limit stress.

237. For structural steel, the Poissons ratio lies between
    (a) $\frac{1}{4}$ and $\frac{1}{3}$
    (b) $\frac{1}{6}$ and $\frac{1}{5}$
    (c) $\frac{1}{5}$ and $\frac{1}{4}$
    (d) $\frac{1}{3}$ and $\frac{1}{2}$.

238. Which of the following is not a classification of Indian standard angle sections
    (a) equal angles
    (b) unequal angles
    (c) sharp angles
    (d) bulb angles.

239. Working stress is obtained when......stress is divided by the factor of safety
    (a) breaking stress
    (b) yield stress
    (c) ultimate stress
    (d) none of the above.

240. An imaginary line joining the locations of the rivets is known as
    (a) rivet line
    (b) scrieve line
    (c) gauge line
    (d) all the above.

241. When one member is placed over the other and the two are joined by two rows of rivets, then the joint is known as
    (a) single riveted lap joint
    (b) single riveted butt joint
    (c) double riveted lap joint
    (d) double riveted butt joint.

242. If N is the numbers of rivets, d is the gross diameter of the rivet, τ_{vf} is the maximum permissible shear stress in the rivet, t is the thickness of the member and p is the pitch of the rivets then the strength of the joint against the single shearing of rivets is given by
    (a) $P_s = N \times \dfrac{\pi}{4} d^2 \times \tau_{vf}$
    (b) $P_s = N \times \dfrac{\pi}{2} d^2 \times \tau_{vf}$
    (c) $P_s = N \times \pi dt \times \tau_{vf}$
    (d) $P_s = N \times 2\pi dt \times \tau_{vf}$.

243. The *rivet value R* is taken as
    (a) the strength of the rivet in shearing
    (b) the strength of the rivet in bearing
    (c) the lesser of (a) and (b)
    (d) the larger of (a) and (b).

244. The most economical section for a column is
    (a) square
    (b) circular
    (c) channel
    (d) tubular.

245. When a column is laterally supported on all sides throughout its length, its effective length is equal to
    (a) L
    (b) v
    (c) $\dfrac{L}{2}$
    (d) $\dfrac{L}{4}$.

246. If the length between inner end rivets is s, and t is the thickness of double flat lacing, then t should not be less than

(a) $\dfrac{s}{60}$
(b) $\dfrac{s}{50}$
(c) $\dfrac{s}{40}$
(d) $\dfrac{s}{30}$.

247. A column is subjected to an axial load W in addition to an eccentric load P_E with an eccentricity of e_x. If A and Z_{xx} are the area and the section modulus of the column, the equivalent load is given by

(a) $P_{eq} = P_E\left(1 - \dfrac{Ae_x}{Z_{xx}}\right) + W$
(b) $P_{eq} = P_E\left(1 - \dfrac{Z_{xx}}{A \cdot e_x}\right) + W$
(c) $P_{eq} = P_E\left(1 + \dfrac{Ae_x}{Z_{xx}}\right) + W$
(d) $P_{eq} = P_E\left(1 + \dfrac{Z_{xx}}{A \cdot e_x}\right) + W$.

248. In the above problem if the eccentric load is also having an eccentricity of e_y along the y-axis, and Z_{yy} is the section modulus about y-axis, then the equivalent load is given by

(a) $P_{eq} = P_E\left(1 - \dfrac{Ae_x}{Z_{xx}} - \dfrac{Ae_y}{Z_{yy}}\right) + W$
(b) $P_{eq} = P_E\left(1 + \dfrac{Ae_x}{Z_{xx}} + \dfrac{Ae_y}{Z_{yy}}\right) + W$
(c) $P_{eq} = P_E\left(1 - \dfrac{Z_{xx}}{A \cdot e_x} - \dfrac{Z_{yy}}{A \cdot e_y}\right) - W$
(d) $P_{eq} = P_E\left(1 + \dfrac{Z_{xx}}{A \cdot e_x} + \dfrac{Z_{yy}}{A \cdot e_y}\right) - W$.

249. A column splice is used to increase
(a) the length of the column
(b) the strength of the column
(c) the rigidity of the column
(d) the cross-sectional area of the column.

250. The distance between centroid of compression flange and centroid of tension flange of a plate girder is known as
(a) clear depth
(b) effective depth
(c) overall depth
(d) moment depth.

251. The distance between the outer faces of flanges of a plate girder is known as
(a) clear depth
(b) effective depth
(c) overall depth
(d) moment depth.

252. The steel beams are embedded in concrete for the purpose of
(a) making the building fire resistant
(b) architectural requirements
(c) both (a) and (b)
(d) none of the above.

253. The maximum depth of an encased beam shall not exceed
(a) 450 mm
(b) 600 mm
(c) 750 mm
(d) 900 mm.

254. The purpose of stiffeners in a plate girder is to
(a) prevent buckling of web plate
(b) increase the moment carrying capacity of the girder
(c) reduce the shear stress
(d) take care of bearing stress.

255. In a riveted joint, the shear failure of plates may occur when
    (a) edge distance is inadequate   (b) shear strength of rivets is inadequate
    (c) bearing strength of rivets is inadequate   (d) none of the above.

256. In the design of Gantry girder, the weight of the crab is taken as
    (a) $\frac{1}{5}$ of lift load + 100 kg   (b) $\frac{1}{5}$ of lift load + 500 kg
    (c) $\frac{1}{4}$ of lift load + 400 kg   (d) $\frac{1}{4}$ of lift load + 100 kg.

257. If a member is likely to be subjected to torsion, which of the following sections is preferable?
    (a) angle   (b) channel
    (c) box   (d) none of the above.

258. The economical depth of a plate girder subjected to moment M, with thickness of web plate t_w, and allowable bending stress σ_b is given by
    (a) $1.1\sqrt{\dfrac{M}{\sigma_b \cdot t_w}}$ (b) $2.2\sqrt{\dfrac{M}{\sigma_b \cdot t_w}}$
    (c) $0.55\sqrt{\dfrac{M}{\sigma_b \cdot t_w}}$ (d) $3.3\sqrt{\dfrac{M}{\sigma_b \cdot t_w}}$.

259. The pressure inside a water main is 12 kg/cm². The diameter of the main is 1.0 m. What is the minimum thickness of the pipe if the pipe material should not be stressed beyond 300 kg/cm²
    (a) 0.5 cm   (b) 1.0 cm
    (c) 1.5 cm   (d) 2.0 cm.

260. When the depth of plate girder is less than 750 mm, it is called
    (a) deep plate girder   (b) shallow plate girder
    (c) economical plate girder   (d) box girder.

261. The size of a rivet is identified by
    (a) diameter of head   (b) shape of head
    (c) diameter of shank   (d) length of shank.

262. Which of the following is a strut
    (a) tension member   (b) flexural member
    (c) torsion member   (d) compression member.

263. A round bar with a threaded portion is subjected to axial tensile load. The stress in the bar will be calculated using
    (a) gross area
    (b) area based on root diameter of threads
    (c) area based on average diameter of threads
    (d) none of the above.

264. The inclination of a lacing bar with the axis of the compression member is θ. Then θ shall not be less than
    (a) 30°   (b) 40°
    (c) 50°   (d) 70°.

265. In the above problem, θ shall not be more than
    (a) 30° (b) 40°
    (c) 50° (d) 70°.

266. Which of the following conditions leads to the development of a plastic hinge in a beam
    (a) when all the fibres of the section have yielded
    (b) when the extreme fibre reaches the permissible stress
    (c) when the extreme fibre reaches the yield point
    (d) none of the above.

267. A rectangular beam has a width b and depth d. Its plastic modulus is
    (a) $\dfrac{bd^2}{12}$ (b) $\dfrac{bd^2}{8}$
    (c) $\dfrac{bd^2}{6}$ (d) $\dfrac{bd^2}{4}$.

268. The section modulus and the plastic modulus of a section are Z and S respectively. Then its shape factor is given by
    (a) $\dfrac{Z}{S}$ (b) $\dfrac{S}{Z}$
    (c) $\dfrac{S-Z}{S}$ (d) $\dfrac{S-Z}{Z}$.

269. The approximate value of the shape factor of an I section is equal to
    (a) 2.15 (b) 1.75
    (c) 1.15 (d) 0.75.

270. In the case of an angle section, if t is the thickness of the angle, the area of the leg shall be computed as
    (a) $\left(\text{length of leg} - \dfrac{t}{2}\right)t$ (b) length of leg × t
    (c) (length of leg − t) t (d) 0.9 × length of leg × t.

271. Increase of carbon percentage in steel decreases its
    (a) strength (b) hardness
    (c) brittleness (d) ductility.

272. The assumption involved in the design of rivet joints
    (a) the load is uniformly distributed among all rivets
    (b) shear stress in rivets is uniformly distributed over gross area
    (c) bonding stress in rivets is neglected
    (d) all the above.

273. If d is the diameter of the rivet, then according to IS : 800 the maximum grip length of rivet is equal to
    (a) 4d (b) 6d
    (c) 8d (d) 10d.

274. Which of the following operations cannot be done easily on mild steel?
    (a) drilling □ (b) hardening □
    (c) punching □ (d) machining. □
275. The percentage of carbon in structural steel is
    (a) 0.2 to 0.27 □ (b) 0.7 to 0.83 □
    (c) 1.7 to 1.83 □ (d) 1.2 to 1.27. □
276. The head of the rivet is made by
    (a) machining □ (b) hot or cold forging □
    (c) welding □ (d) pressing. □
277. The percentage of sulphur in structural steel shall not exceed
    (a) 0.75 □ (b) 0.5 □
    (c) 0.055 □ (d) 0.25. □
278. Which information is not included in the design codes?
    (a) guidance about the loads □ (b) design principles □
    (c) quality of the materials □ (d) allowable stresses. □
279. When the pitch adopted in the design exceeds the minimum pitch to be maintained, additional rivets are used. They are known as
    (a) auxiliary rivets □ (b) tacking rivets □
    (c) packing rivets □ (d) booster rivets. □
280. The size of the rivet is denoted by
    (a) diameter of the cap □ (b) diameter of shank □
    (c) length of shank □ (d) radius of shank. □
281. The effective throat thickness in the case of incomplete penetration of butt weld is taken as
    (a) 7/8th of the thickness of the thinner part joined □
    (b) 7/8th of the thickness of the thicker part joined □
    (c) 5/7th of the thickness of the thicker part joined □
    (d) 5/7th of the thickness of the thinner part joined □
282. A completely penetrating butt weld is specified by
    (a) leg length □ (b) plate thickness □
    (c) effective throat thickness □ (d) penetration thickness. □
283. Intermediate butt welds are used to resist
    (a) shear stresses □ (b) dynamic stresses □
    (c) alternate stresses □ (d) all these. □
284. A structural member subjected to tensile force in a direction parallel to its longitudinal axis is called
    (a) tension member □ (b) tie member □
    (c) tie □ (d) any one of these. □
285. Pick the correct statement from the following :
    (a) tension member in the roof trusses is generally called tie □
    (b) when a tie is subjected to axial tensile force, the distribution of stress will be uniform over the cross-sectional area □
    (c) a member subjected to only axial tension is efficient and economical □
    (d) all of the above. □

286. A steel wire when used as a tie, requires
    (a) no prestressing
    (b) nominal prestressing
    (c) pretensioning to its full capacity
    (d) prestressing to half its capacity.
287. Which of the following is not a tension member?
    (a) cable
    (b) bar
    (c) tie
    (d) boom.
288. If the gauge distance is ...... pitch of the rivets in a tension member the failure will occur in zig-zag line.
    (a) equal to
    (b) less than
    (c) more than
    (d) none of the above.
289. The single channel section used as a tension member has
    (a) low rigidity in the direction of web and high rigidity in the direction of flange
    (b) high rigidity in the direction of web and low rigidity in the direction of flange
    (c) equal rigidity both in the direction of web and flange
    (d) none of the above.
290. In a tension member, when one or more rivets are off the line the failure of plate depends upon
    (a) diameter of rivet hole
    (b) pitch of rivets
    (c) gauge of rivets
    (d) all of these.
291. A structural member subjected to compressive force in a direction parallel to its axis is known as
    (a) stanchion
    (b) post
    (c) end post
    (d) none of the above.
292. In overloaded compression members the failure may be due to
    (a) direct compression
    (b) excessive bending
    (c) bending combined with compression
    (d) any of the above.
293. A strut is a
    (a) compression member
    (b) tension member
    (c) flexural member
    (d) torsion member.
294. The axial load which keeps the column in a slight deflected shape is called
    (a) critical load
    (b) crippling load
    (c) buckling load
    (d) any of the above.
295. The formula adopted by Bureau of Indian Standards for the determination of allowable stress in axial compression is
    (a) Rankine-Gordon formula
    (b) Secant formula
    (c) Merchant-Rankine formula
    (d) Jhonson's formula.
296. For a column supported over its entire length the slenderness ratio will be
    (a) infinite
    (b) zero
    (c) very high
    (d) reasonably high.

297. As the slenderness ratio of a column increases, the allowable stress
   (a) increases
   (b) decreases
   (c) does not change
   (d) these two do not have bearing on each other.

298. The inclination of lacing bars with the longitudinal axis of the component member is usually between
   (a) 40° to 70°
   (b) 30° to 40°
   (c) 20° to 30°
   (d) 10° to 20°.

299. A joint in the length of a column is known as
   (a) longitudinal joint
   (b) load bearing joint
   (c) column splice
   (d) shear joint.

300. When the components of a built up column are connected by a lateral system, the reduction in buckling strength due to shear deflection is ...... that of solid built up columns
   (a) equal to
   (b) more than
   (c) less than
   (d) none of the above.

301. The battening is preferred when the
   (a) column is axially loaded
   (b) space between the two main components is not very large
   (c) both (a) and (b)
   (d) none of these.

302. The formula which takes any initial crookedness of the column and imperfectness of axial loading into account is
   (a) Perry-Robertson formula
   (b) Rankines formula
   (c) Euler's formula
   (d) Secant formula.

303. A column base is subjected to moment. If the intensity of bearing pressure due to axial load is equal to the stress due to the moment, then the bearing pressure between the base and the concrete is
   (a) zero at one end and compressive stress at the other end
   (b) tension at one end and compression at the other end
   (c) uniform compression throughout
   (d) uniform tension throughout.

304. What should be the shape of the base of a combined footing to support two columns with equal loads
   (a) circular
   (b) rectangular
   (c) trapezoidal
   (d) triangular.

305. The main beam is a beam which supports
   (a) floor construction
   (b) joists
   (c) secondary beams
   (d) none of these.

306. Any major beam in a structure is known as
   (a) girder
   (b) joist
   (c) king beam
   (d) main beam.

307. The beams supporting the stair steps are called
    (a) spandrel beams
    (b) rafters
    (c) trimmers
    (d) stringers.

308. A rolled *I* section provides
    (a) large moment of inertia about x-axis with lesser cross-sectional area
    (b) large moment of resistance as compared to other sections
    (c) greater lateral stability
    (d) all of the above.

309. When the load does not pass through the shear centre of the beam, it produces
    (a) torsional moment only
    (b) bending moment only
    (c) torsional and bending moments
    (d) none of the above.

310. Large deflection in beams are not desirable because they
    (a) lead to cracking of ceiling, plastering etc.
    (b) indicate lack of rigidity
    (c) result in poor drainage by forming ponding
    (d) all of the above.

311. If b = length of bearing plate and h = depth of root of fillet from the outer surface of the flange, the bearing length of web B under concentrated load is given by
    (a) $B = b + h\sqrt{3}$
    (b) $B = b - h\sqrt{3}$
    (c) $B = b + 2h\sqrt{3}$
    (d) $B = b - 2h\sqrt{3}$.

312. An out of plane web distortion is known as
    (a) web buckling
    (b) vertical web buckling
    (c) column buckling
    (d) all the above.

313. If h = clear depth of web between the root fillets and t_w is the thickness of the web, then the slenderness ratio for the portion of the web acting as a column when the two flanges are restrained against lateral displacement and rotation is given by
    (a) $\dfrac{h}{t_w}$
    (b) $\dfrac{\sqrt{3}h}{t_w}$
    (c) $\dfrac{\sqrt{5}h}{t_w}$
    (d) $\dfrac{\sqrt{6}h}{t_w}$.

314. The maximum bending moment in the gantry girder under moving loads occurs when its axis and the c.g. of loads are
    (a) at any of the extreme corners
    (b) at $L/4$ from any corner
    (c) at $L/3$ from any corner
    (d) at mid span.

315. If L is the span of a plate girder, then its depth is generally taken as
    (a) $\dfrac{L}{5}$ to $\dfrac{L}{8}$
    (b) $\dfrac{L}{8}$ to $\dfrac{L}{10}$
    (c) $\dfrac{L}{10}$ to $\dfrac{L}{12}$
    (d) $\dfrac{L}{12}$ to $\dfrac{L}{16}$.

316. When the depth of plate girder is less than 75 cm then it is known as
    (a) deep plate girder
    (b) shallow plate girder
    (c) economic plate girder
    (d) optimum plate girder.

317. The self weight of the plate girder in terms of its span L and the total superimposed load W is taken as
    (a) $\dfrac{WL}{100}$
    (b) $\dfrac{WL}{200}$
    (c) $\dfrac{WL}{300}$
    (d) $\dfrac{WL}{400}$.

318. If M is the maximum bending moment and σ_b is the allowable bending stress, the economical depth of web plate of a plate girder is approximately given by
    (a) $\left(\dfrac{M}{\sigma_b}\right)^{1/3}$
    (b) $2.5\left(\dfrac{M}{\sigma_b}\right)^{1/3}$
    (c) $3.5\left(\dfrac{M}{\sigma_b}\right)^{1/3}$
    (d) $4.5\left(\dfrac{M}{\sigma_b}\right)^{1/3}$.

319. A steel welded plate girder is subjected to a maximum bending moment of 1500 kNm. The permissible bending stress is 165/mm². The most economical depth of the plate girder would be
    (a) 600 mm
    (b) 800 mm
    (c) 1000 mm
    (d) 1200 mm.

320. Horizontal stiffeners are provided in plate girders to
    (a) increase the flexural strength of the girder
    (b) increase the shear capacity of the web
    (c) prevent the local buckling of the web
    (d) prevent the local buckling of the flange.

321. In a plate girder with depth of web = d, at least one horizontal stiffener shall be provided if the thickness of the web in less than
    (a) $\dfrac{d}{200}$
    (b) $\dfrac{d}{100}$
    (c) $\dfrac{d}{50}$
    (d) $\dfrac{d}{25}$.

322. In a plate girder with thickness of the web t_w, intermediate vertical stiffeners are essential if the depth of the web exceeds
    (a) $55\, t_w$
    (b) $85\, t_w$
    (c) $115\, t_w$
    (d) $145\, t_w$.

323. A welded steel plate girder consisting of two flange plates of size 350 mm × 16 mm and a web plate of 1000 mm × 6 mm requires
    (a) neither vertical nor horizontal stiffeners
    (b) no vertical stiffeners
    (c) intermediate vertical stiffeners
    (d) both horizontal and vertical stiffeners.

324. The most frequently used section in roof trusses is
    (a) two angles placed back to back (forming T)
    (b) two channels placed back to back
    (c) two channels placed at some distance
    (d) four angles.

325. The distance between two adjacent trusses is known as
    (a) span (b) pitch
    (c) bay (d) tray.

326. The ratio of plastic section modulus to the elastic section modulus is always
    (a) equal to 1 (b) less than 1
    (c) more than 1 (d) in the range 0.75 to 1.25.

327. Which section will have more torsional resistance?
    (a) Angle (b) channel
    (c) tubular (d) I.

328. The most commonly adopted pitch for roof trusses subject to snow loads and heavy winds is
    (a) $\frac{1}{3}$ (b) $\frac{1}{4}$
    (c) $\frac{1}{5}$ (d) $\frac{1}{6}$.

329. The horizontal beams spanning between two adjacent roof trusses are known as
    (a) stringers (b) rafters
    (c) trimmers (d) spandrils.

330. The moment-curvature relation at a plastic hinge is
    (a) linear (b) parabolic
    (c) constant moment for all curvatures (d) same curvature for all moments.

ANSWERS
Objective Types Questions

1. (a)   2. (c)   3. (d)   4. (b)   5. (a)   6. (b)
7. (c)   8. (a)   9. (b)   10. (c) 11. (b) 12. (b)
13. (d) 14. (c) 15. (b) 16. (c) 17. (c) 18. (a)
19. (b) 20. (a) 21. (c) 22. (a) 23. (b) 24. (c)
25. (a) 26. (c) 27. (b) 28. (d) 29. (d) 30. (a)
31. (b) 32. (b) 33. (d) 34. (c) 35. (a) 36. (d)
37. (c) 38. (d) 39. (b) 40. (a) 41. (a) 42. (c)
43. (d) 44. (b) 45. (a) 46. (d) 47. (c) 48. (b)
49. (d) 50. (a) 51. (a) 52. (c) 53. (a) 54. (d)
55. (b) 56. (c) 57. (b) 58. (b) 59. (c) 60. (c)
61. (c) 62. (b) 63. (b) 64. (a) 65. (d) 66. (d)
67. (c) 68. (d) 69. (c) 70. (b) 71. (d) 72. (a)
73. (b) 74. (d) 75. (c) 76. (b) 77. (c) 78. (d)
79. (a) 80. (c) 81. (d) 82. (a) 83. (b) 84. (c)
85. (d) 86. (a) 87. (b) 88. (c) 89. (d) 90. (d)
91. (a) 92. (b) 93. (c) 94. (a) 95. (d) 96. (d)
97. (b) 98. (c) 99. (b) 100. (a) 101. (b) 102. (c)

103. (d)	104. (a)	105. (d)	106. (c)	107. (b)	108. (a)
109. (a)	110. (b)	111. (d)	112. (c)	113. (c)	114. (d)
115. (d)	116. (a)	117. (b)	118. (c)	119. (d)	120. (b)
121. (c)	122. (a)	123. (d)	124. (d)	125. (b)	126. (c)
127. (d)	128. (b)	129. (a)	130. (c)	131. (b)	132. (a)
133. (c)	134. (d)	135. (a)	136. (b)	137. (b)	138. (d)
139. (c)	140. (b)	141. (b)	142. (c)	143. (b)	144. (c)
145. (d)	146. (a)	147. (b)	148. (a)	149. (c)	150. (d)
151. (a)	152. (c)	153. (a)	154. (c)	155. (c)	156. (b)
157. (d)	158. (c)	159. (a)	160. (d)	161. (b)	162. (b)
163. (b)	164. (d)	165. (c)	166. (c)	167. (c)	168. (d)
169. (a)	170. (a)	171. (b)	172. (b)	173. (c)	174. (b)
175. (d)	176. (d)	177. (c)	178. (c)	179. (a)	180. (c)
181. (b)	182. (c)	183. (d)	184. (b)	185. (b)	186. (d)
187. (d)	188. (c)	189. (a)	190. (b)	191. (c)	192. (d)
193. (b)	194. (d)	195. (c)	196. (a)	197. (c)	198. (d)
199. (b)	200. (a)	201. (a)	202. (c)	203. (b)	204. (d)
205. (b)	206. (d)	207. (a)	208. (d)	209. (a)	210. (c)
211. (d)	212. (b)	213. (a)	214. (a)	215. (b)	216. (d)
217. (c)	218. (d)	219. (a)	220. (c)	221. (b)	222. (a)
223. (a)	224. (b)	225. (a)	226. (b)	227. (b)	228. (a)
229. (a)	230. (b)	231. (c)	232. (d)	233. (b)	234. (c)
235. (a)	236. (a)	237. (a)	238. (c)	239. (b)	240. (d)
241. (c)	242. (a)	243. (c)	244. (d)	245. (b)	246. (a)
247. (c)	248. (b)	249. (a)	250. (b)	251. (c)	252. (c)
253. (c)	254. (a)	255. (a)	256. (b)	257. (c)	258. (a)
259. (d)	260. (b)	261. (c)	262. (d)	263. (b)	264. (b)
265. (d)	266. (a)	267. (d)	268. (b)	269. (c)	270. (a)
271. (d)	272. (d)	273. (c)	274. (b)	275. (a)	276. (b)
277. (c)	278. (c)	279. (b)	280. (b)	281. (a)	282. (c)
283. (a)	284. (d)	285. (d)	286. (b)	287. (d)	288. (c)
289. (b)	290. (d)	291. (d)	292. (d)	293. (a)	294. (d)
295. (c)	296. (b)	297. (b)	298. (a)	299. (c)	300. (b)
301. (c)	302. (d)	303. (a)	304. (b)	305. (c)	306. (a)
307. (d)	308. (d)	309. (c)	310. (d)	311. (c)	312. (d)
313. (b)	314. (d)	315. (c)	316. (b)	317. (c)	318. (d)
319. (c)	320. (c)	321. (a)	322. (b)	323. (c)	324. (a)
325. (c)	326. (b)	327. (c)	328. (a)	329. (b)	330. (c).

Chapter 16 SURVEYING

I. INTRODUCTION

Surveying is the art of establishing relative positions of stations both in horizontal and vertical directions.

Levelling is the art of establishment of relative elevations of stations.

The fundamental principle of surveying is to

1. Proceed from whole to part.

2. Establish the position of a station from two independent (linear/and/or angular) measurements.

The earth is a sphere of a diameter of 12742 km. It is taken into account in Geodetic surveying.

In plane surveying we ignore the curvature of earth and take it to be plane.

CHAIN SURVEY

Engineer's Chain	—F.P.S. Chain—100 feet long—divided into 100 links.
Gunter's Chain	—F.P.S. Chain—66 feet long—divided into 100 links.
Revenue Chain	—F.P.S. Chain—33 feet long—divided into 16 links.
Metric Chain	—20 metres—divided into 100 links or 30 metres—divided into 150 links.
"Main Survey Stations"	—Prominent stations.
Main Survey lines	—Chain lines joining these prominent stations.

Number of main survey lines should be as few as possible.

Three main survey lines form a triangle.

As far as possible these triangles should be nearly equilateral triangles.

A well conditioned triangle is one in which no angle is less than 30° nor greater than 120°.

Base line is a prominent line passing through the heart of the area.

Check lines run in between the main lines to test their accurate positions. Each triangle should have at least one check line.

A tie line is a subsidiary line mainly run to get more details within the area.

Offset is the measurement in lateral direction while chaining is done in linear direction.

Perpendicular offsets run at right angles to the chain line.

Oblique offsets make an angle other than 90° with chain line.

Offsets should be as short as possible.

An offset is said to be a long offset when its length is more than 15 m.

Plotting can be done accurately to a minimum length of 0.25 mm.

Limiting length of offset = $\dfrac{0.25}{1000} \times \dfrac{n}{\sin \theta}$ metres

where $\dfrac{1}{n}$ is the scale to which plotting is done.

θ is the deviation of the perpendicular offset in degrees.

Cross staff, optical square and prism square are the instruments to set perpendicular offsets.

Obstacles

1. Obstacles to ranging but not to chaining *e.g.*, a hillock. Reciprocal ranging is adopted then.
2. Obstacles to chaining but not to ranging *e.g.*, a river and a pond.
3. Obstacles to both chaining and ranging *e.g.*, a tall building.

Minor instrument used for ranging is "Line Ranger".

Minor instruments used for setting perpendicular offsets are "Optical square" and "Prism square".

True length = Measured length $\times \dfrac{\text{Incorrect length of chain}}{\text{Supposed length of chain}}$

Similarly,

Correct area = Measured area $\times \left[\dfrac{\text{Incorrect length of chain}}{\text{Supposed length of chain}}\right]^2$

'Planimeter' is an instrument to measure areas.

Area = $M[\text{F.R.} - \text{I.R.} \pm 10n + C]$

I.R. = Initial reading

F.R. = Final reading

n = Number of times the zero mark of the dail passes the fixed index mark + clockwise
− anti-clockwise

C = Additive constant

M = Multiplying constant.

ERRORS IN CHAIN SURVEY

I. Cumulative Errors

1. Length of the chain is shorter than the standard one due to (*i*) kinks, (*ii*) loss of links and (*iii*) knots in links.
2. Slope correction (when > 4°) is not applied.
3. Ranging is not in a straight line.
4. Length of the chain may be more than the standard one due to (*i*) flattening of ring joints, and (*ii*) opening of rings.

II. Compensating Errors

1. Incorrect holding of chain.
2. Chain not uniformly calibrated.

If the total length measured = L, then

$$\text{Cumulative errors} \propto L$$

$$\text{Compensating errors} \propto \sqrt{L}$$

Correction for temperature, $\quad C_t = \alpha(T_m - T_0)L$

Correction for pull, $\quad C_v = \dfrac{(P - P_0)L}{AE}$

Correction for sag, $\quad C_s = \dfrac{l(wl)^2}{24P^2}$

Correction for slope, $\quad C_{sl} = \dfrac{h^2}{2l}$.

COMPASS SURVEY

Bearing of a line is its direction with respect to a reference line *i.e.*, meridian.

True meridian is the line passing through North pole and South pole.

Magnetic meridian is the standard reference line shown by a freely floating balanced magnetic needle free from any external influence.

Arbitrary meridian is any other line taken for reference.

Whole circle bearings are always measured in the clockwise direction with reference to the Magnetic North.

Quadrantal or reduced bearings are taken with respect to Magnetic North or Magnetic South towards East or West *i.e.*, they are measured clockwise in North-East and South-West Quadrants and anti-clockwise in South-East and North-West Quadrants.

Back bearing is the bearing of a line in the opposite direction.

Deflection angle is the included angle between the prolongation of the previous line and the succeeding line.

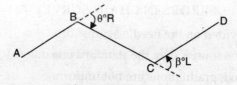

FIGURE 16.1

Clockwise deflection is called right deflection and it is additive.

Anticlockwise deflection is left deflection and is subtractive.

In a closed traverse

Sum of internal angles $= (2n - 4) \times 90°$

Sum of external angles $= (2n + 4) \times 90°$

Algebraic sum of deflection angles $= 360°$.

Dip is the inclination of the magnetic needle with the horizontal.

Northern end is deflected down in the 'Northern hemisphere' while the Southern end is deflected down in the 'Southern hemisphere'. At the equator the dip is zero.

Magnetic declination = True bearing – Magnetic bearing.

When the magnetic meridian is to the East of true meridian, true bearing of a line is greater than its magnetic bearing and therefore magnetic declination is +ve and designated as 'E'.

When the magnetic meridian is to the West of true meridian, true bearing of a line is smaller than its magnetic bearing and therefore magnetic declination is –ve and designated as 'W'.

Magnetic declination at a place is not constant but varies from time to time.

1. *Diurnal variation*—within the same day—more during day time and less during night time.

2. *Annual variation*—within a year—more during summer and less in winter.

3. *Secular variation*—variation over a very long period.

The variation is more near poles and less on the equator.

Isogonic lines are the imaginary lines passing through the stations of same Magnetic Declination.

Agonic line is an Isogonic line of zero magnetic declination.

"Local attraction" at a place is the influence of magnetic materials as steel and nickel objects, iron ore, electric poles and current carrying conductors on the magnetic needle.

Local attraction is the same for all the bearings taken at the place.

ERRORS IN COMPASS SURVEY

I. Instrumental Errors

1. Needle—bent.
2. Pivot—bent, eccentric.
3. Pivot edge—blunt.

4. Needle—sluggish.

5. No counter weight provided on the needle against dip.

6. Sight vanes—bent.

7. Graduated ring—twisted, graduations are not uniform.

8. Sight vanes and pivot—not in one line.

9. Horse hair—too thick.

II. Personal Errors

1. Inaccurate centring of the compass over the station.

2. Improper levelling of aluminium ring.

3. Imperfect bisection of ranging rod.

4. Confused reading in the wrong direction.

5. Careless recording.

Table 16.1

Item	Prismatic compass	Surveyor's compass
Bearing	W.C.B. 0° at South 90° at West 180° at North 270° at East **FIGURE 16.2**	R.B. 0° at North and South 90° at East and West **FIGURE 16.3** East and West interchanged
Graduations	Inverted because we have to see them through prism	Erect
Needle	Broad type—fitted to the bottom of aluminium ring (∴ cannot be seen)	Edge bar type needle—also acts as an index
Scale	Free to float along with the broad type magnet	Attached to the box
Sighting at object and taking bearings	Can be done simultaneously	Sighting is to be done first and then the surveyor has to read the Northern end of the needle
Tripod	Not essential	A must.

PLANE TABLE SURVEY

Field work and plotting are done simultaneously in a plane table survey.

1. Drawing board—400 mm × 300 mm or 750 mm × 600 mm.
2. Alidade—approximately 500 mm long, plain alidade—2 sight vanes, telescopic alidade—telescope mounted on scale.
3. Plumbing fork and plumb bob—to transfer station on to the drawing sheet or *vice-versa*.
4. Trough compass—to mark 'North'.
5. Spirit level—for levelling the board.

Orientation
1. By magnetic needle
2. By back sighting.

Methods of Plane Tabling
1. Radiation
2. Intersection
3. Traversing
4. Resection.

Radiation. From a well commanded single station the details are plotted.

Best suited for smaller lengths and when all the points are accessible. It is very accurate.

Intersection. Plotting a base line *AB*, *A* and *B* being mutually visible and accessible. Draw rays from the stations to all the other points, where these rays meet give the positions of all the other points.

It involves measurement of only one length *i.e.*, that of the base line. Even if the other points are accessible it does not matter.

This is less accurate than radiation.

Traversing. It is locating the plane table almost over all the stations. It may be regarding as a combination of both radiation and intersection.

Resection. It is the location of the instrument station with reference to "two" or "three" already plotted stations.

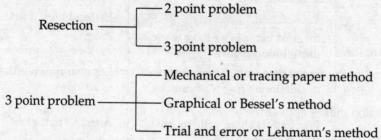

Advantages of Plane Table Survey
1. It is rapid.
2. Area to be surveyed and plotted is in front of the surveyor and hence less possibility exists to miss some details.
3. It is advantageously taken up in magnetic areas.

Disadvantages

1. Heavy, cumbersome and awkward to carry.
2. It has too many accessories.
3. It is difficult to redraw the plotted drawing to a different scale.

Suitability. It is best suited to prepare small scale maps of smaller areas where high precision is not required.

LEVELLING

It is the process of determining relative elevations of places with respect to a datum line.

A 'level surface' is a surface parallel to the mean spheroidal surface of earth.

Mean Sea Level at Karachi was taken as datum by the Great Trigonometrical Survey of India.

Bench Mark is a permanent mark of known elevation.

Height of the instrument is the elevation of the line of sight of the instrument.

Reduction of Levels

1. *Collimation system.* It is rapid but had no check over R.Ls of intermediate sights. It is adopted where more number of "intermediate sights" exist.
2. *Rise and fall method.* It is a slow process but absolute check exists over the computed R.L.s of intermediate sight. It is preferred where less number of intermediate stations do exist.

Curvature. Curvature makes the objects appear 'Lower' than they really are

AB is a horizontal line

AC is a level line.

$$AB^2 = OB^2 - OA^2$$
$$= (OC + BC)^2 - OA^2 = (R + C_c)^2 - R^2$$
$$\simeq 2RC_c \qquad [C_c^2 \text{ is too small}]$$

$$\therefore \quad C_c = \frac{d^2}{2R} = \frac{d^2}{12742} \text{ km} = 0.0785 \, d^2 \text{ metres.}$$

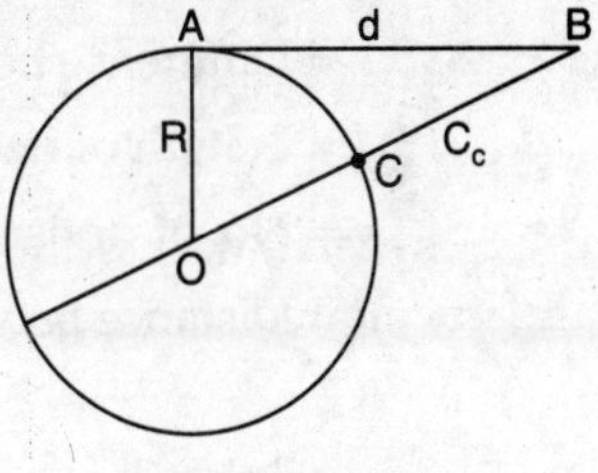

FIGURE 16.4

Refraction. Rays of light passing through different layers of air are refracted down.

The curved path of refracted rays forms an arc of a circle of radius approximately seven times that of the earth.

∴ Refraction makes appear the objects 'Higher' than they really are

$$C_R = 0.0112 \, d^2$$

∴ Combined correction for curvature and refraction $\Big\}$ $C = 0.0673 \, d^2$

Distance of Visible Horizon:

$$d = \sqrt{\frac{h}{0.0673}} = 3.853 \sqrt{h}$$

Reciprocal Levelling. It is a very precise levelling of finding R.L.s between two stations which are well apart by only two settings of levels, one nearer to each station.

Errors due to 'Collimation, curvature and partly due to refraction' are eliminated.

Errors in Levelling

I. Instrumental Errors

1. Collimation error—line of collimation not parallel to the bubble line. 2 peg test is conducted to set it right.
2. Object glass moving in inclined direction while focussing.
3. Sluggish bubble.

II. Personal Errors

1. Improper levelling of telescope.
2. Improper holding of level staff.
3. Imperfect sighting.
4. Settlement of level staff and level.

III. Natural Errors *due to Wind and Sun*

Sensitiveness of Bubble Tube

$$R = \frac{nld}{s}$$

R = Radius of curvature of bubble tube
s = Difference between two staff readings
l = Length of one division of bubble tube
n = No. of divisions
d = Distance between staff and instrument.

CONTOURING

A Contour line is an imaginary line joining the stations of equal elevation.

Contour Interval is the vertical distance between any two Consecutive Contours.

Horizontal Equivalent is the horizontal distance between any two points on two Consecutive Contours.

Characteristics of Contours

1. Contour is a closed line, within the map or outside it, *i.e.*, no Contour shall abruptly end.
2. No two Contours Cross each other. They seem to merge in case of a vertical retaining wall.

    They may appear to cross each other in case of an overhanging cliff.
3. Increased values inside a loop represent a hill. Reduced values inside represent a pond.

4. Contour lines close to each other represent steep slope.

5. U shaped Contours with falling values towards the bend represent Ridge line.

6. V shaped Contours with rising values towards the bend indicate Valley line.

7. Ridge or Valley lines cross Contours at right angles.

Uses of Contours

1. By drawing a section across Contours, we get the profile of land.

2. One can assess the intervisibility between two points.

3. By tracing Contour gradients we align roads, rail roads, canals and pipe lines.

4. One can assess catchment on drainage area of a river.

5. Capacity of a reservoir can be computed.

THEODOLITE SURVEY

Fundamental axes of theodolite:

1. Vertical axis

2. Horizontal or trunnion axis

3. Line of collimation

4. Axis of telescope

5. Axis of plate levels

6. Axis of altitude levels.

Relationship between the Axes

1. Axis of plate levels is perpendicular to the vertical axis.

2. Line of collimation is perpendicular to horizontal axis.

3. Horizontal axis is perpendicular to vertical axis.

4. Axis of altitude levels is parallel to line of collimation.

5. Vertical circle reads zero when line of collimation is horizontal.

Latitude.   '$l \cos \theta$' $\begin{cases} \text{Northing +ve} \\ \text{Southing –ve} \end{cases}$

Departure.   '$l \sin \theta$' $\begin{cases} \text{Easting +ve} \\ \text{Westing –ve.} \end{cases}$

Bowditch Rule:

$$\text{Correction to any side} = \text{Total error} \times \frac{\text{Length of the side}}{\text{Perimeter of transvere}}$$

Transit Rule:

$$\text{Correction to latitude of any side} = \text{Total error in latitude} \times \frac{\text{Latitude of that side}}{\text{Arithmetic sum of all latitudes}}$$

Similarly, correction for departure.

TACHEOMETRIC SURVEYING

It is angular surveying in which horizontal and vertical distances are computed without direct measurement.

Stadia System
1. Fixed hair method
2. Movable hair or Subtense method.

Fixed Hair Method

Horizontal distance $\quad D = \dfrac{f}{i} s + (f + d)$

For inclined sights—Staff held *vertical*

Vertical distance $\quad V = \left[\dfrac{f}{i} s \cos\theta\, (f+d)\right] \sin\theta$

Horizontal distance $\quad D = \left[\dfrac{f}{i} s \cos\theta\, (f+d)\right] \cos\theta$

Staff held *normal* $\quad V = \left[\dfrac{f}{i} \times s + (f+d)\right] \sin\theta$

$$D = \left[\dfrac{f}{i} \times s + (f+d)\right] \cos\theta + h \sin\theta.$$

Anallactic lens is provided in a tacheometer between the diaphragm and the object glass to nullify the additive constant. It is a convex lens.

The distance between the anallactic lens and object glass is given by $= f' + \dfrac{fd}{f+d}$

where f = Focal length of objective, f' = Focal length of the anallatic lens.

Movable Hair Method

$$D = \dfrac{f}{mp} s + (f+d)$$

For inclined sights $\quad D = \dfrac{Ks}{m-e} \cos^2\theta + C \cos\theta$

$$V = \dfrac{Ks}{m-e} \cos\theta \sin\theta + C \sin\theta$$

$K = \dfrac{f}{p}$ constant for an instrument

$C = f + d$ = additive constant

e = index error.

Horizontal Base Subtense Measurement:

Horizontal distance, $d = \dfrac{s \times 206265}{\beta}$

where β = Horizontal angle subtended by the theodolite in seconds between the two ends of the subtense bar.

Tangential Method:

$$D = \dfrac{s}{\tan \beta - \tan \alpha}$$

$$V = \dfrac{s}{\tan \beta - \tan \alpha} \times \tan \beta.$$

CURVES

For a 20 m arc, $\qquad R = \dfrac{1146}{D}$ m

For a 30 m arc, $\qquad R = \dfrac{1718.87}{D}$ m, where D = Degree of curve

ELEMENTS OF A SIMPLE CIRCULAR CURVE

V = Vertex or Point of Intersection it is the starting point. Indicating the necessity of the curve.

ϕ = Deflection angle.

It is Right deflection when the turning is Clockwise and Left deflection when the turning is Anti-clockwise.

θ = Angle of Intersection.

$\theta + \phi = 180°$

T_1 = Beginning of the curve.

= Point of curve (P.C.)

M = Mid-point of curve

= Apex or summit

T_2 = End of the curve

= Point of Tangency (P.T.)

VT_1 = Back tangent

$= R \tan \left(\dfrac{\phi}{2}\right)$

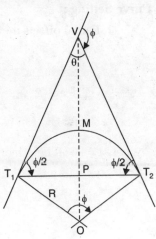

FIGURE 16.5

Length of curve = Arc $T_1 M T_2$

$$= \frac{\phi}{360°} \times 2\pi R$$

Chainage of T_1 = Chainage of vertex – Length of back tangent

Chainage of T_2 = Chainage of T_1 + Length of curve $T_1 M T_2$

VT_2 = Forward tangent = $R \tan\left(\dfrac{\phi}{2}\right)$ No where it is significant.

$T_1 P T_2$ = Long chord = $2 R \sin\left(\dfrac{\phi}{2}\right)$

PM = Mid ordinate = $R - R \cos\left(\dfrac{\phi}{2}\right)$

MV = External distance = Apex distance
= Distance between Apex and Vertex

$= R \sec\left(\dfrac{\phi}{2}\right) - R$

Curve Setting :

1. Radial offsets method:

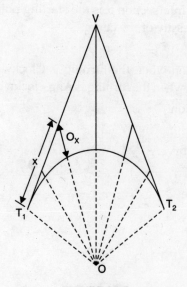

FIGURE 16.6

$$O_x = \frac{x^2}{2R}$$

2. Perpendicular offsets from tangents:

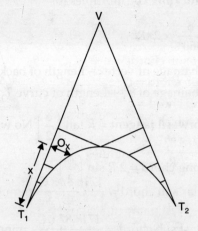

FIGURE 16.7

$$O_x = \frac{x^2}{2R}$$

3. Perpendicular offsets from the long chord:

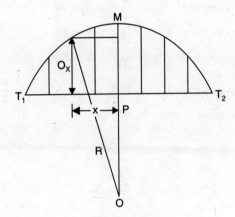

FIGURE 16.8

$$O_x = \sqrt{R^2 - x^2} - OP$$

4. Rankine's method:

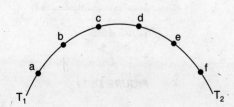

FIGURE 16.9

Chainages of a, b, c, d, e and f must be multiples (of whole numbers) of peg interval.

Peg interval = $ab = bc = cd = de = ef$
= regular sub chords = c
$T_1 a$ = First sub chord = c_1
fT_2 = Last sub chord = c_l

Deflection angle for first sub chord

$$= \delta_1 = \frac{1718.87 \times c_1}{R} \text{ minutes}$$

Diflection angle for regular sub chord = $\dfrac{1718.87 \times C}{R}$ minutes = δ

Deflection angle for the last sub chord = $\dfrac{1718.87 \times C_l}{R}$ minutes = δ_l

Cumulative deflection angle at 'a' = $\lfloor VT_1 a = \delta_1$
Cumulative deflection angle at 'b' = $\lfloor VT_1 b = \delta_1 + \delta$
Cumulative deflection angle at 'c' = $\lfloor VT_1 c = \delta_1 + 2\delta$
Cumulative deflection angle at 'd' = $\lfloor VT_1 d = \delta_1 + 3\delta$
Cumulative deflection angle at 'e' = $\lfloor VT_1 e = \delta_1 + 4\delta$
Cumulative deflection angle at 'f' = $\lfloor VT_1 f = \delta_1 + 5\delta$

Cumulative deflection angle at T_2 = $\lfloor VT_1 T_2 = \delta_1 + 5\delta + \delta_l = \dfrac{\phi}{2}$

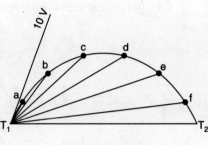

FIGURE 16.10

5. **Two Theodolites method :**

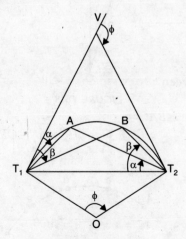

FIGURE 16.11

Angles α and β can be set anywhere from 0 to $\dfrac{\phi}{2}$.

Compound Curve

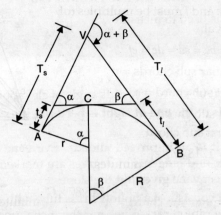

FIGURE 16.12

$$T_s = t_s + (t_s + t_l) \times \frac{\sin \beta}{\sin (\alpha + \beta)}$$

$$T_l = t_l + (t_s + t_l) \cdot \frac{\sin \alpha}{\sin (\alpha + \beta)}$$

Reverse Curve

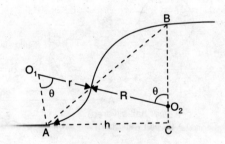

FIGURE 16.13

For a reverse curve between two parallel straights

$$\angle BAC = \frac{\theta}{2}$$

$$AB = 2(R + r) \sin \frac{\theta}{2} = 2(R + r) \times \frac{v}{AB}$$

$$\therefore \quad AB = \sqrt{2v(R + r)}$$

$$AC = h = (R + r) \sin \theta$$

$$BC = v = (R + r)(1 - \cos \theta).$$

MINOR INSTRUMENTS

Abney Clinometer. It is used to measure
1. Vertical angles.
2. Slopes of ground.
3. Grade contours.

Graduated circular arc (90° – 0° – 90°) reads angle of elevation or depression by the vernier.

When the vernier reads 0° the line of sight is perfectly horizontal, it is used as an "Abney Level".

Tangent Clinometer. It is an improved alidade. Eye vane has a peep hole. Object vane consists of a long slit. Degrees and tangents of degrees are marked on either side of the slit. The movable cross hair on the object vane gives out the slopes.

Ceylon Ghat Tracer. It consists of a hollow brass tube with a peep hole and cross hairs. A moving weight sliding on the pinion rack gives out the inclination of the line of sight—sighted towards the sight vane. It reads slope from $\frac{1}{120}$ to $\frac{1}{6}$ upwards or downwards.

Box Sextant

Horizontal and vertical angles can be found by box sextant.

Index glass is fully silvered.

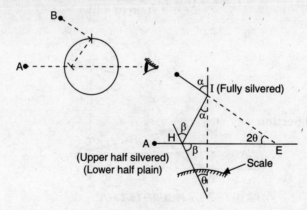

FIGURE 16.14

Horizon glass—upper half silvered, lower half plain.

The angle between the objects = 2 × angle between the mirrors.

It becomes an optical square when the vernier reads zero and the angle between the two mirrors is 45°.

Pantagraph

For reducing maps. Pointer at A, pencil at B.

For enlarging. Pointer at B, and pencil at A.

Weight should be in the same line as B and A.

D is the position of pencil to draw the plan to the same scale.

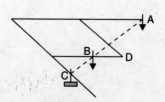

FIGURE 16.15

AREAS AND VOLUMES

1. Mid-Ordinate Rule

$$\text{Area} = \frac{O_1 + O_2 + O_3 + \ldots + O_n}{n} \times L$$

where O_1, O_2 etc., are the mid-ordinates of the trapezium formed.

2. Average Ordinate Rule

$$\text{Area} = \frac{O_0 + O_1 + \ldots + O_n}{n+1} \times L$$

where $O_0, O_1, \ldots$ are the ordinates of each trapezium.

3. Trapezoidal Rule

$$A = d\left[\frac{O_0 + O_n}{2} + O_1 \ldots + O_{n-1}\right].$$

4. Simpson's Rule

$$A = \frac{d}{3}\left[(O_0 + O_n) + 4(O_1 + O_3 + \ldots) + 2(O_2 + O_4 + \ldots)\right].$$

5. Level Section

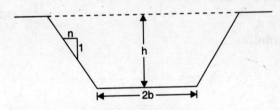

FIGURE 16.16

$$A = (2b + nh)h$$

6. Two Level Section

$$\text{Area in fill} = \frac{(b + rh)^2}{2(r - n)}$$

$$\text{Area in cutting} = \frac{(b + rh)^2}{2(r + n)}$$

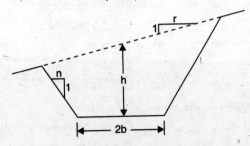

FIGURE 16.17

$$A = \frac{nb^2 \times nh^2 r^2 + 2bhr^2}{r^2 - n^2}$$

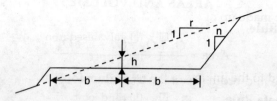

FIGURE 16.18

7. Three Level Section

$$A = \frac{b}{2}(h_1 + h_2) + \frac{1}{2}h(s_1 + s_2)$$

$$s_1 = b + \frac{nr_1}{n + r_1}\left(h - \frac{b}{r_1}\right)$$

$$s_2 = b + \frac{nr_2}{r_2 - n}\left(h - \frac{b}{r_2}\right)$$

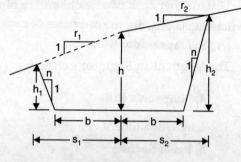

FIGURE 16.19

8. Trapezoidal Formula

$$\text{Volume} = d\left[\frac{A_1 + A_n}{2} + A_2 + \dots + A_{n-1}\right]$$

9. Prismoidal Formula

$$\text{Volume} = \frac{d}{3}[A_1 + A_n + 2\Sigma A_{odd} + 4\Sigma A_{even}].$$

II. OBJECTIVE TYPE QUESTIONS

1. An instrument used for ranging is
   - (a) optical square ☐
   - (b) line ranger ☐
   - (c) clinometer ☐
   - (d) pedometer. ☐
2. Survey plotting can be done with an accuracy of
   - (a) 0.25 mm ☐
   - (b) 0.5 mm ☐
   - (c) 1 mm ☐
   - (d) 1 cm. ☐
3. A chain may get elongated due to
   - (a) change in temperature ☐
   - (b) difference in pull ☐
   - (c) opening of rings ☐
   - (d) kinks in links. ☐
4. A chain is made up of mild steel or galvanised iron wire of diameter
   - (a) 1 mm ☐
   - (b) 4 mm ☐
   - (c) 5 mm ☐
   - (d) 1 cm. ☐

5. Handles of chains are made up of
   (a) mild steel
   (b) galvanised iron
   (c) brass
   (d) copper.
6. Handles are connected to the link by
   (a) flexible joint
   (b) rigid joint
   (c) ball and socket joint
   (d) swivel joint.
7. Distance between two neighbouring brass rings is
   (a) less than 20 cm
   (b) 20 cm
   (c) 1 m
   (d) 5 m.
8. The length of an Engineer's chain is
   (a) 20 m
   (b) 33 feet
   (c) 66 feet
   (d) 100 feet.
9. The length of a link of Gunter's chain is
   (a) 20 cm
   (b) 1′
   (c) 0.66′
   (d) 2.065′.
10. Indirect ranging is adopted when the two ends of chain line are
    (a) mutually invisible
    (b) too distant
    (c) on a sloping ground
    (d) separated by a valley.
11. A 30 m chain after measuring a distance of 6000 m was found to be 10 cm more than the designated length. If the chain was standardised before the commencement of survey then the true length is
    (a) 6020 m
    (b) 6010 m
    (c) 5990 m
    (d) 5980 m.
12. A 20 m long chain when tested should not show an error exceeding
    (a) 2 mm per metre length and 5 mm in the overall length
    (b) 2 mm per metre length and 8 mm in the overall length
    (c) 2 mm per metre length and 20 mm in the overall length
    (d) 2 mm per metre length and 40 mm in the overall length.
13. Drop arrow is used in
    (a) conventional chain survey
    (b) measurements along slopes
    (c) measurement by method of stepping
    (d) measuring with tape.
14. Hypotenusal allowance is
    (a) $(\sec \theta - 1) \times$ measured distance
    (b) $(1 - \sec \theta) \times$ measured distance
    (c) measured distance $\times (1 - \cos \theta)$
    (d) measured distance $\times \cos \theta$.
15. Hypotenusal allowance for a length of 50 m when the slope is 60° is
    (a) 17.32 m
    (b) 25 m
    (c) 37.5 m
    (d) 50 m.

16. Correction for slope is
    (a) $\dfrac{h^2}{2l}$
    (b) $\dfrac{h}{2l^2}$
    (c) $\dfrac{2h}{l^2}$
    (d) $\dfrac{h}{l}$.

17. Distance between two stations A and B is 200 m whereas their difference in elevations is 2 m. Hence horizontal distance between A and B is
    (a) 199 m
    (b) 199.9 m
    (c) 199.99 m
    (d) 199.999 m.

18. Correct length of a 50 m tape, weighing 1.2 kg when a pull of 12 kg is applied at the ends and is freely suspended is
    (a) 49.98 m
    (b) 49.96 m
    (c) 50.02 m
    (d) 50.04.

19. Survey is conducted at 40°C and the measured length was 5 km. If the chain was standardised at 20°C and co-efficient of linear expansion of chain material is 0.3×10^{-5}/°C, then the correct length is
    (a) 5000.15 m
    (b) 5000.30 m
    (c) 4999.70 m
    (d) 4999.85 m.

20. Correction for pull is
    (a) $\dfrac{(P - P_0)l}{AE}$
    (b) $\dfrac{(P - P_0) A \times l}{E}$
    (c) $(P - P_0)EAl$
    (d) $\dfrac{(P - P_0)A}{lE}$

21. Pick up the most accurate statement from the following:
    (a) survey lines in an area should be as many as possible
    (b) number of base lines in an area is limited to one
    (c) main chain lines should form well conditioned triangles
    (d) oblique offsets are inferior to perpendicular offsets.

22. When a chain line encounters a river
    (a) chaining is obstructed but ranging is free
    (b) ranging is obstructed but chaining is free
    (c) both ranging and chaining are obstructed
    (d) both ranging and chaining are free.

23. Reciprocal ranging is adopted when the following is encountered :
    (a) a dense forest
    (b) a hillock
    (c) a river
    (d) a tall building.

24. Convention for a telegraphic line is
    (a) —•——•——•—
    (b) – – – – – – – – – – – –
    (c)  —■——■——■—
    (d) T   T   T   T

FIGURE 16.20

25. Convention for an embankment is

(a)  ☐ (b)  ☐

(c)  ☐ (d)  ☐

FIGURE 16.21

26. In Fig. 16.22, if chainage of A is 400 m, chainage of D is
    (a) 360 m ☐ (b) 377 m ☐
    (c) 423 m   (d) 440 m. ☐

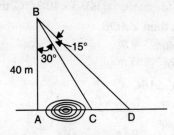

FIGURE 16.22

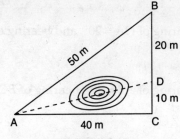

FIGURE 16.23

27. In Fig. 16.23, AD =
    (a) $\sqrt{1700}$ ☐ (b) $\sqrt{2100}$ ☐
    (c) $\sqrt{2900}$ (d) $\sqrt{1500}$. ☐

28. In Fig. 16.24 if bearing of PQ is 287° and QR is 62°, then PR =
    (a) 40 m ☐ (b) 80 m ☐
    (c) 120 m   (d) 160 m. ☐

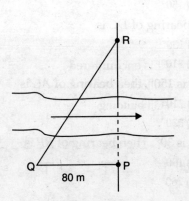

FIGURE 16.24

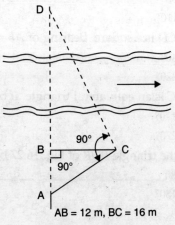

FIGURE 16.25

29. In Fig. 16.25 if chainage of A is 500.000, chainage of D is
    (a) 507.200
    (b) 512.800
    (c) 533.333
    (d) 525.000.

30. In Fig. 16.26 width of the river (CD) is
    (a) 57.74 m
    (b) 63.39 m
    (c) 66.67 m
    (d) 36.60 m.

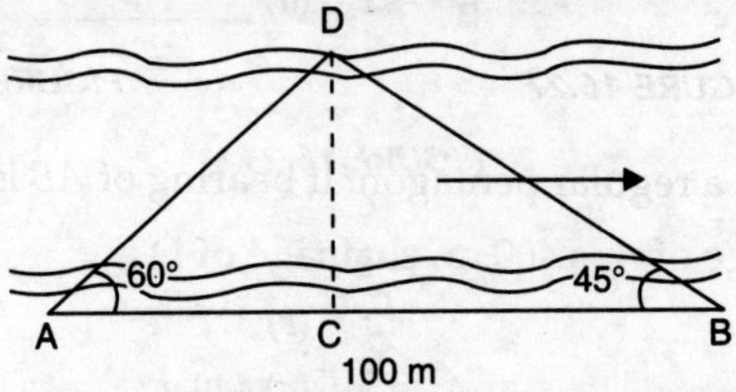

FIGURE 16.26

31. If bearing of OA = 20°, and bearing of OB = 120°, then ∠AOB
    (a) 100°
    (b) 140°
    (c) 280°
    (d) 80°.

32. If bearing of AB = 40°, bearing of BC = 300°, then ∠ABC =
    (a) 80°
    (b) 100°
    (c) 260°
    (d) 180°.

33. If bearing of AB = N 10° W, bearing of BC = N 80° W, then ∠ABC =
    (a) 70°
    (b) 90°
    (c) 110°
    (d) 250°.

34. If bearing of AB = N 30° W, bearing of BC = N 40° E, then ∠ABC =
    (a) 10°
    (b) 70°
    (c) 110°
    (d) 170°.

35. If the bearing of AB = N 40° E, and bearing of BC = S 70° E, then ∠ABC =
    (a) 30°
    (b) 70°
    (c) 110°
    (d) 150°.

36. ABCD is a square. Bearing of AB = 40°. Hence bearing of DC is
    (a) 40°
    (b) 130°
    (c) 220°
    (d) 310°.

37. ABC is an equilateral triangle. If bearing of AB is 150°, then bearing of AC is
    (a) 120°
    (b) 210°
    (c) 270°
    (d) 330°.

38. In the triangle PQR of Fig. 16.27 bearing of PQ is 30°. Then bearing of RP is
    (a) 80°
    (b) 100°
    (c) 150°
    (d) 260°.

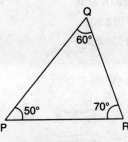

FIGURE 16.27

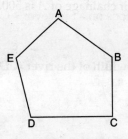

FIGURE 16.28

39. ABCDE in Fig. 16.28 is a regular pentagon. If bearing of AB is 100°, then deflection angle at B is
   (a) 72°R
   (b) 108°R
   (c) 120°R
   (d) 172°R.

40. In Fig. 16.29, if bearing of AB is 190°, bearing of DE

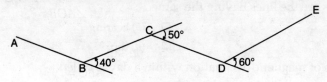

FIGURE 16.29

   (a) 40°
   (b) 140°
   (c) 240°
   (d) 340°.

41. In a closed traverse ABC, the following readings were taken

Line	Fore bearing	Back bearing
AB	19°	200°
BC	100°	277°
CA	277°	49°

If station A is free from local attraction, correct bearing of CB is
   (a) 275°
   (b) 276°
   (c) 277°
   (d) 279°.

42. The magnetic bearing of a line is N 88° E. Its true bearing is S 89° E. Therefore, its magnetic declination is
   (a) 2°W
   (b) 3° W
   (c) 3° E
   (d) 91°.

43. A freely floating needle slightly gets inclined to the horizontal anywhere except on the equator. It is called
   (a) declination
   (b) dip
   (c) local attraction
   (d) secular variation.

44. The graduations on a Surveyor's Compass are

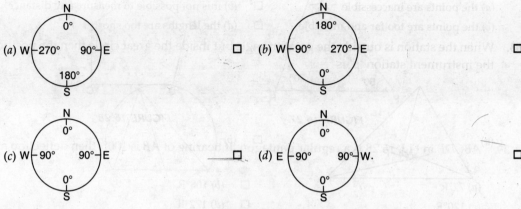

FIGURE 16.30

45. Isogonic lines are the lines having the same
    - (a) elevation ☐
    - (b) bearing ☐
    - (c) declination ☐
    - (d) dip. ☐

46. The variation of magnetic declination within a day is called
    - (a) diurnal variation ☐
    - (b) irregular variation ☐
    - (c) annual variation ☐
    - (d) secular variation. ☐

47. Local attraction at a place may be due to
    - (a) key bunches ☐
    - (b) steel buttons ☐
    - (c) current carrying bare wire ☐
    - (d) electric storm. ☐

48. The amount of correction due to local attraction at a place
    - (a) is a constant for all bearings ☐
    - (b) varies with the bearing ☐
    - (c) changes from time to time ☐
    - (d) sometimes additive and sometimes subtractive. ☐

49. An instrumental error in compass survey is because of
    - (a) inaccurate levelling ☐
    - (b) variation in declination ☐
    - (c) no counter weight provision to counteract dip ☐
    - (d) local attraction due to bare current carrying conductors. ☐

50. The technique of plotting all the accessible stations with a single set up of plane table is called
    - (a) radiation ☐
    - (b) intersection ☐
    - (c) resection ☐
    - (d) traversing. ☐

51. Radiation plane table survey is the best suited when
    - (a) distances are long but accessible ☐
    - (b) distances are short and accessible ☐
    - (c) distances are long and inaccessible ☐
    - (d) distances are short but inaccessible. ☐

52. Intersection is preferred to radiation when
    (a) the points are inaccessible
    (b) it is not possible to measure any distance
    (c) the points are too far and invisible
    (d) the lengths are too short.
53. When the station is outside the great triangle but inside the great circle then the position of the instrument station 'p' is

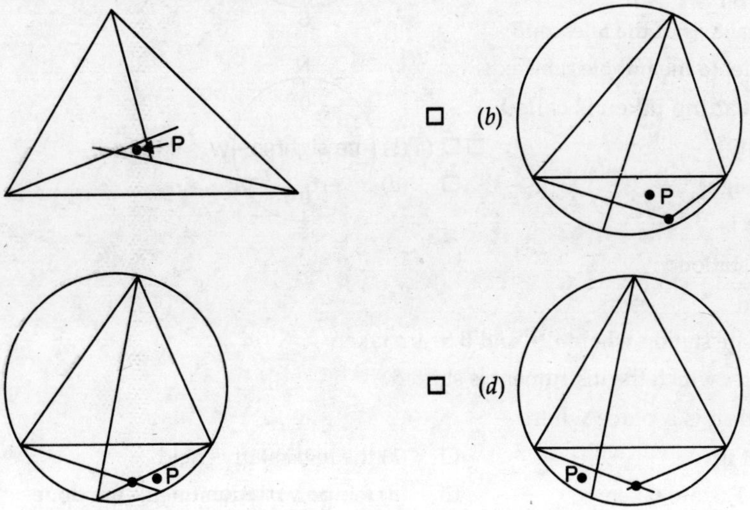

FIGURE 16.31

54. 'The strength of fix' is poor when
    (a) the station is within the great triangle
    (b) the station is outside the great circle
    (c) the station is within the great circle but outside the great triangle
    (d) the station is on the great circle.
55. An advantage of plane tabling is
    (a) it is a tropical instrument
    (b) it has many accessories
    (c) plotting is done out-door
    (d) chances to miss details are less.
56. A disadvantage of plane table survey is
    (a) it is heavy, cumbersome and awkward to carry
    (b) it cannot be used in wet climate
    (c) details may not be available while redrawing to a different scale
    (d) accessories are likely to be lost.
57. An example for a level surface is
    (a) surface of earth
    (b) surface of sea
    (c) surface of a reservoir
    (d) surface of a still lake.
58. Level line and horizontal line are
    (a) the same for longer distances
    (b) both straight lines
    (c) never the same
    (d) same for smaller lengths.

59. A plumb line is
   (a) a vertical line
   (b) a line parallel to a vertical line
   (c) a line perpendicular to level line
   (d) a line perpendicular to the horizontal line.

60. Line of collimation
   (a) is the same as line of sight
   (b) the line joining point of intersection of cross hairs and optical centre of object glass
   (c) the geometrical axis of the telescope
   (d) the line parallel to the bubble tube axis.

61. The very first reading taken is called
   (a) back sight
   (b) fore sight
   (c) intermediate sight
   (d) invert.

62. A change point is
   (a) the very first station
   (b) the last station
   (c) the intermediate station where F.S. and B.S. are taken
   (d) the station after which the instrument is shifted.

63. A levelling station is a place where
   (a) the level is set up
   (b) the level staff is held
   (c) both B.S. and F.S. are taken
   (d) temporary adjustments are done.

64. The telescope of a Dumpy level
   (a) is rigidly fixed to the levelling head
   (b) can be tilted in a vertical plane
   (c) can be taken out of its supports and reversed
   (d) permits interchange of eye piece and object glass.

65. Pick up the odd statement
   (a) temporary adjustments of the Dumpy level are to be performed at every set up
   (b) the eye piece need not be adjusted after the first set up when the same surveyor is taking readings
   (c) parallax error is completely eliminated when there is no change in the staff reading when the eye is moved up and down
   (d) focus the objective towards a white or bright background for the clear visibility of cross hairs.

66. A Bench mark is a
   (a) reference point
   (b) the very first station
   (c) the last station where the survey closes
   (d) point of known elevation.

67. The correct position of holding staff is
   (a) held vertical
   (b) held vertically and swung to left and right and the least reading is recorded
   (c) held vertically and swung towards and away from the person holding and the highest reading is recorded
   (d) held vertically and swung towards and away by the person holding it and the least reading is recorded.

68. "Cross-section" and "Longitudinal sectioning" is
    (a) simple levelling   (b) differential levelling
    (c) profile levelling   (d) check levelling.
69. Height of instrument method of booking readings is adopted
    (a) when less number of intermediate sights exist
    (b) in profile levelling
    (c) in reciprocal levelling
    (d) in differential levelling.
70. An invert is taken when the point is
    (a) having high elevation   (b) above the line of sights
    (c) below the line of sight   (d) below ground level.
71. In a survey it was recorded that Σ Rise = 0, then
    (a) the ground is sloping   (b) it is continuously rising
    (c) it is continuously falling   (d) the survey had many invert readings.
72. When the staff is held on a B.M. of RL 100.000, the staff reading was 2.000. When the staff is held on station P, the reading was 3.000. Hence height of the instrument is
    (a) 100.000   (b) 102.000
    (c) 103.000   (d) 99.000.
73. The following readings correspond to the check at the end of a page of level field book but they are not given in the order. They represent a continuously rising ground
    100.000, 6.000, 0.000, 106.000, 13.000, 7.000.
    Hence in the above ΣBS =
    (a) 0.000   (b) 6.000
    (c) 7.000   (d) 13.000.
74. In the above Σ Rise =
    (a) 0.000   (b) 6.000
    (c) 7.000   (d) 13.000.
75. Ten readings were recorded in a level field book. If the instrument was shifted after 2nd and 6th readings, then the fore sights are
    (a) 1st, 3rd and 7th readings   (b) 2nd, 6th and 10th readings
    (c) 2nd and 6th readings   (d) 3rd, 4th, 5th, 7th, 8th, 9th and 10th readings.
76. The following readings were taken on a uniformly sloping ground
    0.500, 1.000, 1.500, 2.000, 1.2000, 1.700, 2.200, 2.700.
    Hence difference in elevation between the first and last station is
    (a) 1.700 (fall)   (b) 2.200 (fall)
    (c) 2.800   (d) 3.000 (fall).
77. In the above case Σ Rise =
    (a) 0.000   (b) 2.500
    (c) 3.000   (d) 0.800.

78. Due to curvature of earth the object
    (a) looks higher than it is   □   (b) looks lower than it is   □
    (c) looks as it is   □   (d) looks curved.   □
79. Correction for curvature for a distance of 1 km =
    (a) 0.0112 m   □   (b) 0.0673 m   □
    (c) 0.0785 m   □   (d) 0.0673 km.   □
80. Correction for refraction for a distance of 1 km =
    (a) 0.0112 m   □   (b) 0.0673 m   □
    (c) 0.0785 m   □   (d) 0.0673 km.   □
81. In Fig. 16.32, "dip of horizon" is

    (a) $\dfrac{AC}{OB}$ □

    (b) $\dfrac{OA}{OB}$ □

    (c) $\dfrac{CA}{OB}$ □

    (d) $\dfrac{AB}{OB}$ □

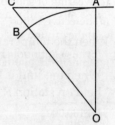

FIGURE 16.32

82. A luminous object on the top of a hill 100 m high is just visible above the horizontal at a certain station at the sea level. The distance between the station and the hill is
    (a) 3.853 km   □   (b) $3.853 \times \sqrt{0.1}$ km   □
    (c) 38.53 km   □   (d) 385.3 km.   □
83. Correct staff readings at A and B in the following cases are

Level at	Staff at A	Staff at B	Remarks
C	4.000	2.000	AC = CB, AB = 100 m
D	3.000	1.050	D is 20 m along BA produced ∴ AB = 20 m, BD = 120 m.

    (a) 3.050 and 1.050   □   (b) 2.990 and 0.990   □
    (c) 3.000 and 1.000   □   (d) 3.060 and 1.060.   □
84. In a reciprocal levelling the following readings were taken :

Instrument nearer to	Staff Readings	
	at P	at Q
P	1.200	1.000
Q	3.000	3.400

Given RL of P = 50.000, then R.L. of Q is
(a) 49.800
(b) 49.900
(c) 50.100
(d) 50.200.

85. In profile levelling, staff readings on two neighbouring pegs 20 m apart are 1.200 and 1.000 respectively. Therefore, the proposed road had a
(a) rising gradient of 1 in 100
(b) falling gradient of 1 in 100
(c) rising gradient of 1 in 20
(d) falling gradient of 1 in 20.

86. An example for instrumental error in levelling is
(a) earth's curvature and atmospheric refraction
(b) collimation error
(c) wearing of shoe of level staff
(d) defective tripod.

87. When the temperature rises, length of bubble
(a) remains unaltered
(b) decreases
(c) increases
(d) sometimes increases and sometimes decreases.

88. Sensitiveness of bubble tube can be increased by
(a) using viscous liquid
(b) reducing length of tube
(c) increasing diameter of the tube
(d) reducing internal radius of the tube.

89. Sensitiveness of Bubble tube
(a) $206265 \times \dfrac{s}{nD}$
(b) $\dfrac{s}{nD}$
(c) $\dfrac{nlD}{s}$
(d) $\dfrac{Rs}{nl}$.

90. A contour map of the area is essential before proceeding with the construction of
(a) a building
(b) a swimming pool
(c) a dam
(d) a bridge.

91. Reciprocal levelling eliminates
(a) collimation error
(b) collimation, curvature and refraction error
(c) curvature and refraction error
(d) collimation and curvature error fully and refraction error partly.

92. Contour lines
(a) end abruptly
(b) cross each other
(c) are uniformly spaced
(d) close somewhere.

93. Contour lines look to cross each other in case of
(a) an overhanging cliff
(b) a dam of vertical face
(c) a steep hill
(d) a deep valley.

94. Section AA indicates

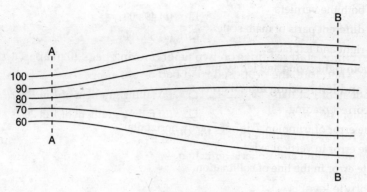

FIGURE 16.33

   (a) steep slope   (b) flat slope
   (c) uniform slope   (d) ridge.

95. These V-shaped contours represent (Fig. 16.34)
   (a) a ridge   (b) a valley
   (c) an overhanging cliff (d) nothing.

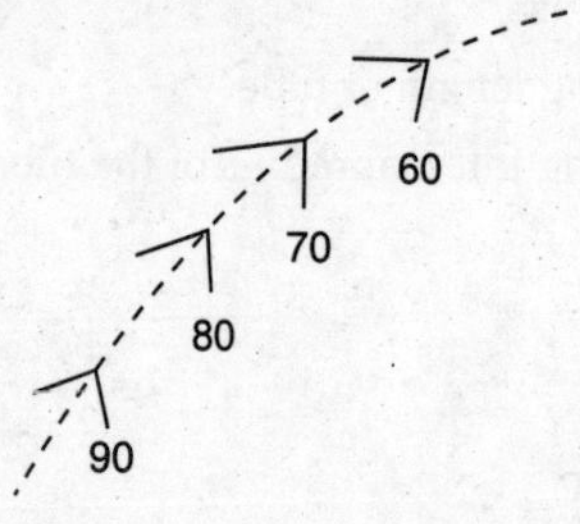

FIGURE 16.34 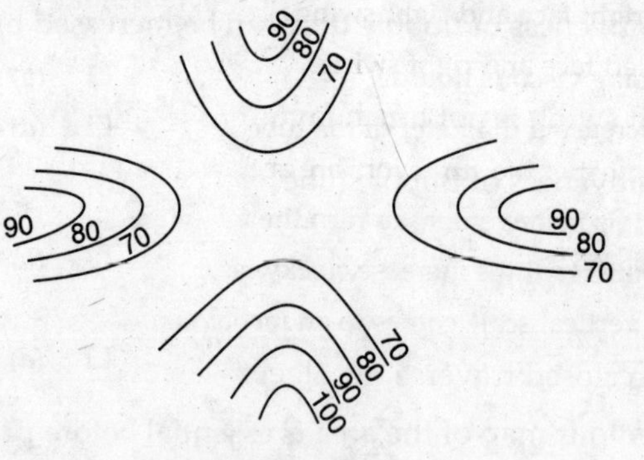   FIGURE 16.34A

96. The above figure represents (Fig. 16.34A)
   (a) a hillock   (b) a valley
   (c) a saddle   (d) a reservoir.

97. Pick up the odd statement
   (a) contours give the topography of the area
   (b) intervisibility between two points can be judged from a contour map
   (c) quantities of earthwork can be computed from the contour map
   (d) for a vertical cliff contours seem to cross each other.

98. In a theodolite the line of collimation is
   (a) parallel to axis of plate levels   (b) parallel to the vertical axis
   (c) perpendicular to the trunnion axis   (d) parallel to the horizontal axis.

99. In a transit theodolite error due to eccentricity of verniers is counteracted by
    (a) reading both the verniers
    (b) reading different parts of main scale
    (c) reading right and left faces
    (d) taking both right swing and left swing readings.

100. Right face and left reading are taken to
    (a) get the correct reading
    (b) eliminate error of trunnion axis not exactly horizontal
    (c) eliminate error in vertical axis
    (d) eliminate error in the line of collimation.

101. Axis of altitude level is
    (a) parallel to trunnion axis
    (b) perpendicular to the vertical axis
    (c) parallel to the line of collimation
    (d) perpendicular to plate levels.

102. A 15 cm theodolite means
    (a) length of the telescope is 15 cm
    (b) height of standards is 15 cm
    (c) diameter of lower plate is 15 cm
    (d) radius of upper plate is 15 cm.

103. The most commonly used set of theodolite operations are
    (a) right face and right swing
    (b) right face and left swing
    (c) left face and right swing
    (d) left face and left swing.

104. Left swing is not much favoured in theodolite survey because
    (a) most of the surveyors are accustomed to right hand
    (b) it is inconvenient to turn the telescope anti-clockwise
    (c) the readings increase clockwise
    (d) vertical scale comes to an inconvenient position to be read.

105. In a closed traverse the algebraic sum of deflection angle is
    (a) 0°
    (b) 360°
    (c) $(2n - 4) \times 90°$
    (d) $(2n + 4) \times 90°$.

106. When internal angles are to be subtended by a theodolite survey for closed traverse
    (a) traverse should run clockwise
    (b) traverse should go counter clockwise
    (c) traverse may go either way
    (d) exterior angles are measured when the traverse goes anti-clockwise.

107. When the first object is to be focussed, the scale is set to 0° 00′ 00″ and then
    (a) the upper screw is fixed, lower screw released, the object sighted, lower screw fixed and lower tangent screw operated
    (b) both the clamp screw fixed, top tangent screw operated
    (c) bottom clamp screw fixed, top screw released, object sighted and upper tangent screw operated

(d) object is exactly bisected by operating both the upper and lower tangent screws whichever is convenient. ☐

108. Can the vertical axis coincide with any other axis in any position of telescope?
(a) Yes, with the geometric centre of level head ☐
(b) Yes, with the axis of altitude levels ☐
(c) Yes, with line of collimation ☐
(d) No impossible. ☐

109. For a transit theodolite the least reading on the main scale is 20′. If its vernier scale should read 20″, then
(a) 59° on the main scale are divided into 60 equal parts ☐
(b) 20° on the main scale are divided into 60 equal divisions ☐
(c) 19° 40′ on the main scale are divided into 60 equal parts ☐
(d) 20° 20′ on the main scale are divided into 60 equal parts. ☐

110. Tacheometry is best suited
(a) where chaining is impossible ☐ (b) for populous areas ☐
(c) in broken grounds ☐ (d) for extremely accurate survey. ☐

111. When the line of sight is inclined at θ to the horizontal and staff is held vertical then horizontal distance is
(a) $\frac{f}{i}s + (f+d)$ ☐ (b) $\frac{f}{i}s\cos\theta + (f+d)$ ☐
(c) $\frac{f}{i}s\cos^2\theta + (f+d)\cos\theta$ ☐ (d) $\left[\frac{f}{i}s + (f+d)\right]\cos\theta + h\sin\theta$. ☐

112. The desirable multiplying and additive constants for a tacheometer
(a) 100 and 0.3 m ☐ (b) 50 and 0.5 m ☐
(c) 100 and 0 m ☐ (d) 200 and 0.15 m. ☐

113. For tacheometer focal length of object glass is 20 cm, distance between object glass and trunnion axis is 10 cm. Spacing between the outer lines of diaphragm = 4 mm. If the staff intercepts are 1.000 (top) and 2.500 (middle) when the line of collimation is perfectly horizontal then, horizontal distance between the staff station and instrument station is
(a) 75.3 m ☐ (b) 78 m ☐
(c) 150.3 m ☐ (d) 153 m. ☐

114. Annallactic lens is provided
(a) between diaphragm and object glass ☐
(b) exactly at the line of intersection of vertical and horizontal axes ☐
(c) just before objective ☐
(d) between eye-piece and diaphragm. ☐

115. Anallactic lens is provided to
(a) nullify both the constants of tacheometer ☐
(b) render additive constant zero ☐

(c) make multiplying constant as 100 and additive constant as zero

(d) improve visibility.

116. Anallactic lens is a
    (a) convex lens
    (b) concave lens
    (c) compound concave and convex lens
    (d) plain lens.

117. In a tacheometer provided with anallactic lens distance between object glass and vertical axis is 15 cm. Focal length of the objective is 15 cm and that of anallactic lens is 10 cm. Hence distance between the two lenses is
    (a) 17.5 cm
    (b) 25 cm
    (c) 30 cm
    (d) 40 cm.

118. For a tacheometer focal length of object glasses = 20 cm. Focal length of anallactic lens = 10 cm. Distance between objective and vertical axis = 15 cm. Spacing between outer lines of diaphragm = 2 mm. If the intercept is 2 m when the line of sight is horizontal, horizontal distance between the instrument and staff is
    (a) 100 m
    (b) 100.3 m
    (c) 200 m
    (d) 200.3 m.

119. The stadia intercept on a fixed hair instrument vertically held is 1 m. If its constants are 800 and 0.5 and total number of turns are 16, the horizontal distance between the instrument and target is
    (a) 5.5 m
    (b) 50.5 m
    (c) 55 m
    (d) 800.5 m.

120. If the length of subtense bar between the extreme targets is 3 m and the angle measured is 3", then the horizontal distance between the instrument and staff is
    (a) 1 m
    (b) 3 m
    (c) $\frac{206265}{3}$ m
    (d) 206265 m.

121. In a tacheometer the central hair reading was 2 m when the angle of elevation was $\tan^{-1} 0.2$ and 3 m when the angle of elevation was $\tan^{-1} 0.25$. Hence the horizontal distance between the instrument and staff is
    (a) 4 m
    (b) 5 m
    (c) 20 m
    (d) 22.22 m.

122. Point of tangency is the
    (a) beginning of the curve
    (b) end of the curve
    (c) common point where the radius changes
    (d) common point where the radius and direction changes.

123. In the above question (121) case, difference in elevation between the instrument and staff is (Fig. 16.35)
    (a) 4 m
    (b) 4.44 m
    (c) 5 m
    (d) 5.55 m.

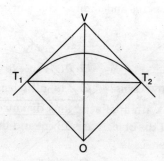

FIGURE 16.35

124. In a simple curve, 'External distance' is the distance between
    (a) vertex and middle point of the curve ☐ (b) vertex and the centre of the curve ☐
    (c) vertex and point of curve ☐ (d) point of curve and point of tangency. ☐

125. If the curve in Fig. 16.35 is a right hand curve, then T_2 is called
    (a) point of intersection ☐ (b) point of tangency ☐
    (c) point of curve ☐ (d) forward tangent point. ☐

126. The angle subtended by a 20 m arc at the centre is (in metric system)
    (a) $\dfrac{1146}{R}$ ☐ (b) $\dfrac{5730}{R}$ ☐
    (c) $\dfrac{573}{R}$ ☐ (d) $\dfrac{1718.88}{R}$ ☐

127. If the radius of a circular curve is 100 m, deflection angle is 90°, then the length of tangent is
    (a) 0 ☐ (b) 70.7 m ☐
    (c) 100 m ☐ (d) ∞. ☐

128. The length of curve in the problem 127 is
    (a) 50 m ☐ (b) 157 m ☐
    (c) 100 m ☐ (d) 314 m. ☐

129. The length of the long chord in problem 127 is
    (a) 50 m ☐ (b) 70.7 m ☐
    (c) 100 m ☐ (d) 141.4 m. ☐

130. Apex distance in problem 127 is
    (a) 29.29 m ☐ (b) 41.40 m ☐
    (c) 70.70 m ☐ (d) 141.40 m. ☐

131. Mid-ordinate for the curve in Fig. 16.35 is
    (a) 29.29 m ☐ (b) 41.40 m ☐
    (c) 70.70 m ☐ (d) 157.00 m. ☐

132. In Fig. 16.36, O_x =
    (a) $-\sqrt{R^2 - l^2}$ ☐ (b) $\sqrt{R^2 - x^2} - \sqrt{R^2 - l^2}$ ☐
    (c) $\sqrt{R^2 - l^2} - \sqrt{R^2 - x^2}$ ☐ (d) $\sqrt{R^2 - (l-x)^2} - (R - O_0)$ ☐

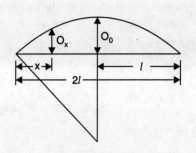

FIGURE 16.36   FIGURE 16.37

133. In Fig. 16.37, $O_x =$
 (a) $\dfrac{x(2l-x)}{R2}$ ☐   (b) $\dfrac{2x(l-2x)}{R}$ ☐
 (c) $R - \sqrt{R^2 - x^2}$ ☐   (d) $\dfrac{x^2}{2R}$. ☐

134. To set a right hand compound curve subtending angles α and β at the centre, the initial reading of the theodolite at the point of common tangency is

 (a) 0° ☐   (b) $\dfrac{\alpha}{2}$ ☐
 (c) $360° - \dfrac{\alpha}{2}$ ☐   (d) $180° - \dfrac{\alpha}{2}$. ☐

135. By Rankine's method deflection angle δ =
 (a) $\dfrac{1718.88\,C}{R}$ degrees ☐   (b) $\dfrac{1817.9\,C}{R}$ degrees ☐
 (c) $\dfrac{90 \times C}{\pi R}$ degrees ☐   (d) $\dfrac{\pi R}{90 \times C}$ degrees. ☐

136. In Fig. 16.38, $T_s =$
 (a) $t_l + (t_s + t_l)\,\dfrac{\sin\alpha}{\sin(\alpha+\beta)}$ ☐   (b) $t_s + (t_s + t_l)\,\dfrac{\sin\beta}{\sin(\alpha+\beta)}$ ☐
 (c) $t_s + \dfrac{(t_s + t_l)\cdot\sin\alpha}{\sin(\alpha+\beta)}$ ☐   (d) $t_l + \dfrac{(t_s + t_l)\cdot\sin\beta}{\sin(\alpha+\beta)}$ ☐

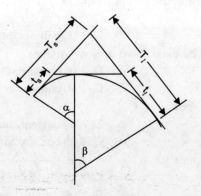

FIGURE 16.38

137. A reverse curve of radii R and r is to be set in between two parallel straights separated by a distance 'y'. If 'α' is the angle subtended at the centre then $y =$
    (a) $(R + r)(1 - \cos \alpha)$ ☐
    (b) $(R - r) \cos \alpha$ ☐
    (c) $(R + r) \cos \alpha$ ☐
    (d) $(R - r)(1 + \cos \alpha)$. ☐

138. In a reverse curve of Fig. 16.39, $h =$
    (a) $(R + r)(1 - \cos \theta)$ ☐
    (b) $(R + r) \tan \theta$ ☐
    (c) $2(R + r) \sin \dfrac{\theta}{2}$ ☐
    (d) $(R + r) \sin \theta$. ☐

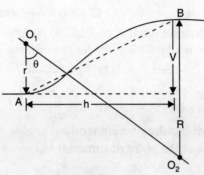

FIGURE 16.39

139. In Fig. 16.39, $AB =$
    (a) $(R + r)(1 - \cos \theta)$ ☐
    (b) $(R + r)(\tan \theta)$ ☐
    (c) $2(R + r) \sin \dfrac{\theta}{2}$ ☐
    (d) $(R + r) \sin \theta$. ☐

140. For a reverse curve between two parallel straights separated by 30 m and having a central angle of 60°, radii are equal. They are each
    (a) 15 m ☐
    (b) 30 m ☐
    (c) 60 m ☐
    (d) 51.96 m. ☐

141. If a 3% downgrade curve is followed by a 1% upgrade curve and the rate of change of grade is 0.1% per 20 m length, then the length of vertical curve is
    (a) 100 m ☐
    (b) 200 m ☐
    (c) 400 m ☐
    (d) 800 m. ☐

142. An instrument used to find slopes of ground is
    (a) planimeter ☐
    (b) clinometer ☐
    (c) box sextant ☐
    (d) pantograph. ☐

143. An instrument that can be used as an optical square
    (a) Ceylon ghat tracer ☐
    (b) box sextant ☐
    (c) tangent clinometer ☐
    (d) clinometer. ☐

144. Pick up the odd instrument
    (a) Abney clinometer ☐
    (b) tangent clinometer ☐
    (c) Ceylon ghat tracer ☐
    (d) box sextant. ☐

145. Pick up the odd instrument
    (a) prismatic compass ☐
    (b) theodolite ☐
    (c) box sextant ☐
    (d) dumpy level. ☐

146. For the Abney clinometer the semi-circular arc is graduated as
    (a) 0° to 180°
    (b) 0° to 90° to 0°
    (c) 90° to 0° to 90°
    (d) 0° to 140°.
147. To use the Abney clinometer as Abney level the vernier should read
    (a) 0°
    (b) 90°
    (c) 0° or 180°
    (d) it is not possible to use clinometer as hand level.
148. To instrument resembling an alidade is
    (a) box sextant
    (b) Ghat tracer
    (c) tangent clinometer
    (d) Abney level.
149. A tangent clinometer was set up at A of R.L. 100.00. When it was sighted to B 200 m away from A, the object vane read − 0.05. If the height of instrument (at A) is 1.2 m, R.L. of B is
    (a) 91.20
    (b) 101.15
    (c) 101.25
    (d) 111.20.
150. Ceylon Ghat tracer can read slopes up to
    (a) $\frac{1}{120}$ to $\frac{1}{6}$ either way
    (b) $\frac{1}{6}$ to $\frac{1}{30}$ either elevation or depression
    (c) $\frac{1}{300}$ to $\frac{1}{2}$ rising or falling
    (d) $\frac{1}{500}$ to $\frac{1}{10}$ upwards or downwards.
151. The horizon glass of the box sextant is
    (a) fully silvered
    (b) full plane
    (c) half silvered and half plane
    (d) concave in shape to form sharp image.
152. If the angle between index glass and horizon glasses of a box sextant is 40° the horizontal angle between A and B sighted by it is
    (a) 20°
    (b) 40°
    (c) 60°
    (d) 80°.
153. The instrument used to reproduce plans to a different scale is called
    (a) planimeter
    (b) clinometer
    (c) Ghat tracer
    (d) pantograph.
154. To reproduce the given map to a small scale, positions of pointer and pencil in a pantograph are
    (a) pointer is to the longer arm and pencil to the shorter arm
    (b) pointer to the shorter arm and pencil to the longer arm
    (c) both pointer and pencil are to the two different short arms
    (d) both the pointer and pencil are to the two different long arms, pencil being fixed and pointer adjusted to the scale.
155. An important rule in using pantograph is
    (a) pointer, pencil and weight should fall in a straight line
    (b) pointer should never be put on the short arm
    (c) the arm on to which the weight is attached should never move
    (d) pencil is always attached to the arm on which the scale is marked.

156. In the mid-ordinate rule, Area =
   (a) $\left[\dfrac{O_1 + O_2 + O_3 + \ldots + O_n}{n}\right] \times nd$
   (b) $\left(\dfrac{O_0 + O_1 + \ldots + O_n}{n+1}\right) \times$ total length
   (c) $\left(\dfrac{O_0 + O_n}{2} + O_1 + O_2 + \ldots + O_{n-1}\right) \times d$
   (d) $[(O_0 + O_n) + 4(O_1 + O_3 + \ldots) + 2(O_2 + O_4 + \ldots)]\dfrac{d}{3}$.

157. Simpson's rule states that Area =
   (a) $(O_1 + O_2 + \ldots + O_{n-1} + O_n) \times d$
   (b) $\left(\dfrac{O_0 + O_1 + \ldots + O_{n-1} + O_n}{n+1}\right) \times$ overall length
   (c) $\left(\dfrac{O_0 + O_n}{2} + O_1 + O_2 + \ldots + O_{n-1}\right) \times d$
   (d) $[(O_0 + O_n) + 4(O_1 + O_3 + \ldots) + 2(O_2 + O_4 + \ldots)]\dfrac{d}{3}$.

158. For a level section area of a trapezoidal cut when base width is '$2b$', side slopes are $1\,V : nH$ and vertical depth is h, is
   (a) $(b + nh)h$
   (b) $(b + 2nh)h$
   (c) $(2b + nh)h$
   (d) $(b + nh) \times 2h$.

159. Area of the section in Fig. 16.40 is
   (a) $(2b + nh + bhr) \times h$
   (b) $(n^2 b + nhr + 2bhr^2) \times (n^2 - r^2)$
   (c) $\dfrac{(nb^2 + nhr^2 + bh^2 r^2)}{n^2 - r^2}$
   (d) $\dfrac{nb^2 + nh^2 r^2 + 2bhr}{r^2 - n^2}$

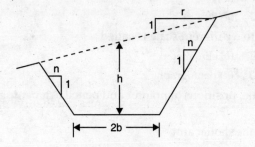
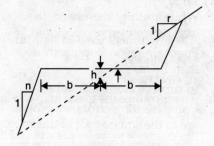

FIGURE 16.40   FIGURE 16.41

160. Area of fill in Fig. 16.41 is
   (a) $\dfrac{nr}{r-n}\left(\dfrac{b}{r} - h\right)$
   (b) $\dfrac{(b - rh)^2}{2(r - n)}$
   (c) $\dfrac{(b + rh)^2}{2(r - n)}$
   (d) $(b + rh)^2 + 2(r + n)$.

161. Area of the section in Fig. 16.42 is
   (a) $b(h_1 + h_2) + h(b + nh_1 + nh_2)$
   (b) $2b(h_1 + h_2) + 2h(s_1 + s_2)$
   (c) $b(h_1 + h_2) + \left(\dfrac{h_1 + h_2}{2}\right)(s_1 + s_2)$
   (d) $\dfrac{b}{2}(h_1 + h_2) + \dfrac{1}{2}h(s_1 + s_2)$.

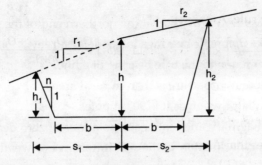

FIGURE 16.42

162. According to prismoidal formula, volume $V =$

(a) $d[A_0 + A_1 + ... + A_n]$ ☐

(b) $\dfrac{d}{3}\left[A_1 + A_n + 4\sum A_{odd} + 2\sum A_{even}\right]$ ☐

(c) $d\left[\dfrac{A_1}{2} + A_2 ..... A_{n\_1} + \dfrac{A_n}{2}\right]$ ☐

(d) $\dfrac{d}{3}\left[A_1 + A_n + 2\sum A_{odd} + 4\sum A_{even}\right]$. ☐

III. TRUE/FALSE

1. In a spherical triangle the sum of the three angles need not be equal to two right angles.
2. Geodetic surveying takes into account the curvature of the earth.
3. Cadastral survey is the same as topographical survey where the boundaries of features as fields, houses etc., are demarked.
4. Mud clogging of links renders the chain longer.
5. Invar is an alloy of steel and nickel which has very low co-efficient of expansion.
6. While pulling the chain, the less experienced man shall be the leader while the more experienced surveyor should be at the rear end of the chain.
7. For chaining on sloping grounds, "stepping" method is adopted when the fall is not greater than 1.8 m.
8. In surveys along hill slopes, it is desirable to proceed up the hill rather than down the hill.
9. Compensating errors are proportional to the distance measured whereas cumulative errors are to the square root of distance measured.
10. Chain survey is based on the principle of triangulation.
11. Chain survey is the best suited when the plane is to be plotted to a small scale.
12. A triangle is said to be well conditioned when any of its angles is greater than 30° but less than 120°.
13. An offset is called long when its length exceeds 15 m.
14. When a map is drawn to a smaller scale more number of offsets are to be taken.
15. The needle of prismatic compass is called a "fast needle".
16. In either prismatic compass or surveyor's compass, the eye vane and sight vane move effectively but not the needle.

17. In a prismatic compass zero is marked at the northern end of the graduated ring.
18. It is customary to take magnetic bearings in geodetic survey.
19. "Azimuth" is another name to the true bearing of a line.
20. Local attraction is more in urban areas than in rural areas.
21. "Dip" is nil on the equator whereas it is 90° at poles.
22. "Agonic lines" are imaginary lines connecting stations of zero declination.
23. Variation in magnetic declination is the maximum in hot weather and the least in cold weather.
24. Amount of magnetic declination is the maximum due to diurnal variation.
25. In plane table surveying orientation by magnetic needle is more accurate than orientation by back sighting.
26. Radiation method of plane table survey is the best suited when a single station with a commanding view is available and the area to be surveyed is small.
27. Two point problem gives more accurate results than the three point problem.
28. In a three point problem when the station is situated on the circumference of the great circle, the rays Aa, Bb and Cc will intersect at one point even when the plane table is not oriented.
29. "Strength of fix" is good when the instrument station is nearer to the periphery of the great circle.
30. "Height of the instrument" is the height of axis of the telescope above ground level.
31. The eye piece need not be adjusted more than once when the same person is operating the dumpy level.
32. Rise and fall system is adopted for profile levelling.
33. Any page of a level field book should begin with a back sight and end with a fore sight irrespective of setting up of a level.
34. By equalising, fore sight and back sight distances, error due to collimation curvature and refraction are nullified.
35. Because of the effect of refraction, objects appear lower than they really are.
36. While levelling across a hill it is advantageous to set the level on the peak as to complete levelling quickly.
37. Length of bubble increases with temperature.
38. Contour is the imaginary line joining points of equal elevation.
39. Horizontal equivalent for any two consecutive contours shall be the same throughout the map.
40. For a ramp sloping in the same direction with no break of slope, contour lines are uniformly spaced parallel straight lines.
41. Contour lines merge in case of a vertical (retaining) wall or a dam across a reservoir.
42. Contour lines cross each other in case of an overhanging cliff.
43. In any level the line of sight should be parallel to the bubble axis.
44. In a dumpy level line of collimation must coincide with the axis of the telescope.
45. A box sextant becomes an optical square when the angle between the two mirrors is 45°.

46. Pantagraph may be highly effective in reducing maps and reproducing them to the same scale but less efficient in enlarging them.
47. "Alidade" of a theodolite is nothing but the upper plate containing standards supporting the telescope.
48. The azimuthal bubble tube can be levelled by the clip screw.
49. When the telescope is to the right of the vertical circle—it is called face right.
50. By observing face right and face left readings error due to imperfect graduations on plate scales is easily overcome.
51. Vertical axis of a theodolite should be at right angles to the trunnion axis.
52. The angle between line of collimation and vertical axis can be anywhere between 0° and 90°.
53. The line of collimation is always perpendicular to the horizontal axis.
54. The internal focussing telescope has a double concave lens which moves forward and backward when focussing is done.
55. The eye piece of a telescope consists of two plano-convex lenses with convex faces towards each other.
56. While measuring a number of horizontal angles the telescope is further swung in the same direction to read the first station again.
57. When the traverse runs clockwise interior angles are measured and when it runs counterclockwise exterior angles are measured.
58. In a closed traverse sum of the exterior angles = $(2n - 4) \times$ right angles.
59. Prismoidal correction = Volume obtained by Trapezoidal formula
    − Volume obtained by Prismoidal formula.
60. The angle subtended by a 30 m arc is $\frac{1146}{R}$.
61. Length of the long chord of a simple circular curve is $2R \sin \frac{\delta}{2}$.
62. The maximum deflection angle for a circular curve = intersection angle.
63. For a simple curve deflection angle of the long chord = $\frac{1}{2} \times$ intersection angle.
64. A transition curve begins with zero radius and ends with the radius of circular curve.
65. The radius of curvature of a transition curve varies inversely as the distance from the beginning of the curve.
66. Tacheometry is ideally suited when very high accuracy is required.
67. Subtense method is a movable hair method of tacheometry.
68. Anallactic lens is not required for the internal focussing telescopes.
69. An anallactic lens absorbs some incident light on it.
70. In a stadia tacheometer, the tacheometric angle is always constant irrespective of the distance.
71. Subtense method is more accurate than stadia method for longer distances.

ANSWERS
Objective Type Questions

1. (b)	2. (a)	3. (c)	4. (c)	5. (c)	6. (d)
7. (c)	8. (d)	9. (c)	10. (a)	11. (b)	12. (a)
13. (c)	14. (a)	15. (d)	16. (a)	17. (c)	18. (a)
19. (b)	20. (a)	21. (c)	22. (a)	23. (b)	24. (d)
25. (b)	26. (d)	27. (a)	28. (b)	29. (c)	30. (d)
31. (a)	32. (a)	33. (c)	34. (c)	35. (c)	36. (a)
37. (b)	38. (d)	39. (a)	40. (b)	41. (d)	42. (c)
43. (b)	44. (d)	45. (c)	46. (a)	47. (c)	48. (a)
49. (c)	50. (a)	51. (b)	52. (a)	53. (d)	54. (d)
55. (d)	56. (c)	57. (d)	58. (d)	59. (c)	60. (b)
61. (a)	62. (c)	63. (b)	64. (a)	65. (d)	66. (d)
67. (d)	68. (c)	69. (b)	70. (b)	71. (c)	72. (b)
73. (d)	74. (c)	75. (b)	76. (d)	77. (a)	78. (b)
79. (c)	80. (a)	81. (d)	82. (c)	83. (b)	84. (b)
85. (a)	86. (b)	87. (b)	88. (c)	89. (c)	90. (c)
91. (d)	92. (d)	93. (a)	94. (a)	95. (d)	96. (c)
97. (d)	98. (c)	99. (a)	100. (b)	101. (c)	102. (c)
103. (c)	104. (c)	105. (b)	106. (b)	107. (a)	108. (c)
109. (c)	110. (c)	111. (b)	112. (c)	113. (c)	114. (a)
115. (b)	116. (a)	117. (a)	118. (c)	119. (b)	120. (d)
121. (c)	122. (b)	123. (c)	124. (a)	125. (b)	126. (a)
127. (c)	128. (b)	129. (d)	130. (b)	131. (a)	132. (d)
133. (d)	134. (c)	135. (c)	136. (b)	137. (a)	138. (d)
139. (c)	140. (b)	141. (d)	142. (b)	143. (b)	144. (d)
145. (d)	146. (c)	147. (a)	148. (c)	149. (a)	150. (a)
151. (c)	152. (d)	153. (d)	154. (a)	155. (a)	156. (a)
157. (d)	158. (c)	159. (d)	160. (b)	161. (d)	162. (d).

True/False

1. T	2. T	3. T	4. F	5. T	6. T
7. T	8. F	9. T	10. T	11. F	12. T
13. T	14. F	15. F	16. T	17. F	18. T
19. T	20. T	21. T	22. T	23. T	24. F
25. F	26. T	27. F	28. T	29. F	30. F
31. T	32. F	33. T	34. T	35. F	36. F
37. F	38. T	39. F	40. T	41. T	42. T
43. T	44. F	45. T	46. T	47. T	48. T
49. F	50. F	51. T	52. T	53. T	54. T
55. T	56. T	57. F	58. F	59. T	60. F
61. T	62. F	63. T	64. F	65. T	66. F
67. T	68. T	69. T	70. T	71. T.	

Chapter 17 — Construction Planning, PERT, CPM

1. INTRODUCTION

Effective and efficient utilization of resources (materials, machinery, manpower, money and time) in an optimal manner is essential for the implementation of any project which is composed of many activities or tasks. The interrelationships among various activities of the project arise out of physical, technological, financial and managerial considerations. Good project management should consider these interrelationships correctly during the planning, scheduling and controlling phases of the project.

Bar charts was probably one of the earliest techniques of project management. This technique was developed by Henry Gantt around 1900 A.D. In a bar chart each activity is represented as a bar, the length of bar being proportional to the time required for the completion of the activity. All the bars are drawn parallel and the time is counted from left to right. Each bar is divided into two halves longitudinally. The progress of the completion of the activity is shown by hatching the top half. The concurrent activities are represented by the appropriate overlap timewise.

Though it is easy to draw a bar chart and compare the progress of the completed work with the original schedule it has certain inherent disadvantages. The bar chart cannot depict the interrelationships among the activities clearly. It is difficult to incorporate the uncertainties in the time schedules. Milestone chart is a modification over the bar chart where in the main activity is subdivided into subactivities. The beginning and end of these sub-divided activities are called milestones.

Network techniques such as critical path method (CPM) and the programme evaluation and review technique (PERT) have been found to be very useful in accounting for the complex inter-relationships that exist among the activities.

A well-defined job or task is called an activity. The beginning or end of each such activity is called an event. A network is a pictorial representation of activities and events. The activities are denoted by arrows while the events are denoted by circles. Events are numbered for identification.

The beginning of an activity is called a tail event. If the tail event happens to represent the commencement of the project, then it is called the initial event. A tail event representing the beginning of more than one activity is said to occur when the first activity starts from it. An event denoting the completion of an activity is called head event. If the head event happens to represent the completion of the project, then it is called the final event or end event. When a head event occurs at the end of more than one activity, the event is said to have occurred when all activities leading to it are completed.

An event which is a head event for some activity and at the same time tail event for some other activity is called a dual role event. In a network all events except initial and final events are dual role events. The two conventions followed in drawing a network are : (1) the time flows from left to right and (2) the head events always have a number larger than that of a tail event.

The activities which can be performed simultaneously and independently of each other are called the parallel activities. The activities which can be performed one after another in succession are known as serial activities. Activity or activities that are to be necessarily completed before another activity commences are called predecessor activities to that activity. Similarly, the activity or activities which can be performed after the completion of other activities are called the successor activities.

An activity which does not require any time or resource is called a dummy activity. A dummy is used as a device to identify a dependence among activities. It is usually denoted by a broken line arrow.

Two errors in logic may occur while drawing a network. They are looping and dangling. In a looping the path of activities leads back into itself. In dangling an activity is drawn with no interconnection.

An event into which a number of activities enter and from which one or several activities leave is called a Merge node. Similarly, events which have one or several entering activities generating a number of emerging activities are known as Burst nodes.

Once the network of the project is drawn it can be analysed by using either PERT or CPM techniques. In PERT the time required for completion of any activity is treated as a random variable. In other words PERT uses the probabilistic approach. It is event oriented. In CPM the time required for the completion of the activity is estimated with fair degree of accuracy. That means CPM uses determinisitic approach and it is activity oriented.

The shortest possible time in which an activity can be completed under ideal conditions is known as the optimistic time estimate (t_0).

The maximum time that would be required to complete the activity under adverse conditions is known as the pessimistic time estimate (t_p).

The time required to complete the activity under normal conditions is known as the most likely time estimate (t_L).

The probability distribution used to describe the time estimate in PERT analysis is Beta distribution. According to this distribution the expected time (also known as the average time or the mean time) of completion of the activity t_E is given by

$$t_E = \frac{t_0 + 4t_L + t_p}{6}$$

The variance of the time of completion is given by $\sigma^2 = \left(\dfrac{t_p - t_0}{6}\right)^2$ and the standard deviation σ is the square root of variance.

The expected time for the path consisting of number of activities in series is obtained as the arithmetic mean of the expected times of the activities in the path. That is

$$(t_E) \text{ for the path} = \frac{t_{E_1} + t_{E_2} + \ldots + t_{E_n}}{n}.$$

Similarly, the standard deviation of the time of completion for the network ending event σ_{t_E} is given by

$$\sigma_{t_E} = \sqrt{\frac{\sigma_1^2 + \sigma_1^2 + \ldots + \sigma_n^2}{n}}.$$

In a network, the critical path is one for which t_E is maximum.

The earliest expected time (T_E) is the time when an event can be expected to occur. This may be indicated above or below the node (that is the event circle) in a network. If more than one parth lead to an event, then the earliest expected time of that event is taken as maximum of t_E's computed along various paths.

The latest time by which an event must occur, to keep the project on schedule is known as the latest allowable occurrence time (T_L).

Slack is defined as the difference between the latest allowable time and the earliest expected time of an event.

$$S = T_L - T_E$$

Positive, zero and negative slacks respectively represent the time of completion ahead of schedule, on schedule and behined schedule.

A critical path is the path which connects the initial and final events through those events for which the slack is either zero or minimum. It represents the path for which the earliest expected time is maximum. All events along the critical path are considered to be critical since any delay in their occurence will lead to delay in the completion of the project.

Since the expected time of completion of the project is the sum of the individual expected times of all the activities along the critical path, it is supposed to follow the normal distribution, according to central limit theorem. Therefore, the chances for completing the project any time between $(t_E - \sigma_{t_E})$ and $(t_E + \sigma_{t_E})$ are 68% and for $(t_E - 2\sigma_{t_E})$ and $(t_E + 2\sigma_{t_E})$ it is 95%.

As already pointed out CPM is activity oriented and it uses the deterministic approach.

The earliest event time (T_E) is the earlist time at which an event can occur. This term is analogous to the earliest expected time used in the PERT analysis except that the degree of uncertainty associated with the word "expected" is now not present. The latest allowable occurrence time (T_L) of an event in CPM is defined in the same way as in PERT.

The earliest time by which an activity can commence is known as the earliest start time (EST) of that activity. So we have

EST of an activity = Earliest event time (T_E) at its tail.

The earliest time by which an activity can be completed is known as the earliest finish time (EFT) of that activity. That means

EFT = EST + activity duration

The latest time by which the activity can be completed is known as the latest finish time (LFT). In other words,

LFT = Lastest event time at the head event of the activity.

The latest start time (LST) of an activity is the latest time by which it can be started without delaying the completion of the project. For no delay condition to be satisfied, it should be obviously equal to the latest finish time minus the activity of the duration. That means

LST = LFT − activity duration.

The term float in CPM is analogous to the term slack in PERT. Float is the range in which the start time or finish time of an activity can fluctuate without affecting the completion time of the project.

Total float (F_T) is the difference between the maximum time available and the actual time required to complete an acitivity.

Free float (F_F) is that portion of the total float that can be consumed by any activity without delaying any succeeding activity. It can also be expressed as the difference between total float and the slack of the head event.

Independent float (F_{ID}) is defined as the minimum available time in excess of the activity duration. It is also equal to difference between free float and slack of the tail event.

Interfering float (F_{IT}) is the difference between total float and free float. It is also equal to slack of the head event.

Total float of an activity is an important concept in CPM. If the total float is negative, the activity becomes supercritical and it requires very special attention and action. If the total float is zero, the activity becomes critical and it demands above normal attention with no freedom of action. When the float is positive, the activity becomes sub-critical allowing some freedom of action.

The critical path consists of those activities for which the total float is zero and which form a continuous chain starting at the first node and ending with the last node.

CPM can make use of cost estimates along with the time estimates and provides a schedule for completing the activities at the minimum total cost.

The standard time that is usually allowed to an estimater to complete an activity is known as the normal time.

The minimum possible time in which an activity can be completed by employing extra resources is called the crash time. It is not possible to complete the activity earlier than crash time by any amount of increase in resources.

Normal cost is the direct cost required to complete the activity in normal time duration. Crash cost is the direct cost required to complete the activity in crash time.

$$\text{The cost slope of direct cost curve} = \frac{\text{Crash cost} - \text{normal cost}}{\text{Normal time} - \text{crash time}}.$$

The minimum time the project takes for its completion will be the sum of the crash times of all the activities along the critical path. The crashing of critical activities may result in some of the non-critical activities becoming critical. It is better to start with crashing that critical activity first which has the lowest cost slope. Sometimes it may not be advisable to crash the critical activities fully. Then they may be crashed in stages.

II. OBJECTIVE TYPE QUESTIONS

1. A project is
    (a) a large dam constructed across a river for a single or multi-purpose ☐
    (b) any job involving many people and large money ☐
    (c) a work of major importance involving huge men and material ☐
    (d) an organised team work to achieve a set task within the time limit. ☐
2. Technology is the
    (a) study of techniques ☐
    (b) study of machines and their operation ☐
    (c) study of industrial relations ☐
    (d) study of behaviour of men and machines. ☐
3. An activity is
    (a) an element of work entailed in the project ☐
    (b) the movement of heavy vehicles from one place to another ☐
    (c) the beginning or the end of a specified job ☐
    (d) the progress of work upto centrain limit. ☐
4. An event is
    (a) a dummy activity ☐
    (b) the useful criterion ☐
    (c) the start and/or finish of an activity or the group of activities ☐
    (d) a definite time interval. ☐
5. The pictorial representation of activities and events of a project is known as the
    (a) Flow chart ☐ (b) Flow net ☐
    (c) Algorithm ☐ (d) Net work. ☐

6. The flow of time in a net work is generally
   (a) from left to right
   (b) from right to left
   (c) both ways
   (d) of no importance.
7. Bar chart is also known as
   (a) Histogram chart
   (b) Flow chart
   (c) Time chart
   (d) Gantt chart.
8. Bar chart is drawn for
   (a) time versus activity
   (b) activity versus resources
   (c) resources versus progress
   (d) progress versus time.
9. Policy is
   (a) a rule or a set of rules never to be violated
   (b) a definite route along which one has to proceed
   (c) a procedure for preparing the project report
   (d) a set of broad guidelines stipulated by management.
10. An activity requires
    (a) events
    (b) resources
    (c) time and resources
    (d) energy and vigour.
11. A drawback of the bar chart is that
    (a) it is difficult to judge whether an activity is completed or not
    (b) all the activities represented are independent of each other
    (c) sequence of activities is not clearly defined
    (d) it is not possible to judge whether the activity is ahead or behined schedule.
12. Milestones are
    (a) the events specifying the beginning and/or the end of subdivided activities
    (b) the stones indicating the distance along a highway
    (c) dates of important happenings in the history of any nation
    (d) none of the above.
13. Henry Gantt developed the bar chart technique around
    (a) 1800
    (b) 1900
    (c) 1850
    (d) 1950.
14. Milestone charts are developed around
    (a) 1800
    (b) 1840
    (c) 1900
    (d) 1940.
15. The earliest method used for planning of projects was
    (a) PERT
    (b) CPM
    (c) Bar chart
    (d) Milestone chart.
16. Bar charts are suitable only for
    (a) minor projects
    (b) medium projects
    (c) major projects
    (d) all the above.

17. PERT is an abbreviation for
    (a) pertinent equation related to a task
    (b) performance, evaluation, rating and timing
    (c) programme evaluation and review techniqe
    (d) periodical estimation of resource treasure.

18. CPM stands for
    (a) computer programming mode
    (b) critical project management
    (c) controlling, planning and maintenace
    (d) critical path method.

19. PERT is
    (a) activity oriented
    (b) event oriented
    (c) time oriented
    (d) resource oriented.

20. CPM is
    (a) activity oriented
    (b) event oriented
    (c) time oriented
    (d) resource oriented.

21. PERT adopts
    (a) deterministic approach
    (b) probabilistic approach
    (c) stochastic approach
    (d) none of the above.

22. CPM adopts
    (a) deterministic approach
    (b) probabilistic approach
    (c) stochastic approach
    (d) none of the above.

23. Which of the following statements is true ?
    (a) PERT is activity oriented and adopts deterministic approach.
    (b) CPM is event oriented and adopts probabilistic approach.
    (c) CPM is activity oriented and adopts probabilistic approach.
    (d) PERT is event oriented and adopts probabilistic approach.

24. When the events are numbered in a network then the number of the head event of an activity is
    (a) always larger than the number of the tail event of the same activity
    (b) always smaller than the number of the tail event of the same activity
    (c) always equal to the number of the tail event of the same activity
    (d) none of the above.

25. Which of the following is an example of parallel activities ?
    (a) construction of walls and casting of roof
    (b) construction of walls and carpentry work of doors and windows
    (c) casting of roof and construction of parapet wall
    (d) digging of well and construction of septic tank.

26. The occurrence of the completion of an activity is called its
    (a) head event
    (b) tail event
    (c) dual role event
    (d) none of the above.

27. The occurrence of the starting of an activity is called its
    (a) head event
    (b) tail event
    (c) dual role event
    (d) none of the above.
28. A dual role event
    (a) is the head event as well as a tail event
    (b) consumes no time
    (c) is the beginning of one event and the end of another
    (d) is any event other than the initial and final events.
29. Pick up the correct statement from the following :
    (a) In a network activities are represented by arrows and events are circled.
    (b) Events into which a number of activities enter and from which one or several activities emerge are called the merge nodes.
    (c) Events into which one or several activities enter and from which a number of emerging activities are generated are called burst nodes.
    (d) all the above.
30. An important principle in drawing a network is
    (a) no activity can start until all the previous activities in the same chain are completed
    (b) parallel activities should begin and end at the same time
    (c) between two events there should not be more than two activities
    (d) the number of dummy activities in a network should not exceed 4.
31. A dummy activity
    (a) has no tail event but only a head event
    (b) has only head event but no tail event
    (c) does not require any resources or any time
    (d) has no bearing on the network and can appear anywhere.
32. At an event other than final event if no activity emerges, it results in an error called
    (a) looping
    (b) dangling
    (c) interfacing
    (d) splicing.
33. If the path of activities leads back into itself, the resulting error in the network is known as
    (a) looping
    (b) dangling
    (c) interfacing
    (d) splicing.
34. In the figure given below, identify the network with the error looping

35. In the figure of the problem 34 identify the network with dangling.
36. Which of the following denotes a dummy activity P?
   (a) ⑨ —P→ ⑩ ☐   (b) ⑨ --P-- ⑩ ☐
   (c) ⑨ --P--→ ⑩ ☐   (d) ⑨ ---P--- ⑩ ☐
37. In the network shown in the figure aside, which of the following statement is true.
   (a) activities A and B should be completed before C commences ☐
   (b) activities A and B should be completed before D commences ☐
   (c) activities A and B should be completed before E commences ☐
   (d) the sequence of the activities is $A \to B \to C \to D \to E$. ☐

   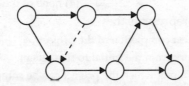

38. If the activity A preceeds B but succeeds C, then network is
   (a) ⑥ —A→ ⑦ —B→ ⑧ —C→ ⑨ ☐
   (b) ⑥ —B→ ⑦ —C→ ⑧ —A→ ⑨ ☐
   (c) ⑥ —C→ ⑦ —A→ ⑧ —B→ ⑨ ☐
   (d) ⑥ —C→ ⑦ —B→ ⑧ —C→ ⑨ ☐

39. A network is given below :

   The correct numbering of the events is given by

   (a) 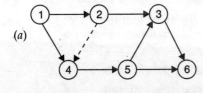 ☐   (b) 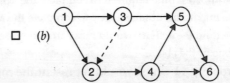 ☐

   (c) 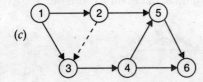 ☐   (d) 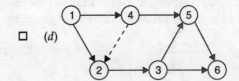 ☐

40. In which of the following are the arrows correctly indicated ?

(a) 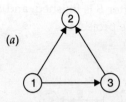   (b) 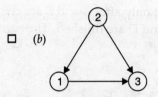 □

(c) 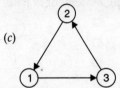 □   (d) □

41. The following figure indicates
(a) a merge □   (b) a burst □
(c) an activity □   (d) none of the above. □

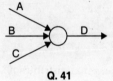
Q. 41

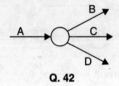
Q. 42

42. What does the figure question 42 indicate ?
(a) a merge □   (b) a burst □
(c) an activity □   (d) none of the above. □

43. In PERT, the critical path represents the
(a) shortest path for the earliest completion of the project □
(b) longent path of the network from the initial to final event □
(c) ideal path by proceeding along which the project can be completed as per schedule □
(d) path which takes into account the completion of the parallel activities. □

44. The optimistic time estimate represents
(a) the shortest time required to complete the activity under favourable conditions □
(b) the maximum time required to complete the activity under adverse conditions □
(c) the time the activity would most often require if normal conditions prevail □
(d) none of the above. □

45. In problem 44 which would represent the most likely time estimate.

46. In problem 44 which would represent the pessimistic time estimate.

47. If t_0, t_P and t_L represent the optimistic, pessimistic and most likely time estimates, the expected time of completion of the activity is given by

(a) $t_E = \dfrac{t_0 + t_L + t_P}{3}$ □   (b) $t_E = \dfrac{t_0 + 2t_L + t_P}{4}$ □

(c) $t_E = \dfrac{t_0 + 3t_L + t_P}{5}$ □   (d) $t_E = \dfrac{t_0 + 4t_L + t_P}{6}$. □

48. A job consists of five activities A, B, C, D and E. Activities A and B can start concurrently. Activity C can start only after A and B are completed, D starts after B is finished, and E can start only after C and D are finished. The correct network for the project is

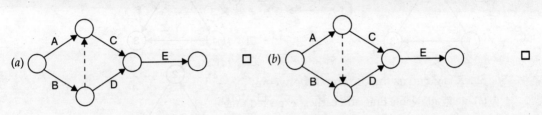

49. The probability distribution taken to represent the completion time in PERT analysis is
    (a) Gamma distribution
    (b) Normal distribution
    (c) Beta distribution
    (d) Log-normal distribution.

50. According to Beta distribution, the standard deviation of the time of completion is given by
    (a) $\sigma = \dfrac{t_p - t_0}{3}$
    (b) $\dfrac{t_p - t_0}{4}$
    (c) $\dfrac{t_p - t_0}{5}$
    (d) $\dfrac{t_p - t_0}{6}$.

51. The optimistic, most likely and pessimistic time estimates of an activity are 5, 10, 21 days. What are the expected time and standard deviation ?
    (a) 12, 3
    (b) 11, 4
    (c) 11, 2.67
    (d) 10, 16.

52. Higher standard deviations means
    (a) higher uncertainity
    (b) lower uncertainity
    (c) nothing to do with uncertainity
    (d) none of the above.

53. If $\sigma_1, \sigma_2, \ldots, \sigma_n$ are the standard deviations of time of completion for the activities along the critical path, the standard deviations of the time of completion of the project is given by
    (a) $\sigma_{t_E} = \dfrac{\sigma_1 + \sigma_2 + \ldots + \sigma_n}{n}$
    (b) $\sigma_{t_E} = \sqrt{\dfrac{\sigma_1 + \sigma_2 + \ldots + \sigma_n}{n}}$
    (c) $\sigma_{t_E} = \sqrt{\dfrac{\sigma_1^2 + \sigma_2^2 + \ldots + \sigma_n^2}{n}}$
    (d) $\sigma_{t_E} = \dfrac{\sqrt{\sigma_1} + \sqrt{\sigma_2} + \ldots + \sqrt{\sigma_n}}{n}$.

54. What is the standard deviation for the network given below:

   $t_0 = 1$, $t_p = 7$ (1→2); $t_0 = 2$, $t_p = 8$ (3→4)

   (a) 1
   (b) $\sqrt{2}$
   (c) 2
   (d) 2^2.

55. Slack is given as the difference between
   (a) latest allowable time and earliest expected time
   (b) latest allowable time and the pessimistic time estimate
   (c) earliest expected time and latest allowable time
   (d) final event time and initial event time.

56. Negative slack indicates
   (a) ahead of schedule condition
   (b) behind the schedule condition
   (c) on schedule condition
   (d) none of the above.

57. For all the events along the critical path, the slack is
   (a) maximum
   (b) negative
   (c) minimum or zero
   (d) none of the above.

58. The probability factor Z in PERT is given by

   (a) $\dfrac{\text{scheduled time} - \text{earliest expected time}}{\text{standard deviation}}$

   (b) $\dfrac{\text{scheduled time} - \text{latest allowable time}}{\text{variance}}$

   (c) $\dfrac{\text{latest allowable time} - \text{scheduled time}}{\text{standard deviation}}$

   (d) $\dfrac{\text{latest allowable time} - \text{earliest expected time}}{\text{variance}}$

59. If the probability factor is zero, the chances of completing the project in scheduled time are
   (a) 0%
   (b) 50%
   (c) 75%
   (d) 100%.

60. Float may be defined as the difference between
   (a) latest start time and the earliest start time
   (b) latest finish time and the earliest finish time
   (c) time available and the time required to complete the activity
   (d) all the above.

61. Interfering float is the difference between
   (a) Total float and independent float
   (b) Total float and free float
   (c) Free float and independent float
   (d) Independent float and free float.

62. Negative float for any activity means that the activity is
    (a) super-critical
    (b) sub-critical
    (c) critical
    (d) none of the above.
63. Zero float for any activity means that the activity is
    (a) super-critical
    (b) sub-critical
    (c) critical
    (d) none of the above.
64. Positive float for any activity means that the activity is
    (a) super-critical
    (b) sub-critical
    (c) critical
    (d) none of the above.
65. The total float for any activity along the critical path in CPM is
    (a) positive
    (b) negative
    (c) zero
    (d) any value.
66. Direct cost of a project is due to
    (a) establishment charges as salaries to administrative staff
    (b) loss or gain of revenue
    (c) penalty imposed
    (d) cost of materials and wages of labour.
67. Total project cost
    (a) increases with increase in time
    (b) reduces with increase in time
    (c) initially reduces and then increases with increase in time
    (d) initially increases and then reduces with increase in time.
68. The total cost versus time curve is
    (a) a straight line
    (b) ∪-shaped curve
    (c) a parabola
    (d) ∩-shaped curve.
69. Crashing to the project means
    (a) reducing the time of completion by spending more resources
    (b) reducing the cost of project by delaying the time of completion
    (c) reducing the project size to save the resources
    (d) all the above.
70. Resource smoothening means
    (a) gradual increase in resources
    (b) adjustment of resources to have the least variations
    (c) complete revamping of resources to suit the requirement
    (d) optimisation and economical utilisation of resources.
71. Cost slope of the direct cost curve is given by
    (a) $\dfrac{\text{crash cost} - \text{normal cost}}{\text{normal time} - \text{crash time}}$
    (b) $\dfrac{\text{crash cost} - \text{normal cost}}{\text{crash time}}$
    (c) $\dfrac{\text{crash cost} - \text{normal cost}}{\text{normal time}}$
    (d) $\dfrac{\text{normal cost} - \text{crash cost}}{\text{crash time}}$.

72. In the time-cost optimization using CPM, the crashing of the activities along the critical path is done starting with the activity having
    (a) least cost slope ☐   (b) highest cost slope ☐
    (c) moderate cost slope ☐   (d) none of the above. ☐

73. When all the paths are arranged according to the descending order of the float, the path which is next to critical path is known as
    (a) under-critical path ☐   (b) sub-critical path ☐
    (c) semi-critical path ☐   (d) secondary critical path. ☐

74. Which of the following is a dummy activity?
    (a) Excavate the foundations ☐   (b) Await the arrival of concrete material ☐
    (c) Lay the foundation concrete ☐   (d) Cure the foundation concrete. ☐

75. The time by which the completion of an activity can be delayed without affecting the start of succeeding activities is called
    (a) total float ☐   (b) interfering float ☐
    (c) independent float ☐   (d) free float. ☐

76. Fulkerson's rule is used for numbering the
    (a) events ☐   (b) activities ☐
    (c) initial events ☐   (d) dummies. ☐

77. Which of the following is a sequential?

    (a) ○—B→○—C→○ ☐   (b) ○—C→○ with branches to D and E ☐

    (c) A and B converging to ○—C→○ ☐   (d) △------○ ☐

78. In the problem 77, which is converging

79. In the problem 77, which is diverging

80. Total cost is
    (a) direct cost ☐   (b) direct cost + indirect cost ☐
    (c) overheads ☐   (d) outage loss. ☐

81. Outage loss plus overheads is equal to
    (a) direct cost ☐   (b) indirect cost ☐
    (c) total cost ☐   (d) penalty. ☐

82. Identify the odd man out
    (a) Mix concrete □　　(b) Excavate trench □
    (c) Assemble parts □　　(d) Design completed. □

83. Which of the following is the correct form of representation of an event ?
    (a) ○ □　　(b) ✕ □
    (c) + □　　(d) ⬡ □

84. Which of the following is not related to cost slope ?
    (a) crash cost □　　(b) normal cost □
    (c) crash time □　　(d) slack. □

85. Use of dummies is required for
    (a) grammatical purpose only □　　(b) logical purpose only □
    (c) both (a) and (b) □　　(d) none of the above. □

ANSWERS
Objective Type Questions

1. (d)	2. (b)	3. (a)	4. (c)	5. (d)	6. (a)
7. (d)	8. (a)	9. (d)	10. (c)	11. (c)	12. (a)
13. (b)	14. (d)	15. (c)	16. (a)	17. (c)	18. (d)
19. (b)	20. (a)	21. (b)	22. (a)	23. (d)	24. (a)
25. (b)	26. (a)	27. (b)	28. (d)	29. (d)	30. (a)
31. (c)	32. (b)	33. (a)	34. (b)	35. (c)	36. (c)
37. (b)	38. (c)	39. (c)	40. (d)	41. (a)	42. (b)
43. (b)	44. (a)	45. (c)	46. (b)	47. (d)	48. (a)
49. (c)	50. (d)	51. (c)	52. (a)	53. (c)	54. (b)
55. (a)	56. (b)	57. (c)	58. (a)	59. (b)	60. (d)
61. (b)	62. (a)	63. (c)	64. (b)	65. (c)	66. (d)
67. (c)	68. (b)	69. (a)	70. (b)	71. (a)	72. (a)
73. (c)	74. (b)	75. (d)	76. (a)	77. (a)	78. (c)
79. (b)	80. (b)	81. (a)	82. (d)	83. (a)	84. (d)
85. (c)					

Chapter 18: Estimating and Quantity Surveying

I. INTRODUCTION

Estimating may be defined as the forecast of the probable cost of the building or a project prior to the commencement of the job. It requires skill, experience, foresight and good judgement on the part of the person making the estimate besides a sound knowledge of cost of materials and labour involved in the work. A good estimate should not differ from the actual cost of the project after completion by more than 15% provided the cost of labour and material remain unaltered. Guess-work means to make a random judgement not based on all the factors concerned. It may result in a too high or too low a value.

Estimating helps in altering or modifying the project to suit to the given budget. It is useful in formulating the tenders for the works and to check the work carried out by contractors for the purpose of interim and final payments. It also provides a basis for fixing the standard rent of buildings.

A fully dimensioned drawing to scale along with sectional views, detailed specifications about the materials to be used, and a schedule of rates of all the items of construction are the essential requirements to proceed with estimating.

Quantity surveying may be defined as the working out or measuring the quantities of a work based on standard method of measurement in a systematic and scientific manner, which when priced give the estimated cost to a reasonable degree of accuracy.

The different types of estimate are
   (*i*) Detailed estimate or Item Rate estimate or Intensive estimate.
   (*ii*) Plinth area estimate.
   (*iii*) Cubicle content estimate or the cube rate estimate.
   (*iv*) Revised estimate.
   (*v*) Supplementary estimate.

Detailed estimate is an accurate estimate and is prepared in two stages. In the first stage the details of measurement of each item are taken out correctly from the drawings and the quantities under each item are computed. In the second stage the cost of each item of work is calculated and all the costs are added to give the total cost. A certain percentage of the estimated cost is added to take care of certain items, which do not come under any head of items and also some unforeseen items. This is known as contingencies. The contingencies is usually taken as 5% to 10%. Some percentages may also be added against workcharged establishment.

In plinth area estimate the plinth area rate is arrived at by dividing the cost of existing structures with similar specifications and construction in that locality by the plinth area. Now the estimated cost of the proposed structure is calculated by multiplying the plinth area of the structure with the plinth area rate. Thus for example if the plinth area rate is ₹ 1500 per sq. m, the plinth area estimate of a building with a plinth area of 100 sq. m will be ₹ 1,50,000. The plinth area is computed as the covered area by taking the external dimensions at the floor level. Courtyard and other open areas are excluded.

The cube rate estimate is made by multiplying the cubical contents of the proposed building by the cube rate which is deduced from the cost of the similar buildings having similar specifications and construction in the locality.

Both plinth area and cube rate estimates are approximate.

Revised estimate is a detailed estimate which is prepared when

(i) there are major deviations from the original proposal, or

(ii) when the original sanctioned estimate is likely to exceed by 5%, or

(iii) when the expenditure exceeds the administrative sanction by 10%.

Supplementary estimate is a detailed estimate which is prepared when additional works are added to the original works.

When the engineering departments take up the works of other departments some amount is charged to meet the expenses of establishment, designing, planning, supervision etc. This is called the centage charges and it varies from 10% to 15% of the estimated cost.

The main items of work in the detailed estimate are

(i)	Earthwork	—both in excavation and filling, the quantities are in m^3.
(ii)	Concrete in foundations	—quantity in m^3.
(iii)	Soling	—quantity in m^2, this is nothing but one layer of bricks put below the foundation concrete.
(iv)	Damp proof course	—quantity in m^2.
(v)	Masonry	—quantity in m^3.
(vi)	Lintels over openings	—quantity in m^3.
(vii)	R.C.C. work	—quantity in m^3.
(viii)	Flooring and roofing	—quantity in m^2.
(ix)	Plastering and pointing	—quantity in m^2.
(x)	Pillars	—quantity in m^3.

(xi) Doors and windows — wood for frames and trusses in m³ wood for door and window shutters in m².
(xii) Iron and steel — quantity in quintals or tons.
(xiii) White-washing or colour-washing — quantity in m².
(xiv) Painting — quantity in m².
(xv) Electrification — about 8% of the estimated cost.
(xvi) Sanitary and water supply — about 8% of the estimated cost.

While working out the quantities of masonry work the deductions are made for the openings bearings, etc. as follows:

Openings of less than 0.1 m² area — no deduction
Bearings of floors and roofs slabs — no deduction
For other openings — full deduction.

In R.C.C. works, if details are not given the volume of steel may be taken as 0.6% to 1.0% of R.C.C. volume. No deduction for steel is made in the volume of concrete.

Deductions allowed in the case of plastering and painting are as follows:

Openings of less than 0.5 m² area — no deduction.
Openings of area between 0.5 to 3 m² — deduction is made for one face only and the other face is allowed for jambs, soffits, sills etc.
Openings of area more than 3 m² — deductions is made for both faces and the jambs, soffits and sills are accounted for.

Weight of iron hold fast is usually taken as one kg. When not specified, 4 hold fasts for window and 6 hold fasts for doors are taken.

For painting works the outer dimensions of doors and windows are taken for computing areas.

The standard size of modular brick is 12 cm × 9 cm × 9 cm. The thickness of wall for the purpose of estimate is taken as an integral multiple of the width of the brick. For example for one brick wall it is 20 cm, for one-and-a half brick wall it is 30 cm, for two brick wall it is 40 cm etc. But, for the 3 brick walls and above the thickness of the wall is actually measured after construction for the purpose of payment.

The price of an item of work is made up of the following components:

(i) Cost of the material
(ii) Cost of the labour
(iii) Cost of tools and plants
(iv) Cost of the overheads
(v) The profit.

The determination of rate per unit of a particular item of work, from the cost of quantities of material, the cost of labour and other expenditure required for its completion is known as the analysis of rate. The profit for the contractor is usually taken as 10%.

In earthwork the nominal lead and lift for the disposal of the excavated soil are 30 m and 1.5 m respectively.

The number of standard modular bricks required for one cubic metre of masonry is 500.

The constituent materials required to make concretes and mortars depend on the mix design. For example to prepare 100 m³ of 1 : 2 : 4 cement concrete, we require 84 m³ of coarse aggregate 42 m³ of sand and 21 m³ of cement. Similarly to prepare 1 : 5 : 10 concrete, we require 92 m³ of coarse aggregate, 46 m³ of sand and 9.2 m³ of cement.

The information which cannot be shown on the drawings is conveyed through specifications. For example, we cannot show the quality of the material to be used or the method of applying paint (2 coats, 3 coats etc.) on a drawing. Without this information we cannot workout the quantities and prepare the estimate. Whenever possible, for the more usual materials, Indian Standards and codes of practice are to be referred.

The technique of determining the fair price or value of a property such as a building, factory, land etc., is known as the valuation. There is a clear distinction between cost and value. Cost means original cost of construction or purchase while value means the present saleable value which may be higher than or lower than the cost.

Book value is the original investment less the depreciation for the period of existence. Salvage value is the value at the end of useful life period without being dismantled. Scrap value is the value of the dismantled material less the cost of dismantling.

Annuity is the annual periodic payments for repayment of the capital amount invested. The gradual accumulation of amount by way of periodic annual deposits meant for the replacement of the structure at the end of its useful life period is known as the sinking fund. Depreciation is the decrease in the value of the property due to structural deterioration, use, wear tear, decay and obsolescence.

II. OBJECTIVE TYPE QUESTIONS

1. Estimate is
   (a) the actual cost of construction of a structure
   (b) the probable cost arrived at before commencement of the structure
   (c) a random guess of the cost of the structure
   (d) none of the above.

2. Which of the following is the most correct estimate?
   (a) plinth area estimate
   (b) cube rate estimate
   (c) detailed estimate
   (d) building cost index estimate.

3. A document containing detailed description of all the items of work (but their quantities are not mentioned) together with their current rates is called
   (a) tender
   (b) schedule of rates
   (c) analysis of rate
   (d) abstract estimate.

4. The approximate cost of the complete labour as a percentage of the total cost of the building is
   (a) 10%
   (b) 25%
   (c) 40%
   (d) 5%.

5. Working out the exact quantities of various items of work is known as
   - (a) estimating
   - (b) mensuration
   - (c) quantity surveying
   - (d) valuation.

6. The essential requirements to prepare a good estimate are
   - (a) a full dimensioned drawing to scale
   - (b) detailed specifications
   - (c) schedule of rates
   - (d) all the above.

7. The covered area of a proposed building is 150 m² and it includes a rear courtyard of 5 m × 4 m. If the prevailing plinth area rate for similar buildings is ₹ 1250/m², what is its cost?
   - (a) ₹ 1,87,500
   - (b) ₹ 2,12,500
   - (c) ₹ 1,62,500
   - (d) ₹ 3,75,000.

8. A layer of dry bricks put below the foundation concrete, in the case of soft soils, is called
   - (a) soling
   - (b) shoring
   - (c) D.P.C.
   - (d) none of the above.

9. The quantity of soling is obtained in
   - (a) m³
   - (b) m
   - (c) lump-sum
   - (d) m².

10. The quantity of wood for the shutters of doors and windows is calculated in
    - (a) m²
    - (b) m³
    - (c) lump-sum
    - (d) m.

11. In the detailed estimate the areas are worked out to the nearest
    - (a) 0.001 m²
    - (b) 0.01 m²
    - (c) 0.05 m²
    - (d) 0.005 m².

12. In the detailed estimate the volumes are worked out to the nearest
    - (a) 0.001 m³
    - (b) 0.005 m³
    - (c) 0.01 m³
    - (d) 0.05 m³.

13. The expenses of items which do not come under any regular head of items and the cost of unforeseen items are called
    - (a) lump-sum
    - (b) extras
    - (c) customary charges
    - (d) contingencies.

14. The size of the standard modular brick is
    - (a) 19 cm × 9 cm × 9 cm
    - (b) 20 cm × 9 cm × 9 cm
    - (c) 20 cm × 10 cm × 9 cm
    - (d) 20 cm × 10 cm × 10 cm.

15. The number of standard modular bricks required to make 1 m³ of masonry is
    - (a) 480
    - (b) 500
    - (c) 520
    - (d) 540.

16. The nominal lead and lift allowed for the earthwork in the excavations of the foundations are
    - (a) 50 m and 2 m
    - (b) 30 m and 2 m
    - (c) 30 m and 1.5 m
    - (d) 20 m and 1 m.

17. The quantity of damp proof course (D.P.C.) is worked out in
    (a) m³
    (b) m²
    (c) m
    (d) lump-sum.

18. No deductions is made in the masonry for the openings if the area of the opening does not exceed
    (a) 0.5 m²
    (b) 0.25 m²
    (c) 0.15 m²
    (d) 0.10 m².

19. When not specified, the volume of steel in R.C.C. work is taken as
    (a) 1% to 1.6% of R.C.C. volume
    (b) 2% to 4% of R.C.C. volume
    (c) 4% to 6% of R.C.C. volume
    (d) 0.6% to 1.0% of R.C.C. volume.

20. The volume of coarse aggregate required to make 100 m³ of 1 : 2 : 4 concrete is
    (a) 84 m³
    (b) 88 m³
    (c) 92 m³
    (d) 96 m³.

21. In the analysis of rates, the profit for the contractor is generally taken as
    (a) 20%
    (b) 15%
    (c) 10%
    (d) 5%.

22. The technique of finding the fair price of an existing building or property is known as
    (a) estimation
    (b) valuation
    (c) pricing
    (d) costing.

23. The value of the property (without being dismantled) at the end of the useful life period is known as
    (a) scrap value
    (b) salvage value
    (c) junk value
    (d) book value.

24. The annual periodic payments made for the repayment of the capital invested is known as
    (a) annuity
    (b) depreciation
    (c) sinking fund
    (d) solatium.

25. The gradual accumulation of amount by way of annual periodic deposits which is meant for the replacement of the structure at the end of its useful life period is known as
    (a) annuity
    (b) depreciation
    (c) sinking fund
    (d) solatium.

26. The thickness of slabs and beams must be measured to the nearest
    (a) 0.001 m
    (b) 0.005 m
    (c) 0.01 m
    (d) 0.05 m.

27. The value of the dismantled material less the cost of dismantling is called
    (a) the scrap value
    (b) salvation value
    (c) rateable value
    (d) none of the above.

28. The quantity of partition walls and honey-comb walls are worked out in
    (a) m
    (b) m²
    (c) m³
    (d) lump-sum.

29. The portion which is wrongly excavated by the contractor is to be filled by
    (a) the earth excavated from the same pit ☐
    (b) the earth excavated from a borrow pit far away from the site ☐
    (c) concrete as specified ☐
    (d) brick masonry. ☐

30. What is the minimum period for which the lime concrete in foundation be left wet without the construction of masonry over it?
    (a) 3 days ☐   (b) 5 days ☐
    (c) 7 days ☐   (d) 15 days. ☐

31. In the centre line method of working out volumes, for cross walls, what deductions must be made from the centre line length at each junction?
    (a) twice the breadth ☐   (b) breadth ☐
    (c) 1.5 breadth ☐   (d) half the breadth. ☐

32. In what units are the quantities for the frames of doors and windows computed
    (a) m ☐   (b) m^2 ☐
    (c) m^3 ☐   (d) lump-sum. ☐

33. When engineering departments undertake the works of other departments the amount charged towards design, supervision and execution etc., is called
    (a) work charged establishment ☐   (b) contingencies ☐
    (c) service charges ☐   (d) centage charges. ☐

34. The plan of a building is in the form of a rectangle with centre line dimensions of outer walls as 9.7 m × 14.7 m. The thickness of the wall in super structure is 0.30 m. Then its plinth area is
    (a) 150 m^2 ☐   (b) 147 m^2 ☐
    (c) 145.5 m^2 ☐   (d) 135.36 m^2. ☐

35. In question 34, what is the carpet area of the building?

36. The quantum of work of any item a skilled labour is supposed to turnout in a day is known as
    (a) unit work ☐   (b) task work ☐
    (c) target work ☐   (d) basic work. ☐

37. In the analysis of rates what percentage of total cost is provided towards sanitary and water supply charges
    (a) 5% ☐   (b) $7\frac{1}{2}$% ☐
    (c) $1\frac{1}{2}$% ☐   (d) 10%. ☐

38. The density of cement is generally taken as
    (a) 1500 kg/m^3 ☐   (b) 1750 kg/m^3 ☐
    (c) 1250 kg/m^3 ☐   (d) 2000 kg/m^3. ☐

39. The weight of cement in one bag is
    (a) 45 kg
    (b) 50 kg
    (c) 60 kg
    (d) 65 kg.

40. The volume of cement in one bag is
    (a) 0.067 m³
    (b) 0.050 m³
    (c) 0.035 m³
    (d) 0.025 m³.

41. The concealed faces of the frames of doors and windows are painted with
    (a) two coats of primer
    (b) two coats of same enamel paint which is applied for the rest of the frame
    (c) varnish
    (d) two coats of coaltar.

42. The information which cannot be included in drawings is conveyed to the estimator through
    (a) specifications
    (b) cover note
    (c) progress chart
    (d) none of the above.

43. The area of the segmental portion of an arch with span l and riser r is approximately given by
    (a) $\frac{2}{3} lr$
    (b) $\frac{1}{2} lr$
    (c) $\frac{3}{4} lr$
    (d) $\frac{5}{8} lr$.

44. The quantity for expansion joint in buildings is worked out in
    (a) m³
    (b) m²
    (c) m
    (d) lump-sum.

45. The volume of cement required for 10 m³ of brickwork in 1 : 6 cement mortar is approximately equal to
    (a) $\frac{3}{7}$ m³
    (b) $\frac{3}{6}$ m³
    (c) $\frac{3}{4}$ m³
    (d) $\frac{5}{8}$ m³.

46. If the bearing is not specified for the lintel, in the estimation it is usually taken as
    (a) thickness of lintel subjected to a minimum value of 12 cm
    (b) $\frac{3}{4}$ of lintel thickness or 12 cm whichever is larger
    (c) $\frac{1}{2}$ of lintel thickness
    (d) 15 cm.

47. In the estimation of plastering surface the deductions are not made for
    (a) ends of beams
    (b) end of rafters
    (c) small openings upto 0.5 m²
    (d) all the above.

48. When it is not specified, the number of hold fasts for a door is usually taken as
    (a) 2
    (b) 4
    (c) 6
    (d) 8.
49. The approximate weight of one cubic metre of mild steel is
    (a) 1000 kg
    (b) 2400 kg
    (c) 14000 kg
    (d) 7850 kg.
50. In the case of roof truss made of steel, rivets, bolts and nuts usually account for
    (a) 1%
    (b) 5%
    (c) 10%
    (d) 15%.
51. Which of the following is not a common size of reinforcement bars
    (a) 16 mm
    (b) 20 mm
    (c) 25 mm
    (d) 28 mm.
52. The number of corrugations in a galvanised corrugated sheet of standard width is usually
    (a) 6
    (b) 8
    (c) 10
    (d) 12.
53. Of the total estimated cost of a building, the cost of electrification usually accounts for
    (a) 1%
    (b) 5%
    (c) 8%
    (d) 20%.
54. For electric wiring such as fan, light, plug etc., the estimate is made in terms of
    (a) type of point
    (b) number of points
    (c) total load at main in kW
    (d) total length of wiring in metres.
55. The explosive for blasting is usually expressed in terms of
    (a) explosive power
    (b) volume of earthwork that can be blasted
    (c) kilograms
    (d) none of the above.
56. The standard width of asbestos cement corrugated sheet is
    (a) 0.9 m
    (b) 1.05 m
    (c) 1.2 m
    (d) 1.35 m.
57. Minimum side lap required for asbestos cement sheets is
    (a) 4 cm
    (b) 10 cm
    (c) 15 cm
    (d) 25 cm.
58. End lap provided in asbestos cement sheets is equal to
    (a) 5 cm
    (b) 10 cm
    (c) 20 cm
    (d) 15 cm.
59. In the case of unsewered areas, an additional provision for septic tank is usually
    (a) 1% of the building cost
    (b) 3 to 4% of the building cost
    (c) 10% of the building cost
    (d) 12 to 15% of the building cost.
60. The capacity of a flushing cistern is normally
    (a) 12 to 15 litres
    (b) 20 to 25 litres
    (c) 30 to 40 litres
    (d) 1 to 5 litres.

61. The minimum size of the pipe connected to septic tank is
    (a) 50 mm
    (b) 100 mm
    (c) 150 mm
    (d) 200 mm.

62. Which of the following is known as job overhead?
    (a) Stationery
    (b) Postage
    (c) Workman's compensation insurance etc.
    (d) None of the above.

63. In which of the following works, the work turned in cu. m per mason per day will be the least
    (a) random rubble masonry in lime mortar
    (b) stone arch work
    (c) brick masonry in super structure
    (d) brick masonry in parapet wall.

64. The volume of cement required to prepare 100 cu. m of 1 : 2 : 4 concrete is
    (a) 16 m^3
    (b) 32 m^3
    (c) 25 m^3
    (d) 21 m^3.

65. Which of the following is known as general overhead?
    (a) Losses on advance
    (b) Interest on investment
    (c) Travelling expenses
    (d) Amenities to the labour.

66. Indicating works left in excavated trenches to facilitate the measurement of borrow pits are known as
    (a) jambs
    (b) posts
    (c) tell-tale
    (d) none of the above.

67. The measurement of steel grills is taken in terms of
    (a) area
    (b) volume
    (c) weight
    (d) none of the above.

68. Whenever the whitewashing or distempering is done on corrugated iron sheets, in the estimation the plan area of the sheets is increased by
    (a) 2%
    (b) 7%
    (c) 10%
    (d) 14%.

69. Whenever colour washing on a.c. corrugated sheets is done, in the estimation the plan area of the sheets is increased by
    (a) 5%
    (b) 10%
    (c) 15%
    (d) 20%.

70. In case of grills, for the estimation of painted area, the flat area is multiplied by
    (a) $\frac{1}{2}$
    (b) 1
    (c) $1\frac{1}{2}$
    (d) 2.

71. In case of steel rolling shutters, for the estimation of painted area, the plain area is multiplied by
    (a) $\frac{3}{4}$
    (b) 1
    (c) $1\frac{1}{4}$
    (d) $1\frac{1}{2}$.

72. The amount required to be deposited by a contractor while submitting a tender is known as
    (a) fixed deposit ☐ (b) caution deposit ☐
    (c) security deposit ☐ (d) earnest money deposit. ☐
73. Actual cost of construction plus certain profit is paid to the contractor. Such a contract is known as
    (a) unscheduled contract ☐ (b) nominated contract ☐
    (c) cost + percentage contract ☐ (d) work order. ☐
74. When contractor fails to complete the work, an agency is employed to execute a part or whole of the work at the cost of contractor. Such an agency is known as
    (a) substitute agency ☐ (b) debitable agency ☐
    (c) secondary agency ☐ (d) creditable agency. ☐
75. Approximate weight of 1 cubic metre of sand is
    (a) 800 kg ☐ (b) 1600 kg ☐
    (c) 2400 kg ☐ (d) 3200 kg. ☐
76. Arrange the specific weights of wood (W), cement (C), steel (S) and coarse aggregate (A) in the increasing order
    (a) W—C—S—A ☐ (b) W—C—A—S ☐
    (c) A—S—C—W ☐ (d) A—C—S—W. ☐
77. An area of one hectare is equal to
    (a) 10 m^2 ☐ (b) 100 m^2 ☐
    (c) 10000 m^2 ☐ (d) 1000 m^2. ☐
78. One hectare-metre represents a volume of
    (a) 1000 m^3 ☐ (b) 10000 m^3 ☐
    (c) 100000 m^3 ☐ (d) 1000000 m^3. ☐
79. The weight of 10 mm diameter mild steel rod per metre length is equal to
    (a) 0.22 kg ☐ (b) 0.32 kg ☐
    (c) 0.42 kg ☐ (d) 0.62 kg. ☐
80. The cross-sectional area of 16 mm steel bar is
    (a) 200 mm^2 ☐ (b) 100 mm^2 ☐
    (c) 256 mm^2 ☐ (d) 64 mm^2. ☐
81. How many kilo-litres are there in one cubic metre?
    (a) 0.5 ☐ (b) 1 ☐
    (c) 10 ☐ (d) 100. ☐
82. Which department looks after the execution and maintenance of water supply and sanitary works?
    (a) public works department ☐ (b) social welfare department ☐
    (c) public health department ☐ (d) none of the above. ☐
83. If the payment of annuity begins after some years in future, it is known as
    (a) Deferred annuity ☐ (b) Delayed annuity ☐
    (c) Readjusted annuity ☐ (d) Regulated annuity. ☐

84. To account for the corrugations, the plain area of the semi-corrugated asbestos sheet is increased by (for white washing)
    (a) 1%
    (b) 10%
    (c) 20%
    (d) 30%.
85. Specifications for the hold fasts are given in terms of
    (a) number
    (b) weight
    (c) volume
    (d) length.
86. If valuable properties are found during excavation, it becomes the property of
    (a) contractor
    (b) labourers
    (c) owner of the site
    (d) government.
87. The volume of sand a normal truck can carry per trip is approximately
    (a) 20–25 m³
    (b) 15–20 m³
    (c) 10–15 m³
    (d) 3–5 m³.
88. The depth of a shallow manhole should not be more than
    (a) 2.1 m
    (b) 3.1 m
    (c) 4.1 m
    (d) 5.1 m.
89. The life of teakwood doors and windows is usually taken to be
    (a) 80 year
    (b) 60 year
    (c) 40 year
    (d) 20 year.
90. Normally the ratio of span to effective depth of a beam is
    (a) 10 to 12
    (b) 4 to 8
    (c) 15 to 20
    (d) 8 to 10.
91. The unit weight of R.C.C. in kg/m³ is
    (a) 1200
    (b) 1800
    (c) 2400
    (d) 3000.
92. One metric horse power is equal to
    (a) 1.36 kW
    (b) 0.736 kW
    (c) 1.736 kW
    (d) 0.75 kW.
93. The head of the division of public works department is
    (a) chief engineer
    (b) superintending engineer
    (c) executive engineer
    (d) divisional engineer.
94. The head of the circle of public works department is
    (a) assistant engineer
    (b) circle engineer
    (c) executive engineer
    (d) superintending engineer.
95. The quantity of cement concrete damp-proofing course is measured in terms of
    (a) m
    (b) m²
    (c) m³
    (d) lump-sum.
96. The total cost of construction including cost of land is termed as
    (a) Market value
    (b) Capital cost
    (c) Book value
    (d) Rateable value.

97. The ...... value of a property is the amount of money whose annual interest at the highest prevailing rate of interest will be equal to the net income from the property.
   (a) book
   (b) salvage
   (c) capitalised
   (d) market.

98. The amount of annuity paid for a definite number of years is known as
   (a) deferred annuity
   (b) annuity certain
   (c) annuity due
   (d) perpetual annuity.

99. The continued payment of annuity for indefinite period is known as
   (a) deferred annuity
   (b) annuity due
   (c) annuity certain
   (d) perpetual annuity.

100. The decrease in the value of the property due to structural deterioration, continuous use, wear and tear, decay and obsolescence is termed as
   (a) years purchase
   (b) salvage
   (c) depreciation
   (d) sinking.

101. The quantity of metal required, in cubic metres, for a 3.7 m wide road of one km long for one layer of compacted thickness 8 cm is
   (a) 296
   (b) 444
   (c) 370
   (d) 222.

102. In standard notation which of the following is prismoidal formula for earthwork calculations
   (a) $V = \frac{L}{6}(A_1 + A_2 + 4A_m)$
   (b) $V = \frac{L}{4}(A_1 + A_2 + 2A_m)$
   (c) $V = \frac{L}{3}(A_1 + A_2 + A_m)$
   (d) $V = \frac{L}{2}(A_1 + A_2)$.

103. Wood work for doors and windows is measured in
   (a) cubic metres
   (b) square metres
   (c) metres
   (d) lump-sum.

104. The unit of measurement of steel work involving flats, angles, channels etc. is
   (a) metre
   (b) cubic metre
   (c) quintal
   (d) kilogram.

105. Revised estimate is warranted when the sanctioned estimate exceeds, for whatever reason, by more than
   (a) 50%
   (b) 25%
   (c) 10%
   (d) 5%.

106. Filling work in trenches shall be carried out in layers of thickness not more than
   (a) 500 mm
   (b) 200 mm
   (c) 300 mm
   (d) 400 mm.

107. The camber provided in long horizontal members to counteract the effects of deflection is
   (a) 1 in 10
   (b) 1 in 100
   (c) 1 in 250
   (d) 1 in 500.

ESTIMATING AND QUANTITY SURVEYING 925

108. The quantity of stone required to produce 10 m³ of rubble stone masonry will be
    (a) 7.5 m³   ☐   (b) 10 m³   ☐
    (c) 12.5 m³   ☐   (d) 15 m³.   ☐

109. The quantity of lime required for one coat of whitewashing of plastered surface of 100 m² will be
    (a) 10 kg   ☐   (b) 7.5 kg   ☐
    (c) 5 kg   ☐   (d) 2.5 kg.   ☐

110. Which of the following comes under general overhead ?
    (a) amenities of workers   ☐   (b) travelling expenses   ☐
    (c) interest on investment   (d) none of the above.

ANSWERS
Objective Type Questions

1. (b)	2. (c)	3. (b)	4. (b)	5. (c)	6. (d)
7. (c)	8. (a)	9. (d)	10. (a)	11. (b)	12. (c)
13. (d)	14. (a)	15. (b)	16. (c)	17. (b)	18. (d)
19. (d)	20. (a)	21. (c)	22. (b)	23. (b)	24. (a)
25. (c)	26. (b)	27. (a)	28. (b)	29. (c)	30. (c)
31. (d)	32. (c)	33. (d)	34. (a)	35. (d)	36. (b)
37. (c)	38. (a)	39. (b)	40. (c)	41. (d)	42. (a)
43. (a)	44. (c)	45. (a)	46. (a)	47. (d)	48. (c)
49. (d)	50. (b)	51. (b)	52. (c)	53. (c)	54. (a)
55. (c)	56. (b)	57. (a)	58. (d)	59. (b)	60. (a)
61. (b)	62. (c)	63. (b)	64. (d)	65. (c)	66. (c)
67. (c)	68. (d)	69. (d)	70. (b)	71. (c)	72. (d)
73. (c)	74. (b)	75. (b)	76. (b)	77. (c)	78. (b)
79. (d)	80. (a)	81. (b)	82. (c)	83. (a)	84. (b)
85. (b)	86. (d)	87. (d)	88. (a)	89. (c)	90. (a)
91. (c)	92. (b)	93. (b)	94. (d)	95. (a)	96. (b)
97. (c)	98. (b)	99. (d)	100. (c)	101. (b)	102. (a)
103. (b)	104. (c)	105. (d)	106. (b)	107. (c)	108. (c)
109. (a)	110. (b).				

Chapter 19 AIR POLLUTION

I. INTRODUCTION

Clean air costs money.

Dirty air costs more !

(When the damage to the individual and property is accounted for)

Pollutant is a foreign element or a natural element in excess concentration whose presence affects the comfort, well being and economy of life, the existing conditions and materials.

Solid waste remain at the same place and its influence is highly "*Local*".

Liquid waste flows along the course of the stream which is definite—thus covering more areas.

Air pollutant spreads over a wider area and may reach even far off places from its origin. It had no definite direction and the direction may not remain the same for a long time.

While solid and liquid wastes can be avoided at a particular place—air pollutant cannot be controlled when once released into the atmosphere. Its influence is "*Global*" unlike solid and liquid pollutants, air pollutants can spread to higher altitudes.

COMPOSITION OF GASES IN THE ATMOSPHERE

Nitrogen   78.1%
Oxygen   20.95%
Argon   0.93%
CO_2 0.03% by volume.

AIR POLLUTANTS

Solids: Suspended particulate matter: dusts, fumes, smokes, bioparticles.

Liquids: Droplets as sulphuric acid, mists.

Gases: CO, SO_2, H_2S, NO_x.

SOLIDS

Particulate is an airborne solid or liquid whose individual particle is greater than 0.0002 µm.

0.0002 µm < Fume < 1 µm < Dust < 75 µm < Grit < 500 µm.

Dusts: They are inorganic or organic solid particles with a regular shape and 1 to 75 µm in size.

Particles of size less than 10 µm remain suspended in the atmosphere. Particles of size less than 1 µm may not settle unless washed down by rain.

Dusts are produced from many operation as spraying, chipping, grinding, crushing, blasting, drilling or pulverising.

Sources: Forest fires, volcanic eruptions, dust storms, ocean spray aerosols, steel mills, cement plants, emissions from cars and trucks. Fly ash, coal dust, cement dust and pollen are the common dusts.

Smoke: It is the product of incomplete combustion of coal, wood, tobacco or a similar solid containing carbon.

It is produced because of

1. inadequate air to achieve perfect (stoichiometric) combustion,

2. inadequate mixing of air with the fuel, and

3. the temperature being too low (below combustion).

Size of smoke particle is less than 0.5 µm.

Sources: Transportation (automobiles), stationary sources as thermal power plants, agricultural or municipal combustion.

Fumes: It is the process of combustion, sublimation and condensation of solids to vapour state during which stage fumes are formed. *e.g.*, lead.

Fume is smoke with wider range of composition.

Their size varies from 0.1 µm to 1 µm.

Mist: Liquid droplets in air.

Fog: Waterdroplets in air.

Intense mist which prevents visibility is termed fog.

Smog: Smoke + fog.

Aerosols: They are solid or liquid particles in finely dispersed gaseous medium. *e.g.*, Insecticides as flit, hair dye, spray of scents.

Bio Particles:

Flowers and crops contribute pollen 5 to 50 µm in size. Feathers and hair are bigger in size.

Micro-organisms as fungi, yeasts, moulds, bacteria are transported by wind.

SOURCES OF SUSPENDED PARTICULATE MATTER

1. **Natural:** Dust storms, forest fires, volcanic eruptions, ocean spray aerosols.

2. **Man made:** In the various operations as

 (*i*) Combustion—Smoke, flyash, soot, cinders.

(ii) **Industrial operations**—Metal powders, metal oxides, cement manufacture, paint spray, saw dust.

Foundary emissions as

1. Arsenic from Copper smelters,
2. Fluorides in phosphate fertilisers
3. Lead from automobiles

Flyash: It is partly burnt fuel, unfused and fused ash of size 1 to 10 μm.

Soot: It is obtained from oil fired engines.

Foundry emissions:

Sand + coke + ash + metallic particles as arsenic from copper smelters, fluorides in phosphate fertilisers and lead in automobiles.

Fibrous Wastes:

Vegetable fibre as cotton and jute,

Animal fibre as silk and wool,

Synthetic fibre as rayon and nylon,

Mineral fibre as asbestos.

EFFECTS OF PARTICULATES

Particulates particularly of size 0.38 μm to 0.76 μm (called Haze particles) are the most effective in the visibility reduction. They absorb and scatter light. The finely divided material is a source of dirt and is responsible for soiling exposed portions.

Particulate of smaller size act as a nuclei inducing the formation of rain drop and hence increased precipitation.

They affect the earth's reflectivity, *i.e.*, the solar radiation reaching the earth and also the terrestrial infrared radiation from earth to the space.

Because of the black smoke canopy, sun's rays are prevented from reaching the earth depriving children of vitamin D, which deficiency leads to '*Rickets*'.

Particulates cause eye irritation.

Inert particles as dust can cause respiratory infections as '*asthma*' and '*bronchitis*' particularly in old people and children.

Particles of lead are *Neurotoxic*.

Inhaled lead is more serious than ingested lead.

Cadmium particulates cause cardiovascular effects.

Nickel causes lung cancer.

Mercury affects nerves, kidney and brain.

Beryllium is lethal.

Spores of moulds of hay cause '*farmer's lung*' infection.

Particles of size 5 to 10 μm are filtered by the respiratory tract and have no effect on human health.

Particles of size 0.5 to 5 μm are collected in lungs and particles smaller than 0.5 μm in alveoli of lungs. They interfere with the cleaning mechanism of lungs.

Inhalation of asbestos dust causes a disease called *"pneumo-coniosis"* also called *'asbestosis'*.

Pollen cause *'hay fever'*, *'bronchial asthma'* and *'dermatitis'*.

Virus infections as *'influenza'* and *'common cold'* are air borne.

Disease producing fungi are responsible for *'coccidioidomycosis'*—a type of *'micosis'*.

Metal	Source	Effects
Lead	Paints, batteries, pipes, gasoline.	Oxygen starvation of haemoglobin of blood, neurosis, brain damage, behavioural disorders anemia, death.
Cadmium	Mining of coal, zinc other metals, burning of plastics, tobacco smoke.	Cardiovascular diseases, hyper tension, kindney damage, zinc and copper metabolism affected.
Nickel	Alloy manufacture, smoke of coal and diesel.	Nickel + CO $\longrightarrow$ Nickel carbonyl. Causes alveoli changes and lung cancer.
Mercury	Extraction and refining of mercury, paints, pesticides, pharmaceuticals.	Blurred vision, brain and kidney damage, neurosis, congenital birth defects.
Beryllium	Ceramics, fluoroscent lamp manfacture, fuels.	'Berylliosis' which affects eyes, lungs and heart.
Asbestos	Extraction of asbestos from mines, manufacture of AC pipes, sheets, fire proof garments, insulators, brake lining	Silicosis (fibrosis), lung cancer (8 times higher in smokers)
Pollen	Flowers, plants.	Asthma, allergy, dermatitis, hay fever.

Smoke spoils the appearance of animal and birds. More areas are required (as their shelter) than in an unpolluted area.

Particulates deposited on vegetation make it unpalatable for animals. Asbestos fibres when consumed resulted in tumuors in birds. Cement dust coated fodder was not accepted by cattle.

Particulates deposited on leaves reduce the intensity of light reaching the interior, interfere with gaseous exchange and damage the outer wax coating.

Clothes, walls and metallic surfaces get dull when soiled. They loose their lustre and brightness.

GASEOUS POLLUTANTS

CARBON MONOXIDE

It is an odourless, colourless, tasteless, inert, toxic gas found less than 0.1 ppm in unpolluted atmosphere.

Sources: It is formed because of incomplete combustion of coal, coke, oil and gas due to **insufficient supply of air** *i.e.*, when fuel to oxygen ratio is more and when the temperature of combustion is low.

Oxidation of methane produces vast amounts of CO in nature.

CO is also produced because of the smoking of tobacco.

Automobile exhausts, mainly gasoline contributes CO (59%). Diesel exhausts contribute relatively less amount (0.2%).

Forest fires, agricultural burning and various industrial processes contribute CO.

Garages, tunnels, and heavy traffic lanes have greater concentration (more than 100 ppm) of CO.

CO had an atmospheric mean life of 75 days.

Effects: CO readily reacts with *haemoglobin* (Hb) forming *carboxy haemoglobin* (COHb) characterised by bright pink coloured flesh. Its affinity for *haemoglobin* is 210 times that of oxygen.

% COHb in blood	Effects
2%	Central nervous system affected
5%	Rate of blood circulation increases to counteract deficiency in oxygen, cardiac and pulmonary functional changes occur.
10%	Sight is blurred. More light required. Chest pain occurs for heart patients.
20%	Laboured respiration. Critical for heart patients
30%	Headache, Nausea
40%	Vomiting, Dizziness
50%	Slurred Speech
60%	Convulsions, Coma
70%	Fatal

When COHb level exceeds 2%, the central nervous system is affected. The person loses senses and cannot discriminate either the 'time interval' or 'the intensity of brightness'.

When COHb exceeds 5%, cardiac and pulmonary functional changes are caused.

When COHb exceeds 10% it causes *coma* and respiratory failure.

Cigarette smokers have COHb level of 5% and chain smokers more than 10%.

CO is equally harmful for warm blooded animals. It had no effect on insects which have no red blood cells.

CO had no effect on vegetation, property or materials.

Its threshold limits are: 10 mg/m^3 – 8 hr. average

40 mg/m^3 – 1 hr. average.

SULPHUR DIOXIDE

It is a non-inflammable, colourless gas of pungent smell and is soluble in water.

It is twice as heavy as air.

It remains airborne for 48 to 96 hrs

Sources: 1. *Natural sources* as

(i) Volcanic bursts,

(ii) Oxidation of H_2S produced by decaying organic matter of land and oceans.

2. *Man made:*

(i) By burning of coal with $\frac{1}{2}$ to 4% of sulphur (as in thermal power plants),

(1% sulphur coal produces 8 kg of sulphur/tonne and 16 kg of SO_2/tonne),

(ii) By burning oil fuel,

(iii) Industrial processes as smelting of metal sulphides.

Effects: SO_2 irritates respiratory tract and injure delicate tissues. *'Bronchitis'* and *'pulmonary emphysema'* are developed. Nose bleeding and eye irritation are caused. It affects people of cardiac disorders.

People suffering from *bronchitis* cannot tolerate even low concentrations of SO_2. Its effect is aggravated when in combination with particulates (*synergistic effect*).

Synergistic effect: When two substances react with each other to form a new substance whose impact is greater than the combined influence of the two individual substances put together (1 + 1 > 2), it is synergistic effect.

SO_2 is highly soluble (in saliva) and get absorbed in the upper part of the respiratory tract. But fine particulates into which SO_2 got impregnated can reach lower part of respiratory tract and alveoli causing immense damage to the sensitive air sacks.

Small particles of Mn, Fe and Vanadium into which SO_2 is absorbed reach alveoli and affect them.

Concentrion of SO_2	Effects
6 ppm	Threshold for brancho-constriction of healthy individuals
10 ppm	Throat irritation, Eye irritation
20 ppm	Spontaneous coughing

SO_2 is readily absorbed by vegetation, soil and water. It combines with ammonia to form ammonium sulphates.

While an amount of SO_2 of 15 µg/m³ (0.005 ppm) is not adequate for the yield of good crops, excesses concentrations are detrimental to the plants like *alfalfa, apple, beans, cotton, lettuce, mulberry, pine* and *soyabean*.

pH of lakes and tanks was reduced to as low as 4 because of SO_2 solution and acid rain reaching the water body nullifying all the aquatic life.

Building materials as limestone, marble and metals as iron, steel, zinc, copper, aluminium and nickel are corroded. Paper becomes yellow and brittle. Cotton, rayon, nylon, linen and leather are weakened.

Visibility of the atmosphere is also reduced because of SO_2.

Threshold limits:

80 μg/m^3 (0.03 ppm)/24 hr Annual average

365 μg/m^3 (0.14 ppm)—24 hr average not to be exceeded more than once in a year.

OXIDES OF NITROGEN

N_2O is *nitrous oxide*—laughing gas.

It is a colourless gas formed because of bacterial decomposition. It is anesthetic and commonly present in the atmosphere.

NO is *nitric oxide*.

It is a colourless gas. It is formed by burning at 1600° to 3000°C.

It is relatively inert and readily gets converted to NO_2.

NO_2 is *nitrogen dioxide*.

It is a suffocating reddish brown gas, heavier than air and readily soluble in water.

It is a strong absorbent of ultraviolet rays from the sun and forms *smog*.

Sources: Naturally occurring nitrogen oxides are more in quantity compared to the man-made sources.

Fuel combustion releases the maximum amount of NO_x.

Effects: NO_x of a concentration greater than 7 mg/m^3 (3.75 ppm) for a period of 10 minutes causes irritation of respiratory tract, headache, loss of appetite and corrosion of teeth.

Concentrations greater than 50 mg/m^3 (27 ppm) cause *pneumonia* and greater than 290 mg/m^3 (155 ppm) are fatal to human beings.

They cause eye and nasal irritation and pulmonary discomfort. They penetrate deep and reach alveoli of lungs reducing body resistance to respiratory infections.

In the presence of hydrocarbons, oxygen and light, NO_2 gives out secondary pollutants as 'peroxyacetyl nitrates' (PAN) and *ozone*. Like CO, NO also combines with *haemoglobin* restricting the transport of oxygen to the tissues.

While smaller amounts of nitrates are helpful for the plant growth, higher concentrations are injurious to plants.

Though not as high as SO_2, NO_2 also dissolves in water, causing acidity and lowering the pH of water bodies.

NO_x cause fading of the textile dyes, yellowing of white fabrics and oxidizing and corrosion of metals.

Brown discoloration of sky occurs due to excessive NO_2 concentration in the atmosphere.

Threshold limits of concentration NO_x 100 μg/m^3 (0.05 ppm).

SECONDARY POLLUTANTS

Secondary pollutants are those which had their origin because of the primary pollutants.

PAN and OZONE

Hydrocarbons + O_2 + NO_2 + Light $\longrightarrow$ Peroxyacetyl nitrate + Ozone

Effects: PAN and ozone are strong oxidizing agents.

PAN irritates eyes at a concentration of 0.1 ppm.

Ozone impairs vision and lowers body temperature.

Photo chemical smog having PAN and *ozone* cause difficult breathing, irritate mucous membrane, cause coughing, choking, severe fatigue eye, nose and throat irritation.

They affect plant cell metabolism, leaves are discoloured and injured. Cotton, grapes, tobacco, cereal crops, vegetables and citrus are worst hit.

Paints and dyes are oxidised, rubber gets cracked, textile fibre lose strength.

OTHER POLLUTANTS

Lead: *Tetramethyl lead* and *tetraethyl lead* are commonly added to petrol to improve its performance as a fuel, *i.e.*, as an anti-knocking compound. Hence the automobile emission contributes the lion share of the lead.

Cigarette smoking produces 0.5 µg of lead per cigarette.

Effects: Weakness, headache, lassitude, blue lining developed along the gums of teeth, damage to the central nervous system, liver and kidney damage, abnormalities in fertility and pregnancy and ill *mental* health of children.

Man have greater level of lead in their blood than women, urban residents than rural folk, and smokers than non-smokers.

FLUORINE COMPOUNDS

These are confined to some local areas.

Sources: Occurs in geological deposits and fluorine compounds are evolved when processed at high temperature.

In glass making, aluminium smelting, fertilizer production, phosphate rock processing, iron and steel industry and coal burning.

Aerosols of fluorocarbons are given out when used as propellants.

Effect: Domestic animals develop *'Fluorosis'* when they consume fodder coated with fluorides.

Mottled enamel of teeth, softening of teeth, bones losing elasticity and heavily loaded bones as the pedal bone being broken.

Some sensitive plants are affected even at low concentrations of fluorides. Irregular and premature ripening and reduction in yield of the crop are noticed.

Fluorine separated out of fluorocarbon aerosols reacts with and destroys the ozone layer. Thus the ultraviolet radiation is directly received by the earth.

Pollutant	Source	Effects
Arsenic	Copper smelting	Carcinogenous
Asbestos	Mineral extraction	Lung cancer (*pneumoconiosis*)
Beryllium	Metallic alloys, fuel for rocket engine	Tumours
Cadmium	Paints, pesticides (cadmium arsenate)	Respiratory poison
3 : 4 benzo (*a*) pyrene	Industrial operations as combustion of coal	Lung cancer
Ethylene	Automobile exhausts	Abnormalities of tomato, pepper, damage to orchids.

AIR POLLUTION EPISODES

While air pollutants are released into the atmosphere because of man made and industrial operations, disasters occur only under certain atmospheric conditions.

Chill nights of winter with no wind currents favour episodes.

Valleys and temperate countries are more vulnerable for air pollution episodes.

Greater solar insolation, greater (> 20 kmph) wind velocity, elevated areas and areas nearer to water bodies (lakes, rivers) experience least air pollution.

Place	Cause	Effects
Meuse valley (Belgium), December, 1930	Inversion lasted for 3 d, $SO_2 + SO_3$ mist, SPM, metallic fumes	Irritation of eyes, nose and throat, breathlessness, constriction of chest 60 excess deaths.
Donora (USA) Oct., 1948	Inversion for 4 d, SO_2	Irritation of eyes, nose and respiratory tract, breathlessness, nausea, cardio respiratory problems 20 excess deaths.
Poza Rica (Mexico) Nov., 1952	Inversion, leakage of H_2S for 25 minutes	Irritation of respiratory tract, 22 excess deaths.
London (England) Dec, 1952	Inversion for 4 d, SO_2	Bronchitis, pneumonia, heart diseases, atmosphere became opaque 4000 more deaths.
London (England) Jan., 1956	Inversion, SO_2	Respiratory and heart diseases, 1000 more deaths.
New York (USA) Nov., 1966	SO_2 particulates	168 excess deaths.
Seveso (Italy) July, 1976	Reactor explosion releasing Dioxin (Tetrachlorodibenzo dioxin)—a herbicide.	Dizziness, diarrhoea, abortions, premature births, skin irritation.

Bhopal (India) Dec., 1984	Leakage of MIC—a pesticide	Choking, asphyxiation, vomiting, violent coughing, irritation of sensory organisms, suffocation, still births, abortions, cardiac arrest, blindness, paralysis, damage to liver, kidney and central nervous system 2,500 people dead >2 lakh were affected.
Chernobyl (Ukraine)	Nuclear disaster Radioactive gases and dust released over 10 d.	Nausea, anemia, skin ulcers, loss of hair, chronic health hazards as aberrations in blood, bones, skin, eyes and other organs 2000 dead.

METEOROLOGY

Meteorology is the study of space. Atmosphere is a layer surrounding the earth and is responsible for the life on earth.

Atmosphere may be subdivided as

(i) Troposphere

(ii) Stratosphere

(iii) Mesosphere

and (iv) Thermosphere.

TROPOSPHERE

It is the layer immediately above the ground extending to an average height of 11 km above Ground level, lesser at poles and highest at the equator (18 km above ground level). Troposphere is the densest of the 4 layers and 70% of the mass of atmosphere is in it. Nitrogen, oxygen, carbon dioxide and water vapour are present in this zone. Cloud formation occurs in this zone.

In Troposphere temperature decreases with increase in altitude uniformly from around 20°C near ground level to – 56°C at the top of Troposphere.

Lapse rate. It is the temperature gradient or rate of fall in temperature with increase in altitude.

In the dry atmosphere of the troposphere temperature falls at the rate of 9.81°C for km rise or roughly at 1°C/100 m rise.

$$\frac{dT}{dZ} = -1°C/100 \text{ m}$$

This is called Dry Adiabatic Lapse Rate (DALR).

When the rate of fall of temperature is more than 9.81°C per km rise in altitude it is called **super adiabatic lapse rate.**

When the rate of fall of temperature is less than 9.81°C per km rise in altitude it is **subadiabatic lapse rate.**

When temperature remain constant with increase in altitude it is **isothermal.** When the temperature increases with rise in altitude it is called **inversion.**

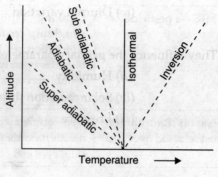

FIGURE 19.1

Tropopause is the transition zone in between Troposphere and Stratosphere where the lapse rate changes direction (*i.e.*, from fall in temperature with rise in altitude to rise in temperature with rise in altitude).

STRATOSPHERE

It is the layer extending above Tropopause but below Stratopause to an altitude of 50 km above ground level. Temperature increases with increase in altitude *i.e.*, from $-56°C$ to $-2°C$ at a uniform rate. Ozone is the only gas present which absorbs ultraviolet rays from the sun shielding the earth. Jet planes prefer to travel in this zone.

Stratopause is the transition zone between the Stratosphere and Mesosphere.

MESOSPHERE

It extends from 50 km to 85 km above ground level. Oxygen and nitric oxide are the gases present. The temperature decreases with increase in altitude from $-2°C$ to $-92°C$.

Mesopause is the transition layer between Mesosphere and Thermosphere.

THERMOSPHERE

It is upper most layer from 85 km to more than 500 km above ground level. Temperature increases with altitude from $-92°C$ to $1200°C$. Oxygen, nascent oxygen and nitric oxide are the gases present.

Based on the extent of spread meteorology can be classified as

1. Macrometeorology spreading over tens and thousands of square kilometres of area of Troposphere, Stratosphere, Mesosphere and Thermosphere over a duration of months.

2. Mesometeorology spreads over tens and hundreds of square km of area of Troposphere and Stratosphere for a duration of hours and days.

3. Micrometeorology spreading over an area of less than 2 sq. km for a height of 1 km above ground level, over a duration of seconds, minutes and hours.

Micrometeorology involves the study of properties of the atmosphere which govern the spread, diffusion and concentration of pollutants at a place and time.

METEOROLOGICAL PARAMETERS

1. Primary parameters: (They control the dispersion and dilution of the pollutants):

(*i*) Wind speed   (*ii*) Wind direction

(iii) Atmospheric stability (iv) Diurnal variation
(v) Mixing depth.

2. Secondary parameters: (They influence the primary parameters):

(i) Temperature (ii) Humidity
(iii) Precipitation (iv) Solar radiation (insolation)
(v) Cloud cover (vi) Pressure
(vii) Topography.

WIND SPEED

Greater the wind speed, more is the 'mechanical turbulence' and hence more is the dilution.

i.e., If the concentration is 4 g/m^3 when the wind speed is 1 m/s, then it is 1 g/m^3 when the wind speed is 4 m/s.

WIND DIRECTION

It is the direction FROM which the wind is blowing, *i.e.*, wind direction is said to be 'East' when it is blowing from 'East' to 'West'.

The direction of the wind will not be the same but will be frequently changing.

The predominant directions and less prevailing directions along which the wind flows and the velocities of flow shall be represented by the 'wind rose'.

Wind rose is a circular diagram representing usually 8 or 16 spokes radiating along the different directions. The proportionate lengths and thicknesses of different conventions represent the frequencies of wind in that particular direction and also the wind speed over a period of time.

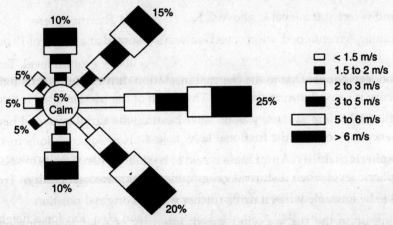

FIGURE 19.2

While N (North), NE (North East), E (East), SE (South East), S (South), SW (South West), W (West) and NW (North West) represent the 8 directions such that the angle between any two neighbouring directions is 45°, N, NNE (North of North East), NE, ENE (East of North East), E, ESE (East of South East), SE, SSE (South of South East), S, SSW (South of South West), SW, WSW

(West of South West), W, WNW (West of North West), NW, NNW (North of North West) are the 16 directions such that the included angle between any two neighbouring directions is $22\frac{1}{2}°$.

A velocity less than 1.5 m/s (6 kmph) is considered as "calm" for which direction has no significance and shall be represented in the central circle.

FACTORS AFFECTING WIND SPEED AND DIRECTION

1. **Surface Roughness:** Smooth surfaces as those of water-bodies as oceans and paved surfaces offer practically no friction and hence no *veering* (veering is clockwise turn of rotation).

Trees and shrubs offer resistance to flow and the veering may be upto 20°.

Hilly areas, urban township with tall buildings and congested dwellings offer more friction and the veering may be even 45° in these cases.

Mechanical turbulence increases with the surface roughness.

2. **Height:** Within the friction layer (*i.e.*, about 700 to 1000 m above ground level), wind velocity changes with the height.

$$\frac{v}{vo} = \left(\frac{h}{h_0}\right)^{1/n}$$

$n = 3$ in stable condition
$\quad = 7$ in neutral condition
$\quad = 9$ in unstable condition

where v = Wind velocity at a level h above G.L.

v_0 = Wind velocity at a level h_0 above G.L.

When nothing is mentioned, wind speed is always assumed at a height of 10 m above ground level.

3. **Diurnal Variations:** Due to the thermal insolation during the bright sunny day, greater thermal turbulence is caused during day time. The depth of the friction layer is more.

During the night time as there was no solar heating, the turbulence and hence dispersion are the least. Hence the depth of the frictional layer is less.

4. **Atmospheric Stability:** An air mass is said to be stable when it tends to subside back to its original atmospheric level when disturbed geographically or meteorologically.

It is said to be unstable when it drifts further from its original position.

Depending upon the surface wind speed, intensity of solar insolation and cloudiness, atmospheric stability is classified as

A—Extremely unstable, B—Moderately unstable
C—Slightly unstable, D—Neutral
E—Slightly stable, and F—Moderately stable.

Thus unstable atmospheric condition is highly desirable to disperse and dilute the gaseous pollutants whereas under stable conditions the gaseous pollutants do not disperse vertically readily but tend to accumulate raising their concentration.

A superadiabatic atmospheric condition causes the air pollutants to rise vertically and to get diluted effectively.

INVERSION

In Troposphere atmospheric temperature falls with increase in altitude. But sometimes due to topographical and meteorological factors, temperature increases with increase in altitude. This atmospheric phenomena is called an **"Inversion"**.

Inversion is a stable atmospheric condition which prevents vertical dispersion of the air pollutants. The inversion layer acts as a lid and prevents any rise of the plume above it.

Types of Inversions

Radiation Inversion:

During the day time the ground gets heated due to solar insolation and gets cooled during night. At the day break (dawn) the temperature gradient (Lapse rate) is A B C as the ground is cool but the layers above are warm and warmer, *i.e.*, temperature increases with altitude uptil the height B. After sunrise the inversion layer AB burns off and the Lapse Rate tends to become a straight line DBC after a few hours of solar insolation. It is called "Radiation or Temperature inversion".

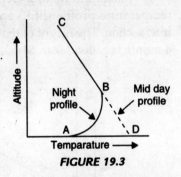

FIGURE 19.3

Solar insolation and strong winds dissipate Radiation inversion. Clouds also reduce the intensity of the Radiation Inversion. Valleys are more vulnerable for this inversion. It is the most common for most nights of winter.

Subsidence Inversion:

It is formed because of descending air mass at a slow rate of about 1 km per day. The sinking layer acts as a lid of the layer below it and prevent vertical dispersion of air pollutants through it. With time the height of the sinking layer gets reduced and hence concentration of air pollutants trapped between the ground level and bottom of inversion layer goes on increasing. Subsidence inversion may occur anytime. Coastal Valleys are more vulnerable for subsidence inversion. It may last for 5 to 6 days.

5. **Mixing Depth:** It is the vertical column of the Super Adiabatic atmospheric layer above the ground level in which free vertical dispersion of the air pollutants takes place and the air pollutant gets diluted.

Mixing depth varies from place to place and from time to time. Mixing Depth is more during day time than at night, more in the afternoon than forenoon, more during summer than winter, is mainly dependent on ground surface temperature and roughness of the ground.

It varies from less than 200 m to more than 4,000 m at a particular place.

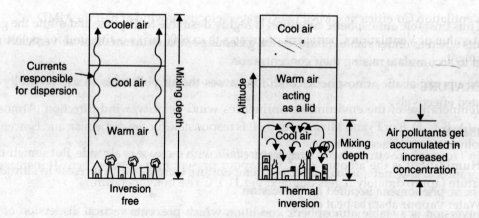

FIGURE 19.4

Maximum Mixing Depth (MMD) at a place can be obtained by plotting early morning temperature profile (ELR) and Dry Adiabatic Lapse Rate (DALR) and locating their point of intersection. The height of that point above ground level gives MMD. The average of MMDs over a month is called Mean Max. Mixing Depth (MMMD).

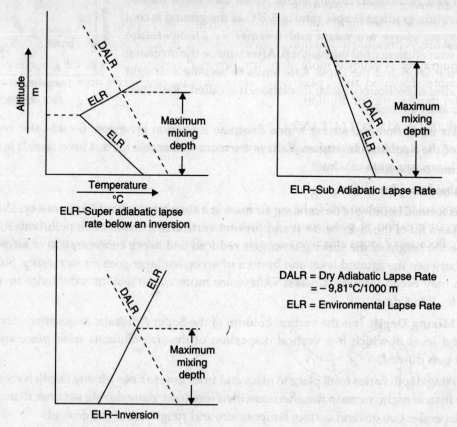

DALR = Dry Adiabatic Lapse Rate
    = $-9.81°C/1000$ m
ELR = Environmental Lapse Rate

FIGURE 19.5

Ventilation Co-efficient = MMD × Average velocity of wind at half the MMD

A value of Ventilation Coefficient of greater than 6000 m²/s is the limit for pollution free environment.

TEMPERATURE

Temperature of the environment influences wind velocity, wind direction, Atmospheric stability (A, B, C, D, E or F) and Mixing Depth. It is responsible for the dispersion and concentration of air pollutants at a place and time.

Humidity: It is the amount of water vapour present in the atmosphere. It increases with temperature (approximately it gets doubled per 11.1°C rise in temperature).

Water vapour absorbs heat energy.

Precipitation: It cleans the atmosphere as the suspended particulate matter is washed down and soluble gases get dissolved in rain water. But in turn the SPM may reach the ground and the dissolved gases may cause water pollution and soil pollution.

Solar radiation: Insolation destroys inversions. It causes unstable conditions even when stable conditions are existing earlier.

Cloud cover: Clouds in sky affect heating and cooling of the lower layers. Cloudy sky presents a cooler day and warmer night compared to clear sky day and night.

Pressure: A pressure of 760 mm of mercury is the standard atmosphere = 1103 millibars.

Topography: A valley is more prone for increased concentrations of the gaseous pollutants and inversions. Elevated areas, plains and areas abutting plains and areas abutting vast water bodies offer more scope for dilution.

DISPERSION OF AIR POLLUTANTS

Plume is a continuous stream of air pollutants ejected from a stack.

Plume behaviour:

1. Looping plume: Looping plume occurs during superadiabatic lapse rate conditions *i.e.*, when strong solar radiation and light winds exist as on the afternoon of a summer with clear sky, which favour creation of good loops. Cloudiness and high winds are unfavourable for loop formation. Because of loops effective dispersion of pollutant takes place and highly get diluted.

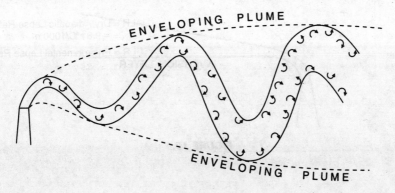

FIGURE 19.6

2. Neutral plume: It is a vertically rising plume occurring under subadiabatic conditions *i.e.*, when the ELR is almost very close to DALR *i.e.*, the stability is neutral and velocity of wind is less than 1.5 m/s (CALM).

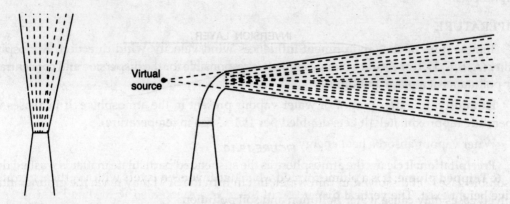

FIGURE 19.7

When the same plume is subjected to a velocity >1.5 m/s it becomes **a Coning Plume.** The dispersion is moderate. Any equation connected with dilution and concentration is ideally suited for this plume rather than for any other plume.

3. Fanning plume: Fanning plume occurs during an inversion *i.e.*, when the temperature increases with altitude. Virtually no dispersion occurs in a fanning plume in the vertical plane.

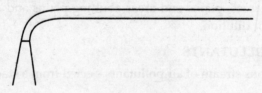

FIGURE 19.8

4. Lofting plume: When the plume is formed above an inversion layer extending uptil the effective height of the stack, vertical spread is prevented downwards into the inversion layer but upward spread can go unlimited.

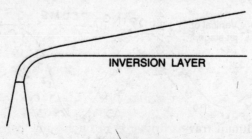

FIGURE 19.9

5. Fumigation plume: When the bottom of the inversion layer is just above the effective height of the stack it acts as a lid and prevent any vertical dispersion above it. There is no bar on the downward spread and the plume may touch the ground at a small horizontal distance from the stack.

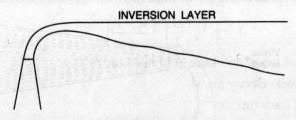

FIGURE 19.10

6. Trapped plume: It is a plume formed when two inversion layers with a vertical gap at the effective height exist. The vertical dispersion of the plume is confined to the band (vertical gap) between the two inversions and the horizontal spread is guided by the bottom of top inversion and top of lower inversion.

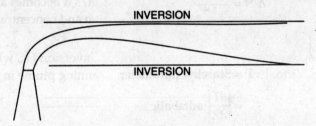

FIGURE 19.11

PLUME RISE

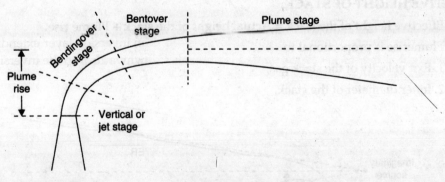

FIGURE 19.12

The hot gaseous pollutant travels upwards through some height ΔH before assuming a horizontal plume profile. This extent beyond the chimney height is called **plume rise**.

For unstable and neutral conditions

$$\Delta H = \frac{150 \, F}{U^3}$$

where U = Cross wind speed at the top of stack (m/s)

$$F = g \cdot V_s \left(\frac{D}{2}\right)^2 \left(\frac{T_s - T}{T}\right)$$

where F = Buoyant force/unit time/unit mass (m^4/s^3)

V_s = Stack exit velocity (m/s)

D = Dia. of stack (m)

T_s = Exit gas temp (Å)

T = Gas temp. at infinite dist.

   = Atmospheric temp. (Å)

For stable and calm conditions

1. $$\Delta H = \frac{5 F^{1/4}}{s^{3/8}}$$

where $s = \dfrac{g}{T}\left(\dfrac{dT}{dZ} + \Gamma\right)$

Γ = Stability parameter

   $= \left|\dfrac{dT}{dZ}\right|$ adiabatic

2. $$\Delta H = 2.3 \left(\frac{F}{U_s}\right)^{1/3}$$

EFFECTIVE HEIGHT OF STACK

Effective height of the stack = Actual height of the stack + Plume rise.

Plumerise is proportional to

1. Exit velocity of the stack gas,
2. Inner diameter of the stack,

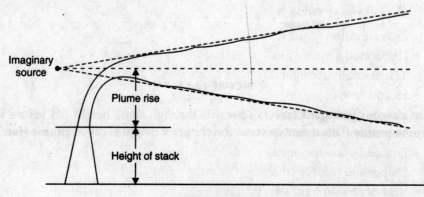

FIGURE 19.13

3. Inversely proportional to the difference in temperature of stack gas and ambient (atmospheric) temperature.

The buoyancy of the plume is more when the stack gas is hotter than the ambient air temperature.

4. Wind speed.

If the wind is still, the plume may vertically rise upwards.

Light winds may deflect the smoke as a gentle curve before assuming into a horizontal stream of the plume.

Strong winds may quickly turn the plume direction to horizontal *i.e.*, without any plume rise or even the smoke is deflected downwards. To prevent such a down draught the exit velocity of the stack gas should not be less than $1\frac{1}{2}$ × wind speed.

MINIMUM HEIGHT OF THE STACK

The height of the stack should be the maximum of

1. $H = 11.5 \times Q^{0.27}$, $\quad Q$ = Particulates emitted kg/h
2. $H = 14 \times Q^{0.3}$, $\quad Q$ = SO$_2$ emission kg/h
3. $H = \dfrac{2.71 \times Q}{U \times C}$, $\quad U$ = Velocity of wind

$\quad C$ = Concentration of the pollutant desired.

4. $H \not< 2.5$ × Height of tallest structure in its vicinity
5. $H \not< 30$ m

THE GAUSSIAN PLUME

$$C_{(x, y, z)} = \underbrace{\frac{Q}{U}}_{\text{(Source strength)}} \times \underbrace{\frac{1}{\sqrt{2\pi} \times \sigma_Y} \cdot e^{-\frac{1}{2}\left(\frac{y}{\sigma_Y}\right)^2}}_{\text{(Diffusion along Y-direction)}} \times \underbrace{\frac{1}{\sqrt{2\pi} \times \sigma_Z} \left[e^{-\frac{1}{2}\left(\frac{Z-H}{\sigma_Z}\right)^2} + e^{-\frac{1}{2}\left(\frac{Z+H}{\sigma_Z}\right)^2} \right]}_{\text{(Diffusion in Z-direction)}}$$

$$= \frac{Q}{2p\,\sigma_Y\,\sigma_Z.U} e^{-\frac{1}{2}\left(\frac{y}{\sigma_Y}\right)^2} \times \left[e^{-\frac{1}{2}\left(\frac{Z-H}{\sigma_Z}\right)^2} + e^{-\frac{1}{2}\left(\frac{Z+H}{\sigma_Z}\right)^2} \right]$$

where C = Concentration (g/m^3)

X = Horizontal distance along the direction of wind (m)

Y = Transverse horizontal distance from the axis of plume (X-axis) (m)

Z = Height above g.l. (m)

H = Height of the stack (m)

U = Velocity of wind (m/s)

σ_Y = Dispersion coefficient along Y-direction (m)

σ_Z = Dispersion coefficient along Z-direction (m)

Q = Rate of emission (g/s)

GLOBAL EFFECTS OF AIR POLLUTION

Global Warming: (Green House Effect):

Solar radiation in the form of short wave light rays reach the earth and get reflected back into the space as long wave infrared (heat) radiation. Atmospheric gases as CO_2, CH_4, N_2O, O_3, CFC and water vapour permit the incoming short wave light rays but prevent the reradiation of the heat from the earth back into the space. Hence the temperature of earth is increasing with time. It is called "Global Warming."

CO_2: It is responsible for 50% of Global Warming.

1. Fossil fuel burning for Thermal power production.
2. Industries, mainly the cement industry.
3. Changed agricultural practices.
4. Deforestation

Deforestation are responsible for increased CO_2 content. It was only 290 ppm a century ago but today more than 355 ppm.

CH_4: 16% of Global heating is because of it.

Cowdung, paddy fields, deforestation, coal mining and natural gas extraction are the sources of Methane.

N_2O: 6% of Global Warming is because of it.

It is the product of Combustion (N_2 and O_2 of the atmosphere) and bacterial decomposition of soils.

O_3: 8% of Global Warming is because of it—which is a secondary pollutant.

CFCs: 20% of Global Warming is because of them.

CFCs are manmade emissions from aerosol propellants, refrigerants, solvents and cleaning agents in Electronics, foam blowing agents in plastics and fire extinguishers.

Global temperatures are rising at 1.5° C near equator and 4.5°C near poles per century.

EFFECTS OF GREEN HOUSE GASES

1. Sea levels may rise from 0.5 m to 1.5 m by 2100 because of ice melting in polar ice caps. Low lying areas of Maldives, Bangladesh and Netherland may get submerged.

   At river mouths salt water may get intruded at 1 m horizontal spread per 1 cm rise.

2. Rate of evaporation may increase reducing the soil moisture content.
3. Frequency of heat waves and droughts may increase affecting crops.
4. Rainfall pattern may change.
5. Floods and forest fires may occur more frequently.
6. Eco systems get affected and some species may become extinct.

URBAN HEAT ISLAND EFFECT

Suspended particulate matter of industries prevent 20% of the solar light rays reaching the earth and also the reradiation of the heat rays from the earth into the space. Also concrete and Asphalt roads, Industrial furnaces, tall concrete structures at closer intervals, busy roads emitting automobile fumes, greater population density (Warm blooded animals), thin vegetation and absence of water bodies as lakes warmup cities and at times they maintain a temperature of 4 to 7°C higher than the rural neighbourhood. Hence, the city is called a 'Heat Island'. Due to the thermal gradient so created airpollutants, dust and smoke from the peripheral area get sucked towards the heart of the city forming a 'DUST DOME'.

ACID RAIN

Pure rain water shall be having a pH of 7. When the pH of the rain water is less than 5.6 it is called 'ACID RAIN'. Crude oil shall be having a sulphur content of 0.1 to 3% whereas it is 0.5 to 10% in Coal. When they are burnt SO_2 is produced which is responsible for 60% of acid rain. SO_2 gets oxidized to SO_3 which forms H_2SO_4 i.e., dilute Sulphuric acid with water. Similarly NO_x emitted out of motor exhaust fumes react with atmospheric moisture to form Nitric acid. 30% of the acid rains are due to the formation of Nitric acid. Hydrochloric acid and phosphoric acid contribute to about 10% of the acid rain.

EFFECTS OF ACID RAIN

1. Water bodies, is particular stagnant pools as reservoirs and lakes get acidified and become dead lakes as their flora and fauna get perished.
2. As the soil is acidified, its nutrients are leached out, fertility and buffering capacity get reduced.
3. Earth worms and other soil organisms get eliminated.
4. Microorganisms helping the plant growth are killed.
5. Wild life destroyed.
6. Lime stone, Marble, cement, concrete and metallic surfaces get corroded. Paints lose their lustre and shade.

OZONE LAYER DEPLETION

Ozone may be a pollutant in Troposphere but is a saviour in Stratosphere.

Ozone layer of the stratosphere absorbs ultraviolet radiation emitted from the Sun.

Ultraviolet rays—Ultraviolet rays are 313 to 400 nm of wavelength and are not so dangerous. *U V-B rays* are of a wavelength of 280 to 313 nm are the most dangerous.

U V-C rays of a wavelength less than 280 nm are the most destructive but are of limited content and are completely absorbed by the atmosphere.

SO_2 and NH_3 were used as refrigerants earlier. Both are reactive, corrosive and toxic. They were replaced by non-toxic, non inflammable, transparent, colourless and odourless CFC 11 and CFC 12.

Besides CFCs, HBFCs, Methyl bromide carbon tetrachloride, Methyl chloroform and Nitrous oxide used as fire extinguishers, cleaning solvents, dry cleaning sprays, adhesives and soil fumigants were found to harm the ozone layer.

Also underground nuclear tests, supersonic air planes travelling in stratosphere are also responsible for the depletion of ozone layer.

CO_2 and CH_4 tend to increase the thickness of ozone layer and are helpful.

Instead of CFCs, HFCs, HCFCs, Hydrocarbons, Terpenes and Helium are used as refrigerants.

EFFECTS OF OZONE LAYER DEPLETION

1. *UV-B rays* induce changes in DNA of blood and may cause skin cancer and ocular damage (Cataract and retinal degeneration).

2. Immune system suppressed.
3. Infections of Bacterial (Tuberculosis, Leprosy), Protozoal (Malaria, Filaria) and Viral (Chicken pox, Measles) origins get easily spread.
4. Terrestrial and aquatic ecosystem get affected.
5. Plants grow slowly. Crops affected and forests get damaged.
6. Paints and Plastics get affected.

MONITORING AIR POLLUTION

AIR POLLUTION SURVEY

It is conducted to
1. identify the sources of pollution,
2. estimate the concentrations of specific pollutants,
3. assess pollutional levels of vehicular emission,
4. assess the extent of damage occurring because of the pollutants,
5. interpret the available data to the city—and to classify it as sensitive areas—schools, sanctuaries and wild life abodes, and
6. propose sites for future industries.

Sensitive areas are to be well separated from industrial areas by providing buffer zones of green belts in between.

Industries are to be located on the leeward side of the city.

LOCATION OF SAMPLING STATIONS

1. For a specific purpose
   (*i*) **Health hazard of humans:** Locate the station at the nose level of an average man.
   (*ii*) **Damage to vegetation:** Locate the station at the foliage level of trees or plants.
   (*iii*) **Corrosion of power lines:** Locate the station at the same level as the wires.

The area to be studied is to be covered by the sampling stations formed by rectangular or circular grids around the source.

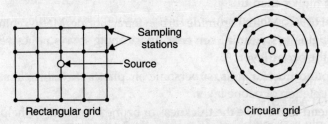

FIGURE 19.14

AIR POLLUTION MONITORING EQUIPMENT

1. Dust fall jar   2. High volume sampler   3. Multigas sampling kit.

Besides:
1. **Anemometer:** to measure wind speed.
2. **Wind vane:** to indicate the horizontal direction of wind.
3. **Bivane:** to indicate the vertical direction of wind.
4. **Temperature sensors:** to measure temperature.
5. **Polarigraph or Pyrheliograph:** to measure global radiation (total incoming solar radiation).
6. **Barometer:** to measure pressure.
7. **Rain gauge:** to record rainfall.

The sampling may be done round the clock which is ideal or for 2 h in the forenoon (between 5 to 11 a.m.) and for 2 h in the afternoon (between 6 to 10 p.m.) for a minimum of two and preferably 3 years.

Observations need be made more frequently during winter months as the chances for dispersion are less and chances for accumulation are more. It can be less for summer when the dispersion is fairly high.

MEASUREMENT OF SUSPENDED PARTICULATE MATTER

Deposit gauges are similar to the rain gauges. In addition bird guards are provided. 10 ml of 0.02% $CuSO_4 \cdot 5H_2O$ is added to prevent the growth of algae.

It can be sometimes as simple as cylindrical jar of about 150 mm in dia. and 280 mm high.

If the exposed surface area of the sampler (*i.e.* the collector) is A and the amount of solids collected in a period of time (may be 1 month or 1 year) is B, then the dust fall rate is B/A tonnes/sq. km/month (or year).

Particles of size greater than 100 μm can be determined by this method. High volume sampler fitted with a filter paper can be used to find the amount of particles of size 1 μm to 100 μm which may not settle down and hence are sucked and retained on the filter paper of the sampler.

SELECTION OF STATIONS

Just as in case of rain gauges the sampling sites should be located as follows:

1. It should be fairly elevated land as the terrace of a building which is atleast 3 to 15 m above the ground level.

2. Sufficient open area should exist around it so as not to obstruct the currents settling.

3. Buildings higher than the site where the gauge is located should maintain a minimum horizontal distance of 1.75 × h, where '*h*' is the maximum height of the building above the gauge.

4. The gauge should be away from smoke tubes and chimneys of residential or commercial centres influencing the readings.

5. Easy collection and evaluation of the results should be possible from the site.

GASEOUS POLLUTANTS

The air sample is passed through a filter paper to exclude suspended particulate matter. Then it is passed through sulphuric acid chamber to absorb CO_2.

SULPHUR DIOXIDE

1. Hydrogen peroxide method: Pass the air sample through 50 ml of 0.05% H_2O_2 solution.

$$H_2O_2 + SO_2 \longrightarrow H_2SO_4$$

Find out the amount of H_2SO_4 formed by titrating the sample with 0.002 NaOH solution.

2. Lead peroxide candle method: SO_2 in the atmosphere reacts with the lead oxide to form lead sulphate.

$$PbO_2 + SO_2 \longrightarrow PbSO_4$$

Lead sulphate reacts with the barium chloride forming insoluble barium sulphate which is precipitated and gravimetrically determined.

$$PbSO_4 + BaCl_2 \longrightarrow BaSO_4 + PbCl_2$$

3. Electrical conductivity method: SO_2 absorbed by a sample of water can be estimated by knowing the conductivity of the water sample before and after the 'absorption' which gives a direct measure of SO_2 concentration.

4. Spectrophotometer (West and Gaeke) method: After absorbing the SO_2, dilute *sodium tetra chloromercurate* aqueous solution forms *dichlorosulphito mercurate*. It reacts with *formaldehyde* and *pararosaniline* to form *pararosaniline methyl sulphonic acid* whose colour intensity is proportional to the SO_2 concentration.

This method is very accurate.

NITROGEN DIOXIDE

1. Greiss Saltzman method: Air samples are collected in evacuated bottles containing 10 ml of a mixture of *sulphanilic acid*, *acetic acid* and N-(1-napthyl) *ethylene diamine dihydrochloride*. Bottle is shaken for 15 minutes to develop colour. Intensity of the colour depends on the concentration of NO_2. The absorbance is measured on the spectrophotometer.

2. Jacobs-Hochheiser method: The air sample is passed through *sodium hydroxide* solution to form nitrous acid. The solution is treated with H_2O_2 to remove SO_2 if any. The amount of *nitrous acid* formed gives an idea of NO_2.

NITRIC OXIDE

Nitric oxide which is a colourless gas gets oxidised to NO_2 which is a reddish brown gas.

Hence the sample is oxidised by passing through acidified *potassium permanganate* solution. But the reaction shall never be complete and only 70% of the NO is converted to NO_2 and hence a correction factor of 0.7 is applied.

OZONE

1. Hydrogen peroxide method: The air solution is absorbing solution and is treated with H_2O_2, boiled, cooled and then 3 N *acetic acid* is added, to have a pH of 3.8. Spectrophotometer is calibrated with *potassium iodate* to read zero for pure water.

2. Chemi-illuminescent technique: Ozone exhibits chemilluminiscent reaction with rhodamin B coated disc absorbed on silica gel. The resulting current flow per unit time indicates the concentration of ozone.

CARBON MONOXIDE

The air sample is passed through *iodine pentoxide* which liberates free *iodine* after the formation of CO_2.

A second method is by infrared analyser of photometric system.

CONTROL OF AIR POLLUTION

SETTLING CHAMBER

Gravitational settling chambers separate particles greater than 50 μm.

Velocity of flow of the gas through the chamber is 0.5 to 3 m/s.

They are ideally suited for the removal of large and heavy solid particles as those from kilns, natural draft furnaces etc.

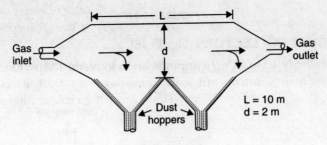

FIGURE 19.15

HOWARD SETTLING CHAMBER

It is a box like chamber with a number of horizontal trays to split the stream and to reduce the depth of settling.

The velocity of flow is 0.3 to 3 m/s. Particles of size greater than 10 μm are trapped. Longer trays are needed to trap finer particles. Their efficiency is less than 50%.

They pose cleaning problems. Higher temperatures tend to bend the plates.

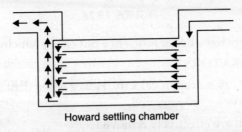

Howard settling chamber

FIGURE 19.16

DUST TRAP (Fig. 19.17)

Heavy dust loaded gas (dust > 100 g/m³ of gas) is rushed into the chamber at a velocity greater than 10 m/s. Its velocity gets reduced to 1 m/s inside the chamber due to increase in the area of cross-section.

Particles bigger than 30 μm are separated.

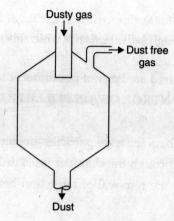

FIGURE 19.17

LOUVERED TYPE DUST COLLECTORS (Fig. 19.18)

Jets of gas of velocity 10 to 15 m/s impinge on to louvers. Their velocity is controlled and particles of size greater than 30 μm are efficiently removed.

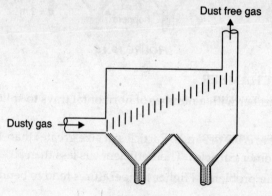

FIGURE 19.18

Closely spaced louvers have better efficiency but easily get clogged.

INERTIAL IMPACT SEPARATORS

Change in direction causes drop in velocity. Hence when dust laden gas has to move along the tortuous course dust particles are removed.

Particles of 5 to 20 μm are effectively removed.

FIGURE 19.19

CYCLONE SEPARATOR

Centrifugal force is used to separate particulates from gas stream.

Dusty gas is tangentially applied at a velocity of 10 to 25 m/s.

The dusty air forms the main vortex whose centrifugal force increases with depth due to reduction in cross-section. Particles are thrust on the walls of the cylinder.

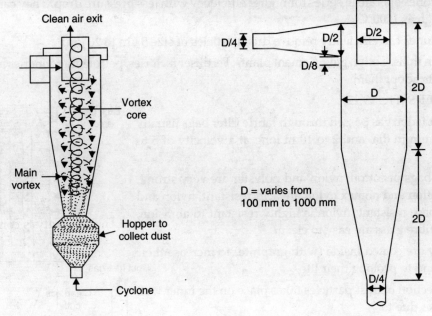

FIGURE 19.20   **FIGURE 19.21**

The upward pure air vortex (vortex core) emanates from the lowest point and pure air goes out.

Collection efficiency increases with

(i) High inlet velocity (v)

Force of separation, $$F = \frac{2mv^2}{D}$$

where F = Centrifugal force (N)
   m = Mass of particulates (kg)
   v = Gas inlet velocity, m/s
   D = Dia. of cylinder, m

(ii) Decrease in cylinder dia. (D)

(iii) Increase of body length

(iv) Increase in particle size and density

(v) Decrease in gas density and viscosity

(vi) Increase with the number of revolutions of the main vortex

(vii) Increase in the ratio of body dia. to outlet dia.

(viii) Increase in the smoothness of the inner cyclone wall.

High efficiency cylones adopt a dia. of 250 mm or less to separate particles of size less than 15 μm.

Cyclones operate in series for higher efficiency with less pressure drop. They can operate at temp. as high as 1000°C.

Cyclones can efficiently remove dust particles of size 5 μm to 20 μm.

Grain mills, Cotton gins, Cement plants, Fertiliser factories, petroleum refineries and Asphalt mixing units adopt them.

BAG FILTERS (Fig. 19.22)

Dust laden gas passed through fabric filter bags usually 100 to 400 mm in dia. and 2 to 10 m long at a velocity of 5 to 15 mm/s.

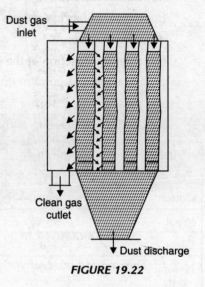

FIGURE 19.22

The bags of cotton, nylon and polyester are very strong, polyester, teflon and nomex nylon are acid resistant, nylon and teflon are base resistant, nylon is highly resistant to abrasion, teflon and fibre glass are easy to clean.

They are coated inside (with graphite) to increase their efficiency and to prolong their life.

Collection of dust particles takes place on the inner wall of the fabrics due to

    interception,

    inertial impaction,

    gravity settling,

    Brownian diffusion,

and   Electrostatic attraction.

The efficiency of the filter increases with the formation of dust cake on the fabric which helps in the straining and finer particles.

Periodic cleaning of the bags is done by:

(i) shaking of the bags,

(ii) pumping dust free air in the reverse direction of filtration, or

(iii) pulse jet cleaning, *i.e.*, subjecting the bags to a pressure of 700 to 850 kPa for 0.1 to 0.2 s in the same direction of filtration.

The "*rate of filtration* is the quantity of air filtered per sq. m surface area of filter per minute" ($m^3/m^2/min$).

In other words it is the quantity of (dusty) air filtered per unit surface area of the bag filters per minute. Hence it is known as '*Air to Cloth Ratio*'.

A value of 0.5 to 5 $m^3/m^2/min$ is usually adopted.

Air to cloth ratio (A/C) depends on:

(i) size of particles,

(ii) density of dust in air,

(iii) type of the fabric of filter bag, and

(iv) method of cleaning of bags.

Bag filters are highly efficient (> 99%) in the removal of even very fine particles (0.3 μm). Their efficiency depends on:

(i) size of the particles,

(ii) dust penetration through the fabric,

(iii) cake release,

(iv) nature of fabric—woven or felted,

(v) temperature,

(vi) chemical reaction of the gas with the filter bag,

(vii) abrasion resistance of the fabric,

(viii) pressure drop during filtration,

(ix) method of cleaning of the bags.

Fibre of bag	Max. Temp. able to withstand	Strength	Weakness
Cotton	80°C	Cheap	Poor resistance to acids
Wool	95°C	Resists abrasion	Poor resistance to bases
Glass	280°C	Resists acids and bases	Poor resistance to abrasion
Nylon	100°C	Resists bases and acids, Easy to clean, Resists abrasion	—
Polyester	135°C	Easy to clean	—
Teflon	260°C	—	Costly
Nomex Nylon	230°C	—	Poor resistance to moisture

SCRUBBERS

These are wet separators of particulates and gases where the downward trickling liquid (usually water) intercepts the rising dust or gas from the low level dirty gas inlet and carrying away the particles along with the effluent water.

Impaction and interception are the predominant forces removing particles of size > 0.3 μm while diffusion is the primary force for particles of size < 0.3 μm. *Diffusiophoretic Deposition, Electrostatic precipitation* and *Agglomeration* are also induced in wet scrubbing. Gas absorption takes place besides the removal of suspended particulate matter.

Hot, explosive and corrosive gases can be passed through them with no adverse effects.

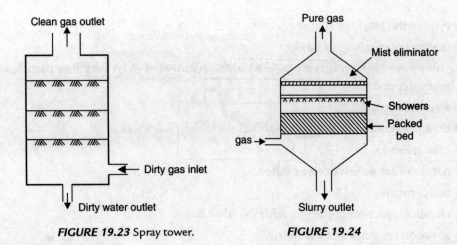

FIGURE 19.23 Spray tower.   FIGURE 19.24

PACKED BED SCRUBBER

The bed may be fixed or floating. It is usually composed of coke, crushed rock or fibre glass.

More than one bed may be provided at times.

Particles of 1 μm size are collected.

Merits of Scrubbers

1. Compact unit requiring less area.
2. Cheaper.
3. Low operating and maintenance cost.
4. Even gases are absorbed.
5. Hazardous wastes can be handled.
6. Not affected by moisture content.

Demerits

1. Not efficient at higher temperatures.
2. Corrosion of the scrubber container may take place.
3. Condensate plume of pollutional characteristics may emerge out.
4. Waste scrubber liquid is to be handled carefully.

ELECTROSTATIC PRECIPITATORS

D.C. voltage of is passed through discharge electrode wires. (Low voltage ESP operates at 6,000 to 8,000 V while high voltage ESP operates at 30,000 to 1,00,000 V). –ve polarity is developed around the wires. Electrostatic field is created in the gap between collection electrode and discharge electrode. The gas stream passing through the electrostatic field charges the dust particles –vely compelling them to get attracted by the +vely charged solid plate (collection) electrodes those are grounded.

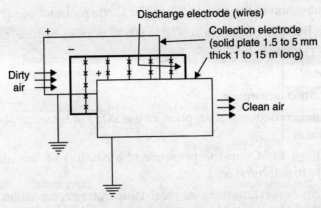

FIGURE 19.25

Whenever the thickness of dusty layer deposited is more than 5 mm thick, the collection electrode is rapped to dislodge the same.

In wet precipitators the cleaning liquid sliding down the electrode clears the dust.

Particles bigger than 1 μm are removed with an efficiency more than 99%.

Hot gases (upto 700°C) can be treated.

Acidic and tarry mists are easily treated whereas their treatment is difficult in any other process.

The pressure drop is small.

ESP can handle 25 to 1000 m³ of gas per second.

$$\text{Efficiency} = 1 - e^{-\left(\frac{AW}{Q}\right)}$$

where A = Area of collection plates (m²)

W = Drift velocity of charged particles (m/s)

Q = Flow rate (m³/s)

However very high voltage is required and particles those are not charged are not removed. More space is required and the equipment and its operation are all costly. They cannot treat explosive gases.

GAS CONTROLS

1. **Fuel substitution:** Coal having less amount of sulphur used.

Coal is crushed and sulphur segregated.

2. **Increasing the height of stack:** By increasing the height of the stack gaseous pollutants can be effectively dispersed. It can be as high as 450 m.

3. **Absorption:** On bed of lime, soda, sodium hydroxide, ammonium hydroxide, citric acid and magnesium oxide.

4. **Adsorption:** through Activated carbon, Allumina, Bauxite and Silica gel.

5. **Condensation:** Vapours are condensed as liquids either by increasing the pressure or by lowering the temperature.

6. **Flaring:** Combustion of the gas at 400° to 850°C when cannot be salvaged.

7. **Incineration:** Burning of the gas at 500° to 900°C to yield a small content of ash.

GAS REMOVAL METHODS

Sulphur dioxide

1. Activated carbon adsorption.
2. Scrubbing and consequent absorption of the SO_2 gas by water, converting the gas to sulphurous acid.
3. SO_2 is oxidized to SO_3 in the presence of a catalyst of vanadium pentoxide and subsequently to sulphuric acid.
4. Injection of dry or wet limestone into high temperature combustion zone along with SO_2 and then followed by scrubbing with liquid lime.
5. Scrubbing with sodium sulphite.
6. Reaction with H_2S to Produce elemental sulphur.

Oxides of Nitrogen

1. Scrubbing with limestone solution.
2. Adsorption on activated charcoal.
3. Flue gas recirculation.
4. Controlling the amount of excess air used for combustion. However, if the air is too less there is a possibility of under burning and CO may be given out.
5. Adopting tangential burning than the central burning flame.
6. Reduction to elemental nitrogen using methane as fuel and platinum vanadium as the catalyst.

Carbon monoxide

1. It has high calorific value. It is a cleaned and used as a fuel.
2. It is burnt in the presence of air.

II. OBJECTIVE TYPE QUESTIONS

1. An average adult at rest inhales about......kg of air per day.
   (a) 8 ☐ (b) 16 ☐
   (c) 24 ☐ (d) 32. ☐

2. Particulate is an airbone
   (a) solid of size greater than 75 μm ☐
   (b) solid of size greater than 10 μm ☐
   (c) liquid of size greater than 1 μm ☐
   (d) solid or liquid of size greater than 0.0002 μm. ☐

3. Smoke is produced out of coal, wood, tobacco or other products containing carbon because of
   (a) inadequate amount of oxygen to achieve stoichiometric combustion ☐
   (b) inadequate mixing of air with the fuel ☐
   (c) temperature of combustion far below the required one ☐
   (d) any one or combinations of the above. ☐

4. Fume is one which has a particle size
   (a) less than 0.0002 μm ☐ (b) between 0.0002 and 1 μm ☐
   (c) 1 to 75 μm ☐ (d) 75 to 500 μm. ☐

5. Haze particles have a size
   (a) less than 0.0002 μm ☐ (b) 0.38 to 0.76 μm ☐
   (c) 1 to 75 μm ☐ (d) greater than 75 μm. ☐

6. Soot is
   (a) vapour generated by volatalization of solids ☐
   (b) finely dispersed solid particles of microscopic size in gaseous medium ☐
   (c) products of incomplete combustion causing finely divided aerosol particles ☐
   (d) finely divided carbon particles containing tar. ☐

7. Haze particles are the most effective in causing
   (a) rickets ☐ (b) farmer's lung ☐
   (c) tumour ☐ (d) visibility reduction. ☐

8. Farmer's lung is caused because of
   (a) spores of moulds on hay ☐ (b) continuous application of insecticides ☐
   (c) spread of pollen ☐ (d) vegetable fibres as cotton and jute. ☐

9. The colourless, odourless, tasteless gas which affects only warm blooded animals and which had no effect on insects, vegetation or metallic surface is
   (a) O_3 ☐ (b) SO_2 ☐
   (c) CO ☐ (d) NO_2. ☐

10. The main source of CO is
    (a) oxidation of methane ☐ (b) smoking of tobacco ☐
    (c) diesel exhaust ☐ (d) gasoline exhaust. ☐

11. Most of the CO is given out (> 60%) by
    (a) agricultural burning ☐ (b) forest fires ☐
    (c) thermal power plants ☐ (d) automobile exhausts. ☐

12. Atmospheric temperature rises because of the release of
    (a) suspended particulate matter ☐ (b) CO_2 ☐
    (c) O_3 ☐ (d) aerosols. ☐

13. Atmospheric temperature decreases because of the release of
    (a) suspended particulate matter ☐ (b) CO_2 ☐
    (c) O_3 ☐ (d) aerosols. ☐

14. Pinkish colour of flesh is due to
    (a) Methenomoglobinemia
    (b) Carboxyhaemoglobin
    (c) Oxyhaemoglobin
    (d) Nitroxihaemoglobin.

15. CO had an atmospheric mean life of
    (a) $2\frac{1}{2}$ hours
    (b) $2\frac{1}{2}$ days
    (c) $2\frac{1}{2}$ months
    (d) $2\frac{1}{2}$ years.

16. The gas that occupies the place after nitrogen and oxygen in the atmosphere is
    (a) Argon
    (b) CO_2
    (c) N_2O
    (d) H.

17. Aerosol is
    (a) air in solution in a liquid medium
    (b) finely dispersed liquid or solid in a gas
    (c) air saturated with a liquid droplets
    (d) air blended with grit.

18. Thermal power plants mostly produce
    (a) CO
    (b) SO_2
    (c) NO_2
    (d) PAN.

19. The air pollutant that causes emphysema is
    (a) CO
    (b) SPM
    (c) SO_2
    (d) NO_2.

20. Paper becomes yellow and brittle because of
    (a) CO
    (b) SPM
    (c) SO_2
    (d) NO_2.

21. The anaesthetic gas is
    (a) N_2O_2
    (b) NO_2
    (c) N_2O
    (d) N_2O_5.

22. The mostly present oxide of nitrogen in the atmosphere is
    (a) N_2O_2
    (b) N_2O
    (c) NO_2
    (d) N_2O_5.

23. The oxide of nitrogen formed at very high temperature is
    (a) N_2O
    (b) NO
    (c) NO_2
    (d) N_2O_5.

24. The gas that forms smog is
    (a) CO
    (b) SO_2
    (c) NO
    (d) NO_2.

25. The major sources of hydrocarbons is
    (a) gasoline exhaust
    (b) diesel exhaust
    (c) forest fires
    (d) agricultural burning.

26. Photo chemical oxidants are produced because of
    (a) NO_2
    (b) NO_2 + Hydrocarbons
    (c) NO_2 + Hydrocarbons + Oxygen
    (d) NO_2 + Hydrocarbons + Oxygen + Light.

27. Sky appears blue because of
    (a) nitrogen in the atmosphere
    (b) the effect of earth's reflectivity
    (c) the terrestrial infra-red radiation
    (d) presence of excess NO_2 in the atmosphere.

28. Ammonia of the atmosphere
    (a) is an inert gas
    (b) neutralises acids
    (c) combines with hydrocarbons to produce ozone
    (d) readily gets stabilised as nitrates.

29. Sky appears brownish because of
    (a) presence of nitrogen in the atmosphere
    (b) the effect of earth's reflectivity
    (c) the terrestrial infra-red radiation
    (d) presence of excess NO_2 in the atmosphere.

30. Rubber loses elasticity and gets cracked because of
    (a) SO_2
    (b) NO_2
    (c) O_3
    (d) PAN.

31. Abnormalities in fertility and pregenancy and ill mental health of children are because of
    (a) fluorine
    (b) lead
    (c) cadmium
    (d) arsenic.

32. Ozone layer is being destroyed by
    (a) CO_x
    (b) SO_x
    (c) NO_x
    (d) F.

33. Pedal bone fracture in cattle occurs mainly because of grazing on fodder coated with
    (a) PAN
    (b) cement dust
    (c) fluorides
    (d) zinc.

34. Tetramethyl and tetraethyl lead are commonly found in
    (a) petroleum
    (b) paints
    (c) zinc refining
    (d) cigarettes.

35. Coal having 1% sulphur produces about....kg of SO_2 per tonne.
    (a) $\frac{1}{2}$
    (b) 4
    (c) 8
    (d) 16.

36. Synergistic effect is
    (a) affecting nervous system
    (b) causing bronchitis
    (c) aggravating the effect of SO_2 in combination with particulates
    (d) formation of ammonium sulphate with ammonia.

37. Acid rain is because of
    (a) SO_x
    (b) CO_x
    (c) COH
    (d) H_2S.

38. Which of the following gases is not an air pollutant?
    (a) CO
    (b) CO_2
    (c) SO_x
    (d) NO_x.

39. The wind flowing towards South is said to be
    (a) North wind
    (b) East wind
    (c) South wind
    (d) West wind.

40. 'Calm' represents a condition when the velocity of the wind is less than
    (a) 10 kmph
    (b) 6 kmph
    (c) 3 kmph
    (d) 1 kmph.

41. As the surface roughness increases
    (a) veering decreases
    (b) thermal turbulence increases
    (c) wind speed decreases
    (d) mechanical turbulence increases.

42. The depth of friction layer is less during the night time because of
    (a) surface roughness
    (b) lack of thermal insolation
    (c) increased mechanical turbulence
    (d) unstable atmospheric condition.

43. Because of heat island effect draft is induced which results in
    (a) pollutants being dispersed uniformly
    (b) pollutants being dispelled away from the heart of the city
    (c) pollutants being sucked towards the heart of the city
    (d) pollutants travelling from higher altitude to lower altitude.

44. Greater mixing depth results in
    (a) greater dispersion of the pollutant
    (b) uniform concentration of the pollutant
    (c) inversion
    (d) green house effect.

45. When the plume rises vertically upwards it is
    (a) looping plume
    (b) coning plume
    (c) fanning plume
    (d) neutral plume.

46. Fumigating plume results in when
    (a) temperature remains unaltered with the height
    (b) an inversion results
    (c) atmosphere is super-adiabatic below the stack height but the lapse rate is negative above
    (d) ambient lapse rate is neutral and a wind velocity of more than 9 m/s results.

47. Plume rise is more when
    (a) wind velocity is more
    (b) less is the linear diameter of the stack
    (c) less is the exit velocity of the effluent
    (d) more is the difference in temperature between **exit gas and air temperature.**

48. To prevent the down draught of the pollutants the minimum exit velocity of the waste gases should be

   (a) $1\frac{1}{2}$ × wind speed ☐
   (b) $1\frac{1}{2}$ = wind speed ☐
   (c) less than wind speed ☐
   (d) more than 3 × wind speed. ☐

49. Minimum stack height should be greater than......times the neighbouring structures to prevent the mechanical turbulence.

   (a) $1\frac{1}{2}$ ☐
   (b) $2\frac{1}{2}$ ☐
   (c) 4 ☐
   (d) 5. ☐

50. Green house effect of CO_2 is

   (a) permitting the outside solar radiation to reach the ground but preventing terrestrial radiation from the ground into the space ☐
   (b) permitting the solar radiation of short length and reradiated terrestrial heat of long wave length ☐
   (c) reflecting the heat rays into the space thereby keeping the temperature of earth unaffected ☐
   (d) causing absorption of heat from troposphere and thereby decreasing the temperature of earth with increase in CO_2 concentration. ☐

51. When the temperature falls at a rate less than dry adiabatic lapse rate then it is

   (a) isothermal ☐
   (b) neutral ☐
   (c) inversion ☐
   (d) stable. ☐

52. Dry adiabatic lapse rate in the troposphere is

   (a) 0 ☐
   (b) – 9.8°C/km ☐
   (c) – 6°C/km ☐
   (d) + 9.8°C/km. ☐

53. Atmospheric stability is reached when

   (a) solar radiation is strong ☐
   (b) velocity of wind is greater than 6 m/s ☐
   (c) sky is cloudy at night ☐
   (d) clear sky exists during the day ☐

54. Plume rise increases the effective height of the stack and thereby

   (a) reduces the ground level concentration of the pollutant ☐
   (b) increases the ground level concentration of the pollutant ☐
   (c) concentration at ground level varies with the temperature of exhaust gases ☐
   (d) concentration at ground level varies with the velocity of exhaust gas. ☐

55. Atmospheric stability increases

   (a) when the lapse rate is less than dry adiabatic lapse rate ☐
   (b) when the lapse rate is more than the dry adiabatic lapse rate ☐
   (c) when the lapse rate equals the dry adiabatic lapse rate ☐
   (d) when vertical motions are enhanced. ☐

56. Rapid diffusion of the pollutant occurs when the atmosphere is
    (a) stable
    (b) unstable
    (c) neutral
    (d) experiencing an inversion.

57. Cloudy skies promote an atmospheric condition of
    (a) stability
    (b) unstability
    (c) neutrality
    (d) inversion.

58. When the fall in temperature per 100 m rise in altitude is more than 0.98°C, it is
    (a) moderately stable
    (b) slightly stable
    (c) moderately unstable
    (d) extremely unstable.

59. Because of heat island effect
    (a) unstable conditions may be caused
    (b) atmosphere is heated
    (c) pollution levels increase during day time
    (d) increase in temperature difference between rural and urban areas more during day time is caused.

60. When the lapse rate is exactly equals adiabatic lapse rate, then the stability is
    (a) stable
    (b) slightly stable
    (c) highly unstable
    (d) neutral.

61. The velocity of exit waste gases should be a minimum of............. **wind speed to prevent down draught.**
    (a) $\frac{1}{2}$
    (b) $1\frac{1}{2}$
    (c) $2\frac{1}{2}$
    (d) $3\frac{1}{2}$

62. Minimum stack height should be............. height of the nearby structure.
    (a) $\frac{1}{2}$
    (b) $2\frac{1}{2}$
    (c) 5
    (d) 10.

63. Deposit gauges are provided with copper sulphate solution
    (a) to prevent the growth of bacteria
    (b) to prevent the growth of algae
    (c) to scare birds
    (d) to prevent the decomposition of SPM.

64. To device used for the easy separation of dry dust of 10 to 100 µm size is
    (a) cyclone
    (b) gravity settling chamber
    (c) bag filter
    (d) scrubber.

65. The unit that cleans dust because of the mechanical straining, sedimentation, inertial force, electrostatic attraction and diffusion is
    (a) electrostatic precipitator
    (b) scrubber
    (c) cyclone
    (d) bag filter.

66. The unit that had the heighest efficiently and which can remove even the finest particle is
    (a) electrostatic precipitator
    (b) scrubber
    (c) cyclone
    (d) bag filter.
67. An air pollutant is
    (a) solid
    (b) liquid
    (c) gas
    (d) solid, liquid or gas.
68. A factor minimising air pollution is
    (a) still air
    (b) chill night
    (c) valley
    (d) rain.
69. The gas that is more than 50% of air pollutants because of anthropogenic source is
    (a) CO
    (b) NO_x
    (c) SO_2
    (d) HC.
70. The air pollutant mostly emitted through natural sources than man made sources
    (a) CO
    (b) NO_x
    (c) SO_2
    (d) HC.
71. Bhopal Gas Tragedy is the result of
    (a) SO_2
    (b) Phosgene
    (c) MIC
    (d) Dioxin.
72. Which of the following is an air pollutant?
    (a) N_2O
    (b) CO_2
    (c) Tropospheric Ozone
    (d) Stratospheric Ozone.
73. Fume is the same as
    (a) Grit
    (b) Dust
    (c) Mist
    (d) Smoke.
74. Fog is responsible for
    (a) Bronchitis
    (b) Opacity
    (c) Dizziness
    (d) Crop failure.
75. Haze particles have a size
    (a) < 0.01 µm
    (b) > 0.375 µm but < 0.75 µm
    (c) > 1 µm
    (d) > 10 µm.
76. A particle of size greater than .... can be seen with naked eye whereas that lesser than .... can be seen only through an electronic microscope
    (a) 1 µm, 1 nm
    (b) 5 µm, 5 nm
    (c) 10 µm, 10 nm
    (d) 50 µm, 5 nm.
77. Which of the following is not a Green House Gas?
    (a) N_2O
    (b) CFC
    (c) O_3
    (d) CO.

78. More than 50% of the Green House Effect is because of
    (a) CO_2 □ (b) CH_4 □
    (c) CFCs □ (d) N_2O. □
79. A gas having more affinity to Haemoglobin as CO
    (a) CH_4 □ (b) SO □
    (c) NO □ (d) SO_2. □
80. Emphysema is
    (a) Throat irritation □ (b) Break down of alveoli □
    (c) Skin allergy □ (d) a type of cancer. □
81. Acid rain results because of
    (a) N_2O □ (b) NO □
    (c) NO_2 □ (d) N_2O_5. □
82. Chlorosis is
    (a) Downward curvature of leaves □ (b) Dropping of leaves □
    (c) Collapse of tissues □ (d) Loss of chlorophyll. □
83. Dust Domes are formed because of
    (a) Global warming □ (b) Acid Rains □
    (c) Heat Islands □ (d) Ozone depletion. □

III. TRUE /FALSE

1. Intense mist which prevents visibility is called FOG.
2. Smoke is given out of solids containing carbon because of incomplete combustion.
3. Particle of size greater than 10 μm are filtered by the respiratory tract of human beings.
4. Particulates in excess concentration reduce the precipitation.
5. Domestic animals in industrial areas require more living space.
6. Particulates are responsible for cooling of the atmosphere.
7. When the level of carboxyhaemoglobin is more in blood the person cannot discriminate the 'time interval'.
8. Men have more lead in blood than women, urban residents than rural folk and smokers than non-smokers.
9. The intensity of rain, fog and warmth are more in urban areas than the surrounding rural areas.
10. When the atmosphere is extremely stable, the directions of wind at two different height may be even opposite to each other.
11. Calm clear nights may present stable condition over rural areas and unstable condition in urban areas.
12. For the thermal power plant, the dominant factor causing the plume rise is the buoyant force but not the momentum of the exit effluents.

13. Smaller the mixing depth, greater is the concentration of the accumulation of the air pollutant.
14. In temperate zones with less solar heating and more cooling 'inversions' are more common at higher altitudes.
15. Strong solar radiation breaks the stable layers formed at night and rapidly becomes unstable. Pollutants are rapidly dispersed.
16. High volume samplers are used to collect finer particles which may not settle in deposit gauges.
17. Within reasonable limits, more is the velocity of wind more is the dilution and hence more is the visibility.
18. An industrial city shall be about 10 per cent more cloudy and may have 10 per cent more precipitation than the adjoining villages.
19. Coal or oil fuels can be made to burn without smoke but they produce more quantities of oxides of nitrogen.
20. Stationary sources as thermal power plants contribute less amount CO than mobile sources as automobiles due to efficient combustion.
21. Wind speed remains constant irrespective of the altitude.
22. In a stable atmosphere vertical motions of the pollutants rapidly take place.
23. When the ambient lapse rate exactly equals the dry adiabatic lapse rate (*i.e.* 0.98°C fall per 100 m rise in altitude), then atmospheric stability is termed as 'neutral'.
24. Vertical motions are not influenced when the atmospheric stability is neutral.
25. When the rate of decrease in temperature with increase in altitude is less than dry adiabatic lapse rate then an inversion results.
26. Vertical motions (of the pollutants) are dampened when the temperature rises with altitude.
27. More is the stability of the environment, more is the turbulence caused.
28. Cloud cover in the sky prevents cooling or heating of the area below it.
29. More is the wind velocity (greater than 6 m/s), higher is the mixing or dilution and hence the stability is neutral.
30. Low wind velocity (less than 2 m/s) with clear sky during the day time may cause 'instability' but 'stability' at night.
31. Under stable atmospheric conditions horizontal dispersion of the pollutants is more.
32. Strong insolation and less wind velocity promote unstability.
33. Unstable atmosphere never results at night irrespective of any wind velocity.
34. The height of the base of stable layer above ground level is called 'Depth (or height) of mixing layer'.
35. More is the wind velocity, more is the plume rise.
36. Greater is the exit diameter of the stack less is the plume rise.
37. Plume rise is independent of stack height.
38. A hotter effluent from a stack may have more plume rise than a colder effluent.
39. Because of precipitation air pollution is rendered water pollution and subsequently land pollution.

40. Most fabric filters cannot withstand a temperature beyond 120°C.
41. In wet type of electrostatic precipitator a film of water flows over the collection electrodes.
42. Scrubbing liquid in a packed tower can be water, sulphuric acid, sodium hypochlorite or caustic soda.
43. Inflammable organic vapours are subjected to incineration when no other method of disposal is economical.
44. To collect hazardous gaseous vapours and pathogenic bacteria, disposable filters are employed.
45. Slow moving motor vehicles contribute more CO than fast moving vehicles.
46. Petrol engines contribute more than 250 time the CO produced by the diesel engine for the same quantity of fuel.
47. CO affects warm-blooded and cold-blooded animals alike.
48. CO has no influence on vegetation and property.
49. Hard coals as anthracite have less S compared to soft coals as lignite.
50. SO_2 is both an oxidizing as well as reducing agent.
51. Gasoline engines contribute more particulates and NO_2 compared to Diesel engines.
52. Ozone and PAN are stronger oxidizing agents than oxygen.
53. Ozone may be a pollutant in Troposphere but is a savious in stratosphere.
54. Methane is a Green House Gas that retards stratospheric ozone depletion.
55. Fly ash can be classified as dust, smoke or fume.
56. Fume is non carbonaceous smoke.
57. Concentrations of CO and HC emissions of automobiles are reduced by supplying more oxygen and heat but concentrations of NO_x increase.
58. Hydrocarbons of smoke react with oxides of Nitrogen in the presence of sunlight to form SMOG.
59. Though a hydrocarbon, Methane will not form SMOG with NO_x and sunlight.
60. Cloudy days are cool and cloudy nights are warm.
61. On a clear summer day an industrial city presents clear and blue air above mixing height while brown and hazy in the mixing height.
62. In haled lead is more harmful than ingested lead.

ANSWERS
Objective Type Questions

1. (b)	2. (d)	3. (d)	4. (b)	5. (b)	6. (d)
7. (d)	8. (a)	9. (c)	10. (d)	11. (d)	12. (b)
13. (a)	14. (c)	15. (c)	16. (a)	17. (b)	18. (b)
19. (c)	20. (c)	21. (c)	22. (b)	23. (b)	24. (d)
25. (a)	26. (d)	27. (b)	28. (b)	29. (d)	30. (c)
31. (b)	32. (d)	33. (c)	34. (a)	35. (d)	36. (c)

AIR POLLUTION 969

37. (a)	38. (b)	39. (a)	40. (b)	41. (d)	42. (b)
43. (c)	44. (a)	45. (d)	46. (c)	47. (d)	48. (a)
49. (b)	50. (a)	51. (d)	52. (b)	53. (c)	54. (a)
55. (a)	56. (b)	57. (c)	58. (d)	59. (c)	60. (d)
61. (c)	62. (b)	63. (b)	64. (a)	65. (c)	66. (d)
67. (d)	68. (d)	69. (a)	70. (b)	71. (c)	72. (c)
73. (d)	74. (b)	75. (b)	76. (d)	77. (d)	78. (a)
79. (c)	80. (b)	81. (c)	82. (d)	83. (c)	

True/False

1. T	2. T	3. T	4. T	5. T	6. F
7. T	8. T	9. T	10. T	11. T	12. T
13. T	14. T	15. T	16. T	17. T	18. T
19. T	20. T	21. F	22. F	23. T	24. T
25. F	26. T	27. F	28. T	29. T	30. T
31. T	32. T	33. T	34. T	35. F	36. F
37. F	38. T	39. T	40. T	41. T	42. T
43. T	44. T	45. T	46. T	47. F	48. T
49. T	50. T	51. F	52. T	53. T	54. T
55. T	56. T	57. T	58. T	59. T	60. T
61. T	62. T				

Chapter 20 WATERSHED MANAGEMENT

I. INTRODUCTION

The area from which all the runoff generated passes through one point on the stream at its outlet is variously known as the catchment, drainage basin, watershed and simply the basin. The catchment area for any point on a stream can be easily obtained from the topographic maps. Generally if the area of the basin is small it is called watershed otherwise the catchment or the drainage basin.

Watershed management is an approach for harnessing the natural resources of the watershed namely the soil, water, vegetation etc. Wherever watershed management practices are not implemented the forests are destroyed, rich and fertile top soil is eroded, reservoirs are silted rapidly, and environmental degradation resulted. Good watershed management practices can help in controlling global warming, greenhouse effect, and levels of chlorofluorocarbons in the atmosphere. It is estimated that 90% of land in India is susceptible for degradation. Watershed management practices aim to control the indiscriminate exploitation of natural resources and try to preserve them.

The general characteristics to be considered are (*i*) size (*ii*) shape (*iii*) slope (*iv*) drainage network (*v*) vegetation (*vi*) soil type (*vii*) climate and (*viii*) land use. Based on the area the watersheds are classified as sub-watersheds (10 – 500 km^2), Micro-watersheds (1 – 10 km^2) and Mini watersheds (< 1 km^2). The area is one of the key input parameters for designing the conservation structures. The watershed may be pear shaped, elongated, triangular, circular, or may resemble a sector of a circle. The runoff characteristics are affected to some extent by the shape of the watershed. The slope of the watershed influences the velocity of overland flow and runoff and also the infiltration. The drainage network of the basin may also influence the runoff characteristics and the erosion pattern in the watershed. The information on vegetation and climate is crucial for planning the type of conservation method. The type of conservation measure depends on land use. Land use also helps in land classification.

The objectives of watershed management are: conservation of soil and water, Improve the water holding capacity of soil, Harvesting the rain water and using it to recharge groundwater, promoting greenery. Proper implementation of watershed management results in increased farm production and higher per capita income.

The land management consists of land surveying and land development. The land development methods include contouring, jungle clearance, uprooting, levelling, shaping, ploughing etc.

The conservation methods to control the soil erosion include ploughing, furrowing, trenching, contour bunding, hedging, terracing, wattling (fencing) etc.

Ploughing along the slope increases runoff, erosion and 50–80% of runoff drains away without giving opportunity for the soil to absorb moisture. Instead ploughing along the contours reduces the erosion by more than 50% and soil retains more moisture.

Continuous furrowing along the contours and ridge development across the furrows is a recommended practice.

Excavating the trench along the contour helps in controlling erosion.

Hedges are grown either in furrows, trenches or on bunds. Almost all rainwater can be harvested by growing hedges.

Check dams are constructed to effectively control the gully erosion.

Even in drought prone areas with scanty rainfall, the rainwater is considerable and of utmost importance and therefore it has to be harvested to the extent possible. Farm ponds and cross bunds are commonly used as rainwater harvesting structures. Farm ponds vary in size. In general they are of 10–30 m long, 5–20 m wide and 2 m deep. They have a harvesting capacity of 100–500 m^3. The efficiency of farm ponds can be enhanced considerably by utilising the harvested water in between rains. Evaporation loss in large ponds can be reduced by spreading retardants.

II. OBJECTIVE TYPE QUESTIONS

1. The area from which all the rainwater drains into a stream and flows through a single point is called
   - (a) Drainage Basin
   - (b) The Catchment
   - (c) Watershed
   - (d) all the above.
2. The crop yield is influenced by
   - (a) depth of farm pond
   - (b) depth of tillage
   - (c) height of check dam
   - (d) none of the above.
3. Implementation of proper watershed management practices will improve
   - (a) the farm production
   - (b) the percapita income
   - (c) both (a) and (b)
   - (d) none of the above.

4. The ideal measurements of a farm pond are
    (a) 5–15 m L 1–5 m W and 1 m D
    (b) 10–30 m L 5–20 m W 2 m D
    (c) 10–50 m L 5–20 m W and 6 m D
    (d) 10–30 m L 5–20 m W and 2 m D.
5. The retardants are used to
    (a) reduce evaporation
    (b) reduce transpiration
    (c) increase infiltration
    (d) to reduce percolation.
6. Groundwater can be protected against pollution by
    (a) locating well at a safe distance from the source of pollution
    (b) grouting and sealing the well casing
    (c) both (a) and (b)
    (d) none of the above.
7. Groundwater legislation is required to
    (a) avoid mining
    (b) protect groundwater against pollution
    (c) regulate the groundwater exploitation
    (d) all the above.
8. Watershed with an area of 1 to 10 km^2 is known as
    (a) Subwatershed
    (b) Mini watershed
    (c) Micro watershed
    (d) Milli watershed.
9. The detachment of soil particle from its present location is called
    (a) sedimentation
    (b) saltation
    (c) silting
    (d) erosion.
10. The best orientation of ploughing is
    (a) along the contours
    (b) across the contours
    (c) at 45° to the contours
    (d) any direction.
11. Narrow excavation along the contours is known as
    (a) Bunding
    (b) Trenching
    (c) Furrowing
    (d) Hedging.
12. Which method is commonly used for soil conservation on steep slopes ?
    (a) Terracing
    (b) Hedging
    (c) Wattling and staking
    (d) Levelling.
13. Which of the following is a saline soil ?
    (a) Alkali soils
    (b) Acidic soils
    (c) Sulfide soils
    (d) All of the above.
14. The water harvesting capacity of farm pond varies from
    (a) 1 to 10 cum
    (b) 10 to 100 cum
    (c) 100 to 500 cum
    (d) none of the above.
15. Water that can be withdrawn from a basin without causing adverse effects to the ground water regime is known as
    (a) maximum perennial yield
    (b) deferred perennial yield
    (c) perennial yield
    (d) none of the above.

WATERSHED MANAGEMENT 973

16. Greening in watershed management involves
    (a) Agriculture
    (b) Horticulture
    (c) Social forestry
    (d) All of the above.

17. Which of the following is benefit of crop rotation ?
    (a) Conservation of soil fertility
    (b) Decrease in soil erosion
    (c) Improving crop yield
    (d) All of the above.

18. Use of harvested rainwater runoff for crop production is known as
    (a) Micro catchment agriculture
    (b) Runoff agriculture
    (c) Dry land agriculture
    (d) None of the above.

19. Ratio of crop yield to evapotranspiration is called
    (a) Water use efficiency
    (b) Water conveyance efficiency
    (c) Water distribution efficiency
    (d) None of the above.

20. Growing fodder trees in pastures is known as
    (a) Silvipastoral practice
    (b) Tree culture
    (c) Farm Forestry
    (d) None of the above.

21. Solar energy potential is maximum in which state of our country
    (a) Andhra Pradesh
    (b) Rajasthan
    (c) Madhya Pradesh
    (d) Uttar Pradesh.

22. Biomass is used for
    (a) Heating
    (b) Production of gobar gas
    (c) Production of electricity and oil
    (d) All of the above.

23. The common sizes of the windmills used for wind power generation, ranges between
    (a) 1.5 to 9.0 m
    (b) 1 to 2 m
    (c) 3.5 to 6.5 m
    (d) 1 to 5 m.

24. Of the following, which zone of the earth stores widespread geothermal energy resources
    (a) Hydrothermal convection zone
    (b) Superheated water pressure zone
    (c) Hot rock zone
    (d) None of the above.

25. 'Pisciculture' is concerned with
    (a) Habitat of fish
    (b) Habitat of cows
    (c) Habitat of sheep
    (d) Habitat of pigs.

26. NWDP Stands for
    (a) Natural Watershed Development Program
    (b) National Watershed Development Program
    (c) Nationalized Watershed Development Program
    (d) Non Watershed Development Program.

27. Buffer irrigation which is advocated for widely spaced orchards is designed for functioning
    (a) at high pressure heads
    (b) at low pressure heads
    (c) at normal pressure heads
    (d) none of the above.

28. A process, in which an unpopular variety plant is cut partially and bound together with a popular plant is called
    (a) Trimming
    (b) Fencing
    (c) Grafting
    (d) Cloning.

29. Chipko is associated with
    (a) environmental movement
    (b) water resources projects
    (c) poverty removal
    (d) literacy.

30. SAT Stands for
    (a) Social Agricultural Terrain
    (b) Semi Arid Tropics
    (c) Semi Agricultural Terrain
    (d) Semi Arid Tree.

31. The geographical area of India is about
    (a) 100 million hectares
    (b) 329 million hectares
    (c) 280 million hectares
    (d) 509 million hectares.

32. DPAP was launched in our country in the
    (a) 1st Five Year Plan
    (b) 5th Five Year Plan
    (c) 3rd Five Year Plan
    (d) 2nd Five Year Plan.

33. In the case of small farms (upto 2 Hectares area) what is the scale adopted for preparation of land map?
    (a) 1 : 1000
    (b) 1 : 10
    (c) 1 : 100
    (d) 1 : 10000.

34. Narrow trenches built along contours for collecting overland flow and increasing soil moisture are known as
    (a) Terraces
    (b) Gradoni
    (c) Hedges
    (d) Barrows.

35. Which of the following is a pervious check dam?
    (a) Rock fill dam
    (b) Concrete gravity dam
    (c) Concrete arch dam
    (d) None of the above.

36. Alkaline Lands are classified as
    (a) red and white
    (b) red and black
    (c) black and white
    (d) none of the above.

37. Which of the following is related to sustainable agriculture?
    (a) Molecular biology
    (b) Genetic Engineering
    (c) Water stress plants
    (d) All of the above.

38. Spider Farming is a term related to
    (a) use of chemical pesticides
    (b) avoiding the usage of chemical pesticides
    (c) use of green manures
    (d) use of organic fertilizers.

39. IRDP Stands for
    (a) Integrated Rural Development Program ☐ (b) International Rural Development Program ☐
    (c) Integrated Rural Drought Program ☐ (d) International Rural Drought Program. ☐
40. "Depression Harvesting" is a method for harvesting which of the following
    (a) rainwater ☐ (b) groundwater ☐
    (c) sub surface water ☐ (d) none of the above. ☐

ANSWERS
Objective Type Questions

1. (d)	2. (b)	3. (c)	4. (d)	5. (a)	6. (c)
7. (d)	8. (c)	9. (d)	10. (a)	11. (b)	12. (a)
13. (d)	14. (c)	15. (c)	16. (d)	17. (d)	18. (b)
19. (a)	20. (a)	21. (b)	22. (d)	23. (a)	24. (c)
25. (a)	26. (b)	27. (b)	28. (c)	29. (a)	30. (b)
31. (b)	32. (b)	33. (a)	34. (b)	35. (a)	36. (c)
37. (d)	38. (b)	39. (a)	40. (a).		

STAGE BY STAGE
DRAMATIS PERSONAE

BY THE SAME AUTHOR

Novels
The Volcano God
The Zoltans: A Trilogy
The Dark Shore
The Evening Heron
Dreams of Youth
Easter Island
Searching

Short Stories
The Devious Ways
The Beholder
The Snow
Three Exotic Tales
A Man of Taste
The Young Greek and the Creole
The Spymaster
The Young Artists

Fantasy
The Merry Communist

Plays
Prince Hamlet
Mario's Well
Black Velvet
Simon Simon
Three Off-Broadway Plays
More Off-Broadway Plays
Three Poetic Plays

Criticism
Myths of Creation
The Art of Reading the Novel
Preface to Otto Rank's The Myth of the Birth of the Hero and Other Essays
Preface to Kimi Gengo's To One Who Mourns at the Death of the Emperor
Some Notes on Tragedy
Preface to Joseph Conrad's Lord Jim
Stage by Stage: The Birth of Theatre
Stage by Stage: Oriental Theatre

Poetry
Private Speech

Philip Freund

STAGE BY STAGE
DRAMATIS PERSONAE
The Rise of Medieval and Renaissance Theatre

PETER OWEN
London and Chester Springs

PETER OWEN PUBLISHERS
73 Kenway Road, London SW5 0RE

Peter Owen books are distributed in the USA by
Dufour Editions Inc., Chester Springs, PA 19425-0007

First published in Great Britain 2006 by
Peter Owen Publishers

© Philip Freund 2006

All Rights Reserved.
No part of this publication may be reproduced in any form or by any means
without the written permission of the publishers.

ISBN 0 7206 1245 4

A catalogue record for this book is available from the British Library

Printed and bound in Croatia by
Zrinksi d.d.

For Robert H. Heiser
A friend indeed

CONTENTS

List of Illustrations	13
Preface	17

1 The Christian Drama — 21
- Early Church drama — 23
- Passion Plays and Mysteries — 36
- Miracle Plays — 59
- Morality Plays — 69

2 Medieval Farces — 87
- The Feast of Asses and the Feast of Fools — 87
- The French *Farceurs* — 91
- German *Fastnachtspiel* — 99
- Medieval Dutch and Italian farces — 105
- English Interludes — 106

3 The Rebirth: Italy — 113
- Pseudo-classical drama — 113
  - Revivals of Seneca — 115
  - Cinthio — 116
- Early Italian theatres and stagecraft — 119
- Renaissance playhouses — 125
- Pastoral dramas — 129
- *Commedia Erudita* — 133
  - Lodovico Ariosto — 134
  - Niccolò Machiavelli — 137
  - Pietro Aretino — 143
  - Angelo Beolco — 158
  - Annibal Caro — 162
  - Alessandro Piccolomini — 168

	Giovan Maria Cecchi	174
	Leone De' Sommi	203
Commedia grava		213
	La Pellegrina	214
	Gian Lorenzo Bernini	226
Commedia dell'arte		233
	The characters	234
	The troupes	238
	Lazzi, stagecraft and scenarios	244
	French *commedia*	250
	English mime: Joseph Grimaldi	252
	In Germany and Austria	256
	Modern evolutions	258

4 Italy: Words and Music — 275

- The Camerata — 277
- Claudio Monteverdi — 281
- Francesco Cavalli — 292
- Marc Antonio Cesti — 297
- Francesco Sacrati — 301
- Alessandro Stradella — 301
- Venetian playhouses — 303
- Alessandro Scarlatti and the Neapolitan School — 316
- *Opera buffa* — 324
  - Pergolesi — 327
- Antonio Vivaldi and Venetian musical theatre — 334
- The librettists — 339
- Stage designers — 341

5 Italy: The Dance — 347
- Origins — 347
- Court dances — 349

6 Drama in Pre-Baroque Germany — 353
- Fifteenth- and early sixteenth-century drama — 353
- Philipp Frischlin — 358
- Heinrich Julius and the transition to German dialogue — 362

7 Spain: A Glorious Era — 365
- Medieval playlets — 366
- *La Celestina* — 367
- Gil Vicente — 368
- Bartolomé de Torres Naharro — 371
- *Commedia dell'arte* in Spain — 372
- Juan de la Cueva — 377
- Miguel Cervantes — 379
- Lope Felix de Vega Carpio — 381
- Guillén de Castro y Bellvis — 393
- Augustín Moreto y Cavana — 398
- Tirso de Molina — 398
- Pedro Calderón de la Barca — 399

8 Spain: Music and Dance — 419
- The operas of Lope and Calderón — 419
- Regional, school, flamenco and court dances — 422

9 France: Its Renaissance — 429
- The slow approach to theatre — 429
- The Hôtel de Bourgogne and the Théâtre du Marais — 436
- Alexandre Hardy and contemporaries — 441
- Ballet — 446

10 To German Baroque: Holland — 451
- The Rederijkers — 451
- Stock drama in seventeenth-century Germany — 456
- Early German opera — 462

11 Leading to Shakespeare — 465
- Nicholas Udall — 466
- Norton and Sackville — 469
- George Gascoigne — 471
- Robert Greene — 472
- Thomas Lodge — 476
- George Peele — 478
- John Lyly — 480

Thomas Nash	485
London's theatres, stagecraft and actors	486

12 England: Kyd and Marlowe — 497

Thomas Kyd	497
The Spanish Tragedy	498
Christopher Marlowe	500
Tamburlaine	509
The Jew of Malta	527
Dr Faustus	533
Edward II	546

13 Shakespeare: The Enigma — 563

Family	563
Schooling	566
Marriage	567
The move to London	569
Narrative poems and sonnets	572
Last years	578
Images of Shakespeare	581

14 Bardolatry — 587

15 Staging Shakespeare's Plays — 599

16 The Bard: Early Plays — 603

Henry VI	603
Richard III	607
The Comedy of Errors	609
Titus Andronicus	612
The Taming of the Shrew	613
Two Gentlemen of Verona	615
Love's Labour's Lost	617
Romeo and Juliet	620
A Midsummer Night's Dream	628
Richard II	631
King John	637

The Merchant of Venice	639
Henry IV, Parts I and II	644
Much Ado About Nothing	646
Henry V	649
Julius Caesar	653
As You Like It	658

17 Shakespeare: Major Phase — 665

Hamlet	665
Twelfth Night	691
The Merry Wives of Windsor	695
Troilus and Cressida	697
All's Well That Ends Well	700
Measure for Measure	701
Othello	704
King Lear	712
Macbeth	726
Antony and Cleopatra	734

18 Shakespeare: The Last Phase — 743

Coriolanus	743
Timon of Athens	745
Pericles	749
Cymbeline	751
The Winter's Tale	760
The Tempest	764
Henry VIII	773
The Two Noble Kinsmen	777

19 Around Shakespeare — 779

Arden of Feversham	779
Thomas Heywood	780
Thomas Dekker	783
George Chapman	787
John Marston	793
Thomas Middleton	799
Philip Massinger	808

Beaumont and Fletcher	813
Cyril Tourneur	827
John Ford	836
James Shirley and Richard Brome	843

20 Jonson and Webster — **847**

Ben Jonson	847
Every Man in His Humour	848
Every Man Out of His Humour	850
Cynthia's Revels	851
The Poetaster	851
Sejanus	853
Eastward Ho!	854
Volpone	856
Epicene	861
The Alchemist	861
Bartholomew Fair	863
Catiline	864
The Devil Is an Ass	865
Collaboration with Inigo Jones	866
The Staple of News	868
The New Inn	869
The Magnetic Lady	869
A Tale of a Tub	870
John Webster	871
Collaboration with Dekker and others	872
The White Devil	873
The Duchess of Malfi	877
Lesser plays	889

Index	893
Index of Producing Companies/Venues	926
Index of Cited Authors	928

ILLUSTRATIONS

Between pages 320 and 321

Illustration to *Roman de Gauvel* by Gervais du Bus depicting a medieval carnival, 1310–14; Bibliothèque Nationale, Paris (© AKG)

Moses and the Golden Calf; Passion Play performed at Oberammagau, Germany (© Eugene G. Shultz)

English Mystery Plays, York (© Manders and Mitchenson Photo Library)

An old witch, Fasching Carnival Parade, Groetzingen near Karlsruhe, Germany, 2004 (© Digital Exposures Picture Library/Alamy)

A red devil, Fasching Carnival Parade, Groetzingen near Karlsruhe, Germany, 2004 (© Digital Exposures Picture Library/Alamy)

English Mystery Plays, York (© Manders and Mitchenson Photo Library)

Frontispiece to an early edition of the English Morality Play *Everyman*, late fifteenth century (© Manders and Mitchenson Photo Library)

Frontispiece to Hans Sachs's play *A Fight Between a Woman and Her Housemaid*; facsimile of a woodcut title page, c. 1556, reproduced in G. Könnecke, *Bilderatlas zur Geschichte der Nationalliteratur*, Marburg, 1895 (© AKG)

Hans Sachs, German playwright, 1494–1576; portrait, c. 1574, by Andreas Herneisen (1538–1610), Germanisches Nationalmuseum, Nuremburg (© AKG)

Passion Plays of Valenciennes depicting the healing miracles of Christ, c. 1500; book illumination from *Mystères de la Passion de Valenciennes*, Bibliothèque Nationale, Paris (© AKG)

Stages for the Valenciennes Passion Plays; coloured drawing, 1547 (© AKG)

Holy Week parade at Sentenil near Ronda in Andalucia, Spain (© Gavin Mather/Alamy)

Teatro Olimpico, designed by Andrea Palladio and completed by Vincenzo Scamozzi in 1580, Vincenza, Italy (© Manders and Mitchenson Photo Library)

Study for a tragic scene; engraving by Sebastiano Serlio (1475–1554); Bibliothèque de l'Arsenal, Paris (© Lauros/Giraudon/Bridgeman Art Library)

Stage machinery for *Germanico sul Reno* by Giovanni Legrenzi, performed in Ferrara in 1676; washed pen and ink drawing, Bibliothèque de l'Opéra, Paris (© AKG)

Stage machinery for the whale that swallows Jonas; engraving from Joseph Furttenbach, *Mannhaffter Kunst-Spiegel*, 1663 (© AKG)

Stage machinery for a sailing ship with rocking device; engraving from Joseph Furttenbach, *Mannhaffter Kunst-Spiegel*, 1663 (© AKG)

Torquato Tasso, Italian poet (1544–95); engraving by Raphael Morghen after a drawing by Ermini Raphael after Raphael (© AKG)

Pietro Aretino, Italian writer (1492–1556); painting by Titian, 1545; Palazzo Pitti, Florence (© AKG)

Niccolò Machiavelli, Italian writer (1469–1527); portrait by Santi Di Tito, 1510; Palazzo Vecchio, Florence (© AKG)

Self-portrait of Gian Lorenzo Bernini (1598–1680); coloured chalk on paper; Ashmolean Museum, University of Oxford (© Bridgeman Art Library)

Performance of a comedy and portrait of Latin writer Terence from a fifteenth-century French manuscript; Bibliothèque Nationale, Paris (© AKG)

Italian comic actors at the Hotel de Bourgogne, Paris; engraving by Pierre Landry from *Grand Almanach Historique*, 1689 (© AKG)

Commedia dell'arte performance in St Mark's Square, Venice, 1614; engraving by Giacomo Franco (© AKG)

Jester in Love by Giovanni Domenico Tiepolo, c. 1793; fresco transferred on to canvas; Ca'Rezzonico Collection, Venice (© AKG)

Nineteenth-century clown Joey Grimaldi (© Manders and Mitchenson Photo Library)

Twentieth-century mime artist Marcel Marceau (© Manders and Mitchenson Photo Library)

Corral de Comedias, Almagro, Spain (© Turespaña)

Fuente Ovejuna by Lope de Vega at the Royal National Theatre, London, 1992 (© Robert Workman)

Procession during the Pia Del Pilar Festival to celebrate the Virgin Mary's visit to St James the Apostle, Saragossa, Spain (© Jeremy Hunter/Impact Photos)

Lope de Vega, Spanish playwright (1562–1635), c. 1630, attributed to Eugenio Caxes (1577–1642); Museo Lázaro Galdiano, Madrid (© Bridgeman Art Library)

Calderón de la Barca, Spanish playwright (1600–81) (© Manders and Mitchenson Photo Library)

Miguel Cervantes, Spanish author and playwright (1547–1616); portrait by Juan de Jauregui y Aguilar, 1600; Royal Academy of Languages, Madrid (© AKG)

Between pages 640 and 641

The 1984 Glyndebourne Festival Opera production of *L'Incoronazione di Poppea* by Claudio Monteverdi (1567–1643), directed by Sir Peter Hall (© Guy Gravett)

The Return of Ulysses by Claudio Monteverdi at English National Opera in 1992, directed by David Freeman (© Richard Mildenhall)

The Banquet of Herod, depicting the beheading of John the Baptist, with Salome holding his head; engraving by Israel Meckenem, c. 1490 (© AKG)

Court Ball in the Louvre on the Occasion of the Wedding of the Duchess Anne de Joyeuse, 24 September 1581; Flemish School, Musée du Louvre, Paris (© AKG)

The Old Horse Market in Brussels, c. 1666, by Adam Frans van der Meulen; Museum of Fine Art, Vaduz, Liechtenstein (© AKG)

Costume design for a ballet performance at the court of Louis XIV, c. 1655; pen and ink drawing ascribed to Henry Gissey, Victoria and Albert Museum, London (© AKG)

The Swan Theatre, London, in 1596; drawing by Johannes de Wit (© AKG)

Engraving of the Globe Theatre, London, c. 1612 (© Manders and Mitchenson Photo Library)

Twentieth-century reconstruction of the interior of Shakespeare's Globe Theatre, London (© Eugene G. Shultz)

Francis Beaumont, English playwright (1548–1616) (© Manders and Mitchenson Photo Library)

Ben Jonson, English playwright (1572–1637); engraving by William Cameron Edwards, 1840, after a contemporary portrait, c. 1610 (© AKG)

Christopher Marlowe, English playwright (1654–93), engraving, c. 1810, after a contemporary portrait (© AKG)

Production of *Doctor Faustus* with Michael Goodliffe as Mephistopheles and Paul Daneman as Faustus, Edinburgh Festival, 1961 (© AKG)

Sir Ian McKellan as the King in Christopher Marlowe's *Edward II*, produced by Prospect Theatre Company, 1969 (© Manders and Mitchenson Photo Library)

David Garrick (1717–79) as King Lear (© Manders and Mitchenson Photo Library)

Laurence Olivier and Ralph Richardson in the film of *Richard III*, 1956 (© Manders and Mitchenson Photo Library)

The Royal Ballet's production of *Romeo and Juliet*, choreographed by Frederick Ashton, with Tamara Rojo and Inaki Urlezaga in the principal roles, London, 2000 (© Linda Rich, Dance Photo Library)

Director Max Rheinhardt surveying a set model during the filming of *A Midsummer Night's Dream* in 1935 (© AKG)

Portrait of Edmund Kean (1787–1833) as Hamlet (© Manders and Mitchenson Photo Library)

Sarah Siddons (1727–88) as Isabella in *Measure for Measure* (© Manders and Mitchenson Photo Library)

Ellen Terry as Lady Macbeth, c. 1888 (© Manders and Mitchenson Photo Library)

Adalbert Matkowsky as Macbeth, Berlin, 1905 (© AKG)

Henry Irving as Cardinal Wolsey in Shakespeare's *Henry VIII*, c. 1892 (© AKG)

Paul Robeson and Peggy Ashcroft in *Othello*, London, 1930 (© Manders and Mitchenson Photo Library)

Maurice Evans as Richard II, New York, 1940 (© Manders and Mitchenson Photo Library)
Bruno Ganz as Hamlet and Jutta Lampe as Ophelia, Berlin, 1982 (© AKG)
Jean-Louis Barrault as Hamlet, Paris, 1946 (© Manders and Mitchenson Photo Library)
John Barrymore as Hamlet, *c.* 1923 (© Manders and Mitchenson Photo Library)
Alexander Moissi as Hamlet, Vienna, 1922 (© Manders and Mitchenson Photo Library)
William Shakespeare (1564–1616); engraving by Martin Droeshout on the title page to the First Folio of the plays, 1610 (© AKG)
Shakespeare's *A Midsummer Night's Dream* performed by the Royal Shakespeare Company at Stratford-upon-Avon, with Charles Laughton as Bottom, Mary Ure as Titania and Vanessa Redgrave as Helena, 1959 (photograph by Brian Seed; © AKG)
Tamburlaine the Great, Royal National Theatre production with Albert Finney as Tamburlaine, directed by Sir Peter Hall, London, 1976 (© Nobby Clark)

PREFACE

Much of this book owes its existence to the inborn, compulsive curiosity of others and their incessant prying. Writers of my ilk, who attempt surveys of epochs or chronicles of a nation's literature, are largely dependent on zealous scholars, the true historians, who master archaic and exotic languages and pore over incunabula and palimpsests in ancient Rhineland and Tuscan monasteries; in hollow, echoing cathedral archives; neglected, dusty libraries in old English manor houses; faded birth, baptismal, marriage, death entries in parish churches and borough registries; crumbling notations of long-forgotten litigation. History is a plodding relay, with the baton passing from hand to hand at very short intervals, each participant relying on those just ahead in the race. It is mostly fact-finding, a small one here, a small one there, and then assembling them. The generalist seeks to arrive at a mosaic, a panorama depicting a period or even an epoch, the evolution of a particular style or an aesthetic movement, Realism, Romanticism, neo-classicism. That is my ambitious attempt here; it has entailed research among the many researchers.

Customarily one's factual sources are acknowledged in footnotes, appendices and lengthy bibliographies, but – as I have said elsewhere – I prefer not to clutter my pages and weigh down my over-long book with such addenda, of concern mostly to other scholars, few in number, not of interest to the general reader. Instead, I cite my debt parenthetically in the text where it is due. I have, naturally, been forced to borrow considerably from others; much of my narrative is simply paraphrased from earlier studies by others; that is unavoidable. In some instances, where an event or an accomplishment has been summarized brilliantly and the phrasing deserves to be shared and enjoyed by lay readers who do not have access to it, I quote directly. I would do more of that if the copyright laws allowed it. The purpose of literature is not the glorification of the author but the enrichment of all of us. There can be no permanent ownership of facts and ideas. Whoever wants to copy from my book is free to do so. (But please mention my name.)

Properly, I should single out major sources of great help to me throughout my task, publications not critiques or biographies of this or that playwright, actor or scene designer, but of broad surveys like my own. My important predecessors – whom I have frequently consulted – include Sheldon Cheney, Allardyce Nicholl, John Gassner, Edward Quinn, George Freedley, John A. Reeves, Kenneth McGowan, Vera Mowry Roberts, Oscar C. Brockett, Margot Berthold and others. I must say that some of them tell the story in elegant prose that I envy, and all possess truly awesome erudition. In addition, I have used several encyclopaedias of the drama: those offered by the Oxford University Press, along with its *Oxford Companion to English Literature*; the *Britannica*; *Collier's*; the *McGraw-Hill Encyclopedia of World Drama*; *The Penguin Dictionary of the Theatre*, edited by Bernard Sobel and John Russell Taylor; *The Dance Encyclopedia*, edited by Anatole Chujoy; the *Simon and Schuster Book of Opera*; and, on music, the *Groves* dictionaries and more.

For decades I have collected articles from academic journals devoted to the theatre and the notices by drama reviewers in newspapers, most often the *New York Times* – delivered each morning to my door – but also the *New York Post*, *New York Daily News*, *Wall Street Journal*, *New York Observer*, *Village Voice*, *The Times* of London, the *Observer*, the *Guardian*, the *Telegraph* and others – bins-full. From magazines: *Time*, *Newsweek*, the *New Yorker*, *America*, *New York* and so on. I have gone often to the splendid New York Public Library and have had access to the stacks in university libraries. By now I find on my desk and shelves a private library of books on Shakespeare that pour unstoppably from the printing presses: and here I am adding my own. I do believe that this volume contains much material not hitherto available in the general histories of the sixteen centuries that are covered.

I must especially acknowledge the contribution of my friend Robert Heiser. When my advance in years made it difficult for me to decipher my always illegible handwriting, he – without being asked – offered with remarkable generosity to copy my entire manuscript before it went to the publisher: to date that comprises this book and two others, 4,000 pages. A deed of friendship that has little equal.

P.F.

doned, and in desperation resorts to prostitution. To redeem her, Abraham takes on the disguise of a lustful customer. Kissing him, she suddenly recognizes him and is overcome by shame. Beseeching her, tenderly and poetically, he persuades her to quit her sinful life and return with him to their home, her future dedicated to prayer and good works. It is not known whether these short plays were meant to be acted or merely read. (The six scripts of Hrosvitha have been translated into English by Christopher St John.)

Services in the Church were conducted in Latin; they were scarcely intelligible to the laity. Gradually, to render portions of the mass meaningful, bits of pantomime were added to the solemn ceremony, and particularly to the wordless passages. This began as early as the ninth century in St Gall, in the German sector of what is now Switzerland. Legend attributes the addition of fresh Latin lines at these heretofore silent moments, among them the *Quem Quaeritis*, to Notker Balbulus, the "Stammerer" (840–912), a monk and musician eminent in his day, who as a choirboy had found it difficult to learn the wordless sections of the mass. He, in turn, generously acknowledged that the idea for his innovation came from a nameless monk of Jumièges, a refugee from the Normans. A further alteration of the liturgy was contributed by Tutilo (850–915), a fellow monk of Balbulus, who inserted prose exchanges, which came to be known as tropes, "turns of dialogue", chanted by celebrants at the altar.

The basic musical scheme was soon one note for each syllable, quite easy to sing. In time the development also became antiphonal; units of the choir sang some lines, and other units responded with subsequent lines, alternately.

A tenth-century text from Limoges in France (923) is a somewhat augmented version of this new treatment of the *Quem Quaeritis*. In this Scriptural incident, appropriately chosen for dramatization at Easter time, three Marys, bearing spices, set out to visit the tomb and anoint the crucified Jesus, and they find the enclosure empty. Alarmed by their discovery, they are confronted by an angel, seated on the rim of the void sarcophagus that holds only the heaped, discarded white grave cloth. Around him lie the sleeping guards. These words ensue:

> ANGEL: Whom seek you in the sepulchre, O followers of Christ?
> MARYS: Jesus of Nazareth, which was crucified, O celestial ones.
> ANGEL: He is not here; He is risen, just as He foretold. Go, announce that He is risen from the sepulchre.

After this the congregation raised an exultant *Te Deum Laudamus*. The episode is taken from St Mark 16:1, 8, and it was sung in Latin. Priests and monks represented the angels, while others – priests, perhaps with their heads cloaked, and gifted with falsetto voices – were the Marys. The action encompassed the raising of the cloth, symbolic of the abandoned muslin shroud left

urged that a "Christian" theatre be used to counteract the lascivious entertainments of the pagans (much as Japanese Buddhist priests introduced Shrine dances and playlets to combat the lewdness of the popular "monkey plays"). His suggestion, both shrewd and hopeful, was premature and discredited, along with his own person, when his arguments failed at the Council of Nicaea (325). Exiled to Gaul for two years, he feigned repentance and was received back and pardoned by Emperor Constantine, but his strange, sudden collapse and death while awaiting communion was looked upon by his contemporaries as retribution by God. Several centuries elapsed before his foresighted idea was realized, when the Church hierarchy exploited for its own ends people's undying fondness for drama and staged presentations.

In the interim, lesser kinds of "entertainment" persisted and even matured: minstrelsy, the telling of stories in song with stringed music as an accessory, was widely practised, especially by German scop and gleemen and French and Italian troubadours. Recounting heroic deeds, romantic attachments – the love of chivalric knights and their chaste, fair ladies – and sometimes adding amusing ditties for variety's sake, they assembled the material out of which "farces" were to reappear. Many of these medieval ballads and romances had an inherent dramatic construction and even passages of dialogue, and doubtless the singer enhanced the effectiveness of his recital with changes of facial expression and bits of pantomime to enforce the sung words, a further preparation for the revival of acting. A pair of wanderers, a singer and his lutanist, came to a castle, wherein the dwellers had many long, lonely and often tedious nights. The talented visitors earned a lodging, meals and perhaps a purse by offering two or three evenings of musical pleasure, part well rehearsed, part improvised.

Nor had the plays of Seneca, Plautus and Terence been forgotten, though known to only a few scholarly souls. In monasteries, the precious Latin manuscripts were carefully copied and recopied, shuddered at or chuckled over, covertly passed from one cloistered generation of monks to the next.

One such recipient was a learned Saxon nun, Hrosvitha (*c.* 932–1001), closeted in a Benedictine abbey at Gandersheim. Into her pious hands came the works of Terence. Admiring and disapproving of them, she set about converting them freely into moral allegories, six in a "wretched Latin" and given new titles: *Gallicanus, Dulcitius, Callimachus, Abraham, Paphnutius* and *Sapientia*. No longer humorous, they are still called "comedies" because they have "happy" endings, their saintly heroines attaining salvation through harsh ordeals and even a blest martyrdom. Her avowed purpose was "to make the small talent vouchsafed to me by Heaven give forth, under the hammer of devotion, a faint sound to the praise of God". Actually she was fluent in Latin; her texts betray a familiarity not only with Terence but Plautus, Horace, Virgil and Ovid. To sample *Abraham*: A virtuous recluse, this elderly man leaves his hermitage to care for an orphaned niece. His concern and efforts are in vain: the girl is seduced, elopes, and is soon aban-

so that they became almost independent works. Even the *Quem Quaeritis* might be given some limited, more realistic "effects", a more solid and ornate sepulchre with a heavy rock beside it, and the priests and boys attired in Biblical dress for authenticity. Among the additions, in later versions, was a visit of shepherds to the Manger. The Christmas plays were similarly enlarged to include the Annunciation, the birth of John the Baptist and – a highly popular subject – Joseph's troubled questioning of his pregnant wife concerning her fidelity.

Still more themes were introduced: Herod, raging, appeared in several little dramas, the archetypal villain, evoking an emphatic response from the naïve, devout spectators; and on 28 December his Slaughter of the Innocents was presented with the choirboys serving as the victims. At the service on Monday after Easter the Disciples' encounter with the risen but as yet unrecognized Christ on the road to Emmaus was portrayed. Eventually all the events in the Nazarene's career were presented chronologically.

Along with these, a variety of other stories from the Bible and the Apocrypha were added, written in Latin by talented priests, the most scholarly and articulate element of medieval society, who had little other outlet for their literary impulses. New music was composed for them, besides that already part of the mass and other liturgical services.

For instance, around 1100 a *Mercator* scene was inserted before the *Quem Quaeritis* trope. This showed the Three Marys buying sweet spices as they approached the tomb to lave the body of Jesus. They are accosted by a gesticulating apothecary eager to sell them his wares. Something of a quacksalver, he is both a familiar character drawn from life and the prototype of a stock figure, the glib vendor of panaceas. His over-zealous merchandising, his display of table, scales, boxes and pots of ointments and powders, contributed a worldly touch and could be amusing, a contrast to the solemn episode that came next. Later he is given a wife. Details of this scene are preserved on a medallion by the Master of the Uta Gospel in Regensburg, Germany, and on bas-reliefs in the French cathedrals of Beaucaire and St Giles. In Constance the apothecary is shown as more dignified, wearing a scholar's cap and grasping a magnifying glass as he peers at the ingredients in his mortar and carefully pounds and grinds them. A thirteenth-century text in Prague has him say: "The best unguents I shall give you, to anoint the Redeemer's wounds, in memory of His burial and to the glory of His name." This is spoken in Latin.

In 1887 Carl Lange collected and published 224 Latin dramatizations of the Three Marys' visit to the sepulchre that were staged during the Easter service. Their brief, poignant verbal exchange with the Angels was enacted in St Gall, Vienna, Strasbourg, Prague, Cracow and south in Italy and Spain, in Padua and the monasteries of Sutri and Silos, and also north in Linköping, Sweden, as well as in Litchfield Cathedral in England.

Another sketch inserted before the *Quem Quaeritis* was the "race to the sepulchre" which had the Disciples Peter and John learning of the Three Marys' discovery of the empty tomb and rushing to behold it for themselves. Following John, who is younger, Peter limps, huffs and puffs, arrives out of breath, but the respectful John steps back, allowing his companion the first chance

behind in the tomb. This trope, the *Quem Quaeritis*, became highly popular and was similarly enacted in other churches throughout Western and Eastern Europe.

Other parts of the mass lent themselves quite as effectively to this dramatic heightening: the *Depositio Crucis*, projected by lowering the veiled Cross on Good Friday and laying it below the altar, and then the *Elevatio Crucis*, uplifting and unwrapping it triumphantly on Easter Sunday; and Christmas-time depictions of the Three Kings, the turbaned Magi bringing gifts to the crèche in the Manger, with ever-growing and more elaborate stage directions; and additional characters, among them the robed Disciples.

Books were exceedingly rare; there were only a few, precious illuminated manuscripts, so the illiterate had to learn elsewhere. Pictorial representations of Scriptural incidents had long been used to instruct the otherwise untaught common laity: bas-reliefs and other carvings on the outer doors and throughout the richly adorned interior of the church, along with images in stained-glass windows, paintings, wood sculptures and mosaics in chapels. So now a scenic element, too, was gradually added to the explanatory pantomime at the altar, and it slowly became more fulsomely theatrical. The performances were surrounded and sustained by anthems, the chanting choir, the music of great organs and pipes and tabors and other medieval instruments, all this lit by torches and candles and framed by rich canopies.

In England, Ethelwold, Bishop of Winchester, issued exact guidance on how the *Visitatio Sepulchri*, the Easter trope, was to be performed (c. 970). The rules were included in the *Regularis Concordia* of Winchester, his diocese, a compilation under way since the seventh century. Even the proper costuming is detailed, together with the equivalent of stage instructions: where to stand, how to gesture, when the chanter's voice should be modulated. At a certain moment the censers are to be taken up, then put aside as the grave cloth is displayed to the choir to indicate that the Lord is risen; at the exultant climax the bells will start ringing, a joyful clamour. The "script" served as a basic pattern that was widely accepted.

One is struck by the resemblance all this bears to the origins of Greek and Oriental drama; in religious ceremony and ritual, and especially its principal occurrence at Easter, springtime, the season of rebirth, of renewal, the sun's reinvigorated warmth, the fresh rains bringing forth a green earth, upthrusting narcissi and buds on branches of yew trees, the vernal change was transposed and symbolized by the epic story of a miraculously resurrected Saviour and His pledge of Eternal Life, or spiritual and physical immortality; this to a frightened humanity that in the Middle Ages was surrounded by decimating plagues and demons, threatening devils and Hell-fire.

(The Christian holiday has many pagan associations. According to the Venerable Bede, the word "Easter" derives from "Eostre", the name of a northern Goddess of Spring and the Dawn. To the Anglo-Saxons, April was a time of her sway, the "Eastur-monath".)

Step by step, the embellished tropes and playlets with music were moved from the choir stalls to be sung only before the altar. The action was sustained to a length of twenty or thirty minutes,

— 1 —
THE CHRISTIAN DRAMA

An esteemed historian, Preserved Smith, has said of Europe's so-called "Dark Ages" – the epoch from approximately the fifth to the tenth centuries – that to us they only seem "dark" because we project our ignorance of them on to the screen of the past. After the barbaric onslaught from the North was over, men began to build new structures, physical and cultural. In truth, much was brilliant and aesthetically exciting during that lengthy span of time. The fancy of man was not asleep. His imagination had not atrophied. His creative impulse was still at work. He lit his squat Romanesque fortress-like churches with red and purple light from stained-glass windows, domed them with luminous golden mosaics in Near Eastern patterns, a Byzantine splendour. He adorned his knight's armour with gaudy embossing, and costumed himself and his lady in attire unprecedentedly witty and unique in design. His cold castle halls and winding stone stairways were warmed with richly coloured banners and tapestries. Sacred books, written on vellum, were delicately inscribed and illuminated, bound in carved ivory and bejewelled panels.

But these were also times of anarchy and the risk of dangerous assault, and most aspects of life were dominated by an ascetic Church. Throughout this period public theatre was almost non-existent. The Christian establishment was its implacable antagonist, and for good reasons. The late Roman stage was depraved, its farces licentious, its melodramas hideously gory, its actors held in ill repute.

Such entertainments did not cease abruptly. In one form or another they lingered on, at first proscribed on Sundays and special holy days. In the sixth century Emperor Justinian married a dancer, an indication that theatrical activity continued there, though the emperor was officially hostile to it. As late as AD 647 performances were given at Rome, cited by Sidonius. In every century thereafter, while the grandiose theatres fell into ruins, there are scattered references to "players" of some sort and their evil impact.

In Rome all mention of theatre vanishes after the close of the seventh century. But there are hints of it at subsequent dates in the Eastern Empire. Presumably what survived, and was deplored and condemned by ecclesiastical authorities, were marketplace offerings by vagabond jugglers, laughing tumblers, dancers, fearless animal trainers and bear-baiters, acrobats, tightrope walkers, amazing knife-throwers, astonishing fire-eaters and puppet-masters. Such performances lent noise and joy to a marriage party hosted by rich and noble families, or would be invited to enliven a holiday festivity. The performers, however, were relegated to a very low caste: their activities were illegal in many towns, they might be arrested. The clergy railed against them; Church councils outlawed them. But, like troupers in every age, they were never permanently discouraged; humanity's universal histrionic tendencies could not be wholly repressed.

The average person's love for spectacle and diversion found expression in other ways. Christianity, spreading over Europe, took over and adapted many pagan festivals that now featured May dances, sword dances, mummers' plays, some of which might include little dramas, such as those in England depicting the deeds of Robin Hood and his band, or the heroism of St George the blessed dragon-slayer. A good majority of these ceremonies and playlets descended from earlier vegetation, fertility and harvest rites, especially those once observed by northern Teutonic tribes. This is recalled in the symbolic display of pine cones and holly wreaths at Christmas, and white lilies at Easter, the lingering tokens of heathen beliefs and seasonal invocatory practices. The pageantry that characterized many special occasions, an assembly of guilds, a gathering to celebrate new wine, a procession to honour a patron saint, revealed the dramatic instinct inherent not only in for-hire performers but also among the common folk. Such events were welcome pretexts for putting on fancy costumes and to indulge in a spate of innocent or mischievous miming. Life, in many ways, was publicly theatrical, even if no formal stages existed.

A paradox: it was in the Church itself that a strange, impressive genre of serious drama eventually developed. In a scathing attack on the Roman theatre, after his conversion to Christianity in *c.* 198, Tertullian had not only decried its flagrant immorality but had also contrasted its feeble, degenerate mimicry with the great, exalting and deeply moving story of the Messiah and of the sufferings of His faithful, martyred followers. Another subject for a dramatist's imagination might be the Last Judgement as set forth with theatrical vividness in the Scriptures, a dread warning of how the earth would gape and the heavens resound, a call for men to return to virtue, urgently to attain redemption.

> Dost thou breathe me a sigh for gaols and theatres? . . . Behold uncleanness thrown down by chastity, perfidiousness slain by faithfulness, cruelty beaten by mercy, wantonness overlaid by modesty. . . . But what sort of show is that near at hand? The coming of the Lord, now confessed, now glorious, now triumphant. What is that joy of the Angels? what the glory of the rising saints? . . . There remain other shows: that last and eternal day of judgment . . . the persecutors of the Name of the Lord, melting amid insulting fires more raging than those wherewith themselves raged against the Christians; those wise philosophers, moreover, reddening before their own disciples, now burning together with them.
> [Translation: C. Dodgson]

As Tertullian evidences early in the Christian Era, a fiery vision obsessed the minds of the medieval faithful: the drama of damnation and salvation; a histrionic vision.

In 380 Gregory Nazianzen, Patriarch of Constantinople, is said to have written a Euripidean tragedy about the Crucifixion of the Redeemer. Also, in the late fourth century, the heretic Arius

to gaze down upon the bare enclosure. This episode, the impetuous "race", is known to have been used in Zurich and St Gall and in the monasteries of St Martial in Limoges and St Florian in Austria, as well as in Helmstedt, Germany. In Dublin there is a revealing note about how the two Apostles are to be dressed: John in a white tunic, Peter in a red garment, and both actors are to be barefoot, with John bearing a palm and Peter holding the keys to Heaven.

When the Crusaders reached Jerusalem, in their quest to regain the Holy Land from the Saracen Infidels, they brought back drawings and precise measurements of the supposed actual tomb so that its representation in the *Quem Quaeritis* and "race" would be true to size, a faithful copy. Emulating them, Conrad, the Bishop of Constance, made three journeys to Palestine to study the Jerusalem model (1280) and had one of similar design and dimensions placed in the Chapel of St Maurice. Earlier, in 1147, Walbrun, the provost of Eichstätt Cathedral, had fulfilled a like mission, returning with a splinter of the Cross and having an exact replica of the "original" crypt installed in a monastery and church, Romanesque in style, dedicated to "the Holy Cross and the Holy Sepulchre", today presided over by Capuchins. There are similar "correct" tombs in San Sepolcro at Bologna, St Michael at Fulda, St Benigne at Dijon and the Holy Sepulchre at Germode in the Hartz Mountains.

Typical of later sketches in the expanding Church dramas is *The Play of Daniel* (1150). It is attributed to Hilarius, possibly English, though it is written in Latin and French. Fortunately, the score and dialogue have been preserved. In 1958 the playlet was revived by Noah Greenberg's New York Pro Musica at the Cloisters, with a collection of ancient musical instruments borrowed from several museums or reproduced from the originals. Spectators were astonished by its visual richness, its gorgeous costumes, the wit and humour of its animal imitators, the delicate lyrical accompaniment of hand bells, triangles, tambourines, trumpets, harps, small and finger cymbals, large and small drums. The Cloisters, a branch of the Metropolitan Museum, holds a rare collection of Romanesque and Gothic art, and New York's Riverside Church, where the presentation was shortly moved to draw larger audiences, is also Gothic in design, so the settings were quite appropriate. In later years the work was repeated in the vast Cathedral of St John the Divine (1977 *et seq.*), becoming a feature of New York's holiday season, besides being taken on tour about the United States and Europe.

In the *Daniel*, Belshazzar is portrayed as he beholds the ominous handwriting on a wall of his palace. He summons the wise, captive Jewish prophet Daniel, who translates the mysterious words, foretelling the king's imminent doom. For this Belshazzar's successor has Daniel cast into the lions' den. An Angel intervenes and with flaming sword protects him, and the lions prove to be not ferocious but friendly. The playlet ends with a *Conductus*, a procession of all the participants, including the amiable beasts. The crowning musical moment is a chanting of the Gregorian *Te Deum*. The marchers carry banners, sacred vessels, swinging fragrant incense-burners. This small play, serious yet touched by whimsy, is essentially a mixture of pagan and Christian elements, providing an occasion for splendid pageantry, its spirit almost one of *faux-naïf*, for there are also hints of deft artistic control and sophistication.

For Noah Greenberg's lavishly furbished revival of the *Daniel* much of the necessary research had been done by a scholarly Benedictine monk of highly diverse talents, Rembert G. Weakland. Musically gifted, he had first considered a career as a concert pianist but instead decided to enter the monastic order. In Italy he studied theology but then chose to pursue his previous inherent interest and sought a degree from New York's Juilliard School of Music. Discovering the thirteenth-century manuscript of *Daniel* in the British Museum, he translated its score into modern notation. He identified it as the devout work of students in the Cathedral of Beauvais, where it had been performed annually at the New Year from 1150 to 1250. Subsequently, when he reached forty, he was elected Chief Abbot of all Benedictine monasteries throughout the world. Recognizing his special abilities as an administrator, Pope Paul VI named him the Archbishop of Milwaukee. His tenure of that post was thereafter characterized by the outspoken liberalism of his doctrinal and political views, which, too, often put him at odds with most others in the Church hierarchy. While carrying out his diocesan duties, he also lectured at times for the British Broadcasting Corporation (BBC) on "The History of Ambrosian Chants", and before the Modern Language Association on "The *Planctus* in Medieval Drama", and to a very different audience at the University of Wisconsin on "Marx and the Bible in Latin America".

In a brief preface to the work, Weakland remarks:

Plays on Old Testament subjects were rare in the Middle Ages, since the drama grew out of liturgical texts which were based on the major events in the life of Christ. The *Play of Daniel* itself, concerned with one of the great prophets of the coming of the Messiah, probably developed from a particular reading in the night office of Christmas which in turn had derived from a sixth-century sermon. The play was composed at a time when the subtleties of music, text and symbol had reached their peak in liturgical drama, and the dramatic aspect had become independent of the liturgical content, asserting itself as a unity of its own. This play marks a turning point, and shows great dramatic advance in its delineation of character and its expressivity. Only the chanting of the Gregorian *Te Deum* at the end betrays it as part of a devotional *Daniel* service.

The *Play of Daniel* was a great favorite, with a highly successful mixture of pagan and religious elements. It provided many occasions for the splendor and display of one of the popular new devices of the time, the *Conductus* or courtly procession. The figure of the queen, of not much importance in the Biblical narrative, was easily enlarged, and the text clearly indicates a great array of costumes, banners, sacred vessels and musical instruments – all the effects of pageantry. But it is the music itself that ensured the play's popularity, and in the tunefulness of its melodies and piquancy of its rhythms we come perhaps as close as we ever shall to Medieval "folk song."

After a lapse of years a different version of this cherished *Daniel* was brought to New York by the Boston Camerata for staging in the Metropolitan Museum's large Medieval Court in 1982.

This production had been playing in Boston's Jordan Hall. Founded in that city in 1954, the well-established Camerata was one of America's thriving early-music ensembles. For the *Daniel* the troupe enlisted Andrea von Ramm, a German medieval-music specialist, to direct and take the title role. Ms von Ramm and the company's leader, Joel Cohen, made no attempt to emulate Pro Musica's opulent offering. As Janet Tassel described it in the *New York Times*:

> The Noah Greenberg production was predicated on its being a period piece, the recreation of a Medieval spectacle. It was a resplendent mounting supervised by Lincoln Kirstein, a sumptuous "thirteenth-century opera," as Nikos Psacharopoulios the stage director called it.
>
> Camerata's, on the other hand, is a scaled-down version, with but five on stage instrumentalists, who slip in and out of their personae, becoming actors and singers as readily as they exchange instruments; in addition, most of the twenty-five member cast are called to percussion duty at one time or another. Mr Cohen, using his own facsimile of the manuscript, developed a different instrumentation from Pro Musica's (the scoring can only be inferred from the manuscript). Though Mr Cohen, like Mr Greenberg, has included drones and doubling, he has made free use, as Mr Greenberg did not, of his own counter-melodies.
>
> The Camerata's vocal sound is not operatic, but rather it is "early;" in the last twenty-four years, says Mr Cohen, scholarship has taught us much about the "white" sound suitable to Medieval music, the voice matching the light, straight tones of the period instruments. Miss von Ramm will sing the young Daniel in the countertenor range, shifting to tenor for the aged prophet. (Pro Musica's Daniel was a tenor, Charles Bressler.)
>
> The setting for Camerata's *Daniel* is Babylonian or "Biblical" rather than Medieval. Camerata's stage designer, Donald Beaman, and costumer, Ronald Guidry, have turned to Persian and Byzantine sources: against a glazed sand-colored background representing the tiled Babylonian palace walls, the figures, draped in brilliantly hued, hand-painted silk, stand forth like *bas-reliefs*.
>
> In the scene during which Belshazzar's people profane the Jewish sacred vessels, Miss von Ramm has her three dancers perform what can only be described as a Middle Eastern belly dance. There is other comic relief in *Daniel*, but this scene, according to Miss von Ramm, should be "the hammiest," an Eastern bacchanal, "with all stops out."

To a question whether such touches were authentic, Ms von Ramm replied: "Oh, how I hate that word these days. . . . If you insist on authenticity, whatever that is, let me remind you that in Medieval days the line between the sublime and the rowdy was very slim – even in a church setting."

Other new motifs were inserted:

> Against the setting of Ancient Babylonia, Miss von Ramm and Mr Cohen have counterpointed a series of specifically Jewish symbols. The writing on the wall, for instance, is in Hebrew, its

appearance accompanied by ominous blasts on the shofar, the ritual ram's horn. When Daniel is found praying by the scheming counselors, he is singing a Sephardic chant from "The Lamentations of Jeremiah".

To signify the universality of the *Daniel* motifs (and to fill out a concert, inasmuch as the play itself takes only about forty-five minutes), Mr Cohen has interpolated "tropes" (elaborations or additions) consisting of varied texts and nineteenth-century American hymns. The interpolations relate specifically to the Biblical Daniel or to the larger themes of the play, and as such explain, interpret, and connect for a modern audience.

Pro Musica, too, used a unifying and explanatory device. Called *Daniel, a Sermon*, it was a narrative composed in the Medieval style of alliteration and stressed verse by W.H. Auden and spoken by an actor dressed as a monk. The spoken texts in Camerata's *Daniel* are also delivered much like a sermon, but in this case the church figure is an early American preacher. To close the program, the Camerata performers sing the American hymn *Wondrous Love* as a recessional.

"We don't know what the medieval productions achieved," says Miss von Ramm, "but we know that they aimed at moving their audiences. Look at the specific rubrics in the *Daniel* manuscript; where Darius comes to see Daniel in the lions' den, for example, it says he is 'in tears.' This is theatre; we must move our audience."

Ms von Ramm had already sung *Daniel* in Europe. She had very fixed ideas about how it should be presented. For instance, she disagreed with how Robert Fletcher, Pro Musica's designer, had conceived the lions that awaited Daniel. Fletcher imagined them as "something between a St Bernard and an enlarged kitten". This hardly consorted with Ms von Ramm's concept. "No! No pussycats! . . . Daniel's aria is the most beautiful thing in the play. We want to be deeply touched, not giggling. We want the *essence* of the lion, the menace and danger, no kitty cats." Accordingly she, Mr Guidry and the choreographer, Carol Pharo, had their three dancers become six lions by borrowing from Oriental tradition and employing "masks, silhouettes, and movement".

Joel Cohen explained: "This production, I hope, is a testimonial to the transferability of art. Whatever century or area it comes from, if it's good art, it breathes and lives in every change of setting and for every new generation. Noah Greenberg knew that. Like Daniel, he was a prophet, too. To my way of thinking, Camerata's production, and every one that comes after, should be considered a memorial to that great man."

Encouraged by the success of the *Daniel*, Pro Musica undertook the mounting of another liturgical drama, and again chose the Fuentiduena Chapel of the Cloisters in New York as the best setting for introducing it. This was *The Play of Herod* (1963), a single piece combining two shorter twelfth-century works, the *Representations of Herod* and the *Slaughter of the Innocents*, originally performed near Orleans in the church belonging to the Benedictine monastery at St-

Benoit-sur-Loire. Most likely the participants were the men and boys of the choir school attached to the monastic edifice. The scripts of both episodes, known as the "Fleury Playbook", now repose in the Municipal Library of Orleans. At the Cloisters this newly created piece, to some degree a pastiche, was staged with the same gilding and glinting regal opulence as was the *Daniel*, and again as a drama meant to have the same popular appeal that it had exerted centuries earlier.

The despotic Herod is the dominant figure in both scripts. Here is an anonymous translation of the instructions given in the text: "When Herod and the other persons are ready, an Angel shall appear with a multitude of the Heavenly host. Seeing this, the shepherds are sore afraid. He shall announce salvation unto them, while the rest keep silent." The shepherds set out for Bethlehem, guided by a mesmerically bright star overhead. Arriving at the manger, they are confronted by a pair of cautious midwives. Paying homage to the Child, they invite the spectators to join in adoration of Him.

The Magi, too, are journeying to Bethlehem, also sighting and being oriented by the gleaming star. Herod, learning of their approach, sends men of his entourage to discover who are the strangers and what is their purpose. The Three Kings proclaim that they have travelled from the East to find and hail a newborn king. Disturbed by word of this, Herod orders his scribes to scan their books to determine if the coming of such a king has been prophesied. A book containing just such a prediction is brought to him. Alarmed and furious, he casts away the book, but his son and heir, Archelaus, tries to reassure him. Herod sends a message to the Magi: they shall discover the Child and bring him to the court.

Having offered their rich gifts to the Child, the Magi rest. Asleep, in a fateful dream they behold the Angel who warns them to choose another route for their homeward journey. Heedful and doing this, they avoid being taken to Herod's palace.

The second part of the play begins with a procession of the Innocents led by a symbolic Lamb of God. Meanwhile, in a grandiose ceremony, Armiger presents Herod with his sceptre. Alerted by the Angel, the Holy Family hastily flees into Egypt to protect the precious Child. Told that the Magi have escaped his control, Herod is seized by panic and rage. He tries to kill himself with his sword but is restrained by Armiger, who suggests that a better move would be to command the slaying of all young boys in the kingdom. Probably the just born king-to-be would be among them. Though their mothers try to save them, the hapless children are massacred. After this horrible deed Rachel is heard lamenting her losses. Her friends and neighbours seek to console her. The Angel appears and sings "Suffer the little children to come unto me", and miraculously the slain little ones "rise up and enter the choir singing".

This leads to Herod's demise; Archelaus replaces him on the throne. The Angel tells Joseph that he can safely return from Egypt. He brings his family back to Galilee. The play concludes with a *Te Deum*.

To round out the score found in the Fleury manuscript, the adapters, Noah Greenberg and

William L. Smoldon, included *Orientis Partibus*, an early processional song, and several thirteenth-century *estamples* and a three-part motet, *Allelya Psallite*. The narration was by Archibald MacLeish, the vestments, from designs by Rouben Ter-Arutunian, were executed by Karinska, the trio a formidable array of artists.

When *The Play of Daniel* was revived at New York's St John the Divine in 1977 the work's brevity presented a problem. It was coupled with *The Play of Herod* to lengthen the offering. The director was Nikos Psacharapoulos. The Boys' Choir was from the Church of the Transfiguration, let by Stuart Gardner; the producers were Toni Greenberg and LaNoue Davenport (the untimely death of Noah Greenberg was a major loss to America's musical life). The sponsors of the event were the Friends of Music in the Cathedral and Stewart R. Mott.

Further re-creations of these works ensued: one was a presentation of *The Play of Daniel* by the Ensemble for Early Music, led by Frederick Renz, given for five performances at the Cathedral of St John the Divine in 1983. According to Bernard Holland in the *New York Times*: "It is a smoothly attractive blending of research and guesswork, mixing ritualistic gesture with more naturalistic acting and featuring singing styles which, in much the same way, lay somewhere between modern vocal technique and the flat primary colors we associate with Medieval music. Faulting the Ensemble for Early Music over matters of authenticity is really beside the point since none of us really knows how these dramas were played and sung. We should be happy for *Daniel*'s bright yet tasteful staging, its stately and ceremonial sense of pace, and for the high quality of its singers. Mark Bleeke was Daniel. . . . Both the instrumentalists and the vocalists were enriched by the Cathedral's gently reinforcing echo. Ralph Lee's lion was an attractive giant-sized puppet with separate players for each paw, and Karen Matthews's costumes created a nice contrast between sobriety and luxury."

The Mannes Camerata, associated with New York's Mannes College of Music, entered the lists with a somewhat changed interpretation of the *Daniel* and *Herod* at the city's Christ Church United Methodist in 1985. This was directed by Paul C. Echols, the group's founder, who was also responsible for the re-edited and restaged version of the playlets. He restored their original titles. John Rockwell, also for the *New York Times*, commented:

> This was, among other things, a performance with a gimmick, but a gimmick that one would like to see become the norm for the presentation of such works. It has long been a puzzlement that otherwise conscientiously authentic or at least plausibly Medieval performances of these dramas have been marred by bright, anachronistic electric lighting and follow spots.
>
> The Mannes Camerata solved that problem by illuminating nearly everything with candles – lots of them; the main exception was for the musicians' scores, unobtrusively lighted electrically up in the choir loft in the rear. The result was not just aptly mysterious and flickering; the lighting lent everything a warm, mystical glow.
>
> In other ways, as well, this was a most pleasing evocation of these works sung in French-

accented Latin and neatly treading the line between scholarly correctness (in so far as we can know what is correct in the first place) and judicious updating for the modern audience.

The updatings came mainly in the form of richer harmonies and more realistic acting than one might have encountered some eight hundred years ago. The score is notated in a monophonic line, but we can assume some octave doublings, descants and drones. Mr Echols has added considerably spicier harmonies here and there than that, yet managed to retain an overall aura of monkish austerity.

Similarly, the staging sometimes approached the lurid, especially in the slaughter of the innocents, complete with smoke effects and full-throated screams. Yet most of these choices made sense, and certainly the resultant musical and dramatic impact was very telling.

At St John the Divine, again, *Herod* was enacted in 1989, this time by the Cathedral's own resident Ensemble for Early Music, headed by Frederick Renz, who prepared the somewhat altered text. Bernard Holland, senior music critic of the *New York Times*, gave this report:

So ancient are the origins of these two *Herods* that its players have not so much re-created an art form as reinvented one. . . . Texts and music survive, and even a few basic stage directions; but when so many centuries have passed, tradition can no longer accumulate and solidify, only crumble. Vagueness like this gives the twentieth-century performer unprecedented freedom – and the accompanying opportunities either to do good or commit abuse. . . . In creating the *Herods*, Gregorian melody was often borrowed or else copied, we are told. Christmas music familiar to the time appears intermittently, presumably to give the thirteenth-century listener a better sense of recognition.

This performance showed the mobility of Mr Renz's imagination. The tone of his production – its sense of time and history – is anchored by the gloomy vastness of this church. A handful of musicians strengthen the illusion of a past revisited by playing copies of bells, winds, brass, the hurdy-gurdy and the like. The ensemble singing is chaste and relatively untainted by current operatic technique.

Under the direction of Philip Burton, players on the raised platform before the altar strike poses in tableaux-like configurations, aping the composition of contemporary painting. Mary and Jesus are represented by gold masks, as if they were Medieval church statuary come to life.

The rest of this production, however, has a way of veering toward up-to-date melodrama. Broad gesticulation, pantomime and ensemble choreography are the order of the day. Herod's rolling of the eyes is reminiscent of a silent screen villain. . . . The slaughter itself is acted out to the furious beating of drums. It is a scene not without effect but one curiously at odds with the restrained and dignified movement practised elsewhere in this production.

Anti-history? Well, maybe history is, after all, what we at this moment choose to make it. With so few facts, with such dim ideas about style, with so much forgotten, the Ensemble for

Early Music can really go pretty much its own modern way. Some of the contrasts jar, not only in the staging but in the occasional juxtaposition of "old" and current singing techniques. Yet the music is carefully prepared and the production professional, and much is done with taste.

Later in the same season (1989/90) the Ensemble for Early Music performed its version of *Daniel and the Lions* at the Cloisters, as well as its interpretation of a secular composition, *Le Roman de Fauvel*, at Florence Gould Hall, the works comprising what the group called the Premier Festival de Théâtre Médiéval.

In various interviews with writers for the *New York Times*, while discussing these productions and other forays into reconstituting very early liturgical drama, Frederick Renz shared his thinking about just how it should be done. He insisted that scholars knew a good deal more about the medieval church plays than when Noah Greenberg had resurrected the *Daniel* two decades before.

Pro Musica was an important factor in congealing the whole movement toward the authentic performance of early music and making it attractive to a popular audience. I find it astounding how much the quality of early-music performance has improved. For instance, the whole style of singing is different from what it was fifteen years ago. A quasi-operatic approach, singing long lines and slurring to the high point, is not necessarily suited to Medieval music, especially in a cathedral, where there's so much echo. So we're working for less vibrato, for a more "*secco*" sound – something that approaches a laser beam as opposed to a floodlight.

He alluded to Greenberg's casting a soprano as Belshazzar's queen, whereas he himself had used only male singers for the adult roles – tenors, baritones and, for all the female characters, counter-tenors.

I feel that it is important that these plays were performed by monks. One doesn't have the whole spectrum from basso profundo to lyric soprano. If you look at the original manuscript, you'll find that most of the melodies fall within the range of an octave, and that it's the same octave. And I should add that our counter-tenors produce a more masculine sound than in the old Pro Musica days. The pioneering counter-tenors, lacking obvious models, emulated the female sound, voluptuous, sensuous, covered. We explore a sound that's brighter, more invasive.

The *Daniel* costumes were very opulent – they were supposed to be imitations of courtly costumes. But the Benedictine monks in the Beauvais Cathedral would never have had that. They wore the basic black robes, plus various vestments to conjure up different characters. And that's what we're going to do.

Generally, we're planning to do the plays in as concentrated a way as possible. *Daniel* was given a lot of stuffing – you see, one can spin the melodies out, interpolate repetitions and make the whole thing really austere and spacey. We're planning to be more direct. We're going where the action wants to take us instead of making it a sort of oratorio event, as *Daniel* was.

Also, we'll be doing more mime, which, combined with a synopsis beforehand, will show people what's being sung. Back in the twelfth century, you know, there was evidently a whole vocabulary of hand and finger symbolisms, especially in various monastic orders where the monks weren't allowed to speak.

He returned to the question of costuming. Pro Musica had based the singers' attire on period models. "The soldiers looked like Medieval soldiers, and not like Benedictine monks playing the roles of soldiers, which made it all look much more secular than I think it was."

About how the music should be performed, Renz observed: "There is really not enough evidence to support any of the proposed theories. I find that in each of the Medieval plays I look at, a different set of rules apply. *Herod* uses a lot of liturgical melodies, so we have a good idea of how those should be done. And in the sections that are set to poetry, there's no problem: you get the rhythm from the emphasis of the text."

In other interviews published in the *New York Times*, Paul C. Echols, of the Mannes Camerata, was asked about his ideas on this subject. For him it had not been a simple task of studying the "Fleury Playbook" and learning the parts. "The scribe wrote on a four-line musical staff with C at the top. Now by C today we mean a pitch location, but it is clear that for them it was a convenience to keep the music contained on the staff so they didn't have to write ledger lines. We infer that the higher voices pitched their melodies up and the lower ones down from this central, baritonish register where everything is written." On this assumption, Echols had carefully worked out transpositions to place the chant-like lines for the lead singers, Herod and the Angel Messenger, as well as the lamenting Rachel.

"We have hewed to authenticity in some areas. For instance, I see no point in singing in modern church Latin, which is basically Latin with an Italian accent. So I had John Collis look into the research that linguists have been doing on Medieval French pronunciation of Latin, and he came back with a set of rules for us." Elsewhere, though, Echols had introduced changes to enhance the solo chanting for modern audiences. "I have choral drones for some sections, and for Herod's call to arms against the newborn King of the Jews, an instrumental drone." Besides this, in the sung and acted passages, he interspersed "'instrumental improvisations and percussion ostinatos based on medieval rhythmic licks' . . . played by an ensemble of vieilles, Medieval harp, Medieval lute, hammered dulcimers, psaltery, bells, small organ and 'lots of primitive percussion.' Now what do you do about costumes? The option is to recreate the twelfth-century context. This is what Noah Greenberg did with the Pro Musica Antiqua twenty years ago when he gave the first modern performances of these dramas. Noah went to the art historians and tried to recreate from the iconographical evidence. He did what I'd call a 'Monks-and-Nuns Show' – the poses and gestures all come from paintings. Another option would be to do it in Biblical dress, which means an ancient Mideastern ambience. I call that the 'Bibles-and-Bathrobes Show,' and I don't want to do it since Hollywood does it so much better. In our production I'm trying to suggest not specifically the twelfth century, nor Biblical times, but just another time, a primitive mysterious ambience, vaguely ancient, like Celtic France. I told the costume designers not to tie me to a certain period." For the same reason, he had ordered that the setting should be without electric lights but instead softly illuminated by 1,000 candles.

Another company appearing on the scene and devoting itself to the music of the Middle Ages and Renaissance was the Waverly Consort, headed by Michael Jaffee. At its inception it drew many members from the former Pro Musica Antiqua. Highly professional and long-lasting, it toured widely, performing a broad variety of ancient works at museums, festivals and universities. Often it did not present the liturgical dramas in their entirety, but rather excerpts from them, as at a Christmas concert in 1974 at the Metropolitan Museum in New York City at which passages from the two Fleury plays about Herod were part of a longer programme on which were sections of a thirteenth-century *Shepherd's Play* from Rouen and a twelfth-century *Journey of the Magi* from Beauvais.

A Czech Music Festival in 1984 included a twelfth-century *Play of Herod* and a thirteenth-century liturgical drama from Poland called *Passion Play* – attesting to the ever-growing interest in these works.

Soon after the thirteenth century, or even earlier in some locations, the enactments at the altar became yet more ambitious; they were moved to roomier space in the vestibule of the church, and afterwards further to the main West Door, the portal and steps, and finally outdoors into the town square or marketplace to accommodate the swelling crowds. They grew into great cycles of short plays, deriving from the Old and New Testaments and dramatizing the Scriptural story from the Day of Creation to the fearful, unsparing Last Judgement. The responsibility for producing the Biblical sketches passed from the priesthood to the guilds, the corporate groups of craftsmen and tradesmen which had arisen in medieval towns to regulate commerce being carried on by an growing middle class. The Fathers of the Church had come to feel that some of the clergy were taking too much pleasure in play-acting, though there were other reasons for this significant shift.

In time, too, more of the playlets began to turn to the Old Testament for their source. Adam and Noah were the principal figures in some of these new vignettes; passages of broad humour entered, with the inclusion of such characters as Noah's stubborn, shrewish wife, a tireless gossip, and Balaam astride a braying ass (enacted by a fellow monk) which he beat until the ireful creature kicked and unseated him. The adventures of Jonah in the Whale's belly comprised another light episode.

During the thirteenth century, Jesus, hitherto unseen in the plays – His existence suggested only by a grave-cloth or in some other symbolic fashion – was given physical form, incarnated in a carefully chosen actor, speaking and participating in relevant scenes. He appeared in the Garden before Mary Magdalene, confronted the Doubting Thomas, walked towards Emmaus, supped with the Apostles; or He descended to face Satan, the Gate of Hell represented by the agape mouth of a beast belching fire and smoke, His purpose being to free the suffering souls condemned to dwell there, among them Adam and Eve and the Patriarchs. With His visible presence, the plays were even more portentous.

The long-accepted belief is that the outdoor theatre of the Middle Ages was an outgrowth of the liturgical drama that began in the choir stalls and at the high altar within the church. This view is challenged by Glynne Wickham in an essay, "Play, Player, Public and Place", in an anthology, *English Drama to 1710*. He cites the existence of religious street plays as early as the tenth century, considerably before the Biblical sketches moved out from the church, and asserts this was a form of theatre that developed independently. Such street plays were part of agricultural and religious festivals, celebrating sowing and harvesting and alluding to the martyrdoms of patron saints. Wickham argues that there was a link, an affinity or overlap, between the sketch as performed at the altar and those in the open air, because they often drew on the same subject-matter; indeed, religious ideas and Biblical lore were as pervasive in the mind of Middle Ages folk when they were outside the church as when they were in it, so it was almost inevitable that they would infuse every thought and expression. These preoccupations were deepened by the thoroughly inculcated sense of sin and the fear of early death in a world beset by plagues.

But if the outdoor plays had an air of piety, they were far less reverential than those seen earlier in hallowed incense-filled precincts when staged for an awed congregation, and they contained increasing elements of robust humour as a spontaneous result of their rude folk origin. Besides, they had a different purpose: mere entertainment. Their authors, if they had any, were not priests who used the altar sketches as "instructional sermons"; instead, their dialogue was conceived by university scholars who were the epoch's other literate class and whose views were not always orthodox. Eventually, as these other sketches became more secular in tone and sometimes echoed voices of social unrest, the conservative clergy became wary and critical of the laity's open-air theatrical offerings and more and more looked on them with disfavour. Among the objects of protest in the plays were not only high living, greedy rulers and unscrupulous merchants but also the Church itself, which – in the instance of the monasteries – was the most important of landlords and was frequently accused of exacting rents that were too high as well as of collecting them too harshly. In *The Second Shepherds' Play* of the Wakefield Cycle, the herdsmen complain, as they beat their hands numbed by cold:

> We poor shepherds that walk on the moor,
> In faith, we are near-hands out of the door;

> No wonder, if it stands, if we be poor,
> For the tilth of our lands lies fallow as the floor,
> > As ye ken,
> We are so hamyd,
> For-taxed and ramyd,
> We are made hand-tamed
> > By these gentlery-men.

Here they are asserting that the landed gentry are heartlessly over-taxing them. Nor dare they deny their lords anything, lest they be made to suffer for it. All this woe is visited upon them by "men that are greater". Ultimately, displeased by their often rowdy aspects and none too hidden notes of dissidence, the Church largely dissociated itself from street performances.

In 1264 Pope Urban IV instituted the Feast of Corpus Christi, a "Eucharistic solemnity", for which the liturgy was composed by St Thomas Aquinas. This Feast of the Blessed Sacrament, on the Thursday following the first Sunday after Pentecost, was also made the occasion for an annual cycle of religious sketches mounted by each town's guilds. The mild weather usually prevalent at this season was an added advantage for the outdoor productions.

These arrays of little plays came to be known as "Passion Plays", because they revolved around depictions of the Passion and Death of Christ, or as "Mystery Plays", construing the word in the way that the Greek religious rites were also termed "mysteries". (A further explanation is that at one time "mystery" had a connotation of "trade", and these plays were put on by guilds, each according to its craft or trade. Also, there were "secrets" connected with them – that is, how certain of their amazing stage effects were accomplished. Again, "mystery" – said to come from "ministerium" – suggests "service" or "office", as in a religious service or office, so that when the word is applied to the playlets it retains more than an echo of a liturgical origin.)

The sketches were in the vernacular, the colloquial speech of the location at which they were staged. In England such cycles were called "Shepherds' Plays". Depending where they were put on, the cycles also borrowed national characteristics: the English the most poetic, the dialogue often lyrical; the Irish more mystical and fantastic; the German more grotesque and satiric, with cruder farcical elements, a fondness for stressing the cruel horseplay of devils and already a pronounced anti-Semitism; the French more rational and rhetorical, and tending to be prolix in form. In all parts of Europe, every town of size celebrated in this fashion, with a day-long pageant of plays; the event might even last two or three days.

The guilds competed with one another to put on the most lavish, most prestigious offerings; they assumed the role of the *choragus*, or wealthy patron of Ancient Greece, who bore the cost of productions. The choice of episode staged by a sponsoring group was related to its craft. From the Old Testament, the Plasterers undertook *The Creation of the World*; the Grocers, *The Birth of Eve* and *The Expulsion from Eden*, perhaps because those stories dealt with "ribs" and "apples".

The Fishmongers and Sailors chose *Noah* and the Water-drawers *The Flood*. The Tanners were assigned *The Fall of Lucifer*, the jealous and rebellious Angel scorched by Hell-fire. From the New Testament, the Shearmen, having to do with lambs, selected *The Annunciation*, and the Tailors, Jewellers and Goldsmiths *The Adoration of the Magi*, for they could provide the most regal raiment. The Barbers acted out *The Baptism of Christ*; the Scriveners, *The Disputation in the Temple*; the Fishermen, *The Draught of Fishes*. To the Butchers and Bakers were given *The Wedding at Cana*, *The Feeding of the Five Thousand* and *The Last Supper*. The Smiths, with their hammers, prepared for *The Crucifixion*. The Cooks and Innkeepers, who presided over roasting spits and ovens, *The Last Judgement* (or *The Harrowing*). To have one's guild excel was a point of honour. The settings and costumes were worked on, and the action and lines rehearsed all year round so that each entry would hold its own and win praise.

A determination of how the Feast of Corpus Christi would be observed was left to the bishop of each diocese. In many places, especially England, the Host was borne aloft in a solemn procession throughout the town, and among the ranks of marchers would be the proud members of all guilds in their fine distinguishing garb. When a gradual opportunity arose to supplant the clergy in offering Biblical plays, some means of moving the performers and the often elaborate stage settings was required. An answer was to use two-storey pageant-wagons, which could stand in the marketplace, with the playing-platform elevated enough to afford all in the crowd to see the action clearly, and also allow the work to travel through the streets to other squares and market centres. Such wagons, six-wheeled, were also employed in Flanders and Spain. (It is from "pagyn", meaning "wagon", that the word "pageant" derives.)

The wagons, a separate one for each Scriptural episode, were also known as "stations". Usually the first stop was in front of the principal or local church, where a mass was celebrated, or else before the town hall or mayor's house; then, one by one, the horse-drawn wagons proceeded, to halt at other appointed "stations", staked out by flags, where additional spectators were gathered. Some might watch the goings-on from the windows or rooftops of neighbouring buildings or other lucky points of vantage. The performance was repeated, at the successive stopping-places, throughout the day, before new audiences. Or each wagon might occupy a permanent post, with scheduled presentations, and people would move about the town, going from one wagon to another, from one "stage" to the next, to take in the whole monumental story. Either way, the entire town became an exciting theatre for one, two or three days, depending on the duration of the festivities. In London the pageant lasted more than a week.

In some places, in England and on the Continent and in particular in France at Valenciennes, the "stations" were not temporary but fixed and architecturally ornate, hence called "mansions". Resembling booths, they were set up in a straight line or in a semi-circle, or it could be a rectangle. Open space between the "mansions", the *platea*, allowed more playing room for the actors. Sometimes rows of seats or scaffolding were put up for better viewing. The spectators turned to look in one direction, then in another, or, if standing, might press forward for a closer view of the exhibit in this mansion or the next. The façade of the church was a background and might be incorporated in the action. Preserving the traditional setting of the playlets when they had taken place around the altar, the booth designated as Hell was at the left, Heaven at the right. An instance of the exploitation of the church's exterior was a play featuring St Augustine, whose chair was set in the very centre, in an open space, the noble House of God appropriately framing him. Some mansions were lavishly canopied and screened off with velvet or silk hangings and tapestries. By reflecting the sun's rays on polished metallic discs, an effect like that of "spotlighting" might be added.

As in earlier Byzantine painting, with its tendency towards semi-abstraction, certain combinations of colours were prescribed for the costumes of the characters: symbolically, the immaculate Mary was delicately yet regally draped in white and gold, her tunic pleated; her mantle was of white silk or some other light, threaded fabric. The King of Hell, Lucifer, boasted horns, bared fang-like teeth and bore a wool-stuffed leather club with which he pummelled his prey – and frequently many hapless onlookers as well. He was masked; his expression was sinister and grotesque. His jacket was hairy. He had in hand a staff, and from his wrist dangled a stout metal chain with which to bind his many victims. Judas's red wig and yellow garment made him easy to identify; Herod and Pilate were opulently robed (which is why they were sponsored by the Tailors) and also had clubs to swing, and were gloved, as were God and the Angels – the gloves being a token of high rank. God was dressed like a high-placed prelate. Angels were gorgeously caparisoned, with gold-leaf overlaying their sheepskin coverings – again, as in Byzantine art; not only their diadems and wings, but sometimes even their features were gilded. The devils, horned and beaked, and with cloven feet and swishing, forked tails, were also hair-covered; their apparel was black. Those whose souls had been redeemed were all in white, and those for ever damned were fittingly in black. Chins might have beards added to them when suitable, and for the roles of domesticated animals – when real sheep or donkeys were not employed – an actor might don an appropriate skin and leap about. To portray a lion or other wild beast he would resort to the same device. To populate Noah's Ark, a motley herd of painted cut-outs was assembled. Many minor characters wore plain medieval attire: they seemed to have wandered from their own times into the Biblical age, which was so very present to the more simple beholders. Ecclesiastical robes might be lent by the parish authorities, but mostly the actors paid for their own costumes – though for unusual requirements the guild or confraternity footed the bill from its treasury. The records they kept are a major source of historical data about how the plays were staged.

In England the upper level of each wagon served chiefly as the actors' platform, while the lower, curtained, was the dressing-room and the place where requisite stage effects were prepared. Trap doors allowed the sudden appearance and exit of devils, with puffs of smoke and flame emanating from "below", and a huge Hell-mouth, or dragon-mouth, might be erected, so that unruly spectators, dragged up from the crowd, might be pushed into it and punished for their infraction, to the hilarity of other onlookers. At Valenciennes the Hell-mouth opened and shut ominously. The Devil and his prankish lesser cohorts were the most popular characters and usually were presented in a comic spirit. But much enjoyment also came from watching the tortures of the damned; they were graphically depicted, the victims lashed to a revolving wheel, or immersed in cauldrons of boiling oil. In time such over-horrendous spectacles diluted the spiritual message of the Scriptures, which sought to do more than instil fear.

The guilds, contesting with one another, spent large sums to mount elaborate and ingenious presentations. In France similar groups were called the *Confrères*; in Florence companies of youthful males organized for this special purpose were known as *del Vangelista*, and in Rome *Gonfalone*. Because a production might take weeks or months to make ready, some of the guilds hired non-member designers, builders and actors to assume the entire responsibility for a playlet; and an experienced director might be put in charge and also take the leading role or double as the prompter, who had an important task always. Contractual terms for such undertakings are preserved in ancient guild ledgers and attest to a medieval love of splendour. Here is one account:

> To Godfrey du Pont for five and one half days of his time employed by him for the fitting of serpents with pipes for throwing flames, at 8s. a day: 44s.
> To Master Jehan du Fayt and his assistants numbering 17 persons for having helped in Hell for nine days during the said Mystery at 6s. a day to each, £45 18s.

In another list:

> A pair of gloves for God . . . four pairs of angels' wings . . . a link for setting the world on fire.

And again, for a production at Mons, France, a third list calls for:

> 18 wagonloads of grass . . . birds, rabbits, lambs, fish and other animals, all alive. (For the Creation, Abraham's sacrifice, etc.) Various trees . . . apples, cherries, fig leaves, various fruits, real and imitation. (To be tied on trees for the Garden of Eden.)

Other such properties included these items:

> False body of St John. Imitation sword for Herodias. Pulleys with which Judas hangs himself. The soft batons with which Jesus is beaten. Roots of trees (To embellish Hell).

Obviously if Hell was terrible, its opposite dwelling-places, Eden and Paradise, had to be shown as abundantly beautiful. Elsewhere:

> A rib coloured red (To become Eve).

Also, all sorts of daring "magical" effects were sought after and achieved. Water was astonishingly changed into wine. The earth quaked. Rocks were terrifyingly split asunder, with devils leaping from them, while thunder shook and rent the air. Cannon exploded. Fountains gushed up from pipes that ran down from overhead. Downpours accompanied the Deluge surrounding Noah and his Ark. Moses' dry staff burst into flower. The miracle of the loaves and fishes was re-enacted and a multitude fed. The Devil visibly metamorphosed into new shapes. He rode a dragon's scaly back. Herod and Judas, the most frequent villains, were swept up into the air and borne off by demons. Occasionally the fantastically attired actors descended to the streets, and the action took place in the very midst of the crowds.

Though the directors might be professional or semi-professional, or still members of the minor clergy, and perhaps trained jugglers and acrobats were brought in to help out, the actors were most likely to be amateurs. A century or so later, in *A Midsummer Night's Dream*, Shakespeare has an amusing picture of a rehearsal by a group of bumbling but eager rustics – caricatured, of course – venturing into a form of theatre. And there were contemporary complaints about the incompetence of the players: they spoke badly, had poor voices, were clumsy, crude. The cast might number hundreds – this was a popular or folk theatre in more than one sense. A prompt-book at Mons reveals that 317 "actors" participated in its pageant, essaying nearly 1,000 roles. They rehearsed long hours tirelessly, from early morning to nightfall, for forty-eight days. Some of these performers got a bit of recompense; the rank of the character to be portrayed might be a factor in how much was paid: God, 3 shillings and sixpence; Noah, 1 shilling; Noah's wife, 8 pence; Saved Souls, 20 pence; Damned Souls, 10 pence. But elsewhere the length of the part determined the wage. Here Noah might earn more than God the Father. A very strenuous assignment was that of Christ, who had to hang on the Cross, with simulated nails through his body. Many attempting this role actually fainted.

In some English towns aspirants for roles were interviewed by a committee appointed by the Mayor. If the festival was at Christmas or some other sacred day other than Corpus Christi, the town council chose the date and laid down rules governing the procession. Town-criers or standard bearers and trumpeters circulated to proclaim the event; they might also invite visitors from nearby towns, in hopes of enlarging the body of spectators. That would bring shoppers, good for the local tradesmen.

Actors who missed rehearsals, drank too much, defied the director's authority, "spoiled a performance" or otherwise misbehaved might be severely fined. They should lead exemplary lives, at least while they were portraying Biblical characters.

As has been remarked, some players undertook multiple roles, but the opposite was also true: a role might be embodied by several players in turn, to present the character at different ages: Christ might be shown as a child, as a young man beginning his evangel, and then as the wracked, dying figure on the Cross. At Mons, it is said, three actors spoke together as the Voice of God, to give his divine pronouncements a weightier timbre.

Another perilous role was that of Judas, who was called upon to hang himself and to be cut down just in time. In some instances female roles were assumed by women, presumably because not enough young males volunteered for the cross-dressing or could create the desired illusion. The performers declaimed their lines and might be asked to sing; the music prepared for the angels was often difficult and demanded much "trilling"; echoes of the mysteries' origin in Church liturgy lingered here. Gestures were broad. A widespread tendency to rant, especially by those chosen to portray evil King Herod, was afterwards deplored by Shakespeare (who saw the Shepherds' Plays in his boyhood) and had his Hamlet counsel the itinerant troupe arriving at Elsinor to avoid "noise that out-herods Herod".

Most depictions were taken very literally by some of the more credulous spectators, to whom Biblical figures and demons were joyfully or fearfully real. The plays were vitally necessary. The medieval world, it must be said again, was crowded with superstitions and menacing delusions, when stalking, early death was omnipresent in a time of plagues, with a dread of Hell-fire and other unworldly punishments always pressing on weak-willed men's minds. Belief in resurrection and a physical immortality in Heaven was imperative to balance the dread that could drive mad the ordinary person whose imagination was lacking in any other resource, and who was confronted by *memento mori* at every hand.

(Much of the specific detail in this chapter is owed to the impressive erudition of Vera Roberts and Margot Berthold in their histories of the era, as well as to the general chronicles by Cheney, Nicoll, Gassner, Durant and Brockett, already acknowledged, and special studies such as Charles Mills Gayley's *Plays of Our Forefathers*, Christopher Ricks's collection of essays by several scholars, *English Drama to 1710*, along with John Russell Brown's performance version of *The Complete Plays of the Wakefield Master*, Ruth H. Blackburn's *Biblical Drama Under the Tudors*, Peter J. Houle's *The English Morality and Related Drama*, and by three authors, E. Catherine Dunn, Tatiana Potich and Bernard M. Peebles, *The Medieval Drama and Its Claudian Revival*.)

Of still extant scripts for the Biblical episodes, the English are deemed the best. This is partly because the Black Death had swept the Continent, decimating its populations, but the epidemics were relatively less devastating in England, allowing the pageants to flourish there more widely and for a longer time. But also the literary quality of the Shepherds' Plays tends to be higher than in most sketches elsewhere; they are naïve, but charming and effective and often poignant. It may be partly for that reason that so many of the texts have survived. The number is surprisingly large. There are four complete cycles: from York, with forty-eight episodes; from Chester, with twenty-five; from Coventry, with forty-two; from Wakefield, with thirty-two. (The Wakefield is also called the Townley, after the Lancashire family that long possessed the manuscript, now in the Huntington Library at Pasadena, California.) Those at York and Chester are still staged today as successful and lucrative tourist attractions in those two towns also filled with historic monuments.

A beautiful, oft-extolled example of the brief dramas in the Wakefield pageant is *The Second Shepherds' Play*, replete not only with pathetic complaints and bitterness but also a robust peasant humour and not a little irony. In its middle scenes the hard-pressed, impoverished Mak and his wife pretend that they are parents of a newborn child, when in fact they are hiding a stolen sheep, which they swaddle with bed-sheets in a cradle. But Mak's sympathetic fellow shepherds, seeking the lost creature, peer with interest at the "infant", whereat the rascally Mak's deception is exposed, after which the perpetrator of the hoax is roughly tossed in a blanket. The play, however, ends with a sudden shift from the English moor to the manger in Bethlehem, where Mary has just given birth to the radiant Christ-child. This concluding scene is alight with mystical joy.

> ANGEL [*sings*]:
>   Gloria in excelsis
> [*To Shepherds*]
>   Rise herdsmen kind, for now is He born
>   That shall take from the fiend what Adam did lorn;
>   That warlock to confound, this night is He born.
>   God is made your friend, now on this morn
>   He behests.
>   At Beth'lem go see;
>   There lies that Lord free,
>   In a crib full poorly.
> [Translation: John Russell Brown]

It is believed that this piece was presented on a pageant wagon that was taken about the town. The author, unknown, is referred to as the Wakefield Master and is thought to have written five other scripts for the cycle: *The Killing of Abel*, *Noah and His Sons*, *The First Shepherd's Play*, *Herod the Great* and *The Buffeting*, the last of these concerned with the torment of Jesus at the hands of Caiaphas despite the persistent intervention of the more lenient Annas. Each of the six playlets lasts from thirty to fifty minutes, depending on how much music and pantomime are inserted, and it has been shown that they can be enacted not only on a movable wagon but on any sort of platform or area, which has prompted modern adaptations and revivals. The archaic dialect is a handicap, however, the speeches needing to be amended.

An affecting, touching episode from another cycle is the Brome *Abraham and Isaac*, which is also cherished for its literary merit, its conveyance of direct emotion. Some scholars believe that it was performed independently, not as part of a series. The incident presented is from Genesis, the widely familiar account of Abraham being ordered by the voice of the Lord to sacrifice his best beloved son. The father's devotion to God is sorely tested by this dreadful command, and moving indeed is the child's plea to be spared:

> Now I would my mother were here on this hill!
> She would kneel for me on both her knees
> To save my life.

And when he appreciates that his father, grasping the knife, must act in accordance with God's will, the loyal and soon submissive boy adds:

> When I am dead, then pray for me.
> But, good father, tell ye me mother nothing;
> Say that I am in another country dwelling.

The boy asks for his father's blessing. Cries the heartbroken father:

> Isaac, Isaac, son, up thou stand,
> Thy fair, sweet mouth that I may kiss.

The boy begs that his death be swift, that his father not tarry. At the last moment the Angel stops the plunging weapon: God is satisfied that Abraham's love for Him is true, constant. At the Angel's prompting a ram is slain and accepted as a burnt offering in place of the exceedingly obedient child. Yet after this reprieve the boy is fearful of his father – a touch of psychological realism.

In an epilogue a "doctor of theology" replaces the actors and lectures the spectators, elucidating the lesson of the play: all should bow to God's will without question, and particularly women should not lament the loss of children, who perhaps have been spared greater ills they might have suffered later in life.

Some of the lively humour of *The Second Shepherds' Play* is also found in *Noah's Flood*, an episode in the Chester cycle. It features the builder's obstinate wife who will not heed her husband's bids and pleas that she embark on his ship. Her objections are various: she prefers the company of her friends, with whom she drinks and gossips, and if they are to drown, naturally she wishes to save them. The pair argue about this and their sons join in the family dispute. Cries the harried Noah, as his perverse, cantankerous wife defies him:

> Lord, that women be crabbed ay,
> And never are meek, that dare I say.
> This is well seen by me to-day.

In this play, too, God is a speaking character: He instructs Noah in his duties. The beasts, according to instructions by the anonymous author, are to be painted images and moved aboard the ark in synchronization with descriptive phrases uttered by Shem, Ham and Japheth:

> Sir, here are lions, leopards, in,
> Horses, mares, oxen and swine:
> Goats, calves, sheep, and kine.

This farcical piece, like many others for the pageants, is in rhymed couplets.

When the shrewish lady finally assents to board, after much shouting, she boxes poor Noah's ears. The vignette ends with the dispatch of a raven and a dove to bring back signs of the flood's ebbing. God accepts Noah's prayer of thanksgiving and promises him that he shall be His agent in repopulating the stricken earth. A rainbow is set in the sky as a pledge of peace between the forgiving Deity and man:

> My blessing now I give thee here,
> To thee, Noah, my servant dear,
> For vengeance shall no more appear;
> And now farewell, my darling dear.

Also from Chester is *The Harrowing of Hell*, similarly combining low comedy and high solemnity. Sponsored by the Guild of Innkeepers and Cooks, it depicts Jesus' descent to the infernal regions, where he confronts Satan, enthroned, surrounded by demons. The King of Hell is overthrown by his followers when he fails to stand up to the Prince of Peace. Jesus redeems a host of spirits – Adam, Isaiah, David, Simeon and others – who have been denied heavenly bliss because they lived before His coming. Among them, too, is the thief who died alongside Him on the Cross. As the Archangel Michael leads the saints and damned to Paradise, Satan laments:

> Out, alas! Now goeth away
> My prisoners and all my prey;
> And I might not stir one stray,
> I am so straitly dight.

In Paradise the new arrivals are joyfully welcomed with much hymning by other saints and prophets. But now the scene shifts back to Hell, where a drunken ale-wife is just making her tipsy entrance, doubtless come "from a seamy corner of medieval life". She is greeted, but in a different way, by the prancing, taunting demons, who delight in beating pots and kettles to make an "infernal" racket. One of her sins has been that of diluting and spoiling the drinks she sold. Satan happily vows to keep her with him eternally. This rowdy conclusion is perhaps a later addition, but provides yet another insight into the daily routine of the city's inhabitants.

The Chester cycle is attributed to Ralph Higden, a monk of that district. Rarely is the authorship of the medieval plays known; here the identification is tentative. The vignettes have a consistency of tone, an admixture of devoutness and good humour, that argues for their coming from a single pen, and it is possible that the same writer contributed episodes performed in neighbouring towns – either that or he has been paid the flattery of imitation, for they bear a marked similarity of style to the excellent Chester sketches. Of course, all have the same Biblical source, and perhaps follow patterns established earlier on the Continent.

The names of a number of English and French authors *have* been recorded, among them Robert Croo, Eustache Marcadé, Arnoul Gréban, Johan Michel, together with that of a nun, Katherine Bourlet. Their episodes are specified. For the most part, they are original only in so far as they expanded on the Scriptural stories, or gave them a personal interpretation, which could only be done within narrow limits.

In a process towards secularization, human values were increasingly stressed, along with the aggrandizement of sheer spectacle. A fondness for dwelling on the misdeeds and personality not only of Lucifer but also of such characters as Salome and the Magdalene is equally marked; then, as now, sin was theatrically more interesting than virtue.

Elsewhere, an intermingling of national characteristics is observed. A token of such melding is a twelfth-century Anglo-Norman *Adam*, which has as its scene Paradise. It is believed to be the oldest drama in French. The stage-instructions are still spelled out and tell exactly how it was performed. A platform was built on the steps leading to the west door of the cathedral, with curtains and cloths of silk hung round it "at such a height that persons" in the cast "may be visible from the shoulders upwards". Presumably this covering was intended to preserve modesty, since Adam and Eve should be naked.

God the Father is a character in this first drama. He is attired in a deacon's wide-sleeved, loose-fitting vestment. When Adam appears before Him, he is now decorously dressed in a red garment, and Eve in a virginally white robe, over which hangs a silken cloak. The actors are told exactly how to read their lines: "neither too quick nor too slow". Indeed, all the cast is told "to speak composedly, and to fit convenient gesture to the matter of their speech. Nor must they foist in a syllable or clip one of the verses, but must enounce firmly and repeat what is set down for them in due order." Furthermore, "Whoever names Paradise is to look and point towards it." No author before or since has ever been more precise. When God need not be in view, he enters the cathedral and stays there, symbolically occupying his earthly home.

Demons invade Eden and seek to tempt Adam, who rebuffs them. The chief devil retreats to the gates of Hell, where he holds a council with his wicked companions, and they concoct a new plan. He approaches Eve, "addressing her with joyful countenance and insinuating manner". The space assigned to Hell is the open area in front of the platform, very near to where spectators are standing; the devils, at times, sally among the people, so that the audience is almost participating in the action and is intimately involved.

The dialogue leading to Eve's seduction is truly dramatic. The devil flatters her, assuring her that she is far more intelligent than Adam:

> Thou art a delicate, tender thing,
> Thou'rt fresher than the rose in Spring;
> Thou'rt whiter than the crystal pale,
> Than snow that falls in the icy vale.
> An ill-matched pair did God create!
> Too tender thou, too hard thy mate.
> But thou're the wiser, I confess;
> Thy heart is full of cleverness.
> [Translation: E.K. Chambers]

The woman is pleased, but hesitant. He promises her that the forbidden apple is "heavenly food!" that will lend her an added grace. She is finally persuaded when the serpent arrives and joins in beguiling her. When she begs Adam to taste the fruit with her, he refuses, until she charges him with unmanly cowardice. He consents, and is delighted with the fruit at first bite; but almost at once is overwhelmed by consciousness of his error and sin. In a trice he disappears behind the curtain and returns to sight, his rich costume changed to a bedraggled one of "fig-leaves sewn together". He expresses his woe most eloquently:

> An evil counsel gave she me!
> Alas! O Eve!

The benighted pair hide in a corner of Paradise, to escape from God's view. But they cannot elude the All-Knowing One, who upbraids them mercilessly. "Didst think through him to be my peer?" He tells them: "How quickly thou lost thy crown!"

Expelled from Eden, they begin an existence of back-breaking toil. Adam blames his mate for all this:

> O wretched Eve! How seemeth it to thee?
> This hast thou gained thee as thy dowry –

She seeks to explain her deed:

> I gave it thee, to serve thee was my aim....
> The fruit was sweet, bitter will be the pain!

The devils re-enter and punish them, shackling them with chains and irons, pushing them about, dragging them towards Hell. A great smoke arises from the fiery pit, while the dancing demons exult and shout together at their triumph, dashing together all sorts of metallic utensils.

The author asks for a comparatively elaborate presentation: "Fragrant flowers and leaves are to be set about, and divers trees put therein with hanging fruit, so as to give the likeness of a most delicate spot." The richly sculptured façade of the cathedral, busy with niches, bas-reliefs and colourful rose window, would provide a very impressive background. Awesome music emanated from its recesses, and the actors – during this early phase – were priests, His devout servants. Of the stately Passion Plays, Sheldon Cheney has said admiringly: "How then doubt that this is one of the noblest theatres the world has known?"

Further episodes from Genesis follow this introductory play, as do vignettes from other books of the Old Testament. A stage instruction for a scene in which Cain beats Abel to death is noteworthy. To ward off the blows and protect the actor, the author proposes: "Abel shall have a saucepan beneath his clothes." In the manuscript all the directions are inserted with red ink.

Though few if any of the host of episodes in the English pageants have the universal humanity of the Brome *Abraham and Isaac*, or attain the level of its plain lyricism together with the matching *naïveté* that enhances its appeal, many do contain memorable passages, and it is obvious that such was their realism that "The shepherds are not of Israel but the English countryside, and Judea is 'not far' from England." In the same way, many of the Passion Plays on the Continent were often folk art of a major sort. From place to place, however, there were some differences of format.

In Spain, at first, there might be enactments of Scriptural stories on a *roca*, a platform borne on the shoulders of twelve sturdy men from station to station. Atop each *roca* was a *tableau*, illustrating a sacred event, that became animated when the procession reached a designated halt. Eventually, as the stage sets grew more ambitious and heavier, the platforms were mounted and travelled on carts.

The Dominican monks of Milan celebrated the exotic tale of the Magi with a wagon-pageant, and in Florence the history of John the Baptist was illustrated on twenty-two scaffolds – *edifizi* – that were moved throughout the city.

In fifteenth-century Flanders, and other parts of the Netherlands, especially Brussels and Bruges, *Gesellen van de Spele*, or craftsmen's acting groups, used little movable stages for dramatized sermons concerning the importance of virtuous living habits. In the small town of Nijmwegen, on Corpus Christi Day, a stage-cart was brought into the marketplace, and actors employed it to offer a trial in which the contestants were the Devil and the Virgin Mary, each seeking possession of the souls of hapless human sinners – needless to say, the compassionate Holy Mother prevailed against the Evil One.

The German artist Albrecht Dürer wrote a description of a pageant he saw while visiting Antwerp (1520) that gives a good idea of its considerable scope:

> The whole city was assembled, all crafts and all trades, each in his best clothes according to his estate. . . . Twenty persons carried the Virgin Mary with Our Lord Jesus, in the most sumptuous finery in honour of God. And in this procession there were many pleasurable things done

and most splendidly arranged. For there were many wagons, plays in ships and other bulwarks. Among them was the host of the Prophets in their proper order, and after that the New Testament, such as the angels' salutations, the Three Kings riding on big camels and other strange, miraculous animals, most prettily decked out, and how Our Lady flees into Egypt, very devout, and many other things here omitted for lack of space. . . . And at the end came a large dragon; he was led by St Margaret and her maids on a leash, that was particularly pretty. She was followed by St George with his knights, a very handsome cuirassier. And in all this crowd there also rode boys and girls, dressed most prettily and splendidly, according to various local customs, to take the place of many saints. This procession, from beginning to end, before it was past our house, took more than two hours.
[An excerpt from his diary, quoted by Berthold]

Passion Plays in France reached their apex in the fifteenth and sixteenth centuries. One, for which Arnoul Gréban prepared a script of almost 35,000 lines, took four days to perform, mixing episodes that were by turns bizarrely and coarsely humorous, serious and sentimental, as it progressed from the birth of Adam, through the earthly career of Jesus, to His suffering, harsh death and Resurrection, to an uplifting climax in the Pentecost. Scores of actors were required to enact the many incidents that Gréban arranged for high theatrical effectiveness. A later and lengthier version of this text was prepared by a younger contemporary, Jean Michel, for staging in his home city of Angers (1486), where it bore the title *Mistère de la Passion de nostre Saulveur Jhesu Christ*. When he was not editing Biblical sketches and composing plays, Michel was a practising physician. An interesting study has been made of how he – and other script-writers – were influenced by paintings and scriptures of their day when depicting the Scriptural characters in their brief dramas.

Even more pretentious were the later pageants in Valenciennes, Paris, Rouen and Bourges, four that have been described as "gigantic" and "swollen". In the sixteenth century stage effects were in the hands of *conducteurs de secret*, master technicians who could truly work marvels. At Valenciennes the mansion housing the infernal regions boasted a fortified tower "surrounded by flames and crowded with wildly gesticulating devils . . . and supplemented by a well into which Satan is hurled after Christ has opened the gates of Hell". God the Father, enthroned with a halo, floated down on a cloud (a decorated platform) to retrieve His beloved Son and return Him to high Heaven. At Rouen the grotesque Hell-mouth was constructed so as to open and shut on cue by means of a mechanism. A remarkable illusion is that called for by Jean Michel for his *Mystère de la Résurrection* (1491): the Holy Ghost is made visible to the Apostles "by tongues of flame lit 'artificially,' with the help of brandy above their heads".

At Mons (Belgium) and Rouen there were huge stages: the platform at Mons was 120 feet long and twenty-four feet deep, and at Rouen 180 feet long. In 1411 the Confrèrie de la Passion of Paris moved its religious production inside to the Hôpital de la Trinité, then to the Hôtel de

Flandre and finally to the Hôtel de Bourgogne, which was later to serve Molière and other repertory companies when at last classic French theatre was developed and permanently established.

Throughout the German-speaking countries all sorts of sacred dramas were popular: Mystery Plays, Prophet Plays, allegories, vignettes celebrating the Passion, the Nativity, the Martyrdom of Saints. There were pageants in the streets of Innsbruck, Bolzano (in the Tyrol), Künzelsau and Freiburg im Breisgau, incorporating the multitude of incidents in the Biblical epic. In Limburg Mary Magdalene was presented as a beauteous courtesan, playing the lute seductively, singing a profane song, enjoying a wild life. Much of the dialogue assigned to devils contained blasphemies and obscenities, additions that horrified some clerics who thought they sprang from an unbridled pagan impulse.

Frankfurt am Main devoted two days to tracing Christ's story from His baptism to His ascension, beginning in 1350. The text included hot-headed disputes based on "Ecclesia and Syngoga and prophets and Jews", and there were topical allusions to the Black Death which had visited and punished the city just a year earlier. At the end St Augustine was represented, giving a lecture on the "true faith" and serving to convert ten Jews. (The "Director's Scroll" for this offering is still extant; it is around fourteen feet long; even so, the instructions for speeches and music are abbreviated, though the actor's cues are clearly marked and the physical action fully written out in red ink.) The staging occurred on a sloping square before the Nikolaikirche: three crosses were set up at the opening's highest elevation; to the east, "against the fine old patrician houses", stood the Throne of Heaven, and at the base of the slope was arranged the Garden of Gethsemane, near enough so that the Angel bearing the cup of bitterness could quickly reach the kneeling Jesus. The playing space was an oval, 120 feet in length, along which were ranged the houses of Mary, Martha and Lazarus, together with Herod's and Pilate's palaces. The Gates of Hell were at the lower west corner of the palace square, where a half-hooded fountain was available for baptismal scenes, and where Satan could make an effective entrance by suddenly rising from an ancient moat, the Wassergraben. In the centre of the cobbled area was the table at which the fateful Last Supper would be served, and alongside it suggestions of the Temple, and a short column with a cock atop it that crowed mockingly as Peter, affrightedly, repeatedly denied knowing anything about the Master. Most likely the actors came forth and afterwards exited back into the church and used it for a convenient dressing-room, much as its priests used its vestry.

Alfeld was notorious for the horrible realism with which Jesus' ordeal on the Cross was portrayed – no easy task for the player given the role – and also for the substantial length of its pageant, the script for it measuring over 8,000 lines. On the same large scale were performances in St Gall and Lucerne, Switzerland; in Lucerne the site for the enactment was its picturesque wine-market. The two-day cycles, produced by religious fraternities, were given until late in the sixteenth century and were appraised as "sumptuous". A city clerk, Renward Cysat, was charged with readying and editing the playbooks, negotiating with the craftsmen hired to erect the scaffolds that served as a stage, and his texts, contracts and floor plans still exist. He rehearsed the actors and personally assumed the role of the Virgin Mary. His records are an important source for historians, and models have been constructed based on his meticulous drawings of how the sets for each scene were arranged: the diagonal strip that became the river Jordan for the baptism of Jesus; the *Haus zur Sonne* (Heaven); Golgotha, with its three stark crosses; the Temple of Jerusalem; Herod's dwelling; and so on.

To the south, in Vienna, stagings of the pageants reached their height under the guidance of Wilhelm Rollinger. A famed sculptor, whose bas-reliefs of Scriptural scenes adorned the "old choir stalls" in the city's St Stephen's Cathedral, he was also a member of the local Corpus Christi fraternity, and by it given the responsibility of designing and mounting an Easter presentation in 1505 that had a cast exceeding 200 actors, which topped all such preceding annual offerings. (Rollinger's famed reliefs in the cathedral were lost in a fire-bombing during the Second World War.) The Viennese Easter Play, at first concerned with the visit to the tomb and the discovery of the Resurrection, is said to have had its origin in a monastery of Augustinian Hermits in 1472. In a later episode, *The Harrowing of Hell*, Adam and Eve, Abraham and Isaac and the Archangel Gabriel are shown supplicating the risen Jesus to grant them salvation and help them escape the flames. The language and actions have a hearty, simple folk flavour, and some passages descend to robust, sensual farce. Yet most of the sketches are found to have a "sturdy piety".

In smaller towns of the South Tyrolean region surrounding the Austrian capital, and particularly those in Bohemia, the elements of peasant humour were similarly emphasized. In the province of Bolzano (by turns Austrian and Italian), rivalry between such thriving trading towns as Bozen (Bolzano), Brixen (Brescia), Stirzing and Eger arose from their citizens' aggressive competitive spirit and burgeoning civic pride. In each of these places the pageants grew longer and longer, year by year, as more episodes were added, until in Bozen the performances went on for seven days, opening with a prologue that showed Christ's entry into Jerusalem (set on Palm Sunday) and comprising thereafter such incidents as the Last Supper, the happenings on the Mount of Olives, the Scourging and the Crucifixion on a gloomy, overcast Good Friday, followed by the Marys' mourning, the awe-inspiring Resurrection, His appearance to the Apostles on the road to Emmaus and, at the climax, His glorious Ascension. This mammoth work was directed by Vigil Raber, also a painter, who dwelled in Sterzing and whose added talents as a writer, stage and costume designer as well as actor were much sought after throughout the Tyrol. Some of his sketches remain and testify to his taste, ingenuity and skill. The week-long play was put on not in the market square but in the local Gothic parish church, with the actors making their first entrance in a solemn file through the main portal from the outside. The nave and transept housed the scene in which Caiaphas and Annas interrogated Christ, and the dwelling of Simon Leprosus; the Mount of Olives was opposite on the right; beyond the altar and choir were Hell, Heaven and the Synagogue, with Solomon's temple rising in the centre. So in Bolzen, as elsewhere, the physical production of the Passion Play was adapted to fit local circumstances, including the amount of funding and the number of would-be players available.

The chosen site had to have room enough to permit a large audience and afford those in it a clear view of the action and a fair chance to hear the spoken lines. Introductory remarks by a *praecursor* helped to explain the story, and his subsequent comments spelled out the moral message, which the more simple-minded might otherwise fail to understand. Outdoors, spectators either stood or brought folding chairs, though those might be difficult to move through the dense throng from mansion to mansion. Inside the churches, of course, some spectators stood but many occupied pews; the acoustics were better, but sight-lines sometimes awkward. An announcement of a play brought flocks of people from neighbouring towns and villages. "The craftsmen bolted their workshops and the gatekeepers barred further entry into the town by closing the heavy gates. All work came to a standstill. . . . Often, each day's performance ended on a deliberately didactic or utilitarian note, as when the faint-hearted were urged to be converted from their *compassio* to a new *promissio*; or, on an altogether worldly level, they were asked to reward the 'poor scholars' with food and drink for their exertions in the play; or, indeed, the highly satisfactory announcement was made that it was time to go 'for a good beer'."

The semi-amateur folk enactments are of lasting significance; they introduced a measure of realism in staging and in many instances had dialogue in the vernacular, so that drama became an even more important instrument for instruction – many ordinary, unlettered persons learned more about the Scriptural stories by seeing the Biblical vignettes than by any other means.

This genre of theatre has not wholly vanished. In 1633 the residents of Oberammergau, a village in Upper Bavaria, were beset by a plague and vowed to perform a Passion Play every decade as a sign of their gratitude if they were spared from death. Once every ten years, faithful to their word, they continue to mount a mass spectacle. Originally they combined a fifteenth-century text in Swabian dialect and a play by Sebastian Wild, an Augsburg *Meistersinger*. Since then the script has been subjected to several revisions, including one in 1810 by P. Ottmar Weiss. Most recently, after the Second World War, objections made to anti-Semitic passages in the presentation led to further rewriting. (There are still protests on this score.) Drawing huge numbers of tourists to their village, the inhabitants have erected a large festival theatre to house their now wide-famed production. All roles in the plays are taken by farmers and craftsmen and other townspeople. Cast members and their parts are chosen by a local council. The dramatic setting, in an Alpine valley, adds to the attraction of the event.

In Chester and York, the Shepherds' Plays are also restaged, again largely to attract tourists who crowd into those picturesque English cities for the rare pleasure of beholding the beautiful, moving and sometimes amusing works of a past age. Going further, in 1977, London's National Theatre commissioned the poet Tony Harrison to edit and adapt episodes from the York Cycle for a compressed offering called *The Passion*; it opened in the Cottesloe. Later, the director, Bill Bryden, expanding on the concept, added sections from the Chester and Wakefield pageants. He returned to the original texts, so that the dialogue was authentic. The work was divided into two parts, the first half, entitled *The Nativity*, started with *The Creation* and extended to the *Massacre of the Innocents*. This segment had its début at the Edinburgh Festival in 1980 then went back to the Cottesloe in London and next to Cologne and Rome.

Subsequent use of material from the Shepherds' Plays, including some from the fragmentary scripts in Coventry and Cornwall, was seen in London at the Mermaid Theatre and elsewhere in Britain. In the early 1990s a new grouping of *The Mysteries* was brought to London's West End and won critical favour.

In America a number of far-spread communities have featured twentieth-century Passion Plays based on new texts: one, given at regular intervals, was staged in the Black Hills of South Dakota; others in like style in Florida and in appropriately named Bethlehem, Pennsylvania; and several, on various occasions, on the campuses of Jesuit universities, among them one in New York City.

The Hartford State Company (Connecticut) chose to do six vignettes from the York and Wakefield Cycles for its Christmas programme in 1984. Mel Gussow, in the *New York Times*, describes how a proper setting was achieved:

> Traditionally these plays are presented in outdoor pageants enacted by local citizens as well as professionals. With the help of John Conklin's stage design, the director Mary B. Robinson skilfully brings a country square indoors. Surrounding a rough-hewn platform is a field of straw and wood shavings. The six actors are dressed in homespun garments. The effect is that of an animated primitive painting.
>
> As adapted by John Russell Brown (an associate director of England's National Theatre), the plays have a conversational and lyrical felicity. Though the stories are filled with portent – including the *Creation* and the *Nativity* – the storytelling is unpretentious. The language is plain, with the occasional interpolation of a colorful archaic phrase.
>
> Each act centers on a sustained story. In the first we see Noah as a man beset even before the Flood by "mickle woe" in the person of his shrewish wife. In the second half there is a rustic folk comedy about an amiable scoundrel who tries to steal a sheep from not-so-gullible shepherds. The plays demand – and receive – a natural, almost naïve, style of performance. No one can overplay his hand, not even Lucifer, who is a kind of everyday devil. The objective is to maintain a ready point of identification with the audience.
>
> John Rensenhouse is especially evocative as he moves from Adam to Joseph to the would-be sheep thief, and Peter Crook undertakes villainous assignments with an insidious ease. Though Christopher McCann is a bit unprepossessing even for an earthly depiction of God, he does have a benign dignity. It is Angela Bassett as Noah's scolding wife who becomes the evening's more assertive figure.
>
> Noah's voyage and the other journeys are threaded together with music by Amy Rubin,

played on the psaltery and other suitable instruments. The Hartford Cycle of *Mystery Plays* is marked by simplicity and bucolic charm.

In England, in the courtyard and nave of Lincoln Cathedral, it has become customary to enact a Passion Play every four years; here the script is based on a 1486 manuscript, and the performance is meant to reflect how the plays were seen by Lincoln's citizens in the sixteenth century until most of such processions were gradually prohibited or simply disappeared with the onset of the Reformation and as a consequence of the Church of England's harsh break from Rome. This modern festival has come about through the exertions of Keith Ramsay, the director of the company and head of the Drama Department at Bishop Grosseteste College in Lincoln. The group, made up of amateurs and professionals, has taken the plays on tour to France, Italy and Germany. In 1989, sponsored by the Oregon Committee for the Humanities, the troupe visited five cities in that western US state and presented a selection of thirteen of the forty-two episodes belonging to the Lincoln pageant, which spans Biblical history from the Creation to the Resurrection. (The 1468 manuscript from which these brief dramas were taken is called the N-Town Plays.)

Twentieth-century playwrights have also been inspired by the innocent spirit of the plays and have attempted to capture it in such *faux-naïf* dramatic works as André Obey's *Noé* (1931), about life afloat on the Ark during the Flood; Marc Connelly's *The Green Pastures* (1930), derived from Roark Bradford's retelling of the Biblical stories in humorous and lyrical African-American dialect; and Clifford Odets' *The Flowering Peach* (1954), another meant-to-be-charming version of the tale of the seafaring Noah. Also, there is a short opera for children, *Noye's Fludde* (1958), by Benjamin Britten, that is pure in heart and similar in tone. A more recent attempt has been Arthur Miller's *The Creation of the World and Other Business* (1972), described by its author as a "reconsideration" of Genesis.

At the urging of the noted choreographer George Balanchine, music for a ballet, *The Flood* (1962), was composed by the equally eminent Igor Stravinsky and aired on a Columbia Broadcasting Company programme. Two decades later Balanchine collaborated with Jacques d'Amboise to remake the piece into a full stage-work. Renamed *Noah and the Flood*, and with costumes and sets based on designs by Rouben Ter-Arutunian for the 1962 presentation, the new version had its début at the State Theater at the Lincoln Center (1982) as part of the New York City Ballet's Stravinsky's Centennial Celebration. Robert Craft was guest conductor, the on-stage narrator was John Houseman, and members of the New York City Opera Chorus, occupying the pit, lifted their voices in the singing passages. In the *New York Times* Anna Kisselgoff commented:

> Only two sophisticated artists such as Mr Balanchine and Mr Stravinsky could have agreed on a theater piece of such apparent simplicity to carry a timeless message of theological import.

Theologians of all religions have, in fact, known this for centuries. *The Flood* is a modern morality play, its text largely drawn from Medieval pageant plays. Moral and religious instruction through popular entertainment is an age-old device of use to all religions: Witness India's Kathakali plays.

What more popular medium could Stravinsky and Mr Balanchine have hit upon than television when they agreed to collaborate on *The Flood*? Conceivably this is not the kind of work that has wide appeal. Yet its mix of deliberate stage-naïveté – complete with masks and puppet-like devices – and deeper allegory has a penetrating resonance. We know the story and yet need to have it told again.

This is a collage. God creates the world with Adam and Eve, Lucifer turns into Satan, God drives Adam and Eve out of Eden, Noah and his family gather the animals into the Ark, the flood subsides and God blesses Noah and his sons. Ingeniously, the spoken part appears totally integrated with the choral and solo singing as well as the orchestra.

Mr Houseman, on stage in a gold chair, narrated and spoke for the mortals, including Noah. John Lankston sang the part of Satan while Robert Brubaker and Barry Carl represented the "Voices of God" from the pit. Led by Lloyd Walser, the chorus opened with a *Te Deum* and closed with another hymn. The idea that Satan is ever-present after the Flood was part of the original but somehow this listener missed his aria at the end and this version seemed more optimistic than its predecessor.

The occasional naïveté of the words derives from the excerpts from the York and Chester Mystery Plays, which Mr Craft chose in 1962 along with passages from the Book of Genesis and contemporary texts. The style of the staging itself derives from this Medieval Mystery Play idea.

From a golden mound, framed by the plastic-tube set used in the Tchaikovsky Festival last year, Adam and Eve emerge in the form of Adam Luders and Nina Fedorova. A stylized tree follows them, a puppet-stage fragment of the set. They strike a pose reminiscent of a Flemish painting. The backdrop, orchid-like angels, is suddenly obscured by a dazzling representation – a coup by Mr Ter-Arutunian – of Lucifer as a golden angel on a white panel. The cloth falls, and Lucifer falls to earth and turns into Bruce Padgett, a writhing black figure ablaze in red light and then in a Muppet-like snake costume manipulated by four dancer-stagehands.

Francisco Moncion's Noah, portrayed in mime, is a masked stationary patriarch, with Delia Peters as the shrewish wife. The only danced section in the conventional sense is an ingeniously abstract view of ark-building for a male-female ensemble in white tights. Geometric shapes and figureheads are discernible through the pure-dance surfaces.

The great delight is the winding procession of animal cut-outs – from peacocks to rhinos and the ultimate doves, paraded by young girls. And when the flood – tarpaulin billowing around furies who slither down covered shapes – comes, the entrance of a child with a dove on a stick seems just right.

An oft-chosen subject in the pageants was the parable of the Prodigal Son. A very early example, one of the first in the French vernacular, is *Courtois d'Arras*, given six performances in modernized English at the Cloisters in New York City in 1980 by the Actors' Guild, a company supervised by Stephen de Pietri. Nan Robertson of the *New York Times* found it to be a "lusty, bawdy affair". The Apse of the Spanish chapel at the Cloisters is "of golden, pitted stone, with a glorious fresco curving high above the altar. The mood is hushed, awed. And then suddenly, a cascade of snores rips the silence." The irreverent disturbance betokens a night of carousing by Courtois d'Arras, the errant son.

> Fleeing home, he literally loses his shirt and the money his father gave him to two beautiful harlots he encounters in a wayside tavern. Misery leads to repentance and final forgiveness at his father's knees.
>
> William Callum, the actor who plays Courtois, manages to swashbuckle and bumble at the same time. William Pritz is the stately father; Christine Mower and Jessica Regelson are the sumptuously gowned harlots, and Cameron Duncan is a sprightly innkeeper.
>
> The same little troupe switches mood for the more solemn and religious *The Woman Taken in Adultery*, another Medieval piece on the same program.
>
> The Consort to the Queen group, directed by Rhea Schnurman, performs Medieval music to accompany both plays. The antique instruments include cornetto, recorder, krummhorn, *rausche pfeife* and wooden flute. The background they provide enhances the atmosphere, including a twelfth-century troubadour song and a love ballad by Richard the Lionheart.
>
> It is as if the art of the Cloisters had come to life – as indeed it has. Every costume in *Courtois d'Arras* and *The Woman Taken in Adultery* has been copied from or inspired by sculpture, tapestries and paintings in the Cloisters, mostly of the fifteenth century. They were designed by Elain Grynkewich Drew.

This verisimilitude of garb was owing to the director's special bent. Though his beliefs were now "unorthodox", he told Nan Robertson, "as a student in parochial schools, I was more Catholic than some of my religious friends. Those early images of the saints were engrained in my mind. They were very important to me." Later he found the parochial school training repressive and rebelled. After study at Tufts University in Boston, he spent a year abroad in London where he enrolled in classes with E. Martin Brown, who was researching and producing medieval plays. On their own, de Pietri and some fellow students staged a play in that style.

Back home, he gained further practical experience in off-Broadway and regional theatres, and also assisted in mounting some of the series of spectacular costume exhibitions at the Metropolitan Museum. "For this play, Elaine and I prowled all through the Lower East Side, the garment district and lower Broadway, buying trims and textiles and old drapes." The materials included fur and gold cloth to trim Courtois' broad-brimmed hat. He also involved himself in details of

Miss Schnurman's arrangements of medieval music for the play, a field of study that she described as having been her "obsession" for years.

Starting in 1975, the Actors' Guild had been putting on a medieval play every year as a component of the Cloister's Summer Festival. "When I was a boy," de Pietri said, "I used to come here and think, 'Wouldn't it be nice to see the paintings and sculpture come to life?' I think we've done it."

(The Cloisters, a gift by John D. Rockefeller Jr, is on a hill overlooking the Hudson River. It was brought over from France and contains the donor's superb collection of Gothic art, including most of the panels of the world-famed Unicorn tapestries.)

Another attempt to recapture the spirit of medieval drama is *The Shepherds' Christmas* (1988), an hour-long chamber opera by William Russo, drawn from the ever-attractive *Second Shepherds' Play*. Among its productions was one by the New York Art Theater Institute at Christ Church United Methodist, an evening attended by Will Crutchfield of the *New York Times*, who wrote:

> The libretto, by Jon Swan, is a pleasing adaptation with some felicity of tone in the dialogue. Mr Russo remarks in a program note that he had set out "to write a work for the voice, a work that would suit the voice and make the voice sound good," and on the whole he has succeeded, notwithstanding the strained sound of some of the actual singing in this performance. Both the tale and the music struck this listener as a little glum in tone (and it might be felt that the discussion of labor and birthing pains is a little graphic for young children), but there are moments of humor and poignancy to be enjoyed and some tuneful, suitably Medieval sounding music.
>
> The best member of the cast was Peter Stewart, who sang mellifluously as the Third Shepherd; Gary Giradina was an oily and slack-lidded villain; Sally Williams gave a lusty account of his irate but complicit wife, and the others held up their responsibilities throughout. Paul C. Echols directed the small instrumental ensemble; Donald T. Sanders was the stage director, and the costumes, sets and lighting were by Vanessa James.

Soon after the end of apartheid, a troupe of black South Africans shared with Londoners a fresh view of the inspired Scriptural tales. In the *New York Times* Alan Riding recounted:

> Vumile Nomanyama was understandably alarmed that his first-ever appearance on stage should be in the role of God. He was told to practice by standing in front of a mirror and repeating "I am God" until it became second nature. And by opening night, he felt ready. He moved to the front of the stage, his muscular chest bared, an African cloth wrapped around his waist, and introduced himself in a confident and clear voice: "I am God."
>
> Several members of the audience immediately rose from their seats and walked out of the theatre.

Six years after the end of Apartheid, the idea of a black God still upset some white South Africans. Yet others saw cause for celebration in the multiracial production of *Yiimimangaliso: The Mysteries*, a song-and-dance adaptation of the Biblical stories as told in Medieval England by the Chester Mystery Plays. The production, presented at the Spier Summer Festival in Stellenbosch near Cape Town in late 2000, showed that on stage – if not yet throughout society – black, white and "colored," or mixed-race, South Africans could work together on an equal footing.

It was also evidence of the wealth of artistic talent that had long been suppressed by Apartheid.

Still, if in South Africa the show was initially viewed as a sociopolitical phenomenon, in London it became an artistic happening. Both the public and the critics here responded enthusiastically.

Writing in *The Guardian* after seeing *The Mysteries*, Michael Billington noted: "This is an event that makes London theatre an infinitely brighter, better place and quite simply raises the spirits."

After playing for a month to full houses on London's West End, the company went on tour about Britain and travelled further abroad.

The medieval appetite for the theatrical was hardly satisfied by the pageantry of the Mysteries, which, though exciting and edifying, came at much too long intervals, usually being annual events. If anything, the popular craving was only whetted by the celebratory and carnival-like Easter processions that enlivened the holiday with dramas that were serio-comic, expensively invested and replete with displays of baffling magic and stark grotesquerie.

The answer was to extend such dramatizations by telling about the extraordinary lives, miraculous acts and martyrdoms of the saints, and to give an appropriate playlet on the day dedicated to each particular saint successively. Miracle Plays, as they were soon called, were sometimes included in the Mystery Cycles; to a degree, the pageants had started as loose collections of Miracle Plays, such as the *Quem Quaeritis* and *Stella*. But those treating with figures who, though religious, were not Biblical were likely to be staged separately, a custom that began in about the twelfth century. Eventually these dramas, too, passed from the hands of the clergy to the guilds; in France, the Confrèrie were responsible for mounting them. The guilds, in general, elected to stage a play paying honour to its patron saint by depicting his or her supernatural deeds or alleviation and healing. Accordingly, the carters and draymen might portray the miracles achieved by St Christopher. But the martyrdoms, the hideous torments, the flayings, burnings, crucifixions to which sainted persons had been subjected made far livelier and compelling fare. Or else the patron saint of a town or village might be duly celebrated. A favourite on the Continent was St

Nicholas, everywhere beloved for his many benefactions; and another, in England, was the courageous St George, the dragon-slayer.

Since the roster of saints was lengthy and ever-growing, and their special days occurred frequently, the townspeople were assured of a steady stream of entertainment week after week. Theatre was no longer a rare once-a-year happening. Furthermore, with works about men and women who had performed miracles and endured martyrdom in post-Scriptural days, fresh subject-matter was introduced, topics and settings that were more recent, even comparatively modern. The scope of theatre was significantly widened, and more realism might be required in the telling, inasmuch as most of the stories did not take place in long-ago Biblical times.

When the plays were staged independently of the cycles, they were likely to be presented on stationary platforms erected for the purpose in the town square or in front of the local cathedral or guild hall; though it might happen, as in some places in England, that the sponsoring guild's pageant-wagon was readily available and made to serve at a fixed spot conveniently open to public view.

Most of the texts of these saints' plays are lost, but their authors are somewhat better known than the nearly anonymous composers of the pageant sketches. In some instances they were the same writers who provided episodes for the Mysteries. One was Hilarius, already mentioned, the English-born cleric and wandering scholar who spent most of his adult life in France as a disciple of the acclaimed teacher and philosopher Peter Abélard, too well remembered for his unfortunate love of Héloïse. A much-liked play about St Nicholas is credited to Hilarius. Another such writer is Jean Bodel, a veteran of the Crusades and later a town official at Arras, where he was a member of the Confrèrie des Jongleurs. His *Jeu de Saint Nicolas* (*c.* 1200) contributed to the body of work about that kindly and pious figure, while also leaving as its legacy a vivid picture of the people – knights, tradesmen, horny-handed peasants – of his day. A pagan king, robbed of his treasures, is helped to recover them, and an honest Christian's life is spared. The play, in which is mixed a good deal of secular adventure, abounds with genre scenes: hard combat between Crusaders and infidels in the Holy Land, riotous excesses in unholy taverns and brothels, and dialogue that gains colour by the early and shrewd use of French argot.

In *The Miracle of Theophilus*, a thirteenth-century Parisian minstrel, Rutebeuf, tells of a clerk who – long before Faust – sells his soul to the Devil, only to repent. Affrighted, he beseeches the help of the Virgin and is ultimately succoured by Her. As in a great many other saints' plays, there is a leavening of humour and adventure; one feels that its spirit is as much secular as religious.

Rutebeuf's play is an example of a myriad of others about the Virgin Mary and the host of miracles she wrought as the Intercessor. These were the most popular of all, a reflection of the Mariolatry that characterized the *Moyen Age*. If anything, adoration of the Virgin was more pronounced than that for the remote, awesome figures of the Father and Son, who were not primarily figures extending love to mankind but rather threatening punishment; it was Mary, the compassionate Queen of Heaven, to whom one prayed for aid and forgiveness. She, generous, ever merciful, responded to the pleas of suffering humanity. Consequently, scores – if not hundreds

– of miraculous interventions were attributed to Her. The art of the Middle Ages is largely one devoted to the Heavenly Mother; from praise of Her, too, sprang much of the cult of chivalry, a new respect for womanhood that was hardly dreamed of in classical times.

Many elements of Mariolatry are exposed in a French cycle of plays, *Les Miracles de Notre-Dame*, each segment of it a complete entity, relating in two parts the heinous deeds of a sinner and his salvation – with the aid of Our Lady, a manifestation of Her divine pity. Needless to say, the first half of each of these little plays – the part depicting, humorously and vividly, the behaviour of the miscreant – is invariably the more interesting, and again a secular tone prevails throughout, until almost the last moment, when each tale winds up on a properly pious note. Among the best known playlets in this cycle of forty-odd episodes is *Jehan le Palu*, in which the ubiquitous Devil tempts an ascetic recluse. Encountering the king's daughter, who has lost her path during a hunt, he rapes and kills her. At the Devil's prompting, he hides her corpse in a pit. Remorseful, he vows to punish himself for his horrible act by henceforth going on all fours and eating nothing but roots. The king visits the glade in the forest where his daughter had disappeared. In a vision, Jehan is told that he is pardoned by Heaven; he confesses his crime, beseeching the Virgin that the princess be brought back from death. This happens, and all praise the Queen of Heaven for her benevolent intervention.

Yet another short drama tells of a woman charged with the murder of her son-in-law and condemned to be burned at the stake. She implores the Virgin to rescue her. The flames consume the cords about her wrists, but she herself is spared.

Perhaps the best known of Miracle Plays is the thirteenth-century *Le Tombeur Notre-Dame* about a wandering tumbler who, finding his life unrewarding, empties his purse and gives away his horse and tawdry garments. He retreats to work in the kitchen of a monastery. He is ignorant, knows no Latin, can recite no prescribed prayers, so is not able to enter the order and has nothing to offer the Virgin, whom he adores. Every night, while the monks sleep, he slips down to the chapel. Taking off his robe and cowl and tightening the belt about his flimsy shirt, he performs his acrobatic tricks before the altar and statue of Our Lady – his humble, instinctive tribute, until he lies exhausted. Told of the newcomer's strange nocturnal movements, the Prior follows to watch him and beholds a marvel: Mary, becoming animated and descending from Her pedestal, leans over and cools her prostrate worshipper's brow; She fully appreciates his fervent, wordless devotion. Jules Massenet, in his opera *Le Jongleur de Notre Dame* (1902), returns to this story, though making the protagonist not an acrobat but a juggler whose manual feats would be easier for an operatic tenor to master.

A lively German playlet, Dietrich Schernberg's *Play of Frau Jutten* (*Ein schön Spiel von Frau Jutten*, c. 1480), has a female Faustus – Lady Jutten – who, at the Devil's prompting, dresses in masculine clothes in order to have the privileged status of a scholar. Her rise is spectacular; she finally bows her head to receive the triple crown and mounts the throne as Pope. But, while studying in Paris, she has taken a lover, Clericus, and become pregnant; now, during a papal procession she gives birth, exposing her true sex and shame. Dying, she is carried off by devils to Hell. In some versions the Virgin rescues her; in others she appeals to St Nicholas, and God sends St Michael to bring her to Heaven after she has exhibited repentance. The language throughout is unusually poetic. Schernberg, a priest of Muhlhausen, in Thuringia, draws on the legend of Pope Joan, who, the story goes, became Pope John VII in 855.

Along with the blessed Virgin Mary, as has been said, a universally favourite figure is St Nicholas. His good deeds, celebrated in Jean Bodel's *Jeu de Saint Nicolas*, are also shown in an amusing sketch by Hilarius. Together with plays depicting the Nativity, those concerning St Nicholas were appropriate offerings on or near the Christmas holiday. In the instance of St Nicholas that was most likely on 6 December, when many schools and church choirs elected a Boy Bishop in anticipation of the lad's mock rule at the Feast of the Holy Innocents, 28 December. On this date the boys would enact short dramatic works in their patron saint's honour. Intercession on behalf of children, travellers and merchants was attributed to this ever-kindly saint, whose benign interest in the young finally led, over the years, to his name being reduced to the childish diminutive Santa Claus, the traditional jolly bearer of gifts, much as were the regal Magi, while rejoicing in the birth of the Infant Jesus.

In the twelfth-century script by Hilarius a wealthy Jew, Barbarus, sets off on a journey and, before his departure, turns over his treasure for safekeeping at a shrine named after the saint. Somewhat arrogantly, he bids an icon of St Nicholas to guard it well, but no sooner has he left than robbers descend on the shrine and loot it. Barbarus, who has not gone far, returns as abruptly and finds that his possessions have disappeared. He is furious at Nicholas, accuses him of stealing and demands the retrieval of the riches in lieu of a whipping.

> By God, I swear to you
> Unless you "cough up" true,
> You thief, I'll beat you blue,
> I will, no fear!
> So hand me back my stuff that I put here!
> [Translation: Charles Mills Gayley]

Nicholas, pursuing the robbers, catches up with them. He knows them to be guilty and threatens dire punishment, at very least hanging, unless they yield their booty. Frightened, they hand back the filched goods. Barbarus, delighted, gives thanks to the image of the saint, who tells him instead to express his gratitude to God. The travelling Jew, overwhelmed, repents of his faults and immediately converts to Christianity. The play is in Hilarius' usual mixture of Latin and "undigested French". The simplicity of the plot suggests that it was probably written for performance by children.

In the thirteenth-century Fleury Playbook, the source of the Benedictine short dramas concerning Herod, there are also four scripts having to do with St Nicholas. The lengthiest of them, *Kidnapped*, informs the spectator, "Tomorrow will be St Nicholas' Day, whom all Christians should devoutly cherish, venerate, and bless, *In crastino erit festivitas Nicolai*." A stage-work could not be more emphatically instructive. The play is about a child, Deodatus, who is taken by his parents to the church of St Nicholas, to whom the boy's father is deeply partial, on that saint's feast day. The celebration is harshly interrupted by infidel soldiers; they rip little Deodatus from his parents' clutches and carry him off. In their foreign land he is pressed into service as cupbearer to the pagan King Marmorinus, who discovers that the boy steadfastly believes his Christian God will rescue him, a hope voiced in simple prayers that the king scorns and ridicules.

A year elapses, while the parents desperately seek to learn the whereabouts of their lost son. Euphrosyne, the mother, beseeches the aid of Nicholas, knowing him to be truly beloved of God. On the anniversary of the boy's abduction she prepares a bounteous feast in honour of the compassionate saint; she and her husband, Getron, are at the table with their guests. On the same day Marmorinus is also feasting, but in an ungodly style. The unhappy Deodatus, serving him, is remembering his distant home and family. Seeing the child's tears, the king berates him, again saying that all his prayers are futile. At this point, Gayley's translation comes to hand:

> Then enters One in the likeness of St Nicholas and whisks away the little cupbearer with his cup of spiced wine in his hand (*scyphum cum recentario vino tenentem*) and setting him down outside his father's house, mysteriously disappears. "Lad, whither away," says a citizen passing by, "and who gave thee that gorgeous cup all filled with wine?" "This is whither away," replies the lad, "and no farther do I go. Praise and glory to St Nicholas, who hath restored me." Out from the table spread with bread and wine that clerks and paupers might refresh themselves, runs the mother to her child, and hugs and kisses him – *quem saepius deosculatum amplexetur* – and returns thanks.

The drama ends with a Latin hymn and a choral chant. It, too, was most likely meant for enactment by boys.

Much the same are the fabulous deeds performed by St Nicholas in *The Three Schoolboys*, a fortunate trio brought back to life, and *The Three Daughters*, by his generous instrumentality spared from careers of impending shame. These two works, also in the Fleury Playbook, are wholly in Latin, yet in phrases plain enough so that the lines could be easily memorized and spoken by young actors in a monastery school. In the first of these two pieces the students on a journey choose to spend a night at an inn; they are about to be turned away, but the innkeeper's wife persuades her husband to let them stay. She has taken note that one of the youths has a hefty purse. At her urging her husband waits until they are asleep and cuts their throats. But Nicholas, in the guise of a pilgrim, appears and seeks provisions. He asks, "Give me meat." The innkeeper replies, "I have none." Says Nicholas, "But you do, in the back room. Bring me the children." As the innkeeper and wife watch, the saint resurrects the slaughtered boys. Awestruck, the guilty pair repent, and all ends happily. In the second play a virtuous man is cruelly beset by poverty. His three daughters are about to be forced into prostitution, but that fate is averted when a bag of gold is tossed through the window of their dwelling, followed by two even larger ones. (This story is said to be the origin of the pawnbroker's insignia, three golden balls.)

Many of the tales are rewritings of far earlier classical plays, and some have been appropriated and boldly adapted from Provençal romances of the period. Incest is a frequent subject. As has also been remarked, acts of harrowing cruelty abound in them when they depict the hideous martyrdoms of the Saints. In the ever-proliferating Mary plays, a major theme is sexual transgressions, which the Virgin forgives. If a weak-willed nun becomes pregnant, the merciful Our Lady on the pedestal becomes animated and helps to conceal the young woman's misstep. Easing the woes of the afflicted, She is endlessly resourceful. A citizen's wife, long infertile, asks for the Virgin's assistance and has her prayers rewarded. Not fully recovered from the physical strain of childbirth, she falls into a doze while bathing the babe, who drowns. The bereaved mother, charged with murder, is sentenced to burn at the stake. Her husband pleads before an image of the Queen of Heaven, who thereupon descends from Her celestial realm to console him. His wife's last request is to gaze upon her child once more, and – behold! – in her arms it comes to life again.

In a desert, to which a Bishop has been banished for his over-zealous faith in the Virgin, he is tormented by devils and left for dead; in truth, he is dying of thirst. She appears, bearing in Her hands a golden vessel filled with refreshing milk from Her own breasts. Or there is the drama of Princess Isabel, too fond of dressing like a man and often donning armour. With Mary's help she is saved from an embarrassing situation by being temporarily changed into a male. The myths and legends go on and on.

Scholars have pointed out that a major theme of the Nicholas plays is money, stolen or recovered, or perhaps given to alleviate the problems of the dispossessed and worthy poor, of whom medieval society had no lack.

The four scripts about St Nicholas found in the Fleury Playbook at the Benedictine abbey at Fleury were themselves "miraculously" resurrected: in the twentieth century by the Ensemble for Early Music at New York's Cathedral of St John the Divine (1979), when performed at the instigation of Frederick Renz, a specialist in all such period works and an advocate of flexibility in interpreting past styles. As had been done with the *Herod* sketches, the most dramatic incidents, those chronicling the saint's kindly interventions, were culled from *The Icon of St Nicholas* about the conversion of the Jewish merchant, Barbaras, together with *The Three Schoolboys*, *The Three Daughters*, *Kidnapped* (here titled *The Son of Getron*), melded into a single composite offering, *The Play of St Nicholas*. Becoming a Christmas event, the programme was repeated in 1980, 1981 and 1993, on the last two occasions with the over-long presentation shortened by omission of the Barbaras episode. The stage director was Peter Klein, and the cast consisted of ten adult singers, a boy soprano, five instrumentalists and a children's chorus. A feature of the production were the processionals that took in the full 200-foot length of the cathedral's lofty nave. Still mindful that the original works had been written, sung and enacted by monks and students, Renz sought to keep the performance simple and seemingly semi-spontaneous, in a fashion that he deemed authentic yet eloquent. About the rightness of this approach, Renz said again: "There are lots of lacunae, lots of things we don't know. And that's where there's an opportunity to be creative. So you wind up with the best of two worlds: you have the discipline of applying scholarly knowledge, plus the freedom to fill in the blanks in a way that really brings the music to life. . . . The music itself consists merely of square notes on a single staff; there are no rhythms, nor is the instrumentation specified. What we're doing here is using folk-instruments to conjure up a sort of street element, and also to conjure up what people in medieval Beauvais would have considered a typically Jewish sound. So we're introducing a Spanish Sephardic influence, using a lot of drones and microtones, plus ethnic instruments." In addition, the Fleury stage instructions are scant.

Renz espoused a performance ideal somewhat parallel to the current trend in playing baroque scores, with a free attitude towards ornamentation and an attempt at purer vocal timbres and a pursuit of instrumental textures that were both varied and transparent, thereby differing from the blended sonority and richness of a twentieth-century symphony orchestra.

The role of the narrator was greatly reduced. "On the one hand, having him was helpful, but it also chopped up and slowed down the action. In our production we haven't interpolated a great deal of narration or instrumental 'travelling' music to get a character from one place to another. We simply let the piece go."

In 1981 the drama was staged in the Great Choir in place of the Nave. "This will make the audience area more limited, and the playing area more intimate, if you can say that about a cathedral. The seeing and the hearing will be quite comfortable. St John is somewhat resonant, but for this music it works well enough, since there's no complicated polyphony. I was pleased, though, when we decided to take the play into the Great Choir, because it sounds more contained there."

The revival of these scripts prompted several articles about the historical Nicholas, a fourth-century Bishop of Myra, a town in Turkey but near the Greek island of Rhodes and at that time a part of the eastern half of the Roman Empire and in a region largely sharing a Greek culture. Uncommonly generous, the good Bishop gave away all his inherited wealth to the poor. After his death in 343 he was not particularly singled out for veneration until a biography attributing to him a list of attractive and endearing miracles was issued in the ninth century. Clearly he was a figure of much appeal. By the eleventh century his cult was widespread throughout Europe; by then he was the patron saint of Greece, Russia and the Kingdom of Naples, and not only children and merchants but also mariners looked to him for semi-divine protection. He became "Santa

Claus" by the gradual elimination of letters from the German spelling of his name, "San(k)t (Nik)klaus".

The Romanesque Chapel of St Martin at the Cloisters in New York City harboured three Lenten performances of a *Planctus Mariae* (*Laments of Our Lady*) arranged and performed by the Early Music Institute of Indiana University in Bloomington in 1983. This, too, is apparently a compilation of songs – some devout, some charmingly or grossly secular – and religious music-dramas. They were from the thirteenth-century Benediktheuern manuscript known as the *Carmina Burana*, a facsimile of which had been published in the United States in 1967. The Bloomington offering was partly made up of the longer of the two Passion Plays included in the ancient collection, both wholly in Latin, together with episodes from a vernacular medieval drama, and the *planctus* cited in the title, where the words of the mourners are set down in both German and Latin. A scholarly critic in the *New Yorker* observed that the score consisted of "liturgical plainchant, the '*quasi*-plainchant' of liturgical drama, 'art song,' popular song, and *planctus*. . . . In the manuscript, the music is notated in unpitched neumes – runelike squiggles above the words, described by John Stevens in *Groves* as 'a mnemonic aid' to long-vanished memories. But, because the play is an assemblage, a fair amount of its music can be found in other, more precisely notated sources, and words and music for narrative chants of which only the opening phrases are given can be supplied from elsewhere." This task was carried out by Thomas Binkley, at one time with the Studio der Frühen Musik, and Clifford Flanigan. The staging at the Cloisters was by Ross Allen, with costumes and scenery planned by Max Röthlisberger, and the lighting by Allen White.

The Bloomington group's offering carried strong conviction.

Not even the Metropolitan's *Parsifal* was more gripping or more moving. How the play was first presented no one is sure. (Does it, indeed, represent any play that was actually performed or some monkish scholar's "grand composite" of available material?) Generally, it seems, liturgical drama was played in church, and vernacular drama out-of-doors. The Benediktheuern play combines elements of both. There is nothing quite like it, though there are parallels for its various components. The preface to a *planctus* from Bordesholm, in Saxony, specifies performance on a platform in the church, "or outside the church if the weather is fine," by a cast of five: a devout priest as Christ, a priest as John, and three youths as the Virgin, the Magdalen, and the mother of John. "When it is done by good and sincere men . . . it truly arouses the bystanders to genuine tears and commission." It has elaborate performance instructions. A fourteenth-century *planctus* from Cividale del Friuli has phrase-by-phrase acting directions: "Here shall She turn to the men with arms outstretched;" "Here shall She point to Christ with open Palms;" "Here with bowed head shall She throw Herself at Christ"s feet." A fourteenth-century Visitation play from Essen, whose collegiate church had both canons and canonesses, was played by a mixed cast – the three Marys by women, the angels by men. This seems to have been an uncommon practice, but in the fourteenth-century Avignon Presentation Play "a very beautiful little girl" took the part of the Virgin, and two young men played instruments. . . . Apropos of the acting directions in the Cividale *planctus*, John Stevens says, "The gestures and movements must surely have been . . . not subtle, shaded, infinitely expressive, but immediately recognizable, demonstrative, and impressive." Apropos of the vocal directions and the "unexpected dimension of psychological realism" they introduce, he says, "The problem is how the implications of these rubrics can be squared with other indications in the plays of formality, impersonality, and emotional restraint." By extension, that "problem" concerns all aspects of performing a Medieval religious drama today. Bloomington's solutions to it were carefully judged, precise, and convincing. But those are cool epithets. What we heard and saw was a drama that "truly aroused the bystanders to genuine tears and compassion." It was passionate, piercing, overwhelming – one of those performances not often encountered that pour the past into the present, that leave listeners for hours afterward all but speechless and make hard the return to mundane life.

Of course, the subject-matter has much to do with it . . . the death of a god, crucified by the men among whom He has come to dwell. Whatever a Westerner's personal beliefs may be, he is conditioned from childhood to respond to this Passion history. His art is founded on it. His music has its roots in the movements through pitches, time, and space of plainchant. These are deep matters.

To the music-and-drama groups in New York City specializing in medieval and Early Renaissance studies and revivals was added the Bon Accord Theater Company in 1991. The founder was Dr B.D. Bills, a faculty member of Lehman College, together with Steve Lavin, an actor, and Ken Ross, a scene and lighting designer, also associated with Lehman College. Bills served as director. The casts were mostly assembled from young professionals. Among the group's productions, taken from the accessible literature of the Mysteries and Miracle Plays put on at the Cloisters, were *The Second Shepherds' Play* (1992) and *The Death of Pilate* (1955), a fourteenth-century text. The material for the latter work is retained in the *Cornish Ordinalia*, or *Official Book*, a manuscript ascribed to the fifteenth century and of unknown authorship. In this book are the texts of several of the short dramatic pieces that were included in the monumentally scaled pageant mounted at Penryn and first enacted by local amateurs along with others recruited from the surrounding region, near present-day Falmouth. It may have been inspired and guided by a monastic community, the Collegiate Church at Glasney, established there until the Reformation brought about its close. Taking three days to unfold in full, the cycle had 170 speaking parts. The dialogue is a mixture of Middle Cornish with a sprinkling of Latin, French and English.

Apparently *The Death of Pilate* was the second in a section of the pageant consisting of three linked plays; the first was *Resurrection* and the third *Ascension*, the whole covering the life of Christ from the Temptation to the Crucifixion and afterwards. The account of Pilate's death is not from the Scriptures but from a much later period, possibly having its source in this instance in the thirteenth-century *Legenda Aurea* (*The Golden Legend*) by Jacobus de Voragine.

It is believed that the Cornish Cycle was performed in the open air in a circular, scraped earthen space referred to as a "round", forty or fifty feet across, with eight platforms or "mansions" about its circumference, and the actors facing inwards or sideways towards the spectators. For *Pilate* the eight scaffolds were assigned, one each, to the emperor, Joseph, Pilate, Nicodemus, Heaven, Hell, Soldiers and Torturers, as the action shifted from one incident and place and group of actors to the next, much as was done in France.

The Death of Pilate is faster paced than most examples of its genre, partly because no narrator, nor any homilies, impede its dramatic progress. Writing in the seventeenth century, long after the event, Richard Carew remarked that a prompter had an important role in the initial stagings: "The players conne not their parts withoute booke, but are prompted by one called the Ordinary, who followeth at their backs with the book in his hand, and telleth them softly what they must pronounce aloud." Carew takes notice of the great popularity of this cycle: "The Country people flock from all sides, many miles off, to hear and see it: for they have therein devils and devices, to delight as well the eye as the ear." (Dr Bills's programme notes offer much of this detail.)

Emperor Tiberius is stricken with leprosy, and is desperate for a cure. He hears of a wonder-working physician, Jesus of Nazareth, who might heal him, and urgently sends for the man, only to learn from Veronica that Pilate, governor of the province, has had the sought-for physician executed. The always unseen Veronica informs the emperor by a messenger of a kerchief on which the likeness of the crucified Jesus is imprinted; by kissing it, the suffering Tiberius might be freed of his loathsome disease. This miracle comes to pass, and at Veronica's instigation the heartless Pilate is ordered to court where he is to be punished for his grossly cruel and over-hasty deed. Appearing before the emperor, Pilate is garbed in Jesus' seamless robe; the sight so moves Tiberius that his wrath vanishes and Pilate is spared, until, at Veronica's participation, the sacred garment is stripped from the unjust Governor. Condemned again, he commits suicide. So repulsive and accursed is his corpse that land and water refuse to accept it.

At the Cloisters, in the apse of a chapel seating 200 spectators, a huge twelfth-century cross was suspended above a platform on which the play was performed. At one side was a Hell's-mouth – an authentic one, imported from France – and in the centre was the emperor's throne. Steps ran down on three sides of the platform, and much of the action transpired on them. Incidental music, arranged by LaNoue Davenport, marked entrances and exits; it was provided by four players with ancient instruments. The set, mostly bare, was designed by Ken Ross, and the costumes were by Susan Soetaert.

Works such as these on apocalyptic events in the Scriptures or legends could bring sublimity to medieval drama, while the writers and performers of saints' plays about post-Biblical and semi-secular subjects learned how to fashion stageworthy pictures of the ordeals of martyrs – evoking a sado-masochistic response from spectators – or else how to contrive roguish adventures, or perhaps expose family melodramas, along with affirming the superior rewards of virtuous love. Year by year, the beginnings of a non-religious theatre became more evident.

The often eloquent Mysteries and always ingenious Miracle Plays did not exhaust the creative imaginations of pious authors of this God-possessed epoch. One of the most original forms of drama, with little or no precedent in classical times, and scant if any parallel in the Orient, is the quite unique Morality Play, which had a strong – at first glance almost inexplicable – appeal in medieval times. In such a presentation, abstract forces and ideas – Evil, Good, Faith, Doubt, Chastity, Lust – are embodied in symbolic or allegorical figures. The conflicts are between half-real, schematic characters, barely humanized, each of them having only a single, obsessive trait, a dominant fault or virtue. For the most part they lack air in their lungs, blood in their veins, flesh on their gaunt bones, though there are some exceptions. The author's sole purpose is didactic: he is dramatizing a sermon or a theological argument. All that he asks of his characters is that each one represent a separate concept, a way of life, a different temperament. The compulsive trait is unmistakably identified by a categorical name: Vanity, Modesty, Recklessness, Prudence and the like. Moved about by the playwright, his people are like blankly carved chess-pieces in a deadly game between God the Father and Lucifer, or the many other metamorphoses of the Devil. The prize is man's soul; the goal is redemption. Each play carries a warning against temptation, error, the danger of sin, and holds up the Church's prescription for due ethical conduct.

The Morality Play is generally thought to have developed following the evolution of the Mysteries and Miracles, but in truth some of its devices preceded those dramatic genres by several hundred years. One antecedent is thought to be the *Psychomachia* of Aurelius Prudentius Clemens, a Spanish poet and orator (*c.* 348–410), who lived in a pagan Roman world but was dedicated to the fervent praise of Christianity. In this work he depicts a "soul-fight" between virtue and vice in the human heart. Still familiar, *Psychomachia* was a favourite of the writers of Moralities ten centuries later.

In the tenth century, as mentioned earlier, the Saxon canoness Hrosvitha of Gandersheim, who has been called the first German poetess, adapted six of the nimble, refined farces of the Roman playwright Terence, determinedly purging them of their mild hints of bawdiness and almost all of their humour. Some historians say that her Latin is fluent, others that it is bad. Terence's scripts were a strange choice. Exceedingly devout, Hrosvitha's aim was to glorify Christ and the miraculous acts of His followers, though in the main she succeeded only – as Gayley put it – "at pruning Terence of his pagan charm and naughtiness, and planting six persimmons". They are "comedies" only in the sense that they end happily in redemptions and conversions to lives of Christian righteousness. Elements of later Moralities are in them, too. In particular, her *Sapientia* is inhabited with prototypically abstract characters named Wisdom, Faith, Hope and Charity. It could also be said that she anticipated the Miracle Plays in other works.

Also in Germany, a Benedictine nun, Hildegard, Abbess of Bingen (1098–1179), composed a liturgical allegory, *Ordo Virtutum* (*c.* 1152), that fully answers the definition of a Morality Play. Belonging to a noble family, she had taken the veil at fifteen and displayed such talent that at thirty-eight she succeeded her teacher as the superior of her nunnery, and thereafter had an ever more active role in the non-Church politics of the day, so that she became known as the "Sybil of the Rhine", dispensing sage and welcome counsel to kings while fulfilling diplomatic missions on their behalf. An independent spirit, she was once taken to task by another abbess for the shapely festive attire worn by her nuns on holy days, and replied tartly: "The feminine form flashes and radiates sublime beauty." Throughout Germany her enviable reputation was that of a prophet, mystic, scientist and theologian.

From her youth onwards she experienced apocalyptic visions of great vividness and was reputed to have been inspired by them to perform miracles. Setting forth many of her transcendent visions in poems and prose works of considerable merit, she often fashioned musical accompaniments for them. The *Ordo Virtutum* is her most ambitious venture in this intensely personal form of expression. A century earlier than other Morality Plays – with the exception of those by Hrosvitha – it is also seen by scholars as one of the earliest attempts at an opera.

Conceived for a special occasion – the Archbishop of Mainz's dedication of Hildegard's own separate conventual order at Rupertsberg – it is in Latin, with all the dialogue sung, save for some brief expostulations by the quick-talking Devil. The libretto outlines the Soul's confused struggle to attain Heavenly perfection. The Devil – the only role assigned to a man – presents a train of obstacles, proffering fiendish temptations, leading the Soul to worldly dalliances that stain her youthful innocence and birthright of purity. Headed by Humility, the Virtues come forward and introduce themselves one by one, a shining contrast to the Devil's minions. Chastened, retreating from the pervasive evils of the world, the Soul elects the company of the Virtues and is permitted to share in the bliss of Heaven, much to the chagrin and frustration of the tireless Tempter.

The *Ordo Virtutum*, hitherto scarcely known in America, but now being discovered, was staged in the noble Sculpture Court of the Metropolitan Museum of Art in New York in 1986 by Sequentia, a medieval music ensemble based in Cologne; the adaptation used was conceived and directed by Barbara Thornton, an American, the group's principal vocalist, who was accompanied by two instrumentalists, Margriet Tindemans and Benjamin Bagby.

John Rockwell, the critic from the *New York Times*, paid tribute to the music's "striking originality, both as to motivic material and formal development". Further:

> Any performance of music and musical theatre this old involves as much re-creation by the performers as creation by the composer. What survives of Hildegard's intentions consists of elegantly notated monody, or single-line chant. Performers must decide what the pitches

mean, how they are to be ornamented, how many singers should sing them, how they should be accompanied and a host of other strictly musical questions – and then they are faced with a whole new set of theatrical issues in the staging.

Musically, Sequentia's performance enlisted a third instrumentalist, Shira Kammen, plus nine additional singers (Mr Bagby was the shouting Devil, complete with rattling chains). . . . There was also an unnamed woman among the singers in a mute attendant's role.

The group's musical arrangement offered mostly sustaining drones underneath the unison melodic lines, sometimes supported by vocal drones. In addition, there were livelier instrumental interludes, usually accompanied by stately dancing (Roberta Lasnik was credited for "movement").

What was most striking, however, was the uniform strength and beauty of the women's voices, headed by Miss Thornton as the Soul. Several were operatically trained, which lent their singing volume and tonal roundness. But they avoided anachronism by keeping their vibratos narrow, favoring a tight, almost pressed top that hinted at Arabic music. Their distant processional entrance, from the far, echoing rear of the space, sounded downright ethereal.

Rockwell was not persuaded by Miss Thornton's dramatic gifts, however, finding her awkward, and he passed a captious judgement on the costumes and lighting, which were unflattering to the cast. "Still, for anyone with an ear and a heart for such music, this was a rich occasion. And the recording, on the German Harmonia label, is strongly recommended."

Four years later, when a new recording of the play – by the same firm – was reviewed, Rockwell had an added comment about it. "Some would argue that the performance of Hildegard's twelfth-century liturgical drama by Sequentia is as much a contemporary as a Medieval work; given the amount of imaginative re-creation that must go into the enactment of any old score so vaguely notated. Especially the pungency of the instrumentation may raise some purist eyebrows, despite the small number of players actually employed. Nevertheless, the performance is a delight, with some particularly strong solo singing from Barbara Thornton, Guillemette Laurens and Jill Feldman." He predicted that this second recording would win a wider audience than its predecessor.

In 1995 a *New York Times* critic was still mindful and admiring of the *Ordo Virtutum*, especially while participating in a feminist symposium, where one of Hildegard's letters was read and discussed. In it the abbess "argues inspiringly that all music praised God and that only the Devil's helpers seek to silence it". An opinion was voiced by panel members that she was a true pathfinder, her output "sublime". Present at the seminar were Clifford Davidson and his wife Audrey Eckdahl Davidson, who together had also edited and staged a version of *Ordo Virtutum*.

A highlight of the Lincoln Center Festival in New York in 1998 was yet another visit and presentation by Sequentia of the *Ordum Virtutum*. By now, as a result of even more recordings, any performance of Hildegard's music was a "hot ticket", and a further edge was added by this year being her 900th birthday. The queue from the box-office of people eager to attend was remarkably extensive. The performance was moved to the altar of the huge Church of St Paul the Apostle, a well-chosen setting. The Gothic edifice was filled to capacity, and there was an overflow of spectators standing at the sides and in the vestibule at the back. As before, there was praise for the singing and disparagement of the staging – "static" and "unimaginative" – and the costuming of plain white gowns. The score was of great interest to music historians, teachers and students of composition, Hildegard having used many original and unusual technical devices at a time of transition to polyphony. To some listeners the sounds were seraphic, "a feather on the breath of God", but to others the chanting gradually became monotonous.

The principal sources of the Morality Play, however, were the intense theological controversies heard in church pulpits and universities throughout the Middle Ages, when religion was overwhelmingly important to the intelligentsia and ordinary folk alike, ruling their lives as never before or since. The debates were fierce, truly intemperate, and to lose could lead to charges of heresy, excommunication, eternal damnation. The Morality was a propaganda tool in such controversies, its simplistic arguments concocted by one side or the other to persuade the average believer who was probably unable to grasp the subtleties, historical citations and paradoxes used by the learned disputants, the earnest, clever scholars and clerics speaking to one another in Latin. In sum, the Morality was a dramatized sermon for the uneducated.

From New York Sequentia went to London, where it was a feature of the Henry Wood Promenade Concerts at the Royal Albert Hall. At about the same time, in the Clark Studio Theater of the Lincoln Center, the Festival continued with an updated, somewhat less reverential version of the *Ordo Virtutum* composed, played and sung by a group of four calling themselves the Hildegurls, who fused the melodies and words of the Latin text with the resonant peal and outpouring of a computer-generated electronic organ. Paul Griffiths wrote in the *New York Times*:

> Where the Sequentia production was stiff and pretentious, the Hildegurls were fluid and direct. Their unmatching combinations of pants and skirts gave them a faintly Indian look, and if the uniform white suggested innocence, it was an assertive innocence. These were Virtues who know about Hell. They were at its rim. Red light and smoke belched up from a hole at the center of the platforms on which they walked, platforms chalked with the names of the Virtues: Victory, Modesty, Hope, Fear of God and so on.
>
> Under the direction of Grethe Barrett Holby, they came up with some striking stage images. In the first scene, Lisa Bielawa as the straying Soul was tied standing to the floor, then Kitty Brazelton sat down in a corner to play a scarlet recorder. In the second scene, the Virtues' bold statements about themselves were sung by recorded voices while strong beams of light shined down on each of their names in turn. And in the third scene, Eve Beglarian, who now took the role of the Soul, gave a powerful impression of remorse.

If the music had been as strong, fresh and unabashed as the staging, all would have been well. But unfortunately, the Hildegurls limited their opportunities – binding themselves as firmly as Ms Bielawa – by preserving the original chant, which was cut and simplified but present all through. It soon became odd to see these vivid contemporary people sweetly singing ancient chant, albeit into microphones. And the discrepancy did not communicate anything very interesting. Occasionally the Hildegurls went in for raw tone or close harmony, and the whole final scene, where Elaine Kaplinsky was in charge, had a quasi-rock sound. But there was altogether too much respect.

The electronic music, created by a different Hildegurl for each of the four scenes, was generally a background wash. It could have been more.

If the Hildegurls had gone a lot further in recomposing the chant and in creating extensions, transformations and counterpoints by means of their sound system, this could have been a real musical adventure.

New York had yet another *Ordo Virtutum*, an orthodox one, staged by students at the Mannes School of Music later that year.

Elsewhere there were recitals of shorter pieces by Hildegard celebrating her birthday – hymns, chants, sequences – and a conference at the University of Vermont drawing scholars to examine the nuances of her works. She was also honoured by feminist groups.

Described as a "Renaissance woman before there was a Renaissance", she turned out heaps of poetry, a treatise on astrology, advice on diet. Indeed, a religious movement that has many followers in Germany, Creation Spontaneously, heeds her preachments. Many claims are made for her views about the preparation of food and its relation to healing – several popular cookbooks containing her recipes have been published.

She herself suffered from migraine headaches and a range of other ills, and referred to semen as "a poisonous foam".

To the populace gathering in the street to watch, every type of religious play had excitement and relevance, even those that now seem hopelessly remote, dull. The English historian G. Wilson Knight remarks that not only was the material offered by the Old and New Testament instinct with drama – and especially for medieval men and women – with their stirring accounts of David's victory over Goliath, Samson's revenge on the wicked Philistines, Christ's trial, death and triumphant resurrection – but so was the theology of the day, and the vituperative style of expression chosen by opponents in heated moments. "Christianity is saturated in the dramatic: its great theologians, Abelard and Aquinas, work, as did Plato, in terms of 'disputation', or dialogue, posing the contrapuntal." One can hardly appreciate today how seriously the hoarse, fervid debates were taken by their hearers, how strongly held were the beliefs. If the Moralities lacked the impact of works that dealt with more perceptibly breathing characters, they nevertheless touched on feelings that deeply moved the spectators, even though they grasped only the barest outlines of complex dogmas and crucial eschatology being expounded.

Another aspect of medieval life reflected in the Moralities was a predilection for the semi-abstract in picturing the Divine, as is seen in Byzantine mosaics and early Sienese and Florentine paintings that shun perspective and depth – and realistic proportion – and are instead flat-surfaced, the backgrounds gilded, the garb of the figures coloured in compliance with a code of prescribed hues symbolizing traits such as purity and the like; Biblical persons and events are not shown too specifically, nor is the external world, which, to some degree, is shut out, so that the mind is largely focused on the Ineffable. Indeed, to some ascetic visionaries, in their narrow monastic cells, the Holy existed in a luminous haze. The half-real characters in the age's dramatic preachments, the Moralities, are of a piece with an often pervasive mystical approach to portrayals of the sacred. Such art has a share of autonomy, a measure of independence from the naturalistic.

Slowly taking shape, this type of play became conspicuous in the fourteenth century and flourished through the fifteenth and sixteenth, while religious zealotry also reached peaks, the doctrinal schisms and antagonisms growing periodically more angry and shrill, the college-and-pulpit wars sharply intimidating, with opponents convinced that the soul's afterlife was at stake. In places, too, there were echoes of the political struggles aroused by papal claims to power over Europe's resistant monarchs.

Yet, whatever their changing social and political context, the authors of the Moralities did not lose sight of their chief aim, which was to inculcate in each spectator a sense of his own moral responsibility, his personal obligation to have a wholesome life. He was warned by vivid examples that spelled out the horrible costs of sin and guided to a direct path to God, Jesus and the compassionate Virgin, who ultimately awaited him and were the dispensers of infinite mercy and salvation.

If the characters in a Morality were conceived without great difficulty, usually being stereotypical, the plots were no longer borrowed from the Scriptures and had to be freshly invented, which might tax the author's resources. The plays also tended to be lengthier than the Miracles or separate episodes in the Mysteries, challenging the writer's ability to sustain the spectator's interest. Some are written in prose, others in rhymed stanzas.

The Moralities did not travel about in wagons but were performed on stationary platforms in public squares. They might compete there with the Miracles, and in some instances both Moralities and Miracles might be components of a Mystery pageant. Their sponsors might be a guild, a town or a university; if the last, they might be enacted by students. In a first phase the scenery and costumes were unpretentious. But since the plays were "talky", the actors had to declaim their lines with full clarity, displaying a mastery of rhetorical speech.

Though religious in essence, the Moralities tended to have contemporary settings and deal with mundane topics from everyday life; in this way they, too, helped to broaden the scope of medieval drama and prepare for the gradual advent of a more popular and secular stage.

The allegorical plays seem to have gained their earliest foothold in England and France. One such staging was that of the *Play of the Lord's Prayer* at York in 1384, with strong evidence that the script had been in existence some years before that date. It continued to be performed until 1582, just about all of two centuries. Though the text is lost, it is believed that it was concerned with "vices for scorn and virtues for praise", illustrated by separate pageants in which the characters were Viciousness, Pride, Lust, Sloth, Gluttony, Hatred, Anger and Avarice, a parade of something of the kind that was sure to include the Seven Deadly Sins. Other such Paternoster Plays, as they came to be called, were put on by a guild in Yorkshire every year thereafter, and in Lincoln and Beverly as well, according to extant records. In these works the Virtues are Good Fame, Wit, Justice, Peace, Art and Science, and other admirable qualities and pursuits, boldly pitted against Disgrace, Idleness, Slander, Sedition and other deplorable psychological and behavioural attributes, with the outcome always satisfying to the State, Church and most proper folk.

The apogee of the English Morality is *The Castle of Perseverance*, variously dated from *c.* 1405 to 1425, the author unhappily anonymous. The material is so well handled and polished that it would seem to be the culmination of the long experience gained by many earlier writers of the genre. (Gassner does not hold this high opinion of it.)

The script belongs to a collection of three plays known as the Macro Moralities because it was once in the possession of Mr Macro. Its separate – yet linked – parts are titled *Mind*, *Will* and *Understanding*. The action carries the protagonist, Humanum Genus (Mankind), throughout his life from birth to death.

By a rare good chance, the manuscript contains the original stage instructions and even a rough drawing of exactly how the site for the play was laid out: it was given outdoors, "on the green", with five scaffolds – or mansions – standing on the perimeter of a circle. In the centre of the large ring was the castellated "Tower of Mankind", surrounded by a water-filled trench. The pretext for having this watery barrier is not clear; perhaps it was meant to designate a space reserved for paying spectators, or it may have merely been put there to suggest a protective moat, an added scenic effect. An enigmatic legend scribbled on the map says: "This is the water about the place, if any ditch be made where it shall be played, or else let it be strongly barred all about." It may have been intended to keep eager spectators from heedlessly pushing forward and intruding on the actors' area. On each of the five platforms was a throne, severally occupied by God the Father, Belial (Satan), Mundus (the corrupt World), Caro (Flesh) and Avaricia (Covetousness), each with an entourage of kin and disciples. The interior of each mansion was hidden by curtains that parted only when the performance was occurring there. This allowed for changes of scene as the audience's attention was directed from one scaffold to another in rapid succession.

All was done on a large scale: the play consisted of 3,650 lines, with thirty-six speaking roles, the leads backed by a swarm of supernumeraries. A Prologue brings two Heralds (Banner-men, or Vexillatores) who salute the audience, identify the Vices and Virtues influencing man's earthly life and offer a brief explanation of the story to come.

The potentate Mundus boasts of the vast reach of his domain; he acknowledges his debt to his minion, Covetousness. Belial and Caro introduce themselves and tell how they prevail over humankind; in Belial's train are Superbia (Pride), Ira (Wrath) and Invidia (Envy); among Flesh's attendants are Gula (Gluttony), Luxuria (Lechery) and Accidia (Sloth). Closely allied to Mundus are Garcio, his boy servant, together with Stulticia (Folly) and Voluptas (Pleasure). In God's company are the Four Daughters, Mercy, Truth, Justice and Peace, each in a garment of a symbolic hue: Mercy in white, Justice (or Righteousness) in red, Truth in green, Peace in black. They are with Mors (Death). Other allegorical characters are Penitencia (Penance), Confessio (Confession), Humilitas (Humility), Paciencia (Patience). Caritas (Love), Abstinencia (Abstinence), Solicitudo (Industry) and Largitas (Generosity) are at odds with Detraccio (Backbiting), as is the Bonus Angelus (Good Angel) opposed to the Malus Angelus (Bad Angel), to both of whom Humanum Genus complains of the harshness of his existence.

Humanum Genus faces a crucial decision: whose counsel shall he accept, that of Bonus Angelus, who points him to a life all too sensible and sedate, or that of Malus Angelus, who opens a far more enticing path? Humanum Genus is seduced by and falls hapless victim to the Seven Deadly Sins, each of whom recites a litany of her inducements. He also swears allegiance to the three Evil Powers. Eventually he realizes, to his cost, his fateful error. Good Angel, assisted by Confession and Penance, persuade him to repent; by way of a sacrament he is reconciled with God.

The ensuing scene is overtly derived from the *Psychomachia* of Prudentius. Learning that Humanum Genus has taken up residence in the Castle of Perseverance, the Evil Powers gather their army and attack to regain his soul. The forces of good – Love, Abstinence, Chastity, Industry, Generosity, Humility – are mustered to defend him. Belial leads two strong assaults on the castle and both times is repulsed by means of roses, "the emblems of the Passion of Christ". Perceiving that a direct attack is futile, Mundus enlists the aid of Avarice, who deviously instils a covetous streak in the now aged Humanum Genus. Enriched by a thousand marks, given him by Avarice, the elderly Humanum Genus prepares to enjoy himself, but is visited by Mors (Death), sent with a dread summons from God. Garcio, the boy servant of Mundus, confiscates the marks for his malevolent master: all that Humanum Genus has left are his sins. Mocked by Bad Angel, Humanum Genus dies while beseeching Mercy.

Standing before the throne of God, the Four Daughters debate the fate of Humanum Genus. Justice and Truth cite the "letter of the law" that has decreed his damnation. Mercy and Peace beg with emotion and eloquence that he be forgiven. God decides in favour of Humanum Genus, and he is saved. The spectators are chided to think on death, and the play ends with a *Te Deum*. Virtually all the major themes of these allegorical offerings are found in *The Castle of Perseverance*.

To the same period belongs *Pride of Life* (1400/1425; here, as elsewhere, the double dates refer first to the time of composition and then to that of publication), its author also anonymous; it is thought to be the earliest English Morality extant – the manuscript, composed of 500 lines with the last part missing, was lost in a fire in Dublin in 1922. It, too, seeks to encompass in an allegory the whole of a man's life, but it differs from most others of its kind in that it has an attractive leaven of humour, the Prologue vowing explicitly that the actors will produce a work "of mirth and eke of kare" arising from "this our game". Luckily, the content of the whole play is summarized by the Prologue. Rex Vivus, the overly proud King of Life, reigns at his court in a circle of handsome knights. Regina, his intelligent and learned wife, warns him to stay aware of his mortality – even though a king, he will not be spared by death. When he continues heedless, she sends for a bishop to repeat her unwanted advice, but this expedient, too, is unavailing. The boastful ruler says that he expects to live for ever. His self-confidence is bolstered by two knights, Strength and Health, who provide him with effective weapons; he is quite without fear. His spirits are also buoyed by Mirth, the jocular royal messenger, upon whom he lavishes the Castle of Galispire and the Earldom of Kent. He sends Mirth abroad to seek anyone reckless enough to challenge him in a passage at arms, a dare extended even to the King of Death. Needless to say, it will be a joust in which the King of Life is struck down. In the missing fragment of the play, one assumes, he is granted salvation by God the Father after the intercession of the always kindly Virgin Mary.

Gayley writes of the play: "It is interesting, not so much for its lofty and ideal conception as for the excellence with which it portrays ingenuous and fundamental types of character, and conducts a plot straightforward, the natural outgrowth of premises common to the play and to a contemporary view of life. In place of the comic in character and episode, the play presents us with a Nuncius, called Mirth or Solas, who sits upon the king's knee, flatters him and sings. While this figure bears a resemblance, indeed, to the court fool, as Professor Brandl has said, he appears to me more nearly related to the herald of the Miracle plays."

Also in the well-preserved Macro manuscript collection are *Mind, Will and Understanding* and *Mankind (Mankynd)*. In Gassner's view the attributes cited in the first work's title – *Mind, Will and Understanding* – are "conspicuous by their absence"; its author is unknown, and its date uncertain. The writer of *Mankind* (c. 1462–85) is also unidentified. (The three ancient scripts are now in the Folger Museum in Washington, DC.) Despite its title, this allegory is not broad in scope in presenting an average man's life story, but it does depict the everlasting struggle in his conscience between his good and evil impulses. Here the three rascals who besiege and tempt him are Nought, New-guise and Nowadays, minions of Mischief. Mercy begins the play by urging the spectators to seek salvation so that on Judgement Day they will be included with the "corn" that shall be saved rather than with the "chaff" destined for burning. Seeking to explain his failings, Mankind describes himself as "a composite of body and soul continually at odds". Mercy, whose help is besought, advises him to live moderately and avoid wicked enticements. Mankind resumes his humble tasks as a farmer, but the three tempters – who have not been to mass for a year – return to bother him once more, until he desperately wields a spade to chase them away.

Titivillus, a devil, joins the trio in their quest for revenge on the would-be virtuous Mankind. They concoct a cruel trick, placing a board beneath the earth where he is trying to dig, weed and plant. Frustrated and exhausted, he rests and falls asleep. While he slumbers, Mischief whispers in his ear that Mercy has been hanged for stealing a horse. If that has been Mercy's conduct and reward, a godly way of life is hardly warranted. Mankind is drawn back to drinking and wenching, a round of loud carousing with his three wild companions. He resolves to cease his churchgoing and even to steal and kill. Mercy, learning of Mankind's new fall from grace, tries to find and rescue him. The evil three impede his search. Ultimately, Mankind is saved, told, in the words of the Saviour: "Go and sin no more." Mercy entreats the spectators to examine their own lives and think on their own need to work towards salvation.

What differentiates *Mankind* is that it abounds with prankish humour, though no advance hint of this is given in the Prologue. All five rogues are comic characters, constantly playing tricks more likely to amuse the spectator than to frighten him. Sir Ifor Evans says: "The three villains are real and contemporary figures. They have comedy and coarseness, and though the play may have little construction, it has a vivid quality in its language." At one point in the action, a moment of keen suspense, a member of the cast stepped forward to announce that Titivillus, the principal devil, could not appear unless the collection already started in the audience brought in a sufficient amount. To Berthold this suggests that this Morality was not sponsored by a guild or other public organization, but instead was a venture by a small group of strolling actors – the cast is limited to a mere five to seven players – performing for their own hoped-for profit. They were taking no chance that the audience would disperse before the piece was over and the money-gathering was completed. Another strange touch: Mercy, usually portrayed as a woman, is here a masculine character.

Though a vast number of Moralities have vanished, Peter J. Houle, in his *The English Morality and Related Drama* (1972), has gathered the titles and plot synopses of fifty-seven of them from the fifteenth and sixteenth centuries, during which they enjoyed a great vogue – indeed, in the sixteenth century becoming the most popular form of theatre. In his list, far from fully comprehensive, are such scripts as *All for Money, Impatient Poverty, Like Will to Like, Longer Thou Livest the More Fool Thou Art, Mundus et Infans*; many of them are fragmentary, and in his Preface Houle confesses that even the summaries of them make tedious reading. The same characters – such as Mundus, Mercy, Humility, Lechery, the Evil Spirit and Justice – appear in play after play, standard figures, their intentions and actions wholly predictable, which, for the unlettered, may have been part of their long-lived attraction.

Surveying the English Moralities in his bibliography, Houle finds that they are likely to fit into one of five categories. Many embrace the "full scope" of the typical man's life, his progression from birth to a major culmination at Judgment Day. Here the main character lacks individual lineaments; the guidelines for appraising him are provided by Christian doctrine. In a second group are "estates", plays in which persons of different classes or professions are set forth and judged by how they demonstrate the health of the state or relate to members of other diverse classes comprising the society. Again, a third kind of Morality encourages a particular virtue – Patience, Charity, Honesty – by presenting a positive example in a character who earnestly essays to acquire that quality. The virtue is always cast as a feminine role. The fourth, a "hybrid" play, portrays a Biblical, or legendary, or historical person – as in *Mary Magdalene*, *Cambises*, *King Darius*, *King John* – with allegorical figures added to the drama. And finally, "youth" plays, perhaps written to be acted by students, proffered advice to parents on the strict upbringing of children.

In France an early instance of the genre is *Le Concile de Bâle* (1431) by Georges Chastellain, a diplomat and historian attached to the court of Philip the Good, Duke of Burgundy. In his play, Ecclesia, Heresy, Peace and Justice are active participants in the Council of Basel, where church practices and articles of faith were in contention. A few years before (1426), students at the Collège de Navarre in Paris had put together a dramatization of a sermon delivered by the university's chancellor, Jean de Gerson. In it, at a trial presided over by Reason, the human sense organs were held accountable for resisting temptations and abiding by firm Christian precepts.

Soon afterwards, at Rennes in 1439, a good amount of money and thought was expended to put on *Bien Avisé, mal avisé*, an allegorical contest between Well-advised and Ill-advised. Ingenious stage effects, borrowed from the Mystery pageants, were required: a Wheel of Fortune revolved, and, at his death, Well-advised was borne aloft by a flock of angels. The text ran to 8,000 lines, spoken and embodied by an ensemble of sixty.

Whereas in such pieces as the students' dramatization of Jean de Gerson's sermon only simple costuming was in order – a long scholar's gown for Reason, a crown for Ecclesia, a blindfold to indicate Synagoga – more original and costly attire was doubtless needed at Tarascon to help identify the figures in Simon Bougoin's *L'Homme juste et l'homme mondain* (1476). The author, a valet to Louis XII, called for a "veritable *carnaval d'allégories*". Cynical and worldly, Mondain busies himself with every sort of depravity; quite his opposite, Juste persists in a blameless existence. Perhaps because the variety of vices is almost boundless – and interesting – the performance of this play was of several days' duration.

The townspeople of Tours were privileged to view *L'Homme pécheur* (1494), written and produced by Nicolas de la Chesnaye, well acquainted with the medical art of his day and ready to exploit his knowledge of it, in this script and again in another of his plays, *Condamnation de Banquet* (c. 1507), which caused a considerable stir and is still carefully studied. In this second work, Rabelaisian in spirit, replete with profanity, Diner, Souper and Banquet personify high living; they are joined in over-indulgence by Bonne-Compagnie, Gourmandise and Passe-temps, their gluttony leading to a pack of ills, among them Gout, Colic, Jaundice, Dropsy and Apoplexy. Brought before a court, in which Galen and Hippocrates are the assessors, Souper is sentenced to wear lead cuffs that will hamper his food-orgies, and Banquet is condemned to death. His hangman: Diet.

The style is lively, with a patina of sophistication and stabs at wit. For historians the script has another dimension, as explained by Berthold, it provides "a wealth of information about table manners, the art of serving and dishing up, as well as about table music. He describes in detail in what costumes the characters are to appear. Moderation, Diet, and all other servants of Dame Experience wear men's dress and speak with men's voices, because they have executive functions at the court of law and 'concern themselves with things to which men are more prone than women. The fool wears his traditional cap with asses' ears, a suit of many colors, bells on his doublet and on his shoes.'"

In *L'Homme pécheur*, the author makes his point with a grisly impact: the allegory concludes with the repentant sinner's soul rising Heavenwards, and his rotting corpse lying on the ground below.

Similarly mirroring the age, showing waste and dissipation, a host of other Moralities projected semi-abstract figures and caricatures of those afflicted by the decadence and evils of the day. A good example is *Les Enfants de maintenant*, describing the problems that lower-middle-class parents had with their children. Written by a teacher for enactment by students, it tells of the misdeeds of Finet and Malduict, the unruly sons of a baker. Luckily, Malduict is not spared the rod; a well-merited caning restores him to society; alas, Finet goes to the gallows.

Moralities flourished in Germany – an instance is *Henselyn* in Lübeck – and in large part they reflect a world in which religious faith was losing its hold and a degree of scepticism was contagious. A remarkably persuasive response in England to such doubts was *Everyman* (c. 1500), which scholars are convinced is closely modelled on a Dutch play, *Spyeghel der Salicheyt van Elckerlijk* (1495), both the original author and the adapter nameless. It could be that they had a common source. After 500 years this work is still viable, always widely read and given productions at universities and by professional companies. Of all the Moralities, it alone is recognized as timeless in its appeal and quite worthy of its enduring fame and stature.

In this stark and beautiful allegory, translated by Petrus van Dienst, Everyman is summoned by Death. He is told to bring his book of reckoning and to prepare for a long journey from which there will be no return. Frightened, he confesses himself unready to render an account. He vainly tries to bribe Death, to win a delay. His tears and entreaties, too, are unavailing: Death is adamant. Everyman begs to be allowed to have companions on this dread journey, and is granted permission, if he can persuade anyone to accompany him. Desperately, he seeks the help of Fellowship, Kindred and Cousin, Worldly Goods, but all fail him; repulsed by his fate, they flee him. He turns next to Good Deeds, who is sympathetic but – in this instance, regrettably – very weak. Good Deeds takes him to Knowledge, who is ready to go with him and leads him further to Confession. Says Confession:

> I know your sorrow well, Everyman.
> Because with Knowledge ye come to me,

> I will you comfort as well I can
> And a precious jewel I will give thee
> Called penance, wise voider of adversity.
> Therewith shall thy body chastised be,
> With abstinence and perseverance in God's service.

Scourged, Everyman is joined on his pilgrimage by his more faithful friends: Discretion, Strength, Five-Wits and Beauty, as well as Knowledge and Good Deeds. With their consolation to sustain him, he advances towards his end, his spirits enforced. Knowledge tells him:

> Be no more sad, but ever rejoice,
> God seeth thy living from His throne aloft.
> Put on this garment to thy behoof,
> Which is wet with your tears,
> Or else before God you may it miss,
> When you to your journey's end come shall.

Everyman asks:

> Gentle Knowledge, what do you it call?

Knowledge replies:

> It is a garment of sorrow:
> From pain it will you borrow;
> Contrition it is,
> That getteth forgiveness;
> It pleaseth God passing well.

In the remaining scenes Everyman quietly disposes of his worldly possessions, distributing half to charity and making restitution to those at whose expense he had profited, and takes the last sacrament. Some criticism of the corrupt clergy is hinted at, and then, in a serene conclusion, after some momentary fear and hesitation, he reaches and enters his grave. Beauty and Strength forsake him, followed by Discretion and Five-Wits. Knowledge stays by him almost to the last instant, but holds back beyond that. Only Good Deeds goes down into the pit with him, promising him inseparable support when he shall face God. An angel appears and tells the spectators that the soul of Everyman will in time ascend to Heaven.

A work of noble simplicity, *Everyman* is a *danse macabre*. Ifor Evans writes of it: "Though the characters are abstract, they have more variety than many of the individual figures derived from Biblical narrative, and the Morality method gives more opportunity for independent treatment by the dramatist. Everyman is a man of his time, close to the audience, and in this strange way the Morality play gains a realism of its own. The strength of *Everyman* lies also in the skill with which the scenes are developed. There is not the stale obviousness which marks some of the other less competent Morality plays.... At each stage of Everyman's journey the abstract is made concrete by lively figures and human situations."

In some ways, it has been said, *Everyman* is atypical of works in which the subject is religious. The protagonist is not faced with temptations or having to make a choice of a path through life; for him it is already too late for that. He has already received the dreaded summons and confronts the summing-up. But it does share a principal motif of so many Morality Plays, a visceral fear of death, probably all the more pressing because life-expectancy for most of the population was several decades shorter than it is today. Like other such plays, it is also intended to prod spectators urgently to take account of their daily habits and set them in order so that they are prepared to meet their Maker at any moment, which is best assured by regular church-going and adherence to the highest Christian principles.

Through the early sixteenth century in England the Morality Plays swerved ever more towards non-religious subjects, or took sides in current battles. At the same time their numbers proliferated. After all, there was no fixed day when by tradition one had to be performed, then and only then – any holiday could serve as a pretext – nor, in most instances, were they costly to stage. An exception to this was *Magnificence* (c. 1516) by the poet John Skelton, once a tutor of Henry VII. The increasingly ostentatious lifestyle at the Tudor court was the object of this self-described "merry" satiric piece, particularly aimed at Cardinal Wolsey. John Bale of Ossory, a Protestant bishop, entered the lists against the Vatican with his polemic *The Treachery of Papists: The Three Laws of Christ*. He was also the author of *God's Promises* and *John the Baptist*, both presented in Ireland by non-Catholics, and *King Johan* (1538; *King John*), in which that monarch of ill-repute – in most chronicles of his reign – is instead portrayed as having been an upright and well-intentioned ruler who, for ungodly reasons, had been beset and cruelly slandered by the Church. Converts to Protestantism are the featured figures in *Lusty Juventus* by R. Wever and *Satire of the Three Estates* (c. 1540) by Sir David Lyndsay; in the latter work, of which the author is a Scot, Pauper rails against landlords and churchmen, with a comic element salient. To these plays the Catholics had ripostes, such as *Interlude of Youth* and *Hyck-Scorner*. John Roos's *Lord Governance*, which incautious young lawyers at Gray's Inn performed before Cardinal Wolsey, so offended the Lord Chancellor that the author and a number of the actors were imprisoned for their rudeness and daring.

The teaching of the young and the advancement of education continued to be favourite topics. John Rastell, a printer and brother-in-law of Sir Thomas More, penned *The Four Elements* (1517), in which he included the heretical suggestion that the earth is round, among the characters' other discussions on scientific matters. He argues that Theology must accept the findings of Science. Similarly up to date was Henry Medwell, whose *Nature* (1486), otherwise deemed dry and inordinately dull, is brightened by a panegyric to "Enlightenment", a foreshadow of the new Humanism.

In John Redmon's *Wit and Science* – he was master of the choirboys at St Paul's Cathedral – a student becomes enamoured of Lady Science. The butts of Redmond's shafts are Idleness and her son Ignorance, along with the outworn teaching methods of the Scholastics. Most of these plays fit into Houle's "historical" and "hybrid" categories of the Morality, intermingling realistic and allegorical characters.

The semi-abstract figures, mixed among the individualized and realistic ones, were to be recognized not only by their generic names and perhaps by the symbolic colour of their attire – as mentioned earlier – but also by certain distinguishing aspects of their make-up and garments. Just as a traditional court jester's costume would be donned by Folly, the one draping Fame would have paintings of eyes, ears and tongues on it – truly a visual comment – and Vanity would be adorned with fluttering vari-hued feathers. Scarcely less subtle was Wealth, whose covering would have gold and silver coins attached to it. In consequence, the psychological qualities represented by figures so aptly garbed could hardly be mistaken.

Moralities persisted in England until 1558, when Elizabeth I – her legitimacy questioned – ascended the contested Tudor throne and banned all religious plays, seeing that the combined political and theological climate was too heated and dangerous, with these forms of popular but argumentative dramas contributing to excessive public emotion and division in the war for souls. Here and there, on scattered stages, such a work might linger, but by 1570 the Morality had, to all intents and purposes, vanished. By then the allegories had been prohibited in the Netherlands, too, after the Council of Trent (1545–63) decreed that the Catholic Church should withdraw from sponsoring them. In Italy they had gone from view by 1547, and in Paris they were under edict by 1548. Only in Germany and Spain did they continue into the seventeenth century.

Some elements of the Morality, especially the humorous ones, were to have a longer life. They are to be found in the Elizabethan theatre, conspicuously in the output of Christopher Marlowe, and above all in his *Dr Faustus*. Later the Morality's legacy is slight, though a few traces of it do remain. They are evident to twentieth-century audiences in any Expressionistic or Proletarian or Absurdist play in which the characters do not have personal names but are designated merely as Man, Woman or the ilk, or when each is obviously a representative of a whole social class or political point of view – as in John Galsworthy's *Strife*, where the protagonist speaks for Labour and his antagonist for Capital, or in George Bernard Shaw's *Major Barbara*, where the munitions manufacturer, Undershaft, is the Realist, and his daughter, the Salvation Army commander, the impractical Idealist – and wherever an author's obvious purpose is to advocate a cause. Other examples are Gerhart Hauptmann's *The Weavers*, another portrayal of class struggle; Ernst Toller's *Masse Mensch*; Eugene O'Neill's *The Hairy Ape* and faith-questioning *Days Without End*.

Some commentators discern the ghost of the Morality in Luigi Pirandello's *Six Characters in Search of an Author*, his people passionately alive yet shadowy and semi-abstract; and in Clifford Odets's *Waiting for Lefty*, with its rousing cry for militant action. Clearly in the vein are Thornton Wilder's *Our Town*, a village located at the centre of the universe, and *The Skin of Our Teeth*, in which Mr and Mrs Antrobus attend a convention of the Ancient and Honorable Order of Mammals, Subdivision Humans. John Steinbeck's *Burning Bright*, similarly allegorical, led one critic to describe the resort to this genre as a "stab at importance" on an author's part, an attempt to inflate his message, prompted by a didactic impulse. Certainly there are parallels to the Morality in the output of Bertolt Brecht and his theory of Epic Theatre, his use of the stage to preach and persuade, to force the spectator to "think", the characters simplistic archetypes, verging on those in cartoons, the aim blunt propaganda. And Samuel Beckett's *Waiting for Godot* and his works that follow might be deemed a new species of Morality Play.

One factor enabling *Everyman* to enjoy a remarkably long life was the move by Max Reinhardt, the noted German producer, to have the Symbolist poet Hugo von Hofmannsthal adapt it somewhat freely for a new staging, first in Berlin in 1911 and next as an annual feature of the Salzburg Festival, beginning in 1920. Here called *Jederman*, it was performed in the great square before the Cathedral – promoted with Reinhardt's usual flair and theatricality, it won worldwide attention and drew large crowds.

The action takes place on a wooden platform built over the Cathedral's steps, but most of the old part of Salzburg serves to expand the *mise-en-scène*, when Death's call to Everyman from a high-perched fortress echoes over the houses and towers and finally resounds hollowly from the maw-like interior of the church, as if sending forth a warning to all mankind. At the end all the bells of Salzburg ring out, "until the air is filled with the deafening jubilation of redemption. This immensely theatrical experience provides the contemporary spectator with a vivid idea of the effect the dramas of the Middle Ages had on the masses who attended. . . . In striving to achieve a simplicity and honesty in this play, Hofmannsthal drew much of his language and imagery from another great writer of Moralities, Hans Sachs (1494–1576). The result is a rich piousness that is at once natural, exalted, and deeply moving." (The quotations are from the McGraw-Hill *Encyclopedia of World Drama*.)

New York's Cathedral of St John the Divine was selected by James Keeler and his Classical Theater Ensemble as appropriate for his version of *Everyman* in 1982 that played for several weeks – three nights weekly – in its vast recess. Eleanor Blau reported in the *New York Times*:

> Here the Cathedral's acoustics and size are used to advantage. In a balcony high above the floor of the cavernous edifice, amid the huge trumpet pipes of the organ, an imposing figure in robes, his unamplified voice carrying with surprising volume, declares, "Drowned in sin, they know me not for their God. . . ." In another scene, Everyman, played by Vince Irizarry, beats himself in penance and utters a prayer, "Hear my clamorous complaint, though late it be." His

words are repeated by actors dispersed all over the Cathedral, seen dimly or not at all in the darkness, in balconies and across the 600-foot space. The spectator cannot tell where the sepulchral voices are coming from.

Diction is crucial among the echoes of St John the Divine, whose dwarfing size presents many theatrical problems. Mr Keeler set out not only to solve them but also to exploit them. The production is geared to take advantage of the stark vastness of the place. The players revel in its reverberations and spread out into its space.

"It's the largest cathedral in the world," said Eliott Sroka, the Cathedral's drama director, "and that can be a liability – or a tremendous asset, if you find a work that fits it. *Everyman* fits it splendidly. For one thing, the loneliness and bleakness of the Cathedral suit this anonymous drama, in which Death tells Everyman – who has not been living virtuously – that he must hold his life before God for judgment."

To preserve a feeling of intimacy in much of the play, the audience, whose size is limited to 100 by Equity show-case regulations, sits near most of the action. Seating is on either side of the main playing area, near the end of the Cathedral that is opposite the altar. The altar is thus to the left or right of the audience members down the length of the Cathedral.

Unlit, this long space seems a dark tunnel out of which actors emerge and return. "When Everyman converts, it's a miracle," and out of the blackness, through lighting effects, the altar at the far end slowly appears, as Mr Keeler sees this happening. Amid now-visible 55-foot marble columns, Everyman begins a joyous procession toward the audience, followed by peasants and knights and the like, accompanied by the organ, recorders and taped music of the period.

Keeler explained to Nan Robertson how he read *Everyman*:

It's a story of people – not bad people, but who in living have had to compromise again and again and again. Suddenly Everyman asks them, "Will you go with me to Death?"

The playwright had great insight into how one rationalizes. Everyman's friends try to avoid direct confrontation, they try to avoid saying no. This one says, "Oh, well, that doesn't sound like much fun," and another says, "Well, I have a child to take care of," and so on. These are people who previously have professed love and loyalty. He shows how really easy it is to leave this bond behind. The difficulty is getting out gracefully – how not to face themselves, and not be caught.

The Everyman legend exists in every language. . . . The latter part of it reflects resistance to the oncoming Reformation. The play shows nostalgia for the Medieval simplicity that was synonymous with the Roman Catholic Church. It views this life as a vale of tears, and distrusts its pleasures, which distract from essential values. There's even a brief section that was probably added on ten or fifteen years later, when, in order to perform the play, it would have been necessary to reconcile it with anti-Roman Catholic sentiment. This section includes, for instance, a character who earlier told Everyman to go to confession, saying now that priests are made blind by their sins."

Keeler felt that he had found a way to reconcile this section with the rest of the script. "At the end, when the now virtuous Everyman dies, he ascended Heavenward – sixty-five feet to the dark heights of the Cathedral. Mr Keeler declined to say how this was accomplished, only that the rising figure was 'more than a scarecrow but less than our player.'" (Ms Robertson saw this *Everyman* while it was still in rehearsal.)

The plaza at the Lincoln Center, with its gushing fountain, held an updated *Everyman* in 1989, an adaptation by Jonathan Ringcamp and the well-known actress Geraldine Fitzgerald. This was a revival of an earlier open-air production (1971) of the ageless allegory by the Everyman Street Theater Company of Washington, performed at Lincoln Center during the first year of a Summer Festival there. The public was invited to see the play without charge. The action was brought to the present day, the scene changed to the nation's capital, the main character identified as the owner of a dance hall. Directed and choreographed by Mike Malone, a cast and crew of 100 were involved. Participation by the audience was besought, with Everyman's fate at the end decided by a poll of the onlookers.

Everyman has been telecast in still other free versions, and it has been chosen for treatment by several dance companies. In one broadcast interpretation it was again done in modern dress and had a jazz score by tenor saxophonist Tubby Hayes.

More recently, the Fuentidueña Chapel of the Cloisters was taken over for a more conventional presentation of the script by the Bon Accord Theater Company in 1993 as one of a series of medieval works to which the troupe was dedicated, another put on there being *Mankind* in 1994.

The Royal Shakespeare Company included an *Everyman* in its offering at the Brooklyn Academy of Music in 1998.

Though scholars long assumed that Hrosvitha's six scripts were meant only to be pored over by pious fellow monastics, but never to be acted, by 1970 there had been no fewer than fifty productions of them in the twentieth century, mostly in Germany – Berlin, Heidelberg, Munich, Halberstadt, Göttingen, Hanover, Nuremberg – but, also in Zurich and London, as well as in New York City, Ann Arbor (Michigan) and Bryn Mawr (Pennsylvania), proving their stageworthiness – and one of the plays was broadcast on Danish radio. Most of these stagings had university or church auspices.

Occasionally the theme of the saints' plays has been replicated in the current theatre, as instanced by T.S. Eliot's *Murder in the Cathedral,* Jean Anouilh's *Becket, or the Honour of God,* Robert Bolt's *A Man for All Seasons* (all these dealing with the martyrdoms of more recent saints), and even – it might be ventured – in Shaw's *St Joan,* which proclaims man's need to believe in miracles.

— 2 —
MEDIEVAL FARCES

Only those having a highly simplified view of humankind would suppose medieval people to have been entirely obsessed by holy thoughts and aspirations, dwelling day by day with ingrained piety. Theirs was truly an all-encompassing religious environment; they were ruled by the Church and its arrogantly dogmatic hierarchy. It warned them not to stray lest they suffer eternal damnation, torment that ended in all-consuming Hell-fire. But the average person has another, less compliant side to their nature. Submissive to civil and divine ordinances, they are also possessed by straining animal spirits, illicit sexual drives, anarchic impulses. If anything, this was probably even more so in the long-past centuries, when the world was narrower and the imposed moral codes overly restrictive.

The clergy, themselves professedly celibate, recognized this repressed ferment among their believers. At intervals, given sanctioned pretexts, there were outbreaks of their congregants' primitive energy in which riot was freely allowed, as later depicted in paintings of alcoholic *kermesses* by Brueghel the Elder and Jan Steen of otherwise staid Dutch peasants and burghers as they loosely and even orgiastically disported themselves. Some of this ungirt liberty is clearly detectable in the horseplay and comic skits on the pageant wagons on Corpus Christi – as seen in *The Second Shepherds' Play*, and in independent sketches such as the robust *Mankind* – but it is far more manifest in a ritual, widespread in Europe for centuries, designated as the *Festa Asinaria (*Feast of Asses) or the Feast of Fools.

This wild holiday, celebrated at first by the minor clergy, mostly the sub-deacons, occurred on the 1 January, coinciding with the Feast of Circumcision. Here the outrageous, iconoclastic spirit that had broken through restraints in the Dionysian revels, or *komos*, of Athens, and similarly in the Saturnalia of the Romans – in which servants exchanged places with their masters – reasserted itself, found a cathartic outlet and brought about an extraordinary reversal: sacred ceremonies of the Church were burlesqued by none other than its purportedly dedicated servants, for whom no doubt the daily Latin chantings and rigid, solemn rites had long since grown boring.

The behaviour of the ordained, usually disciplined participants was sacrilegious indeed. Young priests, disguised as women, acted licentiously, to provoke laughter and indecency reigned, much of it within the sanctuary and during the observance of the mass, which was mocked; the choir brayed like donkeys: there might be dice-playing at the altar; instead of incense, foul-smelling shoes were burned; the congregation might join in, carrying the bawdy disturbance out into the streets.

The Feast of Asses was so called because of the prominent role taken by Balaam and his donkey in the Scriptural tales. How this bizarre pair rose to this destructive peak is traced by Gayley in his *Plays of Our Forefathers*, where he describes their creating

> sad havoc with certain sacred festivals in which at first they had played an innocent and even laudable part. Once the donkey thrust his head within the church door, liturgy, festival, and drama were lost in the stupor of his ears or the bathos of his braying. He began, I think, with Balaam and the processing of the Prophets, proceeded with the Magi, and then with the Virgin, who unwarily rode him into Egypt, and ended with Christ himself in the once solemn, nay even triumphant, Palm Sunday approach to Jerusalem. The *Prophets*, the *Flight into Egypt*, and the *Entry*, he turned into festivals of his own, variously denominated, but always feasts of parody, irreverence, frequently of drunkenness and obscenity. Without doubt some of the profanity and pagan practice which characterized these orgies was a survival of pre-Christian rites by which Teutons and Celts, even Romans as well, had been wont to welcome the approach of Spring or propitiate that of Winter; but the favouring occasion in lands and among peoples called Christian was the appearance of a donkey in the church.

Gayley recounts the long history of this misbehaviour through the ensuing ages:

> We are told that under the dissolute Michael III of Constantinople (842–867), one Theophilus, a buffoon of the court, was invested in the robes of the patriarch; and that, attended by twelve roisterers whom he called his metropolitans, clad also in ecclesiastical vestments, he desecrated the sacred vessels of the altar and parodied the holy communion. Then that, mounted on a white ass, on the day of a solemn festival, and with his train, in which the Emperor himself figures, he met the true patriarch at the head of the clergy and by licentious shouts and obscene gestures disordered that procession. This escapade may have been a burlesque of the *processio prophetarum* conducted by the true patriarch, or of the Entry into Jerusalem, or it may have been unpremeditated devilry. At any rate, the ass is in evidence; and also the revulsion against the straitness of religious ceremony.

In France, in the library of the Cathedral of Beauvais, a twelfth-century manuscript records the ritual of this *Festa Asinaria* as it was observed on New Year's Day – the Feast of Circumcision – a document available for perusal as recently as the eighteenth century, after which it vanished. (Another copy of this script exists in the British Museum.) Gayley summarizes its content:

> At the first vespers the *Cantor* intoned in the middle of the nave a hymn of the day of gladness: 'Let no sour-faced person stay within the church; away on this day with envy and heartache, let all be cheerful who would celebrate the feast of the ass.' . . . After lauds all marched from the

cathedral to welcome the ass which stood in waiting at the great door. The door being then shut, each of the canons stood with bottle of wine and glass in hand while the *Cantor* chanted the Processional of Drink, *Conductus ad Poculum*: 'Solemnize, O Christ, the Kalends of January, and as King acknowledged, receive us at thy nuptials'.... One may picture the pause, the beast in his priestly trappings encircled by hilarious celebrants, the popping of corks and gurgling of wine, the toasting of 'my lord, the Ass,' the quaffing of deep draughts. Suddenly the door is thrown open, and up the aisle the procession streams, conducting the Ass with song:

> Out of the regions of the East
> The Ass arrives, most potent beast,
> Piercing our hearts with his pulchritude,
> And for our burdens, well endued.
>
> Hez, Sire Asnes, come sing and say,
> Open your gorgeous mouth and bray:
> You shall have hay, yur fill alway.
> You shall have oats, to boot, to-day!

In all, there are nine stanzas in Latin, here translated and rhymed by Gayley, that conclude:

> Barley grinds he with its beard,
> Feasts on thistles, purple-speared,
> Thrashes in his own back-yard
> Corn from stubble, chewing hard.
>   Hez, Sire Asnes, etc.
> Say Amen, most reverend Ass, [they kneel]
> Now your belly's full of grass.
> Bray Amen, again, and bray;
> Spurn old customs down the way.
>   Hez va! hez va! hez va! hez!
>   Open your beautiful mouth and bray;
>   A bottle o' hay, and the devil to pay,
>   And oats a-plenty for you, to-day.

Donkey-ears were attached to the cowls of the supposed bishops and other high authorities. With "hee-haws" instead of "amens", the besotted clergy and lay congregation responded to the wickedly distorted chant of the mass. The wine, flowing ever more freely, was used to toast the stupid, humble animal. The young priests, venting their resentment against their strict elders, and reversing their own vestments, wearing them inside out, sometimes donned masks. Some even ventured to take on the roles of cardinal and Pope, their scandalous japery sparing nothing. Processions went noisily through the town, the participants dancing drunkenly and shouting wanton songs.

Another version of this dissolute ceremony, believed to have been set down in 1199 or soon thereafter by Pierre de Corbeil, Archbishop of Sens, incorporates certain aspects of the ritual that are borrowed from the Feast of Fools, the mocking revel usually observed during the Christmas and New Year season. Later Corbeil became Bishop Coadjutor of Lincoln. His purpose was to set bounds to the yearly outburst of debauchery that was permitted at Notre Dame in Paris, and afterwards at Sens. He sought to limit the amount of drinking called for by tradition, and apparently favoured keeping the donkey outside the church, thereby reducing the amount of impious ridicule and tomfoolery. He may have succeeded in accomplishing this at Sens, at least.

The Feast of Fools reached its peak in England, where it acquired traits that differed from those of the Feast of Asses. Its date of origin uncertain, it was well established by the end of the twelfth century and survived into the sixteenth. The riotous proceedings might extend over several days: the deacons took possession of it on St Stephen's Day (26 December), and the choirboys on the Feast of the Holy Innocents (29 December), the sub-deacons on the Feast of the Circumcision (1 January, or alternatively 6 or 13 January). On these occasions church services were conducted by the minor clergy, leading horseback processions through the town and collecting gifts or demanding token payments *en route*. The events were preluded by a harsh, irregular jangling of church bells, followed by out-of-tune singing, the donning of odd, inappropriate garments and grotesque masks. A distinctive feature was the election of a Boy Bishop to preside with complete authority over the ceremonies. This "prelate" was chosen by the choirboys and invested with mock solemnity, after which he was surrounded by other boys also impersonating ecclesiastical dignitaries, while perpetrating gross improprieties. Having taken office, the young "Bishop" was in charge of religious services, delivering hilarious sermons, even taking the real local bishop to task for actual or imaginary shortcomings, and ending his own brief tenure with a raucous banquet. All during the celebrations there were enormous suppers of meat and beer, and the tribute – or loot – they exacted was by no means insubstantial. One accounting, for the year 1396 in York, shows that by today's rates about £650 was collected, "along with two or three gold rings, silk purses, and silver spoons to boot", leaving the Boy Bishop £275 richer after all outlays for the plentiful food and drink.

School holidays were unknown in the Middle Ages; the Feast of Fools was the only holiday allowed the young. As Gayley says, for people at large this was "an opportunity to behold the rising generation in the trappings of maturity and dignity, or in the performance of more or less amusing buffoonery. It also afforded an outlet for the play-acting instinct, natural to the young of all species." This being the Christmas season, the Miracle Plays about St Nicholas might be enacted along with the lads' broad or satiric antics, and in some places the boy elected to rule over the diocese might be known as the "Nicholas Bishop" (and also as the "Bishop of Nothing").

Sometimes the lads, putting on girls' attire, boldly invaded convents. It has been suggested that a strain of repressed bisexuality in the clergy – and, indeed, in the feudal bond of knight and squire – found an oblique expression here. Or the "Bishop" might venture to other sees, his whole ribald, turbulent entourage in train, composed of chaplain, steward, shouting choristers.

This strangely inverted event, overtly anti-religious, essentially a protest, persisted in England until the time of Queen Elizabeth I, with conspicuous elements of it emulated not only in Paris, Sens, Toul, Rouen and other places in France, but also in Italy, where there is a record from the thirteenth century of a boy officiating at vespers on St John's Day.

Initially, it seems, these deviant ceremonies and defamatory caricatures had been undertaken with a serious intent, as a well-deserved tribute to youth and particularly to those who obediently served in the church at the altar and in the arduously trained and disciplined choirs. But they had been misjudged by their elders. As always, through a thin, sedate façade there soon bubbled the innate sense of mischief and impishness that eternally inhabits and possesses the rebellious young, and the hierarchy's control of the planned holiday was all too soon lost, its participants quickly out of hand.

Appalled, Church elders began to rail against the impious, impudent antics, especially the celebrants' shocking fondness for excremental humour. High-ranking clerics denounced such conduct as "execrable" and "defilement", an enormity. The attempts to suppress the mock ceremonies were in vain. The minor clergy were defiant, the festivities went on.

Like the welcome bits of fun that sprang up spontaneously in skits on the Corpus Christi pageant-wagons and occasionally, if somewhat incongruously, in the Miracle Plays, these pseudo-rituals, with their uninhibited mockery and broad japery, were to be another source of impetus and inspiration for the rise of secular comedy, a form for which after a long hiatus Europe was once again ready.

To appease those seeking ever more entertainment or else an opportunity to perform before an audience, a desire that burns in not a few breasts, numerous amateur and semi-professional acting troupes had begun to make an appearance. Such groups had no link to the Church and were largely beyond its purview; what is more, they were emboldened by the laxity shown by the clergy towards the Feast of the Asses and the Feast of Fools. They hardly breached decorum as far, yet they had a degree of freedom in the choice of subject-matter. When they pushed beyond a sharply drawn line, as they were for ever tempted to do, they had trouble. Still, to gain hearers they often did venture across that line to exploit the amusing happenings of ordinary life and eventually won broad clerical displeasure, especially when they aimed barbs at the ubiquitous Church itself. Their scripts were also increasingly infused with hints of social criticism of the lay establishment. As early as 1236, seeing the arrival of a stage of this ilk, which pandered to the unwise, unkempt and unlearned, Bishop Grosseteste of Lincoln excommunicated players who took part in indecent performances and forbade Christians to attend them. In contrast, his Italian contemporary, St Thomas Aquinas, was more indulgent, declaring that the profession of actors had a worthy function, the "solace of humanity"; therefore an actor might, by God's dispensation, escape perdition.

The semi-professional troupes were mostly comprised of mummers and wandering *jongleurs*. An actors' guild was organized in England and recognized by Edward IV, who in 1469 put his own household minstrels in control of it, demonstrating a due measure of prudence. Henry VII had four players always at court, led there by one John English. Henry VIII doubled the number, not to be outdone by his father, and Elizabeth increased the palace's resident company's roster to twelve.

The amateurs and wandering mimes were likely to be seen at revels held periodically at court, or by students at universities, or by the folk in scattered towns and villages; these equally rowdy affairs are believed to have been of very ancient origin, going back to agricultural festivals that over the course of time had also borrowed inspiration from impulsive French observances of the same sort. But these were distinctly non-clerical events. They were sometimes referred to as *ludi*, borrowing the Roman term for public merry-making, occurring in a carnival setting and carefree atmosphere. The participants might emulate, to some extent, the flagrantly unconventional behaviour displayed at the Feast of Asses and Feast of Fools. Each such revel or rampage took place under the guidance of a Lord of Misrule, a leader chosen by members of "orders of fools", variously self-named cliques of madcap young men-about-town or restless, fun-seeking academics. In 1348 their excesses led a shocked Bishop Grandison to prohibit "the proceedings of a certain 'sect of malign men' who call themselves the 'Order of Brothelngyham'". As he depicts them, they wear "a monkish habit, choose a lunatic fellow as abbot, set him up in the theatre, blow horns, and for day after day, beset in a great company the streets and places of the city, capturing laity and clergy, and exacting ransom from them 'in lieu of sacrifice'. This they call a *ludus*, but it is sheer rapine."

The Lord of Misrule might also be called the Abbot of Unreason, or the Abbot of Bon Accord (in Aberdeen), or anything from a broad assortment of fanciful titles, while he was projecting and promoting a reign of utter nonsense, to which his "subjects" enthusiastically subscribed.

Mostly the short plays in these *ludi* were composed by the same authors who wrote and staged sketches for the guilds in the Mysteries, Miracles and Moralities. But here the writers enjoyed their freedom from any obligation to deal with the Scriptures and other facets of religion. As put by Frederick Hawkins they held up a mirror to everyday life and character. Henpecked husbands, imperious wives, exasperating mothers-in-law, good-for-nothing monks, lip-valorous soldiers (one of whom, by the way, is seized with abject terror on beholding a scarecrow, which he takes to be an enemy), these and other personages were connected with more or less whimsical adventures, the dialogue being often lit up by a flash of wit, a hit at common foibles or some pleasantry at the expense of the young *procureurs*.

Secular farce of this kind, portraying domestic intrigue, had vanished from view for centuries, since the days of Plautus and Terence. Medieval works in this genre, exposing marital squabbles and lusty, unsanctioned amours, began to appear earliest in France and Germany, where the liveliest examples of it are to be found, though England was also to yield several happy samples.

The forerunner of the *farceur* was the troubadour – in France the *jongleur*, in Germany the *spielman* – the merry bard, an individual performer, who at each start of summer put on his multicoloured apparel, slung a harp or species of three-stringed violin over his shoulder, then attached a dangling purse to his belt and set out on a gaily-bedecked mule for his annual journey, from town to town, lonely castle to lonely castle, singing to earn his keep, a filling supper and, belatedly, a bed. Gradually he added to his repertoire, telling jokes and reciting anecdotes and stories to round out his offering. His lyrics became sharper, his jests and stories edged with satire, his tone critical of those who were the heartless masters of the state and society. Especially in towns, in the marketplace, this change of emphasis was well received by the populace.

One of the most notable figures in the French literature of this period is Rutebeuf (*c*. 1230–*c*. 1285). The details of his life are mostly vague – even his first name is missing. This much is known: he was of humble birth and uneducated, always poor, enjoyed little recognition or success, had a plain, elderly wife. But he ranked above other *trouvères* in that he possessed a horse instead of a lowly mule. As an entertainer he was much favoured by the feudal barons in whose draughty stone halls he sang and jested. Apparently he had a good voice and an innate skill at rhyming. He also wrote plays, one of which, *Le Miracle de Théophile*, still holds interest. Devoted to exaltation of the Virgin Mary, it tells how a priest sells his soul to a rapacious Devil to gain worldly honours, repents and is saved from a fiery ending by Mary's merciful intercession. In 1913 students in Harvard's 47 Workshop, taught by Professor George Pierce Baker, staged the piece, and twenty years later in Paris Gustave Cohen, forming an experimental theatre group, elected to name his troupe the Théophiliens and launched his venture with Rutebeuf's little drama.

In a different vein is *Dit de l'Erberi* in which a quack physician boastfully reveals all the worthless medications he had induced the Sultan of Egypt to swallow and ingest, the charlatan a portrait drawn by Rutebeuf in a vinegary spirit reminiscent of Menander's pictures of unworthy types in his world, and anticipating Molière's scathing images of fawning, dishonest doctors encountered in his day, and at whose hands he personally suffered.

The *jongleurs* and proto-*farceurs* grew so boldly outspoken that as usual the Church moved to censor them, especially since the clergy were not spared in the acidulous ballads and skits. A Spanish ruler, Alfonso X of Castile (1252–84), was petitioned by a troubadour at his court, Guiraut Riquier, for a precise edict defining what was permissible and what was forbidden, so as to separate "noble entertainers" from those of a baser, wanton, decidedly vulgar sort. The king obliged Riquier and even couched his views in verse, though it is suspected that Riquier himself composed the felicitously worded ruling. Unfortunately, Alfonso would countenance only sanctimonious works, those respectful of religious faith. Despite such censorship, the minstrels, court poets and actors thrived as public favourites, and at some courts were tolerated and even generously welcomed.

At the start of the fifteenth century, in France, several non-clerical "mirthful fellowships" – *Sociétés Joyeuses* – were formed, their members attracted to the idea of extending the revelry of the cherished Feast of Fools. These were the groups shortly emulated by the "orders of fools" in England. The most inspired and active were L'Infanterie Dijonnaise, who celebrated while wearing fools' caps in place of cowls, such as the clergy might have worn when rollicking, the Connards of Rouen and the Enfants sans Souci (Carefree Children) of Paris. For a similar purpose the law clerks of Paris had already founded a guild, the Basoche (1303), partly to be of practical help to one another but also to have good times together, which included putting on buffoonish plays. As already remarked the scripts, sometimes called *soties*, or "foolish pieces", might be provided by the writers who turned out the Mysteries and Miracles. In fact these simple skits, mostly plotless, often served as curtain-raisers to those more sober and moralistic offerings. In the *soties* (variously spelled "*sotties*") the performers – mostly amateurs, with perhaps a professional or two – might be attired in the by then traditional fool's costume, wearing caps with long ears (the donkey's ears), and very likely wielding a bladder on a stick, much as did the clowns in the Mysteries and Moralities, and in particular the "vice", the Devil's helpmate, who added comic relief to the didactic dramas. With this bladder the actor might excite laughter by harmlessly beating a spectator or two, as well as his antagonist in the skit.

Doubtless the farce of the *Moyen Age* germinated here, gradually shaping plots and structure, however slight, based on folklore and incidents in the daily round of life. Short, very light, but persistently vital, these often crude, exuberant works reflect and mix the whimsy of the learned and more sophisticated – that is, literate – citizens of the medieval social order together with the bitter and resentful humour of those toiling and harshly deprived. Most of these little plays had satiric content, ridiculing the Church and the pompous and often predatory figures abounding in the dominant professions – the *avocat*, with his wily speech; the Latin-spouting, bespectacled notary; the apothecary; the physician, ready with his leech for blood-letting, a panacea; and the avaricious money-lender and other constituents of the mercantile class. Still others were concerned with the noisier and rowdier aspects of domesticity, the fractious relationships between man and wife, and through many of them runs a vein of peasant shrewdness and cunning, a delight in trickery.

An exception, a very special one, to the rough French medieval farce is the output of the crabbed hunchback of Arras, Adam de la Hale (or Halle, *c.* 1240–88), who composed a handful of light fantasies. His name is the first recorded in the history of his nation's poet-playwrights who espoused comedy. Supposedly he intended his works to be only for the delectation of his friends, but they won a far wider and lasting audience. In *Jeu de la Feuillée* (1262) he introduced a prototypical Harlequin character, Croquesot, grotesque with a "shaggy, big-mouthed devil's mask". As bells tinkle and a horde of his fellow "herlekins" rush past, he enters and abruptly asks the spectators: "Doesn't it suit me well, this mask, this devil's mug?" The clergy held that such figures were "damned souls", demonic, "sons of Satan", closely akin to werewolves, though within a few hundred years they were to be refined into the merely mischievous Arlecchino of the *commedia dell'arte*. Scholars, quoted by Berthold, conjecture that Croquesot was garbed in "a bright-red cowled cloak", the usual attire of devils and "herlekins".

The play is a free-flowing medley of fairytale motifs, other popular superstitions and cult-lore, wittily and imaginatively commingled, the incidents filled with topical allusions – some obvious, some subtle and many stinging – that the knowing and offended citizens of Arras could readily catch. The supernatural elements were acceptable to them because they were beliefs that they themselves largely shared. *Jeu de la Feuillée* is credited with being the oldest French secular drama, and – in Berthold's words – a script possessed of "satire . . . rudeness and charm, malice and magic spells". It also reveals its author's lifelong passion for seeing justice ever rightly dispensed.

The work had an angry reception in Arras. Young Adam de la Hale left that city, perhaps by compulsion, and enrolled at the university in Paris. He was interested in theology, though he later gave up his study of it. He reverted to writing theatrical pieces, his *Play of the List* (c. 1264) a clear example of a farce. In a later script, *Lijas Adam*, also known as the *Play of Adam* or the *Play of the Greenwood*, he himself appears as a character, relating how he first meant to become a priest, only to change his mind after falling in love with "sweet Marie". He has since come to regret his decision, because he married Marie, and, alas, she no longer seems fair to him, but "pale and yellow . . . My hunger for her is satisfied." The play, strange in other ways, introduces the author's father, too, and an odd variety of irrelevant figures: a madman, a mendicant monk seeking alms in return for miracles and a band of singing fairies.

Two decades later, his notable *Robin and Marion* (1283) is truly lyrical. Written for the entertainment of guests at the court of Robert II, Count of Artois, it anticipates the pastoral drama that was to be favoured by the élite during the Italian and English Renaissance. A pretty shepherdess is urgently wooed by a knight over the objections of her lover, a stout-hearted shepherd. After the amorous conflict is resolved there are dances and games. Often referred to as the first comic opera or operetta, the claim overlooks a similar and apt use of music and dance by the Greeks and Romans. Yet nothing of its kind had been seen in France up to that day. With his original cast of mind and odd fancy, Adam de la Hale is something of a thirteenth-century Offenbach or Gilbert and Sullivan, a startling forerunner.

More typical of medieval exercises of robust humour – and far more famed – is the *Farce of the Worthy Master Pierre Patelin* (c. 1470), author unknown, as are most of the composers of these short plays. Feigning madness, a poor but dishonest lawyer evades paying his debt to a merchant. Then he is hired to defend a shepherd brought to trial for stealing from a neighbour's flock. The chicane lawyer advises him to reply to all questions with a simple "baa!", as though too long association with sheep has reduced him to their limited brain-level. In court this trick works very well. The guilty man is acquitted, but when the lawyer asks for his fee his request is met with a similar series of "baas!", and he is unable to collect any payment. He has met his match.... It is a neat, amusing instance of the cheater being cheated, as well as of a simple man – a peasant – outwitting his putative masters.

(The joke here has the same basic pattern as one in the Talmud, where a scheming Egyptian asks a Jew to teach him the Hebrew alphabet, offering to pay on every letter he masters. "The first is *Aleph*," says the Jew, holding out his hand. But the Egyptian demands, offering no coin, "How do I know it's *Aleph*?" The Jew, taken aback, submissively continues the lesson: "The second letter is *Beth*." The Egyptian, again refusing payment, has the same reply: "How do I know it's *Beth*?" Now furious, the Jew grabs the man's beard and yanks it. "Let go of my beard," cries the Egyptian in pain. "How do I *know* it's your beard?" the Jew asks.)

Another farcical turnabout of this sort is found in *The Man Who Married a Dumb Wife*, which Rabelais (1483–1533) in his immortal, scabrous novel *Gargantua and Pantagruel* tells of having seen performed by a troupe in the marketplace. An old man chooses as his wife a beautiful young woman who has only one lack: she is mute. Finally, frustrated by her perpetual silence, the husband pays a high fee to a surgeon, who renders her able to speak, but she proves so talkative, fatuous and nagging that her distracted and disenchanted husband engages the surgeon to deafen him. A balance reached, the marriage is truly happy. After reading an account of the play in Rabelais, the eminent twentieth-century satirist Anatole France undertook to return it to the stage, and his one-act version is now widely enacted. It has also served as a libretto for a ballet, the setting changed to Spain, the dancers effectively carrying on their concluding conversation by means of clacking castanets.

Into the popular fancy, and rapidly embedding itself in folklore, came a strange creature, two-thirds donkey (the body) and one-third human (the head), a figure possibly seeded in the imagination by the ubiquitous and intrusive presence of Balaam and his cantankerous steed. The pranks and mischievous adventures of this semi-beast, semi-man were soon the subject of hundreds of short verses by legions of unknown fourteenth-century authors. Some of the doggerel pieces were set to music and became fodder for storytellers and street-singers of the day. A collection of 6,000 of the catches is preserved in the Bibliothèque Nationale in Paris. Eventually a few of the poems were linked and took shape as a dramatic ballad, *Le Roman de Fauvel*. "Fauvel" is a name coined by taking the first letter of six of the Seven Deadly Sins: Falsehood, Avarice, Unction, Villainy, Envy and Lechery, all traits exhibited by the trouble-making focal character. The manuscript was discovered in the early 1950s; in 1970 the touring Waverly Consort added it to its repertory. Later, a competing production of it was embraced by the Ensemble for Early Music. In 1990 both productions were on view in New York barely a season apart.

This evoked anger between rival troupes and charges of trespass. The critics, however, looked closely and declared that the two versions were not made up of the same poems and, in fact, were substantially different.

For the most part, royal indulgence gave the French *farceurs* a good deal of liberty; they were even encouraged to be unsparing in their criticism of the regime. The Basoche were allowed to use the hall of the Palais de Justice as a theatre when the weather was inclement; what they did was adapt as a stage a great marble table on which state banquets had formerly been served. The troupe, and also their rivals the Enfants sans Souci, took full advantage of this privilege. Louis XII declared: "My courtiers never tell me the unvarnished truth, and as long as the truth is withheld from me I cannot know how the kingdom is governed. The troupes of the Roi de la Basoche and the Prince des Sots have my authority to expose any abuse they may discover, whether at court or in the town, and to ridicule whom they please. I do not wish to be exempt from their attacks; but if they say a word against the Queen, I will hang them all." He did become a butt of their sharp wit because of his notorious avarice.

On one occasion, somewhat earlier (1486), the eager players had gone too far, likening the young King Charles VIII to "a clear spring muddied by his courtiers, so that they could better fish in troubled waters". This insult raised the hackles of the ruler's councillors, who ordered the script's author and producer, Henri Baude, and the entire cast to be clapped in prison. Soon after, given a discreet hearing in Parliament, they were judged blameless and quietly released. But the incident was a warning: there were always some limits to what they dared to say and do.

Louis XII, truly tolerant, was especially pleased with the offerings of Pierre Gringoire, a member of the Enfants sans Souci and a skilful creator of Mysteries and *soties*. At a moment when the King of France and Pope Julius II were at odds, their feud profoundly serious, Gringoire enlisted his talents and his fellow players on the side of their sovereign with a satiric piece, *Jeu du Prince des Sots et de Mère Sotte*, which he and the troupe mounted at Les Halles in Paris on Shrove Tuesday 1511. Frederick Hawkins, in his *Annals of the French Stage*, synopsizes this madcap and brazenly polemic work, to the staging of which "the highest and the lowest" were invited.

> The author, dressed in the petticoats of Holy Church, disported himself for an hour or two as the warlike Pontiff, who was represented as disguising unbounded hypocrisy and libertinism under the cloak of religion, as seeking to increase his temporal power at the expense of the French, and as obtaining support among bishops and *abbés* by offering them rich benefices and other bribes. At the end, King Louis begins to suspect that his Holiness is not the Church – a suspicion not limited to Paris at this time – and in point of fact is only a sort of *Mère-Sotte* (Fools' Mother). The *Jeu* was followed by a *Moralité* and a *Sottise à Huit Personnages*, in the

former of which the Pope appears as an obstinate and confessedly immoral person. One of the characters was filled by a clever hunchback, Jean du Pontalais, now chiefly remembered in connexion with a piece of amusing boldness.

(To give notice of a production, he had played his drum loudly in a public square that was adjacent to one in which a priest was delivering an outdoor sermon, for which he was rebuked by the indignant cleric: "It is like your impudence to make your announcements while I am preaching." To this, the irrepressible Pontalais retorted: "It is like your impudence to preach while I am making my announcements." The exchange was reported to the magistrates, and Pontalais was imprisoned for six months, a period during which he could learn to improve his scandalous manners.)

As propaganda against the Papacy, Gringoire's play was a marked success. The Parliament, though not wholly sympathetic to the king's hostile policy towards Rome, was moved to contribute to the cost of the openly biased entertainment concocted by Gringoire and the troupe. (Idealized, Pierre Gringoire lives on as a character in Victor Hugo's historical novel *The Hunchback of Notre Dame*.) Henceforth neither the Basoche nor the Enfants sans Souci were to put on works that contained mention of royalty or other important persons. The companies, in their theatres – the Basoche in the Palais de Justice, the Enfants in the Hôpital de la Trinité – avoided politics for a time and were more popular than ever.

The Enfants' fortunes were enhanced by the addition to the company of Jean Serre, who surpassed in the stock roles of a waggish rogue and a drunkard.

Before long, players in the Basoche were in trouble again. They returned to testing the ban against politics and other matters deemed sacrosanct, their irreverence eliciting a lengthy series of parliamentary decrees aimed at them. They conceived an ingenious means of going around the prohibition against representing real persons on the stage by donning masks that transformed them into ugly caricatures of certain obnoxious public figures. Audiences, recognizing the troupes' objects of ridicule, were convulsed with laughter. For this effrontery, in 1536 the actors were threatened with prison sentences and banishment; their licence was withdrawn. Two years later they were back, but soon offended the authorities so gravely that they were in peril of the gallows (1540). They escaped that fate and continued to be in and out of royal and parliamentary favour, their better luck often due to the influence of Marguérite de Valois, who herself had a fondness for writing pastorals and comedies that she had enacted by the ladies of her court.

In 1539 the Enfants had to find new quarters, the Hôpital de la Trinité being rededicated to its original charitable purpose. The company moved to a Hôtel de Flandres, where it was established for a mere four years, after which the house was razed by the king's order. For a time homeless, the actors finally purchased space in the Hôtel de Bourgogne. Seeking to renew their licence, they met with parliamentary opposition, and even intercession by the Court on their behalf did not avail. The space in the Hôtel de Bourgogne had been occupied by the Confrères of the Passion who were also confronted by a hostile Parliament. The Church felt it was time to put an end to all forms of religious drama; indeed, after the staging of such pieces as Gringoire's heretical *Jeu du Prince des Sots et de Mère Sotte*, it frowned on any kind of theatre. The Parliament hearkened to the Church but finally relented, though merely to the extent of permitting only farces to be produced. The members of the Confrères had no interest in devoting their talents to such raffish and even vulgar fare, and so sold their theatre to the Enfants, and then dispersed, bringing to a stop in France a centuries-old tradition of a theatre that was primarily in service to Christianity and its teaching. By 1548 there were no more Mysteries and Moralities. Farce had outlived the liturgical drama. In time, too, the Enfants outlasted the Basoche. But the Church's dislike of the theatre expressed itself in other ways: actors were now religious outlaws; unless they renounced their profession they could not receive Holy Communion or be given Christian burial.

These interdictions did not dissuade the players; audiences still flocked to their jolly offerings. What is more, farces were easy to stage, requiring little or no scenery, very few if any props, a platform with openings at side or back for entrances and exits. All that counted was a comic premise, exaggerated characterization, witty dialogue. Someone was duped by a rogue, but the tables would be turned, with a happy ending. Mistaken identity might be a factor. Emphasis was placed on masks and costumes. Berthold: "The carefully combed beard of the pompous philistine, the lawyer striking attitudes in wig and gown, the bold headdress of the cocotte, the dandy attire of the courtier, and the bell cap of the fool. . . ." At a glance the spectator could recognize the persons in the play and from the familiar garb know what habitual behaviour to expect from the characters. Berthold draws a faint line or distinction between the farce and the *sotie*: "The heroes of the farce are fools in common or court dress – the heroes of the *sotie*, common folk and courtiers in fools' dress." But mostly the terms "farce" and "*sotie*" are used interchangeably.

Throughout the German-speaking countries, from the Baltic south to Bavaria, the Tyrol, Austria and other Alpine regions, pre-Lenten revels much like those observed in England and France were annual happenings. Freed from long seasons of gripping winds and ice, people broke loose from their Catholic moral restraints. Indulging in wild, nocturnal outdoor masquerades, they wore grotesque disguises to appear like demons or fools; some also dressed in animal skins, such primitive, atavistic garb seeming to permit their lewd misbehaviour. Accompanying these revels were playlets that gradually evolved into indigenous kinds of comedy.

Near the mid thirteenth century (1230) a Bavarian knight, Neidhart von Reuenthal, who was also a *Minnesinger*, had a falling out with his feudal lord, Duke Otto II. Taking refuge in Austria, he expressed his long-brewing dissatisfaction with the overly fixed textual conventions of the *Minnesang* and joined those who made use of *höfische Dorfpoesie*, or "courtly village poetry", a peasant speech infused with words and phrases frequently spoken at court. Increasingly, this blended language took over the dialogue in German farce.

The mixture of elevated court humour and its wording and coarser peasant jocularity and argot is epitomized in an early Neidhart play that dramatizes a pretty, local, annual ceremony called "violet picking". The gallant Knight of Reuenthal has a vow from the Duchess of Austria that she will choose him as her "May Lover" if he helps her find the first violet. Preceded by pipers in a festive procession, the knight leads the duchess through fragrant fields along the banks of the Danube. He has already located the precious flower and covered it with his hat. But peasants, offended by mockery of them in his verses, have played a scurrilous trick on him. When he lifts his hat for the noble lady's pleasure, what is revealed is not a sweet-scented violet but instead an evil-smelling, faecal mess. At the end, however, all take part in a round dance, and the episode winds up in a full holiday spirit.

The first fragmentary script for this playlet is preserved in the Benedictine monastery of St Paul in Carinthia and is thought to date from about 1350. Scholars believe that it was to be recited by two minstrels who took all the roles. It is theatrical to the extent that it has an implicit conflict and an elementary plot-turn. Despite its disgusting joke, its dialogue is still mostly courtly.

In later Tyrolean versions, in the fifteenth century, the play is greatly expanded, its setting transferred to a city. As many as 103 actors participate in the added scenes, their colourful costumes identifying their trades and high or low station in life. Each has a personal quirk or obsession. The action is loud and broad, the humour in the new incidents often earthy and exceedingly physical: a dance by peasants on wooden legs, an ear-splitting scuffle between some harridans and an angry innkeeper. The knights and their ladies still speak and dress elegantly. The discovery that the violet has been replaced lets loose an infernal row, and the stage is invaded by argumentative devils.

Such pre-Lenten entertainments, as they grew more sophisticated, came to be known as *Fastnachtspiel*, or Shrovetide Plays. They might be given outdoors on an improvised stage or within a large enough open area of a house or inn. If a platform was desired, flat boards laid atop a row of stout barrels sufficed, with perhaps a wall at the back and ready means of access and departure for the amateur actors. The contrasting and antagonistic characters might be an emperor and prelate, or disputatious abbot; or two rival knights; or a Jew and a zealous priest; a physician and a sceptical patient; a mother superior and a depraved procuress; neighbouring peasants quarrelling over the borders of their lands; a plaintiff and a defendant hotly engaged in a lawsuit, both of them devious and lacking scruples. As elsewhere, the brief, slapdash playlets give lively pictures of the times.

Foreign influences helped to shape the German farce. As early as the tenth century, as remarked, the nun Hrosvitha von Gandersheim had adapted the comedies of Terence (after purging them of what she conceived to be their pagan errors). Her versions, discovered in 1493, reawoke great interest in the Latin originals and their author. A train of studies of Terence's writing by erudite commentators ensued. Terence's well-structured works, therefore, were accessible to serve as models.

In 1480 a "neo-Latin" comedy, *Stylpho*, by Jakob Wimpheling, an Alsatian, was performed at Heidelberg as part of a graduation ceremony. It is about two students, Vicentius, attracted to the new movement called Humanism and striving to embrace its principles, and Stylpho, a shallow opportunist. While Vicentius rises to become a bishop, the foolish Stylpho tries too hard to attain a rank for which he has not adequately prepared himself, falls from an important post in the Roman curia, and ends as a pig-farmer. His chief fault is his poor grasp of Latin, the language of the Humanist. The script has many echoes of Terence, with whose works Wimpheling was obviously already familiar.

Another handy model was the French piece, the *Farce of the Worthy Maître Patelin*, which was much admired in Germany. In 1498 Johannes Reuchlin published *Henno*, its plot closely parallel to that of *Patelin*. A difference is that the thief is Dromo – his name taken from Latin comedy, as is that of Petrucius, the chicane lawyer. Dromo, not a shepherd but a servant of Henno, is the debauched husband of Elsa, who is tricked out of her illicitly gained wealth by her conscienceless spouse. At Henno's instruction Dromo buys filched goods from Danista, a moneylender, with a promise to pay later; he sells the merchandise elsewhere and pockets the proceeds. He later denies having got the cloth from Danista or any money from Henno. Taken to court, his repeated response to all questions is *Ble!*, as suggested by Petrucius, which leads to an acquittal. (He does not say *Baa!*, since he has not been associating daily with sheep.) Thus a set of dishonest rogues – Elsa, Henno, Danista, Petrucius – are cheated, just as they have cheated others, and this act of justice is accomplished by Dromo, himself no paragon of virtue.

In the ensuing twenty-five years *Henno* had thirty-one printings, proof of its wide popularity far beyond the confines of the university. The odd mixture of German and Latin names for his characters is evidence of Reuchlin's acquaintance with the scripts of Plautus and Terence, in addition to his awareness of the French farce from which he freely borrows his nicely constructed plot. The several neat and ironic reversals of fortune in it are an advance for German writers of comedy over what had hitherto been their elementary linear dramaturgy. (Richard Erich Schade's *Studies in Early German Comedy, 1500–1650* offers an erudite survey of the theatre of this historic period.) It is doubtless relevant that Reuchlin had twice visited Rome.

A lesser, cruder variety of *Fastnachtspiel* was fashioned – if not initiated – by Hans Rosenplüt and referred to as *Schwank*. Then as now few writers could earn a living solely by their literary efforts: Rosenplüt was a brazier and gunsmith, practising his trade in Nuremberg. The *Schwank* sketches were grotesque anecdotes in slipshod verse and were mostly involved with exploiting the old device of mistaken identities – accidental or purposeful – and the amusing complications that follow exposure.

Better scripts were those of Hans Folz, a native of Worms, whose other crafts were those of master surgeon and barber. Settling in Nuremberg (1478), he gained a reputation as a reliable and respectable author of *Fastnachtspiel*, winning in seven years approval that prompted the town councillors to issue a document in 1486 – "signed and sealed" – declaring that "Master Hans, the barber, and the rest of his kin" could stage a suitable Shrovetide Play of his own composition, the only proviso being that it be "seemly" and that he be given no financial recompense for it.

Folz, called the *Schnepperer* by his friends and fellow craftsmen, instilled political commentary in his plays, where he spoke out on behalf of the citizens of the Free City in their quarrel with the Margrave of Brandenburg and their struggle against the knights who represented a by now onerous and decadent feudal order. His transparently allegorical *Des Turken Vasnachtspil* contrasts the corruption to be seen about him to a semi-imagined, ideal Oriental country in which the sun rises each day on a landscape where all is blissful. This is Grand Turkey, which a herald proclaims to be a place "where no one has to pay interest". The ruler of this idyllic land, the Grand Turk, pays a visit to Nuremburg, an event that proves to be turbulent, with his Christian hosts disgracing themselves and mistreating their exalted Muslim guest. There are hostile exchanges between papal delegates, imperial envoys and knights. The Grand Turk berates the Christians for their flagrant "arrogance, usury and adultery". They, in turn, threaten to shave off his beard with their sickles and to lave his face with vinegar. Only the intervention of two responsible Nuremberg burghers eases the troubled situation. Departing, the Grand Turk bestows his thanks and blessing on them. His herald declares that the royal procession will move on to somewhere more hospitable. This is a typical ending for a Shrovetide Play, very likely derived from the custom of the Corpus Christi pageants where a sketch would break off abruptly with an announcement that it would be continued at the next "station" as the wagon-train wended its way through town.

In particular, *Fastnachtspiel* flourished in Nuremburg, a centre of prosperous commerce and progressive thought. Its town councillors, however, kept a watch on all forms of entertainment, in the interest of "propriety and public order". One such happening that might become unruly was the *Schembartlauf*, or *Schönbartlauf*, a masquerade annually sponsored by the local guilds, beginning in 1449 during the Shrovetide season, and sometimes offering serious competition to the Plays. The participants, mostly apprentices, were apt to wear bearded masks, which accounts for the name. After 1516, to mitigate excesses, the *Schembart* was officially licensed to last two days only, "lest it discredit itself". Elements of the masquerade linger even today in the folk traditions of Bavaria, the Tyrol and Austria, and there is also a nostalgic reference to it in Goethe's *Faust*.

Unexpectedly, a frequent subject in the *Fastnachtspiel* is King Arthur's court and the brave, chivalric deeds of its knights, the lore of that fabled contingent having been spread throughout Germany and Switzerland by British and Breton minstrels during the eleventh and twelfth centuries.

In the north of Germany, in Lübeck, the Shrovetide Plays were far more circumspect. Presented by clubs, the *Zirkelgesellschaften*, whose members were recruited from the city's patricians, the dialogue avoided any hint of indecencies. The plays were given on a scaffold mounted on a wagon, much as were the Moralities, which eventually they came to resemble.

The pre-eminent German dramatist in the Middle Ages is a cobbler, Hans Sachs (1494–1576) of Nuremberg. Prodigiously energetic, while pursuing his trade to sustain himself, he completed reams of epigrammatic poetry. A *Meistersinger*, he composed 4,000 songs, along with plays to the number of 200. He himself methodically classified his stage-works, listing sixty-three tragedies, sixty-five comedies and sixty-four *Fastnachtspiels* – these last the scripts for which he is best remembered and most applauded. He was a moralist – he wrote Moralities – and evangelistic and didactic. The usual bawdiness is absent from his Shrovetide works – he had in mind that they were to be presented in the pre-Lenten season, a time when his hearers should be preparing themselves for a spiritual state of being.

Travelling widely, he learned much about the world and gathered helpful ideas about theatrical production. For his plays he drew on the Bible and themes from classical antiquity, along with historical chronicles and popular stories, adapting them for staging. He formed a company of amateur players, who in 1550 occupied Nuremberg's disused Marthakirche, which he converted to his use, erecting a simple platform – or trestle – in the nave, and adding a backdrop and side curtains. Stairs on either flank gave access to the bare performance space.

The actors wore contemporary dress, save when portraying Turks or other infidels, or angels, or devils, whose attire could be altogether fantastic. Their skills were by no means rudimentary, for some would be innately gifted. The stage conventions then prevalent were respected: a person in a seated posture was taken to be sleeping, anyone lying prone was considered to have expired, and so forth. The plays were in what is called *Knittelvers* (rhyming four-beat doggerel) and were put on twice weekly in the span between Twelfth Night and Lent.

Among his lengthy list of serious works, most of them no longer held in high regard, though in the past they were often greatly valued, are *The Tyrant Herod* (1552), *The Tragedy of King Saul* (1557) and – dramas rather than tragedies – *Siegfried with the Horny Skin* (1537) and *The Unequal Children of Eve* (1553). As an author of farces, both *Schwank* and *Fastnachtspiel*, his name retains its lustre. Some of his prized scripts are *The Land of Milk and Honey* (1530), *The Cutting Down of Fools* (1543), *The Pregnant Peasant* (1544), *The Devil and the Old Woman* (1545), *The Wandering Scholar from Paradise* (1550) and *St Peter with the Goat* (1555). He composed an original version of *Alcestis* (1555).

His creative impulse was to attack personal evil and preach individual virtue. Several of his lighter plays deal with some form of exorcism of the Seven Deadly Sins – even physically, by surgery. Thus in *Das Narren Scheyden* (*To Cut Fools*; 1557) a sick and stupid patient appeals to a doctor to relieve him of severe constipation. He has already tried a diet of plums, for their emetic qualities, but to no lasting avail. His illness apparently has a deeper cause than at first supposed. After various diagnoses, the doctor prescribes an operation, to which the fearful patient objects, whereupon he is ordered to drink a sample of his own urine. This he meekly does, but again to no avail. He is told to view himself in a mirror; in doing so he must recognize and acknowledge what a fool he is. At last he surrenders to the doctor's prescription and affrightedly goes under the knife. After a deep incision, the physician pulls out the sources of the fool's disease – the

Seven Mortal Sins that have been pervasively festering within him, represented by suitably garbed living figures. With these "extracted", the patient is healed. The infectious Sins, in a ship-like nest that had been lodged in his gut, are sent floating down the river that flows through Nuremberg. Schade interprets this as an allegory applicable to Sachs's native city, which might similarly be cleansed of its avid materialism and growing corruption. "The cure is portrayed in a *Fastnachtspiel* by a writer known to conceive of himself as a selfless wiseman conveying essential moral truths by means of his works."

Considerably more amusing is *The Wandering Scholar from Paradise*, telling of an unscrupulous young man who prevails upon a stupid but kindly peasant woman to pile him high with gifts for her late husband, whom he claims to have seen recently in the cloudy realm, and to which the conniving visitor asserts he himself shall directly return. The wandering spirit of the dead man is in urgent straits, is barefoot, has no breeches, has only a blue hat and the winding-sheet in which he was laid to rest. He is dependent on alms from other inhabitants of Heaven. The woman's second husband, always ill-tempered, is even more angry when he learns of her gullibility. He pursues and overtakes the miscreant, but the quick-thinking, smooth-talking scholar actually succeeds in stealing the man's horse. To cover up his own credulity, the second husband tells his wife that on a generous impulse he gave the horse to the scholar as an added present. He wants her to keep the affair a secret, but alas she has already told the whole parish, and the pair become the laughing-stock of their neighbours.

In *The Hot Iron* (1551) a husband and wife make a trial of each other's fidelity. Unfortunately, neither can honestly pass the test, but he gets by with a deception. Vowing that he is faithful, he "proves" it by pressing a hot iron to his hand, in which he has first hidden a piece of wood in his palm. The wife, not as clever, confesses to at least seven adulteries, one with no less a seducer than their chaplain. At the end, profiting by his ruse, the husband's wholly undeserved moral authority over his wife is established, and she becomes utterly submissive to his rule.

Three years later, in *The Horse Thief*, Sachs tells of another rascal who outwits some angry villagers who have planned to hang him. He gulls these corn-farmers mercilessly, and leaves them all at odds and furiously shouting at one another, as he races away.

In his later years Sachs involved himself wholeheartedly and vociferously in the religious conflict initiated by the obdurate Martin Luther, with whom he took sides against what he agreed was the tyranny and decadence of the Papacy then enthroned in Rome. Engaged in the bitter and risky polemics of the struggling Reformation, his creative energies were diverted: the drama was no longer his foremost concern. He wrote poems in praise of the eloquent Luther, whom he hailed as "The Wittenberg Nightingale", a term that clung to the rebellious theologian.

In succeeding centuries, during the Renaissance and baroque eras, Sachs and his offerings largely passed into oblivion, but in the nineteenth century he and they were discovered by Goethe and the Romantics, who greatly admired him and enjoyed his farces. Like Pierre Gringoire, too, he has gained further immortality by being inserted as a character in Richard Wagner's comic opera, *Die Meistersinger von Nürnberg*, where the famous cobbler has a major and controlling role, if also more than a little idealized. In a scene that is a highlight, he provides disdainful criticism of the fatuous Beckmesser's rehearsal performance by hammering on the soles of the shoes that he is shaping; and, finally, he himself is unexpectedly crowned king of the festival, proclaimed to be the one most worthy of the honours bestowed by his fellow master singers.

The medieval German farce hardly flourished in the period after Sachs. The religious strife with its sharp debates and angry passions that overtook the country soon resulted in a climate that discouraged light popular theatre, though Luther himself was not wholly against the stage, saying: "Playing comedies should not be forbidden, but for the sake of the boys at school allowed and suffered. First, because it is good practice for them in the Latin language; second, because in comedies there are artfully invented, described and represented such persons as instruct people and remind each of his station and office, admonishing what is fit for a servant, a master, a young fellow, and an old man, and what he should do. Verily, they make plain and evident as in a mirror all dignitaries' rank, office, and due, and how each should behave and conduct his outward life in his station." In his remarks, however, a salient word is "suffered"; obviously he had in mind using even comedy for pedagogic reasons – the plays to be in Latin – and to inculcate ideas of public discipline and decorum, not the usual goals of earthy farce.

In the early Renaissance the names of a few comic writers stand out, but for almost two centuries the Germans were to look elsewhere, principally to England and Italy, for their stage-fare, mostly provided by touring foreign troupes, who brought works by their own authors.

The Shrovetide Plays had clones of a sort in the hearty Dutch *vastenavondgrappen*, given on carts during the pre-Lenten season of revels. In 1413 these were described by municipal officials of the town of Dendermonde in the local tradition of offering "amusement" for the young. They were rough-and-ready sketches, featuring mocking mummery and the staple laughable consequences of mistaken identity. The pranks with which they were replete were like those in the Corpus Christi vignettes.

Earlier in origin, and destined to have a sturdier and longer life among the Dutch, were the *Sotterniëen* (the name derived from the *sotie* of the neighbouring French) – or, as they were equally well-known, *Klucht* plays, short burlesques attached as after-pieces to the *Abelespele*, indigenous scholarly dramas, a genre that took form in Brabant around 1350. The purpose of the *Klucht* skits, like that of the Ancient Greek satyr plays, was to engender comic relief after the emotionally draining serious work immediately preceding it. They usually did this by parodying the plot and characters of the sober *Abelespele*, at the conclusion of which spectators were invited to stay for the irreverent comic take-off.

These short farces were produced by the Rederijkers, societies like the *Meistersingers* in Germany, formed by the guilds throughout the Netherlands, beginning in the fifteenth century.

(Again their name has a French source, *rhétoriqueur*, fitting because they also adapted and extended to far greater lengths the theological and ethical debates in semi-abstract Morality Plays.)

The *Klucht* plays were notably robust and colourful, reflecting daily life much as depicted in paintings by the elder Brueghel and Teniers, with rustic folk eating and drinking well beyond satiety, while also making love and energetically dancing. Also appearing in one of these sketches is Mijnheer Werrenbracht, one of the many prototypes of Shakespeare's Falstaff. Here he is a respectable citizen. Suspecting his wife of near infidelity with a priest, he hides in a basket and has himself borne into his house in order to spy on her and take the guilty pair by surprise.

As shall be seen, the activities of the Rederijkers lasted well into the Renaissance, when their contribution became even more important.

South of the German states, students at the University of Pavia staged topical and satirical skits, abounding with collegiate humour: in 1427 they produced *Ianos Sacerdos* and in 1437 *Commedia del Falso Ypocrito*. The Congresa dei Rozzi, an acting group in hilly Siena, made such a happy impression with its high-spirited offerings of the farcical doings of Italian peasants that it got a bid to repeat them at the Papal court and also for the pleasure of the Roman populace. A leading member of this company was Niccolo Campani, to whose excelling gifts as author and actor-manager it owed much of its success. Adopting the stage-name "Strascino", which was also that of one of the characters that he enjoyed portraying, he won high favour with Pope Leo X, as well as with the Roman nobility and crowds of excited, admiring commoners. He sometimes gave solo performances in roles from his own works. After a recital at an Orsini wedding party, where he competed with other superior comedians, he was declared very clearly the best (1518). But nothing of his work, other than his bright, tantalizing legend, has survived.

More fortunate has been Angelo Beolco (1502?–43), of Padua, more familiarly known as Ruzante (or Ruzzante), head of a strolling company that gave plays with dialogue in the dialect of their native town. Though they were not published, several of his scripts are at hand. He typifies the many actors and authors who made the transition from the Middle Ages to the Renaissance. For that reason, more shall be written about him in an ensuing chapter.

In Italy, as elsewhere on the Continent, the plays were enacted in much the same way: mere boards, any platform sufficed. What counted was the learned or improvisatory deftness of the actors, the good humour and wit of the author, his wry observation of character. The readiness of spectators to laugh was also a factor.

What is thought to be the first "wholly secular" play in English is *Fulgens and Lucrece*, which dates from somewhere near 1497; that is, it is the earliest of its kind of which an entire script exists. Its author is Henry Medwall, who composed it while chaplain to Cardinal John Morton, Archbishop of Canterbury. It was commissioned by his employer, the prelate, for performance at a Christmas banquet where several Flemish envoys were guests. Medwall takes his plot and characters from *De Vera Nobilitate*, a Renaissance Latin "discourse". As its title indicates, Medwall's sketch concerns a Roman lady, a senator's daughter, loved by two men, one of noble rank, one a plebeian. The play goes unmentioned by many historians, or is alluded to scantily, because they deem it no longer theatrically viable. A dissenting opinion is that of Sir Ifor Evans in his *A Short History of English Drama*:

> [Lucrece] asks her father what to do, and her father asks the Senate. The two suitors plead before the Senate and nothing is decided. This unpromising plot is handled rather deftly. Two men converse while watching the entertainment at a banquet. They suggest that England has companies of players so well dressed that the gallants emulate their fashions. The suitors plead not to the Senate but by turns to the lady herself, who chooses the poor and virtuous one. In a comic sub-plot, the two conversationalists take sides, and their rivalry finally enters the main theme as they become farcical suitors for Lucrece's maid. This play indicates what could be done in the fifteenth century, independent of Italian models. It reveals how incomplete is the data on which the history of drama is based, for the text of this play was unknown until a copy of it appeared in the Mostyn sale in 1919.

Another of Medwall's advocates is Glynne Wickham in *English Drama to 1710*, a collection of essays brought together by Christopher Ricks. He refers to English borrowings from classical mythology and Roman comedy. "An exceptionally interesting play in this context is Henry Medwall's *Fulgens and Lucrece*, for not only does it contain a parody of a Tournament, a dumb-show and a lengthy moral debate, but it sets out by satiric means to compare the newly emerging social values of England under Henry VII with those of Republican Rome. Granted this kind of treatment, both the lengthy Morality Play and the shorter Moral Interlude could readily be adapted a generation later to serve the cause of Reformation polemic. The initiative for this departure appears to have been taken in Germany in the early 1530s, but it was swiftly copied in England at the instigation of Lord Chancellor Cromwell and Cranmer." This immersion of plays in politics and religious controversy was to lead to the intervention of censorship of such works and their total suppression during the reign of Elizabeth I.

Finally, T.W. Craik, in *The Tudor Interlude*, suggests: "Medwall can claim to be the earliest English dramatist known by name, but even without this accidental distinction his Interlude *Fulgens and Lucrece* entitles him to serious attention as a dramatist. Its theme, birth versus merit, is conventional enough, but Medwall's handling of it shows a fine sense of the resources of the Tudor banqueting hall in which it was staged, its audience and the convivial occasion. While unsuitable for performance in a large public theatre, *Fulgens and Lucrece* still shows its dramatic vitality in more intimate productions. Medwall's other known play, *Nature*, is a Morality drama with a clumsy plot but some lively dialogue."

What has led many historians to Medwall, however, is that *Fulgens and Lucrece* is described as an "Interlude", a term they still find it hard to define. It is not the first play to be so designated: there are theatrical offerings as much as two centuries earlier with the same name, but their scripts are fragmentary or else totally lost. Wickham cites three: *The Interlude of the Student and the Girl* (*c.*1300); a piece about illicit overcharging by leatherworkers, banned by the Bishop of Exeter, who feared that it might start riots – nothing of the text, nor even the title, remains, merely the knowledge that it was too provocative (1352); and John Lydgate's *Mumming at Hartford*, an "entertainment" given at the court of Henry VI at Christmas (*c.* 1425) – its subject comprising an amusing debate between a half-dozen hardened tradesmen and their nagging, sharp-tongued wives.

Some scholars believe that an "Interlude" was at first merely a short, light work given between courses at a banquet on a festive occasion, as probably was *Fulgens and Lucrece*. It is akin to the Morality in that it presents a debate – indeed, that is its principal and distinguishing feature, but its content is wholly secular and possibly humanistic: it tended to be directed at an aristocratic audience, the intelligentsia at court or in the manor houses, and it emphasized the humorous, the fantastic, the somewhat abstract. It had to be witty, in effect a kind of farce. Intended to please and to be instructive, while avoiding overt moralizing, it dared to invite a strictly intellectual controversy, often in those tense times putting its actors and author at risk.

Others hold that the appellation "Interlude" was originally applied to the diverting, mirthful sketches between the serious episodes in the Corpus Christi processions, but Gayley in his *Plays of Our Forefathers* doubts this, and argues that Interludes actually differed little if at all from Moralities, which sometimes had merry passages, nor are all Interludes jocular and fanciful, and he lists a number that are decidedly on the sober side. Length and depth are not discriminating factors, either, though some have contended that they are. Both Moralities and Interludes may vary in each of these respects. A precise definition eludes the erudite, and the term continues to be a loose one.

The best-known writer of Interludes is John Heywood (*c.* 1497–1580), talented as a poet, singer and musician. He was married to a daughter of John Rastell (1475–1536), a printer, member of Parliament for a Cornish constituency, and brother-in-law of the all-powerful Sir Thomas More, the Lord High Chancellor. A literary figure, Rastell translated a comedy of Terence into English and was an earnest author of Moralities such as *The Nature of the Four Elements* (1517/1527?) and *Gentleness and Nobility* (1527?), in which once very certain religious concepts are shown to be in process of secularization. A play of his, *Calisto and Melibes* (1527?), based on a Spanish rogue story, has flesh-and-blood characters, but a pervasive and overwhelming didacticism weakens all his work for theatrical purposes, to judge from the three scripts that have survived and been authenticated as his. To some commentators, Rastell's Moralities verge on being classified as Interludes. (In *Four Elements* he puts forth the novel idea that the earth is not flat but round.)

Sir Thomas More enjoyed aspects of theatre, and when young had once taken part in a court production. Much discussion of intellectual subjects went on in his presence. Heywood, in the midst of the Lord Chancellor's circle, had the career advantage of high connections; besides this, his wife's brother, William Rastell, continued in his father's trade as a printer, so Heywood's compositions readily found their way into print. Six of them are still available.

Converted to Protestantism, the elder Rastell moved deeply into the maelstrom of the Reformation and was active in polemics in its service. While Henry VIII held the throne, this was acceptable.

When not writing, Heywood taught singing and acting to the boys of St Paul's Cathedral in London; despite this churchly occupation and his profession of orthodox religious views, he could write in a Rabelesian vein, some of his lines quite salacious.

An early venture, *The Pardoner and the Friar, the Curate and Neighbour Pratte* (1513/1521) is not ranked as the equal of his next offering, *The Four Ps* (c. 1520/1522), which by many is considered his finest. Three "Chaucerian" characters, a Palmer, a Pardoner and a 'Pothecary, are brought together with a Pedlar; they enter a contest as to who can tell the best lie. No little satire enters into the portrait of each of the contestants, some of whom are purportedly pious, and all of whom are very practised at deception. All are drawn vivaciously: the Palmer (or Pilgrim) views himself as purged of faults by his seemingly faithful journeying from one holy place to another; the Pardoner is selling fake relics, the "big toe of the Trinity" and the "buttock-bone of the Pentecost"; the 'Pothecary is a total charlatan, dispensing worthless nostrums. Some echoes of the *Canterbury Tales* resound at moments. The Pedlar acts as judge, as an acknowledged expert at lying, and the victory finally goes to the Palmer, who claims to have travelled near and far throughout Christendom, and to have encountered no less than 500,000 women of all sorts, but never to have met one who was impatient. The four also explain the nature of their trades and why each is superior to all others.

The Four Ps repeats and elaborates the pattern of the earlier *The Pardoner and the Friar, the Curate and Neighbour Pratte*, a debate among the first pair as to which is the worthiest and deserving of recognition from the parish. Both desire to preach at the same time. Their argument grows so hot and abusive that it leads to a scuffle. The curate enters and, his wrath aroused, is about to drive them out, aided by neighbour Pratte, but the miscreants are clever enough to turn the tables and escape their likely punishment. As in *The Four Ps*, the characters are realistic, and in both stories the wicked figures triumph; Heywood is not trying to make a moral point but merely to fashion works for sheer fun.

A group of scholars argue that the Interlude's debt to the Morality is overstated and suggest that as a fresh genre it owes more to French farce, as clearly evidenced in Heywood's six plays. He borrows the form and spirit from comic writers in France and naturalizes both by using English subject-matter and the broadest possible character-types, rendering them as thoroughly native and thus more easily identified by his contemporaries. French influence had been strengthened by the visit of Henry VIII and his sister Mary to the court of Louis XII and the subsequent marriage of Louis XII and Mary. Also, in the late fifteenth century French minstrels and players had come to England and performed there. Heywood was assuredly familiar with what was already happening across the Channel.

The Play of the Weather (1523/1533) is almost plotless yet was a major success for its author. Jupiter is troubled by all the complaints reaching him from mortals dissatisfied by their daily climate. He assigns Merry-report, a comic rascal, to look into this perpetual disquiet, but whoever is interviewed makes a request for a different kind of sky: a merchant wants a brisk, fair wind for his ships; a Water Miller seeks rain; a Wind Miller no clouds or precipitation; a launderer asks for hot sun to dry the clothes he has scrubbed; and a small boy hopes for frost and snow, so that he can have sport with snowballs; and so on. It is impossible to please everyone, and Jupiter sagely elects to make no changes.

When this offering was produced at the court, a two-storey stage was erected, with Jupiter at the top, flanked there by Aeolus and Phoebe, the deities of wind and sun. The performers were Heywood's "children" – that is, his singing and acting students from St Paul's. At this time it was increasingly common for university and preparatory students to make up a large segment of the casts of semi-professional presentations, and often courtiers might also join them.

A Play of Love (1528/1529?) brings together four assorted characters whose names are as abstract as those of the Seven Deadly Sins or Seven Virtues: these generalized figures are Loved-not-Loving, Loving-not-Loved, Neither-Loved-nor-Loving and Both-Loved-and-Loving. It is somewhat less well constructed than the *The Play of the Weather*, with the diverse speakers not always entering and exiting on cue, but instead staying in view to intervene in the ongoing discussion. Yet Heywood's dialogue is always at least mildly witty and holds attention, overcoming the lack of action and the absence of an engrossing plot. The "pains and pleasures of love" are its only subject, bandied about in an airy manner.

Left unpublished, *Witty and Witless* (or *Wit and Folly*) whimsically raises the question as to whether it is better to be a wise man or a fool. The suggestion of a cross-Channel influence is strengthened here, because the idea and its development closely resemble those in a French piece, *Dyalogue du Pol et du Sage*, which in turn – as Clarence Griffin Child points out – descends from Erasmus' *Encomium Moriae*.

Heywood's debt to a French source is clearly overt in his most admired venture, *The Merry Play Between Johan, Tyb and Sir John* (1533), which neatly parallels a Continental farce, *De Pernet qui va au vin*. In having a motivational plot and ample physical action, this script no longer matches the definition of an Interlude but instead that of a full-fledged play. A henpecked husband, John, growing irate when his wife, Tyb, temporarily vanishes, vows to beat her should she return; he is less firm when she reappears and threatens him. She makes eyes at a priest, Sir John, who has been invited for supper and for whom she bakes a handsome pie. Johan, hungry, hopes to share in the meal but is given no chance: whenever he sits down to eat his wife busies him with a distracting errand. Meanwhile the priest and the wife are gluttonously indulging themselves at the board. Finally the frustrated husband can stand no more. He declares himself master of his house and violently chases both wife and priest from it. As soon as they are out of sight, however, he begins to worry and fret as he imagines what amorous mischief they might now be up to.

While Queen Mary reigned and Catholicism was the state religion, Heywood was the recipient of royal favours. He participated in the pageant that marked her coronation, had his works presented at her court and was granted leases of land in Yorkshire. He is recorded as having been given fees by both Henry VIII and his daughter for his song recitals and playing the virginal. The shifting political and religious winds brought him harsh treatment in his later decades, his position weakened by the demise of Rastell and the displacement from power and execution of their kinsman Sir Thomas More, who heroically opposed Henry VIII's divorce from his unhappy first wife, while her side was taken by the pope and the King of Spain. Following the accession of Elizabeth I, Heywood was charged with involvement in a conspiracy against Archbishop Cranmer, More's successor, who had obligingly called Henry's marriage to Katherine of Aragon invalid. Enduring lengthy imprisonment, the playwright finally gained a pardon. Though he had written improper secular farces, he was still devout enough to accept fourteen years of exile on the Continent for holding fast to his now proscribed faith, spending the years 1575 to 1578 at a Jesuit college in Antwerp. He left its shelter when the members of that evangelizing order were driven out, and with some others went next to Louvain. He died, aged eighty-three, in Belgium. (Some details of his life, and the dates and sequence of his plays, are not established to everyone's satisfaction.)

Though a deeply sincere Catholic, Heywood is paradoxically a voice of the Reformation, for he repeatedly portrays the clergy of his time as corrupt and hypocritical. What is more, he does not moralize, and draws nothing from Scriptures or hagiology; his aim was merely to entertain. He said that he desired "not to teach but to touch".

Other English playwrights, a considerable troop of them and most of them cautiously anonymous, made use of the Interlude and for the same simple purpose, not to edify but to satirize and elicit chuckles and guffaws by their light observations and shafts. No longer in thrall to the censorious, repressive Church, theatre was returning to the people.

— 3 —
THE REBIRTH: ITALY

To later generations, the epoch called the "Renaissance" brings the stirring image of psychological rejuvenation. Too long submissive, having been restricted by theological edicts and dogmas, people's minds regained freedom. The pulse of their intellect quickened. This was an era of exploration, inward and outward, reshaping their view of the world and giving them very different values to live by. After Copernicus Europeans thought of the earth as circling about the sun; after Columbus, their world became round and suddenly more diverse; after Giotto, human likenesses were no longer rendered flat but glowed in more natural hues and in realistic perspective; after Machiavelli, a Christian leader felt justified in being cynical and ruthless in his quest for power. People sloughed off the cowls and robes of medievalism and strode forth into a dazzling, fresh light, dynamically reinvigorated.

No date can be fixed for the beginning of this brilliant span of time: conveniently, and reasonably, it is set as having been engendered by the fall of Constantinople in 1453, when monks and scholars fled westwards to Italy, carrying their treasured copies of classical writings, some still on scrolls of papyrus, endowing thinkers in other European countries with insights into the superior learning of the nearly forgotten Periclean Greeks. Their flight added strong impetus to the process of change. Actually, the advance occurred slowly, with much confusion and conflict due to the boldness of the new concepts and the stubborn lingering of Gothic ideas and traditions.

Ethical, aesthetic and social innovations, starting in Italy, spread gradually northwards and westwards, being observed in and then penetrating France, Spain, Germany and, finally, England, where the new outlook coincided with the Reformation, which was acquiring ever more momentum. In each country the cultural changes took on a different coloration, lent by racial temperament and peculiar historical circumstances. In all instances, however, the new dominant spirit was brought about not only by the excited rediscovery of the classics – they had never been wholly lost, but for the most part had long been accessible only to the very erudite in monastic and academic circles – but also by an outburst of individualism, another eruptive prompting.

With its liberalization of the individual will and imagination, the Renaissance was an age stressing invention, ever more iconoclastic. In the graphic arts it inspired experiments with new materials and techniques – the discovery of oil paints, the addition of *sfumato* and *chiaroscuro*, the play of light and dark – and startling theories of composition and colour-blending. Progress was made, too, in medicine, in physics, as well as in astronomy, enabled by the invention of the telescope, and especially in geography, helped significantly by the introduction of the mariner's compass. Guttenberg found how to print with movable type, which accelerated the distribution of knowledge. From the Far East came gunpowder, and cannon-fire was more destructive, razing the thick castle walls that no longer protected feudal lords. A middle class began to burgeon. Nationalism became an energizing factor.

Oddly enough, the arts in this new period advanced by looking backwards, towards the long-vanished Hellenic and Roman civilizations that fifteenth-century intellectuals idealized and now eagerly sought to emulate.

In this dawning society, ambitiously undergoing transformation, religion had an ever smaller role. The Church, still powerful, was a lavish patron of the arts but no longer a prominent sponsor of the drama, which took on an aspect that has ever since been primarily secular and worldly. If anything, the Church continued to be hostile to the stage. As has already been noted, in 1548 an edict forbade the Confrèrie de la Passion of Paris from producing any work of a sacred nature. For different reasons, Elizabeth I in England proscribed the staging of propagandistic Moralities; and in Germany, for the most part, the Protestant establishment frowned on play-giving. With the loss of this often sublime inspiration a degree of importance and stature was also gone, something luminous, that even today playwrights struggle to recapture, with only limited success.

In Greece and medieval Europe, and in the far-flung Orient, the theatre was a temple, a cathedral, a place for solemn ritual; but now it came to be chiefly dedicated to pleasure, to artificial charades, often as a refuge from clamorous reality. Over the succeeding centuries there are to be a few exceptions to this, but at long intervals: only in rare instances – contributed by Marlowe (an atheist), Shakespeare, Goethe, Eugene O'Neill (a lapsed Catholic), Paul Claudel – is there to be true elevation in theatre again, with humanity viewed from a religious perspective, realizing a spiritual dimension or a troubled lack of it.

Though the Italian Renaissance is one of the most illustrious periods in mankind's history, its playwrights are minor figures, and their output no longer considered stage-worthy. But this may be an unfair judgement. Plays are not adequately tested by whether they read well, but only whether they work well when competently acted before spectators, who follow the action as if hypnotized – the author having induced in them a "suspension of disbelief". The theatre has its own unique attraction; what an audience experiences is unlike what a solitary reader enjoys. There have been few, if any, recent opportunities to assess these Italian Renaissance scripts in performances that respected their conventions and their authors' commendable aim, which was to resurrect Greek tragedy. If mounted today, with the elaborate sets and opulent attire that once adorned them, they might prove to be rewarding and fascinating for curious, open-minded drama-goers.

True, by comparison to the dazzling achievements of Raphael, Michelangelo, da Vinci and others in the visual arts, and to Galileo and fellow delvers into the cosmic sciences, the accomplishments of Italy's Renaissance playwrights seem meagre; three centuries of their often florid works yielded no scripts that have earned immortality outside the peninsula; even in Italy the scripts are mostly antique oddities in mouldering libraries rather than living drama. Yet greatly important steps forward, of lasting and salutary influence, are credited to this epoch. The progress was made in the realms of staging, theatre architecture and acting, rather than in writing. What is more, this is the time and place that saw the birth of opera and formal ballet, and the evolution of a charming and endearing genre of improvisational performance, the *commedia dell'arte*. In all these directions began the types of theatre still familiar and even predominant in our day.

Most of these advances would not have emerged without the help of scripts, which, however deficient they might be as poetry and drama, were still pretexts for staging. A very early entry to the ranks of would-be classical playwrights is Albertino Mussato, who composed at least two tragedies in Latin, modelled on those of Seneca, that master of rhetorical rant and violence. *Eccerinis* (1315), a short work, has an Italian subject: Ezzelino III, a Paduan tyrant. A second play, *Achilleis*, concerns the mythical Greek hero. Though *Eccerinis* is often named as the first secular tragedy, it is impregnated with Christian doctrines. Another treatment of the Homeric story, *Achilles* (1390), by Antonio Laschi, more effectively blends classical form and martial topic.

After a long lapse, other names and personalities appear. A catalyst was the publication of *Arte Poetica*, an essay on Aristotle's *Poetics* by Giangiorgio (Giovanni?) Trissino (1478–1550). An abridged Latin translation of Aristotele's treatise had been brought out by a German scholar a century earlier, and a similar task had been carried out in Spain by Mantinus of Tortus a good many decades subsequently. In 1498 a Venetian printer, Georgio Valla, followed suit with a new Latin version. Finally, the Greek text became available in 1508, but the Italians found it difficult to cope with its archaisms until Trissino's venture. He courageously proceeded to match his creative efforts to Aristotle's precepts, turning away from Senecan models, in *Sofonisba* (1515). Possessed of both considerable erudition and a lyric gift, he relates how the lady of the title, Sofonisba, daughter of Hasdrubal, a Carthaginian general, is pledged to Masinissa, an East Numidian prince, whom she loves. Instead, she is married to the King of West Numidia and travels with him to Rome. She escapes her fate when Masinissa sends her poison in a cup, from which she drinks and expires. Trissino writes with restraint, displaying good taste and eschewing the excessive violence found in the works of Seneca and those who chose him as an exemplar, and the characters are moved by sincere-sounding passion. The tragedy suffers, however, from being somewhat static, too much of the action reported by a series of five messengers, abetted by an over-intrusive fifteen-man chorus that ever more dilutes the physical enactment. Trissino also initiates the use of blank verse, electing "Italian rhymeless hendecasyllables", abandoning classic prosody. Further, he avoids act divisions. His contemporaries revered this poet for his vast scholarship, his meticulous syntax. He is considered one of those who helped elevate Italian to a literary language. Among his other stage-works is *Italia Liberata*.

Even before Trissino's advent the courts of Italian princes were regaled with theatrical entertainments that they sponsored and lavishly subsidized. A good many were comedies, but a substantial share were tragedies by other versifiers of the day. More and still more, so that the spectators might better comprehend the dialogue, the dramatists shifted from Latin to the vernacular.

By now the seven surviving plays of Sophocles had been recovered (1502) and made widely accessible by another eminent Venetian printer, Aldus, who the very next year brought forth an edition of Euripides' extant works, and lastly those of Aeschylus, the oldest and most resistant to translation. The plays offered new models and richer subject-matter, though the gory melodramas of Seneca had by far the strongest appeal. Renaissance courts were hardly ever serene; courtiers were usually busy with fatal intrigues and with exacting vengeance from foes and rivals. The world of power-struggles as seen by Seneca was not too different from that taking place in the ornate palaces of the Medicis and Gonzagas.

A Spanish prince of the Church, Cardinal Riario, was especially fond of Seneca's savage dramas and supported productions of them. He read the script of a play, *Historia Baetica* (1492), by a compatriot, Carlo Vergardi, with which he was so taken that he persuaded Pomponius Laetus, a great scholar, Humanist and headmaster, to have it staged by the students of his Pomponian Academy. To gratify the influential, wealthy prelate, this was done. The subject was the recent defeat and ejection of the Moors who had for too long ruled Granada. The turbulent drama, quite lacking in literary merit, was presented before an élite private audience at a gathering in the Palazzo Riario, as was becoming a current custom.

One important centre of the revival of classical art and encouragement of new works was Ferrara, where in richly decorated halls an enlightened ruler presided over a throng of highly talented painters, musicians, poets, philosophers, all truly devoted to the fresh cult of the ancient. Many of its playwrights aspired to equal the standards set by the Athenian masters whom they revered, though they proved hardly adequate to doing so. Yet they persevered. In Ferrara, in 1499, a breakthrough occurred when Antonio Commelli's *Filostrato e Panfila*, written in Italian, conjoined echoes of Senecan sensationalism with a theme from Boccaccio, to which were added touches from the Mysteries, the result being a tale of unsanctioned love and its dreadful consequences.

Most conspicuous among the Ferrarese dramatists, however, was Giovanni Battista Giraldi (1504–73), better known as "Cinthio", otherwise an eloquent professor of philosophy and rhetoric, whose scripts gained him a popularity that far exceeded that of Trissino and all others. He greatly outdid them – as well as Seneca – in heaping up unending horrors. His *Orbecche* (1541), a drama of unsparing revenge, is written in Italian, as was Commelli's play, but it is the first work in that language to have actually been staged, when resourcefully he arranged to have it performed in his own dwelling. He later declared in an essay on Aristotle, *Discorso delle commedie e delle tragedie* (1543), that what should be sought in a play is the "horrible". Here the characters behave like those distraught and berserk in the House of Atreus. In a preface to *Orbecche* he vows that he will give his listeners "tears, sighs, anguish, terrors, and frightsome death". He crowds into five acts the history of Selina, executed for having committed incest with her husband's son. Returning as a ghost, she promises to revenge herself on Orbecche, the young man's sister, who had revealed their guilt. The ghostly Selina exacts a hideous reprisal: Orbecche's husband has his hands chopped off, his head cut off, his children slain. The severed head and corpses are stuffed in vases that are sent to Orbecche. She, in turn, has her father decapitated and finally kills herself. In addition, Cinthio introduces the relentless Furies and Nemesis. The audience, which obviously had strong stomachs, found the action gripping.

He composed two more tragedies, *Dido* and *Cleopatra*, about loving, unfortunate women, but he appreciated that people really liked happy endings. He turned to offering serious works with pleasing resolutions, as typified by his romantic *Arrenopia*, in which the Queen of Ireland escapes death at the hands of her faithless mate, and all concludes well. In *The Interchanged*, a much lighter piece, girls are disguised as boys, and the changeling theme is rung.

Cinthio originated his plots, the first Italian playwright to do so. Though echoes may be heard here and there, he did not draw on the classics, nor on history, but invented his stories, a move towards modern drama. In his influential *Discorso delle commedie e delle tragedie* he claimed that Seneca was a better guide than the Greeks, because the Roman statesman was graver in tone, wrote more majestic rhetoric, and was more didactic and moral. He favours Horace's division of scripts into five acts, each serving a different purpose. He defends Aristotle's first unity, "time", leaving it to his fellow Italian theoretician Ludovico Castelvetro (1501–71) to add "place". Unity of action was never much stressed by playwrights of this period, though most latter-day critics view it as the most important, perhaps the only true imperative of effective drama. In *Orbecche* all the visible action occurs before a palace, a relay of messengers bringing word of the murders perpetrated within it, much as in the plays of Ancient Greece.

Cinthio has a special place in the history of theatre because his considerable success inspired many other enthusiastic Italian playwrights to follow his lead. His influence extended much further than the peninsula to elsewhere in Europe. The Elizabethan "revenge tragedy" has its prototype in *Orbecche*: the poisoned and skewered bodies that strew and clutter the stage in the final scenes of Shakespeare's *Hamlet* and Webster's *The Duchess of Malfi* offered a picture like one often envisaged by Seneca and replicated by Cinthio. Similarly, the gender-disguises assumed by the young women in Shakespeare's *As You Like It* and *Twelfth Night* are foreshadowed in *The Interchanged*, though the imprint of the Bard's genius is not there. Yet Cinthio's light play is the perfect model of the romantic comedy of intrigue, with complications arising from mistaken identity and abounding with startling recognition scenes. His prose works were also a source of many plots used by others, including Shakespeare for *Othello*.

Among the many reasons for Seneca being chosen as a prime model by the Renaissance Italians is that, contrary to Cinthio's pronouncement, Senecan rhetoric was easier to imitate, while these fifteenth- and sixteenth-century poets were unable to recapture the splendour of Athenian tragic verse.

Even so, the flow of over-ambitious scripts was unstaunched. Giovanni Rucellai (1475–1525), a friend of Trissino, fashioned an *Orestes* dealing with the stranded hero's misadventures in Taurus; he is better known, however, for his *Rosamunda*, a piece that is oddly static though choked with chilling instances of a tyrant's vindictiveness.

From Ludovice Dolce (1506–68), otherwise an educator, came a *Giocasta* (1559), based on Euripides' *The Phoenicians* and treating with the Theban conflict between the mutually envious and hateful sons of the doomed Oedipus and his queen. An English translation of it by George Gascoigne and Francis Kinwelmarsh was enacted at Gray's Inn in London in 1559. In Dolce's *Marianna* the depths of Herod's jealousy of his wife are explored in a script a bit less static than are most turned out in this verbose epoch and filled to brimming with no fewer than six bloody murders. A member of a noble but impoverished family, Dolce was dauntingly prolific, penning at least eight heated tragedies and five light-footed comedies, among other literary works. His *Didone* (*Dido*; 1547) is drawn from Virgil's *Aeneid*, as is Cinthio's.

A professor of literature and philosophy at Padua, Sperone Speroni (1500–1588) was responsible for the gruesome *Canace* (1542), a drama also in the manner of Cinthio and again involving incest, a subject of which these playwrights never seemed to tire. Its publication caused a great stir, one that was long-lasting among serious intellectuals and feuding academicians. What was in contention was a clearer definition of artistic principles.

Besides Trissino, Cinthio and Castelvetro, a number of other theoreticians, Antonio Minturno, Julius Caesar Scaliger, Marco Girolamo Vida and Piero Vettori, wrote essays on the rules articulated by Aristotle and Horace for the proper structuring of a tragedy. Above all, it should have unity of time – "one revolution of the sun" – and place – "one set" – to be plausible, and have five acts. These requirements were accepted as incontrovertible, and they put a yoke on playwrights for centuries to come.

In the second half of the sixteenth century Luigi Groto (1541–85), an actor popularly known as Il Cieco d'Adria ("The Blind Man of Adria"), has a sizeable list of plays and is remembered for three verse comedies – *Emilia* (1579), *Il Tesoro* (*The Treasure*; 1583) and *Alateria* (1587) – as well as two pastorals – *Calisto* (1561) and *Pentimento Amoroso* (*The Lover's Repentance*; 1575). Two tragedies survive: *Adriana*, a romance doused with pathos, and *Dalida*, believed by some historians to have been a source of Shakespeare's *Romeo and Juliet*.

Federico Della Valle (1560–1625) extracted a heroine from the Bible for *Judith* (1590/1600) and borrowed from the climax of a dynastic and religious conflict that had just occurred for his *La Regina de Scozia* (*The Queen of Scotland*; 1591/1600), depicting the last hours of the imprisoned Mary awaiting her execution, decreed by Elizabeth, her rival. Also by Della Valle are *Adelond di Frigia* (*Adelond of Phrygia*; 1595) and *Esther* (1590/1600). Choosing to write about women profoundly beset, his tone is sombre, his judgements stern and moralistic, his overall mood inclined to be infused with melancholy. Of his tragedies, only *Adelond* was performed during his lifetime.

A fuller list of the poet-dramatists of the period would include more names – such as Zinano, Tebaldeo, Cammelli – but some are far better known as authors of comedies and will be discussed later.

When classical subjects and themes from other genres of Italian literature were chosen – such as folk-tales and ballads – the poets took considerable licence in changing the storyline. Historical facts might also be altered. For the most part, however, originality was lacking. If this large, earnest body of work failed, it may have been because its inspiration was artificial, its form too rigidly prescribed. Its subject-matter was almost wholly borrowed from the past; for its audience it had scant immediacy. The playwrights were enamoured of displaying classical learning; their spirit was too academic. It was a theatre created by scholarly philosophers and rhetoricians, their imaginations bound to antiquity, and it was something less than an art-form for full-blooded dramatists. Indeed, it was mostly academies of various sorts – in Naples, Rome, Siena, Florence, Ferrara, Venice – that sponsored these Italian productions.

Many of the scripts were published or enacted belatedly, if at all. In most instances they were given staged readings for the aristocracy and intelligentsia as, it was assumed, took place in Seneca's time. Indeed, the initial purpose was to recreate as far as possible the conditions of that classical setting. In Rome, the Pope and cardinals – and at his academy the eminent scholar and teacher Pomponius Laetus – invited select audiences to performances offered in just that spirit and manner. In Ferrara, where artistic and literary activity especially flourished, Duke Ercole of the d'Este clan took the lead in encouraging theatre for his court. Florence saw the "Magnificent" Lorenzo de Medici try his hand at play writing: his *L'Aridosia* (1536), about a lying, pusillanimous miser, was put on as part of the nuptial festivities for Duke Alessandro. The Sforza dynasty in Milan, the Gonzagas in Mantua and the Della Rovere family in Urbino were similarly enlightened patrons. It may be taken for granted that there was emulation and competition about such matters in the capitals of these small, often hostile states of fragmented Italy.

At first, in many of the cities, theatrical production meant no more than setting up a platform four or five feet above the marble floor at one end of a large, elaborately frescoed room in a princely or ducal palace. (Renaissance rulers, like the Church Fathers, liberally commissioned the best muralists to make sure that not a square foot of wall-space was left bare.) This low platform, atop trestles, was thought to copy the Roman stage as it had been in the days of Terence. At the rear were decorated arches held up by pillars with gilded capitals and bases, and with curtains between them. Some influence of the medieval "stations" serving the Mystery Plays lingered here: one of the curtains might be opened to reveal the "interior" of the house of the leading character, and then the adjacent curtains drawn to expose the next compartment, the dwelling-place of another character; and usually there was a third room. All this resembled the "mansion" theatre – the line of "boxes" – indicating that something similar to what was utilized much earlier at Valenciennes or Lucerne was still being employed. Gradually, painted scenery was added, and in some instances, if performances were outdoors, on terraces or in gardens, the audience sat under awnings.

In 1414, however, occurred an event parallel in significance to that of the publication of Aristotle's *Poetics*: the text of Vitruvius, the first-century BC Augustan architect and theatre historian, came to light. When finally printed in 1486, his *De Architectura* immediately excited Renaissance designers, for it described in very specific detail how Roman auditoriums and stages had been arranged. Whatever was not clear on his pages was shortly subjected to hot debate, for beyond them Vitruvius' laws were now wholly authoritative. In his academy Pomponius Laetus made a close study of them, and students from all over Europe flocked to his lectures, and then took many of his interpretations back to their own countries. Thus all Europe became acquainted with Vitruvius' ideas, to some extent incorporated in the academy's productions.

Berthold suggests that Hans Sachs built the first permanent theatre in Nuremberg's Marthakirche in 1550, but it was at best a modest affair. In fact the Confrèrie de la Passion in Paris had established itself for a very long stay at the Hôtel de Bourgogne in 1548; and when the group was no longer offering plays, until almost the end of the next century, it retained its monopoly of the space and rented it to other companies.

As early as 1435, in Italy, Leon Battista Alberti had published his *Della Pittura*, a treatise in which he gave practical instructions for attaining perspective in drawings, an artistic technique that could also be applied to overcoming the effect of flat planes in painted scenery, and in 1454 he was commissioned by Pope Nicholas V to design a playhouse, though his vision for it was soon outstripped by the efforts of others elsewhere. Experiments by Leonardo da Vinci and Filippo Brunelleschi did much to further and perfect the tricks of perspective whereby the glorious mural and easel painting of the period gained an illusion of depth and more logical spatial relationships; almost at once this knowledge was acquired by scene designers, who frequently were also the same artists as those who laboured at adorning the walls of palaces, churches, royal chapels and convents.

The combination of Vitruvius' helpful guidelines – to which were duly added commentaries by Godocus Badius (1493), Jocundus (1511) and Philander (1544) – and the introduction of perspective painting fascinated and stimulated theatre architects and directors to ambitiously bold attempts at creating spatial illusions, some of them highly extravagant, because the artists were richly subsidized. Renaissance princes had large purses.

An early instance, perhaps the very first, of "perspective painting" being installed as a focal element of a stage-setting is on record in Ferrara for a production of Ariosto's comic *Cassaria* (*The Strong Box*; 1508), where the spectators beheld some free-standing houses in front of a seeming panorama of gardens, dwellings, church spires and towers receding from a near foreground into the hazy blue distance, all on a backdrop. This startlingly fresh theatrical experience was the achievement of Pellegrino da San Daniele. What is more, it accorded with Vitruvius' careful prescription: "Tragic scenes are delineated with columns, pediments, statues, and other objects suitable to kings; comic scenes exhibit private dwellings with balconies and views representing rows of windows after the manner of ordinary dwellings; satyric scenes are decorated with trees, caverns, mountains, and other rustic objects rendered in landscape style." (Some historians claim that a demonstration of perspective scene painting on stage had taken place somewhat previously in the rival city of Mantua.)

In calling for quite different settings for tragedy and comedy, Vitruvius is said to have had in mind a further purpose, a visual projection of his concept of the ideal city, a concern that also captured the imaginations of Italian Renaissance architects and artists. By showing a schematic realization of one, in which a regal, sumptuous area populated with heroic statuary and pillared temples was reserved for an aristocracy, and a less imposing neighbourhood was lived in by common folk, a plan so ordered could be communicated from the stage to spectators who might carry a vivid memory of it from the theatre.

Once started in this direction, the sets grew very elaborate, all three categories of plays – tragic, comic, satiric – richly presented by the eager Renaissance designers making resourceful use of wood, plaster and canvas, filling the stage with fluted columns, pediments and pedestals, or – for a pastoral scene – an "outdoor" landscape filled with trees, rocky caves, grottoes, bushes and other appropriate objects.

Accompanying these solid and dimensional effects, as the artisans raised their sights and sharpened their skills, were all sorts of ingenious, animated marvels to amaze the spectator. Once more – as in the late Roman theatre – fountains shimmered and played; galleons sailed the waves under floating, fleecy clouds; gods flew on high or suddenly vanished; volcanoes spewed smoke and erupted; cannons roared, and battles wavered to and fro on "fields" with pennons flying, and castle walls crumbled under siege; towers fell in flames; chariots raced. Overhead, in the "sky", sun, planets and stars revolved, affixed to a wheel.

Artists no less famed than Mantegna and Raphael were enlisted to furnish the most lavish décor. The stage-machinery was intricate and very costly. Fortunes were spent. But there were also simple devices, such as trap-doors. The resources for staging were soon much more abundant than anything the Greeks or Romans had conceived. (Two greatly helpful sources of information here are Giacomo de Vignola's *The Two Rules of Perspective Practice* (1583) and Nicola Sabbatini's *Manual for Constructing of Theatrical Scenes and Machines* (1638).)

Most often the stage was markedly sloped or raked, to aid the perspective and heighten the illusion of distance as well as to assure the spectator a clear view. The action, however, mostly took place downstage, on a forestage or flat area that corresponded to the *platea* of the medieval theatre. Sometimes steps at either side led from this front platform to the auditorium. Scenery was changed, at first, by resort to a classical *periaktoi*, or revolving prism – its use brought back from antiquity and initiated by Aristotile da San Gallo (1481–1551) at Castro in 1543 – but

Vignola recommended as a better method the sliding in of a series of stiff canvas panels or "flats" with new settings painted on them. These proved so easy to handle that the device was adopted more and more. With them the stage was no longer a "box" – it had no side walls but was open at both right and left, the gaps concealed by intruding "wings" – and there were no doors – the player entered and exited from behind one of these side-frames or "wings" as if he had passed through an unseen portal. The "wings", usually numbering four, might be decorated with seeming three-dimensional detail, especially those nearest the spectators, perhaps suggesting the city streets or forested landscape that extended outside and beyond the palace room or convent cell or dungeon that narrowly housed the violent, plotted action. The first wholly movable sets, it is stated, were adopted at the Teatro di San Cassiano at Venice near the start of the seventeenth century.

Since natural light no longer served, with the theatres totally enclosed, artificial illumination was needed. It came from a multitude of candles or oil lamps with thick wicks hanging above the stage or strategically placed at the sides, and these lights could be altered to create mood – bright gaiety, dim gloom. This was achieved by having the candles stand behind bottles of tinted liquids or even wine; in time, footlights were added, as were reflectors to heighten the candles' gleam and the lamps' glow.

Though a concern with lighting effects was not yet widespread, there were some interesting experiments as this fruitful century unfolded. Leone de' Sommi, in his *Dialogues* (*c.* 1556), suggested that the auditorium should be dimmed, which would result in having the stage appear correspondingly brighter. And in 1598 Angelo Ingegneri ventured further, recommending that the spectators be left in *complete* darkness while the play was in progress, with not even a spill of light from the illuminated stage. These radical ideas were not widely accepted, however, until still another century passed. A German essayist, Joseph Furttenbach the Elder (1591–1667), after spending ten years observing the triumphant advances of the Italian Renaissance, went back to Ulm in his native land and published three treatises on how the many mood-changing lighting effects were accomplished.

Sabbatini spells out, with diagrams and verbal descriptions, exactly how the three-sided revolving prism was to be utilized to project the illusion of a river constantly flowing, or of dolphins and other sea monsters swimming and spouting water, or of clouds traversing the sky, or of larger clouds bearing divinities descending vertically to the stage boards, this last particularly important because so many plays on classical subjects had a *deus ex machina* resolution, a god's pronouncement of absolution solving the hero's predicament.

The sets, though obviously stylized, were increasingly realistic, at least in contrast to the almost total absence of scenery and properties on the hastily put together medieval platform or in the small mansion box. For example, a prince of the Church, Cardinal Bibbiena, wrote a comedy, *La Calandria* (1508), that was staged for the amusement of the Duke of Urbino and his court in 1513. Among those on hand was Count Baldassare Castiglione, author of *The Courtier*, the memorable book on proper conduct for the nobility; he was greatly impressed. "The streets looked as if they were real, and everything was done in relief, and made even more striking through the art of painting and well-conceived perspective." Much of the decoration on the altars, columns and temple friezes was executed in stucco applied to wood. A year later, for the pleasure of Pope Leo X, the play was given in Rome, with sets by Baldasarre Peruzzi (1481–1587) who strongly embraced the new art of perspective. Giorgio Vasari, the painter and biographer, was moved to praise the fidelity of the scene, with "its palaces and curious temples, loggias and cornices, all appearing to be just what they represented".

Few writers on set-design were to have an influence as long-lasting and far-flung as that of Sebastiano Serlio (1475–1554), a former pupil of Peruzzi. His lengthy *Architettura* (1545), containing many sketches and diagrams, is the first work in the Renaissance to devote considerable space to the problems of stage-scenery. Published in Paris, it was circulated throughout the Continent and England, where it was very carefully and eagerly studied, and where its ideas were approved and much borrowed. Serlio wrote his valuable six chapters at a time when the tragedies and comedies were still being put on in the ornate halls of palaces and stately rooms of academies, prompting him to suggest that the platform erected as a temporary stage should be at a height accommodated to the eye-level of the seated duke or prince, or whoever the actors' generous patron might be. The chief preoccupation of Serlio was enhancing the illusion of depth. One means of achieving this was to have the first three wings on either side of the stage ranged to stand at an angle that slants away from the audience, drawing each spectator's gaze in that direction, culminating at a focal point on the backdrop. A broad open door or high, arched window located there, with a vista through it of receding streets, high embattlements or perhaps a forested landscape, would increase the impression of distance. Only the fourth, last wing should face frontally, framing the imagery artfully painted on the back curtain. In addition, the rear section of the raked stage, little or never used by the actors, should slope up abruptly, contributing to the sought-for illusion of far-off objects diminishing, in an inhabited world outside the stage enclosure. Again, the perspective must be conceived and constructed as ideally viewed by the very important person watching from the patron's chair. All solid scenery should be arrayed on this rear, steeply rising part of the stage, permitting the players more space and freedom of physical action on the more level foresection.

Serlio subscribes to Vitruvius' rule with respect to differentiating between the kinds of scenery appropriate to tragic, comic and satyric scripts. He believed that just three designs, one for each of the disparate genres, were all that a well-equipped theatre might require; consequently, the basic floor plan of the three should be the same and match the dimensions of the stage on which they would be installed on repeated occasions. Essentially, they were permanent and unchanging, available for any number of plays, new and old. Later, as audiences grew over-fond of spectacle and variety, the revolving *periaktoi* – two-, three- or six-sided – became a complementary device to indicate a shift of place, as proposed by Vignola and da San Gallo. Sabbatini's ingenious suggestions were also heeded: by inserting grooves in the floor, new flats could be smoothly slid on stage along them; and by quickly covering the painted pictures on the wings with canvases on which were new images, a fresh location was established. As yet there were no front curtains: all these scene changes and others were carried out in full view of spectators.

Serlio, inserting his own drawings in a new edition of Vitruvius' compilation, added perspective, though his Roman preceptor was never a master of that technique; however, this led many to believe that in fact Vitruvius had endorsed it, lending the practice yet more authority. Over the next decades further developments and refinements of the art ensued, and in 1600 Guido Ubaldus brought out his *Six Books of Perspective*, which led to no longer placing the wings at an angle; with more knowledge and skill at perspective, painting on wings set frontally was adequate to create a sense of depth, as was demonstrated by Giovan Battista Aleotti (1546–1636) at Ferrara in 1606. By mid-century the reliance on angled wings was almost entirely outmoded. Instead they were put up parallel to the footlights, one behind the other, each with a different scene on it – it was very simple merely to slide off the one in front to reveal the one behind it, in bare moments altering the location of the action. Any number of rapid site-shifts were possible. At the back of the stage, shutters might be hung over the centred, focal windows, archway or gateway: the horizontal slats could be rolled down while the scene was changed, then rolled up again to show another panoramic vista. The same effect was gained by using curtains that gathered together and opened again, or dropped and then lifted, permitting a different outlook at that point.

The demand for more stage-machinery and astonishing spectacle necessitated overhead scene-masking. Aspiring to be faithful to Ancient Greek and Roman traditions, many plays had outdoor settings, which called for a sky and floating clouds and sometimes the sun – rising or setting – and the moon and stars. A bow-shaped border was the answer: in addition to concealing the mechanism, it could be painted appropriately, and if desired quickly replaced with another kind of sky. For interior sets, an overhang could suggest and illustrate a ceiling or dome. These moves would be coordinated with those of the sliding wings and back shutters.

As the sets and effects grew ever more elaborate, they required precise timing: at the sides, at the back and overhead, the changes had to occur simultaneously; this was difficult and called for many stage-hands and rehearsals for perfect synchronization, taxing the producer with heavy expenditures. Obviating much of all this was a *modus operandi* worked out by Giacomo Torelli (1608–78) known as the "chariot and pole" system. From 1641 to 1645, at the Teatro Novissimo in Venice, Torelli cut slots in the stage-boards and had poles pushed upwards through them to which, above the floor, the wings and other flats were attached. Underneath the stage the poles – or supports – were mounted on carts – chariots – that travelled on tracks at intervals along straight lines parallel to the footlights. The carts were hooked by a maze of ropes and pulleys with weights and a chain of winches to a main winch that, when given a single vigorous turn, impelled all the carts and flats to move at the same time. To the spectator the sight of the fixtures on the stage leaving and being replaced all at once at the behest of an invisible force, as the scene was effortlessly transformed, seemed truly magical. Torelli's efficient solution was soon copied throughout most of Europe and continued to be used almost to the close of the nineteenth century, except in Holland, England and across the Atlantic in the United States, where the older and less intricate groove system was favoured.

At last the front curtain was introduced, but it was on view only before and after the performance. It disappeared into a trough behind the footlights, or rose and finally lowered, or parted at the middle and elegantly drawn into a "festoon drape". Throughout the play's action, however, the stage was curtainless as it had been earlier, the scene changes always visible, along with usually being cued to music, another feature of these productions. In one of his treatises, Furttenbach the Elder urges preparing a special curtain for each play, the decoration suitable for it alone.

Costuming was rich, as was the attire of the upper classes of the period, and who constituted the audience. Members of the cast, portraying heroic figures of the past, such as the gods and goddesses of classical mythology, and love-plagued lords and ladies, needed to be at least as well garbed as the wealthy noble spectators. Also, many of the players were amateurs who were enrolled from the men and women in court circles, delighting in the sport of acting but hardly prepared to humiliate themselves by "dressing down". Even those assigned the roles of servants wore "silks and velvets so long as their masters were clothed in more costly embroidery, laces, and glittering jewels. The aim was to make every actor as unique as possible in style and color, with emphasis on light and definite colors which would enhance the stage picture." This is from Roberts, who also quotes from the *Dialogues* of Leone de' Sommi (1528–92) and his acknowledgement that his players might be performing before friends and acquaintances: "Not only do I try to vary the actors' costumes, but I strive as much as I can to transform each one from his usual appearance, so that he will not readily be recognized by the audience, which sees him daily." On stage, too, might be a titled personage, though most of those actively participating were likely to be chosen from the "dilettantes and scholars", of whom there were always plenty.

For plays set in antiquity, the actor might don a garment loosely adapted from the Greek tunic or Roman toga, but otherwise there was little attempt at historical authenticity, and for most dramas and comedies the performer's own dress was acceptable – providing that it was sumptuous, and regardless of excessive outlay. Each court, on festive occasions – marriages, heralded state visits – sought to outdo the other in lavishness.

De' Sommi allots some pages in his *Dialogues* (1556) to the craft of acting, imparting instruction on effective speech, gesture, posture and physical movement. At all these the player should be as natural as possible. On some days the cast would be augmented by professionals from travelling *commedia dell'arte* troupes.

The Italian Renaissance is acclaimed as an age of great architecture, with some of the world's most gifted artists – Michelangelo, Bramante, da Vinci – lending their genius to creating monumental edifices. Now public theatres, away from the courts, banquet rooms and university halls, finally came to be built, to allow for larger audiences and to facilitate designers in ever more ambitious productions.

The Italian Humanists were somewhat laggard in doing this. Paris already had a permanent theatre – the Hôtel de Bourgogne (1548) – as had Munich; Madrid had two, and London three; but once started, those in various towns of the peninsula far surpassed them in effulgence and capacity.

The first Italian Renaissance playhouse shaped in response to the demand for more space – or at least the oldest surviving example – is the Teatro Olimpico, commissioned in 1580 by the Olympic Academy in Vicenza, a small but handsome city near Venice. The Academy had launched its own company of players and wished to give them a stage. The award to design the theatre, which was to be situated inside the shell of an existing building, went to Andrea Palladio (1518–80). He belonged to the Academy, and with time's passing was to be recognized as one of the world's most influential architects, leaving Vicenza and its verdant hilly environs dotted with neo-classical palaces – now administrative buildings – and large villas in the style that became known as "Palladian", emphasizing a formal balance such as is found in the inspiring and exalted temples of Greece and Rome, their evenly spaced columns, sculptured friezes and architraves empathetically imparting to most beholders an uplifting feeling of dignity and serenity.

Palladio, having pored over the drawings of Vitruvius and thoroughly studied indications in Roman ruins, was strongly guided by them in his plans for the theatre. Unfortunately he died before the task was finished, but left his overpowering impression on it. The architect succeeding him was Vicenzo Scamozzi (1552–1616), who altered some of the scheme for the stage.

Palladio's auditorium is a semi-ellipse, like one illustrated by Vitruvius, with very steep elevation, each row of seats higher than the one before, assuring excellent sightlines. Around the top of this half-circle are columns bearing the ceiling. The broad stage, seventy feet across and only eighteen deep, is of marble. Instead of wings, it has five doors piercing the solid walls, which are also of marble; hence the set is mostly inflexible, permitting very few scene changes. Each of the five doors, when open, yields a vista of city streets "outside": the centre or "royal" door, much the highest and largest, offers a view of *three* streets. (In the spirit of a Renaissance lover of antiquity, Scamozzi likened these openings to the seven gates and extended roadways of Thebes.) Making use of perspective, he did not have these "streets" drawn but instead constructed of plaster and lath in full relief. The illusion is surprisingly real. Above the tiers of seats and columns, and also over the front of the stage, are two rows of statues, numbering eighty, in arched niches. The fore-stage is curved. The whole effect is rich, noble, beautiful.

The inaugural offering was a version of Sophocles' *Oedipus the King* as faithful as possible to the original text. A front curtain, rising and falling, was introduced; wine and fruit were served before the play began; and the ladies in the audience were further regaled and sprinkled with perfume and rose petals, a practice reminiscent of the late decadent Roman theatre.

After more than 400 years the Teatro Olimpico is still intact and is frequently used for concerts and drama festivals, and much visited by sightseers.

Scamozzi was retained by the despotic Duke of Mantua, Vespasiano Gonzaga, to design and oversee the building of another but smaller-scale theatre at Sabbioneta, a village in Lombardy newly acquired as part of his realm. This was completed in 1588. It, too, is contained within a plain, weathered outer structure. Where the Olimpico seats 400, here the horseshoe-shaped arena holds only 300 spectators. The stage is without a proscenium arch and provides at the back only one opening with a vista of streets. Scamozzi himself created the dozen full-length nude classical figures, niched wall paintings and an ornate Corinthian colonnade that enrich the interior. The result, remarkably unified as a consequence of Scamozzi's closely adhering to the spatial proportions suggested by Vitruvius, and of his being in sole charge from start to finish, is of a miniature, oval private theatre in a palace. It, too, is meticulously preserved.

Scamozzi was lucky at having satisfied his patron. The cruel duke was ill-famed for supposedly having poisoned his first wife and her lover, as well as killing his only son, who died from having been kicked in the groin by his angry father. His second wife took refuge in a convent. But the artist's reward for his excellent accomplishment was 110 crowns, a solid gold chain, a silver chalice, and then a safe departure.

Twenty miles from Sabbioneta, the small town of Parma had grandiose dreams. To realize them, in 1618–19 they chose the Ferrarese architect and designer Giovan Battista Aleotti, nicknamed L'Argenta (1546–1636), an early advocate of the use of wings, which he had installed in the Teatro dell'Accademia degli Intrepidi in his native city. He was to give the citizens of Parma and its prince an edifice for the production of drama and music surpassing all others. The result is the Teatro Farnese, so ornate that the spectator's eye is dazzled and confused by its profuse decoration. Painting and statues cover every inch of its ceiling, walls – in arched niches rising row above row – fabulous proscenium and curtain. What is more, it is huge, seating 3,500. By contrast, the Teatro Olimpico seems diminutive and austere. Another difference is that in the Teatro Farnese, named after Parma's ruling family, both stage and orchestra – that is, auditorium – are rectangular, and the floor of the auditorium completely free, so that if desired it can be used for dancing and water ballets – like the elaborately simulated Roman sea battles, *naumachia*. This open space is half-encircled around the sides and rear by twelve mounting, continuous rows of seats, an arrangement that greatly increases the hall's capacity. In the centre is the ducal box, situated above the high, large, main entry. The vast, bare space, of course, is available for added benches and chairs.

Aleotti, abandoning the curvilinear shape of the auditorium favoured by Vitruvius, Palladio and Scamozzi, and replacing the broad, shallow stage seen in the Teatro Olimpico, as well as having a proscenium arch that is both solid and permanent, left as his legacy a prototypical design for theatres that was to be adopted almost everywhere in Europe and America over the ensuing three centuries. His proscenium arch, adorned with sculptures, is credited with being the first of its kind, and associated with it is a rising–falling curtain that conceals scene-changes or marks time-lapses. Behind this front picture frame, at a short distance, is a second one, and a third, constituting a series of inner arches, and between them intruding wings – permitting the players easy access and convenient exits – and finally a painted backdrop, with an illusion of streets or cloud-studded sky and green fields extending further, conjured up by clever perspective. In other theatres to come, the proportions of this narrow, rectangular stage might be greatly widened and deepened to house more spectacular, immense-scaled scenic effects.

Largely of wooden construction, the Teatro Farnese occupied an upper storey of the grim, dark royal palace. It was heavily damaged during the Second World War. In carrying out repairs, the restorers made use of every remaining piece of tinted wood from the original interior.

Venice had a public theatre by 1565, though it was not elaborate; a second one, the San Cassiano, with thirty-one boxes and five balconies, did not arrive until almost three-quarters of a century later (1637), after opera was introduced and had grown very popular. Within a mere four years, by 1641, three other theatres sprang up in that wealthy, art-loving Adriatic city. A republic, the only one among the city-states of Italy, it was a logical place for public theatres to flourish. But here class lines were defined: seats in the boxes and on the first two levels were the most expensive and *de facto* for richer patrons, while the pit – open space on the ground floor – and the top three galleries were the best that the less affluent populace could afford. Here, too, a long-lasting pattern for theatre design was established.

In Florence the Medicis, renowned for fostering magnificence, had Bernardo Buontalenti (1536–1608) convert a section of the east wing of the Pitti Palace into a space for theatrical activities (1585–6); he built a very large stage facing a hall that measured 150 feet by 60, with its floor sloping to allow everybody an unobstructed view of the players. At the front of the hall, at a spot nearest the stage, he sited a dais on which stood a row of thrones reserved for the reigning family. Buontalenti, a master of special effects, put on comedies and other forms of resplendent entertainment with such success that after three years he was able to remodel the hall to resemble a late Roman amphitheatre, with aisles of steps dividing five concentric tiers of seats (1589). Here his productions, and those of Giulio Parigi (1570–1636) and Alfonso Parigi (d. 1656), father and son, attained new heights of scenic grandeur, studied by and a source of inspiration for Furttenbach in Germany and Inigo Jones in England. These spectacles were forerunners of the opulent baroque theatres that came later.

Such grandiose theatres were themselves transplanted palaces. Though larger, they were still not truly "public", being dedicated to an élite: on many occasions the audience was invited, carefully selected and comprised of aristocrats, high clergy, academics, literati, members of scholarly clubs. Again, many factors – the eclectic and erudite subject-matter of the plays, replete with mythological allusions, weighted by extravagant literary language and adherence to strict classical form, and even being displayed in these overwhelmingly splendid playhouses – combined to exclude ordinary folk. The productions were generously subsidized by an indulgent prince or duke or, at his order, by the state's treasury, perhaps to serve a political aim. There was no need to truckle to common tastes, since all expenses were met. Some element of all this is to continue through the ensuing centuries: theatre is to be a cherished but special entertainment attracting a privileged class, the discerning and the rich – though other types of a lively reborn popular theatre now sprang up in Italy and elsewhere.

Besides the pseudo-classical tragedies of Trissino and Cinthio and their learned tribe, another genre of Italian Renaissance drama had a narrow but long-enduring vogue, the pastoral. This is a rather unusual type of play, indeed unique, frequently and appropriately enacted in the open air, on the terraces or in the gardens of the hillside villas of the nobility. It, too, was for the élite and seems to have a remote origin in the poetry of Theocritus (*c.* 380 BC) and other Ancient Greek writers, and in the *Eclogues* of Virgil and his Roman peers, all extolling the beauty of Nature, especially the simple, idyllic pursuits and loves of those living nearest it, shepherds and innocent, virginal shepherdesses. Though sometimes beset by jealousies and more external woes, their troubles prove to be transitory and are resolved happily.

The natural world envisaged in these poetic dramas is never wild and savage, but gentle, a landscape peopled not only with a romanticized peasantry but also with classical mythological figures and benign divinities, laughing fauns and satyrs, dryads, all part of the eternal legend of the lost Golden Age, a past utopia. It was an "escape", a tranquil, pleasing fantasy welcomed by the feuding and often war-weary, over-sophisticated, jaded Renaissance rulers and their courtiers – a healing vision of a world that would eventually be restored to harmony, as in far earlier days.

Italianate pastoral dramas were welcomed in sixteenth-century England, where they were brought to an aesthetic height by a collaboration between the poet-playwright Ben Jonson and the gifted, innovative scenic designer Inigo Jones, and in such interesting ventures as Sir Philip Sidney's *Arcadia* and *The Lady of May*, and by a further host of rivals. Also belonging to this genre are Shakespeare's *As You Like It*, *A Midsummer Night's Dream* and *The Tempest*, fantastic comedies set in the Forest of Arden, in a moon-lit sylvan glade – supernaturally haunted – near Athens, or on a far-off, paradisiacal island, under the spell of a sagacious ruler with magical powers. John Milton entered the list with his *Comus*. In France and Germany, too, such plays were shortly to make an appearance and win a following.

Though the pastoral needed scarcely any stage-setting when presented in the open air, the Italian love of spectacle manifested itself in the actors' costumes, hardly simple even when worn by shepherds and their charming naïve feminine companions. Alois Nagler, in his *Sources of Theatrical History* (a translation of the de' Sommi manuscript in the Biblioteca Palatina, Parma), and in turn quoted by Vera Roberts, describes shepherds wearing animal skins over white silk sleeveless shirts.

Unless they were young and handsome, they also wore a type of fleshings over arms and legs and always some kind of soft shoes. The nymphs were clothed in long-sleeved ladies' shirts, with embroidery and colored ribbons, over which went a long, rich mantle falling from one shoulder and forming the skirt. The animal skins, the mantles, and the ribbons were varied in color and arrangement so that interest would be maintained in the stage picture. Variety in coiffures, some curly, some smooth, some ivy-crowned, was prescribed for the shepherds. The nymphs were uniformly blonde (wigs, of course) and golden fillets or colored ribbons and flimsy, floating veils adorned their heads. These visions carried golden darts and bows, while the shepherds bore leafy branches or sticks.

Hardly realistic! But again, largely the actors were amateurs, not herders of sheep, or, as visitors from a fairy world, they were nymphs who tempted valiant but guileless young men. Actually, they were nobles indulging in the sport of play-acting, while never losing sight of the perquisites of their rank.

When pastorals were enacted in ducal halls, as they were at first, a platform was improvised, but when they were transferred for performance outdoors – the seasonal weather permitting – to the beautifully manicured grounds of summer estates, on hill-slopes with cypress-framed views, landscaped terraces, grottos, fountains, they gained a perfect, appropriate background for their kind of sylvan fantasy: here the semi-permanent stage was surrounded or defined by clipped boxwood or some other high, ambitiously or grotesquely trimmed shrubbery; accordingly they came to be known as "hedge-theatres". Stage-machinery might be added for special effects, often extravagant. Music, pageantry and dance were also conspicuous ingredients; it is generally held that in these aspects the pastoral was a forerunner and contributor to opera, another new Italian Renaissance art-form.

The typical pastoral had very little narrative content and scant dramatic suspense. It was mostly linear, hardly more than an evocation of a wished-for realm, allegorical, extremely stylized and gloriously ideal. In Italy interest in them was short-lived, a span of merely two or three decades. Few of their scripts remain or are remembered, but two of their authors, Torquato Tasso (1544–95) and Giambattista Guarini (1538–1612) were to achieve immortality.

Born in Sorrento, Torquato Tasso, one of Italy's most revered literary figures, very likely inherited much of his creative gift, as his father, Bernardo, was also a poet. Tasso was educated by his father, and at a Jesuit school, and then at the University of Bologna. As early as 1569, when he was twenty-five, his distinctive talent was recognized and he was chosen to be poet laureate. For five years, from 1572 to 1577, he was at the congenial court of the Duke of Ferrara, after which time he set off on lengthy wanderings. He is chiefly known for his epic about the Crusades, *Jerusalem Delivered* (1575). This poem's curious mixture of Christian history and classical and mythological allusions initially stirred clerical hostility.

While he was at Ferrara, the epic was preceded by a very different work, a pastoral, *L'Aminta*

(1573), given its first staging at the Estes family's summer retreat on Belvedere Island in the river Po. Tasso personally supervised the rehearsals. It proved an unquestioned masterpiece, the peak of offerings of this species of fantasy. Its hero, the shepherd Aminta, is enamoured of lovely Sylvia, a devotee of chaste Diana, deity of the hunt. A satyr captures the young woman, binds her to a tree; Aminta and his friends free her. But she, virginal and adamantly true to the dictates of the goddess, will not yield to him. In despair, he twice tries to kill himself, prompted by false reports of her demise. The goddess Diana averts his suicidal efforts, and the ending is joyful.

As in a Greek tragedy, a chorus comments on the action. The work's remarkable appeal arises from its fresh poetry, in tributes to the burgeoning green natural scene – reminiscent, it has been said, of the pristine lyricism of Kalidasa's *Shakuntula*. Its dawn-like quality exerted a spell that was deeply appreciated by the harried, cynical habitués at Ferrara's court. The corrupt present was contrasted to a far better time now gone. In a much-quoted passage, a character, Daphne, laments the New Age's loss of innocence and wonder:

> The world, I think, grows old,
> And, growing old, grows sad . . .
> [Translation: Leigh Hunt]

A complaint voiced more than four hundred years ago!

In addition to its poetic excellence, situations in the story deftly exploit the emotions; there is no lack of heart-rending sentimentality. When Sylvia discovers the inert body of Aminta, who has hurled himself from a cliff in one of his attempts to kill himself, the Chorus relates that:

> It is a feeble love that shame restrains. . . .
> Her eyes appeared a fountain of sweet waters
> With which she bathed his cold cheeks, moaningly,
> Waters so sweet that he came back to life,
> And opening his dim eyes, sent from his soul
> A dolorous "Ah me!"
> [Translation: Leigh Hunt]

The play abounds with topical and local allusions; the mythological characters could be recognized for their likeness to contemporaries in the illustrious circle at Ferrara. Especially complimentary and idealized was the figure patterned on the duke, Tasso's generous patron.

So delighted was its elegant audience with *L'Aminta* that word of its merit quickly spread. In a short period its text had two hundred printings in Italian, nine versions of it in English, along with translations elsewhere. It was ceaselessly imitated by far lesser poets. To the pastoral form Shakespeare and Milton were to add philosophical depth, along with gentle irony, as would Jonson.

Tasso undertook two other dramatic works: an unfinished comedy, *Intrighi d'Amore*, and a pseudo-classical tragedy, *Il Re Torrismundo*, published in 1587 and dedicated to his patron, Vincent di Gonzaga, but deemed less than a success. It explores an incestuous relationship between two sisters.

In his later years actual tragedy befell the poet: his mind clouded and disturbed, he suffered paranoia. Two centuries later Goethe chose him as a subject for psychological study in his play *Torquato Tasso* (1790).

Giambattista Guarini, whose inspiration derived from Tasso, was his successor as reigning poet at the Ferrarese court. His *Il Pastor Fido*, composed between 1580 and 1583, made its stage début in Cremona in 1595, the year of Tasso's death. It was published that year.

This pastoral was destined to have almost as much acclaim and lasting influence as the cherished, enchanting *L'Aminta* – perhaps it was to be held in even more popular affection.

Guarini occupied the post of Professor of Literature and Philosophy at the University of Ferrara. He also served the ruling Estes family on diplomatic missions that had him travelling to Turin, Venice, Rome and even to Cracow, Poland. Eventually, disillusioned by the selfish human behaviour and moral corruption that he encountered, he withdrew from life at the busy court and his role as an envoy. Like Tasso, he was sensitive and scholarly; both had romantic temperaments, easily bruised.

Guarini stated that his only purpose was "to drive away melancholy". Set in Arcadia, a land of woods, fountains and flowers, *Il Pastor Fido* is the light, cheerful story of the shepherd-hero Myrtillus who is passionately drawn to Amaryllis, a nymph. She responds but is compelled to appear indifferent, since her hand has been promised to Silvius, chosen for her by the goddess Diana, the chaste huntress. But Silvius is not interested in Amaryllis, having determined to abandon henceforth the snares of love. Unwarily, Myrtillus confides in Eurilla, hoping that she will intercede on his behalf with Amaryllis. What he does not know is that Eurilla herself is smitten with *him*; she is quite prepared to work against her rival. To add to the amatory complications, Dorinda is infatuated with the would-be celibate Silvius. Declaring her love for him, she, too, is rejected. While Myrtillus sleeps, the repentant Eurilla lays a flowery wreath beside him. As she has anticipated, he thinks the token is from Amaryllis, which restores and raises his hopes in her direction but simultaneously makes her jealous, sensing that he has a competing feminine suitor. Two villains round out the cast. Ambitious in scope, three times as long as *L'Aminta*, the play is a plot-maze, marked by a tangle of emotions, sudden reversals, mistaken judgements, resolved when Silvius changes his mind, marries Dorinda, and Myrtillus and Amaryllis are at last blissfully together, and Diana's high priest, Tyrrhenius, presides over double nuptials. Throughout, explication of the action is helped by messengers and a chorus.

Widely read and admired across Europe, it prompted a host of minor poets to borrow from it. Some elements from it are to be seen in the beautiful *The Faithful Shepherdess* (note the change of gender) by the English playwright John Fletcher in 1610. Handel's opera, *The Faithful Shepherd*, heard in London in 1712, is based on Guarini's play. With a libretto by Giacomo Rossi that was considered to be inferior, it hardly approached the triumphant welcome that a year earlier had greeted the composer's *Rinaldo*, which had a text derived by the same Rossi from an episode in Tasso's *Jerusalem Delivered*. Because Guarini's *Il Pastor Fido* had incidental music by several composers, one of whom is believed to have been Monteverdi, it, too, is singled out as an important precursor of what is to be called "opera", a fully sung dramatic work.

A later pastoral, *L'Amoroso Sdegno* (*The Disdain Caused by Love*; 1597), is perhaps worthy of mention. But scarcely any others. Save for the offerings of Tasso and Guarini, characters in the pastorals are superficial, thinly motivated, too coy and sentimental, too unreal. But the absurdly artificial stories did afford their over-civilized, over-wealthy sponsors a way to support pageantry, while they and their guests indulged in wish-fulfilment – a similar cry for a "return to Nature" is heard at regular intervals in every epoch. These poetic and dramatic fantasies permitted the nobles themselves to display their finery and have innocent fun while participating and acting in them.

Some critics see a Freudian explanation for the fascination by a decadent aristocracy with the idealized "young", "innocent" and "pure" representations of the lower classes, here well scrubbed, conveniently immaculate and beautiful.

Italian comedy of literary quality followed much the same course as the attempt at tragedy, but it had earlier and longer-lasting acceptance and success. One reason for this is that it had fuller and deeper sources. As has been seen, in Germany the tenth-century Benedictine nun Hrosvitha was acquainted with the works of Terence and translated and circulated her bowdlerized versions of them. Nicholas of Cusa discovered twelve comedies by Plautus in 1429, and shortly afterwards a commentary on Terence by Donatus was found – also in Germany – by Cardinal Giovanni Auspira of Mainz. Terence's *Adelphi* was the subject of a lecture by Magister Johann Mandel, of Amberg, at the University of Vienna in 1455. The invention of movable type helped to distribute the texts widely. Soon after the turn of the new century the great scholar Erasmus of Rotterdam was firmly of the opinion that "Without Terence, no one has yet become a good Latinist." Plautus' *Menaechmi* was enacted at the Ferrarese court for the theatre-loving Estes family by 1486. So Italian poets aspiring to write comedies had excellent models to copy, for the Roman playwrights – who, in turn, had borrowed extensively from the Greek Menander – were superior at weaving efficient plots, providing their works with effective dramatic structure, while creating lively, plausible stock characters. They were also a bank of humorous ideas, risible contrivances and situations. The Italian authors of comedy set about robbing that bank.

Besides these, there were also the remembered noisy, capering skits on the Corpus Christi pageant-wagons to feed them bits of comic "business", as well as the impudent and insolent pranks of the uninhibited Feast of Fools, followed by the French *sotie*, the German *Schwank* and other medieval farces. There was no lack of precedents and material, all of which – something from here and there – could be incorporated in a supposedly new offering.

Further, the Italian language and temperament were a good match, truly attuned, to the mellifluous Latin and witty imaginations of their Roman forebears. It was a direct legacy. Indeed, the sway of Terence – and, to only a slightly lesser degree, that of Plautus – over the creative impulses of Italian farce writers cannot be exaggerated. They played the same part in comedy as did Seneca in tragedy, evoking countless translations, adaptations and often slavish imitations.

Acknowledging its classical roots, this new light genre was called the *commedia erudita* – "erudite comedy" – and one of the very earliest to enter the field was the major poet Petrarch, the "father of Humanism", who based his *Philologia* (1349) on Terence. Unfortunately, it is a lost work. In turn, Plautus was the inspiration of Tito Livio dei Frulovisi, of Ferrara, who turned out a series of similar humorous pieces, also in Latin. Piero Paolo Vergerio contributed a *Paulus*, Humanist in spirit, before this crucial century's end. Owing much to Terence, too, the author ingeniously added a few contemporary touches.

A necessary step for the popular appreciation of comedy was a transition from Latin into the vernacular. Even at courts as sophisticated as those of Ferrara and Mantua, not all in attendance – perhaps not even a majority – were conversant enough in that ancient tongue to comprehend the action and catch the jokes that spark a quick-moving farce, nor were many if any of the actors likely to be. These early literary efforts were usually given intimate readings, hardly ever performed. A good many decades elapsed before another important Italian poet, Lodovico Ariosto, yet one more beneficiary of the generous Estes family, ran up the curtain – so to speak – with his *La Cassaria* (*The Strong Box,* sometimes known as *The Casket* in English; 1508) and his *I Suppositi* (*The Counterfeits*; 1509) at the always receptive Ferrara court. It was *La Cassaria* that Pope Leo X had repeated at the Castel Sant'Angelo in Rome in a production boasting striking designs by Raphael. With the great success of these two plays, amusing and readily understood, the *commedia erudita* was merrily on its way.

Three authors of this new type of comedy are still widely known: Ariosto, Niccolò Machiavelli and Pietro Aretino, but they were surrounded and followed by a covey of others whose names and often creditable works cry to be heard. Some were responsible for pseudo-tragedies and have already been mentioned: Giangiorgio Trissino, Cinthio, Ludovico Dolci, Luigi Groto.

World-honoured for his poetic epic, *Orlando Furioso*, which some declare to be the finest work of its kind ever written, Lodovico Ariosto (1474–1533) was a native son of Reggio, and by good fortune was of noble lineage. Residing in Ferrara, the foremost cultural capital of fragmented Renaissance Italy, he was chosen as court poet and spent most of his career there. Given his aristocratic background, remarkable gifts and accomplishments, he could mingle at ease in its social life. He was all too well acquainted with the elegant men and women of his time, and looked on his world with a sharply critical eye. He was well aware of human peccadilloes and hypocrisies. Though he borrowed his plots and devices from the Romans, he changed his characters to have them more representative of contemporary follies. At the same time, he could count on an intelligent, worldly audience closely akin to the courtiers and intellectuals whom he was cavalierly depicting – a curious paradox, but one that assured that the spectators could identify with caricatures of themselves or their friends and acquaintances.

La Cassaria relates in a series of loosely linked episodes how two young men eagerly seek enough money to purchase a pair of slave girls in service to Lucrano, a bawd. With the help of their servants, they succeed. Ariosto wrote this play twice, first in prose (1508) and again in verse (1529), when more practised at stagecraft. What is immediately distinctive in the earlier script is the language, remarkably compact and straightforward, together with initial signs of his talent for creating genuinely individualized characters.

He is chiefly famed, however, for his second venture, *I Suppositi*. He composed this in prose, in his first handling, and two decades later (1529) alchemically rewrote it in verse. He combined plot elements from both Plautus (*The Captives*) and Terence (*The Eunuch*), much as those Roman writers themselves blended ideas from several earlier Greek New Comedies. Ariosto's work, in turn, inspired bits of Shakespeare's *The Taming of the Shrew* (1593–4) and much more of William Wycherley's *The Country Wife* (1675), by way of George Gascoigne's *Supposes* (1566), adapted directly from Ariosto – all of which confirms that one should not expect much true originality in comedy.

In this work Ariosto again uses Ferrara as a setting, and tells how an ardent student, Erostrato, poses as his own servant to win entrance to the house of his beloved, whose father wishes to marry her off to the elderly, wealthy Cleandro. Again, a conniving servant is the spring of the complications, persuading a traveller to impersonate the student's father, lending respectability to the young man's suit, and to offset the dowry of 2,000 ducats promised by Cleandro. The young man's real father inopportunely arrives; troubles and misunderstandings ensue and multiply. The play is distinguished for its fine style, clear and fluent, its successful characterizations, its always bright dialogue. For its presentation in Rome, Raphael introduced a front curtain that was lowered into a trench before the play started, as in late Roman arena theatres. Since much of the action borders on the salacious, or goes somewhat beyond it, one wonders that the Pope and his Vatican court found it so amusing. Its acclaim spread, and it was soon performed at other ducal courts, and as school offerings, and was adopted even by touring *commedia dell'arte* troupes. On the festive occasion of the marriage of the Infanta Maria, daughter of Charles V, Holy Roman Emperor to Maximilian of Austria, the Accademia degli Intronati of Siena chose to honour the royal couple by staging Ariosto's comedy for them at Valladolid in 1548.

Il Negromante (*The Charlatan*; 1520) has a good many touches that seemingly are of Ariosto's invention, as he moves with growing boldness from his dependence on the remote Plautus and Terence. Astrology is one of the butts of his satire here, and the leading character, Jachelino, who gulls everybody, may well have been the model for Ben Jonson's Volpone, that most monumental of rogues. The young hero of this piece must keep secret his having a wife, while he disports in plain sight and seeming flagrance with a mistress. To free himself of this entanglement, he complains of certain physical ailments, which Jachelino purports himself able to cure. Here are foreshadowings of Molière, who was quite as zestful in poking rude fun at mercenary quack doctors.

In 1528 Ariosto was appointed by the Estes family court to organize fêtes and dramatic entertainments; a theatre was built to meet his specifications. *Il Negromante* was produced in it, as well as plays by others which gave him substantial influence over the development of the *commedia erudita*. He himself took roles in the play, which gave him vital experience as an actor. Not only as a writer but as a theatre manager and director, he established modes for staging that were looked to for guidance by less knowledgeable Renaissance impresarios throughout Europe.

In classical style, though a wooden structure, Ariosto's theatre had a raised platform with a painted backdrop, depicting a street façade of five houses, each with a pointed roof and just one door and one window, not unlike the scene – *scaenae frons* – in the day of Plautus and Terence, when the windows allowed characters to lean out and add their voices to a comic tumult below, and the doors permitted them hurried entrances and exits, the accelerated pace abetting the farce. Sets similar to this were to exist almost to the twentieth century for certain kinds of plays.

The licentious courtiers around Ariosto saw themselves mirrored in his *Lena* (*The Bawd*; 1529), which adds up to a bright and keen-edged social documentary. The same has been said of his *Scolastica*, left unfinished at his death and completed by his brother Gabriele, after which a twenty-year-old Christoph Stummel at Wittenberg successfully adapted it into German as *Studentes* (1545) and precociously made a name for himself. As portrayed by Ariosto, two Parisian students fall in love with a pair of girls in Ferrara and are helped in their courtship by a wily servant, Accursio, and by a friendly lodgings-keeper, the principal comic foil. Here are very familiar devices carried over from Plautus and Terence, but a good deal that is new, including harsh bolts of satire against easily bribed churchmen, in the person of a corrupt friar whose itchy palm is ever open, while he piously remarks that a bit of "charity" can always buy forbearance or forgiveness. (One can see why Stummel's version was loudly applauded at Wittenberg, where Melanchthon, the noted Humanist scholar and Martin Luther's sturdy friend and fellow leader in the struggle for Church reform, was among the very academic spectators.) Interesting vignettes of university life are provided.

In all, only these five scripts by Ariosto are extant. How far his example and writing reached is shown by the careful study by John Dryden of *Orlando Furioso* – in Italian – when he and Robert Howard sat down to sketch out their grand poetic and "heroic" drama *The Indian Queen* (1633). What impressed Dryden was the epic's firm design and variety of characters.

Ariosto is an agent of transition in the history of comedy – especially Italian comedy – still making use of the formulas of the past, but creating and anticipating a new spirit, in capturing the voice of his own times, bringing a needed and revealing contemporaneity to the stage.

Somewhat out of character – one might say – is Bernardo Dovizi da Bibbiena (1470–1520), who became Cardinal Bibbiena in 1513, and whose *La Calandria*, offered the same year, takes off somewhat loosely from the *Menaechmi* of Plautus while following Ariosto's lead in blending Roman and contemporary elements. His adaptation is bolder in its immorality than is the original script. Nevertheless, after its initial staging at Urbino, it delighted Pope Leo X, who frequently had it produced at the Vatican for the enjoyment of his guests and the Curia. His Holiness, obviously possessed of hedonistic impulses and worldly tastes, took unbounded advantage of his exalted rank, and was markedly fond of this form of non-ecclesiastical theatre.

A sub-plot for *La calandria* is borrowed from Boccaccio, no exponent of decorum. Two pairs of twins, of opposite sexes, are confusingly involved because of their strong resemblances and reckless disguises when engaging in illicit liaisons. In 1518, paying tribute to Leo X, the newly appointed papal envoy to France, also a Cardinal and a Bibbiena, arranged an expensive revival in Rome of the comedy, with the innovative sets in deep perspective by Baldassarre Peruzzi; they occasioned amazement and excited praise from Vasari, as already mentioned. Again, in 1548, Florentine expatriates in a self-constituted colony in Lyon mounted *La Calandria* as part of a grand reception honouring Henri II of France and his Italian bride, Catherine de' Medici.

(That a Cardinal wrote plays was not altogether exceptional. A century earlier, in 1444, Aeneas Sylvius Piccolomini, later to be enthroned as Pius II, fashioned a comedy, *Chrysis*, its plot and characters largely hearkening back to ever-reliable Terence, on whose works the Pope-to-be had doted in his youth.)

Men of distinction in every field of endeavour entered the play writing lists. It was the fashion, and encouraged by, among others, the indulgent Leo X, as well as by the highly literate, open-minded rulers of Ferrara, the free-spending, art-loving Medicis of Florence, the competitive court of Mantua (where Isabela d'Este resided after a dynastically arranged marriage to a Gonzaga and where she enthusiastically lent a hand to a staging of Terence's *Adelphi* in 1501 and other dramatic works) and by the warring, civilized Montefeltros of Urbino.

So the playwright who now joined the eager throng did not seem wholly out of place, though he was neither a poet nor a conventional scholar. He was none other than Niccolò Machiavelli (1469–1557), son of a middle-class Florentine lawyer and minor government functionary, who was also owner of a small villa and farm ten miles from the city. As a youth Niccolò, like his father, studied law, but dropped it; however, he was well versed in Latin and thoroughly familiar with Roman history. He had scant interest in the arts; his bent was for politics. As had his father, again, he took lowly official posts. Ambitious and alert, delighted and excited by the prospects this world offered, he climbed rapidly. By the age of thirty he had risen to be Chancellor of the Florentine Republic and, admired and respected for his brilliant, incisive mind, served as its able and zealous ambassador to neighbouring petty duchies, city-states and France. Closely associated with Cesare Borgia, he participated in the endless, Byzantine political feuds and intrigues of the day, the internecine struggles for power at home and abroad, that bred in him cynicism and distrust. At the same time his dream, quite visionary, was to achieve the unification of Italy, gathering together its many small principalities, an idealist goal he sought centuries before its actual realization.

He had an unhappy marriage, his domestic frustrations deepening the harsher side of his nature. He was hardly monogamous; having an unusually strong sexual drive he frequented brothels. He reported his exploits there in coarse letters, writing boastfully to one friend, as he neared his fiftieth year: "Cupid's nets still enthrall me. Bad roads cannot exhaust my patience, nor dark nights daunt my courage. . . . My whole mind is bent on love, for which I give Venus thanks" (quoted by Ludwig von Pastor, *History of the Popes*).

When the Medicis superseded the Florentine Republic, Machiavelli faced persecution, was put to the rack four times, and was banished to brooding, bitter exile on the family farm outside San Casciano. There he considered the infamy and treachery of mankind as he had experienced them. As the ensuing fifteen years passed for him, he drew on his unsparing observations of men and their affairs to compose a short, shrewd treatise, *The Prince* (1513), one of the most famous books ever written. It expounds and analyses the unscrupulous means by which a ruler seizes, holds on to, and perpetuates his rank over all rivals, a handbook for those who are self-serving, cunning, devious: it led to "Machiavellian" becoming a pejorative term. Much of the instruction he gives is based on his having watched from near advantage the despotic, manipulative Cesare Borgia at work, though later portraying him as an exemplary statesman, a patriot such as he also deemed himself to have been. During this time he also wrote an essay on *The Art of War*, and over five years dedicated himself to a *History of Florence*, but at the suggestion of Leo X a chronicle containing more than its share of errant details slanted to his own purposes, and to some extent plagiarized.

His other enterprise, in his often tedious and impoverished exile, was putting to pen several stage-scripts, though he never considered it his principal vocation; so it is ironic that of all the authors of *commedia erudita* he is by far the most famous, and the only one of that period who has left a play still intermittently revived, to applause and hearty laughter after the passage of more than four hundred years.

He translated a Latin work and brought out three or four scripts of his own, of which two remain. Aristotle defines comedy as exposing the worst side of human nature, the universal frailties to which envious mind and weak flesh are heirs. The baulked and misanthropic Machiavelli, disenchanted with his fellow citizens, was well equipped by temperament and personal history to spin just that kind of satiric picture of their goings-on.

In *Clizia* (1506) folly reigns in a Florentine household. The husband, Messer Nicomaco, seeks to marry off his ward to a servant for most ignoble reasons: he wishes to cover up an adulterous affair of his own. His wife, in turn, plans to have the girl wed to a more appropriate suitor, a bailiff. On the wedding night, a lad – a page – is substituted for the unwilling bride, and at the climax there is a recognition scene – the long-absent father of the girl happily arrives to retrieve her – and Messer Nicomaco is revealed as the rascal he has certainly been. But he is not really all bad. In the opening moments he has been seen as a decent, respectable merchant, and now at the end he reverts to his inherently good self; the point made about him is that his lustful impulses have temporarily been depriving him of his senses. This resolution strikes many critics as quite unconvincing, far too easy and pat. Apart from that, however,

Machiavelli draws an excellent, rounded portrait of him, and the dialogue is exceptionally deft.

That even this distanced and sophisticated author resorts to the old plot-tricks of implausible disguises and the changeling motif is further evidence of how firmly fixed on the comic writers was the undying formula of Menander and, after him, Plautus and Terence. It is scant wonder that a sensible commentator, Antonfrancesco Grazzini, called Il Lasca, finding fault with the farces of his day, offered a strong complaint: Italy was not Greece or Rome and deserved a stage truly reflecting its own culture. "We have other manners, another religion, another way of life. . . . There are no slaves here; it is not customary to adopt children; our pimps do not put up girls for auction; nor do the soldiers of the present century carry long-clothed babies off in the sack of cities, to educate them as their daughters and give them dowries; nowadays they take as much booty as they can, and should girls or married women fall into their hands they either look for a large ransom or rob them of their maidenhood and honour" (quoted by John Addington Symonds, *Renaissance in Italy*).

Machiavelli, at least, carried Italian farce a bit forward from this outworn if not wholly outgrown pattern. He demonstrated this more surely in his singular masterpiece, *La Mandragola* (*The Mandrake*; 1514/1520). Here is a world of "fools and rogues".

A fatuous lawyer, Messer Nicia Calfucci, has a resolutely virtuous, beautiful wife, Lucrezia. Travelling from Paris to Florence, the lecherous Callimaco hears of her charms and vows to possess her. He enlists the aid of a notorious parasite, Ligurio. They learn that the elderly Calfucci eagerly hopes for a son and heir, but is unable to beget one, and they concoct a preposterous plan. The insidious Ligurio tells the childless lawyer of a doctor who has worked magical feats, and introduces the credulous husband to Callimaco, now disguised and dressed in the robes of a medico. He is informed that in France the king and his nobles are making use of the juice of the mandrake, a narcotic plant, as an aid to fertility; however, the bed-partner who dares to partake of it is doomed to a quick death after a single night of love-making. Where to find such a heedless, impassioned, sexually obsessive performer? It is suggested that some unwary young man be kidnapped and be lured to his death after having fulfilled his purpose as a captive if enraptured lover. With some difficulty, at her husband's urging, and the added persuasion of her mother and her confessor, the reluctant, tearful – if somewhat sex-starved – Lucrezia is brought to consent to the scheme. The lusty Callimaco, now his own true self, allows himself to be kidnapped, spends the assigned night with Lucrezia, who has downed the harmless potion, gains her passionate affections, and joins her in laughter at her husband's absurd credulity.

As can be seen, no one is spared in this farce; all the characters are painted with black edges, even Lucrezia's mother and the friar in whose counsel she meekly trusts. But there is strength in Machiavelli's invention, and wit in his writing, and sharpness in his characterizations. Besides, *La Mandragola* is wholly original – it is adapted from no previous script. In view of this, most critics find it regrettable that Machiavelli did not devote himself to providing more such stage pieces.

London enjoyed a restaging of the play, adapted into English by Ashley Dukes (1885–1959), at the Mercury Theatre in 1940; and it was lucratively brought back to life in Milan in 1965. The season of 1977/8 saw off-Broadway productions of both of Machiavelli's extant scripts. The rarely given *Clizia* was put on at the Nameless Theater, where it was directed by Catherine Ruello and had music by Lynda Kaplan. It caused little stir. But Wallace Shawn's treatment of *La Mandragola* had better auspices at the Public Theater and was a decided hit. As viewed by Richard Eder in the *New York Times*:

> The Public Theater is calling this new version of *Mandragola* a "workshop." This means that it is not a finished production, but still under development.
>
> If *Mandrake* – its translated title – is a "workshop," it is one in which silver hammers are beating out a marvellous contraption: all springs, odd gobbling noises, utterly fresh clowning and enchanting end-runs.
>
> Directed and designed by Willard Leach, *Mandrake* has rough spots. Some need fixing; others seem intentional and can be oddly effective. It has dead moments, and sometimes its talents are applied too heavily and without letup. But even as it stands it provides some of the most stimulating theatre in town, and its further possibilities are very bright.
>
> The play is broad and cynical. . . . Mr Leach has turned it into a circus of incongruities, using buffoonery akin to that of the classic *commedia dell'arte*, but it is governed by a delicate musical rhythm.
>
> A singer, Thelma Nevitt, appears at intervals to deliver a series of lofty Italian lyrics to music – by Richard Weinstock – that range from imitations of classic Italian song to gutter Neapolitan. Wearing ridiculous paper wings and an expression of besotted idiocy – and singing extremely well – she cuts across the stage action with angelic inappropriateness.
>
> Her first song, for example, a fluttery, flowery number, is followed by the appearance on stage of Wallace Shawn, the translator. He is playing the role of Callimaco's servant. He is not a polished actor, but he is terribly funny. He comes on looking like somebody who has lost his way on Canal Street and delivers a vehement and essentially meaningless prologue.
>
> Now Mr Shawn is not only vehement, but also tiny – surely not more than about four foot ten – plump and pale. And when Callimaco appears, played by James Lally, who seems to be about six foot four, the effect of their furiously paced exchange is of a mouse arguing with a moose.
>
> Mr Lally alternates between catatonia and hysteria. He does not simply want Lucrezia; he wants her urgently, immediately, as a three-month-old baby wants its early-morning feeding. His servant and his scheming friend Ligurio try to calm him down. They get him to dress as a doctor so that he can give the necessary advice to Nicia. He is calm and judicial for a moment, but breaks out into a howl the instant Nicia shows signs of hesitation.
>
> The performance needs polishing. Mr Lally's facial gestures are splendid; he is a bit forced and awkward when he is capering around.

Three virtually perfect performances provide the heart of a comedy that is both broad and nuanced. Larry Pine brings all his caterpillar talents to the part of the scheming but absent-minded friar. Angela Pietropinto is infinitely unctuous and greedy as Lucrezia's mother.

Corinne Fischer, playing the virtuous Lucrezia, is the most dazzling of all. She is pale and impassive as a wax doll. Counselled on all sides to betray her husband – with his consent – to promote fertility, she breaks out into an extraordinary array of sobs, squeaks, yips and gibbers. But her mother's qualities are in her: the next morning, the deed done, she is all demure lasciviousness.

Mr Leach has brought a rhythm to the production that makes it a dance of comedy and passion. He has devised a ring-shaped turntable: rooms, doorways appear from the left side and circle, clockwise and stately, to disappear on the right.

There is an extraordinary scene, dappled by moonlight, and accompanied by a quavering song, after all the scheming is done. First the servant and the friend appear, seated, chatting. Then come Nicia and Lucrezia's mother. The turntable carries them off. It brings on the lovers, in bed, frozen in argumentative exploration. And they disappear as well, while the song goes on. It is haunting and pure magic.

A later notice of the play by Mel Gussow, also in the *New York Times*, described Shawn's translation as "felicitous", the work a "malicious sixteenth-century study of sin and corruption", and the result "a campy, rib-nudging farce". For Gussow: "The production is elevated by some of the performances, particularly Angela Pietropinto's as a manipulative mother, and Mr Shawn's own as an effusive lackey, impishly bowing his body to the groveling position."

Two conversions of the play to a musical comedy were also exhibited off-Broadway. The first, *Mandrake* (1984), was brought over from Britain and presented by the SoHo Rep on Mercer Street. In the *New York Times* Stephen Holden found most but not all of the offering to be persuasive:

Mandrake is the kind of musical that Cole Porter might have written had he survived into the age when actors could casually toss around four-letter words. Based on Machiavelli's sixteenth-century satirical comedy, it portrays a Florentine society obsessed with sex, money and "image." Though the church is still powerful, in the words of Fra Timoteo, the local priest: "Attendance has dropped from its medieval peak. Devotion is no longer chic."

By adapting it into a kind of wised-up Gilbert and Sullivan pageant, Michael Alfreds and Anthony Bowles, the Englishmen who created *Mandrake*, have softened the harsher implications of Machiavelli's play. We are meant to enjoy this sensual romp more than we are asked to condemn it. And Mr Bowles, the show's composer, who also directed its American première, has staged it as an odd but engaging mixture of quaintness and titillation.

Mr Bowles's score embraces divergent musical styles with a shrewd sense of the ironic relationship between musical idiom and character, and the background instrumentation, consisting of piano and harp, allows some clever ancient-modern musical juxtapositions.

Mr Bowles's music helps to humanize characters who might easily have been treated as monsters. When Fra Timoteo, having given his approval to a sordid seduction in exchange for alms, confides his venality, it is in a wistful *bel canto* aria that underscores his genuine pangs of conscience. Ligurio, the bluff potion seller whose greedy machinations set the events in motion, has songs that come close to rock. Andrew Barnicle plays Ligurio as a Florentine Dr Feelgood, plying aphrodisiacs, chastity belts and assorted erotic paraphernalia with an unflappable charm and self-assurance.

But the idiom to which the score returns the most frequently is Cole Porter. Porteresque beguines and tangos propel several elaborate production numbers whose intricate part singing is matched by profuse internal rhymes and double-entendres. But the changes in tone from ultrasophisticated patter to streetwise obscenities are so jarring here and there that the mood gets broken. In general, however, the show succeeds in maintaining a tone of elegantly hip irreverence. The best songs – *She Never Knew What Hit Her*, which suggests a hilarious, smutty update of Porter's *Let's Do It* – and *The Waters of the Spa*, which extends the ambiance of Stephen Sondheim's *Weekend in the Country* into an orgiastic free-for-all – have the grand panache of the most accomplished show tunes.

From his strong cast of twelve, Mr Bowles has elicited ensemble performances that have a joyously warm-blooded physicality. Besides Mr Barnicle's Ligurio and Tony Alexander's Fra Timoteo, Mary Testa is particularly outstanding in the role of Dora, a Sicilian servant girl who is the only character not wandering about in a self-deluded haze.

But with all its delights, *Mandrake* is not a wholly successful production. After building a ribald head of steam in a first act that culminates with three successive showstoppers, the second act rushes to tie up the loose ends of a very involved plot and never regains the same momentum of bursting effervescence. And the tiny, cramped stage of the SoHo Rep is precisely the wrong kind of space for a show that wants to spill in all directions with the lavish pageantry of a full-scale comic opera.

(The book and lyrics were by Michael Alfreds; the set design by Joseph A. Verga; the costumes by Steven L. Birnbaum.)

A further update was *Whores of Heaven* (1987) by Michael Wright. Opening at the American Folk Theater, where it was presented by the I Comici Confidanti Company, it transferred Machiavelli's scene from sixteenth-century Florence to the West Village of Manhattan in the 1980s. Again, Stephen Holden of the *New York Times* was on assignment there.

Whores of Heaven, a chaotic musical farce, is an extreme example of the Off-Off-Broadway musical as a frantic, eclectic junk heap of styles thrown together under the scatter-shot philosophy that if the last joke didn't knock 'em dead, the next one will. . . . Its characters include a sexually impotent Mafia don running for mayor, a corrupt priest who has a secret career as an investment banker and other assorted nasty hybrids of classic *commedia dell'arte* and contemporary wheeler-dealers.

Directed and choreographed by Kim Johnson, the production bursts across the stage with the wild careering energy of a two-and-a-half hour burlesque show transported to a contemporary Italian street festival. Mr Wright's digressive score encompasses at least a dozen different idioms, from fake Renaissance chorales to smutty rap songs to knowing pastiches of Mediterranean pop ballads. The songs often run together in extended musical set pieces during which the cast, interacting with a puckish narrator (Suzanne Parke) speaking in mock-Shakespearian couplets, dashes around the stage in masks and disguises that range from the Statue of Liberty to Groucho Marx to the Arab chieftain.

Though *Whores of Heaven* displays a good deal of talent and energy, the production plunges too far over the line separating farce from anarchy. And the show's young, frisky cast of actor-singers lacks the vocal wherewithal to give Mr Wright's elaborate, witty score its due. Crucial duets and ensemble numbers are sung off-key with many of the pungent rhymes not properly enunciated. If *Whores of Heaven* is to have a future – and its premise is intriguing – its current production should be seen as an early stop on the road to a shorter, more disciplined, better-sung show.

Somerset Maugham raided *La Mandragola* for the substance of his short novel, *Then and Now* (1946), and a film version of the play has won praise for its artistry.

After the Medicis were overthrown and the Florentine Republic was restored, and the Papacy had undergone momentous changes, the playwright-philosopher sought to regain his former government posts. His every bid was rejected. Shortly afterwards he died of a painful stomach ailment, knowing that he left his wife and children utterly penniless.

The third of Renaissance Italy's important comic dramatists is emblazoned in history more for his odd and vibrant personality than for his written works. Pietro Aretino, born on Good Friday, 1492, in Arezzo (hence "Aretino", the name by which he is best known), always acknowledged that he was illegitimate, purportedly the son of a poor shoemaker, Luca, and a beautiful if amoral young woman, Tita (Margherita) Bonci, who had sometimes been chosen by local painters to model for their portraits of the Virgin Mary. Yet he laid no claim to having been immaculately conceived; on the contrary, he often asserted that his actual father was an Aretine nobleman, Luigi Bacci, who was notorious as Tita's casual lover – and in later years, when Aretino had won great fame, Luigi Bacci's other sons openly accepted him as their half-brother. Typically, he was inconsistent in putting forth this story of semi-aristocratic birth, at some moments electing to boast of the very humble origin from which he had risen to extraordinary heights. So the facts are uncertain.

Though his life began in straitened circumstances, he quickly showed signs of unusual intelligence; while still a small child he astounded his elders by his precocious ability to read, write and cite passages from the Bible. Biographers disagree as to whether he had any early schooling. If so, it was very elementary and of scant scope and duration.

For some reason, he left home at a surprisingly tender age, somewhere between twelve and fourteen. One explanation given is that he was expelled from his Arezzo school for having written and passed around a sacrilegious poem. Fearing that Luca would severely punish him, he did not return to his house with word of his offence, but instead took to the road with commingling apprehension, rebellion and bravado. He began his lifelong wanderings. His first destination, fifty miles away, was Perugia, where he gained employment as an assistant to a bookbinder. In that city, the birthplace of Perugino and Raphael, he studied art and absorbed a great deal of knowledge from the books he helped to enclose. His stay in Perugia came to an abrupt end, however, with another expulsion, this time for "defacing" an image of the Madonna that piously graced the main piazza: the good citizens of the town discovered that he had changed the sacred painting to have her frivolously holding a guitar in her hands. (Again, there are variants of this account: some say the painting was of the Magdalene kneeling at the feet of the Redeemer; the alteration made it look as though she was not praying but serenading Him.)

He found work as a household servant in Rome, until he was discharged for the theft of a silver cup. From there he went to Vicenza, the showplace of Palladio, and supported himself as a street singer; then he found it more profitable to pose as a wandering friar, seeking alms. On from there to Bologna, his occupation now that of a hosteller, and next that of a moneylender – or, more likely, a collector for one, since he himself had no funds to speak of. (The principal source here is Thomas Caldecot Chubb, editor of *The Letters of Pietro Aretino*, from which quotations will shortly follow.) Many of the subsequent and brief careers ascribed to him may have sprung from the imaginations of his enemies, of whom he was to have many, some quite venomous. But scholars believe the index is mostly accurate: by turns, and always proving himself resourceful and versatile, he was a tax-collector, a mule-skinner and a hangman's helper, a grisly task from which he did not shrink. To this list must be added his having spent a term in the galleys, for some misdeed or other, after which he became "a miller, a courtier, a pimp, a scholar's groom, and a courtier's lackey". Even if only half of this résumé is factual, it is quite daunting. As Chubb says, "With the most generous allowances for malice, it is clear that he did not live a very edifying life."

In 1516, still in his mid-twenties and having been ousted for lechery while working in a monastery, the youthful Aretino returned to Rome. This time his menial post was in the house of a Sienese banker, Agostino Chigi. He was treated well, but was not content to remain a mere servant; in his spare moments, protesting, he scribbled satiric verses. He showed these to Chigi's guests, who recognized the sharpness of his wit. One of his bitter poems described the harsh, demeaning life of a scullion, a lowly dishwasher like himself, "cleaning privies, polishing chamber pots . . . lighting candles, performing lewd offices for cooks and stewards who soon see to it that he is all pricked out and embroidered with the French disease" (quoted from his own *Dialogues*).

Then a trivial piece of doggerel brought about a significant enhancement of his fortunes. Leo X had received a present, a huge elephant named Hanno, from King Manuel of Portugal. The Pope grew exceedingly fond of the beast, as did the citizens of Rome; and when, after two years, Hanno died of angina, the Pontiff was deeply bereaved and disconsolate. Ordering Raphael to paint a life-size portrait of the lost creature, he had the vast canvas hung above the tomb of the beloved animal. Not sharing Leo X's grief, the young Aretino was instead amused. He dashed off a brief, mocking pamphlet, *The Last Will and Testament of the Elephant*, replete with gibes, many of them obscene, targeted at almost every person of dignity and lofty status in Rome, not excluding the Holy Father himself. The jeering pamphlet was hawked throughout Rome and set its citizens laughing disrespectfully at the mournful Pontiff, his Curia and the official entourage. Though the *Will* was unsigned, its authorship was readily traced and exposed. Soon the brazen writer was summoned to the Vatican. Very nervous, he responded and confronted Leo X, who was not angry but, always tolerant, had joined in the laughter evoked by the parody. To prove that he appreciated good fun, and was not offended by rough humour, he offered the rootless Aretino a place in his household, where his duties would be something between those of court poet and candid-tongued jester.

Established as a writer, Aretino had three comfortable years, during which – as one biographer puts it – he ate well. The sudden death of Leo X, his protector, changed that. Most impolitic and imprudent, he wrote a series of satires on the electors gathered in conclave to choose Leo's successor, also verses ridiculing the leading candidates. He had an alert ear for clerical gossip and swirling political ploys and strategems at the Roman court. He made his information and insinuations public in insulting verses that he attached on sheets of paper to a statue of Pasquino, as was a local custom (and from which derives the term *pasquinade*). Though this was done furtively, their authorship was all too clear. They were in the form of *sonetti codati* (that is, "sonnets with a tail to them", so poisonous that their sting hurt like a scorpion's). Aretino supported Cardinal Giulio de' Medici, the late Leo's cousin, but erred. Chosen as the new Pope was Adrian VI, who announced that he would institute wide moral reforms. Recognizing that many at the Vatican and elsewhere in Rome were murderously hostile to him, the hitherto heedless satirist and lampooner hastily left the city.

He took flight to Florence, then moved on to Mantua, where he was able to obtain another post with Federico Gonzaga as court poet, though one much less lucrative (1523). But his literary output increased and became more mature. He gained an important patron in the person of Giovanni dalle Bande Nere, a Medici *condottiere*, whose camp he joined and with whom he formed a close friendship.

After reigning barely more than a year, Adrian VI died, leaving the Papacy vacant, much to the relief of the Romans who had chafed under his ascetic rule. His successor was a rich Medici,

Clement VII. Aretino, like a host of other artists, poets and connivers, hurried back to the "liberated" Holy City, eager to participate in a fresh regime and share in the bounties a wealthy Medici might confer. But Aretino's welcome and stay were short. The talented Guido Romano had painted a series of twenty illustrations of positions that might be used in performing sexual intercourse; engravings were made from them by Marcantonio Raimundi. For sixteen of these Aretino created lewd sonnets, which inspired Vasari to object: "I cannot say which is worse, the drawings or the words." The pictures and texts, circulated among the Roman intelligentsia, caused a scandal.

They reached officials in the Vatican, in particular the influential Bishop Giovanni Matteo Giberti, whom Aretino had already satirized; apprised that Giberti, his enemy, had seen the engravings, Aretino again took his leave.

He sought shelter again with Giovanni dalle Bande Nere, who chided him not for having written pornographic verse but for his lack of discretion. His next destination was Pavia, to fill a place in the retinue of Francis I, whom he successfully charmed, but whose grasp on power was weakening. Extremely pragmatic, and to the considerable surprise of those who knew him best, Aretino composed three laudatory poems, one about Pope Clement, a second about Federico of Mantua and a third about Bishop Giberti. At the same time Federico intervened on his behalf, which led the Pope to forgive the miscreant and permit his return to Rome. Going further, Clement designated him a pensioned Knight of Rhodes. For Aretino this was another very good period, typically enjoyed. This is how a contemporary, a rival, described him:

> He walks through Rome dressed like a duke. He takes part in all the wild doings of the lords. He pays his way with insults couched in tricked up words. He talks well, and he knows every libellous anecdote in the city. The Estes and the Gonzagas walk arm in arm with him, and listen to his prattle. He treats them with respect, and is haughty to everyone else. His gifts as a satirist make people afraid of him, and he revels in hearing himself called a cynical, impudent slanderer. All that he needed was a fixed pension. He got one by dedicating to the Pope a second-rate poem.
> [Quoted from Chubb]

A good example of his arrogance is a request from him to the Mantuan envoy: Would Federico kindly send "two pairs of shirts worked with gold . . . two pairs worked in silk, together with two golden caps". When the response was tardy, Aretino warned that he would make Federico the subject of a diatribe, which prompted the ambassador to write: "Your excellency knows his tongue, therefore I will say no more." The gold shirts and caps arrived, and double in number, which enabled the envoy to report: "Aretino is satisfied."

One who was not so easily intimidated was Bishop Giberti, the Vatican's datary, the official who oversaw the issuance of papal bulls. His enmity was deep and strong. It increased when

Aretino wrote a rude poem about a young woman who worked in the datary's kitchen and was having an illicit affair. As the author of this defamatory piece was walking at night in a dark street near the Tiber, he was set upon by assassins, one of whom was Achille della Volta, an employee of Giberti. In the mêlée Aretino was stabbed in the chest and had two fingers severed from his right hand. He recovered from his wounds and demanded that Achille be taken into custody and charged with the attack, but the Pope took no action. All too aware of his danger and clouded future, Aretino left Rome and thereafter kept it at a considerable distance.

He rejoined his friend Giovanni dalla Bande Nere. Word came that German mercenaries were preparing to invade central Italy, and the warrior Giovanni formed an army to oppose them. Aretino, too, responding to the martial call, donned a helmet. In a battle at Governolo near Mantua, Giovanni was hit by a cannon shot and died soon afterwards, with Aretino at his side and attending him. Two of his letters, one to Giovanni's widow, one to a friend, movingly describe the young Medici's last moments.

With Giovanni no longer at hand to shield him, Aretino went back to Federico and Mantua, hoping to carry out his erstwhile duties at its court. He composed *Pronostici* (1526), a sort of mock almanac, predicting that during the following year Rome would be visited by a major disaster, with calamities befalling certain guilty persons – including Clement VII, whom he now strongly disliked. He felt that the Pope had given Giovanni only wavering support against the German invaders. Clement, in turn, wrote indignantly to Federico for harbouring anyone so notoriously immoral and flagrantly irreverent. The Mantuan solved his dilemma in his own way. He gave Aretino 100 gold crowns and told him to take refuge further north in Venice; then, with the offender safely gone, he replied to Clement, asking whether His Holiness wished him to have the author of the scorching satire done away with.

"Only in Venice does justice hold the scales with an even balance," Aretino declared, feeling secure and well beyond the Pope's lethal reach. He was to remain there the rest of his life. He enjoyed a career of affluence and the utmost splendour. He rented a palace with a wide view of the Grand Canal not far from the Rialto; before long it became a gathering-place for the most eminent citizens of the Republic on the sunny, shining Adriatic. Part of his income came from a pension granted by Doge Andrea Gritti, to whom he addressed "lordly compliments" and exaggerated praise of the "majestic beauty" of the city – he knew when to flatter and when to be vituperative. For his wit, charm and unprecedented hospitality he was admired, sought after, and in equal measure feared. Another part of his funds originated from a very personal kind of extortion and blackmail – in his epistle to the Doge he spoke of himself as "I, who have stricken terror into kings . . . give myself to you, father of your people". What inspired his friends and mere acquaintances to provide him with an endless flow of gifts (bribes) was their well-founded belief that in so doing they could silence his tongue and pen. Nothing was more hurtful than his waspish gossip and flaying satire. No less a contemporary than Ariosto added two lines to his *Orlando Furioso* in tribute to him: "Behold the Scourge of Princes, the divine Pietro Aretino."

To be seen in his company was a privilege much envied. He grew famed far beyond Italy; part of his celebrity arose because of his scathing exposure of corruption (not his own, but that of others, his superiors in exalted places). His writing had a great vogue in Germany, Hungary, Poland, France and England. The quick sale of his books was another source of his ever ample means, permitting his unstinted expenditures. It helped, too, that much of his work was salacious, even by the standards of that somewhat amoral time. Nor was he ever discomposed by being deemed by some a libertine, a scoundrel, a cruel slanderer. "I enjoy being abused by dull people," he rejoined, "for if they praised me it would look as if I were like them." He wrote imaginary dialogues with harlots who described what they claimed were the secret practices of nuns in convents, of respectable girls and wives, and of lustfully accommodating courtesans everywhere.

Even Emperor Charles V, worried lest the Venetian journalist and poet might choose him as an object for his viperish scorn, courted him, seating him at his right at a state banquet in Padua, and bestowing jewels and other gifts on him; as did Francis I of France and Philip II of Spain. At one time he came close to being named a cardinal. From the almost fawning Charles V he received a collar valued at 300 crowns; from Francis I an even more expensive one – according to Durant – and both sovereigns tempted him with promises of large pensions. He was offered another knighthood, but refused it because it would bring no extra recompense. "A knighthood without revenue is like a wall without *Forbidden* signs; everybody commits nuisances there," was his hard-headed comment.

Delighting in perversity, he dedicated his *Dialogues* to his pet monkey, appropriately named Capricio. This is the book in which he discusses the three classes of women and awards the palm to courtesans because they alone give "an honest night's labour for their pay", a conclusion that tickled a multitude of ardent Italian men and set them rocking with sympathetic laughter.

Along with this steamy literature he produced a series of religious books – *The Humanity of Christ, The Seven Penitential Psalms, The Life of the Virgin Catherine,* and pot-boiler accounts of a half-dozen other saints – filled with inaccuracies, stories of miracles that had never happened, fictional passages of his own devising. Privately he called them "political lies", but they won him the admiration of pious readers who looked upon him as a defender of the faith and a pillar of virtue – that is why he was considered by them to be worthy of a cardinal's scarlet robe and hat.

Among his familiars were the magnificent Venetian painters, whose glowing canvases adorned his museum-like house: Bronzino, Vasari, Giulio Romano, Sebastiano del Piombo, along with marbles and bronzes by Jacopo Sansovino and Alessandro Vittoria. Most of these were gifts from the artists, or had been exacted from suppliants of his unreliable favour. In his personal apartments the scene on the ceiling was by Tintoretto. Titian, who thought the burly, heavily bearded Aretino a striking subject, has left two portraits of him, immortalizing an impressive man, with a large head above a fur collar and velvet robe and silk shirt, his features virile and sensual; his gaze direct, unusually vigorous, his thick, wavy beard cascading to his broad shoulders and chest, his excessive girth not only richly caparisoned, but further enhanced by a costly jewel glittering on a chain about his neck. There is also a bust of him by Sansovino. These two, Titian and Sansovino, were his closest male friends: he called them his brothers.

Michelangelo did not take kindly the writer's advice on how to rearrange and improve the composition of the *Last Judgement* in the Sistine Chapel; and Tintoretto, once maligned by Aretino, gave him a true fright by drawing a pistol when his new subject came to his studio to pose. "What's that for?" demanded the poet. "To take your measure, my friend," replied the artist, who thereupon proceeded to count off the trembling Pietro's robust figure with this improvised rule.

He was an indefatigable letter writer. Now collected in six volumes, they comprise about 4,000 pages, from which Chubb has culled his anthology of 262 choice examples. All too often, the bulk of the notes and messages, simply dashed off, reveal his threatening, begging or else cajoling and seducing presents from influential and powerful men, physical favours from women. At times his language is ribald, scatological, brutal; at other moments his prose possesses – as Durant appraises it – "an originality, vivacity, and force unequalled by any other writer of the day. Aretino had style without seeking it. He laughed at those who polished their stanzas into perfect lifelessness; he ended the Humanist idolatry of Latin, of correctness and grace. Pretending to be ignorant of literature, he felt free from cramping exemplars; he accepted in his writing one overriding rule, to enounce spontaneously, in direct and simple language, his experience and criticism of life, and the needs of his wardrobe and larder." Cited by Durant are "tender epistles to a favorite ailing harlot, lusty accounts of his domestic history, a sunset described in a letter to Titian almost as brilliantly as Titian or Turner could paint it". He was energetic in promoting Titian's career, commending him to prospective patrons who might wish to emulate him and sit for a portrait. And after Michelangelo wrote a note rejecting his suggestions on the design of the *Last Judgement*, Aretino, undeterred, continued the exchange, this time soliciting one of the preliminary sketches for the ambitious fresco: the artist should not burn all of them but instead send at least one to Aretino as a token of their true and ever-lasting friendship. "Does not my devotion deserve obtaining from the prince of sculpture and of painters, just one fragment of those cartoons which you toss into the fire, so that it may bring me joy while I am alive, and when I die I may take it with me to the grave?" Nowhere in his plea does he speak of *buying* the sketch. He was a shrewd collector, indeed.

So many of his letters are extant because he kept copies of them and then had them published almost as soon as they were posted; his voluminous compilations of them were printed at comparatively short intervals and eagerly purchased by the public – they gave insight into the gaudy doings of the profligate *haut-monde* and contemporary celebrities, the equivalent of the latter-day gossip column. Aretino is worthy of being profiled this fully, because – though somewhat larger than life-sized – he typifies the active, bold Italian Renaissance artist, his mind restless and questioning, his instincts propulsive, protean – in Spengler's phrase, Faustian – a figure such as is also seen in Benvenuto Cellini's *Autobiography*, representative of his period much as might be said of James Boswell and the England that he graphically and intimately exposed in his *London Journal* two centuries later.

Much practical wisdom is put forth in Aretino's letters. Here is his pithy and worldly advice to a young man thinking about entering military service:

> You are a large fool, Messer Ambrogio, for it was only a few days ago that I had to knock out of your head your unholy notion to take a wife, and I now have to line up my arguments to cure you of your idea of becoming a soldier. Isn't it a known truth that soldiers, like bread, end up mouldy and worthless – even though someone might reply to this: "What would you do without one and the other in times of war?"
>
> It seems to me that you are crazy even to think of enlisting, and stark mad if you actually sign the roster, for the soldier's life is so much like that of the courtier that they might well be called twin sisters. They are both handmaidens of the desperate, and step-daughters of swinish Fortune. Indeed, the camp and the court should always be spoken of in the same breath, for in one you find want, envy and old age, and end up in the poor house; while in the other you earn wounds, a prison camp and starvation.

In his fancy, or in "high-sounding chatter" at a dinner table, Aretino grants, a young person might entertain glorious images of a fighting man's attire and life:

> I am all for talking this fantastic kind of balderdash. It makes a man a Trojan hero in his own mind. But I say a pox on turning words to action. If you do, you will wear out your clothes, your servant and your palfrey in two months. Your patron will become your enemy as will Paradise itself if you should ever get there.
>
> In the same way, the man who fulminates with martial fury makes a bestial fool of himself. Talking big of what you said or did when you served in France, giving yourself a thousand foot soldiers and two hundred men in armour, you invest castles, burn villages, take prisoners, and amass treasure. But if you want to caracole on your steed a couple of times under the window of your doxy, you can do it just as bravely if you stay at home.
>
> For a "hip, hip, hurrah, boys!" in front of some rustic lout's chicken yard, you will go without bread for supper for at least ten weeks. For a bundle of rags – which is all your booty amounts to – and for the dungeon in which you will end up whenever God wills it, you will be rewarded by coming home limping on a cane, and by selling your vineyard so as to keep out of the hands of the loan-sharks.
>
> This is my answer to all your tall talk about the insignia and the medals you have seen worn, forsooth, by those who came back from Piedmont: "*If you had seen those who did not come*

back, if you had seen those who could not keep a groat, you would feel as great pity for them as we all feel for those wretches who go to, and cannot escape, the knaveries of the court."

Come now, change your mind. You can write a sonnet much better than you could go through a drill.... For few people ever get their clutches on the tickets that win the big prizes in the lottery.... My Lord Giovanni de' Medici once said upon a similar occasion: "They babble that I am a valiant man, but that has never kept me from being poor."

Though Aretino had himself put on a helmet and had a bold mind, his general reputation was that of a physical coward. The young man, a minor poet, to whom his advice was addressed, had served as one of Aretino's secretaries. He did not enter the army but even so went on to many stark adventures. He wounded a man in a quarrel, fled to France, stole money from his master and lost it gambling, then sought refuge in England. From Thomas Cromwell he received eighty crowns that slipped through his fingers while he engaged in speculation in the Netherlands – the perceptive Aretino had warned him of the folly of his habitually trying his luck at games of chance. Seeking employment, the young man went to Portugal and joined Cabeza de Vaca's unfortunate expedition to the New World. The last word of him came from somewhere near the headwaters of the Rio de la Plata in what is now Argentina.

In another letter, Aretino expresses his views on censorship and defends the erotic sonnets he had appended to the engravings derived from Giulio Romano's frank paintings. He writes to Messer Battista Zatti of Brescia:

No sooner had I persuaded Pope Clement to set free Messer Marcantonio of Bologna who had been imprisoned for having engraved on copper the sixteen methods, etc., than I had a sudden desire to see those pictures which had caused tattle-tale Giberti to insist that the worthy artist ought to be hung and drawn.

When I saw them I had the same kind of impulse which made Giulio Romano do the original paintings, and inasmuch as the poets and the sculptors, both ancient and modern, have often written or carved – for their own amusement only – such trifles as the marble satyr in the Chigi Palace who is trying to assault a boy, I scribbled off the sonnets which you find underneath each one. The sensual thoughts which they call to mind I dedicate to you, saying a fig for hypocrites. I am all out of patience with their scurvy strictures and their dirty-minded laws which forbid the eyes to see the very things which delight them most.

What wrong is there in beholding a man possess a woman? It would seem to me that the thing which is given to us by nature to preserve the race should be worn around the neck as a pendant, or pinned on to the cap like a brooch, for it is the spring which feeds all the rivers of the people, and the ambrosia on which the world delights in its happiest days.

It is what made you, who are one of the greatest living physicians. It is what had produced all the Bembos, Molzas, Fortunios, Varchis, Ugolin Martellis, Lorenzo Lenzis, Dolces, Fra

Sebastianos, Titians and Michelangelos, and after them all the popes and emperors and kings. It has begotten the loveliest of children, the most beautiful of women, and holiest of saints.

Hence one should order holidays and vigils and feasts in its honour, and not shut it up in a bit of serge or silk. The hands indeed might be hidden since they gamble away money, sign false testimony, make lewd gestures, snatch, tug, rain down fisticuffs, wound and slay. As for the mouth, it spits in the face, gluttonizes, makes you drunk and vomits.

To sum up, lawyers would do themselves honour if they added a clause about it in their fat volumes with something too about my verses and his attitudes.

When you write to Frosinone, greet him in my name.

(Among the many charges of personal immorality brought against Aretino was one of sodomy, which drove him into temporary hiding. On the other hand, he was much praised for his extreme, unmatched generosity. "Everyone comes to me as if I were a custodian of the royal treasury." Often his house was open to the poor and homeless, who loved and respected him.)

He had a succession of mistresses; of their number, no fewer than six of his courtesans proudly and publicly identified themselves as having shared his broad bed. When one of them decamped, a ruffian, Captain Ad, wrote to him volunteering to retrieve her. This earned only a dismissive reply: "My dear brother, in the matter of your wishing to bring back the lady friend and chastise the rogue who ran off with her, thank you in the same spirit that you made the offer, but I do not want you to do either one or the other. The first is not appropriate and the second does not comport with my desires. The first is not appropriate, because when that inconstant she-tramp left me, she gave me back to myself. The second I do not wish, because in taking her for himself, he freed me not merely from a trollop and a loose woman and a cheat, but from expense and from shame and from sin."

The scoundrel who has stolen his mistress – "a living example of every vice" – is "wicked and mad" but not worth being punished, though he has acted with malice and ill will. "I will turn over my revenge to the disease which racks him and tortures him so often and so cruelly that I almost have pity on him. Now God be with you, and may you rejoice in the way things have turned out, just as you would have lamented if I had been robbed of the light of my intellect, my gift of genius and the rewards of my hard work, instead of the vile, vain, pestiferous delights of lust. For when you lose the latter you are the winner, but if you win them, then you lose."

At times he had twenty-two women dwelling in his commodious house, many of them pregnant – they were not there to satisfy his boundless sexual appetite but to find the succour he offered them when they were in dire need of it. He fathered two children, Adria and Austria, whom he treated kindly, as he did their mother, Caterina Sandella, to whom he behaved as though they were married.

A fabulous person! The talk of Italy, he had little or no education yet knew everything and everybody. His plays, which earned fame and good sums, were erotic but droll and easily aroused the mirth of his audiences; he began to write them mostly during his enforced stay at Mantua. By his own account he turned them out at incredible speed, and he was characteristically prolific. He claimed to have written *Il Filosofo* (*The Philospher*; 1546) in "ten mornings", and the earlier *Lo Ipocrito* (*The Hypocrite*; 1541) in "ten sleepless nights"; possibly a boastful understatement, but there is no doubt that his mind and imagination were always racing. Since he continuously revised his scripts, it is difficult to fix the exact order of their initial composition, publication and – in some instances – staging. This especially applies to his two most esteemed comedies, *Il Cortigiano* and *Il Marescalco*.

Il Cortigiano (*The Courtier*, or *The Play of the Court*) is usually ascribed to 1545, the year of its printing, but there was an earlier version a decade previously (1534–5), and most likely one even before that. Set in Rome, it offers a harsh picture of daily life in the environs of the Papacy. Messer Maco, a pious, innocent little man, comes to the Holy City. He wishes to improve his manners and modest station in the world. Unfortunately he falls into the hands of a quick-witted rascal who heartlessly misleads him, taking him into the *demi-monde* while purporting to introduce him to people of high fashion. Actually, in tavern and brothel he learns from these underworld figures – courtesans, cutthroats – how best to comport himself in the realms of the papal nobility and their smooth-tongued hangers-on to which he aspires. That is, he is advised to dissemble, avoid truthful speech and to acquire all other vices practised by Roman aristocrats and their entourages. It is Aretino's savage revenge on his ex-foes in and around the Vatican, a frank, jaundiced image of their thorough corruption. In *Il Cortigiano* Aretino went further than any other Italian of his period in freeing comedy from its classical shackles.

Il Marescalco (*The Stablemaster*, or *The Blacksmith*) is believed to have been undertaken as early as 1526 in Mantua, a year before his move to Venice. Its premise is that the prank-loving Duke of Mantua has a stablemaster who is exceedingly shy of the other sex and apparently holds its members in contempt. Though the cause of his misogyny is never made explicit, it is subtly implied throughout – in a stream of *double entendres* – that he is homosexual. As a character says of him, he has "never seen a woman's nightgown". To tease and test him, the duke orders him to marry and chooses a bride for him, though the stablemaster is not to see her before the wedding. She is described by a string of intermediaries as young, beautiful and bringing with her an enriching dowry of 4,000 *scudi*, a small fortune. The wedding is to be celebrated on the evening of the very day the duke's command reaches him. He is thrown into panic and fury, offering an endless inventory of objections to his entering a wedded state, though never the true one – his sexual orientation. At first he refuses to believe the news, but after it is repeatedly confirmed by representatives from the court he eagerly seeks ways to escape from being afflicted with the "cancer of a wife", even falsely asserting that he is physically unable to perform as a husband, stating he is ruptured. He questions friends and acquaintances as to what married life is really like and receives quite contradictory accounts, some praising it inordinately, others denigrating it as hellish. He vainly resorts to magical incantations. Protesting to the end, he is led to a mock ceremony, only to discover that the disguised "bride" is a boy, Carlos, a handsome page, which leaves the Marescalco deliriously happy.

The script reveals Aretino's considerable deficiencies as a playwright: it is far too long and at moments tediously repetitive; some scenes are totally irrelevant; without exception, the characters are wafer-thin; and the dialogue is so studded with very topical allusions and local argot that latter-day listeners will find much of it incomprehensible. (Along with their English translation and edition of it, Leonard G. Sbrocchi and J. Douglas Campbell include a twenty-page explanatory preface, numerous very necessary footnotes and a forty-page appendix and glossary to clarify the text.) One of the characters, a pedantic schoolteacher, ceaselessly spouts Latin phrases which Aretino's hearers may have understood but which would baffle an audience today – in addition, it is bad Latin, filled with egregious errors, Aretino's point being to prick the pretensions of the half-learned. Much of the humour consists of malapropisms and mispronunciation, puns and word-plays, all favourite devices in farces of the period. Also typical of the age, a Jew is held up to ridicule.

Though the Duke of Mantua is the motivating agent, he does not appear in the play, but it is assumed that he was present at the first performance, and there are frequent flattering references to him and other members of the beneficent Gonzaga family, as well as to other wealthy patrons of the author; and, not least, a few words about Pietro Aretino himself, tongue in cheek, but highly commendatory. Throughout, vulgarity alternates with poetry, and there is a remarkable display of classical erudition by the "uneducated" dramatist.

The special merit of the play, in the view of Sbrocchi and Campbell, is that it sharply mirrors its time. "Just to recall the cast is to find oneself in a typical Renaissance town. All the ranks of the court are present, from the lowest, the maidservant, through the various levels of courtier to the *signore*, the duke, not seen on stagebut frequently mentioned. . . . In addition, one meets two merchants, the Jew and the jeweller, a wet-nurse, and a schoolmaster. Incidental references to the town fool, Ser Paolo, the Nurse's parish priest, and the barber evoke a sense of a living community, as do the Marescalco's nostalgic and wistful reference to the life he might have led as an independent blacksmith if he had not been tempted by the life of a courtier, and the descriptions by the Nurse, Ambrogio and Messer Jacopo of three very different households, with three different kinds of marriage bonds."

The actor who strides on stage to speak the Prologue angrily announces that he is a last-minute replacement for the player who had been assigned the task – the audience gathered on this festive occasion is so exalted and imposing that the weak-kneed fellow is overawed and seized with stage-fright – simply unable to go on. Singled out by name in the assembly of aristocratic guests are the Cardinal of Lorraine, in whose honour the comedy is being presented, and his close friend the illustrious young Ippolito de' Medici, and several noted musicians as well as important visitors from the courts of Ferrara and Bologna. It is Aretino's way of boasting that *he* has been chosen to provide the entertainment at this special event. The speaker says that he is the best performer in the company and, if necessary, could assume any role, or even enact the whole play entirely on his own, taking all the parts, an echo of the author's inherent, unfailing self-confidence.

The fawning compliments paid to the Duke of Mantua, in both the Prologue and the play itself, suggest "the absolute power possessed by a *signore*", a wilful and capricious lord and ruler such as the one seated in this audience. The translators believe that Aretino's deference to this petty despot was edged. "His attack is indirect, almost coy, but his inability to say frankly what he is thinking is as eloquent as any explicit declaration concerning freedom and tyranny might be." His effrontery in the closing lines of the Prologue is quite open, not without risk.

Seeking to persuade the Marescalco to marry and have children, the schoolmaster reels off a list of the great men and women born into the world as a consequence of a coupling of the two sexes. "The parade of statesmen, poets, scholars, artists and courtiers plainly demonstrates Aretino's conviction that he lives in a period of immense achievement." In his roster of toweringly gifted offspring, he inserts the names of his friends Titian, Sansovino, Giulio Romano and Michelangelo, always quick to promote their careers and heighten their reputations.

In several places in the text there is mention of Aristotle; the play answers to his prescription for the classical unities, the incidents happening in less than a single day, within the precincts of one town, and, though at moments diffuse, more or less focusing on one dramatic action, the Marescalco's attempts to avoid having to acquire a wife. Aretino had absorbed the Aristotelian and Horatian rules for sound dramaturgy so prevalent in the Renaissance.

Campbell remarks:

Aretino wrote out of a wide acquaintance with the ways of men. As a self-made man he had gone through the whole gamut of social experience and had observed the behaviour of individuals at all levels. He knew precisely what made men tick, and he presented them directly to his readers in language that is natural, spontaneous and alive. As Pinchera says, "he uses words to capture objects, gestures and attitudes that are characteristic, in the precise instant in which they uncover and reveal a character, or give meaning to a scent; . . . he makes his words participate in the creative process".

The humour of *The Marescalco* arises much more from Aretino's fascination with the real world and the interplay of personalities within it than it does from his manipulation of the stock devices of Renaissance comedy. True, the usual plays on words, misunderstandings, disguises, and the like do appear, but the central device of most Renaissance comedies, the love intrigue, is notably absent. Aretino's is essentially a comedy of character. . . . The central situation, although it is clearly an artificial contrivance, allows Aretino to explore human behaviour in a relatively realistic way. What we enjoy in the play is primarily the vigour, the accuracy, the psychological awareness (it would be an exaggeration to call it subtlety) of Aretino's observation of the characters as they respond to the situation. . . .

In the last analysis, Aretino neither exploits nor explores the Marescalco's homosexuality; given the situation Aretino has chosen, the Marescalco's homosexuality can be seen simply as a device to focus the attack on him with particular intensity. To force the Marescalco to marry a woman would be to force him to cease to be himself, to betray his own identity. It is this radical threat to the Marescalco's being, a threat imposed by the Duke, that gives the character and the play an intensity that once or twice threatens to strain the limits of the comic.

Here and there in the published text are complimentary references to the beauty and charm of Venetian women, passages doubtless added after he left Mantua and obviously meant to ingratiate him with his new audience, though these new lines are fitted in somewhat awkwardly.

While in Venice he revised his earlier works and participated in the staging of two of them. He composed another comedy, *La Talanta*, about the raffish life of courtesans – a subject about which he was well informed – and a tragedy, *La Orazia*, of which Chubb says: "For all its stiffness, it is at least as actable as any other blank verse tragedy of the age, and remember that it was written before Shakespeare, or Marlowe, or even Greene were born, and that the Italians do not have outstanding gifts for this kind of theatre. Regardless of what John Addington Symonds has to say on this matter, the comedies are lively and entertaining, and one of them, *La Cortegiana*, could perhaps be acted today." In recent times the leading Italian philosopher and aesthetician, Benedetto Croce, voiced an even higher view of *La Orazia*, alluding to it as "the most beautiful tragedy written in the sixteenth century".

In his final years Aretino moved from his grandiose house on the Grand Canal to more modest accommodations belonging to a friend, Leonardo Dandolo, on the Riva del Carbon – by some accounts he was evicted from his palatial quarters for not paying his rent. Perhaps his assets were merely illiquid; one biographer says that he had amassed such heaps of golden objects and other valuable works of art that at his death he was one of the richest men in Europe.

He grew very fat, and increasingly developed a religious turn of mind. Earlier, when someone observed that he had spoken ill of everyone but God, he had replied that it was because he did not know God. Apparently he was at last cultivating an acquaintance with Him. It was at this time that there was talk of designating him to be a cardinal, but the Papacy did not look with favour on the startling idea, though the ever aggressive Aretino paid a brief visit to Rome to solicit support for it.

Two years after his enforced move to the Riva del Carbon, apoplexy carried him off at the age of sixty-four, doubtless from over-feasting, over-drinking and incessant love-making. He was buried, as a repentant sinner, in the Venetian Church of San Luca.

He was not without literary influence. It is the belief of many scholars that the English poet-dramatist Ben Jonson's *Epicoene* (or *The Silent Woman*), in which a wealthy old man is tricked into "marrying" a boy disguised as a girl, had *Il Marescalco* as one of its inspirations; and the Homeric catalogue of gourmet dishes in Aretino's play, when Messer Jacopo lauds his wife's cooking, has an echo in Jonson's *Volpone*, in which the eloquent, cunning seducer rattles off a paean to the excellent foods to be served to the beset lady if she will comply with his desires.

While parading his self-acquired erudition, Aretino, the triumphant autodidact, derides classical pedantry such as has taken possession of the Latin-mangling schoolmaster and many of his well-tutored "superiors" in the excited Renaissance world of court and academy that swirled about him and which he himself partly dominated. Alberto Moravia, the Italian novelist, in a *New York Times* review of a new translation of the *Dialogues*, quotes Raymond Rosenthal's comment in the preface to his rendering that "Aretino is a complicated writer." To this, Moravia adds an emendation:

> In my view, Aretino is a simple writer who lived in a complicated period. Or, if you prefer, Aretino is indeed a complicated writer, but unconsciously so, the way a lake is unaware of the "complicated" mountains reflected in its waters. Thus there are two ways of reading Aretino: in terms of the things he wanted to say and in terms of the things he did not want to say but nonetheless said.
>
> To read Aretino in the first way is not difficult. When everything has been said about his acrimony, his licentiousness and his cynicism, we must finally agree that above all else Aretino is a literary man. Not, let it be noted, a poet, not a creator, but a man of letters who shared an ancient and illustrious tradition.
>
> What's more, he was a highly refined literary man, enamored of his craft and indifferent to everything that is not literature. Thus Aretino's great originality – his literary originality, that is – seems to have been his preference for low, popular speech rather than the lofty and noble diction that was usual in writing of the period. Today we would call Aretino an esthete of proletarian language forms, a sensual fancier of plebeian speech. But in the sixteenth century estheticism was still a thing of the future. There was, quite simply, literature, with its conventions of detachment and decorum. In adopting the vulgar language of the people, Aretino was the first to perform a literary feat that became quite commonplace in the following century. That is, he chose an "ignoble" reality that was to be described in ignoble language, all with a view of creating a "miracle" of virtuosity, a miracle endowed with unquestionable animal vitality. And so, for primarily literary motives Aretino seems to be a satirist, a realist, and even at times a moralist – and all without wishing to be or realizing that he was. . . .
>
> Now if we read Aretino in terms of the things "he did not want to say" but unconsciously did say, we must grant Rosenthal his point. Aretino is indeed a very complicated writer. Without his knowing it, he is the perfect son of his century; therefore, we must read between the lines with the eye of the historian, the sociologist, and the moralist. Aretino is naked but believes he is clothed. He is a sleepwalker who strolls over rooftops, eyes closed, never falling, believing he is awake. Notice, for example, the absolutely correct solution he gives to the problem of how to deal with sex. Following the great Greek and Roman tradition, he makes

sex an object of laughter. Here, in the absence of any creative consciousness of his own, it is literature that sets him on the right road. Sex is a venerable, supreme mystery. The sole way to respect this mystery is to laugh at it.

In fact, only the pornographers take sex seriously and analyze and dramatize it. Instead of holding it at arm's length by dint of laughing at it, as Aretino does, they approach it with sentiment, which precisely because it has to do with sex cannot be lewd. But of all this Aretino knew nothing. He wanted only to win the bet that he could produce good literature using sex as the theme. Without intending to do so, he supplied us with a description – and perhaps with something more than a description – of what today we might term the permissiveness of the Italian Renaissance.

The same must be said about his satirizing social customs. As painted by Aretino, the portrait of Italian society of the period is, without question, catastrophic. Was Aretino, then, as some people contend, a scourger of Renaissance corruption, a moralistic judge of Italian decadence? Very likely he was neither. Robustly, sensationally impenitent, Aretino shared too much of that corruption and decadence to treat it as satirist and moralist. One even suspects that the more important indication that the society and the Italy Aretino described were indeed corrupt and decadent is the painful but not pained self-assurance with which the writer tackled his tenebrous themes. . . .

There are two Renaissances, one sumptuous and aristocratic, the other sordid and plebeian. Aretino left us unchallengeable testimony about the second. He was an involuntary witness, one who was incapable of passing judgment – for he himself was often compromised or actively conspiring – but he possessed an exceptionally quick, clear, sharp and precise eye.

Moravia also sees Aretino as alienated, a neurotic.

Another who affected the language of plays, simplifying it, coarsening it by injecting vulgarity and outright obscenities with marked frequency, is Angelo Beolco (1502?–43), who in mid-life – after he became an actor – was better known as Ruzzante. Born in Padua, Giotto's city near Venice, he, too, was illegitimate, his mother believed to have been a maidservant in his father's wealthy family, whose origins were Milanese. He was raised by his paternal grandmother, after his father, Giovan Francesco, chose to marry someone more suitable and had six more children. Little is known about his schooling, but both his father and an uncle were associated with the University of Padua and his work bespeaks a measure of literary sophistication. It is clear that he stayed on good and friendly terms with his father's kin throughout his life. At times he shared in the administration of the family's large country estate and other properties – they were landlords and oversaw various farm holdings – which meant that he was in close touch with the peasants, their lifestyle, their speech and its caustic idioms, their cynical proverbs, their calculating thought-processes, and their abundant superstitions and mocking folklore, which constituted much of the material he later used in his stage-offerings. He moved in two worlds. Through his early interest in theatre he formed an alliance with Alvise Cornaro, a rich Venetian of aristocratic birth, in whose house he sometimes lived. Cornaro also controlled and managed landed estates, while sharing with young Beolco a fervent liking for literature and art. His house in Padua was a gathering place for scholars, architects, artists and craftsmen, philosophers, in every way a cultural centre. It also contained a private theatre.

After some early, tentative ventures as an actor, Beolco established a semi-professional troupe, recruiting its members from youthful Paduan noblemen, for whom he zestfully wrote scripts, usually reserving the leading role for himself. Beginning in 1520, the troupe performed in Padua and gradually expanded its activities to Venice and Ferrara. The company limited itself to this northern region since the plays were in Paduan dialect, not easily comprehended elsewhere. Apparently Beolco never aspired to a national hearing and approval; he published nothing during his lifetime and never translated his scripts into a more literary Italian. Indeed, for the most part they depended on the actors' gifts for improvisation and cannot be fully appreciated by a mere reading of a printed text.

In Ferrara he learned much as a collaborator with Ariosto, then supervising entertainment at the ducal court. Acclaim for the young Beolco as a comic actor began to spread, and eventually he was a popular hero. He and his company worked intensively, producing an impressively long list of scripts. Since he had none of these printed, a good many are lost. As a practical theatre man he constantly revised them for new stagings, so that scholars are often confused as to which surviving text represents his final version and is to be used.

Versatile, he turned out plays of several kinds: some retain elements of the pastoral, and others match the conventional five-act *commedia erudita* in form, observing the three unities, an obligation for which he had little liking; and late in his career he contented himself with two adaptations from Plautus – *La Piovana* and *La Vaccarua* – his purpose, as he put it, to have them "tailored to fit the living", which doubtless meant to cut and update them. His resort to Plautus may have signalled that, though still young, he had run short of invention, perhaps temporarily.

His basic preference, however, was for rowdy peasant farces, at which he excelled and which had overwhelming success. In a series of them, beginning with *La Betìa* (1524–7) and continuing with *Parlamento de Ruzzante che l'eEa Vegnù de Campo* (in essence, Ruzzante's speech after leaving the militia), *Bilora*, *La Moschetta*, *La Fiorina* (*The Little Flower*) and *L'Anconitana* (*The Girl of Ancona*), he introduced a group of characters and repeatedly brought them back, from offering to offering, as he followed their foolish adventures and caprices. It is not possible to fix the exact chronology of these short plays.

The outstanding figure in them was Ruzzante himself, portrayed with such exuberance and canny skill that he became known and referred to by that name, so closely was he identified with the part. He was no longer Angelo Beolco when the public spoke or thought of him: broadly the character took over the man. As the peasant Ruzzante, he is shrewd and scheming in some aspects of his nature, aggressive, rebellious, outspoken, undisciplined and surprisingly stupid in others, by turns bold on impulse and craven on second thought: dishonest, untruthful, selfish, prone to violence, a woman-hater – unpredictable and vital, hence always interesting.

As has been remarked, the language, Paduan dialect, is a distinctive feature of these plays, scatological to a degree never descended to by the profane Aretino. The characters indulge in litanies of what in English would be "four-letter" words; these pour forth unceasingly, present in every light or angry exchange. Yet audiences of nobles, academics and other intellectuals seemed to find this highly amusing, taking no exception to its rawness, possibly recognizing its "truth to life", a realistic echo of the chatter heard daily and everywhere on streets of towns and villages, and especially around vineyards and farms. Like Aretino, Beolco had an acute ear and few if any inhibitions. He may have perceived that the peasants' resort to a narrow range of sexual, excremental and other foul epithets arose from their lack of vocabulary and an inborn inarticulateness that restricted them to voicing feelings in earthy phrases and vile, demeaning insults, some jocular, some irate.

Furthermore, Ruzzante and his fellows reflected the political situation in Padua and the region near the city, which were under the commercial domination of neighbouring Venetians and being exploited by its merchants who exercised a harsh monopoly of purchases. The grain farmers were impoverished and suffering hardships. Beolco, as a manager of family landed estates, was sympathetic to the unhappy peasants and has Ruzzante voice his resentment at the unfair treatment to which they have to yield, as well as at the depredations they endure from marauding mercenaries. He is loudly defiant of authority of any kind, which added to his popularity. Also, as a centuries-old university city, Padua had an enviable tradition of freedom of expression – especially at carnival time – and a dramatist could expect some tolerance for his bluntness and frankness.

For a sampling, one looks at *La moschetta*, translated into English by Antonio Franceschetti and Kenneth R. Bartlett, who rank it as Beolco's "masterpiece" – a pardonable exaggeration. It is hardly a play, but rather an overlong sketch, almost plotless. (The title eludes a clear English equivalent; the translators write: "With the word 'moschetto' Beolco indicates a language spoken by uncultured people trying to express themselves in proper Italian – that is, the Florentine or Tuscan dialect in his time – but making several mistakes in grammar, syntax, diction from their Paduan dialect." So, he is not only showing sympathy for the peasants but in a sense also patronizing them.)

The vituperative Betìa, wife of Ruzzante, is at odds with him and keeps him from her bed, denying his every importunity. She is also eagerly desired by a neighbour, Tonin, a dim-witted, boastful yet timid soldier, who makes advances when Ruzzante is away. The third person beseeching her favours is Menato, her former lover and now a close family friend – he is Ruzzante's *compare* and Betìa's *comare*, to use the local terms for their mutual relationships. In a weak sub-plot, Ruzzante has stolen money from Tonin, who wants it back, being aware of Ruzzante's guilt. At one point Betìa deserts her house and moves in with Tonin, whose dwelling is immediately adjacent. Much of the action consists of little more than clumsy disguises and lengthy soliloquies in which the three individually explain their thoughts and feelings or direct tirades at one another, torrents of invectives. In these, Beolco appropriately inserts domestic and barnyard similes and metaphors: "My love is boiling like a tub of new wine in August," exclaims Menato. And Ruzzante exults, having learned that he can bluff others: "I am so happy that my shirt is that high on my ass. I've made so much money that I could buy half an ox." Betia complains of his laziness: "I'm always pushing you, always shaking you, the way you do with a fish in a frying pan." All sorts of folk ways are evoked in the endless speeches.

Far from admirable, all four characters lie, cheat, are sexually promiscuous: these faults appear in them because they are poor, miserable, frustrated but also because they typify human nature as Beolco assesses it. Delivering the Prologue, a peasant farmer says that he has never found a woman "rotten to the core", nor a man, but if they get into mischief it is because of their "female nature" and their "masculine nature", which is what makes people behave as badly as they do. In brief, they cannot help themselves. This gives his people a considerable licence for misdeeds. They speak of their love for each other, but their emotion is really prompted by a tormenting physical pressure that, once released, lets their thoughts leap wildly elsewhere. Much of their talk of friendship is a sham. Their only basic concern is getting on, regardless of others, which in their bleak, impoverished circumstances is pretty much as it must be.

Certainly Ruzzante, played to great applause by Beolco himself, is no exemplar of virtue; rather he is a blundering survivor, hindered by no moral considerations but – again – not without good qualities, as is true of his companions. Saliently, he is never as brave or wily as he sometimes persuades himself he might actually be or become.

Menato is aware of his ethical shortcomings, describing himself as having been born when "Satan was combing out his tail". Though supposedly a country bumpkin, he is innately manipulative and a double dealer. The object of his desire, Betia, is a tease, constantly obsessed by sexual longings yet anxious to keep a public reputation for modesty. She is guided to her trysts by opportunity, ready to embrace any man who happens to be nearest at hand. Luckily for Tonin, he lives directly next door. At one point Ruzzante recalls nostalgically the idyllic days they enjoyed early in their wooing and marriage, but that relationship has long since vanished.

The play is at its height in a scene that Beolco more or less borrows from Dovizi's *La Calandria*. Ruzzante is knocking on Tonin's door, demanding to know if his wife is there. Inside, Tonin and Betia are bedded, and the enraptured soldier is excitedly shouting aloud the ecstatic progress of their intercourse, reported in a string of *double entendres* which Ruzzante completely misunderstands, thinking his betrayer is breaking in and saddling a she-mule. Here the words are chosen with remarkable adeptness.

With such long wordy passages and often feeble plot complications, much would depend on the actors' mimetic skills. It was Beolco's mastery of these gifts that won him renown.

He was preparing a tragic role when his early death overtook him. For a few decades after that sad event several editions of his works were printed. One facet of a dark farce like *La*

Moschetta, harshly realistic, is that it dispels the romantic image of "love" shaped by Italian Renaissance poets influenced by Dante and subsequently shaped by Petrarch. The exercise of that emotion is much less spiritual and ethereal, and heatedly more glandular and fleshy, downright energetic and breathless, as depicted by Beolco. His contemporaries recognized that and were delighted at having a human balance restored, which assured his plays' continuing appeal.

Observing the wide and lasting popularity of Ruzzante, the *commedia dell'arte* troupes appropriated him, placing him in a different context, inserting him as another stock figure in their repertory. This is one reason that Beolco is seen as a bridge to that genre of theatre, the most enduring form of stage-work to originate in the Italian Renaissance.

By the century's end Beolco's name and reputation faded, his plays lost from sight. Three hundred years later, in the mid nineteenth century (1862), they were rediscovered by Maurice Sand, a French critic, son of the novelist George Sand, who proclaimed their importance. Since then a number of French and Italian scholars have made him a subject of their studies.

Antonfrancesco Grazzini (1503–83), heard earlier finding fault with the clichés of the Italian comedy of his day, failed in his own attempts to provide any new form or fresh subject-matter. His *nom de plume* was Il Lasca, that is, the Roach. A noted writer of comic stories, introducing Pilucca, a prankster, he had a perverse liking for gore. In one tale a husband, finding his wife and his son in *flagrante delicto*, avenges himself by hacking off their hands and feet, gouging out their eyes and severing their tongues, and then letting the guilty pair lie in their blood-soaked bed. His comedy *La Gelosia* (*Jealousy*; 1550) is in a much lighter vein, but even so lacks true vivacity; it is chiefly remembered today for having supplied a plot-turn neatly lifted by Shakespeare for *Much Ado About Nothing*. To this he added *La Spirata* (*The Possessed*; 1560), which also exploits a plot idea later made use of by several lesser Elizabethan writers; and *La Pinzochera* (*The Devotee*; 1582), in which Gerozzo, an old man long cowed by his wife, becomes enamoured of Diamante, a married lady, and is persuaded by an unscrupulous servant to swallow two cherrystones which will make him invisible: he then sets out to complete his conquest, and is of course roundly trounced by the lady's provoked husband. Adultery, crotchety and duped old men and conscienceless servants were still the standard ingredients of this century's farces, and despite his protests Grazzini was not ready to dispense with them.

As a playwright, no one was less prolific than Annibal Caro (1507–66), author of a single script, *Gli Straccioni* (*The Scruffy Scoundrels*; 1543), which he wrote unwillingly – it was only by order of his patron – and did not wish to see produced or ever printed. What is more, the two characters identified in the title are not really scoundrels, nor are they truly scruffy – merely dressed eccentrically, for a purpose – and they are minor figures in the play, taking little part in the action. Yet Caro and *Gli Straccioni* are celebrated: it is a *commedia erudita* that violated old conventions and boldly established new ones.

Born on the Adriatic coast in the small town of Civitanova, Caro was to have a career filled with melodramatic incidents. His father was a prosperous grocer. The son chose to eschew that staple trade, and at eighteen he left the family eaves to complete his education in Florence and make his own fortune. He displayed enough intelligence, charm and wit to win good connections and helpful friends, which resulted in his becoming a tutor in the household of Monsignor Giovanni Gaddi, Clerk of the Apostolic Chamber; his pupil was the Monsignor's nephew. After a time he was chosen to be the lofty churchman's private secretary. This led to his moving to Rome, where he found himself mingling with the entourage of Paul III, the newly enthroned Farnese Pope. Monsignor Gaddi was greatly interested in the arts and especially literature and surrounded himself with their able practitioners, among them classical scholars of high repute, and greatly talented sculptors and painters, including Benvenuto Cellini (who mentions Caro in his famous *Autobiography*) and Sebastiano del Piombo. In this Vatican coterie the young Caro distinguished himself by writing salacious stories. He also proved himself as a poet, turning out felicitous sonnets, and as a sly parodist, an elegant letter-writer and a capable translator from Greek and Latin works, among them his version of Virgil's *Aeneid*.

With the Reformation menacingly under way outside Italy, Paul III was determined to cleanse the Church hierarchy and its bureaucratic establishment of the decadence and corruption that had infected the reigns of his immediate predecessors. He also sought to renovate the city of Rome, many sections of it showing signs of great age and decay. He ordered that whole areas be razed and the buildings on those sites replaced by handsome edifices, a principal one a magnificent Farnese palace. The Holy City was to be superbly embellished, to testify in stone and marble to the grandeur of the Church that he now headed. Titian has bequeathed two virile portraits of him, among the artist's best works – Paul, chosen as Pope because he was in his sixties and not expected to live long, instead frustrated the intriguers and ruled for fifteen years.

At Monsignor Gaddi's death in 1542 the thirty-five-year-old Caro had to find new employment. Once more he was assisted by influential friends, who enabled him to gain a post in the secretariat of Pierluigi Farnese, the elder of Paul III's two illegitimate sons. (They had been fathered when the Pope, then Alessandro Farnese, was a young man, conducting himself as did many of the aristocratic youths of his day.) On behalf of Pierluigi, Caro undertook several diplomatic missions to Lombardy, France and Flanders. Three years later Pierluigi was named to govern the Duchies of Parma and Piacenza, and Caro was appointed an Administrator of Justice. But not for long. The reckless Pierluigi imposed harsh taxes, and his discontented subjects rebelled in 1547 and he was assassinated. Caro was charged with having been involved in the conspiracy, as well as of grand larceny, and immediately fled to save his skin.

Eventually he was back in Rome, and again attached to a powerful Farnese, another Cardinal Alessandro, a nephew of the Pope. Caro stayed with him until retiring in 1563.

Gli Straccioni was commissioned by Pierluigi shortly after Caro was hired; possibly getting the post was contingent on the secretary's preparing the script, a task that both kept secret. Essentially it was to be propaganda, extolling Pierluigi's father, the Pope, and what he was doing to beautify Rome and restore order in the turbulent, crime-ridden capital that only recently had witnessed outright rioting. This "political" aspect of the script is seen by Massimo Ciavolella and Donald Beecher, who have translated it into English, as adding a new dimension to the genre of the *commedia erudita*. If so, it is done obliquely, and is more implicit than explicit.

Aristotle, revered by Italian Renaissance playwrights, stressed that a stage-work should have unity of action, a focal plot and perhaps a single sub-plot. Caro violates that rule – his play has three plots. In the prologue he offers a defence for doing this; further, he asks praise for his innovation, warding off what might be a critical battery. "The constipated traditionalist may take offence at the triple plot, since the ancients never went beyond a single or a double one. But don't forget that though there are no existing precedents for our procedure, neither are there prohibitions against it. . . . Note that the author has followed the traditions in other areas. And in any case, change is inevitable since actions and laws of actions depend on the times and fashions, and these change with every age."

The translators, in their preface, suggest that Caro had deeper motivations than he bares in this speech. Was he merely trying to expand the comic form, providing a more complex set of dramatic intrigues and a greater variety of characters, or was he looking to offer a novelty and stir a controversy? He may have felt, they say, as he wrote to a friend, that existing comedies were tiring to compose because they clung so closely to an outdated formula. They should be made more contemporary in treatment and subject-matter. The triple plot led in that direction, widening the play's scope. "One of Caro's plots is drawn from a little-known Greek romance, a second is taken from an event in contemporary life, while the third is a familiar narrative turn from the traditional theatre. By such means he was able to satisfy the conventions of learned comedy, yet extend the play in terms of its representation of current affairs." Put another way, Caro was impelled to multiply the number and kinds of plot elements because "he was attempting to satisfy two masters, his patron and the rules of regular comedy".

To make the play contemporary, he portrays several readily identifiable Roman citizens; the audience would have no trouble at all in identifying them and would be curious to have the dramatists' explanation of their strikingly odd behaviour. The scattered references to the changes in streets and buildings, owing to Pope Paul's ambitious edict, would also update the scene, making it very current. A few allusions to topical political events would similarly contribute to this end. The third-century Greek romance from which Caro borrows threads of his story is *Leucippe and Cleitophon* by Achilles Tatius, certain proof of Caro's classical erudition. The third segment of the action is compounded of long over-used stock plot-devices from Plautus and Terence.

Ciavolella and Beecher grant that the play is diffuse, at times seeming little more than a series of unlinked sketches, with some scenes, such as Mirandola's fit of madness, quite irrelevant. All the same, it holds interest: What will be the outcome? How will the dilemmas be solved? That is any play's first requisite.

In fact the two brothers, Giovanni and Battista, are Romans of substantial wealth and position, who have suddenly taken to dressing and parading through the capital. Why are they behaving this way? Caro's answer, kept until late in the play, is that they have lodged a lawsuit to regain jewels and funds due them and are angry at court officials who have neglected to hear their case. By their attire, a suggestion that they are impoverished by the prolonged delay, they are hoping to publicize their frustration and call attention to their claim. This has little to do with anything else in the play. What is important, however, is that Battista's daughter, the comely Giulietta, has disappeared, and the brothers believe that she has been abducted by Tindaro, a young man to whom she was attracted and who had sworn love to her. Though he is of good family, he is considered unacceptable by her father, who is anxiously searching for her.

The truth is, Tindaro and Giulietta have eloped and been properly wed. Unhappily, the ship on which they are voyaging is belaid by Turkish pirates and sunk. Tindaro escapes, but Giulietta is taken captive, ravished and killed. Under an assumed name, Gisippo, the grieving young husband returns to Rome, where he is confused by the physical alterations under way in response to the new Pope's command. A friend, Demetrio, meets him and leads him about, pointing out what is being accomplished.

Minor characters are a pair of scoundrels, Marabeo and Pilucca, respectively a steward and a servant in the household of a wealthy young widow, Madonna Argentina, whose husband, Il Cavalier Jordano, has also been reported lost at sea. In collusion, the unscrupulous pair of menials have been regularly mulcting the lady's money accounts.

The widow, lonely, is eager to remarry. The two servants see how this could be to their advantage. To abet their scheming they enlist Barabagia, a well-meaning friend of Madonna Argentina, who is also acquainted with the just-arrived widower Gisippo (Tindaro), an obvious choice to become the lady's second spouse. But Gisippo, still deeply mourning his mutilated and murdered Giulietta, rejects the idea. Every sort of verbal pressure is applied to him until he finally capitulates. The marriage party is about to be held.

Battista and Giovanni encounter Gisippo and learn of Giulietta's death. They berate him, holding him responsible for her unhappy fate.

As might be expected by those familiar with Renaissance comedy, Giulietta was not lost at sea, nor was Il Cavalier Jordano. With much duress she, too, reaches Rome, as does he. Using a false name, Agata, because she fears her father's wrath and believes Tindaro to have drowned, she is in hiding and harassed by Marabeo, who has purchased her from the slaver Turks and is holding her a prisoner, and also by Jordano. Throughout she has steadfastly defended her virtue.

In a bold move, Agata (Giulietta) eludes Marabeo's clutches and flees to safety. Meanwhile, after allowing himself time to recover from his fearful ordeal at sea, Il Cavalier Jordano seeks out his wife and learns that he is about to be superseded as her husband. This leads to his drawing a sword against the reluctant Gisippo (Tindaro), who has no stomach for either matrimony or perilous combat, especially since a rumour has been spread that the "widow" is pregnant.

All matters are straightened out by the intervention of the Pope's attorney, who symbolizes Paul III's intent to restore order and defeat all vestiges of lawlessness in the Holy City. There are recognition scenes and reconciliations. Giulietta is returned to Tindaro's arms, and Battista, delighted that his daughter is alive, relents and sanctions the marriage. The promiscuous and predatory Il Cavalier Jordano reclaims his wife, and the "scruffy brothers" win their case, getting back their jewels and 300 ducats; after donning more fashionable attire they hand over 100,000 ducats to Giulietta as a dowry. Surprisingly, it develops that Madonna Argentina is the daughter of a third brother, hence their niece – she and Giulietta are cousins. Marabeo and Pilucca promise to behave honestly from now on, but instead of offering restitution for their filchings, impudently seem to be contemplating more thievery.

Mirandola, the madman, was a conspicuous figure in the byways of Rome – his presence in the play adds to its topicality, though he has no connection to the plot or any of the characters. The friendship between Tindaro and Demetrio, who acts as his guide, confidant and intermediary, is very close and highly sentimental, and anticipates the depiction of similar male bondings in a host of plays to follow, among them those of Shakespeare. This play, it might be said, has no stock characters; all are individualized.

Throughout the script Caro is heard deploring the decadence and extravagance that are being flaunted by rich people in the city. Demetrio asks: "Why are they parading in those motley trappings? Are they off their rockers? Not that it makes any difference in Rome." Caro wrote in a letter, concerning comedy, that its attraction lay not in the plot or characters but in the verbal exchanges. "The subject-matter is ordinary, since subject-matters of comedy cannot be otherwise, and the rarity of the episodes does not make them any better. What makes them better is the beauty of the sentences and of the style." It was this aspect of the play that earned the most applause. The question of whether comedies should be in prose, and even in the vernacular, was still being debated; *Gli Straccioni* demonstrated the "suppleness and range" of everyday speech on the stage, the skill with which Caro had met a challenge and set an important precedent. He proved that prose and the vernacular could be used with great success. The translators suggest: "Worth noting is the high theatrical quality of the dialogue, generally bright, terse and fast moving. There is a staccato effect in the exchanges with the servants, full of irony and allusions, which is particularly lively." To quote from the text:

> MARABEO: Master Jordano is dead?
> PILUCCA: Master Jordano.
> MARABEO: At sea?
> PILUCCA: At sea.
> MARABEO: The sea beheld him and rolled back her waves. Jordano passed through, just like the Scriptures said.
> PILUCCA: If so, he deserved it.
> MARABEO: And you deserve your money for such good news, Pilucca. And now I'll give you even better news.

PILUCCA: What can top the news of master's death?

MARABEO: I'll tell you the mistress is in love.

PILUCCA: That's a good one! O ho, I get you; my news secures all our past gains, yours sets up for the future.

MARABEO: You've got it. The mistress will have her love and we two will look after her wealth. That business of conscience and loyalty, it's fine for those who don't mind dying of hunger and cold. Riches, Pilucca, wealth, if it's a gentleman you want to be. Since our parents left us nothing, and these others haven't enough sense to give it to us, since we don't know the art of earning it, and hard work is bad for our health, is it any surprise that we make use of our hands? Anyway, the rope's a better way to go than starvation. Are you with me, Pilucca?

Looking ahead, Marabeo warns that they must always be alert and plan carefully. "We should think of something before it happens. Only the virtuous can be heedless. Half a brain is good enough for an honest man, but for a crook even a whole brain isn't enough." When they change their minds and want to prevent the marriage, they invent the story that Madonna Argentina is pregnant. He tells his fellow conniver, "In this sort of affair a little smudge blackens the lot; a tidbit of truth and they'll swallow it all." At a confusing moment, Demetrio cries: "What a catastrophe! Two husbands of one wife and two wives of one husband under the same roof." Declares the indignant Papal Attorney: "To think of such assaults on a virgin, in Rome, in the Piazza Farnese, in the time of Paul III, and with the Pope here. . . ." And it is stressed that it was the Pope's galleys that rescued Giulietta from the Turkish pirates.

Pithiest of all, and very Italianate, is the Attorney's summary of the complex plot: "This comes together perfectly, like cheese on macaroni."

It is believed that Pierluigi Farnese participated with Caro in writing the play and manifested great interest in it. But it was not produced up to the time of his assassination. Afterwards Caro took steps to make sure that very few persons should read it; his reasons for keeping its existence and subject-matter largely a secret are not entirely clear. He gave as an excuse that it was too topical, referring to persons and events now mostly forgotten, and so would hold little general interest. It may have been that with the death of Pierluigi the political situation had changed, and Caro chose to be very cautious. By now he was truly experienced about the risk in such affairs. When Pierluigi's daughter Vittoria sought a copy of the script, he at first refused her request, waiting eight years to send one to Urbino where she resided as the wife of Guidobaldo della Rovere. A stipulation was that the text should not be circulated. Again, after his retirement at the age of sixty from the secretariat of Cardinal Alessandro Farnese, he withheld permission from the Cardinal, his highly placed former patron, to stage the comedy. It was finally published in 1582, sixteen years after his death. The first edition was replete with errors.

Annibal Caro spent his last three years with his family in a small villa in Frascati, well away from the seats of power and crafty intrigue.

Of noble lineage, from an ancient and even famous Sienese family, Alessandro Piccolomini (1508–79) was a figure of wide-ranging interests and talents: science, religion, philosophy, literature. His life reflects the diversity prompted by his open and active intelligence.

Details of his early education are not clear, but Siena boasted a good university, and it is most likely that he first studied there. Later he joined the city's newly founded Accademia degli Intronati and steadily had a leading role in its cultural activities. When he was about thirty he left Siena to enrol in the University of Padua, attending lectures by eminent scholars. There he became a member of another cultural circle, the Accademia degli Infiammati, largely devoted to literature and led by Sperone Speroni, the playwright, an advocate of the use of the Tuscan vernacular in stage-works, a cause which Piccolomini had already embraced, and which was now made more attractive. During his stay in Padua, too, he established a friendship with the aggressive and mercurial Pietro Aretino in nearby Venice.

Alessandro Piccolomini's writing career was a very mixed one. Througout hs life he brought out a series of earnest treatises on every sort of subject, whatever engaged his acute and protean intellect. This led him to discourse on mathematics, astronomy, meteorology, geography – he computed the relative measurements of the land mass and oceans with surprising accuracy – philosophy, ethics and politics – he was a conservative, holding that rule of Siena by Cosimo de' Medici and his Florentine troops would be the best solution of its prolonged factional strife. He was truly a polymath. Strongly drawn to synthesizing the thoughts of Aristotle, he published commentaries on them, including an influential essay on the *Poetics*. Torquato Tasso sought him out for advice on poetic theory. He assisted Pope Gregory XIII in the reform of the Julian calendar, and he is credited with having been the first to develop a vernacular scientific language, a glossary apart from Latin.

Because of the unending turbulence in his native city he moved restlessly about Italy: to Bologna, where he found further insight into Aristotle's concepts under the guidance of a noted scholar, Lodovico Boccadiferro; then back to Siena, where he briefly taught moral philosophy at the university; then to Rome, to become oppressed by unrevealed experiences that brought on a severe mental depression. Deprived of financial independence by events at home, he was obliged to take a secretarial position on the household staff of Francisco de Mendoza, Cardinal of Coria and later of Burgos. In time he returned to Siena for a short stay, and again to Rome. He never stopped writing and publishing. His next employer was Giacomo Cocco, Archbishop of Corfu. By now his thoughts were turning ever more to problems of religious faith. Once more he was back in Siena, where intellectual activity had suffered and almost perished as a result of the city's internecine struggles. He was chosen to be head of the Accademia degli Intronati, an effort to restore its prestige. Soon the Academy incurred the suspicions of both the Jesuits and secular officials as a subversive group, and Piccolomini had to ward off charges of heresy. He was not wholly successful at this, and before long the resurgent Academy was closed. Over the ensuing years he earned a doctorate in theology and became Coadjutor of Siena and Archbishop of Patras.

Amidst all these many activities and preoccupations he also found time to write sonnets and plays, not serious dramas but comedies. A facet that distinguishes them is the prominence awarded to sentiment and the visible role of women, who in Caro's *Gli Straccioni* are endlessly talked about but seldom seen, indeed making an appearance only in the last minutes. Piccolomini's stage-works, like those by other members of the Academy, were written and produced as that organization's contribution to the city's annual festivities at Carnival time. The Academy, advanced and aspiring to be élite, fostered a chivalrous attitude towards women, striving to treat them after the gracious fashion endorsed by Castiglione in *The Courtier*. The plays were largely put on for their entertainment, and a majority of the spectators were women, all invited guests, none being expected to purchase a ticket. Accordingly, as Piccolomini saw it, an increase of delicate feeling and romantic interest was in order, and he supplies it. Yet irony and cynicism are far from absent: he knew that the ladies were not all sugar and spice. Nor is the action without racey incidents; a nice balance is a requisite. This duality of refinement and coarseness in women had also been observed by Castiglione. Piccolomini's unwittingly prepared the way for light works characteristic of the second half of the sixteenth century, that is, sentimental comedies, and he is called the creator of the form.

That few women spoke Latin is another reason for him to use literary Tuscan in his plays. He displays, avows his English translator, Rita Belladonna, "great linguistic sensitivity". The characters speak quite differently, one from the other, each having his or her own idiom and rhythm, proof of Piccolomini's stylistic flexibility. Some, especially the young and the servants, express themselves at a lively pace; their elders, at a lower tempo and perhaps woefully; still others, indignant, in plaintive or furious tirades; and often, when a bawd vents her views, in a racy argot. When lovers revert to the stilted phrases frequently come upon in *commedia erudita*, Piccolomini is clearly writing tongue-in-cheek. A fault, though, is that many speeches are overly long and too expository, hence untheatrical, lacking the briskness needed by verbal exchanges on stage.

His first play, *L'Amor Costante* (1531), satirizes followers of the current cult of Petrarchism and earned plaudits as "an entertaining comedy of romantic conjugal love", one of its peaks occurring when a tryst between a youthful pair is interrupted by a curmudgeonly old man, a bit of action that, having demonstrated its effectiveness, is repeated in *L'Alessandro*. Another early try, *Dialogo della Bella Creanza delle Donne* (*A Dialogue about the Best Behaviour for Women*; 1538), dramatizes a wry, somewhat jaundiced conversation between Raffaella, a procuress, and her prey, Margherita, a young, already bored wife whose compliance she is eager to obtain on behalf of a would-be adulterous suitor. The bawd, in this instance, is almost a clone of Alviga in Aretino's *La Cortigiana*.

L'Alessandro (1544) was also meant to be a Carnival offering to members of the Academy. In its prologue Piccolomini addresses the ladies with fulsome flattery and promises that the play will offer models of morality that they might profitably emulate (though it can hardly be said that it really does). In a second prologue, written for a subsequent performance of the work, he takes his feminine spectators to task for having preferred the episodes of "vulgar buffoonery" in the story over its lessons in propriety. Both prologues are nicely ironical in tone.

Like *Gli Straccioni*, the script has a triple plot, something of a surprise coming from a dedicated Aristotelian, especially one who had published an authoritative commentary on the *Poetics*. The three plot-lines are quite integrated, however, none being independent and each enmeshed with the others, and all three are of equal weight. Piccolomini controls this balancing with considerable aplomb.

The scene is Pisa, an orderly university town, well governed by Cosimo de' Medici. This is meant to suggest that Siena, too, would benefit by submitting to rule by the well-disciplined Florentines. A dynastic *coup* in Sicily has sent two families in flight from Palermo. To hide her identity and ensure her safety, the daughter of one refugee family is disguised as a boy and grows up to young maturity always wearing masculine garb and passing as a male. The son of the other family is given feminine attire and is accepted as a nubile pretty girl, much commended for her modesty. They are strangers to each other. Both have survived their first protectors and have come to live in Pisa, far from their native Sicily and with guardians who are unaware of the young people's true sex. Lampridia (born Aloisio) dwells as the "niece" of old Vincenzio, a wealthy citizen of Pisa; "she" has been put in his charge by his late brother. Fortunio (born Lucrezia) continues in "his" semblance as wholly masculine but finds "himself" hopelessly smitten with Lampridia, an obsessive physical desire that convinces him (her) of being a lesbian.

Another old man, Gostanzo, a friend of Vincenzio and filled with senile lust, is in pursuit of an affair with Brigida, the amoral wife of the foolish, noisy Captain Malagigi, a boastful coward. In his quest of bedding with Brigida, Gostanza is abetted by his rascally servant, Ruzza, who delights in mischief and playing pranks on him.

A third amatory intrigue has to do with Cornelio, son of Vincenzio, so beguiled by Lucilla, daughter of Gostanzo, that he neglects his studies and can only think how to win her affections, which she seems unwilling to grant him. It turns out, however, that she does return his passionate interest and has merely been testing the depth of his feeling for her. Learning this from a daring letter that she sends him, he conspires with his clever servant, Querciuola, how to gain access to her. He and Querciuola resort to a rope ladder by which in hours of darkness he can mount to the window of her bedchamber.

Lucilla, chaste, is not about to yield her maidenhood. Brigida has only scorn for the overheated Gostanzo; her ever-wandering gaze is fixed elsewhere. Fortunio/Lucrezia is devastated at thinking his/her desires are deviant and beyond restraint. The Captain senses that someone is trying to cuckold him and has his sword unsheathed and at the ready.

With all these complications arising from the characters' conflicting goals, endless confusion develops, inevitably brought about by the many disguises.

Old Gostanzo dirties his face and tries to pass himself off as a locksmith to gain entrance to the absent Captain's house. With mistaken identities, logical new obstacles, providential narrow escapes, abrupt reversals and well-placed recognition scenes, the plot unrolls until everything is made clear and an equilibrium once more established. Then ensue the usual grudging reconciliations and happy marriages permitted or announced. The action progresses rapidly, brief episode following brief episode, as the participants' dilemmas pile up crushingly and are ingeniously resolved.

Some of the ideas used here are lifted from Boccaccio's *Decameron* and Ariosto's *I Suppositi*, as well as from Aretino's *Il Marescalco*, which does not signal a lack of originality; like any good scientist, Piccolomini was ready to go forward from facts already proved and accepted: this is what worked. In the Italian Renaissance, of course, there was absolutely no such thing as literary ownership or copyright, and all the playwrights sucked dry whatever material they could from their Greek and Roman predecessors and not a little from one another. Besides, the author probably did not set out to write an immortal comic script but only a trivial diversion suitable to the Carnival, a *pièce d'occasion*.

The Captain, full of bluster but quick to turn tail when an actual threat impends, is the traditional figure of the *miles gloriosus*, traceable to the earliest Roman farce and destined to find an apotheosis in Shakespeare's robust, bellowing, heavy-drinking Falstaff, though more of a caricature and hardly – in more ways than one – as fully rounded.

The exchanges in the dialogue are enlivened by vivid phrases. Piccolomini is fond of likening his people to the subhuman: the Captain has a "frog's face"; Gostanzo walks on the tips of his toes "like a parrot" – he is also toothless; again the Captain is compared to an ox and a jackass; and Gostanzo, erotically aroused, is a "hungry bloodhound".

The play leans towards the romantic and sentimental when the young couples overcome lust to experience ideal love, their personalities maturing, and in such passages between the pairs the writing is lyrical. Elsewhere, in conversations between the servants, and in Gostanzo's Freudian slips and outbursts, and in confidences between two bawds – Niccoletta and Angela, plying their trade – the language descends to the gutter with such frankness that the lady spectators, guests of the Academy, must have felt their cheeks blush and their delicate ears burn.

Considering Piccolomini's elevated status as head of the Academy and a lecturer on moral philosophy, and later as Archbishop of Patras, it is somewhat surprising that his script contains several derogatory remarks about churchmen. "Ruzza, so red and flushed in the face that he looks like a cardinal" certainly implies that the high-living Princes of the Curia tend to imbibe too much. Niccoletta, the bawd, tells Angela how she spied on a friar and a nun in the vestry of a convent, where they were separated by a grating. "Whenever they wanted to kiss each other, they had to squeeze their lips through holes. Their faces were so much like snouts, you've no idea how funny they looked. Eventually, they were caught unawares by the abbess, but she just grinned and left." Angela in turn relates that she has been hired by a wealthy canon of Pisa who wants to be set up with a chosen married woman. "He's promised to give me all the income from his abbey, his parish church, his prebend – everything." It could be a gold mine for her. But Angela has found the woman to be hard and sly and her favours not easily obtained, so seeks Niccoletta's advice on how to approach the affair. Strange? But for some time, Rita Belladonna explains, Siena had been "a focus of religious reform", with some proponents of change having a "vast circle of heretical friends in that city". Apparently Piccolomini belonged to such a group. His friend Pietro Aretino, whom he much admired, may also have helped to instil in him a critical attitude towards the overbearing, free-spending clerics.

While praising womankind in his prologues and in the ensuing dialogue, Piccolomini also has the characters list the general faults: they are unpredictable, fickle and perverse, says Niccoletta:

> It used to be that, if a young man wished to obtain a lady's favour, he had to tell her he was loyal, wise, learned, knew how to write verses that would exalt her to Heaven, and similar wonderful virtues. Now one must make sure one doesn't mention such things. Rather one should tell her that he can play foolish jokes, tell lies, speak nonsense, and such like accomplishments. So be careful, the more so because women aren't any longer friendly towards each other, but full of envy and ill-will. Though you may see them kissing and hugging each other and smiling a lot when they meet, yet if they can start a scandal without being found out, they'll stab each other in the back. They're made happy just by hearing that some woman they know has done something shameful or foolish.

Elsewhere, Ruzzo exclaims: "Women's whims! Don't you know how erratic their brains are apt to be all of a sudden?" Alessandro advises his friend Cornelio:

> I've experienced what women are like today, particularly women of this sort. I know all about it. A lover's virtues, learning and modest ways no longer count for anything. Now girls want something different. They take greater delight in gross pranks and in bragging than in anything else. Just look at the kind of entertainment which is normally offered to ladies these days and compare it with the way in which they used to be diverted only a few years ago. In those days, the wit, wisdom and virtue of ladies and their admirers could shine in a thousand different ways. Now just try saying something brilliant or making a witty remark and they fall asleep. On the contrary, try vulgar horseplay or some saucy remark and they all come back to life and are as merry as crickets. I remember that not long ago a virtuous gentlewoman asked one of these young men why they resorted to such vulgar tricks and rebuked him for knowing practically nothing about courtly love. He just answered that his kind of behavior ensured

success and that was enough. So it's the ladies' fault if they're so little esteemed. You're a gentle young man, so don't expect anything that really matters from a woman.

And Querciuola urges Cornelio to press himself upon Lucilla despite her protestations:

> QUERCIUOLA: Do you mean to tell me you got nothing? Shame on you!
> CORNELIO: She just refused.
> QUERCIUOLA: She shouldn't have imposed her will on you. As for you, you couldn't have tried hard enough. Why didn't you use your hands?
> CORNELIO: What do you mean? God save me from such a thing! I wanted to persuade her by love, not by force.
> QUERCIUOLA: Women say no in order to be conquered.

In all this, Piccolomini was teasing his feminine hearers, and doubtless they were titillated and amused rather than offended.

The old men, too, deplore the changes in manners and the new, cruder social climate. There was much more civility in former times, more value placed on good breeding, mutual respect, honesty, restraint, chastity. In the current era people even eat too much, sitting at table for hours. But Messer Fabrizio, a visitor from Sicily, hearing Vincenzio's querulous lament that "the world grows worse as it grows older", dissents: "We are the ones who grow old, my dear Vincenzio, and the world lags behind us, as safe and sound as it has ever been. . . . Do you really want to know why things appear changed to us? It is merely because we ourselves have changed; we don't look at things or listen to them the way we did when we were young." He says there have always been innocent and shameless lovers, and for them it is always springtime.

The play contains several self-complimentary references to the Accademia degli Intronati, its high-minded aims and its handsome new quarters on a fashionable street, and there are also a number of in-jokes.

Quite baffling at first glance is the comedy's title, *L'Alessandro,* since the character having that name, an emotionally close friend of Cornelio, has little to do; he is almost non-essential to the action. Rita Belladonna suggests that he is a surrogate of the author, who has the same first name. A detached observer, for the most part, Alessandro is for ever lecturing Cornelio, warning him, counselling prudence, discoursing on love and its many facets. He is the personification of common sense and reason. It may also be that Piccolomini wrote *L'Alessandro* atop his script as a sort of signature, a self-identification: here I am, and this is my voice, my view of matters.

L'Alessandro's "run" was not limited to two performances at Carnival in Siena. Over the next fifty years or so there were twelve printings of its text, as well as one in the nineteenth century (1864) and at least two in the twentieth, including Rita Belladonna's English version, quoted here, richly and fascinatingly annotated. She notes that in 1548 Ferrante Gonzaga, governor of

Milan, had it performed in the presence of Philip of Spain. It inspired several imitations, among them *I Trasformati* by Scipio Ammirato and *Les Contens* by Odet Turnèbe. In England many elements of it reappeared in George Chapman's *May-Day* (1602), in which the plot is simplified and many of the lesser characters are eliminated, and there is even more emphasis on satire, and more realism, with scantier attention to romance and the Sienese philosopher's idealization of love.

At his death, Alessandro Piccolomini was entombed in Siena's magnificent cathedral.

A most unlikely background for a prolific dramatist is that of Giovan Maria Cecchi (1518–87), born into a "well-established" and distinguished Florentine family, tracing its lineage back at least 300 years. Members of it had been prominent in public service and politically active, their fortunes rising and falling with those of the intermittently ruling Medicis, to whose unstable cause they were faithful adherents. Indeed, Giovan's birth coincided with the overthrow of the Medicean reign and the start of the brief-lived Republic. When the boy was twelve, his father, Ser Bartolomeo, a notary, was assassinated – "Ser" was a title applied to those in the legal profession. The motive for the killing was both personal and political, and the murderer's identity was known. But despite complaints to Duke Alessandro and Duke Cosimo, no punishment was meted out, and this was considered a setback and loss of prestige for the Cecchis. Another blot on the family's name occurred when Giovan's uncle, his father's brother, Matteo Cecchi, died after being poisoned by his disaffected wife, Prudenza, who protested that he had been abusive to her and their children. She paid for her crime by having her head chopped off at a public execution.

Giovan's mother died a few years after her husband's killing. At sixteen the youth found himself head of the family, in charge of and responsible for his two younger brothers. Besides this, a cousin, son of the late Matteo and Prudenza, was found guilty of several thefts and ordered into exile; he ventured to return and was arrested, held in Florence's chief prison for common criminals. This brought further disgrace and obloquy to the Cecchi clan, and no little added to Giovan's personal burdens and outlook: at an early age he knew all too well what weighty and unjust troubles could be visited on an honest man.

None the less he prospered and rose to a measure of public eminence, highly respected. About his early education little is known, except that he was familiar with the Greco-Roman classics and very fond of them. He embraced his family's long tradition and became a notary, as had his father. He was appointed to various offices, serving the affairs of the Furriers' Guild, and being twice named Proconsul, and Chancellor of the Magistrates of Contracts, as well as Procurator at the Tribunal of Commerce; in all instances he was praised for carrying out his duties efficiently. His tasks required authenticating legal documents and commercial agreements, and he was regarded as the honourable public servant *par excellence*.

At thirty-five he took a wife, Marietta Pagni, by whom he fathered three children. Later in life he was wealthy enough to enter into a commercial venture, with three partners drawn from outstanding Florentine families; they formed a wool trading firm, and though business conditions in Italy were not at their best, the new company did well. By now financially secure, he purchased a villa in the hilly Tuscan countryside to which he retreated at frequent intervals, and to which he eventually retired. He was to do much of his literary work at the villa; its location brought him in touch with the peasantry, whose beliefs, habits and language he studied closely. After Beolco (Ruzzante), no Italian Renaissance playwright was as intimately acquainted with rustic life and country folk as Cecchi, and this was to be echoed in his farces. In his last years he experienced a religious conversion, becoming quite pious. He paid to have the decaying local church, Santa Lucia, properly restored, and also had a monastery built for the Augustinian Fathers of San Francesco di Paola.

How did he find the time, the mental energy, the seclusion, to write twenty-one comedies, as well as a range of serious dramas and works in other genres, while otherwise so busily engaged? He composed poems for special occasions, together with sonnets and pastoral eclogues; compiled a dictionary of local proverbs; translated sermons and parts of the Gospels and Epistles into the Tuscan vernacular, which he espoused; and fashioned mock literary critiques, parodies of learned commentaries written by contemporaries and peers whom he deemed pompous and pedantic. He even turned out a spoof travel book for travellers to Germany, Flanders, Spain and the Kingdom of Naples, though he himself had never gone many leagues beyond his native, sunny, dun-hued Florence. (He makes a point that German Catholics were aware of only four deadly sins, rather than seven, since they considered gluttony, perpetual drinking and greed to be virtues.) Obviously he gathered his material from reading rather than first-hand observation. He compiled guidelines on Medicean bureaucratic functions and procedures, a subject on which he was an unquestioned expert, covering a vast array of topics from the exact duties of magistrates to the most effective organization of charitable societies and the arrangement of honourable marriages for orphaned girls.

His career as a playwright began with *La Dote* (1544) and continued to *Il Debito* (1587), a long stretch. His works fall into four groups: the *commedie osservate*, in five acts, more or less the equivalent of the *commedia erudita*, with secular subject and based on Plautine and Terentian models; the *commedie spirituali*, also in five acts, and with religious themes; the *drami sacri*, similarly touching on matters of faith but more variable in length, sometimes five acts, sometimes only three; and the farces, just three acts, and perhaps mixing "the profane and the spiritual and the commonplace with the grandiose". Some critics, impatient with these nice distinctions, reduce the classifications to three, and even two, since there are frequent overlappings and divergences in the shape and matter of the scripts. Other ways of categorizing them – this is important to some scholars – could be by whether they are in a classical, medieval or contemporary tradition; or simply by their chronology – whether they belong to the first or second phase of his activity as a steadily productive dramatist.

Most of his plays, especially the early ones, the farces, were composed for the Carnival season and convey its light-hearted spirit. He wrote rapidly and stated that none of his scripts took him more than ten days to complete, and some a mere four days. Since many of his plots and characters are borrowed, his speed of composition is credible. Several companies of amateur players had been set up in Florence and were eager to obtain suitable vehicles; besides Carnival, other holidays – Epiphany and St John's Day – were also pretexts for stagings. It is believed that not all his works were enacted; even so, some were eventually published. Presentations might take place in gardens and courtyards, if the weather was clement; or in a spacious ballroom, together with a ballet, concert or opera; or at a banquet in the hall of an academy devoted to study of the classics; or on state occasions, in the grandiose Salone dei Cinquecento of the towering Palazzo Vecchio; in the Pitti Palace, the Sala delle Commedie was also reserved for improvised entertainments; and, finally, in 1586, a permanent theatre for the pleasure of the Medicis and their guests was installed in the Uffizi Palace, available for an ostentatious royal wedding or a welcome to foreign dignitaries.

Unless the production was subsidized by the Medicis or the city – if so it might be spectacular – most offerings by the amateur troupes betrayed limited financial resources. But by claiming to follow the tradition of the ancients they had a handy excuse for having a single set, all the action occurring on a street before a crowded row of houses, with doors and windows flying open and shut, noisily suggesting busy life within. The Florentine troupes had a rich resource, however, a covey of masterful artists for scene-painting, properties and costume designs, as already mentioned: among others Raphael, Cellini, Bronzino and Vasari.

In Florence the spectators were apt to be less patrician than elsewhere in Italy; for the earlier part of Cecchi's life, Florence, a mercantile centre, reached the height of its considerable prosperity; in his audience were clever traders and members of the professions, a new, thriving upper middle class, along with a large number of scholars speaking Latin and Greek and well acquainted with what was as yet known of the classics. Belonging simultaneously to all these intermingling groups, Cecchi aimed his humour at them; hence his plays are described as "bourgeois comedies". They are accurate snapshots of his society, exhibiting characters with whom the spectators could identify and empathize. Yet these are the same characters, given fresh names and an updated setting, who appear in Plautus and Terence. Cecchi succeeds in establishing their timeless universality. "It was his avowed intention," says Douglas Radcliff-Umstead, "to render the ancient comic material a living experience for contemporary Italian audiences."

This is very clear in his first play, *La Dote* (*The Dowry*), which deals with the financial arrangements that precede a marriage and how much the girl's family will settle on the plighted couple. This age-old custom was prevalent in the days of the pre-Christian Roman Republic; indeed, Cecchi takes most of his plot and characters directly from Plautus, as he acknowledges in his prologue, and proudly, for it shows how conversant he is with Latin and the classics. He makes no secret of his literary debt – he boasts of it. Later, in the prologue to another comedy, he speaks of Plautus as his "good companion" and "very dear friend", an affinity real and deep.

His "friendly borrowing" is from Plautus' *Trinummus* (*The Threepenny Day*), which the Roman author derived from Philemon's *Thesauros*. To this, Cecchi adds a scene from Plautus' *Mostellaria*, about a supposedly haunted house, an idea based on the earlier Greek *Phasma*, thought to be by Theognetus. (Such is the scarcity of plots for comedies.) However, Cecchi does not merely change the setting from Ancient Rome to his own-day Florence and give his characters Italian names in place of Greek ones he also eliminates some of the minor persons employed by Plautus, adds a few new roles, excises a number of episodes and rearranges the order of those remaining, to increase the suspense by delaying revelations. From the start he proves that he had a sound dramatic sense. He does not simply copy Plautus' play; he freely adapts it and improves on it.

A wealthy Florentine merchant, Filippo Ravignani, sets out on a long journey; he leaves his two grown children, Federigo and Camilla, in the care of a friend, Manno. Once beyond his father's care, Federigo becomes a profligate, squandering his funds and what would be his inheritance. At one point he is about to sell his family's house to a stranger, but Manno intervenes and buys it at a price far below its true worth, thereby seeming to exploit the young man's weaknesses. Federigo's prudent, idealistic friend, Ippolito is deeply in love with Camilla and wants to wed her, and she desires him. Word comes that her father has been lost at sea in a shipwreck. Federigo is now responsible for Camilla's future. Ippolito's miserly father, Fazio, is adamantly opposed to a marriage between his son and Camilla because Federigo is now impoverished and cannot provide his sister with an adequate dowry. In vain, Ippolito pleads that all that counts with him is the young lady's many good qualities, her intelligence, and her modesty.

Manno alone knows that Filippo, before leaving and to assure a dowry for Camilla, concealed 3,000 ducats in his house; that is one reason for the guardian's haste to buy it before the feckless Federigo sells it to anyone else. But now Manno's search for the hidden treasure is frustrated. His neighbours, not knowing what are his benign motives or about the existence of the hidden hoard, blame him for having betrayed a trust and mulcting the young man. At last Manno discovers the 3,000 ducats. To enable Camilla to marry, but anxious not to reveal the money's source, he has a servant pose as a Levantine merchant newly arrived with letters and the recovered funds. Unexpectedly the real Levantine reaches Florence and encounters the hapless servant who is impersonating him, whereat the usual range of complications arise from the scheming of a large cast – there are many minor characters and several sub-plots – and there is the traditional quota of confusion begotten by mistaken identities and misdirected over-zealousness, much of the trouble being caused by the conniving of Federigo's glib, malicious servant, Moro. Federigo, humiliated at being unable to provide his sister's dowry, is about to enlist as a mercenary in the military. His desperate move is interrupted by the surprising arrival of his father who, of course, has not drowned. To prevent Filippo from entering his house and becoming apprised that his son's extravagance has forced its sale to Manno, Moro and Federigo spread the story that it is haunted. It takes a while for all this to be clarified and the union of Ippolito and Camilla to be finally consummated.

The ending differs in one facet from that of most plays in this genre: Federigo gives no sign that he will change his spendthrift ways; apparently he is unrepentant.

Though Cecchi's purpose is not too overt, *La Dote* is offered as a critical comment on the materialism growing rampant in Florence, where getting and keeping money has become the chief aim of too many of its citizens, so that frequently a marriage is arranged only to enrich the bridegroom and his family, with little concern for the virtues and other attributes of the two people involved in it; and social status and wealth are synonymous. A man's or woman's spiritual worth are of lesser account, and cash ever more important than grace and gentle manners.

Another of Cecchi's innovations is his statement in the prologue that he will offer no advance exposition of his plot. He deems his audience intelligent enough to grasp every turn of the action without help. Like Grazzini, he was in the camp of those who fully accepted Aristotle's precepts, by now canonized by the majority of Italian Renaissance playwrights. If anything, he was even more submissive to and guided by those rules than was Grazzini, who stopped a slight but perceptible distance short of "servile imitation". For the most part, Cecchi's output conforms to what had come to be called "regular theatre".

As a proponent of dialogue in the Tuscan vernacular instead of the then more esteemed Latin, he wrote his earliest plays in ordinary prose that effectively gains colour and impact from his liberal sprinkling throughout of native expletives and idioms; often the lines sound epigrammatic from his deft and appropriate insertion of the local proverbs he so assiduously compiled. For example Moro expresses his disdain for Filippo's lack of foresight by sying "He left ducks to protect the lettuce"; and of one of the tricks, a misrepresentation, played on an unwary soul, he says "He showed him the moon's reflection in a well"; and when Moro has to act quickly to extricate himself from a dilemma, he exclaims "The ball has bounced on my roof." (The quotations are from Douglas Radcliff-Umstead's *Carnival and Sacred Play: The Renaissance Dramas of Giovan Maria Cecchi.*)

This vivacious comedy, mounted by the company of San Bastiano de' Fanciulli, had an excellent public reception. Four decades later, in 1585, following the example of Ariosto, whom he greatly admired, Cecchi revised his script, further rearranging incidents, removing some characters and transposing the dialogue in verse. Though his poetry is adequate, it is hardly inspired; most modern critics prefer the prose version.

The following year (1545) he brought forth *La Moglie* (*The Wife*), also accepted by the Bastiano de' Fanciulli, a group that functioned as his producer for many years. Here mistaken identities abound. Once more he returns to Plautus – *Menaechmi* (*The Twins*) – adding a few devices from Terence's *Andria*. (By commentators of the period, such commingling of classical sources was termed "*contaminatio*", intended as a pejorative and implying that it was offensive to purists.) The same Plautine farce was to serve Shakespeare for his *Comedy of Errors*; however, the English bard was to top his predecessors and double the confusion by having not one pair of twins but two, by chance the brothers' servants also bearing identical features, making it difficult to tell them apart. From Terence, Cecchi borrows elements of romance – the "love interest" – and occasional and discreet dollops of sentimentality.

The scene is shifted from Rome to Florence, the time from antiquity to the present during Carnival, which provides a background of communal chaos – even anarchy – inducive of temporary madness in the characters. It is a season of masking and practical jokes, of exposed feelings and over-reactions. Some details of the plot are standard and all too familiar. Silvano de' Silvani is a rich merchant, has fathered two sons and a daughter. Sailing from Alexandria to Marseilles, he is believed to be lost when the ship is wrecked near Corsica. His sister, married to Alberto Spinola, also an Italian tradesman prospering in Egypt, assumes charge of the three orphaned children. She dies, and her husband decides to return to his native land, taking his nephews and niece with him; *en route* the vessel is waylaid by Turkish pirates. (This was, at the time, a very real threat, hence deemed plausible by the spectators.) Spinola and the three children are separated: he and one of the twins, Alfonso, become slaves in Ragus. Another Italian merchant, Roberto Amidei, buys Alfonso. Recognizing the boy's intelligence, he trains him to become an apprentice and later a manager in the business. Grown to young manhood, Alfonso marries Margherita, Amidei's daughter, and is put in charge of his father-in-law's branch firm in Florence.

Alfonso and his wife become friendly with a young neighbour, Ridolfo, son of a wealthy, politically and socially ambitious merchant, Cambio. On a mission to Ragusa for his father, Ridolfo becomes attached to a slave-girl, called simply Spinola. Impulsively, he marries her. Aware that Cambio would never approve of the union, which brings with it neither riches nor social prestige, he determines to keep it a secret. From what Spinola tells him of her parentage, he guesses that she might be Alfonso's long-lost sister. On the pretext that Alfonso, to be helpful, is taking her with him to Florence to enter a convent, the couple returns there. Then, spinning another tale – that she is the daughter of a prominent Ragusan businessman – Ridolfo asks Alfonso and his wife to shelter Spinola temporarily, suggesting that this might win the goodwill of her "father" and lead to a profitable commercial tie. (This thread of the story derives from Terence.) Since she is now lodged so conveniently near, Ridolfo is able to pay her frequent conjugal visits. His father observes his son's behaviour and grows suspicious – Ridolfo must be toying with a mistress.

To block any mis-step or transgression by his son, Cambio arranges a marriage for him, selecting as a suitable partner a daughter of Pandolfo Agolanti, a prominent Florentine financier. This creates a dilemma for the already wedded Ridolfo. As rumours spread that the reckless young man is having a liaison with a former slave-girl, Agolanti breaks off the marriage compact. Cambio moves aggressively to restore it, exerting pressure on the harassed Ridolfo.

All the action takes place in a single day, with much of the exposition of past events provided in the dialogue. Another serious complication arises with the arrival in Florence of Alfonso's twin, Ricciardo, who after his rescue from the sea has been living with his uncle in Siena. As a token of his mourning and respect for his supposedly demised brother, whom he exactly resembles, he has assumed his name and now is also called Alfonso. This engenders so much confusion, so many people reporting the second Alfonso's doings to Margherita, the true Alfonso's wife, that she is hopelessly bewildered and finally convinced her husband has lost his mind. Active in the misunderstandings and thwarted manoeuvres is Fulgino, servant to Ridolfo, who ceaselessly concocts far-fetched stratagems to foil Cambio's equally devious plans for his son's wedding. The mix-up of identities helps to roil these.

A satisfactory and harmonious resolution of this farrago is only achieved when Alberto Spinola comes from Siena and discovers his long-vanished nephew Alfonso, straightens out the misapprehension about the second "Alfonso" – that is, Ricciardo – and joyfully recognizes his niece, the former slave-girl, Ridolfo's wife. It is further disclosed that Pandolfo Agolanti is actually Alberto's brother-in-law, Silvano de' Silvani. Fleeing from Bologna and a murder charge, he has been forced to hide under a new name. And he, of course, is the father of the twins – the two Alfonsos – and their sister. To even the equation, his daughter by a second marriage, the girl sought by Cambio for Ridolfo, is given instead to Alessandro Rusticelli, an ardent, rival suitor.

One must wonder if, by now, any Italian Renaissance spectator would be taken by surprise at the last-minute reappearance of a father, uncle or other kinsfolk lost at sea or captured by Turks, but apparently this ancient plot device still worked.

Again, Cecchi is holding up to view with disapproval the Florentine emphasis on the monetary facets of marriages that are not made in Heaven but by ambitious fathers or avaricious family elders, with little consideration of the compatibility, mutual affection, physical attraction and strong romantic feelings of the young people facing life-long bonding.

The marriage between Alfonso and the nagging, greedy, jealous Margherita, an arranged union by which Amidei hoped to hold on to a promising assistant and Alfonso sought to advance his career, has long since gone sour. Nor does Alfonso show much generosity of spirit: he is almost certain that Spinola is his sister but does not openly declare it, lest he have to provide her with a sizeable dowry. His motive for helping her escape from Ragusa and sheltering her in Florence is the prospect, conjured up by Ridolfo, of gaining a well-paying customer, though in fact her prosperous "father" is a fiction. Ridolfo knows his friend all too well. Self-interest rules in Florence.

As Radcliff-Umstead remarks, Cecchi has difficulty fusing the Plautine and Terentian elements of the story, and what ultimately serves to unify them is the close biological tie of the leading characters.

The swarm of minor figures offer a picture of workaday Florence: a neighbourhood baker; a tapestry-weaver; several drunken housemen seeking a tavern; a Latin-spouting doctor summoned by Margherita to diagnose the cause of her "husband's" odd behaviour. An abusive customs official, who fines Ridolfo for bringing in his personal silverware, is no doubt intended to add a note of social criticism. All these contending persons contribute to the "world of disorder" brought about by the hapless machinations of Fulgino and the baffling likeness of the twins, who hurry about the city unaware of each other's existence, without ever coming face to face. That they are never seen together mitigates a likely casting problem.

(A local touch: Sienese by the boastful Florentines were regarded as less intelligent and worldly. That Ricciardo and his uncle, Alberto Spinola, seem to come from that nearby city is initially a handicap.)

Marriage still preoccupied Cecchi as he contemplated his next script, *Il Corredo* (*The Trousseau*; 1545–6), though he was to wait another full decade before choosing a wife of his own. Some critics discern a strain of misogyny in his work.

A fourth play, however, finds him going in another direction. *La Stiava* (*The Slave Girl*; 1546) has to do with a strange rivalry between father and son. In an essay already quoted, Antoniofrancesco Grazzini, a friend and fellow dramatist, had objected that there were no slaves in sixteenth-century Italy – making use of them as characters was only a handy and familiar plot-ingredient carried over from the Greeks, as well as from Plautus and Terence. But the incongruity gave Cecchi no pause. In *La Moglie* both Alfonso and his sister Spinola are captured and then sold, after experiencing a disaster at sea. Radcliff-Umstead suggests that in fact there was a class of household workers, some of Tartar extraction, so poor and lowly that theirs was virtually a condition of involuntary servitude. Or they might be bought when Italian ships put in at Black Sea ports.

In *La Stiava* a young Genoese, again Alfonso by name, buys such a girl, Adelfia, when he is on a trading journey to Turkey, where there was a slave market. A significant fact is that she is of Italian birth. Essentially, Cecchi's script is based on the *Mercator* of Plautus, which a decade earlier had also been appropriated by Donato Giannotti, a Florentine statesman, for a play, *Il Vecchio Amoroso* (1533–6). Both works relate how the youthful merchant returns to Genoa with Adelfia and soon learns that his miserly father, Filippo, is so smitten with the girl that he is anxious to supplant his son and take possession of her. In his scheming, the father is abetted by his close friend and neighbour, Nastagio, like him a wealthy merchant, and who also has a son, Ippolito. The two young men, having strong bonds of friendship, are ever ready to help each other. Nastagio, sagacious and better balanced than Filippo, disapproves of the frenzied, senile lust that drives him, yet is himself aged enough to feel sympathy for him. He is a reluctant conspirator, constantly urging restraint in the impetuous attempt to wrest the girl away from her rightful owner – Alfonso speaks of her as his "possession", in sum "property bought and paid for". Filippo quotes a Biblical injunction that a son's goods also belong to the father, which gives him sanction to lay claim to the beautiful Adelfia. Amidst all the loud quarrelling and moves and counter-moves, the girl herself is never seen – oddly, she remains off stage throughout, merely the subject of a prolonged, heated dispute, an inter-generational struggle.

In seeking the girl, Filippo is desperately trying to recapture his youth, though at seventy his physical powers have considerably weakened, until he verges on impotence, a fact that he does not like to admit even to himself. For Alfonso a sense of proprietorship has gradually changed into love. At one point he tells his father that he will sell Adelfia to another bidder, but he has already married her, thereby placing yet another obstacle in his crafty, tyrannical father's way.

Cecchi greatly enriches Plautus' treatment, and his script differs in many respects from Giannotti's handling of the story. The life of Genoa, a busy seaport, is evoked, with references to many comings and goings to and from the crowded harbour. Filippo has the girl abducted, hidden in Nastagio's house, where he wishes to regale her at a banquet and then have her transported to a ship. The plan is botched by Gorgolio, the inept, well-meaning servant to Filippo. While Adelfia is held at Nastagio's, his wife, Giovanna, returns unexpectedly and suspects that in her absence her husband has been entertaining himself with this voluptuous girl. She rails at the bewildered cooks and berates her guileless mate, who has a difficult time exculpating himself. Ippolito, too, fails to be as helpful as wished by Alfonso, and cross words are exchanged: friendship takes its toll.

The outstanding figure in the play is Filippo, doddering yet excessively sensual, his carnal appetite never sated. Grown old himself, he still cannot bear to look at the withered face of his wife, and has a restless, wandering eye. Hypocritical, he masks his rascality, puts on a mien of propriety. Stopping for a short spell during a heinous act, he announces that he is going to attend mass. Well portrayed, too, are his servant, the bumbler Gorgolio, and Nua, a maid in Nastagio's house. They contribute much to the comedy.

Adelfia is rescued during an attempt to transfer her to a bordello presided over by Apollonia, a procuress. In a small box she possesses are trinkets – rings, bracelets – that establish her true identity. She is none other than Nastagio's daughter, who long since had been lost to Turkish pirates. With this recognition her situation is wholly altered: she will have a substantial dowry that will legitimize her middle-class social standing as Alfonso's wife. Her husband and Ippolito, discovering that they are brothers-in-law, are reconciled as calm descends on the two families.

Usually in a farce of this sort, a servant ends the action with an address to the audience, a summary, and perhaps a disarming plea for applause. In this instance the task falls to Filippo, who delivers the farewell and spells out the lesson that the play is meant to convey.

Keeping up his rapid pace, Cecchi turned out *Gli Incantesimi* (*The Enchantments*; 1547) and then *I Dissimili* (*Rival Brothers*, or *Two Unlike Brothers*; 1548). With changes and additions, the latter is Cecchi's adaptation of the *Adelphoe* of Terence. What is the best way to bring up one's sons? In the prologue he states that his purpose is didactic, as was Terence's. The answer should be of interest to fathers, guardians and their charges; that is, young men of marriageable age. The action unfolds in present-day Florence. The dissimilar brothers, Filippo and Simone, now of advanced years, have inherited means enough to live in comfort. Filippo, preferring urban life in expensive courtly circles, has never married. Simone has withdrawn to a farm, his daily routine austere and frugal. He has two sons: Alessandro, who remains with him, and Federigo, who has been permitted to enjoy his uncle Filippo's tutelage and fond care. Alessandro has been raised liberally and indulgently; Federigo has been guided by overly strict rules.

Federigo loves Fiammetta, but is so timid in his approach that his suit seems futile. On his behalf, Alessandro, undisciplined and headstrong, always used to having his way, tries to break down the door to her house and abduct her, a wild deed that creates a great scandal. It is mistakenly assumed that Alessandro was seeking possession of Fiammetta for himself, but he is much in love with and secretly married to Ginevra, who is the daughter of an impoverished widow and brought him no dowry; further, she has recently borne his child. Given very different upbringings, the two young men, one unworldly, the other propelled by heedless and too-generous impulses, have brought disgrace on their elders and themselves are facing disaster. Filippo and Simone are taken aback by this outcome of their theories on the proper education of the young.

Sfavilla, a roguish servant of Filippo, intervenes to no avail. He creates a fictional long-lost brother of Fiammetta in Lucca, a Roberto Burlamatti, who claims custody of his sister; but her guardians, Pietro dall'Acquila and his wife Dorotea, are not persuaded by the story, and Simone's sharp interrogation unmasks the impostor. A forged letter sent to Dorotea, asserting that her husband is dangerously ill in Pisa and summoning her to him, also fails to fulfil its purpose. Before leaving Florence, Dorotea places Fiammetta in the safety of a convent, to remain there until her return to the city. (These plot turns are Cecchi's inventions. He is making the point that duplicitous ploys of this sort often mire those who conceive them. The literal meaning of "Sfavilla" is the "Spark".)

Eventually a remorseful Alessandro makes a full confession to his uncle Filippo and is perhaps all too readily forgiven. Simone recognizes that his governance of Federigo has been too severe. He astonishes everyone by a sudden display of generosity and extravagance, distributing lavish gifts to his sons and servants. Fiammetta's mother, who since her husband's death has dwelled in a convent, comes forth and is reunited with her daughter, for whom a family friend, newly discovered as the girl's uncle, provides a requisite dowry, permitting her union with the timid, infatuated Federigo. A double wedding banquet leads to the comedy's final bows.

Its message is that in rearing children one should not use excessive kindness nor perpetually submit them to a daily, authoritarian routine of close scrutiny and harsh discipline; the constant attempt should be to arrive at a golden mean. But the moral is not too pat: a neat ironic touch is that Simone's open-handed allocation of gifts and grants is actually at his elder brother Filippo's expense, depriving him of much of his inherited property.

Foremost and compelling is the portrait of Filippo, hedonistic, urbane, a worldly courtier seemingly gentle and tactful, yet subtly imperious and something of a snob. He is a superior but flawed person, buying the affection and gratitude of others and not sure of obtaining those emotional responses in any other way. Simone, serious and well-intentioned, avoids being a comic butt by reaching self-understanding and taking steps to correct his faults. Alessandro, inherently bold and impetuous, but now a husband and father, comes to an early maturity. All the leading characters undergo perceptible change.

Once more the question of how important the size of a dowry should be is openly discussed.

Filippo tells his greedy brother that beauty, good breeding and an amiable nature are what really count in a match, not money and social position – though he himself is often condescending to people of lesser wealth, seeming to regard them as inferior to anyone of his own fine qualities.

With *L'Assiuolo* (*The Horned Owl*; 1549) Cecchi's art and reputation as a playwright took a major leap upwards. It is generally considered his best work – even, by some, a comic "masterpiece" – and had a long-enduring popularity matched by no other script of the period, rivalling Machiavelli's *Mandragola*. In the prologue Cecchi announces proudly that this is an original play, one that owes nothing to the classic Greco-Roman farces:

> It's taken neither from Terence nor from Plautus, but as you'll hear, from something which happened recently in Pisa between some young students and gentlewomen. In fact the event is such that, if I'm not mistaken, it will seem pleasant and worthy of your attention. And don't anyone think that this comedy originates from the Sack of Rome or from the Siege of Florence, or because persons became displaced or families had to flee, or some other such event. Nor does it finish with marriages, as the majority of comedies usually do. In our comedy you will not hear anyone complain that he's lost sons or daughters because, as I've told you, no one has lost anybody. No one will be married off, because one of this group's good points, or rather happy rules, is that they cannot get mixed up in marriage, be it their own or that of others. If then you ask what does this comedy deal with, I'll tell you: an event that occurred in ten hours or less, one which you'll hear about right away, if you will give us that welcome attention which one seeks at shows and which you have granted to other comedies by this same author. And, if by chance, it seems slightly more licentious either in words or in action than his other plays, you must excuse him because, wanting for once to get away from marriages and the discoveries of long-lost children, there was nothing he could do about it.
>
> [Translation: Konrad Eisenbichler]

An introduction that was bound to put the spectators in good humour and expectant of enjoyment.

That *L'Assiuolo* is different from works by most of Cecchi's contemporaries is certainly true, and its claim to originality has long been supported by critics and scholars. In truth, though, it is "original" only to a limited degree. Radcliff-Umstead cites at least a half-dozen borrowings of incidents and characters, and lines of dialogue – salacious ones – from the collection of wicked tales in the *Decameron* of Boccaccio (1313–75), who was a copious fount much dipped into by Italian Renaissance playwrights. That earlier Italian's lively wit and spirit, and his concentration on descriptions of erotic diversions and fantasies, are broadly heard and reflected in the output of a host of other writers of Cecchi's century. Furthermore, in *L'Assiuolo* there are light but clear echoes of *Mandragola* in plot-turns, and a decided likeness of attitude in the characters about matters having to do with sexual morality – it might better be said that the bent of nearly all those involved is amoral. (Cecchi's assertion that the play is based on an actual event is probably not to be taken seriously.) To define the script as wholly original, one would have to redefine the term.

Much of the action of *L'Assiuolo* is concerned with putting on disguises and cross-dressing, so it was a singularly appropriate piece for a Carnival offering. It is a celebration of adultery, of sexual indulgence without stint or hesitation, which probably also fitted the exuberant festive season, and its candid sensuality no doubt accounted in considerable part for its long-lived success. Its characters are not driven by romantic love but simply by carnal impulse, and the principal reason the play does not end with marriages is that many of the figures in it are already husbands and wives, narrow roles from which they would very much like to be set free; or, if they are single, marriage is definitely not their intention. All they are seeking is fresh adventure and physical satisfaction.

A pretty widow, Anfrosina, lives with her son, Rinuccio, a student at the University of Pisa. A tenant in their house is Giulio, also a student; the two virile young men are close friends. Near by is the house of Ambrogio, an elderly – almost senile – miserly lawyer who is handling a legal suit for the widow. He has a young wife, Oretta, of whom he is obsessively jealous; he keeps her closely guarded, never allowing her to go out unaccompanied. Rinuccio has eyes for Oretta and, though they have never met, conceives a yearning passion for her; he is determined to seduce her. He seeks the assistance of his friend Giulio to this end and is given a fraternal pledge by him; but, secretly, Giulio is also attracted to Ambrogio's wife, wishes to sleep with her and lays plans to do so, an enterprise in which he is abetted by his self-assured servant, Giorgetto. What complicates matters, however, is that the aged Ambrogio is not content to have a lovely young wife, whom he does not fully satisfy sexually – instead, he is madly possessed by a desire for the sprightly widow Anfrosina and is taking every possible measure to share her bed.

One afternoon Oretta, flanked by the lawyer's dim-witted servant and bodyguard, Giannella, goes to a convent to see a sacred play being enacted there. Giannella cannot enter the cloistered precincts and waits outside, while by chance Oretta finds herself seated alongside Anfrosina and hears her complaint of an elderly neighbour, a lawyer, who is forcing his annoying and unceasing attentions on her. Oretta, outraged, recognizes that the would-be adulterer is her doddering husband. She suggests to the widow that they join to trick and expose him by having Oretta take him by surprise in Anfrosina's room. Rinuccio learns of their intention from his mother's maid-servant, Agnola, who is always ready to garner a tip or other reward.

The two students grasp the opportunity this opens. They spread a false story that they will be away from Pisa for several days. Then they send a letter to Ambrogio, copying Anfrosina's hand-writing and signing her name, saying she will be alone in her house and would welcome him under cover of night. Eagerly responding to this lure, the sex-hungry Ambrogio orders Giannella to post himself on guard alternately before both almost adjacent houses and, in the event anything goes awry, to answer immediately to a pre-arranged signal, the horned owl's cry, *Whooo, Whooo, Whooo* (hence the play's title).

It is Carnival. To make sure that she will not be noticed, Oretta dons male garb. For the same reason, as they approach their neighbour's house, Ambrogio and Gianella dress themselves as pages and look ridiculous. Giulio, to obtain entrance to the lawyer's well-locked home, disguises himself as a maidservant.

In the maelstrom that ensues, Ambrogio is trapped and, inadequately clothed, has to spend the night in a cold courtyard, his urgent *Whoos* unheard by the cowardly Giannella who has been attacked and beaten by Rinuccio and has fled – he later exaggerates the number of the unknowns who assaulted him. The various couplings take place off stage, of course – and in darkened rooms, with every sort of misapprehension as to who are the persons involved. Oretta, into whose bed Giulio has stolen, believes that she is being embraced by her husband and at first is astonished at what seems to be his renewed sexual vigour, and Rinuccio is under the happy illusion that he holds Oretta in his arms, when actually it is her married sister Violante, who has come as a visitor and is napping while awaiting Oretta's return home.

The only one chastened by these few hours of nocturnal mischief and priapic caperings is the aged, chilled Lothario. Both Oretta and her sister are so delighted by the young men's prowess as bed-partners that they express a firm wish to see them soon again – indeed, Oretta even suggests that Rinuccio, who was deprived of her favours, may take Giulio's place on occasion. The moral of the play, if it has one, is that old men should not choose young wives, and certainly, if they have long been physically frustrated, should not try to keep them locked away.

Two of the minor figures are conspicuous, the trickster-servant Giorgetto and the hypocritical Verdiana, a prim-looking lady who between busily going to masses and reciting prayers keeps herself in funds as a semi-professional procuress, and proves herself capable of bargaining successfully with the tight-fisted Ambrogio whenever he has need of her as a go-between. To cap the triumph of his machinations, Giorgetto takes himself off to a brothel.

The action is "fast and furious", with doors being broken in and people climbing through windows to make their escape, walls scaled, and the scuffle in which Rinuccio and Giannella engage. Seeing an aggressor, Ambrogio, duped when he undertakes an illicit liaison, is sure to provoke laughter: it is an ingredient, dubbed the *beffe*, that appears in many of the comedies of this period – a trick is played on someone who, overconfident, is trying to put over a trick on others, a rebound, a form of poetic justice. Also sure to amuse on stage is transvestism, of which there are not one but four practitioners here, since Violante, heading for the Carnival, dresses as a boy and wears a false beard.

The dialogue, in prose, is terse and salty. Cecchi retains the local dialect. The effect is one of "Florentinity", to use a phrase applied to it. The lines are brightened by a prodigal insertion of idioms and popular sayings heard daily in that city's streets and now borrowed from his published collection of proverbs. Expressing contempt for any young man who adheres too closely to conventional rules of good behaviour, Giorgetto warns, "In the end you aren't worth a handful of peanuts." When Rinuccio seeks to justify his spying on Oretta, since it abates his lusting for her, Agnola tells him: "But to put out a fire, you throw on water, not sulphur." Ambrogio declares that he cannot depend on his servants to ward off admirers of his beautiful wife: "It'd be like letting the geese look after the lettuce." Yet he relies on the loyalty of Giannella: "He is more trustworthy than the Lord's Prayer." Bargaining with the sharp-tongued Verdiana, he protests: "You're scalping me like a bad barber." Explaining that her mistress Oretta has no thoughts of seeking a liaison, her servant Agnola asserts: "I found her to be further from these matters than January from roses." Ambrogio lays claim: "I've paid the piper, Madonna Verdiana. I want you to see to it that I also get to dance." Rinuccio wants to know if Giannella can be bribed to keep silent and is told: "I think we'd have problems shutting him up with a knife, even if we shoved it down his throat." Violante laments: "Now we're stuck here like flies in amber." Giving very detailed instructions to the slow-thinking Giannella, Ambrogio is guided by the maxim: "To a thick skull one should feed thick soup." Somewhat shocked by his elderly master's lechery, Giannella muses: "Say what you want, when it comes to sex you can never be sure. When you get the itch, you'll find a way to scratch it or have it scratched for you." Assuring Rinuccio that Oretta's caresses will be distributed in equal measure between the two of them, Giulio promises: "The merchandise will be mutually owned." Of her initial objection that she did not want Giulio to touch her, and the later physical revelation of her ample sexual appetite, Rinuccio comments: "She wouldn't have been the first one to say she wasn't hungry and ended up eating enough supper for seven."

Both Violante and Oretta voice anger at what fate has brought them. In her disguise as a man, Oretta ruefully soliloquizes:

Anyone can come to know how poor and unhappy a woman's lot is if he considers how many inconveniences we are subjected to, how many pleasures we are deprived of, and under what cruel tyranny we must live our lives. When men have to take a wife, they nearly always take what they please. We, on the contrary, must take whoever is given us. Sometimes, and I for one can vouch for it, poor me, we must take one who, to say nothing of his age which would make him our father rather than our husband, is so rough and inhuman that one could sooner call him a two-legged beast than a man. But let's not talk about the bad luck other girls have had, and let's speak of me, the most unlucky one of all. I find myself married to Messer Ambrogio, who could be my grandfather! He's rich, yes. But this doesn't mean I eat any better! Besides having a husband who's old, there's the problem of having one who's jealous, wrongly jealous. And there's no one who's more jealous than him. So, because of his jealousy I'm deprived of pleasure outside the house, and because of his age, of pleasure inside the house.... Poor Oretta, stuck for life with a husband who's old, jealous, philandering and senile. And so, to bring him back to his senses, I have to scale the garden wall, dressed as a man at ten o'clock at night in order to get out, go through Pisa in disguise, enter into other people's homes and perhaps have myself labelled something I've never been nor ever had the intention of being.

Elsewhere, the more independent Violante exclaims: "The first girl who wants to marry an old man should be hanged. Old men are like a gardener's dog: it never eats the lettuce nor lets anyone else eat it" (translation: Konrad Eisenbichler).

In these speeches there are no hints of the author's imputed misogyny. Perhaps he had undergone a change of heart since writing *I Dissimili*.

L'Assiuolo was produced by the Company of the Monsignori and the Fantastichi and was quickly recognized as a hit. Radcliff-Umstead postulates that it struck a taut chord, "expressing a general longing for release from the constraints of the bourgeois marriage". As a vehicle for that wish it was truly a companion piece to Machiavelli's *Mandragola*, which was demonstrated in a most unusual way when a programme consisting of both plays was performed at a gala in the Sala del Papa at the Palazzo Vecchio. One play did not follow the other; instead they were combined, the first act of *Mandragola* preceding the first act of *L'Assiuolo*, and then Machiavelli's second act offered ahead of Cecchi's second, and so on, sections of scripts alternated throughout. The audience welcomed this odd approach to staging a pair of hilarious farces. Scene designs for the plays, a most elaborate row of houses and a courtyard, were by Agnolo Bronzino and Cecchino Salviati.

In *Lo Spirito* (*The Ghost*), written the same year (1549), Cecchi reverts to ending his play with marriages; indeed, four of them. Two of the romantically involved couples are elderly, and two are young, so the complications are intergenerational. Again he partakes of story ideas found in Boccaccio and several other writers, and though he spoke scornfully of plots introducing parents and siblings lost at sea or mistakenly assumed dead in battles with pirates or other alien invaders, he falls back on that device quite shamelessly. (In his prologue he states that – at thirty-one – he is not a professional dramatist but only offers comedies to please his friends.) *Lo Spirito* concerns the troubled relationship between Emilia, a penniless former slave-girl, and Napoleone, a middle-class youth, who must keep their legal bond a secret because it would not be approved of by their guardian. Unaware that she is already married, Neri is preparing to have her wedded to Aldebrando, also a former slave, but now the adopted son and heir to a rich merchant, Anselmo. Away on a business journey, Napoleone vanishes and is reported to have died. He returns before the union of Emilia and Aldebrando is consummated, and convinces the new "husband" that all conjugal rights are still his. Aldebrando consents to this, since his passionate interest lies elsewhere, specifically in the person of the very attractive niece of a foreign physician who is residing in the city before taking a post at the University of Pisa. With Aldebrando's contrivance, Napoleone pays covert visits to his Emilia.

On the elderly Anselmo's conscience is the guilty knowledge that he and Laura, Neri's sister, had once been lovers, until her brother interfered and forced her to marry another man. A child born of the illicit liaison had died. Laura is now a widow, and to ease himself of sinful memories Anselmo openly seeks her hand, but this arouses family opposition. Angered, Anselmo shuts the door of his house to anyone having a link to Neri, with the result that Napoleone can no longer meet Emilia there.

There is pressure on Neri to marry, but he explains that he already has a wife – or has he? While a refugee abroad after the revolt against the Medicis, he had wedded a poor but aristocratic widow in Romania. Later, named a papal envoy to France, he had been separated from her. In an invasion of Romania by Barbarossa's forces she had been taken prisoner and carried off – at that time she was seven months pregnant. He has never learned what became of her and the child.

A solution to the frustrations of Napoleone, Aldebrando and Anselmo appears to be at hand with the arrival of a Greek necromancer, Aristone, who exploits the superstitious nature of all three while glibly charging them exorbitant fees. He instructs them to hide in a trio of chests in which each of the three can be transported to the bedchambers of the ladies they so ardently desire. Also, he has Emilia speak only in Latin, having her claim that she is possessed by a spirit. Hence she must be confined to her room. But all the ruses with the chests go awry when some are delivered to the wrong destinations, one even ending up at the Customs Office, leading to the unhappy Aldebrando's arrest. Officers who come to Neri's house to collect debts left by Laura's late husband block the deposit of Anselmo's coffer there. A piling up of more such mishaps leads to the denunciation of Aristone as a fraud.

It is ultimately established that Aldebrando and Emilia can boast of wealthy and noble lineage – as the audience doubtless expected – and that the foreign physician's sister, Maria, is Anselmo's long-lost wife, and the niece, Laldomine, his daughter. Neri agrees to let Anselmo marry Laura; and Aldebrando gains the hand of Laldomine, and the openly proclaimed bond between Napoleone and Emilia gets their guardian's delayed blessing.

All the characters blame their problems on Fortune – good or bad, everything is at the mercy of chance. Radcliff-Umstead observes that this is the typical outlook of Renaissance playwrights; it is an attitude to be perceived in Ariosto and Machiavelli, too. It differs from the view of medieval dramatists who believed that cosmic interference in human affairs was impartial, at the very least. Later, after his religious conversion, Cecchi would become wholly fatalistic.

A necromancer, pure charlatan, is a figure who appears in other plays of the period. What is surprising is how sympathetically and even respectfully Cecchi portrays this Aristone, a Greek. He does not exhibit vulgar magical tricks but rather, by his urbane and portentous manner, tries to create an image of himself as a lofty master of the fearsome black arts. He asserts that he has degrees from several universities, is an astrologer, a herbalist, a visionary and a descendant – no less – of Aristotle. At the conclusion Neri invites him to stay on as his guest; there is nothing shoddy or common about him, though his aims are wholly mercenary. He is given the honour of delivering the play's last speech.

In the next year and a half, though distracted by starting his notarial career, the indefatigable Cecchi provided the Carnival players with *Il Donzello* (*The Manservant*; 1550) and *La Maiana* (*The Girl from Maiana*; 1550–51), the second of these handily adapted from the ever-inventive Terence, and marked by his growing interest in showing various sorts of ambivalence, the difficulty one experiences in determining what is real in human relationships and the seemingly changing natural world. The task set his characters is to see through deceptive appearances when – after Carnival time – the masks are dropped, along with the "revelry and trickery". Life is suddenly serious once more, moral dictates and practical concerns return and take over. The duality of daily existence, its alternating frivolity and harshness, is displayed.

After a lapse of four years, during which he was preoccupied with his family's dire troubles, he ended his silence with *L'Ammalata* (*The Sick Girl*; 1555), in which a self-sacrificing nurse is an appealing figure, having an aura of kindness and warmth. He draws other such favourable images of family nurses and retainers; though, in more instances, disloyal household staff-members are apt to be shown avid to have their outstretched palms weighted with scudos, susceptible to bribes.

The insidious effects of greed are Cecchi's subject yet again in *Il Servigale* (*The Servant*; 1556), a piece staged by the Company of San Bastiano de' Fanciulli during the Carnival of that year. Arranged by brokers, marriage is a way to increase a family's cash assets. Everyone is in pursuit of money: even the priests want gifts – shirts and handkerchiefs – and the nuns are so zealous in seeking donations that they are dubbed "house-emptiers". Cecchi was not yet quit of his anti-clericalism.

Domenico Ciuffagni, the central figure, is a wealthy businessman, avaricious, a miser, a type frequently created by Cecchi in his picture of Florentine life, a city dominated by mercantile values. But he takes pains to explain what has instilled these unpleasant traits in this wretched old man: born poor, of humble family, he has risen with difficulty to his present status. Having long concentrated on gaining money and position, working hours at hard tasks that he would not pay others to perform for him, he has injured his health, narrowing his world, until piling up scudos and ducats has come to seem to him the end-all of his barren existence. He does not stop short of dishonesty.

Entrusted to his care are two wards, a nephew, Neri, and an impoverished girl, Ermellina, with whom Neri is in love and whom he wishes to marry. But the uncle will not allow the match: he insists that Neri must choose a wife who is of high social rank and will bring a good-sized dowry.

Domenico has a rich friend, Lamberto Lamberteschi, whose son Gentile is also taken with Ermellina. His father similarly objects to his son marrying anyone lacking a fortune and not belonging to the merchant class. The rival young men, balked by their elders, make bids for the girl through third parties – who in turn recruit "brokers" – and who pretend they are acting for themselves, not revealing for whom they are actually speaking. Neri's helper in this deception is Benuccio, a cobbler; while Gentile makes his move through Geppo. As might be expected, both surrogates are men of ill repute.

The broker retained by Benuccio is Agabuto. When Geppo bids 100 ducats on Gentile's behalf, the frantic Neri has Agabito counter with an offer of 300, though he has no such resource.

Domenico is delighted with his success at what is virtually an auctioning off of penniless Ermellina. But Neri has the task of finding the ducats he has pledged. He enlists another ally, who is also a broker, Travaglio, like the others quite unscrupulous.

Travaglio suggests that Neri obtain the money he urgently needs from the tight-fisted Domenico himself and concocts an elaborate plan requiring several disguises. Domenico is about to set off on a business mission to Bologna; Travaglio presents himself as an envoy from a convent and requests that the merchant take with him some valuable wares belonging to the nuns, which Domenico consents to do. A Venetian confederate, in Travaglio's pay, assumes the role of a German commercial traveller who is also going to Bologna. He proposes that for their mutual convenience and safety they undertake the journey together, to which Domenico agrees. To cover expenses, they should mingle their funds. The Florentine entrepreneur places 400 ducats in the valise in which the German has already deposited his share. Even before the departure the carrying case is switched, without Domenico's knowledge of it.

Now Neri has enough to meet his obligation. But Fortune is hostile, as so often in a Cecchi farce. Valentino Renzon da Crema, a military adventurer long unheard from, suddenly reappears in Florence. He is the brother of Antonello, to whose widow Domenico is married; and it is also he who entrusted Ermellina, a war orphan, to his late brother's care, after which she became Domenico's ward when he in turn won the widow, who brought with her a large estate. To Domenico's distress, Valentino is entitled to half of his brother's substantial legacy, with which the old merchant is loath to part.

Facing the loss of such a large sum, Domenico postpones the journey to Bologna in order to cope with the threat of losing half of the financial endowment that had led him to marry. Searching for the "German" to inform him of the change, he cannot find him: then, opening the carrying case, he discovers his loss: it contains only valueless scraps of metal. He is distraught.

With the corrupt Geppo's help, Gentile schemes to gain access to Ermellina's bedchamber, hoping to seduce her. Cross-dressing, he disguises himself as a maidservant, but he is tricked by the unreliable Geppo and Agata, an employee in the house: they lead him, instead, to a darkened room which is occupied by the willing Violante, Domenico's widowed daughter, who has her eye on the rich, young Gentile and believes that if he is caught compromising her he will yield to becoming her second husband. Domenico seizes the intruder, vents his fury on him, and hands him over to Lamberto, his embarrassed father.

Pleased at having cheated Domenico so easily, the brash Travaglio returns with a new proposal, a second ingenious swindle: they should jointly buy a huge quantity of gold thread, to be resold at an amazing profit. But this time Domenico is not so readily deceived. He shakes the truth out of Travaglio and learns of Neri's participation in the theft of the 400 ducats. Confronted by his infuriated uncle, Neri confesses his complicity and avows his remorse.

Lamberto recognizes Ermellina as his daughter, who as a child had been left behind during the combat and chaos in Florence when Cosimo had seized power. Since Ermellina is his sister,

Gentile looks on Violante more favourably and assents to a union. Generously, Valentino relinquishes his claim to his brother's estate, saying that it should serve as Ermellina's dowry if Domenico permits her marriage to Neri. Having recovered his precious store of ducats, the old man is in a rare complaisant humour and consents; he even spares Travaglio from having to face criminal prosecution.

At two points in the play, so strong is Domenico's acquisitive impulse, he engages in fantasies of how he might gain more wealth: he will make prodigiously lucrative investments with the money paid over to him by Benuccio; or he could rob the German while journeying to Bologna, or his fellow traveller might die on the way, whereupon all the contents of the heavily laden carrying case could be his, since no receipts had been exchanged and the German's far-off heirs would have no knowledge of cash money being pooled. His thoughts are even more dishonest than his behaviour. His outcry, when discovering that his golden coins are missing, has a painful edge later echoed by Shakespeare's victimized Shylock:

> Oh God!
> Oh my five hundred ducats! Mine!
> Mine! Because the money has been here in my family
> And in my house, I had directed my love
> To it as something that is mine. . . .
> [Translation: Radcliff-Umstead]

"Mine" is repeated three times because his possessive instinct is so dominant. The four hundred ducats have already become five hundred in his mind, where in his imaginings he had stealthily got hold of the German's share. At the conclusion, one reason for his forgiving Travaglio is that the Venetian rascal has taught him a few new tricks. He will know better next time.

Few of the other people in this play are likeable. Neri abets the robbery of his uncle. Gentile is a braggart, spoiled and unprincipled. Violante heartlessly schemes to entrap him as her husband. The domestic servants are chicane, speaking derisively of their masters behind their backs, and always ready to outwit and prey on them. The cluster of rogues, the crooked Benuccio, the sly brokers Agabito, Geppo and Travaglio, who surround Neri and Gentile, represent a fringe of parasites and sharpers infesting Florentine and Venetian society, seeking to live at the expense of the nobility, the rich merchants and the thriving professional class, if often precariously. It is a broad and distressing panorama, but a colourful one.

Il Servigale is in verse. Each of its five acts is preceded by an allegorical playlet and a madrigal – such *intermezzi*, as they are called, are a series of inserted diversions, perhaps musical, a frequent feature of Italian Renaissance stage-works. Often they are light and frivolous, a form of comic relief. But here Cecchi chose to fill his five interludes with thoughtful musings, hoping to impart moral lessons. A long roster of abstract figures – Purity, Memory, Intellect, Will, Genius, Greed,

Ambition and more – report progress towards the soul's salvation after the final defeat of ugly Greed. Intermingled with these embodied concepts are allusions to a host of historical and heroic mythical characters: Hercules, Achilles, Solomon, Semiramis, Cyrus, Aristotle, Alexander the Great, who in one way or another had been overcome and governed by Love or by all-inspiring Ambition, a gripping desire for immortal fame. And there are also references to the fostering seasons in which these differing and controlling emotions flourish: Purity in spring, Love in summer, Ambition in autumn, Greed in winter. In the last of these interpolations Reason effectively intervenes, dispelling the insatiable drive for earthy satisfactions. The exemplars cited here are Joseph, who chastely rejected his Egyptian master's unfaithful wife; Lycurgus of Sparta, refusing a crown; Fabricus, the Roman, proving himself beyond subornation by bribes; and Emperor Titus, resolved to rule justly and ever with a gentle hand. A closing madrigal exalts the peace that is achieved when worldly aims are put aside and each man's eyes are lifted to God. Radcliff-Umstead says: "All these *intermezzi* function to juxtapose the particular comedy of middle-class Florence in ducal Medicean times against an eternal drama of the soul's triumphant movement toward redemption."

Fortunato Rizzi suggests that *Il Servigale* is the source of Lope de Rueda's contemporary play *Aramellina*.

For *I Rivali* (*The Rivals*; also 1556) Cecchi went back to Plautus' *Casina* and Machiavelli's already current version of it, *La Clizia*; almost without pause, this was followed by *Il Medico* (*The Doctor*; 1557), a caprice about physicians and their treatment of illness. When reprinted in 1585 it bore a changed title, *Il Diamante*. Keeping up the pace, a new play or even two annually, he was ready with *Gli Sciamiti* (*The Coffers*) in 1558.

Il Martello (*The Hammer*; 1561), a comedy, was delayed while Cecchi was preoccupied with more serious endeavours, *Ragionamenti Spirituali* (1558), his rendering of the Gospels and Epistles into the Tuscan vernacular, betokening the beginning of his religious conversion; and *La Morte del Re Acab* (*King Acab's Death*; 1559), a sacred drama.

His growing piety also casts a shadow on *Il Martello*, which is filled with immoral characters engaged in unpardonable acts, but who are finally redeemed, a conclusion that may have been his principal reason for writing the play.

Its plot is comparatively simple, and he states in the verse prologue that it derives from Plautus' *Asinaria*; what he does not acknowledge is that there are borrowings from Terence's *Eunuchus* as well, and from another comedy by Plautus, *Truculentus*. But in transposing the scene and participants to sixteenth-century Florence and making the people resemble his fellow citizens, he adds many significant ideas, devices and touches of his own, creating another lively picture of daily life in that thriving mercantile centre.

Though the literal meaning of *martello* is "hammer", it also connotes a tormenting sexual jealousy, an obsessive, throbbing and beating sensation. Girolamo, elderly but still lecherous, is exceedingly fond of his son, Fabio, and tends to indulge his every wish. The young man, inheriting his father's lustful nature, is a regular patron of Angelica, who presides over a brothel. But she has another, far wealthier client, Captain Lanfranco Cacciadiavoli, who bids her to put a stop to Fabio's visits. Angelica partly accedes to the Captain's demand by putting a price of thirty ducats for her favours, a sum that Fabio cannot meet.

Girolamo moves to help Fabio. Both father and son have almost empty pockets because the widowed Madonna Papera, Girolamo's second wife, Fabio's stepmother, has a firm clutch on the family's funds.

What is more, Girolamo himself is desirous of sharing assaults on Angelica's curvaceous beauty. The father, the son and the Captain are all possessed by jealousy. Accordingly, Girolamo asks his artful servant, Nebbia, to conceive some means, however devious, to raise the money. He exacts from his son an agreement that, with the cash in hand, he may spend two weeks enjoying Angelica's caresses.

Nebbia learns that a peasant, Tognon Di Bartolo, owes Madonna Papera eighty ducats for a recent grain transaction. He has Girolamo pose as his wife's chief steward, and for Angelica to impersonate Madonna Papera, to collect the payment. With the ducats, Fabio replaces the Captain in Angelica's good graces. By now, however, Fabio is infatuated with Emilia, a ward of Angelica, and seeks to elope with her.

The Captain has provided Angelica with rare wines and bounteous fine foodstuffs. She elects to celebrate her well-bought liaison with Giralomo by setting a lavish banquet. The raucous, vinous occasion is interrupted when Madonna Papera, her real steward and the swindled Tognon break in and turbulence ensues. The intruders have been apprised of Nebbia's ploy by Sparecchia, the Captain's parasitic ally, who has been spying on the heedless revellers.

Fabio jubilantly discovers that Emilia is his stepmother's daughter, Selvaggia, fathered by the first husband. As an infant she had been carried off to Genoa by a wet-nurse; on the nurse's death the little girl had been passed along to Angelica's sister, and then later to the watchful charge of Angelica herself. Madonna Papera is joyous at recovering her daughter, issues pardons to all, and opens her purse to spread bounteous gifts, among them a dowry for the new-found Selvaggia. Overcome at witnessing so much domestic bliss, Angelica relinquishes her trade as a courtesan and brothel-keeper, vows to embrace churchgoing and consents to wed the Captain.

Though he professed to disdain the over-used changeling theme, Cecchi did not refrain from exploiting it once more. Over-familiar, too, is the character of the "braggart warrior", the Roman *miles gloriosus*, here represented by the Captain, who is vain, swaggering, boastful but essentially timid. It has been suggested that in Italy during the Renaissance, with the many states and principalities endlessly warring and employing mercenaries, there were doubtless many such men. Cecchi was not only excerpting a popular figure from Plautus and Terence, but also portraying a living example of the species he frequently met.

A fresher role is that of Tognon, the farmer, who is no easy mark, though initially derided as a fool. He does not leave Florence until he has a receipt for his 80 ducats. He is the first of a number of peasants subsequently appearing in Cecchi's comedies who contribute a touch of earthiness. Moreover, this peasant comports himself with reserve and dignity. In this play Cecchi uses the same Tuscan dialect as the others; in later works the rustic characters have a distinctive country vocabulary, emulating those so authentically recorded by Beolco.

Once more, with Girolamo, the playwright offers a crabbed man in the grip of senile lust and not afraid to seem foolish because of it. What is unusual is that this time the elderly husband is not the miser, the wife is.

The utterly frank carnality of the script was probably acceptable in a wild, uninhibited Carnival setting, but to have both father and son bonding to share Angelica's bed on alternate days was probably deemed shocking.

After *Le Pellegrine* (*The Pilgrim Women*; 1567) Cecchi was ever more concerned with writing sacred dramas, a series of four, as his religious crisis deepened. But in 1574–9 he applied his talents to comedy again with *Le Cedole* (*The Promissory Notes*). An ardent Florentine youth, Emilio de' Giuochi, a student at the University of Bologna, is gripped by love for an orphan girl, Angelica. His affair does not progress far because Veronica, the guardian, watches her closely. Emilio gets on good terms with Ramaglia, Veronica's second husband, of noble Florentine lineage but much reduced financially and in spirit; indeed, to earn a living he sometimes works as a humble wool carder or plies the trade of a marriage broker. When Veronica and Ramaglia find themselves almost penniless, Emilio obtains a loan for them by signing a promissory note for 100 ducats from Eustachio Gambale, a Bolognese usurer. As collateral, he hands over the costly textbooks he needs for his studies.

Emilio's desire to marry Angelica seems to be beyond hope of realization: he fears that his wealthy father, Tegghiaio, would disinherit him. The girl he chooses as his wife should be of social station equal to his, or better. None the less he returns to Florence, bringing Angelica and her impoverished guardians with him, and finds lodgings for them in a house near his own. His monetary state continues to worsen, as the time to repay the loan approaches and the feckless Ramaglia's many debts rapidly accumulate.

As do so many of Cecchi's distressed young heroes, Emilio calls on his servant for help with his problem. The ingenious Monello comes up with a list of schemes. One is to have Angelica marry a "front" man, which will enable her to escape from the strict supervision of her two guardians. For this purpose Ghianda, a coachman, is recruited. All of Monello's pleas to Tegghiaio to settle his son's obligation fail, as do a succession of ill-thought-out ruses and gambits, including several disguises. Monello's shortcoming is that he acts too hastily; he does not pause to question and calculate, but spontaneously extemporizes, whereas Tegghiaio is extraordinarily shrewd and wary, never likely to be deceived or taken by surprise. Even the threat that Emilio might be summoned to court and sued does not move him. A good deal of legal phraseology is scattered throughout the dialogue, reflecting Cecchi's ready acquaintance with it as a notary.

As Ramaglia's creditors besiege him and close in, he flees from them, seeking sanctuary in a church to avoid a grim term in debtors' prison. To get Angelica away from her guardians, Ghianda poses as her non-existent brother, Astolfo di Messer Landolfo d'Andal d'Asti; the girl is a war orphan of noble Piedmontese lineage. The "brother" says that he has arranged a marriage for her to the Fantassino (impersonated by Ghianda), but a woman of the neighbourhood recognizes him and exposes his disguise. Veronica is furious at being the object of a cruel trick.

Donning another disguise, with similar lack of success, Ghianda pretends to be Eustachio Gambale and approaches Tegghiaio to obtain the money to settle Ramaglia's sorry affairs and free Emilio from his pressing worries. Suspicious, Tegghiaio refuses to hand over any cash: he will clear the loan only through an intermediary, in Bologna, and only after Emilio's textbooks are given back – and, of course, Ghianda does not have them and cannot get them.

The real Eustachio Gambale reaches Florence and calls on Tegghiaio to redeem Emilio's loan, but the father is convinced that this is an attempt to fleece him; he has already been visited by and negotiated with the actual Bolognese usurer. Arguing, and with accusations flying back and forth, the two old men come to blows. Seeing Eustachio's fear that he, an outsider, might be prosecuted for assaulting a Florentine citizen, Monello offers him refuge in Ramaglia's house, where he passes himself off as Angelica's father. Questioning soon reveals that, though certainly not her father, he is in fact her uncle.

Though seventy, Tegghiaio has developed an amorous interest in young Angelica. To protect the girl from the aged, would-be philanderer, Veronica puts her under the wing of Emilio's mother. The two old ladies plot to punish the wayward Tegghiaio. Meanwhile the lovers find an opportunity to be together.

Anxious to pacify Tegghiaio and the others, Eustachio announces that Angelica will inherit a large estate and offers a dowry if she marries Emilio, whose promissory note he cancels. He assumes responsibility for freeing Ramaglia from his onerous debts. Ghianda is rewarded for his efforts to help Emilio and Angelica.

What concerns still govern Cecchi's characters? Tegghiaio is a shrewd, hard bargainer, Eustachio a grasping usurer, Emilio will not marry his adored one, fearing that he might forfeit his considerable inheritance. Money is the subject most on the irresponsible Ramaglia's mind. Impoverished and pretentious, ineffectual, he represents the end-point of a once aristocratic Florentine family. Commentators note that at the play's conclusion a strong link has been forged between the Tuscan upper middle class and the Piedmontese nobility, but Eustachio Gambale, the moneylender, hardly seems to bring much lustre with him.

Again, Fortune's intervention is decisive, as when Ghianda is recognized by a neighbour, and Eustachio proves to be Angelica's kinsman. Cecchi's religious preoccupation, his growing belief in a Divine power determining the outcome of mankind's endeavours, may account for his introducing and emphasizing the element of seeming chance in his writings.

Dates and chronology are in dispute, but perhaps a half-dozen or more years went by, productive of as many as eight solemn religious dramas, before Cecchi ventured another comedy, some say *Li Contrassegni* (*The Tokens*; 1583 or 1585) or else *Le Maschere* (*The Masks*; 1585). In the prologue to the latter work he describes himself as growing old and in ill health. He asks forgiveness for its having taken sixty days to write, whereas formerly he had spent less than a fortnight on a script.

As in several other Cecchi works, Livia, the young woman around whom the action revolves, is talked about and never seen. She is the beloved of Fabrizio Dal Boschetto, a young Florentine of wealthy family. He seeks to marry her, but his father, Manente, a physician, steadfastly opposes the match – though her background is respectable, her mother, Madonna Clemenza, a widow, has only modest means. After months of frustration, Fabrizio flees Florence and his father's domination. These events occur twelve months before the comedy's opening.

Word comes from Orvieto that young Fabrizio has died there. This awakens in his father, Manente, and his maternal uncle, Baldo dell'Arca, a new regard for the beautiful Livia, and the two elderly men become rivals for her hand. A third contestant appears, an eager young man, Vettorio Ormanni. Suddenly the news is that Fabrizio is not dead; he has been on a journey to the West Indies, trading, and has returned to Genoa with a large fortune. The three contenders in Florence are under pressure to have one of them sign a marriage contract for Livia before Fabrizio reaches home.

The girl's mother, Madonna Clemenza, is away in Pisa to settle a family business affair; the girl has been temporarily entrusted to the care of an aunt, Adriana, who is guiding her in the bewildering choice of a husband among the ardent rivals. Much of their wooing is carried out by a campaign of letter-writing, some of the epistles authentic, some of them false, meant to mislead and defame, as each correspondent seeks to gain an advantage. In addition, letters are stolen or fall into the wrong hands. Among the writers is Fabrizio, still not back in Florence, who depends on a friend, Attilio, to whom he confides his future intentions, and on whom he relies for news. Naïvely, Attilio shows these letters to Vettorio, not realizing that the young man is one of the over-zealous competitors. Each of the rival suitors has enlisted help in his quest: Manente is aided by his clever servant, Catacchio; Baldo by a sharp-witted marriage broker, Chima; Vettorio by his well-named retainer, Imbroglio. As these lesser figures see it, the struggle to capture Livia is a war, one justifying any means to attain victory. Shown a genuine letter from Fabrizio to Attilio, Manente rejects it as a mere trick. He refuses to believe his son is still alive. Adriana's letter to her sister Clemenza is delivered by a dishonest maid to Vettorio, whose Imbroglio tears it up, replacing it with one stating that his should be favoured in the contest. But the forged epistle comes into the hands of Manente, who has Chima fabricate a reply in which Clemenza instructs Adriana to have Livia marry Manente without delay.

It is Carnival. Seizing an opportunity to mask himself, Vettorio plans to abduct the girl and flee with her to Lucca. He relies on Livia's maid, Crezia, to assist him in this bold attempt. But Fabrizio, finally in Florence, gets wind of Vettorio's venture and also puts on a mask. With

Attilio's help he is carrying off Livia when they bump into Vettorio. A scuffle results. The noise of the combat alerts the police, and all the participants are arrested.

Learning that Fabrizio is in prison, proof that he is indeed alive, Manente seeks a reconciliation and consents to the match with Livia, all the more readily since his son has acquired wealth of his own and has little need of a large dowry.

But there are surprising new complications. Having settled matters concerning her late husband's estate, Clemenza returns from Pisa, hears of the impending marriage, and puts a stop to it by revealing that Livia is not her daughter but Manente's, stolen as an infant by gypsies. Fabrizio and Livia are brother and sister. The girl's real name is Porzia. Baldo is her uncle, her mother's brother. This clears the way for Vettorio. Three marriages ensue, with Manente taking Clemenza as his wife, and Baldo joining in matrimony with Adriana, to whom he had at one time been betrothed. Fabrizio is left to be "the most eligible and sought-after bachelor in Florence".

Though Cecchi had spent the intervening years composing sacred plays, he reverted to salacious exchanges on going back to a lively comedy, especially when Manente and Baldo, advanced in years, wonder aloud if they have the sexual potency to satisfy an amorous young wife. The language, physically explicit, crosses the line to the obscene, ranging from the *double entendre* to the utterly lewd.

The cross-section of Florentine society exposed here is unsavoury. With the exception of Fabrizio, none of the leading male characters is a really decent person. Manente is so self-deluding, miserly and heartless that he does not hesitate at trying to outwit and cheat everybody, even his next of kin. When ordering what he thinks will be his wedding banquet he chooses the least expensive dishes, such as crows and his compensation to Chima is incredibly niggardly. He has scant concern for his son's happiness, instead remarking to Baldo that "a man's closest relative is himself". His vanity equals his cupidity, the first quality accounting for his inability to admit defeat, the latter trait evidenced again when he becomes responsible for providing Livia/Porzia's dowry, over which he haggles with Vettorio's father. The generous Fabrizio makes up an appropriate sum for his new-found sister.

Baldo had broken off his affair with Adriana, giving as his pretext a need to marry a younger woman and father an heir. A hunter and sportsman, he, too, is greedy and stingy. His reluctance to pay Chimo a proper fee costs him the broker's services. He falls victim to a swindle concocted by Imbroglio, who leads him to sign two bills of credit in Vettorio's name; they will be shown to Adriana to persuade her to oppose the marriage contract with Manente. The bills supply the funds for Vettorio's thwarted plan to abduct Livia. At the end, however, the money is restored to Baldo.

Vettorio betrays Attilio's confidence, when allowed to read Fabrizio's correspondence. He concurs in Imbroglio's defrauding the elderly Baldo. He also lies to Livia, insisting that Fabrizio is no longer alive, though he has knowledge to the contrary. Once he has gained his prize, by extraordinary good luck – the elimination of Manente, Baldo and Fabrizio by the discovery of their blood ties to Livia – he is belatedly remorseful and rewards those who helped him, notably the maid Crezia, but he is hardly a youth of exemplary virtues.

The lesser figures – Chima, Imbroglio, Crezia – are a raffish lot and once more suggest the dark underside of mercantile Florence, a flourishing social world served and supported by scroungers and their ilk, almost wholly dependent on others for their sustenance and prepared to cadge and steal wherever they are able to do so.

Almost all of Cecchi's usual themes are restated here: the corrupting influence of materialism and the ceaseless quest for affluence by any illicit means except violence; the intervention of Fortune, setting the best-laid schemes awry. Another is what might be seen as Cecchi's persistent note of proto-feminism: nearly all his women are passive characters, their fates determined by the edicts of their fathers, husbands, brothers; they have no voice of their own, and scarcely any legal rights. Many of them are poorly educated, almost illiterate; so they must be submissive, in a kind of life-long bondage to a succession of men around them. This was their status in that day, even in the highest social circles and among the most intelligent. That is why their honour and reputation were all-important, one of the few defensive weapons they possessed.

As yet, Cecchi's last two plays, *I Contrassegni* and *Il Debito* (*c.* 1587), remain unpublished. Most of his works were printed during his lifetime or scattered in libraries and private collections throughout Italy. Because of his abundant use of street argot, the dialect and pithy idioms of sixteenth-century Florence, modern translators face extra difficulty with the texts; furthermore, some exist in as many as four or five different versions, which daunts scholars intent on an infallible rendering of them. *The Horned Owl* is his only play to have been issued in English.

He was even more prolific when writing dramas of spiritual inspiration, *sacre rappresentazioni*, of which some thirty-odd flowed from his pen in the three decades between 1558 and 1587. These sprang out of his religious conversion, and for the most part were composed at the request of Church institutions, celebrating events on the calendar of saints, or for major festivals, perhaps by members of holy orders or confraternities. Some were staged outdoors, others in churches or chapels, or in meeting halls, or in schools and mansions. Florence boasted a strong tradition of such dramas. They evolved into having a fixed form, the lines cast in "hendecasyllabic octaves with a rhyme scheme in abababcc".

Intended to convey edifying lessons, many of them were acted by students. Some were based on episodes in the Scriptures, or described the miracles and martyrdoms of saints and evangelists. For a time, near the start of the sixteenth century, they tended to incorporate over-dramatized and over-romanticized stories. A princess or other chaste damsel in peril was saved by the benign intercession of the Virgin. Plays like these were not approved by the Church hierarchy, with the Catholic Reformation under way, and in 1558 steps were taken to prohibit them; in 1574 Pope Gregory XIII banned such presentations, particularizing that the laity should not stage them in the privacy of their homes; the Archbishop of Florence, Alessandro de' Medici, was increasingly censorious of them, prohibiting them to be offered in places of worship near where masses and other liturgical rites were recited.

Cecchi, as did some others, devoted himself to restoring the *sacre rappresentazioni* to acceptance, a drive led by the Jesuits, and in concert these believers gradually made headway. The Jesuits added baroque spectacle to their offerings, evoking a strong response in an age that loved splendour, especially when it was sanctioned by being dedicated to the glory of God. In convents, nuns and friars revived fifteenth-century scripts, mounting them on a more modest scale.

The Morality-type script, with abstract and allegorical figures, had never been very popular in Italy. But now the Protestants were using dramas of that didactic sort to good effect, and Cecchi adopted and craftily reworked the genre, while also couching some of his output in semi-melodramas, comedies and farces. Shrewdly, he did not hesitate to infuse his preachments with humour, often contributing his native wit to lighten the dialogue. He replaced the traditional octaves with blank verse and added allegorical *intermezzi* to make his message clearer for the younger spectators.

In these works he is a precursor, if not the inventor, of what might now be called "tragicomedy". With his Renaissance veneration for the classics, he reasoned that tragedies required large companies of actors in grandiose settings, and comedies only three or four performers in an intimate *mise-en-scène*, but the offerings he had in mind would fall midway between the two. Tragedies dealt with exalted figures brought low, and comedies with persons who might be considered decidedly ignoble; he wished to write about people with whom those in youthful audiences could readily identify. His preference seems to have been for open-air productions.

His leading characters might belong to royalty, but they would be surrounded by ordinary citizens, and even by riff-raff. In this mixture he often inserted the farcical stereotypes he had previously used, the miser, the braggart captain, the corrupt cleric, the parasite, the swindler, carried over from Plautus and Terence. Blank verse substituted for rhymed lines also helped to make the dialogue sound more natural and modern. He did not feel himself bound by the unities of time and place. He continued to rely on madrigals to enhance his explanatory *intermezzi*.

The subject-matter of his *sacre rappresentazioni* ranges from *La Morte del Re Acab* (*The Death of King Acab*; 1559) – a drama about a regal couple, Ahab and Jezebel, who break from orthodox Judaism and set up a pagan cult worshipping Baal, whereupon the errors of vainglorious rulers are visited with dire consequences on their whole, mostly innocent people – to *La Coronazione del Re Saul* (*The Coronation of King Saul*; 1569) and *Il Figliuol Prodigo* (*The Prodigal Son*; 1569–70), in which a penitent triumphs after his encounters with the Seven Deadly Sins have been illustrated, not with labelled abstractions but embodied in specific contemporary persons. In this the action contains frequent comic touches, capped with a banquet where the father welcomes the errant youth's return; the setting of the Biblical story is updated to Cecchi's middle-class Florence – essentially, he reduces the great parable to a *commedia erudita*. Indeed, some critics object that the original tale grows so faint that it almost fades away. But Cecchi does make his moral point, that contrition can lead to the Father's divine forgiveness.

A further scanning of the titles of these more than thirty sacred dramas is an index of the breadth and depth of his imagination and the boldness of his experimentation; such a list, at random, takes in *Atto Recitabile per alla Capannuccia* (*One-Act Play at the Manger*; 1573); *La Serpe Ovvera* (*The Snake or the Evil Daughter-in-Law*; 1574); *Il Tobia* (*Tobias*; 1580); *La Conversione della Scozia* (*The Conversion of Scotland*; 1581); *Santa Cecilia* (*Saint Cecilia*; 1583); progressing to *Cleofas e Luca* (*Cleofas and Luke*; 1580–7) and *Il Cieco Noto* (*The Man Born Blind*; 1580–7). He tried new form after new form; he even chose outright farce as a vehicle for his moral lessons, as in his final play, *La Romanesca* (*The Roman Woman*; 1585).

Parts of the story of the last work are from Boccaccio. Cecchi advances the scene from ancient times to contemporary Rome. English-born Princess Isabella is wed to the King of France. While he is away the princess learns that her mother-in-law is plotting to have her and her child killed. Isabella cannot take refuge with her father, the King of England, fearing his incestuous desire for her. She flees to Rome for safety and there assumes the humble guise of a nurse.

It is the Year of the Jubilee. The King of France comes to Rome to search for his vanished queen. Another French pilgrim is Claudio, in quest of a long-ago close friend, Sempronio, to whom he owes a profound, never-forgotten debt: when they were fellow students in Paris, Sempronio had pleaded guilty to a murder for which Claudio was wrongfully charged, his gesture springing from a wish to repay Claudio for having earlier yielded his claim to the hand of a young woman whom Sempronio also loved, an act that brought shame, exile and poverty on the self-sacrificing Claudio. The vagaries of chance – again, Fortune – a lengthy chain of coincidences, lucky and unlucky, serve to impede and delay the successful resolutions of the two missions but finally bring them about, ending the persecution of the princess and uniting the long-separated friends, both of them now freed of the accusation of murder. The French king restores Claudio's noble title and estates.

This three-act verse play is compact: Cecchi expertly integrates the two plot-lines, and he goes back to observing the classical unities of time and place. Aristotle had not clearly defined farce as a genre, nor had he ever stated his approbation of it. But Cecchi had finally formulated his own aesthetic theory concerning it, which he voices in the prologue to *La Romanesca*:

> Farce is a third new subject
> Between tragedy and comedy: it enjoys
> The breadth of both of them,
> And it flees their narrowness, because
> It takes unto itself great lords and princes,
> Which comedy does not do. It also takes in,
> As if it were either an inn or a hospital,
> The masses, both the low and plebeian,
> Which Lady Tragedy never wants to do.

> Its actions are not restricted: it accepts
> Merry events and sad ones, secular and ecclesiastical,
> Urban, rural, dreadful and pleasant.
> It takes no account of place: it forms its stage
> Both inside a church and on a square or anywhere.
> Nor does it take account of time. Therefore if it does
> Not finish in one day, it will take two or three.
> [Translation: Radcliff-Umstead]

Farces had flourished for centuries, especially in medieval times, but were looked upon as an inferior and outcast form of drama; Cecchi sought to make the genre respectable. To give it a classical provenance he cited one that reportedly was performed at the court of Caligula; however, that most evil emperor apparently had not cared greatly for it.

The humour in *La Romanesca* is engendered by a nurse, a tailor, beggars, house-servants, guards and other ordinary citizens of Rome who fill out the crowded scene around the leading upper-class characters, comprised of royalty, administrators, court officials and members of the diplomatic corps. Cecchi follows their actions in contrapuntal episodes. He carefully chooses their language to show the differences in vocabulary and inflexion between speakers from a variety of social strata. The Governor of Rome uses sober quotations from the Bible; a tavern patron heightens his exchanges with crude anecdotes and spicy proverbs. The diplomats are eloquent, ambiguous, cloudily verbose.

His purpose was to teach by entertaining; his humour was the bait, a way to lure audiences, especially the youthful. He reveals his aim in his prologues, which are informal and chatty, often whimsical, addressed directly to the spectators. There is little or no didacticism in the action of the plays; his moralizing tends to occur in the allegorical *intermezzi*. He succeeded in fusing the two kinds of major stage-works that dominated the Italian Renaissance, secular comedy and reborn *sacre rappresentazioni*, and was remarkably adept at adapting and modernizing to satisfy the tastes of his contemporaries, which makes him historically important. A shrewd craftsman, he was sure in his timing and had a feeling for what worked on stage. His plays are both thoughtful and amusing.

In his last years he was acclaimed throughout Italy, his country's most popular playwright, partly owing to his unfailing productivity. He influenced the *commedia dell'arte* by showing diverse ways that masking and stock characters could evoke laughter.

His fame began to fade shortly after his death, in his sixty-ninth year (1587), from a brief bout of "catarrh". He was mourned by his peasant neighbours in the village of Gangalandi, where he had contributed to local charities, paying to have the church restored and a small monastery erected.

At the end of the nineteenth century and early in the twentieth, Cecchi – or il Comico, as he was widely known – was discovered by critics and his works scrutinized by scholars, with several books of commentary inspired by them. Today his name and scripts are still largely unrecognized outside Italy and academic realms, and even there appreciated by only a narrow circle.

Unique among the playwrights of the Italian Renaissance is Leone de' Sommi (c. 1526–92), about whom there is not much biographical detail nor means to measure his literary accomplishments. Of his fourteen play-scripts, twelve were consumed in a fire that devastated the National Library of Turin in 1904, along with most of his voluminous output of poems and essays. He was unusual in his day because he was a deeply religious Jew, a leader of his community in Mantua, and wrote plays in both Hebrew and Italian, and is perhaps the first Jewish dramatist so identified in the history of the European theatre. He is noted, too, as the author of a guide to acting techniques and staging that today is of great value to scholars seeking to learn exactly how plays were given professional mountings in the last half of the sixteenth century in Italy.

He belonged to the Portaleone family, many of whose members were eminent physicians whose services were sought by the nobility in various regions of the country. His name at birth was Yehudah Sommo ben-Yitsh; that is, Judah the son of Isaac. But the Italian equivalent of Yehudah is Giuda, or Judas, an unfortunate appellation for him to bear; he chose to be known as Leone, instead. In Genesis 49:9 Yehudah is likened to a young lion, which provided him and other Yehudahs in Italy with a pretext to call themselves Leone. Besides, the lion is also the symbol of the tribe of Judah.

About his education, little is recorded. He studied with Rabbi David Provenzal, a revered teacher. He must have had a gift for languages, because somewhere between his eleventh and thirteenth years he is said to have translated his complex Hebrew grammar into Italian, a bilingual mastery he exhibited throughout his life. Later he showed himself to be broadly knowledgeable, a man of learning and truly sophisticated.

It is not clearly known what was his trade or profession. With only vague clues, his biographers think that he was either a "scribe" or a teacher. A "scribe" might be one who copied holy texts – passages from the Torah – or who assisted with commercial and personal correspondence. One thing is certain: he was dextrous at communicating sacred ideas in a paraphrastic Italian vernacular; his doing this, in a long poem, *Defence of Women*, led to controversy in the Jewish community. Examples of these translations no longer exist, but excerpts of the criticism they aroused still do. He was accused of desecration, of being unfaithful to his religious and cultural heritage.

Living in two very different worlds, as he did, he was fortunate to have been born in Mantua and to have spent most of his life there. Prejudice against Jews was strong in large parts of Italy, especially in Rome, from which Pius V ordered them expelled in 1569. Forced out of the Papal States, they scattered to northern cities, especially Mantua, where the duke, heedless of the Pope's animus towards these people, was notably tolerant. Because of his benign attitude, the Jewish colony there grew, by 1591, to 1,600, eight times its former size, a group that contributed richly to the city's culture and commerce. Here they enjoyed a high degree of autonomy, so that in his daily affairs de' Sommi passed from one busy social stratum to another.

A gap of twenty-two years, from 1538 to 1556, exists in any account of de' Sommi's whereabouts and activities, but he may have been for a time in Ferrara, and have seen and been influenced by presentations given under court auspices, and particularly the plays of Cinthio and Beccari, from which he later borrowed ideas found in one of his own scripts, the pastoral *Hirifile*. It is also possible that he took part in the productions.

Rabbi David Provenzal preached against Jewish involvement in secular court plays, but his former pupil seems to have not absorbed the austere lesson. The Jewish community did foster enactments by Jewish performers on relevant subjects, perhaps based on Biblical incidents, such as Judith's seducion and murder of Holofernes. In Mantua a permanent troupe of Jewish actors was established in 1520. Its accomplishments impressed the duke, who frequently employed their talents on festive occasions. He did not hesitate in ordering them to prepare special works when royal birthdays and weddings were celebrated or foreign sovereigns fêted. The Jews had to bear all the burdensome costs of each production, which could be quite elaborate, competing with the semi-professional Carnival and the omnipresent *commedia dell'arte* ensembles.

The Mantuan company was governed by a three-man committee. The performances at court were a sort of taxation and a way of winning favour. An annual "tribute" – a stage offering – *La Commedia degli Hebrei* (*The Play by the Jews*) – grew to be traditional. Scripts for it and other comedies and dramas commissioned by the ruling Gonzagas soon became de' Sommi's obligation and opportunity: he poured forth a stream of farces, pastorals and *intermezzi*, some intended to please and flatter the court, others for the entertainment of his own community and himself.

One of his two surviving scripts, not counting the badly damaged pastoral *Hirifile*, is *Zahud Bedihuta de-Qiddushim* (*A Comedy of Betrothal*); it exists in four versions, none bearing his name, but many aspects point to his authorship. It is in Hebrew. Tentatively dated 1550, it would have come from de' Sommi when he was in his mid-twenties. It belongs to a category known as *Purimspiel*, a work given at Purim, a religious holiday recalling the valour of Esther, of Jewish faith and married to Ahasuerus, a Persian king, who prevailed against the evil minister Haman and happily saved her imperilled people.

As the Jewish acting company flourished, and de' Sommi's multiple gifts were ever more recognized, he was put in charge of theatrical events at the ducal court, to write and direct not only his plays but also works by others. He is referred to as a choreographer as well, suggesting that he staged the dances; indeed, he was considered to be an expert on all phases of the productions. In his official post he had to observe caution always, since he was a Jew. Of the Italian Renaissance it has been said that it was a "risky and exhilarating" time, and this was especially true for him.

Outlays for entertainments at Mantua were somewhat limited during the reign of Duke Guglielmo, but on the accession of his son Vincenzo the picture underwent a change. The new ruler was passionately fond of all kinds of theatre, and especially of spectacle, and his purse opened wide for many extravagant events. He made use not only of Mantuan troupes but also invited companies from abroad. de' Sommi was commissioned to write a play on the occasion of Vincenzo's second marriage in 1584, and another during the same year for his birthday, the lost *Il Giannizzero*.

De' Sommi's reputation as a playwright and producer was carried far outside Mantua, and he got foreign assignments, but at home two of his personal ambitions were thwarted. He petitioned for a licence to build a public theatre that he would operate free of the court's supervision, but his request was denied. He also nurtured an unspoken wish to be elected to the recently founded Accademia degli Invaghiti, whose membership was made up mostly of the aristocracy and high-ranking clergy, an erudite club that had become the intellectual centre of Mantua. He had been commissioned to write plays for the Accademia, but he was a Jew, so there was an obstacle to his name being entered on its prestigious roster. In one instance he is referred to in an Accademia document as "*nostro scrittore*" – he was being paid for a script – which leads some biographers to speculate that this title was a way of admitting him to a kind of associate membership, getting around the circumstance of his being an infidel.

Though twelve of his scripts vanished irretrievably in the blaze at the Turin library, their titles are known and something about their content. Among them were a verse comedy, *Il Tamburo*; a pair of full-length prose comedies, *La Diletta* and *L'Adelfa*, and a gay *La Fortunata*, which was written for performance at the wedding of Carlo Emanuele I of Savoy to whom he also dedicated his *Le Nozze de Mercurio et di Philologia da Martiano Capella*. Besides these, he provided *intermezzi* for plays by other writers, such as his *Gli Onesti Amori* for Bernardo Pino de Cagli's *Gli Ingiusti Sdegni*.

(Of his surviving works, *A Comedy of Betrothal* has been translated into English by Alfred Golding, as has recently *Le Tre Sorelle* (*The Three Sisters*), de' Sommi's last play, collaboratively by Donald Beecher and Massimo Ciavolella (1992) from whose lengthy preface to their version has been gathered much of the detail here. The illustrative essay on stage production, "*Quattro Dialoghi*", which will be alluded to later, was much earlier rendered into English by Allardyce Nicoll and included in his *Development of the Theatre* (1937).)

Le Tre Sorelle (1588) is dedicated to his Serene Highness, Duke Vincenzo Gonzaga; the date of its first performance is uncertain, scholars wavering between Carnival 1589 and the festive season of 1598, possibly a revival, since the work had a strong popular appeal. The cost of the 1598 production, elaborately mounted, was met by a heavy tax imposed on the community for the duke's many celebrations that year, a levy of about 250 gold scudi, indicating that each of the plays offered was budgeted at about 100 scudi, no small amount. A justification for the tax was that the Mantuan court could not be outshone by those at Venice, Ferrara and Florence, all political and cultural rivals. In any event, Vincenzo was not one to be easily bested in such matters; he excelled all others in his manic extravagance.

Le Tre Sorelle has three major plot-lines, necessitating a triple structure. It deviates widely from the rule laid down by Aristotle and Horace that for simplicity and clarity of dramatic action a play should have at most a main plot and a single sub-plot. The multiplicity here may have been suggested to de' Sommi by Annibal Caro's *Gli Straccioni*, an experimental work that de' Sommi had already helped to revive in Mantua. To handle such intricate plotting required sure craftsmanship, and he was quite equal to the challenge.

A young gentleman of Mantua, Nardino, has inherited a house and a substantial estate from a kinsman. He plans to wed the pretty widow, Olimpia, who had been his late brother's wife for only a year. By doing so he will gain control of the balance of the estate. But because of the fraternal link he must first obtain a dispensation from the Vatican, and he is preparing to go to Rome for that purpose. Knowing that there is a rival for Olimpia's hand, he insists that she be placed in the custody of her mother, the elderly widow Euronica, and that she take up residence in his house, which is to be closely guarded by his servants Gisippo and Capone and by Euronica's maid Lisetta.

The watchful rival, Cavalier Fulvio, is aware of Nardino's impending departure and seeks to take every advantage of it. He is aided by his canny servant, Tansillo, hampered by no scruples. He quickly suborns the simple-minded Capone, and also Lisetta, whose loyalty is ambiguous; she has a habit of saying one thing and meaning another.

On a second plot-track is Captain Frangiferro (Fracasso) Spianamonte, an impoverished braggart warrior, lusting after Lucrezia – she is a sister of Olimpia and wife of Messer Pacifico, a wool merchant. Her husband's amorous gaze is fixed on his mother-in-law's maid, the elusive Lisetta. The Captain has enlisted his servant, Zarda, and a sharp-eyed parasite, Stragualcia, in his manoeuvres to attack Lucrezia's virtue, while her husband, the adulterous Messer Pacifico, is relying on his young helper, Balbino, in his siege of Lisetta.

To this crowded cast is added Suevia, discarded lover of Nardino. Enraged at learning of his intention to marry Olimpia, she dresses as a peasant fruit-seller and rushes to his house to charge him with having betrayed her, but is turned away at the door. To avenge herself on Nardino, she pays Melite, a procuress, to cast a diabolical spell on him, which Melite, in return for some scudos, agrees to do by tormenting a waxen image of him. After that, Suevia wants him back.

Olimpia's other sister, Diletta, still unmarried, is beloved by young Carino, but he knows that his father will oppose a match with anyone having only a small dowry. To win his parents' consent Carino feigns madness, babbling a farrago of wild phrases and images – but he lets the willing Diletta and his good friend Fedele in on his secret. His strategy is to have his distraught father convinced by Fedele that Carino's best hope for a cure is to have a loving woman always by his side. After that, Diletta's mother, Euronica, must be persuaded that special medications have restored Carino to sanity and it is really safe to have him as a son-in-law.

Zarda and the Captain are lodged in Melite's house. Catching sight of Suevia there, he is struck by her. Told that Suevia thinks only of Nardino's return to her, for which she is resorting to spells, Zarda asks Melite to assure Suevia that the spell has been effective and Nardino will be hers again that very night – and then he, Zarda, will slip into the bed in the dark room to accept Suevia's caresses. Melite's reply to this proposal is indecisive.

Intent on gulling the foolish Messer Pacifico, Stragualcia, the parasite, exploits a dream the wool merchant relates to him. In it, he has overwhelmed and impregnated Lisetta. Stragualcia informs him that it is not a dream – he, Pacifico, has actually raped Lisetta while she slept and got her with child. Infuriated, she is going to tell his wife, Lucrezia, who is also pregnant. To quiet Pacifico's panic at being exposed as a philanderer, Stragualcia claims that he has given Lisetta twenty *reali* for a vow of silence – he is repaid by the grateful Pacifico, though the story is wholly untrue.

Ever in quest of illicit gains, Stragualcia imparts to the foolish Pacifico news of the arrival in Mantua of an eminent physician who for a fee can ensure that Lucrezia's child will be the male heir that the wool merchant ardently desires and has heretofore been unable to sire. Taken to the great doctor, who is purportedly a friend of the Captain but is actually Zarda in disguise, Pacifico begs for his help. The infallible prescription, he is told, is that Lucrezia must spend a night with a man who is not her husband. The idea of this shocks Pacifico and initially he rejects it, not wishing to be made a cuckold. His objections are overcome when Zarda and the Captain suggest that the man could be an utter stranger, someone picked out on the street, and that he depart early the very next morning, and that his face be covered all the while he is with Lucrezia. Nor will he ever know in what house he has been or with whom. But will the modest Lucrezia consent to all this? They will tell her that the fertility potion she has taken causes her to have a venomous breath and that to protect himself from it her husband must mask his face, and completely unawares she will think that she is lying with him.

Carino's gout-plagued father is delighted when Fedele conveys to him word that his son can be freed from his madness by a union with Diletta and grants his permission, with a promise of funds, but he insists that the marriage should last only two or three weeks, a concession hardly acceptable to the enamoured young man. Fedele has difficulty calming him. In the end, progressing step by step, they will succeed.

On an errand for Olimpia, Lisetta is carrying a pendant to be altered by a jeweller. In the street she encounters Carino, who is annoyed with her for having been an obstacle to his getting to Diletta. Seeing the pendant, and still feigning madness, he grabs it from her hand and pretends to throw it down a well. He dashes off, leaving Lisetta distraught. What shall she tell her mistress? Zarda appears, still in his guise as a physician. He learns the cause of her distress and offers to help her. "If it's only a pendant thing you're looking for, come right this way and I'll willingly give you one." (The play abounds in *double entendres*: the "pendant" to which he refers is his male organ.) Deceived as to his intentions, Lisetta enters Melite's house, his dwelling-place.

Capone, bribed by Tansillo, arranges for Fulvio to meet and converse with the cloistered Olimpia at the portal of Nardino's house, where he begs her not to rush into marriage with her brother-in-law. She should take into consideration his, Fulvio's, deep love for her. Olimpia informs him that she has no intention of accepting Nardino's offer. She will remain in full control of her own affairs and fortune. Fulvio eloquently and passionately begs her to choose him instead. She grants that she looks on him favourably, and adds that she is quite cognizant of Nardino's persistent trickery.

Lisetta and Suevia chance upon each other in Melite's house. Both reveal that they have been attacked and pawed over by the aggressive and offensive Zarda (in his disguise as a physician) and have been able to repulse him violently and free themselves from his clutches. Suevia also tells Lisetta about her ill-treatment by Nardino, an account of which Lisetta hastens to give her mistress, who is shocked and outraged at further proof of his harshness and duplicity.

Proceeding with their scheme to put the sex-frenzied Captain in Lucrezia's bed and in her unwitting marital embrace, Stragualcia, Zarda and Messer Pacifico exchange clothes and identities – for some reasons too complicated to synopsize – and go through the charade of "kidnapping" a stranger (actually, the transformed Captain) and carrying him, dressed to look like Pacifico, into the wool merchant's darkened house.

Lisetta returns to the vicinity of the well, near Melite's dwelling, seeking assistance in finding the pendant that she mistakenly believes was thrown into it. But the well is being used by Melite to create the illusion that she is working the magic spell paid for by Suevia to punish Nardino and rekindle his ardour for her. Zarda is posted in the cellar at an opening to the dry well, a hollow shaft through which – as instructed by Melite – he issues ghostly sounds. Suevia is duly impressed, but Lisetta is frightened away by them. Melite, enacting the characteristic role of a sorceress, recites a litany of abracadabra, while from far below Zarda also hoarsely intones gibberish, foretelling that Nardino will return "in the form of a Saracen dressed like a Turk".

Carino has snatched the pendant for a purpose: he will return it to Olimpia, using it as a pretext for entering her house to meet her sister, Diletta, his inamorata. But first he must bring her mother, Euronica, to believe that he is back in his right mind. He sends Fedele to the old lady to say that a new medication has restored her daughter's eccentric suitor to his senses. This nostrum, too, is to be prepared by Melite, but she pleads that she has no knowledge of any such prescription. The resourceful Fedele tells her that he is a master of black magic; intimidated, she accepts his dictation and mixes a nonsensical concoction from far-fetched ingredients.

With a clamour in the street and a loud knocking on the door, Nardino and a band of men and women signal their return from Rome, his mission there a disappointment and a shambles. Customs agents at the border of the Holy City had seized several of Nardino's travelling companions as well-known smugglers and imprisoned them; he and a few others had barely made their escape and have hurriedly journeyed back to Mantua. Suevia, recognizing the weary fugitive, greets him vituperatively, upbraiding him for his treachery to her. He is taken aback by her irate welcome.

Lucrezia, awaking, realizes that the man in her bed is not her elderly husband and desperately fights him off. The cowardly Captain, hearing the uproar outside and below in the street, assumes that it is an avenging mob seeking him for his attempted rape; he grabs up his clothes and takes to his heels, making an ignominious retreat.

Olimpia and Fulvio, closeted for an intimate conversation in Nardino's house, are also alarmed by the noise. She asks him to hide, to protect her reputation. He makes his way out through a back door but inadvertently leaves behind an article of masculine apparel. When Nardino breaks in, he notices it and demands an explanation from Capone, whose duty it has been to guard Olimpia. Tansillo and Capone, with a spontaneous falsehood, glibly overcome Nardino's suspicions, then retrieve the tell-tale article and whisk it back to Fulvio.

Chastened by the failure of his venture into Messer Pacifico's house and conjugal bedchamber, the frustrated Captain must still account to the wool merchant for not having accomplished what was necessary to assure him a male heir. His performance was adequate, he insists: "Two or three kisses will do the job." Pacifico is relieved: "I questioned her on this very matter and I found out the knave kissed her at least seven times." The Captain: "That's more than enough." Zarda volunteers to complete the task, but is rudely told to keep quiet.

The denouement has Messer Pompilio, father of Carino, calling on Signora Euronica to offer amends for his son's conduct and to arrange for the betrothal of her daughter Diletta. Fedele is startled to behold Suevia – she is his cousin, a member of the Fedeli family of Porto. She has not been washed overboard at sea, nor left an orphan in the civil wars at home; no, to the disgrace of her clan, she had run away with a very importunate suitor, Nardino. Fedele immediately rebukes her, but Suevia, in tears, protests that Nardino had promised to wed her and that they had gone through a marriage ceremony, though without the proper witnesses, so that their vows are not wholly valid, as a consequence of which she is being cast aside for someone else. Catching sight of Nardino, who has just appeared in the doorway, Fedele draws his sword against him for bringing shame on the Fedeli family, but Fulvio and Messer Pompilio intervene to avert violence. Melite comes forwards. She testifies that she had been present at the troth between Suevia and Nardino, who admits that the ceremony had occurred. He states the he was long loyal to Suevia, but she had been ever more flirtatious with other men, especially when he was away. Feeling that she wanted to be free of him, he, too, had turned elsewhere. Hearing this, Messer Pompilio readily effects a reconciliation between the discordant pair, who reaffirm a matrimonial bond.

Euronica consents not only to Diletta's marriage to Carino but also to the union of Olimpia and Fulvio. Carino then asks his father to permit the shy Fedele, who lacks a fortune, to claim the hand of Urania, Messer Pompilio's other child, Carino's nubile sister. Out of gratitude to Fedele, the old man agrees to this and promises Urania a full dowry. All retire to the house to celebrate the forthcoming nuptials.

Only in the last few lines of the play is there any mention of Urania or Fedele's silent affection for her. Olimpia is seldom in view and is given very few speeches, and Lucrezia and Diletta, though there are many references to them, are never seen on stage.

Some minor characters and details of the action are not in this outline, but it should suffice to show its unusual complexity. Yet with ease the spectator can grasp its development, and it progresses quickly, never failing to sustain interest, all this a proof of de' Sommi's astonishing skill at organization and his innate architectonic sense, his firm control of his intricate story. He felt that a triple plot added richness to his work.

Some elements of the plot are derived from other plays of the period and particularly from Publio Filippo Mantovani's *Formicone* (given in Mantua in 1503) and Machiavelli's *Mandragola*; so obvious are his borrowings from these two long-popular comedies that *Le tre sorelle* must have been seen as openly parodic; de' Sommi expected the knowledgeable members of his audience to perceive the patent similarities. But what he does is to give these familiar plot devices a sudden, wry twist, a surprising fresh alteration. That sort of surprise is flattering, giving the spectator a special pleasure, like the enjoyment of an in-joke, of being let in on a secret.

De' Sommi's handling of bawdy subject-matter like that found in *Mandragola* is far less cynical than Machiavelli's and therefore perhaps more acceptable, especially because the moral climate in Italy had changed since the onset of the Counter-Reformation. (In both *Mandragola* and *Le Tre Sorelle* the woman whose virtue is to be violated is named Lucrezia, and the means of her seduction is the swallowing of a potion distilled from the mandrake root, prescribed by a charlatan; but the Lucrezia in de' Sommi's script drives off her would-be lover, she does not arrange future trysts with him, and her suitor is the threadbare caricature of a warrior, not a virile, handsome young man. In any event, the outcome of *Le Tre Sorelle* preserves "the dignity of marriage".)

Donald Beecher, specialist in Italian Renaissance drama at Carleton University, discusses this in a second preface to *Le Tre Sorelle*, of which he is the co-translator into English.

> Paralleling the stylized use of the setting is de' Sommi's stylized conception of the intrigue, for not only has he fashioned the action into a triple plot, but he has fashioned his intrigues out of extensive references to well-known earlier plays. The procedure, in principle, is not remarkable, for in a sense imitation was essential to the genre. Plautus and Terence were pillaged without apology, and contemporary playwrights preyed upon one another for plot particles and stock characters as though all such elements were held in common and appropriated from a central repository. Moreover, insofar as the conventions of the genre were under constant discussion in the courts and academies, the connoisseur spectators carried to each production a set of generic criteria and expectations against which each production was measured and by which it was held accountable. Hence, memory played its part in the reception of any new work, and the playwright who catered to that memory – an accumulation of prior theatrical experiences based on the same common repertoire of rules and plot motifs – might be looked upon with particular favour. Yet it is one step to imitate plot particles and character types from the ancients, and quite another step to parody entire plots from well-known plays, to interweave

them as parts of a compound design, while calling upon the memories of audience members to identify and compare the originals to their reincarnations. This, again, is the bravura of the artificer replacing the mimetic with the play of formal devices. Such extensive citations remind the reader of the degree to which *The Three Sisters* is a play built up from literary texts as opposed to an imitation of nature. . . . He makes specific new demands on his audience because memory and the identification of analogous forms must play a part in the reception of the play.

In other words, *Le Tre Sorelle* would amuse a newcomer to theatre, but it would entertain a sophisticated play-goer far more.

De' Sommi, while hewing to the conventions of the *commedia erudita*, broadened its scope. He departs from the rule of a single, focused action, introducing a triple plot and moving his people about the traditional stage-setting then available to him. As described earlier, this would consist of a row of houses and a small piazza, with all the incidents occurring in the street and square. This play would require neighbouring houses for Nardino, where Olimpia is under Capone's uncertain guard; for Melite, where she practises her trade and rents lodgings to the Captain and Zarda, as well as to the vengeful Suevia, and where there is a well; for Madonna Euronica, which she shares with her daughter Diletta; for Messer Pacifico and his wife Lucrezia; for Messer Pompilio and his son Carino and daughter Urania; and perhaps less necessary dwellings for Fulvio and Fedele, or one for both. The six houses might be arranged in a "V" formation, facing one another, three on one side, three on the other, a gap at the centre leading the eye to a far-off Mantuan vista.

The characters cross the piazza, entering the half-dozen houses or appearing at upper windows or possibly on balconies, to survey the active goings-on below. In this "claustrophobic acting space" the characters proceed with "choreographic precision", attesting to de' Sommi's adeptness and professionalism, his unfailing grasp of his medium.

Beecher argues that *Le Tre Sorelle*, with its unreal characters and their implausible antics – the Captain assuming various disguises, Pacifico assenting to the despoliation of his wife, Carino feigning madness, Melite concocting absurd spells – is an example of an art-work linked to the then rising Mannerist movement, typified by the "hyperbolic" products of such late sixteenth- and early seventeenth-century painters as Zuccari, Caravaggio and Reni, who sought to startle and shock the viewer, to surpass their predecessors – even the inimitable Michelangelo – by the excess of piety and emotionalism in their portraits of saints casting their eyes upward in ecstasy, though often looking somewhat demented. The spirit of Mannerism was one of artifice, an emphasis on physical straining and grotesquerie. Similarly, de' Sommi's comedy is frankly an artifice, its foolish characters over-sized, too broad, yet compelling if only because of the sheer exaggeration. If people were ready to accept unreality in serious subjects, depictions of saints in eye-popping moments of adoration, why not the totally absurd and fantastic in lighter theatre fare? Lovers like Carino and the egregiously appetent Captain are possessed in much the same way as the mystics and martyrs by their erotic impulses and carried away, behaving like idiots. (A belief that love-sickness could cause insanity was widely held in the sixteenth century.)

Of interest in *Le Tre Sorelle* is de' Sommi's ambivalent use of black magic and witchcraft to further the action. Melite performs a mockery of the prescribed rites, chanting what she knows to be nonsense, abetted by moaning and groaning from Zarda through the sunken air-shaft, and in a soliloquy she bemusedly reveals her lack of faith in what she is doing. She is concerned solely in bilking Suevia to earn a very small fee. But when Fedele, faking a verbal formula, provides her with what sounds like a necromancer's words, she seems to be convinced that they have the power to lift an evil spell. Doubtless many in de' Sommi's audience shared her divided attitude; they were sceptical, but only half-sceptical, in a world in which the popular outlook was still largely pre-scientific.

Twenty-five years earlier, in his *Four Dialogues* – the guide-book to stage production – de' Sommi explained how he set about preparing a script. "I note down the scenes in order, with the names of the characters appearing in them, together with an indication of the house or street by which they are to enter." He has the physical layout clearly in mind, the spatial limitations, and assigns each character to a specific house in the row, allowing the spectator to assume where each person is – i.e., most probably within the house – when not visible on stage. This established, the characters hustle or scurry from door to door, possibly to adjacent houses or across the piazza, in pursuit of their frenzied desires and to attain their legitimate or illicit aims. Most of the episodes are short, the action progressing at an ever-accelerated pace, giving the effect of a rapidly turning kaleidoscope. The characters and tone change from scene to scene; one moment a lover – Fulvio, Carino – is pouring out his heartbreak and frustration, and the next a group of rogues and roués – the Captain, Pacifico, Zarda – are planning sly amorous conquests, adding variety to the action and allowing ironic juxtapositions, while expanding the scope of the play, the sense that a whole cross-section of a community – Mantua – is involved. What is remarkable is how de' Sommi accomplishes the integration of the plot-threads and the actors' on-stage paths so tightly, working not with a single, focused story-line and a small cast but with three basic actions unfolding simultaneously and an ensemble of seventeen dynamic players. Outstanding "choreography", indeed.

A man of the theatre, de' Sommi has Veridico, his surrogate in the *Dialogues*, offer this practical advice: "You may be surprised to hear me say – indeed, I should set it forth as a fundamental principle – that it is far more essential to get good actors than a good play." He does not rule out the importance of having a worthwhile script, but at very least the text must permit the performers to display their gifts as nimble thespians, to exhibit – as Beecher puts it – "their repertory of gestures, expressions, moods, and registers". Elsewhere, Beecher commends *Le Tre Sorelle* for the exceptional opportunities it affords its cast: "Considered in terms of versatile, stylized, and contrasting acting styles, no 'idea' for a play could be better calculated to regale an audience with all the tricks, mimes, and gestures in the repertory of an experienced acting troupe." Inasmuch as the Jewish company was competing with the highly professional and ebullient *commedia dell'arte* groups during the bustle of Carnival, its members had to be considerably skilled to hold their own. Since the *commedia erudita* was burdened by now with over-familiar and repetitive stories and characters, a successful playwright had to seek every kind of novelty – such as the triple plot – and intensify the emotions and drives of the overwrought people engaged in every kind of incident – the Captain, Carino, who go to extreme lengths to attain their goals – giving a looser rein than usual to the actors.

To judge from what de' Sommi writes in his *Dialogues*, the conventions for acting were hardly restrained. As detailed by Beecher: "'To manifest joy the actor may break into a lively dance,' to reveal grief he may 'tear his handkerchief with his teeth,' and to suggest despair he may 'pull his cap to the back of his head.' Those who play fools must maintain a whole vocabulary of gestures, such as 'catching flies' or 'searching for fleas.' If the player is taking 'the part of a waiting maid he must learn to make an exit by tossing up his skirts in a vulgar manner or biting his thumb and so on – actions which the author has not been able explicitly to indicate in his script.'" No subtlety there.

Yet there should be a delving beneath the surface for the motivation. "'Granted the performer has a good accent, good voice, suitable presence, whether natural or achieved by art, it will be his object to vary his gestures according to the variety of moods and to imitate not only the character he represents but also the state in which that character is supposed to be at that moment'" (translation: Allardyce Nicoll).

If there are too many servant-roles in a play – in *Le Tre Sorelle* seven – de' Sommi suggests that each should be dressed in a different colour, to be more readily told apart, with an indication to which household each belongs.

Even with a good script and able actors, a play needs an appropriate stage-setting. Much of de' Sommi's enthusiasm is aroused by the effects and illusions created by scene-painting on backdrops, vistas of fields, forests and turreted cities with houses, streets and squares.

De' Sommi was lucky to be working during Duke Vincenzo's reign, while that Gonzaga ruler was emptying the state's treasury to indulge his taste for ostentatious entertainments, bringing in a host of companies and having a theatre designed by Viani. But the duchy was fast approaching bankruptcy, and there were attempts to curb Vincenzo's spending. Within a few years Mantua began a long commercial decline.

In Naples, far south, another dramatist of many accomplishments – as philosopher and scientist – Giambattista della Porta (1535–1615) led an advance to a final phase of the *commedia erudita*, well categorized as *commedia grava*, scripts overly sentimental. Though making use of stock *commedia dell'arte* characters, his plays are serious to the verge of tragicomedy. For subject-matter he returned to Plautus and other classical writers, and the much nearer-at-hand and always accessible Boccaccio. His *I Due Fratelli Rivali* (*The Two Rival Brothers*; 1601), based on a story by Bandello, is clearly echoed in Shakespeare's *Much Ado About Nothing*. Altogether, writing in prose, he turned out some thirty plays, of which fourteen survive. Among the favourites are *L'Astrologo* (*The Astrologer*; 1570) and *La Fantesca* (*The Maid*; 1592), along with *La Trappolaria* (*The Trick*). In an early venture, *La Cintia* (1560), like so many Italian Renaissance playwrights, della Porta relies on transvestism – a boy garbed as a girl, a girl attired as a young man – for dramatic complications. But in subsequent attempts, such as *La Furiosa* (*The Desperate Girl*; 1580) and *Il More* (*The Moor*; 1607), he manifests more originality. Throughout his work is found an ever stronger accent on romance, inevitably infused by excessive emotion.

In developing this type of drama, which was to grow into what the French later called *comédie larmoyante* – "comedy with tears" – della Porta was joined in Naples by the even more famed philosopher-theologian Giordano Bruno (1548?–1600), best remembered as a playwright for his Humanist satire of alchemy, *Il Candelaio* (*The Candle Maker*; 1582). Alas, this writer of comedy was to fall a victim of the Inquisition, charged with heresy, and with unwanted applause burned at the stake.

The play by this radical Dominican, who perished for his dissent from Roman Catholic orthodoxy, expresses his disgust at the degradation and corruption he saw in the society everywhere around him. In five acts of prose, preceded by three prologues, he depicts the undoing of Bonifacio, who is bisexual, ruthlessly preying on women and boys. ("Candle Maker" was a common pejorative term for a sodomite.) To sate his lust for a courtesan, Vittoria, he solicits the help of a sorcerer to gain entrance to her costly bed. Tricked, he finds himself lying instead with Carubina, his own wife, who takes her revenge by flying off to the arms of her lover. Paralleling these flagrant misadventures are those of Manfredo, a pedant, and the gullible Bartolomeo, misled by another heartless trickster, Ottaviano, and a conniving alchemist. While fake conjurations distract the befuddled victims, Bartolomeo's adulterous wife, too, dallies elsewhere. These three actions unfold simultaneously, after the model of the triple plot in de' Sommi's *Le Tre Sorelle*, but the irate Dominican's tone is quite different, shot through with lashing sarcasm, his aim an exposure of the vacuous credulity, hobbling superstition and moral indifference of his contemporaries. The dialogue is witty, though the play has been faulted for its "gratuitous vulgarity". *Il Candelaio* has had occasional revivals, but usually after having been freely altered to make the language and story complications clearer and less shocking. It is a truly dark comedy.

Like Giordano Bruno, the Sienese-born Girolamo Bargagli (1537–86) is recalled as a playwright for a single script, *La Pellegrina* (*The Female Pilgrim*; 1565–8), which he did not live to see produced. He was the eldest of three brothers – the others, Scipione and Celso – all of whom were highly intelligent and accomplished, excelling in the fields of literature and legal studies. Scipione was prominent in the academic discipline of linguistics, and Celso held chairs in the study of law at two universities. After Girolamo's early death at forty-nine, the younger brothers took care of the financial affairs of his widow and her two sons, the second of whom she bore a few months after her husband's passing.

Biographers are frustrated in attempts to learn much about Girolamo Bargagli's life. During his student years he was oriented towards Humanism, a progressive bias that he subsequently retained. Because he showed literary promise, he was admitted at twenty to Siena's intellectually élite Accademia degli Intronati and shortly afterwards justified his election by publishing a collection of fifty sonnets and two madrigals, as well as a prose essay, *Dialogo de' Giuochi*, the latter dedicated to Isabella de' Medici Orsini. The poems allude to the civil strife that not long before had roiled Florence and Siena and voice hope for a yearned-for return to peace and political and social stability. The *Dialogo* displays his early acquired erudition and his disappointment that the Accademia was no longer holding to the highest classical traditions. Apparently this feeling led him to abandon his youthful thoughts of a literary career. He turned to the legal profession, to which he had also shown a definite inclination.

Another factor in his leaving a writer's life was the forced exile of his friend and mentor, Fausto Sozzini, for making public his strongly held Humanistic and heretical views. The Counter-Reformation had bred an ever stricter censorship in Siena exercised by clerics, and authors had to be especially careful. Along with a zealous theological oversight was a tighter political control. Though he was pro-Medici, Girolamo was hardly likely to feel at ease in having his risky opinions openly circulated.

Where he pored over books filled with the intricate legal procedures and codes of the day is not known, but that he quickly gained a mastery of them is indicated by his being appointed to a lectureship on such Byzantine matters at the University of Siena (1653–4). Then, a bare three years later, he was chosen for a post in Florence as a judge in its Civil Court for several sessions, after which he returned to the University of Siena and a full professorship.

For the rest of his comparatively short life he frequently changed positions, rising in rank and prominence, as he moved about Italy. A lengthy list: from 1568 on he was back in Florence; and next, from 1574, he was Auditor of the Civil Court in Genoa, and briefly Chief Magistrate; after that he was back in Siena again from 1582 to 1586; and he was about to assume the office of Auditor in the Criminal Court of Genoa had not his last illness and death occurred.

The date of his composition of *La Pellegrina* is uncertain; if it was some time between 1564 and 1568, as is conjectured, it was while he was still a lecturer at the University of Siena. The initiative to write it was not his. Cardinal Ferdinando de' Medici, later to be the Grand Duke of Tuscany, wanted a play to grace a royal wedding. The commission for a script was offered to the highly esteemed dramatist Alessandro Piccolomini, who was reluctant to accept it. He had returned to Siena from service with the Archbishop of Corfu (1558), under whose sway he had been ordained, and had been named Archintronato (leader) of the by now apathetic Accademia degli Intronati and was striving to reinvigorate it. Perhaps his wish to avoid a tie to the undertaking was owing to his having become a churchman and ambitious to attain a higher place in its hierarchy. After further correspondence (the record here is confused) a division of labour was suggested: Piccolomini would oversee the project, Girolamo Bargagli would undertake the plotting, Fausto Sozzini would polish and approve the dialogue – he was able to insert anti-clerical gibes in it. It would seem that Piccolomini and Sozzini submitted their ideas, but Girolamo did the actual writing. A good many tokens of Piccolomini's influence are seen in the work, but scholars give credit to Girolamo for the well-constructed comedy that resulted. It is not clear why he was chosen for the task, since he had no theatre experience. Both he and Sozzini were members of the Academy, so Piccolomini was acquainted with them. The Accademia was involved in annual Carnival productions, and many fervent debates on the aesthetics of classical drama were held at its gatherings, and Girolamo's *Dialogo* is filled with allusions to theatre; at least, he had some knowledge of the problems of staging and a degree of interest in them.

Cardinal Ferdinando received the play-script, then returned it with a letter of thanks. No use was made of it; for years, perhaps two decades or more, it lay on a shelf out of sight among Girolamo's other yellowing papers.

In 1582 one Belisario Bulgarini's attempt to engage a cast and stage the play ended in failure. However, a half-dozen years later Ferdinando, now the grand duke – no longer a Cardinal – was preparing to marry Cristina of Lorraine. Invitations were issued to playwrights to submit appropriate scripts for the festivities. Scipione was prompted to send in his brother's comedy. (By now Piccolomini, too, was dead, and Sozzini still in exile.) Court officials asked Scipione to revise the work to make it more contemporary; in particular, Sozzini's anti-clerical tirades should be excised. Scipione complied with these official requests, and *La Pellegrina* had its first performance in 1589, its *intermezzi* sumptuously mounted with singers and dancers, along with a score by Emilio de' Cavalieri and several collaborators, the offering under the direction of Bernardo Buontalenti. Adding to the éclat of the event, the play was put on in the magnificent new theatre just installed in the Medici palace, and the guests were curious and quite eager to take in its ornate décor.

The festivities lasted a fortnight; a repeat enactment of *La Pellegrina* was given during the second week for the benefit of the Venetian ambassadors whose arrival had been delayed.

The play's text was published shortly afterwards, followed by the score two years later. But despite its initial success there is no notation of further staged productions of it.

Along with borrowings from Piccolomini, *La Pellegrina* reveals characters, motifs and bits of action garnered from the works of other well-known predecessors – Ariosto, Grazzini, Cecchi – but this practice was so universal that Bargagli was hardly to be taken to task for it. It flourished because the stock characters were familiar to the audience and endlessly amusing, as were the plots. A playwright displayed his cleverness in fresh, witty dialogue, in a novel if minor twist to the story, perhaps in a small but surprising deviation from the formula that the spectators had repeatedly shown pleasure in accepting: with a few surface embellishments, and even a minimum of competence, a script like Bargagli's was sure-fire. The audience knew a considerable amount of what to expect and was in a good mood and receptive.

La Pellegrina is almost unique for a comedy of this period in having no prologue. Scholars speculate that this may be because it has six *intermezzi* which require so much time that none is left for an introduction. As a result much of the opening scene is taken up with some rather clumsy exposition, the characters telling each other things of which they must already be well aware but about which the audience has to be informed in order to understand the action's complex premise.

An elderly Pisan, Cassandro, anxious to marry off his daughter, Lepida, arranges her betrothal to Lucrezio. But Lepida is secretly married and pregnant, so to delay the imminent ceremony she feigns madness. Cassandro, distressed and fearful that word of her mental state will get about, wants to call in doctors and priests to exorcise evil spirits that may be oppressing her. Lepida is determined to avoid their scrutiny; in this she is abetted by her quick-thinking maid, Giglietta. Her tutor, Terenzio, really her husband using a false name and only pretending to be teaching her, has adopted this ruse to be near her. Far from home, he has experienced many vicissitudes: taken prisoner and enslaved, he had at long last made his escape. On his way back to Germany, and passing through Pisa, he had caught sight of Lepida and fallen in love with her. Putting on what he considered to be a despicable disguise as a pedant, he had wooed and won her. The wedding, at which he had given Lepida a ring, was witnessed only by Giglietta. Now he has written to his father asking assent to the marriage and is awaiting a reply. Meanwhile he and Lepida feel forced to keep silence about their semi-legitimate union.

Lucrezio, too, is reluctant to solemnize his engagement to Lepida, having heard rumours of her mental instability, which Cassandro seeks to dismiss as a temporary attack of nerves at her approaching nuptials. But Lucrezio's strong hesitance has another cause. He is guilt-ridden. While on a business journey to Spain he had given his heart to Drusilla, as she had surrendered hers to him. He had entered into a marriage with her that was kept secret (for reasons which he says are "too complicated to explain"); it remained unconsummated, with Drusilla insisting that they wait until their bond be made public. A single kiss was exchanged as a pledge of their love. But scarcely a day later, recalled to Italy by his employers, he had given Drusilla his pledge to return in a year. In Pisa his affairs had gone catastrophically awry: his business associates had gone bankrupt, and one had died. Two years passed while he struggled to rearrange matters. Then a friend, Fabrizio, brought tragic news: Drusilla was dead. Fabrizio was sure; he had seen her on her funeral bier. Lucrezio is convinced that the fault was wholly his; believing he had abandoned her, she had died of humiliation and grief. He still cannot absolve himself of blame. Only at his family's urging is he now contemplating marriage to Cassandro's daughter, a match that has a commercial aspect. He has a materialistic side.

As the audience no doubt realized at once, Drusilla is not dead. Stricken by the thought that a heedless Lucrezio had forgotten her, she had declined and for several hours' duration had fallen into a coma that resembled death, but at the last moment had revived. Restored to health and her resolve strengthened, she dons the garb of a pilgrim and sets out for Italy to learn the truth about the long-gone but still beloved Lucrezio. Accompanying her, as chaperones and protectors, are two elderly persons, Ricciardo, a lifetime family friend, and the ailing Madam

Tommasa. The pretext that Drusilla has invented for the pilgrimage is that she is fulfilling a vow made during her almost fatal illness.

Arriving in the handsome cathedral city of Pisa, the travellers take lodgings in a house belonging to Violante; they do not know that she is something of a procuress and bawd. (A gentlewoman practising that trade appears in a great many Italian Renaissance comedies.)

Drusilla, making enquiries, learns Lucrezio is about to be married, but the ceremony may be called off because the bride-to-be is rumoured to be deranged. She is intent on encountering both Lepida and Lucrezio face to face. A story is spread that the Spanish pilgrim, briefly in Pisa while *en route* to Loreto, works miraculous cures. Violante is peddling this seductive tale, hoping to profit by drawing all sorts of gullible people to her house. Cassandro, having taken Lepida to a monk, who firmly declared her not possessed by any malignant spirits, hears of the pilgrim's marvellous gift. He decides to have his mad daughter pay her a visit. At the same time, hearing that the pilgrim is from Spain, Lucrezio asks to meet her, believing that she might be able to tell him something more about Drusilla's last hours. He also wants to discuss Lepida's baffling illness with her. Is there a chance that because of it he could break the marriage contract?

A new person enters the story: Federigo, like Terenzio, a student from Germany. Smitten with Lepida, he is bribing Targhetta, a scheming servant of Cassandro, to keep him informed of her condition, the apparent disarray of her wits.

In the encounter with the pilgrim, whose face is half hidden and averted, Lucrezio fails to recognize her, to Drusilla's profound disappointment. He bares his unwillingness to be bound to Lepida, if she cannot be cured of her insanity. He never alludes to an earlier, more passionate relationship he had once experienced in Spain. The pilgrim advises him not to wed.

Fearing that Lepida's pregnancy will soon become embarrassingly obvious, Giglietta urges her to delay no longer, but instead take Lucrezio and covertly resume her liaison with her true love, Terenzio. Lepida, virtuous, loyal, rejects this very immoral suggestion. She continues her charade in public, looking wildly distracted and uttering gibberish.

Federigo is convinced by Targhetta that Lepida, in love with him, is only pretending to be out of her wits in order to avoid the oncoming marriage. He has been sending letters and gifts to her through Targhetta (which the servant has delightedly been pocketing and never delivering). Impatient, Federigo resolves to steal into her house and upstairs to her bedchamber, and there force himself upon her. He believes that she will welcome the assault. He enlists the somewhat dubious and mercenary Targhetta to assist him in carrying out the plan.

Giglietta and Violante are well acquainted and often confide in each other. Federigo is boarding in Volante's house; the long arm of coincidence is considerably extended here – but such fateful chances have proved handy to playwrights since the day of Sophocles, who has Oedipus unknowingly slay his father in an argument at a crossroads, and then marry his mother without the least mutual awareness of their intimate kinship. Bargagli could cite classical precedents, if called upon to do so.

At the separate requests of Cassandro and Ludovico, Drusilla, as a self-announced pilgrim healer, has a session with Lepida alone in her room. Throwing herself at Drusilla's feet, the unhappy girl confesses her well-enacted ruse and the reasons for it. Touched, Drusilla offers to assist her. She orders that Lepida be given a herbal bath, the water perfumed with rare therapeutic ingredients not easy to procure. What Drusilla has in mind is that the bath will fail to ease Lepida's troubled thoughts, thus proving that her seeming madness is incurable, thereby halting her father's plans for her.

Targhetta, eavesdropping on the two, discovers that the bizarre babbling is a pretence, and that Lepida is pregnant; she misconstrues her situation, assuming that Ludovico is her lover and the begetter of the unborn child. Excitedly, hoping to earn a good tip, Targhetta hastens to inform Cassandro. The old man is shocked, angry and confused, not knowing what to make of it. He speculates that Ludovico has conspired in the trick, seeking an even larger marriage settlement.

Cassandro instructs Targhetta to run to the pilgrim and cancel the prescribed bath – he sees no need to waste money on it, and clearly the pilgrim has failed to diagnose the actual state of Lepida's health. Word of the trick's exposure greatly upsets Drusilla, believing Targhetta's account that Ludovico was a participant in the scheme in an effort to extract money from his future father-in-law. In her eyes Ludovico has again proved himself a heartless betrayer of women, a predatory fortune-hunter.

At a meeting in the street, Cassandro berates Ludovico for impregnating Lepida and trying to get a greater dowry. At first Ludovico fails to grasp the old man's accusations, but then furiously denies the charges. They part, both of them speechless with indignation.

Told by Targhetta of Ludovico's "infamous" conduct, Drusilla is also ignited by fury, mixed with despair, at this confirmation of Ludovico's wickedness. Her outcries resound in the street. But Targhetta is now anxious to reach Federigo in time to head off his proposed attack on the pregnant Lepida.

Too late. Federigo has already stolen into Cassandro's dwelling by an unlocked back door, climbed a winding staircase and peered into Lepida's room, and to his consternation beheld her in Terenzio's embrace – the two of them in her bed. He is agitated and insulted that she prefers another – a lowly tutor – to him. He cannot wait to find Cassandro and impart the truth about his duplicitous wayward daughter and the lecherous tutor residing in his house. Federigo claims that he is impelled to tell Cassandro this from a sense of honour, even though Terenzio, a fellow German, is a good friend. At first Cassandro cannot believe this report, then is half persuaded by it. Dumbfounded and shaken, feeling utterly disgraced, he rushes off to ascertain the facts for himself. Gratuitously, Federigo volunteers to accompany him.

Overwhelmed by her bruised emotions, Drusilla sinks into a faint and once again, cold to the touch, appears nearly lifeless. Her companions are alarmed, as is Violante, her landlady, who appeals to Giglietta for advice. But the busy Giglietta, hearing Cassandro's angry voice coming from his house, hurries back there.

Taking Terenzio by surprise, *in flagrante delicto*, Cassandro and Federigo seize him and lock him in a room. They debate what to do next. Federigo offers to tie him in a sack and drop him into the Arno under cover of night. But Cassandro thinks this too dangerous; their vengeful act would eventually be disclosed. He decides to go to the prince, a just ruler, and have Terenzio charged with criminal behaviour, though this might lead to a dreadful public scandal.

Once more herself, Drusilla seeks Ludovico to deliver a final rebuke to him; instead she sees Terenzio in fetters, held by the police, who are explaining to Cassandro and Federigo the legalities of anyone in the tutor's plight – if convicted, he will be sent to row an oar in the galleys. Terenzio pleads his innocence, while the hypocritical Federigo addresses him with jealous scorn and even wilder charges. Cassandro says that he will force Lepida to become a nun.

When Terenzio tries to establish who he is – his family background, his noble origin, his real name – he is recognized with a gasp by Federigo as none other than his long-lost brother, abducted when a child from the family castle in Austria by the marauding Turks. Now, in a complete change of manner, Federigo pleads with Cassandro for his brother's release, and the law officers concur, their behaviour also suddenly turned respectful. But the old man, offended, having suffered a loss of dignity, is adamant that the brazen tutor be punished. Federigo kneels before Cassandro, saying that his brother had been driven by love to take desperate measures of concealment, because he could not as yet prove his high rank, so he should be absolved. Drusilla, watching from a distance, intervenes. She reminds Cassandro: "Man is closest to resembling God when he forgives." After she reasons further with him, the old man yields: he will grant the young couple the same dowry and permit them to have a formal ring-ceremony.

Ludovici and Drusilla meet. She is still in her pilgrim's garb. In an over-long and exceedingly sentimental colloquy, they clear away their fateful misunderstandings; she casts aside her hood and robe and, as befits a comedy, the ending is happy.

But little that is truly or even passingly comic occurs in the story unfolded by the troubled leading characters. This is no longer *commedia erudita* but another example of *commedia grava*. The differences between the two stories are summed up by Bruno Ferraro, of the University of New England in Australia, in a lengthy preface to his English version of Bargagli's script:

> At first there seems to be no sharp dividing line between the Intronati's plays and normal learned comedy as in sixteenth-century Italian theatre, but on closer scrutiny the Italian theatrical production in the second half of the century (particularly in Siena) is characterized by having more romantic and pathetic elements, with serious and almost tragic implications. From a comedy of intrigue where the *beffa* or trickery is prominent, the emphasis is now on sentiments and events bringing anguish, suffering and threats of death; all, however, is resolved in a happy outcome and tragedy is thus avoided. This type of comedy, distinct from tragedy or tragicomedy, can aptly be called serious comedy and will eventually lead to the *commedia dell'arte* and to other theatrical

genres: melodrama, tragicomic plays of the Baroque period, and the *comédie larmoyante* (*teatro lacrimoso*) of the mid eighteenth century.

In serious comedy love is the mechanism which makes characters take on disguises and travel to faraway places to seek out their lovers; it is love which creates a series of misunderstandings and cases of mistaken identity. The characters involved in such situations usually discourse on their fate and suffering, on the uniqueness and nobility of their love and on their impeccable code of honor. . . . According to the canons of serious comedy, Act IV should bring the heroine and hero to the lowest depth of misery and to the highest level of drama. These circumstances are, in turn, reversed in the final act by the customary agnition (recognition of a character thought to be dead or lost forever) and by the happy ending. . . . Unlike learned comedy (especially Florentine) in which much of the action revolves around a central *beffa*, in serious comedy there is no major trickery (Lepida's feigned sickness can be considered only a half-hearted trick and not a central device), hence the minor characters never rise to the status of *mattatori*, inventors of machinations or intelligent solutions which have great bearings on the play. It is the lovers who exercise control over the action of the play since they, and not the servants, devise their own ruses.

The humorous antics in *La Pellegrina* are assigned to minor figures, for the most part the servants: devious Giglietta and gluttonous Targhetta, attached to Cassandro's disarrayed household; Cavicchia, in service to Federigo; Carletto, employed by Lucrezio. All brawl, scuffle, gossip, complain, hurl insults. Especially loud is Violante, engaging in verbal duels with them, some her erstwhile customers. She empties sacks of flour on them. They use street argot, sprinkled with sexual innuendoes. These scenes are not well integrated into the plot; if they were omitted the main action would still advance without impediment and be clear. *La Pellegrina* has been accurately described as a "two-tier play", serious on one plane, rowdy and jokey on another, the passages and moods alternating, with most of the engendering farce occurring in the body of Act III.

The play yields fascinating insights into social values and customs governing daily life. The servants endlessly fulminate at the treatment meted out to them – low wages, stingy table, unfair rebukes. Some are strongly loyal to their employers, others resentful and censorious. Cavicchia quotes Targhetta's account of the upper classes' lifestyle: "Last Sunday, when we were about to go to church, he was complaining about how badly our masters divided up the time set aside for our bodily pleasures. Sometimes, he was saying, they spend four or five hours listening to world news, to music, and to stories, and as much time looking at medals, studying paintings, watching plays, gaping at some woman – all things which aren't worth a cent. They also want to devote time to their noses and they'll spend three hours in a perfume shop sniffing waters, oils, powders, scenting their gloves and other similar stupidities. They don't even devote one hour in the whole day to the mouth, the source of life. But Targhetta talks about it all the time." As portrayed, Targhetta is always hungry and ready to sell his soul for a succulent dish.

A good deal of jeering is directed at the clergy, who are depicted as lecherous and corrupt. When the apparently mad Lepida is brought to be examined by a monk, to determine if she should undergo the harsh rite of exorcism, she is surrounded by a host of young seminarians taking an indecent interest in her. The supposed ascetics live luxuriously. Perhaps these were the speeches and passages the Medici officials stipulated must be removed before the work was staged in the royal palace. Even so, some of the language is so scatological, one wonders how it was received by the exalted and élite spectators.

Bargagli looks about him with a critical eye. He has Cassandro direct a tirade against lawyers and their habits, even though he himself was just embarking on a career in the legal profession. What happened when Cassandro had sought advice from an advocate? Endless delays: "There were twenty-five people around him, one to make a protest against a bill, another to take out a libel action, others to produce documents, to take out a summons, and so on. These lawyers are a devilish torment, by God! Small wonder they cost people their wealth and even their lives, not to mention their wits and their souls." Ferraro remarks that this playwright had a gift for self-satire.

The ongoing debate about language is reflected in exchanges between the two students from Germany, Federigo and Terenzio. When Federigo wishes to retrieve a copy of Petrarch's sonnets from Terenzio's place, the latter says: "I must entreat you not to leave such books in my house. If some student of liberal studies should find it in my rooms and think it was mine he'd defame my reputation and my good name." Federigo replies: "I beg your pardon? Is there any book which exalts our vulgar language more than Petrarch does?" To which Terenzio takes quick exception: "It's called vulgar because it's spoken by the people who know no better. What's this idea of speaking vulgar Italian? We should speak Latin, Latin; Ciceronian, Ciceronian!" Federigo is not persuaded: "I'll say this, Messer Terenzio: though I've come to Italy to learn Latin, I must say that in my country Tuscan is highly valued, especially by any person who intends to become a courtier, as is my intention." Ferraro comments: "Terenzio is a particularly complex figure. He represents the pedant, a traditional figure of derision and laughter in sixteenth-century Italian comedy, who often quotes Latin words and uses a language dear to the *accademici*. . . . Bargagli's caricature of the pedant in this play could also be seen as the author's caricature of intellectuals in general and, since he is an intellectual himself, as a sort of self-caricature. But Terenzio is also one of the lovers and, in this dual role of lover and pedant, he raises the interesting problem of identity when he analyzes his unfortunate state of affairs." In a soliloquy Terenzio complains at how painful it is to masquerade as a pedant: "To wear these clothes, to check my gait, to put on a façade, to say things worthy only of Polyphilus, in brief to lose my identity. But what am I saying? Didn't Jupiter turn into a bull and a swan to assuage his love? Now I find myself even muttering some pedantries – so much am I getting used to playing this role." (Polyphilus: a character in a play by Castiglione whose speeches are a mixture of Latin and the vernacular.)

Though the playwrights of the Accademia degli Intronati made a point of trying to please the ladies in the audience, *La Pellegrina* sounds at times like an anti-feminist tract. Perhaps Bargagli inserts these lines tongue-in-cheek, meaning to tease, and the ladies accepted them in that spirit. The moral standards of the women are depicted in accordance with their social class: Drusilla and Lepida are highly virtuous, Giglietta and Violante far less so – in fact, the ageing, blowzy landlady has been a whore and would still like to be one, though her numerous solicitations meet rebuffs. When Lucrezio protests, "Do you want me to put up with a mad wife?" his servant Carletto asks, "Where will you find any woman who's not a bit daft or light-headed?" And further along, Carletto expounds: "Do you think you'll ever meet a woman who's not somehow devilish? Let's not mention the ones who are so ugly that they actually look like the devil. Even the beautiful ones have a bit of the devil either in their eyes, or on their cheeks, or in their bosoms, or on their mouths, or in their hands, or in the way they dance or in the way they sing. Tell me a gesture, a movement which doesn't reveal the presence of a tempting devil! I think Hell is full of them. I won't mention those who are a bit devilish in their heads or in their brains. Others still, like your future wife, are devilish in another manner and perhaps in a more excusable one since, in this case, they are the tormented ones while in the other cases they are the tormentors."

Later Cavicchia tells his master, Federigo, that he is lucky not to have won Lepida. He explains, "Had you pursued her with the intention of marrying her, as they do in Germany, you would consider yourself fortunate that, seeing that she has turned out to be mad, she has married someone else. If you pursued her as your mistress, in the Italian manner, you'll be better off now that she's going mad than you were when she was in control of herself. To tell you the truth, you only get headaches, problems and delays from a woman in her senses; you're well off only with the crazy ones, for only the mad ones let themselves be plucked."

To this, Federigo retorts: "Just listen to the ravings of this ass!"

And still later this chorus of complaints is joined by an impatient Targhetta. "How true it is that when women have to go somewhere they keep you waiting for a year! It takes them such an effort to tear themselves away from the mirror. . . . Instead of wasting their time with little cream vases, jars and combs, which are worth nothing, how much better they'd spend their time if they attended to cooking-pots, frying pans and roasting-spits, which are important."

Apart from his being always hungry, and having that as his priority, Targhetta deems himself an authority on feminine behaviour. He instructs Federigo: "I'll tell you, my lord, out of the experience I've had of these things after having served many women, one must realize that there are several types of women who want to please their lovers. Some provide your opportunity themselves; others expect you to look for it. Others try to please you in their own way, and you won't achieve anything by being importunate or using force; others, on the contrary, behave like the besieged inhabitants of a castle! They think before they can surrender with honour, they must stand an assault or two. There are others still who are so irresolute, shy, and devoid of willpower that they don't dare do anything, even if they'd like to; the only way with them is to force them. You can be sure that my little mistress belongs to this last type."

These passages do not exhaust the playwright's deprecatory remarks about womankind. Probably the ladies – especially those belonging to the court – looked upon them as mere badinage and were amused. Or were they?

The play is wordy, burdened by too many lengthy speeches, some of them soliloquies and "asides", confidential remarks directed at the audience and theoretically not heard by the other characters. At times the servants use the argot, but at other points in the dialogue their language is formal and they sound like the author – Bargagli – or his collaborator Sozzini.

Foreigners are another butt of jokes and disparagement – Germans and Spaniards, but not the French, since the grand duke was marrying a princess from Lorraine.

The imagery in the dialogue is often pungent – again, Sozzini's contribution? Giglietta, fussing to have Lepida look neatly powdered and coiffed, though she is affecting madness, reminds her: "Even women on their way to the grave are properly made-up and have their hair curled!" When Lepida tells Giglietta that acting half-crazed is difficult, the maid observes: "There are plenty of mad people who try to pass for sane persons, which is much harder!" Elsewhere, an irate Federigo is waiting for Targhetta: "He's like the quintessence of the alchemists: he's never to be found." Soon thereafter Targhetta says in an aside, "It's easy to fool people in love." Carletto discourses on the hazards of seeking to make one's fortune at court, which usually ends badly. Targhetta protests: "You also see a few who have grown rich and have done well for themselves." Carletto rejoins: "They are the exceptions, like white crows."

Violante and Giglietta exchange scandalous confidences. The landlady would like to act as a procuress for Drusilla, who shows no interest. Learning of this, Giglietta declares: "I thought you were an old hand in your profession.... Don't you know that women are like birds? They can all be caught in the end if you use the right approach. Vain women are captured with flattery, miserly women with gifts, haughty women by kissing their feet and simple women with cajolery. Leave it to me, I'll know exactly what she's worth after I've looked her over." But nothing comes of this. Violante confesses her preference for students as lovers; but Giglietta voices her doubts about them. She cites a folk-saying: "Studies and pleasures don't mix," and adds her own impression: "These students look pale, tired, melancholic and quite unsuited to women." She is mistaken, insists Violante, who claims that she has tried all types. But none equals a whole year she once spent with a versatile student. "He never made love to me twice in the same way: a man full of imagination, of new ways. Only those who read books find out about these things, and if I know anything, I owe it to him."

Far more explicit is a passage in which Cavicchia lectures Federigo on the difference between ladies of the nobility and whores and courtesans when men engage them in sexual intimacies, with Cavicchia casting his vote for a paid bed-companion in every instance. Was this not strange fare for a gathering of royalty (including an ex-Cardinal)?

If *La Pellegrina* did go through six more printings yet was not performed again, it must have been widely read, because it served as a model for many later plays in Italy and France, its influence discernible throughout the following century and beyond in the works of such writers as Basilio Locatelli (*La Forestiera*), Carlo Goldoni (*Il Servitore di Due Padroni* and *La Bottega del Caffè*), Jean de Rotrou (*La Pélerine amoureuse, Tragicomédie*), as well as Molière's *Les Femmes savantes*, Regnard's *Les Folies amoureuses* and Destouches' *La Fausse Agnes*, some of which employ the motif of the pilgrimage, others having a character closely resembling the aggressive Drusilla. What is more, as has been stated, Bargagli's script helped to set the pattern of the long-lingering *commedia grava*, a genre into which *La Pellegrina* fits, in Bruno Ferraro's words, as "a comedy of sentiment and manners in which the scenes dealing with love, honor, faith, loyalty and obedience testify to the theatrical taste of the times and to a morality which could exist only in a period of religious and political stability", though that "morality" appears to have been more flexible than one might have expected.

While preparing his copious notes and English translation of the play – from which generous excerpts have been quoted here – Professor Ferraro was surprised to learn that the musical *intermezzi* of *La Pellegrina* were staged at the climax of the forty-fourth Settimana Musicale Senes by the Accademia Musicale Chigiana in 1987. On this occasion the orchestra was conducted by René Clemencic, the Coro Polifonico della Toscano was led by Roberto Sabbiani and the Coro d'Opera della Chigiana was guided by the baton of Lajos Kozma. The 1589 version of the play was the one chosen, though of necessity somewhat adapted. "A narrator read Bastiano de' Rossi's description, the acts of *La Pellegrina* were read in an abbreviated form and during the performance of the *intermezzi* slides of Buontalenti's costumes were projected on a screen with the commentary of Elvira Zorzi." Another performance of the piece was scheduled for the 1989 Maggio Fiorentino. It is interesting to note the massed musical forces used for this rare presentation. (Bastiano de' Rossi was a contemporary who wrote of the offering of *La Pellegrina* shortly after the Medicean festivities and described its opulent production.)

The attraction that writing comedies held for the intelligentsia of the Italian Renaissance is evidenced by the host of names that bid for attention in an adequate survey of this epoch, one in which iconoclastic wit and bold curiosity were highly valued, and the intellectual climate one of swirling change. Gigio Artemio Ciancarli is remembered for his *Capraria* (*The Comedy of the Goats*; 1540), which owes far too much to its classic Roman forerunners; however, his *La Zingana* (*The Gypsy*; 1550) has a contemporary subject and atmosphere. To the list of the overlooked might be added Niccolo Secchi, Girolamo Razzi, Raffaello Borghini, Frances d'Ambra, Luigi Alamanni, Giambattista Celli, Agnolo Firenzuola, Lionardo Salviatti. The roster, only a partial one, does honour to the diligence of scholars, but does it include unrecognized comic dramatists of stature? Most likely not.

Too much of the content of the plays was repetitive and artificial. It has been ventured that besides the enervation induced by the habit of copying so closely from the past – imitation being a form of artistic decadence – another reason for the prevalent lifelessness or quick demise of many of these overly literary farces was that their audience was too small, merely a court élite, with narrow views and interests, Italy's parochialism a consequence of its being divided into so many vying minor principalities and duchies.

Yet these proliferating comedies had a significant influence on the Elizabethans in England and somewhat later Molière and his rivals in France, who were to raid the Italians and make off with plots and characters that they endowed with new life. One cannot have a true perspective of the history of comedy without some acquaintance with the prolific offerings of this period. Ariosto, Machiavelli, Cecchi and their sixteenth-century compeers are vital branches on the family tree of theatre.

In the age that immediately followed the High Renaissance, a dominating figure in the art world is the great baroque sculptor and architect Gian Lorenzo Bernini (1598–1680). His birthplace was Naples, to which his father, Pietro, a Tuscan and a sculptor of substantial reputation, had journeyed to fulfil a commission in 1562–9. Seeing that his son was unusually gifted, Pietro taught him the craft of cutting and shaping marble. In time the father was summoned to Rome to be of service to the powerful Borghese and Barberini families, which enabled the young prodigy to gain access to them and win their notice. Shortly he also gained their patronage, his talent so manifest that his fame quickly far surpassed that of Pietro. As a mere youth he had already been praised by the painter Annibale Carracci, and he became a protégé of Pope Paul V at a young age. In all, during his long life he received work orders successively from eight Popes, and particularly from Urban VIII and Innocent X.

After diligently studying classic statuary, he moved away from its principles. He is credited with having created the elaborate, dynamic baroque style and being incontestably its foremost exemplar, whom no other could match. As an architect he designed much of the vast, ornate basilica of St Peter's and the magnificent colonnade encircling the wide piazza on which it fronts, consisting of 284 granite columns and eighty-eight pilasters. As his legacy he left Rome a city of noble tombs – especially that of the martyred St Peter, with its lofty, gilt-bronze baldachin, and that of Urban VIII – and splashing, graceful, architectonic fountains – the Barcaccia, the Triton – that still adorn its sunlit, broad squares.

He was a painter, though a minor one. As a sculptor he had few equals, ranking close to Michelangelo. In *The Story of Art* E.H. Gombrich writes:

> [the] supreme art of theatrical decoration had mainly been developed by one artist, Lorenzo Bernini. . . . He was a consummate portraitist . . . his bust of a young woman has all the freshness and unconventionality of his best work. When I saw it last in the museum in Florence, a ray of sunlight was playing on it and the whole figure seemed to breathe and come to life. He has caught a transient expression which we are sure must have been most characteristic of his sitter. In the rendering of facial expression, Bernini was perhaps unsurpassed. He used it, as Rembrandt uses his profound knowledge of human behaviour, to give visual form to his religious experience.

And, of a statue of St Teresa of Avila in a moment of uplifted, overpowering, other-worldly bliss:

> It is this vision that Bernini has dared to represent. We see the saint carried Heavenwards on a cloud, towards streams of light which pour down from above in the form of golden rays. We see the angel gently approaching her, and the saint swooning in ecstasy. The group is so placed that it seems to hover without support in the magnificent frame provided by the altar, and to receive its light from an invisible window above. A northern visitor may be inclined, at first, to find the whole arrangement too reminiscent of stage effects, and the group over-emotional. This, of course, is a matter of taste and upbringing about which it is useless to argue. But if we grant that a work of religious art like Bernini's may legitimately be used to arouse the feelings of fervid exaltation and mystic transport at which the artists of the Baroque were aiming, we must admit that Bernini has achieved this aim in a masterly fashion. He had deliberately cast aside all restraint, and carried us to a pitch of emotion which artists had so far shunned. If we compare the face of his swooning saint with any work done in previous centuries, we find that he achieved an intensity of facial expression which until then was never attempted in art. Looking from the head of Laocoön, or of Michelangelo's "Dying Slave", we realize the difference. Even the draperies are completely new. Instead of letting them fall in dignified folds in the approved classical manner, he made them writhe and whirl to add to the effect of excitement and movement. In all these effects he was soon imitated all over Europe.

Intensely religious and austere, Bernini attended mass daily and took communion twice a week. In the spirit of the Counter-Reformation, he sought to inspire the faithful through his fervent art. Details of his life and daily routine are sketchy. He spent most of his years in Rome, venturing out of Italy for any length of time only once. At forty-six he chose a wife who at twenty or twenty-one was less than half his age and by whom he had eleven children. He is described as possessing a fiery temper and preferring a "semi-ascetic existence, on a diet consisting largely of fruit". It is hardly to be supposed that he lived serenely, busy with large projects, as he always was, and dedicating himself to them in a household with distractions provided by eleven active youngsters.

His one journey outside Italy was to Paris, to which he was invited by Louis XIV, who wished his advice on the design of a new palace and its surrounding gardens – what was to be the Louvre and the Tuileries (1665). Arrogant and tactless, the Italian visitor dismissed with disdain all the plans that had been prepared by esteemed French architects and landscapers. He made himself so unpleasant that Louis XIV finally rejected the famed artist's drawings and sent him packing. But by now Bernini's reputation was so widespread that the streets of each city through which he passed were lined with crowds that gathered to catch a mere glimpse of him.

Such was the acclaim he earned that when he died he was considered to have been "not only Europe's greatest artist but also one of its greatest men".

Little known today is that for a good many years Bernini was constantly employed in creating scenery and stage effects for acting troupes, a task for which his masterly skills and the innate theatricality of his personality well suited him. He was deemed to be truly superb at this, and his acceptance of a commission was considered a coup by a producer or director or whoever might be subsidizing a dramatic enactment. He was noted for favouring startling, spectacular effects, magical ones, and needless to say was unceasingly innovative. What is more, the very wealthy sponsors of the presentations – Popes, cardinals, princes – allowed him to be heedless of costs.

Far in advance of his time, he sought to narrow the "distance" between the actors and the audience, to have the spectators feel themselves participants in the story, sharing the same emotions. In one notable instance, to instil fear in the gasping onlookers he flooded the stage and had the turbulent waters roll towards them; at the very last moment a barrier rose to hold back the waves that threatened to inundate the theatre. But the spectators had a shock, a sense of what the unnerved characters had experienced when confronting a deluge caused by the overflowing Tiber. On another occasion he had a torchbearer in a procession "accidentally" set the stage afire, causing the audience to flee in terror, before a providential "rainfall" doused the flames. Spectators, returning, found the scene had changed to "a noble and beautiful garden"; by this device he had let them know what it was like to face a consuming fire. Such tricks required remarkable ingenuity and precise timing. Again, experimenting, he designed a play (1637–8) in a comic vein, for which there were two audiences, two casts and two theatres, for all sorts of interweavings and plot complications, the details of which are unfortunately lost. In this phase of his career (1639–41) he contributed settings for two operas. John Evelyn, the English diarist, saw one of these productions and was deeply impressed.

Even less known now is that the great sculptor also wrote no fewer than twenty plays, all but one of which have vanished. His grounding in practical theatre-craft prepared him to undertake writing scripts, and his genius and originality held a good deal of promise that he might be a dramatist of importance. The one remaining play, short, incomplete and untitled, gives some indication of this. A partial account of what the other scripts contained was provided by a son, Domenico, after his father's death, but it is not explicit enough. The true measure of Bernini's vision and accomplishment in the drama can never be ascertained.

He began writing plays during an illness when he was thirty-six. Despite his piety, many were comedies – as his son confirms – of an obscene and scatological kind, such as might fit into a Carnival context. He did not hand them out to professional groups but had all of them put on under his supervision in his own capacious house for the entertainment of his friends and peers, a decidedly select coterie. Often the cast was drafted from members of his immediate family, who endured long rehearsals, an ideal practice most directors would envy, as Bernini must have known all too well. His improvised stage was small, about twenty-four feet deep. He mounted the plays "on the cheap" out of his own pocket – most of the actors unpaid, the settings quite simple. It was a way of expressing his dislike and impatience with the lavishness, so wasteful, required by his high-placed sponsors, who called for the use of intricate, difficult and cumbersome machinery to create spectacular tricks; he was offering proof that plays could be mounted at far less cost yet hold interest and win admiration. Some of his comedies were then repeated for similar private audiences in the homes of the Roman aristocracy. The simplicity of the settings made this easy to do. He escaped censure for the frequent gross indecency of his dialogue and for satiric gibes aimed at prominent guests, who might even be present, since his art was so revered by powerful patrons.

How to put on a play, indeed, is the subject of his sole surviving and all too fragmentary script. It was discovered as late as 1963 by an Italian scholar, Cesare d'Onofrio, while engaged in research at the Bibliothèque Nationale in Paris. He was poring over a collection of Bernini documents in a ledger relating to repairs on the Fontana di Trevi undertaken by the sculptor, a plan to remove the famous monument from its ancient site. To his great surprise d'Onofrio came upon the long overlooked manuscript, in twenty-five folios that had been folded among the other yellowed papers. He edited the play and had it published shortly afterwards. He hazarded that it was written some time between 1642 and 1644. Since the piece lacked a title, d'Onofrio chose to refer to it merely as *Fontana di Trevi*, alluding to the file in which it was found, though that has no relevance at all to its content. Donald Beecher and Massimo Ciavolella, who brought out the first English translation of the play, have elected to retitle it *The Impresario* (1985).

Beecher and Ciavolella, faculty members of Ottawa's Carleton University, analyse the text in depth in an eighteen-page preface and follow it with thirteen pages of notes that are truly thorough and informative, in fact fascinating, though their comments fill more pages than does the brief truncated play. They find in *The Impresario*, after what they admit were repeated delvings into it, many profundities and subtleties that an ordinary reader might not ever grasp. To such plain readers the script is skeletal, moves awkwardly, and single lines of dialogue to which these scholars attribute major adumbrations may not have all the serious significance the translators find in them. But that is a matter to be decided by each person approaching this unusual text.

Most certainly, *The Impresario* is tantalizing, affording glancing insights into the mind and imagination of a man of unquestioned creative genius and originality, who also possessed a strange personality. Like his theories of stage design, the play is far in advance of its time. It might have been conceived by the twentieth-century master of paradox and proponent of the subjectivity and relativity of truth, Luigi Pirandello, exemplified in such works as *Six Characters in Search of an Author* and *Tonight We Improvise*, which explore the very nature of theatre and the processes of creativity. *The Impresario*, what little there is of it, also has a loose affinity with the exceptional plays of Bernini's Spanish contemporary, Calderón, who established a genre of psychological fantasies with his *Life Is a Dream* and *The Theatre of the World*. And it anticipates by centuries Nikolai Evreinov's advocacy of a stage that acknowledges its intent to be thoroughly artificial, as are his *The Chief Thing* and *The Theatre of the Soul*, embodying the theory the Russian playwright and director put forth in his book *The Theatre for Oneself*. Bernini's piece – it is scarcely more than an unfinished sketch – is a musing on illusion and reality in the theatre, primarily as they are perceived and experienced by a scene designer, one charged with conceiving and executing astonishing stage effects.

Here an artist is essentially a trickster. *The Impresario* is about a covey of tricksters, all trying to outdo or outwit one another. The story is framed about the problems of putting on a play within a play.

Cinthio, a penniless courtier of good family, is deeply in love with Angelica, the yielding daughter of Graziano, famed for his success at producing spectacular theatrical effects. But Graziano will not relinquish his daughter's hand to a suitor who has no money. To solve his dilemma Cinthio turns to his servant Coviello, who is a boastful trickster of a lesser kind. (In these many Italian comedies the impecunious young man always has a personal servant.) Coviello quickly concocts a plan to raise 1,000 scudi, which Cinthio needs not only to marry Angelica but also to pay off his borrowing from some Jews who are pressing their claims. Alidoro, a frustrated rival of Graziano, will give Cinthio the 1,000 scudi to learn the master's secrets in creating the effects that have brought him so much acclaim. So Alidoro, too, is resorting to trickery. At the moment Graziano is busy with other projects and not interested in engaging in stagecraft, so there is not much chance for Cinthio to discover his methods. The young courtier, however, is looked upon with favour by the prince, to whom he has ready access. Coviello prompts Cinthio to approach the prince with word that the hitherto unavailable Graziano is anxious to write and produce a play as a tribute to him, the benign ruler of the state, a prospect that elates that royal person. Next, all too willing to perpetrate a trick on Angelica's obdurate father, Cinthio tells Graziano that the prince commands him to provide a comedy for presentation at court immediately. Reluctant to put aside his other concerns, but flattered by the extravagant praise that Cinthio reports was uttered by the prince at the very mention of Graziano's name, the scene designer impulsively takes steps to fulfil the royal order. Angelica, meanwhile, is to feign illness so as not to be forced to marry anyone else until Cinthio has the necessary funds. With so much duplicity set in motion, Beecher and Ciavolella aptly describe the play as a Hall of Mirrors.

Graziano issues a peremptory call to his usual technical assistants: Iacaccia, a head stage carpenter, and several other handy wielders of nails, hammers, saws, squares and levels; Sepio and Moretto, next in charge of the craftsmen; and Cochetto, a deft French painter. They hastily assemble and under Graziano's direction begin to fashion an ambitious "floating cloud" effect.

The psychological insight here is credible and interesting. Graziano, a scene designer, envisages producing an appealing stage-effect before he even asks himself what the play is about. He will compose a dramatic work, chiefly to reveal his basic talent. Sepio asks just what sort of "cloud" the master has in mind. Graziano replies: "I want it to appear completely natural."

> SEPIO: How do you mean, natural?
> GRAZIANO: By natural I don't mean a cloud stuck in place up there. I want my cloud

standing out, detached against the blue, and visible in all its dimensions like a real cloud in the air.

SEPIO: Up in the air, eh? That's nothing but doubletalk. Detach it from up there, you'll more likely see a cloud on the floor than up in the air – unless you suspend it by magic.

GRAZIANO: Ingenuity and design constitute the Magic Art by whose means you deceive the eye and make your audience gaze in wonder; make a cloud stand out against the horizon, then float downstage, still free, with a natural motion. Gradually approaching the viewer, it will seem to dilate, to grow larger and larger. The wind will seem to waft it, waveringly, here and there, then up, higher and higher – not just haul it in place, bang, with a counterweight.

SEPIO: Well, Messer Graziano, you can do these things with words but not with hands.

GRAZIANO: Now look here. Before we're through, I'd like you to see what the hand can accomplish. Follow me, I'll explain how to go about it. [*Exeunt*]

But Sepio's dire prediction is borne out. When the cloud, painted on canvas, is hung from the flies, it collapses and falls to the stage floor where it lies in an ungainly heap. How shall one interpret this setback? The failure of the "effect" places an obstacle in Cinthio's path in his attempt to steal Graziano's invaluable secrets. This is hardly a trick worth learning about. Or else Bernini is showing what happens – embarrassingly – when stage machinery fails, and also how difficult it is to bring off scenic illusions. One supposes that eventually Graziano will achieve his magical cloud effect, but the script breaks off before that is accomplished.

Though belatedly, Graziano chooses a subject for the play: it will be about an artist who has to write a play and hurriedly prepare its scenic investiture. The writer-designer's name is Graziano, and the challenge facing him is to develop a theme that will fully display his unique power to create magical illusions. Thus the play-within-the-play veers towards autobiography, and Bernini's "second Graziano" is a self-portrait, but in most ways not a complimentary one and at moments almost a caricature. He is in love with Rosetta, a servant-maid in his house – the "first Graziano" also employs a maid by that name towards whom he shows an obvious physical interest. This exchange of dialogue ensues:

GRAZIANO: Let me start writing. Graziano is in love with Rosetta. He loves her so much that he can't wait for his wife to die so that he can marry Rosetta. Got it?

ROSETTA: Will this Graziano have a wife, then?

GRAZIANO: He'll have a wife, yes, but she'll be an old piece of rancid meat. Let me write now. Rosetta has no idea how lucky she is. If she had, she'd show Graziano a bit more . . . tenderness.

ROSETTA: Let me take this Rosetta's part for a minute. If Graziano's wife lives

> another twelve or fifteen years, then Graziano will be a senile old fool, and Rosetta will have wasted the flower of her youth.
>
> GRAZIANO: Not so fast; hear me out. Graziano knows how long the old girl could live, and forestalls that problem by helping himself to Rosetta in advance. He gives her every consideration, like his own wife, so that she'll give him a son.
>
> ROSETTA: Oh no! People won't put up with smut like that. Besides, it wouldn't look natural for a married man to start making children with his maid.
>
> GRAZIANO: Wouldn't look natural? Ha, ha, ha! But that's all you ever hear about – with the maid, the wet nurse, the girl in the pantry – that's all you ever hear!
>
> ROSETTA: Ladies are foolish to keep servants like that in the house.
>
> GRAZIANO: Ha, ha, ha! Lots of women behave that way not because they're fools, no, but because they're so good-hearted – then everybody's happy. Live and let live, right?

At this point Graziano's double – that is, the actor who has the role in the play-within-the-play – breaks in: "Ah, what a pleasure! Don't leave! I just want to know, who is this Graziano who's in love with his maid? Who is he?" (Is this not the actor in every age who seeks to learn the playwright's intention? The better to interpret the character.) The first Graziano rejoins: "Who is he? Why, he's the fool of this play, that's who he is." And, then, the second Graziano: "I see. And if the world's nothing but a play, then Graziano's the biggest fool in the world. Poh, shame on him, one foot in the grave and lechery in his heart. Hasn't he had enough, the hog? That old piece of rancid meat isn't about to die, because Heaven isn't about to grant dismal dodderers the realization of their dirty dreams."

If the first and second Grazianos, to even some degree, represent aspects of the author – Bernini – the depiction is hardly self-congratulatory. The great sculptor was viewing himself with a wry, mocking humour and detachment, with utter irony. But perhaps the two images should not be taken too literally: at the time *The Impresario* was written Bernini was not a dismal dodderer with one foot in the grave; he was in his mid-forties. With his large brood of children, he had no need of another son. It is quite possible, too, than Rosetta and Zanni, Graziano's outspoken servant, had no real-life counterparts. In many respects they are characters borrowed from the *commedia dell'arte*, as their names imply, though they do not always behave as tradition would have those stock figures do. One of Bernini's "tricks" in this play, in fact, is to frustrate the spectators' expectation: the characters do not follow the conventional guidelines – hardly anything in the script conforms to the theatrical conventions of the day. The young lovers, Cinthio and Angelica, are minor participants in the action, whereas in most comedies they are the focus of it. Graziano, though a dupe – everyone is playing tricks on him – never loses his dignity or conducts himself foolishly; he remains in control, even though he is the butt of so many kinds of deceptions. He is tyrannical, with an indecent tongue – many of his remarks about how to construct his "effects" can be construed as sexual *double entendres*, and he makes insulting allusions to his rival Alidero, whom he castigates as "ladylike".

The craftsmen, too, try to ascertain some of Graziano's technical secrets by offering bribes to Rosetta, but she knows nothing, or at least so she claims. Zanni, disloyal to his master, joins Angelica in carrying out the plot against her father; he, too, is a trickster. Whom should one believe and trust in this world? And especially in the realm of the theatre, where all is illusion, though the effort and hope is to have it seem real . . . yet not too real, for the artisans want their accomplishments recognized and appreciated, the actor his skill at acting, the author his gift for wit and plotting, the designer his sure control of effects, all bidding for applause.

The story is never resolved, and the philosophical questions never answered. The play goes nowhere, ending abruptly. Was this because Bernini lost interest, or did not know where to take it next? Beecher and Ciavolella argue that it was deliberately left incomplete, so that the coterie of élite spectators would withdraw to debate the many points it raises. One "fact" that favours the English translators' contention is that Bernini had forty years more in which to round off his comedy if that was his intention, but he apparently never chose to do so.

Was *The Impresario* ever acted? That is not certain.

It is both a trifle and a manifesto. It does not fit into the category of the *commedia erudita*, nor of the *commedia grava*, nor of the *commedia dell'arte* – it admits no improvisation – nor any other traditional form, though it does bear some resemblance to a lesser seventeenth-century genre dubbed the *commedia ridicolosa* which, unlike the *commedia dell'arte*, transferred its situations, byplay and dialogue to print, hence was also not spontaneous.

(Fragmentary though it is, *The Impresario* is significant and belongs in a history of world theatre because it is one more token that in any period – as in the Renaissance, the baroque – some works of art are created that do not accept or adhere to the dominant aesthetic theory and style of the time but are brought forth in defiance of them by a quester whose spirit is independent – a Blake, Melville, Kafka – and whose imagination is prescient of a new and different style that will be prevalent in the future, perhaps one that is far off. This whimsical sketch is also an opening – a peephole – into the thoughts of an artist of genius and profound eccentricity and originality.)

The theatre described thus far is one largely dedicated to the tastes of the aristocracy, looking backwards and admiring the classical arts of Greece and Rome. But Renaissance Italy also contributed a bawdier, more popular form of entertainment, similarly ancient in its origins yet at a later phase – the sixteenth and seventeenth centuries – unique to its time and place and destined to outlast more serious offerings and occupying stages even today. This is the *commedia dell'arte*, the appellation suggesting at first not that it was "artistic" but merely that its players were thoroughly professional, virtuosos in all the arts and tricks of performance. The literary comedies had been fostered by the academies, by princely and ducal courts, by coteries of intellectuals together with dilettantes, and the participants were mostly amateurs. They worked from written scripts, often by such learned authors as Petrarch, Ariosto, Machiavelli. But the boast of the members of the *commedia dell'arte* was that their productions were improvisatory, relying on no set guidelines. The actors made up their verbal exchanges – supposedly – as the action unfolded, choosing their material and inspiration on the spur of the moment, addressing themselves to topical events and conspicuous personages wherever the troupe stopped for a short stay on its endless travels. Their jests and send-ups might be tinged with satiric malice, arousing a jocular response and gibes from the gathered spectators, who easily identified the local dignitaries being ridiculed and boldly caricatured. The targets of such shafts of derision might be a pompous governor, a prefect or a wealthy gentleman engaged in dalliance with a notorious lady of the town. With only a wisp of a plot – a broad yet tenuous dramatic situation borrowed from the Romans – the actors served up lively and apparently spontaneous repartee night after night, salting it with up-to-the-minute allusions. Acrobatics and feats of contortion, juggling and sleight-of-hand were also part of their stock-in-trade, much as with far earlier performers at festivals in Plautus' Rome and during noisy holidays in the Middle Ages. This was a true people's theatre.

Actually, though there were no bound prompt books, long years of training and practice yielded a repertory of useful plots and gags, simple, universal jokes – reflecting minor vices and quirks of personality, human attributes familiar to everyone – a rich bank from which could be drawn ripostes to every turn in the action or challenge or catcall from an onlooker, however unexpected. The performer had these well in mind before venturing on to the open-air platform. That the play was impromptu was mostly not true, save for a degree of ad libbing. Certain jokes were inexhaustibly serviceable. A taunt or sharp comment about adultery, avarice or plain stupidity could be aimed at an unwary spectator in every place the company visited. In the curtained area below the stage, the space utilized as the actors' dressing-room, was posted a list, a reminder of the titles and perhaps brief scenarios or synopses of the sketches to be presented this day; the players would consult it before assembling on the platform for their seemingly unrehearsed appearance.

Furthermore, the nomadic troupes gradually developed a gallery of stereotypical, and even archetypal, characters, masked and semi-abstract, each one embodying a single, unchanging trait; these characters populated almost every sketch. What individualized them was their always having the same name and consistently having a distinctive, traditional costume, his or hers alone, highly familiar to and easily identified by the most illiterate spectator.

In every company there was usually a middle-aged or elderly Pantaleone (or Pantalon, Pantalone or Pantaloon), a merchant with a Venetian dialect and clad in a constricting red vest under a lengthy red or black coat, above red breeches and stockings to match and soft Turkish slippers. His features were hidden by a brown mask, marked by a large curved nose, giving him a predatory aspect; his beard was grey; and untidy wisps of white hair jutted from under his soft, brimless cap. Forever spouting proverbs, he was tight-fisted and greatly concerned about keeping his purse filled, and also had an acquisitive eye for comely young women.

His frequent companion and foil, the Dottore (Doctor), a lawyer, was also attired in black, his cloak resembling a professor's gown with a white ruff, his academic hat similarly black. His mask was dark-hued, relieved by red cheeks, and sported a short beard. He was pedantic, mixing Latin words and phrases in his pronouncements, incongruously delivered with a Bolognese accent. He was gullible, easily taken in by tricksters. Married to a faithless wife, he was jealous and often cuckolded.

The Captain, a descendant of the Roman *miles gloriosus* and a forerunner of Falstaff, had a Spanish look. A blustering swaggart, but a coward at heart, he also had a mask with a long nose. He had a feather in his hat and flaunted a cape. He had a fierce black moustache and brandished a wooden sword. While not altering essentially, he underwent many surface mutations, and also appeared in the *commedia erudita*, where he might be the rival lover who was bested at the conclusion. In such instances his name did change: he might be called Coccodrillo, Matamoros, Rinocorente or Spavento da Vall' Inferno. He was usually short of funds but put up a bold front, and mistakenly considered himself irresistible to the opposite sex. His discovery that he was in error was often a highpoint of the farce.

Pulchinello (or Pulchinella, Pulcinello, Pulcinella, Punchinello) a Neapolitan type, had varied roles. He could be a servant, a merchant, an innkeeper. He was sage and foolish, kindly and merciless, witty and boring. He was truly ugly: his hooked nose was enormous, his back humped, his face spotted with a large wart. From a long, pointed cap drooped a huge feather, and he evolved decades later into the English figure Punch, of the combative pair of marionettes Punch and Judy.

More sympathetic was mischievous Arlecchino (later Harlequin), a prankster, at first a minor character who became increasingly popular. Adventurous, he was cunning and stupid, and was called upon to be an acrobat, agile dancer and the engendering participant in any intrigue. Quite irresponsible, he was a tease and charmingly seductive. His was a black half-mask, through which his eyes gleamed mockingly and merrily. At first his costume was "beggarly", a collection of patches, but over the years refined and formalized into a checkered design of vari-coloured diamond shapes deemed the logo of the *commedia dell'arte*, as Harlequin became the figure most often portrayed by artists. His scalp, shaved, was covered with a rakish hat. He carried a slapstick – or harmless sword – with which he roundly beat his scampering foes.

His boon companion and abettor in escapades was Brighella (sometimes called Buffetto, Scapino, Flautino or Mezzetino – minor roles less clearly defined; mostly but not always they were servants). In contrast to incorrigible but attractive Arlecchino, his friend and co-conspirator was cynical, lascivious, having a sharp-edged verbal wit. His full mask, too, had a hooked nose and a broad moustache, and his garments – wide trousers and short jacket – were decorated with fancy green braid. The impression he gave was evil and grotesque.

In place of a mask, Pedrolino's face was powdered white, stark above a loose white blouse and flowing white pantaloons. His was another small part that grew into a larger, more appealing one, in France as the melancholy, lovelorn, hapless Pierrot (in Russia, Petrouchka). These male minor characters or servants were known as *zanni* – each troupe developed its own variation of them, and these special related characters might be called Truffaldino, Trivellino or Scaramuccia, the last of these a somewhat amorphous figure on occasion, possibly resembling Arlecchino, or else Brighella or even the fatuous, boastful Captain. This was a decided factor with companies based in separate cities where as a consequence of distance the lesser characters might have slightly altered names and traits, and use local dialects.

Mezzetino, who might double for either Brighella or Scapino, stood out as a singer and dancer. He is portrayed holding a musical instrument. His manner was gentler than that of the rough-styled Brighella. By the end of the seventeenth century Constantini, a French actor, provided him with a new costume – that of a valet – and a new personality to go with it. Formerly he was dressed like Scapino in a loose outfit of green-and-white stripes, together with a short cape, the *tabaro* worn by all male players in servant roles, and thought to be a relic of one that had originally draped the shoulders of slaves in Roman farces. Like the Captain and the *zanni*, as well, he had a handy wooden sword; the playlets were always filled with scufflings, burlesqued duellings and slapstick, kicks and beatings. Constantini changed the green-and-white to brighter red-and-white stripes. In his fresh guise Mezzetino might be the betrayed husband or the persistent adulterer, a loyal servant of a double-crosser, readily bribed. (His name means the "half-measure".)

Scapino, another version of Brighella, was mercurial, unable to control his unremitting sexual urges. In a confrontation he quickly took to his heels. Lacking the ability to think logically, he was in a constant state of confusion and further afflicted by a strangely short memory, though never forgetting to seek a monetary tip. He combined the qualities of Ariel and Harpo Marx, his mind and emotions flitting from subject to subject, object to object. In due time he was to be celebrated by Molière.

On the distaff side, the most established figure was Columbine, shallow, flirtatious, heartless, a fit partner for the fey Arlecchino; like him, she wore only a half-mask. Often she was the cunning, scheming maidservant. Pasquella, the old woman, had an ugly full mask, indicating her evil nature. There might also be Harlequina and Pierretta, female counterparts to those rascally *zanni*. It was customary for an actor to play the identical role – or roles – throughout his whole career, until it became second nature. How many of these stock characters were included in a troupe's offerings varied with its financial resources.

The young lovers, the *innamorati*, usually appeared without masks. They were good-looking, well educated and articulate, and by behaving in a normal manner presented an effective contrast to the bizarre and eccentric figures surrounding them. They might be cast as the son or daughter of the greedy Pantaleone or the spuriously learned Dottore, the dull-witted parent who for several acts frustrated the couple's natural desire to mate, though of course the ending was joyous. These young players were instructed to read poetry and the texts of published love-letters, so that their speech would be more heartfelt and lyrical at intimate, sentimental moments. They used their own names, and their apparel was not traditional but of the fashion of their day, as expensively ornamented as the company's coffers could afford.

(Here let it be said that much of this detail is from Pierre Louis Duchartre's rich and exhaustive *The Italian Comedy*, as well as from general histories already cited.)

At a loss for words, the player could resort to stage-tricks, falling back on long-mastered "business". These were known as *lazzi*. For example, Arlecchino might pirouette, or mime catching an annoying fly and swallowing it with a grimace and a gulp, a sure-fire bid for a laugh. He could suddenly insert a cartwheel, upend himself and walk on his hands, or do a "split". At eighty-three, as Scaramuccia (Scaramouche), the famed Fiorilli was still nimble enough to bang his toe against his own ear or the ear of an opponent. Another noted comedian, Tomasso Antonio Visentini, as Trivellino, his hand grasping a glass of water, could somersault without spilling a single drop. The agility to perform such physical feats was a requisite.

Full-length robes or capes were assigned to nobles and elderly men. Their hair was hidden by head-bands or false scalps, so that they seemed to have shaven crowns, as had the ancient mimes, and some of the lower-caste figures sported stuffed phalluses, as had the buffoons in Greek Old Comedy.

The Comedians' acting style has been described by many contemporary admirers. The seventeenth-century player Gherardi, who himself often essayed the antic, mindless role of Arlecchino, gave this personal account and appraisal of their methods (quoted by Duchartre, translated by R.T. Weaver):

The Italian comedians learn nothing by heart; they need but to glance at a play a moment or two before going upon the stage. It is this very ability to play at a moment's notice which makes a good Italian actor so difficult to replace. Anyone can learn a part and recite it on the stage, but something else is required for Italian comedy. For a good Italian actor is a man of infinite resources and resourcefulness, a man who plays more from imagination than from memory; he matches his words and actions so perfectly with those of his colleague on the stage that he enters instantly into whatever acting and movements are required of him in such a manner as to give the impression that all they do has been prearranged.

Still another author-actor in this genre, Riccoboni, a century later, testifies: "Impromptu comedy throws the whole weight of the performance on the acting, with the result that the same scenario may be treated in various ways and seem to be a different play each time. The actor who improvises plays in a much livelier and more natural manner than one who learns his role by heart. People feel better, and therefore say better, what they invent than what they borrow from others with the aid of memory." But Riccoboni acknowledges that such improvisation on stage also has its drawbacks: not all actors are equally talented, and one who is apt and quick at quips might find himself paired with another who cannot hold up his end as cleverly. Then "his own discourse falters and the liveliness of his wit is extinguished".

The major eighteenth-century critic and playwright Count Carlo Gozzi pays tribute in his *Mémoires* to the intelligence, dexterity and versatility of the diligent performers:

> These plays are never withdrawn on account of illness among the actors or because of newly recruited talent. An impromptu parley before going on the stage, as regards both the plot and the way in which it is to be played, is sufficient to insure a smooth performance. It often happens that in special circumstances, or because of the relative importance and skill of certain actors, a change in the distribution of roles is made on the spur of the moment just as the curtain is rising. Yet the comedy is borne along to a gay and sprightly conclusion. It is apparent that these actors penetrate to the very core of their subjects, establishing their scenes on different bases with so many varieties of dialogue that, with each performance, the interpretation seems to be quite new, yet inevitable and permanent.

The Comedians were almost the only Italian Renaissance actors who for the most part still used masks. These devices were of thin leather, moistened and subtly moulded, and had an astonishing vitality and universality; they helped the actor greatly in finding and fixing his character, assuring that it would be long-lasting; the mask truly transforming the person behind it.

Every leading actor had his personal, idiomatic assortment of appropriate movements, gestures and poses: one Arlecchino might bound about the stage exuberantly, indulge in frequent pratfalls or flail impulsively at his fellows with an inflated bladder, or slap them, and another might mince gracefully; a Pulchinello might display his cupidity by over-eating and over-drinking at someone else's table, until his excesses make him ill, or he might serve his guests the most meagre fare and watch anxiously lest they consume too much or ask for more; the bombastic Captain, bellowing as he recounts his past valorous deeds, might at times still have something of an aristocratic aspect, being handsomely accoutred, but in other interpretations appear to be merely a penurious, out-of-work mercenary, at his wits' end and quite bedraggled. Inevitably, one company would steal ideas from its rivals.

As was true of the Japanese Kabuki, the *commedia dell'arte* troupes were most likely to be constituted of family members, one generation teaching the next the fine and even secret details of their craft and traditions. Like medieval mountebanks they travelled by cart from town to town, choosing to arrive – if possible – on the date of a fair or festival, much as had their vagabond predecessors. As had long been the practice, a platform was set up at eye level for standing audience. The townspeople, learning of the Comedians' arrival, flocked to the piazza where bright entertainment was to be presented, a change from the arduous routine of the day. Most of the spectators stood in a semi-circle; a few were seated on benches they brought with them or else rented.

The stage was mostly bare, though in later years some very astonishing effects were introduced: white marble statues sneezed and surprisingly came to life: a temptingly laden banquet table was whisked away just as Mezzetino and Pulcinello were about to treat themselves to a feast. A painted backcloth suggesting a street, square or Arcadian forest would suffice, but if elaborate scenery was available it was welcomed and exploited. On many occasions the performance was enhanced by a fireworks display. Trained animal acts might also be added, such as monkeys, bears, rare and exotic birds.

The humour was vulgar, mostly addressed to a peasant crowd. Every kind of physical assault was demonstrated, the players for ever pummelling one another. Garbage and excrement might be hurled in a coarse fashion reminiscent of the medieval street orgies that often ended the Feast of Fools. A lute-player might accompany the more violent action, helping to stylize it and making it semi-real as in a silent film or animated cartoon. Despite all the slappings, kickings and beatings, no character was ever seriously hurt or felt more than a moment of pain, which might be evinced by a brief cry of protest or outrage, nothing more. Such primitive knock-about behaviour was essentially cathartic for those spectators repressing angry emotions. The impulse to attack in retaliation to an annoyance was covered by laughter and hence made socially acceptable. The punishments inflicted by Arlecchino on his rowdy companions were justified.

The average troupe consisted of seven men, three women. Their lives were hard and insecure, as has always been true of touring actors. They might give offence, be deemed impertinent or scandalous by local or court officials, receive short shrift or be driven out of town. In rural areas they were suspected – with good reason – of immorality and greeted with hostility. The actresses had poor reputations, were looked upon as prostitutes, the men as their procurers. Some companies required their members to sign contracts requiring them to go to confession, refrain from profanity, avoid "moral turpitude".

To this condition there were prominent exceptions, groups that were large, successful, even prosperous. One of the first was the I Comici Gelosi (the "Anxious to Please" troupe), who won prestige and a faithful audience for the *commedia dell'arte*. It was formed in 1552 by performers who remained in Italy when a preceding company, that of Alberto Ganassa, based in Mantua, left for a tour of Spain. The new group was directed by Flaminio Scala (also called Flavio) who created many of its scenarios, mere outlines of the action, leaving the actors to provide spoken lines. Of noble birth, and a very cultured man, he was a highly gifted interpreter of a variety of stage roles. I Comici Gelosi appeared frequently in Venice, Florence, Genoa, Milan, Mantua, and the troupe was twice invited to Paris by Henri III, whose mother enjoyed their performances. However, both times, despite having royal patronage, they made hasty departures from that capital after harassment by rival French troupes and sharp denunciations and threats of fines imposed by critics in the Parliament, where they were charged with offering impious, indecent fare. The court was accused of an inclination towards Italian decadence, the queen having come from Florence. While in France, too, the troupe was kidnapped by the Huguenots and held for ransom, paid by the king. Earlier they had been more welcome in Vienna, favoured there by the emperor. In 1577 they performed in England for Queen Elizabeth. These journeys meant that the actors, whose speech was rapid, had to be fluent in several languages, as well as in local dialects.

About 1600 Scala shifted his allegiance elsewhere and was succeeded as director and leading man by Francesco Andreini (1548–1624), who often took the role of the Lover and, garnering broad acclaim, did much to embellish the character of the Captain, adding further quirks and noisy idiosyncrasies to the part, in a sketch titled *Bravure del Capitano Spavente*. He was able to enrich this portrait of a blustering mercenary by drawing on personal experience and observation. Earlier he himself had served as a soldier on a Tuscan war galley. Captured by the Turks, he was imprisoned for seven years and forced into slavery, finally escaping back to Italy where, master of five languages and every kind of musical instrument, he found in theatre his true *métier*. He and the troupe gained further renown when he married the sixteen-year-old Isabella Canali (1562–1604), whose extraordinary beauty, intelligence and talent were the subject of songs and toasts by princes and commoners wherever the company appeared throughout Europe. The happy pair had seven sons, which further enlarged the troupe. He was a poet and belonged to a literary society.

The most popular of I Comici Gelosi's many splendid presentations was *La Princesse qui a perdu l'esprit*, introducing many novel mechanical devices and featuring a grand sea battle on stage.

The company's emblem was Janus, the two-faced Roman god, and its inscribed motto "*Virtù, fama ed onor ne ser gelosi*", or "They were jealous of attaining virtue, fame and honour". But neither the household deity nor the quest for virtue and honour spared them further trouble. In Mantua short-tempered Duke Guglielmo ordered them to enact a piece of which he was especially fond, in which all the characters were humpbacked. Approving of their rendering of the work, he bade the author and some of the players to approach him. When the entire cast pushed forward, arguing among themselves who should be rewarded, he lost patience and proclaimed that they should all be incarcerated and put to death. They were tortured, but eventually, on second thought, spared by the duke. But he warned that he wanted "works composed by good performers, not by this band of rascals". Escaping the gallows, they were banned from Mantuan territory.

Before Guglielmo's death, however, they were summoned back by his son and successor, Vincenzo, who had a compulsive passion for theatre. He aspired to form a group bringing together the best actors of three competing companies, all resident in his city: the Gelosi, the Confidenti and the Accesi. This led to contradictory petitions to the duke, and much disputation, spurred by personal conflicts, and Vincenzo's effort failed – his father, annoyed, took away his licence to bring actors to the court. Subsequently, when the younger man's reign began, he set up the Duke of Mantua's Servants, of which Isabella Andreini, of the Gelosi, and the temperamental Vittoria, *prima donna* and director of the Confidenti, were members.

Many roles were fashioned to fit the personality, talents and charm of Isabella, whose beauty was admired not only by men but also by imperious Queen Marie of France, later Regent. The Medicean princess was married to Henri IV, an alliance of considerable historical consequence. At age forty, on a subsequent visit to Lyons, Isabella suffered a miscarriage and died. The whole city mourned. Francesco, grieved by her loss, quit theatre work altogether. I Comici Gelosi was disbanded. Strangely, portraits supposedly of Isabella show her as scarcely attractive but instead heavy-featured, with a "bovine expression". This may be due to misattributions or the artists' shortcomings. Of her it was said that she was "as celebrated for her virtue as for her beauty, and she brought such glory to the profession of acting that her name will be held in veneration as long as the world shall endure and unto the end of the ages". Ariosto, Aretino, Giraldi, Guarini and Marino wrote sonnets and other rhapsodic verses in her praise. The great Tasso's hyperbolic lines: "When fostering Mother Nature fashioned the fair veil of her physical graces she sought out beauty and gathered as a flower, taking jewels out of the earth and stars from the heavens." And concluding: "Happy those souls and blessed those hearts in which Love has been stamped in letters of gold after your image." So lasting was the impression she made that the character of the Inamorata, the young, idealized beloved, was hereafter most likely to be named Isabella by an author.

She spoke Latin perfectly, was unquestionably cultivated, was a member of several academies, and wrote poetry. In retirement, Franceso devoted himself to collecting and publishing her dispersed writings.

Though Francesco withdrew from the comedy stage, his children did not. One son, Giovanni Battista Andreini (familiarly known as Lelio (1578–1654), later associated with Comici Confidenti, I Accesi and I Fedeli), inherited his parents' histrionic gifts and was the author of eighteen plays deemed obscene, along with a series of twenty-three sonnets about actors who, having undergone religious conversions, suffered martyrdom for their newly acquired beliefs. The salacious content of his plays did not lessen their popularity – on the contrary. His usual parts were as Arlecchino, Lelio and the Lover.

Alberto Ganassa's troupe, many of whose members were absorbed by I Comici Gelosi when he left for Spain, also visited Paris in 1571. A year earlier it had participated in the festivities in Ferrara attending the marriage of Lucrezia d'Este. In France the company also met obstacles, the Parliament forbidding it – and others – to perform, asserting that entrance fees were too high. Tickets were 5 and 6 sols, "an excessive sum never before levied for such purpose, and an imposition upon the poor". The king sought to intervene but could not prevail over Parliament, which inopportunely recessed. The troupe did appear during a celebration at court of a wedding between the King of Navarre and Marguerite de Valois in 1572, for which they were well paid.

Several historians believe that Ganassa was the first Arlecchino – Duchartre says that the name derives from the French *herlequin*, which in the Middle Ages connoted an "airy sprite, a will-o'-the-wisp, a dramatic character". From France the name went to Italy and changed its spelling. The character was one that Ganassa preferred to enact, and in any event he is the earliest comedian known to have done so. He is also credited with having invented the role of Baron de Guenesche, whose patter combined a mixture of Spanish and the unusual dialect of Bergamo.

Ganassa's company, which had originated in Mantua, remained a good length of time in Spain, given sufficient free rein by Philip II, at whose court it was installed until 1577. The foreign travels of troupes like I Comici Gelosi and Ganassa led to the *commedia dell'arte* becoming a familiar and widely hailed genre outside Italy, eventually becoming more popular abroad than at home.

I Accesi, a group formed and sponsored by the Duke of Mantua, numbered among its players not only Giovanni Battista Andreini but also Virginia Andreini and Flaminio Scala, formerly of I Gelosi. Its leader and star was Tristano Martinelli, who was its Arlecchino. After the company played in Paris and Lyons (1599–1601) Henri IV wrote to his cousin Ferdinand de Gonzaga asking if he might persuade the duke to allow a return tour to France. The duke refused the request, partly for political reasons, but a decade later (1610) relented. While the company was at Henri's court the king and queen consented to be godparents to one of Martinelli's sons, and more stays in France ensued until 1621. As with the other companies appearing in Paris, I Accesi was at the court in the Louvre but also at Fontainebleau and the Hôtel de Bourgogne. From 1623 on Martinelli was with I Comici Fedeli in Venice, where he died in 1630.

I Fedeli (the Faithful Ones) was a troupe assembled by Giovanni Battista Andreini shortly after the death of Isabella and the withdrawal of Francesco in 1605. It enlisted many from the dispersed Gelosi and initially served to display Giovanni Battista's considerable performing skills and enabled him to have his own plays staged. At intervals he left the group – the comedians seemed to have been a restless lot – and it underwent several reorganizations, but never ceased to prosper and continued active for a half-century, until 1651. I Comici Fedeli was invited to assist in the festivities of royal and ducal marriages over the years, journeyed throughout Italy and was on view in Paris, sometimes for extended periods, on four occasions (1613, 1623, 1624, 1625). Like many of the other companies it included serious works in its repertoire. The five or six such pieces by Giovanni Battista Andreini had a limited appeal because the dialogue tended to be a baffling mixture of "Venetian, Bergamask, and Genoese dialects, interspersed with Castillian, French, and German".

From 1615 onwards Andreini shared the directorship with Nicolo Barbieri, who was otherwise identified with the role of Beltrame of Milan. The Inamorata (Florinda) was G.B. Andreini's first wife, and following her death the portrayal fell to his second wife, Lidia, whom he wed when he was seventy.

Revealing is a letter from Marie de' Medici to Giovanni Battista bidding him to hurry to France. She addresses him as "Harlequin", his customary assignment.

> I see from the letter which you wrote me anent the company of comedians that they have finally come to a decision to begin the journey hither, albeit they have waited a long while, and

I had almost lost hope of seeing them. They will, however, be well received, and each and all, I trust, will enjoy the journey. This will suffice to serve you as a Passport. . . . Hasten then, as quickly as you may upon this assurance and dispose yourselves to maintain the high reputation of Harlequin and his troupe, together with the other good rôles which you have recently added to it. The King, my son, and I await the pleasure and diversion that you always provide.

Subsequently, in a second letter from Fontainebleau (1613), she holds out a promise that at the birth of a son – Giovanni Battista's wife was pregnant – the infant will receive a golden chain as a baptismal gift. And ends: "The sooner you set forth, the greater will be your welcome. Come, then, with all speed . . ." (from Duchartre, translated by Weaver).

In yet another reappearance, Flaminio Scala was the director of I Comici Confidenti (the Confident Ones), a group that toured France in 1571, after thoroughly covering the large Italian cities. Besides entertaining French audiences in several provincial towns, they were on show at the Hôtel de Cluny in Paris, with programmes of pastorals, tragedies and *commedie sostenute*, but local rivals finally had them expelled. They won the patronage of the Medici ruler to whom the Duke of Mantua sent a letter commending them. I Confidenti was one of several groups that later were headed by a woman, in this instance Vittoria, using just her first name. Thus it came to be referred to as the Company of Signora Vittoria. This was the singing actress with whom the Gelosi had refused to collaborate when Duke Vincenzo of Mantua had tried to merge Italy's three most brilliant troupes in 1580. Indeed, Vittoria invariably took on the same role, the Inamorata, that Isabella Andreini graced, and the two vied for an equally fervid measure of public adoration. When both companies were at the beck of the Mantuan court at the same time, and the two women competed at alternating performances, drawing comparisons between them was considered to be a rare, exciting privilege. Just as "Isabella" came to be affixed as the perpetual and conventional name of the Inamorata, so quite frequently was that of "Vittoria" chosen to identify the leading female character.

These groups were acknowledged as the most brilliant. Among others of lesser renown, but not unimportant, was I Comici Uniti, playing in Genoa and Padua and other cities. Its roster during the second half of the sixteenth century included the highly rated Silvio Fiorillo (*c*. 1560–1630) and his gifted son Giovanni Battista Fiorillo (1614–51). The father was the original Capitano Matamoros (one of the variants of that stock figure), and from some historians he gets credit for having been the first Pulcinello (Naples, 1609), reincarnating an early Roman precursor, again a character of whom there were endless variants, some quite subtle, some very broad. This process of change went on without cease as the companies large and small proliferated and took on local and regional tinges, and individual actors put their stamp on a role. Silvio was the author of a number of comic sketches as well. His son was valued for his Trappolino, a "second *zanni*" part, and also for his Scaramuccia. His wife, Beatrice (d. 1654), was a pleasing Inamorata.

There was also, in the sixteenth century, I Desiosi, which after 1582 was led by Diana Ponti, a prized actress once with I Confidenti. Its patron was Cardinal Montalto. A poetess, Ponti was best known for her Lavinia, an Inamorata role. Hers was another instance of a family perpetuating a link to the stage. Her father, Adriano Valerini, directed a troupe based in Milan (1585). Of noble stock in Verona, he was no mere montebank but a Doctor of Philosophy and wrote poetry in Greek, Latin and the vernacular. It becomes ever more evident that, paradoxically, many of these acclaimed Comedians were possessed of really solid learning and culture.

In a list of seventeenth-century companies is that of Giuseppe Bianchi, which established itself in Paris in 1639, as will be described hereafter. Another was the Troupe of the Duke Modena (1675). All these groups receiving royal applause and generous recompense were less likely to have to undergo the hardships of the many humbler, smaller ones: exhausting travel in their laden carts over rough roads, the hazard of bad weather, brigands and perhaps open local hostility prompted by the Church.

Much of what is known about the *commedia dell'arte* of the Italian Renaissance comes from images captured in paintings and drawings by great artists of the age in both its homeland and France. They were attracted to the players as subjects because of the graphic masks and colourful attire, the general grotesquerie, the exaggerated gestures, the angular kinetic poses, the Carnival setting in which they performed. Duchartre's comparatively short book contains no fewer than 259 illustrations, some of them copies of the canvases, engravings and pages from the sketchbooks of Domenico Tiepolo and Salvator Rosa, and later ones of Claude Gillot, Jean Antoine Watteau and Jacomo Callot, among a legion of others. Precise details of dress and postures are revealed, and sometimes a specific performer's facial features are impressionistically suggested.

A cornucopia of information is found in a handbook of *lazzi* to which an actor could always look for prompting. The literal meaning of the word "*lazzi*" is "knots", but by the Comedians the term was used to indicate a tie-up, a brief interruption in the onward flow of the action, when the player, improvising but at a loss for words, covered the awkward moment by a sudden verbal inconsequence or absurd physical distraction, perhaps a bit of pointless acrobatics. The *lazzi* were also inserted to relieve tension when the plot threatened to become too serious: they served always to keep the sketch farcical. Before going on stage Arlecchino might fill his mouth with water and then spout it over his antagonist at their first encounter to get a laugh, to establish the right mood from the start for what was to follow. All sorts of amusing, clownish stage business was conceived. A striking "entrance" was important, and it might lead to having Arlecchino dressed as a winged Mercury being lowered from the flies astride a cardboard eagle, or initially arriving in Neptune's moss-draped chariot with a clatter, or in a Venetian gondola, or riding a balky donkey. Surprises had to be frequent. The Inamorata, solitary in her garden, is surrounded by white statues. Then one of them sneezes, and all the "statues" come to life, jump down from their pedestals, and prove to be co-actors transformed by thick white make-up. Or Pascariel is brought on concealed in a sack. Arlecchino argues with Scaramucci about what is in it: coal or a bale of cloth. They open the sack and out pops Pascariel disguised as a three-headed Devil, who puts the inquisitive pair into a squeaking panic.

An even fuller source of knowledge about the style and content of the offerings is a collection of scenarios put together and later published (1611) by Flaminio Scala, who was active in several companies by turns. Of the fifty sketches – they are mostly just short synopses, with no dialogue – one is a tragedy, nine are elaborate fantasies, forty are three-act farces. Each of the farces contains a love-intrigue; all are rapidly paced. Sentimentality prevails in some episodes, alternating with passages of romantic feeling and strong expressions of passion, along with unending comic slapstick. These texts give a perfect picture of the repertory of I Comici Gelosi, the most admired of all the troupes. However brief, the scenarios are less concise than the synopses posted backstage at performances, which Carlo Gozzi says were written on small slips of paper and enabled ten or twelve actors to "keep the public in a gale of laughter for three hours or more and bring to a satisfactory close the argument set for them".

It helped that many in the cast were members of the same family with a long tradition of being in the theatre, or had worked together for many years in the same or various troupes. It was not uncommon for a son or daughter to inherit the roles that had habitually been enacted by a parent or grandparent, the player in the younger generation even making use of the identical tricks.

Some plays did not pretend to be wholly improvised; certain passages were written out and required memorization. In a *comédie mixte*, as such offerings were designated, the prepared speeches would occur at a particular spot in bouts of spur-of-the-moment comedy. They might be a monologue, a soliloquy or an equivalent to an "aria". It would be Pantalone's exasperated tirade against a tormentor or the Doctor's show-off recital of legal or medical terms to impress a guileless hearer. In the *Chevalier du soleil* (1680), based on an earlier Bolognese piece, the Physician is explaining his motive and the scope of his special art to the Doctor (here a lawyer):

> Yes, sir, I practise medicine out of pure love for it. I nurse, I purge, I sound, I operate, I saw, I cup, I snip, I slash, I split, I break, I extract, and tear and cut, and dislocate, I dissect, and trim, and slice, and, of course, I show no quarter.

The Doctor exclaims: "You're a veritable avalanche of medicine." The Physician emphatically affirms this:

> I am not only an avalanche of medicine, but the bane of all maladies whatsoever.
> I exterminate all fevers and chills, the itch, gravel, measles, the plague, ringworm,

gout, erysipelas, apoplexy, rheumatism, pleurisy, catarrh, both wind-colic and ordinary colic, without counting those serious and light illnesses which bear the same name. In short, I wage such cruel and relentless warfare against all forms of illness that when I see a disorder becoming ineradicable in a patient I even go so far as to kill the patient in order to relieve him of his disorder.

The Doctor agrees: "That is an excellent cure." And so on. The Physician is by no means finished with his verbose self-description, which rambles on for perhaps too many more lines.

In another play Arlecchino offers advice to the Captain, who suffers from a toothache:

Take a pinch of pepper, some garlic, and vinegar, and rub it into your butt, and you'll forget your pain in no time.

As the Captain is about to depart, Arlecchino adds:

Wait a moment! I know a better remedy than that: Take an apple, cut it into four equal parts; put one of the pieces in your mouth, and hold your head in an oven until the apple is baked. I'll answer for it if that won't cure your toothache.

In one of his set speeches he accounts for the origin of his name, though this is fanciful rather than accurate:

My name is Arlecchino Sbrufadelli. [*Cinthio bursts out laughing.*] Don't make fun of me; my ancestors were people of consequence. The first Sbrufadel was a pork-butcher by profession, but so eminent that Nero refused to eat any other sausages than those he furnished. Sbrufadel sired Fregocola, a great captain. He married a woman of so lively a temperament that she bore me two days after the wedding. My father was delighted, but his joy was short-lived because of certain complaints lodged against him by the minions of the law.

Unhappy is Arlecchino. When Ottavio asks him, "How many fathers have you?" he replies: "I have only one." "But why have you only one father?" "Well, I'm a poor man, and can't afford any more."

In another long speech and exchange Arlecchino elaborates on his trip to the moon, giving an extravagant account of what he saw and did there. But mostly the scenario would be as scanty and spare as this:

[*Pulchinella enters, dripping wet from the ocean; he tells of the shipwreck in a storm that caused the loss of his master and companions. From the other side comes Coviello, who relates the same misadventure; they behold each other and make the* lazzi *of fear. Next they do the* lazzi *of touching each other, but then realize that they are safe and loudly lament the loss of their master and their friends.*]

[Excerpts from Duchartre, translation Weaver]

After quickly perusing the notice, the actors were on their own – all the dialogue was to be ad libbed – a term for extemporizing derived from the Latin *ad libitum*.

Since a majority of the actors wore masks, much depended on vocal inflection, eloquent mime and illustratory gesture. Some of the indications of emotion were codified. There were standard ways of expressing rage, jealousy, sorrow, joy. Once learned and mastered, these were oft-repeated, easing the actor's burden.

Much of the humour was exceedingly coarse. Apart from the slapping and kicking, there was simulation of dousing characters with the contents of chamber pots.

A study of Flaminio Scala's forty synopses and other extant scenarios reveals adultery as the chief subject of the most zestful farces. Old men are held up to contempt as amorous, impotent fools, bordering on senility and easily duped. Married ladies cherish their virtue very lightly. The kindliness found in the plays of Plautus and Terence is absent, though young love is romanticized, even when the hero and heroine are a bit dull. Servants – such as Arlecchino – are more stupid than clever, causing confusion by failing to comprehend their instructions. The changeling theme, of which the Greeks and Romans were so fond – high-born, long-lost children reunited with their parents – almost vanishes. Social satire runs through the sketch-ideas, externalized in the rough "business" and prompting open salaciousness. One reason the Captain is Spanish is that Spain was then militarily supreme in Europe and heartily disliked, hence the demeaning caricature. On the whole the humour tends to be cruel, truly malicious, betraying the sinister side of daily life in Renaissance Italy.

The plays abound with cross-dressing. Scaramuccia is attired as an aristocratic girl, which allows him to flounce about and flutter. Or: In a street, Mezzetino is approached by a courtesan who asks for direction to the Place de Grève, the site of public executions. With a courteous bow, Mezzetino responds: "You have only to go on as you have started, Mademoiselle, and you're bound to get there." But the "courtesan" is actually Arlecchino in drag, and the encounter leads to fists clenched and a mirthful exchange of blows. Again: Pascariel kicks a pregnant woman in the belly, but no real harm is done because the "pregnant woman" proves to be none other than the facetious Arlecchino. Or vice versa. In one episode Arlecchino appears half in masculine apparel, half in feminine dress, and scurries about, mostly back and forth. He presides over a lemon-stall and an adjacent linen-draper's shop, where he sells layettes "for the children of the eunuchs of the Grand Harem". Pascariel is a customer but can never make out whether he is dealing with the woman vendor in the linen shop or the man dispensing the lemonade, as Arlecchino constantly presents a different face and aspect. Which shop does he own, to which sex does he belong? Then again: Arlecchino comes on as a nurse, cradling a child in his arms. He dashes from one male character to another, demanding to know who is the father, and describing the woeful state of "the poor little brat": "He runs up trustingly to every ass, pig, or ox he sees, thinking each time it is his dear dad." So bereft is the child, declares the nurse, "he'd never consent to suckle unless I rubbed my teats with wine". Pascariel, accused of being the guilty sire, kicks the nurse in the bottom, whereupon "she" cries: ". . . and me big with child, already fourteen months gone!"

The jokes are not new – as already stated, they are timeless. Indeed, many facets of the *commedia dell'arte* are immeasurably ancient. Prototypes of its characters are hinted at in the masks and figurines of singing and dancing celebrants in the vernal Dionysian fertility rites and of the actors in Greek satyr plays. Aristotle tells of Susarion's travelling players who, despite having to pass through warring armies, went from hamlet to hamlet in Icaria to perform in the suburbs of Athens and Sparta. At the great drama contests, ancestors of the Comedians were garbed fantastically in Aristophanes' mocking, wild satires. Related to them are the Sicilian and Atellan clowns, of whom there are dim images on mouldering walls of Etruscan ruins. Similar, too, are the amusing, shrewdly observed stock figures giving life to the New Comedies of Menander. Much the same kind of entertainment was offered by the *funambuli* of Rome, the acrobats, tumblers, tightrope-walkers, fire-eaters, who displayed their prowess in the annual *ludi*. There are portraits of comic actors in wall-paintings preserved in the ashes of houses in ill-fated Herculaneum. Story ingredients are ceaselessly borrowed, from ever-yielding Plautus and Terence. The pageant-trains of the Middle Ages, attended by prancing devils and other knock-abouts, entertained crowds, and the same sorts of buffoons and maskers in disguise mingled among the riotous throngs at Carnivals in Venice and other Italian cities. The *commedia dell'arte* inherited elements and ideas from all these predecessors.

Taking form as an integrated, unique genre in Italy about 1550, the *commedia* flourished there for about 200 years. Some of the earliest troupes were those of Anton Maria, a Venetian, and Soldini, a Florentine. Travelling abroad, Maria's company of nine appeared at the courts of Charles IX and Henri III in France from 1572 to 1578, about the same time as Soldini's ensemble of eleven, as attested by records of payments to both "in consideration of the plays and tumbling which they performed daily before his Majesty". The troupes were also seen in Paris and Blois by royal command. Afterwards Soldini went to Vienna. Another comedian on view in Austria was Giovanni Taborino who began his tour in Linz (1568), made a sortie to Paris (1571), then had a more important engagement in Vienna, where he was named "comedian to his Majesty", a post he is believed to have held until about 1574.

As mentioned before, a strong influence on the shaping and personality of the *commedia dell'arte* was Angelo Beolco, the actor and playwright of Padua who in the early decades of the sixteenth century, about 1521, having organized his own nomadic troupe, gradually created Ruzzante as a leading role for himself, a character that the Comedians quickly adapted for their own use under other names. As previously described in this chapter, the protean Beolco was by a magical osmosis almost wholly absorbed into the part, so popular as a stage figure that he was widely known and hailed not by his true name but everywhere as Ruzzante. Taking note of the rascal Ruzzante's success, they did not copy him exactly but did borrow, in many instances, his crude peasant behaviour and speech-pattern, his earthy vitality, his uninhibited lustfulness, his intermittent shrewdness and stupidity, his dishonesty, the many qualities that made him real and contemporary. He was not a fanciful figure from a remote Hellenic past, or a symbolic or semi-abstract shepherd or demi-god in a pastoral, but someone immediately recognizable with whom many spectators could readily identify. From his example the Comedians were stimulated to instil a modern naturalism in their offerings, significantly enriching their appeal.

Jackson T. Cope, in his *Dramaturgy of the Daemonic*, discounts the claim by Luigi Riccoboni, an actor and author of the first history of the Italian theatre, that Beolco introduced masked personages to his country's stage. "This is an historical shortcut to the birth of the *commedia dell'arte*. No one invented the *commedia*; no one introduced masks. Ruzzante had no role in the earliest *commedia* troupes. Yet Riccoboni was right. It is not that the *commedia* would not have existed without Angelo Beolco/Ruzzante. . . . It is that societies inevitably seize upon some men as epitomes of inevitable events. Beolco was one of those men." His special contributions, however, were his injection of a local dialect and strong hints of social protest that echoed many of the spectators' scarcely voiced feelings. The adoption of dialect – rustic, urban – enhanced communications between actors and audience. He also employed dialect to parody the pastoral and *commedia erudita*, while helping to make folk material more acceptable. Of course, much depended on its authenticity; but of the accuracy of his eye and ear there could be no question, since he based his portraits on the peasants who worked on the family estate. For the most part they are completely amoral. The spirit of all this was transplanted into the aborning *commedia dell'arte*.

This nimble, robust, colourful genre of theatre, having enjoyed two centuries of favour and prosperity, began to degenerate and wane about 1750, after which it almost disappeared in Italy. It had become too scurrilous and lewd, too outspoken, too coarse, provoking Church censorship and civic interdiction, and it is also possible that it had grown too repetitive. Yet it kept thriving because almost from the start it was also much cherished abroad, especially in France and England, where troupes constantly visited, applauded not only by rulers but by eager populaces as well. Comedians were greeted with hearty laughter in Germany and Austria, too, and delighted the Tsar of Russia, who for a time maintained a small, lively ensemble at his court. Far from dying, this form of theatre with its unique idioms and traditions lingered on, attaining immortality far beyond Italy's borders.

In each of the foreign countries in which it established itself it was assimilated differently, taking on new coloration and often a fresh character. In Germany, Hanswurst (Jack Sausage) was a Teutonic version of Pulchinello; while in England, in the eighteenth century, the *commedia dell'arte* suddenly "lost its voice" and became quite wordless, a theatre of pure pantomime. In fact the Italian companies, playing in foreign countries with unfamiliar languages – and regional argot and dialects – had always depended extensively on gesture.

In France this new vogue of mute expression and communication was also gradually accepted, and a special form of entertainment came into being – new, yet ancient, for the Roman mimes had long ago perfected it. Here the *commedia dell'arte* was, if anything, over-refined, over-prettified. The evolution of the energetic scamp Pedrolino into grieving, anaemic Pierrot is one instance of this. Gallicized, idealized, this romanticized aspect of the French *commedia* is charmingly represented – as said before – in the engravings of Jacomo Callot and the illuminating drawings and Rococo paintings of Antoine Watteau.

Italy had seen a popular marionette-theatre develop along with the *commedia*. Both theatres borrowed from each other in the use of masks, plots and jokes. The puppet shows survived the *commedia*'s passing. Oddly, too, Pantalone's Venetian costume left its mark in a type of trouser – the pantaloon – that became widely fashionable at periods for both men and women.

The Italian Comedians, as they were known, had a long reign in Paris, where they eventually comprised Molière's principal competition. Often they occupied the same theatre as his troupe for alternate stays. Watching them, he learned much, incorporating it in his own acting style and the pacing of the scripts he wrote for his company. Both groups were recipients of royal subsidies, but were frowned on by the Church as too free-speaking and indecent; and, at death, actors were denied interment in consecrated ground.

Increasingly, the traditionally costumed figures and *esprit* of the *commedia* characters were to be sighted in all the performing arts – ballet, drama, opera – especially Arlecchino (now more often known as Harlequin outside Italy), Columbine, Pierrot. They became, in a sense, iconic. Though this unique theatre left scarcely any written texts, it proved to be far from ephemeral; its influence is paradoxically broad and pervasive. Literary geniuses such as Lope de Vega, Shakespeare, Marivaux and Goldoni were to be its chief heirs, but hundreds of lesser writers have been inspired by its gentle, audacious humour as well as its sheer artificiality. It is now at a remove from life, an escape from reality, though it offers metaphors of it.

The nineteenth century in Paris brought along a series of distinguished, ineffable mimes seeking to replicate and even improve upon the legendary art of the long-cherished Comedians. The most acclaimed of them was Jean-Baptiste Gaspard Deburau (1796–1846), born in Bohemia, a child of touring acrobats. In 1816 he joined the "noisome" Théâtre des Funambules on the Boulevard du Temple, an after-dark promenade of ill repute. Asthmatic, he spent several years behind the scenes as a stagehand exposed to dust. Allowed to appear before the public, he developed what was to be the ultimately prevailing interpretation of Pierrot, the mournful, unrequited lover, finally quite distant from the original figure. This was an innocent. "His white costume without collar, the loose blouse, and the black skull-cap made him as much like Pagliaccio as Pierrot, and unlike either, except that he wore a white make-up as they did." In the part, he was "pale as the moon, mysterious as silence, supple and mute as a serpent, straight and tall as the gallows". Gradually he won popular notice and brought lustre to the seedy "flea-pit" Théâtre des Funambules, before now mostly devoted to circus acts. He was so successful, indeed, that "Pierrot plays" like his were imitated throughout Europe.

Despite his fame, his life was tragic, which may account for the depressed, lonely character he brought into being on stage. His first wife, who was young, died a mere three months after their marriage. His mistress, who bore him four children, was unfaithful to him. By a second wife he had a daughter who did not survive infancy. But more tribulations awaited him. One day in 1836, while he was walking with his wife, he was beset and taunted by an apprentice who shouted: "Pierrot and his whore!" Enraged, Deburau struck at his tormentor with a cane, killing him. Tried for murder, the great actor stood in court, weeping and dressed entirely in black. The magistrates acquitted him. What was not revealed during the trial was that in his youth he had been trained to wield a stick in lethal fashion.

Though he returned to the theatre and resumed the guise of Pierrot, and audiences cheered him, they could not easily see him as an "innocent" as before. At fifty, this pre-eminent mime was dead.

For a span of years his son Charles inherited the role, then was succeeded in the French public's high esteem by Étienne Decroux (1898–1991), mime and actor, a long-time collaborator with Charles Dullin, the avant-garde director and theatre-manager. At various periods Dullin was associated with the Cabaret Lapin Agile, a centre for poetry readings, and Jacques Copeau's Théâtre du Vieux-Colombier. In 1922 Dullin quit Copeau's innovative company to establish a school for actors, the Théâtre de l'Atelier, an enterprise in which for an eight-year spell he was joined by Decroux, who at the time was in his mid-twenties. Thereafter, ambitious, Decroux was able to bring about a surprising revival of pantomime by having his own school and formulating and teaching a "systematic language of physical expression". His brilliant roster of followers and pupils, drawn by his mastery of the art, includes Elyane Guyon, Catherine Toth, Jean-Louis Barrault and Marcel Marceau, a circle who extended his influence to the end of the twentieth century. (Actually, surviving into his nineties, he outlived most of his disciples and students.) In the early 1940s he produced several well-received programmes of sketches. In 1943 he joined the Théâtre Sarah Bernhardt to teach mime, and between 1947 and 1951 performed in Holland, Switzerland, Britain and Israel, garnering praise and adding to his reputation. Back in Paris, he worked at the Cabaret Fontaine des Quatre Saisons and elsewhere. Invited to New York, he taught mime, in accordance with his well-defined principles, at the New School for Social Research. A glimpse of his artistry is preserved in Barrault's much-touted classic film *Les Enfants du Paradis* (1945). His opinions are summed up in his book *Paroles sur le mime* (1963). A son, Maxmilian, also a mime artist, eventually took over the directorship of his father's school.

The late eighteenth and early nineteenth centuries in England saw mime reach a new heights in the eloquent antics of Joseph Grimaldi (1778–1837), described as "one of the most famous clowns of all time". Of Anglo-Italian lineage, he was also a descendant of the *commedia dell'arte*. Indeed his grandfather, popularly dubbed "Iron Legs", was an entertainer in outdoor Parisian marketplaces and fairs, and also worked as a dancer and pantomimist at the Opéra Comique, where – the story goes – he once leaped "so high as to shatter a chandelier over the Turkish ambassador sitting nearby". On tour, he and his wife performed in Harlequinades at London's Drury Lane (1768) and at various British festivals.

It was at the Drury Lane, too, that his son, Giuseppe, elected to begin a thirty-year career as an able Pantaloon, largely under the shrewd management of David Garrick. There he danced and did acrobatics into his sixties, and at times alternated as a Pierrot, interpreting that chalk-faced character not as he was being portrayed in Paris: he translated him into a "rough-and-tumble" comic figure, hardly a grieving lover. Apparently, this was to current British liking. Growing older, Giuseppe choreographed ballets, while instructing younger dancers in the traditions of pantomime.

Giuseppe's illegitimate son, Joseph – an Anglicized version of his father's name – was his most able pupil, but the relationship was a very unhappy one. Joseph and his younger sister made their stage début when the boy was only two years old, carried on at Sadler's Wells in the arms of his applause-seeking father. It was during Giuseppe's spectacular production there of *The Wizard of Silver Rocks*, or *Harlequin's Release*. From his father the growing boy had daily lessons in acrobatics, dancing, sword-play and what was called "skin-work" – that is, donning animal furs to imitate monkeys and other small beasts who supposedly did tricks. By one account – not verified – Giuseppe had the boy in an animal disguise at the end of a chain and swung it so violently that the affrighted lad was hurled into the audience. Later, in Joseph's *Memoirs*, as well as in gossip of the day, Giuseppe is described as innately cruel, often beating his helpless small son and daughter. While in charge of a troupe of sixty children at Drury Lane he behaved so abusively towards them that some of the parents reported him to the authorities, which led to a criminal investigation. The affair was hushed up, but David Garrick had earlier stated his personal opinion that Giuseppe had a dark streak and should be horse-whipped.

Joseph's mother was a small-part Cockney actress. Giuseppe had several other mistresses and a wife, and supporting that many households was burdensome; so the domestic picture was not pretty for the luckless children. Moreover, Giuseppe had a strange obsession about death and how it would befall him; it haunted his private fantasies. On one occasion he feigned being a corpse to learn what his children would have to say about him, then startled them by his "resurrection". He planted false reports in the newspapers of his sudden, accidental demise, by a fall from a cliff or by unequalled over-eating, then expeditiously had them corrected. These served to make him more of a stage celebrity. He exploited the stories, too, by including a "skeleton" episode and enacting other scenes of "slapstick terror" in his repertory. To some degree he diminished or purged himself of his neurotic emotions and fears by his applauded clowning and miming of them.

Giuseppe Grimaldi died just before his son's tenth birthday, putting an end to much of the boy's ordeal. But the father's will contained a macabre request: Joseph's older sister was to have Giuseppe's head cut off before the coffin lid was closed. A surgeon carried out this task in the daughter's presence, "she touching the instrument at the time".

Joseph's childhood had been miserable and painful. He was scarred by it for life, showing lingering traits of hysteria, while often indulging in thoughts of suicide. It is also believed that he had inherited syphilitic symptoms. But at an early age, at least, he was freed from his father's malign dominance.

Growing up in the world of the theatre, at first at its fringes, he was soon at stage-centre. From the age of thirteen he showed unmistakable talent. By eighteen he was a much-admired Pierrot. He was also married and about to be a father, but his young wife and infant died in childbirth. He sought to assuage his grief in fanatical application to work, honing his natural gifts.

Like Deburau's, his further career was marked by tremendous success and a train of dire misfortunes. Some of his personal experiences were bizarre in the extreme. One night in 1807 he was awakened and told of a horrible riot at Sadler's Wells in which people were being trampled to death. His father-in-law lived there. Joseph rushed to the theatre and gained access through a back window. In the dark, he found himself trapped in a room strewn with the corpses of men, women and children, victims of the violent outbreak. He was the only live person in the enclosure – he felt entombed, sharing his father's long nightmare of premature burial. He shouted and banged on the locked door. Aroused by his screams, those on the other side were terrified, believing only the dead were in the adjacent, temporary morgue. Finally, his frantic voice was recognized and he was released.

He transmuted such experiences into his art. In clown's attire, he sometimes portrayed the dagger scene in *Macbeth*, instilling dread into the onlookers. Years later a spectator recalled: "Notwithstanding that he only made audible a few elocutionary sounds of some of the words, a dead silence pervaded the whole house, and I was not the only boy that trembled. Young and old seemed to vibrate with the effect upon the imagination."

His acrobatics took a severe toll. He claimed that in a very few years he had sustained so many injuries that every bone in his body had been chipped or broken, and he was seldom without physical pain. The result of these and other illnesses was that he aged prematurely. (Born in 1779, he died in 1837, at fifty-eight.) To the public he looked frail, and in 1815 an erroneous rumour of his expiration was widely credited though he lived another two decades. Again, a pattern set by his father, death and resurrection, seemed to be repeated, though this time the rumour was not purposefully contrived.

Life for him became more difficult and psychologically unrewarding. Crippled, he was often unable to perform. The weight of debts led to a failed suicide attempt. His second wife died, and he lost old friends the same way. His son, part of the Grimaldis' act, quit the stage; his profligacy caused his father much despair; then the irresponsible young man met an early death.

This was the background of Joey the Clown, the most famous of all such entertainers in British history – his skills were so appreciated that to this day "Joey" is how some of his fellow countrymen identify any clown: simply, "He's a Joey." Cope, from whose study these biographical facts have come, relates: "He was celebrated by the lowly and great alike. Crowds stampeded the theatre for hours before his farewell benefit performance, and the players stopped the pantomime to lead the audience in hailing him when he anonymously attended a performance in 1837, four months before he died. Byron, Hazlitt, the king and his fellow stagers all celebrated his talents in varied homage."

Cope quotes contemporary observers who attributed the success of "Joey" to his consistency: he was always the same character. He was "thoroughly English, neither dupe nor yokel, but a true John Bull". In Cope's view Joey was like Ruzzante as created by Angelo Beolco three centuries earlier, "the countryman alternately, simultaneously dumb and shrewd, butt *and* trickster". The largely universal figures of the *commedia dell'arte* that for a long time had walked the stage in the Italian Renaissance had mutated in England and assumed different personalities – a process that had also happened in France, though with different and more delicate results. This was perhaps less true of Harlequin and Pantaloon, though with that pair, too, there were innovations. For one, Harlequin (now renamed Joey) was a "drinking clown", initially appearing under the name of Guzzle, with Gobble as a partner. The management had let young Grimaldi change from the traditional Harlequin costume for one extravagantly more colourful, allowing him to command the stage. He had no half-mask. His face was painted grotesquely, the eyebrows exaggerated, the cheeks emblazoned with red triangles. This use of make-up was his invention. He excelled at rapidly discarding outer disguises, which were peeled off to reveal "the red, white, and blue jacket, the knock-kneed knickered uniform of Clown Joey beneath". These quick changes let him appear on the same night – or on successive nights – by innumerable transformations into new characters, though *au fond* still the essential Joey, a sociopathic Anarchist.

To Cope, this creation by Grimaldi was of something more than a mere English clown. "The painted face that stares at one from the dozens of depictions, like the energy Joey exhibited on stage, the grotesque absurdity of his violent 'turns', appears to offer us a share in the daemonic. 'Clown's' face is sub- and suprahuman simultaneously, as had been the half-beast devil mask of the original Arlecchino. Like that Italian original, Joey is a thief, a glutton, a coward, and a sadist whose greed is only expressible as a total incarnation of appetite in all things." Here Cope quotes from David Mayer's *Harlequin in His Element: The English Pantomime, 1806–1836*, to wit: "He was a Cockney incarnation of the Saturnalian spirit; a beloved criminal free from guilt, shame, compunction or reverence for age, class or property. That was Joey." To this, Cope adds: "When we view the character in this light we can see the history of Harlequin's decline in the eighteenth-century popular theatre quite differently. The Joey who replaced Harlequin was actually the resurrection of his original form and spirit as the half-man half-daemon who irrupted from the shadowy history of Germanic devils, a carrier of irrepressible irrationality and desire into the fragile forms of social restraint."

One of his major hits was *Harlequin and Mother Goose* (1806) in which much of the laughter is engendered by physical torture and dismemberment. Pantaloon is caught in a steel trap. As it snaps shut on him it sets off a gun that has the Clown's belly as its target. Harlequin slices off the Clown's ear, but it is reattached with glue. The Clown retaliates with a red-hot poker, slaying Harlequin, who promptly returns to life. Indeed, Harlequin simply refuses to die, defying all efforts to dispose of him: he is chopped to pieces in a cauldron, and next impaled by nails and hammered on to a wall, but each time is revived. This formula of death and resurrection was used over and over again. In *Harlequin and Cinderella* (1820) the Clown loses his arm in a scuffle. He laments its loss, but then sees it a way off hopping about the stage. He rushes to retrieve it, but it eludes him; he chases it, it continues hopping and taunts him; he finally catches it, stamps on it, only to scream in pain. This re-employed old *lazzi* of the *commedia*, but while Grimaldi was enacting the skit he was experiencing intense physical agony from his cumulative injuries, his sprains and multiple cracked bones. He also had a memory of an all-too-real dismemberment, his father's gruesome decapitation and the careful restoration of Giuseppe's head above his neck in the coffin.

Another story is told of a sailor, rendered a deaf-mute by sunstroke, who after some years attended a performance by Joey at Islington and was compelled to exclaim, "What a funny fellow!", thereby permanently recovering his power of speech. Immediately rumours took wing through the town that the superb Grimaldi had also given back sight to a long-blinded soldier and even raised a man from the grave, a beshrouded nineteenth-century Lazarus.

Cope likens Joey's achievement to that of Ruzzante because both put off familiar *commedia* garb, mostly shunned hackneyed Plautine and Terentian themes, and when possible chose sketches that had contemporary native settings. Both actors used local dialects – laid emphasis on doing so – and sought material with very topical content. In their plays there are even echoes – if not loud ones – of political events, including, among others, hints of pacificism: Padua and nearby Florence were in martial turmoil in Beolco's day, and Grimaldi's career flourished while Europe was convulsed by the Napoleonic wars. It is suggested, too, that Grimaldi's concerns and observations were narrowed to what was close at hand, because his father was English-born and his mother a Cockney. Further, the wars were an obstacle to travel, and he never crossed the Channel or even once left the British Isles. Though he inherited an Italian tradition, he applied his genius to transforming it into a new and vigorous English genre. Like Beolco, too, he was wholly absorbed into a dynamic character he had slowly fashioned, losing himself in Joey.

Considering his arduous, miserable childhood and often unhappy domestic life, it is appropriate that his *Memoirs* were edited by his friend Charles Dickens.

Through the ensuing century many clowns emulated Ruzzante and Joey Grimaldi in establishing a distinctive stage personality. Among them, top place was accorded to Grock (Adrian Wettach, 1880–1959), who had a truly unique *alter ego*. By birth French–Swiss, a son of Tyrolean singers, he joined a circus when he was twelve (some say seven), already a capable acrobat; afterwards he toured with his family, improving his musicianship. In 1903 he became the partner of Brich, a musical clown, and together they travelled around Europe and Latin America. Forming a new relationship (1907), he teamed up with the famous clown Antonet. Four years later he became a soloist and chose London as his base, where, over a span of seventeen years, he frequently appeared at the Coliseum, each of his "turns" lasting about an hour. He was expert at playing no fewer than twenty-four different musical instruments; he deftly aroused laughter by pretending to fail when performing on one after another of them. At the climax he would prove to be a virtuoso. In 1919 he was a headliner at New York's Palace Theater, where his act was described as "magnificent". He retired in 1932, in his early fifties, after a career of twenty-seven years, largely on view on international variety stages. He wrote several books of an autobiographical nature, including his informative *My Life as a Clown* (1957).

A strikingly talented Italian mime, Totò (Antonio de Curtis Gagliardi, Duca Commeno di Bisanzio, 1898–1967), was also a recipient of wide praise. As his real name and title indicate, he belonged to an ancient aristocratic family but elected to pursue a career in theatre. He not only practised mime but took parts in revues, variety acts, operettas, serious dramas and, after 1950, films. Audiences were startled and admiringly impressed at the contrast between his effectiveness in comic roles and as a tragedian. Among Italians of his day he was considered to have few if any equals as a comedian, so expressive of nuances of emotion were his facial features.

From about the mid sixteenth century the *commedia dell'arte* had prosperously invaded Germanic realms, in particular the Austrian court, which, as has been noted, played always friendly host to I Comici Gelosi, I Confidenti and I Fedeli, especially after 1570. Eventually there was a considerable measure of assimilation: Arlecchino largely became Hanswurst as local troupes sprang up to compete. This irrepressible comic figure was the inspiration of Josef Anton Stranitzky (1676–1726), a dental student at the University of Vienna and also an actor and puppet-master, as well as co-director of a group calling itself the "German Comedians" (1707). The stage introduction of the coarse-mouthed Hanswurst (Johnny Sausage) took place at Ballhaus in the Teinfaltstrasse, a suburb of Vienna. As conceived by Stranitzky, Hanswurst was a resourceful peasant from Salzburg. Invading the capital, he was trying to settle down, getting ahead by any devious means offering itself. His traits were a dynamic amalgam of Arlecchino's and those of characters in Teutonic folk-art, medieval oafs and English clowns seen on tours. In no time he was recognizable: a round, broad-brimmed, green-pointed hat above a ruddy face with a short beard, a fool's white neck-ruff, brief red jacket and long yellow trousers. With a heavy Bavarian accent, he consumed tankards of beer. His language was scabrous in the extreme, hardly printable here. None the less his initial reception at the Ballhaus was rapturous, and he was soon a favourite in most German regions, his vogue enduring for over a half-century; many scholars blame Hanswurst's long popularity for having impeded the evolution of a respectable literary drama in his native land.

Stranitzky, winning the support of Vienna's council, established the Karntnertor (1708), the city's first public theatre away from the court. In it he continued the tradition of improvisation set by the Italian Comedians. Aspects of Hanswurst's personality and attire varied over time, from town to town, but he remained essentially unchanged. From visiting English clowns Austrian performers borrowed fancifully named stock figures such as John Posset, Stockfish and Pickel Herring.

The Italian Comedians were especially liked in Munich. As early as 1568 a troupe was invited to add jollity to the weeks-long festivities for the wedding of Renata of Lorraine and Prince Wilhelm, son of Duke Albrecht V, Bavaria's ruler. The celebration was under the aegis of Orlando di Lasso, who led the court orchestra. He suggested having a *commedia dell'arte* offering as a component of the entertainment, and himself filled the part of Pantaleone. The sketch chosen for the occasion involved a beautiful courtesan left alone when her rich Venetian lover is summoned away on a matter of state. She is besieged by Pantaleone and his servant Zanni. A rival springs up, a Spanish nobleman on whom the courtesan soon looks more kindly. In quick succession there are clashes and duels, instances of mistaken identity and beatings, serenades directed to the wrong person, all concluding in the usual reconciliations, mutual harmony and geniality, with players and spectators jointly participating in a lively Italian dance.

Such fare so amused the pleasure-loving Prince Wilhelm and his bride that they asked the Italian Comedians to reside with them in Trausnitz Castle at Landshut, giving the actors a run that lasted ten years and ended only when Duke Albrecht deemed his high-living son too extravagant and cut off his financial allowance. Trausnitz Castle still has a vivid life-size portrait of the players, masked and in costume, a fresco by Alessandro Scalzi (also known as Padovano) that stretches up four floors along a wall of what is called the "fools' staircase", illustrating not only the actors, their attitudes, gestures and musical instruments but also dramatic situations and bits of telling action in their skits. Other and later such paintings of historical interest by Lederer are preserved in Krumlov Castle (dating from 1748) in Bohemia, along with ingenious wing-scenery executed by Wetschel and Merkel for a theatre in the same castle (*c.* 1768), the latter commissioned by the Schwarzenberg family who had taken possession of that stronghold (*vide* Berthold and Brockett).

The *commedia dell'arte* was influential in shaping Vienna's popular folk theatre in the early and middle nineteenth century, as was quite obvious in pieces put on by J.J. Laroche in his theatre in Leopoldstadt and by J.A. Gleich and Adolf Bäuerle in their theatre in the Josefstadt, and more elegantly in the magical fairy-tales of Ferdinand Raimund and the witty, indigenous satires of J.N. Nestroy. Newly ambitious to equal Paris as a centre of sophisticated culture and the arts, the Viennese intelligentsia might view the loud, loosely improvised sketches with disdain, but Emperor Joseph II grasped that such knock-about humour was decidedly to the taste of his city's middle- and lower-class spectators and rejected the censors' advice that he ban all such offerings. His solution was to establish a national, classical theatre in which the actors were not to depart from firm adherence to the written text, but for the time being an improvisatory comic stage was still exempt from such hindrances.

Laroche's amusing stock character, Kasperl, was in turn derived from Hanswurst and survives in Austria's puppet theatre, and his buffoon companions were clearly descended from the roistering *zanni*.

Also, in the eighteenth and early nineteenth century, three literary giants, Goethe and Tieck in Germany, Grillparzer in Austria, were involved with theatres separately in Weimar, Dresden and Vienna, as managers or dramaturges, and kept the *commedia dell'arte* vital by translating, adapting and even sometimes personally staging the wry, unsentimental plays of Carlo Gozzi, in which the genre's characters and subjects cavorted as before, often in lush fairy-tale settings. E.T.A. Hoffmann, fabulist compiler of the *Tales of Hoffmann*, contributed a ballet suite, *Arlequino*, and the grotesque *Phantasiestücke in Callots Manier*, and the happy *Prinzessin Brambilla*.

Doors to the Comedians and their plays in the "Italian manner" were opened everywhere in Germany – Cologne, Frankfurt, Bamberg, Leipzig – though they had to compete with accomplished and aggressive troupes bringing comedies and tragedies from Britain and Holland. The Silesian poet Andreas Gryphius (1616–64), who had no practical link to a theatre in his day and whose work was appreciated much later, felt an affinity to the *commedia dell'arte*, as seen in his play-scripts. This general affair with the exuberant Italians lingered into the baroque age and far beyond. In the first quarter of the twentieth century Richard Strauss and poet-playwright Hugo von Hofmannsthal built their lovely eighteenth-century *Ariadne auf Naxos* around an unfortunate incursion of a band of obstreperous Comedians during the palace rehearsal of a young composer's *opera seria*. The most prominent of twentieth-century impresarios, Max Reinhardt, demonstrated his great love for the *commedia dell'arte* in productions at Salzburg and elsewhere.

The travelling companies, with Arlecchino, Columbine, Pedrolino, Pantaleone and *zanni*, reached Portugal, and a few stayed on in Spain. They made conquests in Denmark, whose foremost playwright, Ludvig Holberg (1684–1754), was well acquainted with their *lazzi*: on a trip to Rome he had shared living quarters with several *commedia dell'arte* players, observed their stage-work and learned and remembered their jokes, later making profitable use of them in his twenty-eight comedies. He could also get ideas and quips from a helpful book, Evaristo Gherardi's *Le Théâtre italien*, a rich collection of the lines, situations and basic plots repetitively offered by the Comedians in their "improvisations". Holberg openly acknowledged his frequent indebtedness to it.

To the East, a troupe in service to the Duke of Brunswick (and formerly to the Duke of Modena) was in Warsaw, where apparently it pleased and took hold, for that same year (1730) the King of Poland sent a company to St Petersburg to entertain at the Tsar's court during the coronation of Empress Anna. However, the true reign of the *commedia dell'arte* in Russia occurred belatedly in the early decades of the twentieth century. The world-famed novelist Maxim Gorki, while in "voluntary exile" in Capri, became aware of performances by a Neapolitan group that espoused the style and long-cherished tradition of the *commedia*. Fascinated by the concept of a theatre in which "the actors themselves create the plays", he took notes. Returning to Russia, he discussed this approach with his friend Konstantin Stanislavsky, the founder of the Moscow Art Theatre, where his plays were produced, and showed him his jottings. Stanislavsky, a dedicated proponent of Realism, was open-minded and sponsored several workshops in the Moscow Art Theatre where experimentation was encouraged and given remarkable loose rein. Three of the major directors – Vsevolod Meyerhold, Alexander Tairov, Nikolai Evreinov – who started their careers in Stanislavsky's workshops yet were drawn to anti-Realism, found challenging the concept of a stage where the action responded to the impulses and inspirations of the actors, in some instances with only slight guidance from director and script. The three brash radicals plunged into testing the theory, and various new, derivative genres arose from it grouped under the heading of Presentationalism, a theatre opposed to Representationalism or Naturalism; the search was for "pure theatre", one that was wholly artificial and appealed to the imagination and aesthetic sensibilities rather than the intellect. Symbolism and Expressionism were in the air, which gave the experimenters further scope. (Actually, they did not relinquish much control to the actors, but that was scarcely to be expected from them.)

Examples of these fresh forms of drama drawing on the *commedia dell'arte* as a source are Alexander Blok's *The Puppet House* (1906), which has a disheartened and depressed Pierrot as its protagonist, and Evreinov's monodramas *The Theatre of the Soul* (1915) and *The Chief Thing* (1919). The title of the book in which Evreinov sets forth his visionary ideas is *The Theatre for Oneself* (1912).

So anti-intellectual were these iconoclastic directors that they hailed the ballet, puppetry and circus as the most authentic forms of theatre; nothing is more artificial than the use of dancers or marionettes to tell a story, and the settings for ballets and puppets are totally unreal. Watching them, the spectator has a truly *theatrical* experience; this is not life but something else, entrance into another realm. The same feeling applies to an observer of the highly stylized *commedia dell'arte*, with its actors' masks, bizarre costumes, exaggerated gestures, irrational pranks, acrobatics, topped by oddly simplified characterizations. That is pure theatre beyond question, a creation of art for art's sake, primarily intended to impart aesthetic pleasure and little more.

In Paris, associated with Sergei Diaghilev's exiled ballet company, the Russian choreographer Michel Fokine and composer Igor Stravinsky fashioned *Petrushka* (1911), which has clear and resounding echoes of the *commedia dell'arte*. The music employs brilliantly orchestrated, emphatically rhythmic folk-motifs. The scene is a noisy fair during Butter Week, the Russian equivalent of Mardi Gras, to which an aged Showman has brought three life-size puppets, The Moor, a sad, inarticulate Petrushka and the Ballerina, with whom Petrushka is futilely in love. She is shallow and obviously prefers the fierce, mindless Moor. At a signal moment during the festivities the three figures come to life, leap out of their box and chase through the crowd, to the astonishment of the suddenly fearful fair-goers. The Moor is attacked by the weak, ineffectual Petrushka, who is killed by a sweep of the Moor's scimitar, horrifying the merrymakers. To reassure them and restore calm, the Showman holds up the limp, lifeless Petrushka and convinces the stunned onlookers that the body contains only rags and sawdust. But as the crowd departs, and twilight deepens over the snowy square, the Showman sees the ghost of Petrushka rise to haunt him. To commentators, the ballet has many overtones of meaning. It had a great initial success and has remained in the repertoires of companies very widely, in part because of its epoch-making score.

Shortly before *Petrushka*, Fokine was engaged with the plotless *Le Carnaval*, another work adumbrating the *commedia dell'arte*, though a bit more obliquely. It was intended for a special charity performance at Lent in St Petersburg, but later adopted by Diaghilev's company while in Paris (1910). Set to Robert Schumann's piano pieces of the same name, it portrays a costume ball at which the guests arrive dressed as Columbine, Harlequin, Pantelon, Papillon, Estrella, Chiarina, Pierrot, Eusebius: they are not supposed to be actors or the traditional characters, but instead merely an aristocratic circle of friends gathering at a masquerade to dance and enjoy themselves. A subtle mood is created, matching Schumann's music, here orchestrated by four masters: Alexander Glazounov, Nicholas Rimsky-Korsakov, Anatole Liadov and Alexander Tcherepnine. The décor was by Leon Bakst. The piece caught on in Berlin, London, New York, and later wherever classical ballet was appreciated. Many of history's greatest dancers have had roles in it: Nijinsky, Tamara Karsavina, Leonid Leontiev, Vera Fokina, Bronislava Nijinska, Adolph Bolm and even Vsevolod Meyerhold, later to be acknowledged as the most eccentric and avant-garde of directors. It, too, has a permanent place in many repertoires.

In the spirit of the *commedia dell'arte*, too, is Sergei Prokofiev's opera *The Love for Three Oranges*, which had its première in Chicago in 1921, then in rapid succession in Edinburgh, Cologne, Berlin and at last in Prokofiev's native Russia (1927). Its libretto, by the composer, is derived from Carlo Gozzi's fantastic play of the same name, a fairy-tale about a little prince who has never laughed, alas, prompting his father, the king, to seek a variety of entertainers who might adequately amuse the luckless, mirthless child. Besides its sprightly score, the opera benefits from its many opportunities for cheerful, magical stage effects, among them a parade of unusual clown acts.

The Symbolist playwright Leonid Andreyev chose a circus as the *mise-en-scène* for his *He Who Gets Slapped* (1914), which was successful in Russia and produced on many stages elswhere. The hero, a sensitive aristocrat who has been betrayed by his wife and a friend, joins a circus and seeks to hide his identity and deep hurt behind the paint and make-up of a clown. He bears some resemblance to a brooding Pierrot, his "act" calling on him to be punished by his being slapped by his thoughtless fellow citizens. But the play hardly fits into the categories of farce or comedy; rather like so many other works into which the Gallicized, suffering Pierrot is introduced, it veers towards pathos and tragedy.

The legendary circus in Moscow, acclaimed for its trained bears, acrobats and fast-talking clowns, was especially noted as the venue of Popov (Oleg Konstantinovich Popov, born 1930), who, after training at the State School of Circus Art in that city, initiated his illustrious career by becoming a tightrope walker. In 1955 he was accepted into the Moscow State Circus as a clown, his new personality taken from that of "*Auguste de soirée*", a member of the company whose task is to keep the audience entertained between acts. Popov redefined and enlarged this role by assuming a boyish mien and exuding a roguish, endearing charm, while also seeming to be faced with problems that he cannot quite surmount. With the intrusive curiosity of a naughty child, he becomes entangled with curtain ropes, scene-changes, numerous stage props, causing mild havoc, yet always finally retrieving his balance and self-assurance. In addition, he proves himself to be a superb juggler, mime and musical technician. Parodying the more serious acts on the bill, he lends the lengthy programme a feeling of unity. After being seen frequently on tours about the Continent and the United States, he was appraised in the *Encyclopedia of World Theatre* as no less than the "greatest living clown".

In eighteenth-century England, shortly before the arrival of the Grimaldis, the Harlequinade had more or less originated with John Rich (1692–1761). His father, Christopher, a lawyer, also managed the Drury Lane Theatre. Stingy, he treated his actors so meanly that Thomas Betterton, foremost Shakespearian interpreter of the day, and several of his talented colleagues quit the ensemble and went to Lincoln's Inn Fields. Eventually Christopher Rich's notorious mismanagement caused him to lose his patent and lease on the Drury Lane, whereupon he adroitly moved to obtain possession of the Lincoln's Inn Fields, where his former company members had taken refuge. He embarked on refurbishing the playhouse, but this ambitious venture was cut short by his death in 1714, when control of the house fell into his son's hands.

Though he failed at tragedy, John Rich was fully adept at comedy. His limitation as an actor was an inadequate voice. To compensate for this lack he gradually resorted to dumb-show, becoming Harlequin, and appearing under the name of Lun. Other English actors had impersonated Harlequin a decade or two earlier, among them John Weaver, a dancing master, in 1702, but not on the scale of John Rich, and with only a fragment of his success and expanding popularity. He elevated pantomime to a quite new level, his offerings setting a long-lasting model. In them, serious episodes based on those in classical mythology alternated with comic scenes enacted by the familiar *commedia* figures. The serious passages used poetic dialogue, music and song; the farcical sections were mute, wholly in pantomime, though they were also accompanied by music. The most conspicuous element was spectacle, the transformations of persons and places accomplished by a simple wave of Harlequin's ever-reliable wand. John Rich's creations, some twenty in number – such as *Harlequin Executed* (1716–17), *Amadis, or the Loves of Harlequin and Columbine* – were an enduring draw. About half of these continued in his repertory for years, only a few details occasionally retouched.

With London's theatre-goers showing an almost excessive fondness for Rich's elaborate offerings, he soon had competition, notably from David Garrick, the highly admired actor-manager-playwright, who, through the 1740s, had taken over management of the Drury Lane, where Harry Woodward (1717–71), earlier trained by Rich, was featured, as were in succession the Grimaldis, father and charismatic son. The Drury Lane was available because Rich had left it to build the newer, grander Covent Garden (1732), over which he presided until his death. (The love affair of the English public with spectacular pantomimes – and the "squeaky" Italian opera singers imported and fostered by Handel – irked the fledgeling playwright Henry Fielding, who decried this foreign invasion, alluding to it derisively in his farces. Both Rich and Garrick put on other forms of entertainment – Shakespeare's tragedies and comedies among them – and one of Rich's top hits was his production of John Gay's raffish, ironic *The Beggar's Opera* (1728) which ran for an unprecedented number of nights, a phenomenon that led coffee-house wits to say that it made Rich gay and Gay rich. A compilation of ballads of the day, with new lyrics and an effective libretto, its action set in Newgate prison, it rubbed the nerve-ends of corrupt politicians, while delighting spectators. Indeed, it was the profits from staging this satiric piece that made Rich wealthy enough to undertake Covent Garden. Besides this, in 1735 he founded the Sublime Society of Beef Steaks, a club of eminent personages who met in a room at the theatre to dine and converse, perhaps about drama.)

In the nineteenth century lavishly designed presentations like those fashioned decades earlier by Rich began to develop into a rather new form of theatre, Christmas Pantomimes, mounted during the holiday season – and sometimes continuing beyond New Year's Day – entertainments aimed as treats for children, though the stories might contain sexual innuendoes that would pass over the youngsters' heads to amuse any elders accompanying them. The stories are usually taken from familiar fairy-tales or classic children's literature – *Puss-in-Boots, Little Red Riding Hood, Cinderella, Aladdin, Jack and the Beanstalk, Alice in Wonderland, Robinson Crusoe* – unfolding with interruptions for songs and dances, juggling acts, acrobatics, lengthy clown routines, children's ballets, even circus and music-hall turns and contemporary pop music. The scenery is extravagant, meant to dazzle and evoke gasps and sighs of pleasure; the costumes are glittering; and there are magical transformations, often at the climax of scenes. Cross-dressing is mandatory in these offerings, which have fixed conventions, almost "ritualized" by now: the male hero (principal boy) is played by a slender actress, and the leading comic comes on in "drag", for example as Widow Twankey or a Witch; inevitably there is a truly fiendish villain, perhaps in the person of a Demon King, and of course a good Fairy to rescue anyone in serious trouble. There is room for improvisation and considerable miming, and topical references may be introduced. As the December holidays arrive, the annual Christmas Pantomimes accompany them; these stage-works thrive throughout Britain to this day.

Several twentieth-century English and American playwrights have tried their hand at poeticized *commedia dell'arte* pieces, such as Edna St Vincent Millay's delicate one-act *Aria da Capo*, a particular favourite of Little Theaters in the United States during the 1920s, and the British writer Terence Rattigan's *Harlequinade* (1948). In London and New York, Peter Brook's radical production of Shakespeare's *A Midsummer Night's Dream* (1971) had some of his actors clad in traditional *commedia* costumes.

In Copenhagen's Tivoli Gardens a theatrical ensemble that closely replicates the sketches, style and vibrance of the sixteenth- and seventeenth-century Italian companies has been among the park's permanent attractions. Thousands of miles away, on the Pacific Coast, R.G. Davis founded the San Francisco Mime Troupe to perform silent plays in 1959. But with a change in the *Zeitgeist* (1966), the group adopted new, Leftist principles, abandoned its sole reliance on mime and dedicated itself to outspoken agit-prop works.

An American comic, Jimmy Savo, who had lead roles in vaudeville and Broadway musical comedies, displayed unusual mimetic gifts. When he appeared as Dromio in *The Boys from Syracuse* (1938), an updated, bawdy, loose adaptation of Shakespeare's *The Comedy of Errors*, the *New York Times*'s Brooks Atkinson hailed the offering and gave as one of his prime reasons for doing so the lucky presence of this inspired actor: "First of all, there is Savo, the pantomimic genius, whose humorous gleams and fairy-tale capers have never been so delightful and disarming." But before Savo could fully realize his special talent, his career was untimely ended.

For pure mime, few in the twentieth century have been the artistic peer of Marcel Marceau (b. 1923), a former student of Decroux. After a year with Compagnie Renaud-Barrault (1945), as Arlequin in Prévert's *Baptiste*, a ballet-pantomime about the legendary Deburau, he left to produce only mime-dramas. Before studying with Decroux he had early developed an interest in pantomime while teaching children. Now assembling solo performances of wordless skits, he gradually created the character of Bip, white-faced, with blackened, high-arched eyebrows and scarlet lips, his personality a somewhat romanticized French variant of Pierrot, yet with significant differences. Bip is amiable, foolish, ineffectual, a gentle, downcast vagabond, always chasing butterflies, smelling flowers, trying hopelessly and weakly to do his part in a tug-of-war. Forming a company, Marceau ventured more ambitious playlets, though the Chaplinesque Bip has been his almost constant persona ever since. A high point was a mimed dramatization of Gogol's pathetic short tale *The Overcoat*, staged in 1951. For over five decades he has toured, sometimes having only an assistant who brings to the footlights a placard bearing the name of the sketch to be enacted and perhaps a few props. Some of the most treasured are *Mort avant l'aube* (*Death Before Daybreak*; 1947); *Jardin public* (*Public Garden*; 1949), in which he impersonates a series of ten characters seen in a park; *14 Juillet* (*The Fourteenth of July*, pertaining to France's national holiday; 1956); *Paris qui rit, Paris qui dort* (*Paris Laughing, Paris Sleeping*; 1958); *Le Petit Cirque* (*The Little Circus*; 1958). Haunting is his mute portrayal of the life of a man, from birth to youth, adulthood and death.

Western European and American choreographers have dipped into the *commedia dell'arte*, among them Glen Tetley, one of the first to blend the techniques of modern dance and classical ballet, whose *Pierrot Lunaire* (1962) uses a score by Arnold Schoenberg. Three archetypical *commedia* figures engage in varied sex games, some ribald, some wistful, some verging on the sadistic.

Two Russian expatriates, who worked in Paris and New York, were responsible for differing treatments of the theme and figure of Pulcinella. Leonide Massine, with Diaghilev's Ballets Russes, matched his version (1920) to music by Igor Stravinsky and a scenario based on an old Italian sketch, *The Four Pulcinellas* (c. 1700); at the Paris première Massine himself had the title role. Revived and danced by others, it was renamed *The Two Polichinelles* in London (1935). Yvonne Georgi, a disciple of the German modern dancer Mary Wigman, presented *Pulcinella* while she was ballet mistress of the Municipal Opera of Amsterdam (1941); she toured widely. For a fête celebrating the compositional accomplishments of Stravinsky, who had recently died, George Balanchine and Jerome Robbins joined forces for yet another handling of this *commedia dell'arte* subject for the New York City Ballet, ensconced for the summer at Saratoga Springs, an upstate spa to which it regularly retreated (1972). As a young man Balanchine had seen Massine's *Pulcinella* in Paris and found it "long and boring" (according to Richard Buckle, *George Balanchine, Ballet Master*). He resolved to improve on it. To embellish Stravinsky's score he retained Eugene Berman to design the sets. The curtain, décor and costumes of Massine's piece had been by Pablo Picasso. It might be noted, too, that much of Stravinsky's music for the work consisted of rearranged, charming tunes by the short-lived G.B. Pergolesi (1710–36), which assured it an eighteenth-century ambience. Richard Buckle relates:

> *Pulcinella*, choreographed jointly by Balanchine and Robbins, had so many ideas, so many scene changes, such a large cast and such a plethora of mime, jokes and gambols, that there was no way to get it in shape in time for its Friday-night performance. Balanchine and Robbins, befeathered beggars, appeared briefly toward the end to belabor each other with sticks. Villella mimed furiously throughout – this time (as opposed to *Harlequinade*) in what he called a "down low, gross" genuine Italian comedy style. Even though much of the work that had gone into *Pulcinella* was lost on stage, Violette Verdy remembered her joy at the sight of Balanchine demonstrating passages of broad mime to Villella at rehearsal: "I was there watching with Lincoln Kirstein. We turned to each other, practically with tears in our eyes because it was so touching. I said: 'It's incredible – it's everything that one knows about the theatre.'"

Some of the ballets that bloomed for the festival were destined to fade after one exposure, leaving no trace behind. A few survived for a season or two.

So far, *Pulcinella* seems not to have been among those so chosen.

The *Harlequinade* (1965), alluded to by the ballerina Violette Verdy, was another foray by Balanchine into a work in this tradition. Its predecessor, and partial source, was Marius Petipa's *Les Millions d'Arlequin* in which Balanchine had danced when a child in St Petersburg. (He frequently borrowed from Petipa throughout his career, as he openly acknowledged. Petipa had also created a *Harlequinade*, though possibly this is *Les Millions d'Arlequin* under a different title – the record is unclear.) Petipa, having delved thoroughly into the manners of the *commedia dell'arte*, made considerable use of mime, and Balanchine retained this feature. The score, appropriately "tinkling", was by Riccardo Drigo, an Italian conductor at the Maryinsky.

Edward Villella, who had the leading role, had little knowledge of the traditions of seventeenth-century miming. The dancer told Buckle how Balanchine showed him the proper way to perform it. "I was doing it rough and gruff in a – I thought – comical manner. But he said, 'No, dear, it's *commedia dell'arte*, but it's French, not Italian.' I still didn't understand. Then he turned around and said, 'Harlequin is *premier danseur*,' and walked away. There was this elegant confidence this rogue of a Harlequin has. Harlequin is very pulled-up, not crouching. It was a point of departure."

The set was by Rouben Ter-Arutunian, adapted from one that he had designed for Rossini's comic opera *La Cenerentola* (*Cinderella*) – a bit of artistic economy not unlike Balanchine's propensity to borrow, but perhaps made necessary by budget constraints. The story recounts in two acts how Columbine's wedding to Leandre, an elderly suitor, is prevented when the Good Fairy provides Harlequin with millions of francs, which changes him into a suitable husband in the eyes of Columbine's father, Cassandre. Balanchine seldom composed narrative ballets; this is an exception.

John Cranko, born in South Africa, made his career as a dancer and choreographer in Britain and Germany (Stuttgart) and to a lesser extent also contributed works to companies in Cape Town, Paris, New York, Milan, Edinburgh and a number of cities in Canada. His *Harlequin in April* (1951), in two acts with a prologue and epilogue, was first seen in London at Sadler's Wells, where it was danced by the theatre's resident troupe. The commissioned score was by Robert Arnell, and the décor by John Piper. The title is linked to the opening lines of T.S. Eliot's poem *The Waste Land*. "April is the cruellest month, breeding / Lilacs out of the dead land . . .", a sombre, ironic, sometimes mocking and despairing response to the decline of contemporary civilization, a bleak view of the world apparently shared by Cranko. A synopsis reads: "Man is born from the barren earth, struggles to maturity, loves, triumphs, fails, grows old, and returns to the earth. The fumbling, bumbling Pierrot looks on and comments without in any way being able to help."

One of the most popular operas ever written, Ruggiero Leoncavallo's two-act-with-prologue *I pagliacci* (*The Clowns*; 1892) was first staged at Milan's Teatro dal Verme. Its libretto, by the composer, is about an actor in a troupe of *commedia dell'arte* players. Travelling by cart about

Italy, from town to town, they reach a new destination and set up a platform with curtains for the evening's performance. The principal role is to be taken by Canio, jealously possessive of his wife, Nedda, a company member. She is loved by others: Tonio, a misshapen roustabout, and good-looking Silvio, to whom she is drawn and who stirs her to dream of being freed from her husband's obsessive suspicions. She consents to run off with Silvio. They are overheard by the skulking Tonio, who alerts Canio to the affair, though he cannot identify the lover. Twilight falls. Canio puts on his clown's white mask, while baring in an aria his heartbreak, his having to play the theatrical fool and amuse others while feeling total, personal devastation. The crowd gathers, chatting and eager to be entertained. The trifling play-within-the-play, a light *commedia dell'arte* sketch, begins. Nedda is Columbine, flirting with Harlequin, who withdraws as Columbine's husband enters: it is Canio, cast as Pagliaccio, the clown, Columbine's husband, unable to control his rage. Seizing Nedda, he demands to know her lover's name. She refuses to tell him. Infuriated, he stabs her. The spectators think this is part of the *commedia* sketch, but seeing Nedda's danger, Silvio rushes from the audience to assist her. Canio fatally stabs him. Turning to the audience, he cries pathetically: "The comedy is finished," and collapses.

Leoncavallo, the son of a judge in Calabria, had heard that his father had presided over a trial involving such an incident of tragic jealousy and murder in Montalo. The passionate music is doubtless what accounts for the endless success of this powerful *verismo* opera. The role of Canio has been an inexhaustible favourite of many of history's greatest tenors. In its day *I pagliacci* was deemed shocking, and particularly by some important critics, to touch on subject-matter that was ugly and in poor taste, a judgement that now seems to make no sense.

Besides *Ariadne auf Naxos* and *I Pagliacci*, other operas have *commedia* situations and characters, but they belong to later periods and consequently appear in updated settings and attire, with the people also having different names. Peter Wynn's "Zanies, Lovers, Scoundrels, Fools", an essay in *Opera News*, traces the evolution of such works. An obvious instance is Gaetano Donizetti's much-treasured *opera buffa*, *L'Elisir d'Amore* (*The Elixir of Love*; 1832). The libretto, by Felice Romani, is based on a script, *Le Philtre*, written a year earlier in Paris by Eugène Scribe for the French composer Daniel-François Auber. The competing version by Donizetti for the opera company in Milan was hurriedly prepared when a commissioned work by another composer was not delivered in time. As Wynn points out, the standard figures here are Dr Dulcimara, the fast-talking quack who sells magical potions, and his eager assistant, a *zanni*. The love-hungry, simple-minded Nemorino is the inevitable dupe, a forlorn, rustic Pierrot; his rival, Belcore, is another incarnation of the blustering Captain. Their beloved, Adina, is the Inamorata, and her saucy maidservant Giannetta is an Arlecchina. The scene is the Basque country in the early nineteenth century, but change the place, the garbs, the names of the characters, the time of the action, and *L'Elisir d'Amore* fits neatly into the category of sixteenth- or seventeenth-century *commedia dell'arte*. A similar transformation may be glimpsed in Donizetti's melodious, risible *Don Pasquale*, but it owes a degree of debt to the *commedia erudita* as well. Who is the Don, avid for a young wife, but Pantalone who is deprived of his prey by his nephew's trickery, and a resort to a faked nuptial contract drawn up by a false notary?

Wynn attributes the same provenance to the comic operas of Mozart and Rossini, singling out the high-spirited, pert maidservants who have motivational roles in them and linking them to the flirtatious, conspiratorial Columbine. He sees a close resemblance between the Figaro figure and a Brighella or Mezzotino in the familiar Mozart and Rossini works, and likens Don Basilio – in *The Marriage of Figaro* – to Dr Dulcimara (who in turn descends from the *commedia*'s Latin-spouting Dottore), and declares the manservant Leoporello in *Don Giovanni* to be clearly akin to the typical *zanni*. His list of such perpetuations of characters and their salient traits in other *opera buffe* is lengthy. Another eighteenth-century example: in G.B. Pergolesi's *La Serva Padrona* (*The Maid-Mistress*; 1732) Serpina, the maidservant, craftily inveigles her master, Uberto, into making her his wife. She is assisted by another servant, Vespone, who pretends to be a fierce military man, and who threatens Uberto. "In other words, Columbina tricks Pantalone with the help of Pedrolino, who disguises himself as Capitano Fracasse." Wynn also cites Domenico Cimarosa's *Il Matrimonio Segreto* (1792), its libretto based on the Colman-Garrick *The Clandestine Marriage*, and his *Le Donne Rivali* (*The Lady Rivals*). Of Rossini's many other comic forays, in addition to his *Barbiere di Sivigila* (*The Barber of Seville*; 1816), characters borrowed from the *commedia* abound in *La Scala di Seta* (*The Silken Ladder*; 1812), *L'Italiana in Algeri* (*The Italian Girl in Algiers*; 1813) and the one-act *Signor Bruschino* (1813).

Further on in Wynn's roster are Ferruccio Busoni's curtain-raiser *Arlecchino* (*Harlequin, or The Window*; 1917) and *Turandot* (1917). The *Arlecchino* is a caper fully exemplifying its sixteenth-century models with a scheming hero, his fickle wife (Columbina), her seductive suitor (Leandro), the usual dupe (Matteo) and the other woman (Annunziata) with whom the betrayed Arlecchino finally consoles himself. The tone is wry, ironic. The *Turandot*, differing from Puccini's adaptation of Gozzi's play, retains much more of the original's whimsical, fairy-tale atmosphere and, though once more ironical, is far less serious. The strange, chanting trio of courtiers, in Puccini's version designated as Ping, Pang and Pong, are here named Pantalone, Tartiglia and Brighella and wear their traditional masks. Busoni provided his own libretti for these two works. Still later operatic reincarnations of this kind pointed out by Wynn are Vittorio Giannini's *The Servant of Two Masters* (1967), making expert use of Goldoni's long-lived comedy, and John Corigliano's *The Ghosts of Versailles* (1991), with a fanciful text by William H. Hoffman, partly suggested by the Figaro scripts of Beaumarchais. In conclusion, Wynn argues:

> The *commedia* influence has even been felt in American musical comedy. The book for Tom Jones's and Harvey Schmidt's *The Fantasticks* (1960) was an adaptation of Rostand's *commedia*-based *Les Romanesques* (1894, *The Romancers*), while the staging – still on view after thirty-five years at New York's Sullivan Street Playhouse – was inspired by a Giorgio Strehler production of *The Servant of Two Masters* that Schmidt and Jones had seen in New York. Stephen Sondheim,

Hugh Wheeler and Harold Prince peopled their 1973 *A Little Night Music* with a perfect set of *commedia* characters: the lawyer Egerman is a classic Pantalone, his teenage wife Anne the archetypal *inamorata*. Egerman's son, Henrik, is an *inamorato*, perhaps with a bit of Pierrot mixed in. Count Carl-Magnus is the bragging Capitano; the maid is a Columbina.

At first thought, it's astonishing that characters nearly five hundred years old would still turn up with such frequency, but on reflection the reason is clear: the masks of the *commedia dell'arte* captured mankind's many faces with a vividness the years can't fade. The old fool, the miser, the addle-brained pedant, the bragging bully, the scheming underling are in no shorter supply today than ever.

Again, most of these works owe as much to the *commedia erudita* as to the *commedia dell'arte*, but, as has already been remarked, the two genres were very often intermingled.

An opera not mentioned by Wynn is *Basi e Bote* (1927, with a score by Riccardo Pick-Mangiagalli [1882–1949] and a plot from the well-known Arrigo Boito [1842–1918]; its first performance was in Rome). The setting is Venice: the characters are masked. Pantaleone dei Bisognosi is anxious to wed Rosaura, his young ward, in order to acquire her ample dowry. She and Florindo are in love with one another. Their servants, who also love each other, help them to evade Pantaleone's gimlet eye. They are exposed, but get away in the general uproar and confusion. Pantaleone's servant, Pierrot, whose role is wholly mimed, is wrongly arrested as a thief. As a result of a beating, Pantaleone is invalided. He is attended by Arlecchino, disguised as a physician, who takes away the old man's glasses and stuffs his ears with cotton, so that Pantaleone is no longer aware of what is happening. A notary, Tartaglia, presents him with a marriage contract, which Pantaleone signs, supposing it will unite him with his ward; instead it joins together Florindo and Rosaura. In the same way, Arlecchino and Columbina are wed. Discovering that he has been deceived, Pantaleone is beside himself, but he is placated when Florindo presents him with a heaping casket of ducats. The unlucky Pierrot, dressed as a chimney-sweep, escapes from prison, and there is much rejoicing.

Robert Ward (b. 1917), in collaboration with Bernard Stambler, composed *Pantaloon* (1956) while they were students at the Juilliard in New York; it won considerable praise and notice, leading to a commission from the New York City Opera for another work. The libretto is based on Leonid Andreyev's circus play, *He Who Gets Slapped*. The opera was revived in 1959, this time with the title changed to that of Andreyev's drama, and in 1972 it was staged at the Encompass Theatre, now bearing a combined title: *Pantaloon, He Who Gets Slapped*.

Perhaps the *commedia*-born character most widely found in Italian *opera buffe*, Viennese operettas and American musical comedies is Columbine: she is the soubrette in countless such offerings, flippantly copying the brazen parlour maid Adele in Johann Strauss II's *Die Fledermaus* (*The Bat*; 1874), an undying stock figure, for ever stealing scenes, diverting the audience's attention from her mistress, eliciting laughter and almost as much applause. By whatever name, she enlivens the stage in the light-hearted, lyrical concoctions of Lehar, Kalman, Oskar Straus, Victor Herbert, Friml, Romberg.

I Guillari de Piazza, an Italian folk music and performing ensemble headed by Alessandra Belloni, participated in an annual Greenwich Village pageant by re-staging Silvio Fiorillo's *L'Amor Folle* (1605) at New York University's Tishman Auditorium in 1984. The action in this farce, which is about the birth and death of Pulcinella, was described as "boisterous", with its musical content rendered on period instruments such as the chitarra battente (Renaissance guitar), harp, piccolo, mandolin and recorder, accompanying songs and Neapolitan tarantellas. The leading role was taken by Ms Belloni. A narrator kept the audience abreast of what was going on, since few among the spectators were fluent in this particular Italian dialect of the seventeenth century. The visiting company took the play to other cities.

In the early twentieth century, when the vulgar burlesque theatre flourished in America, drawing mostly men to its programmes of stripteases and slapstick, its comics with their baggy trousers and red rubber noses belonged to a lower stratum of the *commedia* tradition. When the burlesque theatre perished, however, many of these comics moved on to Broadway musical comedy stages and eventually to Hollywood and films.

A special group, who began their careers in London and Paris music-halls, as well as on the Broadway stage, depended more on their artistic miming than on verbal humour, and many of them achieved great prosperity and acclaim in films, especially those who started when motion pictures were silent and expressive facial features and hand gestures were very important. Among such heirs of a purer *commedia* discipline have been Max Linder, Charlie Chaplin, Buster Keaton, Harpo Marx, Jacques Tati, Bert Lahr and Will Irwin.

A British-born actor, Geoff Hoyle, sharing a programme during Lincoln Center's Serious Fun Festival in 1986 in New York City, presented two skits that he called *commedia*. In these he demonstrated two *lazzi*. The first of them had him as a "cringingly hungry Arlecchino catching, salting and consuming a buzzing fly", an oft-repeated bit of business, as has been noted. In the second, *Pantaleone's Purse*, Pantaleone, the ancient miser, and Arlecchino, his wittily predatory servant, vie for the master's gold. In the *New York Times* Jennifer Dunning, a dance critic, commented: "The pieces were performed with impressive skill and offered a gratifyingly unadulterated look into the *commedia dell'arte* tradition. But the relentlessly bawdy tone of the second piece soon palled. As Mr Hoyle rightly said at the start, "I have to remember there may be parents in the audience." The two pieces did suggest the actor's remarkable physical agility, seen most clearly in his rubber-footed negotiation of space."

A few years later Geoff Hoyle returned to New York to appear in *A Feast of Fools* at the Westside Arts Theater (1990). This was an offering of which he himself was the author. Timed to coincide with the work's première, an article by Hoyle was published in the "Arts and Leisure" section of the *New York Times*. In it he discusses his exploration of the *commedia* genre. These are some excerpts:

A wheezy old geezer, more a threat to himself than to the ladies he imagines he attracts, fingers the well-worn purse that hangs from his belt and talks of "*amore*."

A would-be thief tries to convince an investigating officer that the suspicious-looking wire-cutter, crowbar and rope he carries are tools of an honest trade – as is the silverware in his bag.

A spiffily dressed trickster at a busy intersection accosts the unwary with "Whaddya need? Booze, girls, music?"

Characters seen on a modern city street? But these wear masks that cover the upper part of their faces, with exaggerated noses, hairy lips, warts and bushy eyebrows. They look more like conscripts in a strutting parade of humanoid barnyard animals. Yet they wear sixteenth-century clothing. The old man's purse is wrinkled red leather. The thief would have the officer believe that his rope is really a long strand of macaroni. And the spiffy trickster carries a guitar. These are in fact the comic characters of the *commedia dell'arte*.

The old man's name is Pantaleone or Pantaloon. The trickster is Brighella and the thief is Arlecchino, better known as Harlequin and familiar if only as a spangle-suited, bat-wielding symbol on a greeting card.

One of Western theatre's great movements, the *commedia dell'arte* (or Comedy of Skill), cannily frustrates theatre historians and academics because, though it was universally popular in Italy, France and Spain from about 1550 to 1750, it left behind no scripts. It cannot be studied as literature like Shakespeare, Molière, or Lope de Vega, who so profoundly felt its influence.

All we have are descriptions, lists of comic bits known as *lazzi*, some engravings and scenarios. No recordings, no videotapes, none of the modern visual aids we take for granted.

A basic outline pinned up in the wings (from which the phrase "winging it," meaning to improvise, is said to derive) would serve as a springboard for the actors, all skilled improvisers. It is as if what remained of Shakespeare's *Twelfth Night* was a scenario that began, "Duke Orsino speaks of love. He goes out. At this moment, Viola, shipwrecked, disguises herself as a man. She leaves. At this moment, Sir Toby Belch enters, drunk. . . ."

No literature then, but we do have the evidence of the masks. The comic characters, Pantaleone, Arlecchino, Dottore, Brighella, Pulcinella . . . wore masks, specifically half-masks that ended at the upper lip. Actors could therefore speak and vary the changing expressions of the masks with facial gesture.

Audiences flocked to see their favorite characters rather than individual plays. Actors became famous, notorious even, for one particular role. It was as if audiences were tuning in to their favorite characters embroiled in yet another fine mess.

Lazzi were always among the audiences' favorite moments. These comic turns, paradoxically well rehearsed, were used to weave the plot together. They appeared on cue whenever the improvisation got sluggish (and probably whenever an entrance was late, a prop was mislaid, or a fight broke out). There are hundreds of these *lazzi*, many of a sexual or scatological nature, scattered throughout scenarios that read in a collection like a pratfaller's gag book.

These are not "scripts." Comedians don't write their moves down (either it's a clever method of copyright, or they just don't think of it); they pass them on. The *lazzi* are only descriptions. They need the physical comedian's skill to bring them to life. In his book *Lazzi: The Comic Routines of the Commedia Dell'Arte*, Mel Gordon, a professor of theatre at New York University, describes more than two hundred of them.

Hoyle details some of the better-known turns and asks his readers to imagine what Chaplin or the Marx Brothers would do while enacting them, perhaps the *lazzo* of eating the fly, or the *lazzo* of the false arm, in which Gratiano escapes from suspicious characters who are holding and beating him, by freeing himself and taking off, leaving his bewildered captors grasping only a wooden arm (this was performed with many variations, always leading to a surprise).

The *lazzo* of making the bed is spelled out:

Arlecchino and Tartaglia are assigned to make a bed. Seeing wrinkles on the sheets, Tartaglia commands Arlecchino to go under the covers to investigate. Arlecchino does so and announces that there are no wrinkles. Seeing the outline of Arlecchino under the sheets, Tartaglia says it is worse than ever. They trade positions and point-of-view.

Ribald and scathingly satirical, *commedia* was the sitcom of the Renaissance. Its popular appeal was insured by the comic representation of all social classes. Even the "serious" characters were notable for their excesses – the swooning lovers, for example, the vain nobleman later seized on by Molière.

This uniquely *theatrical* of theatre forms has the resonance of myth. The characters penetrate like Jungian archetypes; they are uncompromisingly human, devastatingly satirical populators of "this great stage of fools." Their vulgarity is wholesome and true, and they make you laugh.

As a student actor I had little experience with the form. But I became enthralled by descriptions of *commedia* shows. I saw Giorgio Strehler's famous Piccolo Teatro perform *Harlequin – Servant of Two Masters* in Paris. I pored over Jacques Callot's ironic engravings of *commedia* characters in *I Balli di Sfessania*. And I saw the San Francisco Mime Troupe use classical and half-masks to produce larger-than-life but utterly realistic caricatures – walking social commentaries.

A few years ago I got hold of some masks – Pantaleone, Arlecchino, Dottore – and improvised cautiously in front of a mirror. . . . Putting the masks on one by one, they seemed to me to have magical powers. They were more than just catalysts in a scientific experiment. I felt as though I were dabbling in alchemy.

The masks and all I knew about the characters they represented became a sort of philosopher's stone of physical comedy. The characters and their dilemmas vibrantly sprang to life.

I tried re-creating some of Arlecchino's *lazzi*. It was like finding a trunk full of old toys in the attic. Winding them up carefully, I found they still worked. The masks made me want to attempt impossible gestures. Each mask seemed to make my whole body into a much larger mask....

Today, *commedia* is alive in the compassionate and zany characterizations of artists like Robin Williams and Lily Tomlin. The Italian comedian-playwright Dario Fo, a true inheritor of the spirit of "*improvviso*," has also demonstrated the timeliness of *commedia*. In his signature piece, *Mistero Buffo (Comic Mystery)*, Mr Fo introduces folk tales and molds them into bravura sketches in which, playing all the characters, he skewers those in power and exalts the common folk.

For me it was a logical step to Mr Sniff, the wicked buffoon with the elongated nose – a sort of *commedia* mini-mask – and the weirdly transforming half-mask of the befuddling Fundraiser, both now characters in my current show, *Feast of Fools*. But I don't think these types would have come into being without their ancestors, Arlecchino and Pantaleone.

And theirs is the final irony. As I walk to the theatre each night, these characters still enliven the city streets. I have seen Pantaleone spitting outside the betting shop, a young Arlecchino pretending to be tripped by a passer-by, performing a full forward roll on the sidewalk and demanding compensation in specific dollar amounts. Brighella sells sidewalk bric-à-brac, illuminated by electricity illicitly tapped from a disemboweled street lamp, and offers me his services as a go-between. And the harried waitress Franceschina skillfully nags her impatient flock, including me, in the diner. These *commedia dell'arte* will never die, nor, I believe, would we ever want them to.

A rare event was a loan exhibition at the Frick Collection in New York of fifty-one Domenico Tiepolo drawings illustrating tricks by *commedia dell'arte* players. To round out the showing, which was organized by Adelheld M. Gealt, there were two related drawings by Domenico's even more highly ranked father, Giovanni Battista Tiepolo, and a selection of eighteenth-century fans and porcelain figures bearing *commedia dell'arte* motifs from the Metropolitan Museum of Art and seldom on display there. Also on view was a single "marvelous painting" by Domenico Tiepolo – *The Country Dance* – borrowed from a private owner. Wrote a critic in the *New York Times*:

New York is unlikely to see many exhibitions this year more entertaining – or, indeed, more deeply moving than this.

Punchinello is an extravagant figure of "low" and robustious appetites – earthy, gluttonous, ribald and irreverent, the very model of human excess carried to grotesque extremes. With his tall, cylindrical hat, his beaked nose, bulging belly, he is at once comic and endearing. In the *Divertimento per li ragazzi*, this hero of the popular imagination is depicted in an immense variety of dramatic situations that embrace virtually every event in the cycle of human existence from birth to death.

Scenes of courtship and marriage, glimpses of work and play (the drawings devoted to games and entertainments are particularly wonderful) and scenes of terrible punishments, too (the firing squad and hanging) – these are but the outstanding of the many subjects that inspired the artist's hand in these drawings. By creating whole troupes of Punchinellos to re-enact the cycle of earthly life, the artist gave us, in effect, a comprehensive Human Comedy.

Thus, the drawings fairly teem with anecdotal detail. This is narrative art carried to a rare pitch of perfection. The sheer energy and drive of the draftsmanship are a miracle still, for everything is quickened by a comic impulse that is always alert to the finer shades of feeling even when – as often happens – they are mercilessly mocked. What a marvellous thing the narrative impulse in art is in the hands of a master like Domenico, and how much has been lost to us in the denigration of narrative art in our time.

For the eighteenth century, however, this narrative impulse counted for a great deal, and the *Divertimento* drawings must be counted among the greatest achievements to result from it. Something of their quality is very nicely stated by Mr Gealt in a passage from the introduction to the catalogue accompanying the exhibition:

"Like Mozart's nearly contemporaneous piano sonatas, Domenico's depictions convey at once a boundless energy and an almost enervating but elusive melancholy. In the best tradition of the great clowns, Punchinello bears a streak of tragedy, a constant reminder of the freakishness and irony that make up life itself."

Tiepolo's illustrations, along with other artworks, were one of the sources for Martha Clarke's *Miracolo d'Amore* (1988), highly stylized, dramatic tableaux exploring aspects of eroticism, even brutal encounters between Harlequins and naked women. Mounted off-Broadway, with music by Richard Peasles – echoing Monteverdian melodies – it stressed spectacular visual effects, a parade of other-worldly images: in their episodes, the Harlequins were garbed wholly in white. The impression was one of "formal elegance". Michael Kimmelman in the *New York Times* found that despite the earthiness of its subject the piece tended to be too detached, at moments projecting charm rather than all-consuming passion, the cruel reality lurking beneath the surface.

Analysing the pervasive influence of the *commedia dell'arte* and detecting its ubiquitous presence in modern theatre, Martin Green and John Swan see its starting-point as Paris in the 1830s and 1840s. In their book, *The Triumph of Pierrot*, they assert: "The tawdry Théâtre des Funambules soon became a favorite haunt of the literary *avant-garde*, including Flaubert and Gautier, both of whom wrote scripts for Deburau. Their enthusiasm was inherited by the later generation of Verlaine, Rimbaud and Mallarmé, who turned cheap theatre into poetry. By this devious route the *commedia dell'arte* came to the attention of Western artists in all media, and they took it up with gusto. Between about 1890 and 1930, 'Commedic' images were everywhere, from painting to dance to drama. They declined in prominence thereafter, but we can still find them all around us if we know where to do that." Green and Swan proceed to prove that, perhaps making exaggerated claims to uphold their insight. But they are doubtless right in perceiving an essential affinity between the bold eccentricity and seeming spontaneity of the *commedia* and the ballets and plays of Jean Cocteau (*Parade*), Eugene Ionesco (*The Chairs, The Lesson, The Bald Soprano*), Friedrich Dürrenmatt (*The Marriage of Mr Mississippi*), Bertolt Brecht (*The Resistible Rise of Arturo Ui*), Samuel Beckett (*Waiting for Godot, Endgame*), as well as the whole Theatre of the Absurd. It is to be found in Luigi Pirandello's *Tonight We Improvise* and *Right You Are (If You Think You Are)*, and in many of the paintings of Picasso and Rouault – whether deconstructed women or rueful clowns – and even in the poetry of T.S. Eliot (*The Love Song of J. Alfred Prufrock*) and the mocking, ironic Halloween parades in bohemian Greenwich Village, most often anywhere that public demonstrations and private art are infused with wild fun and personal disenchantment.

Apart from the circus with its clowns, exhibiting their broad antics and preposterous costumes, however, nowhere is the *commedia dell'arte* still to be seen more conspicuously than in the busy parks of London, Paris and other large cities, where the Punch and Judy marionette shows in their gaudy weathered booths continue to evoke outbursts of excited high-pitched laughter from the very young. Punch, the ugly, cantankerous, selfish, violent husband, always berating and striking with a stick at his poor Judy, is the same old, ill-tempered if not ferocious Pulchinello, here an animated, wooden-leather puppet, but hardly lacking verisimilitude.

— 4 —
ITALY: WORDS AND MUSIC

Italian Renaissance writers and scholars were aware that music and dance had accompanied Greek and Roman theatre. With an idolatrous regard for all things classical and a reverent and noble effort to revive the artistic practices of a distant, idealized past – even if it involved delving anew into pagan mythology – the foremost intellectuals of the day debated in their academies and cenacles just how the plays had been staged. Of the nature of Hellenic music, no certain knowledge existed. Illustrations on vases and bas-reliefs on ruined temple friezes depict figures, often only barely outlined, with a variety of instruments – pipes, lyres, tabors – but of course they give off no sound nor indicate rhythm. To Italian scholars then, as to learned classicists now, the chanting and instrumental enforcement heard along with the spoken lines and fervent arias of the dismayed characters in Sophocles' and Euripides' plays have been for ever muted.

The comedies of Aristophanes, too, had interpolated patter songs and lively dances, as in Roman times did the farces of Plautus and Terence and their imitators, but no helpful description of them is available.

In ensuing centuries, especially during the Middle Ages, an increasingly pleasing and expressive art of music had evolved, with new instruments and new theories, some of ever-growing complexity. (It is not within the scope of this chronicle to include the technical aspects of music, but only its application to theatre works, along with some account of the most prominent composers who above all had in mind writing for the stage, creating characters and dramatic situations that were to be interpreted not with music but *through* music.) Among the new forms were simple ballads, the narrative stock-in-trade of minstrels and troubadours, and motets, madrigals and liturgical plainchant – the Easter sketch *Quem Quaeritis* was recited in unison plainchant, as were other scenes in the re-enactment of Christ's death and resurrection. Later this was true of dramatizations of the poignant, miraculous Christmas story; the Church recognized the necromantic, hypnotic power of music. In Winchester Cathedral a tenth-century offering of the *Visit to the Tomb* was backed by a 400-pipe organ that did much to enlarge an emotional response in stirred, uplifted, humble worshippers. Everywhere staged portrayals of the Stations of the Cross and the Passion were strengthened by the insertion of music at the most important moments.

Gregorian chant, unaccompanied single-line melodies, was next superseded by ingenious and inspired combinations of two or more melodies, an interweaving of them that led to polyphony, or singing in parts, a many-voiced music. Eventually, over the long stretch from the tenth to the twelfth century, such contrapuntal writing became so elaborate and busy that listeners could scarcely catch the words to which the music was set. Much later the reformer Giovanni Pierluigi da Palestrina (1526–94) set an example by restoring a necessary transparency in the splendid masses, breviaries and chants he composed for the organ in Rome's glorious Santa Maria Maggiore. This new clarity prepared the way for the human voice to take the foreground once again, rising above the instruments, even dominating them and making some kind of musical drama (or melodrama) more feasible.

Music was already employed to heighten the effect and appeal of semi-liturgical works, often delicately with bells and softly struck drums as in the *Play of Daniel* (Cathedral of Beauvais, thirteenth century). Far earlier, the mystic Abbess Hildegard of Bingen (1098–1179), having written and produced a morality play, *Ordo Virtutum*, at her Rhine Valley monastery near Rupertsberg, gave the nuns performing the allegory over eighty distinct melodic motifs in depicting a battle for the human soul between the sly, heartless Devil and the Sixteen Virtues. Music was intrinsic to *The Play of Herod*, and to the Spanish *Representación del Nacimiento Nuestra Señor*, a fifteenth-century Nativity piece by Gomez Manrique, and to many others detailed on earlier pages, but it was intermittent, a song or psalm here and there, not continuous or all-encompassing.

Secular playlets, such as the contrived *Le Roman de Fauvel*, also made generous use of music. This farce was put together by choosing and combining in a single narrative text incidents from a collection of over 6,000 medieval "fauvels" (verses). Music by various hands was affixed to them soon after they first appeared; much of it has been found on manuscripts in scattered places, some segments monophonic, some polyphonic. Returning from obscurity in the early 1950s, the manuscripts prompted stage productions. In the view of Michael Jaffee, leader of the Waverly Consort, "the music goes beautifully with the Fauvel verses. It has some light moments, but much of it is bound up with heavy metaphysical meaning and is deeply emotional." Most medieval farces, well exemplified by the *Le Roman de Fauvel*, paused at intervals to allow passages of singing and dancing.

Established by the Athenians, this ancient, sprightly tradition was perpetuated during the Italian Renaissance by the ebullient *commedia dell'arte* troupes who had repertoires of songs and proved themselves agile dancers. This had been true of actors in the days of Aristophanes and Menander, of Plautus and Terence. The intelligentsia's desire was to emulate faithfully classical theatre.

But serious drama? Tragedy? What sort of music had the Greeks developed? That was an important question. Pastorals, such as Angelo Poliziano's *Fabula di Orfeo* (1480), given in Mantua, further stimulated interest in the problem by including songs, as did Guarini's cherished *Il Pastor Fido* (1584), a lighter work embellished with music contributed by a profusion of contemporary composers, as were two Florentine productions, Emilio de Cavalieri's *La Disperazione di Fileno* and *Satiro*, both 1590, and both no longer known to exist. There was also Adriano Banchieri's *Prudenze Giovanile* (1607) in which the performers mimed the story, while singers and instruments stayed at a remove behind them, perhaps out of sight – hardly an ideal or assured solution to staging and adding music in an authentic Greek manner!

The speculation about this was lively in Florence. One who had daring theories on the subject and about what forms music should take was Vincenzo Galilei, father of the bold, world-famed astronomer Galileo Galilei, who later barely escaped the Inquisition. A shopkeeper and lute player, Vincenzo Galilei was also a very able mathematician and a member of an intellectual circle, the Camerata (or People of the Chamber), which held frequent meetings in the ornate palace of Count Giovanni Bardi (1534–1612), the group's enlightened sponsor.

Members of the circle engaged in lengthy debates about Aristotle's explicit rule that a tragedy must include a musical element. The great dramatists were composers. Vincenzo gained prestige by publishing an esoteric essay, *Dialogo della Musica Antica e della Moderna*, in which he propounded that by proceeding "from the logic of numbers" one could discover the "calculable secret of musical notes". Displeased with the "courtly music" of the day, he deplored its impropriety and castigated it as "a depraved, impudent whore". He attacked the over-use of counterpoint. He insisted that a musical accompaniment should be less prominent than the words, the poetry, to which it was set; they should be heard more distinctly; and to show exactly what he meant, he composed music – subordinated – to passages from Dante's *Divine Comedy* and Jeremiah's doleful Lamentations as examples of how it should be done. His fellow club members were more or less persuaded to his view.

The earnest participants of the Camerata held meetings for about twenty years (1573–92). Of Count Bardi, his admiring son Pietro later wrote:

> He always had about him the most celebrated men of the city, learned in this profession (music), and inviting them to his house, he formed a sort of delightful and continual academy from which vice and every kind of gambling were absent. To this the young nobility of Florence were attracted with great profit to themselves, passing their time not only in pursuit of music, but also in discussing and receiving instructions in poetry, astrology, and those other areas of knowledge that in turn lent value to this pleasant converse.... Giulio Caccini, considered a rare singer and a man of taste although very young, was at this time in my father's Camerata, and feeling inclined toward this new music he began, solely under my father's instructions, to sing *ariettas* (short songs), sonnets and other poems suitable for declamation, accompanied by a single instrument, in a way that astonished his hearers. Also at this time in Florence was Jacopo Peri ... like Giulio, he sweetened his style and rendered it capable of moving the passions in a rare manner.

(This quotation is from *Opera: A History* by Christopher Headington, Roy Westbrook and Terry Barfoot. Other books that have been very helpful here are *The Stream of Music* by Richard Anthony Leonard; *The World of Opera* by Wallace Brockway and Herbert Weinstock; *The*

Understanding of Music by Charles R. Hoffer; *Grove's Dictionary of Opera*; and several more to be cited later.)

As a result of their discussions, theories and experiments, the members of the Camerata, though most were amateur musicians at best, achieved a breakthrough. Here is how Ethel Peyser and Marion Bauer put it rather simply in *How Opera Grew*: "In passionately trying to imitate the Greek drama with its differently inflected language from the beautiful Italian, they invented song speech or recitative *(stile parlante)*, speech half sung, half declaimed, with more or less musical accompaniment. It is well that they were brave enough to try the impossible, for in doing so they fell upon this device to connect song, scene and dance, which in the hands of future professionals led to the building of a new and beloved art – opera."

The use of recitative kept music flowing throughout a stage-work. The next advance would be to have the words more fully backed by sound, the music continuously blending with and rounding out and increasing the emotional force of the action and dialogue, while suggesting greater, unspoken depths in the characters. At last, a work of this sort was attempted by Jacopo Peri (1561–1633).

In 1592 Count Bardi moved to Rome but the Florentine Camerata carried on, assembling in the house of a younger man, Jacopo Corsi (1561–1604), a munificent patron of music. The contentious Vincenzo Galilei had died in 1591. Somewhat before Peri's initial effort at a full-fledged musical drama, Emilio de' Cavalieri (1550–1602) brought out an oratorio, *La Rappressentazione di Anima e di Corpo* (*The Story of the Body and of the Soul*), in essence a Morality, combining music and scenery and resorting to monody; that is, a tune and accompaniment having a text largely connected by recitative; in fact, a too extensive amount of it. The words were by his sister Laura, active in the Camerata. In a preface, de' Cavalieri conveyed his ideas as to how a composer should go about fashioning a drama consisting largely of music: "The instrumentations should change according to the emotion expressed. An overture or instrumental and vocal introduction are of good effect before the curtain rises. The *ritorno* and *sinfonia* (used as the opener or overture to the drama with music) should be played by many instruments. A ballet, or better a singing ballet, should close the performance. The actor must seek to acquire absolute perfection in his voice . . . he must pronounce the words distinctly. . . . The performance should not exceed two hours. . . . Three acts suffice and one must be careful to infuse variety, not only into the music, but also the poem and even the costumes." These prescriptions were both practical and visionary.

But Peri was a few years and steps behind him. For his venture he chose a mythological subject, *Dafne*, and a libretto prepared for him by the poet Ottavio Rinuccini (1562–1621), taken from Ovid's *Metamorphoses*. In it, Apollo, having slain the Pythian dragon, boasts of it to Cupid, who is offended at the god's display of pride. To humble him and assert his own power, Cupid acts to have Apollo fall hopelessly in love with the river nymph Dafne (Daphne), while – to escape the god's beckonings and advances – she is changed by Zeus into a laurel tree (or perhaps the flowering sweet-scented shrub known today as Daphne). To unfold this pretty tale, a typical pastoral, Peri had his lead singers augmented by a chorus and a good-sized orchestra. Pietro Bardi was present, naturally, and recalls: "I was left speechless with amazement." Actually, some of the songs were not by Peri but by his fellow Camerata member, Giulio Caccini, Count Bardi's erstwhile protégé; and by one account a further contribution to the score came from the group's host, Jacopo Corsi.

The first staging of *Dafne* took place in 1598 before a small, very select audience. One scholar holds that the true date is 1594. Peri declared that his aim was "to imitate speech in song" and that he had accomplished this with "elegance and graces that cannot be notated". In addition, he himself took the role of Apollo, a part for which he deemed himself quite suited: he was still young, twenty-eight, slender, with long blond (some say red) hair. The cast was properly costumed for the period.

Dafne, a slight but epochal first opera, was exceedingly well received and given a second performance, and possibly even more hearings thereafter. Unfortunately most of the score and text have long since vanished, only a few fragments remaining. Better luck has attended Peri's next offering, which, encouraged by the praise he had won with his first effort, followed shortly. Once again he turned to Ottavio Rinuccini and Greek myth for his verbal text. *Euridice* (1600) lends itself nicely to musical treatment since it revolves about the desperate, plaintive search by Orpheus, the wandering, charming poet–singer, through the fearsome depths of the underworld to which a serpent's fatal bite has harshly consigned his adored bride. By the irresistible seduction of his music he hopes to free her from Pluto's realm and lead her back to life in his own bright, upper space. It is so logical and appropriate a subject for opera that many other composers have also elected to elaborate on it through the centuries. The role of Orpheus was claimed by Peri himself.

For this enterprise Caccini seems to have been a full collaborator; the occasion for which the opera was undertaken was the historically important marriage of Maria de' Medici to Henri IV of France, an event that had many major consequences for the arts of Western Europe, since Maria travelled to Paris with troupes of Italian singers and dancers as part of her entourage.

More fluent and melodious, *Euridice* is scored for a harpsichord, a bass viol, two lutes and three flutes, an ensemble not much like a modern orchestra, which would probably sound strange today. Like *Dafne* it belongs to the genre of the pastoral, with bucolic scenes of shepherds and nymphs, telling its fanciful story in ten episodes. In the myth, Orpheus fails to succour Eurydice except fleetingly; warned not to glance back at her as he leads her ephemeral shade from the infernal regions, he cannot resist doing so and consequently loses her for ever. Not so in the Peri–Rinuccini version: Orpheus and his beloved reunited, the ending is joyous. A melancholy curtain would hardly have been welcomed by aristocratic guests at a wedding festivity; the composer and librettist knew how to cut their cloth to fit the royal event. In more detail:

> [The] prologue tells us that fear and sorrow yield to the sweeter emotions evoked by music. *Euridice* features solo recitative, in other words free and declamatory vocal writing, and there are also actual songs, for example in strophic form with succeeding verses sung to the same melody, and choral writing in four or five parts. As for the recitative style, Peri declared in his preface that he had aimed at "an intermediate course, lying between the slow and suspended movements of song and the swift and rapid movements of speech. . . . I judged that the ancient Greeks and Romans (who in the opinion of many sang their staged tragedies throughout in representing them upon the stage) had used a harmony surpassing that of ordinary speech but falling so far from the melody of song as to take on an intermediate form."
>
> [From *Opera: A History*]

It should be added that pretexts are found to include dances, and Eurydice rather than Orpheus has the stellar part; his name is omitted from the title.

Euridice's complete score is extant, revealing Peri's first draft and Caccini's additions. The opera was given beyond Florence in Bologna in 1616, and there have been revivals of it in the twentieth century in Italy and Germany. (A special performance was mounted – most idiomatically – in Florence's Boboli Gardens (1960) near where it was first staged. Bruno Rigacci conducted an orchestra of twelve instruments: lute, viola de gamba, viola da braccio, basso de viola, flauto diritto and spinet.)

The *mise-en-scène* for the gala was entrusted to Bernardo Buontalenti (1536–1608) who had designed and built the great court stage in the Uffizi and had frequently arranged other sumptuous fêtes. He created "wonderful woods", a pastoral setting meant to be a contrast – specified by Rinuccini – to the grim horrors of Hades from which the heroine is fleeing. It is assumed that the suggestion of a change of place was effected by revolving wooden prisms, much like those used in Ancient Greece – a device already employed a decade and a half earlier in a production of Bardi's comedy, *Amico Fido* (1585), which the resourceful Buontalenti had also mounted.

Three days later the stage in the Uffizi was occupied by *Il Rapimento di Cefalo*, an entrant by Caccini, with a libretto by Gabriele Chiabrera. It was designed by Buontalenti, the expenses borne by the city of Florence as a gift to the happy couple. (Henri IV was not actually present at the wedding; his surrogate was the bride's uncle, Ferdinando, Grand Duke of Tuscany.) By official count, 4,000 guests were in attendance, though some historians are sceptical of that figure. As related by Berthold: "Scenic miracles were revealed when the ornamented red silk curtain dropped: the gold chariot of Helios, the magnificent throne of Jupiter, mountains vanishing into the ground, whales popping up and down, frightening earthquakes, and lovely fields scented with perfume."

Apparently Caccini was not satisfied with the work he and Peri had fashioned collaboratively. Within two years he brought out a second *Euridice* (1602); this time, as with the *Cefalo*, he was sole composer, with passages that he had contributed to its predecessor freely lifted and re-employed.

He retains Rinuccini's poetic libretto but dispenses with the overture. He recruits a harpsichord, a lyre (perhaps with twenty-four strings) and a large lute to serve as the orchestra and support the recitative, and gives instructions that the musicians be out of sight, hidden behind the scenes. Again, more detail: "The action in *Euridice*, carried by the solo voice in recitative, was slow and steady. It was poetry little more than chanted. Caccini used mainly background music. At the end of every verse, he wrote a cadence, a 'falling' or ending. These cadences became monotonous. Occasionally there are songs or dances (a forerunner of Italian ballet) at the ends of the scenes. This early 'try' at opera had little expression. But it was dressed in rich costumes and scenery, because of which opera has always been the 'sport' of the wealthy." (From *How Opera Grew*.) This *Euridice*, too, was exhibited in the Medicis' huge Pitti Palace, where the other two works had been staged. Nevertheless, just by being more simply written, it seemed to gain intimacy.

That Caccini gave emphasis to the voice, more so than Peri, was to be expected. He himself was pre-eminently a singer, as was his daughter, Francesca, later alluded to as the first *prima donna*. For her, in most of his other works, he wrote roulades, cadenzas and other fioritura, founding a tradition that opera is for spectacular vocal display. The daughter, too, was an aspiring composer.

Peri's subsequent experiments with the operatic form, mostly lost, included *La Flora* (1628), which survives; but he provided only the music for Clori, the leading female character. Her second-act aria calls for virtuosity, indicating that after three decades he had progressed towards a more free and ornamented vocal and instrumental style.

Caccini won considerable respect with his *Le Nuove Musiche* (*The New Music*), a collection of monodies – one-line melodies – which were viewed as the "modern" music of the day. In addition, as one critic remarks: "He also started the solo song on its enticing way." He excelled at turning out madrigals and, while visiting France, supplied scores for ballets.

Yet these two men, though immensely influential – they introduced recitative in the theatre and a continuous flow of music to augment the dramatic action – had limited talents; paradoxically, their music is thin and repetitive. They launched a great art form but hardly realized the full heights it could attain. To achieve that, a man of genius was needed, who fortunately was at hand, though not a member of the Camerata and not a Florentine.

Claudio Monteverdi (1567–1643) was born in Cremona, a small medieval city and an important centre of music-making, since there were the workshops of the foremost violin-makers of all time: Stradivarius, Guarini, the Amati. Other stringed instruments were crafted there, too. Claudio's father, a barber–surgeon, was noted in his dual professions. The boy's mother died when he was ten. Under his father's guidance he was well schooled, eventually attending university. His home was near Cremona Cathedral; there, at an early age, he took his first music lessons with Marc Antonio Ingegneri, choirmaster and composer, a man of considerable reputation who inducted his gifted pupil into the difficult art of polyphony, which the youth readily mastered. It was soon apparent that this impetuous young Monteverdi had an innate feel for the dramatic and a propensity for breaking rules almost before he fully possessed them. Precocious, as are most master composers, he had his efforts at sacred music for three voices issued by a press in Venice when he was only fifteen (1582).

At twenty he published a book of madrigals (1587) that clearly showed theatrical flair. They combined aspects of polyphony and monody, a new style and genre that he is credited with having originated. In such a piece, dubbed a chamber cantata (*cantata da camera*), one person recited a complete story of a strong dramatic incident with an instrumental accompaniment. This was an excellent preliminary exercise for a future composer of opera. The young man's offering met harsh criticism, however. A monk, Giovanni Maria Artusi, in a diatribe *Imperfections of Modern Music*, charged him with being heedless of "natural laws . . . Though I am glad to hear of a new manner of composition, it would be more edifying to find in these madrigals reasonable *passagi*, but these kinds of air castles . . . deserve the severest reproof. . . . Behold, for instance, the rough and uncouth . . .". It was young Monteverdi's conviction that in dramatic storytelling all music should not be unalterably sweet and melodious that equipped him to write powerful works (quoted from *How Opera Grew*).

His exciting madrigals won him the favour of Mantua's theatre-patron Duke Vincenzo Gonzaga, who, around 1590, hired Monteverdi to entertain at his art-loving court. His first post there was very likely in the string section of the court orchestra for the extravagant ballets sponsored by the duke, but his talent very rapidly gained him higher rank. As the duke travelled about Europe, visiting other courts and capitals, he included musicians – among them Monteverdi – in his retinue. This exposed them to lavish events abroad: pageants, fêtes, regal weddings, mock battles and – at moments – the noise of nearby real ones in the war raging between Turkish invaders and defending Austrians. He also made the acquaintance of a throng of composers and poets in France and Flanders, where musical theory and accomplishment were at a higher point than in Italy, and he was able to learn much.

In 1599 Monteverdi married a woman named Claudia. Their first son was born in 1601, the same year Claudio was promoted to be *maestro di cappella* at the Mantuan court. He continued to display boldness and originality in his musical output, still mostly madrigals. In response to a critic who faulted him for unconventionality (1605), he stated that "with regard to consonances and dissonances there is yet another consideration different from those usually held, which defends the modern style of composition while satisfying reason and senses" (quoted from *Opera: A History*). He was not afraid to write "unbeautiful" music if he deemed it appropriate and necessary to do so. He grew in skill, providing dance music and studying balletic structure and what expertise was required of a choreographer.

One of Duke Vincenzo's sons, Francesco, was educated abroad, at Innsbruck and Pisa. He had genuine aesthetic tastes, especially for music, and himself dabbled in writing poetry and songs. In Florence he acted as "protector" of the Accademia degli Elevati (Academy of the

Elevated Ones), to which Peri and Caccini belonged. His brother Ferdinando also had artistic ambitions. He created the text and music for a ballet put on at the Medici court in 1606 while the imperious rulers of Florence were at their colony, Pisa, for Carnival. A year later Ferdinando followed this first venture with a second theatre work, a comedy with music, now lost. He sent a copy of it to Francesco. The two brothers' enthusiasm for these endeavours led them to compete. Francesco was inspired with the idea of having an opera like those of Peri and Caccini produced at their own Gonzagan court in Mantua. The thought was first bruited in 1602, then dropped for some reason, but in 1607 it came to fruition. It was not difficult to persuade Duke Vincenzo, always engaged in rivalry with the Medicis in artistic matters. What is more, the Mantuan court had all the requisite resources constantly employed there: musicians, designers, performers. Duke Vincenzo had seen Peri's *Euridice* in Florence. His extravagant court was not to be outdone.

Monteverdi, now a ripe forty, was chosen to compose an opera. It is likely that at some time he had visited the Camerata and listened to the debates – there is one listing of his name – and also possible that he had been an invited guest at the élite premières of *Dafne* and *Euridice*, or at least at one of them. The court chancellor (or secretary), Alessandro Striggio, himself a musician, was assigned as librettist. The predictable subject was another treatment of the Orpheus and Eurydice myth, but Monteverdi far outstripped the previous versions of Peri and Caccini, and his *Orfeo* is the first of his enduring masterpieces.

It was meant to enrich Carnival. To avoid its having a depressing ending, Eurydice is not spared her wrenching return to Hades, but Apollo appears to his offspring Orpheus to promise the pair will be together once more in the Elysian afterworld, a joy deferred, not irrevocably denied. Letters between Francesco and Ferdinando tell of the stresses and difficulties encountered during rehearsals, especially in casting the singers, most of whom were from the Mantuan troupe, but a few of whom – among them the young castrato Giovanni Gualberto Magli, a Caccini pupil – were borrowed from those serving at the pleasure of the Grand Duke of Tuscany.

The text of the opera, printed in advance, was distributed to the duke's guests to aid them in following the story. It is believed that performance took place in a room in the Gonzaga palace, presenting a problem, as the stage would not be large enough to accommodate the *deus ex machina*, the machinery for the ultimate ascent of Orpheus and Apollo, son and father, to a heavenly realm. Anticipating this, two versions of the ending were shrewdly provided by Monteverdi and Striggio, based on Latin variations of the Greek myth. As was Peri's opera, this new treatment is in five acts, with a prologue, each act concluding with a chorus. This accorded, more or less, with the structure of both an Athenian tragedy and the Italian pastoral form, such as *Il Pastor Fido*, then so popular. It is suggested that the work was played without pauses between the acts, the action and music flowing on continuously. (This is helpfully discussed by John Whenham, in an essay, "Orfeo: A Masterpiece for a Court", included in *Monteverdi*, a collection of articles edited by Nicholas John, to which there will be a number of future references.) The libretto contains explicit stage-instructions, and probably sliding flats and back-shutters were installed to facilitate the visible changes of scene from earthly, sunny meadows to the dismal underworld and back again.

The choruses at the act-endings are used to comment and pass judgement on the characters and their deeds and decisions, the final moral being that "only the man who can control his passions is worthy of eternal glory". This would satisfy (or pacify) any representatives of the Church in the audience as this celebration of pagan virtues and deities unfolded.

Musically and dramatically, the scope of *Orfeo* far exceeds anything reached or dreamed of by Peri and Caccini. Monteverdi has an orchestra of thirty-eight instrumentalists, comprised of fifteen string players, along with those competent at the keyboard and with brass and winds. The writing for voices is sensitive and varied, truly expressive of the feelings engendered by the changing situations. At moments, there were ballets for nymphs and shepherds, as well as Bacchic and Moorish dances. From Whenham:

> The musical fabric of *Orfeo* is a mixture of traditional and innovatory elements. Its rich instrumentation, with groups of "pastoral" and "underworld" instruments, was derived from the sixteenth-century *intermedio* tradition, as was its choral writing. The most striking new element was, of course, the one which made it possible to set a whole play to music: continuo-accompanied recitative.... The voice could be allotted a number of notes against a sustained bass, freeing the singer to declaim the text in a rhetorical manner.... In the hands of Peri and Monteverdi, "recitative" became a medium capable of conveying a whole gamut of emotions. ... The concept of "recitative" also embraces that of "aria" in early opera, for "aria" here means not only "tunefulness," but also, specifically, the setting of a strophic text.... Technical explanation of the musical styles of *Orfeo* is, in a sense, unnecessary. Given a sympathetic performance, the power of Monteverdi's musical language speaks directly across the centuries without the need for apologists. It is for this reason that his first opera has enjoyed nearly a century of revivals and it also explains why, though the courtly society for which it was written has long since disappeared, *Orfeo* retains a place in the repertory.

Immediately successful, *Orfeo* had two more performances at Duke Vincenzo's order. Monteverdi's name began to spread beyond Mantua's narrow borders. Pleased, the duke commissioned him to write another music-drama to be staged the next year for the festivities at the marriage of Ferdinando to Princess Margherita of Savoy. But Monteverdi's gratification at this was tempered by the serious illness of his wife during his composition of *Orfeo* and his work on its successor. Before his completion of the new project, Claudia died. His grief was profound. But he was called back from Cremona, where he had gone to bury her.

Duke Vincenzo was anxious to impress and outdo the Medicis. He commanded Monteverdi to double the size of his orchestra for this next opera and personally train its members. The duke wanted the new piece to be filled with "furore, pomp, and magnificence". This put an added strain on the weary, grieving composer. Striggio was again the librettist. *Arianna* (1608) borrowed the somewhat tragic legend of Ariadne and Theseus. The Cretan princess, daughter of Minos, after giving her ruthless foreign suitor the thread leading him through the labyrinth and enabling him to slay the Minotaur, sails away with him, only to be abandoned on the island of Naxos. Solitary, she bewails her fate; until at last she is succoured and marvellously uplifted to dwell among the gods, and there becomes Dionysus' bride.

The audience for *Arianna* could hardly have been the most select, for it said to have numbered at least 6,000, surely more than might be properly designated as Mantua's blooded or intellectual aristocracy. Among them was Federigo Follino, the court chronicler, who left this account, possibly somewhat exaggerated:

> The work was very beautiful both because of those taking part, dressed in clothes no less appropriate than splendid, and because of the scenery, showing a wild rocky place among the waves, which in the further distance could be seen continually in motion, giving a charming effect. But since to this was joined the force of the music, by Signor Claudio Monteverdi, maestro di cappella to the Duke, a man whose worth is known to all the world, and who in this work excelled himself, combining with the blend of voices a harmony of instruments behind the scene which always accompanied the voices, and as the mood changed so was the instrumental sound varied; and as it was acted by men and women who were excellent singers, every part succeeded well and most especially wonderfully in the lament which Ariadne sings on the rock when abandoned by Theseus, acted with much emotion and so piteously that no one hearing it was left unmoved, nor among the ladies was there one who did not shed a few tears at her plaint.[Quoted from *Opera: A History*.]

That the orchestra played "behind the scene" indicates that Monteverdi was following Caccini's example in concealing it. It is interesting that Monteverdi is described this early as being known "to all the world".

It is thought that *Arianna* was mostly monodic, though obviously there were some arias. By costly mischance the text survives but the score has not been preserved – only Ariadne's heart-rending plaint, to which Follino alludes, remains. Indeed, the spectators were so haunted by it that it was soon to be heard everywhere: it is now spoken of as opera's first hit song. In it, Monteverdi proved himself able "to move the passions". Many have guessed that this affecting aria reflects his sorrow at his wife's death. The aria was eventually adapted as a popular song, 'The Flowers of May'.

Before the opera was presented, a kinsman of Duke Vincenzo declared it was too "heavy" to be enacted at a wedding celebration; it should have some comic relief. The duke agreed, and Monteverdi was forced to accept the suggestion. He turned not to Striggio the librettist, but Rinuccini, Peri's collaborator, to insert a humorous dance piece, *Il Ballo dell'Ingrate* (*The Prude's*

Ball), which was performed elegantly in "the French style", with a sumptuous *mise-en-scène*, the dancers lavishly clothed in silks and satins, the gods and goddesses descending and ascending "garishly arrayed ... in glorious sheen".

Such light interludes were hardly Monteverdi's *forte*. Over the next few years he started work on other operas but abruptly or reluctantly dropped them; they were not concerned with human beings. One was about the sea, *Le Nozze di Tetide*, a subject that he found inadequate and left unfinished. In a letter he wrote (1616): "I see the characters are winds ... how can I imitate their speech and stir the passions? Ariadne moved the audience because she was a woman, and Orpheus because he was a man and not a wind. ... I find that this tale does not move me at all." He tried again and again, creating a variety of characters, Mercury, Aeneas, Mars, Prosperine, Andromeda, Lavinia, figures from myths and epics mostly for projected theatre pieces now vanished like so much else of his superb output.

His remaining years in Mantua were unhappy. His health never fully recovered from the insidious toll taken of it during the arduous year in which he composed *Arianna* and lost his wife. The city's dank climate and marshy environs constantly oppressed him. In 1612 Duke Vincenzo died and was succeeded by his son, who chose to dismiss Monteverdi from his long-secure position at court. It was a fortunate turn in his life and career.

Opportunely, he received an offer to be *maestro di cappella* at San Marco, Venice's vast opulently Byzantine *duomo*. In the city on the Adriatic, where he spent the rest of his days, he had far more creative freedom, no longer obliged to comply with the whims of often opinionated rulers; though the times in the Republic were hardly serene, its air echoing and tainted with feuds, wars and political intrigues, famines and plagues. Over the decades he was occupied with composing exquisite madrigals, sacred music and more operas – most of these aborted. He wrote a highly dramatic cantata based on verses by Tasso, *Il combattimento di Tancredi e Clorinda* (*The Duel between Tancredi and Clorinda*; 1624) – during the Crusades, a Christian knight seeks to slay his Saracen opponent, unaware that the weak but valiant foe is his beloved in male disguise. It was not intended to be staged, and much later the subject has appealed to several twentieth-century choreographers who have chosen less effective music to accompany their dancers.

A plague took the life of Jacopo Peri. Monteverdi, too, fell ill and vowed to embark on a pilgrimage to the Holy Land if he recovered. Instead, when well again, he returned to composing operas, very often a vain endeavour. These later works were *Andromeda* (1618–20, unfinished), *Apollo* (1620, unfinished), *La Finta Pazza Licori* (1627, commissioned for Mantua, unfinished). *Gli Amori di Diana e di Endimione* (1628, commissioned for Parma), *Mercurio e Marte* (1628, commissioned for Parma), *Prosperina Rapita* (1630), *Le Nozze d'Enea con Lavinia* (1641), *La Vittoria d'Amore* (1641, commissioned for Piacenza). Scores and texts of all these have perished, save for a single excerpt, a trio belonging to *Prosperina Rapita*. That he received a steady flow of commissions throughout these years, with most of the contracts never met, evidences that he was very highly esteemed and remained unsparingly self-critical. He also contributed a ballet for Vienna, *Volgendo il Ciel* (1636?).

The largely intact scores and librettos for two of his later operas have been saved, enriching the world's repertory. They must have set an almost insuperable standard for his contemporaries, showing how dramatic and emotionally stirring an opera could be, and this pair are still frequently performed. *Il Ritorno d'Ulisse in Patria* (*The Return of Ulysses*; 1640) turns to a climactic episode in Homer's *Odyssey*. During Ulysses' long absence, while he participates in the ten-year siege of Troy and then meets endless obstacles throughout his stormy decade-long journey homewards, his faithful Penelope is beset by would-be suitors whom she wards off with desperate and ingenious ploys.

Ulysses' torments are visited on him by Neptune, whom he has offended. Finally Minerva comes to his assistance. (The deities – Jove, Neptune, Mercury, Juno, Minerva – mix with humans in the prologue and main story.) The Goddess of Wisdom, Minerva, has long favoured the wily Ulysses, by far the craftiest of all Homeric heroes. Now she miraculously changes Ulysses into a seeming lowly shepherd so that he can proceed to Ithaca without being recognized and take his wife's wooers by surprise; even his loyal bondsman, humble Eumaeus, does not know him. To prepare for the father's arrival, Minerva also brings Telemachus, Ulysses' son, back from Sparta to his island home.

After revealing his identity to Eumaeus, who joyfully welcomes him, Ulysses undergoes another magical change, now transformed into a beggar. This is wrought by a fire-bolt from Heaven. He seeks bread and alms from the usurpers of his palace, who brusquely deny his plea. Penelope, however, is generous in her response. Though he drops his ragged disguise, he has difficulty convincing her that he is in fact Ulysses, her mate – away for twenty years. She habitually fears a trick. To delay her suitors – Antinous, Peisander and Amphinomous – from progress in their relentless siege, Penelope has proposed that they demonstrate their prowess with Ulysses' huge huntsman's bow. They lack the strength to wield it . . . its strings are taut . . . whereupon Ulysses steps forward, seizes it, fits it with arrows and without hesitation kills his three rivals.

Act Three finds Penelope unpersuaded by Telemachus' assertion that the beggar is most certainly his father. Minerva and Juno appeal to Jove on Ulysses' behalf, but Penelope is obdurate in her disbelief; until the old nurse, having seen the stranger in his bath, notes an old hunting scar. In addition, Ulysses describes an embroidered quilt that once covered their nuptial bed. Happily, the pair embrace, and the opera concludes with a love-duet.

The librettist was Giacomo Badoaro (1602–54), a Venetian nobleman, amateur poet and member of the Accademia degli Incogniti, a literary circle reputed to have been somewhat "libertine". It had a significant influence on the first operas chosen to be heard in Venice, since its members had strong opinions and funds to support what they liked. The inspiration for a work about Homer's hero seems to have been Badoaro's. He submitted the scenario to Monteverdi with a letter saying that he had written this text for a drama with music hoping "to excite

Your Lordship's skill to make known to this city that in the warmth of emotions there is a big difference between a real sun and a painted one" (quoted by Tim Carter in his *Monteverdi Returns to His Homeland*). Assuredly, the poet had found the right composer, though by now Monteverdi was in his early seventies. In the interim since his *Arianna*, still given and admired, many others had written operas, the form had changed and developed in many ways.

The music of *Il Ritorno* is not pretty or refined, like that in the pastorals of Peri and Caccini, but frequently gruff and "rough-hewn" as befits its subject, an evocation of Homer's semi-primitive world. The leading character is strongly masculine, and the shepherds are not cavorting with dainty nymphs and dryads, for they are aged workers. There is a degree of realism. The score has more arias and less recitative, so that despite the harsher story it carries it is more melodious; instead of depending on speech accompanied instrumentally to advance the plot, Monteverdi has more of both the action and the characters' emotions depicted and given expression in full-throated song. He is very deft, too, at shifting from the passages of recitative to the arias, which are well placed. The arias, shaped by their structured verses, lend the work more cohesion, instead of an impression of looseness caused by lengthy stretches of recitative. A sense of cohesion is also achieved by melodic repetition. He adds a number of orchestral devices that suggest or aurally imitate on precise cue what is happening on stage – perhaps changes of rhythm that echo sharply the gluttonous Irus' nervous stutter, or augmenting the sounds of combat, to give two simple examples. He created what he termed the *stile concitato*, or "aroused style" (a clear technical explanation of which will be found in Tim Carter's essay cited above). To be remarked, *Il Ritorno* calls for a small orchestra – fewer strings – and barely uses a chorus. Possibly his sensible reason for this was to lessen costs of production. Venice, a proud Republic, boasted no prince or duke enamoured of theatre and ready to pour forth his own and the state's treasure to subsidize it.

A regard for expenses might also account for the comparative lack of spectacular stage effects in the work, though there are what might today be considered a good many: when Minerva brings Telemachus through the sky in her chariot to his native Ithaca, the usurpers plan to waylay and kill him, but he is shielded by an eagle sent by Jove; a Phoenician ship metamorphoses into a rock; Ulysses, struck by a lightning flash, vanishes below the earth; the gods emerge from the waves and spray and are seated in Olympus – but these were not enough to satisfy the appetites of Venetian audiences accustomed by now to behold staged wonders. Badoaro, criticized for this flaw in his libretto, offered an apology in a preface to a later work, saying that the want of such astonishing effects was most likely due to his "poverty of invention" – was he serious or ironic?

Il Ritorno d'Ulisse in Patria was well appreciated not only in Venice but Bologna, after which it was presented in Venice again the following year (1641). One perceptive critic wrote: "[The story] was described poetically and represented musically with that splendour that will make it remembered in every century" (quoted by Carter). Shortly afterwards a spate of other operas based on the fantastic adventures of the nettlesome Ulysses came forth from other composers.

Badoaro himself provided the libretto for *La Finta Pazza* (*The Pretended Fool*; 1641) and *L'Ulisse Errante* (*Ulysses the Wanderer*; 1644), both with music by Francesco Sacrati (1605–50), who also won applause, not only in Italy but also in France, where, to appease French taste, a ballet was added by Sacrati to *La Finta Pazza*.

(It is puzzling how addicted to selecting the same narrow range of subjects for operas were these Renaissance and early baroque playwrights and composers. Though innovators and experimenters, in choosing what to write about they were decidedly unadventurous. A conspicuous instance: only a few years after Peri's *Dafne*, a Florentine *maestro di cappella*, Marco da Gagliano, offered another *Dafne* (1608). He regularly attended the gathering of the Camerata and was recognized as the most musically gifted member of that intellectually lofty circle. His *Dafne* used mostly the same libretto earlier fashioned by Rinuccini, the poet who so ably served Peri, Caccini and Monteverdi. Accepted to be part of the festivities during a Gonzaga marriage ceremony that was postponed – for political reasons – this *Dafne* was performed instead at Carnival, under the auspices of Duke Vincenzo. It had substantial virtues and was hailed. By nature Gagliano was overly modest, compulsively finding fault with his own compositions. He never again wrote for the stage. (It is suggested that additions to the score were made by Ferdinando Gonzaga, the prince about to be installed as Archbishop.)

Even superior to *Il Ritorno*, in the view of most critics, is Monteverdi's ninth and next to last opera, *L'Incoronazione di Poppea* (*The Coronation of Poppea*; 1643). The composer was seventy-six. Here, too, it is argued that there are additions by other hands, possibly those of Francesco Sacrati and Benedetto Ferrari, who may have rewritten the music for one role – that of Otho – and the whole of the final scene, but this is not agreed upon by all. There are two manuscript versions of the work; they differ in places, which leads to some confusion for scholars seeking the authentic text and score.

Giovanni Francesco Busenello, the librettist, absorbed his material from the annals of Tacitus, among Italian Renaissance scholars the most hearkened-to of Roman historians. In addition, he inserted as a major character the ill-fated, neo-Stoic philosopher-playwright Seneca, to whose contemplative writings many seventeenth-century intellectuals looked for solace and guidance in coping with the daunting burdens and perils of daily existence, especially in periods of oppression and turbulence. Some of the questions raised by Tacitus, how to seize and hold the effective implements of power, were very like those currently posed in Venice, its local scene for ever one of agitation and ugly struggle, as had been all too true in first-century Rome during Nero's sensational reign. For the *Coronation*'s audience the work had political overtones, applicable to a contemporary situation. It was both topical and the first historical opera.

Like Badoaro, Busenello joined convenings of the Accademia degli Incogniti where such matters were intensely argued. Many of the conclusions reached in those earnest discussions affected the way the characters in the *Coronation* are presented. The consensus arrived at in that circle was that no one is to be trusted. Suspicion and scepticism are the best protective devices.

Especially to be viewed warily are any like Nero who possess absolute power over others, and also intense lovers, who are blindingly in the grip of strong feelings. Beneath the surface they are swayed by conflicting emotions. They are deeply complex, and to understand them one must strip away the outer face – or "mask" – they wear. Very often they lack insight into the sources of their dominating impulses.

Such is Nero as shown by Busenello, whose portrait of him is very different from almost all others by whom he is recalled as an inhuman monster, a psychotic despot. Here he is not crazed, rather, he is deeply and sincerely in love with Poppea, overwhelmed by his passion. But is it love, or is it lust? Is his a profound self-deception?

In an allegorical prologue, Fortune, Virtue and Love enter a contest for supremacy, explicating the thesis of the opera: love triumphs over its two other contenders, as to which has the most control over fallible mankind.

Otho, returning to Rome, finds the emperor's guards outside the house of his mistress, Poppea; he realizes that Nero has claimed her. For consolation, he shifts his affections to Drusilla, a lady of the court. Nero pours out to Poppea his urgent desire; she almost persuades him to rid himself of Ottavia, whom she wishes to supplant. Discarded by Nero, who seeks a divorce, the Empress, desperate and jealous, asks for the advice of Seneca, formerly Nero's tutor and now principal counsellor. She beseeches the gods to punish the faithless Nero. The philosophical Seneca tells Ottavia to bear her lot with restraint and stoic dignity, but she cannot accept that course. Similarly, he fails in his attempt to change Nero from being driven by an irrational passion. Beauty is the "mask" behind which Poppea is possessed by an insidious ceaseless ambition. She convinces Nero that the temperate Seneca prevents his having a successful reign. Nero orders that his counsellor perish, but his death is to be self-inflicted.

Heeding the emperor's sentence, Seneca bids farewell to his pupils. Nero praises Poppea for having rid him of his ex-teacher. Ottavia, in a counter-move, enlists Otho to kill her rival, Poppea. He disguises himself in clothes borrowed from Drusilla to gain entrance to the palace. But Poppea is shielded by the Goddess of Love and the attack fails. Her clothes recognized, Drusilla is arrested and charged as the assailant. She asserts her innocence but is about to be executed – Otho steps forward and rightly assumes the guilt. Nero, surprisingly lenient, banishes him from Rome; Drusilla chooses to share his exile. Suspected of having instigated the plot, Ottavia is also banished and must leave at once. In an eloquent aria she laments her departure from her cherished Rome. Poppea becomes Empress; she and Nero sing of their bliss in an exultant duet.

The unusual moral is that Virtue gives way to the pressures of Love. That is, Nero and Poppea personify political ambition and sexual affinity, and victory goes to them; Otho and Ottavia symbolize the "traditional ethical values" which are the losers. This is in accord with a seventeenth-century rebellion against a stifling morality that had too long prevailed under the Church and its strict teachings, a weight of piety from which Italian Renaissance and now baroque intellectuals were trying to emancipate themselves, while they advanced towards a fresh realism. (This interpretation of the text comes from an essay, "Public Vice, Private Virtue", by Iain Fenlon and Peter Miller, and from a brief, unsigned summary of the work in the *Simon & Schuster Book of the Opera* which is translated from a voluminous Italian chronicle compiled by thirteen scholars.)

Unlike the creations of earlier composers, the characters in the *Coronation* are real people. They are flesh and blood, not the airy creatures that inhabited the *Dafnes*. This is true even of minor figures, such as the teasing, flirtatious page boy Valetto, whose efforts to cadge a kiss from ladies-in-waiting provide a leavening touch, a moment of relief, after the scene in which Seneca nobly, even serenely, bids his disciples goodbye. The action is enveloped in music that at appropriate points is opulent and sublime, achieved by a masterly mind and sure hand. Almost at the start, opera reached a high plateau, one that has seldom been exceeded.

Write Brockway and Weinstock: "His innovations were not the fumbling, graceless thinking aloud of a parched theorist, but the rapid, expressive strokes of a creative genius blocking out the architecture of a new medium, and hampered only by the flimsiest of precedents." Peyser and Bauer explain this more specifically:

> His was the first attempt at building the orchestra into a group of mutually assisting instruments that helped balance each other in tone and volume. Before his day, instruments had been used willy-nilly. No one seemed to know or care whether the lute was drowned out by the bass viol or whether the trumpet drowned out the *gravicembalo*, or indeed, whether a trumpet pictured Pluto, the God of the Underworld, or a delicate maiden like Dafne. But Monteverdi used the soft viol and the quaint and plaintive woodwinds to paint in musical tones the shepherdesses or tender maidens and their sweet songs of love. He used drums and flutes and trombones to paint the fury of battle or storm or struggle. He made experiments in painting with music – and in expressing heart quality.
>
> With him instruments began to live for themselves rather than just to prop up the voice. Their days of slavery to song were almost over. Now they were freed and had a life of their own.
>
> He gave to the stringed instruments . . . a new importance, and to the orchestra a new vitality, tenderness, and usefulness.

The *Coronation* was well applauded at its première and repeated throughout Carnival the ensuing year.

In his last years Monteverdi was severely crippled by rheumatism and further handicapped by failing sight that almost blinded him. In 1643 he gained permission from St Mark's to leave his musical duties for a nostalgic journey to his birthplace, Cremona, and then to Mantua, where he had first won fame. Falling seriously ill, he was conveyed back to Venice where shortly afterwards he died.

As a consequence of Monteverdi's long presence in Venice, the thronging city on the Adriatic became Italy's centre for the production of operas. A covey of his pupils and imitators kept busy grinding them out; some of these works had true distinction, though certainly most did not. Foremost among the truly qualified artistic descendants of Monteverdi whose names retain a patina of lustre for their achievements are Francesco Sacrati – who participated to some extent in fashioning or revising the *Coronation* – Francesco Cavalli, Marc Antonio Cesti and Alessandro Stradella, all four of whom assimilated and continued to work in the "Venetian style" in preparing their scores.

By far the best known composer of this period in Venice is Cavalli, whose actual name was Piero Francesco Caletti-Bruni (1602–76), and whose operatic works gained such popularity that before very long he was Monteverdi's chief competitor. (As was a frequent custom, he adopted the name "Cavalli" in a tribute to a generous patron – his earliest – who bore that patronym and was the Venetian governor of Crema, the town from which the young man came, and where his father had been *maestro di cappella* of the local cathedral and his first teacher.) The youthful Cavalli shared with Monteverdi, his subsequent mentor, an inborn understanding of what is dramatic and an innate ability to exploit it. He was good at colourful orchestration, had a sensitive ear for varied rhythms and pleasing harmonies, and could create convincing and sustained moods. He placed more emphasis on the choral aspects of a work than did any of his rivals and compeers, and – as is explained by Peyser and Bauer – "he added a quicker second section in triple time to his first, majestic eight-measure section in duple time, thus establishing the importance of the prelude, or overture". His arias were not spaced at regular intervals, as was the later, more formal practice of many baroque contributors in the period that followed, but were placed where they best and most logically fitted the action.

He was amazingly prolific, in all producing forty-one operas (some say a mere thirty) over a span of three decades, as many as four in two years. His reputation grew throughout Italy and beyond its borders, especially to France. He had other demands on his time and energies. At the beginning of his career he was a treble singer, which was what had attracted the notice of the Governor of Crema who found the fourteen-year-old boy a job in Venice's San Marco (1616), where he stayed for over fifty years, exceeding his father's accomplishment by finally attaining the title of *maestro di cappella* (1668), previously held there by Monteverdi, which heightened its prestige – hardly any other in the musical life of Italy was more important. At the height of his manifold activities, a shrewdly drawn contract with an impresario, Marco Faustini, stipulated that his earnings from his operas would equal his salary at San Marco, which assured him of comfort, though probably not actual wealth. On two occasions, as shall be noted, he himself travelled to France. In Italy one of his contractual requirements was that he not only compose the operas – one or two yearly – but also personally supervise their first two or three presentations. Once again it was an odd alliance: Church – he was a superb organist – and theatre.

He married a wealthy widow, a niece of the Bishop of Pula (1630). At her death in 1652 he inherited most of her estate, as they were childless. He invested funds in the theatres where his operas made up most of the repertory.

A number of his works are still available, and from time to time and here and there in the twentieth century they have been resurrected and have fresh life breathed into their lungs, so that the voices of their characters are heard again – initially a rare event, it has been happening ever more frequently as interest in early Italian and French music has steadily increased.

His first offering, *Le Nozze di Teti e Pelio* (*The Marriage of Thetis and Peleus*; 1639), is full length – a prologue and three acts with a libretto by Orazio Persiani, who chose familiar content from mythology: Jupiter seeks to have Teti wed the nymph Peleo, but the ruler's Olympian consort, Juno, is jealous and intervenes, harassing the lovers. To accomplish this, "Hell arouses Discordia against the pair but is finally conquered by Amore." Abounding with choral passages, and staged during Monteverdi's lifetime, it is the oldest preserved score of the Venetian school.

His second venture, *La Didone* (*Dido*; 1641), indicated how truly endowed with talent he was and superior to almost all others around him. He had the advantage of an excellent libretto by Busenello, one that is ranked by some as the best of the many from that classic-minded poet-playwright. It is described as a "decorative melodrama". In it, Virgil's hero, Aeneas, leaves Troy, sails to Carthage, seduces and abandons its queen, Dido, who after experiencing rapture is left desolate. In the Latin epic she kills herself, but in Busenello's more accommodating ending she is dissuaded from the suicidal act by Iarbas, a Moorish king who has earlier laid suit to her and been rejected. Consoling her, he finally wins her consent to a troth, allowing the opera to wind up with a love-duet.

Cavalli evades any shortcoming here, proves able to handle a large number of characters, provides close links between scenes and furnishes plausible opportunities for more than enough lavish staging. Again, he inserts effective choruses of broad scope. This type of "decorative melodrama" served as a model – almost generic – that was soon accepted and perpetuated for decades thereafter by a host of other composers.

Another librettist, Giovanni Faustini (1619–51), supplied Cavalli with the plot and words for his seventh opera, *Egisto* (*Aegisthus*; 1643), a piece that goes in a contrary direction by dispensing with the chorus, being written for soloists alone, which started a new trend in Venice. At the request of the powerful Italian-born Cardinal Mazarin, Louis XIV's chief minister of state, *Egisto* was one of the first operas staged in Paris (1646). A point of controversy is whether it actually had its première at the Hoftheater in Vienna (1642) and was dedicated to the imperial Habsburgs – scholars exchange harsh words about this. If it was chosen for a first performance by the Austrian court, it would be clear testimony to how early, how far and how rapidly his repute was gaining momentum.

Cavalli's unstinted, exuberant output is thought to have reached its peak with *Il Giasone* (*Jason*; 1649), a retelling of the journey of the Argonauts in pursuit of the fabulous Golden Fleece, the major episode involving the tragic affair of the Greek hero and Medea, but with its horrible climax altered. Medea does not irretrievably wreak vengeance and fly off in a heaven-bound chariot; instead, she is saved from death by an old suitor, previously rejected – and Jason marries Hypsipyle, Queen of Lemnos. The libretto, by Giacinto Andrea Cicognini (1606–60), is deemed quite inferior: a number of scenes are irrelevant or make little sense, and stabs at "wit" are tasteless. None the less the public took to the work; it was the most frequently performed of all Cavalli's operas, and for over two decades was to be heard on stages in most of the leading European cities. The libretto, published, went through six editions in response to continued demand. The score intermingles recitative, arias and brief orchestral interludes. In 1671, while Cavalli was still alive, Stradella revised *Il Giasone*, adding a prologue and inserting three more arias.

L'Hipermestra (1658), sub-titled a three-part "glittering theatrical entertainment", had its début at the Teatro degli Immobile in Florence, where it pleased throngs. Commissioned by Cardinal Giovanni de' Medici, and with a libretto by Giovanni Andrea Moniglia (1614–1700), it was given to celebrate the birth of a son to King Philip IV of Spain.

Obviously Cardinal Mazarin approved of *Egisto*, as he had Cavalli stay on for two years and compose another work, *Ercole Amante* (*Hercules in Love*). A temptingly high fee was offered. The Cardinal was predisposed to like this new genre, dramas with continuous music and amazing stage effects, coming from his native land. He was loudly criticized, however, for financial extravagance of this sort at a time when much of the French population was suffering in abject poverty and oppression. To house *Ercole* a special building was being constructed by royal command, Le Théâtre des Machines, designed by Gaspari Vigarani, to facilitate the creation of scenic illusions, such tricks as the frequent enskyment and lowering of the gods, the multiplying employment of *deus ex machina*. The astute Mazarin's pretext for his considerable outlay was the marriage of Louis XIV to the Infanta Maria Teresa and festivities marking it (1660). Alas, as the date neared, Le Théâtre des Machines was not ready, and the influential Cardinal was ill. At Cavalli's last-minute suggestion, another of his operas was done instead, *Il Xerxes*. It had first been performed in Venice in 1654. This second staging took place in the picture gallery, the Salle des Caryatides, of the Louvre.

The libretto, penned by Niccolo Minato (d. 1698), called for fewer and less difficult illusions, the most taxing one a scene in which Xerxes and his Persian warriors cross the Hellespont over a string of boats that become an improvised bridge. Though introducing historical persons, many components of the plot are fantasy.

Grandson of Cyrus the Great, Othar is heir to the Persian sceptre. To reward him for sustained loyalty, Darius, ruler of the Persian empire, makes him King of Sufia, a small country. Angered that Othar refuses to let his daughter, Amastris, marry their monarch, the Moors attack the capital of Sufia. Unable to resist them, Othar seeks aid from the heroic Xerxes, who arrives with an army, leading eventually to his bold sweep across the watery straits. The Moors are hurled back. In the course of events, Xerxes and Amastris fall in love. (Amastris was sung by a castrato, a male soprano, which did not appeal to the Parisians.) Complications pile up, but in the end the two are wed. This fable was to serve other composers, including Bononcini and Handel, but, once staged at the Louvre, Cavalli's version was never seen again.

Ercole Amante was finally presented in Paris at the Théâtre des Tuileries, as Le Théâtre des Machines had been renamed, two years later (1662), a year after the death of Cardinal Mazarin. The text, by Abbot Francesco Buti, was somewhat altered, again to accommodate French taste: ballets, with music by Jean Baptiste Lully, were inserted. This allowed the dance-loving Louis XIV, elaborately dressed as the Sun King, to have an active role. The opera, consisting of an allegorical prologue and five acts, went on for six hours; to cite one costly effect, the fourth act occurred on the high seas. The work was sung in Italian; the hall was large and acoustics poor; many in the audience could not grasp what was going on. The prologue was not Cavalli's invention, but added for this sole event by Camille Lilius. In it, fulsome compliments are paid to the fifteen ranking royal dynasties of Western Europe, the top place of course awarded to the French sovereigns. The chosen regal families were named by the goddess Diana, singing, looking down from Olympus. The ladies of Louis XIV's court, each representing one of the fifteen crowned lines, participated in the prologue, together with Louis and his queen, Maria Teresa.

In ensuing acts Diana commands Hercules to carry on with his prodigious labours, in return for which his bride shall be none other than Beauty. Hercules loves Iole, as does his son, Hyllo. In the end, however, despite the intrigues of Giunone, Beauty and Hercules are wed, as are Hyllo and Iole.

After the two-year delay and high anticipation, and the large sums laid out, *Ercole*'s reception was disappointing. Applause went to Lully for his ballets, and to Vigarani for his ingenious staging. Cavalli's score was not too persuasive, but much of the muted hostility the work encountered is thought to have been due to resentment against the increasing flow of operas by Italians – as yet, the French had none of their own – and also to an animus still in the minds of many at the court against the recently deceased, firm-handed, foreign-born Cardinal Mazarin. For whatever reason, *Ercole Amante* was never heard again, much as happened to *Xerxes*. Quite disenchanted, Cavalli left France for ever.

Of his forty-one operas, two that had twentieth-century revivals are *La Calisto* and *L'Ormindo*. In *La Calisto* (1651), a "sophisticated tale of gods and mortals", the characters are borrowed from Ovid's *Metamorphoses*. The libretto is by Giovanni Faustini. The gods and goddesses are indulging themselves in erotic, Arcadian capers, paying scant heed to differentiations of gender. Jupiter, infatuated with chaste Calisto, transforms himself into the huntress deity Diana to strengthen his pursuit of the elusive object of his desire. Calisto is a steadfast devotee of the supposedly virginal goddess, whom she wishes to emulate. At the same time, Diana is enamoured of the innocent, all-too-handsome youth Endymion, and seeks to seduce him. Juno, ever jealous of her Olympian mate and cognizant of his undivine philandering, angrily takes a hand to frustrate the goings-on: she changes Calisto into a bear, and the bear is enskyed, the elevation to the heavens unfolding to music that is "a delicate blend of real emotions" (the words are those of Jane Glover, a Cavalli scholar). In this work, with its cavorting nymphs, satyrs, its mischievous use of disguises, humans are innately virtuous, immortals hardly so.

L'Ormindo (1644) tracks the entangled fortunes of two pairs of lovers as they put on a variety of disguises, practise risky deceits and suffer imprisonment and a harsh sentence, struggling to a hard-won happy ending. Once more, Faustini is Cavalli's librettist, concocting a plot that shifts his characters and action from Mauretania (now Algeria) to nearby Morocco and Fesso. The Moorish king claims a young bride; she is in love with his son, Ormindo. They elope and are pursued across North Africa. Overtaken and condemned, they swallow a potion that results in apparent death. The most powerful passage in the opera is this prison scene, for which Cavalli wrote music he liked so much that he used it again in a later venture, *Erismena*. At the close, the king reluctantly yields the girl to his son, while other affairs are concluded satisfactorily. The score is often attractive. Though not revived in the seventeenth century, the work has had staging in the twentieth.

A spirited revival of *Giasone* was a feature of the Spoleto Festival USA at the Dock Street Theater, Charleston, South Carolina, in 1998. *Opera News* carried this report (signed by P.J.S.):

> Produced in 1649, Francesco Cavalli's *Giasone* was the operatic hit of its century (far more popular than Monteverdi), and for good reason, since this send-up of the Jason/Medea myth involved all the features of seventeenth-century opera: cross-dressing, high moral tone, and low comic turns – the cheeky maid, old nurse and overweening soldiers – plus holding the gods up to ridicule. It was a three-ring circus of operatic possibilities, and director David Alden took full advantage of this to update it into an over-the-over-the-top funhouse of madness. It was as if he had a huge wardrobe of costumes (designer Jon Morrell) and told his cast to pick any they liked. Thus, the stuttering servant Demo (Jerold Siena) appeared first as a vacuum-cleaner salesman dressed as a penguin, then as a pizza chef, while Giasone, having won the Golden Fleece, wore it as a cowboy outfit.
>
> The continual inventiveness involved witty sets (by Gideon Davey) and all sorts of props (inflatable boats, windup toys). Though Alden's ideas ran riot, not all of them worked, and he seems not to be a classic farceur, in that he cannot build on humor to arrive at hilarity. The aim here was cruder and more enervating, so that the show wore out somewhere before the opera did.
>
> The producing schizophrenia typical of this kind of opera in performance was in evidence: the musical portions were "straight," in a performing edition of scrupulous "authenticity," played on period instruments by the ensemble La Stravaganza Köln under Harry Bicket's direction. The voices managed the singing and clowning with remarkable aplomb, led by Lawrence Zazzo's pleasing countertenor (Giasone) and his two inamoratas – the fiery, unpredictable Medea of Natascha Petrinsky (Callas-like, lithe and lissome) and the tender, wistful Isifile of Alexandra Coku (Tebaldi-like, her first-act lament is one of the musical highlights of the score). Guy de Mey's outrageous hermaphroditic Apollo was the epitome of camp (con-

trasted to his marvelously henpecked King Egeo), while Herbert Perry's absurd Ercole (Hercules) postured and preened with vocal characterizational distinction. Howard Bender as Medea's nurse, Delfa, did the cross-dressing bit, but the best single scene in this gallimaufry of sight gags and horseplay came at the end of Act II, with the duet of the soldier Besso (Philip Skinner) and the soubrette maid Alinda (Constance Hauman). This combination of singing and rough-and-tumble acting – at one point Hauman rolled across the entire stage – brought down the house, as it would have done on any vaudeville circuit. That the two could sing at all while performing this was amazing enough, but to sing as well as Hauman did was extraordinary. If there were still an *Ed Sullivan Show*, this turn would be a natural.

As stated, some believe that Francesco Sacrati and others stepped forward to write the final act of Monteverdi's *L'Incoronazione* because the master, old and fatally ill, lacked the strength to complete it. More recently Cavalli's handwriting has been identified on one of two surviving and divergent copies of the score – it is not clear which of the pair is Monteverdi's working manuscript or more closely conforms to it. In a good many passages that are most certainly Monteverdi's there are indications of revisions: the notation is more "modern", there are unfinished transpositions and stylistic inconsistencies. Wolfgang Osterhof suggests that these are Cavalli's handiwork, a less than successful intervention on behalf of his revered teacher.

At very least, Cavalli is an important figure. He helped to shape the forms and sustain the traditions that mark opera at the early-flourishing Venetian school of which he was the acknowledged head after Monteverdi's death. His lucrative successes at home and abroad were an influential factor in that. Many others were eager to follow his lead into prosperity by relying on a degree of imitation.

Marc Antonio Cesti (1604–74) came to opera by a different route. His true first name was Pietro; he changed it to Marc Antonio or plain Antonio on becoming a Franciscan monk in 1637. A native of Arezzo, he was inclined to a musical career because he had a fine tenor voice that he controlled with the utmost skill. His training began in Arezzo in a choir and was continued with several other teachers elsewhere – just who, exactly, is disputed. In Volterra, where he entered a monastery, he was already knowledgeable and competent enough to be elected its organist, and from there moved on to Santa Croce in Florence (1664), then returned to Volterra a few years later. He had come to the attention of the Medicis and enjoyed their patronage. In 1647 he sang in an opera in Siena at the launching of a theatre sponsored by Prince Matthias and soon afterwards asked the prince's help in obtaining the post of *maestro di cappella* at Pisa Cathedral – though a monk, he was not averse to seeking material advancement. His application was rejected. But his unusual talent was soon recognized by Cardinal Giovanni Carlo de' Medici and the Grand Duke of Tuscany Fernando II. Through them, it is likely, he became acquainted with a circle of artists and intellectuals who met as the literary Accademia dei Percossi. The group included the famous

Mannerist painter and writer, Salvator Rosa, who became the Franciscan monk's close friend and to whose letters are now owed informative details of Cesti's unusual life and personality. Also in the circle were poets and dramatists who later provided him with librettos.

Earlier, during a stay in Rome, Cesti may have studied under Giacomo Carissimi (1604–74), a noted composer of church music whose oratorios and cantatas – narrative and dramatic songs linked by both speech and recitative – are viewed as having been a direct, preliminary step to the uninterrupted song-form that is opera. Like that master, Cesti had certainly acquired the art of writing effective oratorios and cantatas. That would be helpful when he attempted to fashion his operas. But that he was ever a pupil of Carissimi is not fully authenticated.

In any event, this Franciscan monk had become closely acquainted with the requirements of an opera by having sung in one, a somewhat unique situation (though Peri actually had taken part in his stage-works *after* he had written them – obviously he could not have done so *before*). It was perhaps inevitable, if somewhat incongruous, that the ambitious Father Cesti was moved to embrace the popular new genre; in the exciting climate of the day it would be the logical next step. Once started, he could not stop; he had found his *métier*, and he is credited with having turned out 100 operas and plays with music. Finally, they absorbed so much of his thoughts and daily exertions that he could no longer carry out his ecclesiastical duties and petitioned the Vatican to release him from his vows.

He was no longer in favour with his order or the Holy Father. In a gossipy letter (1650) Rosa alludes to a romance between Father Cesti and a lady, Anna Maria Sardelli, a soprano in a touring company in which the monk also had singing roles. The monastery at Arezzo received an official rebuke from the Superior-General for allowing one of their order to behave in such a "dishonourable and irregular manner". Cited was a particular incident at Lucca.

The Vatican was chilly, haughtily indifferent and deaf to Cesti's pleading. According to Rosa, the restless Franciscan won freedom only after giving four recitals before Pope Alexander VII, who was persuaded that the brown-robed, tonsured singer's proper future might well lie elsewhere – the Pontiff might have been more concerned about preserving decorum in the Establishment than assuring the suppliant's temperamental happiness. Defrocked, though for ever a secular priest, Cesti remained with the papal choir for a few months, then quit it for a more profitable career.

Some years before he had obtained a congenial appointment at the court of Archduke Ferdinand Karl at Innsbruck and had held it for half a decade (1652–7). The post, created especially for him, allowed him to recruit and supervise a select group of musicians to be always available to him. A new theatre, the Komödienhaus, was completed in the city (1654). Its first offering was a revival of Cesti's *La Cleopatra*, now with a prologue and ballets contributed by a fellow composer, G.G. Apolloni. In its previous incarnation it had been put on in Venice (1651–2) and was titled *Il Cesare Amante* (*Caesar in Love*); its libretto, by M. Bisaccioni, had been versified by D. Varotari. A more prestigious event for Cesti, however, was the presentation of his *L'Argia* (1654) on the occasion of a visit to Innsbruck by the just abdicated Queen Christina of Sweden on her journey to lengthy exile in Rome. This was another six-hour performance, but one that "her majesty beheld ... with great pleasures, and attention" (quoted from *Grove's*). It was a lavish affair, employing a great deal of stage machinery, four ballet troupes, a host of supernumeraries and – at outstanding moments – ensembles of four or five voices, given many arias. It was intended to be not really an opera but rather a "private entertainment", a diversion for royalty, "a stylistic anomaly". Unusual is the structure of the strophic arias, many of which often change meter, and some of which run to as long as eighty or more lines, with the lines perhaps of unequal length.

Cesti's major triumphs took place in Venice, which suggests that he had ties to its busy school of opera composers. Indeed, he was very shortly Cavalli's chief rival there and elsewhere. His first submission was *Alessandro Vincitor di se Stesso* (1651); it was followed by *Il Cesare Amante* (later *La Cleopatra*). After the *L'Argia*, in Innsbruck, he made a successful bid for immortality with his *Orontea* (1656), also in Innsbruck. The libretto, by Cicognini – who had provided Cavalli with the scenario for *Il Giasone* – had been used twice before by others; here it is adapted to some extent. A durable success, *Orontea* was frequently revived through the closing decades of the seventeenth century. The very next year he again captured enduring popularity with *La Dori* (1657). As might be expected of one of his temperament, Cesti had an eye for the comic side of life, and this is manifested in amusing, sly portrayals of the servant characters in these two works; it probably accounts for some of their widespread, lasting appeal, not only in Venice but far beyond the buoyant, light-hearted city along the lapping Adriatic, whose waters invaded its dark-green canals. Besides that, his music is "flowing, lyrical and vocally elegant". All these qualities highly suited them for stagings at Carnival and other joyous occasions.

Further cherished works on his long roster – most of them still viable, though now hardly familiar – are *La Magnanimita d'Alessandro* (1662), *Il Pomo d'Oro* (1666–7), *Nettuno e Flora Festeggianti* (1666), *Le Disgrazie d'Amore* (1667), and *La Semirami* (1667); the crowded dates show his fecundity – he was seemingly inexhaustible. The first of these offerings, *La Magnanimita d'Alessandro*, was yet another piece composed in honour of Queen Christina, who was paying Innsbruck a second visit while travelling to Rome from Sweden and her father's funeral. Cesti himself had returned to Innsbruck and a haven assured by the Archduke; in doing so he incurred the Pope's ire and was threatened with excommunication. During the interim period, on leave from the Vatican choir, he had been in Florence taking part in festivities attendant on the marriage of the future Grand Duke Cosimo III and Marguerite Louise d'Orléans; he had singing roles in operas by others staged for the grandiose occasion, but one of the operas put on was his much-praised *Orontea* for which he was able to have the Archduke Ferdinand Karl's musicians, lent by that ruler as a gift to the nuptial couple. There were six performances, with Cesti, so it is said, appearing as the character Alidoro. It is possible that his *Dori* was also presented.

By now Cesti had been ennobled, the generous Archduke having named him "Cavaliere di S. Spirito in Sassia", and through the intervention of Ferdinand Karl, Cosimo III and Emperor

Leopold I, the Holy Father was appeased and the Vatican's threats and claims were waived. He was also granted property and a handsome sum of money.

Ferdinand Karl died in 1652. Three years later Sigismund Francis, his successor, was about to be married. Cesti was engaged in a work for the accompanying festivities, *La Semirami*, which tragically did not take place owing to the death of the new Archduke – the last of the Tyrolean Habsburgs. Soon afterwards the Innsbruck musical establishment was moved to the imperial court in Vienna; the group included Cesti's carefully chosen instrumentalists and his prized Italian singers. At the same time he began a steady correspondence with an impresario, Marco Faustini, and a librettist, Nicolò Beregan, both in Venice.

In Vienna, Cesti's new title was "Honorary Chaplain and Director of Theatrical Music". His creative activity in this new locale was intense. He continued to perform at the Hoftheater in singing roles in works by rivals as well as in those flowing so quickly from his own pen, notably *Nettuno e Flora Festeggianti* and several of the others mentioned in the cluster just above. His *Il Pomo d'Oro* (*The Golden Apple*) was intended to be a contribution to the marriage of Leopold I to the Infanta Margherita of Spain, but a series of problems delayed it, though its lengthy score was completed almost on time; some accounts say it was presented at Carnival in 1666–7, others set the date at 1668. The libretto, by Francesco Sbarra, turned to the Greek legend of the three goddesses who compete for the prize, a golden apple, to be awarded to the fairest of the trio, the unlucky judge the simple-minded, handsome Trojan prince Paris. His choice of Aphrodite, who as a bribe promises him the hand of the most beautiful woman in the world, ultimately leads to his unscrupulous seduction of Helen of Troy and the prolonged, bitter war that follows. Two acts – the third and fifth – are lost, but what is left of the text and score bear out that they called for and were given the most "elaborate and sumptuous" staging ever. The twenty-three settings, rich attire, properties and machinery alone cost 100,000 thalers to construct and gather, and there was also the outlay for the virtuosic singers and added instrumentalists, dancers and a truly great host of extras. The baroque scenery was designed by Ludovico Burnacini, and the added ballet music by Johann Heinrich Schmelzer, a Viennese composer. Comedy, in a broadly popular folk vein, alternated with solemnly enacted allegory. The language, not quite poetic, abounded with erudite mythological allusions. Apparently all this was ephemeral; as far as is known, after a few showings the extravagant *Il Pomo d'Oro* was never seen again, but recollections and praise of its "glittering and gaudy" production long persisted. Engravings of the stage sets were published and remain, as well.

While the stalled *Il Pomo d'Oro* was awaited, the copiously gifted Cesti came forth with *Le Disgrazie d'Amore* to fill the gap; it, too, had interpolated music by Schmelzer, and – more importantly – a prologue composed by none other than Emperor Leopold himself.

La Semirami, which had been meant to grace the Archduke's marriage festivities in Innsbruck but had unfortunately been cancelled, was finally mounted in Vienna at the Hoftheater. The libretto, by the Florentine court poet Giovanni Andrea Moniglia, introduces a mythical Babylonian queen, beautiful and terrifying, to whom are attributed supernal and frightening deeds. It was put on as a tribute to Leopold's birthday. Again, its ballet music is by Schmelzer. Later, in Italy, this work was revised, its title altered to *La Schiava Fortunata*, and staged in Venice (1674) and Modena (1674). Another of Cesti's endeavours for Leopold while the composer was attached to the imperial court was providing music for an equestrian ballet, *La Germania Esultante* (1657), which had a scenario by Sbarra, and again some dance music by Schmelzer.

The stress of all this toil was being felt by Cesti. He gave notice to Emperor Leopold of a wish to depart, and the next year left for Venice and Florence (1658); in the latter city his merits were recognized by the grant of a pension, twenty-five scudi annually. His last known connection with opera is his participation in a revival of *L'Argia* in Siena in 1669. That same year he died, aged forty-six, in Florence. Rumours took wing of poison administered by jealous rivals, but they have never been substantiated.

Less is told of Francesco Sacrati (1605–50), a son of Parma. He is best remembered for supposedly having rewritten the role of Otho and for having completed the last act of *L'Incoronazione di Poppea*, with the assistance of Benedetto Ferrari. If they did, both were truly talented, since the love-duet at the act's end is generally acknowledged to be surpassingly beautiful, possibly the finest segment of *L'Incoronazione*. For certain, Sacrati was highly esteemed by his contemporaries, one of whom – Giacomo Badoaro, the librettist of *L'Ulisse Errante* – said that the younger man, a former pupil of Monteverdi, stood in a relation to his teacher like that of the moon to the sun, his a reflected glory. Another with a generous regard for Sacrati was Prince Mattias de' Medici, who became a good friend. Most of Sacrati's work, alas, is known only from libretti, though there are fragments of a few scores.

He was active as a composer of operas mostly in the 1640s, in northern Italy, and usually in collaboration with the scene designer Giacomo Torelli. At one time he may have belonged to a travelling troupe calling itself the Accademici Discordati, which produced one of his works in Bologna and elsewhere (1641). This was his *La Finta Pazza* (1641), of which only a later, shortened version survives. In 1645 it was taken to Paris by Torelli and the ballet master G. Balbi, who used its music principally to match his choreography and exhibit his dancers. Even so, it was the first Italian opera performed in France not at the court but for the public.

In 1648 Sacrati was *maestro di cappella* of the "musici di Bologna"; the ensuing year he held a similar post at Modena Cathedral. The list of other operas is comparatively brief: *Il Bellerofonte* (1642), *Venere Gelosa* (1643), *L'Ulisse Errante* (1644), *La Semiramide in India* (1648), *L'Isola di Alcina* (1648). In some instances his authorship of these pieces is not certain.

Alessandro Stradella (1639–82) was of aristocratic birth. His father, Marcantonio, of a Tuscan family, belonged to the Cavalieri di S. Stefano, a Pisan order of knights established by the

Medici. The Stradellas lived first at Fivizzano, and then at Nepi, where Alessandro was born. At fourteen he was chosen to be page to the Lante family in Rome, serving there for seven years. Later accounts place him in Bologna. By 1667 there are first mentions of him as a musician. The next year he started his career in opera by composing a prologue to Jacopo Melani's *Il Girello*, a comedy presented in the Palazzo Colonna in Rome (1668); a success, *Il Girello* was shortly exhibited at a substantial profit throughout Italy. Stradella became associated with the capital's first public theatre, the Tordinona, his task being to provide it with more prologues, additional *intermezzi* and arias for the thriving operatic works brought from Venice, which meant that he had to conform to the Monteverdian style, and even more to that of Cesti. He was facile at this. He also composed sacred music and sought commissions of every sort for incidental music, vocal and instrumental, for plays and for private performances – marriages, birthdays – in the palatial dwellings of Roman aristocrats, which were readily open to him. Though he bore a patrician name and comported himself with the assurance of an aristocrat, he had no patrimony and was often hard pressed, haplessly finding himself without money. In 1675 his compositions of liturgical music, and especially of an oratorio for the Church of S. Giovanni Battista, earned him an award, the honorary post of *cameriere extra* at the papal court.

By nature something of a scapegrace, always in need of funds, he soon got into trouble and thereafter was for ever on the move around Italy, mostly because he had to make an escape here and there. An "old and ugly" woman paid him and his friend Giovan Battista Vulpio a tempting large fee to bring about her marriage to a nephew of Cardinal Cibo. The Cibo family was enraged on discovering this conniving, and Stradella prudently took off for refuge in Venice.

In that city of canals he had a patron, Polo Michiel, who greeted him warmly. While in the fostering cultural climate of Venice, Stradella may have written an opera, using a libretto by G.-F. Saliti, but there is no indication anywhere of a performance. In any event, after a few months he was on his way again to Turin for another short stay. His next destination was Genoa; here, too, he had highly placed friends: Duke Giovanni Andrea Doria and his wife Anna Pamphili, whom he had known in Rome. Through them he found a position with the local Teatro Falcone as continuo player in its orchestra and singers' coach. By the very next season he had become the theatre's impresario.

One of his duties was to compose an opera, and he brought out *La Forza dell'Amor Paterno* (1678), and a bare few months later *Le Gare dell'Amor Eroica* (1679), and after a sparse breathing space, *Il Trespolo Tutore* (1679), in an amazing burst of creativity. The first two use reworkings of libretti by the Venetian writer N. Minato. *Il Trespolo* is based on a Tuscan comedy in prose by Giovanni Battista Ricciardi, adapted by Giovanni Cosimo Villifranchi, and it boasts what was becoming a requisite, two mad scenes, on which Venetian audiences doted. There is one in *La Forza dell'Amor Paterno*, too. Stradella knew what would meet the public taste, and perhaps his own likings coincided with it. These works enjoyed overwhelming popularity: *La Forza* had a run of fifteen performances.

Poets dedicated sonnets to him. As a token of their admiration, leading citizens presented him with a golden tray valued at more than his wages for an entire season; and to make sure that he would remain in Genoa, a group of music-addicted noblemen guaranteed him 200 Spanish doubloons annually, a well-furbished house complete with servant and ample provision of food.

A fourth opera, *Moro per Amore* (1681), with a libretto by an illustrious titled Roman, Flavio Orsini, Duke of Bracciano, seems not to have been enacted publicly, though maybe privately.

Another opera score was left unfinished, and others have perhaps been lost. He composed in many other genres, notably cantatas, two hundred of them, for which he is even more highly esteemed. He was kept busy with commissions from patrons for ceremonious affairs.

His remarkable streak of good fortune was short-lived, however. Two attempts were made on his life, for a reason that is not clear. The first attack failed – it is said that the unidentified hired assassins were dissuaded from their murderous intent by the insinuating charms of their victim's music. Not very likely? The instigator – thought to be a jealous rival in a love triangle – obtained the more reliable services of a killer less fond of music. Waylaid, Stradella, just forty-two, was stabbed in the back, a penetrating mortal blow, the attacker still unknown (1682). This incident is the subject of a nineteenth-century opera by the German composer Friedrich Freiherr von Flotow, *Stradella* (1844).

The slain musician was given a solemn burial in the Chiesa delle Vigne, one of Genoa's most fashionable parish churches, and decorous eulogies were paid, along with genuine regret.

Despite the disarray of his private life, he was firmly disciplined in his work, with splendid gifts. "Stradella's style typically features fluent melodic lines which pay careful attention to textual form and underlay: it is also characterized by a frequent and delightful use of counterpoint, and of basses which are either strictly or freely *ostinato*. The language is tonal, but unusual progressions may occasionally surprise the listener. The music constitutes an important link in the period between Cesti and Handel" (quoted from *Grove's*).

Most of the prologues and *intermezzi* he added to the works of others for the Falcone and Tordinona theatres have neat comic touches; he excelled at that. *Il Trespolo Tutore* is a light-hearted account of a guardian who is smitten by love for his pretty ward. Stradella makes him a comic bass, a model for the many other guardians who will similarly be infatuated with attractive, unyielding young wards in a legion of *opere buffe* to follow. Stradella was admired and emulated in this respect not only by Alessandro Scarlatti in Rome but even in distant England by Henry Purcell, who praised the Italian composer's "wit and invention". Though four operas are not a substantial legacy, his example has proved to be a singularly important one.

Of lesser stature in the second and third generation of composers of the Venetian school are a goodly number who were professionally accomplished, their scores revealing strains of originality and personal idioms. Many had artistic links to Venice and its operatic traditions but travelled outside the city to work elsewhere in Italy, as did Stradella, and also abroad, as did Cavalli and

Cesti. So many manuscripts are lost that it is hard to evaluate the quality of this really vast output, but doubtless it was not all journeyman achievement – a considerable amount of valid musical treasure has simply vanished, which is to be greatly regretted. By the century's end three hundred new operas had been brought to Venice's stages, a busy scene.

A priest, Giovanni Legrenzi (1626–90), stood out among a throng of competitors. Of his seventeen operas, a mere four survive. Born poor, he died rich, which suggests that his efforts at composition paid off handsomely. Besides, he had a post as *vice-maestro di cappella* at San Marco, another instance of the strange union of the Church and the commercial theatre that the Renaissance hierarchy and laity did not seem to deem inappropriate. Of course, he wrote sacred music, too. To fill Father Legrenzi's coffers further, the cathedral authorities took an unusual step to reward him, giving him an annual bonus in addition to his stipend after he was promoted to the top as *maestro* in 1685, the grant carrying with it a stipulation that the extra sum was "for the person and not the office" (quoted from Weinstock and Brockway). He was somewhat iconoclastic in his *Giustino* (1683), an *opera seria* in which a brave, loyal peasant lad rises to a throne as emperor. Here was a striking departure from the classical rule that the characters be of noble birth if their tortured emotions and fall were to be considered tragic. (Legrenzi himself, once poor, had risen very high.) *Giustino*, breaking long-emplaced barriers and moving towards realism, was an immense success, having productions not only in Venice but in eight other cities in Italy.

Though a native of Venice, Antonio Sartorio (1630–80) spent much of his career as a composer in Germany, where he was director of music at the Hanoverian court. He made frequent return trips to his birthplace on the Adriatic, however, to take in what was happening, so as to keep abreast of changes there and also to recruit proficient musicians to accompany him back to a colder Hanover. For his operas he tended to choose heroic subjects: these are found in his *L'Adelaide*, *Giulio Cesare in Egito* and *Antonio e Pompeiano* concerning Ancient Rome and its turbulent rulers, in three works produced between 1672 and 1677. His *L'Orfeo* (1672) should be added to the roster of offerings on that myth; it is mostly remembered for an aria, "E Mora Euridice". To establish an almost strident heroic aura, the arias given to certain characters are apt to be accompanied by a trumpet *obbligato*, "the *obbligato* being a kind of secondary solo to complement the singer's melodic line" (quoted from Headington, Barfoot, Westbrook). In *L'Adelaide* he displays a sure skill at choruses for crowd scenes. His arias are bold, even florid, and he has a flair for the comic, such lighter episodes often contrasting sharply with those that aspire to the heroic or to a depiction of the everyday mundane.

Pietro Andrea Ziani (*c*. 1616–84), too, was a priest who injected humour into his scores to match his characters' amusing behaviour and the jocularity in their recitatives, which move at a merry pace. *Annibale in Capua* (1661) is given over to the erotic campaigns of the great Carthaginian general; they are celebrated more than his awesome military triumphs. Again, one wonders at this ordained cleric's choice of avocation and subject, in an age when the Church was especially skittish about impropriety. In his *La Semiramide* (1670), as well as in the *Annibale*,

Ziani mixes popular songs and dance elements in his amiable scores, and both works are precursors of the succeeding century's even more lively *opera buffa*.

Carlo Pallavicino (*c.* 1635–88) satirized Venetian morality – or lack of it – in his *Messalina* (1679). His protagonist, the Roman empress who resorted to poisoning her lame and stuttering elderly husband, is here shown guilty of no crime against the Emperor Claudius worse than her flagrantly deceiving him. To make sure that the spectators would appreciate that the target of his satire was contemporary Venice, not first-century Rome, an Italian Carnival is replicated, a dancing riot of masqueraders, albeit a somewhat anachronistic one.

Talent running in families, Marc'Antonio Ziani (*c.* 1653–1715), a nephew of Father Pietro Andrea Ziani, joined the cluster of industrious Venetian opera composers, keeping up with them in productivity. He was another who transferred to Vienna (1700), accepting an appointment at the court there. Austrians and Germans looked to Italy, and particularly Venice, for musical leadership. In both Venice and Vienna Marc'Antonio was successful. He was excellingly competent at mating text to music and at plausible characterization. He increased the role of the orchestra, augmenting the wind players and giving solo passages to the cello, trombone and chalumeau (a prototype of the clarinet), often to novel effect.

Yet another composer who left Venice for Vienna and a post at the court of Leopold I – formerly held by Cesti and, in the musical realm, the highest of any in Austria – Antonio Draghi (*c.* 1615–1700) had 170 operas to his credit (some say only 118). It should be noted, however, that many were short, perhaps one-act, and others were comprised more of slight tunes for the ballet than sustained melodies for the singers, and were often intended to supplement the offerings of *commedia dell'arte* troupes and amateur or semi-professional skits for Carnival. He also contributed incidental music for secular plays and turned out soaring, dramatic oratorios on sacred themes. Originally from Rimini, the seaside city famed for its churches and tombs adorned with Byzantine mosaics, he brought from Venice to Vienna his feeling and fondness for virtuosic vocal displays, especially by coloraturas. Some hearers rated him higher than Cavalli.

Among Draghi's full-scale works are *Sulpita* (1692, revised 1697), *L'Arsace* (1698) and *La Forza dell'Amor* (1698). In his later years the composer sometimes had a collaborator, his son, Carlo Domenico Draghi (1669–1711). Composers such as Cavalli, Cesti, Ziani, Sartorio and Draghi, who took posts far from home, did much to propagate a growing liking for Italian opera abroad.

As the century neared its end, a prominent figure in Venice was Legrenzi's former pupil, Carlo Francesco Pollarolo (1653–1722), who added eighty-five operas to the city's astonishing output. Much praised by his contemporaries, he experimented boldly: in some instances he divided his orchestra, part on stage, part off stage, obtaining echo effects; he elevated the role of the oboe in his search for ways to extend the expressive range of his instrumental ensembles, and thereby assured the oboe's place in any orchestra playing for an opera, and strove for a harmonic richness and more exactly woven vocal–instrumental textures. He ventured bravely towards more flexibility in form and structure, and favoured ornate vocalization, in a style anticipating the Neapolitan *bel canto* operas of the succeeding century. But few now are aware of what Pollarolo wrote or his technical innovations, which were strong determinants in the development of opera as a lasting genre.

Crossing well over the divide between the seventeenth and eighteenth centuries, Tomasso Albinoni (1671–1751) offered Venetian audiences eighty-one operas (some put the figure at forty-eight). Today his instrumental works are still played and at least vaguely familiar, but without exception his operas are forgotten. Since he was a composer of stature, they could not have been wholly without merit. The titles stir curiosity: *Tigrane, King of Armenia* (1697); *Primislao I, King of Bohemia* (1698) – historical subjects, to some degree – and *Griselda* (1722) and *Aminta* (1722) – from a later phase in his career when apparently he tended towards the pastoral. Anyone interested in the history of opera would like to know more about them.

The importance of Venice as a resounding centre for opera productions more or less ended about the year 1700; their prolixity was sharply diminished in the ensuing century. The shift was southwards to Florence again and even more to Naples.

An inestimable debt is owed to the small circles of sixteenth- and seventeenth-century serious-minded intellectuals and aesthetes in Florence, Mantua and Venice who, caught up in the fevered excitement of unearthing and rebuilding the glories and grandeur of a long-buried classical age, sought to revive Athenian tragedy, its sublime height and depth. They hoped to do this by encouraging the writing of plays on heroic and mythological subjects that were also embellished by verse. Almost at once they recognized the magic that could be exerted by well-chosen words, and nearly as quickly discovered that when the words were enforced by music the magic was even more emotionally compelling. Gratitude is owed, too, to the princely and ducal patrons whose purses were opened to subsidize the costly experiments.

Very soon it was found that when the two arts were conjoined, the music was more important than the words. No one went to an opera to hear the libretto, nor even to pay respectful attention to the story. The composer was to be sovereign in the world of opera. The principal burden and responsibility for a success was to be for ever his.

In the twentieth century the partial discovery of these long-obscure Italian Renaissance operas began in 1904 when the French Impressionist composer Vincent d'Indy translated the text and arranged Monteverdi's *Orfeo* for a drastically cut concert performance in Paris; d'Indy's enthusiasm had been aroused by reading the scores. A staged production followed in 1911. Earlier, in the nineteenth century, several articles that recognized Monteverdi's historical importance had been published by German musicologists; the Italian composer was also allotted three pages in the first edition of the all-comprehensive *Grove's Dictionary of Music and Musicians* (1880), though the author of the entry had doubts about the Renaissance musician's full mastery of his craft, being put off by bold harmonic progressions in the scores; surely they were "careless mistakes . . . it would be absurd to suppose that such evil-sounding combinations could have been introduced deliberately". D'Indy persisted in translating Monteverdi's texts into French and arranging shortened versions of them – the cuts were deep. His reduction of *L'Incoronazione di Poppea* was offered in 1908, and of *Il Ritorno d'Ulisse in Patria* in 1925, both merely at concerts.

In a preface to his edition of *L'Incoronazione* (1908) d'Indy took the Italians to task for neglecting their great composer. How could such magnificent works be ignored by Monteverdi's own countrymen? The next year a concert rendition of *Orfeo* was given in Milan, with many cuts by Giacomo Orefice. This offering was considered unsatisfactory, and it led to further efforts to recapture – or at least closely approximate – what had been Monteverdi's true intentions, by a list of fellow Italians, themselves composers of note. This brought a new edition of the vocal score of *Orfeo* by Gian Francesco Malipiero (1923), followed by more authentic full scores for all three of the surviving operas. Ottorone Respighi issued his scrutiny and interpretation of the *Orfeo* (1935); Luigi Dallapiccola submitted a more modern, updated reading of *Il Ritorno* (1942); and Bruno Maderna, yet another *Orfeo* (1967). In 1984 Luciano Berio gathered five young Italian composers to restudy the *Orfeo*, with results that were not staged publicly. Meanwhile, in Germany, where Monteverdi now stirred much interest, Ernst Krenek was attracted to *L'Incoronazione* (1935), Paul Hindemith applied himself to a personal understanding of *Orfeo* (1943) and Hans Werner Henze chose *Il ritorno* as a project (1984). Hindemith, whose version was finally presented in Vienna in 1954, called for having it played entirely with period instruments. In addition to these interpretations by composers, there have been analyses of the scores by a small army of dedicated musicologists. (A lengthy list of their names is included in an essay, "The Revival of Monteverdi's Operas in the Twentieth Century", by Jeremy Barlow, on whose research this part of this chapter depends.)

The British first heard *Orfeo* in a concert performance in London in 1924 that used d'Indy's French version; but, catching up, it was sung in English by the Oxford University Opera Club, availing itself of a more complete edition by Jack Westrup (1925).

Many conductors relied on their own adaptations, largely unpublished. Besides Maderna and Berio; these include Wenzinger, Goehr, Leppard, Haroncourt, Glover, Norrington, Daniel, Pickett, some of them specialists in medieval, Renaissance and baroque music – this list is not complete.

In New York *Orfeo* was first heard in concert form on a Sunday night at the Metropolitan Opera House in 1912, Josef Pasternack conducting. Seventeen years later it was staged at Northampton, Massachusetts. Werner Josten, the conductor, used the Malipiero version of the score.

In Italy it was revived at the Florentine Maggio Musicale in 1949 in an arrangement by Vito Frazzi. Numerous other revivals occurred elsewhere in Italy, some introducing still newer revisions by Giacomo Benvenuti, Hans Erdmann-Guckel, Carl Orff and Ferdinand Redlich.

In 1960 *Orfeo* was finally staged by the New York City Opera at the City Center on a double bill with Dallapiccola's *Il Prigioniero*; it opened the company's season. Possibly to fit it into the time-limits of the evening's programme, Leopold Stokowski, a conductor noted for his wilfulness, greatly shortened the work and effected many changes in it. It was deemed a considerable success.

An ensemble calling itself Artek put on a semi-staged *Orfeo* in St Joseph's Hall of the Church of St Mary the Virgin in New York in 1993. Seven singers made up the chorus and alternated in solo roles; an orchestra of eighteen players performed on period instruments. A painted backdrop was derived from Maxwell Parrish's familiar *Daybreak*. The production was conceived and conducted by Gwendolyn Toth; the stage direction and choreography were by Alan Tjaarda Jones. Grants from the New York State Council on the Arts and several corporations partially supported the production. (Listeners reported that the work was charming but that the wooden seats grew exceedingly hard over five acts.)

Also in 1993, an *Orfeo* with an attempt at seventeenth-century staging was a feature of the Boston Early Music Festival, a biennial event in that city. The choice of a Monteverdi opera was to commemorate the 350th anniversary of his death. The effort to recapture the original style of performance was chiefly that of James Middleton, who sought to overcome the paucity of information about how the opera was first done in Mantua. Most probably it was given in two fair-sized rooms – adjacent and open to each other – in the Gonzaga ducal palace, which did not boast a theatre. Macmillan relied on drawings and verbal descriptions of playhouses and halls that accommodated works contemporaneous with the *Orfeo*, as well as diagrams of the baroque stage machinery then available. To change scenes he used *telarri*, rectangular rotating units, on each side of which was depicted part of a panorama. As would be true in Monteverdi's day, the costumes – very ornate – were not intended to recreate those of Homer's Greece but instead displayed an Italian Renaissance mode of perhaps fifty years earlier, which was considered sufficient to suggest antiquity. Besides being richly adorned, in Florentine fashion, the performers wore huge wigs. Of the period, too, was their make-up: "white face powder, bright red lipstick and blush – almost garish by today's lights. And there were stylized stage gestures that while neither incomprehensible nor overdone were chiefly remnants of a lost language." Allan Kozin, who was on hand for the *New York Times*, found this insistence on "authenticity" – or something like it – a distraction, often "drawing a veil" over the drama. The stage direction was by Simon Target.

An elaborately costumed *Orfeo*, conducted by Haroncourt and mounted by the avant-garde French director Jean-Pierre Ponnelli and with décor by Pet Halmen, was put on in Zurich in 1979.

Fragments in French excerpted from *Il Ritorno d'Ulisse in Patria* were presented under the baton of Charles van den Borren in Brussels in 1925, shortly before d'Indy's full three-act version was staged the very same year in Paris. Dallapiccola's variation of the text and score was included at the Florentine Maggio Musicale (1942), its singers and orchestra led by Mario Rossi.

England slowly took to the work, introducing it to twentieth-century listeners of a broadcast in 1928 and unveiling it on stage at St Pancras Town Hall in London in 1965. Subsequent performances, usually greeted with accolades, occurred at Glyndebourne (1962, 1972 and 1984), at Kent Opera (1978) and again in London at the English National Opera (1989).

Il Ritorno was presented at Salzburg (1985), in the Henze version, enacted in wildly imaginative sets. At the Netherlands Opera, too, the staging was somewhat bizarre, certainly not realistic.

In the final decades of the century, New York City was treated to two productions of *Il Ritorno*. The first, by the Waverly Consort, conceptualized the Homeric epic, with evocative, impressionistic sets, at the Kathryn Bache Miller Theater on the campus of Columbia University. The group, dedicated to reviving half-forgotten works of early music, designed the production to be handily taken on tour.

The Netherlands Opera's interpretation of the piece, mentioned just above, was brought to the Majestic Theater, under the auspices of the Brooklyn Academy of Music, in 1993. Glen Wilson, the conductor, used his own version of the score and text. Here, too, the setting was simplified, markedly less spectacular, to lessen travel costs. Though the company came from Amsterdam, the opera's interpreters were a polyglot group: German, American, British, Lebanese, Chilean, and the singers were assembled from Holland, America, Chile, England, Wales and Scotland. Dispersed, they carried with them an acquaintance with Monteverdi's long-dormant masterpiece.

Several critics were awed by the emotional depth of the work. Excerpts from Paul Griffiths's lengthy review in the *New Yorker*:

> The musical news in town last month came in the shape of an opera three hundred and fifty years old. . . . Not only did this introduce a new company but it was by far the most candid and expressive piece of operatic staging seen in this city during the entire season.
>
> That this should have happened to Monteverdi may be no accident. Because seventeenth-century opera is still infrequent in the theatre, every performance becomes a kind of première. . . . Monteverdi is still virgin land – and prime land, for his operas aren't just specialty interests but compelling musical dramas, in which worlds hang on the inflection of a singer's line. We may be used to thinking of this composer as standing at the beginning of opera, but his focus on the singer – on musical declamation that sets its own pace and shape, which the instruments must follow – is also an ideal that died with him. Opera went in other ways, toward formal aria, continuous orchestral support, and the rhythm of the mass rather than that of the individual. Only in rare instances is the Monteverdian spirit retained or restored: in French Baroque opera, which maintains the independence of the solitary singing voice, or in Debussy, where the vocal rhythm is often re-separated from that of the orchestra.
>
> *The Return of Ulysses* is, on the face of it, the least bankable of Monteverdi's three surviving

operas. *Orfeo* has advantages as a prototype and as a link in a chain – as one of the first all-sung secular dramas in the Western tradition and as part of the continuing history of the Orpheus myth. *The Coronation of Poppaea* has the voyeur's allure of historical fiction and a richly colorful set of characters. Both works have clear identities. *Ulysses* is altogether more uncertain, perhaps because it's concerned with characters in states of uncertainty....

The gods argue over human fates. Other human characters provide amusing, ironic contrast with the constancy of the principals by pursuing immediate desire: Iris, follower of the suitors, in his ludicrous gluttony, Melantho and Eurymachus in their nimble love affair, which we might catch sight of in corridors as we move from one room to another of the opera's spacious design. Amid all this, there's only one important event, when, near the end of the opera, Ulysses declares his presence by stringing his bow and felling the suitors.

However, such a drama of inaction and postponement – of waiting not for the plot, which barely exists, but for the meaning – may seem even closer to us than the masque of *Orfeo* or the solidly sensual, sensuous world of *Poppaea*. Pierre Audi, the Netherlands Opera's artistic chief and the director of this production, has spoken of *The Return of Ulysses* as Shakespearian, and the notion has some point: Monteverdi was born only three years after the dramatist, and the weave of grand characters has obvious parallels in nearly every Shakespeare play....

Mr Audi thus disposed of the distracting idea – distracting from his own purpose, of presenting a drama of human beings – that people are pieces in celestial games of chance. It was the strength of what resulted that justified the cuts – though justification may be redundant in the case of a work for which there is no intact source, which may include portions by other composers, and which the Venetian audience of 1640 would have thought to be primarily the property of its gentleman poet, Giacomo Badoaro. The very idea of there being such a thing as Monteverdi's *The Return of Ulysses* is a modern anachronism, and if we're interested in that anachronism, rather than in complete justice to Badoaro, then cutting is unexceptionable, even necessary.

The cuts were made by Glen Wilson, who was at the keyboard as he led his small group of musicians and their period instruments: lutes, recorders, bowed strings, occasional recorders. They were seated close to the stage, enabling them to follow the singers.

The sound was quick, lean, and elegant: precisely no more than was necessary in order to provide a harmonic context. There are arguments that Venetian operas of this period would have been given under more opulent musical circumstances when revived at foreign courts, or that modern listeners require more variety of instrumental texture, but those contentions are always confounded when, as here, the barest resources are used with imagination and life. As with the cutting, that's what matters: not whether a performance accords with original practice as we understand it but whether it works.

Griffiths agreed with Glen Wilson, who in a programme note had little good to say about purists, those who seek to "resuscitate baroque manners in set design, costume, and acting". He saw scarce gain in studying and copying the costume drawings of Inigo Jones and others to achieve "authenticity", as had been done by those presenting *Orfeo* at the Boston Early Music Festival under way at almost the same time as *Il Ritorno* in New York. He felt that Monteverdi's powerful music did not need that visual support. Griffiths enlarges on this:

> Part of the problem must be that such drawings presume luxury materials and an unlimited supply of seamstresses; rushed up on the cheap, they look ghastly. But I'd guess the effect wouldn't be much better if some opera company's wardrobe department were to have the means of the Gonzagas.
>
> This is odd. Period-style musical performance is now so much a part of life that we'll probably have to wait another generation before Monteverdi realizations for modern orchestra – such as Hans Werner Henze's version of *The Return of Ulysses* – can be put on again without fear of ridicule. But we don't want Baroque ways of decorating the stage and acting on it. Possibly, the reason is that we live in a visual age, and false historicism is unacceptable to the eyes, though it may get past the ears. That seems too jaundiced a view. Perhaps it's that the theatre is always happening in the present tense: we want to see Ulysses and Penelope now, not a recreation of how they might have looked in theatres long demolished. Music, on the other hand, is always bringing us stories from the past.
>
> That the presentness of theatre and the pastness of music can coincide is one of the miracles of opera, and was one of the miracles of the Netherlands Opera production....
>
> You have to do something special, because these people are expressing themselves vocally at such hypernatural intensity and length. Mr Audi's answer in many cases was to find a single stance or action that could be held throughout a musical number, and to root the acting in the body.

Griffiths cites examples of this: Penelope seated and leaning forward on her throne with "vehement dignity"; Ulysses and Telemachus leaping with joy, yet competing for a staff as in a game, at discovering each other's identity; Melantho cradling her mistress's head in her arms, and the physicality of her lovemaking with Eurymachus – "bringing us close to the characters as human bodies, and thereby as human hearts". Griffiths adds:

> Formality can be expressive, too. The prologue looked beautiful, with Human Mortality a destitute male figure singing against an emblematic tableau of his three eternal rulers: Fate, a lady with a wheel; Love, a woman accompanied by a little winged Cupid; and Time, an old warrior bent over. The production also formalized – spectacularly – the opera's climactic moment: as Ulysses aimed his bow, a blaze of fire erupted across the stage from a row of

torches rising through the floor – visible on the stage, feelable as heat on the face – was an instant reminder in one's own body of being there with Ulysses and the rest of them.

Griffiths paid praise unstintedly to the singers. They were perfect in their roles. He concluded: "Many of the same team will be going forward with Mr Audi to *Poppaea* in Amsterdam next season. I mean to be there, too."

This confirmation of Monteverdi's greatness has been more than borne out by many twentieth-century productions of *L'Incoronazione*, which has been the most frequently revived of his three surviving works. Early French and German interpretations have already been mentioned, and those in modern Italy. In 1937, as well, a performance of the piece was given in the Boboli Gardens, Florence, the version that of Giacomo Benvenuti. Again, in 1937, it was in the repertoire of the Opéra Comique in Paris – in Malipiero's arrangement. Later stagings in England occurred at the English National Opera (1955, 1975), Sadler's Wells (1971), Glyndebourne (1984), Kent Opera (1986). Zurich staged the opera in 1977. This by no means exhausts the count, and one notes Griffiths's allusion to a planned engagement of the music drama in Amsterdam in 1994.

It was introduced to America at Northampton, Massachusetts, in 1926, and next to New York audiences at the Juilliard School of Music in 1933. After a lapse of two decades it returned in concert form to New York, this time offered by the American Opera Society, a chamber group interested in restoring less-known works at the small theatre in the Metropolitan Museum (1953). The conductor, Arnold Gamson, reduced the score and text of the drama to two acts. The English translation was by Chester Kallman. Gamson's abbreviated treatment was repeated at the museum in 1958. Both times it made a very strong impression, partly because the leading roles were sung by the internationally renowned Leontyne Price (Poppea) and Robert Rounseville (Nero). The Caramoor Festival at Katonah, New York, brought out another version (1967 and 1972). A full-scale presentation was that of the Glimmerglass Opera Company at upstate Cooperstown, New York, during its annual summer music festival in that historic town in 1995. The conductor, Jane Glover, chose her own carefully studied reading. The stage director was Jonathan Miller. The orchestra was an intimate ensemble of period instrument players. The seventeenth-century costumes were by Judy Levin. The formal setting, monumental, showed the façade of a palace, guarded by Nero's soldiers. On its steps stood white-clad goddesses, with white wigs and chalk-white faces, resembling antique marble sculptures, though at appropriate moments they moved about and participated in the action. In repose they appeared to be otherworldly, brooding observers. This production was in Italian, with English surtitles. It was put on in Glimmerglass's handsome new Alice Busch Theater. The enactment caused such a stir that it and its cast were transferred in January 1996 to the Brooklyn Academy of Music, where it enjoyed a major critical success. Furthermore, the company's general director, Paul Kellogg, was appointed to head the New York City Opera.

By now there were a good number of recordings of the three extant Monteverdi operas. Barlow cites twelve of *Orfeo*, three of *Il Ritorno*, eleven of *L'Incoronazione*.

After hundreds of years, Monteverdi's hardly known operas have become staples for repertory companies throughout Europe and America, and even in South Africa.

Other Italian Renaissance operas have been less fortunate at finding revival. The New York Pro Musica participated in a music festival in Corfu in 1973 with Marco da Gagliano's *Dafne* as its entry, after which the troupe brought its production from that Mediterranean island to the open-air Caramoor Music Festival. (The ship's journey was stormy, delaying the staging at Caramoor; this proved to be lucky, as the evening originally set was lashed by heavy rains that would have meant a cancellation.)

Apart from Monteverdi, the only seventeenth-century Venetian opera composer to have his works resurrected at least at odd intervals and gain much notice in the twentieth century has been Cavalli. In New York his *Ormindo* was staged at the Juilliard School of Music in 1975, and only a few months later at the Venetian Theater at the Caramoor Music Festival, the second of these an independent production, though the Caramoor staff borrowed a few costumes from their friends at Juilliard. *La Calisto* was a midsummer offering at Glimmerglass in 1996. Scattered enactments of these two Cavalli operas and others have taken place at music conservatories and universities, but these have been rather fugitive, consequently difficult to track and include in a census. They are mostly "in-house", given for the instruction of fellow students, not for the general public. Still, his work has proved to be enduringly vital, however quaint it may seem at times.

This was demonstrated in Britain where, beginning in 1967, Raymond Leppard led performances of *Ormindo*, *Calisto*, *Egisto* and *Orione*, the first pair at Glyndebourne. In Florence, on the 350th anniversary of the composer's birth, his *Didone* was presented.

Venice, a republic, had no ducal court nor a by-divine-right ruler to expend princely grants to encourage its musicians. (It did provide church and palace walls for glorious paintings.) Opera was almost solely a public entertainment and a commercial venture. To satisfy the inordinate demand for it, theatres were needed. This entailed substantial financial outlays and risks, but entrepreneurs, of whom Venice had many, were attracted. Several wealthy families, the Tron, Michiel, Giustinian, Grimani, unhesitatingly opened their coffers, hurrying to share the opportunity. Among those eagerly participating was the alert Cavalli, joining librettist Orazio Persani, choreographer Giovanni Battista Balbi and soprano Felicità Uga to organize an ensemble to occupy the Teatro Tron di S. Cassiano, the first house exclusively dedicated to opera in the city (1641). To keep it busy, Cavalli had to turn out a spate of new works. The enterprise, faltering, eventually failed. Not daunted, he tried again, choosing librettist Giovanni Faustini as his partner. They collaborated as artists and investors, taking over the S. Apollinaire (1650), and were briefly successful. Following Faustini's death, Cavalli was associated with the Teatro Giovanni e Paolo; but when Marco Faustini, the late librettist's elder brother, inaugurated seasons of opera at the Teatro Tron di S. Cassiano in 1558 and 1559 Cavalli once again had his offerings there. He and Balbi also helped in founding a permanent opera company in Naples.

At first the new works by Cavalli and his rivals were put on in already existing theatres or abandoned churches that were renovated or adapted to meet the needs of dramas with continuous music. Perhaps some had housed *commedia dell'arte* troupes, so it was not too difficult to reshape them, as was quickly done. Many of the businessmen entering into this field prospered greatly. Consequently the theatres were larger and larger and multiplied in number; by century's end Venice boasted ten of them, and soon even more – up to seventeen – energetically competing, their stages echoing with arias, recitatives, choruses, soft or swelling orchestral accompaniments. Ticket prices were low, which helped to draw ever larger audiences, as the appetite for this fresh musical genre sharpened. Those spectators who were titled, members of the dominant oligarchy, powerful officials of the Republic, or merely *nouveau riche*, could rent a box, invite important guests and be the generous patron, which was apt to happen during Carnival – the peak season – always a gaudy, riotous spell in the city of busy, gondola-thronged canals.

Typical of the new Venetian playhouses was SS. Giovanni e Paolo, of which a drawing is in London's Soane Museum (whose founder, Sir John Soane, was himself an architect of note). The interior, totally of wood, was somewhat more decorated than were most others. Begun in 1638, its doors were opened the next year for *La Delia* (1639), music by Strozzi and Sacrati. Built solely to mount operas, the stage had a proscenium arch and commodious space for the machinery that helped to create opulent spectacles and astonishing illusions. Beneath the floor was an area accessible by trap-doors for sudden, wondrous appearances and vanishings, and on the stage itself grooves for sliding wings and flats on and off when rapid changes of scene were desired. Other sets of parallel grooves, somewhat slanted, allowed the shifting of flats forwards and backwards, to widen or narrow the space for the performers, when the setting altered from a street with a broad vista to an intimate but confining room, or from a mossy cave in a leafy grove to a deep, high-arched palace hall. Perspective scenery and painted backdrops were used in combination.

The auditorium, an elongated U, was encircled by five tiers of boxes. Well lit, they permitted the occupants to play cards and gamble, or eat and drink, with the singers and orchestra supplying background music. It was true then, as now, that many went to the opera not to listen, but snobbishly to be seen. (Many other playhouses had oval interiors rather than U-shaped ones.)

Even more imposing in size was Marco Faustini's Teatro Tron di S. Cassiano, accommodating 6,000 spectators. It, too, had five tiers of boxes, with thirty-one boxes on each level. The lowest two were the most expensive, the price declining inversely as the tiers rose. Those spectators who defrayed the least were jammed together in the pit. This significant pattern of humble pit, lofty galleries and patrician boxes, a separation by social class and size of purse, was hardly new and was to be long-lasting. For a good while the S. Cassiano was successful. Monteverdi's *Il Ritorno* had its première there, as did Cavalli's *Le Nozze di Teti e Peleo*, *La Didone*, *Egisto* and *Il Giasone*. (*L'Incoronazione* was put on at the Teatro Giovanni e Paolo.) Venice was enjoying a buoyant economy, which may be why Faustini took a chance on leasing a theatre so ambitious in scale.

On average the playhouses in Venice for works without music were far from resplendent; many were "dark and ill-smelling, with seats of hard wood", the luxurious fittings limited to the highest-priced *loges*, and to the rich costumes and elaborate scene-settings found on the stage. The Teatro Tron di S. Cassiano, in particular, was not well ventilated, its benches crowded together and uncomfortable, the chairs in the boxes rickety, and the interior of the house often unswept and dirty. Venice's theatres, bare, environed and pervaded by a dank climate, hardly matched the sumptuous palace rooms and halls in which operas had first been performed. Eventually the S. Cassiano was razed as "too vast".

Nor were casts always equal to their task. An English tourist, the informative diarist Sir John Evelyn, had praise for what he saw, a production in 1645 of a new, now lost opera, *Ercole in Lidia*, by Giovanni Rovetta. He noted that it had "excellent musicians, vocal and instrumental, and machines for flying in the air, and other wonderful motions . . . magnificent and expensive" (quoted from Headington, Westbrook, Barfoot). Kind words were also set down by a French visitor, who in 1680 found "the theatres large and stately, the stage decorations noble but very badly lit", at the same time complaining that the works themselves were overly long, "the ballets or dances between the acts generally so pitiful that they were better omitted . . . [however] the vocal charms amend all imperfections, the beardless men (*castrati*) have silver voices, the women are the best in all Italy" (*ibid.*). Another Frenchman (1699) was wholly displeased, remarking on "a certain confusion and unpleasantness in many aspects of the singing: they dwell many times longer on one quavering than in singing four whole lines, and sometimes run so quickly that it is hard to tell whether they are singing or speaking" (*ibid.*). Other foreign observers reported that the pit was filled with "workers and gondoliers: There is the constant noise of people laughing, drinking and joking, while sellers of baked goods and fruit cry their wares aloud from box to box" (*ibid.*).

The behaviour of the audience was hardly better at a slightly later date in Rome. There, people of rank and fashion, though proudly arrayed in their boxes, were fond of playing chess as the opera unfolded: that game, remarked the sardonic French writer Charles de Brosses (1709–77), provided an escape from the monotony of the dry recitatives; while a melodious aria diverted a contestant from concentrating too tediously on how to make the next move. Dr Charles Burney, the English savant, in Milan during the 1770s, singled out high-stakes *faro* as the favourite distraction of such unheeding opera box-holders.

Some singers stood out from the rest, were highly paid, had faithful and often zealous admirers.

The names of two castrati still resound, Antonio Pistocchi (1659–1756) and Antonio Bernacchi (c. 1690–1756). Apparently they had well-trained voices and interesting personalities. What is more, in Venice female characters were portrayed by women. It was otherwise in Rome, the Holy City governed by celibates, where there was insistence on the presumed inferior, morally weaker sex being kept from the stage; their appearance on it would be improper and sinful. In consequence the female roles were sung by castrati, their high clear voices enthralling Roman listeners.

It is hardly to be wondered that there are controversies over how the Venetian operas are to be interpreted and presented in the twentieth century: often the scripts offer few stage instructions, if any, and the scoring is skeletal or incomplete, and some passages very difficult to decipher. Also, as the reviews of Kozin and Griffiths indicate, historicism raises questions: should not advantage be taken of modern instruments and orchestras to contribute an enriched sound and volume, or does not the attempt to achieve authenticity add an imaginative dimension, evocative of a past epoch, its people's aural likings and cherished fantasies?

As Richard Anthony Leonard puts it, in *The Stream of Music*:

> The rise of this new form of entertainment is remindful of the strange crazes that seized the populations of Europe at various times. . . . Opera spread through Italy like wild fire, until there was hardly a city or town that did not support a number of operatic theatres (either public or in private homes), while a small army of composers was kept busy turning out operatic scores with the speed of commercial artisans. Soon Italian opera became so conventionalized that its whole point and purpose was to show off a new and splendid vocal art. Here again was clear manifestation of the spirit of the Baroque, for the more ornamental, florid, extravagant the aria the more wildly the singer would be acclaimed by the audience. . . . The *castrati*, or male sopranos, created by a revolting practice of mutilation when they were boys, were the most popular performers of all. The schools of singing that grew up in Italy were unsurpassed in thoroughness and excellence.
>
> Within a century Italian opera had swamped the music of the rest of Europe. Its domination was so complete that to this day the nomenclature and terminology of music are still totally Italian.

In Florence the vogue for opera was sustained for about three decades, its fault being that it was "at once too lofty and too precious for long life".

A more varied and durable successor to the Venetian school arose in Naples, partly owing to another man of musical genius, Alessandro Scarlatti (1659–1725), whose work embodied the innate gift of his fellow Neapolitans, dwelling as they do in the sunniest section of Italy. J.J. Lalande, an early traveller, wrote in 1769: "Music is a special triumph of the Neapolitans. It seems as if in that country the membranes of the eardrums are more taut, more harmonious, more sonorous than elsewhere in Europe. The whole nation sings: gesture, tone of voice, rhythm of syllables, the very conversation all breathe music and harmony. Thus Naples is the principal source of Italian music, of great composers and excellent operas" (quoted from Brockway, Weinstock).

Actually, Alessandro Scarlatti was of Sicilian birth, possibly a native of Palermo – there is a touch of mystery about this. Biographers are also unclear about his apprentice years. His father was most likely his early teacher. He may have studied in Rome for two years with the renowned Giacomo Carissimi, master of the oratorio, with whom Marc Antonio Cesti had been a student. Scholars also detect a possible influence of Alessandro Stradella on the output of the young Scarlatti, borrowing a "way with melody", though he avoided Stradella's conspicuous failing, an over-emphasis on repetition. That Scarlatti was a prodigy was evidenced when, somewhere between eighteen and twenty, he wrote an opera, *Gli Equivoci nel Sembiante* (*Misunderstandings and Appearances*; 1679 or 1680); it was staged in the palace of Christina, the exiled Queen of Sweden, who had become a Catholic and taken up residence in the Holy City; she named the young man her *maestro di cappella*, a post he held for five years (to 1684). She also commissioned a second opera from him, *L'Honesta negli Amori* (*Sincerity in Love*; 1680). By now he was married (1678), a union that was to yield him a son, Domenico, who was to exceed him in fame.

To be in service to Queen Christina entailed a certain degree of peril. A patron of the arts, and as devoted to music as she was to her adopted Catholicism, she established an academy to foster it in the Palazzo Farnese where she lived, and she built a private theatre for the presentation of operas. This is where Scarlatti's next works were enacted, rather than publicly. But the eccentric Christina's friendship often ended in disaster for those on whom she bestowed it. In one instance she had her secretary killed. So firm was her grasp that she broke the finger-bones of Pope Alexander VIII in a handshake, and she engaged in violent quarrels with Pope Innocent XI, by most considered to be saintly. It was from the hostility of Innocent XI, who sought to put an end to public showings of operas, that she boldly defended Scarlatti and other composers.

To whatever other dangers he might have been exposed during his employment in Queen Christina's household, the young composer ended there untouched. Critics hold the opinion, however, that his work and career suffered from his not having direct exposure during his formative years to the body of operas by thriving and dynamic composers of the Venetian School, where the art and craft of writing dramas with music was far more advanced. The Venetians had a precious advantage, too, in having to aim their offerings at a broad audience, rather than a private, élite one. His scores and subject-matter might have gained more vigour, been more dramatic, if it had been necessary to please a restless pit filled with ticket-buyers instead of a small palatial hall of polite guests, chosen from the aristocracy and literati. Nevertheless, by his election of subjects and owing to the strict format that he unwaveringly accepted, he happened to create a long-lasting model of what an *opera seria* should be, one that endured at least a half-century, from the baroque era to the classical, until the advent of Gluck, the reformer.

Scarlatti's emancipation from his bonds to the strange Queen Christina – he also composed sacred music for her – came about when the Marquis de Carpio, the Spanish ambassador to the Vatican, was appointed Viceroy of Naples, a possession of Spain, and recruited the talented musician to go there with him. Scarlatti remained in Naples for thirty-five years, in two periods, from 1684 to 1702 and 1708 to 1725. He soon became a dominant figure in the city's musical realm, writing more than half of the operas produced there, with his work being widely heard elsewhere in Italy and far abroad. In all, he composed 115 operas, of which about half the texts and scores survive, though it would be a rare event, indeed, if any were staged today. Greatly influential, his output is quite forgotten, studied by musicologists but never actually played. His son, Domenico, though considered by critics to have been somewhat less gifted, performed and composed for the harpsichord and is now far better known, his works still frequently heard. He also wrote operas that in places like Lisbon and London were more favourably received than his father's, though the elder Scarlatti's *Il Pirro e Demetrio* (1694) had sixty performances in the English capital, where some passages in it were translated from the Italian. In Brunswick and Leipzig a German version of the work was available, enabling it to be sung in either language.

The *opera seria*, as conceived by Scarlatti, largely finds a theme in the loves and obligations of royal personages or other heroic, exalted figures taken from mythology or ancient history, whose dilemmas – of honour, dynastic claims – are ultimately resolved happily or at least satisfactorily; hence the stories are often remote. The composer's attention is focused mostly on the score, as might be expected, but he seems to have had little feeling and certainly no flair for what is compellingly suspenseful or dramatic, nor was he much interested in balancing the story and music, to have the one precisely complement the other, as so ably do Monteverdi and Cavalli. His operas are static, lack unity; mostly they are a lengthy string of arias, often as many as fifty or sixty in a single work, so that a Scarlatti opera is likely to resemble a recital. Each does contain a few duets, trios, ensemble numbers and choruses, the ensembles and choral outbursts usually placed at act-endings, which is effective and became a tradition. He grew into a meticulous craftsman and was fecund with melodies; many of his arias are quite beautiful and to be cherished. To support them he gives prominence to the oboe, and he enhances recitative by having it accompanied by more instruments, perhaps strings, rather than just a solo harpsichord. But the successive arias – and even the ensembles – tend to dilute or wear out their appeal by being *da capo*, that is, each is structured in three parts: the first section is followed by one that more or less contrasts to it, and the third section merely repeats the first. As a result the overall impression sounds less than emotionally spontaneous. An aura of artificiality reigns. Generally, the vocal writing is florid – what is later called *bel canto* – to permit the artists a display of virtuosity, though such high-altitude flights may not always be wholly appropriate to the plot. He makes only a slight bow towards the humorous: not more than one of his 115 "mature" operas could be categorized as comic.

Instead of independent overtures to his operas, Scarlatti began to create each one by interweaving motifs from the music that followed more closely, integrating the prelude to the body of what is to ensue, forecasting the mood and leading melodies, as has been the practice ever since. Furthermore, his overtures have a regular progression: to seize the spectator's interest the first segment is lively, the second is slow and songful to induce more careful listening, and the third, building to a climax, the most robust, to ensure that the audience is fully alert and anticipatory. What will now unfold? The curtain goes up!

Scarlatti's *opera seria*, soon to be designated the classic opera, is usually in three acts, though his *Mitridate Eupatore* (1707) stretches to five, with a ballet – it is one of his major accomplishments. Paradoxically, though he might have been in many ways an innovator, his fate was to be considered by his successors as an arch-conservative, looked upon as old-fashioned even before his death. That is because the type of operas he turned out, and which were much emulated, were so unchanging in their design. They were cut to the same pattern, answering to an admirable formula but one that in time became monotonous.

He was not too happy in Naples; he was overworked and underpaid. As previously in Rome, he had to provide not only an endless stream of fresh operas, but also church music and serenades. His works were staged in Venice, Milan, Rome, Siena, Florence, but this added little or nothing to his sparse income. His salary was not always paid when due. The Neapolitan taste for comedy forced him to insert humorous elements in his pieces, though often he was loath to do so. At long last he decided to leave Naples to freelance and possibly find a better post elsewhere. He made the rounds of other cities; in them a number of patrons and civic officials were ready to offer him commissions for single, particular works, but he sought the security of a permanent position. Finally, back in Rome, he was glad to accept the responsibilities of assistant *maestro di capella* at the church of S. Maria Maggiore (1703–8). But he was not contented there either. To his friend and benefactor, Prince Ferdinando de' Medici, he wrote: "Rome has no shelter for music, which lives here as a beggar" (quoted from Headington, Westbrook, Barfoot).

A political event brought him back to Naples: sovereignty over the hilly city on the Bay and its environs under Vesuvius had recently passed from Spain to Austria as part of the expanding Habsburg Empire. The new Viceroy, Cardinal Grimani, invited Scarlatti to return. This period in his career (1708–25) was once more a fertile one, yielding *Il Tigrane* (1715), *Il Trionfo dell'Onore* (1718, a comedy); *Il Cambise* (1719) and *Telemaco* (1718), among his successes – the last two had their débuts in Rome, where he journeyed to oversee their performances. Rome was also the first to behold two more of his stage-works, *Marco Attilio Regolo* (1719) and finally *La Griselda* (1721). Despite the encroachments of age and ill health, he kept busy.

Handel met Scarlatti and was obviously influenced by his work. Another who was swayed by him was a young German, Johann Adolf Hasse, a pupil of Nicola Porpora, and now anxious to learn from the Italian master. It was, however, the last year of Scarlatti's life; there was not time for many lessons. Hasse was to attain considerable stature in his native land, founding a tradition there, especially in Dresden, of *opera seria* in the Scarlatti style.

Though respected by high-ranking patrons and officially honoured – he was elected to the Arcadian Academy (1706) and the recipient of a papal knighthood conferred by Clement XI (*c.* 1715) – his road was seldom easy. He was the second of eight children; it is believed that conditions in Sicily – famine and political unrest – imposed comparative poverty on his parents and led them to send him and his two sisters to study in Rome, where resources and prospects might be brighter. Alessandro himself was later to have a large family to support – he was fecund in that matter, too. Poorly paid, he was often hard pressed for funds, especially when travelling about Italy with a band of children and servants while job-hunting. He faced many disappointments: for example, shortly after his marriage to Antonia Anzalone he was anticipating the première of his first opera, but at the very last minute the performance was called off because Pope Innocent XI had cancelled all entertainments during Carnival. Similar bans on other productions came about, also by papal edicts, in times of plague, which inspired the Church to proclaim an urgent need for prayers and public renunciations. These were no occasions for frivolity. This further deprived Scarlatti of income. In his attempts to launch his operas in Venice and find permanent employment there, he encountered successive defeats, open hostility and humiliation. His peers in that city did not welcome "outsiders". His manner was somewhat arrogant and condescending; heedless, he made a good many enemies wherever he went. His music, moreover, was more complex and demanding than Venetian audiences tended to like, the instrumentation – to them – "unusual", as was its "contrapuntal texture", and the subject-matter doleful, too serious, despite the happy wind-ups. The Venetians preferred music to be "transparent and tuneful". A local satirist, Bartolomeo Dotti, stated maliciously that the effect of *Mitridate* was soporific, as could have been observed by a glance at spectators in the Teatro S. Giovanni Grisostomo.

The truth was, Scarlatti did not keep up with changing times or shifts in public taste. This was the opinion of Francesco Maria Zambeccari, who in a letter described Scarlatti as "a great man, so good indeed that he succeeds ill because his compositions are extremely difficult and in the chamber style, and so do not succeed in the theatre. *In primis*, those who understand counterpoint will admire him, but in a theatre audience of a thousand people there are not twenty who do understand it, and the rest, not hearing cheerful and theatrical things, are bored. Also, the music being so difficult, the singer has to be extremely careful not to make a slip, and therefore unable to make the gestures he is used to making and becomes too tired. Thus, Scarlatti's theatre style is not pleasing to most audiences, who want cheerful things and *saltarelli* such as they get at Venice" (quoted from *Grove's*).

His most enthusiastic patron, Prince Ferdinando de' Medici, who for several years subsidized productions of the operas in his private theatre at Pratolino near Florence, was limited financially, dependent on his father, Grand Duke Cosimo III, and could not assist Scarlatti as much as he might have wished, though he often approached other possible patrons on the composer's behalf.

Though often spoken of as the "founder of the Neapolitan school of eighteenth-century

l sont en bonne garde mis   pedat lege tienne
our garde de leur anemis   a ui a est faite pour memoire

Previous page An illustration to *Roman de Gauvel* by Gervais du Bus depicting a medieval carnival, 1310–14
Above *Moses and the Golden Calf*; Passion Play performed at Oberammagau, Germany
Right A contemporary production of an English Mystery Play, York

Above An old witch, Fasching Carnival Parade, Groetzingen near Karlsruhe, Germany, 2004

Above A red devil, Fasching Carnival Parade, Groetzingen near Karlsruhe, Germany, 2004

Below A contemporary production of an English Mystery Play, York

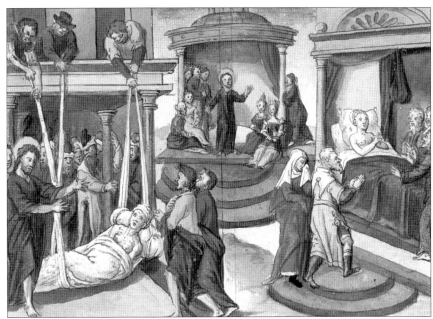

Left Passion Plays of Valenciennes, France, depicting the healing miracles of Christ, *c.* 1500
Below Stages for the Valenciennes Passion Plays, 1547
Right Holy Week parade at Sentenil near Ronda in Andalucia, Spain

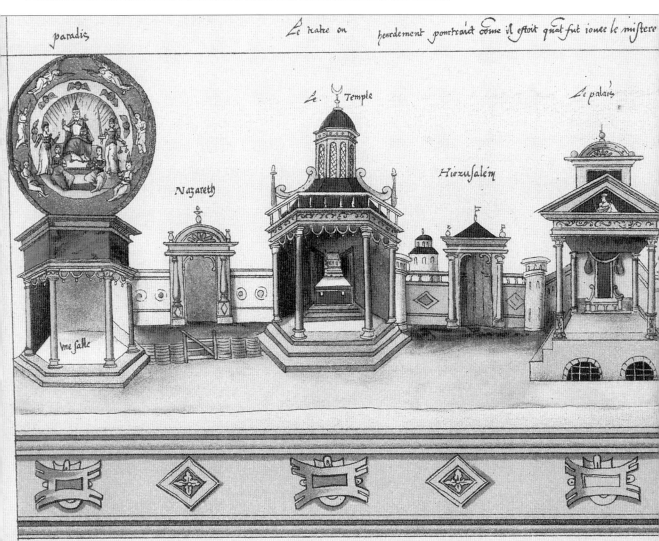

¶ Here begynneth a treatyse how ẏ hye fader of heuen sendeth dethe to somon euery creature to come and gyue a counte of theyr lyues in this worlde / and is in maner of a morall playe.

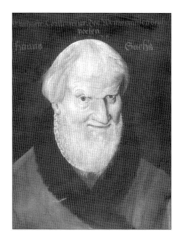

Far left Frontispiece to an early edition of the English Morality Play *Everyman*, late fifteen century
Left Frontispiece to Hans Sachs's play *A Fight Between a Woman And Her Housemaid*; woodcut, *c.* 1556
Above Hans Sachs, German playwright, 1494–1576; portrait, *c.* 1574 by Andreas Herneisen

Above Teatro Olimpico, designed by Andrea Palladio and completed by Vincenzo Scamozzi in 1580, Vincenza, Italy
Left Study for a tragic scene; engraving by Sebastiano Serlio (1475–1554)

Right Stage machinery for *Germanico sul Reno* by Giovanni Legrenzi, performed in Ferrara in 1676
Below left Stage machinery for the whale that swallows Jonas; engraving, 1663
Below right Stage machinery for a sailing ship with rocking device; engraving, 1663

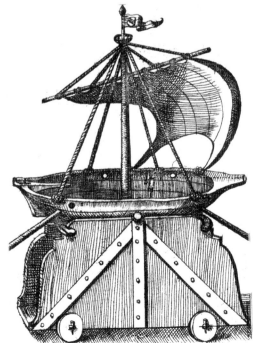

Far left Torquato Tasso, Italian poet (1544–95); engraving by Raphael Morghen after a drawing by Ermini Raphael after the painter Raphael
Left Pietro Aretino, Italian writer (1492–1556); portrait by Titian, 1545

Far left Self-portrait of Gian Lorenzo Bernini (1598–1680); coloured chalk on paper
Left Niccolò Machiavelli, Italian writer (1469–1527); portrait by Santi Di Tito, 1510

Right Performance of a comedy and portrait of Latin writer Terence from a fifteenth-century French manuscript

Left Italian comic actors at the Hotel de Bourgogne, Paris; engraving by Pierre Landry, 1689
Above *Commedia dell'arte* performance in St Mark's Square, Venice, 1614; engraving by Giacomo Franco

Top left *Jester in Love* by Giovanni Domenico Tiepolo, c. 1793; Ca' Rezzonico, Venice
Above The nineteenth-century clown Joey Grimaldi
Left The twentieth-century mime artist Marcel Marceau

Above Corral de Comedias, Almagro, Spain
Left *Fuente Ovejuna* by Lope de Vega at the Royal National Theatre, London, 1992
Right Procession during the Pia Del Pilar Festival, Saragossa, Spain

Above left Lope de Vega, Spanish playwright (1562–1635), c. 1630; portrait attributed to Eugenio Caxes
Above right Calderón de la Barca, Spanish playwright (1600–81)
Left Miguel Cervantes, Spanish author and playwright (1547–1616); portrait by Juan de Jauregui y Aguilar, 1600

opera", and though he was clearly superior to most of his fellow composers in the city, he lacked a faithful public among its exuberant music-lovers, perhaps for the same reasons that had caused his works to fail in Venice. A good number of his offerings were first presented elsewhere, especially – as has been noted – in Rome, where he felt most at home, despite his complaint that he was poorly rewarded for his labours there.

The author of the biographical sketch in *Grove's* provides this summary:

> In the final analysis, Scarlatti occupies an equivocal position in the history of opera. No one did more to establish Naples as a leading operatic centre during the last two decades of the seventeenth century; no Italian during his lifetime wrote operatic music of such quality and depth; and few composers of any age have enjoyed greater respect from their patrons (as evidenced by . . . the epitaph Cardinal Ottoboni wrote for his tomb in the church of Montesanta, Naples, lauding him as "dear to princes and kings"). And yet Scarlatti seems hardly to have influenced the course of operatic history. Most of the "innovations" with which he is sometimes credited . . . can be shown to predate him, while the music itself emerges more as a refinement of seventeenth-century styles than as a harbinger of the classical period. As his art developed, so it became more remote from the audiences whose approval he sought to win. Possibly Zambeccari spoke truer than he intended when he said that by composing so well Scarlatti succeeded ill. Certainly those of his compositions that have been most admired by posterity do not belong in the opera house.

Many others are more generous in their appraisals, suggesting that if Scarlatti was not primarily an innovator, he is to be remembered as one who epitomized an age, his prolific output serving as the best example of his generation's musical attainments and artistic excellences, thereby meriting Cardinal Ottoboni's further phrases on his tomb in which the composer is described as "the greatest 'renewer' of music". Of Scarlatti it has also been said elsewhere that he was hardly a rebel, but that he was a "standardizer" of already existing forms, some that before his contribution were still not completely realized and consolidated. "He invented little, established much." But again, he is blamed for favouring the preening, posturing singers, while overlooking an obligation to tell a dramatic, compact, unified story. He is faulted, too, for having unwittingly calcified the shape of the *opera seria*, denying others after him the freedom to alter it, setting a precedent that restricted even the towering Handel. Yet is he to be held responsible if most younger composers lacked the originality and courage to explore newer forms? That judgement hardly seems to be fair.

(The remarkable history of the Scarlatti family makes a good case for the widely held belief that in many instances musical talent is inherited. Alessandro's father, Pietro, was a professional musician. On the maternal side, an uncle, Vincenzo Amato, the mother's brother, was *maestro di cappella* at Palermo Cathedral; he is remembered for having composed an opera. He may also have given Alessandro early lessons. Alessandro's two sisters, Anna Maria and Melchiorra Brigidam, earned notice as opera singers. They also acquired scandalous reputations. Anna Maria performed in Venice, Rome and Naples; eventually she married a shipowner who became a theatre impresario. Melchiorra's loss of virtue, her liaison with a court official, may have worked to Alessandro's advantage, helping him to obtain an appointment as royal *maestro di cappella*, the premise being that one needs to have the right connections. Her on-stage appearances were in Rome and Naples; later she married a double bass player, who managed the Teatro dei Fiorentini, which might also have been helpful. Of Alessandro's brothers, Francesco was a violinist who had engagements in Naples and Palermo; he composed a dialect comedy, possibly in collaboration with Nicola Pagano, his brother-in-law, the double-bass player; the piece *Lo Petracchio* was staged in Aversa in 1711, and subsequently in London and possibly Dublin. Francesco also turned out cantatas and oratorios. His gifts were matched in the family by Tomasso, a tenor, who like his sisters was an opera singer. Trained at the Conservatorio S. Onofrio in Naples, he was later appointed to the royal chapel for a spell and afterwards took part in *opera buffa*. He had a role in his nephew Domenico's early *Giustino* in 1703. It is hazarded that the Neapolitan opera singer Rosa Scarlatti was his daughter. Besides Domenico, another of Alessandro's sons, Pietro Filippo, was a composer. He held positions in Urbino, as well as in the royal chapel in Naples; his opera *Clitarco* was presented in the latter city at the Teatro S. Bartolomeo in 1728. An impressive assembly!)

A portrait of Scarlatti, by Francesco Solimena, shows a "sad, appealing, and noble-looking man who watches us gravely from the cracked and faded canvas".

Others of consequence in the eighteenth-century Neapolitan School are Leonardo Vinci, Leonardo Leo and Nicola Porpora, who were Scarlatti's successors, though by many critics they might not be regarded as his heirs. Indeed, they differed from him and one another even more than they had traits in common, apart from their living in the same thriving sunny city at much the same time. Dr Burney, the English historian, wrote kindly about Leonardo Vinci (c. 1690–1730), who in his short life composed thirty-five operas, of which two dozen were for production in Naples, the rest for staging elsewhere. Eleven of those first done in Naples were *opere buffe*, the popular new species. What Dr Burney admired in Vinci's output was how aptly the music accommodated both the meaning and the singer. "Vinci seems to have been the first composer who . . . without degrading his art, rendered it the friend, though not the slave, of poetry, by simplifying and polishing melody, and calling the attention of the audience chiefly to the voice part, by disentangling it from fugue, complication, and laboured contrivance." Quoting this, the authors of *Opera: A History* add: "As Burney suggests, Vinci developed vocal writing that is closely responsive to the text, and often intensely expressive in its use of *appoggiaturas* and dotted rhythms." He was fortunate, too, in having an exceedingly capable librettist, the poet Pietro Trapassi Metastasio (1698–1782), for many of his *opere serie*. The texts for such of his works as *Siroe, Re di Persia* (1726), *Alessandro nell' Inde* (1729) and *Artaserse* (1730) are by Metastasio, a name that hereafter is to be cited repeatedly. These three operas had their débuts away from Naples, the first an offering in Venice, the latter two in Rome. Vinci's death, at forty, occurred abruptly: he was stricken by "a sudden colic pain" and was gone before he could even recite his confession or receive last rites. In Naples, as might be expected, some said that he had been poisoned, doubtless because of an adulterous intrigue, but there was never any proof of it. What makes the explanation an unlikely one is a caricature of Vinci by Pier Leone Ghezzi, in which he appears not as a handsome, seductive gallant but as quite middle-aged and ugly, "squat and hook-nosed", scarcely a figure for legend and romance.

Like many other composers of his day, Leonardo Leo (1694–1744) held a post as *maestro di cappella*, in his case at the Neapolitan court chapel, where he wrote sacred music and taught, a task in which he was considered distinguished and was much sought after. He was notoriously slow at composition – on one occasion he was locked in his room with a guard outside to ensure that he would meet a deadline, enabling the singers to learn their parts and hurriedly rehearse them. Even so, more than fifty operas bear his name as their creator. Despite his laggard pace, his inspiration flowed copiously. He was particularly effective at writing choral passages, as shown in *Il Ciro Riconosciuto*, enacted in Turin in 1739, and *Olimpiade*, mounted in Naples in 1742–3; both productions were revivals in which changes from earlier productions had been made. *Il Ciro*, which depicts the distress of a father at the suffering of his child, has a highlight, an aria – "Non So con Dolce Moto" – that has been acclaimed as "sublime, irresistible", the praise coming a half-century later from the German composer and writer Johann Friedrich Reichardt. *Andromaca*, heard in Naples in 1742, retells the story of Hector's distraught wife, widowed in the Trojan siege, whose little son perishes at the hands of the victorious, heartless Greeks. Several scenes have compelling emotional power and have been favourably compared to the tragic versions of the harrowing incident in dramas by Euripides and much later Racine.

Broader is Nicola Porpora's niche in history. He filled his years (1686–1766 or 1767) studying music, teaching it, writing operas. It was as a teacher, especially of singers, that he was most widely celebrated. Among his pupils were Hasse, the German afterwards best known as a composer, and a cluster of male sopranos, including Farinelli and Caffarelli, the most astonishing performers of the day, eagerly bid for by impresarios hoping to sell tickets and pack their theatres. Porpora was truly a task-master. One story has it that he required of the aspiring Detto Caffarelli a daily repetition of exhausting vocal exercises, inscribed on a single sheet of paper, for five years. That one page comprised the basic tricks and techniques of the trade. At Caffarelli's final visit Porpora bade him goodbye, saying: "Go, my son, you are the greatest singer in the world." And many spectators were soon noisily inclined to agree. Porpora's operas were not always greeted as enthusiastically. Of an early work, *Temistocle* (1718), Dr Burney recalled disdainfully, "I never saw music in which shakes were so lavished; Porpora seems to have composed the air *Contrasto assai* in a shivering fit." Other critics have taken issue with that harsh judgement. Frederick Westlake: "He always wrote beautifully for voices." Certainly Porpora's works were greatly popular. In

1733 members of the British royal family and aristocracy – the Prince of Wales and the Duke of Marlborough – formed a London company, commonly alluded to as the Opera of the Nobility, with Porpora as its official composer. Its purpose was to compete with Handel's successful operatic enterprise. The Neapolitan showed that he had shrewd management skills, fortitude, patience and talent to spare. He hired away several of Handel's top singers, and added to them the incredibly gifted Farinelli, a former pupil whose voice was described as "the most beautiful ever heard", and temporarily ruined Handel, whose rival theatre was left half empty. Perhaps it is for this regrettable accomplishment that Porpora is still best remembered. Nor were his successes limited to Italy and England: his *Arianna e Teseo* was enacted in Vienna in 1714. For some of his operas, too, he had the advantage of librettos by Metastasio. A depiction of Porpora in old age is found in George Sand's novel *Consuelo*, and he is also a minor character in the brilliant twentieth-century film, *Farinelli*. An unusual tribute to him – it could be the best of all – was that of a one-time pupil, the famed castrato and teacher Antonio Uberti, who changed his name to Porporino to profit by a tie to his illustrious master.

The Neapolitan School is not of lasting historical significance simply because of the contributions of Scarlatti and his seventeenth- and eighteenth-century successors, but more for having brought into being an allied genre, the *opera buffa*, which seemed to spring inevitably and naturally from the bright weather and high-spirited temperament of that city on the bay and its garrulous, inhabitants. Almost from the beginning, comic elements were to be found in *opera seria*, even in those by Monteverdi, who had reluctantly included them, acceding to the taste of the dilettante, shallow courtiers whom he had to please. Humour had also been inserted into medieval pageant-plays, saints' plays and Moralities. Most often it appeared in *opera seria* as *intermezzi*, slipped in between the acts to afford the spectators a burst of comic relief. Light plays, such as the *commedia erudita*, also added music, but as a subordinate aspect of the presentation. Indeed, there had been patter songs and dances in the farces of Aristophanes, Plautus, Terence, but *opera buffa* was something more, it was all-out farce in which lively, tuneful music carried the principal burden of eliciting laughter, the plot and characterizations being requirements but of less importance. In the birth of this new genre, the nimble Italian language – and sometimes its colourful dialects – were a considerable help, permitting the rapid exchanges and banter that farce must have.

The term "*opera buffa*" became popularly accepted at the very end of the seventeenth century. Earlier, works that were wholly musical and in which continuous provoking stabs at laughter were dominant might bear various casual designations: "*dramma bernesco*", "*dramma comico*", "*divertimento giocoso*", "*commedia pe'museca*" (this last, in Neapolitan dialect) and doubtless in other ways (*Grove's*). A scattering of good-natured musical offerings in the late seventeenth century were more liberally laced with humour than was customary, echoing the *commedia erudita* and looking back at Terence and borrowing his amusing situations and age-old stock characters. This was true in Rome, where Giulio Rospigliosi enjoyed runs of two plays in the "erudite" tradition, with a considerable amount of music added: *Chi Soffre Speari* (or *Il Falcone*; 1637) and *Dal Male il Bene* (1653), though the plot of one is not from Terence but from Boccaccio, and of the other from a little-known Spanish piece. Also in this vein, in Florence, was G.A. Moniglia's rustic *Il Posto di Colognole* with appropriate music by Jacopa Melani (1657), a lively staging chosen to inaugurate the Teatro della Pergola; it was followed by ten or twelve similar offerings at public playhouses and Medici private theatres, until the last year of the century. After that, such works disappeared from Florence for at least two decades. Several of these presentations were taken north to Bologna, where they prepared the way for a future efflorescence of *opera buffa* in that city of arcades, sober professors, lecturers and gowned students. To the works imported from Florence were added eight by local poets and composers (1669–98). But there, too, after a brief spell of prosperity, the budding species vanished for twenty years or more: the Church was raising objections, and the Inquisition was engaged in a close scrutiny; it was prudent to desist.

At last, the scene moved south: *La Cilla* (1706), the score written by a lawyer, Michelangelo Faggioli, to a text by F.A. Tullio, a practised librettist, was put on for an audience of liberated, tolerant members of the legal profession, at the residence of the Neapolitan Minister of Justice, to whom the authors had wisely taken care to dedicate it. Later it was twice repeated in the palace of the Prince of Chiusiano (1707, 1708); obviously it pleased him. The score is lost, but the text has been preserved. It is prototypical of the kind of light musical entertainment that was soon to occupy the stages of Naples' smaller theatres and draw a stream of spectators, as the genre took on viable and independent form.

In the beginning these musical sketches were in Neapolitan dialect and their stories had settings that were a copy of the very district of the city in which the diminutive theatres were located: as a result they had an air of realism despite an essential artificiality; at very least, people readily identified with what they heard and saw in them. The characters, though lifted from the *commedia erudita*, seemed more plausible, no longer almost wholly caricatures. The *innamorati*, the roles always assumed to be played seriously, sounded more natural when the actors used everyday language rather than a poetic vocabulary. A fresh convention was arrived at: the stock characters were now cast according to their vocal range, the young lovers a tenor and lyric soprano, the old man – a stingy, greedy father or guardian, who is an obstacle to the young couple's marriage – a bass, as was the flamboyant boastful captain, and so on. (In not a few instances the male lover might be sung by a soprano, in a reversion to cross-dressing, always a favourite farcical device in Renaissance and baroque theatre.) Gradually this nascent genre evolved beyond these early local boundaries and became more cosmopolitan.

The Teatro dei Fiorentini, full-sized, already a century old, became a venue for these sprightly works – at that time, though briefly, Scarlatti's brother-in-law, Nicola Pagano, was its impresario. The first *opera buffa* staged in its antique precincts was *Patro Calienno de la Costa*

(1709), the score by Antonio Orefice, the libretto by Mercotellis (a *nom de plume*). It was in Neapolitan dialect, as was the next production, *La Spellecchia*, with music by Tomaso de Mauro, text by C. de Petris. As a train of similar works followed, pleasure-seekers gathered at them, at the expense of the Teatro S. Bartolomeo, a home of *opera seria*. In a pragmatic response, the S. Bartolomeo shifted its fare, booking compositions that were in the hilarious new style. Soon other large theatres in Naples, the Pace and the Nuovo, did the same, and more gifted writers were commissioned to provide the offerings – much to the distress of serious music-lovers, faithful to *opera seria*, their numbers painfully reduced. Even Scarlatti acknowledged the profitable vogue and tried his hand at an *opera buffa*, *Il Trionfo dell'Onore*, which seems to have been his sole effort at one. It was done at the Fiorentini (1718), and differed from the others exhibited there, since its young lovers were portrayed as aristocrats, and the libretto was in Italian, not the local dialect. As has also been mentioned, Vinci kept up with the fad by writing eleven comic pieces, all given their premières in Naples; and Hasse, while there, mastered the craft of fashioning comedies that were to be sung throughout.

At this point the singers were apt to be local talent and not too well trained. To compensate for this lack of technique and discipline, the composers avoided making demands that might be vocally taxing; the sopranos were asked to perform few *bravura* passages. The acting was broad; the influence of the robust *commedia dell'arte*, its Carnival heartiness and improvisatory spirit, ever more discernible. Many aspects of the *opera seria* lingered: the arias mostly retained the *da capo* form, and generally the works are restricted to three acts: each of the first two tends to end in a "short brawl", engaged in by three or four of the characters, and the finale of the last act was often the extended outcome of that conflict, a loud if melodic ensemble number. The overture, in three movements, kept the structure that prefaced an *opera seria*.

The enthusiastic reception in Naples of *opera buffa* inevitably prompted a spread northwards. Rome was the next conquest: a well-recognized librettist, Bernardo Saddumene, gathering some singers and enlisting a composer, Giovanni Fischietti, travelled to the Holy City in 1722. His company included castrati to take the female roles, since women were prohibited from appearing on stage in the Holy City. The introductory offering was *Li Zite*, for which he had written the text and which had been staged in Naples with music by Vinci seven years earlier; now Saddumene changed its title to *La Costanza* and replaced Vinci's score with a fresh one by Fischietti. A new touch: the lovers' roles were sung in Italian, the comics' parts in Neapolitan dialect, which had a degree of logic. Saddumene was later to take this practice back to Naples, where it became standard for a stretch of many decades.

He rapidly mounted a second work in Rome, *La Somiglianza* (1723), which confirmed the attraction of his fare. Shortly, Rome, too, was a centre for staging the happy, frolicsome presentations. More productions were brought from Naples, if they had proved their appeal in that city, and Roman composers and scriptwriters saw a chance to reap a good living from commissions to create a new supply. Near the Vatican's precincts, with a more literate audience, reliance on

Neapolitan dialect was soon dropped, and the works were sung only in Italian, the lines in dialect "translated" if need be.

The principal venue in Rome was the Teatro Valle, where was seen Gaetano Latilla's *La Finta Cameriera* (1738, staged a year earlier in Naples as *Il Gismondo*), and also his *Madama Ciana* (1738). This pair had librettos by G. Barlocci, as did two other *opere buffe* presented there in the same season, Rinaldo di Capua's *La Commedia in Commedia* (1738) and his *La Libertà Nociva* (1740). Both composers were Neapolitans, but the singers were Roman. Successful, these works were taken on tour, or performed by local companies in other cities. Within a span of five years *La Finta Cameriera* journeyed to a round dozen Italian cities, beginning in Faenza and Modena, skipping from there to Florence and Genoa and ending in Verona and Parma, with stops at Bologna, Venice, Milan along the way; afterwards it went to Graz, Leipzig and Hamburg. In each city, expectations and welcomes for visits by new *opere buffe* resulted.

The fancy of the capricious Venetians was especially captured. Soon the Teatro S. Angelo, Teatro S. Moise and the vast Teatro S. Cassiano were occupied by *opere buffe*, often displacing the *opera seria*. Initially the composers and librettists were almost all from Naples, but after mid-century this was no longer so, and a new era commanded by northern artists was started, bringing significant changes in this effervescent genre.

Early *opera buffa* reached its peak in the very small output of a young man whose years were also brief, tragically so. This ill-fated, brilliant composer, whose music is immortally witty, was the grandson of a shoemaker, Cruciano Draghi. The family, from Pergola, were referred to as the Pergolesi, which is how the grandson is known. He was Giovanni Battista, born in 1710. His father, Francesco Andrea Draghi, was a surveyor. He and his wife had four children, of whom Giovanni Battista was the third; the other three, two boys and a girl, died in childhood, leaving Giovanni Battista the only survivor. He was frail, probably consumptive from an early age, and – as depicted later by the caricaturist Leone Ghezzi – limped, a leg seriously deformed.

The boy, manifesting talent, probably had his first musical lessons from Francesco Santi, *maestro di cappella* at Iesi, the small town that was home to the maternal branch of the family, where his father, of a respectable profession, had ill-defined links to local nobility; one member of it was the boy's godfather, a lucky circumstance. The youth's next teacher was Francesco Mondini, a violinist and public music master; after which, with the help of the Marquis Cardolo Maria Pianetti of Iesi, Giovanni Battista entered the Conservatorio dei Gesù Cristo in Naples for further study (*c.* 1722–3). His instructors were Gaetano Greco, Leonardo Vinci and Francesco Durante. He remained there as a student until 1731, a spell of eight or nine years; he was exempt from having to pay fees since he participated in public performances, first by singing in the choir and later as a leading violinist in the conservatory's orchestra when it gave concerts in Naples and nearby towns.

At twenty-one he left the conservatory to follow a freelance career and almost at once received a commission to write an *opera seria*, an indication that he was judged to have remarkable gifts – or perhaps had ingratiated himself with indulgent patrons. He set to work, making use of a libretto by Apostolo Zeno, a distinguished Venetian poet and intellectual, impressively versed in Roman history. Originally titled *Alessandro Severo*, the work was renamed *Salustia*; it was not unusual for a libretto to be given more than one musical setting – copyright was non-existent. The prospects for success were high: the young composer had a libretto by a top dramatist, and the great castrato, Nicolini (a favourite of Handel), was engaged for the leading role; his singing and acting drove audiences to raptures. Before the opening night Nicolini died; a replacement was hurriedly obtained, but with little time to spare. Pergolesi had to rewrite many passages to suit the new performer's range and vocal qualities. The score is said to show clear marks of haste. The première was delayed, but the opera was not a success.

The young man's life eased and brightened considerably when he was appointed *maestro di cappella* to Prince Ferdinando Colonna Stigliano, equerry to the Viceroy of Naples. In 1732 Pergolesi composed a comic opera, *Lo Frate 'Nnamorato*, to a libretto by G.A. Federico, a lawyer and Neapolitan comedy writer. It was performed at the Teatro Fiorentini to laughter and shouts of applause. Its run was extended. Revised and with a new cast, it was brought back for Carnival in 1734. Again, it was revived at the Teatro Nuova in 1748. In the intervening years its joyous tunes were heard at all hours sung and whistled in the streets of Naples.

Life was not quiet there; in 1732 the city, lying beneath Vesuvius, was rocked by earthquakes and the theatres were closed. Later that year Pergolesi was commissioned to write another *opera seria* as a birthday tribute to the Empress Elisabeth Christina, consort of Charles VI. For this occasion he composed *Il Prigioniero Superbo*, the libretto based on an adaptation of F. Silvani's *La Fede Tradita e Vendicata*. An unusual piece, it had a small cast, no male soprano, and an alto in the leading feminine role. What was more significant, though hardly noticed at first, was the two-part *intermezzo* between its acts, the brief *La Serva Padrona* (*The Maid as Mistress*), a captivating comic trifle set to a libretto by Federico. He in turn based it on a play by Jacopo Angelo Nelli (1673–1767), a Sienese poet, whose plot mostly employs *commedia dell'arte* stock characters. About this time Pergolesi, as a reward for his having composed liturgical music in honour of St Emidius, was appointed deputy to the *maestro di cappella* of the city, with the right to become his eventual successor, which, of course, never happened.

The following year Naples was besieged by the troops of Charles Bourbon, claimant to its throne. Finally they occupied the city's centre and established sovereignty over it. Pergolesi's patron, Prince Stigliano, had left for Rome. He was replaced, however, by an appreciative local nobleman, the Duke of Maddaloni, who commissioned a performance by Pergolesi of the *Mass in F* at a celebratory service in S. Lorenzo in Rome, as a tribute to St John Neponuk, which attracted much notice. Thereafter Pergolesi became *maestro di cappella* in the Duke of Maddaloni's household and travelled as a member of his entourage.

The young composer was commissioned to write an *opera seria* in honour of the birthday of King Charles's mother in 1734; for this event he wrote *Adriano in Siria*, to a text by Metastasio. It also contained an *intermezzo*, *Livietta e Tracollo*, set to a libretto supplied by T. Mariani. The lead part in *Adriano* was assigned to the great Caffarelli, to whom Pergolesi had to defer, making changes to satisfy the singer's demands; indeed, Metastasio's text, too, was much rewritten to conform with the castrato's wishes. (Adriano was the Emperor Hadrian.) This marked the end of Pergolesi's comparatively short career in writing serious works for the Neapolitan theatre. One local impresario submitted a note to the supervisor of Naples' opera houses omitting Pergolesi's name from the list who might be regularly called upon, and another sent a comment that the young man was "esteemed as a musician, but his last works had failed to please".

The reason for this, doubtless, was Pergolesi's rapidly deteriorating health. None the less, his *Mass in F* greatly impressed the Roman public, which led to his receiving a commission to write the score for Metastasio's *L'Olimpiade*, a production scheduled for Carnival at the Teatro Tordinona in the Holy City in 1735. This did not go well. Metastasio, in Vienna, grew indignant at reports that the size of the chorus had been reduced and the lead singers were of merely so-so calibre. Pergolesi was forced to make endless alterations for them. Perhaps because of his waning strength and creative resources, he inserted four arias used earlier in *Adriano in Siria*. Shortly after the opening, Roman theatres were closed for several days due to the death of a royal personage, and soon again out of respect for the Candlemas festival. By the best account, *L'Olimpiade* was a failure, before long superseded by the theatre's next offering. (A legend has it that a displeased spectator threw an orange that hit Pergolesi on the head.)

The Romans' poor reception of *L'Olimpiade* did not cause its demise. It was staged again as a Carnival feature in Perugia in 1738, and then in Cortona the same year, and subsequently in numerous versions and adaptations in many parts of continental Europe, as well as London at the King's Theatre, where it was titled *Meraspe* (1742). This popularity lasted to at least the century's end, but the unlucky Pergolesi was not on hand to enjoy any of it. Few *opere serie* of the period endured longer.

Pergolesi's lasting fame is hardly owing to his *opere serie*, all undertaken while he was in ill health and other difficult circumstances, especially his having to placate his aristocratic patrons, together with materialistic impresarios and dictatorial castrati, whose whims were often hard to meet. Even musicologists are unlikely to be familiar with these works, yet in places they are filled with truly noble music. *Grove's* remarks that the character of Marziano in *Salustia* has "grandeur and pathos. . . . Intended for Nicolini, his utterances are like an echo of the music of the high Baroque era." In *Il Prigioniero Superbo* "the style is more polished. Pathos is replaced by sentimentality and gallantry, and formal accompanying figures become more prominent in the orchestral writing." The demands of Caffarelli are reflected in *Adriano in Siria*: "His three arias are extended beyond anything else in Pergolesi's music up to that date; they are veritable concert pieces and the focal points of the opera. Each of these expresses a different 'affection', but the expression is subordinated to the need for allowing Caffarelli the opportunity to shine vocally; this is done differently each time and with new effects." That Pergolesi transferred no arias to *L'Olimpiade* from any of his works earlier than *Adriano in Siria* suggests that he recognized his recent advances in accomplishment as a composer. "*L'Olimpiade* is characterized by idyllic and delicate tone-colours, smooth, expressive melodies which exclude virtuosity, free treatment of the text (for example with verbal repetitions of the kind used in *opera buffa*), and a greater intensity of feeling. To some extent the arias have a new function, for instead of summing up the content of a scene and rounding it off, they bring the action to a climax.... In the well-known aria '*Se cerca, se dice*' this causes the *da capo* form to be broken."

Twentieth-century revivals and explorations of Pergolesi's *opere serie* are exceedingly rare, though *Adriano in Siria* was done with little fanfare at the Maggio Musicale Fiorentino (Florence May Festival) in 1985. Along with it was given the *Intermezzo Livietta e Tracollo*, both works directed by Roberto de Simone with the conductor's baton wielded by Marcello Panni. Later in the year, *Livietta e Tracollo* was brought back, coupled with Georges Bizet's *Il Dotter Miracolo*, the director Roberto de Simone again, the conductor now Gyorgy Gyorivanyi Rath. Earlier in 1985, under the same Festival auspices, Pergolesi's *La Vedova Ingegnosa* was performed, the stage action conceived by Talmage F. Fauntleroy, the music led by Marcello Guerrini. Tickets for this event were reserved for schoolchildren.

In his last months, as his strength diminished, Pergolesi completed another comedy, *Flaminio* (1735), again with a libretto by the able Federico. It was produced at the Teatro Nuovo in Naples, and granted a second run at the Teatro Fiorentini in 1737. Siena saw it during Carnival in 1743; it was described there as a *divertimento giocoso*. In 1748 it was combined with *Lo Frate 'Nnamorato* on a double bill given at the Teatro Nuovo in Naples and stayed until the beginning of the following year.

As 1735 wore on, Pergolesi was commissioned to write a *serenata* for the nuptials of the Prince of Sansevero, but he was unable to finish it; the task of providing the second part was relinquished to another composer, Nicola Sabatino.

Aware that his condition was perilous, the almost penniless Pergolesi took up quarters in a Franciscan monastery in Pozzuoli; it had been established by ancestors of the Duke of Maddoloni. He instructed an aunt from Iesi, who had been his housekeeper in Naples, to keep for herself or distribute to the poor and worthy most of his worldly possessions. He was still capable of composing several liturgical pieces, a greatly admired *Stabat Mater* and a *Salve Regina in C minor*, and a secular cantata, *Orfeo*, for strings and soprano.

Dying at twenty-six, he was buried in the common pit next to the outer wall of Pozzuoli's cathedral. Later, the Marquis Comenico Corigliano di Rignano, an admirer of his music and possessor of the *Stabat Mater* manuscript, had a memorial tablet affixed inside the cathedral. In 1890 Pergolesi's remains were transferred to a side chapel within the edifice and the tablet, too, was moved there.

The spread of Pergolesi's fame was an unprecedented phenomenon: there had never been anything like it in the history of music. It began shortly after his death. First came the publication of four of his cantatas in 1735; three years later a second edition was printed. Meanwhile his *intermezzi* caught the public fancy with amazing rapidity. *Lo Frate 'Nnamorato* had already proved itself a popular favourite. Along with his other amusing playlets it was appropriated by travelling *commedia dell'arte* troupes who fitted it into their repertoires. It served them very well and helped to make *opera buffa* a fervently cherished genre.

Federico's plot is unusually intricate. The elderly Marcaniello and his friend Carlo have made wedding plans: Marcaniello's son, Don Pietro, is to marry Carlo's niece, Nena. The young lady, of course, has other ideas; in Nena's eyes Don Pietro is vain and stupid. Another complication is added by Lucrezia, who is pledged to the aged Carlo, a prospect she dreads and resists. All three girls are secretly enamoured of Ascanio, a young man who has been raised in Marcaniello's home. When Ascanio, at his benefactor Marcaniello's request, intervenes with Lucrezia on Carlo's behalf, she openly confesses her preference for him instead. This causes him torment. Don Pietro is really attracted to Vannella, with whom he flirts. And then there is young Nina, whose imminent destiny is to marry Marcaniello, another generational mismatch. She, too, is smitten with the good-looking Ascanio, who is torn between an interest in both Nena and Nina. Marcaniello, aware of Don Pietro's pursuit of Vannella, rebukes him. Nena takes advantage of this to question her husband-to-be's feelings for her. In a ruse, Nina blatantly ignores Marcaniello and feigns interest in the fatuous Don Pietro instead.

By the opening of Act Two all nuptial plans have dissipated: nothing is as it once was or should be. Don Pietro confides in Ascanio that Nena is beside herself with jealousy over his approaches to Vannella and that he also believes Nina to be in love with him. When Ascanio next encounters Nena, she draws him aside and asks whom he likes best, her or her sister Nina, a dilemma that disturbs him more than ever. Nena tells Don Pietro that she will never marry him. Both girls are relieved that the previously arranged matches have been called off, but they are mutually uneasy and suspicious as to which might be Ascanio's choice. He acknowledges that he is fond of both of them, an admission overheard by Lucrezia, infuriated by it. In a tantrum, inspired by her disappointment and frustration, she batters the windows with her fists; her outburst is heard and observed by Carlo, passing by. He tells her that unless she displays better manners, she will never be his wife.

From Nena, Nina and Lucrezia, Marcaniello learns of their contagious passion for Ascanio. Angered at the fuss and confusion this is causing, Marcaniello threatens to kill the much-besought young man. But, a father, he yields to paternal sympathy and agrees not to oppose Lucrezia's efforts to win Ascanio. Recognizing a formidable rival, Don Pietro challenges Ascanio to a duel and wounds him, slightly pricking his arm. Tending the injury, Carlo takes note of a mark like that of a nephew who at four years of age had been abducted long ago. Now Ascanio is discovered to be none other than the lost brother of Nena and Nina, which narrows his selection of a wife to the adoring Lucrezia. All this is unfolded to the accompaniment of "delightfully fresh and expressive" music, composed by the precocious Pergolesi in his twenty-second year.

Livietta e Tracollo, two short pieces that served as *intermezzi* in *Adriano in Siria*, also won considerable favour when put on independently, without the encompassing *opera seria*. A vagabond, Tracollo, dons drag, impersonating a pregnant woman, as a disguise while perpetrating petty thefts. Facenda, his companion, assists him. Livietta vows to apprehend the two rogues. She attires herself as a man, and is joined by Fulvia, who dresses as a peasant girl. They pretend to fall asleep. Seeking easy prey, Tracollo enters the room and tries to make off with Fulvia's necklace. Caught in the act by Livietta, he and the two young women exchange noisy threats and cringing pleas. To escape arrest, Tracollo offers to marry Livietta. She is hardly ready to accept him. In the second short farce, Tracollo pretends to be an astrologer. He feigns madness, hoping to win Livietta's pity and thereafter her hand. In turn, Livietta collapses and seems to be dead. His laments are so loud and apparently real that she is touched by them. Reviving and rising from her death-bed, she welcomes his offer if he will change his ways, which he ardently declares he will. The score for these two slight sketches boasts "happy, melodious arias", and Tracollo is given frequent engaging comic business.

Similarly, Pergolesi's last caprice, *Flaminio*, was soon much staged by the nomadic troupes. Here the plot is more sentimental, and comedy scenes alternate with serious ones, with a perceptible differentiation between the two moods, which is not altogether fortunate. Pergolesi does succeed, however, in characterizing the people in the play by his music, giving them sharply individual traits, a considerable accomplishment.

But his major legacy, a lasting triumph and confirmation of his genius, is *La Serva Padrona*, which made its inconspicuous début as an *intermezzo* in *Il Prigioniero Superbo* and now suddenly began to sweep across all Europe, invariably hailed with applause and laughter. What is more, Federico's libretto and Pergolesi's score went almost unaltered for two decades, an uncommon show of respect in that period when later hands usually had no hesitancy about slicing or expanding the works of predecessors.

Serpina is a servant in the household of the elderly Uberto, a bachelor. Tired of her strict rule of the premises, and her independent airs, he announces that he is going to marry and dispatches Vespone, his valet, to find him a wife, no matter how ugly. The point is, she must be submissive to his whims. Serpina determines that she herself will be that lady; she suspects that her master has a secret fondness for her. Vespone agrees to help her in return for a promise of future rewards. As arranged between the two of them, he informs his master that a certain Captain Tempesta is seeking Serpina's hand. She describes her imaginary suitor as a man of terrifying temper, quick to brandish a sword. Concerned about Serpina's future, since he is also her guardian, Uberto asks to meet Captain Tempesta. The military man arrives – Vespone in disguise. Deliberately, he does not make a good impression. Serpina tells Uberto that the marriage can only take place if he provides her with the very large dowry that the Captain demands as the price of letting her assume his illustrious name. Furthermore, Tempesta insists that if no dowry is paid Uberto himself should marry Serpina. The beset master sees this as the best solution and joyfully accedes to it. As she intended, Serpina will no longer be a mere servant but instead become the absolute mistress of the house.

The piece has only three characters, one of them a silent role, and is scored for a chamber orchestra of strings and continuo, so it is an ideal choice for touring. Also, it was not in Neapolitan dialect, which made it more comprehensible elsewhere.

In 1738 Queen Maria Amalia of Naples called for a revival of *La Serva Padrona* and *Livietta e Tracollo* and spoke of the deceased composer as a "great man". In the ensuing ten years there were at least twenty-four new productions in Italian cities from Rome, Parma and Milan to Venice and abroad in Graz, Munich, Dresden and Hamburg. By the end of the second decade the number had mounted to sixty. Most remarkable was the work's history in Paris, where it was first staged in 1746, an introduction barely heeded. But a half-dozen years later (1752) Eustachio Bambini brought a troupe with a repertoire of fourteen *opere buffe* to the French capital; among the offerings were six that contained excerpts and pastiches of music by Pergolesi, including *La Serva Padrona*. In Paris, a clique of aesthetes and intellectuals nurtured resentment at the steady and growing invasion by Italian music theatre, and the style in which these pieces were enacted, new to the French public. A controversy erupted, mostly carried on in pamphlets and diatribes in newspapers by those who decried the Italian innovations, chiefly written by native composers. Among the latter were Rameau and Lully. An opposing camp, more progressive in outlook, was headed by a distinguished group known variously as the Philosophes or Encyclopaedists. It comprised such sharply intelligent men as Jean-Jacques Rousseau, the progenitor of the Romantic movement, Denis Diderot, the essayist and drama theorist, and Friedrich Grimm, a young German diplomat, who supported the Italians, and in particular described Pergolesi's music as "divine". Even the king and queen took sides: the queen – unpatriotically – favouring the Italians; the king the French. This odd battle was dubbed the Querelle des Bouffons. In the end the progressives prevailed, and the Italian comic idiom was generally accepted. Indeed, Rousseau himself yielded to the influence of *La Serva Padrona* in fashioning his one attempt at a short humorous opera, *Le Devin du village* (1752).

Translated into English, *La Serva Padrona* was performed in open air at Marylebone Gardens in 1759 in a version by Stephen Storace, Sr and J. Oswald; and again, somewhat further revised, successively at the King's Theatre, Drury Lane, Ranelagh Gardens and Covent Garden, manifesting its inexhaustible appeal. Other English adaptations, some quite free, followed. While in Russia, Giovanni Paisiello borrowed Federico's libretto and bravely contributed a new score.

It is not Federico's script, thin and implausible, that accounts for *La Serva Padrona*'s immense success but almost wholly the quality of the young Pergolesi's music, sparkling, neatly satirical, exhilarating and fast-paced. Until the comparatively recent discovery of Monteverdi's and Cavalli's *opere serie*, Pergolesi's little playlet was the only Renaissance musical work for theatre to hold on to a place in the staple nineteenth- and twentieth-century repertoires. What makes it more important is that it was a model and inspiration to the next generation of Italian composers of Italian *opera buffa*, a genre at which they excel, and who looked to it for guidance, if only because of its amazing endurance.

Venice continued to be a scene of busy operatic activity. As many as nineteen theatres were kept occupied. One of the most talented newcomers there was Antonio Vivaldi (1675–1740), nicknamed "the red priest" (he had been ordained in 1693). His orchestral works, especially his violin concertos, enjoy abundant revival, but his thirty-eight or forty operas are quite forgotten, with about half of them lost and the scores of others somewhat fragmented. Among them was a *L'Olimpiade* (*The Olympiad*; 1734), making use of the same libretto by Metastasio that Pergolesi had set a short time before. Brought to Venice, Pergolesi's work, after his death, had a marvellous reception, sharpening among Venetian musicians both indignation and envy. The libretto was obviously attractive; several other composers, including Antonio Caldara, Leonardo Leo, Johann Adolf Hasse and Domenico Cimarosa, also chose it, one soon after the other. Vivaldi's pretext was that Metastasio had been openly irked at the changes made in his scenario by Pergolesi at the behest of the impresario and the castrato Caffarelli, the temperamental lead singer; actually Vivaldi intended his version to be a riposte to the Neapolitan's, which was earning large sums for its Venetian producers, if not for its author.

Metastasio's plot: Clisthenes, King of Sicyon, promises the hand in marriage of his daughter Aristea to the winner of the Olympic Games. Her secret choice would be Megacles, by whom she knows herself to be loved. Lycidas is a suitor, ardently desiring her; he knows that he cannot compete successfully in the games. He turns to Megacles, a friend just returned from Crete, asking him to participate in his place. Megacles, who does not know what the prize is to be, generously agrees to Lycidas' request. Later, learning the facts, he is deeply divided between the strong emotions of friendship and love. He is driven to think of suicide. He urges Aristea to consider Lycidas as a suitable husband. She is puzzled and distressed by his speech and behaviour. He departs and makes ready to kill himself. Lycidas has earlier pledged to marry Argene, who arrives from Crete and discovers that he has transferred his affections to another. Indeed, he has never been faithful. She exposes Lycidas' deception to Clisthenes, who angrily orders the young man's banishment. Seeking revenge, Lycidas fails in an attempt to assassinate the king and is himself sentenced to die. At the scene of the execution the repentant Argene begs to be allowed to perish in his stead. She shows Clisthenes a necklace Lycidas had once given her. The king, startled, recognizes it: it had been worn by an infant son who long past had been thrown into the sea because an oracle had predicted that the offspring would grow up and seek to slay his father. After this revelation, a double wedding follows: Megacles and Aristea, Lycidas and Argene.

Earlier Vivaldi had written *La Fida Ninfa* (*The Faithful Nymph*; 1732), produced under unusual circumstances. The commission to write its score, to a libretto by Scipione Maffei, had gone to another composer, a Florentine, Giuseppe Maria Orlandini. For some unknown reason it was withdrawn and given to the Venetian. By the time Vivaldi completed work on it, Venice was menaced by German troops encamped on its borders. The better part of discretion was to have the opera's première in Verona instead. Learning of this decision, a number of German officers sought permission to go there for the event. To refuse their request was probably undiplomatic, at a time of tension, and it might reveal to the Germans how limited was the number of Venetian troops on the adjacent mainland. Prudently the producers delayed the opera's much-anticipated début. Two years passed; a new opera house was inaugurated in Verona, with Vivaldi's work its first offering. It was elaborately staged, at an expenditure of 20,000 ducats. Vivaldi seems to have followed classic precedents established by *Aminta* and *Il Pastor Fido* in fashioning a bucolic tale of the tribulations of two daughters of an old shepherd in Skyros whose love affairs are exceedingly complicated. They are abducted by Oralto, who transports them to Naxos, where he tries to force himself on one of them, Licori. Oralto's lieutenant, Morasto, once went under the name Osmini and had been Licori's intended husband before he, too, was taken captive and brought to Naxos and involuntarily enrolled in Oralto's service. Attracted to Licori, and seeking to win her affections by arousing jealousy in her, Oralto's brother Tirsi pays exaggerated attention to her sister Elpina. At the right time Osmini identifies himself to Licori, and the happily reunited pair escape from Naxos, along with her sister and aged father. At sea a storm threatens to engulf their ship; they are spared by the intervention of the goddess Juno, who has watched with favour the young couple's long-proven fidelity to each other. Arias and recitative continuo alternate throughout the score, and Vivaldi deftly suits his melodies to the characters' sentiments. A different style of music is affixed to each of the leads, affording each one individuality, and making each one more dramatically interesting.

Tamerlano (1735) deals with Genghis Khan's successor, Tamburlaine, and his complex relationship with the Emperor Bajazet, one of his many victims. Taken prisoner, the deposed ruler and his wife are tortured by the conqueror. In despair Bajazet ends his life by hitting his head ceaselessly against the iron bars of the cage in which he is cruelly pent. The subject of Tamburlaine was popular in the eighteenth century and was taken on by Handel (1720) and Porpora (1730) and in a play by Giacomelli. Vivaldi's score is lost. Later he returned to the theme in his opera *Bajazet*. The identity of the librettist is uncertain, but he is thought to be Augustino Piovene: a far earlier source of the story might be Ruiz Gonzalez de Clavijo (d. 1412), in his day ambassador from Castile. The most powerful depiction of Tamburlaine, of course, strides through Christopher Marlowe's epic script (1590), but apparently the Italian composers and librettists were not aware of its existence. What fascinated them and the public was the figure of a strong man, a plunderer, ruthless in his grasp of sovereignty and wealth. An interesting touch in the opera is Tamburlaine's reliance on a ring inset with a precious stone that changes colour if anyone dares to lie to him. The work ended with a chorus, its music recently discovered.

Vivaldi's *Catone in Utica* (1737) has a libretto by Metastasio to which a score had already been set by Leonardo Vinci (1728) and by Johann Adolf Hasse (1731), but that did not deter the indefatigable "red priest" from using it again, in the spirit of economizing or freebooting that prevailed at that time. It was composed during a spell when he had many distractions and shows signs of carelessness. The opera, like many of his other works, had its première in Verona, with Vivaldi himself conducting, in the presence of Charles Albert, Elector of Bavaria. The libretto is somewhat weak, revolving about wars and alliances, endless intrigues, ill-fated romances. Marcia, Catone's daughter, is secretly in love with Cesare, her father's foe. Cesare offers a peace treaty, but Catone rejects it, insisting that Rome must be freed of every form of dictatorship. Marcia seeks to intervene, and Catone is shocked to discover what are to him his daughter's treasonable sympathies. Pompey's widow, Emilia, using her sway over Fulvio, a general in Cesare's ranks who is infatuated with her, seeks to persuade him to assassinate his leader, but Fulvio stays loyal to Cesare and exposes the plot. The Roman troops, ending their siege of Utica, enter the city. Rather than surrender to Cesare, Catone commits suicide. Considering the price of his victory too high, Cesare does not celebrate a triumph. Instead, he publicly recognizes Catone as a last representative of what were "the traditional virtues of a Roman citizen". The world in Vivaldi's operas could be quite violent.

The critic and biographer Michael Talbott remarks, "Despite their many beautiful moments, Vivaldi's operas cannot lay claim to the historical importance of his concertos. Yet the vigour, complexity and variety of their instrumental writing, especially in the works of the first decade (1713–1723), set a fashion for his older contemporaries. Had he begun to write his operas earlier, or had the rise of the Neapolitans occurred later, their orchestra-dominated style might have established itself more firmly. Though he continued after the critical period around 1725 to produce innovations, these never became consolidated into a 'late-period' style. A work like *Catone in Utica* betrays a self-consciousness foreign to the early operas . . . the malaise of a composer whose ambition has outlasted his capacity for self-renewal" (quoted from Headington, Westbrook, Barfoot).

In recent years there have been recordings of several Vivaldi operas – *Orlando Furioso* (1727), *L'Incoronazione di Dario* (1717), *Catone in Utica* (1737) – and more may follow, so his contribution in this genre may not have wholly vanished.

Semi-staged versions of his *L'Olimpiade* (1999) and *Tamerlano* (2000), both American premières, were included in a Vivaldi's Venice festival by the Little Orchestra Society at Alice Tully Hall in New York. In the *New York Times* Paul Griffiths was not kind:

> In the early eighteenth century, composers bothered to compose only when they could not steal, from their own previous works or from those of colleagues. Singers, too, might insist on thrusting favorite arias into the mix. Since the same situations reappeared in libretto after libretto – fury at an unwanted proposition of love, confusion at the perplexing behavior of a lover – arias could be shuffled to fit any number of stories.

The idea of the inviolable work of art had not penetrated the feverishly busy opera houses of Europe at the time. And, as Dino Anagnost explained before conducting this performance, the composer was near the bottom of the operatic pyramid, with the leading singers right up close to the top.

In some ways this hierarchy has not changed; in other ways it has. The only reason to revive *Tamerlano* now is out of a fascination with Vivaldi: great singers are not clamoring to perform it. Indeed it may be hard to find singers adequate to the demands of arias written as display pieces for stars – and stars with the very special vocal accomplishments of the period.

This performance showed the problem. All the soloists, singing from the book but costumed and in a rudimentary staging, addressed themselves honorably to the task. Many displayed useful virtues. Ellen Rabiner as Tamerlano and Melanie Sonnenberg as Asteria were mezzos of creamy tone and forceful low register. Julia Anne Wolf, assigned the fiercely virtuoso role of Irene, found moments of vividness. Tambra King (Idaspe) came out with a striking cadenza in the *da capo* of her last aria. Philip Cutlip, as the captive Ottoman emperor Bajazette, projected frustrated strength. But only Jane Dutton, as Andronico, sounded free, noble and at ease in this extremely difficult music.

Mr Anagnost and his players were satisfactory in an orchestral score that has few treasures, outside Asteria's recitative monologue on death, which is marked by a very present and ominous bass line. In an unfortunate effort at education, the performance was encumbered with a part for the composer as narrator. This Tim Jerome did, minimizing the embarrassment. Griffiths pointed out that much of the music in *Tamerlano* was not by Vivaldi but borrowed by him from Hasse and other contemporaries.

Vivaldi's considerable output was topped by his fellow Venetian, Francesco Gasparini (1668–1727), whose list of works is in excess of sixty. Two dozen were produced in Venice, the rest in Rome, where he later chose to move and take up residence. His *Ambleto* (*Hamlet*; 1705) was one of the first Italian operas to seek a public in London (1712); perhaps the hope was that the English, generally familiar with Shakespeare's play, would be curious as to how it fared set to music. He, too, wrote a *Tamerlano* (1710). In this he departed from his earlier style – arias accompanied by continuo bass and a clinging to the *da capo* form – to introduce a fuller dependence on the orchestra's strings.

Antonio Caldara (1670–1736) was another prolific figure on the Venetian scene and elsewhere, especially Vienna. He had an obvious fondness for Metastasio's texts, availing himself of a goodly number of them, even though several other composers had also selected the same ones. Typical of his work is *Achille in Sciro*, which had its first performance at the Court Theatre of the Habsburgs in his last year. Teti, the mother of Achilles, foresees his death in the Trojan War. Seeking to help him evade his dire fate, she asks Chirone to hide him away on the island of Skyros, where young Achilles is disguised as a girl and renamed Pirra. He becomes very friendly with Deidamia, daughter of the island's king, Lycomede. Because he has a hot temper, Achilles is always at substantial risk of betraying himself. Deidamia is promised to Theagene, who is discouraged by her coldness towards him; he turns his attention to Pirra instead. A delegation of Greeks headed by Ulysses reaches the island, seeking to enlist Lycomede's support in the expedition against Troy and also to find Achilles, whom they sorely need to strengthen their attacking force. Ulysses employs all his trickery to get Achilles to reveal his true identity, but fails. In this difficult situation, though, Achilles is confused and wracked by emotions of shame, ambition and love, and does not have any idea of how possibly to reconcile them. A banquet is held to mark the departure of Ulysses and the impending wedding of Deidamia and Theagene. Achilles can restrain himself no longer: the wily Ulysses provokes a brawl among the soldiers, and Achilles throws off his disguise and joins in. Lycomede now agrees to give Deidamia to Achilles as his future wife, and all conclude with a chorus of praise, during which Glory, Love and Time make an appearance. (In this opera Caldara was paying tribute to a newly married royal couple, Archduchess Maria Teresa and Francis, Duke of Loraine and future Emperor of Germany. The work was commissioned to be performed during the nuptial festivities.)

In addition to a *L'Olimpiade*, Caldera's lengthy list of works include *Il Demetrio* (1732) and *Demoönate*, both with texts by Metastasio, and both dealing with political conspiracies, one of the librettist's favourite themes. One or the other or both of these two plays were also chosen by Leonardo Leo, Johann Adolf Hasse, Christoph Willibald Gluck, Nicolò Piccinni and Giovanni Paisiello, all within a few years of one another, evidence of how highly was evaluated Metastasio's instinct and skill as a dramatist, particularly in laying out stories calling for musical enhancement and providing frequent opportunities for exciting vocal display.

Antonio Lotti (1678–1741), former pupil of Giovanni Legrenzi, tended to emulate his teacher in his less than three decades' career as a composer of operas. Beginning with *Il Trionfo dell'Innocenza* (1692), he continued to write in Venice until 1717, when he received an invitation to Dresden. Two of his works were presented at its court, *Giove in Argo* (*Jove in Argos*) and *Ascanio* (1718). He was paid a further honour when his *Teofane* was elected to open the city's Neues Opernhaus in 1719. Soon afterwards he went back to Venice and wrote no more operas, devoting himself to turning out reams of sacred music.

A more exhaustive study of the period in Venice would explore the output of numerous other talented composers of the mid eighteenth century who were keeping its stages resounding with song and compellingly busy with intrigues and romances, among them Giovanni Maria Ruggieri, Luca Antonio Predieri, Giuseppi Maria Orlandini and Antonio Zanettini. Unfortunately, not only their music but even their names are now unfamiliar. As a measure of the popularity of musical theatre, it is said that by the century's end almost 2,000 *opere buffe* had been staged in Italy, besides a profusion of *opere serie*.

But the popularity of opera did not mean that it was always profitable. Surprisingly it was not. The competition was likely to be fierce, very cut-throat. The cost of costumes and scenic investiture kept rising, as the public expected ever more elaborate spectacle and astonishing stage effects. The fees exacted by the famed castrati and prized women lead singers never stopped climbing, yet their presence in a cast was an absolute necessity. At the same time ticket prices had to stay low, in view of the rivalry between the many theatres and companies. The best time at the box-office was at Carnival, the height of the opera season. Other months of the year the singers might be idle. Most of the theatres were owned by members of the nobility, who could afford to take losses, though certainly they did not enjoy doing so. Their recompense was an enhancement of social status; they were viewed as public benefactors. Few composers, if any, grew rich. But there were private performances of operas in the palatial halls and gardens of the aristocracy at all seasons, which somewhat increased the musicians' often paltry income.

In histories of opera scant attention is paid to librettists, perhaps a mere sentence or two for each of them. It was very different in the mid and late eighteenth century in Italy, when the poets and dramatists who provided composers with texts were decidedly prominent figures.

In the preceding century the plots supplied for musical settings tended to be too complex, overloaded with characters and story-lines, as were those by Nicolò Minato and Silvio Stampigli where progression of the action was often choked by too many confusing intrigues and dramatic reversals. But gradually the scripts became simple, the action easier to grasp and follow, helping the average spectator to grow absorbed in what was happening.

To some degree this advance towards simplicity and clarity reflected the rising Arcadian movement. About 1690 societies were founded in Rome and other Italian cities aimed at bringing new life to vernacular poetry depicting pastoral idylls, a trend that in a later era might be called a "return to Nature". This movement was widespread and long lasting. Its members put their stamp of approval on "simplicity" for its own sake, and, of course, many of the early operas had pastoral themes, prettily celebrating the customs and loves of nymphs, valiant shepherds and coy, innocent shepherdesses.

Also greatly favoured, as repeatedly noted, were operas that portrayed heroic rulers and military leaders who strode into history from the annals of Ancient Greece and Rome, or from subsequent periods. This arose out of the continuing infatuation with all things classical. The eighteenth-century stage was a revitalized Pantheon of such personages, as seen in the plays of Corneille and Racine, as well as in melodramatic and tragic Italian and French *opere serie*.

Brought to the fore by this interest was Apostolo Zeno (1668–1750), a Venetian playwright who became court poet to the Emperor Charles VI in Vienna. Mention has already been made of the respect in which he was held for his thorough acquaintance with the works of Thucydides and Herodotus, the classical historians, who were storehouses of "biographies" of great men and women of the past (many of the stories possibly apocryphal but none the less fascinating and laden with strong dramatic dilemmas and situations).

Because he was a scholar along with being a sound craftsman, Zeno was in steady demand by composers searching for scripts. He fell in step with the trend towards "simplicity", reducing the number of characters. They had fewer arias, too, though the solos outnumbered the duets and ensembles. He did away with comic scenes, his intent being to maintain the dignity that he felt should be inherent in a classical tragedy. Personal motivations were explored. He had another serious purpose: to have his focal characters serve as moral examples of how sovereigns and the aristocracy should govern and conduct themselves, exhibiting in moments of exceptional crisis courage, loyalty, compassion. This was doubtless of sharp interest to the royal spectators and their courtiers who often made up a good part of his audience. Almost half of Zeno's thirty-five librettos were collaborations; he conceived and worked out his strong plots, while Pietro Pariati applied himself to versifying the dialogue. Zeno wrote his first script, *Gl'Inganni Felici* (*The Fortunate Deceptions*; 1695), when he was about twenty-seven, and Pergolesi's earliest *opera seria* – *Salustia*, originally titled *Alessandro Severo* – was set to a Zeno text.

Zeno's successor as court poet to the imperial Habsburgs was Metastasio (1698–1782, *nom de plume* for Pietro Trapassi); he assumed the post in 1729, not far along in his almost eight decades' career. The son (by adoption) of a rich, scholarly Roman jurist, he was heir to a fortune large enough to let him quit studies in the law and instead take up a poet's life. In Italy he grew to be regarded as a literary figure of major stature, an estimate of him still held. In his mid-twenties he was diverted to undertaking opera scripts by his love of music, and he is said to have written most of his librettos while seated at a keyboard. The first of his texts – twenty-seven in all – was *Didone Abbandonata* (1724), for which Domenico Sarro provided a score. So esteemed was Metastasio's handling of his subjects that some of his librettos were successively adopted by sixty or more composers, some of whom did not hesitate to use the same text twice, appending different music to a second version.

He was hardly an innovator, though he occasionally struck out in new directions, as when he has Queen Dido's love for Aeneas end tragically, as does his scenario for Vivaldi's *Catone in Utica*. Otherwise he stays with the three-act structure, and in virtually every instance he observes the conventions of *opera seria* that by now had become traditional. His skill lay in his ability to stir an audience while the plot and actions were kept within the genre's fixed, formal limits. He knew where best to place arias, before a singer's exit, and how to indicate the rank of the always self-important soloists by the order in which their arias occurred. He varied the metre in his recitative, which was mostly unrhymed. The words for the arias were usually confined to two short-lined quatrains, precisely fitting into the *da capo* form. Here the poetry was of high quality, which was not always true of his phrasing of the recitative. A habitual device of his was the so-called "simile aria", in which a character likens his dilemma to some natural phenomenon, as when a distressed man compares his fate to a ship tossed by a storm, or his happiness at the conclusion to the dawn of a new day.

Even Handel, Gluck and Mozart – in his early *Il Pastore* and *La Clemenza di Tito* – availed themselves of texts by the long-unchallenged Metastasio.

Quite as important as the librettists were the scene designers, perhaps even more so. Opera was increasingly grandiose spectacle. Though the leading scenic artists of the day were Italian, they worked at their height in Vienna, where royal patrons could indulge a shared taste for ostentation, which led to stagings on a remarkably vast scale. Such presentations were in keeping, too, with the newly born baroque age, which architecturally featured asymmetrical monumentality for churches and palaces, a style reminiscent of the best of Augustan Rome, but more richly ornamented. Such settings, replicating the streets and interiors in which the action of historical dramas were supposedly taking place, were likely to be magnificent. Sketches of them by the designers are themselves works of art and today are collected at great cost. Most certainly such investiture was unrealistic.

For Vienna's opera house the splendid stage designs were the work of Giovanni Burnacini (1600–1665), who, beginning in 1652, served Ferdinand III after having demonstrated his bold talent with similar commissions in Venice and Mantua. He had also impressed the Diet of Regensburg with a hasty, temporary structure assembled with what seemed miraculous aptitude for a festival a year earlier.

Burnacini's principal assistant was his son, Ludovici (1636–1707), who was equally gifted and in time his father's successor in the Habsburg capital. He was charged with building the city's new opera house. He also sketched and installed the settings for its first offering (1668), an opera by Cesti, with a libretto by the Jesuit Francesco Sbarra, on the theme of Paris and his golden apple.

Berthold relates: "On this occasion Ludovici outdid himself – and the musical work as well. He presented a gigantic display of picturesquely grouped choruses of gods; massed clouds receding into infinite depth and finally gliding out sideways to reveal Jupiter enthroned; wave upon wave of sea plied by ships; fearful marine monsters and dainty nymphs – all these doubtless held the attention of the admiring gala audience more than did the comparatively modest efforts of singers and orchestra. The actor who was Paris had the honor of descending from the stage in the final apotheosis and handing the golden apple to the young Empress Margareta. She accepted it with a smile. . . ." (Her predecessor, Empress Eleonore, wife of Ferdinand II, came from the ducal house of Mantua, which partly explains the affinity in sophisticated musical matters between the Viennese court and Italy and its artists.) Berthold comments: "Opera had developed to the point where the theatre itself, intended to be its servant, became its master. Opera was a means for the display of the Baroque magic of decoration and machinery. *Il Pomo d'Oro* took second place to the sumptuous scenery."

The illusion of extra width of the setting depended on a new trick of perspective: the vanishing point was no longer at the centre-back, providing the semblance of a distant vista there; instead a significant discovery led the designer to make use of two vanishing points, one at each edge of the stage, or he might have more than two, placed variously. The focus might be on an object located in the middle, a statue, a house, a courtyard, which appeared to be thrust forward, with retreating views flanking it. This caused the stage to look wider, and in some instances deeper as well.

Illusions of scale were also introduced. As Brockett puts it: "The wings near the front were painted as though they were merely the lower portion of a building too large to be contained in the narrow confines of the proscenium. In addition, vanishing points were placed extremely low so as to increase the apparent size of the settings. As a consequence, the designs often seemed so vast that they created a mood of fantasy."

Using this "angle perspective", the designer did not require as much space; he could make even a small stage look impressively large. Though long credited to others, "angle perspective" is now thought to have been developed by Filippo Juvarra (1676–1736) while he was active in Naples about 1706. He later worked as an architect and designer in Rome and Turin, letting others exploit his helpful findings, while he continued his experimentation, sometimes using just draperies or a unit set, sometimes fixed foregrounds with changing backgrounds. He liked circular lines, the spectator's gaze led back to the centre front of the stage rather than off to the wings. Nor, at other moments, was he averse to resorting to a rich *mise-en-scène*.

Ludovici Burnacini's opera house in Vienna – earlier ones in much the same style had been built in Munich (1654) and elsewhere in Germany and Italy – abounded with baroque details: arches; classical pediments; twisted, garlanded columns; faux-marble statues; curvilinear space; a high ceiling and walls covered with frescos of mythological subjects, including a host of frolicsome, flying cherubs. It had four tiers of boxes, elaborately adorned, and a stage framed with pilasters, more statuary in niches, and atop the proscenium an escalloped valance to crown a heavy draped curtain. (Much of the crowded, swirling ornamentation had recently been made possible by the invention of new, more pliable construction materials: plaster and stucco.) The opera house stood for 115 years, until destroyed in 1783.

An even more exceptional family was the Galli-Bibienas, headed by Ferdinando (1655–1743), and also made up of his brother Francesco (1659–1739) and four sons, Alessandro (1687–1769), Antonio (1700–1774), Giuseppe (1696–1757) and Giovanni Maria (1704–69), and Giuseppe's son Carlo (1728–87), a fertile tribe, all with inherent talent. They started and gained notice in Bologna and Parma. Sometimes working singly, sometimes together, they supplied stage designs for formal ceremonies, plays and operas in cities throughout Europe. They were innovative, interested in acoustics as well as stage settings, and the first to use scrims – that is, transparent scenery. Ferdinando was long believed to have originated the *scena per angolo*, but it should rather be said that he mastered it thoroughly and popularized it. After having won special notice in Barcelona, he was summoned to Vienna to mount festive décor for the marriage of the future Emperor Charles VI (1708), who three years later named him successor to Burnacini as court architect (1711). Before long, members of the family were similarly employed in Paris, Lisbon, London, Stockholm, Berlin, Prague and St Petersburg, almost everywhere leaving some kind of lasting mark.

It was Ferdinando who altered the scale of settings, to make the stage seem larger. They had a liking for platforms and stairways as well. The Bibienas also encouraged the trend towards excessive ornamentation, so that a theatrical style came to be named after them, denoting the baroque at its utmost extravagant. Ferdinando also designed the court theatre in Mantua and was the author of a book, *L'Architettura Civile*, widely studied and influential.

Francesco oversaw the erection of theatres in Vienna, Verona and Nancy. From 1726 onwards, Alessandro was attached to the Mannheim Court Theatre, for which he prepared stage sets.

Giuseppe, perhaps the most gifted of the younger generation, first assisted his father in Vienna, and then was his successor there (1727). Also to his credit were the Castle Garden Theatre in Prague (1723) and the still standing Margrave's Opera House in Bayreuth (1744–8), highly admired because its auditorium and stage, harmoniously conceived, attain a perfect unity. His contributions to the Dresden Court Theatre, where he was active for six years (1748–54), also elicited praise. Here he was helped by his son Carlo. He worked next for two years at the Royal Theatres in Berlin at the behest of Frederick the Great; luckily some of his accomplishments are preserved in his book of copper engravings, *Architettura e Prospettiva*. Venice was another location at which he won applause. He excelled in staging open-air offerings in gardens. Berthold: "He remodeled the given garden of architectural setting into an apotheosis of perspective in which reality and illusion totally fused. There exists a series of engravings of his designs for the opera *Costanza e Fortezza* which was performed in 1723 at the park of the imperial castle in Prague, in honor of Emperor Charles VI. They form an optical polyphony so grandly self-sufficient that it seems almost paradoxical to expect an orchestra and singers to make any impact in such a setting."

Some historians assert that with Giuseppe's death (1757) the dominance of the baroque in design and architecture came to an end. But Antonio was still working in Vienna, where he had been responsible for the Redoutensal in the Hofburg (1743), later taking on other commissions in Italy, to which he returned in 1750. Similarly, Giovanni Maria was employed for another fifteen years in Austria, Italy, Spain and Portugal, assuring that the Bibiena tradition of monumentality and over-ornamentation was still alive and to be met at every hand on the Continent, a peculiar legacy that Carlo served for almost two decades more.

The Bibienas had to contend with other remarkable families. One was fathered by Gaspare Mauro (*fl.* 1657–1719) and was engaged in architecture and stage design in Venice, Turin, Monaco, Milan, Dresden, Vienna and other cities, a succession of heirs that lasted until about 1820. An eminent member was Alessandro Mauro, who arranged spectacular galas in Dresden, with waterworks, giant pyrotechnical displays and shifting illuminations.

Enduring even longer were the Quaglios, of whom the first to take up stagecraft was Giulio (1601–58), his loyal progeny staying in the profession for six generations until the mid twentieth century. In all, fifteen members practised as theatre architects and decorators. Italians, they were most often given posts in German-speaking cities. Lorenzo I (1730–1804) was named Court Architect of the Mannheim Theatre. He had the task of enlarging the Schlosstheater and later built the National Theatre, in its day hailed as Germany's most modern in décor and equipment. In Mannheim he created sets for works by Salieri, Galuppi, Hasse and Paisiello. He went next to Dresden, and then back to Italy to study and become more cognizant of recent trends in his field; then on to Zweibrucken. As Court Architect in Dresden he designed settings for the première of Mozart's *Idomeneo* (1781). Domenico II (1787–1837) was prominent as a painter, decorator and lithographer. Simon (1795–1837) was overseer of "scenic art" at the Munich Court Theatre (1828). Angelo II (1828–90) collaborated with Richard Wagner, who had his own revolutionary ideas as to how his myth-laden operas should be mounted. On his own, Angelo became an important figure in Munich's theatre world. Eugen (1857–1942) preserved the family's illustrious tradition into another new century by being active as a stage designer in Munich, Berlin, Stuttgart and Prague. In Monaco, which the Quaglios more or less made their headquarters, a museum contains many of their sketches and plans.

Primarily located in Turin and Milan, though they worked elsewhere as well, the Galliari family boasted several highly capable members, the most esteemed of whom were Bernadino (1707–94) and Fabrizio (1709–90). Others in the clan were productive as designers until about 1823.

Two outstanding designers of the period who did not belong to historically important families are Francesco Santurini and Gian Battista Piranesi. A Venetian, Santurini was chosen to oversee construction of an opera house in Munich in 1650. As mentioned earlier, it was completed in 1654, somewhat before Ludovici Burnacini's new baroque theatre in Vienna and in very much the same fanciful, opulent style. Afterwards, installed as its artistic director, Santurini put on lavish spectacles on its huge stage. His were the scene designs on paper, but their execution was delegated to an assistant, Francesco Mauro, whose title was "master of the machinery". Later two other Mauros, Domenico and Gasparo, were retained in turn at this royal Munich theatre.

Gian Battista Piranesi (1720–78), a Venetian and one of Italy's major artists, designed a considerable number of stage settings, but it was through his prodigious output of more than 1,000 engravings that he exerted influence on his contemporaries and subsequent generations. His preferred subjects, gloomy and morbid, were ruins and vast subterranean prison interiors. He put solid, dark objects in the foreground, light forms behind them. He awoke an interest in "ruins", overgrown, moss-hung relics of antiquity. Designers who heretofore sought to have the details of their stage bright and clear were quickly fascinated by Piranesi's evocations of a sombre mood, his manipulations of perspective and soft hues to create atmosphere. They moved to emulate him. Brockett: "They began to depict picturesque places as seen by moonlight or interiors illuminated by a few shafts of light. Color played only a minor role in this trend, for the palette was limited. Settings were painted in sepia or pastel shades of green, yellow, and lavender. Mood, therefore, was achieved primarily through the juxtaposition of light and shadow." This was a lasting innovation. Centuries later, with the advantage of electric lights and a backstage switchboard, dependence on frequent, subtle changes of the lighting to support and enhance the mood and atmosphere became a dominant factor in mounting any dramatic work.

By the end of the eighteenth century there was a proliferation of baroque opera houses, their interiors gilded and carved, cluttered and over-decorated, across Europe. Some were large, some were small; they might be utilized for plays and dance as well. In addition to those already referred to, there was one in Lisbon, and in St Petersburg and Kharkov (temporarily); in Germany by 1776 as many as fourteen, with the following year's count twenty, rising the following decade to well over thirty – in Weimar, Gotha, Hamburg; the list is too lengthy to give in full.

Such theatres in the "Bibiena style", characterized by scenery of colossal baroque proportions, majestic columns, heavy decoration, bold perspective – which gave one a sense of open space and freedom – were truly imposing, affording the spectator a feeling of an uplifting escape from the everyday world, an enrichment of his life.

—5—
ITALY: THE DANCE

The Middle Ages had folk-dances, "rounds" – lively and sometimes even rowdy – and court dances, very formal, such as the Basse-Dance, "in which the gentleman held his partner by her little finger, and moved to music which was invariably out of step with the verses that sometimes accompanied it". But there was little or nothing of dancing that told a story or was theatrical in any way, except perhaps for some acrobatic or amusing steps performed by a lone troubadour or by a duo of wandering entertainers in a castle hall or marketplace.

Ferdinando Reyna, in *A Concise History of Ballet*, from which the above quotationis taken, asserts that choreography "began" in the fourteenth century in Italy, a claim that will surprise most students of dance – it was long assumed that its inception was much later. Further, according to Reyna, it was not invented by courtiers and other members of the aristocracy, as heretofore supposed; instead it was the creation of men of humble rank, among them a number of Jews. In many countries, and especially Spain, the Jews would have had to conceal as best they could their race, faith and talents, but in Italy they were accorded a considerable degree of freedom. In Mantua, as has been seen, they had provided a playwright and court stage director. If they so wished, and had the gift, they could also embrace careers as dancing masters. One who did and was renowned in the early fifteenth century was Domenico da Piacenza (or Domenico da Ferrara) who published a treatise, *De Arte Saltandi et Ducendi* (*On the Art of Dancing and Conducting Dances*). It can still be perused in the Bibliothèque Nationale in Paris. In it he prescribes what should be the dancer's proper bearing, how much of the floor may be occupied, how best to memorize steps, moves and gestures, how to heed the tempo – fast or slow, as dictated by the music's rhythm – how to leap, how gracefully to lift one's partner, and how, if necessary, to disregard the rules. The dances that he describes are not traditional but are original, his own compositions. His writing is poetic, as the following excerpts demonstrate:

> I am the Basse-Dance, the Queen of Measure, and worthy to wear the crown. With me but few are successful, and he who dances and plays me well must surely have received a gift from Heaven.
>
> I am the Quadernaria, and if musicians treat me wisely they find me one-sixth faster than the Queen of Measures. Played rightly, by a good musician, I will be midway between the Basse-Dance and the Saltarella.

And so on at length. He distinguishes nine "basic" or "natural" steps, and three that are "artificial". Says Reyna: "There had now been created, for the first time, a repertory of movements that were independent of their context; choreography was born. Domenico signed his dances, and was at pains to indicate that they were his 'invention,' both the music and the choreography." The author of the treatise also suggests that there is another dimension to dance besides the overt physical movement: it has a mysticism, a deeper impulse, from which springs inspiration.

Another early theoretician, Messer Ambrosio de Pesaro, described dance as "a demonstrative action, in harmony with the melody of a few voices". He lists fifteen different sorts of Saltarelli, Balli and Basse-Dances, such as a French Ballade, and even a Ballo composed by Lorenzo de' Medici. He also includes dances from other countries – Moorish, Turkish, German.

A more important figure in this early, precursive era was Guglielmo Ebreo (William the Jew), a disciple of Domenico da Piacenza and a much-sought-after dancing master and choreographer. In 1463, after thirty years of experience in his field, he brought forth his richly illustrated *Treatise on the Art of Dancing*, with a flattering dedication to Gian Galeazzo, Duke of Milan. It is a chronicle of the elaborate fêtes and great names of recent years, together with detailed accounts of the mimed masquerades that were featured on such occasions. He gives instructions, "down to the smallest button", on how each dancer should be attired when participating in such entertainments. He also advised his pupils to perform "contra tempo, if they wish to show themselves scientific and intelligent in the art". In his description of a composition, *The Ungracious Lady*, the term "*balleto*" is used – as yet the word "ballet" has not been met.

It is much in evidence two years later, however, in the writings of a nobleman, Antonio Cornazzano, a gentleman who could turn out pious works and coin obscene proverbs with equal aplomb. He comments (1465) on the latest developments in his social realm: "In Italy, ballets are now very much the fashion. They are compositions of several measures which may include all the nine movements of the dance, even those which in appearance are contradictory, such as those of the Coquette and those of the Constant Lover, fittingly combined." This indicates that mime – though not yet drama – had been added to dance.

Reyna summarizes: "Technique had evolved, and new steps had been added to the repertory introduced by Domenico. Dancing was being stylized, refined, and codified, while at the same time there was a growing awareness of the immense possibilities of the human body. Music – which at this period consisted almost entirely of music for dancing – was enriched by a new instrument, the lute. To music and steps were added, as Guglielmo so discerningly puts it, 'certain sweet emotions which, pent up within us contrary to nature, strive to be expressed in movement.'"

As Reyna stresses, ballet did not suddenly arise as the creation of Renaissance Humanists, as frequently propounded, nor as the invention of a single man of genius. Rather it drew on "the rich material of past centuries, sometimes deriving from beyond Italy – from as far away as Arabia, the land of the Moors".

In the Middle Ages, the epoch of the Crusades and the of Moorish settlement in Spain, the drawn-out conflict between Christians and Infidels was echoed and memorialized in a symbolic dance, the Moresque (or Moresca), an enactment of a combat between these implacable foes. At festivals, in villages along the Mediterranean coast, players would sweep into the main plaza, one of them, with blackened face, impersonating a Moor (the "Mattacino"). A mock battle would ensue, the Moor dancing and gesturing in an Oriental manner, a swarm of his enemies gathering menacingly about him. He would be overcome and perish.

In time the Moresque was no longer as warlike; it devolved to a simple entertainment. Singers were added, and a female character (played by a man); drums and fifes provided a rhythmic accompaniment guiding the dancers through complicated movements that called for defiant gestures, thrusts and parries, daring leaps over unsheathed swords. Finally came the "*rosa*", the Moor tossed into the air, the bells attached to him clanging, an eagerly anticipated climax.

Gradually, the Moresque was absorbed into the religious dramas, performed on the pageant carts that wended through the town and set up in a marketplace or before the portals of the local cathedral or parish church. This popular danced pantomime was also an early approach to what was to grow into ballet, a theatrical offering.

More conventional histories of ballet place its origins far later, mostly in the fifteenth and sixteenth centuries. Certainly, dance in Italy in the Renaissance took on a very different aspect from its earliest manifestations in primitive rites and in Greek and Roman theatres. Now it was primarily a component of the spectacles put on in ducal hall or at a garden fête; very often such an event honoured a coronation, or was an "entry" at a princely wedding, or a feature of a royal procession, or was given at an assembly to welcome a foreign dignitary, and on such an occasion the invited, aristocratic audience joined in the dancing – hence its style was stately, courtly.

One such affair was held for the Duke of Milan and his bride on their triumphal visit to Bergonzio di Botta at Tortona in 1489. In large measure the festivity consisted of a banquet, followed by entertainment. Mythological figures appeared in a danced pantomime representing Jason and his Argonauts, who at the conclusion presented a gift to the noble guests, the long-sought Golden Fleece (in this instance a gilt-covered roasted lamb!). Mercury, Diana, Atalanta, Isis, Hebe were other participants, each offering a similarly succulent tribute: a calf, a boar, roasted birds with their feathers still on, a flagon of choice wine – the performers being members of di Botta's family and entourage. Flattery was usually the purpose of these fêtes, as has been said, the goal being political gain. Sometimes the intention of these extravagant "entries" was to impress rulers or envoys from rival duchies or city-states with the wealth and resources of their host, a kind of subtle intimidation.

Poems were recited, perhaps accompanying or interspersed in the pantomimes, as well as songs; and, as had Raphael and Mantegna before him, no less a person than Leonardo da Vinci might be commissioned to supply gorgeous costumes, scene designs and stunning special effects. To fulfil one commission, *Il Paradiso*, the ingenious artist had a mountain split open, revealing Heaven within.

None staged more lavish triumphs that Lorenzo de' Medici (the Magnifico), including balls at which the dancers were masked. He might have Botticelli design the costumes. But the Gonzagas were not to be outshone: when Elizabetta Gonzaga married Guidobaldo di Montefeltro, the father of Raphael organized the entertainment in Mantua: one hundred dancers, directed by the esteemed choreographer Lavagnolo, took part. To help celebrate the wedding of Lucrezia d'Este to Annibale Bentivoglio, a group of Diana's delicate nymphs entered in response to seductive music, then took flight when almost encircled by savages, until finally all were reconciled by the benign intervention of Venus.

The favoured social dances were the Saltarello, the Gaillarde and the Pavane, but the most popular was still the Moresque. As now presented at court it was quite changed, its dancers clad in glinting cloths of gold and silver, its original purpose and meaning hardly discernible; instead it was now filled with classical allusions, which most delighted Renaissance spectators; or it might be concerned with the doings of peasant girls, satyrs and nymphs, perhaps timidly retreating when threatened by savages, from whose gruesome clutches they would be timely rescued. The Venetians excelled at creating such fantasies on commission. (It is said that in England 'Moresque' came to be translated as 'morris' and attached to the ancient folk-dance still bearing that name.)

Deriving from court dances, the role of the male principal in the ballet-to-be became chivalric. He was the cavalier; the lady received his devoted attention. At first the choreography at the fêtes was only an adaptation of the social dances of the day. Slowly the performances detached themselves from that context and took on an expanding theatricality. The dancers, to meet the demands of choreographers, became more and more professional: their technique grew increasingly difficult to master, outstripping the skills and agility of amateurs, however noble their blood. With this new and extended range of steps and leaps, and of expressive miming, the telling of stories by dance alone steadily advanced, until dance itself was recognized as independently dramatic.

It was theatricalized, too, by the players in the *commedia dell'arte* who were trained in the rudiments of dancing at very least, and sometimes also in its intricacies. The *intermezzi* of *opere serie* began to include dances – reluctantly, to satisfy the Mantuan courtiers. Claudio Monteverdi had been obliged to add dance music to the scores of some of his sober, passionate stage-works; the guests of his patron, the Gonzagan duke, insisted on it. For centuries interludes of dancing were generally deemed to be indispensable to any opera production. (In a notorious instance, box-holders, members of the Jockey Club, rioted at a performance in Paris of *Tannhäuser* because they came late and missed the Bacchanale that he had placed in the first act, thoughtlessly inconveniencing them.)

A famed Italian dancer, teacher and choreographer was Cesare Negri. His birthdate is 1530. By that time Milan, where he chose to work, was the "European capital of ballet", but he also moved about the Continent ceaselessly. He was popularly dubbed the Trombone, though just why is not clear. By his own account, in an autobiography and treatise on dance, *Le Grazie d'Amore*, published in his old age, he was somehow present at a good many of the crucial events of the day. He danced on the deck of the admiral's galley, before an assemblage of ships' captains, when Don Juan of Austria sank the Turkish fleet at Lepanto in 1571, and danced again in Milan when Don Juan, the heroic rescuer, was welcomed to Milan. Earlier, in 1564, Negri had established his credentials by staging a grand masquerade: he used twenty-five chariots on which were figures representing every human emotion – Desire, Fear, Suspicion, Anger, Hope and so on. On reaching what is now the Piazza della Scala, the procession halted, permitting the symbolic characters to descend from their garlanded chariots to perform a gigantic Brando (a newly conceived dance). In this very year Henri III hurriedly departed from Poland, summoned to ascend the French throne. *En route* he tarried briefly in Venice to take in the always festive activities there, and then travelled on to Cremona, where he halted to observe the dancing of Negri and a disciple, Farrufino. A further distance along the journey, Henri III and Negri again crossed paths. After that, at each stop along the way, until the Alps were reached, Negri performed for the French sovereign. This greatly increased the repute in which he was now held everywhere.

In 1604 he brought out a revision of *Le Grazie d'Amore*, retitled *Nuove Inventione di Balli*. He was regarded as a teacher without a peer; his influence was vast, indeed. More than forty of his pupils later held posts at the far-flung courts of Europe, engaged by Flemish princes, Bavarian and Polish dukes, French and Spanish kings, an Emperor of Austria, the rulers of Cologne, Parma, Bologna. As a result the principles he laid down became, in a sense, almost universal.

In the revision of his book he specified fifty-five technical rules to be heeded by every dancer. They constitute the fundamentals of classical dance to the present day. He also describes the variety of dances known at that time, and the steps and movements they required. There were already ten different kinds of pirouettes, a half-dozen *cabrioles* and *tour sauté*, and the like. Negri favoured dancing *demi-pointe*, not yet on the toes but half-way there. Feet were turned out, knees bent a little apart when landing as softly as possible from a jump, an elbow lifted slightly during a turn to add grace. He was deeply concerned with imparting elegance with every gesture and move.

Many of the same instructions had been compiled some years earlier by Fabritio Caroso in his treatise *Il Ballarina* (1577 or 1581), which suggests that Italian choreographers and their dancers had shaped a school with defined principles to which they regularly adhered: for them, *this* was how it was to be done. Dedicated to Bianca Cappello de' Medici, Caroso's book contains fifty-four rules. A native of Sermometa, between Florence and Rome, he undertook his choreographic tasks meticulously, sketching steps and placements on paper before mounting an offering. He tried to have his students match their movements to metres of classical verse. He worked in Paris around 1590 and is particularly noted for his *Ballo di Fiore*.

When Catherine de' Medici travelled to France to marry the Dauphin in 1533 her entourage included a troupe of Tuscan dancers to participate in the nuptial festivities. This introduced the Italian style of dance entertainment to Paris and the lavish-spending French court, where the next, more significant evolution of "ballet" was to happen.

(Other sources of the material in this chapter, apart from Reyna, have been *The Dance Encyclopedia*, edited by Anatole Chujoy and P.W. Manchester; *Dance*, by Lincoln Kirstein; *A Short History of Ballet*, by Cyril W. Beaumont; and *The Book of the Dance*, by Agnes de Mille.)

— 6 —
DRAMA IN PRE-BAROQUE GERMANY

Though play-acting had Martin Luther's blessing – if it preached virtuous behaviour and promulgated his heretical new doctrines – the various forms of theatre hardly flourished in Germany after the bold efforts of Hans Sachs. Almost three barren centuries followed during which the stage-arts faltered. At times, the ferment of the Renaissance spilled over from Venice and other Italian eager intellectual centres, and especially from the imperial Habsburg court in Vienna, where by happy chance an Empress was a Gonzaga, and there were incursions by troupes of actors from England who kept a scattering of German poets and liberal rulers abreast of recent developments in theatrical craft elsewhere. Overall, however, these had little effect.

Much of the initial blight that descended on the native output was due to Luther's expectation that the drama be used in his battle against what he proclaimed was a decadent, corrupt Papacy in Rome; plays were to be engaged in vilifying attacks on the dictates of the Vatican, or in bitter counter-attacks as zealous Jesuits took up their quill-pens to defend an orthodox interpretation of the faith. This led to a plethora of scripts that were overly harsh and didactic on both sides of the divide. Such works doubtless held interest for some, the hot-headed partisans, but had little artistic merit and only limited appeal.

Another major obstacle to the growth of audiences was Luther's proposal that the scripts be in Latin and seek to rejuvenate ancient forms, largely Senecan; or, if comedies, they should hearken back to Plautus and Terence. Scholastics, and the Jesuits, welcomed this suggestion, since it helped their students master and speak Latin, enabling them to attain ease and often eloquence in that language. Much of German theatre during this era took place in seminaries and lesser schools. The great scholar and reformer Phillip Melanchthon, in his private academy, had his pupils enact plays in Latin by Plautus and Terence. Melanchthon himself wrote prologues to the comedies. For the most part, though, new scripts in Latin were read and discussed, not actually performed.

Other than the English visitors, the only professional bands of entertainers were travelling groups of jugglers, acrobats and buffoons, offering broad farce in the tradition of the robust, vulgar Hanswurst skits. Actors in serious plays were amateurs, drawn from the nobility and middle class, or else students.

Scripts in English were, in general, as inaccessible as those in Latin, perhaps more so. The touring groups of players from Britain compensated by stressing mime and emphatic physical action, such as swordplay and "cloak-and-dagger" deeds of violence, stage effects picturing the horrifying intervention of the supernatural, abounding in plays supplied to them by Thomas Kyd, Christopher Marlowe, George Peele, Robert Greene, the early Elizabethans whose works tended to pay homage to Seneca's compelling, blood-soaked "revenge tragedies" – that is, by way of close imitation.

An early comedy in Latin style is *Stylpho* (1480) by Jakob Wimpheling, an Alsatian, to whom reference has already been made. His play, merely a dialogue between two former university students, was a feature of graduation ceremonies at Heidelberg. One of the pair, Stylpho, has neglected his study of Latin and other Humanist subjects, and lost the many advantages that he believed were in his grasp; he ends up not in a high post in the Roman curia, as he had anticipated, but instead as a lowly sowherd. His opposite, the earnest, diligent Vigilantius, born a Swabian peasant, has risen from his humble rank to a bishopric. Wimpheling affirms the importance of hard work to acquire knowledge of the classic disciplines.

This simple, homiletic piece inspired a rather similar dialogue by Heinrich Bebel, professor at the newly established university at Tübingen, who scripted a *Comoedia de Optimo Studio Iuvenum* (1501). Here the irresponsible Stylpho is matched against the earnest and dedicated Vigilantius while both are still students. Vigilantius, as yet an imperfectly educated Swabian youth, is clearly an autobiographical figure, since much in his life accords with details of Bebel's own successful career. The rewards of studying liberal arts, especially of becoming an elegant Latin rhetorician, and superior as a poet, are set forth. (Bebel himself, a Swabian, had been poor, and had ascended to being named a poet laureate.) The play offers advice on the academic subjects to which one should apply oneself, and whom a prudent student should select as his mentors. Though designated as "comedies", the plays are only intermittently humorous, granted that they are marked by frequent word-plays and that the authors' favoured characters are destined to fulfil their highest aspirations. If anything, the well-worn term "comedy" was used somewhat loosely at that time.

From German-speaking Switzerland, Johannes Agricola entered the religious controversy with his strident *Tragedia Johannes Huss* (1537), a drama – or better, a tract – so accusatory that he was rebuked by Luther, who complained that it was too angry to serve as a school play.

Thomas Naogeorgus, in *Pammachius* (1538), a highly intellectual and elaborate script covering a thousand years of church history, explored the theme of the Antichrist. The play's title refers to Bishop Pammachius, who lived in the reign of Emperor Julian the Apostate (331–63). In hell, in a weird episode, Satan accords Pammachius, the Antichrist, the papal tiara. At a raucous banquet Pammachius, Satan and his attendants celebrate their brief triumph, which comes to an abrupt halt when word is brought that Luther has affixed his Theses to the church door at Wittenberg. The epilogue has the Antichrist vowing to carry on the battle against Luther, the terrible struggle to continue until the Day of Judgement, with no final decision until then. This ambitious drama was dedicated to Thomas Cranmer, Archbishop of Canterbury, cited by Naogeorgus as the "highest antipapist of the Church of England". (Cranmer had made a trip to Germany and met some of the ardent reformers. He had *Pammachius* staged in Canterbury, possibly in his own palace. Subsequently it was performed at Christ's College, Cambridge, in 1545. Cranmer persistently urged English dramatists to follow Naogeorgus's example in the bitter confrontation with the Vatican.)

In Zurich, Jakob Ruoff wrote *Weingartenspiel* (1539). Besides being a playwright, Ruoff practised as a surgeon and stone-cutter, to some no doubt a daunting combination. His script depicts the vintners as Papists, guilty of having killed God's beloved son.

At twenty, Christoph Stummel of Frankfurt an der Oder was a student of theology and philosophy. Among his professors was Jodocus Willich, who was an authority on Terence and lectured on dramatic technique. Absorbing much from his teacher, young Stummel composed *Studentes* (1545), the title borrowed from Ariosto's comedy and much of the script's content influenced by the Italian poet's work. Faculty members were greatly pleased with Stummel's venture. Acclaim for it rapidly spread beyond the school, making Stummel a celebrity in a very short time. Two performances of his comedy were given at Wittenberg. In the audience, an honoured guest, was the eminent Melanchthon, who declared that the dialogue – "*elegantissima*" – assuredly proved that the youthful playwright was well read. The characters' exchanges resembled those in Plautus and Terence. The script also included references to the Greek comic dramatist Eleutheria, some of the characters bore Greek names, and Stummel displayed a broad acquaintance with Greek mythology. *Studentes* fitted very well in the era's expanding trend to Humanism.

The play, surprisingly realistic, describes daily incidents in the lives of university students, their pleasures and hardships, their delight in learning, their dissipations, distracting carousings, brawls, pub-crawlings, pursuits of loose-virtued young women and true loves. At each act's ending a chorus enters to dispense warnings and good counsel, and at the finale the students' fathers hasten to the scene to dip into their pockets and rescue their reckless sons, as well as to arrange some haphazard marriages.

Best known of the dramatists choosing Latin and spanning the turn of the fifteenth to sixteenth century is Johannes Reuchlin (1455–1522), an outstanding Renaissance scholar. In his works he blended elements from Roman comedy and *Fastnachtspiele*, borrowing plots and mimed "business" from both. He also reached back to French peasant farce in his much-cherished *Henno* (1497), where, as has already been remarked, he gives an updated German setting to the medieval tale of Master Patelin, about how a conniving lawyer is outwitted by a thieving shepherd who responds with *Baa-baa* to all demands for restitution and a fee, feigning to be as stupid as the sheep in his flock, that being what the lawyer has instructed him to do in court. German historians consider Reuchlin to have created their country's first *Lustspiel* (light comedy).

Throughout the first half of the sixteenth century plays in Latin and weightedly didactic – mainly espousing Lutheranism, but some maligning it – were in the forefront, but they were not the only theatre fare. The English troupes brought dramas and comedies on other subjects –

most often wildly melodramatic, secular historical themes – that inspired local poets to write in their native German. So a list of works from this period is a mixed bag. Some are in Latin, but not all are; some grapple with theological problems and proselytize, but not all do. Sebastian Brant's *Tugent Soyl* (1513) and Pangratz's *Die Historie Herculis* (1515) treat allegorically with the youthful Hercules' moral dilemma, having to choose between good and evil ways, his hesitation at an ethical crossroads, a problem also probed by Hans Sachs.

Another favourite topic, taken from the Apocrypha, was that of Susanna, who bathing nude in a stream was spied on by the lecherous elders. At Luther's urging, Paul Rebhuhn used the story for his *Susannen* (1536); it had already attracted the notice of Sixt Birck of Augsburg, who wrote a German version of it and then five years later translated his account of it into Latin (1532, 1537). Other soundings of the deeper meaning of Susanna's ordeal – wholly innocent, she was a victim of slander – were offered in nearby Hungary by Leonhard Stockel, who intended his play to be "an exercise in public speaking and moral behaviour" for the impressionable young, and in neighbouring Denmark by Peder Jensen Hegelund, who modelled his script on that of Sixt Birck but added to it an interlude, *Calumnia*, in which a symbolic figure, Fama Mala, appeared "in a costume picturesquely sewn with tongues of cloth". (All this, and more in this chapter, is gathered from an essay by Richard Erich Schade in *Studies in Early German Comedy*; from Berthold; and the *McGraw-Hill Encyclopedia of the Drama*.)

In Prague, a decade later, a former student of Melanchthon's, Matthias Collin, now a professor of classical philosophy, brought out yet another depiction of the unhappy bather, *History of Susanna* (1543). Produced first at a college, it was repeated at court at the command of the king, who greatly enjoyed it and wished to share his pleasure in it with the whole royal family and his entourage.

Other Old and New Testament figures who were frequently portrayed in school and court plays of the period were Jakob, Joseph, Lazarus and Tobias. To choose subjects from the Scriptures or classical antiquity was a way to evade controversy and the possibility of clerical and political disapproval; it was a measure of safety, when a hint of dissent might be dangerous.

School plays might be put on anywhere, in college courtyards, lecture halls, town halls, guild rooms, ballrooms, public squares. A stage, holding a single set, might be improvised on sturdy boards supported by crossed beams or upstanding, empty wine barrels. If the façades of houses were suggested, they might have their tenants' names attached above the entrance or to a gable in large, printed letters to clarify the action: who belongs where, who is coming from or going to or where; as far as possible, all was spelled out. Many in the audience did not know Latin and needed help to understand just what was happening. For those who did comprehend, the visual scene was created mostly in words, challenging the poet's ability to evoke a room, street or landscape none too obtrusively in the dialogue or more plainly by means of a prologue, narrator or chorus. Stage props, perhaps contributed by generous local craftsmen or obliging guild members, could be shifted about to imply a change of scene, a new place. If outdoors, the stage might be framed in canvas or open to the sky and the inconstant elements.

Notes on props and costumes for a drama in Latin written by Stephan Broelman and presented by his students in Cologne (1581) have been discovered and are illuminating. The performance was given four times in honour of Laurentius, the school's patron saint. A sketch attached to the manuscript shows that the play was enacted in the school yard and that the setting cleverly incorporated two trees standing and arching there. The props consisted of unattached doors that could be moved about, an obelisk, a throne, an altar, "a barred prison, a praetor's curule chair". These were simultaneously visible throughout. Berthold provides details of the players' attire: "The hero-martyr wears a long, ample tunic and a yellow cloak ornamented with plant forms. Faustina appears in a black palla and with a high head-dress; her name is affixed in silver letters to her shoulder." On more modest occasions, we are told that "a cloth thrown over one's shoulder had to do as a Roman toga, some obvious attribute identified the gods or allegorical figures, and a guild emblem served as indication of professional status. A plume of feathers on the hat signified a nobleman, a cudgel a trooper, a white beard an old man, a towel wound around the head a Turk."

There was also a manual on acting, compiled by Jodocus Willich from lectures delivered at Basel and Frankfurt an der Oder. Conveying emotion, every physical part should be used: the angle of the head, furrowing of the forehead, arching of the eyebrows, compression or twitching of the lips, lowering, widening or side-glancing of the eyes, and so on, a Homeric catalogue that extended to the manipulation of the hands, fingertips, stiffening of the knees and placement of the feet, even to how the nape and neck might be held, all enabling the performer to fulfil his role most effectively. Berthold likens the instructions to those given to a Kathak Indian dancer.

The ongoing, never-ending turbulence of these days, which encompassed the Thirty Years' War, is reflected in the more secular plays; the strife and the general misery it brought, was responsible to an appreciable degree for much of the period's artistic sterility. A profusion of writers' names survives, but hardly any of the scripts bearing them are still viable. For instance, Pamphilus Gengenbach's *Der Nollhart* (1517) is a wide-ranging dialogue between Nollhart and ten figures who represent various and divergent aspects of society, from Pope to Jew. Gengenbach sees only chaos lying ahead, unless there is a consensus on the urgent need for a sweeping Reformation – in other words, a justification for Luther's revolt. The play is of special interest to a cluster of historians but would hardly appeal to today's spectator. That might be less true of Willibald Pirckheimer's *Eccius Dedolatus* (1520), described as "a small satirical comedy – an Aristophanic comedy in prose", which has Johann Eck, a Roman Catholic foe of Luther, suffering from a fever and futilely trying to mitigate it by swallowing too much wine. He sends for a doctor, whose initial treatment is to give Eck a sound drubbing – "to trim him down and remove his rough edges". (The name Eck means "corner" or "edge".) He is purged; his faults are "excised", the doctor and his assistants exclaiming "Immortal gods! What a big carbuncle! It's vainglory. And this carcinoma! Slander." At very last, they castrate him. This gruesome bit of "business" has its origin in Ovid and has also been used tellingly by Hans Sachs. But modern audiences would probably not recognize it as a polemic against an enemy of Luther and consequently miss its point. They would wonder why Eck is so roughly handled; nor are they apt to find the "surgery" really amusing.

In German-speaking Switzerland, again, Niklaus Manuel satirizes the sale of papal indulgences in his *Der Ablafskrämer* (1525). An angry band of peasant women sets upon an unscrupulous peddler, Ricardus Hinderlist ("Deviousness"), berate him, force him to confess his wickedness and hang him by his wrists in a manner much like that of the thieves alongside the crucified Christ. He is released after he promises to change his ways. He departs the village for ever, and the money he has collected is promptly given to an honest beggar. This is drama as Luther would have it, as is another of Manuel's plays, *Von Papsts und Christi Gegensatz* (1523) which carries a warning against the omnipresent Antichrist.

The Apocrypha continued to be a rich source of acceptable – sanctioned – plots and characters. In particular, there was a flood of plays about the fateful encounter between the fiercely patriotic Judith and the hated, lustful Holofernes, including one by Sixt Birck (1539). In these scripts her triumph in slaying her country's foe is reinterpreted as symbolizing either the victory of Lutheranism over Roman Catholicism, or else of Christianity over the warriors of Islam, at that time still threatening Europe. Another very popular theme was that of the prodigal son, a leading example being G. Gnapheus' *Acolastus* (1529). This script was considered especially appropriate for students to enact; they could readily identify with its youthful wayward leading figure.

In the second half of the sixteenth century the most important German dramatist was Philipp Nicodemus Frischlin (1545–90), in whose disastrous career is seen the peril that might await anyone then writing for the stage. Born in the village of Erzingen, in the Duchy of Württemberg, he was the eldest son of a Lutheran pastor whose profession and orthodoxy he seemed destined to emulate. He became a student of religion in nearby Balingen; in 1560, entering his teens, he enrolled in a secularized monastery at Königsbronn, where his courses were intended to prepare him to be a clergyman. He mastered Greek, Latin and Hebrew, to be able to read and quote from the Old and New Testaments and the Vulgate. His gifts for rhetoric and exposition were further sharpened by a close study of Terence and Cicero. He went next to Bebenhausen to another *Klosterschule* to equip himself for a pulpit, and finally matriculated in the renowned Evangelical Academy at Tübingen (1562), Württemberg's top-ranking institute for pastoral training. While there, he continued his careful analysis of Cicero's oratorical devices and delved into texts by Homer, Sophocles and Virgil, as well as those by Plato and Aristotle, and some illustrated tomes on astronomy. In 1563 he was granted his baccalaureate degree, the following year a master's degree. Though barely twenty, he was already entitled to take his place in élite Humanist intellectual circles.

Three years later he was chosen for an extraordinary professorship at Tübingen University where he lectured on poetry. Much had been expected of him, and much had already been attained. But he had turned away from theology. Instead, he discoursed to his classes on Caesar's *Gallic Wars* and, on his own initiative, altered his title to *professor poetices et historium*, displaying a high-handedness that affronted some of his elder colleagues. He was quickly drawing attention to himself by his brilliance and self-assurance.

His marriage to the well-connected Margarethe Brenz further assured his worldly advance, while he took care to compose laudatory verses that were sent to Ludwig, Württemberg's youthful duke, to whom they were dedicated. Encouraged by their reception, young Frischlin ventured beyond the duchy to Speyer, where he submitted even more flattering poetry to the Emperor Maximilian II, who had Lutheran leanings (1570). He kept up a flow of laudatory verse in the style of Virgil on the occasion of several royal marriages and other celebratory events. In 1576, when he was just past thirty, he was rewarded by having Emperor Rudolf II crown him *poeta laureatus* and a year later further ennoble him. He had risen very quickly by unquestioned merit and well-directed skill at blandishment.

His decline was to be even more precipitate. He delivered a lecture on Virgil's pastoral verse in which he declared that tilling the soil and living in harmony with Nature was man's ordained fate, decreed by God. For man to live in cities, engage in money-grubbing commerce, and to erect harsh authoritarian structures, establishing a ruling aristocracy, and to deny that all men are equal is contrary to the divine design. Such views were subversive, but he sought to have his lecture published, which required the permission of the ducal chancellery, a request that quickly got him into trouble. As a professor, a civil servant, he was on the state payroll. His tenure was fiercely protested by aroused members of Württemberg's landed gentry, and he received scant support from faculty colleagues, many of whom were his antagonists, perhaps resentful of his leap to prominence. To resolve the threatening situation, the duke, his protector, gave him a safe-conduct pass from the duchy in 1582.

He spent two years "exiled" as a schoolmaster in Laibach, in Carinthia, quietly pursuing his linguistic interests. He went to Italy, where he could have his studies published and read by Humanist scholars in Padua and Venice. Next he returned to Tübingen, hoping to reclaim his university post, only to learn it had been given to Martin Crusis, his arch-enemy, who dominated the faculty senate. Unemployed, he journeyed to the free city of Strasbourg, seeking a job and a safe place to reside; while there, he arranged for the publication of a group of Latin plays (1584). Soon he was summoned back to Württemberg to contribute his talent to a celebration of the duke's second marriage. As part of the festivities, his best-known comedy, *Julius Redivivus*, was performed in Stuttgart in 1585. He also wrote congratulatory verse to mark the happy day.

He hoped to recover his professorial chair at Tübingen, but even the duke's intervention was futile. His efforts obstructed by his enemies in the university, he turned his back on it for ever and for the next several years became a wandering scholar, moving from Prague to Wittenberg, to Brunswick, to Kassel, Marburg, Frankfurt and Mainz, sometimes making do briefly as a schoolmaster, but often finding himself unwelcome, his reputation as being "difficult", an outspoken trouble-maker, having preceded him. Frustrating, too, was that he could not get published. Some monies owing to him were on deposit in the Duchy of Württemberg; he tried to regain them, intending to set up his own print shop in Frankfurt am Main. The officials in the ducal chancellery, fearful of diatribes he might issue, would not release his funds. He responded so vituperatively that the duke, his patience exhausted, finally ordered him seized and imprisoned "for continuous insult to the authorities". Pent too long in a dank cell, in a grim castle, the Hohenurbach, high above the city of Urbach in Swabia, he attempted to escape. Scaling a steep wall, he lost his grip and fell to his death. His rival, Crusis, wrote this cruel obituary: "Frischlinus lieth here, dashed badly by the fall; / A mind so good, misused after all" (translation: Schade).

Seen as a rebel, a non-conformist, this sixteenth-century poet and playwright has elicited interest and sympathy from young mid-twentieth-century students and intellectuals; they have come to feel a deepening rapport with him and his struggle, seeing in him a kindred spirit. This is particularly true among the Swabian intelligentsia. A number of biographies of the unfortunate, put-upon Frischlin and commentaries on his work have appeared in modern Germany. This happened before in the eighteenth century, when Humanism was also in the ascendance. It had not taken long, either, for Frischlin's plays and poetry to become popular, because useful, in schools. Some of the scripts are in Latin, but some in German; for this reason he is cited as one of those who helped in the transition from plays in an ancient, classical tongue, clear to only a few, to those in his native speech, easily understood by all.

Though very little is known about the critical reception of Frischlin's plays in his own day, he himself set down a fairly complete description of how his *Julius Redivivus* was staged in the Long Ballroom of the newly renovated ducal castle. There was an earnest effort by the author to integrate his comedy into the other activities of the wedding festival. He was filling the role, more or less, of court poet, which he had also been at the duke's first marriage. The play had not been written specifically for this purpose; it had been produced some years earlier at Tübingen, but he adapted it to the occasion, though not with complete success – it might be said that his choice was somewhat less than suitable.

The leading characters are Julius Caesar and Marcus Tullius Cicero, with whose lives and writings he was fully familiar. He uses them to exemplify Humanist ideals: one is a military genius, the other a master of strikingly eloquent oratory and broad learning. This famed pair appear from the Underworld to look at sixteenth-century Germany.

They meet and engage in dialogue with two citizens of Württemberg, an imaginary General Hermannus and the erudite Eobanus Hessus – the latter an actual person but by then several decades gone (1488–1540), which allowed Frischlin to make use of him and the lasting esteem in which he was recalled. The subject of the colloquy is the changes that have occurred there since the long-ago days of the Roman Republic. Mostly the tone is boastful – many aspects of life in Germany are marked improvements over those in ancient times. There are some intimations, barbed but quite oblique as caution dictated, that not all at the ducal court and in Stuttgart is perfect – too much alcohol is consumed there (especially by Duke Ludwig) – but this is glossed over by Hermannus, who attributes that fault and others to the insidious presence of foreigners in court circles. Indeed, the author is quite chauvinistic, proclaiming the superiority of Germans over all other Europeans. Partly he does this by inserting two comic characters who derive from the *commedia dell'arte*, semi-clowns, Allobrox, a crafty French merchant, and Caminarius, a lecherous Italian chimney-sweep. Their misdeeds are unmasked, and they are expelled by the Germans. (Caminarius likens sexual intercourse to chimney-sweeping.) Is there not reason for Germans to take pride in their many inventions, such as gunpowder and the printing press, which have led to their highly developed skills at gunsmithing and the outpouring of learned books. Consider the large numbers of eminent Humanist scholars of the classics, among them Erasmus, Melanchthon, Johannes Sturm, the poet Eobanus Hessus himself. This is the race that Caesar had once described as barbarians. To drive home the point that the assured, articulate Germans outreach all others, the French Allobrox and Italian Caminarius speak woeful Latin, much to Cicero's distress. "This is *our* linguistic descendant?" he cries. Everywhere there are handsome cities, where once were Roman camps. Cicero equates them with Athens. At the end, he and Caesar return to the Underworld, truly impressed with what they have seen.

The play was preceded by much elaborate pageantry. Frischlin petitioned the duke for extra funds for sumptuous costumes to garb his actors, and the request was granted. Each member of the cast had the duke's heraldic insignia attached to his tunic.

By Frischlin's own account, *Julius Redivivus* was well received by the nuptial couple and their noble guests, but his report cannot be fully trusted. The court calendar carries praise for the work, but that summary, too, was not without bias. What mattered most was that the author did not achieve his goal, to rehabilitate himself and his career and regain his professorship at Tübingen. He had missed an opportunity, and before the year was out he quit Stuttgart.

The play outlived Frischlin. It had a later production in Kassel (1593), and on subsequent occasions in Stuttgart, and was studied in German schools. It was hardly the first comedy to bring back famous historical persons from their deathly sleep and confront them with a changed world, but it proved one of the most influential, prompting imitations.

Frischlin was predisposed to no one genre, creating a series of spiritual plays, court comedies and tragedies. Much might depend on who was the sponsor and for what occasion. Six of his works were staged under the aegis of Duke Ludwig, his indulgent patron; one was written at the behest of the first duchess, and another as a feature of ceremonies at the centennial of Tübingen University. Some of the titles: *Rebecca* (1576), *Susanna* (1578), *Priscianus Vapulans* (1579),

Comoedie Hildegardis Magna (1579), *Frau Wendelgard* (1579). Considering the often hostile pressures under which he had to compose, his output was substantial.

The transition to drama in German, rejecting Latin, was strongly abetted by Heinrich Julius von Braunschweig, Duke of Brunswick (1564–1613). Wed to a Danish princess, the Protestant duke, well educated and interested in the arts, learned from her of English players who had entertained in Copenhagen. His curiosity aroused, he invited a company headed by Thomas Sackeville, an adept clown, to appear before him at his seat in Wölfenbuttel in 1592. Pleased by their first performances, the duke had them remain for another five years (1593–8). He himself wrote ten German prose plays for them to enact. Primarily intended to amuse a court audience, the scripts were also heavily laden with lessons in decorum. He desired that his throng of sometimes unruly courtiers observe the proprieties and exemplify Christian virtues.

Though eschewing Latin and choosing to write in German, Heinrich Julius was much under the sway of Plautus and Terence, from whom he borrowed characters and what the duke perceived, a bit oddly, to have been a moralizing trait in those much-admired authors of farces. This is well illustrated in his satiric *Von Vincentius Ladislaus* (1594), which centres on the eternal *miles gloriosus*, the craven, braggart warrior. Here he is Vincentius Ladislaus Sacrapa von Mantua Kempfer zu Ross und Fues, "a blustering loudmouth", a self-deluded fantast and inept, foolish adventurer. Learning that an eccentric stranger, purporting to be a gentleman, is staying at the nearby Golden Crown Inn, the fictive local duke, Silvester, bids him to a court banquet, thinking to have some fun. Ladislaus, not happy with the inn's facilities, accepts gladly. At the duke's well-attended table he soon exposes his crass stupidity, offering to enter the duke's forces, after claiming that he is a redoubtable swordsman, having in his youth slain no less than 7,000 foes. He likens himself to Alexander the Great. Put to the test in a fencing match, he fails miserably. He has already recited lengthy, boring tales of his prowess in other fields, led on by the duke, but no one believes him. Convinced that he is making a fine impression, he is sure that a female guest seated next to him is infatuated with him. He takes her laughter at his anecdotes as proof of this, but actually she is secretly derisive. As a dancer and at other courtly arts he lacks measure and grace. Silvester and his jester concoct tricks to be played on this fatuous fellow, promising him a rendezvous and then quick consummation of his erotic passion; instead he ends up most ignominiously in a tub of water, after which he is cast out, a self-described hero in inglorious retreat, though still loudly boasting and vainly threatening.

Heinrich Julius was not merely holding up to ridicule a standard theatrical character to evoke laughs; rather, he was delineating faults – albeit exaggeratedly – perceptible in many at court: overbearing vanity, self-deception along with unrestrained ambition and self-promotion, a lack of proportion, a tendency to tell untruths. The comedy is innately didactic. It asks the spectator, "Do you see anything of yourself in this absurd fellow and his many follies?"

Ladislaus has embellished his name to suggest that he is of noble birth; in fact, he is the son of a cooper. He demands that the innkeeper treat him with uncommon deference. How many others there are like him, even at court! The flagrance and egregiousness of the lies are what make him a comic figure. It is a matter of scale.

Another of his affectations is that he attempts to speak Latin, but his handling of that language is so inaccurate that a priest who hears him finds it unbearable and takes flight from his presence. He dresses extravagantly, exotically, sporting a Hungarian hat that sprouts a feather, twice changing his suit, betraying an obsessive concern with his appearance. Dropping his fork, he dives under the table to retrieve it, showing a lack of assurance by not having a servant do it for him.

Heinrich Julius's list of how not to behave is lengthy. To his courtiers the comedy was clearly a warning. His interdictions accorded with those recently issued governing social and moral conduct by uncouth citizens of his austerely ruled domain. The play was a pleasanter and subtler way of imparting the same strictures to his entourage.

In the works of Jakob Ayrer (c. 1550–1648) linger elements of the *Fastnachtspiel* of which he had earlier been a prominent exponent. Successor to Hans Sachs in Nuremberg, he also drew on the use of mimicry and the greater emphasis on theme to be found in presentations by English troupes visiting Germany. One of his special inspirations was Shakespeare's *The Tempest*, or so many historians believe. (Ayrer's script, *Comedia von der Schönen Sidea*, was published – posthumously – in 1618, Shakespeare's – also posthumously – in 1623. So Ayrer might have seen *The Tempest*, but it is unlikely that he had read it, unless a translation of an actor's playbook came into his hands. From merely being at a performance he would not exactly remember the phrasing but might carry away an overall impression.) Similarities between the two plays are not in plot-details but in the poetic conception of a world fantastically invaded by magic, by the inexplicable.

The Kings of England and Scotland, rivals, are fierce and adamant in their enmity. The English prince shares his father's feeling about Scottish, until he beholds and instantly falls in love with a shy, virginal princess, daughter of the Scottish sovereign, an attraction that is mutual. To evade his father, and be near his newly beloved, the prince announces a forthcoming journey to the Continent. It is a ruse. Once beyond his native land, he dons the habiliment of a jester, slips across the border into Scotland and obtains a position at its court. There he can feast his eyes on the princess. Her father is warned by a magician that the jocular "fool" poses a danger to the daughter, but the prince, given prescience, is alerted and escapes. Shortly he returns in a different disguise, as a Moor, though somewhat unconvincing in his dark-skinned role. Once again the Scottish king is advised by the magician of the identity of the foreign visitor, and the prince is forced to take flight. Pursued, he is captured. His father intervenes; the two kings exchange threats, and a confrontation ensues. The prince is forced to swallow poison and expires. In retaliation, the English ruler orders that the princess, who has come into his hands, be put to death within sight of all. Her radiant innocence moves him, however, and at the last anguished moment he stays the execution. She kneels beside the sprawling body of the fallen prince, and kisses him on the lips, whereupon he springs up; he was not fatally poisoned, merely drugged. The long-hostile kings, now joyful fathers, are reconciled; a wedding between their children is to follow, and the tale is concluded. Towards the end of this "comedy" the stage is incongruously strewn with corpses and gore has been spilled, but the many fantastic components seem to offset that – nothing is to be taken too seriously. Plays of this sort, full of action – "noisy commotion" – and with touches of the supernatural and marvellous, became very popular offerings during festivals at ducal courts. By now, at the start of the seventeenth century, almost all were being given in German.

— 7 —
SPAIN: A GLORIOUS ERA

In Spain, chroniclers speak of the years from the close of the fifteenth century, through the sixteenth and seventeenth, as their nation's Golden Age. Everything was transformed. Though skies were often pure blue and cloudless, particularly in Andalucia, the preceding centuries had witnessed endless strife, with Christians ranged against the victorious Moors, and of one Iberian kingdom – now Granada, now Catalonia, now Galicia – and its predatory monarch in arms against another. A significant turn was the partial unification of these contending regions at the end of the fifteenth century, a design accomplished by the providential marriage of Ferdinand of Aragon and Isabella of Castile, around the time that the Moors were finally expelled from the south. Accompanying this fateful event was the discovery of the New World, from which were shortly brought to Spain heaps of plundered gold and precious artefacts, vastly enriching its treasuries. The spirit of Spain, now bent on world conquest, was adventurous, energetic.

Spain's dominion leaped far north, where the luckless Netherlands was subjugated. The often dour royal family, the Habsburgs, reigned in Vienna and Naples as well as Madrid; Philip II wed Mary Tudor and later laid claim to England, though without success. Yet despite this one setback, Spain, with so many lands under its sway, was the most powerful country on the three Western continents, its empire expanding across an ocean to both Americas.

Along with wealth and strengthened political grasp came a remarkable flowering of the arts. Excellent painters, trained in Italy and preparing the way for El Greco, Ribera, Murillo, the surpassing Velázquez, decorated the court, the homes of nobles, the aisles and walls of magnificent Gothic cathedrals. Poetry flourished, the first novel to be written anywhere was conceived and a number of playwrights yielded works that were to become immortal, founding and perpetuating a vigorous spate of theatre.

The ideological climate was one of intense, narrow-minded religiosity: Spain was Europe's most Catholic country, where the Reformation was met by the fearsome Inquisition and by the establishment of the order of Jesuits, dedicated to a Counter-Reformation, in good part by means of intellectual argument. The rigid orthodoxy cast a dark shadow over what was otherwise a brilliant epoch; besides this, the governance by the Habsburgs was absolutist, permitting no dissent. Still, for literary creativity, this was truly El Siglo de Oro, an age of striving, bright talents.

Here the medieval theatre had consisted of many of the same elements as elsewhere in Europe, though with a unique Iberian quality: sombre, mystical, passionate. This strong religious strain was a continuing factor in the drama of subsequent centuries; nearly all the later playwrights, even the most ribald and secular, fashioned works for sacred occasions, on holy subjects.

During the closing phases of the Middle Ages, Mystery Plays were performed on pageant carts that travelled through the streets on Corpus Christi Day, though perhaps in lesser numbers than in England, Germany or France. That crucial holiday came to be known locally as the "Festival of the Carts (*Carros*)". Well into the early decades of the Spanish Renaissance, Lope de Rueda (*c.* 1510–65), a playwright and leader of a band of professional actors, staged such offerings at the behest of the guilds and other town officials, in addition to his other demanding duties and ventures.

Two-storeyed, the carts were provided by the cities in which they were employed (after 1647 some of these vehicles were four-storeyed). On them were wooden frames dressed and concealed by painted canvas; they were rectangular, about three yards high and sixteen feet long. The top storey might have hinged doors that could be swung open to disclose the inside, where was perhaps a sudden change of scene or some other surprise. The lower storey might be equipped with various stage machinery possibly enabling angels to fly or to achieve a different-seeming magical effect.

An indigenous form of sacred play grew up, the *auto sacramental,* in which were joined features of the pageant playlets and the Moralities. Especially pertaining to the Eucharist, and for didactic purposes, they mixed human and supernatural characters as well as easily recognized allegorical figures such as Grace and Beauty, Sin, Pleasure and Grief, in slightly dramatized preachments. They taught "the efficacy of the sacraments and the validity of Church dogma", and were intended to abet the Counter-Reformation, a mode of propaganda echoing its fervour. They were given on a portable stage, not mounted on a cart but borne on men's shoulders to an appointed station, then on to the next.

As has been noted earlier, both types of sacred plays persisted well beyond the Renaissance, becoming ever more elaborate. Each Mystery was now allotted two linked wagons, one serving to house the dressing-rooms and to facilitate stage entrances (up to 1647), after which, as more space was required, four wagons were assigned to each production: two arranged to stand at the back and one at right-angles at either end. This enlarged stage was bare but had trap-doors for abrupt appearances and disappearances if they were called for in the text. To remove four wagons and replace them with another four for the next playlet was an awkward manoeuvre, so after 1691 the eight wagons were arrayed to create an open square, with a raised platform in the centre, that was fixed and available for all the succeeding playlets. The acting space, now increased, might measure fifty feet wide and nine feet deep. A further embellishment: the carts in the ever more lengthy procession were pulled by bullocks with gilded horns.

To make sure that an offering was appropriate to the solemn religious occasion, a preview of it was staged for a committee of the City Council a month before Corpus Christi Day. Some playlets were new, the rest old and already cherished. The first performance might occur in front of or in a cathedral or church, though this seems not to have been the practice in Madrid. The outdoor stations were specified by the City Council and might be so numerous that the presentations went on for several days. Special showings for the royal family took place in a palace courtyard, and on another day in a plaza facing the City Hall for the Council members, and again for the august Council of Castile, the regional governing body. But all segments of the public were assured of at least two exhibitions of the pietistic works.

Surprisingly, very few scenarios of the medieval playlets have survived, and those remaining are largely from Castile, where literacy was highest. One preserved script, the earliest known, is *Representación del Nacimiento de Nuestro Señor* (c. 1467–81, a *Play of the Birth of Our Lord*), by the aristocrat and poet Gómez Manrique (c. 1412–90); his sister was an abbess, and it is supposed that he composed it for enactment at Christmas by nuns in her convent. A scene in this Nativity piece has the child in the manger beholding the dreadful instruments of his predestined passion, and each stanza of a cradle song, couched in psalmody, ending with a redoubled refrain of "*¡Ay dolor!*" an anticipatory cry of protest and pain. Here, as so often is true, a predilection for the macabre is evident, tingeing a picture that might otherwise be shown as joyous. Of Manrique's three other short works, two are on secular subjects and believed intended for reading, not for performance.

A transitional event of importance was the publication in 1499 of *La Celestina* (*The Spanish Bawd*). It is attributed to Fernando de Rojas (d. 1541), a Christian convert from Judaism about whose life there are few details. It was first issued consisting of sixteen acts of dialogue, then brought out again lengthened to twenty-one acts (1502), possibly by other authors. In either version, it is not a feasible work for the stage, though numerous attempts have been made to adapt it for enactment. Acknowledging the difficulties a performance would offer, the printer's editor (*corrector*) added some final verses advising the purchaser that the work is meant for the pleasure of being read aloud, and he graciously offers suggestions on how to do this to attain the best effects. *La Celestina* is variously categorized as a "dramatic novel" and a "dialogue novel", and is indisputably something of each, a conjoined or hybrid literary genre, dubbed the "*celestinesque*". (Such works, to be read rather than acted, were turned out by scholarly poets in medieval academies and were thought to adhere to a Roman tradition – as exampled by Terence – of passing around scripts and having them perused in private or in small, élite groups, rather than exposed by staging them for the noisy, inattentive, vulgar public.)

Rojas's conversion to Catholicism must not have been complete. Nothing could be further from the pieties of the Moralities and *auto sacramental* than this account of the shrewd and beset Celestina busying herself as a go-between on behalf of the lustful Calisto and his prey, and Rojas shows considerable acquaintance with the practical problems of bawds, even those most discreet and resourceful. (The work is also known as *The Comedy of Calisto and Melibea*.) By any standard, even our own, it is salacious. With it, a flood of secular, scandalous subject-matter was swept into Spanish literature. It contains credible situations and characters, some of them lowlife. In the words of one critic, "As Greek tragedy was composed from the crumbs that fell from Homer's table, so the Spanish drama owed its earliest forms to *La Celestina*." Its cynicism echoes that of Aretino and Machiavelli, and its style is pithy, as it deals with its indecent tale. But then, in a sudden shift of mood, a trio of comic characters are murdered or punished by hanging for their crimes. The hero is killed in a fall from a ladder as he flees a tryst, and the heroine throws herself from a tower to join her beloved in death, a surprising, moralistic ending. Widely popular, the book was translated into Italian, French, English and German.

La Celestina's lasting relevance to historians of Spanish theatre lies in its having proved an immediate source of subject-matter to others more apt in shaping stage-works. Juan del Encina (1469–1529) was one of them. Humanist, a former student of the sage Nebrija, his early inclination was to fashion eclogues in the Italian style, but he was also religious, and his sincere faith entered his output. It was only after he went to live in Italy that he embraced secular themes, as in his pastoral *The Eclogue of Placida and Victoriano* (1513). Some of his plays pre-dated the publication of *La Celestina*, but his later texts betray its influence. Quite simple in form and content, his were the first secular compositions staged in his native country, and for that reason he is often called the founder of Spanish drama. In Rome his patron was the Duke of Alba for whom Encina wrote a series of fifteen eclogues. *Placida and Victoriano* runs to a length of thirteen scenes, is sentimental, and at its climax veers into tragedy. It features a suicide, which is probably why it was later banned by the Inquisition in 1559, three decades after the poet's death. The whole is loosely structured, episodic, "lyrical rather than dramatic in tone". Here, Encina helped to create a stock character, the comic shepherd, who speaks in a local sub-dialect termed *sayaqués*, the vocabulary of the peasantry near Salamanca.

Encina liked to have his pastorals staged outdoors, making use of arbours, woodlands, encircling hills for a setting. His actors mostly wore shepherds' garb. *Placida y Victoriano* was performed in the gardens of Cardinal Arborea in Rome. By this time the playwright had strong church ties, having been appointed Archdeacon at Malaga, often visiting the Holy City. Shepherds also populate his *Egloga del Amor* (1497), commissioned to be a high point of a gala honouring Juan of Castile and Margaret of Austria during their nuptials – she was the daughter of Emperor Maximilian.

Lucas Fernández (1474–1542) began as a disciple of Encina but eventually chose an opposite direction, turning out amorous comedies, some revealing an impulse to experiment with dramatic form. In another vein, his *Auto de la Pasion* (*Passion Play*) is prototypical of the seventeenth-century *comedias de santos* (saints' plays) that were to become a staple of that later period.

The spirit of Italy's Renaissance was gradually seeping into Spain; the two regions had dynastic and political links, and there were other foreign influences. Gil Vicente (c. 1465–c. 1537), a bilingual Portuguese attached to Spain's court, wrote most of his plays in his native tongue, others in

Spanish, and some containing passages in both languages. Abundantly talented, he surpassed both Encino and Fernández in scope and craftsmanship, except that he was weak at plotting. The begetter of forty-four poetic scripts, he is regarded by many as Portugal's greatest playwright.

His biographers have too few facts about him; even the dates of his birth and death are uncertain. His parents were humble folk, yet in his early adulthood he is thought by some to have been a goldsmith and banker, though some surmise that he attended the University of Lisbon as a law student. Perhaps both suppositions are right. He married Bianca Becerra in 1500, fathered a daughter, Paula, and a son, Luis, who inherited his father's poetic talent.

Eventually Gil Vicente decided that his true *métier* was verse and the drama. He had an exceptional lyrical gift. Demonstrating this, he won royal favour and a place at the Portuguese court and held the position for the remainder of his life. His was the task of organizing entertainment on gala occasions, of which there was an endless succession.

His first play, staged in the Lisbon apartments of Queen Maria, was a dialogue in Castilian verse, the language preferred by the queen, whose origin was Spanish. A stream of other eclogues followed, along with pageants, comedies, tragicomedies, farces. He was most fond of pastorals. Under his personal supervision, these were produced with much éclat at the court. More and more, as he matured and widened his audience, his plays tended to be in Portuguese. From the short, simple eclogues with which he began, he advanced to complex, ambitious offerings, handling them well, save that their dramatic conflicts are soft. He held spectators by the beauty of his language. He also has a nice perception of the comic. Because his plots do not build and sustain suspense, he is deemed to be much at his best in the shorter works. Critics suggest that had he been engaged in Italy, rather than in Portugal, which lacked a tradition of theatre – at the time, he was his native land's only dramatist – he might have learned more from able peers and risen to even greater heights.

Among his briefer pieces the most acclaimed are *Auto de la Sibila Casandra* (*Cassandra the Sibyl*; 1513–14); *Auto da Barca do Inferno* (*The Ship of Hell*; 1516); *Auto da Barca do Purgatório* (*The Ship of Purgatory*; 1518); and *Auto da Barca da Glória* (*The Ship of Heaven*; 1519). The first of these, a religious play in one act, relates how Solomon seeks the hand of Cassandra, a shepherdess, whom he is eager to marry. She rejects him; her explanation is that much earlier she had vowed never to wed, yet not to be overly pious. Her aversion to marriage arises from her besetting fears of childbirth, jealousy and possible abandonment. To Solomon, these motivations are irrational. He gathers her aunts, Erythraea, Persica and Cimmeria, to question and reason with her. Cassandra claims that men only show their harshness after the wedding. Her uncles, Isaiah, Moses and Abraham, are summoned to argue with her, but their effort at persuasion fails. She finally reveals her true reason for refusing Solomon's offer: she is convinced that she has been chosen to give birth to the promised Messiah and hence must remain virginal. Hearing this, Solomon believes that she has completely lost her senses.

Suddenly, as all are praising the Virgin and the child-deity it is prophesied that she will bear, a vision of the Nativity appears. Beholding it, the assembly grasps that the Holy Birth has already taken place. They gather about the manger to exalt Mary, begging that she intercede with the Omnipotent Father as they ask pardon for their many sins. Next, they raise their voices in a song of Adoration. Either Vicente was a profound believer or deemed it expedient to seem so. Commentators debate this.

The other three one-act plays comprise the "*barca*" trilogy. *The Ship of Hell* has the Devil and a henchman setting sail to their fiery dwelling-place. Who will accompany them? A trembling line of sinners who have just expired are here to face an eternal, irrevocable Judgement. One by one, they step forward. First is a Gentleman who turns beseechingly to an Angel, the helmsman of the Ship of Heaven anchored close by; his plea for exemption from the flames is denied, and he is sent back to the impatient Satan. The same fate befalls a Usurer and a dishonest Cobbler, as well as a Friar notorious for his fencing and love-making, and his frivolous Mistress. Also seeking help from the Angel is a Procuress, who has ruined the lives of many young girls, and lied, and cheated at cards. Even so, she claims to have been defamed and merits salvation. The Angel does not agree. More are condemned to Hell: a Jew, a dishonest Judge, and Hanged Man, a fearful Miscreant. Spared from damnation are a Fool, who erred due to his essential innocence, and at the last moment Four Knights of the Order of Christ, who were cruel warriors but have given up their lives in the service of the Blessed Saviour. With his motley collection of the un-Christian, the Devil casts off and heads for their final destination.

The Ship of Purgatory utilizes much the same pattern. The souls of ordinary folk are sorted out: on which of the three ships shall they embark? Or will they be delayed? To Hell should go a Farmer, whose hard life led him to be guilty of petty crimes; but since he repented in time, his sentence is lightened. A Market Woman is so garrulous that the Devil can scarcely get in a word. She contends that inasmuch as the Ship of Hell is temporarily aground, she must be permitted to board the Ship of Heaven. Her brazenness and sophistry amuse the Devil, who momentarily defers his decision concerning her. A Shepherd and Shepherdess, along with the others, are told to wait on Purgatory's shore until they are considered truly ready for the Ship of Heaven's next departure. The only soul quickly welcomed by the Angel is that of a little boy who will be immediately awarded God's grace. A blaspheming Gambler is the Devil's prey and hustled aboard the Ship of Hell.

The Ship of Heaven portrays Five Angels watching as Satan takes issue with Death: only lowly creatures are assigned to his vessel. To quiet him, Death allots him several noblemen and high-ranking clerics. They include a shallow-spirited, irresponsible count, a duke who has sold his soul to the Devil, a king who has waged devastating wars, and an emperor who has demanded that his subjects view him as a divinity and abjectly worship him in heathenish rituals. Among these élite sinners, too, are a Bishop, an Archbishop and a Cardinal. Their fault has been that of ceaseless scheming for advancement in the hierarchy. Even a Pope has committed sins of "lust, pride, and simony". Sentenced to perdition, they fall to their knees and pray for absolution, but to no avail. Christ appears. He hands each of the Five Angels an oar symbolizing one of his wounds. The Devil's ship sails off, bearing its cargo of passengers to their doom.

An early vignette, *Auto de los Reyes Magos* (*The Three Wise Men*; 1503), depicts a pilgrimage by a shepherd to pay homage to the Christ Child. *Quem tem Farelos?* (*Serenade*; 1508–9) narrates the ingenious efforts of a destitute courtier to woo a fickle young lady. In *Auto da Índia* (*The Sailor's Wife*; 1509), another comedy, a woman takes advantage of her husband's prolonged absence in the Orient to be unfaithful. *Exhortaçao da Guerra* (*Exhortation to War*; 1513) pays tribute to Portugal's martial achievements. *Auto da Alma* (*The Soul's Journey*; 1518?) is an allegory in which the soul journeys onwards to salvation, with the Church a sheltering inn along the road. *Comedia del Viudo* (*The Widower*; c. 1521) is about an aristocrat, disguised as a servant, who conceals his identity and rank in an attempt to marry the widower's daughter. *Farso dos Almocreves* (*The Carriers*; 1527) presents another nobleman, this one pressed by his creditors and surrounded by rascally servants, while he is trying to extricate himself from his unpleasant dilemma. *Tragicomedia da Serra da Estrella* (1527) places troubled lovers in a pastoral setting. In another, much earlier pastoral, *Auto da Fama* (*Goddess of Fame*; 1510), the feminine deity takes on the shape of a whimsical shepherdess.

Perhaps the most prized of his longer works is *Don Duardos*, a tragicomedy written in 1521 (?) and staged in 1525. The hero, an Englishman (despite his Spanish-sounding name), travels to Constantinople to even matters, an affair of honour, with Primaleon, son of Emperor Palmerin. Instead, he is instantly captivated by the beauty of his enemy's sister, Princess Flérida. To be near her he takes employment as a gardener at the palace, where he can gaze at her. She finds herself oddly attracted to this lowly person. He concocts an aphrodisiac which she accepts not knowing its nature and purpose; her passion for him is aroused, but she is still aware of her rank and the great gap in their social positions. Consequently she fights her growing feeling towards him. He is determined that she love him only for himself and is reluctant to reveal his noble lineage. She finally overcomes her doubts, accepts him though he is a mere gardener, whereupon he tells her the truth and all ends happily as the couple sail for far-off England.

The development of a secular, commercial drama got under way in Spain through the efforts of Bartolomé de Torres Naharro (c. 1465–c. 1537). For a time he was clearly imitating Encina and like him lived and worked for a number of years in Italy, chiefly in Naples, where he was exposed to its increasingly sophisticated theatre. Emancipating himself from Encina's influence, he began to attempt bolder comedies and farces. Initially they were produced in Italy. Having dwelt in the house of a cardinal, he had become intimately acquainted with lesser officials and servants and witnessed their sly machinations in a ceaseless rivalry to obtain influence and sinecures. His *Tinelaria* (1517) reports on the backbiting and tugs-of-war, the unedifying intrigues going on in such putatively respectable precincts. (*Tinelo* is the designation of a hall occupied by administrative officials in a cardinal's palace.) In the prologue, Naharro announces: "What makes you laugh here, you may punish at home." To underline this message, he states in a closing speech that such unscrupulous conduct under the vaulted ceiling of an ecclesiast's house reflects ill on a Prince of the Church supposedly in charge there.

In a gesture of daring, *Tinelaria* was given before Pope Leo X and Cardinal Giulio de' Medici, the latter a future Pope. The lofty guests were not offended, but took the gibes good-naturedly; if anything, Leo X put a high premium on wit. Both their Eminences could attest to the veracity of the picture. They were particularly amused at a raucous drinking scene in which Naharro had unintelligible exchanges in gibberish, arising from the palace being staffed with secretaries and servants from many parts of Europe chatting incessantly in Italian, French, German, Spanish, a veritable Babel.

As a reward, the good-hearted Leo X granted the author licence to publish his scripts over the next ten years. Falling into step, Cardinal Bernadino Carvajal, whose palace was the obvious target of this sharp, unflattering comedy, took no overt exception and accepted Naharro's dedication to him of a printed edition of the work.

Naharro took quick advantage of the Pope's mood of leniency. Collecting a number of his plays, he issued them that very year, 1517, titling the volume *Propaladia*. In a preface he asserted that classical drama, from Ancient Greece and Rome, should no longer be the only model for sixteenth-century scripts. He had already demonstrated his adherence to this radical concept in his first venture, *Comedia Serafina* (1508–9), which he had elongated to five acts, the earliest Spanish play in that full-bodied form. Romantic comedies of the kind he fashioned were new to the Spanish stage, as were "cloak and sword" melodramas which he introduced, filled with violent action and resonating with the themes of honour and its accompanying emphasis on punctilio that were to be frequent motifs in his country's subsequent theatre. In the preface he classified his scripts as either romantic or realistic, defining the two genres. He is not only considered to be Spain's first "modern" playwright, but also its first theorist about the drama.

Six of his plays remain. His bawdiness displeased Spanish Church dignitaries who were less tolerant than the two Italian Popes. Eventually, Naharro's *Propaladia* was put on the Index in 1559 – he himself was no longer alive, and so safely out of reach – but an expurgated version was brought out in Madrid in 1575 and was closely read by younger playwrights, among them, doubtless, the very alert Lope de Vega.

The religious plays were meant to be accessible to the public everywhere, since they served the aims of the omnipotent Church, especially when it was under attack elsewhere in Europe; but attendance at secular dramas was a privilege reserved for the court, the Church hierarchy and the educated nobility. This began to change during the last quarter of the sixteenth century. A precipitating event was a visit to Spain by an Italian *commedia dell'arte* troupe headed by Alberto Ganassa in 1574; it was made so welcome that it chose to remain for several years and built a temporary theatre, the Corral de Pachea, in a style that was new to both Spain and Europe, and that served as a prototype for Spanish theatres for centuries to follow.

The Corral utilized an open space between the walls of neighbouring houses, a patio or area resembling the large courtyard that might stand in front of a sizeable English hostelry and, too, would be pressed into service for early forms of that northern country's dramatic performances. (*corral* is a Spanish term for courtyard.) At one end was a roofless platform; the façades of the surrounding houses would be the backdrop, and their balconies could be borrowed to substitute as an overhanging upper stage on which, if need be, some bit of the play's action took place. The audience stood or was seated on rows of benches rising in tiers. At one side of the podium was a gallery reserved for women of the lower classes, here segregated for modesty's sake. Some windows in nearby dwellings were appropriated for privileged onlookers, and the nobility might have chairs on stage, sharing it with the actors.

Completely absent was a front curtain, but at the back was an inner stage – curtained off – that when opened could provide an interior setting, a study or bedroom. Above, the upper stage, reached by a ladder, might be asked to represent the bastion of a town under siege, or a strategically located window or balcony from which a lovers' meeting is spied on, or it could be assumed to be a cliff, or a sharp precipice. Since the platform was mostly bare, the setting usually suggested by mere allusion in the lines, the action in the plays was allowed to be flexible and fast-moving: a few words could change the place. This encouraged the playwrights to construct scripts loosely, in short scenes or episodes, rapid shifts of location, heedless of the classical unities – a self-indulgence probably not good, as it could result in a loss of the dramatic impact achieved by Aristotelian compression.

The dressing-room was at the back, out of sight, as always. In many respects, this theatre was much like the Elizabethan playhouse, with which it was contemporary, and which was also evolving at this time.

Though intended to be temporary, Ganassa's Corral de Pachea was a marked advance over anything up to that time. Recalling the improvisatory primitive stage employed before, the famed novelist-playwright Miguel de Cervantes had no kind words: he found it crude and elementary, "composed of four boards, arranged in a square, with five or six boards laid across them providing a platform a few feet from the ground". The remaining equipment was "an old blanket drawn aside by two cords, making what they call a tiring-room", behind which were concealed the musicians, contributing old ballads without even the accompaniment of a guitar. He was referring to a performance that he had seen as a boy fifty years earlier.

Shortly after the Corral de Pachea opened its doors, permanent theatres that incorporated many of the same features sprang up in Madrid and Valencia, responding to the public's sudden and ever-growing fondness for entertainment of the Italian sort. They were built by monks. The first two in Madrid were the Corral de la Cruz (1579) and the Corral del Principe, and then a third was added. Soon more theatres were mushrooming in other larger cities, such as Seville,

Burgos, Barcelona, Cordoba, Granada. By royal decree, the privilege of operating a theatre and sponsoring its troupes of actors in Madrid – the visiting companies rotated – was given to three charitable organizations, one the city's General Hospital, the other two dedicated to distributing food and clothing to the desperately needy, from the profits earned by ticket sales. This benign arrangement was in force until 1615, when responsibility for supporting the Hospital and other Church-led charities was shifted to the city government, which in turn taxed the generally thriving theatres for the same purposes. A salutary result of this set-up was that the theatres had less interference than otherwise likely from vigilant Church and civic officials, as the charities and city's treasury were dependent on funds from the troupes' successes and takings. Possibly for this reason, playwrights and actors, too, seemed relatively immune from the Inquisition, though never entirely so.

From this point on, theatres were leased to entrepreneurs, and charities (*cofradias*) had less control over them. The usual term of a lease was four years. In 1638 ownership of the playhouses was transferred to the city; two commissioners kept watch over them. The layout of stages and auditoriums in some cities was altered in a number of respects: a few benches at the front might be installed in semi-circular rows; a tavern might be added at the rear of the patio, with three galleries above it, the first for ladies, the second divided into boxes to accommodate council members and bureaucrats; and the top – the "attic" (*desvánes*) – usually occupied by the clergy, students and assorted intellectuals. If rooms in adjacent houses from which spectators could overlook the stage were used as boxes – even more private and commodious – a fee to allow access had to be paid to their owners.

If, from city to city, there were architectural differences, the general courtyard design was much the same everywhere. At each of the several entrances would be two money-takers, one collecting on behalf of the entrepreneur, the other present for the charity; this ensured that the *cofradias* got their share – about two-fifths – without risk. Extra fees were paid by those who sought seating on stage or in galleries and boxes. At all times, men and women were separated, a rule firmly enforced by the authorities; the sole exception might be in the *aposentos* – private rooms – and then only if the pair or group were known to be legitimately close akin.

For a while, performances took place only on Sundays and feast days. Then Ganassa was permitted to make appearances on a weekday, and soon other companies were given the same right. On Saturdays, however, all theatres were shut. The season for plays began in September and lasted until Lent. The new season started after Easter and continued to July, then summer heat closed them. Special circumstances, a death in the royal family, a plague, might also suspend their activity.

Even before Ganassa's arrival, bands of professional actors were beginning to be formed. They participated in the *autos sacramentales* and religious plays honouring the saints on their special days, helpfully fleshing out the casts of zealous amateurs drawn from seminaries, *cofradias* and guilds. An early leader of such an acting troupe was Lope de Rueda (1510?–1565), under whose guidance the members were trained, well disciplined and enabled to project their individual talents. Born in Seville, his first career – an unlikely choice – was that of a gold-beater, flattening sheets of the rare metal. He turned from that to the stage. He and his well-regarded company were kept busy in presentations of sacred works, a steady enough source of employment, until he saw performances by Ganassa's *commedia dell'arte* ensemble, whose exuberant style of theatre deeply impressed him and moved him to emulate it. Adapting plays from the Italian, and creating new pieces in this foreign mode, he was quickly and warmly embraced by the public.

He was the epitome of the "actor-manager", taking roles, directing and assigning parts to others and providing a good many of the troupe's scripts. He was fastidious about the physical details of his productions. As a dramatist, his chief contribution is the *paso*, a short one-acter exploiting an amusing or surprising incident; it was inserted into a full-length play, though their plots were in no way connected. (The *paso* was to evolve into a more important form later.) He was especially fond of pastorals, appearing in them in shepherd's garb, and might dress after that fashion in other kinds of plays as well.

He entertained by royal command at the court of Philip II, and with his company travelled indefatigably to the far corners of Spain, staging dramas for less noble spectators, thus doing much to "democratize" secular theatre in his native country, where heretofore it had barely existed, with the Church so omnipresent.

As a dramatist, he excelled Encina at plotting – at least, he conceived fuller, more complex plots – and he further developed the character of the simpleton (*bobo*), a favourite figure in Encina's plays. He wrote only in prose. His dialogue is characterized as "earthy" and "outspoken", yet is not without a lyrical strain and a pulse of romantic feeling, and he freely mixed moods. Most familiar of his one-acters is *El Paso de las Aceitunas*. Critics deem the full-length *Eufemia* his best work. It revolves about a "point of honour", which, as has been remarked, has been a constant preoccupation of Spanish playwrights through more than five centuries to the present day.

As theatres proliferated, so did professional acting companies. Some were short-lived, but soon there were so many that the authorities intervened to limit their number, licensing a mere eight (1603), and raising the number of legitimate troupes to twelve (1615). Apparently a good many unregulated furtive bands of actors were active despite the restrictions, as indicated by records of a plethora of productions.

The actors signed contracts with a manager for their services, to last a year or two, and were salaried; or else they shared in the hoped-for proceeds from entrance fees, a sum calculated after deductions for expenses of mounting the play; these troupes were called *compañias de parte*. Though a few players appeared solo, reciting stirring monologues or excerpts and highlights from favourite comedies and dramas, the usual count of performers in a company ranged from sixteen to twenty. A scattering of women might be among them, and also minors of both sexes, these youthful participants the apprentices still learning the tricks of the trade. The question of actresses was likely at all times to ignite sharp controversy, and until 1587 female roles were often assumed by men and boys. The Church, outraged from the start, obtained a royal ban against the presence of women on stage in 1596, but it seems to have been generally ignored. Aroused by this, and after a prolonged and furious debate, the clerics got a stricter edict from the Royal Council: by its terms, no woman could belong to a theatrical company unless her husband or father was also in it (1598–9). In 1608 another rule was promulgated: only players could be backstage – this would reduce opportunities for assignations – and friars were forbidden to attend performances; further, secular plays could not be put on in churches, convents or any other religious establishment. Censorship, quite austere, was imposed on scripts, and dancing proscribed – it was the particular target of many complaints. (Much of this interesting detail comes from Brockett's *The Essential Theatre*.)

The objection to dancing, especially when performed by women, was not fully relaxed for centuries. The actors, like those in the *commedia dell'arte*, were trained to sing and dance; music was an important aspect of their lighter offerings, but the truly pious in the audience viewed the women's graceful movements as "voluptuous" and "licentious". After the edict of 1615, these exhibitions became more decorous.

Players were finally permitted to set up and join a guild, the Cofradia de la Novena, which gave them a status equal to that of persons in other professions. The guild still exists. From the lustrous era known as the Golden Age, a roster bearing the names of 2,000 members has been preserved.

As elsewhere, actors were frowned on by the Church, and they could not receive the sacraments. As early as the thirteenth century they had been stigmatized as "infamous" by Alfonso X, a designation that remained in force until the twentieth century. Yet the condemnation was pragmatically overlooked. Professional actors were required for effective presentations of the *autos sacramentales*, and now were a fount of monies for public charities, the good works carried out by the Church. So they were quietly tolerated.

An actor had to arise early to study and memorize his scripts at about five o'clock, by candlelight or lamplight in pre-dawn darkness. At nine, after breakfasting, he reported to the theatre for rehearsals. Performances were in the afternoon. By seven in the evening he might be free, unless summoned to fill a role in an entertainment at the brightly illumined house of a nobleman. Sudden assignments to take part in a fête at court might also disrupt his schedule and even lead to the cancellation of a regular public performance, a change of plans that annoyed ordinary spectators; indeed, a frequent affront that eventually turned many people away from theatregoing.

Defrayment for costumes might be up to the manager. If the actor brought suitable attire from his personal wardrobe – perhaps having had the same or similar role previously – he got an extra payment. This was spelled out in the contract. Apprentices, of course, were not asked to bear the cost of their stage-garb, nor were they likely to have anything in their possession that would be appropriate. In contrast to the bareness of the stage, garments were colourful and rich: silks, exquisitely stamped velvets, brocades, thick fur-trimming. Troupes were always petitioning civic and Church sponsors for additional grants to cover the cost of the extravagant display intended to dazzle and woo audiences, and in some towns prizes were awarded based not only on the script and acting but on the players' lavish dress as well. In 1534 Charles V tried to curb the excessive outlays by a decree, and subsequent administrations continued to enforce more limits. Further, in 1653 women were singled out for admonitions on this score. They could not appear on stage sporting "strange headdresses, décolleté neck-lines, wide-hooped skirts, or dresses not reaching the floor". They were not allowed costume-changes but had to keep on the same garment throughout, unless the script logically required they do otherwise.

In presentations of historical dramas there was no attempt at authentic dress. Contemporary attire was donned, though occasionally the character's belonging to a past era might be suggested by an eccentric touch in his apparel, possibly outmoded in some way.

According to Brockett: "In 1589, an actor paid 1,100 reals and in 1619 another paid 2,400 reals for a single costume, sums equivalent to about one third of a typical actor's annual income. . . . The actor's wardrobe was considered his greatest financial asset, for it helped him secure employment and could be pawned in bad times."

On tour, which was arduous, troupe members received added compensation for travel expenses. (Brockett draws on Augustín de Rojas Villandrando's *Entertaining Journey* (1604).)

Besides Ganassa's, other *commedia dell'arte* companies visited Spain during these years. From Italy, Maximiliano Milanino led one (1581), as did Tristano Martinelli (1587-8), helping to provide competition and to enliven the scene in Madrid and elsewhere, and finding it quite profitable.

At least 30,000 scripts were written by busy playwrights during the Siglo de Oro, presenting the literary historian with a daunting task. One assumes that many of the plays were short and not all were meritorious. The authors of these works are so numerous that only a few can be recalled here, those acknowledged to have had a special talent or genius, with the significant accomplishments of each described sketchily, since their individual output is also copious. The pens of Spain's greatest dramatists apparently moved quickly, filling hundreds of pages of manuscript.

An overwhelming group of plays continued to have religious content. Secular offerings were of various kinds: *auto* was the term for any script meant to be enacted and often precedes its title; *comedia* indicates that the work is full-length, whether humorous or serious. Further, *comedias* might be categorized as *capa y espada* ("cape and sword"), or *teatro ruido* (that is, "noisy", resounding with combats), or *cuerpo* (succinctly put, "corpse", a designation for dramas treating with events distant in time and place, such as the martyrdoms of saints, the heroic struggles and malevolent assassinations of historical persons, the fantastic powers and magical deeds of mythological figures). Through the first decade and a half of the seventeenth century, and even later, every performance began with a *loa*, a complimentary or dedicatory speech in a prologue, singling out a royal patron, or a grandee whose purse had provided funds for the production, or else with flattery of the audience for having the good sense and good fortune to be in attendance. The *loa* customarily featured songs and dances. Later, as the play unfolded, there might be *entremeses* ("interludes"), extraneous topical sketches between the acts, spoken or sung, or both, to assure a measure of diversity.

Juan de la Cueva (1550?–1610) is another who at an initial phase of his career was strongly influenced by Encina. Not too much is known about his personal life. He was born in Seville and claimed noble descent, with seeming justification – he published a history of the Cueva family line. A venturesome spirit, in his twenties he voyaged with his brother Claudio to Mexico on a hunting expedition and chose to stay in that remote land for three years (1575–7). Missing his homeland, he began to express his nostalgia in verse. Returned to Seville, he kept on with poetry and had the courage to emulate Encina by composing dramatic works. His spell as a playwright was very short, consisting mostly of one overflow of scripts between 1579 and 1581. He saw some of them produced in Seville within that time, and published one more script later in the same decade (1588); another three plays from his writing-table are undated.

He never quite mastered dramatic form, and his plays are flawed, but he described himself as an innovator and had a streak of originality. He soon began to experiment in a direction that took him away from Encina; his importance lies in this, rather than in the stageworthiness of his scripts, which lack cues for physical action, have an over-abundance of declamation, and a tendency to introduce irrelevant characters and detail. From Seneca, too, he borrows the technique of injecting elements of the horrifying to arouse the interest of the audience and get an unnerving, visceral response. This does not save his plays from seeming to be "poetic narratives in dialogue" rather than vehicles for stirring performances.

But he was the first to select incidents from Spanish history for dramatization, and to delve into ballads and folk legend for subject-matter, and at times to capture a ballad-rhythm in his dialogue. All his plays are in verse throughout, and some make use of Italian metrical patterns, which again he was the first to do. In his wake came a host of historical dramas, would-be classical tragedies and indigenous comedies of manners, from his peers and from younger writers, some with more verve, to whom his offerings showed what could be done in secular theatre.

Los Siete Infantes de Lara (*The Seven Princes of Lara*; 1579) his first play, was conceived as a tragedy. At his wife's relentless urging Ruy Velázquez surrenders his brother-in-law Gonzalo to the custody of Almanzor, the Moorish king. His seven sons are murdered. Apprised of this, their father is bitterly shaken by grief; moved by his captive's despair, Almanzor orders him freed. During his detention, Gonzalo has had a secret affair with Princess Zayda, the Moor's sister; she is now pregnant. Before leaving, Gonzalo gives her a ring; their child, when sixteen and of age, should seek him out and be identified by displaying the token.

After the specified lapse of years, Zayda sends their son to his father. Learning how his seven half-brothers perished, the youth, Gonzalo Mudarra, goes forth honour-bound to exact vengeance. In combat, he slays Velázquez, whose evil wife, Doña Lambra, escapes by locking herself in her house. Gonzalo Mudarra sets the edifice afire and, dying, she is reduced to ashes in the flames – a Senecan touch!

Quite different in tone is *El Infamador* (*The Defamer*; 1581), listed as a comedy of manners. The cynical Leucino is convinced that any woman will yield her virtue in exchange for money. But Eliodor remains chaste in spite of all temptations. He persuades the goddess Venus to assist him, but even the Olympian deity's intervention is to no avail: Eliodor is resistant to all seductions. Leucino attempts to abduct her, and in fighting off her kidnappers she inadvertently kills Ortelio, his servant. His master takes advantage of this unforeseen turn of events: he charges that the slaying was deliberate, inspired by Eliodor's infatuation with Ortelio, who spurned her and threatened to tell Leucino about her importunate advances. Arrested, Eliodor receives a death sentence. Another deity steps in, Diana, goddess of chastity, who testifies to Eliodor's courageously guarded purity. Leucino, confronted by Eliodor, confesses to his wicked strategems; it is decreed that he shall be buried alive. (This is not exactly what later generations of theatre-goers would consider a "comedy of manners".)

Other historical dramas by Cueva are: *La Libertad de España por Bernardo de Carpio* (*The Liberation of Spain by Bernardo de Carpio*; 1579?); *El Saco de Roma y Muerte de Borbón* (*The Sack of Rome and Death of Borbón*; 1579?); *La Muerte del Rey Don Sancho y Reto de Zamora* (*The Death of King Don Sancho and Challenge of Zamora*; 1579); *El Principe Tirano* (*The Tyrant Prince*; 1581); and *La Libertad de Roma por Mucius Scévola* (*The Liberation of Rome by Mucius Scaevola*; 1588).

His plays in other veins are: *La Constancia de Arcelina* (*The Constancy of Arcelina*; 1581); *La Comedia del Viejo Enamorado* (*The Comedy of the Infatuated Old Man*; 1579–81?); *El Tutor* (*The Tutor*; 1581?); *La Tragedia de Ayax Telemón* (*The Tragedy of Ajax Telemon*; 1588); *La Muerte de Virginia* (*The Death of Virginia*; 1588); *El Degollado* (*The Man Beheaded*); and *El Vil Amador* (*The Base Lover*). Some of these were never staged; the dates are those of their publication.

Later he wrote a treatise on drama, *Exemplar Poetica* (1606) which did not find its way into print until 1774. In it he commended the newer forms fashioned by younger writers, Lope de Vega and his followers, yet insisted that he himself be recognized as having before them been on the cusp and brought changes and advanced concepts to Spain's rapidly developing theatre. (An article by Andrés Franco has been especially helpful in this account of Juan de la Cueva.)

During this period the attention of the great novelist Miguel de Cervantes Saavedra (1547–1616) was also drawn to the stage. World-famed as the author of *Don Quixote*, he is little known outside Spain as a dramatist, but in his preface to his collection *Ocho Comedias y Ocho Entremeses* (*Eight Plays and Eight Interludes*; 1615), near the end of his life, he states that in the 1580s he had aspired to provide Madrid audiences with stage-works, between twenty and thirty in number, many never produced, and only about half of them now surviving.

He was a son of a proud but impoverished Hildalgo family, a native of Alcalá de Henares. In his early manhood he became a soldier serving under Marco Colonna, a papal commander. At the fateful battle of Lepanto (1571), when he was twenty-four, he was seriously wounded, costing him the use of his left arm. Later he commented that for a writer a hand was less important than understanding, which he well proved. Four years after Lepanto he was captured and sold into slavery in Algiers (1575), a harsh experience he underwent for about a half-dozen years, until finally he was ransomed. He spent some time in Constantinople. Back in Spain, he married and turned to literature, writing poetry and plays, eking out a very scant living.

His early plays are considered to be his best, though only two remain. One is *El Cerco de Numancia* (*The Siege of Numantia*; 1587), about an assault by Roman forces on a Spanish town, a script described by some as the finest tragedy written in his country's language.

Besieged by the remorseless Scipio Africanus, the people within the walled city are outnumbered and slowly starving, but rather than surrender they choose to reduce their town to ashes and slaughter themselves until the streets are strewn with 10,000 corpses. Only one boy survives, until he – with true Spanish pride and patriotism – hurls himself to his death from a turret after a rhetorical outburst. His words and suicidal deed earn the admiration of the Roman conqueror. The play, calling for a huge cast, is filled with ringing speeches of infinite length, allegorical scenes and characters, religious and magical ceremonies, incantations and ghosts, and could scarcely be more static and unfit for holding the attention of a semi-literate audience. It is fiercely animated, however, with praise for the courage of the ancient Spaniards and expresses Cervantes's passionate love for his native land.

In these early works he respected classical principles, but there are soon signs of an independent mind. Departing from the format used by Cueva, he shortened his scripts to four acts, instead of the heretofore accepted five. As in the *Numancia*, he brought over semi-abstract, allegorical characters from the Moralities, placing them on the secular stage. (Encina had been a partial forerunner of this bold shift, having in 1497 mounted a Nativity Play in which contemporary and local peasant types replaced Biblical figures.) Cervantes was too slow, however, to catch up with Lope de Vega, a rival enjoying far more success. By comparison with the racy vernacular introduced by de Vega, the dialogue penned by Cervantes was stilted, outdated. He tried to adapt to the many new, looser practices flaunted by the younger man, but not too ably, which is probably one reason his later offerings are viewed as of lesser worth and too often did not reach the stage for which he optimistically destined them. He simply was not an adept playwright, certainly not the master in that discipline he proved himself as a satirical novelist. Not effective at writing full-length plays, he was at his best with comic *entremeses*, in which the freshness of the *pasos* of Lope de Rueda and the mimetic exuberance of the imported *commedia dell'arte* live on.

Valued among his contributions to the stage-boards are *The Traffic of Argel*, about men held captive by Algerian pirates, for which he drew on his personal experience, and *The Fortunate Ruffian*, which pictured life in the Spain of his own day.

He writes about the theatre in one chapter in *Don Quixote*, describing what the stage was like in his boyhood – a vivid passage already quoted here – and also gives a first-hand report there of a performance by the de Rueda troupe and its versatile leader, whose costumes and properties were merely "four white shepherds' smocks edged with gilded leather, four beards and wigs, and four staves – more or less. The plays were colloquies or eclogues between two or three shepherds and a shepherdess; they were adorned and expanded with two or three interludes, about a Negro woman, a pimp, a fool, or a Basque; for all these four characters and many others were acted by the said Lope with the greatest skill and excellence that can be imagined."

In 1597 he was excommunicated, accused of "offences against His Majesty's Most Catholic Church", and barely escaped the Inquisition. On various other charges he was imprisoned three, four and possibly five times, an experience to which he was probably fully inured.

This most important of writers, Miguel de Cervantes Saavedra lived and died in poverty and was buried in an unmarked grave, leaving a huge legacy of laughter and sly observations of man's many delusions.

It is Cervantes who called Lope Félix de Vega Carpio (1562–1635) a "monster of nature", an encapsulated summary of his rival that is highly accurate and not meant to be pejorative but rather an expression of admiration and wonder. Was there ever anyone else like this Lope? Fathomlessly inventive and immensely energetic, he turned out 2,200 plays by some estimates, 483 of these *comedias* – that is, full-length. By more modest counts, his total was only 1,800 scripts, or even a mere 1,500, some written in one, two or three days, and encompassing every genre then known. Even in his last years – vigorous, he lived to his early seventies – he regularly provided theatre managers with two stage-works a week, so that he earned and keeps another title, "the world's most prolific playwright", a claim not likely to be contested. Along with this ceaseless outpouring he also wrote novels, stories, poems – 1,600 sonnets, Italianate epics, narrative verse on historical and mythological subjects, and lyrics – articles on religion, an essay on the drama, and an autobiography. Of his vast repertory of plays, 450–500 still exist – the statistics about Lope vary wildly after the passing of centuries, and verification is difficult – and are proof of his versatility, fecundity and remarkable competence. The further marvel is that he did this in a life that was incredibly crowded with soldiering, sexual conquests, troubled marriages, illicit escapades. He defended a lawsuit over slander, endured exile, and bore the intermittent responsibility of supporting and bringing up a large brood of children sired with successive wives and mistresses. He was fertile in more ways than one. With so much to tell, it is not easy to write about Lope.

Son of Asturian peasants from Corriedo, he had the advantage of a birth in Madrid, and was spared a quiet life in the country. Instead he entered early into the hurly-burly and mental stimulation of daily experience in Spain's throbbing capital, an appropriate setting for one of hot-blooded, volatile temperament and hyperactive intelligence; yet, of lowly descent, he never lost his understanding of the peasantry, and he was to create striking pictures of their characteristics and habits.

He was a prodigy. At five, he not only read and wrote in Spanish but had already mastered Latin and composed verses. His father, having learned a craft, had risen to be a master embroiderer. At his death (1578), he left an impoverished family, which dispersed. The teenage Lope would have been in straits, but fortunately he became the protégé of a highly influential uncle, Don Miguel de Carpio, the Inquisitor, a grim worthy who doubtless recognized the boy's precocity and got him entrance into good schools. His special training may have started with two years at a recently opened Jesuit academy where he could better his Latin and further an acquaintance with classical authors, and perhaps also take part in theatricals, which the Jesuits used as a teaching tool. (He claims that he wrote a four-act play at twelve.) At the academy he demonstrated adeptness at fencing, dancing and music, and excelled in "ethics" (1572–4?). Now fifteen, however, the restless youth ran away to explore north-western Spain with a classmate. Still fifteen, he enlisted and experienced one fiery battle in an expedition against the Portuguese, or so he asserts. Afterwards he was accepted as a page in the household of the Bishop of Ávila, under whose aegis he was soon enrolled for four years at the University of Alcalá de Henares, where he was awarded a degree. He may also have studied briefly at the University of Salamanca – the record of his schooling is confused, perhaps he intended it to be. By his own admission he was an indifferent student, his mind already settled on becoming a writer, and his attention constantly distracted by his erotic inclinations. In later years, though, he boasted – somewhat defensively – that he was "educated".

Upon his graduation from the University of Alcalá de Henares at seventeen, he chose for himself a path towards ordination as a priest; but at the last moment he fell in love, and his impulse to embrace celibacy abruptly ended. So in time did the infatuation.

In a letter, he lamented having "fallen blindly in love.... God forgive it, I am married now, and he that is so ill fears nothing." As physically bold as he was imaginative, he took part in Philip II's victorious naval sortie against the Azores (1583), where again he saw combat. In the autumn of that year he was back in Madrid. Drawn to the theatre, he got employment with a director-producer, Jerónimo Velázquez, for whom he provided *comedias*, and with whose married daughter, Elena Osorio, he carried on a prolonged liaison (1583–7). When it broke up, largely through the angered father's intervention, Lope wrote insulting satirical verses aimed at Velázquez and sundry actors in his troupe, who found them so provocative that they filed a suit for libel against him, taking Lope to trial (1587–8). He lost the case, his punishment being imprisonment and eight years' banishment from Castile. This incident, which apparently was traumatic, dealing him the harshest of blows, was to be the basis of *Dorotea* (1632), a major five-act play, designated by some as his masterpiece. It is really a long novel in dialogue form, much like *La Celestina*. (Concerning *Dorotea*, Alan S. Trueblood has contributed a scholarly book, *Experience and Artistic Expression in Lope da Vega*, that in 768 pages analyses facets of the play in the finest detail, and that is one origin of material for this chapter, as have been a short biography of Lope by Angel Flores, and summaries in the chronicles of Gassner and Brockett, and references in Bernard Sobel's *New Theatre Handbook,* Martin Esslin's *The Encyclopedia of World Theatre,* McGraw-Hill's *Encyclopedia of World Drama* and *Dent's Companion to the Theatre.*)

After rescuing a scapegrace friend from gaol, he went to Valencia. Fretting in exile there, he kept busily pouring out plays. On one occasion, at considerable risk of being discovered and sent to the galleys, he slipped back into Madrid and, his affair with Elena well behind him, eloped with a nobleman's daughter, Isabel de Urbina, and they were married.

He became secretary to the Duke of Alba. By the time his banishment was finally lifted, he had already established himself as someone whose reliable, unstaunched flow of scripts could help managers in Madrid, Valencia and elsewhere keep their stages steadily occupied. In Madrid again, and rapidly gaining a reputation, he took on the airs of a cavalier, and his plays won applause for their variety, wit and vivacity.

In 1588 his patriotism led him to enlist in the intimidating Armada that Philip II was sending against England, an over-ambitious foray that ended in ignominious disaster after the defeat inflicted on it by a terrible sea-storm and by the harassment of Sir Francis Drake's doughty, darting English fleet. The *San Juan,* the vessel on which Lope served, was one of the few fortunate enough to escape, and on its eventful, six months' homeward voyage he worked devotedly on a lengthy narrative poem, *The Beauty of Angelica.*

Once more in Valencia, he resumed his play-writing. When his young wife died – he had been unfaithful to her and deserted her – he found a new love, Micaela de Luxon, an actress, by whom he had four illegitimate children. He wrote love sonnets to her, but also enjoyed the favours of a number of other ladies, and in 1590 married again, to the daughter of a rich pork merchant who brought him a welcome dowry. He continued his liaison with Micaela, however, and alternately had children by his wife and his mistress, three more in all.

By now his prestige as an unprecedentedly successful author was such that several wealthy noblemen vied in turn to be his patron; he was moderately affluent. Among his sponsors was the generous young Duke of Lessa.

The year 1610 saw Lope re-established in Madrid, and three years later his third wife died. He had begun to take his duties as a father ever more seriously – he had always manifested a fondness for them – and now, his wife no longer on the premises, he had his illegitimate offspring move in to share the house with those born in wedlock. A year later a much earlier impulse returned: he took "minor orders" and was named a "familiar" of the Inquisition, as if following in some measure his zealous uncle's example. But he did not stop writing plays or pursuing women. One of his inamoratas was an actress, Lucia de Salcedo, whom he called *la loca* – "the mad one" – because of her frequent tantrums and unbridled temperament. Another of his autumnal loves was Doña Marta, or "Amaryllis", who, though she had a husband, gave Lope yet another child. He was now almost sixty, and she was decades younger than he. Their sensual attachment gradually diminished, but they remained close friends. At the same time, taking his religious vocation in earnest, he often scourged himself in his cell until the walls were bloodstained. Apparently he saw no contradiction between his sincere piety and the conventional immorality of his life. What is more, he was honoured by the Pope.

His later years were not the happiest: Doña Marta, after becoming blind, died – early deaths took so many women from his life; One of his sons, Lopito, a talented poet, was lost at sea; a daughter, illegitimate, ran off with a cavalier, to her father's deep distress. At last his astounding vitality failed him; and, at seventy-three he followed his many loves into darkness.

So unquestioned was his popularity as a dramatist that a sign at a theatre, "*Es de Lope*" – "It is by Lope" – sufficed: people did not bother to ask the title of the work, simply whether he was its author. His range was limitless. A programme largely from his writing table might consist of a *comedia* or an *auto sacramental*, one secular, the other religious, and both with vulgar *entremeses*, the overall effect like that of a vaudeville show, a mixture of the sacred and the profane. In Lope's eyes this was quite defensible. The tragic and the comic, the divine and the ribald and the lewd, could be offered together. In his essay, *New Art of Making Plays in this Epoch* (1609), he declares that he is conversant with classical theories, having enjoyed a sound education – he is fully aware that tragedy and comedy were looked upon by the Ancient Greeks as quite separate genres and should not be commingled; he himself has composed a half-dozen stage-works that conform to those strictures. But that is not what pleases Spanish audiences, and he is determined for the future to give them what they desire "in defiance of art". He has a thoroughly independent mind-set: "When I have to write a comedy I lock in the precepts with six keys. I banish Plautus and Terence from my study . . . and I write in accordance with that art which they devised who aspired to the applause of the crowd, whom it is but just to honour in their folly, since it is they who pay for it."

He was ready to dispense with the classical unities of time, place and action in favour of a looser form. Cervantes, who never experienced commercial success, criticized Lope's high-handed disregard of Aristotle's revered advice on how best to construct a script. In *Don Quixote*, in a disparaging tone, there is a reference to a work in which "the first act . . . was laid in Europe, the second in Asia, and the third in Africa; and had there been four acts, the fourth would doubtless have been in America". But Lope, at the time he wrote his self-justifying essay, had been marketing and selling scripts for two decades and was well attuned to the public's taste.

The audience, if the king and courtiers were at the theatre, was likely to be somewhat restrained; at such times, too, entrance fees for the groundlings were higher. But if the king and his entourage were away from Madrid, more space in the playhouse was available and fees were lowered, attracting more lower-class spectators, who tended to be unruly. To express their displeasure at a plot-turn or imperfect performance, they might hurl cucumbers or anything that could be thrown at the stage and the actors. The players also had to work against the bustle and noise of vendors who elbowed their way about the pit while hawking fruit, water and sweets, a substantial source of income to the management. Lope knew how to capture and hold the attention of the crowd by fast action, suspense, humour, surprises.

As said before, Ganassa was the first to roof the stage as protection against bad weather, but standees in the pit might be pelted by a summer rain. Later, a protective awning was extended over most of the courtyard. By now the balcony reserved for less affluent ladies had come to be known as the "*cazuela*", or the "stew-pan"; partitioned off, it had separate access. The boxes along the side walls were grated like the balconies of private dwellings: secluded, spectators of both sexes looked down without being clearly seen.

Lope disliked having the actors in historical dramas wear contemporary attire, however lavish, rather than costumes appropriate to the period in which the action was set. He complains of the incongruity and asks why a Roman should sport tight breeches, or a Turk "the neck-gear of a Christian". But the spectators, mostly semi-literate at best, did not object to seeing Aristotle in "a curled periwig and buckles on his shoes". One reason there was little or no attempt at authenticity, of course, was that the actors provided most of their wardrobe. With scenery mostly lacking, the costumes indicated to the audience where the characters were and what they might be about: if wearing a plumed hat and burnished armour, they were probably in a camp of attackers or among those besieged; if in hunting clothes, they were most likely in a leafy forest; and more gorgeous garb might suggest their presence in a richly decorated anteroom at court. Actors portraying the peasantry dressed simply: class distinctions were not overlooked in an excessively caste-conscious society. Male spectators particularly liked to see actresses in breeches-roles, when cross-dressing – as a disguise – was essential to the plot, leading to mistaken identity, a prevalent comic device. The Church, strongly opposed to women participating on stage in the first place, was even more vehemently scandalized by their appearing in breeches, revealing slender ankles and calves, but the hierarchy's specific edict against the titillating practice was weakly enforced and largely unheeded. One reason was the competition of the Italian *commedia dell'arte* troupes with their distaff casting; Spanish companies could not afford to be less attractive, and took chances.

By Lope's day the actors' lot had scarcely improved. There were still long rehearsals, over-frequent performances, the hardships of vagabondage and travel. The programmes were lengthy, especially the stagings of the *autos sacramentales*, which might start at six in the morning and last two days, with exhausting demands on the players. Besides this over-taxing routine, there were also the payless spells of being "at liberty". Many cast members spent a bit of time imprisoned for debt.

None the less the ranks of actors were constantly filled. It is believed that during this era over 2,000 aspirants to the craft of performer were heard spouting verse, singing, playing a musical instrument, and seen dancing or ostentatiously wielding a sword. Over the centuries the many repeated and serious discouragements innate in their beloved profession (or addiction) have never turned its members away from it.

A few stars, such as La Calderone, Charles V's mistress, fared comparatively well, but there is no sign that any ever became wealthy. (It is interesting that La Calderone was the mother of Don John of Austria, the victor in the great naval battle over the Turks at Lepanto; he was the consequence of her notorious liaison with that Spanish king, who commanded in his will that his legitimate son, Philip II, acknowledge his half-brother Don John's royal birth.)

That La Calderone might portray the Virgin Mary or a martyred saint, and a few minutes later appear in a bawdy role, was also anathema to the clergy and pious laymen; they thought the sacred plays should not be tainted by contact with stage folk, whose private lives were too often disreputable.

A successful play might have five or six showings, seldom more. Scripts were sold outright: the fee might enable the lucky author to live comfortably a full year. A *comedia* brought much more than a religious play. A writer as popular as Lope, working without pause to answer the unceasing cry for his works, profited accordingly.

If the public theatres used little scenery, there were exceptions when *autos sacramentales* and entertainments were given indoors at court or outdoors in palace gardens. As were the lavish Italian presentations on which they were modelled, these might be magnificent, rivalling any by the Medicis and Gonzagas. One, ordered by Philip IV, who was truly stage-struck, cost over a half-million reales, a sum defying modern computation. He brought companies to his court from all over Spain, and hired Italian architects to build and design fabulous settings in his private theatres. Stage machinery of every sort was installed – trap-doors, water-tanks for simulated sea-battles, painted backdrops, multi-levels, cranes and pulleys. A reliance on such apparatus provoked Lope. He stated scornfully: "Since good actors are lacking, or our poets are untalented, or our audiences fail of comprehension . . . producers resort to machinery, poets rely on carpenters, and the spectators feast only with their eyes." All that a good script and an inspired cast requires are "four trestles, four boards, two actors, and a passion". A memorable definition that has been much quoted.

Yet costly scenery and startling tricks and stage effects were sometimes added to presentations of his plays, too. For example, a quartet of *autos sacramentales* from his pen, when they were staged in Madrid in 1609, called for a star-filled sky, dragons spitting flames, an ornate palace interior, magically vanishing dinner plates, and a large ensemble in regal attire. But he, for one, was quite able to enthrall an audience without depending on such expensive appurtenances.

If Lope "banished Plautus and Terence" from "the locked room" in which he worked (his mind), as he confidently asserts, he was not uninfluenced by the lengthening line of theatre craftsmen who had preceded him. In boyhood he had seen and responded with glee to the repertory brought to Spain by Alberto Ganassa, and he remembered how compelling was the perpetual liveliness of the *commedia dell'arte*. The scripts of Encina and Cueva brought him practical lessons; at the very least, they told him what to avoid. Overall, though, his fancy and methods were predominantly original. In his hands the cape-and-sword melodrama arrived at its lasting form, more or less – no one was better at it. He excelled, as well, at establishing the mood of a light, romantic comedy and manipulated a series of unexpected dilemmas in a rapid-paced farce. He is also the progenitor of what is to be known, centuries later, as "proletarian theatre", plays in which his sympathy for the lower classes, and especially the stalwart, rebellious peasantry, finds passionate voice. He had a command of rich language. Cervantes praises in particular his "purity of style". At moments, when they are needed, a script is likely to have luminous lyrical passages, with intense emotion throbbing through them.

When drawing on history, Lope theatricalizes it, romanticizing the people and events, giving the incidents a colour and the characters a glamour they never actually had. Yet, in contrast with those given life by the inept tribe of writers before him, his portraits are effective and provide good roles. If they are seldom if ever verifiable, they are credible within the framework of his drama; he brought a great deal of verisimilitude to the stage, where it had been largely lacking. It cannot be said that his people are profoundly probed or fully rounded, but they are heroic and appealing and some have an imposing stature.

His comedies do him proud. He is credited with the invention of the *gracioso*, or buffoon, who – though he had certainly appeared long before – had never previously been given such delightful embodiment. A version of this amusing figure is in *The Gardener's Dog* (1615), one of Lope's best-known and frequently revived works. The affections of a lady of noble birth, Diana, are sought by her handsome secretary, Teodoro. Only when she learns that he is also paying attention to her maid, Marcella, does she ardently respond to him; a repeated motif in Lope's output is that love springs from jealousy and possessiveness. (The title is an allusion to the selfishness of the dog in the manger, who, though not hungry, will not spare a bone to any other dog.) But Diana is acutely aware of her aristocratic rank, so Teodoro's humble origin is a barrier. A solution eventuates when the young man, at the prompting of a wily *gracioso*, claims that he is the son of a count. Even though Diana is aware that this is a falsehood, she prefers to accept it as the truth, since it enables her to wed him. It is enough for her that he can pass as a nobleman in her friends' eyes. In this there is mockery of class distinctions and pretensions, and a delicate juggling of questions of useful illusion and unpalatable truth – here Lope might be said to have anticipated by four centuries Luigi Pirandello's *Right You Are (If You Think You Are)*. It was not new to celebrate the wit, resourcefulness and common sense of the low-born at the expense of their "betters", but few authors handle the subject as well as Lope, himself a son of peasants.

In *Madrid Steel*, another comedy in which the characters display gallantry and fine manners, and unhesitatingly engage in pretence, Lisardo wins the heart of Belisa, who is guarded by her watchful duenna, Teodora. The girl feigns an illness; a doctor is summoned – none other than her lover's servant in disguise. Meanwhile the duenna is lured away by Lisardo's friend and accomplice, Riselo, who astounds the prim lady by his eloquent, if insincere, professions of interest in her. (The play's title derives from a potion, *agua de acero*, water with a taste of iron, prescribed by the "doctor" for the supposedly ailing Belisa.) Complications follow when Belisa's father announces his intention of betrothing her to someone else. This precipitates an elopement. At the same time Riselo marries his true love, who is *not* the duenna. *Madrid Steel* provides the plot-device for Molière's equally popular *Love Is the Best Doctor*, and to a lesser extent his *The Physician in Spite of Himself*.

The Capricious Lady (printed 1617) has two lovers masquerade as servants in a wealthy house, and describes the errors and mischief that result from the caper. It is a comic situation that will serve the authors of ensuing farces over the decades and centuries, among them Oliver Goldsmith for *She Stoops to Conquer* and Jacques Deval for *Tovaritch*. Similarly frolicsome in touch and tone are *The Greatest Impossibility* (1615) and *If Women Didn't See* (c. 1631), both of them concerning love-intrigues and conflicts of honour. The first expresses what was apparently Lope's belief that once a woman sets out to ensnare a man it is highly unlikely that anything will stop her, a premise shared by George Bernard Shaw. *Belisa's Tricks* (1634) also toys with this idea, and with Lope's recurrent preoccupation with the baffling overlapping of illusion and fact in emotional relationships, especially amorous ones.

His interest in this subject had already led him to *The Lunatics of Valencia* (c. 1600), where he examines and contrasts actual madmen, feigned madmen and eccentrics (looked upon as "mad"), and later to *There's Method In 't* (c. 1634), whose clever heroine behaves like a simpleton to attain her ends, no longer a novel idea but well handled.

Lope's seeming personal love of combat, of reckless martial adventures, and his first-hand experiences of them are reflected in his *comedias heróicas* and *comedias historiáles*, taken from history or legend, or perhaps a mingling of both. In this category belong *The Crown of Otun*, relating the fall of Oktokar, the King of Bohemia, and *Rome in Ashes*, a wildly gory work in which Nero is the protagonist. Among his ultra-romantic works is one that will astonish British readers: *Castelvines y Montesses* (c. 1608). The impulsive Roselo Montes encounters Julia at a ball. She is already loved by her cousin, Otavio, and for this reason, though attracted by Roselo, hesitates at accepting his attentions. His passionate pleading wins her heart! The young pair are secretly married by a friendly, well-intentioned friar. Unluckily, in a subsequent quarrel Roselo kills Otavio and is exiled. Bereft, Julia attempts suicide, but the philtre she drinks proves to be ineffective – a trick of the wily friar. She sinks into a coma, is presumed dead, and mourned by her noble parents. Aroused from her death-like sleep, she is rejoined in the tomb by Roselo. Completely disguised, they go back to her family's house and are retained as servants. Concealing herself, Julia speaks to her father, who mistakes her voice for an angel's. This "visitation" leads him to bless the marriage and wins Roselo a reprieve from banishment, so that all concludes happily. Obviously, Lope is exploiting the same legend to which William Shakespeare had turned two decades before for his *Romeo and Juliet* (c. 1595), but there is no evidence that the two playwrights knew each other's works or were even aware that each other existed. *Castelvines y Montesses* has clown scenes, and rejects a tragic ending, as Lope always preferred to do, but his version of the story lacks the bawdy high wit and sardonic humour of Shakespeare's Mercutio and Old Nurse, and is never infused with the superb musical poetry that the English dramatist gives it. (The tale had appeared somewhat earlier in a novel by Bandello and a narrative poem by Arthur Brooke.)

A Certainty for a Doubt (printed 1625) tells of Pedro the Cruel and Don Enrique, brothers competing for the hand of Doña Juana. Pedro, the elder, wields a royal prerogative, orders Enrique into exile and seeks his assassination, an attempt that fails. As a further complication in his harried life, Enrique is besought by two other women besides Doña Juana, whose choice he is. Pedro, exerting force, sets a date for his marriage to her, against her expressed will. In effective disguise, redoubtable Enrique slips into the ceremony, contrives to replace his brother, and weds the surprised, delighted Juana. In an outburst of uncharacteristic generosity, Pedro blesses the happy lovers' union: "What is done admits no remedy." His driving motive has not been true love but a lifelong envy of his high-spirited younger sibling. The plot is far-fetched, but the action is fast and grips attention, and the portrayals of the ambivalent Pedro, ruled by jealousy, and the bravely resistant Juana, who though not attracted to Pedro is not insensitive to the joys of being a queen, lend a strange vitality to the script.

The ever-compelling and recurrent Hippolytus–Phaedra theme is given a Spanish resurrection in *Punishment Without Revenge* (printed 1635). Both the Duke of Ferrara and his illegitimate son love the duke's young wife. The duke discovers an adulterous affair between the two; he tricks the son into slaying his distraught stepmother, then has him killed for the deed. The harsh story is unveiled without the redeeming psychological depth of the great treatments of this ageless dramatic subject by Euripides and Racine. But it does have strength, and the people are drawn boldly.

(Long considered the best of these sensational melodramas and tragicomedies is *The Star of Seville* (c. 1615), acclaimed for its directness and simple power, but recent scholarship tends to attribute it to a hand other than Lope's.)

La Dorotea (printed 1632), referred to previously, has autobiographical overtones. A young poet is gripped by a turbulent affair with a married woman, but she is faithful neither to him nor her husband. He tries to end his infatuation by finding haven in another city, then by joining the Spanish Armada on its misguided expedition. Clearly this love-obsessed poet is Lope himself. The play contains a subtle analysis of human motives, which shows that the author has arrived at a keen self-knowledge, together with insight into others, to a degree that is quite modern.

Even more modern, and proven to be of rousing interest to mid-twentieth-century audiences, are Lope's dramas of social protest. Two plays in this "agitprop" genre fitted the mood of a troubled period, the 1930s, the years of the Great Depression in Europe and America, marked by an upsurge of radical political activity. Enthusiastically rediscovered and widely revived by leftist intellectuals were *Fuente Ovejuna* (*The Sheep Well*; c. 1614) and *El Mejor Alcalde, el Rey* (*The King the Greatest Judge*; printed 1635).

Lope is in his special domain when writing about the Spanish peasantry; he knew them well.

His was an innate affinity; he shared their stubborn, heroic qualities. *The Sheep Well* is not about one man's courageous stand, but that of a whole community of farmers outraged by the conduct of the local Commander, Fernán Gómez de Guzmán, Master of the Order of Calatrava. He lusts after a village girl, Laurencia. She shrugs off his advances, saying, "Though toothsome, I'm just too tough to serve his Reverence for a feast." But it is hardly marriage that the licentious Commander has in mind. When he tries to carry her off, her humble lover, Frondoso, dares to obstruct him by seizing his crossbow and threatening to let fly an arrow at him. Soon afterwards Frondoso and Laurencia are about to be married. The Commander interrupts the wedding rites, has the young man arrested, the girl's aged father beaten; then he kidnaps and rapes the girl. This is but another of the Commander's long succession of infamies, among them the deflowering of other local girls. The put-upon villagers gather; they are angry but hesitant to act against the high-ranking oppressor. The ravished Laurencia cries for vengeance:

> Some of you are fathers, some have daughters. Do your hearts sink within you, supine and cowardly crew? You are sheep, sheep! Oh, well-named, Village of Fuente Ovejuna, the Sheep Well! Sheep, sheep, sheep!

Wild beasts, she tells them, have more sense of honour, and she calls on them to share "the tiger's heart that follows him who steals its young, rending the hunter limb from limb." Aroused by her furious words, her bruised and bloody face, the crowd rushes to the Commander's palace; in the mêlée the vicious nobleman is killed by Esteban, Laurencia's father, mayor of the village.

The peasants, frightened, agree that no one shall reveal who struck the fatal blow. "Fuente Ovejuna shall plead guilty to the crime." They steel themselves by rehearsal against the torture that will be applied to force confessions from them.

King Ferdinand and Queen Isabella hear of the riot and the death of the Commander, the region's hereditary feudal lord. Incensed, they order that the rebellion be quelled. As the villagers foresaw, some of those imprisoned are put to the rack. Even women and children undergo this hideous ordeal, but all stay mute, save for answering that "Fuente Ovejuna" is the culprit. Three hundred are tested but keep silent, as they vowed to do. Finally, impressed by such valour and unable to fix the guilt precisely on anyone, the king discreetly issues a pardon.

Perhaps the merits of this play were somewhat exaggerated by moderns of liberal sympathies, reading virtues in it that it does not wholly possess. It is simplistic in the extreme. The Commander is psychotically evil, and Lope sycophantically praises the king for a wisdom and clemency that he does not really display. Yet for its time, in autocratic Spain, it was daring and original, and its stage-worthiness not to be denied; even now it whips up sympathy and excitement. It was widely popular in, among other places, Soviet Russia, where its celebration of peasant revolt was useful to the Stalinist regime.

The script is based on a once-familiar event in 1476 in Fuente Abejuna, in the province of

Cordoba. Lope learned of it from Rades y Andrada's *A Chronicle of the Three Military Orders*, and was immediately captivated by it. He also included in his text allusions to the broader political background of the period, involving the King of Portugal's claim to the Spanish sceptre. The peasants support Ferdinand of Aragon, while the Commander embraces the cause of the Portuguese interloper.

The King the Greatest Judge also concerns a feudal lord, Don Tellon, rapaciously drawn to a farmer's daughter, Elvira, betrothed to a young peasant. Don Tellon and a follower raid the farmhouse and bear her off. Her father and Sancho, the bridegroom to be, pleading for her release, are rudely rebuffed. They appeal to the king, who promptly signs an order that the girl be freed. Don Tellon brazenly disobeys. The king decides upon a personal visit to the abductor. The father, overwhelmed, is reluctant to have the king, mighty and august, intervene:

> Consider, Sire, a peasant's humble honour
> Touch you not so near. Despatch some judge,
> Some just Alcalde to Galicia
> To do your will.

His sovereign replies:

> The King the greatest Alcalde.

Arrived at the scene, the king orders the execution of the recreant Don Tellon, but commands that first he make Elvira his wife, to restore her honesty, and also to enable her to inherit half her unwilling husband's wealth to bring as a dower to faithful Sancho. Here, again, Lope makes no secret of his sympathy for the lower classes and openly preaches kingship's responsibility for their welfare. The play, distilled from historical fact, is also a tribute to Alfonso VII of Castile. As a stage-work, however, the script has less emotional impact than *Fuente Ovejuna*.

Another side of Lope's protean nature is exposed in his many dramas on Biblical subjects, which are more attuned to his "inconsistent pietism", to borrow Gassner's apt phrase. An example of works of this sort is *The Beautiful Esther*. His searching through the Old Testament for suggestions to keep his pen moving fleetly across blank paper, his many scripts depicting the lives of saints, their sufferings and miraculous, healing deeds – there was always a ready market for such fare, and he provided it with at least forty-four known *autos sacramentales*.

Some of his highly ranked cape-and-sword pieces and plays of romantic intrigue are *The Goblin Lady*, *The Discreet Mistress*, *The Sword of Madrid*, *The Night of Toledo*, *The Countryman in His Homestead*. A historical drama of note is *The Knight of Olmedo*. He ranged far afield from his native Spain for *The Grand Duke of Muscovy*, and to Ancient Greece and Rome for his classical plays, and to mythological realms for his charming pastorals.

Because he was so prolific and versatile, and in view of his unquestioned supremacy in the history of Spanish theatre, he has been likened to Shakespeare. He was born two years before the playwright from Stratford-on-Avon and outlived him by almost two decades. Their careers ran remarkably parallel courses, though Shakespeare was no swordsman or adventurer, or – as far as is known – an incessant womanizer. Both had an amazing perception of human impulses and common motives, an all-encompassing gift of empathy. The English poet did not depend on his personal experience, as his work is largely an imaginative projection; he had probably never encountered an Othello or a Shylock, nor any other Moor or Jew, yet he knew instinctively how they felt and thought. Lope's knowledge of the world and the people in it is wider and more tangible, yet paradoxically never as deep. As a psychologist he is not Shakespeare's equal. His work does not contain the metaphysical questioning that is voiced by an alienated Hamlet and stricken, stumbling Lear, or the flippant mockery of a Mercutio, who challenges a serious appraisal of all life and loves. Nor does Lope have the poetic reach of the author of *Antony and Cleopatra* and *The Tempest* and the immortal sonnets. Then again, a fault of both writers: too much rhetoric.

Composing his myriad scripts so rapidly, Lope is often careless in his plotting, especially when bending likelihood to arrive at a happy ending. His "tragedies" are too fast-moving, or too romantic, or too lightened with comic touches to attain the highest stature; he is more a master of tragicomedy than "tragedy". And he had another shortcoming: like Shakespeare, who unwittingly gave Bohemia a sea coast, he had an inadequate knowledge of geography.

Shakespeare's plays have crossed every boundary. After more than four centuries they are enacted not only wherever English is understood, but in France, Germany, Russia, and on stages as far away as the Orient and Africa; Lope's works are scarcely performed or even read outside Spain, yet his influence on other playwrights has been great everywhere, though many of those affected are unaware of it. The immense number of his scripts, which are infinite variations on the few basic dramatic situations, quickly became an ever-yielding, inexhaustible source of plots for later playwrights, often at very distant removes and not knowing from whom they were actually borrowing. At conceiving effective ideas for stage-works he left to his successors no new worlds to conquer.

Some commentators see in de Vega's work an embodiment of the baroque art-style, flushed with expressive energy and with an inclination towards self-aggrandizement and the ostentatious and grandiose, even if a good part of his writing is direct and earthy. He is hard to evaluate, since he possessed such contradictory traits, which also applied to his overwhelming output. Indisputably true is A.F.G. Bell's summing up of Lope's contribution to his country's evolving theatre: "For a half-century his sensitive impressionable genius was the ready channel through which the spirit of the triumphant Spanish nation could flow and the classic drama become finally nationalized. He added play to play; no single play represents his genius, but in their entirety they fully represent the Spanish nation."

Around the tireless Lope de Vega there were a legion of other playwrights. Since a successful script had a run of only around five productions, managers in Madrid, Seville, Barcelona and other large cities had to be constantly on the alert to find new works.

One resource was Guillén de Castro y Bellvis (1569–1631), about a decade younger than Lope; the two were friends, and Castro was intelligent enough to model his work on that of the professional and demonstrably successful writer from the capital. A native of Valencia, the scion of a distinguished family, the youthful Castro had early shown a marked flair for versifying. At twenty-three, for his accomplishments to that year – two dozen poems and four discourses – he was admitted to the local literary society, Nocturnos. In 1595 he exchanged marriage vows with the aristocratic Doña Marquesa Girón de Rebollada, in what proved to be a deplorable mismatch. That same year he became acquainted with Lope, banished from Madrid and just arrived in Valencia. In the early phase of Castro's career, Lope often spoke openly in praise of the younger man's talents, a commendation that helped considerably.

Five years later he again emulated Lope and entered military service, which eventually led him to a post in Italy, where he was appointed Governor of Stignano, an honorary post conferred on him by the Count of Benavente, Viceroy of Naples. In 1609 he was back in Valencia, and from there moved on to settle in Madrid, where he joined the staff of the Marquis of Penafil (1619), who endowed him with "the house and lands of Casablanca". This gift he rather quickly passed on to his sister Magdalena (1620). By now his marked gifts as an author were recognized and he was elected to the Academy of Poets, permitting him to mingle at ease with the leading dramatists and practitioners of belles-lettres of the day. Another significant honour was paid to him four years later: he became a Knight of the Military Order of Santiago.

A few months later his quick rise to celebrity came up against a sudden obstacle: he was charged as an accomplice in the violent death of a rival. Investigators could not gather specific evidence, and the accusation was finally withdrawn. In 1626 he married again, to Doña Angela Maria Salgado, a second unfortunate choice. The kindly fates that up to now had attended his career did not stay with him. Impoverished, enduring pain and too weak to write, neglected by his influential friends, though he was still rather famous, his life ended comparatively early at sixty-two.

Castro is ranked slightly below Spain's greatest playwrights, but throughout the seventeenth century his work was much applauded, enjoying broad popular success, and is still respected. He was not an innovator, but, though he had chosen Lope as his mentor, he was more than a mere imitator. He had insight into people's motives and created believable characters. Like Lope's they always speak in verse, declaiming about honour or pouring out their hearts and hopes in language that is harmonious and frequently enforced with graphic imagery. His plots are usually well constructed, demanding and holding the spectator's unswerving concentration. His choice of themes is diverse, and like several of his predecessors he found much of his subject-matter in ballads, folklore, mythology, history, bringing fresh ideas and settings to Spain's ever-changing stage. His treatment of them, too, was increasingly individualized. To these were added dramas based on incidents from his own life, which had not been uneventful – this group of scripts portray the society of his day, which lends them extra interest.

His comedies of manners are his most realistic contribution. The title of his *The Ill-Wed of Valencia* (1595?–1604?) suggests that he might have drawn some of its content from his own two failed entrances into matrimony; and *The Man Who Thought He Was Narcissus* (1610?–1620?) is an early, amusing portrait of a fop, who by the next century was to become a stock character. Another of his popular offerings in this category is *The Force of Habit* (1610?–1620? – in Castro's instance, too, datings of his works are not certain.)

From classical mythology and Greek and Roman legends he extracted material for *Procne and Philomena* (1608?–1612?) and *Dido and Aeneas* (1613?–1616?), echoing Virgil. Dramatizing novels by others, he offered stage versions of their stories in his *The Impertinent Meddler* (1605?–1608?) and *The Power of Blood* (1613–14), and he daringly adapted Cervantes's *Don Quixote* for enactment, showing no lack of self-confidence (1605–6?). In his melodramas his characters have exceedingly strong emotions and are seized by unbridled passions, a trait approved and appreciated even today by Spanish audiences, but doubtless even more so in that period, though other European spectators might be somewhat taken aback by such open and turbulent displays of feeling – they are probably not to the taste of many twentieth-century readers. Such scenes occur in *Count Alarcos* (1600?–1602?) and *The Hostile Brothers* (1615?–1620?), among others. In his melodramas his characters are sometimes capable of horrifying brutality. This has diminished his current acceptability, a loss of esteem in recent decades compared with the enthusiasm with which his plays were viewed earlier.

The list of scripts attributed to him is lengthy; only seven have been verified as truly his, though much probability clings to a score or more of others. Oddly, chance, which was so kind and harsh to him at different phases of his life, was to make his lasting fame dependent on a play that he did not write, the immortal *Le Cid* (1636–7) by the French poet-dramatist Pierre Corneille. Two decades earlier Castro brought to the stage a drama about that great eleventh-century hero, sainted in folklore, who with surpassing courage and military prowess had defeated the invading Moors and afterwards become an invincible soldier of fortune. So persuasive and well received was Castro's handling of his subject that he composed a sequel. Corneille, becoming familiar with Castro's success, was prompted to emulate and top him, and accomplished his goal. Today, however, scarcely any chronicle of world theatre – and in particular of French literature – fails to mention the debt owed to Castro, who first recognized the Cid's appeal as the focal figure of a stirring historical play, one that transcends the usual trappings of the *comedia heróica* and *capa y espada*.

Castro's *Las Mocedades del Cid* (*The Youthful Adventures of the Cid*; 1612?–1618?) tells of Diego Laínez and his three sons. When the father is insulted by Count Lozano, he looks to see which of the young men is most ready to defend the family's honour. Rodrigo (later the Cid) is the first to step forward, being the ablest, but he is caught in a tragic vice. He is deeply in love with the count's lovely daughter, Jimena, and she with him. Personal feelings must be set aside: honour comes before all else. In the ensuing duel, Lozano is killed. Summoned to the palace, Don Diego explains that his son has only acted to preserve his father's reputation and dignity.

Rodrigo is advised to join King Ferdinand's troops confronting the Moors. He does so, and one day in camp he and his fellow soldiers are approached by a beggar in tattered garments who asks for sustenance. The men turn away from him in disgust, but Rodrigo, more compassionate, gives him food and assists him. Throwing off his cloak and disguise, the beggar reveals himself as St Lazarus, guardian patron of Spain. He pledges to help Rodrigo in his endeavours.

For the extraordinary valour in battle that he has already shown, a Moorish king has designated him *mio Cide* ("my lord"), the name by which he is hereafter best known. Rodrigo leads a large force against the infidels and achieves a major victory. He returns home.

Now, though still harbouring love for him, it is Jimena's obligation to avenge *her* father's death. Despairing and desperate yet obeying the code, she petitions King Ferdinand to name a champion finally to right her cause. At her pleas the ruler chooses Don Martin, an eager suitor, whom Jimena promises to marry if he brings her Rodrigo's severed head. In the encounter and sword-play, however, it is Don Martin who falls and is beheaded.

Jimena, awaiting the outcome, is told that Rodrigo has perished and voices her dismay. But after the false report, Rodrigo appears bearing the gory token of his triumph. He demands Jimena's acceptance of his offer of marriage. She hesitates, then acquiesces. To reward Rodrigo for his outstanding services, King Ferdinand gives his royal consent to the union.

In the second play, *Las Hazanas del Cid* (*The Exploits of El Cid*; 1610?–1615), Don Sancho, King of Castile, is quarrelling with his brother Don Alfonso and his sister Urarca. Rodrigo seeks to stand aside, since he is uncertain as to which of the disputants is in the right. Unless to protect the king, he will not intervene. Urarca's loyal retainers, Arias Gonzalo and his five sons, are ready to protect her. She is threatened by Don Sancho's plan to grab Zamora, the small domain over which she rules; he is warned by Gonzalo not to make the attempt. The ghost of King Ferdinand appears to his successor, the ambitious Don Sancho, and also warns him to avoid a conflict. But a dissident Zamoran subject, Bellido de Olfor, urges the Castilian ruler to launch an attack. Doing so, Don Sancho loses his life. In a counter-thrust, Don Diego Ordóñez, a faithful adherent of the slain king, joins the fray, as does the fabled Rodrigo. Three of Arias Gonzalo's sons are killed in the combat. Meanwhile, Don Alfonso has been captive of the Moors. A Moorish woman, Zayda, moved by his plight, helps to free him. Don Alfonso hurries back and takes the throne left empty at his brother's death. He pardons the well-meaning Arias Gonzalo. Zayda, converted to Christianity, becomes Don Alfonso's queen, and Rodrigo, the great warrior, vows to serve them both.

The values articulated in these plays and throughout Castro's work – honour, fidelity – are invariably conservative ones. Andrés Franco says that this is a reflection of "the courtly ideal" prevalent in Castro's world, as in Corneille's, and which had great force and meaning for them.

Corneille's version of the Cid's legend is hobbled – or helped – by his having to conform to the rule of the Aristotelian three unities. His drama is far more compact and dispenses with practically all the extra characters and complications added by Castro to stretch out and fill his sequel.

(This hero is also at the centre of another work, Jules Massenet's opera, *Le Cid* (1885), with a libretto by Adolphe d'Ennery, Louis Gallet and Edouard Blau, which largely follows Corneille's treatment.)

Among the many other plays thought to be by Castro are: *Constant Love* (1596?–1599?); *The Birth of Montesinos* (1595?–1602?); *The Happy Ending* (c. 1599); *The Foolish Gentleman* (1595?–1605?); *Proud Humility* (1595?–1605?); *The Count of Irlos* (1600?–1610?); *The Truth Investigated, or The Deceitful Marriage* (1608?–1612?); *The Perfect Gentleman* (1610?–1615?); *The Hostile Brothers* (1615?–1620?) – this list is incomplete. The titles indicate that a goodly number of these scripts are comedies. Most were published while he was alive, which strengthens the claim that they are truly his. But unlike the seven plays firmly identified as by him, and commended for being well constructed, a majority of these just mentioned are loosely shaped, have implausible turns of plot, too many involvements and characters and illogical happy endings, which is partly why their attribution has been questioned.

But there is no doubt that he fashioned vigorous scenes, powerful and gripping moments, of the sort that remain ineradicably in the spectator's memory, and that he was – in the words of Andrés Franco – a "poetic and dramatic genius".

Juan Ruiz de Alarcón (c. 1581–1639), of Mexican birth but high-born Spanish stock, was a lawyer, educated in Spain and holding a government post. Like a surprising number of others in that profession, he turned to play writing and was responsible for some twenty-seven or thirty scripts – in this prolific company, a small body of work! Perhaps one reason his output was comparatively limited is that he was a perfectionist, meticulously reviewing and polishing details, a trait often inherent in those whose minds are concerned with legalities. In this regard for fine points and strict logic he tended to differ from his peers, whose approach was often slapdash. For subject-matter he chose to portray aspects of court life in Madrid, mostly its amusing side.

He, too, was a source of ideas for Corneille, whose *Le Menteur* (*The Liar*) is based on Alarcón's *La Verdad Sospechosa* (*Truth Suspected*; 1628). This comedy has an unexpected ending, not quite a happy one, for the hero, who lies glibly, outdoes himself, and at the close finds himself married to a woman whom he does not love. In time it became a perennial stage success in Spain and, after Goldoni's adaptation of it, in Italy as well. It, and other scripts like it, led to Alarcón being thought of as the originator of character comedy in his adopted country, though due appreciation was to come to him belatedly, for his work was not particularly popular during his lifetime.

In *Walls Have Ears* (printed 1628) Don Meno, too, is punished for not being steadfast in love and for talking too much about the ladies to whom he is paying court. *The Husband's Examination* (printed 1632) offers a deft sketch of a proud young lady who proclaims exacting requirements for any husband-to-be of hers, only to regret it when she becomes enamoured of a gentleman perhaps not fully answering to her original prescription.

Far more sober is *Cruelty for Honour* (c. 1625) in which Alarcón deals with the darker side of human passions. The mood is also serious in *The Weaver of Segovia*, which anticipated Friedrich Schiller's youthful *The Robbers*. Here, a young man turns brigand when treated harshly by his king, though later he is pardoned for having performed unexpectedly generous deeds. Another drama in this heroic mode, which also endorses generosity, is *Gaining Friends* (c. 1630), wherein the altruism of the leading character is fully repaid. The roster includes *The Favours of the World*, *The Proof of the Promises* and *The Anti-Christ*, this last his sole religious work. His plots are economical, his language has a precision abespeaking a lawyer's pen, his people offer contrasting types, their well-defined diversity of response enhancing interest. A footnote: Alarcón and Lope de Vega were bitter foes.

Perhaps of less weight as a dramatist, because lacking originality, is the poet Álvaro Cubillo de Aragón (1596?–1661), though he was neither modest nor unpretentious. But he was an excellent craftsman and very adept at shaping comedies of manners, notable for their refinement and charm. Prized among his nimble works is *Las Muñecas de Marcela* (*Marcela's Dolls*; 1634), a rare depiction of adolescent love. Also esteemed are his *El Señor de las Noches Buenas* (*The Man of Merry Nights*; 1635), as well as a historically based play about a relationship between Infante de Lara and the fabled Mudarra, who bears the burden of his illegitimate birth. His other works include *The Count of Irlos*, *The Commanders of Cordoba* and *The Invisible Prince of Baúl*.

In *Marcela's Dolls* Carlos fatally wounds Valerio's son in a duel. He flees to escape the father's vengeance and, having found refuge in Marcela's home, inevitably falls in love with her. She shares his youthful passion. Her sister, Victoria, warns him that Marcela is betrothed to Octavio. Leaving the house, he encounters Octavio, who is misled into thinking that Carlos is Marcela's brother, Luis. Octavio informs him that Feliciana, Carlos's unmarried sister, has just borne a child which Luis has begotten. In the end, of course, everything is nicely straightened out, with Marcela plighted to Carlos, Feliciana to Luis, and the very willing Victoria to – who else? – Octavio. Pleasantly entertained, audiences were not too critical or demanding, however familiar the plot's resolution.

Between 1640 and 1654 Francisco de Rojas Zorrilla brought forth and had published an exciting series of melodramas concerned with conflicts between honour and love, among them *All Equal Below the King* and *What Women Are*. The subject endlessly gripped the thoughts of Spanish spectators of this difficult transition period, when religious fervour infused daily life, while moral values were challenged and changing.

Three more ornaments of the Golden Age are still to be extolled here. Augustín Moreto y Cavana (1618–69) enjoyed a wide following, nearly equal to that of Lope de Vega, from whom he borrowed no little inspiration. He was considerably better at comedy than serious drama, and again no innovator; but he was proficient at writing for the stage and actors and knew how to characterize quickly. His dialogue has freshness and vivacity. This is evident in *Scorn for Scorn* (printed 1654) and *The Cuckoo in the Nest* (also printed in 1654), as well as *You Can't* (c. 1660). Few of his scripts have new plots, but he reworks the themes used by others and usually to better effect. In turn, he became a source from whom others appropriated ideas.

In a chronicle of world drama, a special niche is filled by Tirso de Molina (Gabriel Téllez, 1583–1648). It is he who first brought the character of the bold, suave, energetically promiscuous lover, Don Juan, to eternal life on the stage. A native of Madrid, he studied at the University of Alcalá de Henares; later, religiously inclined, he took orders in 1613 and became a Mercedarian monk. Much of his time and effort was devoted to travel and military service. But composing plays was also an absorbing occupation for him: he authored over 400 of them, at the apex a purling stream of clever comedies. About eighty of his scripts are extant, among them *The Timid Courtier* (printed 1621). *The Blessed Mistress* (printed 1634), and *Don Gil of the Green Trousers* (1635). This last work is generally considered to be his best: its "hero", Don Gil, is actually a girl, Doña Diana, cross-dressed, daringly resolved to defy conventions if needs be to regain the affections of her fickle husband-to-be. Like Lope de Vega, whom he much admired and by whom he was no little influenced, Tirso de Molina was especially skilled at portraying women. It is said that, though celibate, he learned much about the opposite sex, and about fallible human nature in general, from his long hours listening to troubled penitents in the confessional. *Don Gil* is still frequently revived, almost alone of his vast outpouring.

Though comedy was his true *métier*, he ventured into other genres, historical plays such as *Prudence in the Woman* and the even more sombre *The Doubter Damned*, which asserts that faith in God and the Church is a profound mortal requirement. Despite such pious sentiments, his censorious superiors were not too pleased by his often frivolous stage-writings, which is why his scripts were published under a *nom de plume*; eventually – about 1626 – the Council of Castile elected to rebuke him for his theatrical activities. He ceased his literary endeavours, falling into lengthy silence, a retirement lasting to his seventy-seventh year.

El Burlador de Sevilla y Convidado de Piedra (*The Trickster of Seville and the Stone Guest*) is where he shrewdly introduced the iconic figure of Don Juan Tenorio, taking him from two old Spanish folk-tales, various details of which he neatly combined. The libertine Don, sexually insatiable, seeks to seduce the chaste Doña Ana. Drawing his sword, he kills her aged father, trying to prevent the unsanctioned elopement. Later he slips into the gloomy sepulchre where the dead man is entombed, mocks Don Gonzalo's sculptured effigy, laughingly invites it to dine with him. In the legendary version that originally guided Tirso de Molina, Franciscan friars slay the heartless nobleman; they cover the deed by asserting that he has been transported to the underworld for his rapacious, malicious acts, which is why he has vanished. But in a brilliant *coup de théâtre* Tirso de Molina changes the ending, injecting a strong supernatural element: the murdered man's stone effigy actually appears at the Don's dwelling, knocks ominously at the portal and hoarsely calls out his name. He has accepted the dinner invitation. Pushing into the room, the frightful guest strikes the startled, cowering aristocrat and, with a single blow of its heavy marble arm, crushes him to death. Flames leap up encircling the Don who, with his servant and cynical abettor Catalinon, is consumed in fires rising from Hell.

The critical consensus is that Tirso de Molina wrote better plays than *The Trickster of Seville*. His scope is wide, his work not without power, his fancy tirelessly fertile. His dialogue is witty, ironic. He has sophisticated insight. He has a unique ability to create the comic servant, whom the Spanish of the period dubbed the *gracioso*, always fresh and with endless variety, and each one certain to be truly amusing. Like Lope de Vega, he boldly disregarded the classical unities, deeming them hindrances.

But it is almost solely for having been the first to recognize the dramatic appeal of Don Juan Tenorio as a character that Tirso de Molina is now remembered. The sly, insinuating Don, for ever seeking conquests, compulsively ruining women, soon became an archetype: he inhabits a procession of plays after Tirso de Molina's. Down the years he is portrayed over and over in vibrant scripts and poems by Molière, Pushkin, Rostand and George Bernard Shaw, as well as in Da Ponte's libretto for Mozart's great opera, and is conjured up in Richard Strauss's stormy tone-poem. In the eighteenth century Gluck composed incidental music for Molière's play, and Gaspero Angiolini used it for a ballet on the Don Juan theme, as did Fokine early in the twentieth; followed by Frederick Ashton, who sets his choreography to Strauss's "cyclonic" and "poignant" score. (Don Juan's name has become attached to a description of the neurotic pursuit of women, who are quickly betrayed and abandoned, a pattern of behaviour sometimes seen in men. Sigmund Freud ascribes it to the frustration felt by those who do not openly acknowledge to themselves that they are, in fact, homosexual.)

Lope de Vega has really only one peer in the theatre of Spain's Golden Age, and some would rank him higher than the "monster" for artistry and literary quality. He is also the dramatist who brought down the curtain on this fabulous era: Pedro Calderón de la Barca (1600–81). His talent differs significantly from Lope's, for he is notably refined and sensitive, inclined to the philosophical. He spent much time polishing and perfecting his work, and apparently sought the approval of small aristocratic coteries rather than that of mass audiences. Early in his career he was praised by Lope for his excellence as a poet.

A son of Madrid, he was born to a father of patrician Castilian stock and a mother of Flemish ancestry. The father, Diego, had a position at court as secretary to the Treasury. The boy early manifested a high intelligence. At eight he was enrolled at a Jesuit school, and exposure to discipline that left a durable imprint. His mother died when he was ten, and in time Diego remarried, only to die the following year (1615); his death led to a harsh dispute over the division of his estate between the stepmother, Pedro's two brothers and his sister, with an eventual monetary settlement that left the children an ample provision. This allowed Pedro to enter the University of Alcalá de Henares, and soon afterwards move on to the University of Salamanca in his fifteenth year; he chose to study canon law and theology, while he also involved himself writing verse and plays. He is said to have completed his first stage-script, *El Carro del Cielo* (*The Chariot of Heaven*), when he was thirteen, but no trace of this precocious offering remains. Around his twentieth year he decided that his true bent was literary and he abruptly quit his other preoccupations. Before two years passed he won a verse competition given in honour of St Isidore, patron saint of Madrid – it was this accomplishment that evoked Lope's generous salute to him.

Biographers believe that about this time he collaborated with two other authors, Juan Perez de Montalbán and Antonio Coello, on a commercial script. On his own, the first work of his to appear was *Amor, Honor y Poder* (*Love, Honour and Power*; 1623), based on a story by Matteo Bandello and produced in Madrid.

His life took a turn, however, two years later when he voluntarily embarked on adventurous intermittent service with Spanish troops in Italy and Flanders, as well as later and nearer to home in Catalonia when that stubborn province revolted against Castilian rule. He suffered a minor wound to his hand. He is thought to have been present at the important surrender at Breda, an event he records in his drama *The Siege of Breda* (1625), written soon after the battle ended.

Between these campaigns he was in and out of Madrid, where he had begun the life of a typical young blade of the day, indulging in amorous affairs and even attracting scandal. His brother Diego was wounded by an actor in an encounter. In pursuit of the attacker, who took refuge in a convent, Pedro burst into the sanctuary in order to exact due reprisal. The nuns asserted that they had been "molested", and Pedro was subjected to a public reproof by the court preacher. He responded to this by ably if succinctly parodying the sermonizer's bombastic manner of speech in his subsequent *El Príncipe Constante* (*The Constant Prince*; 1629).

Of upper-class birth and upbringing, and well educated, he fitted in easily with the nobility, several of whom became his patrons; he was attached to a succession of patrician households. He had spurts of creativity that were exceedingly productive. Though he hardly matched Lope de Vega's output, he is credited with having written about three hundred plays of varied categories. Of these, there remain one hundred *comedias*, eighty *autos sacramentales*, twenty *entremeses*, as well as *zaruzelas* – a form of musical theatre that Calderón did not himself originate but elaborated. It may be said that his best works have not dated; a goodly number of them are still revived by repertory companies, especially in Europe but also worldwide.

He began with cape-and-sword melodramas, then so dear to audiences, heated romances filled with exciting physical action, provoked by tyrannical fathers, brothers jealous of the family honour and a sister's good repute, virtuous, ingenious young ladies determined to have the mates of their choice whatever the obstacles, well-intentioned couples at cross-purposes and thwarted by a complex of misunderstandings and calling for ever more desperate intrigues, happily resolved at the last minute, with love and morality triumphant. Of such plays, of which he fashioned a great many, outstanding examples are *La Dama Duende* (*The Phantom Lady*; 1629) and *Casa con Dos Puertas, Mala Es da Guarda* (*A House with Two Doors Is Difficult to Guard*; 1629). His works in this genre are far superior to most of those by his professional contemporaries. He developed the form to its height.

He was already presenting attractive comedies, too, displaying an innate gift for them. For instance, a vein of fun runs through *The Phantom Lady*, along with its melodrama. A capricious young widow, Angela, shares a house in Madrid with her two brothers. Intent on preserving her chastity and the family's good name, they tell no one where she lives. Bored at having to seem prim, she steals out into the town to attend entertainments taking place at the palace grounds. She is disguised, and her brother Luis does not recognize her; he merely sights a comely young lady and, on the prowl with amorous intent, trails her. Affrighted, Angela appeals to a passing stranger for succour. This precipitates a duel between the "rescuer" and the impudent Luis, which is halted by the arrival of the other brother, Juan. Angela has fled, still not identified. To the mutual surprise of the three men, the chivalrous stranger is discovered to be Don Manuel, an old friend, and he is invited to stay in the brothers' dwelling. His room there is adjacent to Angela's and connected to it by a door hidden by a false wall of mirrors. Curious about this handsome, brave young man, Angela slips into his room when it is empty. She mischievously leaves tokens there of her entries, which bewilder his servant Cosme, who is superstitiously persuaded that a goblin is responsible for them. As the situation develops, prompting arguments and concerns about propriety and honour, the brothers' friendship with Don Manuel is seriously strained. Ultimately, Angela and Manuel are to marry, as are Juan and a friend of Angela's. Left disconsolate is the brash Luis, for whom no nubile young lady is on hand at this time. (A lively German revival of *The Phantom Lady* won plaudits in war-ruined Berlin in 1948, where Max Reinhardt had also staged a version of the play some years before.)

Another group of plays have to do with historical subjects, beyond *The Siege of Breda*. Some are accurate, some take necessary liberties with fact and some are wholly drawn from his imagination but convince the trusting spectator that they actually occurred as Calderón describes them. In *The Constant Prince*, given a semi-Oriental setting near Tangiers in North Africa, the Moorish defenders defeat a Portuguese army and capture its leader, Prince Ferdinand. The King of Fez demands that his foes yield the city of Ceuta as ransom, but Ferdinand adamantly refuses to let a Christian city surrender to the infidels. Furthermore, his own father had previously conquered it. To break his spirit, the prince is forced into slavery, starved, ordered to do taxing menial labour. By the time the Portuguese king arrives with his troops to free the suffering prisoner, the emaciated Ferdinand has died, but his indomitable spirit leads the Portuguese avengers to a decisive victory. To punish the Moors, the King of Fez's daughter is seized, and he must yield Ferdinand's body in order to win her release. This drama derives from a popular legend.

The Mayor of Zalamea (1640–4) is unusual for its day because its focal figure is a proud, strong-willed peasant, Pedro Crespo. His daughter is ravished by the captain of a troop of soldiers lodged in the village. During the violation of the hapless girl, the anguished Crespo is tied to a tree in the forest. The arrogant captain is of noble birth. When Crespo demands that the rapist be punished to satisfy a father's code of honour, he is laughed at, regarded as foolishly pretentious. The villagers elect Crespo to be their *alcalde* – chief magistrate – an indication of their sympathy with his cause and anger at the outrage. He orders the captain to be caught and brought to trial before him. The prisoner's superior officer demands that the bound man be set free in the king's name. Crespo defies the request. The king himself reaches the village and asks for the miscreant's liberty, but it is too late. After holding a trial, Crespo has decreed that the captain be garroted; he has the culprit's corpse delivered to the sovereign. The king is impressed by this peasant's courage, his determined pursuit of justice. He declares that Crespo shall be *alcalde* here for life. This work, also frequently revived, is believed to have been inspired by an actual incident. It is notable for Calderón's striking portrait of the bold, independent Crespo. (Critics hear in it echoes of Lope de Vega's *Fuente Ovejuna*.) The picture of village life is "spirited", truly evocative.

Calderón's works are most distinguished and for his time almost unique when they embrace philosophical themes. Foremost among these, and perhaps the script for which he is most renowned, is *La Vida Es Sueño* (*Life Is a Dream*; 1631–2); among its many facets it raises questions about how one should wield power and govern a realm, and what is real, what is illusory, and to what degree does a man have free will, if at all. By astrology, Basilio, King of Poland, learns that his new-born son Segismundo, who will succeed him, will be a despot. To prevent this, he has him raised in solitary confinement, pent in a mountain-top tower, seeing no one but his gaoler and counsellor, Clotaldo. At twenty, and of royal birth, the young prisoner is half-beast and half-man. Gazing from a narrow window, he envies the animals whom he beholds roaming freely.

Ageing, King Basilio worries about the future of his subjects. He decides to bring Segismundo to the castle and find out more about him. By exerting his will, can he overcome the nature given him by fate and the heavenly stars before his conception? Drugged, the comatose Segismundo is brought to his father's court. If he learns to control himself, curb his impulses and passions, he will inevitably reign.

Awaking from his sleep, Segismundo is informed of his regal status as heir to the throne. But, having been raised as little better than a beast, he continues to act like one. Irate, he tries to kill Clotaldo, hurls a servant from a palace window. A humble young lady, Rosaura, has come to court to seek the Duke of Muscovy, who has seduced her. The prince seizes hold of her, and only with force is she freed from his grasp. It is sadly clear to King Basilio that Segismundo should never hold the sceptre.

Once again the young man is drugged. Aroused in his cell, he is told that he has never left it –

the episode at his father's court has merely occurred to him in a dream. But he still has vivid memories of the court. He is deeply confused. Is his life now in this prison a dream, and were his experiences in the castle actual; or is the contrary true? Which is it? Where is he in reality? What is real? Will he ever know? Will he awake to find himself in the castle again?

It does happen. King Basilio chooses the Duke of Muscovy as heir to the throne, but his soldiers revolt against ever having a Russian ruler, a foreigner. They retrieve Segismundo and hail him as their rightful sovereign-to-be. Tempted to punish his father, the prince has finally learned enough to practise restraint. He reproves Basilio, but then kneels before him and beseeches his forgiveness. Next he moves to dispense due justice: he requires the Duke of Muscovy to marry Rosaura, whom he has betrayed – she proves to be Clotaldo's illegitimate daughter. For himself, Segismundo takes the hand of Princess Estrella, who shall be his queen. Though he should now be completely happy, he is not. Is this once again a dream? Will he awake to find himself in his tower cell as before?

This pre-Pirandellian fantasy has evoked many interpretations. Calderón is seemingly explicit yet teasingly ambiguous. He has King Basilio say to Clotaldo:

> In this world,
> All who live are only dreaming.

This casts doubt on all certainty, the solidity and tangible proof of human existence, the worth of mankind's every hope and aspiration.

But this does not lead to utter nihilism. Afterwards, Clotaldo tells Segismundo:

> Even in dreams, I warn you
> Nothing is lost by trying to do good.

Prince Segismundo has gained enough practical wisdom to agree with this:

> That's true, and therefore let us subjugate
> The bestial side, this fury and ambition,
> Against the time when we may dream once more,
> As certainly we shall, for this strange world
> Is such that but to live here is to dream.
> The king dreams he's a king and in this fiction
> Lives, rules, administers with royal pomp.
> Yet all the borrowed praises that he earns
> Are written in the wind, and he is changed
> (How sad a fate!) by death to dust and ashes.

And:

> The rich man dreams his wealth which is his care
> And woe. The poor man dreams his sufferings.
> He dreams who thrives and prospers in this life.
> He dreams who toils and thrives. He dreams who injures,
> Offends, and insults. So that in this world
> Everyone dreams the thing he is, though no one
> Can understand it. I dream I am here,
> Chained in these fetters. Yet I dreamed just now
> I was in a more flattering, lofty station.
> What is this life? A frenzy, an illusion,
> A shadow, a delirium . . .
> The greatest good's but little, and this life
> Is but a dream, and dreams are only dreams.
> [Translation: Roy Campbell]

At the end, he resolves that he wants "no more feigned majesty, fantastic display, nor void illusions, that one gust can scatter like the almond tree in flower". He promises to rule mercifully, justly.

One sees prefigured here Calderón's own eventual withdrawal from the everyday world into the priesthood and other-worldly shadows of the Church; but that decision was not reached until two more decades had elapsed.

Life Is a Dream is not unflawed. In stretches it is overly didactic (but stage directors are never hesitant to cut lines, so that need not be an irremediable fault). The sub-plot, which probes the relationship between the Duke of Muscovy and his former mistress Rosaura, brings in the age-old, over-familiar changeling theme, but does yield several very suspenseful scenes. Segismundo's change of heart and arrival at mature wisdom are very sudden, though plausibly motivated; still, they are not fully persuasive. But this *is* a fantasy.

Gassner sums up the reservations shared by many about this strange and highly original play. He balances them this way: "If *Life Is a Dream* seems too given to moralizing and is not wholly convincing in its psychology, it is nevertheless to be accounted as ingeniously contrived and following a logic of its own. Its subdued speculation is even supported by modern psychiatry; a dream, psychoanalysts assure us, can provide a discharge for destructive impulses. Segismundo's drama, moreover, is to be gauged not by its adherence to strict probability and psychological motivation, but by its power of suggestion and its reflective fantasy. Calderón's verse, besides, is altogether equal to the demands of this philosophical fable; it is marred only by passages that followed the fashionable affectations of language known as euphuism in England and Gongorism in Spain."

Much of the poetry is of a high order, indeed. Spoken aloud it rejoices the ear. In addition, some strands of humour are woven through it. *Life Is a Dream* has proved endlessly attractive to later generations and is probably the most admired work in Calderón's large and varied canon.

El Médico de su Honor (*The Physician of His Honour*; 1635) is another play of considerable moral depth, one of a series in which the central characters are beset by the problem of what is their right course of action. The dilemmas arise from the rigid Spanish code of honour, especially in marital relationships. Don Gutierre Alfonso de Solis, pledged to marry Doña Leonor, withdraws his promise and marries Doña Mencia instead. Though reluctant, she accepts his offer in compliance with her father's wishes. In truth, she is in love with Prince Enrique, the king's brother, away on military service. When the prince returns he obsessively renews his suit, but she virtuously rejects him. He is so persistent that Don Gutierre's suspicions are stoked; he resolves on surveillance. In disguise, he attempts to conceal himself in her bedchamber. In the darkened room she hears a sound and mistakes the intruder for Prince Enrique – who else would be so audacious? She cries out, "Your Highness." Her husband is now certain of her infidelity, though he is quite wrong. How shall he punish her and adhere to his code, yet not evoke a public scandal? In secret he will be "the surgeon of his own honour".

A "bloodletter" is seized, blindfolded and brought to the house. Under duress, a threat to his life if he disobeys, the surgeon bleeds Doña Mencia until she weakens, collapses and dies. Afterwards Don Gutierre tells people that his wife's untimely death was caused by an accident.

The surgeon, troubled, has reported the gruesome incident to the king. He is able to identify the house to which he was taken, having left a clue, the bloody imprint of his hand, on the door; shortly afterwards he returned to take note of it. Don Gutierre is summoned to court. The king, hearing the "betrayed" husband's explanation, pardons him. Their revealing exchange goes this way:

> THE KING: Cleanse straight your doors,
>   A bloody hand is on them.
> DON GUTIERRE: My Lord, when men
>   In any business and its duties deal,
>   They place their arms escutcheoned on their doors.
>   I deal, My Lord, in honour, and so place
>   A bloody hand upon my door to mark
>   My honour is my blood made good.

He is commanded to marry Doña Leonor, to whom he had formerly been betrothed and to whom he has a moral obligation. He asks what he should do if this lady, too, should be unfaithful to him with the same importunate royal brother. The king replies that in such an event murder is indeed justified. Warning Doña Leonor of the risk, Don Gutierre takes the lady's hand, but she – confident of her future fidelity – is unafraid and happily accepts him.

This play has been compared – and unfavourably – to Shakespeare's *Othello*, but perhaps that is unfair. It is not, like Shakespeare's work, a psychological study of one beset man, but rather a reflection of the mores of Calderón's day and place – English society held very different values from those prevailing in intensely Catholic Spain, and Don Gutierre has hardly the complex motivations and emotional tensions of the Moor, in whom are half-hidden insecurities arising from an awareness of racial differences.

Quite similar in plot and development is *A Secreto Agravio, Secreta Venganza* (*Secret Vengeance for a Secret Insult*, also 1635). The heroine, again named Doña Leonor, and Castilian, assents to a marriage by proxy to a Portuguese nobleman, Don Lope de Almeida. She mistakenly believes that her true love, Don Luis, has been slain in battle. But reaching Portugal she discovers that Don Luis is very much alive and passionate to reclaim her. While her husband is absent, she meets Don Luis and tries to persuade him to leave Portugal. Don Lope has grown suspicious. He comes back unexpectedly and discovers the unhappy lovers together; he misinterprets what has been taking place. Greatly disturbed, for after his own fashion he, too, loves Doña Leonor, he acts on a friend's advice and feigns belief in Don Luis's explanation of why the pair were closeted. Don Lope yields to the guidance of the established code of honour but feels that to avoid open scandal he must work in secret. Leonor realizes that she must dismiss Luis from her life, but before she can accomplish this Luis is drowned in what appears to be a fatal mishap – actually the seeming "accident" has been arranged by the vengeful Lope. He stabs Leonor to death and sets his house ablaze to conceal his deed. Her end, too, should appear to have been due to a dreadful mischance. Apprised of the truth and the reason for Don Lope's deeds, the king expresses his approval of them.

Calderón continued address with this theme for many years. Another instance is *El Pintor de su Deshonra* (*The Painter of His Own Dishonour*; c. 1648–50) – such dramas came to be known as *comedias de pundonor*. Here a highly talented artist, Don Juan Roca of Barcelona, is wedded to Serafina, his much younger cousin. He is deeply enamoured of her, but she does not return his feelings. She has accepted his offer of marriage only after the death of Don Alvaro, to whom she was secretly engaged. The report of his demise is false. On a visit to the Governor of Naples, who is Don Alvaro's father, she suddenly comes face to face with Don Alvaro himself, who has survived a shipwreck. He is accompanied by the Prince of Ursino who promptly falls in love with Serafina, too. To Porcia, her intimate friend, Serafina has previously confided that she has never been able to forget or put aside her desire for the supposedly lost Alvaro. Now she is overjoyed and overcome. He misreads her tears and display of emotion as signs that she might still be his.

Returning home, Serafina reconciles herself to continuing as Juan's dutiful wife. But Alvaro has followed her there. While Juan is away, Alvaro pays her overt attention. He barely avoids being caught doing so by Juan, who suddenly returns. At a carnival party, Alvaro in disguise dances with Serafina, but she will not hearken to his erotic pleading. The party is interrupted by a fire. In the excitement Alvaro rescues Serafina and carries her off to his family's country villa. The Prince of Ursino accidentally encounters her there. His interest in Serafina flares up again. He conceives of having an artist furtively paint her portrait. By ill-chance, the man he commissions for this task is none other than Don Juan Roca, who is instructed to hide in the bushes and study a young woman as a subject. In a place of leafy concealment, Don Juan is startled to recognize his vanished wife and watches as Alvaro embraces and kisses her. Drawing a weapon, the infuriated Don Juan fires and kills both of them. Pointing to the blood-spattered corpses, he describes them as offering "a painting of his own dishonour". The grieving fathers of the murdered couple voice no rancour against a man "who defends his honour". The code is inflexible.

Another play along these lines is the earlier *No Monster Like Jealousy* (1634), which treats with the Biblical story of Herod and Mariamne. Determined that his queen shall be faithful to him even beyond his death, Herod plans to have her die, too. All these dramas of possessive lovers are written with sombre power. What is not clear is whether Calderón himself endorsed this adamant and violent code or was presenting images of its workings to stress how insanely costly it was in terms of human life, and how cruel. Most of the young women who are accused in these dramas are in fact innocent.

The death of the indefatigable Lope de Vega in 1635 left vacant the post of director of entertainments at the court. Such was the prestige of the comparatively young Calderón by now that Philip IV appointed him to fill the all-important post. The very next year he was further rewarded by being made a Knight of Santiago. About this time the new Buen Retiro Palace was near completion, an oncoming occasion calling for a gala welcoming party, for which Calderón was ready to contribute *El Mayor Encanto, Amor* (*Love, the Greatest Enchantment*), which eventually was spectacularly enacted on a floating stage in the lake, with the delighted king and his entourage watching it from the shore. This *comedia mitológica* is concerned with the prolonged dalliance between Circe and Ulysses and his stranded crew, all of whom have been ensnared by the siren on their Homeric storm-tossed journey from Troy to their distant home in Ithaca. Ruefully, because at last they must escape, they outwit her. In her effort to entrap them again by her wiles, she and her magical island are brought to ruin.

Calderón was working on this script when, in 1640, as a Knight of Santiago, he was called to arms to participate in the expedition against rebellious Catalonia. Wishing to have the play grace the gala, Philip IV forbade his court poet's departure. Deeply loyal and patriotic, Calderón took only a week to dash off an ending and hurriedly left to assume his place in dangerous combat.

A superb poet, he was well equipped to conceive and write fairy-tales, pastorals and works based on classical mythology – some of them *pièces d'occasion* – and turned out a good many of them. But with time tastes change, and such offerings are of little interest to audiences today. That is also true of his short *autos sacramentales*; they comprise the largest share of his output after he was named court poet, for it was a very religious court, a special audience that he had to please. At least eighty of these pious one-act scripts are still on hand. It was not burdensome for him to dedicate his talents to them: he was drawn to the Church, had delved into theology while in college and had already written a large number of devout dramas. He is no longer famed for them, but some do represent him at his very best. Away from the palace, they would also be presented in streets and plazas and at open-air fairs, especially during the Corpus Christi observance. They had developed quite far beyond the simple medieval Mysteries from which they came. Lope de Vega had helped them to become complex allegories and to be far more elaborately staged, sponsored by civic administrations at great expense. Calderón finally perfected this form of didactic theatre. His playlets have lyrical dialogue and are peopled with ably drawn allegorical figures such as Faith, Folly, Humility, Pride, Charity, Arrogance; he preaches by using aptly chosen symbols to explicate Roman Catholic dogma. He appropriates incidents from the Old and New Testaments, legends of martyrs and saints, adumbrates historical events, and wherever possible links them to the transforming Eucharist, striving to decipher its ultimate significance. His will-to-believe was deeper and stronger than that of Lope.

Though the day of this sort of play is largely gone, several of Calderón's early and more extended offerings in the genre are occasionally brought to the stage. *La Devoción de la Cruz* (*Devotion to the Cross*; *c*. 1633) is one. Categorized as a *comedia de santos*, it has Eusebio in love with Julia but unable to marry her because he is not of noble birth. He continues to court her, which leads to his being challenged to a duel by Lisardo, her vigilant brother. Eusebio tries to avoid the encounter, revealing that he had been found, an abandoned child, at the foot of a cross and bears on his flesh a symbolic image of it. Because of his devotion to the symbol and what it represents, his life has been miraculously spared on several occasions when he was in great peril. Consequently he deems himself to be not ordinary but someone of special rank and merit, inherently noble. Lisardo dismisses this claim, insists on proceeding with the duel and is fatally wounded. In secret, Eusebio gets in touch with Julia, but is forced to conceal himself as her father, Curcio, approaches.

Curcio takes Julia to task, angrily likening her conduct to that of her late mother. He relates how once, in a jealous rage, he had struck his wife, after which she had borne Julia and a long-lost brother at the root of a cross.

Lisardo's corpse is carried in; Julia now hears that he has been slain by Eusebio. Overwhelmed by shock and grief, she tells her lover that they can never be united or see each other again. Her father, blaming her, orders her to enter a convent.

A fugitive, Eusebio resorts to banditry. In one incident he rescues a priest from grave danger; grateful, the man of God promises that, if summoned, he will hear Eusebio's confession wherever he might be at the moment of death.

Julia's image still haunts Eusebio. Finally he breaks into the convent, intent on ravishing her. Beholding the crucifix that hangs at her breast, he is horrified, recoils and flees.

Julia feels she must find him. She dons male garb as a disguise and joins his band of outlaws.

Guilt-stricken, Eusebio begs her to return to her shelter in the convent. Searching for the pair, Curcio discovers them and inflicts a mortal blow on Eusebio, then sees the cross on his bared chest: this must be his long-lost son!

True to his vow, the priest arrives in time to shrive the expiring Eusebio, who miraculously is able to confess his sins even though by now he has been considered quite lifeless.

Told that Eusebio is actually her brother, Julia also confesses her faults. The unforgiving Curcio is about to strike her, but she takes flight from him to the base of a cross marking Eusebio's fresh grave. Another miracle occurs: she and the cross are wafted to Heaven.

Soon afterwards Calderón wrote *La Cena del Rey Baltasar* (*Belshazzar's Feast*; 1634), dramatizing the Biblical story. God sends the loud-voiced, irate prophet Daniel to the hedonistic Babylonian potentate with a warning to repent of his having wed both Idolatry and Vanity. Though the message is given three times over, and Belshazzar half-heartedly tries to heed it, he repeatedly falls short of doing so. At a feast given by the sinful king, Daniel is among the guests. He comes with Death, who purports to be an attendant, and who passes a poisoned drink to the besotted ruler. It is meant to kill his soul, after which Death lifts a knife to dispatch his victim physically. Though Belshazzar frantically grapples with Death, his struggle is futile. As he is dragged away he calls on Idolatry and Vanity to help him, but they have no strength and merely look on.

Quite fanciful is *El Purgatorio de San Patricio* (*The Purgatory of St Patrick*; *c.* 1634), in which Calderón goes far afield for his subject and setting. Shipwrecked on the Irish coast, Patrick and his evil companion Ludovico are seized by the natives, who are pagans. King Egorio has the two imprisoned, but his daughter Polonia is captivated by Patrick and helps to have him freed. He escapes back to Rome, but soon returns to Ireland to convert its un-Christian ruler and his subjects. Meanwhile Polonia has been seduced and killed by Ludovico. Learning of this, Patrick resurrects her, a miracle that leads the wicked Ludovico to repent and change his way of life. Though King Egorio has witnessed the cruel criminal's conversion, he is still not persuaded of the reality of life after death. An Angel leads him through both Purgatory and Hell – the latter an underground cavern – and he and his nation, too, are won over to Christianity.

Especially cherished is *El Gran Teatro del Mundo* (*The Great Theatre of the World*; *c.* 1635), its tone truly exalted. Robed in a mantle of stars, with nine rays of light emanating from His crown, God, the Divine Playwright, apostrophizes His latest creation, Earth, in effulgent phrases: "Behold the windy air where little feathered galleons do sail their singing flight." He names it the "World" and speaks to it, commanding that specimens of mankind – the actors in the play He has written – shall appear before Him. Sun and Moon are set in the sky – "without light there is no play" – to illumine the stage. The cast comes, representative of humanity, heralded by the Law of Grace: a king, a brave warrior, a churchman, a scholar, a rich man, a court lady, a peasant, a beggar, a child and so on. To each is given appropriate raiment and material properties befitting his or her station and duties: for the king, purple and laurel; for the warrior, weapons; for the peasant, rough tools. At first they are subservient – says the king: "Mindful of our obedience, though yet unborn, we come to await Thy pleasure; for in our Sovereign Author's mind we have existed for ever, though never given form and breath. We are but formless dust beneath Thy feet. Breathe Thou upon this dust, and we shall live and set the parts assigned to us in the Great Theatre of This World." God tells them, however, that when the drama is ended He will reward each according to his merits. "So, Beggar, soften thy affliction for the while," He says to the ragged, half-naked cripple, "for I have willed that if the Beggar plays his part effectively and if the king shall do less well, there shall the Beggar have the prize. . . . We do not care what part a man has played; We only care how well he played it." And the title of the play, He reveals to them, is "Do Good, for God is God."

The Law of Grace serves as prompter: "Ever wakeful, a luminous and everlasting light, it will guide your faltering steps." At this point, the play within the play begins:

> While music plays, two globes open up. Within one is the throne of glory, where sits the Author. In the other there is a stage with two doors – and on one is painted a cradle, and on the other a coffin.
>
> [Translation: M.H. Singleton]

One by one the players enter from "birth" and, having enacted their parts, exit to "death", summoned by a singing Voice. What is unfolded is a more mature and sophisticated *Everyman*: each departing spirit is forced to leave behind the "properties" lent him for his role, the king his magnificent robes, the court beauty her garlands and youthful loveliness, the rich man his treasure, the peasant his spade.

Many of the spoken lines are sharply edged, critical of human folly and pretension. The pageant ended, the two globes, earthly and celestial, close. The actors are recompensed, each as he or she deserves. Chosen as worthy, a few sup with the Lord. Others are consigned to Purgatory, or, like the proud and vainglorious rich man, to the Infernal Pit. A still-born child goes to Limbo, where it will feel neither pain nor glory. (This play, too, has had a twentieth-century reincarnation under the auspices of the German producer-director Max Reinhardt.)

Yet another much-prized *comedia de santos* by Calderón is *El Magico Prodigioso* (*The Wonder-Working Magician*; 1637), which is somewhat reminiscent of Christopher Marlowe's *Dr Faustus* – that English play was written three decades earlier (1604), but it is unlikely that Calderón knew of it, and the theme had been a frequent favourite in the Middle Ages, which for both authors was still the comparatively recent past.

In Antioch, Cyprian, a reclusive and celibate philosophy student who rejects Christianity, catches the eye of the Devil and, as they walk together conversing, engages him in a theological debate: "What is God's essential nature?" From the start, Cyprian is suspicious of this well-garbed, articulate stranger.

> 'Tis singular that even within the sight
> Of the high towers of Antioch you could lose
> Your way. Of all the avenues and green paths
> Of this wild wood there is not one but leads,
> As to its centre, to the walls of Antioch;
> Take what you will, you cannot miss your road.
> [Translation: Percy Bysshe Shelley]

But the Devil is not easily put off and replies evasively, while further acting ingratiatingly and insinuating himself into Cyprian's thoughts.

In answer to the subject of their exchange, the Devil argues that God is merely a reflection of man's own nature and a deification of natural forces.

Though not a Christian, and living in Muslim Antioch, Cyprian holds a different view, derived from his reading and meditation:

> . . . There must be a mighty God
> Of supreme goodness, and of highest grace
> All sight, all hands, all truth, infallible.
> Without an equal and without a rival,
> The cause of all things and the effect of nothing,
> One power, one will, one substance, and one essence.
> And, in whatever persons, one or two,
> His attributes may be distinguished, one
> Sovereign power, one solitary essence,
> The cause of all cause.

This is hardly what the Devil wants to hear. Cyprian obviously feels that he has arrived at a better concept of what the Deity must in fact be. But he has come to this conclusion only as a consequence of his reliance on books and metaphysical speculations, which assures the Devil that he can score against him, and makes him ever more eager to do so. (Cyprian's reasonings are much too difficult and abstruse for most audiences to follow and comprehend.)

Residing among the Muslim infidels in Antioch is Justina, a young girl renowned for her chaste Christian life. The Devil is determined to seduce her and attempts to climb to the balcony of her house. Two of her suitors, Lelius and Florus, observe his effort. Mistakenly each thinks that the intruder is his rival, which angrily leads to a duel. The Devil sees his opportunity: he persuades Cyprian to step in between the combatants to mediate the quarrel and stop it.

Cyprian catches a glimpse of Justina and, at the Devil's prompting, is instantly infatuated. Shy, and convinced that the Devil is a magician, he beseeches his help in winning this beautiful young woman. The Devil sets his price: Cyprian's soul. The now vulnerable Cyprian assents. The Faustian bargain is sealed.

Having prevailed over Cyprian, the Devil next approaches Justina, but the purity of her nature defeats him – none of his wiles and spells ever induces her to visit Cyprian's room, where the corrupted young man impatiently awaits her. At last the frustrated Devil sends a phantom in her place. Though the figure looks like Justina, when Cyprian embraces it he is shocked to find himself holding nothing but a skeleton.

Seeking solace, Cyprian finally discovers truth and consolation in Christianity, the faith that has fortified the steadfast Justina. A year later he learns that she has been imprisoned as a non-Muslim. He, too, is brought before the Governor of Antioch and boldly defends Christian doctrine. The Governor orders the execution of both Cyprian and Justina as non-believers, and they are beheaded. An irate God commands the Devil to restore Justina's good reputation and proclaim that the defiant Cyprian's sacrifice testifies to his redemption. (He is later sanctified and becomes St Cyprian.)

El Gran Principe de Fez (*The Great Prince of Fez*; 1668) is a later example of these plays dealing with men and women wrestling with problems of faith, and with supernatural interventions and manifestations in human affairs. Reading the Koran the prince, Muley Mahomet, comes upon phrases that instil serious doubts in his mind. He turns to his teacher, Cide Hamet, to elucidate them but is not satisfied with his response. During the young prince's sleep an Evil Genie and a Good Genie launch a contest over who shall dominate the royal heir. On the next day, in combat with the King of Morocco, the prince tumbles from his horse and lies stunned. His resourceful wife, Zara, enters the fray and takes charge, achieving victory over the Moroccan forces. Recovering from the injuries sustained in his disastrous fall, the grateful prince sets out on a pilgrimage to Mecca. *En route* he is captured by the Christian warrior Don Baltasar and carried off to Malta. On that island-stronghold the prince finds and reads an exciting and enthralling book in Don Baltasar's library – a biographical summary of St Ignatius of Loyola who established the Order of Jesuits, the militant defenders of Christianity.

After his wife, Zara, ransoms him, the prince resumes his sea-voyage to Mecca. During a storm he beholds a vision of the Virgin Mary, who tells him to return to Malta. He does so, and there he accepts baptism. He alters his destination, sailing for Rome instead.

Apprised of his conversion, Zara is enraged. The King of Morocco lays suit to her, hoping to replace the prince and with her help rule Fez. Cide Hamet, feeling humiliated, sends his former pupil a poisoned bouquet, but this attempt to kill the prince is unsuccessful. At the end, the Good Genie has triumphed and the saintly prince attains canonization, his truly earned apotheosis.

Much later Calderón rewrote a number of his secular plays as *autos sacramentales* – among them *Life Is a Dream* – metamorphosing them into allegories, replacing their real-life characters with abstractions, such as Man, Free Will, Grace.

If his cape-and-sword melodramas are preferred by some critics to those by Lope de Vega, it is chiefly because of his more meticulous plotting and determinedly neat craftsmanship. But a philosophical element is nearly always present and lends depth to the characters and action. Among his darker works of this kind is *Amor Después de la Muerte* (*Love After Death*; 1633); it has for its background an insurrection by Moors remaining in Spain after the expulsion of their forefathers by the forces of Ferdinand and Isabella. Some have adapted to their changed circumstances and become "New Christians", but they are still persecuted by harsh royal edicts that seek to extirpate the last traces of the North African invaders and lingering hints of their culture. Goaded too far, they revolt. They retreat from their stronghold, Granada, to the seemingly impregnable Alpujarra range of mountains. A Spanish army, under Don John of Austria, marches against them. At a bloody cost the revolt is finally put down, and the play ends with the grant of an amnesty.

The salient characters are Maltecca, a Moorish girl of great beauty, wealth – in jewels – and courage, and her lover, Don Alvaro Tuzaní, a "New Christian", who seeks to avenge her death at the hands of a Spanish looter.

Over the corpse of his Maltecca, Alvaro swears:

> That the fire that desolates this region
> May see it, and the rolling world may know,
> And all the winds may trumpet it that blow,
> That Fate may help, that Heaven may allow it
> That men, beasts, birds, fish, sun, stars, moon, air, fire,
> Take cognizance of, publish, and resound
> That in a Moorish breast can still be found
> A Moorish heart that's loyal, firm, and sound,
> And, in that heart, a love that outlives death –

A secondary storyline follows the hopeless love between Alvaro's sister, Isabel, and a merciful Spanish courtier, Don John of Mendoza. The plot takes sudden twists and turns that are a tribute to Calderón's ingenuity, as one plausible coincidence is piled upon another suspensefully. Of interest, the Moors are shown with great sympathy by this profoundly Christian dramatist. They are noble, brave – and their revolt is so well motivated that the spectator is likely to side with them and wish them victorious. Humour, some of it coarse or appropriately robust, runs through the play, providing much-needed relief, though at times it is a bit disconcerting. The poetry is exquisite. The atmosphere is richly exotic, especially in the feeling of a Moorish *milieu*. The drama is thoroughly romantic, seeming more like one written in the early nineteenth century than of the mid seventeenth.

Calderón is singularly adept in a semi-Oriental ambience, as here and in the opening scene of

The Constant Prince laid in the gardens of the King of Fez that echo with haunting birdsongs.

Those who are turned off by such dark fare as *Love After Death* will find pleasure in a raft of light comedies that he offered at the very same time or in the following decade – such works were decidedly his forte, too, and showing his amazing versatility. A representative example of such pieces is *Guárdate del Agua Mansa* (*Beware of Still Water*; 1649). A widower, Don Alonso, has two daughters, Clara and Eugenia, who have quite different personalities: Clara is serious and very quiet; Eugenia loquacious and worldly. Their father wishes to get a husband for at least one of them and chooses a plain, rustic hidalgo, who – having a choice – finally selects the younger, Eugenia, though she is obviously flighty. Desperate, she tries to divert his interest to Clara, her habitually silent elder sister. At the same time she flirtatiously encourages the interest of Don Juan and Don Pedro, two cavaliers, to sue for her hand. A friend of the pair, Don Felix, errs in delivering a letter, giving it to Clara instead of Eugenia; it asks that she declare her preference between these two young men so as to avert trouble. Clara is fixed on learning more about Don Felix himself. She pretends that he had not really made a mistake in handing her the letter, that he knows who she is but seeks a pretext for meeting her. Eugenia, annoyed at the overly aggressive Juan and Pedro, sends both of them away. They believe that Felix is courting her on his own. Clara, still passing herself off as Eugenia, arranges a rendezvous with Don Felix; suspicious, Juan and Pedro intrude on them, and a scuffle ensues, which promptly also involves Don Alonso and Eugenia's intended, the simple-minded hidalgo (who is presented as scarcely more than a caricature). Matters calm down when Clara explains what has been her plan. Before long, Don Felix and Clara are married, as are Eugenia and Don Juan.

Among other rollicking scripts by Calderón are *La Banda y la Flor* (*The Scarf and the Flower*; c. 1631); *Antes Que Todo Es Mi Dama* (*My Lady First of All*; c. 1636,); and *El Astrólogo Fingido* (*The False Astrologer*; 1624–5). Unfortunately, space is lacking even to outline them. (Much reliance has been put on Andrés Franco for some of these synopses.) The poet's ventures at writing librettos for operas and *zarzuelas* are briefly discussed later.

More than any other of his many scripts, it is the comedies with their unforced humour that still hold the stage, not only in Spain but Germany and elsewhere.

As has often been stated, what sustains and elevates all the plays, sombre and light, is Calderón's splendid verse – he is deemed a far better poet than Lope de Vega. But much of this cannot be appreciated by those to whom he writes in a foreign language. His style is somewhat blemished by his embrace of "Gongorism", the fad for over-elaborate diction and excessive display of erudition that prevailed in England and Spain at that period, but which is hardly to anyone's liking today, when simplicity and directness – even minimalism – are highly commended as proofs of a poet's sincerity and his firm command of an "economy of means". Despite the difficulties his works present, a number of eminent writers published versions of them in their own tongues. In Britain the most iconic of Romantics, the genius of soaring phrases, Percy Bysshe Shelley (1792–1822), who aspired to be a dramatist (*The Cenci*), offered Englished scenes from

Calderón's scripts and fervently praised his *Absalom's Hair*, the tragedy of King David's errant son. Edward Fitzgerald (1809–93), best known for his rather free adaptation of *The Rubáiyát of Omar Khayyám*, brought out a collection of six of Calderón's plays in translation (1853). German-speaking audiences had access to the plays through the efforts of Franz Grillparzer (1791–1872), the Austrian dramatist who was much influenced by works from Spain's Golden Age, especially those by Lope de Vega and Calderón, the latter's *Life Is a Dream* having inspired Grillparzer's own variation on that drama's basic idea, *The Dream Is a Life* (1834). In France, Albert Camus (1913–60) dedicated his enviable talents to a translation and adaptation of *The Devotion to the Cross* (1953), produced in Paris at the Deuxième Festival d'Art Dramatique. Other translators have been cited along with quotations in this chapter.

Calderón's limitations arise from his zealous and ultra-rigid interpretation of Roman Catholic dogma, which some feel narrows his view of life by its hard-and-fast precepts and in their view do not fully take into account the complexity of human beings – his was the intolerant age of the Inquisition. That allows him to arrive at easy resolutions of his characters' dilemmas – that is, by conversion to the "true faith", or by an epiphany of divine intervention, another form of reliance on *deus ex machina*. He is also too ready to accept the fanatically strict code of honour that dominated the social order of his day and that took a dread toll.

But to many his stature is awesome. Goethe, a poet and playwright, but also a practical theatre manager, said of him: "His plays are entirely stage-worthy; there is not a trait in them that is not calculated for deliberate effect. Calderón was a man of genius who at the same time possessed a superior intelligence."

Alluding to this assessment from the most admired of German writers, Margot Berthold adds enthusiastically:

> Calderón's aristocratic origin gave its imprint to his life, his personality, and his dramatic works. He was not in need of stage mechanisms, but did not despise them. In the productions of his great *autos sacramentales* – with their ceremonial solemnity, their sublimation of matter on the one hand and their personification of abstract concepts on the other – he readily enough availed himself of the technical accessories of stage magic, without becoming dependent on them.
>
> Calderón indeed had more than great intelligence: he had the power of a superb, creative imagination through which he captured the transcendental, and "from the platform of eternity reflected life as a dream before man's awakening in God. *La vida es sueño* – Life is a Dream." Calderón saw the meaning and purpose of his own life as a service of honor to the church, the nation, the king.

Berthold quotes a German critic, Adolf Friedrich von Schack, who granted that Calderón had a strait range of themes.

But he was unmatched in the flawless precision with which the cog wheels of his plots mesh together. Their motive power is the inexhaustible device of disguises and mistaken identities that are the hallmarks of the cloak-and-dagger comedy, together with the witty little interludes known as "*lances de Calderón.*"

But beyond the fine net of intrigues in *The Phantom Lady*, the stiff code of honor of *The Mayor of Zalamea*, and the melancholy self-sacrifice of *The Constant Prince*, Calderón poured his full creative powers into the *autos sacramentales*, the theatrical celebration of man's repatriation into the divine world order. He sublimates and stylizes emotions and reduces human destiny to the underlying conception of God and man.

El gran teatro del mundo is Calderón's great metaphor for the "*maguina de los cielos,*" the divine dispensation that rules the orbits of the stars and the distribution of men's lots. . . . The work was performed in 1675 at the royal palace theatre of Buen Retiro, with Calderón personally supervising the production. The words of the play itself occasionally suggest some of the Baroque sumptuousness lavished on the scenery and costumes for this gala performance. "Fit ornaments and array. . . ."

And when we are told that the creation is set in "a garden of the loveliest design with ingenious perspectives," we can readily imagine the Baroque stage machinery springing into action to open out the view into the green palace gardens of Buen Retiro.

It is obvious that much of Calderón's appeal to Berthold is his religiosity. A different appraisal of him is that of Gassner: "[His] influence on European drama was tremendous, and it was not always good. If it encouraged thoughtfulness, it also propagated artifice. He lacked the breezy health and the rich theatricality of [Lope de Vega] and his character drawing was vitiated by the growing artificiality of Spanish life during the seventeenth century." To Gassner, in a summary: "He was a true son of Spain's rapid decline. Empire was passing from the Iberian peninsula." That change in Spain's fortunes was due to the depletion of the treasure that had been plundered from the New World, and by the expulsion of the Moors and Jews who had propelled the nation's economy. "The lessened energy of the times is reflected in Calderón's generally tame plays, and the decadence observable in his own works grows apace in that of his successors." Describing such scripts as *A Painter of His Own Dishonour* and *Secret Vengeance for a Secret Insult* as "tame" may evoke surprise in some spectators and readers, but Gassner has a deserved reputation as a discerning and judicious critic.

An anonymous contributor to *Companion to the Theatre* refers to Calderón as "one of the great poetic dramatists", and in *The Encyclopedia of World Theatre*, again in an unsigned article, he is singled out as "a supreme poet and master of dramatic situation, one of the great playwrights of the Western world".

After the Catalonian expedition, Calderón never fully recovered his health and strength; he was excused from further military duties and awarded a royal pension – it was not always paid on time, if at all, but he did not lack funds, and he entered the service of the Duke of Alba. But for him these were clouded years: Madrid's theatres were closed, owing to deaths in the ruling family, and two of his brothers were killed. A love affair with an unknown woman – possibly an actress – made him the father of a son, but she died in childbirth (1647). In 1650 or 1651 he was ordained a priest, much as were Lope de Vega and Tirso de Molina, and devoted the last three decades of his long life to the service of God – he was far more orthodox and conventional in Holy Orders than the freewheeling Lope had been. He confined his writing to turning out plays for very important occasions at court and accepted a life-long contract to provide the city of Madrid with two *autos sacramentales* annually to be given at the sacred Corpus Christi festival. Philip IV designated him chaplain of Los Reyes Nuevos in Toledo (1653). Lamenting the poet's decade-long absence from court, the king recalled him to Madrid to be his private chaplain (1663). He died at eighty-one while preparing two new *autos* for that year, *El Cordero de Isaías* (*Isaiah's Lamb*) and *Amar y Ser Amado y Divina Filotea* (*To Love and Be Loved, or The Divine Philothea*). Grown wealthy, he willed his fortune to the Church.

Whether Lope de Vega was this metaphysical poet's superior, his equal or his inferior is a question that embroils Spanish scholars and critics. But there is no doubt that elsewhere in Western Europe Calderón has been the one much more highly esteemed and had his works far more widely performed. That later writers as gifted and lofty as Shelley, Grillparzer and Camus were drawn to translate his plays is itself a special tribute. Where the hundreds of scripts of Lope de Vega, who was a more natural and energetic dramatist, provided a very rich and overflowing warehouse of plot ideas to playwrights everywhere, Calderón's plays themselves – intact, or only slightly adapted – have continued to be viable on stages not only at home but also in foreign languages far outside Spain. They are not only still read, they are still effectively acted.

He was destined to be the last great playwright of his country's dynamic Golden Age. As Spain itself began its material and political retreat, its transcendent literary era was closing down. It is frequently pointed out that Calderón's death and the final days of his nation's artistic supremacy coincided. Spain's stage was to be one of scant distinction for several centuries to follow. All the same, the theatres continued to be crowded, and actors still carried on their hazardous, flamboyant profession with old scripts and thin new ones that changed with ever-successive fashions imported from abroad.

One hundred miles south of Madrid lies the small, sun-laved city of Almagro, very prosperous and an administrative centre in the sixteenth century. Today it boasts eighty noble houses and palaces from medieval times and the *Siglo de Oro*, among them a Corral de Comedias standing just as in the Middle Ages, in the Plaza Mayor. Discovered a few decades ago, it was restored and is now used to produce plays of all origins and eras. In recent years it has been the site each July of the International Festival of Classic Theatre that gathers students and drama-lovers from every part of the world. The year 2000 was the 400th anniversary of Calderón's birth. Seminars and celebrations were held throughout Spain, and the theatre in Almagro revived ten of his works. Among them were the comedies *The Phantom Lady* and *Mornings of April and May*; their leading ladies, when interviewed, expressed pleasure and excitement at Calderón's eloquent defence of feminism, his outspoken pleas for women's rights, hardly a popular attitude in his day. Also offered during the festival was *Life Is a Dream*, Calixto Beito's English-language version that had just been seen in London and New York. Hanging over the mostly bare stage was a huge baroque mirror reflecting the confusing events below. After its visit to Almagro, *Life* was to tour throughout Spain; *The Phantom Lady* was also to be taken about Spain and then go to Panama and Mexico.

— 8 —
SPAIN: MUSIC AND DANCE

Since Naples was claimed and governed by Spain for two centuries, the rulers and intelligentsia of Madrid were well informed of what was happening in that always exciting Italian city on the bay below smoking, cloudy Mount Vesuvius. Those members of the Spanish élite who were interested in the arts quickly became aware that the Neapolitans were developing a new form of musical entertainment, dramatic works in which dialogue was sung throughout with orchestral accompaniment – a type of theatre that was enjoying great popularity and was profitable. Soon such offerings were transported to Madrid, just as was the raffish *commedia dell'arte* with which lighter examples of these operas had an affinity. During the Golden Age, along with expressing pride in its painters and novelists, Spain could boast of and was graced by a number of outstanding composers – Victoria, Guerrero, Gomez and lesser spirits – but they do not seem to have been drawn to this fresh genre, perhaps because much of the music of the day was liturgical, as might be expected in that very Roman Catholic state. Rather, it was the poet-playwrights – Lope de Vega, Tirso de Molina and Calderón – who made the most significant contribution to the newly arrived operas by providing them with librettos and helping to fix the shape of such works for all time.

Earlier, in the final decade of the fifteenth century, the guitar made its appearance in Spain, having developed from the lute and perhaps, over time, having travelled from the Orient by way of North Africa, and then having been brought over by the Moors. It was a handy instrument that almost anyone could play, and it was soon heard in streets and noisy taverns where it added its colour, percussive rhythm and emotional timbre to popular songs and folk-dancing. More cultured composers learned how to inject its vitality into their serious works, and playwrights found it an effective adjunct to scenes, heightening audience appeal. Music became an element of many stage productions. So there was a precedent for accepting operas, innovative offerings that were far more dependent on music, requiring not merely the guitar but other more harmonious and resonant instruments.

Wealthy Spanish families sent their children to Italy to study and absorb its more advanced cultural atmosphere. After all, Rome was the capital of Christianity. Returned to Spain, these young travellers were prepared to accept what Italy sent to Madrid and were already somewhat familiar with such exports. On the other hand, the Italians were impressed by the skill and daring of the leading Spanish playwrights. A lively exchange flourished between the two countries: Italian musicians and scene designers were hired to embellish fêtes at the Spanish court, while the plays of Lope de Vega and Calderón – which those same musicians and designers helped to mount and produce at Philip IV's palace – were admired and deemed superior to what the Italians were writing. As the reputation of the Spanish dramatists spread, Italian opera composers began to look to them for advice on how best to construct plots and handle passages of recitative.

Lope de Vega had already set forth his theories quite explicitly. The Aristotelian unities, so binding, should henceforth be rejected; they resulted in plays that were too artificial. Compressing the action into a single day, having it all occur in one place, put an unbearable strain on plausibility, and restricting the number of sub-plots impoverished the story-line and intrigue, the added suspense that arose when many characters were simultaneously confused and at cross-purposes. Lope, in *New Art for Writing Plays in Our Time* (1609), had expounded these ideas and suggested guidelines for his younger contemporaries. His chief emphasis was on what succeeded in the commercial theatre where he was steadily employed. Of course, he was also justifying his violation of the classical rules, a stance for which he was sharply criticized in some quarters. But he was a thorough pragmatist. What is more, he did not believe a tragic story must be without comic relief; he did not hesitate to mix moods. In his plays, tragicomedies, life was presented more realistically. As he saw it, the world offered a comic-serious picture. If it was the writer's duty to teach, as moralists insisted, he did it best by entertaining while he sermonized, to beguile, "to instruct through pleasure". (This is paraphrased from "The Spanish Contribution to the Birth of Opera", an essay by Jack Sage.)

The eager acceptance of Lope de Vega's argument, and of his violent dramas as models, is why the librettos of most Italian operas are lengthy, loose and sprawling, almost invariably episodic, overladen with complications, filled with far too many characters, to say nothing of choruses of thirty, forty or fifty, as many as the budget permits. Where, and how often, does one find an operatic work with a text that is spare, neat and trim, adhering to the austere, long-revered prescriptions inscribed in Aristotle's *Poetics*?

Sage quotes poet and playwright Jacopo Cicognini (1577–1633) who in a foreword to his *Il Trionfo di David* (1633) stated that he was "imitating" Spanish drama, in particular that of "the famous Lope de Vega", who recommended that he set himself free from traditional classic form, indeed had "beseeched" him to do so, thereby enabling the spectator to "delight in the accidents of history, not just by relating to him what has gone before but by representing these diverse actions in sequence . . . weaving the grave with the ridiculous, pleasure with profit, history with invention". Said Cicognini: The "modern writers" – that is, the Spanish ones, Lope and his followers – were first of all intent on pleasing the listener, and consequently were broadening "the narrow and severe rules of art".

Lope and his disciples further left their impress on Italian opera by their wide and eclectic choice of subject-matter, siphoning personages and outsized incidents from myth and folklore, ancient history and tales from the Scriptures, legendary romances and more recent political events. In Lope's particular case, too, there were tense exposures to domestic crises – jealous husbands, misunderstood wives, the defence of honour – and various "catastrophes of everyday life".

Not only Cicognini but other librettists in Italy, as well as the composers who set music to their words, looked to Spain for lessons in making stronger plots and multiple plots. Among them were Rospigliosi, Busenello and Badoaro, who collaborated with Monteverdi, Cavalli, Castro and others. They were frank about their debt to the Spaniards. In France, too, Rotrou, Racine and Corneille, whose heroic tragic dramas were raided by librettists, had earlier been guided by Lope, Tirso de Molina and Calderón, their acclaimed predecessors.

Lope himself provided the text for Spain's "first opera", *La Selva Sin Amor* (1627), a one-act pastoral, which has a score by two Italian minor composers, Filippo Piccinini and Bernardo Monanni; a very slight offering, it failed. Interestingly, Lope did not apply his own theory in preparing it; it is very much in the old tradition which he now abjured. Perhaps because it was meant to be staged at court, he did not have a free hand.

Calderón's two opera librettos, *La Púrpura de la Rosa* (*The Purple of the Rose*, 1659) and *Celos Aún del Aire Matan* (1660), do have scores by a native composer – the first probably by Juan Hidalgo, and the second certainly by him – and are the first such works sung wholly in Spanish. The music, though, is Italianate, echoing Cavalli. This was a novelty at Madrid's court and hardly a welcome one. In the prologue to *La Púrpura de la Rosa* Calderón admitted being unsure whether Spaniards, whose temperament inclined them to be impatient, would tolerate the opera's "vocal monotony". He was perceptive: the royal audience did not much care for this new genre. Only these two offerings were performed, and afterwards no more in that form for the next three centuries.

Calderón's response was to develop a different kind of musical work that was more to the court's liking, the *zarzuela*, or what in later days would be called an operetta, half-spoken and half-sung, sentimental and abounding in good humour, the vocalized segments appropriately light and tuneful. It also included dancing and elaborate scenic effects, the usual imperative.

They were an outgrowth of ballads that had been partly enacted, and then more fully so. Appropriated by Calderón, they evolved under his contrivance into a highly popular art-form and thrive to the present day. Since 1857 Madrid has had a theatre that houses only *zarzuelas*, drawn from a very large repertory of them by now assembled during more than three centuries.

(The word *zarzuela* means bramble. One of the royal rustic retreats, a former hunting lodge, was in a forest and surrounded by thick undergrowth and brambles; hence it was known as the Zarzuela Palace; it had a theatre chosen by the king to put on this preferred kind of entertainment, most likely to mark gala occasions.)

Calderón's taste was literary and refined. He tended to base his musical plays on classical legends, after adapting them freely. An example is *Eco y Narciso* (*Echo and Narcissus*; 1661), the characters borrowed from Greek mythology. An eclogue, it was most suitable to the isolated rustic setting.

On a hunt, Antaeus, a shepherd, comes upon Liriope, a frantic, wild creature clad in animal skins. In love with Echo, a beautiful maiden, Antaeus hands over to her his strange, grotesque trophy. To Echo, the captive Liriope reveals that she has a son, Narcissus, at whose birth a magician has prophesied "an early death caused by a voice and a beauty". To avert this, she has kept him to herself and raised him in a cave apart from the encroaching world. But now she has lost track of him; she was searching everywhere for him when caught by Antaeus. She begs Echo to help her find the vanished youth. But how? Narcissus adores music. Liriope pleads with everyone to sing, hoping that the young man will hear them and be attracted to the source. They yield to her entreaty, joining voices in song. Above all floats Echo's irresistible voice.

The lure succeeds. Narcissus appears, drawn by the melodious air. Before long he is enamoured of Echo, to the distress of his mother, who orders him to return to the mountains; he does so, reluctantly. Liriope acts to alter Echo's enthralling voice, diminishing it. Echo follows Narcissus to the mountains; after wandering for a time, she discovers him gazing into a pool, speaking to the image he beholds in it. Echo replies, her voice softened, her range limited, so that she can only repeat the last few words of his phrases, as though mocking him. It is as though his reflection is conversing with him, yet not conversing with him. Finally he realizes there is no nymph in the water, he is only infatuated with his own likeness. He himself is ineffably beautiful. Frustrated, infused by knowledge of his mesmerizing self-love, hating the voice that whispers his own words to him, he dies. The gods change Narcissus into a flower. A shepherd seeks to embrace the still lovely Echo, but in vain – she is metamorphosed into air.

Also making use of classical myth is Calderón's *El Laurel de Apolo* (*The Laurel of Apollo*; 1658). Though Sage describes *La Púrpura de la Rosa* as an opera libretto, Andreas Franco considers it better categorized as a *zarzuela*. Commenting on Calderón's use of ancient myths, Sage says that they are subtly integrated – as are the allegorical subjects of his *autos sacramentales* – with contemporary issues, so that the values they embody are surprisingly topical and relevant to his own time, and for some sympathetic spectators to all times.

Spanish dances are varied and unique; there are no others like them. They differ from all other European dance forms; indeed, from any sort of native dance elsewhere. They are as local – indigenous – as Spanish music, to the percussive rhythm of which they owe many of their characteristics, and which in turn responded to the demands of the dances, developing to meet their needs, the two becoming a perfect match. To hear Spanish music, not least the strummed and plucked guitar, is to feel an impulse to dance, but only in a certain manner. No other dances call for finger-snapping, castanet-clicking, heel-tapping, the swooping or proud, rigid posture. Even the use of hands and arms in classic Spanish dance is unlike that employed by performers in other nations. For that, too, the traditional costumes of the male and female participants are *sui generis*: flouncy skirt; for the man, a flat, round-brimmed hat, a short jacket, very tight black trousers.

The origins of the various Spanish dances are far from clear; they derive from many sources, just as the people of Spain are a blend of races. A component is owed to Gypsies, many of whom inhabit Spain, their origins not absolutely certain – though many ethnologists conjecture that their perpetual wanderings began in India in a distant past, as their dark complexions suggest. But the Moors also significantly influenced the culture of Spain, leaving lasting traces of their semi-Oriental arts – architecture, music – on the peninsula they once occupied and ruled; there are strong hints of the Arabic in the idioms of Spanish dances. Since there was an enriching exchange with southern Italy, especially with Naples when it was under Spanish sovereignty, some features of the high-spirited folk dances of the Italians have been absorbed. In the disparate and even antagonistic regions of Spain itself – in Andalusia, Castile, Catalonia, Aragon – distinctive folk dances took shape; some kept their individuality, while aspects of others were gradually assimilated into the major kinds of Spanish dance seen today. Finally, the formal court dances of Italian and French royalty and aristocracy during the Renaissance were adopted and gracefully enacted in the palaces of Habsburgs and Bourbons and the circles of grandees in Madrid. The court dances were truly "courtly" and touches of patrician reticence and dignity still linger in what is now considered to be the classic Spanish dance, which has an air of hauteur.

A professional Spanish dancer's or troupe's offerings will usually consist of works from one of four categories: regional, school, flamenco and court. Or an evening's programme in a café or theatre might present examples of two or three styles, but scarcely ever all four, if only because flamenco – Gypsy – requires dancers trained in a different way. The regional – or folk – selections are drawn from any of the country's forty-three provinces and could include a Jota of Valencia, or a Sardana of Catalonia, a Sequidillas of Seville, a Sevillanas of Andalusia, or – again – a Jota of Aragon, or some of the thirty different dances popular in the Basque country – all these a mere handful of the hundreds that survive and flourish. For each there is an elaborate traditional costume peculiar to its locality, assuring a colourful miscellany. Accompanying each number is music typical of the village or province. The performers may be soloists, duos or larger groups for squares and rounds. As might be expected, many are no longer truly authentic but have been theatricalized. Probably they are seen at their best and are most charming when danced in the region where they originated.

So-called "school dances" are those taught to the young in dance academies and often have a folk basis which has been extensively stylized. One of the most popular is the Bolero, created in 1790 by Cerezo, a dancer at court, who added to a folk-pattern many intricacies borrowed from French ballet, by then considerably advanced. The Bolero caught the fancy of dancers all over Europe, for a time having an astonishing vogue. It has Moorish elements. Other local inventions taken up by the schools are the Malguena and the Jota of Valencia, both freely adapted; and a number of variations on the Sevillanas – El Ole, Las Manchegas, El Vito, Las Panaderos. In the following century, the nineteenth, the list of new dances endlessly contrived by the schools grew very long, largely inspired and guided by a master of classical style, Otero. In some instances, gypsy heel-work was added to them. But by now most of these school dances have been relegated to being period pieces.

The term "flamenco" is applied to Romany dances quite unlike all the others. They are fiery, unrestrained, an expression of real passion. For untold decades a band of Gypsies has made a home in caves in the Sacre-Monte hills outside Granada, and another tribe is located in the Triana quarter in Seville. Both troupes regularly exhibit their wild, breathless flamenco for hordes of curious tourists – needless to say, the dancing is not spontaneous. Their repertory includes such rhythmic pieces as the Alegrias, Soleares, Bulerias, Farruca, Zapateado, Tango and Zambra, staged with nostrils flaring, and usually requiring violent agility and physical strength.

Flamenco is not as prescribed or disciplined as the other types: much in it is supposed to be improvisational, in theory prompted by abrupt emotional impulses or sharp shifts in the music. It tends to be noisy, the guitar accompaniment strident, while onlookers clap hands and shout encouragement. The dancers stamp, and revolve with a flourish, or the male and female participants confront each other with eyes flashing a bold, sensual challenge or invitation.

The male Gypsy dancer (*Gitano*) does not use castanets; many deem them effeminate. He puts emphasis on his arrogant bearing and his exceedingly rapid and exacting heel-work (*tappeno*), at which no other performer is said to be a Gypsy's equal. Here is a technical analysis: "The three basic sounds (that is, the striking of the half-toes, the striking of the heel, and the striking of the full sole) are capable of producing endless varieties of tone and rhythmic combination. Any non-flamenco dance which uses heel-work is called '*agitanado*.'" This feature is believed to have come down from the Near and Middle East.

To manipulate the castanets is surprisingly difficult, requiring many months of practice before one is even competent, let alone has a compelling artistry with them. The best ones are expensive and tuned. Again: "The right is a third higher than the left." An inherent musicality is needed.

In the twentieth century flamenco was effectively transposed to the commercial stage by Carmen Amaya (1913–63). Born in Barcelona, and the daughter of a guitarist, Juan Antonio Aguero, she was also the granddaughter of dancers. She began to perform in public when she was seven, and made her début in Paris at eight. Hers was a train of almost continuous successes. At the onset of Spain's Civil War she and her family left for Mexico and later went to Argentina, where she aroused such enthusiasm that a theatre in Buenos Aires was given her name. In 1941 she and her company – which included her father as guitarist, and two sisters – reached New York, appearing in cabarets. At the end of the Second World War she returned to Argentina (1945) and then to Europe (1948) with her always exuberant troupe, and was seen again in the United States (1955), as before in cabarets but sometimes in large concert halls during tours across the continent. Her unusual talents were caught in a film, *Los Taranatos* (1963), which was also her last year. A bare week before her death, which occurred in her home in Bagur, the Spanish government awarded her the Medal of Isabela la Catolica. (Much of the foregoing and following detail is from an entry by La Meri in *The Dance Encyclopedia*.) "Maya" and "Amaya", Gypsy tribal names, stem from Sanskrit, another clue to a probable origin in India.

Court dances, which without loss of dignity survive from the Renaissance, are the Sarabande, Pavane and Pasacalle, to cite the most familiar. The Sarabande (*La Zarabanda*) lays claim to pure Moorish ancestry and dates from the twelfth century; in Arabic its name means "noise". Initially it was performed only by groups of women to the sound of bells and castanets, in a spirit of abandon. Later, in France, it became more subdued and was enacted by a soloist, male or female. "It was in three-quarters time and its chief step consisted of a quick shift from toe-out to toe-in, characteristic movements of Oriental dancing; the rest was slow glides." The Pavane (Pavin, Panicin) owes its title either to Padua, the Italian city, or more likely to "*pavo*", a word for "peacock". Ladies of the seventeenth century taking part in this dance swept the long trains of their dresses behind them much as a male peacock parades his iridescent tail. The partners curtsy, retreat and advance, with the lady's hand resting on the man's back. As the music to two-four time permits, poses are held, with the dancers maintaining an aloof air. This dance was favoured not only in Italy and Spain but also in France and elsewhere. The stately paced Pasacalle in slow triple-time was matched to a passacaglia, an old Italian or Spanish dance-tune. In France, at the court of Louis XIV, this became the Passacaille, its principal movements long *glissés* with arms held to the sides, to music in three-quarters time.

In Roman days lithe, seductive "girls from Cadiz" might be sent for to entertain in the houses of the wealthy. During the reign of the Moors the imperious Caliphs had dancers perform for their pleasure and when receiving their most important guests. Ferdinand and Isabella, after finally driving out the Moors, welcomed dancers at their court, which before long led to the performers' acceptance at theatres that were beginning to offer light dramatic fare in Madrid and other large cities, as in Naples from which the alert Spanish playwrights took cues. The fantasies staged for Philip IV by Calderón at the Zarzuela Palace had music and dance episodes. Far from disapproving, the Church allowed the addition of choreographic elements to the solemn panoply of its rituals.

"The Spirit of the Spanish Dance" (1925), an eloquent and perceptive essay by the Russian-born critic André Levinson in *Theatre Arts Monthly*, surveyed the state of the Spanish legacy at that time – it was not then at its best and has recovered greatly in the many decades since. But Levinson's remarkably sensitive *aperçus* are helpful for anyone seeking to appreciate this hermetic yet varied dance-form. He saw in the "wild taut savagery" of flamenco "an almost pagan quality of fresh vigor and unrestraint that creates in the spectator the profoundly exciting effect of an elemental manifestation of nature. The rhythm of these dances is a physiological phenomenon; it tightens the throat and lashes the nerves." The school and formal Renaissance dances conjured up for him images of how the "*boleros* and *fandangos* were danced of old at the Escurial, as were the *menuets* and *gavottes* at Versailles, by the ladies of the court, their lace mantillas pinned to their powdered wigs". He continued: "Some of these dances were picked up and stylized by the theatres and thus passed into the traditional repertory of the dance schools. At the time a certain Carlos van Loo, painter to King Louis the Well-Beloved of France, painted his *Conversations Espagnoles*, both the court and the city of Madrid adopted the actual step of the *menuet*, embellishing it with ornaments *à l'espagnole* and substituting for its low-sweeping reverences the arrogant nod of the head, with hand on hip, characteristic of the national *fandango*. This blending of two styles is called *menuet afandangado*. At other times they danced the famous *folies espagnoles* to the altogether Italian music of Corelli. In this way, the idiom of a Scarlatti came to dominate the musical production of Goya's Spain."

He traced the rising interest and fascination with all things Iberian throughout Europe during the Romantic Age, and how Spanish dance was widely imitated, along with an obsession with plays and operas on Spanish characters and subjects – Victor Hugo's *Hernani*, and Verdi's wrenching music drama based on it, to which might be added his more thoughtful *Don Carlos*, its libretto adapted from Friedrich Schiller's stirring script; and later Prosper Merimée's story of a Gypsy cigarette girl, *Carmen*, which gave birth to Bizet's endlessly revived musical version of her unhappy fate. Most of these operas contain Spanish dances. Besides, there was also a flood of symphonic music resounding with *faux*-Spanish idioms – by Chabrier, Rimsky-Korsakov, Debussy, Ravel – that prompted the listener to *imagine* dances set to their compelling rhythms.

Over the years, Spanish choreographers borrowed innovations from abroad – Paris, Milan – to further adorn "their own rich inheritance of Castilian pomp . . . and the Oriental sinuosity of the Moors". Asked Levinson:

What then are the essential dynamic and plastic characteristics of the Spanish dance – the actual origins of its peculiar style? It seems clear that the splendors and the miseries, the glory and the decadence of the Spanish dance have been determined by the uninterrupted duel between the East and West, characteristic of the country itself. The burning ardors of the Moor have been grafted on to the austere race of Castile. An exceptionally violent and sensual nature is held to the rules of an unusually strict society. Andalusia has remained, as far as the dance is concerned, an eternal battlefield. In the realm of art the Cid Campeador has never sheathed his sword and Almançor has never surrendered. Is not the answer to be found . . . in this interpenetration of races, this dual nature of Semite and Aryan, this hot blood of the conquered that beats feverishly in the veins of the conqueror?

But to return to choreography. In a recent article I remarked on the fact that the movement of the Oriental dance is concentric – the knees bent in, the arms embracing the body, everything converging to the centre – while the movement of the dance of Western Europe (the most perfect expression of which is to be found in the traditional classic dance) is just the opposite – the body of the dancer being extended, the arms and legs turned outward, the entire being dilated to its extreme capacity. The European dancer, be she a ballerina or a peasant, moves about freely, she leaps in the air and glides and runs, while the Oriental dancer turns in upon herself, like a statue, turning upon a pedestal. She crouches and stretches. She is a plastic form, coherent but changing. The back hollows, the abdomen swells and sinks. The almost fluid muscles ripple under the glossy skin of the arms, whose sinuous curves undulate in an internal rhythm.

Now if we analyze the most important steps and characteristic attitudes of the Spanish dance (or, to narrow the field of investigation, of the Andalusian dance alone), as well as to compare these with the traditional school steps and positions, we find the same antimony. The diagram of any attitude or pose of classic equilibrium can always be drawn by means of straight lines that cut the vertical at different angles. The decorative curves play about this rectilinear frame. The arms are held parallel to the legs or else in balanced opposition to them. For example, the formula of an arabesque position is angular – an arrangement of crossed straight lines. The Andalusian dance, on the other hand, can best be expressed by a curved line, like the line followed by the stone cutters of the Alhambra or the painters of Ajanta. In a pirouette the leg of the classic dancer describes a periphery around a central point – she radiates outward – while La Argentina in making a turn follows the line of a continuous spiral – like a capital S. That capital S is the image and emblem of Oriental beauty and is depicted quite as lovingly in the Persian miniature as in the Japanese wood carving. In preparing for a dance figure or in an attitude, the French ballet dancer raises one arm, framing the head, while the other arm moves on a level with the shoulder. With the Andalusian dancer the second arm would quite naturally come to cross the body in front – a position that is nothing less than an abbreviation of the famous S. The classic dancer either faces the audience full or else obliquely according to the shoulder line, while the head is held at three-quarters. The fact that a straight line could be drawn from the shoulder blades to the small of the back emphasizes the impression of fixed quantities, definitely proportioned.... With the Spaniard, on the contrary, every detail of the body follows or plays some variation on an outline of curves. The eyebrows that arch, the lashes that hide the glance into the wings, the eye cast upward or downward or fixed in space as the case may be, are as much an integral part of the whole dance movement as are the flexible wrists and the arching insteps. And I am speaking now merely of the purely formal role that these details play in the dance technique and not at all of their expressive, figurative values.

The same antithesis is evident in the traditional costumes of these two schools. Whether at a social ball or on the stage, the Western dancer seeks a fashion that will permit free motion of the arms and legs. The ballet dancer has adopted the décolleté bodice that leaves shoulders and arms uncovered, while the tulle flounces about the waist, a, so to speak, mere abstraction of a skirt, interfere not the least with the action of the legs. The *ghawazi* or Egyptian street dancer will leave the abdomen uncovered but prefers to veil the shoulders and face. The Spanish dancer tends to swathe herself. Over the low-cut neck of her bodice she loves to throw a fringed scarf. She accentuates the mystery of her body. In this connection it should be remarked in passing that what one might almost term the semi-nudity of the classic dancer is purely functional, serving to facilitate the mechanism of her art. Hence its complete absence of sexual significance. What the Oriental withholds from view, on the other hand, no less than the fold of the shawl draping the figure of the Spanish dancer and held up by the shoulder or by the crooked elbow, are replete with erotic suggestion. And here I would add another passing allusion to the symbolism of the Spanish dancer's costume, which tradition and instinct cause her to choose. For each dance there is a certain comb, a certain shawl, a certain flower, rose or carnation, the mantilla worn or discarded, the scarf tied or allowed to float free. Very often the dancer herself could not tell you the reason for her choice, but she never makes a mistake.

An exile from Russia, after the Revolution, André Levinson settled in Paris and viewed Spanish dance from that excellent vantage. The decline in its art that he deplored was considerably reversed in the middle and later stretches of the twentieth century by the appearance of a number of very accomplished performers and their well-trained troupes, among them – apart from Carmen Amaya – Vincente Escudero, La Argentina, La Argentinita, Antonio (of Rosario and Antonio), Carla Goya, José Greco, Roberto Inglesias; theirs are the names that most quickly come to mind, all of them having toured and been acclaimed worldwide. (Ironically, several of these remarkable artists were not born in Spain but in South and North America.)

In 1919 Spanish dance was basic to a story-ballet, *The Three-Cornered Hat*, created for Sergei Diaghilev's Ballets Russes. Though it has music by Manuel de Falla and décor by Pablo Picasso, the choreography is by a Russian, Léonide Massine, who initially had the leading role. He and the Russian company settled in Madrid for a period to study the necessary technique. Among its highlights are a fandango and a farucca. The scenario is derived from a play by Pedro Antonio Alarçón. This semi-farcical work has taken its place in the repertoires of companies on four continents. It always pleases audiences.

In the realm of the fine arts, Spain is probably most popularly admired for its dances.

— 9 —
FRANCE: ITS RENAISSANCE

The spill-over of new new ideas and art-forms from the south – Italy, Spain – heralded the Renaissance in France. These three Mediterranean countries had much in common at that time. But they had many antagonisms, too, which led to frequent military aggressions. Along with their likenesses, the French, Spanish and Italians had different temperaments and sensibilities.

Their ministers engaged in shifting foreign alliances, often seeking to accomplish them by dynastic marriages. Culturally, one was a precipitating event of great and lasting significance, the union in 1533 of the French Dauphin (later Henri II) and Catherine de' Medici (1519–89), daughter of Florentine rulers and bankers who were immensely wealthy, given to ostentation and, more important, close relatives to the Pope. To impress the French and adorn her extravagant nuptials, Catherine brought along her family's gifted troupes of Italian musicians and dancers, some of whom stayed on in Paris on their own or were permanently attached to her court. They displayed what were the latest advances in the staging of fêtes, including operas and ballets.

By now, too, the *commedia dell'arte* was carrying out successful incursions throughout the country, helping to update and enrich the local actors' by-play and theatrical conceits and raising audiences' expectations. From Spain came examples of effective play-making and fresh dance inventions.

Yet the French Renaissance – comparatively brief, from the middle of the sixteenth century to the mid-point of the seventeenth – is viewed as mostly a period of transition in which its theatre moved rather abruptly from the spiritual and elegant Mysteries of the late Middle Ages to what were to be the triumphs of the resplendent baroque or neo-classical age in the second half of the seventeenth.

Life in France during Catherine's lengthy presence there was turbulent. The marriage to Henri II was hardly a happy one, since he was much in love with his long-time mistress, the beauteous Diane de Poitiers. For ten years Catherine, sterile, bore him no heir; she was ugly, large and mannish, but fortunately strong-willed and intelligent. Finally, the athletic Henri insisted on taking part in a tournament (1559). He was mortally struck by his opponent, a spear piercing eyeball and skull. Comatose for nine days, he died at forty-one, leaving Catherine – as she put it – a widow with "three small children and a wholly divided kingdom". Her eldest son, Francis II, was fifteen and physically weak. Her task was dire, but she was equal to it.

France was being torn by civil discord and savage religious feuds between Catholics and

Huguenots, the Calvinist reform faction. Catherine, a Catholic, was beset on all sides. Many of the nobility resented that they were being ruled by a foreigner, especially an Italian descended from a mercantile family. Surrounded by intrigues and plots against her, and at times open rebellion, she held tightly to the reins of power. But the country, unsettled, was devastated, nearing ruin. The culmination of the intense struggle was the dreadful St Bartholomew's Day Massacre in 1572, in which the leading Huguenots were surprised and slaughtered. Catherine and her second son, who by now held the sceptre as Charles IX, had joined in planning and ordering the deed. Shortly afterwards the young king, plunged into melancholy and spitting blood, followed his brother Francis II to an early death in the Queen Mother's arms – he was twenty-four (1574). The French treasury was so bare that Catherine had difficulty paying for his state funeral and the expenses for the journey home of her third son, whom she had made King of Poland. This young man, now Henri III, offended his subjects by his effeminate mannerisms: he was fastidious, witty, intelligent, with liberal tendencies, but he was given to transvestitism and overfond of perfume and jewellery, attributes which led to his being called the Prince of Sodom. His reign, too, was very troubled and concluded when he was stabbed to death in 1589 by a fanatical Dominican who accused him of being partial to the heretical Huguenots. The Queen Mother had taken to her bed and died a short time before him, after saying wearily that she was "at the end of her tether". (The story has strange parallels to that of the last Dowager Empress of China.)

A Medici, Catherine had taste and an innate interest in the arts, seeking to encourage and protect them. Despite her many distractions, she collected treasures – tapestries, figurines, rare books, porcelains, and especially luminous Limoges enamels, jewels, many of the precious objects suggesting a happy blend of French and Italian styles; today they are kept in the Louvre. She gave elaborate revels, conceived and executed by the best craftsmen and artisans. The court was the scene of fine concerts, and the royal clan was deeply concerned with architecture and the aesthetic theories and controversies it inspired. Entertaining lavishly, she had a second motive: she hoped to divert and win over her guests, the nobles who contended against her. In any event, while she was Regent – she was so designated for three spells following the death of her husband and her two eldest sons – the court, with its spectacular galas, was the font from which flowed the new and changing French traditions of drama, opera and ballet. She and her sons saw to it that money was available for such presentations.

At court, only one thing was lacking where theatre was concerned: a playwright of outstanding talent. Another handicap was that Paris had only one licensed playhouse open to the public, the Hôtel de Bourgogne, long controlled by the Confrèrie de la Passion, a semi-religious organization. Beginning in 1599, the group, no longer able to keep its stage perpetually occupied, began to rent it out to other troupes, most of whom had been hopefully touring outside the capital. But there was not much opportunity for a commercial theatre to establish itself and thrive, as was happening in Venice, Rome, Naples and Madrid.

Catholic churchmen and austere Calvinists were censorious, as was the Parlement, and dramatists had to be cautious. On occasion the Parlement criticized the royal family for its lavish expenditures on frivolous affairs. With this background, it is understandable that the chronicle of drama in the French Renaissance is bleak.

The slow approach towards full-blown theatre was by way of translation of Greek and Roman tragedies – above all, Seneca – and comedies – essentially Plautus and Terence – which began in the schools and often led to presentations by students, particularly in the Jesuit colleges, the evangelical order that emphasized play-acting as a teaching tool. The Aristotelian unities were also studied and accepted as imperatives. As might be expected, Biblical subjects were sought; the Jesuits themselves composed such scripts, now treated not as in the Mystery and Miracle Plays, but rather after a Senecan model. In 1546 the production of Mystery Plays was prohibited.

Secular drama owed its introduction to a twenty-year-old youth, Étienne Jodelle, Sieur de Lymodin (1532–73), one of many students in love with classical lore, as was the fashion of the day. He first thought to enter a military career but soon drifted towards indulgence in literary efforts, while also pleasure-seeking. He joined a group of six other young poets in boldly calling themselves the Pléiade; they were led to Pierre de Ronsard and dedicated to furthering the cause of Humanism by praising classical works. However, they strove to embellish their native speech by inserting Greek and Latin words and phrases wherever possible into their essays and verses. They had been stirred to do this by the poet Joachim du Bellay in his treatise *Defence and Illustration of the French Language* (1549). He also cited a need for a French drama that would "return to the dignity of the Greek and Roman works". In response to this plea, Jodelle's fellow members urged him to undertake the task..

He set about it forthwith, with the impetuosity of youth, coming up with *Cléopatre captive* (*Captive Cleopatra*; 1552), based on Plutarch's account of Marc Antony's affair with the Egyptian queen. The play opens after Antony's death. His ghost appears, prophesying future woes. The victorious Octavius and Agrippa engage in an exchange in which they describe the grandeur of Rome. Next comes Cleopatra herself to bewail her overthrow and complain of Octavius' cruelty towards her, for which Antony's spectre rebukes his living Roman foe. In the fifth act, the script's close, a chorus of young Egyptians announce their queen's suicide.

The play was put on, with untrained actors, in the quadrangle of the Hôtel de Rheims for the edification of Henri II and his entourage. Jodelle himself had the role of Cleopatra, and the king was so pleased – apparently – that he opened the royal purse to give him 500 crowns, while declaring that the work was filled with many graces, very rare, very beautiful, a true novelty. *Captive Cleopatra* was given again at the Collège de Boncourt, once more in the presence of the king, shortly afterwards. Frederick Hawkins in *Annals of the French Stage* relates that all the windows of the college, "like the court itself, were choked with persons eager to witness the spectacle".

Though the young author was instantly hailed as "a modern Sophocles", the subsequent critical consensus rates him as at best a mediocre poet with scant flair for play-writing. *Captive Cleopatra* is almost wholly without action, consisting mostly of expositional dialogue, the unfolding far more narrative than dramatic. In the general opinion it is "dull". Going further, Allardyce Nicoll dismisses it as "worthless". The single setting, though, is said to have been magnificent, a palace façade, with Antony's tomb at one side where it might also serve as the enslaved, humiliated Egyptian ruler's refuge and last resting place when she has killed herself. Here the unity of place was ingeniously observed. Time was compressed to just one day by having Antony as a spirit looking back at past events, the focal content of the story occurring in Rome within twenty-four hours.

Emboldened by the acclaim that greeted his initial venture, Jodelle brought out another script, in a sense his country's first comedy, *Eugéne, ou La Rencontre* (1552), a depiction of contemporary French life, revolving about a libertine *abbé*. It, too, was staged with success before the amused king and his courtiers at the Hôtel de Rheims and Collège de Boncourt, and Jodelle was still in high royal favour. But his depiction of a lewd *abbé* provoked the ire of clerics and supplied jealous rivals with occasions to denounce him. Jubilant that his second play, written with "happy courage", had been well received, his fellow members of the Pléiade had a party, a Bacchanalian procession and revel, at which he was crowned with laurel and proffered a goat garlanded with ivy, as was the custom to honour a winner at the Dionysian rites. Word of this pagan event soon spread and prompted further criticism of Jodelle and his bohemian friends; his popularity ebbed away discernibly. Three years later his third play, *Didon se sacrifiant* (1555), met only coldness. Either because his inspiration was exhausted or he was hurt by the rebuff, or he sensed it advisable to stop attracting so much public notice, he withdrew. Even when impoverished in later years, he could not be persuaded to take up his quill and again write for the stage. In a last poem he laments his fate. He died in obscurity at forty.

Yet despite a meagre output, and scant admiration for his work, he is a very important figure in the history of French drama. He set a long-enduring precedent, followed by France's greatest playwrights over the next centuries. A serious play, a tragedy, must respect Aristotle's rules. The three unities are an inflexible requirement. The hero must be larger than life. The dialogue must be poetic. Jodelle had chosen alexandrines, a metre that was adopted by generations of dramatists who succeeded him.

This early regard for classical strictures – they were held almost in awe – is seen in the plays of Jean de la Taille, who came close on Jodelle's heels. A prime example is his *Saul le furieux* (*Saul Enraged*; 1565), for which he chose not a Greek myth but a Biblical subject, handling it somewhat as Seneca might have. Ten years later, when his script was published, he prefaced it with an essay, "L'Art de la tragédie", extolling adherence to Aristotle's tight-binding precepts. To achieve unity of action all the qualities of human nature and conduct must be fused – "good and evil, passion and sentiment" – in a single, clearly defined and crucial deed that faithfully reflects people as they truly are. In practice it was more difficult for de la Taille to obtain unity of place. All that was available to him was a multiple-level medieval platform-stage. He stipulated that his disparate locales – Mount Gilboa, the Cave of Endor, and more – should be juxtaposed as closely as possible. The Greeks, whom he profoundly admired, might have thought him cheating not a little.

In the medieval theatre, mostly the shortcomings and errors of lowly persons were exhibited. De la Taille held that the new tragedy should deal solely with high-born, heroic personages. This upscaling reflected not only Aristotle's psychologically astute suggestion but also an attitude endemic among the arrogant French aristocracy, intellectuals, clergy and academics of the era.

De la Taille had studied at Orleans to be a lawyer before electing instead to be a poet in Paris. He wrote another tragedy that Hawkins describes as "execrable", an appraisal extended to *Saul Enraged*, but he did better with a prose comedy, *Les Corrivaux*, an amiable piece.

The theatre of Jodelle and de la Taille was seen only at court or in the halls and courtyards of the colleges. This continued to be true for about two decades. In those venues a frequent choice would be a pastoral of the Italian sort, a fanciful portrayal, perhaps with elaborate décor, of the loves of flirtatious shepherds and shepherdesses, attended by sprites and other flitting woodland creatures – water-nymphs, dryads – belonging to a sylvan world half-real, half-imagined. Some of these scripts were provided by Nicholas Filleul, who had come to Paris from Rouen; the king found diversion in such bland, light offerings. The Italian comedians were ever more often on hand to entertain; they earned loud applause everywhere in France.

Early in his life, in his teens and disappointed in love, Jacques Grévin, later physician to the Duchesse de Savoie, consoled himself by composing two comedies, *La Trésorie* and *Les Esbahis*; and next, more ambitiously, a tragedy, *La Mort de César*. They were staged at the Collège de Beauvais, where he studied. The tragedy proved to be a bit beyond his grasp, but the comedies capture the fun and vivacity of life in and around the ancient Place Maubert, a popular gathering place; they satirize the manners of the day. The plays are well written, the precocious Grévin having a pure style.

Lazare de Baïf, ambassador to Venice, translated *Hecuba* and *Electra* into French verse and had them produced at the Collège de Coquerel, where his natural son, Jean Antoine de Baïf, became a student a few years later. The younger Baïf was friendly with future members of the Pléiade; sympathetic to their ideals, he joined them. His other passion was music, and he sponsored concerts at his house in the Faubourg Saint Marceau, to which flocked the intellectual élite of Paris. The king granted him the right to establish an Académie de Musique. Feeling that the new semi-professional playwrights were departing too far from the classical norms that the Pléiade espoused, he translated the *Miles Gloriosus*, retitled *Le Brave, ou Taillebras*, and was rewarded by seeing it enacted at the Hôtel de Guise before Charles IX and Catherine de' Medici (1567). Sharing Baïf's conviction that the classical limits set by the Pléiade were being transgressed, another member of the group, Remi-Belleau, drew on a current event as the subject of a five-act verse-comedy, *La Reconnue*, constructing it as it were by a Greek or Roman.

Taking note of Étienne Jodelle's youthful and profitable labours, the royal court librarian, Mellin de Saint-Gelais, translated Trissino's *Sofonisba* from Italian verse to French prose (1556). Henri II consented to a gala production at the Château de Blois (1559). His daughters, all sumptuously attired, took speaking roles in it; in addition, a leading part was assumed by Mary Stuart, who had come to France from Scotland to wed the Dauphin. Staged with "great pomp" and supported by musical interludes, the play was intended to brighten the festivities that marked Mary's forthcoming marriage to the short-lived Dauphin. In some ways the tragedy foreshadowed her own unhappy fate; nor was Mellin de Saint-Gelais much luckier, having died a year before the script was chosen and mounted. Apart from his scholarly duties, the librarian had been known for his scathing wit, a banter before which even Ronsard quailed.

Seeking to write a tragedy on a modern theme, Gabriel Bounyn, Master of Petitions to the Duc d'Alençon, came forth with *La Sultane*. But all of these literary contributions, as has been indicated, were works by poets and aristocratic dilettantes, hardly by professional dramatists, and few are memorable save as historical footnotes. That is probably also true of Robert Garnier (1544–90), who is considered to be the most distinguished of these early Renaissance aspirants to fame as playwrights. He was a native of Ferté-Bernard. Like de la Taille he studied law but in Toulouse, far to the south. He rose in his profession, in 1566 he was admitted to the Paris bar, and three years later was given the post of king's counsel and judge in the criminal court at Le Mans, from which he went higher to lieutenant criminal judge (1574) and next to be a member of the Grand Council (1586), a heady climb. Literary recognition came to him as early as 1565, when he wrote and recited a prize-winning eclogue to Charles IX, who was visiting Toulouse. The poem saw print the following year. That he went on to writing plays while practising law was hardly unusual – many French dramatists were taking just that route. Perhaps they sharpened their skills by learning how to probe and handle an adversarial situation, how to phrase their thoughts clearly – even in verse – as well as how to appeal to magistrates and a jury – an audience – by rising to eloquence at a climax, playing on the hearers' emotions.

Over a span of fifteen years – 1568 to 1583 – Garnier wrote seven tragedies and one work classifiable as a tragicomedy. It is not certain that he ever saw any of them staged, but they were printed and circulated and even rendered into English during his lifetime. It is taken for granted that producers and actors – amateur and professional – possessed and were familiar with the scripts. Six were based on Greek or Roman subjects, one on a Biblical incident and one on an Italian poem, Ariosto's *Orlando Furioso*. He adhered for the most part to a Senecan model. His first, *Porcie* (*Portia*; 1564–5), seems to have been put on at Toulouse when he was about twenty, but he may not have been on hand for the occasion. It was followed at erratic intervals by *Hippolyte* (*Hippolytus*; 1567–9); *Cornélia* (1574); *Marc-Antoine* (1575); *Antigone, ou La Piété* (*Antigone, or Piety*; 1580); *La Troade* (*The Trojan Women*; 1581); *Bradamante* (1581); *Sédécie, ou Les Juives* (*The Jewesses*; 1583). Of these, it is believed that *Hippolyte* was acted in Paris (1567), *Marc-Antoine* in Saint-Maixent (1578), *La Troade* and *Bradamante* also in Saint-Maixent (1581), though the record for *Bradamante* is unsure.

Even though he looked to Seneca, Garnier did not write exciting plays in most instances; his works tend to be static, having far too little action. They were greeted with praise by his contemporaries, however, perhaps because the poetry in them is sometimes sweepingly lyrical. At such moments he does not sound like one who daily prepared dry-as-dust legal documents or handed down weighty, precisely worded judgments.

In a category by itself is *Bradamante*, the tragicomedy adapted from *Orlando Furioso*, Ariosto's verse romance, laid in the time of Charlemagne. Here Garnier puts aside the Senecan model and displays a good measure of originality, even though he borrows much from the great Italian poet. Indeed, as he treats the legend it becomes thoroughly French. Her ambitious parents have promised their daughter, Bradamante, a capable warrior maiden, to be the wife of Leon, the Emperor of Byzantium's heir, a brilliant match; but she loves Roger, which leads to a dispute. Asked to mediate, Charlemagne decrees that it should be resolved by a duel, one of the participants being Bradamante herself. Disguised as a knight, a stranger, Roger is imperilled but rescued by Leon, his friend, who asks for his reward that the "knight" replace him in the encounter with the girl. Charlemagne insists that Leon and Roger must also fight each other, whereupon the disguised "knight" is forced to reveal his identity. Recognizing his friend, Leon generously agrees to relinquish his claim, choosing Charlemagne's daughter Eleanor instead. Roger is named King of Bulgaria, which satisfied Bradamante's parents, and they are persuaded to sanction her union with him.

The Biblical play, *The Jewesses*, combines incidents recounted in Kings, Chronicles and Jeremiah concerning the revolt led by King Zedekiah of Judea against Nebuchadnezzar. The rebellion has failed, and the Jews are besieged in Jerusalem, because Zedekiah has failed to hearken to God's warnings. Amital, the king's mother, pleads with Nebuchadnezzar's wife that her son's life be spared. Compassionate, the wife does intervene successfully, but it proves to be at the expense of Zedekiah's children and the High Priest, who are held as hostages. Learning that the prisoners have been summoned forth, Amital and her daughter-in-law bewail the news, fearing treachery. Their concern is well founded: the children and High Priest are summarily executed, and the rebel Zedekiah, blinded, is led off to his Babylonian captivity. A chorus narrates much of the Old Testament story, and at the opening and close a prophet foretells the impending arrival of the Messiah, and, after that, of the divine Nazarene.

However, Henri III did not much care for such plays, and this royal indifference caused the monarchist Garnier to lay down his pen and never take it up again for endeavours of this sort.

Joseph E. Garreau, in the *Encyclopedia of World Drama* refers to both *Bradamante* and *The Jewesses* as "masterpieces", though he does not spell out why anyone should hold that view – he merely cites it as the opinion voiced by Garnier's peers and contemporaries. (Much of this summary is taken from Garreau's article.) Hawkins is more specific:

In speaking of his theatre, I find it necessary to guard against a temptation to overpraise. To far-reaching scholarship, the fruit of a liberal education at Toulouse, he united a fervid imagination, rare delicacy of thought and feeling, and a fine sense of moral dignity. His style, though originally formed upon that of Seneca and the Pléiade, is often characterized by majestic simplicity, notably in the choruses. Nor did he fall upon an unappreciative age. In the words of Pasquier, he "was allowed on all hands to have eclipsed his predecessors" in France. His success only stimulated him to expend increased care upon his work; the *Juives* . . . marks the culminating point in a course of progressive improvement, and prouder laurels would probably have fallen to his lot if death had not carried him off in his fifty-sixth year. His faith in the antique model did not prevent him from making a few important innovations. He regularly alternated masculine and feminine rhymes; *Bradamente*, in addition to containing the first confidant, was virtually the first tragicomedy written in French. Each of these innovations was adopted by other dramatists, who, indeed, appear to have generally profited by his example.

A play like *The Jewesses* was welcomed by Catholic colleges, since their antagonists, the Huguenots, were also exploiting Scriptural material. Théodore de Bèze, among the foremost Humanists, had much earlier provided them with *Abraham Sacrifiant* (*Abraham Sacrificing*; 1550), performed at a college in Lausanne. It deals with Abraham's doubts and hesitation when having to yield his son to the knife in response to God's dire command. Garreau says of the now ongoing clash of religious dramas used to proselytize: "Whereas the Protestants chose such characters as Abraham, Jephté, and Judith, encouraging active faith and trust in God, the Catholics preferred David or Esther."

In the comic sphere, Pierre Larivey, active at the same time as Garnier, served as a guide to his peers in writing farces. He turned out a half-dozen of them in the late 1570s, their situations and devices more or less appropriated from the ancient classics and current Italians; his initial stimulation to attempt this came from having observed the antics of the *commedia dell'arte* troupes, who were prospering so consistently on their tours. Larivey, a native of Champagne, had a link of some sort to the court. Of Italian parentage, his original family name was Giunti, but they had adopted the more Gallic Larivey. His six scripts have verve; the best of them, perhaps, is *Le Laquais* (*The Lackey*).

Paris needed more than one public theatre, but the Confrèrie de la Passion clung jealously to its monopoly. Besides the Italian comedians, who rented space at the Hôtel de Bourgogne when it was otherwise unoccupied, other nomadic companies of French actors had been formed outside the capital and were anxious to perform for the Parisian populace; unreasonably, they were not permitted to do so. They made various attempts to invade the city. Some players from Bordeaux, wishing to perform the classical works in their repertory, leased a hall in the Hôtel de Cluny; they were threatened with a fine and gaol unless they dismantled their improvised stage before twenty-four hours elapsed. Many forces approved of the imposed limitation: both the Catholic and Protestant clergy looked on the hitherto staid Hôtel de Bourgogne as an edifice of iniquity, a cesspool dominated by Satan, a convenient gathering-place where players and spectators met to arrange scandalous assignations. Yet there was considerable popular demand for dramas; the mere announcement by a flyer affixed to a lamp-post drew knots of citizens who scanned it to learn what play might be coming, and more often than not the stuffy house was filled to capacity. The performances were at irregular intervals, perhaps two or three times a week.

The doors usually opened at one o'clock, an hour before the equivalent of curtain-rise; in winter even a bit earlier, so that apprehensive spectators could head home before twilight and darkness cloaked the streets where criminals might lurk – Paris was hardly well policed. The Hôtel de Bourgogne was small, its dimensions conjectural and theories as to its size ranging from eighteen by sixty feet – hardly comparable to the huge, ornate Italian playhouses – to forty by 110 feet, and able to accommodate as many as 1,600 spectators, if the latter guess is correct. The stage was twenty-five feet wide and had no proscenium, and it was quite deep. It was elevated five or six feet. At its back was a tapestry curtain before which the performers might take an impressive stance or move while declaiming in an elocutionary manner "the sonorous alexandrines of Garnier and the lively dialogue of Larivey". The lighting was dim, emanating from oil-lamps, yet adequately exposed the actors' expressive features. The costumes, even when the subject had to do with ancient history, were apt to be more or less contemporary.

The auditorium had no tiers or seats, except close to the stage, where were loges, and overhanging at back were several galleries, one of which was nicknamed *paradis*. The main floor was carpeted with rushes. Spectators stood – it might be more comfortable with rushes underfoot – or, if monied, enjoyed chairs in the loges. At times there might be benches along the walls on either side of the main floor (*parterre*).

On the stage the settings, if any, were scanty, most likely still relying on a row of "mansions", each indicating a different locale, the actors simply walking from one to the other, giving the script a fluidity characteristic of the medieval practice. This arrangement, around the periphery of the platform, was kept until 1625. Invoices from scene painters have been found, dated after that, suggesting a trend to more costly presentations.

Henri of Navarre, who was the successor to the throne after the assassination of Henri III and who energetically sought to unify his nation, was a not infrequent visitor at the Hôtel de Bourgogne; he did not confine his interest in the drama to what could be seen at court.

The companies banned from Paris finally found a way to enter it. The exclusive licence held by the Hôtel de Bourgogne did not extend to the various kinds of entertainment at the fairs regularly held at the city's outskirts. A group of comedians from the provinces set up a stage at the Foire Saint-Germain, the most important of these frolicsome events. Immediately, the Confrèrie de la Passion lodged a complaint and had the interlopers stopped, though in fact the law was in their favour. At the next performance of the ensemble at the Hôtel de Bourgogne, the actors were hooted and the targets of missiles. To quiet the protesters, the civic authorities issued an edict that anyone causing a disturbance in a theatre faced corporal punishment; at the same time it allowed the troupe at the fair to resume its activities, on a payment to the proprietors of the Hôtel de Bourgogne of two *écus* a year. The Confrères were not satisfied with this; they complained that such competition reduced the value of their holding; they turned to the king, asking him to confirm their right as absolute, and also to permit the return of the Mysteries and other religious playlets, long their chief fare. Though he himself enjoyed fair-going, Henri IV, shrewd and amiable, agreed to this, but Parlement did not wholly concur, determined to retain its independence. Space should not be leased by anyone else in the city for a theatrical enterprise, and to the Confrères only for the presentation of "secular pieces of an inoffensive nature". This meant that the Mysteries were still prohibited in spite of the king's leniency. Henri IV was probably not disturbed by this: he was a Huguenot, but not deeply religious – politic, his impulse had been to placate the Catholics, the faith to which he expediently converted.

The Confrères did not prevail, even so. That very year (1599) Mathieu Lefevre (stage-name Laporte) and his wife, Marie Vernier, together heading a provincial troupe, ignored the edict and performed at the Foire Saint Germain; they dutifully paid the Confrères a fee for doing so, and a few months afterwards obtained permission to establish themselves as a resident ensemble at the Hôtel d'Argent, again sending *écus* to the Confrères. The law still obtained, but for some reason it was not enforced. This new – second – playhouse soon changed its name and was known as the Théâtre du Marais. Though Marie Vernier was not the first woman to appear on a French stage, she is the first in her nation who can be firmly identified as a professional actress.

Actors were not greatly impeded in the provinces; there were organized troupes in Rouen (1556), Amiens (1559), Dijon (1557), Agen (1585) and elsewhere, wherever players gave vent to their irresistible drive to exhibit their skills before others willing to pay to watch them. Unable to offer the sacred dramas that had long been the prime motive for its existence and made up most of its repertory, the Confrèrie de la Passion finally ceased to produce anything of its own; it was reduced to being simply a landlord, collecting rents and fees (*c.* 1600). This was the opportunity itinerant and resident out-of-town companies had been fretfully awaiting. The stages of the Hôtel de Bourgogne and Théâtre du Marais were open to them at last, and the frequency of theatrical activity was accelerated. Given the benign laxity of the authorities, there was soon a surfeit of comedies, which, as everywhere, audiences preferred over serious drama. Among the visitors was a troupe from England (1598).

Beginning in the Middle Ages, tennis was a favourite sport in much of Europe, and Paris had a number of buildings with courts where the game could be played indoors during winter or grey, rainy spells; the dimensions of the courts were almost the same as those of the interior of the Hôtel de Bourgogne. Such a structure could be easily redesigned as a theatre, and this began to happen, as provincial troupes demanded a hearing in the capital. All that was needed was to insert a platform at one end of the court, which could readily be done for even a short stay; if for more permanent occupancy, the spectators' galleries – already in place – could be extended and perhaps divided into loges. So simple and comparatively inexpensive was it to do this, the practice was copied elsewhere throughout Europe, wherever an empty indoor tennis court caught the eye of a resourceful travelling group. With this, the French Renaissance stage was still rather crude, as were most of the scripts available to the players, which – apart from those by Garnier and Larivey – lacked variety and polish.

One new playwright whose work was acceptable to players at the Hôtel de Bourgogne and by other groups was André Montchrétien, who earned a reputation during the last decade of the century, and whose literary gifts at times seemed superior even to those of Garnier. His personal history was decidedly unusual. Born at Falaise, son of a Huguenot apothecary, he was left an orphan while a child. He had the good fortune of getting an education that was paid for by wealthy friends. Supposedly he was headed for a military career, instead, he soon set himself up in Paris as a man of letters and had three of his scripts performed by professional actors, another *Sophonisbe*, a *Les Lacènes* and a *David*, all drawing on familiar sources. Subsequently came an *Aman, ou La Vanité* and a *Hector*, both again derived by his looking back to mythology and the Old Testament. Flushed with success, he besieged and won the hand of a widow, wealthy and aristocratic, whose riches far exceeded his expectations. Accordingly he burnished and expanded his surname, now calling himself Montchrétien de Vasteville.

He consistently displayed bravery. Having offended the Baron de Gourville, he stood off a ferocious attack made upon him by the angry nobleman and two hired henchmen. He sued them and was awarded 12,000 *livres* as compensation for the assault.

His luck changed. Charged with being responsible for the violent death of a gentleman of Bayeux over a breach of trust, he took flight across the Channel. Shortly afterwards, Henri IV allowed his return. While composing his next play, *L'Ecossaise* (*The Plaid*), he developed a profitable sideline: deep in the Orleans forest he installed a forge and produced steel cutlery that was marketed at a shop in the rue de la Harpe in Paris, though he was still quite affluent, with aspirations of belonging to the élite who were disdainful of anyone in trade. Born a Huguenot, he allied himself with their cause at La Rochelle and set out to aid them. *En route*, at a hostelry in Tourvilles, near his birthplace Falaise, he was overtaken and discovered by a squad of Royalist troops and shot dead.

A tragedy, *L'Ecossaise* is about the bitter life and death of Mary Queen of Scots and is dedicated to James I of England, her son. The subject may have commended itself to Montechrétien during his brief exile across the Channel; it was also of poignant interest to the French because of Mary's prolonged and unhappy wait at their court before her arranged marriage to the Dauphin.

In Hawkins's opinion, Montechrétien often surpasses Garnier, notably in his handling of choric passages.

With an overall dearth of truly good tragedies and sprightly comedies, the troupes learned that their best-paying offerings were farces, for which there was always a popular appetite and at which the French, with their tradition of *soties*, generally excelled. The group holding forth at the Hôtel de Bourgogne was well equipped to satisfy this demand by being possessed of three versatile comedians, Henri Legrand, Robert Guérin and Hugues Guéru. This trio had been co-workers in a bakery in the Faubourg Saint Laurent where their daily by-play summoned up so much laughter that they resolved to change their trade. Their first engagement took place near the Estrapade in a renovated tennis court, whence they moved on to another in the rue Mauconseil, where they were recruited to take serious roles – quite a change of casting – until they reached the Hôtel de Bourgogne, where the richness of their talents was fully realized. When asked to portray earnest characters, they gave themselves appropriate new names, Belleville, Lafleur and Fléchelles, and their interpretations were adequate if not outstanding. Then, reverting to comic parts at the Hôtel de Bourgogne, they altered their names once more: Legrand was now Turlupin, Guérin was Gros-Guillaume (Fat William) and Guéru was Gaultier-Garguille, and they perfected three distinct stock roles that they could be depended upon to repeat and embellish with minor variations in farce after farce: Turlupin was the rascally, conniving valet, Gros-Guillaume the insistent pedant, Guéru a confused, doddering old man. Gros-Guillaume wore two belts, one above and the other below his paunch.

Hawkins gives more explanatory detail:

Turlupin, in addition to being of good presence, had *élan* in a very high degree, and in the domain of broad comedy was held to be unapproachable. Gros-Guillaume, as may be inferred from his cognomen, was enormously fat. . . . He had a fund of rich humour, with large black eyes and strangely mobile features. He kept the audience in a continuous ripple of laughter, even when, as was not unusually the case, he suffered so acutely from an internal malady that tears ran down his face. Gaultier-Garguille was hardly less popular, though in a different way. Norman by birth, he could yet imitate the Gascon to perfection, and was dryly funny. . . . His success may have been favoured by peculiarly thin and bandy legs, but few things gave him greater pleasure than to hide this defect under the robe – the stage robe – of a king. Most of the songs and prologues attached to the farces were of his composition. For the rest, he married a daughter of Tabarin, the buffoon who disported himself on a scaffold by the Pont Neuf to attract attention to the remedies devised by the charlatan Mondor for the ills that flesh is heir to. . . . The Trois Farceurs became staunch friends, were never so happy as in appearing with one another, and opposed the introduction of an actress into the troupe upon the ground that they might all fall in love with her.

It is suggested that they learned much from their more experienced competitors, the *commedia dell'arte* troupes who were journeying about France and enjoying opportunities to appear at the

Hôtel d'Argent, which Henri IV had made available to them on certain days of each week – the king was especially fond of the Italian comics' brand of humour. Like the foreign visitors, Turlupin and Gaultier-Garguille frequently wore masks – Gros-Guillaume's broad, jowly face was likely to be whitened with flour – and Turlupin's customary attire resembled that of Brighella. It was through such emulation that the stock figures of the *commedia dell'arte* entered into and were assimilated by French farce, though under new names.

The Théâtre du Marais, lacking such an able trio, lagged in the more lucrative realm of farce. The troupe there retained an author-performer from Toulouse, Jean Deslauriers – on stage, Brascambille – whose material and style were somewhat more refined than that of the Trois Farceurs, which perhaps is why his offerings drew smaller audiences. *Fantasies*, two collections of his scripts – discourses, paradoxes, harangues, prologues and jokes – were published during his lifetime. The theatre's relegation to secondary status was to last for several decades.

But the Marais was a better choice if one preferred a poetic tragedy or an artful comedy, as differentiated from a noisy farce. The commercial theatre was soon changed, however, by the arrival in Paris at about the same time of two remarkable men. Valéran le Conte (or variously Valéran-Lecomte), himself an accomplished actor, had been leading a superior touring company around the provinces for a few years (1595–9), with fitful visits to the capital, appearing briefly at the Marais and elsewhere. In the last year of the century he joined his troupe to that of Benoit Petit, which was occupying the Hôtel de Bourgogne – the very group in which the Trois Farceurs were a prominent attraction. Superseding Petit, Valéran le Conte took a long lease on the premises, a contract that promised to give the city's stage a much-needed stability. In fact the troupe remained there for eighty years.

It did not work out well at first. His repertory consisted of the verse dramas of Jodelle, along with Biblical tragedies and a handful of comedies. For a time he was forced to take his combined company back on tour. By 1610 he was at the Bourgogne again, his troupe permanently established. His presence is not confirmed after 1613, though; what subsequently became of him is not known.

Seeking a reliable supply of new scripts to keep his company busy, he formed a tie with a newcomer to Paris, Alexandre Hardy (or Hardi; *c*. 1570–1632), who became in effect the troupe's resident dramatist, and whose work was destined to be even more conspicuous and personal history even more elusive, since not much is known about him, either. Actually, he was a Parisian by birth (some say in 1560, not 1570), but for a long spell there is no record of where he was and what he was doing. Obviously he had a sound education, as his scripts show easy familiarity with the Greek and Roman classics and contemporary foreign drama. He may have spent some months or years in Madrid. He finally came more clearly into sight as a strolling player going about his own country. From about 1622 on he began writing plays for the rival companies at the Marais and Bourgogne, before reaching a firm alliance with Valéran le Conte's group. He travelled with it on tour and later settled in Paris again when it did, and this time stayed there.

By count, Hardy's literary output can only be compared with that of Lope de Vega and Calderón, with whose works he displays a thorough acquaintance, which is one reason it is thought he may have been in Spain for a while. Estimates of the number of his scripts vary from 300 to 800, of which a mere thirty-four remain. He wholly justifies the claim made for him that he is France's first professional playwright. He had no patron; he earned his living by his pen and had to please his public, which for the most part was not made up of the aristocracy but of middle- and working-class spectators – he was never a court poet.

His works encompass every genre, from pastoral and dramatic poem to comedy and cape-and-sword to tragicomedy and sombre tragedy. He is most concerned with pace and physical action; he realized that his plays had to move fast. Exploiting the medieval platform stage, he sometimes had two and even more actions going on simultaneously: in one compartment or "mansion" might be a throne, and in another compartment next to it a scene occurring on the deck of a ship, and so on, presenting the audience with a multiplicity of events that compelled unswerving attention. He trimmed the rhetoric and declamation, the bombast, that was characteristic of plays of the period, often eliminating them altogether. He dispensed with choruses, cut the length of monologues, had the actors deliver their speeches naturally, with far fewer gestures, avoiding many poses and signs that were traditional and stylized. Verbal exchanges were reduced to the barest number of lines. He subscribed to Lope de Vega's tenet that the Aristotelian unities were no longer applicable. His incidents occur in many locations, often with substantial leaps in time, the events hardly compressed into twenty-four hours. Violent deeds were performed in full view, on stage rather than off. He intermingled comedy and tragedy. He made his plots even more complicated, with twists and turns for striking effects, always adding to the suspense, concentrating on that at the expense of everything else. Audacious, he allowed his lovers to embrace and kiss, hitherto in French theatres considered a gross impropriety.

A good many of his plays are hurriedly tossed off translations or adaptations, in whole or in part, from Spanish and Italian works – his apparent command of Spanish also argues for his having resided in that country for some length of time, as does his acquaintance with Lope de Vega's theories, and his acceptance of them. He must have seen how well they worked, as proved by Lope's amazing success.

Sometimes, like Lope, he would require only a week to turn out a play. He wrote in alexandrines, often inserting rhymed couplets, but just as often they are really meaningless. He is judged to have been a "facile and shallow" poet.

Seeking the public's approval, he won it. Consequently the troupe prospered; an announcement bearing his name almost ensured a full house. He had no equal in popularity. Even so, he remained poor, though on average he received three *écus* for a full-length script, which was unprecedented. This prompted him to write ever more rapidly and carelessly, lessening the value of his work by his hyper-facility. Combining ideas and incidents from scattered sources, he heaped them together so that they had an edged dramatic impact without much regard for plausibility. He is quoted as having said, "Heaven be praised, I can subordinate all loftier aspirations to the demands of my trade." The cost to him was that, an innovator, his slapdash output brought his innovations into disrepute and shortly inspired a reaction against them. Without meaning to, he prepared the way for France's neo-classical drama, a respect – even a reverence – for the unities he had discarded.

A tragicomic cycle of eight plays, *Théagène et Chariclée* (*Theagenes and Chariclea*; c. 1612), is considered to be among the best of his extant works; they are based on a Greek romance by Heliodorus covering incidents occurring over eight days. From Cervantes he took the ideas for *La Force du sang* (*Force of Blood*; 1615–25) and *Cornélie* (*Cornelia*; 1615–25).

His most admired script is *Mariamne* (1605–15), a tragedy on a Biblical theme first told by Josephus; it is essentially a psychological study of jealousy. Herod, now King of Palestine, is deeply frustrated by the attitude of Mariamne, his wife. She looks on him as a usurper responsible for the deaths of her grandfather and brother. She is denounced for having plotted to poison her husband, a charge brought by Herod's brother and sister, Pheroras and Salome, an envious pair who resent her power over the ruler. Mariamne, suicidal, offers no defence, nor does she deny that she has taken a lover, Soemus. Herod, his jealousy exacerbated, commands the immediate execution of the hapless Soemus; Mariamne's end is set for the next day. Herod, still inordinately in love with her, offers a pardon if she will acknowlege guilt. She remains silent, unmoved. Herod hesitantly assents to her death. Afterwards, desolate, he is overwhelmed with grief and regret and makes plans to raise a monument to her. This story was also used by Tristan L'Hermite (1636) and Voltaire, who in 1724 contrived not one but two versions of it.

Hardy, voracious for plots and characters, depended heavily on Greek and Roman myth and history, his long list of scripts showing an *Alcmaeon*, a *Coriolanus*, a *Lucretia*, a *Timocleus*, an *Alceste*, along with depictions of many spectacular demises: *The Death of Achilles*, *The Death of Darius*, *The Death of Alexander*, and also a drama about the suicide of Dido, the Carthaginian queen whose woes seemed endlessly to engross Renaissance spectators. Among his pastorals are *The Rape of Proserpine* and *The Triumph of Love*.

The acclaim and profit accruing to Hardy impressed other playwrights, drawing them to the Hôtel de Bourgogne, but their number included some who did not agree with him that the Aristotelian precepts should no longer prevail. Their aims were more aesthetic. Théophile Viaud, whose father was an advocate in Bordeaux, created a stir with his *Pirame et Thisbé* (1617) when he was twenty-six, but his career as a stage-writer was all too quickly curtailed. Overly fond of writing licentious verse, and with it courting considerable notoriety, he gave offence to those in authority and was sent into exile. He did not refrain from the risky practice, however, and next was suspected of having had a hand in a scurrilous piece, *Parnasse satirique*, which brought even heavier official wrath on his head; the Parlement found him guilty of *lèse majesté Divine* and decreed that he should be burned alive in the Place de Grève, an edict that prompted his renewed flight. In his absence he was incinerated in effigy, a torch put to a bundle of rags intended to represent him. Eventually he was caught in Picardy, brought back and thrown into a dungeon at the Conciergerie, and told that now his sentence would be carried out for certain. During a prolonged second debate in Parlement several powerful voices were raised on his behalf while he waited, agonized; at last it was decided that he should spend another season in exile. Apparently chastened, he changed his ways, won forgiveness from the Church, did lead an exemplary life, which unfortunately was to be a brief one, being taken by death while still quite young. It is not certain that he ever wrote another play, though a tragedy, *Pasiphae*, unacted, is tentatively ascribed to him.

Another poet who adhered to the classical norms was the Marquis de Racan, who contributed a pastoral, *Les Bergeries, ou Arthenice*, that was much cherished by the better-educated spectators. Though inadequately tutored while young, he was apparently acquainted with Tasso's *Aminta* and Guarini's *Pastor Fido*. In this work, as judged by Hawkins, "comparative regularity is allied to refinement, elegance, and tenderness". It is the Marquis's only known play-script.

Very good-looking, Jean Gombaud won the attention and favour of Catherine de' Medici by submitting to her a poem lamenting the death of Henri II, her husband, and rivalled the Marquis of Racan in fashioning pastorals rather like his.

The fourth of these poets, and truly precocious, was Jean Mairet (1604–86). Belonging to an old Roman Catholic family, he was born at Besançon, where his grandfather had sought a haven at the start of the Reformation and the advent elsewhere of the militant and threatening Huguenots. The plague having taken away the youth's parents, he became a student-boarder at the Collège des Grassins in Paris, and while there wrote his first play at the age of sixteen, *Chriséide et Arimand* (performed 1626), a tragicomedy. He followed it with *La Silvie* (1628), a pastoral, also in a tragicomic vein. He went on to enjoy a prestigious literary career, having the doors of the most influential Parisian society opened to him while he was still very young. In particular, he was persuaded by two dedicated classicists, the Cardinal de la Valette and the Comte de Carmail, to pay close heed to the ever more sacred three unities, of which he became an ardent champion. Among his later plays are *La Sylvanire* (1631) and a *Sophonisba* (1634), which firmly embody Aristotelian concepts about form.

The issue was hardly settled as yet. The French intelligentsia has always engaged with fervour in sharp controversies and debates about political, theological, philosophical and aesthetic ideas. Hardy was gone, but there were vocal partisans who defended his emancipating views about dramatic structure and agreed with him (and Lope de Vega) that the classical unities put a hindering clamp on the playwright's freedom and his decision about how best to tell a story on stage. One was Balthasar Baro, at one time secretary to the Marquis d'Urfé, who made use of the Biblical tale of Judith and Holofernes in his *Celinde*, termed a *poème héroïque*. He also completed *Astrée*, a piece left unfinished by his former employer, the Marquis. Both of these are meritorious works, though *Astrée* abounds in "unreal shepherds and shepherdesses", as pastorals of the period were too apt to do.

Well educated, but sparsely paid, Pierre du Ryer (*c.* 1600–1658), who filled a government post, hoped to augment his income by writing loosely constructed tragicomedies – *Clitophon* (1629) and *Argenis and Poliarque* (1631), and their ilk – as well as farces that served as handy vehicles for Gros-Guillaume. But he was persuaded by the example of Mairet, who was becoming the dominant playwright of the moment, and swung over to neo-classicism; the consequence was a series of properly put-together tragedies in the mode of his *Scévole* (1644) which stubbornly held a place in the repertory for over a century. Du Ryer is credited with having helped to make tragedies more attractive to the theatre-going public, essentially creating a new French audience for them.

The conflict of ideas reached its climax at the close of the first quarter of the new century, the seventeenth, with the classics getting the upper hand. Even then there was an overlapping of styles, especially when a major playwright, Jean de Rotrou (1609–50), made his appearance and added to the theatre of the day his ever-growing pile of scripts as non-conforming and free-ranging in structure as Hardy's. But most chroniclers hit on 1625 as the date that marks the end of the French Renaissance. Soon afterwards a half-dozen significant dramatists arrive, men of genius, and their nation's stage has its own delayed Golden Age.

After Valéran's departure (1612), the troupe at the Hôtel de Bourgogne had a series of managers. He had given the company an impressive name, Les Comédiens du Roi (The King's Players), possibly because it had performed for the sovereign on some occasion; actually, no French ruler had granted to it the right to that title, nor any subsidy or exclusive licence. Still, no objection seems to have been raised to the use of that pretentious self-identification. The membership was for ever in flux, some actors leaving for any number of reasons, others replacing them, especially after the deaths of the Trois Farceurs, who had been the star attraction. The players usually signed two- or three-year contracts for their services. The day's receipts were divided after each performance, each actor possessed of a share, and the director of two shares. Those holding lesser rank, who had smaller parts, might own only a fraction of a share. The troupe usually consisted of eight to twelve; if need be, the cast enlarged by outsiders and apprentices; and by 1607 it was commonplace for women to take roles where it was appropriate. A stigma was still attached to being a player, which is why it was customary for those in the profession to adopt a stage-name, particularly if one had a family inclined to be quite religious or socially ambitious. It was nothing to boast of, but instead a scandal, having an actor in the fold.

Because many spectators arrived very early to assure themselves of good places, the manager might assign a player to serve as a "Prologuist" to divert the restive crowd. He would stand alone on the empty stage, telling jokes and seeking to entertain those impatiently awaiting "curtain time". (Actually, there might not be a curtain, since the stage lacked a proscenium.) Standing, and perhaps jostling, in the pit, the crowd might start a commotion. Since the cavaliers in the throng wore swords, and others might have daggers, there was the risk of angry responses leading to quarrels and outright violence, even fatalities. Adding to the bustle and confusion were vendors, elbowing and pushing through the throng to sell food and vinous drinks.

At the time of Valéran's death Guillaume des Gilleberts (1594–1654), one of his leading actors whose stage-name was Montdory (or Mondori, or even Mondory), quit the King's Players and entered the Prince of Orange's Players, where he stayed for many years. Eventually, in the late 1620s, he joined forces with Charles LeNoir (*fl.* 1610–37), also an actor-manager who had been active in Paris. Together they headed a touring troupe, which at the decade's end returned to the capital. After moving from one temporary venue to the next, they took over the Marais and became very serious competitors to the long-resident group at the Hôtel de Bourgogne and finally surpassed it, being recognized as Paris's best group. LeNoir had to leave. Montdory was deemed a great performer, superlative in roles as a tragic hero, imparting conviction to his characters, though tending to be over-declamatory. He helped in the rise of the neo-classical faction, which he favoured, producing and acting in their scripts. This won him the approval of Cardinal Richelieu, who awarded him a pension in 1634. In 1637, unfortunately, Montdory suffered a partial paralysis, ending his career. The cause of his illness is said to have been the extraordinary force and emotion with which he portrayed the jealous Herod in Hardy's *Mariamne*, losing the use of his tongue.

In France the development of opera was preceded by that of ballet; both were imports from Italy, and their evolution took place in the baroque era rather than the Renaissance. The link of music with plays, however, was established in France in medieval times, in the Corpus Christi vignettes at the altar and later on the ranged mansions in the square, illustrating in physical action the Scriptural stories, and music was always a bright component of court entertainments. The troubadours and balladeers depended on music to heighten the dramatic and emotional effect of their narrative songs. At court, such works as Adam de la Halle's *Le Jeu de Robin et de Marion* were enhanced by music, but they were nothing like an opera, where the full burden of revealing the story is borne by song and the music is dominant.

What the French discovered almost by themselves was dignified, highly stylized forms of dancing at elaborate fêtes, staged and led by professionals, and very often with royalty and habitués of the court themselves participating. Such entertainments became very much the vogue with the arrival in 1533 of Catherine de' Medici and her Florentine entourage, including a troupe of well-trained dancers, a keen choreographer and set-designers whose flair and skills wholly captivated the French. But this signal event, though all-important, had been prefigured by a long train of local advances in finding new postures, rhythms and graces. This was not exuberant folk-dance. As put by Ferdinando Reyna: "The rich disorder of the public squares had to be organized and reduced to a scale proper for a palace hall. An end had to be brought to the wearisome solemnity of the *Basse-Dance*. What had until now been a public festivity had to be converted into a private entertainment, of a prescribed length, coloured by individual tastes and ambitions. In short, the semi-improvised had now to be made a work of art." Reyna also recounts that at the end of the fifteenth century, when Charles VIII and his troops returned from Ferrara, he described himself as having been enchanted at what he saw there. "He who wants to be transported from this world to the next, let him listen to Pietrobono playing; he who wants the heavens to open, let him experience the liberality of Duke Borso; and he who desires to see Paradise on earth, let him watch madonna Beatrice dancing in a festival." The impulse to imitate the Italians was born.

The poet Jean Antoine de Baïf, who had belonged to the Pléiades, had quite radical ideas about not only using classical metres in his verses, but also combining his lines with music and dancing, a synthesis like that accomplished by the Ancient Greeks. But he was never to see this realized in his lifetime. He founded an Academy of French Music and Poetry, but it was short-lived. Still, his concepts were implanted in minds of others who followed him.

Louis XII (1462–1515) warred in Italy and was defeated; he made peace and brought back Italian actors and musicians, who had much to teach the French court, patricians and intellectuals not altogether ready to accept foreign ideas or to deem them better than their own. This attitude implanted a continuous tension between admiring the Italians and rejecting them. Henri II shared his wife Catherine's taste for lavish entertainment, and he returned from a protocol visit to Italy with the Gelosi, the famed *commedia dell'arte* players, in tow. Catherine's sponsorship of remarkably costly fêtes at court had a political aim to charm and disarm her foes by her hospitality, as has been said, but she also had an inherent delight in such displays – it was a family trait. Who had ever outshone her father, Lorenzo the Magnificent? It was also a matter of pride: to the French the Medicis were *nouveaux riches*, wealthy bankers from whom French royalty borrowed huge sums they had much difficulty repaying. She was proving that a Medici could outdo everybody in a display of regal sumptuosity. Besides, this was not at her own expense but rather at that of the state treasury, a *modus operandi* more royal than mercantile.

Many Italian performers and artists who had come in Catherine's retinue stayed on, finding a welcome in her court. In 1567 she retained the services of Baldassariono di Belgiojoso (c. 1535–87), a violinist who had come to Paris on his own a dozen years earlier (and who had duly changed his name to Balthasar de Beaujoyeux). He became Catherine's unofficial organizer of festive events after he had been introduced to her by the Duc de Brissac, and was subsequently awarded the title of Intendant of Music and Court Valet (1567) and put in charge of all royal entertainments by Charles IX, her second son. Though by profession a violinist, he had a good eye for colour and design, to which he paid close attention in his court offerings.

Some of his presentations were enacted inside the palace, others in the open. On one occasion delegates arrived from Poland, their mission to offer their country's crown to Catherine's son, the Duc of Anjou, who had made a name for himself in military conflicts. To greet the envoys with éclat, Beaujoyeux had a simulated gigantic rock erected in a palace hall. He had the artificial rock coated with silver paint and sixteen niches hollowed in its sides and decorated to look like clouds. In each of these was hidden a young woman symbolizing provinces of France. On cue to music, they stepped forth, descended, and accompanied by a full orchestra – thirty violins – executed with astonishing precision a series of "*tours mêlés, contre-tours, détours et arrêts entrelacés*" that mesmerized the Poles.

Such a work, which might be flatteringly allegorical but told no specific story, and intermingled music, poetry and dance, was called a *ballet de cour* (the term "ballet" was applied to any kind of stage offering); Beaujoyeux excelled at assembling them. Most such offerings were staged in a hall, the Salle du Petit Bourbon, in a palace alongside the Louvre. Its dimensions were ample, fifty feet by 180, with a deep, rounded apse that added another fifty feet and provided space for the king, the courtiers and other guests, almost seeming to enshrine them. Overhead, balconies jutted out from side walls and accommodated other spectators.

Beaujoyeux's great opportunity came in 1581 when he was asked to prepare a gala to honour the king's brother, the Duc de Joyeuse, and Mademoiselle de Vaudemont, the queen's sister, who were about to marry, strengthening familial royal ties. He withdrew to the country to contemplate what this very special, resplendent offering should be. He chose Olympian mythology, a world peopled by the enchantress Circe and her circle of bizarre creatures; a scenario depicting this domain by Agrippa d'Aubigné had been turned down as too difficult to mount, but Beaujoyeux determined to do it. He gathered four talented helpers: a lyricist, Chesnay; a set- and costume-designer, Jacques Patin, the king's painter; and musicians for the score, Lambert de Beaulieu and Jacques Salmon. Beaujoyeux added some songs that he himself composed – an admirer, Brantome, declared him to be "the best violinist in Christendom". The gifts of these discrete artists were fused, resulting in a nicely balanced work.

The cost of the production rose precipitously to an estimated $800,000 in today's currency. Catherine, the Queen Mother, who had commissioned the extravagant entertainment, seems not to have blanched. It was called the *Ballet comique de la Royne* (or *Reine*). What was truly exceptional about it was Beaujoyeux's hitherto unique prescription: the steps were to be very closely dictated and guided by the musical notes, and the notes were to match as exactly as possible the syllables in the lyrics. Furthermore, the scenario (or "book") comprised a fully developed drama, and the dancers had to "act" their roles, to portray convincing characters.

Here, not in paraphrase but in full, is Reyna's synopsis of this historically significant advance:

The princes and courtiers assembled around Henri III were seated on a platform covered by a canopy around Henri III in the Great Hall of the Petit-Bourbon Palace to watch the opening of the *Ballet comique de la Reine*. The plot begins with the escape of a prisoner of Circe, come to seek assistance of the King in a long tirade full of political allusions. Circe, vexed at the loss of one of her captives, returns haughtily to her cast. At this point comes the first interlude, with an entry of three sirens and a singing triton. A dozen naiads, seated on a chariot, were played by the Queen and ladies of the Court, their faces unmasked. Peleus and Thetis sing a duet, while, from the golden ceiling, an invisible choir responds; twelve pages and twelve naiads dance elegant geometrical figures; suddenly Circe's wand strikes everyone motionless. Mercury then descends from a cloud and frees the prisoners, and the dance of the nymphs begins again – only to stop once more, for now the goddess has cast her spell over Mercury and lured him into her enchanted garden. There then appear a stag, a dog, an elephant and a lion, Circe's rejected suitors, whom she had bewitched into the shapes of animals.

In the second act come eight satyrs playing the flute, who approach four naiads, bow in hand. The satyrs intone a hymn, followed by a maiden who sings homage to the King. All of them go to the grotto of Pan, where the nymph Opis entreats the gods to break the spell. In immediate response to this prayer Minerva appears on a triumphal chariot and delivers a speech before the King, followed by a chorus of six voices. There is a thunderclap, and behold Jupiter on a cloud, draped in a golden tunic. Pan, at the head of eight satyrs, leads the attack on Circe's garden; there is a show of resistance; Circe falls, struck down by a thunderbolt from Jupiter. And all the propagandists march round the hall of the Petit-Bourbon Palace paying homage to the King. This was the great moment, the moment of the *Grand Ballet*, described by Beaujoyeux in his memoirs.

"It was then that the violins changed key and began to play the entrée of the *Grand Ballet*, composed of fifteen passages so devised that at the end of each one all turned their heads toward the King. Having arrived before His Majesty, they danced the *Grand Ballet* of forty passages of geometric figures, some diametrically, some in a square, some in a circle, in many and various fashions, and also in a triangle, accompanied by a few other little squares and other figures. . . . These geometric evolutions sometimes took the form of a triangle with the Queen at the top of it; they revolved in a circle, interwoven like a chain, tracing various figures with a cohesion and accuracy which astonished those present."

The evening's entertainment ended at half-past three in the morning. It had gone on for ten and a half hours, but allowing for court ceremonial, refreshments, dancing during the intervals, and the still rudimentary "machines" which had hardly changed since the days of the Mysteries, the show had lasted barely three hours.

In Italy there were theatres, an appreciative public, a tradition of art as part of civic life. In France – the *Ballet comique de la Reine*, the first truly French "Court ballet," since it had been organized, danced, sung (with the exception of one professional singer) and financed by courtiers and royalty. Its chief importance lay in the way the whole work was co-ordinated. It was, in fact, closer to light opera than ballet. Beaujoyeux, satisfied with his work, wrote: "I think I may claim to have pleased, with a well-balanced production, the eye, the ear and the understanding."

For some years Beaujoyeux's productions were almost mechanically imitated by others. Their stagings grew ever more literary, verbal rather than visual and physical, making use of a

weakening formula. At the instance of Sully, his finance minister, Henri IV, the erstwhile Huguenot, kept a careful eye on his purse and the state budget, and after 1620 no more expensive spectacles were put on at court, though the king did remain fond of fancy-dress balls (*mas gerades*). A lull in experiment, in the discovery of new choreographic concepts, in the improvement of performance techniques, was fated to last until the advent of the young, dance-loving and free-spending Louis XIV.

— 10 —
TO GERMAN BAROQUE: HOLLAND

The havoc of the Thirty Years' War (1618–48), during which German principalities and duchies were wracked by civil strife, was hardly a setting in which the arts could be expected to flourish. The Netherlands, too, was sharply divided by religious and fierce ideological differences, having a predominance of staunchly faithful Catholics in its southern province – Flanders, now part of Belgium, allied to Spain – and obdurate Protestants in the north, who ferociously resisted foreign occupation and oppression. Yet in this small country, beset by the sea, hovering behind dykes, and often under dark skies before, during and after the horrible years of struggle, an efflorescence of artistic creativity yielded a Golden Age in painting and drama. Partly this was possible because an expanding, well-to-do middle class already existed there, brought about by astute and energetic trade at home and abroad; its members were willing and able, and even eager, to support and patronize artists. This was a long era of ballooning prosperity, the consequence of which was a market for its wonderful painters – Rubens, Van Dyck, Hals, van Goyen, Rembrandt, to list only a few of their inspired number – and this period also witnessed the career of the Netherlands' greatest playwright. Most of the successful painters were Catholic and lived and worked in the region where their faith was not likely to be challenged; it was in the north that for dissenters there was constant fear of the Inquisition, and where too many true believers in religious reform knew suffering and exile. In sum, still having a region where peace and normal life prevailed, some fortunate Hollanders could continue to cultivate the arts, whereas most besieged and impoverished Germans, local rulers and intellectuals, could not.

The Rederijkers (Rhetoricians), a writers' circle founded by Flemish aristocrats at the end of the fourteenth century to enact Mysteries and argumentative Morality Plays – and likened to the Meistersingers in neighbouring Germany – was still active during these days and indeed reaching a peak. By now it had many branches and imitators in towns throughout Flanders. Over the years the status of its members was more likely to be middle class, though each of the dispersed circles had a titled patron and a hierarchical set of leaders: a dean, a standard-bearer and so on, much as in the tradesmen's and craftsmen's guilds. To gain and hold prestige, honorary memberships were offered to princes and dukes, too, and the Amsterdam chapter (or "chamber") had received a royal nod of approval from an august personage, Emperor Charles V. As noted earlier, each circle adopted a bloom as its symbol – for example, Amsterdam had the eglantine and Antwerp the violet – covering the range of species familiar to these stalwart, flower-loving citizens.

An annual tradition was a gathering, in one city or another, of all these groups, a very festive occasion that came to be called the *landjuweel* and might go on for several days. It might feature an allegorical procession, *tableaux vivants* and finally a drama contest, which was usually the high point of the much-anticipated convening. This provided a stimulus and opportunity for authors and kept the stage alive. Most of the plays were on religious subjects, still the most relevant topic, but a measure of secularization inevitably crept in. Invitations to participate in the competition often suggested several themes, put forth by an eminent or regal guest, to be taken up by that year's entries; for instance, in 1561 Margaret of Austria, Duchess of Parma, studied a list of two dozen titles tendered to her and chose three: "Is wisdom fostered more by experience or by learning?", "Why does a miser desire more riches?", "What can best awaken man to appreciate the arts?" (quoted from Berthold). The "Violets" of Amsterdam took on the last of these. The programmes, too, were sometimes preserved in unconventional ways, as when a year later (1621) a chapter at Malines had all the spoken and musical contributions set down in print and illustrated with woodcuts, the whole designated rather immodestly as *A Treasure Chest of Philosophers and Poets*.

One of the playwrights who shone at the *landjuweel* was Pieter Corneliszoon Hooft, whose father was the mayor of Amsterdam. His *Achilles en Polyxena*, enacted in 1614, is credited with having brought the classical revival to the Netherlands. He was further influential when his *Granida*, a pastoral bearing traces of the universally admired *Pastor Fido* of Guarino, added this particular new genre to the repertoire embraced by his countrymen; and his Senecan tragedy, *Geeraerd van Velsen*, demonstrated how effective Aristotle's three unities could be – it was the first Dutch play to follow those cardinal rules so carefully. Its subject-matter, too, drawing on history, was Dutch.

G.A. Brendero, belonging to the Amsterdam Eglantines and a contemporary of Hooft, won a popular following with hearty farces and comedies, highly realistic, abounding with characters who might have been lifted from the rowdy stage-works of Plautus or taken from the painted riotous *kermesse* scenes of the elder Brueghel who had depicted such crude peasant merriment a few years earlier. Most of these works were produced on *Klucht* stages, but others, like his *Spaanschen Brabander* (1617), were also performed by fellow members of his chapter of Rhetoricians.

Having gained respect and eminence, the plays given at the *landjuweel* needed to have better settings, and eventually this came about. "An architectural rear stage was closed off to limit the acting space; the partition accomplishing this was decorated with columns and the illusion of arcades, some of them as much as two storeys in height, and an excellent backing for the *tableaux vivants* and serious dramas. To lend a further air of importance to the presentations, they would be preceded by the processions of allegorical figures mentioned before. At their apex, the Rederijker gatherings echoed with Humanist ideas and classical erudition. They also borrowed some of the craft of English companies visiting the Netherlands, resulting in a setting in which relics of the ancient *scenae frons* merged with elements of the Elizabethan stage" (quoted from Berthold).

The voice of a major poet was heard at some of these competitions. Joost van dem Vondel (1587–1679), the most celebrated of Dutch playwrights, was born in Germany. His parents, Anabaptists fleeing Antwerp and the Inquisition, settled in the cathedral city of Cologne. Later, when the boy was twelve, they moved to Amsterdam, where the father established a thriving hosiery workshop. At about seventeen the younger Vondel began to write poems (1604). In 1610 the father died, and Joost, the eldest son, inherited the hosiery business. Shortly after, he married Maria de Wolff, a Fleming who had also been born in Cologne. He let her attend to the workshop while he set about a literary career. His talent was already so evident that he was accepted into the Amsterdam chapter of the Rederijkers – there were several in the city – called Her Wit Lavendel (The White Lavender). In 1610 his first play was published: *Het Pascha ofte De Verlossinge der Kind'ren Israels wt Egypten* (*Passover, or The Deliverance of the Sons of Israel from Egypt*). It was natural that he chose a Biblical subject, for religious questions were always very much on his mind. The mode and text of his initial work also make it clear that he had been closely studying French neo-classical dramas, while Dutch stage traditions lingered – and had some force in his imagination, suggesting with no little potency the shape that his scripts were to take.

That he wrote for the stage was frowned upon by fellow Anabaptists, to whom most worldly activities were sinful and abhorrent. In time he broke with them, deeming them too fanatical. He attached himself for short periods to other sects, searching, shifting uncertainly from one to another.

To enrich his mind and his work, he studied Latin, Italian and French literature, while he poured forth verses. In 1620 he was in serious trouble and went into hiding after writing *Hierusalem Verwoest* (*The Destruction of Jerusalem*). Prince Maurice of Orange had decreed the execution of Johan van Oldenbarneveldt, a statesman who outspokenly advocated political and religious freedoms, and of whom Vondel was a strong adherent. His *Jerusalem* play was a scarcely veiled protest against the prince's too drastic penalty. After a spell of concealment, however, Vondel reappeared and was let off with a fine. He then vented his anger against the conservative Calvinists who were partly responsible for the death sentence imposed on Oldenbarneveldt by the prince. Vondel's allegiance, at this point, was with the Arminian Remonstrants, a more tolerant faction of the Calvinists.

The years from 1632 to 1635 were tragic ones for him, with his wife, a daughter and a son taken from him by death in rapid order. During this dark period he wrote hardly anything. He was already an eminent figure, however, and in 1637, when Holland's first civic theatre was built and about to be opened, he was commissioned to provide the inaugural play, a decided honour. He responded with *Gysbreght van Aemstel*, an epic paying tribute to the city of Amsterdam. The première took place on 3 January 1658; afterwards it became a tradition to produce it on New Year's Day, an annual event observed until 1968, a stretch of more than three centuries. His doubts and questions about religious faith, implanted by the loss of his wife and children, are revealed in it. To the considerable regret of his friends, he converted to Roman Catholicism in 1641, still seeking a firm and lasting faith.

While turning out many more plays, he was not spared further harsh personal blows. After his wife's death he had entrusted the family business to another son, who handled affairs so badly that he finally took flight from the angry creditors to the East Indies, leaving his father to make good on the piled-up debts. Impoverished, the aged Vondel – now in his seventies – had to take a position as a clerk for ten years (1658–68). Yet during this wearisome decade he wrote "some of the greatest poetry in the Dutch language". (This is the critical judgement of H.H. De Leeuve, whose article on Vondel has been helpful here.) Finally, as he was nearing his eighties, the city of Amsterdam took notice of his plight and awarded him a pension. He never ceased his literary tasks, at eighty-four completing a translation of the lively Latin poet Ovid.

In all, he is the author of thirty-two plays, their content progressing steadily from the secular to the ever more religious, as he sought to adapt Biblical matter and Christian doctrines to dramatic forms dictated by the classical theorists and masters: Sophocles, Euripides, Seneca. He embraced the three unities wholeheartedly. In his acceptance of neo-classicism he was following the Italian and French writers whom he had carefully perused, especially Tasso and the Seigneur du Bartas. But despite this infusion of aesthetic concepts from antiquity and foreign sources, his works are considered "thoroughly Dutch". His success depends largely on his superiority as a poet and his earnestness. His plays lack powerful dramatic action, but achieve an effect – asserts De Leeuve – because the dialogue has a "hypnotic rhyme and a rhythm that captivates audiences. They have been called semi-tragic, and they focus primarily on man's struggle to control his rebellious nature and submit to God."

He ceaselessly struggled within and without to define and direct his faith – a gradual approach to Catholicism is perceptible in his work. He also became quite rigid about obeying Aristotle's precepts; displeased that it did not embody the three unities or present a character "who is more than a tool in the hands of God", he subsequently tore up and threw away his first script, *Passover, or The Deliverance of the Sons of Israel from Egypt*, which obliquely referred to the persecution of the Anabaptists and their flight into Germany, a bitter experience shared by his parents. The attraction the classics exerted for him is evidenced by his adaptation of Seneca's *Phaedra*, in his version retitled *Hippolytus* (1628); two years earlier he had similarly borrowed from the *Troades* of Seneca, labelling his offering *De Amsteldamsche Hecuba* (1626), endowing the story of the weeping women of Troy with more local relevance. Thence he burrowed into the Scriptures for a lengthy chain of scripts about the dominant figures found in them. They are brought to life in a trilogy about Joseph – *Joseph at the Court, Joseph in Dothan, Joseph in Egypt*, all in 1640 – after which he added, in the year of his acceptance of Roman Catholicism, *Peter and Paul* (1641), in which he sought to explain and justify the intolerant zeal displayed by that Church in enforcing its dogmas. He wrote of Solomon in a tragic light, and of the burly Samson – *Samson, or The Holy Revenge* (1660), portrayed with an epic sweep – and a *King David in Exile* (1660), coupled with a *King David Restored* (1660). Among his Biblical heroes are an Adam and a Noah: *Adam in Exile, or The Tragedy of Tragedies* (1664) and *Noah, or The Fall of the First World*

(1667). Along with these dramas on religious themes he included a *Mary Stuart, or Martyred Majesty* (1646).

He did not wholly abandon subjects from Hellenic antiquity, however, but paused from time to time to offer more adaptations, selecting for his own interpretations Euripides' *The Phoenician Women* (1660) and Sophocles' *Trachiniae* (1660). In his works are combined Christian and pagan concerns and motifs. Somewhat unexpected in his output is a very late tragedy, *Ch'ung-chen, or The Fall of the Chinese Empire* (1667). (Many of these dates apply to publication rather than enactment.)

In his celebratory *Gysbreght van Aemstel*, composed for the inauguration of the national theatre, Vondel returns to an important historical event, the siege and destruction of Amsterdam in the fourteenth century. Striking out from his castle on the banks of the Amstel river, Gysbreght drives off vicious attackers and rushes out in pursuit of them. A foe, Vosmeer, is taken captive, apparently of his own volition. Recounting mistreatment by his companions-at-arms, he is viewed by Gysbreght as a deserter and welcome source of information about the enemy. In the castle the defenders of the city give thanks to God for the lifting of the siege and the rout of its attackers. Vosmer tells Gysbreght that the enemy had been gathering a huge pile of firewood; he suggests that it be brought inside the walls to deny any further use of it by the besiegers, if they return and renew their assault. When Gysbreght acts on this, the enemy spring forth from their hiding-place beneath the pile and set fire to the city. Beholding it in ashes, Gysbreght asks to perish with it. He is prevented by the Archangel Raphael and told to go with his family to Prussia, to establish a town there to be called New Holland. The Archangel declares that the destruction of Amsterdam was ordained, but he promises that it will some day be restored and shine among the world's cities.

Vondel considered *Jeptha of Offerbelofte* (*Jeptha, or The Promised Sacrifice*; 1659) his best play. Excessive and blind religious zeal and a failure to heed and act as God wills are responsible for Jeptha's downfall – this reiterates Vondel's thesis that man's sufferings result not from external forces but from inner flaws. In the view of many critics this is even better exhibited in *Lucifer* (1654), a fine example of baroque drama. In five acts, couched in alexandrines, this tragedy shows Lucifer eaten by jealousy of man, favoured by God, who has given him an Eden, a paradise, in which to dwell, and elevated him to a rank even higher than that of the Angels. Others in the heavenly company are envious and rebellious, too, and seek Lucifer's allegiance, offering him leadership. He should command them in an assault on Michael, God's chief lieutenant. Hesitant, he is torn between the emotions of loyalty and defection. He finally submits to their flattering offer. Tragic is his belated recognition of his profound error, that his inordinate pride has led him astray, for he no longer has the strength to beg forgiveness and a reconciliation with God. Defeated, he is metamorphosed into a bestial creature and plunged into darkness, though not before he has caused the expulsion of Adam and Eve from the lush and lovely garden prepared for them. It is believed that John Milton was acquainted with *Lucifer* and was influenced by it.

Vondel's plays found audiences during his lifetime and afterwards outside the Netherlands, most especially on German stages, and often there because they were in the repertoires of strolling Dutch acting troupes that gained rising popularity abroad, rivalling English companies and ensembles native to Germany itself.

But the Rederijkers, having the peak of their powers, quickly began to decline; membership dropped, and chapters that had been proliferating started to vanish. In Amsterdam, where there had been several, their number shrank to two, and those two finally merged, bearing a new name, the Academy – their performances took place in a theatre given to them by the city in 1618, called, in Dutch, Schouwburg, which translates as Playhouse. Contemporary engravings show what it was like in 1638, the year in which it was officially taken over by Amsterdam as a municipal theatre, the first such in modern Europe.

The Schouwburg's architect, Jacob van Campen, had travelled about Italy studying theatre design. His stage combined features of those utilized by the Rederijkers and Italian actors, with no proscenium arch or front curtain, and along both sides and the rear rows of pilasters, between which there was space to insert painted panels to indicate changes of location. The panels, reversible, might have different pictures on the front and back, to speed the scene-change even more by simply swinging them about. Instead of panels between the pilasters at the back of the stage, there could be openings to allow deep vistas or create the illusion of receding arcades by rather modest tricks of perspective. A curtain midway to the rear and across the stage also achieved quick changes of place by dividing the stage in two, front and back; it concealed behind it the shifting of furniture and props to prepare a fresh location. If preferred and helpful, old-style medieval mansions might be put up on the stage to facilitate multiple settings for simultaneous or alternating or even ironic contrasting action. Over much of the stage hung a balcony which provided extra space for the actors, or an extra location; and, if needs be, an inner stage – perhaps for a bedchamber – would be fashioned at the middle of the back façade where the opening to a distant vista might otherwise have been.

The stage was elevated seven feet above the floor of the auditorium, which – a novelty – was ovoid in shape. Spectators stood in the pit, which was forty-six by twenty-three feet, encircled by two tiers of boxes – ten of them – reserved for the wealthy and titled; above them was the usual open gallery.

The ceiling was vaulted, and both stage and auditorium were partially gilded and ornate; the theatre was torn down in 1664 and replaced by a more Italianate edifice known as the Nieuwe Schouwburg.

In early seventeenth-century Germany drama did not fare as well. No one of Vondel's stature appeared, and the one playwright of distinction, Andreas Gryphius, had no contact with a theatre. Foreign players continued to visit courts and colleges, and perhaps public fairs, or perform outside a village inn, when and where travel was safe, as the murderous war between the obsessive religious sectarians swept across and cruelly eddied in regions near by. First from England came a troupe headed by Thomas Sackville, as previously noted, and afterwards a venturesome company led by Robert Browne (from about 1590 to 1606); others were from Holland and France – and a wandering Italian *commedia dell'arte* group was apt to show up sporadically somewhere or other, by invitation or on its own initiative. The English were particularly encouraged to go abroad by difficulties with licences at home, and later because during Cromwell's Puritan regime theatres in England were closed completely.

Serious-minded, Robert Browne had elevated literary tastes and brought from London a repertory of plays by Kyd, Marlowe, Shakespeare, Lope de Vega, Calderón and other esteemed and weighty dramatists. He was able to promise the town councillors of Frankfurt, when applying for leave to perform in their city, offerings to inspire in honest spectators "cause and occasion to pursue propriety and virtue". If anything, his choices were too earnest for many German theatregoers, and too different from the knockabout material of the Italians. Discouraged, he returned to England, leaving his troupe in the charge of John Green, an ambitious and enterprising member of the group who was outstanding in comic roles (from 1606 to 1628). Inevitably, John Green changed the group's image, and it thrived. Applied to entertainers, the word "English" came to imply "excellent".

Subsequently, Browne formed another company that he took to Prague, where they found a welcome at the court of the Elector Palatine Frederick and his queen Elizabeth, who was of English birth.

Another nomadic troupe was that of John Spencer, who first gained attention in Leiden and The Hague, after which he and his players roamed over the German states to appear in Stettin, Dresden, Königsberg, Danzig. In 1615, for reasons that were honourable or expedient, he converted to Catholicism, an act that permitted him to perform in more venues and even during the Lenten season (from 1605 to 1623).

The actors in these companies had to master some degree of ease in German, if not fluency; shortly, some local players were added to their rosters, and after the troupes dispersed a number of English members stayed on, recruited into mostly native groups that were eventually organized in the second half of the century.

A late-arriving company was that of George Jolly, making its mark in Germany during the 1650s. After 1680 such incursions by groups from England were ended.

Successful competitors, because they presented Vondel's plays, were invaders from Holland, especially a troupe from Brussels under the patronage of the Archduke Leopold Wilhelm of Austria. After an engagement in Amsterdam it held forth to applause at Gottorp Castle in Holstein (1649) and toured on to Flensburg, Copenhagen and Hamburg, before returning to Amsterdam in 1653. Their leader was Jan Baptista Fornenberg, whose thespian skills were much lauded.

Only rarely were the foreign scripts translated; to help spectators comprehend them, the actors would baldly and off-handedly interject German phrases and idioms here and there, and perhaps even vulgarisms in hit-and-miss fashion, and resort to broad pantomime and gestures. It was also necessary to simplify plots and motivations, as well as add extraneous passages of uncouth humour. So it can hardly be said that this string of strolling players did much to create audiences that were sophisticated and ready to appreciate the refinements of theatrical art.

The Jesuits still put on dramas in their seminaries and colleges, enlisting students for roles in them, choosing versions of the ancient classics – in Greek and Latin, or in translation – or equally erudite offerings composed by the parochial teacher of rhetoric. The audiences would consist of faculty, fellow students, parents and, on ambitious occasions, Church dignitaries, municipal authorities and guests from the duke's or prince's court. The most pretentious ventures were those in schools in southern Germany and Austria. Of course, plays in Greek or Latin would have little appeal to a broad public, and the dramas in German, penned by the Jesuits themselves, were meant to inculcate humility and piety as well as a rejection of gross material pursuits. Works of this sort were Jakob Bidermann's *Cenodoxus* (1602), Jacob Balde's *Jephtias* (1637) and Jakob Masen's *Androphilus* (1647). Gradually there appeared a trend to broaden the overly narrow subject-matter, possibly making use of the vernacular, introducing leavening elements of low comedy and adding music and dance. The collegiate stages and halls were improved, getting proscenium frames and perspective scenery, along with machinery to supply "miracles" attributed to martyrs and saints and other marvels. The seventeenth century saw Jesuit theatre at its apogee in Europe, especially in Vienna. Its productions were far superior to those of the strolling players, which probably did much to deter the better-educated spectators from going to see the foreign actors, who were more professional but too often slapdash and improvisatory, and whose fare was insubstantial. Indeed, the Jesuits sought to dissuade the faithful from attending commercial theatre, which might offer heretical ideas and lead the credulous astray.

Brockett describes as the climax of the Jesuits' activity in this realm a presentation in the Viennese court of *Pietas Victrix* (1659), by Nikolaus of Avancini, given before Leopold I, and with nice flattery associating him with the victory scored by Constantine, the Byzantine Christian emperor, over the pagan Emperor Maxentius. "This production included battles on land and sea, visions, angels and spirits of Hell, and the eventual enthronement of Constantine as an angel hovered overhead on a cloud. On the frame of the proscenium the Habsburg emblem was prominently displayed, making it clear that the Austrian Empire rested on Constantine's victory. *Pietas Victrix* was staged with money provided by the court and was designed by Giovanni Burnacini, the court architect and one of the leading scenic designers of Europe."

In response to Jesuit plays, which were a form of propaganda, Protestant dramatists entered the bitter fray. Only one of them, the poet Andreas Gryphius (1616–64), merits critical scrutiny. Son of a Lutheran minister, he was born in Silesia, in those days a region in the vanguard of cultural activity in Germany, and also the site of the fiercest ideological and military conflicts between the opposing faiths. He studied in Leiden, where he had an opportunity to see stage productions and was lastingly influenced by the works of Vondel. He was not able to experience much more, if any, theatre after a return to his homeland, where the ravaging war swirled about him and left ruins throughout most of his young adult and middle years. Intensely religious, he railed against the barbarism of the fighting and insistently propounded that the cause of such disasters was not external factors but man's inherent sinfulness. Only adamant faith, and possibly martyrdom, could lead him to redemption and an ultimate reconciliation with God. His plays are written in rather bombastic alexandrines. *Cardinio and Celindo* (1646) is innovative; it deals with the problems of bourgeois characters, rather than with those of towering noble stature, as Aristotelian conventions still dicated at that period. This drama is sometimes credited with being the first "bourgeois domestic tragedy". His other serious works include *Carolus Stuardus* (1649–63) and *Catharina von Georgien* (1657). But this share of his output is rarely revived.

More fortunate have been his comedies, the best of which exhibit his considerable erudition. *Herr Peter Squentz* (published 1658) particularly interests scholars, who pay tribute over all to his "intellectual refinement and creative finesse". (A dispute over whether this work is actually by Gryphius seems to have been settled in his favour; it has been determined that "Philip Gregorio Riesentod" was a pseudonym he used.) Critics find many subtleties and ambiguities in it, along with Shakespearian echoes and "irreverent sexual innuendoes". His full title for it was *Absurd Comic Things, or Herr Peter Squentz*, which gives due warning of the amusing commotion about to occur on stage. Squentz is the head of a troupe seeking to present a play at court, a process that is parodied and allows for a "play-within-a-play, in turn a piece not unlike *Pyramus and Thisbe*", hence the suggestion of a borrowing from Shakespeare, though the tale derives from Ovid. Given to hyperbole, Squentz claims to be a subject of a totally fictitious kingdom, purportedly larger than nine European countries, and also to be a fervent monarchist – as was Gryphius himself – and above all a citizen of the Republic of Letters. (Just so might the leader of a foreign troupe present himself when soliciting a chance to perform at any German ruler's court and perhaps earn a permanent appointment there.) He vows to dedicate the printed text of the company's offering to its generous patron. In many respects, despite his grand pretension, his approach to this ruler is quite indecorous. He is heckled by the sceptical court clown, Pickelherring (Pückelhäring), but not disconcerted, so secure is his unjustified self-esteem.

What follows is not only "commotion" and "absurd comic things", but also a display of exuberant word-play – if it be somewhat esoteric – as well as sharp satire aimed at both actors and spectators, the latter (like the troupe's director) supposedly highly learned and given to preposterous intellectual affectations. The cast, who only recently were crude artisans – one, Krick, a blacksmith – are incompetent. Exposed to the critical gaze of a monarch, King Theodorus of Oberland, and his rude, outspoken courtiers, they become nervous and quarrelsome, and finally begin roughhousing, while Squentz vainly tries to restore order.

Much fun is poked at those who believe in astrology, the arcane science to which the characters of the play-within-the-play ascribe predictive and controlling power, and there is amusing talk about the sun, the moon, the awesome conjunctions of the planets. (In Gryphius's day, astrology was a respectable and intimidating cult, though one which Shakespeare, too, touched on facetiously.)

Needless to say, the début of Squentz at the court of Oberland is an utter failure. Most historians view this farce as a good-natured parody of the offerings of the English and Dutch strolling players of the day, or, looking back, those by Hans Sachs and his semi-professional troupe.

In his *Studies in Early German Comedy*, Richard Erich Schade goes further, arguing that Gryphius intended to offer a political allegory, an allusion to the miserable folly of the Thirty Years' War that had recently ended, and Squentz, attempting to quiet his fractious cast, an example of an inept peace-maker. As other scholars disputed his interpretation, however, Schade labelled it as merely tentative, not one on which he insisted.

Horribilicribrifax (published 1663), another comedy, has also warranted revival at intervals since Gryphius's time. It suggests that he was much impressed by the *commedia dell'arte*.

Shortly after his death his tragedies were sometimes staged by the Dutch players in a repertory including Vondel's works. But he reached audiences at home long before that: archives in Cologne yield an announcement of forthcoming presentations of his *Catharina von Georgien* and a martyr's tragedy, *Leo Armenius* (1651). And there is an account of another enactment of the *Catharina* in Wohlau (1654), in which the players wore Oriental costumes. The Bavarian court at Schleissheim Castle witnessed his patriotic political drama *Papinianus* (1685), which was offered by a German troupe and greatly approved.

His Catholic counterpart was Daniel Kaspar von Lohenstein (1635–83), who was also a Silesian. Less religious by far, he wrote heroic tragedies in language even more bombastic and with violent action that emphasized "bloodthirstiness and eroticism, subject-matter much to the taste of the absolutist rulers of the warring German states. His settings were Ancient Egypt, Rome and the Orient – Turkey – where despotic and depraved sovereigns exercised no restraints; they were guilty of monstrous deeds. Lohenstein's pretext for showing such perverse characters was that they demonstrated the depths to which men fell from God when they gave free rein to their essentially flawed nature, a message like that of Gryphius, but one from which the spectator was apt to be distracted by the excessive cruelty and sensuality laid bare in the portrayals."

Very little can be told about Ludwig Hollonius (1570–1621), save that he had once studied at Rostock and then became a Lutheran pastor and liked writing poetry. A fantastic play from his desk, *Somnium Vitae Humanae* (1605), bears an interesting resemblance to Calderón's *Life Is a Dream*, which dates from a quarter of a century later. In it are also elements of a Hans Sachs farce that, of course, appeared much earlier. All this may be coincidence. In any event, the play distilled from Hollonius's imagination is finally quite different. One day Duke Philipp (Philippus Bonus, Philip the Good), walking with his retinue, comes upon a drunken peasant, Ian, sprawled in a gutter in a stupor. Seemingly on impulse, he orders that the comatose lout be carried into the ducal castle. Further, when Ian awakes and is sober, though still disoriented, he is informed that he is the duchy's ruler. This procedure – a proposed charade – surprises and mystifies everyone in the duke's entourage, including Prince Ludovicus, the Dauphin, in disgrace for his misconduct and placed in the charge of the wise, benign Duke Philipp, who is seeing to his proper education. The five-act comedy tracks the startled Ian through his day as a duke: he dons fine attire; he is surrounded by servants and a sycophantic entourage that he leads in an elegant procession to a church, where they attend a special mass and hear a boring sermon; afterwards, as the host, he presides over an extravagant banquet, all the while assured and coming to believe that this is absolutely real. Parallel to the lowly Ian's adventures, in alternating episodes, Hollonius inserts scenes happening in the peasant world. A sick old man, Tytke, brings a flask of his urine to be analysed by Leuthülff (People-helper), a quack doctor, who exacts his fee before delivering his specious diagnosis and irrelevant prognosis, since Tytke's advanced age and deteriorated condition make certain what soon lies ahead for him. It is here that there are echoes of Hans Sachs, whose farces were still enduringly popular in Germany.

The lascivious Leuthülff suggests that Tytke bring his young, childless wife for an examination, but the old man is wary of doing so. Another peasant who is given to excessive drinking, Schmeckebier ("Taste Beer"), tells Tytke that he is going to the duke with a complaint about a landlord who has seized his horses. Later he reports that at court he met only abuse from the duke's underlings, who would not let him have a fair hearing. It is such presumptive courtiers, he states, who are reducing the peasants to pauperism. Obviously Hollonius is sounding a note of social protest. As the ducal procession is on its way from the church, the persistent Schmeckebier hands a petition setting forth his grievances to a minister in the retinue.

The preparation for the banquet is another effective scene. The cook is bothered by the high cost of provender. Plumpert ("Clumsy King"), a peasant, speaks belittlingly of the new priest in the parish: he is remote and overly strict. On the other side, a diatribe comes from Warner, the confessor attached to the court, who has little good to say about the disruptive peasantry.

At the banquet the diners display gross appetites. There is also back-biting and scheming by those resident at the palace, along with impatience at having the besotted Ian present as the putative duke. A half-intoxicated court preacher flirts openly and coarsely with Trine, old Tytke's young wife, and is rebuked by Warner, the confessor, for mixing socially with a woman of her low social status. A parasite, his eye always on food, discloses his tactics for survival. The picture is one of indecorum and unleashed immorality.

The next day hangdog Ian, a peasant again, is retrieved by his family who have been worriedly searching for him. He is easily convinced that all that seemed to have been happening to him was no more than a drunken dream. Duke Philipp has employed this charade to show Prince Ludovicus that power and wealth are fleeting, often mere illusions. Life itself, as instanced by the terminally ill Tytke, is transitory.

It appears that Hollonius had other aims beyond making this didactic point. He defended his play by defining comedy as "an image of truth" and a "mirror of life". *Somnium Vitae Humanae* is a scarcely exaggerated reflection of social conditions in the duchy where the good pastor lived and wrote: landlords were exploiting the peasants, as were professionals – the quack doctor pretending to treat the ailing Tytke – and the complaints of frustrated citizens were brushed aside or blocked by uncaring, rude court officials; the clergy was far from celibate, and corruption rampant; intellectual clarity was beclouded by belief in such nonsense as astrology. All this criticism is offered by Hollonius without rancour, and frequent moments of high comedy.

There is another facet to the play. Apparently it was destined for production during a *Festnachtspiel* at the court of Philipp II, the actual ruler of Pommern-Stettin. He was known to look with favour on theatrical offerings on such occasions. The benign, sage duke in Hollonius's script is also named Philipp, so obviously this was meant to be a compliment to him. Further, Prince Ludovicus bears the same first name as the author, Ludwig. The inference is that the poet-playwright should look to his wise sovereign for mature practical advice and moral guidance, just as the young man in the play does to his kindly appointed guardian. As has been true throughout much of this history of drama, such praise of a wealthy, important patron was almost mandatory. Hollonius rings a clever change on the form it takes.

A scholar of the classics, Martin Opitz, in his *Book of German Poetry* (1624) called for comedies that met the by-now traditional requirements established by Aristotle and Horace and represented in the scripts of Plautus and Terence. He proved his dedication to the Ancient Greeks and Romans by translating Sophocles' *Antigone* and Seneca's *Trojan Women*. His position was an authoritative one in literary circles, but not all of his contemporaries agreed with him. A contrary view was held by Johann Rist, another Lutheran pastor, who preferred the free-for-all, rowdy, often shapeless farces, filled with hilarious bustle and noise, brought over by the itinerant companies from England; such works won far greater popularity at the courts and public venues, anywhere outside hierarchical Church and academic enclaves. Rist contended that just as much praise could be awarded the English offerings. It was low comedy, as opposed to high comedy, but it was what the German folk liked far more than correct imitations of works fashioned in remote pre-Christian epochs. Even royalty regularly chose such rough, hearty fare.

In his summary of German drama during the first half of the seventeenth century, Schade says that the plays are largely derivative, the characters mostly stock.

Music crosses borders more easily than play-scripts since its language is universal and needs little or no translation. Even so, the initial passage of Italian opera, the newest theatrical form, to the courts of German princes was beset by linguistic difficulties. The Thirty Years' War also delayed the flourishing of an indigenous German operatic genre; there was too little money to support it properly.

From the thirteenth century onwards, secular folk-plays often included simple songs and dances; such works were called *Singspiele* and quite popular. They had many scenes of spoken dialogue without musical accompaniment, so they were hardly operas. Hans Sachs composed music of this sort for some of his farces. There was a good deal of music, too, in sacred and liturgical works, chorales and motets, often considerably theatricalized. Martin Luther approved of increasing the employment of hymns and augmented instrumental music at religious services. With these, there was a foundation on which German opera could be built.

In 1609 the Landgrave of Marburg singled out an obviously gifted Saxon youth, Heinrich Schütz (1585–1672), who from boyhood onwards sang in that ruler's chapel choir. In due time, the youth having reached early manhood, the Landgrave granted him a stipend for three years of study with Giovanni Gabrieli at San Marco in Venice. Two decades later, to keep in touch with further advances in musical art in Italy, Schütz returned to Venice (1628), in particular to meet Monteverdi, whose towering greatness he recognized and respected. Also, the year before, Elector Johann George I of Saxony, hearing much about the production in Rome of Peri's *Dafne* (1598), desired something similar for festivities at the marriage of his daughter Princess Louise to the Landgrave of Hesse-Darmstadt. (Some say the lady's name was Sophie.) The same subject was chosen, *Dafne*, and the original libretto by Rinuccini was to be used, with Peri's score adapted by Schütz, by now an acclaimed organist and the court conductor at Dresden. The translator was the ardent classicist – "poet and philosopher" – the aforementioned Martin Opitz,. Unfortunately, he had trouble finding equivalent German phrases for Rinuccini's "honeyed Italian syllables" and fitting his lines to the musical notes; consequently Schütz altered the score to match Opitz's efforts at "liquid German verse", a problem bedevilling composers, librettists and translators in all languages through centuries, and for which there is probably no satisfactory resolution.

Schütz's music for *Dafne* is lost, so the result of their collaboration cannot be judged. It is believed that, influenced by Monteverdi, Schütz changed the music to accord to some extent with the Venetian's dramatic concepts, and also betrayed in it some of the impress Gabrieli's teaching and the output of Alessandro Grandi had on him. Schütz himself is a major composer, so the disappearance of this *Dafne* – "the first German opera" – must be highly regrettable. Alfred Einstein writes in *A Short History of Music*:

This man of deep inwardness of character and tender sensibility, destined to live in a stormy age and isolated in his career, is of all artists one of the most moving, most ideal figures. He was well aware of the danger of dilettantism inherent in monody and issued warnings against it in three great publications of different character, in which he pointed to the permanent worth of the pure acappella style, and thus helped to ensure that the German composers down to Bach should never, like the Italians, have to keep it artificially alive.

Otherwise he accepted the new forms both enthusiastically and deliberately and sought to mould them to the character of the German language. . . . He perfected for Germany the model of the great free church cantata. . . . Out of the spirit of monodic song, the sphere of small vocal and instrumental media and their combination with choir and orchestra, there grew a wealth of ingenious, expressive forms. There is in this music a sense of Springlike awakening, a continuous stirring of dramatic life. From the motet grew the oratorio *scena*; and in late old age, after many other attempts in the oratorio style, Schütz wrote down his conception of the dramatic rendering of the Passion in three separate settings. Here, as always, his expression was of the keenest, not shrinking from the most daring resources of harmony and melody, yet always very simple and genuine. In his urge towards the utmost truth of expression, in his aversion to all surface polish for the sake of mere formal beauty, and at the same time in the instinctive sureness of his construction, Schütz, here at his greatest, is comparable only to the great German painter – Albrecht Dürer.

It is a pity, beyond all question, that his *Dafne* has not survived. He did not write another opera: there was scarcely a prince or duke in Germany, in those stressful years, who could afford to commission and present one, nor did Schütz have truly capable followers or disciples. As Einstein describes him, he was isolated. Harassed by the ceaseless internecine strife that surrounded him, he took refuge at intervals – three – in Copenhagen, but always returned to his post in Dresden. He died at eighty-seven, wearied and dispirited, though in his later decades peace was finally restored.

After him, Sigmund Theophilus Staden (1607–55), a resident of Nuremberg, published *Seelewig* (*Eternal Soul*; 1644), its libretto an odd mixture of a Morality Play and a pastoral drama, its purpose overtly didactic. He explained that it was in the "Italian style". Insecure at writing recitative, he resorted to stringing together a series of many-versed songs, a form that handicapped actors trying to project suspenseful conflict. The day of German opera had not yet arrived, though it was soon to occur.

— 11 —
LEADING TO SHAKESPEARE

It is a cause for wonder that the English Renaissance, almost alone, yielded truly great drama, the finest since the exalted, resplendent days of Pericles in Greece. Many writers have speculated on what brought this about and offer a variety of hypotheses. England lies far north; the Renaissance, with its rich hoard of unearthed ancient art and literature, reached it late, when it coincided with the fecundating Reformation, challenging Christians' profoundly held beliefs, and stimulating new thoughts of every kind. The cross-Atlantic explorations also did much to stir European imaginations, as did startling discoveries in the physical sciences. The English fleet's striking naval successes, its defeat of the Spanish Armada, gave a fresh sense of security to the doughty defenders of the island bastion, invigorating national pride. It also left open to attack the Spanish ships bearing golden treasure from the New World; the Spaniards, having looted there, were in turn looted by English pirates, their prizes adding to wealth at home. At home, too, after the divisive War of the Roses, there was a feeling of political unity, though that was all too transient, since religious differences were sharply alive. Further, it is said, the English theatre flourished like no other because it was already well established from the medieval epoch onwards as a popular form of entertainment. Though it began in the Church and schools, it escaped the piety and academicism that generally blighted the theatre elsewhere. Its known appeal to a wide audience helped it to avoid an élitism that caused it to be a trivial, pleasant, lavish time-killer in the palaces of rulers and rich nobles, where the subject-matter responded only to the patron's whims. Or the rise of English drama to unsurpassed heights may have come about solely by lucky chance, the appearance of several men of indisputable genius, all contemporaries, who set high standards to which other and lesser writers aspired, prompting a daunting artistic competition. "A race of giants" is how they are described: Marlowe, Shakespeare, Jonson, Webster, surrounded by others of lesser yet still considerable stature. It is a period when individualism became a phenomenon, men seeking to realize their talents and add immortal lustre to their names – the day of the anonymous medieval artist was ending. All these and many other explanations have been adduced to account for the intellectual wealth and creative glory of the Elizabethan Age.

(One group of historians holds that climate is a significant factor; a lengthy spell of grey, overcast skies leads artists to melancholy and brooding, their thoughts going deeper, questioning the meaning of life more seriously.)

The nascent English theatre had, at first, little encouragement from the throne. Henry VIII was interested in Italian art. He brought famous painters from all over the Continent to work for him, and he had masques staged at his court in emulation of the ornate Italian fêtes, but seldom had plays performed. Elizabeth I, too, paid scant attention at first to theatricals – she was, if anything, too parsimonious to pay for them. Her tastes seem not to have been aesthetic. So royalty was not the stage's prime sponsor. The birth of English theatre occurred, instead, at universities and schools. Independent scholars and college masters, travelling abroad, studied and dwelt at Ferrara, Florence, Mantua, Urbino and elsewhere on the Continent, and reported what they had seen while there. Foreign diplomats at the Tudor courts, as well as Continental scholars and patrons of the arts, visited England and boasted of the stage-offerings in their own, more "advanced" countries.

In England, of course, the Mysteries and Moralities, along with the wholly secularized Interludes and noisy, sly folk-farces, still flourished – elements from them were to be retained for some time, incorporated in the imported new forms of drama.

Now that books were being printed with movable type, they were far more accessible, cheaper to come by, and the spread of worldly knowledge increased rapidly. The study of Greek and Latin was prescribed in the schools, and Terence, Plautus and Seneca were among the most popular and familiar classic writers. Eventually it became the custom for students to stage the plays as well as read them; as might be expected, they greatly enjoyed being actors. In 1482 King's College, Cambridge, was putting on such performances; and shortly afterwards Magdalen College, Oxford, was doing the same. The dialogue was always in Greek or Latin, and participation in these ventures was deemed so respectable academically that a degree was awarded to one candidate for having written a comedy that followed a classical model. Fifty years later Queens' College and Trinity College, both at Cambridge, record expenditures for several annual productions of comedies and tragedies in Greek. At the Inns of Court, in London, law students also acted, not as a diversion but to meet a curriculum requirement; presumably the training would prepare them to put on a better show of eloquence and manner when arguing a case.

In grammar schools and colleges such as Eton, play-acting was similarly prevalent. Terence's *Phormio* was presented by the choirboys of St Paul's in 1528 for the edification of King Henry VIII and his Chancellor, Cardinal Wolsey. On another occasion Queen Elizabeth I was a spectator. Not only did royalty visit the schools on these occasions, but the collegiate players were sometimes invited to display their acting skills at one of Her Majesty's scattered palaces, of which there were several. Besides classical works, the students also mounted jovial Interludes, including those of John Heywood, noted earlier.

Gradually, schoolmasters and talented students were writing their own scripts, and in English rather than Latin. Among the first known to have done this is Nicholas Udall (also Uvedale, 1504–56), a graduate of Winchester and Corpus Christi College, Oxford, who went on to serve as headmaster at Eton and then at Westminster. He put together a rollicking farce, *Ralph Roister*

Doister, perhaps the earliest sustained English script of its kind. Written some time between 1534 and 1554, with the date of its initial performance equally uncertain, it reached print in 1566. There is no doubt of its immediate success.

Ralph Roister Doister is smitten with love for Dame Christian Custance. To further his faltering importunate suit, he relies on Matthew Merygreeke, a cunning rascal, and in the ensuing complications the stage-tricks dear to Plautus and Terence are quite discernible, while other bits filched from the traditional English Morality crop up here and there. But Udall does well in depicting native types; there is a good deal of comic bustle, and a fondness for impudent word-coinage. English proverbs and songs enliven the noisy exchanges between the characters. Besides his amatory impulse, Ralph is much attracted by the lady's wealth, which Merygreeke calculates for him, but he is undone by his pride and stupidity. Dame Christian marries elsewhere at the end, as she has always intended, becoming the bride of the rather nettled Gawain Goodluck, an estimable merchant. At the tumultuous climax the Dame's stalwart housemaids and Ralph's entourage come to actual blows, the domestics lustily wielding frying-pans as their weapons. "Ralph Roister will no more wooing begin," is the resolve. The boastful, cowardly, discomfited Ralph is one more reincarnation of the eternal *miles gloriosus*; he is matched in effectiveness as a comic figure by the mischief-making Merygreeke. The play is in rhymed doggerel, often quite salty. Here is Dame Custance's maid, Tibet Talkapace, praising her favourite beverage:

> Old brown bread crusts must have good mumbling
> But good ale down your throat hath good easy tumbling.

The songs have a hearty tilt: "Whoso to marry a minion wife" and "I mun be married a Sunday".

Udall's other works are lost or exist only in fragments. He translated selections from Terence, wrote dialogues, sacred plays in Latin scenarios for pageants, and assisted in rendering Erasmus's paraphrase of the New Testament into English. As headmaster, he was apparently a stern disciplinarian. A former student, Tusser, complains of having been flogged by him "for fault but small, or none at all". (Thomas Tusser was a poet of sorts, a collector of maxims, and a writer on agricultural matters. He introduced the culture of barley to England.)

In addition to his headmasterships, Udall held a number of other posts: he was vicar of Braintree, prebendary of Windsor, rector of Calbourne, playwright to Queen Mary. While at Eton he got into trouble that brought him to the attention of the Privy Council; he was sent for a time to the Marshalsea prison.

An even more bumptious and authentic picture of the village life of the time is *Gammer Gurton's Needle*, given at Christ's College, Cambridge, in 1566 and printed in 1575. The author's identity is not firmly determined: he is believed to be either J. Still or William Stevenson, both of whom were fellows at the College – Stevenson was especially active in theatricals there. The manuscript is signed pseudonymously "Mr S. M[aste]r of Art". The guessing does not stop there; also suggested is John Bridges, Bishop of Oxford. (Apparently Still, too, became a bishop.) Whoever the discreet playwright may have been, he was obviously knowledgeable about facets of dramatic structure exemplified in the Latin comedies.

Gammer (Grandmother) Gurton is mending a rip in the nether garment of Hodge, the family's houseman. She misplaces her needle. Everybody searches for it. Diccon, the "Bedlam", a born troublemaker, prods Gammer Gurton to accuse Dame Chat, her neighbour and "gossip" (female friend), of having found and kept it. Dame Chat is innocent, however. The quarrel grows to ridiculous proportions, in a mock-heroic spirit, involving Dame Chat – who is deeply affronted – her maid, Doll; Dr Rat, the curate; Master Baylye and his servant Scapethryft. Summoned to the cottage, Dr Rat gets caught up in the proceedings and ends with his head cracked. In the prologue, in rhymed long doggerel, Diccon sums up the plot:

> As Gammer Gurton with many a wide stich
> Sat piecing and patching of Hodge her man's breech,
> By chance or misfortune, as she her gear toss'd,
> In Hodge's leather breeches her needle she lost.
> When Diccon the Bedlam had heard by report
> That good Gammer Gurton was robbed in this sort,
> He quietly persuaded with her in that stound
> Dame Chat, her dear gossip, this needle had found;
> Yet knew she no more of this matter, alas,
> Than knoweth Tom, our clerk, what the priest said at mass!
> Hereof there ensued so fearful a fray,
> Mas. Doctor was sent for, these gossips to stay,
> Because he was curate, and esteemed full wise . . .

("Stound" means moment.)

Alas, the curate himself is a sot. Befuddled, intervening in the hurly-burly, he only makes things worse. The conclusion is also revealed in advance by Diccon:

> When all things were tumbled and clean out of fashion
> Whether it were by fortune or some other constellation,
> Suddenly the needle Hodge found by the pricking
> And drew it out of his buttock, where he felt it sticking.

Gummer Gurton had left it in the breeches she was mending, and when Hodge sat himself down he discovered its whereabouts painfully. So the problem is resolved, and peace is restored.

The language is far from genteel; rather, it is vituperative and in places more vulgar than that of Rabelais. Not only the prologue but the entire script is in jogging rhymed couplets, which sometimes go lame. Included is a drinking song at the opening of the second act:

> Back and syde go bare, go bare,
> booth foote and hande go colde:
> But Bellye god sende thee good ale ynoughe,
> whether it be newe or olde.

The song, with its exuberant refrain, was taken up widely for generations. The farcical doings end with "a pot of good ale" all around. It is notable that this rowdy little play has a five-act form, borrowing the structure of Roman comedies, seemingly known to the author.

Even before native-flavoured comedies such as these were come upon, serious dramas on secular subjects began to appear. For the most part, they were of two types: plays more or less based on historical characters and events, and imitations of Senecan tragedies. An early "chronicle play", as those of the historical genre were soon called, was the somewhat primitive *King Johan* (*King John*; 1536), from John Bale (1495–1536), Bishop of Ossory, who constantly turned out religious polemics and sacred stage-pieces. A convert from Catholicism to Protestantism, he chose to present and defend King John, the thirteenth-century Plantagenet, a figure to some much maligned. He is seen by the earnest Bale to have been a forerunning proponent of the Reformation, a hero of the new faith. The script, laden with propaganda, is still much like a Morality and is hardly sound history. It is a very effective vehicle for acting when the English ruler defies the "evil" Pope, the Italian antagonist who contradicts the rebel's birthright and seeks to depose him as a usurper. With Bale the English theatre moved towards the chronicle plays that were to be one of its superb achievements.

Two and a half decades later two young law students, Thomas Norton (for whom scholars seem to have no dates) and Thomas Sackville (1536–1608), writing in tandem, tried their hands at a tragedy – it might better be designated a melodrama – in the Senecan style. Collaboration is a tendency quite marked in the Elizabethan period, for reasons not easy to explain. But this joint venture is more unusual than most, since it is thought that Sackville took part in writing only the last two acts, after Norton wrote the first three on his own. Besides being at the Inner Temple, both young men were members of Parliament, and it is believed that Sackville had earlier attended Oxford. The result of the literary partnership is *Gorboduc* (1562), a stage-work described as at the same time violent and dull. It is chiefly remembered in the history of the theatre because it shows how close was the acquaintance by educated Elizabethans with Seneca's bombastic and gory scripts and how pervasive was his influence among playwrights in general during the English Renaissance.

The enterprising pair dipped into Geoffrey of Monmouth's compilation of equally bloody native legends for the plot, adding new details and shaping it like a Senecan verse-drama, establishing a pattern followed by scores of playwrights after them, even by Marlowe and Shakespeare. *Gorboduc* was enacted at the Hall of the Inner Temple, when Sackville was about twenty-six.

King Gorboduc and Queen Videna have two sons, Ferrex and Porrex. The king reaches an unfortunate decision: he will divide his domain between them. Porrex, jealous, is not satisfied with this prospect, and he kills his brother. To avenge her loss of Ferrex, who was her favourite, the queen murders Porrex. At this moment of dire disorder the Duke of Albany tries to seize the throne, a move leading to civil war. In a popular uprising, the queen and her husband themselves are slain. The highly respected poet Sir Philip Sidney liked *Gorboduc* very much; for him, it was "full of stately speeches and well-sounding phrases". By 1590 the text had gone through five printings.

The elegant speeches are its serious fault: nearly all the decisive action in this tale of fratricidal strife – so very Senecan, but now given a turbulent English background – is recounted by messengers, as was done in Ancient Athens. To compensate for the lack of physical encounters, off-stage events are illustrated in dumb show at the start of each act, an expository device borrowed from the medieval stage, and one that suggests a healthy respect by the two authors for the theatre's dependence on perceived gestures and movement, "body language". Even with the interpolated dumb shows, however, the play is far too talky. The dialogue is in blank verse – pentameters without rhymes – setting another precedent, and the classic unities are ignored.

Gorboduc had another facet. With the War of the Roses as its implied context, the play was topical and preached the desirabilty of order and legitimacy in the regal succession. The spectator was most likely to see in it clear allusions to England's recent troubled history, the dynastic quarrels only a short while past – and those that continued.

Besides being a lawyer and member of Parliament, Sackville, who became the first Earl of Dorset and Baron Buckhurst, held the posts of Treasurer and Chancellor of Oxford, as well as a number of other high offices. He wrote poems and prose pieces but no other plays. He was given the task of journeying to Fotheringay to notify the imprisoned Mary Stuart of her sentence of execution (1568). It was a duty in keeping with the events described a few years before in *Gorboduc*.

This awkward play, an opening wedge for the Senecan style, with inflated rhetoric and over-emotionalism, is the prototype of the "revenge tragedy", its structure and idiom that in which most serious English stage-works were to be offered for decades thereafter. It is odd, as has been remarked, that a dramatist of Nero's time, whose works were seldom if ever produced, should have such a wide and lasting influence, even over men of far superior talents, not only in England but in Italy and Spain. As yet, the scripts of the great Athenian writers of tragedies were scarcely accessible, which meant that Seneca alone had the aura of being an exemplar from the rediscovered classical world. Further, his tone appealed because it was moral, while the terrible violence he depicted was being repeated on every side in the sixteenth century. (Mary, Queen of Scots, was to be decapitated, and the heads of felons and men of great rank who had fallen from power were displayed on pikes as warnings along the flanks of London Bridge.) Translations of Seneca's works made even more readers familiar with him. His plays had yet another attraction: while books were becoming increasingly cheap and available in ever larger numbers, most people could not read. History was an engrossing subject to them; there was a taste and demand for it, to which were added illicit romance and skilfully contrived suspense, along with breathtaking intrigue and violence.

Chronicle plays such as *Gorboduc* continued to be written after the enthusiastic initiative taken by Norton and Sackville. So were pastorals, the English nobility copying the Italian vogue for them, together with adaptations of other Italian and Spanish dramas and comedies, and devout scripts on Biblical subjects, in an ever widening choice of themes. This colourful period has been exhaustively researched by scholars with a passion for minutiae. Here it is not possible to do justice to many of the minor yet often gifted precursors and contemporaries of the titans of Elizabeth's day.

However, even in a superficial survey, attention must be directed to George Gascoigne (1525–77 – elsewhere said to have been born in 1534), descended from a Bedfordshire family of repute. He entered Trinity College, Cambridge, and afterwards spent ten years at Gray's Inn from 1555 onwards. In 1556 he married Elizabeth Boyes, mother of the poet Nicholas Breton. This brought problems, since the lady already had a husband; as a result Gascoigne got into legal and financial embarrassments that led him to Bedford's debtors' gaol (1570), coming out from it penniless. To restore his depleted fortunes, he entered military service on the Protestant side in the Netherlands (1572–4), where he was held in prison by the Spanish for four months. While he was absent, a book of poems and plays from his pen appeared, the grouping titled *A Hundred Sundrie Flowres*, though quite without his consent. On his release and return to England he corrected and added new material to what was already in the book. In the short two years remaining to his life he finally enjoyed a measure of recognition and success as a court poet.

Much of his work is now in a single volume, *Posies*. It contains a miscellany of verse, some religious, some not. The secular poems include "The Delectable History of Dan Bartholmew of Bathe", and there are reminiscences in verse of his exploits and misadventures in the Netherlands, "The Fruites of Warre". Of more relevance here are the two plays for which he is best remembered as a dramatist, *Supposes*, based on Ariosto's comedy *I Suppositi*, and *Jocasta*, a blank-verse tragedy claimed to be derived from a script by Euripides. The second of these attributions is invalid. *Jocasta* is translated in collaboration with Francis Kenwelmarsh from an Italian version of the Oedipus myth by Ludovico Dolce, *Giocasta* (1559). Also in the book is a novella, "The Adventures of Master F.J.", described as Chaucerian and supposedly transcribed from an Italian work by one "Bartello" but probably wholly Gascoigne's. While living in Lombardy, the hero – designated merely F.J. – has an affair with a Venetian lady, Leonora, in whose house he dwells. On his return to England they exchange poems and love letters for a time, until the lady's secretary, who is more available, supplants him in her affections. Unhappy at having been dismissed by her, F.J. goes back to Venice, "spending there the rest of his dayes in a dissolute kind of lyfe". Candidly erotic, the novella had first appeared in the *Sundrie Flowres*; for *Poesies* Gascoigne rewrote it, shifting the setting from northern England to Italy throughout; it was also somewhat expurgated. "The Adventures" was suspected of being a *roman à clef* – its characters based on actual persons – and was closely scanned by its more sophisticated readers. It still holds interest now for much the same reasons.

In addition, *Poesies* includes an essay, "Certayne notes of Instruction concerning the making of verse", "a pithy but pioneering account of English versification". (This summary is from *The Oxford Companion to English Literature*, a bountiful source of biographical detail for this chapter.) Later works are *The Glasse of Governement, A Tragicall Comedie* (1575), *The Droomme of Doomes Day* (1576), and *The Steele Glas. A Satyre* (1576), showing that he was still writing shortly before his early death at fifty-one. His *Tale of Hametes the Heremyte* (1575), published after his passing, is a piece he intended, as court poet, for the pleasure of Queen Elizabeth.

He was innovative. In translating Ariosto's *I Suppositi* he turned away from verse and rendered it in prose, the first English playwright to do anything of the sort. This experiment led to his successors having freedom in adapting foreign comedies and also in offerings of their own. The use of the vernacular added vigour, colour and spice to dialogue.

Robert Greene (1558–92) had an even more disruptive life. Born in Norwich, he studied at St John's College and Clare Hall, Cambridge (1575–83), earning an MA, and later was at Oxford. From 1583 onwards, however, he mostly spent his time in London. He was fond of mentioning, boastfully, the extent of his acquired learning at both universities, but in the judgement of many he was "a feckless drunkard, who abandoned his wife and children to throw himself on the mercies of tavern hostesses and courtesans; writing pamphlets and plays was supposedly a last resort when his credit failed". Indeed, that is how he is depicted in *The Oxford Companion to English Literature*, which adds that after squandering his wife's dowry: "He is said to have died of a surfeit of Rhenish wine and pickled herrings, though it may more likely have been plague, of which there was a severe outbreak in 1592." Alas, most other biographical accounts of him are no more kindly. Outside of London he travelled widely through England and the Continent. In the year of his death, as Greene lay ill, the poet–satirist Gabriel Harvey published an attack on him, calling him "the Ape of Euphues" and "Patriarch of shifters". This invective prompted a fierce response by Thomas Nash, a fellow playwright and literary light of the day, who acknowledged that Greene was in truth an inebriate and debtor, but asserted in his defence that "Hee inherited more vertues than vices."

Among those "vertues" were ample talents and enough application to his tasks to be the author of thirty-seven publications, among them moral dialogues, romantic novels and equally romantic plays, together with tellingly realistic pictures of underworld life, all filling fifteen volumes. They sold so well that, as Nash remarks, printers were ready "to pay him deare for the very dregs of his wit". It is possible that Greene's life was neither as vicious, nor the pious sentiments in his moral dialogues as sincere, as he professed. Quite as suspect to some is his death-bed repentance. He may have perversely enjoyed creating a reputation as a wild carouser and bohemian, and pretending at last to be sorry for it, keeping up an ambivalent pose to the final moment. Before he expired, he unloosed a flow of pamphlets on the subject of death-bed conversions, which suggests that his own was well rehearsed.

His eight plays, some of them collaborations, were published over a stretch of several years after his death. Some of his pamphlets and novels, thought to be autobiographical, are scurrilous. He changed styles, however, after the triumphal appearance of Christopher Marlowe's *Tamburlaine*; a shrewd professional writer, studying his market, he saw that this was the new direction. This resolve prompted, in a similar vein to Marlowe's work, *Alphonsus of Aragon* (1588–91), a comparatively weak effort – he was not a Marlowe.

A second attempt, undertaken with Thomas Lodge, *A Looking Glass for London and England* (*c.* 1590), mingles – as do many plays of the period – elements of Moralities, Miracles and a kind of rough-edged satire fashionable in Elizabeth's day. It seeks to hold up an allegorical mirror to the woes of poverty and the ceaseless exploitation of the lowly, by contrasting the lives of the wretched, driven to theft, with the sumptuousness of Ninevah's court, ruled by evil King Rasni. A chorus points out the applicability of this two-sided, ill-matched picture to contemporary London. Incest between brother and sister, another favourite Elizabethan theme, is present here. The language is rich, again reminiscent of Marlowe and anticipating Jonson. The play is also lit with flashes of gutter and alehouse humour. The occult has a role, together with a full measure of conventional moralizing. In most of Greene's tragicomedies Christianity (or, in this instance, Jonas, the Old Testament prophet) prevails over godless magic-working. Again, it is hard to know how sincere was this sentiment on the author's part or whether he was paying lip-service to the orthodox religion of his times.

Far better is *Friar Bacon and Friar Bungay* (1589 or 1591, or perhaps 1594), written quite solo. With all its faults, it is a major achievement. Once more it seems to borrow its verse style from *Tamburlaine* and a fondness for pranks and magic from Marlowe's *Dr Faustus*, but many of its other qualities appear to have been Greene's own. Notable is his celebration of the English setting, idealized, and his tender portrait of Margaret, the Keeper's daughter, loved by Prince Edward, heir of Henry III, while she gives all her bounteous affection to Lacy, the Earl of Lincoln. Margaret is the very embodiment of "warm womanliness" – and described elsewhere as "the country girl, all English" – which leads the prince to exclaim, when first beholding her:

> Into the milk-house went I with the maid,
> And there amongst the cream-bowls did she shine . . .
> She turned her smock over her lily arms,
> And dived them into milk to run her cheese;
> But whiter than the milk her crystal skin,
> Checked with lines of azure . . .

To win this priceless girl, "the lovely maid of Fressingfield", the prince enlists the aid of Friar Bacon, "a brave necromancer . . . A man supposed the wonder of the world". (The assumption is that Greene is referring here to Roger Bacon, the alchemist. A likeness of the prince reappears in plays by Shakespeare and others. Greene's own source was a prose pamphlet, *The Famous Historie of Fryar Bungay*, a fruitful collection of legends.)

Assisted by Friar Bungay, Friar Bacon constructs a head of brass. Next he conjures up the Devil, who instructs him how to make the brass image speak, though only for a month: after that it will fall silent. Someone must always be at hand to listen to it. A servant, Miles, is appointed to the task. Says the head, "Time is." Friar Bacon is asleep; Miles hesitates to awaken him to hear two words. After a lapse of moments, the brass head speaks again: "Time was," and finally, "Time is past." It collapses and breaks apart. Aroused, Friar Bacon rains reproaches on the hapless Miles. A rare opportunity to tap into supernatural wisdom has been missed.

The ardent prince sends his friend Lacy in disguise to a fair. He is to find Margaret and woo her on the prince's behalf, but Lacy loses his heart to her. With Friar Bacon's magical help, Prince Edward beholds in a vision Lacy's courtship of the girl, a highly effective scene on stage. Margaret's adviser is the jolly Friar Bungay, who has seen through Lacy's masquerade and warns her of the young man's true identity. When the Earl presses his impetuous suit, Margaret replies:

> You are very hasty; for to garden well
> Seeds must have time to sprout before they spring:
> Love ought to creep as doth the dial's shade,
> For untimely ripe is rotten all too soon.

The dialogue is light and clever, laced with poetry and often tart, though at moments the farm-maid disconcertingly decorates her otherwise simple speech with classical allusions.

The Earl offers to wed the girl, who joyfully accepts, but when Father Bungay is about to unite the eager pair, Friar Bacon, at the prince's behest, deprives the cleric of a voice and halts the ceremony. Then Friar Bacon has devils spirit Friar Bungay away to Oxford, where he converses with him.

The intrigues multiply, the fantasy grows wilder, with ever more magical abracadabra – in a three-way contest between the occult powers of Friar Bacon, Friar Bungay and a German rival, Vandermast – that exceeds *Dr Faustus* in an amusing and awesome abundance of stage-tricks. This exhibition takes place at a gathering attended by the Emperor of Germany and the Kings of England and Castile.

Subsequent happenings of a serious nature disabuse Friar Bacon of the wisdom of his course; he quits dabbling in necromancy.

Some of the best scenes are pastoral, including that at the country fair. A passage that stands out is this complimentary description of an English seat of learning, by the visiting and admiring German emperor, who says to King Henry:

> Trust me, Plantagenet, these Oxford schools
> Are richly seated near the river-side;
> The mountains full of fat and fallow deer,
> The battling [fertile] pastures lade with kine and flocks,
> The town gorgeous with high-built colleges,
> And scholars seemly in their grave attire,
> Learned in searching principles of art.

No little nostalgia rings in those lines.

The ending is neat and happy. The prince relents, recognizing that Lacy's intentions towards the girl are more honourable than his own. Friar Bacon's "poor scholar" rides off to Hell on the back of a demon. The play has been aptly summed up as "a weird concoction" of many elements: poetry, whimsy, clownish humour. Sir Ifor Evans remarks: "Greene has devised ways of keeping the whole dramatic action together, the Court and the countryside and the world of necromancy. Out of much that was hurried and incongruous in his drama there had emerged something new, whose warm attractiveness is to be found again in Shakespeare's comedies." *Friar Bacon and Friar Bungay*, running along many lines parallel to *Dr Faustus*, is a lighter, merrier handling of a similar topic. And if Greene was influenced by Marlowe, he in turn was to have his ideas and characters appropriated by Shakespeare. The Elizabethan period was one of just such creative interaction.

Even more skilful is Greene's handling of *James IV* (c. 1593), about a feckless Scottish king who is almost seduced and ruined by a favourite. He contemplates the murder of his wife, in order to replace her by another love. But the crime is not consummated, and the contrite king eventually rues having considered it. This action is presented as a play-within-a-play, mounted before the King of the Fairies, Oberon, which permits a good deal of poeticizing, some of it highly worthy. The characterizations are excellent, particularly those of the two women, the queen, Dorothea, and the king's paramour, Ida. Both the wickedly attractive favourite, Ateukin, and the Bishop of St Andrew are strongly imagined persons. It is odd that Greene, who treated his own deserted wife so badly and took up with a mistress in London, could portray good women so well and sympathetically.

Among his other plays are *The History of Orlando Furioso* and two that are less firmly attributed to him, *George-a-Green* and *Pinner of Wakefield*. He is believed to have had a hand as a collaborator in the composition of still more scripts, as was the custom of the day. His novel, *Pandosto; or, The Triumph of Time* (1588), gave Shakespeare the plot he used for *A Winter's Tale*. Greene did not care much for his fellow dramatist. In his pamphlet *Greene's Groatsworth of Witte* (1592) he alludes to

Shakespeare as "an upstart Crow, beautified with our feathers". It is the first known reference in print to the man from Stratford as a newly established London playwright, and obviously Greene resented the young author's borrowings. Nor had he admiration for him as a poet, but denigrates him as one who "supposes he is as well able as most to bombast out a blanke verse as the best of you . . . that is in his own conceit the onely Shakescene in a countrey". So it would also seem that Greene did not welcome rivals. In addition, Shakespeare did not have a university degree, in contrast to Greene himself, Peele, Nash, Lodge, Marlowe, Jonson and other well-educated "University Wits" of the town who were fashioning plays, that being the current term usually applied to those busy script-writers. (Henry Chettle, the publisher of *Greenes Groatsworth of Witte*, in his own *Kind Heart's Dream*, sought to offset Greene's slurs on Shakespeare and give a fairer characterization of him: "Myself have seen his demeanour no less civil than he excellent in the quality he professes. Besides, divers of worship have reported his uprightness of dealing, which argues his honesty, and his facetious grace in writing, that approves his art.")

In all, Greene was a dramatist of amazing if erratic merits and deserves to be more widely read and known, and perhaps to have these works revived on the stage, the medium for which he wrote them – though performing his plays would present many difficulties. He had the ill-luck of sharing the era of Marlowe and Shakespeare, which led to his being overshadowed by them at mastering poetic expression and creating plausible, over-sized characters. To gain eternal fame it is important to have the right contemporaries. Greene is hardly as deep or sure a thinker or craftsman as some of his period, or as unerring a poet, but his is an undeniable flair for theatre, his imagination striking, his phrase-making and gift for fantasy often compelling and bold.

Thomas Lodge (1558–1625) had an initial advantage: his father was Lord Mayor of London. The son was a student at Merchant Taylors' School in that thriving city, and then, like his peers, the "Wits", he entered Oxford. He was at Trinity College there, after which he went on to Lincoln's Inn, implying that he was freighted with much knowledge. At twenty-one he published an essay, *Defence of Poetry, Music and Stage Plays* (1579). It was followed by *An Alarum Against Usurers* (1584), a warning against the dangers facing young men who resorted to borrowing from avaricious money-lenders. Another of his prose works during this period was a prose romance, *Forbonius and Priscaria* (1584). In an adventurous spirit, Lodge set sail on a privateer intending to raid treasure-laden Spanish galleons in waters near the Canaries and Terceras (1586), and five years later, the urge for unusual experiences prompting him again, he embarked on a ship sent to explore South America, a long voyage that extended to the Straits of Magellan (1591–3). During the first of these voyages he composed a romantic novel, *Rosalynde*, which he said was "hacht in the stormes of the Ocean, and feathered in the surges of many perillous seas". He brought out an Ovidian verse fable, *Scillaes Metamorphosis* (1589), about the courtship of a nymph, Scilla, by the sea-god Glaucis; she is calcified into a lonely rock in the sea, punished for cruelty to him. A "minor epic", it may have inspired Shakespeare's *Venus and Adonis*; a few lines of Shakespeare's version have a marked resemblance to those by Lodge, and the rhyme scheme – *sesta rima* – is the same in both poems. Lodge continued to write prose romances, four of them, and poems that conformed to French and Italian models.

Only twice did he venture into play writing, as far as is known. His first effort, the only one on his own, *The Wounds of Civil War* (1594), is about the dictator Sulla's struggle with the general, tribune and oft-chosen consul Gaius Marius. It is of particular interest as an early example of an English play based on Roman history, subject-matter that Shakespeare was also to find quite interesting. It was performed by the Lord Admiral's Men, a professional acting group. He also collaborated with Greene on the eccentric *A Looking Glass for London and England*, as noted.

His subsequent list of publications is lengthy and attests to his remarkable versatility; he contributed satirical poetry, more novels, major works of translation, which reflect the many changes in his personal life. He was converted to Roman Catholicism, studied medicine in Avignon and Oxford, became a physician, issued a treatise on the plague. He typifies the Elizabethan author whose existence was hardly serene and contemplative but crowded with action, risk and incredible zest.

The universities and Inns of Court were an unending source of bright young men descending on London to study law while alternatively dashing off plays and acting in them and displaying a wide range of talents, some quite idiosyncratic. That might also be true of those who lectured them. Thomas Preston (1537–99), a Fellow of King's College, Cambridge, Master of Trinity Hall, and briefly Vice-Chancellor of the university (1589–90), is thought to have written *A Lamentable Tragedie, Mixed Full of Plesant Mirth, Containing the Lfe of Cambises King of Percia, from the Beginning of His Kingdom, Unto his Death, His One Good Deed of Execution, After That Many Wicked Deeds Committed By and Through Him, and Last of All, His Odious Death By God's Justice Appointed* (c. 1561, or 1569?), derived from Herodotus. It is a peculiar *mélange* of tragedy and farce, with mythological (Cupid and Venus), allegorical (Shame, Diligence, Trial, Proof) and comical figures (the well-meaning Hob, Lob and Marian-May-Be-Good, and the villainous Ruff, Huff and Snuff), jumbled together in a crowded text conjoining horrific acts and chamber-pot humour. The dialogue is so bombastic and grandiloquent that it was often quoted derisively by Shakespeare among others, who in *Henry IV* has a character threaten: "I must speak in passion, and I will do in King Cambyses's vein." The attribution to Preston is not definite. Scholars have many problems tracing and dating works of these decades; records were scanty and at best ill-kept, and spellings loose and sometimes bafflingly varied. It should be added that *Cambises*, with its on-stage flayings, stabbings and beheadings, was exceedingly popular, indicating what much of the public liked to see performed.

Robert Wilmot (*fl.* 1566–1608) went to Boccaccio for his *The Tragedie of Tancred and Gismund* (c. 1566–8). Here is a typical instance of collaboration: Act Two is by Henry Noel, and Act Four by Sir Henry Hatton (1540–91), the latter gentleman a favourite of Queen Elizabeth, who admired his graceful dancing. From her he received grants of offices and estates, becoming both Lord Chancellor and Chancellor of Oxford University. There is less to report about Wilmot and Noel. In Boccaccio's tale Sigismonda, the daughter of Tancred, Prince of Salerno, falls in love with his squire, Guiscardo. Becoming aware of the unsanctioned liaison, her father slays Guiscardo, puts the young man's heart in a golden flagon, and sends it to her. She adds poison to the cup, swallows the contents, dies. Remorseful, Tancred has the pair entombed together, much like the frustrated lovers in *Romeo and Juliet*. The story had a sure attraction for the Elizabethans: they had strong stomachs.

There is no limit to the gushing gore and heinous deeds unveiled in Thomas Hughes' Senecan-influenced *Misfortunes of Arthur* (1588), indebted, as was *Gorboduc*, to the often violent legends compiled by Geoffrey of Monmouth. It belongs in the category of historical chronicles that steadily drew spectators, as long as the enactments included unusual thrills and shudders, which Hughes provided in overflowing measure. The same was true of Thomas Legge's *Ricardo Tertius* (1579), which some years earlier had exploited the career of Richard III, a play worthy of Seneca but on a stirring English theme. Both of these works, replete with simulated cruelties, were put on at the Inns of Court, where one assumes that the audience was not made up only of the rude common folk but also of the authors' classmates and grave, learned members of the faculty, authorities on jurisprudence. But perhaps one had to be hardened in order to pass harsh judgements on subversives and malefactors, those robbing purses or plotting to seize the crown and overthrow the currently established religion.

At the end of the sixteenth century London was a small city and the world of its theatre a narrow one. The exchange of ideas among its eager playwrights, whether voluntary or involuntary, was inevitable, and their scripts were certain to bear many resemblances. The haste with which many were prepared also explains an overall sameness. One whose outpouring was more original, as well as prolific, is George Peele (1558–97). He was London-born. His father was Clerk of Christ's Hospital, and some also say that he was a silversmith. The elder Peele had literary leanings, too, providing scripts for city pageants and writing books on accountancy. The son was schooled at Christ's Hospital, Broadgates Hall (Pembroke College) and Christ Church, Oxford. Young Peele, very much a bohemian like the others, settled again in London in 1581, regularly disporting himself in the company of the reckless Greene and Thomas Watson, the latter a poet who had translated Sophocles' *Antigone* from Greek to Latin, an erudite feat, and also rendered Italian sonnets and the lyrics of madrigals. Watson was a close friend of raffish Marlowe, and his investigations into the metrics of poetry and experiments with them influenced Shakespeare. Peele's career as a freelance writer, and the requirements of his wild style of living, took him into three fields: plays, pageants – following a "trade" pursued by his father – and poetry. His verse was of the sort designated as "gratulatory", paying an egregious birthday tribute to a high-born patron, a duke or an earl, commissioned or unsolicited; or hailing a wealthy couple's nuptials; or extending blessings at the birth of a child; or adding to the praise of a national hero upon his victory. A strain of seemingly genuine patriotism runs through Peele's work, but of necessity he was a poet for hire, the fees enabling him to enjoy bouts of debauchery.

His collective works fill three volumes, and they include five extant verse plays – others may have been lost. *The Araygnement of Paris* (printed 1584), a pastoral or graceful court masque, shows that when a work called for it he could write with delicacy and prove himself a poet of true distinction. Young, handsome Paris, a prince of Troy, is watching over his flocks on the verdant slopes of Mt Ida. Along with him is his adoring wife, Oenone. Word comes that he has been appointed to determine which of three goddesses, Juno, Pallas and Venus, deserves the golden apple, symbolizing that she is the most beautiful. He chooses Venus, who wafts him away, leaving Oenone bereft and despairing. Continuing the quarrel, the embittered Pallas and Juno have Paris hauled before judges in Olympus; he is accused of bias in his choice. The final decision about the apple is passed to Diana, who with infinite tact awards the prize to a nymph, "Our Zabeta fayre", which is to say Elizabeth. The play was presented before the plain-faced, red-haired Queen, who appreciated the egregious compliment, now having been declared "fairer than all the goddesses".

His next two scripts are historical chronicles. *Edward I* (printed 1593) yields signs of hurried composition, has a plethora of flaws and marks no progress in that genre. It is scarcely authentic history. Much better, though extravagantly romantic, is *The Battle of Alcazar* (1588–9). The opposing forces are those of Portugal and Morocco. Abdelmec, the Moroccan king, has regained his realm from a usurper, Muly Mahamet, who now turns to Sebastian, ruler of Portugal, to sue for his assistance. Sebastian assents. The Portuguese fleet sails to attack the Moroccans; in the sea-encounter Sebastian and Abdelmec are killed, and Muly Mahamet is drowned taking flight from the scene. Another fatality is that of Thomas Stukely, an English adventurer, erstwhile double-agent in France and his homeland, and later in Ireland and Spain. After having been engaged in privateering, he had finally sold his services to King Sebastian. (An actual person, he was rumoured to be an illegitimate son of Henry VIII.) Peele may have thought the subject was appropriate and timely, since the savaging of the Spanish Armada had recently occurred – a decade after the naval clash at Alcazar (1578) – and the English were basking with pride at their own daring, unpredictable deed. Despite imperfections that prompted one critic to call it "dull, windy stuff", the play has a number of highly effective episodes and passages of excelling poetry. Again, it has inflated echoes of Marlowe's resonant lines in *Tamburlaine*.

The Old Wives' Tale (1592) is usually referred to as Peele's best work, and is certainly the one in which his intrinsic originality is overwhelmingly in evidence. In prose, it is a departure from the style accepted by most of his peers. It opens on a rustic setting, where Madge, the old wife of the title, haltingly seeks to recount a strange story. She interrupts herself several times. Finally she breaks off, while performers enter to enact the tale long trembling on her lips. The tone of her come-to-life narrative changes from the realistic to the utterly romantic, conjuring up fabulous happenings in the search by two brothers for their missing sister, Delia. They learn that she is a captive of a magician, Sacrapant, into whose clutches they also fall. Their rescue is accomplished by Eumenides, a knight, assisted by a ghost, Jack. This Jack is indebted to the knight for having paid the expenses of his funeral, for which he returns to show his gratitude. To enhance this fantasy, there are songs and acts of magic. Peele's lyrics for these songs tend to be exquisite. All the while the spectator is aware that this "tale" is unfolding solely in the mind of Madge, the tremulous ancient. The play has been variously interpreted: some take it as pure, whimsical comedy, a very special one, the first of its kind on the English stage; others construe it as a satire, a parodic thrust at the fanciful and even preposterous offerings of many of his young contemporaries who are aspiring to become well-paid dramatists. Whatever his intention, this venture has an undying charm, tying up in one package "romantic adventure, classical legend, folklore and homespun humour, all combined within the framework of a play-within-a-play". It is intermittently revived, though hardly ever in the commercial theatre; instead it is rediscovered within the bounds of academe by delighted scholars and their students. Critics suggest, too, that Milton may have studied the script before writing his masque *Comus*, in which much of the action is decidedly similar.

Like his fellows, Peele saw Biblical incidents as safe material while having a sure popular appeal. This led him to *David and Fair Bathsabe* (c. 1593), concerning the adulterous seduction of Bathsheba by the Jewish hero-king David, as well as the ruler's struggle with his son Absalom, and the young rebel's death. The play is not ranked high, though its poetry is meritorious.

Peele is said to have been an accomplished actor. In and out of the theatre, he was an intoxicated madcap, given to playing pranks on his bohemian companions. He describes some of them in a book, *The Merrie Conceited Jests of George Peele* (1607), but some of the practical jokes he reports are deemed mere fiction; he was never quite that outrageous. One of his boasts is that, in dire need of trousers, he stole a pair from a friend, causing him considerable embarrassment. Throughout his work are indications of his probable bisexuality, a trait shared by many of the more exuberant, facetious University Wits. Among his other scripts is *Sir Clyomon and Sir Clamides* (1599). He died sometime near his fortieth birthday.

A very different sort of person and playwright was John Lyly (c. 1554–1606). A minor talent, he was nevertheless considerably influential. His paternal grandfather, William, who spelled his last name Lily, was the first high master of St Paul's School and in the forefront of those who brought about a revival of Greek studies in England. He was an authority on Latin syntax and the author of *Lily's Grammar*, a textbook long pored over by British schoolboys, but probably not with glee. (There is a reference to it in Shakespeare's *The Merry Wives of Windsor*.) So the grandson's interest in language *per se* may have been inherent. His education is thought to have included a spell at King's School, Canterbury, and stays at Magdalen College, Oxford, and at Cambridge, which meant that it ran along lines more or less parallel to those of the University

Wits. Somewhat belatedly, at forty-four, he chose to have a political career and was elected to Parliament successively as the member for Hindon, Aylesbury and Appleby (1598–1601).

Despite being born into what might be considered the Elizabethan Establishment, which was certainly helpful, all did not go smoothly for him; perhaps it was because he aimed too high. At Oxford his patron was no less a personage than Elizabeth's wily and powerful chief minister, Lord Burleigh, and later the youthful Lyly held a post as secretary to Burleigh's son-in-law, the Earl of Oxford – the aspiring author's social connections were the very best. In his early and middle thirties he won sensational acclaim and firm assurance that his name would endure to posterity by having published two pastoral prose romances, *Euphues* (1578) and a sequel, *Euphues and His England* (1580). In these slender novels what little there is of plot serves as a pretext for witty and learned discourses between the characters on matters of love and literature, sometimes in exchanges of letters, or one to one, or in groups, in which everyone is ostentatiously erudite and makes a deliberate display of being so. Their speech in these colloquies is ornate to the extreme – over-decorated – akin to the Gongorism to which Calderón was converted.

A young Athenian, Euphues, visits Naples and forms a friendship with an Italian, Philautus. He steals the affections of the Italian's beloved, Lucilla, but all too soon she turns her gaze to Curio, stating her preference for him rather than either of them. After having quarrelled over Lucilla, the two young men feel mistreated by her and resume their ties of friendship. Back in Greece, Euphues directs Philautus to a pamphlet of counsel on affairs of the heart.

The sequel has Euphues and Philautus together on a journey to England, where, heedless of Euphues' advice, the Italian pays suit to more than one local young lady. Though cautioned by Euphues to use caution, Philautus is overly ardent, which puts him at a disadvantage. The question debated between the two friends is whether discretion – even secrecy – or constancy is the quality more requisite. Recalled to Greece, Euphues writes a letter from there addressed to "the ladies of Italy", giving a full description of English society – its queen and institutions, its ladies and gentlemen – and follows this with another letter to Philautus containing further instruction.

So great was the admiration for these short novels, and particularly Lyly's verbal elegance, that "Euphues" has been implanted for centuries in the English vocabulary as the term for excessively fancy prose. *The Pocket Oxford Dictionary*: "*Euphuism*. Affected or high-flown style of writing." The *Oxford Companion to English Literature* suggests that usually in prose of this sort there is "overuse of antithesis, which is pursued regardless of sense, and emphasized by alliteration and other devices, and of allusions to historical and mythical personages and to natural history drawn from such writers as Plutarch, Pliny and Erasmus".

His prestige enhanced by his extraordinary success, Lyly became assistant master of St Paul's School, where his grandfather had reigned in the school attached to the cathedral; in particular, he was put in charge of the company of boy actors at the instigation of his employer, the Earl of Oxford, a financial supporter of its theatrical activities. This gave Lyly, interested in writing plays, an opportunity to have them performed before Queen Elizabeth, frequently in attendance there. Soon afterwards he won a post in the Revels Office, which arranged and oversaw court entertainments. Here he hoped to enlarge his scope and authority.

All of Lyly's plays were put on for special audiences gathered at St Paul's. The boys, turned actors and singers, and also drawn from the Children's Chapel, were not mere amateurs; they had semi-professional skills, but they lent many scenes an artificial air. Yet this consorted nicely with an androgynous quality in Lyly's work, a strain, as has been remarked, that is frequently observed in Elizabethan drama. That young girls were portrayed by handsome boys or graceful, effeminate young men, led to transvestitism becoming an overworked plot device, much used by Greene, Marlowe, Shakespeare, Jonson. It is also likely that bisexuality was somewhat encouraged by the atmosphere of Elizabeth's court. The "Virgin Queen" had a rather masculine character, and she was surrounded by comely, fawning courtiers who flaunted themselves there "in silks and satins". Her intellect was "male", and the powers she wielded usually belonged to a man. In any event, Lyly's plays were well suited to the "boyish treble" and ephebic gestures of his youthful cast.

His chief contribution was in showing how singleness of mood was essential to a work of stage art. He created atmosphere with a sure hand. His careful and deft craftsmanship also set a good, much-needed example. Delicate, witty, charming, filled with philosophical dialectics and allusions to contemporary situations, his plays won and held favour in a restricted circle.

His first, a prose comedy *Campaspe* (or *Alexander and Campaspe*, or *Alexander, Compaspe, and Diogenes*; 1580), relates how a love-triangle between Alexander, the Greek conqueror, Campaspe, his Theban captive, and Apelles, a painter, is resolved. Enamoured of the beauteous Campaspe, Alexander sets her free and engages Apelles to paint her likeness. The artist and his yielding Theban model fall in love. Apelles destroys the finished portrait to have a pretext for more sittings. Suspicious, Alexander tricks Apelles into revealing his ploy, then graciously surrenders the girl to the artist, saying: "It were a shame Alexander should desire to command the world, if he could not command himself." The philosopher Diogenes has a minor role. Pliny is Lyly's source. The script is noted for its songs, especially one that begins, "Cupid and my Campaspe playd, / At Carde for kisses . . ."

Though the dialogue is very much in the style identified with Lyly, some exchanges are effectively succinct:

Is love a vice?
It is no virtue.

More typical is this passage, which occurs between Apelles at his easel as he gazes at Campaspe, who answers him banteringly:

APELLES: I shall never draw your eyes well, because they blind mine.
CAMPASPE: Why then, paint me without eyes, for I am blind.
APELLES: Were you ever shadowed before of any?
CAMPASPE: No: and would you could so now shadow me that I might not be perceivable of any.

Here are two servants talking, their sentences incongruously balanced, alliterative, cultivated:

MANES: I serve instead of a master a mouse, whose house is a tub, whose dinner is a crust, and whose bed is a board.
PSYLLUS: Then art thou in a state of life which philosophers commend, crumb for thy supper, a hand for thy cup, and thy clothes for thy sheets – for "Natura paucis contenta".

Lyly sought to depart from his verbal manner somewhat and simplify his characters' utterances, but in most instances only barely succeeded.

Always the courtier seeking advancement, Lyly intended to have his play seen as a flattering reference to Elizabeth and her entourage. Alexander, the benign ruler, practises self-denial in renouncing his love for Campaspe. The brilliant retinue over which he presides is also like Elizabeth's in its dedication to art and love, though never at the expense of martial courage.

Sapho and Phaon (1583–4) repeats this pattern: a mythological queen, Sapho is a surrogate for England's illustrious Elizabeth I. At Venus' doing, Sapho falls in love with a handsome boatman, Phaon, but the foam-born goddess is also smitten with the young man, and Sapho is forced to yield her claim. In retaliation, she boldly affronts Venus by taking her son, Cupid, as her favourite, thereby herself assuming the role of "Queen of love". Other strands of myth are used to embroider the tale.

Endimion, The Man in the Moon (1585) is another story of a beautiful youth, possibly an allusion to Elizabeth's noble suitor, the Duke d'Alençon, or to another aggressive courtier, the Earl of Leicester. Elizabeth is personified as Cynthia, the chaste lunar goddess, Moon, who is "always one, yet never the same – still inconstant yet never wavering"; a rather apt description of the reigning red-haired Tudor. She is opposed by Tellus, the Earth, perhaps meant to be Mary, Queen of Scots, Elizabeth's cousin and lifelong rival for the sceptre. Tellus joins forces with a witch, Dipsas, by whose evil magic Endimion is enfolded in a sleep that is to last forty years, but he is awakened from the spell by Cynthia's kiss, after which she kindly sets him free. At the end, the references to real-life persons become more specific.

All this flattery was to no avail, however, and Lyly went unrewarded by the tight-fisted, Machiavellian queen. His hope to become Master of the Revels was never requited. He took subtle revenge in later works, *Midas* (1589) and *The Woman in the Moon* (c. 1591), in which his tone grows satirical and his mood cynical and misogynistic. He seems thoroughly disenchanted with womankind. Midas, the Phrygian ruler, is none other than Philip of Spain, who claimed the English throne – this was the time of the Armada, and Lyly does pay high tribute to the heroism of "a heaven-guarded nation", while Midas is portrayed as a monstrous tyrant. *The Woman in the Moon* is in blank verse, its language surprisingly plain.

Gallathaea (1588) ingeniously transforms a serious classical myth into a charming piece. For safety's sake, two girls are separately disguised as boys by their over-protective fathers. Meeting by chance, they become enamoured of each other. One is later changed into a boy, so that their love may be fulfilled. In the realms of myth and fantasy many strange things are possible. The motif of sex-change is sounded again in *The Maid's Metamorphosis* (*c.* 1586–8 – all dates here are problematical). Posthumously attributed to Lyly, this play is also in verse. *Mother Bombie* (1589–90) is Terentian in design: its intricate plot is well handled, making excellent use of the age-old devices of mistaken identity, surprises, reversals, recognitions and reconciliations. Mother Bombie, on whom the title character is based, was an actual citizen of the time, a fortune-teller often mentioned by Elizabethan playwrights. The script treats of incest: a brother and sister desire each other, but at the end prove not to be related. William Hazlitt, the critic, is quoted as saying three centuries later that it is a work "very much what its name would import, old, quaint, and vulgar . . . little else than a tissue of absurd mistakes, arising from the confusion of the different characters one with another, like another *Comedy of Errors*, and ends in their being (most of them) married . . . to the persons they particularly dislike". But not all spectators and readers have held the same opinion of it, and it is a bit startling to find Lyly being categorized as "vulgar". But, incontestably, *Mother Bombie* does preserve a panorama of everyday London life at the time.

In general, apart from *Mother Bombie*, Lyly's plays closely resemble masques. Slighted and disparaged by some later and unappreciative critics, such as Hazlitt, who find him too effete and precious, he is not to be overlooked as an important examplar. He was erudite, yet wore his learning lightly, as other Elizabethan playwrights perceived and began to do. Along with Gascoigne, he helped to establish prose as the natural speech for English comedy. Unfortunately his over-elegant diction tended to cast a baneful spell over many of his contemporaries, with Marlowe and Shakespeare no exceptions, the Bard's *Love's Labour's Lost* showing a first-hand and harmful acquaintance with Lyly's recherché language. Peele's pastoral *Arraignment of Paris* was also written in a manner paradigmatic of Lyly's sophisticated yet sentimental comedies. His own models were Roman – Plautus and Terence, not Seneca – and in this, too, he led the way for some of his immediate successors. Evans comments: "Lyly pleased the courtly audience to which his comedies were addressed; it was the same audience that had welcomed his novels . . . . So topical was Lyly, so neatly adjusted to the age, that much of the light and colour has not disappeared. Shakespeare obviously knew his work and benefited from his study, and though he soon outgrew the clever euphuistic prose, he is in the early plays deeply under his influence. Shakespeare's comedies are at once more romantic and human but they retain much of Lyly's ingenuity. Further, in some more detailed ways Lyly's example was potent. Particularly do the witty servants of Lyly reappear in Shakespeare's comedies." Gassner has praise for the songs in Lyly's plays and their "lovely lyrics, and it must be conceded that he had a 'pretty talent'". (Though the lyrics are by Lyly, it is generally thought that the music itself was not.) But Gassner is less admiring of him as a dramatist: "Shakespeare was indebted to him, not always to advantage, in his romantic comedies for suggestions and style. . . . Lyly's vein, which was also tapped by other writers, added at most a number of divertissements to the theatre." The author of the article on Lyly in *The Oxford Companion* remarks that the plays are now admired "for their flexible use of dramatic prose and the elegant patterning of their construction".

His biographers relate that, having never fulfilled his goal of being Master of the Revels at Elizabeth's court, John Lyly died "an embittered man".

Another University Wit who sought a career in the world of the stage is Thomas Nash (or Nashe, 1567–1601), who was a sizar at St John's College, Cambridge – that is, he received an allowance towards his living expenses, in return for which he was a servant to wealthier fellow students. Like the other Wits, he settled in London after his graduation and was closely linked there with Robert Greene and Christopher Marlowe in their riotous life and borderless debauchery in the city's *demimonde*; the trio knew each other from Cambridge days. His subsequent comments on his fellow writers are illuminating, a help to scholars groping to document more about this talented fraternity. His first published work was a preface to Greene's prose-and-verse romance *Menaphon* (1589), in which Nash took advantage of the forum to vent his satirical view of other aspiring authors of the day. He also took sides in a battle of pamphleteers going on for a time over the liturgical practices and disciplines advocated by certain bishops, and responded to a verbal attack on the mortally ill Greene by the poet-encyclopaedist Gabriel Harvey – Nash's defence of his friend Greene was stinging and triumphant. He wrote a religious tract, as well as a disquisition on dreams and nightmares, *The Terrors of the Night*. For a while he may have served as tutor to the daughter of Lady Elizabeth Carey, a task for which he would seem to have been a very strange choice.

His lingering reputation is chiefly owed to his parodistic, picaresque novel, *The Unfortunate Traveller, or The Life of Jacke* (1594). He continued to carry on his literary feud with Gabriel Harvey, the antagonists writing in a harsh satirical vein, until finally Archbishop Whitgift decreed that the effusions of both writers should be taken out of circulation.

In collaboration with Ben Jonson, Nash turned out a comedy, *The Isle of Dogs* (1597, now lost) a work so critical and licentious in tone that they were ordered to prison for it. Jonson served his sentence; it is not clear whether Nash ever did. In any event, Nash fled from London, perhaps across the Channel to France. As a consequence of this incident a censorship was imposed and all London theatres were shut down for a brief spell.

His pen never ceased to keep him very busy. In *Nashes Lenten Stuffe* (1599) he proffers "a mock encomium of the kipper herring" together with a burlesque treatment of the Hero and Leander tale, which had been celebrated in a splendid poem by his boon companion Christopher Marlowe. It is believed that he also collaborated with Marlowe on several plays, in particular the early *Dido, Queen of Carthage*, composed while they were fellow students at Cambridge; it was produced with a cast made up of the Children of the Queen's Chapel and subsequently published in 1594.

A "disciple" of the great, rather than their equal, Nash is one of the vivid and oft-mentioned figures of London's wild literary scene.

Writing for child actors and seeking to entertain royalty and aristocratic spectators, John Lyly appeared not to be interested in augmenting his income but rather in advancing his status. The contrary was apt to be true of the University Wits, some of whom were poor and constantly in need of money. Their drinking and wenching were costly; when the roisterers did have shillings in their purses, they spent their earnings recklessly. They wrote in order to keep alive, either by attracting the generosity of a noble patron or by selling their works to players and publishers, an uncertain and hazardous way of life.

Alongside the infrequent, semi-professional stagings at Elizabeth's court, and performances at the universities, Inns of Court and Church schools, a commercial theatre was abruptly springing up in London and soon flourishing. At first the wandering bands of ragtag players, jesters and musicians utilized space at fairs, but by now that was too transitory. Next, having to improvise as did their counterparts in Spain, they selected the yards of inns, took over a bear-garden for longer stays, a few days or weeks. There, platforms were set up on barrels; people stood to watch, or gazed down from windows in houses surrounding the available site, and in some places from balconies. A huge number of such inns were scattered about England, as many as six hundred or seven hundred in London alone. Taking longer leases, the troupes installed booths on the platforms and made them permanent. Benches were intended for the gentry. As of 1570 at least six such inn-yards in London were converted after this fashion, and the troupes occupying them were becoming more ambitious.

Soon it became apparent that Londoners had a growing appetite for such entertainment and that investments in enterprises of this sort could yield an entrepreneur quick profits. In 1576 James Burbage, afterwards to be one of Shakespeare's partners, built a structure designed solely for offerings by professional actors; it had no name but was called simply the Theatre. There, in its final years, some of Shakespeare's early works were mounted.

Though there was little opposition to plays given under royal auspices, or those with academic sponsorship, Puritan elements in the city and country were critical of this newly conceived commercial stage; one reason was that performances were given on Sundays. The plays, and the actors, were considered, and perhaps with some justification, to be vulgar and immoral. Indeed, an actor here – as elsewhere in Europe at the time – had no legal status. He was deemed a vagabond, and therefore outside the law, even as he was proscribed by the Church.

The actors, ingenious, found an answer to their dilemma. By attaching themselves to a noble patron they gained considerable protection. Each of the first professional companies, accordingly, sought and gained an alliance with an heir to an illustrious name. Many aristocrats had a passionate liking for theatre, and some, as did Sackville and Sir Philip Sidney, even wrote plays and were willing to lend their moral support and perhaps financial aid. Boasting some such eminent shield, a company came to be known as the Earl of Leicester's Men, or the Lord Chamberlain's Man, and the like. In some instances, too, their elegant sponsors gave the actors their cast-off garments, providing them with expensive, ornate costumes. To be a sponsor lent one prestige. Queen Elizabeth encouraged this practice.

Even so, the theatre and the play remained under the jurisdiction of local justices of the peace, and also of the London municipality. Because of laws against vagabondage, the touring troupes were ever more inclined to limit their activities to the capital, but here, too, they were denounced. Charges of sedition or heresy, loosely defined, always threatened. The hostile Puritans used as a pretext that large gatherings of spectators were a danger to public health in a frequently plague-visited city. To avoid this problem the producers elected to build theatres and stage their shows in outlying areas; in Burbage's instance, across the Thames in a slum or red-light district. Even there he was subjected to harassment by officials, prodded by easily shocked and aggressively vocal churchmen.

Civic authorities had other complaints: the crowds tended to be unruly; pickpockets and other thieves were busy in them. On occasion small riots broke out. Clerics spoke of the damnation to which decent people were exposed when flocking to lurid entertainments. "Whoever shall visit the chapel of Satan, I mean the Theatre, shall find there no want of young ruffians." It was a place, its outraged critics warned, where young ladies might lose their virtue, tempted by masculine spectators who "give them pippins" and "dally with their garments". Even more horrendous is a picture published by a scandalized writer of the time: actors were "crocodiles which devour the pure chastity both of single and married persons". Did they not keep apprentices away from work? Were not the plays lewd? From time to time specific excuses were found to close the theatres; but, struggling on, they reopened again. The queen formed her own company of players.

Plays were put on in daylight, at two or three in the afternoon. The actors, in costume, paraded in the morning to the sound of trumpets and drums to attract an audience. Some time later the custom of holding a street parade gave way to the less noisy passing out of handbills. A flag flapped or drooped over the building to signal that a performance was to be offered: white for a comedy, black for a tragedy. To gain entrance, the spectator had to drop a penny into an iron strong-box (from this we get the term "box-office"). That allowed him standing-room. For seats in the covered gallery or on the stage the price was double or higher. As in Greek theatres, there were no reserved seats: servants or wives were sent ahead to obtain and hold the best places.

In 1581, through the intervention of powerful courtiers, the Master of Revels, who directed the queen's fêtes and other entertainments, was invested with the right to exercise a general censorship. He could grant or rescind a licence, and this afforded the players a measure of security, since what had been approved by an officer of the court was less likely to be attacked by others. What is more, the aristocracy attended the commercial plays with enthusiasm, giving the playhouses a patina of respectability, albeit a thin one. The acting companies had to pay for the licence and post a bond. After any day on which fifty deaths from the plague were listed, the theatres had to shut their gates.

Burbage had his original Theatre torn down in 1597, the ground-lease having expired. Two years later a new building, the Globe, larger and with better stage-machinery, was put up in Southwark by two of his sons. Thriftily, timbers from the original playhouse had been saved and used again. By now the elder Burbage – the "carpenter turned actor" – had contracted to share proceeds from a second venue, the Curtain (1577) in Shoreditch, just outside the City's northern boundary. More exactly, this theatre was at Curten Close (hence, its name), and the new Globe at Bankside; the stage-boards of both houses were to resound with the voices of actors reciting blank verse and the echoes of their restless tread, as they sought to bring to life a young Shakespeare's tentative efforts.

Controlling both the Theatre and the Curtain, Burbage's was a short-lived monopoly. A canny speculator, Philip Henslowe, soon outmatched the Theatre with a finer structure, the Rose, opened in 1587. Starting as a servant to a Mr Woodward of Bankside, Henslowe was clever enough to take to himself his master's wealth and divers kinds of property by winning the hand of Mr Woodward's widow. He had been successively a dyer's apprentice, a bailiff's assistant, a pawnbroker, and a trader in leather and lime. Now he grasped that money was to be made in theatres. Henslowe, almost illiterate, maintained very careful records. His accounts of the dramatic programmes, bear-baiting and bull-baiting spectacles, the casts and sundry costs, and weekly schedules in his several playhouses, are a boon to historians. (Unscrupulous, he lent money to his actors and kept them in debt to him, so as to have a tighter grip on them.)

The Rose was modelled on Burbage's two playhouses, as were other theatres subsequently erected in London to meet the growing popularity of dramatic entertainment, all on the less regulated south bank of the Thames, with the exception of the Blackfriars, a "private" theatre. To the Rose the shrewd Henslowe added the Fortune, operating it with Edward Alleyn as a partner, and the Hope, and there were also more "private" theatres under other auspices: the Whitefriars; Butts; the Salisbury Court; and the Phoenix in Drury Lane, converted by Inigo Jones from the Cockpit-in-Court. Two other public theatres joining the ever fiercer competition were the Swan and the Red Bull, the latter exhibiting boisterous works for the lowest classes. The distinction between a "public" theatre and one designated "private" is not entirely clear. The Blackfriars was in a priory once belonging to that holy order and had long been used by a company of boys who rehearsed for performances at court. Very often, before or after a play was given by child actors to amuse the queen, "private" month-long showings of the work at high prices were added for select, noble audiences, and this could be very profitable. In that sense, this was a "private" theatre, one possible explanation of the term.

The alert Burbage rented the space in the priory, hoping to utilize it for his adult companies, but for a long time was not permitted to do so. Ultimately Burbage's sons succeeded in achieving his aim, and the Blackfriars was also to see Shakespeare's plays staged by his companies. The Blackfriars and Whitefriars stood on land nearer the city's heart and had once belonged to those orders until both had been confiscated during the religious feuding that marked the Tudors' reign. The boys who put on plays there were much acclaimed, popular – they came to be called Children of the Revels of the Queen – but they rapidly lost favour with James I, Elizabeth's successor, when the troupe's offerings took on a political tinge offensive to the new crown. The hypersensitive, Catholic James ordered the troupe disbanded. After all, he was the son of Mary, Queen of Scots, who had a threatening claim to England's throne. The emptying of the priory allowed Burbage to gain possession of it in 1606, but his neighbours obtained an injunction against his using the building for theatrical purposes. They were willing to let the boys perform in it, but not have professional actors occupy it. Eventually, James's irritation having been smoothed, the ban on the Blackfriars troupe was lifted, but the old priory was now under Burbage's control and he certainly did not encourage their competition, so they chose to resume their programmes in Whitefriars instead, where they stayed until 1616, though without official permission. In 1617 the company of boy actors was finally dissolved.

The Blackfriars was much prized; it was more than twice as large as the Globe and could reap more income if a play was heartily welcomed. Besides, it was roofed and performances were not contingent on the weather. The King's Men spent seven months of the year there – from autumn to spring – and the remaining five months of summer at the Globe, which had no roof. Most of the "private" theatres were indoors, and productions at court were similarly sheltered, as might be expected, as was also true of presentations at the universities, the Inns of Court – notably Gray's Inn, among the complex of law schools – and in banquet halls, where plays were still regularly produced, much to the annoyance of the commercial theatres. The bear-baitings and bull-baitings also drew a segment of the paying audiences, but the two circular arenas for crude exhibitions of combats with savage beasts were open to the skies, hence not always available. Of the other public theatres, it is of interest that the Phoenix was so named because it was built on the ashes of an earlier playhouse at the same site. It served a long succession of companies until the Restoration.

What is known about the design of the public playhouses mostly derives from a contract with building specifications drawn up and signed by Henslowe for the construction of the Fortune. Another source of detail is a description of the Swan by a Dutch priest, Johannes de Witt, who in 1596 visited that theatre, was awed by it and wrote about it to a friend, who in turn drew a sketch of it. But much is conjecture, and the result is an endless dispute among the "experts" who seek to envisage the typical playing conditions of the time. Sometimes references are found in the dialogue, and stage-directions in the scripts also give helpful clues.

It is certain that the exteriors of most of the playhouses were of wood and stucco, the shape octagonal, with a turret on one side; most likely the interior was circular – the "wooden O", in Shakespeare's phrase. The inside of the Fortune was square, an exception. This interior design copied that of the inn-yards from which the Spanish *corrales* and the earliest London theatres evolved. In some of the theatres the platform could be removed so that the space could also be used for other purposes – bull- and bear-baiting spectacles, in particular. With most of the houses lacking any cover over the pit – or standing-room – artificial light was unnecessary. Since few people bathed regularly, a bit of fresh air in an enclosure was not unwelcome. The galleries were roofed with thatch.

The stage had no front curtain, but most likely there were some at the back and sides, to be opened for a view of a second room or inner stage at the rear, and at either side to mask the entrances and exits of the players. In some instances the back wall had two or three doors in it. The platform, jutting halfway into the pit, was surrounded by three tiers of galleries; a roof over the projecting stage – and often alluded to as the "Heavens" – was upheld by two visible pillars. The location of the dressing-rooms is harder to determine; the assumption is that they were beneath the stage, where props might also be stored. One of the galleries, in the sketch of the Swan, runs wholly around the playhouse, which meant that it enabled some spectators to look down on the stage action from an unusual angle, the rear, above the back wall or curtain. On some occasions, though, a section of this balcony behind the stage might be used by the musicians or by the actors, or one "box" could serve as a window from which Juliet gazes down at the enraptured Romeo. Others say that her vantage point was more likely in the first balcony. In the canopy over the stage, a "trap-door" allowed the descent of angels or other supernatural beings. Beneath the platform was "Hell", also accessible through a trap-door or from behind the scenes. Through the floor "trap" might rise the ghost of Banquo or Hamlet's father.

The Swan could accommodate 3,000 spectators, twice as many as Burbage's Theatre, and the Globe could also hold 3,000 but only with standees packed together in the pit. De Witt and other witnesses speak of the buildings' interiors as "sumptuous", the wooden pillars painted to look like marble, prompting some visitors to compare them with Ancient Roman models, an appraisal hardly justified. Indeed, these English playhouses in no way equalled those in Italy, especially those with magnificent decoration in Venice. The London theatres doubtless varied from one another in size and shape, and certainly in minor details, yet shared much the same overall aspect.

The Blackfriars, like other "private" theatres, was smaller and had benches for all. It was rectangular, with a better-equipped stage on which elaborate scenic effects were possible. Since it was indoors, artificial lighting was required: it is believed this need was met by a candelabra overhead and elementary footlights – a line of candles or lanterns? Three galleries, some divided into boxes for the rich and noble, ran along both walls of the hall, over which – in the second Blackfriars which replaced the original priory in 1596 – was a vaulted ceiling. The rows of benches for spectators extended on to both sides of the stage. In these playhouses, both public and "private", the actors and spectators were close, instilling a feeling of intimacy, of compelling participation in the emotional excitement engendered by the play. In the "private" theatres the price of admission was considerably dearer, the wealthy and patrician audience more sophisticated and accustomed to seeing royal masques, so a more lavish stage investiture was expected.

In the public theatres the sets were never as bare as has been widely thought. More recent scholarship argues that all the stage-devices used by the Ancient Greeks and Romans, and by the master technicians of the Italian Renaissance, were known to designers of the Elizabethan theatre and employed if needed. Still, the chief appeal was to the spectator's imagination, as in Spain, by way of verbal descriptions of great splendour. But if Robert Greene calls for magical effects and transformations, and similar ones are demanded by Marlowe for his *Dr Faustus*, along with amazing conjuring feats, they were provided. Sound effects were also to be had: martial trumpets, clanging alarum bells, cannon shots, thunderclaps. It was during a realistic presentation of Shakespeare's *Henry VIII* in 1613 that sparks from an exploding cannon set ablaze the thatched roof of the Globe; the building was gutted by the flames, after which a second Globe was put up on the same site, much as had earlier happened to the Phoenix.

If a play needed it, there was a crane – again a *deus ex machina* – to raise and lower winged creatures, demons, divinities. The trap-door, opened, might serve for Ophelia's and Yorick's graves, and so on. It was not that the professional companies were unaware of the theatre's age-old resources, but that they had less money to lay out, and simplicity and plain sets meant larger profits. Often, to help the ausience and set the scene, an actor or stage-hand entered and held up a sign that read "The Forest of Arden" or "Another part of the Forest" to indicate a change of place. The gain in this was fluidity, and a cost was the episodic structure of the plays, the multiplicity of story-lines and scenes in them. Some aristocratic, erudite critics objected to this, arguing for retention of the classical unities of time, place and action, and asking for more credible scenery. But the larger public accepted what was merely suggested by an actor in a line or two, or by the printed signs or a few relevant props. The Elizabethan stage was by choice not literal but conventionalized, like those of Greece and the Orient. Besides, with audiences so near at hand, surrounding the stage on three or even four sides, attempts at realistic settings would hardly have been convincing.

Though Elizabeth hated to spend money, she did like spectacle, and the Master of Revels paid sizeable sums for lavish scenery and costumes for masques and fêtes at court, in imitation of the gorgeously attired amateurs in the Mysteries, as well as of the triumphal processions and entertainments mounted by the Italian nobility and the Habsburgs in Vienna. Here the performances were often in the evening, so that emblazoning artificial lights were added to the glitter. Competing with other European royalty, Elizabeth dared not be too parsimonious. With a royal subsidy, designers such as Inigo Jones introduced illusionary clouds and scene-dissolves, gilding and painted canvas, three-sided prisms, flats, clever tricks of perspective, and a raked (slanted) stage.

Again, as on the Spanish stage, what lent the public theatre presentations colour were the actors' costumes, the bedecked cast-offs and hand-me-downs from their blue-blooded sponsors. It was an age when the audience itself – at least, the richer segment – was adorned like peacocks. In probably no other period have English courtiers put on such finery, been so elaborately coiffed, embellished their features with so much make-up. Only in France and Austria was there an equal personal enhancement. Further, if a company was invited to perform at court for a gala occasion, the Master of Revels might re-outfit the actors, and these new costumes might be kept by them and frequently re-used. Or, when noblemen died, their servants might inherit their clothes and sell them as second-hand ornaments to company members.

This fine stage array would seem incongruous to modern eyes: Julius Caesar might wear the plumed cap, silken blouse and velvet jerkin of an Elizabethan dandy. Though some attempt was made to achieve authenticity, the difficulty lay not only in lack of funds, but also in ignorance of how Greeks and Romans dressed. Historical research was not yet adequate to the task. Still, inventories of wardrobes belonging to the groups and costume sketches by designers show that the "right" sort of attire was often sought; sometimes it was arrived at, but more often it was not. The audience did not insist on accuracy, even if it knew what to look for – and the likelihood was that it did not. If Bottom was to be transformed into an ass, he did don an ass's head, and if a character was frightened by a bear in the woods, the actor impersonating the ferocious beast was convincingly concealed in a bearskin. Some obvious distinctions in contemporary dress were made, too: a priest would wear a black cassock; a cardinal, a red hat; a soldier would appear in a shining metal breastplate and carry a pike; a Turk would boast a turban and scimitar; a king would be more regally apparelled than any of his entourage, and this would also be true of a queen or princess. But servants, in accordance with the general opulence, might be inappropriately garbed in satin and lace, and the absurdity of this did invite a measure of criticism. Make-up was applied sparingly; in daylight, not much was needed. But a ghost would whiten his face, and a Moor – Othello – would resort to burnt cork to darken his lineaments. Wigs and paste-on beards were added where called for, especially when disguises were requisite, as was frequently the case.

With the success of the commercial theatre, the social and economic status of actors began to rise. A few consorted with the great. Some of them became comfortably well-off, substantial middle-class citizens: Shakespeare, as is well known, was able to retire to Stratford, his birthplace, while he was far from old, and there be a moneylender and country squire. Edward Alleyn, one of the most noted actors of the day and step-son-in-law of Philip Henslowe, quit his profession in the mid-thirties and at his death left enough to endow a college at Dulwich and fund other philanthropies. Burbage, too, waxed affluent. Such wealthy actors were exceptions, of course; they grew rich not just by acting but as partners in the companies in which they participated. Both

Burbage and Henslowe offered their players a chance to share in the profits earned by their troupes, perhaps with each investor putting in as much as seventy pounds, a goodly amount. From this capital, playwrights and costumers were recompensed and other expenses met. Lesser actors got meagre wages, perhaps sixpence a week, and boys who took female roles were deemed apprentices whose masters collected a few shillings for their services. For the most part these lads were recruited from the excellently schooled child companies at the Royal Chapel and St Paul's installed at Blackfriars. The box-office "take" was split in various ways: the shareholders divided the gate-receipts; the theatre-owner took what was paid for gallery-seats, that income constituting his "rent" for the house. Other arrangements obtained, however. Also, there were several different formulas for meeting the cost of hiring the box-office men, backstage helpers and musicians. An added source of income was the flat fees doled out for a command performance at court. One explanation of Shakespeare's prosperity was that he fulfilled three functions: he was a shareholder, an actor, a prolific and popular playwright. A special performance, a "benefit", was customarily given for an author at the third offering of his play, and from this he received all the profits. Or scripts might be sold outright to the companies, though they brought little. Finally, the actors did not cavil at spectators being on the stage with them: the extra receipts for those choice seats went to the performers.

By now, in the roofed and candle-lit "private" theatres, plays might be presented at night. This kept the actors very busy. They might rehearse or parade in the morning, perform in a public theatre in the afternoon, and then present a different script at a "private" theatre at night. In between, or when their days and nights were free, they had to learn lines from a new offering, for an ever-changing repertory, and the plays tended to be verbose and excessively long. The companies also toured the provinces, journeying as was the immemorial custom of wandering thespians by horse and wagon, the vehicles piled high with apparel, props, tags-and-ends of scenery. Tours were often resorted to when plague festered in the city and official fiats closed the theatres to limit the infection's spread.

At a remove of centuries it is impossible to make judgement the quality of the acting, and tastes and standards have probably altered considerably since the age of Elizabeth I. Perhaps the most esteemed of interpreters then was Edward Alleyn (1566–1626), a towering six feet, or seeming to be that tall. At seventeen he started his career by joining Worcester's Men, then moved to the Admiral's Men, for many years serving as actor-manager. He married Henslowe's stepdaughter, became his business associate in building the Fortune and with him later owned the Curtain, the Rose and the Hope. He grew very rich, and aspired to a knighthood, but was disappointed in that respect. His second wife, whom he married two years before his death, was the daughter of the great preacher and poet John Donne, Dean of St Paul's. At the Admiral's Men the chief playwright was Christopher Marlowe, and Alleyn's claim to fame rests largely on his interpretations of Marlowe's outsized dominating heroes, Tamburlaine, in the young poet's epic drama of that name; Barabas, the miser and avid gem collector in *The Jew of Malta*; and in the frenzied title role of *Dr Faustus* as well as Greene's *Orlando Furioso*. Incongruously, though a churchwarden of St Saviour's, Southwark, he had financial interests in bear-baiting rings and brothels in London's notorious Bankside area near his theatre. His letters and diaries are preserved in the library of Dulwich College, which was his main beneficiary; they show him as a good husband and responsible householder and a very astute businessman, traits rarely found in the acting profession. His reputation is bolstered by shrewd observers of his performances, among them Nash, Jonson and Heywood, all of whom wrote in praise of him. He was a generous patron of writers.

At the Globe the star was Richard Burbage (1567–1619), Alleyn's most prominent rival. One of the two sons of James, he and his brother Cuthbert had inherited shares in the Blackfriars and he had a half proprietorship of the Globe. It was they who prudently saved the timbers of the Theatre and replaced it with the Globe, where Shakespeare was later a company partner and himself had roles in many of his plays and some by others. The leading parts, however – Richard III, Hamlet, Lear, Othello, Shylock, Romeo – usually fell to Richard Burbage, who was widely hailed for having first portrayed them and to great effect. He also appeared in Jonson's classical tragedies and comedies, *Sejanus*, *Volpone*, *The Alchemist*, *Catiline*; Kyd's *The Spanish Tragedy*; and in works by Webster and by Beaumont and Fletcher, attesting to his astonishing versatility. Shakespeare was obviously grateful to him: he bequeathed him a memorial ring in his will. From most reports of his style, he avoided the flamboyant, declamatory manner common to most of his peers; instead, he was "natural, refined, and sensitive". He was also a talented painter; the Felton portrait of Shakespeare is credited to him. His stay with the Chamberlain's Men at the Globe continued for twenty-five years, and his epitaph, by William Camden, antiquary and historian, is remarkably short and telling: "Exit Burbage."

Of Shakespeare's acting skills very little has been written. He seems to have taken mostly character roles. He is almost unique among actors who also write in that he did not fashion his scripts conspicuously to exhibit his own mimetic gifts. His much-quoted speech of instruction to the players in *Hamlet* makes explicit his bias against over-acting: Hamlet urges them to speak naturally – as Burbage did – and to abjure gesturing too emphatically. His warning suggests that many incompetent actors trod the Elizabethan stage. In their plays, other authors of the day utter disparaging remarks about them. But playwrights often blame others rather than themselves for their failures, so they are not to be relied on for wholly objective views on that subject.

Of note in London then, apart from Alleyn and Burbage, were Richard Tarlton (d. 1588) and Will Kemp (c. 1550–1607), both much appreciated for their art at comedy. Ill-educated, of humble birth, Tarlton was London's most popular clown and a favourite of Elizabeth, to whose attention he was brought by her fond intimate, the Earl of Leicester. In 1583 Tarlton joined the Queen's Men and was adept at ad libs and other forms of extemporization, as well as at composing "jigs" – "rhymed farces sung and danced to traditional airs". He originated a good many long-lasting jokes: a collection of them, *Tarlton's Jests* (c. 1599), was published after his death. He is believed to have been the model for Shakespeare's cynical clown Yorick. The many amusing stories about him are most likely apocryphal. A host of contemporary playwrights speak of him flatteringly, including Nash, Dekker, Harrington, Jonson and Heywood.

Will Kemp (or Kempe), after Tarlton's passing from the scene, succeeded to his very high place in public approbation. After having been a member of the Earl of Leicester's Company, he became the leading comic in the Lord Chamberlain's Men, to which Shakespeare belonged, and had roles in the Bard's plays, such as the malapropist constable Dogberry in *Much Ado About Nothing* and Peter in *Romeo and Juliet*. An agile dancer, he amazed all England when, to settle a wager, he performed the "morris" the whole distance from London to Norwich, a feat which he later boastfully described in his *Kemps Nine Daies Wonder* (1600). He also acted in Jonson's plays, and in the closing phase of his career shifted to the Earl of Worcester's Men, the last three years of his life. The speculation is that he sold his shares and abruptly left the Globe because Shakespeare frowned on Kemp's impudent habit of replacing the author's lines with his own broader quips. To any playwright, for an actor to improvise new wording is *lèse-majesté*. Kemp's end was not a happy one, especially for a player: he died in obscurity.

The plays called for a variety of skills: fencing, dancing, singing. The circumstances in which the performers carried out their tasks – the distracting nearness of the audience, and the rudeness of the groundlings, weary from having to stand so long – were perpetually taxing. An actor had to be self-possessed, quick to catch and hold the interest of the unlettered, restless onlookers in the pit and also to gratify the knowledgeable courtiers upon whose goodwill so much depended. Versatility was needed, for within the short space of a week he might be asked to embody many quite different types of persons, some – in the particular instance of plays by Marlowe and Shakespeare – portrayals of great dimension, most demanding. He could count on being pelted with orange rinds and fruit cores and empty ale-bottles by the noisy tradesmen and arrogant youths – comprising "the many-headed monster of the pit" – if he failed to please. Usually the players underwent long years of apprenticeship. But this was not possible for the adolescent boys who were chosen to be mature, imperious, seductive as Lady Macbeth, Queen Gertrude, Cleopatra, yet there is no word at all of complaint that these characters lacked verisimilitude.

A footnote: amusing and graphic is an account by Sir Henry Wotton (1568–1639), an ambassador and later provost of Eton, of the conflagration that destroyed the first Globe in 1613.

> [It] occurred just as the actor portraying King Henry VIII made his entrance at Cardinal Wolsey's grand mansion. Certain small cannon being shot off . . . some of the paper or other stuff, wherewith one of them was stopped, did light on the thatch, where being thought at first to be but an idle smoke, and their eyes were attentive to the show, it kindled inwardly, and ran around like a train, consuming within less than an hour the whole house, to the very grounds. This was the fatal period of that virtuous fabric, wherein yet nothing did perish but wood and straw, and a few forsaken cloaks; only one man had his breeches set on fire, that would perhaps have broiled him, if he had not by the benefit of a provident wit put it out with a bottle of ale.

— 12 —
ENGLAND: KYD AND MARLOWE

One drama historian describes Thomas Kyd (1558–94) as a "shadowy figure", since little is known about him. In many ways his talent was limited, and today's readers of his plays find his work interesting but scarcely impressive; he wrote only a few scripts, and his life was tragically short – yet hardly any other writer for the theatre has equalled his immense influence. The drama that made him famous when staged, and afterwards went through ten printings in a brief period, did not bear his name: it was attributed to "Anonymous".

By conjecture, his father was a scrivener, acting as a notary, drawing up official contracts and composing letters for others who were less literate. His son was sent to the Merchant Taylors' School, a London institution for the offspring of middle-class tradesmen, founded in 1561. In its third decade, it boasted a headmaster reputed to hold progressive ideas about teaching. Among its other students was the splendid poet-to-be Edmund Spenser. Less certain is whether Kyd continued his studies in Cambridge, and, if so, whether he ever got a university degree there or anywhere else. Whatever the extent of his formal schooling, it is later apparent that he possessed better than average learning for the time, and in particular was conversant in Italian and quite able to correspond in it.

When next located by biographers, he was renting shabby rooms in London and earning a meagre living, probably as a scrivener, his father's often tedious trade. But his life was not dull. His room-mate, six years his junior, was Christopher Marlowe, who had finally come to the city from Oxford and Cambridge, after prolonged and frequently interrupted stays at both venerable universities. His past was somewhat mysterious. The two enjoyed a very bohemian existence, with much drinking and roistering, and with links to the University Wits. Both of them aspired to literary careers – Marlowe was an inspired, precociously gifted poet – and both young men were eager to turn out plays for the suddenly burgeoning and profitable commercial theatre. Besides having Kyd as a companion, Marlowe was a close friend of Thomas Nash.

The members of this group of very talented playwrights, some of them geniuses, who were the founding fathers of English drama, were all remarkably young. Allardyce Nicoll has listed their ages at the beginning of this period (1580): Robert Greene was twenty-two; Thomas Lodge, twenty-three; John Lyly, twenty-six; George Peele, twenty-three; Thomas Kyd, twenty-two. Christopher Marlowe, still at Oxford, was only sixteen but before very long to descend on London and make an early start. William Shakespeare, probably at work with his father in Stratford, was not yet married and probably not ready to launch his career in the playhouses for another five or six years – he was about the same age as Marlowe. This was an astonishing outburst of youthful creativity.

The chronology of Kyd's and Marlowe's early works is uncertain. While at college Marlowe collaborated with Nash on *Dido, Queen of Carthage*; never finished, it was published long afterwards in 1594. In London, in the new realm of theatre in ferment that now surrounded him, a hopeful Kyd kept his pen busy, though scholars cannot determine how many scripts he turned out, nor in what order, nor what were their themes. Here and there are references to his *Hamlet* – today alluded to as the "Ur-Hamlet" – which may well have impregnated Shakespeare's mind when later he fashioned his own version of the melancholy Dane. All that is conjectural. He adapted the Humanist Robert Garnier's French tragedy *Cornelia*, but is scarcely remembered for it. He was probably led to that by having joined a circle of classicists surrounding the Countess of Pembroke and her brother, the distinguished poet Sir Philip Sidney. Some argue that Kyd helped Shakespeare prepare two early plays, the horrendously gory *Titus Andronicus* and the historical chronicle *Henry VI*, when the young actor from Stratford was just getting started as a commercial dramatist and was giving an eager public what it demanded and willingly paid to see. The resourceful Kyd may also have been the author of *Arden of Feversham* (staged some time between 1586 and 1592), which – as the first instance of a bourgeois tragedy – was also greatly influential. Everything concerning this, too, is unproved. Such details of his career remain shadowy, tantalizing those researching the era, and especially those hoping to learn more about the always elusive Marlowe.

About 1581 an English translation of Seneca's plays was published. Almost surely Kyd read them, studied them closely, absorbed many aspects of them thoroughly. Accepting them as models, his handling of Garnier's *Cornelia*, replete with violence, shows his affinity with Seneca. His debt is even more obvious in his *The Spanish Tragedy, or Hieronimo Is Mad Againe* (1588 or 1592), the archetypal example in English of the so-called "revenge" tragedy, an overwrought script that for a long time was the most popular venture on the Elizabethan stage and consequently the forerunner of a legion of melodramas by his contemporaries and immediate successors, among them even Shakespeare and Webster. Several plot ingredients of *Hamlet* are found in it: a drastic deed sought to avenge the murder of one near and dear; feigned madness to achieve that end; ghosts; a play-within-a-play; an initial hesitation finally overcome.

Hieronimo, marshal of Spain, is possessed by tears and driven to fury by the murder of his son, Horatio, by rivals for the love and hand of Bel-imperia. She is the daughter of Don Cyprian, Duke of Castile, brother of Spain's king. The Spanish and Portuguese are at war. Balthazar, son of the Viceroy of Portugal, has been taken prisoner. This does not hinder him from courting Bel-imperia. Her brother, Lorenzo, supports Balthazar's suit, as does the King of Spain, whose reasons are political. Discovering Horatio and Bel-imperia in a garden at night, Lorenzo and Balthazar overwhelm the young man, render him helpless and cause his death by hanging from a tree. The sight of his son's body overthrows Hieronimo's reason. To learn the identity of those guilty of the crime, he makes use of a play, *Solyman and Perseda*, during which Lorenzo and Balthazar are killed. In tracking down those who had been responsible, Hieronimo displays fiendish ingenuity. Scene after scene of intense violence and emotion follow in quick succession: Bel-imperia writes a warning letter in her own blood to Hieronimo, while she herself faces the prospect of being forced into marriage against her will. The mother of the dead Horatio, stricken by her loss, stabs herself. Bel-imperia commits suicide. Rather than confess and betray his confederates, Hieronimo bites out his own tongue, then also ends his life by stabbing, though providentially off stage. At the close the scene is strewn with bodies, a tableau that Shakespeare also borrows.

In an epilogue, poetic justice is meted out by the allegorical figures of Revenge and a Ghost, representing Hieronimo, together with those of his fallen foes, who serve as a chorus: the wicked are to be dreadfully punished in Hell; Hieronimo and Bel-imperia are promised eternal bliss.

The language throughout is monotonously paced and invariably stilted, with only a few striking lines deserving mention. Poetic dialogue is not Kyd's forte. Too often he seems short of breath and has a shaky sense of rhythm. He is fond of inserting Latin phrases and classical allusions that must have been unintelligible to much of his audience. Speeches such as the following surely daunted actors even then:

> ALEXANDRO: Not that I fear the extremity of death
> (For nobles cannot stoop to servile fear)
> Do I, O King, thus discontented live.
> But this, O this, torments my laboring soul,
> That thus I die suspected of a sin
> Whereof, as heavens have known my secret thoughts,
> So am I free from this suggestion.
> VICEROY: No more, I say! To the tortures! When?
> Bind him, and burn his body in those flames,
> That shall prefigure those unquenched fires
> Of Phlegethon, prepared for his soul.

Even so there are patterns in the writing from which Shakespeare learned how to end a scene effectively with an exclamatory rhymed couplet, and there is some gift for word-play and paradox, though little for irony – Hieronimo's feigned or real madness is not used, as in *Hamlet* and *King Lear*, to offer a view of the world from a different angle of vision, yielding a special dark insight. So much of the wording is mere rant and tends to lose force as the lengthy exchanges unwind in bombast, and in a later century it invites Henry Fielding's amusing parody of Elizabethan melodrama when he set forth the mock-heroic exploits of diminutive Tom Thumb.

The motivations of most of the characters, Hieronimo excepted, are only slightly developed. But such artistic shortcomings did not matter for the moment. From its box-office success Kyd's fellow playwrights learned exactly what the uncouth spectators wanted and how to satisfy their already jaded taste. What Kyd brought to the Elizabethan stage was neat craftsmanship; none of the hurried, slapdash script-writers of the time had structured a story quite so well. Kyd, whatever his lack as a poet and psychologist, had an instinct for viable plotting and set a new standard badly needed in the hurly-burly of the theatre just taking form. His sensational play is preposterous and confusing when synopsized, but it can be clearly grasped when enacted, and grips and relentlessly holds attention as the suspense rises, redoubles and accumulates. There is no question of its stage-worthiness. It does much to validate Aristotle's dictum that for a play to be successful far more important than beautiful, eloquent dialogue and profound characterization is the "action" – the plotting.

In the 1602 printed edition of the text, issued eight years after Kyd's death, are several additional passages attributed to Ben Jonson – there is a record of his being paid by a producer to adapt the work, but scholars feel that the new lines do not have an idiomatic Jonsonian sound.

There have been modern stagings of *The Spanish Tragedy*, but they occur rarely, despite proving that the story retains its verve and emotional impact. A disc of the play was issued by Spoken Words in 1962.

To the end of the sixteenth century and through subsequent decades in London, revenge tragedy followed revenge tragedy, all bloodily melodramatic, in a seemingly endless series. They were known as the "tragedies of blood". When the greater playwrights borrowed the form, for *Hamlet*, *King Lear*, *The Duchess of Malfi*, they gave the characters truthful feelings, credible motivations and lyrical, moving, impassioned language that enspelled audiences. Young Thomas Kyd, though at best a mediocre poet, had put his enduring stamp on the theme and shape of England's finest drama.

One difficulty in reconstructing the life of Kyd's friend, the tempestuous Marlowe, is that no one is quite sure how his name was spelled. In some documents there are references to a Marlye, in others to a Kit Marloe, a Marlo, a Marlin, and still again to a Christofer Morley. Are all these the same person? Such variations might easily occur in Elizabethan times when orthography was often dependent on individual choice. (Kyd is sometimes referred to as Kid.) The results can be quite eccentric. In their anxiety to learn more about this mysterious genius, who for good reasons cloaked much of his career in secrecy, scholars grasp at every possible clue, seizing upon the most far-fetched and verbally tortured resemblances. They wish to round out a picture, complete a stubbornly vague biography. But Christopher Marlowe, to use the name with which he was christened, escapes them. A baptismal record in the parish church of St George the Martyr informs posterity that the ceremony took place there on 26 February 1564, just two months before that of William Shakespeare at Stratford-on-Avon. Christopher was the second child and eldest son of John Marlowe, a shoemaker – John's father, in turn, was a prosperous tanner.

Much is made by biographers of the quiet, somnolent but charming atmosphere of walled Canterbury, the cathedral city where the boy grew up; certainly its peacefulness was a contrast to the wild, poetic fevers that burned in him, for his was a heated imagination and sensibility. John Marlowe was no ordinary shoemaker; a person of some substance, he was a freeman of the city, became a churchwarden and died a parish clerk. Apparently he was pious – his respectability is several times attested to – and this might have had its effect on his rebellious son, the only boy surviving of several born to the Marlowes. Having four sisters, Christopher was brought up in a house crowded with women. This might account, too, for certain psychological traits he later developed.

Piety became even more his ambience: at fourteen he was admitted on a scholarship (for "fifty poor boys") to King's School, Canterbury; his day was a round of prayers and psalm-singing, in addition to the study of Latin. After this, winning another scholarship – no one was ever to question his brilliance and learning – he went to Cambridge University as a divinity student, where he gradually changed or else revealed his true, hidden nature. His early choice of a profession may not have been entirely his wish: the scholarships had been set up by the will of Archbishop Parker and were intended to help young men enter the clergy, but in fact the youth had never shown an inclination towards taking holy orders. Quite the contrary.

His university stay lasted six years, with some strange interruptions. It was here that he met the depraved Robert Greene and his disciple, the sharp-tongued Thomas Nash, who were later to be his associates in London's carousing *demi-monde* of literature and theatre.

A water-spout of heretical ideas was shooting up in Cambridge at this time, its students buffeted by a stormy climate of doubts and bold, unsettling questions. Christopher Marlowe was among those who asked them. He was a typical raffish young intellectual of the English Renaissance, but more articulate than others and carelessly unafraid of being outspoken on dangerous topics. Counterbalancing a tradition of faith adhering to medieval orthodoxy was a new scepticism; people were also shocked by revelations about the shape and size of the world that were brought home by daring explorers, who altered the long-established image of it. Political tensions owing to the feuding between Catholicism and Protestantism from reign to reign placed everyone in peril. Scientific theories – a fresh kind of magic – were in the air breathed by young men who were mentally alert. In the university's great, growing library Christopher Marlowe pored over books that later furnished him material for his intellectually well-stocked and often erudite plays. He was particularly interested in astronomy – cosmology – a subject taught at Cambridge. Yet caution was an urgent virtue, and religious scepticism was best kept muted: one university fellow was burned at the stake for his dissident opinions.

The youthful Marlowe's increasing knowledge of the world was hardly confined to books: on several occasions, as noted, he disappeared from the university, and these periods away were so lengthy – often half a term – that the administration was inclined to deny him his degree. Rumours circulated that he had been on the Continent at Rheims in the company of Catholics, and he was suspected of papist leanings. In truth, many students of his age were being won over to Catholicism and leaving England to be trained in preparation for a Spanish invasion that would place Mary, Queen of Scots, a co-religionist, on the throne.

Among Marlowe's classmates was Robert Cecil, son of Lord Burleigh, who was both Lord Treasurer of England and Chancellor of Cambridge and deeply involved in the plots and counter-plots currently under way. Young Cecil and his father might have recognized and exploited Marlowe's reckless nature. It is conjectured that Marlowe, at their behest, pretended to be a convert, went to Rheims, the centre of such subversion, to gather information and report back to Elizabeth's ministers what was afoot there. By his performance the fledgeling poet formed political connections that were both helpful to his advancement and dangerous, for the Elizabethan court was strewn with pitfalls that often proved fatal. Among those whose favour he won now was the powerful Thomas Walsingham, Secretary of the Privy Council, in charge of espionage for the queen. He also met the double-agent Robert Poley, who was to have his hand in Marlowe's violent end.

When Marlowe petitioned to obtain his degree, as yet without success, a message came from the Privy Council on his behalf, instructing the Cambridge authorities that his absences should be excused and the degree awarded, because he had been abroad on "the Queen's business". The message further affirmed that he had conducted himself "discreetelie wherebie he had done her majestie good service, and deserved to be rewarded for his faithfull dealinge". This would seem to say that he had enlisted as a spy for Elizabeth's government. If so, he was probably prompted by a sense of adventure rather than by an allegiance to Protestantism.

When Marlowe was at Cambridge he shared the eager interest in the classics that characterized the Renaissance intelligentsia. He was also beginning to find his voice in poetry, giving vent to his magnificent gift. He combined these two inherent passions in early translations of Ovid's verses and the first book of Lucan's *Pharsalia*, and in doing so formed the style that became uniquely his. His transcription of Ovid was free rather than literal; pedantic academicians might quarrel with it but not true lovers of poetry. Some deem it superior to Ben Jonson's subsequent effort to render the same verses from Latin. Youthful sensuality floods through and colours Marlowe's words:

> Now in her tender armes I sweetly bide,
> If euer, now well lies she by my side.
> The aire is cold, and sleepe is sweetest now
> and birdes send forth shrill notes from euery bough.

And again:

> Loe I confess, I am thy captiue I,
> And holde my conquer'd hands for thee to tie.

But the Lucan translation is even more interesting for another reason. In it he uses, with great effectiveness, blank verse, abandoning traditional rhyme. Though a few other writers, including the Earl of Surrey (in a version of Virgil), Gascoigne and Sackville, had employed it before, they had attained neither grace nor strength with it. Marlowe, with his sweeping run-on lines, is the first to realize how spoken verse of great beauty and rhythmic power could be achieved, and consequently he was already shaping an expressive instrument that would be appropriated by other poets and dramatists for ensuing centuries.

Students at Cambridge frequently flocked to plays. Some were revivals of Latin classics in which they themselves participated, and others were put on by commercial companies from London. That the students took drama seriously – occasionally, over-enthusiastically – is demonstrated by a report that the windows had to be restored after performances at Trinity College during the 1580s, and that some of the youthful ushers wore "visors and steel caps" and were armed with unsheathed swords and daggers to help maintain order. It is not surprising that Marlowe soon sought to write plays, and in a classical style, and definitely and ambitiously undertook one on the subject of Dido, the abandoned, unhappy Queen of Carthage, a theme favoured repeatedly by so many other playwrights from Virgil's epic *Aeneid*. It endlessly fascinated sixteenth- and seventeenth-century dramatists and opera librettists. It is not clear whether this *The Tragedie of Dido, Queene of Carthage*, which was left incomplete, is solely Marlowe's work, or whether being still an apprentice at stagecraft he collaborated with his close friend Nash, as has been mentioned. The confusion arises because Nash's name was appended to it when the piece was published after both had died. But much of the writing is undeniably Marlowe's, including lines and images he thriftily uses again in his more mature works. A matter of dispute is whether *Dido* was staged during his lifetime, a claim being that the child actors of the Chapel put it on in May 1587, at a moment when Marlowe's plea for his degree was under consideration. But interest in it is historical and biographical: it is not one of his viable scripts. It is also thought likely that Marlowe began the composition of *Tamburlaine* and perhaps finished it while he was still at the university.

In July 1587 he left Cambridge with a prized Master's degree and set off for London, where he was quickly absorbed by the more dissolute aspects of life in the city. He became a habitué of its seamier taverns. At the same time he was frequently made welcome at the homes of some the mighty political personages of the day, notably Sir Walter Raleigh, Sir Philip Sidney and his constant friend and patron Thomas Walsingham, all of whom walked in the highest circles of power and were leaders of the English *cognoscenti*.

Marlowe's play *Tamburlaine the Great* was produced by the Lord Admiral's Men some time around 1587, his first year in the capital; it brought him instant celebrity. It rivalled Kyd's *The Spanish Tragedy* in popularity. He was still in his early twenties.

It is hard to grasp the complexity of his existence. He consorted with ruffians and cutpurses, and with political informers; but also with fellow poets, the University Wits – a wild cluster of them, but some of them, such as Greene and Nash, accomplished professionals – and with the most learned and influential nobility. He probably had met and knew Shakespeare, like himself a neophyte, and was seemingly on good terms with Edward Alleyn, a man of intelligence and balanced practicality, who had most of the leading roles in Marlowe's plays, beginning with *Tamburlaine*. All this reflected, perhaps, the contradictions of Marlowe's own divided and perverse nature.

As to the kind of person he was, there are – most naturally – varied opinions. Greene and Nash, who were by fits and starts his friends and foes – for they were a malicious pair, envious of his success – depict him as disagreeable. But their testimony is suspect. Kyd, who knew him most intimately, for they shared rooms, spoke of his belligerence, complaining that his companion was given to "attempting soden pryvie iniuries to me". The accuracy of Kyd's description is borne out by Marlowe's record of having been twice imprisoned at Newgate for public brawling. On one occasion, on 18 September 1589, the gentlemanly poet Thomas Watson, a friend of Marlowe, was summoned by the clash of steel ringing on steel and the cries of a crowd. Hurrying forward, he came upon Marlowe ferociously engaged in a duel with William Bradley, an innkeeper's son. Watson, drawing his sword, rushed into the fray, while Marlowe stopped to catch his breath. Wounded by the twenty-six-year-old Bradley, with whose father Marlowe had been in contention leading to a lawsuit, Watson was forced into a ditch that lined Hog Lane. Fearing he would be killed, Watson finally ran through his antagonist and slew him. In consequence both Marlowe and Watson were held and charged with murder. Their plea was self-defence; luckily for them the late young William Bradley had a bad reputation. After two weeks of immurement Marlowe was released; less fortunate, Watson was kept in durance for three months before winning the queen's pardon. On the second occasion, in May 1592, Marlowe, while reeling home drunk, got into a fight with two constables and was taken into custody for having assaulted them. At a hearing he was reprimanded by the magistrate and freed on probation. All this rounds out the image of his short temper and pugnacity. But he was also sensitive, responsive to physical beauty, enamoured of language and the still occult aspects of science. During this period he also composed some of the finest poetry in English, including "The Passionate Shepherd", the immortal lyric that begins "Come live with me and be my love", and the longer *Hero and Leander*, which inspired the equally youthful Shakespeare to close emulation.

The two sides of Marlowe's temperament are very hard to reconcile. He may have been troubled by his bisexuality – of which there is rather good evidence. He may, unconsciously, have been trying to prove his masculinity to himself and to others by his brawling and his plunges into danger. If his bisexuality – or plain homosexuality – is doubted by any, the rapturous description of Leander's physical splendour is a sufficient answer. It is not the young girl Hero whom the poet adoringly imagines in this work but the handsome young man Leander:

> His body was as straight as Circe's wand;
> Jove might have sipped out nectar from his hand.
> Even as delicious meat is to the taste,
> So was his neck in touching, and surpassed
> The white of Pelop's shoulder: I could tell ye,
> How smooth his breast was, and how white his belly;
> And whose immortal fingers did imprint
> That heavenly path with many a curious dint
> That runs along his back.

A scene of Jupiter dallying with Ganymede is in the first part of *Dido*, and the flagrant love of Edward II for his favourite, Gaveston, is the theme of Marlowe's only historical chronicle play. The roles of women are scant in his plays: his concentration is on the portrayal of strong men, and in this, beyond homoeroticism, there is perhaps no little narcissism and wish-fulfilment.

Coupled with the inner disturbance his sexuality may have caused, Marlowe, a true son of the Renaissance, was fascinated with the phenomenon of power. Though of modest birth, an insurmountable handicap, he consorted with some of England's mightiest figures. Yet, frustratingly, he remained an outsider and would always be one. In this age of fierce individualism, Machiavelli's shrewd and cynical handbook of opportunism was scanned by many with feverish ambitions, intellectual and mercenary as well as political. The desire to climb high, to attain glory, was everywhere more acute than it had been in the more stratified medieval society now rapidly ending. Marlowe surely had this itch; but if he sought fame and wealth for himself, and surely artistic immortality, warranted by his gifts, he was also clear-eyed and intuitive enough to sense the obverse side of the medal: corruption, self-destruction. This is, repeatedly, one of the major themes of his plays, betraying his own compulsive preoccupation with it, and his fear of it. For, in all his dramas, the seeker after power and glory is brought down.

One other habitual component of his behaviour is frequently reported: a former divinity student, he had become a loudly professed "free-thinker". Throughout his work pulses a strong beat of anti-religion, a scorn of Christianity. It is more than mere opposition to a hypocritical, dominating clergy. It appears, however, that his loss of faith left him uneasy: the force of his criticisms and jeering attests to this. In various taverns, and most likely in the drawing-rooms of certain indulgent but more discreet noblemen when the doors were closed, he attacked the truth of the Holy Scriptures, with which of course he was well acquainted, having long studied them in the courses in theology at Cambridge. He pointed out inconsistencies in the Biblical narratives. It was intensely perilous to express "atheistic ideas". As noted, Francis Kett, a Unitarian student in Marlowe's college, was burned at the stake for heresy in 1589 for his deviant views and had died screaming "Blessed by none but God!" The outspoken Marlowe daily confronted the same horrible fate, and that he dared to chance it argues that the subject was one on which he simply could not hold his impetuous tongue. Though he tried to bury the problem of religious belief in his mind, it was too troublesome for him, an uncontrollably articulate artist, to keep silent about it, and drinking also loosened his restraint.

The perils he skirted, for his unorthodoxy and his probable dabbling in political intrigue, were soon to overtake him and lead to his appalling early death, when he had still barely attained his full literary powers. Someone in high councils was offended by Marlowe's rash talk and irresponsible public conduct, or had reason to be apprehensive of what might come of it. His heedlessness made others, more lofty personages, vulnerable, and therefore singled him out. In May 1593 an informer named Richard Baines, a renegade Catholic priest, secretly filed a set of charges against him. He was accused not only of impiety and heresy, in particular of spreading doubts about the literalness of the Scriptures and crude blasphemies about Christ, John the Baptist and the Virgin Mary (he had implied that Christ and John the Baptist were probably lovers and the Virgin Mary had been unchaste), but also of planning to engage with a habitué of Newgate in a brazen counterfeiting scheme. He asserted that he had as much right as the queen to have coins minted. As for his atheistic opinions, it is hinted that he had shared them with other more important men "who in Convenient time shall be named". Mention in the complaint was made of Thomas Harriot, a distinguished mathematician, inventor, New World explorer, "atheist" and close friend and intimate of Sir Walter Raleigh. A promise was appended that witnesses would be produced.

A warrant was issued for the playwright's arrest. Added to the detailed charges was an ominous notation: "He shall be layd for." But the suspect was not in London. To avoid the plague, raging in the capital, he had gone to the tiny nearby village of Deptford. It is suggested that he had been warned of imminent proceedings against him and sought a safe hiding place; or else, possibly, through machinations of some kind or counter-plot that entailed his being secluded and silenced, he was tricked into putting himself at a discreet remove from the City.

The Star Chamber had sent a messenger to the country estate in Kent of his high-placed patron, Thomas Walsingham, ordering that, if there, the blasphemous and criminal Christopher Marlowe be brought for a hearing before it. His rooms in London were entered and searched, and Kyd was taken into custody. Piles of "atheistical" writings and journals were carried off as evidence. When brutally tortured, put to the rack, Kyd insisted on his own innocence but signed a damaging deposition confirming that Marlowe often expressed subversive ideas. It was Kyd's desperate attempt to save his own neck.

In rustic Deptford, on the morning of 30 May 1593, Marlowe and three companions took an early stroll through the countryside. At ten o'clock they stopped at a tavern owned by Eleanor Bull and hired a private room, dined, walked in the garden, supped and retired to their chamber to pass the time playing backgammon or some such game for stakes. Accounts of what ensued are at variance. What were they discussing at such length? Why was Marlowe in the company of a trio with such unsavoury reputations? Ingram Frizer, significantly, was a servant of Thomas

Walsingham. Nicholas Skere, a follower of Frizer, had a history of fraudulent dealings. Robert Poley was regularly engaged in espionage and a notorious *agent provocateur*. It is conjectured that Marlowe had become acquainted with Poley during one of his two stops in Newgate prison. One version of the incident at Deptford is that all four had been drinking too much. Marlowe was sprawled on his bed. When the "reckoning" was presented – for the winnings of the game or the repast – a quarrel erupted. Marlowe, intoxicated and angered, sprang up and attacked Frizer, drawing his dagger, which Frizer deflected, plunging it through Marlowe's eye, killing him. The very next day the playwright's body was interred in a graveyard in Deptford, the exact spot left unmarked and never found. At an inquest Frizer pleaded that he had acted in self-defence and was set free. The affair was quickly dropped; no one openly asked questions about what might really have occurred.

The clergy gratefully pointed to his violent death in a vulgar tavern brawl as a moral lesson. In 1597 Thomas Beard proclaimed: "Here did the justice of God most notably appear, in that hee compelled his owne hand which had written these blasphemies to be the instrument to punish him, and that in his braine, which had devised the same." Also, Marlowe's erstwhile friend Greene had predicted just such an inevitable fate for his rival, saying he himself "preferred to be called an ass rather than to 'dare God out of Heaven with that Atheist Tamburlaine'". And again, in a tract *Diabolical Atheism*, issued a year before the slaying, he had urged Marlowe to seek repentance, "for little knowest thou how in the end thou shall be visited". Marlowe's ignominious end was to be expected if he did not change his sinful ways and curb his treasonable utterances.

A.L. Rowse, the eminent, highly opinionated twentieth-century Elizabethan scholar, dismissed the speculation about Marlowe's murder. "There is no mystery about it: We have all the evidence that makes the matter clear. He would probably have come to some such end sooner or later anyway.... We know that Marlowe was apt to attempt sudden privy injuries to men. He had spent all that day at a tavern with three friends.... No one got into trouble over it.... And they had probably all been drinking. What else?"

The unlucky Kyd was held for a time, thoroughly frightened, and with good cause. He claimed that all "incriminating documents" discovered in their chambers had belonged to Marlowe: he himself was pious. Since Marlowe was dead and the evidence was strong, it did no harm for Kyd to defame him. He asserted that Marlowe was a partisan of Mary, Queen of Scots. He affirmed over and over that Marlowe was godless and of thoroughly bad character. Yet they had shared rooms. He was a suspect. At last Kyd was released, but his health was broken. A marked man, he was shunned. Within a year, on the last day of December 1594, he, too, was dead – some believe from the ill-effects of his imprisonment, fear, prolonged mistreatment; others postulate that he committed suicide.

The picture of the enigmatic Marlowe in the informer Baines's charge is of considerable interest and the best ever available. It is biased, of course, and surely false: they had known each other in Rheims. Yet it yields an image of Marlowe pieced together from tantalizing fragments. Even if it is only half true, it would need to fit the facts to some degree to be convincing in court. In it one seems to hear Marlowe talking, swaggering, somewhat drunkenly boastful, seeking to shock by sweeping over-statements and derisive epithets. He sounds surprisingly young, still a rebellious adolescent determined not to conform to established patterns of belief. Very likely Marlowe actually said the unconventional things that Baines attributes to him – the appearance in court of witnesses was promised – for their underlying defiance was of a piece with his basic character: he was a heretic, independent in thought and deeply concerned about matters that he twisted into jokes. Kyd's charge that his room-mate was dedicated to Mary's cause is illogical, since only a short while earlier Marlowe had spied on the Catholics in the Lowlands, and in any event he hardly leaned towards religious zealotry.

The many ambiguities in the story of Marlowe's slaying have given spring to a throng of fanciful theories. At first the tendency was to malign Marlowe's character, to moralize and say that his was the inevitable and degraded end of a misspent life, as does Rowse. A smirking dissenter was punished for his impiety, debauchery. It was also suggested that the quarrel arose over an accusation of cheating among the gamblers, rather than a settling up of a bill for food and wine as stated by Frizer and his companions, who fortunately were on hand to exculpate him. Later, the fashion was to romanticize the incident: the fight had broken out over the favours of a bawdy barmaid (or prostitute). More recently, biographers have suggested that the murder was politically inspired. Marlowe, knowing too much, was too careless in speech: his arrest was impending. Put to the rack or other torture, he might bear witness against powerful and influential intrigants, Raleigh or Walsingham, bringing about their fall. As is well known, the talented and cultivated Sir Walter Raleigh, who had many enemies at court, subsequently lost his head to the axe on a stained block at the grim Tower of London. Information about the signing of a warrant against Marlowe was sent ahead of time to Walsingham in Kent. Was Marlowe decoyed from London? Was a quarrel with him deliberately provoked by Walsingham's henchmen, to provide a pretext for disposing of him before the queen's constables laid hands on him? Or was there actually an argument? Was Marlowe simply set upon and killed? Raleigh, though literary and certainly a charming gentleman, was known to be ruthless if necessary. A dangerous babbler was removed.

Still another sensational theory put forward is that Marlowe had not been killed at all – a fictitious "murder" was contrived and cunningly staged by Thomas Walsingham, who was his lover and intervened to save him from undergoing long imprisonment and torture. A story was put out, a corpse was wrongly identified as Marlowe; the poet was spirited away to the Continent, where he resumed writing, posting back his scripts to England and having them delivered to William Shakespeare, who obligingly signed his name to them, to the profit of his purse and reputation. According to this hypothesis, offered by Calvin Hoffman in *The Murder of the Man Who Killed "Shakespeare"* (1958), Marlowe is the author of most of the Bard's works. To prove this, Hoffman cites a considerable number of lines from Shakespeare that sound very much like Marlowe. That there are marked parallels and other resemblances is true, and this will be discussed elsewhere. The Hoffman theory, a daring and entertaining one, is not accepted as plausible by Shakespearian authorities. Yet it provoked fervent analysis and debate for a decade or more. It led to permission being obtained to open a Walsingham tomb to learn whether Marlowe's 400-year-old bones and dust might be in it, preserved by a mourning friend. The tomb was empty; but the discovery does not disprove Hoffman's postulation, as it is not central to it.

Another possible account of what happened in the fatal row at Deptford was put forth by Henry Watterson (1840–1921), an esteemed American editor and journalist. Marlowe was not the victim there; his antagonist was, and thereupon the poet fled to the Continent, realizing that his situation was hopeless, in view of his personal history and political ties. He travelled about and finally settled in Padua, where he continued to write plays. He sent them to Francis Bacon, the brilliant essayist and statesman, a nephew of the great Lord Burleigh and a classmate at Cambridge. Bacon polished the scripts and passed them along to Shakespeare, who was assumed to be the author. But all three of Marlowe's companions had appeared at the inquest. Which of the three had fallen, mortally hurt, when struck by the poet's dagger?

In Rowse's study of Marlowe, he repeatedly describes him as schizophrenic, possessed of a shattered personality. His work does not show signs of paranoia, delusions, excessive religiosity, some of the symptoms of that illness, and at twenty-nine he was already past the age at which its onset most often manifests itself. It is possible that his psychological flaw, if he had any, was not as extreme as schizophrenia but a mood disorder that intermittently impelled him to experience semi-psychotic episodes, moments of deep despair and then of physical aggression, during which he struck out at others, fleeting acts of manic violence, an irateness that might also linger. Such rushes of feelings that abruptly lift one to frenetic highs of unrealistic elation, in which a person feels vastly wealthy and powerful, or is too easily provoked to fury, are suffered by uncounted numbers of people who none the less function adequately from day to day. In many instances he would alternate between the two emotional states – as categorized in the profession, his illness was either unipolar or bipolar. No memoir of those who knew him speaks of him as prone to depression; in his plays he chooses central characters of overwhelming drive and influence, world-conquerors. In them he may have been externalizing his own latent manic aspirations.

Incontestably, Kyd's *The Spanish Tragedy* and Marlowe's *Tamburlaine the Great* set the shape of Elizabethan drama. It is not clear which work was the first on the London stage, but some time before 1590 two young men in their shabby rooms were suddenly enjoying amazing success, though their breakthrough from obscurity did little to enrich them. *Tamburlaine* was published in octavo form in that year, after which Part I of the play continued for another fifteen performances by the Admiral's Men (1594–5), with the giant-tall Alleyn in the lead; Part II was presented seven times during the same year. It is believed that the début of the original script – Part I – took place at an inn-yard, rather than in the Theatre or the Curtain where the setting would be more adequate.

While Kyd caught the public's fancy by introducing deep grievances and sensational acts of revenge as a propulsive plot-device, catering to sadistic impulses only half repressed in the spectator, Marlowe brought to the nascent English drama an elevation of language, a heady verbal splendour. It was instantly obvious that he was a supreme poet; this was acknowledged and welcomed, and it was a challenge to many others. Both plays, Kyd's and Marlowe's, were heedless of Aristotelian rules. In England, as nowhere else, the struggle to impose classical form on ambitious plays was futile: romantic, sprawling, episodic narratives were far more to the popular taste. Thereby the Elizabethan theatre gained complete freedom, if not licence, in how to tell a story, and from this came an air of spontaneity.

Yet the attempts of the defeated classicists such as Norton, Sackville and Lyly, to which were added the eloquent arguments of Sir Philip Sidney, were not wholly in vain: their participation in the aesthetic debate at the universities and in the drawing-rooms of the élite and learned led to playwrights observing the need for "decorum" in speech and choosing serious and plausible subject-matter, components that might otherwise have been absent from stage-fare directed only at impatient groundlings. The great Elizabethan plays were written to grip, to stir, to delight the spectators standing in the pit – and whose arches might be aching – but they were also designed to appeal to the intellectual and blooded aristocrats, the truly literate courtiers, who occupied the stage-seats or lent their illustrious names to the troupes. To measure up to that sponsorship, the players had to aim upwards and dared not stoop too low. That was also true of the script-writers. Their stories had philosophical overtones, psychological dimensionality, and language that might include a sprinkling of the vulgar and downright bawdy, and engage in punning and other trivial word-play, but would also soar to unprecedented levels of passionate utterance or breathtaking lyric beauty. A better mixture, a richer combination than this, could hardly be conceived.

All this was not achieved in one stroke by *Tamburlaine*. The play is still journeyman work in many aspects, probably first imagined and begun while Marlowe was resident at Cambridge, and there are even indications that it was completed there, since Greene, in a preface to one of his publications (1588), refers to Marlowe's "daring God out of Heaven with that 'Atheist Tamburlan'", an incident in Part II of the script in which Tamburlaine thrusts a copy of the Koran into a fire and invites the Deity to punish him for his flagrantly sacrilegious deed. The idea for the play came from at least two books: *Magni Tamerlanis Scytharum Imperatoris Vita* (1553), a Latin work by Petrus Perondinus, and a Spanish biography of Timur the Lame by Pedro Mexia (1540) translated into English (1571), but most likely the young Marlowe also drew on other lesser and scattered sources. The figure of the barbaric invader clearly enthralled him. He unrolled maps, fell in love with the names of the exotic countries and capitals that he found on them. A poet, he responded excitedly to their orotund sound, marvellously colourful Oriental cities remote in time and physically far distant. (Like the Ancient Athenians, the Elizabethans were fascinated by place-names, and passages not only in *Tamburlaine* but *Dr Faustus* remind one of the wonderful beacon-speech in Aeschylus' *Agamemnon*, with its Homeric catalogue of Greek peaks and straits across which torchlights convey the news that once-invincible Troy has fallen. Centuries later, in *Remembrance of Things Past*, Marcel Proust silently recites a similar musical litany of village station-signs glimpsed from a train window.)

With confidence, and arrogance, in his brief prologue, the novice dramatist Marlowe promises:

> From jigging veins of rhyming mother wits,
> And such conceits as clownage keeps in pay,
> We'll lead you to the stately tent of war...

And there, he says, one shall hear Tamburlaine:

> Threatening the world with high astounding terms...

This, beyond question, the excessively vocal warrior-hero does. Possibly never in the history of the world has any author written such a torrent of dire warnings and imprecations.

Tamburlaine does not have a conventional plot. It follows the campaigns of Timur the Lame, the fourteenth-century conqueror who rose from being a Scythian shepherd and bandit to end as the merciless lord of the Near East. Taking advantage of the acclaim garnered by his first effort, Marlowe promptly wrote a sequel, the two plays comprising a work ten acts long. Most readers do not appreciate that it is not one but two dramas and, in consequence, find it rambling and repetitive, both its language and its story over-extended. On some occasions the two plays were performed separately, and in modern revivals they are drastically cut and effectively pieced together. Part II describes Tamburlaine's grief at the death of his beloved wife, and finally the fall of his fortunes and his own corruption and death, after winning the crowns of Persia, Turkey and Egypt, and fiercely earning the lasting title of the "Scourge of God".

What fascinated the young author, along with the savage colour, is the sheer intoxication of conquest, the panoply of power. The Scythian victor, yoking captive Oriental kings to his chariot as though they were humble beasts, slashes at them with his whip and shouts:

> Holla, ye pampered jades of Asia!
> What, can ye draw but twenty miles a day,
> And have so proud a chariot at your heels,
> And such a coachman as great Tamburlaine?

And he exults, again and again, in regal ascendancy:

> Is it not passing brave to be a king, Techelles?
> Usumcassane and Theridamas,
> Is it not passing brave to be a king
> And ride in triumph through Persepolis?

To which Theridamas agrees:

> A god is not so glorious as a king.
> I think the pleasure they enjoy in Heaven
> Cannot compare with kingly joys in earth . . .

Indeed, there must have been, as some biographers assume, a sadistic streak in Marlowe himself, which lets him exalt deeds of unbelievable cruelty, as well as an envious and breathless respect and longing for dominion over others. This ruthless thirst for power, however, was not Marlowe's alone; as has been remarked, it was characteristic of Renaissance man, and – closer to home – Elizabethan man. Somewhat later, in subsequent works, one discovers a maturing Marlowe's eventual intuitive rejection of the trait as patently unacceptable.

And another side of Tamburlaine, paradoxically, is his tenderness, his poetic and metaphysical aspiration. These may have matched other qualities in the youthful author. The invader's softer feelings find expression in his outpouring of hyperbolic sorrow at the death of his queen, Zenocrate:

> Black is the beauty of the brightest day . . .

And then:

> What! is she dead? Techelles, draw thy sword
> And wound the earth, that it may cleave in twain,
> And we descend into the infernal vaults,
> To hale the Fatal Sisters by the Hair,
> And throw them in the triple moat of Hell,
> For taking hence my fair Zenocrate.
> Casane and Theridamas, to arms!
> Raise cavalieros higher than the clouds,
> And with the cannon break the frame of Heaven;
> Batter the shining palace of the sun,

> And shiver all the starry firmament,
> For amorous Jove hath snatched my love from hence,
> Meaning to make her stately Queen of Heaven.
> What God soever holds thee in his arms, insatiable
> Giving thee nectar and ambrosia,
> Behold me here, divine Zenocrate,
> Raving, impatient, desperate, and mad,
> Breaking my steelèd lance, with which I burst
> The rusty beams of Janus' temple-doors,
> Letting out Death and Tyrannising War,
> To march with me under this bloody flag!
> And if thou pitiest Tamburlaine the Great,
> Come down from Heaven, and live with me again!

He pleads:

> The Cherubins and holy Seraphim
> That sing and play for the king of kings,
> Use all their voices and their instruments
> To entertain divine Zenocrate.

This is the extravagance of a young writer but one who is immeasurably endowed. It is also the writing of one who is enamoured of classical and other literary allusions: the play abounds with scores of them. And there are also some odd, delicate touches. Describing "milk-white steeds laden with the heads of slain men", he celebrates them with irrepressible delight:

> Besmeared with blood that makes a dainty show.

He also displays an impressive familiarity with military terms, which suggests that in one or more of his absences on the Continent he may have taken part in the religious wars of the Low Countries in the train of Leicester and Sidney.

Some spectators discerned in the play an insistent note of anti-religion. Bold is Tamburlaine's dying threat:

> Come, let us march against the powers of Heaven,
> And set black streamers in the firmament,
> To signify the slaughter of the gods.

The savage conqueror dies unrepentant, bragging and defiant. Greene, always Marlowe's malicious antagonist, was quick to point out how much scarcely concealed impiety is contained in the play.

Another reason that the action of the pageant-like work becomes tediously repetitive is that Tamburlaine's character is unalterable: he learns nothing from setbacks, grief or mortal illness. It may have been that Marlowe, at this early phase of his literary career, was subscribing to the classical view of history that men and events are preordained, unchanging, nothing affects or sways them. This is likely because his sources were largely classical ones. If so, it may explain why Marlowe conceived his work as he did, but the consequence is not good theatre. He learned that, and in his later plays accepts a different, Christian premise, that man does change as a result of suffering and that the attainment of "wisdom" and "enlightenment" is a requisite of Western tragedy – an axiom laid down by Aristotle.

Much cited from the play is Marlowe's tribute to the potential of the human spirit and intellect:

> Nature that framed us of four elements,
> Warring within our breasts for regiment,
> Doth teach us all to have aspiring minds.
> Our souls, whose faculties can comprehend
> The wondrous architecture of the world,
> And measure every wandering planet's course,
> Still climbing after knowledge infinite,
> And always moving as the restless spheres,
> Will us to wear ourselves, and never rest,
> Until we reach the ripest fruit of all –

That ambition prefigures the similar self-immolating quest of Dr Faustus. It also echoes the spirit of the Renaissance, with its emphasis on Humanism and the primary role of the individual.

An even more revealing passage is that in which is described the frustration of every poet who cannot give full utterance to all he feels and searches to say. Something eludes him always:

> What is beauty, saith my sufferings then?
> If all the pens that ever poets held
> Had fed the feeling of their masters' thoughts,
> And every sweetness that inspired their hearts,
> Their minds, and muses on admired themes;
> If all the heavenly quintessence they [di]still
> From their immortal flowers of poesy,
> Wherein as in a mirror, we perceive

> The highest reaches of a human wit;
> If these had made one poem's period,
> And all combined in beauty's worthiness,
> Yet should there hover in their restless heads
> One thought, one grace, one wonder, at the least,
> Which into words no virtue can digest.

Certainly too much of *Tamburlaine* is monotonous, bombastic and ranting. Offsetting these faults, however, is a striking portrait of the towering, ferocious protagonist, to whom is given speeches of vast poetic strength, and an opportunity for a costumer and scene designer to offer a gorgeous spectacle – though it is not known whether the earliest revivals of Parts I and II (1594–5) were ever able to provide the costly settings that doubtless glowed and dazzled in the author's inner vision as he wrote. No stage-directions are in the surviving texts. A Henslowe inventory lists a copper-lace-trimmed coat and crimson velvet breeches for Tamburlaine.

In *Marlowe's "Tamburlaine": The Image and the Stage*, William A. Armstrong surmises that the poet may have acted in student productions while at King's School and Cambridge and later coached Alleyn as Tamburlaine, Barabas and the aspiring Dr Faustus, persuading him to use restraint and refine his interpretations. But the Scythian conqueror is hardly a part that calls for subtlety. (And would Alleyn have been in need of coaching and receptive to advice from a neophyte, a 23-year-old author? Thomas Heywood, a playwright of the day, commends Alleyn as "the best of actors . . . peerless", fit to be ranked with "Proteus for shapes and Roscius for a tongue. So could he speak, so vary.") It is Armstrong's contention that Marlowe was proposing a "historical revolution", ridding the theatre of the acting style then in vogue, such as the crudeness of Tarlton, and replacing it with one in an entirely new mode, proper and dignified. "His second purpose was to revolutionize drama by making it a medium for iconoclasm and for the communication of extremely heterodox ideas about religion and politics. His third revolutionary purpose was to embody in a play an esoteric mythology and symbolism expressive of the contrary impulses at work in his poetic imagination." That seems to be a fair summary of Marlowe's accomplishment, though whether it was his deliberate aim, clearly formulated, must again remain conjectural.

That Tamburlaine, fated by birth to be a mere shepherd, overthrows, cages, humiliates, taunts, slaughters the many who are established sovereigns by "divine right", might be regarded a political statement, implanting in spectators' minds that the exalted privileges granted such rulers were not assured by the Christian God or by heathen deities. If it was Marlowe's intention to say that, he escaped official censure – initially, at least – by choosing for his setting distant Oriental realms, long past and displaying strange ways, events in a barbaric era, peopled by infidels.

In some passages Tamburlaine is boldly disdainful of the Christian God, but in others he seems to look upon himself as serving His goals in destroying weak rulers and kingdoms that are too grossly ill-governed: he is the chosen "Scourge of God".

Though *Tamburlaine* is shapeless, windy, a bit immature, its pageantry and swelling music captivated Elizabethan spectators; they responded to the magic of its language, its unprecedented force, its outrageous goriness, its unending cruelty; the angry Scythian warrior-king, deeming one of his sons weak and cowardly, kills him immediately. A plethora of imitations of this unusual melodrama, plays written to echo its new majestic rhythms, took over London's many stages: Peele's *Edward I*, *The Battle of Alcazar*, *David and Bethsabe*, Greene's *Alphonsus, King of Aragon, Selimus (Part I)*, *The Troublesome Reign of John* and *The Wars of Cyrus* are good instances.

(Borrowings for this chapter are from Nicoll, Gassner, Evans, Armstrong; *The Oxford Companion to English Literature*; George L. Geckle's *"Tamburlaine" and "Edward II": Text and Performance*; Harry Levin's *The Overreacher*; Charles Norman's *The Muses' Darling*; and an essay by Maurice Charney, "Shakespeare and Marlowe as Rivals and Imitators", in *Renaissance Drama*, edited by Leonard Barkan; and in addition, a pamphlet for the Enoch Pratt Library of Baltimore, Marland, prepared by Emily V. Wedge, based on the research of J.L. Hotson, Mark Eccles and F.S. Boas for their various biographies and critical studies of Marlowe's life and poetic and dramatic output.)

After 1595 there are no recorded stagings of *Tamburlaine* for over three hundred years, though it is thought that performances continued at intervals throughout the first half of the seventeenth century. The play's much-belated resurrection occurred in the United States in 1919 under the auspices of the Yale University Dramatic Association at its New Haven campus theatre. The text, much abridged, was subsequently published by the Yale University Press. The student all-male cast was directed by Edmund Montillian Woolley in collaboration with a 21-year-old undergraduate, Stephen Vincent Benét, a youthful member of a distinguished literary family and later a renowned American poet and playwright. The production was highly praised in reviews by English Department faculty, among them the celebrated Professor William Lyons Phelps, and in a critique in the student *Yale Daily News*, signed T.N.W., doubtless another playwright-to-be, Thornton (Niven) Wilder, a senior and twenty-two or so. A very favourable lengthy notice appeared in the *Boston Transcript*, the respected journal issued in that nearby city. Geckle, having obtained the edited script, says that his examination of it reveals, on the part of Woolley and Benét, "little short of the sort of butchery exhibited in Tamburlaine's slaughter of the Virgins of Damascus (which was, in fact, omitted by them). They compressed Part I's five acts into five scenes and Part II's five acts into seven scenes. They cut several minor characters, large parts of speeches and much stage business." Geckle devotes four and a half pages to specifying the discards, many of which he deplores. He particularly disapproves of the excision of symbolic actions and props, such as the succession of looted crowns Tamburlaine loosely handles, wears, rejects. "But one could hardly expect an amateur group with a limited budget to put on a wordy, ten-act melodrama, much of it using an arcane vocabulary, and hope that it would be willingly endured at full-length by a college-age audience . . . and the slaughter of the Virgins of Damascus, if portrayed by fellow students in drag, would far too likely be an episode that evoked hilarity rather than horror."

The play was given on an apron stage. The costumes are described by one reviewer as "absurdly heterogeneous" – which summons up no visual image to today's reader – and by another, Wilder, as having "a touch of fantasy". The invader's role was taken by Louis M. Loeb, who was congratulated by Professor Phelps for "a thoroughly intelligent comprehension of the character" and accorded a similar high appraisal from Wilder. Professor C.F. Tucker Brooke, a Yale scholar who had published an esteemed book on Marlowe, spoke of the evening's offering as "notably conscientious and finished", providing "two and a half hours' entertainment marvelously rich in poetry and in stage action", and he was pleased that "little indeed of Marlowe's grandiose rhetoric failed to get a hearing", though Brooke did have some reservations about how the script had been reshaped. (A measure of courtesy between faculty colleagues, as well as their regard for students' feelings, may have been influential here.) Overall, it would seem that this *Tamburlaine* was deprived of much of its scope, irony and emotional impact. Even so, Yale's was a brave try and certainly one very long overdue.

The next rebirth of the play came about at Oxford in 1933, but it was only of Part II. The Buskins, a student acting group in Worcester College, were directed by Professor Nevill Coghill. Inasmuch as this was the first appearance on native boards in perhaps three hundred years of an English classic, *The Times* saw fit to send a critic, though this was to an amateur production of just the second half of the work. The Zenocrate (Barbara Church) and the Olympia (Judith Masefield) got good notices but not the Tamburlaine (F.B. Hunt). What elicited the reviewer's enthusiasm was the costuming. Coghill had pored over Oriental sketches in the Bodleian, one of which had belonged to the eighteenth-century poet Alexander Pope. It was an Indian miniature of Timur the Lame in his colourful garb. Guided by the picture, Charles Ricketts conceived his designs for Tamburlaine and the others in the cast. The front curtain, painted by the student actors, was a map showing in outline the realms of the lesser kings overtaken in the sweep by the ruthless Scythian warrior.

Almost two decades later, in 1951, a full, professional staging of this long-neglected script occupied the Old Vic in London for five weeks, after which it was to be seen at Stratford-upon-Avon. The director was (Sir) Tyrone Guthrie, and the star was Donald Wolfit, an actor-manager of considerable reputation, noted for excellent interpretations in Shakespearian and Jonsonian plays. Once more, Parts I and II were soldered together, after substantial abridgement, but this was far more deftly accomplished. Many of the same minor characters, lines and incidents were missing here as in the script less ably fitted together by Woolley and Benét, but Geckle testifies that despite the shrinking of the text the essence of Marlowe's drama remained, and this version, too, has been published, with explanatory prefaces by both Guthrie and Wolfit.

Geckle suggests that they were aware of the Woolley–Benét adaptation and followed it closely in some places but, as practical men of the theatre, making more effective changes in other places. He examines and lists all of these, with comments on whether he agrees or disagrees with their efficacy. For the most part he approves of the changes. His study is too detailed to excerpt here.

In his prefatory note Guthrie states his belief that the play has long gone unproduced because of its unwieldy scale and "unpalatable" subject-matter, composed of "an extravagant lyricism . . . heavily laced with cruelty". But the recent world war and the Holocaust, the slaying of millions of Jews, Gypsies and political dissidents, might now have prepared audiences to understand compulsive impulses to slaughter and made the drama more relevant. He blamed the scholarly world for having evaluated *Tamburlaine* only as a work of literature, not appreciating its theatrical virtues, which he sought to realize. In order to achieve this, a tightening and shortening of the script were necessary.

Wolfit wrote that he had initially tested the play as a vehicle for acting by reading it aloud. It satisfied him utterly. "Not only did the verse ring in our ears but characters became clear and distinct, and humour shone here and there." He arrived at a fresh insight. "A long study of the part leads me to think that in the first half there is a great gaiety in the conqueror, a laughing zest for battle and a childish delight in cruelty. To conquer the triple world, to subdue kings and emperors, to win Zenocrate is an intoxicating pastime. Only at the walls of Damascus do we halt for the first time and glimpse the tragedy to come. The insanity and downfall in Part II are as speedy as that of Macbeth, and culminate in the burning of the books and the subsequent challenge to Mahomet." (A somewhat earlier version of the play, edited by Basil Ashmore in 1948, also stressed in inserted stage-directions that Tamburlaine finally descends into a wild madness. Ashmore's adaptation, largely unnoticed, did not reach the stage until a dozen more years went by, but Geckle thinks that Wolfit saw it and gained clues from it.)

Every sort of review – good, bad and indifferent – awaited Guthrie, and Wolfit's bold effort not only to revitalize the play but to do so at considerable cost, in the semi-commercial theatre. The forty performances were well attended. *The Times*'s critic had wondered how modern spectators would react to "a dramatic style separated from our own by three and a half centuries"; the answer: "remarkably well". This was because they looked on Tamburlaine as "an entirely mythical personage . . . the royal god-defying protagonist of a mad dream cast among barbaric splendours and miseries and given theatrical validity by verse which keeps Spenserian melody and informs it with a new driving power". Geckle observes that this includes a borrowing of a phrase by T.S. Eliot, who was present at the first night and is reported to have said: "This . . . makes *King Lear* look as if it had been written by Sir James Barrie." Wolfit, his make-up giving him Mongolian facial characteristics, was applauded for offering "a vibrant figure of pure theatrical flamboyance", and Guthrie's direction, Jill Balcon's Zenocrate, Margaret Rawlings's Zabine and Leo McKern's Bajazeth were also praised. Yet the play itself was not accepted as a successful dramatic vehicle.

The *Daily Telegraph* was represented by W.A. Darlington, to whom it was "a staggering surprise to find that [the play's] vaunted hero was quite such a bestial savage, a mere gangster with, so to speak, knobs on". Walter Hayes, of the *Daily Graphic*, stated that Guthrie, newly appointed as director of the Old Vic, began his tenure there with "a staggering pageant of bestiality that left the audience gasping with horror". (The adjective "staggering" was much invoked.) The *Sunday*

Times's reviewer told its readers that "Wolfit rides hell for leather at Marlowe's high astounding mixture of horror, sadism, magnificence and eroticism, and comes out at the end covered with glory."

The scope and visceral impact of the spectacle merited notice from several critics: John Barber of the *Daily Express* was amazed at beholding "fifty actors dive in and out of 140 costumes, doubling, trebling and quadrupling parts in a production of blazing splendour. The eye is dazzled. The ear is ravished." That must have been as Marlowe dreamed it might be. Philip Hope-Wallace, of the *Manchester Guardian*, was awed by "the magnitude of the carnage . . . The stage is continuously heaped with writhing bodies and the horrors are never shirked – rather the reverse, gloatingly protracted. There is one execution-by-arrows (of the Governor of Babylon) which would have sent the Elizabethans frantic with joy." This incident was obviously quite striking, for Cecil Wilson in the *Daily Mail* also refers to it, incredulous at how realistically it was staged: "The final victim is swung from chains while, by some fiendish trick, his body appears to be riddled with arrows." Wilson was impressed, too, by Guthrie's resourcefulness in having "fifty players share sixty-two speaking parts".

But there were the arrows from the critics, as well. Stinging ones. T.C. Worsley, in the *New Statesman and Nation*, deemed *Tamburlaine*, as mounted at the Old Vic, "a vast, sprawling, truly horrible and, in the last analysis, worthless play". W.J. Igoe, in the *Catholic Herald*, had much the same opinion: the play was bad. Guthrie's himself described it as: "alas, producer's twaddle". Geckle quotes excerpts from more critiques: Ivor Brown, in the *Observer*, lauded Guthrie for his "tempestuously brilliant direction" and thought that Wolfit was "the right actor to marry the ferocity of mood with the finery of speech" and had kind words for others in the cast, but felt that Wolfit's Mongolian make-up was "a handicap, and the emperor's shoddy wardrobe surely all wrong". In the final scenes Wolfit had put off his brighter raiment and donned a shaggy fur robe. Kenneth J. Robinson, in the *Church of England News*, shared Brown's view: Wolfit should not have used make-up that made him look brutish and repulsive, because it raised doubts that Zenocrate would ever have been so readily attracted to him. He also felt that the verse was "deficient in emotional appeal".

The mixed appraisals are typified by Peter Fleming's in the *Spectator*. He admired and was moved by the "poetry which transfigured the ambitions underlying Tamburlaine's megalomania", but he thought mistakes had been made in the costuming. "The tyrant's personal entourage ought, after conquering half the world, to get a little more out of it than they do. His chief of staff, I noticed, acquired a pair of sandals and a sort of peignoir after three or four empires had been subjugated, and at one stage there was an issue of ceremonial headdress. But most of them finished up just as they started, like members of an impecunious water-polo-team." A critic in *Art News and Review* had a quite contrary impression, hailing Leslie Hurry's settings and the costumes as "near masterpieces". Confronted with the Old Vic's framing proscenium arch, which limited the open area for action, Hurry overcame the spatial restriction by "delving into the depths and bounding forward on to a small apron stage". The colours of the set were "subtle" and "barbaric", and there was a sharp visual contrast between "the naked Scythians" and "their resplendent opponents".

The many notices testify to how much attention was paid by the press to the long-delayed return of young Marlowe's drama. J.C. Trewin, a resourceful critic, wrote no less than five reviews concerning it, and placed them in five different journals: *Current Events*, the *Lady*, *John O'London's Weekly*, the *Sketch* and the *Illustrated London News*. In one he revealed that he had been waiting since he was ten years of age for an actual staging of *Tamburlaine*, and on finally beholding it at the Old Vic experienced "the old delight returning, a child's simple wonder at the evocative power of many of the famous sonorous lines. . . . Marlowe had what Milton possessed after him: That love of the proper name, of a rolling swell of syllables. Elizabethan audiences gloried in this." They had not been bothered by the savagery in the work, nor was Trewin. (One supposes that as a child enamoured of *Tamburlaine* he was not only precocious but shockingly bloodthirsty.) "This can sicken, if you allow yourself to be sickened by it. But I doubt whether many people who see *Tamburlaine* will let it eat into their minds. We have supped full with horrors in our own day, horrors compared with which these Marlovian barbarities are shapes in the mist. We must realize that the people of *Tamburlaine the Great* come to life very rarely. They are figures in a monstrous shadow-show against a background lurid with flame." Geckle thinks Trewin's analysis to have been the most perceptive.

On the far Left, the Communist London *Daily Worker* dismissed the whole enterprise as "a rather sadistic circus".

In the last scene, Tamburlaine, near death, calls for a vast map. He "prowls" across it, uttering: "And shall I die, and this unconquered?" Some spectators took this bit of "business" as making visual a boast, the map literally showing all the regions the Scythian has captured, his great accomplishments, or an attempt by the director to win sympathy for a man mortally ill, his goal in life unsatisfied. Geckle believes that Marlowe's intention was "to emphasize the wonder – wonder at the insatiable thirst of Tamburlaine, who, even while dying, yearns for new conquests". Here Geckle quotes Ethel Seaton's essay "Marlowe's Map": "The conqueror's legacy to his sons is the extent of the world yet left for conquest. On the map he traces the five thousand leagues of his journeys, arrogating to himself the campaigns of his under-kings through Africa and beyond *Graecia*. Regretfully he sees world yet to conquer. . . . 'And shall I die, and this unconquered?' It is the cry of Alexander. . . . The play ends on the note of the aspiring motto adapted by Charles V, *Plus ultra*. There is more beyond."

Five years later, in 1956, Guthrie remounted the play with a Canadian cast for a ten-day engagement at the annual Stratford Festival in Ontario. Hurry again provided the sets and costumes, and the lead role was assumed by Anthony Quayle, a tall, burly actor of strong, impressive presence. In his autobiography Guthrie pronounced this to have been his "best production". *The Times* of London had a man at hand who sent a glowing account of the production when it was moved from Stratford to nearby Toronto, where its success was repeated. Its next destination was the Winter Garden, a large house on New York's Broadway. Once again the reviews were mixed, but most were quite negative. Brooks Atkinson, of the *New York Times*, favoured the staging, though he found it "just this side of the egregious". He admired Anthony Quayle's vigorous portrayal and commended others in the visiting company: William Shatner as Usumcasane, William Hutt as Techelles, Barbara Chilcott as Zenocrate (as usual, there were Americans among the Canadians in the company at the Festival). The young Shatner was later to gain an incredible celebrity in the original series of *Star Trek* on television and the later feature-film spin-offs. Strictly opposed to Atkinson's welcoming notice, Eric Bentley in the *New Republic* lashed out at the director, objecting that "while *Tamburlaine* is immoral . . . Mr Guthrie is only amoral, and lightheartedly so. . . . Where Marlowe was defiant, Mr Guthrie is only amused; where Marlowe, in his colossal error, was at least spunky, Mr Guthrie, behind all the false energy, is tired and perhaps even bored." Harold Clurman, in the *Nation*, complained that Guthrie "applied 'effects *to* the text' ". To Wollcott Gibbs, in the *New Yorker*, the offering was "a dark and shapeless mélange of unrelated violence", and Richard Hayes, in the Catholic *Commonweal*, heard little more than "a metallic clangor of rhetoric". Henry Hewes, in the *Saturday Review*, saw Guthrie's approach as lacking focus. But there was general applause for the work as a spectacle. Expected to stay eight weeks, the play closed shortly after a fortnight.

Guthrie's riposte, in his *A Life in the Theatre*: "We came to New York well aware of the difficulty of interesting a public to whom neither the name of the author, nor those of any of the actors, meant anything at all." He was mistaken on that point: in and immediately adjacent to New York City are more than twenty universities, all having English departments and staffed by professors, instructors, and turning out teachers for local high schools, and enrolling flocks of resident graduate students, in a population with a segment of college-educated citizens more numerous than perhaps in any other city in the world. Marlowe's name was hardly unfamiliar, nor regular theatre-goers incurious. What hurt the offering were the bad reviews, their pronouncements that the "the evening was a great, thundering, cavernous bore". (The author of this book was in the audience and thought the presentation was thrilling, a response shared by others who saw it and whose opinions he overheard. The production was so large and expensive that it could not be kept running until such word-of-mouth recommendations could come to its aid, as they might have. The advance announcements that the play would continue for two months may also have caused many to delay buying tickets.)

Interest in *Tamburlaine* was renewed in Britain by Guthrie's venture. Also in 1956, a mere five years after the Old Vic production, the Dramatic Society of St John's College Garden, Oxford, brought forth Parts I and II, combined and abridged, with the sub-plots lopped away. Peter Holmes was the Scythian warlord, and John Duncan the director and responsible for the three-hour adaptation. Once more the critics came from London. Performed in the open air, with a cast of one hundred students in a broad acting arena – "thirty-five by forty *yards*" – the venture suffered from acoustical problems: Holmes spoke well, according to Harold Hobson, in the *Sunday Times*, but could not always be clearly heard. At such distant removes, Hobson remarked, "subtlety becomes impossible". As compensation for that, there were startling visual effects: he was much taken by the spout of blood from the split throat of Tamburlaine's cowardly son, a deed of disdainful punishment inflicted by the unsparing father. Signal, too, was the hanging of one of the Scythian's foes. Kenneth Tynan, in his *Tynan Right and Left* (1967), recalls the staging as "the most accomplished thing the OUDS has done for years" and was beguiled by the "stylized" acting.

The year 1964 was the 400th anniversary of the births of England's two great poet-playwrights: Shakespeare's was widely observed, but Marlowe's was far from overlooked. One salute was offered by the Tavistock Repertory Company at the Tower Theatre, London, which staged *Tamburlaine*, directed by Robert Pennant Jones. It ran for two weeks, playing to "virtually full houses". Malcolm Rutherford, in the *Spectator*, granted it a "lukewarm" notice. The Fletcher Players of Corpus Christi College, Marlowe's *alma mater* in Cambridge, undertook a production of Part I of the taxing melodrama. Geckle was unable to find other critiques of the Tavistock attempt and none at all of the Cambridge venture. By now, perhaps, a revival of *Tamburlaine* was not an extraordinary happening, and the reviewers no longer felt an obligation to visit one. More remarked upon was a radio treatment of the play, guided by Charles Lefeaux and sent over the air by the British Broadcasting Corporation (BBC). Read aloud with fervour, it prompted a commentator in *The Times* to exclaim at the "roll, surge, and clangour of the verse" and declare that: "Grandeur of phrase and sound are no small delights and there is nothing in Shakespeare, or in any other English poet, like the Marlovian purple patch." This is what Gassner felt when he chose for his chapter heading over pages dealing with the poet a descriptive subtitle: "the laureate of magnificence". The usual clash of opinions was voiced after the airing. The *Listener*'s critic faulted Stephen Murray's "bow-wow manner" but liked Joss Ackland as Theridamas, deeming him "a model of how to combine subtle characterization with sensitive verse-speaking". But the Corporation's Audience Research Report gathered figures showing that most listeners thought well of Stephen Murray's vigorous interpretation, found the plot too gory and monotonous but the verse enjoyable. Some protested that the text should not have been abridged, though one wonders if an audience would have stayed attentive for ten hours.

Harvard University marked the quatercentenary by reaching back for Ashmore's adaptation of the play, a version until now not used. The staged concert reading was part of a Shakespeare–Marlowe Festival. Though "lacking visual splendor, the spoken text made a strong impression, much of it down to a splendid delivery of Tamburlaine's lines. Lesser roles were not too well handled," according to the student newspaper, the *Harvard Crimson*, in a review excerpted by Geckle. Somewhat unconvincing was Zenocrate's "sudden transformation from hatred of Tamburlaine to love for him", but this failing is inherent in Marlowe's text. The critic also felt that Ashmore's two-and-a-half-hour script had been cut far too much in order to fit that now conventional length.

A few months later, also in 1964, *Tamburlaine* was on display at the Everyman Theatre in Cheltenham. Derek Malcolm, in the *Guardian*, described this as "a doughty truncated version". He sympathized with Ian Mullins, "the brave producer who tries hard enough to make the audience's imagination work by opting out of the Guthrie spectacle in favour of the pageantry of words". Unfortunately, in Malcolm's opinion, too many of the actors were inadequate, and this was particularly true of Harvey Ashby, as the brutal protagonist, who "hewed his way through the first half like a surprised refugee from Grand Guignol, raping some incomparable verse on the way.... Thereafter he aged twenty years and improved greatly." As seems to happen often with this play, the actress portraying Zenocrate – here, Josephine Tewson – was praised: "She was more certain in effect, well-controlled and mercifully unwooden," though "too much the battered English rose for full conviction".

With the 400-year birthday anniversary scarcely ended, Marlowe's youthful play was on view yet again during "A season of Elizabethan Drama" given by the Canterbury Theatre Trust (1966) at the Marlowe Theatre, in Canterbury. As seen by a critic from *The Times*, this was "a notably robust, straightforward and well-spoken account of the play". The Tamburlaine was Wolfe Morris, "who endows the character with a lethal jocularity that is perfectly justified and highly effective". The adaptor was Professor R.A. Foakes, of the University of Kent, and the director was R.D. Smith.

The 1972 Edinburgh Festival brought *Tamburlaine* to Scotland in a production by the Glasgow Citizens. Keith Hack, the director, chose to have the Scythian represented by not one but three actors, certainly a unique approach. The participants outdid Guthrie's at bloodletting. George Bruce, in his *Festival in the North: The Story of the Edinburgh*, relates: "The arrogance and pride that were fundamental characteristics of Tamburlaine's world were made evident before the play began by actors assembled outside who scrutinized with hostile eyes the audience as it entered, while from the galleries as if over his conquered territories, skeletons were bound to the wooden pillars that support the Assembly Hall, and once the action got going, blood spouted from victims and the Elizabethan motley erupted."

Geckle quotes Irving Wardle of *The Times*: "Philip Prowse, the designer, has transformed the interior into a charnel house that would gladden the spirit of Knox himself. Skeletons stand roped to roof columns, and the stage juts out like the prow of a pirate ship festooned with chains and chariot wheels carrying more grisly cargo." Wardle summarized the production as "an arrogant riot of golden costumes and fountains of blood", and complained that he could not perceive "any consistent purpose underlying the surface show", and felt that "for much of the time the style is decidedly anti-heroic", as when Zenocrate "expresses discontent by throwing herself flat on the floor and kicking her legs". He saw little point in having Tamburlaine portrayed by a different actor in each of the three acts into which the script was handily condensed. In Act One, Rupert Frazer has him starting out as "an erotic adventurer"; in Act Two, Jeffrey Kissoon shows him degenerating into "a brutal dusky killer"; and Act Three calls on Mike Gwilym to lend him "something like Marlovian dignity and sonority". Could not one actor have explored and mastered all three phases? Wardle was also less than happy with the actors' speech, asserting that they did less than justice to the verse.

John Barber, of the *Daily Telegraph*, also deemed the cast's diction somewhat below standard, but he admired the vivid staging. On the other hand, the cast's ability to convey Marlowe's poetry greatly impressed Harold Hobson in the *Sunday Times*, a judgement contradicted by J.C. Trewin, who declared that it was "rather less than good. Marlowe needs splendour, not a sustained shout." Trewin's notice appeared in the *Illustrated London News*; by now he was surely an authority on interpretations of *Tamburlaine*.

A high point in twentieth-century enactments of the play was Peter Hall's, offered by the National Theatre in 1976–7, with cuts of almost 1,000 lines, about a third of the original text. This was a thoroughly professional realization of the drama and probably the fullest version staged since Marlowe's time. Some scenes in Part II were rearranged by Hall. The venture had sixteen performances. Enraptured, Geckle attended *three* of them.

Not all the critics shared his delight in the production. Especially nasty were a number of comments about Albert Finney, who was chosen to be the Scythian leader. Broad-shouldered and husky, he had already won considerable repute for stage and film performances. (As has been noted, most of the players assigned to the role, from Alleyn on, were physically imposing, above average in height. The actual Timur the Lame is reported to have been short, five foot eight, though having a powerful frame. It is said that during rehearsals Hall and Finney practised using a limp, though this notion was finally discarded. Finney more closely resembled the true Timur in size.) John Barber, in the *Daily Telegraph*, had this to say about the lead: "In early scenes, Albert Finney looks extraordinarily handsome with his tousled hair and splendid naked thighs." In the *Daily Mirror*, with a more acidic edge, Arthur Thirkel conceded: "A marathon role . . . But with his full mane of curly hair and dressed in a gold-encrusted tightly fitting miniskirted costume, he struck me [as] more of an overweight elf than the savage conqueror of Asia." Benedict Nightingale, in the *New Statesman*, was quite as unkind: "With his golden curls and beard, he looks rather like Apollo, and acts like him too, swaggering blithely round the stage as if he expected his helmet to sprout wings. There were also times when I was more fancifully reminded of the younger Mohammed Ali, sprawling back and crying, 'I am so *pretty*.' In the programme notes for the production, it was argued that Marlowe 'deliberately makes his hero a figure of Apollonian beauty'." Geckle admits that these observations are accurate. Perhaps some of this criticism should have been aimed at John Bury, the designer, rather than at the actor, who in most instances is apt to have little say in such details as his character's attire. Hall, of course, would have had the final word.

Geckle spells out the varied garb in the scenes, as contrived by Bury: "[He] tried to capture the sense of splendor that the text demands. When he first appeared in Part I, he [Tamburlaine] was wearing a brown blanket, which he flung aside dramatically to reveal his golden armor and impress Zenocrate (played with proper ladylike behavior by Susan Fleetwood). Later in the scene he impressed Theridamas (played by Brian Cox, who looked surprisingly like a young Marlon Brando) by spreading out Zenocrate's captured treasure on a golden cloth. In the banquet scene, Tamburlaine appeared in a bright scarlet robe over his golden armor with, of course, a golden crown, and in Act Five, Scene Two, he wore a black robe signifying his melancholy (before dispatching the Virgins of Damascus, who were clothed in white). The magnificence of the clothing was evident in a £56,000 production in which £27,000 were spent on the costumes."

Despite these negative reviews, Hall and Finney seem to have carried the day. The director lent the production many deft bits of action, some symbolic, some revealing Tamburlaine's evolving character. The only hints of madness are evident when he is stricken with tempestuous grief at Zenocrate's demise, but his bout of frenzied lamentation is transient, as is any further suggestion of the irrational in his behaviour. The passing of time, and its toll on Tamburlaine and those about him, are indicated by their altered appearance at the opening of Part II, which is preceded by an intermission: the Scythian conqueror is now grey-haired; he makes his entrance to the new scene wearily, accompanied by a somewhat careworn Zenocrate and his three teenage sons. His moustache has grown fuller, shaped to give him a more sinister aspect, and his eyes sometimes dart about in his face; there are still traits of wildness, but he lacks the fiery intensity of his former gaze.

If the rhythm flagged and moments of shock were needed, Hall provided them: when Tamburlaine handed over a concubine to reward one of his soldiers, he first stripped her bare to the waist; whereupon other soldiers followed suit, stripping the concubines similarly bestowed on them.

The burning of the Koran and Tamburlaine's defiance of Mahomet is a significant but ambiguous event in the play. He issues a challenge:

> Why sends't thou not a furious whirlwind down ...
> Or vengeance on the head of Tamburlaine
> That shakes his sword against thy majesty,
> And spurns the abstracts of thy foolish laws?

In what follows, he pays tribute to another divine power:

> The God that sits in heaven, if any god,
> For he is God alone, and none but he.

Finney is said to have taken a long pause here and "looked upwards with apprehension and amusement". What did Marlowe intend by these lines? One could hardly expect Tamburlaine to express belief in a *Christian* God. The playwright could not openly voice scepticism about the existence of a benign and watchful Deity. The key phrase may be "if any god", followed by "is God alone", in which he is hedging against a charge of atheism, which would inflict dreadful punishment on an outspoken doubter. Is Tamburlaine the "Scourge of God" in the sense that he is acting on behalf of an ireful supernatural Ruler, or is he God's self-appointed foe? Marlowe's ambiguity is doubtless deliberate. He dare not candidly admit to disbelief, yet he wants to hint at reservations about the stories narrated in the Scriptures and the dogmas and articles of faith expounded from the pulpit.

In an article in *The Times* Hall set forth his understanding of Marlowe's aims. "It is an extraordinary work which uses a Morality play structure to be totally immoral. Indeed it's the most immoral play before Genet. It sets out to prove there is no God, no Jove, no Mohammed, no Nirvana. Man, for all his aspirations, ends up with Hitlers, Mussolinis, Tamburlaines." Finney, interviewed in the *Guardian*, had much the same view. "*Tamburlaine* has primal colours. . . . It's an immorality play. This man does a lot of awful things. He dies quietly. He's not punished. . . . Actually, I believe Marlowe's saying, 'You can do what you like: doesn't matter: you die. Man dies. He can do what he likes, if he can do it, but he dies.'" Guthrie and Wolfit had implied that Tamburlaine's end spelled "some sort of moral retribution", which was their injection of "the Puritan ethic". Hall and Finney did not subscribe to that more readily acceptable interpretation.

The critic John Heilpern – quoted by Geckle – addresses himself in the programme notes to this facet of the work, also dubbing it "the first Immorality Play". He sees Tamburlaine, "the self-proclaimed Scourge of God", as a symbol of the Renaissance spirit: personification of "the age of aspiration and discovery". He assumes that Marlowe was largely sympathetic with the latter concept of him. Heilpern asks: "Are we, then, meant to marvel [at] or condemn him?" Perhaps Marlowe expected the spectator to do both. Says Geckle: "Marlowe enjoyed a strong sense of irony, possessing what T.S. Eliot called a 'savage comic humour.' He offers us a hero designed deliberately to involve us in deep conflicting emotion and judgments. The last two lines of the opening prologue are: 'View but his picture in this tragic glass, / And then applaud his fortunes as you please.' " Heilpern emphasized that the Hall–Finney offering was being done in "a morally neutral way".

Concurring, *The Times*'s Irving Wardle described Finney "treading the action like a tightrope, forbidding you to side for or against him". A like assessment of Finney's performance was that of Robert Cushman in the *Observer Review*: "Marlowe was twenty-three when he wrote the play and English drama was comparably young; at a later stage of development such enthusiastic cruelty would be intolerable. Here it is seductive; the play draws you into Tamburlaine's ruthlessness, makes you one with it. You find yourself hoping he will remain invincible; you do not want so perfect a pattern of victory to be spoiled." And, indeed, only his death cuts it short.

In the Guthrie–Wolfit production the battle scenes were on stage; in this treatment, they were off stage, heard – with much clangour – but not seen. In Part II Tamburlaine has much stronger opponents, and he seems older and somewhat weakened, which increases suspense as to the outcome. To indicate a bloody encounter, Hall had the Olivier Theatre's bare circular stage flooded with intense light from overhead, projecting a red spot at the centre that gradually expanded, accompanied by percussive music. Reconnoitring patrols and refugees from the fighting entered from a surrounding black void, streaming in different directions. With the spectators in a ring around the playing area, a striking intimacy between them and the players was achieved.

Hall and Finney also played a few episodes for laughs, especially in Part I, but the over-arching mood was sombre. The humour came from the foibles and quirks of a few lesser characters among Tamburlaine's train and the captive kings, whose plight at other moments was dire, and throughout the long script are passages of pathos and anguish that were poignantly rendered, swaying the emotions of the audience, as when Olympia kills her son to keep him from mistreatment at the hands of the barbarians and next attempts suicide in his funeral pyre, and when Caliphas is murdered, and when the Captain of Balsera expires. Geckle relates: "When Tamburlaine died, Finney fell over Zenocrate's golden coffin with his head on it facing the audience, as he and the coffin descended via the center-stage trap. The final moment inspired, once again, wonder and, ultimately, relief at the end to the constant sense of limitless energy directed toward carnage." (George L. Geckle is Professor of English at the University of South Carolina. It is not often that one has a first-hand report of this sort from a knowledgeable observer who attended performances of the same play three times in the space of two weeks.)

The number of twentieth-century productions accorded *Tamburlaine* affirm that after hundreds of years it is still an appropriate stage-vehicle, though hardly ever likely to prove a profitable venture.

Possibly *The Jew of Malta* followed *Tamburlaine*, or it might have been a somewhat later work, the experts disagree. It was presented by Lord Strange's Men at the Rose in 1592; the unusually tall Edward Alleyn had the title role. Lord Strange's Men had joined the Lord Admiral's Men to have a stronger company. The play was not published until 1633. The script shows a technical advance over both *Tamburlaine* and *Dr Faustus* and may well have been written after them; it is conceivable, however, that Marlowe worked on his several plays concurrently. Authors do that.

Marlowe chose rich, exotic subjects, strange characters, unlike those favoured by his contemporaries. In that respect his work is distinctive. No other play had a Hebraic miser at its centre, nor Malta as its setting. Barabas, hero-villain, is based on an actual figure, Nassi, a Portuguese Jew, believed to have fled from persecution by Christians to haven in Constantinople at the more tolerant "infidel" court of Sultan Selim. Though earlier he and his family had pragmatically become Christian converts, a strong devotion to his former faith was kept alive in his heart. In Constantinople he flourished, obtaining the grace of the Sultan, who recognized his abilities, named him a duke, and appointed him governor of the flowering island of Naxos, rare honours for a Jew in the Christian and Islamic worlds. In return, like Marlowe's Barabas, he allied himself with the Turks against hostile Catholic Cyprus, a bastion of Crusaders. But in many ways the outline provided by the history of Nassi differs from that given by Marlowe to his arch-manipulator Barabas. Another suggested source is a lost Spanish novel. For the most part, Marlowe did not originate any of his plots; like most other Elizabethans, including Shakespeare, he borrowed ideas for them and then freely adapted them.

In a most unusual prologue, Marlowe replies to critics of *Tamburlaine*, those who, like Greene, had harshly found fault with his frank, large-scaled work for its impiety and presumption. The prologue is delivered by an actor impersonating the great Italian cynic Machiavelli but obviously is Marlowe speaking of himself:

> To some perhaps my name is odious,
> But such as love me guard me from their tongues. . . .
> And weigh not men, and therefore not men's words.
> Admired I am of those that hate me most.
> Though some speak openly against my books,
> Yet will they read me . . .

Doubtless Marlowe was attracted to Machiavelli, admired him and felt an affinity with the Italian's worldly scepticism, and sought to emulate him. "I count religion but a childish toy. . . ." Here, though, the ingenious Marlowe is taking on a disguise, claiming the liberty of letting his characters declare their subversive thoughts, while preserving a right to disclaim them as ever his own.

The Grand Seigneur of Turkey demands that the besieged Maltese pay a ransom, a crushing sum. The island's Governor orders the local Jews to turn over their riches, an edict resisted by the avaricious Barabas, the wealthiest among them. For his temerity his fortune is confiscated and his house seized and turned into a convent. His daughter, Abigail, poses as a nun to gain entrance to her former dwelling and recover a cache of jewels hidden beneath a floorboard, a stratagem that is successful. Barabas then uses her to entice and revenge himself on his enemy, the Governor of Malta, whose son is smitten with the beauteous girl, as is also Don Mathias, an impetuous gentleman. Barabas encourages rivalry between the two young men, hoping it will lead to a quarrel between them. In a duel both youthful suitors are killed. Overtaken by grief, Abigail, who truly loved Don Mathias, recommits herself to the nunnery, now in all sincerity. Her father, infuriated, sends a poisoned gift of food to the convent; all the occupants die, including his apostate daughter. Before her death, however, she reveals her father's malevolent devices to a friar, but he – bound by the inviolable oath of secrecy that rules the confessional – cannot betray what he has been told. Barabas brings about the death of the friar, who is greedy for gold. Blame for the deed is laid on the shoulders of another friar, wholly innocent of any crime. When Ithamore, Barabas's henchman, tries to blackmail his unscrupulous employer, the unfathomably evil Jew retaliates by killing both his treacherous assistant and the fellow's paramour. Finally accused, Barabas feigns death and escapes. He makes his way to the Turks' camp, leads them by a secret path into the stoutly walled city and brings about the destruction of its defenders. He is rewarded by the Turks, who make him Governor of the conquered island. Realizing how much he is hated by its Christian inhabitants, he seeks to change sides again. Plotting a fatal end to the Turkish commander and his troops, he invites them to a banquet in a hall with a collapsible floor. His fiendish design is exposed, and he himself is hurled through the gaping floor into a cauldron of boiling oil: he perishes horribly, as he well deserves, but cursing them all.

The script is mutilated and shows signs of having been edited – or tampered with – by another hand, possibly that of Thomas Heyward, or by Kyd, at what might have been the time of the play's composition.

The general opinion is that *The Jew of Malta* falters badly after the first act. The mistreated Barabas, in the initial scenes, is a human figure and even a sympathetic one; later he is incredibly wicked, even murdering his own daughter without a qualm. But he is a prototype of the "Machiavellian villains" found in subsequent Elizabethan and Jacobean tragedies and melodramas. Rowse likens him to Shakespeare's Aaron in *Titus Andronicus*, Iago in *Othello*, and Richard III. Yet Barabas outmatches them all. He boasts of having poisoned wells and killed unwary Gentiles at random; in all this he fits a dark, frightening stereotype of Jews widely held by the Christian populace of Marlowe's day, most of whom had little or no personal acquaintance with Jews, whose numbers in England at the time were few.

Despite the egregiousness and implausibility of the plot, its development is handled cleverly; it is steadfastly compelling, with neat twists and reversals. Whoever manipulated the plot-line – Marlowe himself, or the more commercial-minded Kyd, or an unknown – was shrewdly inventive. The play was popular for a long time, though often given in versions that did not adhere to the published script. T.S. Eliot called this wildly sensational work "a savage farce". If it has been absent from the recent stage, it is because modern producers, in days too soon after the Holocaust, are sensitive to putting on anything that might have them charged with encouraging anti-Semitism. Barabas is an overwhelmingly defiling caricature.

The play is notable for the vividness of its language, quick-paced, supple, easy for the ear to take in. Marlowe is forging an even better sort of dialogue, using blank verse but having it more conversational and hence more dramatic. An outstanding passage is the very opening one, where the avid Barabas, talking to himself, happily scrutinizes the contents of a small bag of gems, which constitute a large part of his hoarded capital:

> Give me the merchants of the Indian mines
> That trade in metal of the purest mould;
> The wealthy Moor, that in the eastern rocks
> Without control can pick his riches up,

> And in his house heap pearls like pebble-stones,
> Receive them free, and sell them by the weight,
> Bags of fiery opals, sapphires, amethysts,
> Jacthinths, hard topaz, grass-green emeralds,
> Beauteous rubies, sparkling diamonds,
> And seld-seen costly stones of such great price,
> As one of them indifferently rated,
> And of a carat of this quantity,
> May serve in peril of calamity
> To ransom great kings from captivity.
> This is the ware wherein consists my wealth;
> And thus methinks should men of judgment frame
> Their means of traffic from the vulgar trade,
> And as their wealth increaseth, so inclose
> Infinite riches in a little room.

The last line of this soliloquy is among those most frequently quoted from Marlowe's pen. There are also lyric passages as fine as this, given incongruously to Barabas:

> Now Phoebus ope the eyelids of the day,
> And for the raven wake the morning lark,
> That I may hover with her in the air;
> Singing o'er these, as she does o'er her young.

And Ithamore's song, which fittingly embraces rhyme:

> ... But we will leave this paltry land,
> And sail from here to Greece, to lovely Greece.
> I'll be thy Jason, thou my golden fleece;
> Where painted carpets o'er the meads are hurled,
> And Bacchus' vineyards overspread the world;
> Where woods and forests go in goodly green,
> I'll be Adonis, thou shalt be Love's Queen.
> The meads, the orchards, and the prim-rose lanes,
> Instead of sedge and reed, bear sugar-canes:
> Thou in those groves, by Dis above,
> Shalt live with me and be my love.

Occasionally, with fine effect, Barabas's words have an Old Testament assonance and rhythm, making him sound patriarchal. Elsewhere there is a rain of classical allusions that mark the author as a child of the Renaissance.

The Jew of Malta has a contrasting facet: it is outrageously anti-religious. Into the mouths of Barabas and his Turkish slave are put a stream of gibes at Christians, who are depicted as heartless, hypocritical and mercenary. Friars and nuns are imputed to be unchaste. Both of the ill-fated Christian lovers of Abigail are stupid and coarse-grained. Barabas denounces his foes:

> Rather had I a Jew be hated thus,
> Than pitied in a Christian poverty:
> For I can see no fruits in all their faith,
> But malice, falsehood, and excessive pride,
> Which methinks fits not their profession.

And:

> It's no sin to deceive a Christian;
> For they themselves hold it a principle,
> Faith is not to be held with heretics . . .

Today one supposes that this is Marlowe himself speaking, shielded by transposing his views to a Jew and a Turk, who would naturally hold angry and derogatory opinions of Christians, with whom they were at perpetual odds and even war.

Inevitably, *The Jew of Malta* is measured against *The Merchant of Venice*, Shakespeare's portrait of another Jew cruelly isolated, robbed and derided in a hostile environment. Beyond question, *The Merchant* is far superior, but Shakespeare's borrowings from Marlowe are also evident. In the later play Barabas is changed to Shylock, a credible, flesh-and-blood figure. Barabas's daughter Abigail has a parallel in Shylock's Jessica; both young women defect to join Christian lovers; they are cursed by their fathers. *The Merchant*'s plot is better contrived than that of Marlowe's melodrama, with Shakespeare's play a tragicomedy, a refined treatment of the theme of the persecuted Jew and his *hamartia*, greed.

It is interesting to observe the almost direct appropriation by Shakespeare of certain of Marlowe's salient lines and phrases. For example:

> O my girle, my gold . . .

is heard again as:

> My daughter! O my ducats!

Another more significant instance of minor but definite plundering:

> But stay, what starre shines yonder in the East?
> The lodestarre of my life, if Abigail –

sounds once more, though not in *The Merchant* but instead in *Romeo and Juliet*, where the enraptured young lover whispers:

> But, soft! what light through yonder window breaks?
> It is the East, and Juliet is the sun . . .

Such correspondences are too close to be mere chance.

At moments Shakespeare pays Marlowe the flattery of outright imitation. One reason for this may be, as Rowse suggests, that Shakespeare, an actor, having a keen "aural memory", was "drenched with Marlowisms". Such paraphrases occur, too, in Shakespeare's other plays, just as startling echoes of *Hero and Leander* occur in *Venus and Adonis*, the lengthy narrative poems with which both playwrights launched their literary careers. Stylistic effects in *Tamburlaine* are also to be found in *The Merchant*.

It has been argued, convincingly, that Marlowe himself is a character in *Romeo and Juliet*, in the person of the acidulous, witty Mercutio, killed in a street brawl. Also, a line in Shakespeare's *As You Like It* – a "great reckoning in a little room", which is so much like Barabas's summation, "Infinite riches in a little room" – is possibly a deliberate reference to Marlowe's tragic, sordid end in a tavern lodging, as the result of a settling up of a food bill or gambling debt. If so, Shakespeare is making a subtle play here on the word "reckoning".

Shakespeare makes no attempt to conceal his verbal borrowings; he could not have hoped to do so, in a day when the resemblances would be much more apparent to audiences largely familiar with the works of its two most successful writers. Liftings and paraphrases were common then. In *As You Like It* there is an explicit quotation as well from Marlowe:

> Dead Shepherd, now I find thy saw of might:
> "Who ever loved that loved not at first sight?"

All playwrights of the period did this. John Webster helped himself as freely to some of Shakespeare's best metaphors and images, and nowhere was there much striving to fashion new plots, since there is little chance of finding any, and the oldest ones still work very well. It saved effort, and evoked common sense, to make do with those at hand. All that was required was a clever variation.

It is even thought by a few scholars that Marlowe is the "rival poet" mentioned in Shakespeare's sonnets. This might well be, though hard to prove.

Marlowe's masterpiece is *Dr Faustus*. Whatever failings are seen in *Tamburlaine* and *The Jew of Malta* are not found here. It is not unduly lengthy, and it is episodic, hence easy to shorten if a director wishes to snip scenes here and there. It has a subject of magnitude that is relevant in every age and a leading character with whom spectators of any period can readily identify. But in its own day and hour it was particularly important because it reflects perfectly Marlowe's epoch. As do his earlier works, it has a mingling of medieval and Renaissance elements, combining the theatre of the past – Mystery and Morality Plays – and the new popular stage; while within itself, in language and thought, it is replete with loving allusions to the classics. It is, *par excellence*, the *Elizabethan* play, in some respects typical, in others unique.

The date of its first productions, which were by the Lord Admiral's Men at the Rose and the Red Bull, are variously determined to have been in 1588, 1589, 1592 and 1594; that is for scholars to fret about. A text in blank verse and prose was published in 1604 and again in 1616, the second with radical alterations. That is a matter, too, of concern to the seriously erudite. The idea for the play doubtless came to Marlowe while he was reading an English translation of the *Volksbuch*, or *Faustbuch*, issued in Frankfurt am Main (1587 or 1592). The full title of the book was *The Historie of the Damnable Life, and Deserved Death of Doctor John Faustus*. The author, Johann Spies (or Spiess), is credited with being the first to fashion a literary version of a medieval legend, just as Marlowe is believed to have been the first to treat dramatically with the travails and delirious pleasures of a man who sells his soul to the Devil in return for passionately desired but unhallowed powers. But the story was said to have been at least 1,000 years old by the time Marlowe came upon it. Some time in the sixteenth century the legend became attached to Johannes Faustus, well known to fellow alchemists of the high order of Paracelsus and Agrippa, and also to the Humanist Conrad Muth and famed Wittenberg scholar Melanchthon. Marlowe's Dr Faustus, therefore, is compounded of fact and fancy; there is a good deal of confusion as to the model of his wonder-working scientist, an overreacher who ends in catastrophe.

In Marlowe's day "magic" and "science" were not clearly distinguished: hard-headed thinkers in their libraries and laboratories were deeply engaged in trying to transmute base metals into gold, and with delving into numerology, as well as other occult and cabbalistic secrets that might yield supernatural powers to their discoverers, leading them to restore youth, raise the dead, read the future, and even to achieve metamorphosis. In the long-lived legend is incarnated the aspiration of "Faustian" man, ever restless, planning, striving.

The *Volksbuch* immediately fired Marlowe's imagination; it is obvious that he compulsively identified with the heedless, miracle-commanding seeker after power, knowledge, beauty, eternal youth. By contemporary norms this young bohemian had already sold his soul to the Devil – he had renounced his prescribed faith, assailed the pious beliefs of his age, and lusted without inhibitions after power and sensual consummation. He was also poet enough, and possessed enough theatrical flair, to grasp the possibilities of a play on this subject, one that was instinct with colour and invoked the supernatural, allowing him to conjure up the classical past of which he was enamoured, and lending itself to on-stage displays of dazzling trickery: flames, puffs of smoke, inexplicable emanations and evanishments.

The extant text of *Dr Faustus* is, like most of Marlowe's work, fragmented and mutilated; it is not clear whether it was hastily written, but there are signs of speed in its composition, among them its very loose construction. It reads as though it were written by a man literally gripped by a demon, yet it contains the finest poetry in all of Marlowe's work, indeed some of the strongest and most evocative lines of verse in English drama. Small wonder that Ben Jonson was later moved to pay tribute, in a phrase that is so much quoted, to the force of Marlowe's "mighty line".

Though Marlowe follows Spies's book rather closely, with his habitual fidelity to his source, he condenses and transcends his material. Also, he enlarges the character of Faustus so that he is not merely a necromancer who seeks to startle the readily gullible, as does Greene's derivative Friar Bacon, but a true human being, his disturbed inner life painfully revealed. The revelation is stirring, at times shaking – it is an exposure of Marlowe himself, a young man of ambitious intellect, defiance and rash daring, yet still the Canterbury boy, son of a pious father, he has not outgrown the divinity student he once was. Torn by atavistic and childlike fears, he has a psyche that belongs half to the Middle Ages, half to the Renaissance. He is half artist, half materialist; half sceptic, half credulous believer. It must be assumed that Marlowe was not an "atheist" but, rather, like Euripides, a "free-thinker". He rejected the encrusted religious beliefs of his time. But few artists who have a superb poetic response to the scope of the cosmos, an awed and sensuous awareness of its physical grandeur, such as he had, are completely without religious feeling.

Like Oedipus, Hamlet, Lear, Faustus is both real and an archetypal symbol. To create such a dynamic focal figure requires the greatest art. In the opening scene he is in his study. In a long soliloquy he tells of having devoted himself to a thorough absorption of Aristotle's philosophy and of having practised medicine and worked astonishing cures, yet it is not enough. His perusals of law and theology have equally palled. Through his intervention as a physician he has saved whole towns from epidemics, but he has not enabled men to live for ever. He finds theology too full of illogic and has come to look on it as "unpleasant, harsh, contemptible, and vile". He seeks something more. Only "magic" still attracts him, for it promises "a world of profit and delight". His dominant impulse is to make the leap from the natural to the supernatural.

Here Marlowe offers a tragedy whose protagonist is not a king or prince, nor has he achieved his standing by a great military victory. Instead, like Marlowe himself, he is an intellectual aristocrat, a "prince of scholars", an artist-prince, as the young poet – aware of his genius - could have felt himself to be. In the prologue Faustus tells of having been born of parents "base of stock", but this has never deterred him. Again, like the arrogant Marlowe, he has risen to eminence through his own innate gifts and efforts. That the hero of a serious play is not of noble blood is a marked innovation, unprecedented in the history of Western drama to this point. From here on it will no longer be enough for the principal character of a tragedy to be outwardly royal, to bear a proud title or hold a sceptre; he must have kingly attributes of a more inward kind, as in his own opinion does Faustus.

What tempts Faustus above all is the prospect of enormous power:

> All things that move between the quiet poles
> Shall be at my command; emperors and kings
> Are but obeyed in their several provinces,
> Nor can they raise the wind nor rend the clouds;
> But his dominion that exceeds in this
> Stretcheth as far as doth the mind of man,
> A sound magician is a mighty god;
> Here, Faustus, tire thy brains to gain a deity.

Faustian man is no longer content with human endeavours and accomplishments; he dreams of acquiring the prerogatives of God Himself. In this play Marlowe, imaginatively and megalomaniacally lusting after the attainment of power through physical force – as, cruelly and obsessively, does Tamburlaine – or by acquiring vast wealth – as does the unscrupulous Barabas – now with Faust, his surrogate, seeks it through superhuman knowledge, heedless of spiritual cost.

While the closeted scholar contemplates a pause in his career, a Good Angel and an Evil Angel enter his thoughts. Visible, vestigial figures from the Morality Plays, they broaden the drama's significance, lending it metaphysical overtones. But they also have a practical function, letting Marlowe shorten the already over-lengthy expositional soliloquy. (This play is, indeed, the apotheosis of the "Morality", the personifications in concrete images of the struggle between Vice and Virtue. The Angels are an Expressionistic touch, for they are projections of opposite forces in Faustus himself.) The Good Angel bids Faustus to shun the "damnèd book" of spells and conjurations over which he hypnotically pores. He would do better to study the Scriptures instead. But the Evil Angel urges him on:

> Go forward, Faustus, in that famous art,
> Wherein all Nature's treasure is contained;
> Be thou on earth as Jove is in the sky,
> Lord and commander of these elements.

There is a subtle shift in the language from a reference to the Christian God to a pagan deity.

Faustus is weary of pious talk. He is swayed by the vision of what omnipotent magic can assist him to enact:

> Shall I make spirits fetch me what I please,
> Resolve me of all ambiguities,
> Perform what desperate enterprise I will?
> I'll have them fly to India for gold,
> Ransack the ocean for orient pearl,
> And search all corners of the new-found world
> For pleasant fruits and princely delicates;
> I'll have them read me strange philosophy
> And tell the secrets of all foreign kings;
> I'll have them wall all Germany with brass,
> And make swift Rhine circle fair Wittenberg,
> I'll have them fill the public schools with silk,
> Wherewith the students shall be bravely clad;
> I'll levy soldiers with the coins they bring . . .
> And reign sole king of all the provinces;
> Yes, stranger engines for the brunt of war . . .
> I'll make my servile spirits to invent.

Left alone, after a dinner with two magicians who instructed him in their black art, Faustus begins the weird incantations learned from them and is successful. Mephistopheles appears in the guise of a Franciscan friar. At first this surprising visitor seems so compliant and humble that Faustus feels triumphant. But it soon becomes very clear that the messenger from Lucifer demands the scholar's soul in return for being of service. Faustus declares himself unafraid of the threat of damnation and bargains to accept "eternal death" in exchange for being spared for:

> . . . Four and twenty years,
> Letting me live with all voluptuousness . . .
> Having thee ever to attend on me . . .

Intoxicated by the prospect, he pictures himself "great Emperor of the world" who can perform such feats as making "a bridge thorough the moving air" and:

> . . . joining the hills that bind the Afric shore,
> And make that country continent to Spain,
> And both contributory to my crown.

Still, he is not without doubts about his risky course. Mephistopheles reappears; he has gained Lucifer's assent to the compact, and Faustus must sign it in his own blood. The Angels re-enter and try to win him over. When the Good Angel declares that "Contrition, prayer, repentance" will bring Faustus to Heaven, the Evil Angel denounces them as "illusions – fruits of lunacy / That make men foolish that do trust them most".

Accepting the Devil's condition, Faustus pricks a vein:

> I cut my arm, and with my proper blood
> Assure my soul to be great Lucifer's,
> Chief lord and regent of perpetual night!

(The description of Lucifer in one line is quite the equal of anything Milton was to write in *Paradise Lost*.) But when Faustus dips his pen in the blood, it congeals – he cannot write the deed of gift. He is taken aback, frightened by the portent. Mephistopheles brings a chafer of coals, and the heat of its flames clears the blood; it flows again and Faustus can employ it to write *Consummatum est* and sign his name. These are the same words that marked another "deed of blood", the Crucifixion. All this, of course, is a foreshadowing of the last episode in the play, when Faustus will have a glimpse of Christ's blood streaming sacrificially from the Heavens, while the flames of Hell are about to swirl up to engulf him. Again, he is hesitant. Has he gone too far? The Devil acts quickly to distract him and restore his confidence.

Almost all his wishes are fulfilled; one question he asks that goes unanswered is "Who made the world?" He has queries about astronomy – the heavens, the spheres – and about the plants and trees that cover the green earth, and has blind Homer sing to him, and hears the music of Orpheus. In all this he shows his exceptional good taste. But he is never without misgivings, even suicidal impulses.

The Good Angel for ever urges him to repent, the Evil Angel to oppose this counsel, warning him of punishment by imps if he seeks to recant.

Lucifer further rewards him by letting him converse with the Seven Deadly Sins – another reminder of the Moralities. The dialogue is filled with a fabulous account of how the omnipotent Faustus travels by chariot to the peak of Olympus. Afterwards he is observed visiting the Pope in Rome, subsequent to a recounted journey from Trier along the Rhine and on to Paris, Naples, Padua, Venice. Again the Renaissance fascination with geography, the music of far-off names, is exploited for poetic effect, as in *Tamburlaine* and *The Jew of Malta*.

At the Vatican a capricious Faustus and his companion Mephistopheles play irreverent tricks on His Holiness – this is, of course, an echo of the anti-papal feeling in England during the reign of a Protestant queen, whose claim to the throne the Pope refused to acknowledge. At a banquet with the Pontiff meat and dishes vanish, cups are emptied or disappear, and His Holiness has his ears boxed by an invisible hand. The elaborate feast is disrupted by fireworks. This sort of humour, which also includes scenes with devils running about, is broad fun derived from capers in the wagon-mounted Mysteries.

At the request of the Holy Roman Emperor, who has heard of his prodigious feats, Faustus reincarnates the spirits of Alexander the Great and his paramour. He provides grapes in mid-winter for the pregnant Duchess of Vanholt. For these deeds he is richly rewarded. But at moments he still falters, given warnings by some he encounters along his way, and beset by frequent regrets at his perilous bargain. To cajole him further, Mephistopheles grants him another plea: he is permitted to gaze upon the beauteous Helen of Troy, which evokes from Faustus, awed, in an exclamatory whisper, the most quoted lines in the play:

> Was this the face that launched a thousand ships
> And burnt the topless towers of Ilium?
> Sweet Helen, make me immortal with a kiss. [*Kisses her.*]
> Her lips suck forth my soul; see where it flies! –
> Come, Helen, come, give me my soul again,
> Here will I dwell, for Heaven is in these lips,
> And all is dross that is not Helena. . . .
> Oh, thou art fairer than the evening air
> Clad in the beauty of a thousand stars;
> Brighter art thou than flaming Jupiter
> When he appeared to hapless Semele:
> More lovely than the monarch of the sky
> In wanton Arethusa's azured arms:
> And none but thou shalt be my paramour.

This passage clearly prompted Shakespeare's similar, but later, verse in *Troilus and Cressida*:

> Why, she is a pearl
> Whose price hath launched above a thousand ships.

It is probable, too, that Shakespeare learned from it the poetic power of suggestion, which he is to use to excelling advantage in his brief but compelling picture of Cleopatra borne on a gilded barge and "beggaring all description". Marlowe does not offer any specific detail. In the book Spies has a meticulous image of Helen:

This lady appeared before them in a most rich gown of purple velvet, costly imbroidered; her hair hanging down loose, as fair as the beaten gold, and of such length that it reached down to her hams, having most amorous cole-black eyes, a sweet and pleasant round face, with lips as red as any cherry; her cheeks of a rose-colour, her mouth small, her neck white like a swan; tall and slender of personage; in sum, there was no imperfect place in her; she looked about her with a rolling hawk's eye, a smiling and wanton countenance, which near-handed inflamed the hearts of all the students . . . and thus fair Helena and Faustus went out again with one another.

A pretty summation, but never does it dazzle the mind's eye as does Marlowe's sublime evocation, or Shakespeare's tribute to his seductive Egyptian queen, because nothing is left to be freely imagined, nor any allowance set for the individual measures of beauty held by each spectator.

In the same way, Marlowe avoids throughout any references to the burning pitch and other grotesque horrors of a Dantesque or Miltonic inferno. He speaks of Hell only as being "deprived of God's bliss". Hell is the absence of God, exile from His presence. Says Mephistopheles:

> When all the world dissolves,
> And every creature shall be purified,
> All places shall be Hell that is not Heaven.

By spiritualizing – or it might be better to say "dematerializing" – Marlowe heightens and refines his concept of eternal retribution. This is what led the poet Michael Drayton to write of him later that his was "a fine madness", and "his raptures were all air and fire".

As the allotted twenty-four years hurry by – for the spectator, in less than an hour, for this is such a short, episodic work that some prefer to term it not a play but a dramatic poem – Faustus sickens with growing apprehension, anticipating his approaching fate. "The date is expired." The bargain entered into so readily two decades earlier now seems a monstrous folly. The fiends await him. His friends believe him mentally ill, anguished and beleaguered by hideous fancies. He yearns to repent, but he cannot; he is too deeply bound by his compact with evil. It is the evening of his last day.

> Ah, my God, I would weep, but the Devil draws in my tears. Gush forth blood instead of tears! Yea, life and soul! Oh, he stays my tongue! I would lift up my hands, but see, they hold them, they hold them!

Left alone, at his urging, by his friends who have gone to pray for him, he hears the clock strike eleven – such sound-effects, as has been noted, were often employed in the Elizabethan theatre. Their resonance adds to the gravity of the moment. In torment, the doomed scholar cries out:

> Ah, Faustus,
> Now hast thou but one bare hour to live . . .

He is mortally afraid.

Relentlessly, the clock strikes the half-hour. Interestingly, Marlowe achieves a psychologically exact foreshortening of time: as the hours are nightmarishly compressed, he allows himself precisely one line as the equivalent of each remaining minute. In the first half of the great final soliloquy there are thirty-one lines, two of them half-lines. Faustus cries out:

> Ah, the half-hour is past! 'Twill all be past anon!

There are twenty-seven lines in the second half of his self-condemnation, to the climax, ending in the fiery entrance of the Devils, and Faustus's gasping evanishment. Spectators are certainly unaware of how Marlowe paced himself, but knowledge of it affords an insight into how he sometimes worked, at very least in this instance.

(A legend has grown up about the play that at one performance an extra Devil was counted on stage, none other than the Evil One himself, who jumped in to participate, causing widespread consternation.)

The closing soliloquy is one of the supreme moments in theatre, the language matching the torment, the breaking emotion, of the protagonist. Such lines as:

> O, I'll leap up to my God! Who pulls me down?
> See, see where Christ's blood streams in the firmament!
> One drop could save my soul – half a drop: ah, my Christ!

are not to be equalled in religious poetry until the advent of Francis Thompson in *The Hound of Heaven*, and there is good reason to believe that Thompson owed much to Marlowe for his inspiration and imagery. It is all the more difficult to accept the idea that these lines were written by an "atheist". Faustus's last words are: "I'll burn my books."

The Chorus enters, and in a brief epilogue laments that Faustus has perished. His valedictory is one that, ironically and pathetically, applies all too well to the poet himself, tragically and wastefully lost too soon:

> Cut is the branch that might have grown full straight,
> And burnèd is Apollo's laurel bough . . .

Marlowe, as if with an artist's intuitive prescience, had written his own obituary.

Like Oedipus, Faustus has sought too many answers and been too ambitious. That is why the play struck many twentieth-century spectators as highly relevant, even naggingly topical. In an age when man's heedless exploration of the atom and space, his mastery of flight, brought him to the verge of daily dread of extinction, this moral of the play – if it is one that Marlowe intended – was profoundly meaningful. Like Faustus, twentieth-century man was often tempted to wish that he could "burn his books" and turn back to a more modest and blissful past.

Seen in crucial episodes, and under dire pressure and gripped by intense fear, Faustus grows and changes. Not even in the Greek dramas do the characters undergo such a transformation of mood and moral attitude. But despite all the religious allusions in his anguished outcries, he has not achieved redemption – insatiably curious, overly bold, he has gone too far; he cannot be saved.

Was this Marlowe's self-judgement? It could be assumed that in much of this the artist was not working consciously, knowingly writing about himself, but instead was invading his subconscious: he was probing his deepest inner problems through the medium of his art, and the sentence he passed on himself was one that he arrived at obliquely, not yet fully aware that he was formulating it. It has been said of him that his impiety was suspect: he did protest too much. A young man indifferent to religion would be silent about it, especially in that day and age; the man who gibes without cease, who reviles faith and the Church, is beset by continuing doubts and is still possessed by a strong "will to believe" that he is not yet ready to acknowledge.

Yet again, Faustus's submission to the idea of a judgemental God, a bleeding Saviour, a Heaven, a Hell might merely have signified the playwright's willingness to appease the Church's ever-watchful authorities and his audiences' expectations; for purely commercial reasons Marlowe could not have his drama end with the Devil triumphant and Faustus his happy slave. That would certainly bring on the author the contumely of the pious, who far outnumbered the sceptical, along with rejection by a shocked public. In one side of his nature he was cynical enough to make any sort of compromise if need be.

But there are two other possible explanations of Marlowe's consigning Faustus to a horrible, fiery death. One is that he was merely following the outline of the story in Spies's book. The other is that while creating the play, excited by the character he was creating and the magnitude of the subject, his vision of what must be the climax was dictated by a force outside himself, his fancy aroused and heated, incandescent. He provided the ending that his theme demanded: the choice was not his, it was mandatory.

Often asked is whether Marlowe himself thought that "magic" had efficacy, or that Hell or devils truly existed. Some scholars think that he did, but from a reading of his whole body of work the chances are that he did not. Still, most of his audiences did accept the reality of a fiery afterworld to which sinners were sent, and of Lucifer and his malign, cloven-footed agents. The play had a stronger impact on his first audience, therefore, than on latter-day spectators. Marlowe also shows that he was well acquainted with the lore of demonology. As always, he had prepared with care: the conjuring scenes are accurate in all details, as well as theatrically vivid and sometimes amusing.

Frequently it is said that Marlowe's chief lack was a sense of humour. The play has numerous comic episodes, involving Wagner, Faustus's foolish servant, together with a clown who indiscreetly sneaks glimpses into the occult books of their master. Many critics deem these scenes to be blemishes, and some modern directors omit them from their stagings. In fact, some scholars think the comedy is by another hand, possibly Nash. As has been said, the manuscript is very imperfect, and later writers have contributed to it, as was customary in the Elizabethan theatre. But the same comic episodes are in Spies's book, so it is more than likely that Marlowe himself took them from his chief source. Arguably, the scenes do have considerable value within the frame of the drama, providing not only moments of welcome contrast but also an earthier exploitation of its theme. For the most part Faustus makes lofty demands. What he seeks is the extraordinary. The blunt, brutish Wagner, on the other hand, is willing to sell his soul for a much lesser price, "a shoulder of mutton", and, given supernatural powers, his first wish is to change himself into a flea to tickle the girls in their forbidden parts. Is this not an effective bit of counterpoint, the daily "black farce" of men selling their virtue to satisfy simple greed, gluttony, sexual appetite? Throughout, these scenes offer a merry mockery of the exalted central theme and a vulgar development of its possibilities. Faustus and Wagner together are Everyman. It is a superb conception, though the humour might have been even sharper and richer.

And was Marlowe ever wholly without a vein of sardonic levity? Many of the jokes in the play are donnish, academic, the sort with which a University Wit might playfully regale his friends. After Marlowe's death the seldom generous Nash remembered him as "one of the wittiest knaves that ever God made", and said further: "His pen was sharp pointed like a poniard; no leaf he wrote on but was like a burning glass to set on fire all his readers."

Faustus is abjectly and convulsively affrighted, but no physical tortures are shown being inflicted on him, which has led some commentators to remark that Marlowe had turned squeamish, after his depiction of the horrible sufferings of the victims of Tamburlaine and Barabas. Indeed, there are indications that he was beginning to avoid any preoccupation with the ugly and loathsome. (An earlier hint of this trait in him is that there is never any mention that the heroic Tamburlaine is lame. In the conflict between the two aspects of his nature, the fascination with the beautiful was growing ever stronger, infusing his poetry.) To magnify and glorify the past, however, was very much a tendency of the age.

That Marlowe is outstandingly representative of the Renaissance and its ferments is seen not only in the rain of classical references that add so much lustre to his lines, but also in the homage Faustus pays to the Greek philosophers, who attract him far more than do the Christian theologians; and by his interest not only in geography but in all the other new sciences and pseudo-sciences. But if the play expresses the whole sceptical yet excited spirit of the Renaissance, it is at the same time an implicit criticism of the excesses of that epoch: its gross materialism, its arrogance and vaunting ambition, its driven new cult of individualism. Renaissance man, typified by Faustus, wanted to rifle and control all the spirits of the world. In that ever-mounting ambition he invites his eventual self-destruction. It is rare that any single work of art so fully captures and distils the essence of its own time.

From the theologians, though, Marlowe, the former divinity student, probably borrowed his definition of Hell and perdition as "absence from God's bliss". After his years at Cambridge he would have been thoroughly grounded in many current aspects of religious dogma and speculation – there are signs that he never fully discarded or outgrew all of that sober and sometimes mystical indoctrination. It warred with his scepticism, resulting in an ambivalence that complicated his thinking and emotional responses, a dilemma that he did not live long enough or become mature enough to resolve, if that was ever possible.

This dualism is embodied in Faustus, cleft hour by hour by pride and triumphant pleasure, doubt and fear, remorse, all facets of his nature that he cannot reconcile. He is the active designer of his own dismal fate. He is not approached and tempted by Mephistopheles; he himself summons Lucifer's emissary and turns a deaf ear to the Devil's initial warning. True, the Evil Angel prods him, but the two angels are projections of the two most impulsive sides of Faustus's divided personality, not truly external forces. The Evil Angel is no more than the voice of an accommodating self-rationalization such as is possessed by every clever deviate and conventional petty sinner. At the same time, he is a visionary.

Understood as a medieval playwright might, Faustus's "fall" parallels that of Lucifer, the downward swing of "the wheel of fortune" that is the mobile, semi-circular pattern of so many tragic dramas and stories. Renaissance literature did not long retain this concept of "the wheel of fortune"; it is a medieval hangover.

Not all of Faustus's choices are worthy ones. The middle section of the play is taken up with his poor and trivial use of the powers that he has temporarily gained: this could be meant as an ironic commentary. After all the fine, bragging promises he has made himself, he expends a portion of his magical gifts baiting the Pope, outwitting a horse-dealer. He has sold his immortal soul, but what has he purchased at such a terrible price? Was it to have the ability to perform such unseemly tricks? Marlowe, in the mornings after his roistering, might well have asked himself the same question. He was indulging his debased appetites. He was foolishly squandering his time and talents. Given precious opportunities, men do not only seek the answers to cosmic question but waste their strength in the pursuit of what is sordid, mean, of little value. Some critics say that *Dr Faustus* is uneven, these middle scenes sagging, lacking the high accomplishment and tension of the opening and closing ones. Actually, it holds up well throughout.

Goethe was not one of the fault-finders. Far from deeming it badly structured, after studying the script closely, he exclaimed: "How greatly it is all planned." He proposed translating it into German, but then became so taken with the subject that he launched himself into his own play, a different treatment of the same theme, which was to be another major poetic achievement. The Faust legend, of course, is to be told again and again, times hard to number, by lesser writers, down to the present day.

Surely Shakespeare saw this powerful play. It caused such a stir, and London was so small, that he could not have missed hearing about it and most likely seeing it. He could have learned much from it about the craft of writing for the theatre, an art that the young Marlowe seems to have grasped intuitively. One lesson might have been how to repeat, while slightly varying, certain words at intervals throughout a script, so that by their well-spaced recurrence they lend it a feeling of unity. In *Faustus* over and over again there is mention of "fire", "burning", "water", "blood", "Heaven", "Hell", in each instance in a somewhat different context, giving the play a consistent dominant tone, a resonance or ambience. In his best plays Shakespeare also does this, as Mark Van Doren elucidates in his perceptive study of the Bard's lines, but Marlowe had already mastered this subtle technical device.

By instinct, or deliberately, Marlowe works throughout with rare cunning. In particular, the words "despair" and "desperate" are reiterated in thirteen places. To some good Christians, as Helen Gardner remarks in *The Tragedy of Damnation*, Faustus's initial sin has been presumption, and his final sin is despair – he obstinately refuses to believe that he can struggle back to salvation, and he dies with the name not of God but of Mephistopheles on his dry lips and tongue. "Despair" and "desperate" sounded again and again foreshadow his hopeless, unhallowed end.

An eventful revival of the play occurred in New York at the Maxine Elliott Theater in 1937, where it was produced and directed by Orson Welles, who also took the role of Faustus. This offering was sponsored by the Federal Theater, which during the years of the Great Depression in the United States was subsidized by the government to give employment to workers in the theatrical profession, as well to graphic artists and writers. Welles, heading one group of actors in this project, was just twenty-two – or so he claimed, there were some who doubted it – and by his precocious skills as a performer and director had already won considerable celebrity with his fresh, highly imaginative stagings of Shakespeare. Tall and solid, possessed of a sonorous voice, he could have been a stand-in for the fabled Edward Alleyn, the first Faustus.

Among those hailing the presentation was Brooks Atkinson in the *New York Times*:

Although the Federal Theatre has some problem children on its hands, it also has some enterprising artists on its staff. Some of them got together last night and put on a brilliantly original production of Christopher Marlowe's *The Tragical History of Dr Faustus*, which dates from 1589. If that sounds like a schoolboy chore to you, be disabused, for the bigwigs of the Federal Theatre's Project 891 know how absorbing an Elizabethan play can be when it is staged according to the simple unities that obtained in the Elizabethan theatre. Everyone interested in the imaginative power of the theatre will want to see how ably Orson Welles and John Houseman have cleared away all the impedimenta that make most classics forbidding, how skillfully they have left *Dr Faustus*, grim and terrible, on the stage. By being sensible as well as artists, Mr Welles and Mr Houseman have gone a long way toward revolutionizing the staging of Elizabethan plays.

Although *Dr Faustus* is a short play, hardly more than an hour in the telling, it is not a simple play to produce. . . . Like most Elizabethan plays, it has an irresponsible scenario; it moves rapidly from place to place, vexing the story with a great many short scenes; it includes several incidents of supernaturalism and, of course, it is written in verse.

If the directors had tried to stage *Dr Faustus* against descriptive backgrounds it would be intolerably tedious to follow. But they have virtually stripped it of scenery and decoration, relying upon an ingenious use of lights to establish time and place. In the orchestra pit they have built an apron stage where the actors play cheek by jowl with the audience. The vision of the Seven Deadly Sins is shown by puppets in the right-hand box. Up-stage scenes are unmasked by curtained walls that can be lifted swiftly. Entrances are made not only from the wings, but from the orchestra pit and from trap doors that are bursting with light and that make small incidents uncommonly majestical.

The result is a *Dr Faustus* that is physically and imaginatively alive, nimbly active – heady theatre stuff. As the learned doctor of damnation, Orson Welles gives a robust performance that is mobile and commanding, and he speaks verse with a deliberation that clarifies the meaning and invigorates the sound of words. There are excellent performances in most of the parts, notably Jack Carter's Mephistopheles, Bernard Savage's friend to Faustus and Arthur Spencer's impudent servant. There are clowns, church processionals and coarse brawls along the street. Paul Bowles has composed a score which is somewhat undistinguished in itself, although it helps to arouse the illusion of black magic and diabolical conjuration.

Not that Elizabethan dramas have never been staged before under conditions approximating the conventions of Elizabethan theatres. Most of those experiments have a self-conscious and ascetic look to them. But Mr Welles and Mr Houseman have merely looked to the script and staged it naturally. In the first place, it is easy to understand, which is no common virtue. In the second place, it is infernally interesting. *Dr Faustus* has the vitality of a modern play, and the verse sounds like good, forceful writing. For this is an experiment that has succeeded on its merits as frank and sensible theatre, and a good many people will now pay their taxes in a more charitable frame of mind.

Of unusual interest, some decades later, was a production of the drama at Oxford University. Its specific purpose was to raise funds for a new student theatre workshop. The director was Neville Coghill, and its chief draw was the presence of Richard Burton as Faustus, and his wife Elizabeth Taylor as the beauteous Helen. Burton was recognized as an outstanding Shakespearian actor and for his effectiveness in many other roles on stage and screen, and Elizabeth Taylor was perhaps the most famed film actress of the day for her performances and physical attractiveness. The pair had recently married and divorced, and then had married each other again, their tempestuous marital history receiving worldwide attention. During the Second World War Burton had been a Royal Air Force cadet at Oxford, and Coghill had been his tutor. As a gesture of and gratitude to his former mentor, Burton returned to aid the university's cause. He listened carefully to Coghill's suggestions and instructions and followed them, saying: "I think he's going to correct me every second as if I were still writing bad essays for him." Taylor's part in the play, of course, was non-speaking and lasted only two minutes. A critic observed that when Burton intoned "Sweet Helen, make me immortal with a kiss," Taylor obliged, "giving him not just one kiss but four." As expected, the joint appearance of these two players drew throngs of spectators. (Their affiliation was not long-lived; they were soon divorced again.) Mephistopheles was portrayed by Andreas Teuber, an American graduate student at Oxford on a Fulbright scholarship.

Edward II, perhaps Marlowe's last play (1592?), has his thoughts taking a quite different bent. He chose a subject from English history. His principal figure here is, strangely, an obstinate, petulant weakling. To round out an account of events in Edward's troubled reign, he turned to narratives by Raphael Holinshed, John Stowe and – to a lesser extent – Robert Fabyan, all of whom had published chronicles in the late fifteenth or early sixteenth century. But most likely what magnetized his interest in this particular story was Edward II's flagrant bisexuality, his infatuation with a worthless favourite, the low-born Piers Gaveston, whom Edward kept by his side and raised to high rank, infuriating the powerful barons at his court. Marlowe condenses actual time and happenings, accomplished with considerable aplomb. Beyond question, his craftsmanship was improving.

Because of complaints from the barons, Gaveston has been sent away. When Edward ascends the throne after his father's death, he orders his favourite's return. The end of Gaveston's exile is opposed by the jealous lords who detest him as a scheming flatterer. Once more they join with Church hierarchs to drive out this interloper, whose ambitions they fear. Soon after yielding to their demands, Edward arrives to bring him back again. Civil strife erupts; at the play's first climax Gaveston is captured and executed.

Before long he is replaced by a new favourite, Young Spenser (Hugh le Despenser). Caught in these broils is Isabella, Edward's French-born queen, whose love for her regal husband is unrequited. She flees England with her son.

Around Isabella rally the dispossessed nobles, among them the Earl of Kent, Edward's brother, and the elder and younger Mortimers, father and son. The bitter civil war ends when Edward is overthrown. Held a prisoner, he is forced to abdicate in favour of his young son.

The queen has formed an intimate alliance with the younger Mortimer, her chief partisan. The Earl of Kent, having helped to depose his brother, regrets it and seeks to restore him to the throne. To prevent this, Isabella and Mortimer plan to have the imprisoned Edward removed by a professional assassin, Lightborn. Kent is caught, then beheaded.

Edward is held in Berkeley Castle, in a foul-smelling dungeon where he has to stand knee-deep in water. Here Lightborn approaches. Feigning to succour the king, he persuades him to fall asleep and crushes him to death beneath an upturned table. But Lightborn himself is fatally stabbed by the king's warders, recruited into the conspiracy.

When news of the slaying reaches the very youthful Edward III, newly crowned, he surprisingly orders Mortimer's execution, and has the queen, his mother, taken into custody and sent to the Tower.

Of the episode in which the king is murdered, Charles Lamb wrote that it "moves pity and terror beyond any scene, ancient or modern, with which I am acquainted".

In this tragic drama the language is chaste, austere, excellently suited for the stage. In its being largely without the poetic extravagance that coruscatingly enhances Marlowe's other plays, however, it suffers a serious loss. There is little here of "the brave translunary things" he had penned earlier. But a few passages are surpassingly good. In the succession of dungeons in which Edward is kept, he is shaved with gutter-waste and must beg of his captors:

> O, water, gentle friends, to cool my thirst
> And clear my body of foul excrements!

Noteworthy is an earlier speech in which Gaveston tells how he means to beguile his self-indulgent, decadent sovereign:

> I must have wanton poets, pleasant wits,
> Musicians, that with touching of a string
> May draw the pliant king which way I please.
> Music and poetry is his delight;
> Therefore I'll have Italian masks by night.
> Sweet speeches, comedies, and pleasing shows;
> And in the day, when he shall walk abroad,
> Like sylvan nymphs my pages shall be clad;
> My men, like satyrs grazing on the lawn,
> Shall with their goat-feet dance the antic hay.
> Sometimes a lovely boy in Dian's shape,
> With hair that gilds the water as it glides,
> Crownets of pearl about his naked arms,
> And in his sportful hands an olive-tree,
> To hide those parts which men delight to see,
> Shall bathe him in a spring; and there hard by,
> One like Actaeon peeping through the grove,

> Shall by the angry goddess be transformed,
> And running in the likeness of an hart
> By yelping hounds pulled down, shall seem to die;
> Such things as these shall please his majesty.

Marlowe, one observes, writes best on those subjects that excite him most: male comeliness, anti-Church issues and power. Indeed, the anti-religious chord is heard again, though it is made to seem only anti-Catholic, a safe stance. Edward, enraged at the clergy's interference, exclaims:

> Why should a king be subject to a priest?
> Proud Rome! thou hatchest such imperial grooms,
> For these thy superstitious taper-lights,
> Wherewith thy antichristian churches blaze,
> I'll fire thy crazed buildings, and enforce
> The papal towers to kiss the lowly ground!
> With slaughtered priests make Tiber's channel swell,
> And banks raised higher with their sepulchres.

Isabella compares the king's affection for Gaveston to that of Jove for Ganymede. The elder Mortimer comments:

> Thou seest by nature he is mild and calm,
> And, seeing his mind so doats on Gaveston,
> Let him without controulment have his will.
> The mightiest kings have had their minions:
> Great Alexander loved Haphestion;
> The conquering Hercules for Hylas wept;
> And for Patroclus stern Achilles drooped.
> And not kings only, but the wisest men:
> The Roman Tully loved Octavius;
> Grave Socrates wild Alcibiades.
> Then let his grace, whose youth is flexible,
> And promiseth as much as we can wish,
> Freely enjoy that vain, light-hearted earl;
> For riper years will wean him from such toys.

A scholar's life is nicely compared to that of a court intrigant:

> Then, Baldock, you must cast the scholar off,
> And learn to court it like a gentleman.
> 'Tis not a black coat and a little band,
> A velvet-caped coat, faced with serge,
> And smelling to a nosegay all the day,
> Or holding of a napkin in your hand,
> Or saying a long grace at a table's end,
> Or making low legs to a nobleman,
> Or looking downward with your eyelids close,
> And saying, "Truly, an't may please your honour,"
> Can get you any favour with great men;
> You must be proud, bold, pleasant, resolute,
> And now and then stab, as occasion serves.

There are epigrams and neat turns of phrase. About to die, Gaveston demands: "Treacherous earl, shall I not see the king?" To which Warwick replies: "The king of heaven, no other king." And quite resounding is Gaveston's self-epitaph: "I perceive that heading is one, and hanging is the other, and death is all."

Kent's lines, "Fair blows the wind for France; blow gentle gale, / Till Edmund be arrived for England's good!" are famous. In Mortimer's last speech appears the sobering conceit of the "wheel of fortune" which is salient in so many serious plays of this epoch:

> Base Fortune, now I see, that in thy wheel
> There is a point, to which when men aspire,
> They tumble headlong down: that point I touched,
> And, seeing there was no place to mount up higher,
> Why should I grieve at my declining fall? –
> Farewell, fair queen; weep not for Mortimer,
> That scorns the world, and, as a traveller,
> Goes to discover countries yet unknown.

In Renaissance drama, as in the classical and medieval, a tragedy is still the story of a great man's fall.

Edward II is Marlowe's most objective work, an added sign that he was on the verge of maturity as an artist. He was able to write about other people without a clear reference to himself. He had outgrown the narcissism of the young author who projects into his compositions only his self-image or the aggrandizing image – the all-conquering Tamburlaine, the all-knowing Faustus – he half-secretly wishes for himself.

The chief shortcoming of the play is that the central figure wins the sympathy of the audience too late, if at all. He is perverse, as well as perverted. Only when he is brutally imprisoned is his plight likely to elicit a visceral empathic response. The role is not one that has usually attracted great actors, for this protagonist is too effete, too unappealing. In his misfortune Edward pities himself too much. He further degrades himself by his endless complaints. This king is hardly royal. He begs for mercy, though he has never given any. His passions are too histrionic.

The humiliations and hardships to which Edward is submitted indicate anew a curious streak in Marlowe – and again, in most of the Elizabethans – a pleasure in showing and seeing torture and suffering. As has been said, the times were violent, but in the theatre men did not seem to seek an escape from the cruelty that was exercised on every side around them; on the contrary, they apparently sought to behold the violence enlarged upon, heightened.

In *Edward II*, at last, Marlowe brings forth a play that does not revolve entirely around one character: here are a whole parcel of them, many of them depicted with compassion, another quality not too often evidenced in the work of the poet-playwright. Isabella, Kent, the Elder Mortimer, all are sympathetically drawn. The nobles are well individualized.

Inevitably, *Edward II* is compared with Shakespeare's *Richard II*. That chronicle, which came later, resembles Marlowe's closely, with admirers of Marlowe proclaiming his work decidedly the better of the pair. The melancholy music that infuses *Richard II*, sustained throughout, gives the advantage to Shakespeare, but Marlowe's tragedy is better structured and its characters more clearly motivated. Certainly the speech with which Edward resigns his crown is far outmatched by that of the eloquent, self-reproachful Richard when facing abdication, but that does not compensate for other lapses in the script. (Some scholars hazard that Marlowe and Shakespeare collaborated on later chronicles, possibly *Henry VI* and *Richard III*, but no written record substantiates this. Though the two playwrights lived near each other, it is not certain that they actually ever met, as has been stated.)

To some critics *Edward II* is Marlowe's best dramatic work because of its lucid plot, a sign of how much further he might have developed had he lived out his years; his improvement in skill was remarkably rapid. He established a pattern for other chronicle plays, much above the standard set by his predecessors. It surpasses any earlier English example in that genre, in effective characterizations and the variety of them. Other critics see a falling off of his youthful poetic powers. But no one is able to attest that it is his final work. It might have been followed by *Dr Faustus* or *The Jew of Malta*. Like so much else about Marlowe – his life, the probable cause of his death, his promise for the future – the script's chronological order, the date of its composition, defy being definitively fixed.

Geckle has compiled a list of twentieth-century enactments of this oft-neglected play. In what is believed to have been its first professional performance in nearly three hundred years, Marlowe's Edward II once more lived and died on the stage, the place now the New

Theatre in Oxford (1903), with the title role taken by Harley Granville-Barker, the writer-actor-director-producer who is recognized as one of the major innovators of his own century's English drama. In *The Times* an anonymous critic generally approved of Granville-Barker's interpretation of the difficult central character, though it lacked displays of Edward's "kingly temper", while stressing his inherent weariness and vacillation. "The gradual approach of the tragic end was to him only pathetic, and in his dying Edward there was no touch, however faint, of majesty; so that, for example, those lines which, by reason of the sublime imaginative contrast they afford, are perhaps the most deeply charged with tragic intensity in the whole:

> Tell Isabel, the queen, I looked not thus,
> When for her sake I ran full tilt in France
> And there unhorsed the Duke of Cleremont.

brought back none of Edward's faded glory, but were delivered, as was all the rest of the scene, in a uniform voice of quavering despair." Also a note of shocked turn-of-the-century sense of decorum pervaded *The Times*'s review of a play about an unconventional ruler.

In an exchange of letters, Granville-Barker and his friend George Bernard Shaw discussed the production. Shaw felt that the citizens of Oxford should have been offered *Richard II* instead of Marlowe's drama, of which he had a poor opinion. "I have read – or rather re-read – *Edward II*. . . . There is nothing to it – no possibility of success, and the infernal tradition that Marlowe was a great dramatic poet . . . throws all the blame of his wretched half-achievement on the actor. Marlowe had words & a turn for their music, but nothing to say – a barren amateur with a great air."

Stratford-upon-Avon saw the play, given by Frank Benson's Shakespearean Company, two years later (1905). Benson was the unhappy Edward. In the *Manchester Guardian* an unsigned notice found his portrayal "serious, intelligent, well-conceived", but as yet not one that attained greatness. To a reviewer in the *Leamington Courier* the play as a whole was "splendidly presented", though its subject was "not of a character calculated to raise the spirits of the playgoer". Despite that, "there are many beautiful passages, giving ample scope for the elocutionary powers of the actors, and these were keenly enjoyed". In particular, "Benson's interpretation of the headstrong monarch was a clever study. His love for Gaveston and Spenser, while blind to madness politically, was from the personal side generous and kingly." *The Stage*'s critic also praised Benson for endowing Edward with quite worthy impulses towards his two favourites: "His passionate love for Gaveston was clearly to be traced to a generous motive, and it was this love that culminated in the final catastrophe." This critic took note that the actress having Isabella's role strongly implied that she was innocent of conniving at Edward's death, a depiction that caused Geckle to wonder how that impression could be convincingly achieved. As directed by Benson, the murder of Edward was, "thanks to the skill of the dramatist and actors, robbed of its horrors", but again that was a concession to the taste and sensibility of early twentieth-century spectators. But Mortimer's severed head was visible in a basket carried on stage, and this was disturbing to the ladies in the audience. Alluding to this, a commentator in the *Stratford-upon-Avon Herald* described the baring of the head as possibly a mistake, not meant, and one that "certainly lent nothing to the play".

About two decades later *Edward II* was presented at the Regent Theatre in London (1923) and counted as a critical success. Geckle provides no details about this offering, nor of another London production put on by Joan Littlewood in 1956 at the Theatre Royal, Stratford East. He considers more important an amateur staging by the Cambridge University Marlowe Society in 1958, mounted for the club by Toby Robertson, who himself had appeared as Edward in an earlier production at Cambridge in 1951, directed by John Barton. Barton's choice as the principal now was Derek Jacobi, who was not identified at the time, because traditionally performers in Marlowe Society open-air presentations remained anonymous. Of this occasion Gareth Lloyd Evans wrote in the *Manchester Guardian*: "Marlowe's lyric intensity is beautifully conveyed in the verse speaking as is the unsuspected dramatic directness of the language. The nerve-racked, hopeless love of Gaveston and Edward is intensely moving. This Gaveston is live, arrogant, and shifty like a scared terrier, but in Edward's presence he has a fawning humility which striking against the king's obsessive need for him creates a sense of royalty tainted and humanity degraded. He who plays Edward moves superbly from angry, petulant weakness to withered, bent dignity. He gives the king the royalty of pain withstood and the tragedy of constant frustration. Never was anonymity less justified."

Jacobi's picture of Edward was similarly hailed by Laurence Kitchen in the *Observer*: "The crucial relationship between Gaveston and the king is dealt with by giving their rather trite endearments a firm emotional base; it is sincere on both sides, and in the context of baronial persecution has a curious dignity and freshness. Using cadences in the Gielgud tradition and forcing a flexible voice now and then, the anonymous Edward at Stratford goes on to clarify the king's bouts of rhetorical self-assertion. Feeble they are, but never mean; and at times they lead to action on a surge of resentment."

Even more enthusiastic about the direction and acting was Harold Hobson in the *Sunday Times*:

> We have to go back two thousand years into the literature of the Hebrews to find an exposition of unnatural love as unafraid as Marlowe gives us here. Now in general this topic is very boring, principally because it is treated either hypocritically or forensically or with a puritanic shudder. It is astonishing to anyone like me who has been wearied to death by *Cat on a Hot Tin Roof* and *Quaint Honour* and *Tea and Sympathy*, to find how theatrically exciting the condemnable relationship between two young men like Gaveston and Edward can be when the

playwright accepts it as a simple dramatic fact, like the love of Antony and Cleopatra, and not as a matter for argument, dissimulation or moralizing. The hurt neglect of Edward's repulsed queen, the scorn of Mortimer for an ill-born fancy boy, the enmity of the churchmen all become living forces. And every now and again there falls a line that makes the soul shiver with its beauty. . . . This is a play, this is a performance, to strengthen the heart, and to make the senses swim.

From Cambridge, this *Edward II* was taken to Stratford-upon-Avon and then on to the Lyric Opera House in Hammersmith, London, where it earned much the same praise and had a fortnight's run. In most instances the reviewers emulated Hobson in prominently mentioning the homosexual affinities of the king and his favourites. It should be remarked that, though highly lauded, the drama's engagements were comparatively brief.

In 1964, the quatercentenary of Marlowe's birth, John Russell Brown interviewed Robertson at length for the American magazine *Tulane Drama Review*, one topic being his 1958 production of *Edward II*. (The whole issue of the journal was devoted to Marlowe.) Robertson said that what had most gratified him in staging of the play was his having caught "the extraordinary speed of events" in it. Marlowe had given it propulsive action, so that it leaped from incident to incident swiftly. Another facet of his handling of it had been his, Robertson's, concentration on Edward as the dominant figure. For this he had been faulted by Professor Clifford Leech, who saw the play and in a critical essay argued that as much or more attention should be centred on Gaveston. Robertson had not made that wheedling favourite "enough of a Renaissance Machiavellian" who strongly controlled Edward physically, intellectually, and certainly emotionally. Gaveston had been shown as too dependent on Edward. "When Gaveston dies, it is his memory that keeps Edward going, together with those shadows of Gaveston – Spenser and Baldock." Though Brown, the interviewer, did not agree with Leech but instead felt that the production had been marked by "neutrality", Robertson was persuaded that Leech was probably right.

Accordingly, a decade later, when Robertson brought back the play, this time at the 1969 Edinburgh Festival, he shifted various emphases, having Edward "a more deluded character", while stressing that the author's concern had been more than the crippling bond between the king and his love, but as much a search for power, to which was added an appreciation that the work as a whole was meant to be laden with irony, that being one of Marlowe's habitual modes of thought. Robertson felt that ten years earlier his cast at Cambridge, and he himself, had been too young to capture and convey that insinuating quality.

He took a new approach to the casting. "The great thing is to have the right voice, the right organ to play this stuff. Edward needs extraordinary range – from the young man to the mature soldier to the old man. . . . The emotions are minutely and completely expressed, and this needs a voice to express. That's why I was so happy about the bare setting and costumes: the whole thing came back to the words. That is not to say that there wasn't characterization, but it's the language that precisely defines character here."

Sponsored by the Prospect Theatre Company, *Edward II* was acclaimed as the hit of the 1969 festival. Robertson's choice for the lead role was Ian McKellen, who was also handed the title part in Shakespeare's *Richard II*, directed by Richard Cottrell. This gave visitors to the festival a chance to see and compare the two plays in productions that were enacted almost in tandem. The critics were mostly of the opinion that *Richard II* came out ahead, but there were many votes for Marlowe's entry in this unusual contest. *Time* magazine's observer, attending the festival, saw it this way: "For present tastes, honed to instant violence, it is by no means obvious that Shakespeare outwrote Marlowe. McKellen's Richard is Shakespeare's, full-strength and without eccentricity, a prince refined down to holy innocence, so that London critic Harold Hobson could write that 'the ineffable presence of God himself enters into him.' In total contrast, his Marlovian Edward is a performance as hell-inspired as the red-hot poker that, at the conclusion, is used to murder the king by being rammed up his anus."

Benedict Nightingale, in *Plays and Players*, described McKellen as "best actor of 1969" for his accomplishment in the two roles. In a second review, in the *New Statesman*, he attested:

Ian McKellen's performance gives Robertson's reading all the support a director could want. His Edward fairly seethes with repressed energy. He can scarcely keep still. Even when he sits, which tends to be informally on the floor, not formally on any throne, his arms and legs dart and writhe, bent on self-expression and contact. In a court of thick, cold, brooding barons, standing around him like trees in Winter, this king seems to be the only sentient being. No wonder he craves James Laurenson's casual, loose-limbed Gaveston, and no wonder there's such antipathy between him and Timothy West's hard, strong Mortimer. It is a striking performance, particularly in the early and late stages of the play: when McKellen is establishing the character, and when the character begins to disintegrate under the weight of deprivation and suffering. In the abdication scene he's still capable of the large, flamboyant gesture, the arm flung in anger over the head, but it is no longer characteristic. Weariness clogs his movement and his speech, and the creature that Lightborn finally dispatches is a raddled, defeated, pathetic old queer, weakly grappling with his executioner, a parody of his former self.

Effusive with praise was Philip Hope-Wallace, too, in the *Guardian*, singling out McKellen as having:

the right voice ... Edward ... is a part that certain still limited talents can master, but the great thing about this player is that he has presence, height, and control, and in addition the real voice for "Marlowe's mighty line". And he speaks in such a way that the whole constellation moves into place around the bright star in the sky. ...

The neurotic defiance, the unbiddable temper, and the deviant passion are all there, and the rage of grief for Gaveston's death is taken head on with great effect. Only in the later scenes of degradation and despair did Mr McKellen move me less than he had in similar passages in *Richard II*. Here again you have to reckon with the lesser poet and dramatist. But it is a splendid assumption, with the necessary reserves of power, and one which allows the part to be played as it surely must be up to the hilt. No use mumbling Marlowe or trying for a voguish naturalism.

A further sample of the critical reception, this by the much respected J.C. Trewin, in the *Illustrated London News*:

All said, this was Mr McKellen's better performance of the two. . . . He was always within Edward, credibly the hysterically obsessed neurotic, and at the last a man whose suffering would have touched any heart. . . . The death scene conquers all. Just before it, Marlowe had produced, like a spare ace from his sleeve, the murderer Lightborn, one of the most chilling small parts in Elizabethan drama: the man who in the scene with Mortimer introduces himself in the Machiavellian speech on the modes of death, and the line, "Nay, you shall pardon me, none shall know my tricks," Robert Eddison has a terrifying quietness; he governs the stage like an icy emanation. One remembers the conjecture that Lightborn may be, as Professor Merchant has put it, "a surrogate for Lucifer, anticipating in the fate of Edward the tortures of the damned".

The Edinburgh Festival production was transferred to London's Mermaid Theatre. The reviews were in much the same vein. Michael Billington, in *The Times*:

Mr McKellen's Edward . . . is much more than a display of nervous energy. It shows a proper sense of tragic development. In the civil war scenes he moves as if his limbs were suddenly twice as heavy and his body burdened by weighty regalia; and by the end the character has become a worn, ragged shadow, though still with the same insatiable craving for physical contact. Even if packed with slightly too much detail, this is an audacious, powerful and memorable performance.

As a whole, Toby Roberston's production is clear, vigorous, and quicksilver-swift. The homosexuality is handled with justifiable explicitness, the play's emblematic quality is reflected in Kenneth Rowell's stylized designs, and there are strong supporting performances from Timothy West as a cool, guileful Mortimer and from Robert Eddison as Edward's suavely implacable, sinisterly affectionate assassin.

Interviewed by Billington, McKellen shared some of his personal insights into the character; he was not sure he had properly projected the part:

The idea was to show a very young man who suddenly gets the key of the kingdom and who has all the potential to develop into a marvellous person. In fact, because his emotions and desires are thwarted, he develops not into a kind, compassionate man but into a tyrant who, when he defeats Mortimer in battle, feels fulfilled for the first time in his life. The next stage is his degradation and loss of power, and he just becomes a desiccated old shell crawling about waiting to die. That seems the main line. The mistake perhaps was bringing that out in a pictorial way – through changes in make-up, costume, the carriage of the body. When Alleyn played it originally (there is, in fact, no historical proof that Edward Alleyn did play Edward), I'm sure he stayed much the same throughout physically, but a modern audience demands a realism Marlowe did not anticipate. For the speed and continuity of the original production, one substitutes something an audience will readily be able to understand and the kind of attention to detail you're talking about.

Kenneth Rowell's design was doubly innovative. As recalled by John Faulkner, Technical Director of the Prospect Theatre Company, in response to Geckle's query: "The acting area, a disc of brass and aluminium in concentric circles at once brought the strong point of the stage within the same area as the focal point of the auditorium sightlines as well as widening the playing area by six feet. The circular motif was repeated in a curving forestage, sweeping glass fibre ramps from the disc to the gallery and a ring nestling, in the crowded lighting rig, which was on a rake convergent to that of the disc." Geckle adds: "The metallic set design, placed on a tilted stage, produced a strangely alienating effect that was intensified by Edward's and Gaveston's flamboyant behaviour in contrast to that of the dour barons." (To visualize this, the reader might best use a pencil and sketch-pad. What was presented, however, was the tragic history of a fifteenth-century king in and against a twentieth-century avant-garde background.)

In a few reviews there were complaints that the play did not measure up to *Richard II* and that McKellen was too histrionic. But the offering was taken on tour to three cities – Sheffield, Cardiff, Leeds – and then filmed at the Piccadilly Theatre in London for a television showing (1970) not only in Great Britain but also by the Public Broadcasting System in the United States.

Geckle had two conversations with Robertson (1976, 1982) about his lengthy association with *Edward II*; a transcript of the discussions – of considerable interest to scholars of Elizabethan drama – fills over seven pages, too long for this chapter. Touched on again were differing portrayals of Edward, displayed as flagrantly effeminate by McKellen, and as much more reticent by Jacobi, the latter creating a character whose feelings were deeper. McKellen could be perceived, at times, as "acting" rather than always "real". Robertson also suggested that the young player in the role of Edward II should exhibit a streak of toughness, since his avenging murder of his father gives the story's ending a positive upward thrust, which Shakespeare often makes certain to do. The barons, in Robertson's view, are not offended by Edward's sexuality as much as by his raising to power the low-born but disdainful Gaveston – it is in good part a matter of class differences. He believed that Lightborn, the assassin, is homosexual, too, and in love with Edward, whom he bathes and tenderly kisses before killing him, a remarkable bit of business added by the director. Robertson also alluded to Bertolt Brecht's *Edward II* (1924). Freely based on Marlowe's drama, this German adaptation greatly coarsened, distanced, intellectualized the historical incident so that it becomes propaganda justifying the use of torture and illustrating the harsh workings of blind chance.

Apart from Robertson's offering, several other productions of *Edward II* were mounted to mark the quatercentenary. The most noticed of these was that of the Leicester Phoenix Theatre Company, which opened the play in Leicester and then ventured to London, where it was lodged at the Arts Theatre. The director was Clive Perry, the designer was happily named Christopher Morley. The lead, Richard Kay, had played the boy Edward III in Robertson's 1968 staging; John Quentin was Gaveston. Taking three minor parts – Gurney, Lancaster and the Abbot – was Anthony Hopkins, later to win international stardom on the West End, on Broadway and in British and Hollywood films, earning an Academy Award. It was Hopkins alone who won unanimous praise. Otherwise the critical reception varied. In general there was approval for "the rapid and robust pace of the action and the strong sense of physicality in the relationships of the characters". In the *Evening Standard* Milton Shulman gave the nod to Richard Kay for being "impressively authoritarian" as the embattled and angry sovereign; and Jeremy Kingston, in *Punch*, deemed Kay "at his best when petulant and stubborn in the earlier scenes". A number of other reviewers concurred with Bernard Levin, who responded most to Kay "as the wretched king . . . better in his defeat than in his pride, managing the resigned grief to good effect". There was scattered applause from this newspaper covey for Quentin as Gaveston, but little or none for Hilary Hardiman as Isabella, whose transition from loving consort to ruthless foe was not convincing. The fault most mentioned was a lack of eloquence. J. C. Trewin, in the *Birmingham Post*, echoed this complaint: "Marlowe needs vocal splendour. Noise is not enough. Last night, especially before the interval, we had a tumult of bad speaking. Though it improved later, I looked in vain for suppleness or variety."

The Marlowe Society Drama Company – no link to the Marlowe Society at Cambridge University – put on the play at the Commonwealth Institute at Kensington in London. It was also staged by the Bristol Old Vic Theatre School at Berkeley Castle in Gloucester and at the Theatre Royal in Bristol, under the guidance of Glynne Wickham.

Across the Atlantic, a concert reading of *Edward II* was a feature of the Loeb Shakespeare–Marlowe Festival at Harvard University (the year being, and more prominently, Shakespeare's quatercentenary). In 1975, announced in advance as the "First Professional American production of Marlowe's classic about Britain's only known homosexual king", the tragedy finally occupied a New York stage. (Apparently no count was being made of the bisexual Stuarts.) This enactment was directed by Ellis Rabb for the Acting Company, headed by John Houseman, former partner of Orson Welles. Brendan Gill wrote in the *New Yorker*: "The Acting Company took full advantage of the contemporary appetite for sexual revolution; it pursued with relish every clue that the text provided in respect to Edward's irregular personal life." Sylvaine Gold, in the *New York Post*, also felt that the protagonist's womanly attributes were over-emphasized: "Norman Snow's portrayal of the king as a flagrantly effeminate slave of passion was painted with unnecessarily broad strokes, though this is as much the fault of the director as of the actor. Whatever kingly qualities Edward has – and the poor man doesn't have many – were lost behind the posturings of a drag queen." Kinder words were said about other members of the cast: for Peter Dvorsky as Gaveston; for Samuel Tsoutsouvas as Mortimer; and especially for Mary-Joan Negro as a persuasive Isabella. The *New York Times* had its senior critic, Clive Barnes, there: "Mr Snow does not have the range of Mr McKellen – either vocal or physical – but he does have the same air of compelling, even sinister glamour – the sense of a tinsel king with a toy heart." These many differing interpretations of the major characters and of the author's intentions affirm that the script is truly complex, with lifelike ambiguities that ineluctably stimulate thought. Consequently the actors essaying the roles of Edward and Gaveston have an extra burden.

Edward II had another rebirth at the Edinburgh Festival in 1975, this version by the Royal Lyceum Theatre Company. A critic from the *Guardian* dismissed it as "a production of absolutely no subtlety, with declamatory acting pretty well straight from stock". Five years later the Bristol Old Vic had another try at the work; again it was greeted with mixed notices. A bit odd was a production by the Compass Theatre in 1994, unfolded with only six actors. Geckle remarks that this was a new trend in casting, as during this period all three stagings had the same player impersonating Gaveston and later Lightborn.

A month's run was given *Edward II* by the Royal Exchange Theatre Company in Manchester in 1986. Programme notes had an excellent synopsis of Marlowe's personal history, together with "a balanced and judicious analysis" of the play, including its political facets. But the performance itself was chiefly involved with the personal side, a reductive approach. Ian McDiarmid was Edward; Nicholas Hytner, the director. The review in *The Times Literary Supplement* accepted McDiarmid's delineation of Edward as "definitive, for all its unconventionality", showing him as a "nervy, ageing, desperate for affection but irritably aware of his failure to command respect, shifting uncertainly between aggressive bluster, childlike expansiveness and petulant self-pity. Unfortunately, neither of the actors chosen as Gaveston and Young Mortimer lent him adequate support." Another critique, that of Michael Ratcliffe in the *Observer*, reported on the stridence and roughness of the action:

> It is distinguished by physical energy and spectacular gestures that are not afraid to meet Marlowe's excesses by sometimes risking the absurd. The king's spiky toybox crown rolls across the ground as the lovers embrace on the floor. The ominous thrum of plainchant (music, Jeremy Saro) clears for Edward and Gaveston to witness a blasphemous Italianate dance-orgy. . . .

Queen Isabella (Brid Brennan) heaves off the grey cloth of the floor still littered with coins from the orgy like the great burden of her unhappiness itself. Gaveston (Michael Grandage) is trussed up like a piece of meat for the slaughter by the venial barons who – deceptively attired as decent English country accountants, doctors and landowners – have earlier found the king in a tourniquet from the scarlet carpet flung to the floor to welcome Gaveston's return.

The setting, by Tom Cairns, is described by Professor Roger Holdsworth whose appraisal of the production appears in *Research Opportunities in Renaissance Drama*. At the start the stage was "'a sunken, earth-filled circle, covered with a black sheet until the first interval.' The sheet was later removed, and the stage, covered with dirt and with a 'huge cloud-streaked globe above the characters' heads,' gradually turned to mud." Edward, mercurial, was "a bitter mixture of naïvety, tenderness, folly, and corruption", his "dominant quality seeming impulsive immaturity". As opposed to McKellen's picture, McDiarmid never drew out any kingly traits. At the abdication, "Only his own persecuted homosexuality seemed genuinely to possess him."

Geckle's own wry summary of the several versions he witnessed: "What we have in *Edward II* is a play about a homosexual king, to be sure, but those modern productions that stress the adjective and forget the noun are doing a disservice to the complexity of Marlowe's play. As Bernard Levin wittily put it in a review of the 1964 Leicester Phoenix Theatre Company production: '. . . the moral is the one provided by a Mr Krezmer of my acquaintance as we left: "If you want to be a king, don't be a queen."' That holds if you want to portray Marlowe's Edward II with some degree of attention to the text of the play."

Edward II was made into a film in 1992, directed by Derek Jarman, who was afflicted by AIDS. His focus is not on the political aspects of the story but defensively on the homosexual relationship between Gaveston and the weak, dissolute ruler, with explicit scenes of their lovemaking. The settings are bleak and monumental, huge blocks of stone set in sand. Some of the intolerant barons are in period garb, some in twentieth-century attire; Mortimer's henchmen are dressed like Nazi storm-troopers. This mingling of styles follows a fashion prevalent in London at that moment in revivals of classic works to update them only partly. It was to be seen in the new productions of Handel's operas, Benjamin Britten's musical version of *A Midsummer Night's Dream* and Ben Jonson's *The Alchemist*, all running about that time. Apparently the intent is to make the spectator feel that the ancient work is still relevant, that its message also applies to today. The director does not trust the spectator's intelligence and gives crude visual clues to the enduring significance of the offering. (Also the cost of the cast's apparel is reduced.) What is lost is some of the glamour that resplendent costuming adds to the theatrical experience. Men and women in the higher ranks of society never dressed as richly and handsomely as in the Elizabethan and Jacobean decades, and Marlovian and Shakespearian productions often gained greatly from that.

The Edward is Steven Waddington, the insinuating Gaveston is Andrew Tiernan, the Mortimer is Nigel Terry and the emotional Isabella is Tilda Swinton. A number of instances of sado-masochism, sensational episodes of torture, struck some critics as gratuitous. At the film's end Edward's young son is shown gazing into a mirror while putting on earrings and lipstick, implying that he is even more effeminate than his late father.

The Admiral's Men presented another play by Marlowe, *The Massacre at Paris* (c. 1592), on a very topical subject, the St Bartholomew's Day slaughter of the French Huguenots by their Catholic enemies, a horrible deed that sent a compulsive shudder through Protestant England. The only remaining text is short and badly mangled, representing at best a fragment of the original script. The principal character is the Duc de Guise, yet another Machiavellian figure. His language is high-flown, perhaps setting Shakespeare an example of how to write a historical drama. The structure is episodic, covering seventeen years in short scenes leading to the murder of Guise, followed by that of Henri III, and culminating in the accession of Henri of Navarre, for the moment still a Protestant. The killing of Guise is laid to the prompting of Henri III. There is the usual abundance of classical references. It is hard to judge the work's literary merit from the surviving manuscript, but it seems to fall far short of Marlowe's previous accomplishments. It may have been a hurried effort on his part to exploit the emotional reverberations of a sensational current event and thereby fill his purse. Scholars speculate that he wrote it in collaboration with Nash. It loudly rails against intolerance.

Marlowe's plays have had an increasing number of advocates, as proved by these many revivals of scripts long merely read but hardly ever acted, and they have shown continuing stage-worthiness. At the same time, there has been a growing interest for scholars in the scripts. One of the boldest assertions of their lasting excellence is by Maurice Charney in his "Shakespeare and Marlowe as Rivals and Imitators" (1977), a lecture delivered at a gathering of the Modern Language Association and reprinted in an amplified version in *Renaissance Drama New Series X*. Says Charney, Distinguished Professor of English at Rutgers University:

> By the time Marlowe was killed in a tavern at Deptford, his accomplishments at age twenty-nine were considerably greater than those of Shakespeare at age twenty-nine. Had Shakespeare died in the same year, our literary history of Elizabethan drama would need to be radically revised, with Shakespeare among the lesser dramatists like Greene and Peele. Shakespeare not only outperformed his contemporaries, but also outlived most of them. It is not entirely conjectural, therefore, to see Shakespeare in 1593 laboring under the anxiety of influence exerted by Marlowe, his more famous, more energetic, more brilliant and more personally exciting coeval.

Charney joins others in identifying Marlowe as the "Dead shepherd" alluded to in *As You Like It* and, as mentioned, even suggests – and is not the first to do so – that Marlowe might be the Rival Poet of the Sonnets, "who keeps Shakespeare's 'ripe thoughts' from coming to fruition by 'the proud full sail of his great verse' ". If not Marlowe, who else could it have been? What other poet could rival – or outrival him – in Shakespeare's own estimation? Professor Charney also cites the many likenesses between *The Jew of Malta* and *The Merchant of Venice*, mentioned earlier, to which might be added the similarities between *Edward II* and *Richard II*. And between *Hero and Leander* and *Venus and Adonis*, which have also been described on these pages.

But Shakespeare did not simply copy from his prodigiously gifted peer. Warily, "he tried not to compete with Marlowe in his own excellences, but to convert him into a less heroic and more complex mode". Of course, Shakespeare – well after Marlowe's death – did this as he matured. Charney continues:

> There are other ways of phrasing this relation. Marlowe influenced Shakespeare, but Shakespeare also recoiled from that influence. . . . It is as if Shakespeare were trying to conceal the intensity of his own indebtedness to Marlowe, as if he were trying to prove that, although he could not rival Marlowe on his own ground, he could explore areas of dramatic awareness in which Marlowe would never dare to venture.

Analysing *The Jew of Malta* and *The Merchant of Venice*, Charney singles out a significant difference between them. Barabas considers himself far superior to those about him:

> No, Barabas is born to better chance,
> And fram'd of finer mold than common men . . .

whereas Shylock pleads his "common humanity":

> Hath not a Jew eyes? Hath not a Jew hands, organs, dimensions, senses, affection, passions? – fed with the same food, hurt with the same weapons, subject to the same diseases, healed by the same means, warmed and cooled by the same Winter and Summer as a Christian is?

In Shakespeare there is a great deal more fellow feeling and kindness. He did not closely associate with those very high in rank nor walk the corridors of power, as did the brilliant, erratic Marlowe. He was not a wild youth, and he grew in late middle age to the enjoyment of nothing more than retirement and a modest prosperity in a small provincial town where he owned property after prudently saving his theatrical earnings. Charney, further:

There was no chance that the boy from Stratford, not even university trained, could match the boy from Canterbury's soaring and classically derived rhetoric. In an early play like *Titus Andronicus*, for example, Shakespeare may have tried to out-Marlowe Marlowe. The splendid speeches, heavily indebted to Ovid, resemble *Tamburlaine* and *The Jew of Malta*, too. There is no lack of persons with aspiring minds in *Titus*, and the insouciant Machiavellism of Aaron the Moor must be directly indebted to Ithamore and Barabas both. But Shakespeare moved away from grand rhetoric to a more difficult exploration of the details of character and consciousness.

Grandiloquence, such as Marlowe's, is often a phase through which talented young writers make their way.

Though it is probable that Marlowe would have become an even better playwright had he lived as long as Shakespeare, that is not a certainty. In too many instances, youthful poets and dramatists have shown a flash of genius that subsequently dimmed – Tennessee Williams comes to mind; or, in music, Erich Korngold, who never repeated an initial triumph. With Marlowe, dissipation might have taken its toll.

What the world will never know is how much richer might have been the legacy of this poet-dramatist, a question also tantalizing students of Terence, Pergolesi, Keats, Büchner, Mozart, every such artist of dazzling achievements and promise who met premature death. The murder of Marlowe might have been even more costly to the English theatre had not Shakespeare been at hand to carry on from where the slain playwright abruptly left off. After Marlowe, safely entrusted to the quiet pen of his immediate successor, poetic and psychological drama was to mount to an even greater height, with a spate of plays capturing lyricism, beauty, truth and splendour.

— 13 —
SHAKESPEARE: THE ENIGMA

It is a paradox that about the world's most famed author little is truly known. Biographical facts are frustratingly scant, curiosity rife. His name is familiar everywhere, even to the illiterate. His plays are presented almost wherever one finds stages, on at least four continents, but William Shakespeare himself eludes us.

This ambiguity – it might even be called anonymity – surely does not result from scholars not trying to learn more about him. Yet he escapes them, his public persona scarcely related to the dramas he created. Hundreds of eager researchers have fingered through aged, mouldy, yellowed pages for clues, however minute they might be, but the search has been mostly futile. Here and there investigators have come upon his name or his signature in legal documents of his day – a register of his birth or baptism, or a reference to him in a contract, or a listing of him as a juror summoned to decide a petty lawsuit – but these reveal nothing about his physical features, his daily habits, his tastes, his deeper ambitions, quarrels, mature loves. He is nearly faceless. Greene's malicious appraisal of him is not to be trusted. The hearty Ben Jonson was sometimes his drinking companion, but his pithy recollection is unhappily brief.

The following is a short summary of William Shakespeare's life, the facts that have been verified – more or less.

The greatest of playwrights was descended from a family of long-toiling farmers in the English Midlands. (As with "Marlowe", the name Shakespeare had various spellings – "Shakespear", "Shakyspere", "Shaxper", "Shakeshaft" among them – which adds to the difficulty of tracing him and his forebears. It might also be remarked that, in the tavern slang of the day, "Shakespeare" had a ribald connotation.) The Shakespeares lived in Snitterfield, a village four miles outside the market town of Stratford-upon-Avon, in a green area of thick woods and rolling fertile fields that had seen many historic battles during the Wars of the Roses. When William later wrote about those clashes he had a good sense of just where they occurred and the dynastic causes instigating them. Of course, he often altered the details to suit his dramatic purposes as well as pay heed to how the political winds were blowing.

His paternal grandfather, Richard, could have been an eyewitness to some of the actual events, and at the very least had lived through that tumultuous period. Richard Shakespeare enters the village records because he was frequently cited and fined by the local authorities for minor violations, such as letting his cattle roam at large and invade the common, as well as for not trimming his hedges or digging ditches that marked and separated his holding – he rented the land – from those of a fractious neighbour, one Dawkins; apparently the two were always at odds, and after five hundred years their mean, petty quarrels are still remembered. Even so, Richard was a person of some consequence in tiny Snitterfield, undertaking a leading role in communal affairs. In addition, his son John married Mary, the daughter of his father's landlord, Robert Arden of Wilmcote, who owned most of the village. ("Arden" is a name that appears in William's later works.) The Shakespeares had a good-sized house, well located on the High Street, at a corner of a lane running down to a brook that emptied into the ever-flowing Avon.

Richard and his wife had two sons – if not more – the elder of whom, John, was the playwright's father. In turn, John and Mary had eight children, three of them dying while very young. William, third-born, was the eldest survivor. A much younger brother, Edmund, too, was to become an actor. He was sixteen years William's junior. He died at twenty-seven. A histrionic strain may have run in the family. As a clan the Shakespeares were also prone to be in trouble, in particular Henry, a brother of Richard, who was intermittently imprisoned for fighting, trespass, unpaid debts – he was even excommunicated for a time, his sin being that he wore a hat instead of a cap when going into church.

John was apparently no more tidy than his father, and he was fined for not taking proper care of his hedges; he was also fined for not sweeping up a dung-heap on the street in front of his house. Some time after Richard's death John quit Snitterfield and farming and settled in Stratford, at the time a town of 1,500 inhabitants. He had sold his share of the lands he had inherited. He took up a trade, becoming a glover, quickly prospering. He was energetic, ambitious and astute, setting William a good example. Before long, John Shakespeare became a prominent citizen of Stratford and entertained many pretensions. (John Aubrey in his *Brief Lives*, writing some decades later, asserts that John Shakespeare was a butcher. But Aubrey is recognized as an irresponsible gossip, if a diverting one, and his statement about this is disputed. It is true, however, that as a glover he would have tanned and scraped sheep, deer, goat and horse hides, cutting and shaping them to fit hands or to serve as purses, belts or aprons.) Branching out, John also dealt in wools and embraced money-lending. On several occasions he was accused of usury and other illegal transactions.

He set himself up in style. For a while he occupied half a timbered house on Henley Street. Later, having married the comparatively wealthy and much younger Mary Arden, he acquired the other half and added a wing at the rear where it extended into a garden. His shop was in the house. With in time a brood of eight children, doubtless more space was in order.

Despite his minor infractions and some questions about his scrupulousness in business affairs, he was obviously liked and trusted by a majority of his fellow citizens. He was seen to be a born entrepreneur and a risk-taker. Early aspiring to public office, he contrived to have himself appointed an official ale-taster, whose duty it was to assure that the liquor on sale was of acceptable quality; next he rose to be a constable; after that a burgess, which is to say a freeman of the town; and then an alderman. He continued quickly and steadily up the civic ladder, until he was elected bailiff, a post equivalent to deputy mayor – he might be asked to fix the price of corn or other commodities, or to serve writs, or possibly to act as a justice of the peace. He inspected markets and tested the weight of loaves of bread. When travelling about the town from his dwelling to the Guild Hall or with his wife to church on Sundays he would be preceded by a sergeant bearing a silver mace and given the front pew.

His fortune grew apace with his importance. He bought and sold numerous properties, took on mortgages, alert for opportunities to accomplish good deals and flourishing as a landlord. He also engaged in a trail of lawsuits with customers and other tradesmen; a confirmed litigant, he is often referred to in the town's register, though his signature is still unsighted, which raises the question of whether he could spell and write or resorted to making his mark. It is not certain that Mary was literate.

John Shakespeare resolved to apply to the College of Heralds for a coat of arms – he had held high offices and also married the daughter of a gentleman proprietor of large estates, and felt himself entitled to enjoy a like status. He had a sketch of the emblazoned insignia drawn up. But nothing came of this at the time, for reasons that are not clear. Some years later, however, with help from his son William, the arms were granted in 1596, lifting both to the rank of "gentleman". (By this time, following a downturn in his fortunes, John was bankrupt, but William redeemed the family's honour and resurrected its social status in Stratford.) The device: "Gold, on a bend sables, a spear of the first steeled argent, and for his crest or cognizance a falcon, his wings displayed argent, standing on a wreath of his colours, supporting a spear gold, steeled as aforesaid, set upon a helmet with mantles and tassles as hath been accustomed . . ." The motto accompanying this was *Non Sanz Droit* (Not Without Right), hardly a humble boast.

As the eldest son of a prominent man who was waxing richer and richer, the youthful William might have looked forward to an easy and altogether comfortable future. Events turned out differently. Something went wrong with John Shakespeare's calculations; abruptly his fortunes dropped until he was nearly penniless. He tried desperately to retrieve the holdings that rapidly slipped away from him, but he was unable to face up to his ever-multiplying debts. He seemed to lose his grasp of affairs and neglected what remained of his shrinking civic duties. Whereas he had always faithfully attended council meetings, he now never went to them – perhaps he was ashamed of being seen in his publicly reduced state. His colleagues kindly made allowances for his derelictions: he was listed as an alderman for ten more years and exempted from paying the fines that would otherwise be levied for his prolonged absences and from the demands for the higher taxes due from an alderman. That he was spared humiliation suggests that he was still popular and considerably esteemed, or else that he had become an object of his fellow citizens' generosity and pity. He was able to hold on to the Henley Street house, which eventually was to pass to William. But the consequence of the sudden collapse of John's wealth was that it would be up to young William to make his own way in his sixteenth-century world.

How much schooling William Shakespeare had is a major subject of speculation. Education was not mandatory, but the citizenry of Stratford held it in high regard. In 1553 a chartered academy, the King's New School, was established with largesse expended by leading townsmen: the headmaster was granted free occupancy of a house and an annual wage of twenty pounds, a good sum in those days. From it he paid for an assistant – an usher – to look after the younger pupils. This was more than teachers earned at Eton and at many other schools of repute. A number of well-qualified masters filled the Stratford post in succession, two of them holders of Oxford degrees. Most likely a student would be well taught.

As a burgess, an alderman, a bailiff, an aspirant to a coat of arms and a social climber jealous of his station, John Shakespeare probably sent his son to this locally prestigious academy. If so, the boy would have had a grounding in the classics and learned at least some elements of the languages of remote Athens and Rome; brushes with them were prescribed components of the Humanistic curricula of that day and particularly for young persons who might go on to pursue theology. That Ben Jonson, proud of his learning and lamenting the shortcomings of his good-humoured tavern companion Will Shakespeare, complained that Will knew "little Latin and less Greek" is significant, the point being that Jonson's indication that the poet from Stratford *did* possess a smattering of those ancient languages, thus suggesting that he had undergone some years of schooling. No, he had not been to Oxford or Cambridge. That remains a crucial fact. Much is made of his setting one of his plays on "the sea coast of Bohemia", though Bohemia is landlocked, high and dry in Central Europe, far from storm-tossed waves spewing forth half-drowned, shipwrecked youths and maidens. But confusion about geography is common in any age, even today, as anyone who deals even with university students can testify. (Ask your friends exactly where is Bohemia!) One hears of no other serious blunder in Shakespeare's works, consisting of a lengthy narrative poem, a range of sonnets and some thirty-seven dramatic scripts in all. Indeed, there is a continuous and even daunting display of erudition, many mythological and historical references characteristic of the most élite Renaissance literary output. If largely and belatedly self-taught, he must have read extremely widely. But where did he have access to the books, rare and expensive?

One possible answer is that only one of Shakespeare's plays, *The Merry Wives of Windsor*, is wholly original. All the other thirty-six are adaptations, their premises and characters borrowed from earlier plays and tales by other writers, some English, some Italian, some Spanish and sometimes compounded from more than one source. If he progressed from scripts by predecessors who were perhaps better educated, much of the mythological and historical detail was already in hand for him to use and transmute, a task he accomplished with his unique alchemic gift, taking mediocre or merely good material and changing it into something transcendent.

Yet this enablement, a sharing and confiscation of the knowledge and perceptions of lesser-known writers whose scripts he cannibalized, cannot fully account for his becoming a superb play-craftsman and one of the world's great poets, with a remarkable ear for the many accents and nuances of low and high speech, a commanding vocabulary and mesmerizing eloquence hardly acquired in a grammar school – if he attended one – along with a talent for portraying the humorous and tragic facets of life, a memorable flair for the aphoristic, and a sharp, very astute judgement of political moves and often puzzling human behaviour.

One wonders what this Shakespeare might have been if in fact he had gone to university.

A persistent story about the young William is that he was caught stalking deer in a forest near Stratford and brought to court for punishment by an irate Sir Thomas Lucy on whose wooded estate the poaching occurred. This inspired the young man to compose a bitter – no longer extant – ballad concerning his plight and later accounted for his leaving Stratford for London. Alas, this hardy tale has been disproved: it has now been ascertained that Sir Thomas Lucy had not yet purchased the lodge and its park by that time and could not have prosecuted an invader. Proponents of this legend claim that the ballad would have been William's first attempt at poetry.

It is probable that after his schooling – if any – and even during terms at the academy, William worked in his father's shop learning the leather and wool trades, scraping, cutting, dipping his hands in the vats, or hunting and slaughtering the fugitive deer that provided fresh-tasting venison and whose hides could be used for fine, soft gloves.

In later decades it was profitable for citizens of Stratford to preserve and spread such stories and for some of the very oldest inhabitants to assert that they had known the high-spirited youth and recalled him clearly. They might even persuade themselves that this was true, cadging an ounce of celebrity.

Much conjecture is directed to what might have been another occupation while he was in Stratford: perhaps he taught at the school – his education was comparatively scanty, but he could have been a low-ranked assistant. This is mere speculation.

Another theory that has gained currency among a substantial body of historians and literary academics posits that around 1580 William left home to escape accusations of Catholic recusancy – possibly at the suggestion of the schoolmaster John Cottom – and went to work for a year or two at the papist safe haven of Alexander Hoghton in Lancashire as tutor to his family. Through this contact, it has been suggested, he may have come in contact with travelling players and both amateur and professional dramatics.

One fact is sure: in 1582, at the age of eighteen, William Shakespeare was issued a licence to marry Anne Hathaway in Stratford. She was twenty-six and pregnant. John Shakespeare had chosen a wife much younger than he was and William one who was much older than he. A minor, he required his father's consent to wed. Owing to the bride's embarrassing condition the church ceremony took place after only one reading of the banns. The couple's first child, Susanna, was born six months later.

Was this a truly romantic match or a forced marriage, the consequence of an adolescent indiscretion? (The legend also says that at this phase of his minority William engaged in bouts of heavy drinking.) Not too long after his daughter's birth he left Stratford and settled in London, leaving Anne and Susanna behind. From then on, he travelled back to the town intermittently. The Shakespeares had two more children, twins, a girl Judith, a boy Hamnet. Unhappily, William's son died when only eleven. The two daughters outlived him.

Like William, Anne was descended from a family of soil-tillers who dwelt and worked at Slottery, where the parish of Stratford bordered the encroaching Forest of Arden. In Richard Hathaway's will he stipulated that his daughter Anne (Agnes) be given ten marks on her wedding day – this bequest was dated a year before the marriage was actually celebrated, perhaps even before the tie to young Shakespeare was anticipated. He died before the wedding. Twice-wedded himself, Richard Hathaway had seven offspring and owned an ample-sized farm of fifty to ninety acres – his will envisaged dividing them among his children, with his second wife, Joan, presiding over the flock and having the last word in any dispute. The thatched Hathaway cottage still exists, a picturesque tourist shrine. The two-storey exterior, timbered, has a high-pitched roof; latticed, tiny windows let light into its twelve rooms, some panelled. With plaster added to the timbers, the house is surrounded by a garden and embowered with flowers; it offers a vista of an orchard, fertile pasturage and a clear brook near by. In winter the cottage was warmed by a huge fireplace in the master's chamber; this room also contained a capacious bed; below, on the ground floor, was the buttery. All this typifies a Tudor rural habitation. It was partly damaged by fire and restored in 1969.

At the very least, however imprudent, Will had not married out of his class; in many respects the Hathaways were the social equals of the Shakespeares, except that John Shakespeare was more ambitious than Richard Hathaway. Apparently Will deemed Anne reliable and capable: he entrusted her with taking care of whatever of his business affairs might arise in Stratford while he was away at his London pursuits.

Scholars find intimations of what was the ensuing state of the marriage by quoting a passage from a play written much later in his life. In *Twelfth Night* Duke Orsino proffers this advice:

> Let the woman take
> An elder than herself; so wears she to him;
> So sways she level in her husband's heart:
> For, boy, however we do praise ourselves,
> Our fancies are more giddy and infirm,
> More longing, wavering, sooner lost and won,
> Than women's are . . .
> . . . let thy love be younger than thyself,
> Or thy affection cannot hold the bent:
> For women are as roses, whose fair flow'r
> Being once display'd, doth fall the very hour.

If the playwright, with rue, is reflecting here on his own precipitous match, he is at the same time paying tribute to Anne's innate common sense and conceding that it was superior to his own. But one assumes that he did not have her physical image in his thoughts when depicting the beauty of Juliet, Viola and Cleopatra. Most of the couple's lives were lived apart.

Why did William go to London? The special difficulty in answering that inevitable question is that from 1585 to 1592, after the arrival of the twins, his name appears in no document – there is no record of his whereabouts or occupation. Baffled biographers call these the lost years. When last mentioned he was an untested young husband and father, from all the evidence half-educated and unworldly, perhaps engaged in the glover's trade or a clerk in a school – and when next in view an actor and playwright much engrossed in the capital's clamorous theatre world. A remarkable transition. These would be an artist's formative years, hence his biographers try to account for the change in him, but in vain. From this point on, William Shakespeare is an enigma.

Many have put forth possible explanations for his leaving – or fleeing – Stratford. Shame that he had been charged with poaching deer and rabbits – if the incident actually occurred, which is now doubted. Embarrassment at his father's declining fortunes and vanished social status. Anxiety about being arrested for recusancy, if indeed he was Catholic. The natural restlessness of youth, especially in one who was suddenly and prematurely bearing the onerous burdens of marriage and fatherhood. Simple curiosity about large realms beyond provincial Stratford. Any of these reasons may have prompted his departure.

But what led him to a stage career? What had prepared him for it? It was always an uncertain profession, to say the least, and one that tended to be discredited by most clergy and God-fearing citizens, such as had been his neighbours – they were not bohemians, the sort with whom he would associate in London from now on.

It is assumed that in his boyhood William had seen the annual wagon-train Mystery Plays performed to commemorate Corpus Christi Day in Stratford as elsewhere – there are traces of them in his productions as there are in Christopher Marlowe's, lingering elements of pageantry. So he had some notion of "theatre". It is also possible that if he attended King's New School he had participated in enactments of Latin or didactic and moralizing plays there – it was common practice to have students and their preceptors do that. And, as has happened from time immemorial in some young persons, the experience planted in him a love – an infatuation – for performing, which goaded him, and to which he finally yielded in drab Stratford, leading to his impulse to escape and find a new more exciting life.

Another possibility is that if, indeed, he had worked as a tutor at Hoghton Towers in Lancashire he may have come in contact with groups of touring actors and, having developed a taste for drama, made a decision to head for London and to try to make his way in the theatre as a performer.

Certainly several of London's acting companies paid visits to towns outside the capital. The Queen's Men appeared frequently in Stratford. In 1587, on the troupe's way there, William Knell, leading member, was killed in a fight with a fellow player. This created a vacancy in the company's roster. When this was discovered (1961), scholars delightedly leaped at a theory that Shakespeare had been hired to fill the gap – though given minor roles – and accompanied the troupe back to London. He could have cited his background in amateur theatrics as a qualification, saying he was able to take on small parts. Of course, there is no proof that this ever happened.

Concerning his existence in London, unfounded anecdotes have proliferated, while his figure remains vague. One popular tale is that his first employment there was to hold on to the horses outside the theatre while their noble and affluent owners were attending performances. But some scholars hypothesize that he was away at sea during the "lost years" – an idea based on a sprinkling of nautical terms in his scripts. It has also been suggested that like Christopher Marlowe he was spying for Elizabeth in the warring Netherlands. Neither of these contentions has ever been solidly substantiated.

His various lodgings in London have been inscribed in tax-collectors' records; indeed, he was delinquent in paying the taxes due from him over a number of years and was finally tracked down and made to settle up. He was a tenant for a while of a family named Mountjoy in a central area of the city (1604). Their daughter had married an apprentice of theirs, Stephen Belot, who later claimed that he had not received in full the dowry promised him. Master Shakespeare is spoken of, in the suit, as having played the role of matchmaker. The brief filed in the case, brought in 1612, declared that he was no longer a Londoner; by then he had permanently returned to Stratford, his attachment to the theatre ended.

How he broke into the circle of striving and fiercely competitive playwrights is not readily traced. In a recent essay a Cornell University professor, Scott McMillan, conjectures that the young man, having accompanied the Queen's Men to the bustling city, stayed with that esteemed company for a few years. McMillan notes that the plots of six Shakespearian plays have clear parallels to just that many in the Queen's Men's repertory; had they not become familiar to him while he had roles in them? Subsequently he borrowed liberally, adapting and elaborating various strands from them; as noted, such appropriations were common practice. But McMillan concedes that Shakespeare need not have been on the Queen's Men's roster to have been acquainted with those plays: four of them had already appeared in print. Or he might have felt obliged to see performances of scripts by his rivals. He had to know what they were doing and try to equal or top them. It was a busy and demanding market.

Whether he gained entry as a professional actor with the Queen's Men or not, he is eventually counted with the Lord Chamberlain's Men, where he developed as dramatist and shareholder. This group had come from Lord Strange's Men, a nobleman's private troupe. There is a tantalizing hint that "William Shakespeare" had been associated with it before ever setting out for London; this is not widely accepted. Expanding in size, the Lord Chamberlain's Men chose a new name when James I ascended to the throne, succeeding Elizabeth I. The name-change was an opportunistic gesture: James, genuinely interested in theatre, was a more generous patron than his tight-fisted predecessor. The King's Men was a well-regarded company, headed by Burbage and Kempe as chief tragedian and comic respectively. At first they were established at the Theatre owned by Burbage; next, at the Hall of Gray's Inn. Then as now, actors and companies shifted about at short intervals, going wherever their services or special talents were besought. Once he was a shareholder – one of eight "fellows", as they were designated – Shakespeare had sound reasons to stay where he was: the Globe, built from the remnants of the Theatre, became the troupe's permanent home. It mostly flourished and quite often was summoned to offer its presentations from one reign to the next at Elizabeth's and James's courts.

As mentioned in the previous chapter, Shakespeare did not win fame as an actor. He seems to have had lesser parts or character roles: for example, it is believed that he played the Ghost of Hamlet's father or the Player King or both – there was much such "doubling", players appearing in more than one role in a play. Again, as has been suggested, his willingness to subordinate himself testifies to an innate modesty: he did not write *Othello* or *Lear* to test or flaunt his histrionic gifts, to show himself on stage as a commanding figure. It is rare, indeed, for an actor-turned-playwright to conceive scripts in which he will not inevitably have the lead to tumultuous applause. It may also be that he recognized his limitations: he was not a Burbage. But it is obvious that he knew the rules for good acting: he sets them forth eloquently in Hamlet's cogent speech to the players.

The testimony is that in his off-stage deportment, too, he was quiet and gentle. His manner was unlike that of the flamboyant and riotous Marlowe, the bohemian Kyd and Nash. He could quaff ale in sociable circumstances, was witty, congenial, straightforward in speech and a good friend. His colleagues in the troupe later paid fulsome tributes to him (not only for his scripts but for his acting, too), singling out his modesty; and he left remembrances to some of them in his will. By some accounts he did not mingle much or often when outside the theatre; he was too compulsively industrious, always having scripts to write and deliver on time. Also, one of his dominant traits – obviously – was detachment; he stood aside, observing, an onlooker.

To the envious Greene – suitably named – the newcomer was an "upstart crow", an arrant plagiarist, his verse only bombast, his self-conceit grossly inflated. The diatribe was doubtless provoked by the young intruder's having scored a success and won public attention. What was more annoying, he was not one of the University Wits, those who, having been to Oxford or Cambridge, sought to monopolize London's burgeoning stages. Again, as quoted here on earlier pages, the publisher of the dying Greene's pamphlet, Henry Chettle, hastened to issue a correction and apology after his friend Greene's death: the dead man's description was a canard; from what Chettle himself had seen of him, Master Shakespeare's demeanour was always "civil", his manners gentlemanly. And those who did business with him commended him as unfailingly upright, always true to his word.

The highest praise of all came from his fellow poet and playwright, hearty classicist Ben Jonson, an intimate and in some ways his equal, with whom the mature writer from Stratford shared tavern glasses and colloquies fairly often: "I loved the man and do honour his memory (on this side idolatry) as much as any. He was indeed honest, and of an open, and free nature; had an excellent phantasy, brave notions, and gentle expressions," Jonson writes in his pithy *Timbers*; though he goes on to deplore his late friend's unfortunate lack of erudition – that is, his scant knowledge of Greek and Latin – and an unfortunate tendency in his scripts to prolixity. "Would he had blotted a thousand lines," Jonson laments.

The neophyte actor, the young Shakespeare, was fortunate in having one valuable "contact" in London, a prosperous printer, Richard Field, former neighbour in Stratford, where Field's father was a tanner like John Shakespeare – in fact, when the elder Field died, John Shakespeare had helped to appraise his estate. The London printing shop, which Field had inherited by marrying his employer's widow, published works of high quality, and it was there that William brought his very early writing, the narrative poem *Venus and Adonis* (1593) and a second long tale in verse, *The Rape of Lucrece* (written in 1594 but not published until 1609); they received a ready welcome, first from Richard Field and then from the public. The *Venus and Adonis* confirms William's close acquaintance with the Latin poet Ovid, from whom he quotes in Latin on the title page. The work's eroticism had a particular appeal to university students, among whom it became a craze. The skilfully composed dedication of *Lucrece* to Henry Wriothesley, third Earl of Southampton, to whom he had also addressed *Venus*, reveals a growing sophistication. The *Venus* proved to be by far the more popular of the two lengthy poems: Lucrece remains faithful to her vow of chastity, and the lyrics are more sober, whereas Venus as a heroine revels in sexual abandonment. The metaphors are sportive and heated, truly pagan.

Where was Shakespeare when he wrote the sonnets and these narrative poems? Had he brought the manuscripts with him from Stratford? While there was he already a versifier of such gifts and sensibility? It is hard to imagine that to have been so. Once in print, the two slim volumes had a long life; they sold well through many editions and were often pirated in addition. Such literary success provided him with credentials to attempt play-writing.

His sonnets, a far loftier achievement, comprise some of the most original and beautiful imagery to be found in that difficult verse-form. Sonnet-writing was greatly in vogue among poets at that time. The form, conceived by the Italians – Petrarch and others – had now been taken up in England by Sidney and Spenser, by whom Shakespeare was influenced. Though younger and less educated, he quickly surpassed them. His accomplishment in mastering the form has never been equalled. He composed 154 separate sonnets at various moments, apparently while touring with his troupe outside London – there are references in them to spells of travel-weariness. The generally accepted guess is that they date from 1593 to 1594, which would be when he was twenty-eight or twenty-nine, hence not long past his being a half-tutored theatre-aspirant from the Midlands. It is hard to believe that in so few years anyone could reach such heights of expression and control of a restrictive and challenging verse-genre.

Some of his sonnets are so subjective and emotionally revealing that it is thought they were not meant ever to see print, or at least not while their author was living. Only two appeared in anthologies (1599) while he resided in London; the full sequence into which they were finally gathered and arranged, as if each poem of the 154 is causally linked to the next, did not appear until 1609. Now the effect given is of a story being told. The booklet containing them was issued by Thomas Thorpe, almost certainly without Shakespeare's assent – none was needed in the days before copyright. The pamphlet, poorly edited, is replete with inaccuracies; it is unlikely that any poet would have sanctioned his precious lines taking leave of him so carelessly printed. *Venus* and *Lucrece* were still selling well, and Shakespeare had gained a solid reputation. Thorpe probably saw an opportunity to exploit the poet's growing celebrity.

It is widely thought that Thorpe is also responsible for the book's dedication, though this may not be so. In the history of literature no lines have been more fervidly and unrewardingly scrutinized:

> TO THE ONLIE BEGETTER OF
> THESE INSUING SONNETS
> MR W.H. ALL HAPPINESSE
> AND THAT ETERNITE
> PROMISED
> BY
> OUR EVER-LIVING POET
> WISHETH
> THE WELL-WISHING
> ADVENTURER IN
> SETTING
> FORTH
> T.T.

Who is Mr W.H.? What is meant by "begetter"? These questions have vexed and transfixed all self-appointed experts on Shakespeare. The dedication yields no answers, though many are hazarded. The first name that leaps to mind is that of H.W. (Henry Wriothesley), epicene Earl of Southampton to whom the poet had tendered *Venus* and *Lucrece*, seeking his bounty as a patron. This nobleman was young, just eighteen or twenty, having inherited his title while a mere child. He was said to be remarkably handsome – some portraits do not bear this out – with curly hair and soft features, obviously effete in manner, very fond of the arts and profligate in subsidizing artists. His marriage was arranged by his guardian, the formidable William Cecil, Lord Burghley, but at the last moment the designated bridegroom backed away, a retreat for which he was punished by a heavy monetary fine, greatly reducing his substantial patrimony. Eventually he did wed the young noblewoman chosen for him.

Why the avid interest in who Mr W.H. might be? His identity might indicate Shakespeare's sexual orientation. The sonnets in this series chart the progress of a love affair, its ups and downs. One day the poet's hopes are lifted, the next day denied; he is plunged into incessant changes of moods, dependent on how the "beloved" responds to his poetic admirer's addresses. The object of this infatuation – indeed, adoration – is a physically attractive youth of patrician qualities of intellect and feeling. His smile or frown, or frequent show of indifference, never cease to infuse the poet with instant happiness or despair. "When in disgrace with fortune and men's eyes / I all alone beweep my outcast state . . ." The mere mental image of the "beloved" propels the poet's spirit as if on hearing a lark singing hymns "at Heaven's gate". A rival appears on the scene – he is also a poet. Beginning with sonnet 127, the author shifts his affections to a lady who boasts dark eyes and raven hair – the "dark lady" of the sonnets. She is wild, promiscuous, unfaithful. In an odd twist, the young aristocrat lays suit to her and is successful, leaving the poet bereft of both "beloveds". The young man had been his "better angel", the woman had proved to be "The worser spirit . . . coloured ill". The originator of the sonnet cycle resolves to turn away from lovesickness.

The portrayal of the incomparable young man in the sonnets does closely match what is known of Southampton. The poet even encourages this paragon of manhood to enter into a marriage and have a son who will ensure the continuance of all his father's splendid physical and moral attributes. At the same time, Southampton will also be immortalized by the verses dedicated to him. A bold pronouncement!

Yet not all biographers accept Southampton as the inspiration of this cycle of poems. His initials are not "W.H." but "H.W.", and it would have been effrontery – exceedingly tactless – for a mere writer to publicly address a nobleman of such high rank as "Mr" – hardly a way to gain his patronage and affections. To counter this, the proponents of Southampton as the elegant youth in the sonnets have suggested that the poet, in view of the homoerotic tone of the verses, deliberately reversed the initials to shield the earl, already whispered to be homosexual.

Indeed, publication of the sonnets was dangerous, which may have been why Shakespeare withheld them. (It is possible that they were circulated privately, hand to hand, to a limited group of readers.) The penalty for homosexuality was death, though the law was seldom enforced. But to risk exposing the earl at all – and the poet himself – would be folly, and all depictions of Shakespeare that are left – quite scant – are of someone practical, intelligent, cautious.

The Elizabethan court was peopled with a share of somewhat androgynous figures, Southampton among them. The courtiers were so extravagant in greetings that they were spoken of by foreign visitors as the "kissing English". Shakespeare himself, working in a theatrical realm where women's roles were played by attractive boys, was probably surrounded by instances of flagrant homosexuality; and for the greatest part of his time he was away from his wife. That he might have developed an illicit interest in young men is not implausible. If this came to pass, his orientation might more accurately be categorized as bisexual: he was married and father of three. Nowhere in the sonnets is there any intimation that the poet's ardour is reciprocated or that there has ever been any amorous physical encounter, and none is asked for. It would also be very unusual for a homosexual suitor to urge his "beloved" to marry and have children. And both poet and young man abruptly switch their desire to the "dark lady".

In the Elizabethan age, as later in the nineteenth century, elaborate compliments and over-praise of friendship between men would be quite free of overt or half-hidden homosexual content. That may be true of Shakespeare's sonnets. They may have been an overstated attempt by their author to earn again the favour of his wealthy and influential patron. Finding them too explicit, he did not submit them to Southampton or to "Mr W.H.", whoever that person might be. Yet, proud of his skill at writing them, the young poet did allow a circle of friends to read them.

Could it be that the sonnets are the fervent output of a young word-master revelling in poetry for its own sake, on becoming intoxicated with his newfound faculty for writing it? A rush through his mind of apt or witty phrases, images, alliterations, rhymes, verbal music, symbols – the actor joyously and excitedly overwhelmed by all these thronging into his thoughts as if from nowhere? Along with a tendency to exaggerate his emotions, as is true of most actors? Assuredly, his impulse to dramatize was a surpassing one and must have been energetically active in him at almost every moment, though his habitual outward reserve concealed it from the eyes of his incurious, busy colleagues.

That Shakespeare might have been compulsively drawn to someone of his own sex should not be altogether surprising. Today bisexuality is recognized as a likely component of the psyche of many a major artist, accounting for heightened sensibility and often acute insights. It need not ever lead to actual physical expression, instead remaining latent, repressed yet tincturing many feelings.

Southampton is not the only one posited as "Mr W.H."; for a good while William Herbert, the third Earl of Pembroke, was a favourite of scholars delving into Shakespeare's sexual inclinations. Here the initials are a perfect fit. It is to this earl that Shakespeare's First Folio is dedicated by the editors. He was a patron of acting troupes; the family had a connection with the short-lived Lord Pembroke's Men and the much more durable Lord Chamberlain's Men whose players frequently offered works by Marlowe and Shakespeare, and with one or both of which the young man from Stratford was associated as an actor. The earl was equally generous in supporting the arts and their ever-needful practitioners. On four occasions, when appropriate marriages were negotiated on his behalf, he too backed away, emulating the apprehensive and reluctant Southampton. Alas, the zealous advocates of William Herbert as the "beloved" evoking the sonnets lost their seemingly strong case when his birth date was established in 1608, which would have him prepubescent – somewhere between nine and twelve, not even downy-cheeked – at the time it is supposed the poems were conceived.

That does not empty the roster of names, often exceedingly fanciful, put forth as a possible "Mr W.H.". Another is that of the imposing, ill-fated Earl of Essex. (This makes it sound as though the poet had a special fondness for earls!) An ingenious suggestion is that the sonnets voice Shakespeare's hopes and lingering griefs concerning his dead son Hamnet. A less flattering designate is Willie Hughes, a boy actor, a selection that delighted the cynical Oscar Wilde, though he gave scant credence to it.

If their publication was unauthorized, how did the manuscript version of the sonnets reach Thomas Thorpe? Who was their "begetter", to whom the publisher acknowledges his gratitude in the dedication? William Hathaway, Shakespeare's brother-in-law, might have been the person responsible for handing over the poems to Thorpe. Or it could have been William Hart, a son of Shakespeare's sister Joan. Or, with a high degree of plausibility, William Harvey, who became Southampton's stepfather, and who might have come across the verses discarded or mislaid by his dilettante stepson, whereupon he bore them to the printer, both of them hoping to profit thereby. Does the term "begetter" signify no more than that? The "adventure" on which "Mr W.H." was setting forth, with Thorpe's wishes for great happiness, might well have been Harvey's marriage to the Countess of Southampton, the young earl's mother; or more likely to that grand lady's successor, Harvey's second and much younger wife, a plighting, in 1608, consummated barely a year before the book's publication. Again, too little kept in mind is that Shakespeare did not compose the dedication: it is signed T.T., referring to Thomas Thorpe.

What about the "rival poet?" Enthusiastically, some scholars have seized upon Christopher Marlowe for the role. His sexual orientation is really not in doubt. But there is nothing factual to substantiate casting him for the part, though admittedly his presence in them would add glamour and colour to the sonnets.

A busy guessing-game preoccupies the experts about the identity of the faithless "dark lady". Early on, attention centred on Mary Fitton, the Earl of Pembroke's mistress, a convenient choice. Lately the discovery that she was blonde and fair-complexioned tends to disqualify her, unless Shakespeare altered her aspect in a gallant effort at disguise, in doing so protecting her reputation. Or warding off the earl's likely displeasure. Southampton, too, had a mistress, Elizabeth Vernon, whom he made pregnant and wedded. Her portraits show her as blonde as well and round-faced; but how authentic is the likeness? She cannot be ruled in or ruled out. Another competitor weighed by scholars – who themselves are competing in this matter – is Jacqueline Field, the French-born widow who selected Shakespeare's friend and printer the young Richard Field for her second husband. Since she was French, she might have had features somewhat dark of hue and the poet must often have met her. But did she measure up to being the wanton seductress depicted by the aggrieved poet? More recently, with much advanced self-congratulation, the dogmatic and usually authoritative A.L. Rowse proclaimed that he had at last dispelled the mystery: the lady was Emilia Bassano, illegitimate daughter of Baptiste, a court musician; obviously her antecedents were Latin, hence she was most unlikely to have been pale-skinned. Though she was married, she was engaged in a liaison with and had a child by Henry Carey, the first Baron Hunsdon, who as Lord Chamberlain had a clear link to the company that included Shakespeare. Rowse seeks to substantiate his claim by examining some deft, mocking word-play in the sonnets that might refer to Emilia's wronged husband. Other scholars have challenged his interpretation, stating that Rowse gives a wrong first name to the gentleman and mistranslates a key adjective, asserting that the lady was "brown" when the term applicable to her is properly "brave". But perhaps the poet was tempted by someone really "dark", a flirtatious mulatto or quadroon female ale-server in a tavern he habituated, a strain of lively Portuguese blood adding flash to her eyes? Or Lucy Negro, a notorious prostitute of the day?

The explanation that few if any of his biographers wish to accept is that the "dark lady" might be only a fiction, suddenly inserted into the poems for conflict and dramatic effect, by a born master of the theatrical. Or she might be there for another purpose: a belated effort to provide a cover for the poet and his "beloved". It frequently happens that a homosexual marries or takes care that he is often seen in the company of a comely young woman, to ward off suspicion. That might be the expediently invented role of the "dark lady" in the otherwise overtly homoerotic sonnets.

Thorpe's edition having sold out, the sonnets dropped from sight for three decades. In 1640 a fresh printing was brought out by John Benson, who took it upon himself to alter the order of the poems, so that they followed one another in a more narrative fashion, apparently mimicking a troubled love affair and consequently sounding more autobiographical, instead of being mere notations of varying moods. When put in a new place, the sense of some of the sonnets underwent an appreciable change, seeming to convey a different intent. Some of the poems were given titles, quite misleading ones. The dedication to "Mr W.H." was omitted, probably an act of prudence. In addition, Benson forced the "beloved" to undergo a "sex-change": "he" becomes "she" and "him" becomes "her", and elsewhere an endearment is neutered, no specification of gender allowed to remain. By these devices Shakespeare's reputation was cautiously shielded. Subsequent editions, from other hands, emulated Benson's mutilated text. It was assumed that the playwright had written his lyrical and anguished verses about his mistress – or, rather, counting the "dark lady", two mistresses.

Edmond Malone, a later and more responsible scholar, restored the dedication and eliminated Benson's reforms, and thereby brought to light once more the problem of Shakespeare's sexual bent, in an important and correct collection of the poems and plays (1778). He sought to identify "Mr W.H.", dismissed William Hart, the poet's nephew, as an entry – he would have been too young – and pondered the quality of Shakespeare's conjugal relations with the mostly absent Anne, the mother of his children.

With that began an exacting test of lines and words in the poems to discern just what they might tell about William Shakespeare and his secret life. It is safe to say that in the history of the world's literature no verses have ever been so rigorously probed, to no avail.

Most scholars in the nineteenth century were horrified at the thought that the supreme dramatist had ever harboured an "unnatural" interest in a handsome young man. Their response was to reject the idea, some quite violently. The poet-critic Samuel Taylor Coleridge wrote heatedly that he could find nowhere in Shakespeare, whom he wholeheartedly venerated, "even an allusion to that very worst of all possible vices" (1803). Later he declared: "It seems to me that the sonnets could only have come from a man deeply in love, and in love with a woman." This prompts S. Schoenbaum, an eminent twentieth-century authority on all aspects of the playwright, to comment that Coleridge must have read Shakespeare rather carelessly.

The debate goes on. It seems certain that it will never end. Rowse and many others share Coleridge's opinion. But perhaps William Shakespeare is entitled to his privacy. What right have others to ask so many intrusive personal questions about him?

On the other side, there are tales – perhaps made of whole cloth – about his heterosexual proclivities. One is that on his journeys between Stratford and London he was guilty of adultery, his willing abettor in the passionate affair having been Jane Davenant, the young, pretty wife of an innkeeper, at whose hostelry, later called the Cross Inn, the playwright rested a night frequently while *en route*. The source of this story is William Davenant (later Sir William, a prominent figure in the Restoration theatre) who proudly vouched that he was born out of wedlock and that not John Davenant but Shakespeare was his father. It did much to heighten Sir William's stature in stage circles, at least to the extent that his impressive boast was credited by those who heard it.

An anecdote, surely not to be taken at face value, has it that on an occasion when Burbage was playing the lead in *Richard III* a certain lady was overpoweringly moved and attracted by his performance. She sent him a message by her maidservant saying that her rich, aged master was away and giving the address of her house and inviting him to pay her a covert visit. He was to announce his arrival by three discreet knocks on her door. Shakespeare, learning of the contents of the letter and assignation, made his way post-haste to the house, where, upon signalling as instructed with the three knocks, he was promptly admitted. The lady was surprised to behold that her knowledgeable visitor was not Burbage. The eloquent author of *Romeo and Juliet* soon overcame her reluctance to entertain him instead of the tall, burly actor. Well before the anticipatory Burbage reached the dwelling and, in turn, tapped thrice on the door, Master Shakespeare's desires had been fully sated. The message he left for the baffled Burbage was "the reign of William the Conqueror preceded that of Richard III". *Ben trovato*.

As previously stated, in the lawsuit over the unpaid dowry (1612), to which Shakespeare was called as having been a witness to the marriage contract, it is noted that he was no longer a Londoner: he had returned to Stratford-upon-Avon. It is believed that it had always been his intention to go back to his native town, and in 1597, evidencing his prosperity, he had purchased the second largest house there, called New Place. As an actor, dramatist and shareholder in an esteemed troupe he had enjoyed a growing income and obviously had saved much of it. (By one account he had been enabled to make his investment in the company after a grant to him of £1,000, largesse from his acknowledged patron, the Earl of Southampton, but this cannot be verified.) He was shrewd in handling money, as – up to a point – had been his father. And, as his plays surprisingly attest, this poet knew a good deal about writs and contracts and was involved with a number of them. One of his biographers, W. Nicholas Knight, in his *Shakespeare's Hidden Life*, seeks to prove that the poet-playwright had even practised law in some form or other before and during the "lost years" (1585–92). Knight expends 288 pages documenting and presenting his case, though not with total success, many of his fellow experts remaining quite sceptical on the matter. In any event, the author of the sonnets, *Hamlet, King Lear*, became an intelligent businessman.

He had already installed his family in New Place, and in 1611, when he quit acting and writing, except for a few commissions to collaborate on a minor scale or to serve as an occasional "play-doctor", he retired there to spend the rest of his life. The house, built by Sir Hugh Clopton in the fifteenth century, was considered the handsomest in Stratford; it was of brick and timber and had a large garden of mulberry trees and an orchard. (It was destroyed in 1759 by a later owner, irate at the number of visitors who came to see it.) The former poet also acquired two barns in which he stored grain and malt as a hedge against inflated prices. He did a bit of money-lending, speculated in farmland and cottages, and bought a house in the London as another speculation. He displayed remarkable foresight in this, since the site was for many years occupied by *The Times*. It was opposite a theatre in Blackfriars, then a fashionable area.

Apparently he travelled to London on business or pleasure from time to time and met old friends there. Yet he enjoyed the quiet of the small town along the Avon that had been his birthplace.

On 22 March 1616 he made his will, and in another four weeks he was dead. Most likely a bout of illness preceded his passing, which is why he prepared to distribute his estate. But nothing is certain about his final days. A half-century later John Ward, a vicar in Stratford, wrote in his diary a bit of talk he had heard: "Shakespeare, Drayton, and Ben Jonson had a merry meeting and it seems drank too hard, for Shakespeare died of a fever there contracted." But this does not accord with Aubrey's information, which was that "he was not a company-keeper, lived in Shoreditch, wouldn't be debauched, and if invited, write he was in pain".

The last will and testament, appointing three trustees and going through three drafts, has also aroused much curiosity. It indicates that he was comparatively affluent, but the inventory of his assets has vanished. His play-scripts were no longer his property; they belonged to the King's Men. He singled out family members for thoughtful bequests – his sister Joan, her three sons – and provided that Joan could continue to live in one of his houses almost rent-free, for one shilling annually, to the end of her days. To her was to go "all my wearing apparel", which implies that it was valuable and that he dressed well, if not extravagantly. His three nephews got five pounds each; and a grandson, Thomas, was to inherit the house eventually, along with a second Stratford dwelling-place. To his granddaughter, Elizabeth Hall, the only child of Susanna, went all his silver. He earmarked ten pounds for charities in Stratford that served the poor. Some of his friends got small tokens, and to a neighbour's nephew he bestowed a sword. The trustees were to purchase memorial rings to be given to other old acquaintances, among them the former colleagues Richard Burbage, Henry Condell and John Heminges, showing that he still held them in strong affection.

To his daughter Judith, who at thirty-one was about to embark on what her father foresaw was a bad marriage, he left a gold-and-silver bowl and what was equivalent of a trust fund, assuring her, and any children she might have, an income for life, thereby protecting her and the children from her scapegrace husband, Thomas Quiney. This Quiney was four years younger than Judith, so she was following her mother's example in belatedly choosing a mate who was younger than herself. The pair were immediately in trouble; they had not obtained a licence at the proper time. The penalty was excommunication. Besides that, a few months before the wedding Thomas was charged with having had an affair with another young woman and having begotten a girl-child, not yet born when the ceremony with Judith took place. For this sin he later had to do penance in church on three Sundays, appearing there swathed in a stark white sheet – not an auspicious beginning for a new union with the daughter of a leading citizen. He also paid a fine and was required to worship steadfastly in a small chapel that stood amid fields owned by his father-in-law. Judith and Thomas had three sons, the eldest of whom was named Shakespeare Quiney, but all three died in infancy. After her father's death, Judith and Thomas bought a tavern and also undertook a trade as innkeepers. They did not prosper, and lost their lease. In his old age Thomas was reduced to living on twelve pounds a year. Their deaths occurred at nearly the same time, with Judith having reached her seventy-seventh year.

The principal heir was Susanna, who had married more prudently. She had accepted the proposal of John Hall, a physician (or herbalist), eight years her senior and dedicated to his profession, long evincing a serious interest in the "science" of medicine, at least as that pursuit was defined in the sixteenth century. Puritanical, a devout Protestant, he routinely scorned earthly honours, rejecting several awards and honours that came his way, an ascetic stance that led many to praise him. Among his refusals was a knighthood. His patients came to Stratford from everywhere. Growing old, he was seen as "prickly", and was sometimes excluded from Council meetings as too quarrelsome, an obstructionist. In time, his case notes were compiled and published; they were in Latin. It is believed that neither Judith nor Susanna could read or write.

The Halls came into possession of New Place and made it their stately dwelling. Included in the major portion of the estate that accrued to them were Shakespeare's shares in the Blackfriars and Globe if they were still his at the date of his death.

The will displays little concern for Anne, now about sixty. By custom in some localities, a widow was entitled to a lifelong annual income from a third of her husband's estate and also had the right to occupy the family home for her remaining days. If this unwritten tradition applied in

Stratford, she was assured of a comfortable widowhood, and it is suggested that these endowments were not spelled out because they were taken for granted. Otherwise she might have been left dependent on the sense of filial duty and kindness of her daughter Susanna.

Then again, Shakespeare's lack of attention to Anne in the testament might have been a deliberate slight, a final oblique comment on how he looked back at his marriage. One clause that has for centuries aroused much conjecture is his specific bequest to her of their "second-best bed". It might only mean that the "best bed" in the household was reserved for guests, while the "second best" had been shared by husband and wife for years, and she preferred it, and he had obligingly acceded to her wish. Or the "best bed" might be more suitable for a wedded couple, the Halls, and Anne was now destined to sleep alone, for which the "second best" was quite adequate. It has already been established that William Shakespeare was very practical in most matters. He might have been trying to head off the possibility of a family dispute as to who got which of the beds, with Anne demanding the larger and more accustomed one that she really did not need. She lived to 1623, dying at sixty-seven, outlasting her spouse by seven years. Susanna passed on in 1649, and Judith in 1662, neither of them leaving a male heir, nor did the Hall's daughter Elizabeth, so that the Shakespeare line ended, unless Sir William's claim to bastardy and descent from the dramatist is valid. As time proved, what truly survived were the poems and plays, William Shakespeare's uncontested guarantors of immortality.

Burial was in Holy Trinity, a modest yet graceful parish church on the bank of the Avon, always to be linked to his name. Holy Trinity had also witnessed his baptism. Now he was not accorded space in London's Westminster Abbey, where Ben Jonson, his exuberant friend and slightly younger peer, was soon to have a niche and memorial. Shakespeare's reputation as a man of letters had not yet begun to rise; he was by no means the most popular playwright of the day. Jonson, taking the measure of his friend, was shortly moved to write:

> I will not lodge thee by
> Chaucer or Spenser, or bid Beaumont lie
> A little further, to make thee a room:
> Thou art a monument without a tomb
> And art alive still, while thy book doth live
> And we have wits to read and praise to give.

In his will, though, Master Shakespeare had given thought to having his likeness preserved: he set aside money for what turned out to be a sculptured Cotswold limestone half-figure of himself to be placed above his tomb in Holy Trinity. It was to be carved by Gerard Johnson (Garret or Gheerart Jannsen), member of a family of Dutch stonemasons whose chip-strewn yard in London was near to the Globe; how good a resemblance was achieved is now beyond knowledge. The effigy is none too flattering. The face is disproportionately oval, the brow unusually high – the result of balding, which is to be marked in other depictions – the eyes narrow and stonily lifeless. Above the lip, the wide moustache curves up at the ends, and the chin boasts a spade-like beard. The mouth is slightly open, as if – says Professor S. Schoenbaum – he is about to recite a poem of his or declaim a line from a play. To John Dover Wilson – like Schoenbaum, one of the foremost authorities on all aspects of Shakespeare – the face is that of "a self-satisfied pork-butcher". His garment is ornate, indicating that he liked to attire himself opulently. In his right hand he holds a quill pen, in his left hand a sheaf of paper, and both rest on a pillow. Beneath a ledge are Latin inscriptions, unfortunately not error-free: clumsy verses, some also in English, all fulsome with praise, identifying the subject of the statue. (If Anne and her daughters were illiterate, the son-in-law, John Hall, would be the only one able to translate the lines.) The half-figure is journeyman's work; Gheerart Jannsen and his family were stonemasons, and funerary sculpture, unless executed by a Michelangelo or Bernini, is not apt to be skilled or refined. On behalf of the piece, it might be said that at very least the family accepted it as adequate.

When the First Folio, containing the plays, was published in 1623, seven years after Shakespeare's death, the title page bore what was purported to be a correct picture of the author. In it he is younger, perhaps in his forties. On an overleaf, Ben Jonson, who certainly knew the man, attested to the portrait's authenticity. The artist was Martin Droeshout, offspring of a family of Flemish artists. In subsequent editions of the Folio the engraving was subjected to minor alterations, perhaps by other hands, as the copper plate coarsened from over-use. Researchers now state that, as of that time, Droeshout was only a teenager, possibly seventeen, and could not have created the likeness from personal scrutiny but must have referred to a line drawing, perhaps one by Burbage, who was gifted in that medium and frequently did sketches of his colleagues. As is so often the case, this is only a guess. Appraisals of the Droeshout engraving have varied considerably. Dr Rowse is impressed by the "searching look of the eyes understanding everything, what a forehead, what a brain!" Less enthusiastic by far is Professor Schoenbaum who protests that "the engraver has depicted not the brain, only the forehead, described by another observer as that 'horrible hydrocephalus development.' Droeshout's deficiencies are, alas, only too gross. The huge head on the plate of a ruff surmounts a disproportionately small doublet. One eye is lower and larger than the other, the hair does not balance on the sides, light comes from several directions. . . . Still the Folio editors, who undoubtedly knew Shakespeare well, did not reject the likeness." And Schoenbaum detects a "latent irony" in Ben Jonson's endorsement in the concluding sentence of his tribute to his late friend:

> Reader, look
> Not on his picture, but his book.

The poet-playwright is more presentably limned in an oil painting, the Chandos portrait, so called because it long hung in a gallery of the great house of the Duke of Chandos, who chanced to inherit after it passed through several other ownerships. Earlier it is supposed to have been in the possession of the flamboyant Sir William Davenant, from whose bankrupt estate it was purchased at a public auction by the noted player Thomas Betterton, acclaimed on the Restoration stage. He, too, died impoverished. The picture's more or less full provenance has been traced zealously but is not helpful here; suffice that the artist is unknown. On a small, coarse, oval-shaped canvas, a mere eighteen inches by sixteen, the image is of a younger Shakespeare, persuasively in his early forties, with swarthy and Semitic features, which has led to a suggestion that he is dressed and made up to go on stage as Shylock in *The Merchant of Venice*, simultaneously author and actor. (There is speculation that he took that role, having had some experience in money-lending.) Schoenbaum gives this report of the artist's effort to render his subject: "The hair – dark brown verging on black – falls crisply away in a profusion of curls from the massive expanse of forehead. The full, grey-brown eyes, edged with red, gaze thoughtfully out. To one expert the mouth is 'somewhat lubricious, the lips wanton'; another finds the mouth 'combining at once the expression of healthy life with slight melancholy and delicate irony'. The moustache is full, a small beard adorns the pointed chin and ascends in a thick fringe to the hair. In the right ear a gold ring glitters. Could this be Shakespeare of the Midlands?"

The duke proudly and hospitably invited many to view his acquisition, and it has served as a model for many later, more talented craftsmen, among them Sir Godfrey Kneller and Sir Joshua Reynolds. Their appealing conceptions of how Shakespeare must have looked are so widely referred to that today most people simply assume he resembled the Chandos, Kneller and Reynolds versions to some degree. Eventually the Chandos portrait was the first painting of a notable Englishman installed in the National Portrait Gallery (1856), where it remains on display.

In the interim, through the eighteenth and early nineteenth centuries, a string of other images of "William Shakespeare", most of them quickly proved to be mistaken or fraudulent, came to light. The most debatable of these were attributed to Federigo Zuccaro – life-sized, on a wooden panel – Cornelius Jannsen – considerably idealized, quite elegant, and similarly on wood – along with an anonymous study, alluded to as the Felton portrait; and proliferating "copies of copies", as the market for them expanded. A whole literature has grown up, given to arguing about which of them to accept and which to dismiss as true or not true. Much contention greeted the announcement in 1759 by Samuel Felton that he had obtained "a curious portrait of Shakespeare, painted in 1597". He was led to it by a listing in a catalogue of the European Museum (not actually a museum but an art dealer). Others before him had gone to examine it, among them Lord Leicester and Hugh Walpole, but had not ventured to buy it. Felton, more enterprising, sought to know its history. He was told by H.J. Wilson, the shopkeeper, that it had been found in an old house in London's East-chapel, once occupied by an inn, the Boar's Head, that according to legend was frequented by Shakespeare and his theatre friends, and where a fellow actor had dashed off the poet's likeness. (The Boar's Head was not a real tavern but instead a fictional gathering-place for some of Shakespeare's more extroverted characters, such as Falstaff and his convivial companions. What is more, that area of the city had been almost obliterated in the Great Fire of 1666. How had the tavern and the portrait escaped the vast conflagration?) Wilson refused to divulge the picture's route to his hands. Even so, Felton paid five guineas for it, and – through intermediaries – submitted it to a leading authority on Shakespeare for appraisal. That pontifical gentleman, after close investigation, in a subsequent article, declared it to be "the only genuine portrait of Shakespeare" – this set forth by him in italics. The tale that the picture had originated in the "Boar's Head" was thrown out by him as merely an art-dealer's ploy to make a sale, an unworthy ploy not to be encouraged. In return, other experts described the prestigious scholar as gullible. To enforce his claim, the back of the portrait bore the words "Guil. Shakespeare, 1597, R.N.". Some years later, while the back of the painting was being brushed with linseed oil, as a preservative for the wood, the name vanished and beneath it were the initials R.B., immediately suggesting that Richard Burbage was the artist.

The panel is small, eleven by eight inches. The face bears a slight measure of likeness to that etched by Droeshout, but it is a younger man, in his thirties. "The forehead is even larger, incredibly exaggerated, the eyes sorrowful, the nose flat. The moustache, hardly brave, droops at the ends, and the chin is bare." Its advocate George Steevens, the embattled eighteenth-century expert, wrote of it: "There are, indeed, just such marks of a placid and amiable disposition in this resemblance of our Poet, as his admirers would have wished to find." True, there is an element of wishful thinking in identifying this picture, which now hangs in the Folger Library in Washington, DC. X-rays revealed that it had been subjected to considerable over-painting and restoration but is centuries old.

Steevens, for one, disliked the Chandos image, saying of it: "Our author exhibits the complexion of a Jew, or rather that of a chimney-sweeper in the jaundice."

In subsequent decades the limestone effigy in Holy Trinity underwent drastic alterations and other hardships. To counter its natural deterioration it was completely painted over in white, an insult to its surface, however well meant. By 1748 several fingers had broken off and were replaced. In 1861 the white coating was removed, colour was reapplied to the poet's cheeks – they glowed pink – his hair took on an auburn tint, and his doublet became a rich scarlet, which may or may not have been the polychrome hues chosen by Gheerart Jannsen, the sculptor. According to recent reports, the pink of the cheeks has degraded to a deep tan, and the eyes have darkened, so as to seem sightless. A further desecration was perpetrated in 1973 when the half-figure was chipped loose from its stone plinth and slightly damaged; apparently the invaders thought they might find invaluable manuscripts concealed beneath it. But certainly they did not.

Two verbal descriptions from first-hand witnesses yield a half-farthing of guidance as to the physical impression made in encounters with the playwright. As noted earlier, Henry Chettle, apologizing for having published Greene's ugly attack on the "upstart crow" newly come to London theatres, testified that young William Shakespeare was pleasant company, had attractive manners, a good character, and was "handsome and well-shaped". Elsewhere, a spectator at a performance in which the same youthful person undertook a role, remarked that the actor Shakespeare had a "handsome head".

When veneration of Shakespeare began to mount, drawing throngs of awed pilgrims to Holy Trinity, his resting place was moved from the church's north wall. He now sleeps for ever in the railed East Chancel. On his tomb are carved words of a poem – really doggerel – believed to be of his own composition, though, as with so much else about him, that is uncertain:

> Good friend for Jesus sake forbeare,
> To digg the dust enclosed heare:
> Bleste by y man y spares these stones,
> And curst by he y moves my bones.

A modest request from a quiet man respected for his gentleness and reserve. After an interval his wife was laid next to him, a brass plate imparting her name and dates.

(The factual details of this biographical sketch – though not some of the interpretation of them – are culled from many books and essays but principally from studies by Russell Fraser, *The Young Shakespeare*; Stanley Wells, *Shakespeare: A Life in Drama*; S. Schoenbaum, *William Shakespeare: A Collection of Biography and Autobiography*, together with his *Shakespeare's Lives*, an anthology of numerous writings about the poet-playwright over the centuries; Dennis Kay, *Shakespeare: His Life, Work, and Era*; W. Nicholas Knight, *Shakespeare's Hidden Life*; A.L. Rowse, *Shakespeare's Self-Portrait*; Marchetta Clute, *Shakespeare of London*; and Robert Speaight, *Shakespeare: The Man and His Achievement*. These writers, in turn, have absorbed much from distinguished predecessors such as J.K. Adams, E.K. Chambers, Dover Wilson and Leslie Hotson.)

With so little verifiable information about their subject accessible, how have scholars been able to fill volumes, in some instances running to 300 or 400 pages, about his vaguely perceived life? They have done it, after amazing research, delving into yellowed, crumbling documents, and then adding to their discoveries by boldly falling back on rumour and speculations: "he may have", "he could have", "most likely he". From a horn of plenty pour heaps of vividly graphic items that set William Shakespeare in his social context. There are digressions into the histories of people to whom he was related, or with whom he associated, or having to do with the theatre world of the day. It makes interesting reading.

Again the question arises: was this pleasant, ordinary-seeming man the author of *Hamlet, Lear, Midsummer Night's Dream, The Tempest*?

Henry James, who for a time was a drama critic, had doubts. He said that Shakespeare was a fraud. So have many others. One of his early plays, *Titus Andronicus*, a collaborative effort, is exceedingly bad – so bad that his admirers try to deny that he had a hand in writing it. How did he so quickly advance from that to bringing forth a seemingly endless series of masterpieces? If he is not responsible for the plays, who is? The "ghost" of Marlowe in a hideaway on the Continent? Marlowe had talent enough, but the story is too far-fetched to gain credence. There are a dozen reasons why it is implausible, if only because a living Marlowe would not remain anonymous for long – and if safe in Europe would have no need to do so – welcoming obscurity, ignoring the acclaim of which he compulsively dreamed and that would deservedly be his.

Some decades ago a favourite candidate was Sir Francis Bacon, the brilliant statesman – Elizabeth's long-time Chancellor – philosophical essayist and mostly in secret an occasional versifier. No man in England was considered smarter. Fervent Bacon Societies sprang up to promote his cause and sponsor research; a magazine, the *Baconian*, was published. Much of the zeal that inspired these believers has ebbed but is by no means gone. To others it seemed that no one who has read Bacon's leaden-footed poetry could be persuaded for a moment that he fashioned the sonnets or the plays with their exalted lyricism. It should be noted, however, that Shelley for one thought highly of Bacon's poetic gifts.

Several noblemen have had proponents. Perhaps because they were blue-blooded, wealthy, expensively educated, had a liking for the arts and at times dabbled in them. Superior to the plebeian scribbler from Stratford. Leading the list are Edward de Vere, Earl of Oxford, and William Stanley, Earl of Derby; followed by Roger Manners, Earl of Rutland; Sir Walter Raleigh; Robert Devereaux, Earl of Essex; Henry Wriothesley, Earl of Southampton; and Robert Cecil, Earl of Salisbury. Wearing the ermine would seem to endow one with literary skills. In his book *Who Wrote Shakespeare* John Michell examines the claims on behalf of each of them and a half-dozen more and finds that he cannot make a satisfactory case in any instance. But he comes up with a startling suggestion: Sir Francis gathered all the aristocrats in a group, which he personally headed; all contributed plot ideas, characters, passages of dialogue, which Sir Francis turned over to the obliging actor Shakespeare, who winnowed and edited the material, shaping it into his dramas, taking credit for them and never revealing his eminent sources. Classics assembled by committee? A cabal to bring glory to the English stage.

If that is not the answer, what is? Quite often the springs of genius are deeply baffling, attended by mystery.

— 14 —
BARDOLATRY

In fact, Shakespeare's authorship has not been doubted by most readers and theatre-goers, nor the other claimants acknowledged by experts concerned with Elizabethan literature, the scholars who have spent their academic careers scanning what documents there are. The evidence pointing to him, though not as solid as could be desired, is deemed enough to support his credentials. He was recognized as a very talented playwright even in his lifetime; the troupe to which he belonged was regularly at court, and Elizabeth, much amused by his vital portrayal of Falstaff, asked Shakespeare to make the boastful, overweight knight the subject of another script, which he promptly did.

Francis Meres, in 1598, writing about the poets of his day, singled out one above all. "As Plautus and Seneca are accounted the best for comedy and tragedy among the Latins, so Shakespeare among the English is the most excellent in both kinds for the stage." He goes on with quite extraordinary praise, extolling Shakespeare's plays severally, along with his poetry, which he ranks with the best of the Greek and Roman, especially comparing *Venus and Adonis* and *The Rape of Lucrece* to the verses of Ovid. He speaks of Shakespeare as "mellifluous and honey-tongued". This was even before the playwright from Stratford had composed his truly major works.

Soon after Shakespeare's death a collection of thirty-six of his scripts was published; this had a great deal to do with the growth of his reputation. Since plays were hardly esteemed as "literature" it was an honour for a dramatist's work to reach print. But the merit of Shakespeare's contribution to the stage was such that a volume of all his plays was deemed a good risk. That was a shrewd guess. Four editions – "sumptuous" ones – achieved wide reading. In 1623, when he was gone but seven years, appeared this First Folio; in 1632 an expanded Second Folio; and two more had come out by 1685. Besides this, eighteen of the plays had already been printed individually during his lifetime, some – like *Richard III* – as early as 1597; that script went through seven more printings before 1633. Much the same had happened to *Richard II*, *Henry IV*, *Pericles*, *Titus Andronicus*, *Romeo and Juliet*, *The Merry Wives of Windsor* and others. But these were pirated editions, many bearing no author's name and, like the sonnets, issued without his permission or in some instances even his knowledge. This accounts for many inaccuracies and variations in the wordings. Needless to say, from none of these did he or his heirs ever receive any recompense; it is likely that most of the scripts had been purchased outright by his company at the time of the stagings – that was the usual practice. He no longer had ownership.

The company did not want other troupes to have access to the successful plays. But their texts had been peddled to publishers by actors who had used them as prompt books during rehearsals or who had roles in the plays and knew lines and action by heart and were now anxious to make a few pennies more. (Both Henslowe and his son-in-law Alleyn kept detailed records of their financial transactions – wages, cost of costumes and props, setting and the like – but when these ledgers were discovered long after, the excited researchers were puzzled to find no entry of Shakespeare's name anywhere, though as a partner he was entitled to a share of the earnings and payments as a player. The advocates of Bacon and the Earl of Oxford make much of this, saying it indicates that the man from Stratford was not really the source of the plays, nor a shareholder. Admittedly, this absence of any reference to him is strange. Yet he was comparatively well to do when he retired from his activities in the theatre. Again, what is the explanation?)

The same anti-Bard zealots remark that Shakespeare's demise went unmentioned in London; it occasioned notice only in Stratford, marked by his family and local friends, who immediately put up a modest monument to him in the parish church. But news did not spread rapidly in those days; there were few ways for it to do so. Burbage, Condell and Heminges must have learned sooner or later of his passing, because they were included in his will. In any event, it was not from obscurity that this playwright rose rapidly to world fame as is often assumed. His plays in print were best-sellers from the start and even during his lifetime, though other dramatists were more popular.

Ben Jonson, a personal acquaintance and self-reported tavern companion, provided a prologue in verse to the First Folio; it is suspected that he may have had a financial interest in the publishing venture. His tribute was prophetic:

> Soul of the age!
> The applause, delight, the wonder of our stage!
> My Shakespeare rise: I will not lodge thee by
> Chaucer, or Spenser, or bid Beaumont lie
> A little further to make thee a room;
> Thou art a monument without a tomb,
> And art alive still, while thy book doth live,
> And we have wits to read and praise to give.

To which he adds:

> He was not of an age but for all time!

He also addresses his dead friend as "Sweet Swan of Avon". (As quoted earlier, Jonson was also critical of Shakespeare elsewhere, complaining that he was not well educated and his writing over-verbose.)

The First Folio, consisting of five hundred copies at £1 each, was edited with considerable care by two former colleagues, those actors, Henry Condell and John Heminges, to whom he had left bequests to buy rings. They had known him well and seemingly believed him to have written the plays. Surprising numbers of the precious First Folio are extant, 238, and preserved in museums, particularly in the Folger Library in Washington, DC, where there is a trove of 179. At an auction in 2001 a copy in private hands sold for over $40,000. £1 well invested for anyone willing to wait four hundred years. However, there is need to be constantly aware of forgeries, as many skilful ones have been attempted.

The distribution of the Folios proliferated at an incredible rate, along with Shakespeare's reputation. John Milton, in 1630, barely fifteen years after the playwright's death, paid him this impassioned tribute:

> What needs my Shakespeare for his honour'd bones
> The labour of an age in piled stones,
> Or that his hallow'd relics should be hid
> Under a star-ypointing pyramid?
> Dear son of Memory, great heir of Fame,
> What need'st thou such weak witness of thy name?
> Thou in our wonder and astonishment
> Hast built thyself a live-long monument.
> For whilst, to the shame of slow-endeavouring art,
> Thy easy numbers flow, and that each heart
> Hath from the leaves of thy unvalued book
> Those Delphic lines with deep impression took,
> Then thou, our fancy of itself bereaving,
> Dost make us marble with too much conceiving;
> And so sepulchred in such pomp dost lie
> That kings for such a tomb would wish to die.

Much of Shakespeare's increase in stature, even in the succeeding period – the eighteenth century, inimical to Romanticism – is due to the endeavours of the actor and playwright David Garrick, who devoted most of his prized artistry to keeping the poet's work on stage. By this time the plays were freely adapted, shortened, even given new, happier endings. Garrick sought to go against this trend, though he did not always succeed and was sometimes guilty of the same faults. In 1769, chiefly at Garrick's prompting, the anniversary of Shakespeare's birth was celebrated in a huge, 103-day Jubilee at Stratford-upon-Avon. Under the direction of an Italian expert of incendiary effects there was a spectacular fireworks display, abetted by the roar of twelve mortars and thirty cannons emplaced along the river bank. Seven hundred candles shed their flickering light over a temporary wooden amphitheatre erected for the festive occasion, and on to the faces of a thousand spectators – many of them literary celebrities – and an orchestra of a hundred musicians. Garrick and a company of the best actors of the day stood on stage around a statue of the Bard; they were to recite passages from the plays or odes in honour of the dramatist, but unfortunately heavy rain, drumming a tattoo on the roof, drowned them out. Some parts of the jammed amphitheatre collapsed, causing a dangerous situation. The Avon also began to rise, threatening to overflow and flood, and so much of the ceremony was cut short. But the fireworks went off as scheduled, and there was the performance of an Arne oratorio, dancing at night, and a banquet that lasted half a day, at which a turtle weighing 150 pounds was thoroughly consumed. Garrick, as steward of this grand affair, was given a pair of gloves that reputedly had formed part of Shakespeare's attire, and he also received a wand fashioned from a mulberry tree in the orchard formerly belonging to the Bard. It is said that much of the popular image of Shakespeare as a demi-god, a writer without flaw, dates from this much-publicized event.

Similar jubilees in the poet's honour have been held since, the most recent being a year-long celebration in 1964, on what would have been his 400th birthday: to mark it, his plays were revived in many theatres throughout the English-speaking world, special lectures were delivered, commemorative gatherings held, and a torrent of new books on him by professors and Shakespearian authorities came from the ever-busy university presses.

Along with his "love" for the man and dramatist, Ben Jonson – as was just heard – was not above disputing his claims to perfection. With Shakespeare, Fletcher and Webster, one epoch was coming to a close and a new era, the neo-classical, was growing strong. Jonson, as a transitional figure and leader of this literary movement in England, found fault with plays like Shakespeare's not only for prolixity but because they violated the now newly venerated Aristotelian unities and high-handedly mixed the moods of tragedy and comedy, even descending from scenes of elevated passion to low farce in the same work, and doing this repeatedly. Nor did Jonson think a child should be shown growing to manhood or old age within the span of a mere two or three hours, and the frequent shifts of place disturbed him. The classical strictures should be respected. He also wished to avoid the use of fireworks and stage thunder as aids to excitement, and he was weary of the conventions and – by then – clichés of historical chronicles, battle scenes from the Wars of the Roses staged with "rusty swords" and "some few foot-and-half-foot words". Tastes alter, and new preferences prevail in a fresh and ever-changing intellectual climate.

A similar ambivalence of feeling about Shakespeare was shared by John Dryden, the ruling English poet and critic for about three decades during the second half of the seventeenth century. He had much praise for Shakespeare's accomplishments:

> To begin with . . . he was the man who of all modern and perhaps ancient poets had the largest and most comprehensive soul. All the images of nature were still present to him, and he drew them not laboriously but luckily. When he describes anything, you more than see it; you feel it

too. Those who accuse him to have wanted learning, give him the greater commendation: he was naturally learned; he needed not the spectacles of books to read Nature; he looked inwards and found her there.

Furthermore, stated Dryden, Shakespeare is "always great 'when some great occasion is presented to him'". He found Shakespeare remarkably free from affectation. Writers who seek too ambitiously to be elegant:

> forsake the vulgar, when the vulgar is right; but there is a conversation above grossness and below refinement, where propriety resides, and where this poet seems to have gathered his comic dialogue.

Chiefly his approbation is extended to Shakespeare for his "realism". It is truth to nature, especially in characterization, which has kept his work alive from generation to generation:

> Shakespeare is above all writers . . . the poet of nature; the poet that holds up to his readers a faithful mirror of manners and of life. His characters . . . are the genuine progeny of common humanity. . . . His persons act and speak by the influence of those general passions by which all minds are agitated, and the whole system of life is continued in motion.

And again:

> This therefore is the praise of Shakespeare, that his drama is the mirror of life. . . . [The poet], to an unexampled degree, provides the reader with] human sentiments in human language . . . so unusual are the plots and characters that his plays offer us scenes from which a hermit may estimate the transactions of the world, and a confessor predict the progress of the passions.

Therefore in Shakespeare are found no idealized or romanticized "heroes" – no savage, high-sounding Tamburlaines – but men of modest proportions, men just as they are. Though he does summon up fanciful figures – a Caliban, an Ariel, a Puck – he performs the magical feat by which a dramatist of artistic stature "approximates the remote, and familiarizes the wonderful".

Samuel Johnson too pays tributes to Shakespeare's naturalness, which makes even his fantastic and fabulous creatures seem real. "There is a vigilance of observation and accuracy of distinction which books and precepts cannot confer; and from this almost all original and native excellence proceeds." He presents "things as they really exist". In consequence of this faculty, "the ignorant feel his representations to be just, and the learned see that they are complete".

Guided by few rules, for his public was not widely acquainted with them, and there were hardly any critics to enforce them, Shakespeare was free to follow his own temper and talent.

Spontaneity marks his work. He "indulged his natural disposition". At the same time he had the fortunate faculty of using a pure and stable language, so that his writing long remained clear, even to succeeding decades of readers and spectators. Johnson was not too concerned about preserving the sacred three unities, and he defends the poet for having written tragicomedies – that is, for mixing the two genres.

But then Johnson enumerates what are to him the poet's many shortcomings. They are that he seems to have no moral aim in view when he wrote – he is, if anything, amoral. His plots are far from tight. Too many of them are "loosely formed" and "carelessly pursued". He frequently goes off on tangents, forgets to knot together the dangling threads of the story. His endings are often too hurried, too illogical. He is also sometimes inaccurate in matters of historical, geographical and chronological detail. His humour is sometimes gross, and his tragic scenes sometimes tedious, overblown and obscure. His characters tend to be pompous. "His declamations or set speeches are commonly bold and weak." He also indulges in too much punning and word-play; his love for verbal quibbles too often dissipates "terror and pity, as they are rising in the mind". All these charges are substantiated by the formidable Dr Johnson with specific instances taken from the scripts, and he makes each point concretely and forcibly. A distaste for the puerile punning is also voiced by Alexander Pope.

This is the criticism of a different period, however, that demanded that all writers of eminence should have a didactic and moral purpose, that above all valued precision and a nice clarity and decorum in language. But Shakespeare's principal failure was in not helping the spectator distinguish between good and evil. That he was an Elizabethan and lived in a rude period does not excuse him. "He sacrifices virtue to convenience, and is so much more careful to please rather than to instruct. . . . This fault the barbarity of his age cannot extenuate, for it is always a writer's duty to make the world better, and justice is a virtue independent of time and place."

It is chiefly from Johnson's comments, however, that the image of Shakespeare as a "noble savage" is to date, though Milton had expressed much the same estimate of him:

. . . sweetest Shakespear . . .
Warble[d] his native Wood-notes wilde . . .

This was a poet lacking in conscious art, unaware of classical tradition, unlearned, yet a sort of primitive genius.

Voltaire, another giant of the eighteenth century, had much the same view of Shakespeare and spoke of *Hamlet* as "a piece gross and barbarous, that would not be approved by the lowest populace of France or Italy. . . . One would think that this work was the fruit of the imagination of a drunken savage." He objected to all the fighting and eating indulged in by Shakespeare's hearty characters – such behaviour, on the stage, was offensive to a cultivated Frenchman. The

French philosopher and satirist was mindful of Shakespeare's gifts but dismissed him as crude, a "splendid barbarian". It was to take almost two centuries before the French, with their innate respect for order, "reasonableness" and elegance, could embrace Shakespeare with any degree of enthusiasm; and another factor was, and still is in large measure, the lack of good French translations of the plays.

In Germany, in the nineteenth century, Shakespeare's fortunes fared better, especially with the dawn of the new Romantic Age. Gotthold Lessing (1729–81), the distinguished critic and dramatist, took up his cause. In the words of G.G. Gervinus, commenting much later: "The English editors and expositors of Shakespeare's works were yet under the Gallic yoke, when Lessing with one stroke so transformed the age, that we now ridiculed the false sublimity of the French dramas, as they had formerly laughed at the English rudeness." Lessing boldly compared Shakespeare's plays and those of Voltaire himself, and declared the English poet wholly superior. Shakespeare's historical dramas, Lessing stated, "stand to the tragedies of French taste much as a large fresco to a miniature painting, intended to adorn a ring". He concludes, in answer to the question of whether Shakespeare had lacked artistic judgement: "Not every critic is a genius, but every genius is a born critic. He has the proof of all rules within himself." This is in tune with the Romantic exaltation of instinct.

Another German critic, August Wilhelm von Schlegel (1776–1853), was to further Shakespeare's cause on the Continent. He absolved the Bard of nearly all the literary sins instanced by earlier fault-finders. His splendid German translations of the plays soon won them wide popularity. He thought that all the illogicalities or factual errors in the texts were deliberate on Shakespeare's part, purely for artistic effect. For example, Shakespeare had surely been aware that Bohemia had no sea coast, but for the purposes of his fantasy had created one. Similarly, the three unities were not mandatory, and a mixture of moods might be all to the good.

Schlegel was abetted in his advocacy of Shakespeare by Ludwig Tieck (1773–1853) and the even greater German poet and dramatist Friedrich Schiller (1759–1805).

Love of Shakespeare took root in Russia, too; at times there were even more productions of the plays by this English author in German and Russian than in his island homeland.

In actuality, Schlegel argued, Romantic drama offers a different kind of unity:

It delights in indissoluble mixtures; all contrarieties: nature and art, poetry and prose, seriousness and mirth, recollection and anticipation, spirituality and sensuality, terrestrial and celestial, life and death, are by it blended in the most intimate combination.

It embraces at once the whole of the chequered drama of life with all its circumstances; and while it seems only to represent subjects brought accidentally together, it satisfies the unconscious requisitions of fancy, buries us in reflections on the inexpressible signification of the objects which we view blended by order, nearness and distance, light and colour, into one harmonious whole; and thus lends, as it were, a soul to the prospect before us.

All this, asserts Schlegel, is best accomplished by Shakespeare. The "three classical unities", therefore, are of little consequence.

> Why should not the poet be allowed to carry on several, and, for a while, independent streams of human passions and endeavours, down to the moment of their raging junction, if only he can place the spectator on an eminence from whence he may overlook the whole of their course?

Why insist on limiting a play to a supposed few hours' duration, in the hope of making it more credible? Time is a subjective phenomenon. Schlegel insisted: "The intervals of an indifferent activity pass for nothing, and two important moments, though they lie years apart, link themselves immediately to each other." At a play, the spectator's attention "dwells solely on the decisive moments placed before it, by the compression of which the poet gives wings to the lazy course of days and hours". Anyhow, the Greeks themselves had not adhered closely to Aristotle's strictures about time. Shakespeare's detractors should stop cavilling against him on any of these grounds.

With the advent of Romanticism, as expounded and championed by such sponsors as Lessing and Schlegel, Shakespeare's vogue truly began. The English Romantic poet Samuel Taylor Coleridge (1772–1834) also did much to add to his universal fame and almost deify him. It embarrassed Coleridge that German critics had been the first to defend the Elizabethan from this previous belittlement. In lectures and essays, over many decades, Coleridge was Shakespeare's fervent advocate. Though he borrowed much from Schlegel, he protested to his friend Wordsworth that he had held the opinions before the German critic had published them. He assuredly adds many subtle but important insights of his own. Over and over he praises Shakespeare's exquisite judgement; no lack of art or control here – the Bard is infallible.

> Shakespeare knew the human mind and its most minute and intimate working, and he never introduces a word or a thought in vain or out of place: if we do not understand him, it is our own fault or the fault of copyists and typographers; but study, and the possession of some small stock of the knowledge by which he worked, will enable us often to detect and explain his meaning. He never wrote at random, or hit upon points of character and conduct by chance; and the smallest fragment of his mind not infrequently gives a clue to a most perfect, regular and consistent whole.

Goethe contributes his applause, though he prefers Shakespeare as poet to Shakespeare as dramatist. Charles Lamb (1775–1834) thinks Shakespeare too good for the stage – the acting, the scenic investiture, the production are simply inadequate and must be always. To appreciate the plays one must read them. Then one can imagine them as Shakespeare intended them to be, larger, more impressive; no single rendition could possibly do him justice. (Lamb's much-read *Tales from Shakespeare*, written for children's enjoyment, did much to ensure the Bard's future.)

As a result of this noble chorus a cult of a different kind has grown up: it is known as "Bardolatry", a term applied to it by George Bernard Shaw. Shakespeare is not only loved, he is hymned and revered for all sorts of virtues he probably did not have. Most particularly this has occurred in the academies, where scholars devote their hours to his exegesis. Many thousands of books about him moulder on library shelves; hundreds of interpretations of every role have been attempted, and hundreds more are added every year, not only by actors and directors but by psychologists, philosophers, doctoral candidates. The dialogue in his plays has been parsed endlessly, concordances compiled, listing where to find every word he ever used, how often it is repeated and how it was intended. Meanings of every shade have been read into the plots. The plays are constantly analysed from a fresh angle: political, religious, historical, linguistic. The critic's approach may be Freudian, Marxist, Existentialist; Shakespeare's "relevance" is unending, for ever redefined. A quality of the plays is that they are so deep – and ambiguous – that they do lend themselves to this unceasing reinterpretation. In that, they resemble the Sophoclean dramas and novels of Symbolists such as Melville and Kafka. But there is no reason to suppose that Shakespeare ever sought to conceal secret messages in his works. True, his phrasing is not always certain; changes of word-meanings, the presence of many topical allusions that after four hundred years are no longer intelligible to us, altered social values – all these do lend an air of unplumbed profundity to the plays. But most likely far more has been read into them than Shakespeare himself ever dreamed possible. Taken almost literally they are great enough; they do not need to be over-subtilized, over-probed. Nor was the dramatist always the perfect artist. Like Homer, he sometimes nods. Shakespeare was only human – it is his humanity that is one of his chief attractions. He is fallible, like the rest of us.

Whatever the carping of the critics and the shifting intellectual *Zeitgeist*, the public was mostly heedless. The plays were read, and came off the presses in profusion. Many of these were sloppily edited. In 1709 a playwright and scholar, Nicholas Rowe, sought to end this. He prepared a text in which the plays were fitted in a plausible chronology, stage-instructions were inserted, and act divisions indicated – these had been lacking until now. His version became the standard. (By the mid nineteenth century one calculation put the total of different available "collected editions" at 1,250; this did not include the copies of single plays.)

The plays were staged everywhere, even in Californian mining camps by touring troupes. The huge readership, together with the scripts' intrinsic magic, created a universal and perpetual audience. Besides, what actress did not yearn to have a fling at Juliet, Bianca, Lady Macbeth, Cleopatra? What actor could resist the challenge of portraying the eloquent Mark Antony, Shylock, Macbeth, malevolent Iago, tragically confused Othello? Is not interpreting Hamlet the apogee of a player's career?

Teenage schoolchildren, wherever English is spoken, read at least one Shakespeare play and perhaps act in one – *Comedy of Errors, Julius Caesar* are favourites – and university students do the same and many enrol in courses devoted wholly to his works. The count of people familiar with the Bard's works and his supposed physical image adds up to millions. In the late 1890s and early twentieth century companies of English actors performed throughout India, largely for the British colonials; but Indian troupes now produce the plays in their own tongues. Shakespeare is also frequently enacted in Japan and has even been adapted by some African tribal performers. He is put on in South America.

Every season sees at least a half-dozen productions of his works: on- or off-Broadway, in London's West End or at Stratford-upon-Avon, or at the replica of the Globe erected and opened in London on Bankside. Other replicas of the Elizabethan playhouse have been set up in San Diego, California, and in the Folger Library of Washington, DC. Annual Shakespeare festivals are held in Ashland, Oregon, Stratford, Ontario, and in Central Park in New York City. Also in that park is a Shakespeare Garden, in which are all the flowers and plants mentioned in the scripts. A commodious theatre was built in Stratford, Connecticut, to establish a festival there, but the state-subsidized attempt did not prosper.

Paul Levy, of the *Wall Street Journal*, attended the inaugural programmes of the London replica. Some excerpts:

American expatriate actor and director Sam Wanamaker died in 1993, having devoted the last thirty years of his life to his scheme to reconstruct Shakespeare's Globe Theatre – only 150 yards from the site of the first Globe that burned down in 1613. Earlier this month, the theater's première productions opened for previews, and last Thursday the Queen and Prince Philip came here to Bankside, not far away from the site of the new Tate Gallery, to see a special, not too long opening program. It had to be more brief than *Henry V* and *The Winter's Tale*, the plays of the new Globe's opening season, out of respect for the royal backsides, for the seating, I'm sorry to say, is as authentically Elizabethan as the stage.

It will have cost £30 million ($49.2 million) by the time it is finished on September 21, 1999, the 400th anniversary of the first recorded performance at the Globe, and the day when the adjoining year-round Inigo Jones theatre will open. The expense and the toll it took on Mr Wanamaker's health and his family's patience (his daughter Zoë presided over the recent royal event) received their justification when we at the press performance heard the question from the Prologue of *Henry V*: "Can this cockpit hold / the vasty fields of France? Or may we cram / Within this wooden O the very casques / That did afright the air at Agincourt?"

It was spine-tingling. I, for one, had no trouble imagining the French and English armies, though there were hardly ever more than half a dozen people on the stage at once. The Globe really is O-shaped; it has been reconstructed in oak with a thatched roof over the stage; the marbled wooden columns have gilded capitals; the curtainless stage thrusts out into the pit, where the groundlings stand for an admission fee of eight dollars; and most of the rest of us are seated on backless wooden benches.

There are no lighting effects, though there is floodlighting that illuminates the groundlings as well as the actors for the evening performances. The pit is open to the sky, and it rained for both the evening *Winter's Tale* and the matinée of *Henry V*, causing the groundlings to scurry for cover or invest in one of the transparent plastic raincoats being hawked by the vendors who also wandered around the pit with baskets from which they sold small bottles of wine, soda and smoked-salmon bagels. The crowd moved right up to the stage, close enough to see the authentic details of the *Henry V* costumes (natural dyes, some fixed with urine; sleeves and legs tied with bows rather than stitched; even the correct underwear). They booed and hissed the French court, and clapped along with the beat of the period instruments played by the musicians of the Globe.

There were a few concessions in Laurence Olivier's production of *Henry V* (his father was the most famous Harry of them all). The text was shortened by a quarter, and the roles were played, with much doubling up, by a company of fifteen actors (the female parts being taken by men). A program note hints at the changed economic facts: "We believe Burbage's original company would have employed hirelings on a different rate of pay to help cover the forty-five scripted parts. As this is not possible today, some parts have been cut or merged with others."

King Henry V of England was played affectingly by the artistic director of the Globe, Mark Rylance, who shows that his bosses in the Shakespeare Globe Trust (patron HRH Prince Philip) made the right choice. He is the best sort of Shakespearian actor, whose accomplished performance is the result of study as well as instinct. Every gesture is right, because he knows what he is doing and why.

Years ago I was taught Shakespeare by Norman Maclean, one of the Chicago critics. He began his seminar by making us study the conjectured plans and elevations of the Globe, and insisted that every time we read a scene we should mentally stage it there. Many a puzzle disappears when you follow Professor Maclean's injunction; you can imagine how great is the gain in meaning when you actually see a performance staged here.

Scenes flow into each other, as exits and entrances are all made through one of three doors at the back of the stage – the central door or one of the two smaller doors on either side of it. One scene can begin as the other trails off. With no proscenium arch and no curtain, there are no delays for curtains up or down, and no wings for time-consuming lateral entrances or exits – nobody ever has to walk across the width of the stage to get off it. Asides and soliloquies are no longer artificial, but the most natural things an actor can do. With a cock of the head he takes the audience into his confidence – he's standing in the middle of many of them.

Henry V is a politically incorrect play, especially at a time when the new Labour government is thinking of tinkering with the union by holding referendums to allow the Scots and the Welsh to choose a degree of independence from England. This is the most jingoistic of plays, the Bard's paean to Englishness and England, with very non-PC jokes about the French, the

Irish, the Scots and the Welsh. Mr Rylance has the courage to throw himself and his company into the breach and embrace Shakespeare's own view. His performance brought tears to the eyes of the patriotic audience, the majority of whom appeared to me to be Americans. Fearsome as a warrior, Mr Rylance was even better in the comedy scenes where he woos Katherine, the daughter of the French king. That role is charmingly and touchingly played by Toby Cockerell, who is also touching as the Boy.

There are very few props used in *Henry V*, which is just as well. For *The Winter's Tale* suffers from its stage dressing. The production is directed by David Freeman, the Australian founder of Opera Factory, and designed by the artist Tom Phillips, and it is intended to show how the Globe can be used without emphasizing authenticity – or by using some of the conventions of contemporary theatre.

So Mr Phillips, who was responsible for the great Royal Academy Africa exhibition of 1995, has covered the floor of the stage in red earth and made Leontes's throne out of an old rubber tree, while half a used tractor tire marks out a sacred sandpit at the back of the stage, where the actors resort every once in a while for a handful to scatter ritually. Servants and social inferiors use a broom to sweep a path before their betters, and the costumes are drab browns and dull blues.

Mr Phillips explains the Africanization of Shakespeare by saying in the program: "A little knowledge of those parts of the third world where, feeling transported to a timeless place among primal people one suddenly comes across an abandoned tractor, has been a help." Perhaps it was to the performance of Belinda Davison, whose Hermione is warm and moving; to Mark Lewis Jones, whose bluff, butch Leontes is indelicate but thrilling; and to Joy Richardson, whose African medicine-woman Paulina is powerful. But not to Anna-Livia Ryan's shrill Perdita; or to Nicholas Le Prevost's Autolycus, who scores a palpable hit by playing the part as though it has come from some other, slightly better-conceived production.

By now the cast of *The Winter's Tale* will have realized that, despite the all-wood construction and the fact that a mostly roofless building is subject to occasional aircraft noise, it is not necessary to shout. But nothing can be done about the seating on backless benches. The new Globe is not an academic exercise, or a Disney-style tourist trap, but a vibrant contribution to the theater – and a whole chapter in the history of backache.

As of this writing, the plays have served as librettos for sixteen composers for twenty operas; some will be specified on subsequent pages. Very successful musical comedies have been based on *The Comedy of Errors* (*The Boys from Syracuse*) and *The Taming of the Shrew* (*Kiss Me, Kate*). Incidental music for *A Midsummer Night's Dream* by Felix Mendelssohn and *The Tempest* by David Diamond (winner of a Pulitzer Prize) is often heard in orchestral repertories.

Many of the plays have also been made into major films. Choreographers like Shakespeare's plots and characters; they are used for ballets – *The Moor's Pavane*, *Romeo and Juliet*, *Midsummer Night's Dream* – drawing on scores by Purcell, Mendelssohn, Tchaikovsky, Prokofiev and Britten.

— 15 —
STAGING SHAKESPEARE'S PLAYS

In different periods Shakespeare has been performed in a far different manner from that of his own time. It is essential to remember that he wrote for a particular stage, the physical conditions of which – its limitations and resources – inevitably helped to shape his scripts. The simplicity of the productions – a reliance on elaborate costumes, a minimum of scenery – caused the plays to be more verbal, for words were required to evoke the setting; but it also permitted a quick-shifting, episodic, almost cinematic flow of scenes. The platform was thrust into the audience and partially surrounded by spectators, some of whom even shared the stage; this, at the same time, provided both distraction and intimacy.

By the next century the proscenium stage had reached England, and wings and flats replaced the relatively open, bare area on which Hamlet had stalked the turrets and Othello had throttled hapless Desdemona. The plays had begun to retreat from the audience, and intimacy was superseded by "aesthetic distance", to which was added the further distancing of time as Elizabethan drama and its conventions inevitably became more remote. It is to Shakespeare's credit that the vitality of his plays and the vivid reality of the people who lived in them were hardly diminished by these two new obstacles, even though John Evelyn, a seventeenth-century gentleman of neo-classical tastes, relates in his diary how he walked out of a performance of one of the Bard's works because he felt the drama itself too uncouth. His opinion was shared by another candid, lively diarist, his contemporary, the ebullient Samuel Pepys, who recorded: "*Othello* seems a mean thing." (Remarkably, once more, Beaumont and Fletcher were twice as popular as Shakespeare during the Restoration period, and Dryden equated Congreve to the Bard, a critical judgement hard for anyone to comprehend.)

In 1661 a woman – Mrs Colman – assumed the role of Desdemona, and shortly afterwards boy-actors no longer essayed feminine roles, a practice that had already vanished much earlier from Continental stages.

Costuming tended steadily to become more authentic, though not wholly so. In the mid nineteenth century, first in Germany, then in Britain, some characters might wear the ordinary dress of the time, as might Hamlet; but Falstaff, Dromio, Posthumus would have unique, traditional garb long since associated with their roles, and Othello would be attired as a Moor, Mark Antony would be robed in something like a Roman toga. There might be a strange mixture of styles in the same play: the guards in Elizabethan uniforms with breastplates and plumed helmets; while the principals were in garments more nearly appropriate to the period to which they supposedly belonged. Apparently this interfered little with the pleasure the spectators took in the drama, nor in their willing suspension of disbelief. Romeo and Juliet might be in contemporary dress, but the balcony scene was none the less effective. Actresses always wore the latest modes, and usually comedies were given in "modern" attire with no attempt to portray past styles. The effort to do Shakespeare in historically correct costumes began some time after 1760 or 1770. The Bard had a wonderful flair for dazzling display *(vide Pageantry in Shakespearean Theatre,* essays edited by David M. Bergeron) but for a long time this was not exploited by those reviving the plays. (That had to await the arrival of the multi-million-dollar colour film.)

In the nineteenth century "authenticity" finally became highly fashionable, together with a new lavishness of *mise-en-scène*. This trend started in Vienna, about 1810, at the Burgtheater where Josef Schreyvogel was director. A production by Count Brühl of *Henry IV* at the Royal Theatre in Berlin in 1817 also sought to be precisely accurate both in dress and stage-setting. These examples were soon imitated in London by rival actor-managers, Charles Kean and Charles Fechter (the latter French-born, and long a director of the Théâtre de l'Odéon in Paris), in many revivals of Shakespeare with massive stage effects and costumes so right in every detail that they might have been mounted by an antiquarian. Elaborate productions of Shakespeare now became *de rigueur*, and while they may have pleased the eye they had decided disadvantages: they were very costly and clumsy to pack and mount, especially for touring repertory companies, and they necessitated too many intermissions and interruptions while scenery was being shifted for episodic plays like Shakespeare's, so they simply did not work well. Besides they did not create illusion so much as compete with and destroy it. Decades earlier Charles Lamb had complained of being disenchanted by "painted trees and caverns, which we know to be painted" when he attended *The Tempest*; he was similarly displeased by feeble battle scenes portrayed by three or four shouting actors with dull-pointed swords. He protested at the "contemptible machinery by which they mimic the storm" in *King Lear*. The scenery provided by Kean and Fechter and their many competitors was still mostly canvas and paint, and the storm still depended for its effectiveness on the skill of a stage-mechanic.

In the early twentieth century a shift to remove the clutter of props and scenery and strip the stage bare to recapture the fluidity and flexibility of the Elizabethan platform was initiated by designers. They ignored or discarded the proscenium arch. William Poel used only curtains as a background – a shock to those accustomed to the handsome if heavy productions of Henry Irving and his co-star Ellen Terry, who together were the most highly esteemed of Shakespearians. Poel's concepts led to the forming of the Elizabethan Stage Society with Harley Granville-Barker, a notably progressive actor, director and playwright eager to see improvements in staging. Their manifesto read: "The Procrustean methods of a changed theatre deformed the plays and put the art to them to confusion." They advocated a careful restudy of Shakespeare:

Elizabethan stagecraft reflects, variously enough, conditions and obligations (imposed by contemporary circumstances). The forthright telling of the play's story, the freedom with time and place which lets the dramatist rivet each consecutive link in it, the confidences of the soliloquy, the spellbinding rhetoric, the quick alternation of one interest and one group of figures with another – all this is adaptation to environment and the solving of a practical problem. And if the work was rough and ready, well it might be! The Greeks, writing their plays for a yearly festival, could discuss their art and meditate on it. The Elizabethans, one doesn't doubt, discussed furiously, but a hungry public left them little leisure for meditation. If one man could not finish a play quickly enough, two or three more might be called in to help. Old plays must be polished up anew, and if their own authors wouldn't or couldn't do the job, anybody else who was on the spot was expected to oblige. This haste accounts of course, (if nothing else does), for the borrowed and revamped stories. And it is possible that we owe the full version of *Hamlet* to the closing of the Globe during eleven months of plague in London, so that Shakespeare was less rushed than usual. Such a way of working would certainly make more for liveliness than discretion, for impulse and power, but for nothing like perfection of form.

European and American designers and directors began experiments and devised and left as legacies new ways of presenting plays; they were not thinking only of Shakespeare. But their drastic solutions are especially helpful in putting on his often daunting works. Essentially the new styles of production did not differ greatly from those of the Spanish inn-courtyard or the Elizabethan, nor from the Ancient Greek and ascetic Oriental kinds of staging. Or, more recently, how Russian ballet – needing open space – was mounted. Gordon Craig and Adolphe Appia, proceeding quite separately, eliminated most settings – there might be an impressionistic solid object or two – they depended on light-changes and highlighting to create mood and envelope the characters in an hypnotic timelessness on an almost blank stage. The costumes and language alone suggested the time and place. Leopold Jessner chose to have the action occur on flights of steps and intersecting platforms – a throne here, there just a bed suggesting the queen's chamber in which an impassioned Hamlet flails at his errant mother. In Paris, Jean-Louis Barrault did a *Hamlet* on Jessner-like steps; the production was later seen in New York. Overall, in America, Robert Edmond Jones and others absorbed most of these new ideas and eventually altered the look of works on Broadway. No more painted canvas, and a functional minimum of properties. The spectator's imagination was urgently invoked once more. Also available to recapture the freedom of the Elizabethan platform were revolving stages (though not all theatres had such installations) and reversible screens.

All these modifications meant that Shakespeare's works could be produced and toured at far less expense, in an age when the cost of costumes, scenery, large casts, travel and shipping had become forbiddingly high.

Where nineteenth-century directors had emphasized historical accuracy, some of the most progressive in the twentieth century were attracted to a new trend, "updating" and "transporting" the plays. In the mid-1920s a modern-dress *Hamlet* reached Broadway – the unhappy hero in a drab business suit, and those around him also conforming by wearing clothes of that era. Considered an interesting novelty, the production had a short run. But in ensuing decades, the 1930s and 1940s, this visual change became a routine directorial option. *Julius Caesar* was presented by Orson Welles with the surly, shouting mob in black shirts like those of Mussolini's Fascists. The milieu of *Much Ado About Nothing* was transferred to the nineteenth-century American Southwest, and *Twelfth Night* was done with Edwardian appointments and furnishings. Welles also offered a *Macbeth* in New York's Harlem with an all-black cast, the locale of the drama moved to a West Indian island, with a result that was described as "very exciting".

To those critics who objected to this "updating" one reply was that most of the actors who were supposedly Periclean Athenians or Augustan Romans or like Macbeth or Lear belonged to a far earlier period of the island's history were arrayed in Elizabethan attire. At the Globe, too, Shakespeare's audiences also had to make mental allowances and adjustments.

Stranger castings: several actresses have essayed the role of the melancholy prince but none to great applause. Logically, beginning in the nineteenth century, more than one black actor has taken on Othello, the most acclaimed of them the actor-singer Paul Robeson whose deep baritone lent his anguish an even more harrowing resonance. In the later part of the twentieth century a concern for political correctness swept America, especially after the rise of the Civil Rights Movement. Black actors and others belonging to minority groups complained of being shut out of Hollywood and Broadway by a lack of suitable roles. The response in New York was for directors to become "colour-blind" and to ask theatre-goers to be as well. Shakespeare is staged with mixed casts with black courtiers, dukes and earls and even members of the English royal family. Some spectators find this acceptable; to others it is distracting and diminishes plausibility. As yet it has occasioned no public debate.

A long-time tradition after Garrick's day has been that a director could cut Shakespeare's lines, as practically all have done, but never presume to add a word to the texts. In 1938 Maurice Evans, a fine British actor-manager active mostly in the United States, startled Broadway by staging an uncut *Hamlet*, a version no New York play-goer could remember having ever seen. It started at 6.30 and ended at 11.15, six hours with a half-hour break for the actors and audience to catch a snack and a few moments' rest. Brooks Atkinson wrote in the *New York Times*: "It may be a minute or two too long, but no more than that. For the uncut *Hamlet* is a wild and whirling play of exalted sound and tragic grandeur, and Mr Evans acts it as though it were a new text that had not been clapperclawed by generations of actors. Out of the prompt-book he and his director, Margaret Webster, have snatched it and put it on to the modern stage. Apparently Shakespeare was a good writer, because he could put lusty parts into a turbulent play; and this long fiery tale of murder, despair and revenge is the most vivid drama in town today." What is significant is Atkinson's finding that many of the contradictions and ambiguities that perplex interpreters of *Hamlet* vanish when the play is given in full as Shakespeare wrote it.

— 16 —
THE BARD: EARLY PLAYS

A modest disclaimer: it is not possible here to review or analyse all thirty-seven of Shakespeare's extant plays in detail or even adequately. Some must be merely mentioned in passing, with closer attention given to the more important works. But even those cannot be allotted the space they deserve or the scrutiny that might yield a fresh reward. Then again, not every Shakespearian script is a masterpiece; some are trifles that could have been written by anyone. He was not a Midas whose slightest touch turned everything to gold. And the major plays have been probed by thousands of critics and scholars for endless reinterpretation. So it is unlikely that any striking new insights or revelations will be found in these chapters. They are only intended to give a clarifying overview of his incredible career.

The order in which the plays were written is suggested by their sequence in the First Folio and in Rowe's later revision; the accuracy of these is not certain. But no matter; as far as is known he began as a playwright by collaborating on the second and third parts of a massive trilogy, a historical chronicle, about events in the reign of Henry VI. The identity of the author of Part I, a popular but inferior piece of work, cannot even be guessed. In the rush to keep the stage filled, plays were often assembled – slapdash – by several writers as a team, anticipating by four centuries a Hollywood method – to borrow a phrase used earlier, it was "drama by committee".

What was the extent of Shakespeare's participation in Parts II and III (1590–91)? Some experts believe that he wrote only a few scenes in these two sequels; others contend that a substantial part of them is his. The only way to determine that is by a close study of the wording, likened by some researchers to discovering a culprit's fingerprints. Writers tend to have personal vocabularies, to favour particular adjectives, adverbs, idioms and unwittingly – unconsciously – use them repeatedly; it is a factor of what constitutes their style. However, writers in the same era also borrow phrases that are current and oft-spoken by those around them – a *lingua franca* – so that the period to which they belong is more easily recognized than who might be the individual speaker. Much the same casual language is on everybody's lips at the same time. As a test, a compilation of habitual phrases is not a wholly reliable proof of authorship. Characteristic assonance and rhythm in prose and poetry are also components of "style", as are a reliance on sensory details in descriptive passages – some writers are more inclined to dwell on qualities of light or sounds or scents or hues. These are additional clues, but, in sum, there is no DNA that establishes who wrote what.

Was the young, inexperienced William Shakespeare entrusted with revamping most of the earlier *Henry VI*? Was he guided or supervised by more knowledgeable members of the team: were they revising and polishing *his* work? Some scholars hazard that his partners may have been Greene and Marlowe, though no evidence supports this (and again there is no indication that Marlowe and Shakespeare ever met, let alone collaborated). Over 3,000 lines are lifted from two earlier scripts. The scenes usually attributed to Shakespeare are those concerning the valiant English general Talbot and the episode laid in the Temple gardens, a high point of the plays.

Henry VI treats with the bloody, complicated struggles between the houses of Lancaster and York, the so-called Wars of the Roses of which Shakespeare's grandfather will have been a passive witness. It is an often confusing work, crowded with rousing incidents, a large assembly of sharply created characters and not a few passages of excellent verse alternating at places with awkward dialogue and transitions, the whole quite uneven. The plot is too discursive, the overall impression sprawling. The profane Talbot is portrayed with verve, and Jack Cade – whose rebellion is one of the themes of Part II – is an earthy delineation of an agitator. King Henry VI is weak, pathetic, a sovereign whose will falters too often. He wishes only for a simpler destiny. Joan of Arc is looked on harshly, which is only to be expected in a partisan work that sees events solely from an English point of view. (Nor would the spectators have tolerated any other image of her.)

Ascribed to Shakespeare is the much-quoted exchange between Dick and Cade, the rebel leader angry at having signed a legal document:

> DICK: The first thing we do, let's kill all lawyers.
> CADE: Is this not a lamentable thing, that of the skin of an innocent lamb should be made a parchment? That parchment, being scribbled o'er, should undo a man? Some say the bee stings, but I say 'tis the bee's wax, for I did but seal once to a thing, and I was never mine own man since.

This sentiment may have been inspired by Shakespeare's own entanglement in lawsuits.

Coming at a time when the citizens' hearts were swollen with the afflatus that accompanied the victory over the Armada, the trilogy had a public response. Thomas Nash wrote, concerning it: "Brave Talbot should be delighted that in these works he should triumph again and have his bones new-embalmed with the tears of ten thousand spectators (at least several times) who, in the tragedian that represents his person, imagine him fresh bleeding." Professor Alfred Barbage believes that the audience was nearer to 20,000 and further argues that the "groundlings were far more sensitive and intelligent than they have been given credit for being". They differed little from ourselves. How much knowledge of their own national history did Shakespeare's largely illiterate spectators have? Were they partly drawn to the chronicle plays for the information they imparted and to satisfy a deep curiosity? Shakespeare's accounts of past happenings are, of course, quite inaccurate and often biased. Of his own political sympathies and religious beliefs he is unfailingly reticent. Either this was due to a necessary caution – with Elizabeth, a Protestant, on the throne these were perilous times – or he was apolitical as well as an agnostic in matters of faith.

In any event, the young playwright perceived that *Henry VI* had tapped a vein of popular interest, and he quickly followed the trilogy with scripts all his own, constantly improving his art in handling them.

Starting with *Henry VI*, the plays – though fervent in their nationalistic appeal – are by no means uncritical. The horror and waste of civil strife are forever deplored, and battle and dynastic intrigue and ambitions are far from glorified; there are assassinations; son kills father, and father kills son, both for good reasons; a bishop dies of remorse at having conspired in the death of the Duke of Gloucester. York, arms outstretched and bound, succumbs hideously at the stake, defiled by the gibes of his foes and stabbed by Queen Margaret; Mortimer perishes in prison. The anarchy prompted by Cade's revolt is shown as a pervasive, destructive force, yet it is but a corollary of the even more costly regal anarchy that divides England. With all their faults, the three parts of *Henry VI* are exciting theatre in which the action never lags, and the bloodletting between the "Red" and "White" stains the stage unceasingly.

For his first play quite on his own, the neophyte dramatist chose as a protagonist the sinister Plantagenet who grabbed the sceptre as Richard III. As a Duke of Gloucester he is a minor character in *Henry VI*, and Shakespeare's concern with him is already apparent there, but in the trilogy he is simply a villain. In this script he returns with fascinating new dimensions and becomes – though a brazen Machiavellian scoundrel – perversely sympathetic: he is so clever, behaves with so much effrontery, that one grows to like him, however grudgingly, even while perceiving him as loathsome. "A toad" is Queen Margaret's epithet for him in the preceding drama; here he is a limping, crooked-back pretender to the throne – "a bottl'd spider" – and, again, a debt to Marlowe is clear. The scheming Richard is as wicked, nimble and resourceful as Barabas, the unstoppable Jew of Malta. But the verse has an evenness that Marlowe never quite achieved; the plot is far more firm and lucid. However, in comparison to *Henry VI*, the gain in subtlety in Richard's instance is at the expense of the characters around him – they fail to achieve roundness – and the play's scope is far narrower. The lesser figures still tend to explain themselves in words rather than by deeds that let the audience guess at their motivations.

What also makes Richard interesting and wins a reluctant liking for him is his repeated plaint that nature has ill-awarded him: he is seeking to compensate for his physical shortcomings and to avenge himself for the contempt with which he is looked upon by his less intelligent fellow men. A grotesque figure, he is to be forgiven much.

(Outraged by this harsh portrait, a twentieth-century group of interested persons – somewhat anonymous – has formed a worldwide society to obtain redress for the maligned Richard III. Each year, on the anniversary of his death, they publish a kindly obituary notice in the *New York Times* and elsewhere to refute the libel perpetrated by this play.)

Similar gross warpings occur in other of the Folio's scripts, though to a lesser degree: *Richard*

II, *King John*; in fact, they are found in all ten plays dealing with English history. A harried dramaturge, hardly an authority on his country's tumultuous past, he was far too busy for much, if any, obligatory research; his works reflect merely his own not well-founded preconceptions and perhaps even more the prejudices of the chroniclers on whom he drew for effective details. So vivid and ineradicable are his characters that it has been said that as long as Shakespeare is played it is most likely his errant views of events and persons in the Wars of the Roses will be shared by the public at large at the cost of factual truth. If this is unfortunate, it is almost unavoidable. Dramatists reach far more people than do sober authors of textbooks, and no other playwright is as persuasive as this poet whose genius combined so many gifts to convince theatre-goers that what he says is to be believed.

The first lines of Scene One set forth Richard III's case as he sees it. England is at peace; he would wish it otherwise:

> Now is the winter of our discontent
> Made glorious summer by this sun of York . . .

What shall he do, while other men make love?

> But I, – that am not shap'd for sportive tricks,
> Nor made to court an amorous looking-glass;
> I, that am rudely stamp'd, and want love's majesty,
> To strut before a wanton ambling nymph;
> I, that am curtailed of this fair proportion,
> Cheated of feature by dissembling nature,
> Deform'd, unfinished, sent before my time
> Into this breathing world scarce half made up,
> That dogs bark at me as I halt by them; –
> Why, I, in this weak piping time of peace,
> Have no delight to pass away the time,
> Unless to see my shadow in the sun,
> And descant on mine own deformity.
> And therefore, since I cannot prove a lover,
> To entertain these fair well-spoken days,
> I am determined to prove a villain . . .

Arrayed against him, the opposing forces are solid, among them a shrill chorus of bereaved, wailing women, yet he prevails over all, such is his conscienceless wit and deviousness. One of the most brilliant scenes is his wooing of Anne, widow of the Prince of Wales, whom he has slain;

she begins by hating him, and lashing out at him, but finally is seduced by his smooth, serpent's tongue. His is the soft answer that turns away even righteous wrath. This marriage is part of his well-laid plan. He exults:

> Was ever woman in this humour woo'd?
> Was ever woman in this humour won?
> I'll have her, but I will not keep her long.
> What! I, that kill'd her husband, and his father,
> To take her in her heart's extremest hate,
> With curses in her mouth, tears in her eyes,
> The bleeding witness of her hatred by,
> Having God, her conscience, and these bars against me,
> And I no friends to back my suit withal
> But the plain devil and dissembling looks?
> And yet to win her? All the world's to nothing.
> Ha!

He is a consummate actor, like so many of Shakespeare's other heroes: Richard II, Hamlet, Iago. (Is there a touch of the author's innate self-projection in this, to play a role, and to play it well, so that all are deceived? there is great satisfaction in so doing.) Having won Anne, he is soon paying his suit elsewhere, to the Princess Elizabeth, but this time with less luck, though he barely misses his mark.

Step by step he murders nearly everyone who stands in his way – his brother Clarence, his two little nephews, the princes – until the crown is offered to him. He puts on a show of piety, professing a disinclination to take up the heavy, gilded sceptre, and for a short time wins public acceptance, but is finally defeated at Bosworth by the Earl of Richmond, who succeeds him as Henry VII.

Before the end, if Richard does not feel remorse, he does not escape superstitious terror, and, like Macbeth, he is visited by apparitions of his murdered victims. The portrait is sharp, memorable, with the result that the unfortunate actual Richard is today thought of as a monster.

This is an aspect of the play that probably pleased officials at the court and Queen Elizabeth herself. She was the granddaughter of Henry VII, the Tudor who ascended to the throne by overthrowing Richard and the Plantagenets. The Tudor apologists sought every way to discredit the rival line.

The real Richard III had many virtues: he was a good administrator, an outstanding military leader; nor has it ever been proved that he was responsible for the deaths of the two young prisoners in the Tower, though he profited by it.

Richard III, an initial effort, is long but swift-moving, melodramatic, the work of a young man born with a supreme sense of theatre. The limping Richard's desperate cry, "A horse! A horse!

My kingdom for a horse", when he is knocked to the ground and struggles to return to combat at Bosworth, is one of the most quoted lines in all Shakespeare. The sudden advance in craftsmanship is remarkable: the episodes are logically linked, the discursiveness of *Henry VI* is no longer present; throughout the piece it is highly effective on the stage.

The play has been immensely popular; from the very start it caught the public's fancy. As evidence of this, before 1622 six Quartos – that is, individual editions of the script – had been published and snatched up for reading, and it next appeared in the collection that comprised the Folio of 1623. In America, especially, it has been a constant favourite.

An odd interpretation of the script was that of the Romanian *avant-garde* director Andrei Serban at the progressive La Mama Annex in New York. He altered the title to *Richard 3*. To the *New York Times*'s subscribers Neil Genzlinger offered this explanation:

> Mr Serban throws a lot of gimmicks into his production, *Richard 3*, which he created by whittling down *Richard III* and adding some *Henry VI* (its chronological predecessor among Shakespeare's history plays). The most intriguing of these tricks is his triple-casting of the title role.
>
> Three actors step into and out of the character during the evening, sometimes switching off in mid-scene. It makes the production unpredictably entertaining, though frustrating as well.
>
> The Richard-switching sometimes creates great comedy, as when another character starts a conversation with the bald, maniacal Richard (Jason Griffin) only to blink and find that she is now addressing the smooth, long-haired Richard (Chip Persons) or the dark, brooding Richard (Thomas P. Gissendanner). And there are some compelling moments when all three actors join on stage – near the end, for instance, when the ghosts of the king's victims materialize.
>
> But the price of these pluses is that you never get to know Richard or the three young actors completely. Each is good enough to make you wonder what he would do with the entire role; Mr Gissendanner is especially interesting. And, of course, there's not much chance for Richard to transform as he climbs to power when he's been trisected from the start.
>
> That turns *Richard 3* into not much more than a slayfest, which is largely what Shakespeare provided anyway – a lot of killing as Richard strives to get the crown. But Mr Serban and his actors, most of them from the acting program at Columbia University (where Mr Serban teaches), give short shrift to what subtlety Shakespeare did include, and instead go for the gore.
>
> This is literally true: There is so much stage blood that several times in Tuesday's performance actors slipped on it. There are other distractions as well: rope climbing, biohazard suits, intrusive rap music, a video camera.
>
> Still, the show . . . is more invigorating than annoying, made all the more so by the chance

to see emerging actors. Anne Penner as Queen Margaret (sharing the role on alternating nights with Anjali Vashi) had a special intensity.

Be ready to applaud earlier than expected, though. Mr Serban lops off the end of the play, and neither Richard A, B or C gets to offer his kingdom for a horse.

Eight decades earlier Arthur Hopkins produced and directed for Broadway a *Richard III* (1920) in which sixteen abbreviated scenes were transposed from *Henry VI* to afford glimpses of an embittered youthful Richard. The performance lasted until one o'clock in the morning, but the first-night audience stayed long to cheer and demand repeated curtain-calls. (The critics suggested that after the première the play should be shortened.) The mature Richard was John Barrymore, a mercurial member of an eminent American family of actors; it was his first essay of a Shakespearian role, and he was described as "riveting", though working without an adequate supporting cast.

Laurence Olivier sponsored, directed and took the title role in an outstanding film version of *Richard III* (1955), in colour with lavish pageantry.

The Comedy of Errors, written the same year as *Richard III*, is Shakespeare's initial attempt at farce. The plot is largely from Plautus' *The Twins*. Perhaps it is no more than what one critic has called it, "a silly play based on a silly play", an apprentice effort. But it has always worked on the stage, and the young author has made the characters more human.

If he actually did toil as a schoolmaster for some years before undertaking a theatrical career, *The Comedy of Errors* is just such a play, founded on a Latin prototype, as he might be expected to write for his students to enact. His adult audience, partly constituted of those well educated in the classics, would enjoy this rather free adaptation; while for those standing in the pit, the slapstick, the complications provided by the familiar but always effective device of mistaken identity, would surely amuse. It did that, and when revived – as it frequently is, especially in schools – it still does. But no one asserts that *Errors* is anything special, and many another journeyman playwright might have composed it. He does make one ingenious contribution: to the twin masters he adds twin servants, which heightens the comic confusion. And doubtless he learned much from trying his hand at a farce, the most difficult of all dramatic forms, taxing any author.

An Aquila Theater production of the farce in New York in 2002 evoked this perceptive comment from Tom Sellar in the *Village Voice*:

Since Elizabethan times, Western theatre has witnessed four centuries of family dramas – a few great, many dreadful. But still no playwright matches Shakespeare's archetypal but unsentimental visions of that most treacherous human institution. Whether rendering it comic, tragic, or as history, he exposes the absurdities that bring families together and the inevitable rifts that tear them apart (often in the same play). Beyond all the missing lockets, identity thefts, and dynastic struggles, Shakespeare guides us to hauntingly primal scenes of familial suffering uncovering the power battles and psychological anguish of kin desperate to find, recognize, or replace one another.

Most directors treat *Comedy of Errors* as a featherweight variation on this theme, with its double-trouble pairs of unsuspecting twins stirring up spouses, would-be lovers, and impatient creditors. Perhaps because the romp appears light even for a midsummer night's scheme, Aquila Theater sugarblasts their new production with a candy-coated staging. In comic celebration of the quasi-exotic Ephesus setting, director Robert Richmond and costume designer Sarah Hill establish a colorful Tintin-type 1930s adventure look with fezzes and belly dancers aplenty. The seven-actor ensemble plows through a high-octane, shouty, relentlessly goofy gloss on the tale of mistaken identities, Richmond stuffing every scene with pratfalls, lazzi, and double takes.

When smoothly integrated with the language and plot, these antics become enjoyably oversized caricatures of the comedy's growing unrealities and physical confusions – when Antipholus of Ephesus gets arrested, an oddly elaborate scuffle ensues, highlighting a questionable justice system. But more often Aquila's production simply upstages itself, trampling the plot and throwing away lines in a frantic effort to entertain – as in an overworked bit where bookish Luciana compares bust sizes with a courtesan, distracting the audience from some needed plot information.

Richmond's concept tries hard to please but reveals little, missing the opportunity to push the comedy more radically. (Adrian Noble's 1983 RSC production famously exploded the sibling ribaldry through circus clowning.) The boisterous Aquila cast, led by Mark Saturno and Louis Butelli (doubling as both sets of twins), executes the jokiness with exactitude, but the shtick feels imposed and falls back on clichéd business and a lot of mugging.

The larger problem with this approach, however, is that *Comedy* – like many comic strips – has an important darkness at its edges. Egeon's need for ransom money and the Duke's tyrannical threat of execution – desperate dealings launching and driving the plot – get buried by Aquila's fray. Richmond cartoons so much that conflict disappears from most scenes, which rarely feel suspenseful or vertiginous. By starting out so oversized, Aquila's zaniness has nowhere to go; rather than showing how order breaks down and absurdity mounts as the twins intertwine, the production starts silly and stays silly, so the family's realignments at the close only seems superfluous.

As mentioned earlier, *Errors* enjoyed new life in a fresh guise as an immensely successful Broadway musical, *The Boys from Syracuse* (1938), with a fast-paced book by George Abbott – who directed – a score and clever lyrics by Richard Rodgers and Lorenz Hart, "captivating" choreography by George Balanchine (recently arrived in America) and sets by Jo Mielzinner.

The "gorgeous" costumes were designed by Irene Sharaff. Keeping the plotting and characters that had served Shakespeare, this creative team boasted gleefully that they had retained only a single line of his dialogue, replacing the Elizabethan vernacular with that of their own day.

It was briefly revived in 1997 and seen by Donald Lyons for the *Wall Street Journal*:

The Encores series of classic musicals in concert closed the season with *The Boys From Syracuse*, the 1938 version of Shakespeare's *Comedy of Errors*.... Light as a feather and fresh as a spring day, *Boys* looks like an easy achievement – but in fact it stands as an example of complicated nineteen-thirties grace, along with screwball comedies and Astaire musicals. It may not be the best Shakespeare musical, but if Cole Porter's 1948 *Kiss Me Kate* is greater, it is also somehow heavier.

Under director Susan H. Schulman and musical director Rob Fisher, this production breezed through its two-sets-of-unknowing-twins plot as airily as its characters use the two revolving doors on stage. While making for nimble farce, the story served just to hand us on from song to song: a witty farewell to Syracuse ("women don't want divorces there; men are as strong as horses there") sung by the fine Davis Gaines was quickly balanced by romantic tunes like "Falling in Love With Love", sung by the sweet-voiced Rebecca Luker, and "This Can't Be Love" (Mr Gaines and Sarah Uriarte Berry). Later on, the little-known ballad of parting "You Have Cast Your Shadow on the Sea" was counterpointed by a saucy and melodious gold-digging anthem "Sing for Your Supper" terrifically rendered by Ms Luker, Ms Berry and comedienne Debbie Gravitte. The lyrics to this last were wholly inappropriate to their nice-girl characters, even by the loose-book standards of 1938, but who was complaining? Larry Hart did seem to be covertly pushing the envelope in a song about the joys of jail.

John Simon wrote in *New York*:

Few are the musicals with a near-perfect score, and high among these ranks the Rogers-and-Hart adaptation of *The Comedy of Errors* as *The Boys From Syracuse*. Cheek by jowl are ballads and up-tempo numbers, showstoppers and rib-ticklers, each with a musical and verbal seduction that makes them also toe-tappers and breath-takers. This semi-staged revival by Encores! (alas, only for four performances) descends like manna from heaven. Along with a handful of unforgotten golden youngies come also neglected stunners, chief among them a *lied* worthy of Schumann or Hugo Wolf and possibly the most beautiful musical-comedy song ever written, "You Have Cast Your Shadow on the Sea."

To be sure, there is the goofy book, which George Abbott handled skillfully, and which David Ives adapted here with only moderate success. But neither that nor Susan H. Schulman's not-funny-enough staging hurt much: Between the summits of the score, we could use a few valleys to catch our breath in. Kathleen Marshall's choreography, too, is only serviceable; but there is Rob Fisher with his amazing Coffee Club Orchestra, which is to show music what the Vienna Philharmonic is to Mahler and Strauss, John Lee Beatty supplies beguiling bits of scenery, and Peter Kaczorowski's lighting weds the rainbow to the aurora borealis. Last but not least, Toni-Leslie James's "apparel co-ordination" far outclasses the work of her predecessor.

The two supposedly identical sets of twins are easily distinguishable: Davis Gaines is the better-sung Antipholus, Malcolm Gets the better-acted one; Michael McGrath's is the Dromio that's funny, Mario Cantone's the one that isn't. But there is flawless work from Patrick Quinn as the police sergeant, and from three astounding women. As the sisters Adriana and Luciana, Rebecca Luker (tall, willowy, blonde) and Sarah Uriarte Berry (compact, winsome, and brown as a berry) are perfect both as a complementary duo and as unique individuals. Adding the farcical element, the clarion-voiced Debbie Gravitte, as Luce, belts and clowns with equal command. There are other good singers and dancers too numerous to mention, and two delightful spoken bits from Marian Seldes and Tom Aldredge. With a little polishing, this *Syracuse* could transfer to Broadway triumphantly.

Titus Andronicus (1593–4) is as packed with horrors as Thomas Kyd's *The Spanish Tragedy*, which seemingly was its model. By consensus it is Shakespeare's worst play, and his admirers like to think that he did not write it. Some suggest that he merely revised it or touched up minor details in it, but this remains uncertain. A revenge tragedy, it also marks his turning away – as does *The Comedy of Errors* – from English subject-matter to Greek and Roman themes, as he was to do later and far more ably in *Julius Caesar, Coriolanus, Troilus and Cressida* and *A Midsummer Night's Dream*. Titus Andronicus, returning to Rome after triumphs over the Goths, supports Saturninus in his struggle against his brother, Bassianus, to obtain the throne. When Saturninus prevails, Titus expects him to marry his daughter, Lavinia, as had been promised. Instead she elopes with the defeated Bassianus. Enraged at his own son, who has stopped him from overtaking the elopers, Titus slays the young man. Bassianus and Lavinia are found in a forest by a Moor, Aaron, and two sons of Tamora, the captive queen of the Goths, who has married Saturninus, now emperor. Bassianus is slain, Lavinia ravished and left mutilated. Two other sons of Titus are accused of guilt for this deed; Titus, to win a pardon for them, permits his hand to be cut off. His sacrifice is in vain, for the perfidious Moor, Aaron, has the boys executed. This is merely the beginning of a succession of murders, together with instances of adultery coupled with miscegenation. Finally, Titus, to gain revenge, slaughters Tamora's sons. Then, having served their chopped-up bodies in a baked pie to the emperor and empress – Saturninus and Tamora – at a banquet, he kills Tamora and Lavinia, and is in turn killed by Saturninus. (There are echoes here of the legend of Thyestes and Atreus exploited by Aeschylus.) Another son of Titus, Lucius, slays the emperor and himself succeeds to the throne. Of twenty characters, fourteen are dead by the close, all violently. It must be said, however, that while they were alive they were truly vital, though this hardly offsets the many faults of the work.

A few lines have Shakespeare's stamp or seem to be preliminary testings of phrases he will use again, but to better effect. Examples of these are: "Rome is but a wilderness of tigers", which is to be heard, slightly altered, in *The Merchant of Venice*, where Shylock exclaims about "a wilderness of monkeys". "For all the water in the ocean / Can never turn the swan's black legs to white, / Although she lave them hourly in the flood" is a telling image that is much improved upon in *Richard II* and *Macbeth*. An occasional bit of dialogue rings out strongly or finely:

> Some say that ravens foster forlorn children.

And:

> But let her rest in the unrest a while.

And:

> He lives in fame, that died in virtue's cause.

As well as:

> But we worldly men
> Have miserable, mad, mistaking eyes.

This handful of phrases, along with the viability of the cast, argue that Shakespeare did have some part in turning out this horrendous melodrama. One critic, the perceptive Mark Van Doren, has kindly suggested that the play is actually a parody: the still young apprentice Shakespeare was poking fun at the works Kyd fashioned. Though at first blush this seems improbable, Van Doren does submit a good brief for it, pointing out the drama's profusion of anticlimaxes that are wittingly – or unwittingly – comic. In any event, the play was welcomed, won contemporary praise, and appeared in as many as three Quarto editions.

(Quite unexpectedly, in 2001 *Titus Andronicus* provided the material for a film conceived and directed by Julie Taymor, noted for her unusual and uniquely imaginative stagings of out-of-the-ordinary subjects, mingling cut-outs and puppets with live actors, and borrowing elements of Oriental theatre practices. The film won surprisingly little attention.)

Next Shakespeare turned to comedy, composing three works in differing styles seeking to evoke laughter. *The Taming of the Shrew* (1593–4) has as its source Ariosto's *The Counterfeits*, translated into English in 1572 by George Gascoigne as *Supposes*; Ariosto, possibly, had drawn on Plautus and Terence for incidents in the play. Much of the spirit of the *commedia dell'arte* also enters into it. In this robust, hilarious romp, a prosperous gentleman of Padua, Baptista Minola, has two daughters. Gentle Bianca has a trio of suitors, but her father will not let her wed until his elder daughter, Katherina, has found herself a husband. Unfortunately the self-opinionated Katherina has a shrewish disposition and professes to detest men. Bianca's suitors, joined in a common cause, persuade a young man of Verona, Petruchio, to help them out by wooing this termagant. He wishes to marry for money. His desire, as he puts it, is "to wive and thrive . . . To live it wealthily". Baptista gladly gives his consent to the betrothal, but Katherina vows to see Petruchio hanged before she will accept his proposal. The wedding day is set, none the less. To the scandal of everybody, Petruchio arrives in nondescript attire, on a steed with glanders. He misbehaves during the ceremony, curses aloud, and even slaps the priest who is officiating. At the conclusion he refuses to stay for the wedding feast, but instead carries off his ill-tempered bride. At his house the battle between husband and wife goes on, with curses, cuffings and spankings, until Katherina is brought to realize that she has married a masterful man and is happily subdued. To the wonderment of all she becomes a quiet, dutiful mate to her tough but jocular spouse. Bianca chooses herself a husband. The comedy ends with a feast and dance. Strangely, it is a play within a play. In a prologue, a drunken tinker – Christopher Sly – is entertained by a whimsical, amused nobleman who persuades him that he is a lord, and has a troupe of actors perform the noisy farce for him. The play has no epilogue, however; the besotted Christopher Sly is not seen a second time. In modern stagings the irrelevant prologue is usually omitted.

Of all Shakespeare's comedies, *The Taming of the Shrew* – though hardly his best work in a light vein – is perhaps the most often revived. It is a perpetual favourite. The work is most unromantic and has few touches of the poetry and fantasy that infuse his more mature comedies. It is usually played as rough-and-tumble farce. In this piece the female sex is portrayed much less kindly than elsewhere in Shakespeare: a woman who dares to assert her rights or likings is depicted as a nag, a scold, a vixen; a man, on the other hand, no matter how rude or brutal, is merely exercising a natural privilege. A woman is reduced to an item of property, and the high-spirited Katherina is at last brought to submit to this role – and, what is more, she soon seems to like it and lectures other maidens on how to behave in marriage. But that the two principal figures of this farce are witty, arrogant and determined not to yield to each other engenders the fascination and pleasure the spectator takes in watching them, despite themselves, be overtaken by love for each other. They are attracted in a situation that nearly always amuses the onlooker. Shakespeare is to repeat this formula in *Much Ado About Nothing*, in a similar but more subtle, less physical fencing match of stubborn wills between Beatrice and Benedick. The more Petruchio and Katherina rail at and abuse each other, the more certain is the onlooker that they will ultimately surrender to each other completely. The dialogue is vigorous, full of earthy epithets. Both principals, though stock figures, have sharp tongues and gain individuality from that. What the language lacks in poetry is fully made up for by prose that is a display of the vernacular at its most virile and picturesque. As critics have pointed out, the allusions throughout are not to the Italian scene but to the English countryside. Though the setting is Padua, *The Taming of the Shrew* is really an Elizabethan play and might well have been placed in Stratford. Certainly audiences in the Globe could have had little difficulty in identifying with the people, the situation, and the background of this raucous farce.

In 1948 the *Shrew*, too, was converted into a hit Broadway musical, with a "gloriously melodious" score by Cole Porter, perhaps his best ever. The venture was produced by Saint Subber and Lemuel Ayres; the sets designed by Ayres. The choreographer was Hanya Holm. John C. Wilson oversaw the staging. The book, concocted by Bella and Samuel Spewalk, inserted excerpts from *Shrew* into a "play within a play", for an offering on tour, and paralleled the behaviour of Shakespeare's characters with the quarrels backstage between the egotistical leading man and his ex-wife. Needless to say, they are finally reconciled. After its triumph in New York, the musical was seen in London and in an expensive Hollywood film. It was revived on Broadway in 2001 with equal success for a very profitable long stay.

Strangely, the rise of the feminist movement seems not to have affected the enduring attraction of either Shakespeare's comedy or the musical – in the latter case, Cole Porter's irresistible score is doubtless a great help.

Two Gentlemen of Verona (1594–5) is also laid in Italy and is concerned with Valentine and Proteus, who – like Petruchio – are travelling to find themselves wives. Though Proteus has a lady-love in Verona – her name is Julia – he forgets her soon after he follows Valentine to Milan, where he is filled with desire for Silvia, the duke's daughter, who is already beloved by Valentine. But Silvia is betrothed to Thurio, a fatuous courtier. Proteus schemes to remove both Valentine and Thurio as his rivals to Silvia's hand. He causes Valentine to be unjustly banished from the court of Milan. But Silvia, discovering that Proteus has played a trick on his purportedly close "friend", follows Valentine. She is waylaid by polite outlaws in a forest. Their captain is Valentine, now a fugitive. They wish to take her to him, but Proteus rescues her from them; even so, she still refuses to have him. At this he grows angry and threatens to force his attentions on her. Fortunately Valentine is near at hand and saves her from this new peril. After this reversal Proteus repents of his conniving and reveals his true benign personality. Silvia, disguised in male attire, is still pursued by this pair of persistent lovers. When the duke and duchess arrive, affairs are finally cleared up, leading to reconciliations and promises of proper marriages to the tender Julia and wise Silvia.

This is among Shakespeare's slightest works, and possibly written earlier than any of his other comedies – as has been said, their chronology cannot be determined. Its somewhat obscure source is thought to be a Spanish pastoral, *Diana Enamorado* (printed 1542) by Jorge de Montemayor, which could have reached Shakespeare in one of several English translations. If so, the poet borrowed much but added much of his own, including new characters and situations and an enriching lyricism. It has not been performed frequently. Though a trivial piece, *Two Gentlemen of Verona* has some of the romantic tone of the better comedies to come from him. The heroes in all these are well bred, of noble lineage; the heroines are virtuous, witty, gentle, yet resolute; indeed they tend to have independent spirits and intellects. The speech of some is epigrammatic. Declares Lucetta, Julia's waiting-woman: "O, they love least that let men know their love."

The poetry grows more refined. Famous in this play is the song:

> Who is Silvia? What is she,
> That all our swains commend her?
> Holy, fair, and wise is one,
> The heavens such grace did lend her,
> That she might admired be.
>
> Is she kind as she is fair?
> For beauty lives with kindness:
> Love doth to her eyes repair,
> To help him of his blindness;
> And being help'd, inhabits there.
>
> Then to Silvia, let us sing,
> That Silvia is excelling;
> She excels each Mortal thing
> Upon the dull earth dwelling.
> To her let us garlands bring.

Proteus, receiving a letter from his Julia exclaims:

> Sweet love! sweet lines! sweet life!
> Here is her hand, the agent of her heart . . .

But doubtful of the future, adds:

> O, how this Spring of love resembleth
> The uncertain glory of an April day,
> Which now shows all the beauty of the sun,
> And by and by a cloud takes all away!

A minor but persistent stylistic fault is an over-reliance on puns – this grows tedious.

These comedies have a sylvan setting, the green wood. The Forest of Arden that surrounds Stratford-upon-Avon becomes for this series of plays, ostensibly laid in Italy, a magical grove filled with moonlight and grotesqueries, and supernatural creatures. Perhaps the boy William Shakespeare had roamed and day-dreamed in that wood and now in his dramatic fantasies repeopled them with figments half borrowed from his childish fancies. In most of these plays, too, there are clowns – here, it is the servant Launce – who descend from the Roman mimes, the buffoons on the medieval wagons and the cavorting *commedia dell'arte*. Shakespeare was obliged to include them because one segment of his audience demanded crude humour of this sort to balance his rarefied, even airy visions.

The ever-popular device of disguises leading to laughable incidents of mistaken identity is used more and more, and the transvestism of the Elizabethan stage – boys playing the roles of girls – lends itself handily to identity errors and makes them seem plausible. One of Shakespeare's favourite ploys – here as in *As You Like It, Twelfth Night* – is to have a boy actor pretend he is a girl who in turn is pretending "she" is a boy, which provokes much easy confusion. This is a world, too, in which friendship is highly honoured and celebrated, which was an Elizabethan fashion. But it has been justly said that everything Shakespeare does here, he does much better later and somewhere else.

Love's Labour's Lost (1594–5) may have preceded or followed *Two Gentlemen*. Whatever their order of composition, if the dates accorded these and other plays are for the most part correct, Shakespeare was turning out never less than two plays, but sometimes three and even four dramatic works, every year. He wrote quickly. As a young writer he was still under the influence of his predecessors, and here the style he emulates is clearly that of Lyly. Yet one soon discerns that Shakespeare is, in fact, satirizing the over-elegance of Lyly's euphuism. So far as is known, the plot is Shakespeare's own invention, though many of its elements are obvious and familiar. (Meres also mentions a companion play by Shakespeare, *Love's Labour's Won*, but no trace of it can be found, and scholars incline to the idea that it is a reference to *The Taming of the Shrew* under another title.)

The scene of *Love's Labour's Lost* is Navarre, where King Ferdinand has gone into academic retirement for three years. He has resolved to avoid women during that time, and to live austerely and devote his hours to serious study. With him are three young lords, Biron, Longaville and Dumain. Their regimen is to be one meal a day, one full day of fasting each week, and only three hours of sleep at night. All are disturbed when a message comes that the King of France is sending his daughter to them to settle a claim concerning the title to Aquitaine. The princess arrives, attended by three ladies, Rosaline, Maria and Katharine. The resolve of the four gentlemen is considerably diluted by the nearness of these exotic damsels, and they are soon too distracted to continue their bookish pursuits. As might be expected, flirtations and love affairs spring up; there are neat pairings – an attractive lady for each well-spoken lord – and a game which has been likened in its formality to a "minuet". Accompanying the ladies are two French lords, and haunting the court is an eccentric Spaniard, Don de Armado, quaintly extravagant in speech, and his boy-page, Moth, all of whom add their artificial language to this strange pastoral. One of the French courtiers, Boyet, is a mincing poet, described by Biron:

> This fellow pecks up wit as pigeons pease,
> And utters it again when God doth please.
> He is wit's pedler, and retails his wares
> At wakes and wassails, meetings, markets, fairs;
> And we that sell by gross, the Lord doth know,
> Have not the grace to grace it with such show.
> This gallant pins the wenches on his sleeve;
> Had he been Adam, he had tempted Eve.
> 'A can carve too, and lisp; why, this is he
> That kiss'd his hand away in courtesy.
> This is the ape of form, monsieur the mice,
> That, when he plays at tables, chides the dice
> In honourable terms; nay, he can sing
> A mean most meanly; and in ushering
> Mend him who can. The ladies call him sweet;
> The stairs, as he treads on them, kiss his feet.
> This is the flower that smiles on every one,
> To show his teeth as white as whale's bone
> And consciences, that will not die in debt,
> Pay him the due of honey-tongu'd Boyet.

A curate, Nathaniel, and a schoolmaster, Holofernes, also lend themselves to foolery in which fun is poked at their inflated pedantry.

Biron's desperate attempt to escape from love's toils is memorably voiced:

> What! I love! I sue! I seek a wife!
> A woman that is like a German clock,
> Still a-repairing, ever out of frame,
> And never going aright, being a watch,
> But being watched that it may still go right! . . .
> And, among three, to love the worst of all;
> A whitely wanton with a velvet brow,
> With two pitch balls stuck in her face for eyes.

He puts the blame on the little, heartless god of love:

> This wimpled, whining, purblind, wayward boy,
> This senior-junior, giant-dwarf, Dan Cupid;
> Regent of love-rimes, lord of folded arms,
> The anointed sovereign of sighs and groans . . .

King Ferdinand is more patient:

> Come, sir, it wants a twelvemonth and a day,
> And then 'twill end.

But Biron is still not satisfied:

> That's too long for a play.

Even so, no plight is consummated before the ending. Very neatly turned is one lady's rejoinder:

> A jest's prosperity lies in the ear of him that hears it,
> never in the tongue of him that makes it.

In the comedy Biron is the voice of common sense, the one whose feeling for reality keeps the people in it in touch with the world and humanity, however tangentially. He has argued with King Ferdinand against the folly of their studious withdrawal from life. At the conclusion he swears that he will abandon the absurdly pretentious manner of speech to which they have all become addicted:

> Taffeta phrases, silk terms precise,
> Three-piled hyperboles, spruce affectation,
> Figures pedantical; these summer-flies
> Have blown me full of maggot ostentation.
> I do forswear them, and I here protest
> By this white glove – how white the hand, God knows! –
> Henceforth my wooing mind shall be express'd
> In russet yeas and honest kersey noes.

He even pledges himself to give up the practice of using Gallicisms. (One notes that even his renunciation of fine and foreign speech is couched in over-precious language.) The final song of the play, however, does seem to signal the author's own dedication to more homely subjects and a more vigorous and colloquial vocabulary.

Some think *Love's Labour's Lost* is a foolish play; others deem it brilliant, a youthful and impertinent *tour de force* in which Shakespeare not only makes fun of Lyly but shows that he can outdo him at writing in the same coruscating, artificial style, euphuism, and, in the person of Don Armado, Gongorism. Few plays of Shakespeare contain as many oft-quoted lines, and a good part of it is skilfully rhymed, with passages in sonnet form.

The play also has a number of historical allusions to recent events in Navarre – even names of characters – that would have made it topical and hence more interesting to Shakespeare's audience. These references also help scholars to date the work.

As the play unwinds (with only the slightest semblance of a plot) there are a mock play within the play, entertainments and masques, the usual disguises and deceptions, declarations of love inopportunely overheard. King Ferdinand and his three companions lose all further interest in dry scholarship, preferring to pay their court instead. The ladies pretend indifference, however. At last, word is brought of the death of the princess's royal father; she must return home at once. The four men immediately propose marriage, but are told they must wait a year, and during that time demonstrate their worthiness by good deeds. Biron protests at this delay:

> Our wooing doth not end like an old play;
> Jack hath not Jill: these ladies' courtesy
> Might well have made our sport a comedy.

The records show that it was performed at court for Queen Elizabeth's enjoyment in 1597. It remained in the company's repertory for many years. Shakespeare is believed to have revised and augmented it for its presentation before Her Majesty. The supposition is that the immature passages belong to the original, very early version, the better writing to his revision.

The same period in the poet's career yielded a far more notable play, *Romeo and Juliet* (1594–5), assuredly among his best known and best loved. Renaissance Italy – Verona – is the setting; the subject is derived from a story in a collection by Masuccio Salernitano, *Novellino*, issued in Naples in 1476. It was rendered, in turn, into a long, pedestrian poem by Arthur Brooke (1562) that ultimately reached Shakespeare's eye. With his magic, he transfigured the action and characters. Lope de Vega also made use of the tale in his *Catelvines y Monteses*. The plot, though it sounds far-fetched and clumsy in synopsis, moves smoothly on stage and is perhaps too familiar to warrant retelling.

But, briefly, despite the feud between the houses of Montague and Capulet, Romeo, son of the former, meets Juliet, the fourteen-year-old-daughter of the Capulets, at a masked ball, to which he has brazenly ventured in the company of his rakehell friends Mercutio and Benvolio. Infatuated, Romeo later returns, scales the villa's wall, and from below her balcony in a moonlit garden woos the lovely Juliet. When the youth asks the assistance of Friar Laurence to further his suit, the friar assents, hoping an alliance between the two young people might lead to an end of the long feud which, causing street brawls, has disturbed the peace of Verona. He permits the lovers to meet in his cell, where they are secretly wed. Juliet's cousin, the belligerent glowering Tybalt, seeks to pick a public quarrel with Romeo, but he, now hindered by his hidden bond to the Capulets, refuses to be provoked into open fighting. His friend, Mercutio, takes up his cause and in an ensuing duel is killed. As he is obliged to do, Romeo draws his sword and slays Tybalt. For this deed he is banished from the city. A fugitive, he spends a night with Juliet, then flees to Mantua, promising to come back for her.

Juliet's father, unaware that she is married, betroths her to a handsome young nobleman, Paris. In her dilemma she seeks help from Friar Laurence, who gives her a potion. Taking it, she falls into a coma and appears dead. Friar Laurence, meanwhile, has sent a messenger to Mantua to summon Romeo. He is to retrieve her and carry her away with him. Unluckily, before the Friar's message reaches Romeo, he hears news of her death and hastens back to Verona. By now Juliet has been taken to the family tomb by her shocked, bereaved parents. Paris, the would-be bridegroom, goes there to mourn. Romeo, discovering him in the gloomy precincts, runs him through fatally in sword-play. Beholding the "dead" Juliet on her bier, he swallows the poison he has procured and dies. Juliet, reviving, finds Paris slain and the expiring Romeo. She unhesitatingly stabs herself to join him in death. This dreadful consummation leads to a reconciliation of the warring families, both of which are overcome by remorse at the senseless resolution.

This new romantic tragedy of "star-cross'd lovers" exhibits poetry considered to be some of the highest and most beautiful in the world's drama. What educated English-speaking person cannot quote from it? To Charles Lamb, the "love-dialogues" of the hero and heroine are, indeed, as Romeo himself describes them, "silver-sweet sounds"; and A.W. von Schlegel, the German critic, was to write: "All that is most intoxicating in the order of a southern spring – all that is languishing in the song of the nightingale, or voluptuous in the first opening of the rose, all alike breathe forth from this poem. . . . The sweetest and bitterest love and hatred, festive rejoicings and dark forebodings, tender embraces and sepulchral horrors, the fullness of life and self-annihilation, are here all brought close to each other; and yet these contrasts are so blended into a unity of impression, that the echo which the whole leaves behind in the mind resembles a single but endless sight." The remarkable artistic unity of this work has been characterized as "symphonic", everything in it is so well developed and blended.

Caroline Spurgeon and Mark Van Doren, in their perceptive essays on the play, have shown how it is filled throughout with symbolic images of exploding light – and darkness. "The irradiating glory of sunlight and starlight in a dark world . . ." Shakespeare was shaping a new, more flexible style, in which a consistent yet repetitive use of certain colours, images, give each work a unique mood – he will do this again to striking effect in *Macbeth* and *Antony and Cleopatra*. It is a mark, here, of his growing sophistication and craft as a poet. Again, it is like the recurrence of a "motif" in a Wagnerian music-drama.

At their first encounter, the dialogue of the young pair takes the form of a sonnet, the lines assigned to each speaker in almost equal measure.

Glimpsing Juliet at the ball, Romeo exclaims:

> O, she doth teach the torches to burn bright!
> It seems she hangs upon the cheek of night
> As a rich jewel in an Ethiop's ear.

Later she refuses to admit that his being a "Montague" should be a barrier between them. "What's in a name?" she asks herself. "That which we call a rose, by any other name would smell as sweet."

It is her glance, her eyes, that enspell this aroused youth:

> As daylight doth a lamp, her eyes in Heaven
> Would through the airy region stream so bright
> That birds would sing and think it were not night.

Such poetry is literary and hyperbolic, such as is written by a young man; but it befits its subject, the rash and impassioned if immature love of two children, scarcely of an age for wedlock, even at that time.

As Romeo stands in the rose-scented, night-darkened garden beneath the balcony of his beloved, he cries ecstatically:

> But soft, what light through yonder window breaks?
> It is the East, and Juliet is the sun . . .

Optimistic, as he reluctantly tears himself away from beholding the enchanting vision to whom he has addressed himself, he promises that they shall meet once more:

> Parting is such sweet sorrow
> That I shall say goodnight till it be morrow.

Shakespeare prepares deftly for the impending tragedy. Everything has happened in a rush, a mere five days. Friar Laurence, the foolish catalyst, confides his second thoughts:

> These violent delights have violent ends,
> And in their triumphs die, like fire and powder,
> Which as they kiss consume.

Even Juliet has a premonition that their affair is too hasty:

> It is too rash, too unadvis'd, too sudden,
> Too like the lightning, which doth cease to be
> Ere one can say it lightens.

And:

> My only love sprung from my only hate!
> Too early seen unknown, and known too late.

Especially effective are the scenes at the colourful masked ball; in the moonlit garden; along with the street brawl and duel and murder of Mercutio at Tybalt's hand; and the night in Juliet's bedroom, where with her young husband Romeo she awaits the dawn that will part them; and Juliet's fear at taking the potion; and when she expresses her dread at awaking in the tomb surrounded by her mouldering, long-dead kinfolk near the recently slain Tybalt.

The adolescent lovers are convincingly outlined, their impetuosity, their stubbornness, their wrong-headedness, their compulsion and intensity; one does not expect further depth of motivation here than a dawning, overwhelming eroticism and idealization. (They scarcely know each other. Freud was to define such infatuations as an over-valuation of the "love-object".) Strikingly keen characterizations are those of Juliet's old nurse, garrulous and bawdy, who wrongfully and pruriently furthers the girl's illicit affair because of the perverse pleasure she gains from her vicarious participation in it; and the witty, saturnine Mercutio, cynical, reckless, iconoclastic, obscene. (As remarked earlier, some scholars suggest that Mercutio may be Shakespeare's portrait of the turbulent, intellectual Marlowe, who also died from a thrust blade, an event that occurred only a few months before the play's composition, if the script is correctly dated.) For Mercutio is a natural poet, and his exquisite tribute to Queen Mab is quite worthy of Marlowe. Though an Italian cavalier, in the "set speech" – aria – he draws very directly on the English landscape. Excellent, too, is the delineation of Juliet's father, the stern, practical, match-making, plain-speaking parent.

A criticism frequently offered, that the play might well have been a comedy, that only blind chance – the failure of the messenger to reach Romeo, and the inflexibility of the elders – causes the tragedy, is hardly valid: the young lovers are clearly destined to be kept apart by a fierce, bloody vendetta; their parents could scarcely be expected to sanction the union, and the authority of a father in Italian Renaissance times was absolute. The bonding of the youngsters is ineluctably doomed, and not even their unlikely escape from Verona would assure them a happy ending. Friar Laurence's scheme is as harebrained and unrealistic as their own hopeful plans, and they are in more ways than one the victims of their contentious and unwise elders. Have not the young often been made to suffer for that very reason and no other? Wisdom does not always accrue with age.

Apparently Shakespeare had in mind a Juliet of about fourteen. It is hard to imagine a boy actor undertaking the role, though certainly one did. But even more difficult might seem asking spectators to accept a mature woman playing the love-besotted girl. Yet through the centuries a lengthy list of notable actresses have essayed being Juliet, negating plausibility. In the comparatively recent past three took up the challenge on Broadway. Each could justifiably claim in turn to be the "first lady" of the American stage at that time; all three were in mid-life or somewhat past it when they assumed the part. None any longer had a teenage aspect or a girlish figure. All three productions were critically and financially successful and some went on profitable tours.

To be specific: in 1923 John Corbin of the *New York Times* was transported by Jane Cowl's Juliet. "[The] play is a thing of life and beauty in which laughing humor and gaiety of the heart mingle in a multicolored skein with quick human passion tinged with foreboding and despair. . . . Of the Juliet it is difficult to write with moderation. . . . The one thing essential to the part was always there and was denoted by means so simple and true that they defied analysis. There was youth to begin with, touched with the mystery and beauty of great love. The balcony scene was as familiar as a caress, utterly ingenuous and impassioned; yet it positively sang with lyric exaltation." The Romeo was Rollo Peters, no downy-cheeked lad.

Come 1930, Eva La Gallienne, founder and director of the adventurous Civic Repertory Theater, exacted praise from Brooks Atkinson of the *New York Times* for her compelling interpretation. "Her Juliet reveals her as an actress, not merely of intelligence, which she has always been, but of scope and resilience, which she has become this season. Ardently girlish in the balcony scene, her Juliet grows steadily in dignity and command as the tragedy unfolds, and takes the terrors and resolutions of the potion scene with a new fullness of emotion. . . . The clarity and song of her diction are particularly enjoyable." The Romeo: Donald Cameron.

Only a few years later (1934) the third *grande dame* of the American theatre, the tall, statuesque Katharine Cornell, showing herself undeterred by the factor of age and her predecessors' triumphs, enthralled the same admiring Atkinson of the *New York Times*: "Cornell has hung another jewel on the cheek of the theatre's nights. Her *Romeo and Juliet* is on the high plane of modern magnificence. . . . This is an occasion. *Romeo and Juliet* is a drama that drains the playgoer's emotions. In these circumstances all a reviewer can say is 'Bravo!' . . . A performance that will endure in the memory of our theatre. Perhaps vitality is the fundamental motive. Certainly Miss Cornell's Juliet is vital. Looking especially lovely in the flowing vestments of a decorative period, she plays with an all-consuming fervor that takes the big scenes with the little and works them all into a pattern of star-crossed youthfulness. Miss Cornell speaks her lines without declamation, and the singing in them comes more from the heart than the throat.... Here is a complete re-creation – with the suppleness of an actress and the imperious quality of an artist who plays from within. Shakespeare has a vital servant in Miss Cornell." The Romeo was Basil Rathbone. Atkinson's applause was echoed by a chorus of praise from other critics.

A London company, the Old Vic, brought another slant on the play to New York's City Center in 1962 for a limited engagement. The cast had been guided by the internationally active Italian director Franco Zeffirelli, singled out for his opulent – some commentators felt over-lush – productions of plays and operas. Howard Taubman, by then senior drama critic of the *New York Times*, observed that the title roles were given to promising, hardly known young performers, John Stride and Joanna Dunham. As Taubman saw it, Zeffirelli's staging "has the brightness, the gusto, the cutting clarity of the Mediterranean lands". It was an offering of "slashing vigor and wild passion.... Like Shakespeare's cascading images, it is not afraid of abundance, even super-abundance.... This *Romeo* is never content with half-measures." Everything in it was broad, the brawling, the sword-play, the love-making, the humorous episodes. "But Signor Zeffirelli is nearer Shakespeare's mark than the delicate, over-refined *Romeos*. For this tragedy imprisons the overflowing richness and vitality of a young poet who suddenly has full command of the stage.... This production reflects the tumult and fervor of the playwright's unbuttoned epoch."

Subsequently, Zeffirelli put his vision on film, an elaborate work. (Several other cinematic versions of *Romeo and Juliet* preceded it from Hollywood and elsewhere.) Once again he emphasizes the youthfulness of the lovers and their physical beauty, the camera lingering long on close-ups of them quite seductively. His strategy was shrewd, and the film drew large audiences of young people.

If Shakespeare did not originate the story of Romeo and Juliet, his vivid realization of it has given it perpetual life. Its eternal attraction has been attested in a twentieth-century American adaptation with a modern setting, the great musical-comedy success *West Side Story* (1967), with a book by Arthur Laurents and music and lyrics by Leonard Bernstein and Stephen Sondheim, direction and choreography by Jerome Robbins. The updated feud is between a Puerto Rican gang and hostile white neighbours in a Hispanic New York slum. Since the characters are hardly literate and mostly inarticulate, much of the tale is conveyed through dance. The score is eloquent.

The play has also been further embellished in two major operas, Vincenzo Bellini's *The Capulets and Montagues* (1830) and Charles Gounod's *Romeo and Juliet* (1864), and lesser-known works by Nicola Zingarelli (1796), Nicola Vaccai (1823), Riccardo Zavdonai (1922) and Heinrich Sutermeister (1940). Some of these have the title changed to *Juliet and Romeo*, and in Bellini's venture the Romeo is a mezzo-soprano. Needless to say, many of the singers portraying the headstrong pair are neither young nor slender. The story dramatized by Shakespeare has also inspired orchestral suites and tone poems of admirable quality, outstandingly by Tchaikovsky, Berlioz and

Prokofiev, that – along with music by Delius – have been raided by prominent choreographers for ballets using the drama as a scenario. This began in 1811, when the Danish Royal Ballet produced a dance-piece by Vincenzo Galeotti who conceived several works on Shakespearian themes; in this instance he has been followed more recently by Serge Lifar, Birget Bartholin, George Skibine, Frederick Ashton and Kenneth MacMillan. It could be said that every company offering a danced interpretation of the play creates its very own version.

A very different take on the romantic tragedy was that of Joseph Calarco in his off-Broadway *R&J* (1998). He did not originate the project; it was handed over to him by a generous colleague who found herself too busy to complete it. Four good-looking male actors in their mid-twenties constituted the whole cast, filling all the roles, while representing students at an expensive, overly strict private school. They wore grey sweaters and black trousers, their school's uniform. No costume changes ensued, nor did the players resort to cosmetics or use effeminate gestures when portraying Juliet, Lady Capulet, the Nurse. Colarco stressed that his was not an attempt to emulate the style of presentation in Shakespeare's day when boys and graceful youths portrayed girls. There was no scenery and few props.

Calarco wanted to depict Shakespeare's *Romeo and Juliet* as a "dangerous" script, an insight he gained by seeing a film version of Arthur Miller's *The Crucible* in which mass hysteria arises as a result of prolonged sexual frustration. That irrationality is also what afflicts the infatuated young couple in Verona. Italian Renaissance society was also oppressive. This is a summary of what Calarco told a *New York Times* interviewer:

> To contrast the rigidity of the school and nature, he added two passages from *A Midsummer Night's Dream* that evoke the world of magic, as well as several prim pieces of advice from a nineteenth-century etiquette manual entitled *The American Code of Manners*. He cut heavily, especially when Romeo and Juliet profess their love to each other in a way that seemed, as he put it, "melodramatic – over the top."
>
> He also added two Shakespeare sonnets and an erotic wedding scene – "Shakespeare never lets you see the lovers getting married," he said – increasing the play's emotional tension. "This is no sweet romance. The lovers are insane. When Romeo says, 'come death and welcome, for Juliet wills it so,' that's insane."
>
> The ceremony shows two young men getting married and kissing passionately. Is there a homoerotic subtext? "The kiss makes some people uncomfortable, but I never wanted it to be about that," Mr Calarco said. During rehearsals, he urged Mr Shamie (Romeo) and Mr Shore (Juliet) "to make the kisses violent." When they are too gentle, they don't get across the idea of the violence in the play.
>
> For the actors, the female roles have been liberating. Mr Shore said: "It's thrilling to be able to play a woman. It's made me think about a whole new way to approach acting."
>
> Mr Dugan, twenty-five, who transforms himself from a primping, Machiavellian Lady

Capulet to a raucous Mercutio to a one-legged Friar Lawrence, said: "The best thing for me is to turn on a dime from one character to another. Gender is extra."

Mr Calarco agrees: "Inwardly, there's no difference between men and women. Love is a pure thing; it's constant. It's not defined by who's feeling it."

He will have another chance to prove his point. Shooting for the movie version is scheduled to begin in August. Mr Calarco, who is writing the screenplay, is also directing.

Very favourably received by critics and spectators, the play's six-month run exceeded all previous mountings of *Romeo and Juliet* in New York, even that of the much-admired production starring Jane Cowl. (True, the theatre was tiny, seating only seventy-four.)

Donald Lyons, in the *Wall Street Journal*, was among those applauding.

An all-male *Romeo and Juliet*. It sounds like a trendy stunt, something like that *As You Like It* by the English troupe Cheek by Jowl with guys in dresses and beards. But *R&J* at New York's John Houseman turns out to be an electrifying take on the play that "makes it new," as Pound says art must do.

Four students in an all-male prep school are seen performing the right-angled, rigid, unfeeling rites of the school: quick-step marching, recitation of moral dogmas and Latin paradigms (ironically, of *amo, amas* in a loveless world). Then one night the quartet, as in a dream, fall to performing *R&J*. In acting out the most passionate of classics, they cleverly manipulate a strip of shiny red cloth, which becomes, as different times, a banner of femininity, love, blood, death. They perform the play with full-throated, desperate urgency, with no hint of coyness or camp.

Sean Dugan brings a hot-blooded swagger to Mercutio and is later superb as the Friar; these two characters get the tough speeches. Greg Shamie's Romeo is insistent and unironic. The older women (Sean Dugan's Lady Capulet, Danny Gurwin's Nurse) are acted as caricatures of femininity, but Daniel J. Shore plays Juliet with straightforward, bold, unembarrassed simplicity, and it works.

The boys' daytime selves occasionally interrupt their nighttime revels, as when, just after the lovers' first kiss, they suddenly start conjugating that dangerous Latin verb. Adapter-director Joe Calarco sometimes contaminates this play with other Shakespeare texts. By means of robotic movement, Mr Calarco constructs with great ingenuity a contrast between a day world of reason and a night world of emotion; it is this polarity – and not the (quickly forgotten) one-gendered nature of the cast – that *R&J* is really about. It's about the unsettling power of the erotic – not specifically the homoerotic or the heteroerotic but the erotic per se.

The production's viscerally energetic rhythms (half *Stomp*, half ballet) sweep us up afresh into the familiar story. Its whole theatrical vocabulary – that cloth, a symphony of expressive noises, stark lighting, caricatured authority figures – announces at every second that this is artifice, this is play. And yet – such is the nature of theatre – we become all the more eager to enter into the tale. At dawn, the boys wake up to begin again their robotic school routines; but Mr Shamie's Romeo, reciting the epilogue to *A Midsummer Night's Dream*, finds it hard to leave the dream. He has perhaps become an artist. *R&J* is a gem, the most inventive reimagining of a classic in years.

For a fee, visitors to Verona are led by guides (and guide books) to an ancient dwelling that they assert was once occupied by the young Juliet and her intransigent, blue-blooded family. It is rare indeed that a character in a play becomes so alive to spectators and readers that they believe she once actually lived in this noble house.

A Midsummer Night's Dream (1595–6) is another script in a lyrical vein, once again tempered and brought to earth – like *Romeo and Juliet* – with dollops of farce for the groundlings. Here the humour is a neat, sly invention, a satire on amateur players. In all the world's literature there is perhaps not another stage-piece equal to that which Shakespeare accomplished here. Reality is superimposed on fantasy, and the two are blended and flickeringly yet genuinely made integral. It is believed that the play was written for a court revel, perhaps to celebrate the wedding of a fortunate high-born pair or an event at which the royal Elizabeth was a guest.

In a forest glade, Oberon, King of the Fairies, is bickering with Titania, his queen, over a changeling Indian boy whose presence in her train of attendants she obdurately desires. Oberon sends his sprite, Puck (or Robin Goodfellow), to obtain a magic flower whose juice, dropped on the eyelids of a sleeper, causes that person to become infatuated with the first living being glimpsed upon awakening.

Into the fairy-haunted woodland come a quartet of young Athenians: Hermia, temporarily banished from home by the duke, Theseus, because she opposes her father's edict to marry Demetrius – she loves Lysander instead; Helena, unhappy because she wants Demetrius, who professes indifference regarding her; and, of course, the two young men who wish to marry the banished Hermia.

Puck, ordered to place the necromantic juice on Demetrius' eyelids as well as on Titania's, by mistake puts it on Lysander's, so that, when the young man arouses himself, he beholds nearby Helena and redirects his affections to her. To repair his error, Puck applies drops of the potion to Demetrius, who also becomes enamoured of the startled Helena, while the unhappy and confused Hermia is abandoned by both of her former suitors.

A group of dull-witted, simple journeymen, seeking a quiet, secluded spot to rehearse an "interlude", venture into the magical woodland. Their buffoonish number consists of Bottom, the Weaver; Flute, the Bellows-mender; Snout, the Tinker; and Starveling, the Tailor. Their preparations for their proposed play allow Shakespeare to poke rich fun at would-be thespians, who were common enough in his time. (These rustics, though supposedly Athenian, are patently Elizabethan.)

Oberon orders Puck to fasten the head of an ass on Bottom's shoulders, and this ridiculous creature is the first breathing thing seen by Titania when her eyes open. She instantly falls in love with him and, forgetting her regal dignity, pursues him, much to her husband's amusement. Obsessed by her love for Bottom, Titania readily agrees to Oberon's demand for the changeling boy. Her husband then restores her to her senses, and also has Puck stop Demetrius and Lysander, who are about to fight a duel over Helena. The love-tangle between the four young Athenians is easily straightened out. Lysander and Hermia are brought together again and join Demetrius and Helena in double nuptials; while Oberon and his queen, Titania, are also joyfully reunited. Theseus also takes a bride, Hippolyta, the royal Amazon.

Bottom and his group of graceless actors stage a performance of *Pyramis and Thisbe* in honour of these unions, a parody of the ill-fashioned "interludes" then so popular. The *Dream* is enhanced throughout by songs and dances in which the cohorts of Oberon and Titania – such fairies as Peablossom, Cobweb, Moth and Mustardseed – participate.

Perhaps there are touches of Lyly, Lodge and Greene in this remarkable pastoral, with its classical setting and moon-tipped, dew-drenched woodland – there are clear echoes of Ovid in it too– but it is still one of the most original dramatic works ever conceived. Winning the spectator's willing consent, Shakespeare combines Athens, England and fairyland into a single milieu for his sprightly, ensorcelling tale. He does this largely by the marvel of his language, which has enriched English drama with such apt and sparkling lines as Puck's, "Lord, what fools these mortals be!" and such well-phrased promises as his, "I'll put a girdle around the earth / In forty minutes." Over all rules the "watery moon", whose rays gild night-spirits of mischief and fun. And the theme of the work, if it has any, is the inconstancy of love and lovers, whose irrationality seems to be the doing of interfering supernatural forces.

In this play, indeed, as Theseus describes it, the poet "Gives to airy nothing / A local habitation and a name". In light-winged words, a fairy sings:

> Over hill, over dale,
> Thorough bush, thorough brier,
> Over park, over pale,
> Thorough flood, thorough fire;
> I do wander every where,
> Swifter than the moon's sphere;
> And I serve the Fairy Queen,
> To dew her orbs upon the green.
> The cowslips tall her pensioners be:
> In their gold coats spots you see;

> Those be rubies, fairy favours,
> In those freckles live their savours.
> I must go seek some dew-drops here,
> And hang a pearl in every cowslip's ear.

This whole elfin world is ineffably lovely, and – when one remembers the limitations of the Globe's stage – pervasively evocative, so that one almost scents what one is asked to visualize.

Oberon, laying his plot, tells Puck:

> I know a bank whereon the wild thyme blows;
> Where ox-lips and the nodding violet grows;
> Quite over-canopied with lush woodbine,
> With sweet musk roses, and with eglantine:
> There sleeps Titania sometime of the night,
> Lulled in these flowers with dances and delight;
> And there the snake throws her enamelled skin,
> Weed wide enough to wrap a fairy in . . .

Shakespeare, here, has hit the top of his poetic bent. It is witchery inimitable, this amusing fable of wanderers in a vaporous wood where all are "ill-met by moon-light". This gives an insight into the well-stocked imagination of a poet not only in love with words, but with the natural world in all its green, blossoming aspects, and who has an encompassing knowledge of flowers and all growing things, every butterfly and every flying, singing bird. That was the world to which, in his later years, he was to retire – to his garden and orchard and the nearby surrounding forest.

The *Dream* is frequently performed outdoors, in London's and New York's parks. Famed for his production of the play on the Continent was Max Reinhardt who later put his conception on film in Hollywood (1934) after he left Nazi Germany, and also subsequently staged it at the Hollywood Bowl.

The avant-garde director, Peter Brook, was hailed for updating the staging, in effect reducing it to a gymnastic exercise during which well-spoken young actors recited the lines with remarkable clarity. But many spectators felt that the offering was deprived of its other-worldly allure.

Felix Mendelssohn sought to capture the essence of this pastoral in an often-played tone-poem, and Benjamin Britten seized on the Bard's poetic text to create a well-received opera. But *A Midsummer Night's Dream* scarcely needs music to aid in exerting its spell. The English National Opera also staged a modernized version of the piece with the singers in twentieth-century attire. Again, many in the audience found that this was very much at the cost of its unique aura.

With this fantasy behind him, Shakespeare returned to chronicling the deeds of England's kings. As has been remarked, *Richard II* (1595–6) is often compared to Marlowe's *Edward II*, which it certainly resembles. Both plays are about weak, unpopular kings who, for their faults, are forced to abdicate; both monarchs yield their crowns with tearful reluctance, and both are imprisoned and, in dank cells, horribly murdered by agents of the usurpers. Sensitive to his own feelings, insensitive to those of others, proud, intelligent, foolish, egotistical, Shakespeare's Richard is a far more three-dimensional portrait than Marlowe's effeminate Edward. His Richard is a histrionic personality; he is a natural actor, alert to and often delighted by the sound of his own voice, his clever choice of words. This Richard does not merely suffer, he dramatizes his suffering; he poses; he is narcissistic. In adversity his self-pity is endless. He is unstable, his moods quick-changing. He is perceptive, cautious, reckless. He baffles by his inconsistency. He meets every occasion with an apt phase. All these qualities add up to his being real and fascinating.

The play deals again with what is required of "kingship", the need for strong rulers, firm leadership, so that tranquillity and order can prevail. This is a demand that Richard does not meet. The play is filled with metaphors in which court and state are likened to a theatre, the leading personages having roles to enact before the common people.

Bolingbroke, whose father is the rich and powerful John of Gaunt, Duke of Lancaster, charges his rival, Thomas Mowbray, Duke of Norfolk, with treasonable designs. The resolution of the quarrel is an impending trial by combat, which King Richard first commands, then prevents by ordering both antagonists banished. He is pleased to be rid of his cousin, Bolingbroke, whom he fears; he recognizes the strong-minded Lancastrian as ambitious, and observes his growing popularity among the populace.

To put down a rebellion in Ireland, the extravagant Richard must raise new funds. Blind John of Gaunt dies – providentially it seems – and Richard confiscates his property. He equips an army and embarks for the restless island. Bolingbroke, outraged at losing his lawful inheritance, seizes this occasion to return at the head of an army which is partly made up of other malcontent lords, who also dislike the aesthetic, spendthrift king, deeming him an inadequate ruler.

When Richard returns from his expedition and is confronted by this insurgent host, his will collapses. He agrees, at a meeting of Parliament, to yield his crown. Later, held captive in a dungeon at Pomfret, Richard is slain. The new king disowns the murder and vows to make a pilgrimage to the Holy Land to atone for it.

In arranging this historical material Shakespeare was again closely guided by Holinshed's *Chronicles of England, Scotland and Ireland*, but eight other sources have been discerned by scholars, attesting to his thorough research – and theirs. He even went back to a book on which Holinshed had drawn, a work by Edward Hall, and delved into accounts by Sir Johan Froissart and Samuel Daniel, and possibly some French texts. So this time he paid his dues to scholarship. A play by an anonymous author on Richard II had recently been translated into English and may have come into his hands and influenced him, but it is uncertain whether that script was earlier or later than his own. Borrowing here and there, he fashioned the characters and events anew, to suit his purpose, and shifted time and place. He faced the difficult task of making both of the foes, Richard II and Henry IV, sympathetic figures – Queen Elizabeth, a descendant of Henry IV, would not have welcomed an unflattering portrait of her royal ancestor. It is remarkable with what tact Shakespeare succeeds in doing this, a delicate feat of balance. The audience is led to detest Richard for his faults, yet must feel pity for him, for he has many superior traits, including his rare eloquence.

This is not to deny that the verse in the play is not sometimes stiff – Shakespeare is still employing end-stops to his lines and there is an over-frequent and not always felicitous use of rhymes in quatrains and couplets – but Richard's speeches for the most part are exquisitely cadenced and musical.

Most admired is the blind, dying John of Gaunt's superb apostrophe to England, the blessed, proud, green and craggy domain where he has long held lands, tenants and vassal in fief:

> This royal throne of kings, this sceptr'd isle,
> This earth of majesty, this seat of Mars,
> This other Eden, demi-Paradise,
> This fortress built by Nature for herself
> Against infection and the hand of war;
> This happy breed of men, this little world;
> This precious stone set in the silver sea,
> Which serves it in the office of a wall,
> Or as a moat defensive to a house,
> Against the envy of less happier lands;
> This blessed plot, this earth, this realm, this England . . .

It is almost matched by Richard's tribute to his home soil when, landing in Wales, he gratefully kneels to touch his native earth:

> I weep for joy
> To stand upon my kingdom once again.
> Dear earth, I do salute thee with my hand,
> Though rebels wound thee with their horses' hoofs:
> As a long-parted mother with her child
> Plays fondly with her tears and smiles in meeting,
> So, weeping-smiling, greet I thee, my gentle earth . . .

He is especially articulate on the subject of his regal nature, his mystical and still somewhat medieval self-flattering belief in the divine right by which he rules. When he hears that Bolingbroke has amassed a large army against him, he exclaims – apparently fully convinced that what he says is true:

> For every man that Bolingbroke hath press'd
> To lift shrewd steel against our golden crown,
> God for his Richard hath in heavenly pay
> A glorious angel. Then, if angels fight,
> Weak men must fall; for heaven still guards the right.

And again:

> I had forgot myself: am I not King?
> Awake, though coward majesty! Thou sleep'st.
> Is not the King's name twenty thousand names?
> Arm, arm, my name! a puny subject strikes
> At thy great glory.

And yet again:

> Not all the water in the rough rude sea
> Can wash the balm from an anointed King;
> The breath of worldly men cannot depose
> The deputy elected by the Lord . . .

But, as always with him, he lives too much in words, in a public performance:

> We were not born to sue, but to command.

(Though Shakespeare seems in his chronicle plays to lend his emotional support to a similar belief in the divinity of kings and decries sedition – he has been described as probably a conservative in politics – he apparently does not feel that a king divinely appointed should still be allowed to rule if he cannot really govern. A feckless king must give way to an efficient one, though the latter's blood claim to the throne might not be as well founded. It must be remembered, again, that in Queen Elizabeth's day this question of legitimacy was very important, an urgent issue; hence the subject would be of major interest to Shakespeare's audience.)

Later, foreseeing his fate, he intones:

> For God's sake, let us sit upon the ground,
> And tell sad stories of the death of kings:
> How some have been deposed, some slain in war,
> Some haunted by the ghosts they have depos'd,
> Some poisoned by their wives, some sleeping killed,
> All murdered. For within the hollow crown
> That rounds the mortal temples of a king
> Keeps Death his court; and there the antic sits,
> Scoffing his state, and grinning at his pomp;
> Allowing him a breath, a little scene,
> To monarchize, be feared, and kill with looks;
> Infusing him with self and vain conceit,
> As if this flesh, which walls about our life,
> Were brass impregnable; and humoured thus,
> Comes at the last, and with a little pin
> Bores through his castle-wall, and – farewell king!

He confesses:

> For you have but mistook me all this while.
> I live with bread like you, feel want,
> Taste grief, need friends; subjected thus,
> How can you say to me, I am a King?

Soon he pronounces that he is ready to exchange everything he has ever had:

> . . . my large kingdom for a little grave,
> A little little grave, an obscure grave.

He even enjoys his sorrow, if he can have an audience to behold him enacting the outsized role of one flagrantly humiliated and disgraced. In his laments over his lot while a prisoner, there is some self-recognition of his flaws of character, the error of his ways:

> I wasted time, and now doth Time waste me . . .

But in his suffering he also boldly compares himself to Christ, and Christ's foe to Pilate. The scene of the deposition and the one in which Richard tries to defend himself from his murderers are stunning theatre.

By contrast to Richard's musical speeches, Bolingbroke's lines are short and rugged, and characterize him well; he belongs to a new age, a Renaissance man of action, effective and largely sincere. But all the characters talk well. Even the Duke of Norfolk, when sentenced to lifetime exile, protests:

> The language I have learn'd these forty years,
> My native English, now I must forgo;
> And now my tongue's use is to me no more
> Than an unstring'd viol or a harp,
> Or like a cunning instrument cas'd up,
> Or, being open, put into his hands
> That knows no touch to tune the harmony.

Informed of Richard's death, Bolingbroke voices regret. Facing the killer, who has come to claim his reward, the new king utters a harsh rebuke, partly to himself:

> They love not poison that do poison need,
> Nor do I thee. Though I did wish him dead,
> I hate the murderer, love him murdered.

Richard II lends itself to visual magnificence, for this was a monarch famed for the costly brilliance of his heraldic trappings, his slightly effeminate or effete attire and the splendour of the court that surrounded him.

The play disappoints in not always providing the full conflict the story might have offered: Richard surrenders too quickly, bids farewell to his queen too casually; his lamentations go on too long. As has been pointed out by critics, ceremonial phrases and deeds (again in a medieval spirit) often supplant true, naked drama, but, whether read or on stage, for many this historical play is a moving, not easily forgotten experience.

A mellifluous and poignant Richard was that of Maurice Evans in a Broadway revival in 1937. Brooks Atkinson recorded in the *New York Times*:

> Maurice Evans now deserves a sort of reverence for his triumphant performance.... Out of one of the less familiar plays of Shakespeare he has wrought a glorious piece of characterization; his "skipping king" shines through the majesty of inspired acting.... Mr Evans and his colleagues have plucked the heart out of the drama in one of the most thorough, illuminating and vivid productions of Shakespeare we have had in recent memory. When the final curtain descended last evening everyone realized that a play had been honestly presented by one of the finest actors of our time. And all this despite the fact that Richard II is no dominating

hero. According to Shakespeare, he had the air of a king, but he was a popinjay with no mind for authority and no will to rule a state. . . . But now that Mr Evans and his company have pitched their lustrous talents into it, a theatre-goer may well revise his ideas and renovate his mind. Although Richard was no hero, now we know the anguish of his soul is heroic and has the power to make our hearts stand still. Taking Shakespeare at his word, Mr Evans has translated the character of Richard into devastating tragedy. The whole doleful story of weakness in a king is boldly told. Complacent and trivial when his throne is secure, callous and contemptuous in the presence of his betters, he gives way like a sheet of tissue paper at the first opposition. But it is the distinction of Richard that his mind grows keener with destruction before his enemies; although he lacks the power to rule, he has the courage to be his own confessor, and be most kingly when the crown has been snatched from his head. The characterization is strange and progressive. Mr Evans has met it point by point with infinite subtlety and burning emotion. There is not an unstudied corner in any part of this glowing portrait.

There was equal praise for the fine director, Margaret Webster. Before judging a script, it is vital to see what greatly talented stage people can make of it.

Richard II is also the prelude of a tetralogy concerning the fortunes of Henry IV and Henry V. Whether Shakespeare had such an epic canvas in mind when he wrote *Richard II* is not clear, but critics like to think that he did and eagerly perceive architectonic patterns throughout it. Perhaps the design is really there, perhaps not. It could be, however, that Shakespeare developed it retroactively, himself seeing the overarching themes that appear in his works, then wisely developing them further. Any comparison of *Richard II* to his earlier *Richard III* shows his artistic growth to this date, though the earlier play – simple, crowded with action, dominated by a crafty megalomaniac schemer – is more often given and has been better liked by the public. In the subsequent plays of the Henriad (as the tetralogy is sometimes called), the characters often allude ruefully to Richard II with fond respect, which suggests that Shakespeare himself felt no little affection for the "minor poet" he had so ably crafted.

An off-Broadway venture prompted this essay by Donald Lyons in the *Wall Street Journal* (1998):

How to depose an unworthy ruler is the timeless problem faced in Shakespeare's *Richard II*, which is now being done, in tandem with *Richard III*, by the Theater for a New Audience at St Clement's Church.

The problem in staging a meditative, reflective, issue-centered work like *Richard II* is to animate its ideas. Director Ron Daniels takes advantage of the churchly setting to stress the sacred nature of the anointed king. The stage is dominated by a great rose window and filled with candles, altar boys, and bishops surrounding a hieratically posing monarch. It's overdone, this religiosity; the Plantagenets were not, after all, priest-kings, like the Pharaohs. But the visual excess is balanced by some solid playing: Steven Skybell's Richard avoids the traditional mode of languorous mooning; temperamentally unfit for rule and seemingly happier when deposed and free to indulge in eloquent self-pity.

The pairing of the two plays, written in different styles and dealing with different eras, makes little artistic or historical sense, but *Richard III* also offers a terrific title performance. Christopher McCann (he made a hesitant, torn usurper in *Richard II*) gives us a grimly merry demon rejoicing in his deformities (hunched back, withered arm, sideways crab-scuttle) and delighting to share his plans and opinions with the audience. Deriving from the character of Vice in morality plays, this Richard is less a person than a humor. But Mr McCann goes deeper, too, and reveals the perverse child inside the terrorist. In a stirring finale, Richard and Richmond (Henry VII-to-be) share the stage on the night before the battle, during which Richard is stabbed in the back and upended by the leg by Richmond. For the rest, the play's endless curses, lamentations and expositions are staged with commendable, if unexciting, simplicity. We leave the two plays thinking about the knotty problems of power and playful possibilities of language.

Where to place *King John* (1596–7) is a puzzle. Some think it was written before *Richard II*, some afterwards. It is not part of the Henriad but a wholly independent work. And perhaps *The Merchant of Venice* preceded it. Howbeit, Shakespeare probably adapted it from two older plays, author or authors anonymous, *The Troublesome Raigne of King John of England* and *The Death of King John*, anti-papal in tone and intent. Peele may have composed them, but that cannot be determined. In turn they drew on Holinshed, though they handle historical events somewhat cavalierly. Shakespeare also treats them with a very free hand, keeping scarcely a line of the original dialogue, and reconceiving the characters. Again a weak king is portrayed, but he lacks the compensating traits that make Richard dramatically interesting.

The subject is not the Wars of the Roses but England's strife with France and Austria and John's troubles with his nephew, Prince Arthur, a mere boy, and the lad's widowed mother, on whose behalf the French are claiming England's Continental provinces. John and his brave general Faulconbridge hold all English land sacred and refuse to yield a foot of it. The Pope intervenes, having a quarrel with the English monarch over his opposition to the Archbishop of Canterbury, and excommunicates the recalcitrant John. A compromise that has been reached with the French at Angers is broken off in consequence. (How the playwright broached John's defiance of the Pope was of course of lively interest to Elizabethan spectators, who could recall a more recent instance of Tudor insubordination.)

In a battle, John is victorious; young Arthur is taken prisoner to England, where John secretly orders him blinded and murdered. Hubert de Burgh, the prince's warder, spares the boy, having grown fond of him. John is without knowledge of this. The French invade England, and John, learning that his nephew is still alive, is repentant and commands him released in an effort to rally popular support. But the young prince has committed suicide in a leap from the castle's walls. John strategically makes his peace with Rome and defeats the French. But, though victorious, he has been poisoned by a monk and eventually succumbs after much suffering. He is succeeded by his son, Henry III.

The play lacks unity. The language is excessive. An effectively realized character, Constance, has a very stageworthy scene in which she wrangles with Elinor, King John's mother and partisan; but later in the drama her tears and scolding become tedious. She is, like Richard II, too fond of the sound of her own voice and unstoppable once given the least occasion to talk. A medieval Electra, she takes a wild, perverse pleasure in her wrongs and sorrows. The outstanding figure, however, is Philip Faulconbridge (the illegitimate son of King Richard I) whose dialogue is rich and racy with a natural idiom that befits his vigour and bluffness. Through him Shakespeare gives voice once more to his patriotism in the play's only memorable speeches.

It has been suggested that in tenderly creating Prince Arthur, pathetic, vulnerable, the dramatist might have been thinking of his own son, Hamnet, who had just died. The episode in which the boy narrowly escapes blinding is compelling.

Some critics dismiss *King John* as no more than a hurried revision of the earlier and even poorer scripts to which Shakespeare turned at the behest of impatient fellow players. Others find the play important as another transitional piece; in it, even by writing badly at times and hearing his lines spoken, the poet learned more about how *not* to write verse drama – he began to rid himself of a propensity to fustian – and he gradually moves from the form of the simple chronicle play of mass events towards true historical tragedy, plays centring on individual characters delineated and shaped by moments of crisis in the fortunes of a dynasty and state.

Here the spectator's affection goes to Faulconbridge, virile, honest and loyal, with a rough strength and freedom from cant. The Bastard – as he was widely known – is one of the most attractive figures ever drawn by Shakespeare. His are the stalwart words that close the work:

> This England never did, nor never shall,
> Lie at the proud foot of a conqueror,
> But when it first did help to wound itself.
> Now these princes are come home again,
> Come the three corners of the world in arms,
> And we shall shock them: nought shall make us rue,
> If England to itself do rest but true.

Very difficult to classify is *The Merchant of Venice* (also 1596–7). It is hardly realistic. Even as a fantasy it is not easy to place. Both a tragedy and a comedy, it is not a conventional tragicomedy. Its tone is highly romantic. But to categorize it is not important. As has been remarked in the chapter about Marlowe, it owes much to *The Jew of Malta*, but Shakespeare departs considerably from Marlowe's unpleasant melodrama – here much of his tale is light and poetic. A ballad by Gernutus tells the same story, but it is not certain whether it appeared before or after Shakespeare's play. He borrows elsewhere, too– the "pound of flesh" incident is taken from *Il Pecorone*, composed by Giovanni Fiorentine in 1370, published in 1565. So far scholars have found no record of an English translation, but Shakespeare had become acquainted with it somewhere and shrewdly appropriated it. The "caskets" episode is found in the *Gesta Romanorum*, an anthology of fables quite popular in an English translation in Shakespeare's time and included in Boccaccio's *Decameron*. There seems also to have been an earlier Tudor play, *The Jew*, that he may have come across, but this cannot be proved. The way in which Shakespeare's drama differs from these several works from which he compounded his plot lies in the dimensionality of his people and – as always – the musicality of his language, which is at a high level throughout.

Romances with Italian Renaissance settings still fascinated and delighted Elizabethan authors and audiences. Antonio, a wealthy merchant in the city on the Adriatic, offers to help his young friend, Bassanio, who is in love with Portia, a charming and beautiful heiress. Bassanio lacks the money to court her, and Antonio – whose ships are away on trading voyages – also lacks funds. Generously he goes to the Jewish usurer, Shylock, to borrow 3,000 ducats. Shylock resents Antonio, who has berated him for his practices in the past, and lends him the sum for two months on one condition: that if the debt is not repaid when due, he must forfeit a pound of his flesh, from any part of his body that Shylock may decide upon. Antonio, confident his ships will be back a full month before the date set, accepts these very unusual terms.

Portia has promised her late father that she will give her hand only to the man who, choosing among three small caskets, picks out the one in which is enclosed her portrait. One is gold, one is silver, one is lead. She is paid suit by two princes, each of whom – to her secret pleasure – chooses an empty casket. Bassanio, to whom she is really attracted, is more fortunate in his selection, opening the lead box, and Portia is betrothed to him. But their happiness is increasingly clouded by word that Antonio's ships are overdue, and he might have to pay the forfeit to Shylock, which will cost him his life.

Shylock is further embittered by the elopement of his daughter, Jessica, with a Christian, Lorenzo. The girl, running off, has taken with her a large sum of treasure in gold and jewels.

Shylock, at a trial, relentlessly demands payment in full of the obligation entered into by the luckless Antonio. Disguised in masculine attire, Portia appears as her benefactor's lawyer and cleverly argues his case. She admits that Antonio owes the pound of flesh, but points out that the bond says nothing of blood. If, in taking the forfeit, Shylock spills one drop of a Christian's blood, he will be guilty of a crime under Venetian law. Shylock, confronted by this impossible restriction, cannot press his case. The Doge punishes him for his inhuman zeal and, at Antonio's suggestion, sentences him to become a Christian and to pay over to his daughter half his fortune while he is alive and the other half on his death. The news comes that Antonio's ships, surviving storms, have returned belatedly but safely, enriching him once again. Jubilant at the outcome, Portia reveals her identity, and the play concludes on a good-humoured note with marriages.

Both Shylock and Portia, the antagonists, are among Shakespeare's best characterizations. She is charming, gracious, quick-witted, and matures rapidly into a resourceful young woman. Shylock is commanding in his anger, bitterness, vindictiveness. Though he is the villain, his is the leading role.

The Merchant of Venice has evoked much controversy between those who feel that, in effect if not by intent, it is apt to stir anti-Semitism by perpetuating the stereotype of the Jew as an avaricious money-lender; and those who think the portrait of the Jew is actually sympathetic and – for Shakespeare's time – very advanced and compassionate. If the old man has venom in him – he has been cruelly wronged – he also has at times a "prophetic dignity". He is not likeable, but neither is he a monster, for he has been much provoked by these thoughtless and often hypocritical Gentiles:

> If a Jew wrong a Christian, what is his humility? Revenge. If a Christian wrong a Jew, what should his sufferance be by Christian example? Why, revenge. The villainy you teach me I will execute, and it shall go hard but I will better the instruction.

At this point the resemblance to Marlowe's Barabas is strongest. He is fanatical in his insistence, over and again, that the bond must be paid:

> I have sworn an oath that I will have my bond. . . .
> An oath, an oath, I have an oath in Heaven.

And:

> If you deny me, fie upon your law!
> There is no force in the decrees of Venice.
> I stand for judgment. Answer: shall I have it?

He is a legalist, rigid in his interpretation of all things, a close-peering Talmudist, poring over the law, which must be enforced to the letter, and viewing the world around him in the same way. It is not only his desire for vengeance that motivates him, but his inability to accept anything but the most unquestioning obedience of the law, which is his sole protection from his rapacious Christian neighbours and masters, and from his austere and ever-threatening Hebraic God. Most effective, of course, is his superbly impassioned retort to his tormentors:

Above The 1984 Glyndebourne Festival Opera production of *L'Incoronazione di Poppea* by Claudio Monteverdi (1567–1643), directed by Sir Peter Hall
Left *The Return of Ulysses* by Claudio Monteverdi at English National Opera in 1992, directed by David Freeman

Above *The Banquet of Herod*, depicting the beheading of John the Baptist, with Salome holding his head; engraving by Israel Meckenem, *c.* 1490
Left *Court Ball in the Louvre on the Occasion of the Wedding of the Duchess Anne de Joyeuse, 24 September 1581*; Catherine de Medici and her son Henry III are depicted under the canopy to the left.

Above A group of travelling players in *The Old Horse Market in Brussels*, *c.* 1666, painted by Adam Frans van der Meulen

Left Costume design for a ballet performance at the court of Louis XIV, *c.* 1655; drawing ascribed to Henry Gissey

Left The Swan Theatre, London, 1596; drawing by Johannes de Wit
Right The Globe Theatre, London, *c.* 1612
Below right Twentieth-century reconstruction of the interior of Shakespeare's Globe Theatre, London

Above left Francis Beaumont, English playwright (1548–1616)
Above Ben Jonson, English playwright (1572–1637); engraving by William Cameron Edwards, 1840, after a contemporary portrait, *c.* 1610
Left Christopher Marlowe, English playwright (1654–93); engraving, *c.* 1810, after a contemporary portrait

Above Production of *Doctor Faustus* with Michael Goodliffe as Mephistopheles and Paul Daneman as Faustus, Edinburgh Festival, 1961
Right Sir Ian McKellan as the King in *Edward II* in a production by the Prospect Theatre Company, 1969

Above David Garrick (1717–79) as King Lear
Left Laurence Olivier and Ralph Richardson in the film of *Richard III*, 1956
Above right The Royal Ballet's production of *Romeo and Juliet*, choreographed by Frederick Ashton, with Tamara Rojo and Inaki Urlezaga in the principal roles, London, 2000
Below right Director Max Rheinhardt (centre) surveying a set model during the filming of *A Midsummer Night's Dream* in 1935

Left Edmund Kean (1787–1833) as Hamlet
Above Engraving of Sarah Siddons (1727–88) as Isabella in *Measure for Measure*

Left Ellen Terry as Lady Macbeth, *c.* 1888
Above Adalbert Matkowsky as Macbeth, Berlin, 1905

Left Henry Irving as Cardinal Wolsey in Shakespeare's *Henry VIII*, c. 1892
Below Paul Robeson and Peggy Ashcroft in *Othello*, London, 1930
Right Maurice Evans as Richard II, New York, 1940

Top left Bruno Ganz as Hamlet and Jutta Lampe as Ophelia, Berlin, 1982
Top right Jean-Louis Barrault as Hamlet, Paris, 1946
Left Alexander Moissi as Hamlet, Vienna, 1922
Above right John Barrymore as Hamlet, c. 1923
Right William Shakespeare (1564–1616); engraving by Martin Droeshout on the title page to the First Folio of the plays, 1610

Mr. WILLIAM SHAKESPEARES

COMEDIES, HISTORIES, & TRAGEDIES.

Published according to the True Originall Copies.

Martin Droeshout sculpsit London.

LONDON

Printed by Isaac Iaggard, and Ed. Blount. 1623.

Above Shakespeare's *A Midsummer Night's Dream* performed by the Royal Shakespeare Company at Stratford-upon-Avon, with Charles Laughton as Bottom, Mary Ure as Titania and Vanessa Redgrave (extreme right) as Helena, 1959

Right *Tamburlaine the Great*, Royal National Theatre production with Albert Finney as Tamburlaine, directed by Sir Peter Hall, London, 1976

> You call me misbeliever, cut-throat dog,
> And spit upon my Jewish gaberdine,
> And all for use of that which is my own.
> Well then, it now appears you need my help. . . .
> What should I say to you? Should I not say,
> Hath a dog money? Is it possible
> A cur can lend three thousand ducats?

And later:

> I am a Jew.

Still, for all his wonderful humanity and perception, Shakespeare's view of the Jew is Elizabethan: his Shylock remains a hateful alien, a butt for laughter, an outcast whose comeuppance will be loudly applauded. If Shakespeare had any better opinion of Shylock, it is unlikely that he could have persuaded his prejudiced spectators to have shared it with him. At all moments in the play this pragmatic poet anticipates how his audience will respond to his every line and story turn and seeks to please the ticket-buyers. But time has changed the script's values.

There is comedy in the play: the awkward clowning of Launcelot Gobbo.

The language is simpler and more supple than elsewhere, much of it in prose and thronged with lines that ring in memory:

> A Daniel come to judgment . . .

And:

> The Devil can cite scripture for his purpose.

And:

> All that glisters is not gold . . .
> Gilded tombs do worms infold.

As well as:

> How far that little candle throws his beams!
> So shines a good deed in a naughty world.

And:

> How sweet the moonlight sleeps upon this bank!
> Here will we sit, and let the sounds of music
> Creep in our ears; soft stillness and the night
> Become the touches of sweet harmony.

And even better known, of course, is Portia's courtroom speech that begins:

> The quality of mercy is not strained,
> It droppeth as the gentle rain from Heaven,
> Upon the place beneath. It is twice bless'd;
> It blesseth him that gives and him that takes.
> 'Tis mightiest in the mightiest . . .

Less noted are such passages as these given to Gratiano:

> Who riseth from a feast
> With that keen appetite that he sits down? . . .
> All things that are,
> Are with more spirit chased than enjoyed.
> How like a younker or a prodigal
> The scarfed bark puts from her native bay,
> Hugg'd and embraced by the strumpet wind!
> How like the prodigal doth she return,
> With over-weather'd ribs, and ragged sails,
> Lean, rent and beggar'd by the strumpet wind!

And his:

> Let me play the fool,
> With mirth and laughter let old wrinkles come . . .

Portia's:

> The crow doth sing as sweetly as the lark
> When neither is attended; and I think
> The nightingale, if she should sing by day,

> When every goose is cackling, would be thought
> No better a musician than the wren.
> How many things by season season'd are
> To their right praise and true perfection!

If today the play is largely Shylock's, it is because one ambitious and renowned actor after another (some of them Jewish) have sought to essay the part. Calling for a display of applause-getting, virtuosic characterization, it appeals particularly to male stars who are growing too old to be romantic leads – if not Romeo, then Shylock and Lear. Frequently suggested is that had Shakespeare undertaken this script at a later phase of his career he would have written it wholly as a tragedy of the dynamic, ferocious yet pathetic Shylock, dispensing with the trivial sub-plots of the caskets, the disguises and the game with rings that Portia and her maid carry on with their husbands to test their fidelity. It is possible that Shakespeare had never seen a Jew or probably never had dealings with any; there were very few in London in those days. His empathy for them and their burden is amazing. (He did know money-lenders, of course; both he and his father engaged in that trade.)

The inclination of top actors to do *Merchant* may largely explain its long history of revivals; it is another constant favourite. Its major premises are flawed: what intelligent and caring father would stipulate in his will that his daughter choose her husband by taking note of the casket he looks into, and what bright young woman would be persuaded to follow such a course in selecting a husband; what money-lender in his senses would demand the life of a borrower if repayment was delayed, and what made Antonio so desperate for funds to support a young friend's wooing that he would accept such dangerous terms? And was Shylock the only money-lender in Venice? Could not the respectable Antonio have gone elsewhere, and why once more to Shylock whom he had previously charged was an extortionist? Could Portia's male disguise long deceive the Doge and his court? (Portia would have been acted by a young man on stage, which would have helped to sustain the illusion.) Despite these many implausibilities, the play has been accepted by generations of theatre-goers without questioning those facets of it.

John Wain and other critics, seizing upon Antonio's self-description as "a tainted wether of the flock", take it to mean that this elderly Christian "benefactor" is a homosexual, infatuated with the handsome, shallow, fortune-hunting Bassanio and foolishly willing to risk his life to help him. ("Wether": a castrated ram.) Wain further likens Antonio, melancholy without vouchsafing an explanation, to Shakespeare at one period, possibly when he was composing the ambiguous sonnets to an unknown young man. There are seeming parallels. But again, why should an older male lover urge his young companion to marry? It might be a nice way of freeing himself from an outgrown entanglement.

After *Merchant* Shakespeare returned to his historical series. *Henry IV, Part I* and *Part II* (1596–8) are among his finest scripts in any genre. They are packed with breathing characters, among the most vital in all his crowded gallery – impetuous Hotspur, rotund, bohemian Falstaff, mystic Glendower, saturnine King Henry, scapegrace Prince Hal, lean-shanked Justice Shallow – all give these plays teeming life. The source, as often before, is Holinshed, to which was possibly added a dip or two into Samuel Daniel. A slightly earlier, crude script, *The Famous Victories of Henry V*, author unknown, suggested some of the comic incidents. All this Shakespeare wove into a marvellous fabric of brilliant colours and exciting scenes. The first half having been a great success, he promptly followed it with a sequel.

The aged Henry IV finds his kingdom threatened on two flanks. The eccentric, star-gazing Owen Glendower is heading an uprising in Wales, and the Earl of Douglas a rebellion in the north. The latter is put down by Henry Percy (Hotspur), but Glendower defeats and takes prisoner Mortimer, the Earl of March, who has also led a royal army. Hotspur wishes to ransom Mortimer by an exchange of prisoners with Glendower, but the king will not permit it, asserting that Mortimer is to blame for his plight. Infuriated, Hotspur defects to the rebel side, taking a small force of troops. But the Welsh and Scottish dissidents cannot agree on how they will share the kingdom that they have not yet won. A personal dispute between the touchy Hotspur and the vain Glendower is a further divisive complication. Meanwhile King Henry is disturbed by the behaviour of his son, Prince Hal, who spends his time roistering in seamy London taverns with the boastful, dissolute, fat and amusing knight, Sir John Falstaff. The king upbraids his son, who vows to reform. The prince sets out on a campaign against the rebels, though he irresponsibly puts the lying, drunken, plundering Falstaff in command of a company of armed men. With the rebel command split, the royal forces are victorious at Shrewsbury. Hotspur is slain in battle by no less than Prince Hal, and the first play ends with a prospect of peace in the civil war.

In Part II the truce is broken. A new coalition of rebels arises. The leaders of it are the Archbishop of York and Lords Hastings and Mowbray. Hotspur's bereaved father, the Earl of Northumberland, is begged by his wife and widowed daughter-in-law not to join the revolt unless it first gives some sign of winning. Glendower dies, which is a loss to the dissidents. King Henry sends his younger son, John of Lancaster, and the Earl of Westmoreland to counter the forces arrayed against them. The rebels, after a pledge that their complaints will be properly answered, lay down their arms. The royal forces unscrupulously arrest them, and their leaders are executed.

King Henry IV is dying. He is conscience-stricken, and worried about Prince Hal's misconduct, the young man's continued dissipation with Falstaff. The prince, lectured by his father and charged to recognize the duties of the role he must soon play, once again promises to mend his ways.

Falstaff is overjoyed to learn that Prince Hal has succeeded to the throne. The knight assumes that his own tattered fortunes will rise. But when he reaches the court he is rebuffed;

the new king will no longer see him. Prince Hal, now Henry V, has turned over a new leaf at last. Sobered, he intends to rely upon the men who had helped his late father guide the long-troubled kingdom.

The portrait of the proud rebel Hotspur, a high-strung knight with an impatient temperament and a medieval outlook, is one of the most attractive figures in the play; he far outshines the pleasant but comparatively innocuous Prince Hal, his fatal adversary. Hotspur, with his sharp tongue, dominates every scene in which he appears, dedicated as he is to a romantic concept of honour. Even more original is the strange Welshman, Owen Glendower, an heir of the legendary magician Merlin and convinced of supernatural blessings. He lends an almost exotic colour to an already brilliant mural.

Though a panoramic chronicle play on this scale can hardly be expected to have tight unity always, *Henry IV* is lucidly structured, each segment of the plot and the sub-plots clearly placed and followed. By now Shakespeare had developed a sure control of the elements of his stories.

Gravely presented is the calculating usurper, Henry IV, ever insecure about his claim to the throne and haunted by memories of his cousin, the murdered Richard II, for whose death he feels morally responsible. "By what by-paths and indirect crooked ways I met this crown," he confesses. Elsewhere, he cries: "Uneasy lies the head that wears the crown." Prince Hal is strongly praised by the playwright, but does not really earn the spectator's enthusiastic approval. His dual nature and sudden change – wastrel to earnest ruler – are hard to credit. The two sides of his personality do not match

Shakespeare's towering achievement in this play is the genial, besotted Sir John Falstaff, a descendant of Plautus' Braggart Warrior, here enlarged to archetypal proportions, one of the great comic characters of all time. In his person, and in his riotous antics, as has been pointed out, is an anti-heroic treatment of the core events of the play, the revellers in the alehouse mocking all that the court of the remote, lofty and dying sovereign holds solemn. The lascivious, wheedling tongue of Falstaff is also the shrewd voice of common sense, of man in his most unkempt moods interested only in survival and the satisfaction of belly and groin. With his tatterdemalion crew – Poins, Pistol, Bardolf, Mrs Quickly, Dolly Tearsheet – he rules over his own raffish court. Feigning death on the battlefield, he wisely propounds: "The better part of valour is discretion." Much of the spirit of the anti-masque prevails here. Falstaff is a highly talented parodist, a comic mimic of everyone, even of the king and prince. The old man is a "Lord of Misrule" out of the Middle Ages. Thief and wencher – "We have heard the chimes at midnight" – the heavy-bellied Sir John has no inconsiderable role to play. Historical drama is given a new dimension here, when it moves from high events on an exalted plane, to the contrapuntal daily life of tavern, brothel and village, to the Falstaffs, Shallows and Pistols, proving a social picture of broader scope and sharper reality.

The women in the play – high-born Lady Percy, Lady Mortimer, bedraggled Mrs Quickly, Dolly Tearsheet – though they have minor roles, are equally diverse and convincingly animated.

The episode in Wales between the two young warriors, rash, impulsive Percy and Mortimer, and their amorous wives, is a highlight of this many-sided drama.

One theme of the paired plays is the passing of feudalism, with the downfalls and deaths of the chivalric Percy Hotspur and despotic Glendower, representatives of an age in which strong barons held individual sway, and the replacement of that by now anarchic order by a firm, centralized government, embodied by the dying Henry IV and his suddenly thoughtful son, who is to be Henry V.

Another of Shakespeare's frequent concerns, the relationship between father and child, is developed here in the scenes between Henry IV and his wayward, unthrifty royal heir, to whom he is always offering counsel. There are to be similar scenes in *Hamlet* – Polonius and Laertes – and in *The Winter's Tale*, *King Lear* and also in *Romeo and Juliet* and *The Merchant of Venice*. Shakespeare was a parent, and it is believed he did not always enjoy a happy bond with his daughter Judith, whom he was to slight in his will. But the relevance of that to his artistic work is, of course, hypothetical.

The speeches are richly studded with imagery and epigrams, yet have a remarkably natural cadence, which makes them more flexible and easier to mouth. By now, too, Shakespeare has perfected both his prose – as in the instance of Falstaff – and his dramatic poetry, as in the passages of varied tone he assigns to impressively regal King Henry and the tart-worded, over-loquacious Hotspur. They are verse, but also effective dialogue, hitting upon the ear with clarity. Besides, each is in a unique style, fitting each speaker according to his own nature and station.

Next follows *Much Ado About Nothing* (1598–9) – the chronology is, again, speculative, but that is an informed guess. It is counted among his best comedies; it also belongs to a group of three turned out at this period, when his mood was apparently sunny and his wit at its prime. The plot is mostly of his own invention, though traces of a tale by Matteo Banello, published in Italian over a century earlier, are discernible. How – in what translation, if any – it got to Shakespeare's attention has not been discovered; in any event he added many new details to it. A good deal of it – the new characters such as Dogberry and his friends whom Shakespeare introduces – must have been based on his sharp observation of life in his time. By all accounts they are authentic likenesses of Elizabethan constables. But Shakespeare retains an Italian setting, however incongruous it seems at times. A serious streak contrasts throughout to the lighter bands of the story, with the comedy brighter and more dominant – yet the play could easily have been a tragedy, for elements of it put us in mind of *Othello*, with Don John resembling Iago.

Leonato, the Governor of Messina, has a daughter, Hero, who catches the eye of a visitor, Claudio. In the same group of guests is Don John, brother of the Prince of Aragon. A match is promptly arranged between Hero and the infatuated Claudio, but this displeases the villainous Don John, who plots to break up the betrothal. Claudio is accompanied by a friend, Benedick, who has long known Beatrice, Hero's cousin. Between these two, Benedick and Beatrice, there has always been bickering and raillery, which convinces onlookers that *au fond* they are strongly attracted to each other, though strangely unwilling to own it to themselves. Two plots are hatched: in one, Don John seeks to disgrace Hero by making her the victim of calumny; in the other, Beatrice and Benedick are separately allowed to "overhear" how sick with love for the other is each one. Their interests are inevitably stimulated by these pre-arranged conversations staged by their friends to pique their hidden affections. The trick played on Hero is much more serious: Borachio, Don Juan's servant, pretends to be a cavalier carrying on an affair with her – actually it is Hero's waiting-woman who stands by the window accepting Borachio's wanton suit. Claudio, brought to witness this courtship, is deceived by it. At the wedding ceremony he denounces his bride-to-be as unfaithful to him. Startled, Hero faints – and is believed dead. The scene shifts to a prison, where a foolish constable, Dogberry, cross-examines Borachio, who has unwittingly revealed his role in the cruel plot against Hero.

Claudio and Leonato, Hero's father, and Benedick, too, are on the verge of a duel over the "dead" young woman's virtue, when Dogberry arrives with his prisoner, Borachio, who confesses his guilt. Claudio is plunged into remorse and pays heartfelt tribute to his beloved at her "tomb". Leonato learns that his daughter is actually alive and in hiding, sheltered by a friar. He pardons Claudio and urges him to marry Beatrice. The father arranges a masked ceremony, in which Claudio is wed to "Beatrice", only to discover that his bride is his lost Hero, after all. Benedick, much discomfited by the seeming theft of Beatrice by his friend, obtains that lady's hand to his great delight.

If the comedy shines, it is chiefly because of the engaging portraits of the sparring, high-spirited Beatrice and Benedick and their amusing if caustic exchanges of belittling epithets or put-downs. This is the humour of insult at its keenest, and particularly as it may be bandied in skirmishes between a man and a woman. It is a universal fact, too, that very often two such people torment themselves and bait those they love rather than admit that they are drawn to each other. They apparently deem it an indignity, and at the same time they fear to lose their independence by yielding their hearts to the keeping of another. (It is a rather Strindbergian concept, here treated very lightly, entertainingly.) Beatrice is another of Shakespeare's beautiful young women who possesses a masculine cleverness, a strong will and a tongue able to turn each phrase into an apt retort. She belongs to the company of Portia, the Princess of France, Katherine. It would seem that this sort of young lady fascinated Shakespeare himself. The pair of reluctant lovers, Beatrice and Benedick, anticipates those in the sophisticated high comedies of Congreve a century later. Though they belong to the sub-plot, they run away with the play. It has been much debated whether this script, with its serious but ridiculous plot, contrasting merry sub-plot and rustic humour – furnished by Dogberry, Verges and the fumbling Watch – has any semblance of unity. Though the plot-lines are intertwined – Dogberry's intervention leads to a solution of Hero's dilemma – the moods are markedly disparate. But somehow it does work when deftly acted. The scene of the interrupted wedding is much praised for its exemplary craftsmanship. The dialogue is pithy throughout and at its best in the prose passages. It is in verse, though, that Leonato wisely propounds:

> There was never yet a philosopher
> That could endure the toothache patiently.

The brittle prose fits smoothly to the quick action and tone of the play. Admits Benedick, professedly prosaic in speech, "No, I was not born under a rhyming planet, nor I cannot woo in festival terms." His humour, like that of the comedy, is acerbic. Beatrice, his wife-to-be, he rightly calls "my Lady Tongue".

But, for all his frankness and sharpness, Benedick is likeable. He is a lesser Mercutio, a superior Petruchio, another Biron. His friend, Don Pedro, says of him: "He hath a heart as sound as a bell and his tongue is the clapper, for what his heart thinks his heart speaks." This is the very masculine Benedick, an admirable lover.

Though anti-romantic, the play includes two exquisite songs, among the poet's finest: "Sigh no more, ladies . . ." and "Pardon, goddess of the night . . ." The down-to-earth foolery of the idiotic Dogberry provides a good and helpful foil to the ambiance created by such lyrics and by the intrusive, nonsensical plot involving Hero and Claudio. A measure of criticism of minor Elizabethan officialdom might be read into Dogberry's role, for – from all accounts – his ineptitude was a true picture of the local constabulary in many cases.

The Brooklyn Academy of Music was host to a *Much Ado* in 1997 conceived by Cheek by Jowl, a British troupe that had been on Broadway a few seasons previously with an all-male *As You Like It*. For *Much Ado* the cast was made up of both sexes. The décor and costumes were Edwardian. Most of New York's critics endorsed this transfer, though, as always, others objected to any altering of period. The effect at moments was surprisingly Wildean, the frequent hilarity dependent on the humour of insult or on merely flippant, good-natured teasing. The "love-resistant" Benedick (Matthew Maclady) was tall, angular. He explained his sudden change of intention by conceding that "The world must be peopled." Beatrice (Saskia Reeves) was a "bright spitfire", whether expressing her ire or grudging affection. The *Wall Street Journal*'s reviewer observed: "The production (director Declan Donnellan, designer Nick Ormerod) sought a bold, bare, melancholy elegance: the whole cast of courtiers often dancing upstage in slow motion or interposing their frozen presence in scenes where they didn't logically belong. It was as if Ophuls were directing Chekhov." Only the low comic Dogberry failed to amuse.

The French composer Hector Berlioz was a fervent convert to Shakespearian drama. His last opera, *Béatrice et Bénédict*, had its première in Baden-Baden in 1862 with resounding success; since then it has not been conspicuous in worldwide repertories, though it is occasionally staged.

After finishing this frivolous piece, the amazingly versatile, indefatigable Shakespeare returned to weightier labour on his tetralogy – if such it was meant to be. As said before, there is no indication that when he began with *Henry VI* and *Richard III* he had in mind creating a linked sequence of dramas covering a long swathe of English history, the tumultuous reigns of four kings, and perhaps as he proceeded a Henriad never took shape in his imagination. But it seems inevitable that he must have become aware of having brought off just such a feat.

He was guided, as usual, by Holinshed and Hall, and purloined much of the text of an earlier play, *The Famous Victories of Henrye the Fyft* (c. 1588), author unknown, a work from which he had also taken material for his *Henry IV*, featuring a Sir John Oldcastle, who bore a close resemblance to Sir John Falstaff, both of them disporting in some of the same tricks and mischief, incidents obviously lifted by Shakespeare to provide his history with comic relief.

Henry V brings back the once heedless Prince Hal, now an effective leader and sovereign. Quarrels have been renewed with France over provinces there to which England lays claim. At issue is the Salic Law, which prevents inheritance through the female line. When the Dauphin insultingly rejects Henry's bid to bargain, war ensues. To unify his kingdom behind him, the King reaches a pact over clerical rights with the Archbishop of York, who then supports and even sanctifies his cause. The rest of the play occurs in France. At the battle of Agincourt Henry attains a glorious victory. The Duke of Burgundy then brings together the royal contestants. Henry's claims are acknowledged to be just, and a treaty is capped by a marriage between him and the Princess Katherine, the French king's daughter. The final scene, a most charming one, is Henry's courtship of this attractive lady, neither of the pair able to speak the other's language well. (It is hard to tell what are the English king's true sentiments in this meeting, whether he has fallen in love with Katherine or is merely being good-naturedly opportunistic, which would be consistent with his changeful character.)

The play is considerably leavened by comedy. Word is brought of the death of Falstaff, but his erstwhile followers – Pistol, Bardolf, Nym – join King Henry's army. Their antics evoke laughter. Another farcical character, roundly realized, Fluellan, is added to this roster. He is a pedantic schoolmaster who speaks English with an atrocious accent and wishes to fight the war in accordance with classical precepts: he would borrow the strategy of Alexander or Pompey. One of the most affecting speeches in the play, however, is the Hostess's account of Falstaff's passing, which she depicts in her thick, rustic dialect – his last moments are not otherwise shown:

> 'A made a finer end, and went away an it had been any christom child; 'a parted even just before twelve and one, even at the turning o' the tide; for after I saw him fumble with the sheets, and play with flowers, and smile upon his fingers' ends, I knew there was but one way; for his nose was as sharp as a pen, and 'a babbled of green fields. How now, Sir John! quoth I: what man! be o' good cheer. So 'a cried out – God, God, God! three or four times. Now I, to comfort him, bid him 'a should not think of God;

> I hoped there was no need to trouble himself with any such thoughts yet. So 'a bade me lay more clothes on his feet: I put my hand into the bed and felt them, and they were as cold as any stone; then I felt to his knees, and so upward and upward, and all was as cold as any stone.

(Substitute "he" for the "'a" and the sentences become more intelligible.) It is a marvellous prose description and testifies to how vividly Shakespeare saw Falstaff, even to the sick man's "fingers' ends". Oddly, other comic figures in this play die before it ends: Bardolf by hanging for robbing a church; Doll Tearsheet from venereal disease.

In an earlier scene Henry walks unrecognized through his army's camp in the darkness before the crucial battle. He talks to his soldiers and learns much from them, including the not always flattering opinion of him held by various men. The drama is not equal in stature or impact to *Henry IV*, but contains many ringing patriotic exhortations that have made it a long-time favourite of British audiences, especially at times when their nation has been at war and in peril.

Once again a text deals with the themes of the divine authority of kings and the need for a strong monarchy – and it is significant that Henry V expresses regret at the death (at his father's instigation) of ill-fated Richard II, whose claims to sovereignty were beyond doubt the more legitimate. Regicide is not to be easily condoned. If Shakespeare preaches any moral in his historical works, it would seem to be that a consecrated symbol of order is required for national tranquillity, and Henry V finally achieves this: his right to rule is sealed by his decisive victory at Agincourt.

So highly does the dramatist praise this king and celebrate his martial accomplishments that some critics assert that the play glorifies war, exposing an intellectual's typical envy and respect for those less thoughtful who are capable of performing deeds requiring selfless physical daring and courage. The charge ignores the many descriptions of war's horror and gross brutality to be found throughout the script.

Another moral might be that external triumphs, such as Henry V's, need an inner composure in those achieving them. Without such self-control a ruler can hardly hope to command the allegiance and submission of others. Henry V, having an impetuous temperament, does not always govern himself perfectly – his anger rises quickly, and he can be cruel – but on the whole he exercises an iron will and commendable self-restraint at the desperate moments. All the plays of the Henriad treat with problems of kingship: the need for a psychological balance in the king, between active impulse and cautious contemplation, between candour and duplicity, between mercy and sternness, even ruthlessness. If Shakespeare does not – as had Marlowe – applaud Machiavellianism, he is not unaware of the demands of *Realpolitik*. It is no wonder that his kings, put to such endless consuming tasks, often voice a weary sense of their own over-strained humanity and yearn for a simpler life, as do Richard II and Henry V. Speaking in disguise to a soldier, William Bates, in the camp before the battle, he says:

> The king is but a man, as I am; the violet smells to him as it doth to me; . . . all his senses have but human conditions: his ceremonies laid by, in his nakedness he appears but a man; and though his affections are higher mounted than ours, yet when they stoop they stoop with the like wing.

Henry V is not always a good or likeable person, but he seems to be everything a practical ruler ought to be.

The martial note, the brave call to arms, is insistent:

> Now all the youth of England are on fire,
> And silken dalliance in the wardrobe lies;
> Now thrive the armourers, and honour's thought
> Reigns solely in the breast of every man;
> They sell the pasture now to buy the horse . . .

As suggested here, Shakespeare saw a play far larger in scope than the Globe could accommodate. In a prologue, a chorus gives voice to his frustration at being unable to give tangible form and sound to his vision:

> O, for a Muse of fire, that would ascend
> The brightest Heaven of invention!
> A kingdom for a stage, princes to act,
> And monarchs to behold the swelling scene!

He chafes at

> . . . this unworthy scaffold to bring forth
> So great an object. Can this cockpit hold
> The vasty fields of France? Or may we cram
> Within this wooden O the very casques
> That did affright the air at Agincourt?

He appeals to the spectators to use their imagination bountifully:

> Suppose within the girdle of these walls
> Are now confin'd two mighty monarchies. . . .
> Piece out our imperfections with your thoughts;
> Into a thousand parts divide one man. . . .

> Think, when we talk of horses, that you see them,
> Printing their proud hoofs i' th' receiving earth.
> For 'tis your thoughts that now must deck our kings,
> Carry them here and there, jumping o'er times,
> Turning the accomplishment of many years
> Into an hour-glass . . .

He would like to have his stage-army consist of more than a half-dozen actors in the battle scenes.

The chorus is also called upon to describe the opposing camps in the black hours before the conflict:

> Now entertain conjecture of a time
> When creeping murmur and the poring dark
> Fills the wide vessel of the universe.
> From camp to camp through the foul womb of night
> The hum of either army stilly sounds,
> That the fix'd sentinels almost receive
> The secret whispers of each other's watch;
> Fire answers fire, and through their paly flames
> Each battle sees the other's umber'd face;
> Steed threatens steed, in high and boastful neighs
> Piercing the night's dull ear; and from the tents
> The armourers, accomplishing the knights,
> With busy hammers closing rivets up,
> Give dreadful note of preparation.

Such writing, with its catalogue of details, has been compared with that of Homer and Virgil.

In using a chorus, however, to provide commentary and narrative bridges, Shakespeare was reverting to a clumsy device which, though it had its superb virtues in the Greek theatre, had by now become outgrown, and nowhere more so than in his own best dramas.

The epic feeling of national unity promoted by *Henry V* was possibly inspired by England's recent victory over the Spanish Armada. He was pointing back to past similar accomplishments.

Since *Henry V* is so greatly involved with the clash at Agincourt, it is probably the most potentially cinematic of Shakespeare's works, as Laurence Olivier realized when he filmed the play to extensive acclaim (1944) – it was also at a point during the Second World War when Britain was fighting for its existence. It is almost a certainty that the playwright would have enjoyed seeing this grandiose version that has as a highlight a thundering cavalry charge. A half-century later

Kenneth Branagh, a young British actor, assuming the character of Henry and undeterred by Olivier's success, mounted another film version of the play (1989) less opulent, more gritty and gory, that also won wide applause.

A reading of Plutarch's *Lives* may have given Shakespeare the idea for a serious drama taken from events in Roman history. Clearly that is his chief source for *Julius Caesar* (1599), though a number of other Elizabethan authors had preceded him in choosing that subject and had popular success with it, as had the Italian poet, Orlando Pescetti. But Shakespeare's large debt to Plutarch (in Thomas North's translation) outweighs the likelihood of his owing much to anyone else. What is most impressive, however, is his skill in combining and rearranging elements in Plutarch's narrative to shape an effective stage-work. In *Julius Caesar* his craftsmanship throughout is remarkably apt. A comparison with North's version of Plutarch and Shakespeare's deft handling of the same scenes is both illuminating and humbling to anyone who might ever aspire to emulate this playwright's intuitive skill.

Caesar returns in triumph to Rome after defeating his rival Pompey. Seeing him idolized and fearing that he intends to overthrow the Republic and make himself emperor, a hostile group led by Cassius and Casca plan his death. They seek the help of Marcus Brutus, highly respected for his political virtue. The idealistic Brutus hesitates. Three times Caesar is offered the crown by his fellow soldier and admirer Mark Antony, but each time makes a show of refusing to accept it. The conspirators are worried about the approving shouts of the crowd, who favour Caesar's ascendancy. Brutus, though not fully overcoming doubts, finally enlists in the plot. His motives are patriotic; he wishes to save the Republic, which he feels is endangered by Caesar's ambition. His wife, Portia, perceives that some secret is troubling him, but he will not divulge it. Similarly, Caesar's wife, Calpurnia, has a premonition of ill-fortune and pleads with her husband to stay away from the Senate. A soothsayer tells Caesar to "Beware the Ides of March"; but, though superstitious, Caesar finally heeds neither warning. He is welcomed and persuaded by Brutus and his friends, who come to his villa to accompany him to the Capitol. An atmosphere of foreboding and doom has been established.

At the Senate, Metellus Cimber presents a petition for the repeal of his brother's banishment; Caesar's denial of it serves as a signal for the attack. He is stabbed to death by the conspirators. The last and fatal blow is that struck by Brutus. Recognizing him, Caesar gasps, "*Et tu, Brute?*" He falls to the marble floor in a pool of blood. Brutus cries: "People and senators: Be not afrighted; fly not; stand still: ambition's debt is paid."

Panic ensues in Rome. The conspirators are anxious to win Mark Antony's support. He has taken refuge in his home and asks permission to pronounce a funeral oration over Caesar's bloody corpse. His request is granted, but first Brutus mounts the Forum to address the public and explain what has led to the slaying. He ascribes only selfless promptings to the killers. "Not that I loved Caesar less," he says of himself, "but that I loved Rome more." The shocked, appalled crowd seems convinced until Mark Antony takes his turn on the rostrum. With irony and bitterness, and uncanny manipulation of the mob's emotions, he arouses the populace to fury against Caesar's assassins, imputing their designs. The crowd riots, and Brutus and his friends are forced to flee Rome.

A Triumvirate, consisting of Caesar's nephew, Octavius, along with Lepidus and Mark Antony, takes over leadership of Rome. They put to death a hundred senators, including Cicero, in retribution for the slaying. Brutus and Cassius head an army to combat Rome's new leaders. In a tent, at Sardia, Brutus and Cassius fall out bitterly, Brutus accusing Cassius of dishonesty, his taking bribes; they question each other's judgement. At Philippi the armies clash. Octavius is victorious. Word comes of Portia's suicide; Brutus and Cassius also kill themselves. Appreciating that Brutus alone was sincerely motivated in his violent actions, Antony pays just tribute to him:

> This was the noblest Roman of them all.
> All the conspirators save only he
> Did that they did in envy of great Caesar;
> He only, in a general honest thought
> And common good to all, made one of them.
> His life was gentle, and the elements
> So mix'd in him that Nature might stand up
> And say to all the world, "This was a man!"

The play is a keen study of the political breed – power-seeking men whose thoughts and actions are always political; indeed, whose very language invariably is. The dialogue is "oratorical" – all the characters are trained and able at persuasion, at choosing the apt epithet, at evasion of issues, at dissembling, for all their seeming bluntness; at flattery, at the conjuring up and control of emotion. The style is masculine and remarkably monosyllabic – Shakespeare never before or later wrote so plainly. The play does not offer poetry so much as incisive, well-ordered rhetoric – aphorisms, epigrams, so true that again many of them have been absorbed into our everyday language; and when we speak we may be borrowing Shakespeare's words.

Caesar is the shrewd ruler:

> Let me have men about me that are fat;
> Sleek-headed men and such as sleep o' nights.
> Yon Cassius has a lean and hungry look;
> He thinks too much: such men are dangerous.

Calpurnia, fearful of omens, voices her concern:

> When beggars die there are no comets seen;
> The Heavens themselves blaze forth the death of princes.

Caesar replies to her:

> Cowards die many times before their deaths;
> The valiant never taste of death but once.
> Of all the wonders that I yet have heard,
> It seems to me most strange that men should fear;
> Seeing that death, a necessary end,
> Will come when it will come.

Having watched Caesar become a virtual dictator, Cassius' comments about him are caustic:

> Why, man, he doth bestride the narrow world
> Like a Colossus, and we petty men
> Walk under his huge legs and peep about
> To find ourselves dishonourable graves.

Cassius recalls derisively how Caesar had once suffered a fit:

> And when the fit was on him, I did mark
> How he did shake – 'tis true, this god did shake.

And, again, the insistent demand of Cassius:

> Upon what meat doth this our Caesar feed,
> That he is grown so great?

Brutus' speeches tend to be meditative. He urges his co-conspirators to behave with restraint in their attack:

> Let's kill him boldly, but not wrathfully;
> Let's carve him as a dish fit for the gods,
> Not hew him as a carcass fit for hounds.

He miscalculates, which costs him dear:

> And, for Mark Antony, think not of him;
> For he can do no more than Caesar's arm
> When Caesar's head is off.

Thoughtful, when he turns to deeds he is inwardly torn, as he soliloquizes:

> Between the acting of a dreadful thing
> And the first motion, all the interim is
> Like a phantasm or a hideous dream.
> The Genius and the mortal instruments
> Are then in council; and the state of a man,
> Like to a little kingdom, suffers then
> The nature of an insurrection.

Urging decisive action against the army of the Triumvirate, he advises:

> There is a tide in the affairs of men,
> Which, taken at the flood, leads on to fortune;
> Omitted, all the voyage of their life
> Is bound in shallows and in miseries.

Cassius, shrewdly:

> Men at some time are masters of their fate.
> The fault, dear Brutus, is not in our stars
> But in ourselves, that we are underlings.

Brutus' summing-up, before he kills himself:

> My heart doth joy that yet in all my life
> I found no man but he was true to me.

The boy musician, dreaming, awakes and cries out to his master, Brutus:

> The strings, my lord, are false.

Mark Antony is a master of irony and plangent repetition. At first he claims he has only one purpose:

> I come to bury Caesar, not to praise him.
> The evil that men do lives after them,
> The good is oft interred with their bones;
> So be it with Caesar.

But soon he is citing how three times Caesar had refused the imperial crown when it was offered to him. Over and over, Antony asks the crowd: "For this was he ambitious?" He reveals details of Caesar's will, his bequest of park lands for public recreation and of money to the poor of Rome. He recalls Caesar's military accomplishments that enlarged the Roman Empire. He addresses the stiffening corpse:

> O mighty Caesar! dost thou lie so low?
> Are all they conquests, glories, triumphs, spoils,
> Shrunk to this little measure?

His tone changes:

> But yesterday the word of Caesar might
> Have stood against the world; now lies he there,
> And none so poor to do him reverence.

He points to the marks of Brutus' knife-thrust:

> This was the most unkindest cut of all;
> For when the noble Caesar saw him stab,
> Ingratitude, more strong than traitors' arms,
> Quite vanquish'd him: then burst his mighty heart. . . .
> O, what a fall was there, my countrymen!
> Then I, and you, and all of us fell down,
> Whilst bloody treason flourish'd over us.

The speech, inciteful, inflammatory, provokes the shocked mob to its murderous frenzy, turning the rising tide against the conspirators.

Generations of schoolchildren have memorized the funeral oration for their public-speaking classes. Shakespeare proves himself a remarkable speech-writer. Today's heads of state would consider themselves exceedingly fortunate to have his services.

The principal figure in the play is Brutus, the archetypal "liberal" – well-meaning, rational yet easily misled by his feelings, impractically idealistic, made the tool of harder-headed men such as the envious Cassius. Once the rebellion begins to fail, its advocates start to quarrel among themselves, and their campaign falls apart. Brutus is shown confused by his high-mindedness, his search for pure motives; yet he is prone to self-deception, ambitions of his own that he hides from himself, and his error lies in excusing ignoble – even vicious – means to achieve noble ends, the eternal fault of revolutionaries. He is decent, but self-righteous; and soon, like a fanatic Jacobean, his hands are coated with the blood of others. Shakespeare, the psychologist, gives a chilling and truthful portrait of a liberal's self-seduction.

It is said, or hazarded at least, that at this point in his career Shakespeare shifted his settings away from England because he could more freely discuss the political questions that deeply concerned him and other serious-minded Elizabethans: how tyranny was to be overthrown, whether crimes against a government were ever to be condoned. Indeed, there are indications that the Privy Council had grown sensitive to the staging of historical dramas such as those in the Henriad that had a possible allegorical intent, offering subversive parallels to contemporary events. At the request of the Earl of Essex's partisans, *Richard II* was revived and taken as containing allusions to the earl and the queen; the Privy Council banned it. For a long time Shakespeare wrote no more of English dynasties, their quarrels and martial strife. He chose to offer his public either comedies, or tragedies in remote, exotic settings, such as Denmark, far-off Venice, Ancient Troy and Egypt.

Among the comedies he produced in this fertile period is *As You Like It* (1599–1600), with a plot appropriated, as has been remarked, from Thomas Lodge's pastoral novel *Rosalynde*. Lodge, in his turn, had borrowed his plot from a Middle English poem. But, as usual, Shakespeare adds transforming touches and raises the subject to a higher level. Among these are the characters of Touchstone, Jaques and the plain-featured Audrey, who lend a tone of irony and reality to what would otherwise be a simple pastoral. With these contributions and a sharpened characterization of Rosalind, he quite transcends the limited, artificial pastoral form and achieves a romantic comedy counted a masterpiece.

Oliver, who is Orlando's older brother, has deprived him of his inheritance and otherwise cheated and abused him. He now schemes to have him killed in a match with Charles, a dangerous professional wrestler. Witness to the match is to the usurper of the dukedom, Frederick, whose daughter Celia has for her close friend Rosalind, daughter of the banished duke, the rightful ruler. Rosalind is worried about the outcome of the wrestling bout and tries to persuade Orlando not to engage in it; but Orlando is confident and wins. This enrages Oliver, who charges Rosalind with treachery and orders her to leave his palace. She disguises herself as a boy. Her cousin, Celia, is determined to accompany her, and does so dressed as a peasant maid.

They take refuge in the convenient Forest of Arden. Meanwhile Orlando, warned by a loyal servant that Oliver still means to do away with him, also seeks refuge in the woods, his old servant Adam by his side. They are the first to encounter the exiled duke, Rosalind's father, who has established an abode there, a fanciful, rustic court at which Amiens and Jaques are his attendants. This sylvan retreat is a "golden world" in contrast to the intrigue-ridden court from which they have been driven. The duke was a friend of the late Sir Rowland de Bois, Orlando's father, and welcomes the young man to his happy company. Theirs is a sort of Robin Hood existence.

In this crowded woodland Orlando happens upon the wandering Rosalind, now self-named Ganymede. He fails to recognize her in her male disguise of doublet and hose. He tells her of his love for Rosalind, and she vows to bring that lady to him if he will convince him of his sincerity. Led on, Orlando does this by paying court to "him" in a demonstration of how he will woo "her" – Rosalind pretending to be a boy, who is pretending to be Rosalind, this being enacted by a boy passing as a girl. Here Shakespeare ingeniously rings all the changes of a very old device in the requisite circumstance of Elizabethan staging.

The forest is also populated by shepherds, old Corin and young Silvius, and a haughty shepherdess, Phoebe, as well as a lecherous clown, Touchstone, all in the Italian pastoral tradition, and a pair of amusing English tillers, Audrey and her halting William, who is enamoured of her. His love of Audrey is as physical as the feelings of the others are in good part spiritual and romantic.

The burly Oliver comes to the forest in pursuit of his adversary, Orlando; he is attacked by a lioness. (In England? In Italy?) The brave Orlando saves him from the jaws of the savage beast. Oliver, grateful and repentant, begs Orlando's forgiveness. He falls in love with Celia, Rosalind's companion and refugee daughter of Frederick; she grants his suit. Most opportunely word comes that Frederick has undergone a religious conversion and entered a monastery, leaving the exiled duke free to return and rule his full domain. The end is happy for all.

The play could be seen as a parody or burlesque of the typical pastoral. Less obviously structured than some of Shakespeare's other comedies, *As You Like It* still has a subtle design, a mocking juxtaposition of the artificial life of the "envious" court and the Utopian dream of a woodland retreat, an idyllic Eden. It laughs at the same time at the pretensions and self-deceptions of both modes of living. Jaques, in the tradition of Biron, Mercutio and other cynical anti-romantic and even anti-social observers of manners and attitudes, is the detached and disenchanted onlooker, taken in by nothing. (It has been suggested that for the philosophical, melancholy Jaques Shakespeare owes something to Greene.) The odds are in no way stacked in favour of a return to Nature. Touchstone, who is Celia's escort, is none too fond of the forest: "When I was at home, I was in a better place," he laments.

In this supposedly peaceful "golden world" there is really much cruelty, and the Utopians spend a great deal of their time hunting and killing the deer, whose proper habitat is the forest – it is their realm, and, as Jaques points out, men are interlopers here (as, it might be said, is the prowling lioness).

Corin, the aged shepherd, contributes his share of practical, earthy comment: tending sheep is hard work, especially in raw weather. He can only dimly remember that he once suffered love's pangs.

The pastoral elements and the realistic components are nicely blended by Shakespeare's contrapuntal handling, and this is further reflected in the character of Rosalind, who is possessed of witty paradoxes. She is another of his boy-girls, charming, at times saucy, imaginatively playful. She is resourceful, gallant, talkative and intuitively worldly. It is she who says with innate good sense: "Men have died from time to time and worms have eaten them, but not for love." But she herself is not exempt from romantic illusion, and love claims her too as its victim. That she has a doubting mind only deepens her emotion.

As said, in the ancient controversy about the merits of life in the wilds or countryside weighed against those of life at court or in town, Shakespeare takes no sides. He lauds both, and decries both; certainly there is beauty and innocence in the rustic environment, and excitement at court or in the city to which most of the refugees eagerly return at the end. The playwright invites the spectator to choose one or the other, to dwell henceforth "as you like it".

The poetry is extraordinary. Familiar is the duke's tribute to life in a more natural environment:

> Sweet are the uses of adversity,
> Which, like the toad, ugly and venomous,
> Wears yet a precious jewel in his head;
> And this our life, exempt from public haunt,
> Finds tongues in trees, books in the running brooks,
> Sermons in stones, and good in everything.

Most famous is Jaques's summation of the Seven Ages of Man:

> All the world's a stage,
> And all the men and women merely players:
> They have their exits and their entrances,
> And one man in his time plays many parts . . .

In a remarkably brief space and with telling imagery it depicts each human being's progress from birth to death, first the mewling infant, then the reluctant schoolboy, next the strong and ardent youth and on to the hobbling greybeard, and . . .

> Last scene of all
> That ends this strange eventful history,
> Is second childishness and mere oblivion,
> Sans teeth, sans eyes, sans everything.

(Interestingly, Shakespeare's company had just opened its new theatre, the Globe, which chose for its motto, in Latin, the phrase: "All the world plays the player.")

The sadly jesting Jaques confidently believes that he can set all matters straight, if only the troubled populace will hearken to him:

> Give me leave
> To speak my mind, and I will through and through
> Cleanse the foul body of the infected world,
> If they will patiently receive my medicine.

But his offer of "wisdom" is laughingly rejected by the others, nor does Shakespeare ever clearly imply that the saturnine Jaques is in fact as wise as he thinks himself. This is a play which views the world from many angles, and on none does the author ever fully put his stamp of approbation. Could it be that "as you like it" also applies to different kinds of love, from the gross and sensual appetites housed in Touchstone, and in William for his Audrey, to the idealized and poeticized feeling holding Orlando in thrall? He is in the habit of scratching his beloved's name on the bark of trees and tying verses on boughs to vent his rapture; it is when Rosalind comes upon one of these verses that she is aware of his welling emotion:

> From the east to Western Ind,
> No jewel is like Rosalind.

Gibes Touchstone: "Truly, the tree yields bad fruit."

Legend says that Shakespeare himself played the minor role of Orlando's aged servant, and very persuasively.

Amiens has two lovely songs:

> Under the greenwood tree
> Who loves to lie with me,
> And turn his merry note
> Unto the sweet bird's throat,
> Come hither, come hither, come hither!

Another begins:

> Blow, blow, thou Winter wind,
> Thou art not so unkind
> As man's ingratitude;

> Thy tooth is not so keen,
> Because thou art not seen,
> Although thy breath be rude.

A third is sung by two of the banished duke's pages:

> It was the lover and his lass,
> With a hey, and a ho, and a hey nonino,
> That over the green corn-fields did pass
> In the Spring time, the only pretty ring time,
> When birds do sing, hey ding a ding, ding;
> Sweet lovers love the Spring.

As You Like It is frequently revived; that Shakespeare offered such a capricious and whimsical work to his public argues that he sensed a growing refinement and sophistication in his Elizabethan audience.

For John Gassner, *As You Like It* marks one of Shakespeare's climaxes in the field of romantic comedy. "It brings the pastoral form of drama to its highest perfection, and again by virtue of the triple Shakespearian gift of glorious poetry, brilliant characterization, and ingenious mingling of common reality with idyllic grace." Gassner believes that Shakespeare wrote this lark with ease, with "gay abandon".

What would be the ideal site for a production of this play? Daniel Lyons, in the *Wall Street Journal*:

> Arguably, the greatest legacy of Joseph Papp, the founder of the New York Shakespeare Festival, can be expressed in five words: free Shakespeare in a park. Beginning in 1954 in a riverside park on the Lower East Side and moving to Central Park in the late 1950s, Papp pioneered an idea that has spread throughout English-speaking countries. In Central Park itself *Cymbeline* (the only Shakespeare play of the summer) will open next week. From Sydney to London, and in many American cities, Shakespeare on summer evenings has become a civic habit, a cheerfully democratic pastime, a family picnic for the spirit. The comedies and histories are preferred to the tragedies, and the Bard diet is varied, according to local tastes, with modern fare.
>
> Case in point, if a rarefied one: The Hamptons Shakespeare Festival, now in its third season, is presenting *As You Like It* in the Theodore Roosevelt County Park in Montauk at the eastern tip of Long Island through August 23. It's free but there's a "suggested donation of ten dollars" – similarly, the free Sydney, Australia, festival asks for "paper!"
>
> The play, set largely in an idealized forest and featuring in the central character of Rosalind

a strong, witty and winning woman who educates mere men in the ways of love, is an obvious choice for a night outdoors. As skies slowly purple, one sits on a gently sloping hill and faces a stage design suggestive of gnarly wood, behind which rise broad lawns and shapely trees. Director Michael Landman plainly warms to the play's hints of animalistic shape-shifting: Antlered deer and ornery sheep turn up to the glee of the kids in the audience. Nature cooperated spectacularly last Saturday night: Wide V's of cynically cackling geese flew over in tight formation as melancholy Jaques (an eloquent and amusing Josh Gladstone) was orating on folly and the seven ages of man.

When Rosalind (the sharp-featured, amiable and intelligent Amy Prosser) gets to play a boy playing a girl teaching Orlando, her smitten swain, how to make love to a girl, the production comes to life. Remy Auberjonois makes a finely infatuated lover. And when Rosalind takes all the threads of amorous silliness into her hands and smartly unknots them, settling everybody's love-hash with brisk brio, the production gets up and dances. But the good stuff is a long time coming. Mr Landman slows down an already prolix tale with much solemn spectacle: processions, revels, a wrestling bout in slo-mo. Worse are his gratuitous and weird stylizings: Renaissance courtiers look and carry on like Spanish Inquisitors; woodsmen resemble Restoration fops or Ben Franklin. Simple, funny and romantic are best in summer Shakespeare; far-fetched academic concepts can wait for colder times. It was, at last, a merry evening.

— 17 —
SHAKESPEARE: MAJOR PHASE

If Shakespeare's work had come to a halt in 1600, he would have had to his credit some of the finest works in English or any other language. But, amazingly, this sublime plagiarist had scarcely begun his career. His major phase still lay ahead of him. As remarked earlier, one explanation for the greater depth and scope of *Hamlet* (1600) is that the plague had closed London's theatres, and Shakespeare – safer in the country – could write at a slower pace that afforded him more reflection. If, as some scholars contend, the date of its composition were a year or two later, it would also coincide with the death of the poet's father, an event which might have enforced his seriousness and tinged his thoughts a somewhat darker hue. He was now entering into his "tragic period".

Hamlet is based upon another play now called the *Ur-Hamlet*, known to have been produced by the Lord Chamberlain's Men, Shakespeare's company, in 1589; by some it is credited to Kyd, but others hazard that it was by Shakespeare himself, who subsequently revised it. (Textual evidence partly supports this theory: a pirated edition of the play – published in 1603 – is far inferior to today's accepted version, though it claims to be a perfect copy of Shakespeare's manuscript. Which "manuscript" would that be?) Kyd might well have written a still earlier version, however; if not, Shakespeare or the author of the *Ur-Hamlet* was much influenced by him. In Kyd's *Spanish Tragedy* many of the same plot-turns are used: an anguished nobleman, seeking to avenge the death of one near akin, goes mad or feigns madness; a ghost returns to tell how he was murdered; and in both plays the hero has a someone close named Horatio. Still other plays that employ similar devices are John Marston's *History of Antonio and Mellida* (1599), which was followed shortly by a sequel, *Antonio's Revenge*, melodramatic works of crude power wherein a character pretends to be mad, and a ghost of a father appeals to his son to punish a murderer; there is a weak-willed mother and a hero with a hue of melancholy and a feeling of alienation from an ugly world. Plays by three or more playwrights all having the same mood? Some commentators infer from this that the tragic tone of Shakespeare's work during the next few years was not caused by personal experience; he was merely adopting a pose now popular among his contemporaries. Something like this happened in mid-nineteenth-century Russia, when scepticism and world-weariness became fashionable among the intelligentsia, as depicted by Pushkin and other dramatists and novelists. Shakespeare was, aftre all, a practical man of the theatre. It is obvious that in choosing a subject he always asked himself what did ticket-buyers want to see and hear now. He did not truckle to the public, hardly to the pit, but had a keen sense of what would "go", what was likely to be the count at the box-office. He owed that to his company. His ear was attuned to current attitudes; his long file of successes attests to that.

Whatever lay behind Shakespeare's new unhappiness, he makes clear in his play that the provocations facing Hamlet are harsh enough to justify his deep, pervasive, individualized anger and generalized confusion and despair. He is young and sensitive. For him, the moral order is at sixes and sevens. Something is rotten not only in Denmark but in the cosmos.

Another factor in Shakespeare's sudden gloomy outlook might have been the plague, an external force, that was again besetting London and had driven him from the city. Such ills should not prevail in a benignly ruled universe. He was maturing, feeling more mortal, and asking sterner questions. In a man as complex as William Shakespeare there would be no simple pieties.

Today it is known that the "Hamlet" legend goes back much earlier than the *Ur-Hamlet*: towards the end of the twelfth century Saxo Grammaticus mingled historical fact and invention in his *Historica Danica* to set down an account of the events that later comprise the plot of the various plays, including again the ruse of pretended madness and Hamlet's troubled relationship with his mother and Ophelia. Here the hero is called Amlethus. Saxo's chronicle was printed in 1514 and won frequent reprintings. François de Belleforest took the story for his *Histoires Tragiques* (1576), and this French version provides further details that appear in the *Ur-Hamlet*, for which it may have been a second source. A German version of the story, possibly of slightly later date than Shakespeare's drama, is *Der Bestrafte Brudermord* (*The Revenged Fratricide*), apparently based on the *Ur-Hamlet*, which may have been performed in that country by touring English actors. But these French and German versions of uncertain date complicate the problem set to scholars who seek to trace the play's origins. In any event, the plot was largely established before Shakespeare added his handling of it. The advantage of borrowing a pre-existent plot is that a playwright is left free to concentrate on character development, interpretation, and to polish the dialogue; he does not have to think too hard about manipulating his story, though Shakespeare never takes a dramatic idea without significantly improving it, changing it to obtain heightened theatrical effects. Indeed, if *Hamlet* carries on the tradition of the "revenge tragedy", it is at the same time a departure from it. In other plays that follow Seneca's lead it is taken for granted that vengeance is due and should be promptly pursued, but in Shakespeare's treatment no such ready premise is adopted. Hamlet does not know for certain what he should do. He waits and waits, debates, agonizes.

This long and intricate play can be summarized here only in its baldest aspects. The time is the Middle Ages. Denmark's royal castle – at Elsinore – is haunted by the ghost of the late king, named Hamlet like his son. Claudius, the dead king's brother, has married the widowed queen and succeeded to the throne that rightly belongs to his nephew. The young man returns from Germany, where he has been studying at a university. Friends guide him to the night-shrouded parapet of the castle, where he twice beholds the stalking ghost of his father, who tells of being murdered by Claudius. The ghost implores young Hamlet to avenge him, but the son does not know how much faith to put in this accusation from an other-worldly source, a hoarse-voiced apparition. The ghost, indeed, might be the Devil.

To gain time and avoid suspicion, Hamlet makes a harmless show of madness. He evinces interest in Ophelia, daughter of Polonius, the Lord Chamberlain, and the innocent girl returns his seeming affection, but her father and her brother Laertes warn her against him because of his erratic behaviour, which springs not only from his defensive strategy but also from a profound distrust prompted by his father's recent passing and at finding the times "out of joint", with the possibility that his mother has connived with his uncle to kill her former husband.

Claudius and Polonius, wary of Hamlet, perceive a method in his "madness". They enlist two courtiers, Rosencrantz and Guildenstern, to spy on him. They also arrange for Hamlet to encounter Ophelia, as if by chance, and eavesdrop on the pair. Hamlet's outbursts are disturbed, perplexing, at times witty and coherent, but then nonsensical, dense with riddles or abusive. The listeners comprehend nothing; they have no clue to his knowledge or intentions. Polonius is convinced that Hamlet is love-crazed. Only in some remarkably eloquent soliloquies when he is quite alone does he unburden his troubled heart and mind.

A troupe of strolling players comes to the castle. Hamlet engages them, as a purported diversion, to perform at the court. He gives them special instructions as to their choice of a play. He also imparts advice on approaches to good acting, counsel that doubtless sums up Shakespeare's own beliefs on this matter. He rebukes himself when beholding the actors express counterfeit emotions more forcibly than he does while trying to cope with his real grief and increasing outrage.

The play-within-the-play is replete with parallels to the murder of the late king. While he sleeps in his orchard, drops of poison are instilled into his ear by Claudius. At the climax of this enactment the conscience-stricken Claudius betrays deep agitation. The performance is now halted. Even though persuaded of his uncle's guilt, the intellectual and introspective Hamlet hesitates to act. He comes upon Claudius praying in his cabinet and is about to strike him from behind, but draws back lest the usurper's soul rise to Heaven, his having died while in a state of grace.

At her request, Hamlet visits his mother in her bedchamber. He bitterly upbraids her for her hasty remarriage and complicity in the crime of regicide. Noticing that someone is hidden behind the arras (curtain) in the chamber, he plunges his sword at the spot; he finds the body not of Claudius, as he expected, but of Polonius, once more eavesdropping and killed by mischance.

Alarmed by Hamlet's violent behaviour, Claudius orders him to go as envoy to England. He gives secret instructions that the prince should be killed on his arrival there, but with the fortunate intrusion of pirates who board the ship bearing him Hamlet escapes from his would-be slayers and reappears in Denmark. Later, while on the same journey, Rosencrantz and Guildenstern are killed in his stead by his machinations.

Hamlet suspects Ophelia of being in the plot against him. In a confrontation he rails at her, "Get thee to a nunnery" ("nunnery" in this Elizabethan context implying a whorehouse). Ophelia, despondent, her mind unhinged by his insulting rejection of her, as well as the unexplained slaying of her father, sings madly, then drowns herself in a lily pond on the castle grounds.

Laertes, who has been on a mission to France, returns and demands the right to revenge himself on Hamlet who is responsible for the deaths of both Polonius and Ophelia. Though pretending to calm the young man, Claudius actually welcomes his help and abets him. A duel is arranged between Laertes and Hamlet, and Claudius plans to dip Laertes' sword in poison and also to put poison in a goblet from which Hamlet will drink.

Hamlet comes upon the sardonic gravediggers preparing a burial-place for Ophelia and exchanges mordant comments with them. He does not know, however, for whom the grave is being dug. The funeral procession arrives. Laertes, finding his foe at the graveside, immediately quarrels with him. They are parted, each protesting his stronger personal grief at the loss of Ophelia. In the duel that ensues, the rapiers are exchanged – Hamlet does not drink from the poisoned cup, but the queen unwittingly does so and dies. Laertes and Hamlet each wound the other. The poisoned foil kills Laertes, who in a last gasp denounces Claudius, who falls dead when Hamlet's sword runs him through. Hamlet, mortally wounded, asks his loyal friend Horatio to tell his true story to his Danish subjects. He names Fortinbras, Prince of Norway, to be his successor. Fortinbras, who has been contesting with Denmark and threatening war over a small parcel of Polish ground, appears and orders Hamlet borne in state with full honours from the scene, and the play closes with a solemn, noble death march. Though Fortinbras appears very briefly, he is used throughout as a contrast to Hamlet: he is simple, forthright in all his actions, and he also has a father to avenge.

How much Shakespeare was trying to say, and what is his focal theme, in this complex, philosophic drama is uncertain. It has ceaselessly tempted critics to offer tentative guesses or dogmatic explanations. One of the most repeated questions has been whether Hamlet's feigned madness passes over into actual insanity. Another debate has been whether the ghost is merely hallucinatory, a creature of Hamlet's disturbed inner vision, his vexed conscience at not having acted to ferret out the truth about his father's death and his mother's possible culpability and his inability to take action; his indecision and introversion when deeds are urgently required. The play has also been seen as a classic embodiment of an Oedipal relationship, and Hamlet as paralysed and held back by an unnatural bond to his mother. He fears to learn the truth about her, dare not do so. This Freudian interpretation became very fashionable in the twentieth century. Or one beholds Hamlet as an existentialist hero – or, perhaps, anti-hero – in his alienation, his sense of the meaninglessness and absurdity of existence. The play is also an early example of "black humour" – the wit, the high comedy of many of the scenes, the almost camp badinage of the hero, is another strangely modern facet of this timeless drama.

Still others choose to see Hamlet in a Christian framework, a theological probe: the dialogue contains a surprising number of uncharacteristic references to Heaven, angels and even one to Jesus the Saviour. Is it a battle between Good and Evil, with Hamlet aspiring to purity but becoming a victim of incarnate forces of Darkness at the small, rank court where the symbolic action occurs? His fatal error could be ignoring Christian imperatives not to take Mosaic vengeance but to exercise only compassion and forgiveness. As remarked before, Shakespeare's religious inclinations – if any – are not documented; he could see in the fate of Marlowe, Kyd and other imputed atheists that it was best to keep silent in those parlous times, but perhaps there is a clue in his asking in his will for burial in the parish church and for his sculptured likeness to be installed there. Significant, or not?

The drama is also not without a political side. It is a demonstration once again of the dangers of inner corruption and disorder in the state, as personified in its royal family. The dialogue is crowded with images of disease that works from within, spreading its infection throughout the body of Denmark and the world. As in *Oedipus*, a dreadful crime – that of Claudius and Queen Gertrude – has brought on a curse, a plague, that must be cleansed; and it is Hamlet's destiny – by birth – to remove the cause of this malignant ulcerating growth. Like the poison administered to Hamlet's father, the state is infused throughout with a deadly brew. As elsewhere, Shakespeare chooses a few images, repeats and elaborates them in metaphors throughout: disease and poison are signal instances here, and it is from poison that the murdered king, his son Hamlet, the malevolent Claudius, the queen and Laertes all perish – poison dropped in an ear, or quaffed from a cup, or smeared on a foil's tip.

Hamlet may be seen as a scapegoat, whose life – as in Greek tragedy – is a sacrifice, a ritualized offering to appease the offended gods, and the play as a re-enactment of everlasting mythical themes. In it, most interestingly, is a confluence of the Orestes and Oedipus stories, two archetypal motifs reappearing through twenty-five centuries of drama. Here, in powerful combination, they have a strengthened appeal and arouse instinctive responses lurking in the mind, perhaps slightly below the level of verbalized cognition, but there none the less. Shakespeare did not implant them consciously; he was not directly acquainted with either the Orestes or Oedipus legends. But he had inherited them obliquely after many permutations from playwrights at work long before him; they were dormant and implicit in his many sources; it could be said that they come under the heading of the "collective unconscious". *Hamlet* is a script into which much can be read that was not knowingly put there by the author, a genius with an exceedingly fertile imagination that drew on very deep impulses.

All the non-title characters are memorably realized: the Machiavellian Claudius and his errant Queen Gertrude; Laertes; the bumbling, prying, sententious Polonius; the faithful Horatio; the nimble Rosencrantz and Guildenstern; the boisterous, bawdy gravediggers; the polite Osric.

Somewhat enigmatic is the relationship between Ophelia and the prince. Is she quite innocent and virginal, or have they been lovers? Equally unclear is how early her mental instability manifests itself. Polonius may be portrayed as either cunning or stupid, or perhaps both. He is sometimes shown in a comic light and other times as a shrewd hypocrite.

Hamlet is a universal figure because he possesses so many human qualities: mercurial, he is one moment congenial and kind, the next moment ruthless, even savage; his wit is sardonic; he is high-spirited, then despairing – today's psychologist might suspect him of verging on manic-depression. Contemplating suicide, he lists to himself in succinct phrases the many ills that flesh is heir to, the human predicament. He foresees a future in which he will be endlessly called upon to take up arms against a sea of troubles. His sleep will be invaded by disturbing dreams; the nightmares may even visit him after his death – how can he be sure they will not? "Aye, there's the rub." Death is the unknown land from which no traveller returns. To go on – to be or not to be – that is the question. Is life worth the frequent pain of living?

Hamlet has been a challenge to all greatly talented actors (and many who were not adequate to the immensely demanding portrayal). They are drawn to him because Hamlet, like his author, has a histrionic personality, again like Richard II. He puts on a half-dozen roles, altering them to match each person or situation he meets. As a consequence the world of *Hamlet* is in more ways than one a "play-within-a-play". The whole many-faceted world is seen as just that. So many strata of reality and pretence overlap, ambiguity imposed on ambiguity, that nothing is finally knowable. It is Pirandellian. Particularly, it has been said, all the persons at the play-within-the-play that Hamlet has ordered staged are themselves playing roles, wearing "masks", so to speak – Claudius, the queen, Hamlet, Polonius, the courtiers – while the real audience watches them perform, too, putting on the guise of detached observers when all are secretly anxious and scheming. Here is an in-depth embodiment of the idea of how elusive, shifting and relative is truth, how many flashing mirrors are reflected in other mirrors that give each of us our picture of other people and their worlds.

Above all, *Hamlet* is a play about the "mystery of existence". It is filled with more questions than answers; its tone has been described as "interrogative". Its hero is always seeking to know what may be unknowable, the ultimate truth of things, the absolute. The script is about his frustrated metaphysical search.

Hamlet should not be seen as passive or inactive, or too delicate or idealistic, or too good and virtuous to thrive in his environment; he is capable of quick and sudden decisions, as when he attends to the urgent burial of Polonius, and disposes of Rosencrantz and Guildenstern, and leaps into Ophelia's grave to fight Laertes. About only one point is he procrastinating: the proper slaying of Claudius, his stepfather, his sovereign; a difficult affair, surely, and one that demands that he be certain of the facts, and then be convinced of the act's moral justification, and its practical feasibility if he is to inherit the sceptre with his subjects' full consent.

It should also be appreciated that *Hamlet*, like *Oedipus*, is on one level a detective play – the first audience did not know that Claudius had murdered his brother; only slowly was it revealed and confirmed for them. To later generations this aspect of the script has been erased, for they are familiar with the plot. Initially the play had even more stage-suspense. Of course *Hamlet* is a "mystery play" in another and deeper sense; the hero seeks to unravel other secrets: the value of human life, the demands of honour, the daunting paradox of evil flourishing in a universe ruled by a kindly, omnipotent Deity.

Since Aristotle's day the *Oedipus* has been designated by almost all critics as the perfectly constructed play. *Hamlet*, as artfully rearranged by Shakespeare, is steadily exciting, gripping; each scene follows logically, adding to the mounting tension. But several commentators, among them T.S. Eliot and John Wain, feel that it is "imperfect", "an artistic failure". It is too long, too broad in scope. Eliot, never a modest critic, claims that Hamlet over-reacts to the circumstances that he faces; his emotion goes beyond what is called for by external events. Wain largely agrees with Eliot, asserting that while writing the play Shakespeare did not completely master his own feelings about Hamlet – or about the feelings ascribed to Hamlet, the character – and that the resultant script is an inadequate vehicle for them. The difficulty that most other critics have in interpreting the play is the inevitable consequence of that lack of clarification. Ultimately the script is obscure. Shakespeare began to write a revenge tragedy, then he pushed beyond that form, launched on another, deeper study, and the two halves of the drama never fully merge. No one can deny the magnificent theatricality of a rapid succession of vivid events: the encounter with the ghost on the fog-enshrouded parapet, Hamlet's confrontation with his mother in her haunted bedchamber, his tongue-lashing of the startled, bewildered Ophelia, the verbal exchanges with his erstwhile "friends" Rosencrantz and Guildenstern, the rehearsal with the players and his instructions to them on acting, the inadvertent slaying of the furtive Polonius, the "play-within-the-play" at which the guilty Claudius starts up crying "Give me some light! Away!", the near escape from murder of Claudius as he kneels in prayer, and the violent meetings of Hamlet and Laertes, another young man with a father to avenge, to the final duel that leaves corpses strewn about the stage. Harley Granville-Barker calls *Hamlet* "an unsurpassed success – it must be the most successful play ever written – and a convincing evidence that character can be made to tell against every licence and vagary in story and construction". Wain assents that the drama is unfailingly effective on stage; he says that Shakespeare, having found the full strength of his creative powers, suddenly felt that he was without limit, could do everything and attempted too much, was carried away at the expense of true artistic unity. He adds: "But there was no going further on this path; Shakespeare has stretched the capacities of his theatre to their utmost." (Earlier, Goethe – commenting on the play's loose form – placed it "somewhere between a drama and a novel".)

Whatever the merits or demerits of its form, *Hamlet* is a triumph of language. The poet's already remarkable command of words, his style, is now consummate: flexible, epigrammatic, lyric, terse, astonishingly clear, easy to speak. Bartlett's *Quotations* contains twenty pages of memorable and familiar lines from the play. Much as happened after Shakespeare wrote *Romeo and Juliet* and *Julius Caesar*, enriching the everyday vocabulary, a good many English-speaking people now go about their daily affairs easily and repeatedly using phrasing from *Hamlet*; though hardly pausing to be aware of it. At work in Stratford-upon-Avon, the busy, home-bound playwright finds the exact, the perfect, the lasting expression for a host of routine or extraordinary feelings and experiences. Who does not glibly borrow from him and remark: "'Tis a consummation devoutly to be wished", or "There's nothing good or bad but thinking makes it so", or "Brevity is the soul of wit"? Doing so expresses our thought or emotion more pointedly than we could have done on our own and lends a touch of poetry to our habitual speech.

In places the language is extravagant. To Ophelia, the good-looking Hamlet is "the glass of fashion and the mould of form". In a contrite mood, he asks: "Nymph, in thy orisons be all my sins remembered." Reprimanding himself, he declares: "O God, I could be bounded in a nutshell, and count myself king of infinite space, were it not that I have bad dreams." Dying, he begs the loyal Horatio to mourn him: "Absent thee from felicity awhile." And the fond Horatio, as Hamlet's poisoned body is borne off, exclaims: "Now cracks a noble heart. Good night, sweet prince, / And flights of angels sing thee to thy rest."

Lyrical passages:

> This bird of dawning singeth all night long . . .
> But, look, the morn in russet mantle clad
> Walks o'er the dew of yon high eastward hill.

Elsewhere the characters convey their thoughts briefly, encapsulating them in a mere line or two:

> The readiness is all.

> To hold, as 'twere, the mirror up to nature.

Hamlet ruminates:

> I must be cruel, only to be kind.

And:

> There's a divinity that shapes our ends,
> Rough-hew them how we will.

Says Claudius, of the drowned Ophelia:

> Leave her to Heaven.

He has pangs of guilt, haunted by his crime:

> My offence is rank, it smells to Heaven.

And ruefully he seeks to pray:

> My words fly up, my thoughts remain below.
> Words without thoughts never to Heaven go.

Hamlet, plotting to expose Claudius, assures himself:

> The play's the thing
> Wherewith to catch the conscience of the King.

And:

> Thus conscience doth make cowards of us all.

Seeking to learn what has become of his vanished Lord Chamberlain, the worried Claudius cajolingly asks supposedly mad Hamlet:

> Where's Polonius?

The reply:

> At supper.

On being asked by Claudius where Polonius is taking his meal, Hamlet replies with a touch of malice, in an almost camp vein:

> Not where he eats, but where 'a is eaten.

Uncharacteristically, Polonius imparts sound fatherly advice to Laertes, who is leaving on a mission abroad. His attire should be "Rich but not gaudy; / For the apparel oft proclaims the man":

> Be thou familiar, but by no means vulgar . . .
> Give every man thy ear, but few thy voice . . .

> Neither a borrower, nor a lender be . . .
> This above all: to thine own self be true,
> And it must follow, as the night the day,
> Thou canst not then be false to any man.

One might reasonably assume that the courteous, reticent Shakespeare was guided by these prudent maxims in his own conduct, save for the matter of money-lending.

The great soliloquies of the beset Hamlet, too long to be more than cited here, contain some of the profoundest musings on life and morals to be found in scripts produced simply for the theatre. They begin:

> O, what a rogue and peasant slave am I . . .
>
> How all occasions do inform against me . . .
>
> To be or not to be, that is the question. . . .
>
> O that this too, too solid flesh would melt . . .

Hamlet's sharp lecture to the actors is invaluable:

Speak the speech, I pray, as I pronounced it to you, trippingly on the tongue . . . Nor do not saw the air too much with your hand thus, but use all gently . . . O, it offends me to the soul to see a robustious periwig-pated fellow tear a passion to tatters, to very rags, to split the ears of the groundlings . . .

Scattered throughout the text are a score of other aphorisms and excellent phrases that help to define events that almost everyone is likely to encounter. Of his beloved father, Hamlet says: "'A was a man; / Take him for all in all, I shall not look upon his like again." To him, his stepfather is "a little more than kin, and less than kind". In a moment of despair, he tells himself: "How weary, stale, flat and unprofitable / Seem to me all the uses of this world."

And one finds such ringing lines as these:

> I shall talk daggers but use none.
>
> It is a custom
> More honoured in the breach than the observance.

> Angels and ministers of grace defend us!

When he suspects Ophelia of taking part in the cabal against him:

> Frailty, thy name is woman.

And:

> A countenance more in sorrow than in anger . . .

> As if increase in appetite had grown
> By what it fed on . . .

> You would play upon me, you would seem to know my stops,
> You would pluck out the heart of my mystery.

> A king of shreds and patches . . .

Beholding the skull of the deceased court entertainer:

Alas, poor Yorick! I knew him, Horatio; a fellow of infinite jest, of most excellent fancy.

To the manner born . . .

What should such fellows as I do, crawling between Heaven and earth?

And this surprising tribute, when Hamlet is in a better frame of mind:

> What a piece of work is man! How noble in reason! How infinite in faculty! In form and moving how express and admirable! In action how like an angel! In apprehension how like a god! The beauty of the world, the paragon of animals!

Shakespeare conceived Hamlet as young, just home from a university. Most actors who have dared to take the role (except in university productions) have had established reputations and enjoyed major acclaim; they are in or past middle age, like many Romeos – they no longer have a youthful appearance. This somewhat distorts the characterization but seems unavoidable. Many years of experience are needed for the player to encompass all sides of Hamlet's protean nature.

Someone has written a book, *Twenty-five Hamlets,* a descriptive compilation of notable performers who have been the "melancholy Dane". Over four centuries, hundreds have donned Hamlet's traditional garb and mien and conversed with Yorick's hollow skull, holding it in his hand, and affrighted stalked Elsinore's battlements – was it his father's ghost who spoke to him, or an apparition sent by the Evil One?

As might be expected, the first Hamlet was Burbage. After him, during the Restoration, the "authentic" Hamlet was taken to be that of Thomas Betterton (1635–1710). Besides acting, he managed the Lincoln's Fields Theatre, which was owned by Davenant, and later the Dorset Garden Theatre and the Queen's Theatre in Haymarket. (As stated before, Davenant claimed to be the Bard's unacknowledged by-blow.) Betterton asserted that he had learned the role of Hamlet from Joseph Taylor, who had been directed by Shakespeare himself. Few actors have enjoyed the esteem in which Betterton was held by his colleagues for his fine moral character.

In the eighteenth century, the role chiefly belonged to David Garrick (1717–79), Shakespeare's fervent exponent. He produced, directed and sometimes wrote his offerings; his acting was marked by high intelligence and considerable restraint in contrast to often florid enactments by his competitors. Testimony to this comes from a contemporary, Henry Fielding, in whose immortal novel *Tom Jones* an unsophisticated theatre-goer alludes to having seen the Bard's play. He observes that the cast was fine, except for Garrick who was so "natural" that he did not seem to be acting at all. He did surround himself with very talented supporters. Managing the Drury Lane Theatre for a span of twenty-four years, Garrick did much to improve the settings and performances, insisting on adequate rehearsals, better lighting, banishment of spectators from the stage. He consorted with many of the leading intellectuals of his day. (In his youth he had been a pupil of Samuel Johnson.) With his wife, a dancer, he spent some of his later years in France, where his productions were warmly welcomed. He returned to London and the Drury Lane, and at his death was buried in Westminster Abbey, close by a statue of his revered Shakespeare.

Other notable eighteenth-century Hamlets were John Philip Kemble (1757–1823), the scion of an eminent family of actors, whose chief rival was the fiery Edmund Kean (1778–1883). Kemble, educated for the Roman Catholic priesthood, changed his mind and stayed in the theatre instead. One of the twelve children of Roger Kemble, nine of whom became players, he had been on the stage since he was ten. He made an important London début at the Drury Lane as Hamlet at the appropriate age of twenty-six. Afterwards he took over management of the playhouse and subsequently that of Covent Garden. In his troupe was his sister, the famed and adored beauty Mrs Siddons (Sarah Siddons, the "incomparable", who had earlier graced Garrick's company). Kemble favoured historically accurate period costuming.

Kean had a very different temperament. He was the son of an actress and great-grandson of Henry Carey, a playwright and musician who was said to be the illegitimate offspring of the Marquis of Halifax; the identity of Edmund's father was not certain. Neglected in his childhood and adolescence, he grew up wayward and completely irresponsible. Various benefactors tried to help him, offered to adopt him, but he ran away from all of them, in one instance going to sea. He later joined a theatrical troupe, was assigned children's roles, became a ventriloquist, a circus acrobat and tumbler and broke both legs. Yet he advanced enough in his profession that he was asked to recite before George III. By thirty he was acting together with Mrs Siddons. He wed an actress and was offered an engagement at Drury Lane, overcame hostility and won stunning ovations. His life was as melodramatic – and tragic – off stage as on stage. Kean's Hamlet was wildly emotional, passionate in speech and flamboyant in gesture, but he thrilled audiences. According to Coleridge, it was for the spectator "like reading Shakespeare by flashes of lightning". He twice crossed the Atlantic and toured in the United States (1820, 1825). Plays have been written about Kean himself, including one by Alexandre Dumas *père*.

In the nineteenth century, London theatre-goers could look to William Charles Macready (1793–1873), Johnston Forbes-Robertson (1853–1937) and especially Henry Irving (1828–1905) for an interpretation of the aggrieved young Dane. With those three celebrated actor-managers established in the capital, many other British cities had resident provincial troupes and were frequently visited by touring regional companies that tirelessly served up *Hamlet*. The nineteenth century saw Shakespeare as theatre fare at its peak.

Macready contested with Kean for a broad popularity. (More about him later – see *Macbeth*.) Forbes-Robertson made several forays to America. He was versatile, outstanding in a wide range of contrasting roles, and his Hamlet was deemed particularly memorable.

Irving (born John Henry Brodribb), offspring of a small shopkeeper, had his first job at fourteen as a clerk in London. At eighteen he obtained a place with a provincial stock company and at twenty-one with a theatre group in Edinburgh. From there he rose in the ranks to better engagements in Manchester, where he was on view for five years, and finally returned to London (1866) when he was thirty-eight. He met and formed a professional partnership with Ellen Terry that lasted twenty-four years. She was his Ophelia. Together they scored a chain of successes that in time enabled him to take over management of the Lyceum in 1878. (She was the mother of Gordon Craig, the iconoclastic scene designer, and an aunt of John Gielgud, in the ensuing century one of Britain's most distinguished performers.) Irving dedicated his talents to many Shakespearian roles, not the least of them Hamlet. Indeed, for many years the Lyceum presented mostly the Bard's works, though Irving was admired for his parts in scripts by other playwrights. His productions became noted for lavish, cumbersome scenery, a style of presentation against which Gordon Craig, who sometimes had small roles, was to crusade in favour of simpler staging. Highly intellectual, charismatic, Irving's inclination was decidedly romantic. His readings of Shakespeare's characters were often quite original and striking. He was not without shortcomings in the eyes of some critics, one of whom unkindly observed that his was "a mannered elocution and awkward gait". Irving was the first English actor to be knighted in 1895. That honour was later awarded to Forbes-Robertson ,too, in 1913.

In the United States Shakespeare's plays were greatly in demand and displayed the acting gifts of Edwin Forrest (1806–72; again, see *Macbeth*); Junius Brutus Booth (1796–1852); his son, Edwin Booth (1833–93); Maurice Barrymore (1848–1905); James O'Neill (1847–1920); Richard Mansfield (1857–1907); and Robert Mantell (1854–1925), among many others. Touring companies criss-crossed the entire continent and thrived, even – as has been remarked before – in the rude, rough California mining camps where gold had been struck and spending was loose.

Born in London, the elder Booth had a classical education there but was drawn to the stage; he left home for nearby Deptford, where he had his first chance at performing when only seventeen. At a surprisingly early age he was given leading roles in provincial troupes and at twenty-one made his bow as *Richard III* at Covent Garden. He was a tempestuous actor much like Edmund Kean, and it was predicted that he would become Kean's major rival; indeed, he was soon cast as Iago and conspiring against Kean's Othello (1817), at Drury Lane and again at Covent Garden. Spectators took sides; there were riots in the pit every time the pair appeared together in the opposite roles. In 1821 Booth married Mary Anne Holmes, heedless of his already having a wife. Perhaps it was because of the threat of being charged as a bigamist that he moved to the United States, making his début there in Richmond, Virginia, now as the villainous *Richard III*. He and his second wife had several children, who were astonished when the first wife appeared after a lapse of thirty years and sued for a divorce. For the remainder of his career he performed in many cities, except for two seasons when he went back to London. Eccentric and increasingly alcoholic, his behaviour at times bordered on insanity. It was hard to tell whether his brooding and bizarre actions were caused only by his drinking. In any event, audiences found his Hamlet stirring. His sons Edwin and John Wilkes accepted him as an exemplar and became actors, the younger assuring the family an unwelcome place in history. A Southern sympathizer during the Civil War, he was deeply disturbed by its outcome. While President Lincoln sat in a box at Ford's Theater in Washington watching *Our American Cousin*, John Wilkes crept up behind him and fired a shot that killed him. Escaping, his leg broken as he leaped from the box, he was pursued for twelve days, finally being overtaken by soldiers and fatally wounded.

Edwin Booth was more highly regarded than his father and brother; indeed, he became a legendary figure, long revered as the finest American actor of his century. A biography of him by Eleanor Ruggles is appropriately titled *Prince of Players*. Like his father, at sixteen he chose *Richard III* in which to make his first dramatic essay, though in a minor role; it was staged in Boston in 1849. Touring for a dozen years, as did his peers, he achieved marked successes in places as remote as Australia and California, until he returned to New York and assumed management of the Winter Garden Theater (1863–7). He had to overcome the cloud enveloping his family after his brother's horrible deed and withdrew from public sight for almost a year. His admirers insisted on his reappearance. Ambitious, after fire razed the Winter Garden he built Booth's Theater (1869), designed to be the most modern and safe playhouse for actors and spectators; it had every sort of up-to-date stage machinery and required an outlay of a million dollars, straining his financial resources; in 1874 he was bankrupt. Thereafter he resumed touring under other managements, once again visiting scattered American cities as well as London and Germany, his reputation such that he attracted full houses almost everywhere. He and his company were largely devoted to works of Shakespeare, in which his portrayals were considered unequalled, not the least his Hamlet, a part he took at his last performance, at the Brooklyn Academy of Music (1891). New York still has a Booth Theater (not the original), and he left another memorial, a club called the Players, of which he was the founder; his picture hangs near the entrance to his house which he bequeathed to the club, fronting Gramercy Park.

In his youth, accompanying his father to the crude Western towns, he had sometimes put on a black-face and played a banjo to entertain the noisy, drunken pioneers and miners. Of the later career, Howard Taubman writes in *The Making of the American Theater*: "Of Booth's extraordinary qualities there can be no doubt. It is possible to listen to him in a Hamlet soliloquy recorded with primitive equipment in the last years of his life, and to be absorbed by the sensitive and affecting nuances of the reading. There is ample eyewitness testimony to his accomplishments. Small and neat of figure he had the skill, consciously developed, to create the illusion of height.... He served a long, hard and varied apprenticeship.... His *Hamlet* (1864) was a sensation for its day, running an astonishing one hundred performances. In his own theatre, Booth worked to establish high standards. Playing under the management of others, he did not relent in his search for fresh insights. Contemporaries speak of the subtlety of his acting." Taubman quotes a member of Booth's company who recalled that " 'his art of make-up was as ruthless as a modern painter's. It did seem as if he labored to dig out the ugliest traits of his character and paint them into his face.' ... If the American theatre has a patron saint among its actors, it is Booth."

Maurice Barrymore (Herbert Blythe) was another British player who chose to make his way in America. A graduate of Cambridge, he reached New York in 1875 when he was twenty-seven after only brief stage experience in the provinces of his native land. A year later he married Georgiana Drew, a remarkably disciplined actress. As a pair they enjoyed quick acceptance in both America and Britain, he also as leading man for a long line of internationally laurelled ladies, Helena Modjeska, Olga Nethersole, Lily Langtry, Mrs Fiske. The Barrymore–Drew marriage resulted in four children, three of whom – Lionel, Ethel and John – and a granddaughter, Drew Barrymore, were in time to outshine their forebears as stage, film and television luminaries. They were often referred to as "the royal family" of the American theatre and even caricatured for adopting that pose in a hit play bearing that title (1927), by George S. Kaufman and Edna Ferber. Ethel Barrymore was with Henry Irving's company for a short spell. Lionel was dispraised for his *Macbeth*, and John lauded for his *Hamlet* (see below).

Both Robert Mantell and Richard Mansfield were actor-managers, heading touring companies that purveyed the Bard's many plays to receptive theatre-goers nationwide. Mantell was British by birth; Mansfield, the son of a London wine-merchant and renowned German opera singer, was a native of Berlin. Besides playing in Shakespeare, Mantell, too, was associated with Modjeska and later Fanny Davenport. Today it is hard to assess his ability as an interpreter of the major roles. Taubman is deprecatory and describes him as a "ranter". Elsewhere – despite his being outlived considerably by Mansfield – he is more leniently rated as "the last of the great Shakespearian actors in the old tradition".

Richard Mansfield was notoriously unpredictable, often changing his interpretations, stage-business and timing without advance warning to his colleagues, who might be quite disconcerted. In his youth, in England, he sang in Gilbert and Sullivan troupes before abandoning them and sailing to America, where he applied for work, only to be rejected everywhere. He finally got a small part in a current and long-running musical comedy, *The Black Crook* (1882). From there he went to another minor role in a romantic comedy; its lead actor left before the opening and Mansfield replaced him. He did so well that immediately he was hailed as a star. Within a decade his prominence in the Broadway scene was assured. He appeared in hit after hit. The best playwrights of the day wrote effective vehicles for him. He liked roles in which he could be elegantly costumed – Prince Karl, Beau Brummell, Cyrano de Bergerac. He brought Shaw's first play to America, *Arms and the Man*, as well as Ibsen's gnomic fantasy *Peer Gynt*.

Great promise was seen in the early career of James O'Neill, after a bad start. Irish-born, of an impoverished family, he was brought to America as a child of three. His father soon deserted the family, leaving the mother to struggle while raising the boy. He never outgrew his fear of poverty. Giving rein to his instinct to become an actor, he sought employment in the theatre. Eventually he was engaged for parts by Adelaide Neilson and Edwin Booth, scoring as an evil Iago to Booth's gullible Othello. He also took over the title role, prompting Booth to say: "That young man is playing Othello better than I ever did" – or so O'Neill himself asserted afterwards. His future was considered to be an assured one, but then he came across a melodramatic script dramatizing Dumas's novel *The Count of Monte Cristo*. He purchased it, assumed the bravura character of Edmond Dantes, winning so much applause that he appeared in nothing else on stage for the rest of his life. He toured with this popular text year after year, earning large sums, becoming wealthy. His two sons, James Jr and Eugene, unwillingly assisted him. Eugene, rebellious, broke away to become America's foremost playwright, a Nobel Prize recipient. The relationship between the father and his sons was difficult – they considered him miserly, he felt himself to be simply prudent and his hard-drinking sons to be extravagant wastrels. One consequence was a harsh portrait of the senior James O'Neill in Eugene's classic *Long Day's Journey into Night*, but those who knew the O'Neills state that the play's image of him is exaggerated and unjust.

Throughout the twentieth century Britain had an excess of superior male actors, many of whom won wide recognition and knighthoods. Most of them, at one time or another, embodied Shakespearian characters, including the protean Hamlet, each contributing a somewhat different view of him. Opportunities to do so came from the Royal Shakespeare Company in Stratford-upon-Avon and in London, together with the National Theatre established on the South Bank of the Thames, and the sturdy Old Vic, all in ongoing perpetual subservience to putting on the Bard's fare. Finally, the replica of the Globe was added to London's many other stages. The system of actor-managers that prevailed under Irving and that had governed in America more or less vanished, though not entirely so. The national theatres recruited stars and supporting casts when they were needed, after which the player was once more on his or her own. Those aspiring to be Hamlet had a good chance of having a turn at it for part of a season somewhere or some time. It is the most prestigious role of all, hard to refuse. From a roster of illustrious names – an incomplete list – a director could choose for his introverted Dane the talented and resourceful Michael Redgrave, Alec McCowen, John Neville, Anthony Quayle, John Gielgud, Laurence Olivier, Ben Kingsley, Alec Guinness, Albert Finney, Ian McKellen, Derek Jacobi; not all of them actually impersonated Shakespeare's confused hero but a majority did, if not at the South Bank or in the West End, then with a civic repertory company outside London, of which there are many. Alec Guinness startled spectators by offering a Hamlet with a beard. Finney, at the insistence of his director Peter Hall, was Hamlet in an exhausting, uncut version, much as Maurice Evans had ventured to enact some decades earlier in New York. Olivier, in a blond wig – or was his hair bleached? – to look more Scandinavian, filmed the play, directing it; some critics thought that in it he overacted by a considerable margin. (In 1937 he had been Hamlet in a British production put on in the real castle at Elsinore.)

Perhaps the loudest, long-echoing acclaim was reaped by John Gielgud, who twice interpreted the role, the second time self-confidently serving as producer (1934), while keeping the title role. His reward was a near record run in London. The play was then taken to New York, where Gielgud had an unusually brilliant supporting cast that included Dame Judith Anderson as the queen and Lillian Gish – legendarily identified for being a fragile heroine in early silent films – as a nervous Ophelia. Guthrie McClintic was the director. In the *New York Times*, Brooks Atkinson was not wholly persuaded:

> The magnificos of the modern theatre, who latterly were creating a masterly *Romeo and Juliet*, have come to a greater panel in the Shakespearian screen and performed honorably before it. . . .
>
> Mr Gielgud comes now on the clouds of glory that in the last few years have been rising around him in London. He is young, slender and handsome, with a sensitive mobile face and blond hair, and he plays his part with extraordinary grace and winged intelligence. For this is no roaring, robustious Hamlet lost in melancholy, but an appealing young man brimming over with grief. His suffering is that of a cultivated youth whose affections are warm and whose honor is bright. Far from being a traditional Hamlet, beating the bass notes of some mighty lines, Mr Gielgud speaks the lines with the quick spontaneity of a modern man. His emotions are keen. He looks on tragedy with the clarity of the mind's eye.
>
> As the results prove in the theatre, this is one mettlesome way of playing the English stage's most familiar classic. But it is accomplished somewhat at the expense of the full-blooded verse

of Shakespeare. What Mr Gielgud's Hamlet lacks is a solid body of overwhelming emotion, the command, power and storm of Elizabethan tragedy. For it is the paradox of Hamlet that vigorous actors who know a good deal less about the character than Mr Gielgud does can make the horror more harrowing and the tragedy more deeply felt.

Atkinson conceded that the cast and McClintic had studied the play with fresh eyes. "Arthur Byron's Polonius, for example, is no doddering fool but a credible old man with the grooved mind of a trained statesman. As the Queen, Judith Anderson is a woman of strong and bewildered feeling." Fault was found with the settings by Jo Mielzinner, "though his costumes are vivid with beauty. This is an admirable *Hamlet* that requires comparison with the best. For intellectual beauty, in fact, it ranks with the best. But there is a coarser ferocity to Shakespeare's tragedy that is sound theatre and that is wanting in Mr Gielgud's art."

(The author of this book was in the audience and thought the performance was wonderful, among other details leaving for ever in his memory the force and sharpness of Hamlet's instruction to the actors.)

Has there ever been or will there ever be a perfect image of Hamlet, an ultimate explanation of his feelings and emotions? Sensing that people are for ever ambiguous, Shakespeare himself may not have fully defined for himself his hero's character. Overlooked by most stage interpreters is that Hamlet is another Mercutio, Biron, Jaques – an obvious aspect of the playwright's own personality, his natural mischievousness provoked until he retaliates with murderous trickery, while his innate scepticism is enlarged by ill happenings to an existential denial of divine intervention in human affairs.

At the start of the twentieth century Shakespeare was being represented in America by an ingratiating pair, Edward Hugh Sothern (1859–1933) and Julia Marlowe (Sarah Frances Frost; 1866–1950), husband and wife. Both were of British origin, like so many of their fellow actors of that day. Sothern's father, Edward Askew Sothern, was a major figure on the London stage, as was his younger brother George Evelyn Sothern. Beginning his career in the United Kingdom in farces, Edward Hugh mutated to romantic drama after his arrival in the United States, selecting such works as Gerhardt Hauptmann's poetic fantasy *The Sunken Bell* and Justin McCarthy's *If I Were King* (subsequently to be made into the popular operetta *The Vagabond King*), and from there to Shakespeare, becoming in the view of many Americans the best player in the Bard's enduring contributions to theatre literature. In 1899, having reached forty, he formed his own company and began to tour with it, journeying everywhere. A dozen years later he married Julia Marlowe; he was sixty, she was forty-five, an actress. Together, they proved to be an irresistible attraction, in tandem but also separately, she as Rosalind, Beatrice, Juliet, Ophelia, he as one or another of Shakespeare's kings or princes, among them inevitably the troubled Hamlet. In 1926 they presented a series of the Bard's plays at Stratford-upon-Avon.

Walter Hampden (1879–1956) was another of the respected and popular actor-managers much like Mansfield, Mantell and Sothern, but his company stayed mostly in New York, touring only occasionally, and his taste in scripts was more modern than theirs. American-born – Brooklyn – he came to the theatre broadly and diversely educated, having studied at the Brooklyn Polytechnic, at Harvard and in Paris. Crossing to Britain, he joined F.R. Benson's travelling company, a veritable seedbed for able professional British actors, and remained with it for three years (1901–4), participating in staple comedies and Shakespeare. (Back home, the poet Walt Whitman had already protested against the New York stage being taken over by English and English-trained managers, actors and scripts.) Hampden lingered abroad another four years, making London appearances, until returning to New York to support the great Russian star Alla Nazimova in his Broadway début; he was tall, imposing, handsome. Almost two decades later, having acquired more experience and learned the tricks of managing a theatre and troupe, he organized his own company, with himself the producer, director and leading man; he leased a rather seedy playhouse, the Colonial, on Broadway but somewhat to the north of Times Square, and charged low prices for tickets. He changed its name to the Hampden Theater. The interior was a bit dilapidated, but for many eager play-lovers the stage was alight with art. His repertory consisted mostly of Shakespeare – *Othello*, *Hamlet* – Ibsen – *An Enemy of the People* – and Goodrich's romantically florid *Caponsacchi*, and best of all Rostand's *Cyrano de Bergerac*, a role that he made so much his own that no other American actor would dare undertake it. For New Yorkers, among whom was young Howard Taubman, he *was* the long-nosed Cyrano. Whenever one of the company's offerings faltered, Hampden had only to bring back Rostand's play for a limited engagement to recoup the box-office's depleted fortunes. His troupe was always made up of competent actors, though none of them could match the panache of his accurate swordplay or compete vocally with his "sonorous grandiloquence".

His Hamlet, as summed up by Alexander Woollcott in the *New York Times* in 1920, "was a creditable achievement which commands the respect of his generation in the theatre. Both press and public have been responsive enough to it to give him at least a sense of satisfaction." Woollcott felt that the production deserved to be toured and hence more widely seen.

> It approaches *Hamlet* freshly and fearlessly as though Mr Hampden had found the musty script in some delving of his own and staged it as he would any other play, so that it is done without pomposity and with none of the paralysis of ritual. The appropriate and ingenious settings and most of the company are as before – with the addition, notably, of an excellent Polonius contributed by Allen Thomas and a pallid Ophelia by Beatrice Maude.
>
> Mr Hampden's performance as the sweet prince has been lauded to the skies by some of his contemporaries as the best Hamlet of our time. If, in this connection, the undefined phrase "in our time" includes the period in which Forbes-Roberston's performance was visible in the American theater, it would seem to this reviewer that some memories have been ungratefully short. If "our time" means merely yesterday and today, the praise can hardly be said to be

extravagant. In that event, they are not saying much. This performance is unquestionably the best we have – but not quite the best we might have.

Woollcott credited Hampden with "a royal presence, a rich and well-trained voice and a commanding intelligence", but found him without a flair for comedy, a component of Hamlet's volatile nature.

> The humor that makes Hamlet winning, the enriching and deepening humor, is absent. The impishness is there in full force, achieved by a sort of mechanical mockery. But Hamlet's roving, mutinous, lovable sense of fun, his enormous appreciation, are not there. This leaves the play still inexhaustibly interesting but it robs the tragedy of its pathos and poignance, despoils it of its overtones. You could not possibly be bored by Walter Hampden's Hamlet but probably your heart would not be wrung by it either.

Except for Edwin Booth, no American actor has been as loudly and lengthily cheered when venturing a Hamlet (1922) than was John Barrymore (1882–1942), the youngest of Maurice Barrymore's amply talented progeny. He came along a bare two years after Walter Hampden's generally admired portrayal, but John Barrymore belonged to "the royal family" whose every doing commanded attention, so having to follow this closely on Hampden's heels did not deter him. Besides, the producer was Arthur Hopkins, who had sponsored Barrymore's well-greeted *Richard III* and himself was greatly esteemed for a progressive attitude and unerring good taste, a readiness to take chances with serious new dramas. John Barrymore had already proved himself in a series of substantial hits. He was also much talked and gossiped about for his scandalous affairs. He had a quirky sense of humour, a mocking laugh and was decidedly handsome, his noble Roman profile enthralling women spectators – he was *sine qua non*, a matinée idol, engaged in romantic activities on and off stage. Unfortunately he had an over-liking for intoxicating bottled spirits – said to be a family weakness.

Apparently he was meant for the role of Hamlet. John Corbin, in the *New York Times*:

> The atmosphere of historic happening surrounded John Barrymore's appearance last night as the Prince of Denmark; it was unmistakable as it was indefinable. It sprang from the quality and intensity of the applause, from the hushed murmurs that swept the audience at the most unexpected moments, from the silent crowds that all evening long swarmed about the theater entrance. It was nowhere and everywhere. In all likelihood we have a new and a lasting Hamlet.
>
> It was an achievement against obstacles. The setting provided by Robert Edmund Jones, though beautiful as his setting for Lionel Barrymore's *Macbeth* was trivial and grotesque, encroached upon the playing space and introduced incongruities of locale quite unnecessary. Scenically, there was really no atmosphere. Many fine dramatic values went by the board and

the incomparably stirring and dramatic narrative limped. But the all important spark of genius was there.

Mr Barrymore disclosed a new personality and a fitting one. The luminous, decadent profile of his recent Italian and Russian impersonations had vanished, and with it the exotic beauty that etched itself so unforgettably upon the memory, bringing a thrill of admiration that was half pain. This youth was wan and haggard, but right manly and forthright – dark and true and tender as befits the North. The slender figure, with its clean limbs, broad shoulders and massive head "made statues all over the stage," as was once said of Edwin Booth.

Vocally, the performance was keyed low. Deep tones prevailed, tones of a brooding, half-conscious melancholy. The "reading" of the lines was flawless – an art that is said to have been lost. The manner, for the most part, was that of conversation, almost colloquial, but the beauty of rhythm was never lost, the varied, flexible harmonies of Shakespeare's crowning period in metric mastery. Very rarely did speech quicken or the voice rise to the pitch of drama, but when this happened the effect was electric, thrilling.

It is the bad custom to look for "originality" in every successive Hamlet. In a brief and felicitous curtain speech Mr Barrymore remarked that everyone knows just how the part should be acted and he expressed pleasure that, as it seemed, he agreed with them all. The originality of his conception is that of all great Hamlets. Abandoning fine-spun theories and tortured "interpretations" he played the part for its *prima facie* dramatic values – sympathetically and intelligently always, but always simply. When thus rendered, no doubt has ever arisen as to the character, which is as popularly intelligible in the theatre as it has proved mysterious on the critical dissecting table.

Here is a youth of the finest intelligence, the tenderest susceptibility, with a natural vein of gaiety and shrewd native wit, who is caught in the toils of moral horror and barbaric crime. Even as his will struggles impotently to master his external environment, perform the duty enjoined on him by supernatural authority, so his spirit struggles against the overbrooding cloud of melancholy.

If the performance had any major fault it was monotony, and the effect was abetted by the incubus of the scenic investiture. There was simply no room to play in. It may be noted as characteristic that the Ghost was not visible; the majesty of buried Denmark spoke off stage while a vague light wavered fitfully in the centre of the backdrop. In one way or another the play-within-the-play, the scene of the King at prayer and that of Ophelia's burial, all more or less failed to register dramatically.

The production came precious near to qualifying as a platform recitation. But even at that Mr Barrymore might have vitalized more fully many moments. With repetition he will doubtless do so. The important point is that he revealed last night all the requisite potentialities of personality, of intelligence and of histrionic art.

Eight years passed before another Hamlet arrived on Broadway. He was Fritz Leiber (1882–1949), born in Chicago and largely making his stage appearances there, beginning in 1902. For a time he toured with Robert Mantell as well as Julia Marlowe. For thirteen years (1919–32) he directed and acted with the Chicago Civic Repertory Society, then formed his own troupe and briefly toured with it (1934–5). He could boast of solid, extensive familiarity with Shakespeare's characters, having at one time or another assumed many different roles in the Bard's plays.

Confident that its productions were worthy of showing to the country's most knowledgeable audience, the Chicago Civic Shakespeare Society brought its repertory to New York, led by its truly accomplished director. Brooks Atkinson reported for the *New York Times*:

> Fritz Leiber's *Hamlet*, which was acted at the Shubert last evening, has the uncommon quality of being not only a part but a play. It is the tragedy not only of a young man condemned to set an unjointed time to rights, but also of a brother and a sister, a witless old fool, a treacherous king and a faithless queen. For Mr Leiber contends in the face of all stardom that Shakespeare is a dramatic poet telling here a story of great pith and moment, and that the actor's first duty is to make it intelligible. What he has accomplished in the lucidity and rapidity of staging, in the drive of a drama through many scenes and in the interpretation of the lines for their dramatic meaning – is a revelation in the producing of Shakespeare. For the expression of the tragedy as an organic unit, this is the most coherent Hamlet this courier has ever seen.
>
> He approaches the part with the same blunt purpose as the play. His is a straightforward Hamlet – no muddy-mettled spouter of lines, but a character alert to all the relationships of the drama, clarifying the dialogue with simple gestures, waiting his time without slackening pace, keeping his diction clean and his mind on edge. He is, moreover, an actor of sufficient fullness to encompass the gusts of fury that at times sweep through the part, and to rise to the heights of a flagellating passion. Although he cuts through to the likeness of Hamlet he does not pluck out the heart of his mystery. For, after all, Hamlet is no normal person. The paralysis of will that gives the play its story, the hammering of the mind and the malignant brooding are the full measure of the character; and making no pretense to being a virtuoso Mr Leiber does not squeeze the part dry. Applying common sense to the role, Mr Leiber makes Hamlet an understandable character who appeals to the sympathies. But there is still more to this racked and pensive youth; there is a commanding nobility, and also a preternatural sensitivity to the whips and scorns of man's fortunes. Within the dramatic limitations he has set himself, Mr Leiber does extraordinarily well.
>
> Although his company is not all of a piece most of them are equal to their appointed tasks. Marie Carroll's child-like Ophelia puts this much-slandered part in its true perspective. All the attempts to lift this part into a cosmic significance seem as futile as they are in comparison with Miss Carroll's weak, bewildered, pathetic maiden who is caught up in a whirlwind beyond her comprehension. And the mad scene, which tries the souls of even the believers, becomes singu-

larly affecting in the simplicity and the unostentatious design of Miss Carroll's acting. Virginia Bronson is a Queen of such excellent resolution and feeling that William Courtleigh's King seems to lack the royal authority. Nor does Philip Quin get all the bland humor in Polonius. The effeminate mannerisms of Rosencrantz and Guildenstern appear to be nothing but a distasteful truckling after illegitimate laughs. But the Laertes of Lawrence H. Cecil has admirable character, the Horatio of John Burke is to the point, and the First Grave-Digger becomes a hearty low-comedy philosopher in the acting of Robert Strauss.

Mr Leiber comes to New York under the auspices of the Chicago Civic Shakespeare Society with a repertory of seven plays, some of which, like *Twelfth Night*, *As You Like It*, *Richard III* and *King Lear* have almost faded from the New Yorker's memory. After completing a season of twelve weeks in Chicago he is now concluding a brief tour. Next year he will resume his activities in Chicago for the second of five guaranteed seasons.

The significance of his present engagement is, accordingly, more that of a Shakespearian festival than of individual plays. The productions are notably costumed. Herman Rosse has designed sets that make an ingenious compromise between scene suggestion and modernistic stylization, and they can be changed without interrupting the performance. On the whole, Mr Leiber has got a remarkable enterprise well in hand. A man of great energy, he sets a high pace in the performances and whips his company straight through to the end. He stages Shakespeare for those who love the plays and who welcome an unusual opportunity to see them ably acted and produced.

Through the mid decades to the end of the century, the successive Hamlets reaching Broadway continued to be of British origin or trained there, as was true of Evans, whose marathon, uncut version set a new standard for on-stage physical endurance, a nervous and muscular ordeal shared by his engrossed but uncomfortable audience. British, too, was Leslie Howard (1893–1943), who before he tried acting worked as a bank clerk. His first role on the London stage occurred when he was twenty-five (1918), after a mere year's experience on tour. His success was immediate; he was lucky enough to be in several West End hits, and after another two years was brought to New York, where his good fortune continued. He was chosen to play opposite several of Broadway's glamorous leading ladies, among them Katharine Cornell in *The Green Hat*, and had roles in Sutton Vane's allegorical *Outward Bound*, John van Druten's *Berkeley Square* (adapted from Henry James's *A Sense of the Past*), Philip Barry's *The Animal Kingdom* and Robert Sherwood's *The Petrified Forest*, all above-average Broadway offerings. He was invited to Hollywood to repeat his roles on celluloid in several of these dramas, and there as well had the second male lead in the epic *Gone With the Wind*, a film classic. He was firmly established as a performer of assured competence and distinction on stage and screen. He had a good voice, spoke clearly and with upper-class diction, was tall and good-looking, his manner and bearing quietly aristocratic. But there was surprise in theatrical circles when he announced his intention to do *Hamlet* on Broadway, and especially that he dared to do so in the same season in which John Gielgud was to come with his acclaimed London version. Of the two, Gielgud had the longer run and caused more excitement. Howard Taubman reports that Leslie Howard "earned the gratitude of connoisseurs who admired the subtlety of his interpretation". A dozen years later his recurrent good fortune changed abruptly, his end as tragic as Hamlet's. Flying back to Britain to assist in its Second World War effort, his plane crashed into the Atlantic and he perished.

In Minneapolis, in the distinguished regional repertory theatre bearing his name, the British director Tyrone Guthrie staged another updated *Hamlet* – all the players in modern dress – that many people disliked, though some found that for them it cast a new light on facets of the plot and characters. From Britain, too, arrived Richard Burton to present his Hamlet (1964), about which the New York critics were divided. He was robust, strong-voiced, already well recognized for his veritable achievements on stage and screen and also a bit too widely known for his turbulent marriages and addiction to alcohol. His *Hamlet*'s run exceeded that of Barrymore and Gielgud, and certainly of Booth, setting a new record for longevity if not for excellence. A susceptible segment of the public was eager to behold him in person rather than to measure his understanding of the frustrated Dane. Notoriety has its rewards.

An American actor, Richard Chamberlain, who had commended himself to a new broad audience by starring in the television series *Dr Kildare*, made his début as Hamlet in England, after which a film copy of his quite adequate performance was broadcast in the United States. He did not chance a local stage-test. Another American, Mel Gibson, a top-ranker in Hollywood, released his filmed *Hamlet* but avoided a stage-challenge altogether. In the United States, too, the ambitious young British actor Kenneth Branagh, who had made a strong impression as Henry V, had his film venture as Hamlet distributed but again remained away from a flesh-and-blood depiction on Broadway.

In Russia, *Hamlet* has exerted a long and steadfast appeal. Its brooding neurotic hero has a temperament and "soul" akin to one said to be lodged in many thoughtful citizens of that land. The visionary British designer Gordon Craig and Konstantin Stanislavsky, the innovative co-founder and director of the Moscow Art Theatre, had prolonged, exhaustive discussions that led to an experimental production of the play in 1912. A film version has a score by Dmitri Shostakovich, who subsequently shaped the music into a suite.

Germans, too, cherish *Hamlet*, an attitude that has continued through the centuries. An outstanding twentieth-century portrayal of its protagonist was that by Alexander Moissi (1880–1935). He was born in Austria, but his parents were Italian–Albanians; accordingly he spoke with a pronounced Italian accent, handicapping him when he sought a stage career in Berlin. The all-powerful producer-director Max Reinhardt sighted his talent and guided him. He was rewarded when Moissi became one of the country's most admired and prominent stage-personalities. Bilingual, he was almost as popular in Italy. Heading his own company, he visited New York with his *Hamlet* for a limited engagement that elicited a stir among the *cognoscenti*. What the major American critics thought of him cannot be told here, as he performed in German, which most of them could not understand. In those days earphones were not available.

Hamlet is known and staged everywhere. Kawatake Toshio has offered a Japanese translation of it. English-speaking companies on tour there had found that a majority of spectators are familiar with Shakespeare's best works; they read and discuss him in their college literature courses. The avant-garde director Yoshiyuki Fukuda told a *New York Times* interviewer: "*Hamlet* is a very popular play here." Fukuda's "revisionist" version of it (1986) sets the action after the Second World War, before Japan has got over the vast damage and suffering caused by the nuclear attacks. An updated Hamlet has to cope with this widespread despair along with his personal problems and vexing griefs. (A fuller account of the play is included in this author's *Stage by Stage: Oriental Theatre*.) In another instance Hamlet rides a bicycle on stage. During the British colonial period in India *Hamlet* was taken there by venturesome British companies and remained in that country's theatre literature in both English and the many local languages. There are *Hamlets* in Serbian and Spanish. If Robert Frost's definition of "poetry" is true – it is "what is lost in translation" – the endurance of the play in so many languages attests not so much to the poetic sublimity of its dialogue as to its essential strength as sheer drama.

Two operas use Hamlet's story. One is *Amleto* (1715) by the prolific Neapolitan composer Domenico Scarlatti. But scholars believe the librettists Apostolo Zeno and Pietri Pariati drew on the far earlier sources that provided Kyd with ideas for his putative *Ur-Hamlet*, not on Shakespeare's adaptation. In any event nearly all the score is lost, only one aria remaining. In contrast, the libretto for *Hamlet* (1868) by the French composer Ambroise Thomas (1811–96) is directly based on Shakespeare's play, the librettists – Jules Barbier and Michel Carré – having looked closely at its text. Some changes were made – the queen, by mistake, drinks the poison and dies before the duel with Laertes is arranged, and there are interludes for ballet and other kinds of dance, as Parisian audiences were noisily demanding. The libretto is not too effective. Hamlet is not a tenor but a baritone and assigned some highly dramatic arias, and the music is of top quality, especially the parts for soprano – the queen – and coloratura – Ophelia. Rapturously welcomed by opera-goers, the work was considered the best of its kind – that is, lavishly decorated "Grand Opera" – in a good while. Thomas is better known for his *Mignon* (from Goethe's novel, *Wilhelm Meister*) and also made an opera out of *A Midsummer Night's Dream*. His *Hamlet*, despite many virtues, has only intermittent performances.

Shakespeare's tragedy has been an inspiration to choreographers. Perhaps the best received *Hamlet* ballet has been that by the Australian dancer Robert Helpman (1909–86), who also appropriated its title role in 1942. As a youth he studied with Pavlova. Much of his career was spent in London with the Sadler's Wells (renamed the Sadler's Wells Royal) Ballet. He was noted for his unflagging zest. He also acted with the Old Vic and other theatre troupes. His lean, bony features made him an acceptable Hamlet in both the drama and the dance-work fashioned from it. His intellectual *Hamlet* is set to Tchaikovsky's *Fantasy Overture*.

Among others elsewhere is a German balletic version by Victor Gsovsky, with music by Boris Blacher, a scenario by Tatiana Gsovska, that was commissioned by the Bavarian State Opera in Munich in 1950. It adheres more closely to the play than does Helpmann's adaptation. It was moved to Berlin and staged for the Berlin State Opera and subsequently acquired by Berlin Ballett, a company that toured in the United States. Later the work was expanded by Gsovska and performed with a new cast at the Teatro Colon in Buenos Aires in 1951.

Updating the script – a practice by now an unfortunate tradition – Joseph Papp's Shakespeare in the Park, during its marathon run-through of the Bard's complete works, presented a *Hamlet* that was grotesquely antic and distorted. The critics loathed it. As if not to be outdone, London sent a modern facsimile of the great tragedy (1998). Vincent Canby wrote in the *New York Times*:

> The RSC's cinematically inspired *Hamlet* begins promisingly, though irrelevantly, with home movies (Hamlet as a small boy, playing with his father in the family compound). It then jumps to "the present": what looks to be a madcap Long Island evening party, complete with a large, swing band, on the occasion of the marriage of Claudius and Gertrude. Pay no attention to the dialogue, which still contains references to kings and queens and such. Claudius could be the new head of his late brother's brokerage firm, Gertrude a bored, middle-aged socialite who winters in Barbados and Hamlet a Yale student who drinks too much.
>
> As with so many attempts to update and find contemporary associations to the classics, this concept has no place to go once it is established. It simply gets in the way. Reducing the members of this lusting, neurotic royal family to the status of the *haute bourgeoisie*, circa 1950, also reduces the stakes, makes the governing passions seem petty and renders the violence more wanton than inevitable. The concept doesn't seem to have been thought through. When Laertes comes on carrying a rifle and looking for Hamlet, you might think he had been hunting squirrels.
>
> The language is all but lost amid these bizarre images. Alex Jennings's *Hamlet* is likely to be remembered more for his wardrobe, his decadent good looks (and the white-painted face he affects at one point) than for anything he says. This is a *Hamlet* in which the gravedigger sings "September Song;" Ophelia wears red shoes in her mad scene (evoking Judy Garland in *The Wizard of Oz*) and the Ghost turns up in a dinner jacket. Every now and then, when you hear a familiar speech, it seems to be a mistake, as if the soundtrack from another movie were butting in.
>
> Performances are impossible to judge in these circumstances. Susannah York, a most appealing and sexy actress on the screen, appears to be distant, small and unnaturally quiet as Gertrude. This isn't necessarily her fault: the entire play is like a badly edited film composed entirely of long-shots.

If Shakespeare's misanthropy was personal, he gives no sign of it in his *Twelfth Night, or What You Will*. Some editors place it and *The Merry Wives of Windsor*, which also dates from 1600–1, earlier than *Hamlet*. But a clear pattern is discernible in the order of most of the scripts; a serious work is followed by a light one, a pastoral or farce, which in turn is followed by another serious work, and so on. Was this alternation deliberate or unconscious? There could be many psychological explanations for these persistent shifts in mood in the creative process, but who could say what is the true one? It might be that Shakespeare was merely delivering what his company asked for to keep the box-office busy and prosperous.

Twelfth Night is again one of his gayest romances. His source is *The Cheated*, an anonymous Italian work, *Gl'ingannati* (acted 1531, printed 1537). This existed in so many English versions that he was likely to have had ready access to it. Besides, he was often handed material for possible development. Striking parallels to its plot and characters have also been discovered in Sanskrit drama, but this must be coincidence or arise from the worldwide similarity of mythic subject-matter. Indian literature would be far beyond his bounded reach.

Once more the scene is a fictional dukedom, Illyria. Sebastian and Viola, brother and sister, are shipwrecked on its shores. Each thinks the other drowned. Viola disguises herself as a boy and becomes a page of Orsino, Illyria's duke. He is in love with the Countess Olivia, who has recently lost her brother and father and discourages the duke's suit. Orsino asks Viola (whom he thinks a handsome youth) to press his cause with the reluctant countess, though Viola tells him it appears hopeless. The plentiful humour of the play is provided by three members of the Countess Olivia's court: her uncle, Sir Toby Belch, whose demeanour and personality bear a strong likeness to Sir John Falstaff; his friend, Sir Andrew Aguecheek, also in love with the countess; and her steward, Malvolio, who holds a grotesquely exaggerated opinion of his own merits. So far as is known, the sub-plot involving this trio is original with Shakespeare, and it tends to overshadow the main plot of Plautine intrigue and disguise in delighting spectators. Other characters at Olivia's court are: Maria, her maid; Feste, a clown; and Fabian, a servant.

Their betters belong to what might now be called Shakespeare's stock company of young lovers. They are daft, needlessly practising deception, assuming false identities and cross-dressing. They are never shown threatened or endangered; their charade is purposeless. The outworn plot device worked well enough when boys and young men had the roles: it is often quite easy for a man to pass as a girl, but very difficult for a girl to pretend to be a man – in today's theatre few if any of even the most talented actresses do it successfully; the modern audience has to be indulgent, suspending disbelief.

The conceited Malvolio is led by his prankish companions to believe that his countess is smitten with him, and he makes a fool of himself, at the same time raising doubts as to his sanity in Olivia's eyes. The lady, however, is hardly more wise; she has become infatuated with Orsino's young messenger, Viola, who is in male masquerade. Sir Andrew Aguecheek, growing jealous, challenges the "page" to a duel.

Sebastian, the supposedly drowned brother, arrives in Illyria in the company of Antonio, a ship's captain, who rescued him from the sea. The pair separate for safety's sake. Antonio has enemies in Illyria, where he is looked on as a pirate. He comes upon the duellers, and mistaking Viola for her brother Sebastian offers to take "his" place in the sword-exchange. Some officers happen upon the scene and arrest Antonio. He has entrusted his purse to Sebastian and now asks that it be returned to him. Viola, who is astonished by this request, says she does not know what he is talking about, though she offers to share what little she possesses. Taken aback, Antonio accuses "Sebastian" of dishonesty and ingratitude.

Viola goes back to Orsino's palace. Meanwhile Sebastian appears at Olivia's court, and the lady mistakes him for the "page" to whom she is so strongly attracted – the resemblance between brother and sister is so marked that Viola in masculine attire is quickly mistaken for her sibling. When Olivia expresses her overwhelming affection for the young man, he is amazed, but finds that he reciprocates her feeling. An impulsive couple, they are promptly married.

The confusion grows when Olivia sees Orsino and his page together, and addresses Viola as her "husband". The duke is enraged by his page's seeming "treachery". Opportunely Sebastian returns – he has been soundly cuffing Sir Toby and Sir Andrew, who in error have set upon him, thinking him the "page" – and the mystery is clarified. Duke Orsino, learning Viola's true identity, realizes that he prefers her to Olivia, after all. He marries her, pardons Antonio, and all ends genially. As the curtain falls, the clown – Feste – sings one of Shakespeare's most cherished ditties:

> When that I was and a little tiny boy,
> With a hey, ho, the wind and the rain . . .

Viola and Malvolio are the characters who dominate this lilting comedy. The radiance, charm, paradoxical humour of Viola establish her as one of the author's most winning heroines. She is keen, witty, zestful, resourceful, yet in her male guise always delightfully feminine.

Malvolio, blind with self-love, is both mad and pathetic, blissfully unaware of his inadequacies. He is precious yet ponderous, and cold of spirit. Yet he gains and holds the spectator's sympathy. He, more than anyone else, rights the spectator's balance of this fable by his solemnity, his attempt to live by reason in a circle of people who take action headlong, who are in fact as mad as he is supposed to be.

Once again the theme of illusion and reality is stated and explored. This is the irrational world of the young, in which most of the leading players put on "masks": that is, they present themselves as what they are not, either changing sex, or laying claim to a spur-of-the-moment concocted identity, or self-dramatically experiencing an unreal emotion – of love, of grief – that they do not actually feel deeply. Here three over-impulsive pairs of people fall in love at first sight and marry with as little delay as possible. Even Malvolio, when he believes the countess infatuated with him, loses his composure and behaves absurdly. Love, Shakespeare is saying, is the most irrational force of all. Yet, in the end, true passion surmounts barriers of artificial attitudes. But does it? The spectator may wonder whether these hasty marriages will flourish, recalling that young Will Shakespeare was forced to wed at sixteen a woman he apparently did not much cherish.

The language, as always, echoes happily in the ear. As ever, too, there is a host of famed lines. Orsino's:

> If music be the food of love, play on!
> Give me excess of it, that, surfeiting,
> The appetite may sicken, and so die.
> That strain again! It had a dying fall.
> O, it came o'er my ear like the sweet sound
> That breathes upon a bank of violets,
> Stealing, and giving odour.

He is the complete hedonist. But also reflective:

> O, fellow, come, the song we had last night.
> Mark it, Cesario, it is old and plain.
> The spinsters and the knitters in the sun
> And the free maids that weave their thread with bones
> Do use to chant it. It is silly sooth,
> And dallies with the innocence of love,
> Like the old age.

Viola's:

> Make me a willow cabin at your gate,
> And call upon my soul within the house,
> Write loyal cantons of contemned love
> And sing them loud even in the dead of night . . .

To Feste, the ageing, melancholy jester, are given several exquisite songs, including "O, mistress mine, where are you roaming?" as well as sage comment:

> Foolery, sir, doth walk about the orb like the sun;
> It shines everywhere.

Contrast abounds in Sir Toby's unconfined vulgarity. He enters on one occasion with a loud belch: "A plague on these pickled-herring!" and also demands challengingly, "Dost thou think, because thou art virtuous, there shall be no more cakes and ale?"

Michael Feingold, an astute reader of Shakespeare, remarks:

> Because his audience had a short attention span, he always worked with two parallel plots, one high and one low. This was standard Elizabethan practice; his brilliance lay in his ability to link the two thematically and crisscross them dramaturgically. The confusion in Olivia's heart is paralleled by the confusion in her house. As in the main plot, the issue is between two men who don't really love Olivia: Malvolio, who only loves himself, wants her because, like Orsino, he can't have her, while Sir Andrew, as foolish as "Caesario" is unreal, can hardly be called a man at all. Standing in for Olivia herself in this plot, with understandable inadequacy, are the two other people who will marry in the course of the action, Sir Toby and Maria. And where Orsino has two courtiers, Valentine and Curio, Olivia has a professional fool (as distinguished from the twenty-four-hour amateur kind represented by Sir Andrew), Feste, and a sort of associate trickster to Sir Toby, Fabian. The symmetry of the structure, though not easy to make out against the chaos and raging of the production, is fairly dazzling.
>
> Shakespeare's art is never rigid. Symmetry is always set off in two ways, by human individuality and by poetry. The three elements together, you might say, make up the music of the Shakespearian stage – a concept extremely important to *Twelfth Night*, which opens with the act of listening to music and closes with a song. (Another key significator is the affirmation, in the opening speech, that you can't duplicate the effect of a musical performance by repeating it: Once it's done, it's over. "'Tis not so sweet now as it was before.") You can tell almost everything about a director's musical sense by his handling of *Twelfth Night*'s opening speech. Contemporary directors are always in a hurry to start a play, as Shakespeare's audience may well have been; that's precisely why he opened this one with what amounts to a listening session. Orsino isn't going anywhere: A hereditary aristocrat whom others serve, all he has to do is send another love note to Olivia and rack up another rejection. He has plenty of time to brood over the sound of tunes with a dying fall.

This is the last, and perhaps the best, of Shakespeare's sunny or "golden" comedies, in which are the finest elements of pastoral, romance and satire, all expressed with a seeming lyric spontaneity. The play is best summed up in Sebastian's exclamation, when he is being wooed by the impetuous if misguided Olivia, whom he has never seen before: "If it be thus to dream, still let me sleep." This comedy is the happiest of dreams.

A further embodiment of the comedy is found in *Play On!*, a jazz musical with a libretto by Cheryl L. West and melodies borrowed from the late Duke Ellington. Upon its revival at the Goodman Theater in Chicago in 1998 Joel Henning cautioned readers of the *Wall Street Journal* that the "gender-bending" plot of this version needed some "suspension of disbelief". But it was set to Ellington's most inspired music.

> Miss West revised her Broadway book for this production of *Play On!* – as conceived and exuberantly directed by Sheldon Epps. It works exceedingly well here with all the vigor and joy that it lacked when it fizzled last year on Broadway.
>
> Shakespeare's Viola becomes Vy (Natalie Belcon), fresh off the bus from Mississippi and determined to make it as a Harlem songwriter for the melancholic Duke (Charles E. Wallace), who pines for the lost love of Lady Liv (Tonya Pinkins), the premier chanteuse at the Cotton Club. Insisting that a woman has no chance of breaking into the biz, Vy's Uncle Jester (André de Shields) dresses her in men's clothes. The Duke recruits Vy to help him regain the heart of Lady Liv. And, you guessed it, Lady Liv falls for Vy in her male getup, Vy swoons over the Duke, and Lady Liv's rectitudinous manager, the Rev (Paul Oakley Stovall), virtually cross-garters himself in trying to win his boss's heart.
>
> The entire ensemble has its heart and soul in each of the twenty Ellington numbers and pretty much everything separating them. Especially fine are Mr de Shields, Ms Pinkins and Ms Belcon. They display as much talent and energy as a winning World Cup squad. "I Ain't Got Nothin' But the Blues," belted by Ms Pinkins, becomes a classic show-stopper, and the relatively obscure "Rocks in My Bed," sung and danced by Ken Prymus as Sweets and Mr de Shields, gets the whole audience stomping. Ironically, only Mr Wallace's Duke fails to catch the rhythm.

Play On! was shown on US Public Television in 2002.

A much more pedestrian work is *The Merry Wives of Windsor* (1600–1 but perhaps earlier, even before *Hamlet*). Purportedly it was written to order, and Shakespeare's heart was not in the enterprise. As supposed before, Queen Elizabeth, amused by Falstaff's antics in *Henry IV*, expressed a wish to see the bellowing fat knight in the throes of love. Obligingly, Shakespeare resurrected him. Shakespeare is said to have written *The Merry Wives of Windsor* in a mere fourteen days, and the queen was presumably highly pleased with it.

This account of the play's origin may not be true, but it would explain its somewhat mechanical quality. The plot, which is quite serviceable, seems to have been largely the poet's own making, though he may have added details from a tale in William Painter's *Palace of Pleasure* (1566), which in turn draws on Italian stories in print a decade or two earlier. In these, a cavalier pays court to more than one lady simultaneously; they learn of his promiscuity and unite to teach him a lesson. From Tarlton's *Newes out of Purgatorie* (1590), another anthology of stories, Shakespeare may have taken the twist of the lover who unwittingly tells his secret to the lady's jealous husband. Most likely all the devices in the plot had descended from ancient Plautine comedy, and Shakespeare wove them together. Another possible source is an early work of the period, *The Jealous Comedy*. It is his only script dealing wholly with English country life. Elsewhere he treats with that social setting obliquely or gives glimpses only.

Justice Shallow reappears in this farce, and many believe him to be a caricature of Sir Thomas Lucy, of Charlecote, near Stratford, who was said to have accused young Shakespeare of poaching. This is borne out by a reference to Shallow's coat-of-arms, which is like that of the Lucy family. The Justice is having trouble with Falstaff, whom he wishes to prosecute for numerous minor offences, among them robbery, assault on the keepers and libel. The antagonists meet at the house of Page, a gentleman residing at Windsor, who hopes to reconcile them by inviting them to dinner. Falstaff catches sight of Mistress Page and her close friend and neighbour Mistress Ford and becomes enamoured of both ladies. In blatant but clumsy fashion he writes identical love letters to each of them, which they show each other and angrily vow to punish him for his lewd effrontery. He receives word to be at Mrs Ford's house at a certain hour for an assignation. In the meantime Falstaff has met up with Mrs Ford's jealous husband, who under an assumed name wishes to test his wife's fidelity and offers him payment for his assistance. Falstaff gleefully assents, thinking that an ironic joke will be played on Ford. Thus husband and wife are unwittingly working at cross-purposes. But events turn out badly for the amorous knight. Ford returns home inopportunely. Falstaff is forced to hide his portly self under soiled linen in a clothes hamper, which is then thrown into the slimy water of the River Thames.

The remorseless ladies prepare still another trap for him. He keeps a second assignation, is again surprised by the jealous Ford. The fat reprobate is disguised as an old witch, but failing to make his escape receives a beating at Ford's indignant hands. To complete the misfit's discomfiture, the "merry wives" lure Falstaff to meet still more retribution. At Windsor Park, at midnight, he appears masquerading as a deer, wearing a buck's head. A crowd of neighbours, dressed as fairies, descend on him, pinching, cuffing, scorching and otherwise tormenting him. He is taught a lesson, not to lust after or bother respectable married women.

A sub-plot deals with the wooing of Mistress Page's daughter, Anne, by handsome young Fenton, as well as by two comic characters, Slender, a country bumpkin, and Dr Caius, an excitable Frenchman. At Windsor Park, during the night of confusion, Slender and Dr Caius fail in separate plans to make off with Anne; she elopes with the likeable Fenton instead.

If Queen Elizabeth was entertained by the work, her response has not been shared by a parcel of critics. The shortcoming of the play, as a successor to *Henry IV*, is that here Falstaff is the butt of humour rather than the instigator of it. He lacks both his sense of fun and his dignity. "I am not only witty in myself, but the cause of wit in other men," he says of himself boastfully and shrewdly in the earlier, better play. Here he is too foolish, a buffoon, too easily gulled by others. Also there is more than a bit of cruelty in the two "virtuous" ladies who trick and persecute him.

There are those who praise this farce without stint. Revivals of it are frequent. The dialogue is often boisterous and pithy. Much of the action on stage is hilarious. When Falstaff comes back from his dousing, dripping and caked with mud, his authentic voice is heard again. An amusing lesser character, a Welsh parson, affords Shakespeare – perhaps a former country pedagogue – a chance to poke fun at rustic pedantry, and no less at the Welshman's tongue that makes "fritters English". Another well-written minor figure is Mine Host of the Garter. But on the whole this is a comedy without poetry, which is rare for Shakespeare. It has, however, served as the basis for several fine operas, those by Michael Balfe (*Falstaff*, 1838), by Otto Nicolai (*The Merry Wives of Windsor*, 1847) and the late-in-life masterpiece by Giuseppe Verdi (*Falstaff*, 1893), his last work. Verdi's librettist, Arrigo Boito, wisely inserted incidents from *Henry IV* in the text. A more recent entrant is Ralph Vaughan Williams's commendable *Sir John in Love* (1929).

Comedy of a darker hue begins with *Troilus and Cressida* (1601–2), a somewhat sour play. The poet's change of mood is astonishing: what was bright hitherto is now gloomy; disenchantment and anger take over rule of his work.

A comparison to Chaucer's tender handling of the same story is startling; but even more marked, because the time-frame is shorter, is the contrast between this script and that of *Romeo and Juliet*, another drama of unhappily separated lovers, written less than a decade before. Here everything is shown in a jaundiced light.

The exact sources of *Troilus and Cressida* have not been determined. Shakespeare had available to him Chaucer's poem, the plot of which is touched on by Ovid but is largely the invention of medieval poets and, as ever, Boccaccio. Much of the subject, besides that narrated by Chaucer, could have been borrowed from George Chapman's translation of Homer's epic (1598–1616) and from writings by John Lydgate (1370–1449) and William Caxton (1422–91), the latter a translator as well as printer. Greene was probably also looked at. Long before Shakespeare's time the character of Cressida had been debased, as was the conduct of the Greeks, and the Trojans shown somewhat more favourably.

During the siege of Troy one of King Priam's sons, Troilus, falls in love with Cressida, daughter of the priest Calchas. She harbours affection for him but conceals it, especially when her uncle, the lascivious Pandarus, praises Troilus to her. On the Greek side, the Greek "heroes" debate the progress of the war, and in particular a challenge from Hector, brother of Troilus, who wishes to fight in single combat a Greek champion. The Hellenic warriors are jealous and disdainful of one another and anything but "heroic". Achilles, sulking, refuses to confront the valorous Hector. Thereupon, at Hector's own suggestion, Diomedes is chosen to oppose him. The Hellenes also propose to the Trojans that the long and costly war be ended; all that is necessary for this is that Helen, the beauteous Spartan queen stolen by Priam's son Paris, be promptly returned to her husband. Hector favours this, but his fellow Trojans foolishly reject the peace offer.

Through Pandarus' conniving, Troilus and Cressida spend an ecstatic night together in her uncle's house; she has dropped her pretence of not caring for the ardent prince. Meanwhile her father – Calchas – has defected to the Greeks and proposes that she be sent to join him, in exchange for a Trojan prisoner. The Greeks assent to his request.

For a time Cressida refuses to leave Troy, but is finally persuaded. She passionately vows to be true to Troilus, however. Once in the Greek camp, where all the generals make advances to her, she very shortly forgets her yearning Trojan lover. She is spurned only by Ulysses, who deems her a wanton. At last she is awarded as a prize to Diomedes.

It is Ajax who engages Hector in single combat that ends in a draw. Achilles, enraged by the death of his dear friend Patroclus, finally decides to fight Hector. He prevails over the Trojan, but not in a fair struggle: his Myrmidons surround and kill the noble Hector, then tie his corpse to the tail of the Greek champion's horse. Troilus, having learned of the loose Cressida's infidelity, raises his sword against Diomedes, who unseats him, captures his steed and proudly presents it to Cressida.

This "black comedy" is obviously intended as a satiric depiction of the "heroes" of Greek mythology and their fabulous exploits; derisively it seeks to cut them down to size.

The ancient world that evoked adulation in Renaissance – that is, Elizabethan – times was seen differently by Shakespeare. To him it was thoroughly corrupt. In the play all the characters, with the principal exceptions of the keen-witted Ulysses, the idealistic Hector and the betrayed Troilus, are unsympathetic; the tone of the work is cynical, disgusted. Thersites sums it up: "Lechery, lechery, still wars and lechery! Nothing else holds fashion." The women, like all their world, are "fair without, and foul within". Hector, though courageous, will slay to obtain the sumptuous armour of an enemy. One supposes that all this was allegorical and Shakespeare was indicting not only ancient times but, no less, his own.

Love is disparaged. It is the production of "a generation of vipers". Pandarus, the royal pimp, describes it condescendingly, as "hot blood, thoughts, and hot deeds". It is filled with treachery, like the game of war itself.

Explanations for Shakespeare's change of mood from sunny to dark vary from the objective to the subjective: that is, it is ascribed by some critics to his despair at the collapse of the Essex conspiracy, and Elizabeth's order for the beheading of her erstwhile lover, which made her seem to be another inconstant woman; while Shakespeare's long-time patron, Southampton, was imprisoned for his part in the same rebellion. This spelled the loss of an important friend, a protector. Other scholars lay the poet's disgruntlement to personal disappointment in love, a blow reflected in the sonnets. Still others attribute his ill-temper, like that exhibited in *Hamlet*, to a voguish melancholy that was quite impersonal, a mere literary attitude shared by other playwrights of the day. If the plays could be dated more exactly, it would help very much in the effort to interpret them.

The language in this tragicomedy is uneven. Much of the prose is laden with philosophical observations. Most admired is Ulysses' plea for proper measure and order in everything:

> The Heavens themselves, the planets, and this centre
> Observe degree, priority, and place,
> Insisture, course, proportion, in all line of order.

This is the statement of one who believes in hierarchy, moderation and discipline, which is the impression Shakespeare himself seems to have made on others. Another comment has to do with the transience of celebrity and popularity:

> One touch of nature makes the whole world kin,
> That all, with one consent, praise new-born gawds;
> Though they are made and moulded of things past . . .

That sounds like the hurt feelings of a playwright who has received bad reviews and sees rivals with newer ideas rising to outdate him; his day is passing:

> Time hath, my lord, a wallet at his back,
> Wherein he puts alms for oblivion.

Or put another way:

> The present eye praises the present object.

The critics' complaint is that such ideas, though resoundingly true, are merely put into the dialogue: they are not incarnated in the behaviour of the characters or tested by the plot-action. The play can be viewed as essentially a Morality and some of the people in it as scarcely limned personifications of abstract vices. In addition, the script's structure is too diffuse.

Troilus and Cressida voices, among other messages, a strong denunciation of war. It is shown as a sort of game played by capricious personal rules, what is honourable and permissible and what is not. There is a difference between how Hector fights and how Achilles does. But this is not the Shakespeare who patriotically glorified a zestful battle-spirit in *Henry V*.

Coupled with some true poetry in the dialogue, alas, are too much bombast and clumsiness; the poet often seems troubled, unequal to his subject, which in itself – because of his sardonic approach to it – is not wholly worthy of his great talents. Yet a twentieth-century London revival of it in modern dress revealed it to be a rather dazzling and biting satirical fantasy apposite to the arid times after the Second World War.

It may be that after experiencing an emotional *katharsis* of the sort defined by Aristotle – an opinion argued by Freudian analysts such as Ernest Jones in *Hamlet and Oedipus* – the middle-aged Shakespeare was left empty and exhausted but had to deliver yet another play to his company. He could not pause to catch his breath.

The "comedy" could be mounted on a nearly bare stage, just a sparse army camp, the stark walls of Troy; or the Greek tents – the siege having gone on for ten years – might be lavish and gaily decorated. The breastplates of the soldiers could be engraved and embellished and their tunics embroidered and brightly coloured, making a brave show.

The libretto of William Walton's opera *Troilus and Cressida* (1954) relies on Boccaccio's tale and Chaucer's poem rather than on the flawed play.

In the same cluster of "dark" or "problem" comedies is *All's Well That Ends Well* (1602–3). The unequal quality of its dialogue suggests that it might be the rumoured *Love's Labour's Won* – it is one of several candidates – and possibly was rewritten and retitled at a later date. It is another work very difficult to place chronologically. Like *Troilus* it is seldom played, its one period of popularity having been the mid-decades of the eighteenth century, an amoral period when cynicism about upper-class society and marriage was rampant in the theatre. So far is as known, there has never been an American production of it. Indeed there is no record that it was even staged in Shakespeare's lifetime. It, too, is founded upon a story in Boccaccio's lively *Decameron* and includes folkloric elements. Shakespeare, as usual, altered many details.

Helena de Narbona, an orphan, is in love with Bertram, Count of Rousillon, whose mother's ward she is. He scarcely regards her, since she is penniless and far below his station in life. The King of France is desperately ill, and no one can cure his malady. Helena, whose father was a famed physician, believes herself possessed of a secret remedy. She goes to Paris to offer it to the king, but also because Bertram has gone there. When her medicine restores the king to health, he gratefully offers her a choice of a husband from among his faithful young knights. Many wish her hand, but of course her choice is Bertram. He refuses to wed her, until the king, affronted, insists on the marriage. Immediately after the ceremony the hostile bridegroom sets off for the war in Italy and sends Helena back to Rousillon. She receives a message from him that he will not consummate the marriage until she shows him a ring that he wears always on his finger and bears a child that he has begotten by her. Such conditions seem impossible to fulfil. Dauntless, Helena sets out disguised as a pilgrim and this time journeys to Florence.

Bertram, meanwhile, has become infatuated with Diana, daughter of a Florentine widow. He seeks to seduce her. Helena, learning of his design, tells Diana to feign her consent, then takes Diana's place with the help of night's darkness. She persuades Bertram to exchange rings with her, and also conceives his child.

Back in France, Bertram is charged with immorality by Diana. Helena is rumoured dead, and the king suspects that her unloving husband has killed her. Helena frees him by her account of what actually happened. He confesses his cruelty and declares himself truly in love with his wife at last.

Though the plot is obviously absurd, Shakespeare does succeed in making the principal characters surprisingly real: they have a psychological vitality that fits uneasily into the fantastic framework and fairy-tale premise of the plot. The drives of lust and frustrated love are well portrayed. True, none of the people is attractive. Helena's desire for Bertram is too obsessive, and his treatment of her is too rude and unkind; what is it about him that captures and holds her? The spectator may wonder about that, finding him a very unpleasant person.

His conversion at the end is too hurried and implausible. It is impossible to credit that he will now become a loving husband to his obdurately fond wife.

Some humour accrues from a sub-plot concerning the antic doings of Parolles, Bertram's lying, cowardly follower, at some of the lighter moments of this bleak, largely mirthless play. He is hardly to be compared to Shakespeare's more robust comic figures in other works, nor is the clown Lavache, who is made the victim of a heartless practical joke. Parolles proves himself all too ready to betray those dependent on him, and the pragmatic ease with which he recovers from discomfiture is admirably exposed. He has a ready instinct for survival. He is the most believable character on view here. Overall the atmosphere of this strange *comédie noire* is dispiriting: age, disease and loss are among the most prevalent subjects discussed in it.

Measure for Measure (1604–5; perhaps earlier than *All's Well* or later than *Othello*) also has a mood strangely mixed, sombre and at moments gay, a "black comedy". In it, too, the "bed trick" employed in *All's Well* is repeated. (As shown earlier, it was used in Italian Renaissance farces.) In most respects, however, this is a far better work. Shakespeare lifted its plot from George Whetstone's *Promos and Cassandra*, an unacted script, which Whetstone later rewrote and published in a collection of short stories. The tale was not original with Whetstone, however; he borrowed it from a Sicilian collection (printed in 1565) by Battista Giraldi (Cinthio), who had also dramatized it in a play called *Epitia*. Possibly Shakespeare had access to the early Italian version, as well as the two subsequent English ones. He improved upon the plot, making the ending more logical, and deepened the drama's theme, that justice should be tempered by mercy.

Whetstone's play is laid in Hungary; *Measure for Measure* has for its setting Vienna. Vicentio, the duke of that city, takes leave of his duties for a period, appointing Angelo to govern in his place. This young man, of an austere nature, is guided by a rigid morality. He is faced by a corrupt populace. As an initial step towards reform he chooses the death sentence for all seducers. One of the first youths arrested for this crime is Claudio, who has lacked money to let him marry his beloved, Juliet. Isabella, Claudio's sister, a novice in a convent, goes to Angelo to intercede for her brother. The stern young governor rejects her plea, saying that the law must be strictly upheld, but he is so taken with her that he promises her a second audience. When she departs he confesses aloud how strongly this beautiful and eloquent girl disturbs and attracts him. He has power. Isabella returns, and this time he offers to spare her brother if she will become his mistress. Isabella contemptuously refuses his bargain and goes to see her imprisoned brother; she is confident that Claudio will concur with the value she puts on her chastity. But young, trembling Claudio is much more realistic than his righteous sister. He begs her to save his life by yielding to the hypocritical governor. She is appalled by his "cowardice". Quite adamant, Isabella furiously rebukes and quits her brother. She is approached by a friar, who has been counselling Claudio; she does not realize that he is Duke Vicentio in disguise. The helpful "friar" advises her to make believe that she will accept Angelo's vicious offer.

At one time Angelo was betrothed to Mariane, but he had withdrawn from the match when she lost her fortune and dowry. The duke contrives to have Mariane substitute for Isabella in Angelo's quarters, under cover of night. In darkness, Angelo possesses the wrong girl. Though he is unaware of the deception, Angelo treacherously orders Claudio's execution. But the duke returns to his palace, reveals his identity, resumes control, reprieves and sets free the hapless Claudio and has him make proper amends to the girl he had seduced, Juliet. Angelo is accused of having flouted his own laws and told to marry Mariane. Duke Vicentio then takes the fanatically virtuous Isabella as his own duchess. On rather short notice! Several lesser characters are forgiven.

The play is a comedy, then, in the same way as are *Troilus* and *All's Well*: they do not end with deaths, but – in the latter two instances – with reconciliations and neatly paired marriages. But, like those two other plays, it is frequently grim and has rather distasteful characters. It can be said that no happy ending can ask the spectator to overlook the baseness of Angelo, the egotism and selfishness – on each side – of Isabella and Claudio, the often callous deviousness of the duke. A number of the minor people – Lucio, Mistress Overdone, Barnardine, Pompey – illustrate the corruption of the times, too long encouraged by the duke's prolonged lenience. They add to the impression this play gives of having been written by a poet disillusioned and displeased by the world he knows and sees around him. It is good to keep in mind that Shakespeare wrote these so-called comedies at the same time he was composing his greatest tragedies and to assume that perhaps good humour was for this period not his natural bent. Or else the fashionable pessimism of the day – mentioned before – was simply all too pervasive.

The plot, however, is very deftly managed, and many scenes are exceptionally stirring. The action is suspenseful throughout. Though not appealing, the characters are vivid and convincing. In all of them sexual passion is a whip, a drive that almost brings about their undoing; the exception is Isabella, who is cold. The theme of Christian forgiveness is a noble one, and the play does example it well in its unfolding and resolution. The subject is developed with subtle balances, psychological and symbolic touches that mark it as the work of a superior mind and pen. But again there has been a none-too-successful attempt to combine fairy-tale motifs and a harsher realism: the duke in disguise, learning the truth about his subjects, reminds one of the masquerading caliph Harun al-Raschid in *Scheherazade*, and the other disguises and substitutions are similarly fairy-tale materials, highly romantic; but the designs of prison and brothel are too drab, vulgar and evil, and the omnipresent threat of death hanging over Claudio, Barnardine and the others is hard to fit into what is termed a comedy.

The poetry is among Shakespeare's best, again. It rises to and surpasses many occasions in the complicated yet well-ordered plot. Thus Angelo explains why he shall be rigorous in enforcing his edicts:

> We must not make a scarecrow of the law,
> Setting it up to fear the birds of prey,
> And let it keep one shape, till custom make it
> Their perch and not their terror.

In strong language, Isabella protests Angelo's abusive sway and all tyranny:

> But man, proud man,
> Dress'd in a little brief authority,
> Most ignorant of what he's most assured,
> His glassy essence, like an angry ape,
> Plays such fantastic tricks before high Heaven
> As makes the angels weep.

Pleading for her brother, she puts her case very well: Angelo should condemn the sin, but not the sinner. She adds:

> O, it is excellent
> To have a giant's strength, but it is tyrannous
> To use it like a giant!

A lovely song is included:

> Take, O, take those lips away,
> That so sweetly were forsworn;
> And these eyes, the break of day,
> Lights that do mislead the morn.

Outstanding is the passage in which the executioner comes for Barnardine, who is napping:

POMPEY: Your friends, sir: the hangman. You must be so good, sir, to rise and be put to death.
BARNARDINE [*within*]: Away, you rogue, away! I am sleepy.
ABHORSON: Tell him he must awake, and that quickly too.
POMPEY: Pray, Master Barnardine, awake till you are executed, and sleep afterwards.

Richard Wagner transformed *Measure for Measure* into an opera; as always, he wrote the libretto, and he gave it a new title, *The Ban of Love* (*Das Liebesverbot, oder Die Novize von Palermo*; 1836). His book does not follow the play closely. The first and only performance, a failure, took place in provincial Magdeburg's Stadttheater. Wagner and his singers shortly left the city and resettled in Königsberg. The soprano who had Isabella's role, Minna Planer, later became his first wife. Few if any accounts of the composer's life and career refer to this musical work, though it anticipates Wagner's use of leitmotiv, his bold advocacy of free love and his sensitive preoccupation with contrasts of light and darkness.

True tragedy lets Shakespeare achieve grandeur once more, now in *Othello* (1604), and to follow that impassioned drama with three more of his greatest plays, if the best-informed guess as to the chronological order of their composition is accepted. These four, along with *Hamlet*, are the absolute peak of his accomplishment. For this tale of the tormented Moor he was indebted to the same Cinthio to whom he owed the plot of *Measure for Measure*. It is the seventh novel in a collection titled *The Hundred Fables* by this professor of Ferrara who, concerned with morality, examines human virtues and vices. But Shakespeare also had for models two historic persons: Christopher Moro, a Venetian general who served as Lord Lieutenant of Cyprus and while there lost his wife by death; and San Pietro di Bastelica, an Italian adventurer in the pay of France, who murdered his innocent wife, strangling her with her handkerchief, having been misled into believing her unfaithful during his absence on a journey from which he returned abruptly. It is possible that Shakespeare combined these two men, along with the hero of Cinthio's tale, to arrive at his own deeply emotional hero. As usual he tightened the plot, and changed many details of the story to make them more plausible; especially the ending, which he both refined and enhanced.

The Moor, Othello, has won military glory while fighting on behalf of Venice. He woos and elopes with Desdemona, daughter of an aristocratic Venetian senator. Iago, Othello's aide, is jealous and resentful of the Moor's reputation and advancement. He informs Desdemona's father of what has occurred, but despite the father's anger the Duke of Venice permits the mixed marriage, partly because he has immediate need of Othello's services against the Turks. Othello and Desdemona go to Cyprus, of which the Moor is named Governor. With the help of a storm the invading Turkish fleet is sunk and Othello hailed for his decisive victory. Iago, nettled, plots to stir Othello's jealousy against Cassio, the Moor's trusted lieutenant, of whom Iago is also envious. Iago's plan is to convince Othello that Cassio is infatuated with Desdemona and wishes to replace her husband in her affections. Getting Cassio drunk, Iago incites him to engage in a brawl with Roderigo, a Venetian follower of Iago, and also enamoured of Desdemona. The unruly incident results in Cassio's dismissal by Othello. Then Iago suggests that Cassio ask Desdemona to intercede with her husband, asking a pardon for his lieutenant. When the compassionate Desdemona is moved to do this her zeal awakens the Moor's suspicions, already heated by persistent hints from Iago.

Emilia, Iago's wife, has become Desdemona's confidante. She is also used as a dupe by her demonically scheming husband. Iago tells Othello that he has heard Cassio talking in his sleep, expressing his love for Desdemona and his wish to supplant Othello in the marriage bed. As a final trick, Iago obtains from Emilia a distinctly marked handkerchief, a gift from Othello to his wife, which the manipulator plants in Cassio's lodgings. When this is shown to the gullible Othello, his rapidly growing suspicions overwhelm him. He accuses his wife and strikes her. Emilia tries to reassure him of his wife's fidelity, but he is deaf to all such testimony.

Iago, feeling that the intrigue is getting out of hand and endangering him, contrives another duel between Cassio and Roderigo in which Roderigo is killed and Cassio seriously wounded, both by Iago's dagger and furtive intervention. Othello, maddened by jealousy, goes to the frightened Desdemona's bedchamber, where she is asleep, awakes her and smothers her with a pillow. A few moments too late, Emilia enters and reveals Iago's dreadful treachery. Iago follows her into the bedchamber and stabs her. Othello strikes Iago with his dagger, then commits suicide. Iago survives his wound but is held for punishment. Cassio, recovering from his injury, is named Governor of Cyprus to succeed the hapless dead Moor.

This bare outline cannot convey the richness of characterization and the superb poetry that lifts this drama to an exalted height.

Desdemona is shy, gentle. She is well bred, of a noble family; she has been protected from the world. The antithesis in every way of her black, "great-hearted" husband, she shares with him a quality of innocence. A soldier and journeyer, he has had fabulous experiences; it is his vivid recounting of them that has caught and held her interest, stirring wonder in her. She is capable of a far greater love for him than his for her, for his is ultimately possessive and selfish: like Iago, he will ruthlessly destroy what he cannot have. In many ways, the two bound together in this short-lived marriage are badly mismatched.

Othello, proud yet insecure, is powerfully yet poignantly depicted. Conscious of their difference in colour, threatened by the oncoming of age and his imminent loss of virility, uncertain even now of the love of his devoted wife, he falls an easy victim to Iago's machinations. A regal barbarian, of great courage, he has in some ways the simplicity of a child. Once his jealousy is ignited he magnifies every hint into a certainty. His nature is essentially primitive, elemental; he is superstitious, a prey to supernatural fears. He has an innate dignity. He can be tender and trusting, credulous, but also has the ferocity of a soldier, a martial leader. His anger, unleashed, is terrible, savage. He lashes out and hurts what threatens and displeases him. He is wily and wary at times, as befits one of his position; but at other moments is easily deceived. Like other of Shakespeare's heroes he is histrionic, much given to self-dramatization. His overly complex, divided personality dooms him to a tragic fate.

But Othello is not the only one destroyed by jealousy: the devilishly clever, resourceful, envious Iago is led by it to catastrophe as well. He scorns simplicity, honesty and "goodness"; his cynicism is thorough; his dominant attitude is rooted – like Othello's suspicions – in a life-long basic insecurity. He believes in the purity of no one. Lecherous, he is convinced that people can be lured and controlled or enslaved by pandering to their sexual drives. Much has been made of his "motiveless malignity", a phrase applied to his Machiavellianism by Coleridge. Indeed, other commentators have elevated this view to a metaphysical observation, reflecting Shakespeare's conviction that blind evil is a rampant force waiting on every side to waylay man, a concept reinforced by certain statements about the implacable hostility or mere indifference of the gods in *King Lear*, his next play. There it is said that many of the ills that befall man come about through perverse and irrational impulses, seemingly casual ones, inherent in the divine will.

That interpretation may be reaching too far. A mere hearing or reading of the play shows that Shakespeare gives Iago multiple specific promptings for his cruel actions, beyond his being a typical Renaissance villain found in many other Senecan revenge tragedies of the period. As indicated before, Iago, proud, mercenary, is infuriated at having been passed over for promotion by Othello, a black man, in favour of Cassio, a less experienced and possibly less able officer. This has deprived Iago of the added pay and prestige he had expected. He also declares himself persuaded that Othello has, in the past, seduced the apparently vulnerable Emilia:

> I hate the Moor;
> And it is thought abroad that 'twixt my sheets
> Has done my office. I know not if't be true;
> But I, for mere suspicion in that kind,
> Will do as if for surety.

And again:

> For that I do suspect the lustful Moor
> Hath leap'd into my seat; the thought whereof
> Doth like a poisonous mineral gnaw my inwards;
> And nothing can or shall content my soul
> Till I am even'd with him, wife for wife . . .

He questions if Cassio, too, has worn "my nightcap". (Nothing seen of Emilia, despite her light-hearted jest on the subject, justifies this degrading speculation; nor is it in the nature of the large-souled Othello to have violated his wedding vows, though Cassio might be a more likely adulterer.)

No motivation for Iago's quest for "vengeance" could be more explicit. This play is about jealousy, not about the gods or divinely doomed human destiny. If Iago is able to work with such cruel skill on Othello's weakness it is simply because he himself is consumed by it – in one form or another, or in several, and knows all too well its deadly effects. His "vengeance" is disproportionate – and it strikes down the helpless Desdemona, who has never wronged him, and whom he even purports to lust after – but this is because his scheme has got out of hand, has gone further than he expected. Once set in motion, his fiendish intrigue rushes on with its own momentum. Professional soldier and malcontent, coarse yet making a virtue of his seeming candour, jocular, arch dissembler, low-born and envious of his moral and social betters, he seethes with hatred against those more fortunate and illustrious than himself. Critics with a theological bent see in him the personification of Lucifer, or the Vice, a figure out of the Morality Plays that Shakespeare had observed in his youth, and that had so influenced Marlowe's *Dr Faustus*, which in turn had cast a spell over Shakespeare. The tragedy then becomes one of damnation, Iago having sold his soul to the Devil.

In the twentieth century critics having a Freudian perspective have suggested a homosexual strain in Iago, his sharp interest in Othello's love-life, as well as a fugitive desire to replace both Othello and Cassio in the enjoyment of their erotic conquests, and a possible fixation on Roderigo that turns from frustrated love to hate, which is why he plots and brings about his death. These promptings, if there, are harder to discern and substantiate. Modern critics, too, have dismissed Iago as clearly a psychopath.

More narrowly focused on its single theme, jealousy, *Othello* has a sharp dramatic impact; it is not overloaded with ideas, unlike some other of Shakespeare's plays. It also has a classical compression of time and place, a concentration of physical action that was down – perhaps – to the influence of his friendly rival Ben Jonson, who advocated respect for the taut structuring of incidents as opposed to the sprawling story-telling with which most Elizabethan play-craftsmen were content. They worked hurriedly; plays on the board had only brief runs, a few days, and went by in rapid succession; new material was always needed. Seeking to observe the three unities would take time and add difficulties, often perplexing ones – that is, if the teams of journeymen writers were even aware of what Aristotle proposed was the best dramaturgy, for his *Poetics* were still not familiar, known to only a few Greek and Latin scholars like the boastful Ben Jonson, who was temperamentally drawn to attaining an absolute clarity and nice proportion in his work. *Othello* proves that Shakespeare could respond to those demands, if he wished . . . usually he did not care too much about tightness of structure, his imagination was too crowded and fertile, his feelings romantic, his vision too broad, always exceeding limits. Here there are skilfully planted portents of dark events to come, and linking imagery throughout that produces an all-encompassing mood. Frequently pointed out are the numerous places in the dialogue where men are likened to animals, snarling, leaping upon their prey.

The power and beauty of the final scene, the painful encounter that ends in Desdemona's murder and Othello's suicide, arise from some of the most magnificent writing in all dramatic literature, perhaps unequalled since Aeschylus and Euripides and certainly not surpassed by any other theatre poet. Even George Bernard Shaw, who arrogantly disparaged much of Shakespeare's work, paid tribute to the "sublime" music of this play and especially the ultimate scene. But who has not been stirred by such lines as these? Othello, torch in hand, contemplates his sleeping victim:

> It is the cause. Yet I'll not shed her blood,
> Nor scar that whiter skin of hers than snow,
> And smooth as monumental alabaster.
> Yet she must die, else she'll betray more men.
> Put out the light, and then put out the light.
> If I quench thee, thou flaming minister,
> I can again thy former light restore,
> Should I repent me; but once put out thy light,
> Thou cunning'st pattern of excelling nature,
> I know not where is that Promethean heat
> That can thy light relume.

And his dying speech:

> Soft you! A word or two before you go.
> I have done the state some service, and they know't;
> No more of that. I pray you, in your letters,
> When you shall these unlucky deeds relate,
> Speak of them as they are. Nothing extenuate,
> Nor set down aught in malice. Then must you speak
> Of one that lov'd not wisely, but too well;
> Of one not easily jealous, but, being wrought,
> Perplex'd in the extreme; of one whose hand,
> Like the base Indian, threw a pearl away
> Richer than all his tribe . . .
>
> I kiss'd thee ere I kill'd thee. No way but this,
> Killing myself, to die upon a kiss.

It is not only majesty of language, profuse imagery and apt richness of metaphor that mark the dialogue of *Othello*: it has a subtly altered cadence and music everywhere. The poet's control is close to perfect. The Moor's solemn rhythm, his tone often described as organ-like, is varied by the staccato of Iago who pours forth in verse and supple prose his venomous thoughts with the haste of one with an over-stimulated mind propelled by indignation. His thoughts leap ahead of his words, even while he is vaunting his cleverness, which he does unceasingly; his boast is that he works "by wit, and not by witchcraft", and he scorns all around him for being less rational and intelligent than he is. His claim to superiority masks his deeper feelings of inferiority, a trait characteristic of the criminal who finds justification for his malicious acts in his not having been properly regarded and rewarded by the world; it has refused to accept him at his own exaggerated valuation. In retaliation he will go to every extreme. Though critics often complain that Iago's diabolic malevolency is inadequately accounted for, his portrait is in fact one of Shakespeare's most brilliant artistic successes. He is, unfortunately, true to life; others like him abound.

The Venetian setting of *Othello* is seen by some to be analogous to Elizabeth's England. The city-state, fiercely independent, is menaced by an enemy (the Turks instead of the Spanish) whose fleet is lost at sea in a storm. The materialistic corruption of the Venetians, too, had its counterpart in England at this time, and perhaps Shakespeare was making an allegorical reference to that.

In Restoration days Desdemona was the role given to the first actress ever allowed to appear on the English stage. The character of the Moor also prepared the way for black actors to assume serious parts behind the footlights in the United States. Finding his entry to an acting career blocked in his native country because of his race, an African–American, Ira Aldridge (1807–67) went overseas and was welcomed and acclaimed as Othello in Britain and on the Continent, including a performance before the Czar in Russia. The singer-actor Paul Robeson (1898–1976) had the part in less race-conscious Britain in 1930 before, thirteen years later, finally breaking the barrier on Broadway against racially mixed casts in a Theater Guild production. He was strongly assisted by José Ferrer as the insidious Iago and Uta Hagen as the ill-fated Desdemona. Margaret Webster, the director, was Emilia. The actors and producers were dubious and fearful of how the public and critics would respond. Ethan Mardden, in *The American Theater*, sums up the atmosphere backstage and out front on opening night as "charged" and quotes the scene designer Robert Edmund Jones as saying: "If a cat had walked across the footlights it would have been electrocuted." A lengthy ovation rewarded the daring players. Mardden thought that Robeson was really an inept Othello, lacking acting experience and technique in contrast to the well-trained, assured Ferrer and Hagen. But he was physically imposing – a former college athlete – handsome, with a resonant voice; audiences took to him. The Venetian backgrounds were palatial and the costumes colourful and rich. The *Othello* set a new record for a Shakespearian presentation on Broadway: 295 performances. From then on a list of black actors portrayed the

Moor on stage. (Canada Lee reversed the advance: he whitened his face with chalk to appear as Bosola in Webster's *Duchess of Malfi*.) White players continued to darken their features to be the character, however. Olivier, who was an on-stage Othello several times in London, had his enactment preserved on film (1965).

For characterization, Olivier emphasized the use of make-up; experimenting endlessly, he was fascinated with the resemblances and changeful effects he achieved. Donald Spoto, in a biography of Olivier, describes how this very conscientious player sought to become Othello on stage and film. "For the deep vocal volume, he expanded his rib cage by trebling his weight lifting, and he exercised his voice, lowering it a full octave." Spoto quotes from Olivier's diary: "I had to *be* black. I had to feel black down to my soul. I had to look out from a black man's world. . . . External characteristics are to me a shelter – a refuge from having nothing to feel, from finding yourself standing on the stage with just lines to say, without a helpful indication of how to treat them or how to move. I construct my portrait from the outside with little techniques, ideas, images – and once the portrait becomes real, it starts travelling inwards." Spoto relates: "He shaved the hair from his chest, arms, and legs and then applied Max Factor number 2880 over his entire body. When that had dried, he added a black liquid stain and then a third coat to give a mahogany sheen. He and his dresser then used yards of chiffon to polish his skin until it shone (pancake powder would run under perspiration). Then he painted his fingernails with a pale-blue varnish, coated the inside of his mouth with gentian violet, put on a tightly curled black wig, and with pinkish hue polished his palms and the soles of his feet; four hours later, after the performance, almost two hours were required to remove the make-up." Entering the stage, he was greeted by a gasp from the spectators. "Moving like a panther (and sounding more like a West Indian from Notting Hill Gate, he spoke feverishly in a calypso rhythm." But the critics approved. He received mostly his best notices ever from them. His interpretation was, in their eyes, unforgettable, certain to be spoken of with awe for years to come. The film version, sponsored by Otto Preminger, was a hurried production, taking only three weeks, and had a small budget. In the picture Olivier's make-up looks too thick and black, the shine on his face distracting and causing the colour to seem unreal.

Earlier in his career Olivier had been Iago; the Othello was Ralph Richardson. Before rehearsals began, Tyrone Guthrie, the director, arranged for two lengthy discussions between the young man and Ernest Jones, the Freudian analyst, who believed – as remarked before – that Iago, basically homosexual, was in love with the Moor. Convinced of this, Olivier demonstrated Iago's feeling for Othello with such ardour that he was rebuked by Richardson. But on stage, after the première, Olivier had again given expression to this prompting on Iago's part.

Iago, in many aspects, is the tempting role, certainly the more showy and interesting one. This leads some pairs of eminent actors in the same cast to alternate, changing after doing the betrayed Othello one night to impersonating the wily Iago at the next performance, with the previous night's Othello becoming the evil Venetian. If spectators wish, they can see differing enactments of the characters by attending twice, a particular boon to acting students and the box-office.

As with *Hamlet* and *Romeo and Juliet*, there are now *Othellos* with the Moor expressing his grief in a broad range of languages. In Japan, Kawatake Toshio has offered his own idiosyncratic adaptation of the play.

Music adds even more depth and strength of emotion to the tragedy with its engrossing and enspelling power. Francesco Berio di Salsa, librettist of Gioacchino Rossini's *Otello* (1816), took considerable liberties with Shakespeare's treatment of the story. So little is left of the play that it is almost unrecognizable. Rodrigo and Desdemona are about to marry; Otello tries to interrupt the ceremony. Later, Roderigo and Otello fight a duel. But Rossini's is a fine score. Unluckily for him, Giuseppe Verdi's *Otello* (1887), set to the prose and verse of Boito, so far overshadows the earlier opera that Rossini's work is hardly known and scarcely performed. Indeed, Verdi's masterpiece, composed when he was about seventy, also draws attention from the play to which Boito was determinedly faithful. Opera-lovers who go to hear the wonderful opera repeatedly – and it is much the same audience that attends theatre, and likely to choose a Shakespeare offering – tend to stay away from the merely acted *Othello*; the story has become for them over-familiar and less effective without music, despite its intrinsic grandeur. The first singing Othello was Francesco Tamagno, who had "a voice like a trumpet". In recent years at the Metropolitan Opera in New York the principal claimants to the role have been the histrionic Giovanni Martinelli, the fierce Mario Del Monaco and durable Placido Domingo.

Othello, a four-act ballet, has a scenario and choreography by Vakhtang Chabukiani to music by Alexei Machavariani. Originally mounted at the Paliashvili Theatre for Opera and Ballet in Tiflis, Georgia, in 1957, it was later taken with some of the same dancers to Leningrad for performance at the Kirov Theatre (now the Maryinsky). It follows Shakespeare's handling closely but adds several scenes, enacted flashbacks to earlier events that are only verbally alluded to in the play.

The modern dance choreographer, José Limon, who had his own troupe and performed with it, created *The Moor's Pavane* (1949), to a score culled from various works by Henry Purcell. Introduced at the Second American Dance Festival in Connecticut College in New London, it presents the handkerchief episode in an impressionistic evocation of "the love, jealousy, passion and tragedy of the play within the formal framework of a court dance". Winning immediate favour, it is frequently offered by other companies.

The American Ballet Theater and the San Francisco Ballet jointly commissioned choreographer Lar Lubovitch to create a three-act *Othello* that had its début at the Metropolitan Opera House in 1998. Lubovitch was noted mostly for work in modern dance idioms. Jennifer Dunning, in the *New York Times*, felt that his venture emphasized characterization rather than dance; ballet dancers were hardly needed. The title role was filled by Desmond Richardson, an African–American, praised for the mimetic clarity he brought to the part. Dunning deemed him "one of the most majestic dancers ever to tread the Metropolitan's stage. He towered over the dance. His brooding, regal Moor was a man of explosive rage and tenderness, whose powerful and frantic solos suggested a strong man out of control." There is also a hint that Iago's feeling for Othello has an erotic component. Black ballet dancers were few in number in the major troupes. His alternate at some performances was Keith Roberts who struck Dunning as "a more tortured human being than an unstoppable force of nature, making the story smaller-scaled but more poignant". Later the role was assumed by Yuri Possokhov, a Ukrainian – former member of the Bolshoi in Moscow and now with the San Francisco company. He lent the character an animal strength. The score was by Elliot B. Goldenthal; the ambitious and elaborate sets by George Tsypin; and the very effective lighting by Pat Collins.

In the *Wall Street Journal* Joan Acocella responded with stronger dispraise: "Dear God, what a dud!" Of the choreographer: "The smallness of his gift stands horribly exposed." (Lubovitch had a lengthy résumé of successes elsewhere.) Here the music was put down as "largely undanceable". All the characters were "one-note", yet Richardson was a "virtuoso ... noble and high-strung" and should have been perfect for the part. Iago's homosexual feeling for Othello was flaunted. Desdemona's role was reduced. "The dancers don't look tragic; they just look puny and strange. They start; they go nowhere, and then they stop." To mount the ballet had cost a million dollars.

King Lear (1605–6) ascends to even loftier heights; indeed, it is designated by many as Shakespeare's mightiest play, though – exceedingly difficult to stage and act – it is less often given and hence less familiar than are *Hamlet*, *Othello* and *Macbeth*. A tragic drama of old age, it is often thought to be autobiographical, an intensely personal work; but Shakespeare was barely past forty when he wrote it.

What then accounts for its dark outlook, its profoundly morose strain? Beyond the segment of the *cognoscenti* who adopted a pose of nihilism and alienation, the common people of England were suffering; the masses were truly afflicted. The plague was virulent, costing 30,000 lives, including one-seventh of London's citizens. Accordingly, G.B. Harrison says this was known as "the Black Year". The Gunpowder Plot, which was meant to blow up Parliament, took its toll of a good part of the ruling class and shook public confidence in the preservation of due law and national calm. There was a general sense of shock as the identity of the guilty conspirators became known. After a lengthy reign, Queen Elizabeth had died and had been succeeded by a Scot, the still little-known James I, possibly imperilling the political fortunes of many in high places, among them patrons of the theatre. A marked increase of corruption and immorality occurred at court; scandal was rife. That could have been observed by Shakespeare when his company played there. In such circumstances, how could anyone be expected to turn out a bright, optimistic script? It was at this time that Ben Jonson brought out his lacerating *Volpone*, another work expressing disgust at the flagrant avarice in human nature and the decaying social order. Imprisoned for a spell with violent inmates, Jonson had looked on the bare face of depravity.

The plot of *Lear* comes from an earlier minor play, dating from 1594, author anonymous. The story is in Holinshed's *Chronicles* (1567) and in the works of other "historians" and romancers possibly as far back as Geoffrey of Monmouth (1135), and there is a version in John Higgins's *The Mirror for Magistrates* (1574). Edmund Spenser briefly touched on the same material in his *Faerie Queen* (1590). Shakespeare's sub-plot about Gloucester and his sons is to be found in Sir Philip Sidney's *Arcadia* (1590). Weaving together the two stories, Shakespeare – as shall be seen – uses them for "parallelism and contrast", a new unifying device. Once more he alters much of the original tale and provides a new and, in this instance, more tragic ending. Most of the preceding versions conclude happily, with the aged Lear victorious over his enemies and his good daughter Cordelia succeeding to the sceptre. So far as is known, King Lear (or Leir, as earlier spellings have it) has no actual counterpart, his presence in Holinshed's account notwithstanding. He seems to have been, like King Arthur, a figure out of Irish or Welsh mythology – or, if real, he belongs to pre-Roman or pre-Christian Britain. A book by Samuel Harsnett, *Declaration of Popish Impostures* (1603), was the text from which Shakespeare lifted many of the odd phrases and strange names of devils mouthed by his Poor Tom.

In the play, Lear, the venerable and beloved monarch, summons his three daughters and announces his intention to divide his kingdom among them. He bids each of the three to express their fond feeling for him, expecting gratitude and praise. Whoever speaks best shall win the largest share. Goneril, wife of the Duke of Albany, and Regan, married to the Duke of Cornwall, offer him fulsome flattery. The youngest, Cordelia, is embarrassed at being asked for a public protestation of affection. She says merely that she loves her father to the extent a dutiful daughter should. She adds that she must save half of her love for her future husband. The vain old man flies into a rage and abruptly disowns her. One of Cordelia's two suitors, the Duke of Burgundy, drops his negotiation for her hand. But the second, the King of France, impressed by her frankness and honesty, still wishes to wed her and is accepted, though she has forfeited her dowry. Says the king: "She is herself a dower."

King Lear's true friend, the Earl of Kent, tells him that his hasty act is brash and unwise. The earl is rewarded by banishment, but loyally remains in the king's train in disguise under the name of Caius.

Goneril, now in possession of her half of the kingdom, soon reveals her true nature. She finds fault with her father, in his dotage, for all manner of petty annoyances and peremptorily demands that he reduce the number of his quarrelsome, roistering followers from one hundred to fifty. In high dudgeon, Lear leaves her castle and journeys to lodge with his other daughter, Regan, after first sending ahead the disguised Kent as a messenger to announce his coming. He anticipates a proper welcome; but on his arrival learns that Kent has been put into the stocks as the result of a petty altercation with one of Goneril's retainers. Lear, asking shelter for his troop and the immediate freeing of his messenger, is rebuffed by Regan. Instead of redressing her humiliated father's wrongs she insists that he dismiss his whole train of soldiers. Goneril has also arrived at Regan's castle, and the two sisters join to discipline their "mad" father even more. His wits beginning to leave him, the old man flees alone into the stormy night.

Rain and wind are lashing the cold heath that surrounds Regan's castle. The old man, stumbling and near exhaustion, out of his wits, is overtaken and then accompanied by his Fool. In time, as the storm continues, they are also joined by the valiant Kent.

A sub-plot concerns the Earl of Gloucester and his two sons, Edgar and Edmund. The latter, Edmund, is illegitimate but plots to inherit all his father's estate. He discredits his brother with a forged letter written to himself in which – purportedly – Edgar suggests they murder their father. Falsely accused of this design, Edgar flees his father's wrath. He disguises himself as a mad boy, Poor Tom, and is wandering half-naked on the heath in the same storm that encompasses Lear, the Fool and Kent. The four fugitives, joining one another, take haven against the hostile elements in a dark hut. The storm without is but a symbolic reflection of the storm in the hurt, grief-stricken, senile monarch who has lost his mind along with his once flourishing domain. Here the enforcing parallelism, a furious blending of the two storms, is marvellously theatrical. A horde of other dramatists have since borrowed this means of externalizing their characters' inner turmoil.

The Earl of Gloucester confides to his "good" son, the villainous Edmund, a plan to restore the vanished Lear to the throne. Edmund carries this news to the Dukes of Albany and Cornwall, husbands of Lear's elder daughters. He is rewarded by being named Earl of Gloucester in his father's stead, and the old earl is seized, tortured. He is told that Edmund has betrayed him. The cruel Duke of Cornwall gouges out the captive earl's eyes but is himself killed by one of his own serving men, outraged at having witnessed the hideous deed. Regan runs through the servant with a sword.

The Duke of Albany alone has expressed compassion for Lear. He rebukes his heartless wife, Goneril, for her scheming. She, wearied of him, too, has eyes for the dynamic Edmund. But so has the now widowed Regan. The sisters become rivals for Edmund's favour, and he unscrupulously pays court to both.

On the heath, Edgar (Poor Tom) encounters his blinded father. He leads him towards Dover, not betraying his true identity. At one point he pretends to abet his father in a suicide attempt – death by leaping from a cliff – hoping to shock him back to a new interest in life. At last Gloucester recognizes Poor Tom to be his son Edgar. In the same fashion Kent leads the now mad Lear. Their intent is to meet the King of France, who with his queen, Cordelia, has raised a large army and is sailing for England to rescue her deposed father. Cordelia welcomes and cares for the fugitives. But when the battle ensues – Edmund and the Duke of Albany leading their army – the French, headed not by the king but a marshal, are defeated and, unfortunately, Cordelia and Lear are captured. Edmund commands their immediate execution. In another encounter, Edgar and his half-brother duel, and in the exchange Edmund is wounded. Dying, he rues his misdeeds and orders that Cordelia and Lear be spared if possible – but his change of heart is too late, for Cordelia is already dead. Goneril, moved by jealousy and grief, poisons her sister Regan, then stabs herself. Lear, momentarily regaining his senses and discovering that Cordelia, the one daughter who truly loved him, has perished, mourns, collapses and dies across her body. The Duke of Albany, regretting his part in this succession of evil and tragic events, turns over his rule of the kingdom to the always faithful Kent and Edgar.

This melodramatic play, with its complicated action, is filled with some of Shakespeare's most thoughtful observations; at moments it seems meant to have cosmic overtones. There is a lack of clarity – as in *Hamlet*, the motivations of the characters are often ambiguous, the poet's intention oblique. Not only is the storm symbolic but apparently much of the other action, too. This has evoked a variety of interpretations, a whole library of comment and criticism. Like *Hamlet* and *Othello*, the drama has grandeur; the language thunders and soars, the people are vital and have stature, whether they are truly decent or inexplicably evil. The story is gripping, the issues heavy with significance, and spectators are almost always deeply affected. In performance it is poignant, at times terrifying.

The play contains some of the cruellest scenes in all of Shakespeare's work: it is a reversion to *Titus Andronicus*. The episode in which Gloucester's eyes are gouged out, in full view of the audience, and the jelly of the eyeballs stamped underfoot, is shocking, horrible. It must have induced shudders even in Elizabethan spectators long inured to Senecan excesses. The degradation and growing madness of the majestic old king, as he grows feeble and petulant, are almost as painful to behold. He can be identified with too many old men, his fate awaiting a large group. This play, too, has not merely one villain, as has *Othello*, but four: the vilely ungrateful daughters, Goneril and Regan; the malicious Duke of Cornwall; and Edmund, the arch-devil, at dishonest intrigue and roguery quite the equal of Richard III and Iago. It is a harsh appraisal of mankind, again a disenchanted one. Though Lear's doggedly faithful and sometimes serene Fool has a prominent role, segments of humour in the script are largely over-clouded. (That many of the jokes depend on archaic word-plays contributes to this. Few of Shakespeare's texts require so many footnotes or so lengthy a glossary to become intelligible to today's listener and reader.) The prospect of the world it affords is again darkly pessimistic.

Yet, because of the script's power and the splendour of its poetry, its epic scale, it is not depressing. Paradoxically, like other great and massive works of art, it is uplifting. In no other tragic piece by Shakespeare is *katharsis* so surely attained. The theme is universal, and the setting and action have about them a feeling that is primitive, elemental. The play has been likened to the Book of Job with its troubled questioning. The poetry, too, has a different sound and rhythm from that by Shakespeare elsewhere.

It is not a perfect work. The adulatory Coleridge declares: "*Lear* combines length with rapidity, like the hurricane and the whirlpool, absorbing while it advances." But not all would concur. The plot lacks the tautness and symmetry of *Othello*; the story is certainly diffuse, over-long, and filled with wild coincidences, too many meetings on the wind-lashed, rain-swept heath. The play's structure marks a return to that of the chronicle, loose, digressive at times. Many details are inconsistent; there are signs of haste and carelessness. Yet the poet seems to be searching, groping for some new shape, some new vessel for his thought and feeling. If so, he did not succeed in finding it. It is often said that *King Lear* strains the confines of the stage, that it is not a play so much as – like Thomas Hardy's *The Dynasts* – a "dramatic poem". Charles Lamb, for one, and William Hazlitt for another, thought it a violation to put it behind footlights, for the fearful storm, the tempest that engulfs man, and Lear's journey through it, are far better imagined than seen. Some directors argue that *King Lear* fares best when the presentation is highly stylized with little attempt at scenic realism.

The opening scene, the king dividing his domain among his three daughters, two of them wicked – has echoes of a fairy-tale; it is discernibly folkloric. Shakespeare's prose here has the rhythm of a story for children: "Once upon a time . . ." Many in his audience were familiar with the premise of the legend, from Holinshed, Spenser, Higgins and others, so without much thought it was quite acceptable to them, though to a realist the price Lear exacts for his bequest – extravagant expressions of affection – is implausible. As Shakespeare presents the scene, too, it is clearly implied that the division of the kingdom has been fully prepared in advance; the maps are drawn, and to Cordelia is to go the most opulent share. Her suitors have already been apprised of this. The foolish little ceremony in which each of the daughters is asked to profess her love is a sudden, last-minute whim on the part of Lear. When Cordelia, guardedly, refuses to justify her father's award to her of the largest territory, she openly embarrasses and provokes him before his other daughters and the court; his wrath is egotistic but not altogether unbelievable.

The characters learn much during the course of the play. Lear, a heedless autocrat, is taught humility. Accustomed to fawning subservience, he shuts his ears to the honest voices of Cordelia and Kent; he summarily banishes both. It is strange that he does not know the nature of his daughters better; all are mature, and he has lived with them, in turn, for two or more decades. Cordelia, too, should have been aware of his inherent disposition. In fact it would seem that Lear is not very intelligent. Cordelia, like Kent, should have realized by now that there are many moments when it is best to be somewhat discreet. Another point of the play is that Lear's abdication, coupled with distribution of his sovereignty, is foolish, an irresponsible act. It immediately sows discord in his kingdom and provokes a foreign invasion. When a king is weak or incautious, the whole state suffers. Lear has never been self-critical. Regan says of him: "He hath ever but slenderly known himself." At first he is enraged at all who oppose him or deny him anything. He is, in his own view, voiced in a self-righteous phrase, "more sinn'd against than sinning". But gradually he discovers some of his own faults and grows aware of the needs and rights of others. Both old men, Lear and Gloucester, grow in warmth and wisdom. But in many ways it is too late, the lesson too costly. Lear is too aged, his mind too clouded, for him to appreciate the broad folly of his past selfishness, egotism and egregious self-pity. The same is true of the once superficial, hedonistic Gloucester, for he dies shortly after his rescue; his "flaw'd heart" has "'Twixt two extremes of passion, joy and grief, burst smilingly." All the principal figures adopt a stoic attitude before the end.

Miserable in the gale, Lear for once thinks not only of himself as he utters this prayer for all wanderers and rebukes himself for past errors:

> Poor naked wretches, whereso'er you are,
> That bide the pelting of this pitiless storm,
> How shall your houseless heads and unfed sides,
> Your loop'd and window'd raggedness, defend you
> From seasons such as these? O, I have ta'en
> Too little care of this! Take physic, pomp;
> Expose thyself to feel what wretches feel,
> That thou mayst shake the superflux to them,
> And show the Heavens more just.

As has been remarked, the play is the most effective example in all of Shakespeare's works of his use of "parallelism and contrast" – here "reinforcement and expansion" – which he has employed in many of his earlier scripts. W. B. Yeats, whose own search as a playwright was for the "emotion of the multitude", and who saw it provided in Greek tragedy by the voices and responses of the chorus, suggests that Shakespeare attains it by his manipulation of the sub-plot, "which copies the main plot much as a shadow on the wall copies one's body in the firelight. We think of *King Lear* less as the history of one man and his sorrows than as the history of a whole evil time. Lear's shadow is Gloucester, who also has ungrateful children, and the mind goes on imaging other shadows, shadow beyond shadow till it has pictured the world." Such parallels are to be found in *Hamlet*, where the minor figures – Ophelia, Laertes, Fortinbras – also have fathers killed and are called on to avenge them. In *King Lear*, again, Lear's growing insanity is "copied" – or "contrasted" – to two other kinds of madness, the feigned derangement of Edgar in his role of Poor Tom, and the Fool's, his mind being that of the "visionary simpleton". The questions being asked throughout are what is appearance and what is real, what is "sane" and what is "insane". And there is a similar disorder in the state, and in the tempest-wracked cosmos, the natural world. So the play's dimensions expand to where they offer dire perspectives of infinity. Shakespeare ensures this further by having Lear's invectives, his broad curses, include the whole world: he finds it all at fault and would see it all destroyed, the seas overcoming the land. His oaths and apostrophes are aimed at the heavens, the gods, the planet – time and again this is true. Lear thus becomes a symbol of man, and the play in which the forces of Evil and Good are sharply opposed is once more reminiscent of a Morality.

The last scene of *Lear* – the death of Cordelia and the unhappy old king – was too painful for most audiences. It helped to keep the play from the stage. Accordingly, in 1681 a Restoration writer, Nahum Tate, adapted the tragedy and gave it a different ending, one in which Cordelia survives and is soon married to Edgar, who has long loved her. Lear, Kent and Gloucester are also still alive; they retire to a monastery. This version replaced Shakespeare's for a century and a half. It was chosen by such eminent thespians as Betterton, Garrick and Kemble and had the sanction of Dr Johnson, until several critics, notably Lamb and Hazlitt, prevailed upon Kean and Macready to revert to the original text. Since then the true script has been the only one enacted. It gradually won a new and more inured audience, more accepting of a grim resolution.

It is not only Lear who philosophizes in the play. Gloucester, Edgar and the Fool are all given to metaphysical utterances. Most famed of all is Gloucester's biting indictment of a malign universe, the deaf-and-blind or uncaring deities of this timeless, pre-Christian world:

>As flies to wanton boys are we to the gods;
>They kill us for their sport.

Was Shakespeare also applying this to his own day? It is a question much discussed.

As in *Oedipus*, Gloucester is prone to many ironic remarks about his blindness. In response to an old man who exclaims, "You cannot see your way," the sightless earl says:

>I have no way, and therefore want no eyes;
>I stumbled when I saw.

Images having to do with sight are used by all the characters. It is only when Gloucester is blinded that he "sees" his sons and the world in the proper light.

Edgar propounds:

>   Men must endure
>Their going hence, even as their coming hither;
>Ripeness is all . . .

And again, when he first beholds his blinded father:

>   World, world, O world:
>But that thy strange mutations make us hate thee,
>Life would not yield to age.

The Fool, purportedly simple, is forever speaking sagely in epigrams, rhymed adages. He is a one-man chorus, observing and commenting on the errant doings of his betters. (Jesters were still fixtures at the courts of monarchs and noblemen in Shakespeare's time and for some decades afterwards.) He tells Lear: "Thou should'st not have been old till thou hadst been wise." And again, when Lear chides him, "Dost thou call me fool, boy?", he replies with a reminder that the king has surrendered his power and realm: "All thy other titles thou has given away, that thou wast born with."

Edmund offers cynical judgements; they are appropriately shrewd and sharp. Shakespeare implants motivations for this hateful son's villainy: his bastardy, the insensitive levity with which his father refers to it, the secondary status assigned to him, his rights subordinate to Edgar's under the law of primogeniture, his long exile from life under his father's baronial roof. All these help to make his anger, and his urge to strike back, like Iago's, surprisingly credible.

All the characters are concerned with "justice", human and divine. It inspires them to seek vengeance or redress, to get what they consider their "fair" share. They call upon the "gods" to assist them. But no clear answer is ever given to them. Their pleas are desperate, insistent; the silence is obdurate.

Much is made of Shakespeare's use of the words "nothing" and "nothingness" here, since essentially everything that seems important to the people in this story is quickly doomed to disintegrate, reduced to veritable "nothingness": the high rank they held, the love and loyalty they depended on, their sight, their reason and their life itself. Lear, Gloucester, Cordelia, Edgar, Kent lose everything in a world in which only "nothingness" is sure. The play begins with a stately ritual, the characters in their ornate robes of office, a part of the panoply of authority; but some shortly become outcasts, in rags, without shelter. Soon Lear is describing naked man as a "poor, bare, forked animal". He exclaims: "They told me I was everything. 'Tis a lie. I am not ague-proof." A moment later in a valorous effort to re-establish his dignity, he tells himself: "I am every inch a king." But the Fool belittles him: "I am better than thou art now. I am a Fool, thou art nothing." Throughout the play the metaphors and images that spring forth from the dialogue lower man to mere animal form, likening him again to beasts of prey and scavengers, and the point is made that divested of his clothes his ugliness is exposed. This is one reason that Existentialists find *Lear* a play expounding their view of the universe as "absurd", purposeless, meaningless, and therefore a work more welcomed in the mid-twentieth century.

On the other hand, Christian theologians declare that the characters' loss of nearly everything is followed by an ultimate spiritual renewal and restoration of love, faith, property, power, when the people concerned achieve self-knowledge and "illumination". (But by that time most of them are dead.)

The poet Algernon Charles Swinburne called this play "the greatest work of man". A good many other appraisers use the adjective "gigantic". It is an opinion with which Shelley largely concurred: "*King Lear* may be judged the most perfect specimen of the dramatic art existing in the world."

In recent decades British and American actors have taken on the daunting role. Howard Taubman assayed one offering that scored on Broadway in 1964:

> Although Paul Scofield is a Lear you will not forget, he is not merely a star around whom an *ad hoc* production has been reared with faulty underpinning. The *King Lear* that brought the reverberations of great tragedy for the first time last night into the New York State Theater at the Lincoln Center for the Performing Arts is a proud, unified company achievement.
>
> The Royal Shakespeare Company . . . has sent us a great masterpiece produced with regard for the noble arch of its structure, and all the parts are like stones fitted into their niches and carrying their share, however large or modest, of the mighty burden.
>
> Peter Brook has staged this *Lear* with a kind of elemental spareness and simplicity. It is said that he was influenced by the analysis of Jan Kott, the Polish scholar, whose *Shakespeare Our Contemporary* is to be published here in September. In an essay on "*King Lear* or *Endgame*," Mr Kott has attempted to show the parallels between Shakespeare's tragedy and Samuel Beckett's play.
>
> It is Mr Kott's contention that both plays deal with the disintegration of established values. The theme of *Lear*, he insists, is an inquiry into man's journey from the cradle to the grave "into the existence or nonexistence of Heaven and Hell." The theme, he adds, is nothing less than "the decay and fall of the world."
>
> It does not matter whether one accepts Mr Kott's view. The important point is that it apparently has led Mr Brook to conceive of a *Lear* stripped of the panoply of old-fashioned Shakespearian staging. This is a *Lear* whose pertinence to man's predicament yesterday, today and probably always is inescapable.
>
> Mr Brook's design is based on a couple of enormous gray rectangular panels that, standing diagonally at the sides, frame a similar, unornamented gray rear wall. In several places rectangular sheets of metal, looking like copper, are lowered from above; for several scenes two rough-hewn fences are lowered from the sides, and early in the play Lear and his retinue dine at great, rough tables. Otherwise, there is hardly any furniture. The huge stage is like a vast, empty, heartless earth.
>
> The clothes are equally evocative. The dresses and tunics look as though they were made of homespun fabrics and leather. They bring to mind the far past; yet they are simplified and stylized so that they could be of any period.
>
> Mr Brook has composed his scenes in the framework of the open spaces of his deep stage so that man, though often in the foreground, is a small figure against the thick rotundity of the world. Mr Scofield as the mad, spent Lear and the blind Gloucester of John Laurie, consoling each other, form a sculptured pietà in an unfeeling plain. Mr Laurie, seated alone on the bare stage, his body contorted yet limp, while the battle rages unseen, is a harrowing vision.
>
> Mr Scofield's Lear has size, but it remains within the proportions of modern man's

sensibility. In his first scene he allows himself the signs of old age – a cracked, quivering voice and a trembling arm – as he exhorts his daughters to expatiate on the extent of their love. When Goneril later crosses him, he can overturn the table and set his knights roaring in anger. And in the tempest his powerful voice can contend with the elements without resort to ranting.

Although Mr Scofield commands the grand manner, he is at his noblest and most moving as his anguish increases and his mind turns inward and cracks. His scene, after the storm with the Fool, played with poignant lightness by Alec McCowen, with Brian Murray's tender madman of an Edgar and with Tom Fleming's manly, honest Kent has a haunting sense of pity. Mr Brook uses long silences here and elsewhere with remarkable effect.

Mr Scofield finds subtleties of movement, expression and speech to convey the deepening of his awareness and humility. Awakening from his long sleep, he recognizes Cordelia, played with pride and purity by Diana Rigg, with a sense of bewilderment and shame. At the end when he carries in the slain Cordelia, he is like a man who has summoned up unexpected reserves of physical strength and moral courage.

Villainy is stark and brutal in this *Lear*. Irene Worth's Goneril is as stony of voice as she is marble of heart. Pauline Jameson's Regan is vicious and vengeful. Ian Richardson's Edmund has bitterness as well as evil and at the end the necessary touch of gallantry. Tony Church's Cornwall is an eye-gouger to the brutal manner born. Michael Williams's Oswald is foppish and meanly servile.

A former member of the Group Theater had his measure taken by Taubman in the *New York Times* at almost the same time (1965):

There is no end to *King Lear*. Ask Alex Carnovsky; he knows.

Two years ago Mr Carnovsky undertook Shakespeare's most challenging and most profound tragic role at the American Shakespeare Festival Theater here and achieved moments of greatness. Then he went on to play the role in other cities, including Los Angeles. Now he is grappling with Lear again in Connecticut's Stratford, where the production returned last night.

To one who has watched Mr Carnovsky cope with Lear in Connecticut, in California and again in Connecticut, it is clear that he keeps searching for new ways to plumb the overwhelming tragedy's depths. In his latest try he has arrived at his most coherent realization.

Surely and steadily Mr Carnovsky has moved toward a performance that reminds one of a secret of the greatest musicians – their capacity to manage a wealth of subtle inflections within a carefully controlled compass. Mr Carnovsky's current Lear is masterly in the delicacy and penetration of its nuances.

Indeed, it is more poignant than ever because it is so vulnerably and sadly human. But it is no less heroic than before, even if grandeur is not sought for in thunder. For its heroism is firmly rooted in an awareness and acceptance of the human condition.

In Connecticut two years ago Mr Carnovsky managed scenes of surging passion, particularly in the storm, and of ineffable tenderness, particularly in his pathetic encounters with the Fool, Edgar feigning madness, and the blinded Gloucester. But the early scenes were not of the same order, possibly because Mr Carnovsky was saving something for the big one to come.

In Los Angeles last summer Mr Carnovsky succeeded in raising the pitch of the early scenes and maintaining it throughout the play. The result was an interpretation of greater consistency and impressive potency.

For this revival Mr Carnovsky has refined his entire performance. He no longer needs to roar at the extremity of his vocal powers, even in the storm sequence. Because he begins with more restraint, he gets the effects he wants without bellowing.

The risk of this kind of regulated attack is that Lear might become diminished and self-conscious. Nothing of the sort has happened. The reason is simple: Mr Carnovsky's control is not a virtuoso actor's stunt but organic to his conception.

Watch him now as he divides his kingdom. He is old, deliberate and abstracted; the warrants of love he wants from his daughters are routine contracts he expected them to deliver, signed and sealed. The latent fire in the sovereign smoulders as he turns on Kent and as he snarls at Cordelia's suitors. It flares up as he cries out his curse on Goneril. But this is the fierce eruption of a man who is still vain and proud.

As Lear teeters on the edge of madness, there is a delicate balancing between bitterness and tenderness, expressed most movingly in the dialogues with the Fool. The last effort to believe in his stable world is reflected in his desperate embrace of Regan when he hopes that she will be dutiful. The cry, "Reason not the need," becomes a supplication, helpless and grief-stricken.

Like Mr Carnovsky, Allen Fletcher, the director, has subtilized and simplified his production. There is less scenery and there are fewer props than two years ago. The storm scene is played on a bare stage with only a crash of thunder and rays of ominous light to support Mr Carnovsky's defiance of the elements.

Mr Carnovsky's madness is not a performer's effect. It is a release of inhibitions, and it is achingly, piteously gallant. The awakening to sanity and to humility in the final scenes have a wrenching and healing truth.

In the fallible and costly world of the theater it would probably be too much to expect the company to match the maturity and perception of Mr Carnovsky. Those who come nearest to what one would like are Patrick Hines as Gloucester, Roy Poole as Kent and Richard Matthews as the Fool.

John Cunningham plays Edmund with intelligence but the note of commanding authority is not yet there. Patricia Hamilton is a venomous Goneril. Ruby Dee as Cordelia brings warmth to the final scenes. Stephen Joyce as Edgar is pathetic when he feigns madness but needs size at the end.

But *King Lear* as always takes upon itself the mystery of things, and Mr Carnovsky perseveres imaginatively and eloquently in his search for an ultimate Lear, knowing, as we do, that there is triumph in the seeking.

Orson Welles, injured in a fall, was an effective Lear in a wheelchair in New York in 1956. A white-bearded Laurence Olivier production of the play was broadcast on both sides of the Atlantic in 1984.

From Jakarta (1999), Margot Cohen described for readers of the *Wall Street Journal* an odd mutation of the drama:

Shakespeare called Goneril a "thankless child," a "plague-sore," and even an "embossed carbuncle." But the eldest daughter of King Lear appears even more terrifying in the hands of Singaporean director Ong Keng Sen and Japanese playwright Rio Kishida. At the climax of their dazzling new production, simply titled *Lear*, the power-mad dame thrusts a dagger into Daddy's gut. "You created me," she helpfully reminds him.

In probing how despotism breeds violence, *Lear* seems no stranger to Asia. The production certainly struck a chord in Indonesia, a riot-wracked nation struggling to emerge from the dark shadow of former President Suharto. Theatre-goers here gave a standing ovation to the all-Asian cast at the opening-night performance last month, and subsequent shows were sold out.

Fresh from packed theatres in Hong Kong and Singapore in January, the three-million-dollar production moved on to Perth, Australia, and will play in Germany and Denmark in early summer.

Audiences in each country must adjust to the most innovative aspect of *Lear*. Each performer speaks or sings in his own native tongue. Chinese, Japanese, Thai, Indonesian, Javanese, and Minang (which is spoken in west Sumatra), each take a turn, none bowing to the supremacy of the other. To keep Jakarta viewers from drowning in the polyglot, English and Indonesian subtitles were flashed on screens suspended on either side of the stage.

To some extent, the subtitles distract from the majesty of the staging, replete with vivid lighting and elaborate Japanese costumes. Mr Ong's choice stems from the politics of language rather than audience comfort. The director argues that monolingual productions often allow one culture to "appropriate" other cultures, pointing to such works as Peter Brook's *Mahabharata*, which an international cast performed in English.

"They were so anxious to have harmony," says the thirty-six-year-old Mr Ong. "For me, and the whole company, we accepted that harmony is not the most important thing. More often we went for a clash." That might sound like heresy to Singapore's Lee Kuan Yew and other proponents of "Asian values," who maintain that age-old ideas of harmony, family and community are sacrosanct in this part of the world.

Lear offers evidence that a young generation of Asian artists is ready to throw down the gauntlet, without entirely abandoning tradition. Indeed, the tension between traditional and modern art forms provides much of the fascination here. The title role belongs to Naohiko Umewaka, whose voice and gestures are steeped in Japanese Noh theatre. His deep growl and glacial steps contrast sharply with the urbane patter and gamine moves of the Fool as played by Japanese television actress Hairi Katagiri. In sunglasses and a zippered jogging suit, she acts more like a snap-happy Japanese tourist than a fount of Shakespearian wisdom.

The most compelling contrast, however, emerges between Lear and his eldest daughter, played by Beijing Opera veteran Jiang Qihu. ("Goneril" and other familiar Shakespearian monikers have been stripped by Ms Kishida.) Accustomed to male leads at home, Mr Jiang deftly transforms his voice into a shrill soprano worthy of a haughty empress. He gives his wife, an opera teacher, credit for coaching him, though he was not aiming to quash his own gender completely. "The eldest daughter is full of masculine energy," he explains. Mr Jiang's merciless despot displaces Lear as the star of the show.

The gender-bending seems both contemporary and classic, given that men often play women's roles in traditional Asian performance art and did the same in Shakespeare's theatre. For a real shot of male hormones, *Lear* relies on the dynamic choreography of Indonesia's Boi Sakti. Sharp movements drawn from the west Sumatran tradition of martial arts provide an edge to a production that might otherwise drag. The stiff, macho soldiers create a good backdrop for the fluid ballet movements of the king's youngest daughter, delicately played by a male dancer, Thailand's Peeramond Chomdavat.

In Shakespeare's play, the king banishes his youngest daughter, Cordelia, after she fails to provide flamboyant assurances of love. In this version, the youngest daughter not only says the wrong thing, but is struck mute. This plot twist mirrors the obsession of Ms Kishida's script: Language is power.

"Words are weapons! I have won with words," gloats the eldest daughter. The Fool flippantly abandons Lear with the phrase "I'll ramble here, ramble there, till I find a king who knows how to play with words." In essence, the production urges Asians to find their own voice, even if they express themselves in many different tongues.

Ironically, *Lear*'s music often proves far more expressive than the translated dialogue. From the haunting melody of a Japanese lute to a piercing Minang vocal composition, the multilayered accompaniment adds beauty and power to the drama. However, gamelan lovers may find some portions too abrupt, while the synthesized pop music is often sappy.

For all its passion, the production fails to build to a satisfying conclusion. The eldest daughter suddenly finds that it's lonely at the top, after murdering all her relatives and lovers. As the stage darkens, the newly created character of Lear's wife appears. Does the mysterious white-clad figure finally pardon her vicious daughter, or simply extend an embrace of death? The audience is left clueless.

Offstage, in Jakarta, the question also remains unresolved. After enduring thirty-two years of Suharto's harsh regime, Indonesians are wavering between reconciliation and revenge. For better or worse, *Lear* provides no easy answers.

A *Lear* with an African–American cast (2002) was reviewed by Tom Sellar for the *Village Voice*:

Classical Theater of Harlem uses a deceptively pleasant spot to bring home Shakespeare's most crushing tragedy of age and family. Framed by an ivy-covered back wall their courtyard stage consists of various platforms, benches and concrete steps, with only a throne and a few selected set pieces introduced when necessary. Though the company shares this small patch of foliage with July fireflies, it doesn't take long before a balmy evening turns cold, stormy, and dark, as Lear's jolly succession ceremony erupts in family discord.

The most effective scenes – as with most *Lears* – develop from direct confrontations between the conflicting generations, particularly in the first act's falling out and, later, when the rebelling daughters unite to disband their father's retinue. Angela Hughes (Regan) and April Yvette Thompson (Goneril) seize on the sisters' rising intoxication with domestic supremacy, masking their ambitions with dutiful-daughter smiles and issuing threats and ultimatums with hugs. Paul Butler plays the title role with indignation and fortitude; more than a crownless king, Butler shows us a father caught between pride and self-pity, trapped and then destroyed by the choice. When Lear demands respect from his unrelenting daughters, scolding "Art not asham'd to look upon this beard?" Butler's gravelly voice falters, and we understand that the aging man is also asking himself that question.

As if emphasizing the drama's timeless qualities, director Alfred Preisser sets the action in a distant pagan era and puts the company in African robes and capes. The production could use a little visual variety and the verse-speaking needs specificity, but Preisser's largely youthful cast focuses more on character than composition: Hilary Ward plays a disarmingly earnest Cordelia; J. Kyle Manzay turns bastardly Edmund weirdly impish; and Ken Schatz makes an insouciant Fool with a streetwise veneer. When Lear rails against time alone on the rainy heath, however, Preisser arranges a memorable tableau: The others remain on stage and turn their backs, as the courtyard resounds with Shakespeare's cries of protest against filial ingratitude.

Another look at *King Lear* was afforded by Sir Peter Hall's treatment of the work at the Old Vic in 1997 with Alan Howard in the title role. Paul Levy in the *Wall Street Journal*:

The production has been troubled. The August press night was canceled ostensibly because of the death of Princess Diana, though the lukewarm reviews it finally received made it seem that the production simply had not settled down. When I saw it, two performances after its eventual opening, it exceeded its listed running time by forty minutes – all of which had been used to the profit of Mr Howard. He is now a fine, individual Lear, pronouncing all his glorious words with care and obvious concentration. The famous vibrato is still present in his voice, but it's controlled: His execution is a cross between Grand Guignol and method acting.

The text used is the shorter 1623 Folio. In the first scene, Lear's embracing of Goneril and Regan, the kisses, full on the mouth, last a bit too long. Whether the director is suggesting incestuous feelings, or that the sisters are motivated by a history of sexual abuse, is never made plain. This is a seam that is, perhaps mercifully, never unpicked. But maybe it shows what is flawed about this staging. I left the theatre thinking that Mr Howard had convincingly showed the inevitability of the consequences of Lear's division of his kingdom and rejection of Cordelia, and managed the transitions from sanity to madness and back again with consummate skill; but that he, and Sir Peter, had no very clear view of why Lear stooped to such folly in the first place.

Macbeth (1606) has a somewhat fresh setting but is again relegated to a period as remote and primitive as that of *Lear*. It is nearly as dark a tragedy. The choice of a Scottish subject may have been owing to the recent accession of a Stuart, James VI of Scotland, to the English throne as James I, and the forming of the United Kingdom. The Tudor line was ended; Elizabeth was childless. James was known to be fond of theatre and more generous in spending funds to encourage it. Why not write a play that might catch his fancy? Was that Shakespeare's aim? Perhaps. Once more he turned to Holinshed, where there is a lengthy account of the legendary Macbeth, whose character – and that of Lady Macbeth – were, as usual, wholly reconceived by him. In doing this he may have been influenced by another history of Scotland, in Latin, by George Buchanan (1256), but it is not certain that Shakespeare had access to it. Could he read enough Latin? His working habits are unknown.

The play, in Shakespeare's hands, is drenched with supernaturalism – witches, ghosts. In *Hamlet* and *Lear* there are instances of this – Gloucester is much inclined towards reading the stars for portents and is mocked by the sceptical Edmund for doing so. Hamlet is in doubt whether to trust the ghostly voice he hears on the battlement – what if it is not his father's but the Devil's? In *Macbeth* unquestioning belief in other-worldly intervention in human affairs is paramount. And heed to witches' prophecies. This, too, might have been of particular interest to King James, who himself was to write an erudite book on demonology. But Shakespeare had other reasons for placing emphasis on the infernal and diabolic. In doing his "deed of darkness" Macbeth – like Faust – was selling his soul to superhuman Evil. This can be looked upon as merely a symbolic device, but to many in Shakespeare's audience it was doubtless taken at face value, quite literally. The spectators were a mixed lot: some sophisticated; others still unlettered.

Macbeth – Thane of Glamis – and Banquo, two generals in the forces of Duncan, King of

Scotland, have been victorious in a campaign to put down a rebellion. On their way home, on a stormy, lonely heath, in thunder and lightning, they encounter three witches performing their rites. The Weird Sisters hail Macbeth as "Thane of Cawdor" and "King of Scotland" and address Banquo as "the forebear of kings". Macbeth is quickly impressed by this, but Banquo, less credulous, challenges them:

> If you can look into the seeds of Time,
> And say which grain will grow and which will not,
> Speak then to me.

Shortly afterwards Macbeth learns that King Duncan has indeed awarded him the much-sought fiefdom of Cawdor. He cannot forget the second part of the prophecy. His ambitious wife is also consumed by the prospect of the throne but fears that her husband is too irresolute – too full of "the milk of human kindness" – to accomplish the design.

The king, on a progress across his country, comes as a guest to Macbeth's castle. Goaded by his dynamic wife – in this instance it is the woman who is the more Machiavellian figure – Macbeth murders the sleeping Duncan and, with his wife's iron-willed help, smears blood from the dagger on two of the monarch's grooms whose wine has been drugged. Macbeth, accusing the two serving men of the crime, kills them.

Macbeth's crimes begin to haunt him. At a regal feast the ghost of Banquo, whom he also murdered, appears to him. Macbeth, disturbed, cries out; his wife is forced to send away their guests. To learn what awaits him, Macbeth seeks out the three witches. They tell him that "none born of woman" shall ever remove him. What is more, he shall never suffer defeat in battle "until Birnam Wood comes to Dunsinane".

The rightful heir to the throne, Malcolm, son of the slain Duncan, has taken refuge in England. A Scottish nobleman, Macduff, goes there to urge recapture of the kingship from Macbeth, the usurper. While Macduff is in England he hears the dreadful news that his wife and child have been murdered at Macbeth's command. Macduff gathers an army to overthrow Macbeth and avenge his slaughtered family.

Macbeth's wife, her mind overburdened by the guilt of these many accumulated crimes, goes mad and soon dies. The invading forces lay siege to Macbeth's castle at Dunsinane. To hide their advance the soldiers are ordered to cut branches from the trees of Birnam Wood. Seeing the sheltering "forest" marching towards him, the superstitious Macbeth – grief-struck over the death of his wife – is nevertheless prepared to fight on, convinced that "none born of woman" shall prevail over him. He engages in single combat with the vengeful Macduff. To his dismay he hears that Macduff was "from his mother's womb untimely ripped" (that is, by a Caesarian birth). Macduff slays the usurper, fulfilling the witches' warning, and Malcolm assumes the kingship.

Macbeth is one of Shakespeare's shortest tragedies, which suggests that the extant script is fragmented; but, even so, the play flows on smoothly. If it has been cut, by later hands, it was skilfully effected, for the action is clear and rapid. Mostly it delves into a criminal's psyche, the moral deterioration of a gifted man, potentially noble, undermined by ambition, driven on by a ruthless woman who holds him sexually enthralled and demands his bloody deeds as her price. The language again has a dizzying splendour. The haunted Macbeth's soliloquies, when his conscience troubles him, and Lady Macbeth's sleepwalking scene – her mind reliving the horrible details of Duncan's death – are among the most powerful ever seen on the stage. Shakespeare's virtuosity is attested nowhere better than here. As a study of how a criminal justifies to himself his acts, and how he must contend within himself with the gory memory of them, it ranks with Dostoevsky's probing and monumental fiction.

Interestingly, Shakespeare deprives Macbeth of the motivations given in Holinshed for Duncan's murder, a sincere conviction that his claim to the sceptre is more legitimate than that of his rival, Duncan. Furthermore, Duncan's rule is weak and ineffectual. The play also has the killing take place while Duncan is a guest in Macbeth's castle, doubling the horror of the gruesome deed, for Duncan is shown as old, gracious, kind and generous to his hosts. Shakespeare also permits Macbeth to have finer sensibilities than does Holinshed. Thus the ugliness of his career is brought even more into focus, even more heinous than if he were obtuse. Holinshed has Macbeth rule well for ten years; Shakespeare turns him into a short-lived tyrant.

What is amazing is how, despite his vicious acts, Macbeth is still a sympathetic character: he has great physical courage, insight, rare poetic eloquence. Also there is a glimpse of him before his criminal course is started, a man of stature and seeming virtue. And – in partial exculpation – he is shown no little under the influence of the witches and their prophecies. His grasp at power apparently has in his mind the approval of "fate" or "the gods". Much of the blame for his downfall, too, is allotted to his inordinately appetent wife. To soften the impact of their crimes, most of them occur off stage. Finally, the other characters lack the visceral life of the highlighted pair, thane and lady, who dominate the tragedy. The spectator shares their schemes with them, seeing the dangerous task as they do, identifying with them, willingly or unwillingly. But the chief reason they earn and hold sympathy, perhaps, is that both are strongly tormented by their conscience, and the spectator feels pity and terror along with them.

Another outstanding accomplishment is the establishment of an all-encompassing mood for the drama. As Mark Van Doren points out, the adjectives "bloody" and "darkness" recur again and again. Most of the action occurs at night, and the stormy air throughout is overcharged with recriminations and rumours of witchcraft and tokens of fear at imagined as well as real perils, such as might be found in a Gothic novel. Macbeth, his hands still feeling gory, is never at peace or relaxed. The stench of sticky blood seems always to linger in his nostrils.

The castle is on a lonely heather-clad Scottish moor.

> Light thickens, and the crow
> Makes wing to th' rooky wood;
> Good things of day begin to droop and drowse,
> Whiles night's dark agents to their preys do rouse.

The superb "porter scene", an injection of grisly humour a moment after the murder of the sleeping Duncan, is another remarkable theatrical effect that relieves the tension of the audience at the same time that it deepens by acute contrast the horror of the deed just completed. Afterwards, in a vision, a procession of eight kings of Scotland, Banquo's descendants, marches before Macbeth's staring, frightened gaze, yet one more striking bit of stagecraft. He is haunted by phantoms. But it is the vividness of the language that most accounts for the eternal success of this drama. Sleepwalking, and re-enacting the slaying, Lady Macbeth moans: "Who would have thought the old man had so much blood in him?" Wringing her hands, believing them to be ineradicably sanguine, she cries: "Out, damned spot!" and desperately, hopelessly seeks to purge them. "Here's the smell of blood still; all the perfumes of Araby will not sweeten this little hand." Even while participating in the killing, Macbeth himself cannot credit the sight of his hand. Shuddering:

> Will all great Neptune's ocean wash this blood
> Clean from my hand? No, this hand will rather
> The multitudinous seas incarnadine,
> Making the green one red.

This is hyperbole, yet it matches the dimension of the crime that has been committed against the state itself, which might be plunged into anarchy: Macbeth cannot know for certain that he will succeed to the throne without struggling against competitors.

Later, under the burden of his guilt, he groans:

> I am in blood
> Stepp'd in so far, that should I wade no more,
> Returning were as tedious as go o'er.

Guilt and regret have never been given a more eloquent utterance, yet once set on a course of murders he feels that he cannot ever stop in his rise; he has already made too many sworn enemies.

When his wife dies he is isolated, his ultimate despair epitomized in a soliloquy of chilling relevance and universality, especially to listeners in the twentieth century, when many in the intelligentsia reluctantly embraced nihilism and existentialism:

> She should have died hereafter;
> There would have been a time for such a word.
> Tomorrow, and tomorrow, and tomorrow,
> Creeps in this petty pace from day to day,
> To the last syllable of recorded time;
> And all our yesterdays have lighted fools
> The way to dusty death. Out, out, brief candle!
> Life's but a walking shadow, a poor player
> That struts and frets his hour upon the stage,
> And then is heard no more. It is a tale
> Told by an idiot, full of sound and fury,
> Signifying nothing.

An appalling statement! Is it meant to apply just to Macbeth and his misguided ilk, or does it express a deeply held belief – or lack of belief – in Shakespeare himself?

Quite a different set of adjectives (and especially colours) dominates the dialogue in the episodes that involve Duncan, Banquo, Macduff and the other "good" characters: here the allusions are to things alive, bright green and growing. On arriving as a guest, the gracious and unsuspecting Duncan beholds Macbeth's castle as "a pleasant seat":

> This guest of summer,
> The temple-haunting martlet, does approve,
> By his loved masonry, that the heavens' breath
> Smells wooingly here; no jutty, frieze,
> Buttress, nor coign of vantage, but this bird
> Hath made his pendent bed and procreant cradle.
> Where they most breed and haunt, I have observ'd
> The air is delicate.

Beauty is in the eye of the beholder! This alternation of mood is sustained, so that even as the king lies murdered he has a certain sanctity and beauty: "His silver skin lac'd with his golden blood."

Yet another series of images here, as in *Lear*, has to do with attire, the garments that symbolically or metaphorically Macbeth puts on or takes off. Says Angus:

> Now does he feel his title
> Hang loose about him, like a giant's robe
> Upon a dwarfish thief.

And Banquo:

> New honours come upon him,
> Like our strange garments, cleave not to their mould,
> But with the aid of use.

Small details such as these have a cumulative effect. Shakespeare, the poet, had learned how to repeat carefully chosen images, almost imperceptibly ringing changes on them, so that they grow stronger and persist in the mind and ear, much as a Mozart or Beethoven composes variations on a small brace of notes in the development of quartets and titanic symphonies.

> Methought I heard a voice cry, "Sleep no more!
> Macbeth does murder sleep" . . .
> "Glamis hath murder'd sleep, and therefore Cawdor
> Shall sleep no more; Macbeth shall sleep no more."

Macbeth, like the spectator, can truly say: "I have supp'd full with horrors." Gone for him is "Sleep that knits up the ravell'd sleeve of care". When his eyes are closed he is shaken by terrible dreams, nightmares. By contrast, of Duncan it is said that "after life's fitful fever he sleeps well".

Lady Macbeth walks in her sleep, reliving her crime. Sleep is no longer a refuge for her. Her days are filled with perpetual disquiet.

Goaded by ambition, this pair have given to forces of Evil their "eternal jewels", their souls.

Compact, swift, yet philosophically and psychologically profound, *Macbeth* is another demonstration of Shakespeare's occult power as a dramatist, poet and moralist. A terrible retribution is visited on the anti-hero and his wife in this grim story of regicide and its consequence. The play is an analysis without equal of the human conscience sick to death.

Despite its harrowing subject and overall sombre mood, *Macbeth* has been among the half-dozen works by Shakespeare most often staged. Many leading actors aspire to the role, but not many have proved to be successful; they lack the physical and psychological heft the characterization seems to require. Olivier triumphed in the part at Stratford in 1937, though he later acknowledged that his make-up was excessive. One London critic wrote that the portrayal "shook hands with greatness" and another that this was "the best Macbeth since Macbeth", which covers a considerable span of time. He appeared as the Scottish king elsewhere on several other occasions, changing his interpretation somewhat after further probing. For a number of years he tried to raise funds to film his performance but failed and finally had to abandon the project. (Eventually he was able to make a television film of his *King Lear* that was broadcast in America.)

A novel *Macbeth*, with a cast of black actors, was put on by Orson Welles and John Houseman at New York's Lafayette Theater in 1936; it was also referred to as the *Voodoo Macbeth*. This was an undertaking of the Federal Theater, a government agency established by the Franklin D. Roosevelt administration to support stage people during the Depression of the 1930s when there was widespread unemployment. The setting of the play was shifted to the luxuriant jungles of Haiti, the costumes were bizarre and brightly coloured, and there were witch-doctors, chanting, dancing. The critics were of two minds, some put off by the liberties taken with the background and text, others jubilant. The *cognoscenti* and Broadway folk flocked to Harlem to take in the exuberant production, which had its run extended. One objection was that a good many in the company, mostly inexperienced, could not cope with Shakespeare's verse and that Jack Carter, the Macbeth, was hardly adequate, though physically imposing.

Welles later made a film of *Macbeth*, which he directed and in which he himself assumed the title role (1947). He chose to shoot it in black and white, to abet and preserve the shadowy ambience of Glamis castle. (Later, rehearsing *King Lear* for an engagement at New York's City Center in 1955, he broke his leg. He insisted on performing, all the same, making use of a wheelchair. In what proved to be a *tour de force*, he was compelling as Lear, with spectators finding nothing inappropriate about the befuddled old king's inability to walk.)

The Australian actress, Judith Anderson, was Lady Macbeth in Olivier's production at Stratford and two seasons afterwards in a New York staging of the play with Maurice Evans. Her sleep-walking scene was blood-chilling. She had visited a hospital for the mentally ill to observe the halting gait and awkward gestures of chronic somnambulists.

Akira Kurosawa, the master film director, drew on *Macbeth* for his *Throne of Blood* (1957), the Scottish setting transposed to sixteenth-century Japan and enacted with a scattering of stylistic touches taken from the Noh theatre. The Macbeth is the director's favourite actor, Toshiro Mifune. Stanley Kauffman, in *A World on Film*, found that Kurosawa had strayed far from Shakespeare's clear intention. The result is fascinating and exotic but distorts the play on which it is very loosely based. Many of the scenes are powerful, but a Western spectator cannot readily identify with this Macbeth and Lady Macbeth, whose reserve as a Japanese wife ordains that she be seated on a pillow on the floor, her eyes downcast, as she urges her hesitant husband to kill Duncan. (Much later Kurosawa also made *Ran*, his version of *Lear*, with much of the aged man's wanderings shot on the barren slopes of Mount Fuji.)

Verdi, admiring Shakespeare and often looking to him for a subject, settled on *Macbeth*; his opera had its première in Florence in 1847 and was put on in St Petersburg in 1855. The libretto, by Francesco Maria Piave, stayed fairly close to the play though there were naturally elisions and alterations. For a production in Paris, a decade later, Verdi revised the music and had the work considerably reshaped. Piave had oversimplified the story to a point where it was not as dramatic as it might be. Verdi's uneven score reflects this; it is expressive and forceful for some episodes, in other passages mediocre, weak. For a long period of time the work was overlooked, but then it was discovered and hailed excitedly. It now holds a firm place, though not a prime one, in the repertory of a number of major companies.

Twenty-five years old, Ernest Bloch (1880–1959) wrote a *Macbeth*; he worked on it for five years, after which he went on to compose some of the twentieth century's finest orchestral music – quartets, suites. He held back from ever creating another opera, complaining that it was too difficult to obtain a production of any such work. His librettist was Edmond Fleg (Flegenheimer). The action adheres to Shakespeare's for the most part, with some inevitable compression, but more attention is given to the witches and their wild rituals. Lady Macbeth, recalling and recoiling, exhausted, falls dead at the end of her sleepwalking scene. First produced at the Opéra-Comique in Paris in 1910, after a six-year delay, it was revived at the San Carlo in Naples in 1918. Decades later, New York opera-goers had an opportunity to see and hear it when graduate students at the Juilliard School of Music presented three effective stagings of it at their theatre in Lincoln Center. The opera is now little known. Bloch was disappointed at its being largely passed by. In *The Book of the Opera* an anonymous musicologist says of it: "Bloch proved a magnificent Shakespearian interpreter. His music, neither forced nor rhetorical, gives powerful expression to the passion and drama of the original, and he shows with the utmost clarity the sombre figure of Macbeth."

Going beyond Welles is an even more "savage" rendition of the drama, an import from South Africa. Donald Lyons saw it for the *Wall Street Journal* in 1997:

> It was in 1969 that the young South African playwright-director Welcome Msomi got the idea for *Umabatha: The Zulu Macbeth*, a recasting of the Shakespeare play into the context of early-nineteenth-century Zulu history, the time of the great unifying warrior-king Shaka. Unable to study at the University of Natal because he was Black, Mr Msomi composed an epic piece of stagecraft that, in the irony-filled world of Apartheid, had its 1971 première at that university's Amphitheatre. The work, which played in New York in the nineteen-seventies, has now been revived and expanded and is being performed at the Lincoln Center Festival prior to a national tour.
>
> "What inspired me to choose *Macbeth*," Mr Msomi has written, "is that the intrigue, plots and counterplots of the Scottish clans were almost a carbon copy of the drama that took place with the early nations of Africa. In *Macbeth*, Shakespeare had said it all." *Umabatha* does not, though, set out to translate Shakespeare's words; it reimagines the tale as, essentially, a sequence of electrifying communal rituals. There are supertitles that clarify the settings and translate some of the spoken Zulu, but even a high-school memory of the story is enough to render them unnecessary.
>
> After beaded-haired witches utter their prophecies about Mabatha (the *U* in *Umabatha* is a prefix for direct address or reference), his wife Ka Madonsela (a flamboyantly energetic Dieketseng Mnisi) urges her household to cook and weave in preparation for the visit of King Dangane (a splendid Lawrence Masondo). The throbbing ceremony of royal welcome, danced to five big drums and the pounding of spears, is the first of the play's great set pieces.

The stiller, more interior moments – such as Mabatha's soliloquy wondering how he can shed "my father's blood" – are given short shrift; a gleaming dagger rules the night. After the porter's drunken and funny antics comes the next grand occasion: the formal funeral dirge for Dangane by the army. Two lines of warriors arrayed in beautifully patterned brown-and-white shields, neckpieces and loincloths rhythmically squat and kneel about the corpse of the king.

Although Mabatha himself is vividly and sympathetically presented by Thabani Patrick Tshanini, Mr Msomi's Zulu world seems uncomfortable with a hesitant, reluctant, uxorious hero (the play's real Shaka-figure is the marginal Dangane). But this remains an exciting, alive and powerful work in which a sophisticated modern artist uses an old story-form to imagine his people's epic past. It's at its best in the great rites of joy and grief and state occasion where dance is shown as formal expression of a communal soul.

Very different from *Lear* and *Macbeth*, with their pervasive grey climate, blindings and murders, is the next major tragedy: it seems like the work of quite another playwright. It is as if a beclouding mist was lifted from Shakespeare's spirit and imagination. In *Antony and Cleopatra* (1606–7) the action occurs under an airy dome of bright Egyptian sunlight, the characters responding to the heat arising from the vast surrounding, glaring desert. It is a complete change of atmosphere, the sweltering environment contributing to the passions, the reckless sensuality, ruling them.

As before, when approaching a classical subject, Shakespeare resorted to Plutarch's *Lives*. It was available to him in Sir Thomas North's (1535–1601) admirable translation, which in turn was not from the Latin but from Amyot's French version (1579), a work so excellently written that Shakespeare took lines from it intact, including some for the famous description of Cleopatra's gleaming, lavish barge. Other English playwrights, Mary Herbert, Countess of Pembroke and her contemporary Samuel Daniel (1594), had used the ill-fated lovers as a subject; Shakespeare may have had access to their plays, but if so he borrowed little from them.

He departs hardly at all from Plutarch's graphic account except that he elevates Cleopatra's role until it is equal to that of Antony and of the same tragic weight.

The triumvirate, Mark Antony, Octavius Caesar and Lepidus, are governing Rome. Antony goes to Egypt, and there is enthralled by the siren queen, Cleopatra. He neglects his duties, engaged in dalliance with her. Only when he hears of the death of his wife, Fulvia, and civil war and other troubles at home, can he bring himself to part from his regal mistress and return to Rome, though to some extent the affair has begun to pall.

Away from Cleopatra he is momentarily freed from her spell. His political acumen is restored. He confesses his faults to Caesar, and opportunistically strengthens the alliance by a marriage to Octavia, sister of Caesar. When news of this is brought to Cleopatra she is so furious that she almost kills the hapless messenger. Meanwhile in Rome a military dispute with Pompey is peaceably resolved by the triumvirate; order and calm are seemingly re-established.

Antony and Octavia go to Athens, where he takes charge of Rome's eastern army. His generals lead successful campaigns in his name. But Antony is disturbed to learn that Caesar has removed from power the third and weakest triumvir, Lepidus, and has resumed the war against Pompey. Octavia goes to Rome in the hope of smoothing matters between Antony and her brother; but Antony, without waiting longer, prepares an attack upon Caesar. Cleopatra sends a strong naval force to assist her lover, but at a crucial moment in the battle her fleet turns tail; Caesar is victorious. Antony retreats to Alexandria. He is pursued by Caesar, who sends an ultimatum to Cleopatra that she yield Antony to him. Caesar's messenger is whipped by Antony and sent back, his scars a token reply.

Again the forces of Caesar and Antony meet, near Alexandria, and this time Antony is near a triumph, until Cleopatra's forces fail him. He harshly denounces her for this perfidy, and to silence his wrath she has word sent to him that she has died of grief. The still enamoured Antony is moved by this report to commit suicide and runs upon his sword. Mortally hurt, he is borne to Cleopatra's place of refuge and, after a passionate farewell, dies.

Caesar, anxious to obtain Cleopatra's surrender, promises that she will be treated with the respect owed to her sovereign rank. She does not believe him and, foreseeing the humiliation that awaits her in Rome, holds asps to her body, letting the poisonous creatures sting and kill her.

This exotic play offers a challenge that Shakespeare meets surpassingly well. Some of the most glowing dialogue ever set down in English or any language adorns it. The characterizations, too, are fascinating, and they are people new to Shakespeare. Nowhere else has he protagonists like these. Romeo and Juliet are young lovers. Antony and Cleopatra are worldly, mature, keenly intelligent. He is a conqueror, a shrewd and realistic statesman, a master of intrigue. His best qualities are courage and generosity. He is large-souled. She is equally sophisticated and wily. "She is cunning past man's thought," her lover says wryly, and he calls her "this enchanting queen"; she is "the serpent of old Nile". They play at love, like lions; yet despite their games they are deeply serious, emotionally committed. This prompts the hyperbolic phrases that spring to their lips when they reveal their true feelings about each other. They quarrel and make up, laughingly; they enjoy their fallings out. They curse and insult each other, heroically, wittily. Both are actors, histrionic, as are all major Shakespearian characters. Both are liars, and each knows that the other is – they even banter about it. Neither trusts the other. At many moments their games are such that the play skirts high comedy, as do some of the best scenes of *Hamlet*. Both are cynical and alert and devious, and Antony is ageing, weary and all too aware that this prolonged and lustful affair is weakening him, robbing him of his will, his ability to make decisions and take action, causing him to lose an empire. Yet he can never finally abandon her, and ultimately does not even try to do so, and dies partly because he thinks her already dead – the last of the many deceptions she unhesitatingly practises on him.

In young and ruthless Octavius Caesar, Shakespeare presents another Machiavellian figure: worldly, quiet, cold. Implacable, cautious, pragmatic, rational always, he is destined to be a firm ruler, but he is without a heart. This Caesar, step by step disposing of his rivals and taking over for himself the sole imperial power, is a portrait of a familiar type. Strikingly, in Cleopatra this suave careerist is to meet his match, for she, too, is Machiavellian. In the last act, stroke for stroke she cleverly outwits him. Her death is his defeat: she robs him of a spectacular feature of his promised triumph in Rome.

The play has been seen as a Morality: love (Antony and Cleopatra) against the world (Octavius Caesar). This view of the drama is reflected in John Dryden's later rewriting of it, *All for Love, or The World Well Lost*. Antony possesses power, riches, glory, yet relinquishes them for the spiritual and sensual delights of union with the woman who is right for him. He is looked upon with derision, contempt and pity by the practical Caesar and other hard-headed Romans for his folly and decadence, yet Shakespeare surrounds the love affair with an ambiance of glory such that it can hardly be said that the aim of the drama is to condemn Antony; rather, here – as elsewhere – the playwright is clearly on the side of the lovers, though he emphasizes the high cost they must pay. When Antony is given a choice, he elects to remain the lover above all:

> Let Rome in Tiber melt, and the wide arc
> Of the rang'd empire fall. Here is my space.
> Kingdoms are clay. . . .
> . . . the nobleness of life
> Is to do thus, when such a mutual pair
> And such a twain, can do't . . .

Similarly, in response to Cleopatra's repeated query of how much he loves her, he replies: "There's beggary in the love that can be reckon'd." His has been beyond all material measure. Her words are: "Eternity was in our lips and eyes . . ."

The action in the Eastern scenes seems to take place in a nimbus of dazzling, tropical air, and the perspectives are wide open. Cleopatra moves in a world that is barbaric, opulent, semi-Oriental. Enobarbus' famed description of her first appearance is a signal instance of this:

> The barge she sat in, like a burnish'd throne,
> Burnt on the water: the poop was beaten gold,
> Purple the sails, and so perfumed that
> The winds were lovesick with them; the oars were silver,
> Which to the tune of flutes kept stroke, and made
> The water which they beat to follow faster,
> As amorous of their strokes. For her own person,
> It beggar'd all description; she did lie
> In her pavilion, cloth-of-gold, of tissue,

> O'erpicturing that Venus where we see
> The fancy outwork nature. On each side her
> Stood pretty dimpled boys, like smiling Cupids,
> With divers-colour'd fans, whose wind did seem
> To glow the delicate cheeks which they did cool,
> And what they undid did.

It is interesting and instructive to compare this passage to the very similar one in North: it is the small alterations that Shakespeare effected that transmute the words into sublime poetry. For example, in both versions the barge's sails are perfumed, and the oars keep stroke to the music being played; but the playwright adds that the winds are "lovesick" and the waves "amorous" of the silver oars. These are the truly vivifying touches. In an ensuing passage Enobarbus recounts how Antony was abruptly deserted in public, while the populace rushed to observe the queen's approach. Says North, after Plutarch: "Antonius was left post alone in the market place to give audience." In the drama it is:

> ... and Antony
> Enthron'd i' th' market-place, did sit alone,
> Whistling to the air.

No less deft is Shakespeare's giving these speeches to the rough, hard-bitten Roman soldier, Enobarbus, whose grudging admiration and unaccustomed eloquence cause them to sound even more impressive. They are also inserted in a long episode from which Cleopatra is absent and thus serve to remind the audience of her allure and quality of enchantment. Finally, the stage role was to be enacted by a "squeaking boy", and it was incumbent on Shakespeare to endow the performer with this sort of verbal assistance to sustain the illusion.

Not only is sunlight conjured throughout but also moonlight, silvery and golden, in this luxuriant, glistening world of the Mediterranean and Near East. The epithets this pair of mature lovers use to each other, and that others apply to them, also make them important and reveal them as self-important. They are "pillars" of the empire. In Cleopatra's tribute to the dead Antony:

> His legs bestrid the ocean; his rear'd arm
> Created the world; his voice was propertied
> As all the tuned spheres, and that to friends;
> But when he meant to quail and shake the orb,
> He was as rattling thunder. For his bounty,
> There was no winter in't; an autumn 't was
> That grew the more by reaping. His delights

> Were dolphin-like, they show'd his back above
> The element they liv'd in. In his livery
> Walk'd crowns and crownets; realms and islands were
> As plates dropp'd from his pocket.

Caesar, too, lauds his fallen adversary:

> The breaking of so great a thing should make
> A greater crack. The round world
> Should have shook lions into civil streets,
> And citizens to their dens. The death of Antony
> Is not a single doom; in the name lay
> A moiety of the world.

And again Cleopatra cries out:

> O, see, my women,
> The crown o' the earth doth melt. My lord!
> O, withered is the garland of the war,
> The soldier's pole is fall'n! Young boys and girls
> Are level now with men; the odds is gone,
> And there is nothing left remarkable
> Beneath the visiting moon.

Indeed, Antony unblushingly speaks of himself as "The greatest prince o' the world. The noblest". Similar superlatives are lavished on Cleopatra. Enobarbus hears that Antony will leave her, and protests:

> Never, he will not:
> Age cannot wither her, nor custom stale
> Her infinite variety: other women cloy
> The appetites they feed, but she makes hungry
> Where most she satisfies. For vilest things
> Become themselves in her, that the holy priests
> Bless her when she is riggish.

Notably, when Antony speaks his last words to her, he addresses her as if she were more than a mere woman:

> I am dying, Egypt, dying . . .

To him, she is also a domain. And, histrionic to the end, he adds:

> I here importune death awhile, until
> Of many thousand kisses the poor last
> I lay upon thy lips.

In the concluding scene, Cleopatra's death, Shakespeare outdoes himself in poetic evocation, from its first notes – when Iras, her handmaiden, says, "the bright day is done, and we are for the dark", as earlier Cleopatra herself has proclaimed, "Darkling stand / The varying shore o' the world", to her "I have immortal longings in me" and "Now no more the juice of Egypt's grape shall moist this lip", and finally in even more haunting phrases:

> I am fire and air . . .
> My resolution's plac'd, and I have nothing
> Of woman in me; now from head to foot
> I am marble-constant; now the fleeting moon
> No planet is of mine.

Gazing at her rigid form, the momentarily shocked Caesar can only murmur:

> . . . she looks like sleep,
> As she would catch another Antony
> In her strong toil of grace.

Shakespeare seems to have learned from Marlowe what not to do. He offers no verbal details of Cleopatra's form, merely that it "beggars all description". Her attributes are left to the imagination, as is the image of Helen of Troy beheld by Marlowe's over-awed Faustus.

The play is rather like a chronicle: it is loose, episodic (it has forty-two scenes, and hence it is not easy to stage), thronged with well-drawn minor characters – Pompeius, Enobarbus, Eros, Philo, Agrippa, Dolabella. Though he stays close to historical fact, Shakespeare's minor characters are at times very much his own. The play is also virtually a sequel to *Julius Caesar*.

A somewhat curious but well-documented interpretation of the play has been offered by John Wain, who finds it filled throughout with symbolism in which the masculine principle is represented by earth (Rome, Antony) and the feminine principle by water (Egypt, Cleopatra), the former gradually dissolving into the latter. Wain cites a legion of metaphors and images that support his reading and further suggests that Antony, a superb and much-feared general, is defeated by

Caesar because the infatuated Antony chooses to fight at sea, with the unstable assistance of Egypt's fleet, rather than on land, where his Roman forces had always been superior.

Perhaps the play's chief critic has been John Dryden (1631–1700), the neo-classical poet and playwright – he was unquestionably accomplished in both genres – who in his day was a sort of literary dictator, influential, presiding over a flock of lesser aspirants much as had Ben Jonson before him and Samuel Johnson in the years following. He was a dominating figure, though not beyond his own circle. Confirming his importance, he was named Poet Laureate in 1668 and Historiographer Royal in 1670. Involved in many unsparing disputes, he later lost those court offices. He wrote many contentious essays on aesthetic subjects. He translated and adapted *Oedipus*, Ovid and Virgil, and reams of works from foreign languages. His farces, such as *Marriage à la Mode*, are still laugh-getters. As quoted before, Dryden thought Shakespeare was "wonderful" but also to be faulted for a lack of proper – that is, Aristotelian – form and frequent lapses of decorum and artistic good taste. In his view the Bard's writing was "many times flat, insipid, his comic wit degenerating into clenches, his serious swelling into bombast". Also: "The fury of his fancy often transported him beyond the bounds of judgment, either in coining of new words and phrases, or racking words which were in use, into the violence of a catachresis." (A "catachresis" is a far-fetched or improper use of a word.) Dryden revised *The Tempest* (1667) as he thought it should be presented. He disapproved of *Troilus and Cressida* because Cressida was not adequately punished for her infidelity. He liked the opening scenes but found that the treatment soon fell off; and, to him, "the latter part of the tragedy is nothing but a confusion of drums, excursions and alarms". He turned out a variation of that play with justice quite duly meted out (1679). He was moved by *Antony and Cleopatra* to come forth with his own version of that story, *All for Love* (1678), a sign of his self-confidence and plain courage. It is in heroic blank verse; the time frame is limited to seventy-two hours, more or less as recommended by the Greek philosopher and long ago self-appointed, analytical drama-critic. All the action occurs in one place, and allusions to politics, the ponderable links between the lovers and the fate of empires, are mostly eliminated. It is a smaller drama, but by no means without substantial merit. It is respectfully studied in college literary courses and occasionally performed, though mostly by national theatre groups as a curiosity or as a tribute to theatre tradition. *All for Love* illustrates only too well the difference between a first-rate dramatic poet guided by hallowed rules and a truly great one who follows only his intuition.

Antony and Cleopatra is not given often; it calls for an absolutely beguiling actress as the Egyptian queen, a large cast, and many changes of scene, making it expensive to mount and tour. Best forgotten is the attempt of the American star Tallulah Bankhead to fulfil the role. More imperious than seductive, she was described by a cruel reviewer in the *New York Times* as floating down the Nile on a barge that sank. The production promptly went under with it. Katharine Cornell fared better but not well enough in a visually "gorgeous" New York offering in 1947. The reviewers deemed her majestic but too placid and well-behaved.

In 1951 Laurence Olivier and his wife Vivien Leigh were preparing to visit New York with the Shakespeare work. A colleague suggested that, since they had the needed opulent costumes and scenery, they could add Shaw's witty, wise comedy *Caesar and Cleopatra* to alternate at very little added cost. The double package was a strong attraction in both London's West End and on Broadway; in the latter instance it was referred to as "two on the Nile". (Shaw's Cleopatra is a "kittenish sixteen-year-old".)

An event of social and cultural magnitude in New York was the opening of a handsome new theatre to house the Metropolitan Opera Company at Lincoln Center in 1966. A top feature of this gala occasion was the première of an opera for which the American composer Samuel Barber had received a commission. He had already provided the company with a work that had entered its repertory, *Vanessa* (1958), and had won a Pulitzer Prize. For this gala evening he elected to base his offering on *Antony and Cleopatra*. The resplendent sets and costumes were by Franco Zefferelli, the Italian director and designer noted for his elaborate – or, according to some critics, over-elaborate – productions. A black soprano, the cherished Leontyne Price, was Cleopatra; her Antony was Justin Diaz. The choreography was by Alvin Ailey, also black, another signal of the relaxing of racial barriers at the Metropolitan. Alas, the much-anticipated occasion was largely a fiasco when Zefferelli's overweight, monumental installations, erected on a revolving stage, caused the overburdened machinery to break down, to remain fixed, resisting any change of scene, and at one point imprisoning Ms Price, reducing her to near hysteria. Subsequent presentations of the opera inspired little critical enthusiasm for the score. Disappointed and humiliated, Barber fled to Europe for several years. The music was still heard on recordings, and in 2001 the work was finally revived by Chicago's Lyric Opera Company but again failed to persuade listeners of its durability.

— 18 —
SHAKESPEARE: THE LAST PHASE

Three more plays with classical settings – Rome and Athens – followed the imaginative excursion to Egypt by Stratford's busy dramatist. For *Coriolanus* (1608) he continued to rifle the English translation of Plutarch's *Lives*. The verbal correspondences to North's declamatory prose, by Shakespeare often transposed into stirring poetry, are even closer and more numerous than with *Julius Caesar* and *Antony and Cleopatra*. As always, the people are more fully developed and sometimes given new traits. This is particularly true in *Coriolanus*, where, during the course of the story, the title character even changes his name.

The story is also rearranged and simplified, as a playwright must often do, and extra incidents are invented to heighten the stage effect. Other than from Plutarch, details also come from William Camden's *Remaines of a Greater Worke, Concerning Britain* (1605), which had recently been published. These pertain to the role of Menenius, which is enlarged so that he is partly a comic chorus, drunken and garrulous.

The Roman patrician, Caius Marcius, has only disdain for the rabble afoot in the capital. The price of corn is rising, and hungry citizens single out Caius as the leader and spokesman of the aristocrats with whom they contend. His buffoonish friend, Menenius Agrippa, meets a mob and becomes embroiled with some demonstrators, but Caius appears and rebukes the noisy crowd.

The Volscians, a barbaric horde beyond the borders, are planning an attack on Rome, and the aid of the valiant Caius is sought for the city's defence. At Corioli he wins triumphantly, personally scoring a defeat over Tullus Aufidius, his long-time rival-in-arms. In honour of his capture of Corioli, Caius is renamed Coriolanus by the Romans. Meanwhile the Tribunes, weakening before the clamour of the populace, have lowered the price of corn and even distributed some free to the poor. Coriolanus' aristocratic friends urge him to use his popularity to become Consul, but in campaigning for votes he cannot bring himself to plead with or truckle to the mob in the marketplace. Growing impatient, he haughtily insults and dismisses the protests and shouts of those gathered to hear him. His mother, the sagacious Volumnia, tries to persuade him to be more discreet, to dissemble if need be, but he cannot restrain his contempt and anger. The Tribunes, fearful of his growing grasp on power coupled with his unpopularity, contrive to turn the fickle mob against him. His election is annulled and he is exiled.

Brooding, he goes to Antium and offers to join his foe, Aufidius, who is preparing a new thrust at Rome. In the capital the citizens are alarmed by word of this strong hostile alliance. The

Volscians besiege Rome. Coriolanus' patrician friends send emissaries, suing for peace. He is adamant. Only when his mother, Volumnia, together with his wife and his son, beg him to soften does he consent to a truce. He does so not from pity but because he still fears scolding by his domineering mother. He is, in some ways, oddly immature.

The admiration in which the arrogant Coriolanus is held by his soldiers arouses the jealous concern of Aufidius. In Antium he forms a conspiracy against Coriolanus, who is denounced as a traitor for the leniency he has shown his native Rome. He is struck down and stabbed to death. Aufidius immediately pretends to regret the deed he has provoked and orders Coriolanus buried with all honours.

The play is a highly interesting study of a proud man, of aristocratic temperament and ancestry, whose many talents equip him for a statesman's role but who by nature is unable to lower himself in the necessary pursuit of public support. His quick temper also ill suits him for high office, as does his inborn, inflexible conservatism, his refusal at all times to compromise. These faults offset his considerable virtues, including his military prowess. He is, in many aspects, such a cold, disagreeable hero that a play centring on his fall is hardly as moving as it might otherwise be. Confronting his mother's anger, he loses all dignity and manhood. Indeed, some critics believe that Shakespeare was deliberately abandoning the orthodox principles of tragedy: he intended his portrait to be satirical and shared the commoners' dislike of the vitriolic leader whose strength is so inconstant. (Reading the play in this light, George Bernard Shaw dubbed it a comedy.) Others argue that the play reveals Shakespeare's own conservative bias and fear and disapproval of mob rule: he always presents riotous action unfavourably, no matter how justified it might seem; he abhors violent political dissent of any kind. But guessing as to Shakespeare's political leanings is to no avail; he is careful never to reveal them. Here his satire encompasses both sides, the patricians and the plebeians. Both are wrong.

Coriolanus, however, offers a subject that is relevant and topical today, as it was then, an allegorical treatment of a theme that much concerned Shakespeare's contemporaries: political power, how it should be wielded, and by whom. If rule is firm, it must also be compassionate. Coriolanus lacks the common touch. But the mob is also to be despised and put down: "The beast with many heads", Coriolanus terms it. Too much "democracy" and demagoguery can lead to anarchy.

Psychologically, the relationship of mother and son in this play is very modern, almost Freudian. She is barely mentioned in Plutarch; she is largely of Shakespeare's devising. It seems as though Coriolanus is arrogant towards others to compensate for the humiliations inflicted on him by the strong-willed Volumnia. His grown-up petulance is that of a hurt child. She, like him, is a stiff-necked conservative; but she counsels him to compromise, to be evasive, to hide his true feelings. Unfortunately these are lessons he cannot accept. One feels that she would have made a better ruler than he, though she is also ruthless. Overall she is the most arresting figure in the drama.

The poetry is appropriately austere, at times almost stark. It is an excellent, simple rhetoric. A good example is Volumnia's comment to her son:

> You might have been enough the man you are,
> With striving less to be so.

In the late seventeenth century this script was to suffer the same fate as *King Lear*, an adaptation by the resourceful Nahum Tate. The venture was not a success. Another attempt a few decades afterwards by a John Dennis proved no luckier. The poet James Thomson next tried his hand at it (1749); his version was not without virtue, and some elements of it continued to be blended with Shakespeare's original for some while in the following century. It should be noted that the play was popular in Germany during the early twentieth century, the pre-Hitler years.

Another distempered and misanthropic hero appears in *Timon of Athens* (also 1608). If not a historical person, this Timon had at least legendary identity in Ancient Greece and Rome, with references to him appearing in works of Aristophanes, Phrynichus and Plato, as well as subsequently in one of the farces of Plautus. He is mentioned by Plutarch again, in his *Life of Antonius*, and portrayed by the Roman satirist Lucian in an amusing dialogue. Once more, Painter's *Palace of Pleasure* is a possible collateral source, and an earlier Elizabethan comedy (*c.* 1600) is also a likely place from which Shakespeare drew ideas and specific incidents. Two Italian plays, with which he may or may not have been familiar, treated the same subject after the fashion of Lucian; and Timon, as a type of anti-social hermit, had been alluded to by Lyly and Greene, and Dekker and Nash. The outlines of the story were well known to Elizabethan littérateurs.

Wealthy, Timon of Athens is notorious for his extravagance and generosity. He is surrounded by flatterers and parasites. His faithful steward, Flavius, warns him of his reckless waste and mounting debts, but the lavishly living and giving Timon is heedless, until, quite suddenly, he finds himself bankrupt. He is confident that he can collect the money he has so freely lent, and that he will be helped by his many "friends"; but once his impoverishment is known, all these sycophants shun him. Shocked, enraged, Timon bids them to a banquet at which there is no food, berates them, then renounces the company of mankind and retires to a cave in the forest, where he lives on roots and avoids human companionship, a motif of pastoral retreat from the materialistic world often struck in Shakespearian plays, as in *Two Gentlemen of Verona*, *A Midsummer Night's Dream*, *Love's Labour's Lost* and *As You Like It*, though here given a much darker hue. At about the same time his young friend, the soldier and politician Alcibiades, is driven into exile.

A recluse, Timon discovers a vast hoard of gold while digging for roots. It no longer means anything to him. When people come to see him he gives them a part of the gold but hustles them off with harangues and curses. Alcibiades encounters him, and when Timon hears that the banished Athenian wishes to attack their native city he gladly gives him gold to raise and equip an army.

That he has gold still draws people to him, whores, even artists and statesmen – they endure his vituperation in order to share his riches. In a long scene he debates with Apemantus, the cynical philosopher. Thieves approach him; he welcomes them all in the same fashion.

The Athenian rulers, frightened by the threat of Alcibiades' siege, invite him back to the city. He accepts, on condition that his enemies be punished, and Timon's as well. His terms are met. Timon tells the Athenian messengers to hang themselves on his fig tree. He suddenly dies, having first composed his own epitaph, which expresses his hatred and contempt for mankind. His last wish is that he be entombed near the sea, where the tides will twice daily wash over and cleanse his resting-place.

Serious doubts as to the authorship of this script have frequently been expressed. One conjecture is that Shakespeare partially and hastily revised an earlier play by someone less able. A second hypothesis is that Shakespeare himself started the drama, left it unfinished, and someone else of inferior talent – Thomas Middleton, Cyril Tourneur, Thomas Heywood, George Wilkins or John Day – completed it. A third guess is that the extant script, which only came to light in a late Folio (1623), is a version corrupted and altered by actors upon whose collected memories the editors depended. The writing is most uneven, though there are some striking passages attributed to Shakespeare naturally. But which scenes and sections might really be his, which not, is much disputed. It might be said that he did not always compose perfect plays, but he was incapable of bringing forth a thoroughly bad one.

If this harsh "fable" is by Shakespeare, as seemingly the editors took for granted, it is significant that he treats it seriously, not comically, unlike all his predecessors. Is it material for a comedy? Timon behaves foolishly. He is absurdly extravagant. He does not listen to the repeated warnings and good advice of his steward. His companions are grossly ill-chosen. At the last banquet he throws hot water in the faces of his guests. When, by rare luck, he is enriched again, he does not return to Athens to spend his money intelligently; instead he uses it to fund Alcibiades' rebellion. He insults the Athenian emissaries. What accounts for the glum, jaundiced view of the world embodied in this play?

An author's election of a subject is not always voluntary; he is impelled or drawn to it by unconscious forces in himself. Some biographers hazard, with little else to go on, that Shakespeare's spirit was seriously infected by disappointment in love, as revealed in the final sonnets. Or he might have suffered a nervous crisis brought on by overwork and exhaustion. By now he had written over thirty lengthy plays.

The script is too often static, repetitious; in shape it somewhat resembles the *Prometheus* of Aeschylus. Timon is, indeed, almost an allegorical figure in a slightly expanded Morality. It is possible to read the work, as some do, not as a tirade against mankind in general but as a criticism instead of Timon as an archetype of a certain sort of person. As already listed, his faults are many. His excessive generosity has been a form of self-flattery – he admires his own kindness, but it is an attempt to buy friends and to hear praise and to earn binding gratitude. His anger is also excessive, by turns petulant, mean, at moments almost psychotic. He is such a poor judge of character, so unable to tell good men from flagrantly bad ones, that he has mostly himself to blame for his situation. If this interpretation is adopted, the play is not a tragedy but merely a satire.

Still, it is easy to believe that Shakespeare had a part in writing *Timon*. There is a dark, savage power in some of the words the anti-hero hurls at those who have offended him:

> Who dares, who dares,
> In purity of manhood stand upright,
> And say, "This man's a flatterer?" If one be,
> So are they all; for every grize of fortune
> Is smooth'd by that below; the learned pate
> Ducks to the golden fool: all is oblique;
> There's nothing level in our cursed natures
> But direct villainy.

His final advice to the exceptionally loyal Flavius is equally bitter:

> Hate all, curse all, show charity to none,
> But let the famish'd flesh slide from the bone,
> Ere thou relieve the beggar.

His cosmic vision is Swiftian, offered mockingly to the thieves who come to rob him:

> The sun's a thief, and with his great attraction
> Robs the vast sea; the moon's an arrant thief,
> And her pale fire she snatches from the sun;
> The sea's a thief, whose liquid surge resolves
> The moon into salt tears; the earth's a thief,
> That feeds and breeds by a composture stolen
> From general excrement. Each thing's a thief.
> The laws, your curb and whip, in their rough power
> Have uncheck'd theft.

In his withdrawal, his imprecations on materialism, his nihilism, his attempts to commune with nature, Timon almost represents one type of late twentieth-century figure. He does, like a latter-day mystic, a Thoreau, seem to learn how to fare satisfactorily without possessions:

> My long sickness
> Of health and living now begins to mend,
> And nothing brings me all things.

The visual contrast between the opening scenes, in Timon's opulent palace and its bright feasting, and the last two acts, the bleak woods and cave, with the naked self-deprived misanthrope digging for roots in his desolate hermitage near the sea, serves to make this a very stageworthy work; indeed, it acts much better than it reads. Despite its many imperfections, its schematic roughness, some estimable critics rate *Timon* among Shakespeare's great tragedies, but most would say that is going too far.

There is no mention anywhere of a performance of *Timon* in Shakespeare's lifetime, though most probably it was staged then; after all he was a shareholder, doubtless with a voice in what was produced. Thomas Shadwell, in the late seventeenth century, undertook the first of several rather free adaptations of the work by later revisers that did reach the boards. Though never a favoured play, it has been frequently enough revived since then. The roles of Timon and the churlish Apemantus are especially prized.

Updated, the work made its way back to the stage and was viewed by Joel Henning for the *Wall Street Journal* in 1997:

> Spend an evening with Lord Timon of Athens and you suspect that – if he could – Bill Clinton would have him stay overnight in the Lincoln Bedroom. Timon curries favor with important Athenians by lavishing gifts upon them. "Methinks I could deal kingdoms to my friends, and ne'er be weary." Far from being a big-time Shakespearian tragic figure, Lord Timon is to King Lear as Bill Clinton is to Abraham Lincoln.
>
> Interesting but relatively mediocre Shakespeare plays like *Timon of Athens* demand more of their producers than *Othello*, *Lear* and *Macbeth*, which can blow away audiences with no more than a bare stage and one or two transcendent performances. As the final offering of the estimable Chicago Shakespeare Repertory's tenth anniversary season, *Timon of Athens*, which closed last week, got what it needed, especially from its director, Michael Bogdanov.
>
> Timon was an extravagant party-giver, and Mr Bogdanov began his production by inviting audience members to join the reception in Timon's garden and mingle with the sycophants and more notable guests. By the time the audience left the stage and Timon made his grand first entry in a spiffy white suit, surrounded by toadies and bodyguards, the play resonated with twentieth-century significance. When he contributed five talents to redeem Ventidius from the slammer, we could easily imagine him doing the same for a modern cheat.

Timon is a hard character to play, being immoderately generous and in complete denial for one half of the play, and equally immoderate in his misanthropy for the other half. But Larry Yando did as well as he could, keeping the lid on what too easily could have become an intemperate performance. Mr Bogdanov got his best performances, however, from David Darlow as the world-weary Apemantus, who sees it all coming, and Patrick Clear as Flavius, Timon's steward, who cannot get his boss to understand that he's up to his eyeballs in debt and giving away borrowed wealth.

To bring off this unsubtle piece, Mr Bogdanov was greatly aided by designer Ralph Koltai's ingenious use of the cramped playing area. Just as the audience was made party to Timon's hospitality during the play's first half, later we were trapped with him in the filthy squalor of the wasteland in which he rants and then dies alone.

If, as Aristotle said, great tragedy has a soul, *Timon of Athens* doesn't qualify. But perhaps its very limitations make *Timon* resonate for us in this less than heroic age.

Pericles, Prince of Tyre (c. 1606–8) is a turning back from tragedy to romantic drama, but the play is generally deemed not to be his most effective in this genre. Once again Shakespeare's authorship is questioned: is the work wholly his? The names of Thomas Heywood and George Wilkins are put forward as possible co-authors and also that of William Rowley. The consensus credits Shakespeare with major parts of the last three acts, assigns the first two largely to someone else. This conclusion is based on sharply marked differences in style: the dialogue in the earlier acts is clumsy and pedestrian. This script did not appear in the first two Folios. The same theories and arguments put forth concerning *Timon* are applied here: *Pericles* might be a revision of an earlier work by another writer, or a later dramatist bravely took up a script left unfinished by Shakespeare. One reason for adducing Wilkins's collaboration is that in 1608 he published a novel, *The Painful Adventures of Pericles, Prince of Tyre*, with a plot closely paralleling that of the stage piece, but it is not certain whether the novel is based on the play or the play on the novel.

The plot was at very least a thousand years old before Shakespeare and Wilkins utilized it. Many Italian versions of it had been published in the early Renaissance, and the story had been translated into a dozen tongues. The poets John Gower (contemporary of Chaucer) and Laurence Twine (the latter in a novel printed in 1576 and reprinted in 1607) had already reproduced versions of the tale. These two, who acknowledged borrowing much from twelfth-century Godfrey of Viterbo, are doubtless the play's sources, with Gower's influence predominating. Indeed, "Gower" is the name given to the narrator, speaking the Prologue and Epilogue, and at intervals describing the off-stage action – a device Shakespeare had resorted to earlier, in *Henry V*, and would soon employ again.

The venturesome Pericles, Prince of Tyre, goes from one daring exploit to another. In one incident he takes the challenge of the King of Antioch, who has promised the hand of his beautiful daughter to whomever can answer a riddle he propounds. Failure to resolve it means death. Pericles is successful, however: he reveals that the king has committed incest with his daughter. The infamous king plots to poison him to prevent exposure of the secret; but Pericles, after rejecting the hand of the guilty princess, escapes and returns to Tyre. Rightly fearing that envoys from Antioch would pursue and kill him, the prince travels further to elude assassination.

In Tharsus he helps avert a famine by arriving in a vessel laden with corn and gains the lasting gratitude of Cleon, the governor, and Dionyza, his wife. Continuing his journey, because he is warned that he is being sought, Pericles is shipwrecked in a storm and washed ashore not far from Pentapolis. Fishermen, drawing in their nets, find his lost armour in them. The fisher folk befriend him. Putting on the armour, the redoubtable Prince of Tyre participates in a tourney held in honour of the birthday of the Princess Thaisa. He triumphs over adversaries from many countries, and weds lovely Thaisa, daughter of the good King Simonides, who like the princess approves of the courageous Pericles. With his bride he sets sail for Tyre. On the voyage another fearful storm arises. Princess Thaisa seemingly dies in childbirth and is hurriedly buried in the ocean to quiet the superstitious sailors. The prince alters his course and proceeds to Tharsus, where he leaves his infant daughter in the care of Cleon and Dionyza, his sworn friends. Meanwhile Thaisa, not dead but in a deep swoon, is borne by the waves in a floating coffin to Ephesus. A learned doctor restores her numbed senses with fire and music. She is convinced that Pericles has perished in the gale and assumes a veil of chastity in the Temple of Diana.

The incestuous King of Antioch and his daughter have been fatally stricken by a bolt of lightning, a dire punishment from Heaven for their dreadful sin. Word of this reaches Tyre, and the lords of that city spread out in search of their absent prince, to bring him home to rule over them, since he is no longer threatened by murderous hirelings from Antioch. The news is brought to Pericles in Pentapolis.

The years having passed, and Pericles' child – a girl of fourteen, Marina (appropriately named, for she was born at sea) – has grown to young womanhood. Her beauty stirs the jealousy of her guardian Dionyza, whose own daughter is too far outshone. Dionyza plots to have Marina killed. Instead, at the very moment she is to be slain, the girl is kidnapped by pirates and taken to Mitylene where she becomes the property of procurers.

This episode is followed by a sordid scene in a brothel that is so realistic that many prim admirers of Shakespeare refuse to believe he penned it. The procurers make ready to sell the helpless Marina; because she is worth more as a virgin, she has not yet been violated.

Pericles arrives at Tharsus to reclaim his daughter and is told that she is dead. He is shown a monument put up to her memory. Mournfully, he departs.

In the brothel at Mitylene, Marina is about to be put up for auction. Lysimachus, governor of the country, comes to view her. Left alone with him, Marina appeals to be rescued from her captors. Her purity and beauty persuade him of her virtue. He gives her gold with which she buys her freedom.

Marina becomes a teacher of singing, weaving, sewing and dancing. When the winds blow Pericles' ship to Mitylene he is welcomed by Lysimachus, who summons the talented Marina to entertain his profoundly grief-stricken guest. When Marina appears, Pericles recognizes her as his "dead" daughter, because she is so much like his "dead" wife, in a scene written with rare economy and power. Overjoyed, Pericles falls into a faint and in a vision beholds the goddess Diana who tells him to go to Ephesus. Here he finds his queen, the lost Thaisa, and the royal family of Tyre is reunited. Cleon and Dionyza are destroyed and their palace burned by the outraged citizens of Tharsus, and Marina weds Lysimachus.

This long chain of improbable events and coincidences is surprisingly interesting and suspenseful, and handled in a fashion so as to be always lucid. This is high fantasy and fairytale romance; taken on its own terms the work has colour and enchantment, and the verse is often attractive. Pronounces the saintly doctor-magician Cerimon:

> Gentlemen:
> This queen will live.... See how she 'gins to blow
> Into life's flower again! ...
> She is alive, behold....
> Live,
> And make us weak to hear your fate, fair creature,
> Rare as you seem to be!

Marina, recalling her arrival during a storm, says very simply:

> When I was born, the wind was north.

Seldom performed today, *Pericles* is perhaps considerably underestimated. Such works, possessed of a dream-like quality, were now becoming fashionable in the new Jacobean theatre, and Shakespeare, ever responsive to his public's taste, was displaying his skill at the stage's changing fare. Let it be told, offerings of this sort were comparatively easy to spin: episodic, melodramatic, abounding in fabulous deeds, requiring little logic or consistency. Why would the iniquitous King of Antioch invite strangers to guess his secret? Is it not odd that Pericles neglects his beloved daughter for fourteen years? It is not hard, however, to consider many of the incidents as symbolic, to see *Pericles* as richly incorporating mythic and ritualistic motifs.

Cymbeline (1609–10), another romantic fantasy, might be viewed in the same way. It is a work of diverse origins: Holinshed's *Chronicles* and Boccaccio's *Decameron*. An earlier play, author unknown, *The Rare Triumphs of Love and Fortune* (1589), may have contributed some suggestions.

The stylistic influence of Shakespeare's rivals, Beaumont and Fletcher, is no little apparent. Their *Philaster* might have been a model for him. The canny playwright from Stratford, if not in the forefront of the Jacobean theatre's embrace of romantic fantasy, kept apace with it. He knew what would sell tickets. Possibly *Cymbeline* slightly preceded *Philaster*, or the two works were written simultaneously. Some elements of the plot were apparently Shakespeare's own, and he may also have had an assistant while putting together the script.

The time is the first century. Cymbeline, King of Britain – a descendant of King Lear – has a motherless daughter, Imogen. He weds a widow, cruel and intriguing, a trafficker in drugs, who has a scapegrace son, Cloten, to whom Imogen is betrothed. But she elopes with Posthumus Leonatus, son of Sicilius, honoured for the courage he showed in combating Roman invaders. (Born after his father's death, the son was given an unusual first name.) Cymbeline, angered, banishes Posthumus before the marriage is consummated. The bridegroom goes to Rome; but before separating from Imogen he slips a bracelet on her arm, while she gives him a diamond ring, their tokens of mutual constancy while they are kept apart. Abroad, Posthumus proudly affirms his deep trust in Imogen's fidelity. The cynical Iachimo bets 10,000 ducats against Posthumus' precious ring that Imogen can be led astray. Iachimo sets out for Britain to win his wager.

He fails utterly. Frustrated, he conceals himself in Imogen's bedchamber and while she is sleeping carefully observes all the details of her room and even a mole on her breast; he steals the bracelet. He goes back to Rome, and his fictitious but documented account convinces Posthumus of Imogen's promiscuity.

Furious, Posthumus plots his bride's death. He writes her a letter proposing to meet her in Wales, and sends instructions to Pisanio, his servant, in Britain, to accompany Imogen and slay her. But Pisanio, persuaded of Imogen's fidelity, shows her the message. Imogen, as do so many of Shakespeare's heroines, promptly disguises herself in male attire and takes refuge with other hapless characters in a mountain cave. Here she encounters her brothers, Guiderius and Aviragus, two grown sons of Cymbeline, stolen in childhood. But they do not know their true identity, believing themselves the sons of the outcast Belarius, a nobleman banished years before by Cymbeline.

Cloten, the worthless stepson, has pursued Imogen. He is killed by Guiderius whom he has upbraided and threatened. Imogen thinks the headless body is that of Posthumus; she finds it beside her when she arouses from a death-like trance induced by drugs provided by Cloten's mother, the scheming queen. She next falls into the hands of the Romans, who have landed in Britain.

Cymbeline, refusing to pay them tribute, has been defying the invaders, who are led by Caius Lucius. Belarius, Guiderius, Aviragis and Posthumus bravely and loyally join forces with Cymbeline, good Britons all, and hurl back the foreign warriors. Cymbeline, briefly held, is rescued by Belarius and the two inherently noble youths. Posthumus, a prisoner and condemned to death, is freed. Iachimo, who is serving with the Roman legions, is among those captured. He confesses the immoral trick he played to win his wager. He is forgiven. Imogen and Posthumus are joyously reunited, and Cymbeline's two long-lost sons restored to their rightful father and rank. The wicked queen, gone mad, dies.

Once more a preposterous tale is a vehicle for fine characterizations and high poetry, as well as for rich theatrical spectacle; for, by now, the London stage had grown more sophisticated and mounted its scenes lavishly.

In form, *Cymbeline* would seem to be experimental: Shakespeare no longer bothered to present his themes realistically; he was giving loose rein to his poetic instinct and setting free his extravagant imagination, happily creating theatrical illusion for its own sake. Familiar ingredients – disguises, mistaken identity, unexpected returns from "death" – are mixed with a bold hand, a show of bravura. The tendency to do this is marked in mature artists who feel they have exhausted other veins of expression. Having accomplished their work in conventional styles, they now permit themselves a degree of self-indulgence and seek the new. They accept risks; they even go to the brink of self-parody. This applies to all the plays in the last phase of Shakespeare's career and will find its finest embodiment in *The Tempest*.

Certain themes are strongly restated in *Cymbeline*. One is that noble birth will manifest itself regardless of outer circumstances. The lost sons of Cymbeline are brave and good and have most princely impulses, even though they have been reared as outlaws. Belarius comments on this:

> 'Tis wonder
> That an invisible instinct should from them
> To royalty unlearn'd, honour untaught,
> Civility not seen from others, valour
> That wildly grows in them, but yields a crop
> As if it had been sow'd.

It is interesting, too, that the Romans, though the enemy, are presented as courageous and dignified in mortal adversity. This is very much how Ancient Romans were viewed in early Jacobean times: as stoic, gracious, wise. At the end, Cymbeline willingly submits to Augustus Caesar and of his own volition promises to pay the tribute after all. But patriotic sentiment is also appealed to in the play: the Britons fight ferociously to defend their land.

Cymbeline contains two noted songs. One, very widely known, begins:

> Hark! hark! the lark at Heaven's gate sings,
>   And Phoebus 'gins arise,
> His steeds to water at those springs

> On chalic'd flowers that lies;
> And winking Mary-buds begin
>   To ope their golden eyes:
> With every thing that pretty is,
>   My lady sweet, arise;
>   Arise, arise!

The other, a lament for the supposedly dead Imogen, is even more famed and starts:

> Fear no more the heat o' the sun,
>   Nor the furious winter's rages;
> Thou thy worldly task hast done,
>   Home art gone, and ta'en thy wages;
> Golden lads and girls all must,
> As chimney-sweepers, come to dust.

Some describe Shakespeare as growing tired in these last plays, even to have suffered a nervous breakdown, but this does not show in the poetry. His writing, though it is of a different kind, metrically far more complex, is still perfectly controlled. The syntax is sometimes baffling, arbitrary. It is ornate verse, as artificial as the story, but graced with a rococo elegance. At other moments it is very terse. What redeems it from being too artificed is the fervent attention paid to Nature, to the rocky landscape of Wales, the mountain flowers, named – as always with Shakespeare – with an exactness and loving observation testifying to his delight in the world about him. Much has been made of his obvious fondness for his garden and orchard, to which he retired long before he might otherwise have quit his profitable labours.

Making ready to bury the unconscious Imogen, known to him as the boy Fidele, Arvigarus says:

>   With fairest flowers,
> While summer lasts and I live here, Fidele,
> I'll sweeten thy sad grave; thou shalt not lack
> The flower that's like thy face, pale primrose, not
> The azur'd hare-bell, like thy veins, no, nor
> The leaf of eglantine, whom not to slander,
> Out-sweeten'd not thy breath: . . .
> Yea, and furr'd moss besides, when flowers are none,
> To winter-ground thy corse.

Iachimo, claiming to have slept with her, tells Posthumus;

> On her left breast
> A mole cinque-spotted, like the crimson drops
> I' the bottom of a cowslip . . .

He repeats this sensual observation:

> If you seek
> For further satisfying, under her breast,
> Worthy the pressing, like a mole, right proud
> Of that most delicate lodging: by my life,
> I kiss'd it, and it gave me present hunger
> To feed again, though full.

Though he is the villain, Iachimo is given to flights of very delicate eroticism:

> Cytherea!
> How bravely thou becom'st thy bed, fresh lily,
> And whiter than the sheets! That I might touch!
> But kiss: one kiss! . . .
> 'Tis her breathing that
> Perfumes the chamber thus; the flame of the taper
> Bows toward her . . .

Indeed, sexual feeling is given exquisite and refined expression throughout, as when a woman's flesh is described being "as white as unsunn'd snow".

Britain is paid this paean from the queen:

> Remember, sir, my liege,
> The kings your ancestors, together with
> The natural beauty of your isle, which stands
> As Neptune's park, ribbed and paled in
> With oaks unscaleable and roaring waters,
> With sands, that will not bear your enemies' boats,
> But suck them up to the topmast.

Belarius voices his approval of Imogen:

> All of her that is out of door most rich!
> If she be furnish'd with a mind so rare,
> She is alone the Arabian bird . . .

He also contrasts existence on the mountainside to that in the city and at the corrupt court:

> O! this life
> Is nobler than attending for a check,
> Richer than doing nothing for a bribe,
> Prouder than rustling in unpaid-for silk . . .

At the low entrance to the cave:

> Stoop, boys; this gate
> Instructs you how to adore the Heavens, and bows you
> To a morning's holy office; the gates of monarchs
> Are arch'd so high that giants may jet through
> And keep their impious turbans on, without
> Good morrow to the sun.

When Imogen plights her eternal love to Posthumus, he replies:

> Hang there like fruit, my soul,
> Till the tree die.

Many other lines are superbly epigrammatic: "If't be summer news, / Smile to't before . . ."; "Did you but know the city's usuries / And felt them knowingly; the art o' the court, / As hard to leave as keep, whose top to climb / Is certain failing, or so slippery that / The fear's as bad as falling . . ."; "Swift, swift, you dragons of the night, that dawning / May fare the raven's eye!"

Cymbeline has often been described as a play of clouds that part for shafts of sunlight, filled with lovely speeches and ending as it does with widespread if unconvincing reconciliations. Dr Johnson thought it crowded with "imbecilities", its plot is so overladen and far-fetched; but perhaps his literal and logical mind brought to it mistaken criteria.

Hazlitt has discerned a basic unity that combines the many characters and story lines of *Cymbeline*: with all of them Shakespeare is ringing changes on the theme of fidelity. It is also, again, a work that contains many mythical elements: death and renewal, sex changes, separation and rejoining, the legend of Eden – the idyllic natural world – and the classical Golden Age, changelings, the intervention of Divine Providence. The "vision" scene, in which Posthumus is visited by the god

Jupiter descending to converse with him, comprises an effective masque within the play and provides especially an opportunity for dazzling stage effects, but not all critics are persuaded that it represents Shakespeare's handiwork – it is often assumed to be an interpolation by someone later (the cadence is different, much stiffer).

It is easy to see why this melodious play was said to be Tennyson's favourite; he kept a copy alongside him as he lay dying, and it was buried with him in his coffin.

George Bernard Shaw disliked *Cymbeline* but singled out Imogen as a forerunner of the modern woman. She is resolute, commanding and resourceful. He even wrote a variation of the closing scene in which, like an Ibsen heroine, Imogen leaves her distrustful husband.

After directing a production of the play for Shakespeare in the Park in 1998, Andrei Serban engaged in a colloquy about it in the *New York Times* with Adrian Noble, who witnessed his interpretation of it for the Royal Shakespeare Company. With the visit of the British troupe, New York had two major *Cymbelines* in one season, though not at quite the same time. Among their comments:

> NOBLE: It seemed to me that after the very dark plays of *Othello, Hamlet, Coriolanus* – 1605, 1606, 1607 – Shakespeare changed. As an artist, he renewed himself.
> The way he did it was to turn to myth and romance, what we would now call fairy-tales. If you look at that last grouping of plays – *Cymbeline, Pericles, The Winter's Tale* and *The Tempest* – in all of them he sought an exoticism: a desert island, all around the Mediterranean, Bohemia, a sea journey. . . .
>
> In this play, the author is saying that there is a real potential for human beings to live their lives in a more intelligent, more harmonious way, if only they understood themselves, which is the familiar "Know thyself." A good fairy-tale doesn't preach but teaches by example. It's meant to arouse in us an appetite for individual self-development. . . .
>
> The whole idea that by fainting, falling asleep or dying, a part of oneself can be reawakened to another state – all these elements are part of an esoteric fairy-tale, with its deep symbolic allegories. In *Cymbeline*, Shakespeare is much more the Medieval writer than the new rational Renaissance man.
>
> SERBAN: I think young audiences will be drawn to it because there is something mythical in the universal inquiry of "Who are we?" and "What is love?" Which are, in a way, serious moral questions rarely faced in schools or society because they are hard to grasp.
>
> NOBLE: That's absolutely right. The last plays, and *Cymbeline* especially, are attractive to young people because they have no difficulty with something that's not quite naturalistic. The plays are also consonant with the mood of the last years of this century because they are about change and the journey of the spirit. Audiences

that might know other Shakespeare plays usually don't know *Cymbeline* and generally haven't a clue about what's going to happen next. So when Imogen dies, she dies. And then she comes to life again. I think you put your finger on it. That scene is the alpha and omega, the point in the play where, in a way, it starts: the healing process starts. She has to die for her love to live again. This is a play that talks about the creative power of mercy: forgiving somebody is a creative act. It's inspiring, which is why in *Cymbeline* and *The Winter's Tale* audiences are literally elated. In life we are rarely given a second chance. If you do something terrible, your life is damaged for ever. Or, you might do what King Cymbeline does: throw your children out. But suddenly you get a second chance. And it's not sentimental; Shakespeare is telling a story about creativity and how forgiveness can be regenerative.

SERBAN: At the end of the play, Iachimo is forgiven. Whereas in the *Decameron*, from which Shakespeare took the story, the character is killed. Shakespeare saves him and gives him an incredible speech of repentance, which some people find naïve and silly. But I think there is real depth to it because Iachimo, somehow, was made to play his part by divine order, as if he had to play the devil. If it weren't for Iachimo, Posthumus never would have discovered his pride and vanity. Iachimo is almost a devil's advocate to help the hero's self-discovery. Like the Everyman story, this is a tale of redemption: how to attain mastery over the self-destructive impulses in ourselves.

What interests me in *Cymbeline* is the inner spiritual journey; it's about the temptation of the soul. At the beginning is this meeting between Imogen and Posthumus – the quick wedding, which is like a too perfect *Romeo and Juliet*. What Shakespeare seems to be telling us is that marriage is a gift that has to be deserved. In order to be cherished, it first has to disappear. So Posthumus is banished from the court. And suddenly, his pride and vanity take over. In exile, he boasts about how great Imogen is. But when Iachimo lies to him about her, he immediately turns against her. So we see how much his vanity is at stake.

The entire play is about trying to face one's own vanity. It's like a King Arthur story; the hero has to go through the quest for the holy grail in order to find again what he has lost – in this case, sacred marriage, both physical and spiritual.

NOBLE: The artist tries to find a means to show the most challenging choices facing a society. This usually involves holding the mirror up so we see how preposterous we ourselves are.

I often notice, though, that with these last plays of Shakespeare, theatre directors and audiences find them jagged, they're not perfect, they're not finely tuned.

SERBAN: But they're not perfect on purpose.

NOBLE: Exactly.

SERBAN: The whole process of self-discovery is complex. Which is why it's important for audiences now – especially because of what films and television do today – to come to a play like this and discover that the inner movement is sometimes more important than the outer. To become interested, through a fairy-tale, in the interior journey of finding oneself can be unbelievably exciting. And if theatre can do that, then maybe theatre can have a function again. It sounds pompous, but it's true.

NOBLE: It's true.

Above all, to Noble, the play is a "shimmering fantasy". Some scholars may wonder if Shakespeare had these profound didactic and metaphysical intentions when writing it, or whether he was simply trying to take the measure of Francis Beaumont and John Fletcher and their new type of script that was now gaining such wide acceptance.

Donald Lyons, in the *Wall Street Journal*, cast a cold eye at the Royal Shakespeare Company's presentation:

As the last and grandest of its five offerings at the Brooklyn Academy of Music, *Cymbeline* will be played at DC's Kennedy Center. Written in 1609 after *Pericles* and before *The Winter's Tale* and *The Tempest*, the rarely performed *Cymbeline* shares with those other wild, late romances a penchant for the fantastic, the improbable, the unexpectedly redemptive and restorative surprises of life. Compared to the other stories, the plot of *Cymbeline* has the sobriety of a tax return. . . .

Cymbeline is a challenge to stage. Director Adrian Noble and designer Anthony Ward have opted for an abstract look inflected by touches of the Orient and of *Braveheart*. The space is defined by a sky-blue backdrop and an enormous white sheet as carpet, banner or tent: It's as if the action were taking place on a cloud, or in a chi-chi store window, Peter Brook meets Harrods. Mr Ward knows how to make lovely stage pictures: From scene to scene he can evoke the clear palette of Poussin, the hieratic composition of Ingres, the fiery zest of Delacroix. The court of the British king was got up as refugees from *The Mikado*: the queen in black with a tiny parasol on a tall pole; retainers in pajamas, ponytails and black caps. The noble exiles in savage Wales ran to bare chests and hippie hair. A battle resembled a samurai clash in a Kurosawa movie.

Mr Noble appears to believe – as was also evident in his staging of *A Midsummer Night's Dream* on the last RSC visit – that eye-catching décor, plus parodic semi-kitsch, amounts to an interpretation of Shakespeare. But Shakespeare's statements were much more than fashion statements, and all this color cannot mask a superficial and slapdash understanding of the play. (Understanding was a literal problem at BAM: Neither the RSC diction nor the BAM acoustics were what they ought to have been.)

A few performances managed to register amid the décor: Joanne Pearce, deep-voiced and full of no-nonsense energy, made an Imogen full of open friendliness and fun, a Jane Austen heroine plunged into Gothic murk. Guy Henry's Cloten, a hilarious and sinister cross between Richard III and Pee-Wee Herman, stole the play by speaking very, very slowly – as if to idiots – to the audience about his plans. Edward Petherbridge played the king curiously, as a detached and airy exquisite, bored and barely bothering to notice the proceedings. Clad in a monkish white robe, Mr Petherbridge looked like the high priest in *Lost Horizon* and recalled, by his languid delivery, his Lord Peter Wimsey in the TV mysteries.

Exotic and highly emotional, *Cymbeline* lends itself to a blending with music. The librettist Helmina von Chézy extracted elements from it – and from Boccaccio's *Decameron* – for Carl Maria von Weber's opera *Euryanthe* (Vienna, 1823), a much-esteemed and widely performed work. Von Chézy also drew on a medieval romance by Gérard de Nevers. Not too much is left here of Shakespeare's handling of the story.

Yet another of Shakespeare's romantic comedies is *The Winter's Tale* (1610–11). He adapted it from a best-selling short novel, *Pandosto* (1588), by Robert Greene, the dissolute and envious writer who, dying young, in one of his last pieces had referred to London's new playwright, Shakespeare, slightingly. The book was certainly familiar to many in the Globe's audience. The poet made the usual changes: new names for some of the characters and more complex and elevated traits for them. He shifted the scenes back and forth from Bohemia to Sicily, where jealousy is possibly more common and intense. He shortens the time-lapse, introduces new people – Paulina; the roguish Autolycus; Antigonus – and combines others: Camillo conjoins the roles of two other persons in the book. He moderates some of the violence and tragedy, omits a theme of incest – Pandosto's love for his own daughter – and arranges a happy ending. The statue scene, a highlight, is Shakespeare's addition, an instance yet again of his shrewd flair for theatre.

The King of Bohemia, Polixenes, has been the welcome guest of Leontes, King of Sicily, his friend since boyhood. He is about to leave, though Leontes presses him to remain one week more. When Polixenes resolutely declines, Leontes asks his wife, Hermione, to help persuade the visitor to stay. She extends her invitation so charmingly and insistently that Polixenes accedes. Seeing them together, Leontes becomes jealous. He suspects his wife of having been unfaithful; next he even plots to kill Polixenes, who – opportunely warned of his danger by a courtier, Camillo – makes an abrupt departure. This flight deepens Leontes' suspicions: he charges his wife with infidelity. She is with child, and Leontes asserts that it is not his.

A girl is born to Hermione in prison. Leontes declares it has undoubtedly been fathered by Polixenes. He instructs a courtier to take the infant from his sight, into a foreign land, where he will know nothing further of it.

Hermione is forced to stand trial. The Oracle of Apollo at Delphi, to whom Leontes appeals for a verdict, confirms her innocence. Hermione faints. Supine in a trance, she is thought to be dead. At first Leontes refuses to believe the Oracle, but his heir – a little boy – suddenly dies, sorrowing at losing his mother. The stricken Leontes, remorseful and realizing that he is largely to blame, is left to mourn them.

The infant girl, meanwhile, has been taken to Bohemia, where Polixenes rules. (It is here that Shakespeare makes his oft-quoted gaffe about the "sea coast of Bohemia".) The ship is lost in a storm. Ashore, Antigonus, the courtier who has conveyed the child, is chased, killed and eaten by a bear – a woeful fate! A shepherd finds the abandoned babe and adopts her.

Sixteen years go by. King Polixenes is concerned about the frequent disappearances of Florizel, his son. Disguising himself, Polixenes follows the prince. He learns that the youth is entertaining himself in the company of simple people, a shepherd and a peasant girl, Perdita, reputedly his daughter. The king exposes Florizel's true identity to his lowly companions. When ordered to give up Perdita, the infatuated prince refuses to obey his father. Finally the young couple flee. Polixenes follows them to Sicily.

Here the truth of Perdita's royal antecedents is established, qualifying her to become Florizel's bride. Leontes acknowledges his fatherhood. A recluse, long repenting his earlier error, he is taken to see a statue of his "dead" wife; the statue moves, and he discovers that Hermione is alive. The report of her death was a ruse to spare her life. A wholesale reconciliation ensues.

Once more, if one looks upon this play literally – and as synopsized – it is wildly implausible; but if it is considered a fable it has many subtle denotations and obvious excellences.

Its title comes from a story that the little boy, Leontes' son Mamillius, proposes to whisper to his mother:

> A sad tale's best for winter. I have one
> of sprites and goblins. . . .
> There was a man – . . .
> Dwelt by a churchyard. I will tell it softly;
> Yond crickets shall not hear it.

A wide variety of meanings has been read into this play. It is seen by some as a reworking of a fertility myth with much stress in it laid on the opposition of the seasons, winter and spring; or age and youth, death and resurrection. Kindness and harshness, doubt and faith, are also shown at odds. Sometimes these qualities coexist or alternate in the same person, as in Leontes and Polixenes, in their stormy relations with Hermione, Florizel and Perdita. The drama is also filled with lines reflecting the author's worship and delight in beholding pure-hearted youth, the delicate and innocent radiance of Perdita that Florizel prays might be perpetuated eternally, as, of course, it cannot be, save in verse such as Shakespeare's.

Certain themes still preoccupy this dramatist. Jealousy is one. Not only were Othello and Iago obsessed by it but so here is Leontes, as earlier were Richard III, Edmund, Goneril and Regan. It is an emotion that drives people to very cruel acts. It might be sexual jealousy or envy of rank and power. To Shakespeare was this merely a common trait that spectators could readily identify in themselves and others? Hence a good subject to dwell on since it led to violent deeds? Or was it a strain he also perceived in himself? Did he want to achieve the status and popularity as poet and dramatist arrived at by Marlowe, Beaumont and Fletcher? For long stretches of time he lived apart from his wife. Was she faithful to him? He did not seem deeply enamoured of her, or even interested, but as a husband and a prominent citizen of Stratford, aspiring to a coat-of-arms and royal certification as a gentleman, was he concerned about his reputation as conventional head of a family? Did he feel that he ought to be jealous, even if he was not?

Looking back, one notes that he not only repeats certain themes but also particular story-devices. For example, there are a great number of storms at sea and shipwrecks in his plays. Hamlet escapes when pirates attack his ship during one. Othello's decisive victory over the Turks occurs as the seas rage. There are no less than three nautical gales in *Pericles*, and a crucial one brings the heroine to a strange court and awkward dilemma in *Twelfth Night*. In *The Winter's Tale* a wave-tossed ship's sinking strands Antigonus and Perdita on Bohemia's non-existent coast. A storm and shipwreck are important in *The Tempest*. Of course this is a plot incident that hails back to Roman comedy, but Shakespeare was barely acquainted with those scripts. One factor might have been that Britain is an island, a seafaring nation. Storms at sea and the loss of ships and crews would be a constant concern of its people.

In *The Winter's Tale* Shakespeare calls again – as has been remarked – for a reconciliation of the demands of art and nature; that is, between a life in society (and especially in the artificial climate of the court and upper classes) and in a forest glade, the simple huts of somewhat idealized shepherds. Neither style of existence is, in itself, sufficient or best. A middle course might be followed. An exchange of speeches between Polixenes (in disguise) and Perdita at the sheep-shearing festival in Bohemia – a scene actually rich in English folklore – sums up the conflicting attractions of urban and rustic modes of living.

Perdita is another "changeling". Once more Shakespeare suggests that very often nothing can deny the inherent graciousness and beauty of one on the distaff side who is nobly born, and especially if of royal descent. Though raised as a mere shepherdess, Perdita's loveliness causes her to be called "the queen of curds and cream".

As mentioned above, Autolycus, the amusing cutpurse, is one of Shakespeare's additions to the plot. It is thought that he was introduced in order to provide a part for Robert Armin, the Globe's subtle and skilled comic, who also had a good voice. To him are given two fine songs:

> When daffodils begin to peer,
>   With heigh! the doxy, over the dale,

> Why, then comes in the sweet o' the year;
> For the red blood reigns in the winter's pale.
> The white sheet bleaching on the hedge,
> With heigh! the sweet birds, O, how they sing!
> Doth set my pugging tooth on edge;
> For a quart of ale is a dish for a king.

The other is used to sell his wares and starts off:

> Lawn as white as driven snow:
> Cyprus black as e'er was crow;
> Gloves as sweet as damask roses . . .

He deems himself superior to the louts he gulls, terming himself no "simple man". He is an exceedingly well-liked farcical figure, though his link to the plot is very slight. Shakespeare's primary interest always was to hold the attention of the spectators, not to compose an impeccable work of art conforming to neat classical standards.

If the play is taken literally, Leontes' jealousy is too irrational to be viewed seriously, for it is based on virtually no evidence; and Hermione's sixteen years of concealment are difficult to explain, since it is widely known that her husband feels guilt at her supposed "death", leaving no reason for her to keep in hiding. But as a study in neurotic fury, the feverish passions of doubt and anger that seize and hold the unhappy Leontes are remarkably delineated, from his anguished cry, "it is a bawdy planet . . ." to his frenzied retort to Camillo, who has sought to belittle his monarch's suspicions:

> Is whispering nothing?
> Is leaning cheek to cheek? Is meeting noses?
> Kissing with inside lip? Stopping the career
> Of laughter, with a sigh – a note infallible
> Of breaking honest? Horsing foot on foot?
> Skulking in corners? Wishing clocks more swift?
> Hours, minutes? Noon, midnight? And all eyes
> Blind with the pin and web, but theirs; theirs only,
> That would unseen be wicked? Is this nothing?
> Why then the World, and all that's in't, is nothing,
> The covering sky is nothing, Bohemia's nothing,
> My wife is nothing, nor nothing have these nothings,
> If this be nothing.

Hermione is yet another of Shakespeare's resolute women. Paulina, her faithful friend and wife of Antigonus, belongs in the category with Juliet's nurse, with her bold, rough tongue. All these characters are finely realized, however impossible the absurd plot – which was, let it be recalled, borrowed from a novel of which the reading public was extremely fond; the book had gone through at least fourteen editions before Shakespeare latched on to the story.

The play's form, with its broad leaps of time and rapid shiftings of place, while verging on both the realistic and miraculous, as well as the oracular, religious, philosophical, is variously praised as "perfect", a true masterpiece, and condemned as too "irregular", an "artistic failure", depending on each critic's temperament and predetermined point of view. But so much of it is incontestably good, is strongly written, it is hard to believe that Shakespeare was inadequate to fulfilling his purposes; he did what he intended to do. The fabulous, the symbolic and allegorical were his new goal.

This is even better evidenced in *The Tempest* (1611), for no one can question his full mastery in this work, by most considered his last play. Scattered borrowings can be traced – a name of a person here, a minor incident there, a passage of ten or a dozen lines freely translated from Ovid – but these are slight. Mostly the plot seems to be his own. Woven through it, of course, are many motifs, characters and story-devices that he has used before, some of which were stock plot-turns employed by just about all the literary craftsmen of his time; other familiar details embody concepts from myths and ancient folk-tales. Much has been made of the grounding in a hurricane (hence the play's title) of the flagship of an English fleet off the shore of the Bermudas, which occurred in 1609, just a year or so before the composition of *The Tempest*. The disaster received wide public notice, so even oblique allusions to it would be topical. All the crew of the *Sea Venture* had survived, and published accounts of their adventure are somewhat modestly paralleled by actions of the participants in this fantastic comedy.

So frequently does Shakespeare make use of shipwrecks that some critics are led to believe that the storms and natural mischances in his plays are intended to symbolize the intervention of rude external forces or punitive Divine Providence in human affairs. But if any such storm had so deeply impressed him, it was more likely to have been the one that dispersed and swallowed the Spanish Armada two decades earlier, sparing England from invasion.

The Tempest takes place on a subtropical island resembling Bermuda. The unscrupulous Antonio, conspiring with Alonso, the King of Naples, usurps the rule of Prospero, the scholarly Duke of Milan. Prospero, with his three-year-old daughter Miranda, is set adrift in a small boat. Swift currents bear them to an enchanted island. Here the clever, resourceful Prospero becomes sole monarch. He pores over ancient books and learns magic and is able to summon to his aid the sprite Ariel (who is much like Puck in *A Midsummer Night's Dream*) and, as a menial servant, the subhuman, half-bestial, surly Caliban, creatures taken from folklore.

With the passing of thirteen years Miranda develops into a beautiful young woman. Exerting his magical sway, Prospero causes a ship to founder on the island's rock-girt coastline. As the spectators might expect, the survivors of the wreck include the guilty Alonso, King of Naples, the long-ago co-conspirator. Also aboard the demolished ship are his brother, the ambitious Sebastian, and Alonso's good-looking son Ferdinand, as well as Antonio, now Duke of Milan, who had joined with Alonso to dethrone the rightful ruler, Prospero.

Separated from his elders while making his way ashore, Ferdinand assumes that they have drowned, and they believe that has been his unhappy fate. After swimming to land, they are astonished to find their garments are still dry. Wandering about the island, Ferdinand encounters the fair Miranda, and, as so often happens in Shakespeare's romantic comedies, the young couple immediately fall in love. This is Prospero's doing, but he puts on a show of hostility to this youth who so precipitately asks to marry Miranda. Prospero sets young Ferdinand a succession of heavy tasks, such as hauling huge logs.

Elsewhere on the island, Sebastian, wishing to replace his brother Alonso as King of Naples, conspires with Antonio to kill him. In this scheme Antonio is behaving much as he did when Prospero was overthrown, but this time his former confederate is himself to be the victim of the envious, wicked pair. Prospero, learning of the plot, orders Ariel to thwart it, which the sprite succeeds in doing.

The plotters have been prompted by Caliban, whose real wish is to have Prospero murdered and to carry off Miranda. Besides aiming to dispose of Alonso, the schemers have in their sights Gonzalo, a faithful old counsellor in the service of Prospero. The monstrous Caliban also approaches other survivors of the wreck – Trinculo, a jester, and Stefano, a drunken butler – seeking their assistance in ridding the island of the marvel-working Prospero and wins their consent. In honour of Ferdinand and Miranda her father is staging a masque dramatizing the myth of Ceres, Iris and Juno. He interrupts it, as he is warned of his impending danger, which he handily wards off. Famished, some of the courtiers pursue a vanishing feast. Ariel brings the King of Naples and his marooned courtiers before Prospero to be judged; the all-powerful necromancer offers them pardons but demands that his brother, Antonio, yield back to him the dukedom of Milan. They refrain from speaking directly to each other. The repentant King of Naples, Alonso, and his supposedly lost son Ferdinand are reunited. The boatswain and master of the beached ship bring word that it has been mysteriously restored and able to return to sea; the crew is intact and ready to sail. Prospero sets Ariel free, then renounces his magical gifts, and all prepare to embark, returning to their far-off Italian domain. The ugly Caliban is left behind to brood in solitude on the island.

The play's first performance in 1613 was before King James at Whitehall as part of the festivities attendant on the marriage of Princess Elizabeth and the Elector Palatine. The text was published ten years later.

The fairy-tale quality of this work is like that of *A Midsummer Night's Dream*, but it is far more ambitious in intent; it has philosophical connotations quite lacking in the earlier work. In this clear allegory, life is a maze of separations, misunderstandings and reconciliations. Men are both good and evil. But one must learn to live in the world, accept it, be generous; and forgiving. When Prospero has his enemies at his mercy, he forbears to punish them further:

> Though with their high wrongs I am struck to the quick
> Yet with my nobler reason 'gainst my fury
> Do I take part: the rarer action is
> In virtue than in vengeance: they being penitent,
> Not a frown further.

Often cited to confirm the playwright's belief that his career was ended are Prospero's valedictories:

> Our revels now are ended: these our actors –
> As I foretold you – were all spirits and
> Are melted into air, into thin air;
> And like the baseless fabric of this vision
> The cloud-capp'd towers, the gorgeous palaces,
> The solemn temples, the great globe itself,
> Yes, all which it shall inherit, shall dissolve
> And like this insubstantial pageant faded
> Leave not a rack behind: we are such stuff
> As dreams are made on, and our little life
> Is rounded with a sleep.

Again, after describing the miracles he has worked, he resigns his superhuman gift:

> This rough magic
> I here abjure, and, when I have requir'd
> Some heavenly music, which even now I do,
> To work mine end upon their senses that
> This airy charm is for, I'll break my staff,
> Bury it certain fathoms in the earth,
> And deeper than did ever plummet sound
> I'll drown my book.

It is a poet's abdication from the world and from his art. Finally, in the Epilogue the same note is heard – three times reiterated, it cannot have been by mere chance or without significance:

> Now my charms are all o'erthrown,
> And what strength I have's mine own,
> Which is most faint. . . .
>   Now I want
> Spirits to enforce, art to enchant . . .

(Here "want" implies "lack", not "wish".)

Prospero, having mastered the potent magic arts, of his own volition finally returns to humanity – that is, to the everyday world of human duties and responsibilities. He goes back to Milan to take up his former role as a kind, wise ruler over his grateful, happy subjects. He is Shakespeare himself, leaving the enspelled isle – the theatre – where he exerted alchemic power, creating characters and illusionary places, where all appeared or disappeared, lived or died, at his mere whim.

The playwright returns to Stratford-upon-Avon, leaving London and the Globe behind him, and is once more an ordinary citizen. Prospero has found the possession of a godlike control of the world about him too burdensome. Being merely human has rich rewards and makes fewer demands.

If mental and imaginative exhaustion was not a factor in his decision to retire, after writing three dozen lengthy plays on various subjects, some of them emotionally draining, his physical health might account for his quitting. About that nothing is known. Or his age. He was about forty-seven, which today is considered the prime of life but already beyond the years allotted to most in Elizabethan times, when lifespans were much shorter. He would have been looked on as "elderly", if not old. Once back in Milan, Prospero says: "Every third thought shall be my grave." Shakespeare, like most major artists, must have been aware that his span was shortening. As Otto Rank has suggested, a prodding concern about that is what prompts them to seek immortality in the creation of tangible objects that will exist after them and serve as testimony to their having lived, defying oblivion.

But this dramatist's "magic" with words was hardly diminished. He demonstrates that in this semi-gossamer caprice of a play; each of the leading characters has a poet's tongue. Ariel sings, "Come unto these yellow sands . . .", calming the waves as Ferdinand wades ashore. The young prince murmurs in wonder:

> This music crept by me upon the waters,
> Allaying both their fury and my passion
> With its sweet air . . .

He knew that he was, indeed, entering on a fantastic world.

Ariel tells Ferdinand, inaccurately:

> Full fathom five thy father lies;
> Of his bones are coral made;
> Those are pearls that were his eyes:
> Nothing of him that doth fade,
> But doth suffer a sea change
> Into something rich and strange.
> Sea nymphs hourly ring his knell.

As remarked, Ariel resembles Puck. He is ready to obey the behest of Prospero:

> . . . be't to fly,
> To swim, to dive into the fire, to ride
> On the curl'd clouds to thy strong bidding task
> Ariel and all his quality.

His mode of life:

> Where the bee sucks there suck I;
> In a cowslip's bell I lie;
> There I couch when owls do cry;
> On the bat's back I do fly
> After summer merrily.
> Merrily, merrily shall I live now
> Under the blossom that hangs on the bough.

Caliban, grotesque though he be, is allocated some remarkably fine speeches, such as his oath of fealty to the drunken Stefano, whom he mistakes for a god come from the moon:

> I prithee let me bring thee where crabs grow;
> And I with my long nails will dig thee pignuts,
> Show thee a jay's nest, and instruct thee how
> To snare the nimble marmoset; I'll bring thee
> To clust'ring filberts, and sometimes I'll get thee
> Young scamels from the rock. Wilt thou go with me?

Who cannot not but be amazed again at Shakespeare's sharp observation of natural forms? The hulking, shambling Caliban is not a simple embodiment of evil. Though eloquent at moments, for the most part his speech is guttural, sibilant, full of harsh sounds and hisses. Yet he is not without one attractive quality: he responds sensitively to music, and he evokes some sympathy because he is much put upon by his tyrannical master.

Caliban, "this thing of darkness", is the son of the witch Sycorax, who had formerly imprisoned Ariel. Rescued from the clutches of the witch by Prospero, the sprite thereafter owes continued service to the magician, though promised eventual freedom. Caliban is a strangely warped creature:

> A freckled whelp, hag-born – not honour'd with
> A human shape.

As said before, these two odd, contrasting beings are thought to descend from folklore, but it is not yet known just how and where Shakespeare met them, gave them their very specific traits and to what extent they were mostly conceived in his always astonishing imagination. They have kinship to similar fairies and monsters that he might have come upon in his wide reading, including *The Travels of Sir John Mandeville* and Robert Eden's *History of Travaille*, and perhaps in his delvings into Virgil and Pliny. If so, he considerably refashioned what he found there, and, having been raised in a country town, he may also have tapped personally at some point into local beliefs and tales for his future use or only for his pleasure.

Miranda is the last of Shakespeare's young, bewitching heroines. In succession those in these final scripts – Marina, Imogen, Perdita, Miranda – have traits in common that have prompted biographers to suggest that their model might well have been Shakespeare's own elder daughter Susanna, to whom he clearly showed favour in his will, leaving her the major share of his estate.

At beholding human beings, Miranda, who until now has seen none except for her father, exclaims with purity of vision and the insight of innocence,

> O wonder!
> How many goodly creatures are there here;
> How beauteous mankind is! O brave new world
> That hath such people in't.

This kindly affirmation might be taken as Shakespeare's last testament to his fellow beings.

The Tempest abounds with meanings that fascinate and taunt, some of them most likely to elude even hard scrutiny. Of interesting explications there has been no lack. Some of the ideas voiced by Gonzalo, the former counsellor to Prospero, indicate Shakespeare's careful perusal of the writings of the French philosopher Michel de Montaigne (1533–92) and principally his essay *Of Cannibals*. The gist, summed up by Gonzalo, is that a malevolent creature like Caliban represents "the untameable animal nature" innate in the human race and roaming in the world, harbouring impulses that nothing can ultimately mitigate; while opposed is the image of Ferdinand and

Miranda, so transfigured by love that each in the other's eyes is true perfection. The antithesis is for ever ordained. Human nature must be restrained but not too strictly.

Gonzalo also offers his vision of the Utopian state, largely paraphrased from Montaigne's conception in the same essay. In such a place there would be no magistrates, no riches, no poverty and no possession of land, and not even any labour:

> . . . all men idle, all;
> And women too, but innocent and pure;
> No sovereignty.

Gonzalo would also outlaw "treason, felony, sword, pike, knife, gun, or need of any engine". He would restore, in these and other ways, the legendary Golden Age. Overall, however, the play is not an intellectual excursion but an iridescent, lyrical one, with fantastic figures in it.

Still, so much attention is focused on Shakespeare's allegorical farewell in *The Tempest*, his graceful relinquishment of all necromancy, that the script's substantive message is overlooked. Gonzalo is saying that a "middle way" must be embraced, two conflicting instinctive drives moderated by judicious blending for a life invigorated by animal spirits while also uplifted and guided by an Apollonian calm and self-discipline. This idea is explicit in *The Winter's Tale* as well. Most certainly the anarchy and roughness of the brutish Caliban must be tempered, but sensual energy such as his cannot be ruled out of existence, nor should it be, for in each person it strengthens the fibre of life. For wholeness, precious is Ariel, and all that the darting sprite symbolizes, bringing *anima*, soul . . . music.

Gonzalo duly warns that not all magicians are as benign as his Prospero. Steps must be taken to lessen the despotic sway of the interfering "magicians" who order everything and everybody to suit their own wishes and purposes.

A question that arises when one weighs Marlowe's *Dr Faustus* is how much did the playwright himself believe in the efficacy of magic. Was he sceptical? Was he merely using it as a striking theatrical device? Or did he fear and treat with what he imagined were demons, presided over by a Lucifer? That might also be asked of Shakespeare, in whose works the supernatural has so large a part, for in his plays are a profusion of ghosts, witches. In *The Tempest* magic is prevalent. Prospero's knowledge and possession of its practice is the premise of the drama. Never fully explained is just how he has learned the black arts. If merely by poring over old books, why cannot others do so as easily? But it is a power belonging to few. Noel Cobb, in his book *Prospero's Island*, is convinced that Shakespeare was deeply involved and engrossed in demonology and alchemy and clearly well acquainted with its esoteric lore, a believer. Through two hundred pages Cobb seeks to illustrate and prove this, exhibiting remarkable erudition, along with Freudian and Jungian theorizing and occasional mystical flights. He claims, and rightly, that the seventeenth century was when such cults were at their apogee, even subscribed to by King James, as noted before. At almost the same time that Shakespeare was writing *The Tempest*, his friend and rival, the very rational Ben Jonson, was composing his lively satire *The Alchemist* (1610), in which he portrays two fake exponents of the arts, Face and Subtle, selling the Philosopher's Stone – purportedly able to change base metals into gold – and other charms helpful in gambling and courtship, cheating the gullible. The comedy reveals Jonson's profound disbelief in all such miracle-working articles. By contrast, *The Tempest* treats magic-workings with unstinted respect. There is no indication that its force is not actual.

Was Shakespeare superstitious? Even today, long past the Age of Reason, millions of people are instructed by astrology, have their palms read, are wary of walking under ladders, keep their fingers crossed, read tarot cards, have their fortunes told by gypsies and women gazing into crystal balls, start no enterprise on the thirteenth day of the month, believe in angels and the Devil.

In their convivial tavern meetings the two playwrights may well have exchanged views on such matters. Jonson's script might have been his response to the one Shakespeare had under way. (What rich conversations! England's two greatest dramatists. Or it may have been mere backstage gossip, trivial.)

In one aspect, the younger of the two, Jonson, possibly influenced the man from Stratford. The story in *The Tempest* is compact; its action is limited to twenty-four hours and confined to the island. The theme is relatively narrow. Some of Jonson's enthusiasm for observing the Greek unities may have rubbed off on Shakespeare, however belatedly.

Staging *The Tempest* is not easy. It calls for an atmospheric *mise-en-scène*, a large cast – even more than usual for Shakespeare – and the roles of Prospero, Caliban and Ariel are difficult to fill. (Spoto relates that when Laurence Olivier was an acting student he chose to recite a speech by Caliban. He had already developed his keen interest in make-up and dependence on it. "When he entered as the deformed monster, the female students suddenly wanted to leave the room; one fainted, and several required smelling salts, for Olivier had covered himself from head to foot with green slime and had applied coloured plasters to his face and hands with spirit gum, to resemble suppurating carbuncles. He had taken with utter seriousness the playwright's description of the hideous creature. The gasps from his classmates were as heartening as laughter or applause.")

Whatever the problems it presents, *The Tempest* has no lack of repeated offerings. A notable production enhanced the Shakespeare Festival in Stratford, Ontario, in 1962. Lewis Funke was there for the *New York Times*, which sometimes sends its drama critics far afield. William Hutt was the emphatically commended Prospero, stern yet forgiving; Bruno Gerussi the "first-rate" Ariel; John Colicos the Caliban – "an unforgettable creation. Made up as a black monster, in a hornylike forbidding garb, bold and powerful he is gruesome to behold in his semi-animalism. Growling when crossed, his red tongue wigwagging like a serpent's fangs, he brings terror and a tension to his scenes. Nowhere, indeed, is he more frightening than when he finds companions who will aid him in the destruction of Prospero, whom he has come to regard as torturer and usurper of his island. He commences a rhythmic tattoo with his feet, a wail of hysterical joy in a sort of song, his eyes wide and glaring." The interpretations of all in the company pleased Funke, and especially Peter Donat's show of romantic ardour for the pure-eyed Miranda, Martha Henry, "a vision of young beauty". As directed by George McCowan, this play about enchantments was itself "enchanting". The contribution of the scene designer, Desmond Heeley, was also all-important. "He has seized every opportunity for lavishness and color. He has flashed scenes of reds, golds, and blacks, captured the elegance of royalty in satins and brocades against the stage's deep gray. Particularly vivid and breathtaking are the creations for the masque, the delicate, airy entertainment of music and dance." For Funke, this was "a fairy-tale; a dream, full of lovely poetry. . . . Evocation of mood is essential to its fullest appreciation. Only expert minds and expert acting can lift it into its proper realm. The standing ovation at its conclusion was a justified tribute from an audience that had sat spellbound, enveloped by the magic of Prospero's tale."

The Tempest was certain to tempt opera composers. The first to approach it was Henry Purcell in the last year of his all too brief life, 1695. His version, to a libretto by Thomas Shadwell, was staged at London's Drury Lane Theatre. Shadwell, a well-known playwright of the day, had fourteen comedies to his credit, as well as a good many other texts – oratorios, hymns – suited to musical embellishment. Many were "semi-operas" – as was this, *The Tempest* – a hybrid form then much in vogue. Songs alternated with dialogue, the spoken passages much the lengthier, with narrative pauses for ballets. The orchestra was heard more often than the singers, the chorus more than the soloists; there were few arias. Italian influences on Purcell are discernible in the vocal writing, and French in the instrumental parts – the overture, the dance accompaniments. It might be said that a "semi-opera" bears a close affinity to a masque.

Mozart began a work founded on the play. Unhappily, it was left incomplete.

The Romantic German composer Felix Mendelssohn-Bartholdy (1809–47) gave thought to availing himself of at least a portion of Shakespeare's script, translated into Italian by Pietro Giannone. The prolific and very professional French dramatist Eugène Scribe (1791–1861) prepared a libretto freely based on the play. The Mendelssohn-Bartholdy project not having come to fruition, Jacques Fromentin Elie Halévy (1799–1862) turned out a three-act opera (*La Tempesta*) using Scribe's adaptation; it was performed in London at His Majesty's Theatre in 1830, but little is known of it now. (Though Halévy was Jewish, his art was extravagantly lauded by the anti-Semitic Wagner.)

The Tempest (*Der Sturm*), by Frank Martin (1890–1974), is a more recent entry and quite respectful of Shakespeare's treatment, departing from Schlegel's German translation only in minor ways. Speech, *Singspiel* and lyric passages alternate, as do the various comic, elegiac and dramatic episodes. Ariel is mimed by a dancer, accompanied and commented on by the chorus. *Der Sturm* was first performed at the Staatsoper in Vienna in 1956. This work, too, is little known.

Other compositions have had their origin in *The Tempest* and its exceptional characters. Ariel reappears as a rebel angel in Milton's *Paradise Lost*, and *The Tempest* is thought to have inspired him to emulate and perhaps match it with his masque *Comus*. Shelley was similarly moved to write *Ariel to Miranda*, and, an unfortunate coincidence, he drowned when a summer squall upset and sank his small schooner the *Ariel*. Browning brought back the hairy monster for his *Caliban Upon Setebos*. W.H. Auden issued a series of "poetic meditations" that came with his contemplating the allegory. Berlioz and Tchaikovsky wrote orchestral works in which they express their emotional response to the play. For a successful Broadway production of *The Tempest* during the Second World War, David Diamond provided incidental music that won him a Pulitzer Prize.

Covent Garden in London presented an opera by the postmodernist Luciano Berio, *Un Re in Asculo* (*A King Listens*; 1986). It was given by the Lyric Opera Company in Chicago in 1996 and was in both instances much praised. It has roots in Shakespeare's drama. The central figure, Prospero, is a dying impresario who can soon no longer work his marvels. He is besieged by three sopranos competing for his attention and approval and seeking to seduce him. Actually they are hostile to him as well as to one another. The music has been described as "dramatically compelling", in many places reflecting the characters' anger and the chaos of the opera rehearsal.

An opera, *The Tempest*, by the American Lee Holby, had its première in Dallas, Texas, in 1996. It had been commissioned and staged a decade earlier in Des Moines, Iowa, but not by a major company. In Dallas, Willard Spiegelman, for the *Wall Street Journal*, found that the music "sparkled" but faulted the libretto – the words selected by Holby himself and Mark Shulgasser, and taken directly from Shakespeare's text, "got in the way" and were often unintelligible. Archaic, the lyrics' meanings were lost. The critic heard many Mozartian echoes. On the whole, the music was "elegant . . . sumptuous . . . beautiful", capturing the play's "magic and wonder".

This was believed to be the thirty-second adaptation of the Bard's script.

Retired to Stratford, Shakespeare seems to have gone occasionally to London, where he owned property, and while there to have collaborated sporadically on two more scripts: *Henry VIII* (c. 1613) and *The Two Noble Kinsmen* (published 1634). Just how much of *Henry VIII* is his presents another unsolved puzzle, though scholars do not hesitate to offer answers, some tentative, a few dogmatic. Naturally most are inclined – reasonably, if unkindly – to attribute the best scenes to the unquestioned master. Who was his co-worker is another subject for expert guesswork. Here again one procedure is to analyse the dialogue. Shakespeare's line of blank verse usually consists of ten syllables; most lines in the script have eleven syllables and feminine endings. That points to John Fletcher (an identification originally put forth by Tennyson). But some scholars are inclined to select Philip Massinger. Both men were prominent rivals of Shakespeare and now his successors. Again, some authorities rule out Shakespeare altogether, denying his participation; and others insist full credit for the play should go to him regardless of the obvious demerits. After all, Heminges and Condell included it in the First Folio (1623), and who should know better than they, his former colleagues at the Globe? If the play is Shakespeare's, he seems to have imitated Fletcher, whose style is quite distinctive; and if he collaborated with the much younger man he is probably not responsible for the major scenes but, instead, the minor ones, which is hard to explain. One possibility is that Shakespeare, the veteran, was not the author but the director while he was on hand, adding touches and making numerous changes during rehearsals. Then, too, the Globe might have proclaimed that the work was his to increase the ticket sale, since his was a "name", his past plays much admired.

If by the retired Bard, *Henry VIII*, though always popular, is hardly him at his best. Loose, disjointed, overly episodic, its construction is unlike that of *The Tempest*. After the formal neatness and concentration of that fantasy, would the same dramatist have worked as haphazardly as this? The play, indeed, is hardly more than a pageant, emphasizing elaborate ceremony, suited to the new, mechanically equipped Globe that now had more capability for ornamental display. The characters do not develop, nor do they remain consistent. Holinshed's *Chronicles*, that inexhaustible repository of plot material, provides not only much of the action but also much of the dialogue – the words reassembled in blank verse – for the first four acts and the final scene of the fifth. The other scenes in the fifth act come from Foxe's *Book of Martyrs*. The collaborators have chosen only a few highlights from Holinshed, since the narrative spans two dozen years. The chronology is confused, the characters do not age plausibly, the actual sequence of historical events is freely disregarded, but dramatists have always been granted permission to do that by audiences more interested in a good story than a history lesson. That the characters change inexplicably is partly the fault of Holinshed, who portrays them after that fashion, a consequence of his having copied from a variety of earlier texts that presented conflicting points of view. The focus shifts from one character to another: first, Buckingham, then Wolsey, then Katherine and, finally, Cranmer and the infant Elizabeth, to whom lavish flattery is paid. Still, the work purports to be accurate: its alternative title is *All is True*, meaning, however, no more than that it is reasonably close to Holinshed's account, which itself is dramatized and fanciful.

It is surprising that, so soon after Elizabeth's passing, Katherine of Aragon is portrayed so sympathetically. Indeed, as has been said, the play is a treasure trove of what today look like contradictory political impulses. Are the authors pro-Catholic or pro-Protestant?

The opportunities for pageantry offered by *Henry VIII* were to prove excessively costly to the owners of the Globe, however, for it was the firing of the cannons, the sparks touching off the thatched roof – at the end of Act I, when King Henry enters the Cardinal's palace to attend a masque – that set off the bright conflagration, razing the theatre structure in 1613, an event already described. Very likely many of Shakespeare's other scripts, those belonging to his company, perished in this same unlucky blaze, which left standing not a timber of the charred and smoking Globe.

Henry VIII is not without virtues, whatever its shortcomings. Many of the scenes are effective, and the roles – however inconsistent – can become showpieces for good actors. Confirming this was the critical reception of the work in the *New York Times* by Vincent Canby:

> There probably isn't a scholar or critic in the world who would rate *Henry VIII* as one of Shakespeare's great history plays. It possesses no grandly iconic heroes or villains of diabolic ambitions. It contains no patches of soaring verse and commemorates no single splendid event that forever changed the course of the British monarchy. It is essentially a patched-together propaganda piece.
>
> The general belief today is that Shakespeare himself wrote less than half the text. Part pageant, part history, the play seems to have been composed quickly and to order, possibly to celebrate the marriage of James I's daughter in 1613.
>
> Yet as presented by the Royal Shakespeare Company during its recent eighteen-day residency at the Brooklyn Academy of Music, *Henry VIII* is suddenly revealed to be a vivid and compelling theatrical artifact. Here is an emotionally charged, early seventeenth-century docudrama, gorgeously designed and costumed, acted cleanly and without affectation, as engaging to the intellect for the history it omits as for the history it embraces.
>
> This production is the kind of discovery that is expected of the giant, financially besieged RSC, which maintains three theatres in Stratford-on-Avon and two at the Barbican Centre in London, tours regularly in Britain and abroad and has twenty-nine productions in its current repertory. Certainly *Henry VIII* is the most fully realized of the three Shakespeare productions in the company's five-play touring repertory, which, following its Brooklyn engagement, is now at the Kennedy Center in Washington.
>
> The poetry is uneven. Much of it sounds like someone imitating Shakespeare (or worse, Shakespeare imitating himself, but not too badly). Best known are Wolsey's lines, after his fall:
>
>> I have touch'd the highest point of all my greatness. . . .
>> I haste now to my setting. I shall fall
>> Like a bright exhalation in the evening,
>> And no man see me more. . . .
>>   I have ventur'd
>> Like little wanton boys that swim on bladders,
>> This many summers in a sea of glory,
>> But far beyond my depth. . . .
>>   O Cromwell, Cromwell!
>> Had I but serv'd my God with half the zeal
>> I serv'd my King, He would not in mine age
>> Have left me naked to mine enemies.

It still entertains, as attested by Donald Lyons in the *Wall Street Journal*:

For most people, Henry VIII will always be Charles Laughton, tossing chicken bones over his shoulder and harrumphing, "The things I do for England!" as he climbs into bed with a very plain fourth wife. The husband of six, executioner of two, has become, that is, essentially a comic figure. Apparently, Shakespeare had the same problem when, in 1612, toward the end of his career, he was commissioned, probably to celebrate the marriage of one of King James's daughters, to write a serious play about Henry. There had been a recent comedy on the subject, and Shakespeare's Prologue is made to insist that "I come no more to make you laugh," but am "weighty and serious."

Very weighty and very serious, as we can see in the New York Shakespeare Festival production of *Henry VIII* at Central Park's Delacorte Theater. (This is the thirty-sixth, and concluding, play in the festival's ten-year-long Marathon, which has now staged the whole canon.) The play, though late in date, is early in structure, consisting largely of mournful monologues by fallen magnates alternating with the gorgeous pageantry of royal occasions (a masked ball, a coronation, a christening). Director Mary Zimmerman, a thoughtful and inventive experimenter who has long been at Chicago's Goodman Theater, respects the play's difficult ponderousness, its elephantine lurches from lament to ceremony. She creates beautiful patterns on the wide, bare stage and lets us listen and think.

The evening looks splendid: Against a darkening sky and a floodlit Gothic tower rose designer Riccardo Hernandez's Chirico-like diminishing blue colonnade and a huge gold banner. Toni-Leslie James's court costumes run to green and gold. On stage, Henry is as stiff as his clothes, but it's not entirely the fault of the blandly earnest Ruben Santiago-Hudson. It's the character: He can do no wrong; nothing – neither the high taxes imposed by his chancellor Wolsey nor the divorce from the immensely sympathetic Katherine of Aragon – is ever shown to be his fault. Shakespeare had to make the king blameless, but he could not make him interesting.

The playwright's heart belongs to history's losers, to those who had to be got rid of so that Henry could father Elizabeth, whose christening ends the play. First to go is noble Buckingham (a fine Larry Bryggman), who headed for death, says, "the long divorce of steel falls on me" – one of the startling lines that flash across this play like heat lightning. Then at the core is Queen Katherine, that classic victimized first wife who, protesting her love and faithfulness, rejects with contempt the advice of Wolsey to go along with the annulment of her marriage. In vain, she pleads with the king (pictured as anguished and solicitous!); rebuffed, she declines into saintliness, forgiveness and death. It's a big and rather monotonous role, and Jayne Atkinson, clad first in white and then in black, plays it for pathos, plus the odd flicker of rage. She affects, with mixed results, a soft and vaguely Continental accent à la Ingrid Bergman or Elisabeth Bergner.

If the queen is the play's heart, its intellect is Cardinal Wolsey, arrogant and scheming minister who comes a cropper. Josef Sommer, tall, red-cloaked, gaunt as a crow, poised as a cobra, gives us a cold, distanced Wolsey, as unapproachable in his fall from power as he was in the exercise of it. It's a bold take and it works; we're surprisingly moved when this chill, correct man says "I shall fall / Like a bright exhalation in the evening/And no man see me more."

The victors – the ingenuous Anne Boleyn (Marin Hinkle good in a small role); her hilariously disingenuous and greedy maid (a terrific Bette Henritze); the Uriah-Heep-like Cranmer (Michael Stuhlbarg); Henry himself – are less touching than the vanquished. It's to Ms Zimmerman's credit to have given us this static awkward play as it is; this production's vices are, roughly, those of the play. So are its strengths.

Shakespeare's part, if any, in *The Two Noble Kinsmen* is even more conjectural. It is believed by some that he collaborated on this as well with John Fletcher, whose partner – Francis Beaumont – had by now retired. It was the team of Beaumont and Fletcher that became chief playwrights for Shakespeare's company after Stratford's poet withdrew from it.

Derived from Geoffrey Chaucer's *Knight's Tale* – one of the stories told by his pilgrims going to Canterbury (*c.* 1387) and actually borrowed from Boccaccio – this orphaned tragicomedy recounts the rivalry between Palamon and Arcite for the hand of Emilia, sister of Hippolyta, Queen of the Amazons. The two noblemen are being held prisoners by Theseus, who is married to Hippolyta. Theseus allows Palamon and Arcite to contest with each other in a tournament. Arcite, favoured by the god Mars, is about to defeat Palamon, but other deities, Venus and Saturn, intervene at his moment of triumph. Thrown from his horse, Arcite is fatally injured. After lengthy, dignified mourning, Palamon and Emilia are wed. The adaptors added a sub-plot not in Chaucer. The gaoler's daughter, hopelessly infatuated with Palamon, goes mad with frustration and grief. She is cured when a humble suitor presents himself to her as Palamon. To further round out and diversify the offering, giving it a lighter tone, a country festival, presided over by a stuffy schoolmaster, inserts Morris dancing, songs and lyrical flights. *The Two Noble Kinsmen* is not included in most printed collections of Shakespeare's plays, and productions of it are very seldom undertaken, even as a curiosity.

Only a bare handful of men and women, in any art or any language, have equalled the obscure Shakespeare's immense achievement; in English literature he is without peer, his accomplishment is immeasurable. He has given illumination and pleasure to many generations. In addition, he has provided a host of scholars, critics and school lecturers with full-time employment.

— 19 —
AROUND SHAKESPEARE

Time has left Shakespeare in a somewhat lonely eminence. Of all the late Elizabethan and Jacobean dramatists he alone is universally read and staged. At the beginning there were Marlowe, Greene; soon after, Kyd, Nash. He outlived them. Beyond them he was surrounded by and competed with many other writers possessing lively and effective talent, of whom only one, Ben Jonson, is still widely known. Even Jonson is not familiar to most theatre-goers of our time, though his works are beginning to win frequent revival, especially *Volpone* and *The Alchemist.* But in the hurly-burly of the Elizabethan theatre many playwrights, decade after decade, held their own in popular favour with Shakespeare; indeed, some, like Jonson and Beaumont and Fletcher, saw their output even more extensively patronized. Shakespeare's unique merit was appreciated from the start, but his plays did not always draw the most spectators. To students of the period the names of many others, Heywood, Dekker, Chapman, Marston, Middleton, Massinger, Tourneur, Rowley and Webster, stand out. They ought not to be overlooked and nor should Jonson or Beaumont and Fletcher, for all are writers of considerable skill and flair. Had they flourished in a day without Shakespeare they would seem even more impressive; he dwarfs them, of course. On the other hand, if he had not set such an extraordinarily high standard they might not have exerted themselves as strenuously. Nor, as already seen, were they entirely without influence on him.

Historically interesting is an early script, *Arden of Feversham* (1592), by an unknown author, for some time thought to be Shakespeare. But this play is not in his style; neither in spirit nor in language does it sound like him, nor in attainment. It is sometimes ascribed to Kyd or to an imitator of Kyd. Whoever wrote it, the work is important because it is the earliest remaining example of a domestic or middle-class tragedy, as distinct from one employing high-flown verse and dealing with royalty. Of its kind, it is good enough and affords a vivid insight into some aspects of the daily life of the period. It is based on a real murder, a sensational one, that had occurred in 1551, and later was lengthily described in Holinshed. In some ways the script is craftily put together: a lustful, faithless woman in love with a wavering neighbour, Mosbie, conspires with him to kill her wealthy husband so that she can marry her irresolute paramour. But her complicated plan for the killing is delayed by many unexpected obstacles: an accident, a fog, the last-minute intervention of strangers. A considerable degree of suspense is built up by the proliferation of these mishaps, though perhaps this is a bit overdone. The characters are mostly one-dimensional. The best realized are the two cut-throats hired to execute the plan, a pair of robust rogues, Black William and Shakebag. Talking in the argot of the underworld, they are amusing and colourful. Alice Arden has been called "a bourgeoise Clytemnestra", so cunning and determined to dispose of her spouse that at the last moment she herself takes up a dagger and drives in the fatal blow. (The plot anticipates by three hundred years Zola's *Therèse Raquin* and Tolstoy's *Power of Darkness*.) She is soon exposed, together with all her many helpers. Her husband is a hard, grasping, land-hungry exploiter, so that a note of social protest, too, appears in the work. Excessively melodramatic, the dialogue often clumsy, the piece at times barely escapes being a bit absurd, yet mounts in interest and exerts a strong hold on the spectator. It is primitive, however. To say, as some have, that it is "a deeply moving work of art" or "a masterpiece of psychological interpretation which foreshadows *Macbeth*" is to over-praise it greatly.

Such domestic tragedy, often – as with *Arden of Feversham* – having topical reference, became an established genre. Other such plays included *A Warning for Fair Women* (printed 1599), in which the murder of a London merchant, Master George Sanders, is related; *Two Lamentable Tragedies* (printed 1601) and *A Yorkshire Tragedy* (c. 1606). But middle-class dramas of this sort soon fell into better hands, those of Thomas Heywood and Thomas Dekker, who might be called the founders in English of what is later designated the "problem play", dealing with marital infidelity and family conflicts, an approach towards eventual "realism".

Thomas Heywood (*c.* 1570–1641), and not to be confused with John Heywood, the earlier author of such interludes as *The Play of the Weather*, is a writer of awesome versatility. He belonged to the so-called "Henslowe Group", a stable of professional dramatists who turned out scripts on order for Philip Henslowe, the theatre-owner and "angel" and father-in-law of Edward Alleyn, who financed many of the troupes that competed with Shakespeare's, particularly the Admiral's Men. Since Henslowe's diary records his payments and receipts, it comprises – as already noted – an invaluable store of data about theatrical activities for several decades during this bustling period. Sometimes Henslowe's playwrights – as many as twenty available at one time – sent him scenarios or outlines, or the first parts of scripts, and he advanced money to finish them. On occasion he bailed his roistering writers out of gaol, an instance being a discharge for Thomas Dekker; the sum paid as a fine imposed by the magistrate was then deducted from Dekker's fee for a completed work. Henslowe also exerted control over actors, some of whom were under personal contract to him. Most of his stable of dramatists were scarcely more than hacks, but some were highly competent and even true artists, as were Ben Jonson and John Webster. It was their practice of working in groups, two or three writers concocting ideas for a script, that makes specific attributions of authorship so difficult for scholars today.

In Thomas Heywood Henslowe had a most reliable and ready source of material, for he had not only varied gifts but was enviably prolific. He turned out tracts, treatises, histories and biographies, as well as dramas. Educated at Cambridge, he came of a reputable Lincolnshire family. Like Shakespeare, he started his theatre career as an actor, was hired by Henslowe to perform on stage and continued on the boards until he was at least near fifty years old. Accordingly he had a good feeling, again like Shakespeare, for what would work on stage. He was able to gauge, practically, what audiences liked and invariably guided himself by that; he was a crowd-pleaser. He claimed to have written or collaborated in some measure on more than 220 scripts, saying that in all that number he had "a hand or at least a main finger".

His works fall into all the categories of Elizabethan and Jacobean theatre, ranging from *The Four Prentices of London* (c. 1592), a romantic adventure tale – very strange – of four young men in the Middle Ages who combat with the infidel Turks in the Crusades, which still retains many echoes of the Moralities, to *The English Traveller* (c. 1627). He wrote chronicle plays of the Wars of the Roses; *Edward IV* is an example. Classical themes engaged him, too: *The Rape of Lucrece* (1607), and a series dealing with Greek mythology and incidents in the Trojan War. These are episodic; mixing clowning with seriousness, they won wide approval. In other melodramas with a contemporary background he also mingles romantic derring-do with homely realism, his plots illogical but enthralling to the always willingly receptive spectator. His most successful piece, however, is *A Woman Killed by Kindness* (1603). This is in the vein of domestic tragedy, at which he excelled and to which he brought the fullest originality. The plot was not drawn from an actual scandal of his day but borrowed from several authors of Italian stories who also served Shakespeare. But it is far more linear and much less exciting than *Arden of Feversham*.

A wealthy country gentleman, John Frankford, marries an attractive young girl of good family, Anne. She secretly betrays him with his youthful friend and protégé and frequent house-guest, Wendoll. A loyal servant, Nicholas, tells his employer, Frankford, what is happening. At first incredulous, for he deeply loves his wife, Frankford watches them, grows more suspicious, sets a trap and discovers the guilty pair asleep in each other's arms. He does not instantly draw his sword and kill them, which is what usually occurred in like dramas of the period. Instead he spares them. In a rare but somewhat extreme show of Christian charity he commands his wife to bid their two children goodbye, to gather up her possessions and take her leave for ever. He wishes never to see her or hear from her again. He sends her to live in another manor house of his, some miles distant. Repining, the conscience-stricken woman starves herself to death. Her lover, Wendoll, departs for a long tour in foreign lands. Informed of her imminent decease, Frankford arrives just in time to pardon her before she expires, surrounded by her kinfolk in a very tearful final and pathetic scene. Her enlightened husband has not punished her in the conventional way, by murder; instead, as he twice states explicitly, he has killed her by his kindness. Whether his conduct is actually more Christian and charitable, the author does not say. It would be possible to interpret the play's moral as ironical, but this has not been the critical consensus: Heywood is much praised for the "generosity" of spirit his hero displays. Charles Lamb has commended the author for incorporating in his work "courtesy, temperance in the depth of passion, sweetness, in a word, and gentleness". Lamb terms him "a sort of prose Shakespeare". But to others that encomium is a bit unearned. An effective scene is one in which Frankford closely observes his wife and his protégé playing cards and translates each move of the game into terms charged with erotic significance.

Perhaps because he was aware that his drama was a bit tepid at points, Heywood fortified it with a busy sub-plot involving Anne's brother, Sir Francis Acton, and another young nobleman of the neighbourhood, Sir Charles Mountford. Losing a wager to Sir Francis, Mountford is infuriated and kills two of his rival's servants. For this he is imprisoned and ruined, largely at the implacable, angry prompting of Sir Francis.

Susan, Mountford's intensely devoted sister, seeks to free her brother from his durance vile by borrowing money from their richer relatives, but none is willing to help. Encountering her by chance, Sir Francis is immediately possessed by her beauty. He buys her brother's discharge. Humiliated, though glad to be restored to the world, Sir Charles offers his sister in marriage to his former enemy, and a general reconciliation ensues. Once again unwonted "kindness" prevails over the savage impulse for vengeance. The twists and turns of this sub-plot do much to invigorate the play and if anything are more interesting than the principal action. So dominant is the affection between Susan and her brother Charles that the modern reader might perceive it as possibly abnormal, though hitherto this has not been subject to comment.

In *The English Traveller* a young man returns from abroad to find the girl he passionately adores wed to the rich, elderly Wincott. He vows to remain chaste rather than commit the sin of adultery, though he and the wife promise to marry each other when the old man dies. Then the hero discovers that the young woman for whom he had taken the pledge of abstinence is having a secret affair with another gallant. He is outraged but forswears physical revenge on her. Overcome by shame, she dies. This play, unlike *A Woman Killed by Kindness*, has a comic sub-plot which somewhat detracts from it; but the dramatic situation and handling are again quite original. Furthermore, in it Heywood omits the dances and masques which were then very popular – folk-dancing has a part even in *A Woman Killed by Kindness*. He was trying to restore drama to what he deemed was its essential form. In moralizing and lachrymose works like these, Heywood was unwittingly preparing the way for the eighteenth century's sentimental theatre, dealing with intimate marital relationships, offering more personal identification between characters and spectators who now saw themselves, family members, friends replicated on stage. Indeed his choice of domestic topics was deliberate; in the Prologue to *A Woman Killed by Kindness* he declares: "Look for no glorious state, our muse is bent / Upon a barren subject, a bare scene."

His two-part *The Fair Maid of the West* (printed 1631) is a farrago of wild, implausible adventures, following the troubles of a Captain Spencer in love with a tavern maid, Bess Bridges. In its day it delighted packed houses. His realistic comedies included *The Wise Woman of Hogsden* (1604), about a bawdy hag who doubles as a fortune-teller and operator of a baby-farm; set in London's underworld, its rakehell hero is again a prototype of those destined to appear in Restoration comedies. Curiously, Heywood was both old-fashioned and far in advance of his times. His work exposes two veins in his nature: he is pious, yet offers broad humour. He is didactic, yet anxious to please his public at whatever cost. He is an inveterate if excellent punster. At moments, as in Frankford's speech of despair at discovering his wife's infidelity, he writes with poignancy and a touch of poetry. His personality, as it shines through his work, is very much that of the average decent man; modest, decidedly English (and Elizabethan) in his confusing traits.

In his closing years Heywood was appointed City Poet, responsible for staging the Lord Mayor's pageants. Unfortunately for him, only about a tenth of his prodigious output is extant. He is remembered and respected as an innovator at electing crucial new approaches to his subject-matter; but though his work shows national pride, a charitable spirit, an urge to encourage solid virtues in his fellow man, it lacks both profundity and sublimity.

Thomas Dekker (1572?–1632?) had a more cheerful temperament, which must have stood him in good stead during a difficult life; he was fated to spend six years in prison for debt and life-long was harassed by penury. Henslowe was always lending him money as well as having to bail him out. Unabashedly and of necessity a hack writer, he none the less collaborated with some of the giant literary figures of the day, and being considerably talented on his own left to posterity one of the most frequently revisited Elizabethan farces, *The Shoemaker's Holiday*.

Too little is known about his life, but from all indications he had a truly genial personality. Biographers conjecture that he was London-born and raised, and – though he seems not to have attended a university – he was sufficiently conversant with Latin, which argues for his having at least studied at a good grammar school. Besides plays, he wrote satirical books to augment his small income and is well esteemed today for his *The Bellman of London*, a sort of thieves' guidebook to what the city had to offer, and *The Gull's Handbook*, which might be subtitled *A Primer for the Gullible*, a treasury of information about the dark and seedy side of Elizabethan life. One chapter in it, "How a Gallant Should Behave Himself in a Playhouse", portrays how ill-mannered were the rich, noble gentlemen seated on the stage, their comments and gestures humiliating the performers and author. In all, he turned out over twenty prose tracts and pamphlets of all sorts.

His harsh life, and the pressing need to earn money by collaboration, gave him little chance to write his own work. He is identified with some forty stage-pieces, but of these only eight or nine represent what he could do unaided. His early fantasy, the two-part *Old Fortunatus* (initial date unknown but first revived with Henslowe's financial help in 1596), is a strange allegory, based on the old German legend of Fortunatus, his son, a purse of gold and a wishing-cap. It unfolds with often mellifluous verse of unique quality. Dekker has a gift for song, as witness:

> Tomorrow, Shadow, will I give thee gold,
> Tomorrow, pride goes bare, and lust a-cold,
> Tomorrow will the rich man feed the poor,

> And vice tomorrow virtue will adore,
> Tomorrow beggars shall be crownèd Kings.
> This no-time, morrow's time, no sweetness sings.

The play is awkwardly constructed, though it is hard to tell what might have been its original form, since it underwent continual revision, its two parts having been combined into one, some episodes omitted, new ones inserted. At times reminiscent of Marlowe's *Faustus*, it ends tragically: Fortunatus and his son, failing to please the goddess Fortune, who has granted them their desires, must perish for their shortcomings. The play was presented at Elizabeth's court in 1599.

Partnering mostly with younger playwrights, though occasionally with Ben Jonson, Marston, Middleton, Massinger and Rowley, other times with Drayton and Webster, he lent a practised hand to the preparation of scripts for the Admiral's Men and other companies. Many of these dealt with British history, others exploited classical themes – a vanished *Troilus and Cressida*, for example – and he even provided religious dramas, such as *Jeptha*, now lost. *The Whore of Babylon* (1607), an allegory, attacks the Papacy and lauds Queen Elizabeth. He sortied into the field of domestic tragedy, too: *Page of Plymouth*, with Jonson. Another effort in this genre is *The Honest Whore* (1604), obviously influenced by Heywood's success with *A Woman Killed by Kindness* a year before. Possibly Middleton was his collaborator on the first part of it. The script's fault is a constant one in Dekker's work: a weak structure, bordering on formlessness. But some of the scenes have a strong effect, and the characters are sharply alive. A comic sub-plot runs through it, too.

A harlot, Bellafront, loves the false-hearted Count Hippolito. He has persuaded her to change her depraved way of life. By a trick – a death-like swoon much like that experienced by Juliet – he marries Infelice, the daughter of the Duke of Milan. The reformed Bellafront, now an honest woman, weds Matheo, a gambler, who earlier had been responsible for her degradation. Later Hippolito finds her even more attractive and seeks to seduce her, but she repulses him, and vengefully he tries to ruin her but fails. Her husband, needing money, is willing to see her sell herself. She is rescued from her dilemma by her father, Orlando Frescobaldi, a man of true virtue.

The Oxford Companion to English Literature calls the play "painful" but "one of the great dramas of the age". Its dourness is alleviated by the comic sub-plot, which concerns an eccentric linen draper. At first tending to be romantic, Dekker's domestic tragedies grew more realistic as he progressed.

But it is for his comedies that he is best recalled.

The Shoemaker's Holiday (1599) is incontestably topmost. Early in that year Henslowe lent Dekker three pounds to buy a book by Thomas Deloney titled *The Gentle Craft* (1598), and it was the source of his "merry conceited" farce, to borrow his own description of it. It was enacted before Queen Elizabeth on New Year's Day, 1600. Dekker combined three tales by Deloney about cobblers. The protagonist is an actual figure, Simon Eyre, who in the early fifteenth century rose from upholsterer and draper to be Lord Mayor of London. Deloney changed his occupation to shoemaker; Dekker keeps this alteration. The play is a vivacious picture of Elizabethan London, especially the city of busy tradesmen and lively apprentices.

Rowland Lacy, spendthrift nephew and favourite of the Earl of Lincoln, is infatuated with Rose, daughter of Sir Roger Otely, the capital's Lord Mayor. Both his noble uncle and her somewhat prudent father move to impede the affair. Rose is sequestered in the country; Lacy is ordered to France, to lead troops in a military campaign. Pretending to sail, he stays behind disguised as a Flemish shoemaker, Hans, in the employ of the master cobbler, Simon Eyre. Rose's father tries to pair her off with a young London gentleman, Hammon, but she demurs. Meanwhile, the Earl of Lincoln, learning of his nephew's dereliction, is searching for him. Eyre, having scored considerable financial gain in a trade with a Dutch shipmaster, is elected sheriff. At the celebration of this event, at the Lord Mayor's house, Rose encounters her disguised lover, Lacy. Hammon, rejected by Rose, lays suit to the humble Jane, whose husband Ralph has been unwillingly conscripted to fight in France. Hammon, fabricating a report, tells Jane that her husband has been killed. She reluctantly consents to wed him. In reality Ralph has returned wounded and is seeking his lost wife. Rose orders a pair of shoes to be made by "Hans" – her lover Lacy – and by this chance meets him again in London. Hammon sends for a new pair of shoes for his bride-to-be, Jane, and Ralph is the one commanded to make them. Thus shoes reunite the lovelorn pairs. The natural death of several aldermen elevates Eyre to the Lord Mayoralty. Rose and Lacy elope and their marriage is sanctioned by Eyre in his new official capacity. Ralph regains his wife at another church, interrupting the ceremony that would have united her with Hammon. At the request of Eyre, Lacy is pardoned by the king, who is charmed by the shoemaker's frankness and self-confident good humour.

Granted, the plot is laden with coincidences. The outstanding character is Eyre, with his rough-edged tongue, his flow of robust epithets.

The dialogue is earthy, indeed vulgar, replete with London slang, wonderfully vigorous and picturesque, with occasional insertions of poetry and several lovely songs.

> O, the month of May, the merry month of May,
> So frolic, so gay, and so green, so green, so green!
> O, and then did I unto my true love say,
> "Sweet Peg, thou shalt be my summer's queen!"
>
> Now the nightingale, the pretty nightingale,
> The sweetest singer in all the forest's choir,
> Entreats thee, sweet Peggy, to hear thy true love's tale:
> Lo, yonder she sitteth, her breast against a brier.

And so on. Goes another:

> Troll the bowl, the jolly nut-brown bowl,
> And here, kind mate, to thee!
> Let's sing a dirge for St Hugh's soul,
> And down it merrily.

> Down-a-down, hey down-a-down,
> Hey derry derry down-a-down. . . .

The words ring with masculine joy. The play also encompasses a brace of Morris dances by the journeymen. In one instance, it being Shrove Tuesday, the apprentices' holiday, they hold a mad revel in honour of it.

As anyone who has seen it can attest (and it enjoyed a hilarious rebirth in New York City in 1938) *The Shoemaker's Holiday* plays much better than it reads, giving plentiful opportunities for amusing stage "business", outright buffoonery. No one could be fonder of this caper than John Gassner, who writes:

> It is in the frolicsome kindliness of this masterpiece that one finds Dekker's sunny genius at its zenith. Here he is one of several Elizabethans to celebrate the middle class that is rising both politically and socially in some proportion to its prosperity. This endearing romantic comedy combines a "success story" that must have pleased the groundlings with a general picture of artisan life and a romance in which the nephew of an earl disguises himself as a Flemish shoemaker in order to win the Lord Mayor's daughter. Democracy gains a victory when this unequal match is approved by the King. Glorious humor and vigor are supplied by the genial self-reliance of Simon Eyre and by the antics of his mettlesome apprentices who stage the first labor strike ever in a drama; they enable Ralph to win back his wife by flourishing their cudgels. The gaiety of the piece is, moreover, not as "unthinking" as some authorities hold. It is the best social comedy of the age, and its democratic humor has commended it to our time.

From Dekker, later in his career, were to come scenes in *Caesar's Fall* (1602), on which he teamed with Middleton, and *Westward Ho!* (1604) and *Northward Ho!* (1605), with which John Webster was involved. *The Roaring Girl* (1610) is another of his joint efforts with Middleton. It introduces Moll Cutpurse, who dresses like a man, wields a sword and inhabits the underworld, a milieu with which the often-imprisoned Dekker was all too familiar. If apt to be a virago, Moll also has generous impulses and a soft heart and befriends young lovers in distress. (It is from this character that the slang term for a gangster's sweetheart, a "moll", derives. A century later, Defoe made use of it for his *Moll Flanders*.) She is drawn from a young woman, Mary Frith, notorious in those days. In *The Witch of Edmonton* (1621) even the witch, Mother Sawyer, is a sympathetic figure; it is clear that he pitied her. His helpers for this script were John Ford and William Rowley.

Other late scripts were *If It Be Not Good, the Devil Is in It*, making use of a folk-tale about Friar Rush, and *Match Mee in London* (1604–5), the second of these said to have been an utter failure. It was wholly his own. In it Dekker was seeking to emulate the tragicomic form that Beaumont and Fletcher had recently made the fashion. *Keep the Widow Waking* (1624) reunited Dekker's talents with those of Ford, Webster and Rowley. Like many other works of this period, *The Witch of Edmonton* was a dramatization of a current scandal – these plays, in one aspect, were like feature stories in modern newspapers, though they took far more liberty with facts.

Dekker's last known stage contribution is *The Sun's Darling* (1624), capturing the ephemeral spirits of the four seasons. Lacking a firm structure, as always, showing signs of his usual haste and carelessness, it is also marked by his invariable good humour and his gift for delicate lyricism. It was often put on at court and at a small private theatre, the Cockpit.

His pamphlet *The Wonderful Years* (1603), an account of the plague that ravaged London, provided Defoe with poignant details for his subsequent and far more successful *Journal of the Plague Year* (1722).

It is not certain that Dekker died in 1632, but after that date nothing more is heard of him. Some place his death in the year 1641.

He was not a pretentious writer; he sought chiefly to make money he badly needed. But he crowded into his work a command of racy prose and underworld argot and a talent for exquisite poetry – a strange contrast – along with his sharp observations of the swarming life around him and constant evidence of his simple humanity and warm heart. Several of his characterizations are notable, especially those of Bellafront, Orlando Frescobaldi, Simon Eyre, Mother Sawyer and Moll Cutpurse.

George Chapman (*c.* 1559–1634) was also in Henslowe's pressured galaxy. His biographers are vague, and few facts about him are certain. He is, of course, more renowned as a poet, the translator of the *Iliad* and the *Odyssey*. (John Keats entitled one of his greatest sonnets "On First Looking into Chapman's Homer".) As noted earlier, Shakespeare was at times influenced by him. He was truly erudite. His entry into theatre work occurred comparatively late in his career; he was about thirty-six or some say forty. His labours on the Greek epics kept him occupied much of the time before and after. He began his stage-writing with a series of comedies. His first, *The Blind Beggar of Alexandria* (1596), staged at the Rose, met with immediate success and had what was for then a long run of twenty-two performances at intervals over its initial months. The audience was impatient at the play's romantic and semi-tragic elements, so it was expediently cut to preserve chiefly its comic features.

His next, *An Humorous Day's Mirth* (1597), was to show Ben Jonson the way – the two worked together under Henslowe's aegis – by stressing characters dominated by humours, genetic eccentricities, all engaged in a rather simple situation that allowed them to be ruled by an overriding singularity or trait. (Jonson's expanded theory of humours will be discussed later.) In this play the focus is on a jealous husband, a suspicious wife and a scholar of melancholy temperament. Each is wholly motivated by one aspect of his or her personality. At the centre of the action is a dynamic figure, a propulsive mischief-maker who keeps the plot going. Here that responsibility is assigned to Lemet, an intriguing member of the royal household. Chapman takes advantage in this piece of an opportunity to express his fervent dislike of hypocritical Puritanism. He creates the character Forilla, who is given to voicing pious cant while lust fills her heart.

Shortly after this, Chapman shifted his allegiance from Henslowe to a new troupe, the Chapel Children, at the Blackfriars, for whom he wrote a good deal more. He had already completed a new script, *All Fools But the Fool* (1599), which is by far his best in a comic spirit. In this he overcomes his usual failing and fashions a clear though very complicated structure by grafting together two plays by Terence, *The Brothers* and *The Self-Tormentor*, and then adding a new sub-plot, and thoroughly Englishing the whole, finding an acceptable Elizabethan equivalent for each role in the Roman originals. The dialogue is largely in fluent blank verse, with lapses into prose for the more farcical moments. The worldly, sophisticated but strict Gostanzo is outwitted by his rebellious, supposedly very obedient son who surreptitiously marries against his father's wishes. (In Terence's version the son is involved with a courtesan, not a secret bride.) The sub-plot deals once again with the obsessive humour of jealousy. Swinburne has called this one of the best comedies in English. (The datings, here and later, are quite conjectural.)

May Day (c. 1602) is deftly adapted from an Italian farce. In it the hero and heroine put on the dress of each other's sex. The same year brought forth *The Gentleman Usher*, a *mélange* of stately masque, eccentric dance and dramatic action. A duke and his princely son Vicentio are rivals for the hand of Lady Margaret. The son enlists the aid of Bassilio, an effeminate Gentleman Usher, the amusing title character. Good-humoured, self-fancying, Bassilio acts as the son's emissary to Lady Margaret, who secretly marries the young suitor. The bridegroom is forced to flee, pursued by Medic, a wicked favourite of the father. Word comes that the son, Vicentio, is dead. Margaret, rather than marry the father, disfigures herself. But the ending is happy. The son, quite alive, returns; a doctor with magical powers restores the lady's comeliness; the evil favourite, Medic, is banished; and the duke is reconciled with his son and daughter-in-law. Here, as elsewhere, Chapman betrays his interest in the occult. The tone of this confection is also romantic throughout; the possibly tragic dilemmas are miraculously resolved, and thus it presages the work of Beaumont and Fletcher and the last plays – the fantastic tragicomedies, such as *Pericles* and *The Winter's Tale* – of Shakespeare. Chapman's writing, as ever, has an elevation unique to him.

The Old Joiner of Aldgate (performed in 1602) was prompted by a lawsuit then being much talked about, incidents in a marriage gone awry. The play is lost but not the legal records of the sordid scandal on which it was based. *Monsieur d'Olive* (1604) combines romance, comedy and satire and is more pleasing by far, poking neat fun at a foppish courtier.

With Ben Jonson and John Marston, Chapman collaborated on *Eastward Ho!* (1605), which evoked a *cause célèbre* and led to the trio's brief imprisonment, for certain lines in the play offended the new king, James I. They seem not to have been Chapman's but Marston's, but all three authors were condemned. This celebrated affair will be taken up on a later page.

As early as 1598 Chapman began to write poetic tragedies. Gradually this became his preferred genre. The curve of his career, then, is rather like that of Shakespeare's. In some ways he took as his models for these works both Kyd and Marlowe; indeed, he completed Marlowe's unfinished narrative poem *Hero and Leander*. (It has been suggested that Chapman, not Marlowe, is "the rival poet" alluded to in Shakespeare's sonnets.) He had already reached adulthood a decade before *Tamburlaine* set the Elizabethan stage ablaze. It is natural that the first two highly successful writers of that theatre had impressed him. His tragedies, like Kyd's, are Senecan, full of sensational violence. Yet they are often so abstract that they have been described as "dramatic poems" rather than vital stage-works. It is hard to know if this is entirely true, since no one has revived them, and some plays that seem inert on the printed page come surprisingly to life on the boards.

Like Marlowe, Chapman uses his dramas to express his philosophical outlook. As might be expected of one who spent years engrossed with Homer and ancient literature, he is inclined to accept classical precepts almost by assimilation; his views are imbued with Platonism, Stoicism, Plotinianism. He is also preoccupied with the costly and conflicting claims of the state and the individual. He is concerned with the struggle between the mind and the soul against the carnal senses. The body is a prison. The world is to be scorned, for it is rank with depravity. A man is obliged to free his soul and attain Heaven. The plays reveal Chapman's personal synthesis of all these pagan and Christian concepts.

Like Marlowe, too, he builds his drama around an imposing central figure who, like a Tamburlaine, dominates the action. Such is the hero of his most esteemed tragic work, *Bussy d'Ambois* (1604). This is the first of four serious dramas laid in France and depicting actual personages. Chapman had travelled considerably and was well acquainted with foreign parts; but his choice of setting might evince caution on his side. He dared not comment on political problems too near home; his previous unhappy experience with *Eastward Ho!* had confirmed the danger of that. Like Shakespeare, he now resorted to veiled allegory. His Marlovian hero is Louis de Clermont d'Ambois, Seigneur de Bussy, a flamboyant cavalier famed for his feats as a lover, soldier and swordsman at the court of Henry III. Bussy is the Renaissance man in his many attainments, self-sufficient, at odds with his fellow men and society. Excellently constructed after the manner of Kyd, the drama is filled with outsized incidents, including ghostly visitations, tortures, slayings. The demon Behemoth is invoked and appears. The style echoes both Seneca and Marlowe; sometimes the dialogue has been directly translated from Seneca; from Marlowe comes not only the "titanic" protagonist but also the heightened vocabulary and sweeping line. The hero is defeated, suddenly and for reasons he cannot grasp. He dies, after a valiant stand.

Bussy's unique and superior gifts and insolence have incurred the jealousy of the Duke of Guise as well as of Monsieur, the King's Brother, his erstwhile protector. His weakness is his adulterous love of Tamyra, beautiful wife of the Count of Montsurry. Exploiting it, his enemies bring him low. She is also desired by the king's brother, now hostile to Bussy and instrumental in apprising Montsurry of his wife's infidelity. Tortured by her husband, Tamyra yields to his demand that she lure Bussy into a fatal trap.

The play is heavy with metaphysical discussion. Sexual passion is shown to be an irresistible force. Recognizing it as such – and implicitly condoning it – a friar assists Bussy in his guilty liaison. Nature, however wicked, defies man's efforts to rule it and overwhelms him. The poetry has rugged power but little colour. Bussy proclaims his pre-eminence and independence of spirit in words as strong and sober and thoughtful as these:

> When I am wrong'd, and that law fails to right me,
> Let me be king myself (as man was made)
> And do a justice that exceeds the law.

Other passages are vigorously rhetorical. Expiring, yet still on his feet, he announces defiantly:

> Here like a Roman statue I will stand
> Till death hath made me marble.

As in most of his other plays, Chapman prodigally includes many needless characters and too often digresses. None the less, *Bussy d'Ambois* won acclaim at once and brought Chapman his greatest theatrical success. In subsequent years it was produced many times, by various companies. What particularly attracted the public was the boldness of Bussy's fiery temperament.

A later critic who thoroughly disliked the play was John Dryden: "When I had taken up what I supposed was a fallen star, I found I had been cozened with a jelly, nothing but a cold dull mass which glittered no longer than it was shooting; a dwarfish thought dressed up in gigantic words." (Here "jelly" is "jellyfish".)

The Conspiracy and Tragedy of Charles, Duke of Byron (printed 1608) is again set in France. It referred to more recent historical events, the execution for treason in 1602 of the Duke of Byron, friend and companion-in-arms of his king, Henri IV. Everyone in England was familiar with this shocking incident, which also bore an obvious parallel to the death of the traitorous Earl of Essex. A long work, ten acts, it was meant to be performed in two parts. Chapman did not hesitate to have an actor impersonate Henri of Navarre, and to titillate the audience he added a scene in which the queen slaps the face of Henri's mistress. The French ambassador voiced his indignation at this. The English officials duly shut down the play and seized the principal members of the cast. Chapman fled and kept himself in hiding until the affair quieted. Anxious to have the play appear in print, he could not obtain a licence from the Master of Revels. Finally, after much delay, it was brought out, expurgated, with hurtful gaps in the action. "These poor dismembered poems", Chapman ruefully described his two-part play, which he dedicated to Sir Thomas Walsingham, the powerful nobleman who had once been linked to Marlowe.

Chapman's characterization of segments of his drama as "poems" is entirely appropriate. The script consists largely of very long speeches, debates, soliloquies. To this is added choric commentary. Dubbed an "epic" work, it is not dramatic. It again has no pace; it barely moves. Byron, proud warrior, embodies the individualistic spirit of Renaissance man. He is soon at odds with his sovereign, Henri of Navarre, who – now that the wars are ended – seeks to restore peace and order to his kingdom. Henri is the new, absolute monarch of the oncoming neo-classical age, with its firm, hierarchal structure of society. Sent on a diplomatic mission to England and the court of Elizabeth, Byron beholds an exemplar of the "New Monarchy", whose reign – like that of Henry IV – is benevolent, wise. But though deeply impressed by this glimpse of the future, Byron's egotism is too great; he cannot subordinate himself. When he finally wears out Henri's patience by his stubborn defiance, his demands for extended powers, his attacks and conspiring, he is arrested, tried and sentenced to die. After that Byron undergoes terrifying alternations of mood: in consistency with his turbulent nature, his self-aggrandizing aspiration and idealism, he "doubts, storms, threatens, rues, complains, implores". His prophecy is destined to come true:

> By small degrees the kingdoms of the earth
> Decline and wither . . .

Byron is a remarkable portrait of an activist, a revolutionary spirit who cannot accommodate himself to the status quo, and who also cannot separate in himself the twin motives of his personal ambition and his visionary hopes. Henri IV is depicted as rational, astute, objective, burdened by the heavy cares of state.

> He should be born grey-headed that will bear the sword of empire.

He is capable of laughing at Byron at the climax of one encounter. That arrogant egotist, startled, demands of himself:

> What's grave in earth, what awful, what abhorr'd
> If my rage be ridiculous?

This very thoughtful work is probably fated to be closet drama.

Once again the occult is introduced. The risk of second sight is discussed by the hero and by a magician, La Fin. But interestingly, Byron, blindfolded, awaiting death on the gibbet, refuses the consolation offered by the Church. Says the Archbishop:

> My lord, now you are blind to this world's sight,
> Look upward to a world of endless light.

Contemptuously, the rebel retorts:

> Ay, ay, you talk of upward still to others,
> And downwards look with headlong eyes, yourselves.

Byron, like Bussy, has been king in his own eyes and finds it hard to submit even to one divinely appointed. A crabbed play, shot full of insight and fascinating questions, touching on matters of psychology and politics, it repays careful study.

The Revenge of Bussy d'Ambois (1611), a sequel to the earlier play, concerns a moral titan. Clermont is the brother of the murdered hero and, like Hamlet, is urged by the dead man's ghost and by a sister to avenge a slaying. Like Hamlet, he hesitates. He is wholly compounded of good qualities; his is a saintly personality. At times he is apparently possessed of supernatural powers, fending off his enemies by displays of seeming magic, searing the fingers of those who touch him. He is quiet, meditative, led by his sense of duty. He zealously seeks to do only what is good. Though the ghost pleads with him for retribution, Clermont is not convinced that he has the right to enact private justice or make the law his own. He shuns marriage, for he downgrades the sexual impulse, and he forms a platonic relationship – "chaste and masculine" – with the Duke of Guise. (Unlike the principal characters in *Bussy d'Ambois*, who were mostly drawn from life and history, Clermont is Chapman's invention.) When finally he does act and in a "fair duel" accomplishes the death of Montsurry, it is not before he gives and obtains forgiveness from his victim. Next he learns that the Duke of Guise has been murdered at the king's order. To join his lost patron and friend, Clermont commits suicide. None of this follows the usual pattern of the revenge tragedy, but Chapman's originality is purchased at a cost of theatrical force. In this play he prefigures Nietzsche's concept of a man of power whose goal is humane and spiritual, "Caesar with the soul of Christ". The work is again "epic", grave and nobly intentioned, but lacks vitality: Clermont is simply too good, too serious, too stoic and too contemplative to be credible.

Another overly virtuous hero is portrayed in *Chabot, Admiral of France* (c. 1613). It was not printed until 1639, several years after Chapman's death, and then with many alterations effected by James Shirley. The theme is still that of a conflict between a high-minded servant of the state and a king, this time an absolute despot. The idealistic Chabot perishes, but the monarch's triumph has been far too costly to the country and throne. The playwright's idea of the "good" man is also Senecan.

A fervent admirer of Jonson's *Sejanus*, Chapman wrote a *Caesar and Pompey* (*c.* 1613) in which a leading character is Cato, who joins in the chorus of voices scorning the world and who resolves his mortal problems by stoic suicide.

To some extent Chapman shares with Marlowe an acceptance of Machiavelli's pessimistic measure of man and the world, though he also believes in the possibility of transcendence, a sinner's triumphant redemption in the hereafter. His plays are too didactic; they are more intellectualized than Shakespeare's poetic dramas. Indeed they are so freighted with philosophy and allegory that they are at times too weighty for the stage; nor are they easy reading. The thought sometimes is so involuted and paradoxical that obscurity results. Essentially these are Morality Plays, with the characters representing absolute good and evil in opposition, though Byron for one contains both qualities in himself.

In 1612 Chapman returned to comedy with *The Widow's Tears*, an ironical version of the story of the Ephesian widow. Though imperfect, it bespeaks his force and the oddly contrasting trait of cynicism that was ineluctably lodged in his nature. Of all the lines he wrote, perhaps the most quoted is this: "Young men think old men are fools, but old men know young men are fools." A world-weary comment – worthy of Oscar Wilde – that reflects a many-faceted personality.

By now his masques were highly prized, considered to be second in accomplishment only to Ben Jonson's, as stated by Jonson himself. Unluckily, only one survives. His versatility is astonishing, also suggesting his strange complexity. Much more needs to be learned about his life. Did he study at Oxford, as is surmised? When he was young, had he served as a soldier in France and the Netherlands, as had Jonson? Was he a believer in the occult, convinced that he was in intimate and perpetual communion with the spirit of Homer, the long-ago spinner of epics? Did Chapman ever belong to the School of Night, a secret society of "Free Thinkers" of which Marlowe was rumoured to have been a member? Was there, in fact, any such group?

Overshadowed by Shakespeare and Jonson, Chapman is today known only to scholars. He may yet be broadly rediscovered. In the opinion of T.S. Eliot he was "potentially the greatest artist" of the Elizabethan theatre, his genius still unappreciated because in some ways flawed. Though inclined to classicism, he falls sort of adequate discipline.

John Marston (1576–1616) was a slightly later aspirant in the same tradition of satirical and realistic comedy, though he was also responsible for tragic works. Son of a prosperous Shropshire lawyer, he graduated from Oxford, where he studied for the bar and for the next six years was in partnership with his father but then took up the profession of literature, as so many lawyers of that day tended to do in both England and France. On his mother's side his descent was Italian. He, too, had a strange, complex personality, a broad streak of bitterness. He began his career with acerbic topical verse; a sustained attack on Ben Jonson spawned a feud with that formidable man of letters, who in retaliation scathingly portrayed Marston in several scripts. Later they became friendly once more.

About 1599 Marston took up writing plays and was briefly on Henslowe's payroll. Most likely the producer did not find it easy to get along with him. Marston's temper was too savage. Besides, he had independent means; he could be defiant, do as he wished. He soon became yet more involved in the quarrels that embroiled London's theatre, its actors and raffish writers. While with Henslowe he collaborated with Jonson, Chapman and others. During this period he evened the score against Jonson by depicting him as Crispinus, made to swallow a disgusting purgative and vomit his bombast, in *Satiromastix* (1601). Dekker took part in this. Both felt themselves maligned by Jonson's portrait of them in *The Poetaster* (earlier 1601), which has a Roman setting. In it the self-confident Jonson dons the mantle of Horace, the magisterial critic.

Subsequent to his tenure with Henslowe, Marston contributed scripts to the Paul's Boys, who having been banned in 1590 had returned in 1599 and regained popularity. Later he bought a share in another troupe, the Children of the Queen's Revels. It is hard to realize that plays like his, as well as most of the sophisticated farces and heavy philosophical dramas of Chapman and Jonson, were intended to be acted by mere youngsters. Even the controversial *Eastward Ho!*, which caused the imprisonment of its authors, was one put on by these youthful companies.

While with Henslowe, Marston at first used a pseudonym, "Mr Maxton". At other times he called himself "Kinsayder". Some of his satirical thrusts were directed against the Bishop of Exeter, Joseph Hall, who, while a student at Cambridge, had published a "semi-bawdy" novel – in Latin – as well as some plays. He left ten volumes of work, much of it controversial, spent five months in prison, suffered the desecration of his cathedral but afterwards became a favourite of King James. Marston's essays belittling Hall offended the Archbishop of Canterbury, who ordered them burned in 1599.

The serious play by Marston that evoked Jonson's ridicule in *The Poetaster* is *Antonio and Mellida* (probably acted in 1600), followed by *Antonio's Revenge*, the combined scripts totalling ten scenes (the whole printed in 1602). This two-part work, as already mentioned, seems to have had a direct influence on Shakespeare's *Hamlet*, as there are close similarities of plot and incident. More than likely both playwrights owed much in turn to Kyd's *The Spanish Tragedy* and possibly to his rumoured but vanished *Ur-Hamlet*, the phantom predecessor of the Bard's masterpiece. Though Marston's melodrama is far inferior, the atmosphere in both his play and Shakespeare's is much alike, a father's ghost appears to his son, a play-within-a-play is staged and the mood of the hero – melancholic, alienated – resembles the despairing Hamlet's. Others also see in this script certain likenesses to *Romeo and Juliet*, since the noble young lovers face the obstacle of belonging to states that are antagonistic, a circumstance that threatens to separate them for ever.

Venice and Genoa are at war. The Venetians score a victory in battle and demand the heads of Andrugio, Duke of Genoa, and his son Antonio. The young man is enamoured of Mellida, whose father is Piero, the Duke of Venice. To see her, Antonio disguises himself as an Amazon and penetrates Piero's court, where he persuades Mellida to elope with him. He is captured, but her tyrannical father, relenting, permits the marriage, ending the first part on a joyful note.

In *Antonio's Revenge* Piero's dark side is revealed: he has Andrugio killed, causes his daughter's honour to be besmirched to prevent the wedding and seeks to have Antonio murdered. He lays fervent suit to Andrugio's widow. Appalled and broken-hearted, the disgraced Melida dies. Prompted by the ghost of his father to seek revenge, Antonio feigns madness, plays the fool. For self-protection he chooses Machiavelli as his mentor.

Where *Antonio and Mellida* is romantic and aspires to be comic as well, *Antonio's Revenge* is solely, unremittingly melodramatic; indeed, it outdoes Seneca. The hideous action begins with Piero arranging a night-time murder. He enters with his arm smeared with blood, a gore-stained poniard dripping in his hand. To appease the insatiable ghost, the despot's little son is killed. The ghost seeks to prevent the marriage of his widow to his slayer. At the close Piero is tortured, his tongue plucked out and is stabbed by a group of men who hate him for his crimes. Antonio, refusing all reward, elects to withdraw to a religious order to free his heart of hatred – an interesting foreshadowing of what Marston himself was to do.

The verse, sometimes overblown and betraying strain, nevertheless offers initial proof of Marston's authentic poetic gift. It resounds with stinging invective, especially in comic scenes, which ill match the more serious and romantic moments, though the latter are meant to dominate. Piero's court is crowded with characters who typify human faults, after the fashion of Jonson's humours. One courtier, Feliche, serves as Marston's surrogate, heaping scorn on the parasites, venal sycophants and henchmen. Antonio is often frenzied, inarticulate, almost to the point of hysteria, balked in attempting to cope with this dishonest world.

A fault is the play's crude mixing of moods and genres, but it exhibits Marston's flair for theatrical effect, which he was later to carry to sensational heights. It restored the vogue for the "tragedy of blood", a fact of which Shakespeare took notice, being ever sensitive to what thrived at the box-office. John Webster, too, was to use the script as a model for his violent scenarios.

Marston himself calls his drama a "black visage" work. It summed up all the startling and sadistic devices conceived by his predecessors. Though full of bizarre and incredible turns of plot and characters, *Antonio and Mellida* and *Antonio's Revenge* won the applause of audiences, whose appetite for this sort of spectacularly cruel fare was still far from sated. The play's chief artistic virtue is the milieu for dark deeds that it creates, summed up as "tombs and gloom", which makes the implausible acceptable because it appeals to the half-hidden paranoiac fear latent in humankind.

In a somewhat milder vein is *Sophonisba* (c. 1605–6), about the ordeal of the daughter of Hasdrubal, a Carthaginian general. Urged by her lover, Masinissa, she takes poison and dies rather than submit to captivity. The subject had appealed to other Elizabethan and Jacobean playwrights, but Marston was admittedly inspired to choose a theme from Roman history by Jonson's recent *Sejanus*, which he hailed as an incomparable achievement. (As already said, he and

Jonson alternately quarrelled and then settled their differences. Some historians suspect that this "feuding" was a deliberate ploy to draw public attention to the theatre and the dramatists themselves, an early form of advertising not unlike that practised by the Athenian writers of comedies in the Golden Age.) Marston was also impressed by Chapman's treatment of the hero of *Bussy d'Ambois* that had scored so well with London's theatre-goers. Even so he worked largely along different lines, inventing a good many of the play's incidents, proving his decided originality. Little that is ironic tinctures his handling here: his principal characters are truly heroic, especially the doomed lady for whom the drama is named, and the script is moralistic, anti-Machiavellian, an exhortation to preserve virtue whatever the cost. That accorded with the Elizabethan's view of the Romans' stoic code of honour which may or may not have been factual.

In *Sophonisba* the dialogue is often rapid and compelling, and the plot is managed with a better theatrical instinct than was Chapman's. T.S. Eliot discerned in the play hints of "something behind, more real than any of his personages and their actions". This prompts Eliot to pronounce Marston a writer of genius. But not all the critics agree with that very generous appraisal, for even in this more controlled work there are outbreaks of hysteria and moments of obscurity, faults consistently found in Marston's prolific output. Some of these are the consequence of a habitual mishandling of syntax; he is too often careless with it and also handicapped by an insensitivity to English idiom. His innate disgust with the world, his imaginative preoccupation with the darker side of reality, also crops up, despite the more affirmative mood he seeks in his attempt to capture a classical scene as graphically as Jonson had done.

None of the grandeur of Rome is suggested in this passage in which Erichtho describes her visit to the ruins of a once "glorious" temple that had been dedicated to Jove:

> There the daw and crow,
> The ill voic'd Raven, and still chattering Pie
> Send out ungratefull sound, and loathsome filth,
> Where statues and Joves acts were vively lim'd
> Boyes with Blacke Coales draw the valid parts of nature,
> And leacherous actions of imaginde lust,
> Where tombes and beauteous urns of well dead
> Stoode in assured rest, the shepheard now
> Unloads his belly: Corruption most abhord
> Mingling it selfe with their renowned ashes
> Our selfe quakes at it.

Erichtho, incidentally, is a necrophiliac sorceress who is happiest when she can possess herself of a corpse and "gnaw the pale and o'ergrown nails from his dry hand". It is with a trick, by the priapic Erichtho, that Syphax is led to satisfy his lust. How much more debased could these characters be?

Marston's impulse to scourge manifests itself in his lighter works. The most accomplished of these, by far, is *The Malcontent* (1604), which establishes a character and pattern subsequently followed by Molière in *The Misanthrope* and Wycherly in *The Plain Dealer*. Like the bitter heroes of those world-famous (and better) black comedies, his hero is a harsh critic of the social order and humankind in general. The setting is Genoa, where Malevole, actually Giovanni Altofronto, is the legitimate duke. Deposed, he disguises himself and at court takes on the role of a taunter and railer at all around him. He observes much to disgust him in that corrupt realm. Mendoza, the usurper, is carrying on an adulterous affair with the wife of a foolish old man, Pietro. Malevole takes obvious pleasure in apprising the husband of this fact and in a fashion that torments him. He imputes exaggeratedly lascivious impulses to the lady, even suggesting that incest could be the outcome of her illicit liaison, which prods Pietro to exclaim that Malevole suffers from "Hydeous imagination". It is easy to assume that the acidulous-tongued Malevole is an aspect of Marston himself, his mouthpiece, as is Feliche in *Antonio and Mellida*. He is drawn to and angrily sickened by human evil; pruriently, his mind dwells upon it. *The Malcontent*, it has been said, is a "comedy" only in the sense that its conclusion, when Malevole reveals his true identity at a masque and dance, is not tragic. It has been likened, among contemporary works, to Shakespeare's sombre *Measure for Measure* and Malevole to such unhappy, brooding, cynical, carping figures as Iago and Bosola (the latter the somewhat confused villain of Webster's subsequent *Duchess of Malfi*). In fact, Shakespeare's company pirated *The Malcontent* and staged it at the Globe, and Webster wrote a special preamble for it. With his usual inconsistency, Marston later dedicated the printed version of the play to Jonson, who by now was dead. It does attain sounder balance than much of his other work, however, since it has some honest characters, including Malevole's imprisoned wife, whose love for him remains constant even when his successor, Mendoza, seeks to marry her for reasons of state. All the world is not depraved.

His other comedies include *Jack Drum's Entertainment* (*c.* 1600) – which assembles a cross-section of typical English fools and excoriates them – and *What You Will* (1607), which, though a lesser, blander piece than *The Malcontent*, has some exceedingly caustic comment, especially on the fatuity of trying to master facts and gather knowledge:

> I staggered, knew not which was the firmer part;
> But thought, quoted, read, observed and pried,
> Stuffed noting-books, and still my spaniel slept.
> At length he waked and yawned, and by yon sky,
> For aught I know he knew as much as I.

The feeling is one that many a student, cramming for an examination, can share.

The Dutch Courtesan (1605) conjoins comedy and melodrama. The four leading figures in it are not mere humours, as in so many of his other lighter scripts, but full-faced, vital. A reckless young Englishman, Freevill, is discarding his mistress, Franschesina, with the intention of marrying Beatrice. He persuades his cold-veined friend, Malheureux, to visit the "courtesan", and the hitherto frigid Malheureux becomes hopelessly infatuated with her. The former mistress wants him to avenge her and murder Freevill, but even this shocking request does not disenchant him with her. But, after the supposed killing, when Franchesina tries to shift the blame on to the guiltless Malheureux, he finally realizes his mad folly. (He has been arrested and sentenced to death, but it has been a charade and Freevill appears opportunely to clear him.) The wanton Franchesina is perhaps the best drawn of all Marston's characters: vindictive, treacherous, insolent; it suggests a misogynistic strain in the author who conceived her – though by contrast Freevill's betrothed, Beatrice, is most appealing. The sub-plot, concerning the stingy and conniving Mr and Mrs Mulligrub, who are out-cheated by the roguish and debauched Cocledemoy, a master of Billingsgate, is the most realistic in all of Marston's scripts.

All the lighter plays mock a heap of human follies, among them sentimental love, the fickleness of the feminine heart, affectation in manner and speech, foppery in attire, pedantry (as pictured by Marston, Jonson's worst fault). At the same time, in these works Marston portrays himself contrastingly by surrogates who are genial and witty observers of the multiple shortcomings of others, a stab at self-idealization.

Parasitaster, or The Faun (c. 1605) was apparently meant as a rejoinder to Jonson's *The Poetaster*, in which more than anywhere else Marston is taken off. Here the plot is of the slightest. The hypocrisy of court life, where men seek advancement by unctuous flattery, is the target. Lacking is the ingenious intrigue that propels *The Malcontent*. Though the play has its share of barbs, its tone is more uniform and mellow. For reasons of state, Hercules, Duke of Ferrara, decides that his son Tiberio should marry Dulcimel, whose father rules a neighbouring principality. Tiberio rejects the idea; whereupon Duke Hercules, a widower, declares that he himself will take her for his bride. He sends Tiberio to serve as his emissary to negotiate a marriage contract. In disguise, and calling himself Faunus, the duke follows his son to watch his plan unfold. Dulcimel, on meeting Tiberio, is instantly captivated. Resourceful and guileful, she soon wins his affections. As the Duke of Ferrara wished, the young couple are plighted and wed.

During a temporary truce with Jonson, Marston collaborated with that exuberant dramatist and Chapman on the ill-fated *Eastward Ho!* It is believed that he was responsible for the offensive lines, disparaging the Scots, that brought the royal wrath down on the three authors' heads. A character expresses the wish that 100,000 of the king's subjects betake themselves across the ocean to Virginia, "for we should find ten times more comfort of them there than we do here". The remark does not sound too abrasive today, but James I was outraged by it. Jonson and Chapman were incarcerated and in danger of having their ears and noses cut off, but Marston took flight and eluded arrest. (If the "playwrights' war" was serious, and not merely intended to gain the participants steady public attention, this may be why Jonson's anger against Marston returned in full force. He pilloried Marston mercilessly in later works, inviting further ripostes from his well-equipped foe. The hostilities have been a subject of investigation by amused scholars for four centuries.)

Much later, back in action, Marston began a tragedy, *The Insatiable Countess* (1613), with an unidentified collaborator. Left incomplete, the play was subsequently finished by another hand, possibly that of William Barsted. It is a study in bold colours of a nymphomaniac. In the interim, another comedy (*c.* 1608), now lost, got him into trouble with the authorities again. This time he had a brief spell in prison.

Suddenly, in 1616, he withdrew from activity in the theatre, presumably after having undergone a religious conversion. He wed the daughter of a clergyman, took holy orders and for the ensuing eighteen years devoted himself to the study of theology. He became a country vicar and served in a quiet parish.

When he died in 1634 he was interred in Temple Church and on his tombstone was inscribed this epitaph: "*Oblivioni Sacrum*".

Marston was far from alone in writing plays steeped in excessive cruelty: Kyd's *The Spanish Tragedy*, Marlowe's *Tamburlaine*, Shakespeare's *Titus Andronicus* and *King Lear* – in which Gloucester has his eyes gouged out – are a bare few of many examples of works of the period featuring torture. But Marston excelled at depicting the "horrible", the "blood-chilling", the "excruciating"; he was fortunate in writing when he did, when he was rewarded for giving vent to some psychological abnormality, a barely repressed compulsion to imagine and describe hideous deeds with enthusiasm, even a seeming delight. His inclination to this, as betrayed in his work, begs for analysis by critics whose approach is Freudian. His was a split personality, perverse; he was fascinated with yet repelled by sexuality. One strange twist was his turn to the Church after the Archbishop of Canterbury had earlier burned his satirical send-ups of Hall. That other side of his divided self enabled him to compose effective if mordant comedies and pleasing verse. In his last years he attained what was at least outward serenity. He is one of the most baffling and interesting figures in the history of his period, which had a full allotment of odd creative talents, and indeed in the long history of the theatre.

The literary output of Thomas Middleton (1580–1627) was as copious and versatile as that of any of his peers. His father, William, a London bricklayer and builder, had acquired a modest estate and the dignity of a coat-of-arms. Unfortunately he died when the son was only five. His widow took a second husband, with whom she was to be involved in lawsuits over control of the inheritance. Enrolling at Oxford cost Thomas his share of the legacy; consequently he entered the university without sufficient funds and had to cut short his stay there, getting no degree. He took away from it a love of poetry. Not yet twenty-one, penniless, he returned to London, wrote satires, pamphlets and pageants, until he drifted into the theatre.

He was well acquainted with the seamy side of his native city and, like Dekker, he found much of his material there. He worked first for Henslowe but afterwards for a variety of rival troupes – he was a freelance – the Admiral's Men, the Prince's Company, the Paul's Boys, the Queen's Revels Company, the King's Men, and he collaborated with other professional dramatists after the custom of the day. He married an actor's sister and became a father. He did not always sign his scripts and at other times put on only his initials; some of his plays were published as much as a century after his death, so it is hard to know how much he wrote, and exactly what – precisely which parts of the scripts are from his exceedingly able pen. Moreover, his personality was detached, which makes it even harder to identify his contributions. But that he is a writer of stature is beyond question.

His earliest known venture is *Caesar's Fall* (1602), composed with the assistance of Dekker, Webster and others. The script is now lost. *Blurt, Master Constable* (also 1602) is another collaborative effort, most of it probably Dekker's work. The next offering, *Randall, Earl of Chester*, a tragedy, has vanished. *The Honest Whore* (1604–5), described a few pages above, is an early romantic melodrama, once more an undertaking with Dekker, with whom for a time he was clearly a junior partner.

The Phoenix (1607), though it does not bear Middleton's name, is the first surviving play definitely thought to be wholly his own. Laid in Italy, its characters – despite his attempt to have them seen as citizens of Ferrara – are obviously English. The duke's son and heir puts on a disguise and sets out on a journey to observe what is amiss in the realm. He observes miscarriages of justice, the immorality of supposedly respectable women, the making of matches between young couples for monetary gain. The prose dialogue has impact – Middleton has the ear of a born writer for the stage – and poetic passages reveal a talent of a high order. The plot is tenuous, however, too episodic, resembling the looseness of a picaresque novel, and its romantic tone is, for its time, already old-fashioned. Dekker, Heywood, Chapman, Marston and their fellow practitioners had helped to establish the vogue for realistic satire, and this play falls short of fitting neatly into that genre.

In his following work, however, Middleton chooses London as his scene and scores an undoubted success. *A Trick to Catch the Old One* (c. 1605) is truly realistic. Furthermore its edged satire is implicit in the characters and their outrageous behaviour, rather than explicit in sententious dialogue. The plot is nicely contrived, a major advance in craftsmanship. Witgood, an extravagant and lecherous young man, is impoverished as a consequence of his reckless spending and the extortionate terms exacted for loans by his uncle, Lucre, a usurer. To extricate himself the unscrupulous young man has his mistress pose as a rich widow, then spreads the rumour that they are about to marry. His credit is quickly restored. Lucre and a rival moneylender, Master Hoard, descend on the young lady eager to exploit her financially. She plays her role of heiress so well that Master Hoard actually weds her, after paying all of Witgood's pile of debts in return for the young man's forfeiture of his claim to the wealthy "widow's" hand. At this juncture, however, the young courtesan reforms, promising to be a good wife to the old man. Witgood, too, vows to gamble and carouse no more; he chooses to wed Hoard's niece, and the usual reconciliation that ends a comedy takes place, the final tableau being a jovial banquet. Clearly Middleton sympathizes with the amoral Witgood who has handily cozened his avaricious elders. His trick is later to be taken up by Massinger for his not dissimilar farce, *A New Way to Pay Old Debts*. It has been said that at this point in his career Middleton has a sort of Chaucerian attitude: tolerant, slow to censure. He tends to accept life as he finds it, withholding judgement; indeed, avoiding any evaluation of what he beholds around him.

The chronology of Middleton's work is uncertain. The next few years brought forth several more realistic comedies. *Michaelmas Term* (1604–6) makes use of a familiar plot situation, reminiscent of Jonson's *Volpone* and borrowed later for Puccini's *opera buffa Gianni Schicchi* and J.M. Synge's Irish playlet *In the Shadow of the Glen*. A moneylender and wool-draper, Quomodo, pretends to be dead so that he can spy on his wife and son to learn how they will conduct themselves after his passing. The wife quickly marries Master Easy, a country gentleman who has been driven into bankruptcy by the grasping Quomodo. The son as rapidly loses title to the land of which Quomodo had cheated Master Easy, who now recovers it. The tricks, licit and illicit, employed by London tradesmen to enrich themselves and attain a higher social rank at the expense of the less shrewd country gentry are exposed in this lively farce. In particular the merchants are shown to be land-hungry and over-eager to earn enough to send their sons to the universities in order to consolidate their new social status, doubtless a payback for humiliations Middleton had suffered at the hands of such sons of the *nouveaux riches* at Oxford. The young scions of the gentry, on the other hand, are fellows of "great beards, but little wit; great breeches but no money". They belong to a declining species. Plays like this, also written by other London dramatists, about how commerce was carried on by members of a rising, ethically unscrupulous middle class, established a new genre and came to be called "city comedies".

Mad World, My Masters (1606) is a larger-scaled work, combining two plots. In each a courtesan, Gullman, plays a dynamic role. The first concerns Dick Follywit, another pranking young gallant, who enlists Gullman's help in tricking his miserly grandfather, Sir Bounteous Progress. He does this by masquerading under a variety of false names. Finally, as proof of his moral rehabilitation he presents the courtesan as his bride, only to learn that she is the woman who has been kept by his grandsire. The pinching shoe is now on the roguish hero's foot. In the second instance Gullman is hired by a suspicious husband to pose as a pious virgin who will persuade his wife to lead a less promiscuous life. But the courtesan, instead, lends a hand in bringing together the wanton wife and her lover. Later the remorseful lover urges his inamorata to remain faithful to her husband, who renounces his jealousy and shortsightedly hails the erstwhile lover as his best friend. The tone of this comedy is somewhat mixed, some of the scenes being written in a sober spirit. They reveal Middleton changing his direction towards more serious and realistic drama once again.

The Family of Love (1602, printed 1608, written with Dekker) satirizes a religious sect that purportedly sanctioned free love among its zealous converts. The script takes shots at "Puritan" hypocrisy. *Your Five Gallants* (1607) shows a sordid aspect of London in a series of Hogarthian scenes: the appetent, roistering young sons of the gentry explore its lowly taverns, brothels, dark alleys, drab lodgings. A vivid canvas, projected with many hues and shades. Middleton joined Dekker once more for *The Roaring Girl* (c. 1610), the swaggering yet generous and altruistic heroine of which – Moll Cutpurse – has already been met. A tragedy, *Viper and Her Brood*, no longer extant, was probably written also at this time.

A Chaste Maid in Cheapside (c. 1613) also has a somewhat darker mood. Middleton's comedy was becoming more sardonic, the barbs more hooked. Both in construction and in clarity of characterizations there is a further and impressive advance. Three plot-lines are kept going simultaneously, deftly intertwined and then tightly knotted. The chaste maid here is also named Moll. She is beloved by youthful Touchwood, but her parents, Mr and Mrs Yellowhammer, insist that she marry affluent Sir Walter Whorehound. That licentious gentleman has been maintaining a lengthy liaison with Mrs Allwit, by whom he has sired a brood of illegitimate children. The Yellowhammers also seek to wed their son, Tim, a Cambridge student but a dunce, to an alleged niece of Sir Walter, though the "niece" is in fact the "uncle's" Welsh mistress. Many of the minor characters are quickly and excellently realized. They include the cuckold, Mr Allwit, who smiles on his wife's infidelity as long as it provides an income; Tim's pedantic tutor, who debates inanely with his pupil; the childless Mr and Mrs Kix, each blaming the other for their failure to have offspring; and a covey of Puritan gossips, greedy and prurient. Especially amusing is the christening of Mrs Allwit's newest child, an occasion for a noisy feast at which gluttony prevails. In this script Middleton largely gives over prose in favour of an ably handled metrical dialogue that is easy to read and speak. The sum is a racy and ribald picture of the times, funny if at moments a bit harsh. Of all the comedies he conceived in this style, none is better than this.

Soon afterwards Middleton gave up realistic farce. Like Shakespeare he followed the lead of Fletcher into romantic drama. *No Wit, No Help Like a Woman's* (1613) is a transitional piece in which he seeks to blend realism and fantasy, but with little success. The plot could have come from Plautus, though Middleton places his story in London once again. The Twilight family is beset with troubles. The clever and resourceful heroine, disguised in male attire, wishes to help her husband retrieve his lost fortune and finally does so. In the two succeeding years he wrote three more scripts in the same mode: a gruesome and confused tragicomedy, *The Witch* (c. 1615); a comedy, *More Dissemblers Besides Women* (1615); and yet another comedy, *The Widow* (1616), the last of these in helpful collaboration with Fletcher, as well as with Ben Jonson. *More Dissemblers* and *The Widow* are set in Italy and are comprised of various intrigues, over-complex and far-fetched, one of them once again requiring a girl to dress herself in boy's clothing. All are a sharp and unfortunate departure from Middleton's more instinctive and congenial realistic style. In these Fletcher-like concoctions he employs formal verse, but it lacks his earlier sparkle. From *The Witch* was later lifted the figure of Hecate and two witches' songs and their dance for insertion into the corrupt Folio text of Shakespeare's *Macbeth*. This probably occurred at a time when Middleton was asked by the company to enliven that gloomy script for a revival; interpolations and cuts often came about that way.

The Witch is an unrewarding work. Rosamund, Queen of the Lombards, is forced by her bestial husband to offer him a toast by drinking from the skull of her slain father. She schemes to take revenge by a murder. Her husband is still alive at the end, however, as is Rosamund and her accomplice, though all of them escape only narrowly. Even less successful is *Anything for a Quiet Life*, a comedy written with Webster.

With *Women Beware Women* (*c.* 1620) Middleton hit his stride once more. It boasts one of his most complex and compelling portraits, the decadent widow Livia, who seeks to debauch younger women, even her niece Isabella, an intrigant whom she tricks into enjoying an incestuous affair with an uncle. She also acts as pander between the Duke of Florence and a young wife, Bianca Capello, who becomes the duke's mistress.

> Sin tastes at the first draught like wormwood water,
> But drunk again, 'tis nectar ever after.

The atmosphere created is one of a society that is thoroughly tainted, morally rotten. Scarcely a character is wholly decent. In secret, Bianca is wed to Leantio, poor but honest, a merchant's clerk. After his wife is seduced by the duke, Leantio, too, surrenders to the corrupt environment, and tolerates her infidelities for financial reward until an enmity develops between them. She arranges his murder.

Livia helps her uncle Hippolito overcome Isabella's resistance by persuading her that he is not really a blood relation, which is untrue. At the end, after a masque, there is a massacre. Bianca drinks poison and dies. Much in the story is based on a historic scandal, the liaison between Bianca Cappello and Francesco de' Medici (1545–87), Grand Duke of Florence, some decades earlier.

The play, Middleton's only attempt to write a tragicomedy by himself, shows that he had finally mastered the problems of construction. The comic scenes are tinged with irony. Though he tends to include over-long speeches, he contributes much lovely verse and quick-paced colloquial dialogue. The people who throng this work are presented with an exceptional psychological subtlety, much being revealed about them by nuance and sly innuendo. At the end a Cardinal passes judgement, but apart from this there is little didacticism; Middleton retains his accustomed detachment.

Fortunately Middleton now began to team up with William Rowley (*c.* 1585–1626), who proved to be the right partner. All that can be told about Rowley is that he was an actor. Taking up his pen, he had written a few plays of his own: *A Shoemaker a Gentleman* (*c.* 1606); *A New Wonder, a Woman Never Vexed* (*c.* 1610); *All's Lost by Lust* (*c.* 1619) – in this work he provided himself with a fat clownish part. He also turned out *A Match at Midnight* (*c.* 1622) and over the same decade a half-dozen works in tandem with John Day, George Wilkins, Philip Massinger, Thomas Dekker, John Ford, John Fletcher, John Webster – proof that he was most industrious. Joining forces with Middleton, he helped bring out *The Old Law* (*c.* 1616) – Massinger, too, participated in this – and then alone with Middleton *A Fair Quarrel* (1617) and *The World Tossed at Tennis* (1620).

In 1622 Middleton and Rowley offered *The Changeling*, which some consider the finest English tragedy aside from those of Shakespeare. Perhaps that appraisal is over-kind, for the script's premise is wildly improbable and the scope of the story limited: the protagonist has little of the dimensionality with which Shakespeare endows his people, nor is the development of the plot as logical and subsequently as plausible as is required of the action initiated by a fantastic idea or situation; nor is the language ever on a level with that by Shakespeare, Jonson, Webster or several others of this era. Beyond question, though, it is a drama of striking intensity and doubtless acts much better than it reads – it has been revived at intervals over the centuries.

The playwrights made use of a collection of tales by John Reynolds, *God's Revenge Against Murther* (1621), which had just been published. They altered the character of the chief villain, who in Reynolds's book is an attractive gallant. Middleton and Rowley make him instead DeFlores, a servant, physically ugly. He is obsessed with desire for Beatrice, the daughter of his noble, wealthy employer. To her he is hateful. DeFlores dominates the play. Many of the elements combined here had been used by the two authors – and many others – earlier.

The background of *The Changeling* is Alicante, Spain, a good place for a drama of strong passions. Beatrice is betrothed to Alonzo de Piracquo but has fallen in love with Alsemero, a rich young man who after a glimpse of her in church has become infatuated with her. Her father, Vermandero, insists on her marriage to Alonzo, who has an admirable disposition and seems an excellent match. To prevent it Beatrice offers a bribe to DeFlores to kill her fiancé and then to flee. She detests DeFlores and hopes to rid herself of two men she dislikes. He eagerly carries out her bidding, tricks and stabs to death the guileless Alonzo and hides the body. But first he cuts off Alonzo's finger with a ring on it and brings it to Beatrice as proof that he has fulfilled her commission. When Beatrice offers to pay him, however, he refuses the money, demanding instead that she give herself to him. Otherwise he will expose her participation in the crime. She argues with him, offers to increase her bribe but finally is forced to accede to his fierce demand.

Alonzo having disappeared, the newcomer, Alsemero, wins the approval of Beatrice's father, who consents to the young man's becoming the bridegroom. Alsemero, however, seeks assurance that his wife shall be a virgin until their wedding night. With the aid of an arcane treatise he prepares an elixir which will test beforehand the young woman's virtue. Beatrice comes upon the same book, learns what effect the elixir should have on her and when tested by it counterfeits token signs of her chastity.

She still fears that the nuptial night will betray her secret relationship with DeFlores. Accordingly she plays the "bed trick"; her lady-in-waiting, Diaphanta, is paid to substitute herself under cover of darkness for the first night's love-making. So taken is Diaphanta with Alsemero's kisses and embraces that the deceptive consummation of the marriage lasts too long, while the tormented Beatrice jealously and anxiously waits her turn. Fearful that she will now be in Diaphanta's clutches, as she is in DeFlores's, Beatrice expresses her wish for the lady-in-waiting's death, too. The resourceful and implacable DeFlores arranges it; he stabs to death the lady-in-waiting, then sets afire her bedroom in a wing of the palace to conceal the crime. Despite herself, Beatrice is drawn to admire this fiendish lover of hers, whom she has previously loathed. This twist is quite Strindbergian.

Some of the desperate scheming between DeFlores and Beatrice has been overheard by Jasperino, a companion of Alsemero, who warns his friend. At first refusing to believe in the possibility of treachery, Alsemero wavers. Watching Beatrice closely, his suspicions are strengthened. Finally the pair are exposed as double murderers. At the climax DeFlores kills his mistress and then himself.

The sub-plot is a shadow of the main story line, offering parallels to it but also some inversions: Isabella, the young wife of a jealous, elderly doctor, is besought by two reckless gallants, both of whom masquerade as madmen in the doctor's care. Each in this way seeks to have access to her without the other's knowledge. They are assisted discreetly by Lollio, the doctor's servant, who also harbours a wish to claim Isabella. But she proves to be indifferent to all three, and the doctor is assured that his jealousy is without warrant.

The choice of *The Changeling* as a title is somewhat of a puzzle. The term was originally applied to a substitute child, perhaps an elf, or else one whose uncertain parentage causes him to be the victim of unfortunate mistaken identity – a slave who is unaware of his royal lineage is a classic example. Later the word was sometimes applied to one who was "deformed" or "witless". In his adulterous quest Antonio, one of the gallants, seeking to be a patient of Isabella's husband, asserts that he is a "crazed changeling". But why would the authors adopt a title that refers only to a minor character in their sub-plot? Indeed, the complications in the sub-plot are only loosely connected to the major story.

Several critics have discerned in the play variations on the theme of psychological change: Beatrice, like Bianca in *Women Beware Women*, degenerates morally, immersed ever more in crime and depravity. Beatrice is never shown as anything but wicked, for she resolves on the hapless Alonzo's death as soon as she beholds his rival Alsemero. Her evil deeds are compounded as occasions pile up, but perverted impulses have always been in her. DeFlores has been inwardly demonic, an Iago, and he changes for the worse to meet Beatrice's illicit demands. He metamorphoses from her father's servant to her secret master. He is not wholly invulnerable, however. After the first murder he encounters Alonzo's ghost and is daunted by it.

The play does present two very dynamic "lovers", if Beatrice and DeFlores may be described that way, whose savage determination to have what they will possesses them and drives them with fury towards their hideous goals. Beatrice has been likened to Clytemnestra and Lady Macbeth, but she is more like Lavinia in Eugene O'Neill's *Mourning Becomes Electra*, a resolute figure in an intense melodrama but not a heroine of stature in a true tragedy. DeFlores, wanting her, will brook no obstacle:

> I know she hates me,
> Yet cannot choose but love her.
> No matter, if but to vex her, I'll haunt her still;
> Though I get nothing else, I'll have my will.

He is encouraged to think he will have her because he has shrewdly perceived a hidden sluttishness:

> ... if a woman
> Fly from one point, from him she makes a husband,
> She spreads and mounts then like arithmetic,
> One, ten, a hundred, a thousand, ten thousand,
> Proves in time sutler to an army royal.

He is always self-confident. Once having triumphed with her, he is satisfied and even ready to die:

> ... her honour's prize
> Was my reward. I thank life for nothing
> But that pleasure; it was so sweet to me
> That I have drunk up all, left none behind
> For any man to pledge me.

The consensus is that the sub-plot was provided by Rowley, who also wrote the opening and closing scenes, with Middleton contributing the rest. Rowley's influence is thought to have led Middleton to have curbed his usually more verbose style. The dialogue is remarkably taut in many passages. When DeFlores demands that Beatrice submit to him as reward of his stabbing Alonzo, she replies:

> Why, 'tis impossible thou canst be so wicked,
> Or shelter such a cunning cruelty,
> Or make his death the murderer of my honour.

He retorts:

> Push, you forget your selfe!
> A woman dipt in blood, and talk of modesty?

At other moments the language ascends to considerable heights. Alsemero, discovering the truth about the woman to whom he is married, exclaims about all the catastrophic "changes" that have occurred:

> What an opacous body had that moon
> That last chang'd on us? Here's beauty chang'd
> To ugly whoredom: here servant obedience
> To a master-sin, imperious murder:
> I, a suppos'd husband, chang'd embraces
> With wantonness . . .

He continues, speaking to Tomazo, brother of the slain Alonzo, who has challenged him to a duel:

> Your change is come too, from an ignorant wrath
> To knowing friendship. Are there any more on's?

Antonio and Franciscus, the two fatuous "changelings", acknowledge that they, too, have undergone alteration, from their state of foolishness to a condition of more wisdom (after narrowly escaping the gallows on a mistaken charge of having killed Alonzo). At the last Isabella's spouse, the doctor, vows to "change" into a better and more trusting husband.

After the great success of this play, Middleton and Rowley worked together on *The Spanish Gypsy* (1623). But most of this is apparently by Middleton. An exiled nobleman and his friends slip back into Spain by pretending to be gypsies. Love affairs, a rape and a marriage follow. Two of the three neatly intertwined plot strands come from novels by Cervantes, though Middleton effects the usual improvements, softening some of the harsher elements of the original. Yet a firm realism marks this work once more. The characterizations, especially that of anguished and noble Clara, are finely rendered.

Middleton's final, most controversial play is *A Game at Chess* (1624). Seizing on a current political event, the proposed marriage of Prince Charles, heir to the throne, to an Infanta of Spain, the dramatist's offering had an obvious if allegorical reference to the widely unpopular match. The fear was that it might lead to the conversion of the prince to Roman Catholicism, an outcome much dreaded by a very large majority of the populace. In particular, the Spanish ambassador, Gondemar, was ridiculed in Middleton's loosely shaped, complicated script in which the English are the white pieces and the Spanish the black on a huge chess-board. Gondemar, designated the Black Knight, boasts of how he has hoodwinked the English and how Spain plans to establish her hegemony everywhere. At the end, checkmated, he is stowed with his fellow black pieces into a bag. The actor impersonating him was made up to resemble Gondemar closely, even wearing one of the Spanish diplomat's discarded suits, and he was borne on stage on a litter much like that used by the ailing Gondemar. The spirit of the work is Aristophanic, and much of the poetry in it is top flight. Milton praised it. So excited was the public by this sally that seats were almost impossible to obtain; the play ran for nine successive nights – no other play of the period ever did as well, and the theatre profited immensely. The new Spanish ambassador, Gondemar's successor, entered an official protest, and the king had the play shut down. After a time the theatre was allowed to reopen, but further presentations of *A Game at Chess* were forbidden. It was then printed; three editions quickly sold out. They did not bear the names of the author, printer or publisher.

Some years earlier, in 1620, Middleton had been named City Chronologer, a reward for his having designed many of the annual pageants staged for celebrations in honour of the Lord Mayors of London. He retained this post until his death, when he was succeeded in it by the acrimonious Ben Jonson, who for some reason held his predecessor in low regard, dismissing him as "a base fellow". This may have been because apparently Middleton had no aesthetic pretensions; he did not, like Jonson, propound a new *Poetica*. With detachment and tolerance he evokes the Elizabethan and Jacobean world; he finds it materialistic and depraved, yet seldom if ever utters any denunciation. His range – encompassing farce, comedy, fantasy, tragicomedy, psychological melodrama – is broad. Slowly but surely he perfected his craft, observing the conventions prevalent in his day. He mastered a realistic, racy dialogue and at times was capable of lyrical utterance. His characters, though not acutely memorable, are often delightful – as in the instance of Witgood – or horrifying – as are Livia and DeFlores. He looms as a considerable figure in his period, and some of his best work, as evidenced by occasional successful twentieth-century revivals, seems bound to survive.

Henslowe's payroll also lists the name of Philip Massinger (1583–1640). His father had held a "confidential post" with the Earls of Pembroke. The son attended Oxford for three or four years but broke off his studies after his father's death and also left university without obtaining a degree. He was attracted to "poetry and romances" and the close study of Latin verse and prosody at the expense of formal logic. He became adept at reading French and Spanish, which stood him in good stead. Presumably he went directly to London to establish a career, but little is heard of him for his initial half-dozen years there. At next report he was in debtor's prison, along with an actor, Nat Field, and another hack writer on Henslowe's roster, Daborne. All three were begging for an advance payment, five pounds, to set them free. Henslowe obliged them.

Soon after, in 1613, Massinger was collaborating with Fletcher – and sometimes with Field too – on *The Knight of Malta*. At other times he worked with Field only, as on *The Fatal Dowry*. Finally he gained enough confidence to write by himself: *The Duke of Milan* (*c.* 1621–2). From then on he turned out scripts independently or joined forces with Fletcher, Field or Dekker. His output was large, his craftsmanship solid, his bent conservative. He exuded a moral earnestness. It is said that he was somewhat out of touch with his times; he belongs to the Jacobean and early Stuart theatre, though some of his spirit is strongly Elizabethan and that might be why he did not profit much from his labours, having to endure penury on more than one occasion.

A thorough professional, he composed all forms of stage-pieces: realistic and romantic comedies, tragicomedies (especially of the sort Beaumont and Fletcher had made the vogue), tragedies. Though his comedies are only four in number, he is most esteemed for them and especially for *A New Way to Pay Old Debts* (*c.* 1621–2), the core situation of which is borrowed from Middleton's *A Trick to Catch the Old One*. Massinger's treatment of it is much more serious, so that the work verges on drama. The prodigal young Wellborn, financially ruined by his avaricious uncle, announces his impending marriage to a wealthy widow, Lady Allworth, which greatly enhances his prospects and credit. He had once helped her late husband out of a similar dilemma. In Massinger's version, however, the plucking is less sordid than in Middleton's, since the lady is actually of noble lineage, not a courtesan, and most of the people, though shown in a critical light, are of the gentry. The portraits express Massinger's indignation at their failings. He surrounds his hero and villain with a gallery of types – once more incarnations of humours in the Jonsonian sense – whose traits or occupations are revealed by the names he affixes to them: Greedy, Allworth, Overreach, Furnace (the cook), Marrall. This accords, of course, with a custom reaching back to the Greeks. The major role is that of Sir Giles Overreach, who at the end goes mad when his nefarious plans miscarry; the richness of this characterization is the chief cause of the play remaining so long in the active repertory. Overreach is by no means the simple, conventional malefactor. He is based on a contemporary personage, Sir Giles Mompesson, a minion of the unpopular Lord Buckingham. Shortly before, Mompesson had been impeached for his dishonest handling of his office as dispenser of taproom licences. Audiences recognized Sir Giles Overreach as the other, real Sir Giles and were bitterly amused by the picture. Similarly, some of Overreach's suborned helpers, such as Justice, Greedy, Marrall, Topwell and Froth, are stage copies of Mompesson's equally detested minor associates. In his use of humours and his venting of righteous anger Massinger resembles Ben Jonson. It is not established what kind of personal relationship the two men had, but clearly Massinger was closely acquainted with Jonson's new kind of socially conscious comedy and meant to strike out as did Jonson at the callous, mercenary class that was springing up in England. The play is also testimony to the author's sure competence at writing for the stage.

Sir Giles Overreach is remarkably bold, as well as ruthless. He sums up the qualities of his sort in any society. He does not care what it costs others, if he profits. He propounds:

> I would be worldly wise; for the other wisdom
> That does prescribe us a well-govern'd life,
> And to do right to others, as ourselves,
> I value not an atom.

Concerning his fellow conspirator, Justice Greedy, he declares:

> ... so he serve
> My purposes, let him hang, or damn, I care not;
> Friendship is but a word.

His methods, as described by an opponent, are these:

> He frights men out of their estates,
> And breaks through all law-nets, made to curb ill men,
> As they were cobwebs.

Similar in theme is *The City Madam* (c. 1619). Again social climbing is a target, especially those wives of London merchants whose husbands are knighted for having acquired money and who ape their betters in hopes of rising ever further in status. In this piece the haughty, purse-proud wife of Sir John Frugal, a highly successful and prosperous tradesman, aspires to the manner and *milieu* of the ladies at court. She rules her household despotically, tyrannizes her daughters and false, scheming brother-in-law, Sir Luke Frugal, a supposedly reformed rake. He, for a time, is able to obtain a hold on his brother's fortune. But Sir John and two friends, disguised as Indians from the colony of Virginia, return and persuade Sir Luke to hand over to them Lady Frugal and her daughters to be offered as sacrifices to the Devil, purportedly worshipped by the Indians. For this, Sir Luke is promised a gold mine in that distant New World. This is followed by a very dramatic "exposure" scene in which matters are finally set to rights, and a more real scale of values is restored.

Though the plot is certainly extravagant, the play's criticism of materialism and snobbery is delivered forcibly, and it also contains passages that presage the next century, that is, the Restoration and Congreve's witty comedy of manners. This is particularly true of an episode in which Lady Frugal recruits an astrologer, Stargaze, to instruct her free-spending daughters how to treat their suitors and also learn from him how to preserve her youth, for she is inordinately vain. Surprisingly, despite the patent artificiality of the story, the characters and their emotions are valid and breathing. The writing is very pointed and detailed. It has been aptly termed "photographic". The play furnishes a specific image of life in a stratum of London society, upper-middle class, unfortunately corrupted by its accumulating wealth; in brief, the pretentious but indecorous *nouveaux riches*, who provide a spectacle both amusing and saddening. The subject pertains to no single city or epoch but is universal, always repeated. Another point strongly made is that the money has been gained by methods that, though legal, are pitiless and rapacious. This play continued to draw audiences almost as long as *A New Way to Pay Old Debts*, well into the nineteenth century.

A somewhat different kind of comedy, *The Great Duke of Florence* (c. 1627), garners its plot from an old legend, already exploited by earlier writers. Once again Massinger gives his material a fresh turn. An Italian grand duke has heard of the beauty of Livia and sends an envoy to determine the verity of the report; the envoy falls passionately in love with her. Massinger's new twist to this hoary situation – it had long ago contributed the theme of *Tristan and Iseult*, as well as Machiavelli's *Mandragola*, and was later to furnish Henry W. Longfellow the subject of his *Courtship of Miles Standish* – is that the young lady has already lost her heart to the duke's nephew and heir, Giovanni. At the end all is happily resolved, with the duke granting pardon and blessings to all. The poetry in this work is Massinger's best, and the delineation of the young lovers – Livia is charming and Giovanni modest – most engaging. As always, the play's construction is exemplary; no one excels Massinger when it comes to that.

A Very Woman is a romantic comedy (licensed 1634), turned out with the help of John Fletcher, though Massinger is believed to have written most of it. It is probably a revision of a work begun with Fletcher many years earlier. Staged, then preserved in print, it earned high praise from Swinburne in his critical studies of the period's dramas. Its subject is the fickleness of women, and it skirts tragedy at moments – one of the heroine's lovers is nearly killed in a duel. But all concludes well, and the character of the inebriated, amorous Borachia, most probably contributed or heightened by Fletcher, is one of the most amusing in any work to which is attached Massinger's name. A scene in a slave-market also bears Fletcher's lighter touch. The play has gaiety, dignity, restraint and is capably structured. T.S. Eliot, seconding Swinburne's approval of it, has called it "surpassingly well plotted".

Massinger was far more prolific in supplying tragedies to his employers. But most of these have serious faults and were not to survive too long as viable works. Most were fashioned with Fletcher, some with Dekker, as previously mentioned; a few alone. His collaborations include *The Virgin Martyr* (before 1620, with Dekker), a dramatic recitation of the torments suffered in Rome by early Christians during the reign of Diocletian; *Thierry and Theodoret* (with Fletcher; printed 1623), which has a sixth-century setting; *The False One*, telling of the affair between a ruthless Julius Caesar and Egypt's wily Cleopatra and voicing daring questions on political themes. Massinger was often fearless in expressing himself. The justification for the invader's ambitions is challenged:

> Why does this conquering Caesar
> Labour through the world's deep seas of toils and troubles,

> Dangers and desperate hopes? To repent afterwards?
> Why does he slaughter thousands in a battle
> And whip his country with the sword? To cry for't?

On his own, Massinger composed *The Renegado* (c. 1624), the hero, a Jesuit, sympathetically portrayed – a bold gesture at a time of strong anti-papal public emotion. This rather wild adventure story, set in Tunis, testifies not only to Massinger's own strong religious sentiments but also his fondness for exotic *milieux*; most of his serious plays draw on antiquity and have a Roman background. In *The Unnatural Combat* (1623) the Cenci tale, which was later to challenge the talents of Shelley, provides a subject: the actual event, an act of incest followed by parricide, had occurred in Italy only two decades earlier. Here the tyrannical father is called Malefort. He hates his son and nurtures an unnatural desire for his daughter and is murdered by his children.

In the same year (1623) Massinger wrote *The Bondman*, a work long popular, again partly because it contains ample opportunities for acting display in its leading part – in Restoration times it was much admired by that indefatigable theatre-goer Samuel Pepys, who notes in his diary that he went to view it time and again. The play, too, is about Rome (or, more precisely Sicily) and a fictional uprising of slaves. Massinger dares again to frame protest against tyranny through them:

> Equal Nature fashion'd us
> All in one mould . . .

The time is that of the Carthaginian Wars. The nobly born heroine, Cleora, gives her love to the Theban Pisander, believing him to be a slave, the rebel leader "Marullo", though he is counterfeiting that role. This "tragedy" ends with a double marriage and the slaves' emancipation.

A lost play, *The King and the Subject*, offended Charles I, who found a passage "too insolent" and ordered it changed. *The Duke of Milan* (printed 1623) is grisly and sensational, qualities that recur in most of Massinger's melodramas. Yet in it Duke Sforza pleads with the conquering emperor to have his troops avoid rape and slaughter and to show mercy. *The Parliament of Love* (1624) also belongs to this phase of Massinger's career. He termed it "the most perfect birth of my Minerva", an allusion to the goddess of Wisdom expelling children from her brow.

His favourite, though, was *The Roman Actor* (1626). The hero, Paris, a stage performer during the reign of Titus Domitian, is suspected by the emperor of being the wanton empress's lover. Domitian himself intends to punish and kill him. The story is based on the chronicles of Suetonius and Dio Cassius. In one eloquent flight Paris pleads for recognition and understanding of the actor's art. He asserts that the theatre can offer vivid instruction in morals because it teaches virtue by stirring the emotions of onlookers and shows the dangers and consequences of evil actions and the glory of right doing. Accordingly an actor may be more effective than a philosopher who offers only "cold precepts" that are likely to be unreal. Massinger's innate didacticism is well exposed here. In the last act the Empress murders Domitian.

Believe as You List (1631) troubled the authorities and had to undergo substantial change. It included an outright diatribe against Spanish imperialism. Massinger was told to shift the time of the story to antiquity. He chose Ancient Rome again. Flaminius says all must be prepared to sacrifice themselves and others to serve the state and its needs; he is opposed by the liberal Marcellus, who assists unhappy Prince Antiochus, destined victim of Flaminius.

Maid of Honour (printed 1632), a tragicomedy, takes an anecdote, "Camiola and Roland", from Painter's *Palace of Pleasure*. Massinger enlarges the tale, shapes it into an excellent play. In Italy, Camiola has four suitors. Her choice, the inconstant Bertoldo, deserts her when he has a chance of a match with the Duchess of Siena; in doing so he violates the code of the Knights of Malta. After a trial, at which the kind-hearted Camiola pleads his cause, Bertoldo repents. At the end she does not marry him but enters a religious order. Charming and self-reliant, she is cherished by many as Massinger's most attractive heroine.

His major shortcomings in composing tragedies are that his verse is lucid but too generalized and markedly derivative, echoing Shakespeare, Jonson, Webster all too often and suffering severely in comparison; while his characters' minds are not truly plumbed and their motives lack consistency. The tragedies are described as "prosaic". The fault of style in these serious works, their dialogue losing the vigour of his comic writing, has been studied in a disdainful essay by Henry Hallam among others – though earlier Hallam praised Massinger as second only to Shakespeare. Charles Lamb extolled the purity of his English, which he deemed superior to that of any of Massinger's contemporaries. His ethical concerns so enthralled Charles James Fox that, after encountering the plays, for several days the great Whig orator and statesman could speak of nothing else.

After John Fletcher's death Massinger became his successor as playwright-in-chief to the King's Company, a modest-salaried post that he held henceforth. Altogether he was involved in the preparation of a pile of fifty-five works for the stage, of which fifteen are solely by him. Eighteen were penned with the help of Fletcher, the rest along with Dekker, Field and many others. A century later Kemble and Kean liked the outsized roles in his nimble comedies. Some fifteen of his plays are still available; the manuscripts of eight others perished in 1750, when an ignorant domestic used them to light her oven. Dying a pauper, as he lived too much of the time, the luckless dramatist was buried "without memorial" at St Saviour's Church, Southwark. The parish register reads only: "Philip Massinger, a stranger". Much later an inscription was added on a window.

Of the many collaborations in this busy stage epoch, none was as influential as the comparatively brief one between Francis Beaumont and John Fletcher, whose names have already been repeatedly mentioned here. Together they created a new genre of English drama, the romantic tragicomedy.

They wrote chiefly for a new audience that was gathering size during the Stuart reign and was comprised of courtiers and the wealthy – the highly educated weary of gory melodramas – and put on in the small, private theatres now abounding in London, competing with the older, more popular Elizabethan playhouses that still flourished. Both Beaumont and Fletcher differed from most of their literary rivals in having aristocratic antecedents. Beaumont, properly addressed, was "Sir Francis". They were, as Dryden later called them, "gentlemen". Their plays have a special elegance. They wrote for their own kind, perhaps not deliberately but instinctively, and permitted themselves to express innate gifts for sophisticated gaiety, fantasy and wit. Theirs was a coterie theatre. They eschewed realism: the hearty English peasantry, with its Morris dances and Whitsun ale, is not represented here.

Francis Beaumont (1584–1616) was the younger and higher born of the pair, but the shorter-lived, cut down by the plague in his early thirties. His father was Justice of the Court of Common Pleas; his mother was connected with several families of imposing lineage, very blue-blooded. In the footsteps of two elder brothers, Francis attended Oxford but was another dramatist who left without a degree. In 1600 he was admitted to the Inner Temple, which had become a centre for young men interested and active in literature. He was also attracted to the bohemian cenacle of poets and hack writers dominated by the dogmatic Ben Jonson. About this time he met John Fletcher (1579–1625), five years his senior, son of the Bishop of London. The father, now dead, had held many other high posts as scholar and churchman – President of Corpus Christi College, Bishop of Bristol, Mercator, Lord High Almoner to Elizabeth – but his second marriage displeased the queen, and he died penniless and much in debt. Accordingly, his son John, one of eight children, had his own way to make. Of his childhood and early life nothing at all is recorded, but some time after 1600 he met the youthful Beaumont and became his very close friend. They set up housekeeping together. Both were ambitious, needy. It is said that they wore each other's clothes and sat on the same bench when they wrote. It seems to have been complete male bonding. There is talk of a serving wench, Joan, who was kissed by both of them.

Their portraits were painted about this time. This is the young Beaumont as described by Swinburne, after his study of an engraving subsequently derived from the image on canvas: "handsome and significant in feature and expression alike, with clear thoughtful eyes, full arched brows and strong aquiline nose with a little cleft at the tip; a grave and beautiful mouth . . . with full and finely curved lips; the form of face a long pure oval, and the imperial head, with its fair large front and clustering hair, set firm and carried high with an aspect at once of quiet command and kingly observation". A reproduction shows that the picture could pass for that of a Stuart cavalier or courtier by Van Dyck.

Fletcher, also captured by an unknown artist and now in London's National Portrait Gallery, is plain-looking, far less presentable. He has large eyes, a strong nose, a wispy moustache, a pointed, bearded chin. The glance is contemplative, somewhat melancholy, not to be expected from a prolific writer of racy farces. Several other portraits survive and indicate varied aspects of a complex man.

Though poor at the start, Fletcher had rich and well-placed relatives; his uncle had been in the diplomatic service. Two of his cousins were well-regarded poets. Since both Fletcher and Beaumont had access to the right people, when they wished it, and knew the temper and tastes of the aristocracy, Dryden was later to say of them with approbation: "They understood and initiated the conversation of gentlemen, whose wild debauchery and quickness of wit in repartees no poet can ever paint as they have done."

Jonson's initial sway over them is attested by verses they composed to preface the printed text of his *Volpone* (1607). Beaumont was twenty-three, Fletcher twenty-eight. Both were turning out verse of various kinds. When they began writing plays is uncertain. The conventional image of them working on the same scripts and having a clear division of labour is now being discarded. It was long thought that Beaumont contributed plots and Fletcher the poetry and dialogue, their talents complementary. On the evidence, Lawrence B. Wallis argues in his *Beaumont, Fletcher and Company*, they usually undertook individual projects and their gifts overlapped. Each was fully capable of writing a play on his own, and for the most part that happened. Or so it is now believed, though not by all. Living together, they most likely bandied ideas, offered each other criticism and suggestions. Formerly, fifteen scripts bore their names as co-authors. The list is being steadily shortened. Beaumont had less need of money than did Fletcher. It is conjectured that he spent less time at his writing. In instances where they shared the actual task, one or the other took on nearly all of it.

Scholars concur that the first play, *The Woman Hater* (produced by the St Paul's Boys some time before 1606), is a solo effort by Beaumont; if Fletcher added anything, his input is small. The script, too, shows Jonson's lingering influence, for many of its characters embody humours: each one a type, governed by an obsessive trait. The action occurs in Italy, but the people in it are obviously Londoners. Gondarino, a misogynist, is pursued by the mischievous Oriano, who pretends to dislike men. He retaliates by spreading slanders against her, bringing her to the verge of disgrace. Several plots are neatly plaited, and the subsidiary roles – those assigned to represent the figures hopelessly possessed by humours – include Lucio, an epicure, who even weds a courtesan in order to partake of a carefully tended succulent fish, and Lazarillo, who aspires to being a statesman, as well as a variety of merchants, courtiers, fops and political agents.

Much in *The Woman Hater* is parody, cleverly taking off other stage-works of the time, among them Shakespeare's, and this prepares for Beaumont's next and better piece, an outright venture into ridicule, *The Knight of the Burning Pestle* (c. 1607), which was to guarantee him lasting renown. A grocer and his wife (she has never before been in a theatre) attend a "drama". An actor is missing, and they persuade the company to accept their apprentice, Rafe, as a substitute. The youth has dreams of becoming a player. Thereafter they interfere with the performance and keep up a loud stream of comment and criticism and advice on the progress of the script, which relates the wanderings and adventures of a latter-day Don Quixote – the role assumed by the apprentice – who travels about on errands of mercy and chivalry and spouts pseudo-poetic monologues. The free-spending grocer and his talkative, gullible spouse are vigorously drawn and have salty lines. The targets of the author's mirth are many and diverse. One passage sounds cruelly like a high-flown speech in Kyd's *The Spanish Tragedy*, and Dekker's plain, sturdy citizens are lauded – tongue in cheek – in the rough verbal style belonging to Simon Eyre in *The Shoemaker's Holiday*. Shakespeare is not spared: the smitten Rafe's pretence that he is dead, with his beloved Luce preparing to commit suicide and join him in the coffin, is a comic reversal of the macabre situation in *Romeo and Juliet*. The self-confident apprentice could be a character in a Heywood play and resembles all the amateurs taking part in theatricals in those days (similarly pilloried in *A Midsummer Night's Dream* in the scene where the rustics hold a rehearsal in a wooded grove). Basically, Beaumont was poking fun at the fare of the public playhouses and the tastes of the lower-class Londoners who flocked to them, as well as the playwrights who truckled to that segment of ticket-buyers. Besides, this time the play was being done by the Blackfriars' Boys, and no one in the cast was more than fourteen years old; this probably heightened the comic effect, with children pointing out by exaggeration the follies of their elders. Most of the pieces put on by the boys' companies were satirical. This one in particular stresses the materialistic values held by the characters, who are not mere puppets but surprisingly vital.

Though Beaumont's work somewhat inexplicably failed at its first performance, it was to catch on at a revival two decades later and was subsequently to be hailed as "possibly the first and greatest dramatic burlesque". Whether it deserves that much praise readers should decide for themselves. For some it is a one-joke play carried on much too long. It must have amused the spectators, however, to hear Humphrey exclaim:

> . . . although it cost me
> More than I'll speak of now, for love hath tossed me
> In furious blanket like a tennis ball,
> And now I rise aloft, and now I fall.

He exits with this couplet:

> Good night, twenty good nights, and twenty more,
> And twenty more good nights – that makes three score!

The grief-stricken Luce, mourning her seemingly dead lover, Jasper, makes ready to kill herself in this wise:

> First I will sing thy dirge,
> Then kiss thy pale lips, and then die myself,
> And fill one coffin and one grave together.

A sample of Fletcher's unaided early work is *The Woman's Prize or the Tamer Tamed* (acted 1608–9). This ingeniously and rather broadly conceived farce takes up somewhat after where Shakespeare's *Taming of the Shrew* leaves off. Petruchio, the fierce wife-subduer, has married again, and his doughty second wife Maria is determined, for the honour of her sex, to conquer him in turn, reducing him to a more compatible pliability. He is shut out of the bridal chamber on their wedding-night. With the help of a troop of wives, akin to the resolute women in *Lysistrata*, Maria withstands her spouse's prolonged siege. He offers compromises, but she still defies him. All his contrivances are deftly stood off. At last he resorts to a drastic, mean trick: pretending to be dead, he has himself borne to her in a coffin. Her tears are not prompted by his passing but for the numerous follies in which he spent his days. He has to own himself bested by her, and she then vows to behave more affectionately. By no means delicately written, the farce was much liked: given at Court, it was preferred over the Shakespearian work to which it is a sequel. In Restoration times it was loudly applauded, and Pepys admired it greatly.

Another unassisted Fletcher piece, *The Faithful Shepherdess* (1608), presents an opposite facet of his considerable and versatile talent. He was attempting a sophisticated pastoral, modelled on Guarini's *Pastor Fido*. Though there had been earlier English works in this genre, Fletcher brought new elements to it and heightened and refined other aspects. Even his élite audience was not quite prepared for what he was attempting, and initially his offering failed. It lacks action. Quick exchanges of dialogue are replaced by long recitations mostly in rhymed couplets, which adds to the pervading artificiality of this sort of stage-piece. The plot is complicated, running the gamut from idealistic sentiment to brutish passion, all enacted by classical shepherds and shepherdesses taken not from life but from the eclogues of Theocritus and Virgil.

The constant affection of Amoret for her Periget is contrasted to the feelings of Clorin, who having lost her love is tempted by a satyr, and to the lust of the wanton Amaryllis and the Sensual Shepherd. This allows many varieties of love to be exemplified. Besides the Satyr, a river-god has a role – and all this, of course, could be depicted with lavish scenery and stage effects in the private theatres where the price of tickets had steadily risen. (Harbage guesses that such playhouses held about four hundred spectators and gave only one performance weekly.)

Fletcher had his work published and sought and got approving verse prefaces from his friends Jonson and Chapman, who saw the merits of his play. He defended himself, explaining that he had gone back to the subject-matter of ancient poets. He defined his form as "pastoral tragicomedy" and explained what he intended by that: "A tragicomedy is not so called in respect to mirth and comedy, but in respect it wants deaths which is enough to make it no tragedy, yet brings some near it which is enough to make it no comedy." (Here, as elsewhere, "wants" means "lacks" not "wishes".)

From the Middle Ages on, the English drama had freely alternated and juxtaposed tragic and comic episodes. Instead of that practice, which had long ago been denounced by Sir Philip Sidney as resulting in a "mongrel" art-form, Fletcher was proposing a subtler fusion of the disparate genres into a quite new one, in which the hero and heroine face serious peril, yet at the end attain a happy resolution of their dilemma. Even the villain may be spared punishment. In the story itself, many of the devices of comedy – in particular, mistaken identity – are adapted to more sober purposes (as seen in Shakespeare's *Cymbeline*, which shows much of the influence of the vogue soon created by Beaumont and Fletcher's fantastic and escapist tragicomedies). In addition, much symbolism as well as touches of the mythical and supernatural – charms, magic herbs, spells – are introduced. Unfortunately Fletcher's audience was not yet ready for him.

He tried again, this time working with Beaumont. The result was much more successful. *Philaster, or Love Lies a-Bleeding* (1608–10) won immediate acclaim and was frequently restaged. Here the talents of the collaborators seemed to complement each other perfectly. Fletcher, decidedly the better poet, flaunts a unique style, so that now it is seldom difficult to discern his share. His writing is fluent, "easy". As noted earlier, when his likely participation in the script for *Henry VIII* was scrutinized, he favours eleven syllables and feminine (that is, unaccented) endings to his lines. He is fond of alliteration and apposition and frequently repeats words and phrases for sharper emphasis. He excels at capturing elegiac moods, and his lyrics sound spontaneous. He proved to be fertile enough with plot ideas and to have a sure theatrical feel, working up a comic situation to a hilarious climax, and building and sustaining suspense to the last gasp. But his mind is superficial, and he is ready to sacrifice plausibility to move and hold his audience.

Beaumont, a more masculine spirit, is a keener observer of people and their foolish behaviour and in consequence has an outstanding gift for parody and caricature. He seems to have been the dominant partner in their joint task; he was apparently able to chasten and control Fletcher's emotional exuberance and verbosity. He is even the better at plot construction. Together they proved so extraordinarily well matched that their names have been for ever linked, like those of Sir William Gilbert and Sir Arthur Sullivan.

Philaster, or Love Lies a-Bleeding, produced by the King's Company on "divers occasions" in several theatres, is set in Sicily, whose king, a usurper, is worried about his hold on the sceptre and is about to marry his daughter, Arethusa, to a dissolute Spanish prince, Pharamond. She is secretly in love with the rightful heir to the Sicilian throne, Philaster, a young man of incomparable virtues: bravery, honesty, idealism. Though an open foe of the king, he is allowed to move about freely and attend the court because of his popularity with the Sicilian people. The lovers arrange to have as their go-between a handsome boy-page, Bellario. With the aid of a faithful lady-in-waiting, Galatea, Arethusa is able to expose the boastful Spaniard as a philanderer, but Megra, the wanton courtesan with whom he has been having an affair, silences the gossips and critics, including the indignant king, by in turn accusing Arethusa of having lustful relations with the eighteen-year-old page. Friends of Philaster carry the false charge to him, hoping that anger against the king's daughter will inspire him to lead their revolt against the usurper. Torn by doubts, for he is fond of his hitherto loyal page and deeply enamoured of the princess Arethusa, Philaster is immersed in despair. Finding Arethusa and Bellario together in the forest, where the princess has lost her way while on a hunt, the enraged Philaster draws his sword and wounds her. Bellario, already fled, is overtaken by Philaster and also wounded. When both are found by pursuing courtiers, Bellario seeks to take the blame for the attack on Arethusa, and Philaster is much impressed by the boy's proof of his fidelity. He will not accept the youth's self-sacrificial gesture, however, and confesses his guilt. The king condemns him to death. Arethusa secretly weds him, and her furious father threatens her with banishment. The Sicilian people, learning of Philaster's plight, rise to free him. Fearful, the king is forced to ask Philaster's aid in putting down the rebellion. When Philaster prevails over his followers and restores calm, the malicious Megra repeats her slander against Arethusa and the boy-page. Bellario is then ordered to be stripped and killed by the king but reveals "himself" to be Euphrasia, a girl in masculine garb, the daughter of one of Philaster's noble friends at the court. (Oddly, her father has never recognized her, believing her off on a prolonged pilgrimage.) In love with Philaster, she has donned male attire in order to be near him and serve him. The trick of a transvestite disguise had already been used by Shakespeare and Jonson and proved to be acceptable. At the end forgiveness is extended to all. Philaster regains his kingdom and weds Arethusa. Euphrasia (Bellario) announces that she will devote herself to a life of chastity.

This silly story is told so effectively, quick-moving scene after scene of conflict, obstacle and surprise, that even the most patronizing reader will find it hard to put it down. It is not clear how the play would fare on stage today. The lineaments of the Beaumont and Fletcher "tragicomedy" are fully revealed in it: the setting is exotic and remote when it is not fictitious; the plot based on a far-fetched premise, the consequences of which still engage interest by an appeal to the spectator's innate love of melodrama; the characters are simple in outline, easily identified as good, loyal, evil, disloyal; their emotions are rhetorically voiced; the denouement is a neat wrap-up, with everybody quite happy. The play reads effortlessly because the verse is limpid and often has an impact that, though far from Shakespearian, is achieved only by stage poetry at or near its best. For example, Bellario replies to Arethusa's query as to what are Philaster's feelings towards her:

> If it be love
> To forget all respect to his own friends
> With thinking of your face; if it be love
> To sit cross-armed and think away the day,
> Mingled with starts, crying your name as loud
> And hastily as men i' the streets do fire;
> If it be love to weep himself away
> When he but hears of any lady dead
> Or killed because it might have been your chance;
> If, when he goes to rest (which will not be),
> 'Twixt every prayer he says, to name you once,

> As others drop a bead, so to be in love,
> Then, madam, I dare swear he loves you.

There are vivid similes, as this in which we are told of the high regard in which the Sicilian folk hold their prince, Philaster:

> . . . the people . . . are all bent for him:
> And like a field of standing corn, that's moved
> With a stiff gale, their heads bow all one way.

Philaster, believing himself betrayed, uses powerful invective to assail the opposite sex:

> How you may take that little right I have
> To this poor kingdom: give it to your joy,
> For I have no joy in it. Some far place,
> Where never womankind durst set her foot
> For bursting with her poisons, must I seek
> And live to curse you.
> There dig a cave and preach to birds and beasts
> What woman is, and help to save them from you:
> How heaven is in your eyes, but in your hearts
> More hell than hell has; how your tongues, like scorpions,
> Both heal and poison; how our thoughts are woven
> With thousand changes in one subtle web,
> And worn so by you; how that foolish man
> That reads the story of a woman's face
> And dies believing it, is lost forever;
> How all the good you have is but a shadow,
> I'th' morning with you and at night behind you,
> Past and forgotten; how your vows are frosts,
> Fast for a night and with the next sun gone;
> How you are, being taken all together,
> A mere confusion and so dead a chaos
> That love cannot distinguish. These sad texts,
> Till my last hour, I am bound to utter of you.
> So, farewell all my woe, all my delight!

The Simon-pure Philaster has not been staunchly trustful of his beloved Arethusa. He has not penetrated "Bellario's" lengthy masquerade. His conduct is far from chivalrous. He does not seem to qualify as the ideal, intelligent hero. (Again, a boy actor would have been assigned the role of the page, abetting the stage illusion.)

Is there a political side to the play? Some scholars discern in it a clear warning to James I, who was proclaiming the divine right of kings, offending his English subjects. Sicily's ruler, thwarted, exclaims:

> Alas! What are we kings?
> Why do you gods place us above the rest,
> To be served, flattered, and adored till we
> Believe we held within our hands your thunder?
> And, when we have come to try the power we have,
> There's not a leaf shakes at our threat'nings.

For, says the rebel-leader, Dion, even a king's breath will not smell sweet "if once the lungs be but corrupted". This exchange:

> KING: What! Am I not your King:
>   If "ay", then am I not to be obeyed?
> DION: Yes, if you command things possible and honest.

In a final, summary couplet these chastening words are appended:

> Let princes learn
> By this to rule the passions of their blood,
> For what Heaven wills can never be withstood.

Of *Philaster*, Swinburne says it is "the loveliest though not the loftiest of tragic plays which we owe to the comrades or successors of Shakespeare". Interestingly, it has been pointed out that, despite the happy endings, the three principal characters of the play – Philaster, Arethusa, Bellario – all express a death-wish, an escape from life's frustrations. Whether this is another example of the fashionable melancholy of the decade, or a feeling held personally by the still youthful collaborators, cannot be determined.

Having found a formula for success they continued to use it, fabricating tragicomedy after tragicomedy along the same extravagant, fanciful lines. Box-offices, by turns the busiest in London, confirmed their practical wisdom. Other playwrights, seeing what would "go", emulated them.

Not too unlike *Philaster* is *A King and No King* (1611), which raises serious moral questions but evades facing them. Once again most of it is attributed to Beaumont, which may account for its rather darker tone, nearer to pure tragedy, though the artifice of its ending robs it of stature. Arbaces, erratic King of Iberia, defeats Tigranes, King of Armenia, in single combat but treats him with remarkable leniency and offers him the hand of Panthea, his sister. Tigranes is in love with an Iberian woman, Spaconia. (The strange names are typical of dramas of this genre.) When Tigranes meets Panthea, he shifts his affection to her. Arbaces, who has been long away from his native land on foreign campaigns, is also struck with the beauty of his sister, who is almost a stranger to him. He is seized by a guilty passion for her. Growing jealous of Tigranes, Arbaces has him imprisoned. Spaconia visits her faithless lover and forgives him, and his affection for her returns. The more difficult problem of Arbaces and Panthea, who shares his infatuation, is fortunately resolved when it is discovered that he is not her brother after all; indeed, he is only the son of the Lord Protector and not the legitimate heir to Iberia's sceptre. It is properly Panthea's, and she becomes queen. Hitherto proud of his regal status, he is now overjoyed to learn that he has no valid claim to it: *amor omnes vincit*. He and Panthea marry, and he becomes king once more. The story has a fairy-tale quality, save that hints of incest are not usually inserted into children's literature.

The tone of the play is ironic. The prose is strong, and the poetry is Beaumont's best. Arbaces, at the start boastful, is duly chastened. A comic figure, Bessus, a braggart captain in the Iberian army, provides moments of comic relief. He is rather like Parolles in *All's Well That Ends Well*. If Shakespeare owed a debt to *Philaster* for much of what he did with *Cymbeline*, it was repaid by what the young authors now took from him.

Dryden liked *A King and No King*; he was interested in the strange workings of Arbaces' mind, believing him a copy of Alexander the Great, and he deemed the play's climax highly effective. Of the drama overall he declared: "I find it moving when it is read." The passionate drive of the characters, he felt, offset weaknesses in the plot. Other critics objected that the theme was too sensational, a shrewd effort to attract notice. But the script is pale compared with the effusions of some of Beaumont's and Fletcher's contemporaries.

The same year (1611) brought *The Maid's Tragedy*. The heroic soldier Amintor must unwillingly abandon his love, Aspatia. She follows him in boy's attire. He is ordered by the unconscionable King of Rhodes to marry Evadne, who has secretly been the royal mistress and intends to go on being so. The wedding night is unconsummated, with Evadne denying herself to Amintor and revealing that she means to belong only to the king. For a time Amintor accepts the intolerable arrangement, telling no one of it. Eventually he admits the truth of the mock union to his close friend, the grim Melantius, who is Evadne's brother and whose fierce anger is aroused: in his view his family's honour has been blighted by the tyrant from Rhodes. He urges Evadne to kill the king; repentant with a complete change of heart, she is finally convinced and does so. Meanwhile Aspatia, dressed as a man and asserting that she is her brother, challenges the inconstant Amintor to a duel. He is reluctant to fight but feels that he must. He does not recognize his opponent as Aspatia. Deliberately running on to his sword, she is fatally wounded. Overcome by guilt and remorse at her deeds, Evadne comes gory from the place of her crime to where the duel has just occurred and begs Amintor to pardon her. He refuses. She commits suicide and Aspatia, too, dies. Confronted by the demise of the two women, the distraught Amintor stabs himself in a grand denouement.

The play includes poetic passages of high merit. Aspatia bewails Amintor's having deserted her:

> And the trees about me,
> Let them be dry and leafless; let the rocks
> Groan with continual surges; and behind me
> Make all a desolation. Look, look, wenches!

The Maid's Tragedy. Yes, but which maid?

Again, the question of how far a loyal subject must obey the edicts of a villainous king is brought up but hardly probed in earnest: on occasion Amintor does daringly upbraid his monarch, yet acknowledges his divine appointment and submits to all commands.

Though fifty-two plays are in the *Collected Works of Beaumont and Fletcher*, the fewest were written by them in tandem; most were individual ventures or arrived at with a long roster of other writers, which is why Professor Wallis, titling his book *Beaumont and Fletcher*, adds *and Company*. Among the pair's joint works is *Cupid's Revenge* (1607–12), which borrows a bit from Sidney's *Arcadia* and continues in the vein of Fletcher's *The Faithful Shepherdess*. *The Coxcomb* (c. 1609) was later revised by Massinger. Fletcher's contribution to it – about a fatuous gentleman who from an over-exercise of politeness and solicitude invites his being cuckolded by his wife and best friend – is the weakest component of the storyline. Beaumont offers to it the misadventures of Viola, who unwisely elopes late at night, meets her lover only to find that he is so drunk – he has just come from a party – that he does not recognize her. Soon afterwards, while she wanders alone in the darkness, she is robbed and kidnapped by a dishonest tinker, rescued by a man who has lecherous designs on her and sheltered by the milkmaids. Her lover finally discovers her, and all ends happily. This play successfully combines realism – the setting has a rare authenticity – and pleasant and exciting romance, and some of the verse has been termed "exquisite".

The Scornful Lady (c. 1610) was fashioned by the twain for the Queen's Revels and later rewritten by Fletcher for another company, the King's Men. It is mainly Fletcher's work. Mostly the lighter and gayer forms of comedy, the airborne trifles, are from his pen. This work is another attempt at the "comedy of manners", a genre that they did so much to help shape, and that was to flourish some decades later, after the Restoration. The young lady of the title role engages her lover in a gentle duel of wit and words and banishes him from her presence for twelve months because he has flagrantly kissed her in public. He finds himself another lady, whom he proposes to wed, and the alarmed and jealous first miss has to retract her fiat of exile and assent to a hasty marriage. The play is interesting not only for its clever plots and counterplots but as social history, its setting in English manor houses and its collection of minor characters, among them a riotous younger brother and his carousing friends who upset the decorum of the hero's house, despite the efforts of the steward to restrain them. Two other love-matches are patched up before the conclusion. The dialogue, part-verse and part-prose, is vivacious. Given an enthusiastic welcome, the piece was often restaged, both before and after the Civil War and Cromwell's interregnum. Fletcher, in particular, did much to place the witty gallant on the stage, to set comedy in the salons of high society and to match the rakish hero with a high-spirited heroine well able to hold her own in the repartee. Very often the two authors worked together with Massinger or in turn separately, forming a trio or a new pair. In selecting a different hand, Massinger seems to have been their favourite. Some of the many plays that resulted from this mutable combination of talents and practised skills are specified earlier in this chapter. Massinger, the youngest, was the junior partner.

Some time in 1613 Beaumont severed his personal connection with Fletcher and his ties to the bustling theatre world. He made a "profitable" marriage to Ursula Isley, of a suitable family, abandoned London and took up residence in her ancestral manor, Sundridge, in the quiet and green of rustic Kent. They had a daughter.

A portrait of him at about this time, and again from life, shows him somewhat altered. To quote once more from Charles Gayley: "It is of an older man, spade-bearded, of broader brow, higher cheek-bones, and face falling away toward the chin; of the same magnanimity and grace, but with eyes more almond-shaped and sensitive, and eloquent of illness. It is the likeness of Beaumont approaching the portals of death." (The dark hair and beard are luxuriant, indeed!)

Alas, the painting was prophetic. His span of retirement and hoped-for idleness and sport as a country squire was brief; in less than three years after his retreat from London he was dead, the epidemic overtaking him far from the suffering city. Though young, he was so esteemed that he was interred in Westminster Abbey, in a grave next to the even greater poet Edmund Spenser.

What of Fletcher, left to himself? He outlived Beaumont by eleven years. It was long assumed that he never married. But two disparate references – one off-handed in a play by Shadwell – suggest that he, too, may have taken a wife. The other item is in the parish of St Saviour's, Southwark, where is entered the union of "John Fletcher" and "Jone Herring"; the birth of a son is noted three months later. The date is 1612, near that of Beaumont's defection. But several other "John Fletchers" lived in the parish – that name was not an uncommon one – so it may not be the playwright. If he did embrace matrimony, there are signs that he was not happy in it. Several of his poems written after that date speak disparagingly of married life and its much-touted "joys".

Single or wedded, he continued to write; his financial circumstances compelled it. While pouring out comedies he had also been composing tragedies. *Bonduca* (*c.* 1609) is an ill-structured chronicle about Boadicea, Queen of the Britons, and her struggle with the Romans, culminating in her self-slaying and the death of her daughters. In many ways spectacular, and not without theatrical and literary rewards, it is not likely to appeal to readers and audiences today. *Valentinian* (1613–14) tells of a dissolute Roman emperor's attempt to overcome the virtue of Lucina, wife of loyal, honest legionnaire, Maximus. After Lucina dies, the drama – well plotted – devolves to a sensational revenge tragedy, with Maximus prevailing over his foes not by outright courage but by devious means. He contrives the tyrant's murder, having him poisoned by servants. He further intends to kill himself. At the last moment he changes his plan. He seizes the suddenly vacant imperial crown and consolidates his position by wedding Valentinian's widow. When he confesses to her how he cunningly brought about his predecessor's death, the Empress, in turn, kills him by pressing upon him a poisoned wreath. The drama abounds with lavish and sometimes wholly irrelevant stage effects, among them a pageant, a soldiers' dance, the descent of a singing child from fleecy clouds. Some of the characters are firmly delineated – especially the lascivious emperor and the resistant Lucina – and there is plenty of heroic declamation, as well as several charming songs. Fletcher excelled at writing them. This type of inflated "heroic drama" was also to persist into the Restoration decades.

After Beaumont's withdrawal Fletcher attached himself to the King's Men, Shakespeare's former company, and became their chief playwright. Here, as has been seen, he collaborated with all the most able, reliable freelancers. Besides with Massinger, he worked at one time or another with Rowley, Jonson, Chapman, Middleton, Nathan Field, Marston, and possibly Shakespeare and Webster. The list could be lengthened. To determine, or even to put forward an educated guess, as to who wrote what and when, is a scholar's game and likely to involve one in disputes of little worth. Any history of the Elizabethan and Jacobean theatres is partly invalidated by chaotic chronologies and attributions. (Many of the plays that Fletcher wrote with others have been discussed on pages devoted to those writers.)

Nor did he cease undertakings on his own. Generally his comedies were better received than were his serious offerings.

To this period belongs *The Tragedy of Sir Jon Olden Barnavelt* (1619), which concerned a topical event, the recent trial and execution of a Dutch statesman who had opposed Spanish rule there. The censors ordered many lines and episodes excised; its publication was suppressed and the original manuscript not found until two centuries elapsed.

With Massinger Fletcher composed *The False One*, *Thierry and Theodoret*, *The Bloody Brother* and *The Double Marriage*, all melodramas, some enlivened by comic episodes. From 1616 to 1624 he produced by himself seven more plays in a more congenial style, the plots tragicomedies full of good theatre, if often preposterous: *The Mad Lover*, *The Loyal Subject*, *The Humorous Lieutenant*, *The Island Princess*, *The Pilgrim*, *Women Pleased* and *A Wife for a Month*.

Sometimes lyrical and a master of delicate fantasy, sometimes Rabelaisian and downright bawdy, Fletcher pulls all the standard stops employed by the dramatists of his day: murders, magic potions, mortal intrigues, frustrated love affairs and surprises of every kind. The long scroll of joint and solo pieces to which his name is affixed also includes *A Very Woman*, *Four Plays in One* (a group of short works), *The Queen of Corinth*, *The Wild Goose Chase*, *Wit Without Money*, *Rule a Wife and Have a Wife*, *Monsieur Thomas*, *The Spanish Curate*, *The Elder Brother*, *The Little French Lawyer*, *The Chances*, *Henry VIII*, *The Two Noble Kinsmen*. In these Fletcher shows himself by turns to be deft, brilliant, opportunistic, shallow. A writer of genuine but incomplete gifts, he is too pragmatic and lacks artistic integrity; he is inconsistent, concedes too much in order to win the gasps, laughter and cheers of the spectators, who largely – if injudiciously – preferred him to Shakespeare. The two differ in this respect: in Shakespeare's plays the plots are often as fantastic, but the characters are plausible and even universal; in the type of "heroic drama" established by Fletcher – along with Beaumont – little is ultimately credible or significant; the plays are manufactured commodities that exist only to stir and entertain, and this they often did, but with them the theatre began a long-lasting flight from common sense and reality.

Fletcher's very high status – the highest – was perpetuated by critics into the next century, when Restoration audiences were especially fond of his comedies and their tart dialogue. Dryden, elevating him to the very pinnacle, declaring him superior even to the Bard, did so in part because the Beaumont and Fletcher dramas paid homage of the Aristotelian unities, hence were neo-classical in spirit; this accorded with the *Zeitgeist* of the early eighteenth century in England. In anticipating that widespread shift in taste – in redefining what was aesthetically proper – the two fledgeling authors revealed at the very start the deep influence on them of their mentor Ben Jonson, the prophetic, fervent advocate of neo-classicism. One of their prefatory verses to his evangelical *Volpone* hails him for having illustrated and impressed on them

> . . . the art which thou alone
> Hast taught our tongue, the rules of time, of place,
> And other rites, delivered with the grace
> Of comic style, which only is far more
> Than any English stage hath known before.

(Later Dryden had second thoughts and demoted Fletcher to third in rank after Shakespeare and Jonson.)

Extravagant encomiums to Fletcher, the survivor and far more prolific and versatile of the pair, came in heaps from many other sources.

When Fletcher, too, died in 1625 of the ever-winnowing plague he was buried most appropriately in the same Southwark church that was later to receive the body of his other major collaborator, Philip Massinger. (There is no mention of a wife or son.) Today the two dramatists are memorialized there on a painted window.

Approval of the plays and their "gentlemen" authors reached its apogee during the first half of the eighteenth century. The comedies had frequent revivals in London, Bath – the fashionable watering-hole where the nobility and gentry assembled each year to promenade, match-make and gamble for "the season" – and elsewhere in the provinces. "Heroic dramas", such as those composed by Beaumont, Fletcher and Massinger, were enacted by the leading tragedians of the day, including Garrick and William Powell. But most of these imitations were bombastic and hollow; they hardly measured up to the originals. In their insistence on happy endings they laid bare the *deus ex machina* and were not convincing or satisfying.

Beginning just before and after 1750 there was a sudden and sharp decline in acceptance of this repertoire. The number of revivals of the pair's scripts shrank abruptly; their London runs were abbreviated. The works, even Fletcher's comedies, were, overnight, considered out of date, obsolete. The plot complications were outworn. The bright dialogue no longer amused. Who can really say why? Some deemed the plays too salacious. By the century's end only three and then two of the light offerings – *The Woman's Prize, Chances, Rule a Wife and Have a Wife* – were on view anywhere. Shortly into the nineteenth century the reign of Beaumont, Fletcher and company was over. Seldom has there been such a reversal of fortune as in the reputations of these playwrights. They passed into public neglect and almost total oblivion.

Today their names are known but their works not read, except by a few scholars; Shakespeare is everywhere, but no commercial theatre would hazard putting on a play by Beaumont and Fletcher. On very rare occasions a script of theirs might be staged by a civic-subsidized society or group that brings back historical curiosities. From time to time students might stage *The Knight of the Burning Pestle*; a broad parody, filled with mockery, its sophomoric humour has an enduring appeal to collegiate youth.

George Bernard Shaw, whose attitude towards his predecessors was most often grudging, acknowledged the skill of Beaumont and Fletcher in writing for the stage but dismissed their works as having "no depth, no conviction, no religious or philosophical basis, no real power or seriousness". In saying they lack beliefs, however, he overlooks their repeated pleas on behalf of women and their natural rights and the need for equality, a cause in which they were truly in the vanguard, far ahead of their fellow playwrights. (*Vide* Kathleen McLuskie, *Renaissance Dramatists*.)

To the crowded Elizabethan period, too, belongs Cyril Tourneur. His birth date, uncertain, is placed variously between 1570 and 1580. Facts about his life are at best vague. Apparently he attended no university but had a grasp of Latin, which he perhaps acquired at a grammar school. He seems to have come from a respectable family and to have had the powerful Cecils as his patrons. At intervals he published verse and prose tracts of no marked distinction, and it is definite that much of his time and energy were dedicated to military service with the English forces battling the Spaniards in the Lowlands. Some time between 1604 and 1613 he returned to London from the Continent, and about 1606 wrote and saw performed *The Revenger's Tragedy* – however, neither the date nor his authorship can be authenticated. Printed about a year later, the play is unsigned and, though long ascribed to Tourneur, it is the subject of a prime debate; some experts are inclined to credit it to either Middleton or Webster; it contains passages in the style of both those writers.

After *The Revenger's Tragedy* came more poetry, especially verse commemorating the death of Sir Francis Vere, under whom Tourneur had served abroad as secretary. His only signed dramatic work, *The Atheist's Tragedy, or The Honest Man's Revenge*, was printed in 1611, after several performances in unspecified theatres, maybe provincial ones. The next year, 1612, a tragicomedy, *The Nobleman*, was twice acted at Court, but the manuscript has vanished. A few more prose brochures ensued, and in 1613 Tourneur got an assignment to write, along with several collaborators, one act of a proposed dramatization of Dekker's *The Bellman of London*. The indication is that, like many other out-of-muster soldiers, he was desperately impoverished and had to take on hack work. Soon afterwards, however, he was back in military life in the Low Countries. A puzzling incident, in 1617, was his arrest and imprisonment in England on a warrant issued by the Privy Council; the reason for it has not been determined. Within a month his important friend and benefactor Sir Edward Cecil signed his bond and won his release, and the unspecified charge was allowed to lapse or was dropped. There is no account of his activities for the next eight years, but an award to him of a small annual stipend by the Dutch government suggests that he was still in the army.

In 1625 he accompanied Sir Edward Cecil on the disastrous siege of Cadiz. Tourneur had been offered two well-paying posts as Secretary to the Council of War and Secretary to the Marshal's Court as his assignment in this foreign expedition. But at the last moment, before the fleet set sail for Spain, he lost the first of these appointments; a Court favourite replaced him. He stayed with Sir Edward even so. After the Spanish rebuff to the English fleet, the disabled flagship the *Royal Anna*, on board which was Tourneur, limped back to Ireland's port of Kinsale. Half of her crew was dead or wounded, and her riddled hold awash with the intruding sea. Tourneur, mortally ill, was borne ashore. His widow, left penniless, sought financial aid from the ungrateful government of Charles I, but her many pleas were turned down on a technicality.

Of his two extant plays, the earlier – *The Revenger's Tragedy* – is deemed the superior. Both are in the tradition established by Seneca and transposed to England by Kyd. In both, too, Tourneur vents a personal disgust with the world; his loathing is both exacerbating and intense. His two offerings have been described as "two of the most revolting plays in literature". Charles Lamb declared that reading *The Revenger's Tragedy* always caused his ears to burn. All in Tourneur's purview is corrupt, mean, vicious. The world festers with evil.

The setting of *The Revenger's Tragedy* is a sinister Italian court. In the very first scene Vindice (a variant spelling is Vendice) is discovered walking with a skull in his hand, an image like that presented by Hamlet, but it is the bony remains of his beloved, who has died after her ravishment by the old yet lecherous duke. The embittered young man, in some respects the typical "malcontent", is plotting how he shall retaliate on the cruel and degenerate ruling family. His brother Hippolito abets him. Their stratagems bring disaster on everybody, including themselves.

Vindice disguises himself as a pander, is recruited by Lussurioso, the duke's heir, and discovers the venery of his own mother, who for pelf is far too ready to sell her daughter's virtue. He forces the wicked old duke to kiss the gaping, bony mouth of the skull of the young woman for whose suicide he has been responsible. The brothers trick the sexually appetent Lussurioso into an encounter in which he is fatally stabbed. They coldly and craftily arrange the execution by hanging of the duchess's youngest son, guilty of murderous rape. They bring about the downfall of the duchess, the slain duke's second wife, who has been having an illicit affair with her late husband's bastard son, Spurio. Her two other sons by her first husband are trying to get rid of Lussurioso and each other, each hoping to succeed to the ducal throne; both are killed as the outcome of Vindice's scheming. Vindice and Hippolito are undone only when they commit the error of boasting of their successes – the new duke, a bereaved nobleman, Antonio, declares them dangerous to the state and orders their execution. Much of this proceeds in a mood that is grimly amusing. The turns in the action, the abrupt twists of plot, are often ironical; there is much dispensing of poetic justice. Men unwittingly condemn themselves to death by laying traps for others. The dialogue often deliberately invites an appreciative laughter. Errors are compounded, the wrong men murdered. A macabre, grotesque humour informs it all, and before the work is ended the audience is apt to share Vindice's zest for dispatching this huddle of venomous human beings and to lose, as does he, any sense of revulsion at the horrible killings which occur in such sardonic, rapid succession. In fact it is his increasing moral obtuseness that leads Vindice to confess his deeds, expecting to be praised rather than condemned for them.

Like so many other plays this script echoes the didacticism of the Moralities, as it does Jonson's *Volpone*, too, produced a year before. The names of the characters – Vindice (the Avenger), Spurio (the Illegitimate), Lussurioso (the Lascivious), Ambitioso, Supervacuo (the Idle), Dondolo (the Parasite) – are indices of that. Indeed, some have no proper names, though they have leading parts, but are merely designated as the Duke, the Duchess, the Third Son, the Second Noble and so on. They might be viewed as personified abstractions in an allegory. The youthful Tourneur is expressing his bitter revulsion and outrage at the lewdness, greed and heartlessness of the court he pictures, which is an image of a larger world of men and women. It is a fierce indictment. Another touch of the Moralities is the frequency with which the characters speak directly to the audience – the busy resort to "asides" – and the odd mixture of moods, dark and jesting, the lightness with which some of the murders are committed and news of them received by some of the survivors, as when Lussurioso learns that his father has been slain. The stepsons are actually gleeful at the knowledge and can scarcely wait to get rid of the new duke and each other. The play can – and has been – enacted in a spirit of burlesque, as a *reductio ad absurdum* of the revenge drama.

If there are faults in the script, Tourneur's poetry rises above them and it is this that accounts for his odd immortality ("odd" because he may not have written the play and has been famous for something he did not create). Few other dramatic works, aside from those of Marlowe,

Shakespeare, Jonson and Webster, have dialogue of such incisive, flexible and apt excellence. Line after line has a perverse beauty infused by an epigrammatic and a mordant wit. Tourneur is always the moralist, too. Thus Vindice, seeming to feel anger at the death of his beloved but little pity, apostrophizes her skull as he holds it in his hand and bespeaks the vanity of women and the folly of frenzied sexuality:

> Does every proud and self-affecting dame
> Camphire her face for this? And grieve her maker
> In sinful baths of milk – when many an infant starves
> For her superfluous outside, all for this?
> Who now bids twenty pounds a night, prepares
> Music, perfumes and sweet meats? All are hushed:
> Thou may'st lie chaste now! It were fine, methinks,
> To have thee seen at revels, forgetful feasts
> And unclean brothels. Sure, 'twould fright the sinner
> And make him a good coward, put a reveller
> Out of his antic amble
> And cloy an epicure with empty dishes.
> Here might a scornful and ambitious woman
> Look through and through herself. See, ladies, with false forms
> You deceive men, but cannot deceive worms.

But a few minutes earlier, his moods shifting with neurotic suddenness, he has paid the same dead lady a tribute in a quite different key:

> Thou sallow picture of my poisoned love
>   . . . thou shell of Death,
> Once the bright face of my betrothed Lady,
> When life and beauty naturally filled out
> Those ragged imperfections. . . .
>   . . . then 'twas a face
> So far beyond the artificial shine
> Of any woman's bought complexion,
> That the uprightest man, (if such there be,
> That sin but seven times a day), broke custom
> And made up eight with looking after her.
> Oh she was able to ha' made a Usurer's son
> Melt all his patrimony in a kiss . . .

Images are vivid and exact. The Duchess complains of her husband's reluctance to intercede on behalf of the imprisoned rapist:

> One of his single words
> Would quite have freed my youngest, dearest son
> From death or durance, and have made him walk
> With a bold foot upon the thorny law
> Whose prickles should bow under him ...

Lussurioso, planning to have Hippolito replace Vindice as his agent and dispose of him, exclaims:
> I'll employ the brother;
> Slaves are but nails to drive out one another.

Spurio, the bastard, acknowledges the cloud over his birth:

> I'm an uncertain man,
> Of more uncertain woman ...

Unusually evocative is another soliloquy of Spurio's, where he imagines his own begetting as he prepares to have an affair with his father's second wife:

> Duke, thou didst do me wrong, and by thy act
> Adultery is my nature;
> Faith, if the truth were known, I was begot
> After some gluttonous dinner, some stirring dish
> Was my first father, when deep healths went round
> And Ladies' cheeks were painted red with Wine,
> Their tongues as short and nimble as their heels,
> Uttering words sweet and thick; and when they rose,
> Were merrily disposed to fall again.
> In such a whispering and withdrawing hour,
> When base male-Bawds kept sentinel at stair-head
> Was I stolen softly; oh – damnation met
> The sin of feasts, drunken adultery.
> I feel it swell in me; my revenge is just,
> I was begot in impudent Wine and Lust.

Vindice believes in confiding very little to the opposite sex:

> Tell but some woman a secret over night,
> Your doctor may find it in the urinal in the morning.

He can also prescribe the skull as an omnipresent *memento mori* – a reminder of the short span of human life:

> Advance thee, O thou terror to fat folks,
> To have their costly three-piled flesh worn off
> As bare as this; or banquets, ease and laughter
> Can make great men, as greatness goes by clay;
> But wise men little are more great than they.

About the duke he warns:

> O, 'ware an old man hot and vicious!
> Age, as in gold, in lust is covetous.

Of the dishonest scene and people he sees about him he recalls:

> Last revelling night
> When torch-light made an artificial noon
> About the court, some courtiers in the masque,
> Putting on better faces than their own,
> Being full of fraud and flattery . . .

In disguise, as the procurer of Lussurioso, Vindice tempts his own sister with the following vision:

> O, think upon the pleasures of the palace!
> Secured ease and state! the stirring meats,
> Ready to move out of the dishes, that e'en now
> Quicken when they are eaten!
> Banquets abroad by torchlight! music! sports!
> Bareheaded vassals, that had ne're the fortune
> To keep on their own hats, but let horns wear 'em!
> Nine coaches waiting – hurry, hurry, hurry –

Throughout, too, is a hatred of sexuality and materialism:

> Oh,
> Wert not for god and women, there would be no damnation:
> Hell would look like a Lord's Great Kitchen without fire in't.
> But 'twas decreed before the world began,
> That they would be the hooks to catch at man.

Vindice renounces his having been in love:

> I could e'en chide myself
> For doting on her beauty, though her death
> Shall be revenged after no common action;
> Does the silkworm expend her yellow labours
> For thee? For thee does she undo herself?
> Are Lordships sold to maintain Ladyships
> For the poor benefit of a bewitching minute?
> Why does yon fellow falsify his ways
> And put his life between the Judge's lips,
> To refine such a thing, keep horse and men
> To beat their valours for her?
> Surely we're all mad people . . .

Dying, after having kissed the poisoned skull, the duke cries out:

> My teeth are eaten out.

This provokes a grisly jest:

> VINDICE: Hadst any left?
> HIPPOLITO: I think but few.
> VINDICE: Then those that did eat are eaten.

Some lines ring out with a metallic resonance, as Vindice's vow:

> He shall not live
> To see the moon change.

Lussurioso, proclaimed duke after his father's murder, hypocritically proclaims his grief, though told by an eager sycophantic courtier that his new dignity lends him a reassuring brightness.

> Alas, I shine in tears, like the Sun in April.

The disgust felt by the author is taken by some critics as evidence of his immaturity; he was perhaps only in his twenties. Others attribute his attitude to the vogue for melancholy and distaste widespread in his time, as incarnated in the disenchanted Hamlet.

Because *The Atheist's Tragedy* is a lesser work there have been efforts to prove that it is an older script, on the premise that an author writes better as he matures. But most of the evidence is said to be against dating it earlier. Besides, *The Atheist's Tragedy* is a very different kind of play. It glorifies the virtues of Stoic heroism, and possibly this is a reflection of Tourneur's own thoughts about the death of Sir Francis Vere, under whom he had served at Cadiz and elsewhere. During the interval between *The Revenger's Tragedy* and his next play, Tourneur had written a long poem commemorating Vere, and contemplation of that leader's virtues may have wrought a conversion (perhaps religious) in his feelings about life and death. Though the form of his new play is that of a revenge tragedy once again, Charlemont, one of the principals, is convinced that he has no right to kill, even though sorely victimized by his scoundrelly uncle, D'Amville. The ghost of his murdered father appears to him:

> Attend with patience the success of things;
> But leave revenge unto the King of kings.

This dark play, despite its share of horrors, has a great many amusing scenes and bits of action in it, revealing Tourneur's gift for strange and peculiar humour. He seems once more to be burlesquing the conventional revenge tragedy. The scene is France. The uncle, D'Amville, an unbeliever, reasons that man, facing inevitable death, should in the days allotted him experience delight to the full, denying himself nothing. He must obtain wealth that will assure himself an immortality through the happy lives of his children. All methods of obtaining pleasure and riches are justifiable. The play, opposing this philosophy, is another critique of hedonistic materialism. To attain his ends, D'Amville murders his brother (the father of Charlemont), schemes to effect a second killing and even seeks incestuously to rape his hapless daughter-in-law, Castabella. But the sudden demise of both his sons unmans him. His nephew is on trial for a slaying of which he is guiltless. D'Amville seeks to carry out the death sentence personally but instead contrives to brain himself. Before he dies he recants his atheism. He acknowledges the existence of a Divine Power.

In a parallel sub-plot the licentious Levidulcia indulges her physical desires. Her liaison with the attractive, generous Sebastian, one of D'Amville's sons, leads to the quarrel between the young man and her husband that results in their mutual destruction. She laments her forfeited honour, stabs herself to proclaim her remorse and set an example of recovered virtue. The "good" persons, Charlemont and Castabella, are idealizations. Unjustly on trial for his life,

Charlemont makes no defence but anticipates and even welcomes death with Christian fortitude. He accepts the advice of his father's ghost and refuses to exact vengeance by his own efforts. To him is given the summary last line of the play: "Patience is the honest man's revenge." He has trusted in Heaven to right human wrongs, and his faith has been rewarded. Unfortunately he comes off as a cold person. The chaste Castabella is more real and warm, though she, too, is most implausibly willing to die for a deed of which she is innocent.

In style, the play is more sober than *The Revenger's Tragedy*, though there are several fine passages of verse. The symbolism is forced and obvious, the tone didactic. The work, as a whole, wants fire and dramatic vitality. Somewhat incongruous, if theatrically effective, is the mocking comic relief. Tourneur also departs from the conventions of the Senecan revenge tragedy by tempering the passion of his aggrieved hero. He deflates the genre, and this rational attitude was perhaps long overdue.

A most unusual aspect of the play is Charlemont's effort to analyse psychic phenomena: what has caused him to "see" his father's ghost? His tentative answers are surprisingly modern. Has it all been a dream? Has the image risen from his subconscious (though, of course, he does not use that term)? Is the experience a consequence of his having recently been in hand-to-hand combat on the Continent (where his uncle has had him sent to get him out of the way)? In any event, this ghost, counselling moderation, talks quite differently from those in the dramas of Shakespeare.

Tourneur had both versatile talents and a unique intelligence. He might have given much more to the theatre had he chosen to continue in it. His decision to abandon his brief career as a playwright is regrettable. His point of view, like his verse, is highly individual. *The Revenger's Tragedy* is still deemed worthy of revival.

One comparatively recent restaging of the work was undertaken in 1970 by the Yale Repertory Theater in New Haven, Connecticut, a professional group organized and presided over by Robert Brustein, an important American drama critic. He also served as director for this event. Clive Barnes, of the *New York Post*, went to it expecting a stirring evening but was disappointed. Several experienced, well-known actors were in the cast. But for Barnes the pace was too slow. The horrors did not pile up rapidly enough to elicit bitter laughter, which Barnes thought was Tourneur's intent. Such laughter in itself would be cleansing, judgemental. As seen by Barnes:

> The play starts with one of the most famous scenes of Jacobean drama – Vendice caressing the skull of his mistress, who was poisoned when she refused to give herself to the Duke. (The play is set in one of those dark and mysterious Italian courts so beloved of Elizabethan and Jacobean playwrights.) Now he ponders on vengeance – and not just on the lecherous Duke, but on society as a whole. He kills and tricks and he even pretends to pander. He enters the court in disguise, and once there effectively destroys it. Vendice is a Laertes rather than a Hamlet. He would kill in a church if he had to.
>
> The play excels in its theatrical twists – the Duke being slain by being tricked into kissing the

poisoned skull of Vendice's beloved, two brothers sending their third brother to his execution by mistake, and then being given the dead man's head by the executioner, a final masque in which most of the members of the court murder one another. This is a somber and contemptuous view of the human condition.

Mr Brustein's staging has many qualities. I admired his use of the two-level stage that approximated to Jacobean stage conditions and I liked also his use of grotesquerie, including his final invention of a macabre dance of death. Yet the play's purpose still seemed blunted. The horrors were too easily, and too glibly, played for modern laughs, and the terrible speed of the piece was somehow lost.

The acting was variable. Kenneth Haigh, ice-cold, controlled and yet still with a hint of lurking hysteria, was brilliant as Vendice. He at least went through the play like a black-flighted arrow. But his skills were such that he dominated where he occasionally should merely have led.

Of the rest I admired most the strong Lussurioso, offered by David Ackroyd with deep-dyed villainy, and the wavering luxurious Duke of Lee Richardson. The rest tended to be too straight and too bland, or too exaggerated and too absurd. It was here that Mr Brustein failed to draw the playwright's happy medium, and it injured a fragile but still, in the right performance, a darkly effective play.

In contrast to Tourneur, who wrote early in life, John Ford (1586–1637) was a late starter. His father was a Devonshire gentleman. The son may have gone to Oxford, but that is not certain, however, at sixteen he was enrolled at the Middle Temple, where uncle was Lord Chief Justice. He was not a responsible student: he was expelled in 1601 for having fallen behind meeting his bills for food and drink and paid a large fine to win readmittance. He was soon in trouble again for his participation in a "rebellion" over proper dress at dinner and chapel services. When his father died his legacy was tiny in comparison to the larger amounts left to his brothers, indicating his parent's disapproval of John's conduct.

Ford began his career as a poet at about twenty. He followed with prose tracts, and entered the clamorous realm of theatre late. That he chose to compose for the stage was a tribute to the influence of Beaumont and Fletcher, who raised the status of play-writing. (Many courtiers in Stuart times now found that authorship for the theatre could be a respectable profession.) His first work, *The Witch of Edmonton* (1621), was a collaboration with Dekker and Rowley already noted. Stimulated by its success, he devoted most of his remaining literary effort to the drama. With Dekker he turned out four scripts in a single year (1624), of which three have vanished. He also worked with Webster and again with Rowley.

His earliest unaided effort, *The Lover's Melancholy* (1628), met with encouraging applause. This graceful and romantic Arcadian tragicomedy owes something to John Fletcher but also to Burton's recently published *Anatomy of Melancholy* (1621), an imposing essay full of psychological insights, especially into "abnormal" passions. It was, then as now, a widely read work, and of all Jacobean playwrights Ford made the most use of it. *The Lover's Melancholy* is set in Cyprus (but still a fictional island), and its characters have such odd names as Cleophile, Parthenophil (a girl who assumes a masculine disguise) and Palodar. What distinguishes it is not its slight plot but its sustained aristocratic tone, which is disturbed by brief incursions of comic relief. Sometimes too slow, even static, and lacking in lucidity, this is another play redeemed by its poetry. Some scenes also suggest the nascent master – one can already tell that Ford is to take his place among the greatest dramatic craftsmen of his age.

He soon proved this in *The Broken Heart* (c. 1632–3), which evokes antiquity, specifically Sparta. None the less, its chief source is Sidney's *Arcadia*, and for the most part its story is not imagined but taken from life. As Ford intimates in his prologue, it reflects the frustrated passion of Sir Philip Sidney for the "Stella" of his sonnet series. The characters in *The Broken Heart* have names chosen to reflect their dominant qualities: Orgilus is "Angry", Bassanes is so called because he is "Vexation"; Ithocles is no less than "Honour of Lovelinesse", "Amelus" is "Trusty", and so on. "Calantha" is "Flower of Beauty", and the unhappy heroine is given the name of Penthea, "Complaint". Such designations, along with the remoteness of the setting, cast over the play a sense of unreality that the strained plot does little to dissipate. Orgilus has been betrothed to Penthea, but her brother Ithocles, a military hero and the king's favourite, has for political reasons compelled her to marry Bassanes, a fatuous, psychotically jealous older man. Penthea considers herself spiritually wedded to the intense Orgilus, though she refuses to allow her love for him to express itself by physical touch. She is faithful both to her enforced marriage vows and to her repressed feelings. Ithocles himself hopelessly aspires to the hand of Calantha, the King of Sparta's daughter, who is being sought by the Prince of Argos. Though Penthea holds the ambitious Ithocles to blame for her unhappiness and harshly upbraids him, she is forgiving enough to intercede with Calantha on her brother's behalf. The princess returns Ithocles' love and wins from her royal father permission to take him as her husband. Meanwhile Penthea, pining and fasting, sinks into madness and starves herself to death. Orgilus exacts vengeance from Ithocles. He entraps him in a cunningly devised chair and in a sensational moment on stage stabs him to death. During this scene, however, both men address each other courteously and with incongruous respect.

The news of her father's death, Penthea's demise and Ithocles' murder is brought successively to Calantha while she is dancing at a celebration of the nuptials of Orgilus' sister, Euphrania, and his friend, Prophilus. The princess at first seems not to comprehend each new bit of dreadful information and continues to enjoy herself. Suddenly she halts the masque, orders the execution of the guilty Orgilus, who chooses to submit gallantly. "They die too basely who outlive their glories," he declares. Calantha is then "married" in a white-robed ceremony to the corpse of Ithocles, and she, like Penthea before her, dies of a broken heart.

The play has many shortcomings. It shifts focus time and again. The characters behave inconsistently – for example, Bassanes, after hearing his wife merely state her innocence, abruptly ceases to question it, harbouring none of his previous obsessive jealousy but embracing a pious faith in her. Calantha, the major figure in the final scenes, is scarcely on stage in earlier ones.

The language, applauded for its serene and subdued poetry, too often lacks dramatic fire, and not a little of it – despite critics' appraisal to the contrary – is truly stilted and ponderous; it is hard to believe that it sounds as well on stage as has been claimed.

Yet *The Broken Heart* is redeemed from being a typical "horror play" of the period by an air of reserve and gravity. It might be said that in many ways it prefigures Racinian tragedy. Charles Lamb wrote highly of it: "I do not know where to find in any Play a catastrophe so grand, so solemn, and so surprising as this. . . . Ford was of the first order of Poets. He sought for sublimity, not by parcels in metaphors or visible images, but directly where she has her full residence in the heart of man; in the actions and sufferings of the greatest mind." Una Ellis-Fermor declares this play to be "the supreme reach of [Ford's] genius . . . a play . . . which for simplicity and compactness of line, for dignity and compression of emotion, thought and phrase, is unsurpassed in Jacobean drama". Among others who greatly admired *The Broken Heart* is Maurice Maeterlinck, the twentieth-century Flemish playwright, eminent leader of the Symbolist movement. But to some, like Allardyce Nicoll, Ford's writing here and elsewhere is marked by "decadence", though Nicoll grants that it contains "some powerfully affecting situations". Yet on the whole he finds it "fetid and fantastic".

This criticism has been brought even more strongly against Ford's acknowledged masterpiece, *'Tis Pity She's a Whore* (c. 1624). His habitual choice of sensational subjects and abnormal characters is well instanced here. In Italy, Giovanni (no family name) seduces his sister, Annabella. This act of incest accomplished, and brilliantly justified by Giovanni in a much-quoted speech, he permits their father to marry her off to the cruel Soranzo. Pregnant with her brother's child, she soon rues her surrender, but her possessed brother still seeks her embraces and yielding. She is faced with the rage of the suspicious Soranzo, whom she defies by openly admitting a strong feeling for her "angel-like" brother. Her husband threatens to torture her. When Giovanni learns that his brother-in-law has discovered the truth and is about to have revenge, he goes out of his senses, kills Annabella, cuts out her heart, which he wildly exhibits to Soranzo's birthday guests, the bleeding organ held aloft on his sword. He stabs Soranzo and is himself killed by his brother-in-law's four paid assassins.

Anyone who has seen this play on stage can attest to its undoubted power. Suspense is engendered and heated further as the story rushes from climax to climax. The work has been particularly deplored because Ford to a remarkable extent arouses and sustains sympathy for the tainted brother and sister: they are drawn together almost irresistibly, since they feel themselves surrounded by a corrupt world, as represented by her many insistent suitors. The "natural marriage" of these two sensitive young people seems touched with a strange innocence, a peculiar purity of spirit, though as the play progresses Giovanni's lonely, isolated passion hardens and becomes intellectually and selfishly defiant rather than tender as before. He is an existentialist anti-hero, to whom conventional morality is hypocritical and meaningless, and who is seeking to create a more self-satisfying cosmos of his own. At the secret ceremony at which, kneeling, they "wed" each other, they pledge in turn:

> I charge you
> Do not betray me to your mirth and hate:
> Love me or kill me.

This tragic vow is all too soon mortally fulfilled. Appropriate disclaimers are voiced by the author, but he does not sharply condemn the ill-fated young pair. The final words on them are spoken by a cardinal: "'Tis pity." Beckerman observes that these aristocrats do not express moral horror or disgust at a brother–sister coupling, rather, Soranzo is outraged that Annabella is unfaithful to him.

More than anything else, it is the language, both verse and prose, lucid and exact, that lifts the play from the realm of sensational theatre to a higher form of drama, though much must be said for the excellence of the characterizations, both of principals and minor figures, in a well-worn plot. The daily life of Renaissance Italy is created as a busy background for these turbulent events. Besides, Ford does not evade the serious issues he raises: there is no last-minute side-stepping here such as occurs in Beaumont's much weaker story of incest *King and No King*. The guilty lovers pay the full price demanded by the religious ethic that dominates their society. At intervals throughout, Giovanni debates with a friar as to the rightness or wrongness of his unsanctioned passion, and, like Dostoevsky's Raskolnikov, he has the better of the argument in setting forth good reasons for his "crime". Indeed, the friar's answers are feeble by contrast. At the end Giovanni still hopes to be reunited with his beloved sister in a better world.

After the fashion of the day, Ford dedicates the play to a nobleman, John, Earl of Peterborough, a hoped-for patron. He apologizes for the "lightness" of its title; it is really not an appropriate one and at odds with "the gravity of its subject". In fact Annabella is not a whore, a hired sexual partner. But over the centuries the title proved to be a strong box-office draw.

If the play has a source other than Ford's own imagination, it has not been identified; the plot is not derived from any single known predecessor. But Ford does make use of many devices and situations found over and over in works by his rivals, conventional tricks in many Elizabethan and Jacobean scripts. He has a sub-plot in which minor characters are iniquitous, engaged in poisonings and other deadly acts. Even the Friar is morally compromised by practising a deception on the brutal Soranzo. The young couple are in many respects the most virtuous of all those on view in this Italian Renaissance universe.

Bernard Beckerman, in a preface to his *Plays of the English Renaissance*, likens the iconoclastic Giovanni to Marlowe's Faustus and describes *'Tis Pity* as "a throwback to a sophisticated Morality play". He quotes the Friar's first lines, a warning to Giovanni:

> ... wits that presumed
> On wits too much, by striving how to prove
> There is no God, with foolish grounds of art,
> Discovered first the nearest way to Hell,
> And filled the world with devilish atheism.

Love's Sacrifice (printed 1633) shows many of the same motifs and preoccupations as these other Ford tragedies. Carafa, the middle-aged Duke of Pavia, cherishes equally a gallant, scrupulous young man, Fernando, and his own wife, socially inferior Bianca. The young pair become infatuated with each other, and the duke's suspicions and murderous anger are prodded by his envious sister and his evil secretary, D'Avolos. Though the affair is Platonic, despite many temptations, Bianca, when accused by her husband, offers no apology for her feeling for Fernando. She asserts that the young man is "a miracle compounded of flesh and blood", and she is attracted to him by the same inevitable instinct that had inspired Carafa's feeling for her. She is equally candid with Fernando, pleading for their natural right to love:

> Why shouldst thou not be mine? Why should the laws,
> The iron laws of ceremony, bar
> Mutual embraces? What's a vow? A vow?
> Can there be sin in unity?

Her husband imprisons Fernando and kills Bianca. Her words deliberately provoke his deed; denied fulfilment of her love, she wishes to martyr herself. But the duke is soon convinced by the noble-minded Fernando that she was innocent of any sexual infidelity. The remorseful husband commands a grandiose funeral for the wronged lady. The gates of the burial vault are opened to reveal Fernando garbed in the winding-sheet of a corpse: he denies entrance to the murderer, lest the dark, sacred precincts that house the eternally sleeping Bianca be violated. When the duke threatens access by force, Fernando swallows poison and expires. He, too, prefers self-martyrdom. Carafa, overcome by this new catastrophe, also commits suicide, ending a spectacular scene.

A cult of Platonic love had recently grown up at the court of Queen Henrietta Maria, wife of Charles I, and perhaps Ford was inspired by it to compose this warning of the dangers to which such dalliance might lead. By consensus, though, the play is inferior in construction and dialogue to *The Broken Heart* and *'Tis Pity*. The comic relief is particularly distracting, as is a far-fetched subplot. The influence of Burton's *Anatomy of Melancholy* is again discernible, as is that of Sidney's sonnet sequence with its echoes of that poet's real-life frustration in his affair with his self-denying Stella, another instance of psychically destructive Platonism in relations between men and women. Whatever the faults of the play, its psychological insights are often subtle and compel interest.

The fourth of Ford's serious works, *Perkin Warbeck* (printed 1634), is a historical chronicle, a form that as Ford himself remarks in his prologue had fallen from fashion; he hopes to resurrect it. He was attracted to his theme – that of a pretender to the English throne – because he saw the young aspirant as genuinely (if abnormally) self-deluded. In fact, all of Ford's heroes in his major plays – *The Broken Heart*, *'Tis Pity*, *Love's Sacrifice* – suffer from this affliction; they are completely convinced of their own rightness and fail to see themselves as others see them. Thus, whatever the truth might be, the youth is thoroughly persuaded that his claim is rightful, and he lives and dies courageously and always fully immersed in his kingly role. Again, like other Ford heroes, he does not fear death but dismisses it lightly. Indeed, Ford's principal young men seem to welcome death as an opportunity for a grandiose gesture. Once more, as a psychological study *Perkin Warbeck* is a considerable success, and it is also esteemed for the restrained and dignified manner of its telling; its subject, however, an obscure incident during the fifteenth-century reign of Henry VII, limits its appeal outside of England. For his material the playwright made use of Bacon's 1622 *History* of the period.

Besides these serious works, Ford tried his hand at a tragicomedy, *The Queen* (date undetermined), in the style of Beaumont and Fletcher. Probably it is an early work; its double plot is highly artificial and not too well handled. Abnormal jealousy is again one of its themes. The play has been described as a preliminary sketch for *Love's Sacrifice*. Its moral might be "Passions at their best are but sly traitors to ruin honour". Consequently one should make a virtue of one's ruling faults and accept the passions one cannot control. This places one at the mercy of the desired one, whose demands may be capricious, as is taught to Antonio and Valesco in their respective marriages to a submissive queen and a whimsical widow. Also, the point is made that people fear happiness as much as anything else. The attempt in the play is to achieve a sophisticated, fairy-tale atmosphereof the kind that Beaumont and Fletcher as well as Shakespeare (in *Love's Labour's Lost*) attained, but at this time the – presumably – still youthful Ford was hardly their equal.

He closed his career with two fair-to-middling comedies. *The Fancies, Chaste and Noble* (printed 1638) tells of Octavio, Marquis of Siena, who, though elderly and probably sexually impotent, has gathered about himself a covey of young girls, the "Fancies", who to the scandal of onlookers are seen as his harem. The popular opinion wrongs them, however, for the young ladies are his decorous nieces, to whom he is teaching graces and manners that will adapt them to court life. At the end the truth is exposed and marriages result. The lesson drawn is that one should not judge too hastily by superficial appearances. *The Lady's Trial* (1638) has a similar moral. Spinella, the heroine, is slandered, not by direct accusation but by innuendo and rumour, which prove quite as harmful. Auria is another jealous husband, an inevitable figure in a Ford script. But in neither play, though they contain indelicate episodes and occasionally beautifully written passages, does Ford demonstrate himself wholly apt at stage humour. Indeed a famous couplet in *The Time Poets* (1656) attributes to him a saturnine temperament:

> Deep in a dump John Ford was alone got,
> With folded arms and melancholy hat.

He seems to have been lonely, reserved.

After 1639 he vanishes. It is conjectured that he went back to his native Devon and died there. His adult life, apparently, had seen him hard-pressed for money and driven to write for a bare living. Yet he was always something of the "aristocratic amateur" in literature. Several times he refers scornfully to those who "Somewhat of late have made / The noble use of poetry a trade."

He employed the over-familiar theatrical material of his day – the tragedy of blood – and somewhat transmuted it, causing it to seem fresh again. His mind is original, his personality distinct, his approach highly personal. He is certainly a thoughtful writer. Concerning his accomplishment and his artistic stature there is considerable controversy. Some regard him as "the dramatist whose name may follow in honour after Shakespeare, Marlowe, Jonson. . . . He is the last of undoubted genius, and of that peculiar quality that one associates with the age of Sidney, Raleigh, Essex and Shakespeare." His poetic gifts have been likened to those of Shelley. Still others find his psychology and sexual morality surprisingly "modern". In his own day he boldly challenged conventions. But his detractors call him "over-praised". He is charged with being a "sensationalist" who sought too hard "to arouse the jaded appetites of his public. He piled sorrow upon sorrow and assaulted the tear glands with a battering ram." His work is said to lack "vision and ecstasy". T.S. Eliot remarks that Ford's works suffer from "an absence of purpose" and that even his greatest play is "meaningless" because it projects a personal problem of the poet. Allardyce Nicoll decries *'Tis Pity* as "degenerate sensuality, wrapped in an intoxicating riot of words". He sees Ford as a "Mannerist" whose perfervid scripts marked the decline of tragic drama in England after its greatest era. "The worm is upon his pages, the rose is cankered; we are out of the honest breath of Heaven breathing hot-house air." Whatever may be the opinion of today's reader, he is much more likely to be impressed and moved by Ford's plays on the stage than in a book; even when their poetry fails their action captures and rivets attention.

To attempt a summary of a complex dramatist, he liked to explore the minds and feelings of strangely obsessed persons, especially when they are put in extreme and unusual situations. What are their reactions to each other? Such situations are often the consequence of a misalliance, between a man and a woman, or between the real nature of a man and his mistaken image of himself; or both errors. He is curious about them but never without compassion. He is concerned not only with the theme of sexual jealousy but also of honour, the sacredness of vows. His characters are egotistical but self-critically divided, defeatist, too ready to embrace death, extinction, in a search for peace. They have also asked more of life than life can give them.

Beckerman sees Ford, along with Jonson, as a leading transitional figure from the Jacobean theatre to the stage during the reign of Charles I. He gives this picture: "Following the death of James I in 1625, life at court became more rarefied, politics more claustrophobic, and entertainments more idolatrous. Theatrically, the period is considered a ripening to decay of characteristics generated in earlier years. Comedy becomes more flippant, less moral. Tragedy becomes less horrifying, more languid and lush, continuing to mine the traditions of revenge, Italian luxuriousness, and intrigue. To a large extent the theatrical tone of these Caroline years is set by the work of Ford and James Shirley."

James Shirley and Richard Brome are two minor dramatists of the time who deserve some mention. About Brome's background there is more speculation than knowledge. His dates are unrecorded (at a guess 1590–1652/3). His plays competent, if derivative, he laboured as a journeyman in the London script-factories. At first he was attached in some capacity to Ben Jonson; perhaps he was Jonson's secretary. Later he collaborated with Jonson's son on an early work of both. He was soon enjoying immense success on his own with *The Lovesick Maid* (1629). The elder Jonson, who had just experienced a failure, was furious that a former follower had in this instance outdistanced him; he composed a derogatory couplet about the event but soon repented of his ill-temper and softened the lines of his jibing verse. Thereafter Brome was recognized as a true professional, acknowledged and generously praised by Dekker and Ford. He poured out a stream of scripts for several companies successively; no less than nineteen plays are accounted his. His best years coincided with those of Oliver Cromwell's ban on all theatrical activity, and Brome suffered financially in consequence. With the Restoration a handful of his scripts were published and some others revived.

To Jonson-like satire he added a touch of romanticism bearing the stamp of Beaumont and Fletcher. He also reveals the influence of Dekker and Middleton. Indefatigable, modest, good-humoured, he wrote shrewdly and confidently. Mostly his output included realistic comedies of the sort provided by his betters, but he readily supplied tragicomedies on demand. His later offerings contributed to the formation of the comedy of manners with an element of social criticism. His view was expressive of the attitudes of the rising middle class and disdainful of the high-living Cavaliers. He was inventive, often handling plot well. His style, comparatively, was somewhat coarse.

Among his most popular pieces were *The City Wit* (1628), *The Northern Lass* (printed 1629 or 1632) and *The Queen's Exchange* (1631). An ingenious, clever fantasy, *The Antipodes,* makes use of drugs and hallucinations and impersonations to cure several eccentric members of the Joyless family. *The Sparagus Garden* (1635) takes place in a resort where "more or less reputable persons gather to eat asparagus and otherwise amuse themselves", doubtless with gossip and flirtations. Following came *A Mad Couple Well Match'd* (printed 1653). Many of his works were published well ahead of being enacted due to the Lord Protector's prohibition against performances. Brome's last play, *The Jovial Crew* (1641), was later a particular favourite of Pepys; the stage-struck diarist went to take it in on three different occasions and notes that appreciative royalty was also to be seen in the well-entertained audience.

James Shirley (1596–1666) was an even busier writer. His father was a London merchant. The son's seven decades spanned several reigns, including Oliver Cromwell's repressive Commonwealth. He, too, is a transitional figure in the development of Jacobean drama, then on to Restoration farce and the comedy of manners, of which – in such works as *The Witty Fair One* (1628), *Hyde Park* (1632) and *Lady of Pleasure* (1635) – he was an important precursor. He borrowed much from his contemporaries, Jonson, Middleton, Fletcher, Brome and Webster. He first enjoyed a sound education: after four years of classics at the Merchant Taylors' School he attended both Oxford and Cambridge to study theology. Taking orders, he became a schoolmaster and parish priest at or in the vicinity of St Albans in Hertfordshire. (Legend has it that he left Oxford because Laud, the head of St John's College, told him that a man whose face was disfigured by a large mole on his left cheek was hardly the proper person to perform sacramental functions.) Presumably he underwent a second religious conversion, abandoned his parish to become a Roman Catholic and eventually settled in London as a stage-writer, producing not less than two plays a year. He later succeeded Massinger as chief dramatist for the King's Men. For them, and at other times for a number of other companies, he furnished about forty plays – comedies, tragedies, masques – ranging from his farce *Love's Tricks* (1625) to his serious dramas: *The Traitor* (1631) and *The Cardinal* (1641). Some of the more notable, besides those, are *The Wedding* (printed 1629), *The Ball* (1632, with Chapman), *The Bird in the Cage* (printed 1633), *The Gamester* (1634), *The Triumph of Peace* (an elaborate masque, in conjunction with Inigo Jones, who provided the scenery, and Henry Lawes, who composed some of the music). When the plague shut down the London theatres in 1636 Shirley travelled to Dublin to continue his work, contributing to playhouses there, staying abroad until 1640. Then, back to London, he carried on with *The Doubtful Heir*, *The Court Secret* and a score more scripts. When Cromwell and the Puritans closed all London theatres Shirley kept himself alive by taking a post in the royalist forces – and, after their defeat, returned to his former occupation of schoolmaster. With the Restoration he resumed his theatrical activities. Masques such as *Cupid and Death* came from his pen. Establishing himself as a favourite in high circles, he was awarded a lucrative court post, Valet of the Royal Chamber. He prospered, and in his will he left large legacies to his wife and five children. But the lady was not destined to survive him. The Great Fire of London in 1666 enveloped the Shirleys' house, and the shock, exertion and exhaustion caused both the playwright and his wife to die on the same day (most likely the lingering effects of smoke inhalation). They were buried in a single grave.

Shirley's faculty for poetic dialogue, though limited, is often blessed with an aristocratic grace. It is also animated and evocative, helping the spectator to "see" the world surrounding his characters. He can formulate a pithy phrase: to an immoral revel, the gallants and ladies are thronging "by a Subpoena of Venus". He is specific in his verbal depictions of wholesome country life in contrast to an idle, indulgent *haut monde* existence in the decadent city. Explains Lady Bornwell's steward:

> We do not invite the poor o' the parish
> To dinner, keep a table for the tenants;
> Our kitchen does not smell of beef, the cellar
> Defies the price of malt and hops; the footmen
> And coachdrivers may be drunk like gentlemen
> With wine, nor will three Fiddlers upon holidays
> With aid of Bagpipes, that called in the country
> To dance, and plough the hall up with their hobnails,
> Now make my Lady merry. We do feed
> Like princes, and feast nothing but princes . . .

Here is a vivid response concerning an amorous old bawd who proposes an affair to a young man, himself a reprobate:

> Embrace? She has had no teeth
> This twenty years, and the next violent cough
> Brings up her tongue; it cannot possibly
> Be sound at root. I do not think but one
> Strong sneeze upon her – and well meant – would make
> Her quarters fall away; one kick would blow
> Her up like gunpowder, and loose all her limbs.
> She is so cold, an Incubus would not heat her;
> Her phlegm would quench a furnace, and her breath
> Would damp a musket bullet.

Frederick, in *The Lady of Pleasure,* is asked by Kickshaw:

> Does not Jack Littleworth follow?
> FREDERICK: Follow? He fell into the Thames
>   At landing.
> KICKSHAW: The devil shall dive for him
>   Ere I endanger my silk stockings for him.
>   . . . I shall laugh
>   To see him come in pickeld the next tide.
> FREDERICK: He'll never sink, he has such a cork brain.
> KICKSHAW: Let him be hang'd or drown'd, all's one to me;
>   Yet he deserves to die by water, cannot
>   Bear his wine credibly.

His comedies are better than his tragedies and tragicomedies, the latter mostly resembling Fletcher's. Reflecting his years of studying theology, even his comedies of manners have moralistic overtones that will be lacking in the subsequent Restoration theatre: usually his rakes are reformed and marry the innocent young ladies after failing to seduce them. His people are sometimes excellently drawn, though they are not always consistent. In tragedy – *The Cardinal*, reminiscent of Webster's *The Duchess of Malfi*, is deemed his best work – he frequently arrives at effective theatre but not the roundness or intensity of great drama. Like *The Cardinal*, his other well-known tragedy *The Traitor* takes place in Italy; it is about the ruthless Lorenzo de' Medici. As has been seen, Italy was most often chosen by English dramatists of the day for serious plays, and Shirley follows suit. His craftsmanship in these pieces is truly professional and in some aspects brilliant. Both dramas, too, end with scenes of luridly piled-on horror and have considerable power. But Shirley's offerings were not fated to stay in the repertoire. By the time the eighteenth century was out they were no longer to be viewed on stage.

Because he lived long – for that period – and was somewhat younger than most of his more famous rivals, he is often called the last of his kind, the most significant poets and playwrights who belonged to the Elizabethan, Jacobean and Stuart eras.

— 20 —
JONSON AND WEBSTER

The vibrant, amazing theatre of the English Renaissance came to a splendid close with the fortuitous arrival on the scene of two literary giants: John Webster was the son of a tailor; Ben Jonson was raised by a stepfather who was a bricklayer. Genius, *ab ovo*? Neither genes nor early environment offers an adequate explanation. The mystery remains. (But what of the maternal side? Of their mothers little is known.)

Ben Jonson started life in 1572 or 1573 unpropitiously as a posthumous child. His father, who late in life became a zealous Protestant preacher, had died only a month before his son's birth. He had suffered imprisonment for his religious beliefs during the days of the Catholic Mary Tudor, and his estate had been confiscated. Much of the elder Jonson's pulpit fire was later shown by his son in his castigating plays. The Jonsons (originally Johnstone) were of south-west Scottish stock, strong-fibred and austere, and this Puritanism was a strange, conflicting strain in Ben, who was to be not only a moralist but a notorious roisterer.

His widowed mother, left without resources, remarried. Her second husband, the master bricklayer, seemingly failed to recognize his stepson's talents and planned to apprentice him to the bricklaying trade. But the boy had already come to the attention of William Camden, the erudite second master of Westminster, who singled him out for patronage. Early, Jonson got a solid education in the classics, at which he excelled. Afterwards he showed his gratitude to the kind-hearted, discerning Camden by dedicating his first successful play to him. Too poor to attend university, Jonson nevertheless, by his own efforts, became one of the foremost scholars of his time, as was attested in later life when he was awarded honorary degrees by both Oxford and Cambridge. He was proud, even boastful of his learning; indeed, more than a bit pedantic, typical of most autodidacts. It was he who spoke condescendingly of Will Shakespeare's "little Latin and less Greek".

Pushed into bricklaying by his obdurate or unperceptive stepfather, young Jonson soon rebelled. He escaped by running away to join the English forces in the Low Countries where the struggle against the Spanish Catholic oppressors for religious and political freedom exerted a natural appeal to him, the independent-spirited son of a Protestant martyr. It is hard to know the full extent of his participation in the fighting there. He afterwards claimed to have gone ahead of his English companions, to challenge a Spanish foeman to single combat and in full view of all to have slain him and deprived him of his armour in true Greek fashion. But Jonson was never modest, and the story may well have been embellished, if not wholly concocted, by an instinctive playwright. In any event his military service was of short duration.

In 1592 he was back home in his native London, and – though only nineteen or twenty – already wedded. His choice of a wife was not felicitous; the marriage was not to run smoothly. He described his lady succinctly as "a shrew yet honest"; at times he was separated from her but then lived with her again. They had several children, among them a son, born in 1596, of whom Jonson was very proud, speaking of him as "my best piece of poetry". The boy died at the age of seven, taken by the plague. He was badly hit by this loss. Perhaps his wife had much to nag him for during these first years, for what the family lived on is far from clear. Perhaps Jonson had become an actor with a minor troupe of strolling players; it is said that he played the leading role of Hieronimo in a revival of *The Spanish Tragedy*. Later he prepared a new version of this ever-popular melodrama for Henslowe, the money-bag of London's theatre.

In 1597 Henslowe entered in his helpful diary a loan of four pounds to "Bengemen Johnson, player". Very likely this was for Jonson's labour in completing Nash's script *The Isle of Dogs*. It is asserted that Jonson acted in the stage performance of this lost play, which resulted in the arrest of its entire cast and the author. He was to continue at hack work of many sorts, and collaborations with many other writers, throughout most of his long career, having always to make a living. He wrote poems in Latin and English, court masques, occasional pieces and unclassifiable entertainments, as well as comedies and tragedies for Henslowe and others. He never stayed under the aegis of one producer or company of actors but offered his work to several troupes. In October 1597, however, after his release from prison, he did return to Henslowe to request a loan of "twenty s." (shillings, so one pound), which he got, as an advance on a new script he was to prepare. A surviving work of this early period, *The Case Is Altered*, is based on two Latin comedies. In it his style is still not perfected. His name did not long appear on it, and subsequently he denied authorship, but the evidence points to its having come from his pen. What is outstanding about it is his developing craftsmanship, his skill at integrating two borrowed plots into a new, single one.

In 1598 Shakespeare's company, then called the Chamberlain's Men, put on Jonson's first truly noteworthy script, *Every Man in His Humour*, with Shakespeare himself, Condell, Kemp and Burbage acting in it. A distinguished cast! A distinctive work, and a box-office triumph, it afterwards had a prologue added to it in which Jonson issued a "manifesto", his new artistic credo. He has perceived how oft repeated has been the formula of Elizabethan plays up to this time, and he proposes significant changes. In this oft-quoted "manifesto" he calls for more responsible craftsmanship, of which he himself will give a good example. He also thinks the moment has come for much more realism and fresher subject matter on the stage:

> Deeds and language such as men do use,
> And persons such as Comedy would choose
> When she would show an image of the times
> And sport with human follies, not with crimes.

Here and elsewhere, he will search for plots that are topical, "near and familiarly allied to the time". He will eschew the fanciful and far-fetched romantic incidents that filled so many other scripts of the period, situations by now tiresomely overworked:

> As of a Duke to be in love with a Countess, and that Countess to be in love with the Duke's son, and the son to love the Lady's waiting-maid – some such cross-wooing, with a Clown to their serving man.

He is equally weary of historical dramas, such as Marlowe and Shakespeare had often provided, which by now were too much imitated, exploiting a stale format. Therefore he will write in a quite different vein, shunning all such clichés, offering "the groundlings no romantic shudders – no fights between the houses of York and Lancaster or fireworks to please boys and frighten gentlewomen". He will, indeed, compose a comedy "such today as other plays should be". In doing this he intends to follow the best classical models.

What is not fully appreciated today is that in *Every Man in His Humour* Jonson did not break with Elizabethan traditions as fully as he claimed. He more nearly accomplished this in a later revision of the play (*c.* 1612, printed 1616). This first version is still laid in Italy, and much of its comedy derives from the usual plot-materials. What is most original about it is the embodiment of Jonson's theory of humours, that is, the use of characters who are dominated by a single trait, obsessed by one passion in life. As the poet saw it, such an obsession was the result of what we might today call a glandular imbalance. Medieval physiology held that the body contained four kinds of fluids: the blood (which lends sanguinity, or cheerfulness, to a person); the choler (anger); the black choler (melancholy); and the phlegm (apathy). They were also classified as "hot", "cold", "moist" and "dry". Characters so conceived are idiosyncratic; they are like the grotesques – the Uriah Heeps and Pecksniffs – later encountered in the sinister, foggy London depicted by Dickens.

Jonson puts it this way: "When some one peculiar quality doth so possess a man that it doth draw all his affects, his spirits, and his powers . . . all to run one way, they may truly be said to be a humour." Or, as Jonson expressed it to Drummond – as reported in that gentleman's *Timbers* – comedy is concerned with the odd numbers in life, tragedy with the even numbers. The comic figure is abnormal because he allows a single quality to rule him. His every action is fanatically motivated by greed, or gluttony, or lust, or hatred, or miserliness.

In his revision of the play, Jonson shifted the scene to England and gave the characters English names.

Every Man in His Humour is notable, moreover, for Jonson's observation of the three unities – he was dedicated to bringing classical form to the drama. The action unfolds within twelve hours, the only scene is Florence (later changed to London), and the mood is unified throughout; this is comedy and nothing else. Kitely is insanely jealous of his innocent wife. He is surrounded by a group of eccentrics, most of whom, as their names imply, represent different humours: the outspoken Squire Downright, whose forthrightness gets him into trouble; the two Kno'wells, over-curious father and frisky son; the impetuous lover, Well bred; the rascally servant, Brainworm, who stimulates most of the foolish action; Cob, also a jealous husband; Captain Bobadill, a vain braggart and coward – the *miles gloriosus*; a pair of simpletons easily gulled, Stephen and Mathew. The chief virtue of the play is the character portrayals, not its plot, which is very slender, hardly adequate for a work of this length. With the scene transposed to London's bohemian haunts, which Jonson knew only too well, the loosely constructed piece becomes a panorama of Elizabethan low life. A half-dozen human follies are personified and exposed, with laughter arising chiefly from that, hardly from incidents. By far the best passage, in which a father laments the decline of manners in the young but puts the blame for it on too much parental indulgence, as well as on the elders setting their offspring a bad example, is translated directly from Quintilian. The dialogue, vivid, homely, carries the main burden. After a mix-up everyone ends before a magistrate, the whimsical Justice Clement, whose peculiar fixation is quaffing a brimming cup of sack. Jonson was to write much better plays but never to be more acclaimed than for this.

His pleasure was short-lived, however, for soon after his great success he himself was back before a magistrate. Life truly imitated art for him. Having become embroiled with Gabriel Spencer, an actor in a rival company, Jonson killed him in a duel. Partly in self-justification, but also being of a boastful spirit, he claimed that his own sword was almost a foot shorter than his opponent's. He also had his arm slashed. Charged with murder, the young playwright was cast into prison and confronted the gallows. He saved himself by pleading "benefit of clergy" – exemption from capital punishment on the grounds that he was literate, which he proved by reciting Latin verses – and got off with a minor term in gaol and a heavy fine, the loss of all his possessions. His thumb was branded with an "H" for "Homicide". While he was in prison he secretly became a Catholic. In view of the sufferings of his father for the Protestant cause and his own participation in the Low Countries' struggle, this conversion was the more amazing; and it is also testimony to his courage, since to become a Catholic at this time was dangerous. He took the sacrament from a priest who administered it with prison bread. His recusancy lasted twelve years, after which time he reverted to Protestantism.

Henslowe had been grieved by Spencer's death at the hand of "Bengemen Jonson, bricklayer", as he now angrily described him in a letter to Allen. But upon the poet's release the producer was ready to advance him money to obtain his professional services again. Jonson was clearly too good a man to lose. He was commissioned to write a domestic tragedy and a historical chronicle, together with Dekker and others on Henslowe's eminent list.

On his own Jonson readied and had staged a new work, *Every Man Out of His Humour* (1599), which is considerably less effective than his earlier play. Though less popular, it did succeed in catching a measure of public favour. Once again it was put on by Shakespeare's company. Jonson was not satisfied by its reception. He hurriedly had it printed, with a preface in which he insisted that his work had not been properly appreciated. He included lines and passages that had been cut by the performers. He also added a wordy prologue (or "Induction") in which the Author and two "spectators", Mitis and Cordatus, talk about the history and essence of comedy. They continue their discussion throughout the play, acting as a chorus and passing judgement on his people's behaviour in every episode. This lends the work a feeling of heaviness coupled with excessive vindictiveness. But his critical points were accepted by his audience, who tended to agree with him. The plot is certainly less coherent than even the weak one that had served the preceding script. The humours of the characters are even more pronounced; now they are carried to pathological lengths. Their possessors are more openly chided for their varied follies by the two central figures, Macilente, a jealous scholar, and Carlos Buffone, a "cynical jester" and acidulous scold. Among those who come in for tongue-lashing and other forms of punishment are the social climber Sogliardo, the foppish Fastius Brisk, the out-at-elbows conniver Shark, the uxorious husband Delio, the mercenary Sordido. One of the more amusing persons is Fungoso, a law student whose compulsion it is to attire himself after the latest fashion. By the time his tailor finishes a new suit it is already too late, for the mode has changed. At the end most of these characters are shocked out of their fanatical fatuities. But no one is spared. Even the censorious Macilente is given a verbal drubbing, his cankering envy bared. No little personal bitterness entered into Jonson's writing of this play. It is obviously about people he knew at first hand, in particular several of his rivals on Henslowe's payroll. Clef, who spouts fustian, can be recognized as Marston, and the vainglorious Puntarvolo is Sir Walter Raleigh, insisting that he and his dog can journey to Constantinople and back home. Macilente poisons the dog. Brisk's mounting indebtedness lands him in gaol.

Many of Jonson's collaborative works are lost or cannot be identified. In 1600 *Cynthia's Revels, or The Fountain of Self-Love* engaged him further in the War of the Theatres, as the feud came to be called. His chief target was Marston, no mean adversary. In fact Jonson claimed that on one acrimonious occasion the bitter Marston had pointed a pistol at him, which had to be wrested away in a hand-to-hand scuffle that ended with Jonson pummelling his fellow poet. *Cynthia's Revels* was a more literary riposte. Setting out to present Elizabeth's court in an allegory (that the queen was the Goddess of the Moon, Cynthia, was apparent to everybody), Jonson chose for his new victims the posturing, affected courtiers who surrounded her. With this mythological vehicle, in which a mischievous Cupid and Mercury play leading parts, Jonson shot his tart stings and darts. Another butt of his satire was Dekker. At a masque, near the close, the hypocrisies and pretensions of all the characters are stripped away. Though topical allegory of this sort is of scant interest today, it occasioned a considerable scandal at the time, and Jonson had another hit to add lustre to his name.

Marston retaliated with *What You Will*, in which Jonson was again held up to scorn; and Marston joined with Dekker to prepare still another attack in *Satiromastix*. Jonson, having advance word of this, came back with *The Poetaster* (1601). As has already been suggested, to some extent the theatrical companies were probably prompting the feud or allied themselves to it to sell tickets. But there was some actual friction between Jonson and the several troupes that produced his plays; each of them resented his working for competitors.

Opinions about *The Poetaster* vary somewhat among critics, but most agree that it is a better piece of work than *Cynthia's Revels*. Once more using allegory, Jonson resorts to an imaginary Roman scene and introduces such figures as the poet Horace (obviously intended to represent Jonson himself and hence depicted most sympathetically), as well as Virgil and Ovid. At the court of Caesar Augustus the true value of art is acknowledged and rewarded, and once more pettiness and charlatanry are exposed, the principal sufferers being the unfortunate Marston and Dekker, who are thinly disguised by classic names. Marston is Crispinus, who is given a purge by Horace and vomits up chunks of florid and inchoate verse actually lifted from Marston's writing. Dekker appears as Demetrius Fannius.

Much of the rhetoric of this satirical comedy is zestful; there are sharp scenes, though the level of them is uneven. The plot lacks unity; Ovid's affair with the emperor's daughter, Julia, distracts from the Horatian half of the story, which enters too late. The tone is also marred by anachronisms, and that may have been one cause of *The Poetaster*'s failure to win as much approval as its predecessor. Jonson, who did not take a setback lightly, seemed to lose interest in the feud and to realize that the public was weary of it. He announced his decision to withdraw from the absurd contention. He was also ready, he declared, to try his hand at serious drama:

> Since the Comic Muse
> Hath prov'd so ominous, I will try
> If Tragedy hath a more kind aspect.

Another reason for the war's ending was the issuance of the other side's *Satiromastix*, which followed *The Poetaster* and gave his foes the last word, of which they took rich advantage. In it the Jonson character is tossed in a blanket and crowned with nettles – yet the authors grant him his share of virtues. In any event the hostilities were over, and Jonson was soon busily collaborating with Marston again, as though no ill will had ever existed between them.

Accompanying his rise to prominence in the public's eye there had been gathering around him a circle of admiring, talented young writers and outright disciples, among them the youthful Beaumont and Fletcher. They would join for convivial meetings in the Mermaid Tavern and soon elected to call themselves the Sons of Ben. They were fascinated by his uninhibited talk, his store of anecdotes, his iconoclasm, his practical experience and common sense about the theatre world, his energetic intellect. Another quality that drew them to him and held them was his self-assurance, often lacking in the insecure young and by them much sought. He preached neoclassicism and through his own writings and their contributions to the stage he slowly began to dominate the London theatre and the new direction that impended.

He sometimes spoke with a forked tongue. He bragged of his close friendship with Will Shakespeare, loved "just this side of idolatry", but after the Bard's death described him as ill-educated, verbose. He solicited prefatory verse and praise from Beaumont when *Volpone* was printed but commented after the young man's passing that "Beaumont was too full of himself and his poetry".

In his own day Jonson came to overshadow even his "good friend" William Shakespeare, but this was not only because he had a wide-ranging intelligence, as well as skill as a dramaturge and a vigorous eloquence, but also because his was a flamboyant personality. His laurels do not rest on the brow of a quiet, reserved spirit. Ben Jonson was a boisterous, quarrelsome, boastful, witty, shrewd, highly indignant poet.

Sejanus (1603) was Jonson's entry as an author of tragedy; it was staged by Shakespeare's company. He prepared for this serious effort by first fulfilling an assignment to revise *The Spanish Tragedy* and delivering a drama, *Richard Crookback*, now lost, both for Henslowe. The title suggests that the vanished script was about Richard III. A comment by Jonson lends support to the belief that, as a member of the producing company, Shakespeare made certain changes in the acting version of *Sejanus* – in Jonson's words a second pen had a "good share" in it, and the wielder of the second pen was "so happy a genius". But beyond this the reviser is not named; some suggested it was Jonson's close friend Chapman. Whoever he was, he could not save *Sejanus*; though earnest, this ambitious work is dull. It is based on a historical incident during the reign of the Emperor Tiberius, with Jonson gathering his material largely from the satires of Juvenal and the *Annals* of Tacitus. Furthermore, it seems to have been his aim to furnish his tragedy with more factual detail than Shakespeare had used for his Roman drama, which the arrogantly self-confident Jonson may have wished to emulate or outdo.

Sejanus, a Machiavellian, pits himself against the emperor, Tiberius, hoping to replace him as ruler. He first ingratiates himself and seduces the debauched Livia, the Empress, and poisons Drusus, Tiberius' son and heir. Step by step he cunningly removes his rivals by devices and traps and deprives the Roman subjects of their civil rights. Livia plots to kill her husband. Sejanus steadily mounts to power, but Tiberius is his equal in infamy. When the emperor finally grows suspicious he has Sejanus dogged by spies, tricks him into being left unguarded, accuses him in a letter to the Senate. In a popular uprising Sejanus is finally slain. The mob tears him to pieces. Despite all these violent incidents the play is often very static; Jonson succeeds better at the creation of atmosphere than at instilling life into his monstrous contestants, both of whom are vile in their selfishness and cruelty.

Much of this play is meant to be a veiled picture and scarcely muted comment on Elizabeth's court in the declining years of her reign. The queen was old, ill. Corruption was at its height, and struggles for power among her favourites were deadly and multiplying. Essex had been executed for his rebellion. *Sejanus* is Jonson's warning of the crumbling of civic righteousness in England. But despite the drama's bold and admirable purpose it fails to command the stage. Parts of it are translated directly from Jonson's historical sources, which has led it to be described by Hazlitt as a mere "mosaic", but this is unfair as three-quarters of the dialogue is original. (Jonson, proud of his erudition, did append footnotes to his printed version of the work.) Some scholars are enthusiastic about *Sejanus* because certain episodes, such as the book burnings and persecution of liberal intellectuals ordered by the tyrant, could be applied to twentieth-century events.

The play is not without sardonic humour, moments of what has been called "stylized farce", which reflected the general contempt in which Sejanus – and his creator – held the average man. Most of those surrounding Sejanus and Tiberius are typically opportunistic courtiers, servile and fickle. Despite Jonson's long battle for a return to classical form, he violates the cardinal precept that tragedy and comedy should not be mixed, and he dispenses with a chorus and crowds his play with extraneous characters – there are forty roles – and frequently digressive themes. The verse is at moments Marlovian, echoing the cadence and sound of *Tamburlaine* and *The Jew of Malta*, particularly in denunciatory passages:

> On then, my soul, and start not in thy course;
> Though Heaven drop sulphur, and hell belch out fire,
> Laugh at the idle terrors: tell proud Jove,
> Between his power and thine there is no odds:
> 'Twas only fear first in the world made gods.

The poetry, however, has a substance and rigour that is his own, uniquely Jonson's. He sought to develop a style that would be "strict and succinct". He forges lines that are masculine, deliberate and economical, yet without monotony; very different from Shakespeare's, where always fluent and delicate overtones abound.

> These can . . . cut
> Men's throats with whisperings.

That, in *Sejanus*, is the sound of Jonson. He has also a remarkable and continuing fondness for alliteration which marks his verse here and later.

Sejanus, written to conform to what Jonson thought to be classical rules, was an utter failure, though – surprisingly – it won the unstinted praise of Marston, and of others, as has been noted. It took Jonson some time to recover from its harsh reception. He shifted his thoughts to comedy, perhaps realizing he had more innate gifts in that medium, and the result was to be five of the most racy and amusing, full-bodied farcical works in English literature. The first of these, however, was to involve him in serious trouble with the government: *Eastward Ho!* (1605), written with Marston and Chapman.

The title partly echoes that of *Westward Ho!* (1604), by Dekker and Webster, a work that only a year before had helped to introduce realistic comedy to the London stage. This new play was meant to set a new trend, one that was the antithesis of the cynical, immoral farces and romantic scripts that rival companies were offering, while at the same time it would top the piece by Dekker and Webster in its fresh representation of middle-class London life. The titles *Westward Ho!* and *Eastward Ho!* referred only to crossing the bustling Thames to see what was happening on the other side of the river. The triply-authored *Eastward Ho!*, though a rather light script, shines forth as a "citizens' comedy", a lively depiction of the contemporary scene. A moralizing work, its principal figure is a virtuous apprentice, Golding, who marries Mildred, one of his master's daughters, and by honest work rises in status and attains many honours, becoming a Freeman of the City and a deputy alderman. For contrast, another apprentice, Quicksilver, wastes his possessions in riotous living, aping his betters, until he ends in the tight grasp of the constabulary. He repents and is saved from debtors' prison by his kind-hearted former master, Touchstone, a goldsmith. The vices of the "humorous" characters are not as pronounced as in the plays that Jonson composed wholly by himself. A second daughter of Touchstone, Gertrude, wishes only to become a "lady"; she weds a knight, the dissolute Petronel Flash, who recklessly spends her money, then abandons her. She is left in such penury that she must pawn even her petticoat and stockings, until her sister comes to her assistance.

A sub-plot concerns the colonization of the new continent of America. Gertrude's husband has invested all his money to equip a ship that is to voyage to Virginia, where rumour says gold is plentiful. The journey is never begun, however, because the knight gets too drunk. He and his fellow adventurers are wrecked in a storm on the Thames and their ship never actually sets sail. He is arrested but also repents and is released through Golding's generous intercession. It was in connection with this sub-plot that the lines offensive to King James I were inserted, with the consequence that Jonson and Chapman were seized, though the remarks that disparaged the Scots are believed to have been Marston's. As has been noted, Marston seems to have escaped imprisonment – the details of the story are unclear – which raised anew Jonson's always latent ire towards him.

Even before the *Eastward Ho!* incident Jonson had been under suspicion for "popery and treason". Some passages in *Sejanus* sounded subversive, and very likely were. The Earl of Northampton, who was personally hostile to him, summoned him before the dreaded Star Chamber for questioning. The charge had been dismissed, most likely because he had won King James's favour by having collaborated on the formal procession that welcomed the new monarch to London, as well as on several elaborate masques presented at court.

Even royal approval did not spare Jonson from being gaoled for the insult in *Eastward Ho!* (One version of the story, however, says that Jonson had joined Chapman voluntarily in gaol.) The authors faced severe punishment. One possible penalty was that each man might have his nose sliced off. But Jonson had friends at court, noble Englishmen who had secretly enjoyed the jibe at the upstart Scots brought south by the Stuart king. Besides, the writer had composed beautiful masques for them; they knew him well.

To celebrate the playwrights' release a banquet was given them by their fellow writers and friends, the literary world of London. Attending it was Jonson's aged mother, who caused a sensation by exhibiting a packet "full of lusty strong poison" that she had planned to give to her son, mixing it into his drink, if the facial mutilation had been carried out. She added that she, too, to "show she was no churl", expected to partake of the poison, drinking it a few moments before he did. With such heroics on display the banquet was a great success.

In 1605 Webster and Dekker replied with *Northward Ho!*, adding another image of problems, domestic and civic, experienced by ordinary Londoners, many of their conflicts and troubles resulting from their suspicions of one another.

As evidence that Jonson was once again in the good graces of the king, he was shortly afterwards enlisted by the Privy Council to seek for clues in solving the Gunpowder Plot. But his efforts at espionage seem not to have been noteworthy, and he was soon back at his profession as a writer.

His term in prison had, if anything, darkened his mood and deepened an aversion towards his fellow men. He was to demonstrate this in his next and perhaps finest play, *Volpone, or The Fox* (1605–6). Certainly in it his poetic endowment is shown more colourfully than anywhere else in his laughter-provoking stage-works. He had, of course, established himself as a poet in his masques and in short lyrics, the numerous songs mostly translated from the Latin, such as his "To Celia", which begins "Drink to me only with thine eyes . . ." In *Volpone* he brings his superb talent to the composition of a farce in which the language is possessed of a lustrous quality found nowhere else in English comedy save alone in Shakespeare.

Volpone is a scathing play; Jonson's rage against the mercenary temper of his times is given full voice. Ever since the decline and death of Elizabeth and accession of James I, life in England had lost much of its shine. The Stuart court set a bad example by its loose morality. Corruption was prevalent. An aggressive middle class was dispossessing the landed gentry of their estates and influence. If the country gentry had tended to be honest, it was in part because inherited property and ample funds allowed them to live at ease; they did not have to scrounge and scramble to get ahead. But the spirit of the Renaissance was becoming more materialistic. Much of its commerce was shady and illicit. Merchants were increasingly predatory, and too many in the aristocracy and upper classes were hastening their financial ruin by drinking, gambling and extravagance. There was competition among the "best families" in having designed and erected grand manor houses – proclaiming their importance – and collecting fine Italian paintings and French furnishings and engaging in extensive foreign travel, neglecting affairs at home.

Jonson was scarcely the only dramatist to scourge his contemporaries for their derelictions. In medieval times the simplistic Moralities, which his plays resemble in many ways, had made the cost of vice the object of their pious lessons, but no one has ever written of human faults and follies with such fury as his. He has rightly been likened to Aristophanes in his temper.

Venice and its gleaming *palazzos* is the scene of this most savage farce. Petrarch gave Jonson ideas for the plot. He is said to have spent only forty days on it.

An aged sharper, Volpone, lives in sumptuous style on other people's money. He pretends to be fatally ill, a ruse that draws to his bedside numerous dupes, to each of whom he holds out the chance of imminently being his sole heir. They bring him costly gifts, gold and silver plates, jewels, silks. One, the merchant Corvino, overwhelmed by cupidity, even offers his virtuous and carefully sequestered wife Celia as bait, misled into thinking the "ailing" Volpone to be impotent. In his cunning and lecherous schemes the ruthless old scoundrel is abetted by his agile-witted "parasite", Mosca, who if anything is his peer as a cheat. Mosca delights in his skill at mendacity and duplicity. At one point, after yet another coup, he muses exultantly:

> I fear I shall begin to grow in love
> With my dear self and my most prop'rous parts;
> They do so spring and burgeon; I can feel
> A whimsy i' my blood. I know not how,
> Success hath made me wanton. I could skip
> Out of my skin, now like a subtle snake,
> I am so limber. O! your parasite
> Is a most precious thing, dropped from above . . .

He concludes, from this self-congratulatory soliloquy:

> All the wise world is little else in nature
> But parasites or sub-parasites. And yet,
> I mean not those that have your bare town-art,
> To know who's fit to feed 'em; have no house,
> No family, no care, and therefore mould
> Tales for men's ears to bait that sense; or get
> Kitchen-invention, and some stale receipts
> To please the belly, and the groin; nor those,
> With their court-dog tricks, that can fawn and fleer,
> Make their revenue out of legs and faces,
> Echo my lord, and lick away a moth.
> But your fine elegant rascal, that can rise
> And stoop, almost together, like an arrow;
> Shoot through the air as nimble as a star;
> Turn short as doth a swallow; and he here,
> And there, and here, and yonder, all at once;
> Present to any humour, all occasion;
> And change a visor swifter than a thought,
> This is the creature had the art born with him;
> Toils not to learn it, but doth practise it

> Out of most excellent nature, and such sparks
> Are the true parasites, others but their zanies.

As befits a lacerating fable, all the characters are given the names of beasts of prey: Volpone, the Fox; Mosca, the Gadfly; Corvino, the Crow; the doddering Corbaccio, the Raven; the lawyer Voltare, the Vulture; all are avaricious, monomaniacal in their zeal to grow even richer, for none is poor. They try to outbid one another to obtain Volpone's favour. Quite as depraved as Corvino, who would degrade his wife for gain, the aged Corbaccio offers to disinherit his son and name Volpone his presumptive heir, as a ploy. In the end both Volpone and Mosca – who boldly seeks to out-trick his daring master – over-reach themselves. Volpone, wearying of the game, gives out word that he has died and appoints Mosca his heir. Mosca takes advantage of this to seize his master's fortune. The legal complications lead to the intriguants' exposure; they are stripped of all their ill-won possessions. Volpone is sent to prison, Mosca to the galleys. Suitable punishments and harsh disgrace and humiliation are visited on the grasping covey that sought to share most of Volpone's wealth. So, at last, poetic justice is meted out.

Volpone, while in full sway, is properly called a "*magnifico*". Sensual, audacious, deceitful, excited by his power over others, he towers. But the devious Mosca is no less venal and compelling. The role of the half-senile Corbaccio also lends itself to a virtuoso performance on stage. What saves this ugly story from being tragic or repulsive is the sheer effrontery of the two connivers and the egregious avarice of their victims. In addition, the setting is opulent. Even more splendid is the poetry. Volpone's first speech as he catalogues his plunder, with its echoes of Marlowe's *Jew of Malta*, sets the exalted tone. He holds up the gems he hoards in an ornate casket and studies them with intense pleasure; it is a fascinating inventory. Everything is on such an exaggerated scale that the misdeeds of the characters conjure laughter. Spectators begin to sympathize with the extreme stratagems of Volpone and his henchman, and by contrast to their brazen rascality the conventional virtues of Celia and Corbaccio's ill-treated son, Bonario, seem pallid and inadequate, a paradoxical consequence of Jonson's genius for portraying villainy in the grand manner. There is nothing petty about Volpone, yet no one can deny that the portrait is realistic. Every generation has its larger-than-life Volpones.

Nowhere does the poetry reach a more glorious pitch than in Mosca's initial description of Celia, then Volpone's seductive invitation to her after he has cozened her husband into bringing her into the bedchamber of his *palazzo*. Earlier, when Volpone asks, "Has she so rare a face?" Mosca replies:

> O, sir, the wonder,
> The blazing star of Italy! a wench
> O' the first year! a beauty ripe as harvest!

Mosca adds that she has "a soft lip, / Would tempt you to eternity of kissing!"

With these words the shrewd, sybaritic and persistent Volpone seeks to lure the tremulous woman into his bed:

> Thou hast in place of a base husband found
> A worthy lover; use thy fortune well,
> With secrecy and pleasure. See, behold,
> What thou are queen of: not in expectation,
> As I feed others, but possessed and crowned.
> See, here a rope of pearl, and each more orient
> Than the brave Egyptian queen caroused;
> Dissolve and drink 'em. See, a carbuncle
> May put out both the eyes of our St Mark;
> A diamond would have bought Lolia Paulina
> When she came in like starlight, but with jewels
> That were the spoils of provinces; take these
> And wear. and lose 'em; yet remains an earring
> To purchase them again; and this whole state
> A gem but worth a private patrimony
> Is nothing; we will eat such a meal.
> The heads of parrots, tongues of nightingales,
> The brains of peacocks, and of estriches
> Shall be our food, and could we get the phoenix,
> Though nature lost her kind, she were our dish.

This is gorgeous writing, even if the exotic feast probably sounds better than it would taste. Nor does Celia yield, for at a propitious moment she is rescued by Corbaccio's son, Bonario.

A somewhat lengthy extraneous sub-plot revolves about three English tourists: the heartless, youthful Peregrine; the fatuous Sir Politick Would-be (the Parrot); and his loquacious wife, the "madam with the ever-lasting voice", who nearly deafens and drives the impatient Volpone mad. Some of the best scenes are those in which the old Fox, feigning mortal illness, tries to fend off visits from the gossipy, blue-stockinged Englishwoman who is travelling abroad to study fashions and to exhibit her own learning. Her husband, also a chatterbox, plans to enrich himself by selling Venetians a three-year supply of red herrings that he has purchased in Rotterdam. He also claims to have miraculous cures for various physical complaints. Through the manipulations of Peregrine, a cruel jokester, with the assistance of some local merchants, Sir Politick is led to believe he faces incarceration on a charge of high treason against the Venetian state, for trying to sell the city to the Turks. He and Lady Would-be are panicked and prepare for their immediate departure.

Jonson points out in his prologue that his play stays within the bounds of the three unities. Speaking of himself as the "author" he remarks:

> The laws of time, place, persons he observeth:
> From no needful rule he swerveth.

As seen in the instance of *Sejanus*, however, he does not advocate a slavish following of classical form. He looks upon the Greeks and Romans "as guides, not commanders". He adds: "For to all the observations of the ancients, we have our own experience; which, if we will use and apply, we have better means to pronounce." Similarly, he does translate lines from classical poets as much as he paraphrases and embellishes them. In consequence – as with Shakespeare – the borrowed images are often enhanced. Dryden complained that Jonson was a plagiarist whose erudition helped him to steal widely from earlier writers: "You track him everywhere in their snow." This is only partly true. Whatever Ben Jonson filches, he polishes and much improves. He had a fine ear, besides; his is a virile cadence, and his mastery of London's argot – displayed in later works – was certainly his own accomplishment.

The exceedingly clever and high-spirited plot of *Volpone* attracted a number of twentieth-century adaptors. The most notable has been Stefan Zweig, whose freely cut and rearranged version in German had great success in his country, after which it was translated into French by Jules Romains, and then – ironically – into English. Louis Jouvet subsequently made it into a film in France (1941). The English Shakespearian actor-manager Donald Wolfit revived the original play in 1938 and included it in his company's repertory, bringing it to the United States on a visit there. *Volpone* came to life again in a production by the British National Theatre, with Paul Scofield and John Gielgud in the lead roles (1976). In the United States there was a major staging almost every decade, beginning in the 1930s with an opulently mounted production by the Theater Guild, in which Alfred Lunt was an agile limping Mosca and Dudley Digges a wily *magnifico*. Gene Frankel directed Zweig's effective treatment of the shortened script at the Rooftop Theater in the mid-1960s. Brustein, during his first year at Yale, invited an outside troupe to put on a "randy" and "scatological" interpretation of Jonson's masterpiece that provoked protests from subscribers. Indeed there was a cluster of rebirths of the script. Joseph Papp's Shakespeare Festival, making use of a new mobile stage, gave open-air showings of the farce, touring about New York's far-flung boroughs – some of them black and Hispanic ghettos – where the spectators' reaction was unpredictable. (Having the comedy on a truck hearkened back to the medieval pageant-plays.) If onlookers disliked a line or situation they sometimes hurled bottles and fruit at the actors. The cast was racially mixed. The Mosca, Roscoe Lee Browne, was an Afro-American.

In the Midwest, at the Tyrone Guthrie Theater in Minneapolis, the eminent Guthrie personally guided a presentation that, characteristically, emphasized physical humour and by-play at the expense of verbal felicity (1976). The very same year saw the conversion of the play into an even broader offering by Larry Gelbart. The scene was moved to a mining camp in the north-west. George C. Scott was the principal; later, in the long and prosperous run, he was succeeded by Robert Preston. Following came a less ingratiating musical comedy, *Foxy*, headed by the comic – a great clown – Bert Lahr, and inevitably an English-speaking film, *The Honey Pot*. The possibilities of Jonson's story and characters have proved to be inexhaustible. It has been turned into an opera by the American composer George Antheil (1900–1959). But *Volpone* as Jonson himself wrote it has outlasted all variations and never fails to garner shocked laughter and applause.

His enemies now faulted him for working very slowly. He did allow three or more years to pass before his next play made its bow, *Epicene, or the Silent Woman* (c. 1609), a piece with a somewhat brighter and gentler tone. It is the favourite of some Jonson scholars; they prefer it even over *Volpone*, perhaps because it is less strident. It was written during a period when things were going well for him and his mood was quieter, less turbulent.

Even if less angry than *Volpone*, this script has its cruelties. An old man, Morose, a hypochondriac tormented by any kind of noise, is persecuted by his heartless nephew, Dauphine, who wants to extract money from him. He tricks his domineering uncle into wedding an apparently shy, submissive girl, who, once the match is forged, turns into a shrill virago granting him no respite. The rowdy, bad-mannered young wife invites her friends in to plague her husband with wild, loud parties. Dauphine offers to free his uncle of this disastrous marriage and its attendant commotion in exchange for a large sum of money. When Morose, driven half-mad, helplessly agrees to this bargain, the nephew reveals that his new "aunt" is actually a boy in disguise (a fact hitherto withheld from the audience). The whole scheme has been devised by the nephew, who has recruited the boy for this sole purpose. This is one comedy that does not end with a marriage but to the contrary.

The play is brilliant, the story tight and well handled, with many scenes that unfold a panorama of London's most dissolute society, both its high and low strata. Taverns and boudoirs provide the background. *Epicene*, too, verges on the comedy of manners, and when – during the Restoration – Pepys saw it, he wrote enthusiastically: "There is more wit in it than goes to ten new plays." On stage it presents a colourful vista of Jacobean life, though hardly a flattering one. Jonson is still the relentless critic of his age and its many forms of depravity. A host of idiotic minor characters, Sir John Daw, Sir Amorous Foole, the Collegiate Ladies, Captain and Mrs Tom Otter, round out the picture. Set against them, the fun-loving Truewit and sketchier Clerimont somewhat right the balance; they at least are gentlemen and scholars. Here and there, in the dialogue, Jonson borrows lines from Ovid and Juvenal.

Epicene was not immediately successful; its surprise ending apparently disconcerted its first audiences, but before too long it won its way in London, and company after company revived it. A presentation of it at court occurred during Jonson's final year.

In 1610 *The Alchemist* arrived. Some believe that, sobered by the initially poor reception of *Epicene*, Jonson reconsidered the direction his comedies should take. *The Alchemist* is more in the sombre spirit of the immediately acclaimed *Volpone*, a lashing indictment of the follies of the day and mankind's eternal cupidity. The subject was especially topical because "alchemy" had become a craze: the importation of precious metals from America had increased the demand for coins. Even kings – and Queen Elizabeth – lent a willing ear to the promises of charlatans who claimed to be able to transmute lead into gold. The agency for doing this would be the philosopher's stone. Many books were published on the subject, and Jonson had delved deeply into this mystical abracadabra – it was doubtless a theme that appealed to the poet as well as the playwright in him. His play abounds with technical terms drawn from its occult lore. He was also fully aware of the trickery practised by all manner of rogues in London's underworld, with which he maintained an intimate acquaintance. The vogue for dabbling in this pseudo-science was widely exploited by frauds who operated in the city, and Jonson saw the gullibility of the public as another symptom of the avidity of the age: people would do anything to enrich themselves. Outward respectability fell away when any chance to acquire wealth presented itself. Besides alchemy, other kinds of get-rich schemes were plentiful.

The plot of *The Alchemist* is partly owed to the *Mostellaria* of Plautus, specifically the opening and closing scenes, but much of it is Jonson's own, carried out with rare ingenuity. The well-to-do London citizen Lovewit has left the city to avoid the plague; since his elegant town house is standing empty, save for one servant in charge, two rascals – Subtle and his bawdy companion Doll Common – move in and take possession, deceiving others into thinking them to be persons of substance. The servant, Face – afterwards a favourite role for David Garrick – abets their plans. Among their easily fleeced victims are Abel Drugger, a tobacconist; Dapper, a dim-witted law clerk; and the great sensualist, Sir Epicure Mammon. But also they attract and cozen the hypocritical Puritan cult-leaders Ananias and Tribulation Wholesome, who dream of becoming millionaires overnight. Despite their ascetic garb and cant they are as money-hungry as their most acquisitive neighbours. This phase of the play introduces a vein of religious satire that Jonson was to enlarge upon in subsequent works. Eventually, after many complications and crossed lines of story, the farce culminates in an ear-shattering quarrel of vast dimensions. To witness *The Alchemist* is often to spend an excessively noisy evening in the theatre, but all is set to rights by the return of Mr Lovewit. Learning of the deception practised in his absence, he is able to fend off the mob that is angrily besetting his house, and – being a tolerant gentleman – even pardons the errant Face. Lovewit himself marries a widow, Mrs Pliant.

The manner in which Jonson handles this flow of action, with its rapid pace and rising tension, has been widely admired. Coleridge ranks *The Alchemist* along with Sophocles' *Oedipus* and Fielding's novel *Tom Jones* as having the three most perfect plots in literature. Perhaps such praise is a bit overstated, but the play's swift, neat, efficient telling is remarkable, and the classical unities are fully honoured. The dialogue is often coarse, as the subject calls for, but it is also poetic in its raciness, a vivid argot, with passages in a higher stratum. Some of these are given to Sir Epicure Mammon, who reminds one no little of Volpone, though he is less sinister. He is a sybarite of the same stamp. He tells how he will spend the riches he hopes to acquire by his new venture:

> I will have all my beds blown up, not stuft;
> Down is too hard: and then, mine oval room
> Fill'd with such pictures as Tiberius took
> From Elephantis, and dull Aretine
> But coldly imitated. Then, my glasses
> Cut in more subtle angles, to disperse
> And multiply the figures, as I walk
> Naked between my succubae. My mists
> I'll have of perfume, vapoured 'bout the room,
> To lose ourselves in. . . .
> . . .Where I spy
> A wealthy citizen, or a rich lawyer,
> Have a sublimed pure wife, unto that fellow
> I'll send a thousand pound to be my cuckold.
> . . . I'll have no bawds,
> But fathers and mothers; they will do it best,
> Best of all others. And my flatterers
> Shall be the purest and gravest of divines,
> That I can get for money. My mere fools,
> Eloquent burgesses.

This is cynical, extravagant poetry and bespeaks the libidinous and luxurious imagination of its author. The role of Sir Epicure and his quest for an elixir of youth fascinated Charles Dickens, who even gave thought to playing it. The characters, though humours, are always very much alive. To Pepys *The Alchemist* was "most incomparable". He saw it repeatedly over the years. It was as well received as *Volpone* when first produced at the Globe, and then on tour through the provinces by the King's Men. One place where it received much applause was Oxford, where it was staged alternately with Shakespeare's *Othello*.

The fourth of Jonson's great farces is *Bartholomew Fair* (1614). He was prosperous; he was well paid for the court masques he composed – and was able to work more slowly and carefully on his theatre-pieces. Four years elapsed between *The Alchemist* and *Bartholomew Fair*. The preceding spring – in 1613 – he had gone abroad as tutor to the son of Sir Walter Raleigh, though the father was then imprisoned in the Tower. In France, Jonson's noble pupil got him drunk and rode him through the streets of Paris in a wheelbarrow, all the while abusing him with profanity, an incident that Jonson himself took high pleasure in relating afterwards. Such behaviour did indeed comport with that of his riotous characters in *Fair*, written for a new theatre, the Hope, in which plays took turns with bear-baiting. Cognizant of the raucous, uncouth audience apt to gather there, he prepared a comedy that was even noisier than *The Alchemist* and set it at the Smithfield carnival. Here is assembled a broad cross-section of London's lowest life – bawds, pickpockets, ballad-mongers, vendors of roast pig and beer – against a crowded background of puppet shows and the booths of other popular entertainments. The piece has scarcely any formally shaped plot, no discernible beginning, middle, end; it is a series of sometimes uproarious, other times satirical and ironic episodes, betraying in particular the fraudulence of a religious cult leader, a Puritan, self-described as a Rabbi, Zeal-of-the-Land Busy, as well as the vagaries and designs of a dirty pig-woman, Ursula, and a cutpurse, Edgeworth, all of whom weave in and out of the action to provide the successive incidents with a loose connection. The language is again coarse and even brutal but wonderfully colloquial. The Christian "Rabbi" becomes drunk and ends ignominiously in the stocks. Among the faithful followers and dupes are Mistress Purecraft and her pregnant daughter Win-the-Fight, who develops a sinful craving for pork, forbidden by the Old Testament taboo. The way in which she wins a dispensation from Rabbi Busy to eat this delicacy adds laughter to one of the more amusing incidents in this loud and merry caper.

Another amusing person is Troubleall, who is forever demanding "warrants" from the others – what "licence" have they, that justifies their deeds? A former court officer, he demently thinks there should be documentary sanction for everything one does. Such documents should be obtained from Justice Overdo, who is well intentioned but lacks a sense of proportion: he concerns himself with minor miscreants and neglects to punish the major criminals. Or when he does pass sentence on the right rogues it is for the wrong reasons. The first name of the Justice is Adam, and he is finally reminded that he is only human like other men and as liable to error. Then there is Quarlous, who at the end marries the wealthy widow Purecraft. He, though purportedly a gentleman, is actually a knave. His only motive is money and he quickly shifts his suit from one woman to another, only because she is richer. Other rascals include Knockem and Whit, and neat bits of drawing yield the irrational Humphrey Wasp, Cokes the gull, Haggis and Bristle the watchmen, and Mrs Littlewit and Mrs Overdo, whose need to relieve themselves occasions much of the ensuing commotion.

Jonson announced in his prologue that his intention was "to delight all and offend none", and at this he succeeded. It is also a work in which he paid no heed at all to the classical unities he was usually promoting, defining and defending.

In the period between *The Alchemist* and *Bartholomew Fair* Jonson had once more attempted a tragedy. *Catiline* (1611), also produced by the King's Men, was again a failure. An account of Catiline's conspiracy in Ancient Rome, it possesses several merits lacking in *Sejanus*, but that does not spare it from boring its audience as much or even more. Pepys was to write of it that it might read well, but on stage it was at its worst and proved for him a play "the least diverting that I ever saw in my life". The first two acts, at least, were generally said to excel those of *Sejanus*, but in a later preface Jonson declared them to be poorer than the rest of the script, which may have been why – in his opinion – they appealed more to an audience he held in contempt.

The form of *Catiline* is more in accord with that of a Senecan work: in the opening scene a ghost, Sulla's, urges Catiline to accomplish Rome's downfall. Jonson borrows largely from Cicero and Sallust and follows his historical sources too closely. His dialogue is often directly translated from their texts. The play has a chorus which passes moral judgements between the acts. Many passages have a rhetorical splendour, but the weight of declamation is too heavy and at times becomes deadly . Longest of all is a speech by Cicero in the fourth act. The drama has more unity than *Sejanus*, and the plot is more comprehensible.

All the characters are convincing, with Catiline outstandingly so. He is glamorous and dynamic, if evil, and dies courageously. Some of the lesser figures, such as Lentulus and Cethegus, are reduced to being governed by a "humour", and the Roman ladies of light virtue who have political whims give Jonson a chance to display his true talent for scarifying comedy. These episodes, the best in the play, are original interpolations, not found in Cicero and Sallust. One point that Jonson makes with deft irony is that a comparatively minor affair, the intervention of Fulvia, a courtesan who has learned of the plot from a lover, frustrates the conspiracy and fatefully alters the history of the Republic.

Catiline is a born aristocrat whose wish is to overthrow the democratic state. His pretext is that he is acting on behalf of the poor against their oppressive rulers, while he is, in fact, an egocentric revolutionary, fascinated by the idea of destruction for its own sake. He is an insincere and immoral advocate of freedom – really, licence – and dangerous. Cicero, dominating the second half of the drama, is idealized, a model of pragmatic wisdom, altruism and genuine patriotism. But he has to employ ignoble means to attain worthy ends. Caesar vacillates from Catiline's side to the opposition. He is an opportunist, a portrait that departs from Jonson's sources. The picture of all these men grappling for political power is disillusioning.

In 1616 another comedy, *The Devil Is an Ass*, brought Jonson back to practise in that genre. It also signals his decline as a craftsman in that form. Though adroitly worked out in places, it lacks the force of the scripts that had immediately preceded it. In a clever and ingratiating prologue he asks gentry who have bought seats on the stage to leave a bit of elbow room for the actors. In this fable Pug, a lesser devil, gets permission to visit London and finds it more infernal than Hell itself. The "Morality" strain so strong in Jonson's plays is even clearer here. The hapless devil is worsted by the raffish and conniving Londoners he meets, soundly trounced, clapped in gaol as a thief. Satan and his henchman, Vice, wish to spare Hell the humiliation of having one of its minions end on the gallows at Tyburn; so they rescue Pug from Newgate prison, bearing him off concealed by a cloud of smoke and brimstone. Along with this story-line is one about the asinine Fitzdottrel, a country squire, who hires Pug as a servant. Anxious to obtain an estate that would be enjoyed by his heirs for ever, he is totally duped by a scoundrel, Meercraft, to embark on a crazy project, reclaiming property under water. He will then become Duke of Drowndland.

Discovering that he has lost all he owns he pretends to have been bewitched when signing the unfortunate contract, thus making it void. This ploy is not successful, and he is exposed. But his fortune is retrieved by Wittipol and Manly, two would-be seducers of his wife. Her adamant resistance to their advances evokes their pity and admiration; for her sake they step in to rescue her foolish husband from his plight.

Implicit social criticism is introduced here. King James, eager to get hold of funds denied him by Parliament, granted monopolies to exploiters who concocted wild proposals. There was a rising frenzied speculation. Meercraft is a prototype of men of that ilk, indifferent to whom they might ruin. Jonson shows amazing shrewdness and first-hand knowledge of shady business schemes and how they were manipulated, how officials were bribed, shares sold. One of Meercraft's endeavours is to supply toothpicks to the whole kingdom, which is like the harebrained plans of Sir Politick Would-be in *Volpone*. Meercraft's endless flow of ideas are half-fantastic, half-feasible. Jonson himself would have made an able businessman had he turned his talents wholly in that direction. Surrounding and abetting Meercraft are a group of corrupt courtiers, among them Lady Tailbush and Lady Eitherside, and the parasitic Everill.

The Devil Is an Ass is a patchwork of new and old material, hastily put together. Yet much as it is below the four great comedies that came before it, it is superior to those that followed at ever wider intervals. The dialogue is often paraphrastic, this time of Horace. Jonson reapplies and improves what he appropriates from others, however, as when he touches on a familiar theme in his work, intimations of mortality:

> ... think,
> All beauty doth not last until the autumn:
> You grow old while I tell you this.

Fitzdottrel suspects his wife is having an affair with Wittipol and rebukes her:

> O bird,
> Could you do this? 'gainst me! and at this time now!
> When I was so employ'd, wholly for you,
> Drown'd in my care (more than the land, I swear,
> I have hope to win) to make you peerless, studying
> For footmen for you, fine-pace huishers, pages
> To serve you on the knee. . . .
> You've almost turn'd my good affection to you;
> Sour'd my sweet thoughts, all my pure purposes . . .

All this while Jonson was busy with the composition of masques, the form of entertainment at which he was incontestably the finest craftsman in English literary history. The masque, as has been noted, was a further development of the Italian pastoral, with an abundance of mythological material. The genre was imported to the English court during the reign of Henry VIII, who danced in one with Anne Boleyn. Both were in disguise, he as a shepherd, she a shepherdess. William Cornish, Master of the Chapel Royal, offered *The Garden of Esperance* (1517) in which speech was added to the dance and pantomime. He prepared a number of others with singers and music more dominant, such as *The Golden Arbour of Pleasure* and *Love and Duty*. The masque soon became an Arcadian allegory of the sort fashioned by Sir Philip Sidney. Semi-frugal, Queen Elizabeth did not encourage such stagings; they were too costly, and she preferred to have other people present them for her. The Stuarts, more self-indulgent, took delight in them and commissioned many.

Beaumont and Fletcher wrote many masques, as did – on demand – most of the facile and capable dramatists on Henslowe's list. But none of that period compared with Jonson, whose output so pleased King James that eventually he ordered at least two such pieces every year from the robust poet. In 1605 Queen Anne herself performed in Jonson's graceful *Masque of Blackness*. By royal command, too, he had prepared the *Masque of Beauty* and in 1609 *The Masque of Queens*. In all, Jonson was to write at least thirty-two of them. He brought to the creation of these artificial works his classical learning, displayed with lightness and wit, and his effulgent lyrical gift. They continued to be amateur entertainments, mostly enacted by the nobility and their guests, with the help and participation of some professional mimes and dancers, as a private amusement. Jonson's conceptions employed much music. Weak as drama, they were filled with bright poetry. It was Jonson who raised the genre to new heights. Quite as important, however, was the spectacular scenic investiture by the famed designer Inigo Jones.

Son of a Smithfield cloth-worker, Inigo Jones (1573–1652) showed from boyhood a remarkable gift for drawing and draughtsmanship. He capped his apprenticeship at home by richly rewarded studies of painting and architecture in Renaissance Italy, where he meticulously observed ancient monuments and Palladio's revival and pleasing adaptation of that style. In 1615, back in England, he was named the royal Surveyor-General. He proved to be a genius as an architect, improving and adorning much of London as it stands today. Borrowing from the neoclassical inspiration of Palladio, he designed the portico of St Paul's Cathedral and the Banqueting Hall at Whitehall, thus aiding in bringing to a close the age of masculine Tudor beams and gables. But it was in stagecraft that Inigo Jones wrought the most lasting change. He imported the Florentine proscenium arch, dispensing with the plain wooden "O", the constraint and bareness of which had hampered Shakespeare. He introduced perspective scenery, painted on serried ranks of flats, the last presenting the illusion of a distant vista. Admiring this new effect, Jonson described it as "the whole worke shooting downewards, which caught the eye afarre off with a wandring beauty". Jones added the front curtain. He called for many costly stage devices, among them revolving screens. His costumes were particularly elaborate and sometimes scandalous, the gowns of the court ladies having only diaphanous veils inset at the bosom. His Italianate scenery was soon so much in demand that it completely took over the seventeenth-century theatre, revolutionizing the manner in which most plays were mounted.

Everything was going well for Jonson. In 1616 King James conferred a lifetime pension on him, making him in effect England's first Poet Laureate. In the same year the happy writer published his *Workes of Benjamin Jonson*, which occasioned much laughter among his enemies, who deemed it presumptuous of a playwright to call his stage-pieces "workes". But Jonson paid little heed. (By no means were all his plays included in this collection; he told his friend Drummond that not even half of his scripts had been preserved.)

The collaboration between Jonson and Inigo Jones, though it endured for quarter of a century (1605–31), was never easy. Both were strong-willed and, especially on Jonson's part, self-important. Jones spoke of the playwright as "the best of poets, the worst of exasperated men". All too often the splendour of the scenery and costume designs outweighed the lilt of Jonson's speeches and ballads. This led to disputes as to whose contributions were to have priority, culminating in fierce quarrels. Their partnership finally broke up, never to be resumed.

The Sons of Ben moved their gathering place from the Mermaid Tavern to the Apollo Room at the Devil and St Dunstan Tavern. Continuing to hold forth on all matters, he became the archetype of what was to be a long line of English literary dictators (Dryden, Pope, Samuel Johnson *et al.*). Drinking too much, he grew monstrously fat. To shrink his paunch by salutary exercises, he undertook a long walk to Scotland (1618) to visit the poet William Drummond, an excursion that led to Drummond's memoir *Timbers*, a collection of Jonson's pungent observations to his host during his fortnight's stay, including his belittling remarks about Shakespeare and his startling theory – logical to one who had used the trick of transvestism so often in his plays – that Elizabeth, the "Virgin Queen", had never married because Her Majesty was really a boy in "drag". Scholars suspect that Drummond was not wholly sympathetic to Jonson and that some of the comments, set down in highly abbreviated form by the Scottish poet, are not to be taken literally. Basking in fame and prosperity, besides his annual pension of £66 from the king, Jonson usually got generous Christmas and New Year's gifts from such patrons as the Earl of Pembroke. He was granted additional recognition in the form of an honorary Master's degree from Oxford, to which he could add one from Cambridge as well. He was offered a knighthood, which he deemed it the best part of wisdom to refuse. But now ill luck, some of it catastrophic, began to overtake him. His library was destroyed in a fire in 1623. James I, his generous patron, died in 1625. Charles I was not as kindly inclined towards the poet; in theatre he favoured spectacle – as provided by Inigo Jones – to beauty of text. Jonson could no longer count as much on royal support. He turned back from court entertainment to the stage as his principal occupation.

His next venture, *The Staple of News* (1626), is one of the best of his later minor pieces. Taking a plot idea from the *Plutus* of Aristophanes, the ancient dramatist he so much resembled in temperament, he attacked London's first weekly newspaper, *The Courant,* issued by a Nathaniel

Butter. Its content tended to be very unreliable, meant more to excite the reader's interest than to inform him. The newsletter was a fair target, at which other comic dramatists had also taken aim. In one reference to it Jonson describes it as "butter'd news". He imagines a jovial young man, Pennyboy Jr, who is laying suit to the rich Lady Aurelia Clara Pecunia. He accompanies her about London, a tour during which they prodigally dispense kisses and gifts to his hangers-on and flatterers. (Lady Aurelia's wealth has come principally from Spanish mines in the New World.) The couple are beset by every exploiter in the greedy town: lawyers, usurers, fortune-hunters and – again – Puritans. All are satirized.

To watch the young man's behaviour, his father has pretended to be dead and disguised himself as a beggar. At a crucial moment he reappears and takes his son to task for his profligacy. This is a rather overworked device.

As the names of the characters testify, the play is once again much like a Morality. Some of it is lively, but by no means all of it. A good many passages are boring – at least today. The satire is often too broad and simple. The work has a prologue like that of *Knight of the Burning Pestle*, in which four women – Mirth, Tattle, Expectation and Censure – break in upon the action, take stage seats and comment on the play. At one point they call the author "a decayed wit". Much of their dialogue is very racy, packed with allusions to current events.

Studying Jonson from a feminist angle, Professor McLuskie in *Renaissance Dramatists* concludes that he is a thorough misogynist. He fears and despises women most often, portrays them harshly. One does not find in his plays the charming heroines that grace Shakespeare's forested romances and misty pastorals.

In 1628, at fifty-six, Jonson was felled by a paralytic stroke, which he survived but from which he never recovered. He continued to write but was confined to his sick-couch. *The New Inn* (1629) came forth the next year and failed, for which he accepted no responsibility. He blamed poor acting for its weak reception. But this script, like the others that bravely followed it, had innumerable faults.

Luckily he got a grant of £100 from King Charles in return for a poem of praise sent to the monarch. He was also appointed to succeed Middleton as City Chronologer. In 1631 Jonson joined Inigo Jones to shape two masques for the court. This was the occasion for their final quarrel and the severance of their long association. The king valued Inigo Jones's work over Jonson's, and the poet's often profitable connection to royalty was permanently cut off.

His ill luck multiplied. His second son died in 1635. He himself suffered another stroke, but he kept on writing, as he had considerable need to do. Seeking to mend his lot, he turned out *The Magnetic Lady, or Humours Reconciled* (1632), which was somewhat better received, but the frequent use of profanity by the actors in it was deemed offensive.

Lady Loadstone is called the "Magnetic Lady" because, though rich and titled, she invites guests of all sorts and social class. Her niece, the much-sought-after Placentia, is about to be married. It is discovered that she is already pregnant, and during an argument she goes into labour and gives birth to a child. The wedding is cancelled, and this provides a pretext for an uncle, Sir Moth Interest, to gain possession of her large dowry. But next it is disclosed that there has been a mix-up of identities: Placentia is actually Pleasance, Lady Loadstone's waiting-woman, and the waiting-woman is the Lady's niece. (Jonson rings the eternal changeling theme that Gilbert and Sullivan later tunefully burlesque.) Once the two girls are correctly recognized and resituated, they are duly married, Pleasance – now Placentia – to Compass, the hero, who of course recovers the dowry; and Placentia – now Pleasance – to Needle, a steward in the household and the begetter of the out-of-wedlock child. Lady Loadstone herself chooses a husband, Captain Ironside, a brother of Compass.

Unwisely, Jonson perpetuated his feud with Inigo Jones in vituperative poems and by a lampoon of him as Vitruvious Hoop in *A Tale of a Tub* (1633). Audrey has many suitors. She is the daughter of Toby Turf, High Constable of Kentish Town. On St Valentine's Day, when she receives their offers of marriage, her father decides in favour of John Clay, a tile-maker. The wedding party sets off for the church, but the ceremony is delayed by Squire Tub and the vicar Canon Hugh with a baseless story that the bridegroom-to-be is guilty of a highway robbery. The squire desires Audrey to be his wife. His effort is frustrated by Justice Preamble, who enlists the Canon's aid in winning Audrey. Turf takes her off the matrimonial market for the moment, but the Justice nearly succeeds in obtaining her (along with £100). Squire Tub foils this. The ultimately successful contestant, quite unexpectedly, is Pol Martin, usher to Tub's mother. He is an entry on whom no one has counted.

Vitruvious Hoop is not a major figure, but, stung at being caricatured, Jones appealed to the censors; he asserted that being so imaged caused him personal injury. They ruled against Jonson. A year later the expurgated play was acted at court but won little applause.

Jonson was afflicted with another stroke, but despite this added handicap courageously kept on writing. His income ebbed. At times nearly penniless, he lingered several more years, kept abed. He read steadily, his notes later gathered in a posthumous volume, *Discoveries* (1640). He provided a pair of masques for the Duke of Newcastle and worked on another play, *The Sad Shepherd*, left unfinished. It included Robin Hood as a character, along with his merry men in Sherwood Forest, and Puck, and the English witch Maudlin – it was an attempt to anglicize the pastoral, rendering it more colloquial, less artificial in theme and language. The fragments that remain are charming and stir regret that it was never completed. Lonely, and much in debt, he reached the end on 6 August 1637.

(Inigo Jones, after many triumphs, was also destined to experience a spell of obscurity. During Cromwell's austere regime the architect was probably imprisoned, and his last days were clouded.)

Once gone, Jonson's greatness was promptly acknowledged. He was buried in Westminster Abbey, his funeral attended by peers of the realm and a throng. A subscription was started for an imposing monument, but the civil strife that erupted soon after – with the beheading of Charles I –

put a stop to the effort. Accordingly, a friend had carved at the spot a simple but wholly adequate inscription: "O rare Ben Jonson".

Long enjoying only academic esteem and interest for his neo-classicism and influential theory of humours, his work in recent years has had much more frequent revival. *Volpone, The Alchemist* and *Epicene* are widely recognized as still highly vital scripts. More difficult to stage are some of his other major comedies, their dialogue filled with the vivid street-argot he heard and captured and which today's audiences find somewhat unintelligible. A true loss! *Volpone* has been produced more than once on Broadway and *The Alchemist* at the Barbican in London, not as curiosities but as current and relevant offerings.

The output of critical literature about him has also gained momentum. He is seen to be along with Marlowe and Shakespeare at the pinnacle of the English drama and perhaps its greatest writer of comedy. But he is best judged in comparison to other playwrights of the Jacobean age, whom he dwarfs. After Marlowe and Shakespeare, he has no equal to the present day. His accomplishment with language is daunting. He forged a new kind of poetry embracing the vocabulary of London's underworld and the speech of country bumpkins. His energy is to be envied, and he transmits his gusto through the exuberance of his varied characters, who are often created more roundly than some commentators have appreciated. His intellect was formidable, as was his learning for his time, and he synthesized his gifts very well. T.S. Eliot finds him a shrewd appraiser of his world and as a source of social history more important than Shakespeare – the Bard wrote little of life in his own day. Jonson's brutality and vulgarity are hardly as objectionable now as they were in the squeamish nineteenth century when his reputation took a dip. The best explanation as to why his full measure is recognized again is to witness *Volpone*, one of the strongest, richest and most topical comedies ever to come alive on the stage, where it will long remain to dazzle and deeply amuse.

Ben Jonson's life is the best documented of any Elizabethan and Jacobean stage-writer. Of his contemporary John Webster, whose name has frequently been mentioned, almost no biographical facts are available. M.E. Bradbrook, in her *John Webster, Citizen and Dramatist*, says that for long "this darkest of Jacobean poets remained himself almost in complete shadow". Bernard Beckerman, in a preface to his anthology *Five Plays of the English Renaissance*, describes the playwright as "a shadowy figure", adding, "yet seldom has so indistinct a person made so large an impression".

One cause for this confusion is that "John Webster" was a not uncommon name. Scholars seeing it on a church or civic record anywhere – a seventeenth-century birth certificate, baptism entry, marriage licence – seize on it excitedly; it might be a significant discovery bringing academic glory and advancement. Also, it is believed that the playwright's father was "John" and his grandson, the poet's child, was "John" – three "Johns" in one family! That, too, leads to much doubt, questions of identity.

What is certain is that John Webster is one of the finest dramatic poets in English, his talents in many ways ranking him along with Marlowe, Shakespeare and Jonson. His intellect, too, is keen, his theatrical flair sure, his mastery of language breathtaking. Which makes one wonder, futilely, who he was. What inspired his profound morbidity, the macabre turn of his mind? He is a fascinating enigma. It would seem that in *The White Devil* he wrote two lines that might be applied to himself:

> My soul like to a ship in a black storm
> Is driven I know not whither.

Despite the meagre facts, Professor Bradbrook has confidently published a 200-page book tracing much of the playwright's career (and there are other such biographies in the field); but she fleshes it out with detailed lives of four of his neighbours – a truly awesome display of erudition – the text abounding with "perhaps", "possibly", "probably", "may have", "would have", "must have", the words also found in weighty biographies of Shakespeare. It does make for good reading, but were those four persons of importance really the dramatist's neighbours?

From a few references scholars guess that the playwright was born somewhere between 1570 and 1580. His father was a London tradesman, long thought to have been a merchant tailor, since he was a member of the prestigious Merchant Tailors' Guild, and John is supposed to have been enrolled at an elementary level in the guild's reputable school. At the father's death both of his sons, Edward and John, were admitted to the guild "by right", taking over membership from their deceased parent. The latest research, however, suggests that the father was a wealthy coach-maker. His establishment was inherited by his sons. Whether the poet ever practised the trades of tailoring or coach-making is not known. He may have had need for additional income while writing plays. The supposed site of his birth and the business establishment was the corner of Cow Lane and Hosier Lane, West Smithfield, in the parish of St Sepulchre-without-Newgate, a quarter of London that was dotted with playhouses.

John Webster is next sighted at the Middle Temple, studying law and making literary friends. Or such is the assumption. At about twenty-one he married a saddler's daughter and fathered children. Sooner or later, somehow or other, he was drawn to the theatre. He was a born playwright. If he attended the Middle Temple, that could have brought him to Henslowe's attention; other writers on that payroll had come from there. He is clearly in view on the producer's list of reliable collaborators for work with Middleton and two other writers on *Caesar's Fall* (1602). He stayed on the roster but was later engaged by other companies as well, including the three best in London and a boys' troupe.

Chiefly he was paired with Dekker, who was older and decidedly more experienced, though the apprentice was inherently the more gifted. At first his contribution was apt to be slight. Together they turned out three scripts: *Westward Ho!*, *Northward Ho!* and a chronicle play *The*

Famous History of Sir Thomas Wyatt, the last of these a work of little merit. Frederick Frastus Pierce, in his *The Collaboration of Webster and Dekker*, has sought by a line-by-line scrutiny to determine who wrote what. He brings the full academic machinery to the task: word-counts, habitual turns of phrase, the number of three-syllable Latin words, dialectic and metrical tests, feminine endings, incident and character tests, etc. His conclusion is that Webster did little, the burden was largely Dekker's. The younger man may have been helpful in verbal exchanges with his partner.

He also worked with Heywood and Rowley and Ford and was entrusted with adapting Marston's *The Malcontent* for a revival by the King's Men. From close association with these various masters of stagecraft he inevitably learned much of value to him.

Webster is credited with four extant scripts conceived wholly by himself; of these, two are deemed negligible, two have brought him enduring fame. When he worked alone he did so with slow, conscious artistry. His craftsmanship remains uneven, yet he has an instinct for big scenes, an essential requirement for a playwright. It is momentous encounters that endow major stature to a tragedy.

The White Devil (1609) is the first of his two lasting achievements. It and *The Duchess of Malfi* are so superior to everything else tentatively attributed to him that commentators are daunted in trying to explain how he suddenly ascended to such heights of inspiration and afterwards lost it, subsiding to a lower level of accomplishment.

The White Devil was produced at the Red Bull. When it was published a few months later he complained that it was performed "in so dull a time of winter, presented in so open and black a theatre, that it wanted that which is the only grace and setting out of a Tragedy in an open structure, a full and understanding auditory". February was a bleak season for such a grim work, and the Red Bull's usual complement of spectators was apt to favour cruder spectacles of a more sentimental, pious and romantic kind. These factors may well have accounted for the play's initial unsuccess. But Webster had a heartening faith in his composition. He remarks in his preface to the printed version that he has been criticized for having spent such a long time in writing the play, and in reply he cites a tale about Euripides, chided by Alcestides for taking three days to write three verses, while Alcestides had set down three hundred lines in the same span of time. Euripides retorted: "Thou telst the truth . . . but here's the difference, thine shall onely be read for three daies, whereas mine shall continue three ages." In both instances a correct prophecy! In the same preface Webster pays his compliments to Master Beaumont and Master Fletcher, to the "full and heightened style of Master Chapman" and, lastly, to the genius of "Master Shakespeare, Master Dekker, and Master Heywood . . . wishing that I may be read by their light". But, save for Shakespeare, none of these others is really his equal.

The origin of *The White Devil* is a historical incident on which Webster drew freely, altering facts to suit his dramatic purpose. His approach to the material is very much his own and to some eyes perverse. The actual intrigue was engaged in by Vittoria Accoramboni, the daughter of an aristocratic but impoverished mid-sixteenth-century Italian family. At sixteen she married the nephew of a cardinal, who in time became Pope Sixtus V. She later met Paulo Orsini, Duke of Bracciano, who fell in love with her. He murdered his wife, Isabella Medici, for an alleged infidelity with one of his kinsmen. In order to marry Vittoria, Bracciano had her husband killed by musket-fire from an ambush. Their wedding was clandestine and twice celebrated. The slaying of her first husband was investigated, however, with the result that Vittoria was imprisoned for a time in the grim Castel Sant'Angelo in Rome. When Pope Gregory died the pair were openly wed, this being the third time. But the next occupant of the papal throne was that Sixtus V, uncle of Vittoria's assassinated first husband. Once more the guilty couple were persecuted for their crime. They fled to Venice, then Padua. Bracciano expired from natural causes – the travel and harassment had over-taxed his ill health; Vittoria was subsequently murdered by a band of hired killers recruited by the Orsini family, who wished to assure the undisputed inheritance of Virginio, Bracciano's still young son by his earlier marriage to the luckless Isabella de' Medici.

In Webster's drama some names are changed. The cardinal's nephew, Francesco Peretti, becomes Camillo; Bracciano is slightly altered to Brachiano; Pope Sixtus to Paul IV; Virginio to Giovanni. The "White Devil", here applied to Vittoria, is a devil in attractive disguise. Webster makes the hapless Isabella saintly, dying when she kisses a portrait of her husband, who has forsworn her – the glass over the picture is poisoned. Camillo, Vittoria's husband, is a poltroon; his death is caused, deliberately, when he is thrown by a balky horse. Brachiano is a monster of selfishness and cruelty yet admirably defiant of all authority save his own.

Vittoria, charged with complicity in her husband's death, a simulated accident, stands trial and is prosecuted by the slain man's infuriated uncle, Cardinal Monticelso. Her bearing is so composed that he exclaims:

> And look upon this creature was his wife,
> She comes not like a widow; she comes arm'd
> With scorn and impudence: Is this a mourning habit?

When he casts imprecations at her, she boldly replies:

> I am past such needless palsy, for your names
> Whoore and Murdresse they proceed from you.
> As if a man should spit against the wind . . .

She says his "filth" flies back on to his face.

The hot-headed Brachiano stalks from the hearing, and Vittoria is left alone to defend herself. She does so resolutely, parrying word for word. The evidence insufficient, she is acquitted of the murder but condemned for her immorality and ordered to be confined in a house for "penitent whores".

Isabella's brother, the Duke of Florence, prompted by her ghost, plots revenge. He sends a letter to Vittoria in which he professes to be an admirer. He hopes that Brachiano will see it and become jealous. This does happen. Brachiano is smuggled in to visit the captive Vittoria. They have a passionate lovers' quarrel over the letter; she upbraids him for doubting her:

> What have I gain'd by thee but infamie?
> Thou hast stain'd the spotless honour of my house. . . .
> I had a limbe corrupted to an ulcer,
> But I have cut it off; and now Ile goe
> Weeping to Heaven on crutches.

Brachiano is reconciled to her and helps her to escape from the "convent"; they flee the city.

Cardinal Monticelso is elected Pope. His first act is to excommunicate the fugitives. But Isabella's brother wishes to exact a higher price. He continues the poisoning of the hitherto insolent Brachiano, who becomes delirious and expires after crying:

> How miserable a thing it is to die,
> 'Mongst women howling!

The cynical and villainous agent, Flamineo, who is both Vittoria's brother and Brachiano's secretary, is the most dynamic force in the play. He is unfailingly resourceful and philosophically vicious, another Iago. To him are awarded such pithy lines as these:

> For they who sleep with dogs, shall rise with fleas.

and:

> Lovers' oathes are like Mariners' prayers, uttered in extremity.

They may be epigrams from Montaigne. More than once there is an echo of Sidney and Shakespeare, too, and other sharp phrases are proverbs that cropped up in daily speech. Webster knew how to insert them to the best effect, as one brightens a gem by giving it just the right setting. To Flamineo's speeches Webster brings sardonic humour and a disenchanted vision of life. The script contains much direct criticism of the injustice that is engendered at court and spreads outwards from where corruption and flattery prevail. He comments, too, on the brief vainglory of power and wealth and the falsity of women.

Interestingly, it is Flamineo who moralizes the most incisively; he is always conscious of his own evil nature and blaming himself for his dark acts, but so deep is his cynicism that he stops nowhere. He is the Machiavellian figure found so often in plays of this period – after Marlowe – but better motivated than most. He resents his poverty and dependence on others whom he deems his inferiors. He is sceptical of religion, sees himself as having to make his way in a dishonest world where high rank only masks a villainy worse than his own and predicates that any means is justified by his goal. He sometimes has a twinge of compassion, but quickly negates it. His self-confidence leads him to believe that he can accomplish whatever he desires. He is enforced in this conviction by seeing around him others win advancement at a court – extrapolated to his cosmos – compounded of self-interest, decadence, the utmost covetousness. Even the churchmen, among them the bitter, red-hatted Cardinal Monticelso who becomes Pope, are worldly intriguants. Of that merciless foe Vittoria, who shares much of her brother's outlook, he murmurs:

> O poor charity!
> Thou art seldom found in scarlet.

To fulfil his purposes, Flamineo is pander to his sister and wantonly destroys his good, decent younger brother Marcello, a cruel deed that deprives Cornelia, their bereaved mother, of her wits. In a passage like that of the unhinged Ophelia in *Hamlet*, the wailing old woman sings:

> Call for the Robin-Red-brest and the wren,
> Since ore shadie groves they hover,
> And with leaves and flowres doe cover
> The friendless bodies of unburied men.
> Call unto his funerall Dole
> The Ante, the field-mouse, and the mole
> To reare him hillockes, that shall keepe him warme,
> But keepe the wolfe far thence: that's foe to men,
> For with his nailes hee'l dig them up agen.

For all these crimes he reprimands himself, but plunges on from sheer momentum.

At the end, facing disgrace, Flamineo tries to shoot his guilty sister Vittoria and her Moorish maidservant. She is too clever for him, but the duke's co-conspirator, Ludovico, intervenes and brings about the death of his three designated victims. The scene is truly memorable. When her servant, Zanche, is about to be murdered by the avengers, ingeniously disguised as holy friars, Vittoria cries:

> You shall not kill her first. Behould my breast,
> I will be waited on in death; my servant
> Shall never go before mee.

Gaspara, the second assassin demands: "Are you so brave?" Vittoria replies:

> Yes I shall wellcome death
> As princes doe some great Embassadors;
> Ile meete thy weapon halfe way.

She also vows:

> I will not in my death shed one base tear,
> Or if look pale, for want of blood, not fear.

It is no wonder that Carlo, another of the duke's hirelings, calls her "blacke fury". Flamineo is similarly resolute in death, at times even blithe:

> I have caught
> An everlasting cold. I have lost my voice
> Most irrecoverably.

When he complains that he cannot see and asks if he shall have no company on his journey to death, his spirited sister responds:

> O yes thy sinnes
> Do runne before thee to fetch fire from Hell,
> To light thee hither.

All the characters in this uneven work are shaped with power and conviction. If the plot is overladen with bloody deeds, it has the primal appeal of repetitive violence, a magnetic compulsion like that exerted by Tourneur's *The Revenger's Tragedy*. It reaches near Olympian heights in its poetic dialogue. It also has dumb shows, apparitions of those dead by envious hands, duels – not one but two – and much lavish panoply: Webster knew how to make very rich use of stage spectacle. If *The White Devil* is clumsily structured in places, that does not prevent it from being playable – indeed, certain illogicalities may contribute to its effectiveness when it is acted with verve commensurate to its swift and compact style.

The comparative merits of *The White Devil* and *The Duchess of Malfi* (1612–13), Webster's other great play, have been much argued. The majority of critics prefer the later drama, however, possibly because its heroine is more attractive and sympathetic, so that its suspense is considerably stronger. The poetry in it is far superior to that of *The White Devil*, where there are many yet only fitful bright flashes of eloquence; in *The Duchess of Malfi* the incandescence of

Webster's language is constantly aglow. Not even Shakespeare or Marlowe have surpassed the effulgent beauty and painful sharpness of this dialogue.

Once more Webster found his subject in a factual incident. In 1510, a century earlier than the play was written, Joanna of Aragon, granddaughter of King Ferdinand of Naples, was left a widow and Duchess of Amalfi at the age of twenty. She secretly married her steward Antonio Bologna. They kept their relationship hidden for several years, though they had children. But scandalous whispers spread; first Antonio withdrew to Ancona, and later Joanna followed him, renouncing her rank and openly proclaiming her legal connection with her major-domo. Her outraged brothers refused to allow her to settle into private life with the man she loved. The brothers felt humiliated by this misalliance. She fled their vengeance to Siena and then with her three children sought refuge in Venice. *En route*, her brothers' armed hirelings overtook her. Antonio and their eldest son escaped, but Joanna was taken back to her duchy and imprisoned in a castle where she, her other two children and her waiting-maid were murdered. The crime was concealed. Antonio, a year afterwards, tried to learn what had been his wife's fate. He, too, while in Milan, was killed by a professional assassin, one Bozolo. The story was later related in Painter's collection of tales, *Palace of Pleasure* (1566–7), which served many other Elizabethan playwrights.

Webster, as is customary with dramatists – especially those of a romantic temperament – alters details. He adds new aspects and incidents, as was his artistic right. Chiefly he softens and embellishes the characters of the duchess and her Master of the Household. In Painter both are pictured as libertines, she guilty for having stooped so far beneath her in choosing a bedfellow, and he gravely at fault for aspiring to marry too high. Instead Webster makes them both appealing figures, innocently yielding to their natural impulses and affections. They have many fine qualities and are witty and gay. The avenging brothers are ambitious and cruel: one of them, the Cardinal, is crafty and heartless; the other, Ferdinand, Duke of Calabria, is seemingly psychotic and certainly perverse. The Bozolo of the original story becomes Bosola, who plays a leading role not at the end but throughout, tying together many scattered episodes. Another character, named Flamineo, is a faltering Machiavellian and profoundly justified malcontent.

In Webster's version the noble brothers have been visiting their sister, newly widowed. Preparing to leave, they warn her against remarriage or risking her chaste reputation. She makes a pretence of agreeing with them, though she openly suspects them of having rehearsed their farewell speeches to her:

> CARDINAL: You may flatter yourself,
>   And take our own choice; privately be married
>   Under the eaves of night –
> FERDINAND: Think 't the best voyage
>   That e're you made; like the irregular crab
>   Which, though 't goes backward, thinks that it goes right

> Because it goes its own way; but observe,
> Such weddings may more properly be said
> To be executed than celebrated.
> DUCHESS: I think this speech between you both was studied,
> It came so roundly off.

They continue to rail against the possibility of her marriage, Ferdinand stating:

> They are luxurious
> Will wed twice.

Her quick response to such assertions:

> Diamonds are most value,
> They say, that have passed through most jewellers' hands.

His retort is ominous and sets the tone of the ensuing drama:

> Whores by that rule are precious.

She promises:

> I'll never marry.

The Cardinal comments:

> So most widows say;
> But commonly that motion lasts no longer
> Than the turning of an hourglass: the funeral sermon
> And it end both together.

The morbidly sinister characters of these two men have already been described by Antonio in a preliminary conversation with his friend Delio, who asks:

> Now, sir, your promise; what's that Cardinal?
> I mean his temper? they say he's a brave fellow,
> Will play his five thousand crowns at tennis, dance,
> Court ladies, and one that hath fought single combats.

Antonio explains:

Some such flashes superficially hang on him for form; but observe his inward character: he is a melancholy churchman; the spring in his face is nothing but the engendering of toads; where he is jealous of any man, he lays worse plots for them than ever was imposed on Hercules, for he strews in his way flatterers, panders, intelligencers, atheists, and a thousand such political monsters. He should have been Pope; but instead of coming to it by the primitive decency of the Church, he did bestow bribes so largely and so impudently as if he would have carried it away without Heaven's knowledge.

As for the duke:

> . . . a most perverse and turbulent nature:
> What appears in him mirth is merely outside;
> If he laugh heartily, it is to laugh
> All honesty out of fashion. . . .
> He speaks with others' tongues, and hears men's suits
> With others' ears; will seem to sleep o' th' bench
> Only to entrap offenders in their answers;
> Dooms men to death by information;
> Rewards by hearsay.

In quite a different vein, Antonio speaks of their sister:

> . . .the right noble duchess
> You never fixed your eye on three fair medals
> Cast in one figure, of so different temper. . . .
> Her days are practised in such noble virtue
> That sure her nights, nay, more, her very sleeps,
> Are more in Heaven than other ladies' shrifts.
> Let all sweet ladies break their flattering glasses,
> And dress themselves in her.

But his praise is to be expected, since he is in love with her.

When the brothers have departed the duchess immediately declares it her intention to wed her faithful steward. The scene in which she imparts her wish to him is beautifully and wittily written, filled with charming word-play yet touched with poignancy. In view of the danger, and his lowly position, he is scarcely in a position to express his feelings to her. As she herself puts it:

> The misery of us that are born great!
> We are forced to woo, because none dare woo us.

She begins by pretending that she is planning to make her will and asks his advice, as her worthy Master of the Household, what provisions she ought to include in it. When he addresses her as "beauteous excellence" she grasps eagerly at the phrase, a mere woman anxious to be loved by the man who has captured her affections. Most engaging is the moment when she leans forwards and pretends that his eye is inflamed:

> Fie, fie, what's all this?
> One of your eyes is blood-shot; use my ring to 't,
> They say 'tis very sovereign: 'twas my wedding-ring,
> And I did vow never to part with it
> But to my second husband.

He protests, "you have parted with it now". She replies, "Yes, to help your eyesight."

> ANTONIO: You have made me stark blind.
> DUCHESS: How?
> ANTONIO: There is a saucy and ambitious devil
>   Is dancing in this circle.
> DUCHESS: Remove him.
> ANTONIO: How?
> DUCHESS: There needs small conjuration, when your finger
>   May do it: thus, is it fit?

She has slipped the ring on his finger. Even so, he is alarmed:

> Ambition, madam, is a great man's madness . . .
> Conceive not I am so stupid but I aim
> Whereto your favours tend: but he's a fool
> That, being a-cold, would thrust his hands i' the fire
> To warm them.

When, a moment later, he murmurs, "O my unworthiness!" the duchess tells him:

> You were ill to sell yourself:
> This darkening of your worth is not like that

> Which tradesmen use i' th' city, their false lights
> Are to rid bad wares off: and I must tell you,
> If you will know where breathes a complete man
> (I speak it without flattery), turn your eyes,
> And progress through yourself.

She confesses:

> You have left me heartless; mine is on your bosom:
> I hope 'twill multiply love there.

They complete their troth; her maidservant, Cariola, the only witness.

To spy on the duchess and report to them, the brothers have placed in her employ Bosola, an ex-galley slave who has previously worked for the Cardinal. Serving as her Master of the Horse, he lurks about the palace and covertly if belatedly learns that the duchess is pregnant with her third child. His message, when it reaches the short-fused Ferdinand, stirs that strange nobleman to a boundless fury. He does not know of his sister's marriage and loudly suspects her of lewd conduct:

> O confusion seize her!
> She hath had most cunning bawds to serve her turn,
> And more secure conveyances for lust
> Than towns of garrison for service.

He cries out, in his rage:

> Rhubarb, O, for rhubarb
> To purge this choler! here's the cursèd day
> To prompt my memory; and here 't shall stick
> Till of her bleeding heart I make a sponge
> To wipe it out.

The Cardinal asks:

> Why do you make of yourself
> So wild a tempest?

Ferdinand's tumultuous emotion does indeed suggest some deep-lying prod, even a hidden incestuous feelings for his errant sister:

> Methinks I see her laughing –
> Excellent hyena! Talk to me somewhat, quickly,
> Or my imagination will carry me
> To see her in the shameful act of sin.

"With whom?" demands the Cardinal. Ferdinand exclaims:

> Happily with some strong-thighed bargeman,
> Or one o' the woodyard that can quoit the sledge
> Or toss the bar, or else some lovely squire
> That carries coals up to her privy lodgings.

"You fly beyond your reason," chides his calculating, even-tempered churchly brother. Ferdinand's curses are, indeed, among the most fierce and chilling ever penned:

> FERDINAND: Apply desperate physic:
>   We must not now use Balsamum, but fire,
>   The smarting cupping-glass, for that's the mean
>   To purge infected blood, such blood as hers.
>   There is a kind of pity in mine eye, –
>   I'll give it to my handkerchief; and now 'tis here,
>   I'll bequeath this to her bastard.
> CARDINAL: What to do?
> FERDINAND: Why, to make soft lint for his mother's wounds,
>   When I have hewed her to pieces.

And still more:

> I would have their bodies
> Burnt in a coal-pit with the ventage stopped,
> That their cursed smoke might not ascend to heaven;
> Or dip the sheets they lie in in pitch or sulphur,
> Wrap them in't, and then light them like a match;
> Or else to boil their bastard to a cullis,
> And give 't his lecherous father to renew
>   The sin of his back.

The Cardinal finds such passionate anger incredible:

> How idly shows this rage, which carries you
> As men conveyed by witches through the air,
> On violent whirlwinds! this intemperate noise
> Fitly resembles deaf men's shrill discourse,
> Who talk aloud, thinking all other men
> To have their imperfection.

Ferdinand enquires:

> Have not you
> My palsy?

The Cardinal replies:

> Yes, I can be angry, but
> Without this rupture: there is not in nature
> A thing that makes man so deformed, so beastly,
> As doth intemperate anger. Chide yourself.
> You have divers men who never yet expressed
> Their strong desire of rest but by unrest,
> By vexing of themselves. Come, put yourself
> In tune.

But Ferdinand is determined to learn who is the father of his sister's child. He hastens back to her court and feigns ignorance of what has happened, while with Bosola's aid he attempts to ferret out the secret. Their stratagems raise and sustain the tension. The duchess unwittingly betrays herself and, confronted with her brother's vengeance, takes flight with Antonio and their three children, giving as a pretext that they are making a pilgrimage. While they are on the road, footsore and dusty, finding many neighbouring towns have closed their gates to them, they hear the church bells clanging and voices proclaiming their excommunication – the Cardinal's doing. They are overtaken by their pursuers near Loretto. The duchess and two of her children and Cariola are seized, though Antonio and the eldest boy escape. As directed by Bosola, on behalf of Ferdinand, the unhappy woman is imprisoned, pent in a room next to a madhouse, whose shouting occupants are allowed to visit and torment her.

Her immurement is destined to be short; her half-crazed brother has decreed her murder. Bosola returns with four executioners, who prepare to strangle her, the little children, the trembling Cariola. The duchess is undismayed as she faces the black-hooded men who have come to kill her.

"I am Duchess of Malfi still," she tells them, quietly, proudly. Interjects Bosola:

> That makes thy sleep so broken;
> Glories, like glowworms, afar off shine bright,
> But looked too near have neither heat nor light.

(This was an image much favoured by Webster, who used it earlier in *The White Devil*: the theme of reality and illusion, the hollowness of pomp and regality, runs through much of his work. The paradox that grandeur hides misery haunts him.)

Nor is the duchess terrified by the horrible death with which her persecutor threatens her, showing her the cord. Gazing at it, she simply asks:

> What would it pleasure me to have my throat cut
> With diamonds? or to be smothered
> With cassia? or to be shot to death with pearls?
> I know death hath ten thousand several doors
> For men to take their exits; and 'tis found
> They go on such strange geometrical hinges,
> You may open them both ways, – Any way, for Heaven sake,
> So I were out of your whispering. Tell my brothers
> That I perceive death, now I am well awake,
> Best gift is they can give or I can take.

She dies at the hands of the heartless men. The two children are slaughtered. In another few minutes the hysterical Cariola is put to death.

Ferdinand, come to behold the corpses, is overcome by revulsion. "Cover her face, mine eyes dazzle; she died young." (One is reminded of the final scene of *Othello*.) He turns on Bosola, whom he had appointed to be executioner. The pair quarrel. Bosola who has suffered earlier from the Cardinal's treachery, bitterly accuses Ferdinand of an even greater ingratitude. Ferdinand says that he had never really intended to order his sister's death. He reveals that they were twins. He rushes away:

> I'll go hunt the badger by owl-light:
> 'Tis a deed of darkness.

The final episodes show Ferdinand wholly mad, succumbing to lycanthropy, imagining himself changed into a wolf. The Cardinal, having poisoned his mistress, Julia, is stabbed to death in error by his insane brother. This meting out of poetic justice was, of course, necessary to satisfy

Webster's audience; accordingly he provided it. Ferdinand and Bosola dispatch each other. Antonio, too, has been killed by Bosola, and only the duchess's eldest son remains to inherit the title and her fortune.

This encapsulation of the plot hardly does justice to the magnificence of the poetry that enriches every passage of the drama. The play can be dismissed as Grand Guignol, as excessively melodramatic. In all of this Webster was conforming to the taste of his times. In revenge tragedies an author superimposed horror upon horror, slaying upon slaying, carrying the torture and bloodshed to ridiculous lengths. Even Shakespeare, in *Hamlet, Lear, Macbeth*, passes beyond all restraint, as he had to, to meet the bloodthirsty demands of the spectators. The age itself was gory and violent; the stage had to outdo it. Shakespeare, of course, transcends the limitations that the genre imposes; where there is so much violence there is little space or time for contemplation, for subtle portrayals, for grappling with metaphysics. He creates his people so swiftly and deeply and conveys ideas so sharply and succinctly, that he does not seem to need much space or time to do so – though indeed he does add a superfluity of words to keep his actors busy. Webster, accepting the tradition set by Seneca, Kyd and Marlowe, though he possesses scarcely the scope of his exemplar Shakespeare, is here none the less poet and psychologist enough to raise the revenge tragedy to an almost matchless art-form. The passions are intense, the excitement unfailing. Whatever its structural faults, the play shows that he has an inherent feel for what works, as the story develops, blending the real and supernatural, leaping from one stunning climax to the next.

He gives every over-familiar element a new dimension. For example, the visit of the mad and their antic dance is similar to other dumb-shows and mimed scenes staged in madhouses in other plays of this period (an instance, Middleton's *The Changeling*), but Webster transforms his into a symbolic anti-masque, a grim parody of the masques in romantic tragicomedies of the time, a sort of obscene charivari devised by Ferdinand to celebrate the duchess's illicit common-law marriage. In such anti-masques there are totemistic echoes that derive from the Greek Dionysian rites. The duchess's tragic flaw is that with her blue-blooded husband scarcely dead she has hastily remarried, and to her own steward, one so far beneath her in rank. Modern spectators do not realize how daring and heinous this would be considered by an Elizabethan and Jacobean audience, but Webster intended her deed to be a criticism of accepted patrician standards. He sees the relationship between this noble woman and the admirable Antonio as blessed. She boldly refuses to let herself be bound by outmoded strictures and customs:

> Shall this move me? If all my royal kindred
> Lay in my way unto this marriage,
> I'd make them my low footsteps: and even now,
> Even as this hate, as men in some great battles,

> By apprehending danger, have achieved
> Almost impossible actions (I have heard soldiers say so),
> So I through frights and threatenings will assay
> This dangerous venture. Let old wives report
> I winked and chose a husband.

She has Cariola hide behind the arras, to overhear what follows, and begs:

> Wish me good speed;
> For I am going into a wilderness
> Where I shall find nor path nor friendly clue
> To be my guide.

Then she elects Antonio to be her husband, though without benefit of clergy. When he hesitates, she prompts him:

> Sir, be confident:
> What is 't distracts you? This is flesh and blood, sir;
> 'Tis not the figure cut in alabaster
> Kneels at my husband's tomb. Awake, awake, man!
> I do here put off all vain ceremony,
> And only do appear to you a young widow
> That claims you for her husband, and, like a widow,
> I use but half a blush in 't.

She chooses life, not death or death-dealing austerities and conventions. To her, too, the unsanctified marriage is yet a sacred one:

> What can the Church force more? . . .
> How can the Church build faster?
> We now are man and wife, and 'tis the Church
> That must but echo this.

Cariola, the witness, says of this challenge:

> Whether the spirit of greatness or of woman
> Reign most in her, I know not; but it shows
> A fearful madness: I owe her much of pity.

The duchess, a free spirit, is, like Webster himself, a thinker far in advance of her day. Her last words to the stranglers who wind the cord around her frail throat epitomize her remarkable bravery:

> Pull, and pull strongly, for your able strength
> Must pull down Heaven upon me.

But she is also tender, begging before her end that medicine be given to her younger boy, and she is also modest, pleading that after her end her body be bestowed to be cared for by her maidservants.

To actresses the role is a very attractive one. She has a pert tongue, is often sweet, has masculine courage. In some modern offerings the play ends shortly after her death to avoid anti-climax, for she is the focus of most of the major episodes.

Her brothers, the unbalanced Ferdinand and the cold Cardinal, are exceptionally strong and challenging. Ferdinand puts forth many explanations for his acts. At one moment he claims that he opposes his sister's marriage because he wants to be her heir. At another he asserts that his urgent concern is to preserve the family's honour. He himself does not know what his perverse motivation is, what incites his ravings, his rages. His compulsion seems to be a deep-rooted Freudian one, neurotic and finally outright psychotic. Before his demise he declares:

> I do account this world by a dog-kennel:
> I will vault credit and affect high pleasures beyond death.

Antonio is a good man, decent and honest, stoic and cautious. Bosola is complex, morally ambivalent, much more so than either Iago or Flamineo; he is torn between his evil and good impulses and embittered by his lot in life and the unfair treatment he has received at the hands of his social betters. He unwillingly admires the duchess. To him is given the dark song that precedes her murder:

> Hark, now every thing is still,
> The screech owl and the whistler shrill
> Call upon our Dame, aloud,
> And bid her quickly don her shroud.
> Much you had of land and rent,
> Your length in clay's now competent.
> A long war disturb'd your mind.
> Here your perfect peace is sign'd.
> Of what is 't fools make such vain keeping?
> Sin their conception, their birth weeping!
> Their life a general mist of error,
> Their death, a hideous storm of terror:

> Strew your hair with powders sweet:
> Don clean linen, bath your feet,
> And the foul fiend more to check,
> A crucifix let bless your neck.
> 'Tis now full tide 'tween night and day,
> End our groan, and come away.

The atmosphere of the Renaissance court – which even Ferdinand acknowledges is "a rank pasture . . . there is a kind of honeydew that's deadly" – is wonderfully realized; many of the episodes are, appropriately, night scenes, which fits the prevalent darkness of mood. Torches flare, rumours spread, there are alarums and stealthy excursions. Dead hands are pressed into those of the duchess. Such effects would not have been feasible in an open theatre such as the Globe; the small private theatres, with their interior stages, offered new dramatic concepts of which Webster took full advantage, pressing them to extract their utmost artistic possibilities.

Of the three lesser plays thought to have been wholly composed by Webster, one, *The Guise*, is lost and another, *Appius and Virginia*, is now believed to have been written with Heywood.

The plotting of *The Devil's Law Case* (c. 1610 or later) is so labyrinthine as to defy a comprehensible synopsis. At its centre is an evil young man who makes everyone around him his victim, particularly those in his own family, his mother, his sister. The play, subtitled *Or When Women Go to Law the Devil Is Full of Business*, seems to have been intended to inaugurate a new (or restored) private theatre, the Phoenix, to be occupied by the Queen Anne's Men, a company that had gained Webster's favour.

In Naples the anti-hero, a canny merchant, Romelio, decides that he could profit by marrying off his sister Jolenta to a very rich man. Though she is attracted to a blue-blooded wastrel whom Romelio is cheating in rigged business transactions, he chooses an even more affluent dupe to be her mate. However, Jolenta, herself no paragon, after feigning pregnancy is actually about to bear an illegitimate child. Her two lovers have fought a duel and supposedly killed each other, leaving her an heiress. The news of their deaths is false. Meanwhile Romelio has had an affair with a nun that has also resulted in an out-of-wedlock infant. To punish him for his nefarious behaviour his mother, Leonora, decides to accuse a long-vanished "friend" of being Romelio's father, allowing her to disinherit her son. The resourceful Romelio persuades his sister to assert that she has had twins so that he can pass off the nun's child as one of them. By chance the trial instituted by the mother is presided over by the very man whom she has named as the parent. He establishes that he is not the begetter. (It is possible that if Webster did study at the Middle Temple he had become acquainted with legal procedures there.) During the course of the play Romelio engages in "fraud, attempted murder, slander, criminal enticement, fornication, misappropriation – all committed however within the circle of those who should be nearest to him".

Several incidents in the play are edged and ironic. For instance, a sickly man is stabbed by an avenger. He has been suffering from an ailment for which surgery is required but considered too dangerous. The dagger-wound results in his complete cure.

Leonora, unwillingly made an "honest woman" by the verdict, has her eye on one of her daughter's suitors and finally snags him. Jolenta marries the other one.

At the end Romelio comes off rather lightly: he has to pay fines and marry the nun. The happy conclusion is neither logical nor satisfying. Overburdened with irrelevancies and confusingly plotted, *The Devil's Law Case* bears little resemblance to Webster's great tragedies. At intervals, however, there are fine poetic flights that could only have come from him but seem too serious and inappropriate in this context.

The unique quality of Webster's poetry is a masculine compression, a hardness. It is curt, sharply pointed, sometimes piercing, and it is pervasively melancholy. T.S. Eliot, in his much-quoted "Whispers of Immortality", says:

> Webster was much possessed by death
> And saw the skull beneath the skin;
> And breastless creatures under ground
> Leaned backward with a lipless grin.

Addressing this, Bradbrook comments that the poet had reason enough to be "possessed by death", living in London through five years of plague that took away a host of his friends and close associates.

Like Shakespeare, whom he may have been alert enough to study closely and from whom he apparently absorbed much, Webster uses specific words and images repeatedly to create an enveloping mood almost to the border of saturation, and to emphasize a theme. He adroitly coordinates word and line to visual action. As in Shakespeare, too, there are subtle and effective word-plays, even puns. The motivations of his characters are complex and even baffling, which makes them more interesting to modern audiences.

If Heywood partnered Webster in preparing *Appius and Virginia* (1620s?), that may account for its story-line being "clear and simple", in contrast to *The Devil's Law Case*. Another possible explanation is that the script was commissioned by a boys' troupe that also gave performances at the Phoenix. For them the roles would not be too demanding. Its scene is Ancient Rome. The authors (or author) may have had in mind the continuing popular success of Shakespeare's *Julius Caesar*.

The action revolves around the identity of Virginia. Appius, a judge, claims that she is the daughter of a dead slave who belonged to Clodius, one of his associates. Appius' motive is obsessive lust; he wants to gain possession of Virginia with Clodius' assistance. She is, in fact, the child of Virginius, a noble general. According to the plaintiffs his child died and the slave's daughter was substituted for her. The girl asks her father to kill her rather than let her fall into Appius's hands. Virginius, fearing that Appius will prevail, finally does slay her:

> And see, proud Appius, see
> Although not justly, I have made her free.

The moral of the play:

> Better had Appius been an upright judge
> And yet an evil man, than honest man,
> And yet a dissolute judge; for all disgrace
> Lights less upon the person than the place.

A military revolt brings about a change of government. Appius meets his death bravely; Clodius in a craven manner. Bradbrook suggests that the script had an implicit political message, a subversive one, but the boys could hardly be aware of that or face charges for it.

These two last plays are tragicomedies. In her *Tragedy and Tragicomedy in the Plays of John Webster*, Jacqueline Pearson files a brief in their defence; she argues that merely because they are tragicomedies they are dismissed as inferior to *The White Devil* and *The Duchess of Malfi* but that they have many merits and a share of Webster's best poetry. Tragicomedy, with its disjunctions of mood and an addiction to at least partially happy endings, she contends, gives a more accurate picture of life as it is experienced by the majority of mankind than does classical tragedy with its limitations which afford only an unrealistically narrow view. The spectator, leaving the theatre, can more readily identify with what he has just seen and been made to feel. The world is made up of many incompatible elements; a play should not over-simplify its flow in service to art more than to a full image of life.

Through the ensuing centuries Webster's reputation rose and fell. His tragedies were never without sharp detractors. They were viewed with disdain by William Archer. George Bernard Shaw, who had no affinity for plays of this sort, likened the multiple horrors in Webster's scripts to those in Madame Tussaud's waxworks. But such discerning nineteenth- and twentieth-century poets and critics as Algernon Swinburne, Charles Lamb and Rupert Brooke expressed their esteem for him – indeed Brooke wrote his dissertation on Webster and contributed many apt phrases about him.

Webster's vision was macabre: life, as he saw it, was a "deep pit of darkness". His men and women, buffeted by ill chance and evil passions, cannot hope to survive. At best they can meet their fate with the dignity that lends a noble Grecian stature to unhappy human beings, and some do. Above them, as Bosola tells the luckless duchess: "Look you, the stars shine still." But the Heavens are remote and deaf to her pleas. The cosmos is malignly indifferent.

The date of Webster's death, like so much about his life, is unknown. Nothing is heard of him after 1625. Fifteen years later, as a result of the king's beheading and the Civil War, the now reigning Puritans, often scandalized and caricatured, retaliated by closing all English theatres. A glorious era of immeasureable wealth of dramatic literature was cut off. Nor was England ever again to bring forth poet-playwrights of such prodigal genius.

GENERAL INDEX

Notes: Titles of plays and other works which receive frequent mention or detailed analysis have independent main headings; works receiving only passing reference appear as subheadings under the author's name. Detailed analysis is indicated by **bold** type.

Abbott, George, 610
Abélard, Peter, 60, 73
Abraham and Isaac (medieval), 44–5, 49
academies *see under* names of cities
I Accesi, 241, 242
Accoramboni, Vittoria, 873–4
Achille della Volta, 147
Ackland, Joss, 522
Ackroyd, David, 836
actors
    abilities, 42, 276, 495
    companies, 49, 92, 106, 204, 374–7, 436–7, 487 (*see also commedia dell' arte*; guilds)
    court companies, 92
    disciplining, 42, 241
    establishment attitudes to, 21–2, 91–2, 376
    finances, 42, 374, 375, 492–3
    leading performers, **493–6**
    lifestyle, 385
    social status, 21–2, 239, 385–6, 486–7
Adam (twelfth-century Anglo-Norman), **47–9**
Adam de la Hal(l)e, **94–5**
    Jeu de la Feuillée, 95
    Play of Adam (*Lijas Adam*/*Play of the Greenwood*), 95
    Play of the List, 95
    Robin and Marion, 95, 446
Adrian VI, Pope, 145
Aeschylus, 116, 612
    Agamemnon, 511
    Prometheus, 746
Agricola, Johannes, *Tragedia Johannes Huss*, 354
Aguero, Juan Antonio, 424
Ailey, Alvin, 741
Alamanni, Luigi, 225

Alarcón, Juan Ruiz de, **396–79**
    Cruelty for Honour, 397
    Gaining Friends, 397
    The Husband's Examination, 397
    Walls Have Ears, 396–7
    The Weaver of Segovia, 397
Alarcón, Pedro Antonio, 428
Alba, Duke of, 368, 383
Alberti, Leon Battista, *Della Pittura*, 120
Albinoni, Tomasso, 306
Albrecht V, Duke of Bavaria, 257
Alcalá de Henares, University of, 382, 398, 400
The Alchemist (Jonson), 559, 771, **861–3**, 871
    characterization, 862–3
Alden, David, 296
Aldredge, Tom, 612
Aldridge, Ira, 709
Aldus (printer), 116
Alençon, Duc d', 434, 483
Aleotti, Giovan Batista, 124, 127–8
L'Alessandro (Piccolomoni), **169–73**
    characterization, 171–3
    production history, 173–4
Alexander, Tony, 142
Alexander the Great, 822
Alexander VII, Pope, 298
Alexander VIII, Pope, 317
Alfeld, mystery plays, 51
Alfonso X, Pope, 376
Alfonso X of Castile, 93–4
Alfreds, Michael, 141–2
allegorical characters, use of, 69, 83, 192–3, 408, 412, 477
Allen, Ross, 66
Alleyn, Edward, 492, 493–4, 509, 515, 524, 527, 544, 556, 780, 850
All's Well That Ends Well

(Shakespeare), **700–1**, 702, 822
Amati family, 281
Amato, Vincenzo, 321–2
Amaya, Carmen, **424–5**
"Ambrogio, Messer" (correspondent of Aretino), 150–1
Ammirato, Scipio, *I Transformati*, 174
Amor Después de la Muerte (Calderón), **413–14**
L'Amoroso Sdegno (pastoral, anon.), 133
Amsterdam, theatre buildings/companies, 453, 456
Amyot, Jacques, 734
Anabaptists, 453
Anagnost, Dino, 337
Anderson, Judith, 681–2, 732
Andreini, Francesco, 240–1
Andreini, Giovanni Battista (Lelio), 241, 242–3
Andreini, Isabella (née Canali), 240–1, 243
Andreini, Lidia, 242
Andreini, Virginia, 242
Andreyev, Leonid, *He Who Gets Slapped*, 260–1, 268
Angers, mystery plays, 50
Angiloini, Gaspero, 399
Anna, Empress of Russia, 258–9
Anne, Queen (consort of James I), 867
Anouilh, Jean, *Becket or the Honour of God*, 86
Antheil, George, 861
anti-Semitism
    in English society/drama, 529, 531, 640–1
    in Italian comedy, 154
    in Italian society, 203–4

in medieval German drama, 38, 53
Antonet (clown), 256
Antonio (dancer), 428
Antony and Cleopatra (Shakespeare), 622, **734–41**
    characterization, 735–6, 737
    critical commentary, 739–40
    language, 736–9
    modern productions, 740–1
    sources, 734
Antwerp, 49–50
Apolloni, G.G., 298
apothecary, as stock character, 26
Appia, Adolphe, 601
Aquinas, Thomas, St, 38, 92
Aragón, Álvaro Cubillo de, 397
    Marcela's Dolls, 397
Archer, William, 891
Arden of Feversham (Kyd?), 498, **779–80**, 781
Aretino, Pietro, 134, **143–58**, 241
    assassination attempt, 146–7
    biography, 143–50, 156
    correspondence, 150–2
    critical commentary, 156, 157–8
    friendships/acquaintances, 148–9, 155, 168
    influence on later writers, 156–7, 368
    personality, 149–50, 151
    as playwright, 152–8
    pornographic verse, 146
    religious writings, 148
    satirical verse, 144–5, 146–7
    social status/lifestyle, 146, 147–8, 152
    La Cortigiana, 153, 156, 169
    Dialogues, 148, 157–8
    Il Filosofo, 153
    Lo Ipocrito, 153
    Last Will and Testament of the Elephant, 145
    La Orazia, 156
    Pronostici, 147
    La Talanta, 156
"L'Argenta" *see* Aleotti
La Argentina, 428

La Argentinita, 428
Arianna (Monteverdi), **284–6**
Ariosto, Gabriele, 136
Ariosto, Ludovico, **134–6**, 178, 216, 234, 241
    La Cassaria, 120–1, 135
    Lena, 136
    Il Negromante, 135–6
    Orlando Furioso, 134, 136, 147, 434
    Scolastica, 136, 355
    I Suppositi, 135, 171, 471, 472, 613–14
aristocracy, patronage of theatre, 487
Aristophanes, 275, 745
    Plutus, 868–9
Aristotle, 248, 749
    influence on Western drama, 178, 431
    Western commentaries on, 116–17, 118
    Poetics, 115, 120, 164, 420, 707
Arius, 22–3
Armin, Robert, 762–3
Arnell, Robert, 265
Artois, Robert II, Count of, 95
Artusi, Giovanni Maria, *Imperfections of Modern Music*, 282
As You Like It (Shakespeare), 117, 129, 532, 561, 617, **658–63**, 745
    language, 660–1
    modern productions, 662–3
    songs, 661–2
    sources/influences, 658, 659
Ashby, Harvey, 523
Ashmore, Basil, 518
Ashton, Frederick, 399
Asses, Feast of, 87–90
L'Assiuolo (Cecchi), **184–8**, 199
    "originality"/sources, 184–5
The Atheist's Tragedy, or The Honest Man's Revenge (Tourneur), 828, **834–5**
Auber, Daniel-François, 266
Auberjonois, Remy, 663

Aubrey, John, *Brief Lives*, 564, 579
Auden, W.H., 30, 773
Audi, Pierre, 310, 311, 312
audiences
    accommodation, 373
    behaviour, 384–5, 495
    composition/costume, 125, 128, 134–5, 176, 373, 458, 492
Auspira, Giovanni, Cardinal, 133
Austria, 249, 256
autos sacramentales, 366–8, 374, 376, 385, 407–12, 417
    scale of presentation, 386
Avila, Bishop of, 382
Ayrer, Jakob, **363–4**
Ayres, Lemuel, 615

Bacci, Luigi, 143
Bacon, Roger, 474
Bacon, Sir Francis, 586
Badoaro, Giacomo, 287–8, 289, 301, 310, 421
Bagby, Benjamin, 70–1
Baïf, Jean-Antoine de, 433, 447
Baïf, Lazare de, 433
Baines, Richard, 506, 507–8
Baker, George Pierce, Prof., 93
Bakst, Leon, 260
Balaam (biblical character), 36, 88, 96
Balanchine, George, 55–6, 264–5, 610
Balbi, Giovanni Battista, 301, 313–14
Balbulus, Notker "the Stammerer", 24
Balcon, Jill, 518
Balde, Jacob, *Jephtias*, 458
Bale, John, Bishop of Ossory, 82
    King Johan, 469
Balfe, Michael, *Falstaff*, 697
ballet(s), 55–6, 259–60, 264–5, 428
    based on Shakespeare, 626, 690, 711–12
    evolution, **446–50**
    origins, **349–52**

Bambini, Eustachio, 333
Banchieri, Adriano, *Prudenze Giovanile*, 276–7
Bandello, Matteo, 389, 400
Banello, Matteo, 646
Bankhead, Tallulah, 740
Barber, Samuel
  Antony and Cleopatra, 741
  Vanessa, 741
Barbier, Jules, 689
Barbieri, Nicolo, 242
Bardi, Giovanni, Count, 277, 278
  Amico Fido, 280
Bardi, Pietro, 277
"Bardolatry", 595
Bargagli, Celso, 214
Bargagli, Girolamo, **214–25**
  biography, 214–15
  Dialogo de' Guiochi, 215
Bargagli, Scipione, 214, 216
Barlocci, G., 327
Barnicle, Andrew, 142
Baro, Balthasar, 444
Barrault, Jean-Louis, 251, 601
Barry, Philip, *The Animal Kingdom*, 687
Barrymore, Drew, 679
Barrymore, Ethel, 679
Barrymore, John, 609, 679, 684–5
Barrymore, Lionel, 679, 684
Barrymore, Maurice (Herbert Blythe), 678, 679, 684
Barsted, William, 799
Bartholin, Birget, 626
Bartholomew Fair (Jonson), **863–4**
Bartlett, Kenneth R., 160
Barton, John, 552
Basoche (actors' guild), 94, 97, 98
Bassano, Baptiste, 577
Bassano, Emilia, 576–7
Bassett, Angela, 54
Bastelica, San Pietro di, 704
Baude, Henri, 97
Bäuerle, Adolf, 257
Bayreuth, 343
Beaman, Donald, 29–30
Beard, Thomas, 507
Beatty, John Lee, 612

Beaujoyeux, Balthasar de, **447–50**
Beaulieu, Lambert de, 448
Beaumarchais (Pierre-Augustin Caron), 267
Beaumont, Francis, 853
  biography/portraits, 777, 814, 824
  The Woman Hater, 814
  see also Beaumont and Fletcher
Beaumont, Ursula, née Isley, 824
Beaumont and Fletcher, 494, 599, 759, 779, 787, 809, **813–27**, 843
  critical/popular reception, 826–7
  masques, 867
  relationship/collaborative method, 814, 815, 818
  Cupid's Revenge, 823
  The Maid's Tragedy, 822–3
  The Scornful Lady, 823–4
Beauvais Cathedral, 88–9
Bebel, Heinrich, *Comoedia de Optimo Studio Iuventum*, 354
Beccari, Agotino, 204
Beckett, Samuel, 274
  Endgame, 720
Bede, the Venerable, 25
Beglarian, Eve, 72
Beito, Calixto, 418
Belcon, Natalie, 695
Belleforest, François de, *Histoires Tragiques*, 666
Bellini, Vincenzo, *The Capulets and Montagues*, 625
Belloni, Ms, 269
Belot, Stephen, 570
Benavente, Count of, Viceroy of Naples, 393
Bender, Howard, 297
Benét, Stephen Vincent, 516–17
Benson, Frank, 551–2, 683
Benson, John, 577
Bentivoglio, Annibale, 350
Benvenuti, Giacomo, 307, 312
Beolco, Angelo ("Ruzzante"), 106, **158–62**, 175, 248–9, 256
  biography, 158–9, 162

  linguistic/political character, 160, 195
  stage persona, 159–60, 162, 254, 255
Beolco, Giovan Francesco, 158
Beregan, Niccolò, 300
Berio, Luciano, 307
  Un Re in Asculo, 773
Berlioz, Hector, 625–6, 773
  Béatrice et Bénédict, 648
Bernacchi, Antonio, 316
Bernini, Domenico, 228
Bernini, Gian Lorenzo, **226–33**
  biography, 226, 227
  comic idiom, 228
  personality, 227
  sculpture, 226–8
  set designs, 228–9
  surviving script (untitled) see *The Impresario*
Bernini, Pietro, 226
Bernstein, Leonard, 625
Berry, Sarah Uriarte, 611, 612
Der Bestrafte Brudermord (anon.), 666
Betterton, Thomas, 583, 676, 718
Bèze, Théodore de, *Abraham sacrifiant*, 436
Bianchi, Giuseppe, 244
Bibbiena, Bernardo Dovizi da, Cardinal, *La Calandria*, 122–3, 136–7, 161
Bibiena family (theatre architects), 342–3, 345
Bicket, Harry, 296
Bidermann, Jakob, *Cenodoxus*, 458
Bielawa, Lisa, 72–3
Bien Avisé, mal avisé (fifteenth-century Rennes), 79
Bills, B.D., Dr, 67, 68
Binkley, Thomas, 66
Birck, Sixt, 356, 358
Birnbaum, Steven L., 142
Bisaccioni, M., 298
Bizet, Georges, *Carmen*, 426
Blacher, Boris, 690
The Black Crook (author unknown), 680

Blackfriars Theatre (London), 488–9, 490–1
Blau, Édouard, 396
Bleeke, Mark, 32
Bloch, Ernest, *Macbeth*, 733
Blok, Alexander, *The Puppet House*, 259
Boccaccio, Giovanni, *Decameron*, 137, 171, 184, 201, 325, 477–8, 639, 697, 700, 751, 758, 760, 777
Boccadiferro, Lodovico, 168
Bodel, Jean, 60
    Jeu de Saint Nicolas, 60, 62
Bogdanov, Michael, 748–9
Bohm, Adolph, 260
Boi Sakti, 724
Boito, Arrigo, 268, 697, 711
Bolero, 423
Boleyn, Anne, 867
Bolt, Robert, *A Man for All Seasons*, 86
Bonci, Tita, 143
Bononcini, Giovanni, 295
Booth, Edwin, 678–9, 680, 684
Booth, John Wilkes, 678
Booth, Junius Brutus, 678
Booth, Mary Ann, née Holmes, 678
Borghini, Raffaello, 225
Borgia, Cesare, 137
Boswell, James, *London Journal*, 150
Botticelli, Alessandro, 350
Bougoin, Simon, *L'Homme juste et l'homme mondain*, 79
Bounyn, Gabriel, *La Sultane*, 434
Bourlet, Katherine, 47
Bowles, Anthony, 141–2
Bowles, Paul, 545
Boyes, Elizabeth, 471
The Boys from Syracuse (musical), 263, 598
Bradley, William, 504
Branagh, Kenneth, 652–3, 688
Brant, Sebastian, *Tugent Soyl*, 356
Brauschweig, Heinrich Julius von, Duke of Brunswick, **362–3**
Brazelton, Kitty, 72

Brecht, Bertolt, 84, 274
    Edward II, 557
Brendero, G.A., 452
Brennan, Brid, 559
Bressler, Charles, 29
Breton, Nicholas, 471
Breughel, Pieter, the Elder, 87
Brich (clown), 256
Bridges, John, Bishop of Oxford, 467–8
Brissac, Duc de, 447
Britten, Benjamin, *A Midsummer Night's Dream*, 559, 598, 630
Broelmann, Stephan, 357
The Broken Heart (Ford), **837–8**, 840, 841
Brome, Richard, 843
    The Antipodes, 843
    The City Wit, 843
    The Jovial Crew, 843
    The Lovesick Maid, 843
    A Mad Couple Well Match'd, 843
    The Northern Lass, 843
    The Queen's Exchange, 843
    The Sparagus Garden, 843
Bronson, Virginia, 687
Bronzino, Agnolo, 148, 176, 188
Brook, Peter, 263, 630, 720–1, 723, 759
Brooke, Arthur, 389, 620
Brooke, Rupert, 891
Brosses, Charles de, 315
Brown, E. Martin, 57
Brown, John Russell, 54
    see also Index of Cited Authors
Browne, Robert, 457
Browne, Roscoe Lee, 860
Browning, Robert, *Caliban Upon Setebos*, 773
Brubaker, Robert, 56
Brühl, Count, 600
Brunelleschi, Filippo, 120
Bruno, Giordano, 214
    Il Candelaio, 214
Brustein, Robert, 835–6, 860
Bryden, Bill, 53–4
Bryggman, Larry, 776

Büchner, Georg, 562
Bulgarini, Belisario, 216
Buontalenti, Bernardo, 128, 216, 225, 280
Burbage, Cuthbert, 494
Burbage, James, 486, 488–9, 492–3, 494
Burbage, Richard, 494, 571, 578, 580, 582, 584, 588, 676, 848
Burke, John, 687
Burnacini, Giovanni, 341
Burnacini, Ludovico, 300, **341–2**
Burney, Charles, Dr, 315, 322, 323
Burton, Philip, 33
Burton, Richard, 545–6, 688
Burton, Robert, *Anatomy of Melancholy*, 836–7, 840
Bury, John, 524–5
Busenello, Giovanni Francesco, 289–90, 293, 421
Busoni, Ferruccio
    Arlecchino, 267
    Turandot, 267
Bussy d'Ambois (Chapman), **789–90**, 796
Butelli, Louis, 610
Buti, Francesco, Abbot, 295
Butler, Paul, 725
Butter, Nathaniel, 868–9
Byron, Arthur, 682

Caccini, Francesca, 281
Caccini, Giulio, 277, 279, 280, 283
    Euridice, 280–1
    Le Nuove Musiche, 281
    Il Rapimento di Cefalo, 280
Caffarelli, Detto, 323, 329, 334
Cairns, Tom, 559
Calarco, Joseph, *R&J*, 626–8
Caldara, Antonio, 334, 337–8
    Achille in Sciro, 337–8
Calderón de la Barca, Diego, 399–400
Calderón de la Barca, Pedro, 229, **399–418**, 419, 425, 457
    biography, 399–400, 407, 416–17

chosen genres, 400–2, 407–8,
  413, 414, 417, 421–2
critical commentary, 415–16,
  417
influence on later writers,
  415
international reputation,
  419–20
modern productions, 401,
  408, 410, 417–18
opera libretti, 421–2
quadricentenary celebrations
  (2000), 417–18
qualities as dramatist, 399,
  404–5, 407–8, 414, 415–16
translations/adaptations,
  414–15, 417, 418
*A Secreto Agravia, Secreta
  Venganza*, 406, 416
Amor, Honor y Poder, 400
Antes de Todo Es Mi Dama,
  414
El Astrólogo Fingido, 414
La Banda y la Flor, 414
El Carro del Cielo, 400
*Casa con Dos Puertas, Mala
  Es da Guarda*, 401
La Cena del Rey Baltasar,
  409
La Dama Duende, 401, 416,
  418
Eco y Narciso, 421–2
El Gran Príncipe de Fez, 412
Guárdate del Agua Mansa,
  414
El Laurel del Apolo, 422
El Mayor Encanto, Amor, 407
The Mayor of Zalamea, 402,
  416
Mornings of April and May,
  418
No Monster Like Jealousy,
  406
El Príncipe Constante, 400,
  401–2, 413–14, 416
El Purgatorio de San Patricio,
  409
The Siege of Breda, 400, 401
La Calderone, 386

Callot, Jacomo, 244, 250, 271
Callum, William, 57
Calvinists, 453
Cambridge (University),
  intellectual climate, 501–3
Camden, William, 494, 847
  *Remaines of a Greater Worke,
  Concerning Britain*, 743
Cammelli, Antonio, 119
Campani, Niccolo, 106
Campaspe (Lyly), **482–3**
Campbell, Roy, 404
Camus, Albert, 415, 417
Canali, Isabella *see* Andreini,
  Isabella
Cantone, Mario, 612
Cappello, Bianca, 803
Carew, Richard, 68
Carey, Lady Elizabeth, 485
Carey, Henry, 1st Baron
  Hunsdon, 577, 676
Carissimi, Giacomo, 298, 317
Carl, Barry, 56
Carlo Emanuele I of Savoy, 205
Carmail, Comte de, 444
Carmina Burana, 66
Carnovsky, Alex, 721–3
Caro, Annibal, **162–8**
  biography, 162–3, 168
Caroso, Fabritio, *Il Ballarina*, 351
Carpio, Don Miguel de, 382
Carracci, Annibale, 226
Carré, Michel, 689
Carroll, Marie, 686–7
Carter, Jack, 545
Carvajal, Bernadino, Cardinal,
  372
Castelvetro, Ludovico, 117
Castiglione, Baldassare, Count,
  122–3, 169, 222
The Castle of Perseverance
  (fifteenth-century English),
  75–6
Castro y Bellvis, Guillén de,
  393–6, 421
  qualities as dramatist, 393–4,
  395–6
  Count Alarcos, 394
  Dido and Aeneas, 394

The Force of Habit, 394
The Hostile Brothers, 394
The Ill-Wed of Valencia, 394
The Impertinent Meddler, 394
*The Man Who Thought He
  Was Narcissus*, 394
The Power of Blood, 394
Procne and Philomena, 394
Castro y Bellvis, Magdalena,
  393
Catholic Church/Catholicism
  defences of, 82
  opposition to theatre, 83,
  98–9, 199, 317, 320, 376,
  385
  satirical attacks on, 82, 97–8,
  214, 353, 354–5, 357–8,
  371–2
Catiline (Jonson), **864–5**
Cavalieri, Emilio de', 216
  La Disperazione di Fileno, 276
  *La Representazione di Anima
  e di Corpo*, 278
  Satiro, 276
Cavalieri, Laura de', 278
Cavalli, Francesco (Piero
  Francesco Caletti-Bruni),
  292–7, 299, 318, 333–4, 421
  biography, 292–3, 313–14
  La Calisto, 295–6, 313
  La Didone, 293
  Egisto, 293, 294
  Ercole Amante, 294, 295
  Erismena, 296
  Il Giasone, 293–4, 296–7, 299
  L'Hipermestra, 294
  Le Nozze di Teti e Pelio, 293
  L'Ormindo, 295, 296, 313
  Il Xerxes, 294–5
Caxton, William, 697
Cecchi, Giovan Maria, **174–203**,
  216
  biography, 174–5, 190, 202
  categorization of plays, 175
  characterization, 180–1, 190,
  192, 193, 194–5, 198, 200
  creative method, 175–6
  didactic intent, 192–3,
  199–200, 202

language, 178, 186–7, 195, 198
religious writings, 193, 195, 196–7, **199–201**
social satire, 178, 180
treatment of marriage, 178, 180, 181, 183–4, 187–8
women, attitudes to/portrayals of, 181, 187–8, 197, 199
L'Ammalata, 190
Li Contrassegni, 197, 199
La Coronazione del Re Saul, 200
Il Corredo, 181
Il Debito, 175, 199
Il Donzello, 189
Il Figlio Prodigo, 200
Gli Incantesmi, 181
La Maiana, 189–90
Il Medico, 193
La Morte del Re Acab, 193, 200
Le Pellegrine, 195
Ragionamenti Spirituali, 193
I Rivali, 193
Gli Sciamiti, 193
Cecchi, Marietta, née Pagni, 174–5
Cecchi, Matteo, 174
Cecchi, Prudenza, 174
Cecchi, Ser Bartolomeo, 174
Cecil, Lawrence H., 687
Cecil, Robert, Earl of Salisbury, 502, 586
Cecil, Sir Edward, 828
Cecil, William, Lord Burleigh, 481
Le Cedole (Cecchi), **195–6**
La Celestina (Rojas), **367–8**, 382
Celli, Giambattista, 225
Cellini, Benvenuto, 163, 176
Autobiography, 150
censorship/suppression (of drama), 21–2, 82, 83, 98–9, 891
Cervantes Saavedra, Miguel de, **379–81**, 443, 807
comments on Lope, 381, 384, 387

El Cerco de Numancia, 380
Don Quixote, 379, 381, 384, 394
The Fortunate Ruffian, 380
The Traffic of Argel, 380
Cesti, Marc (Pietro) Antonio, 292, **297–301**, 305, 317
biography, 297–8, 300, 301
L'Argia, 298–9, 301
La Cleopatra, 298–9
Le Disgrazie d'Amore, 299, 300
La Germania Esultante, 301
La Magnanimita d'Alessandro, 299
Orontea, 299
Il Pomo d'Oro, 299, 300, 341
La Semirami, 299, 301
Chabrier, Emmanuel, 426
Chabukiani, Vakhtang, 711
Chamberlain, Richard, 688
Chambers, E.K., 48
see also Index of Cited Authors
"Chandos portrait", 582–3, 584
The Changeling (Middleton/Rowley), **804–7**
Chaplin, Charlie, 269
Chapman, George, 697, 779, **787–93**, 853
collaborations, 789, 825, 844
literary qualities, 789, 793, 796
masques, 793
All Fools But the Fool, 788
The Blind Beggar of Alexandria, 787
Caesar and Pompey, 793
Chabot, Admiral of France, 792
An Humorous Day"s Mirth, 788
May-Day, 174, 788
Monsieur d'Olive, 789
The Old Joiner of Aldgate, 788–9
The Revenge of Bussy d'Ambois, 792
The Widow's Tears, 793

Charles I of England, 828, 842–3, 868, 891
Charles V, Holy Roman Emperor, 135, 148, 377, 386, 451, 520
Charles VI, Holy Roman Emperor, 328, 339, 342–3
Charles VIII of France, 97, 447
Charles IX of France, 248, 430, 433, 434, 447
Charles Albert, Elector of Bavaria, 336
Chastellain, Georges, *Le Concile de Bâle*, 79
Chaucer, Geoffrey, 471
The Canterbury Tales, 109, 777
Troilus and Criseyde, 697, 700
The Cheated (anon.), 691
Chester Mystery Cycle, 43, 45–6
authorship, 46–7
modern productions, 53
Chettle, Henry, 476, 571, 584–5
Chevalier du soleil (commedia dell' arte), 245–6
Chézy, Helmina von, 760
Chiabrera, Gabriele, 280
Chigi, Agostino, 144–5
Chilcott, Barbara, 521
Chiusiano, Prince of, 325
Chomdavat, Peeramond, 724
Christian Church
as landlord, 37–8
mockery of, 87–91
opposition to theatre, 21–2, 36, 38, 91, 93–4, 114
role in development of drama, 22–3, 43
role in society, 21, 114
see also Catholic Church; Protestantism
Christina of Sweden, 299, 317–18
Church, Barbara, 517
Church, Tony, 721
Ciancarli, Gigio Artemio
Capraria, 225
La Zingana, 225
Cibo, Cardinal, 302
Cicero, M. Tullius, 358, 360

Cicognini, Giacinto Andrea, 294, 299
Cicognini, Jacopo, 420–1
Cimarosa, Domenico
   Le Donne Rivale, 267
   Il Matrimonio Segreto, 267
   L'Olimpiade, 334
"Cinthio" (Giovanni Batista Giraldi), **116–18**, 204, 241
   Epitia, 701
   The Hundred Fables, 704
   The Interchanged, 117
   Orbecche, 116–17
The City Madam (Massinger), **810–11**
Clarke, Martha, *Miracolo d'Amore*, 273
Claudel, Paul, 114
Clavijo, Ruiz Gonzalez de, 335
Clear, Patrick, 749
Clemencic, René, 225
Clement VII, Pope (Giulio de' Medici), 145–7, 372
Clement XI, Pope, 320
clergy, as comic characters, 222
Clopton, Sir Hugh, 579
clowns, 252–7, 261
Cocco, Giacomo, Archbishop, 168
Cockerell, Toby, 598
Cocteau, Jean, 274
Coello, Antonio, 400
Coghill, Nevill, Prof., 517, 545–6
Cohen, Gustave, Prof., 93
Cohen, Joel, 29–30
Coku, Alexandra, 296–7
Coleridge, Samuel Taylor, 578, 594, 677, 715, 862
Colicos, John, 771–2
collaborations, 477–8, 603, 784, 825–6
   see also names of playwrights
Collin, Matthias, *History of Susanna*, 356
Collins, Pat, 712
Collis, John, 35
Colman, Mrs, 599
Colonna, Marco, 380
"colour-blind" casting, 602

Columbus, Christopher, 113
Comedia von der schönen Sidea (Ayrer), **363–4**
comedias de santos, 368, 410
Les Comédiens du Roi, 445
comedy, defined, 23, 354, 702
The Comedy of Errors (Shakespeare), 263, 484, 598, **609–12**
   modern productions, 609–10
   musical reworking, 610–12
   sources, 178–9, 609
I Comici Confidenti, 240, 241, 243, 256
I Comici Fedeli, 241, 242–3, 256
I Comici Gelosi, 239–41, 242, 245, 256, 447
I Comici Uniti, 243
commedia dell' arte, 125, 135, **233–50**
   artistic depictions, 244, 250, 257, 272–3
   characters, 95, **234–7**, 241–2, 243, 250
   companies, 238, **239–44**, 245, 248–9
   competition with other genres, 385
   costume/masks, 238, 239, 244, 247–8
   international appeal/ variations, 249–50, 257–9, 265, 372–7, 440–1
   lazzi (comic moments), 244–5, 270–1
   literary/comic legacy, **250–6**, **259–69**, 361, 386
   modern re-creations, **269–74**
   performance style, 233–4, 237–8, 239, **244–50**, 276
   repertoire, 162, 245–7
   scenery, 238–9
commedia erudita, 324
   defined, 134
   new developments, 162, 164, 211
   plays' relationship with genre, 175, 200, 220, 233

   standard characters/ situations, 159, 213
commedia grava, 220–1, 225, 233
commedia osservata, 175
commedia ridicolosa, 233
commedia spirituale, 175
Commelli, Antonio, *Filostrato e Panfila*, 116
Condell, Henry, 580, 588, 589, 773–4, 848
Confrèrie de la Passion, 98–9, 114, 120, 436, 437–8
Congresa dei Rozzi, 106
Congreve, William, 599
Conklin, John, 54
Connelly, Marc, *Green Pastures*, 55
Conrad, Bishop of Constance, 27
The Conspiracy and Tragedy of Charles, Duke of Byron (Chapman), **790–2**
Constantine, Emperor, 23, 458
Copeau, Jacques, 251
Copenhagen, Tivoli Gardens, 263
Copernicus, Nikolaus, 113
Corbeil, Pierre de, Archbishop, 89
Corelli, Arcangelo, 426
Corigliano, John, *The Ghosts of Versailles*, 267
Coriolanus (Shakespeare), **743–5**
   sources, 743, 744
Cornaro, Alvise, 158–9
Cornazzano, Antonio, 348
Corneille, Pierre, 339, 421
   Le Cid, 394, 395–6
   Le Menteur, 396
Cornell, Katherine, 624–5, 687, 740
Cornish, William, 867
Cornish (language), plays in, 67–9
Corpus Christi, Feast of, 38, 133, 569
Corral de Pachea, 372–3

Corsi, Jacopo, 278, 279
costumes
    financial responsibility for, 376–7
    historical authenticity, 377, 385, 492, 599–600
    see also under genre names
Cottom, John, 567
Courtleigh, William, 687
Courtois d'Arras, 57–8
Coventry Mystery Cycle, 43
Cowl, Jane, 624, 627
Cox, Brian, 525
Craft, Robert, 55–6
Craig, Edward Gordon, 601, 677, 688
Cranko, John, 265
Cranmer, Thomas, Archbishop, 107, 111, 354–5
Creation, story of, 36
Creation Spontaneously, 73
Cromwell, Oliver, 843, 844
Cromwell, Thomas, 107, 151
Croo, Robert, 47
Crook, Peter, 54
cross-dressing, role in plot development, 117, 186, 385, 398, 482
    in Shakespeare, 617, 691
Crusades, 27, 60
Crusis, 360
Cueva, Juan de la, **378–9**, 386
    output, 379
    Exemplar Poetica, 379
    El Infamador, 379
    Los Siete Infantes de Lara, 378–9
Cunningham, John, 722
Cutlip, Philip, 337
Cymbeline (Shakespeare), 662, **751–60**, 818, 822
    critical commentary, 756–9
    form/dramatic themes, 753, 756–7
    language, 754–6
    modern productions, 757–60
    songs, 753–4
    sources/influences, 751–2, 758
Cysat, Renward, 51–2
Czechoslovakia, 36

da Ponte, Lorenzo, 399
Daborne, Robert, 808
Dallapiccola, Luigi, 307, 308
    Il Prigioniero, 307
d'Amboise, Jacques, 55
d'Ambra, Frances, 225
damnation
    role in Christian/popular thought, 22, 25, 43
    as theme of drama, 22
dance
    ceremonial/triumphal function, 349–51, 447–50
    folk traditions, **422–8**
    role in drama, 276, 349 (*see also* ballet)
    role in opera, 350
    textbooks, 347–8, 351
    training, 423–4
Dandolo, Leonardo, 157
Daniel, Paul, 307
Daniel, Samuel, 631, 644, 734
Daniels, Ron, 636–7
Danish Royal Ballet, 626
Dante (Alighieri), *Divine Comedy*, 162, 277, 538
"Dark Ages"
    art/architecture, 21
    cultural ethos, 21–3
    early developments in drama, 23–6, 37
Darlow, David, 749
d'Aubigné, Agrippa, 448
Davenant, Jane, 578
Davenant, Sir William, 578, 583, 676
Davenport, Fanny, 679
Davenport, LaNoue, 32, 68
Davey, Gideon, 296
Davidson, Audrey Eckdahl, 71
Davidson, Clifford, 71
Davis, R.G., 263
Davison, Belinda, 598
Day, John, 746, 804

de la Taille, Jean, 432–3, 434
    Saul le furieux, 432–3
de la Valette, Cardinal, 444
de Mey, Guy, 297
De Pernet qui va au vin, 110
de Pietri, Stephen, 57–8
de Shields, André, 695
de' Sommi, Leone, 122, **203–13**
    biography, 203–5
    language, 203
    racial/theological identity, 203–4
    translations, 205
    Defence of Women, 203
    Four Dialogues, 125, 129–30, 205, 212–13
De Vera Nobilitate, 107
de Vere, Edward, Earl of Oxford, 481, 586, 588
de Witt, Johannes, 489
The Death of Pilate (fifteenth-century Cornish), 67–9
Deburau, Jean-Baptiste Gaspard, 250–1, 253, 263
Debussy, Claude, 426
Decroux, Étienne, 251, 263
Decroux, Maximilian, 251
Dee, Ruby, 722
Defoe, Daniel
    Journal of the Plague Year, 787
    Moll Flanders, 786
Dekker, Thomas, 745, 779, 780, **783–7**, 799–800, 843, 851–2
    biography, 783
    collaborations, 784, 786–7, 801–2, 804, 811, 813, 851, 872–3
    literary qualities, 783–4, 786
    The Bellman of London, 783, 828
    The Gull's Handbook, 783
    If It Be Not Good, the Devil Is in It, 787
    Match Mee in London, 787
    Old Fortunatus, 783–4
    The Sun's Darling, 787
    The Whore of Babylon, 784

(with Middleton), *The Family of Love*, 801–2
(with Middleton), *The Honest Whore*, 784, 800
(with Middleton), *The Roaring Girl*, 786–7
(with Rowley/Ford) *The Witch of Edmonton*, 787, 836
(with Webster), *Northward Ho!*, 786, 856, 872–3
(with Webster), *The Famous History of Sir Thomas Wyatt*, 872–3
(with Webster), *Westward Ho!*, 786, 854–5, 872–3
del Frulovisi, Tito Livio, 134
Del Monaco, Mario, 711
del Piombo, Sebastiano, 148, 163
Delius, Frederick, 626
della Porta, Giambattista, 213–14
della Rovere, Guidobaldo, 167
della Rovere, Vittoria (née Farnese), 167
Della Valle, Federico, 118–19
   Adelond, 118–19
   Judith, 118
   La Regina di Scozia, 118
Deloney, Thomas, *The Gentle Craft*, 784–5
Denmark, 258, 263, 356
Dennis, John, 745
Depositio Crucis, 25
I Desiosi, 244
Deslauriers, Jean (Brascambille), 441
d'Este, Ercole, Duke of Ferrara, 119
d'Este, Isabella, 137
d'Este, Lucrezia, 241, 350
Destouches, Philippe, *La Fausse Agnès*, 225
Deval, Jacques, *Tovarich*, 388
Devereux, Robert, Earl of Essex, 576, 586, 658, 698, 853
The Devil Is an Ass (Jonson), **865–6**
The Devil's Law Case (Webster), **889–90**
La Devoción de la Cruz

(Calderón), **408–9**
di Botta, Bergonzio, 349–50
di Capua, Rinaldo, 327
di Lasso, Orlando, 257
Diaghilev, Sergei, 259–60, 428
Diamond, David, 598, 773
Diana, Princess of Wales, 725
Diaz, Justin, 741
Dickens, Charles, 256, 849, 863
Diderot, Denis, 333
Digges, Dudley, 860
Dijon, medieval farceurs, 94
d'Indy, Vincent, 306–7
Diocletion, Emperor, 811
I Dissimuli (Cecchi), **182–4**, 188
Doctor Faustus (Marlowe), 83, 410, 473, 475, 491, 494, 511, 514, **533–46**, 550, 739, 784, 839
   characterization, 533–4, 540–1
   comic elements, 542
   leitmotif words/phrases, 544
   modern productions, 544–6
   religious/supernatural content, 535, 540–1, 542–3, 770
   as self-projection, 533–5, 541, 549
   sources, 533–4
doctors, as comic characters, 93
Dodgson, Charles, 22
Dolce, Ludovice
   Giocasta, 118, 471
   Marianna, 118
Domingo, Placido, 711
Don Juan, treatments of character/story, 398–9
Doña Marta "Amaryllis", 383–4
Donat, Peter, 772
Donatus, 133
Donizetti, Gaetano
   Don Pasquale, 266–7
   L'Elisir d'Amore, 266
Donne, John, 493
Donnellan, Declan, 648
d'Onofrio, Cesare, 229
Dorotea (Lope de Vega), 382–3
Dostoevsky, Fyodor, 839
La Dote (Cecchi), 175, **176–8**
Dotti, Bartolomeo, 320

Draghi, Antonio, 305
Draghi, Carlo Domenico, 305
Draghi, Cruciano, 327
Draghi, Francesco Andrea, 327
Drake, Sir Francis, 383
Drayton, Michael, 538, 579
   collaborations, 784
Drew, Elaine Grynkewich, 57–8
Drew, Georgiana, 679
Drigo, Riccardo, 265
Droeshout, Martin, 582, 584
Drummond, William, *Timbers*, 849, 868
Dryden, John, 590–1, 599, 740, 790, 814, 822, 826, 860, 868
   All for Love, 740
   (and Robert Howard), *The Indian Queen*, 136
du Bellay, Joachim, *Defence and Illustration of the French Language*, 431
du Ryer, Pierre, 445
   Scévole, 445
The Duchess of Malfi (Webster), 117, 797, 846, 873, **877–89**, 891
   characterization, 878, 879–80, 886–9
   historical basis, 878
   tragic atmosphere, 886, 889
Dudley, Robert, Earl of Leicester, 483
Dugan, Sean, 626–7
Dukes, Ashley, 140
Dullin, Charles, 251
Dumas, Alexandre, *The Count of Monte Cristo*, 680
Duncan, Cameron, 57
Duncan, John, 521–2
Dunham, Joanna, 625
Durante, Francesco, 327
Dürer, Albrecht, 49–50, 464
Dürrenmatt, Friedrich, 274
Dutton, Jane, 337
Dvorsky, Peter, 558
Dyalogue du Fol et du Sage, 110

Easter, dramas revolving around, 25
Eastward Ho! (Jonson/

Chapman/Marston), 789,
798–9, **854–5**
Echols, Paul C., 32–3, 35–6, 58
Eddison, Robert, 555
Edinburgh, Philip, Duke of, 597
Edward II (Marlowe), 505,
546–60, 561, 631
    characterization, 549–50
    film (1992), 559–60
    language, 547–9
    modern productions, **550–60**
    sources, 546
    treatment of sexuality, 546,
    552–3, 556–7
Edward IV of England, 92
Eisenbichler, Konrad, 184
Eleonore, Empress, 341
Elevatio Crucis, 25
Eliot, T.S., 518, 526, 529, 671,
793, 796, 811, 842, 871
    *The Love Song of J. Alfred
Prufrock*, 274
    Murder in the Cathedral, 86
    The Waste Land, 265
    "Whispers of Immortality",
890
Elisabeth Christina, Empress,
328
Elizabeth, Electress Palatine, 457
Elizabeth I of England, 91,
477–8, 632, 698, 712, 862
    accession, 83, 111
    alleged maleness, 868
    banning of religious drama,
83, 114
    court acting company, 92
    court customs/atmosphere,
482, 491–2, 574–5
    plays performed in presence
of, 240, 466, 488–9, 587,
620
    works written for/aimed at,
472, 479, 483, 607, 695
Elizabethan Stage Society, 600–1
Ellington, Duke, 694–5
Encina, Juan del, 368, 378, 380,
386
    *The Eclogue of Placida and
Victoriano*, 368

Egloga del Amor, 368
Les Enfants de maintenant
(fifteenth-century French), 80
Enfants Sans Souci, 97–9
England, 37–8
    casting resources, 680–1
    climate, 465
    medieval religious theatre,
37–8, 39, 43–9, 53–4,
75–9
    political conditions, 465
    popular entertainments,
90–1, 252–6, 261–3
    Renaissance drama,
465–86
    theatre buildings/companies,
486–95
English, John, 92
English (language), plays
performed in (outside
England), 353–4, 457–8
Ennery, Adolphe d', 396
Eobanus Hessus, 360, 361
Epicoene, or The Silent Woman
(Jonson), 156, **861**, 871
Erasmus (Reinhold), 133, 361
    Encomium Moriae, 110
Erdmann-Guckel, Hans, 307
Escudero, Vincente, 428
Essex, Earl of *see* Devereux
Ethelwold of Winchester, Bishop,
25
Euripides, 116, 873
    Phaedra, 389
    The Phoenicians, 118, 455
Evans, Maurice, 602, 635–6, 681,
732
Evelyn, Sir John, 228, 315, 599
Every Man in His Humour
(Jonson), **848–50**
    characterization, 849–50
Every Man Out of His Humour
(Jonson), **850–1**
Everyman (*c*. 1500, English),
80–2
    modern performances,
84–6
Evreinov, Nikolai, 229, 259
    The Chief Thing, 259

The Theatre for Oneself, 259
The Theatre of the Soul, 259
Eyre, Simon, 785

Fabyan, Robert, 546
Faggioli, Michelangelo, *La Cilla*,
325
The Faithful Shepherdess
(Fletcher), 132, **817–18**
Falla, Manuel de, *The
Three-Cornered Hat*, 428
*The Famous Victories of Henrye the
Fyft* (anon.), 649
The Fantasticks (musical), 267
farce (medieval), **87–111**
    costume/make-up, 99, 103
    forerunners, 87–91, 93
    influence on later styles,
133–4
    modern productions, 93,
96–7
    satirical content, 94
    stock characters, 92–3, 94
farce (post-medieval), 201–3,
440–1, 445
    stock characters, 440
*Farce of the Worthy Master Pierre
Patelin*, **95–6**
    imitations, 101, 355
Farinelli (Carlo Broschi), 323,
324
Farnese, Alessandro, Cardinal
(junior), 163, 168
Farnese, Alessandro, Cardinal
(senior) *see* Paul III
Farnese, Pierluigi, 163–4, 167
Fastnachtspiele (Shrovetide plays),
100–3, 355, 462
Faulkner, John, 556
Fauntleroy, Talmage F., 330
Faustini, Giovanni, 293, 295,
296, 313–14
Faustini, Marco, 292, 300,
314
Fechter, Charles, 600
Federico, G.A., 328, 330, 331
Fedorova, Nina, 56
Feldman, Jill, 71
Felton, Samuel, 583–4

female characters
    performed by boys, 495, 624
    presentation, 169, 172–3,
        222–3, 869
Ferber, Edna, 679
Ferdinand II, Emperor, 341
Ferdinand III, Emperor, 298
Ferdinand Karl, Archduke, 298,
    299–300
Ferdinand of Aragon, 365, 425
Ferdinand of Naples, 878
Fernández, Lucas, 368
Ferrara
    as cultural centre, 116, 119,
        131, 137
    theatre architecture, 120–1
Ferrari, Benedetto, 289, 301
Ferrer, José, 709
Field, Jacqueline, 576
Field, Nathan, 808–9, 813, 825
Field, Richard, 572, 576
Fielding, Henry, 262, 499
    Tom Jones, 676, 862
Filleul, Nicholas, 433
Finney, Albert, 524–7, 681
Fiorentine, Giovanni, 639
Fiorillo, Beatrice, 243
Fiorillo, Giovanni Battista, 243
Fiorillo, Silvio, 237, 243
    L'Amor Folle, 269
Firenzuola, Agnolo, 225
Fischer, Corinne, 141
Fischietti, Giovanni, 326
    La Costanza, 326
    La Sumiglianza, 326
Fisher, Rob, 611–12
Fiske, Mrs, 679
Fitton, Mary, 576
FitzGerald, Edward, 415
Fitzgerald, Geraldine, 86
flamenco dance, **424–8**
Flanders, mystery plays, 49
Flanigan, Clifford, 66
Flaubert, Gustave, 273
Fleetwood, Susan, 525
Fleg(enheimer), Edmond, 733
Fleming, Tom, 721
Fletcher, Allen, 722
Fletcher, John, 590, 836
    biography, 813, 814–15,
        824–5, 826
    collaboration with Beaumont
        see Beaumont and Fletcher
    (possible) collaboration with
        Shakespeare, 773, 777
    collaborations with others,
        802–3, 804, 809, 811, 813,
        825
    critical commentary, 826
    literary qualities, 825–6
    portrait, 814
    The Chances, 827
    Rule a Wife and Have a Wife,
        827
    *The Tragedy of Sir John Olden
        Barnavelt*, 825
    Valentinian, 825
    The Woman's Prize, 817,
        827
    see also Massinger
Fletcher, Richard, Bishop, 722
Fletcher, Robert, 30
"Fleury Playbook", 31
Florence
    Accademia degli Elevati,
        282–3
    audiences, 176
    Camerata (intellectual
        society), 277–8, 283
    cultural life, 277–8, 282–3,
        325
    mystery plays, 49
    Renaissance drama, 119
    social ethos (satirized), 178,
        180, 198
    theatre architecture, 128
Florence, Alessandro, Duke of,
    174
Flotow, Friedrich Freiherr von,
    Stradella, 303
Fo, Dario, 272
Foakes, R.A., Prof., 523
Fokina, Vera, 260
Fokine, Michael, 259–60, 399
Follino, Federigo, 285
Folz, Hans, 101–2
    Des Turken Vasnachtspil, 102
Fools, Feast of, 87–91
Forbes-Robertson, Johnston, 677,
    683
Ford, John, **836–43**
    biography, 836, 842
    collaborations, 787, 804, 836
    literary qualities, 842–3
    personality, 841–2
    *The Fancies, Chaste and
        Noble*, 841
    The Lady's Trial, 841
    The Lover's Melancholy,
        836–7
    Perkin Warbeck, 841
    The Queen, 841
Fornenberg, Jan Baptista, 457
Forrest, Edwin, 678
Fox, Charles James, 813
Foxe, John, *Actes and Monuments*
    ("Book of Martyrs"), 774
Foxy (musical), 861
France
    Classical drama, evolution,
        430–6, 443–5
    Classical drama, rules, 432
    court culture, 430–1
    medieval farce/popular
        culture, 88–90, 93–9
    medieval religious theatre,
        37, 47, 50–1, 75, 79–80
    mime, 250–1
    political conditions, 429–30,
        447
    popular entertainments, 250,
        440–3, 445–6
France, Anatole, 96
Franceschetti, Antonio, 160
Francis I of France, 148
Francis II of France, 429, 430,
    434
Frankel, Gene, 860
Frankfurt-am-Main, mystery
    plays, 51
Frazer, Rupert, 523
Frazzi, Vito, 307
Frederick, Elector Palatine,
    457
Frederick the Great of Prussia,
    343
Freeman, David, 598

French (language), early use for drama, 47
Freud, Sigmund/Freudian theory, 399, 623, 700, 707
Friar Bacon and Friar Bungay (Greene), **473–5**, 534
Frischlin, Margarethe, née Brenz, 359
Frischlin, Philipp Nicodemus, **358–62**
    biography, 358–60
    Julius Redivivus, 359, **360–1**
Frizer, Ingram, 506–7
Froissart, Johan, 631
Frost, Robert, 689
Fuente Ovejuna (*The Sheep Well*) (Lope de Vega), **389–91**, 402
Fukuda, Yoshiyuki, 689
Furttenbach, Joseph, the Elder, 122, 125, 128

Gabrieli, Giovanni, 463
Gaddi, Giovanni, Monsignor, 163
Gagliano, Marco da, *Dafne*, 289, 313
Gaillarde (dance), 350
Gaines, Davis, 611, 612
Galeazzo, Gian, Duke of Milan, 348
Galeotti, Vincenzo, 626
Galilei, Galileo, 277
Galilei, Vincenzo, 277, 278
Gallet, Louis, 396
Galli-Bibiena *see* Bibiena
Galliari family (theatre architects), 344
Galsworthy, John, *Strife*, 83
Gammer Gurton's Needle (Still/Stevenson?), **467–9**
Gamson, Arnold, 312
Ganassa, Alberto, 241–2, 372–3, 374, 377, 386
Gardner, Stuart, 32
Garnier, Robert, **434–6**, 437, 439
    Bradamante, 434–6
    Cornelia, 498
    Sédécie, ou Les Juives, 434, 435–6
Garrick, David, 252, 262, 267, 589–90, 676, 718, 827, 862
Gascoigne, George, 118, **471–2**, 484, 503
    Jocasta, 471
    Poesies, 471–2
    Supposes, 135, 471, 472, 613–14
    "The Adventures of Master F.J.", 471–2
Gasparini, Francesco, 337
Gautier, Théophile, 273
Gaveston, Piers, 546
Gay, John, *The Beggar's Opera*, 262
Gayley, Charles Mills (as translator), 62, 63
    as author *see* Index of Cited Authors
Gealt, Adelheld M., 272
Gelbart, Larry, 860–1
Genet, Jean, 526
Gengenbach, Pamphilius, *Der Nollhart*, 357
Genoa, Anna Pamphili, Duchess of, 302
Genoa, Giovanni Andrea Doria, Duke of, 302
Geoffrey of Monmouth, *History of the Kings of Britain*, 469–70, 478, 713
George, St, theatrical depictions, 22, 60
George III of England, 324, 677
Georgi, Yvonne, 264
"German Comedians", 256–7
German (language), early use for drama, 355–6, 362
Germany/German-speaking regions
    comic popular entertainments, 93, 99–105, 249–50, 256–7
    medieval religious theatre, 37, 51–4, 80
    opera, forerunners/ development, **462–4**
    pre-Baroque drama, **353–64**, **456–62**
    productions of Shakespeare, 688–9
Gernutus, 639
Gerson, Jean de, 79
Gerussi, Bruno, 771
Gervinus, G.G., 593
Gesta Romanorum, 639
Gets, Malcolm, 612
Gherardi, Evaristo, 237
    Le Théâtre italien, 258
Ghezzi, Pier Leone, 323, 327
Giacommelli, Geminiano, 335
Giannini, Vittorio, *The Servant of Two Masters*, 267
Giannone, Pietro, 772
Gianotti, Donato, *Il Vecchio Amoroso*, 181–2
Giberti, Giovanni Matteo, Bishop, 146–7
Gibson, Mel, 688
Gielgud, John, 677, 681–2, 688, 860
Gilbert and Sullivan, 870
Gillot, Claude, 244
Giotto, 113
Giovanni "dalle Bande Nere" *see* Medici
Giradina, Gary, 58
Giraldi, Giovanni Batista *see* "Cinthio"
Gish, Lillian, 681
Gissendammer, Thomas P., 608
Gladstone, Josh, 663
Glazunov, Alexander, 260
Gleich, J.A., 257
Globe Theatre (London), 488, 490, 494
    destruction by fire, 495
    reconstruction, 596–8
Glover, Jane, 307, 312
Gluck, Christoph Willibald, 338, 340, 399
Gnapheus, G., *Acolastus*, 358
God
    physical depictions, 42, 47, 58–9
    as speaking role, 42, 44, 45–6
Godfrey of Viterbo, 749
Godocus Badius, 120

Goehr, Walter, 307
Goethe, Johann Wolfgang von, 114, 258, 415, 543, 594, 671
    Faust, 102
    Torquato Tasso, 132
    Wilhelm Meisters Lehrjahre, 689
Gogol, Nikolai, *The Overcoat*, 263
Goldenthal, Elliot B., 712
Golding, Alfred, 205
Goldoni, Carlo, 250
    La Bottega del Caffé, 225
    Il Servitore di Due Padroni, 225, 267, 271
Goldsmith, Oliver, *She Stoops to Conquer*, 388
Gombaud. Jean, 444
Gone with the Wind (film, 1939), 687
Gonzaga, Elizabetta, 350
Gonzaga, Federico, Duke of Mantua, 146, 147
Gonzaga, Ferdinando, Archbishop, 242, 283, 284, 289
Gonzaga, Francesco, 282–3
Gonzaga, Guglielmo, Duke of Mantua, 203–4, 240
Gonzaga, Vespasiano, Duke of Mantua, 127
Gonzaga, Vincenzo, Duke of Mantua, 204–5, 213, 240, 243, 289
    as Monteverdi's patron, 282, 283, 284–5, 286, 350
Goodrich, Arthur, *Caponsacchi*, 683
Gorboduc (Norton/Sackville), **469–71**
Gorki, Maxim, 259
Gourville, Baron de, 439
Gower, John, 749
Goya, Carla, 428
Goya, Francisco, 426
Gozzi, Carlo, Count, 238, 245, 258
    The Love for Three Oranges, 260
El Gran Teatro del Mundo (Calderón), **409–10**

Grandage, Michael, 559
Grandi, Alessandro, 463
Grandison, Bishop, 92
Grandisson, John, Bishop of Exeter, 108
Granville-Barker, Harley, 550–1, 600–1
Gravitte, Debbie, 611, 612
Grazzini, Antonfrancesco ("Il Lasca"), 139, 162, 178, 181, 216
    La Gelosia, 162
    La Pinzochera, 162
    La Spirata, 162
Gréban, Arnoul, 47, 50
Greco, Gaetano, 327
Greco, José, 428
Greece (Ancient), 87, 465
    influence on Renaissance drama, 116, 129, 233, 248
    music, 275
Green, John, 457
Greenberg, Noah, 27–8, 29, 30, 31–2, 34, 35
Greenberg, Toni, 32
Greene, Robert, 354, **472–6**, 491, 497, 501, 504, 528, 571, 604, 659, 745, 779
    biography/lifestyle, 472–3
    Alphonsus of Aragon, 473, 516
    Greene's Groatsworth of Wine, 475–6
    The History of Orlando Furioso, 475, 494
    James IV, 475
    Menaphon, 485
    Pandosto, or, the Triumph of Time, 475, 760
    (with Thomas Lodge) *A Looking Glass for London and England*, 473, 477
Gregorian chant, 275
Gregory XIII, Pope, 168, 199
Grévin, Jacques, 433
Griffin, Jason, 608
Grillparzer, Franz, 258, 415, 417
    The Dream Is a Life, 415
Grimaldi, Giuseppe (father), 252–3

Grimaldi, Joseph "Iron Legs" (grandfather), 252
Grimaldi, Joseph "Joey" (son), **252–6**
    biography/traumatic experiences, 252–4
    stage persona, 254–5
    Memoirs, 256
Grimani, Cardinal, 319
Grimm, Friedrich, 333
Gringoire, Philippe, 97–8, 104
Grock (Adrian Wettach), 256
Grosseteste, Bishop, 92
Groto, Luigi ("Il Cieco d'Adria"), 118
Gryphius, Andreas, 258, 456
    Cardinio and Celindo, 459
    Carolus Stuardus, 459
    Catharina von Georgien, 459, 460
    Horribilicribrifax, 460
    Leo Arminius, 460
    Papinianus, 460
Gsovska, Tatiana, 690
Gsovsky, Victor, 690
Guarini, Giambattista, 130, **132–3**, 241
Guarini, Guarino, 281
Guérin, Robert (Gros-Guillaume), 440–1, 445
Guerrero, Francisco, 419
Guéru, Hugues (Gualtier-Garguille), 440–1
Guglielmo Ebreo (William the Jew), *Treatise on the Art of Dancing*, 348
Guidry, Ronald, 29–30
guilds
    actors', 92, 94, 105–6, 376
    professional, performances by, 38–42, 59–60
Guinness, Alec, 681
guitar, development of, 419
Gunpowder Plot, 712, 856
Gurwin, Danny, 626–7
Guthrie, (Sir) Tyrone, 517–21, 526, 710, 860

Guyon, Elyane, 251
Gwilym, Mike, 523–4
gypsies, role in development of dance, 423–4

Hack, Keith, 523
Hagen, Uta, 709
Haigh, Kenneth, 836
Halévy, Jacques, 772
Hall, Edward, 631
Hall, Elizabeth, 580
Hall, John, 580, 582
Hall, Joseph, Bishop, 794, 799
Hall, Peter, 524–7, 681, 725–6
Hallam, Henry, 813
Halmen, Pet, 308
Hals, Frans, 451
Hamilton, Patricia, 722
Hamlet (Shakespeare), 43, 117, 494, 499, 646, **665–90**, 698, 704, 717, 726, 834, 886
    casting, 675–88
    characterization, 607, 669–70, 682, 715
    critical interpretations, 668–9
    film versions (1948/91/96), 681, 688
    language, 671–5
    modern productions, 601–2, 680–90
    operatic/balletic versions, 337, 689–90
    reworkings, 689–90
    sources/influences, 498, 665, 666, 795
    structure, 670–1
Hampden, Walter, 682–4
Handel, Georg Friedrich, 262, 295, 319, 321, 324, 328, 340, 559
    The Faithful Shepherd, 132–3
    Rinaldo, 133
    Tamerlane, 335
"Hanswurst" (German comic figure), 250, 256–7, 258, 353
Hardiman, Hilary, 557
Hardy, Alexandre, **441–3**, 444
    biography, 441
    productivity, 442–3

Cornélie, 443
La Force du sang, 443
Mariamne, 443, 446
Théagène et Chariclée, 443
Hardy, Thomas, *The Dynasts*, 716
Harlequin and Cinderella (clown entertainment), 255
Harlequin and Mother Goose (clown entertainment), 255
Harlequin in April (ballet), 265
Harlequinade (ballet), 265
Harlequinade (popular genre), 261–2
Haroncourt, Nikolaus, 307, 308
Harriot, Thomas, 506
Harrison, Tony, 53–4
The Harrowing of Hell, 46, 52
Harsnett, Samuel, *Declaration of Popish Impostures*, 713
Hart, Joan, née Shakespeare (sister), 576, 579
Hart, Lorenz, 610
Hart, William, 576, 577, 579
Harvey, Gabriel, 472, 485
Harvey, William, 576
Hasse, Johann Adolf, 319, 323, 326, 334, 338
Hathaway, Richard, 568
Hatton, Sir Henry, 477–8
Hauman, Constance, 297
Hauptmann, Gerhart
    The Sunken Bell, 682
    The Weavers, 83
Hayes, Tubby, 86
Hazlitt, William, 716, 718
Heeley, Desmond, 772
Hegelund, Peder Jensen, 356
Heliodorus, 443
Helpmann, Robert, 689–90
Heminges, John, 580, 588, 589, 773–4
Henri II of France, 352, 429, 431, 434, 444, 447
Henri III of France, 239, 248, 351, 430, 435, 448–9
Henri IV of France/III of Navarre, 241, 279, 280, 437–8, 439, 450
Henritze, Bette, 777

Henry VI of England, 108
Henry VII of England, 82, 92, 107, 607, 841
Henry VIII of England, 92, 111, 465–6, 867
Henry IV (Shakespeare), 477, 600, **644–6**, 649, 695
    characterization, 644, 645–6
Henry V (Shakespeare), **649–53**, 699, 749
    characterization, 651–2
    film versions (1944/89), 652–3
    modern productions, 596–8, 652–3
    sources, 649
Henry VI (Shakespeare), 498, **603–5**, 609, 649
    authorship, 550, 604
Henry VIII (Shakespeare/others?), 491, 495, **773–7**
    authorship, 773, 818
    characterization, 774
    critical commentary, 775, 776–7
    language, 775
    modern productions, 775, 776–7
    sources, 774
Henry, Guy, 760
Henry, Martha, 772
Henslowe, Philip, 488, 489, 492–3, 588, 780, 783, 784, 788, 794, 800, 808, 848, 850
Henze, Hans Werner, 307, 311
Herbert, Mary, Countess of Pembroke, 498, 734
Herbert, William, 3rd Earl of Pembroke, 575–6
Hernandez, Riccardo, 776
Herod, as character in dramas, 26
Herodotus, 477
Herr Peter Squentz (Gryphius), **459–60**
Herring, Joan, 824
Hesse-Darmstadt, Landgrave of, 463
Heywood, John, **108–11**, 466
    The Four Ps, 109

The Merry Play between Johan, Tyb and Sir John, 110–11
The Pardoner and the Friar, the Curate and Neighbour Pratte, 109
A Play of Love, 110
The Play of the Weather, 110
Witty and Witless (Wit and Folly), 110
Heywood, Thomas, 515, 529, 746, 749, 779, **780–3**
  biography, 780–1, 783
  collaborations, 889, 890–1
  Edward IV, 781
  The English Traveller, 781, 782
  The Fair Maid of the West, 782
  The Four Prentices of London, 781
  The Rape of Lucrece, 781
  The Wise Woman of Hogsden, 782–3
Hidalgo, Juan
  Celos Aún del Aire Matan, 421
  La Púrpura de la Rosa, 421, 422
Higden, Ralph, 46–7
Higgins, John, *The Mirror for Magistrates*, 716
Hilarius, 27, 60
Hildegard of Bingen, 70, 73
  as composer, 276
  Ordo Virtutum, **70–3**, 276
Hill, Sarah, 610
Hindemith, Paul, 307
Hines, Patrick, 722
Hinkle, Marin, 777
Hoffman, William H., 267
Hoffmann, E.T.A., 258
Hoffmannsthal, Hugo von, 258
  Jedermann, 84
Hoghton, Alexander, 567
Holberg, Ludwig, 258
Holby, Grethe Barrett, 72
Holby, Lee, *The Tempest*, 773
Holinshed, Raphael, *Chronicles*, 546, 631–2, 713, 716, 726, 728, 751, 774, 779
Hollonius, Ludwig, **460–2**

Holm, Hanya, 615
Holmes, Peter, 521–2
Homer
  Iliad, 697
  Odyssey, 287
homosexuality, social acceptability/prominence, 482, 574–5, 577–8
  see also under Marlowe; Shakespeare
The Honey Pot (film, 1966), 861
Hooft, Pieter Corneliszoon
  Achilles en Polyxena, 452
  Geeraerd van Velsen, 452
  Granida, 452
Hopkins, Anthony, 557
Hopkins, Arthur, 609, 684
Horace (Q. Horatius Flaccus), 23, 117, 118, 852
Houseman, John, 55–6, 544–5, 557–8, 731–2
Howard, Alan, 725–6
Howard, Leslie, 687–8
Howard, Robert, 136
Hoyle, Geoff, 269
  A Feast of Fools, **269–72**
Hrosvitha of Gandersheim, 23, 69, 86, 100, 133
  Abraham, 23–4
  Sapientia, 69–70
Hughes, Angela, 725
Hughes, Thomas, *The Misfortunes of Arthur*, 478
Hughes, Willie, 576
Hugo, Victor
  Hernani, 426
  Notre-Dame de Paris, 98
Huguenots, 430, 436, 444, 560
humours, theory of, dramatic applications, 788, 849
Hungary, 356
Hunt, F.B., 517
Hurry, Leslie, 519–20
Hutt, William, 521, 771
Hytner, Nicholas, 558–9

Ibsen, Henrik, 757
  An Enemy of the People, 683
  Peer Gynt, 680

The Impresario (Bernini), **229–33**
  characterization, 232–3
  (lack of) ending, 233
  originality, 229–30, 233
L'Incoronazione di Poppea (Monteverdi), **289–91**
  characterization, 291
  composition, 289, 297, 301
  modern revivals/reinterpretations, 307, 312–13
Ingegneri, Angelo, 122
Ingegneri, Marc Antonio, 281
Inglesias, Roberto, 428
Innocent X, Pope, 226
Innocent XI, Pope, 317, 320
Innsbruck, 298–9
"Interlude" (Tudor theatrical form), 107–11
intermezzi
  in opera, 324, 328, 329
  in Renaissance comedy, 192–3, 205, 216–17, 225
Ionesco, Eugène, 274
Irizarry, Vince, 84–5
Irving, Henry, 600, 677
Irwin, Will, 269
Isabella of Castile, 365, 425
Italian Comedians, 250, 257–8, 430, 433
Italy
  comic theatre, **133–250**
  dance, **347–52**
  influence on other cultures, 368, 419–20, 447, 465–6, 471
  medieval theatre, 49, 106
  opera, **276–345**
  Renaissance drama, **112–19**, **129–33**
  theatre architecture/design, **119–29**
Ives, David, 611

Jacobi, Derek, 552–3, 556, 681
Jaffee, Michael, 36, 276
James, Henry, 585–6
  A Sense of the Past, 687
James, Toni-Leslie, 612, 776

James I of England, 489, 570–1, 712, 726, 775, 789, 798, 855, 868
    court atmosphere, 856
Jameson, Pauline, 721
Jannsen, Cornelius, 583
Jannsen, Gheeraert, 581–2, 584
Japan, productions of Shakespeare, 689
Jarman, Derek, 559–60
The Jealous Comedy (anon.), 696
Jehan le Palu, 61
Jennings, Alex, 690
Jerome, Tim, 337
Jessner, Leopold, 601
Jesuits, 200, 431, 458
Jesus
    dramas based on life of, 24–7
    onstage representations, 36–7, 42
    physical demands of performance, 42, 51
    tomb, reproductions of, 27
Jeu du Prince des Sots et de la Mère Sotte, 97–8
The Jew of Malta (Marlowe), 493, **527–33**, 542, 550
    basis in reality, 527–8
    characterization, 529, 535, 605
    critical commentary, 529
    influence on later works, 531–3, 561, 562, 639, 640, 854, 858
    language, 529–31, 537
Jews
    colonies/acting companies, 203–4, 347
    hostility to *see* anti-Semitism
Jiang Qihu, 724
Joanna of Aragon, Duchess of Amalfi, 878
Jocundus, 120
Jodelle, Étienne, Sieur de Lymodin, **431–2**, 433, 434
    Cléopatre captive, 431–2
    Didon se sacrifiant, 432
    Eugène, ou La Rencontre, 432
"Joey" (clown figure), **254–6**

see also Grimaldi
Johann Georg I, Elector of Saxony, 463
John, Don, of Austria, 351, 386
Johnson, Gerard *see* Jannsen, Gheeraert
Johnson, Kim, 143
Johnson, Samuel, Dr, 591–2, 676, 718, 740, 756, 868
Jolly, George, 457
Jones, Alan Tjaarda, 308
Jones, Ernest, 700, 710
Jones, Inigo, 128, 129, 311, 596, 844, **867–8**
    relationship with Jonson, 868, 869, 870
Jones, Robert Edmond, 601, 684, 709
Jones, Tom, 267
Jonson, Ben, 465, 473, 476, 500, 534, 809, **847–71**
    biography, 465, 847–8, 850, 855–6, 863–4, 869
    collaborations, 784, 789, 802, 825, 848, 851
    comments on Shakespeare, 566, 572, 581, 582, 588, 590, 847
    contemporary performances, 494, 848, 863, 866–8
    contributions to pastoral genre, 129, 131
    critical/popular reception, 779, 826, 870–1
    health, 869, 870
    imprisonment, 848, 850, 855
    influences on, 156–7, 788
    literary qualities, 854, 860
    marriage, 848
    masques, 856, **866–8**, 870
    personality/lifestyle, 814, 852–3, 868
    publication of works, 868
    relationship with Marston, 793–4, 795–6, 797, 798–9, 851–2
    relationship with Shakespeare, 563, 579, 771, 853

    relations with other playwrights, 814, 815, 843, 852–3, 868
    religious standpoint, 850
    satirical depictions, 794, 798, 851
    social satire, 856, 861–2, 864, 869
    translations, 502
    views on drama, 590, 788, 848
    women, attitude to, 869
    The Case Is Altered, 848
    Cynthia's Revels, or The Fountain of Self-Love, 851
    Discoveries, 870
    The Isle of Dogs see under Nash
    The Magnetic Lady, 869–70
    The New Inn, 869
    The Sad Shepherd, 870
    A Tale of a Tub, 870
    "To Celia", 856
Joseph II, Holy Roman Emperor, 258
Josephus, Flavius, 443
Josten, Werner, 307
Journey of the Magi (Beauvais, twelfth century), 36
Jouvet, Louis, 860
Joyce, Stephen, 722
Joyeuse, Duc de, 448
Juan of Castile, 368
Julian the Apostate, Emperor, 354
Julius Caesar (Shakespeare), **653–8**, 890
    characterization, 657–8
    language, 654–7
    modern productions, 602
    sources, 653
Julius II, Pope, 97–8
Justinian, Emperor, 21
Juvenal, *Satires*, 853

Kabuki theatre, 238
Kaczorowski, Peter, 612
Kafka, Franz, 595
Kalidasa, *Shakuntala*, 131

Kallman, Chester, 312
Kaplan, Lynda, 140
Kaplinsky, Elaine, 73
Karinska, 32
Karsavina, Tamara, 260
Katagiri, Hairi, 724
Katherine of Aragon, 111
Kaufman, George S., 679
Kay, Richard, 557
Kean, Charles, 600
Kean, Edmund, 676–7, 678, 718
Keaton, Buster, 269
Keats, John, 562
    "On First Looking into Chapman's Homer", 787
Keller, James, 84–6
Kellogg, Paul, 312
Kemble, John Philip, 676, 718
Kemp(e), Will, 494–5, 571, 848
Kenwelmarsh. Francis, 471
Kett, Francis, 505
Kidnapped (Fleury), 63
King, Tambra, 337
A King and No King (Beaumont and Fletcher), **821–2**, 839
King John (Shakespeare), **637–8**
    characterization, 638
King Lear (Shakespeare), 499, 600, 646, **712–26**, 799, 886
    characterization, 715, 716–17, 719, 762
    language/imagery, 718–19, 730
    modern productions/films, 720–6, 731, 732
    revised versions, 718, 745
    sources, 713
    structure, 715–16
Kingsley, Ben, 681
Kinwelmarsh, Francis, 118
Kirstein, Lincoln, 29, 264
    see also Index of Cited Authors
Kishida, Rio, 723, 724
Kiss Me Kate (musical), 598, 611, 615
Kissoon, Jeffrey, 523
Klein, Peter, 65

Klucht plays (Netherlands), 105–6, 452
Knell, William, 570
Kneller, Sir Godfrey, 583
The Knight of the Burning Pestle (Beaumont), **815–16**, 827, 869
Koltai, Ralph, 749
Korngold, Erich Wolfgang, 562
Kott, Jan, 720
Kozma, Lajos, 225
Krenek, Ernst, 307
Kurosawa, Akira, 732, 759
Kyd, Thomas, 354, 457, **497–500**, 506, 571, 779, 828, 886
    alleged authorship of *Arden of Feversham*, 498, 779
    alleged authorship of "Ur-Hamlet", 498, 665, 689, 794
    relationship with Marlowe, 497, 504, 507, 529

la Chesnaye, Nicolas de, *L'Homme pêcheur*, 79–80
La Gallienne, Eva, 624
Lahr, Bert, 269, 861
Lalande, J.J., 316–17
Lally, James, 140
Lamb, Charles, 547, 594–5, 600, 621, 716, 718, 781–2, 813, 828, 891
landjuweel (Dutch drama festival), 451–3
Landman, Michael, 663
Lange, Carl, 26
Langtry, Lily, 679
Lankston, John, 56
Larivey, Pierre, 436, 437
Laroche, J.J., 257–8
Las Mocedades del Cid (Castro), **394–6**
Laschi, Antonio, *Achilles*, 115
Lasnik, Roberta, 71
Latilla, Gaetano
    La Finta Cameriera, 327
    Madama Ciana, 327
Latin
    as language of Church ritual, 24

    plays written in, 23–4, 353, 354–6
Laud, William, Archbishop, 844
Laughton, Charles, 776
Laurens, Guillemette, 71
Laurenson, James, 554
Laurents, Arthur, 625
Lavin, Steve, 67
Lavognolo, Lorenzo, 350
Lawes, Henry, 844
le Prevost, Nicholas, 598
Leach, Willard, 140–1
Lederer, Josef, 257
Lee, Ralph, 32
Lee Kwan Yu, 723
Leech, Cifford, Prof., 553
Lefeaux, Charles, 522
Lefevre, Mathieu, 438
Legge, Thomas, *Ricardo Tertius*, 478
Legrand, Henri (Turlupin), 440–1
Legrenzi, Giovanni, Father, 304, 338
    Giustino, 304
Leiber, Fritz, 686–7
Leicester, Lord, 583
Leigh, Vivien, 741
LeNoir, Charles, 446
Leo, Leonardo, 322, 323, 338
    Il Ciro Riconosciuto, 323
    Olimpiade, 323, 334
Leo X, Pope, 106, 123, 137, 138, 145, 372
Leonardo da Vinci, 120, 350
Leoncavallo, Ruggiero, *I Pagliacci*, 265–6
Leontiev, Leonid, 260
Leopold I, Holy Roman Emperor, 300–1, 305, 458
Leopold Wilhelm, Archduke, of Austria, 457
Lepanto, battle of, 351, 386
Leppard, Raymond, 307, 313
Lessa, Duke of, 383
Lessing, Gotthold Ephraim, 593
Levin, Judy, 312
Lewis Jones, Mark, 598
Liadov, Anatol, 260

Lifar, Serge, 626
lighting, 122, 344–5, 437, 490
Lilius, Camille, 295
Lily, William, 480
Limon, José, 711
Lincoln, Abraham, President, 678
Linder, Max, 269
L'Infanterie Dijonnaise, 94
Littlewood, Joan, 552
Lo Frate 'Nnamorato (Pergolesi), 328, 330, **331–2**
Locatelli, Basilio, *La Forestiera*, 225
Lodge, Thomas, 473, **476–7**, 497
    essays, 476
    Rosalynde, 476, 658
    Scillaes Metamorposis, 476–7
    The Wounds of Civil War, 477
    see also Greene, Robert
Loeb, Louis M., 517
Lohenstein, Daniel Kaspar von, 460
London
    Great Fire (1666), 844
    Shakespeare's life in, 569–71
    St Paul's School, 481–2
    theatre buildings, **488–93**
Longfellow, Henry Wadsworth, *The Courtship of Miles Standish*, 811
Lope de Vega, Lopito, 384
Lope (Félix) de Vega Carpio, 250, 379, 380, **381–93**, 419, 457
    biography, 381–4, 407, 417
    comparisons with other writers, 392, 393, 398, 399, 402, 413, 414, 417
    influence on later writers, 415, 420–1, 442
    influences on, 386
    international reputation, 419–20
    productivity, 381, 391–2, 442
    qualities as dramatist, 384–5, 386–7, 389–90, 392, 398, 399, 420
    relationships with fellow writers, 393, 397
    views on staging, 385, 386, 420, 442, 444
    The Beautiful Esther, 391
    The Beauty of Angelica, 383
    Belisa's Tricks, 388
    The Capricious Lady, 388
    Castelvines e Monteses, 388–9, 620
    A Certainty for a Doubt, 389
    The Crown of Otun, 388
    La Dorotea, 382–3, 389
    The Gardener's Dog, 387
    The King, the Greatest Judge, 389, 391
    The Lunatics of Valencia, 388
    Madrid Steel, 387–8
    The New Art of Making Plays in This Epoch, 384, 420
    Punishment Without Revenge, 389
    Rome in Ashes, 388
    La Selva sin Amor (opera libretto), 421
    There's Method In 't, 388
Lord Admiral's Men, 493–4, 527, 533, 560, 784
Lord Chamberlain's Men, 487, 494, 495, 570–1, 848
Lord of Misrule, figure of, 92
Lord Strange's Men, 527, 570
Lorraine, Cristina of, 216, 224
Lorraine, Francis of, 338
Lorraine, Renata of, 257
Lotti, Antonio, 338
Louis XII of France, 79, 97–8, 447
Louis XIV of France, 227, 293, 294, 295, 425, 426, 450
Louise, Princess, of Saxony, 463
Love's Labour's Lost (Shakespeare), 484, **617–20**, 745, 841
Love's Sacrifice (Ford), **840**, 841
Lubovitch, Lar, 711–12
Lucan, *Pharsalia*, 502–3
Lucerne, mystery plays, 51–2
Lucian, 745
Lucy, Sir Thomas, 567
Luders, Adam, 56
Luker, Rebecca, 611, 612
Lully, Jean-Baptiste, 295, 333
Lunt, Alfred, 860
lute, as acccompaniment to stage action, 23, 239
Luther, Martin, 104, 136, 353, 354, 356, 358
Luxon, Micaela de, 383
Lydgate, John, 697
    Mumming at Hartford, 108
Lyly, John, **480–5**, 486, 497, 510, 745
    biography, 480–2, 485
    influence, 484–5
    prose style, 481, 484–5
    satirized, 617, 620
    Endimion, the Man in the Moon, 483
    Euphues (and His England), 481
    Gallathea, 484
    The Maid's Metamorphosis, 484
    Midas, 483–4
    Mother Bombie, 484
    Sapho and Phaon, 483
Lyndsay, Sir David, *Satire of the Three Estates*, 82

Macbeth (Shakespeare), 253, 607, 613, 622, **726–34**, 886
    characterization, 728
    language/imagery, 729–31
    modern productions, 602, 679, 684, 731–2
    reworkings, 732–4
    sources/influences, 726, 728, 802–3
Machavariani, Alexei, 711
Machiavelli, Niccolò, 113, 134, **137–43**, 234, 368, 793
    biography, 137–8, 143
    influence on characterization, 528, 853, 876, 878
    The Art of War, 138
    Clizia, 138–9, 140
    History of Florence, 138
    The Prince, 138
Maclady, Matthew, 648

Maclean, Norman, 597
Macmillan, Kenneth, 308, 626
Macready, William Charles, 677, 718
Macro Moralities, 75
Maddaloni, Duke of, 328, 330
Maderna, Bruno, 307
Madrid, theatre buildings, 373–4
Maeterlinck, Maurice, 838
Maffei, Scipione, 335
El Magico Prodigioso (Calderón), **410–12**
Magli, Giovanni Gualberti, 283
Mahabharata, 723
Mairet, Jean, 444, 445
Maître Patelin see Farce of the Worthy...
The Malcontent (Marston), **797**, 873
Malipiero, Gian Francesco, 307, 312
Mallarmé, Stéphane, 273
Malone, Edward, 577
Malone, Mike, 86
The Man Who Married a Dumb Wife, 96
Mandel, Johan, Master, 133
La Mandragola/Mandrake (Machiavelli), **139–43**, 184–5, 188, 210, 811
  modern productions/adaptations, 140–3
Mankind (fifteenth-century English), 77–8, 86
Manners, Roger, Earl of Rutland, 586
Mannheim, 344
Manrique, Gomez, 276
  Representación del Nacimento de Nuestro Señor, 367
Mansfield, Richard, 678, 679–80
Mantegna, Andrea, 121, 350
Mantell, Robert, 678, 679–80, 686
Mantinus of Tortus, 115
Mantovani, Publio Filippo, *Formicone*, 210

Mantua
  Accademia degli Invaghiti, 205
  cultural life, 137, 204–5, 240
  Dukes of *see* Gonzaga
  Jewish colony, 203–4
  theatre buildings, 343
Manuel, Niklaus
  Der Ablafskrämer, 358
  Von Papsts und Christi Gegensatz, 358
Manuel of Portugal, 145
Manzay, J. Kyle, 725
Marcadé, Eustache, 47
Marceau, Marcel, 251, 263–4
Il Marescalco (Aretino), **153–7**, 171
Margaret of Austria, Duchess of Parma, 368, 452
Margareta, Empress, 341
Margharita of Spain, Infanta, 300
Margherita, Princess, of Savoy, 284
Maria, Anton, 248
Maria Amalia of Naples, 333
Maria of Portugal, 369
Maria Teresa, Archduchess (later Empress), 338
Maria Teresa, Queen, 294, 295
Mariani, T., 329
Marie, Queen of France (wife of Henri IV), 241
Marino, Giambattista, 241
Marivaux, Pierre de, 250
Marlborough, Duke of, 324
Marlowe, Christopher, 354, 457, 465, 476, 497, **500–62**, 650, 779, 793, 886
  (alleged) authorship of Shakespeare, 508–9, 586
  biography, 500–2, 500–3, 504
  contemporary interpreters, 493–4
  death, 506–9, 560
  influence on other dramatists, 473, 475, 531–2
  (literary) relationship with Shakespeare, 531–3, 537, 544, 550, 560–2, 576, 604
  personality/lifestyle, 504, 507–8, 533–4, 571
  (possible) depiction in *Romeo and Juliet*, 532, 623
  quatercentenary (1964), 522–3, 553, 557
  religious outlook, 505–6, 507, 531, 534 (*see also under titles of works*)
  sexuality, 504–5
  (suspected) involvement in espionage, 502, 506, 570
  translations, 502–3
  Hero and Leander, 504–5, 789
  The Massacre at Paris, 560
  The Tragedy of Dido, Queen of Carthage, 486, 498, 503, 505
Marlowe, John, 500–1
Marlowe, Julia, 682, 686
Marshall, Kathleen, 611
Marston, John, 779, **793–9**, 851, 854
  biography, 793–4, 795, 799
  collaborations, 784, 789, 794, 798–9, 825
  "feud" with Jonson *see under* Jonson
  personality, 793–4, 797, 799
  Antonio and Mellida, 665, 794–5
  Antonio's Revenge, 794, 795
  The Dutch Courtesan, 797–8
  Eastward Ho! see under Chapman
  The Insatiable Countess, 799
  Jack Drum's Entertainment, 797
  Parasitaster, or the Faun, 798
  Satiromastix, 794, 851
  What You Will, 797, 851
Il Martello (Cecchi), **193–5**
Martin, Frank, *Der Sturm*, 772
Martinelli, Giovanni, 711
Martinelli, Tristano, 242, 377
Marx, Harpo, 236, 269
Mary, Queen of Scots, 434, 470–1, 489, 502, 507

Mary ("Bloody") of England, 111, 365
Le Maschere (Cecchi), **197–9**
Masefield, Judith, 517
Masen, Jakob, *Androphilus*, 458
Masondo, Lawrence, 733
masque (genre), **866–8**
mass, dramatic/musical elements, 24
Massenet, Jules
    Le Cid, 396
    Le Jongleur de Notre Dame, 61
Massine, Leonide, 264, 428
Massinger, Philip, 773, 779, **808–13**
    biography, 808–9, 813
    collaborations, 784, 804, 809, 811, 813, 825, 844
    literary qualities, 813
    Believe as You List, 813
    The Bondman, 812
    The Duke of Milan, 812
    The Great Duke of Florence, 811
    The King and the Subject, 812
    Maid of Honour, 813
    The Parliament of Love, 812
    The Renegado, 812
    The Roman Actor, 812–13
    The Unnatural Combat, 812
    (with Dekker) *The Virgin Martyr*, 811
    (with Fletcher) *The False One*, 811–12, 825
    (with Fletcher) *The Knight of Malta*, 809
    (with Fletcher) *Thierry and Theodora*, 809, 825
    (with Fletcher) *A Very Woman*, 811
    (with Nathan Field) *The Fatal Dowry*, 809
Matthews, Karen, 32
Matthews, Richard, 722
Maude, Beatrice, 683
Maugham, W. Somerset, *Then and Now*, 143
Maurice of Orange, Prince, 453

Mauro, Tomaso de, *La Spellecchia*, 326
Mauro family (theatre architects), 343, 344
Maximilian II, Holy Roman Emperor, 359
Maximilian of Austria, 135
Mazarin, Jules, Cardinal, 293, 294
McCann, Christopher, 54, 637
McCarthy, Justin, *If I Were King*, 682
McClintic, Guthrie, 681–2
McCowan, George, 772
McCowen, Alec, 681, 721
McDiarmid, Ian, 558–9
McGrath, Michael, 612
McKellen, Ian, 554–6, 558, 559, 681
McKern, Leo, 518
Measure for Measure (Shakespeare), **701–4**, 797
    characterization, 702
    language, 703–4
    sources, 701
Medici, Alessandro de', Archbishop, 199
Medici, Alessandro de', Duke of Florence, 119
Medici, Bianca Cappello de', 351
Medici, Catherine de', 352, 429–30, 433, 444, 446
    court entertainments, **447–50**
Medici, Cosimo de', Duke, 168, 174
Medici, Cosimo III de', Grand Duke, 299–300, 320
Medici, Ferdinando de', Cardinal (later Grand Duke), 215–16, 224, 280
Medici, Ferdinando de', Prince, 319, 320
Medici, Fernando II de', Grand Duke, 297
Medici, Francesco de', Grand Duke, 803
Medici, Giovanni Carlo de', Cardinal, 294, 297

Medici, Giovanni "dalle Bande Nere", 145–6, 147
Medici, Giulio de', Cardinal *see* Clement VII
Medici, Isabella de' *see* Orsini
Medici, Lorenzo de' ("Il Magnifico"), 350, 447, 846
    L'Aridosia, 119
Medici, Marie de', 242–3, 279
Medici, Matthias de', Prince, 297, 301
Medici family, 128, 137, 143
El Médico de su Honor (Calderón), **405–6**
Medwall, Henry
    Fulgens and Lucrece, **106–8**
    Nature, 83
Melanchthon, Philipp, 136, 353, 355, 361, 533
Melani, Jacopo, 325
    Il Girello, 302
Melville, Herman, 595
Menander, 93, 133, 139
Mendelssohn-Bartholdy, Felix, 598, 630, 772
Mendoza, Francisco de, Cardinal, 168
The Merchant of Venice (Shakespeare), 583, 613, **639–43**, 646
    characterization, 640–1, 643
    influences/sources, 531–3, 561, 639
    language, 641–3
"Mercotellis", 326
Meres, Francis, 587, 617
Mérimée, Prosper, *Carmen*, 426
Merkel, Leo, 257
The Merry Wives of Windsor (Shakespeare), 480, 566, **695–7**
    dating, 691, 695
    sources, 695–6
Metastasio, Trapassi, 322–3, 324, 337–8, **340**
    L'Olimpiade (libretto), 329, 334
Meyerhold, Vsevolod, 259, 260
Michael III of Constantinople, 88
Michel, Johan/Jean, 47, 50

Mistère de la Passion de notre Saulveur Jhesu Christ, 50
Michelangelo (Buonarotti), 149, 155, 226
Michiel, Polo, 302
Middle Ages, popular culture, 43
  theatre *see* Miracle Plays; Morality Plays; mystery plays
Middleton, James, 308
Middleton, Thomas, 746, 779, **799–808**, 843
  biography, 799–800, 808
  collaboration with William Rowley, **803–7**
  collaborations with others, 784, 800, 801–3, 825
  A Chaste Maid in Cheapside, 801
  A Game at Chess, 807–8
  A Mad World My Masters, 801
  Michaelmas Term, 801
  More Dissemblers Besides Women, 802
  No Wit, No Help Like a Woman's, 802
  The Phoenix, 800
  A Trick to Catch the Old One, 800–1, 809
  The Witch, 802–3
  Women Beware Women, 803, 805
  Your Five Gallants, 802
  (with Dekker/Webster) *Caesar's Fall*, 800, 872
  (with Rowley) *God's Revenge against Murther*, 804
  (with Rowley) *The Spanish Gypsy*, 807
  (with Webster) *Anything for a Quiet Life*, 803
  (with Webster) *The Widow*, 802
  see also under Dekker, Thomas
Middleton, William, 799
A Midsummer Night's Dream (Shakespeare), 42, 129, 492,

626, **628–30**, 745, 764, 816
  modern productions, 263, 630, 759
  reworkings, 559, 598, 630, 689
Mielzinner, Jo, 610, 682
Mifune, Toshiro, 732
Milan
  as cultural centre, 351
  mystery plays, 49
Milan, Dukes of, 348, 349–50
Milanino, Maximiliano, 377
Millay, Edna St Vincent, *Aria da Capo*, 263
Miller, Arthur
  The Creation of the World and Other Business, 55
  The Crucible, 626
Miller, Jonathan, 312
Les Millions d'Arlequin (ballet), 265
Milton, John, 455, 520, 589, 592
  Comus, 129, 131, 773
  Paradise Lost, 537, 538, 772–3
mime, 250–1, 263–4, 276–7
Minato, Niccolo, 294, 302, 339
Mind, Will and Understanding (fifteenth-century English), 77
Minnesinger, 99
minstrels *see* troubadours
Minturno, Antonio, 118
Miracle Plays, **59–69**
  humour, 91
  modern productions, 64–7
  sensationalism of subject-matter, 59, 64, 68–9
  staging, 60, 68
Les Miracles de Notre-Dame, 61
Mnisi, Dieketeng, 733
Modjeska, Helena, 679
La Moglie (Cecchi), **178–81**
  characterization, 180–1
Moissi, Alexander, 688–9
Molière (Jean-Baptiste Poquelin), 51, 93, 136
  influences/sources, 226, 236
  Love Is the Best Doctor, 388
  Don Juan, 399

Les Femmes savantes, 225
Le Médecin malgré lui, 388
Le Misanthrope, 797
Mompesson, Sir Giles, 809
Monanni, Bernardo, 421
Moncion, Francisco, 56
Mondini, Francesco, 327
Moniglia, Giovanni Andrea, 294, 301
  Il Posto di Colognole, 325
Mons, mystery cycle, 42
Montaigne, Michel de, 875
  Of Cannibals, 769–70
Montalbán, Juan Perez de, 400
Montalto, Cardinal, 244
Montchrétien, André, 439
  L'Écossaisse, 439
Montdory (Guillaume des Gilleberts), 446
Montefeltro, Guidobaldo de, 350
Montemayor, Jorge de, *Diana Enamorado*, 615–16
Monteverdi, Claudia, 282, 284
Monteverdi, Claudio, 133, **281–92**, 350, 421
  biography, 281–3, 284, 286, 291–2
  madrigals, 282
  modern revivals/ reinterpretations, **306–13**, 333–4
  qualities as composer, 284, 288, 291, 318
  unfinished/unpublished works, 286–7
  Il Combattimenti di Tancredi e Clorinda, 286
  Le Nozze de Tetide, 286
The Moor's Pavane (ballet), 711
Morality Plays, **69–86**, 366–7
  categories, 78–9, 83
  characters, 69, 74, 79, 83
  costume, 79, 83
  forerunners/early examples, 69–73
  humour, 78
  influence on later drama, 82–4, 86, 107, 108,

199–200, 380, 464, 793, 829, 869
    modern productions/ adaptations, 84–6
    moral landscape/purpose, 73–4, 83
    staging, 74, 75
    suppression, 83, 99, 114
More, Sir Thomas, 82, 108–9, 111
Moresque (dance), 349, 350
Moreto y Cavana, Augustín, 398
Morley, Christopher, 557
Moro, Christopher, 704
Morrell, Jon, 296
Morris, Wolfe, 523
Morton, John, Cardinal, Archbishop, 106
La Moschetta (Beolco), **160–2**
Moscow State Circus, 261
Mott, Stewart R., 32
Mountjoy family, 570
Mower, Christine, 57
Mozart, Wolfgang Amadeus, 562, 772, 773
    La Clemenza di Tito, 340
    Don Giovanni, 267, 399
    The Marriage of Figaro, 267
    Il Pastor, 340
Msomi, Welcome, *Umabatha*, 733–4
Much Ado About Nothing (Shakespeare), 495, 614
    influences, 162
    modern productions, 602
Mullins, Ian, 523
mummers' plays, 22
Murray, Brian, 721
Murray, Stephen, 522
musical accompaniment to drama, 275–7, 324, 419, 446
    see also guitar; lute; opera
Mussato, Albertino, 115
Mussolini, Benito, 602
Muth, Conrad, 533
mystery plays, **36–59**
    choice of subject (craft-related), 38–9
    costume/make-up, 40
    cycles, 36, 43, 59

humour, 36, 37–8, 44, 45, 87
influence on later writers, 569
modern productions/ adaptations, 53–9
moral/ethical themes, 47
national characteristics, 38, 43, 47, 49
origins of name, 38
props/expenses, 41
special effects, 42, 50
stage directions, 49
staging, 39–43, 47, 49
suppression, 431

Naharro, Bartolomé de Torres, **371–2**
    Comedia Serafina, 372
    Tinelaria, 371–2
Naogeorgius, Thomas, *Pammachius*, 354–5
Naples
    as cultural centre, 316–17, 319, 325–7, 419, 425
    political situation, 319, 328
Nash, Thomas, 472, 476, **485–6**, 497, 501, 504, 571, 745, 779
    (possible) collaborations with Marlowe, 486, 498, 503, 542, 560
    Nashes Lenten Stuffe, 485–6
    The Unfortunate Traveller, 485
    (with Ben Jonson), *The Isle of Dogs*, 485, 848
Nassi (prototype for *The Jew of Malta*), 527–8
naturalism, opposition to, 259
Nazianzen, Gregory, Patriarch of Constantinople, 22
Nazimova, Alla, 683
necromancer, as stock character, 189
Negri, Cesare, 351
    Le Grazie d'Amore, 351
Negro, Lucy, 577
Negro, Mary-Joan, 558
Neidhart von Reuenthal, 99–100
Neilson, Adelaide, 680
Nelli, Jacopo Angelo, 328
Nestroy, J.N., 257–8

Netherlands
    medieval theatre, 49–50, 80, 105–6
    popular traditions, 451–2
    religious/political situation, 451, 453
Nethersole, Olga, 679
Nevers, Gérard de, 760
Neville, John, 681
Nevitt, Thelma, 140
A New Way to Pay Old Debts (Massinger), 801, **809–10**
Newcastle, Duke of, 870
Nicholas, St
    historical figure, 65–6
    theatrical depictions, 60, 62–5, 90–1
Nicholas of Cusa, 133
Nicholas V, Pope, 120
Nicolai, Otto, *The Merry Wives of Windsor*, 697
Nicolini (Nicolo Grimaldi), 328
Nicoll, Allardyce (as translator), 205, 213
    as author *see* Index of Cited Authors
Nijinska, Bronislava, 260
Nikolaus of Avancini, *Pietas Victix*, 458
Noah, story of, dramatic versions, 36, 40, 45–6, 55
Noah and the Flood (ballet), 55–6
Noah's Flood (Chester), 45–6
Noble, Adrian, 610, 757–60
Noel, Henry, 477–8
Noh theatre, 724
Nomanyama, Vumile, 58–9
Norrington, Roger, 307
North, Thomas, 653, 734
Northampton, Earl of, 855
Norton, Thomas, 469–71, 510
Nuremberg, 120

Oberammergau Passion Play, 53
Obey, André, *Noé*, 55
Odets, Clifford
    The Flowering Peach, 55
    Waiting for Lofty, 84
Old Testament, as source of

material, 36, 38–9
Oldenbarneveldt, Johan van, 453, 825
Olivier, Laurence, 597, 609, 681, 710, 723, 731, 741, 771
Olivier, Laurence, junior, 597
O'Neill, Eugene, 114, 680
   Days Without End, 83
   The Hairy Ape, 83
   Long Day's Journey into Night, 680
   Mourning Becomes Electra, 806
O'Neill, James, 678, 680
O'Neill, James, junior, 680
Ong Keng Sen, 723
opera buffa, **324–34**
opera(s), **276–345**, 861
   based on Shakespeare, 598, 625–6, 648, 689–90, 697, 704, 711, 732–3, 741, 760
   debt to *commedia dell' arte*, 265–8
   forerunners, 276–8, 462–3
   libretti, 339–40, 419, 420–2
   overtures, 318–19
   profitability, 338–9
   scene design, **341–5**
operetta, forerunners/early forms, 95, 421–2
Opitz, Martin, 463
   Book of German Poetry, 462
Orefice, Giacomo, 307
   Patro Calienno de la Costa, 325–6
Orfeo (Monteverdi), **283–4**
   modern revivals/reinterpretations, 306–8
Orff, Carl, 307
Orlandini, Giuseppe Maria, 335, 338
Orleans, 30–1
Orléans, Marguerite Louise d', 299
Ormerod, Nick, 648
Orsini, Flavio, Duke of Bracciano, 303
Orsini, Isabella de' Medici, 215, 874
Orsini, Paulo, Duke of Bracciano, 874
Osorio, Elena, 382–3
Oswald, J., 333
Othello (Shakespeare), 406, 492, 646, 678, **704–12**, 715, 863, 885
   casting, 599, 602, 680, 709–10
   characterization, 529, 607, 705–7
   language, 708–9
   modern productions, 709–11
   operatic/balletic reworkings, 711–12
   sources, 704
   structure, 707–8
   translations, 711
Otto II, Duke of Bavaria, 99
Ottoboni, Cardinal, 321
Ovid (P. Ovidius Naso), 23, 358, 562, 697, 764, 852
   Metamorphoses, 295

Padgett, Bruce, 56
Padovano *see* Scalzi
Pagano, Nicola, 322, 325
Painter, William, *The Palace of Pleasures*, 695, 813, 878
Paisiello, Giovanni, 333, 338
Palestrina, Giovanni Pierluigi da, 276
Palladio, Andrea, 126–7, 867
Pallavicino, Carlo, *Messalina*, 305
Pangratz, Bernhaubt, *Die Historie Herculis*, 356
pantomimes, 262–3
Papp, Joseph, 662–3, 690, 860
Pariati, Pietro, 340, 689
Parigi, Alfonso, 128
Parigi, Giulio, 128
Paris
   cultural life, 429–34
   Hôtel de Bourgogne, 98–9, 120, 126, 430, 436, 437–8, 439
   other theatre buildings/companies, 436–9, 440–1
   medieval farceurs, 94
Parke, Suzanne, 143

Parma, theatre architecture, 127–8
Parrisk, Maxwell, *Daybreak*, 308
Pasquier, Étienne, 436
Passaca(i)lle (dance), 425
Passion Play (Poland, thirteenth century), 36
Passion Plays, 38, 52–3
   modern productions, 53–5
Il Pastor Fido (Guarini), **132–3**, 276, 283, 335, 444, 452, 817
pastoral drama, **129–33**, 276–7, 335, 444, 452, 471, 481, 659–60, 745
Patin, Jacques, 448
Paul III, Pope, 163, 165, 167
Paul V, Pope, 226
Paul VI, Pope, 28
Pavane (dance), 350, 425
Pavia, University of, 106
Pearce, Joanne, 760
Peasles, Richard, 273
Peele, George, 354, 476, **478–80**, 497
   The Araygnement of Paris, 479, 484
   The Battle of Alcazar, 479, 516
   David and Fair Bathsabe, 480, 516
   Edward I, 479, 516
   The Merry Conceited Jests o fGeorge Peele, 480
   The Old Wives' Tale, 479–80
   Sir Clyomon and Sir Clamides, 480
La Pellegrina (Bargagli), **215–25**
   academic debate, 222–3
   characterization, 216, 221–3
   comic/social ethos, 220–4
   composition/première, 215–16, 224
   critical commentary, 216–17, 220–1
   influence on later works, 224–5
   intermezzi, 216–17, 225
   language, 224
   modern productions, 225

sources, 216
structure, 216–17, 221
Penafil, Marquis of, 393
Pennant Jones, Robert, 522
Penner, Anne, 609
Pepys, Samuel, 599, 812, 843, 863
Pergolesi, Giovanni Battista
(G.B. Draghi), 264, **327–34**,
562
   biography, 327–9, 330–1
   qualities as composer,
329–30
   Adriano in Siria, 329–30
   Alessandro Severo (Salustia),
328, 329, 340
   Flaminio, 330, 332
   Livietta e Tracollo, 330, 332,
333
   Mass in F, 329
   L'Olimpiade, 329, 330
   Orfeo, 330
   Il Prigioniero Superbo, 328,
329
   Salve Regina in C minor, 330
   Stabat Mater, 330
   La Vedova Ingegnosa, 330
Peri, Jacopo, 279, 283, 286
   Dafne, 278–9, 289, 463
   Euridice, 279–80, 283
   La Flora, 281
Pericles, Prince of Tyre
(Shakespeare/others?), **749–51**,
757, 762, 788
   authorship, 749
   sources, 749
Perry, Clive, 557
Perry, Herbert, 297
Persiani, Orazio, 293, 313
Persons, Chip, 608
Peruzzi, Baldassare, 123, 137
Pesaro, Ambrosio de, Messer,
348
Pescetti, Orlando, 653
Peterborough, John, Earl of, 839
Peters, Rollo, 624
Petherbridge, Edward, 760
Petipa, Marius, 265
Petrarch, 162, 222, 234, 572, 856
   Philologia, 134

Petrinsky, Natascha, 296
Petris, C. de, 326
Petrus Perondinus, *Magni
Tamerlanis Scytharum
Imperatoris Vita*, 510
Pharo, Carol, 30
Phèdre (Racine), 389
Philander, 120
Philaster, or Love Lies a-Bleeding
(Beaumont and Fletcher), 752,
818–21
Philemon, *Thesaurus*, 177
Philip II of Spain, 148, 365, 375,
382, 383
Philip IV of Spain, 294, 386, 407,
417, 425
Philip the Good, Duke of
Burgundy, 79
Philipp II of Pommern-Stettin,
462
Phillips, Tom, 598
Phrynicus, 745
Piacenza, Domenico da, *De Arte
Saltandi e Ducendi*, 347–8
Pianetti, Cardolo Maria, Marquis
of Iesi, 327
Piave, Francesco Maria, 732
Picasso, Pablo, 274, 428
Piccini, Niccolò, 338
Piccinini, Filippo, 421
Piccolomoni, Aeneas Sylvius,
Cardinal, *Chrysis*, 137
Piccolomoni, Alessandro, **168–74**,
215–16
   biography, 168–9
   L'Amor Costante, 169
   *Dialogo della Bella Creanza
delle Donne*, 169
Pick-Mangiagalli, Riccardo, *Basi
e Bote*, 268
Pickett, Philip, 307
Pietropinto, Angela, 141
Pine, Larry, 141
Pinkins, Tonya, 695
Pino de Cagli, Bernardo, 205
El Pintor de su Deshonra
(Calderón), **406–7**, 416
Piovene, Augustino, 335
Piper, John, 265

Pirandello, Luigi, 670
   *Right You Are (If You Think
You Are)*, 274, 387
   *Six Characters in Search of an
Author*, 84, 229
   Tonight We Improvise, 229,
274
Piranesi, Gian Battista, **344–5**
Pirchkeimer, Willibald, *Eccius
Dedolatus*, 357–8
Pistocchi, Antonio, 316
Pius II, Pope, 137
Plato, 745
Plautus, 23, 93, 353, 745
   adaptations/translations, 159,
431, 433
   influence on Cecchi, 176–7,
178, 181, 182, 193
   influence on Renaissance
drama, 101, 133–4, 135,
139, 362, 467, 609, 613–14
   Asinaria, 193
   Menaechmi, 133, 136, 178,
609
   Mercator, 181
   Miles Gloriosus, 433, 645
   Mostellaria, 177, 862
   The Captives, 135
   Trinummus, 177
   Truculentus, 193
Play of Daniel (?Hilarius, tr.
Weakland), **27–30**, 33–6, 276
Play of Herod, **30–6**
Play On! (musical), 694–5
Pléiade (group of poets), 431,
433, 436, 447
Plutarch, 431, 743, 744
   Lives, 653, 734, 745
Poel, William, 600
The Poetaster (Jonson), 794, **851–2**
Poitiers, Diane de, 429
Poland, 258–9
Poley, Robert, 502, 506–7
Polizano, Angelo, *Fabula di Orfeo*,
276
Pollarolo, Carlo Francesco, 305–6
Pomponius Laetus, 116, 120
Ponnelli, Jean-Pierre, 308
Pontalais, Jean du, 98

Ponti, Diana, 244
Poole, Roy, 722
Pope, Alexander, 517, 592, 868
Popov, Oleg Konstantinovich, 261
Porpora, Nicola, 319, 322, 323–4
    Arianna e Teseo, 324
    Tamerlano, 335
    Temistocle, 323
Porter, Cole, 141–2, 611, 615
Portugal, 368–9
Possokhov, Yuri, 712
Powell, William, 827
Prague, theatre buildings, 343
Predieri, Luca Antonio, 338
Preisser, Alfred, 725
Presentationalism, 259
Preston, Robert, 861
Preston, Thomas (attrib.), *Cambises King of Persia*, 477
Prévert, Jacques, *Baptiste*, 263
Price, Leontyne, 312, 741
Pride of Life (fifteenth-century English), **76–7**
La Princesse qui a perdu l'esprit (commedia dell' arte), 240
Pritz, William, 57
Prodigal Son, parable of, 57–8
Prokofiev, Sergei, 598, 625–6
    The Love for Three Oranges, 260
Prosser, Amy, 663
Protestantism, 104
    attitudes to theatre, 114
    theatrical defences/portrayals, 82, 354–6, 357–8, 469
Proust, Marcel, *Remembrance of Things Past*, 511
Provenzal, David, Rabbi, 203, 204
Prowse, Philip, 523
Prudentius Clemens, Aurelius, *Psychomachia*, 69, 76
Prymus, Ken, 695
Psacharopolis, Nikos, 29, 32
Puccini, Giacomo, *Gianni Schicchi*, 801
Pula, Bishop of, 292–3

Punch and Judy, 235, 274
Purcell, Henry, 303, 598, 711
    The Tempest, 772
Puritanism, opposition to theatre, 486–7
Pushkin, Alexander, 399, 665

Quaglio family (theatre architects), 343–4
Quayle, Anthony, 520–1, 681
Queen's Men, 570
Quem Quaeritis, 24–5, 26–7, 59, 275
Quentin, John, 557
"Querelle des Bouffons", 333
Quiney, Thomas, 580
Quinn, Patrick, 612
Quinn, Philip, 687

Rabb, Ellis, 557–8
Rabelais, François, 96, 468–9
Rabiner, Ellen, 337
Racan, Marquis de, *Les Bergeries, ou Arthenice*, 444
"race to the sepulcher", 26–7
Racine, Jean, 339, 421
Rades y Andrada, *A Chronicle of the Three Military Orders*, 391
Raimund, Ferdinand, 257–8
Raimundi, Marcantonio, 146
Raleigh, Sir Walter, 503, 586, 851, 863
Ralph Roister Doister (Udall), **466–7**
Rameau, Jean-Philippe, 333
Ramm, Andrea von, 29–30
Ramsay, Keith, 55
Ran (film, 1985), 732
Raphael, 121, 135, 145, 176, 350
The Rare Triumphs of Love and Fortune (anon.), 751
Rastell, John, 108–9, 111
    The Nature of the Four Elements, 82–3, 108
Rastell, William, 109
Rath, Gyorgy Gyorivanyi, 330
Rathbone, Basil, 625
Rattigan, Terence, *Harlequinade*, 263

Ravel, Maurice, 426
Rawlings, Margaret, 518
Razzi, Girolamo, 225
Rebhuhn, Paul, *Susannen*, 356
Rebollada, Girón de, Doña Marquesa, 393
Rederijkers (Dutch troupes), 105–6, 451, 453, 456
Redgrave, Michael, 681
Redlich, Ferdinand, 307
Redmon, John, *Wit and Science*, 83
Reeves, Saskia, 648
Regelson, Jessica, 57
Regnard, Jean-François, *Les Folies amoureuses*, 225
Reichardt, Johann Friedrich, 323
Reinhardt, Max, 84, 258, 401, 410, 630, 688–9
religious drama
    Dark Age, **21–7**
    medieval, **27–86**
    Protestant, 353–8
    Spanish, 365–8, 377–8, 407–12
    see also Miracle Plays; Morality Plays; mystery plays
Rembrandt van Rijn, 451
Remi-Belleau, 433
Renaissance
    art/architecture, 114–15, 125–6
    dance, 349–52
    defined, 113–14
    English, 465
    French, 429, 445
    see also Italy
Renaissance comedy, **133–226**
    defining characteristics, 134, 225–6
    influence on later ages, 226
    sources, 133–4 (*see also* Plautus; Terence)
    sub-genres *see commedia* . . .
Renaissance drama, **113–274**
    costume, 125
    staging, **119–29**, 136
Rennes, medieval theatre, 79

Rensenhouse, John, 54
Renz, Frederick, 32, 33–5, 64–5
Respighi, Ottorone, 307
Reuchlin, Johannes, 355
　Henno, 101, 355
Revels, Office/Master of, 482, 485, 488, 492
revenge tragedy, as genre, 500, 834
The Revenger's Tragedy (Tourneur), **828–34**, 877
　authorship, 827–8
　language, 829–34
　modern productions, 835–6
　moral universe, 828, 829, 834
Reynolds, John, 804
Reynolds, Sir Joshua, 583
Riario, Cardinal, 116
Ricciardi, Giovanni Battista, 302
Riccoboni, Luigi, 237–8, 249
Rich, Christopher, 261
Rich, John, 261–2
Richard II (Shakespeare), 613, **631–7**, 658
　characterization, 631, 670
　comparisons with other works, 550, 551, 554, 556, 561, 631
　language/imagery, 632–5
　modern productions, 554, 635–7
　sources, 631–2
Richard III (Shakespeare), 578, **605–9**, 636, 649, 678, 762
　authorship, 550
　characterization, 529, 605, 607
　film (1955), 609
　modern productions, 608–9, 636–7, 684
Richard III Society, 605
Richardson, Desmond, 711–12
Richardson, Ian, 721
Richardson, Joy, 598
Richardson, Lee, 836
Richardson, Ralph, 710
Richelieu, Armand Jean du Plessis de, Cardinal, 446

Richmond, Robert, 610
Ricketts, Charles, 517
Rigacci, Bruno, 280
Rigg, Diana, 721
Rimbaud, Arthur, 273
Rimsky-Korsakov, Nikolai, 260, 426
Ringcamp, Jonathan, 86
Rinuccini, Ottavio, 278, 279, 285–6, 289, 463
Riquier, Guiraut, 93–4
Rist, Johann, 462
Il Ritorno d'Ulisse in Patria (Monteverdi), **287–9**
　modern revivals/reinterpretations, 307, 308–12
Rizzi, Fortunato, 193
Robbins, Jerome, 264, 625
Roberts, Keith, 712
Robertson, Toby, 552–7
Robeson, Paul, 602, 709–10
Robin Hood, theatrical depictions, 22
Robinson, Mary B., 54
Rockefeller, John D., Jr, 58
Rodgers, Richard, 610
Rojas, Fernando de, 367–8
Romains, Jules, 860
Le Roman de Fauvel, 96–7, 276
La Romanesca (Cecchi), **201–2**
Romani, Felice, 266
Romano, Giulio, 146, 148, 151, 155
Rome
　ancient culture, 21–3, 87, 114, 248
　architecture, 226
　cultural life, 317–19, 325, 326–7
Romeo and Juliet (Shakespeare), 495, **620–8**, 646, 794
　characterization, 532, 623
　critical commentary, 621–2
　influences/sources, 118, 532, 620
　modern productions, 624–5
　parodied, 816

reworkings/alternative versions, 388–9, 625–8
Ronsard, Pierre, 431
Roos, John, *Lord Governance*, 82
Roosevelt, Franklin D., 732
Rosa, Salvator, 244, 297–8
Rosenplüt, Hans, 101
Roses, Wars of the, 465, 470, 563
Rospigliosi, Giulio, 325, 421
Ross, Ken, 67, 68
Rosse, Herman, 687
Rossi, Bastiano de', 225
Rossi, Giacomo, 132–3
Rossi, Mario, 308
Rossini, Giaocchino
　Il Barbieri di Seviglia, 267
　L'Italiana in Algeri, 267
　Otello, 711
　La Scala di Seta, 267
　Signor Bruschino, 267
Rostand, Edmond, 399
　Cyrano de Bergerac, 683
　Les Romanesques, 267
Röthlisberger, Max, 66
Rotrou, Jean de, 421, 445
　La Pélerine amoureuse, 225
　Tragicomédie, 225
Rouault, Georges, 274
Rouen, medieval farceurs, 94
round, performances in, 68
Rounseville, Robert, 312
Rousseau, Jean-Jacques, 333
　Le Devin du village, 333
Rovetta, Giovanni, *Ercole in Lidia*, 315
Rowe, Nicholas, 595
Rowell, Kenneth, 555, 556
Rowley, William, 749, 779
　biography, 803–4
　collaboration with Thomas Middleton *see under* Middleton
　collaborations with others, 784, 787, 804, 825
Rubens, Peter Paul, 451
Rubin, Amy, 54–5
Rucellai, Giovanni, 118
Rudolf II, Holy Roman Emperor, 359

Rueda, Lope de, 366, 374–5
  Aramellina, 193
  Eufemia, 193
  El Paso de las Aceitunas, 375
Ruello, Catherine, 140
Ruggieri, Giovanni Maria, 338
Ruoff, Jakob, *Weingartenspiel*, 355
Russia, Tsar of, 249
Russian drama, comic traditions, 259–61
Russo, William, 58
Rutebeuf, 93
  Dit de l'Erberi, 93
  The Miracle of Theophilus, 60, 93
"Ruzzante" *see* Beolco, Angelo
Ryan, Anna-Livia, 598
Rylance, Mark, 597–8

Sabatino, Nicola, 330
Sabbatini, Nicola, *Manual for Constructing of Theatrical Scenes and Machines*, 121–2
Sabbiani, Roberto, 225
Sachs, Hans, 84, **102–5**, 120, 353, 358, 460
  The Horse Thief, 104
  The Hot Iron, 104
  Das Narren Schneyden, 103–4
  The Wandering Scholar from Paradise, 104
Sackville, Thomas, 457, 469–71, 487, 503, 510
Sacrati, Francesco, 289, 292, 297, 301
  La Delia, 314
  La Finta Pazza, 288–9, 301
  L'Ulisse Errante, 288–9, 301
Saddumene, Bernardo, 326
Saint-Gelais, Mellin de, 434
saints, lives of *see* Miracle Plays
Salcedo, Lucia de "la Loca", 383
Salernitano, Masuccio, *Novellino*, 620
Salgado, Angela Maria, Doña, 393
Saliti, G.-F., 302
Salmon, Jacques, 448
Salsa, Francesco Berio de, 711
Saltarello (dance), 350
salvation, as theme of drama, 22
Salviati, Cecchino, 188
Salviatti, Lionardo, 225
San Daniele, Pellegrino da, 120–1
San Gallo, Aristotile da, 121, 123
Sand, George, 162
  Consuelo, 324
Sand, Maurice, 162
Sandella, Caterina, 152
Sanders, Donald T., 58
Sanders, George, Master, 780
Sanserevo, Prince of, 330
Sansovino, Jacopo, 148–9, 155
Santi, Francesco, 327
Santiago-Hudson, Ruben, 776
Santurini, Francesco, 344
Sarabande (dance), 425
Sardelli, Anna Maria, 298
Saro, Jeremy, 558
Sarro, Domenico, *Didone Abbandonata*, 340
Sartorio, Antonio, 304
  L'Orfeo, 304
Saturno, Mark, 610
Savage, Bernard, 545
Savo, Jimmy, 263
Saxo Grammaticus, *Historica Danica*, 666
Sbarra, Francesco, 300, 301, 341
Scala, Flaminio (Flavio), 239–40, 242, 243
  commedia scenarios, **245–7**
Scaliger, Julius Caesar, 118
Scalzi, Alessandro (Padovano), 257
Scamozzi, Vincenzo, 126–7
Scarlatti, Alessandro, 303, **316–22**
  biography, 316–18, 319–20
  family connections, 317, 320, 321–2
  qualities as composer, 318–19
  Mitridate Eupatore, 319
  Il Pirro e Demetrio, 318
  Il Trionfo dell'Onore, 326
Scarlatti, Anna Maria, 322
Scarlatti, Antonia (née Anzalone), 320
Scarlatti, Domenico, 317, 318, 322, 426
  Amleto, 689
  Giustino, 322
Scarlatti, Francesco, 322
Scarlatti, Melchiorra Brigida, 322
Scarlatti, Pietro, 321
Scarlatti, Pietro Filippo, 322
Scarlatti, Tomasso, 322
Schack, Adolf Friedrich von, 415–16
Schade, Richard Erich (as translator), 360
  as author *see* Index of Cited Authors
Schatz, Ken, 725
Scheherezade, 703
Schembartlauf, 102
Schernberg, Dietrich, *Play of Frau Jutten*, 61–2
Schiller, Johann Christoph Friedrich von, 593
  Don Carlos, 426
  The Robbers, 397
Schlegel, August Wilhelm von, 593–4, 621
Schmelzer, Johann Heinrich, 300–1
Schmidt, Harvey, 267
Schnurman, Rhea, 57–8
Schoenberg, Arnold, *Pierrot lunaire*, 264
School of Night, 793
schools/colleges
  educational standards, 466–7
  plays intended for, 353–6, 436, 481–2
  staging of plays, **356–7**
Schulman, Susan H., 611
Schumann, Robert, *Carnaval*, 260
Schütz, Heinrich, 463
  Dafne, **463–4**
Schwank sketches, 101, 133
Schwarzenberg family, 257
Scofield, Paul, 720–1, 860
Scott, George C., 861
Scribe, Eugène, 266, 772
Secchi, Niccolo, 225

Sejanus (Jonson), 793, 795, **853–4**, 855, 860, 864
Seldes, Marian, 612
Selim, Sultan, 527
Seneca, L. Annaeus, 23
    as character in drama, 289, 290
    influence on Renaissance drama, 115, 116, 117–18, 134, 353–4, 378, 434–5, 469–70, 478, 498, 794, 828, 865, 886
    translations/adaptations, 431, 462, 471, 498
    The Trojan Women, 462
sentimental comedy, genre, 169
Serban, Andrei, 608–9, 757–9
Serlio, Sebastiano, *Architettura*, 123–4
Serre, Jean, 98
La Serva Padrona (Pergolesi), 328, **332–4**
Il Servigale (Cecchi), **190–3**
    characterization, 192
    intermezzi, 192–3
Shadwell, Thomas, 748, 772
Shakespeare, Anne, née Hathaway, **567–9**, 577, 580–1
Shakespeare, Edmund (brother), 564
Shakespeare, Hamnet (son), 568, 576, 638
Shakespeare, Henry (great-uncle), 564
Shakespeare, John (father), **564–6**, 567, 568, 572
Shakespeare, Judith (daughter), 568, 580, 581, 646
Shakespeare, Mary, née Arden (mother), 564, 565
Shakespeare, Richard (grandfather), 563–4
Shakespeare, Susanna (daughter), 567–8, 580, 581, 769
Shakespeare, William, 106, 114, 250, 457, 465, 490, **563–777**
    academic study, 595–6
    acting career, 569–72, 578, 661, 848
    anniversary celebrations, 522, 557, 589–90
    biography, 497–8, 500, **563–72**, 578–9, 767, 773
    change of creative direction, 753, 757–8
    contemporaries' comments on, 475–6, 571–2, 587, 847
    contemporary interpreters, 494
    contributions to pastoral genre, 129, 131, 659–60, 745
    critical commentary, **590–5**
    death/burial, 579, 581–2, 585
    education, 566–7, 569
    experimentalism, 753
    Folio editions, 587–9
    historical accuracy, 605–6
    influence on contemporaries, 822, 875
    influences/sources, 475, 484–5, 504, 544
    married life, 567–9, 577, 581, 762
    modern productions, 595–6, **599–602**
    musical/operatic adaptations, 598 (see also under ballet; opera)
    parent–child relationships, treatment of, 646, 668
    personality/lifestyle, 561–2, 571, 579, 585
    pessimism of outlook, 665–6, 698, 700, 712–13, 757
    portraits, **581–5**
    repeated themes/plot devices, 762
    romantic/sexual orientation, 574–5, 577–8, 643
    self-depictions, 682, 762
    theories as to identity, 508–9, 585–6, 587
    will, **579–82**
    The Rape of Lucrece, 572, 573, 587
Shamie, Greg, 626–8
Sharaff, Irene, 611
Shatner, William, 521
Shaw, George Bernard, 388, 551, 595, 744, 757, 827, 891
    Arms and the Man, 680
    Caesar and Cleopatra, 741
    Don Juan in Hell, 399
    Major Barbara, 83
    Saint Joan, 86
Shawn, Wallace, 140–1
Shelley, Percy Bysshe, 414–15, 417, 719
    Ariel to Miranda, 773
The Shepherd's Christmas (opera), 58
Shepherd's Play (Rouen, thirteenth century), 36
Shepherds' Plays, 38
    literary qualities, 43
    modern adaptations, 53–4
Sherwood, Robert, *The Petrified Forest*, 687
Shirley, James, 843, **844–6**
    The Cardinal, 844, 846
    Hyde Park, 844
    The Lady of Pleasure, 844–5
    The Triumph of Peace, 844
    The Witty Fair One, 844
The Shoemaker's Holiday (Dekker), 783, **784–6**, 816
    modern productions, 786
Shore, Daniel J., 626–7
Shrovetide plays *see Fastnachtspiel*
Shulgasser, Mark, 773
Siddons, Sarah, 676–7
Sidney, Sir Philip, 487, 498, 503, 510, 572, 875
    Arcadia, 129, 713, 823, 837
    The Lady of May, 129
Sidonius, 21
Siena, 168
    Accademia delle Intronati, 215
Siena, Jerold, 296
Sigismund Francis, Archduke, 300
Silvani, F., *La Fede Tradita e Vendicata*, 328
Simone, Roberto de, 330
Sixtus V, Pope, 873–4
Skelton, John, 82
Skere, Nicholas, 506–7
Skibine, George, 626

Skinner, Philip, 297
Skybell, Steven, 637
slaves, anachronistic use as characters, 181
Smith, Henry Preserved, 21
Smith, R.D., 523
Smoldon, William L, 31–2
Snow, Norman, 558
Soane, Sir John, 314
social comment
    in *commedia dell' arte*, 234
    in Lope de Vega, 389–91
    in medieval farces/popular entertainemnts, 91–2, 94
    in morality plays, 82
    in Renaissance drama, 153, 160, 221–2
    in shepherds' plays, 37–8
    see also Jonson, Ben
Sociétés joyeuses, 94
Soetaert, Susan, 68
Soldini (comedian), 248
Solimena, Francesco, 322
Sommer, Josef, 777
Somnium Vitae Humanae (Hollonius), **460–2**
Sondheim, Stephen, 142, 625
    A Little Night Music, 267–8
Sonnenberg, Melanie, 337
Sonnets (Shakespeare), 533, 561, **572–8**, 643
    "dark lady", identification, 576–7
    dedication, 573–4, 575–6, 577–8
    publication history, 573, 576, 577–8
    "rival poet", identification, 789
Sophocles, 116
    Antigone, 462
    Oedipus Rex, 126, 669, 671, 718, 862
    Trachiniae, 455
Sophonisba (Marston), **795–6**
Sothern, Edward Askew, 682
Sothern, Edward Hugh, 682
Sothern, George Evelyn, 682
soties, 99, 133
    see also farces, medieval

South Africa, all-black companies, 58–9
Southampton, Countess of, 576
Sozzini, Fausto, 215–16, 224
Spain
    dance, **422–8**
    dramatic forms, 377–8
    "Golden Age" drama, **365–418**
    influence on European culture, 426
    opera, 419–22
    political situation, 365
    religious climate, 365, 419
    religious drama, 366–8, 377, 407–12
The Spanish Tragedy (Kyd), 494, **498–500**, 503, 509–10, 612, 665, 794, 799, 848
    parodied, 816
    revised by Jonson, 853
Spencer, Arthur, 545
Spencer, Gabriel, duel with Ben Jonson, 850
Spencer, John, 457
Spenser, Edmund, 497, 572, 713, 716
Speroni, Sperone, 168
    Canace, 118
Spewalk, Bella/Samuel, 615
Spies(s), Johann, *The Historie of the Damnable Life and Deserved Death of Doctor John Faustus*, 533–4, 537–8
Lo Spirito (Cecchi), **188–9**
Spyeghel der Salicheyt van Elkerlijk (fifteenth-century Dutch), 80
Sroka, Elliott, 85
St Bartholomew's Day Massacre, 430, 560
St John, Christopher, 24
Staden, Sigmund Theophilus, *Seelewig*, 464
stages, fixed, early use of, 40–1, 50–1
Stambler, Bernard, 268
Stampigli, Silvio, 339
Stanislavsky, Konstantin, 259, 688

Stanley, William, Earl of Derby, 586
The Staple of News (Jonson), **868–9**
Steen, Jan, 87
Steevens, George, 584
Steinbeck, John, *Burning Bright*, 84
Stella, 59
Stevenson, William, 467
Stewart, Peter, 58
La Stiava (Cecchi), **181–2**
Stigliano, Ferdinando Colonna, Prince, 328
Still, J., 467–8
Stockel, Leonhard, 356
Stokowski, Leopold, 307–8
Storace, Stephen, Sr., 333
Stovall, Paul Oakley, 695
Stowe, John, 546
Gli Straccioni (Caro), 162, **163–7**, 169, 170
Stradella, Alessandro, 292, 294, **301–4**, 317
    La Forza dell'Amor Paterno, 302
    Le Gare dell'Amor Eroica, 302
    Moro per Amore, 303
Stradivarius, Antonius (Antonio Stradivari), 281
Stranitsky, Josef Anton, 256–7
Strauss, Johann II, *Die Fledermaus*, 268
Strauss, Richard, *Ariadne auf Naxos*, 258, 266
Strauss, Robert, 687
Stravinsky, Igor, 55–6
    Petrushka, 259–60
    Pulcinella, 264–5
Strehler, Giorgio, 267, 271
Stride, John, 625
Striggio, Alessandro, 283, 285
Strozzi, Giulio, *La Delia*, 314
Stuhlbarg, Michael, 777
Stukely, Thomas, 479
Stummel, Christoph, *Studentes*, 136, 355
Sturm, Johannes, 361
Subber, Saint, 615
Suharto, President, 723, 725

Sully, Duc de, 450
Surrey, Earl of, 503
Susanna (biblical story), 356
Susarion, 248
Sutermeister, Heinrich, 625
Swan, Jon, 58
Swan Theatre (London), 489–90
Swinburne, Algernon Charles, 719, 788, 814, 891
Swinton, Tilda, 560
Symbolism, 259, 260–1
Synge, J.M., *In the Shadow of the Glen*, 801

Taborino, Giovanni, 248
Tacitus, 289
Tairov, Alexander, 259
Talmud, 96
Tamagno, Francesco, 711
Tamburlaine the Great (Marlowe), 335, 493, 503, **509–27**, 542, 789, 799, 854
    characterization, 511–13, 514–15, 535, 549
    critical commentary, 515, 518, 526, 528
    imitations, 473, 516, 562
    language, 515, 520, 522, 537
    modern productions, **516–27**
    religious content, 507, 510, 513–14, 515–16, 525–6
    sources, 510–11
The Taming of the Shrew (Shakespeare), 598, **613–15**, 617
    influences/sources, 135, 613–14
Los Taranatos (film, 1963), 424
Target, Simon, 308
Tarlton, Richard, 494–5
    Newes out of Purgatorie, 695–6
Tasso, Bernardo, 130
Tasso, Torquato, **130–2**, 168, 241
    Intrighi d'Amore, 132
    Jerusalem Delivered, 130, 133
    L'Aminta, 130–1, 335, 444
    Il Re Torrismondo, 132
    Il Combattimenti di Tancredi e Clorinda, 286
Tate, Nahum, 718, 745

Tati, Jacques, 269
Taylor, Elizabeth, 545–6
Taylor, Joseph, 676
Taymor, Julie, 613
Tchaikovsky, Pyotr, 598, 625–6, 773
    Fantasy Overture, 689
Tcherepnine, Alexander, 260
Te Deum, use in religious dramas, 24, 28, 31
Tebaldeo, Antonio, 119
The Tempest (Shakespeare), 129, 363, 598, 740, 757, **764–73**, 774
    characterization, 768–9
    dramatic/philosophical atmosphere, 765–7, 769–70
    language, 766, 767–8
    literature inspired by, 772–3
    modern productions, 771–2
    operatic adaptations, 772, 773
    supernatural content, 767–8, 770–1
tennis courts, use as theatres, 438–9
Tennyson, Alfred, Lord, 773
Ter-Arutunian, Rouben, 32, 55, 265
Terence (P. Terentius Afer), 93, 100, 353, 358, 562
    Christian adaptations, 23–4, 69–70, 133
    influence on Western drama, 133–4, 135, 139, 179, 182–3, 189–90, 193, 362, 467
    later imitations/translations, 101, 431
    Adelphi, 133, 137, 182–3
    Andria, 133
    The Brothers, 788
    Eunuchus (*The Eunuch*), 135, 193
    Phormio, 466
    The Self-Tormentor, 788
Teresa of Avila, St, 227
Terry, Ellen, 600, 677
Terry, Nigel, 559–60
Tertullian, 22
Testa, Mary, 142
Tetley, Glen, 264

Teuber, Andreas, 546
Tewson, Josephine, 523
theatres
    design/architecture, **119–29**, 136, 228–9, **341–5**, 372–4, **489–93**, 599–601, 867–8
    earliest (fixed), 98–9, 120, 126
    (see also stages)
Theocritus, 129
Theophilus (of Constantinople), 88
Thirty Years' War, 357, 451, 460
Thomas, Allen, 683
Thomas, Ambroise
    Hamlet, 689
    Mignon, 689
Thompson, April Yvette, 725
Thompson, Francis, *The Hound of Heaven*, 540
Thomson, James, 745
Thornton, Barbara, 70–1
Thorpe, Thomas, 573, 576
The Three Daughters (Fleury), 63–4
The Three Schoolboys (Fleury), 63–4
Throne of Blood (film, 1957), 732
Tieck, Ludwig, 258, 593
Tiepolo, Domenico, 244, 272–3
Tiepolo, Giovanni Battista, 272
Tiernan, Andrew, 559
Timon of Athens (Shakespeare/others?), **745–9**
    authorship, 746
    characterization, 746–7
    language, 747–8
    modern productions, 748–9
    sources, 745
Tindemans, Margriet, 70
Tintoretto, 148–9
Tirso de Molina (Gabriel Téllez), **398–9**, 417, 419
    El Burlador de Sevilla, 398–9
    Don Gil of the Green Trousers, 398
    The Doubter Damned, 398
'Tis Pity She's a Whore (Ford), **838–40**, 841
Titian, 148–9, 155
Titus Andronicus (Shakespeare),

498, 562, **612–13**, 715, 799
  characterization, 529
  critical commentary, 585–6
Titus (film, 2001), 613
Toller, Ernst, *Masse Mensch*, 83
Tolstoy, Nikolai, *The Power of Darkness*, 780
Le Tombeur Notre-Dame, 61
Tomlin, Lily, 272
Torelli, Giacomo, 124–5, 301
Toshio, Kawatake, 689
Toth, Catherine, 251
Totò (Antonio de Curtis Gagliardi, Duca Commeno di Bisanzio), 256
Tourneur, Cyril, 746, 779, **827–36**
  biography, 827, 828
  literary qualities, 829–30
  The Nobleman, 828
Tours, medieval theatre, 79–80
Townley Mystery Cycle *see* Wakefield
Le Tre Sorelle (De' Sommi), **205–12**, 214
  critical commentary, 210–12
  sources, 210
Trent, Council of, 83
Trissino, Giangiorgio, 115–16
  Sofonisba, 115, 434
Tristan and Iseult (medieval romance), 811
Tristan L'Hermite, 443
Troilus and Cressida (Shakespeare), 537, 697–700, 702
  characterization, 698
  language, 699
  sources, 697
Les Trois Farceurs, **440–1**, 445
troubadours, 23, 93–4
Tshanini, Thabani Patrick, 734
Tsoutsouvas, Samuel, 558
Tsypin, George, 712
Tübingen, University of, 354, 358, 359–60
Tullio, F.A., 325
Turnèbe, Odet, *Les Contens*, 174
Tusser, Thomas, 467
Tutilo, 24
Twelfth Night (Shakespeare), 117, 270, 568–9, 617, **691–5**, 762
  characterization, 691
  critical commentary, 694
  language, 693–4
  modern productions/reworkings, 602, 694–5
  sources, 691
Twine, Laurence, 749
Two Gentlemen of Verona (Shakespeare), **615–17**, 745
Two Lamentable Tragedies (anon.), 780
The Two Noble Kinsmen (Shakespeare/others?), 773, 777
Tyrol, mystery plays, 52–3

Ubaldus, Guido, *Six Books of Perspective*, 124
Uberti, Antonio, 324
Udall, Nicholas, 466–7
Uga, Felicità, 313
Umewaka, Naohiko, 724
United States
  commedia influence, 263, 267–73
unities (of time/place/action), 117, 164
  dispensed with, 399, 442
  observation, 431, 432, 849
  restrictive effect, 396
"University Wits", 476, 480, 485, 486, 497
"Ur-Hamlet" *see Hamlet*: sources
Urban IV, Pope, 38
Urban VIII, Pope, 226
Urbina, Isabel de, 383
Urbino, as cultural centre, 137
Urfé, Marquis d', 444
Uta Gospel, 26

Vaccai, Nicola, 625
The Vagabond King (musical), 682
Valenciennes, mystery plays, 39, 50
Valéran le Conte, 441, 445
Valerini, Adriano, 244
Valla, Georgio, 115
Valois, Marguerite de, 98, 241
van den Borren, Charles, 308
van Druten, John, *Berkeley Square*, 687
Van Dyck, Anthony, 451, 814
van Goyen, Jan, 451
van Loo, Carlos, 426
Vane, Sutton, *Outward Bound*, 687
Varotari, D., 299
Vasari, Giorgio, 123, 148, 176
Vashi, Anjali, 609
Vaudemont, Mlle de, 448
Vaughan Williams, Ralph, *Sir John in Love*, 697
Velázquez, Jerónimo, 382
Venice
  Accademia degli Incogniti, 287, 289–90
  as cultural centre, 286, 299, 303–6, 313–16, 327, 350
  political life, 286, 289, 313
  theatres, 122, 128, 314–15
Venus and Adonis (Shakespeare), 504, 561, 572, 573, 587
  sources, 476–7
Verdi, Giuseppe
  Don Carlos, 426
  Falstaff, 697
  Hernani, 426
  Macbeth, 732
  Otello, 711
Verdy, Violette, 264–5
Vere, Sir Francis, 828, 834
Verga, Joseph A., 142
Vergardi, Carlo, 116
Vergerio, Piero Paolo, 134
Verlaine, Paul, 273
Vernier, Marie, 438
Vernon, Elizabeth, 576
Vettori, Piero, 118
Viaud, Théophile, 443–4
  Parnasse satirique, 443
  Pirame et Thisbé, 443
Vicente, Bianca, née Becerra, 369
Vicente, Gil, **368–71**
  Auto da Alma, 371
  Auto da Barca da Glória, 369, 370–1
  Auto da Barca do Inferno, 369, 370

Auto da Barca do Purgatório, 369, 370
Auto da Fama, 371
Auto da Índia, 371
Auto de la Sibila Cassandra, 369–70
Auto de los Reyes Magos, 371
Don Duardos, 371
Exhortaçao da Guerra, 371
Farso dos Almocreves, 371
Quem tem Farelos?, 371
Tragicomedia da Serra da Estrella, 371
Vicente, Luis, 369
Vicente, Paula, 369
Vicenza, theatre architecture, 126
Victoria, Tomás Luis de, 419
Vida, Marco Girolamo, 118
La Vida Es Sueño (Calderón), **402–5**, 412, 418, 460
Vienna
  cultural life, 300, 305, 353
  mystery plays, 52
  popular theatre, 257–8
  theatre buildings, 257, **341–2**, 343
Vigarani, Gaspari, 294, 295
Vignola, Giacomo de, *The Two Rules of Perspective Practice*, 121–2, 123
Villandrado, Augustin de Rojas, *Entertaining Journey*, 377
Villella, Edward, 265
Villifranchi, Giovanni Cosimo, 302
Vinci, Leonardo (eighteenth-century composer), 322–3, 326, 327
  Li Zite, 326
Virgil (P. Vergilius Maro), 23
  Aeneid, 163, 293, 394, 503
  Eclogues, 129
Virgin Mary, medieval cult of, 60–1
  plays based on, 61–2, 93, 199
Visentini, Tomasso Antonio, 237
Visitatio Sepulchri, 25
Vitruvius, *De Architectura*, 120–1, 123–4, 126, 127
Vittoria, Alessandro, 148

Vittoria (*commedia* troupe leader), 240, 243
Vivaldi, Antonio, **334–7**
  Bajazet, 335
  Catone in Utica, 336, 340
  La Fida Ninfa, 334–5
  L'Incoronazione de Dario, 336
  L'Olimpiade, 334, 336–7
  Orlando Furioso, 336
  Tamerlano, 335, 336–7
Volpone (Jonson), 135, 156–7, 801, 829, 853, **856–61**, 862, 866
  characterization, 858
  language, 857–9
  modern productions/reworkings, 860–1, 871
  preface (by Beaumont and Fletcher), 815, 826
Voltaire (François-Marie Arouet), 443, 592–3
Von Vincentius Ladislaus (Brauschweig), **362–3**
Vondel, Joost van dem, **453–6**
  Biblical dramas, 454–5
  Ch'ung-chen, or the Fall of the Chinese Empire, 455
  De Aemsteldamsche Hecuba, 454
  Gysbrecht van Aemstel, 453, 455
  Het Pascha . . ., 453, 454
  Hierusalem Verwoest . . ., 453
  Hippolytus, 454
  Jephta of Offerbelofte, 455
  Lucifer, 455
Vondel, Maria, née de Wolff, 453
Voragine, Jacobus de, *The Golden Legend*, 68
Vulpio, Giovanni Battista, 302

Waddington, Steven, 559
Wagner, Richard, 772
  Das Liebesverbot, 704
  Die Meistersinger von Nürnberg, 104–5
  Tannhäuser, 350
wagons, use in staging, 39–40
Wain, John, 643, 671, 739–40
Wakefield Mystery Cycle, 43–4

Second Shepherds' Play, 37–8, 44, 45, 67–9, 87
Walbrun of Eichstätt, 27
Wall, Max, 269
Wallace, Charles E., 695
Walpole, Hugh, 583
Walser, Lloyd, 56
Walsingham, Thomas, 502, 503, 506–7, 508–9
Walton, William, *Troilus and Cressida*, 700
Wanamaker, Sam, 596
Wanamaker, Zoe, 596
Ward, Anthony, 759
Ward, Hilary, 725
Ward, John, Rev., 579
Ward, Robert, *Pantaloon*, 268
A Warning for Fair Women (anon.), 780
Watson, Thomas, 504
Watteau, Jean-Antoine, 244, 250
Watterson, Henry, 509
Weakland, Rembert G., 28
Weaver, John, 261
Weaver, R.T., 237
Weber, Carl Maria von, *Euryanthe*, 760
Webster, John, 465, 494, 532, 590, 779, 795, 797
  biography, 847, 871–2, 891
  collaborations, 784, 786, 787, 802–3, 804, 825, 872–3
  critical/popular reception, 890, 891
  literary qualities/outlook, 877–8, 885, 890, 891
  (with Heywood?), *Appius and Virginia*, 889, 890–1
  The Guise, 889
  see also under Dekker; Middleton
Webster, Margaret, 602, 636, 709
Weinstock, Richard, 140
Weiss, P. Ottmar, 53
Welles, Orson, 544–5, 602, 723, 731–2
Wenzinger, August, 307
West, Cheryl L., 694–5
West, Timothy, 554

GENERAL INDEX | 925

West Side Story (musical), 625
Westrup, Jack, 307
Wetschel, Johann, 257
Wettach, Adrian *see* Grock
Wever, R., *Lusty Juventus*, 82
"W.H., Mr", 573–4, 575–6
Whetstone, George, *Promos and Cassandra*, 701
White, Allen, 66
The White Devil (Webster), **873–7**
  characterization, 875–6
  historical basis, 873–4
  preface, 873
Whitgift, John, Archbishop, 485, 794, 799
Whitman, Walt, 683
Whores of Heaven (*La Mandragola*), 142–3
Wickham, Glynne, 557
  see also Index of Cited Authors
Wigman, Mary, 264
Wild, Sebastian, 53
Wilde, Oscar, 576, 793
Wilder, Thornton
  Our Town, 84
  The Skin of Our Teeth, 84
  see also Index of Cited Authors
Wilhelm of Bavaria, Prince, 257
Wilkins, George, 746, 749, 804
Williams, Michael, 721
Williams, Robin, 272
Williams, Sally, 58

Williams, Tennessee, 562
Willich, Jodocus, 355, 357
Wilmot, Robert (and others), *The Tragedie of Tancred and Gismund*, 477–8
Wilson, Glen, 309–12
Wilson, H.J., 583
Wilson, John C., 615
Wimpheling, Jakob, *Stylpho*, 101, 354
Winchester Cathedral, 275
The Winter's Tale (Shakespeare), 646, **760–4**, 788
  characterization, 762–4
  critical commentary, 757, 758
  modern productions, 596–8
  sources, 475
Wolf, Julia Anne, 337
Wolfit, Donald, 517–19, 526, 860
Wolsey, Thomas, Cardinal, 82, 466
A Woman Killed by Kindness (Heywood), **781–2**, 784
women, as actors, 43, 239, 375–6, 438, 445, 599, 709
  costume, 377, 385
  opposition to, 376, 385
Woodward, Harry, 262
Woolley, Edmund Montillian, 516–17
Wordsworth, William, 594
Worth, Irene, 721
Wotton, Sir Henry, 495
Wright, Michael, 142–3
Wriothesley, Henry, 3rd Earl of

Southampton, 572, **573–4**, 579, 586, 698
Württemberg, Ludwig, Duke of, 359, 361–2
Wycherley, William
  The Country Wife, 135
  The Plain Dealer, 797

Yando, Larry, 749
Yeats, W.B., 717
York, Susannah, 690
York Mystery Cycle, 43
  modern productions, 53–4
A Yorkshire Tragedy (anon.), 780

Zambeccari, Francesco Maria, 320
Zanettini, Antonio, 338
zarzuela, 421–2
Zavdonai, Riccardo, 625
Zeffirelli, Franco, 625, 741
Zeno, Apostolo, 328, **339–40**, 689
Ziani, Marc'Antonio, 305
Ziani, Pietro Andrea, Father, 304–5
  Annibale in Capua, 304
  La Semiramide, 304–5
Zimmerman, Mary, 776
Zinano, Gabriele, 119
Zingarelli, Nicola, 625
Zola, Émile, *Thérèse Raquin*, 780
Zorilla, Francisco de Rojas, 397
Zorzi, Elvira, 225
Zuccaro, Federigo, 583
Zweig, Stefan, 860

INDEX OF PRODUCING COMPANIES/VENUES

This index refers only to only modern productions; contemporary performers are to be found in the General Index.

Accademia Musicale Chigiana, 225
Acting Company, 557–8
Actors' Guild, 57–8
Almagro (Spain), International Festival of Classic Theatre, 417–18
American Ballet Theater, 711
American Opera Company, 312
Aquila Theater, 609–10
Artek, 308

Ballet Russe, 259–60, 264–5, 428
BBC Radio, 522
Berlin State Opera/Ballet, 690
Bon Accord Theater Company, 67–9, 86
Boston
    Early Music Festival, 308, 311
    Jordan Hall, 28–9
Boston Camerata, 28–30
Bristol Old Vic, 557, 558

Cambridge (University)
    Corpus Christi College, Fletcher Players, 522
    Marlowe Society, 552–3
Canterbury Theatre Trust (Marlowe Theatre), 523
Caramoor Music Festival, 312, 313
Charleston, NC, Dock Street Theater, 296–7
Cheek by Jowl, 648
Cheltenham, Everyman Theatre, 523
Chicago, Goodman Theater, 694–5
Chicago Civic Shakespeare Society, 686–7
Chicago Lyric Opera Company, 694–5
Chicago Shakespeare Repertory, 748–9
Classical Theater Ensemble, 84–5
Classical Theater of Harlem, 725
Compass Theatre, 558
Consort to the Queen, 57–8
Cooperstown, NY, 312

Dallas, Texas, 773
Des Moines, Iowa, 773

Edinburgh Festival, 54, 523–4, 553–5, 558
Elsinore Castle, 681
Ensemble for Early Music, 32, 34–5, 64–5, 96–7

Federal Theatre (Project 891), 544–5, 731–2
Florence
    Boboli Gardens, 312
    Maggiore Musicale (May Festival), 330

Glasgow Citizens, 523–4
Glimmerglass Opera Company, 312, 313
Glyndebourne, 309, 312

Hampton Shakespeare Festival, 662–3
Hartford (CT) State Company, 54–5
Harvard University, 522
    47 Workshop, 93
    Loeb Shakespeare–Marlowe Festival, 557
Hildegurls, 72–3

I Comici Confidanti, 142–3

I Guillari de Piazza, 269
Indiana University, Bloomington, Early Music Institute, 66–7

Jakarta, 723–5

Katonah, NY, 312
Kent Opera, 309, 312

Leicester, Phoenix Theatre Company, 557, 559
Leningrad, Kirov Theatre, 711
Lincoln Cathedral, 55
Little Orchestra Society, 336–7
London
    Arts Theatre, 557
    Commonwealth Institute, 557
    Covent Garden Opera House, 557
    English National Opera, 309, 312
    Globe Theatre (reconstructed), 596–8, 681
    Lyric Opera House, Hammersmith, 553
    Mercury Theatre, 140
    Mermaid Theatre, 54, 555–6
    National Theatre, 53–4, 524–7, 860
    Old Vic, 517–20, 680–1, 725–6
    Regent Theatre, 552
    Royal Albert Hall, 72
    Sadler's Wells Theatre, 312, 689
    St Pancras Town Hall, 309

INDEX OF PRODUCING COMPANIES/VENUES | 927

Theatre Royal, Stratford East, 552
Tower Theatre, 522
West End, 58–9

Maggio Fiorentino, 225
Manchester Royal Exchange Theatre, 558–9
Mannes Camerata, 32–3
Marlowe Society Drama Company, 557
Minneapolis, Tyrone Guthrie Theater, 860
Montauk (NY), Theodore Roosevelt County Park, 662–3
Municipal Opera of Amsterdam, 264

Natal, University of, 733–4
Netherlands Opera, 309, 311
New Haven, CT, Yale Repertory Theater, 835–6
New York, 557–8, 681–2
    Alice Busch Theater, 312
    Alice Tully Hall, 336–7
    American Folk Theater, 142–3
    American Shakespeare Festival Theater, 721
    Broadway, 610–12, 635–6, 648, 684–8, 690
    Brooklyn Academy of Music, 86, 309–12, 648, 679, 759–60, 775
    Cathedral of St John the Divine, 27, 32–4, 64–5, 84–5
    Central Park, 662, 690, 757–9
    Christ Church United Methodist, 32–3, 58
    Church of St Mary the Virgin, 308
    City Center, 732
    Cloisters, 27–30, 36, 57–8, 66–9, 86
    Delacorte Theater, 776–7
    Florence Gould Hall, 34
    Frick Collection, 272–3
    Hampden Theater, 683–4
    Juillard School of Music, 312, 313
    Kathryn Bache Miller Theater, 309
    Lafayette Theater, 731–2
    LaMama Theater, 608–9
    Lincoln Center, 55–6, 71–3, 86, 720–1, 733–4
    Majestic Theater, 309–12
    Mannes College of Music, 32–3, 73
    Maxine Elliott Theater, 544–5
    Metropolitan Museum, 27, 70–1, 312 (*see also* Cloisters)
    Metropolitan Opera House, 307, 711
    Nameless Theater, 140
    off-Broadway, 626–8, 636–7
    Public Theater, 140–1
    Rooftop Theater, 860
    SoHo Rep, 141–2
    Tishman Auditorium, 269
New York Art Theater Institute, 58
New York City Ballet, 55–6
New York City Opera, 312
New York Shakespeare Festival, 662–3, 690, 860
Northampton, Massachusetts, 312

Ontario, Stratford Festival, 520–1, 771–2
Oxford (University), 545–6
    New Theatre, 550–1
    St John's College Dramatic Society, 521–2
    Worcester College Buskins, 517

Pro Musica, 27–8, 30–2, 34, 35, 313
Prospect Theatre, 553–6

Royal Lyceum Theatre Company, 558
(Royal) Shakespeare Company, 86, 551–2, 680–1, 720–1, 757–60, 775

San Francisco Ballet, 711
San Francisco Mime Troupe, 263
Sequentia, 70–2
Stratford, Connecticut, 721–3
Stratford-upon-Avon, 517–21, 551–2, 553, 731
La Stravaganza Köln, 296–7

Tavistock Repertory Company, 522
Theater Guild, 860
Théophiliens, 93

Washington, DC, Kennedy Center, 759–60
Waverly Consort, 36, 96–7, 276

Yale University Dramatic Association, 516–17

INDEX OF CITED AUTHORS

Notes: Books are indexed under the author's name, followed by the title where this is mentioned in the text. Reviews are indexed by the title of the periodical, followed by the reviewer's name where given. This index lists only modern authors; ancient commentators are to be found in the General Index.

Adams, J.K., 585
American Theater
    Mardden, Ethan, 709
Armstrong, William A., *Marlowe's "Tamburlaine": The Image and the Stage*, 515, 516
Art News and Review, 519–20

Barfoot, Terry *see* Headingon, Christopher
Barkan, Leonard (ed.), *Renaissance Drama*, 516
    see also Charney, Maurice
Barlow, Jeremy, "The Revival of Monteverdi's Operas in the Twentieth Century", 307, 313
Bartlett, John, *Familiar Quotations*, 671
Bauer, Marion *see* Peyser, Ethel
Beaumont, Cyril W., *A Short History of Ballet*, 352
Beckerman, Bernard, *Plays of the English Renaissance*, 839, 842–3, 871
Beecher, Donald, and Massimo Ciavolella, 164, 205, 210–13, 229
Bell, A.F.G., 392
Belladonna, Rita, 169, 172, 173–4
Bergeron, David M. (ed.), *Pageantry in Shakespearean Theatre*, 600
Berthold, Margot, *A History of World Theatre*, 43, 49–50, 80, 99, 120, 257, 280, 341, 356, 357, 415–16, 452
Birmingham Post, 557
Blackburn, Ruth H., *Biblical Drama Under the Tudors*, 43
Boas, F.S., 516

Boston Transcript, 516
Bradbrook, M.E., *John Webster, Citizen and Dramatist*, 871–2
Brockett, Oscar G., *The Essential Theatre*, 43, 257, 342, 376, 377, 383, 458
Brockway, Wallace, and Herbert Weinstock, *The World of Opera*, 277–8, 291, 304
Brooke, C.F. Tucker, Prof., 517
Brown, John Russell, 553–4
    The Complete Plays of the Wakefield Master (ed.), 43
Bruce, George. *Festival in the North: The Story of the Edinburgh*, 523
Buckle, Richard, *George Balanchine, Ballet Master*, 264–5

Campbell, J. Douglas *see* Sbrocchi, Leonard G.
Carter, Tim, *Monteverdi Returns to His Homeland*, 288
Catholic Herald
    Igoe, W.J., 519
Chambers, E.K., 585
Charney, Maurice, "Shakespeare and Marlowe as Rivals and Imitators" (in Barkan (ed.), *Renaissance Drama)*, 516, 560–2
Cheney, Sheldon, 49
Child, Clarence Griffin, 110
Chubb, Thomas Caldecot, *The Letters of Pietro Aretino*, 144, 146, 150–2, 156
Chujoy, Anatole, and P.W. Manchester (eds.), *The Dance Encyclopedia*, 352
Church of England News
    Robinson, Kenneth J., 519

Ciavolella, Massimo *see* Beecher, Donald
Clute, Marchetta, *Shakespeare of London*, 585
Cobb, Noel, *Prospero's Island*, 770–1
Commonweal
    Hayes, Richard, 521
Cope, Jackson T., *Dramaturgy of the Daemonic*, 249, 254–5
Craik, T.W., *The Tudor Interlude*, 107
Croce, Benedetto, 156

Daily Express
    Barber, John, 519
Daily Graphic
    Hayes, Walter, 518
Daily Mail
    Wilson, Cecil, 519
Daily Mirror
    Thirkel, Arthur, 524
Daily Telegraph
    Barber, John, 524
    Darlington, W.A., 518
Daily Worker, 520
de Mille, Agnes, *The Book of the Dance*, 352
Dent's Companion to the Theatre, 383, 416
Duchartre, Pierre-Louis, *The Italian Comedy*, 237, 241, 243, 244, 247
Dunn, E. Catherine, Tatiana Potich and Bernard M. Peebles, *The Medieval Drama and Its Claudian Revival*, 43
Durant, Will, 43

Eccles, Mark, 516

Einstein, Alfred, *A Short History of Music*, 463–4
Evans, Sir Ifor, 78, 81–2, 107, 475, 484–5, 516
Evening Standard
    Shulman, Milton, 557

Feingold, Michael, 694
Fenlon, Iain, and Peter Miller, "Private Vice, Public Virtue", 291
Ferraro, Bruno, Prof., 220–1, 222, 225
Flores, Angel, 383
Franco, Andrés, 396, 414, 422
Fraser, Russell, *The Young Shakespeare*, 585

Garreau, Joseph E., 435
Gassner, John, 43, 75, 77, 383, 404, 416, 485, 516, 522, 662, 786
Gayley, Charles Mills, 69, 77, 824
    Plays of Our Forefathers, 88–9, 90, 108
Geckle, George L., *"Tamburlaine" and "Edward II": Text and Performance*, 516, 517, 518, 519, 520, 522, 524–5, 526, 527, 550–1, 556–7, 559
Gombrich, E.H., *The Story of Art*, 226–7
Green, Martin, and John Swan, *The Triumph of Pierrot*, 273–4
Grove's Dictionary of Music and Musicians, 306
Grove's Dictionary of Opera, 277–8, 303, 320, 321, 329–30
Guardian, 551, 558
    Billington, Michael, 59
    Evans, Gareth Lloyd, 552
    Hope-Wallace, Philip, 519, 554–5
    Malcolm, Derek, 523

Harrison, G.B., 712
Harvard Crimson, 522
Hawkins, Frederick, *Annals of the French Stage*, 92, 97–8, 431, 433, 435–6, 444
Headington, Christopher, Roy Westbrook and Terry Barfoot, *Opera: A History*, 277–8, 285, 304, 315, 319, 322
Heilpern, John, 526
Hoffer, Charles R., *The Understanding of Music*, 277–8
Hoffman, Calvin, *The Man Who Killed "Shakespeare"*, 508–9
Holdsworth, Roger, *Research Opportunities in Renaissance Drama*, 559
Hotson, J.L., 516, 585
Houle, Peter J., *The English Morality and Related Drama*, 43, 78, 83

Illustrated London News, 524, 555

John, Nicholas (ed.), *Monteverdi*, 283
    see also Whenham, John
Juvarra, Filippo, 342

Kauffman, Stanley, *A World on Film*, 732
Kay, Dennis, *Shakespeare: His Life, Work and Era*, 585
Kirstein, Lincoln, *Dance*, 352
Knight, G. Wilson, 73
Knight, W. Nicholas, *Shakespeare's Hidden Life*, 579, 585

Leamington Courier, 551
Leonard, Richard Anthony, *The Stream of Music*, 277–8, 316
Levin, Bernard, 557, 559
Levin, Harry, *The Overreacher*, 516
Levinson, André, "The Spirit of the Spanish Dance", 425–8
The Listener, 522

Manchester Guardian see Guardian
Mayer, David., *Harlequin in His Element*, 254–5
McGraw-Hill Encyclopedia of World Drama, 84, 261, 356, 383, 435
McLuskie, Kathleen, *Renaissance Dramatists*, 827, 869
McMillan, Scott, 570
Michell, John, *Who Wrote Shakespeare?*, 586
Miller, Peter see Fenlon, Iain
Moravia, Alberto see *New York Times*

Nagler, Alois, *Sources of Theatrical History*, 129–30
Nation
    Clurman, Harold, 521
New Republic
    Bentley, Eric, 521
New Statesman (and Nation)
    Nightingale, Benedict, 524
    Worsley, T.C., 519
New York
    Simon, John, 611–12
New York Post
    Gold, Sylvaine, 558
New York Times
    Atkinson, Brooks, 263, 521, 544–5, 602, 624–5, 635–6, 681–2, 686–7
    Barnes, Clive, 558, 835–6
    Blau, Eleanor, 84–5
    Canby, Vincent, 690, 775
    Corbin, John, 624, 684–5
    Crutchfield, Will, 58
    Dunning, Jennifer, 269, 711–12
    Eder, Richard, 140–1
    Funke, Lewis, 771–2
    Genzlinger, Neil, 608–9
    Griffiths, Paul, 72–3, 309–12, 316, 336–7
    Gussow, Mel, 54–5, 141
    Holden, Stephen, 141–3
    Holland, Bernard, 32, 33–4
    Hoyle, Geoff, 269–72
    Kimmelman, Michael, 273
    Kisselgoff, Anna, 55–6
    Kozin, Allan, 308, 316
    Moravia, Alberto, 157–8
    Riding, Alan, 58–9
    Robertson, Nan, 57, 85–6
    Rockwell, John, 32–3, 70–1

Tassel, Janet, 29–30
Taubman, Howard, 625, 720–3
Woollcott, Alexander, 683–4
uncredited reviews, 71, 272–3, 626–7
New Yorker, 66–7
    Gibbs, Wolcott, 521
    Gill, Brendan, 558
Nicoll, Allardyce, *The Development of the Theatre*, 205, 432, 497, 516, 838, 842
Norman, Charles, *The Muses' Darling*, 516

Observer
    Brown, Ivor, 519
    Cushman, Robert, 526
    Kitchen, Laurence, 552
    Ratcliffe, Michael, 558–9
Opera News
    "P.J.S.", 296–7
Osterhof, Wolfgang, 297
Oxford Companion to English Literature, 485, 516, 784

Pastor, Ludwig von, *History of the Popes*, 138
Pearson, Jacqueline, *Tragedy and Tragicomedy in the Plays of John Webster*, 891
Peebles, Bernard M. *see* Dunn, E. Catherine
Peyser, Ethel, and Marion Bauer, *How Opera Grew*, 278, 291, 292
Phelps, William Lyons, 516–17
Plays and Players
    Nightingale, Benedict, 554
Potich, Tatiana *see* Dunn, E. Catherine
Punch
    Kingston, Jeremy, 557

Radcliff-Umstead, Douglas, *Carnival and Sacred Play*, 176, 178, 180, 181, 184
Rank, Otto, 767
Reyna, Ferdinando, *A Concise History of Ballet*, 347, 348, 446–7, 448–9
Ricks, Christopher (ed.), *English Drama to 1710*, 43, 107
    see also Wickham, Glynne
Roberts, Vera Moury, *On Stage*, 43, 125, 129–30
Rosenthal, Raymond, 157
Rowse, A.L., 507, 509, 529, 532
    Shakespeare's Self-Portrait, 576–7, 582, 585
Ruggles, Eleanor, *Prince of Players*, 678

Sage, Jack, "The Spanish Contribution to the Birth of Opera", 420, 422
Saturday Review
    Hewes, Henry, 521
Sbrocchi, Leonard G., and J. Douglas Campbell, 154
Schade, Richard Erich, *Studies in Early German Comedy, 1500–1650*, 101, 104, 356, 460
Schoenbaum, S., Prof., *William Shakespeare: a Collection of Biography and Autobiography*, 582, 583, 585
Esslin, Martin, *Encyclopedia of World Theatre*, 383, 416
Seaton, Ethel, "Marlowe's Map", 520
Simon and Schuster Book of the Opera, 291, 733
Sobel, Bernard, *New Theatre Handbook*, 383
Speaight, Robert, *Shakespeare: The Man and His Achievement*, 585
The Spectator
    Fleming, Peter, 519
    Rutherford, Malcolm, 522
Spoto, Donald, *Laurence Olivier*, 710, 771
Spurgeon, Caroline, 621–2
The Stage, 551–2
Stevens, John, 66, 67

Stratford-upon-Avon Herald, 552
Sunday Times, 518–19
    Hobson, Harold, 522, 524, 552–3, 554
Swan, John *see* Green, Martin
Symonds, John Addington, *Renaissance in Italy*, 139, 156

Talbott, Michael, 336
Taubman, Howard, *The Making of American Theater*, 679, 680, 683, 688
Theatre Arts Monthly see Levinson
Time magazine, 554
Times Literary Supplement, 558
The Times (London), 517, 518, 520–1, 522, 523, 551
    Billington, Michael, 555–6
    Wardle, Irving, 523–4, 526
Trewin, J.C. (various publications), 520, 524, 555, 557
Trueblood, Alan S., *Experience and Artistic Expression in Lope de Vega*, 382–3
Tulane Drama Review, 553–4
Twenty-Five Hamlets (author unknown), 675–6
Tynan, Kenneth, *Tynan Right and Left*, 522

Van Doren, Mark, 613, 621–2, 728
Village Voice
    Sellar, Tom, 609–10, 725

Wall Street Journal, 648
    Acocella, Joan, 712
    Cohen, Margot, 723–5
    Henning, Joel, 694–5, 748–9
    Levy, Paul, 596–8, 725–6
    Lyons, Donald, 611, 627–8, 636–7, 662–3, 733–4, 759–60, 776–7
    Spiegelman, Willard, 773
Wallis, Lawrence B., *Beaumont, Fletcher and Company*, 815, 823

Wedge, Emily V., 516
Weinstock, Herbert *see* Brockway, Wallace
Wells, Stanley, *Shakespeare: A Life in Drama*, 585
Westbrook, Roy *see* Headingon, Christopher
Westlake, Frederick, 323
Whenham, John, "Orfeo: A Masterpiece for a Court" (in John (ed.), *Monteverdi*), 283, 284
Wickham, Glynne, "Play, Player, Public and Place" (in Ricks (ed.), *English Drama to 1710*), 37, 107–8
Wilder, Thornton *see Yale Daily News*
Wilson, John Dover, 582, 585
Wynn, Peter, "Zanies, Lovers, Scoundrels, Fools", 266–8

Yale Daily News
   "T.N.W." (Thornton Wilder), 516–17